STUDIES IN NEOTROPICAL ORNITHOLOGY HONORING TED PARKER

Cover drawing: *Herpsilochmus parkeri,* by John P. O'Neill

ORNITHOLOGICAL MONOGRAPHS

Edited by

JOHN M. HAGAN
Manomet Center for Conservation Sciences
P.O. Box 1770
Manomet, Massachusetts 02345 USA

Ornithological Monographs, published by the American Ornithologists' Union, has been established for major papers too long for inclusion in the Union's journal, *The Auk.* Publication has been made possible through the generosity of the late Mrs. Carll Tucker and the Marcia Brady Tucker Foundation, Inc.

Copies of *Ornithological Monographs* may be ordered from Max C. Thompson, Assistant to the Treasurer, Department of Biology, Southwestern College, 100 College St., Winfield, Kansas 67156. Communications may also be routed through the AOU's permanent address: Division of Ornithology, National Museum of Natural History, Washington, D.C. 20560.

Editor of this issue, J. V. Remsen, Jr., Museum of Natural Science, Louisiana State University, Baton Rouge, Louisiana 70803 USA

Price of *Ornithological Monograph* 48: $49.95 prepaid. Add $4.00 for handling and shipping charge in U.S., and $7.00 for all other countries. Make checks payable to American Ornithologists' Union.

Library of Congress Catalogue Card Number 97-075147

Printed by Allen Press, Inc., Lawrence, Kansas 66044

Issued December 3, 1997

Ornithological Monographs, No. 48 1–918 pp.

Printed by Allen Press, Inc., Lawrence, Kansas 66044

Copyright © by the American Ornithologists' Union, 1997

ISBN: 0-935868-93-3

STUDIES IN NEOTROPICAL ORNITHOLOGY HONORING TED PARKER

Editor:
J. V. REMSEN, JR.

ORNITHOLOGICAL MONOGRAPHS NO. 48
PUBLISHED BY
THE AMERICAN ORNITHOLOGISTS' UNION
WASHINGTON, D.C.
1997

LIST OF SPONSORS

W. Alton Jones Foundation
Ted and Dorothy Parker
Curtis C. Sorrells
Robert and Gail DeBellevue
Victor Emanuel Nature Tours
James M. Bishop
James R. Stewart, Jr.
Lancaster County Bird Club

(photo by John P. O'Neill)

Theodore A. Parker III (1953–1993)

Here I sit a world away
Among old books that you will never read,
Thinking of you.

When you lie down
Among the night-cries of your brother birds,
Where leaf and cloud and stone
Enfold there intertexture without end,
My prayers go with you
Even so far as Cusco and La Paz,
The peaceful, mountain cradle of the stars.

Blanford Parker

PREFACE

The future of Neotropical ornithology changed forever when Ted Parker died in August 1993. We'll never know exactly how it was changed, but we know it is going to be different without him. Most of his first 40 years were spent developing an unrivaled knowledge of natural history, and he had just begun to acquire the confidence and stature to apply that background to big problems. His death was, therefore, a multidimensional tragedy. The first two papers in this volume provide many details on Ted and his career, thus allowing this preface to be brief.

Soon after Ted's death his legions of friends searched for ways to express their appreciation for his friendship and his influence. This volume is one of those efforts. A primary goal was to complete as many of Ted's unpublished manuscripts as possible. Otherwise, the only initial criteria for submission of manuscripts was that they involve Neotropical ornithology, that at least the lead author was one of Ted's friends, and that Ted would have enjoyed reading the manuscript. We all hope that we succeeded in the latter; certainly Ted would be gratified to see so many of his favorite themes represented. Ted also would have appreciated the range in author experience, from "household names" in Neotropical ornithology to students and non-professionals submitting their very first papers for peer-review. He would also have appreciated the range in nationalities of the authors as well as the number of authors (15) from countries in the Neotropics.

I thank the many people who reviewed the manuscripts; each paper was reviewed by at least two researchers. Lola de Quintela and Manuel Plenge graciously translated many abstracts into Spanish. The Museum of Natural Science, LSU, provided logistic and clerical support for completion of the manuscript; Marilyn Young was especially helpful. John P. O'Neill donated his time to produce the cover artwork. Tom Schulenberg provided much sound advice and encouragement.

A special thanks is due to John M. Hagan III, editor of Ornithological Monographs. John should really be listed as an official co-editor of the volume, but John declined my requests to do this because he said that what he did was just part of his job as editor of the series. But handling the headaches of more than 50 separate manuscripts was far beyond the normal duties of an editor of a monograph series. John's careful editing improved many papers dramatically, and the countless hours of work that he did for this volume merit special recognition. Special thanks also are due to his assistants, Cammy Collins and Rebecca Hagan, for their work on copyediting the final proofs.

Publication of this volume was generously financed by the W. Alton Jones Foundation, through J. P. Myers. Many people and organizations also contributed generously to this project through the Ted Parker Memorial Fund, LSU: ARA Records, R. H. Barth, Jr., Will B. Betchart, James Bishop, Kevin J. Burns, Angela Chapman, Allen T. Chartier, George W. Clayton, Ben and Lula Coffey, Mario Cohn-Haft, John B. Crowell, Jr., Robert and Gail DeBellevue, Jonathan L. Dunn, Paul R. Ehrlich, Ethyl Corporation, Peter and Alice Fogg, Daniel T. Forster, Kimball L. Garrett, William and Marcella Hackney, James C. Hageman, Linda S. Hale, Tom and Jo Heindel, Franklyn K. Hoover, Ned K. Johnson, Jody Kennard, Cecil C. Kersting, Ralph and Gail Kinney, Joseph P. Kleiman, Christopher P. Kofron, Lancaster County Bird Club, Paul Lehman, George A. Loker, David O. Matson, Harold B. Morrin, S. H. Mudd, Ted and Dorothy Parker, David N. Pashley, Robert R. Reid, Jr., Jack Reinoehl, Mark B. Robbins, Scott K. Robinson, Sammie Rodden, Douglas A. Rossman, Judith A. Schiebout, Donna and Greg Schmitt, Peter E. Scott, John P. Sevenair, Frederick H. Sheldon, Alfred E. and Gwen B. Smalley, Peter A. Soderbergh, Curtis C. Sorrells, Ronald J. Stein, James R. Stewart, Jr., John W. Terborgh, Guy A. Tudor, Peter D. Vickery, Victor Emanuel Nature Tours, Carol Walton, David and Melissa Wiedenfeld, Kevin J. Zimmer, and Dale A. Zimmerman.

Long may you run, Ted, long may you run.

J. V. Remsen, Jr.

STUDIES IN NEOTROPICAL ORNITHOLOGY HONORING TED PARKER
J. V. REMSEN, JR., EDITOR

TABLE OF CONTENTS

DEDICATION	vii
PREFACE	ix
TABLE OF CONTENTS	xi
LIST OF COLOR ILLUSTRATIONS	xv

Robbins, Mark B., Gary R. Graves, and J. V. Remsen, Jr. In Memoriam: Theodore A. Parker III, 1953–1993 1

Remsen, Jr., J. V., and Thomas S. Schulenberg. The pervasive influence of Ted Parker on Neotropical field ornithology 7

Graves, Gary R. Colorimetric and morphometric gradients in Colombian populations of Dusky Antbirds (*Cercomacra tyrannina*), with a description of a new species, *Cercomacra parkeri* 21

Fitzpatrick, John W., and Douglas F. Stotz. A new species of tyrannulet (*Phylloscartes*) from the Andean foothills of Peru and Bolivia 37

Krabbe, Niels, and Thomas S. Schulenberg. Species limits and natural history of *Scytalopus* tapaculos (Rhinocryptidae), with descriptions of the Ecuadorian taxa, including three new species 47

Remsen, Jr., J. V. A new genus for the Yellow-shouldered Grosbeak 89

Bates, John M. Distribution and geographic variation in three South American grassquits (Emberizinae, *Tiaris*) 91

Bierregaard, Jr., Richard O., Mario Cohn-Haft, and Douglas F. Stotz. Cryptic biodiversity: an overlooked species and new subspecies of antbird (Aves: Formicariidae) with a revision of *Cercomacra tyrannina* in northeastern South America 111

Brumfield, Robb T., David L. Swofford, and Michael J. Braun. Evolutionary relationships among the potoos (Nyctibiidae) based on isozymes 129

Budney, Gregory F., and Robert W. Grotke. Techniques for audio recording vocalizations of tropical birds 147

Capparella, A. P., Gary H. Rosenberg, and Steven W. Cardiff. A new subspecies of *Percnostola rufifrons* (Formicariidae) from northeastern Amazonian Peru, with a revision of the *rufifrons* complex 165

Chesser, R. Terry. Patterns of seasonal and geographical distribution of austral migrant flycatchers (Tyrannidae) in Bolivia 171

Cohn-Haft, Mario, Andrew Whittaker, and Philip C. Stouffer. A new look at the "species-poor" Central Amazon: the avifauna north of Manaus, Brazil 205

Collar, Nigel J., David C. Wege, and Adrian J. Long. Patterns and causes of endangerment in the New World avifauna 237

Garvin, Mary C., John M. Bates, and J. M. Kinsella. Field techniques for collecting and preserving helminth parasites from birds, with new geographic and host records of parasitic nematodes from Bolivia 261

Hackett, Shannon J., and Cathi A. Lehn. Lack of genetic divergence in a genus (*Pteroglossus*) of Neotropical birds: the connection between life-history characteristics and levels of genetic divergence 267

Haffer, Jürgen. Contact zones between birds of southern Amazonia 281

Hardy, John William, and Theodore A. Parker III. The nature and probable function of vocal copying in Lawrence's Thrush, *Turdus lawrencii* 307

Hilty, Steven L. Seasonal distribution of birds at a cloud-forest locality, the Anchicayá Valley, in western Colombia 321

Isler, Morton L. A sector-based ornithological geographic information system for the Neotropics 345

Isler, Morton L., Phyllis R. Isler, and Bret M. Whitney. Biogeography and systematics of the *Thamnophilus punctatus* (Thamnophilidae) complex 355

Kratter, Andrew W., and Theodore A. Parker III. Relationship of two bamboo-specialized foliage-gleaners: *Automolus dorsalis* and *Anabazenops fuscus* (Furnariidae) 383

Marantz, Curtis A. Geographic variation of plumage patterns in the woodcreeper genus *Dendrocolaptes* (Dendrocolaptidae) 399

Marín, Manuel. Species limits and distribution of some New World spine-tailed swifts (*Chaetura* spp.) 431

Marra, Peter P., and J. V. Remsen, Jr. Insights into the maintenance of high species diversity in the Neotropics: habitat selection and foraging behavior in understory birds of tropical and temperate forests 445

O'Neill, John P., and Theodore A. Parker III. New subspecies of *Myrmoborus leucophrys* (Formicariidae) and *Phrygilus alaudinus* (Emberizidae) from the upper Huallaga Valley, Peru 485

Oren, David C., and Theodore A. Parker III. Avifauna of the Tapajós National Park and vicinity, Amazonian Brazil 493

Parker, III, Theodore A., and Jaqueline M. Goerck. The importance of national parks and biological reserves to bird conservation in the Atlantic forest region of Brazil 527

Parker, III, Theodore A., Douglas F. Stotz, and John W. Fitzpatrick. Notes on avian bamboo specialists in southwestern Amazonian Brazil 543

Parker, III, Theodore A., and Edwin O. Willis. Notes on three tiny grassland flycatchers, with comments on the disappearance of South American fire-diversified savannas 549

Perry, Alan, Michael Kessler, and Nicholas Helme. Birds of the central Río Tuichi Valley, with emphasis on dry forest, *Parque Nacional Madidi*, depto. La Paz, Bolivia 557

Renjifo, Luis Miguel, Grace P. Servat, Jaqueline M. Goerck, Bette A. Loiselle, and John G. Blake. Patterns of species composition and endemism in the northern Neotropics: a case for conservation of montane avifaunas 577

Robbins, Mark B., and Theodore A. Parker III. What is the closest living relative of *Catharopeza* (Parulinae)? 595

Robbins, Mark B., and Theodore A. Parker III. Voice and taxonomy of *Caprimulgus* (*rufus*) *otiosus* (Caprimulgidae), with a reevaluation of *Caprimulgus rufus* subspecies 601

Robbins, Mark B., Gary H. Rosenberg, Francisco Sornoza Molina, and Marco A. Jácome. Taxonomy and nest description of the Tumbes Swallow (*Tachycineta* [*albilinea*] *stolzmanni*) 609

Robinson, Scott K. Birds of a Peruvian oxbow lake: populations, resources, predation, and social behavior 613

Robinson, Scott K., and John Terborgh. Bird community dynamics along primary successional gradients of an Amazonian whitewater river 641

Rosenberg, Kenneth V. Ecology of dead-leaf foraging specialists and their contribution to Amazonian bird diversity 673

Schmitt, C. Gregory, Donna C. Schmitt, and J. V. Remsen, Jr. Birds of the Tambo area, an arid valley in the Bolivian Andes 701

Schulenberg, Thomas S., and Theodore A. Parker III. Notes on the Yellow-browed Toucanet *Aulacorhynchus huallagae* 717

Schulenberg, Thomas S., and Theodore A. Parker III. A new species of tyrant-flycatcher (Tyrannidae: *Tolmomyias*) from the western Amazon Basin 723

Sillett, T. Scott, Anne James, and Kristine B. Sillett. Bromeliad foraging specialization and diet selection of *Pseudocolaptes lawrencii* (Furnariidae) 733

Silva, José Maria Cardoso Da, David Conway Oren, Júlio César Roma, and Luiza Magalli Pinto Henriques. Composition and distribution patterns of the avifauna of an Amazonian upland savanna, Amapá, Brazil 743

Stotz, Douglas F., Scott M. Lanyon, Thomas S. Schulenberg, David E. Willard, A. Townsend Peterson, and John W. Fitzpatrick. An avifaunal survey of two tropical forest localities on the middle Rio Jiparaná, Rondônia, Brazil 763

Tallman, Dan A., and Erika J. Tallman. Timing of breeding by antbirds (Formicariidae) in an aseasonal environment in Amazonian Ecuador 783

Vuilleumier, François. Status and distribution of *Asthenes anthoides* (Furnariidae), a species endemic to Fuego-Patagonia, with notes on its systematic relationships and conservation 791

Whitney, Bret M., and José Fernando Pacheco. Behavior, vocalizations, and relationships of some *Myrmotherula* antwrens (Thamnophilidae) in eastern Brazil, with comments on the "Plain-winged" group 809

Wiedenfeld, David A. Land of magnificent isolation: M. A. Carriker's explorations in Bolivia 821

Zimmer, Kevin J. Species limits in *Cranioleuca vulpina* 849

Zimmer, Kevin J., and Steven L. Hilty. Avifauna of a locality in the upper Orinoco drainage of Amazonas, Venezuela 865

Zimmer, Kevin J., Theodore A. Parker III, Morton L. Isler, and Phyllis R. Isler. Survey of a southern Amazonian avifauna: the Alta Floresta region, Mato Grosso, Brazil 887

LIST OF COLOR ILLUSTRATIONS

FRONTISPIECE: Ted Parker vi
PLATE I. Parker's Antbird (*Cercomacra parkeri* sp. nov.) and Dusky Antbird (*Cercomacra tyrannina*) 20
PLATE II. Cinnamon-faced Tyrannulet (*Phylloscartes parkeri* sp. nov.) 36
PLATE III. Three new species of tapaculo (*Scytalopus*) 46
FIG. 12. Plumages of the Slaty Antshrike (*Thamnophilus punctatus*) 368
PLATE IV. Orange-eyed Flycatcher (*Tolmomyias traylori* sp. nov.) 722
FIG. 4. Ventral views of Rusty-backed Spinetail (*Cranioleuca vulpina*) 858

IN MEMORIAM: THEODORE A. PARKER III, 1953–1993

MARK B. ROBBINS,[1] GARY R. GRAVES,[2] AND J. V. REMSEN, JR.[3]

[1]*Division of Ornithology, University of Kansas Museum of Natural History, University of Kansas, Lawrence, Kansas 66045, USA;*
[2]*Bird Division, National Museum of Natural History, Smithsonian Institution, Washington, D.C. 20560, USA; and*
[3]*Museum of Natural Science, Louisiana State University, Baton Rouge, Louisiana 70803, USA*

We face the impossible task of summarizing in a few pages the brilliant life of our friend, Ted Parker. Those who knew Ted will understand our dilemma. For those who did not, we share some reminiscences of a charismatic and remarkable biologist, one who in our opinion was the most gifted field ornithologist of the 20th century.

Ted was born into a nurturing family in Lancaster, Pennsylvania, on 1 April 1953. One of his earliest memories was of his grandmother taking him to Lancaster's North Museum when he was six years old. Ted immersed himself in natural history and decided at an early age that he wanted to become a naturalist. Birds, reptiles, amphibians, and butterflies of the Lancaster area were all subjected to his penetrating focus, and he spoke fondly of the times he spent on field outings with his younger brother, Blanford. An example of Ted's modus operandi during his middle-school years was his approach to shell collecting. Not content with beachcombing or buying shells from commercial outlets, he spent hours at local fish markets dissecting fish intestines in search of rare deep sea mollusks. Ted was a devoted conservationist from his childhood days, and as a senior in high school he conducted an environmental impact statement for the Lancaster Waterworks. By his own account, he was bored with the educational curriculum, and his predilection for birds came to the forefront as a teenager. His reputation as a "birding phenom" was born during this period.

During the last semester of his senior year in high school, Ted embarked on an attempt to break a bird-listing record, the number of species seen in the United States and Canada in one calendar year, that had stood since the 1950s. With the help of Harold Morrin and other friends he criss-crossed the United States several times in the spring and summer of 1971. During the fall he enrolled in the University of Arizona, strategically chosen for its proximity to good birding localities. Ted smashed the listing record, but wryly recounted that he was forced to drop every course, except golf, during his first two semesters because of his chronic birding. Although his start in college was inauspicious, Ted's birding exploits earned him national recognition among birdwatchers and mention in the *Reader's Digest*. He rapidly became an authority on Arizona birds and within two years assumed the editorship for the southwest region of *American Birds*. Ted majored in biology but later switched to anthropology to avoid organic chemistry and physics. In later years, he often joked that superfluous college course requirements impeded his pursuit of worthwhile knowledge. All too often, extraordinarily gifted but unconventional students become lost in academic gristmills. Fortunately, some of Ted's professors at Arizona encouraged Ted to channel his prodigious birding talent into mainstream ornithology (e.g., with H. R. Pulliam, 1979, Fortschr. Zool. 25:137–147).

Ted made his first trip to Mexico and the Neotropics during his second semester at Arizona. In a sense he never returned. Birding with Ted in Mexico was not for the fainthearted, and sleep was accommodated only after exhaustion had a firm grip. After birding all day and then catching the evening chorus of nightjars and owls, it was back in the car for an all-night drive, dodging animals on the road, with the radio blaring rock-and-roll. The dawn chorus heralded that a new locality had been reached. These extended trips resulted in significant discoveries from Jalisco to Chiapas and some of his first contributions to the scientific literature (1976, Amer. Birds 30:779–782).

During his stint at Arizona, when Ted was not planning his next trip to Mexico, he was devouring literature on the Neotropics. Serious students of any discipline would have been awe-

FIG. 1. Ted examining one of his favorite groups, the cotingids, at LSU in September 1990. He wrote a number of the species accounts in Snow's *The Cotingas* (1982). Courtesy of *The Advocate*. Photo by Stephan Savoia.

struck by the amount of time he spent studying and his degree of retention. Dorm furniture and bed were buried by books and papers, and his room became the focal point for naturalists and birdwatchers. Nightly discourses on birds, cacti, or any of a hundred other natural history topics, were sprinkled with liberal politics.

In spring 1974, George Lowery phoned his former student, Steve Russell, to see if there were any promising students at Arizona interested in helping Louisiana State University (LSU) survey the birds of Peru. Russell knew just the person. Ted's acceptance of LSU's offer was immediate and unqualified, even when told that he would have to cut off his shoulder-length hair to meet Lowery's genteel standards. Unfortunately, Lowery, who died in January 1978, did not live long enough to appreciate fully how fortuitous he was in landing Ted. During that initial trip, Ted spent nearly eight months in Peru, traveling the length and breadth of the country collecting bird specimens. By the end of the expedition, Ted had acquired Huánuco-accented Spanish from his Peruvian field companion, Reyes Rivera, and a keen insight into local customs that would serve him well during the remainder of his life. Ted's passion for taping bird vocalizations blossomed on the 1974 expedition. Afterwards, he was seldom seen in the field without a bulky reel-to-reel tape recorder. Not content with exhausting whirlwind trips and anecdotal observations of rare birds, he also began studying avian communities in single localities for months at a time.

Another maturing event was his marriage to Susan Allen in 1976. However, much to the chagrin of his benefactors, Ted's extended field expeditions interfered with university course work, delaying his graduation until 1977. That he finished at all is a testament to the efforts of Susan, as well as to the encouragement of his friends, especially John O'Neill, then curator of birds at LSU.

Almost immediately after they were married, Ted and Susan departed for a seven-month expedition to Peru. At Explorer's Inn, the future Tambopata Reserved Zone, Ted fine-tuned what was to be one of his most valuable contributions to Neotropical ornithology—the avifaunal site inventory. Until recently, inventories had been largely based on the cumulative record of museum specimens, captured and released birds, and sight records. In the old museum tradition, thorough

Fig. 2. Ted (in center) at El Triunfo, Chiapas, Mexico in April 1973. Results from this trip included the first description of the chick of the Horned Guan (*Oreophasis derbianus*) (1976, Amer. Birds 30:779–782). Photo by Mark B. Robbins.

inventories in species-rich Amazonian sites required thousands of man hours and years to complete. Ted focused on vocalizations. With the use of tape playback, he methodically tracked down almost every avian sound in the rain forest, and more importantly, remembered everything he had heard. Ted demonstrated to his skeptical colleagues at LSU that inventories as he conducted them were an order of magnitude more efficient than those based on traditional methods. In hindsight, we view the "Parker inventory" as a methodological revolution. Although identifying Neotropical birds by voice has a long history, no one had done it as well or applied it as effectively as Ted. By the mid-1980s, he was so proficient that he could inventory 80–90% of any local avifauna from Mexico to southeastern Brazil in a few mornings. By the time of his death he had deposited an astonishing 15,000+ recordings in the Library of Natural Sounds (LNS) at Cornell University. Although Ted championed the use of sound recordings in inventory work (1991, Auk 108:443–444), he continued to collect specimens until the end of his life. Over his 19 years association with LSU, he added over 3,000 specimens to the collection.

Ted's long absences in the field created strains in his marriage and soon he and Susan parted. After the divorce, LSU and Baton Rouge would be his home, at least spiritually, for the remainder of his life.

From the mid-1970s through the 1980s Ted's primary source of income came from leading bird tours, mainly for Victor Emanuel Nature Tours. This gave him the unprecedented opportunity to study bird communities in literally hundreds of locations in the Neotropics. From 1974 through 1993, Ted averaged six months annually in the field. In part through the success of his tours, Ted rapidly achieved the reputation as the foremost authority on the identification and distribution of Neotropical birds. His Peruvian tours quickly gained fame, as over 700 species were routinely recorded. But the number of species recorded was secondary to the enthusiasm and appreciation of the Neotropical avifauna that Ted conveyed to his tour participants. Ted was the consummate tour leader—he combined unsurpassed birding skills, an audiophilic memory, and museum background with a disarming lack of ego. Under the guise of a tour co-leader, MBR witnessed first hand the chemistry between Ted and participants during a West Indian tour in 1988. Most participants, many of whom had booked several previous tours with him, worshipped Ted. His charming personality coupled with his unrivaled knowledge endeared him to a vast community, ranging from the general public to the most intractable of all groups, politicians and the military. Kenn Kaufman (1993, Amer. Birds 47:349–351) effectively related one of Ted's greatest assets, that of conveying information to people without offending them; he made them feel as if they were part of the inner circle.

Ted labored on and off for 15 years on the species accounts for *Birds of Peru*. Given the enormity of the task (1700 + species), and the fact that he spent much of the year in the field obtaining additional data, his colleagues teased Ted that he wouldn't finish the project until he quit field work, which of course everyone knew would never happen. Nevertheless, he was surprisingly productive during his brief stays in Baton Rouge. Between 1977 and 1990, he authored 37 technical papers, 2 audiocassettes, and 1 annotated check-list. This was no easy task for him because attending to his voluminous correspondence and "decompressing" between field trips often required weeks. He would often lose himself for hours playing and watching basketball as well as studying specimens and literature on Neotropical birds. It was a common and predictable event, from his college days up to his last years, to find him sweating profusely in black high-top Converse basketball shoes, with a white towel wrapped around his neck, watching a basketball game after he had just spent a couple of hours playing on the court. His obsession for basketball was second only to his love of birds, and he never missed an LSU game when he was in Baton Rouge. At home he was frequently found reposing on the couch editing his tape-recordings and filling out LNS data forms while watching basketball. One minute he would be shouting "Did you see that play!", and in the next instant he would point out some barely audible vocalization in the background of a recording.

In 1980, Ted met Carol Walton, and in 1985 they were married. Ted's insatiable thirst for field work continued, and his ambivalence toward domestic activities caused their marriage to suffer the same fate as his first. With every passing year, more demands were made of Ted's time, but he made sure that he was with his family at Christmas, and he made every effort to make it back to Lancaster during the sweet-corn harvest. His appreciation for good food was well known.

Ted's presence set LSU apart from other Neotropical programs. He was a beacon for prospective graduate students from around the country even though his official connection with LSU was only as a research associate. Ted's comprehensive knowledge gave him a unique view of the links between a species' vocalizations, foraging behavior, habitat preference and its biogeography and systemic relationships. His accomplishments there drew further attention, when he, John O'Neill, and the LSU program were the focus of Don Stap's book, *A Parrot Without a Name: the Search for the Last Unknown Birds on Earth*. Stap devoted several pages to how Ted discovered a new species of flycatcher by first hearing its song. Ted's contributions to the LSU program were recognized with a posthumous honorary Ph.D.

We are among the legions that acknowledge picking his brain. One only needs to peruse the Neotropical literature for the past two decades to get a feel for how generous he was. It would be difficult to find a serious student of Neotropical ornithology who did not incorporate some of Ted's knowledge in their work. Jon Fjeldså and Niels Krabbe (1990, *Birds of the High Andes*) perhaps said it best: "In particular we thank Theodore A. Parker III for an enormous amount of life history data. As he is by far the greatest capacity on the life histories of Neotropical birds there ever was, his contribution to our knowledge of Neotropical birds can not be stressed strongly enough."

Regardless of his audience, Ted was not timid in conveying the urgency needed to conserve the world's fauna. Brent Bailey, of Conservational International, recounted an unforgettable meeting that CI's Bolivian representative, Guillermo Rioja, and Ted had with the Vice President of Bolivia. At the end of one of Ted's classic conservation diatribes, which Rioja was certain would result in them being asked to leave, the Bolivian official leaned forward at the table and said, "It is people like you that have made your country great." Murray Gell-Mann, the Nobel laureate physicist and avid birdwatcher, was exposed to Ted's haranges during joint field work in South America, and, like so many others before him, was deeply impressed by Ted's field skills. Gell-Mann was also a director of the John D. and Catherine T. MacArthur Foundation, and he helped encourage the MacArthur Foundation to fund Ted's idea of a Rapid Assessment Program (RAP), through Conservation International. The core of the RAP team was formed from a handful of eminent biologists, each of whom brought years of experience in the tropics: Ted, botanists Alwyn Gentry and Robin Foster, and mammalogist Louise Emmons. The premise of the RAP program was to assess quickly, usually in a few weeks, the diversity and uniqueness of an area, and then to transmit the results, via a rapidly published report, to conservationists, biologists, administrators, and politicians. As Ted argued, the RAP protocol was an effective way to obtain sufficient data for guiding conservation priorities in the tropics, where land use changes can occur with frightening rapidity. Nevertheless, he and his RAP colleagues were well aware of the importance, both to science and to conservation, of the more classical, long-term surveys. Ted was instrumental in persuading the MacArthur Foundation to fund some traditional faunal inventories.

Ted's successes and recognition in conservation soothed some of the bitterness that he had felt about being snubbed by some mainstream ornithologists, who in spite of their academic credentials held but a fraction of Ted's knowledge about Neotropical birds. Ted was always amused by, and often galled, by the cadre of "experts" who spent little time in the field. He mellowed a bit by his late-30s, knowing that the "arm-chair" biologists would eventually be swept away by the tidal wave of new information and ideas generated by field people like himself.

Ted was well aware of the extraordinary risks that he took by working in remote areas. Upon recounting his many close calls, ranging from dodging animals and vehicles on Mexican roads to having his and John O'Neill's boat full of specimens sunk by Aguarunas, to being chased by a stone-throwing mob in Peru, even the most seasoned field person would shake their heads. However, Ted had no aspiration of becoming a martyr, and the thing that he feared most about doing field work was having to rely on small planes to get him to many otherwise inaccessible sites. His fear was well-founded, because on at least two prior occasions, once while flying through a severe thunderstorm in the eastern Peruvian Andes and the other during a landing at an abandoned airfield in Bolivia, he was nearly killed.

On 3 August 1993, Ted, Al Gentry, Jaqueline Goerck, one of Ecuador's leading conservationists, Eduardo Aspiazu, and two Ecuadorian biologists, Alfredo Luna and Carmen Bonifaz, left Guayaquil in a small plane on a routine mission to survey the rapidly diminishing forest in southwestern Ecuador. No flight path had been filed, navigational errors were made, and in the late afternoon the plane crashed into a remote mountain cloaked in a cloud bank. The pilot and Eduardo died shortly after impact, and Al passed away during the night. The following morning, Jaqueline, suffering a broken ankle and spinal injuries, and Carmen struggled down the forested mountain side and brought help that afternoon. By then, Ted had died, and Alfredo barely survived. In all likelihood, the knowledge that ebbed away over those few hours will require decades to recover, if ever.

At the time of his death, Ted was probably the happiest he had been during his professional career, as a result of the success of RAP, the recent and imminent publication of several other long-term projects, and his engagement to Jaqueline Goerck. In the months before his death, Ted talked about spending less time in the field, settling down, and having children. One can only speculate what Ted would have accomplished if he had lived. It is certain that his role in the conservation of tropical biotas would have continued to accelerate, and his storehouse of knowledge would have come to fruition through his publications and collaboration with others. Such monumental tasks as finishing *Birds of Peru* will be left to his colleagues, with the impossible burden of attempting to reach the expectations generated by Ted's involvement.

Envious of the amount of time Ted spent in the field, we vicariously lived his adventures and recounted fond memories of distant camps and forest trails. Many of us took for granted that we could ride his train for years to come. In an interview less than a year before his death, Ted related that he wanted the following words from a Robert Frost poem carved on his tombstone: "Two roads diverged in a wood, and I took the one less traveled by." Ted steadfastly took the less traveled road and we were fortunate to have been his friends and colleagues along the way.

Susan Allen Lohr, Harold Morrin, John O'Neill, and Thomas Schulenberg helped illuminate reminiscences of an extraordinary individual.

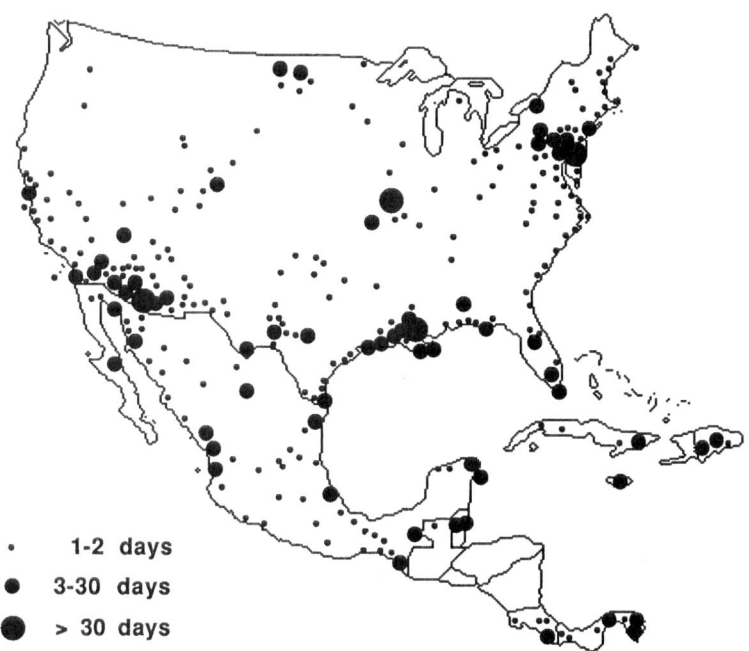

FIG. 1. Geographic distribution of Ted Parker's fieldwork in North America south of Canada.

David N. Pashley, A. Townsend Peterson, H. Douglas Pratt, Carlos E. Quintela, Mark B. Robbins, Gary H. Rosenberg, Kenneth V. Rosenberg, Peter E. Scott, T. Scott Sillett, Daniel A. Tallman, Erika J. Tallman, Francisco J. Vilella, David A. Wiedenfeld, Morris D. Williams, and Robert M. Zink. The list of persons similarly influenced beyond LSU would be enormous and would include, just as a start, all authors in this volume.

The breadth and depth of Ted's experience in the Americas was remarkable for someone so young. Localities from his field trips (Figs. 1 and 2) illustrate his thorough coverage of the hemisphere. From 1974 to 1993, he spent a total of approximately 115 months (9.6 years) in the field in the Neotropics, or roughly six months per year, and visited 22 countries. Thus his field experience was incomparable among contemporaries, and of historic proportions.

Parker had the ability not only to remember facts accumulated during field-work but to integrate them into an overall picture. He knew the habitat, voice, and foraging behavior of nearly all 4,000+ species of New World birds. Particularly after he began collaborating regularly with Robin Foster and Al Gentry during their work for Conservation International, Ted learned a great deal about tropical plant communities. The breadth of Ted's knowledge, combined with a questioning mind and a rare talent for synthesis, allowed him to see patterns that no one else had seen or could detect. His insights often amazed and intimidated academic ornithologists. Original ideas, difficult to generate for most of us, flowed from Ted. When his colleagues discovered something that Ted had not already realized, from natural history anecdote to general pattern, they regarded it as a major triumph. Mixed with the tragedy of his death is that he had just begun to put his ideas in writing, and many of these were buried in papers with titles that did not reflect the ideas contained. Most of the facts are in his book on Neotropical birds (Stotz et al. 1996). Parker was primarily responsible for the database in this book, a massive matrix of 3,751 species by 47 variables (e.g., habitat, elevational range, guild) for each species—or roughly 175,000 data entries. However, most of the original ideas concerning the patterns revealed by these data had not been written down at the time of his death.

Ted had superior hearing abilities. He became the expert on Neotropical bird voices by combining this natural talent with hard work. He recorded thousands of individual birds, lugging heavy recorders and awkward microphones everywhere, often through rugged terrain and in predawn darkness. He spent countless hours studying his recordings of bird vocalizations. Ted's

FIG. 2. Geographic distribution of Ted Parker's fieldwork in South America and the Lesser Antilles.

more than 15,000 cuts represented approximately 25% of all recordings at the Library of Natural Sounds at the Cornell Laboratory of Ornithology as of 1994 (G. F. Budney, pers. comm.). Ted, who served on the Lab's Administrative Board, also recruited many other recordists to contribute their tapes to the LNS. These contributions, both direct and indirect, were instrumental in shaping the LNS into the world's largest sound library.

To put Parker's expertise in proper context, consider that when he started taping to identify South American birds by voice, only a handful of pioneers had begun recording Neotropical bird voices. Few records and no audiotapes of Neotropical bird sounds were then available. Likewise, published descriptions of the voices of most Neotropical species did not exist; indeed, the voices of most Neotropical birds were unknown. Few field guides were available to help even with visual identifications. Ted learned largely by first recording the voice and then collecting voucher specimens to insure that his identification was correct; therefore, his specimens and tape-recordings are doubly valuable.

On LSU expeditions, Parker prepared roughly 2,750 bird specimens and collected many hundreds more that were prepared by others. Almost all his specimens are housed at the Museum of Natural Science, LSU; a few are at the Museo de Historia Natural, Universidad Nacional Mayor de San Marcos, Lima, Peru, and the Museo Nacional de Historia Natural, La Paz, Bolivia. The following taxa have been named for Ted: a genus of cardinaline grosbeak (Remsen 1997); the Ash-throated Antwren, *Herpsilochmus parkeri* (Davis and O'Neill 1986); the Subtropical Pygmy-Owl, *Glaucidium parkeri* (Robbins and Howell 1995); a species of *Cercomacra* antbird (Graves 1997); a species of *Scytalopus* tapaculo (Krabbe and Schulenberg 1997); a species of *Phylloscartes* tyrannulet (Fitzpatrick and Stotz 1997); a subspecies of the Coppery Metaltail, *Metallura theresiae parkeri* (Graves 1981); a species of carabid beetle, *Batesiana parkeri* (Erwin 1994); and a species of chewing louse, *Furnariphilus parkeri* (Price and Clayton 1995). Parker described the following bird taxa: *Schizoeaca fuliginosa plengei* and *Uromyias agraphia squamigera* (O'Neill and Parker 1976); *Pipreola riefferii tallmanorum* and *Chlorospingus ophthalmicus hiaticolus* (O'Neill and Parker 1981); *Grallaricula ochraceifrons* (Graves et al. 1983); *Thryothorus eisenmanni* and *Thryothorus euophrys schulenbergi* (Parker and O'Neill 1985); new subspecies of *Myrmoborus leucophrys* and *Phrygilus alaudinus* (O'Neill and Parker 1997); and a new species of *Tolmomyias* (Schulenberg and Parker 1997).

Ted's influence can be seen in seven major areas:

1. *Importance of thorough knowledge of vocalizations for surveying tropical bird communities.* Everywhere Ted went, he proved that knowledge of bird distribution, particularly tropical bird distribution, was incomplete until sampled by someone skilled in voice identifications. He showed that few tropical bird communities had been thoroughly surveyed and that even at well-studied localities, common species had been overlooked. Others before him had, of course, recognized the importance of vocal identification, but no one else dramatized and communicated more broadly its importance. Perhaps the best example was Ted's visit to the Cocha Cashu study site in Peru's Manu National Park, where John Terborgh's team of competent ornithologists had been studying the birds for more than a decade before Parker's visit. In only a few weeks, Ted added 20 species to what was already the most thoroughly studied tropical forest locality in the world. Among Parker's discoveries was the Rufous-fronted Antthrush (*Formicarius rufifrons*), known from just two female specimens until he found it at Manu by hearing and then tape-recording an unfamiliar antthrush song (Parker 1983).

Parker's knowledge and skill allowed him to detect an undescribed species by voice and to know that it had to be an undescribed species before he collected it. In 1983 at his Sucusari study site in Peru, Parker heard an unfamiliar *Tolmomyias* flycatcher. Knowing the voices of all other species in the genus, Ted knew that this was certainly a new species (Schulenberg and Parker 1997). Adding to the novelty was that it was not discovered at some remote or previously unvisited locality but on the banks of the Napo River near Iquitos, just the kind of river-bank locality most accessible to, and frequently visited by, ornithologists for more than a century. It was the Amazonian equivalent to discovering a new species of bird from the side of an interstate highway. This discovery by Ted was a focal point of Stap's (1990) book, which received many enthusiastic reviews (e.g., Plimpton 1990).

Parker showed repeatedly that by knowing their voices, species thought to be rare and local were often actually common and widespread, from inconspicuous canopy tyrannids such as White-lored Tyrannulet (*Ornithion inerme*; Parker 1982) to large, unmistakable birds like Nocturnal Curassow (*Nothocrax urumutum*; Parker, in press). He teased us all mercilessly when we presented him our tape-recordings from areas we had visited—"Now," he'd say, "I can find out what was *really* there." One of Ted's most frequent sermons was that any tropical bird species not prone to frequent mist-net capture was much more common than recent literature indicated. All this led to his influential commentary on the overwhelming importance of knowledge of voices in surveying tropical birds (Parker 1991).

2. *The importance of voice in determining taxonomic relationships.* As Parker accumulated an encyclopedic knowledge of the songs and calls of Neotropical birds, the patterns of relationship that they often revealed became evident. For example, the plantcutters were considered a separate and distinctive family, the Phytotomidae, until Parker noted how similar the voices of plantcutters were to those of *Ampelion* cotingas. Subsequently, Lanyon and Lanyon (1989), using biochemical and morphological characters, found that the plantcutters and *Ampelion* were sister taxa. Likewise, Parker predicted from voice that "*Synallaxis*" *gularis* did not belong in that

genus, a result confirmed by biochemical data (Braun and Parker 1985). The "species accounts" sections of his many papers on natural history contain many additional hypotheses on relationships based on voice (e.g., Parker et al. 1985).

At a different taxonomic level, Parker frequently pointed out in conversation that defining species limits by abrupt, dramatic shifts in vocal characters would result in the elevation of many hundreds of subspecies of Neotropical birds to species rank. He was frustrated by the amount of time required to document properly and publish all these examples. It seemed futile to him to spend several months on one such project when he knew of literally hundreds of such examples. Although many such examples were mentioned briefly in his papers, the vast majority of them remained in Parker's head, although he alerted many of his colleagues to them and Parker encouraged them to work out the details (e.g., Kratter 1997).

3. *The importance of foraging behavior in taxonomic relationships.* Parker watched birds with an exceptional eye for detail, particularly with respect to foraging behavior. He synthesized the details of foraging behavior into patterns that he believed often revealed phylogenetic relationships. For example, Ted recognized that the antwren genus *Myrmotherula* with its 30 or more species, contained at least two distinctive species groups: those that persistently foraged in dead leaves suspended above ground (Remsen and Parker 1984), and those that did not. Hackett and Rosenberg's (1990) genetic study indicated that these two groups indeed represented not only different lineages but that these lineages were not each others' closest relatives.

4. *The interrelationship between foraging behavior and habitat selection in understanding a species' biogeography.* Ted discovered not only that many bird species were restricted to the vicinity of bamboo thickets but also that these birds often had specialized foraging behaviors associated with features of their special habitat (Parker 1982). Such compound specialization explained their restricted geographic distributions. Similarly, Parker (1981) pointed out that *Polylepis* woodland, which occurs in tiny relictual patches in the Andes high above cloud-forest elevations, supports several bird species restricted to this woodland, and that some of these species have specialized foraging behaviors or diets that explain this restriction.

5. *The importance of general natural history knowledge.* Ted's involvement with professional ornithology coincided with a period of dramatic increase in the emphasis on theoretical and quantitative aspects of avian ecology and systematics. Although Parker's limited background in mathematics prohibited him from active involvement in such analyses, he was nonetheless deeply interested in conceptual issues. Ted was greatly dismayed, however, by his perception that an increasing proportion of researchers and students devalued general natural history knowledge. He was alarmed by encounters with students and professors who were more interested in which statistical tests were appropriate for their data analyses than in the basic biology of their study organisms. He delighted, therefore, in pointing out errors generated by ignorance of natural history that he found in publications, and how better knowledge of the bird's natural history would have changed their design or interpretation. One mistake that he often ridiculed, in part because it was so prevalent in Neotropical ornithology, was equating differences in mist-net capture rates with differences in relative abundance, without taking into account the radically different movement and spacing patterns that make some species much more prone to capture (Remsen and Parker 1983). Fortunately, it is once again becoming a popular theme that extensive, accurate knowledge of natural history is at the core of good science and good scientists (e.g., Wilson 1994, Noss 1996).

Parker's species accounts in his many papers on natural history of Neotropical birds set the standard for a generation of field ornithologists. When composing our own accounts of little-known species, we modeled them after Ted's accounts and revised them from the viewpoint of "what would Ted say?"

6. *The importance of voice, foraging behavior, and habitat in field identification.* Field identification of tropical birds, especially those of forests, is notoriously difficult. Parker greatly influenced the field skills of a generation of Neotropical field ornithologists by insisting that subtle differences in plumage, criteria that have received increasing attention among experts in the Northern Hemisphere, were nearly useless as field marks in most tropical habitats. He preached instead the importance to field identification of knowing not only voice but also foraging behavior, posture, habitat, and microhabitat—in other words, everything about the way a bird behaves in life, not just a narrow view of "what it looks like." His seminal paper on foliage-gleaner identification (Parker 1979) revolutionized the manner in which many of us approached field identification. As the reigning expert on Neotropical field identification, Parker was besieged by people who sent him detailed descriptions of plumages, accompanied by proposed identifications, often of species that would never occur in the regions, habitats, or microhabitats where

sighted. When such descriptions were accompanied by notes on voice and behavior, Parker usually could point out the correct identification.

7. *Tropical conservation requires rapid assessment programs.* Ted left an indelible imprint on Neotropical conservation. He had three special talents that were the key elements of his influence. First, his comprehensive knowledge of Neotropical birds and his genius for synthesizing this knowledge into patterns gave him a unique overview of their conservation status. Second, his extraordinary ability at bird identification allowed him to survey avifaunas more rapidly and more accurately than anyone previously imagined. Third, Ted was charming. From campesinos living next to parks, to Nobel scientists, to movie stars with funds to donate, to presidents of countries, they all loved talking with him, because they could tell he was well-informed and passionate about his beliefs. Finally, unlike many scientists, Ted was willing to become involved.

These ingredients were all essential to Ted's innovative conservation program, the Rapid Assessment Program (RAP), which he designed and directed for Conservation International from 1989 until his death. The RAP design was to assemble a team of world experts on the field identification of organisms, such as birds, whose conservation status could be assessed and then to survey these groups in the target region over a relatively short time, usually about one month. The RAP team members then analyzed their survey data to evaluate the region's importance to Neotropical conservation (Parker and Bailey 1991; Parker and Carr 1992; Foster et al. 1994; Parker et al. 1993a; Parker et al. 1993b; Parker et al. 1993c). Under RAP protocol, it could take less than a year from when the survey was completed to formal conservation recommendations to those in power. This was indeed rapid compared to traditional approaches.

Consequently, Ted's Rapid Assessment Program rankled some mainstream conservation biologists, who preached that a proper inventory should be more thorough and should include the entire biota, not just a few well-known taxa (see Roberts 1991; Abate 1992). Ted privately labeled this criticism the "what-about-the-nematodes" approach. He of course agreed that a more thorough inventory would be better, and nothing would have pleased him, the-all-around-naturalist, more than a complete survey of the biota, nematodes and all. But Ted knew that tropical habitats would be all but gone by the time traditional inventories were conducted, by the time all those species of nematodes were properly identified, and so he argued for concentrating on those indicator organisms that can be identified most efficiently and whose conservation status can be assessed.

So, the RAP team shrugged off criticism, generated tremendous publicity for conservation in the popular press (e.g., Booth 1990, Germani 1990, Conniff 1991, Wolf 1991, Lipske 1992, Reed 1992, Churchman 1993), and accomplished so much so quickly that soon many other conservation organizations were talking about creating their own RAP versions. A photo of Parker in action for the RAP team has already appeared in at least one general biology text (Solomon and Berg 1995). RAP is clearly Ted's most important legacy for conservation.

That Ted accomplished what he did should inspire all young ornithologists. Ted showed what can be accomplished, with minimal academic credentials, by total dedication to learning and by publishing his findings in technical journals. We can only wonder what he would have accomplished had he had 80 years, instead of 40. Ted Parker has already influenced a generation of young ornithologists here and in Latin America. They are now doing fieldwork in the Parker tradition, and hopefully they can help make up for the 40 years we didn't get. We have tried to capture the essence of Ted Parker's influence by quantifying it and reducing it to facts and categories. We believe, however, that this approach of describing the whole by listing its parts fails to portray the true strength and pervasiveness of his influence in Neotropical ornithology, in part because Parker's charisma, so obvious to those who knew him, cannot be captured adequately by words alone.

Although we have attempted to be as clinical as possible in our assessment of Parker's influence, we cannot end without noting what is well known to all who knew Ted, namely that he was loved by many and considered a close friend by hundreds of ornithologists, birders, naturalists, and conservationists. He gave his time enthusiastically to help and encourage hundreds of people, from prominent scientists to beginning students. Now it remains for those of us influenced by Ted to make the most of what he taught us.

ACKNOWLEDGMENTS

We thank Angelo Capparella, Steve Cardiff, Mario Cohn-Haft, John Hagan, Jaqueline Goerck, Gary Graves, Dan Lane, Susan Lohr, Debby Moskovits, Harold Morrin, John O'Neill, Doug

Pratt, Mark Robbins, Dave Wiedenfeld, and Kevin Zimmer for their valuable contributions to the manuscript.

LITERATURE CITED
(see also Parker bibliography below)

ABATE, T. 1992. Environmental rapid-assessment programs have appeal and critics. BioScience 42:486–489.
ANONYMOUS. 1994. In memory of Ted Parker. LSU Tigers Basketball Magazine (February 5):34.
BATES, J. M., AND T. S. SCHULENBERG. 1997. In memoriam: Theodore A. Parker III, 1953–1993. Auk 114: 110.
BOOTH, W. 1990. 'Now or never' surveys of Earth's diversity. Washington Post (October 1, 1990):A3.
BUDNEY, G. 1994. A legacy for conservation. Birdscope 8:1–2.
CHURCHMAN, D. 1993. RAP team to the rescue. Ranger Rick 27:16–23.
COHEN, S. 1994. The collectors. Washington Post Magazine (January 9):8–11, 24–26, 28.
COLLAR, N. 1995. Ted Parker: A personal memoir. Bird Conserv. International 5:141–144.
CONNIFF, R. 1991. RAP: on the fast track in Ecuador's tropical forests. Smithsonian 22:36–49.
DAVIS, T. J., AND J. P. O'NEILL. 1986. A new species of antwren (Formicariidae: *Herpsilochmus*) from Peru, with comments on the systematics of other members of the genus. Wilson Bull. 98:337–352.
EMANUEL, V. 1993. Remembering Ted. Victor Emanuel Nature Tours Newsletter 16:10–12.
ERWIN, T. L. 1994. Arboreal beetles of tropical forests: the Xystosomi group, subtribe Xystosomina (Coleoptera: Carabidae: Bembidini). Part I. Character analysis, taxonomy, and distribution. Can. Entomol. 126:549–666.
FITZPATRICK, J. W., AND D. F. STOTZ. 1997. A new species of tyrannulet (*Phylloscartes*) from the Andean foothills of Peru and Bolivia. Pp. 37–44 *in* Studies in Neotropical Ornithology Honoring Ted Parker (J. V. Remsen, Jr., Ed.). Ornithol. Monogr. No. 48.
FORSYTH, A. 1994. Ted Parker: in memoriam. Conserv. Biol. 8:293–294.
FOSTER, R. 1995. A campside conversation. Conservation International Member's Rep. (Spring 1995):9–12.
GELL-MANN, M. 1994. The Quark and the Jaguar. W. H. Freeman, New York.
GERMANI, C. 1990. Biologists rate the rain forests. Christian Science Monitor (September 24):14.
GRAVES, G. R. 1981. A new subspecies of Coppery Metaltail (*Metallura theresiae*) from northern Peru. Auk 98:382.
GRAVES, G. R. 1997. Colorimetric and morphometric gradients in Colombian populations of Dusky Antbirds (*Cercomacra tyrannina*), with a description of a new species, *Cercomacra parkeri*. Pp. 21–35 *in* Studies in Neotropical Ornithology (J. V. Remsen, Jr., Ed.). Ornithol. Monogr. No. 48.
HACKETT, S. J., AND K. V. ROSENBERG. 1990. Comparison of phenotypic and genetic differentiation in South American antwrens. Auk 107:473–489.
HURLBERT, K. J. 1994. A tribute to Alwyn H. Gentry. Conserv. Biol. 8:291–292.
JAMMES, L. 1994. Una figura de la ornitología boliviana. Domingo (Santa Cruz de la Sierra, 25 Dec. 1994):2.
KAUFMAN, K. 1993. Theodore A. Parker III, 1953-1993. Amer. Birds 47:349–351.
KRABBE, N., AND T. S. SCHULENBERG. 1997. Species limits and natural history of *Scytalopus* tapaculos (Rhinocryptidae), with descriptions of the Ecuadorian taxa, including three new species. Pp. 47–88 *in* Studies in Neotropical Ornithology Honoring Ted Parker (J. V. Remsen, Jr., Ed.). Ornithol. Monogr. No. 48.
LANYON, S. M., AND W. E. LANYON. 1989. The systematic position of the plantcutters, *Phytotoma*. Auk 106: 422–432.
LIPSKE, M. 1992. Racing to save hot spots of life. National Wildlife (April/May):40–48.
MUTH, D. P. 1995. In memory of Theodore A. Parker III. J. Louisiana Ornithol. 3:i–ii.
MYERS, J. P. 1993. Incomparable, irreplaceable, immortal. Amer. Birds 47:346–347.
NOSS, R. F. 1996. The naturalists are dying off. Conserv. Biol. 10:1–3.
O'NEILL, J. P. 1993. Ted Parker's antwren. Amer. Birds 47:348–349.
PLIMPTON, G. 1990. New bird in birdland. New York Review of Books 37(16):3–4.
PRICE, R. D., AND D. H. CLAYTON. 1995. A new genus and three new species of chewing lice (Phthiraptera: Philopteridae) from Peruvian ovenbirds (Passeriformes: Furnariidae). Proc. Entomol. Soc. Washington 97:839–844.
PULLIAM, H. R. 1975. Coexistence of sparrows: a test of community theory. Science 189:474–476.
REED, S. 1992. Where the wild things are. People (January 20):28–33.
REMSEN, V. 1993. Theodore A. Parker III, ornithologist and conservationist, 1953–1993. Louisiana Ornithol. Society News 154:3–4.
REMSEN, J. V, JR. 1997. A new genus for the Yellow-shouldered Grosbeak. Pp. 89–90 *in* Studies in Neotropical Ornithology Honoring Ted Parker (J. V. Remsen, Jr., Ed.). Ornithol. Monogr. No. 48.
ROBBINS, M. B., G. R. GRAVES, AND J. V. REMSEN, JR. In memoriam: Theodore A. Parker III. Pp. 1–5 *in* Studies in Neotropical Ornithology Honoring Ted Parker (J. V. Remsen, Jr., Ed.), Ornithol. Monogr. No. 48.
ROBBINS, M. B., AND S. N. G. HOWELL. 1995. A new species of pygmy-owl (Strigidae: *Glaucidium*) from the Eastern Andes. Wilson Bull. 107:1–6.
ROBERTS, L. 1991. Ranking the rainforests. Science 251:1559–1560.
SCHMIDT-LYNCH, C. 1993. Ecología de duelo. Sí 337:40–41.

SCHULENBERG, T. S. 1995. A tribute to Ted Parker. Bird Conserv. International 5:137–139.
SCHULENBERG, T. S., AND N. J. COLLAR (EDS.). 1995. In memory of Ted Parker. Bird Conserv. International 5:137–439.
SOLOMON, E. P., AND L. R. BERG. 1995. World of Biology. Fifth edition. Saunders College Publ., Philadelphia, Pennsylvania.
STAP, D. 1990. A Parrot Without a Name: The Search for the Last Unknown Birds on Earth. A. A. Knopf, New York.
STEVENS, W. K. 1993. Biologists' deaths set back plan to assess tropical forests. New York Times (Science Times; August 17):B5, B8.
WILSON, E. O. 1994. Naturalist. Island Press, Washington, D.C.
WOLF, E. C. 1991. Survival of the rarest. World Watch (March/April):12–20.
ZIMMER, K. J. 1993. Ted Parker remembered. Birding 25:377–380.

PUBLICATIONS BY T. A. PARKER, III

1972

PARKER, T., III. 1972. 626 species in one year: a new North American record. Birding 4:6–10.

1973

PARKER, T. 1973a. Nesting season, Southwest Region. Amer. Birds 27:902–905.
PARKER, T. A., III. 1973b. First Snow Bunting specimen for Oklahoma. Bull. Oklahoma Ornithol. Soc. 6:33.
SPEICH, S., AND T. A. PARKER III. 1973. Arizona bird records, 1972. West. Birds 4:53–57.

1974

PARKER, T. 1974a. Fall migration, Southwest Region. Amer. Birds 28:87–90.
PARKER, T. 1974b. Winter season, Southwest Region. Amer. Birds 28:672–676.

1976

KAUFMAN, K., T. A. PARKER III, AND M. ROBBINS. 1976. Notes on the birds of Puerto Los Mazos, Sierra de Autlán, Jalisco. Mexican Birds Newsl. 1(3):12–18.
O'NEILL, J. P., AND T. A. PARKER III. 1976. New subspecies of *Schizoeaca fuliginosa* and *Uromyias agraphia* from Peru. Bull. Brit. Ornithol. Club 96:136–141.
PARKER, T. 1976. On the behavior of the Marvelous Spatuletail *Loddigesia mirabilis*. Birding 8:175.
PARKER, T. A., III, S. HILTY, AND M. ROBBINS. 1976. Birds of El Triunfo cloudforest, Mexico, with notes on the Horned Guan and other species. Amer. Birds 30:779–782.
PARKER, T. A., III, AND J. P. O'NEILL. 1976. An introduction to bird-finding in Peru. Part I. The Paracas Peninsula and Central Highway (Lima to Huanuco city). Birding 8:140–144.
PARKER, T. A., III, AND J. P. O'NEILL. 1976. An introduction to bird-finding in Peru. Part II. The Carpish region of the Eastern Andes along the Central Highway. Birding 8:205–216.

1977

O'NEILL, J. P. AND T. A. PARKER III. 1977. Taxonomy and range of *Pionus "seniloides"* in Perú. Condor 79:274.

1978

O'NEILL, J. P., AND T. A. PARKER III. 1978. Responses of birds to a snowstorm in the Andes of southern Peru. Wilson Bull. 90:446–449.
PARKER, T. A., III, S. A. PARKER, AND M. A. PLENGE. 1978. A checklist of Peruvian birds. Tucson, Arizona.
TALLMAN, D. A., T. A. PARKER III, G. D. LESTER, AND R. A. HUGHES. 1978. Notes on two species of birds previously unreported from Peru. Wilson Bull. 90:445–446.

1979

ELEY, J. W., G. R. GRAVES, T. A. PARKER III, AND D. H. HUNTER. 1979. Notes on *Siptornis striaticollis* (Furnariidae) in Peru. Condor 81:319.
HILTY, S. L., T. A. PARKER III, AND J. SILLIMAN. 1979. Observations on Plush-capped Finches in the Andes with a description of the juvenal and immature plumages. Wilson Bull. 91:145–148.
PARKER, T. A., III. 1979. An introduction to foliage-gleaner identification. Continental Birdlife 1:32–37.
PULLIAM, H. R., AND T. A. PARKER III. 1979. Population regulation of sparrows. Fortschr. Zool. 25:137–147.

1980

PARKER, T. A., III. 1980. Birding the selva of southeastern Peru at Explorer's Inn. Birding 12:221–223.
PARKER, T. A., III, AND J. P. O'NEILL. 1980. Notes on little known birds of the upper Urubamba Valley, southern Peru. Auk 97:167–176.

PARKER, T. A., III, AND S. A. PARKER. 1980. Rediscovery of *Xenerpestes singularis* (Furnariidae). Auk 97: 203–205.

PARKER, T. A., III, J. V. REMSEN, JR., AND J. A. HEINDEL. 1980. Seven new bird species for Bolivia. Bull. Brit. Ornithol. Club 100:160–162.

1981

O'NEILL, J. P., AND T. A. PARKER III. 1981. New subspecies of *Pipreola riefferii* and *Chlorospingus ophthalmicus* from Peru. Bull. Brit. Ornithol. Club 101:294–299.

PARKER, T. A., III. 1981. Distribution and biology of the White-cheeked Cotinga *Zaratornis stresemanni*, a high Andean frugivore. Bull. Brit. Ornithol. Club 101:256–265.

PARKER, T. A., III, AND L. J. BARKLEY. 1981. New locality for the Yellow-tailed Woolly Monkey. Oryx 16: 71–72.

PARKER, T. A., III, AND J. P. O'NEILL. 1981. An introduction to bird-finding in Peru. Part III. Tingo Maria and the Divisoria Mountains. Birding 13:100–106.

SCHULENBERG, T. S., AND T. A. PARKER III. 1981. Status and distribution of some northwest Peruvian birds. Condor 83:209–216.

1982

PARKER, T. A., III. 1982. First record of the Chilean Woodstar *Eulidia yarrellii* in Peru. Bull. Brit. Ornithol. Club 102:86.

PARKER, T. A., III. 1982. Observations of some unusual rainforest and marsh birds in southeastern Peru. Wilson Bull. 94:477–493.

PARKER, T. A., III. 1982. (Species accounts for): genus *Ampelion*. Pp. 55–64 *in* The Snow Cotingas (D. W. Snow, Ed.). Cornell University Press, Ithaca, New York.

PARKER, T. A., III, AND S. A. PARKER. 1982. Behavioural and distributional notes on some unusual birds of a lower montane cloud forest in Peru. Bull. Brit. Ornithol. Club 102:63–70.

PARKER, T. A., III, S. A. PARKER, AND M. A. PLENGE. 1982. An Annotated Checklist of Peruvian Birds. Buteo Books, Vermillion, South Dakota.

REMSEN, J. V., JR., T. A. PARKER III, AND R. S. RIDGELY. 1982. Natural history notes on some poorly known Bolivian birds. Gerfaut 72:77–87.

1983

GRAVES, G. R., J. P. O'NEILL, AND T. A. PARKER III. 1983. *Grallaricula ochraceifrons*, a new species of antpitta from northern Peru. Wilson Bull. 95:1–6.

PARKER, T. A., III. 1983. A record of Blackburnian Warbler (*Dendroica fusca*) for southeastern Brazil. Amer. Birds 37:274.

PARKER, T. A., III. 1983. Rediscovery of the Rufous-fronted Antthrush (*Formicarius rufifrons*) in Peru. Gerfaut 73:287–292.

PARKER, T. A., III. 1983. [Review of]: South American Landbirds, by John S. Dunning. Auk 100:774–776.

REMSEN, J. V., JR., AND T. A. PARKER III. 1983. Contribution of river-created habitats to bird species richness in Amazonia. Biotropica 15:223–231.

1984

PARKER, T. A., III. 1984. Notes on the behavior of *Ramphotrigon* flycatchers. Auk 101:186–188.

PARKER, T. A., III, AND R. A. ROWLETT. 1984. Some noteworthy records of birds from Bolivia. Bull. Brit. Ornithol. Club 104:110–113.

REMSEN, J. V., JR. AND T. A. PARKER III. 1984. Arboreal dead-leaf-searching birds of the Neotropics. Condor 86:36–41.

1985

BRAUN, M. J., AND T. A. PARKER III. 1985. Molecular, morphological, and behavioral evidence concerning the taxonomic relationships of "*Synallaxis*" *gularis* and other synallaxines. Pages 333–346 *in* Neotropical Ornithology (P. A. Buckley, M. S. Foster, E. S. Morton, R. S. Ridgely, and F. G. Buckley, Eds.). Ornithol. Monogr. No. 36.

HARDY, J. W., AND T. A. PARKER III. 1985. (Audio cassette): Voices of the New World thrushes. ARA Records, Florida State Museum, University of Florida, Gainesville.

PARKER, T. A., III. 1985. (Audio cassette): Voices of the Peruvian rainforest. Cornell Laboratory of Ornithology, Ithaca, New York.

PARKER, T. A., III, AND J. P. O'NEILL. 1985. A new species and a new subspecies of *Thryothorus* wren from Peru. Pages 9–15 *in* Neotropical Ornithology (P. A. Buckley, M. S. Foster, E. S. Morton, R. S. Ridgely, and F. G. Buckley, Eds.). Ornithol. Monogr. No. 36.

PARKER, T. A., III, T. S. SCHULENBERG, G. R. GRAVES, AND M. J. BRAUN. 1985. The avifauna of the Huancabamba region, northern Peru. Pages 169–197 *in* Neotropical Ornithology (P. A. Buckley, M. S. Foster, E. S. Morton, R. S. Ridgely, and F. G. Buckley, Eds.) Ornithol. Monogr. No. 36.

ROBBINS, M. B., T. A. PARKER III, AND S. A. ALLEN. 1985. The avifauna of Cerro Pirre, Darién, Panama. Pages 198–232 *in* Neotropical Ornithology (P. A. Buckley, M. S. Foster, E. S. Morton, R. S. Ridgely, and F. G. Buckley, Eds.). Ornithol. Monogr. No. 36.

1987

FJELDSÅ, J., N. KRABBE, AND T. A. PARKER III. 1987. Rediscovery of *Cinclodes excelsior aricomae* and notes on the nominate race. Bull. Brit. Ornithol. Club 107:112–114.

PARKER, T. A., III. 1987. [Foreward] Pages 10–11 *in* The Tanagers (M. L. Isler and P. R. Isler). Smithsonian Institution Press, Washington, D.C.

PARKER, T. A., III, AND J. V. REMSEN, JR. 1987. Fifty-two Amazonian bird species new to Bolivia. Bull. Brit. Ornithol. Club 107:94–107.

SCHULENBERG, T. S., T. A. PARKER III, AND R. A. HUGHES. 1987. First records of Least Tern, *Sterna antillarum*, for Peru. Gerfaut 77:271–273.

1989

PLENGE, M. A., T. A. PARKER III, R. A. HUGHES, AND J. P. O'NEILL. 1989. Additional notes on the distribution of birds in west-central Peru. Gerfaut 79:55–68.

1990

PARKER, T. A., III. 1990. La Libertad revisited. Birding 22:16–22.

REMSEN, J. V., JR., AND T. A. PARKER III. 1990. Seasonal distribution of the Azure Gallinule (*Porphyrula flavirostris*), with comments on vagrancy in rails and gallinules. Wilson Bull. 102:380–399.

TERBORGH, J., S. K. ROBINSON, T. A. PARKER III, C. A. MUNN, AND N. PIERPONT. 1990. Structure and organization of an Amazonian forest bird community. Ecolog. Monogr. 60:213–238.

1991

HARDY, J. W., T. A. PARKER III, AND B. B. COFFEY, JR.. 1991. (Audio cassette): Voices of the woodcreepers. ARA records, Florida State Museum, University of Florida, Gainesville.

PARKER, T. A., III. 1991. On the use of tape recorders in avifaunal surveys. Auk 108:443–444.

PARKER, T. A., III, AND B. BAILEY (EDS.). 1991. A biological assessment of the Alto Madidi region and adjacent areas of northwest Bolivia May 18–June 15, 1990. RAP Working Papers No. 1. Conservation International, Washington, D.C.

PARKER, T. A., III, AND O. ROCHA. 1991. Notes on the status and behaviour of the Rusty-necked Piculet *Picumnus fuscus*. Bull. Brit. Ornithol. Club 111:91–92.

PARKER, T. A., III, M. GELL-MANN, A. CASTILLO U., AND O. ROCHA. 1991. Records of new and unusual birds from northern Bolivia. Bull. Brit. Ornithol. Club 111:120–138.

PARKER, T. A., III, AND O. ROCHA O. 1991. La avifauna de Cerro San Simón, una localidad de campo rupestre aislado en el Depto. Beni, noreste boliviano. Ecología en Bolivia No. 17:15–29.

1992

BATES, J. M., T. A. PARKER III, A. P. CAPPARELLA, AND T. J. DAVIS. 1992. Observations of the *campo, cerrado,* and forest avifaunas of eastern Dpto. Santa Cruz, Bolivia, including 21 species new to the country. Bull. Brit. Ornithol. Club 112:86–98.

COLLAR, N. J., L. P. GONZAGA, N. KRABBE, A. MADROÑO NIETO, L. G. NARANJO, T. A. PARKER III, AND D. WEGE. 1992. Threatened Birds of the Americas. The ICBP/IUCN Red Data Book. Third edition, part 2. International Council Bird Preservation, Cambridge, United Kingdom.

COX, G., J. COX, T. A. PARKER III, AND O. ROCHA O. 1992. Las aves y los mammiferos de Perseverancia. Ecología en Bolivia, Documentos, Ser. Zool. 1:1–15.

ENGLISH, P. H., AND T. A. PARKER III. 1992. (Audio cassette): Birds of Eastern Ecuador. Cornell Laboratory of Ornithology, Ithaca, New York.

PARKER, T. A., III, J. M. BATES, AND G. COX. 1992. Rediscovery of the Bolivian Recurvebill with notes on other little-known species of the Bolivian Andes. Wilson Bull. 104:173–178.

PARKER, T. A., III, AND J. L. CARR (EDS.). 1992. Status of forest remnants in the Cordillera de la Costa and adjacent areas of southwestern Ecuador. RAP Working Papers No. 2. Conservation International, Washington, D.C.

1993

KRATTER, A. W., T. S. SILLETT, R. T. CHESSER, J. P. O'NEILL, T. A. PARKER III, AND A. CASTILLO. 1993. Avifauna of a chaco locality in Bolivia. Wilson Bull. 105:114–141.

PARKER, T. A., III, R. B. FOSTER, L. H. EMMONS, P. FREED, A. B. FORSYTH, B. HOFFMAN, AND B. D. GILL. 1993a. A biological assessment of the Kanuku Mountain region of southwestern Guyana. RAP Working Papers No. 5. Conservation International, Washington, D.C.

PARKER, T. A., III, A. H. GENTRY, R. B. FOSTER, L. H. EMMONS, AND J. V. REMSEN, JR. 1993b. The lowland dry forests of Santa Cruz, Bolivia: a global conservation priority. RAP Working Papers No. 4. Conservation International, Washington, D.C.

PARKER, T. A., III., B. K. HOLST, L. H. EMMONS, AND J. R. MEYER. 1993c. A biological assessment of the Columbia River Forest Reserve, Toledo District, Belize. RAP Working Papers No. 3. Conservation International, Washington, D.C.

WILLIS, E. O., D. W. SNOW, D. F. STOTZ, AND T. A. PARKER III. 1993. Olive-sided Flycatchers in southeastern Brazil. Wilson Bull. 105:193–194.

1994

FOSTER, R. B., T. A. PARKER III, A. H. GENTRY, L. H. EMMONS, A. CHICCHÓN, T. SCHULENBERG, L. RODRÍGUEZ, G. LAMAS, H. ORTEGA, J. ICOCHEA, W. WUST, M. ROMO, J. A. CASTILLO, O. PHILLIPS, C. REYNEL, A. KRATTER, P. K. DONAHUE, AND L. J. BARKLEY. 1994. The Tambopata-Candamo Reserved Zone of southeastern Perú: a biological assessment. RAP Working Papers No. 6, Conservation International, Washington, D.C.

PARKER, T. A., III. 1994. Habitat, behavior, and spring migration of Cerulean Warbler in Belize. Amer. Birds 48:70–75.

1995

HARDY, J. W., T. A. PARKER III, AND B. B. COFFEY, JR. (Audio cassette): Voices of the woodcreepers. Second Edition. ARA Records, Florida Museum of Natural History, University of Florida, Gainesville.

PARKER, T. A., III, T. S. SCHULENBERG, M. KESSLER, AND W. H. WUST. 1995. Natural history and conservation of the endemic avifauna of north-west Peru. Bird Conserv. Int. 5:201–231.

REMSEN, J. V., JR., AND T. A. PARKER III. 1995. Bolivia has the opportunity to create the planet's richest park for terrestrial biota. Bird Conserv. Int. 5:261–293.

1996

HARDY, J. W., T. A. PARKER III, AND T. TAYLOR. (Audio cassette): Voices of the toucans. ARA Records, Florida Museum of Natural History, University of Florida, Gainesville.

STOTZ, D. F., J. W. FITZPATRICK, T. A. PARKER III, AND D. K. MOSKOVITS. 1996. Neotropical birds: Ecology and Conservation. University of Chicago Press, Chicago.

1997

HARDY, J. W., AND T. A. PARKER III. 1997. The nature and probable function of vocal copying in Lawrence's Thrush, *Turdus lawrencii*. Pages 307–320 *in* Studies in Neotropical Ornithology honoring Ted Parker (J. V. Remsen, Jr., Ed.). Ornithol. Monogr. No. 48.

KRATTER, A. W., AND T. A. PARKER III. 1997. Relationship of two bamboo-specialized foliage-gleaners: *Automolus dorsalis* and *Anabazenops fuscus* (Furnariidae). Pages 383–397 *in* Studies in Neotropical Ornithology honoring Ted Parker (J. V. Remsen, Jr., Ed.). Ornithol. Monogr. No. 48.

O'NEILL, J. P., AND T. A. PARKER III. 1997. New subspecies of *Myrmoborus leucophrys* (Formicariidae) and *Phrygilus alaudinus* (Emberizidae) from the upper Huallaga Valley, Peru. Pages 485–491 *in* Studies in Neotropical Ornithology honoring Ted Parker (J. V. Remsen, Jr., Ed.). Ornithol. Monogr. No. 48.

OREN, D. C., AND T. A. PARKER III. 1997. Avifauna of the Tapajós National Park and vicinity, Amazonian Brazil. Pages 493–525 *in* Studies in Neotropical Ornithology honoring Ted Parker (J. V. Remsen, Jr., Ed.). Ornithol. Monogr. No. 48.

PARKER, T. A., III. 1997. Bird species recorded at three sites on the northern and western slopes of the Cordillera del Cóndor. Pages 168–179 *in* The Cordillera del Cóndor region of Ecuador and Peru: a biological assessment (T. S. Schulenberg and K. Awbrey, Eds.). Rapid Assessment Program Working Papers No. 7. Conservation International, Washington, D.C.

PARKER, T. A., III, AND J. M. GOERCK. 1997. The importance of national parks and biological reserves to bird conservation in the Atlantic forest region of Brazil. Pages 527–541 *in* Studies in Neotropical Ornithology honoring Ted Parker (J. V. Remsen, Jr., Ed.). Ornithol. Monogr. No. 48.

PARKER, T. A., III, D. F. STOTZ, AND J. W. FITZPATRICK. 1997. Notes on avian bamboo specialists in southwestern Amazonian Brazil. Pages 543–547 *in* Studies in Neotropical Ornithology honoring Ted Parker (J. V. Remsen, Jr., Ed.). Ornithol. Monogr. No. 48.

PARKER, T. A., III, AND E. O. WILLIS. 1997. Notes on three tiny grassland flycatchers, with comments on the disappearance of South American fire-diversified savannas. Pages 549–555 *in* Studies in Neotropical Ornithology honoring Ted Parker (J. V. Remsen, Jr., Ed.). Ornithol. Monogr. No. 48.

ROBBINS, M. B., AND T. A. PARKER III. 1997. What is the closest living relative of *Catharopeza* (Parulinae)? Pages 595–599 *in* Studies in Neotropical Ornithology honoring Ted Parker (J. V. Remsen, Jr., Ed.). Ornithol. Monogr. No. 48.

ROBBINS, M. B., AND T. A. PARKER III. 1997. Voice and taxonomy of *Caprimulgus rufus otiosus* (Caprimulgidae), with a reevaluation of *Caprimulgus rufus* subspecies. Pages 601–607 *in* Studies in Neotropical Ornithology honoring Ted Parker (J. V. Remsen, Jr., Ed.). Ornithol. Monogr. No. 48.

SCHULENBERG, T. S., AND T. A. PARKER III. 1997. Notes on the Yellow-browed Toucanet *Aulacorhynchus huallagae*. Pages 717–720 *in* Studies in Neotropical Ornithology honoring Ted Parker (J. V. Remsen, Jr., Ed.). Ornithol. Monogr. No. 48.

SCHULENBERG, T. S., AND T. A. PARKER III. 1997. A new species of tyrant-flycatcher (Tyrannidae: *Tolmomyias*) from the western Amazon Basin. Pages 723–731 *in* Studies in Neotropical Ornithology honoring Ted Parker (J. V. Remsen, Jr., Ed.). Ornithol. Monogr. No. 48.

ZIMMER, K. J., T. A. PARKER III, M. L. ISLER, AND P. R. ISLER. 1997. Survey of a southern Amazonian avifauna: the Alta Floresta region, Mato Grosso, Brazil. Pages 887–918 *in* Studies in Neotropical Ornithology honoring Ted Parker (J. V. Remsen, Jr., Ed.). Ornithol. Monogr. No. 48.

Female and male (top) of *Cercomacra parkeri* sp. nov. and females of *C. tyrannina crepera* (Panama, lower left) and *C. t. tyrannina* (northern Colombia, lower right). From a painting by Larry McQueen.

COLORIMETRIC AND MORPHOMETRIC GRADIENTS IN COLOMBIAN POPULATIONS OF DUSKY ANTBIRDS (*CERCOMACRA TYRANNINA*), WITH A DESCRIPTION OF A NEW SPECIES, *CERCOMACRA PARKERI*

GARY R. GRAVES

Department of Vertebrate Zoology, National Museum of Natural History, Smithsonian Institution, Washington, D.C. 20560, USA

ABSTRACT.—Two morphologically similar antbirds, *Cercomacra tyrannina* and *C. parkeri* sp. nov., are sympatric in the foothills of the Cordillera Occidental and Cordillera Central of the Colombian Andes. The two taxa appear to replace each other elevationally: *C. tyrannina*—sea level to ca. 730 m; *C. parkeri*—ca. 1,130 to 1,830 m. Colombian populations of *C. tyrannina* exhibit significant clines in plumage color and size that are consistent with the predictions of Gloger's and Bergmann's ecogeographic rules.

RESUMEN.—Dos hormigueros morfológicamente similares, *Cercomacra tyrannina* y *C. parkeri* sp. nov. son sympatrica en las lomas de las Cordilleras Occidentales y Central de las Andes Colombianas. Las dos taxa probablemente se reemplazan a diferentes niveles: *C. tyrannina*—nivel de mar ca 730 m; *C. parkeri*—ca. 1,130 to 1,830 m. La población de *C. tyrannina* Colombiana demuestra significante clines en el color de plumaje y tamaño que es consistente con las reglas ecogeográficas de Gloger y Bergmann.

The Dusky Antbird (*Cercomacra tyrannina*) is a polytypic species that ranges from southern Mexico through Central America to eastern Brazil (Peters 1951). Recently I discovered that the series of *Cercomacra tyrannina* from Colombia examined by Chapman (1917), Meyer de Schauensee (1950), and Wetmore (1972) were actually composed of two regionally sympatric taxa, the widespread *C. tyrannina* (including *C. t. tyrannina* and *C. t. rufiventris*) of the lowlands and a distinctive, undescribed highland taxon.

This paper focuses on the geographic variation, taxonomy, and systematics of *C. tyrannina* species complex in Colombia west of the Andes. First, I present a formal description of the new taxon based on female plumage and morphology. A second objective is to analyze patterns of geographic and elevational variation in plumage color and morphology in the trans-Andean populations of *Cercomacra tyrannina*.

MATERIALS AND METHODS

I quantitatively evaluated the external morphology and plumage color of museum specimens of adults of the *Cercomacra tyrannina* species complex from trans-Andean populations in Colombia. Measurements of wing chord, tail length (from point of insertion of central rectrices to tip of longest rectrix), tarsus length, bill length (from anterior edge of nostril), and bill width (at anterior edge of nostril) were made with digital calipers to the nearest 0.1 mm.

Color of the plumage of the crown, upper breast, and center of back was determined with a reflectance spectrophotometer (Colorscan, Hunter Laboratories) equipped with a 12-mm sample port. Each color measurement was independently replicated three times and averaged for analysis. Specimens with ruffled or worn body plumage were omitted. Geographic variation in plumage color of *Cercomacra tyrannina* is more pronounced in females than in males, a widespread trend among antbirds (Hellmayr 1929). For this reason, the spectrophotometric evaluation of color was restricted to females. Brown and rufescent plumage is susceptible to post-mortem color change over time. However, nearly all females used in this study were collected by M. A. Carriker, Jr., and K. von Sneidern in the 1940's, and so this limits potential biases caused by differences of post-mortem change. [The diagnosis of males is uncertain at present and will be addressed in a future paper.]

Colorimetric characters were described in terms of opponent-color coordinates (L, a, b) (Hunter and Harold 1987). The "LAB" system is based on the hypothesis that signals from the cone receptors in the human eye are coded by the brain as light-dark (L), red-green (a), and yellow-blue (b). The rationale is that a color cannot be red and green or yellow and blue at the same time. Therefore, "redness" and "greenness" can be expressed as a single value a, which is positive if the color is red, negative if green. Likewise, "yellowness" or "blueness" is expressed by b for yellows and $-b$ for blues. The third coordinate L, ranging from 0 to 100, describes the "lightness" of color; low values are dark, high values are light. For example, consider the opponent-color coordinates of breast color for two hypothetical specimens: (1) $L = 40$, $a = 20$, $b = 40$; and (2) $L = 50$, $a = 15$, $b = 35$. The breast of the first specimen is darker and more saturated with red and yellow than that of the second specimen.

Geographic coordinates of collecting localities were taken from Paynter and Traylor (1981). Elevational data were recorded from specimen labels whenever possible, otherwise from Paynter and Traylor.

I used principal components analysis (PCA) to reduce the dimensionality of data and to facilitate the analysis of color and morphology in two dimensions. Color analyses were performed on correlation matrices derived from untransformed variables, whereas morphology was analyzed with log-transformed variables and covariance matrices (Wilkinson 1989). Mensural differences between the taxa were evaluated with t-tests. When appropriate, α-levels (0.05) were adjusted for the number of simultaneous tests. Least-squares regression lines were projected on bivariate scatterplots for heuristic purposes, not for hypothesis testing. Clinal variation of plumage color and morphology in Colombia was illustrated by plots derived from trend-surface analyses of raw and principal components data (SURFER, Golden Software, 1987).

Cercomacra parkeri sp. nov.
Parker's Antbird

Holotype.—National Museum of Natural History (USNM), No. 436469; adult "laying" female from La Bodega, on the north side of the Río Negrito on the road from Sonsón to Nariño, ca. 5,800 ft. [1,768 m], on the eastern slope of the Cordillera Central, depto. Antioquia, Colombia; collected 16 June 1951 by M. A. Carriker, Jr.; original number 20461.

Diagnosis.—*Cercomacra parkeri* females are most similar to those of *C. tyrannina crepera* of Central America, but differ as follows: (1) auriculars of *C. parkeri* brownish-gray instead of rufescent; and (2) eyering and lores gray or brownish-gray rather than brown or rufescent. The white intrascapular patch, variable in size in *C. t. crepera*, is absent or vestigial in female *parkeri*. Male *C. parkeri* tend to have olivaceous rather than gray flanks but cannot be distinguished by plumage color from all specimens of *C. t. tyrannina*. Mean tail length of *C. parkeri* is significantly longer than that of Colombian populations of *C. t. tyrannina* (Table 1).

Description of holotype.—All qualitative color comparisons were made under Examolites (Macbeth Corp.). Crown gray with faint olive-brown tint becoming progressively more olive-brown on the mantle, scapulars, lower back, rump, and upper tail coverts (Table 2). Contrasting intrascapular patch absent. Outer webs of wing coverts, alula, remiges, and rectrices dull olive brown. Concealed inner webs of remiges dark grayish-brown. Lores and anterior half of eyering pale gray. Preorbital bristles black. Sides of neck and distal two-thirds of auricular feathers gray, concolor with the crown; rachi of auriculars basally white or pale buffy-white. Underparts from chin to undertail coverts rich rufescent, darkest and most saturated on the upper breast. Opponent-color coordinates (L, a, b) based on the average of three independent measurements: crown (27.1, 1.3, 7.4); back (27.2, 2.3, 10.7); and breast (46.3, 16.2, 39.4) (compare with Table 2). Soft part colors of dried specimen: upper mandible brownish-black; lower mandible horn-colored, brownish-black near the tomia; feet and tarsi, brownish-black.

Measurements of holotype (mm).—Wing (60.6), tail (61.7), tarsus (21.3), bill length (9.7), bill width (4.6).

Distribution.—Females are known from the western slope of the Cordillera Occidental and northern and eastern slopes of the Cordillera Central of the Colombian Andes (3,700 to 6,000 ft [1,128 to 1,829 m]) (Fig. 1). Males provisionally assigned to *C. parkeri* on the basis of elevation (collected above 1,000 m) occur on the western slope of the Cordillera Oriental, depto. Santander, in addition to known localities for females.

Etymology and dedication.—This taxon is named for my friend and brother-in-arms, Theodore A. Parker III (1953–1993), who put his life on the line for the cause of nature conservation. Those of us privileged to have known him witnessed the ascension of a savant—the most talented field ornithologist of our generation, and perhaps the best in history.

TABLE 1

Ranges, Means (± SD), and Sample Sizes of Selected Measurements (mm) of Colombian Populations of *Cercomacra tyrannina* and *C. parkeri*. Male Specimens Were Assigned to Species Based on Elevation. Significant *P*-values (Two-tailed t-test), by Sex, between *C. parkeri* and *C. tyrannina*, Were Adjusted for the Number of Simultaneous Tests, 0.05/10 = 0.005: * = $P < 0.001$

Characters	*parkeri* ♂	*parkeri* ♀	*tyrannina* ♂	*tyrannina* ♀
Wing chord	61.7–66.8 64.2 ± 1.6 n = 16	58.1–63.0 60.8 ± 1.4 n = 11	60.8–66.5 63.7 ± 1.6 n = 46	56.2–64.4 60.3 ± 1.9 n = 47
Tail length	57.0–69.8 * 63.3 ± 3.9 n = 14	58.9–64.7 * 60.6 ± 2.0 n = 10	51.6–63.6 57.1 ± 2.9 n = 45	50.0–62.6 55.1 ± 3.3 n = 38
Tarsus length	21.1–24.1 22.9 ± 0.9 n = 17	21.1–23.8 22.4 ± 0.8 n = 11	21.0–23.9 22.6 ± 0.8 n = 45	20.6–23.4 22.0 ± 0.6 n = 45
Bill length	9.8–11.7 10.6 ± 0.5 n = 16	9.7–11.2 10.4 ± 0.6 n = 10	10.3–12.5 11.0 ± 0.5 n = 39	9.7–11.4 10.5 ± 0.5 n = 41
Bill width	4.0–5.0 4.7 ± 0.3 n = 17	4.1–4.8 * 4.4 ± 0.2 n = 11	4.3–5.4 4.7 ± 0.3 n = 46	4.3–5.4 4.8 ± 0.2 n = 46

Specimens examined (Colombia).—Cercomacra parkeri (females). Depto. Antioquia: La Bodega (USNM holotype); La Frijolera (AMNH 133482, USNM 256136 [formerly AMNH 133483]); Valdivia (USNM 402341, 402347, 402348). Depto. Bolívar: Volador (USNM 398011). Depto. Caldas: Hacienda Sofía (USNM 436463, 436464); La Selva (ANSP 158180). Depto. Valle del Cauca: "Salencio" on Novita trail (AMNH 111924).

Males provisionally assigned to *C. parkeri* on the basis of elevation (collected above 1,000 m). Depto. Antioquia: Botero (USNM 426347, 426348); La Bodega (USNM 436470); La Frijolera (AMNH 133484); Valdivia (USNM 402342, 402343, 402344, 402345, 402346, 402349). Depto. Caldas: La Selva (ANSP 158181); La Sofía (USNM 436466, 436467, 436468). Depto. Risaralda: Pueblorrico (ANSP 158179). Depto. Santander: Virolín, 6,000 ft. (USNM 373602). Depto. Valle del Cauca: Palmira (AMNH 108924, 108925).

Cercomacra tyrannina (females, sample size in parentheses). Depto. Antioquia: El Pescado

TABLE 2

Spectrophotometric Measurements (Range, Mean, and Standard Deviation) of Opponent-Color Coordinates (*L*, *A*, *B*) of Female *Cercomacra parkeri* and Trans-Andean Colombian Populations of *C. tyrannina*. Significant *P*-values (two-tailed t-test) Were Adjusted for Number of Simultaneous Tests 0.05/9 = 0.0056: * = $P < 0.001$; ** = $P < 0.0001$

		Color characters		
		Lightness *L*	Red *a*	Yellow *b*
parkeri (n = 11)	Crown	25.3–28.9 27.4 ± 1.3	1.3–1.9 ** 1.6 ± 0.2	5.9–8.2 ** 6.8 ± 0.9
	Back	26.3–29.9 * 28.1 ± 1.1	1.7–2.6 2.1 ± 0.3	6.3–10.7 * 8.0 ± 1.3
	Breast	40.0–49.0 45.2 ± 3.0	15.9–18.8 17.3 ± 1.1	36.7–43.0 39.8 ± 1.8
tyrannina (n = 47)	Crown	24.3–32.6 29.1 ± 2.0	1.8–4.3 2.8 ± 0.6	8.4–16.8 11.8 ± 1.6
	Back	26.2–35.0 30.5 ± 2.1	1.7–3.5 2.5 ± 0.5	5.8–15.3 10.7 ± 2.2
	Breast	37.9–55.8 47.5 ± 4.4	12.0–20.1 16.5 ± 2.0	34.2–42.8 39.2 ± 2.1

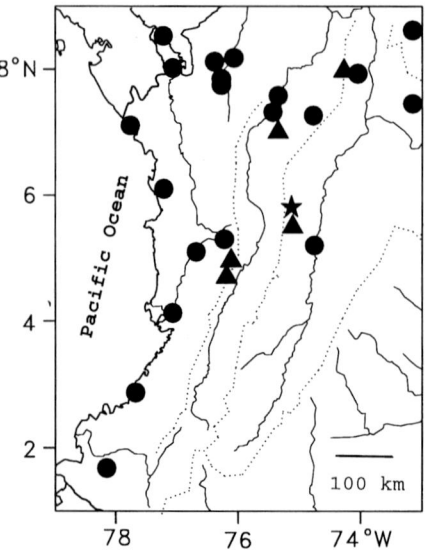

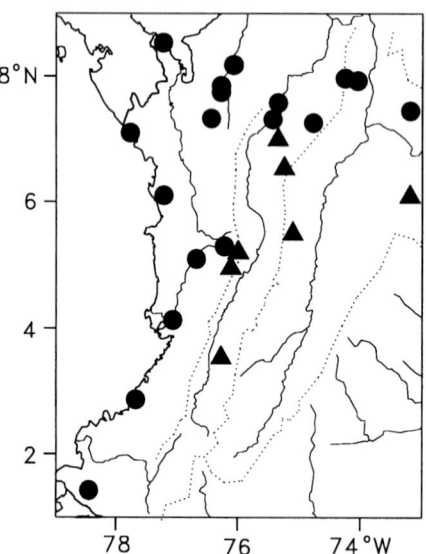

Fig. 1. Distribution of examined specimens of *Cercomacra tyrannina* (circles) and *C. parkeri* (triangles) in Colombia: (top) females; (bottom) males provisionally assigned to species by elevation—*C. tyrannina* (sea level to 1,000 m) and *C. parkeri* (above 1,000 m). The type locality of *C. parkeri*, La Bodega, depto. Antioquia, is indicated by a star. Some symbols represent two localities. Dotted lines approximate the divides of the three Andean cordilleras. Degrees of north latitude and west longitude appear along the side and bottom.

(2); Hacienda Belén (4); Tarazá (1); Puerto Valdivia (2). Depto. Antioquia/Cordoba: Quimarí (1). Depto. Cordoba: Quebrada Salvajín (2); Socorré (3); Tierralta (4). Depto. Bolívar: Santa Rosa (6). Depto. Cauca: Río Saija (1). Depto. Chocó: Acandi (3); Alto del Buey (3); Andagoya (2); Río Jurado (3); Unguía (1). Depto. Nariño: Barbacoas (1). Depto. Norte de Santander: Bellavista (1). Depto. Risaralda: Santa Cecilia (1). Depto. Santander: Hacienda Santana (4). Depto. Tolima: Honda (1). Depto. Valle del Cauca: Punto Muchimbo (2).

Males assigned to *C. tyrannina* by elevation (sea level to 1,000 m). Depto. Antioquia: El

Pescado (1); Hacienda Belén (4); Tarazá (1); Valdivia (3); Villa Arteaga (1). Depto. Bolívar: Puerto Nuevo (2); Santa Rosa (6); Volador (2,500 ft) (1). Depto. Cauca: Rio Saija (1). Depto. Chocó: Acandí (3); Alto del Buey (1); Andagoya (1); Río Jurado (3). Depto. Cordoba: Quebrada Salvajín (3); Socorré (3); Tierralta (2). Depto. Nariño: Guayacana (3). Depto. Risaralda: Santa Cecilia (2). Depto. Santander: Hacienda Santana (5). Depto. Valle del Cauca: Punto Muchimbo (4).

Large series of *C. tyrannina* from Middle America (Mexico, Honduras, Costa Rica, Panama), Ecuador, Venezuela, and Brazil were also examined, as well as smaller series of *C. laeta, C. serva, C. manu, C. cinerascens, C. nigricans,* and *C. nigrescens.*

Behavior, vocalizations, and ecology.—Unknown.

RESULTS

I addressed the relationship of *C. parkeri* and *C. tyrannina* along the lines broached by the following questions.

(1) Are patterns of plumage color and morphological variation within *tyrannina* and *parkeri* different?
(2) Does *parkeri* represent the end point of a phenotypic continuum that corresponds to elevational gradients?
(3) Do environmental or physical barriers separate the elevational ranges of the two taxa?
(4) Should *parkeri* be recognized as a subspecies of *C. tyrannina* or as a separate species?

PLUMAGE COLOR

Color correlates among plumage regions.—Colorimetric values from the three plumage regions (crown, back, breast) of female *C. tyrannina* specimens were significantly correlated (Table 3). For example, lightness of the crown, back, and breast were positively correlated. Redness of plumage regions was also positively correlated. By comparison, only the crown and back exhibited significantly correlated values for yellow. Lightness values were uncorrelated or negatively correlated with the saturation of red and yellow. In other words, darker specimens of *C. tyrannina* were more richly colored.

None of the color correlates for *C. parkeri* (Table 3) were significant when P-values were adjusted for the number of simultaneous tests ($n = 36$). Nevertheless, the relationship among colorimetric characters in *C. parkeri* seems to differ substantially from those exhibited in *C. tyrannina*. For example, when the signs of the correlation coefficients for *C. parkeri* and *C. tyrannina* (Table 3) were compared, only 16 of 36 possible pairwise correlations had the same sign for both species—about the number that would be expected by chance. Most significantly, the lightness of the back and redness of the crown were positively correlated in *C. parkeri,* whereas lightness and redness of the dorsal plumage (crown and back) of *C. tyrannina* were negatively correlated.

Plumage color differences between C. tyrannina *and* C. parkeri.—The crown of *C. parkeri* was significantly less saturated with red and yellow than in *C. tyrannina,* whereas the back of *C. parkeri* was both lighter and less yellow than that of *C. tyrannina* (Table 2). Most significantly, values for yellowness of the crown of the two species did not overlap, unlike breast color. A bivariate plot of factor scores for plumage color confirms the difference between the two taxa (Fig. 2, Table 4).

Geographic variation of C. tyrannina.—Annual rainfall ranges from approximately 1,500 mm in the middle Cauca and Magdalena valleys to more than 8,000 mm in the headwaters of the Atrato and San Juan rivers on the Pacific slope of the Cordillera Occidental (see Haffer 1967, 1975; Fig. 1.3 in Instituto Geográphico "Agustin Codazzi" 1982). In general, annual rainfall increases westward along transects drawn from the lower Río Magladena to the Gulf of Urabá on the Caribbean coast and the depto. Chocó on the Pacific coast.

Female plumage color ($n = 47$ individuals) appeared to vary along rainfall gradients. Trans-Andean populations are currently divided into two subspecies, *C. t. tyrannina* and *C. t. rufiventris.* Wetmore (1972: 190) noted that *C. t. rufiventris* was "merely a transition from the northern *crepera* [which ranges from southern Mexico to central Panama] to typical *C. tyrannina tyrannina,* found through most of Colombia, southern Venezuela and northwestern Brazil . . . occasional specimens in the area assigned to *rufiventris* are as pale as the average *tyrannina.*" Spectrophotometric measurements confirm his observations; *C. t. tyrannina* and *C. t. rufiventris* specimens are combined in the following analyses.

Lightness of the back ($r^2 = 0.18$, $P < 0.01$) and breast ($r^2 = 0.09$, $P < 0.05$), but not the

TABLE 3

Pearson Correlation Coefficients of Opponent-color Coordinate/Plumage Characters of Females, with *Cercomacra parkeri* above the Diagonal, *C. tyrannina* below the Diagonal. Significant P-values for Each Species Were Adjusted for Number of Simultaneous Tests, $0.05/36 = 0.0013$: * < 0.001; ** $P < 0.0001$

		(1)	(2)	(3)	(4)	(5)	(6)	(7)	(8)	(9)
(1) Crown	L	—	0.41	0.01	0.30	0.11	−0.22	0.11	0.23	−0.30
(2) Back	L	0.47*	—	0.53	0.75	0.25	−0.23	0.04	−0.28	0.44
(3) Breast	L	0.47*	0.53*	—	0.36	0.41	−0.55	0.06	0.10	0.73
(4) Crown	a	−0.14	−0.53*	−0.37	—	−0.11	−0.08	0.50	−0.37	0.29
(5) Back	a	−0.27	−0.42	−0.20	0.69**	—	−0.18	−0.45	0.04	0.33
(6) Breast	a	−0.43	−0.45	−0.64**	0.49	0.34	—	−0.13	−0.73	0.13
(7) Crown	b	0.15	−0.26	−0.10	0.77**	0.47*	0.38	—	0.18	−0.09
(8) Back	b	−0.08	−0.52*	−0.02	0.57**	0.76**	0.28	0.58**	—	−0.46
(9) Breast	b	0.02	0.13	0.41	−0.05	−0.03	0.33	0.20	0.20	—

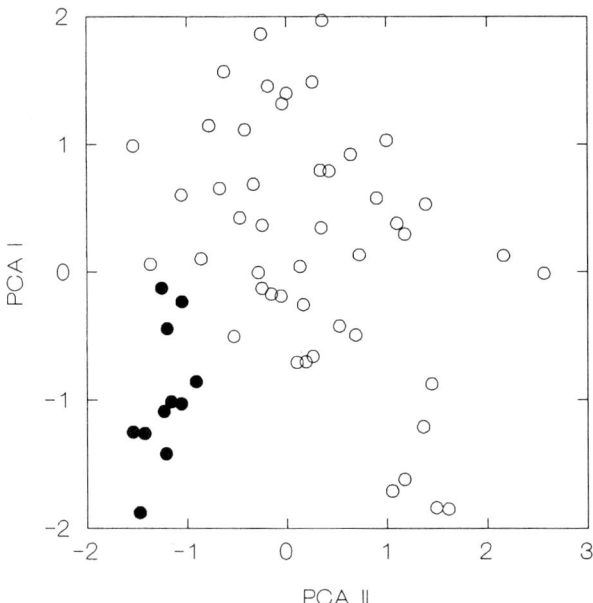

FIG. 2. Bivariate plot of factor scores (PCA I and II) from a principal components analysis of plumage color in female *Cercomacra tyrannina* and *C. parkeri* (see Tables 2, 3, and 4).

crown ($r^2 = 0.06$, $P > 0.10$), was negatively correlated with longitude, which in turn appears to be roughly correlated with rainfall. Saturation of dorsal (crown and back) and ventral (breast) plumage with red was positively correlated with longitude: crown ($r^2 = 0.33$, $P < 0.0001$); back ($r^2 = 0.39$, $P < 0.0001$); and breast ($r^2 = 0.14$, $P < 0.01$). Saturation of yellow in dorsal, but not the ventral plumage, also was positively correlated with longitude: crown ($r^2 = 0.22$, $P < 0.001$); back ($r^2 = 0.22$, $P < 0.001$); and breast ($r^2 = 0.01$, $P > 0.50$).

Specimens from the Pacific coast in the Chocó region and the Gulf of Urabá are darker and significantly more saturated in pigments expressed as red and yellow (Fig. 3). Clinal gradients are steepest near the base of the northern tip of the Central Andean Cordillera (compare Fig. 1 and Fig. 3). These data are conveniently summarized by PCA I (Fig. 4), which was positively correlated with longitude ($r^2 = 0.26$, $P < 0.001$). PCA 2 was uncorrelated with longitude ($r^2 = 0.00$, $P > 0.87$).

Only lightness and yellow of the breast ($r^2 = 0.46$, $P < 0.0001$) were positively correlated

TABLE 4

FACTOR LOADINGS FROM A PRINCIPAL COMPONENTS ANALYSIS OF OPPONENT-COLOR COORDINATES (L, A, B) OF FEMALE *Cercomacra parkeri* AND TRANS-ANDEAN COLOMBIAN POPULATIONS OF *C. tyrannina* (SEE FIG. 2)

		PCA Axes	
		I	II
Crown	lightness L	0.10	0.79
	red a	0.92	0.06
	yellow b	0.84	0.35
Back	lightness L	−0.14	0.80
	red a	0.84	−0.10
	yellow b	0.87	0.07
Breast	lightness L	−0.09	0.83
	red a	0.29	−0.73
	yellow b	−0.00	0.09
Percent variance explained		34.9	29.4

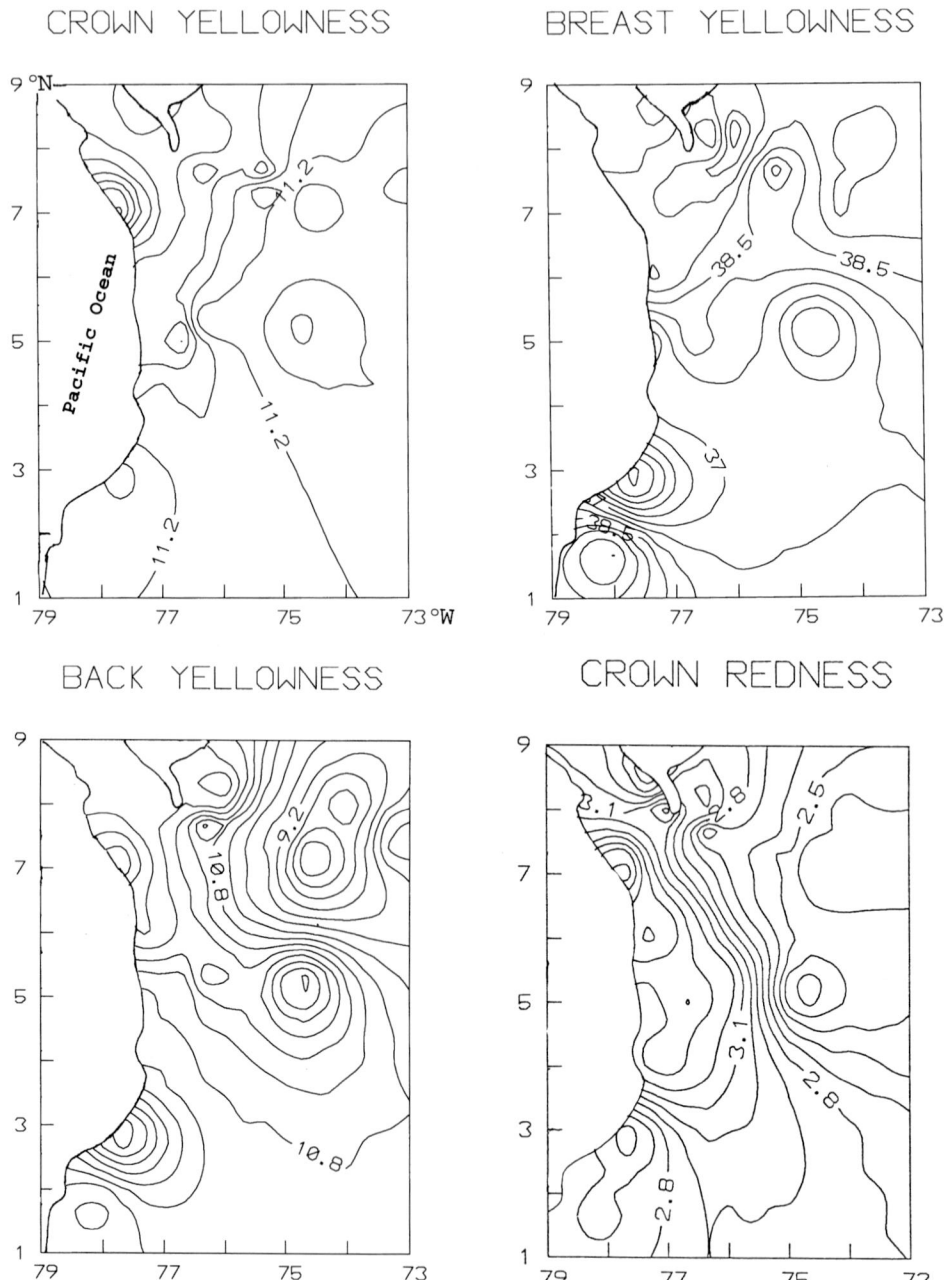

FIG. 3. Geographic variation of crown redness, and yellowness of crown, back, and breast of female *C. tyrannina* in Colombia (see Table 2). Note the difference in isoline orientation between dorsal and ventral yellowness. Degrees of north latitude and west longitude appear along the side and bottom.

with latitude. Color clines of the dorsal and ventral plumage thus appeared to be somewhat orthogonal to one another. PCA 1 ($r^2 = 0.03$, $P > 0.25$) and PCA 2 ($r^2 = 0.02$, $P > 0.42$) were uncorrelated with latitude (Fig. 3). In sum, geographic variation in female plumage of *C. tyrannina* is consistent with the expectations of Gloger's ecogeographic rule, which posits that populations in more humid environments tend to be more heavily pigmented.

Elevational variation of C. tyrannina *and C.* parkeri.—Although the elevational range of *C. tyrannina* is narrow in Colombia (sea level to 915 m in the Serranía del Baudo), significant

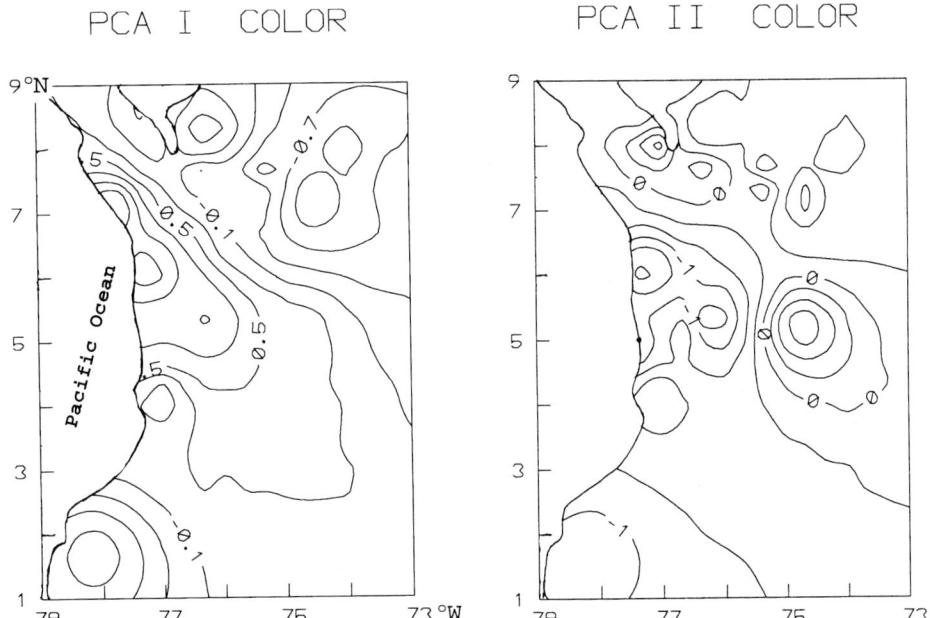

FIG. 4. Trend surface analysis of factor scores from a principal components analysis of plumage color of female *C. tyrannina*. Degrees of north latitude and west longitude appear along the side and bottom.

elevational clines in plumage color were observed. Two of the nine possible color/plumage characters of female *C. tyrannina* were negatively correlated with elevation: yellowness of the crown ($r^2 = 0.11$, $P < 0.03$) and back ($r^2 = 0.10$, $P < 0.03$), although these probabilities fall above the adjusted *P*-value when all characters are considered simultaneously (Table 5). In general (8 of 9 correlation coefficients were negative), plumage color was less rufescent at higher elevations.

Perhaps owing to small sample size ($n = 11$) of female *C. parkeri*, none of the colorimetric variables were significantly correlated with elevation (Table 5). However, the relationship between elevation and plumage coloration in *C. parkeri* differed qualitatively from that in *C. tyrannina* (Fig. 5). Only two of nine correlation coefficients had the same sign for both species (Table 5)—a result not expected by chance (binomial test, $P < 0.01$). Collectively, these data demonstrate the distinctiveness of *C. parkeri*.

TABLE 5

PEARSON CORRELATION COEFFICIENTS OF THE RELATIONSHIP BETWEEN COLORIMETRIC VARIABLES AND ELEVATION FOR FEMALE *C. parkeri* AND *C. tyrannina*. NONE OF THE P-VALUES WAS SIGNIFICANT WHEN ADJUSTED FOR THE NUMBER OF SIMULTANEOUS TESTS FOR EACH SPECIES $\alpha = 0.05/9 = 0.0056$

		parkeri	*tyrannina*
Crown	lightness (*L*)	0.14	−0.14
	red (*a*)	−0.08	−0.19
	yellow (*b*)	−0.20	−0.33
Back	lightness (*L*)	0.21	−0.05
	red (*a*)	0.61	−0.21
	yellow (*b*)	0.53	−0.32
Breast	lightness (*L*)	0.43	−0.11
	red (*a*)	−0.46	0.07
	yellow (*b*)	0.13	−0.07

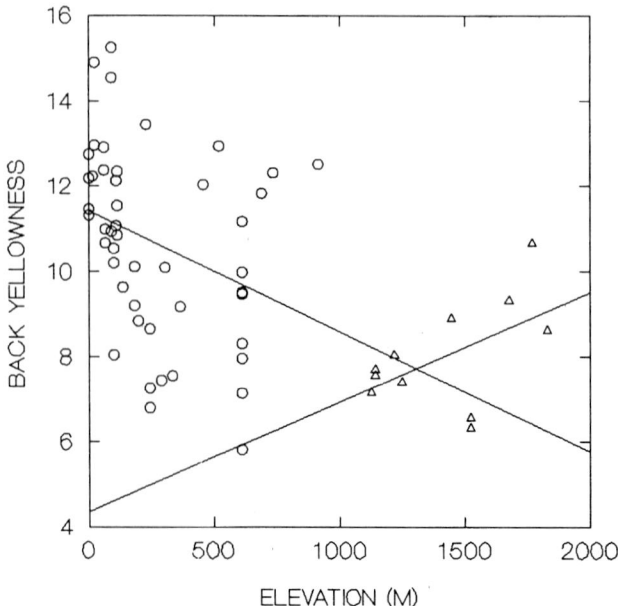

Fig. 5. Relationship of the "yellowness" of back plumage of female *C. tyrannina* (circles) and *C. parkeri* (triangles) with elevation. Least-squares regression lines for the two species are illustrated for comparison.

MORPHOLOGY

Morphological correlates within subspecies.—The correlative relationships among morphological variables of female *C. tyrannina* and *C. parkeri* (Table 6) differ significantly. Seven of ten correlation coefficients for pairs of variables of the two species had the same sign. Wing chord/tarsus and wing chord/bill length correlations were significantly positive for *C. tyrannina* but strongly negative for *C. parkeri*. These differences reflected a fundamental dissimilarity in morphological shape.

Morphological differences between C. tyrannina *and* C. parkeri.—From pooled samples of specimens (Table 1), the tail of male and female *C. parkeri* was significantly longer than that of *C. tyrannina,* and the bill width of female *C. parkeri* was significantly narrower.

Geographic variation of C. tyrannina.—Although this study focused on a fraction of the species' geographic range, females exhibited significant patterns of morphological variation in Colombia (Table 7). Wing chord and tail length were significantly correlated with longitude. Factor scores (PCA 1) derived from a principal components analysis of wing chord, tail length, and tarsus length were significantly correlated with longitude ($n = 36$, $r^2 = 0.34$, $P < 0.0002$) and uncorrelated with latitude. The size of females increased from the Pacific slope eastward to the upper Magdalena Valley (Fig. 6).

PCA I factor scores derived from morphology (wing, tail, tarsus) were significantly correlated with PCA I scores derived from the analysis of nine plumage color variables ($n = 36$, $r^2 = 0.17$,

TABLE 6

PEARSON CORRELATION COEFFICIENTS OF MORPHOLOGICAL MEASUREMENTS OF FEMALES. *C. parkeri* (N = 9–11) ABOVE THE DIAGONAL, *C. tyrannina* (N = 33–46) BELOW THE DIAGONAL: * = $P < 0.005$, ** = $P < 0.0005$

	Wing	Tail	Tarsus	Bill length	Bill width
Wing	—	0.21	−0.57	−0.34	0.02
Tail	0.35	—	−0.41	−0.27	0.18
Tarsus	0.51**	−0.09	—	0.35	0.02
Bill length	0.44*	0.03	0.42	—	0.37
Bill width	0.06	0.22	0.05	0.16	—

TABLE 7
PEARSON CORRELATION COEFFICIENTS FOR THE RELATIONSHIP BETWEEN MORPHOLOGICAL MEASUREMENTS AND LATITUDE AND LONGITUDE FOR FEMALE *C. tyrannina*. SAMPLE SIZE IN PARENTHESES. SIGNIFICANCE: * = $P < 0.01$, ** = $P < 0.005$.

	Latitude	Longitude
Wing	0.03 (47)	* −0.39 (47)
Tail	0.19 (38)	** −0.61 (38)
Tarsus	−0.18 (45)	−0.17 (45)
Bill length	−0.26 (41)	0.11 (41)
Bill width	−0.28 (46)	0.22 (46)

$P < 0.024$). This suggests that both morphology and plumage color are labile and respond adaptively to environmental gradients.

Elevational variation of C. tyrannina *and* C. parkeri.—Two of five morphological characters of female *C. tyrannina* were positively correlated with elevation (Table 8). Populations inhabiting the Andean foothills and the Serranía del Baudo, depto. Chocó, had longer wings and tails and presumably greater body masses than lowland populations. In contrast, body size of *C. parkeri* (wing, tail, tarsus) was uncorrelated with elevation. However, bill length was negatively correlated with elevation (Fig. 7). These data reinforce the conclusions drawn from the analyses of plumage color—that *C. parkeri* was morphologically distinct and did not represent the end point of an elevational cline of *C. tyrannina*.

DISCUSSION

Subspecies of Cercomacra tyrannina *in Colombia*.—Patterns of geographic variation in color and morphology among Colombian females of *C. tyrannina* are complex and defy simple taxonomic categorization. In general, populations from the drier interior were paler in color and larger than those from coastal localities. However, trend-surface projections demonstrated that clinal patterns varied among the chosen color characters, often discordantly. Should the smaller, darker Pacific slope populations, "*T. c. rufiventris*," continue to be recognized taxonomically? Although color clines appear to be steepest near the northern end of the Cordillera Central, where does one draw the dividing line between *rufiventris* and nominate *tyrannina*? Any such division would be arbitrary at best. I agree with Haffer and Fitzpatrick (1985) that subspecies

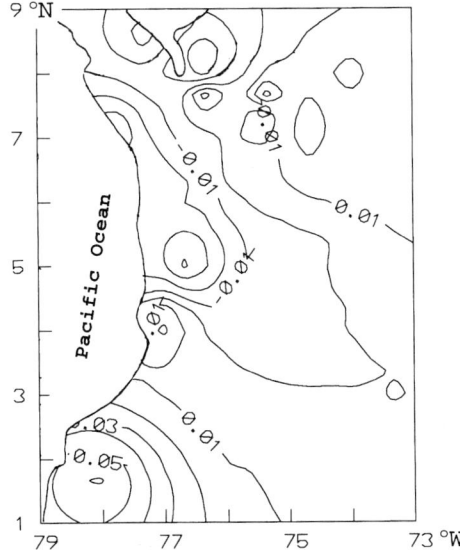

FIG. 6. PCA I factor scores from an analysis of body size of female *C. tyrannina*. Degrees of north latitude and west longitude appear along the side and bottom.

TABLE 8

Pearson Correlation Coefficients for the Relationship between Morphological Measurements and Elevation for Female *C. parkeri* and *C. tyrannina*. Sample Size in Parentheses. Significance: * = $P < 0.05$; ** = $P < 0.005$

	parkeri	tyrannina
Wing	0.03 (11)	** 0.43 (46)
Tail	0.21 (10)	** 0.34 (38)
Tarsus	0.32 (11)	0.02 (44)
Bill length	* −0.74 (10)	0.13 (40)
Bill width	0.21 (11)	0.13 (45)

names should be applied only to populations that exhibit relatively uniform character "plateaus" separated by zones of rapid phenotypic change. Applying trinomials to a series of populations whose phenotypes imperceptibly grade into one another denies statistical reality. In the absence of well-defined subdivision between coastal and interior populations, I recommend that *C. t. rufiventris* (Lawrence) be placed in the synonymy of *C. t. tyrannina* (Sclater).

The roughly parallel clines in size and color of female *C. tyrannina* that corresponded to rainfall gradients suggest ecophenotypic adaptation, although environmental induction also may be partially responsible (James 1983). Color variation in this species is the best-documented case of Gloger's rule in the Neotropics. Size variation is subtle, but the statistically significant trend toward larger birds from drier interior locations and at higher elevations is consistent with the expectations of Bergmann's ecogeographic rule (see James 1970; Murphy 1985).

Is C. parkeri *a species or a subspecies of* C. tyrannina?—In contrast to the clinal variation among lowland populations, the distinction between highland (*C. parkeri*) and lowland (*C. tyrannina*) "dusky" antbirds is discrete. The intraspecific relationship of morphological and color characters differed between the two taxa, as did the relationship between plumage and morphological variables and elevation. No evidence of plumage intermediacy between the two taxa was found.

A narrow elevational hiatus may occur between the two species. *C. tyrannina* has been recorded up to 730 m in the Andean foothills, well below the range of *C. parkeri* (ca. 1,130–1,830 m). Females of both species have been collected along elevational transects at two localities (Table 9). Based on M. A. Carriker's specimens, catalogs (1947–1948), and field maps (deposited in USNM), elevational gaps of 885 m and 518 m, respectively, were recorded between *C. parkeri* and *C. tyrannina* along the El Pescado-Valdivia and Santa Rosa-Volador transects. Carriker did not distinguish between the two taxa in his field catalog and collected specimens of both opportunistically. In other words, he made no special effort to obtain "dusky antbirds" at higher elevations, perhaps because they were relatively common throughout the humid lowlands of northern Colombia. Thus, the actual width of the elevational hiatus at the two localities in the late 1940's could have been much narrower or nonexistent. In any event, there is no evidence of physical or environmental barriers between the two taxa. Ecological, behavioral, and biochemical data are needed to conclusively determine the specific relationship of *parkeri*. However, specimen data and elevational distributions indicate that *parkeri* is a biological species rather than a subspecies of *C. tyrannina*. I cautiously suggest that the burden of proof lies with those who would disagree.

Within the genus *Cercomacra*, at least three taxa appear to be restricted to the lower slopes of the Andes: (1) *C. parkeri* from Colombia; (2) *C. nigrescens aequatorialis* (Zimmer 1931) from eastern Ecuador; and (3) *C. nigrescens notata* (Zimmer 1931), including *C. n. jelashtei* (Carriker 1933), from eastern Peru. The elevational limits of a fourth taxon, *C. tyrannina vicina* (Todd 1927), are imperfectly known, but it may be restricted to the foothills of the eastern slope of the Cordillera Oriental in Colombia and the Andes of Venezuela.

Zimmer (1931) noted that *Cercomacra nigrescens aequatorialis* and *C. n. notata* were more similar in coloration to one another than either taxon was to *C. n. fuscicauda*, of the adjacent lowlands. Whether or not the lowland and highland populations of *C. nigrescens* are in genetic contact is unknown. However, the finding of discrete differences between *C. parkeri* and *C. tyrannina* suggests the possibility that a similar situation may occur between highland and lowland taxa of *C. nigrescens*.

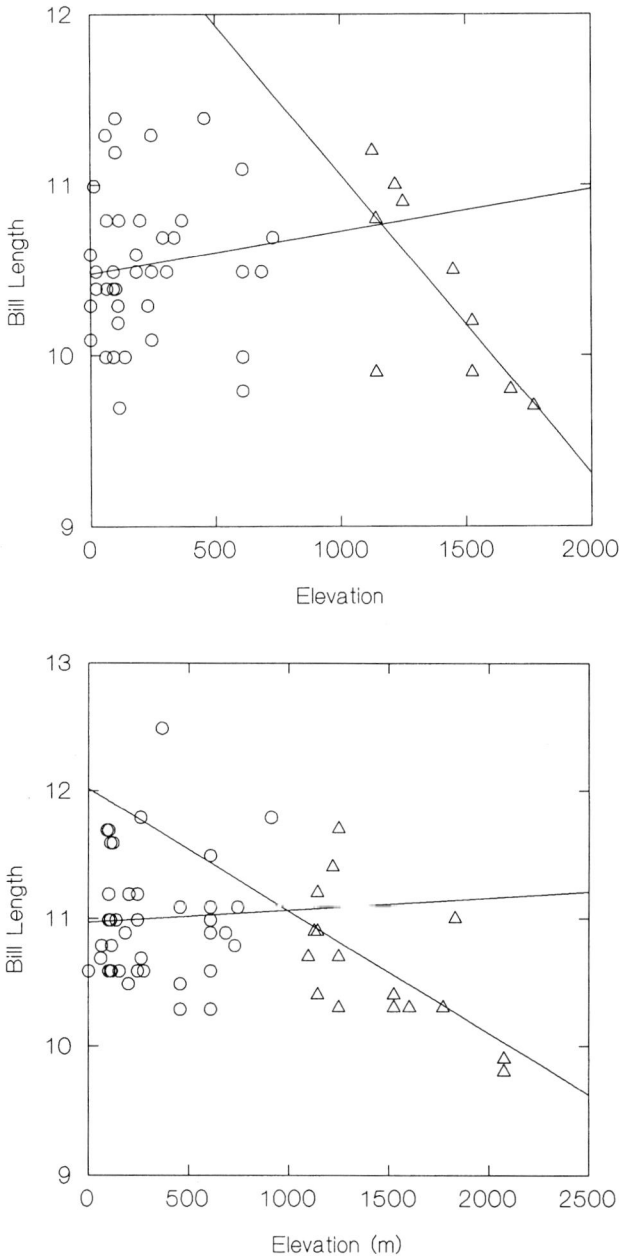

FIG. 7. Relationship of bill length and elevation for *C. tyrannina* (circles) and *C. parkeri* (triangles): females (top); males (bottom). Note that the slopes of the least-squares regression lines were of different sign for the two taxa. Intercepts and slopes for females were $10.48 + 0.00025 \times$ elev. for *tyrannina* and $12.82 - 0.00175 \times$ elev. for *parkeri*; for males, $10.97 + 0.00009 \times$ elev. for *tyrannina* and $12.02 - 0.00096 \times$ elev. for *parkeri*.

EPILOGUE

Speciation.—Although our knowledge of cryptic sibling species in the Andes is rudimentary (e.g., Graves 1987; Arctander and Fjeldså 1994; Whitney 1994), allopatric distributions may be the rule among recently evolved, geminate taxa (Mayr 1963). Because of the relatively short distances between the lowlands and the crest of the Andes, genetic isolation and differentiation

TABLE 9
LOCALITIES (ELEVATIONS) WHERE FEMALE *C. parkeri* AND *C. tyrannina* HAVE BEEN COLLECTED ALONG THE SAME ELEVATIONAL TRANSECT

Transects	
El Pescado-Valdivia, depto. Antioquia	Santa Rosa-Volador, depto. Bolívar
tyrannina	*tyrannina*
El Pescado	Santa Rosa
198 m (USNM 402360)	610 m (USNM 392645)
335 m (USNM 402361)	(USNM 392815)
	(USNM 392875)
Puerto Valdivia	(USNM 392876)
110 m (AMNH 133489)	(USNM 392878)
	(USNM 398012)
parkeri	*parkeri*
Valdivia (above Sevilla)	Volador
1220 m (USNM 402341)	1128 m (USNM 398011)
1250 m (USNM 402347)	
1448 m (USNM 502348)	
La Frijolera	
1524 m (AMNH 133482)	
(USNM 256136)	

of *C. parkeri* (vis-a-vis *C. tyrannina*) is unlikely to have occurred along elevational gradients in the Colombian Andes (Graves 1988). A likely scenario is that *parkeri* evolved in allopatry in the lowlands, perhaps in the Cauca Valley, and was later displaced by *tyrannina* to mid-elevation foothills of the Andes after secondary contact.

Conservation.—*Cercomacra parkeri*, like many other Andean forest birds (Graves 1985, 1988), has a limited geographic distribution and elevational range. These factors translate to small population sizes and vulnerability to extinction. Mid-elevational forest in Colombia is rapidly disappearing (Hilty 1985), and most remaining forest patches are second growth or have been degraded by selective cutting. Because *Cercomacra* antbirds occur frequently in second growth, tangled thickets, forest edge, and stream-side vegetation, *C. parkeri* may be better off than primary forest obligates.

Summation and the future.—Avian speciation in the Andes and adjacent lowlands, especially the ecological and evolutionary processes that permit the co-existence of sister taxa, remain *terra incognita* despite more than a century of study. In a field in which narrative models are the norm, few hypotheses regarding differentiation and speciation have been tested quantitatively. The study of geographic variation in Neotropical birds also has a long tradition, but, again, only a few papers have quantitatively explored variation in morphology or plumage color or the ecological or evolutionary implications of character gradients. Why? Not only are pertinent data difficult to obtain but also the few ornithologists specializing on the systematics, taxonomy, and evolutionary biology of Neotropical birds have been reluctant to embrace quantitative methods. As a consequence, the poverty of our collective knowledge of quantitative patterns of geographic variation and their relationship to environmental gradients and speciation is overwhelming.

Ted Parker did not spend his time performing statistical analyses of character gradients or formulating null models of mixed-species flocks, but more importantly he collected data and observations that made my research on these topics possible. That was part of his genius. In a letter written many years before his death, Ted expressed his philosophy and work ethic: " . . . as long as I can see and hear and write I'll be ready for any kind of field work." He was an incredible resource of unique and unpublished data on Neotropical birds. Ted amazed his colleagues and inspired generations of students. Now, we go it alone.

In homage to his spirit, we must abandon the comfort of our academic offices for the wild parts of the planet while there is still time and rededicate our efforts to discover new facts—new information about the ecology, behavior, genetics, morphology, systematics, and distributions of birds. In final tribute to Ted's life, we must then do meaningful things with what we have learned.

ACKNOWLEDGMENTS

I thank Steve Cardiff, Mario Cohn-Haft, Andy Kratter, Van Remsen, Tom Schulenberg, and Doug Stotz for helpful comments, and Mort Isler for preparing the maps. Mary Baker of the Conservation Analytical Laboratory, Smithsonian Institution, graciously allowed access to the spectrophotometer.

I am grateful to the curators and staff of the Academy of Natural Sciences (ANSP), American Museum of Natural History (AMNH), Carnegie Museum of Natural History, Field Museum of Natural History, and the Museum of Natural Science, Louisiana State University, for the loan of specimens, and to the National Museum of Natural History (USNM) for support. Javier Piedra kindly translated the abstract for this paper.

Lastly, I thank Larry McQueen for his beautiful frontispiece. Ted would have loved it.

LITERATURE CITED

ARCTANDER, P., AND J. FJELDSÅ. 1994. Andean tapaculos of the genus *Scytalopus* (Aves, Rhinocryptidae): a study of speciation using DNA sequence data. Pp. 205–225 *in* Conservation Genetics (V. Loeschcke, J. Tomiuk, and S. K. Jain, Eds.). Birkhauser Verlag, Basel, Switzerland.

CARRIKER, M. A., JR. 1933. Descriptions of new birds from Peru, with notes on other little-known species. Phil. Acad. Nat. Sci. Philadelphia 85:1–38.

CHAPMAN, F. M. 1917. The distribution of bird-life in Colombia. Bull. Amer. Mus. Nat. Hist. 36:1–729.

GRAVES, G. R. 1985. Elevational correlates of speciation and intraspecific variation in plumage in Andean forest birds. Auk 102:556–579.

GRAVES, G. R. 1987. A cryptic new species of antpitta (Formicariidae: *Grallaria*) from the Peruvian Andes. Wilson Bull. 99:313–321.

GRAVES, G. R. 1988. Linearity of geographic range and its possible effect on the population structure of Andean birds. Auk 105:47–52.

HAFFER, J. 1967. Speciation in Colombian forest birds west of the Andes. Amer. Mus. Novitates 2294:1–57.

HAFFER, J. 1975. Avifauna of northwestern Colombia, South America. Bonner Zoologische Monographien No. 7.

HAFFER, J., AND J. W. FITZPATRICK. 1985. Geographic variation in some Amazonian forest birds. Pp. 147–168 *in* Neotropical Ornithology (P. A. Buckley, M. S. Foster, E. S. Morton, R. S. Ridgely, and F. G. Buckley, Eds.), Ornithol. Monogr. No. 36.

HELLMAYR, C. E. 1929. On hetergynism in formicarian birds. Pp. 41–70 *in* Festschrift Ernst Hartert. J. Orn.

HILTY, S. L. 1985. Distributional changes in the Colombian avifauna: a preliminary blue list. Pp. 992–1004 *in* Neotropical Ornithology (P. A. Buckley, M. S. Foster, E. S. Morton, R. S. Ridgely, and F. G. Buckley, Eds.). Ornithological Monogr. No. 36.

HUNTER, R. S., AND R. W. HAROLD. 1987. The Measurement of Appearance. 2nd edition. Wiley, New York.

JAMES, F. C. 1970. Geographic size variation in birds and its relationship to climate. Ecology 51:365–390.

JAMES, F. C. 1983. Environmental component of morphological differences in birds. Science 221:184–186.

MEYER DE SCHAUENSEE, R. 1950. The birds of the Republic of Colombia. Caldasia 5:645–871.

MAYR, E. 1963. Animal Species and Evolution. Harvard University Press, Cambridge, Massachusetts.

MURPHY, E. C. 1985. Bergmann's rule, seasonality, and geographic variation in body size in House Sparrows. Evolution 39:1327–1334.

PAYNTER, R. A., JR., AND M. A. TRAYLOR, JR. 1981. Ornithological Gazetteer of Colombia. Harvard University, Cambridge, Massachusetts.

PETERS, J. L. 1951. Check-list of Birds of the World. Vol. 7. Museum of Comparative Zoology, Cambridge, Massachusetts.

TODD, W. E. C. 1927. New gnateaters and antbirds from tropical America, with a revision of the genus *Myrmeciza* and its allies. Proc. Biol. Soc. Wash. 40:149–177.

WETMORE, A. 1972. The birds of the Republic of Panama. Part 3. Passeriformes: Dendrocolaptidae (Woodcreepers) to Oxyruncidae (Sharpbills). Smithsonian Misc. Coll. Vol 150, Part 3.

WHITNEY, B. 1994. A new *Scytalopus* tapaculo (Rhinocryptidae) from Bolivia, with notes on the other Bolivian members of the genus and the *magellanicus* complex. Wilson Bull. 106:585–614.

WILKINSON, L. 1989. SYSTAT: The system for statistics. Evanston, Illinois, Systat Inc.

ZIMMER, J. T. 1931. Studies of Peruvian birds. I. New and other birds from Peru, Ecuador, and Brazil. Amer. Mus. Notitates 500:1–23.

A NEW SPECIES OF TYRANNULET (*PHYLLOSCARTES*) FROM THE ANDEAN FOOTHILLS OF PERU AND BOLIVIA

JOHN W. FITZPATRICK[1,2] AND DOUGLAS F. STOTZ[1,3]

[1]*Division of Birds, Field Museum of Natural History, Roosevelt Road at Lake Shore Drive, Chicago, Illinois 60605, USA;*
[2]*present address: Cornell Laboratory of Ornithology, 159 Sapsucker Woods Rd., Ithaca, NY 14850, USA;*
[3]*present address: Environmental and Conservation Programs, Field Museum of Natural History, Roosevelt Rd. at Lake Shore Dr., Chicago, IL 60605, USA*

ABSTRACT.—We describe a new species of tyrant flycatcher (Tyrannidae) based on nine specimens from the forested Andean foothills of southeastern Peru and one specimen from adjacent northern Bolivia. The new form appears most closely related to *Phylloscartes flaviventris*. We cite evidence that the latter species is endemic to the coast range of northern Venezuela. We hypothesize that these two sister-species are more distantly related to a third, *P. superciliaris*, which has a relictual distribution in midmontane forests of Middle America and the northern Andes. The new species is an obligate member of mixed-species flocks and forages high in the canopy of intact primary forest. In Peru it inhabits moist forest between 650 and 1,200 m elevation; elevational range may be somewhat lower in Bolivia, but even there the species is confined to the foothills. Conservation of large tracts of these vanishing forests is essential for the persistence of this and many other species narrowly restricted to the Andean foothills.

RESUMEN.—Decribimos una especie nueva de atrapamoscas (Tyrannidae), conocida de nueve ejemplares de los bosques húmedos premontanos del sudeste de Perú, y un ejemplar del norte de Bolivia. El pariente mas cercano de la nueva especie es *Phylloscartes flaviventris*, que es endémico a las montañas de la costa venezolano. Nuestra hipótesis es que *Phylloscartes superciliaris* es el pariente más cercano a esta par de especies. *Phylloscartes superciliaris*, también, tiene una patrón de distribución de tipo-reliquia en los bosques de elevaciones medias en América Central y los Andes al norte. La nueva especie es un miembro obligado de bandadas mixtas, y forrajea en las copas de bosque primario. En Perú, la distribución elevacional está limitada entre 650 y 1,200 m. La distribución elevacional en Bolivia puede ser un poco más baja, pero todavía la especie está limitada a los bosques premontanos. Es esencial la conservación de áreas grandes de estos bosques, que son desapareciendo, para la persistencia de esta especie y muchas otras, que son especialistas dependientes de los bosques premontanos en los Andes.

In 1899 the German collector Otto Garlepp obtained a small, yellowish tyrant flycatcher near the town of Marcapata, in the forested Andean foothills of southeastern Peru (male; depto. Cuzco, 1,000 m; collector's number 1398; Mus. Berlepsch no. 43968). Hellmayr (1927: 349) correctly predicted that this specimen would "doubtless prove to be an undescribed form," noting several ways in which it differed from the Venezuelan endemic, "*Pogonotriccus*" (= *Phylloscartes*) *flaviventris*. In 1972 John Terborgh and John Weske mist-netted, photographed, and released an unidentified tyrannid (see frontispiece), at their "Camp 0" on the Cordillera Vilcabamba, depto. Apurímac, Peru (685 m). They suspected this to be the same form as Garlepp's specimen but did not encounter the species again (Terborgh, pers. comm.).

In 1981 JWF located and collected several specimens of this same species in the lower Cosñipata Valley of southeastern Peru, and in 1982 DFS collected a single male near Puerto Bermudez, depto. Pasco, Peru. We later discovered the species to be a regular member of mixed-species canopy flocks in foothill forests throughout southeastern Peru. Theodore A. Parker III also encountered the species at several localities in the foothills of southern Peru and in depto. La Paz, Bolivia, and in 1995 Bret Whitney, Alan Perry and Marcelo Hinojosa secured one specimen from depto. Beni, Bolivia.

TABLE 1
Mean Linear Measurements[1] (mm ± s.d.) of Three "Rufous-lored" Species in the Genus *Phylloscartes*

	parkeri		*flaviventris*		*superciliaris*	
	Male (*n* = 6)	Female (*n* = 3)[2]	Male (*n* = 2)	Female (*n* = "1")[3]	Male (*n* = 8)	Female (*n* = 4)
Wing	52.9 ± 0.9	48.5 ± 2.3	53.5	47.7	56.8 ± 3.3	53.6 ± 1.3
Tail	48.4 ± 1.1	45.0 ± 1.4	50.8	46.0	53.3 ± 3.5	53.2 ± 1.6
Tarsus	16.2 ± 0.2	15.2 ± 0.8	17.6	16.0	17.1 ± 0.6	17.1 ± 0.3
Culmen	12.1 ± 0.1	11.3 ± 0.3	12.2	11.7	12.7 ± 2.0	12.9 ± 0.3
Bill tip	6.5 ± 0.4	6.0 ± 0.0	6.6	6.3	6.7 ± 0.5	7.6 ± 0.2
Bill width	3.7 ± 0.4	3.9 ± 0.1	4.0	4.0	4.0 ± 0.2	3.7 ± 0.2

[1] Culmen measured from base to tip, bill tip measured from anterior edge of nares to tip, bill width measured at anterior edge of nares.
[2] Measurements of female *parkeri* exclude one immature specimen, noticeably smaller than the others.
[3] An unsexed specimen of *P. flaviventris* (AMNH 498860) presumed female on basis of small size.

In Peru the new species occupies a narrow elevational zone in tall, upper tropical forest, mainly between 650 and 1,200 m. This spectacular ecological setting was among Ted Parker's favorite haunts. We hope that the attention Ted helped bring to the diverse and seriously threatened Andean foothills will continue to further their long-term protection.

Phylloscartes parkeri sp. nov.
Cinnamon-faced Tyrannulet

Holotype.—Field Museum of Natural History (FMNH) 315959; male (skull 20% pneumatized) collected on ridge above Hacienda Amazonia, depto. Madre de Dios, Peru, 12°57'S, 71°11'W, elevation 1,050 m; collected and prepared 23 November 1983 by John W. Fitzpatrick (field catalog # 83-075).

Diagnosis (see frontispiece).—Similar in plumage and size to *P. flaviventris* of northern Venezuela (Table 1), but differs as follows: cinnamon-rufous lores and eyering darker and lores slightly wider; crown grayish-green contrasting with green nape and back, not dark brownish olive nearly matching back; auricular patch pale cinnamon-buff, narrowly bordered with blackish, not extensively black with only a few buffy feathers; chin and throat pale yellowish-white, not bright yellow; yellow breast and flanks suffused and indistinctly streaked with olive, not clear bright yellow throughout; outer webs of greater and median secondary coverts edged and broadly tipped pale yellow forming wingbars with indistinct proximal edges, not wingbars clearly demarcated against all-black coverts; mandible black, not pale flesh at base. Differs from *P. superciliaris* by being slightly smaller (Table 1) and in many details of plumage: in *parkeri*, underparts yellow, not pale gray; tips to secondary coverts and innermost remiges pale yellow contrasting with rest of wing, not light green and concolor with rest of wing; auricular patch cinnamon-buff diffusely edged brownish black, not largely white with solid black border forming a complete and distinct ring connecting with the black malar stripe; crown grayish-green contrasting with green nape and back, not uniform slate gray with the gray extending posteriorly onto nape.

Description of holotype.—Dorsum and rump Olive-Green (Color #46; capitalized color names and numbers from Smithe 1975). Entire crown, except narrow band on lores and forecrown, distinctly grayer than back, closest to Glaucous (#79) but lightly suffused with olive. Lores, eyering, and forecrown rich Cinnamon-Rufous (#40), nasal tufts creamy white. Faint postocular streak white. Distinct "facial patch" consists of broad auricular patch pale cinnamon-buff, nearest Cinnamon (#39), bordered above and posteriorly by broad, dusky band and, along malar region, by narrow and indistinct dusky band. Chin whitish, faintly tinged pale yellow laterally and grading into darker yellow throat. Breast and flanks broadly but indistinctly streaked dull Olive-Green; midabdomen and undertail coverts clear, pale Spectrum Yellow (#55). Thighs olive. Wings generally dusky to blackish, but edges of secondaries bright Yellowish Olive-Green (#50) to Olive-Yellow (#52); greater and median secondary coverts broadly tipped and narrowly edged pale Olive-Yellow forming two distinct wingbars. Primary coverts blackish, narrowly edged Olive-Green matching dorsum. Outer web of innermost secondary broadly edged pale yellowish-white, this broadening distally to form conspicuous pale spot on closed wing; traces of this terminal spot also present on adjacent four secondaries. Rectrices dusky, edged same Yellowish

Olive-Green as on secondaries. Six long rictal bristles on both sides of gape, the longest 8.5 mm. Soft parts in life: iris dark brown, bill entirely black, tarsus and feet medium gray. No molt evident.

Measurements of holotype (mm).—Wing chord 51.5; central rectrices 49.0; tarsus 16.2; culmen, from base 12.0, from anterior edge of nares 6.0; bill width at anterior edge of nares 3.2. Mass 8.1 g. Testes < 1 mm, not enlarged.

Geographic distribution.—Known from foothill forests of the eastern Andes in central and southeastern Peru and adjacent northern Bolivia. The northernmost specimen is from depto. Pasco (Puellas, km 41 on Villa Rica-Puerto Bermudez highway; Louisiana State University Museum of Zoology 106326), where considered uncommon by Schulenberg et al. (1984; listed as *P. flaviventris* in that publication). T. A. Parker also found the species in depto. Huánuco (Parker et al. 1991), probably at La Divisoria (T. S. Schulenberg, pers. comm.), about 180 km NNW of the Pasco locality. Farther south, where primary forest remains at appropriate elevations, we have found *parkeri* common in the foothills of the Peruvian departments of Cuzco and Madre de Dios. Parker found the species in similar foothill forest at Cerros del Távera, depto. Puno (Parker and Wust 1994). The single Bolivian specimen is from depto. La Paz, Prov. Franz Tamayo, near San Jose de Uchupiamonas, above Río Tuichi and below Serranía de Eslabon, in the Parque Nacional Madidi (14°16'S, 68°6'W by gps, fide B. Whitney). The southernmost record comes from the Serranía Pilon, depto. Beni, at 875 m (Parker et al. 1991, uncataloged tape recordings deposited at Cornell Library of Natural Sounds, LNS).

Elevational distribution.—All Peruvian specimens and all sight records by the authors in Peru are between 650 and 1,200 m. However, V. Emmanuel and T.A. Parker III found the new species in forest along the Río Urubamba at 1900 m below Machu Picchu, depto. Cuzco, in July 1985 (TAP unpubl. fieldnotes; V. Emmanuel, pers. com.). The Bolivian specimen is from 570 m (fide B. Whitney; recorded as 535 m on label). Between 1980 and 1985 we conducted thorough inventories along elevational gradients on the foothill ridges of Cerro de Pantiacolla (depto. Madre de Dios) and above Atalaya (near Hacienda Amazonia, depto. Madre de Dios), and along the "Cosñipata Highway" on the main Andean slope (above Pilcopata, and near the Río Tono, both depto. Cuzco). The new species was most reliably encountered from 900 to 1,100 m on the main slope, and from 700 to 900 m on outlying ridges. Although we never encountered the species above 1200 m, the exact upper limits to its distribution remain to be determined.

Specimens.—Peru (n = 9): one male (LSUMZ 106326) from Puellas, km marker 41 on road from Villa Rica to Puerto Bermudez, depto. Pasco; two males (FMNH 315959 [holotype], 323129) and one female (FMNH 323128) from type locality; one immature female (FMNH 323127) from Cerro de Pantiacolla, depto. Madre de Dios; two males (FMNH 311513, 311514) and one female (FMNH 311512) from slopes above Consuelo, km marker 65 on Cosnipata Valley road, depto. Cuzco; one "male" (?; measurements suggest female) from Marcapata, depto. Cuzco (Mus. Berlepsch 43968).

Bolivia (n = 1): one male (Col. Boliviana de Fauna 02949) from depto. La Paz (locality given above).

Etymology.—Named in honor of our friend, the late Theodore A. Parker III, whose skill in the field, unbridled enthusiasm for birds and conservation, and twinkling smile the authors will sorely miss, always.

REMARKS

Juvenal plumage.—One specimen is an immature female (FMNH 323127; skull 10% pneumatized, ovary tiny and translucent) collected 18 November 1985 at 720 m on the Cerro de Pantiacolla. The plumage pattern of this specimen is similar to that of adults, but the plumage is somewhat fluffy and decomposed, especially below. Also, the rufous lores are distinctly paler, the crown greener, underparts paler, and breast slightly more mottled with greenish compared to adults.

Sexual dimorphism.—Males average 7% to 9% larger than females in linear dimensions (Table 1), but the sexes are similar in plumage. Plumage variation among the paratypes is related to age (see above) and molt conditions (see below).

Breeding and molt.—The only direct evidence of breeding was the presence of presumed family groups in several flocks during November 1985, one of which included the immature specimen mentioned above. Enlarged gonads are clearly documented on the labels of five specimens: the single Bolivian male (collected 25 August 1995, no molt), three Peruvian males (4 August 1985, no molt; 11 and 24 November 1981, both in heavy wing, tail, and body molt), and one Peruvian female (9 November 1981, some body molt only). The three specimens without enlarged gonads were collected on 8 April 1982 (light body molt), 13 August 1985 (central rectrices half-grown, no wing molt), and 23 November (no molt). These data suggest a late dry-season breeding period, reaching its peak between August and November, as is typical of most bird species of lowland and foothill forests in southern Peru and northern Bolivia (pers. obs.). Molt appears to coincide with or immediately follow breeding.

Behavior.—Like all other members of its genus, *P. parkeri* virtually always forages in mixed-species flocks. On the same forested mountain slopes where *P. parkeri* occurs in Peru and Bolivia, *P. orbitalis* forages mainly in understory flocks, *P. ventralis* is in both understory and subcanopy flocks, and *P. ophthalmicus* typically accompanies mixed-species flocks in the subcanopy and canopy. *P. parkeri* normally forages much higher than these congeners, often at the highest portions of the forest canopy. In the vicinity of the type locality, where we surveyed mixed-species canopy flocks in detail, about a third of them at appropriate elevations contained *parkeri*, usually a pair or a pair plus one or two presumed immatures. These typically foraged near the treetops, most often among sparse, small-leaved foliage near the outermost branch-tips of taller trees, including emergents.

Treetop habits would make *parkeri* one of the most challenging flock members to spot and identify, but its diagnostic "gnatcatcher-style" foraging behavior facilitates detection. During active foraging the tail is held horizontal or slightly above horizontal and the wings are held out slightly from the body. Often both wings are flashed partially outward simultaneously while the bird hops steadily out along a thin perch, alternately facing one side then the other. We did not detect any of the more deliberate, single-wing elevations with pauses, such as those performed by some other members of the genus *Phylloscartes* (and by other tyrannid genera such as *Leptopogon* and *Mionectes*). The tail rarely is fanned. Sallies for prey mostly consist of upward or outward hover-gleans (Fitzpatrick 1980, Remsen and Robinson 1990) less than 1 m long, taking prey from the undersides of leaves. Similar behavior was described in Parker et al. (1991).

Vocalizations.—During late dry season and early wet season (i.e., July through December) individuals intermittently sang while foraging. In addition, individuals occasionally ceased active foraging and vocalized from a stationary perch (during which time the tail was held outward or slightly downward, but not vertical). We noted and tape-recorded only two types of vocalizations during dozens of encounters with *P. parkeri*. The more common call was a single, sharp note uttered at irregular intervals of 0.5 to 2.0 s (chevron-shaped note in Fig. 1). The second vocalization—a short, sweet trill—usually followed a deliberate series of call notes. The trill lasted about 0.5 s, beginning at the same pitch as the call but rising sharply then falling back down (Fig. 1). These trills were typically separated by up to 20 s, with call notes uttered periodically between them. Both sexes appeared to sing; for example, we recorded one active bout (Library of Natural Sounds cut 76068) which included many trills given in rapid succession by a presumed pair in the close company of at least one immature, presumably their offspring (collected, see above).

Systematics.—As surmised by Hellmayr (1927), *P. parkeri* and *P. flaviventris* very likely are sister-taxa. Within *Phylloscartes*, only these two species and *P. superciliaris* have well-defined rufous lores and eyering, and only *parkeri* and *flaviventris* also have bold facial patterns containing buffy or rufous feathers. Both *parkeri* and *flaviventris* also have bright yellow underparts, bright yellow wingbars, and yellow tips to the inner remiges, resulting in extremely similar overall plumage patterns. As described above (see *Diagnosis*), many plumage differences also exist, however, and we have no hesitation in recognizing *parkeri* at the species level. Species-level treatment is further justified by the considerable difference in the sound and structure of the primary vocalizations of the two forms (Fig. 1).

We hypothesize that *Phylloscartes superciliaris* is a sister taxon to the *parkeri-flaviventris* clade because it is the only other species in the genus with well-marked rufous lores and eyering and a bold facial pattern. It differs considerably from the *parkeri-flaviventris* clade, however, by averaging somewhat larger (Table 1), lacking wingbars and spots on the inner remiges, having grayish white underparts, and having a predominantly white face patch.

The active, "gnatcatcher-style" foraging behavior of *parkeri* also characterizes certain other species of *Phylloscartes*, including *flaviventris* and *superciliaris*. Some members of the genus, however, including *Phylloscartes orbitalis* and *P. ophthalmicus*, forage with more deliberate pauses while searching for prey and with the tail held nearly vertically downward. We strongly suspect that this behavioral dichotomy within *Phylloscartes* reflects two phylogenetic groups within this large tyrannid genus. This hypothesis remains to be investigated.

Two other members of the "gnatcatcher-style" foraging group also have tinges of cinnamon

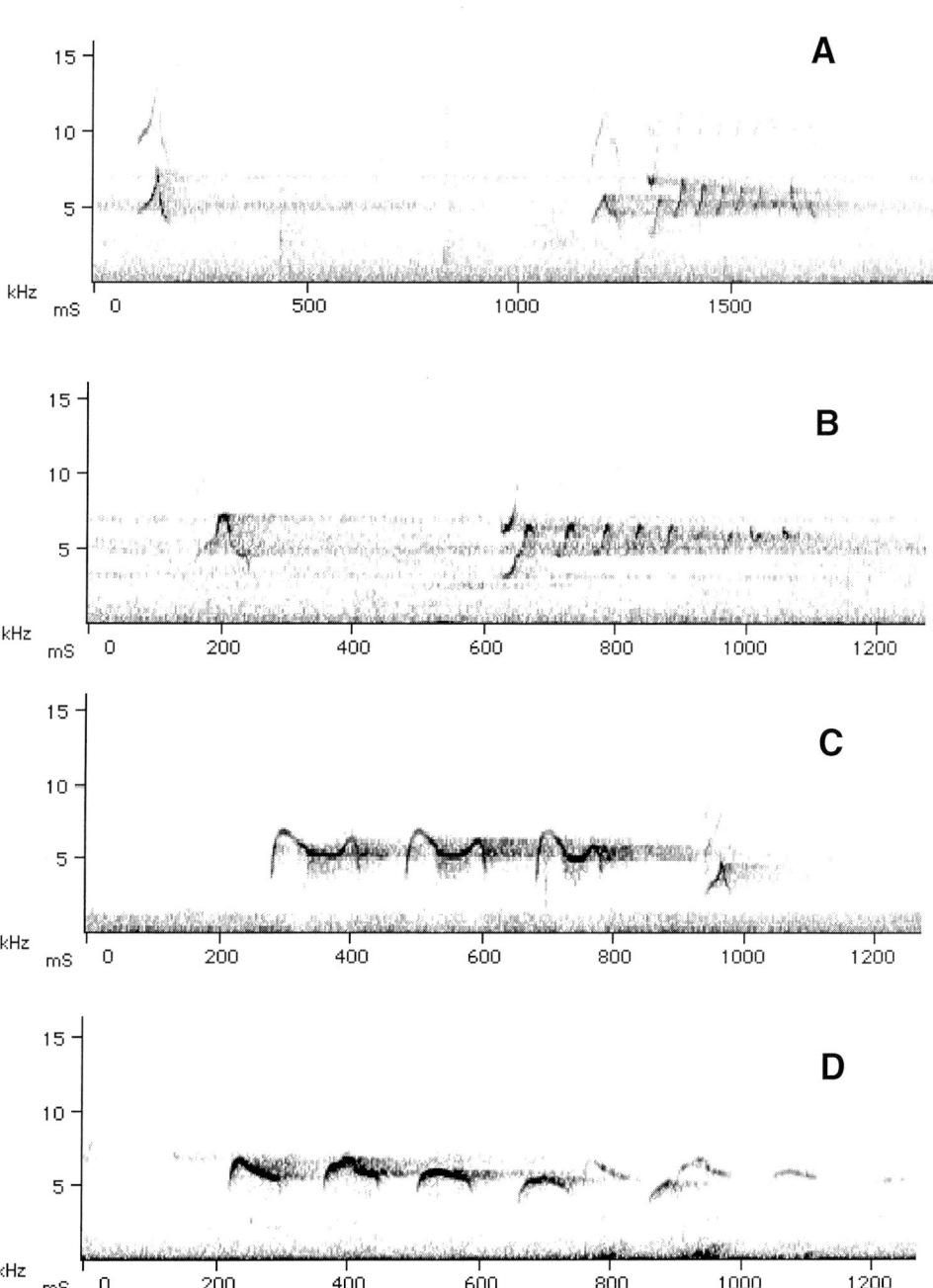

FIG. 1. Spectrograms of typical vocalizations of *Phylloscartes parkeri* (A, B) and *P. flaviventris* (C, D). Call notes of *P. parkeri* are the isolated, chevron-shaped notes, one of which also immediately precedes the trill in A. Both examples of *P. parkeri*, perhaps by different sexes or by an adult and an immature, from recording of actively vocalizing family group made 18 November 1985, Cerro Pantiacolla, depto. Madre de Dios, Peru, by J. W. Fitzpatrick (Library of Natural Sounds 76068). Recordings of spontaneous songs identified as *P. flaviventris* recorded by Steven L. Hilty, 12 January 1990, Parque Nacional Henri Pittier, Maracay, Venezuela (C; LNS 52689) and by John C. Arvin, 20 January 1996, same locality (D; uncataloged).

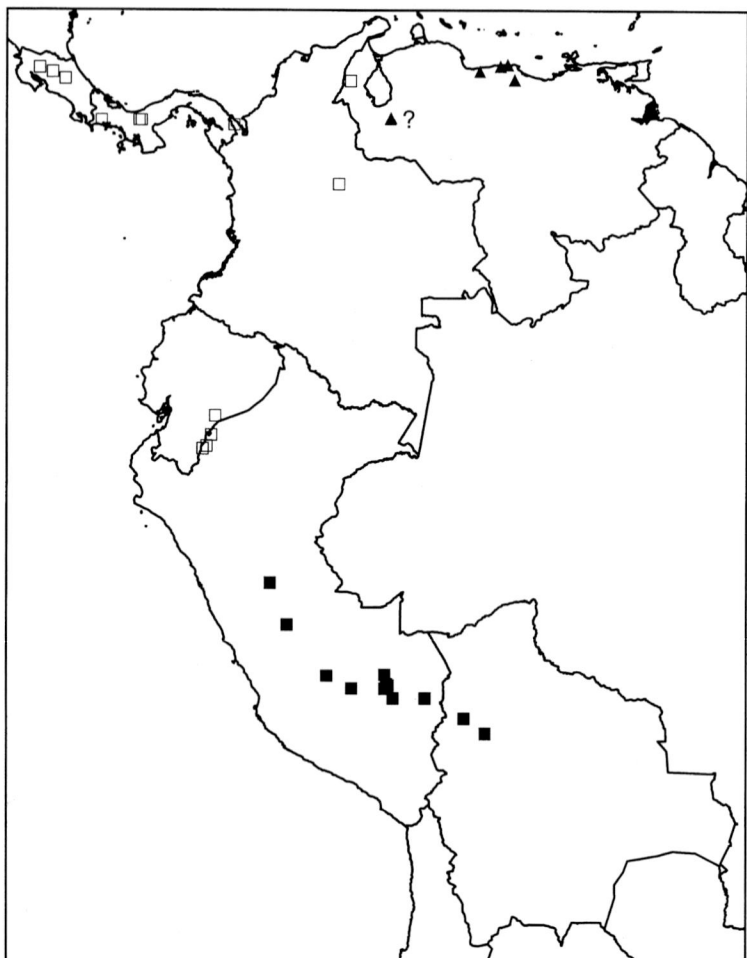

FIG. 2. Map of northeastern South America and adjacent Middle America, showing sight and specimen localities of *Phylloscartes parkeri* (solid squares), plus specimen localities of *P. flaviventris* (solid triangles), and *P. superciliaris* (open squares). Question mark identifies the type locality of *P. flaviventris*, which we believe to be an error on the original labels (see text); we suspect that *flaviventris* is endemic to the Venezuelan coast range.

or rufous about the lores and face. Both species are from the eastern lowlands of Brazil. *Phylloscartes roquettei* is known from the type specimen and a few recent sight records in dry forest bordering the rio São Francisco (Willis and Oniki 1991, incl. sonograms; Collar et al. 1992). *Phylloscartes sylviolus* is more widely distributed in the forests of southeastern Brazil, plus adjacent Paraguay and Argentina. These two species lack the bold face patterns of the three rufous-lored species discussed above, however, and they may not be closely allied with the three Andean, rufous-lored species.

Biogeography.—We suspect that the huge geographic gap between *P. parkeri* and *P. flaviventris* (Fig. 2) is even bigger than published locality data suggest (Traylor 1979). It is likely that *P. flaviventris* actually is restricted to the coastal mountain range of northern Venezuela, even though two localities mentioned in the original description of *flaviventris* (Hartert 1897) are from the Mérida Andes. *Phylloscartes flaviventris* has not been positively recorded outside the Venezuelan coast range despite a century of exploration in the Mérida Andes since the species was described. Furthermore, considerable evidence exists that the collector of the specimens referenced by Hartert, Albert Mocquerys, mislabelled many specimens, and that he definitely collected numerous specimens from the coast range during an appropriate time period prior to Hartert's description (Zimmer and Phelps 1954).

Phylloscartes parkeri, *P. flaviventris*, and *P. superciliaris* may constitute a relictual species-group (Fig. 2). All three species occupy similar, narrow elevational zones in humid forests from lower to middle montane elevations. As mentioned above, *P. flaviventris* apparently occurs only in the coastal mountains from Aragua to Miranda, where it occurs from 750 m to 1,400 m. *Phylloscartes superciliaris* is poorly represented in collections and has a patchy distribution (Fig. 2), seemingly replacing the *flaviventris-parkeri* complex from 600 to 1,200 m. in Costa Rica (Stiles and Skutch 1989) and western Panama (Ridgely and Gwynne 1989), eastern Panama (Cerro Tacarcuna), Venezuela (Sierra de Perija), the eastern Cordillera of Colombia (Río Virolin, 1,700 m.; Hilty and Brown 1984), southern Ecuador (Cordillera Cutucu, 1,700–1,750 m., Robbins et al. 1987; Cordillera del Condor, 1700 m.; Krabbe and Sornoza 1994), and extreme northern Peru (C. del Condor, T. S. Schulenberg, pers. comm.). South of the recently discovered Ecuadorian and Peruvian populations of *P. superciliaris*, a gap exists from the Río Marañon to depto. Pasco in Peru (Fig. 2). Whether this gap is real or a collecting artifact remains unknown. Sight records of *P. parkeri* in central and southern Peru and in Bolivia, and its occurrence on both main slopes and isolated ridges, suggest that the new species is nearly continuously distributed along the Andes from central Peru to northern Bolivia (Fig. 2).

Ecology and conservation.—*Phylloscartes parkeri* joins the considerable list of bird species whose distributions are confined to a narrow elevational range along the base of the eastern Andes. The narrowest of these elevational zones tend to occur in the Upper Tropical Zone, between 500 and 1,500 m (Stotz et al. 1996). In southern Peru, *P. parkeri* is relatively common only from 650 to 1,200 m, making it among the narrowest of all the foothill specialists. This narrow range combined with treetop habits probably help explain why the species went unrecognized for so long.

Phylloscartes parkeri is limited to primary evergreen forest having an intact and relatively tall canopy. Neither of us has recorded the species among the many mixed-species flocks we have censused in second-growth forest, even where the species is common in adjacent, intact forest. For example, JWF and colleagues spent 45 days inventorying birds between 900 and 1400 m on the main Andean slope of the Cosñipata Valley (near Consuelo, depto. Cuzco), during October and November 1981. Selective logging had been underway for several years in this area, producing irregular openings in the forest canopy up to several hundred m across. Older openings were regenerating, and many contained dense stands of bamboo (*Guadua* sp.; see also Fitzpatrick and Willard 1990). Mixed-species flocks were still common throughout this forest, but we found *P. parkeri* only in two canopy flocks (both at 1,000–1,100 m). These two flocks occupied tall, intact forest on the two steep hillsides near our camp that had thus far escaped selective logging. We strongly suspect that persistence of *P. parkeri* (along with many other species occupying this diverse elevational zone) absolutely depends upon protection of large, virgin stands of Upper Tropical wet forest along the Andean foothills. Important national parks that contain complete elevational transects, and good populations of *P. parkeri*, include Parque Nacional Manu in southeastern Peru and the newly established Parque Nacional Madidi in northern Bolivia.

ACKNOWLEDGMENTS

We are grateful to the Ministerio de Agricultura of Peru for permitting our fieldwork and collecting in southeastern Peru during the 1980s. For tireless assistance in the field on Field Museum expeditions in 1980, 1981, 1983, and 1985, we owe special thanks to David E. Willard. We also thank Barbara Clauson, Dale Clayton, Enrique Deza, John Douglass, Molly Fitzpatrick, Robin Foster, Robert Izor, Linda Kinkel, Nina Pierpont, William Southern, Tyana Wachter, and John Weske for assistance on one or more of these expeditions. We are very grateful to Bret Whitney for sending us the specimen, tape recordings, and detailed locality information from Bolivia, for lending us recordings of *Phylloscartes flaviventris* and *P. superciliaris*, and for his careful reading of the manuscript. John C. Arvin, Steve Hilty, and Scott M. Lanyon also contributed recordings of *P. flaviventris*. We thank Phyllis Isler for making preliminary sonograms and Greg Budney and Katherine Brese for making the final sonograms accompanying this article. The manuscript was improved by comments from Laurie Binford, Mario Cohn-Haft, Van Remsen, and Tom Schulenberg. Fieldwork was supported by the H. B. Conover Fund and the Ellen Thorne Smith Fund of the Field Museum of Natural History. The 1985 expedition also was partially supported by the National Science Foundation (BSR-8508361). Fieldwork by Stotz in depto. Pasco in 1982 was part of a LSUMZ expedition supported by John S. McIlhenny, Babette M. Odom, and H. Irving and Laura S. Schweppe. For lending us specimens and access to the collections under their management we thank the curators of the Belepsch Museum, and J. V. Remsen, Jr. (Louisiana State University Museum of Natural Science), Storrs Olson (U.S. National Museum of Natural History), Mary LeCroy (American Museum of Natural History), David Willard (Field Museum of Natural History), David Agro (Academy of Natural Sciences, Philadelphia), and Jaime Sarmiento and Carmen Quiroga (Coleccion Boliviana de Fauna).

LITERATURE CITED

COLLAR, N. J., L. P. GONZAGA, N. KRABBE, A. MADROÑO NIETO, L. G. NARANJO, T. A. PARKER III, AND D. C. WEGE. 1992. Threatened Birds of the Americas: the ICBP/IUCN Red Data Book. Smithsonian Inst. Press, Washington.

FITZPATRICK, J. W. 1980. Foraging behavior of Neotropical tyrant flycatchers. Condor 82:43–57.

FITZPATRICK, J. W., AND D. E. WILLARD. 1990. *Cercomacra manu*, a new species of antbird from southwestern Amazonia. Auk 107:239–245.

HARTERT, E. 1897. [A new species of *Leptotriccus*]. Bull. Brit. Ornithol. Club. 7:5.

HELLMAYR, C. E. 1927. Catalogue of birds of the Americas. Part V. Field Mus. Nat. Hist., Zool. Ser. 13.

HILTY, S. L., AND W. L. BROWN. 1986. A Guide to the Birds of Colombia. Princeton Univ. Press, Princeton, New Jersey.

KRABBE, N., AND F. SORNOZA M. 1994. Avifaunistic results of a subtropical camp in the Cordillera del Condor, southeastern Ecuador. Bull. Brit. Ornithol. Club 114:55–61.

PARKER, T. A., III, A. CASTILLO U., M. GELL-MANN, AND O. ROCHA O. 1991. Records of new and unusual birds from northern Bolivia. Bull. Brit. Ornithol. Club 111:120–138.

PARKER, T. A., III, AND W. WUST. 1994. Birds of the Cerros del Távara (300–900 m). Pp. 83–90 *in* The Tambopata Reserved Zone of Southeastern Perú: A Biological Assessment (R. B. Foster, T. A. Parker III, A. H. Gentry, L. H. Emmons, A. Chicchón, T. Schulenberg, L. Rodríguez, G. Lamas, H. Ortega, J. Icochea, W. Wust, M. Romo, J. A. Castillo, O. Phillips, C. Reynal, A. Kratter, P. K. Donahue, and L. J. Barkley, Eds.). RAP Working Papers 6. Conservation International, Washington, D.C.

REMSEN, J. V., JR, AND S. K. ROBINSON. 1990. A classification scheme for foraging behavior of birds in terrestrial habitats. Pp. 144–160 *in* Avian Foraging: Theory, Methodology, and Applications (M. L. Morrison, C. J. Ralph, J. Verner, and J. R. Jehl, Jr., Eds.). Studies in Avian Biology 13.

RIDGLEY, R. S., AND J. A. GWYNNE. 1989. A Guide to the Birds of Panama with Costa Rica, Nicaragua, and Honduras. Second edition. Princeton Univ. Press, Princeton, New Jersey.

ROBBINS, M. B., R. S. RIDGELY, T. S. SCHULENBERG, AND F. B. GILL. 1987. The avifauna of the Cordillera de Cutucú, Ecuador, with comparisons to other Andean localities. Proc. Acad. Nat. Sci. Philadelphia 139:243–259.

SCHULENBERG, T. S., S. E. ALLEN, D. F. STOTZ, AND D. A. WIEDENFELD. 1984. Distributional records from the Cordillera Yanachaga, central Peru. Gerfaut 74:57–70.

SMITHE, F. B. 1975. Naturalist's Color Guide. American Museum of Natural History, New York, New York.

STILES, F. G., AND A. F. SKUTCH. 1989. A Guide to the Birds of Costa Rica. Cornell University Press, Ithaca, New York.

STOTZ, D. F., J. W. FITZPATRICK, T. A. PARKER, III, AND D. K. MOSKOVITS. 1996. Neotropical Birds: Ecology and Conservation. Univ. of Chicago Press, Chicago, Illinois.

TRAYLOR, M. A., JR. 1979. Subfamily Elaeniinae. Pp. 3–112 *in* Check-List of Birds of the World, Vol. VIII (M. A. Traylor, Jr., Ed.). Museum of Comparative Zoology, Cambridge, Massachusetts.

WILLIS, E. O. AND Y. ONIKI. 1991. Avifaunal transects across the open zones of northern Minas Gerais, Brazil. Ararajuba 2:41–58.

ZIMMER, J. T. AND W. H. PHELPS. 1954. A new flycatcher from Venezuela, with remarks on the Mocquerys collection and the piculet, *Picumnus squamulatus*. Amer. Mus. Novit. 1657:1–7.

Three new species of tapaculo (*Scytalopus*): Chusquea Tapaculo (*S. parkeri*), female upper left, male upper right; Ecuadorian Tapaculo (*S. robbinsi*), male bottom left; Chocó Tapaculo (*S. chocoensis*), bottom right (male above, female below). From a watercolor painting by Jon Fjeldså.

SPECIES LIMITS AND NATURAL HISTORY OF *SCYTALOPUS* TAPACULOS (RHINOCRYPTIDAE), WITH DESCRIPTIONS OF THE ECUADORIAN TAXA, INCLUDING THREE NEW SPECIES

NIELS KRABBE[1,2] AND THOMAS S. SCHULENBERG[3,4]

[1]*Zoological Museum, University of Copenhagen, Universitetsparken 15, DK-2100, Copenhagen, Denmark;*
[2]*present address: Casilla 17-21-791, Quito, Ecuador;*
[3]*Committee on Evolutionary Biology, University of Chicago, Chicago Illinois 60637 and Department of Zoology, Field Museum of Natural History, Chicago, Illinois 60605, USA;*
[4]*present address: Environmental and Conservation Programs, Field Museum of Natural History, Roosevelt Road at Lake Shore Drive, Chicago, Illinois 60605, USA, and Conservation International, 2501 M Street NW, Suite 200, Washington, D.C. 20037, USA*

ABSTRACT.—The taxonomy of the genus *Scytalopus* (Aves; Rhinocryptidae) is revised. Full descriptions of the 10 Ecuadorian taxa are given, with some indication of their annual cycle, and details of their vocalizations, habitats, and geographical distribution. Notes are given on all extralimital forms, and a new taxonomy based on vocal data, and often supported by new distributional and genetic data, is proposed. The number of species in the genus rises from 11 to 37, two or three of which remain unnamed. Two additional populations are identified as being distinct, but their geographical distributions have not been satisfactorily delimited. Sonagrams of most taxa are presented.

Three new species are named and described. Full species status is proposed for most taxa ranked as subspecies by Zimmer (1939) and Peters (1951). Several of these species have sympatric distributions, and others have different songs. Allopatric taxa with the same body mass and elevational distributions, and with relatively similar, but yet distinctive songs are united into superspecies.

RESUMEN.—Se revisa la taxonomía del género *Scytalopus* (Aves; Rhinocryptidae), uno de los géneros más problemáticos del mundo. Se dan descripciones completas de los 10 taxa ecuatorianos, con algunas indicaciones sobre su ciclo anual, y detalles de sus vocalizaciones, habitats, y distribución geográfica. También se dan notas sobre todas las formas extralimitales, y se propone una nueva taxonomía basada en datos de vocalizaciones, en muchos casos respaldados por nuevos registros geográficos y análisis genéticos. El número de especies en este género sube de 11 hasta 37, dos o tres de ellas permanecen innominanadas. Adicionalmente dos poblaciones han sido claramenete identificadas, pero su distribución geográfica todavía no está satisfactoriamente delimitada. Sonogramas de las vocalizaciones conocidas para casi todas las especies se presentan.

Tres especies nuevas han sido nominadas y descritas. Se propone el status de especie para la mayoría de los taxa ubicados como subespecie por Zimmer (1939) y Peters (1951). Algunas de estas especies tienen distribuciones simpátricas, otras tienen cantos diferentes. Taxa allopátricos con el mismo peso corporal y la misma distribución altitudinal, y con cantos relativamente similares pero aún distintos, se agrupan en superespecies.

Taxonomically *Scytalopus* is one of the most complicated of all bird genera. Even the diagnosis of taxa has been problematic, and only recently has headway been made in the study of the relationships among taxa.

The taxonomy of *Scytalopus* long was based on the analysis of Zimmer (1939), which entailed, among other revisions, the description of no fewer than nine new taxa. Even so, Zimmer, working entirely from museum material, was not satisfied with his own results. Little further progress was made until recent years, when museum ornithologists became familiar with these birds in the field. Beginning in the late 1970s, our own experiences with the genus quickly convinced us that the taxonomy of the genus was confused. Working at first independently, and then for many years in collaboration, we have studied these birds in the field and museum, solicited information from other field workers, and attempted to apply our expanding knowledge to the classification of the group. During the course of our work on the genus in the last decade, it has become clear that the traditional morphological approach is not at all sufficient to delimit species in this genus. Allopatric forms inhabiting the same ecological zone are not necessarily conspecific, and genetic data (Arctander and Fjeldså 1994) show that such taxa are often not each other's closest relatives. Species of *Scytalopus* are extremely similar by plumage characters and structure, and differ from each other primarily in vocalizations, in body mass, and in elevational distribution, although some have diagnostic plumage patterns. Many individual specimens cannot be unequivocally assigned to a species because plumage differences often become apparent only when comparing large series of specimens collected within the same decade. The collecting and weighing of specimens of known song types at known elevations, and comparison of material collected in the same decade, has made it possible to clarify some of the confusion.

Our preliminary results have formed the basis for several recent, somewhat differing classifications of the genus (Fjeldså and Krabbe 1990; Ridgely and Tudor 1994; Parker et al. 1996), and have helped inspire independent studies as well (Whitney 1994). To date, however, most of the underlying documentation for our classification has not been published. This paper, then, represents an overdue "progress report" for our ongoing studies of *Scytalopus*. It treats in depth only the taxa occurring in Ecuador. We also include new distributional and vocal data on a number of extralimital forms, and point out additional taxonomic problems. We compare the traditional taxonomy (i.e., that of Peters 1951 and of the describers of new taxa published later) with one that we suggest. We are well aware that we have not "solved" all taxonomic problems in *Scytalopus*, and we regard our classification as provisional on certain points. We outline remaining problems or areas of uncertainty, and encourage additional studies of these poorly known, but fascinating birds.

GENERAL DESCRIPTION OF THE GENUS

Tapaculos of the genus *Scytalopus* (Gould 1837) are small (13 to 43 g), usually dark gray birds. Almost exclusively montane, and primarily Andean, they are found from Costa Rica to Panama, and through the Andes from Colombia and Venezuela to Cape Horn. They also are found in the coastal mountains of Venezuela, and from eastern and southeastern Brazil into adjacent Misiones, Argentina (Peters 1951). A possible *Scytalopus* has been found in Quaternary deposits on Cuba (Olson and Kurochkin 1987).

The foraging behavior described for an individual of *Scytalopus schulenbergi* (Whitney 1994) agrees with our 20 or 30 observations of forest species, and is probably typical for the genus. None was ever seen reaching up, and none gleaned green foliage with leaves more than a few millimeters in size. We can add several observations of birds using tunnels through piles of rocks, tussocks of grass, moss-covered roots, or root tangles on banks or steep slopes while foraging. We have seen *S. unicolor latrans* spend up to an hour in the same thicket and sit on the same perch for up to 30 s, but most foraging birds observed were active and moved along persistently, covering up to 75 m in 20 min. They feed almost entirely on arthropods, usually tiny insect imagos, mostly gleaned from mosses, dead twigs, or grass (pers. obs. both authors; Whitney 1994). One adult among 15 *S. vicinior* had eaten a 4 to 5-cm grasshopper, one a 2-cm imago scarab beetle. Only 4 of over 300 stomachs we examined contained vegetable matter: a single juvenile among 36 *S. canus opacus* had eaten a berry, and three among four *S.* [*superciliaris*] *zimmeri* had their stomachs crammed with tiny seeds. Thus there is little evidence of trophic segregation of the species. Up to 60 individual arthropods were present in some stomachs.

Scytalopus tapaculos typically inhabit dense forest understory, shrubbery, or dense bunch grass or moss. They often seem reluctant to fly, and usually do so for only short distances. We have not seen birds naturally fly more than 3 m and consider them nearly flightless. The longest flight we have witnessed, and which was heavily labored, was of a bird that escaped handling at chest level and flew ca. 20 m across a creek. When they escape handling, they usually flutter to the ground and dart into cover without flying. They could be described as "agoraphobic."

These tapaculos sing and respond to playback of song mainly for two or three hours after dawn and for an hour before dark, but occasionally at other times, if the sky is overcast (pers. obs.). They often are shy of humans. After playback of their voice, they usually approach through dense thickets, crossing small open spaces with great speed, but occasionally (males only) an immediate, direct attack on the tape-recorder takes place; during attacks or when fleeing, they are exceptionally rapid (pers. obs.).

They often occur in high densities, as judged from numbers of birds vocalizing and the rapidity with which collected birds are replaced (pers. obs. N.K.). The aversion of most taxa to open habitats and their weak flying abilities render them poor dispersers. Their narrow elevational montane distributions, which are subject to fragmentation by geological and climatic changes, give them unusual opportunities to differentiate in isolation.

Owing to the dense habitat in which they are found and their active foraging mode, their life history is difficult to study, but apparently they stay in close pairs during breeding. The few known nests (Johnson 1967; Skutch 1972; Rosenberg 1986; Sick 1993; and two nests of *Scytalopus affinis* found by N.K.) were globular structures of grass, moss, and roots, one also with some feathers, with a side entrance, hidden near or just under the ground under cover of moss or bunch grass. As far as is known, clutch size is two, and eggs are white. Eggs of *S. indigoticus* took 15 days to hatch (Sick 1993). A brood patch has so far been noted only in females (pers. obs. N.K.). Males defend their territory vigorously during breeding. They may respond to the songs of other *Scytalopus* species, but do so much more consistently to voices of their own species.

As in several rhynocryptid genera the feathers of the lores and forehead are more or less erect.

Unusual (and possibly unique) among passerine birds is the complete lack of skull ossification in the genus. In many other suboscine passerines the skull apparently only ossifies partly (see Miller 1963; Winkler 1979), but during our own collecting of over 300 specimens of 22 of the taxa that we include in *Scytalopus,* we have never seen signs of even an incipient ossification. We therefore consider reports of specimens with partly or completely ossified skulls in error.

Few plumage features serve to distinguish the taxa. Sexual dimorphism is also weak. Females are usually somewhat smaller and paler than males, and have more brown in their plumage. A large percentage of the individuals have one or two albino feathers in their plumage. Albinism usually occurs on different feathers of different individuals, but occasionally it is inbred. Most of a *Scytalopus* population may show identical partial albinism, but in one sex only. For example, in a series of 15 specimens of *S. canus opacus* from Cordillera las Lagunillas, southern Loja, Ecuador, seven of eight males (and three more seen), but none of four females, have a prominent white patch on the wing-coverts, whereas in a series of *S. parvirostris* from Cordillera Colán, depto. Amazonas, Peru, a similar spot is present in four out of six females, but in none of the six males.

The plumage sequence is not well known, but more or less asymmetrical molt is the rule rather than the exception. Birds generally molt into plumages with less brown. We therefore consider only the grayest birds true adults. The very large proportion of immatures and subadults to adults in series of specimens, despite that mainly territory-holding birds are collected, and the very gradual transformation in each molt, suggest that several molts are needed to attain full adult plumage. Generally the brown, juvenal plumage is more or less barred with dusky or black (the amount of barring being extremely variable even within one population: see under *S. canus opacus* and *S. unicolor subcinereus*). Subadults often have silvery feather-tips on the belly and have variable amounts of brown in the plumage. Adult plumages are predominantly gray or black. In many species the adults have brown-barred flanks; in some, the vent and rump also are barred. Other plumage variations, which may be more or less developed, are: brownish tail, whitish belly, whitish throat, whitish supercilium, silvery fore-crown, a white central crown-patch, and straight bars versus scallops on the flanks. The iris is dark brown. The bill is dark gray to blackish in Andean forms. The feet in all Ecuadorian forms are gray-brown to blackish brown on the frontal and lateral surfaces, light gray-brown to gray-brown on the medial surfaces, generally lightest in *S. parkeri* and *S. robbinsi* (described below, p. 81 and 78). A reversal of the pigmentation of the feet (yellowish or dull reddish, and palest on the frontal and lateral surfaces) is found in a few non-Ecuadorian forms (*S. zimmeri, S. superciliaris,* and *S. indigoticus*).

The plumage colors in *Scytalopus* change ("fox") quickly in dried museum skins (pers. obs. both authors; Whitney 1994). Gray pigments become considerably paler over just a few years and may eventually take on a brownish tone; brown pigments become more reddish but change at a slower rate. This makes it difficult to compare old and recent material. The descriptions in this paper, therefore, are restricted to recent material except when otherwise noted.

TAXONOMY

The morphological similarity of many *Scytalopus* taxa gives few clues about the relationships of allopatric forms. Variation in characters is often subtle, overlapping, and, especially in the case of plumage characters, difficult to quantify. Many taxa merely show lack of any distinctive characters. We also are reluctant to unite taxa on the basis of what few plumage characters as do exist, because we fear that some such characters may be homoplastic. We only feel confident in using plumage characters to unite taxa in cases in which such an interpretation is supported by additional evidence (e.g., similarities in elevational distribution among different taxa).

Although many individual specimens cannot be assigned to a given species by plumage alone, certain plumage differences do become apparent on larger series of recent specimens of known song-type. Each song-type is associated with a distinctive elevational distribution (as was reported for Bolivia by Whitney [1994]). Specimens of birds belonging to a given song-type are of the same size (primarily as indexed by body mass). Measurements of wing, tail, tarsus, and bill (Table 1) also differ among the taxa, but with more overlap than body masses. The number of rectrices ranges from eight to 14, but may vary within at least six taxa (Table 2). No differences in wing formula, tarsal scutellation, feather-shape and rigidity, or microscopic location of feather pigmentation could be discerned. Apparently voice is the major factor in species recognition.

We generally treat as species the allopatric taxa that differ in territorial songs. We are aware of some cases (e.g., between the northern and southern populations of *S. canus opacus*) in which call notes vary between populations that have similar or identical songs. Such variation in vocalizations presumably reflects underlying genetic variation as well, but we are reluctant to recognize populations of taxa purely on the variation in calls. This reluctance stems in part from an assumption, admittedly untested, that the song is more important in species-recognition and pair-formation than is a call. Although we strongly doubt it, the different calls might not be homologous.

Our treatment gives species rank to the majority of the vocally known taxa previously treated as subspecies. We do not change the taxonomic status of vocally unknown taxa. We unite into superspecies the morphologically distinguishable taxa that show some vocal similarities, and that occupy similar habitats at similar elevations.

This approach is based partly on the positive correlation between vocal and genetic differences found in the genus by Arctander and Fjeldså (1994); on the assumption that vocal characters are as important as, or more important than, morphological characters; and on the assumption that vocal characters are entirely inherited in *Scytalopus*. Although the latter assumption has not been demonstrated specifically for tapaculos, it may be true of all sub-oscines (Kroodsma 1982, 1984).

A comparison of the traditional taxonomy and the one that we propose is presented in Table 4.

METHODS

During the last decade we have tape-recorded several hundred individual birds of the genus *Scytalopus* at a number of sites in Ecuador; these data form the core of the distribution sections of the species accounts below. More than 200 individual tapaculos (Appendix), representing all the call-/and song-types heard, were called in with voice playback and then shot or netted, with a blood-sample extracted for DNA-sequencing, and the number of rectrices counted. All specimens were weighed and sexed by gonad inspection. Elevation, locality, and stomach contents were also recorded on the specimen labels. These series of specimens of known song-type form the basis of the descriptions below. We also examined the *Scytalopus* specimens in all major museum collections, to check the previously published distributions of morphologically identifiable species.

Capitalized color names in the descriptions refer to Ridgway (1912).

Only juveniles and females with active gonads (largest ovum over 2 mm and oviduct greatly thickened and curled) or a brood patch have been used to assess timing of breeding. The material is limited, but apart from single August juveniles of *Scytalopus chocoensis* (described below, p. 75) and *S. spillmanni*, there is no sign of breeding of any species in Ecuador between late May and late September, i.e., during the dry season in the highlands and on the Pacific slope. If data of males with enlarged gonads are included, breeding birds have been found throughout the year except in May ($n = 4$) on the east slope, and in August ($n = 14$) (no May and June specimens) on the west slope.

Acronyms of museums and acoustic libraries cited in the text are: American Museum of Natural History, New York (AMNH); Academy of Natural Sciences of Philadelphia, Philadelphia, Pennsylvania (ANSP); British Museum (Natural History), Tring, U.K. (BMNH); Bioakustisk Laboratorium, Aarhus, Denmark (BLA); British Library of Wildlife Sounds, London (BLOWS); Colección Boliviana de Fauna, Museo de Historia Natural, La Paz, Bolivia (CBF); Carnegie Museum of Natural History, Pittsburgh, Pennsylvania (CM); Escuela Politécnica Nacional, Quito, Ecuador (EPN); Fundación Miguel Lillo, Tucumán, Argentina (FML); Field Museum of Natural History, Chicago, Illinois (FMNH); Instituto de Ciencias Naturales, Universidad Nacional de Colombia, Bogotá, Colombia (ICN); Los Angeles County Museum, Los Angeles, California (LACM); Library of Natural Sounds, Cornell Laboratory of Ornithology, Ithaca, New York (LNS); Louisiana State University Museum of Natural Science, Baton Rouge, Louisiana (LSUMZ); Museum of Comparative Zoology, Harvard University, Cambridge, Massachusetts (MCZ); Museo Ecuatoriano de Ciencias Naturales, Quito, Ecuador (MECN); Museo de Historia Natural, Universidad Nacional Mayor de San Marcos, Lima, Peru (MHN); Musée d'Histoire Naturelle, Neuchâtel, Switzerland (MHNN); Museu Nacional, Rio de Janeiro, Brazil (MNRJ); Museu de Zoologia da Universidade de São Paulo, Brazil (MZUSP); Naturhistoriske Riksmuseet, Stockholm (NRS); Phelps Collection, Caracas, Venezuela (PCC); National Museum of Natural History, Washington, D.C. (USNM); Western Foundation of Vertebrate Zoology, Camarillo, California (WFVZ); Zoological Museum, University of Copenhagen, Copenhagen (ZMUC).

VOCALIZATIONS

Sonagrams of most non-Brazilian taxa are shown in Figures 1 to 85. Earlier discussions of *Scytalopus* vocalizations documented by sonagrams are found in Vieillard (1990) (most Brazilian taxa), Fjeldså and Krabbe (1990) (most Andean taxa), and Whitney (1994) (some Bolivian and Peruvian taxa).

Over 600 tape-recordings, representing 35 of the roughly 42 taxa in the genus, were available for this study. Some 350 of our own recordings are from Ecuador, and sonagrams of these recordings form the basis for the descriptions in the following section. Descriptions of pitch (frequency) of vocalizations are complicated by the usual presence of genuine overtones (harmonics). The first overtone (second harmonic) is usually the loudest, but this may vary among species, or even individually (see under vocalizations of *S. unicolor latrans*). The pitch and shape of individual notes, and the rate of delivery are generally distinctive for each taxon. They may vary in one individual in different states of excitement (compare Figs. 11 and 12), but as is typical of most suboscine passerines (see Ridgely and Tudor 1994), vocalizations generally show lack of plasticity over geographic distance. A study of syringeal morphology in *Scytalopus* might reveal variations comparable to those found in some other suboscines (see Prum 1992).

Most *Scytalopus* vocalizations are repetitions of a single note, given from 1 to 34 per second. In some taxa the song consists of a rapid trill. Analyses of mitochondrial DNA sequences suggest that these trilled songs have developed independently several times in the genus. For example, *S. schulenbergi* has a rapid trilled song (Fig. 8) that resembles that of *S. parvirostris* (Fig. 6). The closest known relative of *S. schulenbergi* (J. C. da Silva, pers. comm.), however, is an unnamed taxon from depto. Apurímac, Peru, with a very different song (Fig. 35).

Functions of the different vocalizations are not always easy to infer. We assume that songs are homologous across taxa, but it is possible that there is more than one kind of song. For example, the song of the male *S. parkeri* during duets (pair-formation?) is different from its territorial song (Figs. 82 and 79). Some vocalizations appear to be contact calls (Fig. 3), others distress calls (Fig. 4), and others might serve both as territorial patrol calls and contact calls (Fig. 48).

We have some evidence for seasonal variation in the repertoire of some taxa. For example, only male songs and female (contact?) calls of *Scytalopus robbinsi* were heard in El Oro in September, whereas at the same site in November only high-pitched, descending series from females were heard.

Territorial advertising songs of most species are given only at large intervals (one to three times) during the morning and evening. During sunny days birds normally do not respond to playback of their song between 0800 and 1700 hr. During territorial disputes song may be given nearly continuously for up to 30 min. Male songs during territorial disputes sometimes differ from the usual advertising song (compare Fig. 79 with 80 and 61 with 62). As pointed out by Whitney (1994) scolds (alarms) are often relatively similar in different species, but the territorial songs of *Scytalopus caracae* (Fig. 77) and *S. a. atratus* (Fig. 14) sound much like scolds of other species. Some species appear to have fewer kinds of vocalizations than others. High-pitched descending series of notes by females sometimes initiate duets with males (Figs. 81 and

TABLE 1
Measurements of Body Mass and Length of Wing, Tail, Tarsus, and Bill of Ecuadorian *Scytalopus* Tapaculos. Taxa Arranged by Increasing Mean Values for Males

	Males			Females		
	N	Mean	Range	N	Mean	Range
Body mass (g):						
opacus (north)	23	16.2	13.9–17.9	7	14.8	13.4–15.5
opacus (south)	9	16.5	15.0–17.8	3	15.7	15.0–16.0
latrans (west)	15	18.2	16.8–20.9	14	16.9	15.5–18.0
subcinereus	4	18.8	15.9–20.0	7	16.7	14.2–19.1
latrans (east)	8	18.8	17.5–19.6	5	17.5	15.8–20.0
robbinsi	8	19.6	18.1–21.0	3	19.1	18.7–19.5
chocoensis	17	21.0	19.0–22.5	5	19.0	17.0–20.1
parkeri	13	22.5	21.0–24.4	4	20.1	18.8–22.3
vicinior	14	22.5	20.5–24.6	2	23.6	22.9–24.2
spillmanni	36	25.2	21.0–30.0	5	24.1	20.0–29.5
atratus	7	26.2	24.6–32.5	1	25.3	—
micropterus	9	29.7	27.0–32.5	—	—	—
Wing (flat) (mm):						
robbinsi	8	53.9	52.0–55.0	3	56.5	52.0–59.0
latrans (east)	6	54.7	53.0–56.0	2	58.5	57.0–60.0
chocoensis	20	55.3	52.0–60.0	8	52.6	47.6–58.0
subcinereus	4	56.2	54.0–59.0	7	52.6	49.0–59.0
opacus (north)	22	57.2	52.0–63.0	4	55.3	54.0–56.0
vicinior	13	57.3	55.0–60.0	3	58.5	57.0–60.0
latrans (west)	20	57.5	53.0–63.0	19	56.2	49.0–63.0
opacus (south)	8	57.8	55.0–59.0	1	56.0	—
atratus	8	60.1	55.0–65.0	2	60.5	59.0–62.0
micropterus	15	61.4	59.9–64.0	1	64.0	—
spillmanni	38	61.9	56.0–67.0	7	58.4	56.0–62.0
parkeri	14	62.9	59.0–66.0	5	61.4	58.0–65.0
Tail (mm):						
robbinsi	8	36.0	34.0–39.2	3	37.2	35.7–39.0
chocoensis	19	38.7	33.8–40.7	9	36.9	34.0–38.5
latrans (east)	6	39.3	37.0–41.0	2	40.0	37.0–43.0
latrans (west)	18	39.8	36.0–45.0	18	37.4	34.0–40.1
subcinereus	5	40.4	39.0–42.0	7	37.2	34.0–42.3
opacus (north)	19	42.8	35.4–46.7	4	39.8	39.0–42.0
opacus (south)	8	43.4	39.0–47.0	1	41.7	—
parkeri	13	44.5	42.3–50.0	5	44.3	42.0–48.0
spillmanni	35	45.3	39.0–54.0	7	41.7	38.0–45.0
atratus	8	47.2	44.0–52.0	2	44.7	42.4–47.0
vicinior	11	48.1	46.0–52.0	1	43.5	—
micropterus	12	53.5	48.0–59.0	1	58.0	—
Tarsus (mm):						
latrans (east)	5	21.9	21.0–23.1	2	21.4	21.2–21.6
robbinsi	4	22.2	22.2–22.3	3	21.6	21.2–22.0
chocoensis	8	22.2	21.2–25.0	2	21.4	21.1–21.7
opacus	27	22.5	20.8–24.2	8	20.8	19.8–22.0
latrans (west)	12	22.6	20.5–24.0	12	21.6	20.0–23.1
subcinereus	3	22.9	22.2–23.4	6	21.2	20.6–21.7
vicinior	9	23.5	22.8–24.3	2	23.5	22.8–24.1
spillmanni	18	24.5	22.2–26.0	5	24.0	23.5–24.2
parkeri	14	24.6	23.8–25.3	3	22.0	21.0–23.0
micropterus	6	24.9	24.3–25.4	1	26.0	—
atratus	6	25.5	24.7–26.2	1	22.6	—

TABLE 1
CONTINUED

	Males			Females		
	N	Mean	Range	N	Mean	Range
Bill from fore edge of operculum to tip (mm):						
opacus	19	5.5	5.0–6.1	1	5.1	—
parkeri	8	6.4	6.1–7.0	—	—	—
latrans (west)	10	6.4	5.6–7.0	3	5.7	5.3–6.2
subcinereus	2	6.5	5.8–7.1	—	—	—
latrans (east)	2	6.7	6.3–7.0	2	6.6	6.5–6.6
robbinsi	5	6.9	6.6–7.2	—	—	—
spillmanni	14	7.0	6.1–7.6	2	6.8	6.4–7.1
vicinior	9	7.4	6.6–8.3	—	—	—
atratus	3	7.4	7.3–7.7	—	—	—
chocoensis	7	7.7	7.5–8.1	2	6.6	6.5–6.6
micropterus	4	7.8	7.5–8.2	—	—	—

82), but are sometimes given alone, in which case the function is less clear. Both males and females are often attracted to playback of their various vocalizations, but strong responses usually are elicited only by playback of male territorial song.

DESCRIPTIONS OF ECUADORIAN TAXA WITH NOTES ON EXTRALIMITAL FORMS

In the following section we give accounts of the *Scytalopus* species that are found in Ecuador, presenting for each taxon a brief diagnosis (generally valid only for Ecuador), and detailed data on plumage, voice, habitat, elevational, and geographic distribution in Ecuador. In addition, we briefly discuss extra-limital taxa that have been regarded as conspecific with taxa found in Ecuador, and we provide comments on taxonomy. Measurements of Ecuadorian taxa are given in Table 1. Comparisons of elevational ranges are given in Table 3. A comparison of different classifications of the genus is given in Table 4. Geographical distributions in Ecuador are shown in Figure 86. Elevational distributions in Ecuador are shown in Figures 87 and 88.

Scytalopus unicolor, sensu Zimmer 1939

Zimmer (1939) described three new taxa, *subcinereus*, *intermedius*, and *parvirostris*, which he united, together with *latrans*, as subspecies of *S. unicolor*. That at least two species were included in Zimmer's *S. unicolor* became evident when we by 1983 had learned that the vocalizations of *parvirostris* and *latrans*, as detailed below, are entirely different. Furthermore, the distribution of *parvirostris* may overlap with that of *intermedius* in north-central Peru. Series of

TABLE 2

RECTRIX NUMBER IN ECUADORIAN *Scytalopus* TAPACULOS. ENTRIES IN EACH COLUMN SHOW THE NUMBER OF INDIVIDUAL SPECIMENS PER TAXON THAT POSSESS THE INDICATED NUMBER OF RECTRICES

Taxon	Number of rectrices					
	8	10	11	12	13	14
opacus	3	29				
latrans (west)		12	2	6		
latrans (east)	1	3				
subcinereus	1	8		1		
spillmanni				14	1	2
parkeri				16		
vicinior		2		11		
chocoensis		11				
robbinsi		8				
atratus		2		4		
micropterus				4		

TABLE 3
Elevational Distribution of *Scytalopus* in Ecuador. Taxa Are Arranged by Increasing Mean Elevation (m)

	Mean	Range
Generalized characterization of elevational distribution in Ecuador:		
chocoensis	700	350–1,065[a]
robbinsi	975	700–1,250[b]
atratus	1,250	850–1,650
micropterus	1,725	1,250–2,100
vicinior	1,800	1,250–2,350
parkeri	2,700	2,250–3,150
spillmanni	2,700	1,900–3,500
subcinereus	2,750	1,500–4,000
latrans	2,990	1,975–4,000
opacus	3,465	3,050–3,980
Pacific slope in provs. Carchi and Esmeraldas:		
chocoensis	700	350–1,065[a]
vicinior	2,000	1,650–2,350[c]
spillmanni	2,775	2,350–3,200
latrans	3,225	3,100–3,350
opacus	3,690	3,400–3,980
Pacific slope in prov. Pichincha:		
vicinior	1,675	1,450–1,900[c]
spillmanni	2,625	1,900–3,350
latrans	3,285	2,700–3,870[d]
Pacific slope in prov. Azuay:		
robbinsi	1,070	890–1,250[b,e]
subcinereus	2,750	1,500–4,000
Amazonian slope in prov. Napo:		
atratus	1,275	1,200–1,350[f]
micropterus	1,675	1,250–2,100
latrans	2,225	2,000–2,450[g]
spillmanni	2,800	2,100–3,500
opacus	3,650	3,350–3,950
Amazonian slope in prov. Zamora-Chinchipe:		
atratus	1,150	1,050–1,250[f]
micropterus	1,700	1,400–2,000[h]
latrans	2,200	2,100–2,300
parkeri	2,825	2,300–3,350
opacus	3,350	3,050–3,650

[a] Might range higher and lower. In Colombia known down to 250 m, in Panama up to 1,465 m.
[b] May formerly have ranged lower.
[c] Probably ranges lower. In prov. Cotopaxi known down to 1,250 m.
[d] Might range higher and lower. On Pasochoa, prov. Pichincha, known up to 4,000 m, in prov. Cotopaxi down to 2,300 m along clearings.
[e] Probably ranges lower.
[f] Might range higher and lower. In prov. Zamora-Chinchipe known down to 1,050 m, in prov. Morona-Santiago up to 1,650 m and possibly down to 850 m. In prov. Napo reportedly up to 1,950 m at Cosanga (P. Greenfield, M. Lysinger, both, pers. comm. to N.K.).
[g] Might range higher and lower in disturbed areas. There are old prov. Napo specimens labelled "Baeza, 1830 m" and "Papallacta, 3100".
[h] Might range slightly higher and lower.

two taxa, consistent in morphology with *parvirostris* and *intermedius*, were collected in sympatry on Cordillera Colán, depto. Amazonas (LSUMZ). There also exists a single tape-recording (B. Whitney) of a bird (not collected) with a *parvirostris*-like song, obtained within the range of *intermedius*. Unfortunately vocal and genetic data are lacking for *intermedius*, but we presume that this *parvirostris*-like recording corresponds to the birds with *parvirostris*-like morphology collected nearby, strongly suggesting local sympatry of these taxa.

Likewise, we have no knowledge of the voice of nominate *unicolor*, which, like *intermedius*, has only a small distribution in northern Peru. Consequently, although there is evidence for more than one species within the taxa in Zimmer's polytypic *S. unicolor*, we are not certain of the relationships of nominate *unicolor* to other taxa, and the taxonomy of all these forms cannot be

TABLE 4
Provisional Taxonomy of *Scytalopus* Tapaculos. The Linear Sequence Is Arbitrary

Zimmer 1939; Peters 1951	Krabbe and Schulenberg 1997	Comments[a]
Scytalopus unicolor unicolor	*Scytalopus unicolor unicolor*	a
Scytalopus unicolor subcinereus	*Scytalopus unicolor subcinereus*	
Scytalopus unicolor intermedius	*Scytalopus unicolor intermedius*	a
Scytalopus unicolor latrans	*Scytalopus unicolor latrans*	b
Scytalopus unicolor parvirostris	*Scytalopus parvirostris*	b
Scytalopus speluncae	*Scytalopus speluncae*	
Scytalopus macropus	*Scytalopus macropus*	
	Scytalopus [*femoralis*]:	
Scytalopus femoralis femoralis	*Scytalopus femoralis*	
Scytalopus femoralis micropterus	*Scytalopus micropterus*	
	Scytalopus [*bolivianus*]:	
Scytalopus femoralis bolivianus	*Scytalopus bolivianus*	
Scytalopus femoralis atratus	*Scytalopus atratus atratus*	b, d, f
Scytalopus femoralis confusus	*Scytalopus atratus confusus*	c, f
Scytalopus femoralis nigricans	*Scytalopus atratus nigricans*	f
Scytalopus femoralis sanctaemartae	*Scytalopus sanctaemartae*	e
Scytalopus argentifrons	*Scytalopus argentifrons argentifrons*	
Scytalopus chiriquensis	*Scytalopus argentifrons chiriquensis*	c
	Scytalopus [*panamensis*]:	
Scytalopus panamensis panamensis	*Scytalopus panamensis*	
Scytalopus panamensis vicinior (in part)	*Scytalopus chocoensis*	
	Scytalopus robbinsi	
Scytalopus panamensis vicinior (in part)	*Scytalopus vicinior*	e
	Scytalopus [*latebricola*]	
Scytalopus latebricola latebricola	*Scytalopus latebricola*	
Scytalopus latebricola meridanus (in part)	*Scytalopus meridanus*	b
Scytalopus latebricola meridanus (in part)	*Scytalopus infasciatus* (?) (unnamed ?)	
Scytalopus latebricola caracae	*Scytalopus caracae*	e
Scytalopus latebricola spillmanni	*Scytalopus spillmanni*	
	Scytalopus parkeri	
	Scytalopus [*indigoticus*]:	
Scytalopus indigoticus	*Scytalopus indigoticus*	
	Scytalopus psychopompus	a, c
	Scytalopus novacapitalis	
Scytalopus magellanicus magellanicus	*Scytalopus magellanicus*	e
Scytalopus magellanicus superciliaris	*Scytalopus superciliaris superciliaris*	g
	Scytalopus superciliaris santabarbarae	a, c, g
Scytalopus magellanicus zimmeri	*Scytalopus zimmeri*	g
Scytalopus magellanicus simonsi	*Scytalopus simonsi*	g
Scytalopus magellanicus urubambae	*Scytalopus urubambae*	g
	Scytalopus unnamed species	
	Scytalopus unnamed species	
Scytalopus magellanicus altirostris	*Scytalopus altirostris*	g
Scytalopus magellanicus affinis	*Scytalopus affinis*	g
Scytalopus magellanicus opacus	*Scytalopus canus opacus*	g
Scytalopus magellanicus canus	*Scytalopus canus canus*	a, g
Scytalopus magellanicus griseicollis	*Scytalopus griseicollis griseicollis*	
Scytalopus magellanicus fuscicauda	*Scytalopus griseicollis fuscicauda*	a, g
Scytalopus magellanicus fuscus	*Scytalopus fuscus*	b, e
Scytalopus magellanicus acutirostris	*Scytalopus acutirostris*	
	Scytalopus schulenbergi	

[a] a = voice unknown; b = two or more species may be involved; c = doubtfully valid; d = population south of the Río Marañón vocally distinct and perhaps closer to *Scytalopus nigricans*. Treatment by Ridgely and Tudor (1994) when differing from traditional taxonomy: e = distinct species; f = subspecies of *S. bolivianus*; g = subspecies of *S. griseicollis*.

FIGS. 1–85. Sonagrams of most non-Brazilian taxa of *Scytalopus* tapaculos. Except when otherwise noted recordings are from Ecuador. Last locality modifier is province (Chile, Argentina, Ecuador), department (Bolivia, Peru, Colombia), or state (Venezuela). Dates are denoted day/month/year. PB = after playback, i.e., recorded in response to a tape-recording of the individual's or the species' voice. NV = natural vocalization, i.e., recorded without the stimulus of tape playback. Roman numerals refer to the original N.K. tapes (available from N.K. upon request), following numerals to counter of tape-recorder used.

→

satisfactorily resolved on present knowledge. In their uniform plumage, for what it is worth, *unicolor* and *intermedius* appear more similar to the northern taxa (*latrans* and *subcinereus*) than to *parvirostris*.

We suggest that the best taxonomic solution at the moment is to unite the northern forms into one polytypic species, *S. unicolor* (including *subcinereus*, *intermedius*, *unicolor*, and *latrans*), and to regard *S. parvirostris* as an independent species. This approach is conservative because it masks subtle, but consistent, vocal differences between *subcinereus* and *latrans*, and even between populations of *latrans* east and west of the Andes, as detailed below. We suspect that our *S. unicolor*, like Zimmer's before us, will prove to include more than one species. We continue to investigate this complex, but, because vocal and genetic data are lacking or are incomplete for two of these taxa, we are reluctant to introduce further taxonomic changes, until new information becomes available.

UNICOLORED TAPACULO *Scytalopus unicolor latrans*
(Hellmayr 1924; type in FMNH)

Brief diagnosis.—Relatively small. Depicted in Fjeldså and Krabbe (1990, pl. XLI 3a). The only Ecuadorian tapaculo (except *subcinereus* male) that may lack brown in the plumage. Even when some brown is present, the brown is usually unbarred and less extensive than in congeners, but a few individuals cannot, unless they weight more than 18 g, be separated from dark and fully adult males of *S. canus opacus*.

Plumage.—Adult male ($n = 19$): Fourteen western birds uniform: Chaetura Drab, Dark Mouse Gray, Chaetura Black, or Blackish Mouse Gray. Five eastern birds uniform black and may represent a different subspecies (see voice). *Adult female* ($n = 14$): Twelve western birds between Hair Brown and Chaetura Drab, Chaetura Drab, or Chaetura Black. Ten uniform, two with an almost indiscernible wash of Dresden Brown to Olive Brown on the tips of some flank feathers. One of two eastern birds similar, the other darker (Blackish Mouse Gray), about as dark as western males. *Immature/subadult male* ($n = 2$): Two western birds. One Fuscous Black with a faint wash of Dresden Brown to Olive Brown on thighs. One Dark Mouse Gray, with a wash of Dresden Brown to Mummy Brown on flanks, thighs, tertials, and edges of remiges, and with traces of Mouse Gray barring on flanks, thighs, and tertials; two primaries and a tertial in the left wing fresh or growing, uniform Blackish Mouse Gray. *Immature/subadult female* ($n = 3$): Two western, one eastern bird. One molting from juvenal into fairly uniform plumage, Dark Mouse Gray above, Light Mouse Gray below, distinctly lighter than its fresh-plumaged presumed mother, which is uniform Chaetura Black. One has 5 mm broad Cinnamon tips to the feathers of lowermost belly and vent, and has traces of this color as faint bars on the flanks and undertail coverts. The eastern bird similar to the latter but with even less brown, despite exhibiting juvenal primary-coverts, alula, and flight feathers except tertials. *Juvenal female* ($n = 2$): One western bird, molting. Crown and side of head between Ochraceous-Tawny and Cinnamon-Brown, mantle between Cinnamon-Brown and Brussels Brown, throat between Clay Color and Tawny-Olive, lower belly Cinnamon-Buff. Flanks obscurely barred gray and Tawny-Olive. Axillars Cinnamon-Buff. Wing-coverts and inner tertial Cinnamon-Brown with dusky bases and subterminal bar, remiges Dark Mouse Gray narrowly edged Cinnamon-Brown, inner remiges with narrow Cinnamon-Brown tips and faint indications of a presubterminal bar. One eastern bird (ANSP 176878) similar, but lores and feathering along the mandibular rami near Clay Color, contrasting somewhat with adjacent areas; feathers of lores also narrowly fringed black. Throat near Light Ochraceous-Buff. Rectrices unmarked, dark gray. *Juvenal male* ($n = 1$): One western bird, aberrant (ZMUC 80073). This schizochroic specimen had pinkish bill and feet, and the gray colors replaced with pallid gray, and appears uniform pale brown at a distance. We have seen several old museum specimens (NRS, BMNH) that we believe also represent the juvenal plumage of *latrans*. They all have less barring on the underparts than juvenals of other Ecuadorian species.

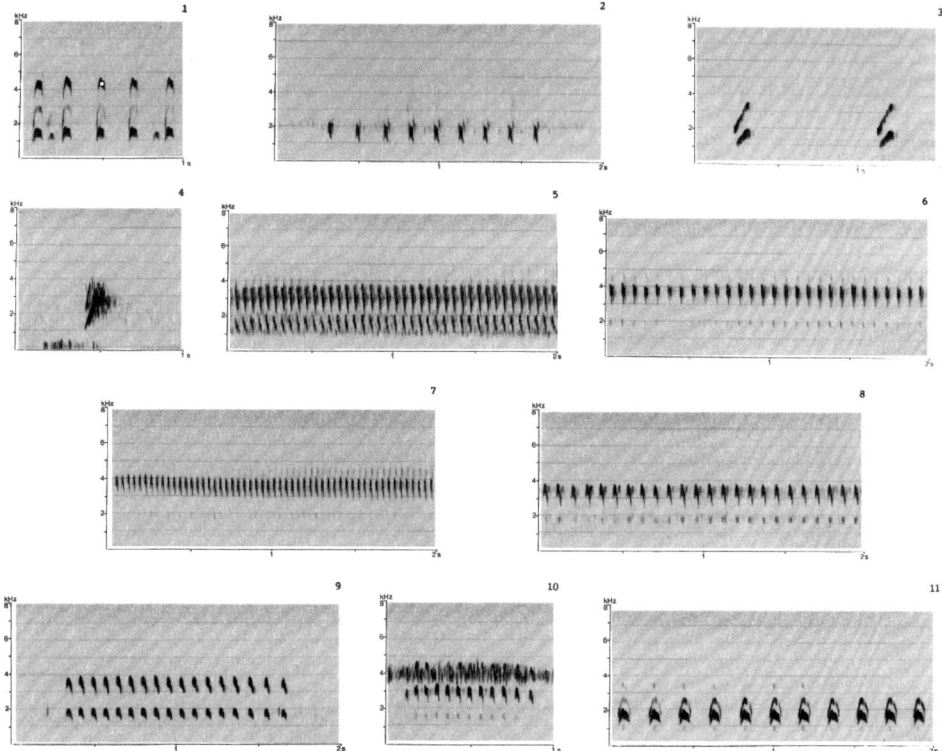

FIGS. 1–11. *Scytalopus unicolor latrans*: 1. Excited song by pair, female fast, male slow. PB. Lloa, Pichincha, 2,700 m, 21/7/84. Both collected. N.K. 2. Song on east slope. PB. Cutucú, Morona-Santiago, 1,525 m, 1/7/84. T.S.S. 3. Call. NV. Yanacocha, Pichincha, 3,400 m, 30/9/83. N.K. 4. "*Brzk*", presumably by a female or young. PB of female. Yanayacu, Pichincha, 3,700 m. 7/3/92. XLVIIA 410–417. N.K. *Scytalopus parvirostris*: 5. Male song. PB. Playa Pampa, Pasco, Peru, 2,325 m, 24/6/85. Collected. T.S.S. 6. Song. PB. Machu Picchu, Cuzco, Peru, 2,600 m, 13/7/83. N.K. 7. Song. PB. Siberia cloud forest, Santa Cruz, Bolivia, October 1983. T. A. Parker. *Scytalopus schulenbergi*: 8. End of male song. NV. Below Abra Malaga, Cuzco, Peru, 3,000 m, 4/12/83. Collected. N.K. 9. Call. PB. Below Abra Malaga, Cuzco, Peru, 3,350 m, 21/7/85. T.S.S. 10. Call. NV. Above Sandia, Puno, Peru, 2,800 m, 25/12/83. N.K. *Scytalopus unicolor subcinereus*: 11. Excited male song. PB. Sural, Azuay, 2,650 m, 4/3/91. Collected. XXIIA 381–385. N.K.

Annual cycle in Ecuador ($n = 13$).—*Juvenals*: 7, 7 October; 7, 19, 20 March; 27 April; 2, 10 May (all prov. Pichincha); 11 July (prov. Morona-Santiago). *Brood patch:* 5 January (female) (west prov. Cotopaxi). *Active ovary:* 7 March (prov. Pichincha); 15 March (southeast prov. Carchi). *Nest:* 8 October (near Baños, prov. Tungurahua) (Skutch 1972).

Vocalizations in Ecuador.—Song a repeated note 2–8 times/s (fastest at high excitement, such as countersinging birds). The song also may be given by both sexes in duet, with the song of the female slightly higher-pitched than that of the male (Fig. 1) (also Fjeldså and Krabbe [1990, p. 425 sonagram 2]). Song on most of the east slope (Papallacta, prov. Napo southwards; Fig. 2) averages faster than in remainder of Ecuadorian range, the notes are sharper, and the song is more frequently rhythmic, reminiscent of that of *subcinereus*. A song often begins with a "stutter," in which notes may be given singly, less often in doublets or triplets. In the song, the fundamental note is frequently loudest, 1.2–1.4 kHz, but first and second overtones almost equally as loud, and occasionally the first overtone is louder than the fundamental. Each note is about 0.06 s long, the pitch rising in the beginning, and then descending smoothly to the end.

The call of both sexes is a single note that resembles song notes in pitch and overtones, but is given singly or in a series of two or three at irregular intervals. Each note is distinctly rising throughout (Fig. 3). Commonly heard in the west, rarely in the east.

An excited "*brzk*" (Fig. 4) is only heard infrequently. On two occasions it was found to be given either by a female or its young, and was interpreted as a distress call.

Habitat in Ecuador.—Humid forest undergrowth, *Chusquea* bamboo or *Neurolepis* cane, hu-

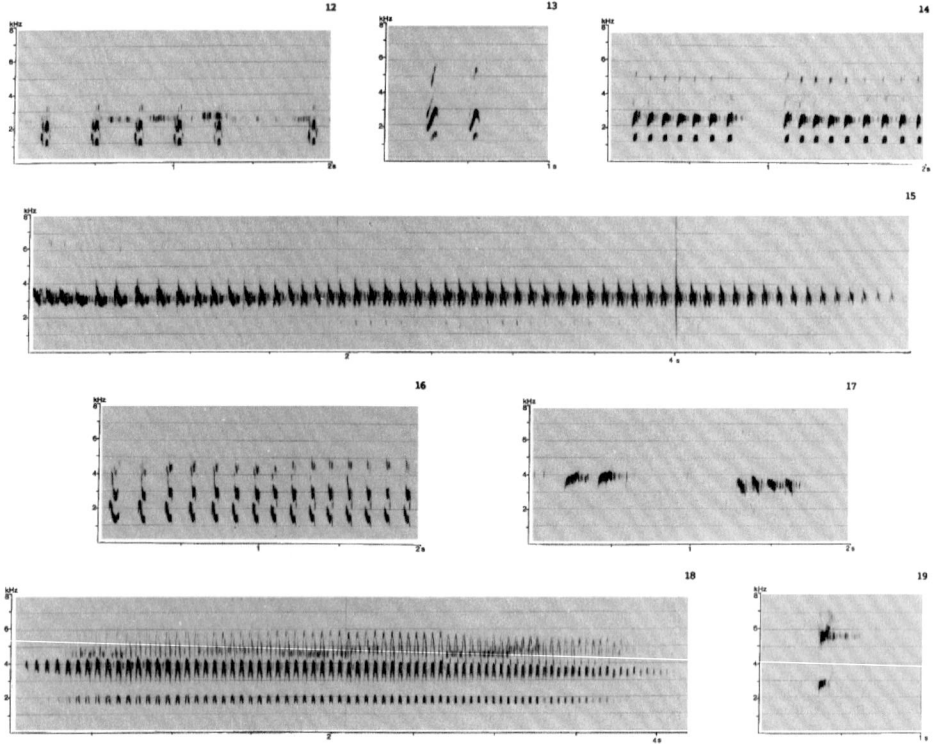

FIGS 12–19. *Scytalopus unicolor subcinereus*: 12. Normal male song, each note with a double quality. NV. Same bird as previous. XXIIA 386–395. N.K. 13. Call of female with young. PB. Sural, Azuay, 2,650 m, 4/3/91. Both collected. XXIIA 306–324. N.K. *Scytalopus atratus atratus*: 14. Male song. PB. Hollín road, Napo, 1,350 m, 11/10/92. Collected. LIXB 216–224. N.K. *Scytalopus bolivianus*: 15. Song. PB. Calabatea, La Paz, Bolivia, 1,400 m, October 1983. T. A. Parker. 16. Song. PB. Chapare road below Miguelito, Cochabamba, Bolivia, 1,800 m, 30/10/83. R. A. Rowlett. 17. Call. PB. Calabatea, La Paz, Bolivia, 1,400 m, October 1983. T. A. Parker. *Scytalopus sanctaemartae*: 18. Song. NV. Below Cincinnati, above Campana, nw-slope of Santa Marta mountains, Colombia, 1,480 m, 6/2/84. S. Hilty. 19. "*Brzk*". PB. Santa Marta mountains, Colombia, 1,700 m, early January 1994. P. Coopmans.

mid *Polylepis* scrub, shrubbery, swampy areas, along ditches. Opportunistic and often numerous in second growth, entering relatively dry regions through riparian shrubbery. Where in contact with *S. canus opacus*, *S. u. latrans* haunts more humid and broad-leaved shrubbery or bamboo rather than ericaceous scrub. Where syntopic with *S. spillmanni*, *S. u. latrans* may be found slightly more at edge and in open or drier understory.

Distribution in Ecuador.—Two nearly isolated populations, differing somewhat vocally and in color, but not notably in measurements. One at 1,975–4,000 m on the Pacific watershed south to northern prov. Cañar, just "spilling over" and down to 2,750 m on the east slope along the Colombian border in the Sucumbíos region, and to 3,000 m in Napo at Oycacachi. Another locally at 1,975–2,450 m on the east slope from Papallacta, prov. Napo, southwards; known also from the outlying ridges of Sumaco (prov. Napo), Cutucú (prov. Morona-Santiago), and the highest parts of Cordillera del Cóndor (prov. Zamora-Chinchipe) (EPN). Birds recorded at 2,600 m on Volcán Tungurahua, prov. Tungurahua, may also belong here. The eastern form may deserve taxonomic recognition. It meets *subcinereus* along the right bank of the Río Paute, prov. Azuay, where N.K. recorded both song-types, but their voices, and the plumage of subadult female (one such collected), are so similar that intermediates are difficult to identify on this basis.

UNICOLORED TAPACULO *Scytalopus unicolor subcinereus*
(Zimmer 1939; type in AMNH)

Brief diagnosis.—Relatively small. Adult males uniform blackish like *intermedius* and east Ecuadorian *latrans*. Adult females considerably paler and with extensively brown flanks that are

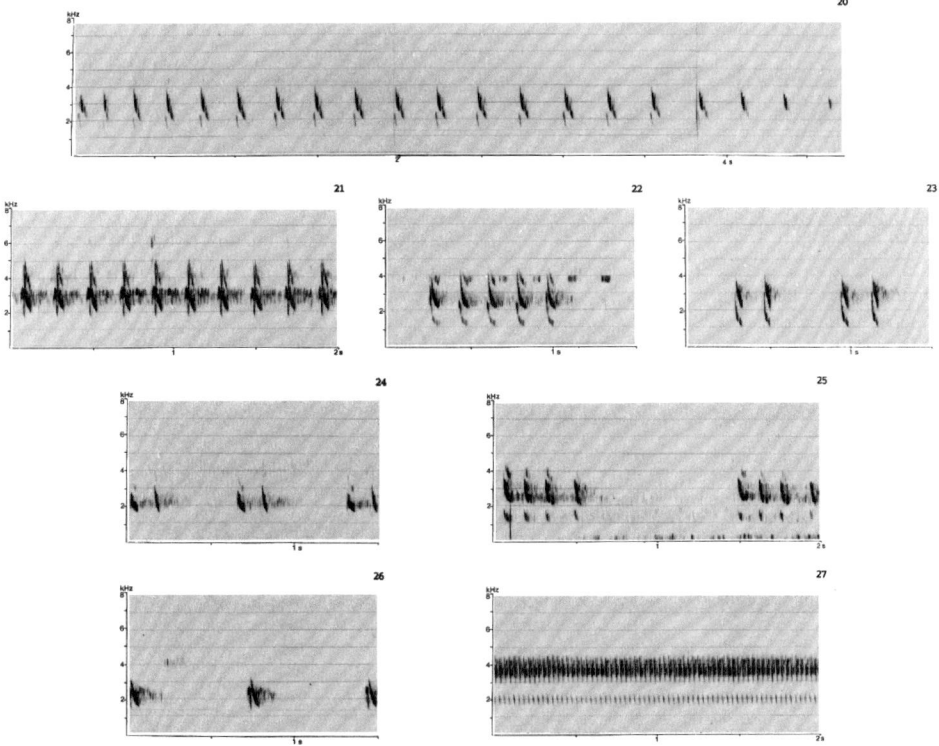

FIGS. 20–27. *Scytalopus atratus nigricans*: 20. Song. NV. Cerro El Teteo, Táchira, Venezuela, 1,050 m. P. Schwartz. *Scytalopus* new species?: 21. Song. PB. Finca Merenberg, Huila, Colombia, 2,200 m, 18/10/86. B. Whitney. *Scytalopus atratus "atratus"*: 22. Part of song. NV. Jirillo, San Martín, Peru, 1,350 m, 26/7/86. T.S.S. 23. Part of song. NV. Jirillo, San Martín, Peru, 1,350 m, 27/7/86. T.S.S. *Scytalopus micropterus*: 24. Song. NV. Cutucú, Morona-Santiago, 1,750 m, 24/6/84. T.S.S. 25. Call/alarm, possibly female. NV. Guacamayos, Napo, 2,035 m, 23/8/91. XXVIIIA 231–232. N.K. *Scytalopus femoralis*: 26. Song. PB. Abra Patricia, Amazonas, Peru, 2,250 m, 7/9/83. N.K. *Scytalopus canus opacus*: 27. Male song. PB. Limón road, Azuay, 3,400 m, 13/12/91. Collected. XLA 12–22. N.K.

uniform or obscurely and faintly barred; paler than female *intermedius*, but usually darker than female *unicolor*. Wings and tail distinctly shorter than in *S. parkeri*. Flanks not distinctly barred as in *S. robbinsi*. Subadult male with barred flanks much paler than *S. robbinsi*. Juveniles variably barred or uniform below, but much paler than *latrans*.

Plumage.—Adult male ($n = 2$): Uniform black, morphologically indistinguishable from males of east Ecuadorian *"latrans"*. Adult female ($n = 5$): Deep Mouse Gray, Dark Mouse Gray, or Blackish Mouse Gray. Flanks, under-tail coverts, and in all but one, also the lower belly Tawny-Olive; usually rump and sometimes upper-tail coverts and lower sides also more or less washed with this color; either uniform or with faint and obscure blackish barring on the flanks. Axillars rather light, between Pinkish-Buff and Cinnamon-Buff. One specimen has nape, back, wings, and upper-tail coverts faintly washed with brown, and has Pallid Gray edges to the feathers of the throat and belly, giving a slightly streaked effect. *Immature/subadult male* ($n = 2$): Both molting from juvenal plumage, one with large bursa, slightly enlarged testes and singing; the other without bursa and with small testes. Wings, upper-tail coverts, most of tail, lower sides and flanks Tawny Olive, barred dusky on wing-coverts, tertials, upper-tail coverts and flanks. Some rectrices brown and dark-barred at their tip, others uniform, dark gray. Lower belly Pinkish-Buff. No older subadults from Ecuador were available, but two males from Porculla, depto. Lambayeque, Peru (ANSP), and Cruz Blanca, depto. Piura, Peru (LSUMZ), have some dull brown mixed into the flanks; the first is paler gray than adults. *Juvenal male* ($n = 1$): Crown, side of head, mantle, and back between Dresden Brown and Mummy Brown, feathers narrowly tipped blackish, wing-coverts Dresden Brown, each with blackish tip and central bar. Remiges Chaetura Drab, narrowly edged Mummy Brown. Inner tertial much like wing-coverts, next two

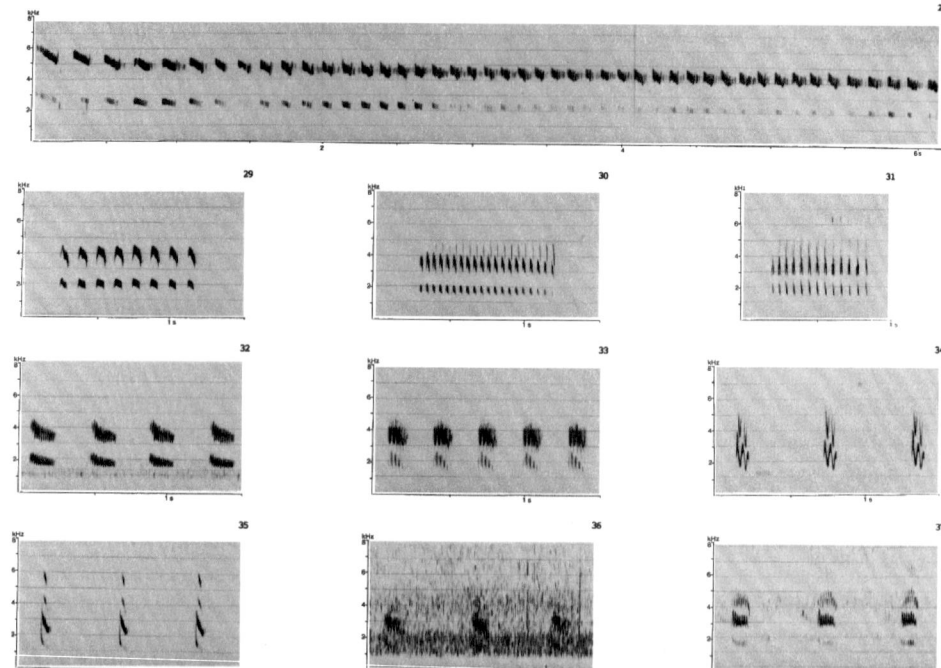

FIGS. 28–37. *Scytalopus canus opacus*: 28. Female part of duet, probably pair-formation. PB. Near Papallacta, Napo, 3,900 m, 14/10/83. N.K. 29. Male call. PB. Cerro Mongus, Carchi, 3,600 m, 22/3/92. Collected. XLVIIIA 61–65. N.K. 30. Call. PB. Cerro Chinguela, Piura, Peru, 1980. T. A. Parker. *Scytalopus affinis*: 31. Female alarm near nest (male collected). NV. Quebrada Pucavado, Ancash, Peru, 4,100 m, 16/2/87. N.K. 32. Male song. PB. Chinancocha, Quebrada Llanganuco, Ancash, Peru, 9/2/86. T.S.S. *Scytalopus altirostris*: 33. Song (?). Not seen. PB. Bosque Unchog, Huánuco, Peru, 3,500 m, 15/11/83. N.K. *Scytalopus* unnamed species: 34. Song. NV. Millpo, Pasco, Peru, 3,650 m, 8/7/85. T.S.S. *Scytalopus* unnamed species: 35. Male song. PB. Nevado Ampay, Apurimac, Peru, 3,500 m, 17/3/87. Collected. N.K. 36. Call. NV. Nevado Ampay, Apurimac, Peru, 3,500 m, 18/3/87. N.K. *Scytalopus urubambae*: 37. Song. NV. Near Totora, Río Ucumare, Santa Teresa valley, Cuzco, Peru, 3,700 m, December 1990. G. Engblom.

like remiges, but with a narrow Dresden Brown subterminal bar. Rump and upper-tail coverts barred blackish and Antique Brown. Chin and throat Cinnamon-Buff, grading to Clay Color on rest of underparts, feathers of the entire underparts narrowly tipped dusky, those of belly, flanks and under-tail coverts also with a presubterminal dusky bar. Rectrices Chaetura Drab, central pair uniform, remainder with faint barring of Dresden Brown. Two birds described by Zimmer (1939) also have barred underparts. *Juvenal female:* Five were described by Zimmer (1939), four nearly uniform ochraceous or dull ochraceous below, one completely barred below. Three of these were re-examined and found to have more or less uniform brown backs. One from Canchaque, depto. Piura, Peru (LSUMZ), similar, but with light-brown feathers of nape scalloped with blackish. Below all with pale ochraceous breasts and bellies, the relatively bright, buffy, lower bellies barred.

Annual cycle in Ecuador ($n = 4$).—*Juvenal:* 4 March. *Active ovary:* 6 February; 5 March; 12 April.

Vocalizations in Ecuador.—The normal song (Fig. 12) is sharper than that of *latrans*, and each note has a double quality, the first part sharp (0.013 s long). The excited song (Fig. 11) is very much like that of *latrans*, but with a different spectrum of harmonics, and of a somewhat sharper quality, as seen by all notes being slightly peaked, rather than most notes smooth and rounded as in latrans (Fig. 1).

The song is also more rhythmic than *latrans*, with a short pause after the slightly lower first note followed by a varied number of 2–5 or more similar notes (Fig. 12); pauses between phrases usually 0.2–1.0 s, excited song continuous (Fig. 11).

The first overtone is loudest, 1.6–2.2 kHz. In low-pitched notes the second overtone is relatively loud, the fundamental and third overtones are barely audible, in higher-pitched notes the

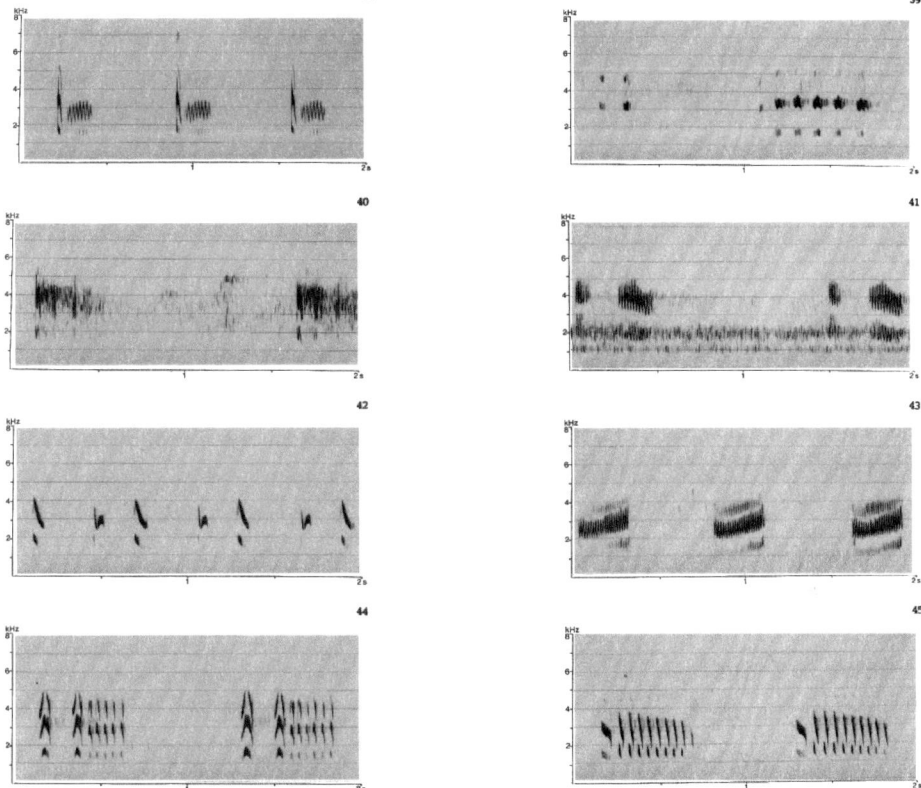

FIGS. 38–45. *Scytalopus simonsi*: 38. Male song. NV. 70 road km west of Cochabamba on Oruro road, Cochabamba, Bolivia, 3,850 m, 14/4/87. N.K. 39. Female call. PB. Above Cuyocuyo, Puno, Peru, 3,600 m, 22/12/83. Collected. N.K. *Scytalopus zimmeri*: 40. Song. NV. Cerro Campamiento (20°48′S, 64°30′W), Chuquisaca, Bolivia, 2,700 m, 23/9/91. J. Fjeldså. *Scytalopus superciliaris superciliaris*: 41. Song. PB. Above Tafi del Valle, Tucumán, Argentina, 2,400 m, ca. 1989. R. S. Ridgely. *Scytalopus magellanicus*: 42. Song. NV. Parque Nacional Nahuelbuto, Malleco, Chile, 1,300 m, 25/8/85. B. M. Whitney. *Scytalopus fuscus*: 43. Song. PB. Parque Nacional Campana, Santiago, Chile, 450 m, 6/12/87. B. Whitney. *Scytalopus acutirostris*: 44. Male song. NV. Carpish tunnel, Huánuco, Peru, 2,760 m, 9/2/87. Collected. N.K. 45. Male song. PB. Same bird as previous cut. N.K.

fundamental is almost as loud as the first overtone, and the second overtone barely audible; apparently any sound between 0.9 and 2.6 kHz is loud irrespective of whether it is an overtone or a fundamental.

Call (Fig. 13; also Fjeldså and Krabbe [1990, p. 425, sonagram 1]) given by both sexes is similar to that of *latrans*, sometimes with a harsher quality, and often given in slow series of 1–3 notes.

As in *latrans* a rarely heard, excited "*brzk*" is given, presumably by the female or young, during distress.

Habitat in Ecuador.—Humid and semi-humid forest undergrowth and shrubbery, including small and secondary patches along streams and ditches into the arid zones. Also mixed *Polylepis-Gynoxys* woodland. Tolerates drier conditions than congeners, but invades humid forest in absence of competitors. Where syntopic with *S. parkeri, S. u. subcinereus* avoids stands of bamboo.

Distribution in Ecuador.—1,500–4,000 m. Southwestern Ecuador from the Pacific slope in prov. Azuay, the Cajas plateau, and the inter-Andean slopes of Cuenca valley south and west on the inter-Andean and Pacific slopes, including all the more-or-less isolated mountains over 2,000 m. Along the right bank of Río Paute, it meets east slope *latrans* near Amaluza, where both song-types can be heard. Here its range also overlaps with that of *S. spillmanni*. Elsewhere on the west slope of the Eastern Andes *subcinereus* is replaced above by *S. canus opacus* with no overlap. On the Páramos de Matanga both occur right to the crest, *S. canus opacus* on the eastern

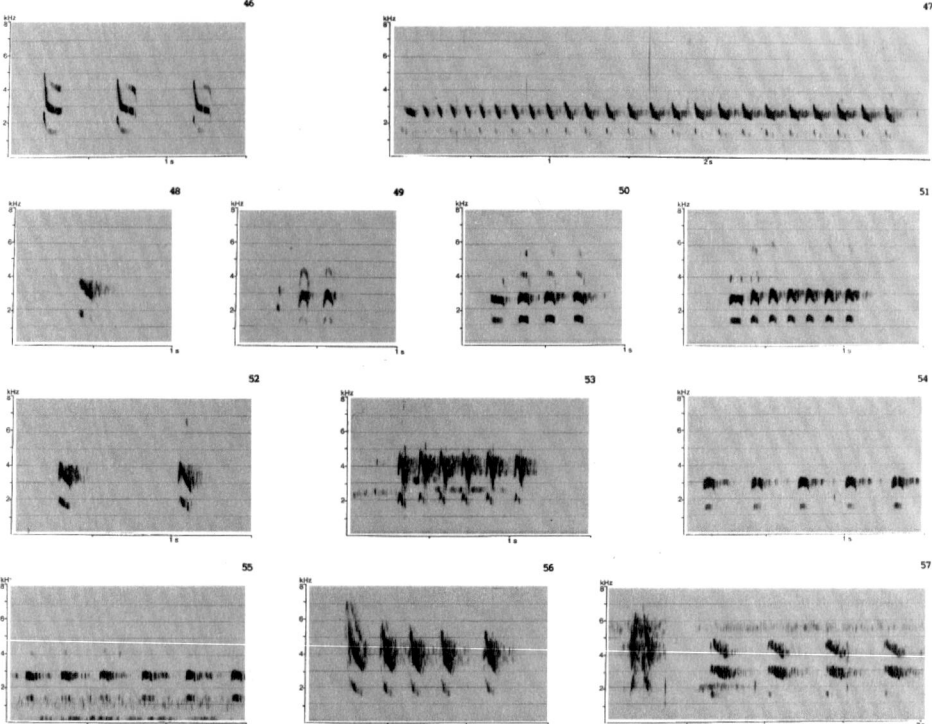

FIGS. 46–57. *Scytalopus acutirostris*: 46. Male call. NV. Same bird as previous cuts. N.K. *Scytalopus vicinior*: 47. Song. NV. La Planada, Nariño, Colombia, 1,800 m, January 1989. F. Lambert. 48. Call. NV. Parque Nacional Farallones, Valle del Cauca, Colombia, 1,950 m. 13/1/83. B. Whitney. 49. Call. NV. Parque Nacional Farallones, Valle del Cauca, Colombia, 1,950 m. 3/1/83. B. Whitney. 50. Female call. PB. Mindo, Pichincha, 1,700 m, 15/11/92. Collected. XLIIIA 340–345. N.K. 51. Female call. PB. Maquipucuna, Pichincha, 1,775 m, 31/3/93. Collected. LXVIIIA 339–343. N.K. 52. Female end of very high-pitched descending series. Mindo, Pichincha, 1,700 m, 2/11/91. XXXVIB 121–122. N.K. *Scytalopus panamensis*: 53. Alarm? Headwaters of Río Tigre, Cerro Tacarcuna, Chocó, Colombia, 1,150 m, October 1990. M. Pearman. *Scytalopus chocoensis*: 54. Song. PB. El Placer, Esmeraldas, 670 m, 13/8/87. T.S.S. Collected. 55. Male song. PB. El Placer, Esmeraldas, 670 m, August 1987. T.S.S. 56. Female call. PB. Alto Tambo, Esmeraldas, 550 m, 16/2/92. Collected. XLVB 201–209. N.K. 57. "*Brzk*" and beginning of song. "*Brzk*" presumably given by a female. NV. El Placer, Esmeraldas, 670 m, 5/7/84. N.K.

slope, *subcinereus* on the western. Near the top they occur in similar habitat within 20 m of one another but on their respective slopes. On the Pacific slope in prov. Azuay, *subcinereus* is replaced below by *S. robbinsi* with little or no overlap.

Distribution beyond Ecuador of S. [*unicolor*].—In Venezuela *S. unicolor latrans* is known from the Andes of edos. Mérida and Táchira at 1,800–2,200 m (AMNH, Berlin Museum; see also Meyer de Schauensee and Phelps [1978]). In Colombia it is known from all slopes of the three main Andean chains, ranging from 1,800 to 3,800 m (AMNH, BMNH, FMNH, LACM, NRS, USNM, WFVZ). Three recently collected specimens, two from the east slope and one from the west slope of the Eastern Andes in depto. Cundinamarca (ICN) show the same black color found in eastern Ecuadorian birds. Birds from depto. Antioquia in the Central Andes sound closest to the west Ecuadorian birds (tape-recordings by N.K.). In Peru birds singing like those in east Ecuador occur east of the Huancabamba valley in northern depto. Cajamarca (Cerro Chinguela, 2,000 m [Parker et al. 1985; T. A. Parker tape-recordings in LNS]). Others collected in Peru without vocal data, but taken nearby at Chaupe, depto. Cajamarca, northeast of Huancabamba, 1860 m [AMNH]; and at Lomo Santo above Jaen, depto. Cajamarca, 1,525 m [AMNH], undoubtedly also belong here. *S. unicolor intermedius* (Zimmer 1939; type in AMNH) occurs in depto. Amazonas in the northern end of the Central Andes, Peru (AMNH, ANSP, BMNH, LSUMZ), and similar specimens have been taken further south in depto. San Martín (Puerto del Monte, 3200 m [LSUMZ]). A male from Chira, 2,290 m (ANSP) was referred to

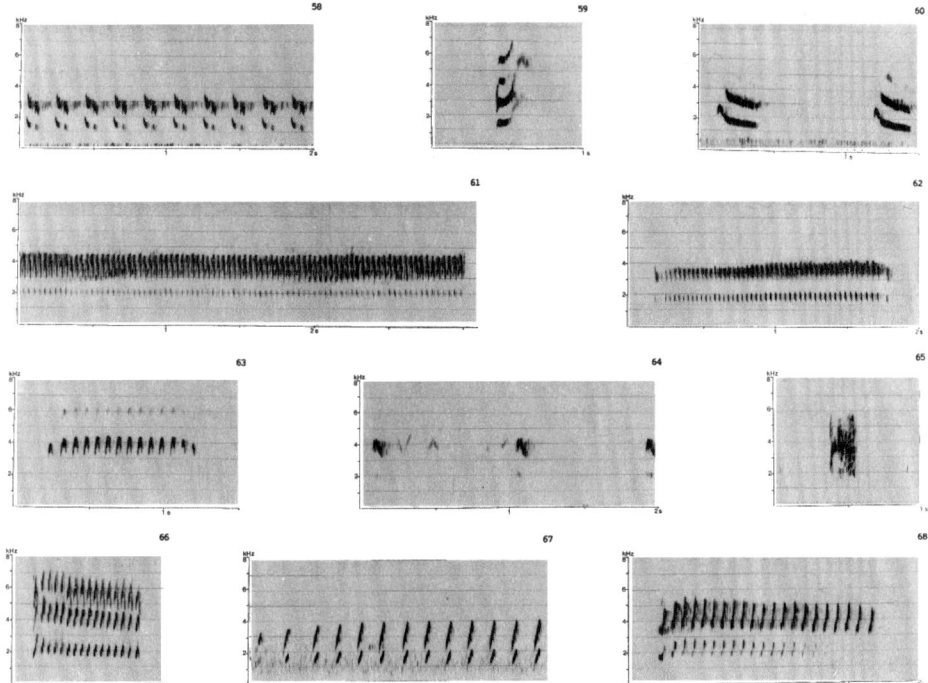

FIGS. 58–68. *Scytalopus robbinsi*: 58. Song. NV. Piñas, El Oro, 900 m, 12/6/85. M. B. Robbins. 59. Female single-note call. NV. Piñas, El Oro, 900 m, 26/9/90. VIIB 221–230. N.K. 60. Female descending series. PB. Piñas, El Oro, 900 m, 16/11/91. Collected. XXXVIIA 45–78. N.K. *Scytalopus spillmanni*: 61. Song. PB. Above Tandayapa, Pichincha, 2,300 m, 29/7/84. T.S.S. 62. Male excited (short rising trills). PB. Maldonado road, Carchi, 2500 m, 20/11/90. Collected. XIIB 195–200. N.K. 63. Alarm. PB. Above Tandayapa, Pichincha, 2,300 m, 29/7/84. T.S.S. 64. Female long, slowly descending series as also given in duet. PB. Apuela road, Imbabura, 2,300 m, 12/6/87. Collected. VIIB 272–274. N.K. 65. "*Brzk*" presumably given by a female. NV. Llanganates, Napo, 3,300 m, 27/5/92. XLIXB 217–219. N.K. *Scytalopus latebricola*: 66. Call. PB. Santa Marta mountains, Colombia, 2,350 m, January 1994. P. Coopmans. *Scytalopus meridanus*: 67. NV. Los Frailes Hotel, Mérida. Venezuela. P. Coopmans. 68. Alarm. NV. Universidad Los Andes (ULA) forest, Mérida, Venezuela, 2,200 m, 18/1/94. G. Engblom.

latrans by Zimmer (1939), who believed the locality to be north or west of Río Marañón. As far as it is now known, however, Chira is south and east of the river, so the specimen is presumably referable to *intermedius*.

S. unicolor subcinereus occurs in northern Peru at 1,220–3,200 m on the Pacific slope of Western Andes in depto. Piura (Palambla, 1,220 m [AMNH, ANSP]; El Tambo, 2,800 m [ANSP, MHN]; Cruz Blanca, 1,700–3,200 m [Parker et al. 1985, tape-recordings by T. A. Parker in LNS]), depto. Lambayeque (Porculla, 1,830 m [ANSP]), and depto. Cajamarca (vicinity of Cascabamba, 2,400–2,900 m [MHN and tape-recordings by N.K.]; Hda. Taulis, 1,700–2,700 m [AMNH, MHN]; Nancho and [=] Montaña de Nancho, 2,350 m [MHN, Warsaw Museum]). It seems possible that *subcinereus* crosses over to the east slope of West Andes in depto. Cajamarca. A male and a female from Cutervo, depto. Cajamarca, 3,000 m (BMNH, Warsaw Museum), were ascribed to *S. magellanicus* by Taczanowski (1884), then to *latrans* by Hellmayr (Cory and Hellmayr [1924]). Zimmer (1939) discussed them and left them in *latrans*, although the description of a dark male and a pale (molting) female with a nearly uniform juvenal tail suggests they belong with *subcinereus*, as might a specimen from Tambillo [1,770–2,440 m] (MHN).

The pale *Scytalopus unicolor unicolor* (Salvin 1895; type in BMNH) is definitely known from farther south on the eastern slope of the Western Andes, Peru, in southern depto. Cajamarca (Cajabamba, 2,750 m [AMNH, BMNH]) and depto. La Libertad (Huamachuco, 3,170 m [AMNH]; Succha [AMNH]; Soquián, 2,000 m [ANSP]), and has been recorded also from the Pacific slope in depto. Cajamarca (Chugur [AMNH]; Sunchubamba, 2,650 m [MHN]) (Koepcke 1961). Chugur is north of the southernmost locality known for *subcinereus*. Zimmer (1939) pointed out that two males from Chugur are no darker than some topotypical specimens of

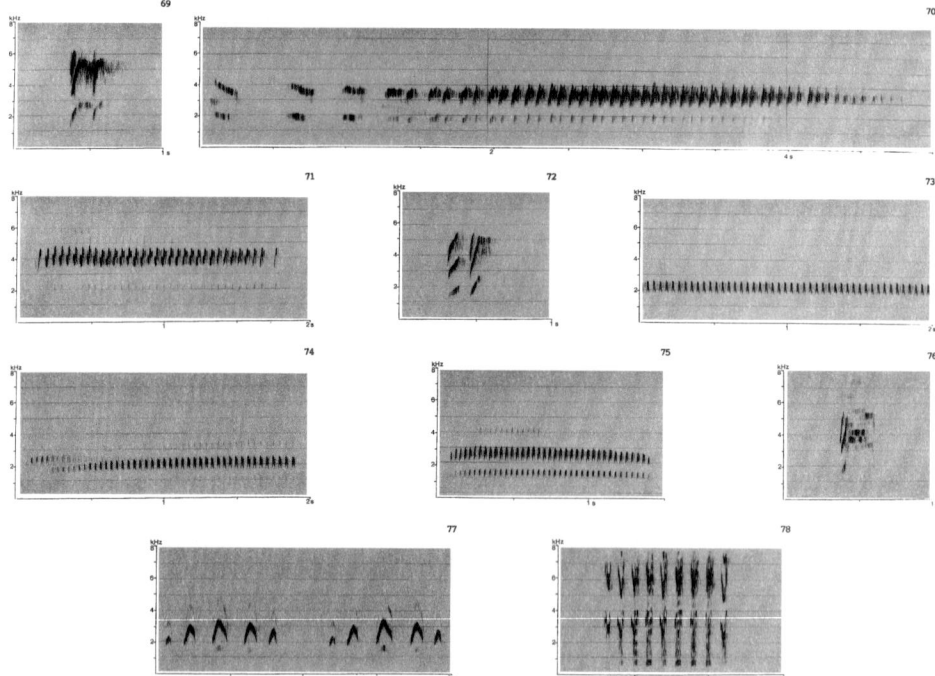

FIGS. 69–78. *Scytalopus meridanus*: 69. "*Bzrk*". NV. ULA forest, Mérida, Venezuela, 2,200 m, 18/1/94. G. Engblom. 70. Song. NV. ULA forest, Mérida, Venezuela, 2,200 m, 18/1/94. G. Engblom. 71. Alarm. NV. Near Lagrita, east Táchira, Venezuela, 2,680 m, 21/10/88. B. Whitney. 72. Call. NV. Near Lagrita, east Táchira, Venezuela, 2,680 m, 21/10/88. B. Whitney. *Scytalopus "meridanus"* (= *S. infasciatus*?): 73. Song. NV. Parque Nacional Chingaza, Cundinamarca, Colombia, in tall mossy forest, 3,100 m, February 1989. F. Lambert. 74. Short rising trills. Same bird as in Fig. 73. NV. F. Lambert. 75. Alarm. East of Bogotá, Cundinamarca, Colombia. P. Coopmans. *Scytalopus caracae*: 76. "*Brzk*". NV. Near Colonia Tovar, coastal mts, Aragua, Venezuela, 1985. J. P. O'Neill. 77. Song of same individual as in Fig. 76. PB. J. P. O'Neill. 78. Alarm. PB. Near Colonia Tovar, coastal mts, Aragua, Venezuela, 26/2/91. D. Fischer.

unicolor. The morphological differences between *unicolor* and *subcinereus* are not great. In view of this similarity, and the geographic proximity of these taxa, it is possible that they intergrade. Such intergrades would be difficult or impossible to detect morphologically, however. Again, we can only lament the paucity of information on nominate *unicolor*.

Populations that currently we treat as *Scytalopus parvirostris* (Zimmer 1939; type in AMNH; the type from La Paz, Bolivia, and a molting juvenal from Machu Picchu, Peru, depicted in Fjeldså and Krabbe [1990, pl. XLI 4c and 4b]), occur in central Peru from Cordillera Colán, depto. Amazonas (LSUMZ; no vocal data but the series is consistent in morphology with *S. parvirostris*) and Florida, depto. Amazonas (one tape-recording by B. M. Whitney), south at least to Machu Picchu, depto. Cuzco, Peru. *Scytalopus parvirostris* is also found from depto. La Paz south to depto. Santa Cruz, Bolivia, in which area it occurs up to 3,200 m (AMNH, ANSP, FMNH, LSUMZ, NRS, ZMUC).

The rate of delivery (notes/s) of the song of *Scytalopus parvirostris* varies from 21 (depto. Pasco, central Peru) (Fig. 5) to 14 (Machu Picchu, depto. Cuzco) (Fig. 6), to 20–28 in deptos. Cochabamba and Santa Cruz, Bolivia (Fig. 7). Also, the elevational distribution of birds that we assign to *S. parvirostris* varies latitudinally. In central Peru (depto. Pasco) they range from 1,800 to ca. 2,400 m, whereas in Bolivia the elevational range is ca. 2,000–3,200 m (Whitney 1994), and the elevation for the type locality of *S. parvirostris* was reported to be 3300 m (Zimmer 1939). As also noted by Whitney (1994), further work might show that more than one species is involved in what we here call *S. parvirostris* (the name *sylvestris* Taczanowski, 1874 may be available for the central Peruvian population).

Some puzzling recordings from the Amazonian slope in depto. Cuzco (below Abra Málaga, 3,000 m, subad. male [ZMUC 80007] tape-recorded by N.K. [Fig. 8] and 3,350 m, tape-recorded by T.S.S. [Fig. 9]), and in depto. Puno (above Sandia, 2,800 m [tape-recording by N.K.; Fig.

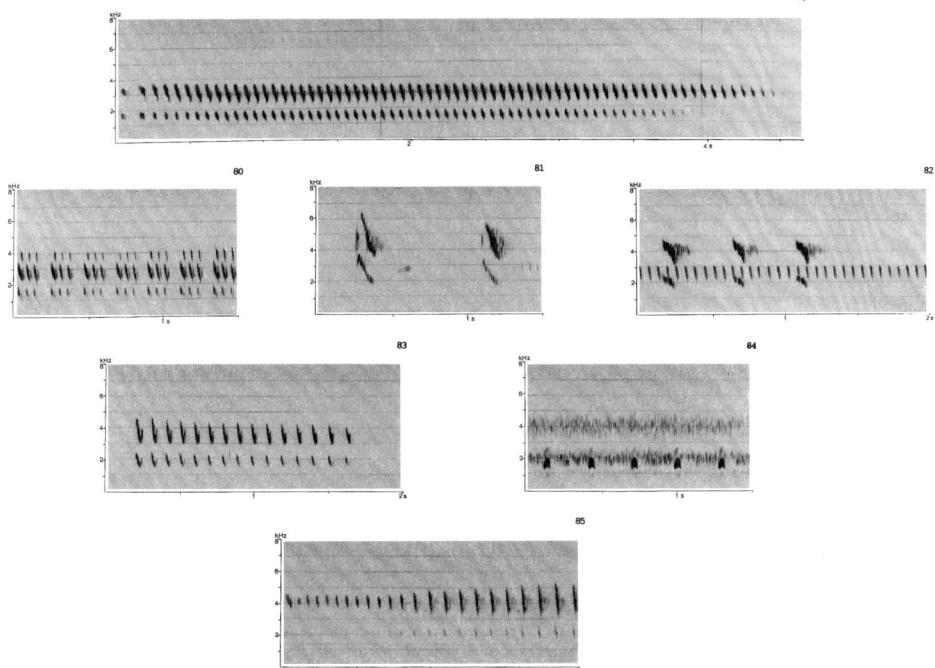

FIGS. 79–85. *Scytalopus parkeri*: 79. Male song. NV. Cord. del Cóndor, Zamora-Chinchipe, 2,100 m, July 1993. T. A. Parker. 80. Male excited song. PB. Limón road, Morona-Santiago, 3,100 m, 9/6/84. Collected. N.K. 81. Beginning in duet, female highest pitched. NV. Limón road, Morona-Santiago, 2,280 m, 18/6/84. Female collected. N.K. 82. Later in same duet as Fig. 81. N.K. 83. Call (typical of male). NV. Acanamá, Loja, 3,100 m, 13/2/91. XXIA 335–367. N.K. *Scytalopus macropus*: 84. Song. NV. Bosque Unchog, Huánuco, Peru, 3,500 m, 17/11/83. N.K. *Scytalopus argentifrons argentifrons*: 85. Song. PB. Parque Nacional Monteverde, Costa Rica, 24/5/79. O. Jakobsen.

10]) referred to under *parvirostris* in Fjeldså and Krabbe (1990) appear to be representatives of a recently discovered species. This species, *Scytalopus schulenbergi*, which is found above *parvirostris*, was described from deptos. Puno, Peru, and La Paz and Cochabamba, Bolivia, by Whitney (1994). Two individuals tape-recorded (and one of them collected) at Abra Málaga sang like Bolivian birds. Scolds recorded at both Abra Málaga and in depto. Puno are slightly softer than scolds of Bolivian birds.

We do not know where to place two specimens with relatively pale gray chests and bright brown, unbarred flanks from depto. Ayacucho (NE Tambo, 2,600 and 3,390 m [LSUMZ]). They might represent an undescribed taxon.

Scytalopus femoralis, sensu Zimmer 1939

Zimmer considered the forms *bolivianus* (Allen 1889; type in AMNH), *femoralis* (Tschudi 1844; type in MNN), *micropterus* (Sclater 1858; type in BMNH), *confusus* (Zimmer 1939; type in AMNH), *atratus* (Hellmayr 1922; type in CM), and *sanctaemartae* (Chapman 1915; type in AMNH) as subspecies of *S. femoralis*. Zimmer noted that some specimens that he included in this polytypic species exhibited a white spot on the forecrown, but evidently considered the presence or absence of this white crown spot to be an individually variable character. Later *nigricans* (Phelps and Phelps 1953; type on deposit to AMNH) also was described as a subspecies of Zimmer's *S. femoralis*.

T.S.S. and others found that in Peru and Ecuador Zimmer's *Scytalopus femoralis* encompassed two sympatric species. Birds with a white crown spot differ from birds without it in vocalizations and by being darker, smaller, relatively shorter-tailed, by often having white tips to the belly feathers, and also by generally occurring at lower elevations, at ca. 1,070–2,030 m, whereas dark-crowned birds occur at ca. 1,300–2,300 m. Specimens (LSUMZ) documenting the co-occurrence of both types along the same elevational transects in Peru were collected on the Cor-

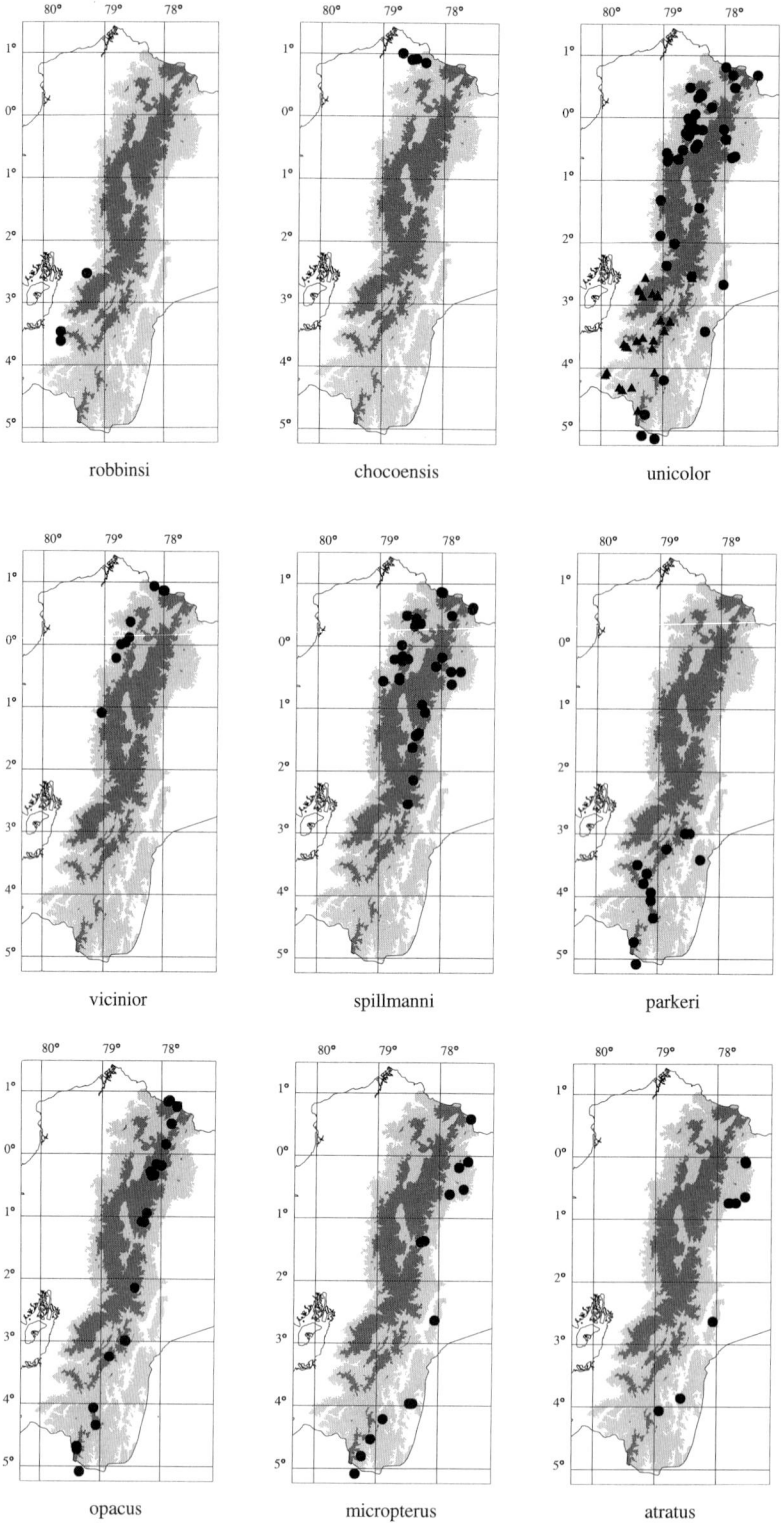

Fig. 86. Geographical distribution of *Scytalopus* tapaculos in Ecuador. Light gray shading indicates areas between 1,200 and 3,000 m elevation, dark gray shading for areas above 3,000 m. For *Scytalopus unicolor*, circles represent the subspecies *latrans*, and triangles represent the subspecies *subcinereus*.

SPECIES LIMITS OF *SCYTALOPUS* TAPACULOS

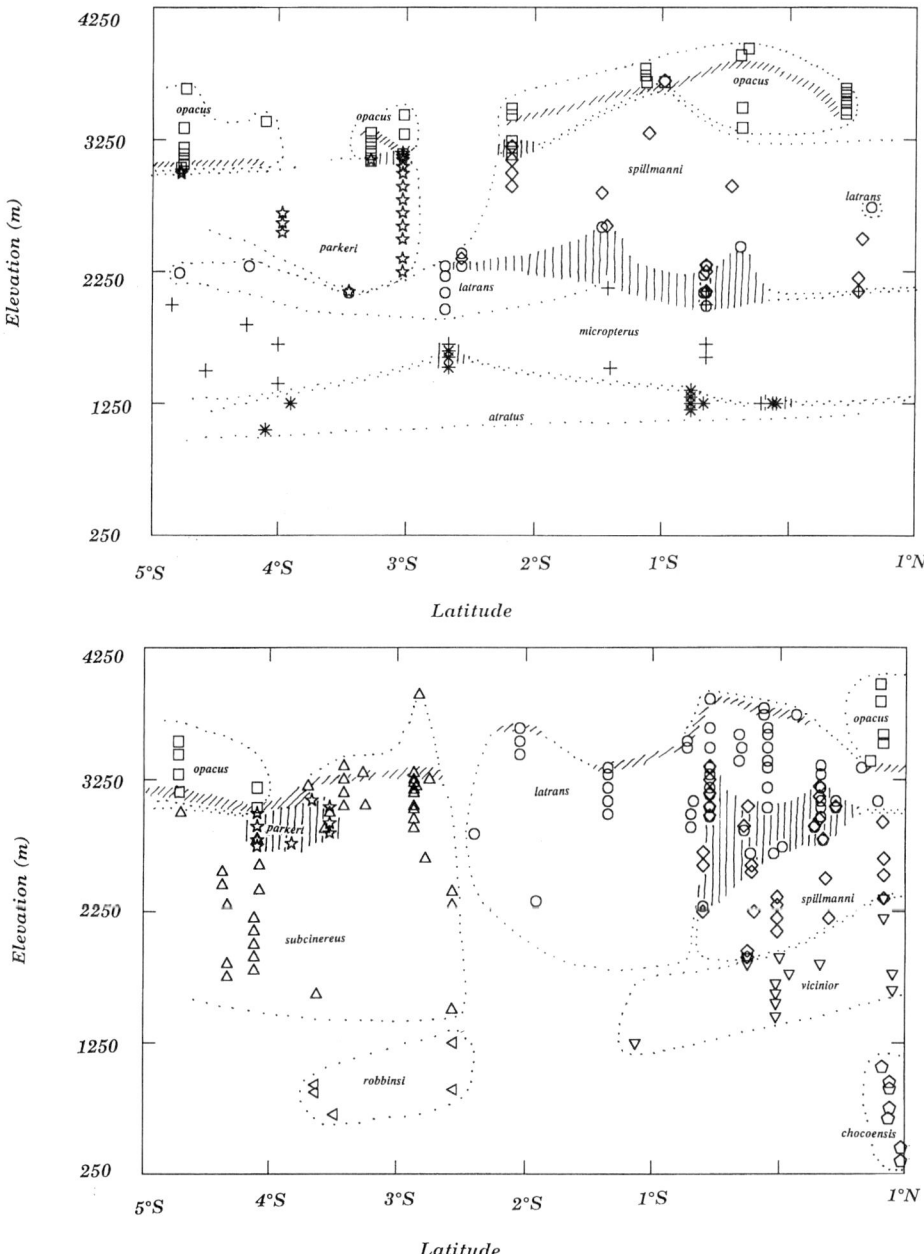

FIGS. 87 and 88. Elevational distribution of *Scytalopus* taxa in Ecuador. X-axis is latitude in degrees, y-axis is elevation (m). Fig. 87. Amazonian slope including Cordillera de Cutucú and Cordillera del Cóndor. Fig. 88. Pacific slope, in the south including the western slope of eastern Andes. Square *Scytalopus canus opacus*, star *S. parkeri*, diamond *S. spillmanni*, circle *S. unicolor latrans*, triangle pointing up *S. unicolor subcinereus*, plus *S. micropterus*, asterisk *S. atratus*, triangle pointing down *S. vicinior*, triangle pointing left *S. robbinsi*, pentangle *S. chocoensis*. Vertical hatching known areas of overlapping ranges, slanting hatching treeline.

dillera Colán, depto. Amazonas; along the trail from Cumpang to Utcubamba, depto. La Libertad; and in the Carpish Mountains, depto. Huánuco. This same pattern was discovered in 1984 in Ecuador in Cordillera de Cutucú, prov. Morona-Santiago, later also in provs. Napo and Zamora-Chinchipe (ANSP, MECN, ZMUC).

Clearly Zimmer's *Scytalopus femoralis* is in need of revision. The nomenclature of the dark-crowned birds is relatively straightforward. Birds along the eastern slope of the Andes from southern Colombia (where crossing over to the head of Magdalena valley) south to northern Peru north of the Río Marañón, are the large, very long-tailed *micropterus*. Those populations from northern Peru, just south and east of the Río Marañón south to central Peru, are what historically have been referred to nominate *femoralis* (but see below). Thus these taxa are allopatric, and under the biological species concept, their taxonomy is another question. The songs of *femoralis* and *micropterus* (Figs. 24 and 26) are similar, but consistently different. Furthermore, there are two fixed allelic differences between these two taxa (T.S.S., unpublished), suggesting long-standing genetic isolation. For these reasons we recommend that *S. femoralis* and *S. micropterus* be treated as allospecies in a *S. [femoralis]* superspecies.

The remaining forms in Zimmer's *Scytalopus femoralis* all have a white crown spot, which we consider to be a possible synapomorphy, but *sanctaemartae* is unlike the others in its pale color and entirely different song (Fig. 18). It is isolated in the Santa Marta mountains, Colombia, at elevations from 900 to 1,700 m or slightly higher (ANSP, AMNH, ICN). The other taxa with a white crown spot (*bolivianus*, *atratus*, *confusus*, and *nigricans*) all appear morphologically very alike, but most differ somewhat vocally (Figs. 14, 15, 16, 17, 20, 22, and 23) and we treat them as allospecies in a *S. [bolivianus]* superspecies.

The nomenclature of the white-crowned birds is particularly problematic. The oldest available name is *bolivianus*, the type of which is from depto. La Paz, Bolivia, and is a small, dark specimen with a well-developed crown spot. Birds assigned to this taxon are found from depto. Chuquisaca, Bolivia (Fjeldså and Mayer 1996;, pers. obs. T.S.S.), north to deptos. of Santa Cruz (tape-recording by T. A. Parker in LNS), Cochabamba (NRS) and La Paz (AMNH, ANSP, LSUMZ), Bolivia, and depto. Puno, Peru (AMNH, ANSP, LSUMZ), at elevations ranging from 1,000 to 2,300 m (and possibly to 2,850 m [Whitney 1994]). Tape-recordings of *bolivianus* show an unusually wide range of variation (Figs. 15, 16, and 17), but all vocalizations are fairly different from the known voices of other white-crowned birds farther north in Peru. Hence, we recommend that *S. bolivianus* be regarded as a separate species, but emphasize that more needs to be learned about its morphological variation and distributional limits. Among the unresolved questions surrounding this treatment are birds from depto. Cuzco, Peru, with a *bolivianus*-like song, but without a white crown spot; and the possibility that the type of *femoralis*, which is very dark and has very little barring on the flanks, may represent the central Peruvian species that usually has a white crown spot.

On the basis of current knowledge, then, available names for the white-crowned birds from central Peru north are *atratus, confusus,* and *nigricans*. All three holotypes are dark specimens with well-developed crown spots. The pale throat on the female and on one of the two males in the type series of *atratus* are matched in the female and (in part) by the male in the type series of *nigricans*, in a female specimen from Ecuador (ZMUC 80142), and is suggested as well by prominent white scalloping on the throat of a male from Ecuador (ANSP 176885); these markings, although distinctive, probably reflect nothing more than individual, or, perhaps, ontogenetic variation.

There is no information on the vocalizations of topotypical populations of either *atratus* (the type of which is from the east slope of Eastern Andes of Colombia) or of *confusus* (type locality in the upper Cauca valley, Colombia). Recent recordings of a white-crowned bird (not collected) from the west slope of the Eastern Andes of Colombia in depto. Cundinamarca (P. Coopmans) are similar to our recordings of white-crowned birds from the east slope of the Andes in Ecuador. Consequently, we provisionally use the name *atratus* for the birds of the east slope of the Andes in Colombia, Ecuador, and Peru. We restrict the name *confusus* to the birds of the Central and Western Andes of Colombia, and, in the absence of firm knowledge of the vocalizations of *confusus*, we provisionally regard this taxon as a subspecies of *atratus*. Clearly, judicious field work, including the collecting of specimens and the preservation of tissue or blood samples, in the Andes of Colombia will be necessary to resolve these nomenclatural issues. Yet another problem is posed by an apparent pattern of geographic variation in the songs of white-crowned birds. Songs recorded in Peru (Figs. 22 and 23) and Venezuela (Fig. 20) are very similar, yet these populations are separated by the different-sounding birds (Fig. 14) of eastern Ecuador and, apparently, Colombia. We can think of several possible explanations for this peculiar situation.

Most likely may be that the white-crowned populations in eastern Peru are a species different from those in Ecuador, either conspecific with, or closely related to, *nigricans* of Venezuela. In view of the surprising biogeographic picture that this paints, and of the already convoluted nomenclatural problems posed by this group, we refrain at this time from naming new taxa, but recommend further studies of these white-crowned populations throughout the Andes.

NORTHERN WHITE-CROWNED TAPACULO *Scytalopus atratus atratus*
(Hellmayr 1922; type in CM)

Brief diagnosis.—Relatively large. Male blackish, with white crown spot and distinct, brown bars on the flanks (Fjeldså and Krabbe [1990, pl. XLI 9 as *confusus*]). Female, at least when immature or subadult, slightly paler, washed with brown on most of the upperparts, with a smaller crown spot and a whitish throat. Both sexes may have whitish tips on the belly feathers.

Plumage.—Male ($n = 9$): All Blackish Mouse Gray, forecrown with a snow white central spot (very small in two), flanks barred Ochraceous Tawny. Two specimens are washed with dark brown and have faint, dark bars on rump and upper-tail coverts; one is also washed with dark brown on terminal half of tertials. Belly variable (age-related ?). Pale, broad tips or subtips to feathers of upper belly silvery white and conspicuous in two specimens, gray and inconspicuous in one, faintly indicated on a few feathers in one, absent in others. One appears uniform blackish on the belly, the others have slight Ochraceous-Orange to Ochraceous-Tawny or more olivaceous barring on the lower belly and under-tail coverts. One has feathers of a restricted area on upper belly with grayish white, subterminal bars, appearing scaled. Axillars gray with a slight olivaceous wash. *Female* ($n = 1$): White crown spot very small, white feathers broadly dark-tipped. Rest of forecrown dark gray, remainder of upperparts between Argus Brown and Brussels Brown, of a very dark shade; rump and upper-tail coverts barred blackish, tertials with 1–2 dark bars near tips and edges. Greater wing-coverts with subterminal Ochraceous bar, bordered with blackish. Throat whitish. Rest of underparts Deep Mouse Gray, 2 mm wide terminal or subterminal bars on upper belly whitish. Lower belly, lower sides, flanks and under-tail coverts Ochraceous-Tawny barred blackish. Axillars Olivaceous. *Juvenal female:* An old specimen (BMNH) taken at an elevation of 850 m in prov. Morona-Santiago, probably of this taxon, shows no trace of a pale crown-spot (the spot is found in juveniles from Peru [LSUMZ]).

Annual cycle in Ecuador ($n = 1$).—*Juvenile:* 6 January (prov. Morona-Santiago).

Vocalizations in Ecuador.—Song (Fig. 14) by both sexes, a series of 5–8 similar notes lasting 0.6–0.7 s, and repeated about once/s. First overtone loudest, 2.5 kHz, the fundamental weaker. Although resembling the scold of some other species, we have never heard other vocalizations from this species that could be interpreted as song. Some high-pitched squeaks have been noted from males after playback of song.

Habitat in Ecuador.—Humid, primary forest undergrowth, at edge and in second growth. In broader-leaved vegetation than *S. micropterus* in the zone of overlap.

Distribution in Ecuador.—850–1,650 m, along the entire east slope of Eastern Andes and on the outlying ridges (Volcán Sumaco-Pan de Azúcar, prov. Napo; Cordillera de Cutucú, prov. Morona-Santiago; Cordillera del Cóndor, prov. Zamora-Chinchipe). Replaced above by *S. micropterus*, but with considerable elevational overlap.

EQUATORIAL RUFOUS-VENTED TAPACULO *Scytalopus micropterus*
(Sclater 1858; type in BMNH)

Brief diagnosis.—Large and heavy, dark gray with long blackish tail composed of 12 rectrices. Flanks always barred. Bill relatively stout. No white crown spot.

Plumage.—Adult male ($n = 7$): Blackish Mouse Gray to Dark Mouse Gray, lower back and rump between Mars Brown and Argus Brown with or without dark bars. In three birds brown also found on the nape (two birds), wing-coverts, edges of remiges, and on upper-tail coverts; one also has tips of innermost remiges with a subterminal light brown bar bordered blackish. Tail blackish. Below Deep Mouse Gray to Dark Mouse Gray; the two birds brownest above with Light Mouse Gray tips to feathers of upper belly. Lower sides, flanks, lower belly and under-tail coverts between Amber Brown, Sudan Brown, Ochraceous-Tawny and Cinnamon-Brown, distinctly barred blackish. Axillars Sayal Brown. *Adult female:* Zimmer (1939) described 14 birds as being duller than males, with a slight tinge of drab in gray of back and anterior underparts, but flanks sometimes more contrastingly barred.

Annual cycle in Ecuador.—Unknown.

Vocalizations in Ecuador.—Song (Fig. 24; also Fjeldså and Krabbe [1990:430, sonagram 2]) given by male consists of two resonant notes 0.2 s apart, second note shortest. Some birds may start with a single note, sometimes with a slight double quality, and later change to the typical double-noted song (occasionally and irregularly triple-noted) *"cu-ock"* at 2.0–2.2 kHz and repeated endlessly at 0.3–0.7 s intervals. An alarm-type call (Fig. 25), and a single *"kick"* note may be given by the female.

Habitat in Ecuador.—Humid shrubbery at forest edge and along streams, frequently in second growth. In the zones of overlap found in more microphyllous vegetation than *S. atratus*. The replacements above by *S. spillmanni* and *S. parkeri* appear to be very abrupt.

Distribution in Ecuador.—1,250–2,100 m along the entire east slope of Eastern Andes and on the outlying ridges (Volcán Sumaco - Pan de Azúcar, prov. Napo; Cordillera de Cutucú, prov. Morona-Santiago; Cordillera del Cóndor, prov. Zamora-Chinchipe; Cordillera de Tzunantza).

Distribution beyond Ecuador of species.—In Colombia *S. micropterus* has been tape-recorded on the west slope of Eastern Andes at the head of Magdalena Valley (Cueva de Los Guácharos; B. M. Whitney recording), and a specimen from the east slope of Central Andes in depto. Huila (La Palma, 1,525 m [BMNH]) appears to be of this form (wing 59.5, tail 50, tarsus 24.5 mm, coloration and bill like *S. micropterus*). Some specimens from the Amazonian slope in southern Colombia, at Llorente, depto. Nariño, 1,800 (?) m (FMNH 292139, female), at Cerro Pax (ANSP), and at 30 km E Cerro Pax (ANSP), all in depto. Nariño, also seem to belong here, but apparently the species does not range further north. In Peru, it occurs north of Río Marañón in depto. Cajamarca (Cerro Chinguela, 1,700–1,950 m [Parker et al. 1985, tape-recordings by T. A. Parker in LNS]; Chaupe, 1,830 m [AMNH, ANSP]; Lomo Santo [ANSP]).

South and east of the Río Marañón in Peru the very similar, but slightly shorter-tailed *S. femoralis* (Tschudi 1844; type in MNN) is widespread, but its exact southern limit (in southern Peru) has not yet been located (see also notes under *S. atratus*).

Scytalopus magellanicus, *sensu* Zimmer 1939

All forms referred to *Scytalopus magellanicus* by Zimmer (1939) are small and are found near or above treeline. In life all have a relatively flat crown that at certain angles appears to have a silvery sheen, forming a contrast with the exposed, dark bases of the feathers of the loral and ocular region. These characteristics could be plesiomorphies, the small size possibly even an adaptation to the dense treeline scrub and tussock grass. The songs of most forms (all but *canus*, *fuscicauda*, and *santabarbarae*) are now known, and they are all different (compare Figs. 27 through 43). Furthermore, there are three cases of elevational parapatry between two forms.

In 1983, N.K. found that birds assigned to *Scytalopus magellanicus* from the Carpish mountains, depto. Huánuco, central Peru were different in plumage, song, and habitat, from birds, also assigned to *S. magellanicus*, occurring at treeline in deptos. Apurímac and Cuzco, south-central Peru. A Louisiana State University expedition (T. J. Davis and G. and K. Rosenberg) independently found two sympatric species in depto. Pasco, central Peru, and later also in the Carpish mountains. A dark species was found in the upper reaches of forest, whereas a more or less white-superciliaried species was at and above treeline. The dark, forest populations in Huánuco and Pasco were morphologically similar to each other, and had similar vocalizations. For reasons outlined below, we apply the name *acutirostris* (Tschudi 1844, type in MHNN) to these birds.

The treeline populations in Huánuco and Pasco, however, differed from each other in vocalizations. The Huánuco treeline form is referable to *altirostris* (Zimmer 1939), whereas the Pasco population remains undescribed (K. Rosenberg and T. Davis, in prep.; specimens LSUMZ). Whitney (1994) reported a parallel situation in south-eastern Peru and northern Bolivia, in which a white-diademed bird was found in the upper reaches of forest (*S. schulenbergi*), and a white-superciliaried bird at and above treeline (*simonsi*).

Farther south, local sympatry has been reported in Chile between two other taxa, *fuscus* and *magellanicus* (Johnson 1969, Riveros and Villegas 1994).

Evidently Zimmer's "*Scytalopus magellanicus*" must be subdivided into several component species. Based on vocalizations, Whitney (1994) proposed to split Zimmer's "*S. magellanicus*" into two groups, *S. griseicollis* and *S. magellanicus*, found north and south of "the North Peruvian Low," respectively, whereas Ridgely and Tudor (1994), also based on vocalizations, treated *fuscus* and *magellanicus* as distinct species and grouped all other taxa as members of a polytypic *S. griseicollis*. Because of their different songs (and the above-mentioned cases of elevational parapatry) we feel obliged to treat all the vocally known forms in *S. magellanicus*, *sensu* Zimmer (1939), as distinct species. Biogeographically and morphologically, *canus* appears to be closest to *opacus*, which we therefore treat as a subspecies of *S. canus*. We treat *fuscicauda* as a subspecies of its nearest neighbor, *S. griseicollis*. Although they appear to be widely separated geographically and could be unrelated, they are very similar in morphology. The newly described *santabarbarae* appears to be a recent isolate of *S. superciliaris*, and we leave it as a subspecies of that species.

PARAMO TAPACULO *Scytalopus canus opacus*
(Zimmer 1939, 1941, type in AMNH)

Brief diagnosis.—Very small. Adult male (depicted in Fjeldså and Krabbe [1990, pl. XLI 1 d]) varies from light to deep gray; usually lighter-colored and with more distinct barring than brown-flanked individuals of *S. unicolor latrans* and *S. unicolor subcinereus*, but old museum specimens not always identifiable. Females and subadults on the other hand, are easily separated by being considerably browner above. Healthy-looking birds (breast muscle bulging above sternum) weighing less than 15.5 g immediately after capture, safely referable to *S. canus opacus*.

Plumage.—*Adult male* ($n = 21$): Above Deep to Dark Mouse Gray, below from between Light Mouse Gray and Mouse Gray to Deep Mouse Gray, belly palest. Birds from prov. Morona-Santiago average lightest. Flanks, under-tail coverts, and sometimes vent Ochraceous-Tawny to Cinnamon-Brown, relatively narrowly and densely barred blackish. One very dark with faint bars on flanks. Crown feathers minutely tipped with Light Mouse Gray and with a darker subterminal band, and in certain lights all show a silvery sheen on the crown, contrasting with the darker loral and ocular region. Axillars gray like rest of underwing. Birds from prov. Zamora-Chinchipe have faint brown and dark bars on tips of tertials. *Adult female* ($n = 5$): Four are Mouse Gray on crown; between Dresden Brown and Sepia on nape, mantle and back; feathers with minute Dark Mouse Gray tips except on back. Rump between Dresden Brown and Sudan Brown, faintly barred dusky; tail Cinnamon-Brown with faint barring at edges and tip. Wings Brussels Brown, inner five remiges with Sudan Brown subterminal band bordered with blackish. Below Mouse Gray, in one grading to Light Mouse Gray on the apical 2–3 mm. Central lower belly Pinkish Buff. Flanks and under-tail coverts like rump or slightly brighter. One much darker and more uniform, like ad. male except for a faint wash of dark brown above. *Immature/subadult male* ($n = 5$): One has crown Deep Mouse Gray; nape, mantle, and wash on central crown between Brussels Brown and Raw Umber, feathers of these parts indistinctly tipped dusky. Back, wing-coverts, and edges of remiges Brussels Brown; wing-coverts and back-feathers with subbasal and subterminal black bar; tertials with subbasal, presubterminal, and narrow, terminal, black bar, and like the inner remiges with narrow Snuff Brown subterminal bar, that is lighter, forming a spot near the shaft on the outer web. Rump like mantle, feathers with narrow subbasal and subterminal black bar. Tail above with regular, 2 mm wide between Brussels Brown and Sudan Brown, and 1 mm wide blackish bars. Sides of head and neck like crown, grading into Mouse Gray of throat and breast. Most of belly Warm Buff to light Buff, almost uniform. Feathers of extreme lower belly with subterminal blackish bar. Sides, flanks, and under-tail coverts between Ochraceous Tawny and Tawny Olive, regularly barred blackish like the upperside of the tail. Underwing dull Light Mouse Gray, greater coverts and axillars with spot-like silvery-white streak along shaft. Underside of tail blackish. Another much like it, but brown colors more olive, belly-feathers broadly tipped whitish rather than Light Buff. One molting into adult plumage still brown on nape, upper mantle, some feathers on central back, some wing-coverts, and edges of the remiges. Tertials and edge of rectrices vermiculated brown and black, rump and upper-tail coverts barred brown and blackish. Two like adults, but one with faint wash of brown on back, wing-coverts, tips of inner remiges, and on tail, one with light brown dots on tips of greater wing-coverts, and with brown vermiculations on tail and tips of inner remiges. *Immature/subadult female* ($n = 2$): Entire upperparts between Dresden Brown and Sepia, greater wing-coverts and tertials blackish subapically and with Ochraceous Buff tips. Lower rump and upper-tail coverts barred, edge of rectrices barred or vermiculated, with blackish. Side of head, throat, and breast Mouse Gray washed with brown, entire belly uniform Ochraceous Buff. Lower sides, flanks, and under-tail coverts Cinnamon to Cinnamon Buff and barred blackish. They thus mainly differ from subadult male by their uniform bright ocher bellies. *Juvenal male* ($n = 2$): One with upperparts including wing-coverts and tail Cinnamon Brown barred blackish, dark bars inconspicuous on crown and sides of head. Edges of remiges Cinnamon Brown, tertials and wing-coverts more evenly barred than in most congeners. Below Tawny Olive, palest on belly, throat and most of breast uniform. Blackish subterminal markings appear as spots on lower breast and widen to become regular bars on lower underparts. Axillars Tawny Olive. One much like it, but darker brown above, more uniform below, spots confined to sides of breast, and bars to sides, flanks and under-tail coverts; belly whitish. *Juvenal female* ($n = 3$): One with upperparts much like the first-mentioned juvenal male, but brown brighter, near Sudan Brown. Underparts also similar, but brighter, near Yellow Ocher, and regular barring confined to sides, flanks, and under-tail coverts. Spots on lower breast continue onto the almost uniform central belly as scattered streaks and spots. One similar above, but unmarked below except for dark bars on flanks and under-tail coverts, and feathers of cheeks, lower throat and breast with extensive, ill-defined, Mouse Gray bases and minute tips, this area from most angles appearing as a gray that grades into the Yellow Ocher chin and belly. One somewhat brighter brown above, and with more conspicuous bars on crown, cheeks, nape, and back. Underparts Yellow Ocher with dark bars, nearly spot-like on throat and breast. All with axillars between Yellow Ocher and Tawny Olive.

Annual cycle in Ecuador ($n = 8$).—*Juveniles:* 16, 20 March (se prov. Carchi); 6, 11, 22 May (prov. Napo); 8 November (prov. Zamora-Chinchipe). *Brood patch:* 9, 14 November (females) (provs. Loja and Morona-Santiago).

Vocalizations in Ecuador.—Song by male (Fig. 11; also Fjeldså and Krabbe [1990:442, sonagram 2]) is a fast, long trill of ca. 34 notes/s, the loudest, first overtone at 3.8 kHz. The beginning of the song is a distinctive "stuttering," usually of 1–3, rarely up to 5, somewhat lower-pitched, slower notes. Simultaneously with the male song the female may break into a usually high-pitched, descending (5.7–4.2 kHz) series of 15–20 notes, 5–8/s (Fig. 28). Call in most of range consists of 5–9 "*kee*" notes, 4.0 kHz in male (Fig. 29), 4.5 kHz in female, lasts ca. 1 s, and is repeated at 2–6 s intervals. In southernmost Ecuador (Cordillera Las Lagunillas) and immediately adjacent Peru (Cerro Chinguela, depto. Piura) the call by both sexes is very different (Fig. 30), and resembles a short burst of male song. It should be noted that calls are given rather frequently, i.e., the noted differences are between homologous calls.

Habitat in Ecuador.—0.2–2.0 m tall, dense, humid to fairly dry shrubbery at treeline, notably *Escallonia myrtilloides* and ericaceous scrub. Locally the upper parts of taller humid forest, mainly in *Chusquea* bamboo. The replacement below by *S. unicolor latrans* and *S. unicolor subcinereus* is everywhere very sharp. There are small zones of overlap with *S. spillmanni* and *S. parkeri* in humid forest, but those species never enter *Escallonia* and ericaceous scrub, even where *S. canus opacus* is absent.

Distribution in Ecuador.—3,050–3,980 m at treeline, lowest in the south. On Páramo El Angel, depto. Carchi, in the northwest, and along the entire Eastern Andes, where the rivers Pastaza, Paute, and Zamora, and the low ridge between Yangana (prov. Loja) and Valladolid (prov. Zamora-Chinchipe) divide it into four populations. Birds of the southernmost population may be subspecifically distinct on basis of their distinctive call-note, and the presence of a white wing spot in most males. Reports from the eastern edge of the Cajas plateau in western prov. Azuay (King 1989) are in error, caused by confusion with the female of *S. unicolor subcinereus*, the only tapaculo inhabiting the plateau and its temperate slopes.

Distribution beyond Ecuador of Zimmer's Scytalopus magellanicus.—In Colombia *S. canus opacus* occurs in depto. Nariño at Puerres (Denumbo), 2,820 (?) m (ANSP), and on the Colombian side of the border at Aguas Hediondas, Páramo El Angel ([tape-recordings by N.K.]). Tape-recordings by F. Lambert and B. M. Whitney from the southern end of Central Andes in depto. Cauca (Volcán Puracé, 3,300–3,400 m) prove indistinguishable from Ecuadorian birds. In Peru *S. canus opacus* is found in northernmost depto. Cajamarca on the border to depto. Piura (Cerro Chinguela, 2,600–3,500 m [Parker et al. 1985; specimen from 3,100 m in LSUMZ; tape-recording by T. Parker in LNS]).

The Colombian form *S. canus canus* (Chapman 1915; type in AMNH; type depicted in Fjeldså and Krabbe [1990, pl. XLI 1a]) occurs in the northern end of Western Andes in depto. Antioquia (Paramillo [type-locality], 3810 m [AMNH, BMNH]; Páramo Frontino [Hilty and Brown 1986]). Birds from Central Andes in depto. Caldas (La Leonera east of Manizales, 3,600 m) were included in this taxon by Peters (1951), but there are no tape-recordings from either region.

Scytalopus griseicollis griseicollis (Lafresnaye 1840; type presumably in Paris Museum, once on loan to MCZ; depicted in Fjeldså and Krabbe [1990, pl. XLI 1c]) is found at 2,600–3,900 m in low, often fairly dry scrub ("mattoral") in, around, and east of Bogotá, depto. Cundinamarca and in depto. Boyacá, in the Eastern Andes, Colombia (ICN, AMNH, ANSP, BMNH). Although its bright, unbarred flanks are matched by south Peruvian *urubambae*, its whitish abdomen differs from any adults of the forms to the west or south of it. Its vocalizations differ distinctly from those of *S. canus opacus* (F. G. Stiles, pers. comm.), and consequently we regard *opacus* and *griseicollis* as separate species.

Scytalopus griseicollis fuscicauda (Hellmayr 1922; type in CM; depicted in Fjeldså and Krabbe [1990, pl. LXIV 31]) occurs at 2,500–3,200 m in south edo. Lara and edo. Trujillo, Venezuela (CM). It resembles *griseicollis* so much that, at least until its voice is known, we treat it as a subspecies of that form.

Scytalopus affinis (Zimmer 1939; type in ANSP; an unusually dark specimen depicted in Fjeldså and Krabbe [1990, pl. XLI 1e, sonagram 1 p. 442]) occurs in the northern Andes of Peru in depto. Cajamarca (Colmena, 2,835 m: LSUMZ; Chota [LSUMZ]) and depto. Ancash (Cordillera Blanca, 3,050–4,100 m [ANSP, LSUMZ, MHN, ZMUC]). It is very pale and lacks a silvery supercilium. It has a rather peculiar call (Fig. 31), and a distinctive song (Fig. 32) with a rapid delivery of three bursts per second, and an equally loud fundamental and first overtone.

Scytalopus altirostris (Zimmer 1939; type in ANSP; imm. female depicted in Fjeldså and Krabbe [1990, pl. XLI 1f, sonagram 2 p. 441]) occurs in deptos. San Martín (Puerto del Monte, 3250 m [MHN]), Amazonas (Atuén [type-locality; ANSP]), La Libertad (Patas [ANSP]), and Huánuco (Bosque Unchog, 3450 m [LSUMZ and tape-recordings by N.K.]), Peru. We only have a single recording of its presumed song (Fig. 33). This song is distinctive, given with the same rapid delivery as the song of *S. affinis*, but rougher, with a broader frequency amplitude, and with a different relative volume of the fundamental and the first overtone.

In depto. Pasco, central Peru (Chipa [AMNH]; near Millpo, 3,450–3,650 m [LSUMZ, MHN]) a rather brownish, yet unnamed form is found (see above). Its song (Fig. 34) is distinctive. Bouts are given at ca. 2 per second, and are composed of a high-pitched, rough, fundamental with a wide frequency amplitude, and an equally loud first overtone.

In Apurímac, south-central Peru (Nevado Ampay, 3,000–4,000 m [ZMUC and tape-recordings by N.K.]; Cerro Queñua Khasa, 4,000–4,600 m [ZMUC]) another, yet unnamed form is found (depicted in Fjeldså and Krabbe [1990, pl. XLI 1h, sonagram p. 440]). It has a distinctive song (Fig. 35), consisting of a single note given twice per second. The fundamental is low-pitched, the first overtone loudest, and both second and third overtone clearly audible. A rarely given vocalization, only documented by a poor recording (Fig. 36), shows some resemblance to the song of *urubambae*.

In the southern Cordillera Vilcabamba, depto. Cuzco, Peru, the enigmatic *Scytalopus urubambae* (Zimmer 1939; type in AMNH) is found at treeline at 3,660–4,170 m. Superficially it resembles *S. griseicollis* of Colombia and differs from adjacent forms by lacking a supercilium and by having no bars on its fairly orange flanks. Hence we doubted its correct systematic position as well as the labelling of the types. However, in 1990 and 1991 G. Engblom (in litt.) obtained photographs, several tape-recordings, and a juvenile specimen near the type locality. The photographs are unmistakably of birds resembling the type of *urubambae*. Its song (Fig. 37) is distinctive. Bouts are given twice per second, and are of a fairly narrow frequency amplitude, in roughness finer than *S. altirostris*, slightly rougher than *S. affinis*, much like *S. superciliaris*. The first overtone is loudest, the second overtone and the fundamental audible.

We refer all (usually white-browed) birds with similar voices (Figs. 38 and 39) at and above treeline from Vilcanota mountains, depto. Cuzco, south through depto. Puno, Peru, and to deptos. La Paz and Cochabamba, Bolivia (specimens in AMNH, ANSP, BMNH, LSUMZ, NRS, ZMUC; tape-recordings by N.K.), to a single form, *S. simonsi* (Chubb 1917; type in BMNH; depicted in Fjeldså and Krabbe [1990, pl. XLI 1i, sonagram p. 439]). Bouts of song are given once to twice per second. Each bout has two components, a single note followed by a short "*churr*", which is about as rough as the "*churr*" of *S. altirostris*.

In deptos Chuquisaca and Tarija, Bolivia, *Scytalopus zimmeri* (Bond and Meyer de Schauensee 1942; type in ANSP; depicted in Fjeldså and Krabbe [1990, pl. XLI 1j]) is found. Its plumage is intermediate between *S. simonsi* and *S. superciliaris* but closest to the latter. Adult males may have yellow feet, which we have not seen in *simonsi*, but which is suggested in dry museum skins of *superciliaris*. The 3-syllabled song of *zimmeri* is distinctive (Fig. 42, clearer recordings by N.K. were obtained too late for inclusion in this paper). Records are from 25 km E Padilla, 2,500 m (ANSP); and Monte Chapeados, 2,500–3,000 m (CBF, ZMUC; tape-recordings by J. Fjeldså and N.K.), both depto. Chuquisaca, and from numerous sites in depto. Tarija (Fjeldså and Mayer 1996).

A few apparent hybrids or intergrades between *simonsi* and *zimmeri* have been reported (Whitney 1994). Although suggestive, these hybrids have not been corroborated by genetic data. One of the putatuve hybrids (ZMUC 80031) gave a typical *simonsi* song, and was collected along steep stream banks in open grassland at 3,600 m. Records of *zimmeri* with the 3-syllabled song are from *Alnus* and *Podocarpus* forest below 3,000 m on the easternmost slopes of the Andes in deptos. Chuquisaca and Tarija. We are not sure at present where *zimmeri* would come into contact with *simonsi.* The two may be separated by the dry valley of the Río Grande and the extensive dry regions in western and central depto. Chuquisaca.

Scytalopus superciliaris (Cabanis 1883; type in Berlin Museum; depicted in Fjeldså and Krabbe [1990, pl. XLI 1k]) (Fig. 41, see also sonagram by Whitney [1994, Fig. 4 C]) is found at 1,500–3,350 m in northwest Argentina in provs. Jujuy, Salta, Tucumán, and Catamarca (specimens in AMNH, Berlin Museum, FML, FMNH, NRS; tape-recordings by R. S. Ridgely and N. Gardner).

The recently described *Scytalopus superciliaris santabarbarae* (Nores 1986; type in FML) from Santa Barbara in prov. Jujuy, Argentina, is similar to, but slightly darker than *S. superciliaris* (specimen in AMNH examined). We do not know of any recordings of its song.

Apparently there is a large distributional gap from prov. Tucumán to prov. Mendoza in north-central Argentina where no *Scytalopus* is found. Farther south nominate *S. magellanicus* (Linnaeus 1789, p. 979; type lost; depicted in Fjeldså and Krabbe [1990, pl. XLI1, sonagram p. 437]) occurs from prov. Río Negro, Argentina, and prov. Valdivia, Chile, south to Cape Horn (AMNH, ANSP, Buenos Aires Museum, BMNH, FMNH, LACM, NRS, WFVZ; tape-recordings by B. M. Whitney). It lacks the brown back and white throat and supercilium of *S. superciliaris,* and has a silvery fore-crown in adult males. Its song (Fig. 42) is distinctive, with bouts consisting of two different single notes given about twice per second.

In central Chile *Scytalopus fuscus* is found (Gould 1837; type lost; depicted in Fjeldså and Krabbe [1990, pl. XLI 1m, sonagram p. 438]). It occurs from sea level to 2,900 m from prov. Coquimbo to the Río Bío Bío (AMNH, Berlin Museum, BMNH, FMNH, LACM, NRS, MCZ, Paris Museum; tape-recordings by B. M. Whitney). Two birds from prov. Mendoza, Argentina (Berlin Museum, BMNH), one taken as high as 3,500 m, might also belong to this taxon. Morphologically *fuscus* differs from *magellanicus* by its larger size, and its uniform black plumage is matched by only some individuals of *magellanicus* (Riveros and Villegas 1994). Its song (Fig. 43) is also very different from that of *magellanicus.* The two occur sympatrically from Valdivia to Río Bío Bío (Johnson 1969; Ridgely and Tudor 1994).

The possible presence of a third taxon (in part "*albifrons*" of Landbeck 1857) resembling *S. fuscus* in large size and blackish coloration, but with a white forecrown in both sexes (specimens from Colchagua in Berlin Museum and BMNH) may have caused some of the earlier confusion in the identification of central Chilean specimens.

As mentioned above, we use the name *Scytalopus acutirostris* (Tschudi 1844; type in MHNN) for the dark-bodied taxon that occurs in central Peru in forest below the treeline taxa. This name has been used by most authors (Hellmayr 1924; Zimmer 1939; Peters 1951; Whitney 1994) for populations of the former polytypic "*magellanicus*" from central Peru, or from there south to western Bolivia.

The type of *acutirostris* was collected by Tschudi at an unspecified locality in central Peru; Hellmayr (1924) restricted the type locality to Maraynioc, depto. Junín, in central Peru, which is where Tschudi made his collection (Tschudi 1844). The type is unsexed, and the tail is missing. Recent field work by T. J. Davis, G. H. Rosenberg, K. V. Rosenberg, G. Engblom, and ourselves has shown that, except for *urubambae*, birds found above treeline in the range ascribed to *acutirostris* by Hellmayr (1924) and Zimmer (1939) are: very small; more or less silvery-/or white-browed in adult males; the rump and flanks are densely barred, the bars forming more or less straight lines; and with a dark mask when viewed at certain angles. Such birds do not agree well with the type of *acutirostris.* The type is washed with brown above, and so is presumably a female or an immature/subadult. The following description is by N.K., based upon his examination of this specimen in 1987. The description of the type by Berlepsch and Hellmayr (1905) is consistent with N.K.'s, but is less detailed: Crown and mantle Dark Mouse Gray with a single white feather over the right eye. Rump Argus Brown very faintly dark-barred. Wings dark with faint wash of Argus Brown, most conspicuous on primaries and tertials, which have a slightly lighter (between Argus and Amber Brown), inconspicuous, subterminal bar. Lores Light to Pale Mouse Gray tipped Dark Mouse Gray. Sides of head and entire throat and breast Mouse Gray, central belly similar, but feathers with 2 mm or longer, relatively well-defined Pale Mouse Gray tips. Flanks Amber Brown with curved, dark bars. Lower central belly to vent Sudan Brown to Ochraceous Tawny with irregular, dark, curved bars. Wing (both) 56 mm, tarsus 21.6 mm, the middle toe with claw 21.5 mm.

The lack of distinct barring on the rump of such a brown individual, the lack of a dark mask, the rather acute bill shape, and the curved bars on the flanks convinces N.K. that the type of *acutirostris* does not represent any of the central Peruvian taxa found above treeline.

So what population, then, does the type of *acutirostris* represent? In the 1980s we independently discovered that two vocally distinct species replace each other elevationally in humid forest of the upper subtropical and temperate zone of Central Peru. One occurs at 1,850–2,500 m, sings much like *S. parvirostris* from Bolivia (as described above), and often has silvery tips to the belly-feathers. Another is found at 2,675–3,500 m from eastern depto. La Libertad (tape-recordings by T. A. Parker, LNS 17224 and LNS 17254) south to deptos. Huánuco (ZMUC, LSUMZ), Pasco (LSUMZ), and apparently Junín (MHNN, Warsaw Museum). This species differs from *S. parvirostris* in that the adult males are darker and more uniform with no pale sheen on the belly, and the vocalizations (Figs. 44, 45 and 46) differ greatly from sympatric forms of *S. parvirostris*, *S. altirostris* and the unnamed, superciliaried form at and above treeline in depto. Pasco. This taxon was referred to as "*Scytalopus* unnamed species" by Fjeldså and Krabbe (1990:427–428). Provisionally we use the name *acutirostris* for this second taxon, because its morphology is consistent with the type of *acutirostris*, and because this taxon otherwise has no name. At the present time we can not rule out the possibility that the name *acutirostris* refers to the populations from central Peru that we are calling *parvirostris*; there is an additional complication, as noted above, that these populations are distinct from typical *parvirostris* from Bolivia, in which case the determination of the identity of this type will become even more important. Or, perhaps, if the type of *acutirostris* can not be identified with certainty, it may become necessary to declare it a *nomen dubium*, and designate a fresh type specimen for this taxon or to rename the population to which we have applied this name.

Scytalopus panamensis, sensu Zimmer 1939

Zimmer (1939) described a new taxon, *vicinior*, as a subspecies of *Scytalopus panamensis*. In the 1980s we discovered that *vicinior* encompassed two sympatric species that replace each other elevationally. They differ in vocalizations, and recognizably in morphometrics. Tape-recordings from the type-locality of *vicinior* (D. Willis and F. Lambert) and the information that only one species of *Scytalopus* occurs there showed that the high-elevation species represents *vicinior*. The low-elevation species thus lacks a name. The call of the low-elevational species (Fig. 56) shows some resemblance to that of *S. panamensis* (Fig. 53) as noted by Pearman (1993). We thus treat the two as allospecies. We see no particular similarity, either in plumage or in voice, between true *S. vicinior* and *S. panamensis*, and recommend that they are treated as genuinely independent species. We propose to describe the new low-elevation form as:

Scytalopus chocoensis sp. nov.
Chocó Tapaculo

Type: ANSP 180144; adult male from El Placer, ca. 670 m, prov. Esmeraldas, Ecuador, 00°52′N, 78°33′W; collected 14 August 1987 by T. S. Schulenberg, original field number 4577; copies of tape-recordings of this individual deposited at LNS, recording LNS 40016.

Paratype: ZMUC 80094; adult male from 18 km north-northwest of Alto Tambo, ca. 450 m, prov. Esmeraldas, Ecuador, ca. 00°58′N, 78°43′W; collected 18 February 1992 by N. Krabbe, blood sample N.K. 29-18.2.92. Copy of a tape-recording of this individual deposited at LNS, recording LNS 65993.

Paratype: MECN 6362; adult male from ca. 10 km west of Lita, ca. 900 m, prov. Esmeraldas, Ecuador, ca. 00°48′N, 78°28′W; collected 23 November 1991 by N. Krabbe, blood-sample no. 2-23.11.91.

Description of type.—Forehead, crown, sides of nape, and feathers around eye near Dark Mouse Gray; feathers narrowly tipped blackish, creating a slightly scaled appearance. Nape and center of upper back between Argus Brown and Brussels Brown, but very dark (in the rest of the descriptions of this species this color is referred to as very dark brown); feathers of these parts tipped blackish. Proximal scapulars and sides of back near Dark Mouse Gray, the (generally concealed) feather-centers very dark brown, tips narrowly blackish. Distal scapulars and lower back very dark brown, feathers narrowly tipped blackish. Lower rump brighter, near Amber Brown, feathers with one or two blackish subterminal bars. Upper-tail coverts very dark brown. Rectrices blackish-gray. Wing-coverts dull brownish-gray, amount of brown increasing medially such that innermost coverts are largely very dark brown, with narrow blackish tips. Primaries brownish-gray. Outermost secondaries brownish-gray with a narrow, very dark brown fringe on outer web, broadening into a narrow subterminal band at tip of feather. Amount of brown in secondaries increases medially, such that innermost secondaries are almost entirely brown, with a small subterminal Amber Brown spot at tip of outer web. Lores near Pale Mouse Gray with darker tips. Chin and throat near Light Mouse Gray, shading darker, near Mouse Gray, on breast and belly. Feathers of central lower belly with broad whitish gray tips, forming an irregular, pale belly patch. Flanks and under-tail coverts Amber Brown, feathers with two or three narrow, blackish bars. Body mass 21.5 g, wing (chord) 53.5 mm, tail 39.8 mm, tarsus 21.3 mm. Iris brown, bill black, feet gray-brown.

Description of paratypes.—Both with 10 rectrices. Coloration of soft parts and plumage like the type, but center of upper back Dark Mouse Gray, nape with only a faint wash of very dark brown, and innermost secondaries without a small, pale spot at tip of outer web. ZMUC specimen: body mass 21.7 g, wing (flat) 55 mm, tail 36.2 mm, tarsus 21.4 mm. MECN specimen: body mass 21.0 g, wing 57 mm, tail 37.0 mm, tarsus 25 mm.

Etymology.—The name refers to the distribution of the taxon, which is restricted to the Chocó region, an important center of endemism, from Cerro Pirre, easternmost Panama, south through western Colombia to extreme north-western Ecuador.

Brief diagnosis.—Small, with 10 rectrices. Wings and tail shorter, underparts paler and more extensively gray, and flanks darker than in *S. vicinior*, which usually has 12 rectrices. Virtually indistinguishable from an allopatric population (p. 78), but *S. chocoensis* averages heavier, the gray of plumage is slightly more towards Neutral Gray and paler below, male has broader silvery feather-tips on the belly when fresh, female distinctly paler throat, brown of plumage is slightly darker, and feet in dried specimens are darker. Lacks the broad silvery supercilium of *S. panamensis*.

Plumage.—*Adult male* ($n = 7$): Includes the paratypes but not the type. Crown, nape and mantle Dark Mouse Gray, feathers narrowly tipped black, feathers of mantle basally Blackish Mouse Gray. Wings Blackish Mouse Gray with faint wash of a very dark brown on edge of remiges and on the entire outer half or third of each of the three tertials, that in three specimens have a subterminal pale dot or bar on the outer web. Back, most of rump, and upper-tail coverts the same very dark brown, grading to Amber Brown on lower rump, feathers narrowly tipped, those of rump barred, with blackish. Tail blackish. Side of head like crown, grading to Mouse Gray on throat, breast and belly. Four specimens have the gray feathers below tipped with Pale Mouse Gray, minutely on the throat and breast, 2 mm wide on the belly. Three others have Light Mouse Gray feather-tips on the belly. Extreme lower belly Ochraceous Tawny barred black. Lower sides, flanks and under-tail coverts Amber Brown with black bars, barring scaly on lower sides and upper flanks. Axillars Ochraceous Tawny. *Adult female* ($n = 3$): Much like male, but brown above extended to nape, mantle and wing-coverts, and more extensive on the remiges. Wing-coverts sometimes with black presubterminal and terminal bar. The tertials may have a small, light spot near tip of outer web, Amber Brown, and encircled with black. Edge of tail with slight, dark brown wash. Below also like male, but throat distinctly paler (Pale Mouse Gray), breast also paler, and entire lower belly Ochraceous Tawny, barred black. All brown colors in plumage slightly brighter than in male. *Subadult/immature male* ($n = 1$): Indistinguishable from adult males with pale feathertips on the belly, but with a bursa. *Juvenal male* ($n = 1$): Above more or less uniform, drab brown, feathers narrowly tipped blackish. Rump and upper-tail coverts as in adult. Rectrices and remiges as in adult, although the primary coverts differ by being broadly edged with dark brown on the outer web and tip of inner web. This brown edging on the outer, middle primary-coverts is paler, brighter brown (near Yellow Ocher) than the brown of the other feathers. Feathers of chin and throat basally light gray, those of breast and belly blackish, all with a paler subterminal band and a narrow, blackish tip, the entire underparts thus appearing barred. The pale, subterminal band is very pale, drab brown on the throat but shades to a deeper brown caudally (to near Tawny Olive). Flanks and under-tail coverts as in adult.

Annual cycle in Ecuador ($n = 1$).—*Juvenile:* 3 August (prov. Esmeraldas). *Active ovary:* 16 February (prov. Esmeraldas).

Vocalizations in Ecuador.—Song by male (Fig. 54) a 5–60 s or longer (longest at high excitement) series of 0.09–0.11 s long notes delivered at 2.6–3.6/s, first overtone loudest, 3 kHz, beginning of each note rising. Frequently, a song begins with two or more slightly faster and lower-pitched notes, and the last note may be given after a short pause. There is some variation in the shape of the notes between different individuals (Fig. 55). Call (Fig. 56) by both sexes a rapid series of 3–8 short, sharp notes, lasting 0.4–1.0 s, and repeated every 3–4 s, first overtone loudest, 2.5 kHz. As pointed out by Pearman (1993), there is some similarity between this call and one given by *S. panamensis* (compare Figs. 53 and 56). Also given, at least by the female *S. chocoensis*, is a sharp, explosive, buzzy "*brzk*" (Fig. 57), with equally loud fundamental and first overtone. The vocalizations of *S. chocoensis* were described by Robbins et al. (1985) under *S. vicinior* and by Hilty and Brown (1986) under *S. femoralis*.

Habitat in Ecuador.—Dense undergrowth of wet, mainly primary forest.

Distribution in Ecuador.—Pacific lowlands north of Río Guayllabamba in prov. Esmeraldas and immediately adjacent prov. Imbabura. Known from along the Ibarra-San Lorenzo railroad and the nearby parallel road, between 350 and 950 m. A specimen from prov. Imbabura (Paramba, 1065 m [AMNH]) is probably also referable to this taxon. The species is replaced at higher elevations by *S. vicinior*, but the extent of overlap or existence of a gap between them still needs to be determined.

Distribution beyond Ecuador.—Cerro Pirre, Panama, where known at 1,340–1,465 m (ANSP; also tape-recordings by T. A. Parker in LNS), south along the Pacific slope of Colombia, where it occurs at elevations between 250 and 1,250 m in deptos. Antioquia (Alto Bonito, 450 m [AMNH]), Valle de Cauca (Río Anchicayá, 700–1,050 m [AMNH]; also tape-recordings by S. Hilty), and Nariño (La Guayacana, 250 m [FMNH, LACM]; below El Diviso, 600 m [tape-recordings by P. Coopmans]; Ricaurte, 1,200 m, [LACM]; Río Nambi, 1,250 m [heard by P. Coopmans]).

NARIÑO TAPACULO *Scytalopus vicinior*
(Zimmer 1939; type in AMNH)

Brief diagnosis.—Medium-sized and long-tailed. Lower belly in both sexes more extensively brown and with feathers more distinctly dark-tipped than in *S. spillmanni*. Individual specimens of the two rarely identifiable with certainty. Longer-tailed than *S. chocoensis*, and usually with 12 instead of 10 rectrices, and without the pale throat of female *S. chocoensis*.

Plumage.—*Adult male* ($n = 12$): Crown Deep Mouse Gray, feathers narrowly tipped blackish. Nape, (mantle), back, upper-tail coverts, wing-coverts and edges of remiges very dark Brussels Brown, feathers of nape, mantle, and back tipped blackish. Seven specimens have some or most of feathers of mantle Dark Mouse Gray, either uniform, with narrow, blackish feather-tips, or with subterminal brown area and blackish tips. Central rump Cinnamon Brown faintly barred dusky, tail blackish. Breast Hair Brown to Deep Mouse Gray, throat only slightly paler except for one bird, which has Light Mouse Gray throat grading to a Mouse Gray breast. Upper or most of belly Light Mouse Gray, feathers very narrowly tipped dusky. Lower belly from between Ochraceous Orange and Cinnamon to between Mars Yellow and Sudan Brown, feathers very narrowly, but invariably tipped dusky. Flanks and under-tail coverts Amber Brown, distinctly barred blackish. Axillars Cinnamon Brown. *Adult female* ($n = 2$): One has crown Deep to Blackish Mouse Gray, feathers tipped blackish. Nape, mantle, back, most of rump, and wings Blackish Mouse Gray very faintly washed with blackish Brussels Brown on edge of remiges, on wing-coverts and broadly subterminally on feathers of nape, mantle, back, and rump; lower rump barred blackish and Brussels Brown, brightest on lowermost rump. Upper-tail coverts and tail blackish. Chin and throat Mouse Gray grading to Deep Mouse Gray on breast, upper sides and upper belly. Feathers of upper and mid belly with 2–3 mm wide Light Mouse Gray terminal or subterminal area, those of mid belly with very narrow, blackish tips. Lower belly between Ochraceous Tawny and Ochraceous Orange, feathers very narrowly tipped blackish. Lower sides and flanks Antique Brown with scaly, black barring; lower flanks brighter, Ochraceous Orange with straighter black bars. Axillars Cinnamon Brown. Another is like it, but with slightly paler throat, heavy Ochraceous-Buff wash on central belly, and with indistinct and smaller dusky markings on rump and flanks, forming mottles rather than bars. *Immature/subadult male* ($n = 1$): Molting from juvenal plumage (only fresh feathers described). Crown Deep Mouse Gray, feathers tipped blackish. Nape, mantle, back, most of rump, and wings Blackish Mouse Gray washed with blackish Brussels Brown on edge of remiges and on wing-coverts. Inner wing-coverts also with this wash subbasally, and broadly subterminally on feathers of nape, mantle, back, and upper rump, feathers of lower rump barred blackish and Brussels Brown. Upper-tail coverts and tail blackish. Chin and throat Light Mouse Gray to Mouse Gray, grading to Deep Mouse Gray on breast, upper sides, and upper belly. Feathers of breast and upper belly with pale shafts basally until 2–3 mm from tips. Feathers of upper and mid belly with 2–3 mm wide Light Mouse Gray terminal or subterminal area. Narrow tips blackish on mid belly and on the Ochraceous Tawny lower belly. Lower sides and flanks as in the first described adult female. Axillars Tawny Olive. *Immature/subadult female* ($n = 1$): Only the type. Type-description by Zimmer (1939): "Upper parts very dark reddish brown with the uropygium brighter (light Auburn) and banded with blackish; forehead and superciliary region slightly tinged with grayish; lores, auriculars, chin, throat and breast light Neutral Gray in strong contrast to the dark brown lower parts; flanks, femoral areas and under-tail coverts deep Argus Brown barred with blackish; middle of belly paler, near Pinkish Cinnamon. Tail and wings like the back, with a slight, pale spot and dusky bar at the tip of the shortest tertial." *Juvenal male* ($n = 1$): In molt (only juvenal feathers described). Feathers of sides of throat, throat, breast and belly with a 1.0–1.5 mm wide Tawny Olive subterminal bar and subbasal area around the shaft, with blackish narrow tip and 1.5–2.0 mm wide presubterminal area, and with gray extreme base.

Three juveniles from southwestern Colombia at Ricaurte, depto. Nariño, 1,200 m (LACM 30865), Ricaurte, depto. Nariño, 2,500 m (LACM 37071), and Cerro Munchique, depto. Cauca, 2,500 m (FMNH 249752) are worth mentioning. They are quite alike, but on distributional grounds the latter two probably belong with *S. spillmanni*, whereas the first may represent the lowest known elevation for *S. vicinior*. They differ from a juvenal *S. chocoensis* in the following features: barring of ventral surface broader on throat; brown subterminal bars of underparts brighter, closer to Yellow Ocher, deepening only slightly or not at all from throat to belly; bases of feathers of forehead brown, not gray as in *S. chocoensis*; barred pattern of underparts extends farther onto sides of throat; a superciliary is present, feathers with the same barred pattern as the underparts.

Annual cycle in Ecuador (n = 2).—Juvenile: 6 November (prov. Pichincha; advanced molt). *Active ovary:* 15 November (prov. Pichincha).

Vocalizations in Ecuador.—Song by male (Fig. 47) a 2.5–20 s long (longest at high excitement), decelerating series of notes given at 13–7/s, and increasing in amplitude (also Fjeldså and Krabbe [1990:432, sonagram 1]). The first few notes are a bit slower and sometimes higher-pitched than the following. Loudest first overtone 3.2 kHz in the first few notes, 2.8 kHz in the following ones. Call by male (Fig. 48) a 3.5 kHz descending "*ki*" given every 6–7 s. Alarms by female and apparently also male include a higher-pitched "*ke ki ki*" (Fig. 49), "*kekikiki*" (Fig. 50), and "*kekikikikike*" (Fig. 51); the first note is slightly higher-/or lower-pitched than the following, which is at 2.8 kHz (3.6 kHz through most of the corresponding call of *S. spillmanni*). Also given by female *S. vicinior* is a 5 s-long series of roughly 10 notes at 3.3 kHz, and a 10 to 15 s-long, descending series of roughly 9 high-pitched, explosive notes (Fig. 52).

Habitat in Ecuador.—Humid forest undergrowth. Sometimes in ferns and broader-leaved, more primary vegetation than *S. spillmanni*.

Distribution in Ecuador.—1,250–2,000 m, locally (prov. Carchi) to 2,350 m. Range abuts that of *S. spillmanni* above, contact with *S. chocoensis* not yet demonstrated. Pacific slope from the Colombian border south at least to southernmost prov. Cotopaxi. A specimen from prov. Chimborazo (Chaguarpata, 1,740 m [Warsaw Museum]) may belong here. P. Coopmans (pers. comm.) was the first to point out the presence of *S. vicinior* in Ecuador, in 1990.

Distribution beyond Ecuador.—1,400–1,950 m, probably both higher and lower, on the Pacific slope of Colombia, where known from depto. Risaralda (Alto de Pisonas, 1,750 m [ICN, specimens and tape-recordings collected by F. G. Stiles]), depto. Valle del Cauca (Las Lomitas, 1,400 and 1,525 m [AMNH, no vocal material]; Parque Nacional Farallones, 1950 m [tape-recording by B. Whitney]) and depto. Nariño (Mayasquer, 1,465 m [ANSP, no vocal material]; Ricaurte, 1500–1,800 m, type-locality, also tape-recordings by F. Lambert, D. Willis and F. G. Stiles). Some subspecific differentiation may have occurred in Colombia, because the two Risaralda specimens are exceptionally dark and heavy (male 24.5 g, female 26.7 g).

ECUADORIAN TAPACULO new species

During field work by ANSP in south-west Ecuador a tapaculo was collected that was thought to represent a southern population of the morphologically similar *Scytalopus chocoensis*. However, N.K. found the southern population to differ in all vocalizations, and it is also genetically distinct (Arctander and Fjeldså 1994, T.S.S. unpub.). We therefore treat it as an independent species, which we propose to name:

Scytalopus robbinsi sp. nov.

Holotype: ZMUC 80102; adult male from 9.5 km west of Piñas, 870 m, prov. El Oro, Ecuador, at 03°40'S, 79°44'W, collected 25 September 1990 by N. Krabbe; blood sample N.K.2–25.9.90. Copy of a tape-recording of this individual deposited at LNS, recording LNS 65994.

Description of type.—10 rectrices. Above Dark Mouse Gray, feathers faintly tipped blackish, notably on crown. Rump, upper-tail coverts and faint wash on nape, lower back, and inner remiges Prout's Brown. Tail blackish. Underparts between Mouse Gray and Deep Mouse Gray, belly with ill-defined, 2 mm wide silvery gray feather-tips. Lower sides, flanks, extreme lower belly, and under-tail coverts Cinnamon Brown barred blackish. Axillars Tawny Olive. Body mass 18.4 g. Wing (flat) 55 mm, tail 36 mm, tarsus 21.5 mm. Iris dark brown, bill blackish, feet gray-brown.

Etymology.—We take the pleasure of naming this bird after Mark B. Robbins, who was the first to tape-record and collect it (recordings in LNS; specimen in ANSP); his tape-recordings greatly facilitated the collecting of further specimens and sound material. We also take the opportunity to acknowledge his substantial contribution to Neotropical ornithology.

Brief diagnosis.—Small. Told from dark females of *S. unicolor subcinereus* by having distinctly barred flanks. For differences from the allopatric *S. chocoensis* see under that species.

Plumage.—*Adult male* ($n = 7$): Includes the type. Above Dark Mouse Gray, feathers tipped blackish. Nape, lower back, upper-tail coverts, sometimes rump, and usually inner remiges Prout's Brown, rump otherwise like flanks. Tail blackish. Below between Mouse Gray and Deep Mouse Gray, belly sometimes with broad and indistinct silvery gray feather-tips. Lower sides, flanks, extreme lower belly, and undertail coverts Cinnamon Brown barred blackish. Axillars Tawny Olive, usually with faint, dusky barring. *Adult female* ($n = 3$): Like male, but brown of nape reaching onto upper mantle; wing-coverts brown with black subterminal dot or bar, tertials with pale, Warm Buff spot at tip of outer web. Entire lower belly Cinnamon Brown barred blackish in two, only extreme lower belly in one. Most of belly in the latter specimen with distinctive whitish feather-tips. *Juvenal:* Unknown.

Annual cycle in Ecuador.—High song activity was noted 25–26 September 1990 and 1 February 1991, only females vocalized 15 November 1991, low or no song activity 9 December 1991, 14–16 April 1991 and 17–19 April 1993.

Vocalizations.—Song by male (Fig. 58) and possibly also female somewhat reminiscent of that of *S. chocoensis*, but considerably faster, 4.4 to 5.3 notes/s, and each note with two distinct components, the latter part lower-pitched than the first, the first overtone loudest, at first 2.7–3.0 kHz, then 2.6–2.8 kHz. The beginning of each note is descending (rising in *S. chocoensis*). Call of female (Fig. 59) an 0.11 s-long single note rising at both beginning and end, centered at 1.4 kHz. The loudest is variably the fundamental, first or second overtone, third overtone sometimes as loud as some of the others. Other female calls include an often slowly descending, 15 to 20 s-long series of some 10 to 20 high-pitched notes, that after the first, somewhat faster 10 notes are given at 1/s (Fig. 60), each note about 0.77 s-long and descending; only the first and the louder second overtones at 3.8–2.4 and 5.7–3.6 kHz are audible.

Habitat.—Undergrowth of wet forest.

Distribution.—Ecuador, where known at 700–1,250 m. Restricted to the Pacific slope in provs. Azuay and El Oro. In prov. Azuay it is replaced at higher elevations by *S. unicolor subcinereus* with little or no overlap.

Discussion.—This species appears to be a southern isolate of *Scytalopus chocoensis*, on the basis of its similar morphology, and geographical and elevational distribution. Our decision not to place them in the same species is based in part on the contrast between the uniformity of the voice of *S. chocoensis* from Panama south to northern Ecuador, and the different vocalizations of *S. robbinsi*. Furthermore T.S.S. (unpubl. data) found several apparently fixed allelic differences between *S. robbinsi* and *S. chocoensis*. The distribution of *S. robbinsi* is similar to that of the recently described parakeet *Pyrrhura orcesi* (Ridgely and Robbins 1988). Both species occupy restricted geographic ranges, within which their habitat is now largely destroyed, and the remnants highly fragmented (Ridgely and Robbins 1988; Collar et al. 1992). Long-term survival of these species, and of the many endemic bird species of the adjacent Tumbesian center of endemism, will depend upon effective measures to maintain these last forests (see Best and Kessler 1995).

Scytalopus latebricola, sensu Zimmer 1939

Zimmer united the forms *caracae*, *meridanus*, *latebricola*, and *spillmanni* in a polytypic *S. latebricola*. These forms, however, share only a lack of diagnostic features. Three of them have different vocalizations. The fourth, nominate *latebricola*, is not well known vocally. The alarm call of a bird presumed to represent it bears some resemblance to the alarm call of *meridanus*. These calls are not entirely alike, however, and as the two are geographically isolated, with *latebricola* found in the upper reaches of the Santa Marta massif, which is known for its many endemic species, we recommend that they are best treated as allospecies of a *S. [latebricola]* superspecies.

The two remaining forms are vocally so distinctive that they are best treated as genuinely independent species: *Scytalopus caracae* and *S. spillmanni*. Birds from the Eastern Andes of Colombia were included in *meridanus* with some reservation by Zimmer. Vocally they appear to be a genuinely independent species, which might be represented by *S. infasciatus* (Chapman 1915), but a comparison of adequate material with the type (in AMNH) has not taken place.

SPILLMANN'S TAPACULO *Scytalopus spillmanni*
(Stresemann 1937; type in Berlin Museum)

Brief diagnosis.—Relatively large (depicted in Fjeldså and Krabbe [1990, pl. XLI 6 as unnamed species]). Tail shorter than in the even heavier *S. micropterus*. Heavier, shorter-tailed, and belly less extensively brown and less dark-scaled, in female also slightly brighter than in *S. vicinior*, but the two not always separable. Female virtually indistinguishable from some specimens of the allopatric *S. parkeri*, but usually heavier.

Plumage.—*Adult male* ($n = 25$): Above Blackish Mouse Gray, feathers indistinctly tipped blackish. Edges of inner remiges, lower back, rump, upper-tail coverts, usually edges of retrices, and faint wash on nape Prout's Brown to Brussels Brown. Rump sometimes barred blackish and then lighter (Cinnamon Brown). Below Deep Mouse Gray, belly sometimes with silvery-gray feather-tips. Extreme lower belly Pinkish Cinnamon to Cinnamon Buff. Flanks and under-tail coverts Ochracous Tawny to Cinnamon Brown, barred blackish. Axillars Cinnamon Buff to Ochraceous Tawny. *Adult female* ($n = 3$): Includes the type. Above much like male but more extensively brown and with somewhat more distinctive dusky feather-tips. Below lighter than male (Mouse Gray), and lower belly extensively Ochraceous Orange to Cinnamon Buff. Lower sides, flanks, and under-tail coverts also lighter than in male (Sudan Brown), barred blackish. *Juvenal female* ($n = 1$): Fore-crown Ochraceous Tawny grading to Brussels Brown on rest of upperparts, and barred blackish throughout. Tail blackish. Primary coverts with Pale Orange-Yellow subterminal bar and blackish vermiculations. Below Pale Pinkish Buff barred blackish, throat almost without bars. Axillars Clay Color. *Juvenal male* ($n = 1$): Similar to juvenal female, but somewhat darker, Pinkish Buff rather than Pale Pinkish Buff below.

Annual cycle in Ecuador ($n = 2$).—*Juveniles:* 5, 6 January (prov. Napo).

Vocalizations in Ecuador.—Song by male (Fig. 61) (also Fjeldså and Krabbe [1990:432, sonagram 2 as "*vicinior*" Pichincha]) a 10 to 20 (rarely 60 or more) s-long, very fast series of notes (25–35/s). Rate of delivery steady at the beginning, the pitch often slightly rising towards the end (whereas the very similar song of *S. canus opacus* begins with a "stutter," never rises, and stops abruptly). First overtone loudest, 4 kHz. At high excitement (Fig. 62) such as after playback of song, during encounters with other males, or in the presence of a female, the male may repeat every 2 s an 0.5 to 1.0 s-long, distinctly rising series of notes delivered at 26–30/s. First overtone loudest, rising from 2.8 to 3.4 kHz (Fjeldså and Krabbe [1990:432, sonagram 3 as "*vicinior*" Pichincha]). A call (Fig. 63), perhaps of an alarm type, is a roughly 1 s long series of 11 to 15 notes, first overtone loudest, 3.6 kHz, first note lower (3.1 kHz), last one or two slightly lower (3.2–3.3 kHz); compare with corresponding, but lower-pitched call of *S. vicinior* (Fig. 51). The female may utter various high-pitched notes, sometimes in a descending series of 5 to 6 such notes (Fig. 64) and at high excitement initiated with a sharp and high-pitched "*brzk*" (Fig. 65). This descending series often triggers male song.

Habitat in Ecuador.—Humid forest undergrowth including *Chusquea* bamboo.

Distribution in Ecuador.—1,900–3,200 m, locally (prov. Napo) to 3,500 m. On the east slope of Eastern Andes from the Colombian border south, only just crossing to the right bank of Río Paute. Not found in Cordillera de Cutucú, prov. Morona-Santiago, or Cordillera del Cóndor, prov. Zamora-Chinchipe, but specimens from Pan de Azúcar (prov. Napo: 00°27'S, 77°43'W, 2,900 m [MECN, WFVZ]) undoubtedly referable here. On the western slope of Western Andes from the Colombian border south at least to west of Sigchos in western prov. Cotopaxi (N.K. tape-recordings). A specimen from Hacienda Porvenir, ca. 2,500 m, prov. Bolívar (BMNH) may be this taxon. Also locally on the upper slopes of the inter-Andean valleys (southeast prov. Carchi, Volcán Tungurahua). Only after obtaining females of the present form in 1990 and after delimiting the ranges of the two species and collecting near the type locality, did we realize that the type of *spillmanni*, which is a female taken on Volcán Iliniza in northwestern Ecuador, represents it and not the morphologically similar *S. parkeri*; hence their distributions were confused in Fjeldså and Krabbe (1990). The elevational ranges of *S. spillmanni*, *S. micropterus*, *S. vicinior*, and *S. canus opacus* barely overlap where they meet. In the lower part of its elevational range on the east slope, *S. spillmanni* co-occurs with *S. unicolor latrans* in the zone where *S. micropterus* replaces *S. spillmanni*.

Distribution beyond Ecuador.—In Colombia *S. spillmanni* is known from the Central Andes in deptos. Antioquia (3 km southeast of the town of Caldas, 2,530–2,650 m [tape-recordings by N.K.]), Cauca (Volcán Puracé [heard by P. Coopmans]), and the eastern slope of Eastern Andes in depto. Nariño (La Victoria, 2,700 m [FMNH]). Undoubtedly also occurs on the Pacific slope in Nariño, whence come a number of specimens (Mayasquer, 2,375 m [ANSP]; Piqualé, [WFVZ,

LSUMZ]; Ricaurte, 2,000 and 2,500 m [LACM]), but a lack of vocal data makes distinction from *S. vicinior* uncertain.

Scytalopus latebricola (Bangs 1899; type in MCZ; depicted in Fjeldså and Krabbe [1990, pl. LXIII 30]) is restricted to the Santa Marta Mountains in northernmost Colombia, where it apparently occurs at 2,000 m (ICN; tape-recordings by P. Coopmans; Fig. 66) and at 3,660 m (AMNH, type in MCZ). There are no known tape-recordings from the elevation where the type was collected, but we consider it likely that the birds at 2000 m represent the same form as the type.

Scytalopus meridanus (Hellmayr 1922; type in AMNH; depicted in Fjeldså and Krabbe [1990, pl. XLI 2a, sonagram p. 434]) was based on a specimen from 4000 m in the Andes of edo. Mérida, Venezuela. Other specimens from the same mountains and edo. Táchira at elevations ranging down to 1980 m (AMNH, ANSP, BMNH, FMNH, PCC) have been referred to this taxon (Hellmayr 1922, 1924; Peters 1951; Phelps and Meyer de Schauensee 1978), as have some from the Central and Eastern Andes of Colombia; also in Fjeldså and Krabbe (1990:434, sonagrams 1 and 2). Zimmer (1939) questioned the allocation of Colombian specimens to *meridanus* and noted that more material was needed to clarify the relationships. Recently *S. spillmanni* was found in the Central Andes of Colombia (tape-recordings by N.K.). Birds from the Eastern Andes of Colombia are vocally distinct (Figs. 73–75) from birds from the upper subtropical zone of Mérida and Táchira (Figs. 67–72). Vocal material from the elevation in Mérida where the type of *meridanus* was collected, and a comparison of material from the Eastern Andes of Colombia with the type of *infasciatus* (Chapman 1915; type in AMNH from Páramo de Beltrán, 2,970 m) is needed before the taxonomy and nomenclature of these forms can be further addressed.

The coastal mountains of Venezuela are inhabited by *S. caracae* (Hellmayr 1922; type in AMNH). Its voice (Figs. 76–78) and plumage differs strongly from those of any other tapaculo.

CHUSQUEA TAPACULO new species

Field work by N.K. has shown that a population at 2,250–3,150 m in south-east Ecuador differs from *S. spillmanni* in voice and habitat, and to some degree in morphology. It also differs genetically (Arctander and Fjeldså 1994). We propose to name this population:

Scytalopus parkeri sp. nov.

Holotype.—ZMUC 80173; subadult male from ca. 20 km south-southwest of San Lucas, 2,770 m, prov. Loja, Ecuador, at 03°50′S, 79°16′W, collected 7 March 1991 by N. Krabbe; blood sample N.K.1–7.3.91. Copy of a tape-recording of this individual deposited at LNS, recording LNS 65995.

Description of type.—12 rectrices. Crown, mantle, upper back, and wings Dark Mouse Gray, feathers narrowly and indistinctly tipped blackish, most conspicuous on crown. Nape, lower back, rump, upper-tail coverts, edge of rectrices, and a faint wash on the wing-coverts and edges of remiges between Dresden Brown and Snuff Brown; tertials wholly this color but for a buff, subterminal bar bordered blackish. Rump and upper-tail coverts faintly barred blackish, most conspicuous on tips of upper-tail coverts. Below Mouse Gray, central belly with 3 mm wide, silvery feather-tips, appearing uniform silvery. Lower belly bright, uniform Ochraceous Buff. Lower sides, flanks, and under-tail coverts between Ochraceous Tawny and Cinnamon Brown barred blackish. Barring distinct on lower sides, otherwise faint. Axillars between Tawny Olive and Cinnamon Buff. Body mass 24.4 g. Wing (flat) 61 mm, tail 43 mm, tarsus 24.5 mm. Iris dark brown, bill blackish, feet light brown.

Etymology.—We take pleasure in naming this species after the late Theodore. A. Parker III (who was the first to tape-record and collect it), in recognition of his emphasis of the importance of field study and the role of vocalizations in bird systematics, and in honor of his vast knowledge and generous heart, which led to outstanding contributions to our knowledge of Neotropical birds.

Brief diagnosis.—12 rectrices. Not identifiable with certainty from female *S. spillmanni*. Medium-sized and long-winged, tail brown or broadly brown-edged (depicted in Fjeldså and Krabbe [1990, pl. XLI 2b as *spillmanni*]). Central belly often silvery white (presumably younger birds) and lower belly then brighter and lighter brown than in the black-tailed *S. micropterus*, and unbarred. Older birds are easily told from adult *S. micropterus* by their virtually unbarred rump, flanks, and lower belly, and from female *S. unicolor subcinereus* by their much longer wings and longer tail, and by 12 rather than usually 10 rectrices. In dry specimens the feet average slightly paler than in Ecuadorian congeners. Largely confined to dense stands of *Chusquea* bamboo.

Plumage.—*Adult male* (*n* = 14): Above Dark Mouse Gray, feathers almost always narrowly, and usually indistinctly, tipped blackish. Nape, lower back, rump, upper-tail coverts, tail, and usually inner (sometimes all) remiges between Dresden Brown and Snuff Brown, lower rump slightly brighter. Upper-tail coverts usually barred blackish. Below Mouse Gray, belly sometimes with 1–3 mm wide, silvery feather-tips. Lower or extreme lower belly from Ochraceous Tawny to Ochraceous Buff. Flanks and under-tail coverts between Ochraceous Tawny and Cinnamon Brown, more or less barred blackish, sometimes (older birds?) virtually unbarred. Axillars Cinnamon to Tawny Olive. *Adult female* (*n* = 3): Similar to male, and like it may lack silvery feather-tips on the belly and have unbarred flanks. *Immature/subadult male* (*n* = 1): The type, described above. *Juvenal male* (*n* = 2): One molting. Feathers of crown and side of head Dark Mouse Gray with (sub)terminal Amber Brown bar and with a narrow, black tip that may wear off. Mantle, wing-coverts, and remiges Argus Brown, feathers vermiculated with black and Cinnamon to Pinkish Buff at their tips. Lower back, rump, upper-tail coverts, and tail Argus Brown barred blackish, tail darkest and only indistinctly barred. Underparts Mouse Gray to blackish, evenly and narrowly barred Pinkish Buff (thus appearing dark with pale bars rather than the opposite as in most or all congeners). Flanks broadly barred Prout's Brown and blackish, under-tail coverts somewhat lighter. Axillars between Tawny Olive and Cinnamon Buff. Bill almost as dark as in adult, feet wholly so.

Annual cycle in Ecuador (*n* = 1).—*Juvenile:* 13 December (prov. Morona-Santiago).

Vocalizations.—Song given by male (Fig. 79) a 1 to 9 s long (up to 15 s or more after playback) series, of initially descending notes delivered at 10 to 12/s and repeated at 1 to 8 s intervals. First overtone loudest, 3.4 to 3.6 kHz (Fjeldså and Krabbe [1990:436, sonagram 2 as *spillmanni*-group]). Each note is relatively short (0.03–0.04 s) with a distinctly descending beginning and a less pronounced, rising end. At excitement (Fig. 80) the delivery of notes may rise up to 19/s, each note may lack the rising end, and each series may be shorter (down to 3 notes: Fjeldså and Krabbe [1990:436, sonagram 3] as *spillmanni*-group) and repeated with hardly any interval; the notes are then lower-pitched, 2.7 kHz. During duets, possibly given at pair-formation (Figs. 81 and 82; also Fjeldså and Krabbe [1990:436, sonagram 1] as *spillmanni*-group) male sustains an even series of 2.8 kHz notes delivered at 19/s, while the female simultaneously gives a long, descending series of 2 notes/s (sometimes faster), the first two or three notes explosive and high-pitched (up to 7 kHz), the following falling from 5 to 4 kHz (first overtone). Call (Fig. 83) much like song, but sharper and composed of only 9 to 12 notes. It lasts ca. 1 s and is repeated every 4 to 7 s. Some of the species's vocalizations were described by Parker et al. (1985) under *S. latebricola*.

Habitat in Ecuador.—Dense stands of *Chusquea* bamboo and adjacent humid forest undergrowth.

Distribution in Ecuador.—2,250–3,350 m. Found south of Río Paute on the east slope of the Eastern Andes and in the highest parts of Cordillera del Cóndor, prov. Zamora-Chinchipe, and from the eastern Chilla mountains south along the west slope of Eastern Andes to the headwaters of Río Catamayo. Does not occur on the Pacific slope of prov. Azuay, in the Celica mountains, or on the eastern rim of the Cajas plateau, and has so far not been found on the west slope of Cordillera de Sabanilla (Río Calvas drainage). The recent discovery of *S. spillmanni* south of the Río Paute, within 50 km of the northernmost specimens of *S. parkeri*, and with no apparent habitat break between them, suggests that the two may be genuinely sympatric, separated by habitat and voice only. Where *S. parkeri* co-occurs with *S. unicolor subcinereus*, *subcinereus* is restricted to drier and more secondary vegetation.

Distribution beyond Ecuador.—Peru in northern depto. Cajamarca and immediately adjacent depto. Piura on Cerro Chinguela, 2,590–2,900 m (LSUMZ and tape-recordings by T. A. Parker in LNS).

NON-ECUADORIAN SPECIES

For completeness we will briefly mention the six known species not discussed above (see also Table 1):

The huge (32–43 g: LSUMZ), dark *Scytalopus macropus* (Berlepsch and Stolzmann 1896; type in Warsaw Museum or lost; species depicted in Fjeldså and Krabbe [1990, pl. XLI 10a,b, sonagram p. 429]) (Fig. 84) is endemic to the Central Andes of Peru, at 2,590–3,500 m from depto. Amazonas south to depto. Junín (ANSP, FMNH, MHN, LSUMZ).

Three forms with white or pale superciliaries are confined to Central America: *Scytalopus argentifrons argentifrons* (Ridgway 1891; type in USNM; tape-recording by O. Jakobsen in BLA; Fig. 85) in Costa Rica and Volcán de Chiriqui in western Panama (AMNH, BMNH, USNM, WFVZ); the closely related (Wetmore 1972) *S. argentifrons chiriquensis* (Griscom 1924; type in AMNH; vocalizations similar to those of nominate *argentifrons* [B. M. Whitney, pers. comm.]) to eastern provs. Chiriqui and Veraguas, western Panama (AMNH, BMNH); and the more distantly related *S. panamensis* (Chapman 1915; type in AMNH) to Cerro Malí (Ridgely 1976) and Cerro Tacarcuna in easternmost Panama (AMNH, BMNH) and immediately adjacent Colombia (tape-recording by M. Pearman in BLOWS; Fig. 53).

Three species are confined to eastern Brazil: *Scytalopus indigoticus* (Neuwied 1831; types in AMNH), *S. speluncae* (Ménétriès 1835; type in St. Petersburg Museum), and *S. novacapitalis* (Sick 1958; type in MHNRJ). Sonagrams of most of their vocalizations were published by Vielliard (1990). Systematics and biology of all the Brazilian rhinocryptids were discussed by Sick (1960). The recently described taxon *S. psychopompus* (Teixeira and Carnevalli 1989; type in MHNRJ; no vocal data available) inhabits a small area in coastal Bahía, Brazil (MHNRJ, MZUSP). Despite close resemblance to *S. indigoticus* (differing only in unbarred flanks and bluish slate thighs), it was given full species status by its describers. This treatment was followed with some reservation by Ridgely and Tudor (1994). We consider the arguments for granting species rank to this taxon to be weak, and its taxonomic status was also doubted by Gonzaga et al. (1995). However, without vocal or genetic data, we are in no position to suggest changes in its taxonomic status.

We have no doubts that further research in the Andes, in particular Peru and Colombia (see e.g., Fig. 21), will reveal the presence of yet undescribed species of *Scytalopus* and will shed further light on the taxonomy on some of the taxa that are now scarcely known in life.

Related genera: Three other rhinocryptid genera seem to be the closest relatives of *Scytalopus*:

The monotypic *Eugralla paradoxa* (Ochre-flanked Tapaculo) of Valdivian forest in Chile and immediately adjacent Argentina is very similar to *Scytalopus* both morphologically and vocally. It mainly differs by its more elevated base of the bill.

The monotypic *Myornis senilis* (Ash-colored Tapaculo) inhabits thickets of *Chusquea* bamboo in the northern half of the Andes. It was considered to be a member of *Scytalopus* by Hilty and Brown (1986), a view with which we disagree. It is relatively light (21 g) and differs from *Scytalopus* by its more slender shape, much longer tail (*Myornis*: tail 58–68, tail/body mass ratio 3.0–3.8 mm/g; *Scytalopus* tail 35.1–57.2 mm, tail/body mass ratio 1.7–2.6 mm/g), more rounded wings and slightly more elevated base of the bill. Its song (described by Fjeldså and Krabbe [1990]) is structured differently from that of any *Scytalopus*.

Merulaxis ater (Slaty Bristlefront) and the similar, but larger *Merulaxis stresemanni* (known from two specimens) inhabit the Atlantic forests of eastern Brazil. They are shaped like *Myornis*, have narrow, pointed, stiffer and longer, erect feathers of the loral region than *Scytalopus* and *Myornis*, and are sexually dimorphic, males being gray, females rather uniform brown, resembling the juvenal of *Myornis*. As pointed out to us by the late T. A. Parker III, the similarity of the songs of *Merulaxis* and *Myornis*, minute-long repetitions of a single note terminating with one or more descending series of "hysterical laughter," suggests that they are each others closest relatives.

ACKNOWLEDGMENTS

Our field work in Ecuador was made possible by the collaboration of the MECN, Quito. The Ministerio de Agricultura, Quito, kindly issued the necessary permits. N. K. acknowledges the support of his field work by ZMUC, and field work by T.S.S. was supported by LSUMZ and ANSP. We would like to thank the following museum curators for loans and for letting us examine the collections: M. LeCroy, L. L. Short, and F. Vuilleumier, AMNH; F. B. Gill, ANSP; P. Colston, BMNH; K. C. Parkes and D. S. Wood, CM; R. Mena, EPN; J. W. Fitzpatrick, S. M. Lanyon, and D. E. Willard, FMNH; F. G. Stiles, ICN; K. Garrett, LACM; J. V. Remsen and S. W. Cardiff, LSUMZ; M. Moreno, MECN; I. Franke, MHN; F. Gehringer, MHNN; G. R. Graves and R. L. Zusi, USNM; O. Grönwall, NRS; and L. Kiff, WFVZ. We would like to thank P. Hansen, BLA, for use of their sonagraph and for help with the time-consuming preparation of the sonagrams; LNS and R. Ranft, BLOWS for use of vocal material from these libraries, and the individual recordists: A. van den Berg, B. Best, C. Carter, C. Clarke, P. Coopmans, G. Engblom, D. Fischer, J. Fjeldså, N. Gardner, P. Greenfield, S. Hilty, O. Jakobsen, F. Lambert, D. McDonald, J. P. O'Neill, T. A. Parker, R. S. Ridgely, R. A. Rowlett, M. Pearman, M. B. Robbins,

F. G. Stiles, R. Templeton, B. M. Whitney, and D. Willis. The field work and museum visits of N.K. were generously funded by H.R.H. Crownprince of Denmarks Foundation, Dr. Bøje Benzons Foundation, Gads Foundation, and a collection study grant from the Chapman Fund (AMNH). J. M. Carrión translated the abstract into Spanish. Bent Otto Paulsen provided needed measurement data on short notice, for which we most grateful. The manuscript greatly benefitted from the criticism of J. Fjeldså, C. Marantz, J. V. Remsen, J. M. C. da Silva, B. M. Whitney, and F. Vuilleumier.

LITERATURE CITED

ALLEN, J. A. 1889. List of birds collected in Bolivia by Dr. H. H. Rusby, with field notes by the collector. Bull. Amer. Mus. Nat. Hist. 2:77–112.

ARCTANDER, P., AND J. FJELDSÅ. 1994. Andean tapaculos of the genus *Scytalopus* (Aves, Rhinocryptidae): a study of modes of differentiation, using DNA sequence data. Pp. 205–225 in Conservation Genetics (V. Loeschcke, J. Tomiuk, and S. K. Jain, Eds.). Birkhauser Verlag, Basel, Switzerland.

BANGS, O. 1899. On some new or rare birds from the Sierra Nevada de Santa Marta, Colombia. Proc. Biol. Soc. Washington 13:91–108.

BEST, B. J., AND M. KESSLER. 1995. Biodiversity and Conservation in Tumbesian Ecuador and Peru. Birdlife International, Cambridge, U.K.

BLOCH, H., M. K. POULSEN, C. RAHBEK, AND J. F. RASMUSSEN. 1991. A survey of the montane forest avifauna of the Loja province, southern Ecuador. International Council for Bird Preservation Study Report No. 49.

BOND, J., AND R. MEYER DE SCHAUENSEE. 1942. The birds of Bolivia. Part 1. Proc. Acad. Nat. Sci. Philadelphia 94:307–391.

CABANIS, J. 1883. Bericht über die December-Sitzung. J. Ornithol. 31:104–106.

CHAPMAN, F. M. 1915. The more northern species of the genus *Scytalopus* Gould. Auk 32:406–423.

CHUBB, C. 1917. *Scytalopus simonsi*; sp. nov. Bull. Brit. Ornithol. Club 38:17.

COLLAR, N., L. P. GONZAGA, N. KRABBE, A. MADROÑO NIETO, L. G. NARANJO, T. A. PARKER III, AND D. C. WEGE. 1992. Threatened Birds of the Americas. International Council for Bird Preservation, Cambridge, U.K.

CORY, C. B., AND C. E. HELLMAYR. 1924. Catalogue of birds of the Americas. Field Mus. Nat. Hist., Zool. series 13, part 3.

DE LAFRESNAYE, F. 1840. Oiseaux noveaux. Rev. Zool. 3:101–106.

FJELDSÅ, J., AND N. KRABBE. 1990. Birds of the High Andes. Zoological Museum, University of Copenhagen, Denmark.

FJELDSÅ, J., AND S. MAYER. 1996. Recent ornithological surveys in the "Valles" region of southern Bolivia and the possible role of Valles for the evolution of the Andean avifauna. Danish Environmental Research (DIVA) Technical Report No. 1.

GONZAGA, L. P., J. F. PACHECO, C. BAUER, AND G. D. A. CASTIGLIONI. 1995. An avifaunal survey of the vanishing montane Atlantic forest of southern Bahia, Brazil. Bird Conserv. International 5:279–290.

GOULD, J. 1837. Exhibition of birds allied to the European Wren, with characters of new species. Proc. Zool. Soc. London Part 4:88–90.

GRISCOM, L. 1924. Descriptions of new birds from Panama and Costa Rica. Amer. Mus. Novitates No. 141.

HELLMAYR, C. E. 1922. Neue Formen der Gattung *Scytalopus*. Ornithol. Monatsber. 30:54–59.

HILTY, S., AND W. L. BROWN. 1986. Birds of Colombia. Princeton University Press, Princeton, New Jersey.

JOHNSON, A. W. 1967. The Birds of Chile and Adjacent Regions of Argentina, Bolivia and Peru. Vol. II. Platt Establecimientos Gráficos, Buenos Aires.

KING, J. R. 1989. Notes on the birds of the Rio Mazan valley, Azuay Province, Ecuador, with special reference to *Leptopsittaca branickii, Hapalopsittaca amazonina pyrrhops* and *Metallura baroni*. Bull. Brit. Ornithol. Club 109:140–147.

KOEPCKE, M. 1961. Birds of the western slope of the Andes of Peru. Amer. Mus. Novitates No. 2028.

KROODSMA, D. E. 1982. Learning and the ontogeny of sound signals in birds. Pp. 1–23 in Acoustic Communication in Birds, vol. 2 (D. E. Kroodsma and E. H. Miller, Eds.). Academic Press, New York.

KROODSMA, D. E. 1984. Songs of the Alder Flycatcher (*Empidonax alnorum*) and Willow Flycatcher (*Empidonax traillii*) are innate. Auk 101:13–24.

LANDBECK, L. 1857. *Pteroptochos albifrons* n. sp. Arch. Naturg. 23:273–275.

MÉNÉTRIÈS, E. 1835. Monographie de la famille des Myiotherinae où sont décrites les espèces qui ornent le Musée de l'Académie Impériale des Sciences. St. Pétersb. Acad. Sci. Mém., (6th ser.) 3, part 2 (Sci. Nat.):443–544.

MEYER DE SCHAUENSEE, R. 1970. A Guide to the Birds of South America. Livingston Pub. Co., Narberth, Pennsylvania.

MEYER DE SCHAUENSEE, R., AND W. H. PHELPS, JR. 1978. A Guide to the Birds of Venezuela. Princeton University Press, Princeton, New Jersey.

MILLER, A. H. 1963. Seasonal activity and ecology of the avifauna of an American equatorial cloud forest. Univ. Calif. Publ. Zool. 66:1–74.

NORES, M. 1986. Diez nuevas subespecies de aves provenientes de islas ecologicas argentinas. Hornero 12: 262–273.
OLSON, S. L., AND E. N. KUROCHKIN. 1987. Fossil evidence of a tapaculo in the Quarternary of Cuba (Aves: Passeriformes: Scytalopodidae). Proc. Biol. Soc. Washington 100:353–357.
PARKER, T. A. III, T. S. SCHULENBERG, G. R. GRAVES, AND M. J. BRAUN. 1985. The avifauna of the Huancabamba region, northern Peru. Pp. 169–197 in Neotropical Ornithology (P. A. Buckley, M. S. Foster, E. S. Morton, R. S. Ridgely, and F. G. Buckley, Eds.). Ornithol. Monogr. No. 36.
PARKER, T. A. III, D. F. STOTZ, AND J. W. FITZPATRICK. 1996. Ecological and distributional databases for Neotropical birds. Pp. 118–436 in Neotropical Birds: Ecology and Conservation (D. F. Stotz, J. W. Fitzpatrick, T. A. Parker III, and D. K. Moskovits, Eds.). University of Chicago Press, Chicago.
PEARMAN, M. 1993. Some range extensions and five new species to Colombia, with notes on some scarce or little known species. Bull. Brit. Ornithol. Club 113:66–75.
PETERS, J. L. 1951. Check-List of Birds of the World, Vol. 7. Museum of Comparative Zoology, Cambridge, Massachusetts.
PHELPS, W. H., AND W. H. PHELPS, JR. 1953. Eight new subspecies of birds from the Perijá Mountains, Venezuela. Proc. Biol. Soc. Washington 66:1–12.
PRUM, R. O. 1992. Syringeal morphology, phylogeny and evolution of the Neotropical manakins (Aves: Pipridae). Amer. Mus. Novitates No. 3043.
RIDGELY, R. S. 1976. A Guide to the Birds of Panama. Princeton University Press, Princeton, New Jersey.
RIDGELY, R. S., AND M. B. ROBBINS. 1988. *Pyrrhura orcesi*, a new parakeet from southwestern Ecuador, with systematic notes on the *P. melanura* complex. Wilson Bull. 100:173–182.
RIDGELY, R. S., AND G. TUDOR. 1994. The Birds of South America, vol 2. University of Texas Press, Austin, Texas.
RIDGWAY, R. 1891. Notes on some Costa Rican birds. Proc. U.S. Nat. Mus. 14:473–478.
RIDGWAY, R. 1912. Color Standards and Color Nomenclature. Published by the author. Washington, D.C.
RIVEROS, G., AND N. VILLEGAS R. 1994. Análisis taxonómico de las subespecies chilenas de *Scytalopus magellanicus* (Fam. Rhinocryptidae, Aves) a través de sus cantos. An. Mus. Hist. Nat. Valparaíso 22: 91–101.
ROBBINS, M. B., T. A. PARKER III, AND S. E. ALLEN. 1985. The avifauna of Cerro Pirre, Darién, Eastern Panama. Pp. 198–207 in Neotropical Ornithology (P. A. Buckley, M. S. Foster, E. S. Morton, R. S. Ridgely, and F. G. Buckley, Eds.). Ornithol. Monogr. No. 36.
ROSENBERG, G. H. 1986. The nest of the Rusty-belted Tapaculo (*Liosceles thoracicus*). Condor 88:98.
SALVIN, O. 1895. On birds collected in Peru by Mr. O.T. Baron. Nov. Zool. 2:1–20.
SCLATER, P. L. 1858. Notes on a collection of birds received by M. Verreaux of Paris from the Río Napo. Proc. Zool. Soc. London Part 26:59–77.
SIBLEY, C. G. AND B. L. MONROE. 1990. Distribution and Taxonomy of Birds of the World. Yale University Press, New Haven.
SICK, H. 1958. Resultados de uma excursão ornitológica do Museu Nacional a Brasília, novo Distrito Federal, Goiás, com a descrição de um novo representante de *Scytalopus* (Rhinocryptidae, Aves). Bol. Mus. Nac., nova série, Zool. 185:1–20.
SICK, H. 1960. Zur Systematik und Biologie der Bürzelstelzer (Rhinocryptidae), speziell Brasiliens. J. Ornithol. 101:141–174.
SICK, H. 1993. Birds in Brazil. Princeton University Press, Princeton, New Jersey.
SKUTCH, A. F. 1972. Studies of tropical American birds. Publ. Nuttall Ornithol. Club No. 10.
STRESEMANN, E. 1937. Vögel vom Monte Illiniza (Central-Ecuador). Ornithol. Monatsber. 45, 3:75–77.
TACZANOWSKI, L. 1874. Liste des oiseaux recueillis par M. Constantin Jelski dans la partie centrale du Pérou occidental. Proc. Zool. Soc. London 1874:501–565.
TACZANOWSKI, L. 1884. Ornithologie du Pérou, vol. 1. Oberthur, Paris.
TEIXEIRA, D. M., AND N. CARNEVALLI. 1989. Nova espécie de *Scytalopus* Gould 1837, do nordeste do Brasil (Passeriformes, Rhinocryptidae). Bol. Mus. Nac., nova série, Río de Janeiro, 331.
VIELLIARD, J. M. E. 1990. Estudio bioacústico das aves do Brasil: o gênero *Scytalopus*. Ararajuba 1:5–18.
VON BERLEPSCH, H. G., AND C. E. HELLMAYR. 1905. Studien über wenig bekannte Typen neotropischer Vögel. J. Ornithol. 53:18–19.
VON BERLEPSCH, H. G., AND M. J. STOLZMANN. 1896. On the ornithological researches of M. Jean Kalinowski in central Peru. Proc. Zool. Soc. London 1896:322–388.
VON LINNAEUS, C. 1789. In Caroli Linnaei Systema naturae sive regna tria naturae systematice proposita per classes, ordines, genera et species. 13 edition: 1 (2) (J. F. Gmelin, Ed.). Leipzig.
VON NEUWIED, M., PRINZ. 1831. Beitrage zur Naturgeschichte von Brasilien. 3 (2). J. B. Delamolliere, Weimar.
VON TSCHUDI, J. J. 1844–1846. Untersuchungen über die Fauna Peruana. Ornithologie. Scheitlin und Zollekofer, St. Gallen, Switzerland.
WETMORE, A. 1972. The Birds of the Republic of Panama. Smiths. Misc. Coll. Vol. 150, part 3.
WHITNEY, B. M. 1994. A new *Scytalopus* tapaculo (Rhinocrytidae) from Bolivia, with notes on other Bolivian members of the genus and the *magellanicus* complex. Wilson Bull. 106:585–614.
WINKLER, R. 1979. Zur Pneumatisation des Schädeldachs der Vögel. Ornithol. Beob. 76:49–118.
ZIMMER, J. T. 1939. Studies of Peruvian birds. No. 32. Amer. Mus. Novitates No. 1044.

ZIMMER, J. T. 1941. Studies of Peruvian birds. No. 38. Amer. Mus. Novitates No. 1126.

APPENDIX

Ecuadorian localities and specimens. Unless otherwise noted the specimens are deposited in ZMUC. Province, coordinates, and elevations are given in parentheses.

Scytalopus canus opacus.—Birds with one call type: 6 ad. males: Páramo El Angel (w. Carchi: 00°43–49'N, 77°47–56'W, 3,400–3,980 m); 4 ad. males (1 in ANSP), 1 subad. male, 1 imm. female, 1 imm. unsexed., 1 juv. female: Cerro Mongus (se. Carchi: 00°27'N, 77°52'W, 3,460–3,650 m); tape-recordings (N.K.; no specimens): Laguna San Marcos (Pichincha: 00°07'N, 77°58'W, 3400 m); 4 ad. males (2 in ANSP), 1 subad. male: Papallacta (Pichincha/Napo: 00°19–23'S, 78°09–13'W, 3,350–3,950 m); 1 male (tape-recording and blood sample only): Río Anatenorio (Napo: 00°59'S, 78°17'W, 3,350–3,700 m); tape-recordings (N.K.; no specimens): around Oyacachi (Napo: 00°11–13'S, 78°03–08'W, 3,400–3,600); 2 ad. males, 2 subad. males, 1 imm. male, 2 juv. females: Cordillera de Los Llanganates (Tungurahua/Napo: 01°07–08'S, 78°19–22'W, 3,700–3,800 m); male type and female topotype (AMNH): "Tambillo, 2440 m" [cf = Tambo de Ashilán, 3,200 m]; tape-recordings and sightings by N.K.: Tambo de Ashilán (Chimborazo: 02°11'S, 78°29'W, 3,150–3,500 m); 2 ad. males (1 in MECN), 1 female: Cordillera Zapote-Najda (Azuay/Morona-Santiago: 03°01–02'S, 78°38–39'W, 3,150–3,450 m); 2 ad. females: Páramos de Matanga (Azuay/Morona-Santiago: 03°17'S, 78°54'W, 3,100–3,350 m).

Birds with another call type: heard by Bloch et al. (1991) and N.K.: Cajanuma (Loja/Zamora-Chinchipe: 04°06'S, 79°09'W, 3050–3400 m); 1 ad. male (MECN) and tape-recordings by N.K.: Cerro Toledo (Zamora-Chinchipe: 04°23'S, 79°07'W, 3150–3350 m); 7 ad. males (1 in ANSP), 3 ad. females (2 in ANSP), 1 juv. male: Cordillera Las Lagunillas (Loja/Zamora-Chinchipe: 04°43–46'S, 79°25–26'W, 3,050–3,650 m).

Scytalopus unicolor latrans.—Dark gray birds with one song type: 1 male (MECN): Pacific slope of Páramo El Angel (Carchi: 00°47'N, 78°01'W, 3,100 m); 1 male: interandean slope of Páramo El Angel (Carchi: 00°39'N, 77°54'W, 3,350 m); 2 males (1 in MECN): near Santa Barbara (Sucumbíos: 00°39'N, 77°30'W, 2,750 m); 1 male, 3 females (1 in ANSP): Cerro Mongus (Carchi: 00°27'N, 77°52'W, 3,220–3,300 m); 1 male: Cordillera de Toisán (w. Imbabura: 00°27'N, 78°36'W, 3,100 m, tape-recordings by N.K. down to 3,050 m); 1 male, 1 female (LSUMZ): Apuela road (w. Imbabura: 00°21'N, 78°26'W, 2,800 m); tape-recordings by N.K.: Apuela road (w. Imbabura: 00°20'N, 78°24–25'W, 2,980–3,365 m); tape-recording by N.K.: Loma Taminanga (Imbabura: 00°17'N, 78°28'W, 2,900 m); tape-recording by N.K.: Lag. Negra, Mojanda Mts. (Imbabura/Pichincha: 00°08'N, 78°15'W, 3,750 m); 5 males, 5 females (1 in MECN), 1 juv. male: Volcán Pichincha (Pichincha: 00°03–15'S, 78°30–38'W, 2,700–3,750 m); heard by N.K.: Mt Ilalo (Pichincha: 00°14'S, 78°24'W, 2,400 m); 2 unsexed (ANSP): Chiriboga road (Pichincha: 00°17–18'S, 78°37–40'W, 2,875–3,500 m); tape-recordings by N.K.: w. slope V. Atacazo (Pichincha: 00°19'S, 78°37'W, 3,400–3,600 m); tape-recordings by N.K.: Pasochoa (Pichincha: 00°27–28'S, 78°29'W, 2,700–4,000 m); tape-recording by O. Jakobsen: 1 km NE Machachi (Pichincha: 00°31'S, 78°31'W, 3,000 m); tape-recordings by N.K.: Corazon (Pichincha: 00°33'S, 78°43'W, 2,980–3,870 m); 3 females: Volcán Iliniza (w. Cotopaxi: 00°42'S, 78°47–48'W, 2,900–3,000 m); 2 males, 1 female: tape-recordings by N.K.: Río Rayo (Cotopaxi: 00°36'S, 78°59'W, 2,250–2,300 m); tape-recordings by N.K.: 2 km SSE Quillotuna (Cotopaxi: 00°41'S, 78°57'W, 3,100 m); 2 males, 1 female: Cerro Parcato (w. Cotopaxi: 00°44'S, 78°58'W, 3,500–3,550 m); tape-recordings by N.K. and blood sample: 10 km NW Salinas (Bolívar: 01°21'S, 79°05'W, 3,000–3,350 m); tape-recording by N.K.: 3.5 km NW Chillanes (Bolívar: 01°55'S, 79°05'W, 2,340 m); heard by N.K.: Loma Totol (Chimborazo: 02°03'S, 78°51'W, 3,450–3,650 m); 1 male: 11 km N Zhud (Cañar: 02°24'S, 78°58–59'W, 2,850 m). Black birds with another song type: 1 imm. female: above Cuyuja (Napo: 00°23'S, 78°01'W, 2,450 m); 3 males (1 in MECN): Cordillera de Guacamayos (Napo: 00°39'S, 77°52'W, 2,000–2,300 m); heard by N.K.: Hacienda Aragón (Napo: 00°40'S, 77°55'W, 2,100–2,235 m); heard by N.K.: east of Oyacachi (Napo: 00°13'S, 78°03'W, 3,020 m); tape-recording by N.K. (calls, song-type?): nw. slope of Volcán Tungurahua (Tungurahua: 01°28'S, 78°27'W, 2,600 m); 6 males (5 in ANSP, 1 in MECN), 1 female (ANSP), 1 juv. female (ANSP): Cordillera de Cutucú (Morona-Santiago: 02°42'S, 78°03'W, 1,975–2,300 m); 1 female and tape-recordings by N.K.: Arenales, right bank of Río Paute (Azuay: 02°34'S, 78°34'W, 2,300–2,400 m); 1 male (EPN) and tape-recordings by T. A. Parker III: Cordillera del Cóndor (Morona-Santiago: 03°27'S, 78°21'W, 2,100 m); tape-recording by N. Flanagan: Cordillera de Tzunantza, Romerillos-San Luis trail (Zamora-Chinchipe: ca. 04°14'S, 79°01'W, 2,300 m); 1 male (ANSP) and tape-recordings by M. B. Robbins: Río Isimanchi (Zamora-Chinchipe: 04°47'S, 79°20'W, 2,250 m).

Two males (BMNH, no vocal data) labeled "Papallacta, 3,100 m" (Napo: ca. 00°22'S, 78°08'W) and "Baeza, 1830 m" (Napo: 00°27'S, 77°53'W) undoubtfully also belong here.

Scytalopus unicolor subcinereus: tape-recordings by N.K.: Arenales, right bank of Río Paute (Azuay: 02°34'S, 78°34'W, 2,300–2,400 m); heard by N.K.: Molleturo road (Azuay: 02°34'S, 79°20'W, 1,500 m); tape-recordings by N.K.: Cerro Paredones (Azuay: 02°45'S, 79°26–27'W, 3250 m); 2 males, 2 females, 1 juv. female: Sural (w. Azuay: 02°47'S, 79°26'W, 2,650 m); 1 female and tape-recordings by N.K.: above Chaucha (Azuay: 02°52'S, 79°23'W, 2,880–3,300 m); tape-recordings by N.K.: Laguna Illincocha (Azuay: 02°50'S, 79°13'W, 3,890 m); tape-recordings by N.K.: Laguna Llaviuco (Azuay: 02°51'S, 79°08'W, 3,200 m); 1 imm. male and tape-recordings by N.K.: Río Mazan (Azuay: 02°52'S, 79°07'W, 2,950–3,300 m); tape-recordings by N.K.: Guagualoma (Azuay: 03°15'S, 79°05'W, 3,055 m); 1 subad. male: Páramos de Matanga (Azuay/Morona-Santiago: ca. 03°16'S, 78°56'W, 3,300 m); old specimen (AMNH) and tape-recordings by

N.K.: Bestión (Azuay: 03°25′S, 79°01′W, 3,050–3,350 m); heard (Bloch et al. 1991): between Selva Alegre and Manu (Loja: 03°32′S, 79°22′W, 2,850–2,950 m); 1 female (MECN): near San Antonio de Cumbe (Loja: 03°34′S, 79°12′W, 2,875 m); tape-recordings by N.K.: Acanamá (Loja: 03°42′S, 79°13′W, 3,200 m); 1 female and tape-recordings by N.K.: Celica Mts. (Loja: 04°05–07′S, 79°57–59′W, 1,800–2,410 m); tape-recordings by N.K.: Cajanuma (Loja: 04°05′S, 79°11′W, 2,600 m); heard (Bloch et al. 1991): Uritusinga (Loja: 04°06′S, 79°09′W, 2,800–2,950 m); tape-recordings by N.K.: 2 km S Cariamanga (Loja: 04°20′S, 79°33′W, 2,300 m); 1 female and tape-recordings by N.K.: Utuana, Sozoranga Mts. (Loja: 04°20–22′S, 79°42–45′W, 1,750–2,550 m); tape-recordings by N.K.: above Jimbura (Loja: 04°42′S, 79°27′W, 3,000 m). AMNH specimens from Taraguacocha (5 males) (El Oro: ca. 03°35′S 79°28′W ?), El Chiral (2 males) (El Oro: ca. 03°38′S, 79°41′W, 1,615 m), Zaruma (3 males) (El Oro: 03°41′S, 79°37′W), and above Zaruma (1 female) (El Oro: ca. 03°40′S, 79°39′W) first mentioned by Chapman (1926) were referred here by Zimmer (1939).

Scytalopus spillmanni.—3 males: Laurel (w. Carchi: 00°49–50′N, 78°01–03′W, 2,350–2,930 m); 1 male: Cordillera de Toisán (w. Imbabura: 00°27′N, 78°36′W, 3,050 m); 4 males: Apuela road (w. Imbabura: 00°19–24′N, 78°23–27′W, 2,200–3,200 m); 3 males, 1 female: Loma Taminanga (w. Imbabura: 00°17′N, 78°28′W, 2,900 m); 11 males (2 in MECN, 7 in ANSP), 1 female (ANSP), 2 unsexed (1 in MECN), 1 imm. male (ANSP), 1 imm. female (ANSP): Volcán Pichincha (Pichincha: 00°01–15′S, 78°35–41′W, 2,100–3,050 m); tape-recordings and heard by N.K.: Chiriboga road (Pichincha: 00°15–17′S, 78°40–48′W, 1,900–2,900 m); tape-recordings by N.K.: Corazón (Pichincha: 00°33′S, 78°43′W, 2,980–3,350 m); 1 female (the type in Berlin Museum): Volcán Iliniza (Cotopaxi/Pichincha: ca. 00°35′S, 78°43′W); tape-recordings by N.K.: Río Rayo (Cotopaxi: 00°36′S, 78°59′W, 2,250, 2,500, and 2,700 m); 3 males: near Santa Barbara (Sucumbíos: 00°33–35′N, 77°31–32′W, 2,100–2,500 m); tape-recording by G. Rosenberg: Cerro Mongus (Carchi: 00°27′N, 77°52′W, 3,200 m); 4 males: Cordillera de Guacamayos (Napo: 00°39′S, 77°52′W, 2,100–2,300 m); 1 female, 1 juv. male, 1 juv. female: Río Anatenorio (Napo: 00°59′S, 78°21′W, 3,000–3,500 m); tape-recordings by N.K.: Cordillera de Los Llanganates (Napo: 01°06′S, 78°18′W, 3,300 m); 2 males: Volcán Tungurahua (Tungurahua: 01°22–28′S, 78°24–29′W, 2,600–2,850 m); tape-recordings by N.K.: Orregán, w. slope Volcán El Altar (Chimborazo: 01°39–40′S, 78°30′W, 3,100–3,300 m); old specimen (AMNH) and tape-recordings by N.K.: upper Río Upano, "Tambillo" [cf. = Tambo de Ashilán] (Chimborazo: 02°11′S, 78°29′W, 2,440–3,200 m); 1 male: Arenales, right bank of Río Paute (Azuay: 02°34′S, 78°34′W, 2,300–2,400 m). A male (MECN, more specimens in WFVZ) from 2 km south of Pan de Azúcar, 15 km east-southeast of Borja: (Napo: 00°27′S, 77°43′W, 2,900 m), a male (MECN) from Papallacta (Napo: ca. 00°22′S, 78°08′W), and a female (AMNH) from above Baeza (Napo: ca. 00°27′S, 77°53′W) look similar and should undoubtedly be referred here.

Scytalopus parkeri.—5 male, 5 females, 2 unsexed, 1 imm. male, 1 imm. unsexed, 1 juv. male: Cordillera Zapote Najda (Morona-Santiago: ca. 03°02′S, 78°31–36′W, 2,250–3,150 m); tape-recordings by N.K.: Páramos de Matanga (Azuay: 03°17′S, 78°54′W, 3,000–3,250 m); tape-recordings by T. A. Parker III: Cordillera del Cóndor (Morona-Santiago: ca. 03°27′S, 78°21′W, ca. 2,300 m); heard (Bloch et al. 1991): between Selva Alegre and Manu (Loja: 03°32′S, 79°22′W, 2,850–2,950 m); 2 males: 7 km NE San Lucas (Loja: 03°40′S, 79°13′W, 3,100 m); 1 subad. male (type): ca. 20 km SSW San Lucas (Loja: 03°50′S, 79°16′W, 2,770 m), heard and blood sample (Bloch et al. 1991): ca. 10 km ENE Loja (Loja: 03°58′S, 79°09′W, 2,550–2,700 m); 2 males (1 in ANSP): Cajanuma (Loja: 04°06′S, 79°09′W, 2,750–2,900 m); tape-recording by M.K. Poulsen: Uritusinga (Loja: 04°06′S, 79°09′W, 2,800–3,000 m); 1 male (MECN), 1 female (ANSP): Río Isimanchi (Zamora-Chinchipe: 04°46′S, 79°25′W, 3,000–3,350 m); tape-recordings by N.K.: Cerro Toledo (Zamora-Chinchipe: 04°23′S, 79°07′W, 3,000–3,275 m).

Scytalopus vicinior.—4 males (1 in ANSP, 1 in MECN): Maldonado road (w. Carchi: 00°50–54′N, 78°03–12′W, 1,650–2,350 m); tape-recordings by N.K.: Cordillera de Toisán (Imbabura: 00°20′N, 78°36′W, 1,850 m); 1 female: Maquipucuna (Pichincha: 00°05′N, 78°37′W, 1,775 m); tape-recordings by N.K.: 1 km above Tandayapa (Pichincha: 00°00′, 78°41′W, 1,900 m); 9 males, 1 female, 1 imm. male: near Mindo (Pichincha: 00°02′S, 78°45–46′W, 1,700 m, heard by N.K. down to 1,450 m); tape-recording by N.K.: Chiriboga road (Pichincha: 00°15′S, 78°50′W, 1,850–1,900 m); heard by N.K.: El Corazón (Cotopaxi: 01°08′S, 79°05′W, 1,250 m).

Scytalopus chocoensis.—17 males (9 including the type in ANSP, 1 in MECN), 5 females (2 in ANSP), 1 juv. male (ANSP): near Ibarra-San Lorenzo railroad (Imbabura and Esmeraldas: 00°52–58′N, 78°30–43′W, 350–950 m). A male (AMNH) from Paramba (Imbabura: 00°49′S, 78°21′W, 1,065 m) undoubtedly also belongs here.

Scytalopus robbinsi.—3 males (1 in ANSP): Molleturo road (w. Azuay: ca. 02°34′S, 79°20′W, 890 m, heard by N.K. to 1,250 m); tape-recording by N.K.: Daucay (El Oro: 03°30′S, 79°45′W, 700 m); 5 males including type (1 in ANSP, 1 in MECN), 3 females (1 in ANSP): 9 km W Piñas (El Oro: 03°39′S, 79°45′W, 870–930 m).

Scytalopus atratus: heard by P. Coopmans: Coca Falls (Napo: 00°06′S, 77°35′W, 1,250 m); 1 male: n-slope of Pan de Azúcar (Napo: ca. 00°08′S, 77°34′W, 1,250 m); 5 males (1 in EPN, 2 MECN), 1 female: Rio Hollín road (Napo: 00°40–46′S, 77°35–51′W, 1,200–1,350 m); 3 males (ANSP): w. slope of Cordillera de Cutucú (Morona-Santiago: ca. 02°40′S, 78°06′W, 1,525–1,650 m); tape-recordings by N.K.: w. slope of Cord. del Cóndor, S Paquisha (Zamora-Chinchipe: ca. 03°58′S, 78°37′W, 1,250 m); tape-recording by N.K.: Río Bombuscaro (Zamora-Chinchipe: 04°06′S, 78°58′W, 1,050 m).

Scytalopus micropterus.—heard by N.K.: La Bonita road (Sucumbíos: 00°33′N, 77°32′W, 2,100 m); 3 males: Pan de Azúcar (Napo: ca. 00°08′S, 77°34′W, 1,250 m); tape-recording by O. Jakobsen: El Salado

(Napo: 00°13'S, 77°43'W, 1,250 m); 11 males, 10 females (AMNH, no vocal data): lower Volcán Sumaco (Napo: ca. 00°34'S, 77°38'W, subtropical); tape-recordings by N.K.: Cordillera de Guacamayos (Napo: 00°39'S, 77°52'W, 1,600–2,100 m); 5 males (1 in MECN, 4 in ANSP), 1 female (BMNH, no vocal data): w. slope of Cordillera de Cutucú (Morona-Santiago: ca. 02°40'S, 78°06'W, 1,700 m); 2 males: w. slope of Cordillera del Cóndor (Zamora-Chinchipe: 04°00'S, 78°27–30'W, 1,400–1,700 m); 1 male (ANSP): Quebrada Avioneta, Cordillera de Tzunantza (Zamora-Chinchipe: 04°15'S, 78°55–56'W, 1850 m); tape-recording by P. Coopmans: near (S) Valladolid (Zamora-Chinchipe: 04°34'S, 79°08'W, 1,500 m); heard by F. Sornoza: San Andrés (Zamora-Chinchipe: 04°50'S, 79°17'W, ca. 2,000 m).

2 old specimens (Warsaw Museum, no vocal data): Mapoto and Machay, Volcán Tungurahua (Tungurahua: 01°24–25'S, 78°16–20'W, 1,521–2,125 m) probably belong here.

A NEW GENUS FOR THE YELLOW-SHOULDERED GROSBEAK

J. V. REMSEN, JR.

Museum of Natural Science, Louisiana State University, Baton Rouge, Louisiana 70803, USA

ABSTRACT.—Recent molecular data have shown that the genus *Caryothraustes* (Cardinalinae) as currently recognized is paraphyletic because one of its member species, *humeralis*, is not the closest relative of the other two species in the genus. Therefore, a new genus is created for this species, the Yellow-shouldered Grosbeak, formerly known as *Caryothraustes humeralis*.

RESUMEN.—Recientes datos moleculares han mostrado que el género *Caryothraustes* (Cardinalinae) como es reconocido actualmente, es parafilético, porque una de sus especies integrantes, *humeralis*, no tiene la relación más próxima con las otras dos especies del género. Por lo tanto, se crea un nuevo género para la especie conocida antiguamente como "*Caryothraustes*" *humeralis*.

Molecular genetics (Tamplin et al. 1993; Demastes and Remsen 1994) have confirmed hypotheses based on morphology and natural history (Hellmayr 1938; Hellack and Schnell 1977; Remsen and Traylor 1989) that the Yellow-shouldered Grosbeak ("*Caryothraustes*" *humeralis*) is not the closest relative of the other two species in the genus *Caryothraustes* (*C. canadensis* and *C. poliogaster*). Thus, its inclusion in *Caryothraustes* would make that genus paraphyletic. To indicate the uncertain affinities of *humeralis* and to remove it from genera for which there is no evidence of sister relationship, I here establish a new genus for it.

The species *humeralis* has been placed in three genera. It was described by Lawrence as a member of the genus *Pitylus* Cuvier. That genus, however, was subsequently restricted (e.g., Ridgway 1901) to just two species, *P. grossus* and *P. fuliginosus*. Demastes and Remsen (1994) showed that recognition of the genus *Pitylus* caused the genus *Saltator* to be paraphyletic, and they recommended placing *Pitylus* in the synonymy of *Saltator*, a recommendation followed by the American Ornithologists' Union (1995) Check-list Committee. Ridgway (1901) treated *humeralis* as a member of the genus *Caryothraustes* Reichenbach. Chapman (1926) treated *humeralis* as a member of the genus *Saltator* Vieillot, but did so reluctantly, stating: "In its rounded, decurved culmen and more pointed wings, it appears to differ generically from *Saltator* though apparently nearer that genus than to *Pitylus*." Hellmayr (1938) reluctantly placed *humeralis* in *Caryothraustes*, and it has been treated as a member of that genus since then (e.g., Paynter 1970; Sibley and Monroe 1990). Demastes and Remsen (1994) found that *humeralis* was not a sister taxon either to *Pitylus* or to *Saltator sensu strictu*.

In plumage color and pattern, *humeralis* shares characters with members of both *Caryothraustes* and *Saltator*, and these shared features were clearly responsible, historically, for the placement of *humeralis* in these two genera. Therefore, naming a new monotypic genus based on plumage characters could be avoided by merging *Caryothraustes* into *Saltator*. However, available molecular data (Demastes and Remsen 1994) show that to combine *Saltator* and *Caryothraustes* and also to avoid a paraphyletic genus would require the merger into one genus of all other cardinaline genera analyzed so far (*Cyanocompsa*, *Cardinalis*, *Pheucticus*, and *Spiza*). Such a genus would be unusually, perhaps uniquely, heterogeneous in birds. Furthermore, many other genera of cardinalines have yet to be analyzed genetically, and so their retention as separate genera or placement in this broad genus (for which *Saltator* Vieillot is the oldest name) would be based on inferences from phenotypic data. Finally, the genus *Saltator* itself is probably paraphyletic (Hellack and Schnell 1977). Therefore, I prefer to keep *humeralis* distinct at the generic level.

Because the type species for *Caryothraustes* is *C. canadensis*, no other generic name is available for *humeralis*. I propose the following:

Parkerthraustes, new genus

Type species.—*Pitylus humeralis* Lawrence, 1867.

Diagnosis.—The evidence for creation of a new genus for *humeralis* is largely molecular and morphometric; see Tamplin et al. (1993) and Demastes and Remsen (1994) for details. Traditional descriptions of genera, however, are based on external phenotypic characters that in themselves may have little phylogenetic significance. Therefore, a diagnosis of the genus based on these characters is somewhat artificial. Nevertheless, I provide the following means for distinguishing *Parkerthraustes* from other genera. *Parkerthraustes* is the only genus of thick-billed, hook-billed, sexually dimorphic passerine in the New World that has a mottled malar streak and throat. It may be distinguished from *Caryothraustes* as follows: (1) yellow coloration in ventral plumage restricted to undertail coverts, with traces on lower flanks and thighs (versus conspicuous yellow or greenish yellow on breast or belly in *Caryothraustes*); (2) chin and throat irregularly mottled with black (male) or gray (female) (versus solid black in *Caryothraustes*); (3) malar feathers mottled black and off-white (male) or gray and off-white (female) (versus solid black in *Caryothraustes*); (4) auriculars black (male) or gray (female) (versus yellow or greenish-yellow in *Caryothraustes*); (5) crown and nape gray (versus greenish-yellow in *Caryothraustes*); (6) forehead gray (versus black in *Caryothraustes*); and (7) *Parkerthraustes* is sexually dichromatic, whereas *Caryothraustes* is not. *Parkerthraustes* can be distinguished from any species in the genus *Saltator* (including *Pitylus*) by (1) undertail coverts: yellow in *Parkerthraustes* versus not yellow in *Saltator*; and (2) tail shape: outer three rectrices in *Parkerthraustes* the same length or slightly longer than the inner two, thus giving the tail a slightly notched shape, versus rectrix 5 shorter than rectrices 4 and 3, giving the tail a more rounded shape in *Saltator*. In *S. rufiventris*, an aberrant *Saltator* in other respects (Hellack and Schnell 1977), the differences between rectrices 5, 4, and 3 are minimal, approaching the condition in *Parkerthraustes*. Also, the mottled center of the chin and throat in *Parkerthraustes* is not found in any species of *Saltator*, although *S. atriceps* and *S. similis* show some approach in this character.

Etymology.—I am pleased to name this genus in honor of my friend and colleague, the late Theodore A. Parker III. Long ago, Ted pointed out to me the distinctive nature of this species and how different it was behaviorally from true *Caryothraustes*. Its generic separation symbolizes one of Ted's favorite themes, namely the importance of detailed natural history observations and the clues that they provide for elucidating the systematic relationships of birds. This and other themes that Ted promoted greatly influenced a generation of Neotropical field ornithologists (Remsen and Schulenberg 1997).

Relationships.—A phenetic analysis of data on allele frequencies indicated that *Parkerthraustes* was closer to the cardinaline grosbeaks than to the saltators, whereas a cladistic analysis of the same data set produced the opposite result (Demastes and Remsen 1994). Without a well-corroborated phylogeny of the cardinalines, I hesitate to speculate on where *Parkerthraustes* fits in the phylogeny of this group. Therefore, I recommend that *Parkerthraustes* be placed incertae sedis at the end of the linear sequence of cardinaline genera.

ACKNOWLEDGMENTS

Richard C. Banks, M. Ralph Browning, Gary R. Graves, and Robert W. Storer made valuable comments on the manuscript.

LITERATURE CITED

AMERICAN ORNITHOLOGISTS' UNION. 1995. Fortieth supplement to the American Ornithologists' Union Checklist of North American Birds. Auk 112:819–830.
CHAPMAN, F. M. 1926. The distribution of bird-life in Ecuador. Bull. Am. Mus. Nat. Hist. 55:1–784.
DEMASTES, J. W., AND J. V. REMSEN, JR. 1994. The genus *Caryothraustes* (Cardinalinae) is not monophyletic. Wilson Bull. 106:733–738.
HELLACK, J. J., AND G. D. SCHNELL. 1977. Phenetic analysis of the subfamily Cardinalinae using external and skeletal characteristics. Wilson Bull. 89:130–148.
HELLMAYR, C. E. 1938. Catalogue of birds of the Americas. Part XI. Field Mus. Natur. Hist. (Zool. Ser.) 8: 1–662.
PAYNTER, R. A., JR. 1970. Subfamily Cardinalinae. Pp. 216–245 in Check-List of Birds of the World (R. A. Paynter, Jr., Ed.). Museum of Comparative Zoology, Cambridge, Massachusetts.
REMSEN, J. V., JR., AND T. S. SCHULENBERG. 1997. The pervasive influence of Ted Parker on Neotropical field ornithology. Pp. 7–18 in Studies in Neotropical Ornithology Honoring Ted Parker (J. V. Remsen, Jr., Ed.), Ornith. Monogr. No. 48.
REMSEN, J. V., JR., AND M. A. TRAYLOR, JR. 1989. An Annotated List of the Birds of Bolivia. Buteo Books, Vermillion, South Dakota.
RIDGWAY, R. 1901. The birds of North and Middle America. Part 1. Family Fringillidae—the finches. Bull. U.S. Nat. Mus. 50:1–715.
TAMPLIN, J. W., J. W. DEMASTES, AND J. V. REMSEN, JR. 1993. Biochemical and morphometric relationships among some members of the Cardinalinae. Wilson Bull. 105:93–113.

DISTRIBUTION AND GEOGRAPHIC VARIATION IN THREE SOUTH AMERICAN GRASSQUITS (EMBERIZINAE, *TIARIS*)

John M. Bates

Department of Ornithology, American Museum of Natural History, Central Park West and 79th Sreet, New York, New York 10024-5192, USA; present address: Zoology Department, Field Museum of Natural History, Roosevelt Road at Lake Shore Drive, Chicago, IL 60605, USA

ABSTRACT.—The taxonomy and South American distribution of three species of *Tiaris* grassquits are reviewed. I focus on the two most widespread species, Dull-colored Grassquit (*T. obscura*) and Sooty Grassquit (*T. fuliginosa*), because together they form a circum-Amazonian distribution. Membership of the Dull-colored Grassquit in *Tiaris* has been proposed only recently. Although only four subspecies of Dull-colored Grassquit are currently recognized, this species contains at least six diagnosable populations that include a "leapfrog" pattern of geographic variation through the Andes. In contrast, insufficient specimen material is available to diagnose apparently isolated populations of the Sooty Grassquit that occur north, east, and south of the Amazon Basin. The three currently recognized subspecies are not separable based on the male plumage characters used to describe them. South American subspecies of the primarily West Indian Black-faced Grassquit (*T. bicolor*) occur syntopically with *T. obscura* and *T. fuliginosa* in northern South America. The similarity of these three species has lead to confusion regarding their South American distributions, and many specimens have been incorrectly identified. I present the first records of Dull-colored Grassquit for Paraguay and possibly Brazil, and of Sooty Grassquit for Bolivia. I also present an unpublished record of Sooty Grassquit for Colombia. Colombian specimens previously referred to this species are actually *T. obscura*.

RESUMEN.—Se hizo una evaluación de la taxonomía y de la distribución de tres especies de Tordillos *Tiaris*. Pongo mayor enfoque en las dos especies que tiene la más amplia distribución en Sudamérica, Tordillo pardo (*T. obscura*) y Tordillo ahumado (*T. fuliginosa*), porque los dos juntas forman una distribución circum-Amazónica. Se ha propuesto solo recientemente que los Tordillos pardo pertenecen a los miembros de *Tiaris*. Aunque solamente se reconocen cuatro subespecies de Tordillo pardo; las especies contienen a lo menos seis poblaciones diagnosticables que incluyen un patrón de "saltacabrilla" de variación geográfica a través de los Andes. A diferencia, aunque hay tres subespecies, aparentemente no hay suficientes características diferentes disponibles para hacer un diagnóstico de las poblaciones aisladas del Tordillo ahumado que se encuentran por la parte norte, este, y sur del Amazonia. Las subespecies sudamericanas principalmente del Tordillo común antillano (*T. bicolor*) ocurren de una manera sintópica con *T. obscura* y *T. fuliginosa* en la parte norte de sudamérica. La similitud entre estas tres especies ha creado mucha confusión en cuanto a su distribución sudamericana y muchos especímenes han sido identificados de una manera incorrecta. Presento los primeros récords del Tordillo pardo para Paraguay y posiblemente de Brasil, y del Tordillo ahumado para Bolivia. También presento un récord no publicado del Tordillo ahumado para Colombia. Otros especímenes colombianos que han sido referidos a esta especie son realmente *T. obscura*.

Distribution patterns of South American birds have fascinated ornithologists since they first began describing the myriad of species found on the continent. Although many Neotropical distribution patterns, such as those occurring among Amazonian birds, have been discussed at length (Haffer 1974, 1987, and references therein), others have only recently been noticed. One such pattern is the circum-Amazonian distribution described by Remsen et al. (1991), in which taxa occur in a ring of populations around the Amazon Basin. In some cases, single species exhibit this distribution, but Remsen et al. (1991) also noted allopatric congeners that, together, showed the pattern. If more than one species is involved, then incorrect taxonomy can obscure a circum-Amazonian distribution pattern. In the genus *Tiaris*, this has occurred because one species (Dull-colored Grassquit, *T. obscura*) has long been placed in *Sporophila* (Paynter 1970). In this paper, I review the characters that support placement of *T. obscura* in *Tiaris* and document the South American distributions of, and geographic variation in, this species and two congeners. The combined distributions of two species (*T. obscura* and *T. fuliginosa*, Sooty Grassquit) appear to represent another example of the circum-Amazonian distribution. In addition, I find that the Andean populations of *T. obscura* represent another example of the "leapfrog" pattern observed in other Andean birds (Remsen 1984).

The genus *Tiaris* is composed of five small emberizine finches, called grassquits, found throughout the Caribbean and in portions of Central and South America. There are no phylogenetic studies of relationships within the genus; however, evidence suggests that the five *Tiaris* species, including *T. obscura*, represent a monophyletic group (Clark 1986). The five may be divided further based on the presence or absence of yellow in male plumage. This character unites *T. olivacea* (Yellow-faced Grassquit) with *T. canora* (Cuban Grassquit) relative to *T. obscura*, *T. fuliginosa* and *T. bicolor* (Black-faced Grassquit), which lack yellow in any plumage. In this paper, I focus on *T. obscura* and *T. fuliginosa*, the species most widespread in South America; however, I also include distribution information on the South American populations of *T. bicolor*. This species is sympatric with *T. obscura* and *T. fuliginosa* in northern Colombia and Venezuela, and plumage similarities and geographic variation in these species have caused confusion regarding their distributions in this region.

METHODS

Locality data for mapping distributions of the taxa were taken from the literature and from specimens housed in the American Museum of Natural History, New York (AMNH); the Academy of Natural Sciences, Philadelphia (ANSP); the Carnegie Museum of Natural History (CM), Pittsburgh; the Field Museum (FMNH), Chicago; the Louisiana State University Museum of Natural Science (LSUMZ), Baton Rouge; the Museum of Vertebrate Zoology, University of California, Berkeley (MVZ), and the National Museum of Natural History, Smithsonian Institution, Washington D.C. (USNM). I measured wing chord and tail length, as well as culmen length (from the distal end of the nares to tip of the bill), width, and depth on specimens with digital calipers. Measurements followed Baldwin et al. (1931).

I also noted coloration of the maxilla and mandible on each specimen. Although bill coloration may change in specimens over time, I believe that these characters can be used with some caution, especially when series of birds are available for comparison to one another. In a few cases, bill colors were recorded on specimen labels by the collectors, and for these specimens (even if collected more than 50 years ago), the overall pattern of color (particularly contrast in color between maxilla and mandible) had not changed from the colors written on the label. In addition to specimens that I examined, curators at other museums provided locality data for holdings in their collections. These institutions include: the Colección Ornitológica Phelps, Caracas, Venezuela (COP); the Natural History Museum of Los Angeles County (LACM), Los Angeles; the Peabody Museum of Natural History, Yale University, New Haven (YPM); and the Western Foundation of Vertebrate Zoology, Camarillo (WFVZ). Two localities were based on tape recordings housed at the Library of Natural Sounds, Cornell Laboratory of Ornithology, Ithaca (LNS). I mapped collecting localities using the Geographic Information System described by Isler (1997).

RESULTS AND DISCUSSION

Tiaris bicolor

This species is widespread in the West Indies, but is limited to northern portions of Colombia (Col.) and Venezuela (Ven.) in South America (Fig. 1). Only two of the eight described subspecies of *T. bicolor* occur on the continent. *Tiaris b. omissa* is distributed primarily along the northern coastal plain, but there are a few records from further inland (Fig. 1). This subspecies also occurs on Isla Margarita, Tobago (but not on Trinidad, where only *T. fuliginosa* occurs), and throughout the Lesser Antilles to Puerto Rico. The other subspecies in South America, *T. b. huilae*, is known only from the upper Magdalena valley in south-central Colombia (Miller 1952).

Adult males of both *T. bicolor* subspecies are distinguishable from *T. obscura* and *T. fuliginosa* by their olive backs and sooty-colored heads. Female and immature *T. bicolor* resemble females and immatures of the other two species, but *T. bicolor* females and immatures are generally

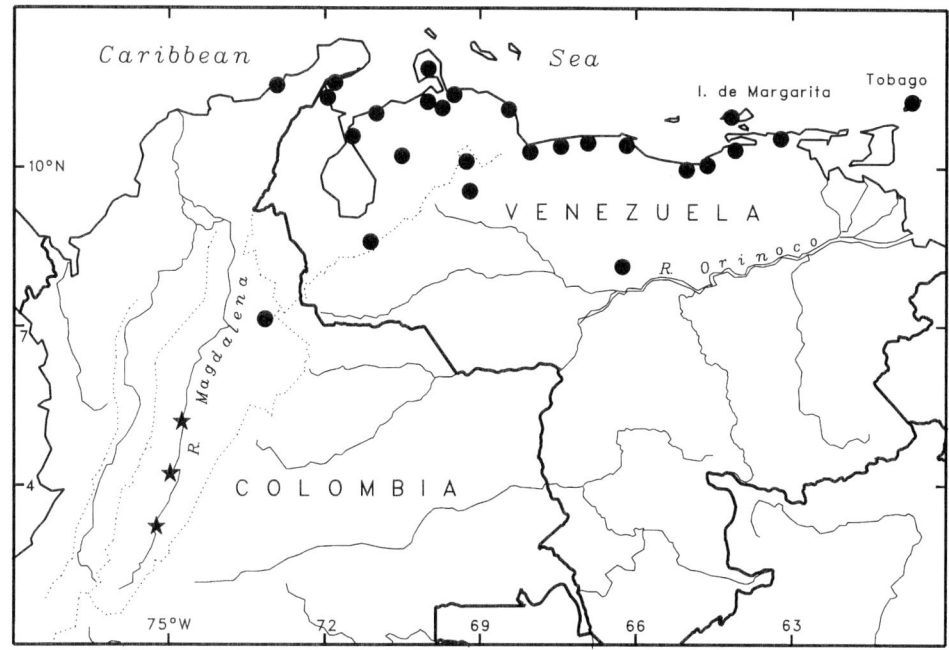

FIG. 1. South American specimen localities for the two subspecies of Black-faced Grassquit (*Tiaris bicolor*) that occur on the continent. Black dots = *T. o. omissa*, black stars = *T. o. huilae*. Dotted lines along the Andes indicate the cordilleran divides.

paler overall with more olive backs than *T. fuliginosa* females and *T. obscura* (both sexes). Both subspecies of *T. bicolor* also have slightly shorter tails than the other species (Table 1).

Tiaris b. huilae differs from *T. b. omissa* in that both sexes are paler brownish-olive on the back, and males have paler and more extensively gray flanks than *omissa* males. Specimens collected in the northern part of the Magdalena valley near Bucaramanga (Col.) have been referred to *T. b. omissa* (Miller 1952). In the vicinity of Bucaramanga and Caracas (Ven.), *T. bicolor*, *T. obscura*, and *T. fuliginosa* all have been collected (Figs. 1, 2); however, both subspecies of *T. bicolor* favor xerophytic lowland habitats more than the other species. Most specimen localities on the mainland are below 500 m, with records from near Bucaramanga, Mérida (Ven.), and Caracas occurring as high as 1,000 m (Appendix).

TIARIS OBSCURA

As the English name (Dull-colored Grassquit) implies, *T. obscura* is a dull brown bird. Unlike the other members of the genus, this species exhibits no obvious sexual dimorphism in plumage (bill color may differ between sexes in some populations). Both sexes resemble the females and immatures of the other two brown species of *Tiaris*, especially *T. fuliginosa*. *Tiaris obscura* occurs in a variety of habitats, from humid and dry forest clearings to xerophytic scrub, from northern Colombia and Venezuela to northern Argentina (Fig. 2). It is distributed throughout the mountain ranges of northern Colombia and Venezuela; however, from extreme southern Colombia through Ecuador, it is known only from the western slope of the Andes. In Peru, populations are found in dry habitats west of the Andes and along both humid and dry slopes of the eastern valleys. The species also occurs on the eastern side of the Andes in Bolivia south to northern Argentina and Paraguay. Before discussing geographic variation in this species, I summarize the evidence leading to its placement in *Tiaris*.

Taxonomic position.—For many years, *T. obscura* was placed in the genus *Sporophila* (e.g., Todd 1912; Chapman 1926; Hellmayr 1938; Meyer de Schauensee 1952, 1966) or *Catamenia* (Hellmayr 1913). Paul Schwartz (1972, unpubl. data) first suggested placing it in *Tiaris* because it shared a buzzy song and construction of a dome-shaped nest (most other South American emberizines, including *Sporophila*, build cup-shaped nests) with other species of *Tiaris*. Paynter (1970) placed *obscura* in *Sporophila*, but acknowledged Schwartz's suggestion in a footnote.

TABLE 1

Specimen Measurements for Three South American *Tiaris* Grassquits. Taxonomic and Geographic Subdivisions of the Specimens Are Explained in the Text. Values in Each Measurement Column Are Mean, Standard Deviation, and Sample Size

	Sex	Wing	Tail	Culmen length	Culmen width	Culmen depth
T. bicolor omissa	Male	51.5 (2.0), 5	37.0 (1.6), 5	6.4 (0.5), 5	5.5 (0.2), 5	6.4 (0.7), 5
T. b. huilae	Male	52.2 (2.3), 2	37.5 (0.7), 2	6.4 (0.7), 2	5.5 (0.1), 2	6.0 (0.3), 2
T. obscura haplochroma (Venezuela)	Male	50.6 (0.3), 2	40.3 (0.2), 2	6.1 (0.3), 2	6.1 (0.1), 2	—
T. o. haplochroma (Santa Marta Mtns.)	Male	51.7 (2.2), 17	41.7 (2.1), 16	6.4 (0.5), 16	5.8 (0.5), 17	6.6 (0.3), 17
T. o. haplochroma (Magdalena Valley)	Male	53.8 (2.2), 19	42.5 (1.9), 19	6.6 (0.3), 18	5.8 (0.4), 19	6.8 (0.2), 12
T. o. haplochroma (central Colombia)	Male	54.0 (2.1), 9	42.5 (1.4), 9	6.3 (0.4), 9	5.9 (0.3), 9	6.6 (0.2), 9
T. obscura (southern Colombia, Ecuador)	Male	50.1 (2.4), 13	39.6 (2.3), 12	5.4 (0.5), 12	5.4 (0.1), 8	5.9 (0.2), 5
T. obscura (northern Peru)	Male	52.6 (1.4), 8	41.4 (2.0), 8	5.6 (0.4), 8	5.2 (0.3), 8	6.2 (0.3), 7
T. obscura pacifica[1]	Male	61.1	47.0	5.5	6.7	—
T. obscura (central Peru, depto. La Paz, Bol.)	Male	55.0 (1.8), 17	41.5 (1.7), 16	6.0 (0.3), 17	5.5 (0.2), 17	6.1 (0.3), 15
T. obscura (central/southern Bolivia)	Male	54.7 (2.2), 11	43.4 (2.3), 11	6.0 (0.3), 11	5.9 (0.2), 11	6.2 (0.2), 11
T. obscura ? (Mato Grosso, Brazil)	Male	—	—	—	—	—
T. obscura (Paraguay)[1]	Male	53.6	43.8	6.1	5.9	6.2
T. obscura (Argentina)	Male	57.3 (2.0), 4	43.5 (1.0), 4	6.2 (0.2), 4	6.1 (0.1), 4	6.7 (0.2), 4
T. fuliginosa (Trinidad)	Male	57.2 (2.3), 9	42.9 (2.0), 9	6.7 (0.4), 9	6.4 (0.2), 9	7.0 (0.3), 9
T. fuliginosa (Venezuela)	Male	59.9 (0.8), 6	44.0 (1.4), 6	7.0 (0.7), 6	6.6 (0.5), 6	6.9 (0.2), 5
T. fuliginosa (Colombia)[1]	Male	59.1	41.2	6.9	6.6	7.1
T. fuliginosa (Maranhão, Brazil)	Male	57.4 (1.6), 6	41.2 (1.7), 6	7.0 (0.2), 6	6.3 (0.3), 6	7.3 (0.2), 6
T. fuliginosa (southeastern Brazil)	Male	57.7 (2.0), 11	42.5 (1.6), 11	7.1 (0.4), 11	6.8 (0.4), 11	7.3 (0.3), 10
T. fuliginosa (Mato Grosso, Brazil)[1]	Male	58.4	42.9	—	6.7	—
T. fuliginosa (Bolivia)[1]	Male	56.2	42.0	6.7	6.1	7.5
T. bicolor omissa	Fem.	49.6 (1.9), 7	36.8 (1.0), 6	6.4 (0.5), 7	5.7 (0.3), 7	6.7 (0.6), 7
T. b. huilae[1]	Fem.	50.4	36.3	7.0	5.6	6.6
T. obscura haplochroma[1] (Venezuela)	Fem.	50.7	38.0	6.4	6.0	6.3
T. o. haplochroma (Santa Marta Mtns.)	Fem.	50.2 (1.6), 15	39.1 (1.3), 15	6.2 (0.4), 15	5.7 (0.3), 15	6.3 (0.3), 12
T. o. haplochroma (Magdalena Valley)	Fem.	51.3 (1.9), 4	41.1 (2.3), 4	6.2 (0.2), 4	6.2 (0.2), 4	6.6 (0.8), 2
T. o. haplochroma (central Colombia)	Fem.	52.4 (1.8), 7	40.3 (1.9), 7	6.3 (0.2), 7	6.1 (0.3), 6	6.5 (0.7), 5
T. obscura (southern Colombia, Ecuador)	Fem.	49.5 (2.3), 5	38.7 (2.6), 4	5.7 (0.2), 5	5.1 (0.3), 4	5.8 (0.2), 4
T. obscura (northern Peru)	Fem.	52.7 (1.6), 11	39.0 (2.1), 11	5.7 (0.5), 10	5.2 (0.6), 11	6.0 (0.3), 9
T. obscura (central Peru, depto. La Paz, Bol.)	Fem.	53.0 (1.9), 17	41.0 (1.2), 16	5.9 (0.3), 17	5.7 (0.3), 16	6.0 (0.2), 15
T. obscura (central/southern Bolivia)	Fem.	53.3 (1.7), 18	43.1 (1.9), 18	5.7 (0.2), 18	5.7 (0.3), 18	6.2 (0.2), 18
T. obscura ? (Mato Grosso, Brazil)	Fem.	55.5 (0.1), 2	45.9 (3.4), 2	6.1 (0.5), 2	5.8 (0.1), 2	6.7 (0.1), 2
T. obscura (Paraguay)	Fem.	—	—	—	—	—
T. obscura (Argentina)	Fem.	—	—	—	—	—
T. fuliginosa (Trinidad)	Fem.	57.5 (1.6), 3	42.3 (0.3), 3	6.5 (0.3), 3	6.4 (0.1), 3	7.0 (0.1), 3
T. fuliginosa (Venezuela)	Fem.	54.5 (0.8), 3	42.0 (3.3), 3	6.7 (0.5), 2	6.0 (0.2), 3	6.9 (0.3), 3
T. fuliginosa (Maranhão, Brazil)	Fem.	51.8 (1.0), 3	40.5 (1.6), 3	6.0 (0.1), 3	6.2 (0.1), 3	6.8 (0.2), 3
T. fuliginosa (Mato Grosso, Brazil)	Fem.	—	—	—	—	—
T. fuliginosa (southeastern Brazil)[1]	Fem.	56.3	41.2	7.3	6.3	7.2

[1] Only one individual was measured.

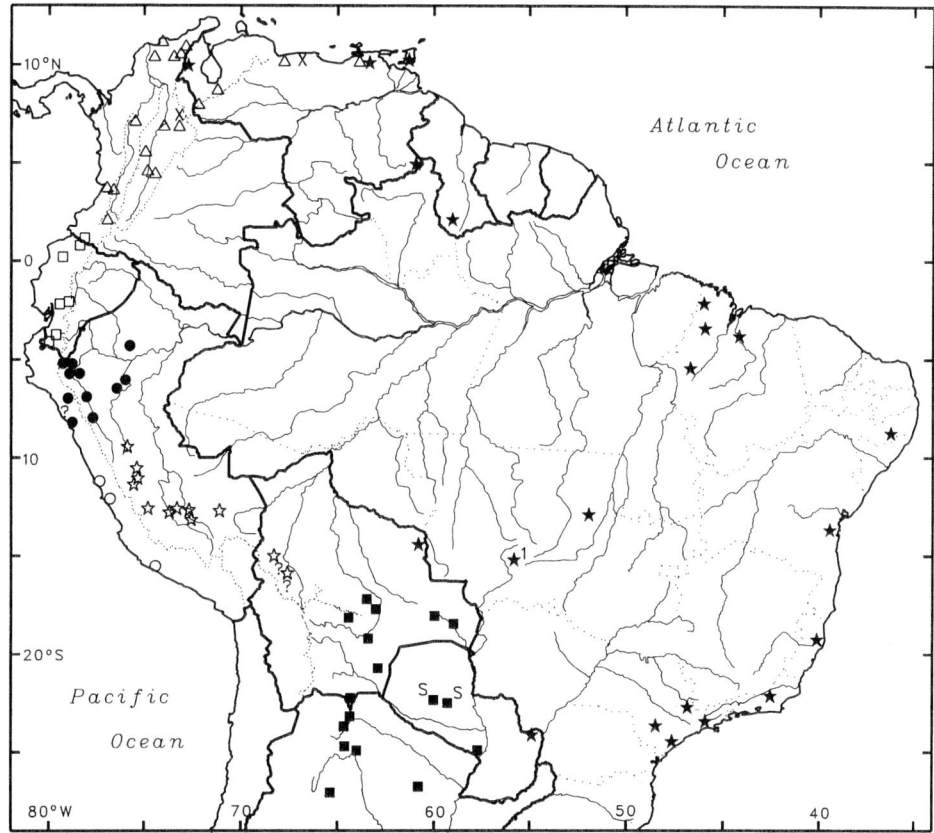

FIG. 2. Specimen localities of Dull-colored Grassquit (*T. obscura*) and Sooty Grassquit (*T. fuliginosa*). Specimens of *T. obscura* are divided into six diagnosable populations based on plumage variation (see text). From south to north, black squares = birds from Paraguay, Argentina; and southern Bolivia (an "s" next to two Paraguayan localities indicates that these are sight records only); open stars = birds from northern Bolivia and central Peru, open dots = *T. o. pacifica*; black dots = birds from northern Peru; open squares = birds from Ecuador and southern Colombia; and open triangles = *T. o. haplochroma*. Black stars = *T. fuliginosa* (a "1" next to Chapada, a site in southwestern Brazil, is to indicate that I believe several specimens from this site, long considered *T. fuliginosa*, may be *T. obscura*. If correct, they represent the only records of *T. obscura* for Brazil, see text). No geographic variation in this species was detected. Two "X"s in Venezuela and Colombia indicate localities where *T. obscura*, *T. fuliginosa*, and *T. bicolor* (see Fig. 1) have been collected at the same site. Three "?" indicate an odd-plumaged specimen from western Peru (a *T. obscura* from Ascope, depto. La Libertad, Peru) and two specimens that I have not examined from two localities in depto. La Paz, Bolivia. The Bolivian sites possibly lie near where two distinct populations come together. Dotted lines along the Andes indicate the cordilleran divides, and those in Brazil delineate Brazilian states.

Domed nests of *T. obscura*, in addition to those observed by Schwartz in Venezuela, have been collected in other parts of its range (Colombia, Peru, and Argentina; nests and eggs at WFVZ, Lloyd Kiff in litt.). Other studies have provided further documentation that *obscura* belongs in *Tiaris*.

In an unpublished manuscript in which they proposed to name a new subspecies of *obscura* (from northern Venezuela) after Schwartz, A. Wetmore, and W. H. Phelps, Jr., wrote the following regarding cranial characters in *obscura*:

> "In the skull of species of the genus *Sporophila*, the prepalantine bar of the palate in ventral view is slender at its central portion, and expands gradually forward to its fusion with the premaxilla. In this genus and its allies, there is no indication of a lateral palatomaxillary on either side. The species of *Tiaris* are distinct in this regard as the prepalantine bar expands only slightly at its anterior end, and there is a clearly marked, slender palato-

maxillary on the other side. The anterior end of this process is fused with the premaxilla, and for somewhat more than half its length it rests against the outer side of the anterior end of the prepalantine bar, from which it is separated by a suture. The distal part projects free as a very slender, pointed spine. (Those not familiar with the terminology of this description will find the details illustrated in Tordoff, 1954, fig. 1). To check on this we had the bones removed from a study skin of *Sporophila obscura haplochroma* (U. S. N. M. No. 25464) from San Lorenzo in the Sierra Nevada de Santa Marta. Though the skull was somewhat damaged, the palato-maxillary is preserved on one side from its attachment to the premaxilla back along the prepalantine bar to the basal part of the free spine (the tip of that process having been broken in cleaning). This verifies the interesting discovery made through the records of the song, so that the species is to be known as *Tiaris obscura*."

Also, Collins and Kemp (1976) found that *obscura* resembled other *Tiaris* rather than *Sporophila* in natal pterylosis, and Clark (1986) found that members of *Tiaris*, including *obscura*, possess a unique foot-scute pattern relative to other Neotropical emberizines.

Distribution and geographic variation.—Paynter (1970) recognized four subspecies of *T. obscura*: 1) *T. o. haplochroma* (Todd 1912) of northern Colombia and Venezuela, 2) *T. o. pauper* (Berlepsch and Taczanowski 1884) of the Colombian Andes south to northwestern Peru, 3) *T. o. obscura* (Lafresnaye and d'Orbigny 1837) of the eastern slopes of the Andes of central Peru to northern Argentina, and 4) *T. o. pacifica* (Koepcke 1963) confined to the western side of the Andes in southern Peru. However, a number of authors (e.g., Chapman 1926; Hellmayr 1938; Meyer de Schauensee 1952; Haffer 1986) have followed Hellmayr's (1913) suggestion that *T. o. pauper* is indistinguishable from nominate *obscura*. In the following paragraphs, I review geographic variation in this species. At the outset, I note that even if *pauper* is considered valid, distinctive geographic variation is obscured by recognizing only the four subspecies listed above. I find at least six diagnosable geographic units.

1) *Tiaris o. haplochroma*, as traditionally recognized, occurs in the Santa Marta Mountains, in the northern portions of the Colombian and Venezuelan Andes, and in the coastal ranges east to Caracas (Paynter 1970; Fig. 2, open triangles). I have not examined the Venezuelan specimens from the Cordillera de Mérida (La Azulita, Mérida and Queniquea, Táchira; COP) that Wetmore and Phelps (unpubl. manuscript) proposed to name after Schwartz, but I include them in *haplochroma* because Venezuelan specimens that I have examined from further east (Picacho de Galipán, Distrito Federal; Sierra de Carabobo, Carabobo; Cerro Turumiquire, Sucre; Río Cocollar, Monagas; Appendix) are not distinguishable. As discussed below, I also include central Colombian specimens of *T. obscura* in this subspecies.

Specimens of *T. o. haplochroma* that I have examined from the type locality (Cincinati, depto. Magdalena, Col.) and other sites in the Santa Marta region (AMNH, CM, FMNH, LSUMZ, MVZ; Appendix) are dark brown or gray-brown (some variation is almost certainly due to foxing), both on the back and on the breast with only the slightest olive tinge. These specimens have darker brown underparts than other *T. obscura* populations, and the belly color contrasts little with breast color. In any plumage, these birds are difficult to distinguish from juveniles and females of *T. fuliginosa*; however, adult *T. o. haplochroma* tend to be duller brown with less buffy tones than *fuliginosa*. In his type description of *T. o. haplochroma*, Todd (1912) described both the back and breast of the type, an adult male, as olive. Although there are slight olive tones to the coloration of *haplochroma*, emphasizing this color over brown may lead to confusion of this taxon with *T. bicolor* which, as noted above, is distinguished from the other two species by its distinctly olive coloration on the back. Hellmayr (1938) and Meyer de Schauensee (1952) both gave more accurate descriptions of the coloration of *T. o. haplochroma*. Collecting localities for *haplochroma* in northern Colombia and Venezuela are generally above 700 m, with records up to 2,000 m (La Africa, depto. La Guajira, Col.; Picacho de Galipán, Cerro Turumiquire, Ven.).

The presence of a bicolored bill often is cited as a character that distinguishes *T. obscura* from *T. fuliginosa* (e.g., Hilty and Brown 1986; Ridgely and Tudor 1989), but variation in this character exists in both taxa. In *T. obscura*, bicolored bills are typical of populations from extreme southern Colombia south, but not of *T. o. haplochroma*. In the majority of *T. o. haplochroma* specimens (exceptions are some females), the lower mandible is light only on the underside; thus, seen from the side, the bill is all dark. Also, many *T. fuliginosa*, including adult males, have light lower mandibles. Reliance on this character has lead to confusion about the identification of specimens from central Colombia.

With one exception, all *Tiaris* specimens from central Colombia outside the Santa Marta region are brown and many have all dark bills. Thus, although Chapman (1926) considered a female *T. o. haplochroma* from La Frijolera in the lower Cauca Valley (depto. Antioquia, AMNH 134144) to be indistinguishable from Ecuadorian specimens that he assigned to *T. o. pauper*, this individual (with a light lower mandible) is not typical of central Colombian specimens. In plumage coloration, central Colombian birds resemble *T. o. haplochroma* from the Santa Marta region in that they are more grayish-brown on the breast than *T. fuliginosa* from Trinidad (where *obscura* does not occur).

Dark-billed *Tiaris* from the upper Patía and middle Magdalena Valleys were identified originally as *T. fuliginosa* (Borrero and Hernandez 1961; Meyer de Schauensee 1966). However, in describing six specimens (two males, three females, one unsexed) from the middle Magdalena Valley, Borrero and Hernandez (1961) quoted a portion of a letter from E. Eisenmann in which, although he agreed that their specimens were *T. fuliginosa*, he cautioned that they appeared to be immature birds because there were no sooty-plumaged males. He further noted that the bill and breast color in these specimens were darker than that of the *T. fuliginosa* specimens used for comparison. I have examined two of the Borrero and Hernandez specimens from the eastern slopes of the middle Magdalena valley (Llana Fría and El Guadual, depto. Santander, AMNH), and I agree with Haffer (1986), who re-identified these birds as *T. obscura* based on plumage color. Only one is a male, but there is no reason to think this specimen or seven other dark-billed males that I have examined from central Colombia (Appendix) are all immatures. Two November males collected by Haffer in the Patía valley (Río Guachicono, AMNH 708660, and Río San Jorge, AMNH 708661, both in depto. Cauca) may have been breeding adults based on their fully ossified skulls and enlarged testes (4.5 × 6.5 mm and 7 × 6 mm, respectively). Also included in this group are three AMNH specimens from the western slope of the western Andes in the vicinity of the Dagua valley (Dagua, AMNH 107359; Jiménez, AMNH 515264; and Primavera, AMNH 515265, all in depto. Valle de Cauca) that had been identified previously as *T. fuliginosa* or *T. olivacea*.

In measurements, *T. obscura* specimens from Venezuela, the Santa Marta Mountains, the Magdalena Valley, and central Colombia are similar (Table 1). In wing chord and the three bill measurements, both sexes from all four regions are smaller on average than *T. fuliginosa*. Average measurements of both sexes from these regions are slightly greater than *T. obscura* from southern Colombia and Ecuador.

The only definite record of *T. fuliginosa* in Colombia is an adult male (in sooty plumage) collected on the eastern side of the Magdalena valley near Bucaramanga (see *T. fuliginosa* below) ca. 100 km north of the Borrero and Hernandez localities. Thus, *T. o. haplochroma* co-occurs with *T. fuliginosa* along the coastal range of northern Venezuela and in the Sierra de Perijá (Fig. 2). In Colombia, both are known from the eastern side of the lower Magdalena valley, but all Colombian records attributed to *T. fuliginosa* from farther south and west appear to be *T. obscura* (Fig. 2). Localities for *T. obscura* in central Colombia range in elevation from 500 m (Jiménez; I have not seen the Barrancabermeja record from 82 m) to 2400 m (El Zancudo, depto. Caldas).

Following Hellmayr's (1913, 1938) belief that *T. obscura* from throughout the Andes were indistinguishable, Haffer (1986) assigned specimens from central Colombia and the Magdalena Valley to *T. o. obscura*. Relative to other *T. obscura* specimens, I consider specimens from these regions to be *T. o. haplochroma* based on their all dark bills (in most specimens), dark gray-brown breasts, and generally larger measurements (Table 1). Additional study may find some of these populations and those from Venezuela to be distinct from one another.

2) Birds from extreme southern Colombia (Ricaurte, depto. Nariño) and Ecuador (open squares, Fig. 2) differ from *T. o. haplochroma* in that all specimens have light-colored lower mandibles. They also are lighter brown on the breast and have light-cream-colored bellies that contrast with brown breasts and buffy flanks. They differ from populations of central Peru by being brownish rather than dark gray-brown on the breast. These Ecuadorian birds also are less pale than birds from western Peru and the dry northern Peruvian valleys (see below). I have not examined the type specimen of *T. o. pauper* from Cayandeled (prov. Chimborazo) in southwestern Ecuador (Berlepsch and Taczanowski 1884); however, of nine specimens from the same region in the AMNH collection (Bucay, Pallatanga, and Chimbo), all but two from Pallatanga (AMNH 173542) and Chimbo (AMNH 514512) fit the description given above for Ecuadorian birds. The Pallatanga and Chimbo specimens are grayer on the breast than other specimens from the region, and therefore they approach the coloration of specimens from central Peru (see below). Specimen records in western Ecuador range from 100 m (Quinindé, prov. Esmeraldas) up to 1,500 m (Pallatanga).

3) Taczanowski (1884, 1886), Koepcke (1961, 1963), and Paynter (1970) placed specimens from the dry western slopes of the Andes in northern Peru and the Marañon and Huallaga valleys (black circles, Fig. 2) with populations to the north in *T. o. pauper*, but the birds from the dry regions of northern Peru are paler gray on the chest than populations to the north, and the flanks are almost as pale as the belly. Some individuals have whitish throats. A few specimens from higher elevations are grayer on the breast than other birds, and the grayest, a bird from Viñas, depto. La Libertad, Peru (AMNH 514509) resembles birds from central Peru (see below). A single specimen from Ascope, depto. La Libertad, in the Chicama Valley (FMNH 399631) is very rusty colored, unlike any other specimens I have examined of this species (the locality is labeled with a "?" in Fig. 2). Northern Peruvian specimens range from 50 m on the coast (Víru, depto. La Libertad; 105 m in the Huallaga valley, 20 km NE of Tarapoto, depto. San Martín) to ca. 2,000 m (Taulís, depto. Cajamarca; I did not examine this specimen, but Koepcke [1963] noted that this bird approached specimens from central Peru in possessing a darker breast).

4) Koepcke (1963) described *T. o. pacifica*, from the "lomas woods" of the southern coast of Peru (open circles, Fig. 2), as larger in measurements than all adjacent populations. My measurements of the wing chord and tail length of a paratype (AMNH 766699) agree with those she reported in being longer than any other *T. obscura* populations (Table 1). In coloration, this bird most resembles the pale birds from northwestern Peru (see group 3, above). I am unaware of additional distribution information for this region since Koepcke's (1963) original description; thus, gaps she described persist between the distribution of *pacifica* and other Peruvian populations (Fig. 2). The three localities for *pacifica* are all below 400 m.

5) All non-coastal specimens from central Peru south are considered to be *T. o. obscura* (e.g., Paynter 1970); however, geographic variation is also evident among these birds. Specimens from central Peru south to depto. La Paz, Bolivia (open stars, Fig. 2), are darker gray-brown on the chest than adjacent populations, and therefore the chest color contrasts strongly with the belly. The flanks are buffy. Records in this region range from 450 m (Kiteni, depto. Cuzco, Peru) to 2,650 m (Huanta, depto. Ayacucho, Peru). As noted above, some specimens from northern Peru and Ecuador approach central Peruvian birds in darkness of breast coloration; thus, variation could be clinal in north-central Peru and southern Ecuador (alternatively, there may be elevational separation). The distribution of these dark-chested birds may extend south of depto. La Paz, Bolivia, but paler birds occur in the Andes of depto. Santa Cruz (Fig. 2, see below).

6) Specimens from the eastern Andean slope south of depto. La Paz, Bolivia, to northern Argentina (black squares, Fig. 2) greatly resemble birds from the dry regions of northern Peru (group 3 above) and are much paler than birds from immediately to the north (La Paz and central Peru, Zimmer 1930). Hellmayr (1913) had specimens only from northern Peru and southern Ecuador for comparison when he re-examined the type of *T. o. obscura* collected at Chiquitos, Bolivia (Lafresnaye and d'Orbigny 1837), so it is not surprising that he concluded *T. o. pauper* was inseparable from *T. o. obscura* (he had no specimens from localities in depto. La Paz or central Peru, where dark-breasted populations occur).

Although there are few data on the pale *T. obscura* from this region, existing specimen material suggests that there may be seasonal movements of these populations. In the Andean foothills near Tambo, depto. Santa Cruz, Bolivia, during January and February 1984 (austral summer), D. C. and C. G. Schmitt collected a series (LSUMZ) of males with enlarged testes (>5 × 7 mm) and females in or nearly in breeding condition (one with a brood patch, LSUMZ 125040) from around 1,500 m. In the lowlands east of the Andes, specimens have been collected only in the austral winter. Remsen (in litt.) suspected that birds observed and collected in prov. Chiquitos, depto. Santa Cruz (LSUMZ, Parker et al. 1993), were nonresidents in this region, which lies some 200 km east of the Andes. Specimens were collected from flocks inter-mixed with small numbers of *Coryphospingus cucullatus* (Red Pileated-Finch), another species that exhibits seasonal movements in this part of Bolivia (Davis 1993; Remsen in litt.). All known collection dates of *T. obscura* from prov. Chiquitos (Lafresnaye and d'Orbigny 1837, CM, FMNH, LSUMZ) and for other localities more than 100 km east of the Andes are for the austral winter (outside dates are 14 May and 21 October). These include the first records of this species for Paraguay (with additional sight records) and possibly the first records for Brazil.

On 9 October 1945, B. Podtiaguin collected a small finch at Chaco-í, depto. Presidente Hayes, Paraguay (USNM 405363). This individual matches pale *T. obscura* specimens from central Bolivia and Argentina in measurements and paler overall coloration relative to specimens from northern Bolivia and central Peru. Hayes (1995) did not list this species for Paraguay; thus, this specimen represents the first record of this species for the country. Chaco-í (75 m) is over 600 km from the Andes (Fig. 2). On 3, 4, 9 and 10 August 1995, two groups of ornithologists, including Floyd Hayes, Bret Whitney, Tom Schulenberg, and the author, observed at least eight individual *T. obscura* among other birds (including *C. cucullatus*, *Polioptila dumicola* Masked Gnatcatcher, *Parula pitayumi* Tropical Parula, and *Poospiza melanoleuca* Black-capped Warbling-Finch) that responded to imitations of *Glaucidium brasilianum* (Ferruginous Pygmy Owl) in Chaco woodland and woodland edge near Chaco Lodge, depto. Presidente Hayes (Lat. 22° 30′S. Long. 59°19′W, 130 m), and near Colonia Filadelfia. This site is 500 km from the Andes. We never observed more than two individuals at one time. These records along with a specimen (AMNH 142183), collected on 16 May 1916 at Avia Terai, prov. Chaco, Argentina (400 km from the Andes) suggest that this species might be widespread in the Chaco at least during the austral winter; however, it has not been recorded at other well-studied Chaco sites (Kratter et al. 1993).

Two old specimens (AMNH 32582–83), identified as females, that H. H. Smith collected at Chapada, Mato Grosso, Brazil (Allen 1891), on 29 July and 30 August 1885 have always been considered to be *T. fuliginosa* (e.g., Naumberg 1930; Hellmayr 1938). This is probably because of the existence of a male *T. fuliginosa* in the same collection. However, these birds were not collected at the same time of year (the male was collected in November), and the two females are much paler than any brown *T. fuliginosa* that I have examined. Although it is conceivable that these specimens are *T. fuliginosa* (they are similar in measurements, Table 1), this would mean that female *fuliginosa* from the cerrado region are distinct from other *fuliginosa* (contrary to the observations of earlier authors). I believe these two females could represent the only records of *T. obscura* for Brazil. They resemble Paraguayan and Argentinean *T. obscura* in color and size (although they are slightly larger than Chiquitos specimens in most measurements). They were collected during the austral winter in the same months as other *T. obscura* specimens from the eastern lowlands. Chapada is approximately 1,000 km east of the Andes.

The available records are consistent with *T. obscura* being a regular visitor to the eastern lowlands of Bolivia, Paraguay, and northern Argentina during the austral winter. If so, the two Chapada specimens could represent extra-limital records in the non-breeding season. Clearly, additional specimens and observations are needed to determine the seasonal status of this species throughout this region.

TIARIS FULIGINOSA

Like *T. bicolor*, this species is sexually dimorphic. Adult males are entirely sooty-black with a slight olive tinge to the back. Females and immatures are brown like *T. obscura*, but generally exhibit richer tones, and they are more buffy below. The species favors forest clearings and edge and does not generally occur in drier habitats, in contrast to *T. bicolor* and *T. obscura*. Local movements are thought to occur in some populations. Seasonal changes in elevation have been reported in Trinidad (Ffrench 1991), and the species is known to invade areas of seeding bamboo in southeastern Brazil (D. Stotz, J. M. Cardoso da Silva, pers. comms.) and Venezuela (M. Lentino, pers. comm.).

Distribution and geographic variation.—Specimen records are scattered. *Tiaris fuliginosa* has been recorded from widely separated sites that form a crescent around the southern, eastern, and northern edges of Amazonia (Fig. 2). Along the southern edge, the species has been collected at the Bolivian site documented below and from two sites in Mato Grosso, Brazil (Chapada, Allen [1891]; Serra do Roncador, Fry [1970]). The species also recently has been recorded for the first time in Paraguay (12 Sept. 1992), from the Atlantic forests on the eastern edge of the country at Lagunita, depto. Canendiyú (Brooks et al. 1993, Hayes 1995, photos of mist-netted male on file at *VIREO*). In eastern Brazil, it is found along the Atlantic coast from the state of São Paulo north to Pernambuco. The species also is known from further north in the state of Maranhão (E. Dente specimens, LSUMZ; Oren 1990). These records are not included in any recent summaries of the species' distribution (e.g., Ridgely and Tudor 1989; Sibley and Monroe 1990; Sick 1993) and include records from farther inland than in other parts of eastern Brazil (Fig. 2).

Along the northern edge of Amazonia, *T. fuliginosa* is known in the Tepui region from a single specimen taken by H. Whitely at approximately 1,200 m on the slopes of Mount Roraima, Venezuela (Salvin and Godman 1884; Phelps 1938). In addition, Peberdy (1939) apparently collected this species from Bat Mountain, Guyana (approx. 350 km SSE of Mt. Roraima); however, Snyder (1966) reported that the specimen was destroyed. In northern Venezuela, *T. fuliginosa* overlaps in distribution with the other two non-yellow *Tiaris* along the slopes of the coastal range from the state of Sucre west to the Sierra de Perijá. It also is the only member of the genus that nests on Trinidad (Ffrench 1991). Although all recent accounts of this species' distribution also include the upper Patía and middle Magdalena valleys of Colombia, I agree with Haffer (1986) that these specimens are *T. obscura* (see *T. obscura* above).

There is at least one specimen of an adult male *T. fuliginosa* from Colombia. This is a bird taken by M. A. Carriker, Jr., on 10 May 1964 at Portugal, prov. Santander (MVZ 35684), a site probably on the eastern side of the northern Magdalena valley. I examined other Carriker specimens from Portugal and Azufrada, prov. Santander (also MVZ), that have been identified as female and immature male *fuliginosa* and believe them to be *T. obscura*. The exact locations of Portugal and Azufrada are not known (R. A. Paynter, Jr., in lit.), but they are likely in the vicinity of Lebrija, a known location where Carriker also collected in May 1964 (specimens at LSUMZ, MVZ, YPM). Carriker, then 84 years old, was living in nearby Bucaramanga (Anonymous 1963).

The first record of T. fuliginosa *for Bolivia.*—On 19 July 1988, at Arroyo del Encanto, depto. Santa Cruz, a male finch (LSUMZ 137740) was netted in tall humid forest and prepared by D. C. Schmitt. The following information was recorded on the specimen label: testes 1×1 mm, skull 100% ossified, no fat, and no molt; the iris was dark brown, the maxilla dusky brown, the mandible buffy with sides dusky brown, and the toes and tarsi were fuscous. The bird was collected in humid tropical forest, interspersed with overgrown clearings resulting from selective logging of mahogany (*Swietenia macrophylla*) and other hardwoods. The site lies along the western base of the Serranía de Huanchaca in Noel Kempff Mercado National Park (Bates et al. 1989) and was 2 km from the nearest open habitats (cerrado or pampa).

Although we subsequently recorded *Sporophila schistacea* (Slate-colored Seedeater) at several forest sites, and we recorded seven other *Sporophila* species in the seasonally flooded grasslands below the plateau (Bates et al. 1992, unpubl. data), the 1988 specimen remained identified only as *Sporophila* sp. (Bates et al. 1989) with the exception of a note by J. V. Remsen pointing out that, of specimens in the LSUMZ collection, the bird most resembled specimens of *T. o. haplochroma*. The Arroyo del Encanto specimen differs from *T. obscura* specimens by its richer brown coloration on the back and breast, and by the reduced pale feathering on the belly. Like many *haplochroma*, it has little contrast between the colors of the maxilla and mandible. The bird was not compared to *T. fuliginosa* until it was realized that this species had been recorded from "only" 550 km to the east in Mato Grosso (Chapada). The Arroyo del Encanto specimen is similar to immature male *T. fuliginosa* from eastern Brazil (LSUMZ) and much more brownish than the two female *Tiaris* collected by Smith at Chapada, Mato Grosso (Allen 1891, see *T. obscura* above). It was the only individual *Tiaris* encountered by LSUMZ personnel despite several field seasons in the Huanchaca region (Bates et al. 1989; Bates et al. 1992; Kratter et al. 1991). As with lowland records of *T. obscura*, this *T. fuliginosa* could represent a wandering individual. The only definite records from this part of South America are the adult male from Chapada collected 100 years ago, and Fry's (1970) record from northeastern Mato Grosso. Whether a stable population exists in southern South America west of the Atlantic forests is unknown.

Three subspecies of *T. fuliginosa* have been described, all based on males in adult (sooty) plumage; however, I have examined 24 adult males from throughout the species' distribution, including the types of *T. f. zuliae* (Aveledo and Ginés 1948, AMNH) and *T. f. fumosa* (Lawrence 1874, AMNH), and intra-population variation in plumage is as great as that among specimens belonging to the proposed subspecies. Measurements of named populations also show no consistent differences. The only apparent distinction in measurements for any population is that females from Maranhão, Brazil, may be smaller than females from other regions (Table 1). Thus, although *T. fuliginosa* is composed of a number of apparently allopatric populations, the current subspecies are not diagnosable by male plumage or measurements of available material, and I recommend that no subspecies be recognized.

CONCLUSIONS

Sympatry of *Tiaris bicolor* (Black-faced Grassquit), *T. obscura* (Dull-colored Grassquit), and *T. fuliginosa* (Sooty Grassquit) in parts of Colombia and Venezuela, has created confusion in the identification of some specimens from this region. This is especially true for *T. obscura* and *T. fuliginosa*.

At least six diagnosable populations of *T. obscura* occur from northern Colombia south to Argentina; only two of which (*T. o. haplochroma* and *T. o. pacifica*) are adequately named. The other four are currently treated as a single taxon, the nominate subspecies. Populations from the central valleys of Colombia that I consider to be *T. o. haplochroma* also may prove to be distinct.

These populations previously have been thought to be *T. fuliginosa*. Andean populations form a "leapfrog" pattern similar to that described for many other Andean birds (Remsen 1984), with dark populations (northern Andes and central Peru/northern Bolivia) and light populations (Southern Col. and Ecuador, northern Peru, and central Bolivia/Paraguay/Argentina) alternating along a north/south gradient.

Based on scattered specimen records, the distribution of *T. fuliginosa* is appears quite disjunct. In several regions (the Tepuis and central Brazil), *T. fuliginosa* is known from fewer than two specimens (taken up to 100 years apart); thus the status of these populations is unknown. The only definite record for Colombia is from the Bucaramanga area on the eastern side of the lower Magdalena Valley. The currently recognized subspecies of *T. fuliginosa* are not diagnosable based on type descriptions of adult males because intra-population variation in adult male plumage is substantial.

For both *T. fuliginosa* and *T. obscura*, additional specimens, data on song and natural history, and molecular studies would help clarify relationships among and distributions of populations, especially in regions where the species co-occur (Colombia, northern Venezuela, and possibly central Brazil). Information on the relationships among the currently recognized *Tiaris* species also is needed to determine if *fuliginosa* and *obscura* are sister taxa, a question whose resolution has bearing on the evolution of the circum-Amazonian distribution discussed below.

These poorly known *Tiaris* species can be grouped with a number of other small finches (e.g., members of *Haplospiza*, *Amaurospiza*, and some *Sporophila*) about which little is known in terms of distribution and status in many parts of the Neotropics (e.g., see Barrajas and Phillips 1994 on *Haplospiza*). Although widespread geographically, these species are considered local and uncommon-to-rare in most of their distributions. For many regions, they are poorly represented in collections. With so little information, we cannot assess whether some populations might be threatened by habitat changes, because we do not know if a population even exists. In the future, efforts should be made to better understand the natural history and seasonal distributions of these often-overlooked species.

THE CIRCUM-AMAZONIAN DISTRIBUTION PATTERN IN BIRDS

Based on the distribution of *Platyrinchus mystaceus* (White-throated Spadebill), Remsen et al. (1991) defined the circum-Amazonian distribution as one in which "sister taxa are found in (a) *montane* forested areas on the humid slopes of the Andes and the Coastal Range of Venezuela or the Tepui region, and (b) *lowland* forested areas south and east of the lowland forests of Amazonia." Phylogenetic relationships among members of *Tiaris* have not been determined; however, *bicolor*, *fuliginosa*, and *obscura* lack the yellow that occurs in the male plumages of the other two species in the genus (*T. canora* and *T. olivacea*). Thus, the three could be sister taxa. For the following discussion, I assume that this is the case. It also might be argued that *Tiaris* grassquits are not strictly forest inhabitants; however, *obscura* and *fuliginosa* are associated with forest edge based on descriptions of most collecting localities. Taken together, the ranges of *T. obscura* and *T. fuliginosa* encircle Amazonia (Fig. 1) in a circum-Amazonian pattern like that described by Remsen et al. (1991) for *P. mystaceus*. *Platyrinchus mystaceus* and *T. fuliginosa* also occur together on Trinidad (on Tobago, *P. mystaceus* occurs as does *T. bicolor*). Remsen et al. (1991) noted the ranges of several species of thraupids that also exhibited a circum-Amazonian distribution, including *Piranga flava* (Hepatic Tanager), *Chlorophonia cyanea* (Blue-naped Chlorophonia), and *Pipraeidea melanonota* (Fawn-breasted Tanager). With such shared distributions among a growing number of taxa, it is tempting to look for a single explanation to explain the pattern. However, a closer examination of these circum-Amazonian distributions is warranted. To do this, I briefly discuss the distributions of these three tanagers, *P. mystaceus*, and the *Tiaris* species at several geographic scales based on an examination of collecting localities for specimens housed at the AMNH and from the literature.

The five circum-Amazonian taxa mentioned above often have been collected at the same sites. For example, *T. fuliginosa* or *T. obscura* has been collected with *P. mystaceus* at 12 AMNH localities from throughout the circum-Amazonian distribution. The three tanager species also have been collected at some of these same localities. *Piranga flava* has been collected with one of the two *Tiaris* species at 22 localities. As localities in common for these circum-Amazonian taxa include such diverse sites as Chapada, Mato Grosso (central Brazil, all but *C. cyanea*), and Mount Roraima (southern Venezuela, all but *P. melanonota*, which is known from the region), some common aspect to the ecology or evolutionary history of these species is possibly supported.

However, at the regional scale these species also have extensive non-overlapping portions to their distributions. For example, *P. mystaceus* occurs in portions of Paraguay, southern Brazil, and northeastern Argentina where *Tiaris* is absent, whereas *Tiaris* occurs along the western Andes in Peru, where *P. mystaceus* is absent. The distributions of the genera to which these species belong suggest differences in evolutionary history or ecology. *Tiaris* occurs throughout the West Indies where *Platyrinchus* is absent, and *P. mystaceus* has congeners throughout Amazonia, whereas *Tiaris* does not. Only three of the five taxa mentioned above occur on the western slopes of the Andes south to Ecuador (*C. cyanea* and *P. melanonota* occur in the western Andes only in western Colombia). The lack of overlap at the regional scale does not refute a common history leading to circum-Amazonian distributions, but it does demonstrate variation exists, similar to what Remsen (1984) found for his Andean "leapfrog" distributions. Clearly, more thorough study of the evolution of these interesting patterns is needed.

ACKNOWLEDGMENTS

The author was supported by a Chapman post-doctoral fellowship from the American Museum of Natural History. This study would not have been possible without the help and information provided by many people including K. Burns, M. LeCroy, J. M. Cardoso da Silva, J. Haffer, F. Hayes, M. Lentino, M. Marín, R. Paynter, Jr., T. Schulenberg, F. G. Stiles, R. Storer, D. Stotz, P. Sweet, and B. Whitney. Mort Isler provided the expertise necessary to construct the distribution maps. Curators at the Academy of Natural Sciences, Philadelphia (F. Gill) and the National Museum of Natural History (G. Graves) provided access to collections in their care. I am grateful to curators and staff at the following institutions for specimen loans: Carnegie Museum of Natural History (K. Parkes), Field Museum (S. Lanyon and D. Willard), Louisiana State University Museum of Natural Science (J. V. Remsen and S. Cardiff), and Museum of Vertebrate Zoology, University of California (C. Cicero). Curators and collection managers at other institutions provided information on collections in their care: Colección Ornitológica Phelps (M. Lentino), Library of Natural Sounds, Laboratory of Ornithology (G. Budney), Natural History Museum of Los Angeles County (K. Garrett), Yale Peabody Museum of Natural History (F. Sibley), and Western Foundation of Vertebrate Zoology (L. Kiff). M. Lentino provided a copy of the unpublished Wetmore and Phelps manuscript. LSUMNS field work in Bolivia was carried out in collaboration with staff of the Parque Nacional Noel Kempff Mercado and the Museo de Historia Natural "Noel Kempff Mercado, Santa Cruz, Bolivia, and funded by grants to J. V. Remsen from J. S. McIlhenny and the National Geographic Society with additional support from P. Machris. The present manuscript was improved by comments from S. Hackett, M. Isler, R. Paynter, Jr., J. V. Remsen, T. Schulenberg, and D. Stotz.

LITERATURE CITED

ALLEN, J. A. 1891. On a collection of birds from Chapada, Mato Grosso, Brazil, made by Mr. Herbert H. Smith. Bull. Amer. Mus. Nat. 3:337–380.
ANONYMOUS. 1963. M. A. Carriker, Jr. Nebraska Bird Review 31:2–7.
AVELEDO H., R., AND H. GINÉS. 1948. Ave nueva de Venezuela. Mem. Soc. Cien. Naturales, La Salle 8: 107–108.
BALDWIN, S. P., H. C. OBERHOLSER, AND L. G. WORLEY. 1931. Measurement of birds. Sci. Publ. Cleveland Mus. Nat Hist. 2:1–165.
BARRAJAS L., F. C., AND A. R. PHILLIPS. 1994. A *Haplospiza* finch in western México; the lessons of an enigma. Bull. Brit. Orn. Club 114:36–46.
BATES, J. M., M. C. GARVIN, D. C. SCHMITT, AND C. G. SCHMITT. 1989. Notes on bird distribution in northeastern depto. Santa Cruz, Bolivia, with 15 species new to Bolivia. Bull. Brit. Orn. Club 109: 236–244.
BATES, J. M., T. A. PARKER III, A. P. CAPPARELLA AND T. J. DAVIS. 1992. Notes on bird distribution in northeastern Santa Cruz, Bolivia II, with 21 species new to the country. Bull. Brit. Orn. Club 112: 86–98.
BERLEPSCH, H. V., AND L. TACZANOWSKI. 1884. Deuxième liste des oiseaux recueillis dans l'Ecuadeur occidental par M. M. Stolzmann et Siemiradski. Proc. Zool. Soc. London 1884:281–293.
BOND, J., AND R. MEYER DE SCHAUENSEE. 1941. Descriptions of new birds from Bolivia. Part IV. Notulae Naturae 93:1–7.
BORRERO, J. I., AND J. H. HERNANDEZ C. 1961. Notas sobre aves de Colombia y descripción de una nueva subespecie de *Forpus conspicillatus*. Novedades Colombianas 1:430–445.
BROOKS, T., R. BARNES, L. BARTRINA, S. H. M. BUTCHART, R. P. CLAY, E. Z. ESQUIVEL, N. I. ETCHEVERRY, J. C. LOWEN, AND J. VINCENT. 1993. Bird surveys and conservation in the Paraguayan Atlantic forest. Birdlife Int. Study Rep. 57:1–145.

CHAPMAN, F. M. 1921. The distribution of bird life in the Urubamba valley of Peru. Bull. U. S. Nat. Mus. 117:1–138.
CHAPMAN, F. M. 1926. The distribution of bird-life in Ecuador. Bull. Amer. Mus. Nat Hist. 55:1–784.
CLARK, G. A., JR. 1986. Systematic interpretations of foot-scute patterns in Neotropical finches. Wilson Bull. 98:594–597.
COLLINS, C. T., AND M. H. KEMP. 1976. Natal pterylosis of *Sporophila* finches. Wilson Bull. 88:154–157.
DAVIS, S. E. 1993. Seasonal status, relative abundance, and behavior of the birds of Concepción, Departamento Santa Cruz, Bolivia. Fieldiana, Zool. No. 71.
FFRENCH, R. 1991. A Guide to the Birds of Trinidad and Tobago, 2nd ed. Cornell Univ. Press, Ithaca, New York.
FRY, C. H. 1970. Ecological distribution of birds in northeastern Mato Grosso State, Brazil. An. Acad. Brasil. Ciênc. 42:275–318.
HAFFER, J. 1974. Avian speciation in tropical South America. Publ. Nuttall Orn. Club 14:1–390.
HAFFER, J. 1986. On the avifauna of the upper Patía Valley, southwestern Colombia. Caldasia 15:533–553.
HAFFER, J. 1987. Biogeography of Neotropical birds. Pp. 105–150 in Biogeography and Quaternary History in tropical America (T. C. Whitmore and G. T. Prance, Eds.) Clarendon and Oxford University Press, New York, New York.
HAYES, F. E. 1995. Status, distribution and biogeography of the birds of Paraguay. Monographs in Field Ornithology 1:1–224. American Birding Association.
HELLMAYR, C. E. 1913. Critical notes on the types of little-known species of Neotropical birds.—Part II. Nov. Zool. 20:227–256.
HELLMAYR, C. E. 1938. Catalogue of birds of the Americas and adjacent islands. Field Mus. Nat. Hist., Zool. Ser. 13, pt. 11.
HILTY, S. L., AND W. L. BROWN. 1986. A Guide to the Birds of Colombia. Princeton University Press, Princeton, New Jersey.
ISLER, M. L. 1997. A sector-based ornithological information system for the Neotropics. Pp. 345–354 in Studies in Neotropical Ornithology Honoring Ted Parker (J. V. Remsen, Jr., Ed.), Ornithol. Monogr. No. 48.
KRATTER, A. W., M. D. CARREÑO, R. T. CHESSER, J. P. O'NEILL, AND T. S. SILLETT. 1992. Further notes on bird distribution in northeastern depto. Santa Cruz, Bolivia, with two species new to Bolivia. Bull. Brit. Orn. Club 112:143–150.
KRATTER, A. W., T. S. SILLETT, R. T. CHESSER, J. P. O'NEILL, T. A. PARKER III, AND A. CASTILLO. 1993. Avifauna of a Chaco locality in Bolivia. Wilson Bull. 105:114–141.
KOEPCKE, M. 1961. Birds of the western slope of the Andes of Peru. Amer. Mus. Novitates, No. 2028:1–31.
KOEPCKE, M. 1963. Zur Kenntnis einiger finken des peruanischen Küstengebietes. Beitr. Neotrop. Fauna 3: 2–19.
LAFRESNAYE, A., AND A. D. D'ORBIGNY. 1837. Synopsis Avium. Mag. Zool. (Paris), 7, cl. 2, p. 81.
LAWRENCE, G. N. 1874. Descriptions of six supposed new species of American birds. Ann. Lyc. Nat. Hist. 10:395–399.
MEYER DE SCHAUENSEE, R. 1952. A review of the genus *Sporophila*. Proc. Acad. Nat Sci., Philadelphia 104: 153–196.
MEYER DE SCHAUENSEE, R. 1966. The species of birds of South America with their distribution. Livingston Press, Narberth, Pennsylvania.
MILLER, A. H. 1952. Two new races of birds from the upper Magdalena valley of Colombia. Proc. Biol. Soc. Wash. 65:13–30.
NAUMBERG, E. 1930. The birds of Mato Grosso, Brazil: a report of the birds secured by the Roosevelt-Rondon expedition. Bull. Amer. Mus. Nat. Hist. 60:1–432.
OREN, D. C. 1990. New and reconfirmed records from the state of Maranhão, Brazil. Goeldiana Zoologia 4:1–13.
PARKER, T. A., III, A. H. GENTRY, R. B. FOSTER, L. H. EMMONS, AND J. V. REMSEN, JR. 1993. The lowland dry forests of Santa Cruz, Bolivia: a global conservation priority. RAP Working Papers 4. Conservation International, Washington, D.C.
PAYNTER, R. A., JR. 1970. Subfamily Emberizinae. Pp. 3–214 in Checklist of Birds of the World, vol. XIII (R. A. Paynter, Jr., Ed.), Museum of Comparative Zoology, Cambridge, Massachusetts.
PAYNTER, R. A., JR. 1982. Ornithological Gazetteer of Venezuela. Museum of Comparative Zoology, Cambridge, Massachusetts.
PAYNTER, R. A., JR. 1985. Ornithological Gazetteer of Argentina. Museum of Comparative Zoology, Cambridge, Massachusetts.
PAYNTER, R. A., JR. 1989. Ornithological Gazetteer of Paraguay, 2nd ed. Museum of Comparative Zoology, Cambridge, Massachusetts.
PAYNTER, R. A., JR. 1993. Ornithological Gazetteer of Ecuador, 2nd ed. Museum of Comparative Zoology, Cambridge, Massachusetts.
PAYNTER, R. A., JR., AND M. A. TRAYLOR, JR. 1981. Ornithological Gazetteer of Colombia. Museum of Comparative Zoology, Cambridge, Massachusetts.
PAYNTER, R. A., JR., AND M. A. TRAYLOR, JR. 1991. Ornithological Gazetteer of Brazil. Museum of Comparative Zoology, Cambridge, Massachusetts.

PAYNTER, R. A., JR., M. A. TRAYLOR, JR., AND B. WINTER. 1975. Ornithological Gazetteer of Bolivia. Museum of Comparative Zoology, Cambridge, Massachusetts.

PEBERDY, P. S. 1939. A Report of the British Guiana Museum. British Guiana Museum, Georgetown, Guyana.

PHELPS, W. H. 1938. The geographical status of the birds collected at Mount Roraima. Bol. Soc. Venez. Cienc. Naturales 36:83–95.

PHELPS, W. H., AND W. H. PHELPS, JR. 1963. Lista de aves de Venezuela con su distribución. Bol. Soc. Venez. Cienc. Naturales 24:1–479.

PINTO, O. 1944. Catálogo das aves do Brasil. Pt. 2. Passeriformes. São Paulo: Publ. Dept. Zool., Sec. Agric. Indús., Comér. 700 p.

REMSEN, J. V., JR. 1984. High incidence of "leapfrog" pattern of geographic variation in Andean birds: implications for the speciation process. Science 224:171–173.

REMSEN, J. V., JR., AND M. A. TRAYLOR, JR. 1989. An Annotated Checklist of the Birds of Bolivia. Buteo Books, Vermillion, South Dakota.

REMSEN, J. V., JR., K. C. PARKES, AND M. A. TRAYLOR, JR. 1987. Range extensions for some Bolivian birds, 3 (Tyrannidae to Passeridae). Bull. Brit. Orn. Club 107:6–16.

REMSEN, J. V., JR., O. ROCHA O., C. G. SCHMITT, AND D. C. SCHMITT. 1991. Zoogeography and geographic variation of *Platyrinchus mystaceus* in Bolivia, Peru, and the circum-Amazonian distribution pattern. Ornithol. Neotropical 2:77–84.

RIDGELY, R., AND G. TUDOR. 1989. The Birds of South America. Vol. 1: The Oscine Passerines. University of Texas Press, Austin, Texas.

SALVIN, O., AND F. D. GODMAN. 1884. Notes on birds from British Guiana. Ibis 1884:443–452.

SCHWARTZ, P. 1972. El sonido como implemento adicional en la taxonomía ornitológia. Acta IV Congr. Latin. Zool. 1:207–217.

SIBLEY, C. G., AND B. L. MONROE, JR. 1990. Distribution and Taxonomy of Birds of the World. Yale Univ. Press, New Haven, Connecticut.

SICK, H. 1993. Birds in Brazil: A Natural History. Princeton University Press, Princeton, New Jersey.

SNYDER, D. The Birds of Guyana. 1966. Peabody Museum, Salem, Massachusetts.

STEPHENS, L., AND M. A. TRAYLOR, JR. 1983. Ornithological Gazetteer of Peru. Museum of Comparative Zoology, Cambridge, Massachusetts.

STEPHENS, L., AND M. A. TRAYLOR, JR. 1985. Ornithological Gazetteer of the Guianas. Museum of Comparative Zoology, Cambridge, Massachusetts.

TACZANOWSKI, L. 1884, 1886. Ornithologie du Perou, vol. 3. Rennes.

TODD, W. E. C. 1912. Descriptions of seventeen new Neotropical birds. Ann. Carnegie Mus. 8:198–214.

TODD, W. E. C., AND M. A. CARRIKER, JR. 1922. Birds of the Santa Marta region of Colombia: a study of altitudinal distribution. Ann. Carnegie Mus. 14:1–611.

TORDOFF, H. B. 1954. A systematic study of the avian family Fringillidae based on the structure of the skull. Univ. Michigan Mus. Zool. Misc. Publ. 81:1–42.

WIED, M., PRINZ ZU. 1830. Beiträgezur Naturgeschichte von Brasilien, von Maximilian, Prinzen zu Wied. Vogel 3:628.

ZIMMER, J. T. 1930. Birds of the Marshall Field Peruvian Expedition, 1922–1923. Field Mus. Nat. Hist., Zool. Ser. 17, pt. 7.

APPENDIX

LOCALITIES OF SPECIMENS IN THE LITERATURE AND SPECIMENS EXAMINED. FOR *T. o. haplochroma*, ABBREVIATIONS FOLLOWING LOCALITIES REPRESENT THE GENERAL REGION IN WHICH BIRDS FROM EACH LOCALITY WERE PLACED IN THE MEASUREMENT DATA SET (TABLE 1, SEE TEXT FOR JUSTIFICATION). SM = SANTA MARTA MOUNTAINS, MV = MAGDALENA VALLEY (INCLUDING THE WEST SLOPE OF THE SERRANÍA DE PERIJÁ), CC = CENTRAL COLOMBIA (CAUCA, PÁTIA, AND DAGUA VALLEYS)

Locality (country: department, province or state, when available; and locality)[1]	Elevation[1]	Specimens measured	Citation[2]
Tiaris bicolor omissa			
Colombia: La Guajira, Ríohacha	S. L.[3]		Todd and Carriker (1922)
Colombia: Santander, Lebrija	1,086 m		YPM
Colombia: Santander, Bucaramanga	1,018 m		LACM, YPM
Colombia: Santander, Chitota	800 m		LACM
Venezuela: Falcón, Sabaneta	60 m		COP
Venezuela: Falcón, Quiuragua	110 m		COP
Venezuela: Falcón, Moruy	110 m		COP
Venezuela: Falcón, San Luis	780 m		COP
Venezuela: Falcón, San Juan de Los Cayos	S. L.		COP
Venezuela: Falcón, Casigua	S. L.		COP
Venezuela: Falcón, La Vela de Coro	S. L.		AMNH
Venezuela: Falcón, Boca del Tocuyo	S. L.		COP
Venezuela: Miranda, Carenero	S. L.		COP
Venezuela: Miranda, Cabo Codera	20 m		COP
Venezuela: Anzoátegui, Barcelona	S. L.		COP
Venezuela: Anzoátegui, Pfritu	S. L.		COP
Venezuela: Zulia, Palmarejo	S. L.		COP
Venezuela: Zulia, Cojoro	S. L.		COP
Venezuela: Zulia, Paraguaipoa	20 m		COP
Venezuela: Portuguesa, Acarigua	S. L.		COP
Venezuela: Gúarico, Santa Rita	120 m		COP
Venezuela: Sucre, Carúpano	S. L.	1M, 1F	AMNH
Venezuela: Sucre, Plains of Cumana	?	2M, 4F	AMNH
Venezuela: Mérida, Mérida	1,000 m		AMNH
Venezuela: Nueva Esparta, Boca del Río	S. L.		AMNH
Venezuela: Lara, El Cují	550 m	2M	AMNH
Venezuela: Lara, Quebrada Arriba	600 m		COP
Venezuela: Lara, Barquisimeto	566 m	1F	AMNH
Venezuela: Distrito Federal, El Valle	850 m		AMNH
Venezuela: Distrito Federal, Cotiza	1,000 m	1F	COP
Venezuela: Distrito Federal, Puerto Cabello	S. L.		AMNH
Venezuela: Aragua, Cata	S. L.		AMNH
Tobago	20 m		Collins and Kemp (1974)
			AMNH

APPENDIX
CONTINUED

Locality (country: department, province or state, when available; and locality)[1]	Elevation[1]	Specimens measured	Citation[2]
Tiaris bicolor huilae			
Colombia: Tolima, Honda	230 m	1M, 1F	AMNH
Colombia: Tolima, Chicoral	500 m	1M	AMNH
Colombia: Huila, Villa Vieja	439 m		Miller (1952), MVZ
Tiaris obscura haplochroma			
Colombia: Cesar, Pueblo Viejo (SM)	?	4M, 3F	ANSP, CM
Colombia: Cesar, San José (SM)	1,500 m	2M, 1F	USNM
Colombia: Cesar, Atanquez (SM)	820 m	1M	USNM
Colombia: Magdalena, Minca (SM)	700 m	3M, 3F	AMNH, ANSP, CM, LSUMZ
Colombia: Magdalena, Vista Nieve (SM)	1,200 m	1M	USNM
Colombia: Magdalena, Cincinati (SM)	1,480 m	6M, 7F	CM, USNM
Colombia: La Guajira, La Africa (MV)	ca 2,000 m	1M	USNM
Colombia: La Guajira, Monte Elias (MV)	ca 1,350 m	1M	USNM
Colombia: Norte de Santander, Ocaña (MV)	1,200 m	4M	CM
Colombia: Santander, Hacienda Santana (MV)	1,150 m	4M	USNM
Colombia: Santander, El Tambor (MV)	ca 500 m	1F	CM
Colombia: Santander, Palo Negro (MV)	1,200 m	1M	LSUMZ, MVZ
Colombia: Santander, Lebrija (MV)	1,086 m	2M, 1F	LSUMZ, MVZ, YPM
Colombia: Santander, Azufrada (MV)	600 m	4M	MVZ
Colombia: Santander, Bucaramanga	1,018 m		YPM
Colombia: Santander, Gómez (MV)	ca 750 m	1F	FMNH
Colombia: Santander, Portugal (MV)	850 m	1M	LSUMZ, MVZ, YPM
Colombia: Santander, Barrancabermeja (MV)	82 m		Borrero and Hernandez (1961)
Colombia: Santander, Las Peñitas	?		Borrero and Hernandez (1961)
Colombia: Santander, El Guadual (MV)	800 m	1M	Borrero and Hernandez (1961), AMNH
Colombia: Santander, Llana Fría (MV)	800 m	1F	Borrero and Hernandez (1961), AMNH
Colombia: Cundinamarca, El Colegio	ca 700 m		LACM
Colombia: Cundinamarca, El Hospicio	1,300 m		Borrero and Hernandez (1961)
Colombia: Antioquia, La Bodego (CC)	1,750 m	1M	USNM
Colombia: Antioquia, La Frijolera (CC)	1,500 m	1F	AMNH
Colombia: Antioquia, La Sofía (CC)	1,150 m	1M	USNM
Colombia: Cauca, Río Guachicono, 7 km NE El Bordo (CC)	700 m	1M	AMNH
Colombia: Cauca, Popayán (CC)	1,760 m	2F	FMNH
Colombia: Cauca, Río San Jorge 16 km S El Bordo (CC)	700 m	1M	AMNH
Colombia: Caldas, El Zancudo (CC)	2,400 M	1F	CM
Colombia: Valle de Cauca, Dagua (Caldas) (CC)	816 m	3M, 1F	AMNH, CM

APPENDIX
CONTINUED

Locality (country: department, province or state, when available; and locality)[1]	Elevation[1]	Specimens measured	Citation[2]
Colombia: Valle de Cauca, Primavera (CC)	1,700 m?	1F	AMNH
Colombia: Valle de Cauca, Jiménez (CC)	500 m	1M	AMNH
Colombia: Valle de Cauca, Yumbo (CC)	985 m	1M	CM
Venezuela: Aragua, El Limón near Maracay	400 m		Collins and Kemp (1974)
Venezuela: Táchira, Queniquea	1,900 m		Phelps and Phelps (1963)
Venezuela: Táchira, Bramón	1,200 m		Phelps and Phelps (1963)
Venezuela: Miranda, Petare	850 m		Phelps and Phelps (1963)
Venezuela: Mérida, La Azulita	1,135 m		Phelps and Phelps (1963)
Venezuela: Carabobo, Sierra de Carabobo	?	1M, 1F[5]	FMNH
Venezuela: Distrito Federal, Caracas	ca 900 m	2M, 2F[5]	FMNH
Venezuela: Distrito Federal, Picacho de Galipán[4]	2,000 m	2M, 1F	CM
Venezuela: Sucre, Cerro Turumiquire	ca 2,000 m	1F[5]	FMNH
Venezuela: Monagas, Río Cocollar	1,180 m	3M[5]	FMNH

Tiaris obscura (southern Colombia, Ecuador)

Colombia: Nariño, Ricaurte	1,250 m	2M, 1F	ANSP
Ecuador: Esmeraldas, Quinindé (Rosa Zárate)	100 m	1F	USNM
Ecuador: Imbabura, Hacienda Paramba	ca 900 m		Hellmayr (1938)
Ecuador: Loja, Las Piñas	1,100 m	1M	AMNH
Ecuador: Loja, Cebollal	1,000 m	2F	AMNH
Ecuador: Chimborazo, Cayandeled	1,500 m		Hellmayr (1938)
Ecuador: Chimborazo, Chimbo	345 m	1M	AMNH
Ecuador: Chimborazo, Pallatanga	1,500 m	1M	AMNH
Ecuador: Guayas, Bucay	100 m	1?	AMNH
Ecuador: Guayas, Naranjo	300 m	6M, 1F	AMNH
Ecuador: El Oro, Porto Velo	640 m	1M	AMNH
Ecuador: El Oro, Zaruma	1,200 m	1M	AMNH

Tiaris obscura (northern Peru)

Peru: Cajamarca, Cabico	800 m	1F	AMNH
Peru: Cajamarca, Callacate	ca 1,500 m		Hellmayr (1938)
Peru: Cajamarca, Perico	200 m	2F	AMNH
Peru: Cajamarca, 9 km S Jaen	900 m	1F	LSUMZ
Peru: Cajamarca, Hacienda Limón	2,011 m	1M	LSUMZ
Peru: Cajamarca, Puerto Tamborapa		1F	CM
Peru: Cajamarca, Taulís	ca 2,000 m		Koepcke (1963)
Peru: Amazonas, Bagua	1,700 m	1M	MVZ
Peru: Amazonas, San Juan (Bagua Grande)	1,000 m	1F	FMNH

APPENDIX
CONTINUED

Locality (country: department, province or state, when available; and locality)[1]	Elevation[1]	Specimens measured	Citation[2]
Peru: Amazonas, Balsas	822 m	1M	LSUMZ
Peru: Piura, Huancabamba	1,930 m	1?	AMNH
Peru: La Libertad, Ascope	230 m	1?	FMNH
Peru: La Libertad, Poroto	ca 500 m	1M, 1F	AMNH
Peru: La Libertad, Soquian	2,000 m	1M	CM
Peru: La Libertad, Trujillo	100 m	1M, 1F	AMNH
Peru: La Libertad, Viñas	1,680 m	2M, 2F	AMNH
Peru: La Libertad, Virú	68 m	1F	AMNH
Peru: San Martín, 20 km NE Tarapoto	105 m	1?	LSUMZ
Tiaris obscura (central Peru and depto. La Paz, Bolivia)			
Peru: Junín, Perené	1,000 m	2M	AMNH
Peru: Junín, Utcuyacu	1,465 m	1M	AMNH
Peru: Junín, La Merced	800 m	1F	AMNH
Peru: Junín, Oreja de Capela	1,600 m	1F	FMNH
Peru: Junín, Huacapistana	1,800 m	1F	AMNH
Peru: Junín, Paltaypampa	1,678 m		Hellmayr (1938)
Peru: Pasco, 9 km SEE Oxapampa	2,050 m	1F	LSUMZ
Peru: Ayacucho, Santa Rosa	800 m	1M, 1F	AMNH
Peru: Ayacucho, Huanta	2,650 m	1M	LSUMZ
Peru: Cuzco, Kiteni	ca 450 m	3?	LSUMZ
Peru: Cuzco, Cordillera Peru, Cuzco, Vilcabamba	ca 1,850 m	2F	AMNH
Peru: Cuzco, Idma	1,524 m	1M, 1F	AMNH
Peru: Cuzco, Santa Ana	1,060 m	3M, 4F, 1?	AMNH
Peru: Cuzco, Foot of Machu Pichu	2,130 m	1M	AMNH
Peru: Cuzco, Río Cosireni	ca 900 m	1? imm.	AMNH
Peru: Cuzco, Chaullay	1,300 m		Chapman (1921)
Peru: Cuzco, Hacienda Villa Carmen	ca 600 m	1F	FMNH
Peru: Cuzco, Río Mapitunari	685 m	1?	AMNH
Peru: Huánuco, Huachipa	ca 1,000 m	1M	FMNH
Peru: Huánuco, Vista Alegre	ca 1,000 m	2M, 1F	FMNH
Bolivia: La Paz, Caranavi	800 m	1F	LSUMZ
Bolivia: La Paz, 47 km N Caranavi	1,350	1M	LSUMZ
Bolivia: La Paz, Cerro Asunta Pata	1,300 m	3M, 2F	LSUMZ
Tiaris obscura (Boliva, Paraguay, Argentina)			
Bolivia: La Paz, Omeja	ca 300 m		Hellmayr (1938)
Bolivia: La paz, Huanay	ca 300 m		Bond and Meyer de Shauensee (1941)

APPENDIX
CONTINUED

Locality (country: department, province or state, when available; and locality)[1]	Elevation[1]	Specimens measured	Citation[2]
Bolivia: Santa Cruz, Buena Vista	400 m	1M imm.	FMNH, Hellmayr (1938)
Bolivia: Santa Cruz, Gutiérrez	900 m	1F	LSUMZ
Bolivia: Santa Cruz, San Carlos	ca 300 m	1M	FMNH
Bolivia: Santa Cruz, Santiago de Chiquitos	ca 300 m	3M, 4F, 1?	FMNH, LSUMZ, Hellmayr (1938)
Bolivia: Santa Cruz, Tambo	1,500 m	4M, 11F	LSUMZ
Bolivia: Santa Cruz, Santa Cruz	480 m	1M, 1F	CM, Hellmayr (1938)
Bolivia: Chuquisaca, Carandaiti	600 m		LACM, Remsen et al. (1987)
Bolivia: Tarija, Yacuiba	ca 500 m	1M, 1F	CM
Bolivia: Tarija, Río Lipeo (Monte Bello)	600 m		Meyer de Schauensee (1952)
Brazil: Mato Grosso, Chapada	800 m	2F	AMNH
Paraguay: Presidente Hayes, Chaco-í	75 m	1M	USMN
Paraguay: Presidente Hayes, Laguna salada	130 m		Sight record (see text)
Paraguay: Borquerón, Colonia Filadelfia	ca 175 m		F. Hayes (pers. comm.)
Argentina: Salta, Embarcación	300 m	2M	AMNH
Argentina: Salta, 40 km SE Parque Nacional el Rey	290 m		LNS (B. Whitney)
Argentina: Salta, 75 km E Joaquin V. Gonzalez	100 m		LNS (B. Whitney)
Argentina: Chaco, Avia Terai	350 m	1M	AMNH
Argentina: Jujuy, Yuto	450 m	1M, 1?	AMNH, FMNH, WFVZ
Argentina: Tucumán, Río Sali			Hellmayr (1938)

Tiaris obscura pacifica

Peru: Lima, Lachay	400 m	1M	AMNH
Peru: Lima, Lima	154 m		Koepcke (1963)
Peru: Arequipa, Atiquipa	325 m		Koepcke (1963)

Tiaris fuliginosa

Trinidad: Caparo	?	3M, 2F	AMNH
Trinidad: Waller Field	?	1M	AMNH
Trinidad: Alexander	?	1M	AMNH
Trinidad: Valencias	?	1F	AMNH
Trinidad: St. Joseph	?	1M	AMNH
Trinidad: Aripo	ca 1,000 m	1M, 2M imm.	AMNH
Venezuela: Zulia, Pejochaina	ca 2,100	1M	Aveledo and Ginés (1948)
Venezuela: Zulia, Kunana	1,150 m	1M	Aveledo and Ginés (1948)
Venezuela: Zulia, Ayapa	1,010 m	1M	AMNH, Aveledo and Ginés (1948)
Venezuela: Carabobo, Sierra de Carabobo	?	1M	CM
Venezuela: Aragua, Pié del Cerro	1,000 m	1M imm.	CM

APPENDIX
CONTINUED

Locality (country; department, province or state, when available; and locality)[1]	Elevation[1]	Specimens measured	Citation[2]
Venezuela: Distrito Federal, Cerro de Avila	2,160 m	1F?	CM, Hellmayr (1938)
Venezuela: Sucre, Quebrada Seca, S of Cumaná	200 m	1M[6], 2F?	AMNH
Venezuela: Sucre, Santa Ana	500 m	1M? imm.	AMNH
Venezuela: Sucre, Los Palmales	450 m	1F?	AMNH
Venezuela: Sucre, Campo Alegre [Valley]	?	1M[6]	AMNH
Venezuela: Sucre, San Rafael	900 m	2M	CM
Venezuela: Sucre, Rincón of San Antonio	500 m	1M	AMNH
Venezuela: Bolívar, Monte Roraima	1,250 m		Phelps (1938)
Colombia: Santander, Portugal	950 m	1M	MVZ
Guyana: Rupununi, Bat Mountain	?		Snyder (1966)
Brazil: Maranhão, Coroatá	?	3M, 3M imm., 3F	LSUMZ, Oren (1990)
Brazil: Maranhão, Serra de Conseição	?		Oren (1990)
Brazil: Maranhão, Pedra Chata	150 m		Oren (1990)
Brazil: Maranhão, Aldeia Uruwati	?		Oren (1990)
Brazil: Pernambuco, Quipapá	100 m		Hellmayr (1938)
Brazil: Bahia, Canmamú	50 m		Weid (1830)
Brazil: Espírito Santo, Rio São José	50 m		Pinto (1944)
Brazil: Rio de Janeiro, Cantagalo	400 m		Pinto (1944)
Brazil: São Paulo, Itapetininga	650 m	1M, 1M imm.	LSUMZ
Brazil: São Paulo, Rio Guaratuba, Santos	50 m	2M, 1M imm.	LSUMZ
Brazil: São Paulo, Monte Alegre	750 m		Pinto (1944)
Brazil: São Paulo, Vitoriana	200 m		YPM
Brazil: São Paulo, Rio Juguia, Poço Grande	?		YPM
Brazil: São Paulo, Boa Vista	?	3M imm.	FMNH
Brazil: São Paulo, Iguapé	?	1M, 1M imm.	FMNH
Brazil: Minas Gerais, Rio Doce	?	1M	AMNH
Brazil: Mato Grosso, Chapada	800 m	1M	AMNH
Brazil: Mato Grosso, Serra do Roncador	400 m		Fry (1970)
Paraguay: Canendiyú, Lagunita			Brooks et al. (1993)
Bolivia: Santa Cruz, Arroyo del Encanto	550 m	1M imm.	LSUMZ

[1] Localities and many elevations determined from Paynter (1982, 1985, 1989, 1993), Paynter and Traylor (1981, 1991), Paynter et al. (1975), Stephens and Traylor (1983, 1985), and Hayes (1995).
[2] Citations include data taken from museum specimens examined by the author (ANSP = Academy of Natural Sciences, Philadelphia; AMNH = American Museum of Natural History; CM = Carnegie Museum of Natural History; LSUMZ = Louisiana State University Museum of Natural Science; MVZ = Museum of Vertebrate Zoology, University of California; USNM = National Museum of Natural History), from specimens in collections not examined by the author (COP = Colección Ornitológica Phelps, Caracas, Venezuela; YPH = Peabody Museum of Natural History, Yale University; LACM = Natural History Museum of Los Angeles County), from the literature and from data sheets at the Library of Natural Sounds (LNS).
[3] S. L. = Sea Level.
[4] Hellmayr (1938) apparently considered two of these specimens to be *T. fuliginosa*.
[5] Specimens not included in the measurement data set.
[6] These specimens, from the Carraciolo collection, are labeled females, but their sooty plumage suggests they were incorrectly sexed. This also calls into question the sex of other specimens in this series from Quebrada Seca, Los Palmales, and Santa Ana (several of these may be *T. obscura*). None of these specimens were included in the measurements.

CRYPTIC BIODIVERSITY: AN OVERLOOKED SPECIES AND NEW SUBSPECIES OF ANTBIRD (AVES: FORMICARIIDAE) WITH A REVISION OF *CERCOMACRA TYRANNINA* IN NORTHEASTERN SOUTH AMERICA

RICHARD O. BIERREGAARD, JR.[1,2], MARIO COHN-HAFT[3], AND DOUGLAS F. STOTZ[4,5]

[1]*Biodiversity Programs, National Museum of Natural History, Smithsonian Institution, Washington, DC 20560, USA;*
[2]*present address: Biology Department, University of North Carolina at Charlotte, 9201 University City Boulevard, Charlotte, North Carolina 28223-0001, USA;*
[3]*Museum of Natural Science, 119 Foster Hall, Louisiana State University, Baton Rouge, Louisiana 70803-3216, USA; and*
[4]*Museu de Zoologia, Universidade de São Paulo, CP 7172, 01051 São Paulo, SP, BRAZIL;*
[5]*present address: Environmental and Conservation Programs, Field Museum of Natural History, 1200 Lake Shore Drive, Chicago, Illinois 60605, USA*

ABSTRACT.—An undescribed population of antbirds in the *Cercomacra tyrannina/nigrescens* complex that is sympatric with *Cercomacra tyrannina* was discovered north of Manaus, Brazil. The new population differs from *Cercomacra tyrannina* in plumage, certain mensural characters, and especially voice. Based on similarity of the song, the undescribed form from Manaus is considered conspecific with *Cercomacra tyrannina laeta*. However, because of the sympatry of the Manaus population with *Cercomacra tyrannina, laeta* must be elevated to full species level, with three subspecies: *Cercomacra laeta laeta* in extreme southeastern Amazonia; *C. laeta sabinoi* in northeastern Brazil; and a new subspecies described in this paper, known only from the vicinity of Manaus and eastern Roraima, approximately 650 km north Manaus. The population of *Cercomacra tyrannina* north of the Amazon River and west of the Rio Negro, formerly assigned to *Cercomacra tyrannina laeta,* is shown to be allied with *Cercomacra tyrannina* rather than *Cercomacra laeta*. In this paper, we recognize a valid species and subspecies from two areas whose avifaunas are among the best known in tropical South America. The cryptic species, *Cercomacra laeta,* was already represented by many dozens of specimens in museum collections. Our findings demonstrate that the accurate cataloging of the planet's biodiversity will rely not only on expeditions to remote mountain valleys but also on a careful analysis of specimens and information already at our disposal.

RESUMO.—Uma população ainda não descrita do complexo *Cercomacra tyrannina/ nigrescens* foi descoberta em simpatria com *C. tyrannina* ao norte de Manaus, Brasil. Esta nova população difere de *Cercomacra tyrannina* pela plumagem, certas medidas, e sobretudo por suas vocalizações. De acordo com as similaridades apresentadas por este último carater, a forma não descrita de Manaus foi considerada coespecífica com *Cercomacra tyrannina laeta*. Em virtude dessa simpatria com *Cercomacra tyrannina, laeta* deve ser elevada ao nível de espécie, estando composta de três subespécies, a saber: *Cercomacra laeta laeta* do sudeste extremo da Amazônia, *C. laeta sabinoi* do nordeste do Brasil, e uma subespécie nova, descrita neste trabalho, conhecida apenas dos arredores de Manaus e leste de Roraima, cerca de 650 km ao norte de Manaus. A população de *Cercomacra tyrannina* encontrada ao norte do Rio Solimões e oeste do Rio Negro, anteriormente atribuida à *Cercomacra tyrannina laeta,* na verdade revelou-se mais relacionada à *Cercomacra tyrannina* que *Cercomacra laeta*. A espécie revalidada no presente trabalho pertence a avifauna de duas áreas sul-americanas relativamente bem trabalhadas desde o ponto de vista ornitológico, estando representada em coleções por várias dúzias de espécimens. Esta constatação demonstra que a precisa catalogação da biodiversidade mundial depende não apenas de expedições a lugares remotos, mas também de uma análise detalhada das informações e exemplares já disponíveis.

Our efforts to stem the tide of extinction brought on by our own ever-hastening incursion into the last wild places on earth will be bolstered by an accurate inventory of the species in those ecosystems. Because birds are so well studied, we expect to discover new species only through expeditions to remote corners of the world. In this study, we demonstrate that a relatively common subspecies of *Cercomacra tyrannina,* the Dusky Antbird, which occurs in one of the better-known regions of Amazonia and is well represented in ornithological collections, is a valid species. This discovery emphasizes that a thorough understanding of the planet's biodiversity will require not only continued efforts such as the Rapid Assessment Program of poorly known areas, championed by the late Ted Parker and Conservation International, but also careful attention to the ecosystems much closer to home.

Cercomacra tyrannina is generally a bird of forest edge, tall second growth, and small clearings. In Amazonia it is generally encountered in such habitats in *"terra firme"* situations, rather than river edge or flooded forests. Seven subspecies were recognized by Peters (1951), ranging from southeastern Mexico to the lowlands of Ecuador west of the Andes and east of the Andes broadly through the tropical lowlands of Amazonia (except southwest Amazonia) and the Guianas. A disjunct population, *C. tyrannina sabinoi,* occurs in northeastern Brazil, several hundred kilometers southeast of the Amazonian forest.

J. Vielliard and E. O. Willis initially drew our attention to a peculiar situation with *Cercomacra* antbirds in Belém and Manaus, Brazil. Vielliard did not recognize a recording of *Cercomacra tyrannina* from Manaus, because he knew the very different voice of *C. t. laeta* south of the Amazon in the vicinity of Belém, near the mouth of the Amazon River 1,500 km east of Manaus. Willis, who was familiar with the voice of *C. tyrannina* from Manaus north into Panama but not *C. t. laeta* south of the Amazon, later confirmed (in litt.) that in Belém and Manaus he had heard a *Cercomacra* that was different from the *C. tyrannina* with which he was familiar. He did not know at the time that the song in Belém was being given by birds considered to be a subspecies of *C. tyrannina* and did not pursue the matter.

In April 1988 Bierregaard and Cohn-Haft found a population of *Cercomacra* antbirds on a small island in the Balbina hydroelectric project reservoir, about 150 km north of Manaus (59°20'W × 2°S; site 2, Fig. 1). Our attention was initially drawn to the birds because we did not recognize their songs, which proved to be the key to resolving the taxonomy of this and related populations. The birds were singing in such close proximity to *C. tyrannina* that our parabolic microphone was occasionally focussed on both songs. The songs of the unknown birds were strikingly different from the typical song of *C. tyrannina* north of the Amazon. As we later discovered, these "new" songs resembled closely the songs of *C. t. laeta,* which Willis had heard in Belém and Manaus. In 1992, Stotz found a population of *Cercomacra* antbirds in Roraima (site 1, Fig. 1) with a song similar to those of the Balbina birds.

Descriptions of new subspecies.—Below we describe the details of comparisons of specimens collected and recordings made at the Balbina site and in Roraima with museum material and recordings of other Brazilian populations of *C. tyrannina* and *C. nigrescens.* These studies have demonstrated that populations placed in *C. tyrannina* in Brazil represent two species.

General similarity of morphology and plumage lead us to believe that the Balbina/Roraima population is closely allied to *C. tyrannina,* although its sympatry with this species and differences in plumage and voice indicate that these populations do not belong to the same species. A comparison of the song of the Manaus birds and reciprocal playback experiments with *C. t. laeta* in Belém lead us to ally these two populations at the subspecific level within the biological species *Cercomacra laeta.* Tapes from the state of Pernambuco indicate that the population in northeastern Brazil, *sabinoi,* is also a subspecies of *C. laeta* rather than *C. tyrannina.* We characterize *C. laeta* as a species distinct from *C. tyrannina* with three subspecies, as follows:

Cercomacra laeta laeta Todd 1920

Holotype.—Carnegie Museum 69242, adult female collected in Benevides, Pará, Brazil, by S. M. Klages on 5 Sep. 1918.

Diagnosis.—Males virtually indistinguishable from *Cercomacra tyrannina saturatior* except for their smaller wing and tail (Table 1) and different voice (Fig. 2). Females distinguished from allopatric *C. t. saturatior* by smaller wing, tail and tarsus, paler underparts, and olive-gray, rather than chestnut-rufous, auricular regions.

Range.—Southeastern Amazonian Brazil, east of the Rio Tocantíns and south of Marajó Bay (Baia de Marajó), southeast through the state of Pará and adjacent Maranhão (Fig. 1). There are no specimens of *laeta* from the island of Marajó in any of the museum collections we studied,

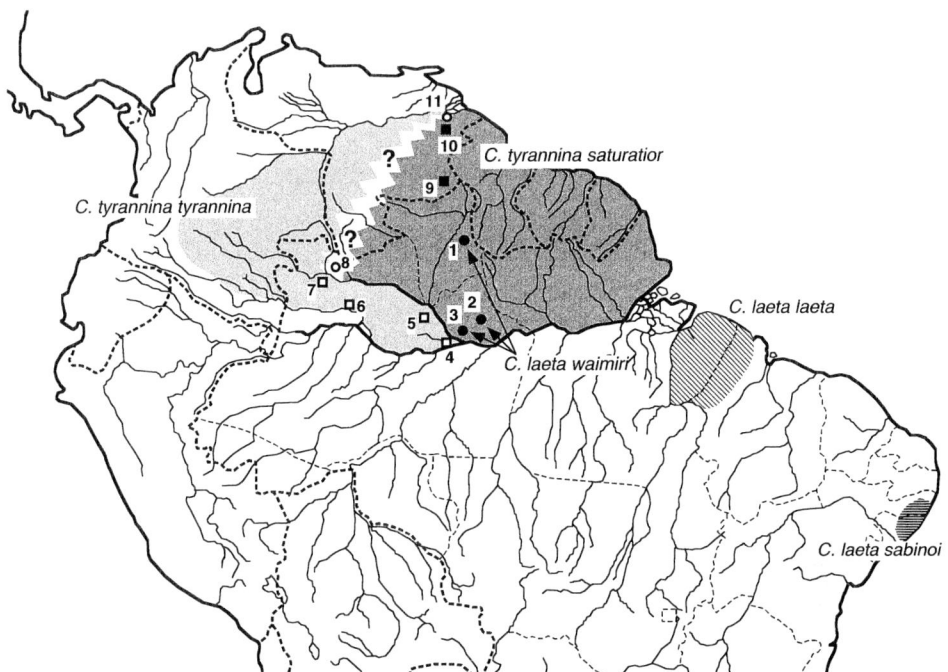

FIG. 1. The ranges of *Cercomacra tyrannina* and *Cercomacra laeta* in northern South America east of the Andes. Solid black circles are known locations for *C. l. waimiri*: 1 = southeast of Boa Vista, Roraima; 2 = Balbina hydroelectric reservoir; 3 = KM 47, Manaus-Boa Vista Highway (BR174). Open squares indicate locations for *C. t. tyrannina* in the southeastern portion of its range: 4 = Manacapurú and Fazenda São Francisco, km 50, AM070; 5 = Jaú National Park; 6 = Mavaã, north bank Rio Japurá; 7 = mouth of the Rio Uaupés. The western limit of the range of *saturatior* is not known. Solid squares indicate sites where songs typical of *saturatior* in Manaus and Suriname have been recorded: 9 = Santa Elena, Venezuela; 10 = El Palmar, Venezuela. Open circles are two recordings with songs noticeably slower than typical *saturatior* (see text): 8 = São Gabriel de Cachoeira, Brazil; 11 = "Campamento Rio Grande," Venezuela. Note that some trans- and cis-Andean subspecies (e.g. *vicina*) of *C. tyrannina* and populations of *C. t. tyrannina* in the Cauca and Sinu valleys are not shown because they are not germane to this paper.

so although range maps of *C. tyrannina* often indicate a continuous distribution for the species from the north bank of the Amazon to the known range of *laeta* east of the Tocantíns (e.g., Ridgely and Tudor 1994), we find no evidence that the form crosses the vast Marajó Bay (Fig. 1).

English name.—We propose to call *Cercomacra laeta* "Willis' Antbird," in recognition of E. O. Willis' indefatigable studies of the Amazonian avifauna, especially members of the Formicariidae, and his early recognition of the unresolved taxonomic problem addressed in this paper.

The males of the population of *C. laeta* in the vicinity of Manaus have longer tails (Table 1) and are darker than the eastern birds from Belém. Additionally, although the songs of the two populations are superficially very similar, there is a consistent difference in the song of the males that further distinguishes the two populations (see Fig. 2). Based on these differences and the geographic barrier (the Amazon River) between the populations, we name the Manaus birds

Cercomacra laeta waimiri, subsp. nov.

Holotype.—Mus. Paraense Emílio Goeldi number 52521, adult male collected on an unnamed island (the site of "Base 2" of the pre-flooding animal rescue effort) in the Balbina Hydroelectric Project Reservoir 150 km northeast of Manaus, Amazonas, Brazil, by R. O. Bierregaard, Jr., on 11 Apr. 1988 (BDFFP capt. 41799).

Description of holotype (male).—(Colors in capitals are from the Munsell soil color charts followed by the Munsell notation for hue, color and chroma [Anon. 1994]). Throat, breast, underwing coverts and upper belly Dark Gray (2.5 Y 4/0); lower flanks, thighs, and lower belly washed with olive (Dark Grayish Brown 2.5 Y 4/2); crown to lower back Very Dark Gray (2.5

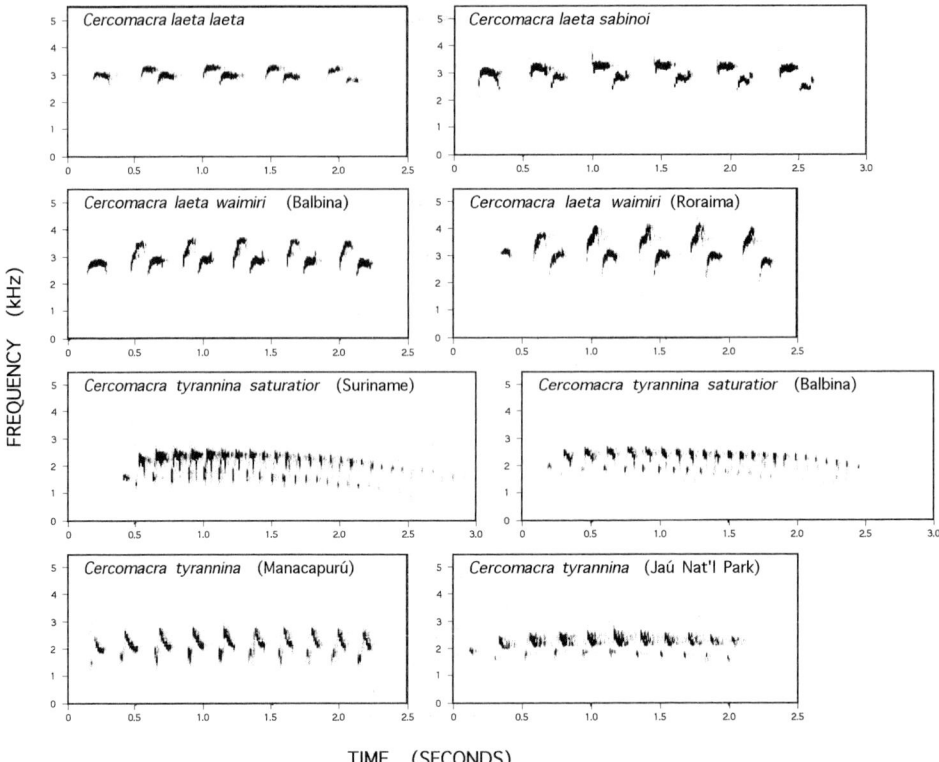

FIG. 2. Male songs of *Cercomacra laeta* and *C. tyrannina*. All three subspecies of *C. laeta* exhibit the distinctive pattern of a single note followed by a series of couplets. Unlike nominate *laeta* or *sabinoi*, the first note of each couplet in male *waimiri* rise sharply in frequency. The rapid trill of the male song in *C. tyrannina saturatior* is similar across its range from Suriname to Manaus. The songs of male *C. tyrannina* on the west bank of the Rio Negro (Manacapurú and Jaú) resemble *saturatior* but are substantially slower. Recording details as follows: *C. l. laeta*, natural song, recorded south of Belém (Rio Guajará), Brazil, LNS 48156; *C. l. sabinoi*, response to playback of *C. l. laeta*, recorded on 18 Nov. 93, in Aldeia, near Pernambuco, Recife, Brazil, by B. M. Whitney; *C. l. waimiri* (Balbina), natural song, LNS 46961; *C. l. waimiri* (Roraima), natural song recorded by Stotz, ASN DFS 09/3; *C. t. saturatior* (Suriname), response to playback, LNS 11530; *C. tyrannina* (Manacapurú), natural song, LNS 46847; *C. tyrannina*, recorded 14 Jul. 1993, in Jaú National Park (west bank, Rio Negro), Amazonas, Brazil, by A. Whittaker.

Y 2/0); upper tail coverts washed with olive, concolor with flanks and lower belly; upper wing coverts black (2.5 Y 2/0) very narrowly (<1 mm) tipped with white; a large, concealed white interscapular patch is present and the leading edge of the wing at the wrist is white; rectrices and remiges blackish (concolor with black), leading edge of primaries narrowly edged in a lighter gray, and trailing edge of proximal primaries edged pale gray. Iris dark brown. Bill black, and legs bluish to blackish gray.

Description of female.—Breast, belly and undertail coverts orange-cinnamon (Strong Brown 10 Y/R 8/8); crown, nape, back, and rump dark olive gray (5 Y 3/2); lores, superciliaries, and auriculars orange-cinnamon, slightly paler than breast (Yellow 10 Y/R 7/8); auriculars distally concolor with back. Rectrices dark grayish brown (2.5 Y 3/2). Leading edge of wing at wrist orange-cinnamon, slightly lighter than breast (Strong Brown 10 Y/R 6/8). Upper wing coverts blackish (Very Dark Gray 10 Y/R 3/1) tipped about 1 mm with orange-cinnamon concolor with breast; primary and secondary coverts edged Very Pale Brown (10 Y/R 7/3); a weak, concealed, white interscapular patch is present. Edge of anterior vane of alular quills creamy buff, paler than the breast and edging to secondary coverts.

Diagnosis.—An antbird of medium size, recognizable as *Cercomacra* by long, slender, and slightly hooked bill, distinct rictal bristles, and the rather long, graduated tail with ten rectrices (Ridgway 1911:90). Within the genus, the new form is distinguished from the "*nigricans* group" (*C. manu, C. melanaria, C. nigricans, C. ferdinandi,* and *C. carbonaria*) by the female's warm

TABLE 1

MORPHOLOGY OF *Cercomacra laeta*, *C. tyrannina*, AND *C. nigrescens* IN SOUTH AMERICA EAST OF THE ANDES. VALUES ARE MEANS (MM), FOLLOWED BY SAMPLE SIZE, STANDARD DEVIATION, AND RANGE

Form	Wing	Tail	Tarsus	Culmen
C. n. nigrescens				
Males	68.6 (15, 1.31, 66.7–70.9)	63.6 (15, 1.84, 61–67.5)	24.6 (14, 1.57, 19.3–25.8)	21.2 (14, 0.87, 19.2–22.1)
Females	65.9 (5, 1.27, 64.3–67.7)	61.7 (5, 4.68, 56.4–67.5)	23.9 (5, 0.71, 23.0–24.8)	20.1 (5, 0.52, 19.5–20.7)
C. n. approximans				
Males	66.7 (12, 1.93, 64.0–71.0)	62.1 (11, 3.21, 54.7–65.7)	23.4 (12, 1.61, 19.2–25.0)	20.6 (12, 1.27, 19.2–24.1)
Females	63.68 (5, 1.61, 61.7–65.5)	62.14 (5, 2.83, 59.0–65.7)	23.0 (5, 0.33, 22.7–23.5)	20.14 (5, 0.54, 19.4–20.9)
C. t. tyrannina				
Males	64.13 (6, 1.38, 62.0–65.5)	60.38 (6, 1.27, 58.6–61.6)	24.16 (5, 0.88, 23.0–25.4)	19.72 (5, 0.62, 18.8–20.3)
Females	61.06 (5, 0.82, 60.4–62.3)	55.56 (5, 1.30, 53.7–56.9)	22.9 (5, 0.95, 22.0–24.2)	18.14 (5, 0.64, 17.3–18.9)
C. t. tyrannina (between Rio Negro and Amazonas)				
Males	63.54 (17, 1.73, 60.4–67.0)	58.62 (17, 2.17, 55.0–62.4)	23.09 (16, 1.19, 20.5–25.9)	19.11 (14, 0.89, 17.0–20.6)
Females	60.82 (14, 1.30, 58.4–62.7)	57.21 (14, 2.36, 53.0–61.8)	22.49 (14, 0.86, 21.3–23.9)	18.62 (13, 0.64, 17.8–19.8)
C. t. saturatior				
Males	62.54 (28, 1.89, 58.8–66.2)	59.65 (29, 1.96, 56.0–63.5)	22.79 (28, 0.62, 21.1–23.9)	19.25 (29, 0.67, 18.1–20.7)
Females	59.14 (23, 1.86, 55.8–63.6)	58.62 (22, 2.59, 55.0–64.9)	22.15 (21, 0.61, 20.9–23.2)	18.80 (23, 0.73, 17.3–20.1)
C. l. laeta				
Males	59.3 (11, 1.80, 57.0–63.7)	55.82 (10, 2.54, 52.7–61.9)	21.97 (11, 0.23, 21.4–22.2)	18.41 (11, 0.76, 17.4–19.6)
Females	56.88 (9, 1.38, 54.3–59.0)	53.0 (9, 1.53, 50.9–54.8)	21.3 (9, 0.60, 20.0–21.9)	18.24 (9, 0.44, 17.6–19.0)
C. l. sabinoi				
Males	61.59 (8, 1.64, 59.7–64.0)	56.04 (8, 1.32, 53.9–57.4)	22.29 (8, 0.66, 21.1–23.2)	14.81 (8, 0.43, 14.1–15.2)[1]
Females	59.55 (6, 1.26, 58.0–61.0)	53.4 (5, 1.24, 50.6–53.4)	21.84 (5, 0.38, 21.6–22.5)	14.72 (6, 0.42, 14.3–15.5)[1]
C. l. waimiri				
Males	61.95 (4, 1.43, 60.0–63.4)	59.3 (4, 1.29, 58.3–61.0)	23.88 (4, 0.25, 23.6–24.2)	19.08 (4, 0.15, 19.0–19.3)
Females (1)	59.2	55.9	23.2	17.8

[1] Measurements from nares, not comparable to other forms (see text).

cinnamon-brown ventral surface, rather than gray or olive gray typical of females in this group (Fitzpatrick and Willard 1990). Within the *"tyrannina* group" (*C. tyrannina, C. laeta, C. nigrescens, C. serva,* and *C. brasiliana*), the new form most closely resembles *tyrannina, laeta,* and *brasiliana.* The new form is readily distinguished from *C. cinerascens* (intermediate between the *tyrannina* and *nigricans* groups) by the lack of conspicuous, broad white (male) or buffy (females) tips to the rectrices.

The new population can be differentiated from *C. nigrescens* and *C. serva* by more prominent white edging to the upper wing coverts (greatly reduced to absent in *approximans* and *serva*) and pronounced olive wash to the flanks, thighs, lower belly, and upper tail coverts in males. The single female collected differs in having cinnamon-buff tips of the upperwing coverts. Both sexes are also substantially smaller than *nigrescens* (see Table 1).

C. l. waimiri from the Balbina site differs from *C. tyrannina saturatior* and *C. l. laeta* in having a darker belly and is also lacking white tips to belly feathers that give a faint, narrow scalloping found in some *C. t. saturatior* and nominate *laeta.* This scalloping may be less pronounced in *C. t. saturatior* specimens with more advanced feather wear, but there is no indication of it at all in the Balbina *C. l. waimiri* specimens, including those that are in fresh plumage. Freshly molted adult male *C. t. saturatior* show narrow (ca. 1 mm) off-white tips to the rectrices (wider in the outer feathers), which do not appear in any of the male *C. l. waimiri* specimens. Some *C. t. saturatior* have browner remiges than our *C. l. waimiri* specimens, but this is not consistent and may be a function of feather weathering or age (M. and P. Isler, pers. comm.).

Males of *C. l. waimiri* are darker gray above and below than nominate *laeta,* which has an olive brownish wash to the remiges contrasting with the rest of the back. Also, nominate *laeta* have very narrow (up to 1 mm) off-white tips to the rectrices, which probably can be lost to feather wear; the tails of all five *C. l. waimiri* specimens lack these white tips.

The strongest character separating female *C. l. waimiri* from nominate *laeta,* sympatric *C. t. saturatior,* nominate *tyrannina,* and *C. tyrannina* north of the Amazon River and east of the Rio Negro is the darker upper wing coverts (darker gray and less olive than rest of back), with the greater and median secondary coverts prominently edged with cinnamon-brown concolor with breast, a trait lacking in all other forms considered here. The female further differs from nominate *laeta* and nominate *tyrannina* in having orange-cinnamon, rather than gray, auriculars.

Because the plumages of *C. tyrannina* and *C. brasiliana* are practically indistinguishable (Ridgely and Tudor 1994), we assume that comparisons between *laeta* and *brasiliana* would be no different from those between *laeta* and *tyrannina*. We did not compare our specimens to *C. brasiliana,* a species restricted to the Atlantic forest region in southeastern Brazil, about 1,000 km south of Amazonia, which has a different voice (Ridgely and Tudor 1994) than any of the populations joined in *C. laeta.*

Variation.—The three remaining males collected at the Balbina site agree closely with the holotype. The one male collected in Roraima, however, is paler on the belly, with faint white edgings to the feathers of the central belly, much like many male *C. t. saturatior* or nominate *tyrannina,* but lacks the white tips to the rectrices, in agreement with the Balbina specimens.

Range.—This form is known from four male specimens collected several km above the Balbina Dam, (59°20'W × 2°S; site 2, Fig. 1), a pair collected 4 km north of the Rio Cachorro, on the road from Cantá to Confiança, Roraima, Brazil (60°40'W × 2°15'S; site 1, Fig. 1), recordings elsewhere in Roraima, and sight records north of Manaus (km 47, BR174) made 10 to 15 years prior to our discovery of the Balbina birds (Willis, in litt.; Oren, pers. comm.) (site 3, Fig. 1). Despite the difference in the ventral plumage of the one male collected in Roraima, we tentatively consider the Roraima birds to represent the same form as the Manaus birds based on the similarity of song and lack of any barrier between eastern Roraima and the Balbina site.

Etymology.—We name this subspecies for the tribe of Amerindians on whose former lands we originally encountered these birds. The Waimiri and neighboring Atroari tribes were displaced by the Balbina Hydroelectric Project.

Based on plumage and voice, we ally the northeastern Brazilian population currently known as *C. tyrannina sabinoi* with the Amazonian populations of *C. laeta* and characterize it as follows:

Cercomacra laeta sabinoi Pinto 1939

Holotype.—Adult male collected by Oliveiro Pinto on 15 Dec. 1938 at Fazenda São Bento, Taperá station, Pernambuco state, Brazil; Museu de Zoologia, USP (São Paulo) 18.122.

Diagnosis.—In the original description of this form, Pinto (1939) effectively only compared *sabinoi* with *laeta*. He noted that males are indistinguishable in plumage, whereas female *sabinoi* are paler overall, especially ventrally where they are light cinnamon with no appreciable rusty suffusion.

Range.—Known only from coastal forests of extreme northeast Brazil, occurring in the states of Pernambuco and Alagoas from sea level to 450 m (Fig. 1) (Teixeira, unpub. data).

METHODS

During three visits to the Balbina island, four adult males of the *C. l. waimiri* were collected. No females were seen or heard at Balbina. In Roraima, Stotz collected a pair of *Cercomacra* antbirds and recorded male and female songs similar to those of the Balbina birds.

We searched for additional populations of *C. l. waimiri* at various sites, including along the road from Manaus to Itacoatiara (AM-010), in the vicinity of Manaus, at the INPA silviculture station (km 47 on BR174; site 3, Fig. 1) where Willis and Oren (pers. comms.) had seen and heard the birds in the 1970s, and west of the Rio Negro along the road (AM070) from Cacau Pireira to Manacapurú, a small village on the north bank of the Amazon (Solimões) River about 75 km west of Manaus. Our efforts west of the Rio Negro were concentrated along the access road to the Fazenda São Francisco, km 50 AM 070 (site 4, Fig. 1). We stopped wherever we saw suitable habitat, i.e., the edge of tall forest skirted by 3- to 10-m-tall second growth, often dominated by *Cecropia* spp., and played an endless loop cassette tape made from J. Vielliard's recording of *C. l. waimiri* as well as our own recordings of *C. tyrannina* from the Manaus vicinity.

We compared specimens from the Museu Paraense Emílio Goeldi, Belém, Brazil (MPEG), the Museu Nacional, Rio de Janeiro, Brazil (MN), the Museu de Zoologia da Universidade de São Paulo, São Paulo, Brazil (MZUSP), the American Museum of Natural History (AMNH), and the Museum of Comparative Zoology, Harvard (MCZ), to our specimens (Appendix 1). We studied skins from all described forms of *C. tyrannina* in Amazonia and northeastern Brazil, as well as two eastern Amazonian subspecies of the Blackish Antbird, *C. nigrescens*: *C. n. nigrescens*, which is sympatric with *C. tyrannina* north of the Amazon; and *C. n. approximans*, which occurs south of the Amazon and west of the east bank of the Rio Tapajós. The populations of *C. tyrannina* examined were: *C. t. tyrannina* from northwestern Amazonia (Venezuela and Colombia); birds east of the Rio Negro and north of the Amazon River through the Guianas, hereafter referred to simply as *C. t. saturatior*; the population west of the Rio Negro and north of the Amazon (Solimões), traditionally considered to be *C. tyrannina laeta*; *C. l. laeta* south of the Amazon delta (Marajó Bay) and east of the Tocantins River to the southeastern limits of the Amazonian forests in the state of Maranhão; *C. l. sabinoi* from Alagoas and Pernambuco in extreme northeastern Brazil; and *C. l. waimiri*.

Field capture data for specimens collected by Bierregaard and Cohn-Haft are archived in the computer data base and original field slips of the Avian Ecology Subproject of the Biological Dynamics of Forest Fragments Project (BDFFP). Capture numbers link specimens described herein to these data.

Wing chord, tail, tarsus, and upper mandible length (from the base of the skull) were measured with vernier calipers on at least five, but often more, specimens of each sex of each subspecies or population, with the exception of *C. l. waimiri,* for which only five males and a single female have been collected. Bierregaard measured all specimens except the specimens of *sabinoi*, which were measured by D. M. Teixeira (see acknowledgments), and the male *C. l. waimiri* collected and measured by Stotz in Roraima. Bill measurements made by Teixeira are not comparable to those made by Bierregaard and were not used in statistical comparisons.

Morphological data were log-transformed prior to statistical analyses. Using the SYSTAT (Wilkinson 1990) statistical software package, a one-way analysis of variance (ANOVA) was performed on the four measurements across the eight study populations for males. For females, only seven populations could be compared, because only one specimen of *C. l. waimiri* was available.

Prior to the ANOVA, Bartlett's test for homogeneity of group variance was performed. For each sex, a post-hoc test of the equality of all pairs of means was then performed using Tukey's HSD procedure (Tukey 1977, as cited in Wilkinson 1990). Additionally, an independent samples t-test was performed to test the equality of means between sexes for each of the four morphological characters in all subspecies except *C. l. waimiri*.

Bierregaard and Cohn-Haft recorded the songs and calls of the Balbina birds and *C. tyrannina* from Manaus and Manacapurú. These recordings are archived at the Library of Natural Sounds (LNS) at the Cornell Laboratory of Ornithology and the Arquivo Sonoro Natural (ASN) at the

TABLE 2

Sexual Dimorphism in *Cercomacra laeta, C. tyrannina,* and *C. nigrescens*: Independent Samples *t*-test. Values Presented are Sample Sizes and Results of Tests, as Follows: ns = H_0 of Equal Means Accepted; * = H_0 of Equal Means Rejected, $0.05 > P > 0.01$; ** = H_0 of Equal Means Rejected, $0.01 > P > 0.001$; *** = H_0 of Equal Means Rejected, $P < 0.001$. There Were Insufficient Data to Perform This Test for *C. l. waimiri*. Pooled Variances Used for Wing, Tail, and Culmen

Population	Wing	Tail	Tarsus	Culmen
C. l. laeta	18,**	17,**	18,*	18, ns
C. l. sabinoi	12,*	11,***	11, ns	12, ns
C. n. nigrescens	18,***	18, ns	17, ns	17, *
C. n. approximans	15,**	14, ns	15, ns	15, ns
C. t. saturatior	49,***	49, ns	47,***	50, *
C. t. tyrannina	9,**	9,***	8, ns	8, **
C. t. tyrannina (west of R. Negro)	29,***	29, ns	28, ns	25, ns

University of Campinas, São Paulo, Brazil. Stotz's recordings from Roraima are archived at the ASN. We obtained recordings of *laeta* from the state of Pará, *sabinoi* from Pernambuco, and of *tyrannina* from the west bank of the Rio Negro in the Jaú National Park from various colleagues (see Acknowledgments). Additional recordings of *C. tyrannina* and *C. nigrescens* were provided by the LNS (see Appendix 2).

Our recordings around Manaus and Manacapurú were made by Bierregaard on a Nagra III and Cohn-Haft on a Sony TCM-5000 tape recorder. Songs were analyzed and spectrograms produced with the Canary software (Charif et al. 1995) from the Cornell Laboratory of Ornithology on a Power Macintosh 7100/80.

In the course of playback experiments, we recognized two types of response. When playback of a test recording was made while a bird was singing, a positive response was for the singing bird to abruptly stop singing, utter a short "jut-ut" call, approach the recorder, eventually resuming singing, coming out of hiding with the white interscapular patch prominently displayed. Both sexes of *C. tyrannina* and *C. laeta* exhibited this behavior. The birds were considered to have shown no response if they continued singing and did not approach the recorder or come out of hiding. For playbacks in the known territories of birds not singing, a positive response was recorded when the birds started singing and aggressively approached the recorder as described above.

RESULTS

Morphology.—All taxa studied are sexually dimorphic. On average, for all taxa females are smaller than males, although there is overlap (Table 1). The independent samples *t*-test rejected the null hypothesis of equal means for all forms for wing length (Table 2). Except for *C. nigrescens approximans,* all female means were less than those of their respective males for tail and tarsus length, although these differences were significant only for *C. l. sabinoi, C. l. laeta,* and *C. t. tyrannina* for tail length, and *C. l. laeta* and *C. tyrannina* north of the Amazon River and east of the Rio Negro for tarsus length (Table 2). Because of this sexual dimorphism, inter-taxon comparisons were made separately for males and females.

Inter-taxon comparisons showed significant (ANOVA, $P < 0.001$) taxon effect for both males and females. For males, variances were homogeneous for wing, tail, and culmen, and for all four traits in females.

Males of *C. l. laeta* are the smallest of the populations studied in all dimensions, although there is overlap and not all comparisons are statistically significant (Tables 1 and 3). Statistically, males of *C. l. laeta* are not different from *C. l. sabinoi* or *C. l. waimiri,* are smaller in all dimensions measured than either form of *C. nigrescens,* have shorter tails and tarsi than *C. t. saturatior* and *C. tyrannina* west of the Negro, and have shorter wings, tails, and tarsi than nominate *tyrannina* (Table 3). No differences were found among *C. l. sabinoi, C. tyrannina* west of the Negro, and the *C. l. waimiri* males. Individuals of *C. l. sabinoi* are smaller than either form of *C. nigrescens* considered here in all but tarsus length and has a shorter tail than nominate *tyrannina* and *C. t. saturatior* (Table 3). The *C. l. waimiri* males show few differences from other forms, having shorter wings than *C. approximans* and *C. nigrescens,* and shorter tails and culmens than *C. nigrescens* (Table 3).

Female nominate *laeta* are smaller than females of both subspecies of *C. nigrescens* for all characters, smaller than *C. t. saturatior* and *C. tyrannina* west of the Negro in all but culmen, and have smaller wings and tarsi than nominate *tyrannina.* Female *C. l. sabinoi* differ from *C. l. laeta* only in having significantly longer wings (Table 3). They are smaller than both subspecies of *C. nigrescens* in wing, tail, and tarsus, except for the comparison of tarsus length with *C. n. approximans,* for which the null hypothesis of equal means was not rejected. Females of *C. l. sabinoi* differed from females of *C. t. saturatior* and *C. tyrannina* west of the Negro only in having shorter tails. No statistical differences were found in wing, tail, or tarsus when the sample was compared to nominate *tyrannina* (Table 3).

Vocalizations.—In suboscine passerines, song is believed to be genetically determined, rather than learned (Kroodsma 1984). Differences in songs, therefore, are probably a phenotypic expression of genetic differences between populations and consequently an important clue to taxonomic relationships. Furthermore, because songs are used in intraspecific territory and in mate advertising, playback experiments may be used to test biological species limits in allopatric forms (e.g., Lanyon 1963).

The genus *Cercomacra* is composed of two species groups, the members of which are allied by both plumage and voice characteristics (Fitzpatrick and Willard 1990). The forms reviewed in this paper are in the "songster" or "whistler" group, characterized by clearer, higher-frequency songs than the "croakers." In both groups members of a pair commonly duet.

Our comparisons of songs are based on three recordings of *C. l. laeta,* at least six individuals of *C. l. waimiri,* one recording of *C. l. sabinoi,* and an ample number of recordings of *C. tyrannina* from Brazil, Suriname, and Venezuela (see Appendix 2).

The spectrograms of the songs of male nominate *laeta, sabinoi,* and *C. l. waimiri* are similar to each other and differ greatly from other forms of the *Cercomacra* "songsters," including *tyrannina* (Fig. 2). Whereas the song of male *C. tyrannina* consists of a fairly rapid (roughly 10/s) series of single notes sometimes descending in frequency, the song of male *laeta* is much slower, consisting of an introductory note followed by a series of three to six (usually four, but as many as eight if birds are responding to playback of conspecific songs) couplets of notes. The typical song (four couplets) lasts between 1.75 and 2.0 seconds.

The male *C. laeta* song begins with a single syllable at around 2.80 to 3.10 kHz, with the first note of each subsequent couplet higher in frequency (ca 3.2 kHz) than the second, which is very similar to the introductory note in shape and frequency (Fig. 2). The last note of each song is often slightly lower in frequency than the preceding ones. We describe the song as "pee pee-ter pee-ter pee-ter . . . ."

The songs of male *C. l. waimiri* consistently differ from those of *C. l. sabinoi* and nominate *laeta* in that the first note of each couplet in *C. l. waimiri* is sharply slurred upward from about 2.6 kHz to 3.7 kHz, whereas the first note in nominate *laeta* and *C. l. sabinoi* after a very brief increase in frequency remains at a constant peak frequency of 3.3 to 3.4 kHz (Fig. 2).

The song of female *laeta* is almost always given in a duet with the male; the female begins singing after the male has started his song (Fig. 3). The songs of male and female often overlap. The male often utters a relatively quiet, "chittery" series of two to five syllables when the female finishes her song. This general pattern is similar to the duets of other species in the *Cercomacra* songster species. However, only in *laeta* does the male's song consist of coupled syllables. In the recordings we examined, male songs are more common than duets, but often alternate with them, a pattern witnessed in *C. tyrannina* as well.

The female song is an ascending series of two to four notes, each of about 0.5 s duration, with all but the ultimate note rising abruptly and then slowly descending in frequency, forming inverted "U"s in a spectrogram. In contrast, the last note uttered ascends in frequency (Fig. 3). This is consistent over all three subspecies.

In *C. tyrannina* duets north of the Amazon River and east of the Rio Negro, the male's song begins with one or two slow notes whistled at about 2 kHz, followed by a series of 15–20 faster, higher notes, each slurred downward from about 2.6 to 2 kHz. The length of the song varies considerably and is often particularly extended by males responding to playback of their own or conspecific songs. The female begins her song about 1.5 seconds after the male begins his. After an introductory note slurred upwards at about 1.6 kHz, the remaining series of 6–10 notes appear as inverted "U"s in a spectrogram (Fig. 4). The peaks of the notes rise in frequency from 2.4 to 2.9 kHz and decrease in duration from about 0.1 to 0.05 seconds and then quietly

TABLE 3

Matrix of Results of Pairwise t-tests (Tukey) between Subspecies of *Cercomacra laeta*, *C. nigrescens*, and *C. tyrannina*, for Wing, Tail, Tarsus and Culmen Length. Results for Males above the Diagonal, Females below. Group Variances Homogeneous for Wing, Tail, and Culmen for Males and for all Characters for Females. Data are not Comparable for Culmen Length in *C. l. sabinoi*, and Sample Size was Insufficient for Female *C. l. waimiri*

Taxon	l. laeta	l. sabin.	l. waimi.	n. appro.	n. nigre.	t. satur.	t. tyran.	t. tyran. (w. of Rio Negro)
C. laeta laeta								
Wing		ns	ns	$P < 0.001$	$P < 0.001$	$P < 0.001$	$P < 0.001$	$P < 0.001$
Tail		ns	ns	$P < 0.001$	$P < 0.001$	$P < 0.001$	$P < 0.01$	$P < 0.05$
Tarsus		ns	ns	$P < 0.05$	$P < 0.001$	ns	$P < 0.01$	ns
Culmen		—	ns	$P < 0.001$	$P < 0.001$	ns	ns	ns
C. laeta sabinoi								
Wing	$P < 0.05$		ns	$P < 0.001$	$P < 0.001$	ns	ns	ns
Tail	ns		ns	$P < 0.001$	$P < 0.001$	$P < 0.01$	$P < 0.01$	ns
Tarsus	ns		ns	ns	$P < 0.001$	ns	ns	ns
Culmen	—		—	—	—	—	—	—
C. laeta waimiri								
Wing	—	—		$P < 0.001$	$P < 0.001$	ns	ns	ns
Tail	—	—		ns	$P < 0.05$	ns	ns	ns
Tarsus	—	—		ns	ns	ns	ns	ns
Culmen	—	—		ns	$P < 0.01$	ns	ns	ns
C. nigrescens approximans								
Wing	$P < 0.001$	$P < 0.01$	—			$P < 0.001$	ns	$P < 0.001$
Tail	$P < 0.001$	$P < 0.001$	—			$P < 0.05$	ns	$P < 0.01$
Tarsus	$P < 0.01$	ns	—			ns	ns	ns
Culmen	$P < 0.001$	—	—			$P < 0.001$	ns	$P < 0.01$

TABLE 3
Continued

Taxon	l. laeta	l. sabin.	l. waimi.	n. appro.	n. nigre.	t. satur.	t. tyran.	t. tyran. (w. of Rio Negro)
C. n. nigrescens								
Wing	$P < 0.001$	$P < 0.001$	—	ns		$P < 0.001$	$P < 0.001$	$P < 0.001$
Tail	$P < 0.001$	$P < 0.001$	—	ns		$P < 0.001$	ns	$P < 0.001$
Tarsus	$P < 0.001$	$P < 0.001$	—	ns		$P < 0.001$	ns	$P < 0.01$
Culmen	$P < 0.001$	—	—	ns		$P < 0.001$	$P < 0.05$	$P < 0.001$
C. tyrannina saturatior								
Wing	$P < 0.01$	ns	—	$P < 0.001$	$P < 0.001$		ns	ns
Tail	ns	$P < 0.001$	—	ns	ns		ns	ns
Tarsus	$P < 0.05$	ns	—	ns	$P < 0.001$		ns	ns
Culmen	ns	—	—	$P < 0.01$	$P < 0.01$		ns	ns
C. t. tyrannina								
Wing	$P < 0.001$	ns	—	ns	$P < 0.001$	ns		ns
Tail	ns	ns	—	$P < 0.01$	$P < 0.01$	ns		ns
Tarsus	$P < 0.01$	ns	—	ns	ns	ns		ns
Culmen	ns	—	—	$P < 0.001$	$P < 0.001$	ns		ns
C. t. tyrannina (w. Rio Negro)								
Wing	$P < 0.001$	ns	—	$P < 0.05$	$P < 0.001$	$P < 0.05$	ns	
Tail	$P < 0.01$	$P < 0.01$	—	$P < 0.01$	$P < 0.05$	ns	ns	
Tarsus	$P < 0.01$	ns	—	ns	$P < 0.01$	ns	ns	
Culmen	ns	—	—	$P < 0.01$	$P < 0.01$	ns	ns	

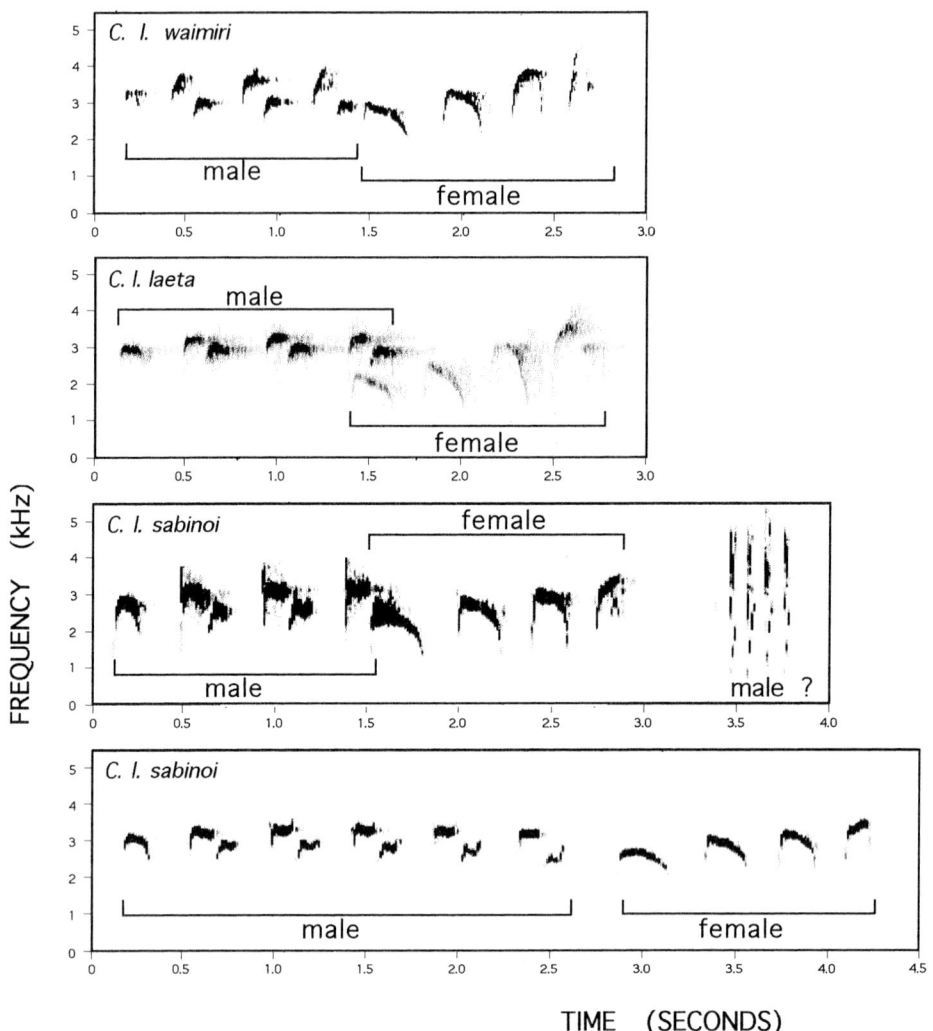

FIG. 3. Male and female duets of *Cercomacra laeta*. Female song is similar to that of *C. tyrannina* (Fig. 4) except that the last note rises in frequency. Male songs are often extended (more couplets) after playback. Recording details as follows: *C. l. laeta*, natural song, recorded south of Belém (Rio Guajará), Brazil, LNS 48156; *C. l. sabinoi*, response to playback of *C. l. laeta*, recorded on 18 Nov. 93, in Aldeia, near Pernambuco, Recife, Brazil, by B. M. Whitney; *C. l. waimiri* (Roraima), natural song ASN DFS 09/3.

taper off. As the female ends her song, the male often utters a brief series of quiet notes. These patterns are consistent in all recordings from Manaus, Suriname, and southeastern Venezuela.

The songs of males and duets of *C. tyrannina* (formerly assigned to *C. t. laeta*, see below) west of the Rio Negro is similar to *C. tyrannina* east of the Rio Negro but is substantially slower (Fig. 4). Most of the difference lies in the male's song, in which the notes are given at a rate of about 4/s west of the Rio Negro and 10–12/s in *C. tyrannina* east of the Rio Negro. The female song is somewhat slower west of the Negro, but the difference is not as pronounced as in the male song (Fig. 4).

Playback experiments.—In reciprocal playback experiments, one male and one female *C. l. laeta* in Belém responded quickly and aggressively to the song of *C. l. waimiri*. The male was collected and tissue preserved. Similarly, a male of *C. l. waimiri* responded aggressively to the *C. l. laeta* song recorded by Vielliard east of the Rio Tocantíns. Likewise, B. M. Whitney (in litt.) used Belém *C. l. laeta* recordings to lure in *C. l. sabinoi* near Recife, Pernambuco. He reported that a pair of *C. l. sabinoi* reacted very aggressively to the playback of the Belém birds.

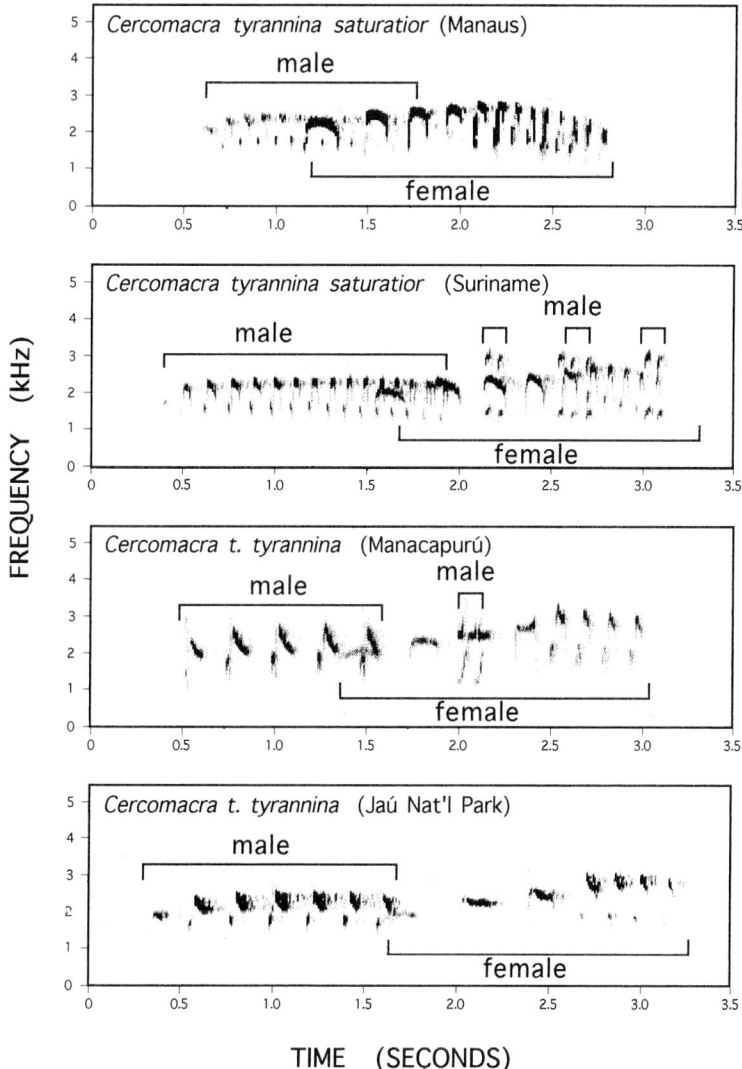

FIG. 4. Male and female duets of *Cercomacra tyrannina*. Female song usually overlaps male song and often ends in a descending trill, unlike female *C. laeta*. Males often utter a two-syllable call, "jut-ut" as the female sings. As with males, female songs of nominate *tyrannina* are generally similar in frequency and pattern to *saturatior* but slower. Recording details as follows: *C. t. saturatior* (Manaus), response to playback, LNS 46965; *C. t. saturatior* (Suriname), response to playback, LNS 25516; *C. t. tyrannina* (Manacapurú), natural song, LNS 46847; *C. t. tyrannina*, recorded 14 Jul. 1993, in Jaú National Park (west bank, Rio Negro), Amazonas, Brazil, by A. Whittaker.

At Balbina, the resident *C. tyrannina* did not respond to the playback of either *C. l. waimiri* or Belém *C. l. laeta*. We did not quantify the number of playback trials, because we were using the recordings to lure birds in the newly found population into nets over a period of several days, often with individuals of *C. tyrannina* in close proximity. Also, during hours of searching for additional populations of *C. l. waimiri* around Manaus and across the Rio Negro on the Fazenda São Francisco, as we repeatedly played the Belém recording, no individual *C. tyrannina* responded to the song of either *C. l. laeta* or *C. l. waimiri*.

Geographic variation in Amazonian C. tyrannina.—*Cercomacra tyrannina* along the north bank of the Amazon from the Rio Jarí west to Manaus have traditionally been placed in *laeta* (Cory and Hellmayr 1924; Naumburg 1939). Peters (1951) extended the range of *laeta* north of the Amazon east to Macapá (in the state of Amapá) and west across the Rio Negro to Manacapurú.

Analysis of the ample collection of Amazonian *C. tyrannina* in the collection of the AMNH revealed that all well-prepared female specimens of *C. tyrannina* north of the Rio Amazonas, from Amapá west to Manaus have a chestnut-rufous auricular region, whereas the auriculars of female *laeta* south of the Amazon and the birds west of the Rio Negro are olive-gray. The gray-cheeked Manacapurú and Belém populations appear, therefore, to be disjunct. Additionally, the Manacapurú birds are larger than those from Belém in wing and tail (males) and wing, tail, and tarsus (females) (Table 1). They differ morphometrically from *C. tyrannina* north of the Amazon River and east of the Rio Negro only in females having longer wings (although with substantial overlap).

Most importantly, the song of the birds west of the Rio Negro, although significantly slower than the typical song of *C. tyrannina* east of the Rio Negro, is clearly much like that of *C. tyrannina* and unlike *C. laeta* (Fig. 2). Whittaker's recording of *C. tyrannina* in the Jaú National Park (about 200 km northwest of Manaus; site 5, Fig. 1) shows that the slow song typical of the Manacapurú population extends up the west bank of the Rio Negro.

Considering these differences in morphology, plumage, and voice between these birds and *C. laeta*, we propose that the population west of the Rio Negro and north of the Amazon River is conspecific with *C. tyrannina*. Based on the four traits we measured, we are unable to distinguish these birds morphometrically from *C. t. saturatior* or nominate *tyrannina*. Nominate *tyrannina*, which ranges from Colombia in the Cauca and upper Sinú valleys into the Amazonian lowlands on the Rio Vaupés and into Venezuela in the states of Bolivar and Amazonas and in northern Brazil on the upper Rio Negro (Peters 1951) (Fig. 1), differs from Guianan *C. t. saturatior* in having grayish rather than reddish brown auriculars in the females (Chubb 1918). Because this characterizes specimens that we examined from the west bank of the Negro from Manacapurú to Jucali and Tatú near the mouth of the Rio Uaupés (site 7, Fig. 1), we tentatively place this population in the nominate subspecies.

Having shown that the population of *C. tyrannina* north of the Rio Amazonas and east of the Rio Negro is not *laeta*, it remains to justify our use of *saturatior* for *C. tyrannina* in this area. *Cercomacra t. saturatior* Chubb 1918 was considered by Cory and Hellmayr (1924) to be limited to British Guiana. Peters (1951) expanded this range to eastern Venezuela on the upper Cuyuni River and to Suriname. Because the Manaus area is floristically (Prance 1990) and faunistically (Bierregaard et al. 1987; Stotz and Bierregaard 1989; Prum 1994; Cohn-Haft et al. 1997) similar to the Guianas, the Brazilian birds north of the Amazon and east of the Rio Negro could be referred to the Guianan *saturatior*.

Chubb's (1918) original diagnosis of *C. t. saturatior* stated that the subspecies differed from nominate *tyrannina* in being "almost black above, and very dark slate-color on the undersurface." In fact, the full series at AMNH of male *C. t. saturatior* from the Guianas, *C. tyrannina* from the north bank of the Amazon, and nominate *tyrannina* broadly overlap in their coloration (which is not very dark slate below, nor as black as *C. nigrescens* ventrally), with the majority of specimens showing a pronounced white wash to the belly and some scalloping. Typically, then, the male plumage does not appear to be useful in identifying the subspecies of the lower Amazonian *C. tyrannina*.

However, female *C. t. saturatior* are distinguished from nominate *tyrannina* by their bright chestnut rufous on the ventral surface extending into the cheeks (Chubb 1918). Our specimens from the Manaus region as well as specimens from eastern Roraima (Ilha de Maracá), other north-bank sites, and the state of Amapá (F. C. Novaes, pers. comm.) agree in this trait with female *saturatior*. Therefore, based on female plumage, we believe the range of *C. t. saturatior* extends into Brazil north of the Rio Amazonas westward to the Rio Negro and north to the Orinoco region of Venezuela. We were unable to find sufficient museum specimens to determine the western extent of the range of *C. t. saturatior*.

Several recordings of *C. tyrannina* from eastern Venezuela and the east bank of the Rio Negro in far western Brazil offer tantalizing clues as to the limits of the ranges of *C. t. saturatior* and nominate *tyrannina*. In northeastern Bolivar, Venezuela, close to the Guyanan border, Ted Parker recorded songs typical of *C. t. saturatior* at El Palmar (LNS 30576) (site 10, Fig. 1) and at Santa Elena (LNS 30576) (site 9, Fig. 1), whereas only a few dozens of km northeast of El Palmar, he recorded a pair of *C. tyrannina* with a noticeably slower song (LNS 34463). Additionally, Cohn-Haft recently recorded *C. tyrannina* on both sides of the Rio Negro near São Gabriel de Cachoeira (site 8, Fig. 1). His recordings from the west bank are similar to other recordings from Jaú National Park and Manacapurú, whereas a single (unarchived) recording from the east bank seems intermediate between the west bank recordings and "typical" *C. t. saturatior*. These recordings suggest that further study is needed to resolve the range limits of these species and that northeastern Bolivar, Venezuela, and the headwaters of the Rio Negro are areas likely to provide important information.

The ecology of C. laeta.—The natural history of *C. laeta* is not well known. Typically, *Cercomacra* antbirds form stable pair bonds, are strongly territorial and do not participate in mixed-species flocks. As do many insectivorous birds of the understory or forest edge, individual *C. tyrannina* will take advantage of army-ant swarms and forage for insects flushed by the ants when they move through the birds' territories but are not "obligate" ant followers (Willis 1985). Except for the Balbina birds, where the males were not obviously associated with females, all of our observations and those reported to us (M. and P. Isler, pers. comm.; A. Whittaker, pers. comm.; B. M. Whitney, pers. comm.) suggest that *C. laeta* is also territorial, not a regular participant in mixed-species flocks, nor an obligate army-ant follower. A more refined portrait of the species' ecology will require detailed field studies

In Belém, *C. laeta* is a relatively common bird of lower stages of várzea (flooded) forest or streamside habitat (D. C. Oren, pers. comm.; M. and P. Isler, pers. comm.). The birds' Portuguese name around Belém is *"mãe do igarapé,"* or "mother of the stream." However, near Belém, Bierregaard collected a male *C. l. laeta* along a dry roadside several hundred meters from any stream, in habitat typical of *C. tyrannina* in Manaus, so the relationship with riparian habitat is not obligatory. There have been no published studies of *C. l. sabinoi*, but, based on anecdotal reports, it seams to inhabit low, scrubby, often disturbed vegetation, at least sometimes near streams (E. O. Willis, in litt.; B. M. Whitney, pers. comm.).

Given its sympatry with *C. tyrannina,* the ecology of *C. l. waimiri* remains somewhat of a mystery. In the Manaus vicinity, *C. tyrannina* is a bird of the forest edge, typically found along road cuts or at the edge of large clearings where tall forest is fringed by moderately dense second growth, often dominated by *Cecropia* spp. about 3–10 m height. We suspect that *C. l. waimiri* may, as does nominate *laeta,* show a preference for stream- or river-edge forest. The clearing where we discovered the population at Balbina was contiguous with the shore of the reservoir, and thus the birds may have been displaced from their normal habitat by rising waters after the dam was closed. Consistent with this hypothesis is the lack of any *C. l. waimiri* around the edge of a hilltop clearing not contiguous with the lake, where *C. tyrannina* was present, about 1 km from the clearing where we discovered the male *C. l. waimiri.*

In three expeditions to the Balbina site, we never saw or heard anything that we could identify as a female *C. l. waimiri.* Most *Cercomacra* form tight pair bonds and are almost always found in pairs, singing together. At Balbina, the male *C. l. waimiri* were moving and singing alone. The lack of females at the Balbina site is further indication that the camp clearing may not have been the typical habitat for *C. l. waimiri.* The only other site near Manaus where *C. l. waimiri* has been seen was at a permanent camp clearing 47 km north of Manaus, near a white-sand "campina" formation. The birds were heard at this site in the late 1970s by both E. O. Willis (in litt.) and D. Oren (pers. comm.), but are no longer present. There is no river nearby from which these birds may have wandered, but a large stream wanders through the area and perhaps the scrubby vegetation of the campina is structurally similar to river-edge habitat.

In Roraima, Stotz found the birds in pairs in disturbed *terra firme* forest and forest edge, not associated with river edge. One area was in forest near a small patch of campina, similar to the Manaus clearing. The forest had been logged, and in all areas agricultural clearings were nearby. *C. tyrannina* was not present at any of the sites where Stotz found *C. l. waimiri.*

DISCUSSION

External characteristics offer no absolute distinction between these taxa. Plumage differences, especially in males, are very subtle to virtually non-existent in some allopatric populations of *Cercomacra tyrannina* and *C. laeta*. Indeed in some cases only the females can be readily distinguished, a phenomenon termed "heterogyny" by Hellmayr (1929). This, in fact, is the norm for the genus *Cercomacra* (e.g., Fitzpatrick and Willard 1990). Intraspecific and interspecific differences in the four morphological traits measured were subtle. Even when differences were statistically significant, the ranges of values often overlapped substantially.

To understand the relationships in this group, close attention to the birds' songs was required. It would have come as no surprise to Ted Parker that vocalizations were the key to unraveling the relationships between the antbird populations discussed in this paper. Ted was keenly aware of the importance of bird song in Neotropical avian systematics (Remsen and Schulenberg 1997).

Careful attention to vocalizations of tropical forest birds has revealed a number of such "new" taxa. Cohn-Haft (1993) discovered a new population of White-winged Potoo (*Nyctibius leucopterus*) by identifying the source of an unfamiliar nocturnal vocalization in the forests north of Manaus. Other cryptic species have been distinguished from very similar populations based on distinctive vocalizations (e.g., *Cymbilaimus sanctaemariae,* Pierpont and Fitzpatrick 1983; *Chamaeza meruloides,* Willis 1992; *Thamnophilus punctatus,* Isler et al. 1997; *Cranioleuca vulpina,* Zimmer 1997).

With increasing human pressure on the environment, it becomes ever more important to catalog accurately the planet's biodiversity. Adequate inventories of the biota will be invaluable as we decide which areas are most important to save and which can best suffer the brunt of our own species' expansion into currently less-disturbed habitats. Our study reminds us that unrecognized taxa occur even in well-studied areas. With so many cryptic species being discovered in birds, perhaps the best described group of organisms on earth, it is apparent that this "cryptic biodiversity" is clearly just the tip of an immense iceberg awaiting discovery.

ACKNOWLEDGMENTS

We thank the staff of Eletronorte and Brazil's National Institute for Research in Amazonia (INPA) for their cooperation and assistance during our expeditions to the Balbina hydroelectric project and Roraima. D. M. Teixeira kindly measured specimens of *Cercomacra laeta sabinoi* in the collection he curates (Setor de Ornitologia, Museu Nacional, Rio de Janeiro). The following curators kindly permitted us access to their collections: D. Oren and F. C. Novaes (Museu Goeldi, Belém), R. Paynter (Museum of Comparative Zoology, Harvard University), F. Gill (Academy of Natural Sciences, Philadelphia), M. LeCroy (American Museum of Natural History, New York). Greg Budney provided many recordings of *Cercomacra* from the Library of Natural Sounds, Cornell Laboratory of Ornithology. We thank the following for generously providing sound recordings: J. Vielliard and M. and P. Isler (*Cercomacra laeta* from near Belém); B. M. Whitney and J. F. Pacheco (*C. l. sabinoi* from northeastern Brazil); and A. Whittaker (*Cercomacra tyrannina* from Jaú National Park). F. C. Novaes was particularly generous with his time, sending photographs of and notes on specimens in his collection. K. Rosenberg focused our attention on the importance of this case as an example of cryptic biodiversity. L. Barden, G. Graves, M. and P. Isler, E. Morton, K. Rosenberg, and B. M. Whitney offered useful comments on earlier drafts. P. Isler shared very useful suggestions for the preparation of spectrograms. Financial support was provided by the Pew Charitable Trusts, the Smithsonian Institution, and many private contributors to the Biological Dynamics of Forest Fragments Project, administered jointly by the Smithsonian Institution and INPA. Stotz thanks P. Vanzolini for help in São Paulo and FAPESP for financial support. Cohn-Haft thanks D. Kroodsma and the Library of Natural Sounds for the loan of recording equipment, and Field Guides, Inc. for the opportunity to visit remote sites in Amazonia. This is publication number 158 in the BDFFP Technical Series.

LITERATURE CITED

ANON. 1994. Munsell Soil Color Charts. Kollmorgan Instruments Corp., New Windsor, New York.

BIERREGAARD, R. O., JR., D. F. STOTZ, L. H. HARPER, AND G. V. N. POWELL. 1987. Observations on the occurrence and behavior of the Crimson Fruit Crow (*Haematoderus militaris*) in central Amazônia. Bull. Brit. Ornith. Club 107:134–137.

CHARIF, R. A., S. MITCHELL, AND C. W. CLARK. 1995. Canary 1.2 User's Manual. Cornell Laboratory of Ornithology, Ithaca, New York.

CHUBB, C. 1918. Description of new forms of *Myrmophila, Rhopias, Cercomacra, Rhopoterpe, Hylopezus, Grallaricula, Furnarius, Lochmias.* Bull. Brit. Ornith. Club 38:83–87.

COHN-HAFT, M. 1993. Rediscovery of the White-winged Potoo (*Nyctibius leucopterus*). Auk 110:391–394.

COHN-HAFT, M., A. WHITTAKER, AND P. C. STOUFFER. 1997. A new look at the "species-poor" central Amazon: The avifauna north of Manaus, Brazil. Pp. 205–236 *in* Studies in Neotropical Ornithology Honoring Ted Parker (J. V. Remsen, Jr., Ed.), Ornithol. Monogr. No. 48.

CORY, C. B., AND C. E. HELLMAYR. 1924. Catalog of Birds of the Americas and Adjacent Islands in Field Museum of Natural History. Part III, Pteroptochidae–Conopophagidae–Formicariidae. Field Museum of Natural History, Zool. Series., Chicago.

FITZPATRICK, J. W., AND D. E. WILLARD. 1990. *Cercomacra manu,* a new species of antbird from southwestern Amazonia. Auk 107:239–245.

HELLMAYR, C. E. 1929. On heterogynism in Formicarian birds. J. Orn. 77(Suppl.):41–70.

ISLER, M. L., P. R. ISLER, AND B. M. WHITNEY. 1997. Biogeography and systematics of the *Thamnophilus punctatus* (Thamnophilidae) complex. Pp. 355–382 *in* Studies in Neotropical Ornithology Honoring Ted Parker (J. V. Remsen, Jr., Ed.), Ornithol Monogr. No. 48.

KROODSMA, D. E. 1984. Songs of the Alder Flycatcher (*Empidonax alnorum*) and Willow Flycatcher (*Empidonax traillii*) are innate. Auk 101:13–24.

LANYON, W. E. 1963. Experiments on species discrimination in *Myiarchus* flycatchers. Amer. Mus. Nov. 2126:1–16.

NAUMBURG, E. M. B. 1939. Studies of birds from eastern Brazil and Paraguay, based on a collection made by Emil Kaempfer. Bull. Amer. Mus. Nat. Hist. 76:231–276.

PETERS, J. L. 1951. Checklist of Birds of the World. Vol. VII. Harvard Univ. Press, Cambridge, Massachusetts.

PIERPONT, N., AND F. W. FITZPATRICK. 1983. Specific status and behavior of *Cymbilaimus sanctaemariae*, the Bamboo Antshrike, from southwestern Amazonia. Auk 100:645–652.

PINTO, O. M. O. 1939. Duas formas novas da avifauna de Pernambuco. Bol. Biol. (Nova série), Clube Zool. do Brasil 4:189–195.

PRANCE, G. T. 1990. The Floristic Composition of the Forests of Central Amazonian Brazil. Pages 112–140 *in* Four Neotropical Rainforests (A. H. Gentry, Ed.). Yale Univ. Press, New Haven, Connecticut.

PRUM, R. O. 1994. Species status of the White-fronted Manakin, *Lepidothrix serena* (Pipridae), with comments on conservation biology. Condor 96:692–702.

REMSEN, J. V., JR., AND T. S. SCHULENBERG. 1997. The pervasive influence of Ted Parker on Neotropical field ornithology. Pp. 7–19 *in* Neotropical Ornithology Honoring Ted Parker (J. V. Remsen, Jr., Ed.), Ornithol. Monogr. No. 48.

RIDGELY, R. S., AND G. TUDOR. 1994. The Birds of South America. Vol. II. Univ. of Texas Press, Austin, Texas.

RIDGWAY, R. 1911. The birds of Middle and North America, Part 5. Bulletin of the United States National Museum. No. 50.

STOTZ, D. F., AND R. O. BIERREGAARD, JR. 1989. The birds of the fazendas Porto Alegre, Dimona and Esteio north of Manaus, Amazonas, Brazil. Revista. Bras. Biol. 49:861–872.

TUKEY, J. W. 1977. Exploratory Data Analysis. Addison-Wesley, Reading, Massachusetts.

WILKINSON, L. 1990. SYSTAT: The System for Statistics. SYSTAT, Inc., Evanston, Illinois.

WILLIS, E. O. 1985. *Cercomacra* and related antbirds (Aves, Formicariidae) as army ant followers. Rev. Bras. Zool. 2:427–432.

WILLIS, E. O. 1992. Three *Chamaeza* antthrushes in eastern Brazil (Formicariidae). Condor 94:110–116.

ZIMMER, K. 1997. Species limits in *Cranioleuca vulpina*. Pp. 849–864 *in* Studies in Neotropical Ornithology Honoring Ted Parker (J. V. Remsen, Jr., Ed.), Ornithol. Monogr. No. 48.

APPENDIX 1
SPECIMENS EXAMINED

Cercomacra l. laeta: MPEG 20581, 22360, 22383, 22725, 22970, 23214, 23428, 29752, 31811, 31813, 33012, 37379, 43190; AMNH 430795, 430797, 430798, 430799, 430801, 430802, 430804, 491087. *Cercomacra laeta sabinoi*: MZUSP 18.122, 18.123, 39.063, 39.064, 39.065, 63.522, MN 32.049, 32.050, 32.051, 33.894, 33.895, 33.896, 33.897, 36.376. *Cercomacra laeta waimiri*: MPEG 52521 (type), 52522, 52523, Museu Nacional (Rio de Janeiro) MN 42847, Museu de Zoologia (São Paulo) MZUSP (DFS 92-252, DFS 92-251).

Cercomacra t. tyrannina: AMNH 121920, 434548, 448302, 448303, 460371, 460372, 460374, 460375, 460376, 460380, 460382. *Cercomacra tyrannina saturatior*: AMNH 125622, 125623, 125629, 125630, 125632, 128474, 176898, 176899, 248556, 248557, 248559, 248563, 248568, 248570, 248571, 309883, 309886, 309889; MPEG 16115, 16118, 16592, 20219, 20289, 24813, 28708, 29180, 29450, 39118, 39119, 39120, 39121, 39122, 40439, MPEG 52528, 52529, 52530, 52531, 52532, 52533, 52534, 52535, 52536; MN 42848, 42849, 42850, 42851. *Cercomacra tyrannina* from the upper Rio Negro: AMNH 310692, 310693, 310694, 310695, 434551, 434554. *Cercomacra tyrannina* from the Orinoco in Venezuela: 491088, 491091, 491092, 491094, 491095, 491097, 491098, 491099. *Cercomacra tyrannina* from n. Roraima: AMNH 824584. *Cercomacra tyrannina* (west of R. Negro): AMNH 312041, 312042, 312044, 312045, 312047, 312048, 312049, 312051, 312052, 312053, 312055, 312056, 312057, 312059, 312060, 312070, MPEG 52524, 52525, 52526, 52527.

Cercomacra nigrescens approximans: MCZ 174923, 174924, 174925, 174927, 174928, 174929, 174930, 174932, 174936; MPEG 39907, 39908.

APPENDIX 2
RECORDINGS EXAMINED

Cercomacra l. laeta: LNS 48156, 48229, 48230, ASN JV 149/1b. *Cercomacra l. sabinoi*: 1 recording by B. M. Whitney. *Cercomacra l. waimiri*: LNS 46949, 46950, 46951, 46952, 46953, 46954, 46957, 46961, 46967, ASN DFS 09/3.

Cercomacra tyrannina: West bank, Rio Negro: LNS 46847, 46893, 46895, 46931, 46941, 46943, 46988, 46991, 47009, 48555, 48556, 48557, 48558, 46847, A. Whittaker (unarchived), M. Cohn-Haft (unarchived); Suriname: LNS 11530, 11622, 25311, 25516, 26593; Manaus: 32413, 46852, 46853, 46855, 46859, 46864, 46865, 46965; São Gabriel (east bank, upper Rio Nego, Brazi); M. Cohn-Haft (unarchived); Bolivar, Venezuela: LNS 30468, 30576, 34463.

Cercomacra nigrescens: LNS 11511, 35618, 37284.

EVOLUTIONARY RELATIONSHIPS AMONG THE POTOOS (NYCTIBIIDAE) BASED ON ISOZYMES

ROBB T. BRUMFIELD,[1,2] DAVID L. SWOFFORD,[1] AND
MICHAEL J. BRAUN[1,2]

[1]*Laboratory of Molecular Systematics, National Museum of Natural History, Smithsonian Institution MSC MRC 534, Washington, D. C. 20560, USA; and*
[2]*Department of Zoology, University of Maryland, College Park, MD 20742, USA*

ABSTRACT.—Isozyme electrophoresis was used to assess genetic variation in potoos (Nyctibiidae), a distinctive Neotropical family of caprimulgiform nightbirds. Interspecific levels of genetic differentiation among potoos are extremely high (range of Nei's D = 0.191–1.172) and are comparable to intergeneric levels of differentiation in other bird families. In addition, levels of genetic differentiation between populations of both *Nyctibius grandis* and *N. griseus* from east and west of the Andes are comparable to the high genetic distances found in cross-Andes comparisons in other isozyme studies of Neotropical birds. These data suggest that extant potoo lineages are quite old, and that substantial genetic diversity exists in potoos that is not conveyed in the current taxonomy, in which most potoo species lack named or described intraspecific variation, and all species share a single genus.

Phylogenetic analysis of isozyme data supports the monophyly of Nyctibiidae through comparisons with outgroups from five other caprimulgiform families. Our results also support the monophyly of a clade composed of *Nyctibius maculosus*, *N. leucopterus*, and *N. griseus*, and confirm *maculosus* and *leucopterus* as sister taxa. The relationships of other potoos remain essentially unresolved, although there is weak support for the placement of *N. bracteatus* as the basal taxon. Relationships among caprimulgiform families are also essentially unresolved by these data, although there is some support for a clade composed of Caprimulgidae, Aegothelidae, and Eurostopodidae. The very high genetic distances from *Steatornis* to all other caprimulgiforms indicate that it represents the earliest branching lineage in the order.

RESUMEN.—Electroforesis de proteínas fue usado para estudiar la variación genética de los nictibios (Nyctibiidae), una familia Neotropical de aves nocturnas. Los niveles de diferencia entre las especies de nictibios son altos (Nei's D = 0.191 hasta 1.172), tan alto como los niveles que se encuentran entre los géneros en otras familias de aves. Los niveles de diferencia entre las dos especies con representantes en los dos lados de los Andes (el occidente y el oriente), *Nyctibius grandis* y *N. griseus*, están de acuerdo con esos observados en otros estudios de aves de quienes los representantes ocurren en los ambos lados de las montañas. Estos resultados sugieren que los lineajes evolucionarios de nictibios son antiguos y que bastante diversidad genetica existe que no se puede reconocer con la taxonomia corriente.

Un análisis filogenético de los datos indica que los nictibios pertenecen a un grupo monofilético y diferente de otros grupos de aves nocturnas. Los resultados tambien apoyan la existencia de un grupo monofilético compuesto de *N. maculosus*, *N. leucopterus* y *N. griseus*, en que las especies *maculosus* y *leucopterus* tienen una relación mas cercana que con *griseus*. Las relaciones evolucionarias de los otros nictibios no fueron resolvadas, sin embargo los datos sugieren que *N. bracteatus* es la especie mas antigua. Tampoco resolvamos las relaciones entre las familias caprimulgiformas, aunque los datos sugieren que Caprimulgidae, Aegothelidae y Eurostopodidae son consanguineos. La grande distancia genética que existe entre *Steatornis* y otras familias en el estudio, indica que *Steatornis* es la familia mas antigua de las aves nocturnas.

The potoos (Nyctibiidae) are an exclusively Neotropical family of nocturnal birds characterized by their distinctive mimicry of vertical tree stubs, upon which they often perch. Mimicry is achieved by pointing their bill upward, closing their eyes, and laying their tail flat along the branch (Sick 1993). Their cryptic behavior, nocturnal habits, and tropical distribution have made them one of the most poorly known groups of birds. However, recent fieldwork in South America has yielded new information on vocalizations and life history of several little-known taxa that clarifies species boundaries and provides new material for anatomical and molecular studies (Remsen and Traylor 1983; Schulenberg et al. 1984; Parker et al. 1985; Cohn-Haft 1993).

All potoos are currently united within the genus *Nyctibius* (Monroe and Sibley 1993). Most treatments of the family recognize five to seven potoo species (Chapman 1926; Peters 1940; Schulenberg et al. 1984; Sibley and Monroe 1990), depending on whether the Middle American form, *jamaicensis* (Northern Potoo), is given specific status or lumped within *griseus* (Common Potoo), and whether *maculosus* (Andean Potoo) is lumped within *leucopterus* (White-winged Potoo) (Schulenberg et al. 1984). Confusion concerning the specific status of *maculosus* began when Peters (1940) reduced it to a subspecies of *leucopterus*. This was presumably based on Chapman's (1926) conclusion that *maculosus* was an Andean representative of *leucopterus*, although Chapman did not explicitly state that he considered them conspecific. More recent analyses of morphology and voice strongly support the species status of *leucopterus* and *maculosus* (Schulenberg et al. 1984; Cohn-Haft 1993).

As with most Neotropical avian taxa, there have been few attempts to elucidate the evolutionary relationships within this family. The Northern Potoo has been hypothesized to be the sister taxon to *grandis* (Great Potoo) based on their vocalizations, rather than to the phenotypically similar Common Potoo (Davis 1978). Schulenberg et al. (1984) proposed that *maculosus* is a highland relative of *griseus*, not *leucopterus*, based on similarities in size and plumage. Mariaux and Braun (1996) recently performed a molecular phylogenetic survey of Nyctibiidae using DNA sequence data from the mitochondrial cytochrome *b* (cyt *b*) gene. They found evidence for a *maculosus-leucopterus* clade, confirming Chapman's (1926) early view. However, they were not able to fully resolve relationships among potoos, partly due to unexpectedly high levels of divergence among potoo cyt *b* sequences. To confirm and extend those observations, we examined nuclear genetic markers generated from electrophoresis of isozymes to perform a phylogenetic analysis of Nyctibiidae.

METHODS

We use the taxonomy of Sibley and Monroe (1993) as the most recent comprehensive treatment of Caprimulgiformes. All available frozen tissue samples of *Nyctibius* were obtained for this study ($n = 14$). Specimens examined in this study included one to three individuals of each currently recognized potoo species (Table 1), aside from *N. [griseus] jamaicensis*, of which no samples were available. One individual from each of five other caprimulgiform families was used as outgroups (Table 1). Because the SOD locus could not be resolved for *Podargus strigoides*, an individual of *Podargus papuensis* was scored for this locus.

Protein electrophoresis was performed on Titan III cellulose acetate plates (Helena Laboratories Inc.) according to methods described by Richardson et al. (1986). Tissue homogenates were prepared by grinding approximately 50 mg of heart, liver, and pectoral muscle in 500 µl of distilled water. The mixture was spun in a Brinkman 5415C Eppendorf centrifuge at 14,000 rpm for 2 min. The resulting supernatant was divided into 20 µl aliquots and frozen ($-80°C$) for subsequent electrophoretic analyses. Running conditions for all loci appear in Table 2. Electromorphs were coded alphabetically in order of relative mobility from the origin with the most anodally migrating allele as "a."

BIOSYS-1 (Swofford and Selander 1981) was used to compute Cavalli-Sforza and Edwards (1967) and Nei (1978) genetic distances (D), and to perform a UPGMA cluster analysis using the Cavalli-Sforza and Edwards (1967) chord distance. Phylogenetic analyses were performed using FREQPARS (available electronically via anonymous ftp from onyx.si.edu; see Swofford and Berlocher 1987) and PAUP (Swofford, 1993; the edition used was a prerelease version of PAUP* 4.0).

RESULTS

GENETIC VARIATION

Interspecific variation.—Levels and patterns of variation at 23 presumed genetic loci from 20 enzyme systems were resolved for all ingroup and outgroup taxa (Table 3). One locus represents an unknown dehydrogenase (UDH) that appeared as a lightly staining but well resolved locus on SORDH. Because three additional loci from three enzyme systems (GPT, ME-1, NP; Table 3) could not be fully resolved for all outgroups, they were not included in the phylogenetic analyses. Twenty-two of the 26 loci (85%) were variable among the potoos; the monomorphic loci were AK, GOT, GPI, and MDH-1. There were no monomorphic loci when outgroups were

TABLE 1

Specimens Examined in This Study. Specimen Numbers Refer to the Tissue Catalog and Not the Voucher Specimen. In Addition to Samples Available at the United States National Museum of Natural History (USNM), Tissue Samples Were Provided by the Frozen Tissue Collections of the Following Institutions (in Decreasing Order of Amount Borrowed): Louisiana State University Museum of Zoology (Baton Rouge; LSUMZ), Museum of Victoria (Australia; MV), Academy of Natural Sciences of Philadelphia (Philadelphia; ANSP)

Taxon/specimen number	Collector	Locality
NYCTIBIIDAE		
Nyctibius aethereus (NAET)		
LSUMZ B10877	A. S. Meyer	PERU: depto. Ucayali: SE. slope Cerro Tahuayo.
LSUMZ B11236	D. C. Schmitt	PERU: depto. Ucayali: SE. slope Cerro Tahuayo.
Nyctibius bracteatus (NBRA)		
LSUMZ B4509	S. W. Cardiff	PERU: depto. Loreto; Lower Río Napo region, E. bank Río Yanayacu, ca. 90 km N Iquitos.
LSUMZ B20270	M. Cohn-Haft	BRAZIL: Amazonas; Munic. Manaus, km 34 ZF-3, FAZ. Esteio, ca. 80 km N Manaus.
LSUMZ B20318	K. V. Rosenberg	BRAZIL: Amazonas; Munic. Manaus, km 41 ZF-3, Faz. Esteio, ca. 80 km N Manaus.
Nyctibius grandis (NGRA)		
USNM B3223	R. T. Brumfield	PANAMA: prov. Bocas del Toro; 6 km E Changuinola on road from Changuinola to Almirante.
LSUMZ B8954	C. G. Schmitt	BOLIVIA: depto. Pando; Nicolás Suarez, 12 km by road S Cobija, 8 km W on road to Mucden.
LSUMZ B15415	J. M. Bates	BOLIVIA: depto. Santa Cruz, Velasco, Pre-Parque Nacional: "Noel Kempff Mercado," 30 km E Aserradero Moira.
Nyctibius griseus (NGRI)		
USNM B3252	M. J. Braun	PANAMA: prov. Panamá; Chiva Chiva Rd.
ANSP B3238	F. Sornoza M.	ECUADOR: prov. Sucumbios; Imuya Cocha.
Nyctibius leucopterus (NLEU)		
LSUMZ B20267	M. Cohn-Haft	BRAZIL: Amazonas: Munic. Manaus; km 34 ZF-3, Faz. Esteio, ca. 80 km N Manaus.
LSUMZ B20315	M. Cohn-Haft	BRAZIL: Amazonas: Munic. Manaus; km 34 ZF-3, Faz. Esteio, ca. 80 km N Manaus.
LSUMZ B20319	M. Cohn-Haft	BRAZIL: Amazonas: Munic. Manaus; km 34 ZF-3, Faz. Esteio, ca. 80 km N Manaus.
Nyctibius maculosus (NMAC)		
LSUMZ B271	M. J. Braun	PERU: depto. Cajamarca; Lucuma on Sapalache-Carmen Trail.
LSUMZ B1825	T. S. Schulenberg	PERU: depto. Pasco; Santa Cruz, about 9 km SSE Oxapampa.
CAPRIMULGIDAE		
Chordeiles minor (CHOR)		
LSUMZ B5279	L. Hale	USA: Louisiana: Cameron Par.; Holly Beach, ¼ mi N Holly Beach Hwy.
EUROSTOPODIDAE		
Eurostopodus mystacalis (EURO)		
MV JWC 129	J. Wombey	AUSTRALIA: Australian Capital Territories; Canberra; 35°17'S, 149°08'E.
AEGOTHELIDAE		
Aegotheles cristatus (AEGO)		
MV C450	J. Wombey	AUSTRALIA: Queensland; Kroombit Tops; 24°26'S, 150°43'E.
PODARGIDAE		
Podargus papuensis		
MV C876	J. Wombey	AUSTRALIA: Queensland; Silver Plains; 13°59'S, 143°33'E.
Podargus strigoides (PODA)		
LSUMZ B8654	A. P. Capparella	AUDUBON ZOO, New Orleans, Louisiana.
STEATORNITHIDAE		
Steatornis caripensis (STEA)		
LSUMZ B7474	D. E. Willard	VENEZUELA: terr. Amazonas; Cerro de la Neblina Camp VII, 1,800 m.

TABLE 2
ENZYMES EXAMINED, BUFFERS USED, AND RUNNING TIME FOR EACH ENZYME

Enzyme (E.C. no.)	Abbreviation	Number of loci	Running buffer[a]	Running time (hr)[b]
Aconitate hydratase (4.2.1.3) (aconitase)	ACON	1 (anodal)[d]	C	1
Adenosine deaminase (3.5.4.4)	ADA	1	B	1
Adenylate kinase[c] (2.7.4.3)	AK	1	C	1
Alanine aminotranserase (2.6.1.2) (glutamate-pyruvate transaminase)	GPT	1	B	1
Aspartate aminotransferase[c] (2.6.1.1) (glutamate-oxaloacetate transaminase)	GOT	1 (anodal)[d]	B	1
Creatine kinase[c] (2.7.3.2)	CK	2	D	1
Esterase (α-napthyl acetate) (3.1.1.1)	EST	1	C	1
Fumarate hydratase[c] (4.2.1.2) (fumarase)	FUM	1	A	2
Glucose-phosphate isomerase[c] (5.3.1.9)	GPI	1	B	1.5
Glutathione reductase (1.6.4.2)	GSR	1	E	1
α-Glycerophosphate dehydrogenase (1.1.1.8) (glycerol-3-phosphate dehydrogenase)	αGPD	1	B	1.5
Guanine deaminase (3.5.4.3)	GDA	1	B	0.75
Isocitrate dehydrogenase[c] (1.1.1.42)	IDH	2	A	2
Lactate dehydrogenase (1.1.1.27)	LDH	2	A	2
Malate dehydrogenase[c] (1.1.1.37)	MDH	2	C	1
Malic enzyme (1.1.1.40) (NADP-malate dehydrogenase)	ME	2	B	1
Mannose phosphate isomerase[c] (5.3.1.8)	MPI	1	B	1
Phosphoglucomutase[c] (2.7.5.1)	PGM	2	B	1
6-Phosphogluconate dehydrogenase (1.1.1.44)	6PGD	1	B	1.5
Purine nucleoside phosphorylase (2.4.2.1)	NP	1	B	0.75
Peptidases (3.4.11)				
Leucine-alanine	LA	1	A	0.75
Leucine-glycine-glycine	LGG	1	A	1
Phenylalanine-proline	Phe-Pro	2	A	.75
Valine-leucine	VL	1	A	1
Pyruvate kinase[c] (2.7.1.40)	PK	1	A	1.5
Sorbitol dehydrogenase (1.1.1.14) (L-iditol dehydrogenase)	SORDH	1	C	1
Superoxide dismutase (1.15.1.1)	SOD	1 (anodal)[d]	A, B	1
Unknown dehydrogenase[c] (1.1.1.?)	UDH	1	C	1

[a] A = 0.01 M Citrate-phosphate (10 mM di-sodium hydrogen orthophosphate, 2.5 mM citric acid), pH 6.4, B = 0.02 M Phosphate (11.6 mM di-sodium hydrogen orthophosphate, 8.4 mM sodium di-hydrogen orthophosphate), pH 7.0, C = 0.05 M Tris-maleate (50 mM Tris, 20 mM maleic acid), pH 7.8, D = 0.015 M Tris-EDTA-borate-$MgCl_2$ (15 mM Tris, 5 mM di-sodium EDTA, 10 mM magnesium chloride, 5.5 mM boric acid), pH 7.8, E = 0.13 M Tris-EDTA-borate (130 mM Tris, 2.2 mM di-sodium EDTA, 6 mM sodium hydroxide, 71.3 mM boric acid), pH 8.9. Recipes found in Richardson et al. (1986).

[b] At 200 V or 7 mA.

[c] Conservative loci suggested for birds by Les Christidis (pers. comm.) with the exception of UDH, which was considered conservative based on the small number of alleles.

[d] Indicates direction of migration of scored locus. Most vertebrates have more than one locus for these enzyme systems.

included. Eleven additional loci (ACON-1&2, EST, LA-1&2, LDH-2, LGG, ME-2, PGM-2, PHEPRO-1, and SORDH) showed variation among the potoos, but either could not be fully resolved or exhibited uninterpretable variation for some species. When these additional loci are considered, 92% of the loci are variable within the potoos.

Genetic distances (Table 4) among potoo species are extremely high (average Nei's $D = 0.655$; range = 0.191–1.172). Likewise, genetic distances between potoo species and the outgroups are high, dramatically illustrated by the absence of shared alleles between *N. bracteatus* (Rufous Potoo) and *Aegotheles cristatus* (Tables 3 and 4). In fact, the smallest average genetic distance between the potoos and an outgroup is 2.031 with *Eurostopodus mystacalis*.

TABLE 3

Matrix of Allele Frequencies Used for Phylogenetic Analysis. Species Acronyms Are as in Table 1. Loci Followed by Asterisks Were Considered Conservative

Locus	Allele	NMAC	NLEU	NGRI	NAET	NGRA	NBRA	CHOR	EURO	AEGO	STEA	PODA
ADA	a	—	—	—	—	—	—	—	—	—	1.0	—
	b	—	—	—	—	—	—	—	1.0	—	—	1.0
	c	—	—	—	—	—	—	—	—	1.0	—	—
	d	1.0	0.833	1.0	1.0	1.0	—	0.5	—	—	—	—
	e	—	0.167	—	—	—	—	—	—	—	—	—
	f	—	—	—	—	—	1.0	0.5	—	—	—	—
AK*	a	—	—	—	—	—	—	—	—	—	—	1.0
	b	—	—	—	—	—	—	1.0	—	1.0	—	—
	c	1.0	1.0	1.0	1.0	1.0	1.0	—	1.0	—	1.0	—
CK-1	a	—	—	—	—	—	—	—	1.0	—	—	—
	b	1.0	—	1.0	—	—	1.0	—	—	—	—	—
	c	—	—	—	—	—	—	—	—	—	1.0	—
	d	—	1.0	—	—	—	—	—	—	—	—	—
	e	—	—	—	—	—	—	1.0	—	—	—	—
	f	—	—	—	—	—	—	—	—	1.0	—	—
	g	—	—	—	1.0	—	—	—	—	—	—	—
	h	—	—	—	—	—	—	—	—	—	—	1.0
	i	—	—	—	—	1.0	—	—	—	—	—	—
CK-2*	a	—	—	—	—	—	—	—	—	—	—	1.0
	b	—	—	—	—	—	—	—	—	1.0	—	—
	c	—	—	—	—	—	—	1.0	1.0	—	1.0	—
	d	—	—	—	—	—	1.0	—	—	—	—	—
	e	—	—	1.0	1.0	1.0	—	—	—	—	—	—
	f	1.0	1.0	—	—	—	—	—	—	—	—	—
FUM*	a	—	—	—	—	0.667	—	—	—	—	—	—
	b	—	—	—	—	0.333	—	—	—	—	—	—
	c	—	—	—	—	—	—	—	—	—	—	1.0
	d	—	—	—	—	—	—	—	—	1.0	—	—
	e	—	—	—	—	—	—	—	1.0	—	—	—
	f	—	—	—	—	—	—	1.0	—	—	—	—
	g	—	—	—	—	—	—	—	—	—	1.0	—
	h	1.0	1.0	1.0	—	—	1.0	—	—	—	—	—
	i	—	—	—	1.0	—	—	—	—	—	—	—
GDA	a	—	—	—	—	1.0	—	—	—	—	—	—
	b	—	—	1.0	—	—	—	—	—	—	—	—
	c	—	—	—	—	—	—	—	—	—	—	1.0
	d	1.0	—	—	—	—	—	—	—	—	—	—
	e	—	1.0	—	—	—	—	—	—	—	—	—
	f	—	—	—	—	—	—	1.0	—	—	—	—
	g	—	—	—	1.0	—	—	—	—	—	—	—
	h	—	—	—	—	—	1.0	—	—	—	—	—
	i	—	—	—	—	—	—	—	—	1.0	—	—
	j	—	—	—	—	—	—	—	—	—	1.0	—
	k	—	—	—	—	—	—	—	—	1.0	—	—
GOT*	a	—	—	—	—	—	—	—	—	—	—	1.0
	b	—	—	—	—	—	—	1.0	1.0	1.0	1.0	—
	c	1.0	1.0	1.0	1.0	1.0	1.0	—	—	—	—	—
αGPD	a	—	—	—	—	—	—	—	1.0	—	—	—
	b	—	—	—	—	—	—	—	—	1.0	—	—
	c	—	—	—	—	—	0.333	—	—	—	—	—
	d	1.0	0.667	1.0	—	1.0	0.667	—	—	—	—	—
	e	—	—	—	—	—	—	—	—	—	1.0	—
	f	—	0.333	—	1.0	—	—	—	—	—	—	—
	g	—	—	—	—	—	—	—	—	—	—	1.0
	h	—	—	—	—	—	—	1.0	—	—	—	—

TABLE 3
Continued

Locus	Allele	NMAC	NLEU	NGRI	NAET	NGRA	NBRA	CHOR	EURO	AEGO	STEA	PODA
GPI*	a	—	—	—	—	—	—	—	—	0.5	—	—
	b	—	—	—	—	—	—	1.0	—	—	—	—
	c	—	—	—	—	—	—	—	1.0	0.5	—	—
	d	—	—	—	—	—	—	—	—	—	1.0	1.0
	e	1.0	1.0	1.0	1.0	1.0	1.0	—	—	—	—	—
GSR	a	—	—	1.0	1.0	1.0	—	—	—	—	—	—
	b	—	1.0	—	—	—	—	—	—	—	—	—
	c	—	—	—	—	—	—	—	—	—	1.0	—
	d	1.0	—	—	—	—	—	—	—	—	—	—
	e	—	—	—	—	—	—	1.0	—	—	—	—
	f	—	—	—	—	—	—	—	1.0	—	—	—
	g	—	—	—	—	—	1.0	—	—	—	—	—
	h	—	—	—	—	—	—	—	—	1.0	—	—
	i	—	—	—	—	—	—	—	—	—	—	1.0
IDH-1	a	—	—	0.25	—	0.667	—	—	—	—	—	—
	b	—	—	—	—	—	0.167	—	—	—	—	—
	c	1.0	1.0	0.75	1.0	0.333	0.833	—	—	—	—	—
	d	—	—	—	—	—	—	1.0	—	—	1.0	—
	e	—	—	—	—	—	—	—	—	0.5	—	—
	f	—	—	—	—	—	—	—	1.0	—	—	—
	g	—	—	—	—	—	—	—	—	—	—	1.0
	h	—	—	—	—	—	—	—	—	0.5	—	—
IDH-2*	a	—	—	—	—	—	—	—	—	0.5	1.0	—
	b	—	—	—	—	—	—	—	0.5	—	—	—
	c	1.0	1.0	1.0	—	—	—	—	—	—	—	—
	d	—	—	—	—	—	—	1.0	—	0.5	—	0.5
	e	—	—	—	—	—	—	—	0.5	—	—	—
	f	—	—	—	1.0	1.0	1.0	—	—	—	—	0.5
LDH-1	a	—	—	—	—	—	—	—	—	1.0	—	—
	b	1.0	1.0	1.0	—	—	—	—	—	—	1.0	—
	c	—	—	—	1.0	—	—	—	1.0	—	—	—
	d	—	—	—	—	1.0	—	—	—	—	—	—
	e	—	—	—	—	—	1.0	1.0	—	—	—	—
	f	—	—	—	—	—	—	—	1.0	—	—	—
	g	—	—	—	—	—	—	—	—	—	—	1.0
MDH-1*	a	—	—	—	—	—	—	—	—	1.0	1.0	1.0
	b	1.0	1.0	1.0	1.0	1.0	1.0	—	—	—	—	—
	c	—	—	—	—	—	—	—	1.0	—	—	—
	d	—	—	—	—	—	—	1.0	—	—	—	—
MDH-2*	a	—	—	—	—	—	—	—	—	—	—	1.0
	b	1.0	1.0	1.0	1.0	0.833	—	1.0	1.0	1.0	1.0	—
	c	—	—	—	—	0.167	1.0	—	—	—	—	—
MPI*	a	—	—	—	—	—	1.0	—	—	—	—	—
	b	—	—	—	—	—	—	—	—	1.0	—	—
	c	—	—	—	0.25	0.667	—	—	—	—	—	—
	d	—	—	0.25	0.75	—	—	—	—	—	—	—
	e	—	—	—	—	0.333	—	—	—	—	—	—
	f	1.0	1.0	0.75	—	—	—	—	—	—	—	—
	g	—	—	—	—	—	—	—	—	—	0.5	—
	h	—	—	—	—	—	—	1.0	1.0	—	—	—
	i	—	—	—	—	—	—	—	—	—	0.5	—
	j	—	—	—	—	—	—	—	—	—	—	1.0
6PGD	a	—	—	—	0.25	—	—	—	—	—	—	—
	b	—	—	—	—	—	—	—	1.0	—	—	—
	c	1.0	—	—	—	—	—	—	—	—	—	—
	d	—	—	—	—	—	—	—	—	—	—	1.0

TABLE 3
Continued

Locus	Allele	NMAC	NLEU	NGRI	NAET	NGRA	NBRA	CHOR	EURO	AEGO	STEA	PODA
	e	—	0.833	1.0	0.75	1.0	—	—	—	—	—	—
	f	—	—	—	—	—	—	1.0	—	1.0	—	—
	g	—	—	—	—	—	1.0	—	—	—	—	—
	h	—	0.167	—	—	—	—	—	—	—	—	—
	i	—	—	—	—	—	—	—	—	—	1.0	—
PGM-1*	a	—	—	—	—	—	1.0	—	—	—	—	1.0
	b	—	—	—	—	—	—	1.0	1.0	1.0	—	—
	c	1.0	1.0	1.0	1.0	1.0	—	—	—	—	—	—
	d	—	—	—	—	—	—	—	—	—	1.0	—
PHE-PRO-2	a	—	—	—	—	—	—	1.0	1.0	—	—	—
	b	—	—	—	—	1.0	—	—	—	—	—	—
	c	—	—	—	—	—	—	—	—	—	—	1.0
	d	—	—	—	—	—	—	—	—	1.0	—	—
	e	—	—	—	—	—	1.0	—	—	—	—	—
	f	1.0	1.0	1.0	1.0	—	—	—	—	—	1.0	—
PK*	a	—	—	—	—	—	—	—	1.0	—	—	—
	b	—	—	—	—	—	—	—	—	—	—	1.0
	c	1.0	1.0	1.0	—	—	—	1.0	—	—	—	—
	d	—	—	—	—	—	1.0	—	—	—	—	—
	e	—	—	—	—	1.0	—	—	—	1.0	1.0	—
	f	—	—	—	1.0	—	—	—	—	—	—	—
SOD	a	—	—	—	—	—	—	—	—	0.5	1.0	—
	b	—	—	—	1.0	—	—	—	—	—	—	—
	c	—	—	—	—	—	—	1.0	—	—	—	—
	d	—	—	—	—	—	—	—	1.0	0.5	—	—
	e	1.0	1.0	1.0	—	1.0	—	—	—	—	—	—
	f	—	—	—	—	—	1.0	—	—	—	—	—
	g	—	—	—	—	—	—	—	—	—	—	1.0
UDH*	a	—	—	—	—	—	—	—	—	—	—	1.0
	b	—	—	—	—	—	—	—	—	1.0	—	—
	c	1.0	1.0	1.0	1.0	1.0	1.0	1.0	1.0	—	—	—
	d	—	—	—	—	—	—	—	—	—	1.0	—
VL	a	—	—	—	—	—	—	—	—	1.0	—	—
	b	—	—	—	—	—	—	1.0	—	—	—	—
	c	—	—	—	—	—	—	—	—	—	1.0	—
	d	—	—	—	—	—	—	—	—	—	—	1.0
	e	—	—	0.25	—	—	—	—	—	—	—	—
	f	—	—	—	1.0	—	—	—	1.0	—	—	—
	g	1.0	1.0	0.75	—	1.0	1.0	—	—	—	—	—

The small sample sizes may have resulted in some inflation of genetic distances because lower frequency alleles shared between taxa were unlikely to be detected. To assess the contribution of small sample size to genetic distance, we calculated pairwise genetic distances between individuals within a species. In this analysis, the average genetic distance among individuals within species in our data set (see Appendix) was 0.04 over all possible comparisons ($n = 12$). If comparisons among populations of *griseus* and *grandis* separated by the Andes are eliminated from the analysis (see *Intraspecific Variation* section below), the average genetic distance is 0.02. These results illustrate that the extremely high genetic distances found among potoo species are not simply an artifact of small sample size.

Examination of the genetic distance matrix reveals two striking patterns (Table 4). First, *griseus* consistently has among the lowest genetic distances to all other potoo taxa, a pattern also revealed in the matrix of raw distances generated from cyt *b* sequences (Mariaux and Braun 1996). Secondly, *bracteatus* consistently has the highest genetic distance to all other potoo taxa. These patterns may reflect lineage-specific reduction or acceleration in evolutionary rate. An alternative explanation for the high distances to *bracteatus* is that it represents the basal potoo

TABLE 4

Genetic Distance Matrix Based on 23 Loci Resolved for All Taxa. Upper Matrix, Nei (1978) Unbiased Genetic Distance; Lower Matrix, Cavalli-Sforza and Edwards (1967) Chord Distance. Nei's Genetic Distance between NBRA and AEGO Was Unobtainable Because They Shared No Alleles. The Cavalli-Sforza and Edwards Genetic Distance between NBRA and AEGO Is Not 1.0 Because of Correction Factors. For Acronyms Refer to Table 1

Taxon	NMAC	NLEU	NGRI	NAET	NGRA	NBRA	CHOR	EURO	AEGO	PODA	STEA
NMAC	******	0.191	0.198	0.811	0.679	0.865	1.872	2.026	3.090	2.026	3.125
NLEU	0.388	******	0.208	0.708	0.622	0.976	1.869	1.999	3.063	1.999	3.098
NGRI	0.394	0.410	******	0.539	0.419	0.881	1.838	1.992	3.056	1.992	3.091
NAET	0.677	0.643	0.589	******	0.578	1.172	2.186	1.716	3.068	2.004	3.795
NGRA	0.637	0.619	0.541	0.601	******	0.979	2.233	2.039	2.440	2.039	3.774
NBRA	0.684	0.713	0.690	0.753	0.709	******	2.189	2.412	∞∞∞∞	3.105	2.700
CHOR	0.825	0.826	0.825	0.846	0.847	0.846	******	1.168	1.374	1.727	3.807
EURO	0.840	0.840	0.840	0.818	0.841	0.860	0.751	******	1.693	1.727	3.114
AEGO	0.881	0.881	0.881	0.881	0.862	0.900	0.781	0.809	******	1.470	2.856
PODA	0.840	0.840	0.840	0.840	0.841	0.881	0.818	0.818	0.787	******	2.420
STEA	0.881	0.881	0.881	0.886	0.886	0.866	0.886	0.881	0.870	0.860	******

lineage, and simply has not shared a recent common ancestor with the other extant potoo taxa. Regardless, these patterns are indicative of rate variation, and the distance-based methods of phylogenetic inference used herein that assume rate constancy (i.e., UPGMA) should be interpreted with caution.

Intraspecific variation.—Nine additional loci (ACON-1, EST, LA-1, LDH-2, LGG, ME-2, PGM-2, PHEPRO-1, and SORDH) were scored for the analysis of intraspecific differentiation within *griseus*, *grandis*, and *bracteatus* for a total of 34 or 35 loci (EST was not resolved for *griseus*). The genetic distance (Nei 1978) between samples of *griseus* on opposite sides of the Andes was 0.131 based on 34 loci (0.034 for the 23 loci used in phylogenetic analyses; Tables 3 and 4). *Nyctibius grandis* had a similar across-Andes genetic distance of 0.188 based on 35 loci, but the distance based on 23 loci (Nei's $D = 0.101$) was considerably higher than that of *griseus*. Levels of genetic differentiation between populations on opposite sides of the Andes are consistently higher than levels of genetic differentiation between populations on the same side of the Andes (0.188 versus 0.010 for *grandis*). These values are quite high for intraspecific comparisons in birds, but are consistent with levels of genetic differentiation across the Andes that have been found in a taxonomically diverse array of avian species (Brumfield and Capparella 1996). It is noteworthy that the magnitude of genetic divergence between cross-Andean populations of *grandis* approaches that separating *maculosus*, *leucopterus*, and *griseus* (Table 4).

The two populations of *bracteatus*, between which the Río Negro represents the largest potential barrier to dispersal, had a genetic distance of 0.035 based on 35 loci. Although this value is relatively low, it suggests that some genetic differentiation exists among populations of *bracteatus* because the genetic distance between two individuals from the same population was 0.010. Analysis of more specimens will be necessary to determine if significant genetic structure exists in *bracteatus*.

PHYLOGENETIC ANALYSIS

Phylogenetic analyses were performed on the allele-frequency data shown in Table 3. Only phylogenetically informative loci consistently scoreable in all taxa were included. We chose the FREQPARS method (Swofford and Berlocher 1987) as our primary method of analysis. FREQPARS implements a parsimony method for polymorphic character data that assigns (for any given tree) a set of ancestral allele-frequency arrays that minimize the total amount of frequency change implied by the reconstruction, with change measured in terms of Manhattan distance between nodes (the "MANAD" criterion). Although our sample sizes are small, methods of analysis that incorporate frequency information are, in fact, less sensitive to sampling error than coarser "presence-absence" coding strategies (Swofford and Berlocher 1987).

The FREQPARS program has limited searching capabilities and is best used by evaluating user-defined trees that cover the range of trees likely to be optimal. Because over 34 million unrooted trees are possible for the 11 taxa included in our study, evaluation of all possible trees was impractical. However, only 99,225 unrooted trees are consistent with monophyly of the six potoo taxa. We evaluated all of these trees using FREQPARS; this strategy guarantees finding the optimal tree(s) assuming only that the potoos are monophyletic (see Potoo monophyly section below). The input treefile for this analysis was constructed by creating a dummy data matrix containing a single uninformative character and performing an exhaustive search using the "topological constraints" feature of PAUP with the "collapse zero-length branches" option deselected. The resulting trees were then exported in FREQPARS format.

Potoo relationships.—A single most-parsimonious tree resulted (219.668 "steps" or units of allele-frequency change) from the FREQPARS analysis, placing *aethereus* as the basal potoo taxon (Fig. 1A). On this tree, a *grandis-bracteatus* clade is sister to a clade composed of *maculosus*, *leucopterus*, and *griseus*, with *maculosus* and *leucopterus* appearing as sister taxa within the latter clade. The cost of rejecting either the *maculosus-leucopterus-griseus* or the *maculosus-leucopterus* clade is $222.268 - 219.668 = 2.500$ steps, equivalent to 1.25 allelic substitutions. The five next-most-parsimonious trees were less than one-half step longer than the most parsimonious (each 220.002 steps) and should probably be treated as equally parsimonious given the small sample sizes. All six trees agree on the ([*maculosus*, *leucopterus*], *griseus*) relationship. However, basal to this clade, all permutations of potoo relationships occur, with the exception that *grandis* never appears as the basal taxon.

The extremely high levels of genetic divergence found among the potoos raise the possibility that the true phylogenetic signal might be obscured by high rates of substitution at loci evolving too rapidly to provide reliable information. Consequently, we performed a second FREQPARS

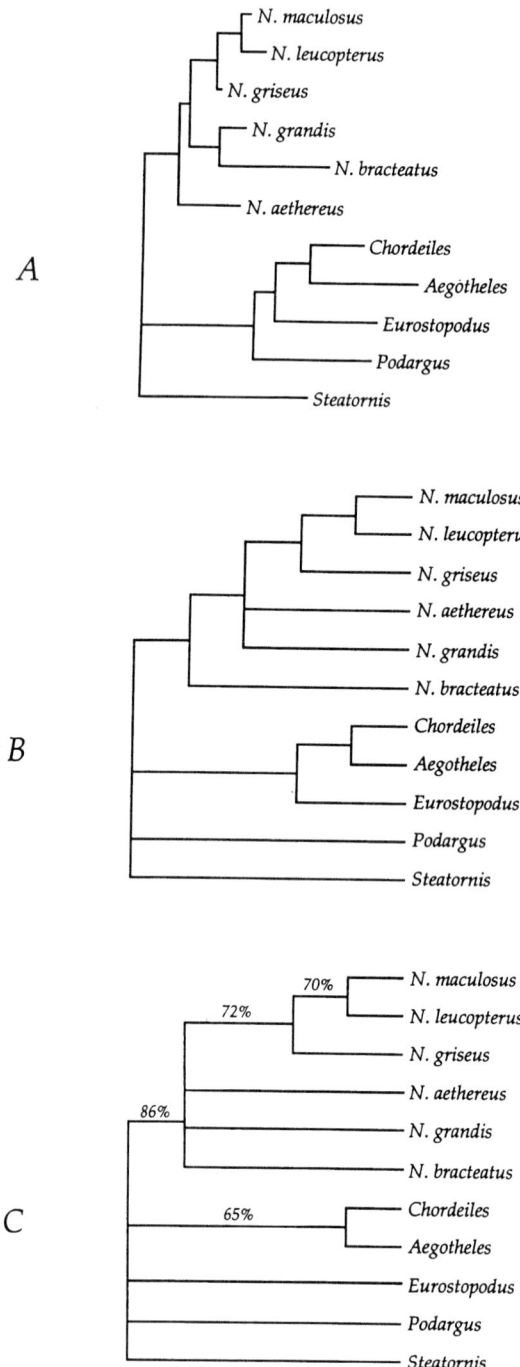

FIG. 1. Cladograms of caprimulgiform relationships based on parsimony (FREQPARS) analysis of allele-frequency data from isozyme loci (Table 3). The potoos were constrained to be monophyletic, and the trees are drawn rooted to *Steatornis* based on its large genetic distance to all other taxa (Table 4). A) Most-parsimonious tree (219.668 steps) of caprimulgiform relationships based on 23 isozyme loci (Table 3). Branch lengths reflect amount of allele-frequency change (measured as Manhattan distance) between each pair of nodes. B) Strict consensus of nine most-parsimonious trees (97.020 steps) based on data from 12 conservative loci (Table 3). Branch lengths do not reflect genetic distance. C) 50% Majority-rule consensus tree resulting from bootstrap analysis under MANOB criterion. Values shown on branches reflect the percentage of 1,000 bootstrap replicates in which the corresponding clade was found.

analysis in which only 12 loci considered to be conservatively evolving in birds (Table 3) were included. Nine equally parsimonious trees (97.020 steps) were found; the strict consensus (also the majority rule consensus) of these nine trees is shown in Figure 1B. The trees from the conserved-loci-only analysis preserved most aspects of the all-loci analysis, except that *bracteatus* consistently appears as the basal potoo taxon, a relationship supported by an apparent PGM-1^a synapomorphy in the remaining five potoo taxa.

Taken jointly, we interpret the FREQPARS analyses as supporting a terminal clade composed of *maculosus*, *leucopterus*, and *griseus* based on the synapomorphic alleles MPI^f, IDH-2^c, LDH-1^b, and PK^c. The presence of some of these alleles in other taxa is most parsimoniously interpreted as homoplasy, perhaps due to coincident migration of nonidentical alleles. The monophyly of *maculosus* plus *leucopterus* is supported by the shared allele CK-2^f. These analyses failed to resolve definitively whether *aethereus* or *bracteatus* is the basal potoo taxon, but suggest that *grandis* is not. Although distance analyses (see below) provide weak support for a *grandis-aethereus* relationship, no allelic synapomorphies for such a grouping are evident in the data matrix, and FREQPARS analysis rejects this relationship. The only unambiguous synapomorphy for the *grandis* plus *bracteatus* clade is the allele MDH-2^c, which occurs at low frequency in *grandis*; evidence for this clade should therefore be regarded as tentative in light of the small sample sizes.

Potoo monophyly.—The analyses described above do not directly address the question of whether the six potoo species constitute a monophyletic group, as potoo monophyly was assumed. To evaluate the evidence for potoo monophyly, we used a technique (Berlocher and Swofford in press; available in PAUP* 4.0) that obtains an exact solution to the "MANOB" criterion of Swofford and Berlocher (1987), which is a good approximation to the MANAD criterion. MANOB requires that allele-frequency arrays assigned to each internal node of the tree (hypothetical ancestral taxa) be chosen from the set of allele-frequency arrays observed in the terminal taxa. In this analysis, each unique allele-frequency array is treated as a character-state, and "stepmatrices" are created in which the cost of transformation between any pair of states is the Manhattan distance between the allele-frequency arrays represented by the two states. The generalized parsimony (Sankoff and Rousseau 1975; Swofford and Maddison 1992) algorithms available in PAUP can then be used for tree searches, overcoming the limitations of FREQPARS.

An unconstrained search using MANOB found the same most-parsimonious tree and tree-length as the exact FREQPARS analysis, demonstrating that in this case MANOB's approximation to MANAD is perfect. By performing a constrained search in PAUP, we found that the shortest tree incompatible with potoo monophyly required 223.668 steps for the full set of loci. Thus, the cost of rejecting potoo monophyly is 223.668–219.668 = 4.0 steps, equivalent to two complete allelic substitutions. The comparable analysis using the conserved-locus set yields a cost of rejecting potoo monophyly of 100.668–97.020 = 3.648 steps. In view of these results and the uniformly greater genetic distances between potoos and non-potoos than within potoos, we believe that potoo monophyly is reasonably well-supported by the allozyme data.

Support for potoo monophyly, as well as for relationships within potoos, was also evaluated with the bootstrap procedure (Felsenstein 1985), using the PAUP* MANOB approximation. The bootstrap proportions shown in Figure 1C reflect the extent to which these groupings might be supported by an independent sample of loci, although in many cases they provide conservative estimates of the probability that each group represents a true phylogenetic clade (Hillis and Bull 1993). The bootstrap results indicate reasonably strong support for potoo monophyly, with somewhat weaker support for the (*maculosus*, *leucopterus*) and (*griseus*, (*maculosus*, *leucopterus*)) clades.

Caprimulgiform relationships.—*Steatornis* was used to root the FREQPARS analysis of all taxa. Although *Steatornis* is not an unequivocal candidate for the basal caprimulgiform, the high genetic distances between it and all other taxa (Table 4) make it the best *ad hoc* outgroup. In addition, a weighted parsimony analysis of cyt *b* sequence data, including *Gallus gallus* as an outgroup, placed *Steatornis* as the basal caprimulgiform (Mariaux and Braun 1996). In the FREQPARS analysis of all loci (Fig. 1A), the most parsimonious tree joins *Chordeiles* and *Aegotheles* as sister taxa based on the shared allele AK^a. This is inconsistent with the UPGMA and minimum evolution analyses (see below), in which *Chordeiles* and *Eurostopodus* are sister taxa. There are no unambiguous synapomorphies that support *Chordeiles* and *Eurostopodus* as sister taxa, and constraining *Chordeiles* and *Eurostopodus* to be monophyletic requires an additional 2.67 steps relative to the most parsimonious FREQPARS topology (222.668 steps). With respect to the non-potoo taxa, the analysis of the conserved-loci-only data set is consistent with the all-loci data

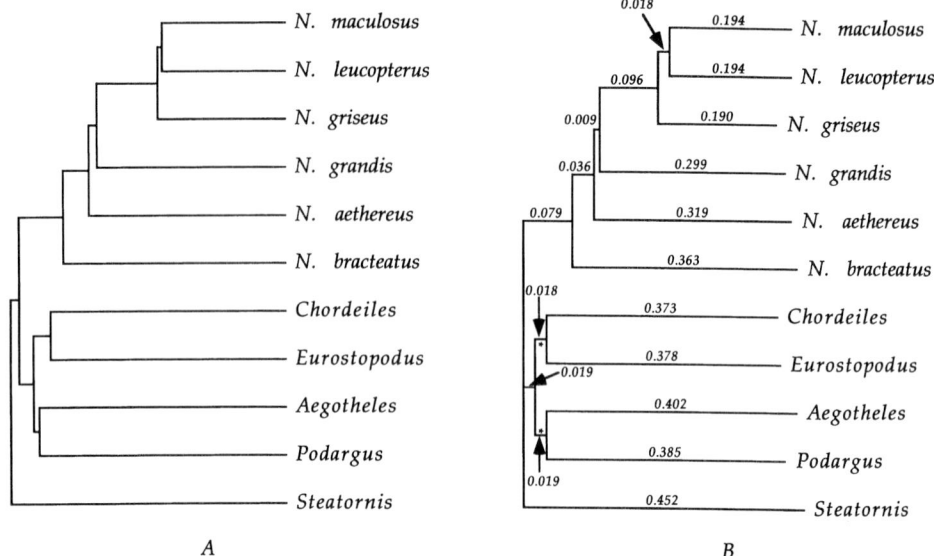

FIG. 2. Distance analyses of caprimulgiform relationships based on chord distances of Cavalli-Sforza and Edwards (1967) derived from allele frequencies at 23 isozyme loci. A) UPGMA phenogram. B) Optimal minimum evolution tree (Kidd and Sgaramella-Zonta 1971, Rzhetsky and Nei 1992) found by branch-and-bound search under the constraint of potoo monophyly.

set. However, the grouping of *Podargus* as a sister taxon to the *Chordeiles-Aegotheles-Eurostopodus* clade is not unequivocally supported, and is therefore absent from the consensus of the nine equally parsimonious trees (Fig. 1B). The *Chordeiles-Aegotheles* clade is the only one for which bootstrap support exceeds 50% (Fig. 1C).

Distance analyses.—The data of Table 3 were also analyzed using several distance methods for comparison. The results of a UPGMA cluster analysis are shown in Figure 2A. Figure 2B shows the optimal tree under the "minimum evolution" criterion of Rzhetsky and Nei (1992) (originally described as "LS-length" by Kidd and Sgaramella-Zonta 1971). This tree was found by a branch-and-bound search under the constraint of potoo monophyly and was also found in every one of 100 unconstrained heuristic searches using the random-addition-sequence option of PAUP*. The score for the minimum-evolution tree is slightly better (3.837 vs. 3.842) than the tree found by the neighbor-joining method (Saitou and Nei 1987) (not shown), and one additional tree had a score better than the neighbor-joining tree. All of these contain a *Chordeiles-Eurostopodus* clade absent from the FREQPARS trees. Except for this discrepancy (addressed above), all trees from the distance analyses are generally consistent with the tree from the conserved-loci FREQPARS analysis, differing only in rearrangements around the short branches indicated by asterisks in Figure 2B. In particular, *bracteatus* is consistently the sister taxon to the remaining potoos, and the (*griseus*, (*maculosus*, *leucopterus*)) clade is consistently present.

DISCUSSION

Intraspecific geographic variation.—Although our sample sizes are small, the genetic distances found between populations of *griseus* and *grandis* from opposite sides of the Andes are clearly large, indicating that significant geographic variation does exist in these taxa, and supporting the notion that the Andes have played an important role in the diversification of Neotropical birds restricted to the humid lowlands (Chapman 1917; Haffer 1967; Brumfield and Capparella 1996). It is unclear from the literature whether the populations of *griseus* we sampled from opposite sides of the Andes are currently united within the same subspecies (Chapman 1926; Peters 1940; Meyer de Schauensee 1964). All populations of *grandis* that we sampled are recognized as the same subspecies (Land and Schultz 1963). The isozyme data suggest that a careful examination of morphological characters and vocalizations may reveal significant differences between cis- and trans-Andean populations of both *griseus* and *grandis*.

Relationships among potoos.—The large mouth and eyes, gray-brown vermiculated plumage, and distinctive body form give most potoos a common general appearance that likely explains why they have traditionally been classified in a single genus. *Nyctibius bracteatus*, however, is strikingly different, with unvermiculated rufous plumage and pronounced white spots on the wing coverts and flanks that make it appear like *Steatornis*. However, the isozyme data demonstrate that not only are the potoos monophyletic, but they have probably not recently shared a common ancestor with another caprimulgiform.

The fossil record provides little help in resolving relationships and divergence dates within Nyctibiidae, although a fossil of *griseus* from the Pleistocene (20,000 years ago) of coastal Brazil confirms that diversification events within *Nyctibius* are at least as old as the late-Pleistocene (Brodkorb 1971). In addition, the high degree of overlap in current potoo distributions obfuscates any insight into evolutionary relationships based on biogeography.

The isozyme data strongly support the monophyly of a terminal clade composed of *maculosus*, *leucopterus*, and *griseus*, based on synapomorphic alleles at MPI, IDH-2, LDH-1, and PK. Genetic distances among these three species are remarkably similar, indicating a relatively rapid radiation among them. Interestingly, cyt *b* sequence divergences among these three taxa were also similar, although the taxonomic relationships of *griseus* remained unresolved by parsimony analyses of those data (Mariaux and Braun 1996). The similarity in genetic distances among these three species highlights the controversy that has surrounded their relationships (Chapman 1926; Peters 1940; Schulenberg et al. 1984; Cohn-Haft 1993; Mariaux and Braun 1996).

All analyses (FREQPARS, UPGMA, minimum evolution, neighbor-joining) performed on the isozyme data indicate that *maculosus* and *leucopterus* are sister taxa (Figs. 1–2). However, close examination of the data reveals that the monophyly of *maculosus* and *leucopterus* rests largely on a single synapomorphy at the CK-2 locus, assuming monophyly of the *maculosus-leucopterus-griseus* clade. Polymorphisms present in *griseus* and not in *maculosus* and *leucopterus* (IDH-1, MPI, and VL) also contribute to the clustering of *maculosus* and *leucopterus* in the analyses, yet the small available sample size renders allele-frequency estimates imprecise. Increased sample sizes of both individuals and loci should help in verifying relationships within this clade. Although the possibility of rate deceleration in *griseus* further complicates interpretation of the genetic distance data, such a deceleration may actually be responsible for the difficulty in placing *griseus*, because it would tend to make *griseus* appear more closely related to *maculosus* and *leucopterus* than it really is. On the whole, the isozymes provide weak support for the sister taxon relationship of *maculosus* and *leucopterus*. This same grouping is strongly supported by cyt *b* sequence data, however (Mariaux and Braun 1996). Taken together, we believe the genetic data provide a reasonably firm resolution of this trichotomy in favor of a *maculosus-leucopterus* clade, as originally proposed by Chapman (1926).

Of the 23 loci in which two or more potoo species shared an allele, *bracteatus* possessed a unique allele at 12 of those loci. This level of divergence suggests that either *bracteatus* represents the basal branch of the potoo lineage or it has undergone an accelerated rate of evolution relative to the other potoos (e.g., Fig. 1A). The presence of a synapomorphic allele for all other potoos at PGM-1 (see also ADA and MDH-2) favors the former interpretation. Although the FREQPARS analysis of all loci was unable to clearly resolve whether *bracteatus* or *aethereus* represents the basal potoo taxon, the FREQPARS analysis of conservative loci and all distance analyses placed *bracteatus* in that position. We conclude that the isozymes provide some evidence for the placement of *bracteatus* as the basal potoo, but consider this conclusion tentative based on these data alone. Unfortunately, the cyt *b* data did not elucidate the placement of *bracteatus* (Mariaux and Braun 1996).

The relationships of *aethereus* and *grandis* also remain problematic. It seems unlikely, however, that *grandis* is the basal potoo. The FREQPARS analysis of all loci suggested that *aethereus* may represent the basal lineage, but this relationship was unsupported in the analysis of conservative loci. Of the 23 loci in which two or more potoo species share an allele, *aethereus* and *grandis* possess unique alleles at seven and five of those loci, respectively. The high number of unique alleles in *bracteatus*, *aethereus*, and *grandis* suggest an ancient divergence among these potoos. Unfortunately, the high number of autapomorphies obfuscates attempts to resolve their relationships. Relationships of these potoos will most easily be resolved through examination of more conservative genetic markers.

Relationships among caprimulgiforms.—The fossil record indicates an early divergence date among caprimulgiforms. Fossils of Nyctibiidae (*Euronyctibius kurochkini*), Aegothelidae, and Podargidae all appear in the fossil record of France by the upper Eocene (Mourer-Chauviré 1982, 1987). These fossils demonstrate that not only was the diversification of caprimulgiforms ancient, but also that current distributions are relicts of once more extensive ranges. The uniformly high isozyme genetic distances among caprimulgiform families are consistent with their apparent age. Sibley and Ahlquist (1990:412, 840) also found high genetic divergences among caprimulgiform families using DNA-DNA hybridization data.

The unvermiculated rufous plumage and aberrant behavior of *Steatornis*, a frugivorous echo-locating troglodyte, make it the most distinctive caprimulgiform. Phylogenetic analysis of all caprimulgiform families based on cyt *b* sequences (including *Gallus gallus* as an outgroup) placed *Steatornis* as the basal caprimulgiform taxon (Mariaux and Braun 1996). The occurrence of a fossil oilbird in the early Eocene (ca. 50 Mya) of Wyoming (Olson 1987) confirms that Steatornithidae has had a long and complex history. Sibley and Ahlquist (1990:412, 840) presented an alternative phylogenetic view based on UPGMA analysis, which placed Nyctibiidae and Steatornithidae as sister taxa in a terminal clade. However, another analysis presented by these authors (1990:819) linked *Steatornis* with *Podargus* in a terminal taxon, and a reanalysis of these data (Harshman 1994) produced a "star" phylogeny of caprimulgiform families, indicating a lack of resolution in the data.

The large isozyme genetic distances to *Steatornis* (Table 4) make it a likely candidate for the earliest branching lineage, as indicated by the UPGMA phenogram and minimum evolution tree (Fig. 2). However, because we did not include any non-caprimulgiform outgroup taxa, we could not determine cladistically whether *Steatornis* is the basal caprimulgiform. In addition, there are no unambiguous isozyme synapomorphies that unite all other caprimulgiform families. In sum, we believe that the available evidence indicates that *Steatornis* is the earliest branching lineage of the caprimulgiforms, although this position cannot be considered incontrovertible.

Our analyses of the isozyme data result in various groupings of the other caprimulgiform families (Figs. 1–2). The high genetic distances involved probably make resolution of their relationships difficult. One interesting linkage that does arise in the FREQPARS analyses is that of *Chordeiles* and *Aegotheles* with *Eurostopodus*. This putative clade is supported by an isozyme synapomorphy at PGM-2. It also received some support from parsimony analysis of the cyt *b* data (Mariaux and Braun 1996), although it did not appear in the phylogenies of Sibley and Ahlquist (1990) or Cracraft (1981). Further data are required to confirm or refute this hypothesis of relationship.

Potoo taxonomy.—The levels of isozyme genetic divergence found among the currently recognized potoo species may be the highest ever found within a single genus of birds. Randi et al. (1991) examined levels of allozyme divergence among genera within Strigiformes (Strigidae and Tytonidae), an order hypothesized to represent the sister taxon to Caprimulgiformes (Sibley et al. 1988; Bleiweiss et al. 1994). Although they did not analyze intrageneric differentiation, they found levels of genetic differentiation among genera within Strigidae (average Nei's D = 0.88) comparable to those found among species within Nyctibiidae (average Nei's D = 0.65). An isozyme analysis of genetic differentiation among the procellariiform families (Barrowclough et al. 1981) found an average interfamilial genetic distance (average Nei's D = 0.68; range = 0.336 to 1.214) less than that found among potoo species.

The amount of genetic variation that exists among potoo species is not adequately conveyed in the current nomenclature, in which all potoos share a single genus. Simply elevating all of the current species to monotypic genera, however, will not greatly improve the utility of the taxonomy, and would, in fact, remove any phylogenetic information. If additional support is found for the *maculosus-leucopterus-griseus* clade, an alternative treatment would be to place these three taxa in one genus, and elevate *aethereus*, *grandis*, and *bracteatus* as monotypic genera. This classification would retain phylogenetic information while also recognizing the high genetic differentiation among the taxa. Another possible arrangement would be to treat *bracteatus* as a monotypic genus, leaving the other species in *Nyctibius*. Again, this treatment can only be recommended if additional support for *bracteatus* as the basal taxon can be mustered. We prefer the conservative approach of retaining the traditional taxonomy until additional data allow more certain resolution of potoo phylogeny.

The paucity of specimens of most potoos in museum collections has prevented detailed analyses of geographic variation and speciation. A conspicuous case in point is that of *leucopterus* and *maculosus*, two dramatically differentiated species that were conflated well into the 1980's due to lack of comparative material (Schulenberg et al. 1984). Conservative morphology, sometimes coupled with intrapopulational plumage variability, has made discerning species and subspecies limits difficult even in well-collected taxa like *griseus* and *grandis*. For example, although the Northern Potoo, *jamaicensis*, has often been treated as a subspecies of *griseus*, it is still uncertain whether the two are even close relatives (Davis 1978). Further analyses of genetics, vocalizations, and morphology will clarify variation within several of the currently recognized species. Such studies may identify other geographically delineated taxa differentiated at or near the species level. Populations of both *griseus* and *grandis* on opposite sides of the Andes seem to be prime candidates, as they appear to be well-differentiated genetically, based on the limited samples available. In particular, the amount of genetic divergence between populations of *grandis* from opposite sides of the Andes is comparable to that found among *maculosus*, *leucopterus*, and *griseus*.

ACKNOWLEDGMENTS

This paper really began on a mountainside in northern Peru on June 30, 1980, when M.J.B. and Theodore A. Parker III recorded the first *Nyctibius maculosus* for the country. Parker and Braun immediately realized that the bird must be specifically distinct from *N. leucopterus*, contrary to the literature of the time. For sharing the thrill of discovery on that day, and much knowledge, insight, and inspiration before and since, M.J.B. and R.T.B. acknowledge a deep debt to Ted Parker. We thank the many skilled field workers named in Table 1 for collecting samples, and Les Christidis, Van Remsen, Diana Reynolds, Mark Robbins, Fred Sheldon, and Bob Zink for providing them to us. Mario Cohn-Haft, Van Remsen, and Tom Schulenberg provided helpful reviews of the manuscript. Fernanda Zermoglio helped with the Spanish translation of the Abstract.

LITERATURE CITED

BARROWCLOUGH, G. F., K. W. CORBIN, AND R. M. ZINK. 1981. Genetic differentiation in the Procellariiformes. Comp. Biochem. Physiol. 69B:629–632.

BERLOCHER, S. H., AND D. L. SWOFFORD. 1997. Searching for phylogenetic trees under the frequency parsimony criterion: an approximation using generalized parsimony. Syst. Biol. 46:209–213.

BLEIWEISS, R., J. A. W. KIRSCH, AND F. LAPOINTE. 1994. DNA-DNA hybridization-based phylogeny for "higher" nonpasserines: reevaluating a key portion of the avian family tree. Mol. Phylogenet. Evol. 3:248–255.

BRODKORB, P. 1971. Catalogue of fossil birds: part 4 (Columbiformes through Piciformes). Bull. Florida State Mus., Biol. Sci. 15:163–266.

BRUMFIELD, R. T., AND A. P. CAPPARELLA. 1996. Historical diversification of birds in northwestern South America: A molecular perspective on the role of vicariant events. Evolution 50:1607–1624.

CAVALLI-SFORZA, L. L., AND A. W. F. EDWARDS. 1967. Phylogenetic analysis: models and estimation procedure. Evolution 21:550–570.

CHAPMAN, F. M. 1917. The distribution of bird-life in Colombia. Bull. Am. Mus. Nat. Hist. 36:1–729.

CHAPMAN, F. M. 1926. The distribution of bird-life in Ecuador. Bull. Am. Mus. Nat. Hist. 55:1–784.

COHN-HAFT, M. 1993. Rediscovery of the White-winged Potoo (*Nyctibius leucopterus*). Auk 110:391–394.

CRACRAFT, J. 1981. Toward a phylogenetic classification of the recent birds of the world (Class Aves). Auk 98:681–714.

DAVIS, L. I. 1978. Acoustic evidence of relationship in potoos. Pan Am. Studies 1:4–21.

FELSENSTEIN, J. 1985. Confidence limits on phylogenies: An approach using the bootstrap. Evolution 39:783–791.

HAFFER, J. 1967. Speciation in Colombian forest birds west of the Andes. Am. Mus. Novit. 294:1–57.

HARSHMAN, J. 1994. Reweaving the tapestry: What can be learned from Sibley and Ahlquist? Auk 111:377–388.

HILLIS, D. M., AND J. J. BULL. 1993. An empirical test of bootstrapping as a method for assessing confidence in phylogenetic analysis. Syst. Biol. 42:182–192.

HILTY, S. L., AND W. L. BROWN. 1986. A Guide to the Birds of Colombia. Princeton Univ. Press, Princeton, New Jersey.

KIDD, K. K., AND L. A. SGARAMELLA-ZONTA. 1971. Phylogenetic analysis: Concepts and methods. Am. J. Human Genetics 23:235–252.

LAND, H. C., AND W. L. SCHULTZ. 1963. A proposed subspecies of the Great Potoo, *Nyctibius grandis*. Auk 80:195–196.

MARIAUX, J., AND M. J. BRAUN. 1996. A molecular phylogenetic survey of the nightjars and allies (Caprimulgiformes) with special emphasis on the potoos (Nyctibiidae). Mol. Phylogenet. Evol. 6:228–244.

MEYER DE SCHAUENSEE, R. 1964. The Birds of Colombia, and Adjacent Areas of South and Central America. Livingston Publishing Company, Narberth, Pennsylvania.

MEYER DE SCHAUENSEE, R. 1970. A Guide to the Birds of South America. Livingston Publishing Company, Narberth, Pennsylvania.

MONROE, B. L., JR., AND C. G. SIBLEY. 1993. A World Checklist of Birds. Yale Univ. Press, New Haven, Connecticut.

MOURER-CHAUVIRÉ, C. 1982. Les oiseaux fossiles des Phosphorites du Quercy (Eocène supérieur a Oligocène supérieur): implications paléobiogéographiques. Geobios, Lyon, mém. spéc. 6:413–426.

MOURER-CHAUVIRÉ, C. 1987. Les Caprimulgiformes et les Coraciiformes de l'Eocene et de l'Oligocéne des phosphorites du Quercy et description de deux genres nouveaux de Podargidae et Nyctibiidae. Pp.

2047–2055 *in* Acta XIX Congressus Internationalis Ornithologicus, 1986. Univ. Ottawa Press, Ottawa, Canada.

NEI, M. 1978. Estimation of average heterozygosity and genetic distance from a small number of individuals. Genetics 89:583–590.

OLSON, S. L. 1987. An early Eocene oilbird from the Green River Formation of Wyoming (Caprimulgiformes: Steatornithidae). Docum. Lab. Géol. Lyon 99:57–69.

PARKER, T. A., III, T. S. SCHULENBERG, G. R. GRAVES, AND M. J. BRAUN. 1985. The avifauna of the Huancabamba region, northern Peru. Pp. 169–197 *in* Neotropical Ornithology (P. A. Buckley, M. S. Foster, E. S. Morton, R. S. Ridgely, and F. G. Buckley, Eds.). Ornithol. Monogr. No. 36.

PETERS, J. C. 1940. Check-List of Birds of the World, vol. 4. Museum of Comparative Zoology, Cambridge, Massachusetts.

RANDI, E., G. FUSCO, R. LORENZINI, AND F. SPINA. 1991. Allozyme divergence and phylogenetic relationships within the Strigiformes. Condor 93:295–301.

REMSEN, J. V., JR. AND M. A. TRAYLOR, JR. 1983. Additions to the avifauna of Bolivia, Part 2. Condor 85: 95–98.

RICHARDSON, B. J., P. R. BAVERSTOCK, AND M. ADAMS. 1986. Allozyme electrophoresis: A handbook for animal systematics and population studies. Academic Press, New York.

RIDGELY, R. S., AND J. A. GWYNNE, JR. 1989. A Guide to the Birds of Panama: With Costa Rica, Nicaragua, and Honduras. Princeton University Press, Princeton, New Jersey.

RZHETSKY, A., AND M. NEI. 1992. A simple method for estimating and testing minimum-evolution trees. Mol. Biol. Evol. 9:945–967.

SAITOU, N., AND M. NEI. 1987. The neighbor-joining method: a new method for reconstructing phylogenetic trees. Mol. Biol. Evol. 4:406–425.

SANKOFF, D., AND P. ROUSSEAU. 1975. Locating the vertices of a Steiner tree in an arbitrary metric space. Math. Prog. 9:240–246.

SCHULENBERG, T. S., S. E. ALLEN, D. F. STOTZ, AND D. A. WIEDENFIELD. 1984. Distributional records from the Cordillera Yanachaga, Central Peru. Gerfaut 74:57–70.

APPENDIX

GENOTYPE FOR EACH INDIVIDUAL AT ALL LOCI. REFER TO TABLE 1 FOR TAXON ACRONYMS. ALLELE DESIGNATIONS IN UPPER CASE ARE REFERABLE ONLY TO VARIATION WITHIN A SINGLE SPECIES.

Individual	ACON	ADA	AK	CK-1	CK-2	EST	FUM	GDA	GOT	αGPD	GPI	GPT	GSR	IDH-1	IDH-2
1825 (NMAC)	S	d	c	b	f	—	h	d	c	d	e	c	d	c	c
271 (NMAC)	F	d	c	b	f	—	h	d	c	d	e	c	d	c	c
20315 (NLEU)	—	d	c	d	f	—	h	e	c	df	e	h	b	c	c
20319 (NLEU)	—	de	c	d	f	—	h	e	c	df	e	h	b	c	c
20267 (NLEU)	—	d	c	d	f	—	h	e	c	d	e	h	b	c	c
3238 (NGRI)	F	d	c	b	e	—	h	b	c	d	e	cg	a	ac	c
3252 (NGRI)	S	d	c	b	e	—	h	b	c	d	e	g	a	c	c
10877 (NAET)	—	d	c	—	e	—	i	g	c	f	e	d	a	c	f
11236 (NAET)	—	d	c	g	e	—	i	g	c	f	e	d	a	c	f
15415 (NGRA)	S	d	c	i	e	S	a	a	c	d	e	i	a	ac	f
8954 (NGRA)	S	d	c	i	e	S	a	a	c	d	e	i	a	ac	f
3223 (NGRA)	F	d	c	i	e	F	b	a	c	d	e	i	a	a	f
20270 (NBRA)	F	f	c	b	d	S	h	h	c	cd	e	b	g	c	f
20318 (NBRA)	F	f	c	b	d	F	h	h	c	cd	e	b	g	bc	f
4509 (NBRA)	S	f	c	b	d	F	h	h	c	d	e	b	g	c	f
5279 (CHOR)	—	df	b	e	c	—	f	f	b	h	b	f	e	d	d
129 (EURO)	—	b	c	a	c	—	e	i	b	a	c	a	f	f	be
450 (AEGO)	—	c	b	f	b	—	d	k	b	b	ac	e	h	eh	ad
7474 (STEA)	—	b	a	h	a	—	c	j	a	g	d	—	i	g	df
8654 (PODA)	—	a	c	c	c	—	g	c	b	e	d	—	c	d	a

SIBLEY, C. G., J. E. AHLQUIST, AND B. L. MONROE, JR. 1988. A classification of the living birds of the world based on DNA-DNA hybridization studies. Auk 105:409–423.

SIBLEY, G. C., AND J. E. AHLQUIST. 1990. Phylogeny and Classification of Birds. A Study in Molecular Evolution. Yale Univ. Press. New Haven, New Jersey.

SICK, H. 1993. Birds in Brazil: A Natural History. Princeton Univ. Press, Princeton, New Jersey.

SWOFFORD, D. L. 1993. Phylogenetic Analysis Using Parsimony, Version 3.1.1. Computer program distributed by the Smithsonian Institution, Washington, D.C.

SWOFFORD, D. L., AND S. H. BERLOCHER. 1987. Inferring evolutionary trees from gene frequencies under the principle of maximum parsimony. Syst. Zool. 36:293–325.

SWOFFORD, D. L., AND W. P. MADDISON. 1992. Parsimony, character-state reconstructions, and evolutionary inferences. Pp. 186–223 in Systematics, Historical Ecology, and North American Freshwater Fishes (R. L. Mayden, Ed.). Stanford Univ. Press, Stanford, California.

SWOFFORD, D. L., AND R. B. SELANDER. 1981. BIOSYS-1: A FORTRAN program for the comprehensive analysis of electrophoretic data in population genetics and systematics. J. Hered. 72:281–283.

APPENDIX
EXTENDED

LA	LDH-1	LDH-2	LGG	MDH-1	MDH-2	ME-1	ME-2	MPI	NP	6PGD	PGM-1	PGM-2	PHE-PRO-1	PHE-PRO-2	PK	SOD	SOR DH	VL	UDH
—	b	—	—	b	b	c	—	f	ad	c	c	—	—	f	c	e	—	g	c
—	b	—	—	b	b	c	—	f	eh	c	c	—	—	f	c	e	—	g	c
—	b	—	—	b	b	c	—	f	g	e	c	—	—	f	c	e	—	g	c
—	b	—	—	b	b	c	—	f	g	e	c	—	—	f	c	e	—	g	c
—	b	—	—	b	b	c	—	f	g	eh	c	—	—	f	c	e	—	g	c
FS	b	b	F	b	b	c	F	f	hi	e	c	F	F	f	c	e	F	g	c
M	b	b	F	b	b	c	S	df	h	e	c	F	F	f	c	e	F	eg	c
—	c	—	—	b	b	c	—	d	f	ae	c	—	—	f	f	b	—	f	c
—	c	—	—	b	b	c	—	cd	f	e	c	—	—	f	f	b	—	f	c
S	d	a	F	b	b	—	S	c	f	e	c	F	S	b	e	e	F	g	c
F	d	a	F	b	bc	a	F	c	f	e	c	F	S	b	e	e	F	g	c
S	d	a	F	b	b	a	F	e	fh	e	c	F	F	b	e	e	F	g	c
BD	e	b	F	b	c	ab	F	a	c	g	a	F	F	e	d	f	F	g	c
DE	e	b	F	b	c	b	F	a	bc	g	a	F	F	e	d	f	F	g	c
AC	e	b	F	b	c	b	F	a	c	g	a	F	F	e	d	f	F	g	c
—	e	—	—	d	b	—	—	h	—	f	b	—	—	a	c	c	—	b	c
—	g	—	—	c	b	—	—	h	—	b	b	—	—	a	a	d	—	f	c
—	a	—	—	a	b	—	—	b	—	f	b	—	—	d	e	ad	—	a	b
—	b	—	—	a	a	—	—	j	—	d	a	—	—	c	b	g	—	d	a
—	h	—	—	a	b	—	—	gi	—	i	d	—	—	f	e	a	—	c	d

TECHNIQUES FOR AUDIO RECORDING VOCALIZATIONS OF TROPICAL BIRDS

GREGORY F. BUDNEY AND ROBERT W. GROTKE
*Library of Natural Sounds, Cornell Laboratory of Ornithology,
Ithaca, New York 14850*

ABSTRACT.—Audio recordings of tropical birds are important tools for biologists involved in the study, management, and conservation of birdlife; the ability to acoustically identify a species in a number of tropical habitats is essential. Recording method, equipment, and the condition of equipment can affect the accuracy and quantity of audio recordings collected. Suitability of currently available analog recording systems, emerging digital recording formats, and differing microphone designs for field work varies. This paper discusses essential and effective criteria that can be used to select a recording system based upon research goals and financial resources.

To survey and study tropical birds, biologists increasingly recognize that it is essential to be able to identify the birds by their sounds and to have the skill to make audio recordings of their voices (Parker 1991). Although several papers have been written on making audio recordings of bird sounds for research (Gulledge 1976; Wickstrom 1982; Ranft 1991; Vielliard 1993), recent advances in recording technology, together with the specific requirements of the tropical researcher, warrant revisiting and updating the information presented in these publications. This paper presents the fundamental technical information required to master the operation of a field recording system and prepares the recordist for situations that may be encountered in the tropics.

RESUMEN.—Las grabaciones del sonido de aves tropicales son herramientas importantes para los biólogos envueltos en el estudio, manejo y conservación de las aves. La habilidad para identificar acústicamente a una especie en varios tipos de hábitats tropicales es esencial. El método de grabación, equipo y la condición del equipo, pueden afectar la exactitud y cantidad de grabaciones coleccionadas. La conveniencia de los sistemas de grabación análogos disponibles al presente, los formatos de grabaciones digitales actuales y los diferentes disenos de micrófonos para trabajo de campo varían. Este artículo discute los criterios esenciales y efectivos que pueden ser utilizados para seleccionar un sistema de grabación basado en las metas de investigación y los recursos financieros disponibles.

Para estudiar y hacer censos de aves tropicales, los biólogos reconocen que es esencial poder identificar los pájaros por sus sonidos y tener las destrezas para hacer grabaciones de sus voces (Parker 1991). Aunque varias publicaciones sobre la preparación de grabaciones de sonidos de aves han sido escritas para la investigación (Gulledge 1976; Wickstrom 1982, Ranft 1991, Vielliard 1993), avances recientes en la tecnología de grabación, junto con los requerimientos específicos del investigador del trópico, ameritan revisar y poner al día la información presentada en estas publicaciones. Este artículo presenta la información técnica fundamental requerida para dominar la operación de un sistema de grabaciones en el campo y prepara al grabador de sonidos de aves para situaciones que se pueden encontrar en los trópicos.

THE IMPORTANCE OF RECORDING

About 3,100 bird species (Ridgely and Tudor 1989), roughly a third of the world's bird species, occur on the continent of South America. Although much of the region's baseline natural history remains undocumented, the rate of habitat destruction is outpacing the ability of scientists to document and study these species and populations. The next decade offers a critical window of opportunity in which to document the voices of wildlife in the Neotropics before many disappear.

Every researcher collecting sound recordings in this region can make a unique contribution to bird research and conservation (Kroodsma et al. 1996a). Relative to temperate latitude birds, the biology of tropical species is little known. Recordings of their voices provide important baseline data for research in the fields of avian systematics, behavior, and bioacoustics, and are essential to conservation initiatives as training and playback tools for surveys and censuses.

The researcher who records in the tropics faces not only unprecedented opportunities, but also unique challenges. Many bird species are elusive, hidden in impenetrable vegetation or high in the canopy. Recording equipment must be transported to remote locations with limited repair facilities and often must function under conditions of extreme humidity. With a combination of good field technique, common sense, and appropriate recording equipment, a recordist can ensure that equipment remains operational and that the objective of recording tropical bird voices is consistently achieved.

TECHNIQUE

Good recording technique combines advance preparation with field savvy, the ability to recognize and create recording opportunities, and the knowledge of how to operate recording equipment properly. Good technique maximizes the recorded level of the target bird's sound compared to the level of background sounds and the recording system's self-noise. It also involves an understanding of animal behavior and using it to one's best advantage. Proper technique will yield superior acoustic data for analysis, experimentation, and publication.

Preparing to Record

Many important recordings have been made as a result of serendipitous encounters in the field (e.g., recordings of the flight calls of parrots). To have the maximum chance of success in unpredictable situations, one must have the recording system in a constant state of readiness. When an opportunity arises, the action of turning on the recorder and aiming the microphone should be reflexive.

Operational readiness begins with the testing of a fully assembled recording system prior to entering the field. Assemble the components of the system, connect them, and make certain the system is capable of recording. If a system is in daily use, check it routinely at the end of each day's work. Prepare the system for the next day's operation the night before, remembering to load new tape and batteries if required.

Setting a Recording Level

To maximize the signal-to-noise ratio, one must set the record level on the tape recorder correctly. The objective is to record as strong a signal (bird sound or other animal sound) as possible on tape without distortion. For analog recorders the signal-to-noise ratio of the recorded sound is maximized with respect to inherent noise produced by the tape and components of the recorder itself. For digital recorders, recording at an optimal level on tape ensures that signals are recorded at or near the maximum bit rate (16-bit in modern 16-bit digital recorders). This practice helps reduce quantizing errors. To properly convert an analog signal to digital, two dimensions must be stored. Sampling implicitly creates a value representing time information, and quantization creates a value for amplitude information. With quantization, as with the measurement of any analog event, accuracy is limited by the system's resolution. Because of a 16-bit word length, the digital system's resolution also is limited, thereby introducing a measuring or quantizing error. Carefully recording at or near the maximum allowable level permits the recordist to use the system's resolution capabilities more fully, thereby minimizing quantizing errors.

Skilled recordists familiar with a particular system will set their recorder's gain control at an intermediate setting before venturing into the field. If the record level is turned completely down or left at the maximum setting, it will certainly require adjustment. With experience, one will have an idea of the appropriate operating range for the record gain setting, depending on the type or class of sound and the distance. Set the record level control at the low end of this range before recording. If the recorder must be turned on quickly to capture a suddenly heard sound, the chance of recording the signal without distortion will be greatly increased.

Audio Recorder Meters

The ideal recording level can be difficult to determine. To understand proper record level setting, one must understand the characteristics of a recorder's meter. Three types of meters are typically found on commercially available audio recorders: VU (Volume Unit), PPM (Peak Program Meter), and bar graph, or LED (Light Emitting Diode) meters.

The integration time of a meter is of paramount interest to those who record wildlife. In simplest terms, a meter's integration time is the time it takes the needle, or LEDs, to go from a position of rest to "0," the point beyond which distortion will occur. The response time or speed of a meter determines whether it can provide accurate readings when a bird sound is recorded. Of the three meter types, VU meters offer the slowest ballistics (response time), with an integration time of 300 msec. Peak Program Meters are considerably faster, with an integration time of 12 msec. Bar graph meters offer the fastest integration time, typically 10 msec. or less.

The slow response time of the VU meter makes it less desirable than the PPM and bar graph meters for recording wildlife sounds. Many of the sounds produced by birds such as warblers, sparrows, and tanagers have such quick onsets or rapid transients that the VU meter's pointer shows little or no response to the sound. An inexperienced recordist who sees no meter movement or only a slight deflection of the meter pointer often reacts by increasing the record gain until the meter indicates a satisfactory level. But this action is a mistake; if the meter's ballistics were capable of indicating the true signal amplitude, then the VU meter pointer would likely register well into the range of distortion.

Despite the disadvantages of VU meters, several popular field cassette recorders are equipped with them (Sony TCM-5000EV, Sony TC-D5 ProII, Marantz PMD221, and PMD222). If a recorder has off-tape monitoring capability (e.g., Marantz PMD221, PMD222, and Sony TCM-5000EV), one can circumvent the VU meter's foible by using headphones to listen carefully for distortion while making the recording. Marantz and Sony, recognizing the limitation of the VU meter alone offer a "Peak" light to supplement the recorder's VU meter in the models PMD222 and TC-D5 ProII respectively. This "Peak" light has the integration time of a Peak Program Meter, 12 versus 300 msec., and is therefore capable of accurately indicating signals with a quick onset or transients (e.g., wood-warbler call notes or songs).

How Long to Record

Turning on a tape recorder at the appropriate time is important when making any recording. Sound archives are filled with "decapitated" recordings–recordings that lack the introductory portion of a song or call. For example, of the nine examples of Nightingale Wren (*Microcerculus marginatus*) from north of the Amazon River archived in the Cornell Laboratory of Ornithology's Library of Natural Sounds, four do not contain a complete song. The problem of decapitation in *M. marginatus* recordings is restricted to the form north of the Amazon River, where song consists of a very brief series of musical rising and falling notes followed by an extended series of single whistled notes descending in scale, each song separated by long periods of silence. Decapitated recordings are the result of waiting until the bird vocalizes before turning on the recording system. The most common rationale for recording in this manner is "to save tape," but tape is the least expensive component of any recording system or recording expedition. The solution to this problem is straightforward: anticipate when a bird will vocalize and start the recorder in advance.

Allowing the tape to run 5–10 seconds after the bird stops vocalizing is also important. It will be invaluable if the recording is used in an audio publication; the brief amount of ambient sound lets the producer insert transitional fades into and out of the recording. Again, tape is the least expensive component in a recording system–don't be afraid to use it.

Assuming that one has anticipated the bird's song and captured the beginning, for how long should one record? The answer depends on one's purpose for recording, of course. The longer a recording is, the broader its potential range of uses and the greater the number of potential users.

If one wants to document the song repertoire of a Sedge Wren (*Cistothorus platensis*) from Brazil, for example, one will need several hours and thousands of songs (Kroodsma, pers. com.). For biological inventory purposes or some types of taxonomic investigations relatively brief recordings will suffice. For example, while conducting biological surveys, Parker (1991) recommended recording at least one minute of "natural" song (barely long enough to record a complete song of a Nightingale Wren). For many suboscines, songs may be very stereotyped, but if one is interested in their singing behavior, a recording of 10–20 minutes in length is not excessive for a bird that sings for extended periods. Similarly, a recording with intact intersong intervals is infinitely more valuable for research than 20 brief song samples from the same song bout of an individual. In contrast, some species may vocalize only sporadically or only for a brief period of the day or season. Most species of woodcreepers (Dendrocolaptidae) sing naturally only in the hours prior to or at dawn. It may be necessary to record for significant stretches of time during that period to obtain just a few vocalizations.

Integrating Technical Knowledge and Field Skills

Although the quality of recording equipment is important, one's recording technique is even more important. In competent hands, a modest cassette system can yield excellent results. Conversely, high-quality gear will yield inferior results if improperly used.

The distance from the microphone to the bird is crucial. One of the most important recording techniques to remember is that one can substantially increase the sound level of a recording by halving the distance between the microphone and the bird. By cutting the distance to the singing bird in half, one can potentially achieve as much as a 6dB increase in the signal going to tape.

When recording a bird, attempt to get as close as possible without causing the bird to flee. Ted Parker maintained that if one heard an unusual sound, one should pursue it, because rarely does the sound come to the recordist. Although not without its risks (remember to carry a compass and note reference points to avoid getting lost), moving toward the sound will increase the probability of obtaining a recording (and visually identifying the bird). Closer proximity to the vocalizing bird also maximizes the signal-to-noise ratio and reduces unwanted interference from other vocalizing animals. Plan an approach accordingly, so that when aiming the microphone at a bird, an unobstructed line of sight can be achieved. Foliage can affect the transmission of sounds, particularly high frequencies, which have shorter wavelengths and are therefore more subject to scattering.

Try to be aware of the noise that you, the recordist, make. Avoid nylon clothing or any other type of apparel that makes a great deal of noise with slight movement. Minimize the movement of feet when standing on a potentially noisy surface, such as leaf litter or gravel. If necessary to turn while recording to keep the microphone focused on a moving bird, pivot the upper body to avoid shifting one's feet.

Because of naturally high levels of ambient sound, many habitats pose inherent challenges to acquiring good quality recordings. The sounds of swift-running mountain streams, for example, are inescapable in many Andean locations. To attenuate background sound relative to the vocalizing bird, take advantage of topography. Look for features that can act as barriers between the microphone and unwanted ambient noise. Retreating just a few meters from the edge of a ravine with a stream below will attenuate higher frequency elements of "white" noise generated by the rushing water; the edge now effectively cuts off the direct reception of those frequencies from the microphone. Moving to a position behind a buttressed tree has a similar effect. In windy, open situations, take cover behind a rock or earthen berm, or construct a wind break to shield against wind that would otherwise vibrate a microphone diaphragm uncontrollably. Conditions will vary and so will the resources available, but in nearly every instance, one can find a way to improve the signal-to-noise ratio through careful positioning.

Another aspect of good recording technique is timing. Be at the research site before first light (Parker 1991). For many species, the pre-dawn and dawn hours represent the period of greatest (and sometimes only) activity. Forest-falcons (*Micrastur* spp.) vocalize in the pre-dawn hours and are easily overlooked during the rest of the day. Some tyrannid flycatchers give vocalizations at dawn that they do not make during any other period of the day.

Learning the habits of individual birds also helps. If one is aware of a bird's habitual behaviors, such as the repetitive use of a particular song perch, attempt to position oneself near the perch and be prepared to record before the bird arrives. One can also use a remote microphone with a long cable to record birds that predictably return to a certain location. For example, lekking manakins use the same display sites for extended periods of time. By placing a microphone near a male's display arena and running a long microphone cable back to a unobtrusive vantage point, one can observe the birds' behavior and make recordings of a high signal-to-noise ratio.

Using Playback

Whether working with a known species or an unidentified one, "playback" (broadcasting a recording of a song or call) is an essential and invaluable tool for detecting and identifying birds in the tropics. In many cases playback is the only practical method for observing a species. An elusive bird hidden in thick foliage will often move into the open to investigate playbacks.

The successful use of playback involves not only operating a recording system, but also understanding bird behavior. For example, when feasible, determine that you are within the territory of the target individual. Also, try to select a location that contains suitable cover through which a "skulking" bird can approach, but also a break in the vegetation where the bird can be lured into view. Next, play one or two songs and wait for a response. Impatience and sudden movement by the recordist can lead to failure. The response time varies among species. Some species, such as wrens, can be attracted with playback of only one or two songs, whereas other species, such as tapaculos, may require additional playback and patience.

Be aware that birds may use different vocalizations or alter the temporal pattern of their song in response to playback. Therefore, we strongly recommend that recordists clearly announce on tape when vocalizations are in response to playback.

Playback can be used in extraordinary circumstances, with seemingly extraordinary results. Ted Parker used playback to call in parrots in flight. As soon as he heard parrots in the distance, he would immediately turn on his recorder and tape their flight calls as they passed nearby. Then, directing his recorder toward the departing parrots, he would play back the tape. Often the parrots would respond by changing their flight direction and moving toward the sound of the playback. A positive identification and an additional recording opportunity were then possible.

Playback is often used as a research tool because it is so effective at eliciting a vocal response from a bird or causing it to reveal itself (Johnson et al. 1981). Birds will respond to signals of varying fidelity, including distorted signals. The equipment used for playback can directly influence the results of one's research, however (Weary and Krebs 1992; Dhondt and Lambrechts 1992). If an investigation involves a quantitative and qualitative study of a species' response to playback, then it is essential to know the frequency response of each component in the playback system, including the system used to record the original source recording.

DOCUMENTATION

To have scientific value, a recording must be accompanied by documentation: species, how identified (sight and/or sound), sex, behavior, habitat, time of day, exact locality (including latitude, longitude, and elevation, if known), and model numbers of recording equipment. These data can be announced directly on tape, at the end of a recording. Announcing this information during a recording makes the recording less useful. If some of the information remains constant for several recordings, one need not announce it following every recording, only when some parameter changes (such as habitat description or the type of equipment used). Taped notes can be brief, just enough to jog the memory when more detailed text records are constructed. Kettle and Vielliard (1991) and Kroodsma et al. (1996b) provide recordists with a detailed text data convention for audio recordings that is accepted by most major sound archives.

Another important aspect of documentation is assigning numbers to the original field tapes. When tapes are numbered, recordings are more easily located within a collection. One commonly used method is a sequential numbering system, akin to the sequential specimen numbering used by collectors of museum specimens. Another scheme involves numbering reels first by year, then by reel within a year. Some far-ranging researchers find a country prefix to be a useful part of a numbering scheme. For more detailed information on the organization of field recordings see Kroodsma et al. (1996b).

EQUIPMENT

A range of analog and digital recorders, directional microphones, shockmounts and windscreens, and related equipment are available to researchers. Selecting the most suitable recording components depends on the frequency range produced by the subject, field conditions, and money available for purchasing equipment.

For work in the tropics, a good portable field recording system should have the following features: (1) resistance to humidity, (2) off-tape monitoring, (3) playback capability, (4) the option to use conventional dry-cell batteries as a power source, (5) a PPM or LED record-level meter, and (6) a three-pin-XLR style microphone input. Every system, regardless of its capabilities, should be calibrated to known standards before taken into the field.

ANALOG RECORDERS

Cassette and reel-to-reel recorders are the two most commonly used types of analog field recorders, with cassette recorders by far the most common. Analog recording systems offer proven reliability to researchers even under the harsh conditions of the tropics. Analog tape transport mechanisms are virtually immune to humidity-related problems.

To record faithfully the vocalizations that birds produce, recordists must be certain that their analog recorders are properly calibrated to a recognized standard (NAB, AES, CCIR, DIN & IEC). These internationally recognized standards were adopted by the audio recording industry to allow accurate interchangability of tapes from machine to machine, i.e., cassette to cassette, open reel to open reel, and R-DAT to R-DAT. Track placement, recording equalization, and reference flux levels are just a few of the many standards that apply to audio recording. Also important is that the recorder be calibrated for the exact tape stock (type of tape). Even newly purchased audio-recorders may not perform to specification because of the rigors of shipment. Head misalignment, speed errors, and improper bias settings are just a few of the problems that plague even new machines.

Cassette recorders.—Cassette recorders have limited-frequency bandwidth and a lower signal-to-noise ratio, but are much lighter and use longer-running tapes than reel-to-reel machines. Cassette recorders are a low-cost alternative to other types of recorders: the highest price for a portable cassette recorder is about $1125. When calibrated and used properly, they can produce excellent results.

Several cassette recorders offer the basic features required of a field recorder for tropical field work: off-tape monitoring, good metering capability, professional microphone connectors, playback speaker, durable construction, and light weight. Although no one recorder has all features desired by field recordists, several offer combinations of most of these features: such as Sony's TCM-5000EV and TC-D5 ProII, and Marantz's PMD221, PMD222, and PMD430 models. All but the Sony TC-D5 ProII has off-tape monitoring. All have VU meters. The Sony TC-D5 ProII and Marantz PMD222 also provide a Peak light in addition to the VU meter.

Noise reduction features such as Dolby or DBX should be not be used. These features may work well for voice and music recording but are detrimental to bird song recording. Similarly, avoid using any limiter or AGC (automatic gain control) feature. Limiters and noise reduction processors all have a response or reaction time that in most cases is slower than most birds produce sounds. This lag in response time creates undesirable artifacts recorded along with the target sound. Once recorded, it is impossible to remove them.

Connectors are the components in a recording system most likely to fail, because they are subject to tugging, flexing, and other forms of strain. The Sony TC-D5 ProII and Marantz PMD222 offer the best type of microphone input connectors, namely professional 3-pin Cannon (also known as XLR) connectors. The Marantz PMD430 uses a less desirable, but durable, ¼" phone connector, whereas the Sony TCM-5000EV and Marantz PMD221 use a 3.5 mm mini-plug connector that is unreliable under difficult field conditions.

Of the machines discussed here, the Sony TCM-5000EV has the loudest playback capability. Many field biologists find the playback volume of other recorders insufficient, where others find the playback adequate. For those who require loud playback, but prefer the performance of recorders other than the TCM-5000EV, a small battery-powered playback speaker usually solves the problem.

Durability of a recorder is determined by the quality of its construction. With one exception, the recorders mentioned are well designed and constructed. With proper care and regular maintenance, a life span of 3–5 years under hard use is reasonable. One recorder stands out with respect to durability among this group of cassette recorders: the Sony TC-D5 ProII. It is very well made and able to withstand field work, maintaining speed and head alignment even under rough conditions. Weight and dimensions are also important considerations for anyone working in remote locations. All of these cassette recorders are of the same approximate weight and dimensions.

Analog reel-to-reel tape recorders.—Reel-to-reel analog recorders, also known as open-reel recorders, have wider tape width and faster linear tape speeds than do cassette recorders; they offer the widest frequency bandwidth, greatest fidelity, and best signal-to-noise ratios of any analog recorder. These machines are capable of accurately recording sonically challenging bird sounds, such as tanager songs and calls, sounds that most cassette recorders cannot record accurately.

At present, the only portable analog reel-to-reel recorder suitable for recording bird sounds is the stereo Nagra IV-S. The IV-S features off-tape monitoring, a Peak Program Meter, professional Cannon-style microphone inputs, and recording speeds of 15 ips (inches per second), 7.5 ips, and 3.75 ips. The Nagra IV-S is extremely durable, has excellent speed stability, and is unaffected by humidity. Power usage by a IV-S is conservative, up to 25 hr on 12 D-cell batteries, a distinct advantage in remote locations.

Other aspects of reel-to-reel tape recorders are potential limitations for field work. The IV-S weighs approximately 22 lbs (8.2 kg) with 12 D-cells installed. At 7.5 ips, a 5-inch reel of 1.5 mil tape provides 16 minutes of recording time, and only 8 minutes at 15 ips. The weight of both the machine and tape present significant concerns with respect to portability in the field and shipment to a field location.

With the price of a new Nagra IV-S exceeding $10,000, a reel-to-reel recorder might not be an option for many researchers. However, used recorders offer an alternative. It is possible to acquire a used stereo Nagra in good condition for $2,500–$3,000; monaural Nagras cost somewhat less and are excellent alternatives.

DIGITAL RECORDERS

Digital audio recording is an extremely promising technology, one that offers great potential as a field recording format for research. Three digital-format portable recorders are commercially available in North America: R-DAT (Rotary Head Digital Audio Tape), Nagra open-reel digital, and MiniDisc (MD). Simply put, digital recorders convert audio signals to binary code stored either on magnetic tape or optical disc. R-DAT and Nagra digital recorders record signals on magnetic tape. The MiniDisc format uses an optical disc as its storage media.

The convenience of digital recorders is seductive. Digital machines are lightweight, typically weighing 2 to 4 lbs (~0.75–1.5 kg), compared to more than 20 lbs (7.5 kg) for portable reel-to-reel analog recorders. They are small, with dimensions comparable to most cassette recorders. The Sony TCD8 is the smallest digital recorder, weighing only 1lb 1 oz (510 g), including battery, and measuring 5¼" × 1½" × 3½" (132 mm × 36.7 mm × 88.2 mm). Furthermore, a single R-DAT cassette has a maximum running time of two hours, far more than the few minutes for some reel-to-reel recorders. R-DAT recorders also feature a convenient track indexing scheme, making it possible to program a maximum of 99 randomly accessible points.

R-DAT recorders.—Of the digital formats mentioned, R-DAT is the one most often used for field recording. To record a signal, the R-DAT recorder uses a helical scanning head-drum assembly, a structure mechanically similar to the head on a video cassette recorder. This head assembly spins rapidly (2,000 rpm) while recording on slow-moving tape (8.15 mm/sec). In the "play" mode, this same head functions as a playback head. Few R-DAT recorders actually have separate record and play heads (discussed in a later paragraph).

R-DAT recorders deliver high-quality audio, comparable to that of a reel-to-reel recorder. The standard professional sampling rate of 48 kHz yields a frequency bandwidth limited to approximately 24 kHz. Although some R-DAT machines are capable of recording at lower sampling rates (44.1 kHz and 32 kHz), recordings should be made at the highest possible sampling rate.

Many recordists have successfully used R-DAT recorders in tropical climates, although others have had problems. The R-DAT recorder's main drawbacks for tropical field use are susceptibility to humidity (the recorder records unreadable data or shuts down), heavy power consumption, and lack of off-tape monitoring capability for most machines. Under humid conditions, condensation can form on the spinning head drum of R-DAT recorders. This moisture creates what audio engineers refer to as "stiction:" as the tape moves over the rapidly spinning head, moisture-induced friction causes the head to slow down or stall completely. The net result is that no readable data are recorded on the tape. If operation of the recorder continues under such conditions, damage to the recorder can result.

R-DAT recorders consume considerably more power than conventional analog recorders. Running out of tape may be the bane of the open-reel recordist, but running out of battery life is the bane of the R-DAT recordist. Most R-DAT recorders are powered by ni-cad (nickel-cadmium) battery packs. Although the running time for a fully charged pack is typically 1.5 to 2 hrs, recharging ni-cad batteries can be difficult under remote field conditions. Furthermore, R-DAT recorders take several seconds to go from a "power off" mode to "record-ready" state. So, a recordist who wants to be ready to record unpredictable vocalizations must keep the recorder in "pause" mode. But, in contrast to reel-to-reel analog recorders, which draw little power when paused, R-DAT recorders consume almost full power, so batteries are depleted quickly.

A final drawback of R-DAT recorders is that most do not have off-tape monitoring. The recordist is therefore unable to listen to what is actually stored on the tape. R-DAT manufacturers Fostex, HHB, and Sonosax offer recorders with off-tape monitoring, a feature reflected in their higher purchase prices. Additionally, Sonosax has engineered "climate-control" on their recorders, resulting in machines virtually unaffected by humidity. The advent of off-tape monitoring and climate-control on portable R-DAT recorders are positive developments toward addressing the requirements of field work. Machines offering one or both of these features should receive careful consideration, in spite of additional cost, by anyone recording in the tropics.

One of the most difficult decisions a recordist makes in assembling a recording system today is whether to invest in a digital recorder. Anecdotal accounts both support and refute the reliability of R-DAT recorders under extremely humid conditions. The lack of off-tape monitoring capability of most R-DAT recorders, together with problems under humid conditions and heavy power consumption, contribute additional risks to the rigors of collecting tropical bird sound recordings. The cost of R-DAT machines is not inconsequential and reflects important distinguishing features. Prices range from $750 for the Sony TCD8 with 3.5 mm microphone input, to $1,500 for the basic professional Tascam DA-P1, to $4,000–6,000 for HHB and Fostex recorders that provide off-tape monitoring. Recordists should weigh the convenience, light weight, and cost of an R-DAT recorder against the possibility of equipment failure in a remote location or at the moment when a rare recording opportunity presents itself. Some recordists address this issue by carrying a back-up recorder, such as a reliable analog cassette recorder.

The Nagra D recorder.—A second digital format was introduced by Kudelski S.A., manufacturer of the renowned Nagra portable reel-to-reel analog tape-recorders. The Nagra D is a digital, open reel, two or four channel, 20 bit, rotary head recorder with a sampling rate of 48 kHz and uses ¼" DASH, PD type tape. It has off-tape monitoring capability, a feature absent from many R-DAT digital cassette recorders. Recording time per 5" reel is 2.0 hrs when recording on two channels and 1 hr when recording on four channels. Tape speeds are 49.6 mm/sec and 99.2 mm/sec respectively. Weight of the recorder is 17.8 lbs (6.64 kg) with the battery.

Nagra tape-recorders are renowned for their reliability under adverse field conditions. The Nagra D's rotary-head configuration and open-reel tape transport are designed with field use in mind. The tape transport is covered by a hermetically sealed, clear, plastic cover, thus minimizing potential damage from dust and moisture. "Confidence monitoring," the ability to listen to the signal recorded on tape during recording, is a standard feature that is invaluable for field recording.

Two aspects of this recorder limit its suitability for field recording, however. The Nagra D has been designed for on-location use for movie shoots and music recording. This is reflected in the placement of meters and many controls (i.e., record level controls) on the recorder's top surface rather than on the front panel as with analog Nagras. Carried "over the shoulder," as required in most wildlife field recording situations, it is virtually impossible to view the meters while recording. With a price of nearly $20,000, cost is also a significant factor. For those who can afford its cost and are not handicapped by the position of its meters, the Nagra D will undoubtedly offer years of reliable and outstanding performance.

Digital recorders with data compression.—Recently introduced, the Mini-Disc recorder outwardly appears that it might have application for field use. Mini-Disc is a compact digital recorder that stores up to 74 minutes of audio on optical disc. Like other digital formats, the MiniDisc offers convenient track numbering along with the rapid accessing of a compact disc.

A significant drawback to this recording system is that it uses data compression algorithms to maximize data storage capacity. These algorithms are based on psychoacoustic research—the study of how humans hear and perceive sound. As a result, when bird sounds are recorded, harmonic content at the upper margin of human hearing can be lost. Furthermore, the "program material" (i.e., the bird song *plus* the background noise) is "reviewed" as a whole. This means that if, in processing the algorithm, the machine determines that the background sounds are masking elements of the signal of principal interest, the "masked" components of the principle signal are not necessarily recorded. Data compression algorithms may also add spurious data during the recording process (Fig. 1).

For the present, we cannot recommend Mini-Disc recorders or other recorders that use data compression, although they do have utility as a sound storage system with quick access capability for listening purposes.

MICROPHONES

Two types of directional microphone systems are typically used to record in the tropics: the parabolic reflector/microphone combination and the shotgun microphone. Both have excellent directional characteristics and minimize interference from other sound sources. Each has advantages and disadvantages, depending on recording circumstances.

The parabolic reflector.—The terms "parabolic reflector" and "parabola" are used interchangeably to mean the combination of a reflector (a mechanical amplifier) and a microphone (which can be either omni-directional or cardioid in pick-up pattern). The reflector's amplification capability is what makes it useful for field work. The parabola's surface area determines the degree of amplification, whereas its diameter determines its low-frequency operating range (Little 1964, Wickstrom 1982). All properly designed parabolas amplify the source signal hundreds of times; furthermore, amplification is achieved without addition of electronic noise that commonly

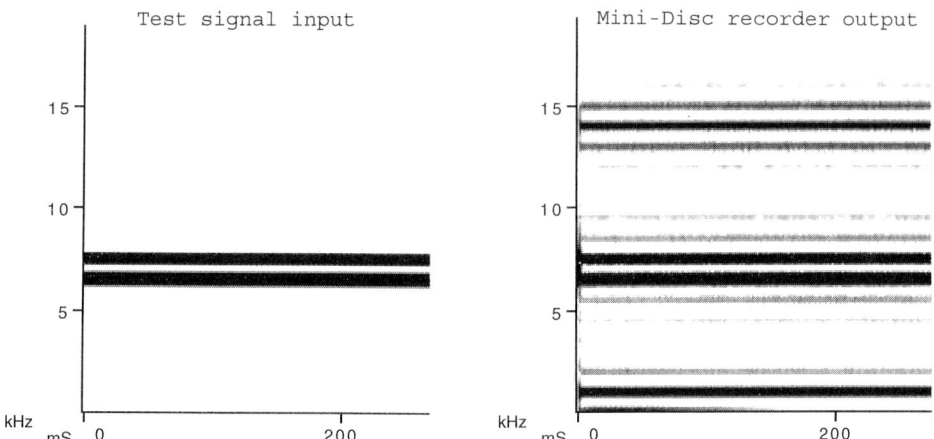

FIG. 1. Recording devices that use data compression schemes or other algorithms designed for human speech or music can severely alter recorded vocalizations of birds. Here an "Audio Precision System One," a computer-based audio test system used in standard tests of audio equipment, is used to generate a test signal of a 6.5 kHz and 7.5 kHz tone. This test signal, when recorded on a Sony portable Mini-Disc recorder, is noticeably distorted, producing strong signal anomalies at 1 kHz, 14 kHz, and a variety of other frequencies as well. How the Mini-Disc will record a bird's signal with background noise, or how the Mini-Disc might record two independent signals produced by the two syringes of a bird, is unpredictable. Abscissa is kiloHertz, ordinate milliseconds.

occurs with in-line preamplifiers. A field recordist using a parabola can make recordings with a high signal-to-noise ratio at a greater operating distance than is possible with other microphone systems. Canopy species, such as tanagers or *Terenura* antwrens, are probably best recorded with a parabola.

A parabolic reflector must be set up properly to achieve the best results. The parabolic reflector can be thought of as a system made up of a microphone and a reflector. The microphone must be accurately placed in the focal point (the area where all the collected sound energy is concentrated or focused) of the reflector. Dan Gibson and Telinga are two parabolic reflector manufacturers that solve this problem by providing a fixed rigid-mounting system for the microphone. When the system is fully assembled the microphone is guaranteed to be in the proper placement. However, if one chooses to assemble a system using discrete parts, e.g., a Roché reflector and a Sennheiser microphone, then it is necessary to determine the correct position. There are two ways to accomplish this.

One such method requires using a small portable FM radio as a fixed sound source. Set up the recording system, complete with parabola, microphone and recorder. Next place the radio on a table or chair. Turn it on and adjust the volume to a comfortable listening level. Now tune the radio between stations so that only noise is heard. Position the recording equipment about 20 feet away from the radio. Turn all equipment on and place the recorder in the record-pause mode. Manually aim the reflector so that it is pointing in the direction of the radio speaker. Adjust the record level on the recorder so that the meter is indicating mid-scale. Be sure to keep the reflector perfectly still. Mounting the reflector to a camera style tripod works well. Now carefully and slowly slide the microphone back and forth in its mount while watching the meter on the recorder. One should be able to see the level rise and fall as one moves the microphone around the focal point. The point at which the highest sound level is attained is the optimal position to place the microphone. One can also monitor the sound level with headphones, although the differences may not be as easy to determine.

The second method uses the sun, but the microphone must be removed before attempting this procedure. The concentrated energy from the sun will create dangerously high temperatures that can easily destroy a microphone. While aiming the reflector at the sun, move a small white card around the area of the microphone mount. As the card is moved in and out of the reflector, the reflected light pattern moves in and out of focus. Find the spot that gives the sharpest focus and note the relative position from the center of the reflector. This dimension is the focal point. The next step is to determine where the diaphragm in the microphone capsule is located. Usually the diaphragm (which looks like a gold or silver disc) can be seen by holding the microphone up to a light and looking through the wire screen mesh of the capsule. Note its position relative to the actual face of the microphone. To complete the installation make sure that the face of the diaphragm is positioned precisely at the focal point determined in the first step.

The optical and acoustic focal points should be the same for a parabola. However, this is not always the case. For this reason the preferred method for establishing the focal point is the acoustic technique. The optical method offers an alternative when resources required for the acoustic method are not available.

Aiming a parabola in the field also requires precision. If the parabola is not aimed accurately at the sound source, the recording will lack the high-frequency content (detail) of the sounds. To aim a parabola correctly, one must listen through headphones while the recording is made. When the bird is not visible, use the following technique to aim the parabola. First, smoothly "pan" the parabola across the horizon, beginning at the extreme right or left of where the bird is believed to be and listen through headphones for the point at which the sound's clarity and loudness, particularly high frequency elements, are the greatest. Once the correct position on the horizontal axis is determined, pan along the vertical axis until the angle of elevation at which the sound is the loudest is determined. At this point, the parabola will be properly aimed.

Parabolas are available in both clear and opaque materials. Clear parabolas are typically constructed of plastic and enable one to see a vocalizing bird through the parabola and to aim the parabola accurately. Opaque parabolas may be made of plastic, fiberglass, or carbon fiber. Many recordists outfit opaque parabolas with a sight, which allows them to use visual cues for accurate aiming.

The parabola is not without its drawbacks. The reflector's unwieldy shape makes it inconvenient in some situations, such as seeking out a bird in dense undergrowth. Furthermore, parabolas are better at capturing high-frequency (short wavelength) sounds than low-frequency (long-wavelength) sounds, so recordings made with a parabola generally have a high-frequency emphasis. A bird song that contains both high-frequency and mid-frequency sounds will be altered when recorded with a parabola, with a relative increase in the amplitude of high frequencies with respect to the source sound. How low of a frequency can be recorded is limited by the diameter of the parabola. Very low frequency (long-wavelength) sounds may be picked up by the microphone but will not be amplified by the reflector at all, because these wavelengths are larger than the reflector's diameter.

The parabola's large size can also make shipping it to remote locations a problem. Shipping cases that will stand up to abuse are available and worth the investment. Such cases not only protect a parabola but often provide ample room to pack a recorder, cables, blank tapes, and other gear. One parabola that is especially easy to pack is the 22-inch (56 cm) Telinga Pro parabolic reflector/microphone system. The dish is made of a clear, thin plastic that can be rolled up into a cylinder 22 inches (56 cm) long and 8 inches (20 cm) in diameter. The microphone and handle apparatus can be removed and conveniently stored inside the rolled dish for transport. The entire system weighs approximately 1.5 lbs (0.56 kg) Although the Telinga Pro's diameter is somewhat smaller than ideal, this system's transportability makes it especially useful for work in remote locations, particularly where there is a need to record canopy species.

Shotgun microphones.—Shotgun microphones are well suited for all-around use in tropical situations. This type of microphone does a good job of selecting a target sound from surrounding noise by rejecting off-axis sound. Shotgun microphones achieve their directionality through the hollow, ported, interference tube that surrounds the microphone; the longer the interference tube, the more directional the microphone. The ports in a shotgun microphone create a time delay for sounds arriving off-axis (sounds not from the direction aimed). This time delay causes certain frequencies to be canceled out, thus giving the shotgun its directionality.

The shotgun microphone has a broader "angle of acceptance" than the parabola and is, therefore, easier to aim. A shotgun microphone is ideal for recording loud, moving signals, such as the flight calls of parrots and the song or call of a moving antbird.

Shotgun microphones have some disadvantages, too. Although shotguns are directional, they are not as directional as parabolic reflectors. Also, shotgun microphones lack the amplifying capability of parabolic reflectors. This can be a problem when recording distant birds or low level signals. Thus, a recordist using a shotgun microphone must be much closer to the bird than a recordist with a parabola to get a recording of the same amplitude.

Condenser and dynamic microphones.—The two microphone designs most commonly used for bird sounds are dynamic and condenser microphones. Both shotgun microphones and the omni-directional or cardioid microphones used in parabolas can be either dynamic or condenser in design. Dynamic microphones are durable and require no electrical power, but are typically low in signal output strength. Condenser microphones typically have greater sensitivity and higher signal output than dynamic microphones, but also require electrical power and can be affected by humidity.

Condenser microphones come in two designs: electret condenser and rf (radio frequency) condenser. An example of an electret condenser microphone is the Sennheiser ME66/K6 "short" shotgun microphone. An example of an rf condenser shotgun microphone is the Sennheiser MKH70. Under humid conditions, electret condenser microphones are susceptible to static discharge, which is manifested in the form of audible crackling and popping on the recording (it is easy to mistake the sound of static discharge for a failed cable or connection). RF condenser microphones are more resistant to humidity. No matter what kind of microphone one selects, it should be stored with desiccant and be serviced routinely to minimize humidity related problems. An alternative to dessicant is to use a "hot closet," a closet with a light bulb or heating element in the bottom. If a "hot closet" is used, avoid placing equipment close to the light bulb or heating element as temperatures may exceed that which is safe for some types of equipment.

Because the microphone is the first stage of the recording system, one should purchase the best microphone affordable. One superb all-purpose shotgun microphone for tropical recording is the Sennheiser MKH70 rf condenser shotgun microphone. The Sennheiser MKH20 rf condenser omnidirectional microphone is the rf condenser equivalent for use with a parabolic reflector.

SHOCKMOUNTS AND WINDSCREENS

With today's sensitive microphones, shockmounts and windscreens are essential components for field recording. Shockmount systems are mechanical devices in which a microphone is typically suspended by elastic cords or bands, effectively isolating it from hand-held vibration. With the sensitive microphones preferred for field research, handling noise can be substantial, and because it is picked up by the microphone, it is recorded directly on tape. It is not uncommon for recordist-generated noise of this type to be as loud as the bird vocalization recorded.

Windscreens are essential, too. Without a windscreen, light breezes of wind can often obscure the bird vocalization on tape. Windscreens perform their isolating function by creating a turbulence-free zone around the microphone itself, while allowing the sought-after sound to arrive effectively at the microphone's diaphragm.

Use of a windscreen and shockmount are relatively inexpensive ways to minimize noise created by wind and by hand-holding a microphone. These two accessories come in many different styles. Most microphone manufacturers offer a range of custom-fit windscreens and shockmounts for their microphones. A few manufacturers also offer universal shockmounts that will accommodate many different styles of microphones. One example of a universal shockmount is Audio Technica's AT-8415, which is popular among researchers who use shotgun microphones. With a little ingenuity, it is also possible to fabricate a shockmount at little cost from a variety of basic materials, such as aluminum strap stock, rubber bands, plastic conduit, and neoprene tubing.

PREAMPLIFIERS, TRANSFORMERS AND ATTENUATORS

Nearly all audio equipment that bird sound recordists use today has been designed with another purpose in mind, such as recording concert or studio music, spoken interviews, or on-location motion picture sound-tracks. Most portable audio equipment comes with enough built-in "gain" or amplification to accommodate these types of recording events, because the target signal is usually strong and the microphone is typically placed close to the subject.

Recording birds in the field is another matter. Many bird sounds are not loud, and it is not always possible to place the microphone close to the bird without disturbing it. The recording of bird sound thus requires more "gain" than most portable recorders typically offer. Furthermore, the dynamic range of wildlife sounds pushes audio equipment to its performance limits.

One way that field recordists can address this problem is by using an in-line preamplifier to achieve a stronger signal on tape. An in-line preamplifier installed between the microphone and tape recorder will boost the signal from the microphone prior to arrival at the tape recorder's internal microphone preamplifier. These devices require power, typically supplied by a dry-cell battery. "Pre-amps" are generally small and relatively inexpensive ($100–600). Unfortunately, most inexpensive preamplifiers add a substantial amount of electronic noise to recordings.

Some recorder/microphone combinations can also benefit by using a matching transformer between the microphone and the recorder. To facilitate the transfer of electrical energy from the microphone to the recorder, it is important to have the microphone's impedance (AC resistance) closely match that of the recorder's input. In the case of the Sony TCM-5000EV cassette recorder and the Sennheiser ME80, one does not have the optimal match. They will function as a system, but the usable gain will be less than ideal.

To solve the problem of a mismatched recorder and microphone, many recordists use an impedance matching transformer between the microphone and recorder. The recordist can realize a 10–12 dB increase in recorder sensitivity by adding the proper matching transformer between the microphone and the recorder. Because these devices are free of electronic noise, they have an advantage over preamplifiers in that they give a similar, but cleaner result. They also do not require electrical power, and so they work well in field conditions. Note, however, that matching transformers will not work with all recorder/microphone combinations; check with a competent audio engineer for recommendations.

Just as the above section addresses ways to increase signal strength, some situations may arise where the target signal is so strong that distortion occurs. This is where the microphone attenuator switch comes into play. The attenuator is a resistive device designed to lower the signal strength. It can be found on some of the more expensive Sennheiser RF condensor microphones, as well as on many of the recording machines. One situation that sometimes causes distortion is having a microphone set up remotely to record a vocalizing bird. When the bird is close to the microphone, it is possible that the signal level is so strong that a condenser microphone's internal preamplifier is overloaded (distorting) or the signal coming from the microphone is overloading the microphone input preamplifier of the recorder. When these situations arise, simply lowering the record level will not solve the problem. An overloaded signal occurring at the microphone cannot be corrected or cleaned up at the recorder. Lowering the record level on the machine simply lowers the signal level, distortion included. A good telltale sign of a possible overload condition (aside from listening) is to note the level control setting on the recording device. Let's say the record level control has a range of 1 to 10. If one is forced to set the control at around 1.5 or 2.0 to obtain proper meter levels, then the signal coming from the microphone is probably too strong. A better operating position for the control would be in the 5 to 7 range. One way to correct this overloaded condition is to switch the attenuator on at the microphone (if it is so equipped). For those using microphones that do not have attenuators or if the microphone attenuator alone proves insufficient, the next option is to use the attenuator switch on the recorder. If neither option is available, simply moving the microphone further away from the bird should correct the problem. Always remember to switch the attenuator off when it is not needed because it will significantly degrade the system's signal-to-noise ratio.

CABLES

The quality of the cables that connect the components of a recording system are critical. Select high-quality cables that are flexible, shielded (braided shield only), and only as long as required. Balanced cables with three-pin XLR-style connectors are preferred whenever possible. (Look for three-pin connectors also when selecting a recorder and microphone.) Use professional-grade connectors that incorporate good strain-relief mechanisms.

No matter how good the cables are, carry spares. No component in a recording system undergoes more physical stress, and failure is always a possibility. It is far easier to attach a new cable to a recording system than to try to repair a damaged one in the field.

AUDIO TAPE

Analog tape-recorder manufacturers design and calibrate their machines to meet "general" technical tolerances (electronic performance specifications) on a variety of different tape stocks (brands). For example a Marantz PMD222 will function satisfactorily with TDK, Scotch, Sony, BASF, and Fuji tapes. Open reel machines are similar. Although these "general" calibrations are fine for music and speech applications, sounds produced by animals are much more complex and difficult to store on tape; thus the calibration requirements for recording these natural sounds must be much more precise. To insure the best possible recordings, it is important to calibrate a recorder for a single tape stock and then to consistently use it for all of one's recording work. When an analog tape-recorder is calibrated for a particular tape stock, it is literally "fine tuned" to optimally match the recorder's electronic performance to the magnetic properties of that tape stock.

Digital tape-recorders are exempt from this fine-tuning. For example, any R-DAT digital tape from any manufacturer can be used on any R-DAT recorder without compromising machine performance.

Cassette tape.—Cassette tape is physically small and delicate. Its thickness is 0.67 mil, 0.47 mil, and 0.33 mil for C-60/C-90/C-120 lengths, respectively. The tape width is 3.81 mm (1/8"). Each manufacturer has its own tape stocks (formulations), each of which offers different performance levels. These performance levels are known as "IEC types" of tape.

Three "IEC types" (not to be confused with brands) of tapes are made for use in analog cassette recorders, and a given cassette recorder may accommodate one or more of these three types. The three types of tape are IEC Type I (also called ferric oxide/normal bias), IEC Type II (chromium dioxide/high bias), and IEC Type IV (metal/high bias). All cassette tape manufacturers make all three types.

Obtaining optimal performance from a cassette recorder demands that one must use the type or types of tape recommended by the manufacturer. For example, a Sony TCM-5000EV is designed only for Type I (ferric oxide) tape. If Type II tape is used in this machine, the resulting recording may contain machine-generated distortion, such as spurious harmonics, and the recording will not be a faithful representation of the original sound (Figs. 2A and 2B). If a cassette recorder accepts more than one tape type (the Marantz PMD222 and Sony TC-D5 ProII are examples), be sure that the tape selector switch is set for the tape used. Type II is the preferred tape for bird sound recording, provided the recorder is designed to accommodate it. Metal tape also offers good recording performance, but battery life is somewhat reduced when this type is used because of the extra demands placed on the recorder electronics.

For best results, use only C-60 (60-min) or shorter cassettes. Long-playing cassettes, the kind that record for 90 or 120 minutes, are made with a thinner backing material than 60-min tapes. These thin-backing tape stocks are more susceptible to stretching, deformation, and print-through. Print-through occurs when high level recorded signals are transferred from one layer of tape to adjacent layers. It manifests itself as a pre- and post-echo of the principal signal when a tape is played back.

To increase recording time and to minimize costs, one might be tempted to record on both "sides" of a cassette or open reel tape. Whenever possible, however, one should record on only one side. Signals on opposite sides of a tape can sometimes bleed across "guard bands" (unrecorded areas that separate tracks) onto adjacent tracks. This is called "cross-talk." The contaminating signal will appear as low level backwards audio in the background of the recording. Once this problem has occurred, the offending signal cannot be removed.

Reel-to-reel tape.—Reel-to-reel or open-reel tape is available from many manufacturers. As with cassettes, each manufacturer has its own various tape stocks that offer varying degrees of performance. These different formulations are also available in different lengths. The length (running time) of the tape is determined by three factors: the size of the reel, 5" or 7"; the thickness of the tape used, 1.0 or 1.5 mil; and the recording speed, 3.75 ips, 7.5 ips, or 15 ips. As mentioned in the "AUDIO TAPE" section, it is imperative to calibrate the reel-to-reel recorder for the specific tape stock being used. Reel-to-reel tape is time proven as a reliable field recording format. When compared to cassette audio tape, open-reel tape's physical proportions of 1.0 and 1.5 mil thickness and its ¼" width make it less susceptible to stretching, damage from dust, and tape-edge damage. Reel-to-reel tape is capable of sustaining considerable damage or contamination and yet remaining playable, whereas an R-DAT tape under the same circumstances would become unplayable.

R-DAT tape.—The DAT cassette, slightly more than half the size of the analog cassette, was developed and standardized exclusively for the R-DAT recording format. As with a video cassette, the tape is protected until the cassette is installed in a recorder. At that time a small flap on the cassette opens, and the tape is extracted and wrapped around the head-drum assembly. R-DAT tapes, which use a metal-powder oxide, come in standard lengths of 15, 30, 46, 60, 90 and 120 minutes. The tapes are 3.81 mm wide, the same as the standard 1/8" analog cassette tape, and 0.47 mils thick, the same as analog cassette C-90 tape stock. Because R-DAT tape is transported past the heads at about one third the speed of an analog cassette, tape length required for 120 minutes is 60 meters with R-DAT and 171.45 meters for analog cassette. The combination of the rapid spinning head assembly, the helical data writing pattern, and the slow linear tape speed of the R-DAT allow tremendous amounts of data to be stored on very little tape. This is a mixed blessing, however. Although R-DAT allows long recording time and the ability to search an entire tape quickly, a small crinkle or even a finite amount of dirt or dust can render significant amounts of R-DAT data useless.

PROTECTIVE FIELD CASES

Protective cases for field recorders are a frequently overlooked but useful accessory. A well-designed field case, constructed of Cordura cloth, will protect a recorder from dust, detritus,

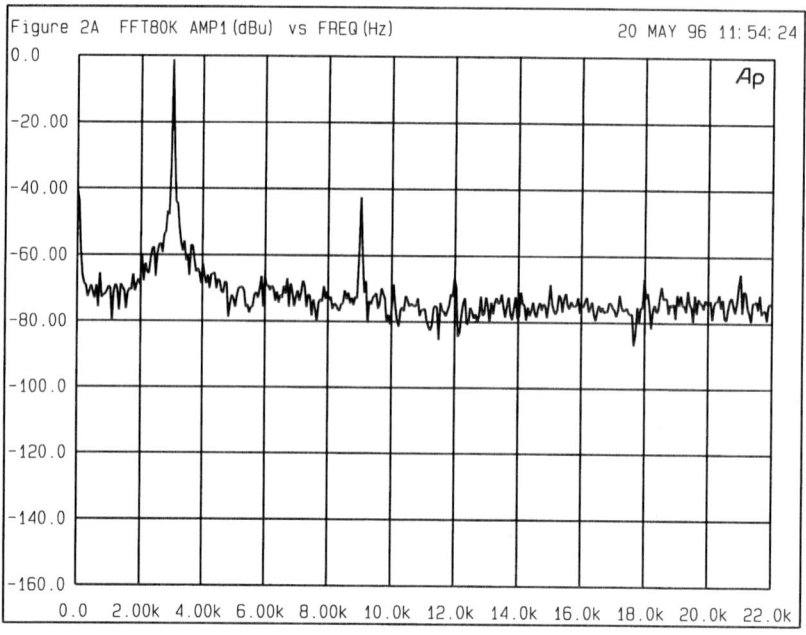

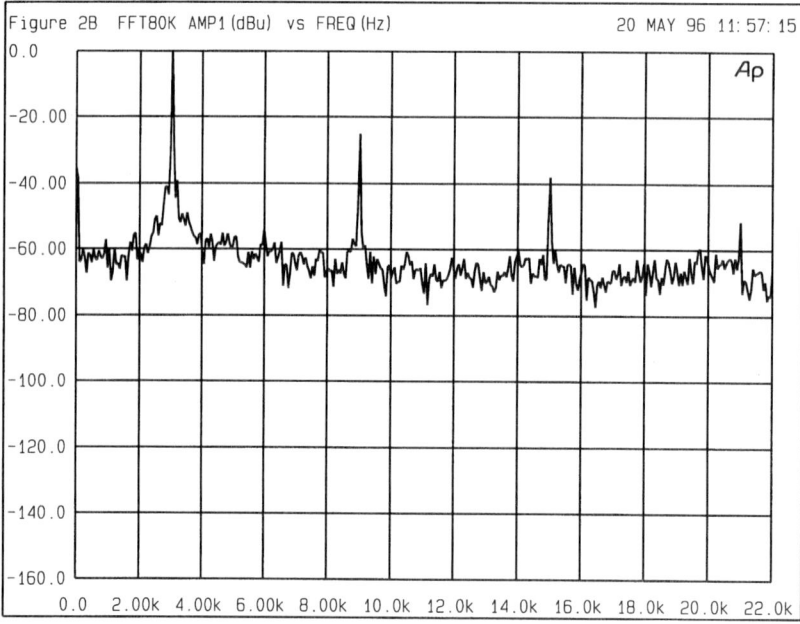

Fig. 2. A seemingly minor mistake, such as using Type II cassette tape on a recorder designed or set for Type I tape, can amplify undesirable harmonics in a recording, thus confounding spectral patterns in complex bird vocalizations. A. Spectral analysis of a 3 kHz test signal recorded at "O" VU on a "modified" Sony TCM-5000EV cassette recorder, which is designed for using IEC normal-bias tape (we used Maxell XLI-S60). Aside from the overall noise and the 3 kHz fundamental test signal, the only harmonic clearly visible is the 3rd harmonic, at 9 kHz. When a cassette recorder is properly calibrated, this 3rd harmonic, an inescapable byproduct of magnetic audio recording, is about 50 dB below that of the test signal; it should not be audible or cause serious problems in any spectral analysis. B. The same test, but this time using IEC Type II high-bias tape (Maxell XLII-60). Note that the 3rd harmonic is now only 20 dB below that of the test signal, and that the 5th harmonic becomes significant, too, at 40 dB below the fundamental. Abscissa, signal amplitude in dBu; ordinate, frequency in Hertz.

and rain, and will cushion it if dropped. A case will also cover and protect cables and connections and will put gear in a tidy package less likely to snag in dense undergrowth. Padded shoulder straps and waist belts, a common feature of high-quality cases, distribute weight and reduce fatigue associated with carrying a heavy recorder. Although the initial investment of $125–200 for a good case may seem high, the protection it provides will add years to a tape recorder's life.

CALIBRATION AND MAINTENANCE

Recorders should receive routine maintenance and be calibrated prior to the start of each field season, as well as before any major expedition. Audio recorders are complex mechanical and electronic devices. All internal moving parts are subject to wear. Many moving parts also have some kind of lubricant coated on them to help overcome friction. This lubricant is like a magnet for any dirt, dust, and small debris. Routine maintenance (changing belts, drive rollers/idlers, and internal cleaning) can greatly reduce the chances of total machine failure in the field.

Calibration should be performed by a professional audio service center or the manufacturer's authorized service facility. When sending a machine in for calibration you should do the following:

1. Supply a new, unused tape from the stock you will be using. The performance of the recorder can then be optimized for your tape stock.
2. Give detailed instructions as to the level of calibration required (see calibration instruction list below).
3. Request proof of performance, i.e., charts, graphs, documentation etc., that states the recorder's performance after calibration.
4. Provide a time-table so the service center is aware of when you need the recorder back. Be aware that the turn-around time for service centers can typically run 4-6 weeks.

Audio recorder calibration instruction list (applies only to analog recorders): check take-up and supply reel tensions; check and set absolute speed; check wow and flutter specification; check speed drift; demagnetize and adjust record/playback head azimuth (head height if needed); calibrate playback level and frequency response to reference standards; adjust bias and record equalization for flattest frequency response using the tape stock provided; adjust recorder calibration for unity gain in record/play mode (if unable to obtain unity gain please note record/playback gain offset); check distortion levels (THD, 3rd and 2nd harmonic); check signal-to-noise ratio (both "A" weighted and flat); print-out calibrated performance data. Once the above is completed, your machine will be calibrated to known reference standards. These procedures will maximize the scientific validity of your recordings.

Clean the recorder routinely, even daily if frequently used. For analog recorders, a good agent for cleaning the tape path (the heads, guides, and capstan) is isopropyl alcohol. (Do not use agents such as rubbing alcohol, which contain lubricants and will leave a residue.) Cotton swabs moistened with alcohol, the excess squeezed out, are effective for reaching and cleaning recorder heads and guides. If an excessive amount of alcohol comes into contact with the recorder's bearings and motor lubricants, it can cause premature failure.

R-DAT recorders should not be cleaned in the same manner as analog recorders. Use only the manufacturer-recommended cleaning tape or have the recorder cleaned by a professional, certified technician.

FIELD REPAIRS

With a modest amount of preparation and a bit of common sense, the recordist can ensure that a recording system will operate. A schematic of a recorder's circuitry will be invaluable if one can locate a good repair facility. Be aware, however, that an unskilled technician can do more damage than good.

Problems such as broken cables can completely disable a recordist. With an assortment of simple tools, however, one can handle most small repairs in the field. We recommend carrying the following:

1. Leatherman® or equivalent multipurpose tool (includes knife, screwdrivers, scissors, and pliers). A Leatherman-type tool together with a Swiss Army knife will cover a wide range of tool requirements.
2. Wire cutters and strippers.
3. Black electrical tape.
4. Portasol® (or equivalent) butane-powered soldering iron and 60/40 rosin core solder. Note that butane is not typically permitted on commercial aircraft. It will, therefore, be necessary to purchase butane in the country of destination.
5. Simple, battery-powered voltage/continuity multi-meter.
6. "Helping hand" holding apparatus (available through most electronic repair shops) or vice.
7. Extra batteries. In remote locations it is not uncommon to discover that newly purchased, name-brand batteries have already been partially or completely discharged.

If soldering is required to repair a broken cable and a suitable electronic repair facility cannot be located, jewelry shops often have soldering equipment.

SOUND ARCHIVES

Once recordings are collected, the recordist must consider how to preserve them for long-term use. Depositing recordings in an established sound archive ensures not only that they will be preserved for others to use in research and conservation, but also that they are protected against damage or destruction. Sound archives provide safe, long-term storage of recordings and also provide centralized access to recordings, which facilitates their use by others (Kroodsma et al. 1996b). Some sound archives extend their service role to include the repair and calibration of field recording equipment.

The five sound archives with the largest collections of Neotropical recordings are the Arquivo Sonoro Neotropical, Universidade Estadual de Campinas, Brazil; Laboratorio de Sonidos Naturales, Museo Argentino de Ciencias Naturales 'Bernardino Rivadavia', Argentina; Library of Natural Sounds at the Cornell Laboratory of Ornithology; Bioacoustics Archives and Library, Florida State Museum, University of Florida; and the Wildlife Section of the British National Sound Archive (formerly the British Library of Wildlife Sounds [BLOWS]).

Most archives provide copies rather than original tapes to those who wish to use archived recordings. Users may range from biologists studying animal communication and taxonomy, to resource managers who use recorded sounds to prepare for and conduct censuses, to commercial users such as television or film producers.

Responsible sound archives take the following steps to preserve original recordings:

1. High-quality copies of the original tapes are generated using rigidly calibrated professional studio recorders. One of these copies is the "working" copy, which is used as the source recording when additional copies are made, thus avoiding wear and tear on the original recording.
2. A good archive will also produce a first-generation safety copy. This copy should be stored off-site.
3. To prevent tape deterioration, original tapes, working copies, and safety copies should be stored under strict climate-controlled conditions.
4. The archive should restrict access to working copies, originals, and safety copies, allowing only authorized personnel to work with them.
5. The archive provides copies to users and does not loan the original on a routine basis.
6. The archive requires users to sign an agreement that outlines the manner in which copies may be used.

SUMMARY STEPS FOR RECORDING

To maximize your recording opportunities, here is a short list of the steps one should follow:

1. Assemble the entire recording system and check the equipment in advance.
2. Be at the investigation site before first light.
3. Position the microphone so that a clear path exists between it and the vocalizing bird.
4. Carefully aim the microphone.
5. Get closer to the bird; remember that halving the distance to the vocalizing bird (repeatedly if necessary) doubles the signal level reaching the microphone.
6. Position the microphone so that interference from background sounds is minimized.

7. Set the record level for the loudest element in the target bird's vocalization, and then leave it there.
8. Record for at least one minute, or longer if the bird allows.
9. Minimize handling and machine noise.
10. Announce basic data at the end of each recording.
11. Review and organize your field tapes at the end of each day.

ACKNOWLEDGMENTS

Among the late Ted Parker's many accomplishments are the recordings he made of more than 1,600 species of Neotropical birds. These recordings are archived in the Cornell Laboratory of Ornithology's Library of Natural Sounds (LNS). LNS is indebted and immensely grateful to Ted, not only for his recordings, but for being an ambassador for LNS—nearly all of the archive's current contributors of Neotropical material can trace their affiliation with LNS back to contact with him. The authors wish to thank Ted for sharing his knowledge, his energy, and his tremendous spirit that has inspired so many of us. Donald Kroodsma kindly applied his significant editorial skill and recording expertise to review this paper, significantly improving the content and presentation of information. We wish to thank Kenneth V. Rosenberg for his comments and encouragement in getting this paper published. We also wish to extend our gratitude to Cynthia Berger, Carol Bloomgarden, Tim Gallagher, Sandra L.L. Gaunt, Morton L. Isler, Phyllis R. Isler, and David L. Ross, Jr., who reviewed this manuscript and provided many insightful comments. We thank Jorge Saliva for translating the abstract to Spanish.

LITERATURE CITED

DHONDT, A. A., AND M. M. LAMBRECHTS. 1992. Individual voice recognition in birds. Trends in Ecology and Evolution 7:178–179.

GULLEDGE, J. L. 1976. Recording bird sounds. Living Bird 15:183–203.

JOHNSON, R. R., B. T. BROWN, L. T. HAIGHT, AND J. M. SIMPSON. 1981. Playback recordings as a special censusing technique. Pp. 68–75 in Estimating Numbers of Terrestrial Birds (C. John Ralph and J. Michael Scott, Eds.). Studies in Avian Biology No. 6.

KETTLE, R. AND J. M. E. VIELLIARD. 1991. Documentation standards for wildlife sound recordings. Bioacoustics 3:235–238.

KROODSMA, D. E., J. M. E. VIELLIARD, AND F. G. STILES. 1996a. Study of bird sounds in the Neotropics: urgency and opportunity. In Ecology and Evolution in Acoustic Communication in Birds (D. E. Kroodsma and E. H. Miller, Eds.). Cornell Univ. Press, Ithaca, New York.

KROODSMA, D. E., G. F. BUDNEY, R. W. GROTKE, S. M. E. VIELLIARD, S. L. L. GAUNT, R. RANFT, AND O. D. VEPRINTSEVA. 1996b. Natural sound archives: guidance for recordists and a request for cooperation. In Ecology and Evolution in Acoustic Communication in Birds (D.E. Kroodsma and E. H. Miller, Eds.). Cornell Univ. Press, Ithaca, New York.

LITTLE, R. S. 1964. Acoustic properties of parabolic reflectors. Bioacoustics Bull. 4:1–3.

PARKER, T. A., III. 1991. On the use of tape recorders in avifaunal surveys. Auk 108:443–444.

RANFT, R. 1991. Equipment for recording the sounds of birds and other animals. Bioacoustics 3:331–334.

RIDGELY, R. S., AND G. TUDOR. 1989. The Birds of South America. Vol. 1. The Oscine Passerines. Univ. of Texas Press, Austin, Texas.

WEARY, D. M., AND J. R. KREBS. 1992. Great Tits classify songs by individual voice characteristics. Anim. Behav. 43:283–287.

WICKSTROM, D. C. 1982. Factors to consider in recording avian sounds. Pp. 1–52 in Acoustic Communication in Birds (D. E. Kroodsma and E. H. Miller, Eds.). Academic Press, New York.

VIELLIARD, J. 1993. Recording wildlife in tropical rainforest. Bioacoustics 4:305–311.

A NEW SUBSPECIES OF *PERCNOSTOLA RUFIFRONS* (FORMICARIIDAE) FROM NORTHEASTERN AMAZONIAN PERU, WITH A REVISION OF THE *RUFIFRONS* COMPLEX

A. P. CAPPARELLA,[1] GARY H. ROSENBERG,[2] AND STEVEN W. CARDIFF

Museum of Natural Science, Foster Hall 119, Louisiana State University, Baton Rouge, Louisiana 70803, USA;
[1]*present address: Department of Biological Sciences, Illinois State University, Normal, Illinois 61790, USA;*
[2]*present address: 8101 North Wheatfield Drive, Tucson, Arizona 85741, USA*

ABSTRACT.—*Percnostola rufifrons jensoni*, a new subspecies of Black-headed Antbird, is described from northeastern Peru on the north bank of the Amazon River east of its confluence with the Napo River. Comparison with the three other subspecies and the historical taxonomic treatment of formicariids suggests this complex consists of two pairs of subspecies that should be ranked as species, *Percnostola rufifrons rufifrons* + *Percnostola rufifrons subcristata* (=*Percnostola rufifrons*, the Black-headed Antbird, with two subspecies *rufifrons* and *subcristata*) and *Percnostola rufifrons minor* + *Percnostola rufifrons jensoni* (= *Percnostola minor*, the Amazonas Antbird, with two subspecies *minor* and *jensoni*).

RESUMEN.—*Percnostola rufifrons jensoni*, una nueva subespecie de Hormiguero Cabecinegro, ha sido descrita desde el noreste del Perú sobre el banco norte del Río Amazonas al este de su confluencia con el Río Napo. Comparaciones hechas con las otras tres subespecies, además del tratamiento histórico del grupo de los formicáridos, sugieren que este complejo consiste de dos pares de subespecies que deben ser jerárquizados como especies, *Pernostola rufifrons rufifrons* + *Percnostola rufifrons subcristata* (= *Percnostola rufifrons*, el Hormiguero Cabecinegro, con dos subespecies *rufifrons* y *subcristata*) y *Percnostola rufifrons minor* + *Percnostola rufifrons jensoni* (= *Percnostola minor*, el Hormiguero Amazonas, con dos subespecies *minor* y *jensoni*).

From 6 June to 3 July 1984, we inventoried the avifauna of a locality in nonflooded (*terra firme*) forest on the north bank of the Amazon River in Peru, approximately 10 km northeast of the mouth of the Napo River, on the west bank of the Quebrada Orán. At this site a series of *Percnostola rufifrons* (Black-headed Antbird) was collected, a species previously known from Peru only from one male taken at Nauta, depto. Loreto, in 1866 by Bartlett (Hellmayr 1924). Nauta is on the north bank of the Marañón River near its confluence with the Ucayali River, approximately 175 km southwest of our Orán camp (Fig. 1). Hellmayr (1924) assigned the Nauta specimen to the subspecies *P. r. minor*. Although the males from Quebrada Orán resemble males of *minor* quite closely, the females differ dramatically from *minor* as well as the other described subspecies of *P. rufifrons*. We propose to name a new subspecies.

Percnostola rufifrons jensoni subsp. nov.

Holotype.—Louisiana State University Museum of Natural Science (LSUMZ) No. 119831; adult female, from Quebrada Orán (local name and name used on specimen labels; also known as "Río Yanayacu de Orán" on the 1967 Mapa Política, 1:1,000,000, Instituto Geográfico Militar del Perú), ca 5 km N of the Amazon River, 85 km NE of Iquitos, 3°25'S, 72°35'W, depto. Loreto, Peru, elevation ca. 110 m; 19 June 1984; prepared by Gary H. Rosenberg (original number 1204).

Paratypes.—Three females: LSUMZ 119829 (skin and tissues), 119832 (skin, complete skeleton, tissues), and 119834 (skin only, head destroyed). One adult male: LSUMZ 119833 (skin and tissues). One immature male: LSUMZ 119830 (skin and tissues).

Diagnosis.—Males are 100% separable from the three described subspecies (*rufifrons*, *sub*-

cristata, and *minor*) in crown coloration. Male *jensoni* have black crown feathers with broad gray edges that give the crown a distinctive, scaled appearance. In contrast, nominate *rufifrons* and *subcristata* have solid black crowns and *minor* has crown feathers that are mostly black with narrow gray edges. The crown feathers of *jensoni* and *minor* are short (ca 10.5 mm), in contrast to the longer crown feathers in *rufifrons* and *subcristata* (ca 14.7 mm).

Differences among the subspecies are more pronounced in females. Female *jensoni* are 100% separable by the (1) more concolor grayish crown and back versus blackish crowns contrasting with grayish-brown backs (*rufifrons* and *subcristata*) or a dull rufous crown contrasting with a grayish-brown back (*minor*), (2) sides of the head gray, not rufous or ferruginous as in the other subspecies, and (3) primaries dark grayish-brown with pale brown edging to the outer webs, not lighter brown with dull rufous edging on the outer webs as in the other subspecies. As in males, the crown feathers of female *jensoni* and *minor* are shorter than in *rufifrons* and *subcristata*. The relatively pale underparts of *jensoni* more closely resembles *minor* than *rufifrons* or *subcristata*.

Iris color for both sexes is gray in *jensoni* (as in *minor*; FMNH specimens), but red in *rufifrons* (Dunning 1982) and *subcristata* (L. Harper, pers. comm.).

Description of holotype.—Forehead and crown feathers with Blackish Slate basal two-thirds, recurved Black subterminal band, and broad, dull Sepia tips, giving the crown a scaled appearance; nape, back, scapulars, rump, and upper-tail coverts Blackish Slate tinged dull Sepia; lores, auriculars, and subocular region predominately Blackish Slate mixed with Light Drab; indistinct Raw Umber posterior supercilium; sides of neck Blackish Slate mixed with Raw Umber; chin Ochraceous Buff; center of breast Amber Brown blending to darker Raw Umber on sides; belly and under-tail coverts Ochraceous Buff blending to darker Raw Umber on flanks and tibiotarsus feathers; remiges Fuscous Black, narrowly edged Raw Umber; lesser, median, and greater upper-wing coverts Black, broadly tipped Amber Brown, forming three distinct wing bands; rectrices Blackish Slate with narrow Pale Olive Buff tips on outer three pairs (capitalized color names from Ridgway 1912). Colors of bare parts (in life): iris gray; maxilla black; mandible, tarsus, and foot gray.

Measurements of holotype.—Wing (chord) 62.7 mm; tail 44.9 mm; tarsus 27.7 mm; culmen from anterior edge of nares 17.5 mm; body mass 21 g.

Variation.—Two males and four females were examined from the type locality. One male is an immature (skull 90% pneumatized, left testis 2 × 1 mm, wing 64.2 mm, tail 50.5 mm, body mass 21.5 g) that differs from the adult male (skull 100% pneumatized, left testis 3 × 1½ mm, wing 69.2 mm, tail 52.5 mm, 22 g) in lacking the black throat, having some brown-edged feathers intermixed in the normally slate-gray body plumage (mainly in the breast, belly, and flanks), and retaining brown-edged juvenal (instead of gray adult) remiges and greater upper-wing coverts.

Crown and ventral coloration varies in females. Crowns vary from Black with feathers edged Blackish Slate (as in 119829), to Black with feathers edged Sepia (as in the holotype). The holotype has relatively dark underparts compared to the other three females, especially on the breast and throat. We suspect that variation in the intensity of the ochraceous coloration of the underparts is age-related. All three females with intact heads had 100% pneumatized skulls. However, the headless female with relatively pale underparts (LSUMZ 119834) was obviously an immature bird with a 3 × 2 mm smooth ovary. The other two relatively pale females had ovaries that were 6 × 3 mm (ova minute, LSUMZ 119829) and 7 × 5 mm (largest ovum 1 mm diameter, LSUMZ 119832). Thus, the holotype has the darkest underparts and the largest ovary (8 × 5 mm, largest ovum 1 mm); we presume it to be more typical of more mature birds, but more specimens will be needed to confirm this. Unfortunately, existing female specimens have no data on bursa condition or oviduct configuration to further aid in age determination.

Soft part colors vary between the sexes only in bill coloration; males have an all black bill, whereas females have a black maxilla and a gray mandible.

Specimens examined.—*Percnostola r. rufifrons* BRAZIL: Amapá, Alto Rio Araguari (LSUMZ 67332-3; 1 ♂, 1 ♀); Amapá, Prosperidade, Rio Maracá (LSUMZ 67330, 1 ♂); Amapá, Macapá, Rio Amapari (LSUMZ 67331, 1 ♀); Amapá, Monte Caiari, Uassa Swamp (CM 68376, 1 ♂); Amapá, Serra do Navio (ANSP 170079-80, 1 ♂, 1 ♀, LACM 59750-3, 2 ♂, 2 ♀); Pará, Obidos (CM 82644, 82660, 82714, 82716-7, 82788-90, 82812, 82841-3, 82971-2, 83044, 83236, 83369, 84153, 11 ♂, 7 ♀, YPM 28916-9, 2 ♂, 2 ♀); FRENCH GUIANA: Cayenne (AMNH 233853, 1 ♂, CM 55569, 55588, 55656, 56180-1, 56310, 56834, 3 ♂, 4 ♀, YPM 28912-4, 2 ♂, 1 ♀); Mana River, Tamanoir (CM 60896, 60935-6, 61060, 61281, 61453, 61530, 62237, 62262, 62340, 62393-4, 8 ♂, 3 ♀, 1 ♀?, YPM 28908-11, 2 ♂, 2 ♀); Oyapock River, Pied Saut (CM 64765, 64957, 64975, 65021, 65123, 65561, 2 ♂, 3 ♀, YPM 28915, 1 ♂); GUYANA: Omai mine, Essequibo River (LSUMZ 64748, 1 ♂); 5 km N Rockstone, E. bank Essequibo River (ANSP 186620, 186622, 1 ♂, 1 ♀); *ca* 5 km NW Mabura Hill, between Essequibo and Demerara rivers

(ANSP 186642-3, 1 ♂, 1 ♀); Waratilla Creek, Demerara River (UMMZ 108584, 1 ♂); SURINAM: "interior" (AMNH 491394, 1 ♀); locality unknown (UMMZ 116714-5, 1 ♂, 1 ♀); vicinity of Paramaribo (MCZ 144852-4, 144856-8, 4 ♂, 2 ♀); Apoera, Courantyne River (YPM 27475, 1 ♀); Paraku, Saramacca (MCZ 199709-10, 1 ♂, 1 ♀); Albina (MVZ 123567-8, 2 ♂); District of Para (MCZ 83324, 1 ♀).

P. r. subcristata BRAZIL: Pará, Faro, Rio Jamundá, São José (AMNH 283926, 1 ♂); Pará, Faro, Rio Jamundá, Castanhal (AMNH 283929, 1 ♀); Amazonas, Itacoatiara, Rio Amazonas (FMNH 50701, 1 ♂).

P. r. minor BRAZIL: Amazonas, Rio Negro, Cucuhy (USNM 316487, 1 ♂); Amazonas, Rio Negro, São Gabriel (USNM 327252-3; 1 ♂, 1 ♀); Amazonas, Serra Imeri, near Salto do Huá (USNM 326271-3, 326397-8; 2 ♂, 3 ♀); Amazonas, Rio Cauabury, Panela de Onca (USNM 327254-5; 1 ♂, 1 ♀); Amazonas, Rio Solimões, Tonantins (CM 96556, 96599-600, 96636-7, 96667, 96703, 96826-7, 96932-3, 97060, 97213-4, 97398, 97430, 97560, 97722, 97728, 97858, 97898, 13 ♂, 8 ♀, YPM 28920-3, 2 ♂, 2 ♀); VENEZUELA: Amazonas, Mt. Duida, Valle de los Monos (AMNH 273304, 1 ♂); Amazonas, confluence of Río Huaynía and Río Casiquiare (AMNH 432651, 1 ♀); Amazonas, Brazo Casiquiare near Caño Duratamoni (USNM 327256, 1 ♀); Amazonas, San Carlos de Río Negro (FMNH 319235-8; 3 ♂, 1 ♀).

P. r. jensoni PERU: Loreto, Quebrada Orán (LSUMZ 119829-34; 2 ♂, 4 ♀); Loreto, Nauta (BMNH 69.5.25.46, 1 ♂).

Etymology.—It is our pleasure to name this distinctive subspecies after Peter Jenson, owner of Exploraciones Amazónicas (Explorama Tours), Iquitos, Peru, in thanks for his generous help and support through the years for Louisiana State University Museum of Natural Science field work in northeastern Peru.

Distribution.—All LSUMZ specimens for this subspecies were collected at the type locality on the north bank of the Amazon River east of the confluence with the Napo River. We have assigned tentatively the only other Peruvian specimen of this species, collected at Nauta, to *P. r. jensoni* (see remarks below).

DISCUSSION

Distribution.—The distributional limits of the four subspecies of *Percnostola rufifrons* are known poorly (Fig. 1). The species is not known from south of the Amazon River (Peters 1951). The boundary between *P. r. rufifrons* and *P. r. subcristata* is given as the Trombetas River in eastern Brazil (Peters 1951). Published locality data (Zimmer 1932) suggest that the Branco River could be the boundary between *subcristata* and *minor*. The species is known from Nappi Creek just east of the upper Branco River (ca. 3°17'N, 59°39'W), a critical locality for determining subspecies distribution, but unfortunately no subspecies identification was given and no specimens were reported (Oniki and Willis 1972). The western boundary of *subcristata* appears to be the lower Negro River, except for one locality, Igarapé Cacão Pereira, on the immediate west bank of the Negro across from Manaus. A male and a female were collected from this locality by the Olalla brothers (Zimmer 1932). The general lack of specimens from north of the Amazon River between the Negro River and our Orán site precludes determination of the boundary between *minor* and *jensoni,* although the series of *minor* from Tonantins is highly suggestive that the Putumayo/Içá River separates these two forms (Fig. 1). It is interesting that two of the four subspecies have at least part of their ranges delimited by rivers, a phenomenon often found among taxa of understory *terra firme* forest suboscines (Capparella 1988, 1991).

It is notable that this species remained unrecorded in eastern Peru between 1866 and 1984, despite intensive collecting at several sites in the Iquitos region in the interim (e.g., Puerto Indiana, Pebas, Río Curaray). Our specimens were collected in *terra firme* forest, but this species was not found by LSUMZ in six other *terra firme* sites with similar coverage in the same region (Fig. 1). This suggests *P. r. jensoni* is patchily distributed in eastern Peru, perhaps because it requires a particular subset of *terra firme* habitat. The species appears to inhabit edges, mainly treefall gaps, in *terra firme* forest around Manaus, Brazil (C. E. Quintela, pers. comm.) and is caught with some degree of regularity in mist nets (Bierregaard 1988); at Orán, five of the six specimens were netted and one was shot. Oniki and Willis (1972) reported that the species is found in second growth and edges of dense forest at their study sites of Nappi Creek, Reserva Ducke (3°08'S, 60°02'W), and Serra do Navio (0°55'N, 52°01'W). Hilty and Brown (1986) remarked that the species is found mainly in white-sand soil forests in Colombia. Tostain et al. (1992) listed understory of primary forest and second growth in addition to treefall gaps as the habitat of *P. rufifrons* in French Guiana. Ridgely and Tudor (1994) described the habitat as undergrowth and treefalls at the edge of humid forest and mature second-growth.

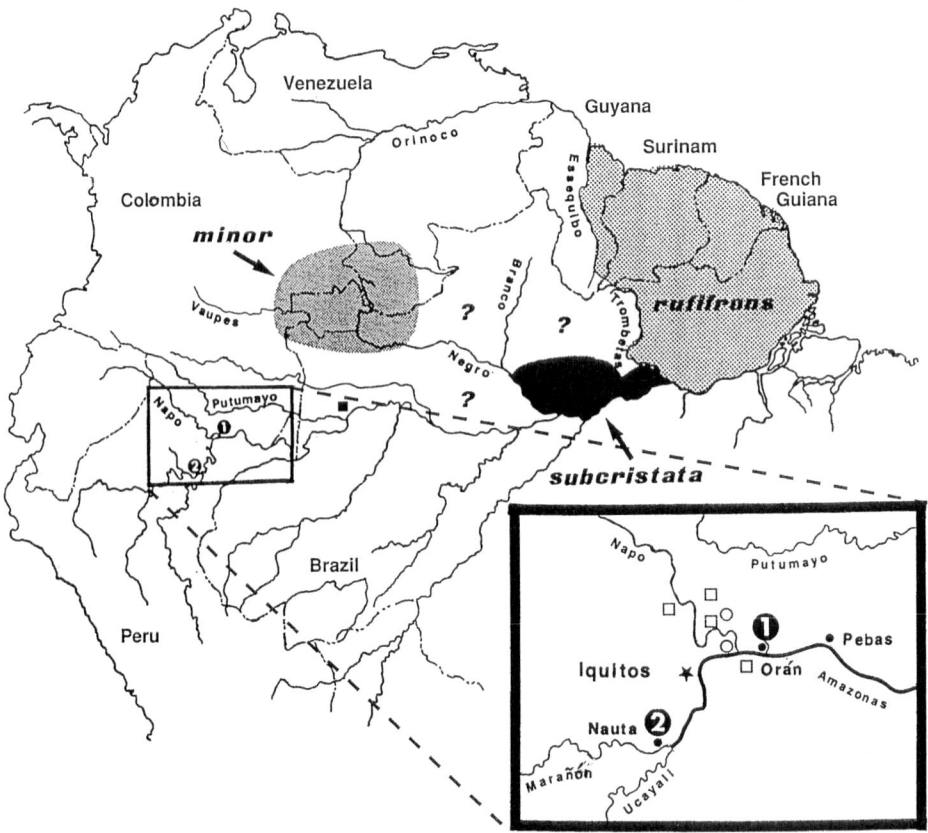

FIG. 1. Range of the taxa comprising the *Percnostola rufifrons* complex based on specimens examined. Within the square are: (1) filled circles showing the two Peruvian localities of *jensoni*, (2) open squares showing four other LSUMZ collecting sites where *jensoni* was not found, and (3) open circles showing two sites worked by Theodore A. Parker III, where *jensoni* was not found. Filled square shows Tonantins locality of *minor*.

Hellmayr (1924) noted that the male from Nauta (a specimen in poor condition, with broken bill, bent tarsi, and missing feathers) is not typical of *minor* because of its lighter, nearly slate-gray pileum. Zimmer (1932) suggested that this characteristic was within the range of variation exhibited by *minor* as seen in males from the upper Negro River. Our examination of the Nauta male revealed that the crown was not concolor but was faded black and broadly edged with Blackish Slate like specimens of *jensoni*. Therefore, based on crown color and locality we assign tentatively the Nauta male to *jensoni* until the more distinctive female is collected from that locality.

Systematic relationships.—There are no recordings of the vocalizations of *jensoni*. Unpublished recordings by S. L. Hilty of *minor* at Mitu, Colombia, and *rufifrons* at the Brownsberg Nature Park, Surinam, were compared by one of us (Rosenberg). Although the sample size is insufficient for definitive conclusions, analysis of sonograms shows subtle but distinct differences in the overall frequency of the song and in the change of frequency within the individual notes of the song. Willis (1982) described in detail the vocalizations of *rufifrons* at Reserva Ducke, Brazil, but made no comparisons to other subspecies. Ridgely and Tudor (1994) reported that the song of *rufifrons* and *subcristata* are the same. The apparently allopatric distribution of the taxa prohibit direct determination of reproductive compatibility.

We believe that recognition of two species is warranted based on the rather broad suite of characters (various aspects of plumage color and pattern discussed above, eye color, crown feather length, non-overlapping measurements of tail [Table 1]) that can be used to separate the subspecies *minor* and *jensoni* from nominate *rufifrons* and *subcristata*. In our opinion, however, morphological differences between *minor* and *jensoni* (males separable by the relative width of gray tips to the crown feathers, females separable by the more rufous versus slaty crown and face) and between *rufifrons* and *subcristata* (females separable by the black versus dark brownish

TABLE 1
Measurements (in mm) of Taxa in the *Percnostola rufifrons* Complex, with Range, Mean, and Sample Size

Taxon (N)		Wing (chord)	Tail	Tarsus	Culmen (from nares)	Body mass (grams)
rufifrons	(51 ♂♂)	69.8–77.3; 73.0 (N = 50)	55.3–69.7; 59.8 (N = 50)	27.1–30.8; 29.0 (N = 51)	11.2–14.0; 12.4 (N = 48)	24–31; 26.8 (N = 5)
	(36 ♀♀)	69.1–75.1; 71.1 (N = 36)	54.2–62.9; 58.1 (N = 34)	27.2–30.6; 28.7 (N = 35)	11.1–13.3; 12.0 (N = 35)	23.5–29; 25.4 (N = 4)
subcristata	(2 ♂♂)	72.1–74.0; 73.1 (N = 2)	57.6–60.8; 59.2 (N = 2)	28.3–28.6; 28.5 (N = 2)	11.5–13.1; 12.3 (N = 2)	
	(1 ♀♀)	69.9 (N = 1)	58.5 (N = 1)	29.3 (N = 1)	12.1 (N = 1)	
minor	(20 ♂♂)	66.3–74.6; 68.7 (N = 20)	44.4–50.8; 47.6 (N = 19)	25.9–30.0; 28.1 (N = 20)	11.2–12.7; 12.0 (N = 19)	23.7 (N = 1)
	(17 ♀♀)	61.7–70.3; 66.8 (N = 16)	42.9–50.0; 47.1 (N = 17)	27.1–29.3; 28.1 (N = 17)	11.0–12.9; 11.8 (N = 16)	26 (N = 1)
jensoni (Orán)	(2 ♂♂)	64.2–69.2; 66.7 (N = 2)	50.5–52.5; 51.5 (N = 2)	26.3–27.4; 26.9 (N = 2)	11.2–12.6; 11.9 (N = 2)	21.5–22; 21.8 (N = 2)
	(4 ♀♀)	62.7–68.3; 65.5 (N = 3)	44.9–49.6; 47.3 (N = 2)	26.8–27.7; 27.1 (N = 3)	10.8–11.1; 11.0 (N = 3*)	21–22.5; 21.8 (N = 3)
jensoni (Nauta)	(1 ♂)	66.6 (N = 1)	51.7 (N = 1)	**	**	

* Culmen measureable on skeleton.
** Not measureable.

black crown) are comparatively slight and consistent with maintenance of these two pairs of taxa as subspecies. In contrast, *rufifrons* + *subcristata* and *minor* + *jensoni* are at least as well-differentiated phenotypically as several other currently recognized pairs or complexes of species within the Formicariidae, most notably in the genera *Thamnomanes* (*caesius* vs. *schistogynus*), *Myrmotherula* (the *brachyura-obscura*, *gutturalis-fulviventris-leucopthalma*, and *haematonota-spodionota-ornata* complexes), and even within the "*Percnostola*" assemblage itself (e.g., *Schistocichla* [*Percnostola*] *schistacea* vs. *leucostigma*). Based on nomenclatural priorities, the two species would be called *Percnostola rufifrons* and *Percnostola minor*. The English name for *Percnostola rufifrons* would remain Black-headed Antbird. We suggest that the English name for *Percnostola minor* be Amazonas Antbird because the major part of the known range is in Amazonas state, Brazil, Amazonas state, Colombia, and Amazonas state, Venezuela.

In summary, our recommended taxonomy, with approximate ranges (all north of the Amazon), is as follows:

Percnostola rufifrons. Black-headed Antbird
  P. r. rufifrons (eastern Guyana, Surinam, French Guiana, northeastern Brazil)
  P. r. subcristata (north-central Brazil)
Percnostola minor. Amazonas Antbird
  P. m. minor (eastern Colombia, northwestern Brazil, southern Venezuela)
  P. m. jensoni (northeastern Peru)

ACKNOWLEDGMENTS

We dedicate this paper to Theodore A. Parker III. His inspirational pursuit of knowledge of Neotropical birds and perpetually insightful guidance were instrumental in the success of the authors' graduate work at LSU. He will be missed greatly.

In addition, we thank the curators and staff of the Academy of Natural Sciences of Philadelphia (ANSP), American Museum of Natural History (AMNH), British Museum (Natural History) (BMNH), Carnegie Museum of Natural History (CM), Field Museum of Natural History (FMNH), Natural History Museum of Los Angeles County (LACM), Louisiana State University Museum of Natural Science (LSUMZ), Museum of Comparative Zoology, Harvard University (MCZ), Museum of Vertebrate Zoology, University of California at Berkeley (MVZ), National Museum of Natural History, Smithsonian Institution (USNM), University of Michigan Museum of Zoology (UMMZ), and the Yale Peabody Museum (YPM) for permission to examine specimens. We thank R. O. Bierregaard, Jr., L. Harper, S. L. Hilty, and C. E. Quintela for critical information. We thank S. M. Lanyon, C. A. Marantz, J. V. Remsen, Jr., R. S. Ridgely, and T. S. Schulenberg for helpful comments on the manuscript. We thank the Dirección General Forestal y de Fauna, Ministerio de Agricultura, Lima, for permission to work in Peru.

LITERATURE CITED

BIERREGAARD, R. O., JR. 1988. Morphological data from understory birds in *terra firme* forest in the central Amazonian basin. Rev. Brasil. Biol. 48:169–178.
CAPPARELLA, A. P. 1988. Genetic variation in Neotropical birds: implications for the speciation process. Acta Congr. Internat. Ornithol. 19:1658–1664.
CAPPARELLA, A. P. 1991. Neotropical avian diversity and riverine barriers. Acta Congr. Internat. Ornithol. 20:307–316.
DUNNING, J. S. 1982. South American Land Birds. Harrowood Books, Newtown Square, Pennsylvania.
HELLMAYR, C. E. 1924. Catalogue of Birds of the Americas and the Adjacent Islands. Field Mus. Nat. Hist., Zoological Series, 23, part III.
HILTY, S. L., AND W. L. BROWN. 1986. A Guide to the Birds of Colombia. Princeton Univ. Press, Princeton, New Jersey.
ONIKI, Y., AND E. O. WILLIS. 1972. Studies of the ant-following birds north of the eastern Amazon. Acta Amazonica 2:127–151.
PETERS, J. L. 1951. Check-List of Birds of the World. Volume 7. Museum of Comparative Zoology, Cambridge, Massachusetts.
RIDGELY, R. S., AND G. Tudor. 1994. The Birds of South America. Volume II. University of Texas Press, Austin, Texas.
RIDGWAY, R. 1912. Color Standards and Color Nomenclature. Published by the author, Washington, D.C.
TOSTAIN, O., J.-L. DUJARDIN, C. ÉRARD, AND J.-M. THIOLLAY. 1992. Oiseaux de Guyane. Société d'Études Ornithologiques, Brunoy, France.
WILLIS, E. O. 1982. The behavior of Black-headed Antbirds (*Percnostola rufifrons*) (Formicariidae). Rev. Brasil. Biol. 42:233–247.
ZIMMER, J. T. 1932. Studies of Peruvian birds. VII. Amer. Mus. Novit. No. 584.

PATTERNS OF SEASONAL AND GEOGRAPHICAL DISTRIBUTION OF AUSTRAL MIGRANT FLYCATCHERS (TYRANNIDAE) IN BOLIVIA

R. TERRY CHESSER[1]

Museum of Natural Science, Foster Hall 119, and Department of Zoology and Physiology, Louisiana State University, Baton Rouge, Louisiana 70803, USA;
[1]*present address: Department of Ornithology, American Museum of Natural History, New York, New York 10024, USA*

ABSTRACT.—Austral migrants are bird species that breed in temperate South America during the southern summer and migrate north for the austral winter. Tyrant flycatchers are the dominant group of austral migrants. Detailed examination of geographical and seasonal distribution of flycatchers in Bolivia reveals that 57 species or subspecies show evidence of seasonal movements. Most of these (33 taxa) are present within the country throughout the year, and many of these have overlapping breeding and wintering ranges, as is typical of austral migrants. Of the taxa present in Bolivia during only part of the year, 18 occur exclusively as winter residents, five as summer residents, and one solely as a transient. Bolivian austral migrant tyrannids tend to occur in the lowland, puna, and valle life zones, where they occupy mainly scrub, woodland, and open habitats. Relatively few breed in humid areas, including the Amazonian lowlands and upper tropical and subtropical zones, but 16 species occur mainly or exclusively in winter in Amazonia, where they tend to occupy secondary and edge habitats. Many austral migrant tyrannids show seasonal elevational differences in Bolivia, in addition to latitudinal differences. *Pseudocolopteryx acutipennis* is the most extreme of the species studied in range of seasonal elevational differences and delayed timing of migration. The easternmost portions of the Bolivian Andes may be important as a migration route for long-distance austral migrants, including tyrannids.

RESUMEN.—Las aves migradoras australes son aquellas especies que se reproducen en América del Sur durante el verano y migran hacia el norte cuando llega el invierno. Los Tyrannidae son el grupo dominante de aves migradoras australes. Un estudio detallado de la distribución estacional y geográfica de los migradores australes de la familia Tyrannidae de Bolivia reveló que 57 especies y subespecies tenían movimientos estacionales. La mayoría de éstos (33 taxones) están presentes en el país durante todo el año, y muchos presentan superposición de sus rangos de apareamiento y de invierno, como es típico de los migradores australes. De los taxones presentes en Bolivia durante sólo una parte del año, 18 residen exclusivamente durante el invierno, cinco son residentes de verano, y uno solo es transitorio. Los tiránidos que son migradores australes tienden a habitar las zonas de vida de áreas bajas, puna, y valles, donde ocupan ambientes de matorral, arbustivos, y abiertos. Algunos pocos se aparean en áreas húmedas, incluidas los bajos del Amazonas y las zonas tropicales y subtropicales superiores, pero 16 especies existen en Amazonia principalmente o exclusivamente en el invierno, donde tienden a ocupar ambientes secundarios y marginales. Muchos de los tiránidos migradores australes de Bolivia muestran tanto diferencias estacionales de altitud como diferencias latitudinales. *Pseudocolopteryx acutipennis* es la especie que presenta un mayor rango de diferencias estacionales de altitud y retardo migratorio. Las áreas de más del este de los Andes bolivianos podrían ser importantes rutas migratorias para los migradores australes de larga distancia, incluyendo los Tyrannidae.

Austral migrant birds are species that breed in south temperate South America during the southern summer, and migrate north for the austral winter. Austral migration is in this respect a mirror image of the north temperate migration systems, such as Nearctic-Neotropical migration, although the number of austral migrant species is lower and distances migrated are generally less, often involving overlapping breeding and wintering ranges (Chesser 1994). Austral migration was recognized as long ago as the time of Azara (1802–1805), and many ornithologists, such as Hudson and Barros, have written extensively about migrants in their countries, but system-wide investigation of austral migration was until recently limited to brief discussions in Zimmer (1938) and Sick (1968). The last decade has seen greatly increased interest in austral migration, however, and has witnessed the publication of a number of works dealing solely or in part with individual migrant species (e.g., Lanyon 1982, Remsen and Parker 1990, Marini and Cavalcanti 1990, Marantz and Remsen 1991, Chesser and Marín 1994), migration in specific regions (e.g., Nores et al. 1983; Belton 1984, 1985; Sick 1984, 1993; Willis 1988; Fjeldså and Krabbe 1990; Narosky and Di Giacomo 1993; Hayes et al. 1994), and austral migration as a whole (Chesser 1994).

In this paper I examine the distributional patterns in Bolivia of austral migrant tyrant-flycatchers (Tyrannidae), the dominant group of austral migrants (Chesser 1994). Most previous discussions of austral migrants have focussed on temperate localities, such as Argentina, Chile, and southeastern Brazil (although see, for example, Zimmer 1938, 1931–1955; Pearson 1980). Bolivia, in contrast, lies wholly within the tropical zone of South America, and patterns of migration there differ considerably from those in more southern localities. In addition, Bolivia is extremely topographically diverse, allowing for elevational, as well as latitudinal, migratory movements.

Bolivia's avifauna is the most diverse of any land-locked country in the world, owing primarily to its Neotropical location and the great topographical diversity (Remsen and Traylor 1989). Because these factors are so important to the distribution of Bolivian birds, an understanding of the results and discussion to follow will be greatly enhanced by an introduction to the geography of the country.

Bolivia is the fourth largest country in South America, extending from approximately 10° to 23° south latitude and 57° to 70° west longitude, and bordered to the west by Chile and Peru, to the north and east by Brazil, and to the south by Paraguay and Argentina (Fig. 1). The country may be divided into two main parts: the lowlands, which encompass the whole of northern and eastern Bolivia, and the Andean highlands, which are found in roughly the southwestern quadrant of the country, although they extend north of this area in depto. La Paz (Fig. 1). Much of the northern part of Bolivia, including all of depto. Pando and portions of depto. El Beni and northern deptos. La Paz, Cochabamba, and Santa Cruz, is composed of lowland tropical forest. Interspersed among these forests, especially in depto. El Beni, but also in deptos. La Paz and Santa Cruz, are seasonally inundated savannas, which also extend along the Brazilian border in depto. Santa Cruz south of the zone of lowland tropical forest. The major part of depto. Santa Cruz, in contrast, is composed of chaco and dry and semi-humid lowland forest, all of which extend south into eastern deptos. Chuquisaca and Tarija.

The highland region of Bolivia is likewise composed of a variety of habitats. Montane forests line the eastern slopes of the easternmost Andes (Cordillera Oriental). From depto. La Paz through depto. Cochabamba and into portions of extreme western depto. Santa Cruz, these are humid or cloud forests; in the southern half of the country, from western depto. Santa Cruz to depto. Tarija, these montane forests are semihumid. West of the semihumid montane forest, between the Cordillera Oriental and the Cordillera Central, is an area of semiarid valleys, extending from depto. Cochabamba south through western deptos. Chuquisaca and Tarija, and eastern depto. Potosí. The southwestern portion of Bolivia, encompassing southern depto. La Paz, most of deptos. Oruro and Potosí, and part of southwestern depto. Cochabamba, is composed of an arid montane plain, the altiplano, surrounded by the high montane puna of the Cordillera Central to the east, and the Cordillera Occidental to the west, along the Chilean and Peruvian borders.

METHODS

The information presented here is based on specimens and, occasionally, published sight records. Although Bolivia lacks the history of local ornithological interest enjoyed by some South American countries, most regions of the country have been studied in sufficient detail for patterns of seasonal distribution to emerge. The major exception to this, and a potential source of bias in the results, is the extreme northern portion of Bolivia, in the lowland rainforests and savannas of depto. Pando and northern deptos. La Paz, El Beni, and Santa Cruz. Relatively few collections have been made in this area, and this collecting activity has been concentrated during the dry season (May to November); there are no records for this region for most of December, January, and February.

Data on known or suspected austral migrant tyrannids were gathered during research trips to major North and South American collections of Bolivian birds, from a thorough search of the literature on Bolivian birds, and occasionally through correspondence with personnel at museums

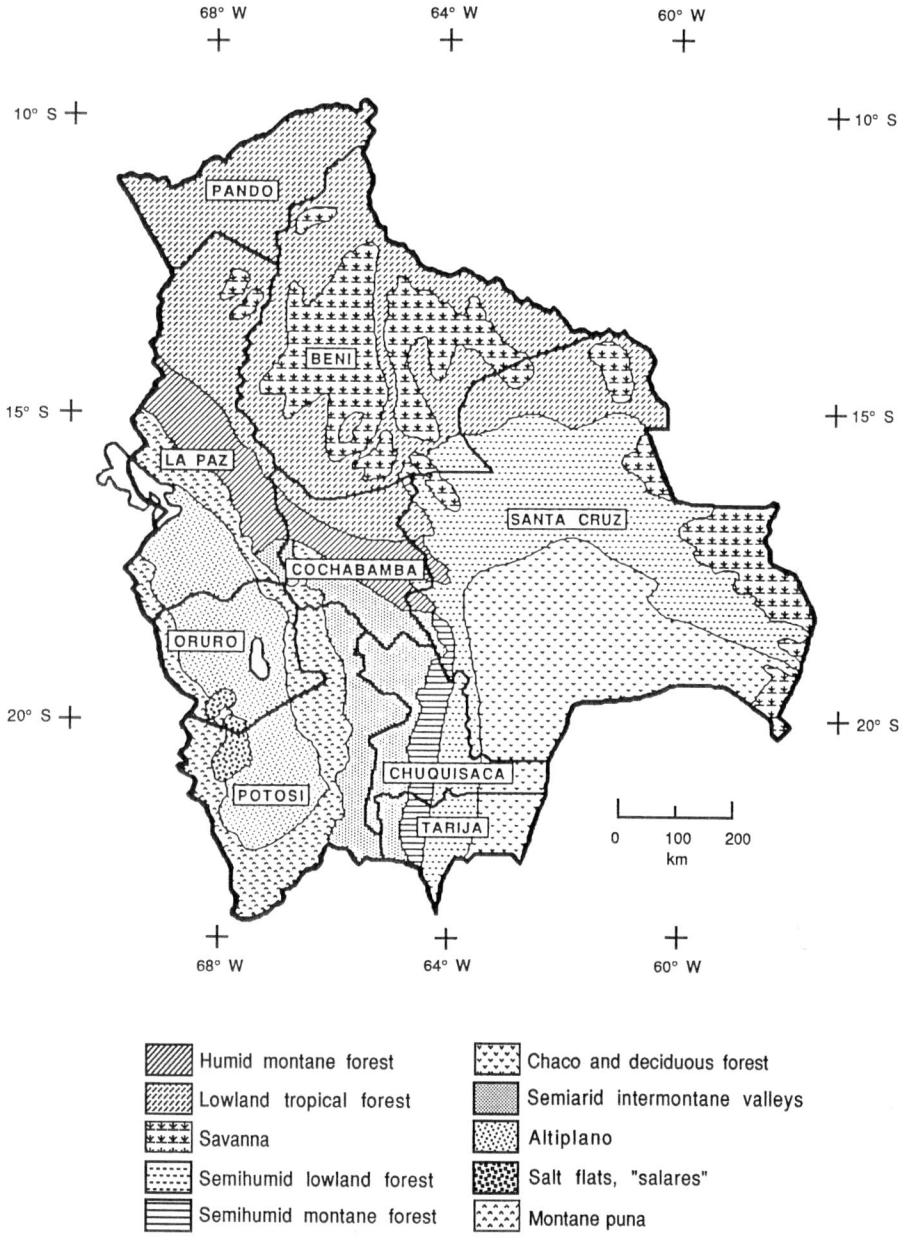

FIG. 1. Map of Bolivia, showing major habitats and departamentos (from Remsen and Traylor 1989).

I was unable to visit personally. Principal information taken from each specimen included museum number, collector, date of collection, locality, elevation, sex, and any supplemental data on subcutaneous fat deposits, habitat, and gonad size or other information on breeding condition. Specimen or literature records lacking date of collection could not be used in most cases. Data on date and locality of collection were then mapped, using geographical coordinates from the Bolivian ornithological gazetteer of Paynter (1992) and coordinates determined from local maps (for localities not covered in the gazetteer), for all species of austral migrant tyrannids known or suspected to be migratory in Bolivia. Patterns of seasonal distribution were assessed and fit into a continental context. For those few taxa whose seasonal distributions were unclear, or for which collecting bias, especially in extreme northern Bolivia, might have played a factor, available distributional data, including specimens and literature records, were examined from neighboring countries.

Geographical patterns of Bolivian austral migrant tyrannids were grouped according to available data into categories of migration (see Table 1 et seq.), and analyses were conducted relating migration patterns to taxonomy, life zone occurrence (following the scheme of Remsen and Traylor [1989]), habitat (following the methods of Chesser [1994]), and foraging (after Fitzpatrick [1980]). Detailed maps of seasonal distribution were drawn for nine flycatchers (Figs. 2–10), illustrating the variety of migration patterns among Bolivian tyrannids. Patterns of seasonal differences in elevation and timing of migration of Bolivian migrant tyrannids were also considered. In four instances, multiple subspecies (of the same species) were found to undergo seasonal movements in Bolivia; in all analyses, these were considered to be separate taxa.

The species accounts follow the taxonomy and sequence of Remsen and Traylor (1989). The accounts contain information concerning the status of each species or subspecies in Bolivia, its winter and summer distribution, months of occurrence (including early and late records) for those parts of the range where the species is not present year-round, and elevational range, as well as any additional relevant data (e.g., specimens with subcutaneous fat deposits). Status for a species in any particular portion of its range is indicated as either permanent resident, summer resident, winter resident, or transient (passage migrant); in many cases, the same species is of different status in different parts of Bolivia. Taxa that are migratory in southern South America, especially in Argentina, but which do not show seasonality of distribution in Bolivia, are not discussed. Taxa present in an area during the austral spring and summer, the breeding season for these taxa in more southerly portions of their range, are assumed to breed in that area.

Specimens of particular interest are identified by museum acronym and collection number. Acronyms used are AMNH (American Museum of Natural History, New York), ANSP (Academy of Natural Sciences of Philadelphia), BMNH (British Museum [Natural History], Tring), CM (Carnegie Museum, Pittsburgh), DMNH (Delaware Museum of Natural History, Greenville), FML (Fundación Miguel Lillo, Tucumán), FMNH (Field Museum of Natural History, Chicago), LACM (Los Angeles County Museum, Los Angeles), LSUMZ (Louisiana State University Museum of Natural Science, Baton Rouge), MACN (Museo Argentina de Ciencias Naturales, Buenos Aires), UMMZ (University of Michigan Museum of Zoology, Ann Arbor), and PMNH (Peabody Museum of Natural History, Yale University, New Haven).

RESULTS AND DISCUSSION

Fifty-seven species or subspecies of austral migrant flycatchers were found to undergo migratory movements in Bolivia; another six are of uncertain status or have been recorded solely as vagrants (see Appendix). Many of these taxa are found in some portion of Bolivia throughout the year, but others are present only during winter or, less frequently, solely as summer residents or transients.

SPECIES ACCOUNTS

Camptostoma obsoletum bolivianum (Southern Beardless-Tyrannulet).—Apparent permanent resident (81 specimens) in most of its range in Bolivia, including deptos. Tarija, Chuquisaca, Cochabamba, Santa Cruz, and possibly El Beni; probably present exclusively as a winter resident (seven specimens) in northwestern depto. Cochabamba and depto. La Paz (and adjacent depto. Madre de Dios, Peru), where there are records only for May, June, August, September, and October (earliest 20 May [CM 119644], latest 11 October [Fjeldså and Krabbe 1989; J. M. C. da Silva, pers. comm.]). There is also a published sight record from depto. Potosí (Nores and Yzurieta 1984). This species is migratory in Argentina (Nores et al. 1983, Capurro and Bucher 1988) and may be migratory elsewhere (Short 1975, Belton 1985). It has been found in Bolivia from the lowlands to over 3,000 m in both the breeding and non-breeding seasons.

Phaeomyias murina ignobilis (Mouse-colored Tyrannulet).—Summer resident (34 specimens) in deptos. Tarija, Chuquisaca, southwestern depto. Santa Cruz, and apparently southern depto. Cochabamba; permanent resident (12 specimens) in deptos. Cochabamba and Santa Cruz north and east of this area. *Phaeomyias m. ignobilis* has been collected in the area of summer residence, along the Andean foothills south of ca. 18°S, in all months from October through March (earliest record 9 October [ANSP 136192], latest 25 March [CM 119535]). There are only eight winter specimens of this subspecies (one from June, seven from August), indicating that it is overlooked in the non-breeding season. B. Whitney (in litt.) has observed individuals of this species in apparent fall migration along the base of the Andes near Santa Cruz (depto. Santa Cruz), and

Hayes et al. (1994) reported *P. murina* as a summer resident in Paraguay. *Phaeomyias m. ignobilis* breeds in Bolivia up to 2,150 m, and possibly as high as 3,085 m (FMNH 181518, collected 20 October), but has been found in winter only at low elevation (800 m and below).

Sublegatus modestus brevirostris (Scrub Flycatcher).—Winter resident (15 specimens) in northern Bolivia; permanent resident (31 specimens) along the Andes from depto. Tarija and western depto. Potosí north to northern depto. Chuquisaca, western depto. Santa Cruz, possibly southeastern depto. Cochabamba, and possibly in eastern depto. Santa Cruz (although it may occur there only as a transient). This subspecies has been recorded from deptos. Pando and La Paz (one record; ANSP 120029), and northern deptos. Cochabamba and Santa Cruz, north of ca. 17°S, only in June, July, August, and September (earliest record 23 June [LSUMZ 133585], latest 22 September [LSUMZ 150870]). *Sublegatus m. brevirostris* has been recorded from eastern depto. Santa Cruz during September, March, and April only, suggesting that the subspecies may occur there exclusively as a passage migrant. Likewise, specimens collected in southern depto. Santa Cruz, near the Paraguayan border, in September 1990 (LSUMZ 153753–55), were probably migrants. This species was not found at this locality in June of that year (T. A. Parker, pers. comm.), and the September specimens all had deposits of subcutaneous fat, ranging from light to heavy. Elevations at which *S. m. brevirostris* has been collected in Bolivia range from 300 m to 2,750 m, with a wide elevational range throughout the year.

Suiriri s. suiriri (Suiriri Flycatcher).—Winter resident (64 specimens) in lowland areas of deptos. Santa Cruz and Chuquisaca; permanent resident (68 specimens) in the montane and semiarid intermontane zone of depto. Cochabamba and extreme western deptos. Santa Cruz and Chuquisaca, and probably most of depto. Tarija. Sixty-eight of the seventy lowland records (below 1,000 m) for this subspecies have been recorded from March through October (earliest record 9 March [FMNH 295414], latest 24 October [Laubmann 1930]); the other two records are a February specimen from depto. Tarija (Salvadori 1897), probably a post-breeding wanderer, and a January specimen from eastern depto. Santa Cruz (FMNH 295413), an apparent non-breeder (skull completely ossified, but testes 1×1 mm). With the exception of these two lowland records, *S. s. suiriri* has been found from 1,525 m to ca. 3,000 m during the breeding season, and from 300 m to ca. 3,000 m in the non-breeding season.

Myiopagis c. caniceps (Gray Elaenia).—Of uncertain status in Bolivia, but possibly an uncommon winter resident. There are only two records from Bolivia, one from depto. Tarija, collected 21 August (ANSP 136150), the other from depto. Santa Cruz, collected 21 July (LSUMZ 124630). *Myiopagis c. caniceps* has also been found near the Bolivian border in the state of Mato Grosso, Brazil (Willis and Oniki 1990) and in Salta and Jujuy, Argentina. This subspecies is considered a possible migrant in southeastern Brazil (Belton 1985). It appears that all records from the westernmost portion of its range, from ca. 58°W to 65°W, are from the non-breeding season (June through October; $n = 8$; Chesser, unpubl. data), such that the Bolivian specimens form part of an unusual pattern. It has been collected only at ca. 700 m in Bolivia.

Myiopagis v. viridicata (Greenish Elaenia).—Summer resident (41 specimens) in southern Bolivia; apparent permanent resident (12 specimens) in much of central Bolivia; apparently occurs only as a winter resident (11 specimens) in extreme northern Bolivia. This subspecies has been recorded in the breeding season from deptos. Tarija and Chuquisaca, and southern depto. Santa Cruz, south of ca. 18°S, in October, November, December, January, February, March, April, and May (earliest record 19 October [LACM 35550]), latest 7 May [DMNH 67075]). North of ca. 17°S, in deptos. Pando, La Paz, El Beni, and northern depto. Santa Cruz, *M. v. viridicata* has been recorded during February, March, July, August, September, October, and November. It seems clear that the species winters in this region, and the relative lack of summer records, combined with the presence of birds with deposits of subcutaneous fat in the early spring (e.g., LSUMZ 150876, collected 4 October with extremely heavy fat), suggests that many individuals vacate the area during the austral summer. Davis (1993) recorded it only in February, March, and November at Concepción, depto. Santa Cruz, and was unsure of its status there. This species apparently occurs only in winter in Peru, and in the states of Amazonas, Acre, and Rondônia, northwestern Brazil (Chesser 1995), coinciding with its period of occurrence in northern Bolivia. *Myiopagis v. viridicata* has been found in Bolivia as high as 1,300 m during breeding, but only below 750 m in the non-breeding season. It has not been recorded from deptos. Oruro and Potosí.

Elaenia s. spectabilis (Large Elaenia).—Summer resident (28 specimens) in southern Bolivia; winter resident and transient (24 specimens) to the north of ca. 18°30′ S. There are breeding season records for this species from southwestern and extreme southeastern depto. Santa Cruz, and depto. Tarija, in all months from October through April (earliest record 30 October [ANSP

136088], latest 22 April [LSUMZ 124558]). *Elaenia spectabilis* has been recorded north of this area, in deptos. Pando, La Paz, El Beni, and Santa Cruz, only in April, May, June, July, September, October, and November (earliest record 15 April [FMNH 181435], latest 20 November [FMNH 334472]). Davis (1993) reported this species as an uncommon transient during October and November at Concepción, depto. Santa Cruz, and late October and November specimens from the non-breeding range had moderate [FMNH 334471] to very heavy [FMNH 335163] deposits of subcutaneous fat. This lowland species has been recorded to 1,125 m in the breeding season, and to 600 m in the non-breeding season. There are no records of *E. spectabilis* from deptos. Cochabamba, Chuquisaca, Oruro, or Potosí.

Elaenia albiceps chilensis (White-crested Elaenia).—Transient (48 specimens) in the lowlands of depto. Santa Cruz and in deptos. Cochabamba and La Paz, and apparent summer resident (11 specimens) in depto. Tarija. Intermediates between this subspecies and *E. a. albiceps* (the resident subspecies in depto. Cochabamba) breed in depto. Chuquisaca, and hybrids between *E. a. chilensis* and *E. parvirostris* breed at intermediate elevations on the eastern slopes of the Andes in deptos. Tarija, Chuquisaca, and Cochabamba (Traylor 1982). Traylor suggested that populations of *chilensis* on the eastern slopes of the Andes are probably sedentary, because of wing shape, but there appear to be no Bolivian specimens of *chilensis* from June and July; it is here considered to be wholly migratory, pending further information. Specimens of pure *E. a. chilensis* have been taken in deptos. Santa Cruz, La Paz, and Cochabamba in March, April, and May (earliest 13 March [FMNH 294294–95], latest 20 May [CM 94751]), and in August, September, October, and November (earliest 6 August [ANSP 120001], latest 14 November [FMNH 181466]), a pattern probably best considered that of a passage migrant. There are records of pure *chilensis* (depto. Tarija) and the intermediate and hybrid birds (deptos. Tarija, Chuquisaca, Cochabamba, and Santa Cruz) from October, November, December, January, and February (earliest record 19 October [ANSP 136098], latest 27 February [FMNH 294304]). Specimens of presumed transient *E. a. chilensis* have been collected from 300 m to 3,300 m; breeding season records (including the intermediates and hybrids) range from 1,400 m to 2,450 m (see Traylor 1982 for further details). The continent-wide seasonal and geographic distribution of this species was reviewed by Marini and Cavalcanti (1990).

Elaenia parvirostris (Small-billed Elaenia).—Winter resident (40 specimens) in northern Bolivia; summer resident (213 specimens) along the eastern slopes of the Andes in deptos. Tarija, Chuquisaca, and southern deptos. Santa Cruz and Cochabamba, and possibly further east in depto. Santa Cruz. There are records from deptos. Pando, El Beni, and northern deptos. Santa Cruz and Cochabamba, north of ca. 17°S, for all months from March through November (earliest record 23 March [FMNH 335168], latest 20 November [FMNH 334473]). The 20 November specimen had heavy deposits of subcutaneous fat, indicating a migrating bird. *Elaenia parvirostris* has been recorded from its breeding area along the base of the Andes in all months from September through May (earliest record 15 September [ANSP 43810], latest 25 May [CM 80219]). This species is of uncertain status in eastern depto. Santa Cruz, where there are specimens only from March, April, and May (although the March specimen is a juvenile [M. A. Traylor, in litt.]), and at Concepción, north-central depto. Santa Cruz, where Davis (1993) reported records from October and January through April, but whose specimens indicate its presence in September and March only (FMNH 335166–68). This species has been reported in January from Estação Ecológica Serra das Araras, Mato Grosso, Brazil (Silva and Oniki 1988). *Elaenia parvirostris* is found up to 1,500 m (intergrades with *E. albiceps chilensis* up to 2,150 m, and possibly to 2,450 m) during the breeding season in Bolivia, and up to ca. 1,000 m in the non-breeding season. See Traylor (1982) for data on hybridization of this species with the higher elevation *E. albiceps*.

Elaenia strepera (Slaty Elaenia).—Summer resident (59 specimens) in deptos. Tarija, Chuquisaca, and western depto. Santa Cruz. There are records from Bolivia for all months from September through March (earliest record 19 September [CM 50855], latest 27 March [CM 81091]). *Elaenia strepera* has been collected in Bolivia only along the eastern slope of the Andes from 500 m to 2,700 m. The seasonal and geographic distribution of this species was reviewed in detail by Marantz and Remsen (1991).

Elaenia chiriquensis albivertex (Lesser Elaenia).—Summer resident (20 specimens) in depto. Santa Cruz; apparent permanent resident (21 specimens) in depto. El Beni; recorded once from depto. La Paz. Specimens have been collected in depto. Santa Cruz during September, October, January, February, March, and April (earliest record 11 September [FMNH 62623], latest 30 April [UMMZ 106860]). There are records for depto. El Beni from March, August, September, October, and December, and a July specimen for eastern depto. La Paz (ANSP 119997). This species has been recorded in the Bolivian lowlands up to 700 m. The continent-wide seasonal and geographic distribution of this species was reviewed by Marini and Cavalcanti (1990).

Serpophaga nigricans (Sooty Tyrannulet).—Of uncertain status in Bolivia. This species has been recorded only in depto. Tarija, and only during September and October (earliest record 17 September [ANSP 136028], latest between 21 and 31 October [Nores and Yzurieta 1984]). Carriker, who collected all seven Bolivian specimens, found this species breeding in late October at Entre Rios (Bond and Meyer de Schauensee 1942). *Serpophaga nigricans* is thought to be migratory (Meyer de Schauensee 1970; Short 1975; Traylor 1979) or partially migratory (Barrows 1883) in at least some portion of its range.

Serpophaga munda (White-bellied Tyrannulet).—Permanent resident (68 specimens) in most of its range in Bolivia, but apparently occurs exclusively as a winter resident (four specimens) in depto. Santa Cruz east of ca. 62°30'W. There are both summer and winter records for *S. munda* from deptos. Tarija, Potosí, Chuquisaca, and Cochabamba, and western depto. Santa Cruz; there is also a single record from La Paz on 11 November (Niethammer 1956). Davis (1993), evidently treating *S. munda* as a subspecies of *S. subcristata*, and not distinguishing between the subspecies, observed "*S. subcristata*" at Concepción, depto. Santa Cruz (16°08'S, 62°02'W), only in April, June, September, and October; her single specimen is an individual of *S. munda*, suggesting that this species is a winter resident in her area. Other evidence of seasonal occurrence comes from eastern depto. Santa Cruz and across the border in western Mato Grosso, Brazil, where *S. munda* has been recorded in May, July, and August (Chesser 1995, unpubl. data). The species is migratory to the south in Argentina (Olrog 1963; Meyer de Schauensee 1966; Traylor 1979; Nores et al. 1983; Fjeldså and Krabbe 1990). This species has been recorded at a wide range of elevation (300 m to 3,500 m) more-or-less throughout the year.

Serpophaga subcristata (White-crested Tyrannulet).—Winter resident (40 specimens) in Bolivia. This species has been recorded from depto. Tarija, eastern depto. Chuquisaca, and depto. Santa Cruz, with single records from deptos. Cochabamba (ANSP 135970), El Beni (Cabot 1990), and La Paz (Remsen, unpubl. data). There are records for all months from April through September (earliest record 13 April [Laubmann 1930], latest 19 September [ANSP 135971]). Curiously, this species was recorded exclusively in summer (September through February) across the Bolivian border in the Pantanal of Mato Grosso, Brazil, by Cintra and Yamashita (1990). *Serpophaga subcristata* has been recorded in a variety of lowland habitats in Bolivia, from 300 m to 850 m.

Inezia inornata (Plain Tyrannulet).—Winter resident (23 specimens) in deptos. Pando, El Beni, and northern depto. Santa Cruz, north of ca. 17°S; apparent summer resident (five specimens) in southwestern depto. Santa Cruz and depto. Tarija; permanent resident (20 specimens) from northern depto. Cochabamba east through eastern depto. Santa Cruz, south of 17°S. Records from the "winter only" range are from May, June, July, August, September, and October (extreme dates are 9 May and 7 October [both from Gyldenstolpe 1945]). The November records from the Río Itenez on the El Beni/Rondônia border (Ridgely and Tudor 1994) are evidently erroneous, the result of transposing IX (September) to XI (November; AMNH 792118–792120). *Inezia inornata* also occurs only as a winter resident in depto. Madre de Dios, Peru, and the Brazilian states of Acre, Rondônia, and Mato Grosso (Chesser 1995; Willis and Oniki 1990). Records for the breeding range are from November and January (earliest record 4 November [ANSP 135980], latest 11 January [DMNH 66039]). This lowland species has been recorded in Bolivia up to 1,100 m in summer, and to 800 m during the non-breeding season.

Culicivora caudacuta (Sharp-tailed Grass-Tyrant).—Probably an uncommon and local permanent resident in Bolivia, but some individuals may migrate south to breed. The 14 Bolivian specimens of this species were collected in deptos. Santa Cruz, El Beni, and La Paz, in March, June, July, August, October, and November (earliest record 16 March [CM 79126], latest 20 November [CM 51046]). Although these records are most consistent with winter resident status, the birds observed in June were singing and in family groups (Parker and Rocha 1991), and may have been territorial and resident. A series of this species (LSUMZ 150981–87, 151790, 151884) was collected atop the Serranía de Huanchaca in northeastern depto. Santa Cruz in early October; to date this is the only month that collections have been made at this locality. One of these birds (LSUMZ 150985) had heavy deposits of subcutaneous fat, possibly indicating pre-migratory fattening, but the others had no fat or light fat deposits. *Culicivora caudacuta* may be partially migratory in Argentina (B. Whitney, in litt.). This grassland species has been found from 350 m to 725 m in Bolivia.

Polystictus p. pectoralis (Bearded Tachuri).—Winter resident (eight specimens) in Bolivia. This species has been collected only in western depto. Santa Cruz, where there are records from

July, August, September, and October (earliest record 30 July [CM 43944–45], latest 18 October [AMNH 498980]). Across the Brazilian border in Mato Grosso, the species is also apparently present only in winter, from May through September (Chesser 1995, unpubl. data). All Bolivian records are from lowland localities (350 m to 700 m).

Pseudocolopteryx sclateri (Crested Doradito).—Of uncertain status in Bolivia, where recorded from only two localities. Three specimens were taken in south-central depto. El Beni from 18–20 October 1988 (LSUMZ 123065, 124487–88), and seven in western depto. El Beni from 30-31 December 1937 (Gyldenstolpe 1945). The two October specimens for which fat and gonad data are available, had no or only a trace of fat, and were probably not breeding (female with ovary 3×2 mm, ova and oviduct not enlarged, although male with largest testis 7 × 5). The species is probably a local permanent resident in Bolivia, although it may be migratory in the southern portion of its range in Argentina and Brazil (Meyer de Schauensee 1970, Short 1975, Belton 1985; but see Ridgely and Tudor 1994).

Pseudocolopteryx dinellianus (Dinelli's Doradito).—Apparently an uncommon winterer in southern Bolivia; there are only three Bolivian records for this marshland species. A recently re-identified specimen was collected from the Sierra de Santa Rosa (Itau), depto. Tarija, 1,600 m, in September 1924 (BMNH 1925.6.5.22), and there are two specimens from Villa Montes, depto. Tarija, ca. 500 m, collected 16 April and 5 May 1926 (Laubmann 1934; elevation from Paynter 1992). This species was considered migratory by Short (1975) and Olrog (1984).

Pseudocolopteryx acutipennis (Subtropical Doradito).—Summer resident (64 specimens) in the highlands of deptos. Tarija, Chuquisaca, Cochabamba, and La Paz; occurs solely as a non-breeder (17 specimens) in the lowlands of deptos. Santa Cruz, Cochabamba, El Beni, and La Paz. This species has been recorded from its breeding area in Bolivia during December, January, February, March, and April (earliest record 17 December [FMNH 181345, but specimen not located], other early 22 December [FMNH 181340]; latest 20 April [LSUMZ 37967]). Records from the wintering range are from May, June, August, October, and December (earliest record 2 May [FMNH 296279], latest 5 December [LSUMZ 37969]). This species is migratory to the south in Argentina (Short 1975; Traylor 1979; Nores et al. 1983; Fjeldså and Krabbe 1990). Elevational range in Bolivia during the breeding season is from 2,200 m to 3,550 m; remarkably, *P. acutipennis* has not been recorded above 700 m outside of the breeding season.

Pseudocolopteryx flaviventris (Warbling Doradito).—Vagrant. There is only one Bolivian record for this migratory species, a specimen taken 15 May 1915 at Buenavista, depto. Santa Cruz, 400 m (CM 51268; Remsen et al. 1987).

Euscarthmus m. meloryphus (Tawny-crowned Pygmy-Tyrant).—Winter resident in eastern, central, and northern Bolivia; of uncertain status in depto. Tarija, where the only record is a specimen collected 7 November (ANSP 135897). This subspecies has been recorded from deptos. Pando (Davis et al. 1994), El Beni, and Santa Cruz in all months from May through November (earliest record 2 May [FMNH 296276; Remsen et al. 1987], latest 6 November [FMNH 334479]). Although the 6 November specimen, collected in eastern depto. Santa Cruz, had only light fat, it was not in breeding condition (testes 4 × 3 mm). Records of *E. m. meloryphus* from western Mato Grosso, Brazil, indicate that it is also present there only during the non-breeding season (Willis and Oniki 1990; Chesser 1995). This species is found at low elevation in Bolivia, from 175 m to 700 m.

Myiophobus fasciatus auriceps (Bran-colored Flycatcher).—Summer resident (79 specimens) along the Andes and Andean foothills from depto. La Paz south through Tarija and into Argentina; winter resident (64 specimens) principally to the north and east of this area; apparently occurs year-round (six specimens) in central depto. Cochabamba (Fig. 2). Records from the area of winter occurrence only (deptos. Pando and El Beni, central and northern depto. La Paz, and all but extreme western depto. Santa Cruz) range from 25 March (FMNH 335212, 335214) to 1 November (PMNH 38669), and include all months in between. Records from the area of summer residence range from 10 October (FMNH 181279) to 3 May (Laubmann 1930). This species breeds from 1,000 m to 2,800 m in southern depto. La Paz and deptos. Cochabamba, Santa Cruz, and Chuquisaca (dropping down to 450 m near the Argentine border in depto. Tarija), but has been found almost exclusively below 1,400 m during the non-breeding season (one record from 2,225 m [FMNH 181275], and one from ca. 2,700 m [Fjeldså and Krabbe 1989], out of 64 total). At least some breeding populations in Bolivia may be subspecifically distinct from *auriceps* (unpublished LSUMZ data).

Lathrotriccus e. euleri and *L. e. argentinus* (Euler's Flycatcher).—The nominate subspecies appears to be a winter resident (36 specimens) in deptos. Santa Cruz, Cochabamba, El Beni, and La Paz; *L. e. argentinus* is an apparent summer resident (39 specimens) in southern and central

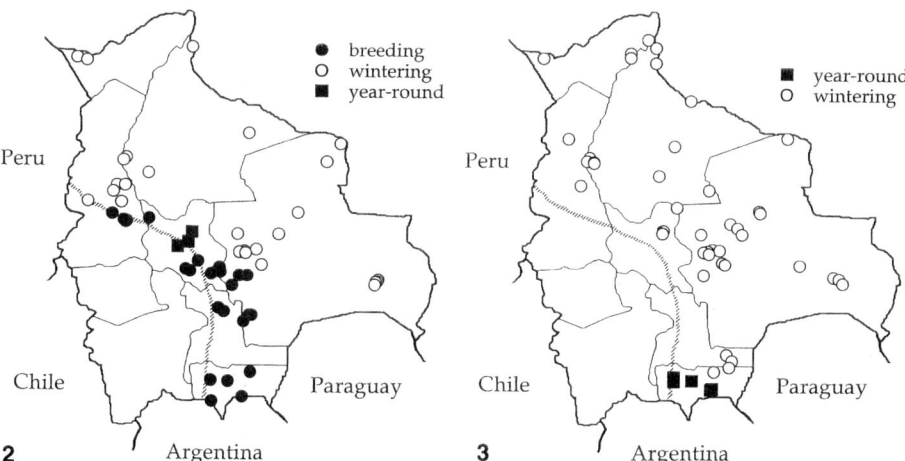

FIG. 2. Seasonal distribution in Bolivia of *Myiophobus fasciatus*, the Bran-colored Flycatcher. In this map and those to follow, each symbol represents one or more specimens collected at the locality indicated, and the lines partitioning off the southwestern portion of the country roughly divide high elevation localities (ca. 2,000 m and above) from low elevation localities (below ca. 2,000 m). *Myiophobus fasciatus* is a summer resident along the Andes, and a winter resident in the lowlands to the north and east, with areas of apparent permanent residency in depto. Cochabamba.

FIG. 3. Seasonal distribution in Bolivia of *Pyrocephalus rubinus*, the Vermilion Flycatcher. This species is a permanent resident in the lowlands of extreme southern Bolivia and a winter resident in the lowlands to the north, including Amazonia.

Bolivia in deptos. Tarija, Chuquisaca, Santa Cruz, and Cochabamba, and an apparent winter resident (25 specimens) in the north in deptos. La Paz and El Beni. There are records for *L. e. euleri* for April, June, July, August, September, and October (earliest record 15 April [FMNH 181242], latest 13 October [FML 11924]). *Lathrotriccus e. argentinus* has been recorded in southern and central Bolivia, south of ca. 16°S, in all months from August through April (earliest record 7 August [ANSP 138726], latest 10 April [CM 81147]). This subspecies has been found in deptos. La Paz and El Beni in April and August only, suggesting that it may be present there only during migration. *Lathrotriccus e. euleri* ranges up to 1,650 m in Bolivia; *L. e. argentinus* has been recorded up to 2,150 m on its Bolivian breeding grounds.

Cnemotriccus fuscatus bimaculatus (Fuscous Flycatcher).—Permanent resident (18 specimens) in southeastern Bolivia; apparent summer resident (13 specimens) along the lower foothills of the Andes from depto. Tarija north to depto. Cochabamba; apparent winter resident (18 specimens) in northeastern depto. Santa Cruz. Records for the area of apparent summer residence are from September, October, and November (earliest 30 September [FMNH 217774], latest 29 November [LSUMZ 36305]), and for the area of apparent winter residence from July and August (earliest 1 July [LSUMZ 137535], latest 8 August [LSUMZ 137552]). This species was found to occur mainly as a summer resident in Paraguay (Hayes et al. 1994) and is present only in winter in western Amazonian Brazil (Chesser 1995). Year-round elevational range for *C. f. bimaculatus* in Bolivia is generally 700 m or below, but there is a single November record from ca. 3,000 m (AMNH 139436). Some migrant *C. f. bimaculatus* may winter within the range of *C. f. beniensis*, but distinguishing migrants is difficult, because of extensive character overlap between these subspecies.

Pyrocephalus r. rubinus (Vermilion Flycatcher).—Winter resident (164 specimens) in Bolivia, except in the extreme south (four specimens from depto. Tarija, where permanently resident), and perhaps in depto. Cochabamba (10 specimens; possible permanent resident); not recorded in depto. Oruro or Potosí (Fig. 3). Collected in deptos. Pando, La Paz, El Beni, Santa Cruz, and Chuquisaca in all months from April through November (earliest record 8 April [FMNH 296226], latest 1 November [FMNH 335226]). Records from depto. Cochabamba include three November specimens and an early December specimen (LSUMZ 36231–33, 37894), indicating that some individuals may stay to breed, but no breeding or fat information was noted, and there are no other austral summer records. Recorded elevations for this species in Bolivia range from 175 m to 2,400 m, with the vast majority below 1,000 m.

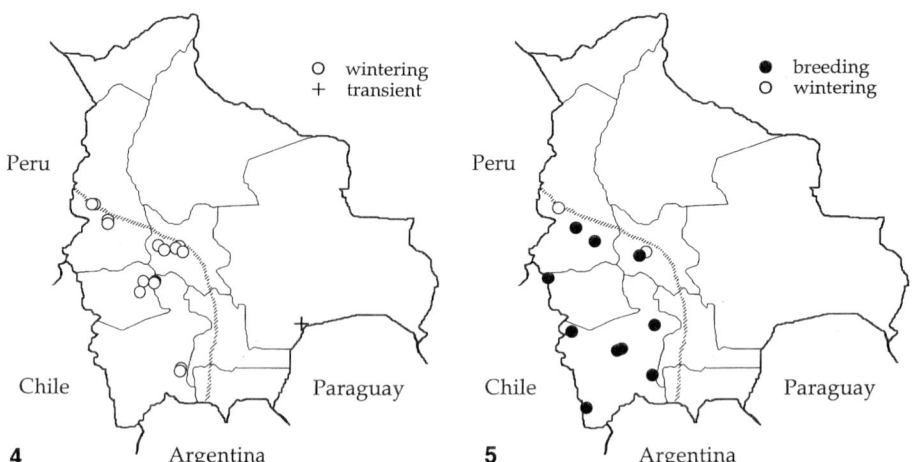

FIG. 4. Seasonal distribution in Bolivia of *Muscisaxicola capistrata*, the Cinnamon-bellied Ground-Tyrant. This species is a winter resident in the high Andes of Bolivia.

FIG. 5. Seasonal distribution in Bolivia of *Muscisaxicola flavinucha*, the Ochre-naped Ground-Tyrant. This highland species is present in Bolivia mainly during the breeding season.

Xolmis coronata (Black-crowned Monjita).—Winter resident (15 specimens) in the chaco and dry forest zone in eastern depto. Tarija and southeastern depto. Chuquisaca, with a single record from western depto. Santa Cruz (CM 32854). This species has been recorded in Bolivia in April, May, July, August, and September (earliest date 24 April [Laubmann 1930], latest 4 September [LACM 35527]). All specimens for which elevation was noted were taken at ca. 500 m.

Agriornis m. microptera (Gray-bellied Shrike-Tyrant).—Uncommon winter resident (six specimens) in Bolivia. There are specimens from deptos. Tarija and Cochabamba, southern depto. La Paz, and western depto. Santa Cruz, collected in May, June, July, and September (earliest record 30 May [LSUMZ 102445], latest 22 September [ANSP 135202]). The depto. La Paz specimen (LSUMZ 102445) is the northernmost record for the subspecies (ca. 16°19'S). This subspecies has been recorded in Bolivia at widely divergent elevations, from 375 m to 3,750 m.

Agriornis murina (Lesser Shrike-Tyrant).—Winter resident (37 specimens) in deptos. Tarija, Chuquisaca, and Cochabamba, and western depto. Santa Cruz. Months during which *A. murina* has been recorded from Bolivia are May, June, July, August, and September (extreme dates are 2 May [CM 80016] and 22 September [ANSP 135257–58]). This species has been found from 375 m to 2,570 m in Bolivia.

Muscisaxicola m. maculirostris (Spot-billed Ground-Tyrant).—Permanent resident (127 specimens) in much of its range in Bolivia, but occurs only as a winter resident (43 specimens) to the north and east. This species occurs in the altiplano and puna regions of deptos. Oruro and Potosí throughout the year, as well as in the semiarid intermontane zone of depto. Cochabamba. It is apparently much more common in depto. Cochabamba in winter than during the breeding season; 86 of 97 specimens were collected from April through September, and only two from November through February. This species has been recorded exclusively in winter in deptos. Chuquisaca, Tarija (a single July record [ANSP 135387]), and La Paz, where specimens have been collected in all months from April through August (earliest record 6 April [ANSP 175121], latest 2 August [AMNH 803450]); and as a presumed non-breeder (vagrant?) in depto. Santa Cruz (a single October record [FMNH 180921]). Elevational range for *M. m. maculirostris* is from ca. 2,500 m to over 4,000 m during both breeding and non-breeding seasons, but at least 39 of the 47 records below 3,000 m are from April through early October (non-breeding season).

Muscisaxicola capistrata (Cinnamon-bellied Ground-Tyrant).—Winter resident (65 specimens) at high altitude in deptos. Potosí, Oruro, Cochabamba, and La Paz; recorded as a vagrant from depto. Santa Cruz (Fig. 4). *Muscisaxicola capistrata* has been recorded from Bolivia during May, June, July, August, and September (earliest record 5 May [ANSP 135316], latest 18 September [UMMZ 212733]). The species has been found wintering from 2,600 m to 4,780 m, but most records (52 of 65) are at elevations of 3,400 m and above. The record from depto. Santa Cruz was at low elevation (520 m) on 13 September, coincident with a strong *surazo*, or southern cold front (Kratter et al. 1993).

Muscisaxicola rufivertex pallidiceps (Rufous-naped Ground-Tyrant).—Apparent permanent resident (29 specimens) in the western portion of its Bolivian range, along the altiplano and puna zone from southeastern depto. Potosí to northwestern depto. Oruro, and possibly into southern depto. La Paz; apparent winter resident (nine specimens) to the northeast, in eastern depto. Oruro, northern depto. Potosí, and in the semiarid intermontane region of depto. Cochabamba, where it overlaps with the resident *M. r. occipitalis*. Records from the area of winter resident status are from May, June, and September (earliest record 7 May [UMMZ 222966], latest 7 September [CM 120414]). This species breeds in Bolivia at elevations from 3,350 m to 4,350 m; it is found in winter down to 2,225 m.

Muscisaxicola albilora (White-browed Ground-Tyrant).—Winter resident (26 specimens) in Bolivia. This species has been recorded from the altiplano and puna zone of southern depto. La Paz and northeastern depto. Potosí, and from the intermontane region of depto. Cochabamba. There are records for April, May, June, July, August, and September (earliest 7 April [ANSP 138789], latest 5 September [ANSP 146681]), at elevations from 2,400 m to 4,000 m.

Muscisaxicola c. cinerea (Cinereous Ground-Tyrant).—Summer resident (15 specimens) in deptos. Cochabamba, Chuquisaca (one September record only), and Potosí, on the altiplano and in the puna and intermontane zones; apparent permanent resident (13 specimens) in eastern depto. La Paz and northern depto. Oruro; possibly only a winter resident (one specimen) at the northern end of its Bolivian range (in western depto. La Paz). There are records from the area of summer residence for September, October, December, January, and February (extreme dates are 21 September [BMNH 1902.3.13.656] and 17 February [ANSP 135360]). Records for the area of apparent permanent residence are from March, August, October, November, and December. The northernmost Bolivian specimen (LSUMZ 124298) was collected on 22 May, in depto. La Paz (ca. 16°19'S, 68°01'W); *M. c. cinerea* may occur in this area solely as a winter resident. Available data indicate that specimens have been collected in Peru only from April through September (Chesser, unpubl. data), although Fjeldså and Krabbe (1990) published a record of an immature from Puno in January. Elevations occupied by *M. c. cinerea* range from 3,600 m to over 4,500 m; elevation is similar throughout the year.

Muscisaxicola flavinucha (Ochre-naped Ground-Tyrant).—Summer resident (18 specimens) in Bolivia in the altiplano and puna zones of deptos. Potosí, Oruro, and southern depto. La Paz; apparent winter resident (20 specimens) north and northeast of this region in deptos. Cochabamba and La Paz (Fig. 5). Specimens from the breeding area have been collected during December, January, February, and March (earliest record 8 December [LSUMZ 124289], latest 25 March [ANSP 135310]). There are records from the apparent non-breeding range for April, May, June, August, and October (earliest 21 April [CM 81234–35], latest 2 October [ANSP 135309]). Available data indicate that *M. flavinucha* occurs only as a winter resident in adjacent Peru, and exclusively as a summer resident in Argentina and Chile (Fjeldså and Krabbe 1990; Chesser 1995). This species has been found in Bolivia from 3,875 m to 4,780 m during the austral summer, and from 3,220 m to 4,300 m in the non-breeding season.

Muscisaxicola frontalis (Black-fronted Ground-Tyrant).—Uncommon winter resident (six specimens) in Bolivia. This species has been recorded from the altiplano and puna zone of deptos. Potosí, Oruro, and La Paz, during June, July, and August (earliest record 5 June [ANSP 135333], latest 15 August [LSUMZ 101480]). It is apparently found only in winter in Peru, as well (Traylor 1979, Fjeldså and Krabbe 1990, Chesser 1995). The elevational range of *M. frontalis* in Bolivia is from ca. 4,000 m to ca. 4,600 m.

Lessonia oreas (Andean Negrito).—Apparent summer resident (19 specimens) on the altiplano of southern depto. La Paz and deptos. Oruro and Potosí; winter resident (65 specimens) to the north and east in central depto. La Paz, eastern depto. Oruro, northeastern depto. Potosí, and deptos. Cochabamba and Tarija. Records from the area of summer residence range from 7 October (FMNH 180926, 180934) to 7 February (ANSP 135397–98). Records for the wintering zone are from March, April, May, June, July, August, and October (earliest date 2 March [CM 94477, 94482–83], latest 26 October [BMNH 1902.3.13.676]). Records of *L. oreas* from the breeding season range from 3,600 m to 4,200 m (17 of 19 records at 3,900 m or above), whereas those from the non-breeding season range from 2,400 m to 4,500 m, with 16 records below 3,000 m. This species also appears to be migratory in Peru (Parker et al. 1982, Chesser 1995), and apparently at the southern end of its range in Argentina (Chesser 1995).

Lessonia rufa (Austral Negrito).—Uncommon winter resident (eight specimens) in depto. Tarija, with a single record from western depto. Santa Cruz (LSUMZ 37859). This species has been

FIG. 6. Seasonal distribution in Bolivia of *Knipolegus hudsoni*, Hudson's Black Tyrant. This species is a winter resident in the lowlands of Bolivia.

FIG. 7. Seasonal distribution in Bolivia of *Myiarchus swainsoni ferocior*, one of the subspecies of Swainson's Flycatcher. This subspecies is a summer resident in the lowlands of southern Bolivia and a winter resident to the north. Breeding and non-breeding ranges do not overlap.

recorded in Bolivia only during May, July, and September (earliest date 10 May [LSUMZ 37859; Remsen et al. 1987], latest 19 September [ANSP 135390, 135392–93, 135395–61]). Bolivian records range from the lowlands to the semiarid montane zone, from 375 m to 2,400 m.

Knipolegus striaticeps (Cinereous Tyrant).—Winter resident in depto. Tarija, southeastern depto. Chuquisaca, and depto. Santa Cruz south of ca. 17°S; a few individuals may breed in the southern portion of the Bolivian range. Of 27 specimens, 25 were collected in April, June, July, August, and September (earliest record 15 April [FMNH 181038], latest 25 September [CM 119406]). The other records are a 20 November specimen from depto. Tarija (LACM 35569), and a 28 January specimen from extreme southwestern depto. Santa Cruz (MACN 41039); although there is no information on the breeding condition of these birds, it is possible that they are permanently resident, breeding individuals. This species is also more common in winter, and a rare breeder, in the chaco of western Paraguay (Hayes et al. 1994), and apparently a rare non-breeder in the state of Mato Grosso, Brazil (Chesser 1995). *Knipolegus striaticeps* has been found in Bolivia only in areas of chaco or deciduous or semihumid forest, at low elevation (175–750 m).

Knipolegus hudsoni (Hudson's Black-Tyrant).—Winter resident (36 specimens) in Bolivia, primarily in depto. Santa Cruz, west of ca. 61°W, and in depto. El Beni; there is also a single record from depto. La Paz (LSUMZ 96312; Remsen et al. 1987) (Fig. 6). This species has been recorded during all months from May through October (earliest date 12 May [Gyldenstolpe 1945], latest 11 October [LSUMZ 124320]), in a variety of lowland habitats up to 800 m. A bird collected 19 September 1990, near the Paraguayan border in the chaco of southern depto. Santa Cruz (LSUMZ 153784; Kratter et al. 1993), had moderate levels of subcutaneous fat and was probably a spring migrant. Records from Paraguay suggest that it is found there only as a southbound migrant (in September and early October; Hayes et al. 1994).

Knipolegus a. aterrimus (White-winged Black-Tyrant).—Permanent resident (220 specimens) along the Andes, from western depto. Tarija north through central depto. Chuquisaca into extreme western depto. Santa Cruz and southern depto. Cochabamba, south of ca. 17°S; apparent winter resident (13 specimens) in the lowlands to the east and north, in eastern deptos. Tarija and Chuquisaca, southwestern depto. Santa Cruz, and northern depto. Cochabamba; apparent summer resident (22 specimens) in eastern depto. Potosí and western depto. Chuquisaca. There are records from the area of apparent winter residence from April, May, July, August, September, and October (earliest 28 April [Laubmann 1930], latest 21 October [CM 51639]). Specimens have been collected in the zone of apparent summer residence in September, November, December, and February (earliest 13 September [BMNH 1902.3.13.625], latest 26 February [ANSP 135532]). *Knipolegus a. aterrimus* has been recorded in Bolivia during the breeding season

(November through February) only from 2,150 m to 3,350 m (additionally at 1,500 m and 3,800 m in March), primarily within areas of montane forest and semiarid intermontane valleys. In the non-breeding months, it can also be found to more than 3,000 m, but many individuals have been recorded at low altitude (down to 250 m) in northern depto. Cochabamba, southwestern depto. Santa Cruz, and eastern deptos. Chuquisaca and Tarija.

Hymenops p. perspicillatus and *Hymenops p. andinus* (Spectacled Tyrant).—Both subspecies occur as winter residents in Bolivia. Females and immatures are not diagnosably different; therefore, the subspecies records below are based on specimens of adult males. The nominate subspecies (14 specimens) has been recorded only from depto. Santa Cruz, in March, May, June, July, August, and November (extreme dates are 22 March [CM 119864] and 17 November [CM 79938]). *Hymenops p. andinus* (five specimens) has been collected in deptos. Santa Cruz, El Beni, and Cochabamba (Niethammer 1956), in April, May, June, and July (earliest date 17 April [LSUMZ 124329], latest 7 July [LSUMZ 124327]). There are records of females and immatures (29 specimens) of this species from deptos. Santa Cruz, El Beni, and Tarija, for March, May, June, July, August, September, October, and November (no specific date for the March specimen [MACN 8979]; latest record 7 November [MACN unnumbered]). *Hymenops p. perspicillatus* has been found up to 1,200 m in Bolivia, and *H. p. andinus* up to ca. 900 m.

Fluvicola pica albiventer (Pied Water Tyrant).—Permanent resident (53 specimens) in most of central and eastern Bolivia; possibly present only as a summer resident (three specimens) in depto. Tarija (records from October, November, and March), and apparently only as a winter resident (four specimens) in northern depto. El Beni and northern depto. Santa Cruz (records from June, August, September, and October; the November specimen published in Peña 1962 is *Fluvicola leucocephala* [PMNH 38627]). This species also breeds in northwestern Argentina, but apparently is not present there during winter (no specimens between 10 May [FML 11372] and 23 August [USNM 284359]). Likewise, this subspecies appears to occur only in winter in the northwestern Brazilian states of Amazonas, Acre, and Rondônia, and in Peru (Traylor 1952; Parker et al. 1994; Chesser 1995). Willis and Oniki (1990) found this species exclusively in winter at two sites in Mato Grosso, Brazil, although Cintra and Yamashita (1990) reported it as common in the Pantanal of that state from August through May. Elevations occupied by *F. pica albiventer* in Bolivia range from 150 m to 1,100 m.

Satrapa icterophrys (Yellow-browed Tyrant).—Permanent resident (90 specimens) in the southern half of Bolivia, from central depto. Cochabamba and western depto. Santa Cruz south; occurs only as a non-breeder (16 specimens) in the northern half of the country. Except for a sight record of a visiting individual in February from Concepción, depto. Santa Cruz (Davis 1993), records from deptos. La Paz and El Beni, and northern deptos. Cochabamba and Santa Cruz (north of ca. 17°S) are from June, July, August, September, and October only (earliest record 1 June [AMNH 792072], latest 3 October [Fjeldså and Krabbe 1989; J. M. C. da Silva, pers. comm.]). The species appears to occur solely in winter in Peru, as well (Parker et al. 1982; Chesser 1995). Its status in eastern depto. Santa Cruz, where there are records only from September and October, is uncertain, as is its status in adjacent Mato Grosso (Chesser, unpubl. data). Willis and Oniki (1990) recorded it exclusively in winter in Mato Grosso, but it was present at only one of their ten sites, whereas Cintra and Yamashita (1990) recorded it from July through January in the Pantanal. *Satrapa icterophrys* ranges in elevation up to ca. 3,300 m in Bolivia, during both breeding and non-breeding seasons. The species has not been recorded from deptos. Oruro and Pando.

Attila phoenicurus (Rufous-tailed Attila).—Apparent winter resident in northeastern depto. Santa Cruz, although it is possible that it is present only as a transient. The only records for Bolivia are a tape-recording from 25 August and a vocal record from late September, both from a forest island atop the Serranía de Huanchaca at 720 m (Bates et al. 1992). This species breeds in southeastern Brazil and northeastern Argentina, and may winter over a broad expanse of Amazonia (Ridgely and Tudor 1994, Chesser 1995). However, these "winter" records are primarily from April, May, and September, and may refer to transient individuals. The actual wintering range of *Attila phoenicurus* may be much more restricted than has been indicated (cf. *Elaenia strepera* [Marantz and Remsen 1991]).

Casiornis rufa (Rufous Casiornis).—Permanent resident (139 specimens) in most of Bolivia, but apparently present in the extreme north only during the austral winter (13 specimens). There are records from all deptos., excluding the southwestern deptos. of Oruro and Potosí, although relatively few from deptos. La Paz and Cochabamba. Probably present only as a winter resident in depto. Pando, northern depto. El Beni, and extreme northern depto. Santa Cruz, where there are records only from June, July, August, September, and October (27 records from 18 June

[LSUMZ 133454] to 14 October [Gyldenstolpe 1945]). Seasonality of occurrence is indicated in neighboring areas by specimen data from southern Peru, and the states of Acre and Rondônia, Brazil, where *C. rufa* appears to occur only during the austral winter (Chesser 1995). Parker et al. (1994) listed this species as a wintering austral migrant at the Tambopata Reserve in southern Peru, and Willis and Oniki (1990) found it only in winter at the three northernmost Mato Grosso sites at which it was present. Olrog (1963) observed migratory behavior by this species in Bolivia. This species occurs up to 1,630 m (ANSP 136366; CM 80995) from October through March, with many records above 1,000 m, but the highest elevational record from the nonbreeding season is a late April specimen from 875 m (LSUMZ 124400).

Myiarchus tuberculifer atriceps (Dusky-capped Flycatcher).—Summer resident (38 specimens) in forests and woodlands of the southern Bolivian Andes; permanent resident (42 specimens) to the north. This subspecies has been recorded in deptos. Tarija, Chuquisaca, and Santa Cruz only in October, November, December, January, February, and March (records range from 19 October [ANSP 135735] through 17 March [CM 81009]). *Myiarchus t. atriceps* is present year-round in depto. Cochabamba and possibly in depto. La Paz (although there are specimen records here only from July and October), as it is also in Peru. The species breeds and is mostly migratory in northwestern Argentina, although there are winter records (only two of 82 available records south of depto. Cochabamba, Bolivia, are from winter). Elevational range in Bolivia, which is similar throughout the year, is from 1,000 m to 3,400 m. This subspecies has not been recorded from deptos. Pando, El Beni, Oruro, or Potosí.

Myiarchus s. swainsoni and *M. s. ferocior* (Swainson's Flycatcher).—The nominate subspecies occurs only as a passage migrant (four specimens) in Bolivia; *M. s. ferocior* is a summer resident (nine specimens) in deptos. Tarija and Chuquisaca and southwestern depto. Santa Cruz, and a winter resident or transient (36 specimens) throughout central, northern, and eastern Bolivia (Fig. 7). Of the four records for *M. s. swainsoni*, three are from the lowlands (two immature birds from eastern Bolivia collected in April [FMNH 296254–55], and one from northern depto. El Beni, collected on 13 October [Gyldenstolpe 1945]), and one is a March specimen from high elevation (3,300 m) in Cochabamba (FMNH 181198). There is also a specimen of an *M. s. swainsoni* × *M. s. ferocior* hybrid from lowland depto. Santa Cruz in March (FMNH 294126). Davis (1993) reported the species as a fairly common spring transient at Concepción, depto. Santa Cruz, and all of her specimens (collected 13 September through 9 November; FMNH 335250–55) are referable to *M. s. ferocior*. There are records for *M. s. ferocior* in southern Bolivia, where a female collected 4 December (FMNH 294124) is noted to have already laid eggs, only from October, November, and December. North of 18°S, this subspecies has been recorded only in March, April, May, July, August, September, October, and November (extremes are 8 March [AMNH 497136] and 9 November [FMNH 335254]). *Myiarchus s. ferocior* has been recorded from low elevation to over 2,000 m during the breeding season, but only below 600 m during winter (although there is a probable transient specimen from 3,300 m [UMMZ 106864] in March).

Myiarchus t. tyrannulus (Brown-crested Flycatcher).—Permanent resident (120 specimens) in most of Bolivia, although not recorded from deptos. Oruro and Potosí; apparently present only in winter (37 specimens) in much of northern Bolivia. There are records of adults for depto. Pando and along the Brazilian border in depto. El Beni and northern depto. Santa Cruz, from June, July, August, September, and October (extreme dates are 20 June [LSUMZ 133427] and 28 October [Gyldenstolpe 1945]); an immature, however, was collected on 19 December in depto. El Beni (PMNH 38664; Peña 1962). Specimen records from southern Peru, and the Brazilian states of Acre and Rondônia, likewise range from June through October, except for a 20 November specimen from depto. Cuzco, Peru (AMNH 822446). Donahue (1987) listed this species as a possible (wintering) migrant at the Tambopata Reserve, Peru, and Willis and Oniki (1990) reported it only in winter at the four northernmost Mato Grosso sites at which they recorded this species. Olrog (1963) observed migratory behavior of this species in Bolivia. Elevational range of *M. t. tyrannulus* in Bolivia is from 175 m to 2800 m, and appears not to change seasonally.

Myiodynastes maculatus solitarius (Streaked Flycatcher).—Summer resident (62 specimens) in eastern and southern Bolivia, south of ca. 18°S; permanent resident (86 specimens) in much of the central and northern regions; possibly present only in winter (10 specimens) in extreme northern Bolivia (depto. Pando, northern depto. El Beni); not recorded from deptos. Oruro and Potosí. Specimens have been collected from deptos. Tarija and Chuquisaca, southern depto. Cochabamba, and southwestern depto. Santa Cruz in October, November, December, January, February, March, and April (earliest record 10 October [ANSP 135636], latest an undated April specimen [Lönnberg 1903]. There are specimens from eastern depto. Santa Cruz (east of ca. 63°W, and south of ca.

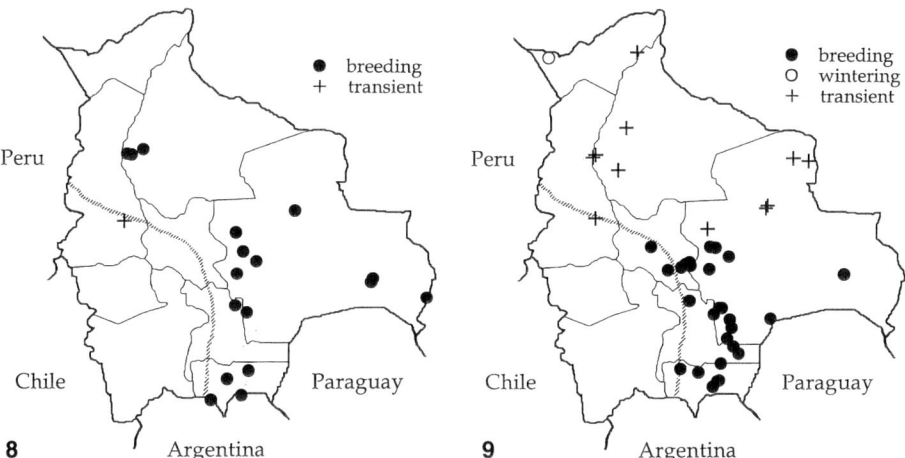

FIG. 8. Seasonal distribution in Bolivia of *Empidonomus varius*, the Variegated Flycatcher. This species is a summer resident in the lowlands of Bolivia.

FIG. 9. Seasonal distribution in Bolivia of *Empidonomus aurantioatrocristatus*, the Crowned Slaty-Flycatcher. This species is a summer resident in lowlands in the southern half of the country, wintering only in depto. Pando in northwestern Bolivia. Note the large number of transient records.

15°S, the southern limit of tropical forest in eastern Bolivia) from September, October, November, December, February, March, and April (extreme dates are 26 September [Laubmann 1930] and 25 April [FMNH 296242]). This species was recorded at Concepción, depto. Santa Cruz, only during these same months in 1985–87 (Davis 1993). *Myiodynastes m. solitarius* is a permanent resident in northern deptos. Santa Cruz, Cochabamba, and La Paz, primarily in areas of humid lowland and montane forest, although also in semihumid lowland forest in western depto. Santa Cruz. It has been recorded in the area of possible winter residence, in depto. Pando and northern depto. El Beni, in July, August, September, October, and November (extreme dates are 11 July [LSUMZ 133410] and 2 November [Gyldenstolpe 1945]). The elevational range for this species in Bolivia is from ca. 150 m to 2,700 m throughout the year.

Legatus l. leucophaius (Piratic Flycatcher).—Summer resident (26 specimens) in most of its range in Bolivia, although it also winters (eight specimens) in depto. Pando. There are multiple breeding-season specimens from deptos. El Beni and Santa Cruz, two specimens from depto. La Paz, and sight records from depto. Tarija (Nores and Yzurieta 1984). These records are from August, September, October, November, December, January, and February (extreme dates are 29 August [BMNH 1902.3.13.810] and 6 February [AMNH 496438]). The single specimen from Cochabamba (ANSP 135633) was collected 8 September and was probably a migrant bird. Records for depto. Pando are from July, August, and October. There are also records of apparent wintering birds from the adjacent Brazilian states of Acre and Rondônia. This species has been found up to 1,500 m during the breeding season, and from 175 m to 325 m in winter in depto. Pando.

Empidonomus v. varius (Variegated Flycatcher).—Present in Bolivia during breeding and migration only (54 specimens; Fig. 8). This species has been recorded in deptos. La Paz (collected 26 August [Niethammer 1956], presumably in migration), El Beni, Santa Cruz, Chuquisaca, and Tarija. Breeding season records are generally north or east of the high Andes, extending as far east as the Brazilian border; however, there are no records for Bolivia north of 14°S. There are specimens for all months from August through April (earliest 26 August [as above], latest 15 April [FMNH 181126, 296237]). Davis (1993) reported *E. varius* as present but uncommon during the wet season (November through April) at Concepción, depto. Santa Cruz. This species breeds in Bolivia from 150 m to 1,775 m, but a migrant individual has been seen in March at 2,850 m (Whitney et al. 1994).

Empidonomus a. aurantioatrocristatus (Crowned Slaty-Flycatcher).—Summer resident (94 specimens) in central, southern, and southeastern Bolivia; apparent winter resident (three specimens) in the extreme northwest (Fig. 9). Present only as a transient (17 specimens) in much of northern Bolivia. There are records for the breeding area in deptos. Tarija and Chuquisaca, southern depto. Cochabamba, and western and southeastern depto. Santa Cruz, for all months

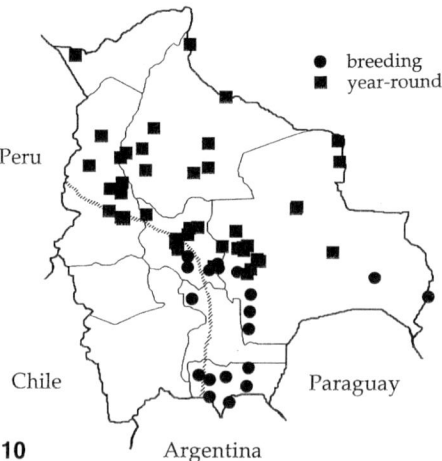

FIG. 10. Seasonal distribution in Bolivia of *Tyrannus melancholicus*, the Tropical Kingbird. This species is present year-round in most of lowland Bolivia, but is found in the southern portion of the country during the breeding season only.

from September through May (earliest record 10 September [FMNH 181135], latest 2 May [an immature; LSUMZ 37993]). Three wintering individuals were collected in July in northwestern depto. Pando (LSUMZ 133403–05); the species is also present in Peru and northwestern Brazil only during winter (Pearson 1980; Parker et al. 1982; Chesser 1995). Records for this species during apparent passage migration, from areas for which summer or winter records are lacking, are from deptos. La Paz and El Beni, eastern depto. Pando, and northeastern depto. Santa Cruz. Except for a sight record from March (Davis 1993), these records ($n = 14$) are exclusively from spring migration (September, October, November). Individuals from these areas, and from which data on subcutaneous fat were taken, had either heavy (LSUMZ 125883, 151153-54) or very heavy (FMNH 335275) fat. Other specimens with heavy fat deposits were taken in southern (LSUMZ 153786; collected 20 September) and eastern (FMNH 334517; collected 9 November) depto. Santa Cruz. *Empidonomus a. aurantioatrocristatus* occurs in the breeding season at elevations up to 2,500 m, with two March specimens (one an immature) at 3,300 m (CM 120285; FMNH 181145); it has been recorded in winter only at 325 m in lowland tropical forest.

Tyrannus m. melancholicus (Tropical Kingbird).—Permanent resident (155 specimens) in northern and central Bolivia; summer resident (35 specimens) south of ca. 18°30'S (not recorded from deptos. Oruro or Potosí) (Fig. 10). There are records from deptos. Tarija, Chuquisaca, and southern depto. Santa Cruz for all months from October through May (ranging from 21 October [ANSP 135601] to 4 May [Laubmann 1930]). Davis (1993) reported *T. melancholicus* as an uncommon or rare permanent resident in Concepción, depto. Santa Cruz, but stated that its numbers were supplemented during migration periods (March–April and September–October), when it became common. This species has been recorded throughout the year at elevations up to 2,500 m.

Tyrannus s. savana (Fork-tailed Flycatcher).—Present in Bolivia as a breeding bird (48 specimens) and passage migrant (27 specimens). *Tyrannus savana* has been recorded from all deptos. except Potosí, but apparently only as a transient in deptos. Oruro (Whitney et al. 1994), Pando, northern depto. El Beni, and northern and central depto. Santa Cruz, where there are specimen records only from September, October, and November (spring migration), and February and March (fall migration). This species was a common transient at Concepción, north-central depto. Santa Cruz, during these same months, and was additionally seen in April (Davis 1993). A bird with heavy subcutaneous fat was collected here 15 September (FMNH 335290). *Tyrannus savana* has been found in deptos. La Paz and Cochabamba only during February and March, indicating that it may occur there only during fall migration. Definite breeding season records occur from western depto. El Beni, western and southeastern depto. Santa Cruz, and depto. Tarija, where the species has been recorded during all months from September through April (earliest record 10 September [CM 32996], latest 1 April [Lönnberg 1903]). According to Pearson (1975, 1980), *T. s. savana* occurs in Tumi Chucua, depto. Beni, from April through November (non-breeding season), but this appears to be an error. Pearson was present at this site only in September, October, and November (spring migration), and the southernmost wintering records for this species appear to be at least several hundred kilometers north of Tumi Chucua. This species evidently breeds only at low elevation (below 1,000 m) in Bolivia, but specimens have been taken as high as 2,570 m during apparent fall migration, and two migrants were sighted on 20 March 1992 at 3,760 m (Whitney et al. 1994).

Tyrannus albogularis (White-throated Kingbird).—Occurs in Bolivia only as a summer resident (15 specimens). Specimen and published sight records indicate that this species is present from August (earliest record 28 August [LSUMZ 151155]) through April (no specific date [Davis 1993]). *Tyrannus albogularis* has been recorded only from the lowlands of depto. El Beni and northern depto. Santa Cruz. Davis (1993) reported it to be a common species, primarily in the cerrado vegetation, in Concepción, depto. Santa Cruz, where nests were found. Remsen (1986) noted that it was fairly common in savanna and around buildings at Estancia Inglaterra, depto. El Beni. Two birds collected in prov. Velasco, extreme northeastern depto. Santa Cruz (LSUMZ 151155–56), were found in open *campo* and open *campos cerrado*, respectively. This lowland species has been found up to 725 m in Bolivia.

Xenopsaris a. albinucha (Xenopsaris).—Apparently a winter resident in Bolivia. The seven records (six specimens, one sight record) are from deptos. El Beni and Santa Cruz, from May, September, and early October (extreme dates are 10 May [Gyldenstolpe 1945] and 3 October [Parker and Rowlett 1984]). Available data indicate that *X. albinucha* has been recorded in nearby Mato Grosso, Brazil only during probable non-breeding times (April and May), and is found to the south in Argentina in the breeding season only, from mid-September through April ($n = 48$ specimens; Chesser 1995, unpubl. data). The lack of Bolivian records for June, July, and August is most likely a result of the uncommonness of the species, rather than its absence during these months, although it is possible that it merely migrates through Bolivia and Mato Grosso towards an unknown wintering area (there is also an apparently disjunct population in northeastern Brazil; Short 1975, Chesser 1995). This species has been found in Bolivia at elevations from 200 m to 600 m.

Pachyramphus polychopterus spixii (White-winged Becard).—Present only during the breeding season (12 specimens) in central, southern, and eastern Bolivia, south of ca. 17°S; winter resident (34 specimens) in northern Bolivia. There is an area of apparent permanent residence (20 specimens) in western depto. Santa Cruz (and no records from depto. El Beni and from the southwestern deptos. Oruro and Potosí). In the area of summer residence, in southwestern and southeastern depto. Santa Cruz, and deptos. Chuquisaca and Tarija, *P. p. spixii* has been recorded in October, November, December, February, March, and April (extreme dates are 29 October [LACM 35749] and 26 April [FMNH 296211]); the species has additionally been collected in January in the area of permanent residence to the north. North of ca. 18°S, in deptos. Cochabamba, La Paz, Pando, and northern depto. Santa Cruz, there are records for this species for all months from February through October (earliest 13 February [FMNH 335295–96], latest 11 October [Gyldenstolpe 1945]). Davis (1993) observed this species only in February, March, and April at her study area near Concepción, depto. Santa Cruz, and could not determine its status, but seven specimens at nearby localities from May and June (CM 79994, 80089, 80101–02, 80222, 80345, and 80432), suggest that it is a winter resident in this region. This subspecies is apparently a winter resident in Peru and the state of Acre, Brazil (Chesser, unpubl data). *Pachyramphus p. spixii* has been collected up to 2,150 m during the breeding season, but only to 1,500 m in winter, when 41 of 43 Bolivian records are below 900 m.

Pachyramphus v. validus and *P. v. audax* (Crested Becard).—The nominate subspecies is uncommon and of uncertain status in Bolivia, but may be a winter resident at the edge of its range; *P. v. audax* is a summer resident (21 specimens) in southern Bolivia, and a probable permanent resident to the north. The three Bolivian records for *P. v. validus* are a specimen of unknown date from eastern depto. Santa Cruz (Hellmayr 1925), and two specimens from northeastern depto. Santa Cruz, a male collected on 4 July (LSUMZ 137484) and a female collected on 12 October (LSUMZ 151166; identified by range); the latter bird had moderate levels of subcutaneous fat. A wintering female (LSUMZ 102284), possibly of this subspecies, was collected in June in depto. La Paz (females of *validus* and *audax* are not distinguishable [Hellmayr 1929]). This subspecies is migratory in the southern part of its range (Willis 1979; Belton 1985; Chesser 1995), and is apparently found only in winter in Rondônia, Brazil, across the border from northeastern depto. Santa Cruz (Chesser 1995). There are also two winter specimens in female plumage from Rondônia, Brazil (FMNH 330650–51), presumably but not certainly this

TABLE 1
NUMBER OF SPECIES OF MIGRANT TYRANNIDS IN BOLIVIA, ARRANGED BY SUB-FAMILY AND MIGRATORY PATTERN. VAGRANTS OR TAXA OF UNCERTAIN MIGRATORY HABITS IN BOLIVIA ARE NOT INCLUDED IN TOTALS

	Elaeniinae	Fluvicolinae	Tyranninae	Tityrinae	Total
Breeding season only	2	0	3	0	5
Wintering only	4	11	1	1	17
Transient only	0	0	1	0	1
Year-round in Bolivia					
Breeds S, perm N	3	0	4	1	8
Perm S, winters N	4	5	2	0	11
Breeds S, winters N	4	8	2	1	15
Uncertain/vagrant	(5)	(0)	(0)	(1)	(6)
TOTAL	17	24	13	3	57

subspecies. Both Bolivia specimens from northeastern depto. Santa Cruz were collected 12–15 m up in mixed-species flocks in lowland tropical forest.

Pachyramphus v. audax breeds in western depto. Santa Cruz, and in deptos. Chuquisaca and Tarija, south of ca. 17°30'S, in the zone of semihumid lowland and montane forest. Records from this area date from October, November, December, January, February, March, and April (earliest 28 October [ANSP 138514], latest 10 April [Laubmann 1930]). There are evidently only three, possibly four, Bolivian records of *P. v. audax* north of this area: an August specimen from Cochabamba (ANSP 138515), a February specimen from depto. El Beni (FML 11410), a December specimen from La Paz (ANSP 120135), and the June specimen mentioned above (LSUMZ 102284). The subspecies appears to be resident but uncommon in this region. The paucity of winter records throughout the range of this subspecies is striking. Available data show no records in Argentina from mid-June until late September, indicating that the species migrates (see also Nores et al. 1983), but there are few winter records from Bolivia or Peru. That one of the Peruvian specimens (AMNH 820184) was collected 40 m up in the lower canopy of cloud forest (at 1,680 m), suggests that this species may be overlooked in winter. Elevation for breeding birds in Bolivia ranges from 650 m to 2,400 m, with a March specimen collected at 2,700 m. The two Bolivian winter specimens were found at 300 m and 600 m, in the zone of lowland tropical forest.

GENERAL CONSIDERATIONS

Geographic patterns.—Many discussions of regional patterns of austral migration (e.g., Hilty and Brown 1986, Hayes et al. 1994) have considered only species whose presence within the region is strictly seasonal (i.e., species that are winter or summer residents). Such focus is warranted at the northern or southern extremes of species' ranges. Bolivia, situated in the southern end of the tropical zone, presents more complicated patterns of distribution. The latitudinal position and great topographic diversity of Bolivia allow species to migrate fairly extensively, yet remain within the borders of the country throughout the year.

Indeed, most tyrannids migratory in Bolivia (34 of 57) are present somewhere within its borders year-round (Table 1), and many show overlapping breeding and wintering ranges, as is typical of austral migrants (Chesser 1994). Most of these taxa are found in the northern (and eastern, for many species that show seasonal elevational differences) portion of their range only during winter, but nine are resident in the northern part of their range, and extend their range southward during the breeding season. These within-Bolivia migrations tend to be geographically independent; predominant patterns of range demarcation do not appear to occur.

Of the 23 migrant flycatchers present in Bolivia only part of the year, the vast majority (17 species or subspecies) are winter residents (Table 1). Five of these are species of the high-elevation genera *Muscisaxicola* and *Agriornis*. Species from eleven other (often dissimilar) genera (e.g., *Serpophaga*, *Lathrotriccus*, *Knipolegus*, *Hymenops*, and *Xenopsaris*) are also represented in this group.

Only one tyrannid, the subspecies *Myiarchus s. swainsoni*, occurs solely as a transient in Bolivia, and five are present in summer only. That these five tyrannids (*Elaenia albiceps chi-*

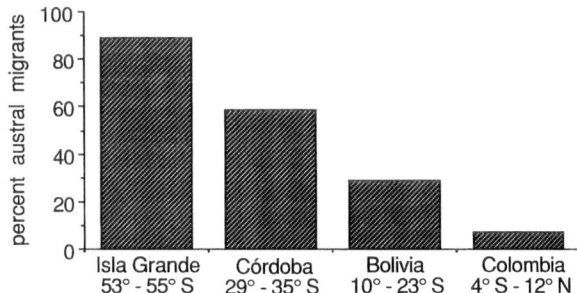

FIG. 11. Percentage of tyrannids that are austral migrants in selected regions of South America. Data for Isla Grande from Humphrey et al. (1970), for Córdoba (Argentina) from Nores et al. (1983), and for Colombia from Hilty and Brown (1986).

lensis, Elaenia strepera, Empidonomus varius, Tyrannus savana, and *Tyrannus albogularis*) completely and two more (*Legatus leucophaius, Empidonomus aurantioatrocristatus*) nearly vacate Bolivia during winter is fairly remarkable, given its location at tropical latitudes and the range of habitats present there. Several species, including *L. leucophaius,* are known to leave at least portions of their range at similar latitudes in the northern Neotropics during winter, apparently in response to resource fluctuations associated with the dry season (Eisenmann 1963; Morton 1977). The dry season in northern Bolivia coincides in part with the absence of the "summer only" species, but not precisely. Wintering of several of these austral migrants (*E. varius, T. albogularis, T. savana*) is associated in part with Amazonian river islands (Rosenberg 1990), a habitat not found in Bolivia. However, these migrants winter in many other habitats, and occur as far north as the Caribbean coast of South America, far from Amazonian river islands. Breeding ranges of three of the summer resident taxa (*Elaenia albiceps chilensis, Empidonomus varius,* and *T. savana*) extend from tropical South America far into the temperate zone, south to central Argentina or beyond. Tropical/temperate overlap in breeding range is fairly unusual in long-distance migrants of other migration systems, but common among austral migrants, owing to greater habitat continuity between temperate and tropical South America, to the small area of the South American temperate region relative to its tropical area, and to the lack of geographic barriers between temperate and tropical areas (Chesser 1994).

Three species of austral migrant tyrannids (*Hymenops perspicillatus, Lathrotriccus euleri, Myiarchus swainsoni*), and possibly a fourth (*Pachyramphus validus*), are represented by more than one migratory subspecies in Bolivia. In one case (*H. perspicillatus*), both subspecies occur exclusively in winter in Bolivia. In the others, the subspecies show different patterns. *Lathrotriccus e. euleri* occurs in Bolivia solely as a winter resident, whereas *L. e. argentinus* is a summer resident in central and southern Bolivia, and an apparent winter resident to the north; range of these subspecies overlap in winter. *Myiarchus s. swainsoni* is present in Bolivia only as a transient, whereas *M. s. ferocior* is a summer resident in southern Bolivia and a winter resident or transient to the north. *Pachyramphus v. audax* is a summer resident in southern Bolivia and a probable permanent resident to the north, whereas the nominate subspecies may occur in Bolivia only as a winter resident.

It is fairly common for austral migrant taxa to winter partially within the range of a resident (usually tropical) subspecies to the north (Chesser 1994). In Bolivia, this occurs in only six instances, probably owing to its proximity to the southern temperate zone: *Sublegatus modestus brevirostris* winters within the range of the nominate subspecies in northern Bolivia, *Lathrotriccus e. euleri* and *L. e. argentinus* winter within the range of *L. e. bolivianus* in northern Bolivia, *Agriornis m. microptera* winters somewhat within the range of *A. m. andecola* in depto. La Paz, *Muscisaxicola rufivertex pallidiceps* winters within the range of *M. r. occipitalis* in depto. Cochabamba, *Myiarchus swainsoni ferocior* winters within the range of *M. s. pelzelni* in northern Bolivia, and *Pachyramphus polychopterus spixii* winters within the range of *P. p. nigriventris* in depto. Pando. Because of the ease of overlooking subspecific differences in specimens or in the field, migration in such cases is difficult to ascertain, and further research may prove that other species show this same migratory pattern.

A small percentage of tyrannids are austral migrants in Bolivia relative to temperate regions of South America, but a high percentage relative to tropical South America north of Bolivia (Fig. 11). As might be expected, the percentage of austral migrant tyrannids declines steadily

TABLE 2
Status of Austral Migrant Tyrannids in Selected Regions of South America. Data for Colombia after Hilty and Brown (1986), for Córdoba after Nores et al. (1983), and for Isla Grande after Humphrey et al. (1970)

	Summer resident	Winter resident	Summer and winter	Transient only
Colombia (4°S–12°N)	0%	84%	8%	8%
Bolivia (10–23°S)	9%	30%	59%	2%
Córdoba, Arg, (29–35°S)	60%	20%	17%	3%
Isla Grande (53–55°S)	100%	0%	0%	0%

from south to north. Twenty-nine percent (57 of 195) of Bolivian flycatchers are migratory to some degree. In contrast, 89% (eight of nine) of tyrant flycatchers occurring regularly on Isla Grande (Tierra del Fuego) are austral migrants, and 60% (30 of 50) of those found in the province of Córdoba (central Argentina), but only 7% (12 of 175) of the tyrant flycatchers of Colombia.

Comparative analysis of the status of austral migrant tyrannids in these four regions reveals that, although the percentage of migrants that are summer residents and winter residents, respectively, increases or decreases with increasing south latitude, the percentage of austral migrant flycatchers present throughout the year within Bolivia is extremely high (Table 2). Fifty-nine percent of these taxa occur year-round in Bolivia, compared to 8 percent for Colombia, 17 percent for Córdoba, and 0 percent for Isla Grande, thus underscoring the "crossroads position" of Bolivia in the geography of austral migration. Although the latitudinal range covered by Isla Grande and Córdoba is less than that of Bolivia, this accounts for only a small portion of the observed pattern. Expanding the latitudinal range of Córdoba south by seven degrees (to equal that of Bolivia), for instance, changes the percentage of "summer and winter" migrants by less than ten percent.

Taxonomic patterns.—Bolivian austral migrant tyrannids are most numerous in the subfamily Fluvicolinae, followed by the Elaeniinae, Tyranninae, and Tityrinae (Table 1). This pattern is the same as that for austral migrant tyrannids as a whole (Chesser 1994). All 24 migrant fluvicolines are present in Bolivia in winter only, or are present in the northern portion of their Bolivian range in winter only; none of these taxa extend their range south in summer from a permanent Bolivian population (although other subspecies of a few of these species [e.g., *Muscisaxicola rufivertex*] are permanent residents north of the range of the migratory subspecies), or occur only as breeding residents that winter further north.

Migrant elaeniines exhibit the widest range of migration patterns in Bolivia, and include all geographic patterns except "transient only" (Table 1). Most of these occur in Bolivia throughout the year, but four species are present in winter only, and two are breeding season residents. The 13 migrant tyrannines likewise exhibit a variety of geographic migratory patterns, although the majority of tyrannines occurs in part or all of their Bolivian range during the breeding season only. Three of the five "breeding season only" migrants are included in this group, and six other taxa that extend their range southward during the breeding season from a permanent Bolivian population. Only one tyrannine flycatcher occurs in Bolivia solely as a winter resident.

Life zones and habitat.—Analysis of life zone preferences of migrant flycatchers in Bolivia reveals that large numbers of migrant tyrannids occur in four life zones or combinations of life zones (non-Amazonian lowlands, lowlands in general, puna, and lowlands + valle zone), and few in the other life zones (Table 3). Of the ten most common life zones occupied by Bolivian flycatchers, these four contain from 43 to 75 percent migratory species, whereas the other six range from 0 to 17 percent migratory species. Life zones containing few tyrannids but high percentages of migrants are the non-Amazonian lowlands + valle zone (three of four taxa migratory), the lowlands in general + upper tropical zone (two of two migratory), and the "widespread" category (two of two migratory). Life zones with unusually low numbers of migrant taxa are the Amazonian lowlands (1 of 52 taxa migratory), the upper tropical zone (0 of 14 migratory), the temperate + temperate/puna transition zone (0 of 10 migratory), and the upper tropical + subtropical zone (1 of 10 migratory). In part this may be explained by the lack or scarcity of these life zones south of Bolivia, and the fact that species of southern South America do not breed in most of these zones, but this does not necessarily preclude species from wintering in these zones (as do some Nearctic-Neotropical migrants; several austral migrants winter in part in Amazonian lowlands [see below]), nor does it apply to occupants of the temperate life zone.

TABLE 3

Number of Species of Migrant Tyrannids in Bolivia, Arranged by Life Zone or Macrohabitat. Vagrants or Taxa of Uncertain Migratory Status in Bolivia Are Excluded from Totals. Life Zone Data Follow Remsen and Traylor (1989), with Slight Modification. Key to Life Zone Abbreviations: A = Amazonian Lowlands; N = Non-Amazonian Lowlands; L = Lowlands in General; L, V = Lowlands + Valle Zone; P = Puna Zone, Including Species that Occupy in Addition to the Puna Zone the Puna/Temperate Transition Zone or the Valle Zone; U = Upper Tropical Zone; U, S = Upper Tropical + Subtropical Zones; T = Temperate Zone; F = Andean Foothills, Including Species that Occupy Other Zones in Addition to the Foothills; S = Subtropical Zone; and Other = Various Life Zones or Life Zone Combinations Occupied by Relatively Few Species, Including N, V (= Non-Amazonian Lowlands + Valle Zone) and L, U (= Lowlands in General + Upper Tropical Zone). See Remsen and Traylor (1989) for More Detail on Life Zone Characteristics

	A	N	L	P	L, V	U	U, S	T	F	S	Other
Breeds only	0	1	1	0	1	0	0	0	0	1	1
Winters only	1	7	2	4	0	0	0	0	0	0	3
Transient	0	0	1	0	0	0	0	0	0	0	0
Year-round in Bolivia											
Breeds S/perm N	0	1	1	0	4	0	1	0	0	0	1
Perm S/winters N	0	1	0	3	5	0	0	0	0	0	2
Breeds S/winters N	0	0	8	2	2	0	0	0	0	0	3
Uncertain/vagrant	(1)	(4)	(1)	(0)	(0)	(0)	(0)	(0)	(0)	(0)	(0)
Total migr. tyr. (Bol.)	1	10	13	9	12	0	1	0	0	1	10
Total tyrannids (Bol.)	52	23	19	17	16	14	10	10	7	6	27
% of Bol. tyr. migr.	1.9%	43.5%	68.4%	52.9%	75.0%	0.0%	10.0%	0.0%	0.0%	16.7%	37.0%

TABLE 4
NUMBER OF SPECIES OF MIGRANT TYRANNIDS IN BOLIVIA, ARRANGED BY BREEDING HABITAT.
TAXA OF UNCERTAIN MIGRATORY STATUS IN BOLIVIA ARE EXCLUDED FROM THE TOTALS

	Open	Marsh/wetland	Open/scrub	Scrub	Scrub/woodland	Woodland	Woodland/forest	Forest
Breeds only	0	0	1	0	2	1	0	1
Winters only	5	3	3	1	3	0	1	1
Transient	0	0	0	0	1	0	0	0
Year-round in Bolivia								
Breeds S/perm N	0	0	0	1	2	3	2	0
Perm S/winters N	3	0	0	1	6	1	0	0
Breeds S/winters N	2	2	1	2	2	2	3	1
Uncertain/vagrant	(1)	(2)	(0)	(1)	(1)	(1)	(0)	(0)
Total migr. tyr. (Bol.)	10	5	5	5	16	7	6	3

Relative temporal stability of food resources and high species diversity of the resident avifauna are often advanced as reasons for low numbers of migrants in humid tropical habitats (e.g., Willis 1966, Karr 1976), and are likely applicable here, as well.

A closer examination of the life zone data indicates that taxa that winter in Bolivia tend to occur in the non-Amazonian lowlands and in the puna zone. The lowland species represent a wide variety of taxonomic and foraging (see below) groups, including *Serpophaga subcristata* and other perch-gleaning elaeniines; *Xolmis coronata*, a perch-to-ground feeding fluvicoline; *Knipolegus hudsoni*, an aerial hawking fluvicoline; and *Xenopsaris albinucha*, a tityrine species (Prum and Lanyon 1989) whose foraging behavior is little known. The puna species include principally ground tyrants of the genus *Muscisaxicola*. In contrast, the five species found in Bolivia only during the breeding season represent five life zones: non-Amazonian lowlands (*Tyrannus albogularis*), lowlands in general (*Tyrannus savana*), lowlands + the valle zone (*Empidonomus varius*), the subtropical zone (*Elaenia strepera*), and the temperate + subtropical zone (*Elaenia albiceps chilensis*).

Broad-scale analysis of breeding habitats of passerine austral migrants has shown a preponderance of migrants in open and scrubby habitats, owing in part to the prevalence of these habitats in temperate South America (Chesser 1994). Austral migrant tyrannids are similar to all migrant passerines in breeding habitat, but somewhat fewer flycatchers are found in open areas (32 percent to 42 percent) and slightly more in various woodland habitats (34 percent to 25 percent), presumably due to the primarily insectivorous habits of tyrannids on their breeding grounds. Bolivian tyrannids occupy habitats in almost exactly the same proportions as austral migrant tyrannids as a whole. However, with the exception of *Elaenia spectabilis, E. strepera*, and *Attila phoenicurus*, the few austral migrant tyrannids that breed only in forest do not occur in Bolivia. One of these species (*Colorhamphus parvirostris*) breeds more-or-less exclusively in *Nothofagus* forest in southern Chile and Argentina, wintering mainly in Chile; others, such as *Phyllomyias fasciatus*, occur only in the Atlantic forests of eastern South America.

Examination of habitat and migratory habits of Bolivian tyrannids on a finer scale shows that most breed in scrub/woodland, woodland, and open habitats (Table 4). Species of open habitat tend to occur in Bolivia as winter residents, but most other habitat types are also represented among wintering tyrannids. Woodland is the most common habitat of summer resident flycatchers in Bolivia; one species (*Empidonomus v. varius*) occupies primarily woodlands, and two others (*Elaenia albiceps chilensis* and *Tyrannus albogularis*) occur in woodland as well as other habitats. Scrub/woodland species exhibit a variety of migration patterns, but many (six of 16) are permanent residents in the southern part of their Bolivian range, and extend their range northward during the non-breeding season.

Although only one migrant flycatcher, *Attila phoenicurus*, is found exclusively in Bolivia's Amazonian lowlands (with the possible addition of *Pachyramphus v. validus*), numerous taxa occur there in part. Most (15 of 24) are present exclusively (or nearly so) in winter. These species tend to breed in scrub or woodland habitats; thus some individuals change macrohabitats during the non-breeding season (Table 5). Six tyrannids are resident in Bolivian Amazonia, but extend their range southward during the breeding season. Two other tyrannids (*Empidonomus varius, Tyrannus savana*) breed in Amazonian Bolivia but move north in winter, and one (*Myiar-*

TABLE 5

HABITAT OF MIGRANT TYRANNIDS PRESENT IN AMAZONIAN LOWLANDS EXCLUSIVELY OR MAINLY DURING WINTER. DATA FROM BIRDS COLLECTED BY LSUMZ EXPEDITIONS IN AMAZONIAN PORTIONS OF DEPTOS. SANTA CRUZ (PROV. VELASCO), PANDO (PROV. NICOLÁS SUAREZ), EL BENI (PROV. GRAL BALLIVIÁN), AND NORTHERN DEPTO. LA PAZ

Species	Predominant breeding habitat	Microhabitat in Amazonia
Sublegatus modestus brevirostris	Scrub/woodland	1–20 m up in tropical hill forest
Myiopagis v. viridicata	Woodland/forest	1–15 m up in edge or understory of tropical hillside or lowland forest, or upland forest adjacent to varzea
Elaenia s. spectabilis	Forest	Low along edge of tropical forest, river-edge second growth
Inezia inornata	Scrub	4–35 m up in tropical hill, lowland, or riverine forest, regularly in canopy or with mixed-species flock, occasionally along forest edge
Euscarthmus m. meloryphus	Scrub	Low in tropical hillside or lowland forest, in bamboo in varzea, or in scrub along forest edge
Myiophobus fasciatus auriceps	Scrub	Low in second-growth of tropical forest, or river-edge forest
Lathrotriccus e. euleri	Woodland/forest	Understory (up to 3 m) of upper tropical forest, tropical hill forest, occasionally in understory of forest islands or along streams in bamboo
Pyrocephalus r. rubinus	Open country/scrub	2–10 m up in second growth, varzea, or forest clearings in tropical hill or lowland forest
Fluvicola pica albiventer	Marsh/wetlands	Low along forest edge near open, wet areas
Satrapa icterophrys	Woodland	Gallery forest (?)
Attila phoenicurus	Forest	Forest canopy
Casiornis rufa	Scrub/woodland	1–15 m up, occasionally to 30 m, in varzea, vine tangles, edge, or understory of tropical lowland or hill forest, occasionally with mixed-species flocks
Myiarchus t. tyrannulus	Scrub/woodland	10–30 m up, sometimes lower, along tropical lowland or hill forest edge, clearings, varzea, roadsides, or in canopy or subcanopy
Myiarchus swainsoni ferocior	Scrub/woodland	Up to 12 m in tropical forest
Empidonomus a. aurantioatrocristatus	Woodland	Up to 20 m along edge, clearings of tropical hill forest
Pachyramphus polychopterus spixii	Woodland	1–15 m up in tropical hill or upland forest, second-growth, forest edge, river-edge, or sub-canopy, often with mixed-species flocks

chus s. swainsoni) is a transient. Of the flycatchers present in Amazonia during winter, all, except *Attila phoenicurus* and *Euscarthmus m. meloryphus*, appear to breed in Bolivia; these taxa are either permanent (e.g., *Myiarchus t. tyrannulus*) or summer (e.g., *Inezia inornata*) residents in southern Bolivia.

Pearson (1980) and Robinson et al. (1988) have argued that a principal strategy of wintering migrant birds in Amazonia, including several austral migrant tyrannids, is use of secondary and edge habitats. Data from specimens in the LSUMZ tend to support this hypothesis (Table 5). Secondary microhabitats regularly occupied by wintering tyrannids in tropical forest include forest edge, second-growth, varzea, forest clearings, and river-edge forest. Thus, despite the macrohabitat differences between breeding and wintering grounds, *micro*habitats of austral migrant tyrannids wintering in Bolivian Amazonia resemble those of their breeding grounds. Canopy has been considered similar to secondary habitats in seasonal resource availability patterns (Stiles 1980, Levey and Stiles 1992), and at least two wintering austral migrant flycatchers (*Inezia inornata* and *Attila phoenicurus*) occur regularly in the canopy of tropical forest. Several species additionally may be found in mixed-species flocks.

Foraging behavior.—The most common characteristic modes of foraging among Bolivian migrant tyrannids are ground-foraging, perch-gleaning, aerial hawking, fruit/upward hover-gleaning, and outward hover-gleaning; eight to twelve migrants use each of these foraging types (Table 6). In contrast, all other foraging modes are characteristic of four or fewer migrant tyrannids, and upward striking and near-ground foraging are characteristic of only one species each.

By percentage, ground foragers are the most migratory (70.6%) group of Bolivian flycatchers. Fewer than one-tenth of Bolivian upward strikers (2.3%) and near-ground foragers (8.3%) are migratory. Species characterized by each of the other seven foraging types range from 25.0% migratory (perch to ground foragers) to 66.7% migratory (enclosed perch hawkers). Thus migrants are not concentrated among only a few foraging types; species using many foraging modes are fairly similar in their tendency to migrate.

Analysis of foraging with migratory pattern reveals that high percentages of ground foraging, perch-to-ground foraging, and perch-gleaning flycatchers occur in Bolivia in winter only. Likewise, only aerial hawkers and fruit/upward hover-gleaners are present in Bolivia exclusively as breeders. Such patterns undoubtedly indicate a degree of seasonality in resources available to species of particular foraging types, but are more properly analyzed on a continental scale, at which foraging of resident species, particularly in temperate South America, can be taken into account. The "crossroads" position of Bolivia, where both wintering, breeding, and year-round migrants occur, makes these patterns particularly difficult to assess. For example, in addition to the aerial hawking summer residents cited previously, an aerial hawking species (*Knipolegus hudsoni*) occurs in Bolivia as a *winter* resident, "replacing," in terms of feeding behavior, the summer resident taxa. The degree to which species are not reliant solely on insects (i.e., take fruit in winter) probably also plays a role in their distribution, and many austral migrant flycatchers are at least somewhat frugivorous (Chesser, unpubl. data).

Seasonal differences in elevation.—Of the 34 migrant tyrannids present within Bolivia throughout the year (hence, those for which both breeding and non-breeding season data are available), a surprising 15 appear to show seasonal differences in elevational range of 500 m or more (Table 7). The breeding and non-breeding ranges of one species, *Pseudocolopteryx acutipennis*, are completely and strikingly separate; breeding-season elevation ranges from 2,200 m to 3,550 m, whereas non-breeding records are all below 700 m (Figure 12). All other tyrannids showing seasonal elevational differences have elevationally overlapping breeding and wintering ranges, and form two groups: taxa that breed at middle or high elevation and extend the lower limit of their ranges downward during winter, and taxa that breed at low to middle or high elevations but are found in winter only at low elevation. In the former group, it is the lower elevational limit that changes, whereas the upper elevational limit shifts seasonally in the latter group. Other species (e.g., *Myiophobus fasciatus auriceps*) probably undergo at least an elevational shift in abundance, with most winter records at the lower end of the breeding range, or most breeding season records at the upper end of the winter range. However, range of elevation does not appear to differ seasonally for such species (see species accounts for other examples).

Excluding the exceptional *P. acutipennis*, those species that undergo a seasonal downward shift of their upper range limit are roughly evenly divided between the 500–1,000 and 1,000+ m groups (five and four species, respectively; Table 7). In contrast, four of five species whose *lower* range limit shifts downward in winter show differences of 1,000+ m. This may result from the fact that the lower range limits of the middle and high elevation species of the latter group ($\bar{x} = 2,900 \pm 1,012$ m) are significantly higher than the upper range limits of the middle

TABLE 6

Number of Species of Bolivian Migrant Tyrannids, Arranged by Predominant Foraging Mode. "Total Migr. Tyr. (Bol.)" is Number of Bolivian Migrant Flycatchers Using a Particular Foraging Mode, "Total Tyrannid (Bol.)" is Number of Bolivian Flycatchers (Migrant and Resident) Using a Particular Foraging Mode, and "% of Bol. Tyr. Migr." is Percent of Bolivian Flycatchers Using a Foraging Mode that Are Migratory. Data on Foraging Taken Principally from Fitzpatrick (1980). Problematic Species *Xenopsaris albinucha* and *Culicivora caudacuata* Have Been Placed in "Outward Hover Glean" and "Perch Glean," Respectively

	Out. hov. glean	Fruit/hawk	Fruit/up. h.-gl.	Perch glean	Upward strike	En. perch hawk	Aerial hawk	Near ground	Perch to ground	Ground
Breeds only	0	0	2	0	0	0	3	0	0	0
Winters only	2	0	0	4	0	1	1	0	3	6
Transient	1	0	0	0	0	0	0	0	0	0
Year-round in Bolivia										
Breeds S/perm N	1	1	4	0	1	0	1	0	0	0
Perm S/winters N	3	1	0	3	0	0	1	0	0	3
Breeds S/winters N	1	0	2	3	0	3	2	1	0	3
Uncertain/vagrant	(0)	(0)	(1)	(5)	(0)	(0)	(0)	(0)	(0)	(0)
Total migr. tyr. (Bol.)	8	2	8	10	1	4	8	1	3	12
Total tyrannids (Bol.)	17	6	24	27	43	6	20	12	12	17
% of Bol. tyr. migr.	47.1%	33.3%	33.3%	37.0%	2.3%	66.7%	40.0%	8.3%	25.0%	70.6%

TABLE 7
Seasonal Differences in Elevational Range of Austral Migrant Tyrannids in Bolivia. "Elevational Difference" Indicates Which End of a Species' Elevational Range Differs Seasonally (See Text for Full Explanation); "Degree of Difference" Is Categorized as 500–1,000 m, 1,000+ m, or 2,000+ m

Species	Elevational difference	Degree of difference (m)
Phaeomyias murina ignobilis	Upper limit	1,000+
Suiriri s. suiriri	Lower limit	1,000+*
Myiopagis v. viridicata	Upper limit	500–1,000
Elaenia s. spectabilis	Upper limit	500–1,000
Pseudocolopteryx acutipennis	Total (no overlap)	2,000+
Muscisaxicola rufivertex pallidiceps	Lower limit	1,000+
Muscisaxicola flavinucha	Lower limit	500–1,000
Lessonia oreas	Lower limit	1,000+
Knipolegus a. aterrimus	Lower limit	1,000+
Casiornis rufa	Upper limit	500–1,000
Myiarchus swainsoni ferocior	Upper limit	1,000+
Legatus l. leucophaius	Upper limit	1,000+
Empidonomus a. aurantioatrocristatus	Upper limit	1,000+
Pachyramphus polychopterus spixii	Upper limit	500–1,000
Pachyramphus validus audax	Upper limit	500–1,000**

* Excludes an obviously non-breeding lowland specimen from January (see species account for details).
** Includes a Peruvian winter record from 1,680 m; for Bolivian specimens only, the degree of difference is 1,000+ m.

and low elevation species of the former ($\bar{x} = 1,862 \pm 491$ m; *t*-test; $p = 0.01$). Thus the latter group has greater potential for elevational range shifting. Differences in degree of seasonal climate change between high and low elevation may also play a factor in this pattern, although, significantly, no high-elevation species appears to vacate the upper portion of its range (with the possible exception of *Muscisaxicola flavinucha*; see species account), as would be expected if harshness of climate were driving the seasonal differences in elevation.

It should be noted that patterns of seasonal differences in elevation among these migrants are not necessarily the result of altitudinal migration. If a species breeds and is migratory in areas to the south of a permanent population in Bolivia, then only in those cases in which the upper portion of the range is vacated during winter can it be ascertained that altitudinal migration is taking place. If the lower limit of such a species' elevational range is adjusted downward during winter, it may merely be the result of latitudinal migrants from the south occupying the lower elevational sites. Such a pattern may occur, for example, in *Suiriri s. suiriri*. Similarly, an upward shift in elevational range in the breeding season may not be an indication of altitudinal migration, if the species winters and is migratory in areas to the north of a permanent Bolivian population. In this instance, the shift may result from latitudinal migrants from the north occupying the higher elevational sites, although, if these migrants winter at lowland sites, their migration is partially altitudinal as well. Most austral migrant tyrannids that exhibit seasonal shifts in elevation, however, do not show these patterns, and appear to undertake actual elevational movements.

That many Bolivian austral migrant tyrannids are both latitudinal and elevational migrants emphasizes the continuity of the two forms of migration. Levey and Stiles (1992) suggested that

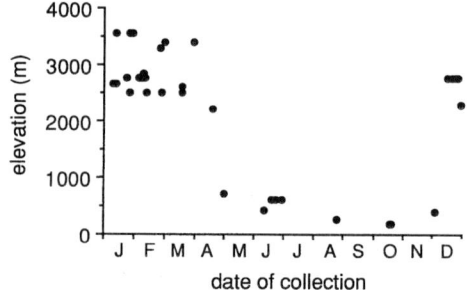

Fig. 12. Seasonal elevational distribution in Bolivia of *Pseudocolopteryx acutipennis*.

seasonal movements of Neotropical birds occur at a variety of scales, from local and elevational to long-distance, and that movements on all levels appear to be related to tracking variable resources. They argued that long-distance migration (from the tropics to the temperate zone) is merely the evolutionary endpoint of a continuum of movements associated with resource fluctuations, including elevational migration. The data on the flycatchers considered here are consistent with this hypothesis.

A second elevational pattern, characteristic of long-distance migrant flycatchers in Bolivia, is the extent to which records of lowland taxa have been recorded at high elevation (3,000+ m) during fall migration periods, particularly in March. There are some nine species whose breeding ranges extend extensively south of Bolivia, in Argentina, and whose wintering ranges extend north of Bolivia in western Amazonia, or in northern portions of Colombia and Venezuela; these taxa vacate large areas of their ranges in both winter and summer. These species (*Elaenia spectabilis*, *E. albiceps chilensis*, *E. parvirostris*, *Pyrocephalus r. rubinus*, *Myiarchus swainsoni*, *Myiodynastes maculatus solitarius*, *Empidonomus v. varius*, *Empidonomus a. aurantioatrocristatus*, and *Tyrannus s. savana*) likely pass through Bolivia in reasonably large numbers during migration, in addition to breeding or wintering there. The highest elevational record for five of these nine species (*E. a. chilensis*, *M. swainsoni*, *E. v. varius*, *E. a. aurantioatrocristatus*, and *T. s. savana*) was recorded from mid to late March. Because these species are present in Bolivia for an average of roughly eight months of the year (either as breeders and transients or as winterers and transients), the expected chance of observing a high elevational record during any particular month is approximately 12.5 percent. That high elevational records for five of nine species are from March is a highly significant result ($\chi^2 = 18$; d.f. = 1; $p < 0.0001$), and suggests either that portions of the Bolivian high Andes may lie along fall migration routes, or that elevational vagrants tend to occur in the Andes during fall migration. T. A. Parker (pers. comm.) and B. M. Whitney (in litt.) have noted "fallouts" or unusual numbers of migrants, especially tyrannids, along the base of the Andes in depto. Santa Cruz in the southern fall, indicating that at least the eastern slopes of the Andes may be important as a fall migration route (or a possible source of vagrants). The geographical position of the central and southern Bolivian (and northern Argentine) Cordillera Oriental, where the Andes reach their easternmost point, places these mountains on the most direct route between much of Argentina and western Amazonia, and provides at least circumstantial geographical support for this hypothesis. That migration routes may differ in spring and fall is supported by the observations of Davis (1993) in lowland depto. Santa Cruz, who noted two of four long-distance transient flycatchers (*Elaenia spectabilis*, *Myiarchus swainsoni*) only during spring migration (September, October, and November).

Timing of migration. Although analysis of timing of migration using specimens is best accomplished with a large data set from the entirety of a species' range, data available from specimens from Bolivia (e.g., specimens with high levels of subcutaneous fat, specimens from areas of passage migration only, and early and late dates of occurrence on breeding and wintering grounds) are potentially useful for examining timing of migration. Indeed, several patterns emerge from analysis of the data for Bolivia alone and are corroborated by available data from other regions.

Data for eight species for which sufficient spring and fall data are available indicate that spring migration of Bolivian flycatchers occurs generally from September to November, and fall migration from March to May (Table 8). However, analysis of timing of migration of individual species suggests that intraspecific variation in timing is extensive. This is particularly clear in the data for spring migrations of *E. parvirostris*, *E. aurantioatrocristatus*, and *T. savana*, which show strong indications of migration occurring over periods of two months or more, from early-to-mid September to early-to-mid November. Given the relatively compressed breeding area of southern South America, relative to the wintering area apparently available to austral migrants (cf. the situation for many Nearctic-Neotropical migrants; Chesser 1994), one might predict that populations of many austral migrants are limited by habitat on their breeding grounds, and that selection for timing of spring migration in the austral system would be relatively strong. Instead, these data suggest that timing is rather relaxed, at least for the species considered here.

Several deviations from the general pattern of timing of migration, or timing that occurs rather early or late within the general period, are evident among austral migrant tyrannids in Bolivia. The most obvious of these is the late migration of *Pseudocolopteryx acutipennis*, for which there are breeding range records from 17 December to 20 April and non-breeding records from 2 May to 5 December (Figure 2). These data are supported by the available specimens from breeding areas in northwestern Argentina, where records extend from December through early May, and with observations that the species arrives in the Cachi area of pcia. Salta (Argentina) in mid-

TABLE 8

Data Related to Timing of Migration in Bolivia for Selected Austral Migrant Tyrannids. Codes for Significant Dates Are: **HF** = Specimen with Heavy or Extremely Heavy Subcutaneous Fat Deposits, **MF** = Specimen with Moderate Fat Deposits (This Category Used Only Occasionally in Support of Other Data), **T** = Transient, Specimen Collected in Area of Apparent Passage Migration (Species Neither Breeds Nor Winters at Geographical or Elevational Locality), **WE** = Early Date of Arrival on Bolivian Wintering Grounds, **WL** = Late Record from Bolivian Wintering Grounds, **BE** = Early Date of Arrival on Bolivian Breeding Grounds, and **BL** = Late Record from Bolivian Breeding Grounds. Records for Two Species from the Wintering Grounds (*S. m. brevirostris*) [WE] and *E. a. aurantioatrocristatus* [WE, WL] Were Seasonally Biased Due to Lack of Collecting Effort in Their Wintering Areas, and Were Not Used (See Species Accounts for these Records)

Species	Significant fall dates	Significant spring dates
Sublegatus modestus brevirostris	29 Mar (T), 16 Apr (T)	15 Sep (T), 16 Sep (HF), 22 Sep (WL)
Myiopagis v. viridicata	7 May (BL)	4 Oct (HF), 19 Oct (BE), 7 Nov (HF)
Elaenia s. spectabilis	15 Apr (WE), 20 Apr (MF), 22 Apr (HF, BL)	30 Oct (HF), 30 Oct (BE), 19 Nov (MF), 20 Nov (WL)
Elaenia parvirostris	23 Mar (WE), 25 May (BL)	15 Sep (BE), 16 Sep (HF), 7 Oct (MF), 9 Oct (HF), 20 Nov. (HF, WL)
Myiarchus s. swainsoni	13 Mar (T), 4 Apr (T), 13 Apr (T)	13 Oct (T)
Empidonomus a. aurantioatrocristatus	18 March (T), 27 March (T), 2 May (BL)	10 Sep (BE), 17 Sep (HF), 20 Sep (HF), 27 Sep (T), 2 Oct (T), 5 Oct (T), 11 Oct (HF, T), 7 Nov (HF), 9 Nov (HF, T)
Tyrannus m. melancholicus	17 Mar (HF), 8 Apr (HF), 10 Apr (HF), 14 Apr (HF), 4 May (BL)	Sep (T)*, 21 Oct (BE), 3 Nov (HF)
Tyrannus s. savana	10 Feb (T), 15 Feb (T), 16 Feb (T), 25 Feb (T), 27 Feb (T), 5 Mar (T), 11 Mar (T), 13 Mar (T), 16 Mar (T), 18 Mar (T), 20 Mar (T), 1 Apr (BL), 5 May (T), 15 May (T), 25 May (T)	10 Sep (BE), 15 Sep (HF, T), 19 Sep (T), 22 Sep (T), 24 Sep (T), 30 Sep (MF, T), Oct (T)*, 1 Oct (T), 25 Oct (T), 5 Nov (T)

* Undated sight records of transient individuals from Davis (1993).

December (B. M. Whitney, in litt.) and that it is still fledging young in early April in depto. Cochabamba, Bolivia (B. M. Whitney, in litt.), although Fjeldså and Krabbe (1990) reported nesting in northwestern Argentina in November. Relatively late timing of spring migration also seems to be characteristic of *Elaenia s. spectabilis* (Table 8), although not to the extent of that of *P. acutipennis*. Spring migration of *E. spectabilis* appears to occur only in late October and November; this is further supported by a series of specimens with no fat from the Bolivian wintering grounds in early- to mid-October (LSUMZ 124540–42, 125913–15).

A second pattern involves early timing of spring migration among the ground-tyrant genera *Muscisaxicola* and *Agriornis*. Late records from the Bolivian non-breeding range of seven of these species, excluding only the uncommon *Muscisaxicola frontalis* and *Agriornis m. microptera*, range from 5 September to 13 October, with only two records after 18 September. It therefore appears that these species are uniformly early spring migrants from Bolivia, with most individuals presumably beginning migration in August and early September. This suggestion is supported for at least one of these species by data from Humphrey et al. (1970), who noted that *Muscisaxicola capistrata* arrives on Isla Grande, Tierra del Fuego, in early September (cf. *Elaenia albiceps chilensis*, for which the earliest record for Isla Grande is 14 October [Humphrey et al. 1970]).

Presumably such deviations are related to timing of breeding (or activities associated with breeding, such as territory settlement), which in temperate areas is apparently closely tied to annual peaks in food availability (Lack 1954; Perrins 1970; Martin 1987). Thus one would predict that the food relied upon by nesting ground tyrannids reaches its most abundant stage earlier than, and that of *Pseudocolopteryx acutipennis* later than, that of most other tyrant-flycatchers, or that early arrival or territory settlement is especially important for ground tyrants. Late migrations in some other austral migrant species (e.g., some *Sporophila* seedeaters) are believed to be related to seasonality of food resources (seeds, for these species) on their breeding grounds (Remsen and Hunn 1979), and high abundance of migrant birds on the Planalto Central of Brazil has been linked to periods of great insect abundance (Negret 1988). The unusual migrations of *P. acutipennis* are particularly intriguing, for other high Andean and temperate marsh-nesting species (e.g., *Tachuris rubrigastra*, *Phleocryptes melanops*) do not appear to share the same annual cycle (see data on earlier nesting in Peña 1987, Fjeldså and Krabbe 1990); this would be a rewarding avenue for future research.

Other issues.—This study was based upon the most traditional and data-rich method of determining migrations of birds: a thorough geographical and seasonal investigation of legitimate records and observations of particular species. Although bird-banding potentially yields valuable information not obtainable through analysis of seasonal distributional records, such data are slow in accumulating and should not be viewed as a panacea. Recovery of a banded bird provides two data points only, and in South America, particularly Amazonia, recoveries of banded migrants, especially small passerines, will likely be rather rare, even if and when banding programs become more common. In contrast, the data points provided by seasonal distributional records of South American birds are many and continue to increase.

Nevertheless, it must be emphasized that, even in areas where collecting or observational activity has been plentiful, absence of records during a particular season does not necessarily indicate absence of the species. Such information is compelling, however, when combined with data indicating presence of the same species in other areas during the same season that the species is absent in a given region (and vice-versa). The only other explanation for this phenomenon would be massive geographical and seasonal sampling error, an extremely unlikely possibility.

The sampling error argument is potentially applicable to the elevational data presented here, as well as the geographical information; that is, that the elevational patterns observed are the result of inadequate collection or observation at particular elevations during certain seasons. If this were the case, one would expect to find few, if any, species resident at the elevations that produced the bias; and one would expect to find patterns of bias such that changes in elevational distribution would be the same for different species at similar elevations. However, species were found to be resident at all elevations considered, from the lowlands to the puna zone, and those species exhibiting seasonal differences in elevation frequently differed in pattern of elevational change (Table 7).

Migration of birds in at least some portions of Bolivia was previously discussed or mentioned by Olrog (1963), Pearson (1980), and Davis (1993). The species considered in this paper as austral migrants include all tyrannids listed as long-distance migrants by Pearson and Davis for their study areas. Two other species designated here as austral migrants (*Myiophobus fasciatus* and *Casiornis rufa*) were considered local migrants at the study site of Davis (1993); investigation at a larger scale reveals these species to be true austral migrants.

Olrog (1963) reported having observed migratory ("migratorio") or displacement ("desplazamiento") behavior in six tyrannids not considered here to be austral migrants (*Tolmomyias sulphurescens grisescens* [*pallescens*?], *Ochthoeca r. rufipectoralis*, *Ochthoeca cinnamomeiventris thoracica*, *Myiotheretes striaticollis pallidus*, *Xolmis cinerea*, and *Knipolegus signatus cabanisi*), as well as four thought to be migratory (*Knipolegus striaticeps*, *Knipolegus hudsoni*, *Myiarchus t. tyrannulus*, and *Myiarchus swainsoni ferocior*). Movements of the former six taxa would appear to be strictly local or elevational movements, and not evidence of austral migration. Significantly, Olrog, in his final checklist of the avifauna of Argentina (1979), mentioned migration of only the latter four tyrannids in describing their continent-wide geographical ranges; the former six were either tacitly considered resident, or had not been recorded in Argentina and thus were not discussed (in the case of the two *Ochthoeca* species).

Nevertheless, Bolivian flycatchers not considered here may eventually prove to be austral migrants, and many species will certainly be shown to undergo elevational or local movements. Moreover, it should be emphasized that the geographical and elevational range boundaries given for migrants are approximations, and that further fieldwork will likely demonstrate that migratory patterns of a number of species differ from those presented here. There are great opportunities for research on migrants in Bolivia and elsewhere in South America, and studies of these taxa are strongly encouraged. For Bolivia, collections in the northern portion of the country from December or January, or year-round fieldwork there, would be particularly helpful in sorting out migration patterns of many species.

ACKNOWLEDGMENTS

I wish to thank the following for use of museum collections under their care, for the loan of specimens, or for information concerning specimens: George Barrowclough, John Bates, Mary LeCroy, and François Vuilleumier (American Museum of Natural History), Peter Colston and Robert Prŷs-Jones (British Museum [Natural History]), David Agro, Mark Robbins, and Christopher Thompson (Academy of Natural Sciences of Philadelphia), Robin Panza and Kenneth Parkes (Carnegie Museum), Gene Hess (Delaware Museum of Natural History), Estela Alabarce and Claudio Laredo (Fundación Miguel Lillo), Scott Lanyon, Tom Schulenberg, Melvin Traylor and David Willard (Field Museum of Natural History), Kimball Garrett (Los Angeles County Museum), Jorge Navas (Museo Argentina de Ciencias Naturales), Nelly Bó and Aníbal Camperi (Museo de La Plata), Douglas Stotz (Museu de Zoologia da Universidade de São Paulo), Dante Teixeira and Jorge Nacinovic (Museu Nacional de Rio de Janeiro), David Oren (Museu Paraense Emílio Goeldi), Raymond Paynter (Museum of Comparative Zoology), Julio Contreras (Programa de Biologia Básica y Aplicada Subtropical), Janet Hinshaw and Robert Storer (University of Michigan Museum of Zoology), and Fred Sibley (Peabody Museum of Natural History). I thank José Maria Cardoso da Silva and Jon Fjeldså for providing data on specimens in the Swedish Museum of Natural History, David Good for writing the mapping program used in Figures 2–10, Luis Chiappe for translating the abstract into Spanish, and Bret Whitney for providing unpublished observations. J. V. Remsen and J. M. C. da Silva made many helpful comments on an earlier draft of this manuscript. This paper is dedicated to the memory of Ted Parker, who encouraged me to study austral migration and was a source of continuous ideas and insight.

LITERATURE CITED

ALLEN, J. A. 1889. List of the birds collected in Bolivia by Dr. H. H. Rusby, with field notes by the collector. Bull. Amer. Mus. Nat. Hist. 2:77–112.

AZARA, F. DE. 1802–1805. Apuntamientos para la Historia Natural de los Paxaros del Paraguay y Río de la Plata. Vols. I–III. Madrid.

BARROWS, W. B. 1883. Birds of the lower Uruguay. Bull. Nuttall Ornith. Club 8:198–212.

BELTON, W. 1984. Birds of Rio Grande do Sul, Brazil, Part I. Rheidae through Furnariidae. Bull. Amer. Mus. Nat. Hist. 178:369–636.

BELTON, W. 1985. Birds of Rio Grande do Sul, Brazil, Part II. Formicariidae through Corvidae. Bull. Amer. Mus. Nat. Hist. 180:1–242.

BOND, J., AND R. MEYER DE SCHAUENSEE. 1942. The birds of Bolivia. Part I. Proc. Acad. Nat. Sci. Philadelphia 94:307–391.

BOND, J., AND R. MEYER DE SCHAUENSEE. 1943. The birds of Bolivia. Part II. Proc. Acad. Nat. Sci. Philadelphia 95:167–221.

CABOT, J. 1990. First record of *Upucerthia validirostris* from Bolivia and new Bolivian distributional data. Bull. Brit. Ornith. Club 110:103–107.

CAPURRO, H. A., AND E. H. BUCHER. 1988. Lista comentada de las aves del bosque chaqueño de Joaquin V. Gonzalez, Salta, Argentina. Hornero 13:39–46.

CHAPMAN, F. M. 1926. The distribution of bird-life in Ecuador. Bull. Amer. Mus. Nat. History 55:1–784.

CHESSER, R. T. 1994. Migration in South America: an overview of the austral system. Bird Conservation Intl. 4:91–107.
CHESSER, R. T. 1995. Biogeographic, ecological, and evolutionary aspects of South American austral migration, with special reference to the family Tyrannidae. Unpublished Ph.D. dissertation, Louisiana State University.
CHESSER, R. T., AND M. MARÍN A. 1994. Seasonal distribution and natural history of the Patagonian Tyrant (*Colorhamphus parvirostris*). Wilson Bull. 106:649–667.
CHUBB, C. 1919. Notes on collections of birds in the British Museum, from Ecuador, Peru, Bolivia, and Argentina. Ibis 1919:1–55, 256–290.
CINTRA, R., AND C. YAMASHITA. 1990. Habitats, abundância e ocorrência das espécies de aves do Pantanal de Poconé, Mato Grosso, Brasil. Pap. Avul. Zool. 37:1–21.
DAVIS, S. E. 1993. Seasonal status, relative abundance, and behavior of the birds of Concepción, departamento Santa Cruz, Bolivia. Fieldiana Zool. N. S. 71:1–33.
D'ORBIGNY, A. 1847. Voyage dans l'Amerique Méridionale. Vol. 4, pt. 3, Oiseaux. Bertrand and Levrault, Paris and Strasbourg.
EISENMANN, E. 1963. Mississippi Kite in Argentina, with comments on migration and plumages in the genus *Ictinia*. Auk 80:74–76.
EISENTRAUT, M. 1935. Biologische Studien im bolivianischen Chaco. VI. Beitrag zur Biologie der Vogelfauna. Mitt. Zool. Mus. Berlin 20:367–443.
FITZPATRICK, J. W. 1980. Foraging behavior of Neotropical tyrant flycatchers. Condor 82:43–57.
FJELDSÅ, J., AND N. KRABBE. 1989. An unpublished major avian collection from the Bolivian highlands. Zool. Scripta 18:321–329.
FJELDSÅ, J., AND N. KRABBE. 1990. Birds of the High Andes. Zoological Museum, University of Copenhagen, and Apollo Books, Svendborg, Denmark.
GYLDENSTOLPE, N. 1945. A contribution to the ornithology of northern Bolivia. Kungl. Svensk. Vet.-Akad. Handl. (ser. 3) 23:1–300.
HAYES, F. E., P. A. SCHARF, AND R. S. RIDGELY. 1994. Austral bird migrants in Paraguay. Condor 96:83–97.
HELLMAYR, C. E. 1925. Review of the birds collected by Alcide d'Orbigny in South America. Parts IV-VI. Novit. Zool. 32:1–30, 175–194, 314–334.
HILTY, S. L., AND W. L. BROWN. 1986. A Field Guide to the Birds of Colombia. Princeton Univ. Press, Princeton, New Jersey.
HUMPHREY, P. S., D. BRIDGE, P. W. REYNOLDS, AND R. T. PETERSON. 1970. Birds of Isla Grange (Tierra del Fuego). University of Kansas Museum of Natural History, Lawrence, Kansas, for the Smithsonian Institution, Washington, D.C.
KRATTER, A. W., T. S. SILLETT, R. T. CHESSER, J. P O'NEILL, T. A. PARKER III, AND A. CASTILLO. 1993. Avifauna of a Chaco locality in Bolivia. Wilson Bull. 105:114–141.
LACK, D. 1954. The Natural Regulation of Animal Numbers. Clarendon, Oxford, England.
LANYON, W. E. 1982. Evidence for wintering and resident populations of Swainson's Flycatcher (*Myiarchus swainsoni*) in northern Surinam. Auk 99:581–582.
LANYON, W. E. 1984. A phylogeny of the kingbirds and their allies. Amer. Mus. Novit. 2797:1–28.
LAUBMANN, A. 1930. Vögel. Wissenschaftliche Ergebnisse der Deutschen Gran Chaco-Expedition. Strecker and Schroder, Stuttgart, Germany.
LAUBMANN, A. 1934. Weitere Beiträge zur Avifauna Argentiniens. Verh. Orn. Ges. Bayern 20:249–336.
LEVEY, D. J., AND F. G. STILES. 1992. Evolutionary precursors of long-distance migration: resource availability and movement patterns in Neotropical landbirds. Amer. Nat. 140:447–476.
LÖNNBERG, E. 1903. On a collection of birds from northwestern Argentina and the Bolivian Chaco. Ibis (1903):441–471.
MARANTZ, C. A., AND J. V. REMSEN, JR. 1991. Seasonal distribution of the Slaty Elaenia, a little-known austral migrant of South America. J. Field Ornith. 62:162–172.
MARINI, M. A. AND R. B. CAVALCANTI. 1990. Migrações de *Elaenia albiceps chilensis* e *Elaenia chiriquensis albivertex* (Aves: Tyrannidae). Bol. Mus. Para. Emilio Goeldi, ser. Zool. 6:59–67.
MARTIN, T. E. 1987. Food as a limit on breeding birds: a life-history perspective. Ann. Rev. Ecol. Syst. 18:453–487.
MEYER DE SCHAUENSEE, R. 1970. A Guide to the Birds of South America. Oliver and Boyd, Edinburgh, Scotland.
MORTON, E. S. 1977. Intratropical migration in the Yellow-green Vireo and Piratic Flycatcher. Auk 94:97–106.
NAROSKY, T. AND A. G. DI GIACOMO. 1993. Las aves de la Provincia de Buenos Aires: distribución y estatus. Asoc. Ornitológica del Plata, Vázquez Mazzini Editores y L. O. L. A., Buenos Aires, Argentina.
NEGRET, A. 1988. Fluxos migratorios na avifauna da Reserva Ecológica do IBGE, Brasília, D. F., Brasil. Rev. Bras. Zool. 5:209–214.
NIETHAMMER, G. 1953. Zur Vogelwelt Boliviens. Bonn. Zool. Beitr. 4:195–303.
NIETHAMMER, G. 1956. Zur Vogelwelt Boliviens (Teil II: Passeres). Bonn. Zool. Beitr. 7:84–150.
NORES, M., AND D. YZURIETA. 1984. Registro de aves en el sur de Bolivia. Doñana, Acta Vert. 11:327–337.
NORES, M., D. YZURIETA, AND R. MIATELLO. 1983. Lista y distribución de las aves de Córdoba, Argentina. Bol. Acad. Nac. Cienc. Córdoba 56:1–114.
OLROG. C. C. 1963. Notas sobre aves bolivianas. Acta Zool. Lilloana 19:407–478.

OLROG. C. C. 1979. Nueva lista de la avifauna Argentina. Opera Lilloana 27:1–324.
OLROG. C. C. 1984. Las Aves Argentinas. Admin. Parq. Nac., Buenos Aires, Argentina.
PARKER. T. A., III, P. K. DONAHUE, AND T. S. SCHULENBERG. 1994. Birds of the Tambopata Reserve (Explorer's Inn Reserve). Pp. 106–124 in The Tambopata-Candamo Reserved Zone of Southeastern Perú: a Biological Assessment (R. B. Foster et al.). RAP Working Papers 6. Conservation International, Washington, D.C.
PARKER. T. A., III, S. A. PARKER, AND M. A. PLENGE. 1982. An Annotated Checklist of Peruvian Birds. Buteo Books, Vermillion, South Dakota.
PARKER. T. A., III AND O. ROCHA O. 1991. La avifauna del Cerro San Simon, una localidad de campo rupestre aislado en el depto. Beni, noreste Boliviano. Ecología en Bolivia 17:15–29.
PARKER. T. A., III AND R. A. ROWLETT. 1984. Some noteworthy records of birds from Bolivia. Bull. Brit. Ornith. Club 104:110–113.
PAYNTER, R. A. 1992. Ornithological Gazetteer of Bolivia. Mus. Comp. Zool., Cambridge, Massachusetts.
PEARSON, D. L. 1975. Un estudio de las aves de Tumi Chucua, departamento del Beni, Bolivia. Pumapunku 8:50–56.
PEARSON, D. L. 1980. Bird migration in Amazonian Ecuador, Peru, and Bolivia. Pp. 273–283 in Migrant Birds in the Neotropics: Ecology, Behavior, Distribution, and Conservation (A. Keast and E. S. Morton, Eds.), Smithsonian Inst. Press, Washington.
PEÑA, L. E. 1962. Anotaciones sobre las aves collectadas en Bolivia. Rev. Univ. Catol., Santiago 47:167–201.
PEÑA, M. DE LA. 1987. Nidos y Huevos de Aves Argentinas. Published by the author, Santa Fe, Argentina.
PERRINS, C. M. 1970. The timing of birds' breeding seasons. Ibis 112:242–255.
PRUM, R. O., AND W. E. LANYON. 1989. Monophyly and phylogeny of the *Schiffornis* group (Tyrannoidea). Condor 91:444–461.
REMSEN, J. V., JR. 1986. Aves de una localidad en la sabana húmeda del norte de Bolivia. Ecología en Bolivia 8:21–35.
REMSEN, J. V., JR., AND E. S. HUNN. 1979. First records of *Sporophila caerulescens* from Colombia; a probable long distance migrant from southern South America. Bull. Brit. Ornith. Club 99:24–26.
REMSEN, J. V., JR., AND T. A. PARKER III. 1990. Seasonal distribution of the Azure Gallinule (*Porphyrula flavirostris*), with comments on vagrancy in rails and gallinules. Wilson Bull. 102:380–399.
REMSEN, J. V., JR., AND M. A. TRAYLOR, JR. 1989. An Annotated List of the Birds of Bolivia. Buteo Books, Vermillion, South Dakota.
RIDGELY, R. S., AND G. TUDOR. 1994. The Birds of South America. Vol. II. The suboscine passerines. Univ. Texas Press, Austin.
ROBINSON, S. K., J. TERBORGH, AND J. W. FITZPATRICK. 1988. Habitat selection and relative abundance of migrants in southeastern Peru. Pp. 2298–2307 in Acta XIX Congressus Internationalis Ornithologici (H. Ouellet, Ed.), University of Ottawa Press, Ottawa, Canada.
ROSENBERG, G. H. 1990. Habitat specialization and foraging behavior by birds of Amazonian river islands in northeastern Peru. Condor 92:427–443.
SALVADORI, T. 1897. Viaggio del Dott. Alfredo Borelli nel Chaco boliviano e nella Republica Argentina. 7. Uccelli. Boll. Mus. Zool. Anat. Comp. Torino 12:1–36.
SCLATER, P. L., AND O. SALVIN. 1879. On the birds collected in Bolivia by Mr. C. Buckley. Proc. Zool. Soc. London (1879):588–645.
SHORT, L. 1975. A zoogeographic analysis of the South American Chaco avifauna. Bull. Amer. Mus. Nat. Hist. 154:165–352.
SICK, H. 1968. Vogelwanderungen im kontinental Südamerika. Vogelwarte 24:217–243.
SICK, H. 1984. Ornitologia Brasileira, uma introdução. 2 vols. Editora Univ. Brasília, Brasília.
SICK, H. 1993. Birds in Brazil, A Natural History. Princeton Univ. Press, Princeton, New Jersey.
SILVA, J. M. C. DA, AND Y. ONIKI. 1988. Lista preliminar da avifauna da Estação Ecológica Serra das Araras, Mato Grosso, Brasil. Bol. Mus. Paraense Emilio Goeldi, sér. Zool. 4:123–143.
STILES, G. 1980. Evolutionary implications of habitat relations between permanent and winter resident landbirds in Costa Rica. Pp. 421–435 in Migrant Birds in the Neotropics (A. Keast and E. S. Morton, Eds.). Smithsonian Institution Press, Washington, D.C.
TRAYLOR, M. A., JR. 1979. Tyrannidae. Pp. 1–245 in Checklist of Birds of the World, Vol. 8 (M. A. Traylor, Jr., Ed.). Harvard Univ. Press, Cambridge, Massachusetts.
TRAYLOR, M. A., JR. 1982. Notes on tyrant flycatchers (Aves: Tyrannidae). Fieldiana Zool. N. S. 13:1–22.
WHITNEY, B. M., J. L. ROWLETT, AND R. A. ROWLETT. 1994. Distributional and other noteworthy records for some Bolivian birds. Bull. Brit. Ornith. Club 114:149–162.
WILLIS, E. O. 1979. The composition of avian communities in remanescent woodlots in southern Brazil. Pap. Avul. Zool. 33:1–25.
WILLIS, E. O. 1988. Land-bird migration in São Paulo, southeastern Brazil. Pp. 754–764 in Acta XIX Congressus Internationalis Ornithologici (H. Ouellet, Ed.), University of Ottawa Press, Ottawa, Canada.
WILLIS, E. O., AND Y. ONIKI. 1990. Levantamento preliminar das aves de inverno em dez áreas do sudoeste de Mato Grosso, Brasil. Ararajuba 1:19–38.
ZIMMER, J. T. 1931–1955. Studies of Peruvian birds, nos. 1–66. Amer. Mus. Novit. (between numbers 500 and 1723 inclusive).
ZIMMER, J. T. 1938. Notes on migrations of South American birds. Auk 55:405–410.

APPENDIX

SUMMARY OF DATA FOR TYRANT-FLYCATCHERS MIGRATORY IN BOLIVIA. LIFE ZONE ADAPTED FROM REMSEN AND TRAYLOR (1989) AND FORAGING MODE FROM FITZPATRICK (1980)

Species or subspecies	Seasonal distributional pattern	Breeding habitat	Life zone	Foraging mode
Camptostoma obsoletum bolivianum	Perm. S, winters N	Scrub/woodland	Lowlands/valle	Perch-glean
Phaeomyias murina ignobilis	Breeds S, perm. N	Scrub	Lowlands/valle	Fruit/upw. hover-glean
Sublegatus modestus brevirostris	Perm. S, winters N	Scrub/woodland	Lowlands/valle	Outward hover-glean
Suiriri s. suiriri	Perm. S, winters N	Scrub/woodland	Non-Amaz. lowlands/valle	Perch-glean
Myiopagis c. caniceps	Uncertain	Woodland	Non-Amaz. lowlands	Perch-glean
Myiopagis v. viridicata	Breeds S, winters N	Woodland/forest	Lowlands	Perch-glean
Elaenia s. spectabilis	Breeds S, winters N	Forest	Lowlands	Fruit/upw. hover-glean
Elaenia albiceps chilensis	Breeds	Scrub/woodland	Temperate/subtropical zones	Fruit/upw. hover-glean
Elaenia parvirostris	Breeds S, winters N	Woodland/forest	Lowlands/valle	Fruit/upw. hover-glean
Elaenia strepera	Breeds	Forest	Subtropical zone	Fruit/upw. hover-glean
Elaenia chiriquensis albivertex	Breeds S, perm. N	Scrub/woodland	Non-Amaz. lowlands	Perch-glean
Serpophaga nigricans	Uncertain	Scrub	Non-Amaz. lowlands	Perch-glean
Serpophaga munda	Perm. S, winters "N"	Scrub	Non-Amaz. lowlands/valle	Perch-glean
Serpophaga subcristata	Winters	Scrub/woodland	Non-Amaz. lowlands	Perch-glean
Inezia inornata	Breeds S, winters N	Scrub	Lowlands	Perch-glean
Culicivora caudacuta	Uncertain	Open	Non-Amaz. lowlands	Perch-glean
Polystictus p. pectoralis	Winters	Open	Non-Amaz. lowlands	Perch-glean
Pseudocolopteryx sclateri	Uncertain	Marsh	Non-Amaz. lowlands	Perch-glean
Pseudocolopteryx dinellianus	Winters	Marsh	Non-Amaz. lowlands	Perch-glean
Pseudocolopteryx acutipennis	Breeds S, winters N	Marsh	Widespread	Perch-glean
Pseudocolopteryx flaviventris	Vagrant	Scrub	Lowlands	Perch-glean
Euscarthmus m. meloryphus	Winters	Scrub	Lowlands	Perch-glean
Myiophobus fasciatus auriceps	Breeds S, winters N	Scrub	Lowlands/valle	Enclosed perch-hawk
Lathrotriccus e. euleri	Winters	Woodland/forest	Lowlands/upper tropical zone	Enclosed perch-hawk
Lathrotriccus e. argentinus	Breeds S, winters N	Woodland/forest	Lowlands/upper tropical zone	Enclosed perch-hawk
Chemotriccus fuscatus bimaculatus	Breeds S, winters N	Woodland/forest	Lowlands	Enclosed perch-hawk
Pyrocephalus r. rubinus	Perm. S, winters N	Open/scrub	Lowlands	Near ground
Xolmis coronata	Winters	Open/scrub	Non-Amaz. lowlands	Perch to ground
Agriornis m. microptera	Winters	Open/scrub	Puna/valle	Perch to ground
Agriornis murina	Winters	Open/scrub	Non-Amaz. lowlands/valle	Perch to ground
Muscisaxicola m. macilirostris	Perm. S, winters N	Open	Puna/valle	Ground
Muscisaxicola capistrata	Winters	Open	Puna	Ground
Muscisaxicola rufivertex pallidiceps	Perm. S, winters N	Open	Puna	Ground
Muscisaxicola albilora	Winters	Open	Puna	Ground
Muscisaxicola c. cinerea	Perm. S, winters N	Open	Puna	Ground

APPENDIX
CONTINUED

Species or subspecies	Seasonal distributional pattern	Breeding habitat	Life zone	Foraging mode
Muscisaxicola flavinucha	Breeds S, winters N	Open	Puna/valle	Ground
Muscisaxicola frontalis	Winters	Open	Puna	Ground
Lessonia oreas	Breeds S, winters N	Open	Puna	Ground
Lessonia rufa	Winters	Open	Valle	Ground
Knipolegus striaticeps	Perm. S, winters N	Scrub/woodland	Non-Amaz. lowlands	Aerial hawk
Knipolegus hudsoni	Winters	Scrub/woodland	Non-Amaz. lowlands	Aerial hawk
Knipolegus a. aterrimus	Breeds S, winters N	Scrub/woodland	Widespread	Aerial hawk
Hymenops p. perspicillatus	Winters	Marsh	Non-Amaz. lowlands	Ground
Hymenops p. andinus	Winters	Marsh	Non-Amaz. lowlands	Ground
Fluvicola pica albiventer	Breeds S, winters N	Marsh	Lowlands	Ground
Satrapa icterophrys	Perm. S, winters N	Woodland	Lowlands/valle	Fruit/hawk
Attila phoenicurus	Winters	Forest	Amaz. lowlands	Outward hover-glean
Casiornis rufa	Perm. S, winters N	Scrub/woodland	Lowlands/valle	Outward hover-glean
Myiarchis tuberculifer atriceps	Breeds S, perm. N	Woodland	Upper tropical/subtropical zones	Outward hover-glean
Myiarchus s. swainsoni	Transient	Scrub/woodland	Lowlands	Outward hover-glean
Myiarchus s. ferocior	Breeds S, winters N	Scrub/woodland	Lowlands/valle	Outward hover-glean
Myiarchus t. tyrannulus	Perm. S, winters N	Scrub/woodland	Lowlands/valle	Outward hover-glean
Myiodynastes maculatus solitarius	Breeds S, perm. N	Woodland/forest	Lowlands/valle	Upward strike
Legatus l. leucophaius	Breeds S, perm. N	Woodland	Lowlands	Fruit/hawk
Empidonomus v. varius	Breeds	Woodland	Lowlands/valle	Aerial hawk
Empidonomus a. aurantioatrocristatus	Breeds S, winters N	Woodland	Lowlands/valle	Aerial hawk
Tyrannus m. melancholicus	Breeds S, perm. N	Woodland	Lowlands/valle	Aerial hawk
Tyrannus s. savana	Breeds	Open/scrub	Lowlands	Aerial hawk
Tyrannus albogularis	Breeds	Scrub/woodland	Non-Amaz. lowlands	Aerial hawk
Xenopsaris a. albinucha	Winters	Scrub/woodland	Non-Amaz. lowlands	Outward hover-glean
Pachyramphus polychopterus spixii	Breeds S, winters N	Woodland	Lowlands	Fruit/upw. hover-glean
Pachyramphus v. validus	Uncertain	Scrub/woodland	Lowlands	Fruit/upw. hover-glean
Pachyramphus v. audax	Breeds S, perm. N	Scrub/woodland	Lowlands/upper tropical zone	Fruit/upw. hover-glean

A NEW LOOK AT THE "SPECIES-POOR" CENTRAL AMAZON: THE AVIFAUNA NORTH OF MANAUS, BRAZIL

MARIO COHN-HAFT,[1,2] ANDREW WHITTAKER,[1,3] AND
PHILIP C. STOUFFER[1,4]

[1]*Projeto Dinâmica Biológica de Fragmentos Florestais, INPA Ecologia,
C. P. 478, Manaus, Amazonas 69011, Brazil;*
[2]*Museum of Natural Science, Foster Hall 119, and Department of Zoology and Physiology,
Louisiana State University, Baton Rouge, Louisiana 70803, USA;*
[3]*Conjunto Acariquara Sul, Rua Samaumas 214, Manaus, Amazonas 69085, Brazil;* and
[4]*Department of Biological Sciences, Southeastern Louisiana University,
Hammond, Louisiana 70402, USA*

ABSTRACT.—The birds of the Fazendas Dimona, Porto Alegre, and Esteio, and adjacent areas, ca. 80 km north of Manaus, Amazonas, Brazil, have been intensively studied since 1979. Stotz and Bierregaard (1989) published a list of 352 species, based on seven years of study. Here we modify that list based on an additional eight years of work at the same sites. We add 49 species, revise the identification of four species, and remove seven species, for a total of 394 species. We also list 12 additional species as "hypothetical," requiring further substantiation, and revise the status of numerous species. Additions include 22 species believed to have been overlooked previously, 15 species believed to have colonized the area recently in response to anthropogenic change, and 12 species that are probably vagrants from habitats not well represented in the study area. Status revisions reflect an increase in availability and diversity of secondary habitats or better knowledge of vocalizations of birds that are much more commonly heard than seen. Specimens indicate that *Campylorhamphus procurvoides* and *Hemitriccus zosterops*, not *C. trochilirostris* and *H. minor*, are present at the site. We also present evidence that *Hemitriccus z. zosterops* and *H. z. griseipectus* are distinct species, and we report the first Brazilian records (including specimens) of *Phylloscartes virescens*.

Comparison with the avifauna of two well-studied sites in Amazonian Peru, Manu National Park and Tambopata Reserve, reveals similar species richness in the *terra firme* forest component of all three sites. The difference in richness between the sites is because of the variety of habitat types present at each. We suggest that habitat heterogeneity, not primary productivity or rainfall as have been proposed elsewhere, is the major determinant of patterns in bird species richness within Amazonia.

RESUMO.—As aves das fazendas Dimona, Porto Alegre, e Esteio, e das áreas adjacentes, a cerca de 80 km ao norte de Manaus, Amazonas, Brasil, foram estudadas intensivamente desde 1979. Stotz e Bierregaard (1989) publicaram uma lista com 352 espécies, baseada em sete anos de estudo. Modificamos essa lista a partir de mais oito anos de trabalhando nas mesmas áreas. Acrescentamos 49 espécies, corrigimos a identificação de quatro, e removemos sete, resultando num total de 394 espécies. Também listamos doze espécies adicionais como "hipotéticas," que necessitam maiores evidências de sua ocorrência, e revisamos o status de várias espécies. Adições incluem 22 espécies provavelmente negligenciadas anteriormente, 15 espécies que acreditamos terem colonizado a área recentemente em resposta a mudanças antropogênicas, e 12 espécies que provavelmente são visitantes irregulares de ambientes pouco representados na área de estudo. As revisões de status refletem um aumento na disponibilidade e diversidade de vegetação secundária e um maior conhecimento das vocalizações de aves que são muito mais comumente ouvidas do que vistas. Espécimes indicam a presença na área de estudo de *Campylorhamphus procurvoides* e *Hemitriccus zosterops*, em vez de *C. trochilirostris* e *H. minor*. Também apresentamos evidência de que *Hemitriccus z. zosterops* e *H. z. griseipectus* são espécies distintas, e confirmamos com exemplares o primeiro registro de *Phylloscartes virescens* para o Brasil.

A comparação desta avifauna com a de duas áreas igualmente bem conhecidas na Amazônia peruana, o Parque Nacional de Manu e a Reserva Tambopata, revela uma riqueza de espécies semelhante nas áreas de floresta de *terra firme* dos três locais. A maior riqueza total nas áreas peruanas é devido à maior variedade de tipos de hábitats presentes. Sugerimos que a heterogeneidade de habitats é o maior fator determinante de padrões de riqueza de espécies de aves dentro da Amazônia, em vez da produtividade primária ou pluviosidade propostas anteriormente.

Accurate and complete avifaunal site lists are a cornerstone in improved understanding of avian biogeography and community ecology. With recent improvement in knowledge of vocalizations and with new, high-quality field guides, skilled observers can produce reasonably accurate species lists for many Neotropical sites. However, considerable research (including taxonomic revision) remains to be done before identification to species will be acceptable without tangible evidence (voucher specimens, tape-recordings, or photographs). Perhaps even more daunting than accuracy, however, in the species-rich Neotropics is completeness of species lists. Most lists are based on short-term samples and qualify only as preliminary. To enable meaningful comparison between sites or over extended time intervals, complete lists are needed; these are labor-intensive.

Until recently, all avifaunal site lists from Amazonia were preliminary in nature. One of the most-studied sites in all of South America is in Manu National Park in Amazonian Peru, where John Terborgh and his students have been working at the Cocha Cashu Research Station since 1973 (Terborgh et al. 1990). The Cocha Cashu bird list (in Karr et al. 1990) is probably the most complete published list for any South American, and certainly any Amazonian, location. The recently published list (Parker et al. 1994) for nearby Tambopata Nature Reserve, also in Peru, may rival Cocha Cashu in completeness.

The subject of our study is the reserves of the Biological Dynamics of Forest Fragments Project (BDFFP, formerly Minimum Critical Size of Ecosystems Project) north of Manaus in central Amazonian Brazil. This is probably the most intensively studied area in central Amazonia. Ecological studies began at the site in 1979, aimed at examining the effects of fragmentation of continuous forest on a wide variety of organisms and abiotic parameters (see Lovejoy and Bierregaard 1990 for a history of the project). Ornithological investigations have been a major component of the project. A program of mist-netting in the understory of continuous forest and fragments of one, 10, and 100 ha began in 1979 and has continued nearly uninterrupted since then (see Lovejoy 1985; Bierregaard and Lovejoy 1988, 1989; Bierregaard 1990a, b; Stouffer and Bierregaard 1995a, b). Other studies have dealt with focal species or groups, especially those not sampled in mist nets (Bierregaard 1984, 1988; Quintela-Almeida 1985; Bierregaard et al. 1987; Harper 1987, 1989; Quintela 1987; Klein and Bierregaard 1988a, b; Klein et al. 1988; Mesquita 1989; Powell 1989; Stotz et al. 1992; Cohn-Haft 1993, 1995; Stouffer and Bierregaard 1993; Whittaker 1993, 1995; Borges 1995), and scores of volunteer banders and visitors have contributed their observations.

Stotz and Bierregaard (1989) listed the bird species known in the BDFFP reserves as of 1987, after seven years of study. Since then, an increased emphasis on vocal recording, auditory and visual censuses, and use of a canopy tower (see Methods) have added numerous species to the project's avifauna. Habitat change within the study area, mostly in the form of abandonment of pasture and subsequent regeneration of second-growth forest, and encroaching deforestation along the road from Manaus have probably led to colonization by some species and changes in abundance of others. Finally, specimen collecting has permitted a closer examination of some difficult species, leading in some cases to reidentification of species previously listed. Here we present a modified list of bird species from the BDFFP site and adjacent areas, based on an additional six years of field work, incorporating increased knowledge of the birds, changes in abundance, and documentation of evidence for inclusion. We discuss the differences between this list and the earlier one and lists from other Amazonian sites. Our approach emphasizes the study area as a whole and any changes on a regional scale. Detailed treatment of the effect of habitat fragmentation on a local scale is covered elsewhere (e.g., Stouffer and Bierregaard 1995a, b; Bierregaard and Stouffer 1997).

STUDY AREA

The BDFFP study site, about 80 km north of Manaus, Brazil, is located on three adjacent 15,000 ha ranches partially deforested for cattle pastures (fazendas Dimona, Porto Alegre, and Esteio, 60°W, 2°20'S; Fig. 1). In addition to the fazendas, the exclusive focus of Stotz and Bierregaard's (1989) list, we include the area surrounding a 40-m tower, located on the ZF-2 road, and the intervening region (Fig. 1). We do not include a large, isolated, white-sand *campina*

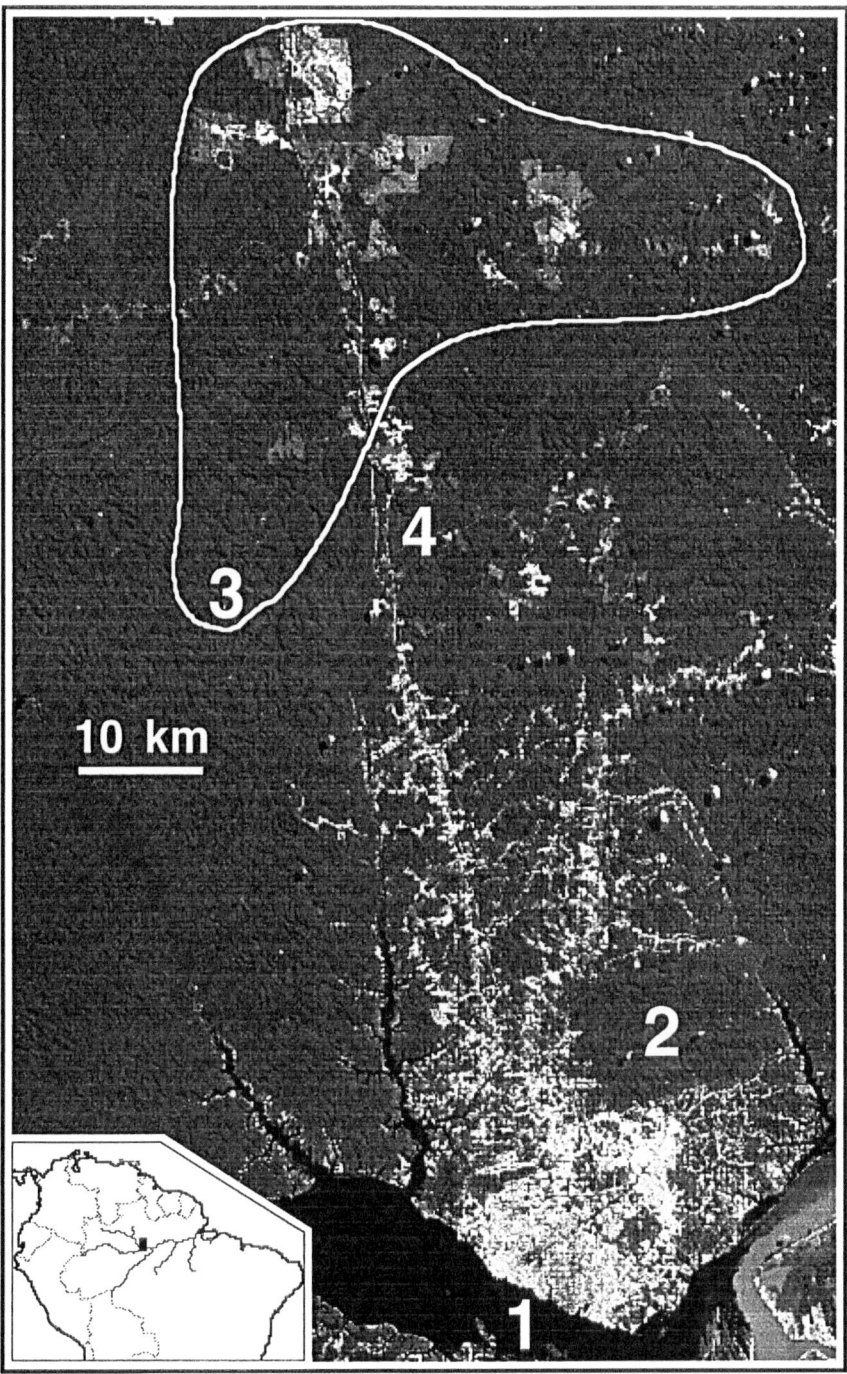

FIG. 1. Map of Manaus (1) and area to north, showing study area (outlined in white), Reserva Ducke (2), and locations of the tower on the ZF-2 road (3) and of the campina (4). Dark gray represents primary forest, light gray is secondary forest, and white represents open areas such as pastures, roads, and urban construction. The fazendas studied by Stotz and Bierregaard (1989) are in the upper part of the study area, centered around the large areas of pasture and second growth. Map based on Landsat TM image; bands 3,4,5; 22 May 1992; source INPE.

(see Fig. 1; described in Anderson 1981) that is roughly the same distance south of the fazendas as the tower, because it represents a distinct habitat, not found on the fazendas (but see *campinarana* below), with an endemic avifauna (see Willis 1977; Oren 1981). The study area does not extend farther south to avoid the direct affects of urban expansion from Manaus and to permit comparison with Reserva Ducke (Fig. 1). Records from north of the fazendas are also excluded because there is a small river (Rio Urubu) and a marked change in terrain (including the presence of rocks, greater relief, and caves) roughly 100 km north of Manaus. Several bird species known from that region that have not been recorded closer to Manaus (see Discussion).

The site was continuous forest until the late 1970s, when some development began, consisting mostly of land clearing for cattle pasture. Stotz and Bierregaard (1989) reported that 10,000 ha had been cleared from the fazendas by the middle 1980s. Since that time almost no new land has been cleared, and much pastureland has been abandoned. The second-growth forest, formed on abandoned clearings, is all less than 15 years old and of known history (R. Mesquita, unpubl. data). Disturbance is still minimal to the north, east, and west of the site, where the forest stretches for hundreds of kilometers. To the south there is increasing disturbance, including some fairly large farms, but even near Manaus clearing is only close to existing roads. Primary forest connecting the study area to Reserva Ducke and the outskirts of Manaus is interrupted in most places by only one or two narrow roads. The main road north out of Manaus, BR-174, is gradually being paved, and this will undoubtedly bring with it increasing settlement and hunting pressure. Game animals (such as monkeys and cracids) and large predators (such as jaguars, *Felis onca*, and pumas, *F. concolor*) persist in the study area in good numbers, although local ranchers and weekend hunters from Manaus exert some pressure on these species.

The study area is roughly 500 km^2, made up mostly of primary *terra firme* forest, but containing several "islands" of disturbed habitats connected to each other (and eventually to a larger region of disturbed habitat, Manaus, and the Rio Negro) by narrow dirt roads (Fig. 1). There are two major types of second growth present, tall second growth dominated by *Cecropia sciadophylla* and a shrubbier vegetation dominated by *Vismia* spp. Both types occur on abandoned pastures, depending on land use before abandonment, and in narrow belts fringing the dirt roads cut through primary forest. Some *Cecropia* stands are as tall as 20 m and provide a dark, open understory much like primary forest. Details of bird use of these two kinds of second growth, which are not distinguished in this study, were studied by Borges (1995). Open water is restricted to four or five (depending on rainfall) small ponds formed in the middle of pasture and one seasonal pond occuring naturally in the midst of undisturbed forest.

The primary forest in the study area has been described in detail elsewhere (Lovejoy and Bierregaard 1990). Important features are a canopy height of ca. 30 m, with occasional taller emergents, open understory dominated by stemless palms, a relatively closed canopy, except for areas of treefall gaps, high tree species diversity, and fairly low epiphyte loads. There is little macro-relief and no major watercourse, but small streams (less than 2 m across and 1 m deep) have eroded many steep valleys 10–30 m deep. Annual rainfall averages about 2,200 mm, with an annual peak in March and April and a pronounced dry season from July through September (MME 1978, Stouffer and Bierregaard 1993).

Relative to many Amazonian sites, the study area has low habitat heterogeneity. There is no equivalent to Manu's "floodplain" or bottomland forest, that is, forest on flat, ancient alluvial plain, not subject to annual flooding. There is also no seasonally flooded forest (*várzea* or *igapó*), no bamboo, nor oxbow lakes—all habitat types associated with specialized fauna. White-sand *campina* scrub and woodland ("Amazonian white-sand caatinga" of Anderson 1981) is an important and locally distributed vegetation type in the Manaus region, with a characteristic avifauna. Only one small patch of this vegetation, surrounded by typical *terra firme* forest, is known within the study area, and it is a closed-canopy woodland (*campinarana*), not a fully developed scrub (*campina*); consequently, it lacks species typical of more open *campinas* (e.g., *Elaenia ruficeps*; Oren 1981).

METHODS

This study includes all bird species observed in the study area during the entire 15-yr history of the BDFFP through 1994. Our own work over an 8-yr period (1987–1994) immediately follows the period (1979–1986) covered by Stotz and Bierregaard's (1989) list, and provides the basis for determination of species status and any changes believed to have occurred in the project's duration. Whittaker and Cohn-Haft began working in the area in 1987, and Stouffer began in 1991. We each worked in various parts of the study area at all times of year, focusing on a variety of research objectives (described elsewhere), during which time the data for this paper were incidentally gathered. Cohn-Haft worked mostly in unbroken primary forest and collected specimens during a study of bird diets (Cohn-Haft 1995). Whittaker and Stouffer focused more on disturbed forest, surveying birds in isolated forest fragments and studying avian use of forest edge and second growth, respectively. All of us ran mist nets in continuous forest and isolated patches at one time or another, as well as conducted general avian surveys using tape recorders, both for playback and for documentation; we have also made over 50 morning bird surveys from the tower. Our field work was facilitated by, but not limited to, grid systems (100 or 200 m) of narrow trails in demarcated study plots and small rustic camps located within the study area. In addition to our own observations, we drew on observations of other project participants over the years, the project's database of more than 50,000 mist net captures in over 150,000 net-hours, and previously published records from the tower (Bierregaard 1982).

We expanded the study area to include the tower because it permitted a better assessment of the abundance of canopy species. Although this extended the region covered by some 15 km to the south beyond that of Stotz and Bierregaard (1989), the overall areal extent of coverage is only minimally increased by that amount visible from the tower itself and from the roads leading to it. Our extensive observations from the tower and mist-netting below it (unpubl. data) reveal that the avifauna is essentially identical to that of the BDFFP reserves. Therefore, we do not attribute any differences between our results and those of Stotz and Bierregaard (1989) to increase in area studied (see Discussion).

Status categories.—Species' abundance was determined subjectively, combining frequency of detection (visual or auditory) and capture rate to reflect our impression of actual population density in preferred habitat. This differs from the categories of Stotz and Bierregaard (1989) in having one fewer category (their "fairly common" is usually subsumed in our "common") and in not being based on quasi-quantitative rates of detection, which we believe are strongly methods-biased. Thus, our ratings represent dimensionless hypotheses of density that can (and should) be tested by quantitative census techniques (e.g., Terborgh et al. 1990). "Common" is used for species believed to occur everywhere in appropriate habitat; for small passerines this probably translates to contiguous territories up to roughly 15 ha each (unpubl. data). Species listed as "uncommon" occur in most, but not all seemingly appropriate habitat and probably have densities roughly an order of magnitude lower than common species. "Rare" species appear to be absent from more appropriate habitat than that in which they occur and probably have densities an order of magnitude lower than uncommon species. These three coarse categories include all species that we consider the "core avifauna" (Remsen 1994) at our site. In addition, we use "casual" (equivalent to "extremely rare" of Stotz and Bierregaard [1989]) for species registered three or fewer times. Unlike the other abundance categories, "casual" is based strictly on number of detections and may include vagrants as well as extremely low-density or sporadic residents, which our limited data are unable to distinguish. The single abundance rating giving to each species refers to its period of greatest abundance. For species with variable annual abundance, seasonal status was rated as "austral migrant" (present April–September), "boreal migrant" (usually October–April, but shorebirds arrive much earlier; see Stotz et al. 1992), or as having "unspecified movements" of undetermined seasonality, possibly at a local scale, leading to periods of lower abundance or absence.

Birds were classed with respect to habitat and microhabitat. We distinguished five major habitat types: primary *terra firme* forest; secondary forest, of which specific types and age classes were not differentiated; pasture, often including some low bushes, solitary tall trees and snags, and fences; open water, in the form of ponds either in pasture or in forest; and *campinarana*. These categories are similar but not identical to those used by Stotz and Bierregaard (1989). We do not consider treefall gaps, edges, small woodland streams (*igarapés*), and overhead airspace as major habitats, but rather as microhabitats within the above habitat types. Other microhabitats or positions (including vertical strata) were also recognized: terrestrial, understory, midstory, canopy, water surface.

Following Stotz and Bierregaard (1989), we use "sociality" to refer to intra- and interspecific associations of each species (i.e., pairs, monospecific flocks, mixed-species flocks, etc.). To these categories, we added "lekking" for species in which males are usually gathered at communal display grounds (e.g., some hummingbirds and manakins).

Evidence documenting each species was ranked heirarchically, and only the highest quality evidence available for each species is listed (Appendix). We consider the best evidence to be a specimen. Next is a permanent record, either a photograph or tape recording, which can be used to confirm identification to species in most cases. Third is a capture record, that is, an in-hand sight record augmented by some morphometrics (usually wing chord, tail, and weight) that could help to confirm identification. The lowest form of evidence are field observations, either sighting or vocalization heard. A few species based only on sight records are listed separately as "hypothetical," meaning that the species was positively identified by the observer, but that such an identification requires a higher category of evidence for inclusion in the main list. This is used for especially difficult field identifications or species deemed very unlikely to occur by range.

The taxonomy used here follows Morony et al. (1975 and corrigenda) except for the Ardeidae and the expanded Emberizidae, which follow the AOU Check-list (1983), and the Tyrannidae, Cotingidae, and Pipridae, which follow Traylor (1979). Other minor exceptions, including any differences in nomenclature from that used by Stotz and Bierregaard (1989), are explained in footnotes to the Appendix.

All specimens were registered at the Museu Paraense Emílio Goeldi (MPEG) in Belém and will be deposited there, at the Louisiana State University Museum of Natural Science (LSUMZ) in Baton Rouge, or at the National Museum of Natural History (USNM) in Washington, D.C. Specimens are referred to by preparator's field number, pending final deposition. Tape recordings are or will be deposited at the Cornell Library of Natural Sounds (LNS) and the Arquivo Sonoro Natural (ASN) in Campinas. Photographs are or will be deposited at Visual Resources in Ornithology (VIREO) in Philadelphia.

RESULTS

As of January 1995, 394 bird species were recorded in the study area (Appendix); in addition to these we list another 12 species as "hypothetical" (see Table 4). Seven species included in Stotz and Bierregaard's (1989) list of 352 species were removed, four were reidentified, and 49 species were added.

REMOVALS

Earlier inclusion of *Tinamus guttatus* was based on vocalizations and two mist net captures. Examination of the capture records revealed a carefully described adult *Crypturellus variegatus*, with appropriate measurements, and a small chick, doubtfully identifiable to species. Identification of tinamous by voice has improved considerably in the decade since *T. guttatus* was placed on the reserve list (although there is still much to learn and recording accompanied by playback and collection is badly needed to resolve remaining problems). We have never heard in the study area the vocalization we attribute to this species elsewhere in Amazonia, based on recordings at LNS: two clear, whistled notes. The vocalization previously ascribed to this species (D. Stotz, in litt.) may be that of *T. major*, which we hear frequently during the rainy season: a single, clear whistle followed by a tremulous descending note. Vocalizations of *C. brevirostris* (see below) may also have been attributed to *T. guttatus*. Considering that *T. guttatus* has never been seen in the reserves (despite frequent sight records of *T. major* and *C. variegatus*) in nearly 15 years of work and that it is unknown north of the Amazon and east of the Rio Negro, we doubt that the species is present in the study area. We have placed it on the hypothetical list.

The identification of *Aramides calopterus*, known in the area from a single sight record (Stotz and Bierregaard 1989), was based on the terse plumage descriptions in Meyer de Schauensee (1970) and may have been erroneous (D. Stotz and R. Bierregaard, pers. comm.). Because the species is otherwise unknown from the region east of the Negro and north of the Amazon rivers, we have listed it as hypothetical from our study area.

Nyctiphrynus ocellatus had been listed on the basis of a single vocal record (D. Stotz, pers comm.). Although the vocalization is fairly distinctive, we suspect that the species was included erroneously, and so we list it as hypothetical. We have never heard the characteristic song despite scores of hours of night survey work in appropriate habitat. In regions where it occurs this species is normally fairly common and readily detected in primary forest, so we doubt we have overlooked it. It is not known from the Guianas or from Brazil north of the Amazon and east of the Negro rivers, so its presence at Manaus would represent a major range extension.

Myrmotherula surinamensis was included earlier based on a single record lacking habitat description (D. Stotz, pers. comm.). Considering recent discoveries of extremely similar-looking *M. klagesi* and *M. cherriei* near Manaus (Cohn-Haft, unpubl. data), and the preference by all three species for seasonally flooded river- or lake-edge habitats, we prefer to list *M. surinamensis* as hypothetical. To our knowledge, there are no specimens of *M. surinamensis* from the Manaus area (see Gyldenstolpe 1930), although its occurrence in appropriate habitat is to be expected.

Hypocnemoides melanopogon was included earlier on the basis of a single record from ca.

km 100 of the BR-174 along the Rio Urubu (D. Stotz, pers. comm.). This is outside of the study area and so we have removed the species entirely from the BDFFP avifauna. There is probably no appropriate habitat for it in the reserves, although it could appear as a vagrant.

Hylophilus brunneiceps is associated with white-sand vegetation, especially along the upper Rio Negro, and is not known from specimens near Manaus (Oren 1981). Like other *campina* specialists, it could certainly appear in our area, although we have never registered it. Given the difficulties in greenlet identification and apparently considerable earlier confusion (see "Reidentifications" below), we prefer to list D. Stotz's single sight record as hypothetical.

Conirostrum bicolor was included earlier based on a single record included among the BDFFP notes from the earliest years (D. Stotz, pers. comm.). In central Amazonia, this species occurs in seasonally flooded vegetation along white-water rivers, so we feel its appearance in the study area is very unlikely. We list it as hypothetical.

REIDENTIFICATIONS

We question the identification of *Forpus passerinus* (including one record of our own). In the city of Manaus and in disturbed and river-edge habitats nearby, we have only positively identified *F. crassirostris*. (Note, as pointed out to us by D. F. Stotz [in litt.], that *F. crassirostris* is the correct name for the species commonly called *F. xanthopterygius*; see Pinto 1978.) Although both species are possible by range (and some authors consider them conspecific), they are difficult to distinguish. We are not convinced that the few sight records of *Forpus* parrotlets from the study area refer to *F. passerinus*, best told from *F. crassirostris* by the absence of a blue rump in the male; we prefer to list "*Forpus* sp." in recognition of the need for better documentation. Bierregaard (1982) listed a single sight record of "*Forpus* cf. *sclateri*" from the tower, which we suspect refers to the same taxon as the other study-area sightings.

The scythebill found at Manaus was identified as *Campylorhamphus trochilirostris* by Stotz and Bierregaard (1989), after Willis (1977) reported *C. procurvoides* from Reserva Ducke. Manaus birds had long been recognized to bear the most conspicuous distinguishing field mark for *C. procurvoides* north of the Amazon, the unstreaked mantle (Meyer de Schauensee 1970; Hilty and Brown 1986; Ridgely and Tudor 1994); nevertheless, this identification was overruled by bill measurements of netted birds and of a single specimen (D. Stotz, pers. comm.), which we were unable to locate. A recent specimen (MCH 225) from the study site, however, is indistinguishable in plumage from a series of nominate *C. p. procurvoides* at the American Museum of Natural History. The unstreaked mantle, sagittate breast streaks (vs. long, linear stripes) not extending to the belly and lacking conspicuous dark borders, and the generally browner, less rufescent ground color clearly distinguish this form from any *C. trochilirostris* subspecies found in nearby regions. The bill chord of our specimen (63 mm, measured from base of bill) and of netted birds (BDFF Project, unpubl. data) is longer than expected for *C. procurvoides* (46–56 mm) and within the range for *C. trochilirostris* (58–64 mm), according to Hilty and Brown (1986). Those measurements, however, may only apply to forms found in Colombia because, among central Amazonian forms of the two species, Zimmer (1934) found a considerable range of overlap (56–61 mm). Zimmer also noted that bill length, as measured along the curve of the culmen, is similar in the two species, the generally shorter chord length of *procurvoides* indicating the deeper curvature of that species' bill. We are unable to place all specimens to species by subjective estimation of bill curvature and, pending larger samples of consistently measured bills, we suggest that neither curvature nor chord length is diagnostic at the species level. Zimmer (1934) proposed that tarsus length might consistently distinguish *procurvoides* (18–20 mm) from *trochilirostris* (21–23 mm); the tarsus of our specimen (18.7 mm) supports that dichotomy.

We are confident that *C. procurvoides* is the only regularly occurring scythebill in the *terra firme* forest north of Manaus. The species has been known from Manaus since Hellmayr's (1925) accurate reidentification as *procurvoides* of a "*trochilirostris*" specimen collected by Natterer in 1833 (Natural History Museum of Vienna 15911, examined by Cohn-Haft). Zimmer (1934) apparently overlooked this but, without access to the specimen, independently suspected it might be *procurvoides*. Subsequently, Peters (1951) did not include Manaus within the range of any scythebill. To our knowledge, the Manaus area represents a gap in the central Amazonian distribution of *trochilirostris*, falling between two distinct subspecies, *C. t. notabilis* to the south and *C. t. snethlageae* occurring as close as Faro, 400 km east of Manaus on the same side of the Amazon and Negro (Zimmer 1934). In general, the distribution of *C. trochilirostris* is mostly peripheral to the Amazon basin, in which *procurvoides* is widespread (see maps in Ridgely and Tudor 1994). In areas of sympatry in central Amazonia, *trochilirostris* is apparently restricted

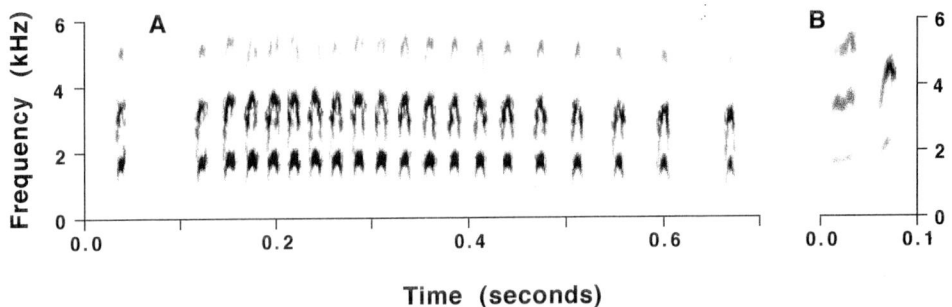

FIG. 2. Song of (A) *Hemitriccus z. zosterops* (from study site north of Manaus, Brazil; recorded by MCH) and (B) *H. z. griseipectus* (Tambopata Reserve, Madre de Dios, Peru; recorded by T. Parker; LNS 12872). Sound spectrograms were produced using Canary software of the Bioacoustics Research Program of the Cornell Laboratory of Ornithology, Ithaca, New York.

to seasonally flooded forest (E. Snethlage in Zimmer 1934). Thus, we predict that (1) if *trochilirostris* is found to occur near Manaus, it will inhabit *várzea* forest (not present at our site), and (2) *procurvoides* will prove to be resident in *terra firme* forest throughout the region between Manaus and Guyana (shown as distributional gap on map in Ridgely and Tudor 1994: 213). Although there is considerable geographic variation in both species and a thorough taxonomic revision is needed, the form occurring at our site, nominate *procurvoides*, is unlikely to be affected by revision.

Earlier identification of *Hemitriccus minor* (Bierregaard 1982, Stotz and Bierregaard 1989) was based on morphometrics of netted birds and similarity of the voice to that of *H. minor* in Rondônia. Specimens collected in the reserves (MCH 14, 56, 96, 97, 109), however, have confirmed the identification as *H. zosterops*, by such diagnostic characters as wing formula and nostril shape (Hellmayr 1927, Zimmer 1940, Cohn-Haft 1996). The form of *H. zosterops* found in the reserves is the nominate subspecies (following the taxonomy of Traylor 1979), widespread in Amazonia north of the Amazon River. This yellow-bellied form looks much like *H. minor* in plumage, but is slightly larger. Because *Hemitriccus* spp. are sexually dimorphic in size (females smaller; Zimmer 1940) and because female *zosterops* overlaps in size with male *minor* (Cohn-Haft, unpubl. data), small, sex-blind samples could easily fail to distinguish the two morphometrically. In the field they are virtually identical, nominate *zosterops* differing subtly from *minor* by more conspicuous wingbars, more distinctly streaked throat, greenish-white as opposed to beige-white iris, white versus white-to-buffy eyering, all-black mandible versus variable amounts of pale coloration at base, and flatter culmen (see also Cohn-Haft 1996). The "tip-trill" song of Manaus birds (Fig. 2A), described by Stotz and Bierregaard (1989), closely resembles songs of *H. minor*, as well as both *Lophotriccus* species found in the study area. The song of *H. zosterops*, however, differs from all of these by always beginning with a "tip" note and by gradually slowing throughout the trill (Cohn-Haft 1996). Near Manaus, *H. minor* occurs only in *igapó* (unpubl. data), and the species is unknown from the region east of the Rio Negro and north of the Amazon, except in French Guiana (Tostain et al. 1992), where we question the identification. A specimen from Surinam (Haverschmidt 1968) proved to refer to nominate *zosterops* (see Ridgely and Tudor 1994).

Proper identification of the Manaus birds is further complicated by the fact that *H. z. griseipectus*, the familiar "*zosterops*" to workers in southern Amazonia, looks and sounds more different from the Manaus form than does *H. minor*. White-bellied, gray-breasted *griseipectus*, found throughout Amazonia south of the Amazon River, was originally described as a full species (Snethlage 1907). It gives a distinctly different song from nominate *zosterops*: a rapid, high-pitched, metallic "ca-DEEK" (Fig. 2B), confirming the full species status of *Hemitriccus griseipectus* (as accorded without justification by Sibley and Monroe [1990]).

Stotz and Bierregaard (1989) listed *Hylophilus pectoralis*, followed by a question mark indicating doubt as to identification, as rare in primary forest. We have found *H. thoracicus* (not listed by them) at only two locations in the study area, including the one where Stotz (pers. comm.) registered the possible *pectoralis*, and we assume that this is the species in question. Our identification is based on the yellowish-white iris, complete yellow pectoral band, head lacking contrasting gray coloration, and the song (recorded) lacking the terminal trill given by

TABLE 1
Species Added that Were Probably Overlooked Previously

Crypturellus brevirostris	*Chaetura chapmani*	*Progne subis*
Leptodon cayanensis	*Avocettula recurvirostris*	*Riparia riparia*
Chondrohierax uncinatus	*Chrysolampis mosquitus*	*Vireo altiloquus*
Accipiter poliogaster	*Pharomachrus pavoninus*	*Piranga rubra*
Coccyzus melacoryphus	*Ornithion inerme*	*Tersina viridis*
Ciccaba huhula	*Phylloscartes virescens*	*Dendroica petechia*
Nyctibius leucopterus	*Tyrannus tyrannus*	
Streptoprocne zonaris	*Progne tapera*	

pectoralis (Ridgely and Tudor 1989). Both species are fairly widespread in Amazonia, but *thoracicus* is typical of forest canopy and edge (Hilty and Brown 1986), whereas in central Amazonia *pectoralis* is known only from natural savannas. Willis (1977) listed both species as occurring just south of our study area, near but not expressly in the *campina* (Fig. 1). We have recorded neither species from the *campina* itself, where *pectoralis* (and *H. brunneiceps*; see above) could theoretically occur. Outside of it, we would not be surprised to find *thoracicus* in tall forest, and similar-looking *H. semicinereus* in second growth. There is also a single record of *H. thoracicus* from the tower by J. Fitzpatrick (Bierregaard 1982).

In addition to these species, Stotz and Bierregaard (1989) questioned their identifications of *Ciccaba virgata* and *Anthracothorax nigricollis*, suggesting that the latter might be *A. viridigula*. We have found both *C. virgata* and *C. huhula* in the study area (see below). The only *Anthracothorax* species we have observed at the site is *A. nigricollis*. In the region, we find *A. viridigula* to prefer late successional vegetation on islands in the Amazon River, although both species are present in planted areas in the city of Manaus.

Also, several species listed in Bierregaard (1982) from the tower would be additions to the list, but were not included because they are surely early misidentifications of similar species now known to be regular in the area. *Trogon collaris*, otherwise unknown in the Manaus area, was reported heard on a single occasion by J. Fitzpatrick; however, this species sounds similar to *T. rufus*. *Polioptila plumbea*, which occurs near Manaus only in *igapó* (pers. obs.), was reported as uncommon in canopy flocks, exactly the role filled by *P. guianensis*. The single record of *Nasica longirostris* from the tower has since been rescinded (R. Bierregaard, pers. comm.).

ADDITIONS

To facilitate discussion of the 49 species not reported by Stotz and Bierregaard (1989), we divided them into three groups (Tables 1–3), emphasizing what we believe are the characteristics most responsible for the species being encountered in the study area.

The largest group of additions are 22 species that have probably always been present in the abundance that we encountered (Table 1). Six of these (*Crypturellus brevirostris*, *Ciccaba huhula*, *Nyctibius leucopterus*, *Chaetura chapmani*, *Ornithion inerme*, and *Phylloscartes virescens*) are relatively common forest birds, present all year. With the exception of *C. chapmani*, all these species are regularly heard and are much easier to detect by voice than by sight. All are either canopy species, nocturnal, or difficult to see or identify. Specimens support identifications of *N. leucopterus* (see Cohn-Haft 1993), *C. chapmani* (MCH 324), and *P. virescens* (MCH 38, 136, 137, 194). Our records and specimens of *P. virescens* are the first for Brazil (see Stotz 1990). *Ciccaba huhula*, first sighted in the reserves by Whittaker, is the more common *Ciccaba* owl in primary forest, judging by its characteristic calls and by sight records. Although we are not able to distinguish all of its varied vocalizations from those of *C. virgata*, the latter is also present, known from one capture and one sight record (Stouffer), and may prefer more disturbed forest habitat. *Ornithion inerme* was first reported in the study area from the ZF-2 tower (Bierregaard 1982) and is easiest to see there, but can be heard throughout the area from primary forest canopy.

Crypturellus brevirostris has been seen twice in the reserves (Whittaker). Before this, in December 1989, T. Parker identified and recorded (LNS) a tinamou vocalization as a *C. bartletti*-type song. Although we have never seen the singer, we believe the song belongs to *C. brevirostris*. This is not a radical suggestion considering that the song of *brevirostris* is undescribed, that *bartletti* is unknown north of the Amazon River, and that the two are considered close

TABLE 2
SPECIES ADDED SINCE 1986 THAT ARE TYPICAL OF OPEN OR SECOND-GROWTH VEGETATION AND
ARE BELIEVED TO BE RECENT ARRIVALS TO THE STUDY AREA (SEE TEXT)

Athene cunicularia	*Thamnophilus punctatus*	*Manacus manacus*
Phaethornis ruber	*Camptostoma obsoletum*	*Sporophila bouvronides*
Polytmus theresiae	*Phaeomyias murina*	*Sporophila lineola*
Amazilia fimbriata	*Elaenia chiriquensis*	*Saltator maximus*
Chelidoptera tenebrosa	*Pachyramphus rufus*	*Euphonia chlorotica*

relatives (Blake 1977), even conspecific by some (Hellmayr and Conover 1942). The type specimen of *Crypturellus brevirostris* was collected in Manaus (see Hellmayr and Conover 1942), and there is one specimen (MPEG 29939, collected in 1964) from Reserva Ducke, where we have since heard the same song (although the species is not included on the Ducke list [Willis 1977]). The song is a prolonged series of evenly spaced pure tones, rising in pitch slightly throughout, and reminiscent of songs of *Microcerculus bambla* in the same region. Prior to Parker's identification we had noted this vocalization frequently in our field notes without identification as to species. Judging by voice, *C. brevirostris* is the least common of the three tinamous found principally in primary *terra firme* forest, where it occurs exclusively, and may be associated with small forest streams. We have heard this vocalization throughout the study area and presume the species to be widespread in the *terra firme* forest north of Manaus. Although larger and longer-billed, *C. variegatus* is extremely similar and would probably be difficult to distinguish from *C. brevirostris* in the field, which may explain the lack of earlier sight records or captures.

The remaining species in Table 1 are probably either resident at very low densities, vagrants from other habitats, or are scarce long-distance migrants. Many of these prefer non-forest habitats, and some (e.g., swallow spp.) are found in considerable numbers outside of our study area (Stotz et al. 1992), but within it are generally seen as they pass overhead. *Tyrannus tyrannus* and *Piranga rubra* breed in North America and winter in greatest numbers in western Amazonia (see Stotz et al. 1992). *Accipiter poliogaster* is known in our area from two sight records from the ZF-2 tower (Bierregaard 1982; Whittaker and K. Zimmer). The second record, of an immature-plumaged individual on 13 January 1995, does not fit the pattern of austral winter records noted in Amazonian Colombia (Hilty and Brown 1986). *Avocettula recurvirostris* and *Chrysolampis mosquitus* are both known at the site from single female individuals, seen on the same day (7 November 1993) from the tower by Whittaker, K. Zimmer, and V. Emanuel. *Pharomachrus pavoninus* appears to be extremely rare in *terra firme* forest north of Manaus; our record is of a single singing male, noted repeatedly over a period of several months near a camp in continuous forest. *Tersina viridis* has been seen from the tower (Whittaker) and in the canopy of undisturbed forest on one of the fazendas (T. Towles, pers. comm.).

Another group of newly added species (15 spp.) are birds of second growth and open or disturbed areas that are common in similar habitats closer to Manaus and appear to have colonized the study area since 1987 (Table 2). Colonization by these species probably reflects gradual northward expansion along roads from areas of more extensive habitat near Manaus. Roads resemble rivers—being long, narrow, and bordered by disturbed habitat—and as such may facilitate the dispersal of riverine species (e.g., *Chelidoptera tenebrosa*, *Camptostoma obsoletum*, *Pachyramphus rufus*) as well as inhabitants of secondary forest (e.g., *Phaethornis ruber*, *Saltator maximus*). During our tenure at Manaus, we have noted *Chelidoptera tenebrosa* progressively farther north along the BR-174 road (see Fig. 1), and Whittaker first noted it on one of the fazendas (Dimona) on 10 September 1991. Similarly, *Phaethornis ruber*, whose earlier absence was "particularly perplexing" to Stotz and Bierregaard (1989), and *Pachyramphus rufus* were first noted on a fazenda in 1988 and have since spread and been confirmed nesting (Whittaker). We suspect that this mode of colonization may be typical of species occupying disturbed and second-growth habitats, and all species listed in Table 2 may be expected to increase gradually in numbers at our site.

One probable exception to the dispersal pattern just described is *Athene cunicularia*. It is widespread in arid and open habitats throughout most of the New World. We are unaware of any other records from rainforest areas in Amazonia. Our record is of a single individual in a

TABLE 3
Species Added that Are Typical of Water-Related and Other Marginally Present Habitats

Butorides striatus	*Pipile cumanensis*	*Celeus flavus*
Oxyura dominica	*Aratinga leucophthalmus*	*Tachyphonus phoenicius*
Pandion haliaetus	*Crotophaga major*	
Nycticorax nycticorax	*Chloroceryle amazona*	
Mesembrinibis cayennensis	*Chloroceryle americana*	

pasture on the Fazenda Dimona, first seen by R. Bierregaard on 4 December 1992 and seen again as recently as June 1994 (Whittaker).

A third group of newly added species (12 spp.) is made up of birds that normally occupy habitats absent or only marginally present in our area, and that appear as vagrants from nearby areas, but are not expected to colonize (Table 3). All are listed as casual (known from three or fewer records), and so none is considered part of the core fauna. One, *Tachyphonus phoenicius*, is a specialist of *campina*. The two sight records, both from the tower (Bierregaard 1982, Stotz, pers. comm.), probably represent wandering individuals. Other species found mostly in *campina* (e.g., *Galbula leucogastra* and *Thamnophilus punctatus*) also occur to a lesser extent in second growth or edge situations in the study area, but *T. phoenicius* is not one of these. Nor has it appeared in the *campinarana*, the only site where the *campina*-specialist *Neopelma chrysocephalum* has been recorded in our area.

The remaining species in Table 3 are associated with aquatic environments, either open water (*Oxyura dominica*, *Pandion haliaetus*), river and lake edge (*Butorides striatus*, *Nycticorax nycticorax*, *Mesembrinibis cayennensis*, *Aratinga leucophthalmus*, *Crotophaga major*, *Chloroceryle amazona*, *C. americana*), or bottomland and flooded forest (*Pipile cumanensis*, *Celeus flavus*). The appearance of *O. dominica*, *P. haliaetus*, and *C. americana* is clearly related to the formation of the ponds in pasture where they were found. Surprisingly, *O. dominica* may have taken up residence in a single pond, where up to two males have been seen repeatedly over several years. Nevertheless, the small extent of habitat seems likely to limit any expansion in the study area by these species. *Aratinga leucophthalmus* might be expected to visit second-growth areas in the reserves, given its abundance in the city of Manaus. However, birds noted in the city seem to prefer wet areas or to be in wandering flocks, perhaps associated with the large rivers nearby. Our two records are of small groups overhead, once over primary forest from the tower and once over pasture.

Finally, the other river- and lake-edge species listed above were all recorded at camps in the middle of unbroken primary forest. *Nycticorax nycticorax* (Whittaker) and *Mesembrinibis cayennensis* (A. Martins, pers. comm.) were each noted once, heard calling at night; the latter bird apparently perched atop a snag at the edge of camp. A single *B. striatus* was found at dawn in one camp, walking nervously on the sandy ground and periodically "freezing" in response to human activity in camp; it was only seen for one day (A. Martins, pers. comm.). The one record of *C. major* (Cohn-Haft) was of an individual skulking low in dense vegetation at the edge of a camp. These records suggest dispersal, in many cases nocturnal, of species resident in aquatic habitats in the general vicinity. Seen from the air, our camps, which are more or less round breaks in the canopy, may resemble ponds, especially at night when sandy ground, camp roofs or tarps, or even pockets of mist probably reflect light differently from the relatively uniform canopy around them. Similar camp records at dawn of single *Podiceps dominicus* (W. Strickland, pers. comm.), *Gallinago gallinago* (Whittaker), an unidentified shorebird (A. Martins, pers, comm.), and a pair of *Ceryle torquata* (Cohn-Haft and Whittaker) also suggest nocturnal movement and mistaken identification of camps as ponds. Barbara Zimmerman (pers. comm.) reports that a species of *Phyllomedusa*, a nocturnal frog that normally vocalizes only around ponds or pig wallows, also commonly calls at the edge of camps with blue tarp roofs.

HYPOTHETICAL

Twelve species reported based on inadequate documentation, but possibly present, are listed as "hypothetical" (Table 4). We believe that substantiation in the form of a specimen, or at least an unambiguous tape recording or photograph, is necessary to confirm the presence of any of these species in the area. *Tinamus guttatus*, *Aramides calopterus*, *Nyctiphrynus ocellatus*, *Myr-*

TABLE 4
Hypothetical Species (See Text)

Tinamus guttatus	Nyctiphrynus ocellatus	Capsiempis flaveola
Accipiter striatus	Chaetura cinereiventris	Pipreola whitelyi
Falco deiroleucus	Popelairia langsdorffi	Hylophilus brunneiceps
Aramides calopterus	Myrmotherula surinamensis	Conirostrum bicolor

motherula surinamensis, Hylophilus brunneiceps, and *Conirostrum bicolor* had been listed by Stotz and Bierregaard (1989). We suspect that they were identified incorrectly for reasons discussed above ("Removals"). Other hypothetical records are as follows:

Accipiter striatus is unknown from the Amazon basin. Our records of unidentified small accipiters include individuals resembling North American (Whittaker), plain-breasted (Cohn-Haft) and rusty-thighed (Cohn-Haft) forms. Unusual plumages of *A. superciliosus* or *Harpagus diodon*, an austral migrant to eastern Amazonia noted once by Willis (1977) just outside of the study area, may be involved in these sightings.

Whittaker observed a possible adult *Falco deiroleucus* at a considerable distance perched on a snag on one of the fazendas. There are several Manaus-area records for this species (Whittaker 1996), including photographs (Cohn-Haft), although the status of the species in Amazonia is poorly known. At the time of the reserve sighting (7 April 1988), the observer was unfamiliar with the species and could not positively rule out the similar *F. rufigularis*. After subsequent comparative experience, Whittaker strongly suspects that the initial identification was correct, but we would prefer better documentation.

Chaetura cinereiventris was seen on one occasion (two individuals on 15 May 1990 by Whittaker); although we are familiar with the species from elsewhere in Amazonia, given the difficulty identifying swifts in the field, we would prefer specimen confirmation.

Whittaker and K. Zimmer observed a female coquette from the tower that had a pure white rump band, suggesting *Popelairia langsdorffi*. Typical buff-rumped females of *Discosura longicauda* were also present; however, the variation in female plumage in these birds appears to be poorly documented, and it is not clear if this mark is sufficient to distinguish the two species. Furthermore, several other coquette species have white rump bands, and *P. langsdorffi* is unknown from east of the Negro and north of the Amazon.

John Fitzpatrick reported seeing an adult *Capsiempis flaveola* feeding young in the canopy of primary *terra firme* forest from the tower (Bierregaard 1982). Considering that this species normally occurs in low, shrubby vegetation or bamboo and that the only Manaus-area record besides this one is from lake-edge bushes south of the Amazon (Jan Pierson and Cohn-Haft), we prefer to list the species as hypothetical in the study area. Perhaps this record refers to *Phylloscartes virescens* (J. Fitzpatrick, pers. comm.).

Pipreola whitelyi, known only from montane forest in the eastern tepuis of Venezuela and adjacent Guyana near the Brazilian border (Traylor 1979), is an unlikely candidate to appear near Manaus. Nevertheless, the male should be entirely distinctive, so we are inclined to recognize here two independent sight records of single males from the Manaus area. Jan P. Smith (pers. comm.) noted one with other canopy frugivores foraging on fruits at the edge of Camp Florestal. Outside of the reserves, in a residential neighborhood of Manaus, Randy Downer (pers. comm.) reports observing this species with other frugivores also in a fruiting tree over a stream.

Status Changes

The following types of status were evaluated for each species and assigned a separate column in the Appendix:

Abundance and seasonality.—Many species were assigned a different abundance code from that in Stotz and Bierregaard (1989). This is in part an inevitable function of the different criteria and number of categories employed. Most such changes are for greater abundance than previously estimated (e.g., most species listed as "fairly common" earlier are simply "common" here). Greater familiarity with species, especially their voices, generally led to upward revisions for a large number of species (e.g., *Micrastur ruficollis, Nyctibius aethereus, N. bracteatus, Frederickena viridis, Zimmerius gracilipes, Myiopagis caniceps*). The longer total period of study reflected here has given us a more accurate picture of the abundance of some rare species. Three species not seen since 1986, *Laterallus melanophaius, Pionites melanocephala*, and *Chordeiles acutipennis* were reclassified as casual. Just as some of the casual species noted for the first time after 1986 may have appeared in response to changing habitats, the lack of subsequent records of these species could signify their disappearance from the area; however, there are simply too few data to know.

In certain cases, our different abundance assignment appears to reflect a real change in abundance since 1986. Many edge and secondary species have probably increased in abundance as time for colonization has elapsed. This group includes several open-country species rarely or never found in primary forest (e.g., *Heterospizias meridionalis, Tyrannus savana,* and *Sturnella militaris*), but is dominated by species that are occasionally encountered in disturbed areas within primary forest, but that thrive along edges and in second growth (e.g., *Melanerpes cruentatus, Cymbilaimus lineatus, Myrmeciza atrothorax, Tolmomyias poliocephalus, Myiarchus tuberculifer, M. ferox* and *Cyclarhis gujanensis*). Other second-growth species appear to have decreased as older and more diverse secondary habitats have developed (e.g., *Troglodytes aedon*). One unusual change is the apparent shift in color-phase predominance in *Buteo brachyura* from light in the earlier years (D. Stotz, pers. comm.) to dark phase in recent years. Perhaps this reflects founder effects.

Species whose numbers vary considerably during the year represent 12% of the avifauna. Eight species are austral migrants, 26 are boreal, and 15 engage in unspecified movements (Appendix). For certain species known to have migrant and resident populations we have not established which form is present in our area (e.g., *Empidonomus varius, Tyrannus savana, T. melancholicus*).

Several relatively common species not known to be migratory appear to vary in abundance at the fazendas, as we have indicated with an "m" after the abundance code (Appendix). Unlike Stotz and Bierregaard (1989), we agree with Willis (1977) that *Pionus* and *Amazona* spp. fluctuate greatly in numbers in the area. In particular, *P. menstruus* and *A. autumnalis* may vary from abundant to virtually absent. Many of these species, including *Geotrygon montana* and several parrots, are frugivores and may make relatively short-scale movements to track food resources. For *G. montana*, capture data indicate that abundance is highly variable and does not follow an annual cycle, although abundance peaks have occurred during the wet season (Stouffer and Bierregaard 1993). Cohn-Haft (unpubl.) noted flocks of thousands of *P. menstruus* in Roraima during August 1991 when the species was absent from the reserves.

Habitat.—Most of the revisions of habitat codes reflect our recognition of use of second growth and edges by forest birds. Some species (e.g., *Phaethornis superciliosus, P. bourcieri, Mionectes macconnelli, Terenotriccus erythrurus,* and *Myiarchus ferox*) have become quite common in both *Vismia-* and *Cecropia*-dominated secondary growth, although *Cecropia* forests appear to be used by more forest birds. We did not list second growth as a habitat for species that we suspect use it only in passing between areas of primary forest (see Borges 1995).

Microhabitat/position.—Although Stotz and Bierregaard (1989) did not list this category, their habitat classification included what we consider microhabitats. The most noteworthy difference in our microhabitat ratings concerns several manakin species (*Tyranneutes virescens, Corapipo gutturalis, Pipra pipra, Pipra erythrocephala*), which we routinely observed foraging at fruiting trees in the forest canopy. Although all of them lek (and may nest) in the understory and midstory of closed forest (Prum 1990; pers. observ.), we believe that they forage mostly in the canopy at our site. This behavior has not been observed elsewhere and could be attributed to the unusual poverty of fruits in forest understory at Manaus (Gentry and Emmons 1987). Nevertheless, the forest canopy is poorly studied at most sites, and where manakin foraging has been observed at lower strata it is most often at treefall gaps (Levey 1988). Thus, we believe that these species are not typical inhabitants of closed forest interior, despite the frequent classification of most manakins as such. Guild studies also routinely class all woodpecker species as "bark-gleaning insectivores," although we have observed several species (especially *Melanerpes cruentatus* and *Celeus grammicus*) regularly consuming large quantities of fruit in the canopy. Mesquita (1989) rated several woodpecker species as the most important dispersers of *Clusia grandiflora* (Clusiaceae) fruits at our site.

Sociality.—We made few revisions in sociality codes, but some of these changes reflect interesting behavior changes associated with use of secondary habitats. *Dendrocincla fuliginosa, Xiphorhynchus pardalotus,* and *Myrmotherula axillaris* are all typically found in mixed-species flocks in primary forest, but appear to persist outside of flocks in secondary areas. This change is most notable for *M. axillaris*, which is almost never found outside of flocks in continuous forest, but has persisted and even reproduced in 1- and 10-ha fragments (Stouffer and Bierregaard 1995a).

We believe that the flycatcher *Platyrinchus coronatus* normally engages in lekking behavior, although this apparently has not been noted elsewhere. We usually encounter several singing individuals (presumably males) at dependable locations and elsewhere may go several hundred meters through forest without encountering any. At these presumed leks, birds often flare their colorful crests and make short flights between perches without foraging, often accompanied by an abrupt "chirr" sound, apparently produced by their wings. We have no evidence of lekking in the other *Platyrinchus* species at our site. Species in another genus of small flycatchers, *Lophotriccus*, are purported to lek elsewhere (Hilty and Brown 1986), but do not do so at our site. Both *L. vitiosus* and *L. galeatus* appear to maintain permanent territories separated by considerable distances in favored habitat.

Evidence.—Seventy percent of all species listed in the Appendix are documented by photograph, tape, or specimen (level 2 or better). Most of the others are not controversial. Species listed as hypothetical (Table 4), as mentioned above, require better documentation and do not appear on the main list (Appendix).

DISCUSSION

A variety of techniques is clearly needed to accurately and completely survey the avifauna of a tropical forest (Terborgh et al. 1990). Our species list represents about a 14% increase in species over the list of Stotz and Bierregaard (1989). Many additions are a result of more night work, more time spent in the canopy (mostly at the tower), more emphasis on vocalizations, more collecting, and more prolonged coverage by the same people, leading inevitably to greater familiarity with the fauna. Extensive work with mist nets prior to our arrival in Manaus had produced a thorough characterization of the understory fauna. Nevertheless, one understory species, *Crypturellus brevirostris*, had gone completely undetected by mist nets (which undersample ground-walking birds like tinamous; Karr 1981) and was only recognized after it was noted by voice. Uniform coverage throughout the year is also important. More than 10% of the avifauna at our site may be seasonal to some extent; these species could be missed by incomplete seasonal coverage.

Time is also a critical component of faunal surveys (Remsen 1994), and high-diversity tropical sites seem to require at least several years to be completely characterized. Like Manaus, Manu National Park in Peru added some 40 species (and removed about five) to its list (Terborgh et al. 1984; and updated version in Karr et al. 1990), despite a much longer history of bird work than at Manaus. Very rare species or vagrants, such as the majority added to this list, may continue to be added at a slow rate nearly indefinitely, especially in the tropics, where a much larger species pool is fairly close at hand. Nevertheless, one important objective of a species list is to characterize the "core" fauna in a region. At our site only one habitat could be said to have a core fauna. This is the primary *terra firme* forest. The other three major habitat types present in some abundance (second growth, pasture, and open water) are of recent anthropogenic origin. We believe that we have characterized the core avifauna in primary *terra firme* forest at our site to greater than 99% accuracy and completeness.

Another objective of our list was to monitor change over time. Our assessments of change suggest the role of changing availability of anthropogenic habitats and varying rates of colonization. A more fine-scale approach involving censusing of bird numbers (e.g., Terborgh et al. 1990) is necessary to quantify change. Standardized mist-netting over long periods detects changing habitat use patterns by birds, although interpretation of capture rates can be complicated (Remsen and Good 1996).

Certain second-growth species were surprisingly slow to colonize the area (e.g., *Phaethornis ruber*, *Chelidoptera tenebrosa*, *Thamnophilus punctatus*, *Camptostoma obsoletum*, *Phaeomyias murina*), but we attribute their eventual arrival to gradual dispersal along roads. We predict the appearance of other species by the same means, for example, *Herpetotheres cachinnans* (seen as far north as km 30 of the BR-174 road in August 1993 by Cohn-Haft), *Megarynchus pitangua* (noted at km 45 by Whittaker), *Todirostrum maculatum*, and *Cacicus cela*—all typically river-edge species that occur along the roads and at fazendas nearer to Manaus. The last three are among the species listed for Ducke Reserve (Willis 1977) that still have not been recorded in our area (see below).

Disturbed areas, embedded in a seemingly endless expanse of forest, still make up only a tiny fraction of the habitat available north of Manaus. Nevertheless, rapid growth of Manaus and gradual paving of the BR-174 leading to Boa Vista and Venezuela will inevitably lead to greater forest fragmentation. Human disturbance processes have probably only increased species richness at our site to date, and apparently no primary forest species have suffered for it. However, levels of habitat disturbance and hunting are extremely low and will probably increase. On a smaller spatial scale, the negative effects of habitat fragmentation on primary forest species is clear (e.g., Stouffer and Bierregaard 1995a).

More importantly, increasing local species richness by increasing habitat diversity in our area is not a meaningful conservation objective. Secondary vegetation is not an endangered habitat, nor are the species that occupy it. On the other hand, large areas of primary forest, with *low* habitat heterogeneity, are increasingly subject to destruction even if soon abandoned. These areas may offer a key to understanding patterns of species richness throughout the Amazon (see below) and may be important population sources for forest species.

AVIFAUNAL AFFINITIES

The avifauna north of Manaus is distinctly Guianan in affinities. All but a handful of species (and subspecies where known) also occur in the Guianas. Those not more widespread in Amazonia represent the Guianan "center of dispersal" (Haffer 1969) or "area of endemism" (Cracraft 1985), which appears to be delimited to the south by the Amazon River and to the southwest by the lower Rio Negro. In general, the pattern of range extensions found at Manaus has been of species previously known only from the Guianas (see Stotz and Bierregaard 1989), and our addition of *Phylloscartes virescens* is consistent with this pattern.

Other typically Guianan species recently found at Manaus and believed to terminate their westward distribution there have been found outside of the Guiana area. *Pachyramphus surinamus* has been found at several sites west of the Negro and south of the Amazon (Whittaker 1995); *Haematoderus militaris* was found south of the Amazon in Rondônia (Stotz et al. 1997) and at the Urucu (Peres and Whittaker 1991); *Polioptila guianensis* occurs at the Urucu (Peres and Whittaker 1991), in Rondônia (Stotz et al. 1997), and at Borba (Cohn-Haft and B. Whitney, unpubl. data); and *Cyanicterus cyanicterus* has been noted at Borba (Silva and Willis 1986; Cohn-Haft and Stouffer, unpubl. data) and the Urucu (Peres and Whittaker 1991). *Nyctibius leucopterus* was entirely unknown in Amazonia before its discovery in the reserves (Cohn-Haft 1993). Ted Parker (in Parker et al. 1993) since found it in Guyana, indicating that it is probably widespread in the Guianas, and recent unpublished records from west of the Rio Negro (Whittaker) and south of the Amazon (Cohn-Haft and Stouffer) suggest that the species may occur throughout Amazonia. This Amazonian form of *N. leucopterus* represents an undescribed taxon (Cohn Haft, in prep.).

One interpretation of these records is that the Negro and Amazon rivers are not insurmountable barriers for these species, all of which inhabit the canopy and lack close relatives elsewhere in the Amazon. They may occur across the great rivers as vagrants and may have established populations at low levels. This is the "center of dispersal" view espoused by Haffer (1969). Alternatively, these records may represent previously undetected established populations of equivalent stature to those in the Guiana region. In this case, these species had merely been incorrectly assigned to their "area of endemism" (*sensu* Cracraft 1985). There is simply not enough distributional information yet to know. The view that species known only from one side of a major river should not occur on the other side may have biased our opinion with respect to some of the species we list as "hypothetical" (Table 4).

The general outlines of Amazonian avian endemism proposed by Haffer (1969) have withstood well the test of time. Nelson et al.'s (1990) warning that apparent areas of endemism can be a simple function of the areas studied does not seem to apply to the broad patterns in bird distributions. Nevertheless, most of the Amazon is barely explored (Oren and Albuquerque 1991), and continued avifaunal surveys will be necessary to fill in the gaps in our understanding.

The closest area studied to ours is Reserva Ducke. Although the published list (Willis 1977) is admittedly preliminary, additions listed by Stotz and Bierregaard (1989) and our own (all included in the Appendix) indicate an avifauna very similar to that of the reserves. Of the 32 species listed for Ducke that Stotz and Bierregaard did not find in the reserves, we failed to find only 16 of them (*Agamia agami, Tapera naevia, Otus choliba, Glaucidium brasilianum, Hylocharis cyanus, Veniliornis passerinus, Nasica longirostris, Megarynchus pitangua, Attila cinnamomeus, Todirostrum maculatum, Sublegatus glaber, Mionectes oleagineus, Turdus leucomelas, T. ignobilis, Cacicus cela, Sporophila americana*). Of these, all appear to be absent due to a shortage of their preferred habitat in the reserves, either extensive secondary or riverine vegetation. Several species (mentioned above) are expected to colonize the reserves soon. More species occur in the reserves that have not been found yet at Ducke; however, we feel this is a function of coverage. Our limited experiences at Ducke have led to the addition of a number of common species, and other workers (especially R. Cintra, R. Ridgely, T. Schulenberg, D. Stotz) have unpublished records that we did not include here. Thus we are confident that the core primary forest avifaunas at the two sites are actually nearly identical, although there may be differences in the relative abundances of some species.

PATTERNS OF SPECIES RICHNESS

It is generally believed that western Amazonia has the highest species richness (for most taxa, including birds) in the basin and in the world. Because species lists form the basis for such statements, care must be taken to ensure the comparability of these lists. Remsen (1994) pointed out several major potential sources of incomparability, including unequal sampling effort. The only Amazonian sites with published avifaunal lists based on surveys as thorough and complete as we believe ours to be are Cocha Cashu Biological Research Station in Manu National Park, studied by Terborgh and colleagues (in Karr et al. 1990), and the Tambopata Reserve (Parker et al. 1994), both in southeastern Peru. Total species richness at Manu is 554 and at Tambopata is 572, both substantially higher than that of our site (394), despite the considerably larger area covered at Manaus. Thus, the two western Amazonian sites clearly contain more species than one central Amazonian site.

This unstartling result does not, however, shed any light on the nature or causes of the difference in species richness between sites. An obvious source of difference is the number of habitats found at each site. The Peruvian sites contain numerous habitats not present in our area, including rivers, lakes, associated successional vegetation, bamboo, *várzea*, and bottomland (floodplain) forest. All three sites contain extensive *terra firme* forest, but at our site *terra firme* is really the only extensive habitat type present; therefore, it is the only habitat type that can be validly compared (Remsen 1994).

If we compare only the *terra firme* component at each site (Table 5), the tendency for increased species richness in the western sites disappears. The *terra firme* total for Manu (271 species) is remarkably similar to that for Manaus (264 species), especially in light of the 40% greater overall site richness at Manu. Furthermore, the Manu total may be an overestimate because no abundance codes were given and so some extremely rare or accidental species may have been included, whereas only "core" species (*sensu* Remsen 1994; excluding those listed as "casual") were included for Manaus. Tambopata, which has the highest overall richness of any site, had a markedly lower richness (200 species) in its *terra firme* component. We suspect that this is a result of inadequate sampling of *terra firme* forest at Tambopata. The overall similarity between the Manu and Tambopata avifaunas and the proximity of the two sites lead us to expect their *terra firme* avifaunas also to be very similar. Support for this interpretation comes from the low totals at Tambopata for families (Table 5) that are nocturnal (Nyctibiidae), tend to stay in the canopy (Psittacidae, Trochilidae), or are otherwise inconspicuous (Bucconidae, Dendrocolaptidae, Tyrannidae); the fact that many of the species at Tambopata not listed for *terra firme* are known to occur in that habitat elsewhere; and the fact that the nearest *terra firme* to the site's lodging is nearly 2 km away (J. V. Remsen, pers. comm.). Alternatively, some species that use *terra firme* at Manu (or Manaus) might actually prefer other habitats to the exclusion of *terra firme* at Tambopata, making the *terra firme* avifauna there genuinely poorer than at these other sites. Although unlikely in our opinion, that possibility highlights the importance of comparing sites by habitat type (Remsen 1994). Whether the low total for Tambopata is real or an artifact, it is clear that the higher total species richness there and at Manu versus Manaus is due entirely to habitats other than *terra firme*. This suggests that the major cause of the difference in total richness between the sites is habitat heterogeneity (beta-diversity).

We cannot rule out the possibility that the much larger area covered at Manaus (ca. 500 km^2) than at the other sites (ca. 10 km^2) is responsible for Manaus not having many fewer species in *terra firme* than the other sites; however, we do not believe that size is a problem. In fact, we predict that the entire core fauna listed for Manaus can be found in an area of just a few hundred hectares. We note that the study area contains a high proportion of disturbed habitats so that a considerably smaller area of primary forest was surveyed than the overall size of the study area suggests. Also, the considerable extension of the study area to include the tower did not add any species to the core fauna. Four species (*Accipiter poliogaster, Avocettula recurvirostris, Chrysolampis mosquitus, Tachyphonus phoenicius*) were found only at the tower, but they are considered casual, and two others (*Leptodon cayanensis* and *Euphonia chlorotica*) seen only along the BR-174 road outside of the fazendas are also casual and so did not enter in the core fauna.

TABLE 5
SPECIES RICHNESS IN PRIMARY *Terra Firme* FOREST AT THREE AMAZONIAN SITES

Family or major subfamily	Manaus[a]	Manu[b]	Tambopata[c]
Tinamidae	4	6	3
Ardeidae	1	2	3
Cathartidae	2	2	1
Accipitridae	11	10	11
Falconidae	7	4	5
Cracidae	3	3	2
Odontophoridae	1	1	1
Psophiidae	1	1	1
Rallidae	0	1	1
Eurypygidae	0	1	1
Columbidae	3	2	2
Psittacidae	12	12	8
Cuculidae	2	5	4
Strigidae	6	6	5
Caprimulgidae	4	2	2
Nyctibiidae	5	3	1
Apodidae	4	0	2
Trochilidae	10	13	5
Trogonidae	4	5	5
Momotidae	1	2	2
Alcedinidae	2	2	3
Bucconidae	7	6	4
Galbulidae	4	3	3
Capitonidae	1	3	1
Ramphastidae	4	4	5
Picidae	10	8	7
Dendrocolaptidae	13	13	10
Furnariidae	11	17	11
Formicariidae	29	37	24
Conopophagidae	1	1	0
Tyrannidae	37	33	27
Pipridae	7	5	6
Cotingidae	6	6	3
Hirundinidae	2	0	0
Troglodytidae	3	4	2
Turdinae	3	4	1
Polioptilinae	3	0	0
Vireonidae	7	5	3
Emberizinae	1	1	1
Cardinalinae	3	4	2
Thraupinae	22	27	19
Parulinae	2	1	1
Icterinae	5	6	2
Total	264	271	200

[a] Data from Appendix: all species containing "1" (primary *terra firme* forest) in habitat column and abundance "rare" or greater.
[b] Data from Karr et al. (1990): all species containing "U" (upland forest) in habitat column.
[c] Data from Parker et al. (1994): all species containing "Fh" (upland forest), "Fsm" (forest stream margins), "Fo" (forest openings), or "Fe" (forest edges), and not marked with asterisk (observed ≤ 3 times).

Another line of evidence is that, if both the Manu and Tambopata faunas were likely to increase a great deal by increasing the geographic extent of the study areas, then the *terra firme* fauna of the two sites, which are only about 100 km apart, should be considerably different, representing different subsamples of the complete fauna of the area. On the contrary, however, the somewhat smaller *terra firme* fauna listed at Tambopata appears simply to be a subset of that at Manu. Unfortunately, we have no concentrated survey data from a restricted subplot of our study area to test our hypothesis, but we look forward to the results of such a study in the future. In the meantime, we assume that habitat heterogeneity is the sole explanation necessary for the results of our intersite comparison.

Near Manaus, habitat heterogeneity is also implicated in increased site richness. Balbina, the site of a hydroelectric plant roughly 120 km north of Manaus, appears to have all of our species plus a number of others, although lists are only preliminary (Willis and Oniki 1988; Bierregaard et al., unpubl. data). This is an area at the edge of the Guiana shield with considerable relief, caves, waterfalls, and distinct soils. There is also a large river, flooded forest, and some bamboo at the site. These different habitats probably account for the addition of numerous species, such as *Zebrilus undulatus, Pyrrhura picta, Automolus rufipileatus, Sakesphorus melanothorax, Thamnophilus amazonicus, Microrhopias quixensis, Cercomacra laeta* (see Bierregaard et al. 1997), *Hemitriccus josephinae, Ochthornis littoralis, Rupicola rupicola, Atticora* spp., *Henicorhina leucosticta*, and *Granatellus pelzelni.*

We propose that the exceptional species richness of western relative to central Amazonia is actually a function of habitat heterogeneity. Ted Parker and J. Haffer (unpubl. data) noted remarkable constancy in *terra firme* species richness at a variety of Amazonian sites. Habitats such as bamboo (Parker et al. 1997; Kratter 1997), *campina* (Oren 1981), Amazonian savanna (Silva et al. 1997) and various river-created environments (Remsen and Parker 1983; Rosenberg 1990) are known to have specialized fauna associated with them and to contribute to species richness. Increased habitat heterogeneity at the periphery of the Amazon basin is probably a natural consequence of the geography of the region, where increased relief and erosion create a more dynamic and fine-grained environment. By contrast, the enormous size of the lower reaches of the basin's rivers and the relatively flat terrain necessarily create large expanses of single habitat types. Furthermore, broad rivers probably limit potential for mixing of forest species, making species distributional limits more clearly defined in central Amazonia.

Factors such as rainfall and primary productivity, which have been shown to correlate with patterns of species richness (e.g., Haffer 1990; reviewed by Rosenzweig and Abramsky 1993), are probably not causally related to richness in birds across Amazonia, but may be merely coincidentally related due to unique aspects of Amazonian geography. The mountains enclosing the basin to the north, west and south are responsible for higher rainfall in the west due to orographic effects and prevailing equatorial easterlies. High primary productivity is associated with high rainfall and richer montane soils. Within-habitat diversity, however, as exemplified by the above discussion of *terra firme* sites, may be independent of these factors. Rainfall probably only affects bird diversity insofar as it determines type of vegetation cover. On the other hand, absolute biomass or number of individuals on a site should be positively related to primary productivity. This says nothing about whether this biomass is distributed among many or few species. The relationship between biomass and species richness could be tested by comparing population densities of the same species between sites of differing productivity (e.g. Stouffer 1997).

CONCLUDING REMARKS

Discussions of regional distribution patterns and conservation decisions depend on accurate identification and taxonomy. Our reidentifications of common and familiar species after collecting should serve as a warning regarding sight records and an example of how much taxonomic work remains to be done in the Amazon. All lists based on sight records alone should be considered preliminary, awaiting specimen confirmation.

Most species probably have diagnostic vocalizations, and so tape recordings should serve someday soon as satisfactory documentation for nearly all species. However, the only way to sort out the relationship between forms and voice is with voucher specimens of recorded individuals. Until these correlations are better established, the only adequate documentation for some species is a specimen. The accumulation of well annotated recordings in sound libraries (such as LNS and ASN) provides a tremendously valuable resource akin to a tissue collection.

ACKNOWLEDGMENTS

This paper is dedicated to the memory of Ted Parker, who we wish were here to read it and set us straight.

We received generous financial and incomparable logistical support from the Biological Dynamics of Forest Fragments Project (BDFFP) in Manaus and from its parent organizations, the Smithsonian Institution and the Instituto Nacional de Pesquisas da Amazônia (INPA); this paper is number 160 in the BDFFP technical series. We also received support or assistance from Louisiana State University's Museum of Natural Science and Board of Regents (to MCH), Field Guides, Inc. (to MCH), and Victor Emanuel Nature Tours (to AW). MCH gratefully acknowledges the Instituto Brasileiro do Meio Ambiente (IBAMA) for collecting permits. Don Kroodsma and the Cornell Laboratory of Ornithology's Library of Natural Sounds (LNS) both made long-term loans of field recording equipment, which provided our biggest window to understanding Amazonian birds, but which we inevitably ended up destroying; to them we extend our sheepish gratitude. LNS also provided use of recordings in its care and Canary software. Museum work was graciously facilitated by the curatorial staffs of the Louisiana State University Museum of Natural Science, American Museum of Natural History, Natural History Museum of Vienna, and Museu Paraense Emílio Goeldi; J. M. C. Silva provided data for specimens we were unable to examine. This study benefited immeasurably from the help of our field companions and the many field assistants of the BDFFP, especially S. Borges, R. Downer, V. Emanuel, A. Martins, K. Rosenberg, J. Smith, T. Towles, B. Whitney, S. Wilson, K. Zimmer, and particularly J. Pierson, who insistently questioned the earlier identifications of *Hemitriccus* and *Campylorhamphus*. Doug Stotz was extraordinarily generous with his help at every stage of this project, especially in reviewing numerous drafts of the manuscript and in answering our desperate pleas for information over the years. John Bates, R. Mesquita, T. Schulenberg, and V. Remsen offered useful criticism of the manuscript. Bruce Nelson provided the satellite image. Finally, we thank Rob Bierregaard for introducing us to Amazonian Brazil, for sharing his knowledge and data, and for encouraging our continued work there.

LITERATURE CITED

AMERICAN ORNITHOLOGISTS' UNION. 1983. Check-list of North American Birds, 6th ed. Am. Ornithol. Union, Washington, D.C.

AMERICAN ORNITHOLOGISTS' UNION. 1995. Fortieth supplement to the American Ornithologists' Union *Checklist of North American Birds*. Auk 112:819–830.

ANDERSON, A. B. 1981. White-sand vegetation of Brazilian Amazonia. Biotropica 13:199–210.

BIERREGAARD, R. O., JR. 1982. Levantamentos ornitológicos no dossel da mata pluvial de *terra firme*. Acta Amazonica 12:107–111.

BIERREGAARD, R. O., JR. 1984. Observations on the nesting biology of the Guiana Crested Eagle (*Morphnus guianensis*). Wilson Bull. 96:1–5.

BIERREGAARD, R. O., JR. 1988. Morphological data from understory birds in *terra firme* forest in the central Amazonian basin. Rev. Bras. Biol. 48:169–178.

BIERREGAARD, R. O., JR. 1990a. Avian communities in the understory of Amazonian forest fragments. Pp. 333–343 *in* Biography and Ecology of Forest Bird Communities (A. Keast and J. Kikkawa, Eds.). SPB Academic, The Hague.

BIERREGAARD, R. O., JR. 1990b. Species composition and trophic organization of the understory bird community in a central Amazonian *terra firme* forest. Pp. 217–236 *in* Four Neotropical Rainforests (A. H. Gentry, Ed.). Yale Univ., New Haven, Connecticut.

BIERREGAARD, R. O., JR., M. COHN-HAFT, AND D. F. STOTZ. 1997. Cryptic biodiversity: An overlooked species and new subspecies of antbird (Aves: Formicariidae) with a revision of *Cercomacra tyrannina* in northeastern South America. Pp. 111–128 *in* Studies in Neotropical Ornithology Honoring Ted Parker (J. V. Remsen, Jr., Ed.), Ornithol. Monogr. No. 48.

BIERREGAARD, R. O., JR., AND T. E. LOVEJOY. 1988. Birds in Amazonian forest fragments: Effects of insularization. Pp. 1564–1579 *in* Acta XIX Congressus Internationalis Ornithologici (H. Ouellet, Ed.). Ottawa, Ontario, 1986. National Museum of Natural Science, Ottawa.

BIERREGAARD, R. O., JR., AND T. E. LOVEJOY. 1989. Effects of forest fragmentation on Amazonian understory bird communities. Acta Amazônica 19:215–241.

BIERREGAARD, R. O., JR., D. F. STOTZ, L. H. HARPER, AND G. V. N. POWELL. 1987. Observations on the occurrence and behaviour of the Crimson Fruitcrow *Haematoderus miltaris* in central Amazonia. Bull. Brit. Ornithol. Club 107:134–137.

BIERREGAARD, R. O., JR., AND P. C. STOUFFER. 1997. Understory birds and dynamic habitat mosaics in Amazonian rainforests. Pp. 138–155 *in* Tropical Forest Remnants: Ecology, Management, and Conservation of Fragmented Communities (W. F. Laurance and R. O. Bierregaard, Jr., Eds.). Univ. of Chicago, Chicago, Illinois.

BLAKE, E. R. 1977. Manual of Neotropical birds. Vol. 1. Univ. of Chicago, Chicago, Illinois.

BORGES, S. H. 1995. Comunidade de aves em dois tipos de vegetação secundária da Amazônia central. M. S. thesis, Instituto Nacional de Pesquisas da Amazônia and Universidade Federal do Amazonas, Manaus, Brazil.

COHN-HAFT, M. 1993. Rediscovery of the White-winged Potoo (*Nyctibius leucopterus*). Auk 110:391–394.

COHN-HAFT, M. 1995. Dietary specialization by lowland tropical rainforest birds: Forest interior versus canopy and edge habitats. M.S. thesis, Tulane Univ., New Orleans, Louisiana.

COHN-HAFT, M. 1996. Why the Yungas Tody-Tyrant (*Hemitriccus spodiops*) is a *Snethlagea*, and why it matters. Auk 113:709–714.

CRACRAFT, J. 1985. Historical biogeography and patterns of differentiation within the South American avifauna: Areas of endemism. Pp. 49–84 *in* Neotropical Ornithology (P. A. Buckley, M. S. Foster, E. S. Morton, R. S. Ridgely, and F. G. Buckley, Eds.). Ornithol. Monogr. No. 36.

GENTRY, A. H., AND L. H. EMMONS. 1987. Geographical variation in fertility and composition of the understory of Neotropical forests. Biotropica 19:216–227.
GYLDENSTOLPE, N. 1930. Notes on ant wrens allied to *Myrmotherula surinamensis* Gmelin, together with the descriptions of two new forms. Arkiv för Zoologi 21A(26):1–38.
HAFFER, J. 1969. Speciation in Amazonian forest birds. Science 165:131–137.
HAFFER, J. 1990. Avian species richness in tropical South America. Studies on Neotropical Fauna and Environment 25:157–183.
HARPER, L. H. 1987. The conservation of ant-following birds in small Amazonian forest fragments. Ph. D. dissertation, State Univ. of New York, Albany, New York.
HARPER, L. H. 1989. The persistence of ant-following birds in small Amazonian forest fragments. Acta Amazônica 19:249–263.
HAVERSCHMIDT, F. 1968. Birds of Surinam. Oliver and Boyd, Edinburgh.
HELLMAYR, C. E. 1925. Catalogue of Birds of the Americas and the Adjacent Islands. Part 4, Furnariidae-Dendrocolaptidae. Field Mus. Nat. Hist. Publ. 234, Zool. Ser., vol. 13.
HELLMAYR, C. E. 1927. Catalogue of Birds of the Americas and the Adjacent Islands. Part 5, Tyrannidae. Field Mus. Nat. Hist. Publ. 242, Zool. Ser., vol. 13.
HELLMAYR, C. E., AND B. CONOVER. 1942. Catalogue of Birds of the Americas and the Adjacent Islands. Part 1, no. 1. Field Mus. Nat. Hist. Publ. 514, Zool. Ser., vol. 13.
HILTY, S. L., AND W. L. BROWN. 1986. A Guide to the Birds of Colombia. Princeton Univ., Princeton, New Jersey.
KARR, J. R. 1981. Surveying birds with mist nets. Pp. 62–67 *in* Estimating Numbers of Terrestrial Birds (C. J. Ralph and J. M. Scott, Eds.). Stud. Avian Biol. 6.
KARR, J. R., S. K. ROBINSON, J. G. BLAKE, AND R. O. BIERREGAARD, JR. 1990. Birds of four Neotropical forests. Pp. 237–269 *in* Four Neotropical Rainforests (A. H. Gentry, Ed.). Yale Univ., New Haven, Connecticut.
KLEIN, B. C., AND R. O. BIERREGAARD. 1988a. Movement and calling behavior of the Lined Forest-Falcon (*Micrastur gilvicollis*) in the Brazilian Amazon. Condor 90:497–499.
KLEIN, B. C., AND R. O. BIERREGAARD. 1988b. Capture and telemetry techniques for the Lined Forest-Falcon (*Micrastur gilvicollis*). J. Raptor Res. 22:29.
KLEIN, B. C., L. H. HARPER, R. O. BIERREGAARD, AND G. V. N. POWELL. 1988. The nesting and feeding behavior of the Ornate Hawk-Eagle near Manaus, Brazil. Condor 90:239–241.
KRATTER, A. W. 1997. Bamboo specialization by Amazonian birds. Biotropica 29:100–110.
LEVEY, D. J. 1988. Spatial and temporal variation in Costa Rican fruit and fruit-eating bird abundance. Ecol. Monogr. 58:251–269.
LOVEJOY, T. E. 1985. Minimum size for birds [sic] species and avian habitats. Pp. 324–327 *in* Acta XVIII Congressus Internationalis Ornithologici (V. D. Ilyichev, and V. M. Gavrilov, Eds.). Nauka, Moscow.
LOVEJOY, T. E., AND R. O. BIERREGAARD, JR. 1990. Central Amazonian forests and the Minimum Critical Size of Ecosystems Project. Pp. 60–71 *in* Four Neotropical Rainforests (A. H. Gentry, Ed.). Yale Univ., New Haven, Connecticut.
MESQUITA, R. C. G. 1989. A biologia reprodutiva de *Clusia grandiflora* Split.: Variação individual e remoção de sementes. M.S. thesis, Instituto Nacional de Pesquisas da Amazônia and Universidade Federal do Amazonas, Manaus, Brazil.
MEYER DE SCHAUENSEE, R. 1970. A Guide to the Birds of South America. Livingston, Wynnewood, Pennsylvania.
MINISTÉRIO DE MINAS E ENERGIA (MME). 1978. Projeto Radam Brasil, Folha SA 20 Manaus. Departamento Nacional de Produção Mineral, Rio de Janeiro.
MORONY, J. J., JR., W. J. BOCK, AND J. FARRAND, JR. 1975. Reference List of the Birds of the World. Am. Mus. Nat. Hist., New York.
NELSON, B. W., C. A. C. FERREIRA, M. F. SILVA, AND M. L. KAWASAKI. 1990. Endemism centres, refugia and botanical collection density in Brazilian Amazonia. Nature 345:714–716.
OREN, D. C. 1981. Zoogeographic analysis of the white sand campina avifauna of Amazonia. Ph.D. dissertation, Harvard Univ., Cambridge, Massachusetts.
OREN, D. C., AND H. G. ALBUQUERQUE. 1991. Priority areas for new avian collections in Brazilian Amazonia. Goeldiana Zool. 6.
PARKER, T. A., III, P. K. DONAHUE, AND T. S. SCHULENBERG. 1994. Birds of the Tambopata Reserve (Explorer's Inn Reserve). Pp. 106–124 *in* The Tambopata-Candamo Reserved Zone of Southeastern Perú: A Biological Assessment (R. B. Foster, T. A. Parker III, A. H. Gentry, L. H. Emmons, A. Chicchón, T. Schulenberg, L. Rodríguez, G. Lamas, H. Ortega, J. Icochea, W. Wust, M. Romo, J. A. Castillo, O. Phillips, C. Reynal, A. Kratter, P. K. Donahue, and L. J. Barkley, Eds.). Rapid Assessment Program Working Papers No. 6. Conservation International, Washington, D.C.
PARKER, T. A., III, R. B. FOSTER, L. H. EMMONS, P. FREED, A. B. FORSYTH, B. HOFFMAN, AND B. D. GILL. 1993. A Biological Assessment of the Kanuku Mountain Region of Southwestern Guyana. Rapid Assessment Program Working Papers No. 5. Conservation International, Washington, D.C.
PARKER, T. A., III, D. F. STOTZ, AND J. W. FITZPATRICK. 1997. Notes on avian bamboo specialists in southwestern Amazonian Brazil. Pp. 543–547 *in* Studies in Neotropical Ornithology Honoring Ted Parker (J. V. Remsen, Jr., Ed.), Ornithol. Monogr. No. 48.

PERES, C. A., AND A. WHITTAKER. 1991. Annotated checklist of the bird species of the upper Rio Urucu, Amazonas, Brazil. Bull. Brit. Ornithol. Club 111:156–171.

PETERS, J. L., JR. 1951. Check-List of Birds of the World. Vol. 7. Mus. Comp. Zool., Cambridge, Massachusetts.

PINTO, O. M. O. 1978. Novo Catálogo das Aves do Brasil. 1ª Parte. Editora Gráfica dos Tribunais, São Paulo.

POWELL, G. V. N. 1989. On the possible contribution of mixed species flocks to species richness in Neotropical avifaunas. Behav. Ecol. Sociobiol. 24:387–393.

PRUM, R. O. 1990. Phylogenetic analysis of the evolution of display behavior in the Neotropical manakins (Aves: Pipridae). Ethology 84:202–231.

QUINTELA, C. E. 1987. First report of the nest and young of the Variegated Antpitta (*Grallaria varia*). Wilson Bull. 99:499–500.

QUINTELA-ALMEIDA, C. E. 1985. Forest fragmentation and differential use of natural and man-made edges by understory birds in central Amazonia. M. S. thesis, Univ. of Illinois, Chicago, Illinois.

REMSEN, J. V., JR. 1994. Use and misuse of bird lists in community ecology and conservation. Auk 111:225–227.

REMSEN, J. V., JR., AND D. A. GOOD. 1996. Misuse of data from mist-net captures to assess relative abundance in bird populations. Auk 113:381–398.

REMSEN, J. V., JR., AND T. A. PARKER III. 1983. Contribution of river-created habitats to bird species richness in Amazonia. Biotropica 15:223–231.

RIDGELY, R. S., AND G. TUDOR. 1989. The Birds of South America. Vol. 1, The Oscine Passerines. Univ. of Texas, Austin, Texas.

RIDGELY, R. S., AND G. TUDOR. 1994. The Birds of South America. Vol. 2, The Suboscine Passerines. Univ. of Texas, Austin, Texas.

ROSENBERG, G. H. 1990. Habitat specialization and foraging behavior by birds of Amazonian river islands in northeastern Peru. Condor 92:427–443.

ROSENZWEIG, M. L., AND Z. ABRAMSKY. 1993. How are diversity and productivity related? Pp. 52–65 *in* Species Diversity in Ecological Communities (R. E. Ricklefs and D. Schluter, Eds.). Univ. of Chicago, Chicago, Illinois.

SCHWARTZ, P. 1975. Solved and unsolved problems in the *Sporophila lineola bouvronides* complex. Ann. Carnegie Mus. 45:277–285.

SIBLEY, C. G., AND B. L. MONROE, JR. 1990. Distribution and Taxonomy of Birds of the World. Yale Univ., New Haven, Connecticut.

SILVA, J. M. C. DA, D. C. OREN, J. C. ROMA, AND L. M. P. HENRIQUES. 1997. Composition and distribution patterns of the avifauna of an Amazonian upland savanna, Amapá, Brazil. Pp. 743–762 *in* Studies in Neotropical Ornithology Honoring Ted Parker (J. V. Remsen, Jr., Ed.), Ornithol. Monogr. No. 48.

SILVA, J. M. C. DA, AND E. O. WILLIS. 1986. Notas sobre a distribuição de aves da Amazônia brasileira. Bol. Mus. Paraense Emílio Goeldi (Ser. Zool.) 2:151–158.

SNETHLAGE, E. 1907. Neue Vogelarten aus Südamerika. Ornithologische Monatsberichte 15:193–196.

STOTZ, D. F. 1990. Corrections and additions to the Brazilian avifauna. Condor 92:1078–1079.

STOTZ, D. F., AND R. O. BIERREGAARD JR. 1989. The birds of the fazendas Porto Alegre, Esteio and Dimona north of Manaus, Amazonas, Brazil. Rev. Brasil. Biol. 49:861–872.

STOTZ, D. F., R. O. BIERREGAARD, M. COHN-HAFT, P. PETERMANN, J. SMITH, A. WHITTAKER, AND S. V. WILSON. 1992. The status of North American migrants in central Amazonian Brazil. Condor 94:608–621.

STOTZ, D. F., S. M. LANYON, T. S. SCHULENBERG, D. E. WILLARD, A. T. PETERSON, AND J. W. FITZPATRICK. 1997. An avifaunal survey of two tropical forest localities on the middle Rio Jiparana, Rondonia, Brazil. Pp. 763–781 *in* Studies in Neotropical Ornithology Honoring Ted Parker (J. V. Remsen, Jr., Ed.), Ornithol. Monogr. No. 48.

STOUFFER, P. C. 1997. Interspecific aggression in *Formicarius* antthrushes? the view from central Amazonian Brazil. In press, *Auk*.

STOUFFER, P. C., AND R. O. BIERREGAARD, JR. 1993. Spatial and temporal abundance patterns of Ruddy Quail-Doves (*Geotrygon montana*) near Manaus, Brazil. Condor 95:896–903.

STOUFFER, P. C., AND R. O. BIERREGAARD, JR. 1995a. Use of Amazonian forest fragments by understory insectivorous birds. Ecology 76:2429–2445.

STOUFFER, P. C., AND R. O. BIERREGAARD, JR. 1995b. Effects of forest fragmentation on understory hummingbirds in Amazonian Brazil. Conservation Biol. 9:1086–1095.

TAMPLIN, J. W., J. W. DEMASTES, AND J. V. REMSEN, JR. 1993. Biochemical and morphometric relationships among some members of the Cardinalinae. Wilson Bull. 105:93–113.

TERBORGH, J. W., J. W. FITZPATRICK, AND L. EMMONS. 1984. Annotated checklist of bird and mammal species of Cocha Cashu Biological Station, Manu National Park, Peru. Fieldiana (Zoology) 21:1–29.

TERBORGH, J., S. K. ROBINSON, T. A. PARKER III, C. A. MUNN, AND N. PIERPONT. 1990. Structure and organization of an Amazonian forest bird community. Ecol. Monogr. 60:213–238.

TOSTAIN, O., J. L. DUJARDIN, C. H. ERARD, AND J. M. THIOLLAY. 1992. Oiseaux de Guyane. Societe d'Etudes Ornithologiques. Maxéville, France.

TRAYLOR, M. A., JR (ED.). 1979. Check-list of Birds of the World, vol. 8. Mus. Comp. Zool., Cambridge, Massachusetts.

VIELLIARD, J. 1989. Uma nova espécie de *Glaucidium* (Aves, Strigidae) da Amazônia. Rev. Bras. Zool. 6: 685–693.
WHITTAKER, A. 1993. Notes on the behaviour of the Crimson Fruitcrow *Haematoderus militaris* near Manaus, Brazil, with the first nesting record for the species. Bull. Brit. Ornithol. Club 113:93–96.
WHITTAKER, A. 1995. Range extensions and nesting of the Glossy-backed Becard *Pachyramphus surinamus* in central Amazonian Brazil. Bull. Brit. Ornithol. Club 113:93–96.
WHITTAKER, A. 1996. First records of the Orange-breasted Falcon *Falco deiroleucus* in central Amazonian Brazil, with short behavioral notes. Cotinga 6:65–68.
WILLIS, E. O. 1977. Lista preliminar das aves da parte noroeste e áreas vizinhas da Reserva Ducke, Amazonas, Brasil. Rev. Bras. Biol. 37:585–601.
WILLIS, E. O., AND Y. ONIKI. 1988. Aves observadas em Balbina, Amazonas e os prováveis efeitos da barragem. Ciência e Cultura 40:280–284.
ZIMMER, J. T. 1934. Studies of Peruvian birds. No. 13. The genera *Dendrexetastes, Campyloramphus,* and *Dendrocincla.* Amer. Mus. Novitates 728:1–20.
ZIMMER, J. T. 1940. Studies of Peruvian birds. No. 34. The genera *Todirostrum, Euscarthmornis, Snethlagea, Poecilotriccus, Lophotriccus, Myiornis, Pseudotriccus,* and *Hemitriccus.* Amer. Mus. Novitates 1066:1–23.

APPENDIX
LIST OF SPECIES OBSERVED IN STUDY AREA FROM 1979–1994

Family (or major subfamily)	Abundance/ seasonality	Habitat	Position/ microhabitat	Sociality	Evidence
Tinamidae					
Tinamus major (D)	c	1	t	s	2t
Crypturellus soui (D)	u	2, 1	t	s	4h
Crypturellus brevirostris (D)	u	1	t	s	4s (2t)[a]
Crypturellus variegatus (D)	c	1	t	s	2pt
Podicipedidae					
Podiceps dominicus	c	w	w	s	2t
Anhingidae					
Anhinga anhinga	x	p	o	s	4s
Ardeidae					
Tigrisoma lineatum	r	l, w	eit	s	4sh
Ardea cocoi	r	w	t	s	4s
Ardea alba[b]	r	w	t	s	4s
Bubulcus ibis	x	p, w, l	to	m	4s
Butorides striatus	x	l	e	s	4s
Pilherodias pileatus	x	w	t	s	4s
Nycticorax nycticorax	x	l	o	s	2t
Cochlearius cochlearius	x	l	i	s	2p
Ciconiidae					
Mycteria americana (D)	x	p	o	s	4s
Threskiornithidae					
Mesembrinibis cayennensis	x	l	o	s	4h
Anatidae					
Oxyura dominica	r	w	w	s	4s
Cathartidae					
Coragyps atratus (D)	u	p	oct	sm	4s
Cathartes aura (D)	u	p, 2	oct	sm	4s
Cathartes melambrotos (D)	c	1, p	oct	sm	2p
Sarcoramphus papa (D)	u	1, p	oct	sm	2p
Accipitridae					
Pandion haliaetus (D)	xb	w	c	s	4s
Leptodon cayanensis	x	1	o	s	4s
Chondrohierax uncinatus	x	1	o	s	4s
Elanoides forficatus (D)	um?	1, 2	oc	sm	2pt
Gampsonyx swainsonii	r	p	c	s	4s
Harpagus bidentatus (D)	u	1	om	s	4s
Ictinia plumbea (D)	um?	1, 2	oc	sm	4s
Accipiter superciliosus (D)	r	1	c	s	4s
Accipiter bicolor (D)	r	1	mu	s	4s
T *Accipiter poliogaster*	x	1	c	s	4s
Leucopternis melanops (D)	r	1	c	s	3
Leucopternis albicolllis (D)	c	1, 2	coe	s	2pt
Buteogallus urubitinga (D)	u	1, 2	coe	s	2pt
Heterospizias meridionalis	u	p	c	s	4s
Buteo nitidus (D)	c	2, p	ec	s	2t
Buteo magnirostris (D)	u	p, 2	ec	s	4sh
Buteo platypterus (D)	ub	2, 1	ec	s	4s
Buteo brachyurus (D)	u	2, p	oc	s	2t
Buteo albicaudatus	r	p	oc	s	4s
Morphnus guianensis	r	1	cm	s	2p
Harpia harpyja (D)	r	1	c	s	2pt
Spizastur melanoleucus	x	1, p	oce	s	2p
Spizaetus tyrannus (D)	x	1, 2	oc	s	4sh
Spizaetus ornatus (D)	u	1	oc	s	2pt

APPENDIX
Continued

Family (or major subfamily)	Abundance/ seasonality	Habitat	Position/ microhabitat	Sociality	Evidence
Falconidae					
Daptrius ater	r	1	o	sm	2t
Daptrius americanus (D)	c	1	c	ms	2t
Polyborus plancus	r	p	ct	s	2t
Milvago chimachima	u	p	ct	s	2t
Micrastur ruficollis (D)	c	1, 2	mu	s	2pt
Micrastur gilvicollis (D)	c	1	mu	s	1
Micrastur mirandollei (D)	u	1, 2	m?	s	2t
Micrastur semitorquatus (D)	u	1, 2	sm	s	2t
Falco rufigularis (D)	c	1, p	co	s	2pt
Cracidae					
Ortalis motmot (D)	c	2	e	ms	2t
Penelope marail (D)	c	1	c	sf	2t
Penelope jacquacu (D)	u	1	cte	sf	2t
Pipile cumanensis[c]	x	1	c	s	4s
Crax alector (D)	u	1	te	sm	2pt
Odontophoridae[d]					
Odontophorus gujanensis (D)	u	1, 2	t	sm	2pt
Psophiidae					
Psophia crepitans (D)	u	1	tm	m	2pt
Rallidae					
Laterallus melanophaius	x	w	t	s	4s
Laterallus viridis (D)	u	p	t	s	2t
Aramides cajanea	r	1, 2	it	s	2t
Heliornithidae					
Heliornis fulica	x	w	w	s	2t
Eurypygidae					
Eurypyga helias	x	1	it	s	2t
Charadriidae					
Pluvialis dominica	rb	w	t	sm	4s
Charadrius collaris	x	w, p	t	s	4s
Jacanidae					
Jacana jacana	c	w	t	s	2t
Scolopacidae					
Tringa melanoleuca	ub	w	t	sm	4s
Tringa flavipes	rb	w	t	sm	4s
Tringa solitaria (D)	ub	w	t	sm	4s
Actitis macularia	ub	w	t	s	4s
Calidris minutilla	xb	w	t	sm	4s
Calidris fuscicollis	ub	w	t	sm	2pt
Calidris melanotos	rb	w	t	sm	2p
Micropalama himantopus	xb	w	t	sm	4s
Gallinago gallinago	x	w, 1	te	s	2p
Columbidae					
Columba plumbea (D)	c	1	c	sf	1
Columba subvinacea (D)	c	1, 2	c	sf	2t
Columbina passerina (D)	u	2, p	t	sm	2p
Columbina talpacoti (D)	u	2, p	t	s	4sh
Leptotila verreauxi (D)	c	2, p	t	s	4sh
Geotrygon montana (D)	cm	1	t	s	2pt
Psittacidae					
Ara ararauna (D)	c	1	co	ms	2t
Ara macao (D)	r	1	co	ms	2p

APPENDIX
Continued

Family (or major subfamily)	Abundance/ seasonality	Habitat	Position/ microhabitat	Sociality	Evidence
Ara chloroptera (D)	c	1, p	co	ms	2pt
Ara manilata	u	p, 1	c	m	4sh
Aratinga leucophthalmus	x	1	o	m	4sh
Forpus sp.[e]	x	1, 2	ce	m	4s
Brotogeris chrysopterus (D)	c	1	co	m	2t
Touit purpurata (D)	u	1	c	m	2t
Pionites melanocephala	x	1	e	m	4s
Pionopsitta caica (D)	u	1	c	m	2t
Pionus menstruus (D)	cm	1	co	ms	2t
Pionus fuscus (D)	um	1	co	ms	2t
Amazona autumnalis (D)	cm	1	co	ms	2t
Amazona farinosa (D)	cm	1	co	ms	2t
Deroptyus accipitrinus (D)	c	1	c	m	2t
Cuculidae					
Coccyzus euleri	ra	1	c	sc	2tp
Coccyzus melacoryphus	xa	2	e	s	4s
Piaya cayana (D)	u	2	ce	s	4sh
Piaya melanogaster (D)	c	1	c	cs	2pt
Crotophaga major	x	1	e	m	4s
Crotophaga ani (D)	c	p, 2	et	m	2pt
Dromococcyx pavoninus	x	1	u	s	2t
Tytonidae					
Tyto alba	r	2, p	e	s	4sh
Strigidae					
Otus watsonii (D)	c	1, 2	m	s	1
Lophostrix cristata (D)	c	1	m	s	2t
Pulsatrix perspicillata (D)	c	1	c	s	2t
Glaucidium hardyi[f] (D)	c	1, 2	c	s	1
Athene cunicularia	x	p	t	s	4s
Ciccaba virgata	r	2, 1	c	s	2p
Ciccaba huhula	u	1, 2	cm	s	2t
Caprimulgidae					
Lurocalis semitorquatus (D)	u	1	oe	s	1
Chordeiles acutipennis	x	p	ot	s	4sh
Chordeiles minor	rb	1, p	o	m	4s
Nyctidromus albicollis (D)	c	2, 1	et	s	2t
Caprimulgus nigrescens (D)	u	1, 2	et	s	1
Nyctibiidae					
Nyctibius grandis (D)	r	2, 1	c	s	4h
Nyctibius aethereus	r	1, 2	me	s	2tp
Nyctibius griseus (D)	u	2, 1	ce	s	2t
Nyctibius leucopterus (D)	u	1	c	s	1
Nyctibius bracteatus	u	1	mu	s	1
Apodidae					
Streptoprocne zonaris	rm	1, p	o	sm	4s
Chaetura spinicauda (D)	c	1, w, p	o	m	2t
Chaetura chapmani (D)	c	1, w	o	m	1
Chaetura brachyura (D)	u	2, w, p	o	m	1
Panyptila cayannensis (D)	r	1	o	s	4s
Tachornis squamata[g] (D)	r	p	o	m	1
Trochilidae					
Phaethornis superciliosus (D)	c	1, 2	ue	sl	2pt
Phaethornis bourcieri (D)	c	1, 2	u	sl	2pt
Phaethornis ruber (D)	r	2	ue	s	2p
Campylopterus largipennis (D)	c	1, 2	ce	s	2pt

APPENDIX
Continued

Family (or major subfamily)	Abundance/ seasonality	Habitat	Position/ microhabitat	Sociality	Evidence
Florisuga mellivora (D)	u	1, 2	c	f	2p
Anthracothorax nigricollis (D)	r	1	cei	s	4s
T *Avocettula recurvirostris*	x	1	c	f	4s
T *Chrysolampis mosquitus*	x	1	c	f	4s
Discosura longicauda (D)	r	1, 2	ce	f	4s
Thalurania furcata (D)	c	1, 2	umc	sf	2pt
Hylocharis sapphirina	u	1, 2	c	f	3
Polytmus theresiae (D)	x	p	c	s	4s
Amazilia versicolor (D)	r	2	ce	s	4s
Amazilia fimbriata	x	1	ce	s	4s
Topaza pella (D)	r	1, 2, w	cei	sl	1
Heliothryx aurita (D)	c	1, 2	cme	sf	2p
Trogonidae					
Trogon melanurus (D)	c	1	c	sc	2pt
Trogon viridis (D)	c	1, 2	cem	s	2t
Trogon rufus (D)	u	1	u	s	2pt
Trogon violaceus (D)	c	1	ce	sc	2t
Pharomachrus pavoninus (D)	x	1	mc	s	4sh
Momotidae					
Momotus momota (D)	c	1	mui	s	2pt
Alcedinidae					
Ceryle torquata (D)	r	w	c	s	4sh
Chloroceryle amazona (D)	x	w	c	s	4s
Chloroceryle americana (D)	x	1	ie	s	2t
Chloroceryle inda (D)	r	1	i	s	2p
Chloroceryle aenea (D)	r	1	i	s	2p
Bucconidae					
Notharchus macrorhynchus (D)	c	1	c	s	2pt
Notharchus tectus (D)	u	1, 2	ce	s	1
Bucco tamatia (D)	u	1, 2	cm	s	1
Bucco capensis (D)	r	1	mu	s	2pt
Malacoptila fusca (D)	u	1	u	s	1
Nonnula rubecula (D)	r	1	m	s	2p
Monasa atra (D)	c	1, 2	ec	m	1
Chelidoptera tenebrosa (D)	r	p, 2	c	sm	4s
Galbulidae					
Galbula albirostris (D)	c	1, 2	ue	s	1
Galbula leucogastra (D)	r	c, 2, 1	ce	s	2t
Galbula dea (D)	c	1, 2	ce	cs	1
Jacamerops aurea (D)	u	1	m	s	2pt
Capitonidae					
Capito niger (D)	c	1	c	c	2pt
Ramphastidae					
Pteroglossus viridis (D)	c	1, 2	c	mf	1
Selenidera culik (D)	c	1	c	sf	2pt
Ramphastos vitellinus (D)	c	1	c	sf	2pt
Ramphastos tucanus (D)	c	1	c	sf	2t
Picidae					
Picumnus exilis (D)	r	1, 2	ce	sc	2p
Melanerpes cruentatus (D)	c	2, 1	c	mf	2t
Veniliornis cassini (D)	c	1	ce	c	1
Piculus flavigula (D)	c	1	cm	uc	2t
Piculus chrysochloros (D)	x	1	ci	s	4s
Celeus undatus (D)	r	2, 1	c	sf	2pt
Celeus grammicus	c	1	ce	sf	2t

APPENDIX
Continued

Family (or major subfamily)	Abundance/seasonality	Habitat	Position/microhabitat	Sociality	Evidence
Celeus elegans (D)	u	1	cm	s	2t
Celeus flavus (D)	x	1	ci	s	4h
Celeus torquatus (D)	u	1	c	s	2t
Dryocopus lineatus (D)	c	2, 1, p	ce	s	2t
Campephilus rubricollis[h] (D)	c	1	cm	s	2pt
Dendrocolaptidae					
Dendrocincla fuliginosa (D)	c	1, 2	u	sua	2pt
Dendrocincla merula (D)	c	1	u	a	1
Deconychura longicauda (D)	c	1	um	uc	1
Deconychura stictolaema (D)	c	1	u	u	1
Sittasomus griseicapillus (D)	c	1, 2	cem	cus	1
Glyphorynchus spirurus (D)	c	1, 2	mue	us	1
Dendrexetastes rufigula (D)	u	1	ce	s	1
Hylexetastes perrotii (D)	u	1	m	sa	2pt
Dendrocolaptes certhia (D)	c	1	m	as	2pt
Dendrocolaptes picumnus (D)	u	1	m	a	2t
Xiphorhynchus pardalotus (D)	c	1	mu	u	1
Lepidocolaptes albolineatus (D)	c	1	c	c	1
Campylorhamphus procurvoides (D)	u	1	m	u	1
Furnariidae					
Synallaxis rutilans	u	1	u	s	2pt
Philydor erythrocercus (D)	c	1	m	u	2pt
Philydor pyrrhodes (D)	u	1	ic	s	1
Automolus infuscatus (D)	c	1	u	u	1
Automolus rubiginosus	u	1, 2	ue	s	2pt
Automolus ochrolaemus (D)	c	2, 1	eu	su	1
Xenops milleri (D)	c	1	c	c	2p
Xenops minutus (D)	c	1	m	u	1
Sclerurus mexicanus	u	1	t	s	2p
Sclerurus rufigularis (D)	c	1	t	s	1
Sclerurus caudacutus (D)	r	1	t	s	2pt
Formicariidae					
Cymbilaimus lineatus (D)	c	2, 1	me	s	2t
Frederickena viridis (D)	r	1, 2	u	s	1
Thamnophilus murinus (D)	c	1, 2	me	su	1
Thamnophilus punctatus (D)	u	2, c	m	s	2pt
Thamnomanes ardesiacus (D)	c	1	u	u	2pt
Thamnomanes caesius (D)	c	1	um	u	2pt
Myrmotherula brachyura (D)	c	1, 2	ce	c	1
Myrmotherula guttata (D)	u	1	u	su	1
Myrmotherula gutturalis (D)	c	1	u	u	1
Myrmotherula axillaris (D)	u	1, 2	em	us	1
Myrmotherula longipennis (D)	c	1	um	u	1
Myrmotherula menetriesii (D)	c	1	mu	u	1
Herpsilochmus dorsimaculatus (D)	c	1	c	c	1
Terenura spodioptila (D)	c	1	c	c	2t
Cercomacra cinerascens (D)	c	1	ce	s	2t
Cercomacra tyrannina (D)	r	2	e	s	2t
Hypocnemis cantator (D)	c	1, 2	eu	s	1
Sclateria naevia	x	1	it	s	2p
Percnostola rufifrons (D)	c	1, 2	ue	sa	1
Schistocichla leucostigma[i] (D)	u	1	it	s	1
Myrmeciza ferruginea (D)	c	1	te	s	1
Myrmeciza atrothorax (D)	u	2, 1	iet	s	2pt
Pithys albifrons (D)	c	1	u	a	2pt
Gymnopithys rufigula (D)	c	1	u	a	1
Hylophylax naevia	r	1, 2	u	s	2pt
Hylophylax poecilinota (D)	c	1	u	sa	1

APPENDIX
Continued

Family (or major subfamily)	Abundance/ seasonality	Habitat	Position/ microhabitat	Sociality	Evidence
Formicarius colma (D)	c	1	t	s	1
Formicarius analis (D)	c	1	t	s	2pt
Myrmornis torquata (D)	u	1	tu	s	1
Grallaria varia (D)	c	1	t	s	2pt
Hylopezus macularius (D)	u	1	t	s	2pt
Myrmothera companisona (D)	c	1, 2	te	s	2t
Conopophagidae					
Conopophaga aurita (D)	u	1	u	s	2pt
Tyrannidae					
Phyllomyias griseiceps (D)	x	2	ce	c	4s
Zimmerius gracilipes (D)	c	1, 2	c	c	2t
Ornithion inerme	u	1	c	sc	2t
Camptostoma obsoletum	x	2	ce	s	2t
Phaeomyias murina (D)	r	2	e	s	2t
Tyrannulus elatus (D)	c	1, 2	ce	sc	2t
Myiopagis gaimardii (D)	c	1	c	c	2t
Myiopagis caniceps (D)	c	1	c	c	1
Elaenia parvirostris (D)	ra	2	e	s	4s
Elaenia chiriquensis	xm	2, p	c	s	1
Mionectes macconnelli (D)	c	1, 2	umec	lf	1
Phylloscartes virescens (D)	c	1	c	c	1
Corythopis torquata (D)	u	1	t	s	2pt
Myiornis ecaudatus (D)	u	1, 2	me	s	2t
Lophotriccus vitiosus (D)	c	1, 2	me	s	1
Lophotriccus galeatus (D)	r	2	e	s	2t
Hemitriccus zosterops (D)	c	1, 2	m	s	1
Todirostrum pictum[j] (D)	c	1, 2	c	s	1
Ramphotrigon ruficauda (D)	r	1	m	s	2pt
Rhynchocyclus olivaceus (D)	u	1	m	u	2pt
Tolmomyias assimilis (D)	c	1	c	c	1
Tolmomyias poliocephalus (D)	c	1, 2	ce	sc	1
Platyrinchus saturatus (D)	u	1	u	s	1
Platyrinchus coronatus	c	1	um	sl	1
Platyrinchus platyrhynchos (D)	u	1	m	s	1
Onychorhynchus coronatus (D)	u	1	u	su	2pt
Terenotriccus erythrurus (D)	c	1, 2	me	su	1
Myiobius barbatus (D)	c	1	m	u	1
Contopus borealis (D)	rb	2, 1	ce	s	4s
Contopus virens (D)	rb	2, 1	ce	s	4sh
Pyrocephalus rubinus (D)	xa	2	ce	s	4s
Attila spadiceus (D)	c	1	c	s	2pt
Rhytipterna simplex (D)	c	1, 2	c	sc	2pt
Laniocera hypopyrra (D)	c	1	m	sl	1
Sirystes sibilator (D)	c	1	c	sc	1
Myiarchus tuberculifer (D)	u	2, 1	e	s	2t
Myiarchus ferox (D)	u	2	e	s	2t
Pitangus sulphuratus (D)	r	2, p	ec	s	4sh
Myiozetetes cayanensis (D)	c	2, p	ec	s	1
Myiozetetes luteiventris[k]	x	2	ec	s	4s
Conopias parva[l] (D)	c	1	c	cm	2t
Myiodynastes maculatus (D)	ra?	2, p	c	sc	4s
Legatus leucophaius (D)	u	2, 1	c	s	4sh
Empidonomus varius (D)	um?	2	e	s	4s
Empidonomus aurantioatrocristatus (D)	ra	1	c	s	4s
Tyrannopsis sulphurea (D)	u	1	ci	s	2t
Tyrannus melancholicus (D)	cm	2, p	e	sm	2t
Tyrannus savana (D)	ua?	2, p	ec	sm	4s
Tyrannus tyrannus	xb	p	c	s	4s
Pachyramphus rufus (D)	x	2	e	s	2t

APPENDIX
CONTINUED

Family (or major subfamily)	Abundance/ seasonality	Habitat	Position/ microhabitat	Sociality	Evidence
Pachyramphus marginatus (D)	c	1	m	c	2t
Pachyramphus surinamus (D)	c	1	c	sc	2t
Pachyramphus minor (D)	u	1	cm	c	2pt
Tityra cayana (D)	c	1, 2	ce	c	2t
Pipridae					
Schiffornis turdinus (D)	c	1, 2	u	s	2pt
Piprites chloris (D)	c	1	c	c	1
Neopipo cinnamomea	x	1, 2	m	s	2p
Tyranneutes virescens (D)	c	1	mc	lf	1
Neopelma chrysocephalum (D)	u	c	m	l	2t
Manacus manacus (D)	x	2	u	l	3
Corapipo gutturalis (D)	c	1	uc	lf	2pt
Pipra pipra (D)	c	1	uc	lf	1
Pipra serena (D)	c	1	u	l	1
Pipra erythrocephala (D)	c	1, 2	mc	lf	2pt
Cotingidae					
Phoenicircus carnifex (D)	u	1	m	sl	2pt
Iodopleura fusca	xm?	1	ce	s	4s
Lipaugus vociferans (D)	c	1	m	lsc	2pt
Cotinga cotinga	x	1	c	f	4s
Cotinga cayana (D)	u	1	c	sf	4s
Xipholena punicea (D)	c	1	c	sfl	1
Haematoderus militaris (D)	u	1, 2	ce	s	2t
Perissocephalus tricolor (D)	u	1	cm	ls	2t
Hirundinidae					
Progne tapera	xa	p	o	m	4s
Progne subis (D)	rb	1, p	o	m	4s
Progne chalybea (D)	u	p, 2	oc	m	4sh
Neochelidon tibialis (D)	u	2, 1	eo	m	2t
Stelgidopteryx ruficollis (D)	u	2, p	e	m	4sh
Riparia riparia	xb	p	o	m	4s
Hirundo rustica (D)	ub	p	o	m	3
Troglodytidae					
Thryothorus coraya (D)	c	2, 1	eu	s	1
Thryothorus leucotis	x	2	e	s	4sh
Troglodytes aedon (D)	c	p, 2	u	s	2pt
Microcerculus bambla (D)	c	1	ut	s	1
Cyphorhinus arada (D)	u	1	tu	sm	2pt
Muscicapidae (Turdinae)					
Catharus fuscescens (D)	rb	1, 2	met	s	2p
Catharus minimus (D)	rb	1	mt	s	2p
Turdus albicollis (D)	c	1	mt	s	1
(Polioptilinae)					
Microbates collaris (D)	c	1	ue	su	1
Ramphocaenus melanurus	c	1	ce	c	1
Polioptila guianensis (D)	u	1	c	c	1
Vireonidae					
Cyclarhis gujanensis (D)	c	2, 1	ce	s	2pt
Vireolanius leucotis (D)	c	1	c	c	2pt
Vireo olivaceus[m] (D)	ub	1, 2	c	c	2p
Vireo altiloquus	rb	1	ce	fc	4s
Hylophilus thoracicus	r	1	c	cs	2t
Hylophilus semicinereus	x	2	me	s	4sh
Hylophilus muscicapinus (D)	c	1	c	c	1
Hylophilus ochraceiceps (D)	c	1	um	u	1

APPENDIX
Continued

Family (or major subfamily)	Abundance/ seasonality	Habitat	Position/ microhabitat	Sociality	Evidence
Emberizidae (Emberizinae)					
Ammodramus aurifrons (D)	c	p, 2	t	s	2pt
Volatinia jacarina (D)	c	p, 2	te	s	2t
Sporophila bouvronides[n]	xm	p	et	sm	4s
Sporophila lineola	xm	p	et	sm	4s
Sporophila castaneiventris (D)	c	p	tc	s	2t
Oryzoborus angolensis (D)	u	2, p	ect	s	2pt
Arremon taciturnis (D)	r	1	te	s	2pt
(Cardinalinae)					
Caryothraustes canadensis (D)	c	1, 2	c	mc	2t
Saltator grossus[o] (D)	c	1, 2	c	cs	2t
Saltator maximus (D)	x	2	ce	sf	2t
Cyanocompsa cyanoides[p] (D)	c	1, 2	e	s	2pt
(Thraupinae)					
Lamprospiza melanoleuca (D)	c	1	c	mc	2t
Hemithraupis flavicollis (D)	c	1	c	cf	1
Lanio fulvus (D)	u	1	m	c	2tp
Tachyphonus cristatus (D)	c	1	c	cf	2pt
Tachyphonus surinamus (D)	c	1, 2	uemc	mc	1
T *Tachyphonus phoenicius*	x	1	c	s	4s
Piranga rubra (D)	xb	2	e	s	4s
Ramphocelus carbo (D)	c	2, p	e	ms	2pt
Thraupis episcopus (D)	c	2, p	ec	sm	2p
Thraupis palmarum (D)	c	2, 1	ec	mc	3
Cyanicterus cyanicterus	r	1	c	c	2t
Euphonia plumbea	x	1, 2	c	c	4s
Euphonia chrysopasta (D)	u	1, 2	c	cs	2t
Euphonia chlorotica (D)	x	2	ce	s	4h
Euphonia minuta (D)	u	1, 2	c	f	2t
Euphonia cayennensis (D)	c	1	c	fs	2t
Tangara mexicana (D)	r	2, 1	c	mfc	4sh
Tangara chilensis (D)	c	1	c	mfc	1
Tangara punctata (D)	c	1, 2	c	cf	2t
Tangara varia (D)	r	1	c	cf	1
Tangara gyrola	r	1	c	c	4sh
Tangara velia (D)	c	1	c	mfc	2t
Dacnis lineata (D)	c	1	c	fc	4sh
Dacnis cayana (D)	c	1	c	fc	1
Chlorophanes spiza (D)	c	1, 2	c	fc	2pt
Cyanerpes nitidus (D)	u	1	c	fc	4sh
Cyanerpes caeruleus (D)	c	1	c	mfc	4sh
Cyanerpes cyaneus (D)	c	1, 2	c	fc	4sh
Tersina viridis	xm	1	c	m	4sh
(Parulinae)[q]					
Dendroica petechia	xb	1	c	c	4s
Dendroica fusca	xb	2	c	c	4s
Dendroica striata (D)	rb	2	c	c	4s
Phaeothlypis rivularis (D)	u	2, 1	tie	s	2t
Conirostrum speciosum	x	2, 1	c	c	4s
Coereba flaveola (D)	c	1, 2	ce	fc	2t
(Icterinae)					
Psarocolius viridis (D)	c	1	c	mf	2t
Cacicus haemorrhous (D)	u	1, 2	ce	mf	2t
Icterus cayanensis	r	1, 2	ce	sf	4s
Icterus chrysocephalus (D)	u	1, 2	ce	sf	2t
Sturnella militaris (D)	c	p	t	s	2t
Molothrus bonariensis (D)	u	p, 2	ec	sm	4s
Scaphidura oryzivora (D)	u	p, 2, 1	coe	msf	4sh

APPENDIX
Continued

^a Tape recording of voice believed to be this species; see text.
^b Called *Casmerodius alba* by Stotz and Bierregaard (1989); see AOU (1995).
^c Sometimes called *Aburria* (=*Pipile*) *pipile*; we follow Sibley and Monroe (1990).
^d Treated as subfamily of Phasianidae by Stotz and Bierregaard (1989); we follow Sibley and Monroe (1990).
^e See text, "Reidentifications."
^f Called *G. minutissimum* by Stotz and Bierregaard (1989); recently split (Vielliard 1989).
^g Called *Reinarda squamata* by Stotz and Bierregaard (1989).
^h Called *Phloeoceastes rubricollis* by Stotz and Bierregaard (1989).
ⁱ Called *Percnostola leucostigma* by Stotz and Bierregaard (1989); see Ridgely and Tudor (1994).
^j Treated as subspecies of *T. chrysocrotaphum* by Stotz and Bierregaard (1989).
^k Sometimes placed in genus *Tyrannopsis*; see Ridgely and Tudor (1994).
^l Treated as subspecies of *Coryphotriccus albovittatus* by Stotz and Bierregaard (1989).
^m Includes both migratory and resident forms (discussed in Stotz et al. 1992).
ⁿ Sometimes considered a subspecies of *S. lineola* (see Schwartz 1975).
^o Called *Pitylus grossus* by Stotz and Bierregaard (1989); see Tamplin et al. (1993), AOU (1995).
^p Sometimes placed in genus *Passerina*; we follow AOU (1983).
^q Including *Conirostrum* and *Coereba* (as in Morony et al. 1975).

Status codes (see Methods for further explanation).
Abundance (in preferred habitat): c—common, u—uncommon, r—rare, x—casual; followed by code for seasonality (if not year-round resident): a—austral migrant, b—boreal migrant, m—unspecified movements.
Habitat (more than one listed in order of decreasing preference): 1—primary *terra firme* forest, 2—secondary forest, p—pasture, w—open water (pond), c—*campinarana*.
Position/microhabitat (more than one listed in order of decreasing preference): t—terrestrial, u—understory, m—midstory, c—canopy, e—edge or treefall gap, i—small woodland stream (*igarapé*), o—overhead airspace, w—water surface.
Sociality: u—accompanies understory mixed-species flocks, c—accompanies canopy mixed-species flocks, m—in monospecific flocks, s—solitary or in pairs, f—in mixed-species assemblages at fruiting or flowering trees but not flocking, a—army-ant follower, l—lekking.
Evidence (number represents quality of evidence in descending order, only highest-quality evidence available listed): 1—specimen; 2 followed by t—tape recording, by p—photograph; 3—mist net capture record; 4 followed by s—sighting, by h—heard.
"(D)" following species—noted at Reserva Ducke by Willis (1977), Stotz and Bierregaard (1989), or by us (unpubl.)
"T" preceding species—registered at ZF-2 tower only.

PATTERNS AND CAUSES OF ENDANGERMENT IN THE NEW WORLD AVIFAUNA

NIGEL J. COLLAR, DAVID C. WEGE, AND ADRIAN J. LONG
BirdLife International, Wellbrook Court, Girton Road, Cambridge CB3 0NA, U.K.

ABSTRACT.—*Threatened birds of the Americas* (1992) detailed 327 species, of which only four had ranges entirely outside the Neotropics, showing how important this latter region is for global bird conservation, contributing 30% of all threatened birds on earth. Brazil had 97 threatened species, Peru 64, and Colombia 56. These countries, plus Mexico, held three-quarters of all threatened birds in the Americas. Over 78% (256) of all threatened bird species possessed ranges of less than 50,000 km^2. Some 57% of all threatened birds were confined to wet forest, 17% to dry forest, and 10% to grasslands, a rapidly disappearing habitat type. Over 76% suffered from loss of habitat (for 49% this is the only threat); 16% and 11% suffered significantly from hunting and trade respectively, and 8% were threatened as a function of their restricted ranges. Roughly 30% (twice as many as in Africa) were Endangered (highest category), another 30% divided equally between Indeterminate and Vulnerable, 30% were Rare, and 10% were Insufficiently Known (lowest). Of 146 species in the two highest categories, only nine were under sufficient management regimes, 23 might already have become extinct, 16 needed immediate intervention, and 42 needed very urgent attention. Parrots (28% of New World species threatened) and cracids (26%) suffered disproportionately through the combination of habitat loss and intensive human exploitation (trade and hunting respectively). A key means of saving threatened species lies in the identification and protection of areas in which they are sympatric.

The New World, and in particular its Neotropical region, has long been recognized as holding a disproportionately large number of species. Of the world's roughly 9,500 bird species, we compute from a variety of sources that 4,130 (43%) occur in the New World (29% of the planet's land area), and 3,800 (40%) occur in the Neotropics (16% of the planet's land area).

The New World's globally threatened bird species, defined according to standard criteria of IUCN (The World Conservation Union), have been listed in six ICBP/BirdLife studies (Anon. 1964; Vincent 1966–1971; King 1978–1979; Collar and Andrew 1988; Collar et al. 1992, 1994). Over the last 30 years the list has expanded five-fold, with most growth in continental South America. It had risen to 360 species by 1988, but with the most detailed and focused review of the situation (Collar et al. 1992), which included the Neotropical Pacific and the Caribbean, the number fell to 327, of which only four occurred entirely outside the Neotropical region. Bibby (1994) showed that the 1988 and 1992 reviews differed by 141 species. Some (24 species) of the discrepancy was because of taxonomic changes or the discovery of new species, but much of it was attributable to precautionary inclusions in the 1988 list (which was in any case preliminary in nature); of 29 species considered threatened for the first time in 1992, 14 had been indicated as "near-threatened" (i.e., subjectively judged as falling close to but outside the boundary for threatened status), and 15 were omitted altogether in the 1988 review.

Eligibility for threatened status in both 1988 and 1992 was measured against presently outdated IUCN criteria, the vagueness and subjectivity of which had already led to a search for new criteria based on broadly applicable numerical thresholds (Mace and Lande 1991; Mace et al. 1993). Under the old criteria, a species was considered threatened if, by virtue of a declining world population or small range, it was somehow deemed to be at imminent or steadily increasing risk of global extinction. Under the new criteria—issued in draft to the IUCN General Assembly in January 1994 and condensed in tabular form in the introduction to Collar et al. (1994)—the same process of analysis was objectified and rendered more rigorous by the introduction of numerical thresholds for population sizes, range sizes, and decline rates. Nevertheless, use of the new criteria in 1994 largely confirmed the number and composition of the New World's threatened species in 1992. (Differences in composition from the 1994 list mostly involved transfer of species between the threatened and near-threatened lists, not movement onto or off the non-threatened list.) In this paper we analyze the extensive data-set on threatened birds presented in the 1992 volume, to identify such fundamental elements as the New World's most important countries, habitats, threats, and taxonomic groups, and hence to establish a general framework and set of priorities for individual and collective conservation efforts.

Three considerations must preface the analysis presented in this paper. First, every effort was made in Collar et al. (1992) to make the data-set as complete as possible. All traceable published and many unpublished references containing information relevant to the conservation of the species under review were consulted (the bibliography, running to 80 pages, lists over 2,600 citations). Over 550 individual correspondents, representing expertise throughout the Americas, were acknowledged. As many as 60 museums in the Americas and Europe were contacted or visited for their unpublished specimen data (a source that yielded new information for virtually every species treated). We therefore believe that the data presented in the species accounts were a reasonably accurate reflection of available knowledge up to mid-1992, and that there was broad agreement among those consulted over the selection of the species in the book.

Second, although biological criteria, as proposed by Mace and Lande (1991) and Mace et al. (1993), are fundamental to a primary assessment of extinction risk, the usefulness of the resulting species categorizations can be enhanced by information on active management or intervention affecting a species. This was recognized by Collar et al. (1992) in Appendix B, where species regarded as "Endangered" (then the highest category of threat) but under active management (e.g., Whooping Crane [*Grus americana*] and Puerto Rican Amazon [*Amazona vittata*]) were regarded as having a different level of need from those that were not, whereas species in lower categories of threat were subdivided on the basis of their presence or absence within protected areas. Species well represented in protected areas were not regarded as seriously at risk of extinction, except for those with very small populations, or which were little known, or which probably possessed some additional management need ("largely protected, but for which vigilance is needed"). Thus, presence or absence of management was a factor for the Collar et al. (1992) classification.

Third, this paper considered only threatened species, not those classified as near-threatened, because the equivalent data-sets on the latter have not been systematically assembled.

METHODS

To identify trends and patterns from various attributes of the New World's threatened birds, we used a comprehensive data-base of these species compiled by BirdLife International researchers. This includes geo-referenced point-locality data stored for all threatened Neotropical species, derived directly from Collar et al. (1992), and for all "restricted-range" species (i.e., those with historic breeding ranges of less than 50,000 km^2), taken from the BirdLife Biodiversity Project (ICBP 1992; Balmford and Long 1994; Crosby 1994; Thirgood and Heath 1994; Long et al. 1996; Stattersfield et al., in press). In the data-base we coded information on the 327 threatened New World bird species in Collar et al. (1992). Each species was scored in various ways geographically, ecologically, by threat ("etiologically"), and taxonomically, and these data were then summed to determine the importance of various elements bearing on the conservation of the species.

Geographically, each species was classified by range state, general region, range size (greater or less than 50,000 km^2), and, where appropriate, Endemic Bird Area (EBA) (areas with two or more restricted-range species entirely confined to them are referred to in the BirdLife Biodiversity Project as EBAs). To represent the patterns graphically we plotted the distribution of point-locality data for all threatened species by 1° squares and graded each square by number of species present (a density grid-cell analysis), using the Geographic Information System, Atlas*GIS.

Ecologically, species were classified by elevation and habitat (see Appendix 1 for codes). A density grid-cell analysis incorporating species habitat codings and distributional data was performed to identify the key areas for particular habitats for threatened birds. We did not attempt here to identify patterns in other ecological aspects of threatened species, such as body weight, population biology, life-history, etc.

Etiologically, species were coded by type of threat and degree of threat. A subjective distinction was made between alteration of habitat and loss of habitat, the former being broadly seen as reducing the abundance of a species and the latter as potentially eliminating it. The degree of threat was derived entirely from Appendix B of Collar et al. (1992).

Taxonomically, the threatened species were examined by family and number of species per family. Collar et al. (1992) accepted the families distinguished by Morony et al. (1975). However, our analyses use the family limits and sequence of Sibley and Monroe (1990), and their table of contents for the number of species per family. Moreover, because our data-set only covers the New World,

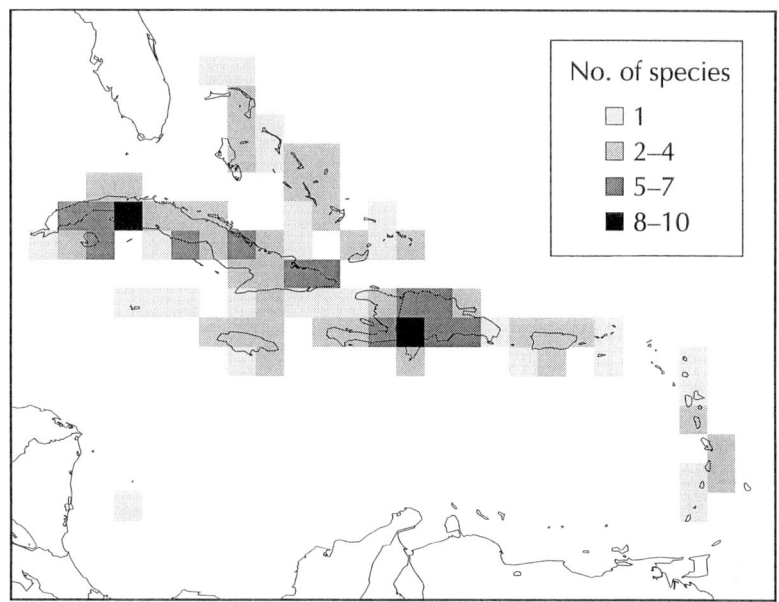

FIG. 1. The density distribution of threatened species within the Caribbean per 1° grid.

we could not—except in the case of parrots—make comparisons involving families shared with the Old World.

The evidence in Collar et al. (1992) on range, habitat and threats, and their assessment of degree of threat, was accepted without modification here, except in the three cases mentioned below. Most other codings for the present analyses required some degree of interpretation of the written information in Collar et al. (1992), sometimes supplemented by inference or assumption.

New information on threatened species steadily accumulates and can modify or render obsolete previous classifications. This has been the case since Collar et al. (1992) was sent to press in August 1992. However, here we have included new information on only three species, Buckley's Forest-falcon (*Micrastur buckleyi*), erroneously reported for Brazil based on a misidentification (Wege and Long 1995: 65); Blue-throated Macaw (*Ara glaucogularis*), rediscovered in Bolivia (Jordan and Munn 1993); and Rufous-sided Pygmy-tyrant (*Euscarthmus rufomarginatus*), recorded in Paraguay (Olrog 1979) but omitted by oversight from Collar et al. (1992). A few other minor adjustments and corrections have been made, which means that several small discrepancies exist in totals and tables in this review when checked against figures in Collar et al. (1992). Updates are given in Collar et al. (1994).

RESULTS

GEOGRAPHY: KEY AREAS

Regional patterns.—The 327 threatened birds of the Americas were distributed in the four regions as follows: North America 12, or 4% (eight shared, with one species migrating to the Pacific); Caribbean (including the Lesser Antilles, Trinidad and Tobago; [Fig. 1]—where, however, Trinidad and Tobago are excluded) 40, or 12% (seven shared); Middle America 29 (Fig. 2), or 9% (six shared); and South America 245 (Fig. 3), or 75% (seven shared). These data emphasize how Neotropical species, and in particular those from South America, dominated the list of threatened species (Fig. 4). The Andes and the relatively restricted regions of the Atlantic Forest and cis-Andean lowlands (Pacific and Caribbean), along with the islands of the Caribbean and eastern Pacific, held a disproportionately large number of threatened species. Those confined to islands represented some 15% of the total (47 species), the majority (32 species) of which were within the Caribbean basin, the rest (15 species) being on the Pacific islands of the Revillagigedo, Guadalupe, Desventuradas, Juan Fernández, and Galápagos groups.

National patterns.—Of the nations of the Americas (Appendix 2, Fig. 5), Brazil clearly contained the highest number of threatened species, with 97 (29.7%), of which 65 were endemic: the three key biomes in the country were the Atlantic forest belt (56 species), the interior dry forests and

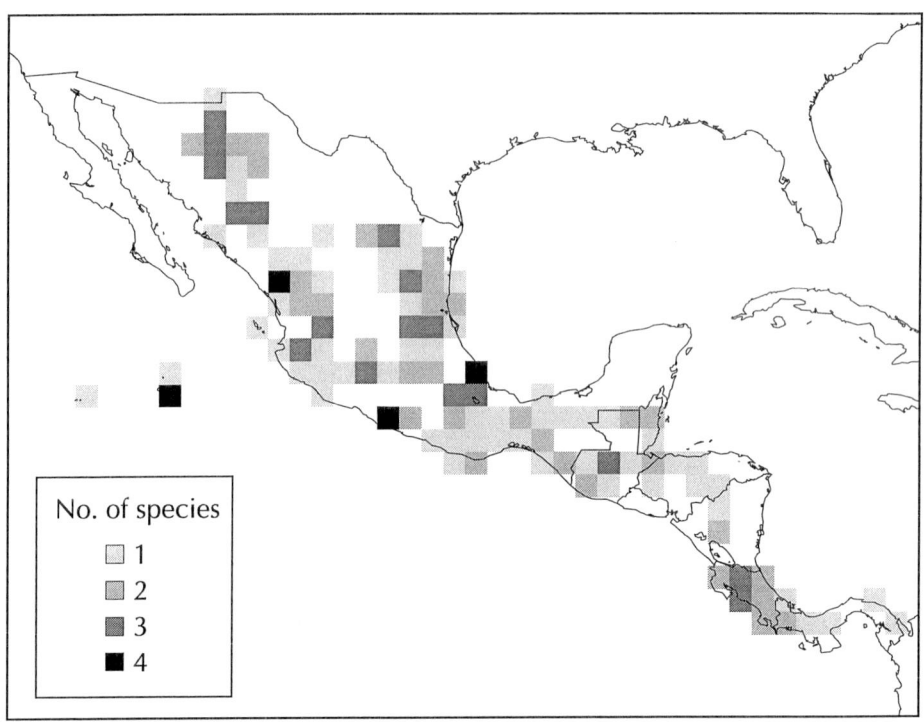

FIG. 2. The density distribution of threatened species within Middle America per 1° grid.

savannas (15 species), and the inland (wet) grasslands (15 species), with few threatened species occurring elsewhere. Only six threatened species were confined to the Amazon basin.

Peru had 64 threatened species (19.4%), of which 31 were endemic. Three main areas of endemism (EBAs) contained at least 45% of these species: the Tumbesian region of northwest Peru (at least 14 threatened species in Peru alone, one extra in Ecuador); the Peruvian High Andes (at least 13); and the Marañón valley (at least nine). Colombia had 56 threatened species, of which 31 were endemic. Ecuador and Argentina were fourth and fifth in terms of numbers of threatened bird species, but they shared substantial proportions of these forms with their various neighbours. A critical faunas analysis, whereby species on a list for one country or area are excluded if they already appear on a larger one for another (Vane-Wright et al. 1991), would take Mexico from sixth place in terms of total numbers of threatened bird species to fourth in terms of complementarity (i.e., absence of overlap) of species. Indeed, there was almost no overlap in species between any of these four countries (Colombia shares seven with Brazil or Peru; Mexico shares none), and their cumulative percentage totals reveal that Brazil alone held just under one-third, Brazil and Peru one-half, Brazil, Peru, and Colombia two-thirds, and these three plus Mexico three-quarters of all the threatened birds of the Americas.

Despite this apparent unevenness in spatial distribution, perhaps the most striking feature of the list was the occurrence of threatened bird species throughout the nations of the New World. The only country without any threatened species was El Salvador. Even the small island states and many dependencies in the Caribbean possessed them. Indeed, threatened birds extended virtually throughout all the islands, whereas substantial areas of Central and South America had none (Figs. 1–3). These trends reflect in part the vulnerability of species confined to small ranges (in particular to islands), and in part the uneven distribution of human colonization of these regions.

Biogeographic patterns.—Of the 2,609 restricted-range bird species on earth identified by ICBP (1992), 1,009 occurred in the Americas as defined by Collar et al. (1992) and were located within 79 distinct areas of endemism (EBAs). Of these 1,009, 256 (25%) were threatened. These 256 species represented 78% of the New World's threatened species (and for some 12% of these 256, the smallness of the range was itself considered a threat). An additional 66 threatened species with ranges greater than 50,000 km^2 also occurred in EBAs. Thus, 93% of the species considered by Collar et al. (1992) were to be found in EBAs (Bibby 1994).

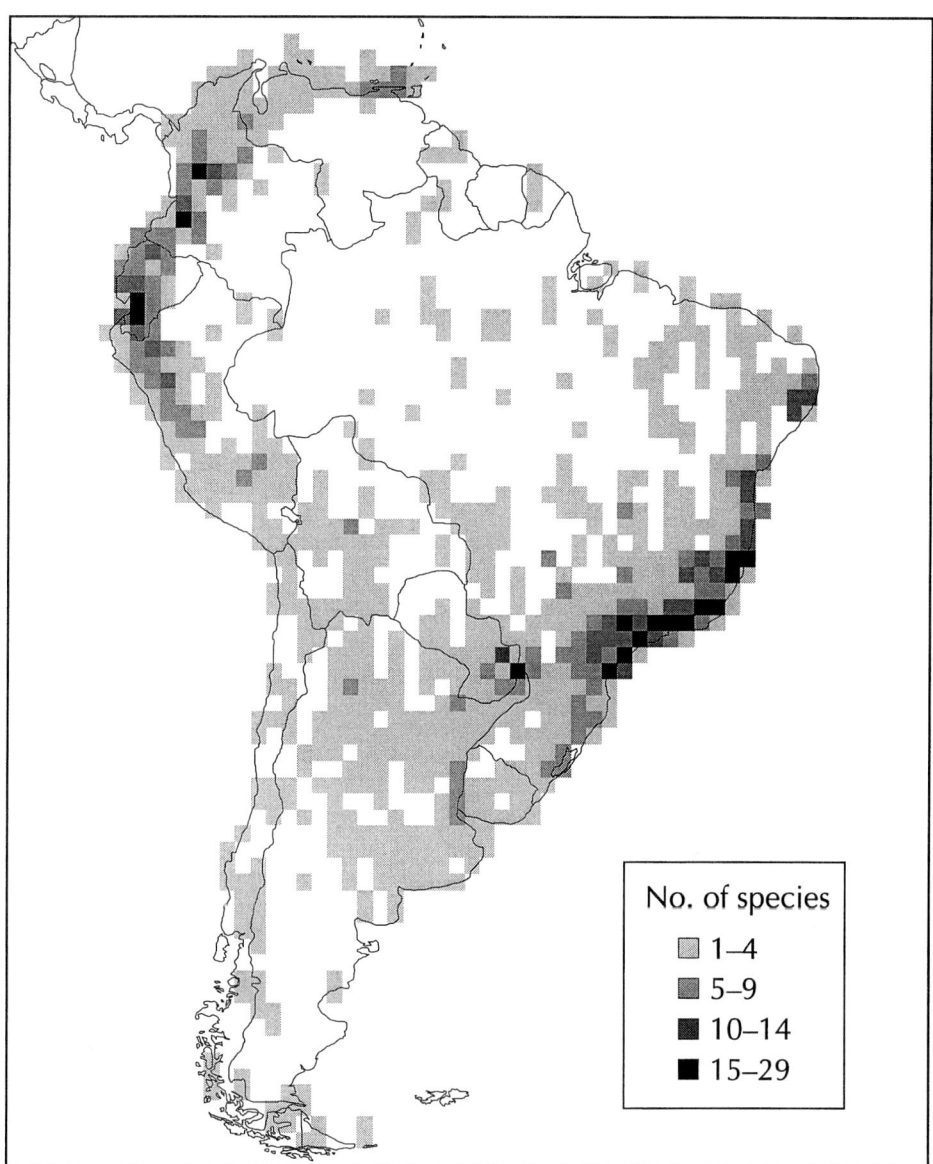

FIG. 3. The density distribution of threatened species within South America per 1° grid.

As threats were not evenly distributed across continents, some EBAs proved to be nearly or entirely composed of non-threatened species. One example of this pattern occurred in the tepuís of southern Venezuela. Another lay in Costa Rica and Panama, where five EBAs contained 100 restricted-range species, the most diverse assemblage of restricted-range species per unit area anywhere in the Neotropics; yet only seven threatened species occur in these EBAs. This disparity is due in part to the confinement of most restricted-range species (there are 52 in the Costa Rica and Panama Highlands EBA) to the relatively well-protected montane regions, where they were classified as safe (see *Etiology* below), the only exceptions being the Glow-throated Hummingbird (*Selasphorus ardens*), restricted (as far as was known in 1992) to two unprotected highland areas in Panama, and an elevational migrant, the Bare-necked Umbrellabird (*Cephalopterus glabricollis*).

There were, however, striking examples of highly threatened areas of avian endemism, most notably: the Atlantic coastal forests of Alagoas state in Brazil, which supported 14 restricted-range species and which have experienced such extensive forest fragmentation and clearance that 12 (86%)

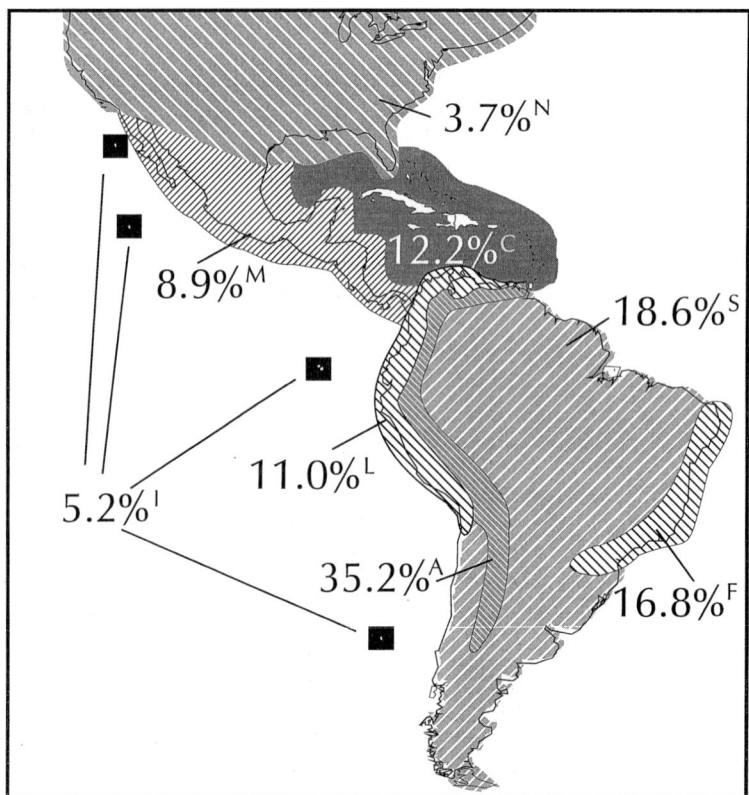

Fig. 4. Percentages of threatened species by general region in the New World (owing to overlap between areas in which the percentages do not add up to 100). The total number of species is 327. Area N, North America, holds 12 species; M, Middle America, 29; C, Caribbean, 40; I, the (Neotropical and other) Pacific islands, 17; L, the cis-Andean lowlands of South America, 36; A, the Andes, 115; S, the South American plateau and lowlands, 61; F, the Atlantic Forest region, 55. These regions correspond to those given in column 2 of Appendix 1.

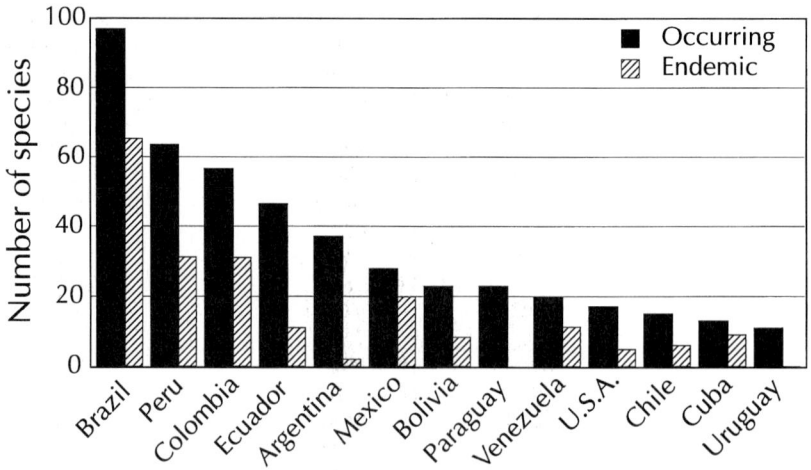

Fig. 5. Priority countries for threatened birds in the Americas. The histogram shows all countries which hold 10 or more threatened species, and indicates the proportion endemic to each country. Precise numbers of species are given in Appendix 2.

TABLE 1
Distribution of Threatened Bird Species by Elevation and Habitat. Habitat: 1 = Wet Forest; 2 = Dry Deciduous Forest (Including Scrub and Caatinga); 3 = Grassland (Including Cerrado, Chaco, Savanna, Páramo, and Montane Scrub); 4 = Wetland; 5 = Riverine (Including Riparian and Gallery Forest); 6 = Mangroves; 7 = Coastal, Marine; ? = Unknown. Elevation: L = Lowland/Tropical/0–500 m; M = Submontane/Subtropical/500–2,000 m; H = Highland/Temperate/2,000–5,000 m; I = Island

Habitat/elevation	1	2	3	4	5	6	7	?	Total
L	43	15	16	6	6	3	2	1	92
LM	33	6	6	1	3	1	2	—	52
LMH	—	2	—	—	1	—	—	—	3
M	35	11	2	—	1	—	—	1	50
MH	26	3	—	—	1	—	—	—	30
H	30	4	8	7	—	—	—	1	50
I	19	13	1	2	—	3	10	—	48
?	1	1	—	—	—	—	—	—	2
Total	187	55	33	16	12	7	14	3	327

were threatened; the Cordillera de Caripe and Paria Peninsula of northern Venezuela, which supported 14 restricted-range species, six (43%) of which were threatened; and the Tumbesian region of southwest Ecuador and northwest Peru (see above), where 55 restricted-range species occurred, of which 15 (27%) were threatened. These must stand among the most critically important targets for bird conservation in the Neotropics.

Ecology: Key Habitats

Analysis of threatened species by elevational distribution and by habitat revealed their predominance in lowland areas and wet forest (Table 1). Discounting islands, 33% of threatened birds (92/280) were tropical lowland species, 18% submontane, and 18% highland temperate forms, with the remainder less clearly defined. The three main habitat types for threatened species were wet forest, dry forest, and grasslands (Fig. 6).

Wet forest.—Over 57% (187 species) of threatened birds were confined to this broad habitat type. In South America, the Atlantic coastal forests of primarily Brazil appeared especially important, with between 15 and 24 threatened species in some of the 1° grid cells in Bahia, Espírito Santo, Rio de Janeiro, Minas Gerais, and São Paulo (Fig. 7). Also important were the humid tropical forest on both slopes of the Andes from Colombia south to central Peru. In comparison to all other habitat types, these areas supported between two and four times as many threatened species. Long (1994) provided a more detailed review of the key cloud-forest areas.

Dry forest.—Nearly 17% (55 species) of threatened birds relied on dry forest habitat types, which included deciduous forest, dry scrub and caatinga (Table 1). The most important 1° grid cells were those with between three and six species present, and were located in southwest Ecuador and northwest Peru (the Tumbesian region), and also the Río Marañón in north-central Peru (Fig. 8). Threatened species were spread thinly in the Caatinga Domain of Brazil; especially important were the dry

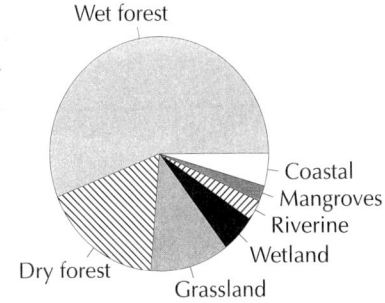

Fig. 6. Key habitats for threatened birds in the Americas. The pie chart expresses percentages based on main habitat preference as given in Appendix 1.

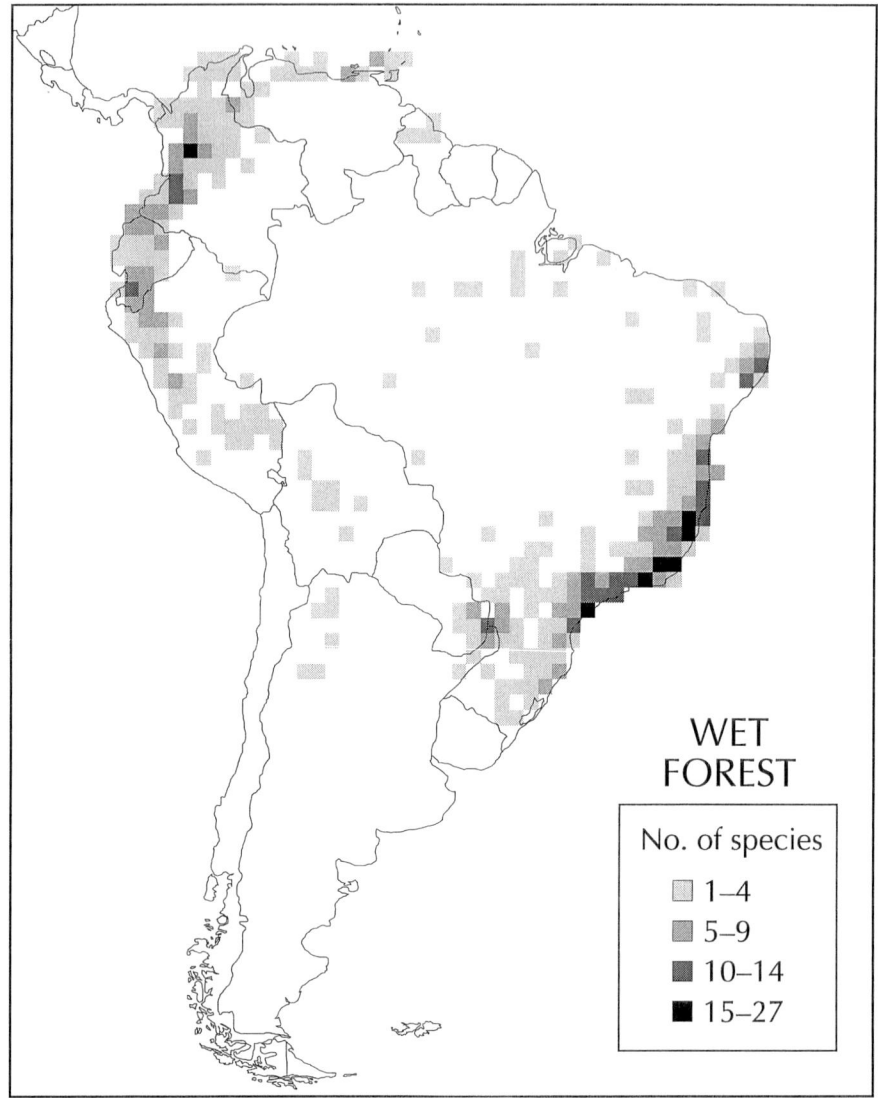

Fig. 7. The density distribution of threatened species restricted mainly to wet forest within South America per 1° grid.

forests in the central Brazilian states of Bahia, Minas Gerais, and Goiás, which have experienced extensive clearance for irrigated and dry-field agriculture, and for charcoal production for domestic and industrial fuel (Silva and Oren 1992).

Grasslands.—With 10% (33 species) of threatened birds confined to grasslands (which include campo limpo, páramo, savanna, and bushier areas such as campo sujo and open cerrado), this habitat contained over twice as many threatened species as any of the remaining non-forest habitat types. The Andean páramo/puna and Patagonian grasslands held a number of these birds (eight species occurred in páramo/puna grasslands and montane scrub: Table 1), but of greatest concern were the grasslands of southern Brazil and northern Argentina (Fig. 9). As a result of agricultural development, nearly all species endemic or near-endemic to the open vegetation of central Brazil have suffered drastic declines (e.g., Parker and Willis 1997), and a few may even be extinct through large parts of their former range. Most of the remaining campo and cerrado habitat specialists of the region would also have been considered threatened if they had not also retained reasonably healthy populations in the largely intact grasslands of north-central Bolivia (Parker et al. 1993). The wet "Mes-

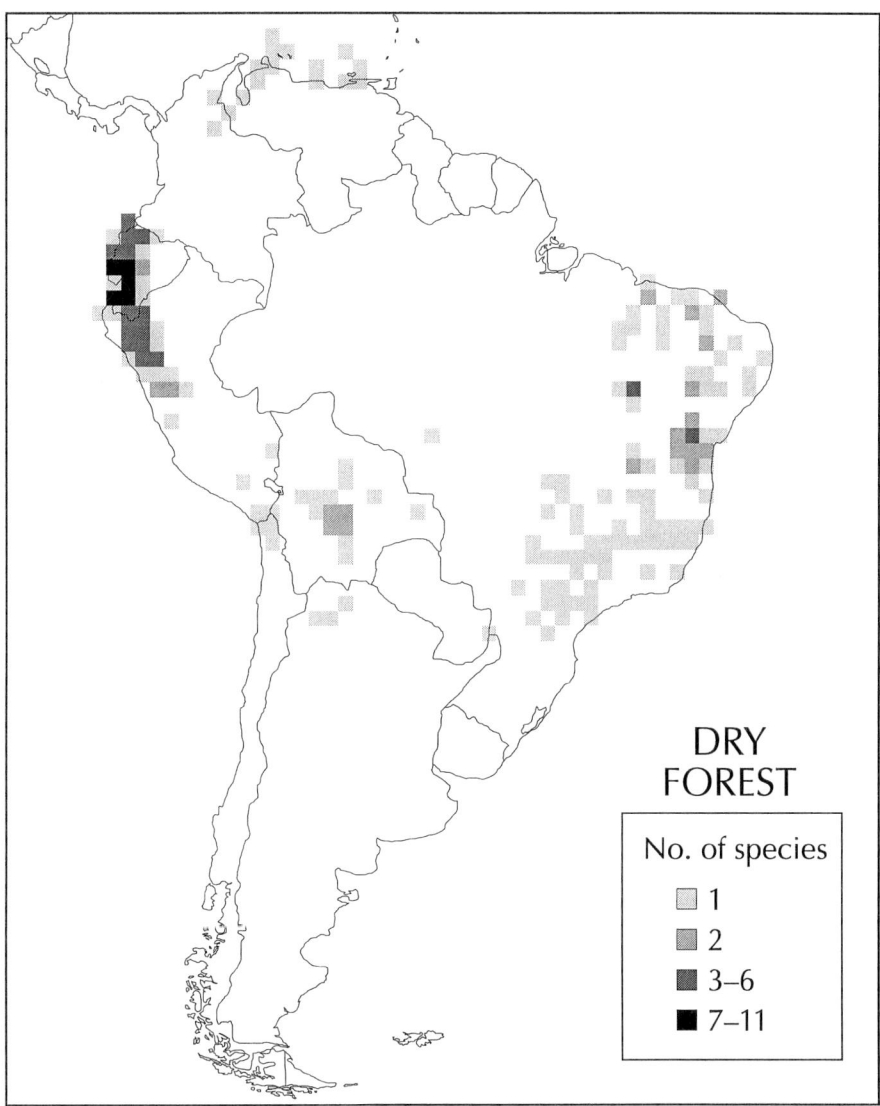

FIG. 8. The density distribution of threatened species restricted mainly to dry forest within South America per 1° grid.

opotamia" grasslands of Entre Ríos and Corrientes provinces, Argentina, held 12 threatened species, and have suffered from overgrazing and uncontrolled annual burning (Pearman and Abadie, unpubl. data). There are few protected areas in this region, all inadequately managed.

ETIOLOGY: KEY PROBLEMS

Throughout the Americas the primary threat to bird species proved to be the destruction and disturbance (or alteration) of the habitats on which their existence depends; over 76% of the threatened species were in part regarded as such because of loss of habitat, and for almost 49% of threatened birds in the New World this factor alone was the main threat (Table 2). Over 8% of bird species were threatened solely as a function of their restricted ranges (the "old" IUCN category Rare allowed for this), although all these species were in some way additionally predisposed to extinction in the face of human pressure within their ranges. Hunting and trade were important components of the threat profile for 16% and 11% of New World species respectively, but virtually

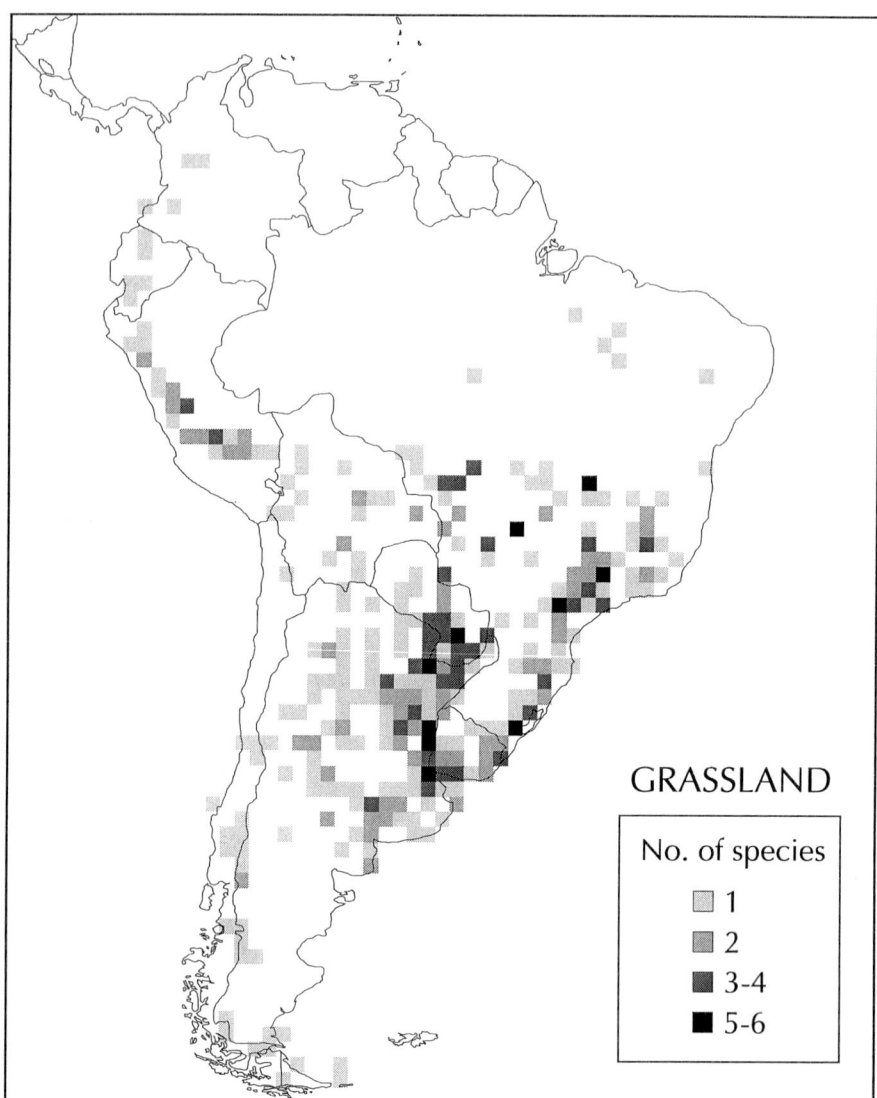

FIG. 9. The density distribution of threatened species restricted mainly to grasslands within South America per 1° grid.

never figured as the sole threat to a species. Trade nearly always occurred alongside habitat loss, whereas hunting in single combination with habitat loss accounted for 6% of all threatened birds.

The distribution of species by category of threat showed an even pattern, with almost 30% Endangered, just over 30% divided equally between Indeterminate and Vulnerable, 30% Rare, and just under 10% Insufficiently Known. The category Indeterminate was intended to register certainty that a species is threatened, but at an uncertain level; Insufficiently Known accepted that the species in question might not be threatened at all. Because a species placed in Indeterminate might therefore be Endangered, Indeterminate stood next in line and potentially nested entirely within Endangered; in other words, a precautionary view would regard as many as 45% (146) of all New World threatened species as at the highest level of risk.

TAXONOMY: KEY FAMILIES

Of 41 families represented on the threatened list, 15 were assessed for the percentage of their members at risk within the Neotropics (Table 3). *Chi*-squared tests performed on each of these 15

TABLE 2

Threatened Bird Species by Type of Threat. Column 1 = Threats in Order of Prevalence (for Explanation See Codes for Column 5 of Appendix 1). Column 2 = Total Number of Species for which the Threat is Registered; Column 3 = Number of Species for which This is the Only Threat; Column 4 = Number of Species for which This Threat Has Been Registered as Minor. "Restricted Range" Includes Species with Other Intrinsic Factors Rendering Them Susceptible to Extinction (See Number of Species)

Threat	With threat	Only threat	Minor threat
L	250 (76.5%)	159 (48.6%)	15 (4.6%)
H	51 (15.6%)	2 (0.6%)	9 (2.8%)
T	36 (11.0%)	1 (0.3%)	7 (2.1%)
R	31 (9.5%)	27 (8.3%)	—
I	21 (6.4%)	8 (2.4%)	11 (3.4%)
U	13 (4.0%)	11 (3.4%)	—
P	12 (3.7%)	1 (0.3%)	4 (1.2%)
A	11 (3.4%)	—	2 (0.6%)
N	10 (3.1%)	—	1 (0.3%)
B	7 (2.1%)	1 (0.3%)	1 (0.3%)
D	7 (2.1%)	—	—
E	4 (1.2%)	1 (0.3%)	1 (0.3%)
C	2 (0.6%)	—	1 (0.3%)
F	1 (0.3%)	—	—
G	1 (0.3%)	—	—
K	1 (0.3%)	—	—
O	1 (0.3%)	—	—

families showed that only three varied significantly from expected (on the basis of 8% of bird species in the New World being threatened: see Discussion). The parrot family had the greatest proportion (28%) of species (39 out of a New World complement of around 140) in the threatened category ($\chi^2 = 39.52$, d.f. = 1, $p < 0.001$). Evaluation elsewhere of virtually the same figures (Collar and Juniper 1992) indicated that this percentage stemmed from the potent combination of habitat loss and direct human exploitation in the form of trade, a trend also observed for Venezuela (Desenne and Strahl 1991). The next most vulnerable group, by percentage, was the Cracidae, with 13 of 50 species (26%) threatened ($\chi^2 = 10.25$, d.f. = 1, $p < 0.005$). Habitat loss and another form of direct human exploitation—hunting—were the key factors in the decline of this group. The family with the next highest percentages of threatened species were the tinamous (15%), which also faced loss of habitat (in many cases, grasslands) and pressure from hunting.

Families with below-normal percentages of threatened species were the Ramphastidae (3.5%) and Vireonidae (4%), although it is clear that if figures for certain taxonomic groupings now regarded as subfamilies were to be included (the "old" Tyrannidae, for example), or if figures were easily available (as for parrots) to compute the New World numbers of widespread families, many other taxonomic groupings would emerge with fewer than expected threatened species. All would tend to be relatively large (e.g., with over 40 members) and characterized by their mobility and wide-ranging behaviour, rendering them less prone to confinement and speciation within small areas (a circumstance already indicated above as a liability).

DISCUSSION

A detailed analysis of Africa and related islands (Collar and Stuart 1985) resulted in the identification of 172 threatened bird species; allowing roughly 1,500 species in Afrotropical Africa and 200 for related islands, and excluding the Palearctic African mainland, some 10% of the region's avifauna thereby emerged as at risk. A general review at the global level (Collar and Andrew 1988) indicated that 1,029 (11%) of all bird species are threatened. If these trends were to be repeated in the New World, then over 400 species should qualify for listing; in fact, Collar et al. (1992) considered only 327 to be threatened, which is 8% of the New World avifauna.

However, although 8% is slightly less than the global level of threatened birds, these 327 threatened species still represent around 30% of the global total. Given that the Nearctic contributes only 12 species (six of which migrate to the Neotropics and two of which are themselves primarily Neotropical), the dimensions of the crisis in bird conservation in the Neotropics could scarcely be more

TABLE 3
NUMBERS OF THREATENED BIRD SPECIES BY FAMILY. NUMBERS IN BRACKETS REFER TO THE TOTAL NUMBER OF SPECIES IN THAT FAMILY, JUDGED (EXCEPT IN THE CASE OF PARROTS) ONLY FOR FAMILIES CONFINED TO THE NEW WORLD (SEE METHODS)

Fringillidae 55	Falconidae 2
Psittacidae 39 (140 in New World) 28%	Galbulidae 2 (18) 11%
Tyrannidae 34 (537) 6.5%	Muscicapidae 2
Trochilidae 29 (319) 9%	Odontophoridae 2 (31) 6.5%
Furnariidae 24 (280) 8.5%	Podicipedidae 2
Thamnophilidae 22 (188) 12%	Ramphastidae 2 (55) 3.5%
Cracidae 13 (50) 26%	Scolopacidae 2
Rallidae 10	Vireonidae 2 (51) 4%
Columbidae 10	Anatidae 1
Passeridae 10	Apodidae 1
Procellariidae 9	Charadriidae 1
Formicariidae 7 (56) 12.5%	Ciconiidae 1
Tinamidae 7 (47) 15%	Cinclidae 1
Accipitridae 6	Dendrocygnidae 1
Caprimulgidae 5	Gruidae 1
Picidae 5	Laridae 1
Certhiidae 4	Momotidae 1 (9) 11%
Rhinocryptidae 3 (28) 11%	Phalacrocoracidae 1
Sturnidae 3	Strigidae 1
Coccyzidae 2 (18) 11%	Trogonidae 1
Corvidae 2	

striking (Collar 1992). This perception is greatly reinforced by the fact that in Africa only 16% of threatened species were placed in the highest category, Endangered (Collar and Stuart 1985), whereas almost 30% of the New World's threatened avifauna were given that status in 1992 (Table 4). Moreover, whereas no extinctions have been recorded in continental Africa since 1800, the three mainland American regions have experienced several each (Labrador Duck [*Camptorhynchus labradorius*], Passenger Pigeon [*Ectopistes migratorius*], and Carolina Parakeet [*Conuropsis carolinensis*] in North America, plus the Ivory-billed Woodpecker [*Campephilus principalis*] shared with Cuba; Atitlán Grebe [*Podilymbus gigas*] and Slender-billed Grackle [*Cassidix palustris*] in Middle America; and almost certainly both Colombian Grebe [*Podiceps andinus*] and Glaucous Macaw [*Anodorhynchus glaucus*] in South America), plus three extinctions in the wild (California Condor [*Gymnogyps californianus*], Socorro Dove [*Zenaida graysoni*], and Alagoas Curassow [*Mitu mitu*]).

Habitat destruction has been repeatedly identified as the cardinal cause of endangerment. King (1978) found that it affected 65.3% of all threatened birds, whereas Temple (1986) estimated the figure to be 82%, so the 75% level found in this study, targeted only on the Americas, indicates the evenness of the effect across the planet. Similarly, King (1978) and Temple (1986) reported substantial percentages of forest-dwellers among birds at risk globally, although (owing to the influence of non-forest oceanic island species) not as high as in the Americas, where the number of threatened bird species confined to wet (187) and dry (55) forest (Table 1) means that 242 (74%) are in forest habitats.

That a high proportion of the New World's threatened birds should occur in forest (the great

TABLE 4
DISTRIBUTION OF THREATENED BIRD SPECIES IN AFRICA AND THE AMERICAS BY PERCENTAGE IN EACH CATEGORY OF THREAT. DATA FOR AFRICA ARE TAKEN FROM STUART AND COLLAR (1988)

Category	Africa (%)	Americas (%)
Endangered	16.0	29.2
Indeterminate	18.0	15.2
Vulnerable	9.0	16.4
Rare	45.5	29.8
Insufficiently Known	10.5	9.4

majority of the Neotropics having originally been forested), and that a high proportion of them should be threatened because the forest is being destroyed (given the degree of endemism within relatively small areas of forest in areas long exploited by human colonists), is unsurprising. More significant is the emergence of Neotropical grasslands as a major conservation issue (Collar 1992; Parker and Willis 1997). With as many as 33 grassland species at risk (10% of all threatened American birds), action to conserve substantial tracts of such habitat is long overdue. Collar et al. (1992: 35) emphasized the extremely rapid (post-1960) and near-total conversion of open Brazilian grasslands to large-scale agriculture, promoted by new techniques such as liming to cure aluminium toxicity and acid soils, and involving the planting of exotic vegetation (eucalyptus, pines) and crops (sugarcane, soybeans). More than 95% of potential arable or stock-raising land has probably already been appropriated or otherwise thoroughly degraded. Relatively pristine tracts of upland grassland in Brazil south of 15°S may now be confined to five national parks and a small number of other reserves, and most natural grassland vegetation elsewhere in the country seems likely to disappear altogether by the end of the century (see also Parker and Willis 1997). The situation in Argentina appears no better.

The spread of agriculture throughout the Americas brings a further threat in the form of increased levels of brood-parasitism from Brown-headed (*Molothrus ater*) and Shiny Cowbirds (*M. bonariensis*). Only seven species were identified as at risk from this threat (Table 2), but many more may be so in the future. Only the U.S.A. has the financial and technical resources to contain the cowbird threat at critical sites. For species like the Montserrat Oriole (*Icterus bonana*) and the Cipó Canastero (*Asthenes luizae*), the chances of such intervention appear negligible. The steady spread of cowbirds through the Caribbean and South America, in the wake of continuing forest clearance, could result in the endangerment of many restricted-range passerines.

That 27 species are threatened only as a function of restriction of range (Table 2) illustrates the inherent susceptibility of such species to events whose occurrence may be unpredictable, but whose impact is predictably serious: all 27 demand constant vigilance as the price of their survival. Their selection was, however, based on the fact (unstated in Collar et al. 1992) that they possessed other features, relating to their biology or the potential of their habitat for human damage or exploitation, which magnified their inherent precariousness. Even so, range-size as a major threat (that is, significant enough to cause a species to be treated as threatened) is not fully reflected in the number of species (31) listed as affected by it because this was a common (again unstated) consideration in the "Threats" paragraphs of the 1992 review. That 256 threatened species also have ranges of less than 50,000 km^2 is a much better indication of the decisive importance of range size in shaping judgment of conservation status.

The 27 species for which range size was sufficient in itself as a threat were in "islands" of habitat, in most cases (24) on continents. However, an important 15% of all threatened New World species were on genuine islands, two-thirds (67%) in the Caribbean, where Cuba and Hispaniola emerged as considerably important (their combined totals surpassing those of both Venezuela and the U.S.A.). The influence of introduced predators and competitors came most strongly into play for island species, but they were also the species to which the largest numbers of threats apply. Although in some cases this latter phenomenon may have been an artifact of the intensity with which island species have been studied, extinction vortices are certainly more pronounced on islands, where restriction of range and susceptibility to introductions compound the more usual pressures applied by human inhabitants, themselves constrained by the area of the island. Important lessons are to be learned from the plight of such species—exemplified by the Puerto Rican Amazon, as definitively documented by Snyder et al. (1987)—and they will increasingly emerge as critical models on which continental conservation efforts must draw.

Perhaps the most disturbing point to emerge from Collar et al. (1992) is simply the degree or potential degree of endangerment in the New World avifauna (Table 5). Of 146 species actually or possibly in the highest rank of threat (i.e., Endangered and Indeterminate combined), only nine were receiving management sufficient to bring them slowly towards recovery; 23 were so rare that they could not be found and may be extinct, 16 were so critical that they needed immediate intervention, 42 were in urgent need of attention, and six, although almost certainly extant, first needed to be found before they could be helped. A few of the 23 that may already be gone (E/Ex[4]) simply required, in the first place, focused effort to determine their status. Prime amongst these was the Imperial Woodpecker (*Campephilus imperialis*) of the Sierra Madre Occidental in Mexico. In the next category down (E[1]), possibly only one bird, the Junín Grebe (*Podiceps taczanowskii*) of one lake in the Peruvian Andes, was slipping beyond the reach of human capability to reclaim; some others may have required little more than solid, sensible study and effort to turn their fortunes around, most

TABLE 5

NUMBERS OF THREATENED BIRD SPECIES BY CATEGORY OF THREAT (IN WHICH V/R IS CLASSIFIED AS A SUBSET OF R). THE FIGURES IN PARENTHESES REPRESENT THE PERCENTAGES OF THE TOTAL NUMBER OF THREATENED SPECIES (327) IN THE AMERICAS. SUPERSCRIPT NUMBERS REPRESENT THE 12 SUBDIVISIONS OF THE OLD IUCN CATEGORIES OF THREAT MADE BY COLLAR ET AL. (1992) TO DISTINGUISH DEGREES AND TYPES OF URGENCY

E	ENDANGERED	96 (29.4%)
E^1	Species at critically low levels that need immediate intervention	16 (4.9%)
E^2	Species in urgent need of attention	42 (12.8%)
E^3	Species in urgent need of attention when found	6 (1.8%)
E/Ex^4	Species in urgent need of attention if found	23 (7.0%)
E^5	Species receiving urgent attention	9 (2.8%)
I	INDETERMINATE	50 (15.3%)
I^6	Species needing urgent attention if taxonomic status confirmed	6 (1.8%)
I^7	Species for which evidence is conflicting: possibly urgent	37 (11.3%)
I^8	Species possibly in need if and when relocated	7 (2.1%)
V	VULNERABLE	52 (15.9%)
V^9	Species largely unprotected and in need of attention	52 (15.9%)
R	RARE	98 (30.0%)
V/R^{10}	Species with populations only partly protected	79 (24.2%)
R^{11}	Species largely protected but vigilance needed	19 (5.8%)
K	INSUFFICIENTLY KNOWN	31 (9.5%)
K^{12}	Species for which further protection is desirable	31 (9.5%)

notably perhaps—if only because of its taxonomic distinctiveness—the White-breasted Thrasher (*Ramphocinclus brachyurus*) of St Lucia and Martinique.

Another analysis of Collar et al. (1992) showed that, despite all the material mustered, remarkably few data existed for many species (Bibby 1994). Distribution tended to be the best known parameter, but many species had been recorded from a relatively small number of localities and could only be presumed to occur in nearby or intervening areas of similar habitat. Less than 25% of threatened species had been subject to any formal population estimate, the majority being assumed to be rare because they were infrequently seen within what were presumed to be their limited ranges. Formal estimates of *trends* in populations were even less common, because even fewer species were censused 20 years ago (trends were almost always informally inferred from qualitative reports of habitat loss). Straightforward surveys of threatened species are therefore critically important first steps towards establishing whether and where populations are viable, with ecological studies likely to enhance the management process by diagnosing causes of decline and proposing remedies.

Brazil, Peru, Colombia, and Mexico have all been identified as "megadiversity" countries that deserve the highest priority for conservation (Mittermeier 1988). Our analysis supports this view from the perspective of threatened species, but greatly refines the focus for future work by highlighting particular areas, habitats, issues and, in some cases, species that need to be targeted within these countries if the deployment of large-scale "biodiversity" funds is to have an appropriate impact. All the evidence indicates that the most frequent pattern for threatened birds in the Neotropics combines restricted range, a degree of habitat loss, and limited representation within protected areas. The effect of hunting and trade is not only to intensify the degree of threat but also to draw in other, often more wide-ranging types of species, notably parrots and cracids, by amplifying the effects of habitat loss on them.

Common sense and economic interests require conservationists to develop initiatives that cater to as many species as possible, chiefly by targeting sites of sympatric occurrence of threatened or locally endemic forms as the units of concern (Collar 1987; Collar et al. 1987; Collar and Stuart 1988; ICBP 1992; Wege and Long 1995). Thus, roughly 90% of threatened birds in the Americas could be largely secured by the maintenance of sufficient natural habitat in EBAs, and indeed site conservation is central to the future of all the New World's threatened species: over 70% of the individual recommendations made in Collar et al. (1992)—1,045 for the 327 species—involved a set of four site-specific proposals, namely, to find new sites, to protect known sites, to manage them, and to enumerate populations within them; only five species accounts made no site-oriented recommendations (Bibby 1994).

The existence of EBAs as targets for conservation does not, however, make further work on documenting threatened species redundant. To begin with, many threatened species by their nature have specific requirements that would not necessarily be satisfied by general habitat or site management within an EBA. Moreover, it is the evaluation of the status of the restricted-range species in the Red Data Book that forms the best guide to the priority order in which EBAs should be addressed. Indeed, it is the establishment of the precise distributions of and threats to the restricted-range species that forms the best guide to the priority areas *within* EBAs to target for survey and protection. Finally, of course, threatened species that do not have restricted ranges (in this case 71, of which 16, marked in Appendix 1, were not recorded within any EBA) cannot simply be ignored.

Nevertheless, conservation of threatened species in areas of their sympatry is an essential way of coming to terms with the myriad numbers of species claiming the attention of financially and logistically constrained environmental managers and decision-makers. An analysis of the distributions of all mainland Neotropical threatened bird species, as given in Collar et al. (1992), has permitted the identification of a suite of such sites, which represent priority areas for the conservation of threatened birds in the region (Wege and Long 1995).

Postscript.—Ted Parker, co-author of *Threatened birds of the Americas*, the book this paper analyzes, was a tireless advocate of action for conservation. He would have been moved to know from the number of papers and authors in this volume just how much he was loved and respected. It is a volume that helps us come to terms with our feelings of loss and grief, and it gives us a chance to add our individual voices in one solid, collective testimonial. We all know, however, that this book is not the monument Ted would have wanted. What he wanted, and with his startling energy and devotion he might have done most of it single-handedly, was to find and save the Imperial Woodpecker, assure the conservation of Murici in northeast Brazil, see gazetted the new key areas he was finding in Bolivia, fight to save the grasslands of the South American planalto, and have a guiding hand in all the other tasks needed to prevent not only the extinction of any species of wildlife but also the loss of any major tracts of wilderness in the New World. Ted's monument, no less, will be the preservation of the Neotropical avifauna. If he is to rest in peace, we have everything yet to do.

ACKNOWLEDGMENTS

F. E. Hayes and R. S. Ridgely combined to point out the Paraguayan record of *Euscarthmus rufomarginatus*. H. van Dijkhuisen designed the software that allowed us to perform the grid-cell analyses; J. H. Fanshawe and A. Payne facilitated the statistical analysis. C. J. Bibby, M. J. Crosby, M. G. Kelsey, and A. J. Stattersfield reviewed the typescript before submission, and A. W. Kratter and D. F. Stotz were constructive referees, the latter contributing many changes to our original codings in Appendix 1. J. V. Remsen and J. M. Hagan greatly improved the text.

LITERATURE CITED

ANON. 1964. List of rare birds, including those thought to be so but of which detailed information is still lacking. International Union for Conservation of Nature and Natural Resources, Morges, Switzerland.

BALMFORD, A., AND A. J. LONG. 1995. Across-country analyses of biodiversity congruence and current conservation effort in the tropics. Conserv. Biol. 9:1539–1547.

BIBBY, C. J. 1994. Recent past and future extinctions in birds. Phil. Trans. Royal Soc. London 343:35–40.

COLLAR, N. J. 1987. Red Data Books and national conservation strategies. World Birdwatch 9(2):6–7.

COLLAR, N. 1992. A Red Data Book for the Americas. World Birdwatch 14(4):8–9.

COLLAR, N. J., AND P. ANDREW. 1988. Birds to watch: the ICBP world list of threatened birds. International Council for Bird Preservation, Techn. Publ. 8, Cambridge, U.K.

COLLAR, N. J., M. J. CROSBY, AND A. J. STATTERSFIELD. 1994. Birds to Watch 2: The World List of Threatened Birds. BirdLife International (BirdLife Conservation Series 4), Cambridge, U.K.

COLLAR, N. J., T. J. DEE, AND P. D. GORIUP. 1987. La conservation de la nature à Madagascar: la perspective du CIPO. Pp. 97–108 in Priorités en matière de conservation des espèces à Madagascar (R. A. Mittermeier, L. H. Rakotovao, V. Randrianasolo, E. J. Sterling, and D. Devitre, Eds.). International Union for Conservation of Nature and Natural Resources, Occas. Pap. 2, Gland, Switzerland.

COLLAR, N. J., L. P. GONZAGA, N. KRABBE, A. MADROÑO NIETO, L. G. NARANJO, T. A. PARKER III, AND D. C. WEGE. 1992. Threatened Birds of the Americas: The ICBP/IUCN Red Data Book, 3rd ed., part 2. International Council for Bird Preservation, Cambridge, U.K.

COLLAR, N. J., AND A. T. JUNIPER. 1992. Dimensions and causes of the parrot conservation crisis. Pp. 1–24 in New World Parrots in Crisis: Solutions from Conservation Biology (S. R. Beissinger and N. F. R. Snyder, Eds.). Smithsonian Institution Press, Washington, D.C.

COLLAR, N. J., AND S. N. STUART. 1985. Threatened Birds of Africa and Related Islands: The ICBP/IUCN Red Data Book, 3rd ed., part 1. International Council for Bird Preservation, and International Union for Conservation of Nature and Natural Resources, Cambridge, U.K.

COLLAR, N. J., AND S. N. STUART. 1988. Key forests for threatened birds in Africa. International Council for Bird Preservation, Monogr. 3, Cambridge, U.K.

CROSBY, M. J. 1994. Mapping the distributions of restricted-range birds to identify global conservation priorities. Pp. 145–154 in Mapping the Diversity of Nature (R. I. Miller, Ed.). Chapman and Hall, London.

DESENNE, P., AND S. D. STRAHL. 1991. Trade and the conservation status of the family Psittacidae in Venezuela. Bird Conserv. International 1:153–169.

ICBP. 1992. Putting Biodiversity on the Map: Priority Areas for Global Conservation. International Council for Bird Preservation, Cambridge, U.K.

JORDAN, O. C., AND C. A. MUNN. 1993. First observations of the Blue-throated Macaw in Bolivia. Wilson Bull. 105:694–695.

KING, W. B. 1978. Endangered birds of the world and current efforts toward managing them. Pp. 9–17 in Endangered Birds: Management Techniques for Preserving Threatened Species (S. A. Temple, Ed.). University of Wisconsin Press, Madison, Wisconsin; and Croom Helm, London.

KING, W. B. 1978–1979. Red Data Book, 2: Aves, 2nd ed. International Union for Conservation of Nature and Natural Resources, Morges, Switzerland. [Reissued as King, W. B. 1981. Endangered Birds of the World: The ICBP bird Red Data Book. Smithsonian Institution Press and International Council for Bird Preservation, Washington, D.C.]

LONG, A. J. 1994. Restricted-range and threatened bird species in tropical montane cloud forests. Pp. 47–65 in Tropical Montane Cloud Forests (L. S. Hamilton, J. O. Juvik, and F. N. Scatena, Eds.). East-West Center, Honolulu.

LONG, A. J., M. J. CROSBY, A. J. STATTERSFIELD, AND D. C. WEGE. 1996. Towards a global map of biodiversity: patterns in the distribution of restricted-range birds. Global Ecol. and Biogeogr. Letters 5:281–304.

MACE, G., N. COLLAR, J. COOKE, K. GASTON, J. GINSBERG, N. LEADER-WILLIAMS, M. MAUNDER, AND E. J. MILNER-GULLAND. 1993. The development of new criteria for listing species on the IUCN Red List. Species 19:16–22.

MACE, G. M., AND R. LANDE. 1991. Assessing extinction threats: toward a reevaluation of IUCN threatened species categories. Conserv. Biol. 5:148–157.

MITTERMEIER, R. A. 1988. Primate diversity and the tropical forest: case studies from Brazil and Madagascar and the importance of megadiversity countries. Pp. 145–154 in Biodiversity (E. O. Wilson, Ed.). National Academy Press, Washington, D.C.

MORONY, J. J., W. J. BOCK, AND J. FARRAND. 1975. Reference List of Birds of the World. American Museum of Natural History, Department of Ornithology, New York.

OLROG, C. C. 1979. Notas ornitológicas, XI: sobre la colección del Instituto Miguel Lillo. Acta Zool. Lilloana 33:5–7.

PARKER, T. A., III, A. H. GENTRY, R. B. FOSTER, L. EMMONS, AND J. V. REMSEN, JR. 1993. The Lowland Dry Forests of Santa Cruz, Bolivia: a Global Conservation Priority. Conservation International, Washington, D.C.

PARKER, T. A., III, AND E. O. WILLIS. 1997. Notes on three tiny grassland flycatchers, with comments on the disappearance of South American fire-diversified savannas. Pp. 549–555 in Studies in Neotropical Ornithology Honoring Ted Parker (J. V. Remsen, Jr., Ed.), Ornithol. Monogr. No. 48.

SIBLEY, C. G., AND B. L. MONROE, JR. 1990. Distribution and Taxonomy of Birds of the World. Yale University Press, New Haven.

SILVA, J. M. C. DA, AND D. C. OREN. 1992. Notes on *Knipolegus franciscianus* Snethlage, 1928 (Aves: Tyrannidae), an endemic of central Brazilian dry forests. Bol. Mus. Paraense Emílio Goeldi 16:1–8.

SNYDER, N. F. R., J. W. WILEY, AND C. B. KEPLER. 1987. The Parrots of Luquillo: Natural History and Conservation of the Puerto Rican Parrot. Western Foundation of Vertebrate Zoology, Los Angeles.

STATTERSFIELD, A. J., M. J. CROSBY, A. J. LONG, AND D. C. WEGE In press. Endemic Bird Areas of the World: Priorities for Biodiversity Conservation. BirdLife International, BirdLife Conservation Series No. 7, Cambridge, U.K.

STUART, S. N., AND N. J. COLLAR. 1988. Birds at risk in Africa and related islands: the causes of their rarity and decline. Proc. VI Pan-Afr. Orn. Congr.: 1–25.

TEMPLE, S. A. 1986. The problem of avian extinctions. Pp. 453–485 in Current Ornithology, 3 (R. F. Johnston, Ed.). Plenum Press, New York.

THIRGOOD, S. J., AND M. F. HEATH. 1994. Global patterns of endemism and the conservation of biodiversity. Pp. 207–227 in Systematics and Conservation Evaluation (P. L. Forey, C. J. Humphries, and R. I. Vane-Wright, Eds.). Clarendon Press, Systematics Association Special Volume 50, Oxford.

VANE-WRIGHT, R. I., C. J. HUMPHRIES, AND P. H. WILLIAMS. 1991. What to protect?—Systematics and the agony of choice. Biol. Conserv. 55:235–254.

VINCENT, J. 1966–1971. Red Data Book, 2. Aves. International Union for Conservation of Nature and Natural Resources, Morges, Switzerland.

WEGE, D. C., and A. J. LONG. 1994. Key Areas for Threatened Birds in the Neotropics. BirdLife International, BirdLife Conservation Series No. 5, Cambridge, U.K.

APPENDIX 1
THREATENED BIRD SPECIES OF THE AMERICAS, ACCORDING TO COLLAR ET AL. (1992)

Species are arranged by family and sequence according to Sibley and Monroe (1990); line spaces within families define previously recognized families. Codes for this appendix are as follows:

An asterisk (*) indicates species without restricted ranges (i.e., greater than 50,000 km^2), including seabirds. A sword (†) indicates restricted-range species that are not in endemic bird areas (EBAs). All other species on this list therefore have restricted ranges and occur in EBAs.

Region (Column 2) (The regions in column 2 correspond to those mapped in Figure 4.): n = North America; M = Middle America; C = Caribbean; I = Neotropical Pacific Islands; L = Pacific/Caribbean Cis-Andean lowlands; A = Andes; S = South American interior lowlands; F = Atlantic Forest.

Habitat (Column 3): 1 = wet forest; 2 = dry deciduous forest (including scrub and caatinga); 3 = grassland (including cerrado, chaco, savanna, páramo, and montane scrub); 4 = wetland; 5 = riverine (including riparian and gallery forest); 6 = mangroves; 7 = coastal, marine; ? = unknown.

Elevation (Column 4): L = lowland/tropical/0–500 m; M = submontane/subtropical/500–2,000 m; H = highland/temperate/2,000–5,000 m; I = island.

Threats (Column 5) (note there is no J, M, Q or S in this list): A = alteration of habitat (including water perturbation); B = brood-parasitism; C = competition (resulting from man-induced imbalances); D = disturbance; E = egg-collecting; F = fish-net accidents; G = global warming; H = hunting (including persecution as pest); I = introduced predators or competitors (incl. fish); K = collision with powerlines; L = loss of habitat; n = natural disasters (eruptions, erosion, storms); O = overexploitation (including of food resource); P = pollution, pesticides; R = range restriction; T = trade; U = unknown.

In the elevation and threat columns, alphabetical order of codes indicates order of importance, and lower case indiciates minor importance.

Species	Region	Habitat	Elevation	Threat
Tinamidae				
Tinamus osgoodi	A	1	M	LH
Crypturellus kerriae	M, L	1	LM	Lh
Crypturellus saltuarius	L	2	L	LH
Nothoprocta taczanowskii	A	3	H	LH
Nothoprocta kalinowskii	A	3	H	Ha
Nothura minor*	S	3	LM	L
Taoniscus nanus	S	3	LM	L
Cracidae				
Penelope barbata	A	1	MH	LH
Penelope perspicax	A	1	MH	LH
Penelope albipennis	L	2	L	LH
Penelope ochrogaster*	S	5	L	HL
Pipile pipile†	C	1	I	LH
Pipile jacutinga*	F	1	LM	LH
Oreophasis derbianus	M	1	H	LH
Mitu mitu	F	1	L	LTHp
Pauxi pauxi	A	1	M	LH
Pauxi unicornis	A	1	M	H
Crax alberti	L	1	LM	LHt
Crax globulosa*	S	5	L	H
Crax blumenbachii	F	1	L	LHT
Odontophoridae				
Dendrortyx barbatus	M	1	MH	LH
Odontophorus strophium	A	1	M	LH
Dendrocygnidae				
Dendrocygna arborea*	C	4	I	LAHPET
Anatidae				
Mergus octosetaceus*	F	5	LM	LAHPT
Picidae				
Picoides borealis*	N	1	L	LN
Colaptes fernandinae	C	2	I	L
Dryocopus galeatus*	F	1	L	L
Campephilus imperialis	M	1	H	LH
Campephilus principalis*	C, N	1	IL	L
Ramphastidae				
Capito hypoleucus	L	1	M	L
Aulacorhynchus huallagae	A	1	H	R
Galbulidae				
Jacamaralcyon tridactyla	F	5	LM	L
Galbula pastazae	A	1	M	L
Trogonidae				
Euptilotis neoxenus	M	1	MH	L

APPENDIX 1
CONTINUED

Species	Region	Habitat	Elevation	Threat
Momotidae				
Electron carinatum*	M	1	L	L
Coccyzidae				
Hyetornis rufigularis	C	2	I	Upl
Neomorphus radiolosus	A	1	M	L
Psittacidae				
Anodorhynchus hyacinthinus*	S	3	L	TLH
Anodorhynchus leari	S	2	L	LTHDN
Anodorhynchus glaucus*	S	5	L	LH
Cyanopsitta spixii	S	2	L	TLRi
Ara glaucogularis	S	3	L	T
Ara rubrogenys	A	2	M	THL
Guaruba guarouba*	S	1	L	LTH
Aratinga brevipes	I	1	I	R
Aratinga euops	C	2	I	LHT
Aratinga auricapilla*	F	2	L	LT
Leptosittaca branickii*	A	1	H	LH
Ognorhynchus icterotis	A	1	MH	L
Rhynchopsitta pachyrhyncha	M, N	1	MH	LTh
Rhynchopsitta terrisi	M	1	MH	L
Pyrrhura cruentata	F	1	L	Lt
Pyrrhura orcesi	A	1	M	L
Pyrrhura albipectus	A	1	M	L
Pyrrhura calliptera	A	1	MH	Lth
Bolborhynchus ferrugineifrons	A	3	H	L
Forpus xanthops	A	2	M	TL
Touit melanonota	F	1	M	L
Touit surda*	F	1	L	L
Touit stictoptera*	A	1	M	l
Hapalopsittaca amazonina	A	1	H	L
Hapalopsittaca fuertesi	A	1	H	L
Hapalopsittaca pyrrhops	A	1	H	L
Amazona vittata	C	1	I	LHTNI
Amazona pretrei	S	1	L	TLh
Amazona viridigenalis	M	2	L	TL
Amazona rhodocorytha*	F	1	L	LT
Amazona brasiliensis	F	1	L	THL
Amazona barbadensis	L, C	2	L	TLn
Amazona oratrix*	M	2	L	TL
Amazona vinacea*	S	1	L	LTh
Amazona versicolor	C	1	I	LHTic
Amazona arausiaca	C	1	I	LTHNi
Amazona guildingii	C	1	I	LTHN
Amazona imperialis	C	1	I	LTHN
Triclaria malachitacea	F	1	LM	LHT
Apodidae				
Cypseloides lemosi	A	?	M	Ulp
Trochilidae				
Glaucis dohrnii	F	1	L	L
Campylopterus ensipennis	A, C	1	M	L
Lophornis brachylopha	M	2	M	LR
Popelairia letitiae†	S	1	?	U
Thalurania ridgwayi	M	1	LM	L
Lepidopyga lilliae	L	6	L	L
Amazilia luciae†	M	2	L	L
Amazilia distans	A	1	L	l
Amazilia boucardi	M	6	L	L
Amazilia castaneiventris	A	1	M	L
Eupherusa poliocerca	M	1	MH	L

APPENDIX 1
Continued

Species	Region	Habitat	Elevation	Threat
Eupherusa cyanophrys	M	1	M	L
Hylonympha macrocerca	A	1	M	LR
Aglaeactis aliciae	A	2	H	R
Coeligena prunellei	A	1	MH	L
Sephanoides fernandensis	I	1	I	I
Heliangelus regalis	A	1	M	R
Eriocnemis nigrivestis	A	1	H	L
Eriocnemis godini	A	?	H	L
Eriocnemis mirabilis	A	1	H	R
Haplophaedia lugens	A	1	MH	LT
Metallura baroni	A	1	H	L
Metallura odomae	A	1	H	L
Taphrolesbia griseiventris	A	2	H	R
Loddigesia mirabilis	A	1	MH	L
Eulidia yarrellii	L, A	5	LM	L
Acestrura bombus*	L, A	2	LMH	L
Acestrura berlepschi	L	1	L	L
Selasphorus ardens	M	1	M	R
Strigidae				
Xenoglaux loweryi	A	1	H	R
Caprimulgidae				
Siphonorhis americanus	C	2	I	Il
Caprimulgus noctitherus	C	2	I	Li
Caprimulgus candicans*	S	3	L	L
Caprimulgus maculosus	S	?	L	U
Eleothreptus anomalus*	F	3	LM	L
Columbidae				
Columba caribaea	C	1	I	LHN
Columba oenops	A	2	MH	L
Columba inornata*	C	2	I	LHNIC
Zenaida graysoni	I	1	I	I
Columbina cyanopis†	S	3	L	L
Claravis godefrida*	F	1	LM	L
Leptotila wellsi	C	2	I	Li
Leptotila ochraceiventris	L, A	2	LMH	L
Leptotila conoveri	A	1	MH	L
Starnoenas cyanocephala	C	1	I	LH
Gruidae				
Grus americana*	N	4	L	PXDNKhl
Rallidae				
Coturnicops notatus*	S	4	L	U
Laterallus levraudi	L	4	L	P
Laterallus tuerosi	A	4	H	PA
Laterallus xenopterus†	S	4	LM	L
Rallus wetmorei	L	6	L	L
Rallus semiplumbeus	A	4	H	LPDH
Rallus antarcticus*	S	4	L	l
Porzana spiloptera†	S	4	L	L
Cyanolimnas cerverai	C	4	I	Li
Fulica cornuta*	A	4	H	heap
Scolopacidae				
Numenius borealis*	N, S	7	L	hl
Numenius tahitiensis	N, I	7	LMI	IH
Charadriidae				
Charadrius melodus*	N, M, C	7	LM	IDL
Laridae				
Larus atlanticus*	S	7	L	E

APPENDIX 1
Continued

Species	Region	Habitat	Elevation	Threat
Accipitridae				
Accipiter gundlachii	C	1	LM	LH
Leucopternis lacernulata*	F	1	L	LH
Leucopternis occidentalis	L, A	2	LM	L
Harpyhaliaetus coronatus*	S	3	L	LH
Buteo ridgwayi	C	1	I	LDH
Buteo galapagoensis	I	2	I	hi
Falconidae				
Micrastur plumbeus	L, A	1	LM	L
Micrastur buckleyi*	S	1	LM	L
Podicipedidae				
Podiceps andinus	A	4	H	LPIH
Podiceps taczanowskii	A	4	H	PA
Phalacrocoracidae				
Nannopterum harrisi	I	7	I	OIDF
Ciconiidae				
Gymnogyps californianus*	N	3	LM	HP
Procellariidae				
Pterodroma defilippiana*	I	7	I	I
Pterodroma phaeopygia*	I	7	I	I
Pterodroma cahow*	C	7	I	LHEICPADNG
Pterodroma hasitata*	C	7	I	IHL
Pterodroma caribbaea*	C	7	I	IH
Puffinus creatopus*	I	7	I	I
Puffinus auricularis*	I	7	I	I
Pelecanoides garnoti*	I, L	7	I	LHEI
Oceanodroma macrodactyla*	I	7	I	I
Tyrannidae				
Hemitriccus cinnamomeipectus	A	1	MH	R
Hemitriccus mirandae	F	2	M	L
Hemitriccus kaempferi	F	1	L	R
Ceratotriccus furcatus	F	1	LM	L
Todirostrum senex	S	1	L	U
Anairetes alpinus	A	1	H	L
Pseudocolopteryx dinellianus†	S	4	L	R
Euscarthmus rufomarginatus*	S	3	L	L
Phylloscartes lanyoni	L	1	LM	L
Phylloscartes roquettei	S	2	L	L
Phylloscartes paulistus*	F	1	L	L
Phylloscartes ceciliae	F	1	LM	L
Platyrinchus leucoryphus*	F	1	LM	L
Onychorhynchus occidentalis	L	1	L	L
Lathrotriccus griseipectus	L, A	2	LM	L
Agriornis andicola*	A	3	H	U
Yetapa risoria*	S	3	L	LA
Attila torridus*	L, A	1	LM	L
Tyrannus cubensis	C	1	I	L
Pachyramphus spodiurus	L, A	2	L	L
Laniisoma elegans*	F	1	LM	L
Tijuca condita	F	1	M	R
Carpornis melanocephalus*	F	1	L	L
Phytotoma raimondii	L	2	L	L
Zaratornis stresemanni	A	1	H	L
Iodopleura pipra	F	1	LM	L
Calyptura cristata	F	1	LM	L
Lipaugus lanioides	F	1	M	L
Cotinga maculata	F	1	L	L

APPENDIX 1
Continued

Species	Region	Habitat	Elevation	Threat
Xipholena atropurpurea	F	1	L	L
Carpodectes antoniae	M	6	LM	L
Cephalopterus glabricollis	M	1	LM	L
Pipra vilasboasi	S	1	L	U
Piprites pileatus	F	1	M	L
Thamnophilidae				
Biatas nigropectus	F	1	LM	L
Clytoctantes alixii	L, A	1	LM	L
Clytoctantes atrogularis	S	1	L	L
Xenornis setifrons	M, L	1	L	R
Thamnomanes plumbeus	F	1	L	L
Dysithamnus occidentalis	A	1	M	L
Myrmotherula snowi	F	1	LM	L
Myrmotherula fluminensis	F	1	L	U
Myrmotherula grisea	A	1	M	L
Herpsilochmus parkeri	A	1	M	L
Herpsilochmus pectoralis*	S	2	L	L
Formicivora iheringi	F	2	M	L
Formicivora erythronotos	F	1	L	L
Formicivora littoralis	F	1	L	L
Terenura sicki	F	1	LM	L
Terenura sharpei	A	1	M	L
Cercomacra carbonaria	S	5	L	R
Pyriglena atra	F	1	L	L
Rhopornis ardesiaca	F	2	M	L
Myrmeciza ruficauda	F	1	L	L
Myrmeciza griseiceps	L, A	1	MH	LA
Pithys castanea	S	1	L	U
Furnariidae				
Cinclodes aricomae	A	1	H	L
Cinclodes palliatus	A	3	H	R
Aphrastura masafuerae	I	1	I	L
Leptasthenura xenothorax	A	1	H	L
Synallaxis courseni	A	1	H	R
Synallaxis infuscata	F	1	LM	L
Synallaxis tithys	L	2	LM	LA
Synallaxis cherriei*	S	1	LM	L
Synallaxis zimmeri	A	2	HM	L
Poecilurus kollari	S	5	L	R
Asthenes luizae	S	3	M	B
Asthenes huancavelicae*	A	3	H	L
Asthenes berlepschi	A	3	H	R
Asthenes anthoides*	A, S	3	LM	L
Thripophaga cherriei	S	5	L	R
Thripophaga macroura	F	1	L	L
Thripophaga berlepschi	A	1	H	LA
Premnoplex tatei	A	1	M	L
Philydor novaesi	F	1	LM	L
Simoxenops striatus	A	1	M	L
Automolus ruficollis	L, A	1	MH	L
Automolus erythrocephalus	L, A	2	M	LA
Megaxenops parnaguae*	S	2	LM	L
Xiphocolaptes falcirostris*	S	2	L	L
Formicariidae				
Formicarius rufifrons	S	1	L	L
Grallaria gigantea	A	1	MH	L
Grallaria alleni	A	1	H	L
Grallaria chthonia	A	1	M	R
Grallaria rufocinerea	A	1	H	L

APPENDIX 1
Continued

Species	Region	Habitat	Elevation	Threat
Grallaria milleri	A	1	H	L
Grallaricula cucullata	A	1	MH	L
Rhinocryptidae				
Merulaxis stresemanni	F	1	L	L
Scytalopus novacapitalis†	S	5	M	L
Scytalopus psychopompus	F	1	L	L
Vireonidae				
*Vireo atricapillus**	N, M	2	M	LB
Vireo caribeus†	C	6	I	L
Corvidae				
Cyanolyca nana	M	1	MH	L
Cyanolyca mirabilis	M	1	MH	L
Cinclidae				
Cinclus schulzii	A	5	MH	U
Muscicapidae				
Turdus haplochrous	S	2	L	R
Turdus swalesi	C	1	I	L
Sturnidae (Mimini)				
Nesomimus trifasciatus	I	2	I	Ri
Mimodes graysoni	I	2	I	I
Ramphocinclus brachyurus	C	2	I	Li
Certhiidae (Troglodytinae)				
Hylorchilus sumichrasti†	M	1	LM	L
Cistothorus apolinari	A	4	H	LP
Ferminia cerverai	C	3	I	Li
Thryothorus nicefori	A	2	M	R
Passeridae				
*Anthus nattereri**	S	3	LM	LA
*Carduelis yarrellii**	S	1	MH	L
*Carduelis cucullata**	L, C	2	LM	TL
Carduelis siemiradzkii	L	2	LM	L
Junco insularis	I	1	I	LI
Xenospiza baileyi	M	3	H	L
Torreornis inexpectata	C	2	I	L
Atlapetes flaviceps	A	1	H	L
Atlapetes pallidiceps	A	2	M	L
*Gubernatrix cristata**	S	3	L	Tl
Fringillidae				
*Vermivora bachmanii**	N, C	1	L	L
Dendroica chrysoparia†	N, M	1	M	LB
Dendroica kirtlandii†	N, C	1	L	LB
Geothlypis speciosa	M	4	H	L
Leucopeza semperi	C	1	I	il
Myioborus pariae	A	1	M	Lt
Basileuterus griseiceps	A	1	MH	L
Xenoligea montana	C	1	I	L
Conirostrum tamarugense	L, A	2	H	L
Conothraupis mesoleuca†	S	2	?	U
Chlorospingus flavovirens	A	1	M	R
Hemispingus goeringi	A	1	H	L
Nemosia rourei	F	1	LM	L
Calyptophilus frugivorus	C	1	I	L
Buthraupis melanochlamys	A	1	MH	L
Buthraupis aureocincta	A	1	H	L

APPENDIX 1
Continued

Species	Region	Habitat	Elevation	Threat
Buthraupis aureodorsalis	A	1	H	R
Buthraupis wetmorei	A	1	H	L
Wetmorethraupis sterrhopteron	A	1	M	L
Chlorochrysa nitidissima	A	1	M	L
Tangara cabanisi	M	1	M	L
Tangara fastuosa	F	1	LM	TL
Tangara peruviana	F	1	L	Lt
Tangara meyerdeschauenseei	A	2	M	R
Tangara phillipsi	A	1	M	R
Dacnis hartlaubi	A	1	M	L
Dacnis nigripes	F	1	LM	lt
Dacnis berlepschi	L, A	1	LM	L
Oreothraupis arremonops	A	1	MH	L
Xenospingus concolor	L, A	5	LMH	L
Incaspiza ortizi	A	2	MH	L
Poospiza alticola	A	1	H	l
Poospiza rubecula	A	1	H	L
Poospiza garleppi	A	1	H	L
Poospiza baeri	A	1	H	l
*Poospiza cinerea**	F	3	M	l
Sporophila frontalis	F	1	LM	LT
Sporophila falcirostris	F	1	LM	LT
Sporophila melanops†	S	3	L	U
Sporophila nigrorufa†	S	3	L	L
Sporophila palustris	S	3	L	LT
*Sporophila hypochroma**	S	3	L	LT
Sporophila zelichi	S	3	L	LT
Sporophila insulata†	L	3	L	L
Diglossa venezuelensis	A	1	MH	L
Camarhynchus heliobates	I	6	I	R
Saltator rufiventris	A	2	H	L
Psarocolius cassini	L	1	L	L
Cacicus koepckeae	S	1	L	L
Icterus bonana	C	2	I	Bl
*Xanthopsar flavus**	S	3	L	LPb
Agelaius xanthomus	C	6	I	LIB
*Sturnella defilippi**	S	3	L	L
Hypopyrrhus pyrohypogaster	A	1	MH	L
Curaeus forbesi	F	1	LM	LBt

APPENDIX 2

THREATENED BIRD SPECIES OCCURRING IN, AND ENDEMIC TO, COUNTRIES AND THEIR DEPENDENCIES IN THE AMERICAS. THE ASTERISKS INDICATE THAT ALL EIGHT SPECIES IN THE DOMINICAN REPUBLIC AND HAITI ARE ENDEMIC TO THE ISLAND OF HISPANIOLA. THE PRESENCE OF FRANCE, U.K., AND THE NETHERLANDS REFLECTS THESE COUNTRIES' VARIOUS TERRITORIAL POSSESSIONS IN THE CARIBBEAN AND NORTHERN SOUTH AMERICA

1	Brazil	97/65	29.7%
2	Peru	64/31	19.4%
3	Colombia	56/31	17.1%
4	Ecuador	46/11	14.0%
5	Argentina	37/2	11.3%
6	Mexico	28/20	8.6%
7=	Bolivia	22/8	7.0%
7=	Paraguay	22/—	7.0%
9	Venezuela	20/11	6.1%
10	U.S.A.	17/5	5.5%
11	Chile	15/6	4.6%
12	Cuba	13/9	4.0%
13	Uruguay	11/—	3.4%
14=	Dominican Republic	8/—*	2.4%
14=	Haiti	8/—*	2.4%
16=	Jamaica	5/3	1.5%
16=	Panama	5/1	1.5%
18=	France	4/2	1.2%
18=	Costa Rica	4/1	1.2%
18=	Canada	4/—	1.2%
18=	Guatemala	4/—	1.2%
22=	Dominica	3/2	0.9%
22=	U.K.	3/1	0.9%
22=	Bahamas	3/—	0.9%
27=	Trinidad and Tobago	2/1	0.6%
27=	Belize	2/—	0.6%
27=	Nicaragua	2/—	0.6%
30=	Grenada	1/1	0.3%
30=	St. Vincent	1/1	0.3%
30=	Antigua and Barbuda	1/—	0.3%
30=	Guyana	1/—	0.3%
30=	Netherlands	1/—	0.3%
30=	St. Kitts-Nevis	1/—	0.3%
30=	Surinam	1/—	0.3%

FIELD TECHNIQUES FOR COLLECTING AND PRESERVING HELMINTH PARASITES FROM BIRDS, WITH NEW GEOGRAPHIC AND HOST RECORDS OF PARASITIC NEMATODES FROM BOLIVIA

MARY C. GARVIN,[1,2,3] JOHN M. BATES,[1,4] AND J. M. KINSELLA[2,5]

[1]*Museum of Natural Science and Department of Zoology and Physiology, Louisiana State University, Baton Rouge, Louisiana 70803, USA;*
[2]*Department of Pathobiology, P.O. Box 110880, University of Florida, Gainesville, Florida 32611-0880, USA;*
[3]*present address: Vector Biology Laboratories, Dept. of Biological Sciences, University of Notre Dame, Notre Dame, Indiana 46556, USA;*
[4]*present address: Dept. of Zoology, Field Museum of Natural History, Roosevelt Rd. at Lake Shore Dr., Chicago, Illinois 60605, USA;*
[5]*mailing address: 2108 Hilda Ave., Missoula, Montana 59801, USA*

ABSTRACT.—The collection and preservation of helminth parasites from birds under field conditions are discussed and suggestions are made for identification and deposition of specimens. We also report on the parasitic helminths observed during study skin preparation during an expedition to Bolivia in 1989. Of the approximately 272 bird species collected (1,110 individuals), nematodes were detected in 43 species (16%). Nematodes from 28 of the bird species could be identified, representing 12 species and 10 genera. This represents the only published report of nematode parasites of birds from Bolivia, including 2 new host and 12 new geographic records. These data provide additional baseline information on the helminth parasites of Neotropical birds that is necessary to understand the consequences of parasitic infections.

RESUMEN.—Se discuten la colección y preservación de parásitos helmintos de las aves bajo condiciones de campo y se hacen sugerencias para la identificación y deposición de los especímenes. También reportamos sobre los helmintos parasíticos observados durante la preparación de especímenes de aves colectados en una expedición a Bolivia en 1989. De las aproximadamente 272 especies de aves colectadas (1,110 individuos), 43 especies (16%) hospedaban nematodos. En 28 de las especies de aves, se pudieron identificar nematodos representando doce especies y 10 géneros. Esto representa el único reporte publicado de parásitos de nematodos para las aves de Bolivia, incluyendo dos nuevos huéspedes y 12 nuevos récords geográficos. Estos datos proporciona información adicional para datos básicos sobre parásitos helmintos de aves neotropicales. Datos como estos son necesarios para poder lograr entender las consecuencias de infecciones parasíticas.

Although birds have been collected in Bolivia for a century, no published records exist of parasitic helminths of birds in the country. This is unfortunate in light of recent implications of the effects of bird parasites on host fitness (Atkinson and van Riper 1991), sexual selection (Hamilton and Zuk 1982), and behavior (van Riper et al. 1986). The implication of parasitic infections for their host community structure and their importance in conservation biology are also becoming more evident (Scott 1988, Minchella and Scott 1991). However, even the most basic data on parasite distribution, prevalence, and life cycle are lacking for common parasite species in Neotropical birds. By examining bird specimens for parasites, ornithologists increase the value of those specimens and provide essential background data for work necessary to evaluate fully the impact of parasites on avian evolution and ecology. Unfortunately, ornithologists on field expeditions typically lack the time and training to properly examine study specimens for parasites, or to collect and preserve specimens.

Here we outline the phyla of helminths that parasitize birds and suggest techniques for collecting and preserving helminth specimens in the field. We also report on nematode parasites acquired from bird specimens collected on an expedition to Bolivia to provide examples of parasites that may be encountered during study skin preparation. This report is not a complete survey of the helminth fauna of the birds collected, but it represents the only published report of nematode parasites of birds from Bolivia, including 2 new host and 12 new geographic records.

Recommendations.—Little information is available on the parasite fauna of Neotropical birds. It is therefore useful to document the prevalence, intensity, and distribution of infections when possible and to deposit voucher specimens in appropriate collections for future taxonomic study. Such information is especially valuable because host voucher specimens and associated data are available for cross-reference. Scientific collecting expeditions provide excellent opportunities to obtain baseline information because helminths may be collected during study skin preparation. We suggest that investigators use the following methods to search for and adequately preserve specimens in an efficient manner under the constraints of field conditions.

The helminth fauna of birds consists of three phyla, which can be distinguished in the field and handled accordingly. Members of the Phylum Nematoda (roundworms) are easily distinguished by their cylindrical bodies and range from several millimeters to several centimeters long. Nematodes may be found in the following locations: under the nictitating membrane or in the eye socket after removal of the eye; between the skin and muscle fascia; in the stomach or crop upon examination of contents; under the Koilon lining of the gizzard; and in the body cavity and fascia of visceral tissue during inspection of gonads, or collection of tissue samples. Larval forms may also be found encysted on organ tissue. Nematodes found alive should be killed in a vial of glacial acetic acid and then transferred to 70% ethanol with 5% glycerine, preferably within one hour. Dead nematodes can be placed directly into the ethanol.

The Phylum Platyhelminthes (flatworms) includes the flukes, Class Trematoda, and the tapeworms, Class Cestoidea. Platyhelminths are flattened dorsoventrally and usually white or cream in color. Trematodes in birds are usually small (<1 mm to several mm long), elongate or nearly spherical, and often have distinct anterior and posterior regions. Although the alimentary tract is the most common site of infection, nearly every organ system in birds has been colonized by at least one family of trematodes. Adult cestodes, in contrast, are almost exclusively found in the lower alimentary tract and can be distinguished by their segmentation and longer bodies (up to several centimeters). The anterior holdfast organ (scolex) is essential for identification, so every effort should be made to obtain the complete worm.

Living trematodes and tapeworms should be relaxed in cool water for an hour or two, fixed in AFA (85% ethanol, 85 parts; concentrated formalin, 10 parts; glacial acetic acid, 5 parts) for 24 to 48 h, then transferred to 70% ethanol for storage. Dead specimens can go directly into AFA, then 70% ethanol. An alternative way to kill very small trematodes is to drop them into hot 10% formalin and then fix them in AFA.

Helminths of the Phylum Acanthocephala (spiny-headed worms) are less common in birds and are usually found in the intestine of their host, with their anterior, spiney, proboscis embedded in the intestinal wall. Although living specimens are typically white, they may take on the color of the intestinal contents and reach two or more cm in length (Rausch 1983). Care should be taken to tease the proboscis from the intestinal wall, because the spines are vital to identification. Live worms should be kept in cool water until relaxed and then fixed in AFA and transferred to ethanol as above.

If time permits, intestinal tracts may be removed, opened longitudinally, and rinsed in a shallow pan with enough water to suspend any contents. Helminths may be recovered from the suspension or, if attached to tissue, removed with forceps. The cecum, lungs, trachea, and esophagus may be examined in this fashion as well. Parasites from each organ should be stored seperately and labelled accordingly. Utensils should be rinsed between each use.

The proper killing, fixing, and preservation of specimens is critical to maintaining the integrity of the helminth for species identification upon return from the field, at which time, specimens should be sent to an expert for identification. Author J. M. Kinsella welcomes parasitic helminths for identification. Specimens should then be deposited in a recognized reference collection. Two such collections in the United States are the National Parasite Collection at Beltsville, Maryland, and the Harold W. Manter Collection at the University of Nebraska, Lincoln. Voucher specimens may then be used as a basis for further study, including the effects of parasites on the ecology and evolution of the host species (May 1988).

Host-parasite records from Bolivia.—From June through October 1989, an expedition from the Louisiana State University Museum of Natural Science continued a survey of the avifauna of the Parque Nacional Noel Kempff Mercado in eastern Bolivia (Bates et al. 1989, 1992).

Methods.—Birds were collected in mist nets or with shotguns at three study sites in prov. Velasco, depto. Santa Cruz, eastern Bolivia: (1) 21 km SE of Catarata Arco Iris (Lat. 13°55′ S, Long. 60°45′W, elev. 670 m); medium-sized forest fragment (600 ha) on top of plateau; most birds netted inside forest; (2) approximately 45 km E of the town of Florida (Lat. 14°34′S, Long. 60°40′W, elev. 720 m); small forest fragment (350 ha) on top of plateau; most birds netted inside forest; (3) Piso Firme (Lat. 13°35′ S, Long 61°55′W, elev. ca 250 m); primary forest with variable canopy height; also some small patches of cerrado in the area. Further descriptions of the study sites may be found in Bates et al. (1992).

Although collecting focussed on forest and forest edge, specimens were also collected in the adjacent cerrado habitats (e.g., *Rynchotus rufescens, Otus choliba, Synallaxis albescens, Formicivora rufa,* and *Cyanocorax cristatellus*). Bird specimens were prepared by Angelo Capparella (APC), Maria Dolores Carreño (MDC), Abel Castillo (AC), Jorge Cayalo (JC), Curtis Marantz (CAM), Tristan Davis (TJD), Hugo Hurtado H. (HHH), Gary Rosenberg (GHR), Manuel Sánchez S. (MSS), and Armando Yépez (AY). Bird specimens are housed at the Museum of Natural Science, Louisiana State University, and the Museo de Historia Natural "Noel Kempff Mercado" (Santa Cruz).

Parasites were collected when observed during bird specimen preparation and placed directly into 10% formalin or into warm glycerine for several hours before being transferred to formalin. Helminths were identified by J. M. Kinsella and deposited in the helminth collection of the Harold W. Manter Laboratory, University of Nebraska.

Identification of parasites to the species level was not always possible for two reasons. First, because only females were collected from some specimens, species-specific taxonomic characters of males could not be used (e.g., *Oxyspirura, Eulimdana*). In other cases, inadequate relaxation of the specimens prior to fixation prevented identification to the species level.

Results and discussion.—Approximately 1,100 bird specimens of 272 species were collected. Parasitic nematodes were found in 43 of these individuals (16%), representing 27 species (10%). The 12 species of parasitic nematodes that were identified came from 28 individuals of 15 species (Table 1).

Bird specimens were not searched systematically and exhaustively for parasites because of time constraints and preparator inexperience. Thus, some parasites were certainly overlooked. Therefore, parasite prevalence and intensity were underestimated.

Eye.—Eye parasites of the family Thelaziidae were found in six species of birds, usually under the nictitating membrane. The two most common genera of thelaziids, *Oxyspirura* and *Thelazia*, can cause conjunctivitis in birds (Brooks et al. 1983). These two genera exhibit dramatically different life cycles. Eggs of species of *Oxyspirura* pass down the lacrimal duct, into the gut, and out of the body in the feces. They are then ingested by an insect intermediate host such as a cockroach (Fielding 1927), which is, in turn, ingested by the next avian definitive host. Pence (1972) found that two species of *Oxyspirura* in North America exhibit ecological, rather than host, specificity. Their occurrence in a variety of unrelated birds in a particular habitat appears to be related to the occurrence of the intermediate host(s). This genus was represented in our sample by *Oxyspirura cruzi* and *Oxyspirura* sp. found in two cerrado-inhabiting birds (Table 1).

In contrast, species of the genus *Thelazia* are transmitted by flies (Muscidae) of the genera *Musca* and *Fannia* that feed on lacrimal secretions (Schmidt and Roberts 1989). Larvae deposited on the conjunctiva are ingested by the flies and, after development, transmitted by the fly to the eye of another host. Host distribution is probably related to the feeding habits of the fly vector. *Thelazia digitata* was the most common eye worm, found in four host species of three different families (Table 1). All bird species infected with this helminth inhabit the forest middle story. *Thelazia campanulata* in the accipitrid *Leptodon cayanenisis* represents a new host record.

Air sacs.—Species of the genus *Diplotriaena* are parasites of the air sacs of birds. They are often found in the body cavity during study skin preparation. *Diplotriaena* was the most common genus of parasite encountered, with three species identified from 13 birds (3 species, Table 1). Both records of *Diplotriaena attenuatoverrucosa* were from *Xiphorhynchus elegans*. One *Diplotriaena tricuspis* infection was reported from *Cyanocorax cristatellus*. The most common helminth encountered was *Diplotriaena henryi*; all specimens were collected from *Turdus amaurochalinus*. Helminth infections have been frequently observed in the body cavity of *T. amaurochalinus* on other expeditions (Bates, pers. obs.). Orthopterans and coleopterans are known intermediate hosts for *D. henryi*. Because birds acquire infection through ingestion of an infected

TABLE 1
HOST SPECIES AND LOCATIONS IN BOLIVIAN BIRD HOSTS WHERE PARASITES WERE FOUND.
NUMBERS FOLLOWING THE COLLECTOR'S NUMBER REFERENCE THE EXACT LOCALITY WHERE EACH
SPECIMEN WAS COLLECTED

Host (Coll. Cat. No.[a], locality[b])	Location in host	Parasite (parasite accession number[c])
Tinamidae		
Rynchotus rufescens		
(AC 575, 2)	Eye	*Oxyspirura cruzi* (Rodriguez 1962) (37907)
(GHR 2484, 2)	Fascia of oviduct	*Tetracheilonema quadrilabiatum* (Molin 1858) (37915)
(AC 582, 2)	Body cavity	*Tetracheilonema quadrilabiatum* (37916)
Accipitridae		
Leptodon cayanensis		
(JMB 910, 2)	Crop	*Physaloptera acuticauda* Molin 1960 (37917)
(JMB 910, 2)	Eye	*Thelazia campanulata* (Molin 1858) (37918)
Cracidae		
Pipile cujubi		
(JMB 589, 1)	Eye	*Thelazia digitata* Travassos 1918 (37908)
Cuculidae		
Piaya cayana		
(AC 512, 3)	Body cavity?	*Cyrnea semilunaris* (Molin 1860)
Strigidae		
Otus choliba		
(JMB 973, 2)	Nasal cavity	*Squamofilaria sicki* (Strachan 1957) (37919)
Trogonidae		
Trogon melanurus		
(JMB 741, 2)	Eye	*Thelazia digitata*
(APC 3519, 3)	Stomach	*Subulura travassosi* Barreto 1919 (37909)
Picidae		
Celeus torquatus		
(AC 518, 2)	Eye	*Thelazia digitata* (37910)
Veniliornis affinis		
(MDC 262, 2)	Eye	*Thelazia digitata*
Dendrocolaptidae		
Xiphorhynchus elegans		
(CAM 224, 3)	Body cavity	*Diplotriaena attenuatoverrucosa* (Molin 1958) (37911)
(APC 3359, 1)	Body cavity	*Diplotriaena attenuatoverrucosa*
Furnariidae		
Synallaxis albescens		
(AC 392, 1)	Eye	*Oxyspirura* sp. (female)
Formicariidae		
Hylophylax poecilinota		
(JMB 580, 1)	Body cavity	*Eulimdana* sp. (female) (37913)
Tyrannidae		
Rhytipterna simplex		
(CAM 215, 3)	Body cavity	*Monopetalonema solitarium* Caballero 1948 (37914)
Casiornis rufa		
(APC 3618, 3)	Body cavity	*Monopetalonema solitarium*
Corvidae		
Cyanocorax cristatellus		
(JMB 992, 2)	Body cavity	*Diplotriaena tricuspis* (Fedtschenko 1874) (37912)
Turdidae		
Turdus amaurochalinus		
(APC 3574, 3)	Esophagus	*Diplotriaena henryi* Blanc 1919 (37920)
(MDC 242, 3)	Body cavity	*Diplotriaena henryi*
(MDC 250, 3)	Body cavity	*Diplotriaena henryi* (37921)

TABLE 1
Continued

Host (Coll. Cat. No.[a], locality[b])	Location in host	Parasite (parasite accession number[c])
(AY 1, 3)	Body cavity	*Diplotriaena henryi*
(JC 22, 3)	Body cavity	*Diplotriaena henryi* (37923)
(MSS 3134, 1)	Body cavity	*Diplotriaena henryi* (37922)
(MDC 264, 2)	Body cavity	*Diplotriaena henryi*
(HHH 2, 2)	Esophagus	*Diplotriaena henryi*
(MDC 249, 3)	Esophagus	*Diplotriaena henryi*
(MDC 92, 1)	Body cavity	*Diplotriaena henryi*

[a] See text for collector names.
[b] See text for localities.
[c] Accession numbers are for the H. W. Manter Laboratory, University of Nebraska, Lincoln.

arthropod intermediate host, the frequency of this host-parasite interaction may be indicative of food preference of *T. amaurochalinus*. It is also of interest that *Turdus amaurochalinus* is found in this part of Bolivia only in winter, and therefore, might have acquired infections of *D. henryi* on the breeding grounds.

Body cavity.—*Eulimdana* sp. was collected from the body cavity of *Hylophylax poecilinota*. Microfilaria, the embryonic stage, are found in the skin. Biting lice (order Mallophaga, suborder Amblycera) have been shown to serve as the intermediate host of species in this genus (Bartlett and Anderson 1987).

Monopetalonema solitarium was found in the body cavity of two species of Tyrannids, *Rhytipterna simplex* and *Cassiornis rufa*. *Tetracheilonema quadrilabiatum* was collected from the fascia of the oviduct and the body cavity of *Rynchotus rufescens*.

Gastrointestinal tract.—A number of helminths that occur in the gastrointestinal tract are often found upon examination of stomach contents. *Cyrnea semilunaris* was collected from *Piaya cayana*. Although exact location of the parasite in the host was not determined, *C. colini*, the only species of *Cyrnea* with an elucidated life cycle, typically is found in the wall of the proventriculus, near its junction with the gizzard (Wehr 1971). Infections are acquired upon ingestion of an infected insect, a cockroach in the case of *C. colini* (Cram 1933). *Physaloptera acuticauda* was found in the crop of *Leptodon cayanensis*. Physalopterids often inhabit the stomachs of their host. *Subulura* spp. are parasites of the caecum. The only specimen collected in this study, however, was found in the stomach of a trogon (*Trogon melanurus*). Both grasshoppers and beetles are known intermediate hosts for this genus (Wehr 1971). Birds are infected when the invertebrate intermediate host is ingested.

Nasal cavity.—In 1957 Strachan (1957) described a species of nematode based on specimens collected from the subcutaneous tissue around the eye of an unidentified species of *Otus* as *Thelazia sicki*. Those same specimens were later redescribed as *Squamofilaria sicki* by Anderson and Chabuad (1958). Our record of *S. sicki* from the nasal cavity of *Otus choliba* is the first from an identified host species and only the second recorded occurrence of this nematode. Other species of Neotropical *Otus* should be examined to determine the host distribution of the parasite.

In conclusion, scientific collecting expeditions provide unique opportunities to help determine patterns of parasite distribution and host-parasite associations. Although a thorough examination of each bird specimen is required to determine parasite prevalence and intensity, casual examination, collection, preservation, and identification of parasites will help to identify host-parasite relationships and increase our understanding of geographic distributions of parasites. Such data provide a springboard for future detailed studies that may fully measure the consequences of parasitic infections on populations of Neotropical birds.

ACKNOWLEDGMENTS

The 1989 Expedition was funded by a National Geographic Society grant to J. V. Remsen, Jr., with additional support from Mrs. Paquita Machris. Field work was done in conjunction with personnel of both the Museo de Historia Natural "Noel Kempff Mercado" (Universidad Autonoma Gabriel Rene Moreno, Santa Cruz, Bolivia) and the Parque Nacional Noel Kempff Mercado (PNNKM). The staff of the PNNKM, under the direction of Ing. Gregorio Cerro Grande, Lic. Arturo Moscoso V. of the depto. de Vida Silvestre, Centro de Desarrollo Forestal (Santa Cruz), and Lic. Teresa R. de Centurion and her staff at the Museo provided advice and help throughout our work in Bolivia. Finally, we thank Hermes Justiniano of the Fundación Amigos de la Naturaleza (Santa Cruz) for additional logistic support. Manuel Marín A., J. V. Remsen, Jr., and Mercedes Foster provided help and comments on the manuscript.

LITERATURE CITED

ANDERSON, R. C., AND A. G. CHABAUD. 1958. Taxonomie de la filaire *Squamofilaria sicki* (Strachan 1957) n. comb. et place du genre *Squamofilaria* Schmerling, 1925 dans la sous-famille Aproctinae. Annales De Parasitologie Humaine et Comparee 33:254–266.

ATKINSON, C. T., AND C. VAN RIPER III. 1991. Pathogenicity and epizootiology of avian haematozoa: *Plasmodium*, *Leucocytozoon*, and *Haemoproteus*. Pp. 19–48 *in* Bird–Parasite Interactions: Ecology, Evolution, and Behavior (J. E. Loye and M. Zuk, Eds.). Oxford Univ. Press, Oxford, U.K.

BARTLETT, C. M., AND R. C. ANDERSON. 1987. *Pelecitus fulicaeatrae* (Nematoda:Filarioidea) of coots (Gruiformes) and grebes (Podicipediformes): skin-inhabiting microfilariae and development in the Mallophaga. Can. J. Zool. 63:2803–2812.

BATES, J. M., M. C. GARVIN, D. C. SCHMITT, AND C. G. SCHMITT. 1989. Notes on bird distribution in northeastern dpto. Santa Cruz, Bolivia, with 15 species new to Bolivia. Bull. Brit. Orn. Club 109:236–244.

BATES, J. M., T. A. PARKER, III, A. P. CAPPARELLA, AND T. J. DAVIS. 1992. Observations on the *campo*, *cerrado*, and forest avifaunas of eastern dpto. Santa Cruz, Bolivia, including 21 species new to the country. Bull. Brit. Orn. Club 112:86–98.

BROOKS, D. E., E. C. GREINER, AND M. T. WALSH. 1983. Conjunctivitis caused by *Thelazia* sp. in a Senegal parrot. J. Amer. Vet. Assoc. 183:1305–1306.

CRAM, E. B. 1933. Observations on the life history of *Seurocyrnea colini*. J. Parasitol. 20:98.

FIELDING, J. W. 1927. Further observations on the life cycle of the eye worm of poultry. Australian J. Exp. Biol. Medical Sci. 4:273–281.

HAMILTON, W. D., AND M. ZUK. 1982. Heritable true fitness and bright birds: A role for parasites? Science 218:384–386.

MAY, R. M. 1988. Conservation and disease. Conserv. Biol. 2:28–30.

MINCHELLA, D. J., AND M. E. SCOTT. 1991. Parasitism: a cryptic determinant of animal community structure. Trends Ecol. Evol. 6:250–254.

PENCE, D. B. 1972. The genus *Oxyspirura* (Nematoda: Thelaziidae) from birds in Louisiana. Proc. Helminthological Soc. Wash. 39:23–28.

RAUSCH, R. L. 1983. The biology of avian parasites. Pp. 367–442 *in* Avian Biology (D. S. Farner, J. R. King, and K. C. Parkes, Eds.). Academic Press, New York, New York.

SCHMIDT, G. D., AND L. S. ROBERTS. 1989. Foundations of Parasitology. Times Mirror Mosby Publishing, New York, New York.

SCOTT, M. E. 1988. The impact of infection and disease on animal populations: implications for conservation biology. Conserv. Biol. 2:40–56.

STRACHAN, A. A. 1957. Eye worms of the family Thelaziidae from Brazilian birds. Can. J. Zool. 35:179–187.

VAN RIPER III, C., S. G. VAN RIPER, M. L. GOFF, AND M. LAIRD. 1986. The epizootiology of and ecological significance of malaria in Hawaiian land birds. Ecol. Monogr. 56:327–344.

WEHR, E. E. 1971. Nematodes. Pp. 185–233 *in* Infectious and Parasitic Diseases of Wild Birds (J. W. Davis, R. C. Anderson, L. Karstad, and D. O. Trainer, Eds.). Iowa State University Press, Ames, Iowa.

LACK OF GENETIC DIVERGENCE IN A GENUS (*PTEROGLOSSUS*) OF NEOTROPICAL BIRDS: THE CONNECTION BETWEEN LIFE-HISTORY CHARACTERISTICS AND LEVELS OF GENETIC DIVERGENCE

SHANNON J. HACKETT[1] AND CATHI A. LEHN[2]

Museum of Natural Science and Department of Zoology and Physiology, Louisiana State University, Baton Rouge, Louisiana 70803, USA;
[1]*present address: Zoology Department, Field Museum of Natural History, Roosevelt Road at Lake Shore Drive, Chicago, Illinois 60605, USA;*
[2]*present address: Animal Science Department, Texas A & M University, College Station, Texas 77843-2471, USA*

ABSTRACT.—Relationships among 14 species of araçaris (toucans) in the genus *Pteroglossus* (Ramphastidae) were addressed using allozyme data. Low levels of genetic differentiation among species in the genus were found in comparison to most other studies of Neotropical birds. A scenario relating life-history characteristics of organisms to possible levels of genetic differentiation is proposed. Analyses of allozyme genetic distance data identified two major groups of species: the first group includes members of the *torquatus* complex (*torquatus, frantzii, sanguineus,* and *erythropygius*), *flavirostris, mariae, castanotis,* and *pluricinctus;* the second group includes *inscriptus, bitorquatus,* and *beauharnaesii.* Of three outgroups (*Ramphastos, Selenidera,* and *Baillonius*), the monotypic Saffron Toucanet, *Baillonius baillonii,* is most allozymically similar to *Pteroglossus.* Mitochondrial DNA sequences of the cytochrome *b* gene confirm low levels of genetic differentiation among members of the *Pteroglossus torquatus* complex. Because genetic distances among most species are low, congruence of these branching patterns with phylogenies based on other character systems is assessed. Biogeographic implications of the phylogenetic pattern are also discussed.

RESUMEN.—Se vieron las relaciones entre 14 especies de araçaris (tucanes) en el género de *Pteroglossus* (Ramphastidae) usando datos de alozimas. Se encontraron niveles bajos de diferenciación genética entre las especies en comparación a otros estudios de aves neotropicales. Se propone un escenario que relaciona las características de la historia de vida de los organismos a posibles niveles de diferenciación genética. El análisis de los datos de la distancia genética de las alozimas identificó dos grupos principales de especies: el primer grupo incluye miembros del complejo *torquatus* (*torquatus, frantzii, sanguineus,* y *erythropygius*), *flavorostris, mariae, castanotis,* y *pluricinctus;* el segundo grupo incluye *inscriptus, bitorquatus,* y *beauharnaesii.* De los tres grupo externo (*Ramphastos, Selenidera,* y *Baillonius*), el Toucanet azafrán monotípico, *Baillonius bailloni,* es más parecido en sus alozimas a *Pteroglossus.* Las secuencias del ADN mitocondrial del gen del citocroma *b* confirman niveles bajos de diferenciación genética entre miembros del complejo de los *Pteroglossus torquatus.* Porque las distancias genéticas entre la mayoría de las especies son bajas, discuto estos patrones de ramificación con filogenias basadas en otros sistemas de caracteres. También se discuten las implicaciones biogeográficas del patrón filogenético.

Araçaris, toucans of the genus *Pteroglossus*, have been long considered a group whose distribution conforms to the "refugia hypothesis" of historical diversification in the Neotropics (Haffer 1969, 1974, 1985, 1987). This nonpasserine genus is composed of some 30 differentiated units (species and subspecies) distributed throughout the Neotropics, and Haffer (1974) has suggested that much of the diversification of *Pteroglossus* has occurred as of the results of Pleistocene climatic fluctuations. In the last 30 years, since Lewontin and Hubby (1966), genetic information from protein electrophoresis and, more recently, DNA sequencing, have been used to document genetic variation both within and among species. Although genetic data-bases for many vertebrates have been developed (Nevo 1978), genetic information on nonpasserine birds remains meager (Barrowclough et al. 1985), especially Neotropical nonpasserine birds. In this paper, results of genetic data on toucans are used to address the relationships within *Pteroglossus* and to assess the role of Pleistocene versus pre-Pleistocene events in shaping the current distribution of taxa.

METHODS

Tissue samples for this analysis were obtained from the Louisiana State University Museum of Natural Science (LSUMNS) Frozen Tissue Collection. Collecting localities for specimens used in the allozyme and DNA analyses are listed in Appendix 1. The allozyme study encompassed all species in the genus *Pteroglossus* for which tissues were available at the time. Not included were *Pteroglossus viridis* and *P. aracari*. Ramphastids chosen as outgroups to *Pteroglossus* were *Baillonius bailloni*, *Selenidera reinwardtii*, and *Ramphastos cuvieri*. Cladistic assessments of allozyme variation were rooted at *Ramphastos* (based on information suggested by Haffer 1974). Because one goal of this research was to assess phylogenetic and biogeographic relationships among the four members of the *P. torquatus* complex, the allozyme data were used to choose the relevant taxa outside the *torquatus* complex for mitochondrial DNA (mtDNA) sequencing. In addition, allozymes demonstrated little differentiation among species (see results), and mtDNA sequence data were used in an attempt to increase the number of phylogenetically informative characters. Finally, with two independent molecular data sets, strength of the phylogenetic signal in the *P. torquatus* complex could be assessed based on topological congruence of branching patterns (Miyamoto and Cracraft 1991, Miyamoto and Fitch 1995).

Protein electrophoresis.—Standard horizontal starch-gel electrophoresis of proteins was performed as outlined in Hackett (1989), Hackett and Rosenberg (1990), and Hackett (1995). Locus names follow Hackett and Rosenberg (1990). Alleles were coded by their relative mobility from the origin; the most anodally migrating allele was coded "a." Isozymes were coded in a similar manner, with a "1" indicating the most anodally migrating isozyme.

BIOSYS-1 (Swofford and Selander 1981) was used to compute genetic distances (Nei 1978, Rogers 1972) and a UPGMA phenogram. The computer program PHYLIP (Felsenstein 1986) was used to construct (from Rogers [1972] genetic distances) a tree that assumes a constant rate of evolution ("KITSCH"), and one that does not ("FITCH").

Cladistic assessment of allelic variation was performed in three ways. First, each locus was coded as a multi-state unordered character and alleles at each locus as character states (Buth 1984). In the second method, phylogenetically informative alleles were considered as characters and coded as present or absent (Rogers and Cashner [1987]; but see also Buth [1984], Mickevich and Mitter [1981], and Swofford and Berlocher [1987], Murphy [1993]). The Branch-and-Bound option of the computer program PAUP 3.1.1 (Swofford 1993) was used for these two cladistic analyses. One hundred bootstrap replicates were performed on the presence/absence coding of alleles to assess confidence in the branching pattern (Felsenstein 1985; Sanderson 1989). None of the above cladistic analyses attempts to take into account frequency differences between taxa. Thus, the third cladistic assessment was a FREQPARS analysis (Swofford and Berlocher 1987) that does take into account frequency information in a parsimony analysis. This analysis was implemented using the pre-release version of Swofford's PAUP*. One hundred bootstrap replications of the FREQPARS data set were also performed in PAUP*.

DNA sequences.—A total nucleic acid preparation was made from 0.1 gram of liver tissue frozen at $-80°C$ [Hillis et al. (1990); see Appendix 1 for specimens sequenced]. Amplifications of a 307 base pair fragment (not including primers) of the mitochondrial cytochrome *b* gene were performed via the polymerase chain reaction (PCR). The primers L14841 (5'-CCATCCAA-CATCTCAGCATGATGAAA-3') and H15149 (5'-CCTCAGAATGATATTTGTCCTCA-3'; Kocher et al. 1989) were used. Double-stranded PCR amplifications were performed in 50 µl total reaction volumes [10 µl of a 10^{-2} dilution of the total DNA preparation, 2.5 µl of a 10 µM solution of each primer, 5 µl of 10× buffer (including $MgCl_2$), 2 µl of a 1.0 mM solution of dNTP's, 0.20 µl *Taq* DNA polymerase (Promega), up to 50 µl with H_2O]. Thirty to 35 cycles were performed using the following cycling parameters: first cycle—denaturation at 94°C for 3 minutes, annealing at 56°C for 1 minute, extension at 72°C for 30 seconds; remaining cycles—denaturation at 94°C for 1 minute, annealing at 56°C for 1 minute, extension at 72°C for 30 seconds.

Single-stranded DNA was generated following the procedure of Allard et al. (1991) in which only one primer is used (no limiting primer). Five µl of the double-stranded product were used to generate single-stranded DNA in 100 μl reactions [5 μl double-stranded DNA, 2 μl of a 10 μM solution of one primer, 10 μl of 10× buffer (including MgCl$_2$), 4 μl dNTP's, 0.40 μl *Taq* DNA polymerase (Promega), up to 100 μl with H$_2$O]. Twenty cycles were performed using the following cycling parameters: first cycle—denaturation at 94°C for 3 minutes, annealing at 56°C for 1 minute, extension at 72°C for 45 seconds; remaining cycles—denaturation at 94°C for 1 minute, annealing at 56°C for 1 minute, extension at 72°C for 45 seconds. Single-stranded DNA was generated for both the heavy and light mtDNA strands, and the products were cleaned by five washings with H$_2$O through Ultrafree®-MC 30,000 NNMWL filters (Millipore Corp., Bedford, MA), and concentrated to a final volume of approximately 30 μl. Seven μl of cleaned single-stranded DNA were used for DNA sequencing using T7 DNA polymerase (Sequenase® version 2.0, United States Biochemical, Cleveland, OH).

The DNA sequence data were analyzed cladistically using PAUP 3.1.1 (Swofford 1993). *Pteroglossus flavirostris* was sequenced as the outgroup to the *Pteroglossus torquatus* complex (see allozyme results), and sequences of other Piciformes (toucans, barbets, and a woodpecker; Lanyon and Hall, 1994) were used as more distantly-related outgroups. All base positions were used in the analysis, and no weighting of substitutions was performed. In addition, the presence/absence allozyme data set and sequence data set were combined and analyzed cladistically (Kluge 1989).

RESULTS

Protein electrophoresis.—Twenty-five loci were resolved for all taxa listed in Appendix 1 (Table 1). Low levels of genetic differentiation were observed among *Pteroglossus* species in the allozyme analysis (Table 2). Genetic distances (Nei 1978) among *Pteroglossus* species averaged 0.095 (±0.045 s.d.); among the four members of the *Pteroglossus torquatus* complex (*frantzii, torquatus, sanguineus,* and *erythropygius*), genetic distances averaged only 0.047. However, heterozygosity within species was typical of birds, averaging roughly 4%.

Distance analyses (Fig. 1) showed many short branch lengths separating most *Pteroglossus* species. There seemed to be two major groups: the first consisting of members of *torquatus* complex, *flavirostris, mariae, castanotis,* and *pluricinctus,* the second consisting of *inscriptus, bitorquatus,* and *beauharnaesii.* Within the *torquatus* complex, *torquatus* and *frantzii* were most similar, followed by *sanguineus* and *erythropygius. Baillonius bailloni* was allozymically more similar to *Pteroglossus* than to *Selenidera.* Because genetic distances were so low among many of the species, congruence of these branching patterns with phylogenies based on other analyses and other character systems should be assessed.

Cladistic analysis of loci with the alleles as unordered character states resulted in at least 200 equally most-parsimonious trees, with a consistency index (C.I.) of 0.923. The strict consensus of these trees (not shown) resulted in little resolution. Two nodes appeared in all of the most-parsimonious trees: one indicating monophyly of the genus *Pteroglossus*, and the second supporting a clade of *inscriptus, bitorquatus,* and *beauharnaesii.* In 80% of the most-parsimonious trees, *Baillonius bailloni* was the sister taxon to *Pteroglossus.* When alleles are coded as present/absent, nine equally most-parsimonious trees resulted (C.I. = 0.61; trees not shown). Monophyly of *Pteroglossus* was again supported, as was the *inscriptus/bitorquatus/beauharnaesii* clade. There was also strong support for *Baillonius* as the sister taxon of *Pteroglossus* (supported by all most-parsimonious trees, and a bootstrap value of 80%). Bootstrap values for relationships within the genus *Pteroglossus* were all less than 60%. The FREQPARS analysis resulted 10 trees (not shown) with equal scores (56.453) and showed the same two basic groups as described above, with the exception that the *inscriptus/bitorquatus/beauharnaesii* group falls inside the other group (species relationships unresolved in this group). There was once again little resolution among *Pteroglossus* species. The bootstrap analysis of the FREQPARS data set resulted in a similar degree of resolution as the cladistic analysis of both loci and alleles. Thus, cladistic analyses of allozyme data provided little support for species-level relationships within *Pteroglossus*; however, there seems to be allozyme support for two groups of *Pteroglossus* toucans and for *Baillonius* as the sister taxon to *Pteroglossus*.

DNA sequences in the Pteroglossus torquatus *complex.*—Only 10 base positions of the 307 sequenced (3.3%) were variable within the *P. torquatus* complex (Fig. 2). Seven of the 10 variable sites were at third codon positions; three were at first codon positions. Within the *P. torquatus* complex, the transition:transversion ratio averaged 9:1. Two substitutions were inferred to cause an amino acid replacement in this region of the cytochrome *b* protein. Including the outgroup to the *torquatus* complex, *P. flavirostris*, an additional 16 positions were variable.

TABLE 1

Allozyme Frequencies for the *Pteroglossus* Species and Outgroups (see text) Analyzed in This Study. The Following Seven Loci Were Monomorphic and Fixed for the Same Allele Across All Species: SOD1, ICD2, MDH1, LAP, LDHA, PGM1, PGM2

	CK2	CK1	LDHB	GPI	MPI	GDH	FUM	NP
P. torquatus	A	B	B	B	C (0.50) D (0.50)	B	C	E
P. frantzii	A	B	B	B	C	B	C	E
P. sanguineus	A	B	B	B	C	B	C	E
P. erythropygius	A	B	B	B	C	B	C	E
P. castanotus	A	B	A (0.25) B (0.75)	A (0.13) B (0.87)	C	B	C	A (0.87) D (0.13)
P. pluricinctus	A	B	B	B	C	B	C	D
P. inscriptus	A	B	B	B	C	B	C	A
P. bitorquatus	A	B	B	B	C	B	C	A
P. flavirostris	A	B	B	B	B (0.25) C (0.75)	B	C	E
P. mariae	A	B	B	B	C	B	C	E
P. beauharnaesii	A	B	B	B	B (0.17) C (0.83)	B	C	A (0.33) B (0.16) D (0.17) E (0.33)
Selenidera	A (0.50) B (0.50)	A	B	B	E	A (0.50) B (0.50)	A	D
Baillonius	A	B	B	B	C	B	B	C
Ramphastos	A	C	B	B	A	B	A	C

Twelve of these were at third positions, one at a second position, and three at first positions. There were 18 to 21 changes between members of the *torquatus* complex and *P. flavirostris*, and percent sequence divergence averaged 6.3%. Between *P. flavirostris* and members of the *torquatus* complex, the transition:transversion ratio averaged 3:1.

Within the *P. torquatus* complex, there was only one phylogenetically informative character in the sequence data set, and it united *P. erythropygius* and *P. sanguineus*. The other characters were autapomorphic. There is obviously insufficient DNA sequence data to support relationships in this complex of species. The bootstrap analysis of the combined presence/absence allozyme data set and sequence data set (not shown) united *P. frantzii* and *P. torquatus* as sister taxa and *P. sanguineus* and *P. erythropygius* as sister taxa. However, the bootstrap values for nodes were only 41% and 45%, respectively.

DISCUSSION

Genetic differentiation and the initial conditions hypothesis.—Both allozyme and sequence data demonstrate little differentiation among taxa in the genus *Pteroglossus*. This is surprising given the high levels of genetic differentiation observed for allozymes among species in many other Neotropical genera (summary in Hackett and Rosenberg 1990; Hackett 1995, 1996). For example, Nei's (1978) genetic distances within the formicariid genus *Gymnopithys*, another Neotropical genus similarly distributed across Central and South America, but only containing four species, averaged 0.173 (Hackett 1993). Within *Gymnopithys leucaspis*, which overlaps in range with the *P. torquatus* complex in Central America and South America west of the Andes (the Chocó region), Nei's (1978) genetic distances averaged 0.053 (Hackett 1993) in comparison to 0.020 among *Pteroglossus*. These low levels of genetic differentiation are even more surprising given the high degree of sympatry of many Amazonian species of *Pteroglossus* (Haffer 1974). One prediction that might be made from the high degree of sympatry is that *Pteroglossus* is old and has had a great deal of time to evolve phenotypic differences and establish large ranges that

TABLE 1
Extended

ICD1	MDH2	GOT1	PGD	ADA	GOT2	GDA	SDH	LA	αGPD
C	B	C	B	A (0.87) B (0.13)	C	A	C	B	B
C	B	C	B	A	C	A	C	B	B
A (0.13) C (0.87)	B	C	B	A (0.87) B (0.13)	C	A	C	B (0.25) D (0.75)	B
C	B	B	B	A	C	A	C	D	B
B (0.13) C (0.87)	B	C	B	A	C	A	B (0.25) C (0.75)	D	B
A (0.50) C (0.50)	B	C	B	A	C	A	C	D	B
C	A	C	B	A	C	A (0.50) B (0.50)	A	D	B
C (0.50) D (0.50)	A	C	B	A	C	A	C	D	B
C	B	C	B	A	C	A	C	C (0.25) D (0.75)	A
A	B	C	A (0.50) B (0.50)	A	C	A	C	D	B
C	A	A (0.17) C (0.83)	B	A	B (0.17) C (0.50) D (0.33)	A	C	D	B
C	B	C	C	A	F	A	C	B	B
C	B	C	B	A	E	A	C	A	B
A (0.50) C (0.50)	C	A	B (0.50) C (0.50)	A	A	A	D	B	B

overlap extensively (Mayr and Short 1970). However, speciation and geographic range overlap in *Pteroglossus* toucans has been accomplished with little genetic differentiation at allozyme loci, and thus are presumably fairly recent.

There are several possible explanations for the low levels of genetic divergence. First, speciation could be extremely recent. However, the pattern of area relationships among the endemic *Pteroglossus torquatus* taxa matches those for a number of other taxa whose genetic distances are much greater (Hackett 1992, 1993, 1996). One tenant of vicariance biogeography is that matching area cladograms of endemic taxa are evidence of a single series of vicariant events. Thus, we doubt that more recent speciation in toucans compared to the other taxa studied is a probable explanation for the low levels of genetic differentiation. Second, rates of molecular change at allozyme loci and mtDNA might be lower in *Pteroglossus* in comparison to the other taxa that show a similar pattern of area relationships. Testing this hypothesis would require work at the molecular level to look at the efficiencies and effectiveness of replication and repair enzymes of toucans in comparison to other birds. However, toucans have the same levels of heterozygosity and average number of alleles per locus as do other birds; thus, it's not a lack of variation within populations that explains the lack of differentiation among populations and species.

Finally, population life-history characteristics may affect molecular divergence among populations and species. A vicariance biogeographic analysis of other taxa co-distributed with the *P. torquatus* complex indicated a single set of area relationships for Pacific and Caribbean Central America, and the Chocó region of western South America (Hackett 1992; Hackett 1993; Hackett 1996; see below). However, these toucans are much less differentiated than other taxa which presumably shared a common series of speciation events. If there were a common series of vicariant events, then the different "initial genetic conditions" that taxa experienced before speciation events influenced the amount of genetic differentiation after speciation events (Fig. 3). This scenario would work as follows: initially, toucan populations are characterized by large population sizes with considerable levels of gene flow among populations (T_1 of Fig. 3). Thus,

TABLE 2
ALLOZYME GENETIC DISTANCES FOR *PTEROGLOSSUS* SPECIES AND OUTGROUPS (SEE TEXT) ANALYZED IN THIS STUDY. NEI (1978) GENETIC DISTANCES BELOW THE DIAGONAL. ROGERS' (1972) GENETIC DISTANCES ABOVE THE DIAGONAL

	1	2	3	4	5	6	7	8	9	10	11	12	13	14
1. *P. torquatus*	0.000	0.025	0.055	0.105	0.133	0.125	0.205	0.165	0.098	0.125	0.149	0.280	0.185	0.360
2. *P. frantzii*	0.009	0.000	0.040	0.080	0.108	0.100	0.180	0.140	0.086	0.100	0.133	0.280	0.160	0.360
3. *P. sanguineus*	0.030	0.022	0.000	0.060	0.083	0.070	0.160	0.113	0.070	0.070	0.113	0.320	0.166	0.390
4. *P. erythropygius*	0.074	0.063	0.033	0.000	0.108	0.100	0.180	0.140	0.100	0.100	0.124	0.360	0.200	0.400
5. *P. castanotis*	0.095	0.083	0.041	0.083	0.000	0.078	0.121	0.088	0.128	0.120	0.120	0.345	0.188	0.407
6. *P. pluricinctus*	0.107	0.095	0.049	0.095	0.043	0.000	0.160	0.100	0.120	0.080	0.118	0.300	0.180	0.380
7. *P. inscriptus*	0.199	0.185	0.142	0.185	0.087	0.151	0.000	0.080	0.200	0.200	0.113	0.420	0.260	0.420
8. *P. bitorquatus*	0.154	0.141	0.096	0.141	0.054	0.096	0.060	0.000	0.160	0.135	0.073	0.380	0.220	0.400
9. *P. flavirostris*	0.079	0.074	0.041	0.085	0.086	0.097	0.189	0.144	0.000	0.120	0.140	0.352	0.206	0.432
10. *P. mariae*	0.107	0.095	0.044	0.095	0.090	0.063	0.200	0.131	0.097	0.000	0.153	0.355	0.220	0.400
11. *P. beauharnaesii*	0.112	0.103	0.061	0.096	0.059	0.078	0.067	0.026	0.102	0.117	0.000	0.345	0.199	0.376
12. *Selenidera*	0.298	0.308	0.346	0.426	0.369	0.326	0.511	0.448	0.405	0.416	0.378	0.000	0.320	0.360
13. *Baillonius*	0.188	0.174	0.167	0.223	0.177	0.188	0.288	0.238	0.215	0.238	0.192	0.365	0.000	0.360
14. *Ramphastos*	0.417	0.426	0.463	0.490	0.487	0.480	0.511	0.497	0.539	0.497	0.432	0.405	0.426	0.000

GENETIC DIFFERENTIATION AND BIOGEOGRAPHY IN *PTEROGLOSSUS*

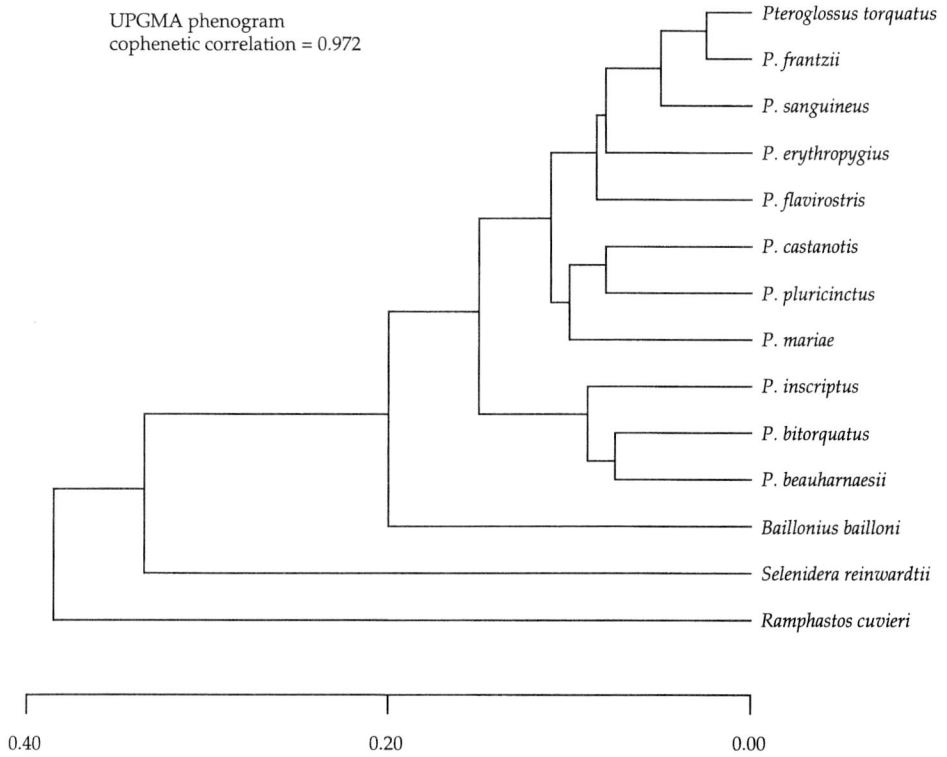

FIG. 1. UPGMA phenogram of Rogers' (1972) genetic distances (Table 2) of *Pteroglossus* species and outgroups. The "KITSCH" and "FITCH" trees of PHYLIP have the same topology.

```
flavirostris    TTTTGGATCCCTCCTAGGCATCTGCCTCGCCACACAAATCATCTCTGGCCTCCTCTTAGCCGCCCATTATACCGCAGACACC
erythropygius   ...C.........................................A.A.........C.....................
frantzii        ...C.........................................A.A.........C...........T.........
sanguineus      ...C.........................................A.A.........C.....................
torquatus       C..C..........................A............G..A.A.........C....T...............

flavirostris    TCCTTAGCCTTCTCATCCGTTGCCCACATATGTCGGAATGTCCAATATGGCTGACTAATCCGCAACCTACATGCTAACGGAG
erythropygius   ...C...............................A..C...............................T....
frantzii        .............................C..A..C..................................T....
sanguineus      ...C...............................A..C...............................T....
torquatus       ...................................A..C.....................................

flavirostris    CCTCATTCTTCTTCATCTGCATCTACCTTCACATCGGACCGAGGTTCTATTACGGATCCTACCTCTTCAAAGAAACCTGAA
erythropygius   ............T.....T..T.....C..........G.G.T.................................
frantzii        ............T.....T..T.....C..........G.G.T.................................
sanguineus      ............T.....T..T.....C..........G.G.T.................................
torquatus       ............T.....T..T.....C..........GTG.T.................................

flavirostris    CATCGGTGTTATCCTCCTCCTAACCCTCATAGCAACACGCTTCGTGGGCTACGTTCTCCCA
erythropygius   ............................G.........A..T.................
frantzii        ............................G........T.A..T................
sanguineus      ............................G.........A..T.................
torquatus       ............................G.........A..T.................
```

FIG. 2. Mitochondrial cytochrome *b* sequences for *Pteroglossus* species. Dots indicate identity to the sequence of *Pteroglossus flavirostris*.

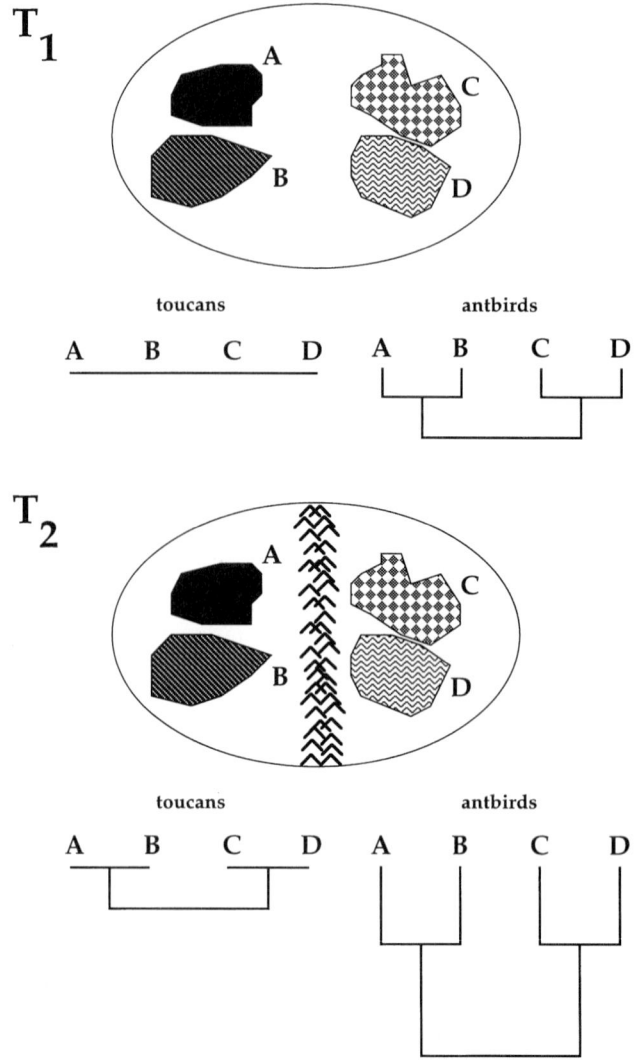

FIG. 3. Initial conditions hypothesis to explain different levels of genetic differentiation among lineages separated by the same vicariance events. At time T_1, there are no major barriers to gene flow. The toucan populations are undifferentiated; the antbird populations are differentiated due to low levels of gene flow among populations. At time T_2, after a major barrier to gene flow has arisen, toucan populations on either side of the barrier are starting to differentiate, and antbirds also continue to differentiate. Even though the same barrier to gene flow has affected these co-distributed taxa, levels of genetic differentiation across the barrier differ because different life-history characteristics caused pre-isolation differences in genetic structure.

there is little differentiation among populations. When strong barriers to gene flow arise (like a mountain range in T_2 of Fig. 3), populations on either side of the barrier are extremely genetically similar. Contrast this with the situation that could arise in a taxon characterized by little gene flow among populations, such as for an understory antbird. In the antbird, populations may be genetically differentiated, even at time T_1, due to isolation by distance. Thus, when strong barriers to gene flow arise, populations on either side of the barrier are already genetically differentiated. Under this scenario, the same pattern of phylogenetic relationships among areas result from the same series of vicariance events for both lineages, but the levels of genetic differentiation differ greatly between the co-distributed lineages. The araçaris are undifferentiated, whereas the antbirds show considerable differentiation.

More lineages need to be examined for the effects of "initial genetic conditions." If our predictions are correct, then one could look for these life-history correlates to genetic divergence in other lineages of birds: other toucan groups, canopy-dwelling tanagers, canopy-dwelling flycatchers, and perhaps some parrot lineages, to name a few. Also, if these initial conditions have widespread influence, then the use of molecular clocks is problematic, especially for fairly recent divergence events. One would expect these initial conditions to become less significant over time because as time since a speciation event increases and genetic divergence increases, the relatively small levels of population differentiation before a speciation event will contribute only a small amount of the total divergence among taxa. In any case, molecular clocks may need to be recalibrated for different lineages of birds with different life-history characteristics before they can be used to date speciation events.

Phylogeny.—Genetic differentiation within *Pteroglossus* is low enough that many species differ only in frequencies of alleles at allozyme loci. In addition, mtDNA sequence data were uninformative with respect to relationships within the *Pteroglossus torquatus* complex. Thus, the phylogenetic content of allozyme frequency differences must be addressed. Congruence of branching diagrams derived from independent character systems is an important measure of the phylogenetic content of data sets (Cracraft and Helm-Bychowski 1991; Kluge 1989; Miyamoto and Cracraft 1991; Miyamoto and Fitch 1995; Zink and Avise 1990). In many cases, congruence of independent data sets is difficult to establish because there are few published phylogenies of Neotropical birds. However, Cracraft and Prum (1988), Haffer (1974), and Prum (1988) have published partial phylogenies of *Pteroglossus* toucans based on different character systems that can be compared to our allozyme results.

Perhaps the only strongly supported relationship in the allozyme data, beyond the monophyly of *Pteroglossus*, is the placement of *Baillonius bailloni* as the sister taxon to *Pteroglossus*. This relationship was suggested by Haffer (1974) and was demonstrated by rigorous phylogenetic analyses of morphological characters by Prum (1988). All allozyme analyses place *Baillonius* as the sister to *Pteroglossus*, and the bootstrap values support this grouping at 80%.

Also, at higher taxonomic levels, the tree in Figure 1 is concordant with allozyme distance trees published by Lanyon and Zink (1987). Lanyon and Zink, in their allozyme analysis of piciform birds, concluded that *Selenidera* and *Pteroglossus* were more closely related to each other than they were to *Ramphastos*. The present allozyme distance analyses also show this. Unfortunately, *Baillonius* was not included in their analyses.

Prum (1988) used cladistic analyses of plumage characters to document monophyly of the *Pteroglossus torquatus* complex, and bill color and patterns to hypothesize relationships within the complex. Prum found the two Chocó taxa, *P. sanguineus* and *P. erythropygius*, to be sister taxa as were *P. frantzii* and *P. torquatus*, the two mostly Central American taxa. This is also the phylogeny found by the combined allozyme/DNA sequence data set. The allozyme data set supported monophyly of the *P. torquatus* complex and sister-taxon relationship of *P. frantzii* and *P. torquatus*; however, the sister-taxon relationship of *P. sanguineus* and *P. erythropygius* was not supported by allozyme genetic distances (Fig. 1).

Cracraft and Prum (1988) addressed relationships among the *P. bitorquatus* and *P. viridis* species-groups of Haffer (1974) using morphological characters. Within the *P. bitorquatus* species-group, they found that *P. flavirostris* and *P. mariae* were sister taxa (these differentiated taxa are often placed as subspecies of a single species [Haffer 1974]). The allozyme data were unable to recover this grouping; however, the two are placed in the same major group in the distance analysis (Fig. 1). The separation of *P. mariae* and *P. flavirostris* definitely needs to be checked with additional molecular data; it is at odds with the plumage similarity of these birds. The UPGMA phenogram of allozyme genetic distances (Fig. 1) differs from the relationships suggested by Cracraft and Prum (1988) in the placement of *P. bitorquatus*. The allozyme data support a clade comprised of *P. inscriptus* (traditionally a member of the *P. viridis* species-group), *P. beauharnaesii*, and *P. bitorquatus*, an hypothesis not suggested by Cracraft and Prum (1988) or Haffer (1974), that results in nonmonophyletic species-groups. However, until a rigorous and complete morphological phylogeny of the genus *Pteroglossus* is published, it is difficult to assess monophyly of the species-groups used by Haffer (1974) and Cracraft and Prum (1988), and therefore, the degree to which there is conflict between allozyme distance data and morphological phylogenies. In addition, our allozyme analysis did not include *P. viridis*, the only other member of the *P. viridis* species-group, or *P. aracari*, the other member of the *P. aracari* species group. Inclusion of these taxa could influence the allozyme trees. Because *Pteroglossus* species differ from each other mostly by allozyme frequency differences and because many of the allozyme relationships should be considered tentative, further genetic data (especially DNA sequence characters) are necessary to obtain a robust genetic estimate of phylogenetic relationships of these toucan species.

Haffer's (1974) tree is the only estimate of relationships in the genus that included all species. He hypothesized species-groups based on a number of characters (for example, plumage similarity, calls, distribution) but was tentative about joining the species-groups into higher-level relationships. He did place the *P. torquatus* complex in the same group as *P. pluricinctus* and *P. castanotis* (the *Pteroglossus aracari* species-group); the grouping of these taxa was also found in the allozyme distance analysis (Fig. 1). To make the tentative phylogeny of Haffer (1974, page 189) conform to the allozyme data would result in an increase of standard deviation of the "KITSCH" tree from 16% to 26%. For the presence/absence allozyme data set, forcing Haffer's tree increased tree length from 18 to 26 steps. Most extra steps are the result of making the *P. aracari* species-group monophyletic, and placing *P. flavirostris* and *P. mariae*, which are members of the *P. bitorquatus* species-group, with *P. bitorquatus*.

Crother (1990) dismissed the phylogenetic information in allozyme frequency differences due to temporal instability of allele frequencies. However, few studies have demonstrated temporal instability of alleles in birds (Burns and Zink 1990; see also discussion by Murphy et al. 1990, pg. 51). Thus, the agreement of different data sets on relationships within the *Pteroglossus torquatus* complex and the placement of *Baillonius* suggests that the allozyme distance data have a phylogenetic component.

Biogeography of the P. torquatus *complex.*—The pattern of phylogeny of the *P. torquatus* complex suggested by the combined allozyme/mtDNA sequence data set and the morphological data set of Prum (1988) supports the following general area cladogram: Pacific and Caribbean Central America as sister areas (*P. frantzii* and *P. torquatus*), Chocó as the sister area to Central America (*P. sanguineus* and *P. erythropygius*), and Amazonia as the sister to the Chocó/Central American area (*P. flavirostris* and the rest of the genus). This same area cladogram was suggested by Cracraft and Prum (1988), Prum (1988), and Hackett (1992, 1993, 1996) for other taxa and its implications are outlined in Cracraft and Prum (1988) and in Prum (1988). The low levels of genetic differentiation among members of the complex suggest that these divergence events are fairly recent and may have occurred at a time consistent with Pleistocene climatic fluctuations, even though direct evidence for climatic fluctuations is limited (Bush 1994; Haffer 1982).

The sister-taxon relationship of *Baillonius* and *Pteroglossus* also has biogeographic implications. *Baillonius* is endemic to southeast Brazil, and Prum (1988) hypothesized that some lineages endemic to southeast Brazil are remnants of an ancient diversification, and are sister taxa to large radiations throughout lowland Neotropics. *Baillonius* appears to be such a lineage, and the high degree of genetic differentiation between *Baillonius* and *Pteroglossus* suggests that it may have been evolving independently for millions of years. Prum also suggested that some diversification in southeastern Brazil was recent. This is probably reflected in the distribution of *P. aracari*, almost certainly the result of a more recent connection between Amazonia and southeastern Brazil.

In conclusion, diversification of lowland Neotropical taxa of birds occurred as the result of many different vicariant events over different time periods. Divergence within the *Pteroglossus torquatus* complex probably occurred during the Pleistocene, whereas *Baillonius* has probably been evolving independently for millions of years. Thus, it is likely that the events of a single time period can not explain the high diversity of forms found in the Neotropics, and the challenge is to figure out which particular geologic events and time periods influenced which particular taxa.

ACKNOWLEDGMENTS

This research was supported by grants from the National Science Foundation (NSF Doctoral Dissertation Improvement Grant, BSR-9101289), the Fugler Fellowship of Louisiana State University Museum of Natural Science, the Chapman Fund of American Museum of Natural History, the American Ornithologists' Union, and National Sigma Xi. We are grateful to J. M. Bates, R. L. Chapman, M. S. Hafner, S. M. Lanyon, D. P. Pashley, R. O. Prum, J. V. Remsen, K. V. Rosenberg, and R. M. Zink for helpful comments on this manuscript. We thank the Audubon Park and Zoological Gardens in New Orleans for the samples of *Pteroglossus sanguineus* and *Baillonius bailloni*. Also, we wish to thank the many institutions and governmental agencies in Costa Rica, Ecuador, Peru, and Bolivia for their assistance in obtaining specimens used in this study.

LITERATURE CITED

ALLARD, M. W., D. L. ELLSWORTH, AND R. L. HONEYCUTT. 1991. The production of single-stranded DNA suitable for sequencing using the polymerase chain reaction. BioTechniques 10:24–26.
BARROWCLOUGH, G. F., N. K. JOHNSON, AND R. M. ZINK. 1985. On the nature of genic variation in birds. Current Ornithol. 2:135–154.
BURNS, K. J., AND R. M. ZINK. 1990. Temporal and geographic homogeneity of gene frequencies in the Fox Sparrow (*Passerella iliaca*). Auk 107:421–425.
BUSH, M. B. 1994. Amazonian speciation: a necessarily complex model. J. Biogeogr. 21:5–17.
BUTH, D. G. 1984. The application of electrophoretic data in systematic studies. Ann. Rev. Ecol. Syst. 15: 501–522.
CRACRAFT, J., AND K. HELM-BYCHOWSKI. 1991. Parsimony and phylogenetic inference using DNA sequences: some methodological strategies. Pp. 184–220 in Phylogenetic Analysis of DNA Sequence Data (M. M. Miyamoto and J. Cracraft, Eds.). Oxford Univ. Press, New York.
CRACRAFT, J., AND R. O. PRUM. 1988. Patterns and processes of diversification: speciation and historical congruence in some Neotropical birds. Evolution 42:603–620.
CROTHER, B. I. 1990. Is "some better than none" or do allele frequencies contain phylogenetically useful information? Cladistics 6:277–281.
FELSENSTEIN, J. 1985. Confidence limits on phylogenies: an approach using the bootstrap. Evolution 79:783–791.
FELSENSTEIN, J. 1986. PHYLIP—Phylogeny inference package, Version 3.0. University of Washington, Department of Genetics, Seattle, Washington.
HACKETT, S. J. 1989. Effects of varied electrophoretic conditions on detection of evolutionary patterns in the Laridae. Condor 91:73–90.
HACKETT, S. J. 1992. Molecular phylogenies and biogeography of Central American birds. Ph.D. dissertation, Louisiana State University, Baton Rouge.
HACKETT, S. J. 1993. Phylogenetic and biogeographic relationships in the Neotropical genus *Gymnopithys* (Formicariidae). Wilson Bull. 105:301–315.
HACKETT, S. J. 1995. Molecular systematics and zoogeography of flowerpiercers in the *Diglossa baritula* complex. Auk 112:156–170.
HACKETT, S. J. 1996. Molecular phylogenetics and biogeography of tanagers in the genus *Ramphocelus* (Aves). Molecular Phylogenetics and Evolution 5:368–382.
HACKETT, S. J., AND K. V. ROSENBERG. 1990. Evolution of South American antwrens (Formicariidae): comparison of phenotypic and genetic differentiation. Auk 107:473–489.
HAFFER, J. 1969. Speciation in Amazonian forest birds. Science 165:131–137.
HAFFER, J. 1974. Avian speciation in tropical South America. Publ. Nuttall Ornithol. Club. 14:1–390.
HAFFER, J. 1982. General aspects of the refuge theory. Pp. 6–24 in Biological Diversification in the Tropics (G. T. Prance, Ed.). Columbia Univ. Press, New York.
HAFFER, J. 1985. Avian zoogeography of the Neotropical lowlands. Ornithol. Monogr. 36:113–146.
HAFFER, J. 1987. Biogeography of Neotropical birds. Pp. 105–150 in Biogeography and Quaternary History in Tropical America (T. C. Whitmore and G. T. Prance, Eds.). Clarendon Press, Oxford.
HILLIS, D. M., A. LARSON, S. K. DAVIS, AND E. A. ZIMMER. 1990. Nucleic acids III: sequencing. Pp. 318–370 in Molecular Systematics (D. M. Hillis and C. Moritz, Eds.). Sinauer Assoc., Inc., Sunderland, Massachusetts.
KLUGE, A. G. 1989. A concern for evidence and a phylogenetic hypothesis of relationships among *Epicrates* (Boidae, Serpentes). Syst. Zool. 38:7–25.
KOCHER, T. D., W. K. THOMAS, A. MEYER, S. V. EDWARDS, S. PAABO, F. X. VILLABLANCA, AND A. C. WILSON. 1989. Dynamics of mitochondrial DNA evolution in animals: amplification and sequencing with conserved primers. Proc. Natl. Acad. Sci. 86:6196–6200.
LANYON, S. M., AND J. G. HALL. 1994. Reexamination of barbet monophyly using mitochondrial-DNA sequence data. Auk 111:389–397.
LANYON, S. M., AND R. M. ZINK. 1987. Genetic relationships in piciform birds: monophyly and generic and familial relationships. Auk 104:724–732.
LEWONTIN, R. C., AND J. L. HUBBY. 1966. A molecular approach to the study of genic heterozygosity in natural populations. II. Amount of variation and degree of heterozygosity in natural populations of *Drosophila pseudoobscura*. Genetics 54:595–609.
MAYR, E., AND L. L. SHORT. 1970. Species taxa of North American birds. Publ. Nuttall Ornithol. Club 9:1–127.
MICKEVICH, M. M., AND C. M. MITTER. 1981. Treating polymorphic characters in systematics: a phylogenetic treatment of electrophoretic data. Pp. 45–58 in Advances in Cladistics, vol. 1 (V. A. Funk and D. R. Brooks, Eds.). New York Botanical Gardens, New York.
MIYAMOTO, M. M., AND J. CRACRAFT. 1991. Phylogenetic inference, DNA sequence analysis, and the future of molecular systematics. Pp. 3–17 in Phylogenetic Analysis of DNA Sequence Data (M. M. Miyamoto and J. Cracraft, Eds.). Oxford Univ. Press, New York.
MIYAMOTO, M. M., AND W. M. FITCH. 1995. Testing species phylogenies and phylogenetic methods with congruence. Syst. Biol. 44:64–76.

Murphy, R. W. 1993. The phylogenetic analysis of allozyme data: invalidity of coding alleles by presence/absence and recommended procedures. Biochem. Syst. Ecol. 21:25–38.
Murphy, R. W., J. W. Sites, Jr., D. G. Buth, and C. H. Haufler. 1990. Proteins 1: isozyme electrophoresis. Pp. 45–126 in Molecular Systematics (D. M. Hillis and C. Moritz, Eds.). Sinauer Assoc., Inc., Sunderland, Massachusetts.
Nei, M. 1978. Estimation of average heterozygosities and genetic distance from a small number of individuals. Genetics 89:583–590.
Nevo, E. 1978. Genetic variation in natural populations: patterns and theory. Theor. Pop. Biol. 13:121–177.
Prum, R. O. 1988. Historical relationships among avian forest areas of endemism in the Neotropics. Proc. Int. Ornithol. Congr. 19:2562–2572.
Rogers, J. S. 1972. Measures of genetic similarity and genetic distance. Studies in Genetics VII. Univ. Texas Publ. No. 7213:145–153.
Rogers, J. S., and R. C. Cashner. 1987. Genetic variation, divergence, and relationships in the subgenus *Xenisma* of the genus *Fundulus*. Pp. 251–264 in Community and Evolutionary Ecology of North American stream fishes (W. J. Matthews and D. C. Heins, Eds.). University of Oklahoma Press, Norman, Oklahoma.
Sanderson, M. J. 1989. Confidence limits on phylogenies: the bootstrap revisited. Cladistics 5:113–130.
Swofford, D. L. 1993. "PAUP: Phylogenetic Analysis Using Parsimony," Version 3.1.1, computer program distributed by the Illinois Natural History Survey, Champaign, Illinois.
Swofford, D. L., and S. H. Berlocher. 1987. Inferring evolutionary trees from gene frequency data under the principle of maximum parsimony. Syst. Zool. 36:293–325.
Swofford, D. L., and R. B. Selander. 1981. BIOSYS-1: a FORTRAN program for the comprehensive analysis of electrophoretic data in population genetics and systematics. J. Hered. 72:281–283.
Zink, R. M., and J. C. Avise. 1990. Patterns of mitochondrial DNA and allozyme evolution in the avian genus *Ammodramus*. Syst. Zool. 39:148–161.

APPENDIX 1

LSUMNS Tissue Numbers (beginning with B) and Collecting Localities for *Pteroglossus* Specimens Analyzed in This Study. Taxonomy Follows Haffer (1974). * Indicates Specimens Sequenced for mtDNA Analyses

P. aracari superspecies		
torquatus	B16280*	Costa Rica: prov. Limón; 11 km by road W Guápiles
	B16284	Costa Rica: prov. Limón; 11 km by road W Guápiles
	B16301	Costa Rica: prov. Limón; 11 km by road W Guápiles
	B16302	Costa Rica: prov. Limón; 11 km by road W Guápiles
frantzii	B16075	Costa Rica: prov. Puntarenas; Río Copey, ca. 4 km E Jacó
	B16076*	Costa Rica: prov. Puntarenas; Río Copey, ca. 4 km E Jacó
sanguineus	B11864	Ecuador: Prov. Esmeraldas, El Placer
	B11783	Ecuador: Prov. Esmeraldas, El Placer
	B11787	Ecuador: Prov. Esmeraldas, El Placer
	B11995*	Ecuador: Prov. Esmeraldas, El Placer
erythropygius	B13479	Audubon Park Zoo, New Orleans, Louisiana
	B16320*	Audubon Park Zoo, New Orleans, Louisiana
castanotis	B18436	Bolivia: depto. Santa Cruz; Parque Nacional Noel Kempff Mercado, 86 km ESE Florida
	B18451	Bolivia: depto. Santa Cruz; Parque Nacional Noel Kempff Mercado, 86 km ESE Florida
	B7636	Bolivia: depto. Beni; 38 km by road W Trinidad
	B12519	Bolivia: depto. Santa Cruz; Parque Nacional Noel Kempff Mercado, 50 km ESE Florida
pluricinctus	B7112	Peru: depto. Loreto; Quebrada Orán, ca. 5 km N Río Amazonas, 85 km NE Iquitos
P. viridis superspecies		
inscriptus	B18300	Bolivia: depto. Santa Cruz; Parque Nacional Noel Kempff Mercado, 86 km ESE Florida
	B18303	Bolivia: depto. Santa Cruz; Parque Nacional Noel Kempff Mercado, 86 km ESE Florida
P. bitorquatus superspecies		
bitorquatus	B18412	Bolivia: depto. Santa Cruz; Parque Nacional Noel Kempff Mercado, 86 km ESE Florida
flavirostris	B3559	Peru: depto. Loreto; S Bank Río Marañón along Río Samiria, Estación Biológico Pithecia
	B10814*	Peru: depto. Ucayali; W bank Río Shesha, ca. 65 km ENE Pucallpa
mariae	B4635	Peru: depto. Loreto; S Río Amazonas, ca. 10 km SSW mouth Río Napo on E bank Quebrada Vainilla
P. beauharnaesii	B10705	Peru: depto. Ucayali; W bank Río Shesha, ca. 65 km ENE Pucallpa
	B4950	Peru: depto. Loreto; S Río Amazonas, ca. 10 km SSW mouth Río Napo on E bank Quebrada Vainilla
	B9295	Bolivia: depto. Pando; ca. 12 km by road S Cobija, ca. 8 km W on road to Mucden
Baillonius bailloni	B19010	Audubon Park Zoo, New Orleans, Louisiana
Selenidera reinwardtii	B4164	Peru: depto. Loreto; lower Río Napo region, E bank Río Yanayacu, ca. 90 km N Iquitos
Ramphastos cuvieri	B7197	Peru: depto. Loreto; Quebrada Orán, ca. 5 km N Río Amazonas, 85 km NE Iquitos

CONTACT ZONES BETWEEN BIRDS OF SOUTHERN AMAZONIA

JÜRGEN HAFFER
Tommesweg 60, D-45149 Essen, Germany

ABSTRACT.—Numerous well-differentiated and geographically representative taxa of birds (subspecies and species) meet in southern Amazonia along zones of secondary contact whose locations are independent of the courses of large and small rivers and their floodplains. Many of these contact zones cross the rivers at right angles, and others are displaced locally. Representative taxa that meet in the Rio Madeira-Rio Tapajós interfluvium inhabit the southern portion in the headwater region and the northern portion in the Amazonian lowlands, respectively. A number of northeastern Amazonian species extend their ranges across the lower Amazon River southward and, in some cases, exhibit secondary contact with relatives of southeastern Amazonia. The representative members of another group of species and subspecies pairs (which, in most cases, are separated by the wide lower Rio Tapajós) are in direct contact in the Rio Teles Pires region, the eastern headwater stream of the Rio Tapajós. Further contact zones cross large rivers in upper Amazonia. Birds that exhibit zones of contact include not only canopy birds but also species of the middle levels of the forest and some that are restricted to the rainforest understory.

Contact zones represent regions of major biogeographic discontinuity in a continuous rainforest environment. An historical interpretation of their origin as zones of secondary contact implies large-scale separation of the respective bird populations during one or more periods of rainforest reduction during the geological past. Numerous indications of climatic-vegetational fluctuations during the Cenozoic, i.e., the Tertiary and Quaternary periods, and before are known from peripheral and central portions of Amazonia and other tropical areas of the world; these favor a refugial interpretation for the origin of the representative taxa of birds (and other animals) and their distribution patterns. Geographic separation of populations to the north and south of the Amazon River was caused by Pleistocene sea-level fluctuations that alternately flooded and exposed the central portions of the Amazon Valley (as well as portions of the continental shelf regions of South America) during interglacial and glacial periods, respectively.

RESUMO.—Várias espécies e subespécies de aves bem diferenciadas e geograficamente representativas encontram-se ao longo de zonas de contato secundário na região sul da Amazônia. Essas zonas não estão relacionadas a cursos de rios grandes ou pequenos, nem a áreas inundáveis. Taxa representativos que se encontram no interflúvio dos rios Madeira-Tapajós habitam, respectivamente, a porção sul na região de cabeceira e a porção norte nas terras baixas Amazonenses. Várias espécies do nordeste da Amazônia têm sua distribuição estendida em direção sul e em alguns casos exibem contato secundário com espécies relacionadas do sudoeste da Amazônia. Outro grupo de pares de espécies e subespécies (que na maioria dos casos estão separadas pelo Rio Tapajós) estão em contato direto na região do Rio Teles Pires, o rio de cabeceira ao leste do Rio Tapajós. Outras zonas de contato atravessam grandes rios na região superior da Amazônia. Aves que exibem zonas de contato incluem espécies de dossel e de nível intermediário, e algumas espécies de sotobosque.

Zonas de contato representam regiões de grande descontinuidade biogeográfica em um ambiente de floresta contínua. Uma interpretação histórica de sua origem como zonas de contato secundário refere-se á separação em larga escala de populações de aves durante um ou mais períodos de contração da mata durante o passado geológico. Durante o Cenozóico (durante os períodos Terciário, Quaternário e anteriores), flutuações climático-vegetacionais são conhecidas de regiões periféricas e centrais da Amazônia, e de outras áreas tropicais do mundo. Essa evidência sugere a hipótese de refúgios para a origem e distribuição de taxa representativos de aves (e outros animais). A separação geográfica de populações ao norte e ao sul do Rio Amazonas deve-se a flutuações do nível do mar que inundavam e expunham, alternadamente, a região central do Vale da Amazônia (e a região das plataformas continentais da América do Sul) durante perídos interglaciais e glaciais, respectivamente.

Early naturalist-explorers who travelled in tropical America during the 19th Century reported that many distinctly different representative species and subspecies of birds, primates and butterflies inhabit opposite banks of large Amazonian rivers or of their floodplains (Wallace 1853, Bates 1863). This "river effect" has been confirmed and discussed on the basis of additional observations and material by many later scientists who usually implied the origin of species in Amazonia through geographic isolation of populations by these river barriers (either through the subdivision of an originally continuous ancestral range by the developing river courses or through the dispersal of founder individuals across the river barriers), e.g., Hellmayr (1910, 1912), Snethlage (1913), Mayr (1942: 228), Sick (1967), Willis (1969), Hershkovitz (1969, 1977). None of these authors, however, took into consideration the diminishing barrier effect of Amazonian rivers toward their headwater regions, where many of the animal taxa inhabiting opposite sides of the broad rivers further downstream are in direct contact. There, they either hybridize, exclude each other geographically without hybridization (parapatry), or overlap their ranges (sympatry) to some extent (Haffer 1974, 1978, 1982, 1987b, 1992b, 1993).

In this paper I illustrate a number of avian contact zones in southern Amazonia whose locations are unrelated to or are merely displaced locally by broad river courses. Well-differentiated taxa of forest birds replace each other geographically along these contact zones, which represent conspicuous biogeographic discontinuities in a continuous rainforest environment. In May 1993, Theodore Parker III and I discussed several maps accompanying this article. He was particularly intrigued by the difficulties that many of these zoogeographic situations pose for the "River hypothesis," i.e., an interpretation of the origin of many Amazonian bird species under presently existing ecological conditions through geographical isolation of populations by the rivers and their floodplains alone, without the effect of certain historical circumstances, such as climatic-vegetational changes. Of course, broad Amazonian rivers do represent important barriers for many birds inhabiting the forest interior, as is again shown by several distribution maps accompanying this article. The significance of the rivers for the speciation process, however, is not immediately obvious. Herein, I illustrate and discuss the complexities in inferring historical processes from patterns of distribution.

METHODS

I visited portions of Colombian and Peruvian Amazonia during 1957–1967 (when stationed in Bogotá, Colombia) and of Brazilian Amazonia during 1990–1994 (regions around Manaus and Belém as well as the upper Rio Tapajós region in northern Mato Grosso). Specimens collected in southern Amazonia in 1993 have been deposited in the E. Goeldi Museum, Belém, Brazil. In 1994, I travelled along the Amazon and Solimões rivers from the mouth upriver to Iquitos in eastern Peru (and back again to the Atlantic Ocean), observing birds on excursions into the forests on both river banks during daily stops of the vessel.

In 1968–1973 and 1992–1995, I studied Amazonian birds, checked specimens, and noted unpublished locality records while visiting the major museums in the United States, Brazil, and Europe to supplement those records available in the published literature (see Acknowledgments and Haffer 1970, 1974, 1978). From this data base I constructed the distribution maps that accompany this article, which use only valid specimen records.

RESULTS

The members of many pairs of avian subspecies and species (not necessarily sister taxa) inhabiting primarily the *terra firme* rainforest meet along contact zones that extend from west to east (or NW to SE) across many rivers in the interfluvium between two northward-flowing rivers: the Madeira in the west and the Tapajós in the east (Fig. 1). Several contact zones even cross the latter river and continue in a southeastern direction toward or across the Rio Xingú; others are displaced for some distance by the Rio Tapajós in the east or the Rio Madeira in the west. Further groups of representative taxa meet in the region of the Rio Teles Pires, the eastern headwater stream of the Rio Tapajós and in southwestern Amazonia, where several contact zones cross the Ucayali, Javarí, Juruá, and Purus rivers. The latter group of contact zones is here only briefly mentioned. Contact zones in southern Amazonia are scattered over wide regions, although they tend to be found between the central portions of areas of endemism (Haffer 1987b). Contact zones in northern Amazonia, especially in the upper Rio Negro region and in southern Venezuela, are more conspicuously clustered than those of southern Amazonia. The distributions and interrelationships (if known) between the conspicuously different members of the subspecies and species pairs in contact in southern Amazonia are briefly discussed below.

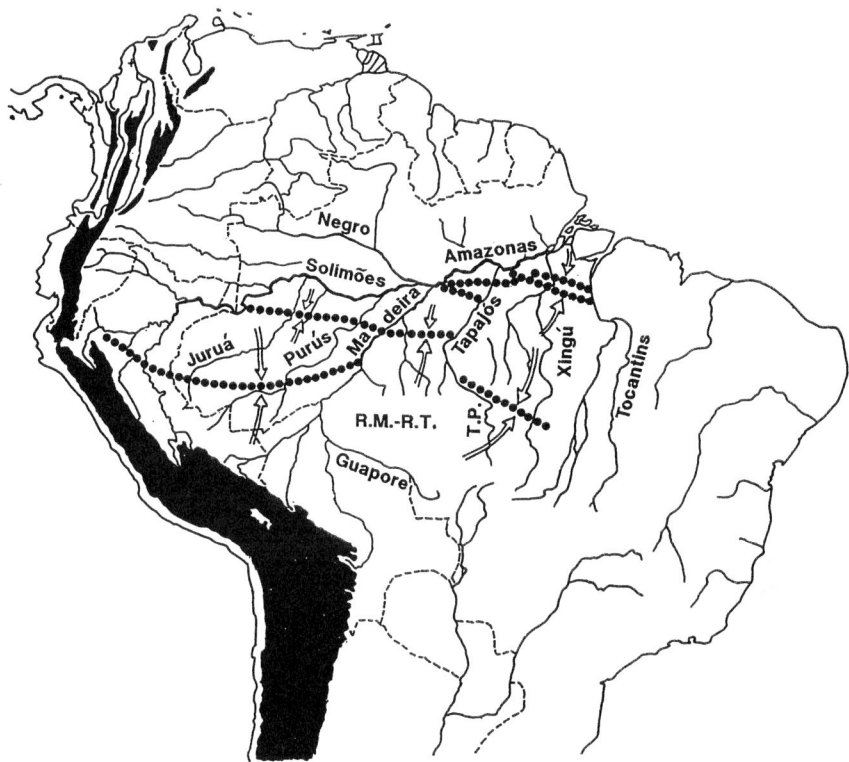

Fig. 1. Location of selected contact zones between subspecies and species of birds in southern Amazonia. These zones of biogeographic discontinuity and geographical replacement are independent of, or are variously displaced by, broad rivers. See Figures 2–12 and text for details. R.M.-R.T. = Rio Madeira-Rio Tapajós interfluvium; T.P. = Rio Teles Pires.

Contact Zones Between Birds in the Rio Madeira-Rio Tapajos Interfluvium

The taxa that established contact in the Rio Madeira-Rio Tapajós (R.M.-R.T.) interfluvium (Table 1) replace each other in the northern and southern portions of this interfluvium.

Parrots (Psittacidae).—The Vulturine Parrot (*Pionopsitta vulturina*), endemic to southeastern Amazonia, ranges, without obvious differentiation, from the Belém region west across the lower Tocantins, Xingú, and Tapajós rivers to the lower Rio Madeira and south to the Rio Cururú and the Serra do Cachimbo (Fig. 2). Its western representative of upper Amazonia is the Orange-cheeked Parrot (*P. barrabandi*), which crosses the upper Rio Madeira in a southeastern direction to occupy a large portion of the states of Rondônia and northern Mato Grosso, including the eastern headwater region of the Rio Tapajós (Rio Aripuanã, Rio Peixoto de Azevedo). Both species evidently meet in the R.M.-R.T. interfluvium along a parapatric contact zone that runs approximately NW-SE, crossing all rivers in this region, which here flow from south to north. Hybridization is not known to occur.

Jacamars (Galbulidae).—Among the member species of the *Galbula galbula* superspecies, the Green-tailed Jacamar (*G. galbula*) inhabits northeastern Amazonia, north of the lower Amazon River. The southern bank is occupied by the widespread Brazilian *G. ruficauda rufoviridis* (probably better treated as a species) except in the area between the lower Rio Madeira and the lower Rio Tapajós, where the northern species *G. galbula* is common (Haffer 1974). Evidently *G. galbula* had crossed the Amazon River from north to south into the northern portion of the R.M.-R.T. interfluvium before *G. r. rufoviridis* occupied this area, spreading from the south. The species probably meet along creeks and streams of the R.M.-R.T. interfluvium along an E-W contact zone at some distance south of the Amazon River. Details remain unknown.

Barbets (Capitonidae).—A distribution pattern and contact zone similar to the jacamars is found in a pair of barbets (Fig. 3), the Brown-breasted Barbet (*Capito brunneipectus*) and the Black-girdled Barbet (*C. dayi*). *Capito brunneipectus* inhabits the northern portion of the

TABLE 1
Rondônia Elements (RO) and Upper Amazonian Elements (A) of the Avifauna in Contact with Their Lower Amazonian Representatives in the *terra firme* Rainforests of the Rio Madeira–Rio Tapajós Interfluvium, Southern Amazonia; See Figures 1–5 for the Location of Some of These Contact Zones

Rondônia (RO) and Upper Amazonian (A) elements	Lower Amazonian elements
Pionopsitta barrabandi (A)	*P. vulturina*
Capito dayi (RO)	*C. brunneipectus*
Pteroglossus beauharnaesii (A)	*P. aracari*
Ramphastos (vitellinus) culminatus-pintoi (A)	*R. (vitellinus) ariel*
Xiphorhynchus (guttatus) guttatoides (A)	*X. (g.) eytoni*
Myrmotherula i. heteroptera (A)	*M. i. iheringi*
M. longipennis transitiva (RO)	*M. l. ochrogyna*
M. l. leucophthalma (A)	*M. l. phaeonota*
Terenura humeralis (A)	*T. spodioptila*
Myrmoborus m. myotherinus (A)	*M. m. ochrolaema*
Formicarius colma nigrifrons (A)	*F. c. amazonicus*
Rhegmatorhina hoffmannsi (RO)	*R. berlepschi*
Xipholena punicea[1]	*X. lamellipennis*
Basileuterus fulvicauda (A)	*B. rivularis*

[1] A species widespread in the Guianas and northeastern Amazonia that expands its range into upper Amazonia.

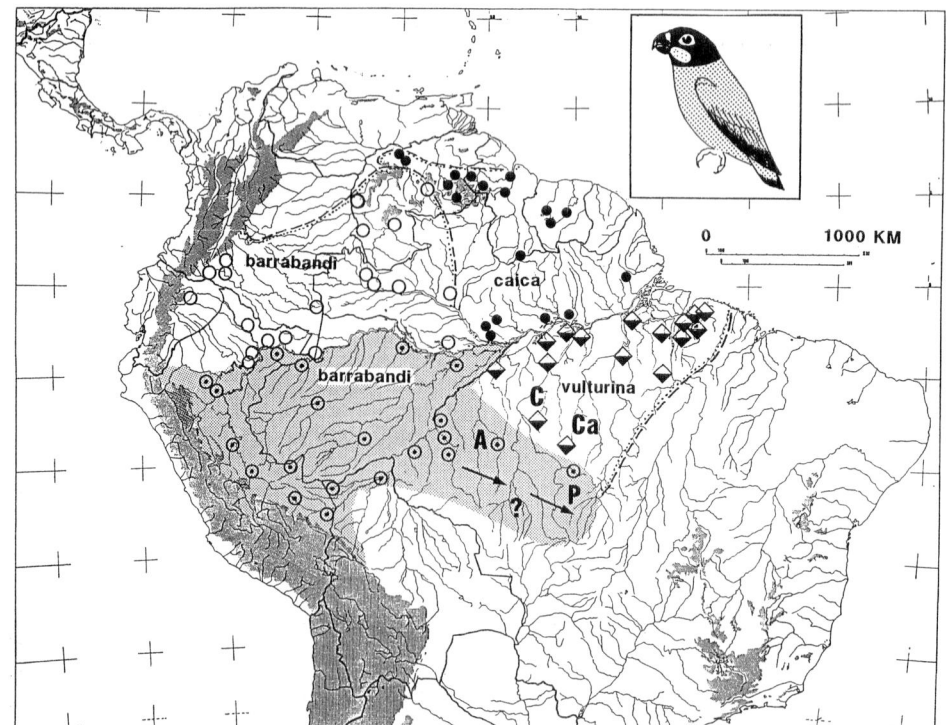

Fig. 2. Distribution of parrots of the *Pionopsitta caica* superspecies (from Haffer 1992b). Locality records (symbols) refer to the following species and subspecies: *P. barrabandi* (bird sketched; *P. b. barrabandi* open circles; *P. b. aurantiigena* circles with center dot); *P. caica* (solid circles); *P. vulturina* (half-solid squares). C = Rio Cururú, Ca = Cachimbo, A = Rio Aripuanã, P = Rio Peixoto de Azevedo. Dashed and dotted lines follow range limits.

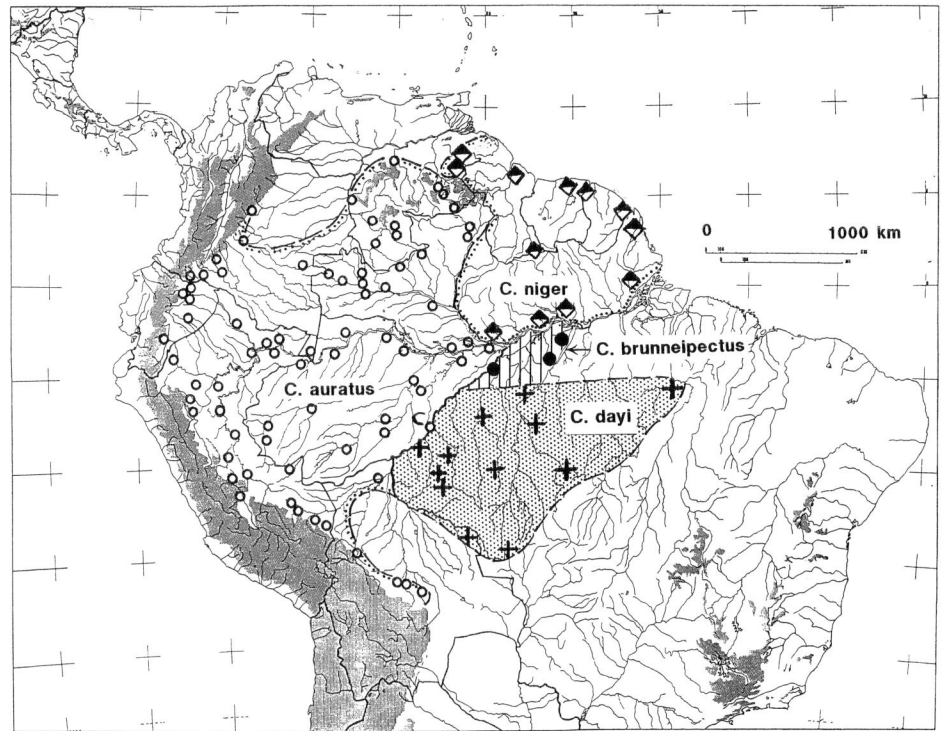

FIG. 3. Distribution of several Amazonian barbets (Capitonidae). Locality records (symbols) refer to the following species: *Capito auratus* (open circles); *C. brunneipectus* (solid circles and vertical hatches); *C. niger* (half-solid squares on edge); *C. dayi* (crosses and stippling). C = opposite Calamá. Dashed and dotted lines follow range limits.

R.M.-R.T. interfluvium, whereas *C. dayi* is found in the hilly area to the south (eastward to the lower Rio Tocantins). Sick (1958) reported a series of nine specimens of *C. dayi* in the Museu Nacional (Rio de Janeiro) from "the Guaporé region (Rio Guaporé, Javarí, Jaurú)." This Javarí River evidently is not identical with the Rio Javarí in the state of Amazonas that forms the Brazilian boundary with Peru and which Sick (1993) erroneously mentioned later as the western range limit of this species. Comments on the rationale for treating these barbets and *C. auratus* as species rather than subspecies are given in Appendix 1.

Toucans (Ramphastidae).—The Curl-crested Aracari (*Pteroglossus beauharnaesii*), a southern Amazonian species, crossed the upper Rio Madeira in a southeastern direction, reaching the headwaters of the Tapajós and Xingú rivers (Fig. 4). The northern portion of the R.M.-R.T. interfluvium is inhabited by the Black-necked Aracari (*P. aracari*), which probably meets its southern representative in the same general area where the parrots *Pionopsitta vulturina/P. barrabandi*, as well as the members of several other species pairs, have established contact. Two well-differentiated forms of the Channel-billed Toucan (*Ramphastos vitellinus*) also meet in the central region of the R.M.-R.T. interfluvium, where they hybridize extensively (Haffer 1974): the northern *R. (v.) ariel* and the southern *R. (v.) culminatus-pintoi*. This hybrid zone extends in an E-W direction across the rivers Juruena, Teles Pires, Xingú, and others.

Woodcreepers (Dendrocolaptidae).—The Ocellated Woodcreeper (*Xiphorhynchus ocellatus*) inhabits upper Amazonia and, in the east, reaches the lower Rio Tocantins. It is missing, however, from the southbank of the Amazon between the lower Rio Tapajós and Xingú where, on the other hand, its northern representative species *X. pardalotus* is common (Zimmer 1934: 14). This is a zoogeographic situation reminiscent of that of the jacamars *Galbula galbula/G. ruficauda rufoviridis* discussed above.

Antbirds (Formicariidae).—The members of several pairs of antbird species or well-differentiated subspecies probably also established contact in the R.M.-R.T. interfluvium along an E-W contact zone, although details of their interrelationships remain unknown: *Myrmotherula i. iher-*

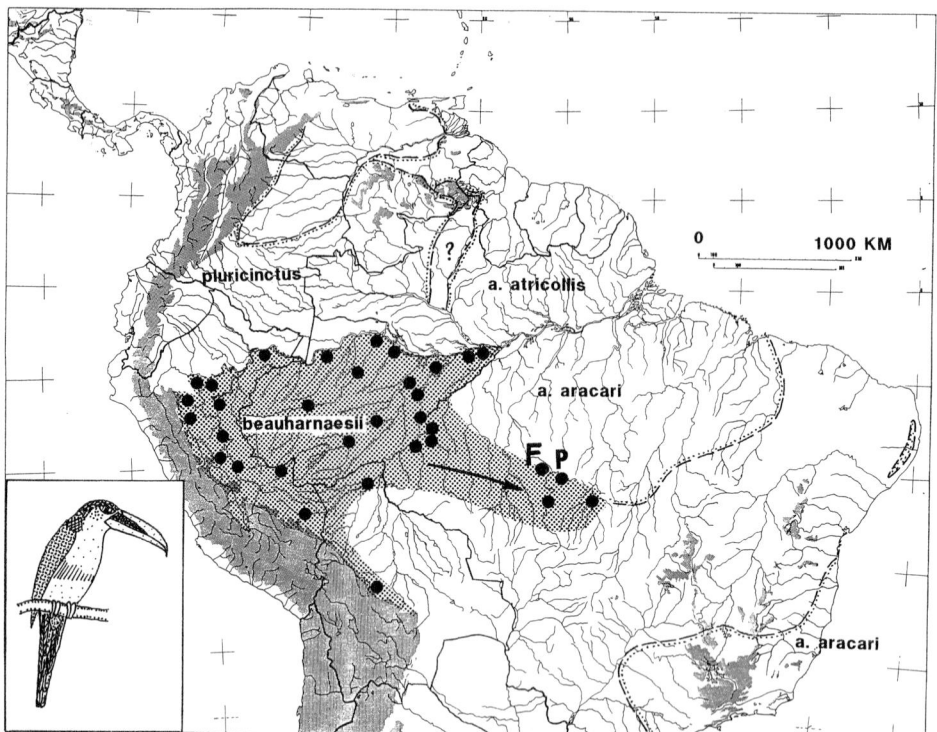

FIG. 4. Distribution of the Curl-crested Aracari (*Pteroglossus beauharnaesii;* from Haffer 1992b). Solid circles denote locality records. F = Alta Floresta, P = Rio Peixoto de Azevedo. Names of representative species are indicated in their respective distribution areas. Dashed and dotted lines follow range limits.

ingi/M. i. heteroptera, M. longipennis ochrogyna/M. l. transitiva, M. l. phaeonota/M. l. leucophthalma, Terenura spodioptila/T. humeralis, Myrmoborus m. ochrolaema/M. m. myotherinus. The first-named representatives inhabit the northern portion of the R.M.-R.T. interfluvium, whereas the last-named representatives occupy the region of Rondônia and northern Mato Grosso to the south (Cory and Hellmayr 1924). The members of the species pair *Hypocnemoides melanopogon/H. maculicauda* also replace each other in the northern and southern portion of this interfluvium, respectively (see Fig. 5.7 in Haffer 1987b). In the case of the subspecies pair *Formicarius colma amazonicus/F. c. nigrifrons*, the latter upper Amazonian form (with a black front) established a rather restricted "bridgehead" east of the upper Rio Madeira in Rondônia, leaving most of the R.M.-R.T. interfluvium to its eastern representative *F. c. amazonicus* (Fig. 5), a member of the *ruficeps*-group of subspecies (upper head entirely rufous brown). Another pair of ant-following antbirds in the R.M.-R.T. interfluvium are *Rhegmatorhina berlepschi/R. hoffmannsi* (Willis 1969; see Fig. 8 in Haffer 1992b); their contact zone, however, is far to the north near the Amazon River.

Cotingas (Cotingidae).—The members of two species pairs established contact in the northern portion of the R.M.-R.T. interfluvium near the Amazon River: *Xipholena lamellipennis/X. punicea* and *Phoenicircus carnifex/P. nigricollis*. In the former pair, the Pompadour Cotinga (*X. punicea*) crossed the Rio Madeira to occupy all of Rondônia and northern Mato Grosso into northernmost Bolivia. Its representative, the White-tailed Cotinga (*X. lamellipennis*), is restricted to southeastern Amazonia, as illustrated by Haffer (1970; see Fig. 11 in Haffer 1992b) and Snow (1982). In the *Phoenicircus* species, the Guianan Red-Cotinga (*P. carnifex*) extended its range southward across the lower Amazon River (Fig. 6). The contact zone with its wide-ranging representative, the Black-necked Red-Cotinga (*P. nigricollis*) of western and southern Amazonia, runs mostly east-west, crossing the southern tributaries of the Amazon River, although it is appreciably displaced by the broad lower Rio Tapajós. Overlap and sympatry of these species between the lower Madeira and Tapajós rivers, as suggested by the distribution maps of Ridgely and Tudor (1994), probably do not exist.

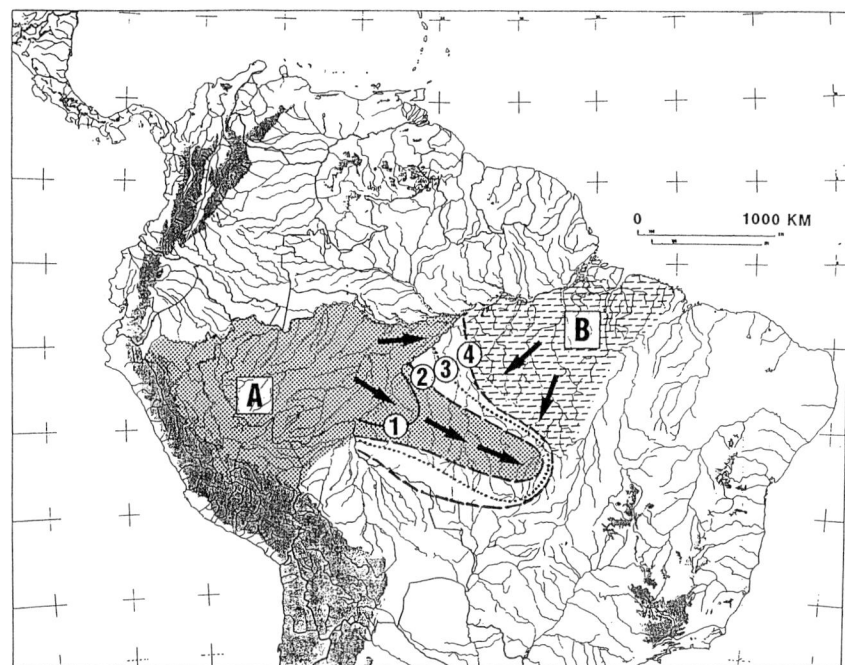

FIG. 5. Geographical replacement of some bird species and subspecies in southern Amazonia (slightly modified from Haffer 1992b). Several taxa of southwestern Amazonia (A; shaded area) and, in part, northern Amazonia, have advanced across the upper Rio Madeira southeastward into Rondônia and northern Mato Grosso, reaching the headwater regions of the Tapajos and Xingu rivers, where they replace their respective geographical representatives of southeastern Amazonia (B; hatched area) along sharply defined contact zones. In the following list the name of the western subspecies/species is given first: 1—*Formicarius colma nigrifrons*/*F. c. amazonicus*; 2—*Pteroglossus beauharnaesii*/*P. aracari*; 3—*Pionopsitta barrabandi*/*P. vulturina*; 4—*Xipholena punicea*/*X. lamellipennis*; see Figures 2 and 4 for details.

Manakins (Pipridae).—The Crimson-hooded Manakin (*Pipra aureola*), which inhabits both banks of the lower Amazon River, and the Band-tailed Manakin (*P. fasciicauda*) are in contact in the valleys of the lower Tocantins, Xingú, and Tapajós rivers, and in the upper Madeira Valley (Fig. 7). This zone of geographical replacement is again independent of the river courses, which are crossed at right angles. The ranges of these manakins are strictly complementary; sympatry of *P. aureola* and *P. fasciicauda* in the area between the lower Xingú and Tocantins rivers as illustrated by Ridgely and Tudor (1994) is not known to exist. Occasional hybrids between these species have been reported from two contact zones: near Santarém at the mouth of the Rio Tapajos between *Pipra aureola* and *P. fasciicauda*, and just west of the lower Rio Madeira between *P. filicauda* and *P. aureola* (*P.* "*heterocerca*"; Haffer 1970, 1974:83).

Miscellaneous.—Scale-breasted Woodpecker (*Celeus grammicus*) and Waved Woodpecker (*C. undatus*) are geographical representatives of the same lineage. They may meet at the lower Rio Xingú. In southern Amazonia, the range of *C. grammicus* extends east to the upper Xingú River (a male from Posto Jacaré preserved in the Goeldi Museum, Belém, may have been collected on either river bank). The distribution pattern of the well-differentiated forms of the tyrannulet *Zimmerius acer* resembles that of the *Phoenicircus* species mentioned above. The Guianan *Z. a. acer* occupies the area to the north of the lower Amazon River as well as extensive portions of the lowlands to the east of the lower Rio Tapajos, where it is in contact with its representative, *Z. a. gracilipes*.

A group of northern Amazonian species (Table 2) crossed the lower Amazon River southward at different points and occupied variously extensive portions of southeastern Amazonia without, however, establishing contact with geographically representative species. The parrot *Touit purpurata* is fairly widespread south of the Amazon River, whereas *Myrmeciza ferruginea* occupied only a small area between the lower Rio Madeira and the Rio Tapajós (see also *Galbula galbula* and *Xiphorhynchus pardalotus* mentioned above). Other species in this group crossed the Am-

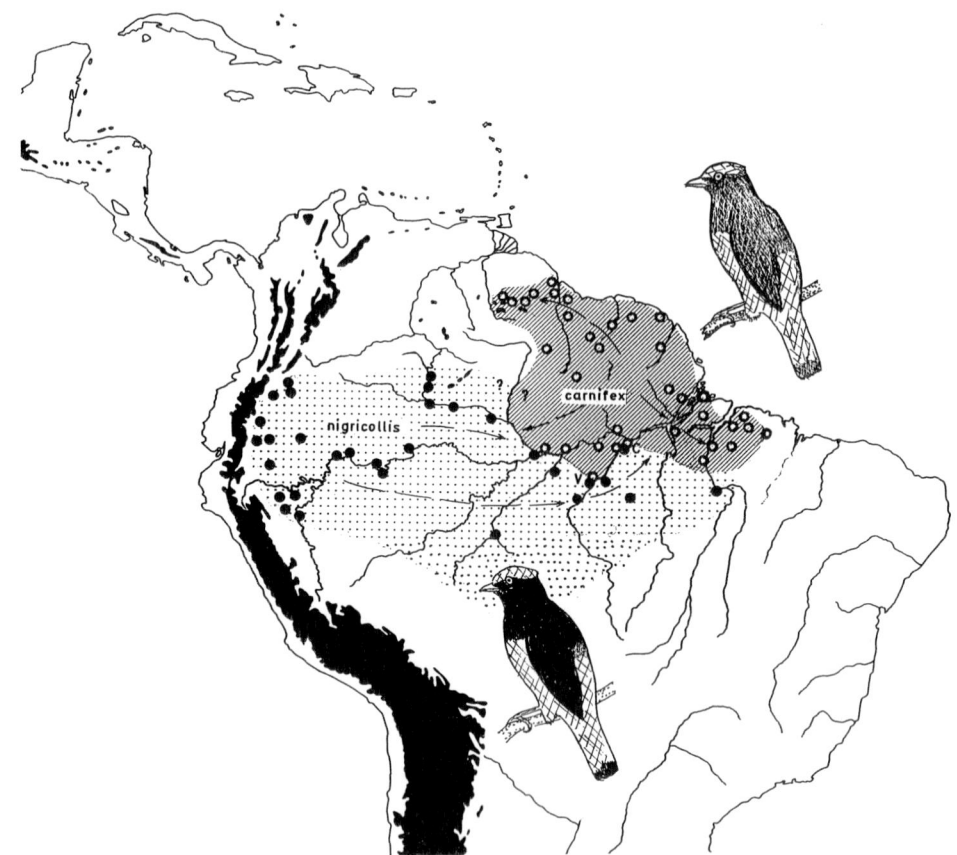

FIG. 6. Distribution of the Red-Cotingas, *Phoenicircus carnifex* superspecies (males illustrated; modified from Haffer 1970 and Snow 1982). Locality records (symbols) refer to *P. carnifex* (open circles and shading) and *P. nigricollis* (solid circles and stippling). The record of *P. nigricollis* from Marabá, Rio Tocantins, requires confirmation.

azon River nearer its mouth, where the river channel is fairly narrow and where numerous islands may have facilitated their transfer from north to south. The White Bellbird (*Procnias alba*) breeds in the hill forests of the Guianan region (Snow 1982) and, to the south of the lower Amazon River, in the Serra dos Carajás (Roth et al. 1984, Oren and Novaes 1985). A vagrant individual (or seasonal migrant) was collected near the city of Belém (A.R. Wallace 1853; see Oren and Novaes 1985). The Red-and-black Grosbeak (*Periporphyrus erythromelas*) inhabits humid forests both to the north and south of the mouth of the Rio Amazonas.

CONTACT ZONES BETWEEN BIRDS IN THE RIO TELES PIRES REGION, THE
EASTERN HEADWATERS OF THE RIO TAPAJOS

A number of representative taxa are in contact in the eastern headwater region of the Rio Tapajós (Fig. 1; Table 3), where they either hybridize extensively (subspecies) or exclude each other geographically along parapatric contact zones (species). Details on the location of these contact zones and on the interrelationships of these taxa, however, are usually lacking. In the area of the lower Rio Tapajós most, although not all, of these birds inhabit the forests on opposite banks of this wide river course.

The populations of the Dark-winged Trumpeter (*Psophia viridis*) are differentiated as subspecies between the wide southern tributary streams of the lower Rio Amazonas. They integrade in the forested headwater regions of the same river that effectively separates two subspecies further downstream. I collected specimens intermediate between *P. v. dextralis* and *P. v. viridis* near Alta Floresta, Mato Grosso. The population of the White-bellied Parrot (*Pionites leucogaster*)

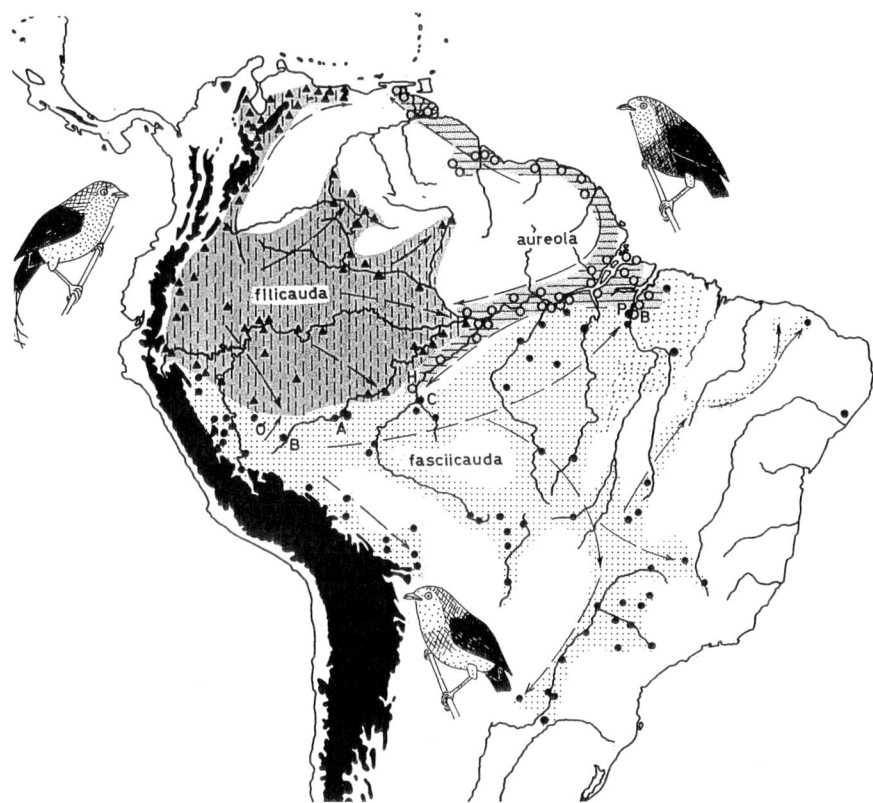

FIG. 7. Distribution of manakins of the *Pipra aureola* superspecies (males illustrated). Locality records (symbols) refer to the following species: *P. aureola* (open circles and horizontal hatching), *P. fasciicauda* (solid circles and stippling), and *P. filicauda* (solid triangles and vertical dashes). An isolated occurrence of *P. fasciicauda* in Alagôas, northeastern Brazil has been reported by Ridgely and Tudor (1994).

inhabiting the Rio Teles region is highly variable and intermediate between *P. l. xanthurus* and the nominate form (Novaes 1981). In the hummingbird species of the genus *Phaethornis* listed in Table 3, nothing is known about geographical exclusion with or without hybridization in the upper Tapajós region, although the representative taxa are probably in contact (*P. philippii* and *P. bourcieri major*; the latter form was described by Hinkelmann 1989). The same may be true for the well-differentiated subspecies of the piculet *Picumnus aurifrons* in this region (*P. a. aurifrons/transfasciatus*). Certain intermediate specimens indicate gene flow, in this area, between the forms of the Red-necked Aracari (*Pteroglossus bitorquatus*; see Fig. 5 in Haffer 1992b).

The medium-sized woodcreepers *Xiphorhynchus elegans* and *X. spixii* are in contact a short distance east of the upper Rio Teles Pires that here flows from south to north (Fig. 8). Typical specimens of each species have been collected in the area of the Rio Peixoto de Azevedo (P of Fig. 8; Novaes and Lima 1992; *X. elegans*) and at Cachimbo (C of Fig. 8; Pinto and Camargo 1957; *X. spixii*). I have examined these birds at the Goeldi Museum in Belém and at the Museu de Zoología in São Paulo, respectively. No river barrier separates the ranges of these species in this region. This is the case, however, a short distance west of the Rio Peixoto de Azevedo, where I collected additional material along the northern (*X. spixii*) and southern bank (*X. elegans*) of the Rio Teles Pires near Alta Floresta, Mato Grosso (specimens in Goeldi Museum, Belém, Brazil). This river is here about 200–300 m wide. None of these birds show any phenotypic indications of hybridization. Accordingly, I treat *X. spixii* and *X. elegans* as specifically distinct (*contra* Ridgely and Tudor 1994; see Appendix 2 for comments on species limits and taxonomy in these woodcreepers).

The geographical representatives (subspecies) of several antbirds (Formicariidae) meet and probably intergrade in the Rio Teles Pires region. Examples (Table 3) include subspecies of

TABLE 2

Northern or Northeastern Amazonian Bird Species that Crossed the Lower Amazon River Southward for Various Distances, in Many Cases Establishing Contact with a Southern or Upper Amazonian Representative; See Figures 6 and 7 for the Location of Two of These Contact Zones. Several Species Have No Obvious Southern or Western Representatives

Upper Amazonian or southcentral Amazonian representative	Northeastern Amazonian species present south of the lower Amazon River
Forpus xanthopterygius	*F. passerinus*
Galbula ruficauda rufoviridis	*G. galbula*
Ramphastos (tucanus) cuvieri	*R. (t.) tucanus*
Celeus grammicus	*C. undatus*
Xiphorhynchus ocellatus	*X. pardalotus*
Phoenicircus nigricollis	*P. carnifex*
Pipra fasciicauda	*P. aureola*
Zimmerius a. gracilipes	*Z. a. acer*
—	*Touit purpurata*
—	*Myrmeciza ferruginea*
—	*Poecilotriccus andrei*
—	*Lophotriccus galeatus*
—	*Cotinga cotinga*
—	*Procnias alba*
—	*Haematoderus militaris*
—	*Polioptila guianensis*
—	*Cyanicterus cyanicterus*[1]
—	*Periporphyrus erythromelas*

[1] Reported to the south of the Amazon River at Borba on the lower Rio Madeira (Silva and Willis 1986) and along the upper Rio Urucú, south of Tefé (Peres and Whittaker 1991).

TABLE 3

Representative Taxa of Birds in Contact in the Region of the Rio Teles Pires, Eastern Headwater Stream of the Rio Tapajós, Southern Amazonia. Most But Not All of These Birds Inhabit Opposite Banks of the Wide Lower Tapajós River; See Figures 8–11 for the Location of Selected Contact Zones

Western representative	Eastern representative
Psophia v. viridis	*P. v. dextralis*
Pionites l. xanthurus	*P. l. leucogaster*
Phaethornis superciliosus insignis	*P. s. muelleri*
Phaethornis philippii	*P. bourcieri major*
Pteroglossus bitorquatus sturmii	*P. b. reichenowi*
Picumnus a. aurifrons	*P. a. transfasciatus*
Xiphorhynchus elegans	*X. spixii*
Myrmotherula longipennis transitiva	*M. l. paraensis*
Myrmotherula menetriesii berlepschi	*M. m. omissa*
Myrmotherula l. leucophthalma	*M. l. sordida*
Microrhopias quixensis bicolor	*M. q. emiliae*
Rhegmatorhina hoffmannsi	*R. gymnops*
Hylophylax poecilinota griseiventris	*H. p. nigrigula*
Phlegopsis nigromaculata subsp.	*P. n. bowmani*
Pipra nattereri	*P. iris*
Chiroxiphia (pareola) regina	*C. (p.) pareola*
Todirostrum chrysocrotaphum similis	*T. c. illigeri*
Thryothorus genibarbis	*T. coraya*
Hylophilus muscicapinus griseifrons	*H. hypoxanthus inornatus*

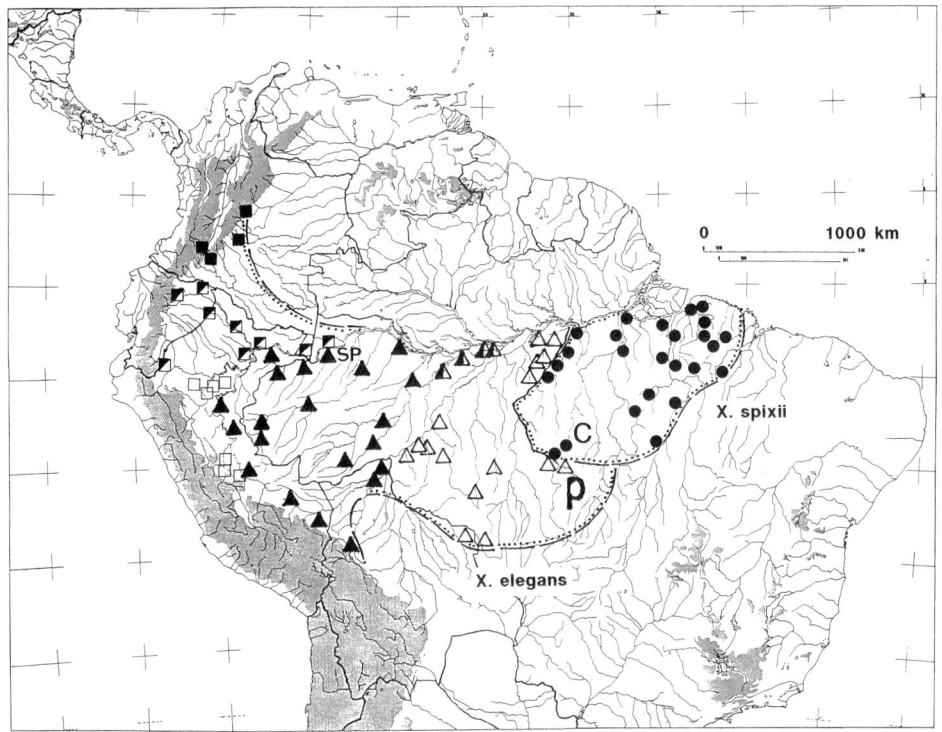

FIG. 8. Distribution of woodcreepers of the *Xiphorhynchus spixii* superspecies. Locality records (symbols) refer to the following taxa: *X. spixii* (solid circles), *X. e. elegans* (open triangles), *X. e. elegans*≷*juruanus* (half-solid triangles), *X. e. juruanus* (solid triangles), *X. e. insignis* (open squares), *X. e. ornatus* (half-solid squares), and *X. e. buenavistae* (solid squares). Dashed and dotted lines follow range limits. C = Cachimbo, P = Rio Peixoto de Azevedo, SP = São Paulo de Olivença.

Myrmotherula longipennis, *M. menetriesii*, *M. leucophthalma*, and *Microrhopias quixensis*. In each species, rather well-differentiated subspecies inhabit large areas of southcentral and southeastern Amazonia, respectively. They intergrade probably along fairly narrow contact zones. In *Hylophylax poecilinota*, the Rio Teles Pires separates the ranges of *H. p. nigrigula* (east bank) and *H. p. griseiventris* (west bank), as I established during recent fieldwork (Haffer, unpubl. data). Farther north, however, the black-throated form *nigrigula* occupies the forests both west and east of the lower Rio Tapajós (Fig. 9). *Hylophylax p. nigrigula* is also known from the Serra do Cachimbo and from the Rio Peixoto de Azevedo (Novaes and Lima 1992). In western Brazil, the upper Rio Juruá separates, at least locally, the ranges of another black-throated subspecies *H. p. gutturalis* (phenotypically, a form of the *lepidonota* group) north of this river (Gyldenstolpe 1945) from *H. p. griseiventris* to the south (Santa Cruz on the Rio Eiru; Pinto 1947, Novaes 1957). Nominate *H. p. poecilinota* inhabits northeastern Amazonia north of the lower Amazon River and east of the Rio Negro, including the surroundings of Manaus and Itacoatiará (Rio Atabany, Igarapé Anibá; material in Museu de Zoología, São Paulo). Specimens of *H. p. duidae* from this same area ("Igarapé Anibá") mentioned by Gyldenstolpe (1945) probably are either mislabelled or misidentified (see also Willis 1982). *Rhegmatorhina hoffmannsi* and *R. gymnops* meet somewhere in the region of the Rio Juruena, because both banks of the Rio Teles Pires near Alta Floresta are inhabited by the latter species (see Fig. 8 in Haffer 1992b). The population of *Phlegopsis nigromaculata bowmani* at the Rio Teles Pires near Alta Floresta shows some intermediacy with an undescribed form of the R.M.-R.T. interfluvium (see Fig. 9 in Haffer 1992b).

Among the manakins (Pipridae), *Pipra nattereri* is a species of the R.M.-R.T. interfluvium that extends its range eastward across the Rio Teles Pires into the headwaters of the Rio Xingú. In that region it is in contact with *P. iris* of southeastern Amazonia (Haffer 1970, 1992b). Several specimens of the supposed species *Pipra "vilasboasi"* and *P. "obscura"* (Sick 1959a, b) are

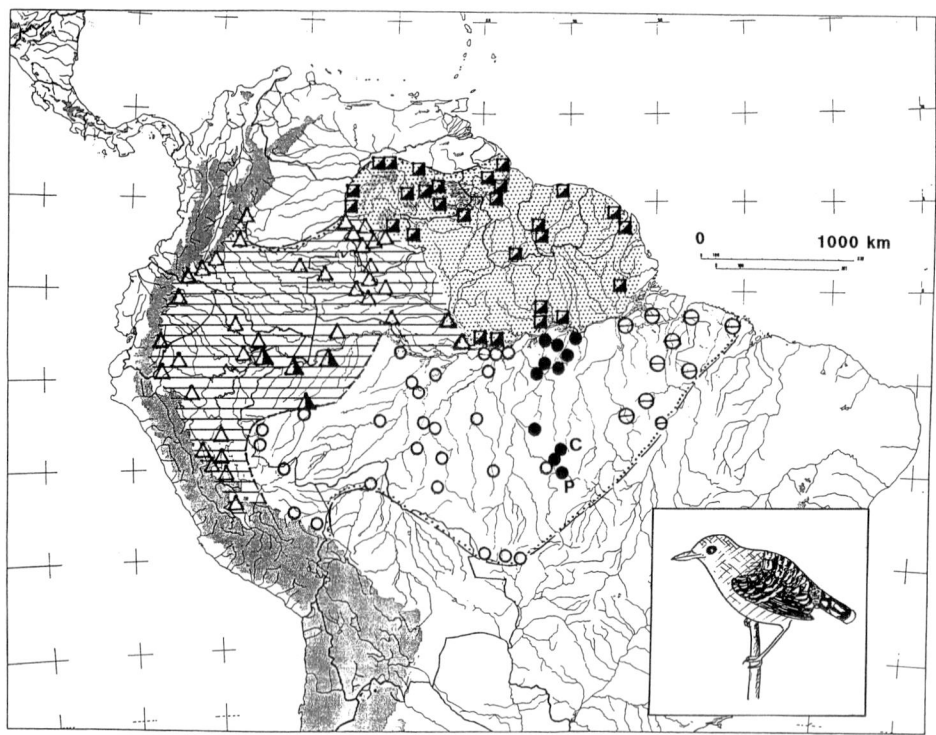

FIG. 9. Distribution of the Scale-backed Antbird (*Hylophylax poecilinota*); male of a plain-throated form illustrated. Locality records (symbols) refer to the following subspecies: *poecilinota* (half-solid squares and stippled area), *lepidonota* group (hatched horizontally): *lepidonota* and *duidae* (open triangles), *gutturalis* (half-solid triangles); *griseiventris* group (blank area): *griseiventris* (open circles), *nigrigula* (solid circles), and *vidua* (open circles with horizontal dash). Dashed line in southeastern Peru and western Brazil indicates the approximate location of the contact zone between *lepidonota-gutturalis* and *griseiventris*. Dashed and dotted lines follow range limits. C = Serra do Cachimbo, P = Rio Peixoto de Azevedo.

known from the small Rio Cururú, a short distance east of the Rio Teles Pires in the general region of contact between *P. nattereri* and *P. iris*. These birds represent probably no more than hybrid phenotypes between the latter two parapatric species (Haffer, in prep.). Future field studies may find that hybrid individuals similar to "*vilasboasi*" and "*obscura*" occasionally occur in the populations along the contact zone between *P. nattereri* and *P. iris* (see also *P.* "*heterocerca*" mentioned above). The lack of hybridization among *P. coronata*, *P. nattereri* and *P. iris* in other regions is due to the fact that the ranges of these species are separated by the large rivers Madeira and Tapajos. However, occasional hybridization between *P. nattereri*, and *P. iris* probably does occur in the headwater region of the Rio Tapajos, where this river ceases to act as an effective barrier permitting direct contact between these species. Parkes (1961) and Stotz (1993) discussed general aspects of hybridization in Amazonian manakins.

Chiroxiphia (*pareola*) *regina* with a yellow cap and *C.* (*p*) *pareola* with a red cap, as well as the two forms of the small flycatcher *Todirostrum chrysocrotaphum* also meet in the Rio Teles Pires region (Table 3). Two closely related and geographically representative species of wrens, Coraya Wren (*Thryothorus coraya*) and Moustached Wren (*T. genibarbis*), inhabit mainly northern and southern Amazonia, respectively (Fig. 10). The Rio Solimões-Amazonas separates their ranges over long distances. However, northern *T. coraya* crossed the Amazon River southward in two areas; first, near the Peruvian Andes and, second, in lower Amazonia between the Rio Tapajós and the Rio Tocantins. The range of *T. genibarbis* is complementary to that of *T. coraya*, and both species probably are in direct contact in many areas (including the Rio Teles Pires region), although no details on their interrelationships are known. The distributional ranges of these species as described and depicted by Ridgely and Tudor (1989) suggest wide zones of overlap south of the Amazon River for which there is no evidence.

Two species of greenlets (*Hylophilus*) also have rather complex complementary distribution

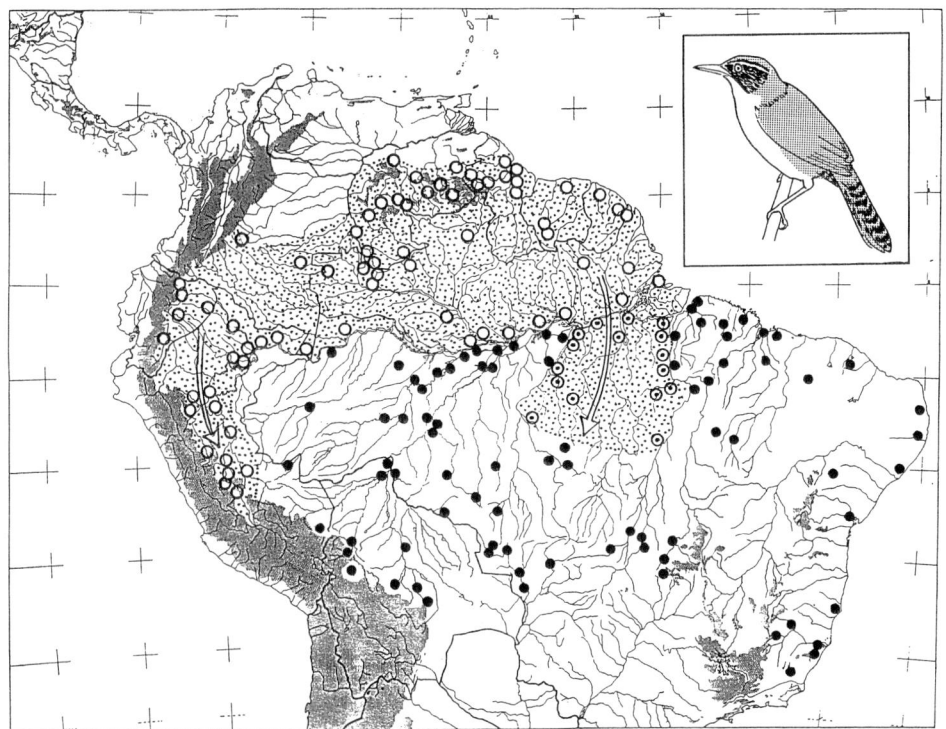

Fig. 10. Distribution of wrens of the *Thryothorus coraya* superspecies. Locality records (symbols) refer to the following taxa: *T. coraya* (various subspecies; open circles, stippling and bird sketched), *T. c. herberti* (open circles with center dot and stippling), and *T. genibarbis* (solid circles).

patterns (Fig. 11). Upper Amazonian Dusky-capped Greenlet (*H. hypoxanthus*) ranges eastward to the Rio Negro and the Rio Madeira beyond which rivers it is replaced by the Buff-cheeked Greenlet (*H. muscicapinus*). However, a geographically isolated form of *H. hypoxanthus* reappears, south of the Amazon River, in the area between the Rio Tapajós and the lower Rio Tocantins (*H. h. inornatus*). Several specimens of *H. muscicapinus griseifrons* preserved in the American Museum of Natural History, New York, and supposedly collected at the lower Rio Xingu, probably are mislabelled (Haffer, in prep.). The distributional ranges of these species as depicted by Ridgely and Tudor (1989) suggest zones of regional overlap for which there is no evidence. A sight record of *H. hypoxanthus* from northeasternmost Bolivia (Bates et al. 1989) is no longer considered valid (Bates, pers. comm.). Chapman's (1921) name "*albigula*" for birds from Ilha de Santa Julia, Rio Iriri (lower Rio Xingú region) and the Rio Jamauchim (Santa Helena), most probably is a synonym of *H. h. inornatus* (type locality Cametá on the lower Rio Tocantins), as suggested by Zimmer (1942). The type localities of *inornatus* and "*albigula*" are only 400 km apart (Fig. 11). Ridgely and Tudor (1989) confirmed the probable identity of these forms by comparing series of both from the lower Rio Tapajos. However, the yellower-bellied specimens from west of the lower Rio Madeira (Caviana, Rio Solimões and lower Rio Purus, at Carnegie Museum, Pittsburgh) which Ridgely and Tudor (1989) accepted as representing "true *albigula*," following Todd (1929) and Blake (1968), actually belong to *H. hypoxanthus ictericus* Bond (1953), as I established when comparing these specimens recently.

DISCUSSION

There is no question that the wide lower portions of many Amazonian rivers effectively separate the ranges of numerous forest birds, thus probably causing or enhancing the development of genetic and morphological differences of the separated populations (Capparella 1988; Haffer 1974, 1992b). However, many of these populations are in broad contact in the headwater regions of Amazonian rivers where rivers cease to be effective barriers, and more or less uninhibited gene flow connects the intergrading populations. Examples are *Pteroglossus bitorquatus*

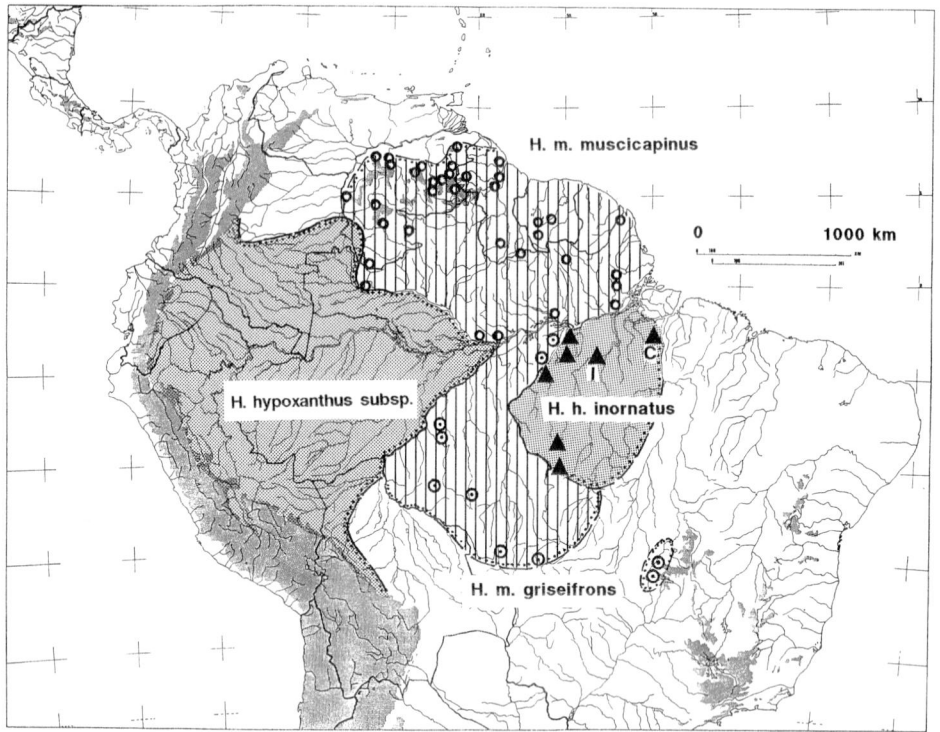

FIG. 11. Distribution of two Amazonian species of greenlets (*Hylophilus*), *H. muscicapinus* (hatched vertically) and *H. hypoxanthus* (shaded). Locality records (symbols) refer to the following taxa: *H. m. muscicapinus* (open circles), *H. m. griseifrons* (open circles with center dot), *H. hypoxanthus* subsp. (upper Amazonia, no records shown; includes *H. h. hypoxanthus, fuscicapillus, flaviventris* and *ictericus*), and *H. h. inornatus* (solid triangles). Dashed and dotted lines follow range limits. C = Cametá, I = Ilha de Santa Julia, Rio Iriri.

sturmii/reichenowi and *Psophia v. viridis/dextralis* that, in both cases, are separated by the broad lower Rio Tapajos, but intergrade in the headwater region, as mentioned above. In other cases, river-separated species exhibit direct contact in the headwater regions and there replace each other geographically with no or only limited hybridization along parapatric contact zones (the first taxon to reach the headwaters where the river was narrow enough to cross being able to "conquer" the opposite bank). This has been illustrated and discussed above on the basis of the distribution patterns of selected bird species and subspecies pairs in southern Amazonia, e.g., *Formicarius colma, Pteroglossus beauharnaesii, Pionopsitta barrabandi* (Fig. 5). Three additional examples may be mentioned of contact zones between birds which, west of the upper Rio Madeira, meet in western Brazil and eastern Peru. Among the ground-cuckoos (*Neomorphus*), *N. geoffroyi* (including *squamiger*) inhabits southern Amazonia east of the Rio Madeira (Fig. 12). To the west of this river, this species meets *N. pucheranii lepidophanes* along a contact zone that crosses the Rio Purus and the Rio Juruá. Similarly, the Band-tailed Manakin (*Pipra fasciicauda*) inhabits most of southeastern Amazonia as well as humid areas of central Brazil. To the west of the Rio Madeira, *P. fasciicauda* meets an upper Amazonian representative, *P. filicauda*, along a contact zone which crosses the Rio Purus, Rio Juruá, and Rio Ucayali (Fig. 7). The contact zone between *Galbula tombacea* and *G. cyanescens*, west of the Rio Madeira, also crosses the lower Rio Juruá and Rio Purús in this region (Haffer 1974).

Among the birds forming contact zones that cross Amazonian rivers are members of diverse ecological groups: (1) birds of the canopy level (*Pionopsitta, Capito, Pteroglossus, Ramphastos*), (2) birds of the middle levels of the rainforest (*Xiphorhynchus, Galbula, Pipra*), and (3) birds restricted to the forest understory and the forest floor (*Neomorphus, Myrmotherula, Hylophylax, Formicarius, Thryothorus*).

Contact zones represent major zones of biogeographic discontinuity. They may indicate the current or former location of barrier zones. The main hypotheses proposed to explain barrier

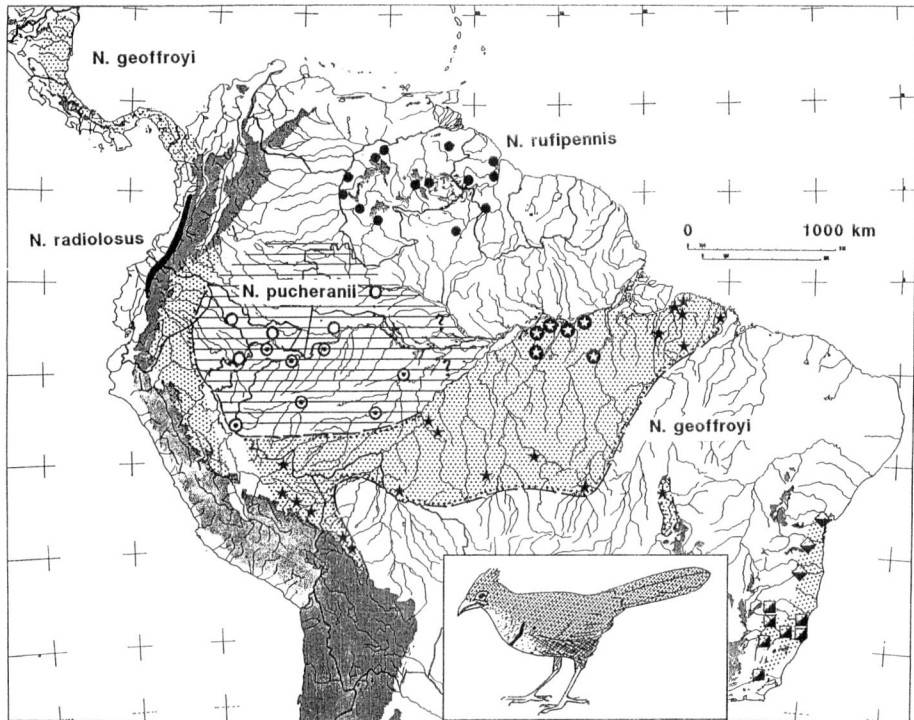

FIG. 12. Distribution of the ground-cuckoos *Neomorphus* (modified from Haffer 1977). *N. geoffroyi* (shaded and bird sketched), *N. pucheranii* (hatched horizontally), and *N. radiolosus* (solid). Locality records (symbols) refer to the following taxa: *N. g. geoffroyi* in southern Amazonia and *N. g. australis* along the base of the Andes in southeastern Peru and Bolivia (solid stars), *N. g. squamiger* (open stars in black circles), *N. g. maximiliani* (semi-solid squares on edge), and *N. g. dulcis* (semi-solid squares). For locality records of *N. g. aequatorialis* near base of Andes in Peru, Ecuador, Colombia, and for *N. g. salvini* in northwestern Colombia and Middle America see Haffer (1977). *N. p. pucheranii* (open circles), *N. p. lepidophanes* (open circles with center dot), *N. rufipennis* (solid circles). Dashed and dotted lines follow range limit. An additional locality of *N. pucheranii* is Araracuara in Amazonian Colombia (not indicated on this map), where Cuadros (1991) observed this unmistakable bird.

formation separating populations and causing speciation in Amazonia are based on different (mostly historical) factors (Table 4). These various hypotheses emphasize the effect of vertical tectonic movements, the development or existence of river barriers, the effect of environmental gradients, the changing community composition due to climatic-vegetational fluctuations during the Cenozoic (Tertiary and Quaternary), or a combination of these factors (Haffer 1993, 1997). Each may be relevant to a different degree for different faunal groups or different levels of faunal differentiation. I recommend distinguishing these models under separate designations, especially hypotheses 3 and 4 (Table 4), to minimize the potential of misunderstanding in discussions of the historical biogeography of Amazonia.

Interpretations of the origin of contact zones based on separation of the respective populations by Amazonian rivers or on ecological differences of the areas occupied by the respective taxa appear difficult. An interpretation of the origin of representative taxa through isolation by rivers under presently existing environmental conditions is difficult even in many forest understory birds when rivers do separate the ranges of the representatives (at least for some distance). Each of these taxa frequently occupy huge areas traversed by large rivers that are broader than the ones that separate the ranges of the representative taxa. An example is *Hylophylax poecilinota griseiventris* of south-central Amazonia (Fig. 9). The Rio Juruá, which may locally separate the ranges of *H. p. griseiventris/gutturalis* in the west, and the Rio Teles Pires, which separates the ranges of *H. p. griseiventris/nigrigula* in the east, are smaller rivers than the Rio Madeira, the Rio Purús, and the Rio Juruena within the range of *H. p. griseiventris*. Moreover, *H. p. nigrigula* inhabits both banks of the broad lower Rio Tapajós. Future analyses will demonstrate whether

TABLE 4
Models of Geographic Speciation in Amazonia During the Tertiary and Quaternary Periods (Cenozoic), as Proposed by Various Authors (From Haffer 1997)

	1 Paleogeography hypothesis	2 River hypothesis	3 River-refuge hypothesis	4 Refuge hypothesis	5 Disturbance-vicariance hypothesis	6 Gradient hypothesis
Forest reduction during dry climatic periods of the Cenozoic	Not considered (irrelevant)	Not considered (irrelevant)	Weak; only peripheral portions in northern and southern Amazonia affected	Strong; peripheral regions and central Amazonia affected	Weak; only peripheral portions in northern and southern Amazonia affected	Not considered (irrelevant)
Barriers separating populations	Continental seas, plateaus, flooded plains	Rivers (and their floodplains)	Broad rivers in central Amazonia and unforested areas in the headwater regions	Open forests and nonforest regions; rivers locally	Ecologically unsuitable forests	Steep environmental gradients
Cause of barrier formation	Tectonic movements and/or sea-level fluctuations	Development of rivers or dispersal of founders across preexisting river barriers	Fairly weak ←→ Humid/dry periods ←→	←→ Climatic fluctuations during the Cenozoic ←→ Strong (also rain shadow effect due to tectonic movements)	Cold/warm periods	Strong ecotones
Authors	Emsley (1965) for butterflies	Sick (1967), Hershkovitz (1977), Capparella (1988)	Ayres (1986), Capparella (1991), Ayres and Clutton-Brock (1992)	Haffer (1969, 1974), Vanzolini and Williams (1970), Vrba (1992)	Colinvaux (1993)	Endler (1982)

or not many of such river-separated and phenotypically uniform populations exhibit strong genetic differences.

An historical interpretation of zones of geographical replacement as zones of secondary contact between the respective taxa is simpler and appears more probable than under the River hypothesis. Such an interpretation implies largescale separation of the respective bird populations during one or more periods of geographical isolation in the geological past (either due to paleogeographic changes in the distribution of land and sea or due to climatic-vegetational fluctuations affecting the peripheral regions of Amazonia only or affecting all of Amazonia differentially). In most cases an interpretation of the origin of the contact zones through climatic-vegetational fluctuations is probable in view of numerous indications from many areas of Amazonia for vegetational changes during the last several million years (summaries by Haffer 1987a, 1993; Ab'Saber 1993). Recent evidence for vegetational changes comes not only from peripheral portions of Amazonia but also from central and upper Amazonia (thus favoring a Refugial hypothesis over the River-refuge hypothesis; see below).

Paleopollen studies indicate the occurrence of several periods of rainforest regression in the region of the Serra dos Carajás in southeastern Amazonia during the last 60,000 years (Absy et al. 1991). A drier climate than today and an open vegetation prevailed repeatedly during certain intervals of the last glacial period in the areas of the Rio Tapajós, in portions of the State of Rondônia as well as north of Manaus in central Amazonia (Bibus 1983; Emmerich 1988; Veiga et al. 1988). Large fossil dune fields in northcentral Amazonia between the Rio Branco and the Rio Negro also document one or more dry phases during the recent geological past (Santos et al. 1993). In upper Amazonia, an extensive fluvial-lacustrine system dried up about 53,000 years ago (Kronberg et al. 1991, Kronberg and Benchmol 1993). Moreover, paleontological studies of fossil vertebrates led Rancy (1991, 1993) to conclude that an open savanna vegetation was widespread in upper Amazonia during long periods of the Pleistocene before the currently existing forests returned and again covered this area completely.

By contrast, Colinvaux (1993: 473, 485) suggested "that the prime environmental forcing of tropical forests in ice age America was cooling rather than aridity. The forests of the central Amazon were probably not markedly fragmented . . . (although) savanna regions at the periphery were probably more extensive than now." This author assumed a different community composition in Amazonian forests during the Quaternary due to cold/warm cycles (rather than dry/wet cycles) leading to intensive species interactions, speciation, and endemism (without, however, discussing any details). Because Colinvaux neglected most or all geoscientific evidence for dry/wet climatic cycles and corresponding vegetational changes recently collected in portions of central and upper Amazonia, it is difficult to follow his reasoning. Additional geological evidence for dry/humid cycles in tropical South America during the Quaternary has been summarized by Haffer (1987a).

Bush (1994) also stated that climatic cooling, rather than aridity, was the factor driving a Pleistocene re-assortment of vegetation in Amazonia, although he did accept climatic drying (by about 20%) over Amazonia during glacial periods, which led to the expansion of dry-adapted vegetation types into the transverse climatic belts crossing lower Amazonia from southeast to northwest in the Manaus-Santarém region, as well as crossing southwestern Amazonia along the border region of Peru and Brazil. In this way, Bush (1994) accepted a separation of humid rainforest blocks in the Guianas and at the mouth of the Amazon River from the upper Amazonian forests (as discussed under the Refuge hypothesis). Moreover, he spoke of species that "only survived in areas that were optimal." Apparently, Bush (1994:13) had species-specific refugia in mind when he stated:

> "If the cooling and drying stressed individual species to the point where they went extinct over parts of their range and only survived in areas that were optimal, a mechanism for allopatric speciation emerges. Each time a population was stressed by climatic change, and this may occur several times for each Quaternary glaciation, the chance of it becoming fragmented increases, especially where it is in competition with species that are better adapted to the prevailing conditions. Species that contained considerable environmentally-related genotypic variation may have contracted into optimal locations for each genotype. . . . The presence of invading, cold-adapted, or dry-adapted taxa could have resulted in local competitive exclusion of some lowland taxa, or genotypes, further increasing the possibility of isolation and allopatric speciation."

It is evident that this generalized interpretation uses arguments derived from both the *Refuge hypothesis* (dry/humid cycles) and the *Disturbance-vicariance hypothesis* (cold/warm cycles). As in the model of Colinvaux (1993), the question arises under Bush's interpretation as to the speciation mechanism during the pre-Quaternary (Tertiary, etc.), when primarily dry/humid cycles (but basically no cold/warm cycles) occurred.

Climatic-vegetational changes due to astronomical Milankovitch cycles influenced the climate of the world during the entire Cenozoic (Tertiary-Quaternary) and before. During dry climatic periods, extensive humid forests probably existed in fairly large regions of the Amazonian lowlands, where enough surface relief was present to create rainfall gradients, e.g., near the Andes, around the mountains of southern Venezuela and the Guianas as well as in Rondônia (to the north of the Parecis Mountains in central Brazil), and in the hilly areas of eastern Pará (Haffer 1969), besides dry forests and humid gallery forests in other portions of Amazonia. These postulated forest refugia were not located in areas "between rivers" but in the lowlands adjacent to hilly and mountainous regions irrespective of the location of rivers. Rancy (1991, 1993) assumed that, during cold-dry periods, most or all of the upper Amazonian forest region was converted to woodland savannas with only gallery forests preserved along the rivers serving as "mini-refugia" for the rainforest faunas. This interpretation appears rather extreme.

According to the refugial interpretation, subspecies and species of birds and other animals originated through geographical isolation in ecologically favorable regions of Amazonia (forest refugia) during adverse (dry) climatic periods (Haffer 1969, 1974; Vanzolini and Williams 1970; Vanzolini 1973, 1992). The effect of the forest refugia, of course, was species-specific in the sense that strictly rainforest-adapted taxa were more affected and refuge populations were more effectively separated from one another than were ecologically more broadly adapted species. Many refuge populations (although not all) extended their ranges during humid periods (like today) and established secondary contact; for general reviews of this macroevolutionary "habitat theory" see also Vrba (1992, 1993). A variant of the "classic" refuge hypothesis assumes the origin of taxa in small remnant forests fringing the main forest refuge areas which themselves supposedly served as "sinks," where the newly originated forms accumulated (Fjeldså 1994). In the course of their repeated range expansions and retreats animal populations, of course, adjusted their movements to the presence of highly species-specific barriers (such as the rivers in Amazonia) whose efficiency depended and still today depend on the varying dispersal capability of individual species of birds, mammals, and insects. Forest birds with poor dispersal capabilities across open water were able to circumvent wide rivers in the headwater regions, at least in those cases where no ecologically equivalent geographical representatives blocked their range expansion. *Ecologically competing representative species (and, of course, subspecies) are more effective zoogeographical barriers than many rivers in Amazonia.* The latter may be crossed or circumvented in the headwater regions, which is not possible if a biological barrier (i.e., a competing taxon) is present. Some examples discussed above document that even the Amazon River has been crossed in cases where no close relatives occupied the opposite bank.

Contact zones with or without hybridization indicate "tension" between the two taxa replacing each other geographically. Very little is known about the processes that maintain the contact itself or the location of the replacement zones in Amazonia. Similarly, nothing can be deduced as to the time when the representative taxa originated and when they established (or re-established) contact. In view of the alarming rate at which the rainforests disappear, I emphasize that contact zones are another phenomenon that must be considered when areas are identified for future parks or forest reserves.

The model that I designated as *River-refuge hypothesis* (Haffer 1992b, 1993) combines aspects of the River hypothesis and of the Refuge hypothesis of faunal differentiation. Animal populations have been presumably isolated in "semi-refugia" separated by a combination of broad rivers (plus their floodplains) and by extensive, ecologically unsuitable terrain in the headwater regions of northern and southern Amazonia that were more or less unforested during dry climatic periods. This hypothesis should not be included under the same designation as the *River hypothesis* (or *Riverine barrier hypothesis*), because the effect of repeated climatic-vegetational changes is not required under the latter hypothesis, whereas the effect of such changes is an essential part of the River-refuge hypothesis (and of the Refuge hypothesis). The Amazon forest region contracted in a north-south direction under the River-refuge hypothesis but did not fragment. By contrast, under the Refuge hypothesis climatic-vegetational changes also affected central Amazonia leading to fragmentation of the forest region. The River-refuge hypothesis was proposed on the basis of the patterns of distribution and differentiation of Amazonian primates by Ayres (1986) and Ayres and Clutton-Brock (1992) and of genetic studies of certain bird species of the rainforest understory by Capparella (1991). The designation "River-forest contraction hypothesis" would also be feasible for this model. However, I prefer the label "River-refuge hypothesis" because this model is somewhat intermediate between the River hypothesis and the Refuge hypothesis and draws arguments from both.

Direct support for the Refuge hypothesis (versus the River-refuge hypothesis) comes from geoscientific evidence collected in central Amazonia as discussed above (vegetational changes north of Manaus and in the lower Rio Tapajós region) and in upper Amazonia. Moreover, large areas of central Amazonia probably were flooded when sea-level was raised somewhat worldwide during interglacial periods. On the other hand, the barrier effect of large Amazonian rivers probably was considerably reduced, when sea-level fell 100 m below the current level during dry glacial periods exposing most of the central Amazon Valley and portions of the continental shelves (Haffer 1987a). During these periods, the narrower rivers of Amazonia flowed in the central, deep portion of their current beds. Also, many or most Amazonian rivers, except the widest ones, can change course quickly, especially in flood season, even locally shifting their main beds several kilometers overnight. In this way, even the most sedentary birds and other animals of the interior of flood plain forests are passively transferred across the rivers and, thereby, continually reintroduce genes from one bank to the other.

In addition, a general reduction of humidity and rainfall over Amazonia during a dry glacial period would not only lead to a contraction of forests on broad latitudinal fronts from the north and south, but would also cause the separation of upper Amazonian forest blocks from lower Amazonian forest blocks along conspicuous "dry" transverse belts crossing southwestern and lower Amazonia from southeast to northwest (Haffer 1969; Van der Hammen and Absy 1994). The geological data on vegetational changes from the Manaus and Tapajós regions appear to corroborate this suggestion.

Arguments from avian distribution patterns in favor of the Refuge hypothesis over the River-refuge hypothesis may be summarized as follows:

(1) Several pairs of upper and lower Amazonian taxa meet along contact zones that extend from north to south across central Amazonia, e.g., species of the *Penelope jacquaçu* group (Haffer 1987b, Fig. 5.11), the *Brotogeris chrysopterus* superspecies (Haffer 1987b, Fig. 5.12), the *Ramphastos* toucans (Haffer 1974), and the *Pipra aureola* superspecies (Fig. 7). This situation suggests that the representative taxa of upper and lower Amazonia were separated during dry periods of the geological past by a (vegetational) barrier zone that crossed central Amazonia from north to south.

(2) Speciation in strong-flying canopy birds which readily cross broad rivers remains unexplained by the River-refuge hypothesis. Ecologically unsuitable areas between forest refugia presumably were more effective barriers.

(3) The differentiation of species that inhabit river-created vegetation zones along floodplains and river banks also remains unexplained by the River-refuge hypothesis. Under the Refuge hypothesis, these birds were also affected by vegetational changes in central Amazonia. Examples of a west-east separation and differentiation along the Amazon River among birds of river vegetation are the two pairs of antbirds *Myrmoborus l. berlepschi/lugubris* and *Thamnophilus cryptoleucus/T. nigrocinereus* (Haffer and Fitzpatrick 1985, Haffer 1987b).

Some birds and other vertebrates may have originated in central Amazonia through geographic isolation of ancestral populations by rivers alone or a combination of isolation by broad river courses and unsuitable ecological conditions in the headwater regions (e.g., *Capito brunneipectus*, Fig. 3; *Rhegmatorhina berlepschi*, see Fig. 8 in Haffer 1992b). However, I doubt that these models explain the origin of a large portion of the Amazonian diversity.

As mentioned above, sea-level fluctuations probably affected the separation and differentiation of certain bird populations. During interglacial periods of raised sea-level, the ancestors of some or all of the representative populations (Table 2) were separated to the north and south of the broad lower Amazonian embayment. They may have established contact in post-Pleistocene time when a lower sea-level and the reappearance of islands in the lower Amazon River and near its mouth (Ilha Marajó, Ilha Mexiana, Ilha Caviana, and others) permitted their crossing the narrower lower Amazon River in a southern direction.

During long periods of the Tertiary, a portion of the present lower Amazon Valley, i.e., the region between Manaus and Obidos, was apparently dry land and permitted a direct faunal exchange between the land areas of the Guianan Shield to the north and the Brazilian Shield to the south. Until the late Miocene, this "bridge" was a gentle divide between the broad upper Amazonian (Solimões) basin to the west and the comparatively small sedimentary basin in the lowermost Amazon Valley including the Marajó trough to the east (Mosmann et al. 1986). This low drainage divide was worn down by erosion and disappeared during the late Miocene tectonic movements leading to the continuity of the Amazon Valley from the upper Solimões region to the Atlantic Ocean.

During the Tertiary, no or very little effect is seen of the Iquitos Arch that crossed the upper Amazon basin from south to north during the previous Mesozoic and Paleozoic eras. It remains unknown whether and when the area over this arch was above sea-level. A slight thinning of the Upper Tertiary Pebas Formation over the former Iquitos Arch (Petri and Fúlfaro 1983, Fig. V-5) appears insufficient to suggest an upland connection between the Guianan and Brazilian shields in this area during the late Miocene. However, the paleogeographic situation is still poorly known. Occasional marine incursions from the Pacific Ocean and the Caribbean Sea reached western Amazonia during the Tertiary, when this region probably was covered with huge lakes, swamps and rivers. This area was closed off from the Pacific Ocean during the Middle Miocene due to continued uplift of the Andes mountains, which led to the reversal of the drainage pattern from a previous western and northwestern direction to an eastern direction. Complete continuity of the Amazon Valley to the Atlantic Ocean developed during the late Miocene when the northern Andes were strongly uplifted (Katzer 1903; Hoorn et al. 1995).

This rather simple paleogeographic setting of the greater Amazon region during the Tertiary period (of ca. 60 million years duration) does not seem to provide a sufficiently complex and changing geological background to have caused the intensive faunal evolution and speciation that certainly took place on the stable land areas of the Guianan shield to the north and the Brazilian shield to the south of the Amazon basin as well as in portions of the Amazon basin itself during the Tertiary. In any case, the geological data are not sufficiently finegrained to permit the details of physiographic changes in central South America during this time interval to be analyzed. Nevertheless, it appears likely that other factors than the paleogeographic development of these regions determined the speciation patterns of the Tertiary faunas in South America more effectively, such as paleoclimatic fluctuations (caused by Milankovitch cycles) leading to repeated ecological vicariance events and the formation of Tertiary forest refugia (Refuge hypothesis).

Recent geological studies in various parts of the world have established the fact that astronomical Milankovitch cycles, causing worldwide climatic-vegetational fluctuations, influenced the continuously changing distribution of forest and nonforest vegetation on earth not only during the Ice Ages of the last 2 million years (Quaternary) but also during the entire Tertiary period and before (Haffer 1993; see also Broecker and Denton 1990). The cyclic origin and disappearance of forest and nonforest refugia, which simultaneously served as "species traps" and "species pumps," probably underlies, as a predictably reversible mechanism, much of organic evolution, in particular the differentiation of vertebrate faunas in Amazonia and other regions of the world during the Cenozoic (Tertiary-Quaternary); see also Terborgh (1992). This statement is not intended to diminish in any way the general biogeographic significance of paleogeographic changes in the distribution of land and sea in South America and other regions due to tectonic movements during the course of the geological history or of the far-reaching effect of continental rifting and subsequent continental drift. An understanding of the significance of the latter vicariance processes as well as of jump dispersal between islands is needed for a complete analysis of the zoogeographical history of the various groups of animals.

ACKNOWLEDGMENTS

I am grateful to the late Theodore Parker III for discussions on several aspects of Neotropical ornithology in letters and during several personal encounters. He was a scientist who, in an innovative way, opened a new promising approach to faunal study as the basis of conservation work in the New World tropics and was just beginning to harvest interesting results on a broad front. J. V. Remsen suggested several useful editorial improvements of the text. I also gratefully acknowledge critical comments by A. Capparella and J. M. Bates.

The curators of the following museums granted me access to the bird collections under their care: American Museum of Natural History (New York), Carnegie Museum (Pittsburgh), National Museum of Natural History (Washington), Field Museum of Natural History (Chicago), Museum of Natural Science (Louisiana State University, Baton Rouge), Museum A. Koenig (Bonn), Zoologisches Museum (Berlin), Museu de Zoología (Sao Paulo), and Museu Paraense E. Goeldi (Belém, Brazil). Some data used in this report were obtained during fieldwork in the headwater region of the Rio Tapajós in 1993; I thank Dr. Edson de Carvalho and Dona Vitoria Riva Carvalho, Instituto Ecológico Cristalino, Alta Floresta, Mato Grosso, Brazil, for the necessary permits as well as Drs. F. C. Novaes and D. C. Oren, Museu E. Goeldi, Belém, Pará, Brazil, for their administrative support. I also gratefully acknowledge travel grants received from the Deutsche Forschungsgemeinschaft (Bonn) and from the Conselho Nacional de Desenvolvimento Cientifico e Tecnológico (CNPq, Brasilia).

LITERATURE CITED

AB'SABER, A. 1993. As bases do conhecimento sôbre os paleoclimas modernos da Amazonia. Ciência Hôje (São Paulo) 16, no. 93:1–3.
ABSY, M. L., A. CLEEF, M. FOURNIER, L. MARTIN, M. SERVANT, A. SIFEDDINE, M. FERREIRA DA SILVA, F. SOUBIES, K. SUGUIO, B. TURCO, AND T. VAN DER HAMMEN. 1991. Mise en évidence de quatre phases d'ouverture de la foret dense dans le sud-est de l'Amazonie au cours des 60,000 dernières années. Première comparaison avec d'autres régions tropicales. Compte-Rendu Acad. Sci. Paris, vol. 312, série II:673–678.
AYRES, J. M. C. 1986. Uakaris and Amazonian flooded forest. Ph.D. dissertation, University of Cambridge, Cambridge, U.K.
AYRES, J. M. C., AND T. H. CLUTTON-BROCK. 1992. River boundaries and species range size in Amazonian primates. Amer. Natur. 140:531–537.
BATES, H. W. 1863. The Naturalist on the River Amazons. Murray, London.
BATES, J. M., M. C. GARVIN, D. C. SCHMITT, AND C. G. SCHMITT. 1989. Notes on bird distribution in northeastern dpto. Santa Cruz, Bolivia, with 15 species new to Bolivia. Bull. Brit. Ornith. Club 109: 236–244.
BIBUS, E. 1983. Die klimamorphologische Bedeutung von stone-lines und Decksedimenten in mehrgliedrigen Bodenprofilen Brasiliens. Z. Geomorph., N. F., Suppl. 48:79–98.
BLAKE, E. R. 1968. Vireonidae. Pp. 103–138 in Checklist of Birds of the World, vol. 14 (R.A. Paynter, Jr., and E. Mayr, Eds.). Harvard University Press, Cambridge, Massachusetts.
BOND, J. 1953. Notes on Peruvian Icteridae, Vireonidae and Parulidae. Notulae Naturae 255, p. 1–15.
BOND, J., AND R. MEYER DE SCHAUENSEE. 1943. The birds of Bolivia. Part II. Proc. Acad. Nat. Sci. Philadelphia 95:167–221.
BROECKER, W. S., AND G. H. DENTON. 1990. What drives glacial ages? Scient. Amer. 262 (January):43–56.
BUSH, M. B. 1994. Amazonian speciation: a necessarily complex model. J. Biogeogr. 21:5–17.
CAPPARELLA, A. P. 1988. Genetic variation in Neotropical birds: implication for the speciation process. Acta XIX Congr. Intern. Ornith. (Ottawa 1986), vol. 2:1658–1664.
CAPPARELLA, A. P. 1991. Neotropical avian diversity and riverine barriers. Acta XX Congr. Intern. Ornith. (Aukland 1990), vol. 1:307–316.
CHAPMAN, F. M. 1921. Descriptions of proposed new birds from Colombia, Ecuador, Peru, and Brazil. Amer. Mus. Novitates 18:1–12.
COLINVAUX, P. 1993. Pleistocene biogeography and diversity in tropical forests of South America. Pp. 473–499 in Biological Relationships Between Africa and South America (P. Goldblatt, Ed.). Yale University Press, New Haven, Connecticut.
CORY, C. B., AND C. E. HELLMAYR. 1924. Catalogue of Birds of the Americas. Part III. Field Mus. Nat. Hist., Chicago, Illinois.
CUADROS, T. 1991. Registro visual del cucu terrestre Piquirrojo (*Neomorphus pucheranii*) en Colombia. Bol. Sociedad Antioqueña Ornitología (Medellín, Colombia) 2:26–27.
EMMERICH, K.-H. 1988. Relief, Böden und Vegetation in Zentral-und Nordwest-Brasilien unter besonderer Berücksichtigung der känozoischen Landschaftsentwicklung. Frankfurter Geowiss. Arbeiten, Ser. D (Phys. Geogr.), no. 8.
EMSLEY, M. G. 1965. Speciation in *Heliconius* (Lep., Nymphalidae): Morphology and geographic distribution. Zoologica (New York) 50:191–254.
ENDLER, J. 1982. Pleistocene forest refuges: fact or fancy? Pp. 179–200 in Biological Diversification in the Tropics (G. T. Prance, Ed.). Columbia University Press, New York.
FJELDSÅ, J. 1994. Geographical patterns for relict and young species in Africa and South America and implications for conservation priorities. Biodiversity and Conservation 3:207–226.
GYLDENSTOLPE, N. 1945. The bird fauna of Rio Juruá in western Brazil. Kungl. Svenska Vetenskapsakademiens Handlingar, 3rd ser., 22:1–338.
GYLDENSTOLPE, N. 1951. The ornithology of the Rio Purús region in western Brazil. Arkiv för Zoologi 2: 1–320.
HAFFER, J. 1969. Speciation in Amazonian forest birds. Science 165:131–137.
HAFFER, J. 1970. Art-Entstehung bei einigen Waldvögeln Amazoniens. J. Ornith. 111:285–331.
HAFFER, J. 1974. Avian speciation in tropical South America. Publ. Nuttall Ornith. Club no. 14.
HAFFER, J. 1977. A systematic review of the Neotropical ground-cuckoos (Aves, *Neomorphus*). Bonn. Zool. Beitr. 28:48–76.
HAFFER, J. 1978. Distribution of Amazon forest birds. Bonn. Zool. Beitr. 29:38–78.
HAFFER, J. 1982. General aspects of the refuge theory. Pp. 6–24 in Biological Diversification in the Tropics (G. T. Prance, Ed.). Columbia University Press, New York.
HAFFER, J. 1987a. Quaternary history of tropical America. Pp. 1–18 in Biogeography and Quaternary History

in Tropical America (T.C. Whitmore and G.T. Prance, Eds.). Clarendon and Oxford University Press, Oxford, U.K.

HAFFER, J. 1987b. Biogeography of Neotropical birds. Pp. 105–150 in Biogeography and Quaternary History in Tropical America (T.C. Whitmore and G.T. Prance, Eds.). Clarendon and Oxford University Press, Oxford, U.K.

HAFFER, J. 1992a. Parapatric species of birds. Bull. Brit. Ornith. Club 112:250–264.

HAFFER, J. 1992b. On the "river effect" in some forest birds of southern Amazonia. Bol. Mus. Paraense E. Goeldi (Brasil), Zoología 8:17–245.

HAFFER, J. 1993. Time's cycle and time's arrow in the history of Amazonia. Biogeographica 69:15–45 (Compte-Rendu des séances de la Societé de Biogéographie, Paris).

HAFFER, J. 1997. Alternative models of vertebrate speciation in Amazonia: An overview. Biodiversity and Conservation, in press.

HAFFER, J., AND J. W. FITZPATRICK. 1985. Geographic variation in some Amazonian forest birds. Ornithol. Monogr. 36:147–168.

HELLMAYR, C. E. 1910. The birds of the Rio Madeira. Novitates Zool. 17:257–428.

HELLMAYR, C. E. 1912. Zoologische Ergebnisse einer Reise in das Mündungsgebiet des Amazonas (L. Müller). II. Vögel. Abhandl. Königl. Bayer. Akad. Wiss., Math.-Phys. Klasse 26:1–142.

HERSHKOVITZ, P. 1969. The Recent mammals of the Neotropical region: a zoogeographic and ecological review. Quart. Rev. Biol. 44:1–70.

HERSHKOVITZ, P. 1977. Living New World Monkeys. Vol. 1. Univ. Chicago Press, Chicago.

HILTY, S. L., AND W. L. BROWN. 1986. A Guide to the Birds of Colombia. Princeton Univ. Press, Princeton, New Jersey.

HINKELMANN, C. 1989. Notes on the taxonomy and geographic variation of Phaethornis bourcieri (Aves: Trochilidae) with the description of a new subspecies. Bonn. Zool. Beitr. 40:99–107.

HOORN, C., J. GUERRERO, G. A. SARMIENTO, AND M. A. LORENTE. 1995. Andean tectonics as a cause for changing drainage patterns in Miocene northern South America. Geology 23:237–240.

KATZER, F. 1903. Grundzüge der Geologie des unteren Amazonasgebietes. Weg, Leipzig.

KRONBERG, B., AND R. BENCHIMOL. 1993. Aridez no Acre. A historia climática de uma região. Ciência Hôje (São Paulo) 16, no. 93:44–47.

KRONBERG, B. I., R. E. BENCHIMOL, AND M. I. BIRD. 1991. Geochemistry of the Acre subbasin: window on ice-age Amazonia. Interciência 16:138–141.

MAYR, E. 1942. Systematics and the Origin of Species. Columbia Univ. Press, New York.

MEYER DE SCHAUENSEE, R. 1966. The Species of Birds of South America and Their Distribution. Livingston Publ. Co., Narberth, Pennsylvania.

MEYER DE SCHAUENSEE, R. 1970. A Guide to the Birds of South America. Livingston Publ. Co., Wynnewood, Pennsylvania.

MOSMANN, R., F. U. H. FALKENHAIN, A. GONZALES, AND F. NEPOMUCENO. 1986. Oil and gas potential of the Amazon Paleozoic basins. Pp. 207–241 in Future Petroleum Provinces of the World (M.T. Halbouty. Ed.). Amer. Assoc. Petrol. Geol., Memoir 40.

NOVAES, F. C. 1957. Contribuição à ornitología do noroeste do Acre. Bol. Mus. Paraense E. Goeldi, nova série, Zoología 9:1–30.

NOVAES, F. C. 1981. A estrutura da espécie nos periquitos do genero Pionites Heine (Psittacidae, Aves). Bol. Mus. Paraense E. Goeldi (Belém, Brasil), nova série, Zoología 106:1–21.

NOVAES, F. C., AND M. F. C. LIMA. 1992. As aves do Rio Peixoto de Azevedo, Mato Grosso, Brasil. Rev. Brasil. Biol. 7:351–381.

OREN, D. C., AND NOVAES, F. C. 1985. A new subspecies of White Bellbird Procnias alba (Herrmann) from southeastern Amazonia. Bull. Brit. Ornith. Club 105:23–25.

PARKES, K. C. 1961. Intergeneric hybrids in the family Pipridae. Condor 63:345–350.

PERES, C. A., AND A. WHITTAKER. 1991. Annotated checklist of the bird species of the upper Rio Urucú, Amazonas, Brazil. Bull. Brit. Orn. Club 111:156–171.

PETERS, J. L. 1948. Check-List of Birds of the World. Vol. 6. Harvard University Press, Cambridge, Massachusetts.

PETRI, S., AND V. J. FULFARO. 1983. Geología do Brasil. São Paulo, Univ. São Paulo.

PHELPS, W. H., AND W. H. PHELPS, JR. 1958. Lista de las aves de Venezuela con su distribución. Part 1 (No Passeriformes). Bol. Sociedad Venez. Ciénc. Naturales 19 (no. 90):1–317.

PINTO, O. 1947. Contribuição à ornitología do baixo Amazonas. Arquivos de Zoología (São Paulo) 5:311–482.

PINTO, O., AND E. A. DE CAMARGO. 1957. Sôbre uma coleção de aves da região da Cachimbo (sul do Estado do Pará). Pap. Avulsos (Mus. Zool., São Paulo) 13:51–69.

RANCY, A. 1991. Pleistocene Mammals and Paleoecology of the Western Amazon. Ph.D. dissertation, Univ. of Florida; Univ. Microfilm Int., Ann Arbor.

RANCY, A. 1993. A paleofauna da Amazonia indica áreas de pastagem com pouca cobertura vegetal. Ciência Hôje 16 (No. 93):48–51 (São Paulo).

RIDGELY, R. S., AND G. TUDOR. 1989. The Birds of South America, vol. 1 (The Oscine Passerines). Oxford Univ. Press, Oxford.

RIDGELY, R. S., AND G. TUDOR. 1994. The Birds of South America, vol. 2 (The Suboscine Passerines). Univ. Texas Press, Austin, Texas.
RIPLEY, D. S. 1945. The barbets. Auk 62:542–563.
RIPLEY, D. S. 1946. The barbets—Errata and addenda. Auk 63:452.
ROTH, P., D. C. OREN, AND F. C. NOVAES. 1984. The White Bellbird (*Procnias alba*) in the Serra dos Carajás, southeastern Pará, Brazil. Condor 86:343–344.
SANTOS, J. O. S. DOS, B. W. NELSON, AND C. A. GIOVANNINI. 1993. Corpos de areia sob leitos abandonados de grandes rios. Ciência Hôje (São Paulo) 16 (no. 93):22–25.
SICK, H. 1958. On the distribution of Day's Barbet. Condor 60:339.
SICK, H. 1959a. Zwei neue Pipriden aus Brasilien. J. Ornith. 100:111–112.
SICK, H. 1959b. Zur Entdeckung von *Pipra vilasboasi*. J. Ornith. 100:404–412.
SICK, H. 1967. Rios e enchentes na Amazônia como obstáculo para a avifauna. Atas Simp. sôbre a Biota Amazônica, vol. 5 (Zoología):495–520.
SICK, H. 1993. Birds in Brazil. Princeton University Press, Princeton, New Jersey.
SILVA, J.M.C. DA, AND E. O. WILLIS. 1986. Notas sôbre a distribução de quatro espécies de aves da Amazônia brasileira. Bol. Mus. Paraense E. Goeldi (Belém, Brasil), sér. Zoologia 2:151–158.
SNETHLAGE, E. 1913. Über die Verbreitung der Vogelarten in Unteramazonien. J. Ornith. 61:469–539.
SNOW, D. 1982. The Cotingas. Bellbirds, Umbrellabirds and Their Allies. Oxford Univ. Press, Oxford.
STOTZ, D.F. 1993. A hybrid manakin (*Pipra*) from Roraima, Brazil, and a phylogenetic perspective on hybridization in the Pipridae. Wilson Bull. 105:348–351.
TEIXEIRA, D. M., J. B. NACINOVIC, AND G. LUIGI. 1988. Notes on some birds of northeastern Brazil (3). Bull. Brit. Ornith. Club 108:75–79.
TEIXEIRA, D. M., R. OTOCH, G. LUIGI, M. A. RAPOSO, AND A. C. C. ALMEIDA. 1993. Notes on some birds of northeastern Brazil (5). Bull. Brit. Ornith. Club 113:48–52.
TERBORGH, J. 1992. Diversity and the Tropical Rainforest. Freeman & Co., New York.
TODD, W.E.C. 1929. A review of the vireonine genus *Pachysylvia*. Proc. Biol. Soc. Washington 42:181–206.
TODD, W. E. C. 1948. Critical remarks on the wood-hewers. Annals Carnegie Mus. 31:5–18.
VAN DER HAMMEN, T., AND M. L. ABSY. 1994. Amazonia during the last glacial. Palaeogeogr. Palaeoclimat. Palaeoecol. 109:247–261.
VANZOLINI, P. E. 1973. Paleoclimates, relief, and species multiplication in equatorial forests. Pp. 255–258 *in* Tropical Forest Ecosystems in Africa and South America: A Comparative Review (B.J. Meggers, E.S. Ayensu, and W.D. Duckworth, Eds.). Smithsonian Press, Washington, D.C.
VANZOLINI, P. E. 1992. Paleoclimas e especiação em animais da América do Sul tropical. Estudos Avançados (São Paulo) 6, no. 15:41–65.
VANZOLINI, P. E., AND E. E. WILLIAMS. 1970. South American anoles: The geographic differentiation and evolution of the *Anolis chrysolepis* species group (Sauria, Iguanidae). Arquivos de Zoología (São Paulo) 19:1–298.
VEIGA, A. T. C., M. A. DARDENNE, AND E. P. SALOMÃO. 1988. Geologia dos aluviões auríferos e estaníferos da Amazônia. Anais XXXV Congr. Brasil. Geol. (Belém, Pará) 1:164–177.
VRBA, E. S. 1992. Mammals as a key to evolutionary theory. J. Mammal. 73:1–28.
VRBA, E. S. 1993. Mammal evolution in the African Neogene and a new look at the Great American interchange. Pp. 393–432 *in* Biological Relationships between Africa and South America (P. Goldblatt, Ed.). Yale University Press, New Haven, Connecticut.
WALLACE, A. R. 1853. A Narrative of Travels on the Amazon and Rio Negro. Reeve, London.
WILLIS, E. O. 1969. On the behavior of five species of *Rhegmatorhina*, ant-following antbirds of the Amazon basin. Wilson Bull. 81:363–395.
WILLIS, E. O. 1982. The behavior of Scale-backed Antbirds. Wilson Bull. 94:447–462.
ZIMMER, J. T. 1934. Studies of Peruvian birds. No. XV. Notes on the genus *Xiphorhynchus*. Amer. Mus. Novitates 756:1–20.
ZIMMER, J. T. 1942. Studies of Peruvian birds. No. XLI. The genera *Hylophilus*, *Smaragdolanius*, and *Cyclarhis*. Amer. Mus. Novitates 1160:1–16.
ZIMMER, J. T. 1948. A new name for *Xiphorhynchus spixii similis* Zimmer. Auk 65:446.

APPENDIX 1

Comments on species limits and taxonomy in the barbets *Capito niger* - *C. auratus* - *C. brunneipectus* - *C. dayi*.

These four Amazonian barbets (Capitonidae) had been treated in the literature as species until Bond and Meyer de Schauensee (1943) included *auratus* as a subspecies of *C. niger*; their only "reasoning" reads: "Clearly all the forms grouped under *auratus* should be considered subspecies of the scarlet-throated *niger* of the Guianas, which in addition, has a red fore-head." In his brief review of the barbet family, Ripley (1945) continued this "lumping" trend, including also the other two representative taxa, *brunneipectus* and *dayi*, as subspecies of the enlarged species *C. niger* without giving any explanation for this rearrangement. The geographic representation of these rather differently colored taxa appears to have been Ripley's only reason, because shortly afterwards he reversed his previous action with respect to *C. dayi* stating: "*Capito dayi* should be maintained as a separate species (fide Peters in litt.) due to its overlapping range with *C. n.*

insperatus" (Ripley 1946). This statement evidently refers to the fact that, although the upper Rio Madeira normally separates the ranges of *C. auratus* and *C. dayi* (Fig. 3), three specimens of *C. auratus insperatus* from "Calamá" (Hellmayr 1910), a town on the right (eastern) bank of the upper Rio Madeira, supposedly came from "within" the range of *C. dayi*. The details of the distribution of these two barbets along the upper Rio Madeira are poorly known, but I doubt that both species occur sympatrically to any extent. The simplest solution is to assume that the critical specimens of *C. auratus* from "Calamá" actually came from the left (western) bank of the Rio Madeira "opposite Calamá" (where I placed this record on the map; C on Fig. 3). No other individual of *C. auratus* has ever been reported from the right bank of the Rio Madeira. Peters (1948) treated *C. dayi* as a monotypic species and, like Ripley (1945, 1946) combined *C. niger, auratus*, and *brunneipectus* as another (polytypic) species.

During recent decades, numerous pairs of non-hybridizing representatives in geographic contact, i.e., parapatric species, have become known from Amazonia and from other parts of the world (Haffer 1992a). Therefore, the uncritical "lumping" of geographic representatives as subspecies simply on the basis of rather similar plumage coloration is no longer acceptable without further evidence (e.g., extensive hybridization and intergradation of taxa in contact or detailed comparison of taxa separated by a barrier with other allies that do hybridize). *Capito dayi* has a totally different plumage coloration from the other three Amazonian relatives; its color pattern somewhat resembles that of the north-Colombian *C. hypoleucos*, which may be the closest ally of *C. dayi*. The upperparts of *C. brunneipectus* are similar to those of *C. auratus*, but the brown breast-band, the pale cinnamon-buff throat (spotted in ♀) and the heavily marked remaining underparts are quite different from *C. auratus* populations to the west. These populations possess an orange-chrome throat and light yellow belly. The wide lower Rio Madeira separates *C. brunneipectus* from *C. auratus*, and the lower Amazon River prevents contact with *C. niger*, a Guianan species with a deep red throat. I doubt that *C. brunneipectus* would hybridize with *C. dayi, C. auratus*, or *C. niger* if in contact. In central Amazonia, the wide lower Rio Negro and Rio Branco separate the ranges of *C. auratus* and *C. niger*; however, they are probably in contact east of the upper Rio Branco in northernmost Brazil and in southeasternmost Venezuela; no intermediate specimens have been collected so far. The calls of *C. niger* also differ appreciably from those of *C. auratus* (D. F. Stotz, pers. comm.). The subspecies of *C. auratus* that is in contact with the red-throated and red-fronted *C. niger* in these areas is the yellow-thoated and yellow-fronted *C. a. aurantiicinctus*. This is a rather common bird in southern Venezuela (more than 100 specimens in the Phelps Collection in Caracas); on the other hand, *C. niger* barely enters southeastern Venezuela from the Guianas (only five specimens in the Phelps Collection; Phelps and Phelps 1958). *Capito niger, C. auratus*, and *C. brunneipectus* are certainly each others' closest relatives and may best be considered to form a superspecies. Pending a more detailed study of these barbets, I maintain species status for the taxa of eastern Amazonia (*C. niger, C. brunneipectus, C. dayi*) and for the upper Amazonian *C. auratus*, which is polytypic.

APPENDIX 2

Comments on species limits and taxonomy in the woodcreepers *Xiphorhynchus elegans - X. spixii*.

Currently, the various forms of the Elegant Woodcreeper (*X. elegans*) are in a state of taxonomic confusion, because Meyer de Schauensee (1966, 1970) treated *X. elegans* and *X. spixii* as two widely sympatric species, each with several geographical subspecies. Actually, only one set of geographically representative taxa inhabit the forests south of the lower Amazon River (monotypic *X. spixii*) and from the Rio Tapajós west to the foothills of the Andes in Peru (polytypic *X. elegans*) (Fig. 8). *Xiphorhynchus elegans* also occurs northward through eastern Ecuador into southeastern Colombia. The subspecies of *X. elegans* are *elegans, juruanus, insignis, ornatus*, and *buenavistae*. Except for the weakly differentiated form *buenavistae* (= *similis*; Zimmer 1934, 1948) in southeastern Colombia and for *insignis* in the Andean foothills of eastern Peru, each of these subspecies inhabits large regions of Amazonia, shows little or no (clinal) geographical variation, and intergrades with neighboring subspecies along fairly narrow, rather localized contact zones. The nominate form (*X. e. elegans*) inhabits the R.M.-R.T. interfluvium. Specimens from the lowlands west of the mouth of the Rio Madeira are intermediate between *elegans* and *juruanus* (Zimmer 1934; specimens examined at AMNH), as are those from the lowermost Rio Purus that Gyldenstolpe (1951) assigned to the nominate form. *Xiphorhynchus e. juruanus* inhabits southwestern Amazonia to the foothills of the Andes, where it intergrades via *X. e. insignis* with *X. e. ornatus*. The latter form inhabits the Amazonian lowlands of eastern Ecuador and southeastern Colombia and resembles distant *X. e. elegans*. The subspecies *ornatus* and *juruanus* are separated by the Rio Solimões (Fig. 8). Todd (1948; also Zimmer 1934) reported specimens of both *ornatus* and *juruanus* from São Paulo de Olivença on the southern bank of this river (SP on Fig. 8). Todd insisted that Samuel Klages had collected the three specimens of *ornatus* at the "same place" and on the "same day" as several specimens of *juruanus*, thereby supposedly proving the specific distinctness of these forms. During my visit to the Carnegie Museum in Pittsburgh in May 1993, K. C. Parkes and I checked the collecting dates of the Klages specimens of these forms from São Paulo de Olivença. They were collected three and more days apart. None of the *ornatus* specimens was collected on the same day as any specimens of *X. e. juruanus*. Moreover, Klages mentioned in his unpublished field report (archives of the Bird Section, Carnegie Museum) that he collected near São Paulo de Olivença not only in the *terra firme* forests on the southern bank of the Rio Solimões but occasionally visited the forested islands in the Rio Solimões as well. I strongly suspect that this is the habitat where Klages obtained his specimens of *X. e. ornatus*. No intergradation is obvious in the specimens of *ornatus* and *juruanus* from "São Paulo de Olivença." However, I suspect that if the populations of these forms would meet in *terra firme* forest without separation by a river, they would intergrade in a manner similar to *juruanus-insignis-ornatus*. In March 1994, I travelled twice along the Rio Solimões and observed from the ship on the south bank the small town of São Paulo de Olivença and, opposite from it, a large uninhabited and densely forested island in the river (ca. 5 km long and up to 2 km wide). This island, where Klages most probably collected his specimens of *X. e. ornatus*, is separated from São Paulo de Olivença by the main channel of the Rio Solimões, which is here about 400–500 m wide. My observations support Zimmer's (1934:9) suspicion that the specimens of *ornatus* from "São Paulo de Olivença" were probably collected in a habitat "connected ecologically, if not otherwise, with the left bank of the Amazon." A specimen of *X. e. ornatus* collected by A. M. Rea on Yanamono Island in the Río Marañón at the mouth of the Río Napo (Dep. Loreto, Peru; Carnegie Museum) indicates a similar distribution pattern of *ornatus* and *juruanus* in this region upriver from São Paulo de Olivença. A supposed "sympatry" of *X. spixii* and *X. elegans* in southeastern Colombia (as suggested tentatively by Hilty and Brown 1986) does not exist either. The Colombian specimens of *"elegans"* turn out to be indistinguishable from *X. "spixii" ornatus* and *buenavistae* (F. G. Stiles, pers. comm.), which are subspecies of *X. elegans* as discussed above.

Since the above comments were submitted for publication, I received a copy of Ridgely and Tudor (1994). These authors treated *X. elegans* and *X. spixii* as conspecific, granting the possibility that *spixii* may deserve "to be separated as a monotypic species." In general, I agree with their taxonomic notes. However, I would like to mention that although the text described the range of *X. spixii* correctly as extending east to Maranhão, the distribution map of this species (p. 200) illustrated an isolated occurrence farther east in the Brazilian state of Ceará. This record should be deleted from the map, because it was based on a misidentified subadult male of *X. picus* from the Serra do Baturité, Ceará (Teixeira et al. 1988, 1993).

THE NATURE AND PROBABLE FUNCTION OF VOCAL COPYING IN LAWRENCE'S THRUSH, *TURDUS LAWRENCII*

JOHN WILLIAM HARDY[1] AND THEODORE A. PARKER III[2,3]

[1]*Florida Museum of Natural History, University of Florida. Gainesville, Florida 32611, USA;*
[2]*Museum of Natural Science, Foster Hall 119, Louisiana State University, Baton Rouge, Louisiana 70803, USA;*
[3]*Deceased*

ABSTRACT.—In this paper we document with analysis of tape recordings the unusual song of the Neotropical Lawrence's Thrush (*Turdus lawrencii*). This song belongs in the category of "vocal appropriation" under the broad heading of "vocal copying" in the sense of Dobkin (1979). Study of recordings of about 30 individuals of Lawrence's Thrush, all probably males, revealed imitations of 173 bird species, as well as those of frogs and insects. Males sing for hours from song posts in the rain-forest canopy, with intermissions of five to 15 minutes, presumably during which they feed. The song is a halting, thrasher-like (Mimidae) rendition that sometimes exceeds 250 phrases without a break. In such a sequence, vocal copying of as many as 51 bird species has been detected. Individual singers almost never respond visibly to playback of such song, suggesting that it is directed to females. Based on its vocal behavior, we predict this species will prove to have an atypical thrush mating system, either promiscuous or polygynous. Copying by three other species of Neotropical thrushes is discussed.

RESUMEN.—En este artículo documentamos el canto excepcional de *Turdus lawrencii* a través de un analysis de grabaciones. Este canto está clasificado dentro de la categoría de "apropriacíon vocal" bajo el concepto general de "copias vocales" en el sentido de Dobkin (1979). El estudio de grabaciones de 30 individuos de *Turdus lawrencii*, probablemente machos en su totalidad, revela imitaciones de 173 especies de aves, tanto como imitaciones de ranas e insectos. Machos cantan por horas desde perchas en el docel del bosque lluvioso, con interrupciones de cinco a 15 minutos, período en el cual supuestamente se alimentan. El canto con sus cortas pausas se parece a las vocalizaciones típicas para la familia Mimidae y a veces excede 250 frases sin descansar. En tales sequencias, se han detectado copias vocales de hasta 51 especies de aves. Los individuos cantantes casi nunca responden de forma obvia a las grabaciones de sus propios cantos, indicando que éstos son dirigidos a las hembras. Basandonos en su comportamiento vocal, predecimos que esta especie tiene un sistema de apareamiento promiscuo o poligínico, cual no es típico para los túrdidos. También discutimos las imitaciones por tres otras especies neotropicales de la familia Turdidae.

Vocal copying, in the broadest sense, is a widespread phenomenon in birds, and as Baylis' (1982) review emphasized, it has raised many questions, few of which have been answered definitively. "Vocal appropriation," a category of "vocal copying," erected and defined by Dobkin (1979), designates the use of calls or songs of other bird and animal species that are neither potential competitors nor predators of the appropriator. It functions intraspecifically either to attract or stimulate a mate or potential mate, or in intraspecific territoriality to facilitate dispersion and individual distinctiveness. (Dobkin reserved "vocal mimicry" for interspecific copying, and "vocal imitation" for intraspecific copying, as, for example, when young males hear and adopt the song dialect of their fathers.) We value the distinction of Dobkin's categories, but find "vocal appropriation" and its variations often unwieldy in narrative use. Thus we herein sometimes use the words "copy," "imitation," and "imitate," in substitution for "vocal appropriation" and "vocally appropriate." Vocal appropriation is proving to be more common than previously believed. Its detection, and that of vocal copying in general, requires that sounds be recognizable to the listener and attributable to the subject being copied. Only in recent years have field students in places such as the range of Lawrence's Thrush (*Turdus lawrencii*), the subject species of this paper, gained expertise with the songs and calls of the native bird species and thus the ability to recognize this thrush's vocal capabilities.

Of the five widely sympatric breeding species of *Turdus* in Amazonia, Lawrence's Thrush is the least known. It is found in tall rain forest from the base of the eastern Andes in southeastern Colombia, Ecuador, Peru, and northern Bolivia east to the rivers Negro and Tapajos in Brazil, and in southern Venezuela (Ridgley and Tudor 1989). The bird is nowhere common, but is most numerous in low-lying, seasonally flooded areas, especially in the narrow transitional zone between *varzea* and *terra firme* (Parker, pers. obs.). Away from this zone, this species usually occurs in wooded swamps or along streams (Parker 1982; Parker and Remsen 1987). Unlike sympatric congeners, such as Hauxwell's Thrush (*T. hauxwelli*) and White-necked Thrush (*T. albicollis*), Lawrence's Thrush is not uniformly distributed, often being absent from seemingly suitable habitat (Parker, pers. obs.).

Until the mid-1970s, Lawrence's Thrush was, to most field and museum students, just one of the tropical "brown thrushes": poorly known, little studied, and seemingly as plain and "ordinary" as all the others. This is no longer the case. In recent years, a new generation of bioacoustically oriented field ornithologists has worked in the Neotropics using tape recorders to become unprecedentedly familiar with the sounds of a thousand or more bird species. Lawrence's Thrush has thus become known as arguably the world's most accomplished avian vocal appropriator. Songs of males are composed almost entirely of imitations of vocalizations of other bird species strung together in a sequence of song phrases that may, with short intermissions, endure for several hours. We use the term sequences, rather than bouts, because all of these sequences are probably mere segments of song-bouts of much greater length.

In this paper we analyze 36 song sequences, of approximately 30 unmarked individuals (probably males), and present detailed analysis of the songs of five of these birds. We also speculate on the function of song in this species, and from this we make a prediction about the character of this thrush's as-yet-unknown mating system. In fact, we believe that the song of Lawrence's Thrush represents the phenomenon of avian mimicry in its pioneer stages. Although the songs of scores of species have been studied, it is remarkable how few Neotropical species have been discovered to use mimicry.

METHODS

Parker spent 3–8 months each year for the last 19 years of his life in some of the most species-rich bird habitats of South America, much of that time in Amazonia. During field work for the Louisiana State University Museum of Natural Science, Parker became thoroughly acquainted with the sounds of South American birds, while making hundreds of hours of tape recordings for the Library of Natural Sounds (LNS) at the Laboratory of Ornithology, Cornell University. As a result, he could listen to a song of a Lawrence's Thrush and quickly recognize most, if not all, imitated phrases of other birds' vocalizations.

Hardy assembled the recordings analyzed here and re-recorded each cut on track one of a two-track stereo tape recorder. While doing so, he lengthened the intervals between successive phrases. Parker then listened to each song and, following each phrase, entered by microphone on track two the name of the imitated species. From these tapes we selected a total of six sequences for more detailed analysis (see Appendix I): three of one Peruvian bird (*#1*), a cut by bird *#5*, a distant neighbor of *#1*, and single cuts of two additional birds from Brazil and Peru, respectively. We then retrieved data on these cuts that allowed comparison of repertoire among song sequences of bird *#1*, between bird *#1* and the others, and among all the birds.

In the survey part of this study, 36 song sequences of approximately 30 birds, recorded at three localities in Peru and Brazil were analyzed. The localities were Tambopata Reserve, near Puerto Maldonado, depto. Madre de Dios, Peru; Quebrada Sucusari Camp, on the north bank of the lower Río Napo, depto. Loreto, Peru; and Humaita do Moa, edo. Acre, Brazil. These recordings were mostly of birds recorded at the Peruvian locations, with only one from Brazil. Parker's identification (see Appendix II) of vocal appropriation by these Lawrence's Thrushes yielded 176 kinds of organisms, mostly species of birds (11 orders, 25 families, and 6 subfamilies), as well as one frog (Dendrobatidae) and one orthopteran or frog sound, otherwise unidentified. A few kinds of birds are identified only to a group such as "hawk" or "hummingbird."

Most of the 36 recorded song sequences that we surveyed were incomplete, interrupted by the recordists or by disturbance to the bird. The six sequences (Appendix I) selected for more detailed analysis were from 51 to 262 phrases long.

RESULTS

SINGING AND RELATED BEHAVIOR OF LAWRENCE'S THRUSH

From a few, often-used song perches high in the canopy, individuals sing a remarkable song comprising long, continuous series of imitations of vocalizations of other birds, and occasionally of frogs and insects. Song perches of four birds observed over a period of eight years at the Tambopata Reserve were ca. 25–35 m above ground in the upper, often sunlit, branches of tall trees. The centers of the "territories" of these individuals were separated by ca. 600–1,000 m; whether these singing birds could hear each other is unknown. In southern Peru, individuals sang almost continuously from dawn until late in the morning, leaving their perches (presumably to forage) for brief periods of five to 15 min. They also sang for long periods during the warmest afternoon hours (14:00–18:00) on clear days when most other forest bird species are silent. At Tambopata, males sang daily during clear, warm weather from April to mid-October (the dry season) and then sporadically during clear periods throughout the intervening rainy season. In contrast, singing individuals studied from 1982–1989 at Quebrada Sucusari camp, in northern Peru, were silent during the drier months of May to September, but characteristically vocal in January and February, when all four local thrushes (*T. albicollis*, *T. hauxwelli*, *T. ignobilis*, and *T. lawrencii*) are breeding (Parker, pers. obs.). At Sucusari Camp, Parker found what were presumably incubating female *T. lawrencii* in late January. The grass and mud cup nests were similar to those of many other *Turdus* species, wedged between bromeliads and the trunks of medium-sized, smooth-barked trees, ca. 10–20 m above water several cm deep, within tall flooded forest. In southern Peru, the breeding season of this and other *Turdus* species may also coincide with the wetter months of November to March.

ANALYSIS OF VOCAL APPROPRIATION BY LAWRENCE'S THRUSHES

Species imitated by approximately 30 different individuals.—Birds imitated approximately 30 species in about 50 song phrases. In the two song sequences that extended for more than 150 phrases (Fig. 1, A and C), a maximum number of imitated species was reached by around 180 phrases. In no sequence did a bird imitate more than 51 species (Figs. 1 and 2), although we know that more species are in some individuals' repertoires (Parker and Remsen 1987). Most unidentified phrases seem to represent original vocalizations, not imitation.

Bird *#1* was resident near the head of a popular nature trail, where tape-recorded often by various people. We chose three long sequences from this bird for comparison. Bird *#5*, also at Tambopata, was a distant neighbor of bird *#1*. The single song-bout sequence analyzed reveals vocal appropriation of sounds of 51 bird species, 21 of which were also in the three cuts analyzed for bird *#1*.

We also analyzed six other 50-song phrase sequences from LNS: cuts 11416, 34215, and 37832 (again, bird *#1*); 28961 (Tambopata); 28539 (Limoncocha, Ecuador); 33781 (Sucusari Camp, N. bank of lower Río Napo, Loreto, Peru). In these 50-song phrase sequences, the males imitated 32, 27, 26, 33, 38, and 33 identifiable species, respectively. The range of imitated species for Peruvian and Ecuadorian Lawrence's Thrushes was thus 26 to 38 in 50 song-phrase sequences, with a mean of 32.7. In Appendix III we present a portion of one of the song sequences of bird *#1* (LNS 12848), shown in Figure 1A, listing the imitated species and unidentified phrases exactly as they were uttered by the bird.

Some general characteristics of Lawrence's Thrush.—Among diurnal predators, only the Black Hawk-Eagle (*Spizaetus tyrannus*) was commonly imitated. It is often vocal in flight. However, another common diurnal predator, the Roadside Hawk (*Buteo magnirostris*), a highly vocal species, was not detected in any Lawrence's Thrush repertoire. No owl (Strigiformes) imitations were recorded, nor were any imitations of nightjars (Caprimulgidae), although these species are often vocal in twilight hours, when many thrushes apparently learn orthopteran and frog sounds. *Pionites* parrots (Psittacidae) were frequently imitated. The White-bellied Parrot (*Pionites leucogaster*) is commonly imitated at Tambopata, whereas its allospecies, the Black-headed Parrot (*P. melanocephala*), is imitated elsewhere. Jacamars (Galbulidae) of various species were imitated frequently. Although not evident from the data presented, the Great Jacamar (*Jacamerops aurea*) is usually imitated poorly but unmistakably. Only once in the scores of renditions did a thrush perform the imitation well (hear on Hardy and Parker 1985, example 1, which is Vielliard's recording at Humaita do Moa, edo. Acre, Brazil, bird *#20* herein). By far the most consistently imitated species in most Lawrence's Thrush repertoires was either the Bluish-slate Antshrike (*Thamnomanes schistogynus*) in s. Peru and Brazil, or its northern replacement, the Cinereous Antshrike (*T. caesius*), which has an essentially identical voice. These

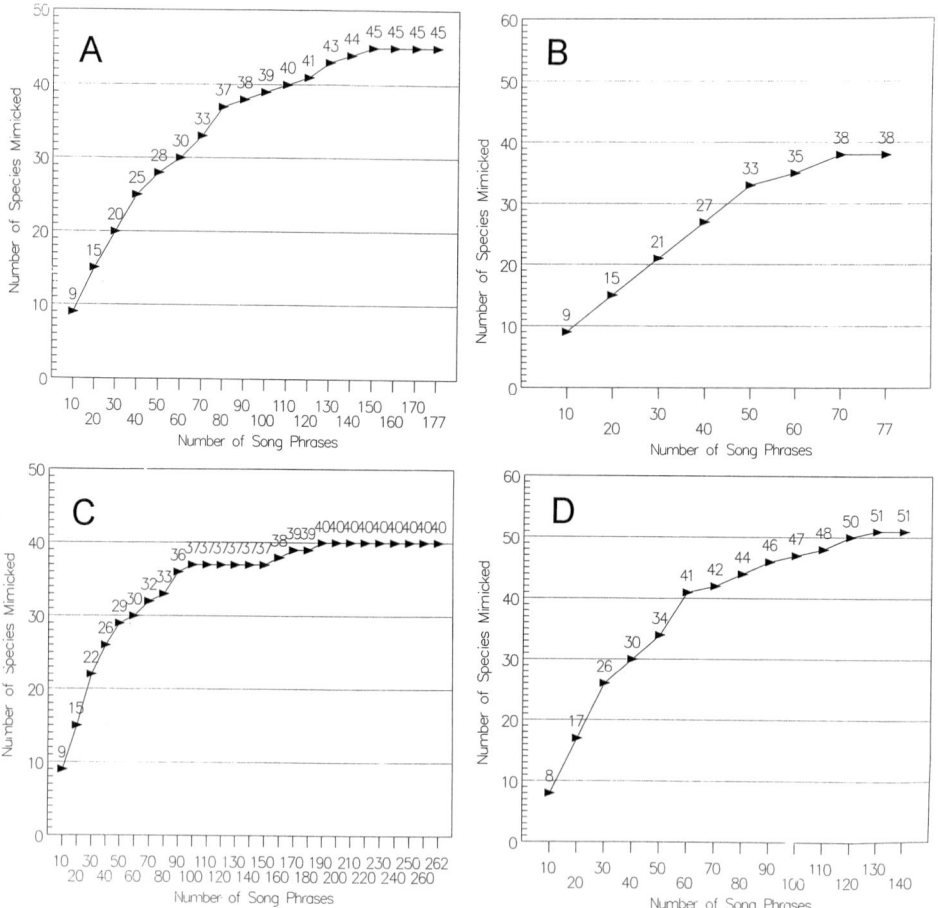

FIG. 1. Number of species copied vs. number of song phrases uttered in four song-phrase sequences by Lawrence's Thrush: (A) LNS 12848, referred to as bird #1 herein; (B) LNS 17825, bird #1; (C) LNS 18189, bird #1; (D) LNS 13648, bird #5 herein. All recordings made at Tambopata Reserve, depto. Madre de Dios, Peru.

antbirds' repertoires are unusually varied and variable, and the thrush's imitations reflect this. Antbirds (Formicariidae) are copied by the thrush, probably because they are numerous and highly vocal throughout the thrush's range. Manakins (Pipridae) and tyrant-flycatchers (Tyrannidae) are also among the suboscines that are widely and richly imitated. Although there are exceptions, most Lawrence's Thrushes included songbirds in their repertoires by imitating only brief segments of the species' full songs. For example, with one exception (Fig. 3C, D), imitation of the song of the Southern Nightingale-Wren (*Microcerculus marginatus*) consisted merely of the opening notes of its song. The same was true of the songs of the *Thryothorus* wren species and of the Slate-colored Grosbeak (*Saltator grossus*), all of which are frequently copied. Although most copying by Lawrence's Thrush is of common birds, sounds of rare birds (e.g., Brown-rumped Foliage-gleaner, *Automolus melanopezus*, and Red-billed Pied Tanager, *Lamprospiza melanoleuca*) are represented fairly consistently in most individual repertoires, and generally silent, retiring migrants such as the Veery (*Catharus fuscescens*) can be represented. Individuals of species imitated by Lawrence's Thrush are not known to react to the copying, probably because imitated phrases are brief and simply passing components of a continuing rendition of utterances. One anonymous referee pointed out that Lawrence's Thrush would be disadvantaged if constantly attacked by every bird it imitates. However, Parker makes no mention of such attacks in his field notes. See Dobkin (1979) for a discussion of this subject.

It seems possible that in no other bird species has imitation been reported to so dominate the

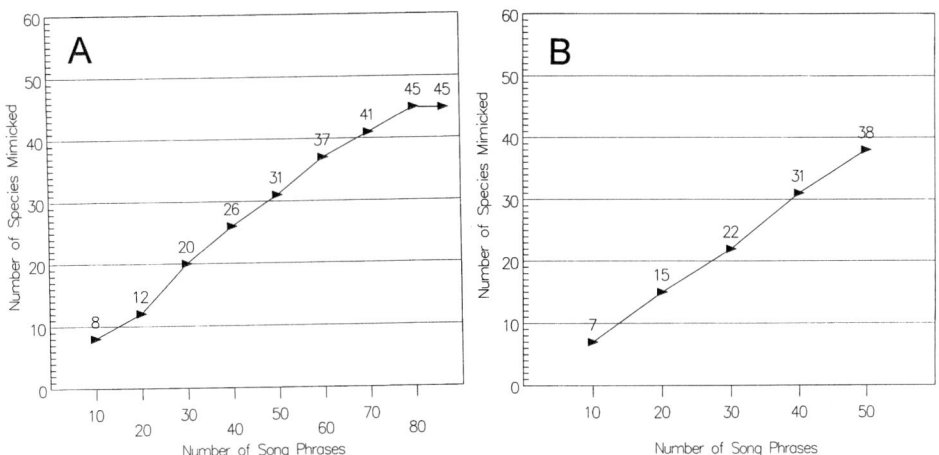

FIG. 2. Number of species copied vs. number of song phrases uttered in two song-phrase sequences by Lawrence's Thrush: (A) LNS 24216, bird *#10* herein, recorded at Quebrada Sucusari Camp, lower Río Napo, depto. Loreto, Peru; (B) bird *#20* herein, recorded by Jacques Vielliard at Humaita do Moa, edo. Acre, Brazil.

song. One other impressive example of such domination has been suggested, however, by Remsen et al. (1982), who proposed that the song of Lawrence's Goldfinch (*Carduelis lawrencei*) is almost completely composed of interspecifically copied notes. As might be expected, some individual Lawrence's Thrushes imitate better than others. Some imitate poorly, but the vocally appropriated species can be recognized even though the renditions are in a different pitch, or are otherwise distorted. Whether this is related to reproductive condition or age is unknown. The easiest performances to follow are those of individual thrushes that copy with each phrase

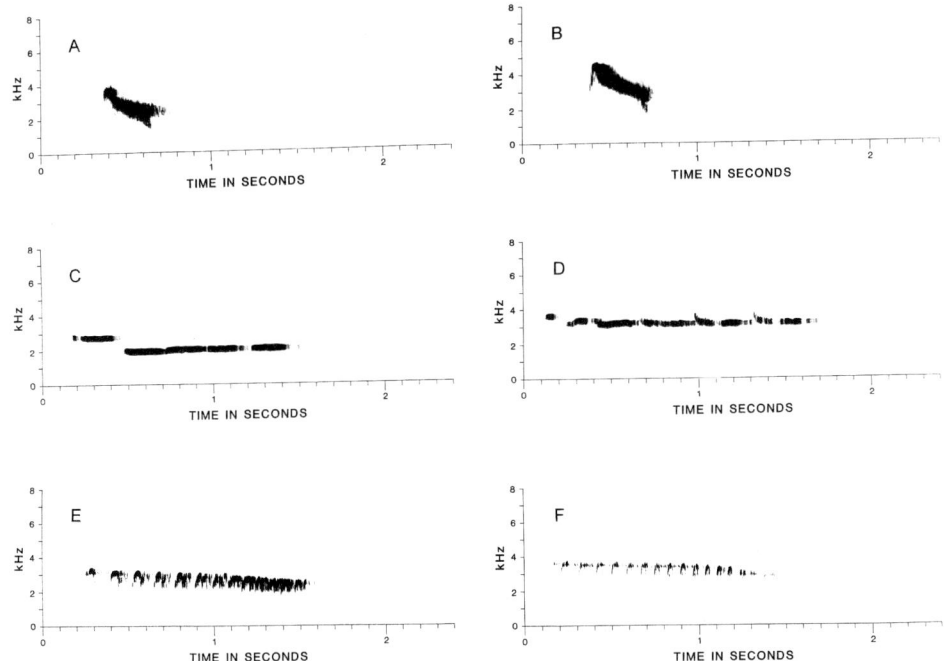

FIG. 3. Comparison of sonograms of characteristic vocalizations of three bird species with sonograms of vocal copying of those species by Lawrence's Thrush: (A) Violaceous Jay (*Cyanocorax violaceus*), call; (B) imitation; (C) Southern Nightingale-Wren (*Microcerculus marginatus*), introductory portion of song; (D) imitation; (E) Lineated Woodcreeper (*Lepidocolaptes albolineatus*), song; (F) imitation.

only one species, followed by a pause, and another species copied, etc. Some birds are, however, highly complex imitators, and thus may imitate two to four species without pause, combining one song component from each of the copied birds into a single phrase (see Discussion). One such utterance was a phrase that included calls of the Black-faced Antthrush (*Formicarius analis*), the Band-tailed Manakin (*Pipra fasciicauda*), and the Bluish-fronted Jacamar (*Galbula cyanescens*).

Lawrence's Thrush also has distinctive call notes. One recording made at Tambopata by Lewis Kibler (LNS 35179) consists of two (alarm?) calls, which are sometimes coupled. One is a soft *kuk* and the other is a piercing, shrill *peer*! On this recording, these calls were usually uttered together as *kuk-cheer*! If one were unfamiliar with Lawrence's Thrush song, these repeated calls could be mistaken for song. On LNS 2539, recorded by Arnoud van den Berg at Limoncocha, Río Napo, pvcia. Loreto, Ecuador, three different calls preceded, but were not incorporated into, the song. They were a plaintive *perwheee*, an abrupt *weecheee*, and an exclamatory *peep peep peep*! Lawrence's Thrushes often respond vigorously to playback of such vocalizations, in contrast to their lack of visible response to playback of song.

ANALYSIS OF SONG OF TWO MALE LAWRENCE'S THRUSHES AT
TAMBOPATA RESERVE

Individual male variation in songs.—Male *#1* (identification as a single individual putative because the bird was *not* marked) copied 50 species in the three song sequences analyzed here. Forty-nine of these species were birds. Of these, 27 (55%) were on all three sequences, 11 (22%) were on two sequences, and 11 on only one sequence. The three sequences were recorded over a two-year period. This individual of Lawrence's Thrush was sedentary, singing on the same song-post day after day—at the same popular trail head. That is why we became acquainted with the bird. The 55% degree of similarity of repertoire on these three recordings suggests that a basic repertoire is established in adult birds, but that considerable latitude for repertoire modification may exist. A slight tendency exists for frequently imitated species to be on all sequences. The Bluish-fronted Jacamar and the Bran-colored Flycatcher (*Myiophobus fasciatus*), copied 11 and 12 times, respectively, were notable exceptions, both being absent from one sequence. The Forest Elaenia (*Myiopagis gaimardii*) was copied five times and was missing from one of the sequences. The Dusky-headed Parakeet (*Aratinga weddellii*) was imitated three times but was on all three sequences. The Epaulet Oriole (*Icterus cayanensis*) was also imitated only three times, but all were on only one sequence. The Chestnut-tailed Antbird (*Myrmeciza hemimelaena*) was copied twice, both on the same sequence. The eight other species imitated three to five times were absent from one of the three sequences.

In the three song sequences of bird *#1*, there were 513 phrases. Of these, 444 (86%) were identified as vocal copying. Twenty (3.9%) of those were classed as either "Orthopteran or frog (= dusk sound)," and 422 (82%) were identified as imitations of particular bird species. Five phrases were identified as "*Myiozetetes* sp." and were not counted in the total number of 50 species copied because all three locally occurring members of that genus, *granadensis, luteiventris,* and *similis,* were identified among the imitations.

Of the 49 bird species imitated by bird *#1*, 17 (35%) accounted for 271 (64%) of the 425 imitative phrases uttered, each of the 17 having been copied at least 10 times (range 10 to 37). The Bluish-slate Antshrike was imitated 37 times, the Slate-colored Grosbeak 28 times, the Black-faced Antbird (*Myrmoborus myotherinus*) 23 times, and both the Striolated Puffbird (*Nystalus striolatus*) and Blue-backed Manakin (*Chiroxiphia pareola*) 20 times.

Differences in repertoire between two males at Tambopata.—Both birds, *#1* and *#5*, were in the Tambopata Reserve, but as previously noted, were not close neighbors. It is doubtful that they could hear each other sing from their primary song stations. It seems likely, however, that the resource of singing and calling bird species from which they could construct their imitative song repertoires was essentially the same. Surprisingly, male *#1*'s repertoire of imitations of 49 bird species in the three sequences analyzed here include mimicry of 28 species not heard in the recorded sequence of male *#5*. This could be because only one song sequence of male *#5* was available for analysis. Yet that song sequence of 142 phrases, in which imitations of 51 bird species are evident, included 30 bird species evidently not in *#1*'s repertoire. The two males' imitative repertoires included 21 bird species that they both copied in the sequences analyzed. The shared species make up less than half (43% of *#1*'s repertoire; 41% of *#5*'s; 27% of the sum) of the bird species imitated in the songs of these two individuals, even though they shared an essentially identical acoustical environment. Selection of material thus appears to differ among

TABLE 1
SUMMARY OF DEGREE OF VOCAL APPROPRIATION IN 25-SONG PHRASE SEQUENCES OF LAWRENCE'S THRUSH

	Total phrases	Number with copying	Percent with copying
Bird #1			
(LNS 11416) Cut 1	72	62	86
(LNS 12848) Cut 4	177	155	86
(LNS 17815) Cut 11	74	64	86
(LNS 18189) Cut 13	262	226	86
(LNS 13621)	22	19	89
(LNS 17829) Cut 12	40	35	88
(LNS 12628)	39	35	90
Bird #2			
(LNS 11482)	34	30	88
Bird #6			
(LNS 13660)	67	54	88
Bird #7			
(LNS 28961)	52	43	82
Bird #5			
(LNS 13948) Cut 9	142	117	82
Bird #10			
(LNS 24216)	86	69	80
Bird #20			
(Vielliard)	51	41	80
(LNS 24193)	48	44	92
(LNS 28539)	56	45	80
(LNS 28931)	29	26	89
(LNS 31914)	27	25	92
(LNS 31917)	19	17	89
(LNS 33781)	82	73	82
(LNS 31293)	15	13	86
(LNS 35114)	48	36	75
(LNS 35160)	24	23	96
(LNS 34215)	66	63	95
(BLA D-23)	23	16	70
(BLA D-73)	26	16	62
(LNS 37391)	17	17	100
Totals: 25	1,598	1,364	85

individual Lawrence's Thrushes. It is fascinating to us, for example, that both males imitated two species of antwrens from the genus *Myrmotherula*, but not the same two! Bird #1 copied the White-flanked (*M. axillaris*) and Plain-throated (*M. hauxwelli*) antwrens, both undergrowth species, whereas bird #5 copied the Pygmy (*M. brachyura*) and Sclater's (*M. sclateri*) antwrens, both canopy species. As D. S. Dobkin (*in litt.*) properly pointed out: "... a serious undersampling error exists here. An obvious variability in sequence organization by individual singers indicates that much greater sampling of individual birds is necessary before definitive statements can be made concerning such things as the lack of extensive overlap in imitated species by males sharing the same acoustic environments (e.g., at Tambopata)."

PERCENTAGE OF SONG PHRASES INCORPORATING OR CONSTITUTING IMITATION

Of the 36 song-phrase sequences analyzed, we chose 25 (that consisted of from 15 to 262 phrases) to determine the frequency of occurrence of imitation (Table 1). We rejected some shorter sequences marred by distortion of recording quality, causing Parker to be uncertain of the species. To be tallied in the non-imitative category, the phrase had to be totally without a component of vocal copying. Many phrases had components of apparent non-copying coupled with copying. These were tallied as copying. A phrase listed as unidentified but as "sounding familiar" was also listed as copying. The 25 sequences analyzed contained 1,598 song phrases, 1,364 (85%) of which included or constituted copying of another bird species or other animal sound. The range of variation was 62% (16/26) to 100% (17/17). Note that the male identified as male #1 had a degree of copying of 86% in *each* of the three song-phrase sequences.

When the 25 song-phrase sequences in Table 1 are arranged from fewest phrases (15) to most (262), no correlation exists between number of phrases in a recorded sequence and degree of vocal appropriation (Product-Moment correlation coefficient $r = -0.044$, $N = 25$, $P > 0.5$). That is, there is no suggestion of effect of observer sampling duration on likelihood of detection of vocal appropriation.

QUALITY OF IMITATION, ORGANIZATION, AND DEGREE OF ABSTRACTNESS IN SONG

Figure 3 presents sonograms that display the accuracy with which an individual Lawrence's Thrush copied the calls and songs of three other bird species. For example, Figure 3A represents the call of Violaceous Jay, which may be compared directly with Figure 3B, a Lawrence's Thrush imitation of that call. Similar comparisons may be made of Lawrence's Thrush imitation of Southern Nightingale-Wren in Figure 3C, D and Lineated Woodcreeper in Figure 3E, F. As these sonograms illustrate, Lawrence's Thrush imitations can show remarkable fidelity to the vocalizations on which they are based (all three may be heard on Hardy and Parker 1985, 51:ex. 2).

For the more detailed analysis, we chose song sequences that were straightforward, well-recorded, with well-spaced phrases, usually involving imitation of one bird or other animal species, or an unidentified sound believed to be an original utterance of Lawrence's Thrush.

As noted above, some males' songs are more abstract, having phrases composed of imitation of two or more species or a species plus an unidentified component, with distortion of pitch or cadence. A 21-phrase portion of an 82-song sequence recording is presented here to illustrate such singing. The male is LNS 33781, recorded by Parker at Sucusari Camp, on 18 January 1984. An approximate transcription of what Parker dictated between this thrush's first 21 song phrases follows: "*Legatus leucophaius* call; *Thryothorus* sp. slow phrase + *Formicarius colma* rising note series; *Myrmotherula hauxwelli*, beginning of song; *Ramphastos cuvieri* + *Spizaetus tyrannus* flight call; *Hylophilus thoracicus* song (distorted); *Cnipodectes subbrunneus*; *Xenops*?; *Phlegopsis nigromaculata* song; *Dendrocolaptes certhia*-like song, *Tyrannopsis sulphurea* calls + *Celeus torquatus* calls (distorted); *Myrmeciza melanoceps* + *Thryothorus coraya* song; *Myrmotherula haematonota*-like song; unidentified + *Schiffornis major*; unidentified; *Spizaetus tyrannus* flight call; *Terenotriccus erythrurus* call; unidentified *Automolus rufipileatus*-like?; unidentified; *Myiarchus ferox* trill; *Piaya cayana* song + *Attila bolivianus* song; *Thryothorus*-like + unidentified; and *Cnipodectes subbrunneus* + *Myrmotherula hauxwelli* calls." It should be clear why the analysis of such song-phrase sequences is difficult.

Questions such as whether old birds mimic more frequently or copy more species than younger individuals, and whether compound phrases and original phrases are characteristic of old vs. young males, must await study of color-banded birds of known age over periods of several years.

DISCUSSION

Lawrence's Thrushes are almost impossible to find when not singing. More than 95% of all our observations ($n > 300$) were of solitary males on high song perches. (The birds were occasionally seen at fruiting trees in the mid-level or canopy of forest and, rarely, were flushed from the forest floor, where they search leaf litter in the manner of other thrushes.) Singing males almost always ignored playbacks of their own songs or those of neighboring males. This suggests to us that their elaborate "advertising" songs are delivered primarily to attract or stimulate females, not to defend territories. In contrast, singing males of some other Amazonian thrushes (e.g., Black-billed Thrush, *T. ignobilis*) are regularly encountered with their presumed mates. This observation leads us to the following prediction: Lawrence's Thrush has an atypical mating system that is either promiscuous or polygynous. The atypical behavior of male Lawrence's Thrushes, namely continuous, nearly year-round vocalizing from a few preferred song-perches, typifies the behavior of many Neotropical species known or assumed to be promiscuous or polygynous, e.g., numerous cotingas (Cotingidae), manakins (Pipridae), and some tyrant-flycatchers (Tyrannidae). We further predict that the elaborate advertising song of Lawrence's Thrush, which differs dramatically from other *Turdus* songs in the degree to which it is based on vocal appropriation, is the product of strong sexual selection.

Only two other New World thrushes, the Pale-eyed Thrush (*Platycichla leucops*) and the

Yellow-legged Thrush (*P. flavipes*), are known to have elaborate advertising songs in which many bird species are vocally copied (Hardy and Parker 1985). Numerous species of New World *Turdus* deliver songs similar in pattern to those of the three known vocal copiers, but none of their sounds are recognizable examples of vocal copying. Until recently, the only other known example of vocal copying among New World thrushes was to be heard on a recording of a peculiar Hauxwell's Thrush (*T. hauxwelli*) song obtained in response to playback of its advertising song (Hardy and Parker 1985). Upon hearing a tape of its own (typical) song, which is rather like that of an American Robin (*T. migratorius*), but slower, the bird flew in from 50 m away and began singing a "whisper song" comprising a succession of perfect imitations of at least 10 bird species found in the same habitat. The function of this song type in Hauxwell's Thrush seems to be territorial rather than mate attraction, in contrast to the structurally similar song of Lawrence's Thrush. Parker recently (pers. obs., 1989) heard a Pale-breasted Thrush (*T. leucomelas*) deliver an even more elaborate imitative song in response to playback of its own slow song, but this performance has not yet been analyzed.

ACKNOWLEDGMENTS

We are grateful to Greg Budney and Jim Gulledge of the Library of Natural Sound, Cornell University, for providing copies of recordings upon which this paper is in large part based, and to the late Ben B. Coffey, Jr., Victor Emanuel, and Jacques Vielliard for other recordings useful in our study. The Coffey recordings are in the Bioacoustic Archives of the Florida Museum of Natural History, and the Vielliard recordings are in the Arquivo Sonoro Neotropical, Universidad Estadual de Campinas, Brazil. We thank Mary V. McDonald and Dianna Carver for computer assistance, and Louisiana State University Museum of Natural Science and Victor Emanuel Nature Tours for providing Parker with recording opportunities in South America. We also thank J. V. Remsen, Jr., Lewis Kibler, and Robert M. Chandler for reading the manuscript, and Arnoud van den Berg and David S. Dobkin for their careful study of the manuscript and many helpful suggestions that improved it. Finally, thanks to Terry Taylor for his assistance in the final revision and preparation of the manuscript and to Adam Kent and Marcus Tellkamp for their help with the Spanish abstract.

LITERATURE CITED

BAYLIS, J. R. 1982. Avian vocal mimicry: its function and evolution. Pp. 1–83 *in* Acoustic Communication in Birds. Vol. 2. (D. E. Kroodsma and E. H. Miller, Eds.). Academic Press, New York.

DOBKIN, D. S. 1979. Functional and evolutionary relationships of vocal copying phenomena in birds. Zeit. für Tierpsych. 50:348–363.

HARDY, J. W., AND T. A. PARKER III. 1985. Voices of the New World Thrushes. Audio-cassette tape with insert notes. ARA Records, Gainesville, Florida.

PARKER, T. A., III. 1982. Observations of some unusual rainforest and marsh birds in southeast Peru. Wilson Bull. 94:477–492.

PARKER, T. A., III, AND J. V. REMSEN, JR. 1987. Fifty-two Amazonian bird species new to Bolivia. Bull. Br. Ornithol. Club 107:94–106.

REMSEN, J. V., JR., K. GARRETT, AND R. A. ERICKSON. 1982. Vocal copying in Lawrence's and Lesser goldfinches. West. Birds 13:29–33.

RIDGELY, R. S., AND G. TUDOR. 1989. The Birds of South America. Vol. I. The Oscine Passerines. Univ. Texas Press, Austin, Texas.

APPENDIX I
Singing Male Lawrence's Thrushes and Their Recorded Song Sequences Analyzed in Greatest Detail in This Paper

Bird	Designation	Sequence length	Locality	Date	Recordist
#1					
A	LNS 1283 (4)	10:45	Tambopata Reserve, Puerto Maldonado, Madre de Dios, Peru	8/7/77	Parker
B	LNS 17825 (11)	6:14	Tambopata Reserve, Puerto Maldonado, Madre de Dios, Peru	?/7/79	V. Emanuel
C	LNS 18189 (13)	24:11	Tambopata Reserve, Puerto Maldonado, Madre de Dios, Peru	9/9/79	M. Palmer
#5	LNS 13648 (5)	8:43	Tambopata Reserve, Puerto Maldonado, Madre de Dios, Peru	23/7/79	Parker
#10	LNS 37832 (10)	8:01	Sucusari Camp, N Bank Rio Napo, Loreto, Peru	5/1/85	G. Budney
#20	ASN, Campinas, Brazil (20)	2:30	Humaita do Moa, Acre, Brazil	19/4/81	J. Vielliard

APPENDIX II
Vocal Appropriation in Lawrence's Thrush (*Turdus lawrencii*): Systematic List of Species Copied by Approximately 30 Males in 36 Song Sequences at Three Localities in Peru, Brazil, and Ecuador

TINAMIDAE
  1. *Crypturellus soui,* Little Tinamou

ACCIPITRIDAE
  2. *Leptodon cayanensis,* Gray-headed Kite
  3. *Spizaetus tyrannus,* Black Hawk-Eagle
  4. "Hawk"

FALCONIDAE
  5. *Daptrius ater,* Black Caracara
  6. *Falco rufigularis,* Bat Falcon

CHARADRIIDAE
  7. *Vanellus cayanus,* Pied Lapwing

PSITTACIDAE
  8. *Pionites leucogaster,* White-bellied Parrot
  9. *Pionites melanocephala,* Black-headed Parrot
  10. *Pionus menstruus,* Blue-headed Parrot
  11. *Aratinga weddellii,* Dusky-headed Parakeet
  12. *Aratinga leucophthalmus,* White-eyed Parakeet
  13. *Pyrrhura rupicola,* Rock Parakeet
  14. *Pyrrhura melanonota,* Maroon-tailed Parakeet
  15. *Brotogeris cyanoptera,* Cobalt-winged Parakeet

CUCULIDAE
  16. *Piaya cayana,* Squirrel Cuckoo
  17. *Dromococcyx pavoninus,* Pavonine Cuckoo

APODIDAE
  18. *Reinarda squamata,* Fork-tailed Palm-Swift

APPENDIX II
Continued

TROCHILIDAE
  19. "Hummingbird"

TROGONIDAE
  20. *Trogon collaris,* Collared Trogon

ALCEDINIDAE
  21. *Chloroceryle amazona,* Amazon Kingfisher

GALBULIDAE
  22. *Galbula cyanescens,* Bluish-fronted Jacamar
  23. *Galbula tombacea,* White-chinned Jacamar
  24. *Galbula dea,* Paradise Jacamar
  25. *Jacamerops aurea,* Great Jacamar

BUCCONIDAE
  26. *Notharchus macrorhynchus,* White-necked Puffbird
  27. *Nystalus striolatus,* Striolated Puffbird
  28. *Nonnula* sp., Nunlet
  29. *Monasa nigrifrons,* Black-fronted Nunbird
  30. *Chelidoptera tenebrosa,* Swallow-wing

RAMPHASTIDAE
  31. *Pteroglossus castanotis,* Chestnut-eared Aracari
  32. *Pteroglossus beauharnaesii,* Curl-crested Aracari
  33. *Pteroglossus flavirostris,* Ivory-billed Aracari
  34. *Ramphastos cuvieri,* Cuvier's Toucan
  35. *Ramphastos culminatus,* Yellow-ridged Toucan

PICIDAE
  36. *Celeus flavus,* Cream-colored Woodpecker
  37. *Celeus torquatus,* Ringed Woodpecker
  38. *Celeus grammicus,* Scale-breasted Woodpecker
  39. *Campephilus rubricollis,* Red-necked Woodpecker

DENDROCOLAPTIDAE
  40. *Sittasomus griseicapillus,* Olivaceous Woodcreeper
  41. *Glyphorynchus spirurus,* Wedge-billed Woodcreeper
  42. *Dendrexetastes rufigula,* Cinnamon-throated Woodcreeper
  43. *Xiphocolaptes promeropirhynchus,* Strong-billed Woodcreeper
  44. *Dendrocolaptes picumnus,* Black-banded Woodcreeper
  45. *Dendrocolaptes certhia,* Amazonian Barred-Woodcreeper
  46. *Lepidocolaptes albolineatus,* Lineated Woodcreeper
  47. *Xiphorhynchus spixii,* Spix's Woodcreeper
  48. *Xiphorhynchus elegans,* Elegant Woodcreeper
  49. *Xiphorhynchus guttatus,* Buff-throated Woodcreeper

FURNARIIDAE
  50. *Ancistrops strigilatus,* Chestnut-winged Hookbill
  51. *Philydor erythropterus,* Chestnut-winged Foliage-gleaner
  52. *Philydor ruficaudatus,* Rufous-tailed Foliage-gleaner
  53. *Automolus infuscatus,* Olive-backed Foliage-gleaner
  54. *Automolus rufipileatus,* Chestnut-crowned Foliage-gleaner
  55. *Automolus melanopezus,* Brown-rumped Foliage-gleaner
  56. *Automolus ochrolaemus,* Buff-throated Foliage-gleaner
  57. *Xenops minutus,* Plain Xenops
  58. *Sclerurus caudacutus,* Black-tailed Leaftosser

FORMICARIIDAE
  59. *Cymbilaimus lineatus,* Fasciated Antshrike
  60. *Pygiptila stellaris,* Spot-winged Antshrike
  61. *Thamnomanes ardesiacus,* Dusky-throated Antshrike
  62. *Thamnomanes caesius,* Cinereous Antshrike
  63. *Thamnomanes schistogynus,* Bluish-slate Antshrike
  64. *Myrmotherula brachyura,* Pygmy Antwren
  65. *Myrmotherula sclateri,* Sclater's Antwren
  66. *Myrmotherula hauxwelli,* Plain-throated Antwren

APPENDIX II
CONTINUED

67. *Myrmotherula haematonota*, Stipple-throated Antwren
68. *Myrmotherula ornata*, Ornate Antwren
69. *Myrmotherula axillaris*, White-flanked Antwren
70. *Myrmotherula longipennis*, Long-winged Antwren
71. *Myrmotherula iheringi*, Ihering's Antwren
72. *Myrmotherula menetriesii*, Gray Antwren
73. *Herpsilochmus sticturus*, Spot-tailed Antbird
74. *Drymophila devillei*, Striated Antbird
75. *Cercomacra cinerascens*, Gray Antbird
76. *Cercomacra serva*, Black Antbird
77. *Myrmoborus myotherinus*, Black-faced Antbird
78. *Myrmoborus leucophrys*, White-browed Antbird
79. *Hypocnemis cantator*, Warbling Antbird
80. *Hypocnemis hypoxantha*, Yellow-browed Antbird
81. *Hypocnemoides maculicauda*, Band-tailed Antbird
82. *Percnostola lophotes*, White-lined Antbird
83. *Schistocichla schistacea*, Slate-colored Antbird
84. *Sclateria naevia*, Silvered Antbird
85. *Myrmeciza hemimelaena*, Chestnut-tailed Antbird
86. *Myrmeciza hyperythra*, Plumbeous Antbird
87. *Myrmeciza melanoceps*, White-shouldered Antbird
88. *Myrmeciza goeldii*, Goeldi's Antbird
89. *Myrmeciza fortis*, Sooty Antbird
90. *Hylophylax naevia*, Spot-backed Antbird
91. *Phlegopsis nigromaculata*, Black-spotted Bare-eye
92. *Chamaeza nobilis*, Striated Antthrush
93. *Formicarius colma*, Rufous-capped Antthrush
94. *Formicarius analis*, Black-faced Antthrush

COTINGIDAE

95. *Lipaugus vociferans*, Screaming Piha
96. *Pachyramphus marginatus*, Black-capped Becard
97. *Pachyramphus polychopterus*, White-winged Becard

PIPRIDAE

98. *Pipra filicauda*, Wire-tailed Manakin
99. *Pipra fasciicauda*, Band-tailed Manakin
100. *Pipra erythrocephala*, Golden-hooded Manakin
101. *Chiroxiphia pareola*, Blue-backed Manakin
102. *Tyranneutes stolzmanni*, Dwarf Manakin
103. *Piprites chloris*, Wing-barred Manakin
104. *Schiffornis major*, Greater Manakin

TYRANNIDAE

105. *Myiopagis gaimardii*, Forest Elaenia
106. *Myiopagis caniceps*, Gray Elaenia
107. *Myiopagis viridicata*, Greenish Elaenia
108. *Tyrannulus elatus*, Yellow-crowned Tyrannulet
109. *Ornithion inerme*, White-lored Tyrannulet
110. *Zimmerius gracilipes*, Slender-footed Tyrannulet
111. *Mionectes oleagineus*, Ochre-bellied Flycatcher
112. *Hemitriccus iohannis*, Johannes' Tody-Tyrant
113. *Hemitriccus flammulatus*, Flammulated Pygmy-Tyrant
114. *Corythopis torquata*, Ringed Antpipit
115. *Platyrinchus platyrhynchos*, White-crested Spadebill
116. *Platyrinchus coronatus*, Golden-crowned Spadebill
117. *Tolmomyias poliocephalus*, Gray-crowned Flycatcher
118. *Tolmomyias assimilis*, Yellow-margined Flycatcher
119. *Ramphotrigon fuscicauda*, Dusky-tailed Flatbill
120. *Cnipodectes subbrunneus*, Brownish Flycatcher
121. *Terenotriccus erythrurus*, Ruddy-tailed Flycatcher
122. *Myiophobus flavicans*, Flavescent Flycatcher
123. *Myiophobus fasciatus*, Bran-colored Flycatcher

APPENDIX II
Continued

124. *Lathrotriccus euleri,* Euler's Flycatcher
125. *Contopus virens,* Eastern Wood-Pewee
126. *Cnemotriccus fuscatus,* Fuscous Flycatcher
127. *Colonia colonus,* Long-tailed Tyrant
128. *Attila cinnamomeus,* Cinnamon Attila
129. *Attila bolivianus,* Dull-capped Attila
130. *Attila spadiceus,* Bright-rumped Attila
131. *Rhytipterna simplex,* Grayish Mourner
132. *Myiarchus ferox,* Short-crested Flycatcher
133. *Sirystes sibilator,* Sirystes
134. *Megarynchus pitangus,* Boat-billed Flycatcher
135. *Pitangus sulphuratus,* Great Kiskadee
136. *Myiozetetes cayanensis,* Rusty-margined Flycatcher
137. *Myiozetetes similis,* Social Flycatcher
138. *Myiozetetes granadensis,* Gray-capped Flycatcher
139. *Myiozetetes luteiventris,* Dusky-chested Flycatcher
140. *Myiozetetes* sp.
141. *Legatus leuceophaius,* Piratic Flycatcher
142. *Tyrannopsis sulphurea,* Sulphury Flycatcher

CORVIDAE

143. *Cyanocorax violaceus,* Violaceous Jay

TROGLODYTIDAE

144. *Thryothorus genibarbis,* Moustached Wren
145. *Thryothorus coraya,* Coraya Wren
146. *Thryothorus leucotis,* Buff-breasted Wren
147. *Henicorhina leucosticta,* White-breasted Wood-Wren
148. *Microcerculus marginatus,* Southern Nightingale Wren
149. *Donacobius atricapillus,* Black-capped Donacobius

MUSCICAPIDAE
    Turdinae

150. *Turdus ignobilis,* Black-billed Thrush
151. *Turdus hauxwelli,* Hauxwell's Thrush
152. *Catharus fuscescens,* Veery

VIREONIDAE

153. *Vireo olivaceus,* Red-eyed Vireo
154. *Hylophilus thoracicus,* Lemon-chested Greenlet
155. *Hylophilus hypoxanthus,* Dusky-capped Greenlet
156. *Smaragdolanius leucotis,* Slaty-capped Shrike-Vireo

EMBERIZIDAE
    Icterinae

157. *Cacicus cela,* Yellow-rumped Cacique
158. *Cacicus solitarius,* Solitary Black Cacique
159. *Icterus chrysocephalus,* Moriche Oriole
160. *Icterus cayanensis,* Epaulet Oriole
    Parulinae
161. *Basileuterus (Phaeothlypis)* sp.
    Thraupinae
162. *Dacnis* sp., Dacnis
163. *Tersina viridis,* Swallow Tanager
164. *Lamprospiza melanoleuca,* Red-billed Pied Tanager
165. *Lanio versicolor,* White-winged Shrike-Tanager
166. *Lanio fulvus,* Fulvous Shrike-Tanager
167. *Tachyphonus luctuosus,* White-shouldered Tanager
168. *Habia rubica,* Red-crowned Ant-Tanager
169. *Euphonia chrysopasta,* Golden-bellied Euphonia
170. *Euphonia minuta,* White-vented Euphonia
171. *Euphonia rufiventris,* Rufous-bellied Euphonia
172. *Tangara chilensis,* Paradise Tanager

APPENDIX II
CONTINUED

Cardinalinae
173. *Saltator grossus,* Slate-colored Grosbeak
174. *Saltator maximus,* Buff-throated Saltator
175. *Cyanocompsa cyanoides,* Blue-black Grosbeak

Insecta
176. Orthopteran or frog (= dusk sound)

Amphibia
177. Dendrobatidae "frog"

APPENDIX III
AN EXAMPLE OF A SONG PHRASE SEQUENCE, CONTINUOUSLY GIVEN, OF LAWRENCE'S THRUSH

Song phrase sequence LNS 12848, Parker Cut 4, male *#1*

1. Buff-throated Woodcreeper; 2. Moustached Wren; 3. Dwarf Tyrant-Manakin; 4. Spix's Woodcreeper; 5. unidentified; 6. unidentified; 7. White-browed Antbird; 8. Black-capped Becard; 9. Golden-bellied Euphonia; 10. Black-faced Antbird; 11. Black Hawk-Eagle; 12. unidentified; 13. Great Kiskadee; 14. Brown-rumped Foliage-gleaner; 15. White-winged Shrike-Tanager; 16. Piratic Flycatcher; 17. Goeldi's Antbird; 18. Sirystes; 19. Bluish-slate Antshrike; 20. Chestnut-eared Aracari; 21. Dull-capped Attila; 22. Amazon Kingfisher; 23. Cream-colored Woodpecker; 24. White-lored Tyrannulet; 25. Wing-barred Manakin; 26. Black-spotted Bare-eye; 27. Black-faced Antbird; 28. Black-faced Antthrush; 29. Euler's Flycatcher; 30. Bluish-slate Antshrike; 31. Epaulet Oriole; 32. Dusky-tailed Flatbill; 33. unidentified; 34. Bluish-slate Antshrike; 35. Southern Nightingale Wren; 36. Bluish-fronted Jacamar; 37. Buff-throated Woodcreeper; 38. unidentified; 39. Spix's Woodcreeper; 40. Bluish-slate Antshrike; 41. unidentified; 42. "hummingbird"; 43. Paradise Jacamar; 44. unidentified; 45. Dwarf Tyrant-Manakin; 46. Brown-rumped Foliage-gleaner; 47. Moustached Wren; 48. Pied Lapwing; 49. unidentified; 50. Black Hawk-Eagle; 51. Chestnut-crowned Foliage-gleaner.

SEASONAL DISTRIBUTION OF BIRDS AT A CLOUD-FOREST LOCALITY, THE ANCHICAYÁ VALLEY, IN WESTERN COLOMBIA

STEVEN L. HILTY

6316 West 102nd Street, Shawnee Mission, Kansas 66212-1718, USA

ABSTRACT.—This study presents information on the seasonal status of a cloud-forest avifauna on Colombia's Pacific Andean slope at 950–1050 m elevation. Two hundred seventy-one species were recorded at the 70–80 ha site. About 64% of these were year-round residents. The remainder (36%) were vagrants and species showing varying degrees of seasonal movements; 22% were short-distance (elevational or local) migrants, 8% were long-distance migrants from North America. Short-distance migrants were predominantly frugivores and nectarivores; over half (58%) of all nectarivores were elevational migrants. Seasonal movements were recorded in all months, but short-distance migrants were most numerous March-June and November-January, periods when migrants from both the highlands and lowlands were present. The abundance of flowers and fruits also showed two small peaks. About twice as many short-distance migrant nectarivores came from high elevations as low elevations. A majority of short-distance migrant frugivores also were from the highlands. Eight *Tangara* tanagers, although resident, showed significant expansions and contractions of foraging range, diversity of fruit eaten, and amount of fruit included in their diet, an indication that these birds track resource levels, and that variation in fruit abundance affects the dynamics of frugivore populations.

RESUMEN.—Este estudio muestra una información preliminar de cambios estacionales de la avifauna de un bosque nublado en la vertiente pacífica de los Andes colombianos. Se registró un total de 271 especies en un área entre 70–80 ha. Alrededor de un 36% de las especies mostraron evidencias de movimientos estacionales; 22% presentaban movimientos altitudinales o locales y 8% eran aves migratorias de larga distancia que se reproducen en Norte América. Las aves migratorias de corta distancia eran principalmente nectívoros y frugíforos; 58% de los nectívoros eran aves migratorios que se presentaban movimientos altitudinales o locales. Los movimientos estacionales fueron complejas y distribuidos a lo largo del año, pero más aves migratorias de corta distancia estuvieron presente durante los períodos de marzo a junio y de noviembre a enero, epocas cuando provenía especies de elevaciónes mayores y menores. La abundancia de flores y fruitas tambien mostraron evidencias de cambios estacionales. Los nectívoros migratorios provenía predominantemente de elevaciónes mayores; tambien la majoría de frugíforos estacionales provenía de regiones altas. Ocho especies de *Tangara* no mostran evidencias de movimientos altitudinales pero mostraron aumentos y reducciónes de sus rangos de alimentaciones, equalmente en la diversidad y cantidad de frutas consumidos.

It is generally acknowledged that populations of many tropical latitude-breeding birds vary seasonally in abundance (Karr 1977; Smythe 1982; Karr and Freemark 1983; Stiles 1983; Levy 1988). Variations in abundance over time may be a result of local movements within a habitat, or between habitats at similar elevations, or between habitats at different elevations (Feinsinger 1978; Stiles 1980; Karr et al. 1982; Levy 1988; Loiselle 1988, 1991; Stiles 1988; Loiselle and Blake 1990, 1991). In southern Central America long-term studies by Skutch (1954, 1967), Karr (1971), Feinsinger (1976), Stiles (1983, 1985a, 1988), Martin and Karr (1986), Levy (1988), and Loiselle and Blake (1990, 1991) have documented extensive local and elevational migration within avian communities. Movements have been broadly linked to temporal and spatial variation in food, especially of fruit and nectar (Stiles 1988; Levy and Stiles 1992; Loiselle and Blake 1991).

No work comparable to that in Costa Rica exists on seasonal movements of birds in the Andes, and the extent of movements within Andean avifaunas is largely unknown. Nevertheless, an understanding of migratory movements within the Andes is of interest because: (1) many migrant birds are nectarivores or frugivores that play important roles in pollination and seed dispersal; and (2) increasingly intensive land use in the Andes makes information on seasonal migratory movements important for conservation-planning and long-term preservation of biotic communities.

Several short-term surveys of avian diversity have been conducted in the Andes (i.e., Terborgh 1971; Ridgely and Gaulin 1980; Parker and Parker 1982; Remsen 1985; Davis 1986; Robbins and Ridgely 1990; Willis and Schuchmann 1993), but none has lasted longer than a few weeks, and these studies are of insufficient duration to detect seasonal movements or seasonal abundances of birds. The present study is the first documentation of seasonal bird distribution throughout a full annual cycle in an Andean avifauna. Because comparable work in montane avifaunas exists mainly for sites in the mountains of southern Central America, this study addresses the question: are there differences in seasonal occurrences of bird species at comparable elevations between western Colombia (3°N) and Costa Rica (10°N)? Relationships between food resources, habitats, and short-distance migratory movements in birds in western Colombia also are compared with studies in Costa Rica. The data were gathered as part of a larger study aimed at identifying seasonal patterns in foraging and breeding behavior of frugivores (Hilty 1977, 1980a, b).

STUDY SITE

Field work was carried out in the upper Anchicayá Valley, depto. de Valle, Colombia, from April 1972 through June 1973, at a locality known as Alto Yunda (3°32″N, 76°48″W). The study site, at 950–1050 m elevation was about 70–80 ha in extent on a long ridge that forms part of the watershed divide separating the eastern side of the upper Anchicayá Valley from the Río Mono Valley to the east.

The area also was visited from 9–18 June 1975. Alto Yunda was originally an overnight waystation for mule trains carrying supplies to the construction site of the Alto Anchicayá hydropower plant. Later, a road at lower elevation alleviated the need for mule trains, and the building was occupied for a number of years by a forest guard employed by the Corporación Autónoma Regional del Cauca (C.V.C.) to prevent settlers from illegally entering and destroying forest in the watershed around the dam. Most trails are now overgrown and completely closed, or have been destroyed by landslides, so fewer guards are required today. The building at Alto Yunda no longer exists, and there is no ready access to the study site.

Alto Yunda lies within a belt of very heavy rainfall that characterizes most of the Pacific slope of the Western Andes southward to the Ecuador boundary. Rainfall at four sites in or near the Anchicayá Valley was slightly biseasonal and averaged 4,000–5,000 mm a year, but data for most years were incomplete. At Danubio (3°38″N, 76°56″W), which is nearby but lower in elevation, rainfall averaged about 5,600 mm (C.V.C., unpublished data; Fig. 1). Mean monthly temperate was 21.5 degrees C. Fog occurred almost daily at Alto Yunda, averaging more than 5 hrs. per day from September–November and in March. The least foggy months, January–February and July–August, average 2–3 hrs. per day.

The vegetation at Alto Yunda is Premontane Rain Forest (Holdridge 1967). In a popular sense it would be called cloud forest. The forest is evergreen, and trees in the understory and the canopy are covered to varying degrees with epiphytes. Large trees generally support heavy loads of herbaceous and woody epiphytes. Because the terrain is steep, numerous small streams, landslides, and treefalls create many light gaps, and the canopy is discontinuous and broken. A dense understory of herbaceous and woody plants makes movement away from established trails difficult.

By 1973, subsistence farmers and squatters to the east of Alto Yunda, in the valley of the Río Mono, had cleared or disturbed perhaps 35–40% of the forest at elevations below 850 m. At higher elevations there were few clearings. The entire upper Anchicayá Valley was undisturbed rainforest except for a small area around the hydroelectric plant, a construction site, housing site, and one road (all located at about 500 m). The Anchicayá Valley remains protected today, and access to the valley is strictly controlled. The construction site and most of the housing area is now closed, and today this wilderness is less disturbed than 20 years ago. At Alto Yunda one large pasture surrounded the house and extended eastward down the valley a distance of approximately 350 m. Except for the rather wide entrance trail (10 m wide in places) from La Cascada to Alto Yunda and continuing on to the dam site, there were no other man-made disturbances to the natural forest vegetation. Use of the trail from Alto Yunda to the dam was discontinued about two years prior to my arrival, and this portion of the trail was already so overgrown and closed by successional plants that it was impassable without use of a machete.

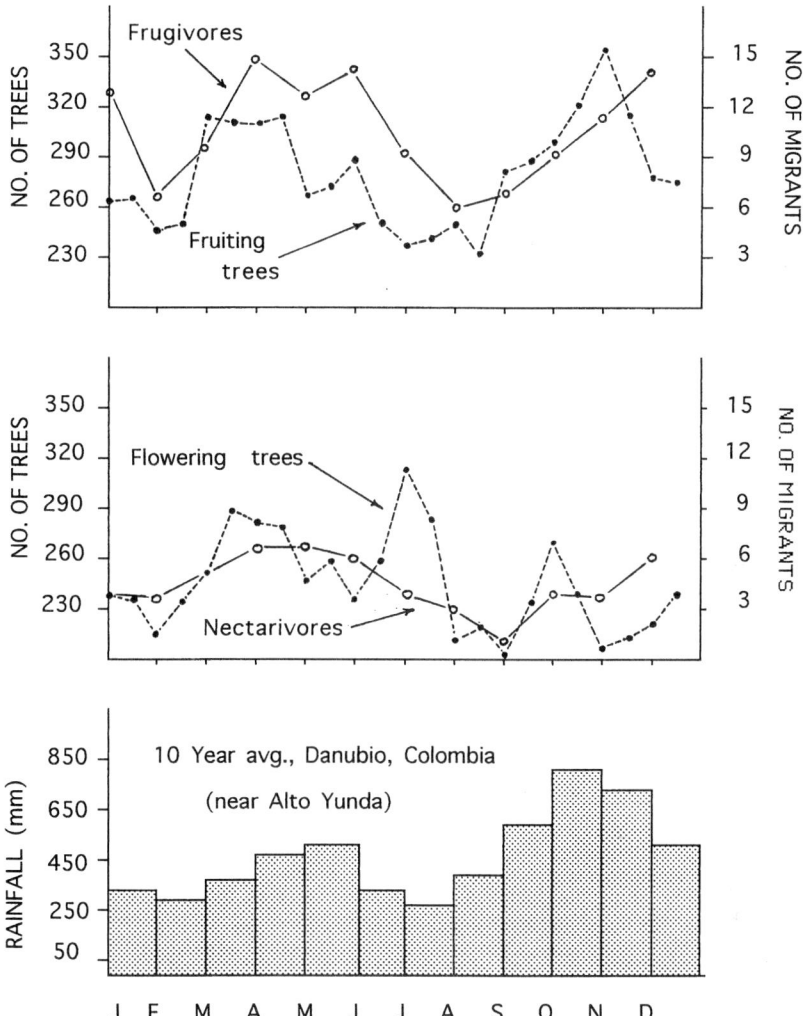

FIG. 1. Number of individual fruiting trees and number of species of short-distance migrant frugivores (top); number of individual flowering trees and number of species of short-distance migrant nectarivores (middle); and 10 year average annual rainfall at Danubio (near Alto Yunda), Colombia, 1963–73.

Because of generally pristine conditions in the Anchicayá Valley, mammals and birds were present in natural numbers, although illegal hunting occurred occasionally. One cracid, the Sickle-winged Guan (*Chamaepetes goudotii*), was common at Alto Yunda, but no others occurred there. The Baudo Guan *(Penelope ortoni)*, although collected at lower elevation in the Anchicayá Valley (Meyer de Schauensee, 1948–1952), was never seen at Alto Yunda. Forest guards and residents of the Río Mono Valley never indicated others occurred here, although they mentioned the presence of a "pajuí" (*Crax rubra*) at lower elevations.

METHODS

Plants.—A total of 621 trees (164 species) greater than 3 m height were marked along a 1 km trail where birds were captured in nets. All tagged trees were examined twice monthly for evidence of flowering and fruiting (see details in Hilty 1980a). Variations in the number of species and individuals flowering and fruiting were taken as broad indications of seasonal change in nectar and fruit resources.

Birds.—An average of 12 mist nets (approx. 3 × 10 m) were operated for 2–4 mornings each month (except September and January) from May 1972 through June 1973, for a total of 1,784 net hours and 1,394 net captures. From this sample, 436 birds, mostly frugivores, and nectarivores, were marked with unique combinations of color bands for later recognition. Net captures, field observations of marked birds, censusing, and other field techniques (below) provided a fairly complete picture of the diversity and seasonal movements of birds at Alto Yunda. Five mornings each month (for a total of 30 hrs each month throughout the study) were spent recording all birds and their activities at a group of 19 *Cecropia reticulata* trees growing in an old regenerating landslide scar inside the forest. Several other species of fruiting trees (*Miconia* sp., *Cecropia burriada*, *Coussapoa oligoneura*) were kept under observation for shorter periods of time during the study. Seven to 10 mornings each month were spent walking trails to record foraging behavior of frugivorous birds. I kept records of the variety of fruit eaten by all frugivores, and I kept detailed foraging observations for many common tanagers including all the *Tangara* tanagers. The locations of all sightings and recaptures of color-marked tanagers were recorded and compared to original capture sites. These combined techniques provide a relatively complete inventory of the avifauna and one that is comparable from month to month.

Species were assigned to diet, movement (migratory, etc.), and habitat categories based on field observations and occasionally also by information from published sources (i.e., Stiles and Skutch 1989; Remsen et al. 1993). The seasonal movements, diets, and habitats of each species are not as rigid and simplistic as categorized here. Diets vary seasonally (Leck 1972; Morton 1980; Loiselle and Blake 1990), habitat preferences vary geographically (Hilty, pers. obs.), and seasonal movements are complex and poorly understood (Levy 1988). Despite these limitations, the categories permit broad-based comparisons within this avifauna, and with studies elsewhere.

Diets.—Bird species were assigned to one of the following diet categories: (1) carrion and scavenging, (2) raptorial species that eat mainly vertebrates, (3) arthropods, (4) fruit, (5) fruit and arthropods, (6) fruit and seeds (for parrots and parakeets) and, (7) nectar. Species were assigned to the category deemed a "best fit" despite the fact that most species consume foods in more than one of these categories. In species in some genera (i.e., *Attila*, *Pachyramphus*, *Cacicus*, *Atlapetes*, and *Arremon*) in which the diet consisted of almost equal proportions of fruit and insects, species were split and assigned to each of the categories. For the analysis some ecologically similar categories (e.g., all categories in which fruit was a major component of the diet) were merged.

Habitats.—Species were assigned to one of three habitats: (1) forest interior, which includes species found primarily inside shady understory of forest, (2) forest canopy and edge, which includes species found primarily in the upper levels of old growth forest, older regrowth forest, or along forest borders, and (3) early successional vegetation and semi-open, which includes birds typically found in pastures or bushy overgrown areas. A single aquatic species was recorded flying over the area, although no aquatic habitats existed at Alto Yundo.

Seasonal status.—All species at Alto Yundo are classified into one or more of five categories. These categories are artifically rigid, and many species may eventually be shown to fit into combinations of these categories. (1) Residents include all species that show no evidence of seasonal movement beyond normal post-breeding dispersal of juveniles. (2) Elevational migrants include species that show some seasonal patterns of abundance, e.g., between higher and lower elevations. Movements are mainly short-distances and may involve all, or only a portion of the species' population. Some elevational migrants may breed at Alto Yunda and then move to higher or lower elevation, although breeding data are insufficient for analysis. (3) Local migrants move seasonally, but mostly within the same elevational zone and region. Some species classified as local migrants may prove only to be post-breeding dispersers, wandering immatures, and unmated individuals of resident species that are searching for appropriate habitat, mates, etc. (4) Long-distance migrants (Nearctic/austral breeding migrants) include those that return, for part of each year, to tropical latitudes. (5) A final category, *vagrants*, comprise a few rare or accidental species observed on only one or a few days (in most cases only a single individual was observed or captured in nets during the 15-month study) and whose movement patterns, if any, are unclear. Additional work may show that a few species breeding in or near Alto Yunda undertake longer-distance intratropical migrations that take them out of Colombia. *Elanoides forficatus* and *Ictinia plumbea* are examples.

To estimate the proportion of migrants at Alto Yunda from highland and lowland areas, I examined the overall geographic distribution of each species (using Hilty and Brown 1986). If a species was found primarily higher than 1,000 m (the elevation of Alto Yunda), then it was considered to belong to a highland community; if it was primarily lower, then it was considered to belong to the lowland community. A local species was one whose center of distribution was close to the elevation of Alto Yunda. For example, *Pharomachrus antisianus* occurs primarily in cloud forest at elevations above 1,400 m but was recorded at Alto Yunda during the months of June through August, suggesting a downslope movement during this time of the year. The reverse was true for lowland and foothill species like *Amazona farinosa* and *Columba goodsoni*, which were common in the Pacific lowland plain but also were present for several months (both at broadly similar times of the year) at Alto Yunda. The general direction of movement of short-distance migrants, therefore, was based on an examination of their broader elevational distributions in Colombia.

Guilds.—The guild categories are similar to those used by other researchers in South America. Assignments are based largely on personal data.

RESULTS

Community composition.—I recorded 271 species of birds at Alto Yunda. This is one of the largest lists reported for a single lower montane wet forest locality in the Neotropics. The number of "core" species (Remsen 1994) at Alto Yunda is higher than ones reported at similar elevations elsewhere in the Andes (e.g., Miller 1963; Remsen 1985; Davis 1986; Parker and Parker 1982). Subtracting vagrants, the core Alto Yunda avifauna is 256; removing vagrants, and all species of the early successional vegetation in the pasture, reduces the core total to 217 species, still substantially above the 152 (which includes some early succession species) reported by Davis (1986). Davis' (1986) lists are some of the highest reported totals in Andean wet forest localities at elevations similar to Alto Yunda.

Several additional species occurred at the same elevation in other parts of the Anchicayá Valley or in adjacent river systems. For example, *Merganetta armata, Sayornis nigricans, Serpophaga cinerea* and *Cinclus leucocephalus* were not recorded because Alto Yunda lacks a permanent stream. *Rupicola peruviana* was virtually absent at Alto Yunda (two vagrant immatures seen in 15 months) because there were no cliffs suitable for nest sites. *Semnornis ramphastinus* and *Tangara parzudakii* occurred on a ridge only 100 m higher and a little over 1 km away, but were never found at Alto Yunda, and *Pipreola jucunda*, common on this same ridge, was recorded only once in 15 months at Alto Yunda. A curious example of a patchy distribution is illustrated by *Boissonneaua jardini*. Fairly common at 500–600 m in the Anchicayá Valley, and as high as 1,200–1,600 m further south in the Colombian Andes of Nariño, it was never recorded at Alto Yunda.

Residents.—About 64% (174 of 271) of the Alto Yunda avifauna is resident. This 271 species total includes vagrants and species of early successional habitat (15 and 39 species respectively), which are not part of the core forest avifauna and may show more seasonal movement than the core avifauna. When these species are removed, the percentage of residents in the core avifauna increases to 71% (154 residents out of 217 core species). When species of the forest interior are considered, the change is even more dramatic; 81 of 94 forest-interior species (86%) are sedentary. The 13 nonsedentary species of the forest interior include four vagrants, a long-distance migrant (*Wilsonia canadensis*) and eight short-distance migrants. The taxonomic composition of these eight short-distance forest migrant species spans six families. They are united, however, by diet—all eight are either nectarivores or frugivores—and include a quail-dove, oilbird, two hummingbirds (which also feed at forest edges), a fruiteater, two thrushes, and a tanager.

Resident species were predominantly insectivores, but also include most raptors and some nectar- and fruit-feeding species. Resident frugivores of the forest interior were mostly small-bodied species, i.e., manakins and tanagers, that ate small fruits and berries, or a mixed diet of fruit and insects. Two species, *Tinamus major* and *Chamaepetes goudotii*, were exceptions, in that they were large-bodied resident frugivores. Other large-bodied forest frugivores—among them a quetzal, fruitcrow, cock-of-the-rock, and umbrellabird—were elevational migrants.

Most resident species probably bred in or near Alto Yunda. For resident species in which part of the population is migratory, e.g., *Elanoides forficatus, Ictinia plumbea* and *Chlorophonia flavirostris*, local breeding was suspected but not proved.

Migrants: community patterns.—In a broad context, migrants should not be viewed as distinct entities but as a continuum of species whose behaviors range from virtually sedentary to intercontinental, and from occasional to regular in occurrence (Lack 1944; Keast and Morton 1980; Levey and Stiles 1992). Species at Alto Yunda displayed this entire spectrum of migratory behavior, including species that (1) transited an area but did not forage, (2) migrated in or out of the region one or more times during the year (ranging from local to intercontinental in distance), and (3) resident species whose populations were augmented or depleted during part of the year by migratory populations.

TABLE 1
NUMBER OF SHORT-DISTANCE MIGRANT (ELEVATIONAL AND LOCAL) BIRD SPECIES BY MONTH AT ALTO YUNDA, ANCHICAYÁ VALLEY, COLOMBIA, APRIL 1972 TO JUNE 1973

Category[1]	J	F	M	A	M	J	J	A	S	O	N	D
Highlands												
Frugivores[2]	5	1	4	4	7	5	3	1	2	3	5	5
Nectarivores	3	1	3	5	4	3	2	1	1	1	3	6
Lowlands												
Frugivores	7	5	5	10	5	8	4	4	4	4	4	8
Nectarivores	1	3	2	2	3	3	2	2	0	3	1	0
Mid-El. (local)												
Frugivores	1	1	1	1	1	1	2	1	1	2	2	1
Summary												
All frugivores	13	7	10	15	13	14	9	6	7	9	11	14
All nectarivores	4	4	5	7	7	6	4	3	1	4	4	6
Total	17	11	15	22	20	20	13	9	8	13	15	20

[1] Highlands = distribution of species mostly above 1,000 m; lowlands = range mostly below 1,000 m; mid-elevation = range centered at about 1,000 m. Includes all species for which fruit is an important component of the diet. Most also consume insects and/or arthropods.
[2] Includes all psittacines (both frugivores and seed predators/granivores).

Two psittacines, *Aratinga wagleri* and *Bolborhynchus lineola,* were seen flying over the area on several occasions but were not observed foraging. In other areas these species are known to move seasonally or erratically as fruit crops vary, but the extent and timing of these migrations are not documented. Sightings of lone individuals of *Pandion haliaetus* in November and December could represent long-distance migrants or individuals wandering from one river system to another.

About 36 percent (97 of 271) of all Alto Yunda species showed some type of seasonal or irregular movement other than normal post-breeding dispersal of young and unmated adults. When vagrants are removed the figure drops to 30%; when both vagrants (6%) and long-distance migrants (8%) are removed the figure drops to 22%. These short-distance migrants (22% of total avifauna) were mostly elevational migrants and were recorded in all months (Table 1). Nearly twice as many were present during March-June and November-January as during February and August-September.

Elevational migrants at Alto Yunda are more likely to depend wholly or partly on fruit or nectar than on any other food resource (Table 2). As a group, nectarivores were the most highly migratory. Fifty-eight percent of all nectarivores were migratory, compared with 44% of the frugivores. Frugivores, however, were numerically the most important group of migrants (58% of all short-distance migrant species were frugivores). Most insectivorous species, on the other hand, were residents (72%) or long-distance migrants (13%). Short-distance migrants (elevational and local) accounted for a mere 7% of all insectivore species (Table 2).

Nectarivores.—Nectar-feeders showed three main patterns. First, the number of elevational migrants reaching Alto Yunda from the highlands was twice that of the lowlands (8 species versus 4). Secondly, the highest number of species of migrant nectarivores was recorded during two periods of the year—March-June and December (Table 1). The first peak generally overlaps an early-year increase in the number of woody plants (individuals and species) in flower (Fig. 1); a small late-year flowering spike in October was accompanied by only a small increase in migrant nectar-feeders. The number of migrant nectarivores was positively, but not significantly, correlated with the number of individual trees and shrubs in flower (Spearman's rank correlation, $z = 1.81$, $P < 0.10$). The lowest number of migrant nectarivores was recorded in August-September when only a single highland hummingbird, *Aglaiocercus coelestis*, was present. The fewest number of individual plants flowered in September, and this also was at or near the time when the fewest species were flowering (Hilty 1980a).

Thirdly, well over half the nectar-feeding community was migratory, and most of these migrants (12 of 15) were hummingbirds. The dominance of the migrant nectar-feeding category by hummingbirds, however, was strongly correlated with the habitat in which these birds spent most of their time. Ten of the 12 migrant hummingbirds inhabited canopy, forest edge, and early successional vegetation; only two were associated with the forest interior (Table 3 and Appendix).

TABLE 2
Association Between Movement Status and Diet of Birds at Alto Yunda, Colombia, April 1972 to June 1973

Diet class	Frugivore[1]	Nectarivore	Insectivore	Granivore	Raptor[2]
Status					
Short-distance migrants					
Elevational[3]	32	15	9	1	
Local	2				
Vagrant	4		9		2
Long-distance migrants[4]	5		16		2
Resident	55	11	87	7	14
Total[5]	98	26	121	8	18

[1] Includes species for which fruit is an important, but not exclusive, component of the diet, i.e., tinamous, parakeets, parrots, trogons, thrushes, dacnises, tanagers, emberizine finches. *Attila, Pachyramphus, Cacicus, Atlapetes,* and *Arremon* are split between frugivore and insectivore categories.
[2] Includes only those whose prey is principally small vertebrates or carrion.
[3] *Cotinga nattererii, Pipreola jacunda,* and *Cephalopterus penduliger,* considered as elevational migrant frugivores, may be only vagrants to Alto Yunda.
[4] All but one species are from north temperate latitudes.
[5] *Elanoides forficatus, Ictinia plumbea, Chlorophonia flavirostris,* and *Heterospingus xanthopygius* were each split into both resident and migrant categories. *Pionopsitta pulchra* was counted only as an elevational migrant, and *Diglossa indigota* as a resident and migrant.

By contrast six of 10 resident hummingbirds fed primarily inside mature forest. Only one resident, *Amazilia tzacatl,* occurred largely in early successional vegetation.

Frugivores.—More short-distance migrant frugivores at Alto Yunda were from the highlands than the lowlands. A total of 19 frugivores was classed as migrants from higher elevations to Alto Yunda, and 13 from lower elevations (Table 4). Migrant frugivores from the lowlands remained at Alto Yunda longer, respectively, than highland species (mean of 4.6 months vs. 2.6 months).

The greatest number of migrant frugivores were present March-June and November-January, periods of time broadly overlapping the migrant nectar-feeding community. These two peaks of frugivore numbers broadly track a twice annual increase in fruit abundance (Hilty 1980a), although they appear to lag it slightly (Fig. 1). Correlation between the number of individual trees in fruit (fruit abundance), and the number of elevational migrants present was positive but not significant (Spearman's Rank Correlation, $z = 1.360$, $P < 0.10$).

Some short-distance movements may overlap resident populations. Populations of *Elanoides forficatus* and *Chlorophonia flavirostris* included both resident and migrant individuals. Individuals of *Elanoides forficatus* were recorded almost daily during every month of the year at Alto Yunda. However, large flocks moving along a forested ridge in August and September (maximum of 31 individuals on 8 September 1972), greatly augmented the small numbers of birds usually present. *Chlorophonia flavirostris* was present all year in small numbers, but during January and February flocks of 18–20 fed in *Miconia* sp. trees at Alto Yunda, and in June, at elevations of 100–300 m in the lower Anchicayá Valley, I encountered flocks of over 80 birds. Groups of up to a dozen also have been seen at 1,800 m, and the origins and seasonal movement of this species remain unclear.

About 23 percent (22 of 97) of the migrant species at Alto Yunda breed in north-temperate latitudes (Table 2). Only one (*Myiopagis caniceps*) may be an austral migrant. Six species were recorded only on one or a few days of the year, i.e., *Buteo platypterus, Chaetura pelagica*

TABLE 3
Hummingbird Communities at Alto Yunda, Colombia and La Montura, Costa Rica

Category	Alto Yunda	La Montura
Resident	10	10
Regular seasonal migrants[1]	7	4
Rare/accidental migrants	5	8

[1] Regular elevational migrants at Alto Yunda include *Doryfera ludoviciae, Androdon aequatorialis, Threnetes ruckeri, Colibri delphinae, Florisuga mellivora, Aglaiocercus coelestis,* and *Philodice mitchellii.*

TABLE 4
NUMBER OF SHORT DISTANCE MIGRANTS REACHING ALTO YUNDA FROM HIGHER, LOWER, AND LOCAL (OR UNKNOWN) SOURCES

	Number of species	Mean number months present
Elevational migrants		
Highland species	19	2.6
Lowland species	13	4.6[1]
Local migrants[2]	2	?

[1] Mean based on 11 species; *Pionopsitta pulchra* and *Heterospingus xanthopygius* not included in calculations because suspected resident and migratory population could not be distinguished.
[2] Origin (highlands or lowlands) unknown.

(identification not certain), *Myiarchus crinitus, Riparia riparia, Hirundo rustica,* and *Piranga olivacea,* whereas the remainder were resident for 5–9 months of the year. The highest number of individuals and species of north-temperate migrants was recorded in October, many arriving around the middle of the month (Hilty 1980b). Most remained only a few days, but one *Dendroica fusca* that I color-marked was with a mixed-species flock until its northward migration the following spring.

Short-distance migrants at Alto Yunda were primarily frugivores and nectarivores (Table 2). Insectivorous elevational migrants were mainly aerial-feeders—four swifts, two swallows and two insectivorous raptors. The remainder consisted of an icterid and a flycatcher. The two raptors, *Elanoides forficatus* and *Ictinia plumbea,* capture large, flying insects (among other things) by snatching them from canopy foliage or grabbing them in the air. *Elanoides forficatus* was present year-round, but its numbers greatly increased in August and early September, suggesting a migratory movement. *Ictinia plumbea* was not present all year but several large migrating groups were noted in August and September. Both species are migratory in parts (all?) of their range (Stiles and Skutch 1989), and southbound flocks have been noted in July and August in Costa Rica (Stiles, pers. comm.).

Four species of swifts, all aerial insectivores, displayed a seasonal distribution that suggests they were elevational migrants. During 1972 and 1973, three species, *Streptoprocne rutila, Chaetura cinereiventris* and *C. spinicauda,* were present mainly during February–May and October–December. *Chaetura cinereiventris* was also recorded during June 1975. *Streptoprocne rutila* was probably present continually from November through May (Appendix) despite the fact that I failed to record it in December or April. This highland species was seen nearby at elevations even lower than Alto Yunda during December. *Panyptila cayennensis* was recorded in August and November–December, thus overlapping in part the pattern shown by the others. Except for *S. rutila,* all of these swifts are believed to be lowland and foothill breeding species, and their occurrence at elevations as high as Alto Yunda during most of the rainiest months of the year may be related to wet season increases in small aerial insects at this elevation or, alternatively, to depressed aerial insect populations at lower elevations. This is the first report of presumed seasonal elevational migration in Neotropical swifts. Marin (1993) summarized patterns of elevational distribution of swifts in the Andes of Ecuador, but did not document or discuss the possibility of seasonal movement.

The late-year, down-slope movement of highland species included several, i.e., *Entomodestes coracinus, Platycichla leucops, Pipraeidea melanonota,* and *Bangsia edwardsi,* that had not previously been in the area, and stayed only a few weeks or a month or two. Lowland species stayed longer—some were present more than half the year—i.e., *Capito maculicoronatus, Pteroglossus sanguineus,* and *Tangara lavinia*—and could have bred there (no breeding evidence was noted). *Columba goodsoni* (from lowlands) and *Ampelioides tschudii* (from highlands) were noted singing for several weeks upon arrival. Except for *Ampelioides tschudii,* whose loud whistles were heard for several months early in the year, it is unlikely that any highland species bred there. The short residency, absence of song, and unpaired behavior of most elevational migrants, such as *Geotrygon frenata* and *Entomodestes coracinus,* suggest that they did not breed at Alto Yunda.

On the other hand *Amazona farinosa* was recorded on the study site in eight different months, but was still not known to breed there (it was known to breed at sites about 500 m lower in elevation). Late in the year it was present in large flocks for several months, but in other months

TABLE 5

Kendall's Rank Correlation of Number of Species of Fruit Taken Each Month by Eight Species of *Tangara* with the Number of Tree Species in Fruit Each Month and Kendall's Rank Correlation of Number of Species of Fruit Taken Each Month by 26 Species of Tanagers with the Number of Tree Species in Fruit Each Month, at Alto Yunda, Colombia

	Month											
	J	F	M	A	M	J	J	A	S	O	N	D
Number of fruiting species	47	45	55	60	61	63	52	53	53	58	66	51
Tangara[1]	22	18	25	29	30	29	25	29	22	34	33	21
26 Tanagers[2]	27	23	33	36	41	36	33	40	33	46	44	30

[1] Kendall's coefficient of rank correlation with ties, $r_s = 0.76$, $P < 0.05$.
[2] Kendall's coefficient of rank correlation with ties, $r = 0.71$, $P < 0.05$.

it was noted only sporadically and in scattered pairs or small groups that flew over the study area but rarely stopped. Records for some presumed elevational migrants, e.g., *Dacnis cayana* and *Carduelis xanthogastra*, were so scattered or spotty that elevational movements or breeding status were unknown.

Vagrants.—I list 15 species as vagrants (Table 2). A little over one-half (nine of 15) were insect-eating species, and the majority (12 of 15) were species of forest canopy or early successional growth. Two raptorial species, *Milvago chimachima* and *Falco sparverius,* are typical of open or partially deforested regions and probably represent wandering individuals from deforested pastureland at lower elevation. Partial clearing and opening of forest habitats at lower elevation probably also opened an avenue of upward expansion for *Synallaxis brachyura* and *Tolmomyias sulphurescens*. Another recent invader from open habitats, *Molothrus bonariensis*, was already established in the single clearing at Alto Yunda, and *Zonotrichia capensis*, although not recorded every month, was classed as a resident.

A pair of antbirds, *Cecromacra tyrannina*, were considered vagrants, although they were heard and seen frequently during a three week period from late November to mid-December. They were never seen thereafter. This species was not found near Alto Yunda—the nearest resident pair known to me was several kilometers away and at least 250 m lower in elevation—and this pair's temporary presence probably represented wandering behavior in search of a suitable territory.

Some insectivores were recorded only irregularly during the study, e.g., several woodpeckers, foliage-gleaners, a motmot, a puffbird, and two flycatchers (*Rhynchocyclus brevirostris* and *Myiobius barbatus*). A combination of foraging areas that lay largely outside the study area, and in some cases low population densities, may account for the inconsistent records of these species.

Without records from many years it is impossible to distinguish some short-distance migrants responding to environmental perturbations from vagrants or normal non-breeding dispersal of immatures and unmated or mated adults. *Pipra pipra* provides an example. It foraged mostly for small berries and fruit in the lower story of the forest and was easily captured in mist nets but infrequently observed. I captured it during six widely scattered months of the year. All were adult females or female-plumaged immatures. The broad scatter of records throughout the year could point to post-breeding dispersal or to diet-related seasonal movement. The nearest lek known to me, apparently occupied more or less year-round, was nearly 1 km away, at 660 m elevation.

Seasonality of foraging and diet.—The number of kinds of fruit eaten each month by eight species of *Tangara* tanagers, and by the entire tanager community (26 species), was positively and significantly correlated with the number of species of trees bearing fruit (Kendall's Rank Correlation, $r_s = 0.76$, and $r_s = 0.71$ respectively, $P < 0.05$, Table 5). Thus, the greatest range of fruits was taken during months when the greatest diversity was available. The amount of fruit in their diets also reached an annual high of 86.3% ± 1.3 S.E. in October and dropped to an annual low of 66% ± 6.4 S.E. in August ($n = 3,849$ foraging records). These months are at or close to the annual highs and lows respectively of general fruit abundance in the community (Hilty 1980a).

None of the eight common *Tangara* was a local or elevational migrant (all were present in approximately the same numbers in all months) but, as a group, they showed significant foraging

TABLE 6

Ranking of Fruit Abundance (Canopy Surface Area) with Mean Monthly Foraging Distance (N = 1,340) of 11 Species of Tanagers[1]

Parameters[2]	J	F	M	A	M	J	J	A	S	O	N	D
Forag. dist. (m)	121	136	146	143	125	112	110	113	127	129	123	119
Fruit abund. (%)	40.4	40.8	32.3	34.2	35.8	49.5	48.0	41.2	37.3	45.9	49.5	38.3

[1] *Tangara cyanicollis, T. arthus, T. nigroviridis, T. palmeri, T. rufigula, T. icterocephala, T. gyrola, T. florida, Euphonia xanthogaster, Chlorospingus flavigularis, Chlorothraupis stolzmanni.*
[2] Kendall's coefficient of rank correlation with ties, $r_s = -0.61$, $P < 0.05$.

responses to seasonally varying fruit resources. The average foraging distance (distance of recapture or sighting away from original capture location) varied from a maximum of 146 and 136 m in July and August to 110 and 112 m in November and December, respectively (Table 6). The greater foraging distances in July and August, compared to later in the year, coincided with a general decrease in the amount of fruit at this time of year (Hilty 1980a); conversely, foraging distances were smallest when fruit spiked to one of its highest levels of the year in November. The foraging distance of eight *Tangara*, and three additional species with sufficient data, are inversely and significantly correlated with fruit abundance (Kendall's Rank Correlation with ties, $r_s = -0.61$, $P < 0.05$). Foraging distances did not contract as markedly during the early year fruiting peak of March–April, a period when many species were breeding (Hilty 1977).

DISCUSSION

Community composition and comparisons.—The Alto Yunda avifauna represents one of the largest single-site, montane avifaunas reported. Aside from possible historical factors, the avifauna at Alto Yunda is not believed to be inherently richer than other Andean montane sites. Rather, it represents the addition of some vagrants, rare or uncommon species, and especially, many short-distance migrants that were recorded over a full annual cycle. Many of these species would not be recorded during the relatively brief visits that have characterized work at other Andean sites.

Few comparable studies exist of seasonal distribution patterns in Andean regions. Miller's (1963) year long survey at San Antonio, Colombia (2,100 m), less than 40 km away, provides a comparison, although his conclusions differ from those of the present study. San Antonio, a region of mixed "cloud forest," disturbed areas, and pasture land, had fewer species (167) than Alto Yunda. Moreover, Miller reported that apart from 15 species that migrate to North America to breed, the avifauna is permanently resident, with only one species (*Elaenia obscura*) [sic = *frantzii*] showing regular seasonal movement.

Despite Miller's (1963) results, dynamic avifaunas similar to Alto Yunda's may characterize much of the Andes, perhaps even more so than the avifauna of my study area. Alto Yunda is an unusually wet mountain region close to the equator, where rainfall is not strongly seasonal. Thus, at Alto Yunda, seasonal variation is minimized.

The large proportion of tropical breeding species that showed evidence of seasonal or irregular movement at Alto Yunda (22%) is comparable to studies by Stiles (1983, 1985b, 1988), Levy (1988), and Loiselle and Blake (1991) in Costa Rica. These studies have demonstrated that both local and migratory movements are characteristic of montane avifaunas in southern Central America. Stiles (1985b) reported that up to half of the avifauna in Costa Rica shows evidence of migratory movement, and that 26% of the species on Costa Rica's wet Caribbean slope show seasonal changes in elevation of at least 500 m. More than 75 species of birds in Parque Nacional Braulio Carrillo, a montane park in Costa Rica, undergo seasonal elevational movements (Stiles 1985b); at Alto Yunda 57 species were classified as elevational migrants.

Most local and elevational migrants are frugivores (ca. 58%) or nectarivores (ca. 25%), proportions similar to those found in Costa Rica (Loiselle and Blake 1991) but higher than the representation of these groups in the avifauna. The high proportion of short-distant migrants that are frugivores and nectarivores in Costa Rica suggests that their movements are related to varying abundances of food (Loiselle and Blake 1991; Levy and Stiles 1992).

The two seasonal increases in numbers of migrant frugivores and nectarivores are of interest because species from both the highlands and lowlands were present. Contributions from species of both higher and lower elevation account for a doubling of the number of short-distance migrants during April–June and November–January (only April–June and December for nectarivores). These movements differ in timing from those reported in Costa Rica, and they also appear more complex. Stiles (1985b) reported that migration in both nectar-/and fruit-eating birds was correlated with flower and fruit production, but the respective elevational movements differed in timing—frugivores moved downslope during the latter part of the year, at about the same time that nectar-feeders moved upslope. The seasonal movement of elevational migrant nectarivores at Alto Yunda is similar to that in Costa Rica, in that more species apparently moved downslope during the first (drier) part of the year, and left (presumably moving back upslope) during the rainiest latter months of the year.

The pattern in fruit-eating birds suggests more complex movements than those reported in Costa Rica. In Costa Rica a major downslope movement of fruit-eating birds characterizes the wet season (last half of year), and a major upslope movement occurs in the dry season (early months of the year) (Stiles 1988; Loiselle and Blake 1991), whereas at Alto Yunda two peaks of elevational migrants were apparent, both approximately coincident with the highest levels of fruit in the environment. Fruit and flower production was not strongly seasonal at Alto Yunda (probably less so than in Costa Rica which lies seven degrees of latitude northward), and some measures of flower and fruit abundance at Alto Yunda showed no significant seasonality. Consequently, elevational migration at Alto Yunda may not be so rigorously constrained by climate and food supply as in Costa Rica, thus permitting a greater variety of seasonal movements to occur.

Another difference between Alto Yunda and the Costa Rican localities is the elevational origin of the species. At Alto Yunda the seasonal increases of elevational migrant frugivores and nectarivores contain species from both high and low elevations. However, at present it is impossible to know the true sources and origins of the species contributing to these seasonal patterns.

Nectarivores.—Proportionately more nectarivores are migrants than any other dietary group of birds at Alto Yunda. Nectar dependence, however, does not necessitate migratory behavior. Stiles (1988), Feinsinger and Colwell (1978) and Levy and Stiles (1992) found for hummingbirds in Costa Rica a positive correlation between seasonal migratory movements and use of canopy and edge habitats. The results of this study are similar to those at La Montura in Parque Nacional Braulio Carillo, Costa Rica (Stiles 1988). The sites are at similar elevations, have similar rainfall, and similar hummingbird communities, each with 10 residents in a community of 22 species (Table 3). However, Alto Yunda hosted more regular seasonal migrants, with fewer migrant species classified as rare or accidental, than La Montura. Most regular migrants at La Montura were immatures that moved downslope during the early- to mid-rainy season, with only one species, *Thalurania furcata*, moving upslope. Alto Yunda's regular migrant hummingbirds were drawn about equally from the highlands and lowlands, although almost all of the rare species (except an immature *Amazilia amabilis*) were from the highlands. Of seven regular migrant hummingbirds at Alto Yunda, five (*Doryfera ludoviciae*, *Androdon aequatorialis*, *Threnetes ruckeri*, *Florisuga mellivora* and *Aglaiocercus coelestis*) were present for 5–9 consecutive months and may have bred, but each was absent for at least two consecutive months. The two remaining species, *Colibri delphinae* and *Philodice mitchellii*, were present only for about three months during the year. The latter's two brief visits coincided with the flowering of several *Hamelia* sp. shrubs.

At La Montura, Stiles (1988) correlated the seasonal presence of several hummingbirds with definite elevational shifts of their populations. At Alto Yunda the seasonal visits of several species of hummingbirds also represent elevational shifts because their populations were derived from elevational zones primarily above or below Alto Yunda. *Threnetes ruckeri* and *Florisuga mellivora*, for example, breed mainly below 900 m, *Aglaiocercus coelestis* mostly from 900–2,000 m, and *Colibri delphinae*, *Haplophaedia aureliae*, and *Ocreatus underwoodi* even higher still. The movements of a few hummingbird species may represent primarily nonelevational shifts to habitats adjacent to, but outside of, the study area. Brief absences, for example, of *Phaethornis syrmatophorus*, *Amazilia tzacatl*, and *Urostice benjamini*, and fluctuating numbers of *Popelairia conversii*, may have been due to local changes in the abundance of food plants. Each of the four latter species was relatively numerous at Alto Yunda, and, although all breed at higher or lower elevations, their relatively brief absences suggest local, rather than longer, elevational movements.

Seasonal foraging behavior and resource abundance.—One response to adverse changes in food resources is to alter the diet and, consequently, foraging behavior. Another response is a local, elevational, or long-distance migration. This study lacks direct evidence linking food supply to migratory movements, but it demonstrates that small tanagers track resources and respond by adjustments in diet and expansion and contraction of foraging range. It suggests that food resources may be limited at some times for some species.

All avifaunas are composed of a variety of species with different life-history strategies, habitat requirements, foods, migratory tendencies, and foraging techniques, but seasonal limitations of food are an important factor in migratory movements (Stiles 1985b, 1988; Loiselle and Blake 1991; Levy and Stiles 1992). The eight species of *Tangara* represent corroborating evidence relating resource abundance to seasonal movements of the Alto Yunda avifauna. Most *Tangara* are not territorial in the strict sense of defending exclusive, rigid territories (pers. obs.). The size of their foraging areas can easily expand or contract as resource abundance varies. Such a response to variation in diversity and abundance of food resources could be a precursor to the evolution of local or elevational migration.

Conservation.—The extent of short-distance migration reported here has important implications for conservation planning. About a third of the avifauna at Alto Yunda showed evidence of seasonal movement. To protect these species, and migratory species like them elsewhere, future research and land planning must focus not only on communities at specific sites, but also on maintaining corridors of habitat to accommodate species' movements within and between communities.

ACKNOWLEDGMENTS

Stephen M. Russell aided this study in many ways, including pointing the way to Colombia. A joint research program with the University of Arizona Department of Ecology and Evolutionary Biology and Peace Corps provided support. Many Colombians, including especially Eliécer Solarte, Hectar Perdomo, and Ernesto Schrimpff, helped with logistical support in the field. I especially thank F. Gary Stiles, J. V. Remsen, John Hagan, and Niels Krabbe for helpful comments on the manuscript, and my wife, Beverly J. Hilty, for help with field work, and for companionship and support throughout.

LITERATURE CITED

DAVIS, T. J. 1986. Distribution and natural history of some birds from the departments of San Martín and Amazonas, northern Peru. Condor 88:50–56.

FEINSINGER, P. 1976. Organization of a tropical guild of nectarivorous birds. Ecol. Monogr. 46:257–291.

FEINSINGER, P. 1978. Ecological interactions between plants and hummingbirds in a successional tropical community. Ecol. Monogr. 48:269–287.

FEINSINGER, P., AND R. K. COLWELL. 1978. Community organization among Neotropical nectar-feeding birds. Amer. Zool. 18:779–795.

HILTY, S. L. 1977. Food supply in a tropical frugivorous bird community. Ph.D. dissertation, Univ. of Arizona, Tucson, Arizona.

HILTY, S. L. 1980a. Flowering and fruiting periodicity in a premontane rain forest in Pacific Colombia. Biotropica 12:292–306.

HILTY, S. L. 1980b. Relative abundance of north temperate zone breeding migrants in western Colombia and their impact at fruiting trees. Pp. 265–272 *in* Migrants Birds in the Neotropics: Ecology, Behavior, Distribution, and Conservation (A. Keast and E. S. Morton, Eds.). Smithsonian Institution Press, Washington, D.C.

HILTY, S. L., AND W. L. BROWN. 1986. A Guide to the Birds of Colombia. Princeton Univ. Press, Princeton, New Jersey.

HOLDRIDGE, L. R. 1967. Life Zone Ecology. Rev. ed. Tropical Science Center, San José, Costa Rica.

KARR, J. R. 1971. Structure of avian communities in selected Panama and Illinois habitats. Ecol. Monogr. 41:207–233.

KARR, J. R. 1977. Ecological correlates of rarity in a tropical forest bird community. Auk 94:240–247.

KARR, J. R., AND K. E. FREEMARK. 1983. Habitat selection and environmental gradients: dynamics in the "stable" tropics. Ecology 64:1481–1494.

KARR, J. R., R. W. SCHEMSKE, AND N. BROKAW. 1982. Temporal variation in the undergrowth bird community of a tropical forest. Pp. 441–453 *in* The Ecology of a Tropical Forest: Seasonal Rhythms and Long-Term Changes (E. G. Leigh, Jr., A. S. Rand, and D. M. Windsor, Eds.). Smithsonian Institution Press, Washington, D.C.

KEAST, A., AND E. S. MORTON. (EDS.). 1980. Migrant Birds in the Neotropics: Ecology, Behavior, Distribution, and Conservation. Smithsonian Institution Press, Washington, D.C.

KRABBE, N., AND T. S. SCHULENBERG. 1997. Species limits and natural history of *Scytalopus* tapaculos (Rhinocryptidae), with descriptions of the Ecuadorean taxa, including three new species. Pp. 47–88 *in* Studies in Neotropical Ornithology Honoring Ted Parker (J. V. Remsen, Jr., Ed.), Ornithol. Monogr. No. 48.

LACK, D. 1944. The problem of partial migration. Brit. Birds 37:122–130.

LECK, C. F. 1972. Seasonal changes in feeding pressures of fruit- and nectar-eating birds in Panama. Condor 74:54–60.
LEVEY, D. J. 1988. Spatial and temporal variation in Costa Rican fruit and fruit-eating bird abundance. Ecol. Monogr. 58:251–269.
LEVEY, D. J., AND F. G. STILES. 1992. Evolutionary precursors of long-distance migration: resource availability and movement patterns in Neotropical landbirds. Amer. Nat. 140:447–476.
LOISELLE, B. A. 1988. Bird abundance and seasonality in a Costa Rican lowland forest canopy. Condor 90: 761–772.
LOISELLE, B. A. 1991. Temporal variation in birds and fruits along an elevational gradient in Costa Rica. Ecology 72:180–193.
LOISELLE, B. A., AND J. G. BLAKE. 1990. Diets of understory fruit-eating birds in Costa Rica. Studies in Avian Biology 13:91–103.
LOISELLE, B. A., AND J. G. BLAKE. 1991. Temporal variation in birds and fruits along an elevational gradient in Costa Rica. Ecology 72:180–193.
MARIN, M. 1993. Patterns of distribution of swifts in the Andes of Ecuador. Pp. 117–123 in Recent Advances in the Study of the Apodidae (G. Malacarne, C. T. Collins and M. Cucco, Eds.). Avocetta 17, Special issue.
MARTIN, T. E., AND J. R. KARR. 1986. Temporal dynamics of Neotropical birds with special reference to frugivores in second-growth woods. Wilson Bulletin 98:3–41.
MEYER DE SCHAUENSEE, R. 1948–52. The birds of the Republic of Colombia. Caldasia (pts. 1–5), nos. 22–26:251–1212.
MILLER, A. 1963. Seasonal activity and ecology of the avifauna of an American equatorial cloud forest. Univ. of Calif. Publ. Zool. 66:1–74.
MORTON, E. S. 1980. Adaptations to seasonal changes by migrant land birds in the Panama Canal Zone. Pp. 437–453 in Migrant Birds in the Neotropics: Ecology, Behavior, Distribution, and Conservation (A. Keast and E. S. Morton, Eds.). Smithsonian Institution Press, Washington, D.C.
PARKER, T. A., III, AND S. A. PARKER. 1982. Behavioural and distributional notes on some unusual birds of a lower montane cloud forest in Peru. Bull. Brit. Ornithol. Club 102:63–70.
REMSEN, J. V., JR. 1985. Community organization and ecology of birds of high elevation humid forest of the Bolivian Andes. Pp. 733–756 in Neotropical Ornithology (P. A. Buckley, M. S. Foster, E. S. Morton, R. S. Ridgely, and F. G. Buckley, Eds.). Ornithol. Monogr. No. 36.
REMSEN, J. V., JR. 1993. The diets of Neotropical trogons, motmots, barbets and toucans. Condor 95:178–192.
REMSEN, J. V., JR. 1994. Use and misuse of bird lists in community ecology and conservation. Auk 111: 225–227.
RIDGELY, R. S., AND S. J. C. GAULIN. 1980. The birds of Finca Merenberg, Huila Department, Colombia. Condor 82:379–391.
ROBBINS, M. B., AND R. S. RIDGELY. 1990. The avifauna of an upper tropical cloud forest in southwestern Ecuador. Proc. Acad. Nat. Sci. Phil. 142:59–71.
SKUTCH, A. F. 1954. Life histories of Central American birds. Pacific Coast Avifauna No. 31.
SKUTCH, A. F. 1967. Life histories of Central American highland birds. Publication of the Nuttall Ornithological Club No. 7.
SMYTHE, N. M. 1982. The seasonal abundance of night-flying insects in a Neotropical forest. Pp. 309–318 in The Ecology of a Tropical Forest. (E. Leigh, Jr., A. S. Rand, and D. Windsor. Eds.). Smithsonian Institution Press, Washington, D.C.
STILES, F. G. 1980. The annual cycle in a tropical wet forest hummingbird community. Ibis 122:322–343.
STILES, F. G. 1983. Birds. Pp. 502–530 in Costa Rican Natural History (D. H. Janzen, Ed.). Univ. of Chicago Press, Chicago, Illinois.
STILES, F. G. 1985a. Seasonal patterns and coevolution in the hummingbird-flower community of a Costa Rican subtropical forest. Pp. 757–787 in Neotropical Ornithology (P. A. Buckley, M. S. Foster, E. S. Morton, R. S. Ridgely, and F. G. Buckley, Eds.). Ornithol. Monogr. 36.
STILES, F. G. 1985b. On the role of birds in the dynamics of Neotropical forests. Pp. 49–59 in Conservation of Tropical Forest Birds (A. W. Diamond and T. E. Lovejoy, Eds.). International Council for Bird Preservation Technical Publ. No. 4.
STILES, F. G. 1988. Altitudinal movements of birds on the Caribbean slope of Costa Rica: implications for conservation. Pp. 243–258 in Tropical Rainforests: Diversity and Conservation (F. Almeda and C. M. Pringle Eds.). California Academy of Sciences, Memoirs No. 12.
STILES, F. G., AND A. F. SKUTCH. 1989. A Guide to the Birds of Costa Rica. Cornell University Press, Ithaca, New York.
TERBORGH, J. 1971. Distribution on an altitudinal gradient: Theory and a preliminary interpretation of distribution patterns in the avifauna of the Cordillera Vilcabamba, Peru. Ecology 52:23–40.
WILLIS, E. O., AND K.-L. SCHUCHMANN. 1993. Comparison of cloud-forest avifaunas in southeastern Brazil and western Colombia. Ornithologia Neotropical 4:55–63.

APPENDIX

Birds of Alto Yunda, Anchicayá Valley, Colombia With Months of Occurrence, Seasonal Status, Habitat, Elevational Distribution, and Guild. April 1972 Through June 1973; ● (Bullet) Indicates 1 or More Rec(s). For the Month

Species	J	F	M	A	M	J	J	A	S	O	N	D*	Status[1]	Habitat[2]	Distrib. elev.[3]	Guild[4]
TINAMIDAE																
Tinamus major	●	●	●	●	●	●	●	●	●	●	●	●	R	FI	L	Fr(Gran?), T
Crypturellus soui	●	●	●	●	●	●	●	●	●	●	●	●	R	ES	L	Fr(Gran?), T
CATHARTIDAE																
Coragyps atratus	●	●	●	●	●	●	●	●	●	●	●	●	R	A	W	Car
Cathartes aura	●	●	●	●	●	●	●	●	●	●	●	●	R	A	W	Car
ACCIPITRIDAE																
Elanoides forficatus	●	●	●	●	●	●	●	●	●				R & SDM	A	L	Ins, Aer/Cnpy
Harpagus bidentatus	●	●	?		●		?	?	?			●	R	FI	L	R, D
Ictinia plumbea	●			●	●		●	●	●				R & SDM	A	L	Ins, Aer/Cnpy
Accipiter superciliosus												●	R	CN/ES	L	R, D
Buteo platypterus			●						●				LDM	CN	H	R, D
Buteo magnirostris		●			●	●	●	●	●	●	●	●	R	ES	L	R, D
Leucopternis princeps	●	●			●	●	●			●			R	FI	C	R, D
Harpia harpyja		●											R	FI	L	R, D
Spizastur melanoleucus				●								●	R	FI	L	R, D
Spizaetus ornatus										●			R	FI	L	R, D
PANDIONIDAE																
Pandion haliaetus											●	●	LDM?/V?	A	W	R, D
FALCONIDAE																
Herpetotheres cachinnans	●	●	●	●	●	●	●	●	●	●	●	●	R	CN/ES	L	R, D
Micrastur ruficollis		●				●	●		●				R	FI	L	R, D
Milvago chimachima								●					V	ES	L	Car
Falco rufigularis					●†								R	CN/A	L	R, D
Falco sparverius											●		V	ES	W	R, D
CRACIDAE																
Chamaepetes goudotii	●	●	●	●	●	●	●	●	●	●	●	●	R	FI	C	Fr, A
ODONTOPHORIDAE																
Odontophorus erythrops	●	●										●	R	FI	C	Gran, T

APPENDIX
CONTINUED

Species	J	F	M	A	M	J	J	A	S	O	N	D	D*	Status[1]	Habitat[2]	Distrib. elev.[3]	Guild[4]
RALLIDAE																	
Laterallus melanophaius	●					●					●			R	ES	L	Ins, T
COLUMBIDAE																	
Columba goodsoni	●	?	●	●		●			●	●	●			SDM	CN	L	Fr, A
Columba plumbea	●	●	●	●	●	●				●	●			R	CN	L	Fr, A
Columba subvinacea	●	●	●			●				●	●		●	R	CN	C	Fr, A
Geotrygon saphirina	?	?									●		●	R	FI	C	Fr, T
Geotrygon frenata							?		●				●	SDM	FI	H	Fr, T
PSITTACIDAE																	
Aratinga wagleri	●													SDM	CN	H	Fr/Gran, A
Bolborhynchus lineola				●										SDM	CN	H	Fr/Gran, A
Touit dilectissima	●					●	●	●	●	●	●		●	R	FI	C	Fr/Gran, A
Pionopsitta pulchra	●	●	●	●	●	●	●	●	●	●	●		●	R & SDM	CN	L	Fr/Gran, A
Pionus menstruus	●	●				●	●	●	●	●	●		●	R	CN	L	Fr/Gran, A
Amazona farinosa	●	●				●			●	●	●		●	SDM	CN	L	Fr/Gran, A
CUCULIDAE																	
Piaya cayana	●	●				●	●	●	●	●	●		●	R	CN	L	Ins, A, G
Crotophaga ani	●	●				●	●	●	●	●	●		●	R	ES	W	Ins, A, G
STRIGIDAE																	
Ciccaba virgata	●		●			●				●			●	R	FI	L	R
Otus columbianus	●					●				●	●		●	R	FI	C	R
STEATORNITHIDAE																	
Steatornis caripensis							●		●		●			SDM	FI	W	Fr, A
NYCTIBIIDAE																	
Nyctibius griseus	●						●							R	CN	L	Ins, A, Sal
APODIDAE																	
Streptoprocne zonaris	●	●									●		●	R	A	L	Ins, Aer
Streptoprocne rutila		●							●	●	●		●	SDM?	A	H	Ins, Aer
Chaetura pelagica *										●				LDM	A	L?	Ins, Aer
Chaetura cinereiventris		●		●	●†						●		●	SDM?	A	L	Ins, Aer
Chaetura spinicauda				●							●			SDM?	A	L	Ins, Aer
Panyptila cayennensis								●					●	SDM?	A	L	Ins, Aer

APPENDIX
CONTINUED

Species	J	F	M	A	M	J	J	A	S	O	N	D*	Status[1]	Habitat[2]	Distrib. elev.[3]	Guild[4]
TROCHILIDAE																
Doryfera ludoviciae		●	●	●	●	●							SDM	CN	H	N
Androdon aequatorialis	●	●	●	●	●	●				●	●		SDM	CN	L	N
Threnetes ruckeri		●	●	●	●	●	●			●	●		SDM	ES	L	N
Phaethornis yaruqui	●	●	●	●	●	●	●	●	●	●	?	●	R	FI	L	N
Phaethornis syrmatophorus							?		?		?	?	R	FI	H	N
Phaethornis longuemareus	●	●	●	●	●	●	●	●	●	●	●	●	R	FI	L	N
Eutoxeres aquila	●	●	●	●	●	●	●	●	●	●	●	●	R	FI	L	N
Florisuga mellivora							●	●	?				SDM	CN	L	N
Colibri delphinae								●	●				SDM	CN	H	N
Popelairia conversii	●	●	●	●	●	●	●	●	●	●	●	●	R	CN	C	N
Thalurania furcata	●	●	●	●	●	●	●	●	●	●	●	●	R	FI	L	N
Amazilia amabilis													SDM	ES	H	N
Amazilia tzacatl	?	?	●	●	●	●		?	?				SDM	CN	L	N
Amazilia franciae	●	●	●	●	●	●	●	●	●				R	ES	C	N
Urosticte benjamini	●	●	●	●	●	●			?	●	●	●	R	CN	C	N
Heliodoxa imperatrix									●				R	FI	H	N
Coeligena wilsoni	●	●	●	●	●	●	●	●	●	●	●	●	SDM	FI	H	N
Haplophaedia aureliae				●	●								SDM	CN	H	N
Ocreatus underwoodii												●	SDM	CN	H	N
Aglaiocercus coelestis	?	●								?	●	●	SDM	FI	H	N
Heliothryx barroti	●	●	●	●	●	●	●	●	●	●	●	●	R	CN	L	N
Philodice mitchellii												●	SDM	ES	H	N
TROGONIDAE																
Pharomachrus antisianus		●			●†							●	SDM	CN	H	Fr, A
Trogon massena	●	?	?	●	?	●	●					●	R	FI/CN	L	FR/A, A, Sal
Trogon collaris	●	●	●	●	●	●	●	●	●	●	●	●	R	FI	C	Fr/A, A, Sal
MOMOTIDAE																
Electron platyrhynchum		●					●	●			●		R	FI	L	Ins, A, Sal
Baryphthengus martii	●											●	R	FI	L	Fr/A, A, Sal
BUCCONIDAE																
Malacoptila mystacalis	●										●	●	R	FI	C	Ins, A, Sal
Micromonacha lanceolata		●				●					●	●	R	FI	C	Ins, A, Sal

APPENDIX
CONTINUED

Species	J	F	M	A	M	J	J	A	S	O	N	D*	Status[1]	Habitat[2]	Distrib. elev.[3]	Guild[4]
CAPITONIDAE																
Capito maculicoronatus	•			•		•	•	?	•	•			SDM	CN	L	Fr, A
Eubucco bourcierii		•	?	•	•	•	•	•	•	•	•		R	CN	H	Fr/A, A, G, DL
RAMPHASTIDAE																
Aulacorhynchus haematopygus	•	•				•	•	•	•	•	•	•	R	CN	H	Fr, A
Pteroglossus sanguineus	•	•		•		•	?	•	•	•	•	•	SDM	CN	L	Fr, A
Ramphastos ambiguus	•	•		•		•	•	•	•	•	•	•	R	CN	L	Fr, A
PICIDAE																
Picumnus olivaceus	•	•		•		•	•	•	•	•	•	•	R	CN	L	Ins, B, I
Piculus rubiginosus	•	•				•	•	•	•	•	•	•	R	CN/ES	H	Ins, B, I
Piculus leucolaemus	•	•				•	•	•	•	•	•	•	R	CN	C	Ins, B, I
Celeus loricatus						•	•		•	•	•	•	R	FI/CN	L	Ins, B, I
Dryocopus lineatus						•	•	•	•	•	•	•	R	CN/ES	L	Ins, B, I
Veniliornis fumigatus	•					†•		•		•	•	•	R	FI	H	Ins, B, I
Veniliornis cassinii		•				•	•	•	•	•	•	•	R?	CN	L	Ins, B, I
Campephilus melanoleucos	•	•		•		•	•	•	•	•	•	•	R	CN/ES	L	Ins, B, I
Campephilus haematogaster				•		•	•	•	•	•	•	•	R	FI	C	Ins, B, I
DENCROCOLAPTIDAE																
Dendrocincla fuliginosa	•	•				•	•	•	•	•	•		R	FI	L	Ins, B, S
Glyphorynchus spirurus	•	•				•	•	•	•	•	•	•	R	FI/ES	L	Ins, B, S
Dendrocolaptes sanctithomae												•	V	FI	L	Ins, B, S
Xiphorhynchus erythropygius						•	•	•	•	•	•	•	R	FI	L	Ins, B, S
Xiphorhynchus triangularis							•	•	•	•	•		R	FI	C	Ins, B, S
Campylorhamphus pusillus									•	•	•	•	R	FI	C	Ins, B, S
FURNARIIDAE																
Synallaxis brachyura	•	•											V	ES	L	Ins, A, G
Cranioleuca erythrops	•	•							•	•	•	•	R	CN	C	Ins, A, G
Premnoplex brunnescens		•		•	?				?	•			R	FI	H	Ins, B, S
Pseudocolaptes lawrencii										•	•	•	R?	CN	C	Ins, A, G
Syndactyla subalaris	•									•			R	FI	H	Ins, A, G
Anabacerthia variegaticeps												•	R/V?	FI	C	Ins, A, G
Anabacerthia striaticollis		•											R	FI	C	Ins, A, G
Philydor rufus		•					•	•				•	R	CN	L	Ins, A, DL
Automolus rubiginosus	•	•										•	R	ES	H	Ins, A, G(?)

APPENDIX
CONTINUED

Species	J	F	M	A	M	J	J	A	S	O	N	D*	Status[1]	Habitat[2]	Distrib. elev.[3]	Guild[4]
Thripadectes ignobilis	•	•	•	•							•	•	R	FI	C	Ins, A, DL
Xenops minutus	•	•								•	•	•	R	CN	L	Ins, B, S
Sclerurus mexicanus	•	•		•						•	•	•	R	FI	L	Ins, T, G
FORMICARIIDAE																
Taraba major	•												R	ES	L	Ins, A, G
Thamnophilus unicolor	•	•		•		•	•	•	•	•	•	•	R	FI	H	Ins, A, G
Thamnistes anabatinus	•		•						•	•	•	•	R	FI	C	Ins, A, G
Dysithamnus puncticeps	•			•		•	•	•	•	•	•	•	R	FI	C	Ins, A, G
Myrmotherula brachyura													V	CN	L	Ins, A, G
Myrmotherula surinamensis	•												R	ES	L	Ins, A, G
Myrmotherula fulviventris											•		V	FI	L	Ins, A, DL
Myrmotherula schisticolor	•												R	FI	C	Ins, A, G(DL?)
Herpsilochmus axillaris							•	•	•	•	•	•	R	CN	H	Ins, A, G
Microrhopias quixensis	•						•						R	CN/ES	L	Ins, A, G
Terenura callinota	•												R	CN	H	Ins, A, G
Cercomacra tyrannina													V	ES	L	Ins, A, G
Myrmeciza immaculata	•	•				•					•	•	R	FI	L	Ins, A, G
Myrmeciza nigricauda	•												R	FI	L	Ins, A, G
Gymnopithys leucaspis													V	FI	L	Ins, AF
Conopophaga castaneiceps	•							•	•	•	•	•	R	FI	C	Ins, A, G(?)
RHINOCRYPTIDAE																
Scytalopus chocoensis[5]	•									•	•	•	R	FI	C	Ins, T, G
COTINGIDAE																
Cotinga nattererii	•	?											SDM/V?	CN	L	Fr, A
Pipreola jacunda	•	•	•	•	•								SDM/V?	FI	H	Fr, A
Ampelioides tschudii	•	•									?	•	SDM	ES	H	Fr, A
Pachyramphus cinnamomeus	•	•	•	•			•	•	•	•	•	•	R	ES	L	Fr/A, A
Pachyramphus polychopterus	•	•	•	•	•	•	•	•	•	•	•	•	R	CN	L	Fr/A, A
Tityra semifasciata	•	?	•					?	•	•	•	•	R	CN	L	Fr, A
Querula purpurata	•											•	SDM/R?	CN	L	Fr, A
Cephalopterus penduliger	•												SDM/V?	CN	C	Fr, A
Rupicola peruviana													V	FI	C	Fr, A

APPENDIX
CONTINUED

Species	J	F	M	A	M	J	J	A	S	O	N	D*	Status[1]	Habitat[2]	Distrib. elev.[3]	Guild[4]
PIPRIDAE																
Pipra pipra	•	•	•	•	•	•	•	•	•	•	•	•	R/V?	FI	C	Fr, A
Masius chrysopterus	•	•	•	•	•	•	•	•	•	•	•	•	R	FI	C	Fr, A
Manacus vitellinus													R	ES	L	Fr, A
Machaeropterus deliciosus													R	FI	L	Fr, A
Schiffornis turdinus													R	FI	L	Fr/A, A
TYRANNIDAE																
Zimmerius viridiflavus	•	•	•	•	•	•	•	•	•	•	•	•	R	CN	C	Fr, A
Ornithion brunneicapillum	•	•											V	CN	L	Fr/A, A
Camptostoma obsoletum	•												R	ES	L	Ins, A, G
Myiopagis caniceps												•	SDM/V?	CN	?	Ins(?), A, G
Elaenia flavogaster									•	•	•	•	SDM	ES	W	Fr/A, A, G
Mionectes olivaceus	•												R	FI	C	Fr, A
Leptopogon superciliaris	•		•	•	•	•	•	•	•	•	•	•	R	FI	C	Ins, A, Sal
Pseudotriccus pelzelni												•	R	FI	H	Ins, A, Sal
Myiornis ecaudatus													V	CN	L	Ins, A, Sal
Lophotriccus pileatus	•										•	•	R	FI	C	Ins, A, Sal
Todirostrum nigriceps	•	•											R	CN	L	Ins, A, Sal
Todirostrum cinereum													R	ES	L	Ins, A, Sal
Rhynchocyclus brevirostris												•	R	FI	L	Ins, A, Sal
Tolmomyias sulphurescens													V	ES	L	Ins, A, Sal
Platyrinchus mystaceus	•										•	•	R	FI	C	Ins, A, Sal
Myiotriccus ornatus	•										•	•	R	FI	C	Ins, A, Sal
Myiobius barbatus	•	•										•	R	FI	C	Ins, A, Sal
Mitrephanes phaeocercus	•							•	•	•	•	•	R	CN	C	Ins, A, Sal
Contopus cooperi							?	?	?	•			LDM	CN	C?	Ins, A, Sal
Contopus sordidulus (or virens)								•	?	•	•	•	LDM	CN	C?	Ins, A, Sal
Contopus fumigatus						•							R	CN	C	Ins, A, Sal
Empidonax virescens	•									•	•	•	LDM	ES	L	Ins, A, Sal
Colonia colonus	•					?	?	?					R	CN/ES	L	Ins, A, Sal
Attila spadiceus	•												R	CN	L	Fr/A, A, Sal
Myiarchus crinitus										•	•	•	LDM	ES	L?	Fr/A, A, Sal
Myiarchus tuberculifer	•	•					•						R?	CN	L	Ins(?), A, Sal
Myiozetetes cayanensis	•	•								•			R	ES	L	Ins, A

APPENDIX
CONTINUED

Species	J	F	M	A	M	J	J	A	S	O	N	D*	Status[1]	Habitat[2]	Distrib. elev.[3]	Guild[4]
Conopias cinchoneti	•	•	•	•						•	•	•	R	CN	C	Ins, A, G
Myiodynastes chrysocephalus	•	•	•	•	•	•	•	•	•	•	•	•	R	CN	C	Ins, A, Sal
Tyrannus melancholicus					•								R	CN/ES	W	Ins, A, Sal
Tyrannus tyrannus										•		•	LDM	CN/ES	?	Fr, A, Sal
HIRUNDINIDAE																
Notiochelidon cyanoleuca				•		•	•	•	•	•		•	SDM	A	H	Ins, Aer
Neochelidon tibialis				•	?	•	•	•	•	•	•	•	SDM	A	L	Ins, Aer
Stelgidopteryx ruficollis	•						?						R	A	W	Ins, Aer
Riparia riparia										•			LDM	A	?	Ins, Aer
Hirundo rustica										•	•		LDM	A	L	Ins, Aer
CORVIDAE																
Cyanolyca pulchra		•											V	CN	H	Fr/A, A
TROGLODYTIDAE																
Campylorhynchus albobrunneus	•	•				•		•	•	•	•	•	R	CN	C	Ins, A, G
Odontorchilus branickii	•	•				•	•	•	•	•	•	•	R	CN	C	Ins, A, G
Thryothorus spadix													R	FI	C	Ins, A, G/DL?
Thryothorus nigricapillus	•	•				•	•	•	•	•		•	R	CN/ES	L	Ins, A, G
Troglodytes aedon													R	ES	W	Ins, A, G
Henicorhina leucophrys	•	•				•	•	•	•	•		•	R	FI	H	Ins, A, G
Microcerculus marginatus	•	•				•	•	•	•	•		•	R	FI	L	Ins, T, G
POLIOPTILINAE																
Microbates cinereiventris	•					•	•	•	•	•		•	R	FI	L	Ins, A, G
TURDINAE																
Myadestes ralloides													R	FI	C	Fr, A
Entomodestes coracinus		•											SDM	FI	H	Fr, A
Catharus ustulatus	•									•	•	•	LDM	CN/ES	C?/W	Fr, A
Platycichla leucops	•												SDM	FI	H	Fr, A
Turdus serranus		•											SDM	CN	H	Fr, A
Turdus ignobilis		•		•	•		•						SDM	CN/ES	H	Fr, A
Turdus obsoletus			•	•	•								R	FI	H	Fr, A
VIREONIDAE																
Vireo leucotis	•												R	CN	L	Ins, A, G
Vireo olivaceus (incl. *flavoviridis*)				•	•				•	•	•	•	SDM?/V?	CN/ES	L	Fr/A, A
Hylophilus semibrunneus					•								R?	CN	C	Ins, A, G

340 ORNITHOLOGICAL MONOGRAPHS NO. 48

APPENDIX
CONTINUED

Species	J	F	M	A	M	J	J	A	S	O	N	D*	Status[1]	Habitat[2]	Distrib. elev.[3]	Guild[4]
PARULINAE																
Parula pitiayumi	•	•		•		•		•				•	R	CN/ES	L	Ins, A, G
Vermivora chrysoptera	•	•	•	•						•	•	•	LDM	CN	C	Ins, A, DL
Vermivora peregrina										•	•	•	LDM	CN	L	Fr/A, A
Dendroica castanea										•	•	•	LDM	CN	C	Ins, A, G
Dendroica cerulea								•	•	•			LDM	CN	C	Ins, A, G
Dendroica fusca	•	•		•	•					•	•	•	LDM	CN	H	Ins, A, G
Mniotilta varia	•	•		•				•	•	•	•	•	LDM	CN	H	Ins, B, S
Setophaga ruticilla	•	•		•						•	•	•	LDM	CN/ES	W	Ins, A, S
Wilsonia canadensis	•	•		•						•	•	•	LDM	FI	H	Ins, A, G
Oporornis philadelphia									•	•		?	LDM	ES	C	Ins, A, G
Myioborus miniatus	•			•	•	•	•	•	•	•	•	•	R	FI/CN	L	Ins, A, G
Geothlypis semiflava	•			•	•	•	•	•	•	•	•	•	R	ES	L	Ins, A, G
Basileuterus tristriatus	•	•		•	•	•	•	•	•	•	•	•	R	FI	H	Ins, A, G
Basileuterus culicivorus	•	•		•	•	•	•	•	•	•	•	•	R	CN/ES	C	Ins, A, G
Basileuterus fulvicauda	•	•		•	•	•	•	•	•	•	•	•	R	FI/AQ	L	Ins, T, G
THRAUPINAE																
Diglossa indigotica	•			?			?		•			•	R & SDM	CN	C	N
Dacnis venusta	•	•							•	•	•	•	SDM	CN	L	Fr/A, A
Dacnis cayana						•							SDM?	CN	L	Fr/A, A
Cyanerpes caeruleus													SDM?	CN	L	N
Cyanerpes cyaneus													SDM?	CN	L	N
Chlorophanes spiza	•					•	•		•	•		•	R	CN	L	Fr/A, A
Iridophanes pulcherrima								•	•		•		SDM	CN	H	Fr/A, A
Coereba flaveola	•	•		•	•	•	•	•	•	•	•	•	R	CN/ES	W	N
Chrysothlypis salmoni	•			•		•			•	•	•	•	R	CN	C	Fr/A, A
Chlorochrysa phoenicotis	•	•		•	•	•					•		R?/V?	FI/CN	H	Fr/A, A
Tangara cyanicollis	•			•	•	•	•		•	•		•	R	CN/ES	C	Fr/A, A
Tangara arthus	•					•						•	R	FI/CN	C	Fr/A, A
Tangara nigroviridis	•	•		•		•		•	•	•	•	•	SDM/V?	CN	H	Fr/A, A
Tangara palmeri	•				•	•			•	•	•	•	R	CN	L	Fr/A, A
Tangara rufigula	•				•						•		R	CN	C	Fr/A, A
Tangara icterocephala	•				•	•		•	•	•		•	R	FI/CN	C	Fr/A, A
Tangara vitriolina	•									•			V	ES	C	Fr/A, A
Tangara gyrola	•	•			•	•		•	•	•	•	•	R	CN/ES	C	Fr/A, A
Tangara lavinia	•	•	•			•			•	•	•	•	SDM	CN	L	Fr/A, A

APPENDIX
CONTINUED

Species	J	F	M	A	M	J	J	A	S	O	N	D*	Status[1]	Habitat[2]	Distrib. elev.[3]	Guild[4]
Tangara florida	•	•	•	•	•	•	•	•	•	•	•	•	R	FI/CN	C	Fr/A, A
Tangara larvata	•	•	•	•	•	•	•	•	•	•	•	•	R	CN/ES	L	Fr/A, A
Euphonia minuta												•	SDM	CN	L	Fr, A
Euphonia xanthogaster	•	•	•	•	•	•	•	•	•	•	•	•	R	FI/CN	C	Fr, A
Euphonia saturata			•										SDM	CN/ES	L	Fr, A
Euphonia fulvicrissa													V	CN?	C	Fr, A
Chlorophonia flavirostris	•								•	•			R & SDM	CN	C	Fr, A
Chlorospingus flavigularis	•	•	•	•	•	•	•	•	•	•	•	•	R	FI/CN	C	Fr/A, A
Chlorospingus flavovirens	?	?											R	FI/CN	C	Fr/A, A
Pipraeidea melanonota												•	SDM	CN/ES	H	Fr/A, A
Bangsia rothschildi	•	•	•	•	•	•	•	•	•	•	•	•	R	FI/CN	C	Fr/A, A
Bangsia edwardsi	•	•	•									•	DM	FI	H	Fr/A, A
Thraupis palmarum	•	•	•	•	•	•	•	•	•	•	•	•	R	CN/ES	W	Fr/A, A
Thraupis episcopus	•	•	•	•	•	•	•	•	•	•	•	•	R	ES	W	Fr/A, A
Piranga rubra										•			LDM	CN/ES	H	Fr/A, A
Piranga olivacea										•			LDM	CN	?	Fr/A, A
Ramphocelus flammigerus												•	SDM	CN/ES	H	Fr/A, A
Ramphocelus icteronotus	•	•	•	•	•	•	•	•	•	•	•	•	R	CN/ES	L	Fr/A, A
Chlorothraupis stolzmanni	•	•	•	•	•	•	•	•	•	•	•	•	R	FI	C	Fr/A, A
Mitrospingus cassinii	•	•	•	•	•	•	•	•	•	•	•	•	R	FI	L	Fr/A, A
Heterospingus xanthopygius	•	•	•	•	•	•	•	•	•	•	•	•	R & SDM	CN	L	Fr/A, A
Tachyphonus rufus	•	•	•	•	•	•	•	•	•	•	•	•	R	ES	W	Fr/A, A
ICTERIDAE																
Molothrus bonariensis												•	R	ES	L	Ins, T
Icterus chrysater	•	•											SDM?	ES	H	N (Fr/A)
Amblycercus holosericeus													SDM?	ES	W	Ins, A
Cacicus uropygialis	•											•	R	CN	L	Fr/A, A
Psarocolius angustifrons													V	CN	L	Fr/A, A
CARDINALINAE																
Saltator grossus	•	•	•	•	•	•	•	•	•	•	•	•	R	CN	L	Fr/A, A
Saltator maximus	•	•	•	•	•	•	•	•	•	•	•	•	R	CN/ES	L	Fr, A
Saltator atripennis	•	•	•	•	•	•	•	•	•	•	•	•	R	CN	C	Fr, A
Pheucticus ludovicianus	•	•	•	•						•	•	•	LDM	CN/ES	C	Fr/A, A
Cyanocompsa cyanoides	•	•	•	•	•	•	•	•	•	•	•	•	R	FI	L	Fr/A, A

APPENDIX
CONTINUED

Species	J	F	M	A	M	J	J	A	S	O	N	D*	Status[1]	Habitat[2]	Distrib. elev.[3]	Guild[4]
EMBERIZINAE																
Oryzoborus angolensis	●	●	●	●	●								R?	ES	L	Gran
Volatinia jacarina	●	●	●	●	●	●	●	●	●	●	●	●	R	ES	L	Gran
Tiaris olivacea	●	●	●	●	●	●	●	●	●	●	●	●	R	ES	C	Gran
Sporophila americana	●	●	●	●	●	●	●	●	●	●	●	●	R	ES	L	Gran
Sporophila nigricollis	●	●	●	●	●	●	●	●	●	●	●	●	R	ES	W	Gran
Sporophila intermedia	●	●	●	●	●	●	●	●	●	●	●	●	R	ES	W	Gran
Atlapetes tricolor	●	●	●	●	●	●							R	CN	C	Fr/A, A
Buarremon brunneinucha	●	●	●	●	●	●		●	●	●	●	●	R	FI	H	Fr/A, A & T
Buarremon (torquatus) atricapillus	●		●	●	●								R	FI	C	Fr/A, A & T
Arremon aurantirostris												●	R	FI	L	Fr/A, T
Zonotrichia capensis	●										●		R	ES	H	Gran, T
FRINGILLIDAE																
Carduelis xanthogastera	●				●	●			●			●	SDM/R?	H	H	Gran, A

* Month abbrev. J (Jan.) through D (Dec.); ? indicates status uncertain.
† Recorded only in June of 1975.
[1] Key to status: LDM = Long-distance migrants from north temperate zone. SDM = short-distant migrant from within tropical latitudes. All short-distance migrants are elevational migrants except *Elaenia flavogaster* and *Tangara vitriolina*, which are classified as local migrants. R = resident throughout year. V = vagrant, including juveniles, immatures, and adults (unmated or mated birds). Some represent normal non-breeding dispersal in search of suitable habitat; others are probably expansions of edge and open terrain birds into newly deforested zones.
[2] Key to habitats: FI = interior of mature forest, mainly shaded lower and middle levels. Cn = canopy and old second growth, principally edge species. ES = early succession, young second growth, open pasture with scattered trees. A = aerial.
[3] Key to elevational distribution: L = species distributed largely below 1,000 m in western Colombian Andes. H = species distributed largely above 1,000 m in western Colombian Andes. C = species distribution centered at about the 1,000 m contour. W = widespread from lowlands through highlands.
[4] Key to guilds: *n* = Nectar (hummingbirds). Fr, T = Fruit, Terrestrial. Fr, A = Fruit, Arboreal. Fr/A, T = Fruit/Arthropods, Terrestrial. Fr/A, A = Fruit/Arthropods, Arboreal. Fr/Gran = Fruit/Granivore (applied to parrots and parakeets). Gran, T = Granivore, Terrestrial. Gran, A = Granivore, Arboreal. Ins, A, G = Insectivore, Arboreal, Gleaning. Ins, A, Sal = Insectivore, Arboreal, Sallying (sally here includes any wing-powered maneuver). Ins, AF = Insectivore, Ant-Following. Ins, DL = Insectivore, Dead Leaf. Ins, Aer = Insectivore, Aerial (swifts, swallows). Ins, Aer/Cnpy = Insectivore, Aerial/Canopy (2 species of kites). Ins, B, S = Insectivore, Bark, Surface (mostly woodcreepers). Ins, B, I = Insectivore, Bark, Interior (woodcreepers, some furnariids). Ins, T = Insectivore, Terrestrial. R, *n* = Raptorial, Nocturnal (owls). R, D = Raptorial, Diurnal. Car = Carrion/scavenging.
[5] See Krabbe and Schulenberg (1997).

A SECTOR-BASED ORNITHOLOGICAL GEOGRAPHIC INFORMATION SYSTEM FOR THE NEOTROPICS

Morton L. Isler

Division of Birds, National Museum of Natural History, Smithsonian Institution, Washington, District of Colombia 20560 USA;

ABSTRACT.—Current efforts to study and conserve Neotropical birds suffer from imprecise delineations of their geographic distributions. By coding geographic data to small sectors, the geographic information system described in this paper provides fine-grained computer mapping of avian ranges that portrays distributions clearly without sacrificing important information such as which side of a river a taxon has been found. The sector approach also supports graphical and statistical analyses that relate the distribution of taxa to one another and to other geographically variable factors such as climate and habitat.

RESUMO.—Esforços atuais em estudo e conservação de aves Neotropicais sofrem com o impreciso delineamento de suas distribuições geográficas. O sistema de informações geográficas descrito neste artigo, através da codificação de dados geográficos em pequenos setores, fornece recursos para mapeamento informatizado de alta definição das distribuições de aves, que as retratem claramente sem prejuízo de informações importantes, tais como qual margem do rio tem sido o táxon encontrado. A abordagem em setores favorece ainda análises gráficas e estatísticas que relacionam a distribuição entre os táxons e entre estes e fatores variáveis geograficamente, como clima e hábitat.

Current geographic range descriptions for many Neotropical birds are typically generalized and vague. Collecting stations tend to be widely dispersed and, in Amazonia, concentrated in riverside locations (Oren and Albuquerque 1991). Range statements (e.g., Meyer de Schauensee 1970) often incorporate assumptions that distributions are continuous between known locations and that allopatric sister taxa will be found to intergrade. Today, rapid expansion of avian field studies and collections provide an opportunity to portray distributions accurately. Avifaunal surveys have attained a new, high level of comprehensiveness, and although the cautions of Remsen (1994) should be observed, a taxon's absence can be noted with some certainty, particularly where vocalization inventories (Parker 1991) are employed. Especially among the often sedentary suboscines, the numerically dominant segment of the Neotropical avifauna, it is becoming apparent that many taxa have complex ranges that include disjunctions or situations in which sister taxa are parapatric without intergradation. Taxa formerly considered representatives of continuously distributed species are increasingly being recognized as specifically distinct (e.g., Isler et al. 1997; Prum 1994).

In addition to studies of species limits, precise distributional knowledge and an ability to interrelate spatial data are vital to a wide range of ornithological studies including: 1) reviews of systematics and phylogeny requiring detailed geographic knowledge of opportunities for gene flow (e.g., Hackett and Rosenberg 1990); 2) definitions of areas of endemism, central to analyses of historical biogeography (e.g., Cracraft 1985, Haffer 1985); 3) examinations of geographic variation in a species' morphology as it relates to habitat, climate, etc. (e.g., Kratter 1993); 4) regional identifications of concentrations of endangered species used to establish priorities for acquisition of conservation areas (e.g., Whitney and Pacheco, in press); and 5) broad-scale analyses that interrelate species distributions to climate, vegetation, etc., in order to establish conservation priorities among biomes (e.g., Stotz et al. 1996).

Because they require detailed and accurate locational data, these types of studies are facilitated by geographic information systems (GIS). A GIS combines computer cartography with analytic tools (Cole 1993) and may be broadly defined as a "computer-based system—used to store and manipulate geographic information" (Aronoff 1989). Geographic information systems vary in their design especially in relation to the spatial scale, spatial density and resolution of the data, and quality of the information they contain. This paper does not seek to review the potential applications of GIS to ornithology (see Shaw and Atkinson 1990 for some other applications) but rather to report on the progress and discuss issues in the development and implementation of one geographic information system designed to map and analyze avian distributions over large regions. Examples of maps produced with this GIS may be found in papers by Bates, Graves, Isler et al., and Zimmer in this volume. Even though its development will not be complete, the author hopes to make this GIS available to researchers by 1998 by installing it at the following institutions: Field Museum of Natural History (Chicago, IL, USA), Museo Paraense Emilio Goeldi (Belém, PA, Brazil), Museum of Natural Science, Louisiana State University (Baton Rouge, LA, USA), and National Museum of Natural History, Smithsonian Institution (Washington, DC, USA).

THE SECTOR APPROACH

The essence of this GIS is that information is coded to sectors, areas approximating 50 kilometers square but individually shaped to reflect geographic features such as rivers. Techniques of sector definition are described in the next section of the paper. Alternatives to the sector approach in mapping distributions across wide regions include 1) using the precise longitude and latitude of each collecting station, and 2) coding locations to a latitude-longitude grid. The first alternative affords greater precision in mapping collecting sites, but the amount of precision lost in implementing the sector approach is trivial in maps that span hundreds or thousands of kilometers. The second alternative permits easier coding of collecting stations, but as this coding is only done once, in the initial programming, it presents no advantage to the user. In comparison to these alternatives, the sector approach has the following advantages:

1. Sectors serve to clarify on which side of a geographic barrier a taxon occurs. For example, major Amazonian rivers mark the outer limits of the ranges of many birds (Capparella 1988, Haffer 1992), and many Neotropical collecting stations are along river banks. Following the first alternative, if one plots the precise coordinates of a riverine location on a map of Amazonia with a large and easily read map symbol, the symbol will overlap the line of the river and cause confusion as to which side of the river the taxon is found. The second alternative, using a geometric grid, ignores topography altogether. In contrast, the sector method displaces mapping symbols to the proper sides of rivers and other topographic features such as cordillera that isolate taxa restricted to lower elevations. In a related vein, the sectors also clarify locations with respect to major political boundaries. Although biogeographically immaterial, political jurisdictions are important from a conservation perspective and have value in written range descriptions.

2. Sectors improve map presentation by reducing the number of symbols. Map simplification is especially important when a large number of collecting stations are clustered about a city, as they are around Belém, Brazil. The mass of overlapping symbols that results when every station is plotted promotes a distorted image of distribution by visually confusing the prevalence of collecting stations with distributional status of a taxon. Moreover, overlap makes the shape of map symbols difficult to discern. The alternative, the geometric grid, also simplifies map presentation but has the danger of consolidating unrelated sites on different sides of rivers, etc.

3. Sectors provide a common unit of analysis. A common unit of analysis allows a variety of techniques to be used in relating avian distributional data to other geographically variable information, such as rainfall or edaphic conditions, as discussed below.

The sector approach is inappropriate, however, for mapping detailed distributions within small regions where precise locations of collecting stations are important. In such situations, it is best to place individual locations on the computer map using exact coordinates.

How sectors were drawn.—To establish the sectors, I first divided the Neotropics into 13 conveniently sized regions (Fig. 1) which were then divided into sectors (e.g., Ecuador, Fig. 2). I drew the size and shape of the sectors to accomodate a variety of shapes (squares, triangles, etc.) of map symbols used to locate collecting stations so that they would be readable on journal-size distribution maps. To help the reader envision the expanse of a sector, a "model" sector size is a square, 50 by 50 km. In reality, sector sizes and shapes differ greatly from this "model" because they are drawn to reflect prominent features such as coastlines, cordilleran divides, and major rivers as exemplified by the sector shapes shown on Figure 2.

The sectors are numbered with five-digit codes on Figure 2. The first two digits are a country code or, in the case of Brazil, a state code. Digits 3–5 are unique to the sector. The codes given to localities also include a sixth digit not shown on this map. The sixth digit is a subsector code,

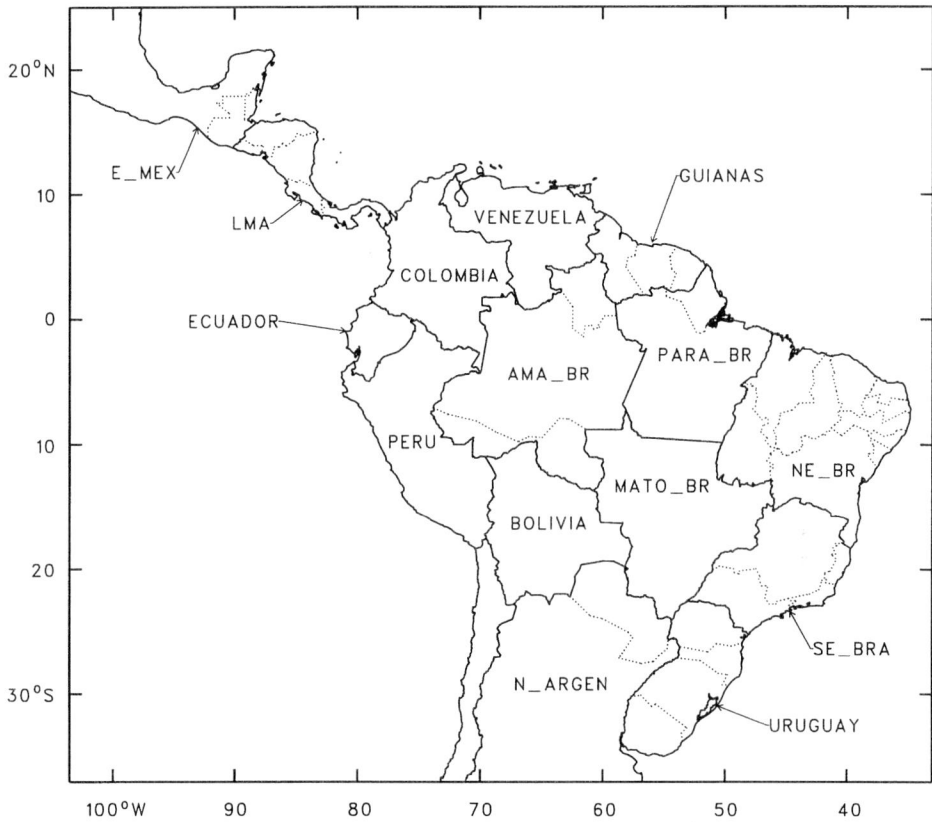

FIG. 1. Key to sector maps. This map delineates the area currently incorporated in the GIS. The heavy lines are the boundaries of computer maps used to define sectors (Fig. 2 is an example). The maps are identified by their file names. The dotted lines are national boundaries (states in Brazil) encompassed within sector maps.

used to make fine physiographic distinctions among locations. Section A of Figure 3 provides an enlargement of a single basic sector and "190290" is its full identification number as subsector codes of "0" are given to basic sectors. Collecting stations located in this sector, such as those labeled a and b, are coded with this identification number and entered into the appropriate computer file, called a *system gazetteer*, for the country or Brazilian state. Figure 3 also shows how a sample map symbol, a square, fits in the sector. The coordinates, longitude and latitude, of the cross in the center of the square are used to place map symbols for that sector on computer maps. Of course, the size of the symbol will be reduced greatly on a map of the Neotropics, but it will be placed clearly on the proper side of the river and will serve for both collecting stations.

Subsectors were established to overcome a shortcoming of the sector approach when applied to regions of rapidly shifting or complex topography and habitat. An example of subsector use involving river island habitats is provided by Section B of Figure 3, where an island and a collecting station, c (double circles), have been added to the river. A subsector has been created, centered on the river and located by the sample map symbol (square) with the cross in the center setting the subsector longitude and latitude. A "1" has been made the sixth digit of the code for the subsector, and collecting station c is assigned sector code 190291. Therefore, whenever a taxon is found on a river island, its location will show up on the map in the middle of the river rather than displaced to one bank or the other, and maps for river island birds (such as those in Rosenberg 1990) will reflect their distribution appropriately.

Subsector codes are used to represent a variety of geographic features. For example, in steep and narrow valleys such as the Magdalena and Cauca valleys in Colombia, five subsectors are typically employed. One is placed on the valley floor, two are used for intermediate elevations on either side of the river, and two for the high elevations on the cordilleran divides bordering

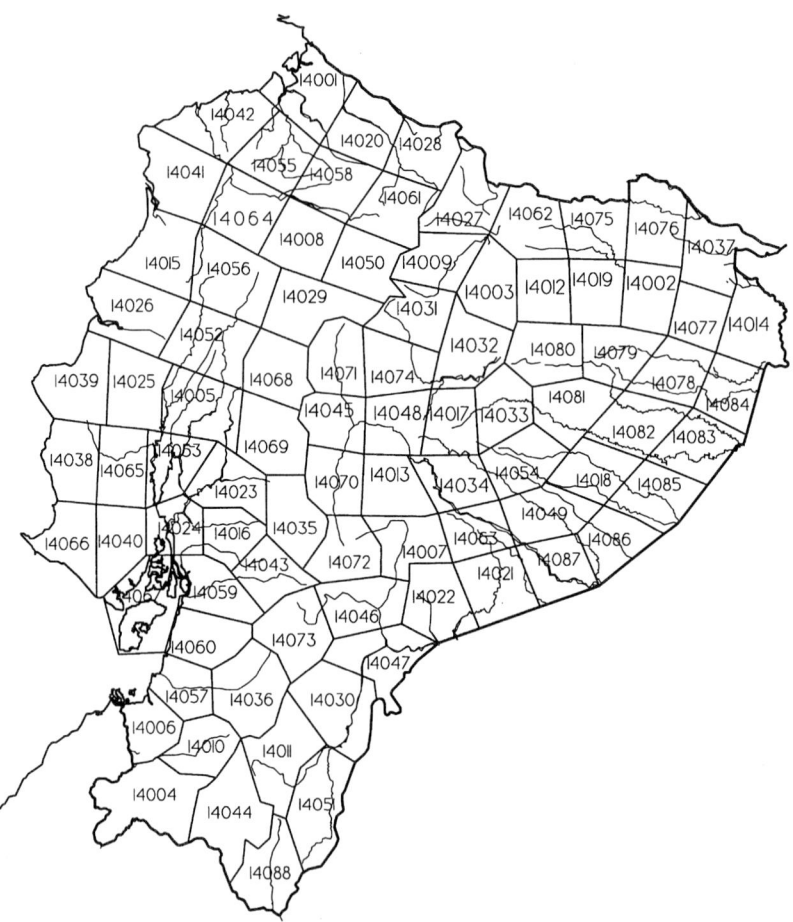

FIG. 2. Sector outlines in Ecuador. Ecuador is the smallest sector map. To simplify presentation, this map does not show subsectors, and sector identification numbers are limited to five digits. Although sectors cross the upper reaches of rivers (such as the Río Napo), subsectors are used to differentiate locations on opposite banks.

the sector. A supplemental file to the system gazetteers identifies the physical features used in subsector codes.

Collecting stations are being coded to sectors and listed in system gazetteers incrementally. As of December 1996, over 4900 subsectors had been defined and over 12,000 locality names coded. Our ability to code localities has been aided greatly by the ornithological gazetteers (e.g., Paynter 1993) prepared under the direction of Raymond A. Paynter, Jr., Museum of Comparative Zoology. In addition to the system gazetteers, the GIS consists of computer files of the 13 sector maps that allow users to place additional localities into the system gazetteers, a computer file (named LONGLAT) that associates points of longitude and latitude with each subsector number, and a package of computer base maps. How these files are employed in cartography and analysis follows.

CARTOGRAPHY

Steps in map-making.—The procedure used to make a distribution map is to: (1) enter sites, taxon names, and associated information in a relational database; (2) transport points of longitude and latitude to a mapping program; and (3) place the points on a base map and add map symbols, text, hatching, etc., that best communicate the distributional information. Step 1, entering the data, is accomplished in an Access (Microsoft Corp., Redmond, Washington) database application that I have written and named NeotropicDB. The application incorporates the aforemen-

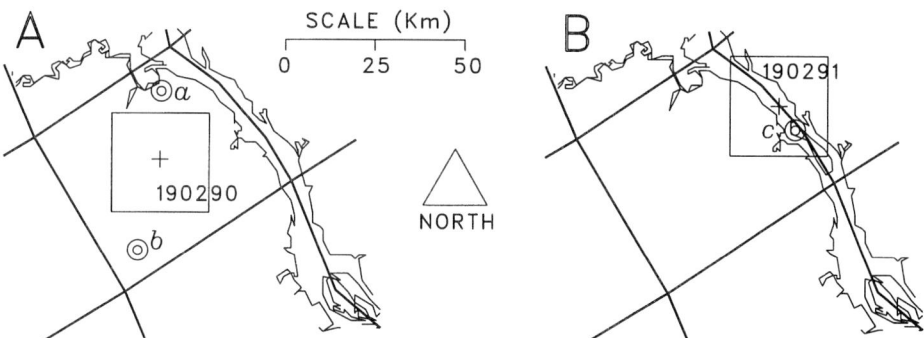

FIG. 3. Methods used in sector and subsector definition. A. *Basic sector definition*. Heavy lines are sector boundaries that follow the center of a major river to the east but cut through a smaller tributary to the north. The square is an example of a map symbol and shows the relationship of the size of the map symbol to the size of the sector. The cross in the center of the square was used to provide the mapping coordinates used to locate map symbols for this sector. The double circles, identified by a and b, exemplify collecting stations coded to this sector. B. *Subsector definition*. An additional collecting station, identified by c, has been placed on a river island. This station and other river island localities are coded to a subsector (identification number 190291) placed in the center of the river. The second square is an example of a map symbol for river localities, and the cross at its center locates the subsector coordinates.

tioned gazetteers. Site data include both exact geographic coordinates and sector coordinates. Lists of source names (e.g., journal articles) and taxon names that will be referenced are entered initially by the user. Then, site, source, and taxon data are selected for each record from dropdown lists, thus insuring consistency of data as well as ease of entry. Additional fields in the record allow entry of a variety of information such as type of documentation, specimen data, links of specimen and tissue samples, and references to observations (e.g., natural history) that may be associated with the record. Records may be cloned to simplify processing repetitive entries such as lists of taxa at the same site by the same collector.

Data in the resultant table can be manipulated in many ways, and for purposes of Step 2, transporting the data to a mapping program, the process is simple. The user selects the taxon and the type of coordinates (exact or sector coordinates), and the database application searches the table of records and exports an ASCHII file of coordinates of sites or sectors at which the taxon has been found. The ASCHII coordinates must then be translated into map points by a CADD (Computer-Aided Design and Drafting) or a GIS program. I have been using ARC/INFO (Environmental Systems Research Institute, Redlands, California), a powerful but expensive program that supports GIS, to make the translation to map (DXF) points. DXF points can then be manipulated in ARC/INFO or in various other CADD or GIS programs in Step 3. Users without access to ARC/INFO can input the ASCHII file of coordinates directly into their mapping program, if enabled by the program, or by keyboard entry.

The final step, the preparation of the maps, begins with a decision on which base map of coasts, rivers, etc., to place the sector points. This GIS includes a variety of base maps that extend from southern Mexico to northern Argentina. They employ the "Plate Carrée" projection, in which each unit of longitude and latitude is the same size square. Although this projection produces large mapping distortions at temperate latitudes, it is satisfactory for maps near the Equator. A major advantage of computer mapping is that the base map can be modified easily to reflect taxa distributions. Moreover, as the scale is "zoomed" in or out, the level of detail, such as the number of rivers shown, can be adjusted. In addition to the base maps provided in this GIS, the user may use world computer maps, such as Mundocart (Petroconsultants, London) and the Digital Chart of the World (U. S. Defense Mapping Agency, Washington, DC) or may scan or manually digitize printed maps. In digitizing, map features are traced by a pointing device on a special table. Unfortunately, at this time, elevational contour data are unavailable (on Mundocart) or portrayed in feet with data omitted in some regions (on Digital Chart of the World) and must be captured from printed base maps by manual digitizing or other means.

Map presentation issues.—Computer cartography permits any researcher to become an excellent draftsman equipped with a wonderful variety of mapping tools. Maps can be manipulated instantly by transforming scales, varying line widths, creating different map symbols, altering text fonts, etc. Options are almost limitless. The choice of mapping conventions shapes the image conveyed to the reader, however, and care must be taken not to distort the information because alternate presentations may convey different messages concerning distributions.

The examples in Figure 4 afford an idea of the versatility of computer cartography and point out the need for careful consideration of decisions involved in map presentation. They portray the distribution of the Tufted Antshrike (*Mackenziaena severa*) of southeast Brazil and adjacent Argentina and Paraguay. Most maps of widespread avian distributions (e.g., Isler and Isler 1987; Ridgely and Tudor 1989) use shading (termed "hatching" in many computer programs) to convey the extent of ranges. An alternative approach is to place symbols for collecting stations without defining outer range boundaries (e.g., Snow 1982). Ideally, when shading is used, its extent should be confined to areas containing types of habitats occupied by the taxon whose range is being portrayed. Currently, this tenet can be followed only in the most general way (such as forested vs. nonforested regions) because we lack knowledge of the habitat preferences of many taxa, and satisfactory maps distinguishing Neotropical microhabitats at the level of detail expressing taxa habitat preferences do not exist. Consequently, ranges are usually portrayed by simply drawing a line around the outermost collecting stations and filling in the space with shading (Fig. 4A). Although providing a strong visual presentation of the outer distributional limits, the shading exaggerates the amount of area occupied by *Mackenziaena severa*, a forest-dwelling species, and provides no information on distribution inside the shaded area.

In the absence of detailed habitat information, the computer may be used to reduce the shaded area to a more limited representation of the species' known distribution (Fig. 4B). To outline the shading on this map, circles with a 50-km diameter were first drawn around the center of each sector in which the species has been found. Then, to complete the boundary, circumferences of circles within 150 km of each other were connected. Although arbitrary, this approach limits the error of extending shading any great distance into areas where the species has not been found while preserving the visual impact of shading.

Figure 4C provides additional information regarding the distribution of *Mackenziaena severa*. Extensive alterations of forest have occurred in southeastern Brazil in recent years, and map symbols are used to differentiate locality records before and after 1960, expressing the possibility that pre-1960 locations may no longer harbor this species. Another map symbol is used to identify records that lack specimen or vocalization documentation. Shading is placed in accord with distance rules as in Figure 4B, but a 75 km, rather than 150 km, rule was used to connect modern locations and two types of shading are used. Dark shading delimits the areas where modern records occur, the more lightly shaded areas encompass pre-1960 records. Compared to Figure 4B, this map is a more conservative representation of areas where the species is known to survive today. As satellite photos become more available, it will be possible to replace mechanical rules for outlining shading with an computer portrait of habitat distribution drawn from photos.

Finally, shading may be eliminated entirely (Fig. 4D). No assumptions are made that extend the range of this species beyond known localities. Although most precise of the four examples, this approach lacks the visual impact of maps with shading. The reader is required to study portions of the map to interpret the distributional status of the species.

ANALYSIS OPTIONS

In addition to supporting cartography, geographic information systems provide a way to link spatial databases analytically. For example, a map of annual average precipitation can be related to data describing avian distributions. Either graphical or numerical (quantitative) methods can be used. Graphical analysis uses the full capability of GIS and the computer screen. Numerical analysis uses ordinary statistical methods to analyze data stored in the GIS database. To date, this GIS has been primarily used for cartography, but the system is designed to support analysis. Descriptions of analytic methods may be found in GIS texts (e.g, Aronoff 1989), and only the barest outline of the analytic options are presented here to illustrate some of the opportunities offered by a sector-based system.

Some types of ornithological questions that may be addressed include: (1) Where does a taxon occur relative to other taxa? (e.g., how much overlap occurs among taxa?) (2) What kinds of spatial patterns (e.g., patterns of endemism and contact zones) are shared by taxa? (3) What common characteristics, such as climate or vegetation, define the set of places in which a taxon occurs? (4) How do avifaunal distribution patterns relate to geographic barriers such as rivers or cordilleras? (5) How do patterns of avian distribution relate to the known history of changes in the geology, climate, or human activity of various regions?

Graphical analysis.—The simplest type of graphical analysis is visual inspection in which the researcher overlays different data sets on the computer screen. For example, distributions of avian species may be superimposed on maps expressing current environmental conditions, reconstructions of the geologic past, distribution patterns of other organisms, etc. The resultant displays provide opportunities to acquire insights regarding the level of association among the spatial patterns of diverse data sets even when the patterns are complex. Flexibility in manipulating and displaying different data sets in the same file stems from a feature of advanced computer graphics programs known as data layers. Different types of information are placed on different layers in the file (as illustrated in Shaw and Atkinson 1990). Layers act like sheets of tracing paper overlaying one another, and they may be displayed and hidden rapidly in an interactive process.

Going beyond visual inspection, a multitude of specialized GIS techniques, even a simple listing of which is beyond the scope of this paper, have been developed to analyze spatial data. Most are designed for applications using much finer (scale, density, resolution and quality) data than is currently available for distributional studies of Neotropical avifauna. Appropriate to the quality of Neotropical data, the sector approach of this GIS provides a simple data framework in which operations interrelating data sets can be implemented easily and efficiently. For example, GIS programs (e.g., ARC/INFO) incorporate "overlay operations" to map regions that meet certain requirements. One may ask the computer to map all sectors where three species occur together or may ask questions of increasing complexity such as where Species 1 AND [Species 2 OR Species 3] occur together. If Species 1, 2, and 3 have been found at three different but geographically proximate localities, their placement in the same sector overcomes problems of their not being found at precisely the same data point. Although GIS programs include proximity techniques for dealing with this problem, the sector approach takes into account topographic barriers as well as distance in its solution.

Numerical analysis.—The same relational database files used to input data for computer mapping, as described earlier, may be used to interrelate data without employing mapping or GIS programs. In a relational database program, every data field can be used as a key, and data files can be linked to allow for searches of common data points, etc. For example, as in the graphical analysis, a search can be made for sectors in which three species have all been found or more complex search requirements can be implemented. A difference is that the graphical approach will produce a map of sectors that meet the requirements of the query, whereas the answer from the relational database program will take the form of a list of sector numbers. In additional to their use in responding to such direct queries, the GIS database files also facilitate conventional statistical analysis, such as computing correlations in the distribution of species. In this regard, the sector approach provides the same advantages as it does in graphical analysis.

GIS AND NEOTROPICAL ORNITHOLOGY

Understanding the distribution and phylogeny of Neotropical birds poses an enormous challenge. The number of taxa is so large and the environmental conditions so complex that powerful tools are needed to sort through vast amounts of data to perform complex spatial analyses. The pioneering biogeographic work of Haffer (1974, 1987, 1993) and others is extraordinary not only for the quality of the original thought and analysis but for the massive volumes of distributional data plotted by hand.

GIS offers the opportunity to build on this work by employing much larger data sets than can be handled manually. By maintaining and manipulating spatial data in a digital format, GIS is speedy and cost-effective in its repetitive use of data files. Above all, these data-handling capabilities provide a way to integrate different types of data within a single analytic framework. Moreover, given the analytic speed afforded by modern computers, refinements can be made quickly through successive analyses.

An understanding of the biogeography of Neotropical avifauna is vital to its conservation. As Ted Parker was fond of pointing out, the conservation decisions made in this decade will determine the future survival of many Neotropical birds. Detailed surveys and analyses of the distribution of taxa, coupled with studies of genetic, morphological, and behavioral variation, are needed to help determine which habitats—at all geographic scales—have the highest priority for conservation. GIS can provide an important tool in this process.

ACKNOWLEDGMENTS

Theodore A. Parker III was head "cheerleader" as this project developed, and his inspiration will always be there as the work proceeds. My wife and coworker, Phyllis R. Isler, and our

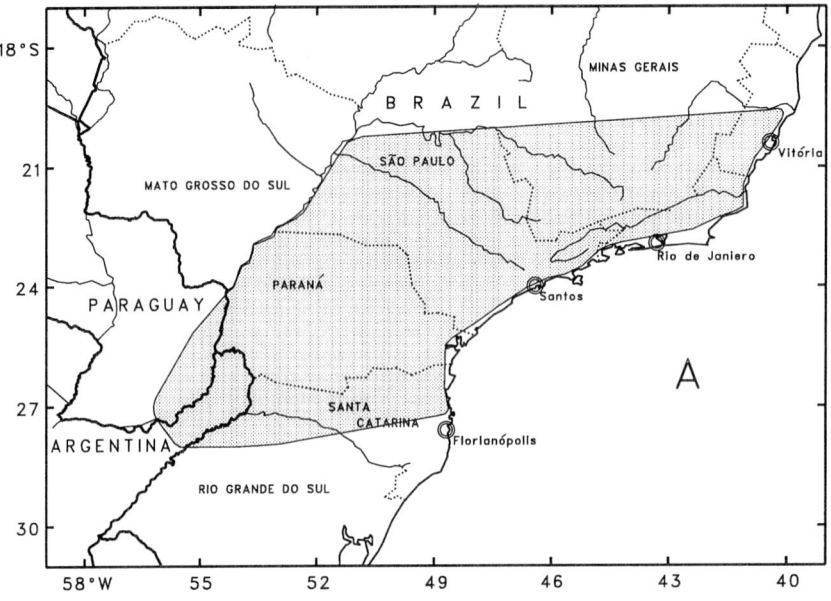
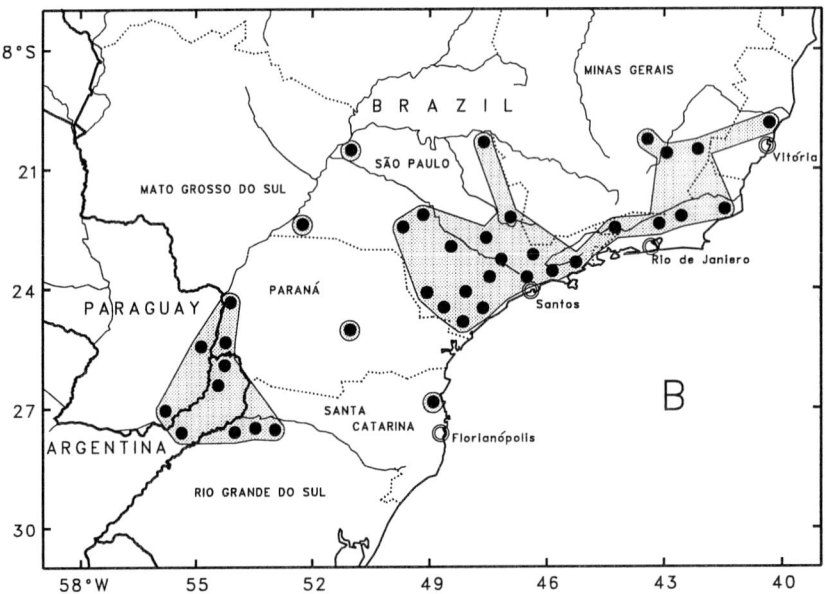

Fig. 4. Alternate map presentations for the Tufted Antshrike, *Mackenziaena severa*. Rules used in creating maps are described in the text. Dotted lines are state boundaries in Brazil. A. Hatching without pinpointing localities. B. Partial hatching with black circles locating all records, documented and undocumented. C. Partial hatching with dark hatching confined to areas with modern records identified by black circles; "H" inside a circle = historical (pre-1960) records; "V" inside a circle = visual, undocumented records. D. Same use of map symbols as in C but no hatching.

A GIS FOR THE NEOTROPICS

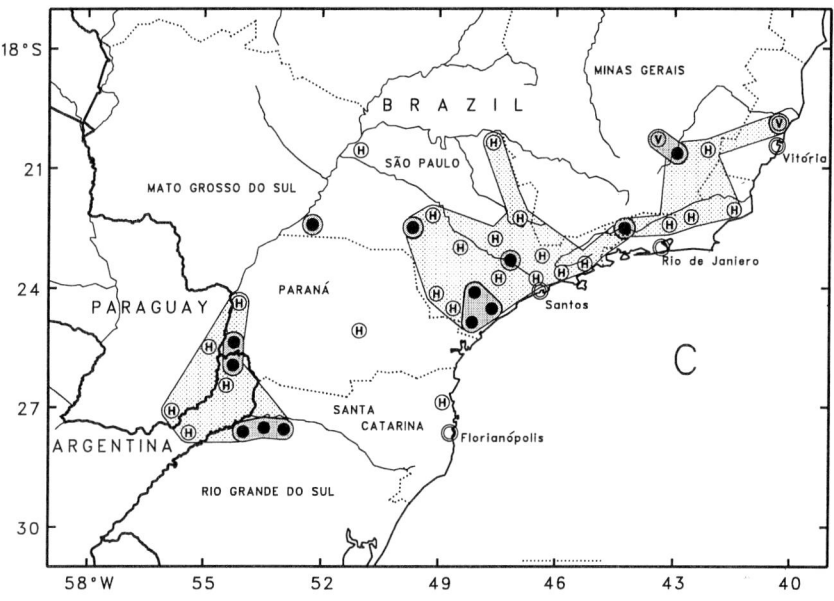

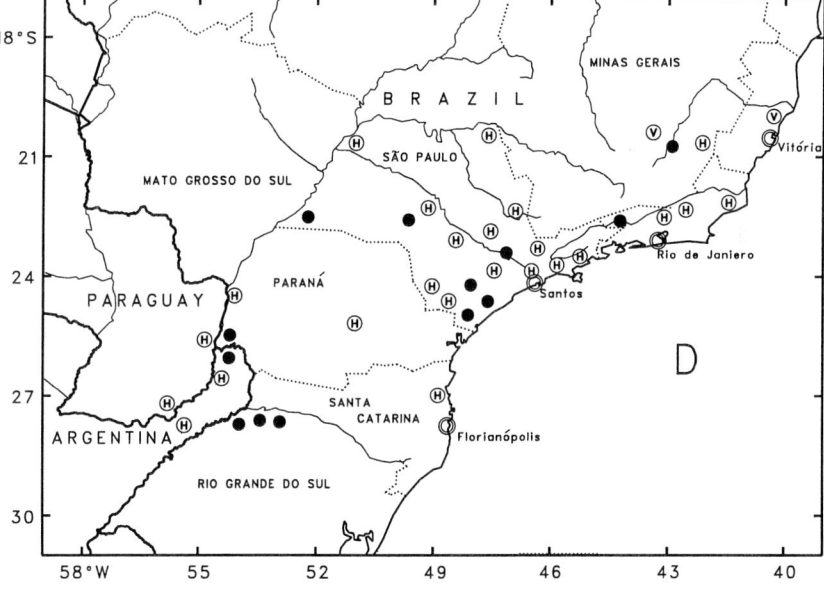

FIG. 4. Continued.

coauthor on an upcoming monograph on Thamnophilidae, Bret M. Whitney, provided support in innumerable ways. John M. Bates made useful suggestions on the organization of this paper, and he, Robert Behrstock, and Robin Panza provided useful comments on an earlier draft. Mario Cohn-Haft and Jane Read reviewed the submitted manuscript and made many helpful suggestions. Dan Cole delivered vital instruction in GIS techniques and implementation of ARC/INFO procedures, and his teachings (Cole 1993) are an important underpinning of this paper. Craig Ludwig provided important programming assistance. Cole, Jürgen Haffer, Luiz P. Gonzaga, José F. Pacheco, and Paulo Sergio Fonseca provided stimulating ideas on issues of map presentation. I am also grateful to Gary Graves who first encouraged me to prepare this paper, and to J. V.

Remsen, Jr., who originally taught me the value of locality-by-locality mapping. Lorna Anderberg has been making a major contribution to this project in her continuing development of the system gazetteers.

LITERATURE CITED

ARONOFF, S. 1989. Geographic Information Systems: A Management Perspective. WDL Publications, Ottawa, Canada.

CAPARRELLA, A. P. 1988. Genetic variation in Neotropical birds: implications for the speciation process. Acta XIX Congr. Intern. Ornith. (Ottawa 1986):1658–1664.

COLE, D. 1993. Introduction to Geographic Information Systems. Unpublished ms. Smithsonian Institution, Washington, D.C.

CRACRAFT, J. 1985. Historical biogeography and patterns of differentiation within the South American avifauna: areas of endemism. Pp. 49–84 in Neotropical Ornithology (P. A. Buckley, M. S. Foster, E. S. Morton, R. S. Ridgely, and F. G. Buckley, Eds.). Ornithol. Monogr. No. 36.

HACKETT, S. J., AND K. V. ROSENBERG. 1990. Comparison of phenotypic and genetic differentiation in South American antwrens. Auk 107:473–489.

HAFFER, J. 1974. Avian speciation in tropical South America. Publ. Nuttall Ornithol. Club, No. 14.

HAFFER, J. 1985. Avian zoogeography of the Neotropical lowlands. Pp. 113–146 in Neotropical Ornithology (P. A. Buckley, M. S. Foster, E. S. Morton, R. S. Ridgely, and F. G. Buckley, Eds.) Ornithol. Monogr. No. 36.

HAFFER, J. 1987. Biogeography of Neotropical birds. Pp. 105–150 in Biogeography and Quaternary History in Tropical America (T. C. Whitmore and G. T. Prance, Eds.). Clarendon and Oxford Univ. Press, NY.

HAFFER, J. 1992. On the "river effect" in some forest birds of southern Amazonia. Bol. Mus. Paraense E. Goeldi, Zoología 8:217–245.

HAFFER, J. 1993. Time's cycle and time's arrow in the history of Amazonia. Biogeographica 69:15–45.

ISLER, M. L., AND P. R. ISLER 1987. The Tanagers. Smithsonian Inst. Press, Washington, D.C.

ISLER, M. L., P. R. ISLER, AND B. M. WHITNEY. 1997. Biogeography and systematics of the *Thamnophilus punctatus* complex. Pp. 355–381 in Studies in Neotropical Ornithology Honoring Ted Parker (J. V. Remsen, Jr., Ed.), Ornithol. Monogr. No. 48.

KRATTER, A. W. 1993. Geographic variation in the Yellow-billed Cacique, *Amblycercus holosericeus*, a partial bamboo specialist. Condor 95:641–651.

MEYER DE SCHAUENSEE, R. 1970. A Guide to the Birds of South America. Livingston Publ. Co., Wynnewood, Pennsylvania.

OREN, D. C., AND H. GUERREIRO DE ALBUQUERQUE. 1991. Priority areas for new avian collections in Brazilian Amazonia. Goeldiana Zoologia 6:1–11.

PARKER, T. A., III. 1991. On the use of tape recorders in avifaunal surveys. Auk 108:443–444.

PAYNTER, R. A., JR. 1993. Ornithological Gazetteer of Ecuador. Museum of Comparative Zoology; Cambridge, Massachusetts.

PRUM, R. O. 1994. Species status of the White-fronted Manakin, *Lepidothrix serena* (Pipridae), with comments on conservation biology. Condor 96:692–702.

REMSEN, J. V., JR. 1994. Use and misuse of bird lists in community ecology and conservation. Auk 111: 225–227.

RIDGELY, R. S., AND G. TUDOR. 1989. The Birds of South America, Vol. I. Univ. Texas Press, Austin, Texas.

ROSENBERG, G. H. 1990. Habitat specialization and foraging behavior by birds of Amazonian river islands in northeastern Peru. Condor 92:427–443.

SHAW, D. M., AND S. F. ATKINSON. 1990. An introduction to the use of geographic information systems for ornithological research. Condor 92:564–570.

SNOW, D. 1982. The Cotingas. Cornell Univ. Press, Ithaca, New York.

STOTZ, D. W., J. W. FITZPATRICK, T. A. PARKER III, AND D. K. MOSCOVITS. 1996. Neotropical birds: ecology and conservation. Univ. of Chicago Press, Chicago, Illinois.

WHITNEY, B. M., AND J. F. PACHECO. 1995. Distribution and conservation status of four *Myrmotherula* antwrens (Formicariidae) in the Atlantic forest of Brazil. Bird Conservation International 5:421–439.

BIOGEOGRAPHY AND SYSTEMATICS OF THE *THAMNOPHILUS PUNCTATUS* (THAMNOPHILIDAE) COMPLEX

MORTON L. ISLER,[1] PHYLLIS R. ISLER,[1] AND BRET M. WHITNEY[2,3]

[1]*Division of Birds, National Museum of Natural History, Smithsonian Institution, Washington, D.C. 20560, USA;*
[2]*% Field Guides Inc., P.O. Box 160723, Austin, Texas 78716, USA;*
[3]*Museum of Natural Science, 119 Foster Hall, Louisiana State University, Baton Rouge, Louisiana 70803, USA;*

ABSTRACT.—The Slaty Antshrike is currently considered a single polytypic species (*Thamnophilus punctatus*) that ranges from Belize to southeastern Brazil. We present evidence that its distribution is highly patchy and that geographic representatives differ in vocalizations and external morphology. As a consequence, we recommend that six species be recognized within the complex and that the taxonomic position of two additional taxa, potentially warranting species status, be reevaluated when their vocalizations and behavior are better known. Recognition at the species level has important conservation implications because most members of the complex are concentrated in tropical dry-forest habitats that are rapidly disappearing in the Neotropics. Moreover, our results suggest that studies of taxa occupying dry forest can provide new insights into the evolution of Neotropical birds.

RESUMO.—A choca-bate-cabo (*Thamnophilus punctatus*) é correntemente considerada uma espécie politípica única que se distribui de Belize até o sudeste do Brasil. Apresentamos evidência que sua distribuição é bem espalhada e que seus representantes geográficos diferem em vocalizações e morfologia externa. Como consequência, recomendamos que sejam reconhecidas seis espécies dentro do complexo e que a posição taxonômica de dois táxons adicionais, potencialmente merecedores do "status" de espécie, sejam reavaliados quando suas vocalizações e comportamento estiverem melhor conhecidas. O reconhecimento ao nível de espécie têm importante implicação na conservação porque a maioria dos membros do complexo estão concentrados em hábitats tropicais de floresta seca que estão rapidamente desaparecendo nos Neotrópicos. Além disso, nossos resultados sugerem que estudos de táxons que ocupam florestas secas podem fornecer novas perspectivas para entender a evolução de aves neotropicais.

Birds restricted to dry forest are among the most threatened in the Neotropics (Stotz et al. 1996). Neotropical dry forests are patchily distributed, and disjunct regions are often home to sister taxa. Such taxa typically have been considered "geographical representatives" and reduced to subspecies rank (Mayr 1982). Biogeographic and systematic studies of these allopatric populations are needed to identify isolated regions that contain strongly differentiated populations as a basis for setting conservation priorities.

Additionally, most studies of patterns of avian speciation in the lowland Neotropics (e.g., Haffer 1987) focus on species inhabiting rain forest or savanna/grassland. Evidence (summarized in Haffer 1993) suggests that tropical dry forests were more widespread during dry climatic-vegetational cycles in the Pleistocene and possibly during earlier climatic oscillations of the Tertiary as well. Reexamination of the systematics and biogeography of the dry forest avifauna offers the potential of a new perspective into avian evolutionary history in the Neotropics.

Taxa in the *Thamnophilus punctatus* complex are an important component of some dry forest avifaunas, but their systematics has received little attention recently. Earlier studies by Hellmayr (1924) and Zimmer (1933) were hampered by a paucity of specimens and behavioral information. Moreover, descriptions of the range of *T. punctatus* (e.g., Meyer de Schauensee 1966; Sibley and Monroe 1990) do not portray the patchy distribution of the complex accurately.

In the past two decades, substantial information has been obtained regarding distribution, morphology, vocalizations, habitat, and foraging behavior that provides a basis for reexamining systematics of the complex. Our revision aims to take into account all of these attributes, to provide objective and replicable classifications of taxa, and to help build a foundation for conservation of dry forest avifauna and examination of avian evolutionary history in the Neotropics from a dry forest perspective.

METHODS

Our analysis consisted of three principal steps: 1) initial diagnoses of taxa by mapping distributions and examining differences in external morphology; 2) comparison of vocalizations, habitat, and behavior among the taxa; and 3) identification of taxa as species or subspecies based on explicit criteria.

Distribution.—Collecting localities were identified in a comprehensive literature search and from inventories of collections at The Academy of Natural Science, Philadelphia (ANSP); American Museum of Natural History, New York (AMNH); Carnegie Museum, Pittsburgh (CM); Field Museum of Natural History, Chicago (FMNH); Louisiana State University Museum of Natural Science, Baton Rouge (LSUMZ); Museo Paraense Emílio Goeldi, Belém (MPEG); and National Museum of Natural History, Washington (USNM). Non-specimen records documented by tape recordings were added to this inventory. Localities were entered into a geographic information system (Isler 1997) and mapped. Sight records by experienced observers were also added, but these were designated as sight records on the distribution maps when they represented a range extension.

Plumage and measurements.—We surveyed collections of the aforementioned museums (except FMNH) to assess the range of plumage variation. To make detailed plumage comparisons and measurements, we randomly selected five males and five females of each currently recognized taxon and, additionally, of populations exhibiting plumages differences not expressed in named taxa. In describing plumages, "median tail spots," an important diagnostic character, refers to a white or light spot halfway down all or some inner rectrices. Alphanumeric color designations were based on Munsell® Soil Color Charts (1994 Edition; Macbeth Division of Kollmorgen Instruments Corp, Baltimore, Maryland, USA). The accompanying descriptive names were hyphenated to distinguish them (e.g., very-dark-brown). Alphanumeric codes, rather than descriptive names, should be used to refer to the Munsell charts because descriptive names were sometimes modified by us to portray the color more completely.

Measurements were made with MAX-CAL electronic digital calipers. Measurement methods followed Baldwin et al. (1931). Bill length was measured from anterior edge of the nares to the tip. Wing measurements are of the natural wing chord. Bill measurements were taken to hundredths of mm (rounded to tenths in tables), and other measurements were taken to tenths (rounded to mm in tables). As no evidence of sexual dimorphism was found in the measurements taken, we combined males and females in the sample for each taxon.

Establishment of taxonomic units.—Taxa were initially diagnosed by external morphology. To be "diagnosable," all individuals (one or both sexes) had to possess a morphological character that was sufficiently distinct to identify the taxon without comparison to other populations. Employing distributional data, we then determined whether taxa were allopatric or parapatric. Populations that represented clinal variants were not considered taxa.

Measurements were used to diagnose differences between pairs of taxa when both of two conditions were met. First, the ranges of measurements of the two samples could not overlap. Second, the means ($\bar{x}$) and standard deviations (SD) of the set of smaller measurements (a) and the set of larger measurements (b) had to meet the requirements of

$$\bar{x}_a + 1.88 SD_a \leq \bar{x}_b - 1.88 SD_b$$

If the distributions of both sets of measurements are normal, 97 percent of the measurements of the entire population of either taxon should lie outside the range of all the measurements of the other taxon (Amadon 1949).

Vocalizations.—We assume that vocalizations of antbirds, like those of other suboscines, are primarily, if not entirely, transmitted genetically (Kroodsma 1984, 1989; Kroodsma and Konishi 1991) and that they provide useful characters for systematic study, as in other suboscines (Lanyon 1978; Kroodsma 1984; Whitney 1994). An acoustic character was used to diagnose differences between taxa only when its occurrence or its measurement on a spectrogram permitted every individual recorded to be identified as one or the other taxon. To be considered diagnosable, measurements of vocal characters had to meet the same criteria, stated in the previous paragraph, as did measurements of morphological characters.

We classified vocalizations into three categories: "Loudsongs," "Rattles" including Rattlelike-calls, and "Calls." A Loudsong is a consistently patterned, multiple-note vocalization that is commonly repeated at regular intervals. A Rattle is a rapidly delivered series of sharp, grating notes lasting ca. 0.5–2.0 s. Rattlelike-calls are sharp, often grating calls given by taxa that do not deliver a typical Rattle. Both Rattles and Rattlelike-calls are usually delivered in repetitive sequences. Based on our field observations, we believe that both are given in the same behavioral context by taxa in the *T. punctatus* complex. They serve as a disturbance call by which we mean a vocalization delivered as the individual stops its normal behavior, such as foraging, and directs the vocalization at the appearance of a potential threat, including humans. In the text and figures, we identify both Rattles and Rattlelike-calls as "Rattles" in order to distinguish them from other Calls.

In describing vocalizations for the *T. punctatus* complex, we use the following terms: a "distinct terminal note" is the final note of a Loudsong that differs from preceding notes by being longer, usually more strongly accented, and sometimes structurally dissimilar; a "roll" is a series of rapidly delivered, similar notes, below 2.0 kHz in frequency; and a "trill" is structurally like a roll but is above 2.0 kHz in frequency and sounds qualitatively more musical.

To identify acoustic characters that serve to distinguish antbird species from one another (species-specific characters, sensu Becker 1982) and therefore are most relevant to studies of antbird systematics, we (Isler, Isler, and Whitney in prep.) are studying: (1) which vocal characters tend to vary individually or clinally, and (2) which characters distinguish vocalizations of closely related, sympatric species-pairs (e.g., *Thamnophilus murinus* and *T. schistaceus*). For example, we are finding that individual variation in Loudsongs usually involves shifts in overall frequency level or song length, i.e., the song pattern is the same but is delivered at a higher frequency or with more notes. On the other hand, species-specific variation typically consists of different patterns of change in frequency, delivery rate, and note shape within the Loudsong. These tentative conclusions are consistent with Becker's (1982) survey of studies of species-specific characteristics in bird sounds, although Becker's review did not incorporate any studies of suboscine species.

We assembled 246 recordings of taxa in the *T. punctatus* complex. Locations and recordists are listed in Appendix 2. Every recording was viewed in its entirety as a real-time spectrogram on a *Uniscan II* (Multigon Industries). Measurements were taken directly from the *Uniscan II*. Printouts were made of each type of vocalization from every locality.

Loudsong recordings normally included a number of song repetitions by the same individual. We typically measured characteristics of every song in the recording that was recorded well enough to analyze. Where the recording included more than five measurable songs, we took measurements on every third song in addition to the first three songs and the last song. The measurements taken always included number of notes and length of song. Other measurements depended on the pattern of the song. For example, if a song type accelerated or decelerated, time intervals between notes were measured. Frequencies were measured where the peaks of notes were most intense in amplitude. Length (s) of first five notes, important in the diagnoses, was measured from the beginning of the initial note to the beginning of the sixth note.

Typically, we found little variation between the Loudsongs of an individual in a recording (data available upon request from the authors). Therefore, the measurements of an individual's Loudsongs within a recording were combined to provide values for the individual used in analyses, and unless otherwise stated, sample sizes reflect numbers of individual birds recorded rather than numbers of songs.

We measured Calls differently depending on whether they were single notes, groups of similar or dissimilar notes, or patterned series of notes such as a rattle. Patterned Calls were measured to determine length, change in frequency, and rate of delivery. We noted whether each Call was repeated regularly or irregularly and obtained the number of Calls per minute for regularly repeated Calls. Based on these analyses, we determined whether Calls for each taxon were diagnosable.

Spectrograms used in illustrations were selected to reflect mean or median (e.g., number of notes) characteristics. These printouts were made using a Macintosh Centris 650 computer, Canary version 1.1 (Bioacoustics Research Program, Cornell Laboratory of Ornithology, Ithaca, New York), Canvas version 3.0.6 graphics software (Deneba Software, Miami, Florida), and a LaserWriter Pro 630 printer.

Habitat and behavior.—Over a 15-year period the authors observed habitat and behavior of all taxa in the *T. punctatus* complex except *huallagae* and *interpositus*. Our observations are supplemented by unpublished observations of Theodore A. Parker III and Kevin J. Zimmer as well as by information from the literature.

Criteria for assigning taxonomic rank.—To be considered species, allopatric populations had to exhibit diagnosable differences in *both* external morphology and vocalizations. This requirement reflects a desire to be conservative, given the current level of uncertainty as to which aspects of plumage and vocalizations indicate sufficient genetic divergence to consider a taxon a species. Allopatric populations diagnosable by either morphology or vocalizations, but not both, were considered subspecies, but no populations were found in the *T. punctatus* complex that were diagnosable by vocalizations and not morphology. Parapatric taxa diagnosable by morphology were assigned species rank even if we could not detect diagnosable vocal differences. This less stringent requirement was adopted because a lack of clinal intermediacy in morphology at the contact zone was taken to indicate that gene flow between the taxa is absent or highly restricted. Although we were also prepared to accept a small, stable cline as evidence for restricted gene flow, this condition was not found. Finally, named subspecies based on variants in an extensive cline were synonomized.

To apply these criteria, we used the comparative method, as exemplified by the foregoing discussion of determination of species-specific characters in vocalizations. If consistent measurable differences in a morphological or vocal character were known to occur in sympatric or non-interbreeding parapatric populations of related thamnophilids, then we accepted comparable differences in the same character in considering species status for allopatric populations. Conversely, we avoided using a character if differences in that character between allopatric populations were found to be within the range of individual or clinal variation of related taxa.

RESULTS

Distribution and morphology.—Eleven allopatric or parapatric (Fig. 1) taxa were found to be diagnosable by morphology as summarized in this section. Complete results of the analysis and detailed maps of geographic distribution are provided in Appendix 1.

Based on three plumage characters, the 11 taxa can be sorted into three groups (Fig. 2). Group I taxa differ from all other taxa in the dull crown color and pale-buff wing spots of the female and from Group III taxa by the absence of median tail spots on inner rectrices of males. Group I consists of two allopatric taxa occurring west of the Andes: *atrinucha* on the mainland and *gorgonae* on Gorgona Island off the coast of Colombia (Fig. 1). We found that *"subcinereus"* of northwestern Colombia represents the pale end of a cline. On the other hand, the reddish-brown plumage coloration of *gorgonae* lies outside the range of variation of the olive-brown coloration of *atrinucha* clines (Fig. 3).

Group II is comprised of six taxa, five of which are named. The sixth, *taxon nov.*, is described in Appendix 1. Morphologically, Group II taxa differ from Group I in that females have brightly colored crowns and white wing spots, and they differ from Group III in that males lack median tail spots on the inner rectrices (Fig. 2). Two of the Group II taxa, *stictocephalus* and *taxon nov.*, are found in close proximity to two Group III taxa, *sticturus* and *pelzelni* (Fig. 4). Although the historic collecting localities cannot be located precisely, type localities for *sticturus* and *stictocephalus* appear to be within 50 km of each other, along or near the Rio Guaporé. Because of the close proximity of these type localities, Hellmayr (1924) thought that *stictocephalus* was an individual variant of *sticturus*. In 1933, Zimmer resurrected the name *stictocephalus* by showing a female from Utiarity (see Fig. 4) resembled the type specimen and shared plumage characters with other specimens from an extensive region further north in southeastern Amazonia. Zimmer, however, was at a loss to explain the geographic proximity of these distinct forms and called for additional collecting near their type localities, a need that persists to this day.

Within the range of *stictocephalus*, we were not able to diagnose topotypical specimens of two named taxa, *"zimmeri"* and *"saturatus,"* from one another nor from specimens of *stictocephalus*. In Appendix 1, we describe a new taxon known only from the Serranía de Huanchaca, Bolivia, which differs from *stictocephalus* by the gray crown streaks and grayer coloration of the male and by the pale wing coverts and less reddish coloration of the upperparts of the female (Fig. 5).

Two other Group II taxa, *leucogaster* and *huallagae*, are geographically isolated in dry, rain-shadow valleys of northern Peru. The black crowns of males of both taxa typically contain concealed white spots, a unique character in the *T. punctatus* complex. Otherwise, *leucogaster* and *huallagae* differ: in particular, in *leucogaster* the male is extensively white below and the female has light-olive-brown upperparts; in *huallagae* the male is dark-gray below and the female

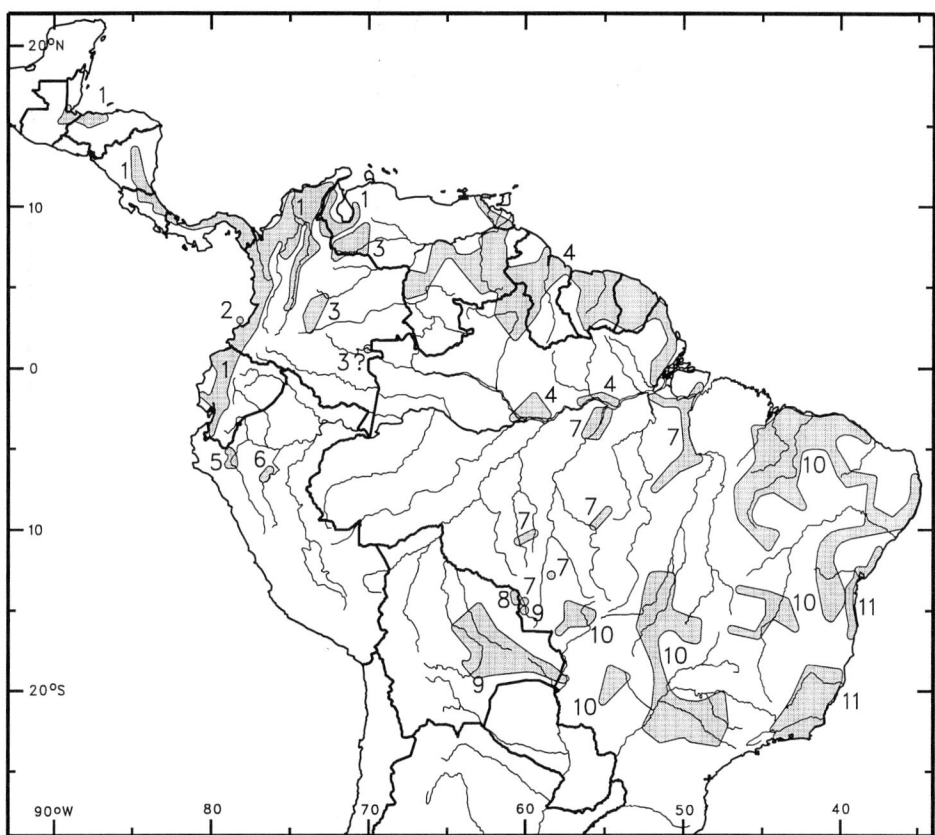

FIG. 1. Overview of the distribution of taxa in the *Thamnophilus punctatus* complex that are diagnosable by plumage: 1 = *atrinucha*. 2 = *gorgonae*. 3 = *interpositus*. 4 = *punctatus*. 5 = *leucogaster*. 6 = *huallagae*. 7 = *stictocephalus*. 8 = *taxon novum*. 9 = *sticturus*. 10 = *pelzelni*. 11 = *ambiguus*. Shading covers 50 km circles centered on sectors containing collecting stations (see Isler 1997) and also includes interstitial areas when circles are within 150 km of each other, but is constrained by elevational limits in the Andes and other mountainous regions. Spaces between shaded areas may be the result of insufficient field work in some regions. Figures 4 and 7 and maps in Appendix 1 provide additional detail.

dark-grayish-brown above (Fig. 5). In addition to these characters, bills are smaller in *leucogaster* than *huallagae*, and the two species can be diagnosed by bill depth, the only morphological measurement difference between geographically proximate taxa in the entire *T. punctatus* complex that met our criteria for diagnosability.

The remaining two taxa of Group II are found east of the Andes and north of the Solimões/Amazon River. In *interpositus*, the male is blacker and the female richer in color than in other Group II taxa (Fig. 5). Nominate *punctatus* is paler than *interpositus* and is most easily identified by the back of the male, which is entirely gray or shows only a few small black spots. The plumage of *punctatus* varies clinally with the palest individuals found to the east, in Amapá, Brazil, and the darkest in the western portion of its range.

Group III comprises three taxa in which males have median tail spots on some or all inner rectrices and females have brightly colored crowns and white wing spots (Fig. 2). Males of two of these taxa, *pelzelni* and *ambiguus*, have patches on both webs of the inner rectrices; males of the third, *sticturus*, have median patches only on the outer webs of some inner rectrices (Fig. 6).

The taxa *ambiguus* and *pelzelni* are most easily distinguished by underpart color of males and upper back color of females (Fig. 6). The two taxa appear to be parapatric in Bahia, Brazil, where *ambiguus* occurs on the coastal plain and *pelzelni* in adjacent uplands (Fig. 7). Zimmer (1933) reported that *pelzelni* and *ambiguus* intergrade in eastern Bahia, Brazil, but did not provide details. Naumburg (1937) examined plumages of *pelzelni* and *ambiguus* from eastern Brazil (including specimens examined by Zimmer) and did not find intermediates, although she found

Group	Taxon	Morphology ♀ crown	Morphology ♀ wing spots	Morphology ♂ inner median tail spots	Vocalizations Loudsong	Vocalizations Rattle or Rattlelike-call	Behavior Tail quiver
Group I	*atrinucha*	dull	Pale-buff	absent	Loudsong 1	Rattle 2	no
	gorgonae	dull	Pale-buff	absent	?	?	no
Group II	*punctatus*	bright	white	absent	Loudsong 2	Rattle 3	yes
	interpositus	bright	white	absent	Loudsong 2	Rattle 3	?
	leucogaster	bright	white	absent	?	Rattle 4	yes
	huallagae	bright	white	absent	?	?	?
	stictocephalus	bright	white	absent	Loudsong 4	Rattle 5	yes
	taxon novum	bright	white	absent	Loudsong 4	?	?
Group III	*sticturus*	bright	white	present	Loudsong 3	Rattle 1	yes
	pelzelni	bright	white	present	Loudsong 5	Rattle 1	yes
	ambiguus	bright	white	present	Loudsong 6	Rattle 1	yes

FIG. 2. Characters defining three groups of taxa within the *T. punctatus* complex. Full description of plumage coloration of female crowns and wing spots may be found in Appendix 1.

two females she could not assign clearly to either taxon. We examined these specimens (AMNH 242900 and 242912) and found them to be *pelzelni* based on underpart coloration (and supported by measurement differences as suggested in Appendix 1), although their upper backs are grayer than typical *pelzelni*. In all our museum visits, we searched for and did not find specimens intermediate between *pelzelni* and *ambiguus*.

Turning to the western edge of the distribution of *pelzelni*, this taxon and *sticturus* are not known to come into contact, although no major physical features separate their ranges (Fig. 4). No intermediates have been found, but it should be noted that this region is poorly studied ornithologically. Plumages of *pelzelni* and *sticturus* differ in extent of median tail spots, forecrown color of males, and female coloration (Fig. 6).

Loudsongs.—We found six types of Loudsongs (Fig. 8) in the *punctatus* complex. Loudsong 1 is an evenly paced or slightly accelerating series of similar notes ending with a distinct terminal note, whereas Loudsongs 2–6 accelerate substantially in pace and lack the distinct terminal note. Loudsong 1 is delivered by *atrinucha* ($n = 53$ individuals) throughout its range including northeastern Colombia (region of the named subspecies, "*subcinereus*") and Venezuela east of the Sierra de Perijá and north of the Andes. Songs from Middle America, northern Colombia, and

	atrinucha	*gorgonae*
♂ forecrown	Typically black spotted gray, but varying from solid black to solid gray (between N5 and N6)	Extensively gray to ca. 10 mm from base of bill. Compared to *atrinucha* specimens with gray forecrowns, those of *gorgonae* are paler (between N6 and N7) and feathers usually show numerous whitish shafts.
♀ upper back	Varying clinally from dark-olive-brown (2.5Y3/3) to olive-brown (2.5Y4/3).	Dark-reddish-yellow-brown (7.5YR3/4), often tinged more reddish.
♀ underparts	Varying clinally from olive-yellow (2.5Y6/6) to pale-yellowish-olive-brown (2.5Y7/4), throat whitish.	Brownish-yellow (10YR6/6) tinged reddish, only slightly paler on throat.

FIG. 3. Plumage characters defining Group I taxa.

	sticturus	*pelzelni*	*ambiguus*
♂ median tail spots	Confined to outer webs of 2-4 outermost rectrices on each side of the tail.	On both webs of all rectrices. (sometimes lacking on inner webs of rectrices of subadults).	
♂ forecrown	Gray to ca. 8 mm from base of bill.	Mixed gray and black; typically mostly gray. More gray in *pelzelni* than *ambiguus*, but difference not diagnosable.	
♂ underparts	Gray (N5 to N6) with center of belly extensively whitish.	Gray (usually N6, sometimes N7) with center of belly extensively whitish; some populations tinged buff.	Gray (N5 to N6) with whitish on belly absent or confined to spots or bars.
♀ upper back	Yellowish-red-brown (7.5YR4/6), almost as bright as crown.	Brown (7.5YR5/4 to 4/4) to reddish-brown (5YR4/4); some populations paler (10YR4/3).	Olive-brown (2.5Y4/3).
♀ underparts	Extensively white; sides yellowish-brown (10YR7/6) fading to very-pale-brownish-yellow (10YR7/4) on breast and crissum.	Yellow-ochre to brownish-yellow (10YR7/6 to 6/6 to 6/8 in some populations); center of underparts paler; some populations extensively white.	Light-yellowish-olive-brown (2.5Y5/4 to 6/4, sometimes paler).

FIG. 6. Plumage characters defining Group III taxa.

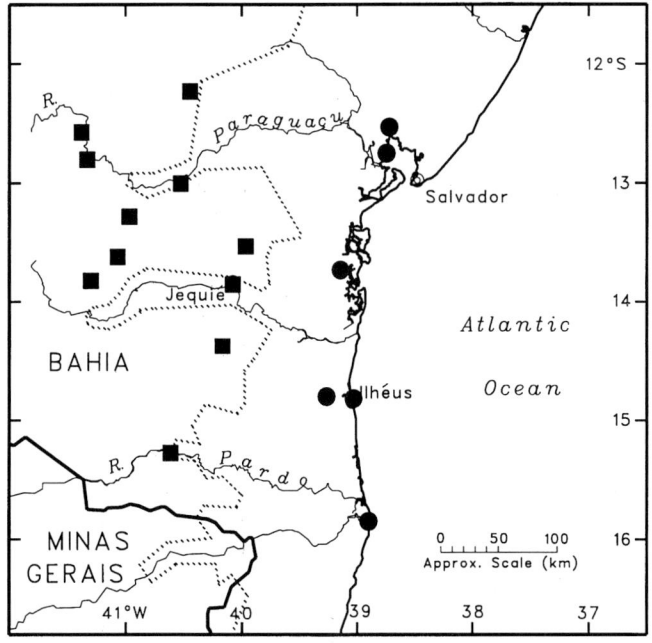

FIG. 7. Region of apparent parapatry between *pelzelni* and *ambiguus*. Black squares locate collecting stations at which *pelzelni* has been found; black circles = *ambiguus*. Dotted line is approximate 400 m contour.

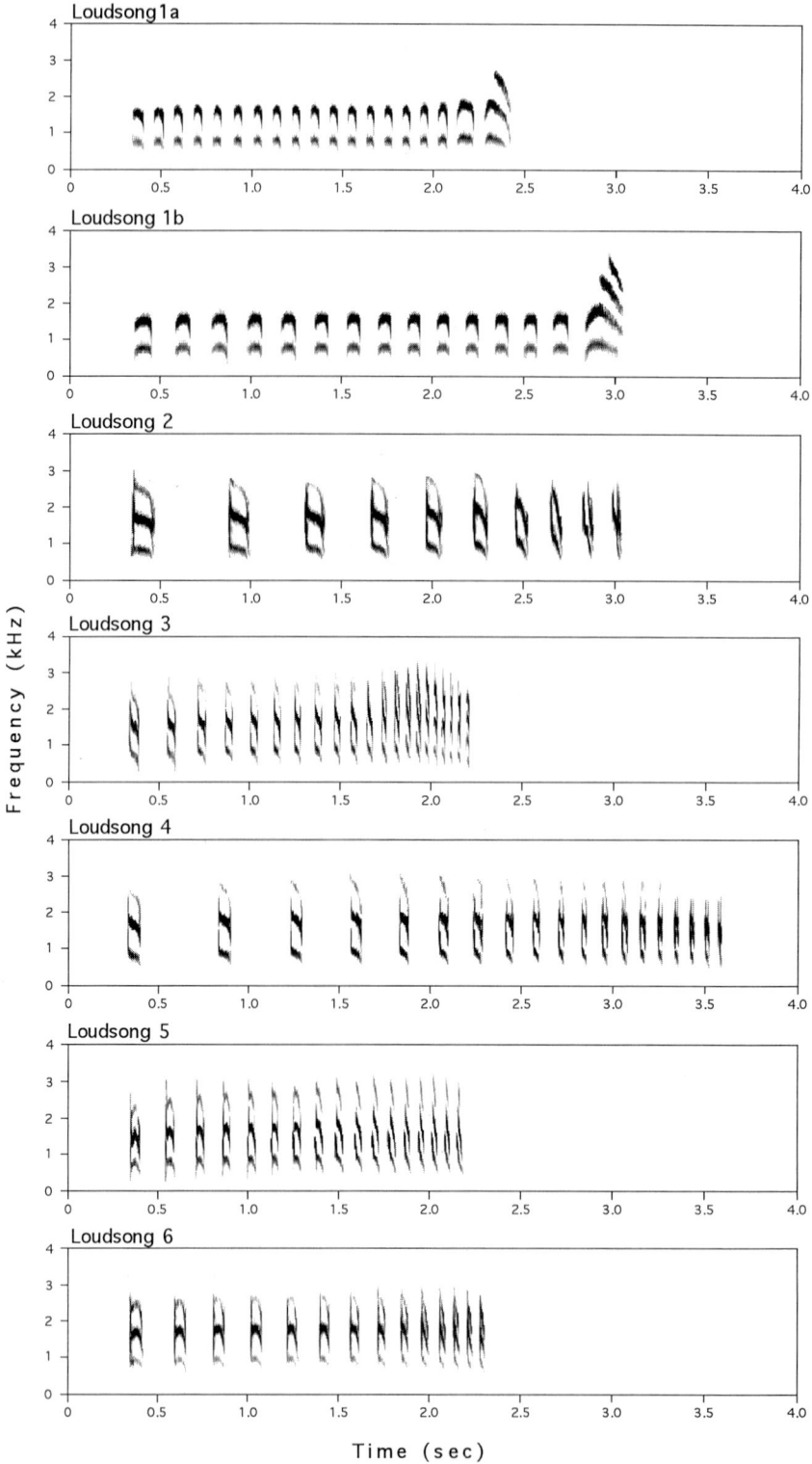

FIG. 8. Types of Loudsongs recorded from taxa in the *T. punctatus* complex.

		stictocephalus (Loudsong 4)	*pelzelni* (Loudsong 5)	*ambiguus* (Loudsong 6)
Sample size	*n* individuals	41	27	18
	n songs	166	88	63
Number notes per song	median	19	15	15
	range	10-34	10-24	10-24
Total length of song in sec	$\bar{x}$	3.1	1.7	2.0
	SD	0.66	0.30	0.26
	range	1.9-4.9	1.2-2.9	1.6-2.6
Total length in sec of first five notes	$\bar{x}$	1.57	0.82	1.04
	SD	0.28	0.08	0.11
	range	1.04-2.13	0.60-0.99	0.79-1.29

FIG. 9. Measurements of selected characteristics of Loudsongs 4, 5, and 6.

of note as delivered by *punctatus* and *interpositus*, has been recorded for *leucogaster* (Rattle 4, $n = 2$). The very distinctive Rattle of *stictocephalus* (Rattle 5; $n = 13$) is a growl extended into a roll that is delivered so rapidly that notes are barely distinguishable. We have no recordings of Rattles or Rattlelike-calls of *gorgonae*, *huallagae*, or *taxon novum*.

Calls.—A "caw" or "cow" Call (Fig. 11; Calls 1a and 1b) has been recorded for most taxa in the *T. punctatus* complex (total *n* for all taxa = 53). This type of call is given by many other *Thamnophilus* species (pers. observ.). Within the *T. punctatus* complex, this Call is individually variable, and the examples (1a and 1b) illustrate two of the forms that it may take.

Sample sizes for recordings of Calls other than the *caw* are currently too small to be evaluated

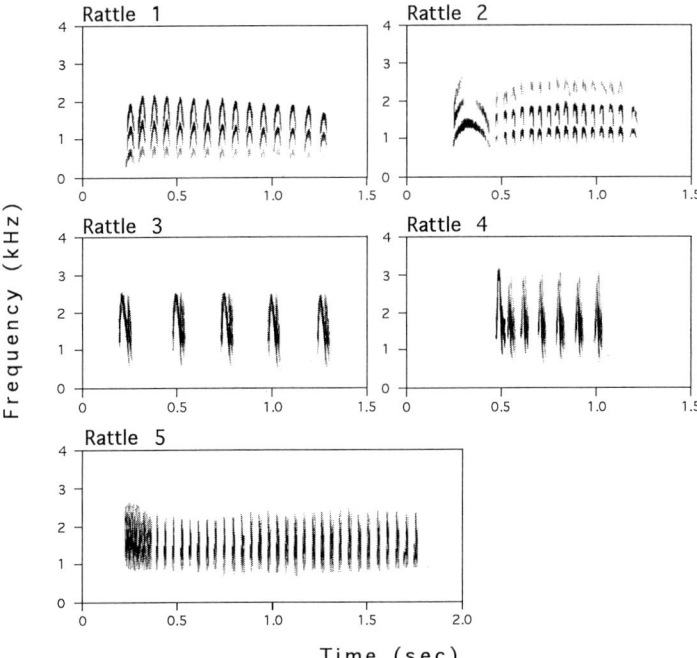

FIG. 10. Types of Rattles recorded from taxa in the *T. pumctatus* complex.

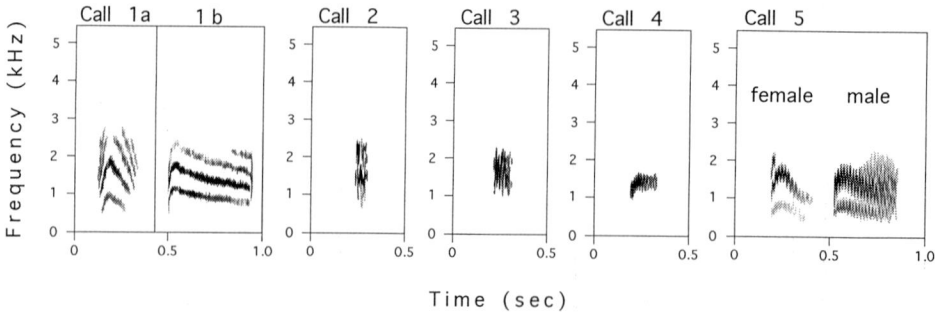

FIG. 11. Types of Calls recorded from taxa in the *T. punctatus* complex.

for taxonomic purposes. Call 2 (Fig. 11) is a short guttural vocalization recorded for *punctatus* (n = 3). The waveform of Call 3 is similar to that of Call 2 and has been recorded for *leucogaster* (n = 2) and *ambiguus* (n = 1). Call 4, given by *stictocephalus* (n = 15), is similar to Calls 2 and 3, but is often a longer note, rising in frequency, and covering a narrower frequency band (0.6 kHz versus 1.4–2.0 kHz). A raspy, guttural *caw*, Call 5, has been recorded for *sticturus* (n = 2) and *pelzelni* (n = 1). When sample sizes are larger, we suspect that Calls other than the typical thamnophiline *caw* will prove to be phylogenetically informative within the *punctatus* complex.

Habitat.—Taxa east of the Andes are most widespread in drier regions and, in more humid regions, they are largely restricted to pockets of vegetation growing on sandy or rocky soils. Taxa inhabiting southern Venezuela (*punctatus*) and southeastern Brazil (*ambiguus*) often are found at forest borders and light gaps. West of the Andes, *atrinucha* appears to inhabit a wider range of vegetation types than do forms occurring east of the Andes. Although the variety of habitats occupied by taxa in the *T. punctatus* complex will be described by the authors in a forthcoming monograph, habitat data are currently insufficient to be used in systematic studies.

Behavior.—The behavioral attributes, including foraging behavior, of taxa in the *T. punctatus* complex appear to be similar. Given current knowledge, only one behavior appears to have phylogenetic significance (Fig. 2). Whitney has observed that all taxa occurring east of the Andes (*huallagae* and *interpositus* have yet to be studied) quiver their tails regularly, especially just after they alight on a new perch, but that birds west of the Andes (*atrinucha*) lack this tail motion. Oniki (1975) first noted differences in call notes and behavior between populations of *T. punctatus* east and west of the Andes, but did not specify their nature. Willis (1984) noted tail quivering in some *punctatus* forms east of the Andes but interpreted it as a reaction to disturbance. Tail motion has been considered to reflect antbird phylogeny by Willis (1967, 1985) and Whitney and Rosenberg (1993).

Evidence from molecular studies.—Using allozyme data, Brumfield (1993) recently examined relationships between populations of 29 species (or species pairs) of Neotropical birds. Values for Nei's (1978) genetic distance (*D*) were obtained using tissue samples collected in 1) Central America west of the Canal, 2) the Chocó region (Darién, Panama, to western Ecuador), and 3) Amazonia. Within the *T. punctatus* complex, the genetic distance between populations of *atrinucha* from the Chocó and Central America was negligible (*D* = 0.003). In contrast, the genetic distances between the Chocó and Central American populations of *atrinucha* and *taxon novum* from extreme eastern Bolivia (*D* = 0.172 and 0.175 respectively) were among the largest reported within any sets of populations studied by Brumfield (although it should be noted that the Amazonian sample used in the *T. punctatus* complex was geographically more remote from the Chocó and Central America than the sample used in most other population sets). The genetic distance between *atrinucha* and *taxon novum* was comparable to mean genetic distances among species within three groups of antwrens (Hackett and Rosenberg 1990), within the antbird genus *Gymnopithys* (Hackett 1993), and within other antbird genera (K. Rosenberg, unpubl. data).

TAXONOMIC CONCLUSIONS

Taxa west of the Andes.—*Thamnophilus atrinucha* is considered specifically distinct from other forms in the complex based on differences of morphology, vocalizations (Loudsongs and Rattles), and tail movement. We consider the Gorgona Island population, *gorgonae*, to be a subspecies of *atrinucha*; the two taxa differ in plumage, but no recordings of the vocalizations of *gorgonae* were available. A currently named taxon, "*subcinereus*," of northeastern Colombia and northwestern Venezuela, could not be diagnosed by either plumage or vocalizations; therefore we synonymize it with *atrinucha*.

Taxa east of the Andes, north of the Amazon.—*Thamnophilus punctatus* is diagnosable by both plumage and vocalizations (Loudsongs and Rattles) from all other taxa except *interpositus*. Although they differ in plumage coloration, *interpositus* and *punctatus* could not be distinguished vocally, and consequently we consider *interpositus* to be a subspecies of *punctatus*. With regard to *interpositus*, it should be noted that all available recordings were made in Venezuela and that all specimens available to us were collected in Meta, Colombia. We were unable, therefore, to confirm that *interpositus* populations in Venezuela and central Colombia constitute a single taxon.

Taxa in northern Peru.—*Thamnophilus leucogaster* is diagnosable by plumage characters. The few recordings that we have of *leucogaster* indicate that its Rattle is distinctive. However, we consider the number of recordings to be insufficient for diagnosis. The coloration of *Thamnophilus huallagae*, especially the upper back of both sexes, is unique in the *punctatus* complex. Although the male's hidden white crown-patch is shared with the geographically proximate *leucogaster*, the external morphology of the two taxa are otherwise quite dissimilar. Without further knowledge of their vocalizations, we cannot recommend that *leucogaster* or *huallagae* be elevated to species rank, nor given the unique qualities of their plumages and geographic isolation, can we recognize them as subspecies of another form. We therefore consider *leucogaster* and *huallagae* to be taxa of uncertain rank.

Taxa in Bolivia and Brazil south of the Amazon.—We consider *Thamnophilus sticturus*, *T. stictocephalus*, *T. pelzelni*, and *T. ambiguus* to be separate species. *T. sticturus* differs from the other three taxa morphologically and vocally (Loudsong); it also appears to be parapatric with *T. stictocephalus* without evidence of intergradation. In addition to its possible parapatry with *sticturus*, *T. stictocephalus* differs in morphology and vocalizations (both Loudsongs and Rattles) from the other three taxa. The apparently parapatric taxa, *T. pelzelni* and *T. ambiguus*, are considered species based on morphological distinctions and the lack of clinal intermediacy in their contact zone. Although the Loudsongs of *pelzelni* and *ambiguus* are almost always identifiable, overlap in their characteristics precluded us from using Loudsongs to distinguish these two taxa.

The new taxon (see Appendix 1) is considered a subspecies of *T. stictocephalus* because although it is diagnosable by plumage characters, its vocalizations cannot be separated from those of *stictocephalus*. We were unable to diagnose either "*saturatus*" or "*zimmeri*" by plumage or vocalizations, and thus we synonymize them with *stictocephalus*.

Summary.—Thus, on the basis of existing information, we submit that the *Thamnophilus punctatus* complex consists of at least six species and possibly eight species as follows.

Thamnophilus atrinucha	Western Slaty-Antshrike
T. a. atrinucha	
T. a. gorgonae	
Thamnophilus punctatus	Guianan Slaty-Antshrike
T. p. interpositus	
T. p. punctatus	
Thamnophilus (p.)? leucogaster	(Maranon Slaty-Antshrike)
Thamnophilus (p.)? huallagae	(Huallaga Slaty-Antshrike)
Thamnophilus stictocephalus	Natterer's Slaty-Antshrike
T. s. stictocephalus	
T. s. taxon novum (described in Appendix A)	
Thamnophilus sticturus	Bolivian Slaty-Antshrike
Thamnophilus pelzelni	Planalto Slaty-Antshrike
Thamnophilus ambiguus	Sooretama Slaty-Antshrike

Note: (*p.*)? denotes a taxon of uncertain taxonomic rank. The English name in parentheses is proposed if the taxon is found to merit species rank.

DISCUSSION

In this section we employ knowledge of the *T. punctatus* complex to consider: 1) geographic variation of color intensity; 2) current maintenance of populations of Neotropical birds in dry forest mini-refugia; 3) future research potential.

Implications of geographic variation in color intensity.—Large differences in plumage color intensity (darkness or saturation) are evident within *T. punctatus* taxa. Male underparts differ

Fig. 12. Color key to plumages of taxa in the *Thamnophilus punctatus* complex. Female *atrinucha* is most easily diagnosed as its crown is olive brown with only a tinge of yellowish red-brown or brownish yellow (as in *n*), presenting little contrast with back, whereas other forms have crown bright red brown (as in *o*). Also, female *atrinucha* has wing and tail markings clearly tinged yellow (palest in *gorgonae* and "*subcinereus*"), whereas in other forms of the *punctatus* complex these markings are white or nearly so (compare *n* and *o*).

Extent of median tail spots distinguishes some taxa in both male and female plumages. Most taxa have median spots on outer rectrices only (*c*) whereas *ambiguus* and *pelzelni* have light median spots on both inner and outer webs of all rectrices creating a double row of spots that may be seen from above and below when tail is open (*d*); *sticturus* has median spots on the outer webs of most rectrices that may only be seen from above (*i*).

Otherwise, taxa are distinguished by a combination of three male characteristics (color of forecrown, back, and underparts) and three female characteristics (color of crown, back, and underparts) as follows. ***atrinucha* dark populations:** Male: forecrown *f* (becoming more heavily spotted with gray on Pacific slope south of Chocó, Col. and more like *e* on the eastern base of the S. Perijá, Ven.); back *u* (*t* east of S. Perijá); underparts like *r* but slightly darker on breast (slightly paler than *q* east of S. Perijá). Female: crown and back *n*; underparts like *k* but more extensively tinged brownish yellow including throat and belly. ***atrinucha* pale populations:** Male: forecrown like *f* but more spotted with gray; back *u*; underparts between *r* and *s*. Female: crown and back like *n* but paler; underparts like *m* but throat and belly very pale brownish yellow. ***gorgonae*:** Male: forecrown *h*; back *u*; underparts *r*. Female: crown and back like *n* but back color more like crown; underparts like *l* but throat and belly brownish yellow. ***punctatus*:** Male: forecrown *g*; back *v*; underparts like *r* but slightly darker, belly paler in Amapá. Female: crown *p* but slightly yellower; back *n* but slightly paler; underparts *k*. ***interpositus*:** Male: forecrown *e*; back *t*; underparts *q*. Female: crown *p*; back *n*; underparts like *k* but more brownish yellow, including throat. ***leucogaster*:** note small size and small bill. Male:

along the gray scale from white to nearly black; females exhibit similar variability in intensity. The plasticity of color intensity is most clearly illustrated in the geographic variation of *atrinucha* which may exemplify Gloger's Rule that populations in humid regions have darker coloration than those of dry regions. Other examples in the complex differing substantially in color intensity (but not in vocalizations) include the subspecies pairs: *T. p. punctatus* and *T. p. interpositus*, and *T. s. stictocephalus* and *T. s. taxon novum*. Such variation suggests that chromatic intensity in the complex is subject to relatively rapid change, probably in response to environmental conditions, and that caution should be employed when considering color intensity as a character in phylogenetic studies of the Thamnophilidae.

Maintenance of populations in mini-refugia.—East of the Andes, a number of taxa are locally distributed within the shaded areas delineated in Fig. 1. For example, along the Amazon River and its major tributaries, *T. stictocephalus* and *T. punctatus* are usually confined to isolated patches of white-sand scrub vegetation known as *campinarana* or *campina*; away from rivers these same species are typically restricted to vegetation bordering rocky outcroppings of hills and tepuis; and *T. punctatus* also occupies the narrow strip of sandy soil forest bordering the Atlantic Ocean in the Guianas. Other species east of the Andes, such as *T. sticturus* are found in tropical dry forests, especially those with an abundance of lianas, in less humid regions surrounding Amazonia.

Tropical dry forests were undoubtedly much more widespread in Amazonia during dry climatic periods (Haffer 1993). It is also reasonable to assume that elements of the *T. punctatus* lineage were more widespread and continuously distributed during these periods than they are today. With increasing rainfall and humidity, tall rain forests expanded until only remnants of dry forest vegetation remained in some regions, especially on pockets of sandy or rocky soil. Currently, in addition to members of the *T. punctatus* complex, other species occupying mini-refugia of relatively dry, non-grassland associations (*campinarana* and *campina*) in Amazonia include *Rhytipterna immunda* (Pale-bellied Mourner), *Elaenia ruficeps* (Rufous-crowned Elaenia), *E. chiriquensis* (Lesser Elaenia), *Hemitriccus margaritaceiventer* (Pearly-vented Tody-Tyrant), *Xenopipo atronitens* (Black Manakin), *Neopelma chrysocephalum* (Saffron-crested Tyrant-Manakin) and *Tachyphonus phoeniceus* (Red-shouldered Tanager), all of which, like taxa in the *T. punctatus* complex, occur in similar habitats beyond the limits of Amazonia, such as the Orinoco basin, the Chaco region centered in western Paraguay, and the *caatingas* of northeastern Brazil. Analysis of the *punctatus* complex suggests that a study of the biogeography of these habitat specialists (and some others that occur in grasslands within and outside of Amazonia, like *Rhynchotus rufescens* [Red-winged Tinamou] and *Chordeiles pusillus* [Least Nighthawk]) would shed new light on the history of Amazonia.

Future research potential.—The current distribution of the *T. punctatus* complex lends itself to speculation regarding its evolutionary history. The majority of characters distinguishing principal groups within the *T. punctatus* complex (Fig. 2) indicate that *T. atrinucha* represents one clade, which presumably was isolated by the Andes possibly in combination with subsequent changes in sea level (Haffer 1975), and that taxa east of the Andes appear to have diverged from one another following separation from the *atrinucha* clade. Beyond these conclusions, we believe that further construction of a phylogenetic tree for the complex requires more research including: 1) obtaining missing vocal and behavioral data, especially for *leucogaster* and *huallagae*; 2) determination of the closest relatives to the *T. punctatus* complex, possibly through molecular studies, and extension of all analyses to those taxa (candidates include taxa currently

---

←

forecrown like *e* but with a few gray feathers adjoining bill; back between *u* and *v*; underparts like *s* but throat grayer. Female: crown and back like *p* but back slightly less reddish; underparts similar to *m* but brownish yellow areas are replaced by light olive brown. **huallagae**: Male: forecrown *e*; back between *u* and *v*; underparts *r*. Female: crown and back *o*; underparts like *j* but slightly paler and tending towards *k* in color. **sticturus**: note small size and small bill. Male: forecrown *h*; back between *u* and *v*; underparts *s*. Female: crown and back like *p* but back less reddish; underparts *m*. **stictocephalus**: Male: forecrown *f*; back varying between *t* and *u*; underparts varying between *q* and *r*. Female: crown and back like *p* but back sometimes darker; underparts varying between *j* and *k*. **parkeri**: Male: forecrown *h* but gray extends though crown in streaks and variably in posterior feather edges; back *v*; underparts between *r* and *s*. Female: crown and back like *p* but paler; underparts like *k* but paler with white center of belly. **pelzelni**: Male: forecrown *g*; back between *u* and *v*; underparts *s* (sometimes tinged brown). Female: crown and back *p*; underparts *l*. **ambiguus**: Male: forecrown *f*; back *u*; underparts like *r* but with some whitish feathers on belly. Female: crown *p*; back *n*; underparts like *k* but more brownish yellow.

included in *Thamnophilus amazonicus*, *T. nigrocinereus*, and *T. insignis*); 3) continuing study (currently underway by the authors) of the vocalizations of thamnophilids to assess the value of various vocal characteristics as species-specific characters; and 4) further analysis of the complex utilizing internal morphological or molecular characters. As suggested in the previous section, the geographic distribution of the *T. punctatus* complex may shrink into refugia in periods when the distribution of humid forest species expands, and the pursuit of research into its evolutionary history has the potential of affording a different perspective of avian evolution in the Neotropics.

ACKNOWLEDGMENTS

This paper is dedicated to the memory of Theodore A. Parker III, with whom we were so fortunate to have shared many years and whose teachings continue to inspire us. It was Ted who initially stimulated our efforts to understand geographic variation in the *T. punctatus* complex after he returned from the Serranía de Huanchaca with the first recordings of what is now *T. stictocephalus taxon novum*.

We are grateful to J. M. Bates, who led the expedition that first collected *taxon novum*, G. R. Graves, S. J. Hackett, T. S. Schulenberg, D. F. Stotz, and K. J. Zimmer, all of whom made useful suggestions as they reviewed various drafts of this paper. During its development, this study also benefitted from discussions of emerging findings with M. J. Braun, R. T. Brumfield, J. Haffer, D. E. Kroodsma, J. F. Pacheco, K. V. Rosenberg, and J. M. C. da Silva. R. C. Banks helped us deal with questions of taxonomic nomenclature. Stotz, J. V. Remsen Jr., and C. A. Marantz made numerous helpful comments in their reviews of the sumitted manuscript.

The development of the criteria used to assign taxonomic rank was stimulated by the many helpful comments of Banks, Bates, Braun, Graves, Haffer, Remsen, Schulenberg, K. Winker, and R. L. Zusi. The criteria are the responsibility of the authors and not necessarily consistent with the views of commentators.

Braun and Brumfield confirmed our behavioral observations of *atrinucha* in Panama. M. Cohn-Haft examined plumages of specimens at LSUMZ. C. W. Thompson helped and provided measurements at ANSP. We thank the curators of AMNH, ANSP, CM, FMNH, LSUMNH, and MPEG for permission to examine specimens in their care, and S. W. Cardiff, Remsen, M. B. Robbins, Thompson, and D. E. Willard for expediting loans of critical specimens. At the USNM, P. Angle, R. Browning, J. Dean, and C. Dove provided important assistance.

We thank the many recordists, identified in Appendix 2, who provided recordings. Many of these are archived at the Library of Natural Sounds, Cornell Laboratory of Ornithology, where G. Budney, R. Grotke, and others on the staff provided important support. Additional recordings are archived at the National Sound Library, British Library (R. Ranft, Curator, Wildlife Section) and at the Florida Museum of Natural History (J. Hardy and T. Webber). E. S. Morton, National Zoological Park, loaned us equipment vital to the analysis of vocalizations.

LITERATURE CITED

AMADON, D. 1949. The seventy-five per cent rule for subspecies. Condor 51:250–258.
BALDWIN, S. P., H. C. OBERHOLSER, AND L. G. WORLEY. 1931. Measurements of birds. Sci. Publ. Cleveland Mus. Nat. Hist. 2:1–165.
BATES, J. M. 1993. The genetic effects of forest fragmentation on Amazonian forest birds. Ph.D. dissertation, Louisiana State University, Baton Rouge, Louisiana.
BATES, J. M., T. A. PARKER III, A. P. CAPPARELLA, AND T. J. DAVIS. 1992. Observations on the *campo*, *cerrado* and forest avifaunas of eastern Depto. Santa Cruz, Bolivia, including 21 species new to the country. Bull. British Ornith. Club 112:86–98.
BECKER, P. H. 1982. Species-specific characteristics of bird sounds. Pages 213–252 *in* Acoustic Communication in Birds, Vol. I. (D. E. Kroodsma and E. H. Miller, Eds.). Academic Press, New York.
BRUMFIELD, R. T. 1993. Historical diversification of birds in the Chocó area of endemism: a molecular perspective. M.S. thesis, Dept. Biol. Sci., Illinois State Univ., Normal, Illinois.
HACKETT, S. J. 1993. Phylogenetic and biogeographic relationships in the Neotropical genus *Gymnopithys* (Formicariidae). Wilson Bull. 105:301–315.
HACKETT, S. J., AND K. V. ROSENBERG. 1990. Comparison of phenotypic and genetic differentiation in South American antwrens (Formicariidae). Auk 107:473–489.
HAFFER, J. 1975. Avifauna of northwestern Colombia, South America. Bonner Zool. Monogr., no. 7. Bonn, Germany.
HAFFER, J. 1987. Biogeography of Neotropical birds. Pp. 105–150 *in* Biogeography and Quaternary History of Tropical America (T. C. Whitmore and G. T. Prance, eds.). Clarendon and Oxford Univ. Press, New York, New York.
HAFFER, J. 1993. Time's cycle and time's arrow in the history of Amazonia. Biogeographica 69:15–45.

HELLMAYR, C. E. 1906. Revision der Spix'schen Typen brasilianischer Vögel. Abhandlungen der Königlich Bayerischen Akademie der Wissenschaften 22:563–726.
HELLMAYR, C. E. 1924. Catalogue of birds of the Americas. Pteroptochidae-Conopophagidae-Formicariidae. Field Mus. Nat. Hist. Zool. Ser. 13, Pt. 3, 1–369.
ISLER, M. L. 1997. A sector-based ornithological geographic information system for the Neotropics. Pp. 345–354 in Studies in Neotropical Ornithology Honoring Ted Parker (J. V. Remsen, Jr., Ed.), Ornithol. Monogr. No. 48.
KROODSMA, D. E. 1984. Songs of the Alder Flycatcher (*Empidonax alnorum*) and Willow Flycatcher (*Empidonax traillii*) are innate. Auk 101:13–24.
KROODSMA, D. E. 1989. Male Eastern Phoebes (*Sayornis phoebe*, Tyrannidae, Passeriformes) fail to imitate songs. J. Comparative Psychology 103:227–232.
KROODSMA, D. E., AND M. KONISHI. 1991. A suboscine bird (eastern phoebe, *Sayornis phoebe*) develops normal song without auditory feedback. Anim. Behav. 42:477–487.
LANYON, W. E. 1978. Revision of the *Myiarchus* flycatchers. Bull. Amer. Mus. Nat. Hist. 161:429–627.
MAYR, E. 1982. Of what use are subspecies? Auk 99:593–595.
MEYER DE SCHAUENSEE, R. 1966. The Species of Birds of South America. Livingston, Wynnewood, Pennsylvania.
MEYER DE SCHAUENSEE, R., AND W. H. PHELPS, JR. 1978. Birds of Venezuela. Princeton University Press, Princeton, New Jersey.
MONROE, JR., B. L. 1989. The correct name of the Terek Sandpiper. Bull. British Ornith. Club 109:106–107.
NAUMBURG, E. M. B. 1937. Studies of birds from eastern Brazil and Paraguay, based on a collection by Emil Kaempfer. Bull. Amer. Mus. Nat. Hist. 74:139–205.
ONIKI, Y. 1975. The behavior and ecology of Slaty Antshrikes (*Thamnophilus punctatus*) on Barro Colorado Island, Panama Canal Zone. Anais Acad. Brasil. Ciênc. 47:477–515.
PACHECO, J. F., AND B. M. WHITNEY. 1997. On the origin of some birds collected by George Such, and the type localities of several forms. Auk 114:303–305.
PHELPS, W. H., AND W. H. PHELPS, JR. 1963. Lista de las aves de Venezuela con su distribución. Part 2. Passeriformes. Bol. Soc. Venez. Cienc. Nat. 24:1–479.
ROCHA O., O. 1990. Lista preliminar de aves de la reserva de la biósfera "Estación Biológica Beni." Ecologia en Bolivia 15:57–68.
SIBLEY, C. G., AND B. L. MONROE, JR. 1990. Distribution and Taxonomy of Birds of the World. Yale Univ. Press, New Haven, Connecticut.
STOTZ, D. F., J. W. FITZPATRICK, T. A. PARKER III, AND D. K. MOSCOVITS. In press. Neotropical Birds: Ecology and Conservation. University of Chicago Press, Chicago, Illinois.
WHITNEY, B. M. 1994. A new *Scytalopus* tapaculo (Rhinocryptidae) from Bolivia, with notes on other Bolivian members of the genus and the *magellanicus* complex. Wilson Bull. 106:585–614.
WHITNEY, B. M., AND G. H. ROSENBERG. 1993. Behavior, vocalizations, and possible relationships of *Xenornis setifrons* (Formicariidae), a little-known Chocó endemic. Condor 95:227–231.
WILLIS, E. O. 1967. The behavior of Bicolored Antbirds. Univ. California Publ. Zool. 79:1–132.
WILLIS, E. O. 1984. Antshrikes (Formicariidae) as army ant followers. Papéis Avulsos Zool. (São Paulo) 35:177–182.
WILLIS, E. O. 1985. *Cercomacra* and related antbirds (Aves, Formicariidae) as army ant followers. Rev. Bras. Zool. 2:427–432.
WILLIS, E. O. 1988. Behavioral notes, breeding records, and range extensions for Colombian birds. Rev. Acad. Colombiana Cienc. Exactas Fisicas Nat. 16:137–150.
ZIMMER, J. T. 1933. Studies of Peruvian Birds. No. 10. The Formicarian genus *Thamnophilus*. Part 2. Amer. Mus. Nov. 647:1–27.

APPENDIX 1

MORPHOLOGICAL CHARACTERS, DISTRIBUTION, AND SYNONYMIES

This appendix provides further information regarding the morphological characters and distributional data that were used in the initial definition of the 11 taxa. Lists of specimens examined only include specimens for which measurements were taken; many other specimens were examined during the course of this study. Plumage characters are defined primarily in charts; only highlights and supplementary information are provided in the text. Plumage comparisons are generally restricted to contrasting taxa within each of the three groups identified in Figure 2. In the synonymies, all taxa are referenced to Hellmayr (1924), and names synonymized by Hellmayr are not listed as they may be found in that volume. Similarly, type localities and sources are given in the synonymies only for post-1924 references.

Six measurements are provided in Figure 13. Because the rules allowing measurements to be used in diagnoses were stringent, measurements were found to differ sufficiently to differentiate only one pair of taxa in the *T. punctatus* complex. However, other size differences among taxa are evident in specimens. In this appendix, in sections identified as "Additional measurement information," we present the outcomes of tests for significance of measurement differences for samples of pairs of closely-related taxa and for geographically-proximate taxa without respect to group. These tests were made using independent samples *t*-tests. Because of the large number of pairwise comparisons undertaken, differences are reported only for probabilities of .01 or less, and because of the small sample sizes, the results are reported as "typically has".

Taxon	Bill width	Bill depth	Bill
Group I			
atrinucha West Andes & Perijá n = 30	4.5-5.6 (5.18, 0.20)	5.7-6.5 (6.05, 0.24)	11.5-13.6 (12.6, 0.55)
East of Perijá n=16	4.6-5.6 (5.21, 0.30)	5.8-6.7 (6.33, 0.28)	11.8-13.1 (12.3, 0.45)
gorgonae n = 9	5.0-5.5 (5.27, 0.17)	6.3-6.9 (6.58, 0.21)	12.3-13.7 (12.9, 0.57)
Group II			
interpositus n = 8	4.8-5.2 (4.98, 0.16)	6.0-6.6 (6.31, 0.18)	9.8-12.0 (10.9, 0.56)
punctatus	4.9-5.7 (5.14, 0.25)	5.9-6.4 (6.17, 0.21)	10.3-12.6 (11.3, 0.75)
leucogaster	4.3-4.7 (4.50, 0.17)	5.3-5.9 (5.60, 0.23)	9.8-10.7 (10.3, 0.30)
huallagae	4.8-5.4 (4.96, 0.18)	6.2-6.9 (6.48, 0.23)	10.3-12.0 (11.0, 0.61)
stictocephalus n = 24	4.4-5.6 (4.98, 0.29)	5.6-6.9 (6.10, 0.27)	10.1-12.3 (11.0, 0.58)
taxon nov. n = 13	4.4-5.4 (5.02, 0.38)	5.5-6.4 (6.06, 0.22)	9.9-11.4 (10.5, 0.62)
Group III			
sticturus	3.9-4.6 (4.23, 0.20)	5.4-6.0 (5.69, 0.20)	8.7-10.5 (9.6, 0.50)
pelzelni	3.9-4.5 (4.27, 0.22)	5.3-6.2 (5.83, 0.31)	9.4-10.5 (9.9, 0.35)
ambiguus	4.7-5.4 (4.92, 0.27)	5.8-6.7 (6.11, 0.34)	10.2-11.7 (10.8, 0.49)

Taxon	Tarsus	Tail	Wing
Group I			
atrinucha West Andes & Perijá n = 30	19-22 (21.1, 0.7)	51-59 (55.4, 1.6)	64-73 (68.2, 2.8)
East of Perijá n=16	20-22 (21.4, 0.6)	52-56 (54.4, 1.3)	64-68 (66.6, 2.0)
gorgonae n = 9	21-24 (22.6, 0.9)	57-60 (57.9, 1.0)	66-72 (69.4, 1.7)
Group II			
interpositus n = 8	21-22 (21.7, 0.4)	56-58 (57.2, 0.8)	66-70 (68.1, 1.6)
punctatus	21-22 (21.7, 0.6)	52-56 (53.9, 1.4)	64-69 (66.2, 1.7)
leucogaster	20-22 (21.3, 0.5)	53-58 (55.6, 1.9)	64-68 (65.2, 1.4)
huallagae	21-23 (22.0, 0.9)	52-58 (55.3, 1.7)	67-70 (68.8, 1.2)
stictocephalus n = 24	20-23 (21.9, 0.7)	55-66 (59.2, 2.8)	66-72 (68.7, 1.3)
tax. nov. n = 13	21-23 (21.6, 0.5)	56-59 (57.6, 1.1)	63-69 (66.3, 2.0)
Group III			
sticturus	21-24 (22.5, 0.6)	51-56 (53.8, 1.4)	62-67 (64.9, 1.6)
pelzelni	22-25 (23.4, 0.9)	50-58 (53.4, 2.7)	63-67 (65.3, 1.7)
ambiguus	22-26 (23.6, 0.9)	54-60 (58.6, 3.1)	66-70 (67.8, 1.4)

FIG. 13. Measurements. Sexes pooled; $n = 10$ unless noted. The large sample sizes obtained for *atrinucha* and *stictocephalus* reflect an earlier concern that measurements might vary geographically within these taxa; no significant differences were found. First number in parentheses = mean; second number in parentheses = standard deviation.

Although not used in the diagnoses, these measurement differences are presented for their supporting value in confirming identification of specimens and as a starting point for future analysis.

Group I.—With respect to plumage, differs from Groups II and III in that crowns of females are dull (dark-yellowish-brown [10YR 3/4] to dark-brown [7.5YR 3/4] slightly tinged reddish-brown) and the bold wing and tail spots of females are pale-buff (10YR8/3–8/4), occasionally lighter (10YR8/2) in pale individuals or darker (10YR7/4) in dark individuals. Also differs from Group III in lacking median tail spots on inner rectrices of males.

Thamnophilus a. atrinucha Salvin and Godman

Thamnophilus atrinucha Salvin and Godman, 1892:200.
Thamnophilus punctatus subcinereus Todd, 1915:80.
Thamnophilus punctatus atrinucha Hellmayr, 1924:90.

Plumage characters.—Figure 3 describes the characters distinguishing *T. a. atrinucha* and *T. a. gorgonae*.

Additional measurement information.—See *T. a. gorgonae* for measurement comparison with Group I. The large bill of *T. atrinucha* is evidenced by the lack of overlap between bill lengths measured for *T. atrinucha* and most taxa in Groups II and III. The bill length of *T. punctatus* is closest, and although the ranges of measurements of bill lengths of *T. atrinucha* and *T. punctatus* overlap, the mean bill length of *T. punctatus* typically is smaller ($t = 5.546$, df $= 30$, $P < .001$).

Distribution.—Figure 14. The Caribbean slope in southeastern Guatemala, extreme southern Belize, northwestern Honduras, Nicaragua, Costa Rica, and Panama. The Pacific slope from Coclé, Panama, eastward and thence southward through Colombia and Ecuador to southwestern El Oro. Northern Colombia eastward to the western slope of the Serranía de Perijá and southward in the Magdalena Valley to Tolima. North of the Andes and east of the Serranía de Perijá in Norte de Santander, Colombia, and in Zulia, Táchira, Mérida, and Trujillo, Venezuela. Elevation: lowlands; to 500 m in Guatemala, Honduras, and Nicaragua; to 1,000 m in Costa Rica and Panama; and to 1,200 m in Venezuela, Colombia, and Ecuador.

Specimens examined.—PANAMA: Cana (USNM 1 ♂).—Cerro Chicanti (USNM 1 ♀).—Chiriqui Lagoon (USNM 1 ♂, 1 ♀).—"Panama" (USNM 1 ♀).—Río Jaque (USNM 3 ♂).—Río Payo (USNM 1 ♀).—Río Tuquesa (USNM 1 ♀). COLOMBIA: Antioquia: Puerto Valdivia (USNM 1 ♂). Bolívar: La Raya (USNM 1 ♂).—Regeneración (USNM 1 ♀).—Santa Rosa (USNM 1 ♂, 1 ♀).—Simití (USNM 1 ♂). Cesar: Ayacucho (USNM 1 ♀).—Caracolicito (USNM 1 ♂).—La Esperanza (USNM 1 ♂). Chocó: Nuquí (USNM 2 ♀)—Río Jurubidá (USNM 1 ♂). Córdoba: Pueblo Nuevo (USNM 1 ♂).—Río Salvajín (USNM 1 ♂).—Socorré (USNM 1 ♂, 1 ♀). La Guajira: Los Gorros (USNM 1 ♀). Santander: Hacienda Santana (USNM 1 ♀). Santander del Norte: Bellavista (USNM 1 ♂, 1 ♀).-Petrolea (USLM 2 ♂, 3 ♀). Sucre: Colosó (USNM 1 ♀). Valle del Cauca: Río San Juan (USNM 1 ♀). VENEZUELA: Mérida: Santa Elena (CM 1 ♂, 2 ♀). Trujillo: Sabana de Mendoza (CM 3 ♂, 1 ♀). Zulia: Guachi (CM 3 ♂).

Remarks.—The extensive geographic variation in color intensity is summarized in Figure 14. Note that the blackest males are found in Santander del Norte, Colombia, on the east side of the Serranía de Perijá, whereas the palest males occur on the west side of the same mountain range. We found "*subcinereus*" to be the extreme pale variant on the cline that stretches across northern Colombia. Gray markings (occasionally found in black crowns of males in Ecuador) and white spots (found in crowns of some specimens in northern Colombia) appear to be associated with immaturity.

Thamnophilus atrinucha gorgonae Thayer and Bangs

Thamnophilus gorgonae Thayer and Bangs, 1905:95.
Thamnophilus punctatus gorgonae Hellmayr, 1924:89.

Plumage characters.—Figure 3 describes the characters distinguishing *T. a. gorgonae* and *T. a. atrinucha*.
Additional measurement information.—Compared to *atrinucha*, *gorgonae* typically has a deeper bill ($t = 5.884$, df $= 37$, $P < .001$), longer tarsi ($t = 5.228$, df $= 37$, $P < .001$), and a longer tail ($t = 4.021$, df $= 37$, $P < .001$). Like *atrinucha*, *gorgonae* typically has a longer bill than other taxa in Groups II and III.
Distribution.—Figure 14. Isla Gorgona, Cauca, Colombia.
Specimens examined.—COLOMBIA: Isla Gorgona (AMNH 1 ♂, 2 ♀♀; CM 1 ♂, 1 ♀; USNM 2 ♂, 2 ♀)
Remarks.—Overall plumage coloration of male resembles *atrinucha* specimens at mid-level of color saturation.

Group II.—With respect to plumage, differs from Group III by males lacking median tail spots except on outer webs of outer rectrices. Differs from Group I by females having the crown bright yellowish-red-brown (7.5YR 4/6) and wing spots white.

Thamnophilus p. punctatus (Shaw)

Lanius punctatus Shaw, 1809:3.27.
Thamnophilus p. punctatus Hellmayr, 1924:92.

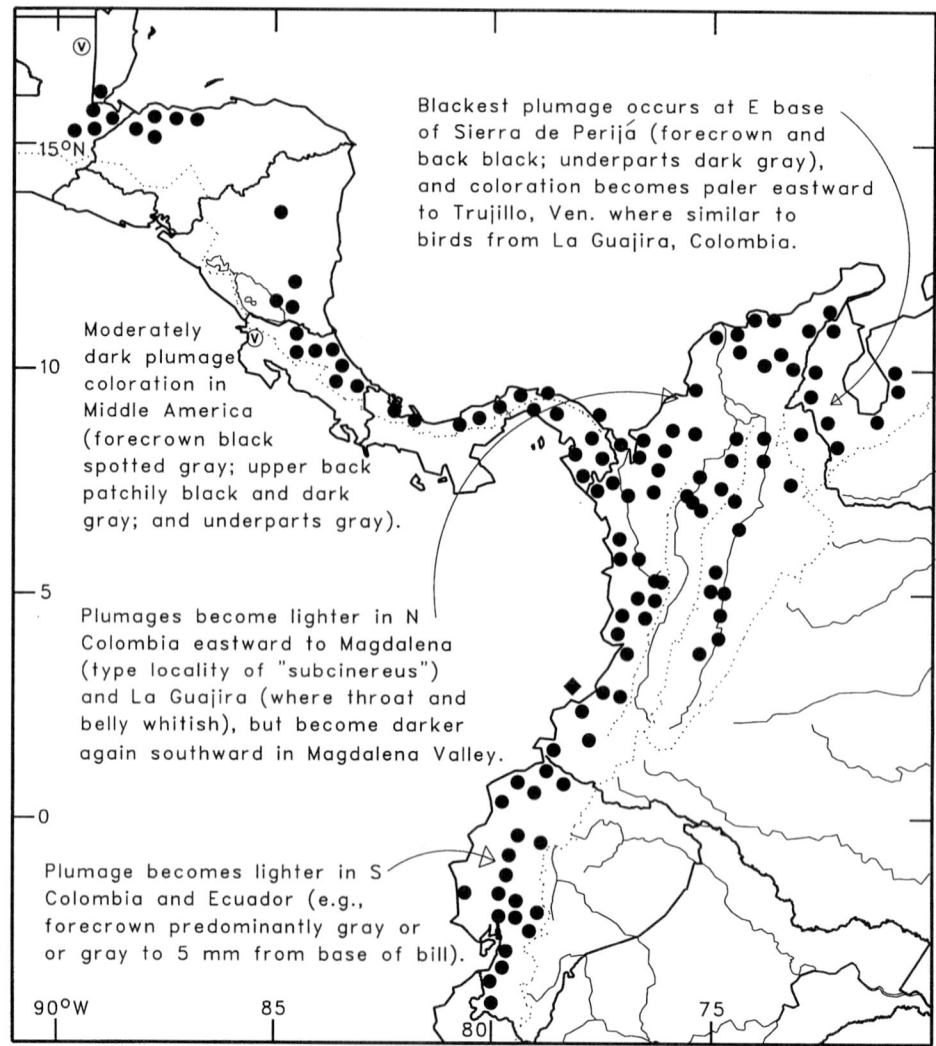

Fig. 14. Distribution of *Thamnophilus atrinucha* with notes on clinal variation of male plumage coloration. Sector locations are designated by: black circle = *T. a. atrinucha*; black diamond = *T. a. gorgonae*; V in a circle = undocumented sight record. The fine lines of dots represent cordilleran divides.

Plumage characters.—Figure 5 describes the characters distinguishing *T. p. punctatus* from other Group II taxa.

Additional measurement information.—Compared to *T. s. stictocephalus*, *T. p. punctatus* (including *interpositus*) typically has a shorter tail ($t = 4.868$, df = 40, $P < .001$) and a shorter wing chord ($t = 3.365$, df = 40, $P = .002$). Also see *T. a. atrinucha* and *T. punctatus interpositus*.

Distribution.—Figure 15. Extreme eastern Venezuela, north of the Orinoco in eastern Sucre and Delta Amacuro; Venezuela south of the Orinoco in northern Amazonas and northern and eastern Bolívar; eastward through the Guianas and thence southward through Amapá, Brazil, to near the mouth of the Amazon, and thence westward along the north bank of the Amazon through Pará and extreme western Amazonas, Brazil, to the west bank of the Rio Negro opposite Manaus. Elevation: mostly lowlands below 1000 m but occurs to 1500 m on Venezuelan tepuis (Phelps and Phelps 1963).

Specimens examined.—BRAZIL: Amapá: Rio Tracajatuba (USNM 2 ♂, 2 ♀). Amazonas: Hacienda Rio Negro (AMNH 1 ♀). Pará: São José, Rio Nhamunda (AMNH 1 ♂). VENEZUELA: Amazonas: Río Cataniapo (AMNH 1 ♂). Bolívar: Río Arabopó (AMNH 1 ♀). Sucre: Ensenada Cariaquito (ANSP 1 ♂, 1 ♀).

Remarks.—Plumage, especially underparts, becomes paler clinally eastward to French Guiana and Amapá, Brazil.

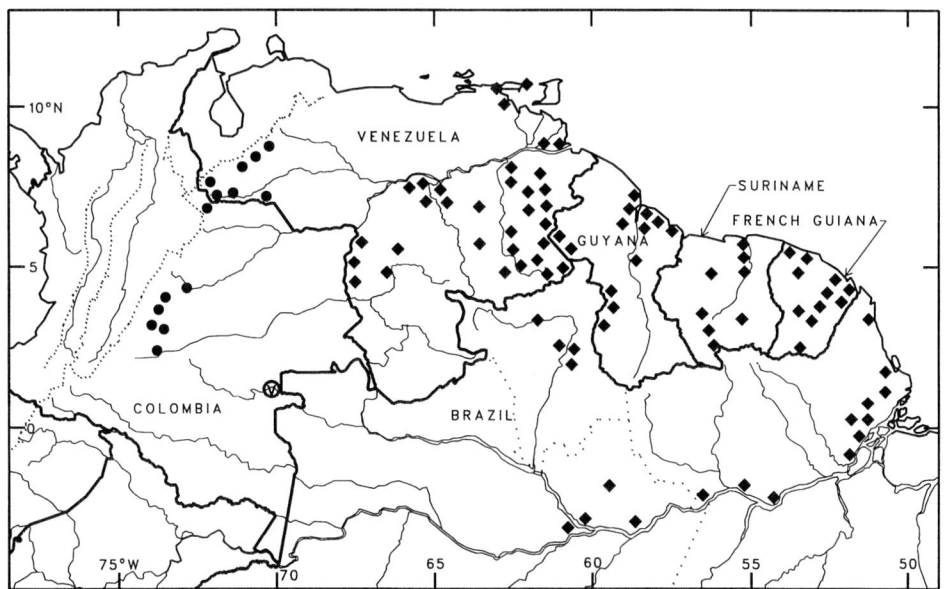

FIG. 15. Distribution of *Thamnophilus punctatus*. Sector locations are designated by: black diamond = *T. p. punctatus*; black circle = *T. p. interpositus*; V in a circle = sight record, probably *interpositus*. The fine line of dots represent cordilleran divides; the more widely spaced lines of dots are Brazilian state boundaries.

Thamnophilus punctatus interpositus Hartert and Goodson

Thamnophilus punctatus interpositus Hartert and Goodson, 1917:496; Hellmayr, 1924:94.

Plumage characters.—Figure 5 describes the characters distinguishing *T. p. interpositus* from other Group II taxa.

Additional measurement information.—Compared with *punctatus*, *interpositus* typically has a longer tail ($t = 5.775$, d.f. = 16, $P < .001$).

Distribution.—Figure 15. South of the Andes in southwestern Táchira, western Apure, and Barinas, Venezuela, and adjacent Arauca, Colombia. Also near the base of the Andes in Meta, Colombia. A sight record in eastern Vaupés, Colombia, appears to be of this subspecies (see Willis 1988). Elevation: 100–500 m.

Specimens examined.—COLOMBIA: Meta: "Meta" (ANSP 1 ♂, 1 ♀).—Río Duda (AMNH 1 ♂, 1 ♀).—Puerto Barrigón (AMNH 1 ♂, 1 ♀).—Villavicencio (AMNH 1 ♂, 1 ♀).

Remarks.—We were unable to examine or measure Venezuelan specimens of *interpositus*. Description of its distribution in Venezuela follows Meyer de Schauensee and Phelps (1978). It is conceivable that plumages of the possibly allopatric populations in Venezuela and Meta, Colombia, differ. Moreover, additional collecting is needed in the region between the ranges of *interpositus* and *punctatus* to ascertain whether these taxa come into contact and whether intermediates occur.

Thamnophilus (p.)? leucogaster Hellmayr

Thamnophilus naevius albiventris Taczanowski, 1884:9.
Thamnophilus punctatus leucogaster Hellmayr, 1924:94.

Plumage characters.—Figure 5 describes the characters distinguishing *T. (p.)? leucogaster* from other Group II taxa. Eleven of fourteen males examined had concealed white spots in the crown. In addition to differences in plumage characters, the bill of *leucogaster* is diagnostically less deep than that of *huallagae*.

Additional measurement information.—Compared to *huallagae*, *leucogaster* typically has a narrower bill ($t = 5.917$, d.f. = 18, $P < .001$), shorter bill ($t = 3.301$, d.f. = 17, $P = .004$), and a smaller wing ($t = 5.989$, df = 18, $P < .001$).

Distribution.—Figure 16. Middle Río Marañón valley, Cajamarca and Amazonas, Peru. Elevation: 200–1200 m.

Specimens examined.—PERU: Cajamarca: Cabico (AMNH 1 ♂).—Jaén (AMNH 1 ♂).—Perico (AMNH 1 ♂).—San José de Lourdes (LSUMZ 2 ♂, 2 ♀).—Santa Cruz, 28 km north of (LSUMZ 3 ♀). Additional males were examined to determined the extent of white spots in the crown.

Remarks.—Resembles *stricturus* of Group III, but in addition to lacking median tail spots, male differs by

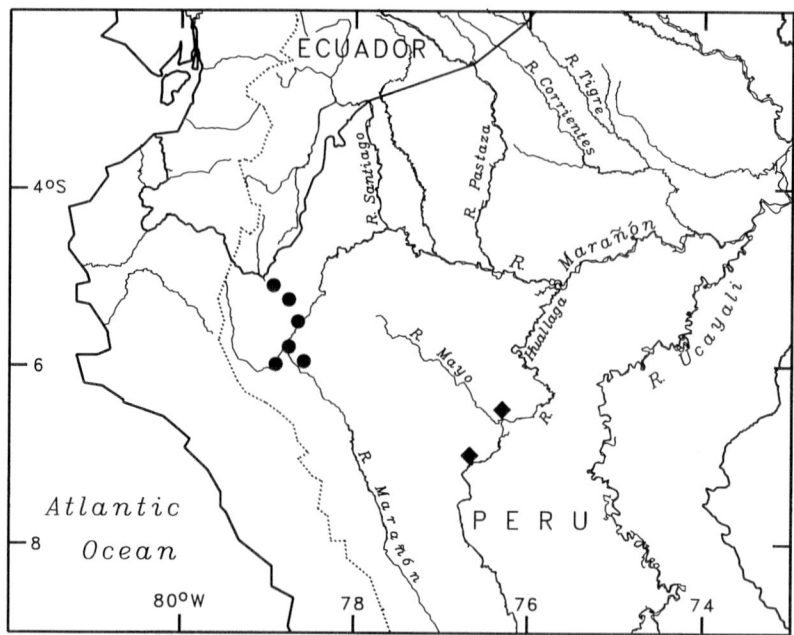

FIG. 16. Distribution of *Thamnophilus (p.)? leucogaster* (black circles) and *T. (p.)? huallagae* (black diamonds). Individual collecting localities are portrayed although some pairs of closely spaced collecting stations of *T. (p.)? leucogaster* are represented by single symbols. The fine line of dots represents the Andean divide.

having hidden crown spots (lacking in *sticturus*) and a black forecrown (gray in *sticturus*); female has less reddish back than *sticturus*, and sides are light-olive-brown (yellowish-brown in *sticturus*).

Zimmer (1933) restored the original name of *albiventris* to this taxon, but we believe that Hellmayr was correct in renaming it. *Thamnophilus albiventris* Taczanowski and *T. albiventer* Spix (= *Taraba major stagurus* [Spix]) were concurrently placed in the same genus (in Hellmayr 1906) and therefore primary homonymy exists (Monroe 1989).

Thamnophilus (p.)? huallagae Carriker

Thamnophilus amazonicus huallagae Carriker, Proc. Acad. Nat. Sci. Philadelphia, 86, 1934:324; El Tingo, San Martín, Peru.
Thamnophilus punctatus huallagae Meyer de Schauensee, Species Birds South Amer., 1966:272.

Plumage characters.—Figure 5 describes the characters distinguishing *T. (p.)? huallagae* from other Group II taxa. Occasionally females of other taxa may appear grayish on underparts, but they lack the dark-grayish-brown back of female *huallagae*. Four of five males examined had concealed white spots in the crown.

Additional measurement information.—See *leucogaster*.

Distribution.—Figure 16. West slope of the middle Río Huallaga valley, San Martín, Peru. Elevation: ca. 200–900 m.

Specimens examined.—PERU: San Martín: Shapaja (ANSP 2 ♀♀).—Tingo de Saposoa (ANSP 5 ♂, 2 ♀; CM 1 ♀).

Thamnophilus s. stictocephalus Pelzeln

Thamnophilus stictocephalus Pelzeln, 1869 (1868):77, 146. (?) *Thamnophilus punctatus sticturus* [in part] Hellmayr, 1924:96.
Thamnophilus punctatus saturatus Todd, Proc. Biol. Soc. Washington, 40, 1927:153; Villa Braga = Vila Braga, left bank Rio Tapajos, Pará, Brazil.
Thamnophilus punctatus stictocephalus Zimmer, Amer. Mus. Novit., 647, 1933:13.
Thamnophilus punctatus zimmeri Pinto, Arq. Zool. Est. São Paulo, 5, 1947:446; Rio Pracupi, tributary to right bank of the Amazon between the Xingu and the Tocantins, Pará, Brazil.

Plumage characters.—Figure 5 describes the characters distinguishing *T. s. stictocephalus* from other Group II taxa.

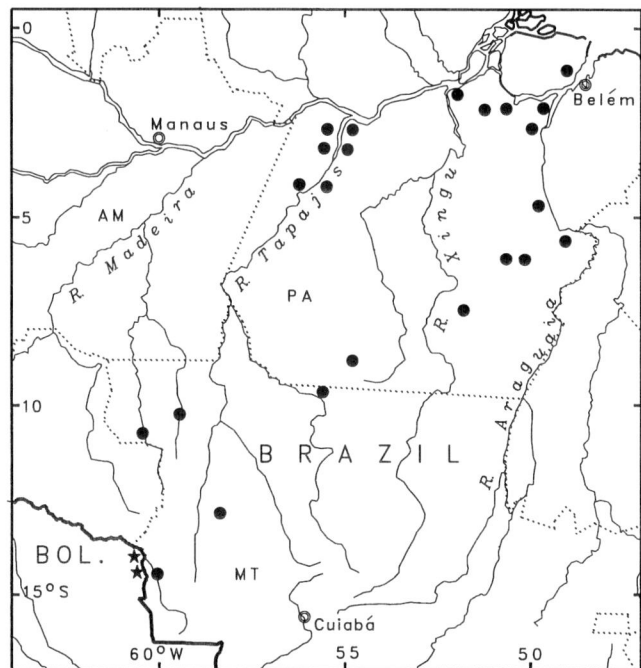

FIG. 17. Distribution of *Thamnophilus stictocephalus*. Sector locations are designated by: black circle = *T. s. stictocephalus*; black star = *T. s. parkeri*. Dotted lines are Brazilian state boundaries; two-letter codes identify Brazilian states.

Additional measurement information.—Compared to *T. p. punctatus*, *T. s. stictocephalus* typically has a longer tail ($t = -4.868$, d.f. = 40, $P < .001$) and a longer wing ($t = -3.365$, d.f. = 40, $P = .002$). Also see the next taxon.

Distribution.—Figure 17. Locally distributed in central Brazil, south of the Amazon, with possibly isolated populations extending from both banks of lower Rio Tapajós eastward to the Ilha de Marajó (MPEG specimens) and the west bank of the lower Rio Tocantins, and thence southward to the Serra dos Carajás (Whitney recording) and Serra do Cachimbo, Pará. In Mato Grosso, from the Rio Cristalino north of Rio Teles Pires opposite Alta Floresta (Whitney recording) westward to the Rio Aripuanã drainage and southward to the Rio Guaporé. Elevation: near sea level along Amazon to ca 700 m in Mato Grosso.

Specimens examined.—BRAZIL: Pará: Aramanaí (AMNH 1 ♂).—Diamantina (USNM 3 ♂, 2 ♀).—Itaituba (ANSP 1 ♀; CM 3 ♂, 1 ♀).—Mocajuba (AMNH 1 ♂, 1 ♀).—Baião (AMNH, 1 ♂, 1 ♀)—Tapará (AMNH 2 ♀).—Vila Braga (ANSP 1 ♂; CM 4 ♀).—Vilarinho do Monte (AMNH 1 ♂, 1 ♀).

Remarks.—The type specimen of *T. s. stictocephalus* is in poor condition (Hellmayr 1924) and lacks the tail feathers necessary to identify the bird with certainty; more collecting is needed from the vicinity of the type locality to confirm that the range of *stictocephalus* extends to the Rio Guaporé. Two named taxa synonymized herein (*"saturatus"* and *"zimmeri"*) were described by their authors in comparison to *T. p. punctatus* and without reference to each other or to specimens of *T. s. stictocephalus*. Our examination of the type specimen of *"saturatus"* and specimens called *"saturatus"* by Zimmer (1933), and topotypical specimens of *"zimmeri"* revealed substantial individual variation in color intensity, and we were unable to diagnose differences between these and other populations of *stictocephalus*. Specimens from Ilha de Marajó in the MPEG, identified by us as *stictocephalus*, show some tendency in plumage coloration towards *punctatus*, and vocal recordings are needed of that apparently isolated population. Some females are entirely yellowish-olive-brown below (lacking the whitish throat and belly); this appears to be a subadult plumage.

In previous discussion in this paper we have referred to *taxon novum*, an undescribed population whose plumage is diagnosably distinct but whose vocalizations could not be separated from those of *stictocephalus*. We are now prepared to describe this form as:

Thamnophilus stictocephalus parkeri
subsp. nov.

Holotype.—LSUMZ 150730, adult male from Serranía de Huanchaca, 21 km southeast of Catarata Arco Iris (see Bates et al. 1992), Santa Cruz, Bolivia, elevation 670 m. Collected on 1 July 1989 by J. M. Bates,

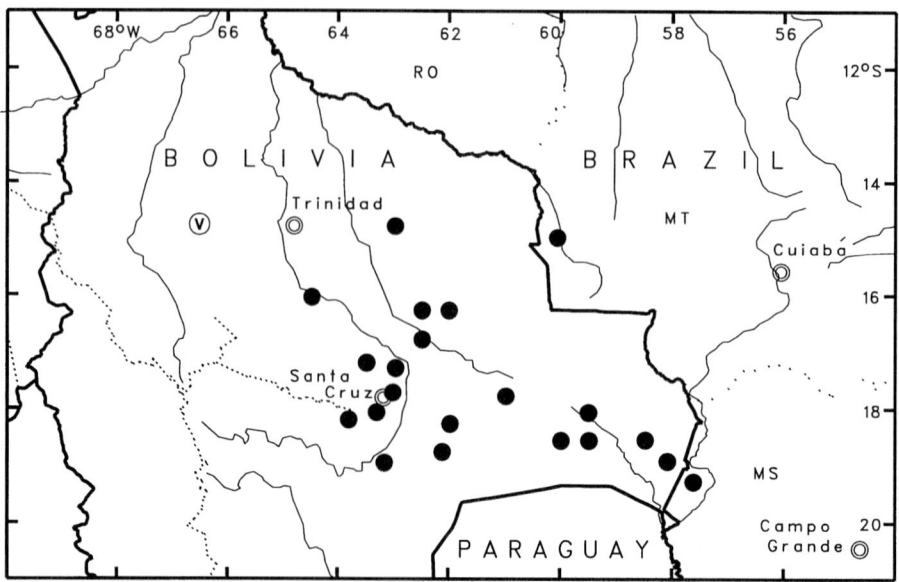

FIG. 18. Distribution of *Thamnophilus sticturus*. Sector locations are designated by: black circle = *T. sticturus*; V in a circle = undocumented record. Lines of dots are Brazilian state boundaries; two-letter codes identify Brazilian states.

original number 626. Skull 100 percent ossified; weight 21.5 g; iris red; toes and tarsi gray; bill black, maxilla with a light base.

Plumage characters.—Figure 5 describes the characters distinguishing *T. stictocephalus parkeri* from other Group II taxa. Of the males examined, two specimens had crown feathers completely edged gray, and six specimens (including the holotype) had crown feathers partially edged gray. In addition to the characters listed in Figure 5, the female differs from female *T. s. stictocephalus* by having paler greater wing coverts. Those of *T. s. parkeri* are dark-yellowish-brown (10YR4/6 to 3/6) with very-dark-brown (approaching blackish-brown) subapical spots, whereas entire greater coverts (except tips) of *T. s. stictocephalus* are very-dark-brown (10YR2/2 to 2/1).

Measurements of holotype.—Bill width 4.3; bill depth 6.4; bill 10.5; tarsus 21.2; tail 59.1; wing 67.1.

Additional measurement information.—Compared to *stictocephalus*, *parkeri* typically has shorter wings ($t = 3.479$, d.f. = 28, $P = .002$). Compared to *sticturus*, *parkeri* typically has a wider bill ($t = 4.233$, d.f. = 16, $P < .001$), deeper bill ($t = 3.589$, d.f. = 16, $P = .002$), longer bill ($t = 9.617$, d.f. = 16, $P = .006$), and a longer tail ($t = 5.716$, d.f. = 16, $P < .001$). Compared to *pelzelni*, *parkeri* typically has a wider bill ($t = 5.076$, d.f. = 14, $P < .001$) and a shorter tarsus ($t = 4.425$, d.f. = 14, $P = .001$).

Distribution.—Figure 17. Known only from top of the plateau of the Serranía de Huanchaca, northeastern Santa Cruz, Bolivia. Elevation: 650–725 m.

Specimens examined.—BOLIVIA: Santa Cruz: 21 km southeast of Catarata Arco Iris (LSUMZ 7 ♂, 3 ♀).—25 km southeast of Catarata Arco Iris (LSUMZ 2 ♀).—45 km east of Florida (LSUMZ 1 ♀). The crown feathers of an additional male specimen was examined.

Etymology.—Named for Theodore A. Parker III, who discovered this population and who first tape-recorded its loudsong.

Remarks.—Specimens of a few obviously subadult males of other taxa in the *punctatus* complex show some gray in the crown. All seven male specimens of *parkeri* were noted as having skulls 100% ossified, although two of them have brownish primaries, an element of pre-definitive plumage in *Thamnophilus* spp. Of the two specimens with brownish primaries, the crown of one had extensive gray feather edgings, and the other had the least gray in the crown of any of the seven males studied. The forest fragments atop the plateau on which *parkeri* was collected are isolated by grassland from continuous forest at the edge of the plateau (see Bates [1993] for description of region). Whitney tape-recorded but did not collect a form of *stictocephalus* in forest at the northern base of the Serranía (south bank of the Río Itenez) whose subspecific identity is unknown.

Group III.—With respect to plumage, differs from Groups I and II by presence of median tail spots on all or some of the inner rectrices of males (absent or reduced in females). Differs from Group I by brightly colored crowns of females (reddish-yellow-brown [7.5YR4/6] in *pelzelni* and *ambiguus*; yellowish-red-brown [5YR4/6] in *sticturus*) and by white wing spots.

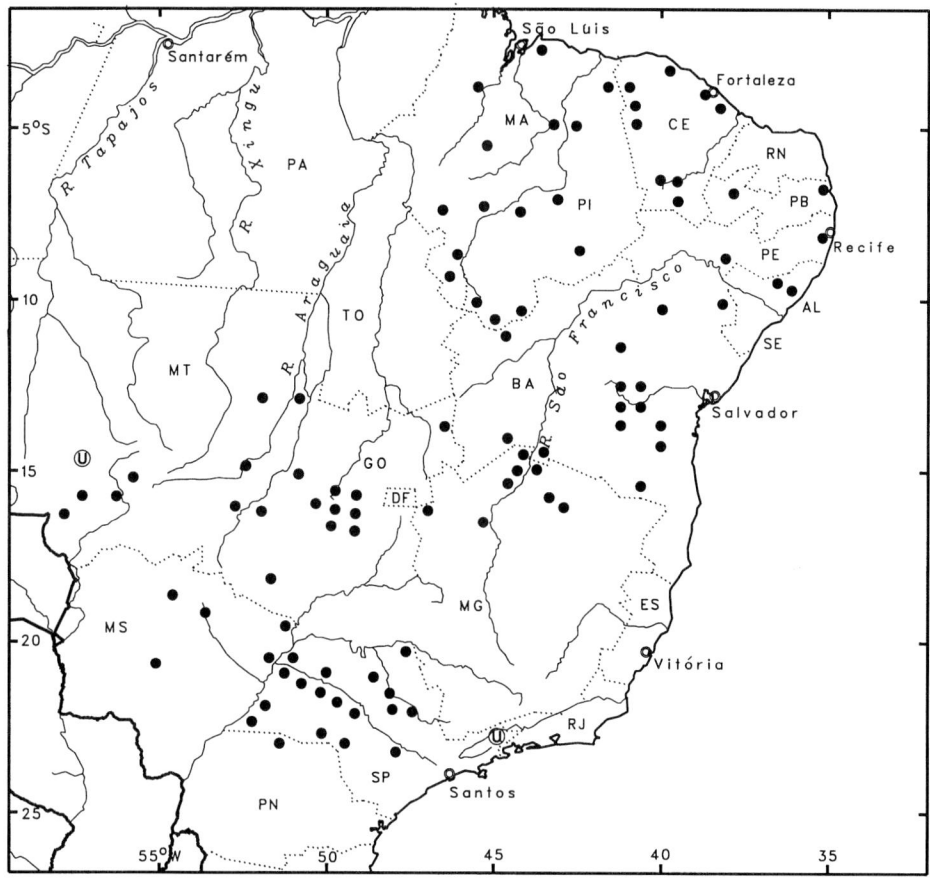

FIG. 19. Distribution of *Thamnophilus pelzelni*. Sector locations are designated by black circles. Lines of dots are Brazilian state boundaries; two-letter codes identify Brazilian states.

Thamnophilus sticturus Pelzeln

Thamnophilus sticturus Pelzeln, 1868: 76, 144.
Thamnophilus punctatus sticturus Hellmayr, 1924: 95.

Plumage characters.—Figure 6 describes the characters distinguishing *T. sticturus* from other Group III taxa.

Additional measurement information.—Compared with *T. pelzelni*, *T. sticturus* typically has shorter tarsi ($t = 2.963$, df = 20, $P = .008$). Also see measurements under *T. stictocephalus parkeri*.

Distribution.—Fig. 18. Extreme eastern Cochabamba and Santa Cruz (except north), Bolivia, eastward to extreme southwestern Mato Grosso, Brazil. One undocumented record for Beni, Bolivia (Rocha 1990). Elevation: 300-900 m.

Specimens examined.—BOLIVIA: Santa Cruz: Río Quizer (LSUMZ 1 ♂, 3 ♀).—Río Tucuvaca (LSUMZ 1 ♂, 1 ♀).-Santiago de Chiquitos (LSUMZ 4 ♂, 2 ♀).

Remarks.—See remarks under *T. leucogaster* (Group II).

Thamnophilus pelzelni Hellmayr

Thamnophilus punctatus pelzelni Hellmayr, 1924: 96.

Plumage characters.—Figure 6 describes the characters distinguishing *T. pelzelni* from other Group III taxa.

Additional measurement information.—Compared with *T. ambiguus*, *T. pelzelni* typically has a narrower bill width ($t = 5.835$, d.f. = 18, $P < .001$; note ranges in our sample do not overlap), a shorter bill ($t = 4.713$, d.f. = 18, $P < .001$), a shorter tail ($t = 3.950$, d.f. = 18, $P = .001$), and a shorter wing chord ($t = 3.630$, d.f. = 18, $P = .002$). Also see *T. stictocephalus parkeri*, *T. sticturus*.

Distribution.—Figure 19. Possibly discontinuously distributed in Brazil from Maranhão, Piauí, Ceará, and

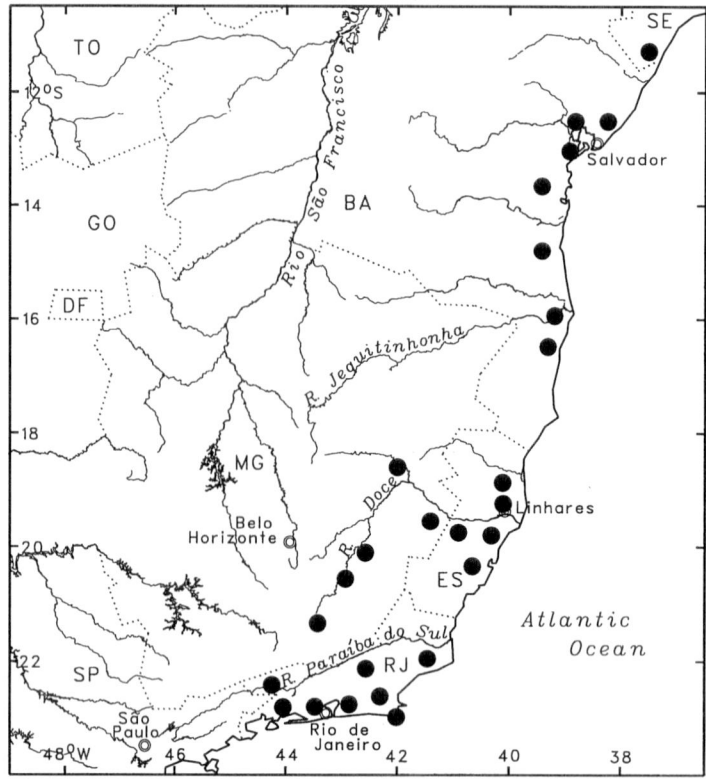

FIG. 20. Distribution of *Thamnophilus ambiguus*. Sector locations are designated by black circles. Lines of dots are Brazilian state boundaries; two-letter codes identify Brazilian states.

Paraíba southward to southern Bahia (except coastal zone in Bahia and southern Sergipe), northwest Minas Gerais, and São Paulo (except along coast) and westward to eastern and southern Mato Grosso and central Mato Grosso do Sul. Elevation: mostly 400–800 m (close to sea level in the northeast and occasionally to 1,100 m in the planalto).

Specimens examined.—BRAZIL: Bahia: Jaguaquara (USNM 1 ♂).—Morro do Chapéu (AMNH 1 ♀). Goiás: Aragarças (USNM 1 ♂, 1 ♀). Mato Grosso: Chapada dos Guimarães (USNM 2 ♂♂, 2 ♀). São Paulo: Rincão (USNM 1 ♂, 1 ♀).

Remarks.—Many questions remain about whether possibly disjunct populations of *T. pelzelni* come into contact, and as noted in Figure 6, populations are variable in coloration (MPEG specimens and J. M. C. da Silva, pers. comm.). Further data collection and a fine-grained analysis of distribution and plumage coloration are needed to understand geographic variation in *T. pelzelni*.

Thamnophilus ambiguus Swainson

[*Thamnophilus naevius*] var. a? *T. ambiguus* Swainson, 1825:91.
Thamnophilus punctatus ambiguus Hellmayr, 1924:97.

Plumage characters.—Figure 6 describes the characters distinguishing *T. ambiguus* from other Group III taxa.

Additional measurement information.—See *T. pelzelni*.

Distribution.—Figure 20. Coastal zone of Brazil from extreme southern Sergipe (Whitney recording) southward through Bahia and Espírito Santo to Rio de Janeiro; inland in the Rio Doce valley to eastern Minas Gerais. Elevation: lowlands below 400 m.

Specimens examined.—BRAZIL: (location unspecified) (USNM 1 ♀). Bahia: (location unspecified) (USNM 1 ♂, 1 ♀). Espírito Santo: Ibiraçu (Pau Gigante) (USNM 2 ♂, 2 ♀).—Lagoa Jupiranã (AMNH 1 ♀). Rio de Janeiro: Floresta Tijuca (CM 1 ♂).—Represa Rio Grande (CM 1 ♂).

APPENDIX 2

VOCALIZATION RECORDINGS EXAMINED

The following list identifies recordings used in the study. The inventory is organized by taxon and by recording location at the level of department or state. Recordist names are provided as an aid in identifying the recording. Numbers following the recordist name identify number of cuts. Archived recordings are identified with the abbreviated names of the archive and the number of recordings accessed at that location. LNS = Library of Natural Sounds, Cornell Laboratory of Ornithology, Ithaca, New York. NSA = National Sound Archive, The British Library, London. FSM = Florida State Museum Sound Archive, Gainesville, Florida. Recordings not yet archived but copied into the sound library of the Islers are identified with the abreviation ISL followed by the number of recordings. The large majority of these unarchived recordings are either in the process of being archived or will be archived by the recordists.

T. a. atrinucha.—COSTA RICA: "Caribbean slope": K. Zimmer 3 (ISL); Heredia: R. Ranft 1 (NSA), G. Stiles 1 (FSM), B. Whitney 1 (ISL). PANAMA: Bocas del Toro: D. Wiedenfeld 1 (ISL); Canal Zone: R. Behrstock 3 (FSM), R. Brown 1 (NSA), G. Clayton 1 (FSM), P. Donahue 1 (NSA), O. Jakobsen 1 (NSA), L. Kibler 2 (LNS), E. Morton 4 (LNS), D. Ross 1 (LNS), B. Whitney 1 (ISL), K. Zimmer 1 (ISL); Darién: E. Morton 1 (LNS), T. Parker 2 (LNS), B. Whitney 3 (ISL); Panamá: L. Davis 1 (LNS), E. Morton 1 (LNS); San Blas: B. Whitney 3 (ISL), K. Zimmer 1 (ISL); location within Panama uncertain: Willis 3 (FSM). COLOMBIA: Chocó: S. Hilty 1 (ISL); Magdalena: B. Coffey 1 (ISL); Caldas: B. Whitney 4 (ISL). ECUADOR: Cañar: T. Parker 1 (ISL); Esmeraldas: O. Jakobsen 1 (NSA), T. Parker 4 (ISL); Los Ríos: T. Parker 2 (ISL); Pichincha: R. Behrstock 1 (ISL), B. Coffey 2 (ISL), G. Rosenberg 1 (ISL), A. van den Berg 1 (LNS). VENEZUELA: ?Depto.: P. Schwartz 1 (LNS); Zulia: P. Schwartz 8 (LNS); Trujillo: P. Schwartz 1 (LNS).

T. a. gorgonae.—no recordings.

T. p. punctatus.—VENEZUELA: ?Depto.: P. Schwartz 2 (LNS); Amazonas: K. Zimmer 1 (ISL); Bolívar: R. Behrstock 1 (ISL), Parish/Altman 1 (ISL), T. Parker 1 (LNS), P. Schwartz 15 (LNS), B. Whitney 6 (ISL), K. Zimmer 4 (ISL). SURINAME: T. Davis 1 (LNS), P. Donahue 1 (NSA), V. Mees-Balchin 3 (NSA), P. Trail 1 (LNS), B. Whitney 1 (ISL). BRAZIL: Amazonas: B. Whitney 3 (ISL).

T. p. interpositus.—VENEZUELA: Apure: P. Schwartz 1 (LNS); Táchira: P. Schwartz 2 (LNS).

T. leucogaster.—PERU: Cajamarca: B. Whitney 2 (ISL).

T. huallagae.—no recordings.

T. s. stictocephalus.—BRAZIL: Pará: T. Schulenberg 1 (ISL), B. Whitney 27 (ISL); Mato Grosso: B. Whitney 2 (ISL), K. Zimmer 3 (ISL).

T. s. parkeri.—BOLIVIA: Santa Cruz: S. Mayer 1 (ISL), T. Parker 1 (ISL), B. Whitney 3 (LNS) 4 (ISL).

T. sticturus.—BOLIVIA: Santa Cruz: T. Parker 2 (LNS) 6 (ISL), B. Whitney 1 (LNS) 2 (ISL).

T. pelzelni.—BRAZIL: Bahia: T. Parker 1 (LNS), B. Whitney 20 (ISL); Ceará: B. Whitney 15 (ISL); Mato Grosso: K. Zimmer 2 (ISL); Mato Grosso do Sul: Whitney 3 (ISL); Minas Gerais: B. Whitney 10 (ISL); Pernambuco: B. Whitney 1 (ISL); São Paulo: Willis 1 (FSM).

T. ambiguus.—BRAZIL: Bahia: B. Whitney 2 (ISL), E. Willis 1 (FSM), K. Zimmer 2 (ISL); Espírito Santo: M. Isler 1 (LNS), P. Isler 3 (LNS), T. Parker 1 (LNS) 2 (ISL), B. Whitney 6 (ISL); Rio de Janeiro: B. Whitney 5 (ISL); Sergipe: B. Whitney 4 (ISL).

RELATIONSHIP OF TWO BAMBOO-SPECIALIZED FOLIAGE-GLEANERS: *AUTOMOLUS DORSALIS* AND *ANABAZENOPS FUSCUS* (FURNARIIDAE)

ANDREW W. KRATTER[1,4] AND THEODORE A. PARKER III[2,3]
Museum of Natural Science, Foster Hall 119,
Louisiana State University, Baton Rouge, Louisiana 70803 USA;
[1]*Department of Zoology and Physiology, Louisiana State University,*
Baton Rouge, Louisiana 70803 USA;
[2]*Conservation International, 1015 18th St. NW., Washington D. C., 20036 USA;*
[3]*Deceased;*
[4]*present address: Florida Museum of Natural History,*
University of Florida, Gainesville, Florida 32611 USA

ABSTRACT.—Analyses of several characters indicate that two bamboo specialist foliage-gleaners in the Furnariidae—*Anabazenops fuscus* (White-collared Foliage-gleaner) of southeastern Brazil and *Automolus dorsalis* (Dusky-cheeked Foliage-gleaner) of western Amazonia—are sister species. The nest site of *Automolus dorsalis*, above-ground in cavities, indicates that this species does not belong with other species in the genus. Similarities in foraging behavior and vocalizations, as well as similarities in plumage, provide evidence for a close relationship between *dorsalis* and *Anabazenops*. Although some morphological similarities are evident, analyses of eight mensural characters were equivocal with regard to a close relationship between these two species. We recommend that *A. dorsalis* be moved to the genus *Anabazenops*. Relationships of these two species with other taxa in the subfamily Philydorinae are investigated.

RESUMEN.—Análisis de varios caracteres indican que dos especies de la familia Furnariidae—*Anabazenops fuscus* (White-collared Foliage-gleaner) del sureste de Brasil y *Automolus dorsalis* (Dusky-cheeked Foliage-gleaner) del oeste de la Amazonía—son especies hermanas. Las dos especies son especialistas de bambú. La ubicación del nido de *Automolus dorsalis* en huecos naturales del bambú indica que la especie no pertenece al género *Automolus*. Los nidos de las otras especies de *Automolus* se ubican en huecos entre las barrancas. Semejanzas entre comportamiento de forrajeo y vocalizaciones, así como semejanzas entre el plumaje, proveen evidencias de una relación entre *dorsalis* y el género *Anabazenops*. Aunque algunas semejanzas morfológicas sean evidentes, análisis de ocho medidas son ambiguas con respecto a una relación entre las dos especies. Recomendamos que *A. dorsalis* sea cambiado al género *Anabazenops*. Relaciones con otros grupos taxonómicos de la subfamilia Philydorinae son investigadas.

Correct taxonomy is essential to the investigation of the evolution of behavioral and ecological characters (Brooks and McClennan 1991). Because ecological specialization helps promote the high diversity of tropical ecosystems, knowledge of taxonomic relationships among specialized taxa can give a foundation to understanding how high-diversity ecosystems arise. Bamboo specialization by birds in the Neotropics provides an excellent opportunity to study the evolution of specialization. Bamboo specialization has not only arisen many times across a number of unrelated taxa, but it is also shared by some species that are presumably closely related (e.g., species in *Claravis*, *Drymophila*, *Ramphotrigon*, *Hemitriccus*, *Haplospiza*). Bamboo specialization also represents both habitat specialization (on bamboo thickets) and foraging specialization (on bamboo substrates). Determining how bamboo specialization is shared among taxa can thus give insight into the evolutionary flexibility of habitat or foraging specialization (e.g., see Richman and Price 1992; Kratter 1993). The biogeography of these species is also of interest as some presumably related bamboo specialists are shared among several important areas of endemism in tropical South America (western Amazonia, Andes, and southeastern Brazil).

While conducting field studies throughout tropical South America, one of us (Parker) was struck by plumage, vocal, and behavioral similarities between two presumably distantly related species, both bamboo specialists in the family Furnariidae: the White-collared Foliage-gleaner (*Anabazenops fuscus*) of southeastern Brazil and the Dusky-cheeked Foliage-gleaner (*Automolus dorsalis*) of western Amazonia. A close relationship between these species had not been suggested previously. The latest revision of the family (Vaurie 1980) not only kept them in their traditional places as separate genera within the foliage-gleaner subfamily (Philydorinae), but also stated that these genera were not closely related. In 1992, Kratter discovered a nest of *Automolus dorsalis* unlike that of any other *Automolus* (Kratter 1994). In this paper we present several lines of evidence that support a close relationship between *Anabazenops fuscus* and *Automolus dorsalis*. This relationship will be discussed in reference to their relationship to other taxa in the Philydorinae and to the evolution of bamboo specialization.

The limits of the subfamily Philydorinae are controversial: Vaurie's (1971, 1980) enlarged Philydorinae included genera placed by Cory and Hellmayr (1925) in the separate subfamilies Sclerurinae and Margarorinae; Vaurie also lumped many genera into the single genus *Philydor*. Herein we accept Cory and Hellmayr's limits for the subfamily. Whereas Cory and Hellmayr's classification grouped the behaviorally and ecologically uniform "foliage-gleaner" genera (see Parker 1979), Vaurie's broadly unaccepted classification (e.g., see Fitzpatrick 1982) included some quite distinct genera (e.g., the terrestrial *Sclerurus*) within the foliage-gleaner assemblage.

In addition, we assume that *Automolus dorsalis* and *Anabazenops fuscus* belong to a group of similar genera within the Philydorinae, most of which are known as "foliage-gleaners" (*Syndactyla, Anabacerthia, Anabazenops, Philydor, Automolus*; English names from Ridgely and Tudor 1994), but also including the Chestnut-winged Hookbill (*Ancistrops strigilatus*), Striped Woodhaunter (*Hyloctistes subulatus*), the two recurvebill (*Simoxenops*) species, and some treehunters (*Cichlocolaptes leucophrus* and the seven *Thripadectes* species). We assume that these ten genera are monophyletic. Hereafter, we call this group of 39 species the "foliage-gleaner clade." Although these ten genera are presented sequentially in Sibley and Monroe's classification (1990), the originating phylogeny for this classification (Sibley and Alquist 1990) does not preclude the possibility that this clade is not monophyletic. Some other genera with unclear affinities (*Berlepschia, Pseudocolaptes, Pseudoseisura, Heliobletus, Xenops, Megaxenops,* and *Pygarrhicus*) were included by Cory and Hellmayr in the Philydorinae. We assume that these taxa lie outside the foliage-gleaner clade and do not consider them further. We follow the species-level classification given by Sibley and Monroe (1990), which included 39 species within the foliage-gleaner clade.

METHODS

Morphology.—The morphological relationships of *Anabazenops fuscus* and *Automolus dorsalis* to each other and to other species in the foliage-gleaner clade were investigated by phenetic analyses of eight characters: 1) length of culmen from the base of the skull (Bill Length), bill width at the distal end of the nares (Bill Width), bill depth at the distal end of the nares (Bill Depth), wind chord (Wing), tail length (Tail), tarsus length (Tarsus), hind toe length without the claw (Hind Toe), and hind claw length (Hind Claw). All measurements were taken with digital calipers and measured to the nearest 1/10th millimeter by Kratter.

Anabazenops fuscus and *Automolus dorsalis* were compared to 16 other foliage-gleaner species that either represent the morphological diversity within the clade (*Hyloctistes subulatus, Ancistrops strigilatus, Philydor erythrocercus, Philydor erythropterus, Philydor rufus, Anabacerthia striaticollis, Syndactyla rufosuperciliata, Cichlocolaptes leucophrus, Automolus ochrolaemus, Automolus infuscatus, Hylocryptus erythrocephalus, Thripadectes holostictus*) or share similar behavior, especially bamboo specialization, or plumage with the two species in question (*Simoxenops ucayalae, Automolus roraimae, Automolus melanopezus, Automolus rufipileatus*). Except for three species (see below), we measured three to five males (five preferred) from one or two nearby sites for each species to reduce sexual dimorphism and geographic variation. Sample size depended on specimens available in the Louisiana State University Museum of Natural Science (= LSUMZ). A sufficient sample of only females (N = 3) was available for *Cichlocolaptes leucophrus*. Larger series (37 male *Automolus dorsalis* and 11 male *Anabazenops fuscus*) of the two main taxa in question were measured to account for any variation across the species' range (specimens from LSUMZ and other museums; see Acknowledgments).

Each character was examined individually using Analysis of Variance (ANOVA). Student-Newman-Keuls (SNK) post-ANOVA tests were used to examine patterns of univariate characters if ANOVAs showed a significant difference. A Principal Component Analysis (PCA) reduced variation in the eight characters to a few orthogonal axes. The covariance matrix of mean values for each character for each species was used for the PCA. Each measure was initially ln-transformed to equalize variances. In addition, distance values in the same matrix (8 character by 18 species) were calculated, and a phenogram of the species was produced using the UPGMA algorithm. All statistical analyses were run using SAS (SAS Institute 1992).

Plumage.—Plumage characters of *Automolus dorsalis* and *Anabazenops fuscus* were compared to 36 other species in the foliage-gleaner clade (mainly LSUMZ specimens; also see Acknowledgments). The plumage of the only species not examined directly by us, *Philydor novaesi*, is described in Ridgely and Tudor (1994).

Vocalizations.—Tape recordings of *Automolus dorsalis* and *Anabazenops fuscus* from the Library of Natural Sounds (LNS) at Cornell University were analyzed with "Canary" software from the Bioacoustics Research Program at the Cornell Laboratory of Ornithology, Ithaca, New York. Seven different *Automolus dorsalis* cuts of separate individuals from Peru (deptos. Huánuco, Madre de Dios, and Cuzco) and Brazil (Mato Grosso) and 10 different *Anabazenops fuscus* cuts of separate individuals from southeastern Brazil (Espíritu Santo and Rio de Janiero) were analyzed.

Songs and calls of these two species were compared with other species in the foliage-gleaner clade. Parker had field experience with vocalizations of every species in this clade, except presumably *Philydor novaesi*. In addition, Kratter analyzed recordings from personal tapes and from catalogued recordings in the Florida Museum of Natural History (FLMNH) for 24 (of 39) species in this clade. These included tapes of every recognized genus in the clade except *Cichlocolaptes*. Vocalizations of most other species in the clade, including *Cichlocolaptes*, are described in Ridgely and Tudor (1994).

Foraging ecology.—The foraging behavior of *Automolus dorsalis* has received almost no attention in the literature. Based on "limited data," Remsen and Parker (1984) classified this species as a probable dead-leaf specialist; dead-leaf searching is the predominant foraging technique in the subfamily Philydorinae (Remsen and Parker 1984; Rosenberg 1990). The foraging behavior of *Anabazenops* was described by Rodrigues et al. (1994). In 1992–1993, Kratter studied the foraging behavior of *Automolus dorsalis* in lowland *Guadua* bamboo thickets along the Río Tambopata in depto. Madre de Dios, Peru (sites described in Kratter, in press). Observations were recorded on a microcassette recorder; variables recorded for each observation followed Remsen and Robinson (1990). We compared the substrate use of *A. dorsalis* with three other species in the foliage-gleaner clade that frequented bamboo at the same site (*Simoxenops ucayalae*, *Automolus melanopezus*, and *A. rufipileatus*) and with observations of *Anabazenops fuscus* and *Philydor rufus* made by Rodrigues et al. (1994) in southeastern Brazil.

We analyzed substrate use of these six species using Chi-square tests for two data sets: first, bamboo vs. nonbamboo substrates (2 substrate by 6 spp. matrix), and second, internode vs. node vs. live-leaf vs. dead-leaf substrates (4 substrate by 6 spp. matrix). These latter four substrates were chosen to provide consistency between the foraging data of Kratter and that of Rodrigues et al. (1994), and to minimize low expected values in the matrix. Post-hoc cell contributions in the contingency tables were analyzed to see which cells contributed to significant chi-square tests. These post-hoc cells are residuals (observed − expected) adjusted to have expected values that are distributed normally with a mean of zero and standard deviation of one (Siegel and Castellan 1988). For the 4 × 6 matrix, a significant level of $P < 0.01$ was used because the number of cells in the table exceeded 20.

RESULTS

Morphology.—All eight characters (Table 1) showed significant differences among the 18 species (one-factor ANOVAs, $P < 0.001$ for all characters). Across all univariate characters, *Anabazenops fuscus* and *Automolus dorsalis* were more similar to one another than to any other species analyzed. In SNK post-ANOVA tests, the two bamboo specialists significantly differed in only one of eight characters (Hind Claw), whereas the average number of characters showing significant differences for each species pair was 3.9 for *Automolus dorsalis* and 3.6 for *Anabazenops fuscus*. Each of the other 16 species analyzed differed from *Automolus dorsalis* and *Anabazenops fuscus* in at least two characters.

In the PCA, the axes that accounted for interpretable variation in the matrix were PCA1, PCA2, and PCA3; these three axes accounted for 86% of the total variation (Table 2). The first PCA axis had high positive loadings for all eight characters (Table 2), and thus represented a size component (Johnston and Selander 1971). Based on this axis, the bamboo specialist *Simoxenops ucayalae* was the largest species, and the epiphyte searching *Hyloctistes subulatus* (pers. obs.) was the smallest (Fig. 1). The two species in question were among the four largest species analyzed (along with *Thripadectes holostictus*). Thus, three of the four largest species were

TABLE 1

Sample Size (N), Mean ($\bar{X}$), and Standard Deviation (S.D.) of Eight Variables for the 18 Species in the Morphological Analyses. See Text for Definitions of Variable Codes. * Sample Size is N-1

Species	N	Bill length $\bar{X} \pm$ s.d.	Bill width $\bar{X} \pm$ s.d.	Bill depth $\bar{X} \pm$ s.d.	Wing $\bar{X} \pm$ s.d.	Tail $\bar{X} \pm$ s.d.	Tarsus $\bar{X} \pm$ s.d.	Hind toe $\bar{X} \pm$ s.d.	Hind claw $\bar{X} \pm$ s.d.
Anabacerthia striaticollis (ANBC)	5	18.1 ± 0.5	3.8 ± 0.1	5.4 ± 0.3	88.4 ± 1.7	91.1 ± 4.4	18.4 ± 0.6	8.5 ± 0.6	7.3 ± 0.1
Anabazenops fuscus (ANBZ)	11	*25.5 ± 0.8	4.8 ± 0.2	*7.2 ± 0.2	92.2 ± 3.3	95.2 ± 5.2	22.5 ± 1.0	12.1 ± 0.5	*13.0 ± 1.0
Ancistrops strigilatus (ANCS)	5	23.5 ± 0.8	5.2 ± 0.4	6.8 ± 0.3	95.0 ± 4.3	86.6 ± 5.0	21.9 ± 0.9	8.6 ± 0.9	7.5 ± 0.4
Automolus dorsalis (AUDO)	37	*24.0 ± 1.1	5.0 ± 0.3	7.4 ± 0.4	93.0 ± 4.3	*88.5 ± 5.1	23.2 ± 1.3	*11.4 ± 0.7	10.5 ± 0.6
Automolus infuscatus (AUIN)	5	24.7 ± 0.9	4.9 ± 0.3	6.5 ± 0.3	95.5 ± 1.2	90.7 ± 4.8	22.2 ± 0.9	9.9 ± 0.5	8.3 ± 0.5
Automolus melanopezus (AUME)	5	22.9 ± 0.8	4.8 ± 0.1	6.3 ± 0.2	84.9 ± 2.0	87.5 ± 4.6	20.3 ± 1.0	6.5 ± 1.1	8.1 ± 0.7
Automolus ochrolaemus (AUOC)	5	24.2 ± 0.4	4.7 ± 0.4	6.3 ± 0.3	89.6 ± 1.0	91.0 ± 4.3	21.6 ± 0.6	10.3 ± 0.9	8.8 ± 0.5
Automolus roraimae (AURO)	3	20.7 ± 1.6	4.2 ± 0.3	5.5 ± 0.5	80.1 ± 2.9	82.0 ± 3.2	21.2 ± 0.3	8.7 ± 0.9	8.1 ± 0.5
Automolus rufipileatus (AURP)	5	24.3 ± 0.9	5.0 ± 0.3	6.3 ± 0.4	87.0 ± 3.6	86.4 ± 3.6	22.9 ± 0.6	8.5 ± 0.8	9.0 ± 0.4
Cichlocolaptes leucophrus (CICH)	3	28.0 ± 0.6	4.9 ± 0.7	6.0 ± 0.1	80.9 ± 1.8	84.8 ± 2.2	22.6 ± 0.5	7.4 ± 0.7	8.4 ± 0.6
Hylocryptus erythrocephalus (HYCR)	5	31.0 ± 0.5	4.8 ± 0.2	6.9 ± 0.3	92.1 ± 3.4	95.3 ± 6.5	27.3 ± 0.6	9.9 ± 1.1	9.9 ± 0.4
Hyloctistes subulatus (HYCT)	5	24.1 ± 0.3	4.4 ± 0.3	5.5 ± 0.3	77.2 ± 0.8	71.2 ± 3.1	18.7 ± 0.6	5.7 ± 0.7	8.1 ± 0.5
Philydor erythrocercus (PECE)	5	19.9 ± 0.7	4.1 ± 0.1	5.1 ± 0.2	85.2 ± 5.5	85.3 ± 3.0	20.2 ± 1.0	8.0 ± 0.6	6.8 ± 0.3
Philydor erythropterus (PEPT)	5	23.5 ± 1.6	5.2 ± 0.3	6.4 ± 0.2	97.0 ± 2.3	84.8 ± 1.6	21.0 ± 0.8	8.2 ± 0.6	7.3 ± 0.5
Philydor rufus (PHRU)	5	20.4 ± 0.9	4.4 ± 0.3	5.5 ± 0.2	88.8 ± 0.7	92.9 ± 3.8	22.5 ± 0.6	6.3 ± 0.4	6.3 ± 0.5
Simoxenops ucayalae (SIMO)	3	25.7 ± 0.2	4.8 ± 0.2	8.5 ± 0.4	99.4 ± 5.7	92.8 ± 1.7	26.7 ± 0.9	14.1 ± 1.8	11.9 ± 0.6
Syndactyla rufosuperciliata (SYRS)	5	20.3 ± 0.5	4.0 ± 0.1	5.6 ± 0.3	77.2 ± 2.4	90.0 ± 4.9	21.5 ± 0.4	11.0 ± 1.3	9.6 ± 0.3
Thripadectes holostictus (THHO)	5	26.4 ± 0.6	5.1 ± 0.4	6.4 ± 0.3	87.5 ± 2.0	99.5 ± 2.0	26.0 ± 1.1	8.8 ± 0.7	9.6 ± 0.2

TABLE 2
Loadings of Characters on PCA Axes, with Eigenvalues and Percent of Total Variance Explained for Each Axis in the PCA

Character	PCA1	PCA2	PCA3
mean ln(Bill Length)	0.338	−0.453	−0.301
mean ln(Bill Width)	0.316	−0.571	0.192
mean ln(Bill Depth)	0.433	−0.054	−0.038
mean ln(Wing)	0.333	0.007	0.616
mean ln(Tail)	0.290	0.416	0.378
mean ln(Tarsus)	0.399	−0.027	−0.023
mean ln(Hind Toe)	0.343	0.493	−0.137
mean ln(Hind Claw)	0.350	0.215	−0.572
Eigenvalue	4.63	1.33	0.97
% total variance	57.9	16.6	12.1
Cumulative %	57.9	74.5	86.6

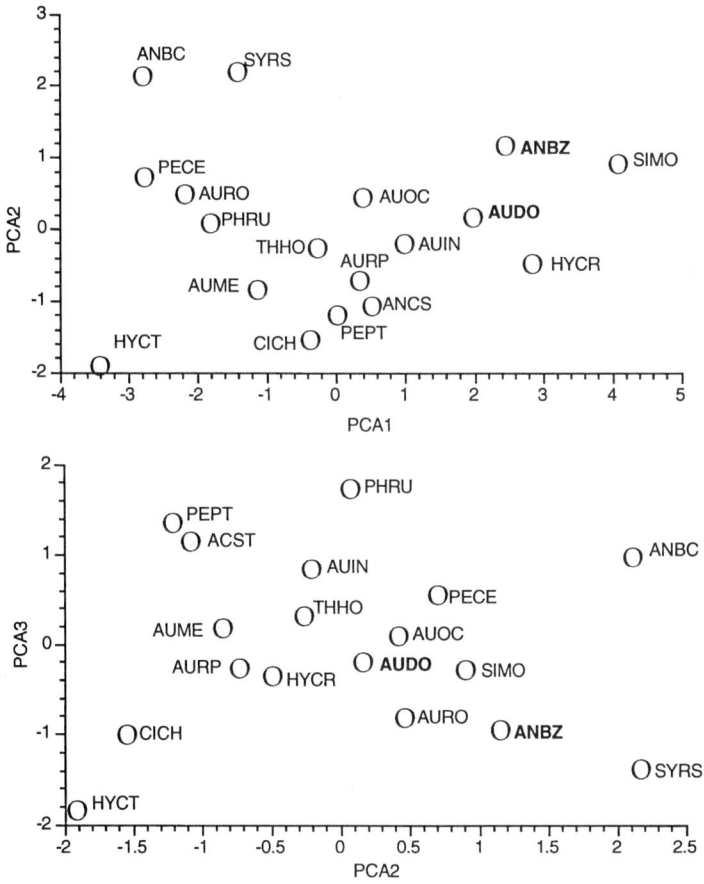

FIG. 1. Biplots of PCA results for eight morphological characters of 18 species in the foliage-gleaner clade. (Upper plot) PCA1 and PCA2. (Lower plot) PCA2 and PCA3. See Table 1 for species codes.

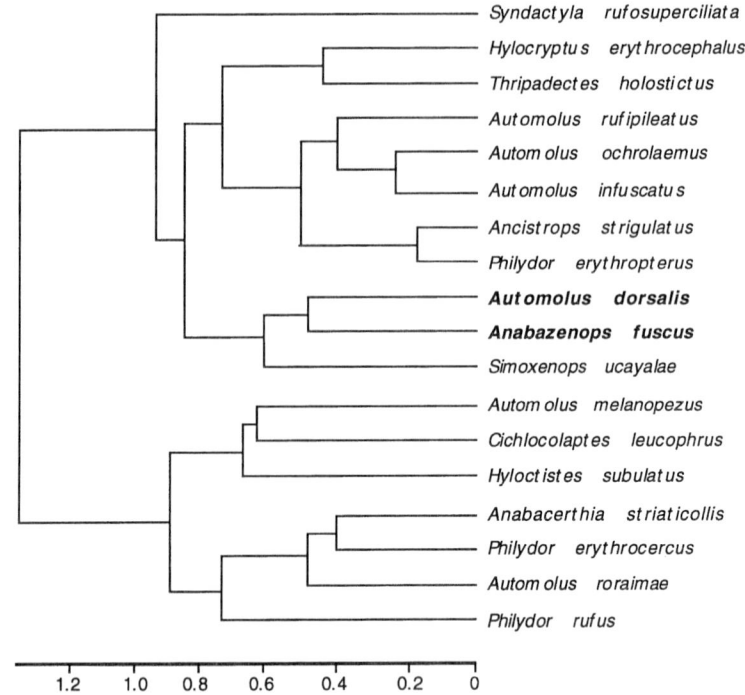

FIG. 2. Phenogram of eight morphological characters for 18 species in the foliage-gleaner clade.

bamboo specialists. The shape components represented by PCA2 and PCA3 were difficult to correlate with any behavioral or ecological attributes. PCA2 had high positive loadings for Tail and Hind Toe and high negative loadings for two bill characters (Culmen Length and Bill Width). Two small montane foliage-gleaners, *Syndactyla rufosuperciliata* and *Anabacerthia striaticollis*, were small-billed and long-tailed; the large-billed, short-tailed *Hyloctistes* was also isolated at the negative end of this axis (Fig. 1). *Anabazenops fuscus* showed a larger score on this axis than *Automolus dorsalis*. The third PCA axis had high positive loadings for Wing and Tail, and high negative loadings for Hind Claw and Culmen Length. Again *Hyloctistes* showed an extreme value, negatively on this axis (Fig. 1). *Philydor rufus* showed a high positive score. Negative scores were shown by *Anabazenops fuscus* and, to a lesser extent, *Automolus dorsalis*, on the this axis (Fig. 1). Although the PCA showed that *Anabazenops fuscus* and *Automolus dorsalis* were somewhat alike morphologically as compared to the 16 other species in the clade that were analyzed, these results have to be considered equivocal with respect to a close relationship.

Groupings in the phenogram (Fig. 2) also do not show much evidence that overall morphology reflects relationships as far as known. For example, the two *Syndactyla* species were placed in different major groupings in the phenogram; none of the three *Philydor* species were placed with one another; and only three of the five *Automolus* species (excluding *dorsalis*) grouped together. At least one of the "misclassified" *Automolus* species (*A. melanopezus*) is quite similar to others in the genus (particularly *A. rufipileatus*) in foraging behavior (Table 3; see also Rosenberg 1990), plumage (Ridgely and Tudor 1994, pers. obs.), and song (pers. obs.). It was placed with the monotypic *Cichlocolaptes*. The other species analyzed in *Automolus* (*A. roraimae*) may not belong with this genus (B. Whitney, pers. comm.; see below). Nonetheless, *Anabazenops fuscus* and *Automolus dorsalis* grouped together (Fig. 2); the bamboo specialist *Simoxenops ucayalae* was then added to this cluster.

Plumage.—Even with the rather conservative plumage diversity in the Philydorinae, the similarity in plumage between *Anabazenops fuscus* and *Automolus dorsalis* is striking, and it remains a mystery why a close relationship between these species was not suggested before based solely on plumage characters. Both species share a distinct white to buff supercilium that starts at the lores and broadens as it continues anteriorly past the eye, although this is much more distinct in *Anabazenops fuscus*. Both species share distinct white throats, contrasting with either plain

TABLE 3
Percentage Substrate Use (of Total Observations in All Cases) by Foraging Foliage-Gleaners. Numbers in Parentheses are Sample Sizes. Data for *Automolus dorsalis, A. melanopezus,* and *Simoxenops ucayalae* are from Río Tambopata in Peru (Kratter, pers. obs.). Data for *Anabazenops fuscus* and *Philydor rufus* are from Rodrigues et al. (1994)

Substrate	*Anabazenops fuscus*	*Philydor rufus*	*Automolus dorsalis*	*Automolus melanopezus*	*Automolus rufipileatus*	*Simoxenops ucayalae*
Total observations	150	239	157	115	35	63
Bamboo	86.7 (130)	1.3 (3)	61.8 (79)	44.3 (51)	40.0 (14)	95.2 (60)
Stems	73.3 (110)	1.3 (3)	35.0 (55)	7.0 (8)	8.6 (3)	95.2 (60)
Nodes[1]	52.7 (79)	3 (1.3)	25.5 (40)	6.1 (7)	5.7 (2)	11.1 (7)
Internodes[2]	20.7 (31)	0	9.6 (15)	0.9 (1)	2.9 (1)	84.1 (53)
Dead leaves	5.3 (8)	0	25.5 (40)	35.7 (41)	31.4 (11)	0
Live leaves	8.0 (12)	0	1.3 (2)	1.7 (2)	0	0
Nonbamboo	13.3 (20)	98.7 (236)	38.2 (60)	55.7 (64)	60.0 (21)	4.8 (3)
Stems	6.7 (10)	0.8 (2)	3.8 (6)	0	0	1.6 (1)
Dead leaves	3.3 (5)	15.9 (38)	34.4 (54)	55.7 (64)	60.0 (21)	3.2 (2)
Live leaves	3.3 (5)	82.0 (196)	0	0	0	0
Total nodes	52.7 (79)	1.3 (3)	25.5 (40)	6.1 (7)	5.7 (2)	11.1 (7)
Total internodes[3]	27.3 (41)	0.8 (2)	13.4 (21)	0.9 (1)	2.9 (1)	85.7 (54)
Total dead leaves	8.7 (13)	15.9 (38)	59.9 (94)	91.3 (105)	91.4 (32)	3.2 (2)
Total live leaves	11.3 (17)	82.0 (196)	1.3 (2)	1.7 (2)	0	0

[1] Nodes include bamboo sheaths, spines, and nodes.
[2] Internodes include stems and broken-off tips of stems.
[3] Total internodes include bamboo stem internodes and all nonbamboo stems.

grayish (*A. dorsalis*) or grayish-brown (*Anabazenops fuscus*) underparts. The crown, face, back and wings are plain rusty brown in both species, the crown being a bit brighter rufous. The tail of both is orange-rufous to deep rufous. The only striking difference is the whitish collar that spans the nape in *Anabazenops fuscus* is missing in *A. dorsalis*.

A major plumage distinction in the subfamily is the presence of streaks on either the upper or underparts. Species of *Ancistrops, Hyloctistes, Syndactyla, Anabacerthia, Cichlocolaptes,* and *Thripadectes* share streaked plumage. *Simoxenops* has both an unstreaked species (*S. ucayalae*) and a streaked species (*S. striatus*). Most of these streaked taxa have distinct light-colored superciliaries. Among the more or less unstreaked taxa, *Simoxenops ucayalae*, the two species of *Hylocryptus*, and most *Automolus* lack a distinct pale supercilium; *Automolus leucophthalmus* and some races of *A. ochrolaemus* have an indistinct supercilium. *Automolus roraimae* closely matches *Anabazenops fuscus* and *Automolus dorsalis* in the contrasting white throat, distinct white supercilium, and unstreaked plumage; its underparts, however, are much darker than either species. It is also a much smaller species (Table 1, Fig. 1). The entire genus *Philydor* (*sensu* Sibley and Monroe 1990) shares with *Anabazenops fuscus* and *Automolus dorsalis* unstreaked plumage and the pale supercilium. Many species of *Philydor* also have distinct throat color, but none are white as in *Anabazenops fuscus* and *Automolus dorsalis*. The white collar is unique to *Anabazenops fuscus* in the foliage-gleaner clade, although *Syndactyla ruficollis* has a rufous collar.

Anabazenops fuscus and *Automolus dorsalis* also have parallels in their juvenal plumages: the underparts and supercilium have a suffusion of bright tawny buff (AMNH 129798, 180958, 234724 for *A. dorsalis* and AMNH 316821 for *Anabazenops*). Vaurie (1980) described "immature" plumages for many species in the foliage-gleaner clade; the only other species that share distinct juvenal plumages characterized by bright buffy or rufous suffusion to the underparts and supercilium are *Philydor erythrocercus, Syndactyla subalaris,* and *Automolus roraimae*. Juveniles of the latter species, initially identified as a separate species (*Philydor hylobius*), also has dusky scalloping edges to the breast feathers (Dickerman et al. 1986), which is apparently absent in juveniles of *Anabazenops fuscus* and *Automolus dorsalis*. Although the buffy suffusion of juveniles may be difficult to detect in those Philydorinae species with bright buff or rufous underparts (e.g., *Philydor pyrrhodes*), the juvenal plumages of most other species in the subfamily tend to be very similar to adults, especially the genus *Philydor* (see Dickerman et al. 1986), or have brownish edges to the contour feathers (Vaurie 1980).

Vocalizations.—The primary songs of *Anabazenops fuscus* and *Automolus dorsalis* are very similar (Fig. 3): each gives a long series of slowly delivered, rough "chuck" notes that show

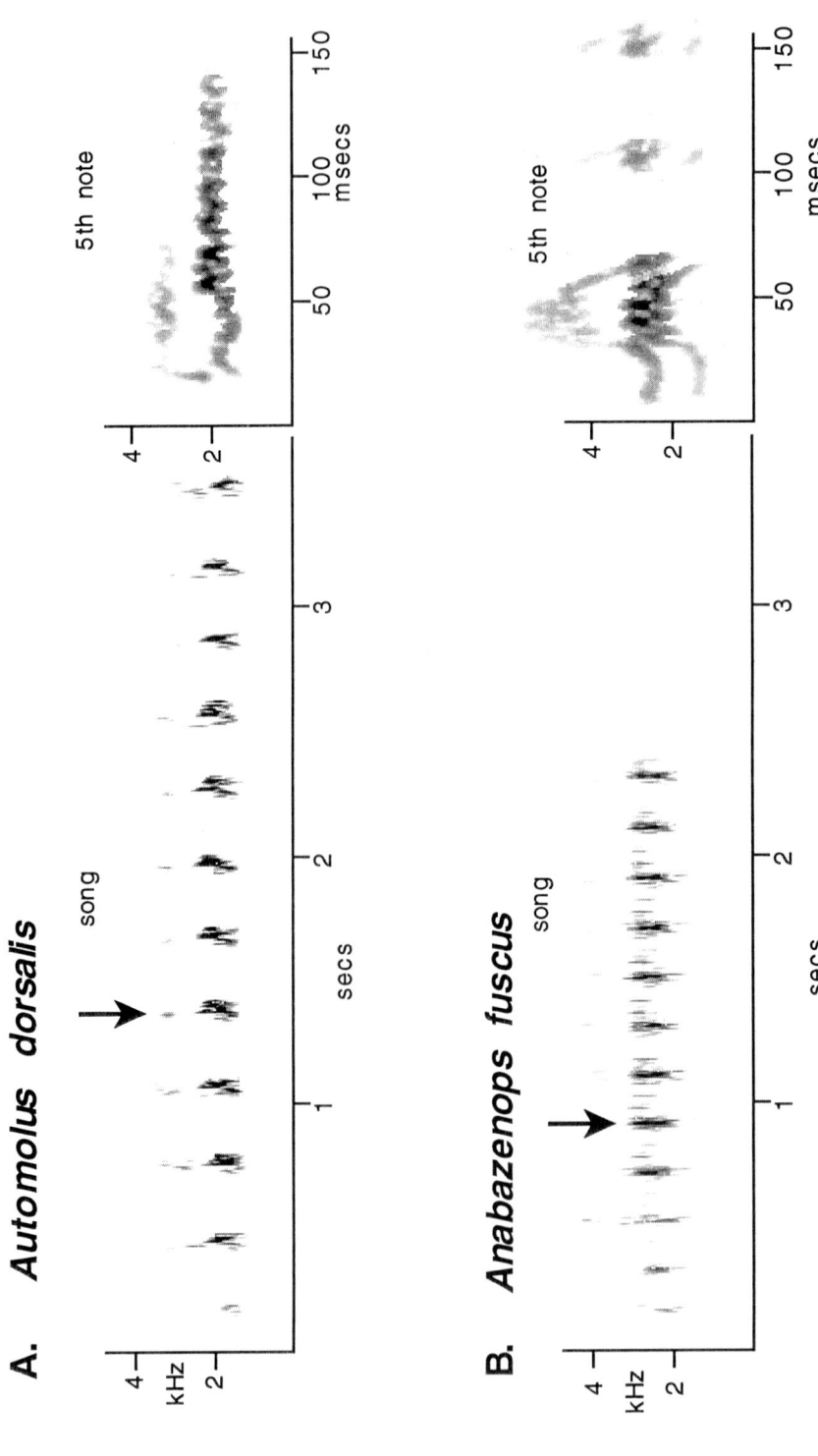

FIG. 3. Representative songs of (a) *Automolus dorsalis* (LNS 48124 from Mato Grosso, Brazil) and (b) *Anabazenops fuscus* (LNS 39082 from Rio de Janeiro, Brazil). The 5th notes from each song (indicating by arrow) are at right; these are at slower speed to show greater detail of individual notes.

little variation in pitch. The songs of both species have a distinct and similar cadence, starting with the notes closer together and then becoming more or less evenly spaced. The notes of both species' songs increase in amplitude over the first 3–5 notes (Fig. 3). The structure of the individual notes is also similar (Fig. 3). The notes of both species are two-part. The first part, with a small range in frequency, is at lower frequency (ca. 1.7 kHz in *Anabazenops fuscus* and 1.9 kHz in *Automolus dorsalis*) and less loud than the second part. This initial part slurs upward in both species, *Automolus dorsalis* at the beginning and *Anabazenops fuscus* at the end. The second part of the note is composed of upside-down "u"-shaped pieces. The frequency of each piece ranges between 1.4 and 2.6 kHz in *Automolus dorsalis* and 1.8 and 3.2 kHz in *Anabazenops fuscus*.

Both species show some variation in this primary vocalization. The number of notes per song varied from 5–24 in *Automolus dorsalis* (N = 35 songs from 4 individuals), and from 5–42 in *Anabazenops fuscus* (N = 38 songs from 6 individuals). Both species had modes of 11 notes/song (N = 8 for both species). *Anabazenops fuscus* will sometimes give its song in paired notes (e.g., paired notes 0.12 sec. apart, separated from next pair by 0.2 sec pause: LNS 39082). *Automolus dorsalis* sometimes begins its song with the notes much closer together, producing a sustained chatter (up to 24 seconds; LNS 30093) that eventually slows to the cadence of its regular song (Kratter, pers. obs). Both species often sing antiphonal duets (Parker pers. obs.). Although the duets consist of a chatter and the primary song in both species, there are some differences. The duet in *Automolus dorsalis* apparently consists of one member giving the song described above with the chatter beginning; the other member gives a quick series of about nine even-pitched notes during the chatter (LNS 30093 and 30012). The duet of *Anabazenops fuscus* apparently consists of one member of the pair giving the song described above, while the other gives a sustained chatter (LNS 48323).

In addition to these songs, both species give other vocalizations. The usual contact call of *Automolus dorsalis* is a harsh "cheff" (LNS 28786), similar to a single note from its song. Also ascribed to this species is a quick, stuttering series of harsh notes (LNS 48130, 48104). Both of these cuts were recorded in Mato Grosso, Brazil. However, during seven months of field work in southeastern Peru, where *Automolus dorsalis* was common, Kratter never learned this vocalization while studying this species. *Anabazenops fuscus* has a distinct vocalization that is quite different from its song (LNS 23773, 32058, 32064). This is a "chec" note, immediately followed by a quick series of 3–5 down-slurred *jéee-ur* notes (but up to 12 notes on occasion: LNS 32058). The notes are distinctively high-pitched and squeaky, and quite unlike other species in the foliage-gleaner clade.

The foliage-gleaner clade shows a wide range of vocalizations, but no species is very similar to *Anabazenops fuscus* or *Automolus dorsalis*. Although antiphonal duetting is widespread in the Furnariidae (Parker pers. obs.), within the foliage-gleaner clade, antiphonal duetting is known for only one other species (*Cichlocolaptes leucophrus*; D. Stotz, pers. comm.). Like *Anabazenops fuscus* or *Automolus dorsalis,* the duet of *Cichlocolaptes* consists, in part, of a chatter (Ridgely and Tudor 1994). Also somewhat similar to the chatter portions of songs given by the two species in question are the harsh chattery songs of the two *Simoxenops* species and three *Syndactyla* (Parker et al. 1985; Parker et al. 1992). B. Whitney (pers. comm.) described the song of *Automolus roraimae* as also being similar to *Syndactyla*. The songs of both these other genera differ from distinct slow cadence of the songs of *Anabazenops fuscus* and *Automolus dorsalis* in that their notes tend to be spaced irregularly, the notes tend to accelerate through the song, and the notes have a distinct "scratchy" quality to the notes (FLMNH recordings of *Simoxenops ucayalae, Syndactyla subalaris, S. ruficollis, S. rufosuperciliata*; see also Parker et al. 1990). These other genera are not known to duet anitphonally (Parker, pers. obs.).

Vocalizations of other species in the foliage-gleaner clade tend to be quite different from *Anabazenops fuscus* and *Automolus dorsalis*. In addition to their duetting, these two species differ especially in their slow delivery and evenly spaced cadence. The songs of species in *Philydor* vary widely, but the notes are never as harsh as the two species in question, and the songs tend to have the notes very close together in chatters or even trills (e.g., the long ascending trill of *P. pyrrhodes* [pers. obs.], the monotonic accelerating song of *P. erythrocercus* [pers. obs.], the high pitched whinny of *P. atricapillus* [FLMNH 1311-26-1], the variable ascending or descending chatter of *P. rufus* [FLMNH 1299-10-3], and the descending trill of *P. erythropterus* [Ridgely and Tudor 1994]). The song of *P. lichensteini* is said to be similar to that of *P. rufus* (Ridgely and Tudor 1994). The song of *P. dimidiatus* (FLMNH 1333-15-1) is quite different than others in the genus, with staccato notes that are spaced further apart. The chattery songs of *Anabacerthia striaticollis* (FLMNH 875A-6-1) and *A. variagaticeps* (FLMNH 1097-23-2) tend to be irregularly paced with high-pitched, squeaky notes. The song of *A. amaurotis* (D. Stotz, pers. comm., in Ridgely and Tudor 1994) is apparently quite different than others in the genus: "stuttering chatter lasting several seconds, followed by 3–4 loud shrieking notes." Although varying among species, the songs of *Automolus* species tend to be much shorter than those of *Anabazenops fuscus* or *Automolus dorsalis* and differ in note structure. These songs include short chattery rattles (*A. rufipileatus, A. infuscatus, A. melanopezus, A. ochrolaemus* call only; pers. obs., Ridgely and Tudor 1994), a descending series of 5–10 nasal notes (*A. ochrolaemus* song; pers. obs., Ridgely and Tudor 1994), a simple two-note nasal song (*A. rubiginosus*; pers. obs., Ridgely and Tudor 1994), and a variably delivered series of two-part, squeaky, high-pitched notes (*A. leucophthalmus:* FLMNH 897-13-1). Vocalizations in the genus *Thripadectes* also vary widely; calls of most species are harsh and grating. *T. ignobilis* has two-note calls that are squeaky and high-pitched (FLMNH 1299-25-2). *Thripadectes rufobrunneus* has harsh grating calls and chatter (FLMNH 1115-15-1); the song of this species is described as a "burry *chi-wóhr, chi-wórh* with the rolling quality of Boat-billed Flycatcher" (Stiles and Skutch 1989). The calls of *T. virgaticeps* (FLMNH 1129-33-2) share the same harsh quality shown in *T. rufobrunneus*, but are higher pitched. Songs of *T. flammulatus* (Ridgely and Tudor 1994) and *T. melanocryphus* are described as accelerating series of grating notes. The songs of *Hyloctistes subulatus* and *Ancistrops strigilatus* are similar to one another (Ridgely and Tudor 1994) and quite different than other species in the clade (pers. obs.): two squeaky introductory notes followed by or interspersed with very dry chatter. The calls (FLMNH 1301-3-1) and songs (Ridgely and Tudor 1994) of *Hylocryptus erythrocephalus* are quite distinctive with short series of ringing and staccato notes that have a much different quality than the two species in question.

Foraging ecology.—Most species in the foliage-gleaner clade search for arthropods and small vertebrates in clumps of dead leaves suspended above-ground (Remsen and Parker 1984, Rosenberg 1990, see below). Although *Anabazenops fuscus* and *Automolus dorsalis* share this general diet (pers. obs.), they stand apart from most other species in this clade by their frequent use of bamboo stem substrates. In their study of *Anabazenops fuscus,* Rodrigues et al. (1994) showed that this species typically searched stem substrates, especially nodes of bamboo stems (Table 3). Although Rodrigues et al. (1994) did not quantify the "attack" maneuvers of this species (see Remsen and Robinson 1990), they described the foraging maneuvers as "usually . . . probing into old and rotten nodes," but sometimes "hammer away at internodes," or "run the bill along internodes and remove sheaths." Judging by their tabulated data presented for substrate use (Table 1 in Rodrigues et al. 1994), these last two behaviors did not total more than 20.7% of foraging maneuvers. Only 8.6% of foraging attempts were directed at dead-leaf substrates. Although *Automolus dorsalis* also frequently searched bamboo stems, particularly nodes, it only did so to half the extent of *A. fuscus* (Table 3). *Automolus dorsalis* particularly differed from *Anabazenops fuscus* in its increased use of dead-leaf substrates (Table 3). In its use of bamboo stem substrates. *Automolus dorsalis* appeared to focus on the small dead sheaths clasping around spines at bamboo nodes and the large leafy sheaths clasping around bamboo nodes. Most bamboo stems inspected were dead. It would most often use hang maneuvers (hang upside-down, hang down, hang-sideways) to probe into or around sheaths, stems, and dead leaves. It was never seen to use "hammer" or "chisel" maneuvers.

Substrate use of *Anabazenops fuscus* and *Automolus dorsalis* were compared to four other species in the foliage-gleaner clade (Table 3). The foraging observations of all species examined were conducted in bamboo habitats, but the percentage of bamboo substrates used differed significantly among species ($\chi^2 = 369$, $P < 0.0001$), and varied from the extremely infrequent use (1.3%) of bamboo by *Philydor rufus* to the near exclusive use (95.7%) of bamboo by *Simoxenops ucayalae* (Table 3). This latter species, along with *Anabazenops fuscus* and *Automolus dorsalis*, used bamboo substrates significantly more than expected (Table 4). *Philydor rufus* was the only species that used nonbamboo substrates significantly less than expected.

The six species also significantly differed in their relative used of stem-internode, stem-node, dead-leaf, and live-leaf substrates ($\chi^2 = 256$, $P < 0.0001$). The three species that used bamboo substrates more often than expected (*S. ucayalae, Anabazenops fuscus,* and *Automolus dorsalis*) also used stems to a much greater degree than the other three species (Table 3). *Anabazenops fuscus*, and especially *S. ucayalae*, used stem-internodes significantly more than expected. *Anabazenops fuscus* and *Automolus dorsalis* differed from *S. ucayalae*, however, in their significantly greater use of stem-nodes (all bamboo) (Table 4). The attack maneuvers used by *Simoxenops ucayalae* also are quite different from other species analyzed. It frequently (32.2% in Kratter's data) used pecking or chiseling maneuvers on bamboo stem internodes (Parker 1982; Remsen and Robinson 1990). It also used a foraging maneuver unknown in other species in the

TABLE 4
Residuals of χ^2 Tests of Substrate Preferences for (a) Bamboo vs Nonbamboo Substrates, and (b) Stem Internode vs Stem-Node vs Dead-Leaf vs Live-Leaf Substrates (See Table 3). Residual Values Greater than 1.96 are Significant at $P < 0.05$ (Appropriate for a), and Values Greater than 2.575 are Significant at $P < 0.01$ (Appropriate for b). Significant Values are Bold-Faced

Substrate	Anabazenops fuscus	Philydor rufus	Automolus dorsalis	Automolus melanopezus	Automolus rufipileatus	Simoxenops ucayalae
a) Bamboo vs nonbamboo substrates						
Bamboo	**10.932**	**−17.038**	**4.233**	−0.566	−0.822	**8.051**
Nonbamboo	**10.932**	**17.038**	**−4.233**	0.566	0.822	**−8.051**
b) Stem internode vs stem-node vs dead-leaf vs live-leaf substrates						
Stem-node	**12.306**	**−8.156**	2.661	**−3.621**	−1.943	−1.495
Stem-internode	**4.255**	**−7.707**	−0.939	**−4.793**	−2.165	**15.798**
Dead-leaf	**−8.124**	**−8.305**	**6.529**	**12.964**	**6.761**	**−5.865**
Live-leaf	**−5.222**	**22.081**	**−8.506**	**−6.918**	**−3.833**	**−5.245**

clade—inserting its bill into holes (almost always bamboo stem-internodes) and enlarging the hole by prying with its strange-shaped bill. These prying maneuvers made up 25.4% of the foraging attempts recorded by Kratter.

The three *Automolus* species (including *dorsalis*) used dead-leaf substrates significantly more than expected (Table 4). *Automolus rufipileatus* and *A. melanopezus* searched dead-leaf substrates in greater than 90% of their foraging attempts; *A. dorsalis* searched these substrates in only 60% of its foraging attempts. *Philydor rufus* was the only species that used live-leaf substrates significantly more than expected, although Rodrigues et al. (1994) stated that this species is a "typical dead-leaf foliage-gleaner."

Nest site.—As far as known, all *Automolus* species, except for *A. dorsalis* (Kratter 1994), nest in burrows dug in banks. Vaurie (1971, 1980) used nest site to place the genus *Automolus* near *Thripadectes* and *Sclerurus*, and not with other "foliage-gleaners" in the Philydorinae, which, as far as known, nest in tree cavities above ground. The single *A. dorsalis* nest described was in a hollow bamboo stem (Kratter 1994). *Anabazenops fuscus* also nests above ground, in tree cavities (Sick 1988).

DISCUSSION

Because of its above-ground nest, Kratter (1994) suggested that *Automolus dorsalis* does not fit in its traditional placement with other *Automolus*, all of which, as far as known, nest in terrestrial burrows. In this paper, we give additional evidence that *A. dorsalis* dramatically differs from its congeners in vocalizations and foraging behavior. The main question thus becomes "What is the taxonomic position of *Automolus dorsalis*?" The evidence presented here indicates that its closest relative is *Anabazenops fuscus*, and we argue that *dorsalis* should be placed in *Anabazenops*. The characters best supporting this arrangement are, in order of decreasing importance, vocalizations, foraging behavior, habitat selection, and plumage. Although none of these characters absolutely indicates a sister relationship in and of itself, the sum total of all these characters provides the most compelling evidence for a close relationship.

In the foliage-gleaner clade, antiphonal duetting appears to be restricted to *Anabazenops fuscus*, *Automolus dorsalis*, and *Cichlocolaptes leucophrus*. This latter species, however, does not show any other evidence (ecological, behavioral, morphological, plumage) of close relationship with either of the former two species. Within the subfamily Philydorinae, antiphonal duetting is more widespread, and is known for *Megaxenops* (Whitney and Pacheco 1994), *Berlepschia* (Whitney, pers. comm.), and *Pseudoseisura* sp. (at least *lophotes*, pers. obs.). Although this suggests the possibility that antiphonal duetting may be a shared-primitive character within the foliage-gleaner clade, a close relationship between *Anabazenops fuscus* and *Automolus dorsalis* is still indicated by similarities in the pattern, cadence, and note structure of their primary vocalizations.

Although *Anabazenops fuscus* and *Automolus dorsalis* have some marked differences in substrate preferences during foraging, their frequent use (53 and 25% for *Anabazenops fuscus* and

Automolus dorsalis, respectively) of bamboo nodes sets them apart from other species in the foliage-gleaner clade. *Simoxenops ucayalae* also frequently searches bamboo stems, but mainly internodes (Table 3). Its attack maneuvers are quite different than the two species in question (see results); its frequent use of a prying attack behavior is unique in the clade. Most other species in the foliage-gleaner clade search dead-leaf substrates, especially those in the genera *Anabacerthia. Automolus, Philydor, Hyloctistes* and *Thripadectes* (Remsen and Parker 1984; Rosenberg 1990; Ridgely and Tudor 1994; pers. obs.). Rosenberg (1990) found that dead-leaf searching dominated the foraging maneuvers of four species of *Automolus* (100% for *A. rufipileatus*, 94% for *A. ochrolaemus*, 88% for *A. infuscatus*, and 97% for *A. melanopezus*), two species of *Philydor* (80% for *P. erythrocercus* and 92% for *P. ruficaudatus*), and *Hyloctistes subulatus* (85%) along the Río Tambopata in southeastern Peru. Remsen and Parker (1984) provided evidence that dead-leaf searching is probably the predominant foraging technique in many species of the Philydorinae (including *Anabacerthia* and several species of *Philydor* and *Automolus*). They hypothesized that most species in the foliage-gleaner genera *Philydor, Automolus, Hyloctistes,* and *Thripadectes* would be dead-leaf specialists.

Several species in the foliage-gleaner clade search bromeliads or other epiphytes, including *Cichlocolaptes leucophrus* (Ridgely and Tudor 1994), *Syndactyla ruficollis* (Parker et al. 1985), *S. subalaris* (Ridgely and Tudor 1994), and *Hyloctistes subulatus* (pers. obs). The two *Hylocryptus* species differ from others in the clade in their somewhat terrestrial habits (Wiedenfeld et al. 1985; Ridgely and Tudor 1994)

The shared juvenal plumage of *Automolus dorsalis* and *Anabazenops fuscus* is found in only a few other species in the foliage-gleaner clade. The combination of the whitish throat, pale underparts, lack of streaks, and distinct pale supercilium of adult birds are unique. For these characters, however, *Automolus roraimae* differs from the pair of bamboo specialists only in the color of the underparts.

The relationship of *Automolus dorsalis* and *Anabazenops fuscus* to other taxa in the Philydorinae is more difficult to ascertain. If nesting site (above ground cavities vs. terrestrial burrows) is a useful character in separating *Automolus* and *Thripadectes* from the other philydorines (see Vaurie 1971), then *Anabazenops fuscus* and *Automolus dorsalis* clearly belong with the other above-ground nesting taxa (as far as known *Philydor, Syndactyla, and Anabacerthia*) and not with the terrestrially nesting *Automolus* and *Thripadectes*. Some species of *Syndactyla* and *Simoxenops* share with the two species in question frequent use of bamboo habitats (especially *Simoxenops ucayalae* and *Syndactyla ruficollis*: Parker 1982; Parker et al. 1985), similar juvenal plumage (*Syndactyla subalaris*), and somewhat similar songs (see Results). *Syndactyla* and *Simoxenops* may be closely related (Parker et al. 1990). *Automolus roraimae* also may be close to the two species in question. This species shares similar plumage, including that of the juvenile. Its natural history is less well-known. Although Willard et al. (1991) described this species as primarily searching bromeliads, it also searches clasping sheaths around bamboo nodes (J. P. O'Neill, pers. comm.), a foraging behavior often observed for *Automolus dorsalis* (Kratter, pers. obs.). B. Whitney (pers. comm.) described the song and foraging behavior of *A. roraimae* as similar to that of *Syndactyla*.

Of course, the characters supporting a sister relationship between *Automolus dorsalis* and *Anabazenops fuscus* could also arise by convergence among species that are not closely related. In this case, convergence may be expected in morphological or foraging characters, because the two species share similar habitat. Both species occur in thickets of *Guadua* bamboo (Rodrigues et al. 1990, pers. obs.), a genus with large diameter stems and prominent spines. In fact, *Simoxenops ucayalae,* the only other species in the clade that frequently searches *Guadua* stems, is morphologically similar to *Automolus dorsalis* and *Anabazenops fuscus* (Figs. 1 and 2). However, nest-site placement and vocalization similarities are less likely to result from convergence; in fact, vocalizations have, by and large, been useful characters in elucidating relationships within the Furnariidae (e.g., Parker et al. 1985; Whitney 1994). The morphological characters formally used to distinguish *Automolus* may not have been phylogenetically useful, given the results here and also those of Parker et al. (1985) for "*Automolus*" *ruficollis*, which has since been placed in the genus *Syndactyla* (e.g., Ridgely and Tudor 1994). This study gives further evidence that behavioral and ecological characters may shed important light on relationships.

Substrate specialization is common within the foliage-gleaner clade: for example, *Cichlocolaptes leucophrus* may be a bromeliad specialist in south-central Brazil (Ridgely and Tudor 1994); *Philydor pyrrhodes* is a palm specialist throughout Amazonia (Parker, pers. obs.); and *Simoxenops ucayalae* is a bamboo stem specialist in southern Amazonia (Parker 1982). The greatest diversity of philydorines occurs in the high-diversity tropical forests in western Amazonia. At a single study site on the Río Tambopata, 15 species in this subfamily have been found (Foster et al. 1994). Although a majority of species search dead leaves, they tend to segregate ecologically by differences in habitat (e.g., river edge forests vs. *terra firme* forests) or substrate preference (e.g., bamboo, palm fronds, canopy trees) (see Parker 1979). The evolutionary history of specialization is important in understanding how this high diversity avifauna arose and how these ecologically similar species can occur sympatrically.

Taxonomies that reflect the true phylogenetic history are critical for assessing the evolution of specific characters, such as bamboo specialization, and the historical biogeography of the taxa in question (Brooks and McClennan 1991). For instance, Kratter (1993) showed that bamboo specialization can be a rather flexible strategy evolutionarily, having probably arisen twice within a single species (*Amblycercus holosericeus*). In traditional classifications (e.g., Cory and Hellmayr 1925; Peters 1951; Vaurie 1980; Sibley and Monroe 1990), "*Automolus*" *dorsalis* is placed next to the almost wholly sympatric *A. infuscatus* (along with *A. leucophthalmus* of southeastern Brazil). *Automolus infuscatus* is a widespread forest-understory species in Amazonia. In southeastern Peru, *A. infuscatus* differs from *A. dorsalis* in habitat selection; it completely avoids the dense thickets of bamboo favored by *A. dorsalis* (pers. obs.). In areas where *A. dorsalis* is absent (e.g., southern Amazonia), *A. infuscatus* apparently uses a wider variety of habitats (B. Whitney, pers. comm.). If *A. infuscatus* is the sister species of *A. dorsalis*, then bamboo specialization is an autapomorphic character for *A. dorsalis*, indicating that bamboo specialization evolved relatively recently in this clade and that the evolutionary split may have occurred as a result of habitat differences between *A. infuscatus* and *A. dorsalis*. A very similar situation can be found in the genus *Cymbilaimus* (Formicariidae): *C. sanctaemariae*, a bamboo specialist of western Amazonia (Pierpont and Fitzpatrick 1983) is clearly the sister species (there are no other species in the genus) of *C. lineatus*, a widespread lowland species in the Neotropics.

If *Automolus dorsalis* and *Anabazenops fuscus* are sister species, then a very different interpretation of the evolution of bamboo specialization is indicated. Bamboo specialization is a synapomorphy in this clade, suggesting that a bamboo-specialized ancestor was widespread from western Amazonia to southeastern Brazil. This biogeographic pattern is shared by other closely related bamboo specialists, including the genus *Drymophila* (*D. devillei* in western Amazonia, *D. caudata* in Andean foothills, and four bamboo specialist *Drymophila* species in southeastern Brazil) and a group of species in *Hemitriccus* (*H. flammulatus* in western Amazonia, and *H. obsoletus* and *H. diops* in southeastern Brazil) (Parker 1982).

TAXONOMIC RECOMMENDATION

We recommend that "*Automolus*" *dorsalis* be placed in the genus *Anabazenops*; the correct name would thus be *Anabazenops dorsalis* Sclater and Salvin. Although *Anabazenops* shows some affinities with *Syndactyla* and *Simoxenops*, without further investigation into the relationships of genera in Philydorinae, we recommend that *Anabazenops* be kept in its traditional position near *Philydor* (e.g., Peters 1951, Vaurie 1980, Sibley and Monroe 1990).

NOTE

Although the untimely death of Parker occurred before this manuscript was begun, Ted had suggested to me and others the possible close relationship between *Anabazenops fuscus* and *Automolus dorsalis*, based on shared vocalizations and bamboo specialization. I was not able to have access to all of Ted's notes, so many data presented in the paper are solely my own. The section on vocalizations especially could have been broadened immeasurably with Ted's input. Andrew W. Kratter.

ACKNOWLEDGMENTS

The Rapid Assessment Program of Conservation International funded Parker and Kratter's research on the Río Tambopata in 1992. Kratter's field work in Peru in 1993 was funded by a Frank M. Chapman grant from the American Museum of Natural History, a Grants-in-Aid research award from Sigma Xi, Alexander Wetmore and AOU Council awards from the American Ornithologists' Union, and a Fugler Fellowship in Tropical Biology from Museum of Natural Science, Louisiana State University. Van Remsen, Doug Stotz, and Bret Whitney made many useful criticisms of this manuscript. We thank M. A. Rodrigues for sending us a pre-publication copy of his manuscript on foraging in *Anabazenops*. The following museums kindly loaned their specimens of *Anabazenops fuscus* (AF) and *Automolus dorsalis* (AD) to us: Field Museum of Natural History (FMNH: 1 AF and 13 AD specimens), American Museum of Natural History

(AMNH: 18 each), and Academy of Natural Sciences in Philadelphia (ANSP: 5 AF and 6 AD specimens). John Bates at the AMNH kindly allowed us to measure *Automolus roraimae* specimens. The Library of Natural Sounds (LNS) at the Cornell Laboratory of Ornithology supplied tape-recordings of these species' songs. Tom Webber and J. W Hardy kindly allowed Kratter access to the Florida Museum of Natural History's Bioacoustic collection and equipment. Amanda Stronza helped translate the abstract.

LITERATURE CITED

BROOKS, D. R., AND D. A. MCCLENNAN. 1991. Phylogeny, Ecology, and Behavior. Univ. Chicago Press, Chicago, Illinois.

CORY, C. B., AND C. E. HELLMAYR. 1925. Catalogue of birds of the Americas and the adjacent islands. Field Mus. Nat Hist. Zool. Ser, 13(14):1–390.

DICKERMAN, R. W., G. F. BARROWCLOUGH, P. F. CANNELL, W. H. PHELPS, JR., AND D. E. WILLARD. 1986. *Philydor hylobius* Wetmore and Phelps is a synonym of *Automolus roraimae* Hellmayr. Auk 103: 431–32.

FITZPATRICK, J. W. 1982. [Review of] Taxonomy and geographical distribution of the Furnariidae (Aves, Passeriformes). Auk 99:810–813.

FOSTER, R. B., T. A. PARKER III, A. H. GENTRY, L. H. EMMONS, A. CHICCHÓN, T. SCHULENBERG, L. RODRÍGUEZ, G. LAMAS, H. ORTEGA, J. ICOCHEA, W. WUST, M. ROMO, J. A. CASTILLO, O. PHILLIPS, C. REYNEL, A. KRATTER, P. K. DONAHUE, AND L. J. BARKLEY. 1994. The Tambopata-Candamo Reserved Zone of Southeastern Perú: A Biological Assessment. RAP Working Papers 6. Conservation International, Washington, D. C.

JOHNSTON, R. F., AND R. K. SELANDER. 1971. Evolution of the House Sparrow. II. Adaptive differentiation in North American populations. Evolution 25:11–28.

KRATTER, A. W. 1993. Geographic variation in the Yellow-billed Cacique (*Cacicus holosericeus*), a partial bamboo specialist. Condor 95:641–651.

KRATTER, A. W. 1994. The nest of the Crested Foliage-gleaner *Automolus dorsalis*. Ornitologia Neotropical 5:105–107.

KRATTER, A. W. 1995. Status, habitat, and conservation of the Rufous-fronted Antthrush *Formicarius rufifrons*. Bird Conservation Intl. 5:391–404.

PARKER, T. A., III. 1979. An introduction to foliage-gleaner identification. Continental Birdlife 1:32–37.

PARKER, T. A., III. 1982. Observations of some unusual rainforest and marsh birds in southeastern Peru. Wilson Bull. 94:477–493.

PARKER, T. A., III, T. S. SCHULENBERG, G. R. GRAVES, AND M. J. BRAUN. 1985. The avifauna of the Huancabamba region, northern Peru. Pp. 169–197 in Neotropical Ornithology (P. A. Buckley, M. S. Foster, E. S. Morton, R. S. Ridgely, and F. G. Buckley, Eds.). Ornithol. Monogr. 36.

PARKER, T. A., III, J. M. BATES, AND G. COX. 1992. Rediscovery of the Bolivian Recurvebill with notes on other little-known species of the Bolivian Andes. Wilson Bull. 104:173–178.

PETERS, J. L. 1951. Check-list of Birds of the World. Vol. VII. Mus. Comp. Zool., Cambridge, Massachusetts.

PIERPONT, N. J., AND J. W. FITZPATRICK. 1983. Specific status and behavior of *Cymbilaimus sanctaemariae*, the Bamboo Antshrike, from southwestern Amazonia. Auk 100:645–652.

REMSEN, J. V., JR., AND T. A. PARKER III. 1984. Arboreal dead-leaf searching birds of the Neotropics. Condor 86:36–41.

REMSEN, J. V., JR., AND S. K. ROBINSON. 1990. A classification scheme for foraging behavior of birds in terrestrial habitats. Pp. 144–160 in Avian Foraging: Theory, Methodology, and Applications (M. L. Morrison, C. J. Ralph, J. Verner, and J. R. Jehl, Eds.). Studies in Avian Biol. 13.

RICHMAN, A. D., AND T. PRICE. 1992. Evolution of ecological differences in the Old World lead warblers. Nature 355:817–821.

RIDGELY, R. S., AND G. TUDOR. 1994. The Birds of South America. Vol. 2. The Suboscine Passerines. University of Texas Press, Austin, Texas.

RODRIGUES, M., A., S. M. R. ALVARES, AND C. G. MACHADO. 1994. Foraging behavior of the White-collared Foliage-gleaner (*Anabazenops fuscus*): a bamboo specialist. Ornitologia Neotropical 5:65–67.

ROSENBERG, K. V. 1990. Dead-leaf foraging specialization in tropical forest birds. Ph.D. dissertation. Louisiana State University, Baton Rouge, Louisiana.

SAS INSTITUTE. 1992. SAS/STAT User's Guide. Version 6. SAS Institute, Inc. Cary, North Carolina.

SIBLEY, C. G., AND J. E. AHLQUIST. 1990. Phylogeny and Classification of Birds. Yale Univ. Press, New Haven, Connecticut.

SIBLEY, C. G., AND B. L. MONROE, JR. 1990. Distribution and Taxonomy of Birds of the World. Yale University Press, New Haven, Connecticut.

SIEGEL, S., AND N. J. CASTELLAN. 1988. Non-parametric Statistics for the Behavioral Sciences, 2nd edition. McGraw-Hill, New York.

SICK, H. 1988. Ornitologia Brasileira, Vol. 2, third ed. Brasília, Brazil.

STILES, F. G., AND A. F. SKUTCH. 1989. A Guide to the Birds of Costa Rica. Comstock, Ithaca, New York.

VAURIE, C. 1971. Classification of the Ovenbirds (Furnariidae). Privately published, London.

VAURIE, C. 1980. Taxonomy and geographical distribution of the Furnariidae (Aves, Passeriformes). Bull. Amer. Mus. Nat. Hist. 166:1–357.
WIEDENFELD, D. A., T. S. SCHULENBERG, AND M. B. ROBBINS. 1985. Birds of a tropical deciduous forest in extreme northwestern Peru. Pp. 305–315 *in* Neotropical Ornithology (P. A. Buckley, M. S. Foster, E. S. Morton, R. S. Ridgely, and F. G. Buckley, Eds.). Ornithol. Monogr. 36.
WHITNEY, B. M., AND J. F. PACHECO. 1994. Behavior and vocalizations of *Gyalophylax* and *Megaxenops* (Furnariidae), two little-known genera endemic to northeastern Brazil. Condor 96:559–565.
WILLARD, D. E., M. S. FOSTER, G. F. BARROWCLOUGH, R. W. DICKERMAN, P. F. CANNELL, S. L. COATS, J. L. CRACRAFT, AND J. P. O'NEILL. 1991. The birds of Cerro de la Neblina, Territorio Federal Amazonas, Venezuela. Fieldiana: Zoology, N. S. 65:1–80.

GEOGRAPHIC VARIATION OF PLUMAGE PATTERNS IN THE WOODCREEPER GENUS *DENDROCOLAPTES* (DENDROCOLAPTIDAE)

CURTIS A. MARANTZ[1,2,3]

[1]*Museum and Natural Science, Foster Hall 119, and* [2]*Department of Zoology and Physiology, Louisiana State University, Baton Rouge, Louisiana 70803, USA;*
[3]*present address: Biology Department, University of Massachusetts, Amherst, Massachusetts 01003, USA*

ABSTRACT.—To reassess taxonomic relationships within *Dendrocolaptes*, I examined plumage variation in the five members of the genus. Following a descriptive analysis of the species, I conducted cladistic and phenetic analyses of a suite of 21 plumage characters along with bill color. Results suggest that approximately one-fifth of the presently recognized subspecies are invalid. Many taxa were based on inadequate sampling at a geographic scale too coarse to expose subtle variation of a clinal nature. A cladistic analysis of plumage characters did not support the monophyly of *Dendrocolaptes*, although the barred species-group was monophyletic. Within the barred species group, trans-Andean and cis-Andean representatives each formed monophyletic subgroups, but *D. concolor* was placed within the Amazonian *D. certhia*. *Dendrocolaptes hoffmannsi* was placed as the sister-group to the barred taxa in contrast to the conclusions of some previous workers. The relationships predicted by phenetic techniques paralleled the cladistic analysis except that the Nechí and Central American populations of *D. certhia* were separated, as were *D. concolor* and the Amazonian *D. certhia*. The inability of the quantitative techniques to support monophyly of *Dendrocolaptes* likely resulted from the fact that the streaked *Dendrocolaptes* exhibit the ancestral character state, and therefore differ little from the outgroups based on plumage patterns. Although subspecies distributions within *Dendrocolaptes* conform to areas of endemism proposed by Haffer (1967, 1985) and Cracraft (1985), relationships among most taxa were too weakly defined to make strong predictions regarding the events responsible for isolating populations. This inability to infer relationships was likely due, in large part, to convergence among taxa occurring sympatrically and in regions within similar climatic conditions.

RESUMEN.—Para re-evaluar los vinculos taxonómicos dentro del género *Dendrocolaptes*, se examinó las variaciones en el plumaje de cinco miembros del género. En base a un análisis descriptivo de las especies se realizaron análisis cladísticos y fenéticos de un grupo de 21 caracteres del plumaje, además del color del pico. Los resultados sugieren que aproximadamente un quinto de las subspecies actualmente reconocidas no son válidas. Muchos grupos taxonómicos se basaron en muestras inadecuadas recolectadas en una escala geográfica muy gruesa como para revelar sutiles variaciones graduales. Un análisis cladístico de los caracteres del plumaje, no respaldó la condición monofilética del género *Dendrocolaptes*, aunque el grupo de especies con bandas era monofilético. Dentro del grupo de especies con bandas, los representantes transandinos y cisandinos forman sub-grupos monofiléticos, respectivamente, pero se colocó a *D. concolor* con la especie amazónica *D. certhia*. *Dendrocolaptes hoffmannsi* se colocó como un grupo hermano al de las especies con bandas, contrario a las conclusiones de algunos trabajos previos. Los vínculos pronosticados por las técnicas fenéticas son paralelos a los obtenidos por el análisis cladístico, excepto que las poblaciones Nechi y centroamericanas de *D. certhia* fueron separadas, al igual que *D. concolor* y la especie amazónica de *D. certhia*. La incapacidad de las técnicas cuantitativas de apoyar la condición monofilética del género *Dendrocolaptes*, probablemente resultó del hecho que los *Dendrocolaptes* con rayas exhiben caracteres del estado ancestral y, por consiguiente, difieren poco de los grupos extremos en base a los patrones de plumaje. Aunque la distribución de subespecies dentro del género *Dendrocolaptes* conforma a las áreas de endemismo propuestas por Haffer (1967, 1985) y Cracraft (1985), los vinculos entre la mayoría de los grupos taxonómicos fueron muy débilmente definidos como para hacer pronósticos firmes respecto a los acontecimientos responsables por el aislamiento de las poblaciones. Esta incapacidad de diferenciar vinculos se debe probablemente, en gran parte, a la convergencia entre los grupos taxonómicos que ocurren en forma simpatrica, y en regiones con condiciones climáticas similares.

Because differentiation of geographically segregated populations is believed to be a primary event in the evolutionary process, studies of geographic variation are of key importance in systematics (Baker 1985; Zink and Remsen 1986). Geographic variation in birds traditionally has been studied through the description and analysis of morphological patterns (Fjeldså 1985). Although geographic variation has been assessed using many new techniques (see reviews by Selander 1971; Barrowclough 1983), most taxa have been defined largely on the basis of morphological (both morphometric and plumage) characters of study skins (Selander 1971). Whereas some taxa are distinguished morphometrically, many subspecies have been defined almost exclusively on the basis of plumage characters. A careful analysis of geographic variation in plumage patterns therefore represents a logical first step toward developing a framework for the later comparison of morphometric and vocal variation. Because many recent studies of geographic variation have found that too many populations have been formally named (Fjeldså (1985), a reassessment of plumage variation and taxonomic limits among group members will likely be necessary before plumage characters can be compared meaningfully with other data sets.

Even though most of the world's biodiversity is found in the tropics, few studies have focused on tropical birds relative to their overwhelming proportion of taxa occurring worldwide. One of the many understudied Neotropical groups is the family Dendrocolaptidae. Few passerine families show as much overall similarity among species, both in morphometrics and in plumage coloration, as does the Dendrocolaptidae (e.g., Vaurie 1980). All woodcreepers are long-tailed, branch-climbing birds with olive and rust colors predominating the plumage. Furthermore, even though size varies substantially among species, all are relatively similar in body shape. As a result of this morphological similarity, most woodcreeper taxa (both species and subspecies) have been characterized on the basis of plumage characters. Plumage patterns result from phylogenetic effects that might include natural selection for cryptic coloration. However, the close similarity of the rusty coloration of dendrocolaptid wings and tail to that found in the philydorine ovenbirds, the group suggested to have given rise to the Dendrocolaptidae (Ihering 1915; Feduccia 1973), may indicate that some aspects of woodcreeper plumage are highly constrained. Apart from the characteristic rusty wings and tail of woodcreepers, the generally brownish plumage of many species contains markings that may form complex patterns, often evident in the form of streaking and barring. The presence of barring, especially on the belly, distinguishes the genera *Xiphocolaptes*, *Hylexetastes*, and *Dendrocolaptes* from the remaining woodcreepers. Furthermore, within *Dendrocolaptes*, barring is found either throughout the body plumage, or on the belly in combination with a streaked crown, throat, and breast.

In this paper I reassess plumage variation in the woodcreeper genus *Dendrocolaptes* and compare patterns of plumage similarity using both phylogenetic and phenetic methods. The genus *Dendrocolaptes* traditionally has contained between four and six species (Cory and Hellmayr 1925; Peters 1951; Raikow 1994), with the most thorough review being that of Peters (1951). Because of its completeness, the taxonomy of Peters (1951) was used as a starting point for this study (Fig. 1). On the basis of plumage patterns, the five recognized species can be divided into groups with streaked plumage (*D. picumnus* and *D. platyrostris*) and those with extensively barred plumage (*D. certhia* and *D. concolor*). A fifth species, *D. hoffmannsi*, has been allied with both the streaked (Willis 1982) and barred groups (Raikow 1994) by past workers. Among these species, two are widespread (*D. picumnus* and *D. certhia*), occurring in both Central America and South America from Mexico south to northern Argentina and Bolivia, respectively (Peters 1951). *Dendrocolaptes picumnus* occurs in the lowlands of the Amazon Basin and the Chaco (eastern Bolivia, southwestern Brazil, and western Paraguay), as well as in montane regions in Central America, the Andes of South America, and the coastal ranges of northern Venezuela and Colombia (Figs. 2 and 3). By contrast, the remaining species are all lowland forms. *Dendrocolaptes certhia* is divided by the northern Andes into Haffer's (1967, 1985) trans-Andean (Central America, Chocó, and Nechí; Figs. 4 and 5) and cis-Andean (Amazonian; Fig. 5) populations. The last three species are confined to South America, with both *D. concolor* and *D. hoffmannsi* occurring exclusively in the southern Amazon Basin (Figs. 6 and 7, respectively), and *D. platyrostris* occurring in central and eastern Brazil and adjacent parts of Paraguay and Argentina (Fig. 8).

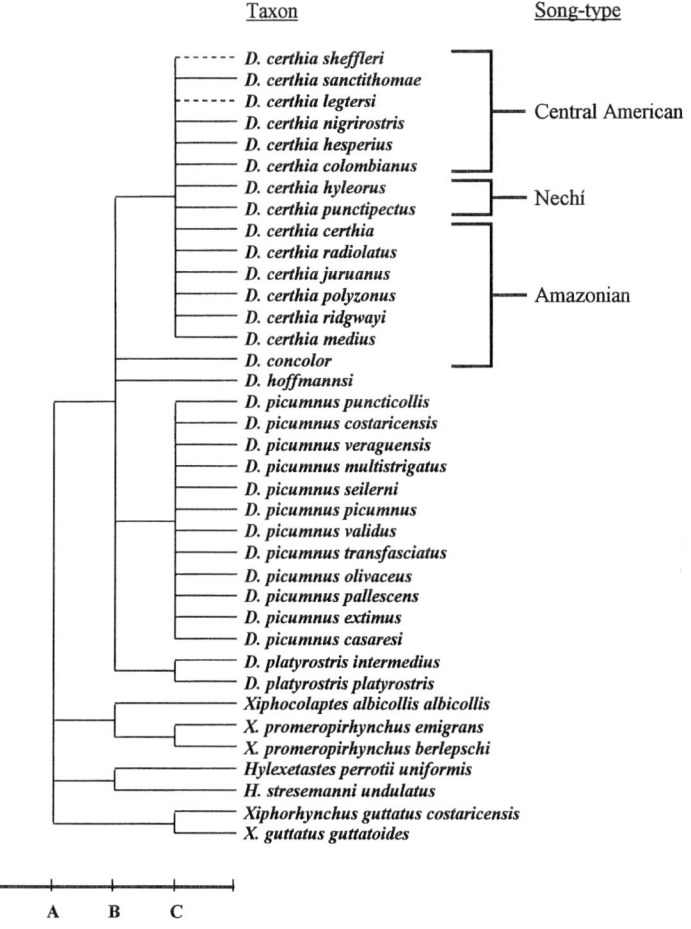

FIG. 1. Taxonomic relationships within *Dendrocolaptes*, and of the outgroup taxa studied, as reflected by Peters (1951). Song-type groups are indicated for the barred woodcreeper complex. The taxonomic levels pictured are A—genus, B—species, and C—subspecies. Note: *D. certhia sheffleri* and *D. c. legtersi* (dashed lines) were described after the publication of Peters (1951).

METHODS

I used two techniques to analyze plumage patterns. To assess the validity of currently recognized taxa, I compared plumage characters from a series of study skins either to the descriptions presented by Cory and Hellmayr (1925) or, for more recently described taxa, the type descriptions. Here I present inconsistencies between past descriptions and patterns I noted in the study skins. Because of complex patterns of variation noted among the Amazonian "barred" woodcreepers (*D. concolor* and *D. certhia*), three ornithologists familiar with the ecology and biogeography of Neotropical birds (J. V. Remsen, Jr., Bret M. Whitney, and David A. Wiedenfeld) independently examined skins of these taxa. A similar method was used previously by Remsen et al. (1991). Rather than assess relationships among each of the subspecies, the primary goal of this consultation was to determine, in a relative sense, how diagnosable the northern and western Amazonian taxa were relative to those occurring in the southeastern part of the basin. Before examining the skins, I provided each observer with a review of the characters used to define the various taxa and a single "type" specimen from each taxon. These "type" specimens were selected from a pool of birds collected from as near as possible to the type locality for each subspecies, and were used as references by which to sort birds into groups. In addition to recognized subspecies, distinct plumage patterns among a series of birds from southeastern Peru warranted their treatment as a separate "taxon" for this analysis. Using the information provided, these observers were asked to divide a randomized series of skins into subspecies groups.

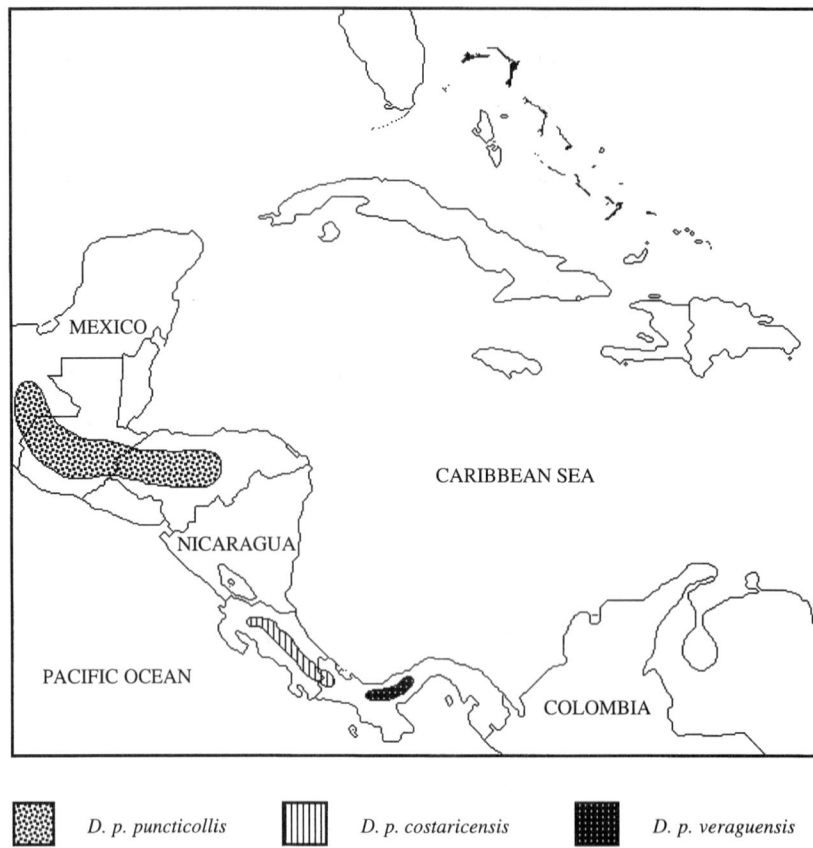

FIG. 2. Distribution of the Central American *Dendrocolaptes picumnus* based on Peters (1951).

In addition to the qualitative analysis of skins, I selected a single specimen from each taxon for comparative analysis. Even though I refer to these taxa using trinomials, the subspecies I studied are comparable to the phylogenetic species used by Cracraft and Prum (1988). A preliminary inspection indicated that use of representative skins was possible because within-population variation in the characters examined was small relative to that among taxa. Although sexual dimorphism in plumage patterns was not evident, adult males were used whenever possible. Skins were chosen at random from among the available series of well-prepared study skins. With the exception of the determination of barring versus streaking on a given region of the body, most plumage characters involved the qualitative assessment of subtle patterns. In addition to 21 plumage characters, bill color was included in the analysis (Table 1). Bill color was assessed from dried study skins because data for soft part colors often were not recorded on specimen labels and, when available, were too variable to be useful. Only crude estimations of bill color were possible due to the variable effects of fading (see Willis 1992). Thus, 22 qualitative binary and multistate characters were analyzed. Abbreviations for museums that housed specimens studied are in the Acknowledgments.

I performed both cladistic and phenetic analyses of the plumage data. Cladistic analysis involved maximum parsimony (Hackett and Rosenberg 1990). For this analysis, heuristic searches were used with the computer program PAUP (Swofford 1991). The resulting cladogram represented a 50% majority-rule consensus of the shortest length trees. To avoid making *a priori* assumptions regarding the evolution of plumage patterns, the characters were treated as unordered, and the trees were rooted using outgroup analysis. These outgroup taxa were chosen from Feduccia's (1973) "strong-billed" woodcreeper group. The placement of outgroups follows the relationships proposed by Raikow (1994), with *Xiphocolaptes* and *Hylexetastes* considered close relatives to *Dendrocolaptes,* and *Xiphorhynchus* treated as a more distant outgroup. The seven outgroups were: *Xiphocolaptes promeropirhynchus emigrans*, *X. p. berlepschi*, *X. albicollis*, *Hy-*

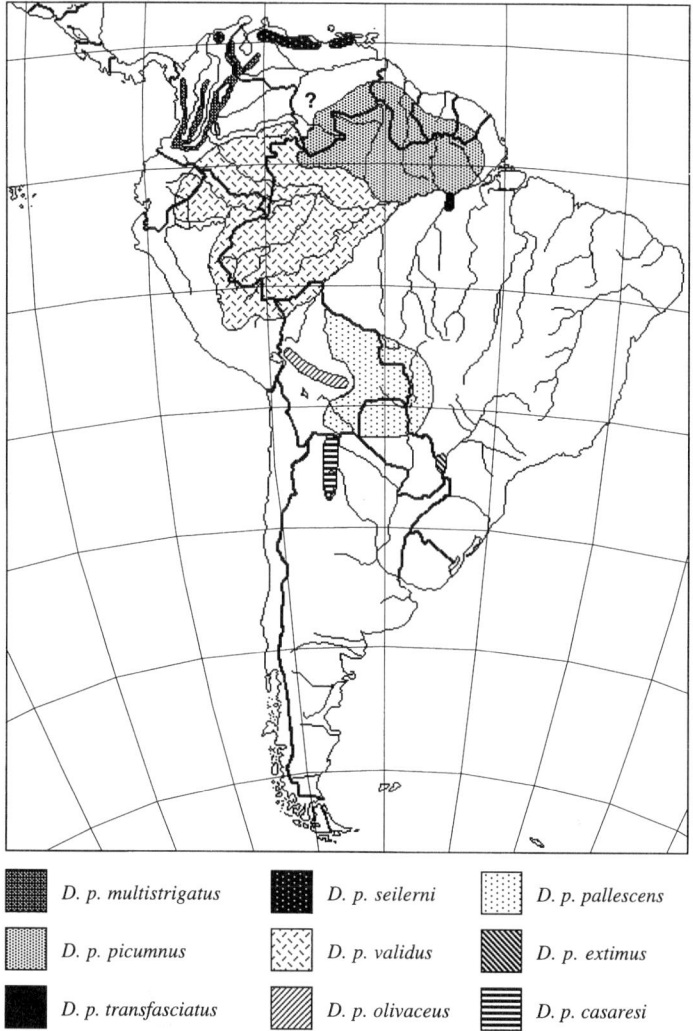

FIG. 3. Distribution of the South American *Dendrocolaptes picumnus* based on Peters (1951). Because the extent of this species' range (and the subspecific identification of birds) in southern Venezuela is uncertain, I have thus used a question mark (?) to denote this region on the map.

lexetastes stresemanni undulatus, *H. perrotii uniformis*, *Xiphorhynchus guttatus costaricensis*, and *X. g. guttatoides*.

For phenetic analyses I constructed a matrix of general similarity coefficients (Gower 1971) and then analyzed this matrix three ways using the software package BioΣtat (Pimentel and Smith 1986). The coefficients were clustered using the UPGMA algorithm (Sneath and Sokal 1973: 230–234) and ordinations were created using both Principal Coordinates Analysis (PCORD) and Non-metric Multi-dimensional Scaling (MDS; Pimentel 1979). I chose these ordination techniques because Pimentel and Smith (1986) cautioned against using multistate characters with Principal Components Analysis. A minimum-spanning tree, constructed from the initial similarity matrix, was superimposed onto the ordinations to examine their distortion of the original similarity values.

RESULTS

Taxonomic limits and the validity of presently recognized taxa.—My analysis of variation within *Dendrocolaptes* (Appendix) suggests that one-fifth of the previously recognized taxa

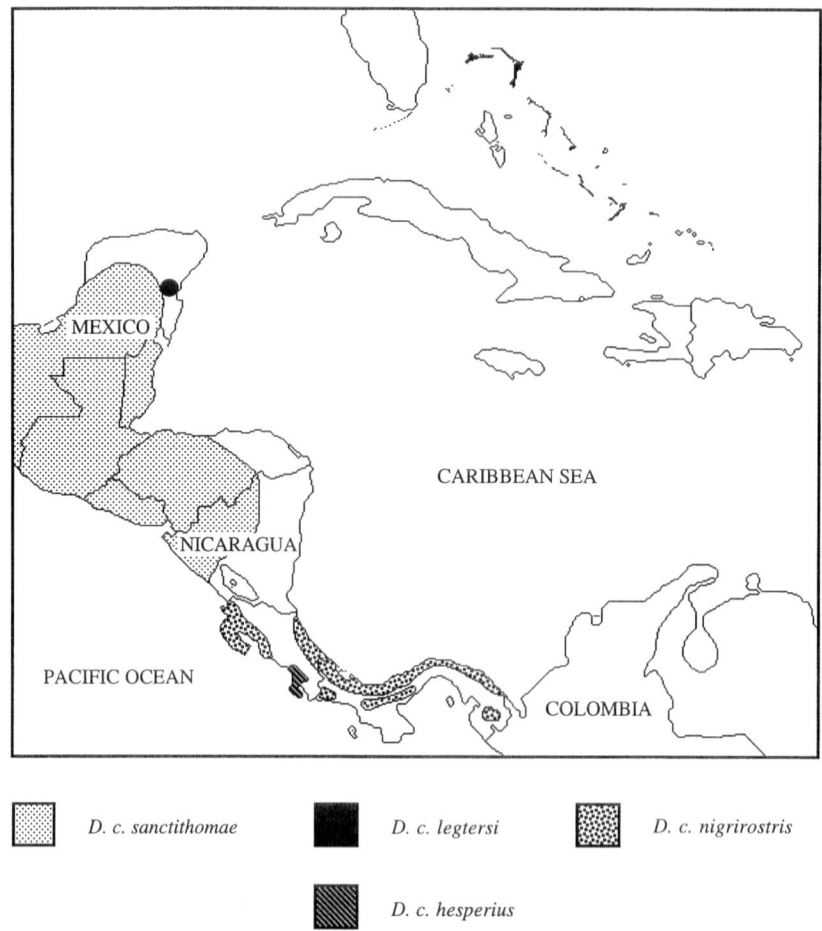

FIG. 4. Distribution of the Central American *Dendrocolaptes certhia* based on Peters (1951). Note that the range of *D. c. sanctithomae* extends off the map to the north and west into Mexico as far as central Veracruz and Campeche (Peters 1951), and that *D. c. sheffleri* (not shown) is restricted to southern Oaxaca (Binford 1965).

should be synonymized (Table 2). This result was not surprising given that Fjeldså (1985) noted that most recent studies of geographic variation have found that too many populations have been formally named. Two factors are likely responsible for the past recognition of invalid taxa within *Dendrocolaptes*: taxa were described on the basis of samples too small to account for intrapopulation variation, and geographic sampling was not at a scale sufficiently fine to expose patterns of clinal variation. Although several apparently valid taxa are recognized from a limited number of specimens (e.g., *D. c. sheffleri*, *D. c. punctipectus*, and *D. p. transfasciatus*), they are quite distinct.

Within the *D. certhia* complex distinct subgroups occupy trans-Andean and cis-Andean regions of the Neotropics. Each group contains several valid subspecies as well as others that seem to reflect artificial breaks in smooth clines (i.e., they appear to have been recognized in large part as a result of collecting bias). In the trans-Andean *D. certhia* there are four or five discrete plumage groups, three (or four, if *D. c. legtersi* is included) within the Central American songtype, and one representing a Nechí population. Although closely allied with the Central American *D. certhia* based on plumage patterns, the Nechí taxa are distinguished by the presence of breast spotting (Phelps and Gilliard 1940; Wetmore 1942). At present, however, sample sizes are too small to recognize the Nechí population as a full species, especially given the apparent introgression between the Nechí *D. c. hyleorus* and the Central American *D. c. colombianus* in central Colombia. Whatever the relationship between the Central American and Nechí populations of

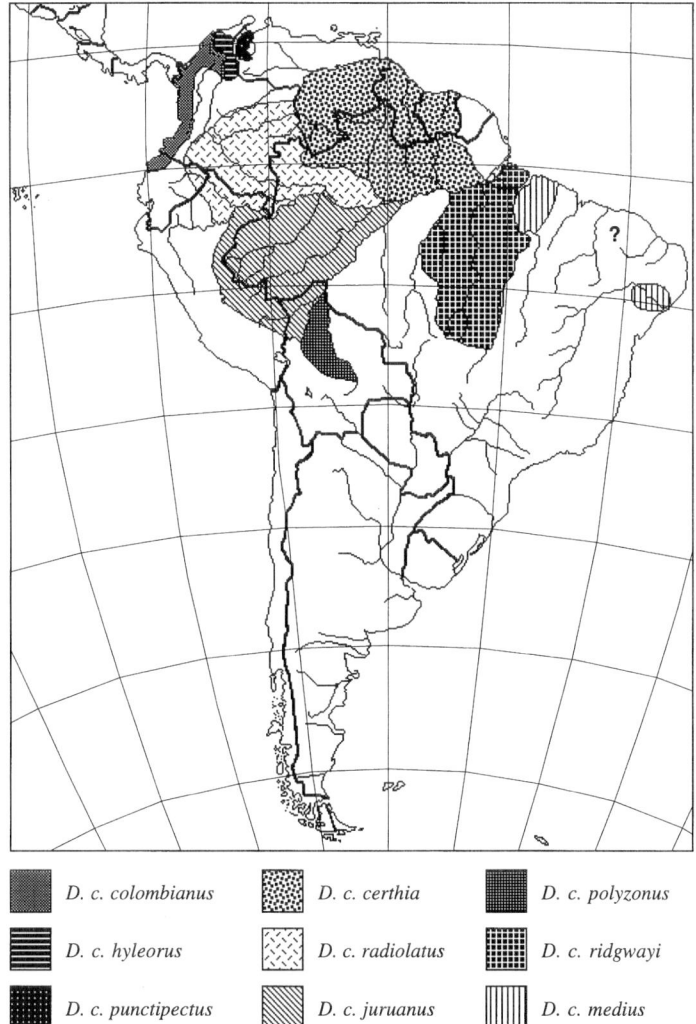

FIG. 5. Distribution of the South American *Dendrocolaptes certhia* based on Peters (1951). Because the extent of this species' range in northeastern Brazil is uncertain, I have used a question mark (?) to denote this region on the map. See text for a clarification of the true extent of the range of *D. c. ridgwayi* and the degree of sympatry between it and *D. concolor*.

D. certhia, the discrete plumage characters separating these trans-Andean forms from cis-Andean taxa (Appendix), in combination with recognized behavioral, vocal, and mensural characters (Marantz 1992; Willis 1992), support the recognition of the trans-Andean populations as a full species under the name *D. sanctithomae* (see also Willis 1992).

In contrast to the relatively linear ranges of the trans-Andean barred woodcreepers, the remaining taxa in this group occupy large areas in the Amazon Basin much like slices in a pie. Most Amazonian taxa are presently separated by major Amazon tributaries (such as the Rio Negro, Rio Madeira, and the Rio Tocantins) that sharply demarcate plumage groups (e.g., Fig. 5). Most populations within this region also seem to lack the clinal variation expected to result from either introgression or subtle variation in climatic or ecological parameters. Only in southeastern Peru and adjacent northwestern Bolivia does plumage variation indicate that some populations have circumvented river systems at their headwaters to come into contact; it is here that introgression between *D. c. radiolatus*, *D. c. juruanus*, and *D. c. polyzonus* is apparent (Fig. 9). Relative to the well-defined *D. certhia* populations in northern and western Amazonia, the southern Amazonian barred woodcreepers are weakly differentiated. Not only are birds from south-

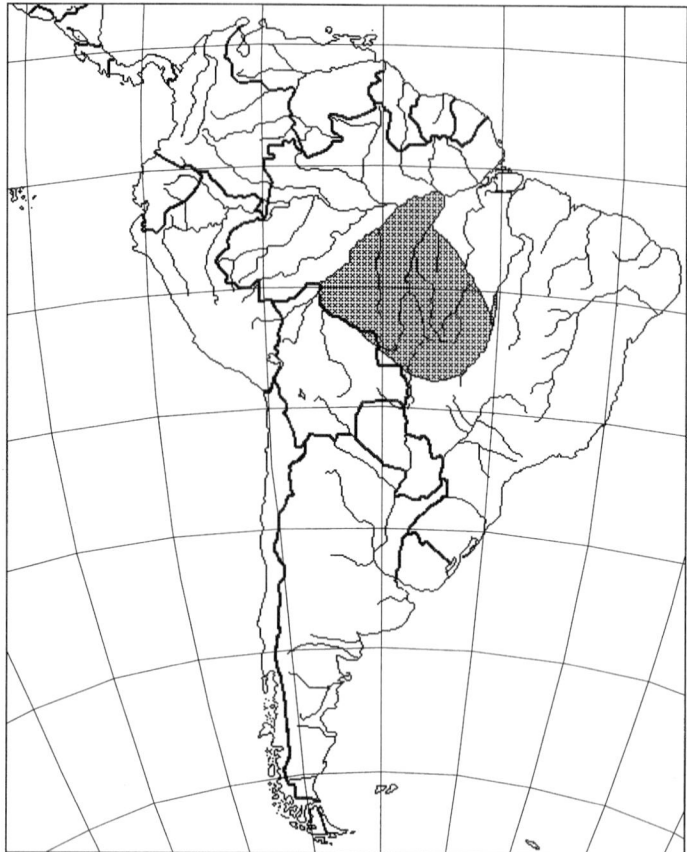

FIG. 6. Distribution of *Dendrocolaptes concolor* based on Peters (1951) and Bates et al. (1989, 1992).

eastern Amazonia weakly differentiated, but plumage patterns appear to vary clinally, suggesting either some form of ecological selection or, more likely, intergradation of two reproductively compatible forms. In either instance, the sample I examined argues against the recognition of *D. c. ridgwayi*. Moreover, the questionable localities of the Olalla specimens (Haffer 1978, 1992), especially with respect to the river bank from which the birds were taken, further complicates the taxonomy of many southern Amazonian species. Because specimens from both sites where *D. concolor* and *D. certhia* reportedly have been collected together were taken by the Olalla brothers, it is possible that birds were collected on both sides of the Rio Tocantins. Before making firm conclusions regarding the specific status of southern Amazonian taxa, an assessment of the true degree of sympatric occurrence and reproductive isolation between *D. certhia* and *D. concolor* should be undertaken. However, plumage data from the specimens I examined do not support the recognition of *D. concolor* as a full species separate from *D. certhia*.

Relative to the biogeographic patterns of the Amazonian *D. certhia*, those of the streaked *Dendrocolaptes* are relatively simple, with discrete plumage variation in many *D. picumnus* taxa likely resulting from their "insular," montane distributions. Among the Middle American *D. picumnus*, I found two discrete plumage groups. Likewise, most of the South American taxa were well-differentiated, even though the southernmost populations of this species appear to be in need of revision. In this case, the two subspecies described most recently (Brodkorb 1941; Steullet and Deautier 1950) were compared with limited series of *D. p. pallescens* that likely did not account for the full range of variation in that subspecies. Although *D. picumnus* was easily differentiated from *D. platyrostris*, the distinction between these two taxa was no more apparent than that between the trans-Andean and cis-Andean populations of *D. certhia*. By contrast, *D. hoffmannsi* was strikingly different from all other *Dendrocolaptes*, based on plumage characters.

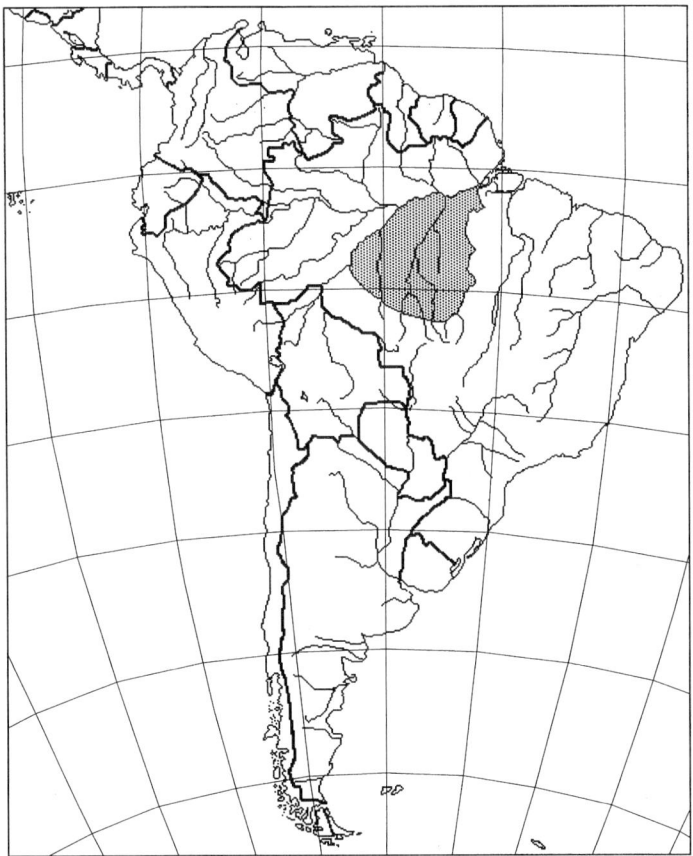

Fig. 7. Distribution of *Dendrocolaptes hoffmannsi* based on Peters (1951).

Comparative analyses.—Monophyly of the *Dendrocolaptes* was not supported by my cladistic analysis of plumage characters because the streaked *Dendrocolaptes* formed a polyphyletic group with respect to the outgroup taxa (Fig. 10). In all trees, however, the barred woodcreepers formed a clade, with *D. concolor* located within the Amazonian *D. certhia*. Although the two Nechí subspecies of *D. certhia* were placed within the Central American song-type, this trans-Andean clade was separated from the Amazonian *D. certhia/D. concolor* clade in a majority of the trees produced. Although 36 most-parsimonious trees were produced (127 steps, Consistency Index = 0.457), the only relationships found in fewer than 100% of the trees related to those among taxa within the Amazonian and Central American barred woodcreeper clades, and in the relationship of these two clades to one another. *Dendrocolaptes hoffmannsi* was the sister-species of the barred woodcreepers in all shortest trees. Beyond this, relationships departed from expectations based on present taxonomy because the streaked *Dendrocolaptes* formed a polyphyletic group.

The UPGMA phenogram reflects patterns of overall similarity based on plumage characters (Fig. 11); the patterns shown were similar but not identical to those of the cladogram. The phenogram produced a clear separation of the Central American and Nechí *D. certhia*, whereas the cladistic analysis placed the Nechí birds among Central American taxa. The plainly marked *Hylexetastes perrotii uniformis* was associated with the *Xiphorhynchus guttatus-Xiphocolaptes albicollis* group rather than with *H. stresemanni*; this placement was most likely a result of the unmarked throat of *H. perrotii*. By contrast, *H. stresemanni* was grouped with *Xiphocolaptes promeropirhynchus* within the streaked *Dendrocolaptes*; unlike *H. perrotii*, *H. stresemanni* has a streaked throat. There was no consistent separation between *D. picumnus* and *D. platyrostris* because *D. picumnus transfasciatus* (from central Amazonia) and nominate *D. platyrostris* (endemic to the Atlantic Coastal forests of southern South America) were allied, probably as a result

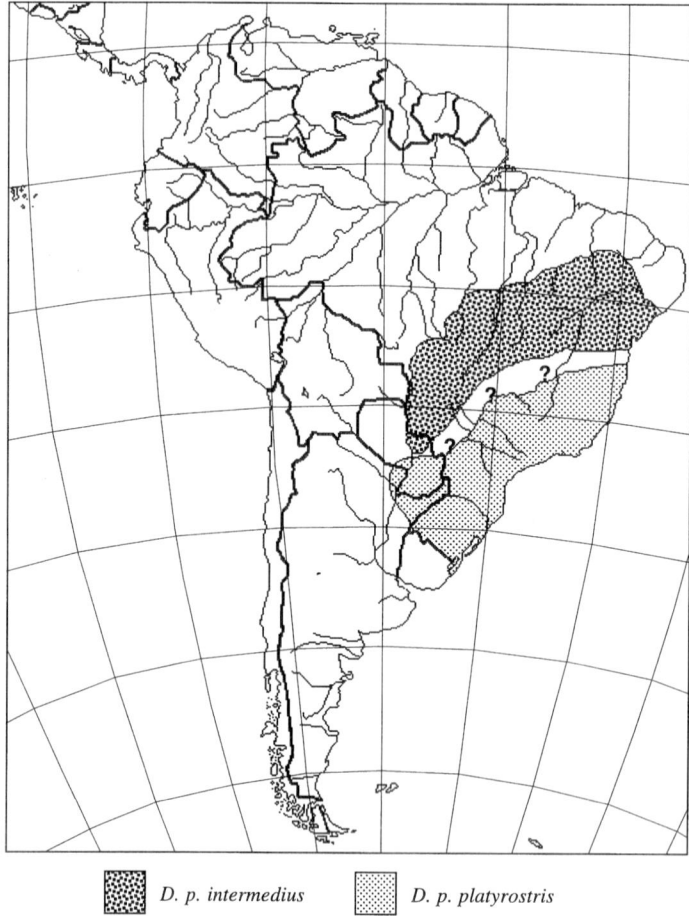

FIG. 8. Distribution of *Dendrocolaptes platyrostris* based on Peters (1951).   D. p. intermedius   D. p. platyrostris

of the shared, blackish crown. Placement of *D. hoffmannsi* in a sister group to a mixed-genus cluster of streaked birds was not surprising given the results of the cladistic analysis; however, the placement of the Middle American *D. picumnus* with this species differed from both my cladistic analysis and the findings of previous workers.

Ordination techniques revealed relationships similar to those suggested by UPGMA cluster analysis. Even though only the first two axes had eigenvalues exceeding 1.0 in the PCORD analysis, the third axis still explained over 10% of the total variance. Because PCORD and MDS produced similar graphical results, however, I discuss only the plots from MDS (Fig. 12). Together, the first three axes separated nearly all of the species groups. The first axis separated barred forms from streaked taxa, with *D. hoffmannsi* and the nearly unmarked *Hylexetastes perrotii* placed centrally along the axis. Among the remaining outgroups, *Xiphorhynchus guttatus*, *H. stresemanni*, and *Xiphocolaptes* were grouped with the streaked *Dendrocolaptes*. The second axis separated the barred *Dendrocolaptes* into three distinct groups. The Amazonian *D. certhia* placed high along the axis (as did *D. concolor*), whereas the two Nechí populations were at the opposite extreme. Placed between these two groups were the Central American *D. certhia*. Although the second axis poorly segregated the streaked birds, the subspecies of *D. picumnus* and *D. platyrostris* were separated marginally along the third axis. The two *Hylexetastes* and the three *Xiphocolaptes* taxa were not well-separated from *D. platyrostris* on either axis. The low scores of *X. guttatus* were sufficient to separate it from the remaining streaked birds along the third axis, whereas the high scores of *D. hoffmannsi* along this axis again segregated it from the remaining taxa.

TABLE 1
Characters and Character States for Plumage Analysis of *Dendrocolaptes* and Selected Outgroup Taxa

Throat		
1.	Chin color:	(a) paler than breast; (b) same shade as breast.
2.	Pattern:	(a) none; (b) barring; (c) streaking; (d) both.
3.	Width of barring:	(a) none; (b) narrow; (c) intermediate; (d) broad.
4.	Width of streaking:	(a) none; (b) narrow; (c) intermediate; (d) broad.
5.	Shade of barring:	(a) none; (b) pale; (c) intermediate; (d) dark.
6.	Shade of streaking:	(a) none; (b) pale; (c) intermediate; (d) dark.
Breast		
7.	Pattern:	(a) none; (b) barring; (c) streaking; (d) both.
8.	Width of barring:	(a) none; (b) narrow; (c) intermediate; (d) broad.
9.	Width of streaking:	(a) none; (b) narrow; (c) intermediate; (d) broad.
10.	Shade of barring:	(a) none; (b) pale; (c) intermediate; (d) dark.
Belly		
11.	Pattern:	(a) none; (b) barring; (c) streaking; (d) both.
12.	Width of barring:	(a) none; (b) narrow; (c) intermediate; (d) broad.
13.	Shade of barring:	(a) none; (b) pale; (c) intermediate; (d) dark.
Crown		
14.	Color:	(a) rusty shades; (b) golden-brown to olive; (c) blackish shades.
15.	Pattern:	(a) none; (b) internal feather barring; (c) narrow, pale shaft-streak; (d) pale shaft-streak rounded as a spot.
16.	Shade of barring:	(a) none; (b) pale; (c) intermediate; (d) dark.
Back		
17.	Pattern:	(a) none; (b) barring; (c) streaking; (d) both.
18.	Shade of barring:	(a) none; (b) pale; (c) intermediate; (d) dark.
19.	Extent of streaking:	(a) none; (b) upper-back; (c) mid-back; (d) throughout.
Bill		
20.	Shade:	(a) pale; (b) intermediate; (c) dark; (d) black.
Other		
21.	Dark loral region:	(a) present; (b) absent.
22.	Barred greater-coverts:	(a) none; (b) indistinct; (c) distinct.

DISCUSSION

The evolution of barring in woodcreepers.—Whereas streaked plumage is shared by nearly all woodcreepers and several philydorine ovenbirds, barring is uncommon in both groups. Although this in itself does not support the derived nature of barring, the only woodcreepers that show any barring whatsoever comprise a single clade that contains the genera *Dendrocolaptes*, *Hylexetastes*, and *Xiphocolaptes* (Raikow 1994). Whereas my data were not sufficient to determine relationships among these three genera, Raikow's results, based on a suite of morphological characters, suggest that barring evolved only once in the family, and that the above genera evolved from a common ancestor with a barred belly. The close similarity in plumage patterns among these three genera is further illustrated by their lack of separation in my phenetic analyses. Two possible explanations might account for the evolution of belly barring in woodcreepers: protective coloration and sexual selection. It seems unlikely, however, that the belly pattern of a woodcreeper that spends most of its life clinging to a tree trunk provides protective coloration (cf. Stiles and Skutch 1989: 259), even though numerous Neotropical woodpeckers are also barred below (Hilty and Brown 1986). Sexual selection likewise appears to be a poor explanation for maintaining such a pattern, especially given that none of the information provided by Willis (1982, 1992) relates to a behavior in which the belly is exposed.

Although belly-barring in *Dendrocolaptes* appears to be a retained, ancestral character, most plumage variation in the genus is consistent with selection for a cryptic lifestyle in birds that spend much of their lives clinging to tree trunks. Outgroup analysis suggests that barring in woodcreepers appeared first on the belly. In the more derived, barred *Dendrocolaptes,* however,

TABLE 2

Revised Taxonomy of the Genus *Dendrocolaptes* Based on Plumage Variation. This Table Contains the Reduced Subset of Discrete Subspecies, the Previously Described Taxa They Include (n = Number of Specimens Examined), Their Geographic Ranges, and Comments on Variation Noted. Specimens of Possible Hybrid Combinations (e.g., the "Balta" Population) Are Not Included Here

Taxon	Geographic range	Comments on plumage variation
D. sanctithomae (= Central American and Nechí populations of *D. certhia*)		
D. s. sheffleri (2)	SW Mexico in vicinity of type locality	none evident in limited sample
D. s. sanctithomae (129) *D. s. nigrirostris* (122) *D. s. colombianus* (45)	S Mexico south through Central America to NW South America (the Chocó region)	brighter plumage and more extensively pale-based bill in northwestern portion of range
D. s. legtersi (4)	E Yucatan Peninsula, Mexico, in vicinity of type locality	none evident in limited sample
D. s. hesperius (25)	SW Costa Rica and adjacent Panama	none evident
D. s. punctipectus (1) *D. s. hyleorus* (14)	NE Colombia and extreme NW Venezuela	Venezuelan birds average brighter overall and more olive than do those from Colombia
D. certhia (= Amazonian populations of *D. certhia* in addition to *D. concolor*)		
D. c. certhia (177)	N South America west to Rio Negro, south to Amazon	worn birds may appear like *D. c. juruanus*
D. c. radiolatus (136)	NW portion of the Amazon basin, east to Rio Negro	none evident
D. c. juruanus (52)	SW portion of the Amazon basin, east to Rio Madeira	see comments under *D. c. polyzonus*
D. c. polyzonus (6)	N Bolivia and SE Peru at base of Andes	none evident, but birds showing characters of this subspecies occur north well into Peru
D. c. medius (34)	NE Brazil, east of Rio Tocantins and south of Amazon	hybridization with *D. concolor* evident at western limit of range
D. c. concolor (57) *D. c. ridgwayi* (26)	NE Bolivia and central Brazil, south of the Amazon between the Rio Madeira and the Rio Tocantins	paler and grayer in SW, more extensively barred towards NE; variation appears to result from hybridization with *D. c. medius*
D. hoffmannsi (22)	Central Brazil, south of the Amazon between the Rio Madeira and the Rio Xingu	immatures appear to be more obviously streaked below than are adults
D. picumnus		
D. p. puncticollis (58)	mountains of S Mexico south into Honduras	none evident
D. p. costaricensis (14) *D. p. veraguensis* (4)	highlands of Costa Rica and western Panama	birds in western Panama are somewhat more rufescent than those in Costa Rica
D. p. multistrigatus (23)	Andes of Colombia and western Venezuela	none evident
D. p. seilerni (23)	Santa Marta region of Colombia, and coastal ranges of northern Venezuela	none evident

TABLE 2
Continued

Taxon	Geographic range	Comments on plumage variation
D. p. picumnus (26)	N South America north of Amazon, east of Rio Negro	none evident
D. p. validus (48)	W Amazonia east to Rio Negro and Rio Madeira	none evident
D. p. transfasciatus (1)	central Brazil, south of Amazon, east of Rio Tapajós	only one specimen examined
D. p. olivaceus (11)	Andean foothills in central Bolivia	northern birds more olivaceous, southern birds more rusty overall
D. p. pallescens (25) *D. p. extimus* (5)	Chaco region of SW Brazil, S Bolivia, Paraguay in addition to adjacent humid forests in E Paraguay	highly variable with respect to intensity of rufous tones and extent of barring
D. p. casaresi (5)	NW Argentina	identical to *D. p. pallescens* based on plumage patterns
D. platyrostris		
D. p. intermedius (57)	central and NE Brazil, and NE Paraguay	crown coloration of birds from C Paraguay suggests intergradation with *D. p. platyrostris*
D. p. platyrostris (280)	SE Brazil, NE Argentina, and adjacent Paraguay	see comments under *D. p. intermedius*

this pattern occurs throughout the body plumage. In this respect, *D. hoffmannsi* is intermediate between the streaked and barred forms in that it shows minimal streaking (which is more prevalent on immatures than on adults) and is weakly, but more extensively, barred below than are other streaked taxa. Whereas it is difficult to provide an adaptive explanation for barring that is restricted to the belly, Willis (1982) proposed that streaking in *Dendrocolaptes* may be an adaptation for blending in with furrowed bark or the spaces between epiphyte leaves. These streaks may also serve to break up the outline of the bird, thus making it more difficult for predators to locate. By contrast, barred birds used trunks that were better illuminated, and they moved more slowly and remained stationary for longer periods of time than did streaked birds (Willis 1982). Even though these observations are anecdotal, they provide the first hypothesis suggesting an adaptive basis for plumage variation within *Dendrocolaptes*.

Plumage characters support two clades within the barred *Dendrocolaptes*. These clades correspond to the trans-Andean and cis-Andean regions of the Neotropics. Although these two groups were easily diagnosable using plumage characters, the differences between them are subtle enough that all barred woodcreepers were considered a single species by some workers. This subtlety makes it difficult to provide an adaptive explanation for the discrete characters separating the two clades.

In the taxa studied, there appear to be three, discrete transitions in plumage patterns: 1) the appearance of barring on the belly that sets the *Xiphocolaptes*, *Hylexetastes*, and *Dendrocolaptes* clade apart from all other woodcreepers; 2) the extensively barred body that lacks streaking, which sets the *D. certhia*/*D. concolor* complex apart from the remaining *Dendrocolaptes* (with *D. hoffmannsi* apparently an intermediate step in this transition); and 3) the separation of the trans-Andean and cis-Andean barred woodcreepers based on crown coloration, the presence of a loral spot, and the character of barring on the underparts. It is these three, well-supported transitions that appear to represent discrete phylogenetic characters. By contrast, subgroups within each of these clades are weakly supported by characters that appear to reflect both convergence and local climatic variation.

Biogeographic patterns within Dendrocolaptes.—Biogeography of the lowland *Dendrocolaptes* follows closely the areas of endemism proposed by Haffer (1967, 1985) and Cracraft (1985). Because few cladistic hypotheses have been developed for Neotropical species, however, the

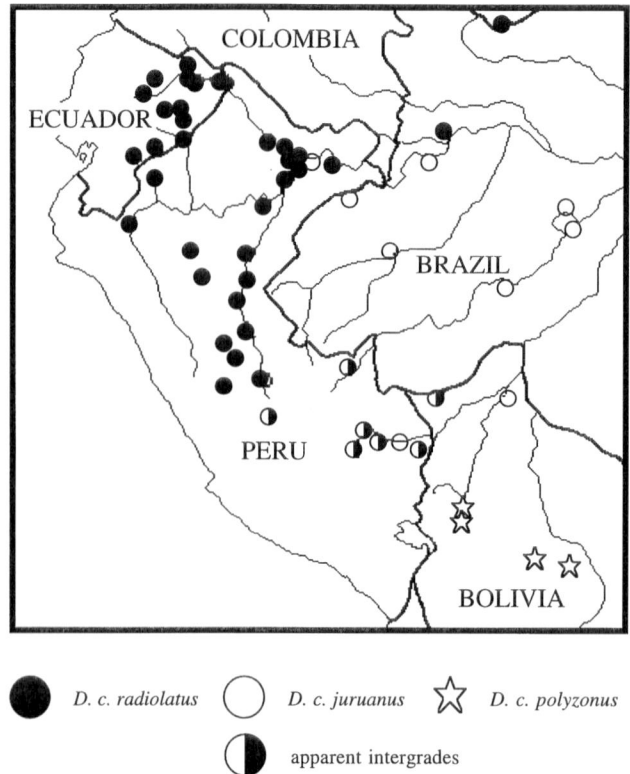

FIG. 9. Distribution of *Dendrocolaptes certhia* in western and southwestern Amazonia based on specimens examined for this study. "Apparent intergrades" refers to birds showing a combination of deep ochraceous coloration and finely barred underparts (suggesting either *D. c. polyzonus* or introgression between *D. c. juruanus* and *D. c. radiolatus*).

congruence of relationships among taxa in this region is difficult to assess. Taxa in the barred woodcreeper complex are separated into groups occupying both trans-Andean and cis-Andean regions of the Neotropics. Among the streaked *Dendrocolaptes*, *D. hoffmannsi* and *D. platyrostris* are exclusively cis-Andean; only *D. picumnus* occurs west of the Andes, where it is restricted to the highlands. All three streaked *Dendrocolaptes* occur in Amazonia, and both *D. picumnus* and *D. platyrostris* are found in Haffer's (1985) "non-forested" regions as well as in forested areas.

Two distinct plumage groups of *D. certhia* occur in Central America. One of these groups corresponds to the region of maximum endemism within Haffer's (1985) Pacific Southern Middle America, whereas the other occurs in lowland forests of Central America on both the Caribbean and Pacific slopes. Like Cory and Hellmayr (1925), I found no discrete break in plumage characters between specimens from this latter group and those occurring in the Chocó region of northwestern South America. Chapman (1917) and Cracraft and Prum (1988) similarly noted the link between Central American taxa and those from northwestern South America, and also that this influence may be traced as far north as southern Mexico (as is the case in *D. certhia*). One possible explanation for the colonization of Central America by barred *Dendrocolaptes* is that the South American forms invaded Central America during a period when humid climates allowed the widespread appearance of lowland forest across the Central American land-bridge (Haffer 1967). Hackett (1993) supported a similar hypothesis for the colonization of Central America by the antbird genus *Gymnopithys*, using molecular data. The mechanism by which trans-Andean and cis-Andean populations were later isolated is less clear (Cracraft and Prum 1988), but likely resulted from either the uplift of the Andes (Chapman 1917) or successive climatic fluctuations (Haffer 1967). A few populations appear to have been isolated in lowland forest refugia in Central America (e.g., the Chiriquí area of endemism (Haffer 1967), south-

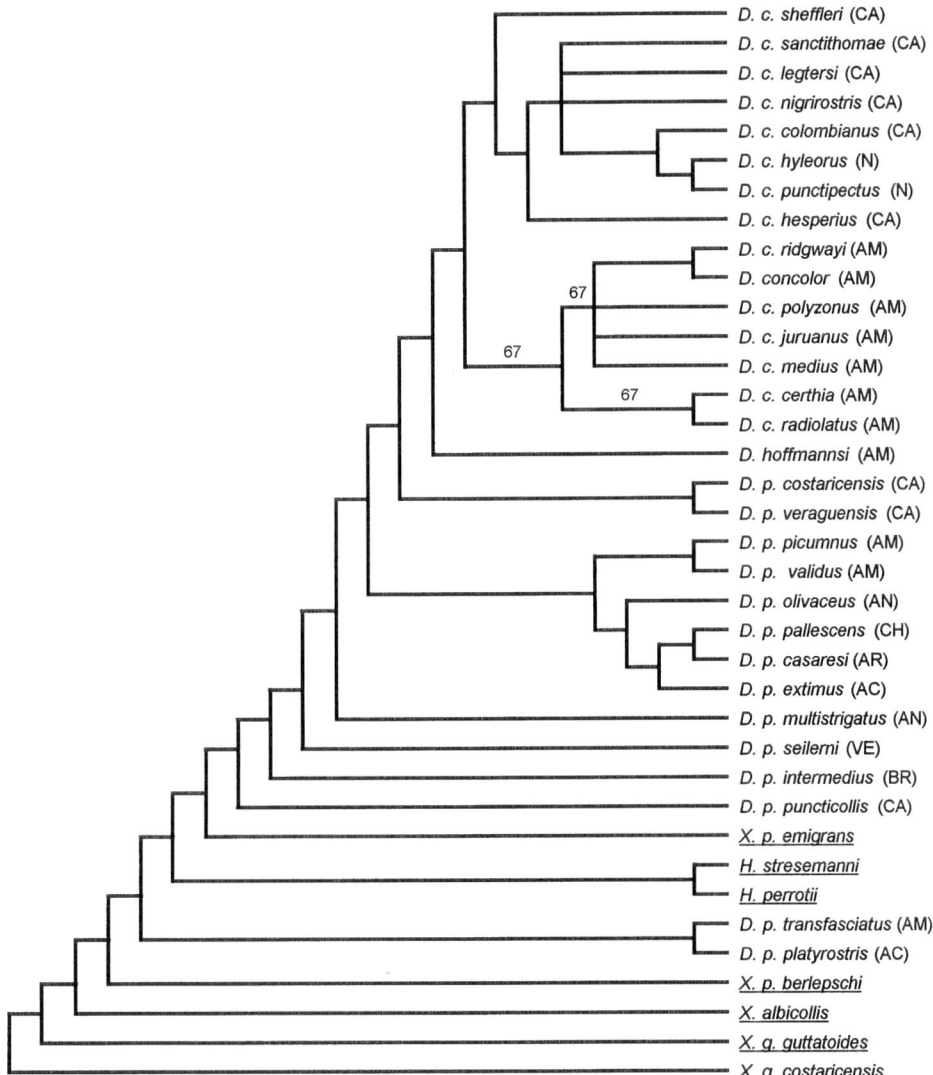

FIG. 10. A 50% majority rule consensus from a cladistic analysis of *Dendrocolaptes* plumage characters. The only differences among the various trees relate to the placement of taxa within both the Amazonian and Central American barred woodcreeper clades, and in the relationship of these two clades. The length of the 36 most-parsimonious trees was 127 steps. Numbers at the nodes correspond to the percentage that a given node was represented in the set of 36 trees; those nodes with no number were represented in all 36 trees. Underlined taxa are outgroups. The abbreviations appended to the species names refer to: Central American (CA), Nechí (N), Amazonian (AM), Andean (AN), coastal ranges of Venezuela (VE), central and eastern Brazil (BR), Chaco (CH), Atlantic Coastal Forest (AC), and northern Argentina (AR).

western Mexico, and the Yucatan Peninsula) while the more widespread population retreated. The final occupation of Central America resulted from either a reinvasion from western Colombia, or from the expansion of a relatively undifferentiated Middle American population to both the north and south. Some Central American populations of *D. certhia* have remained geographically isolated to this day; however, in others that did come into secondary contact, intergradation is apparent. Although the weak step-cline evident in bill coloration in *D. c. sanctithomae* may have resulted from either selection against intermediates or from secondary contact, the presence of a step-cline cannot by itself be used to distinguish the two (Endler 1983). I believe that the generally variable nature of bill coloration in *Dendrocolaptes* argues against selection for this character, and indicates that secondary contact is likely. During the last reinvasion of Central

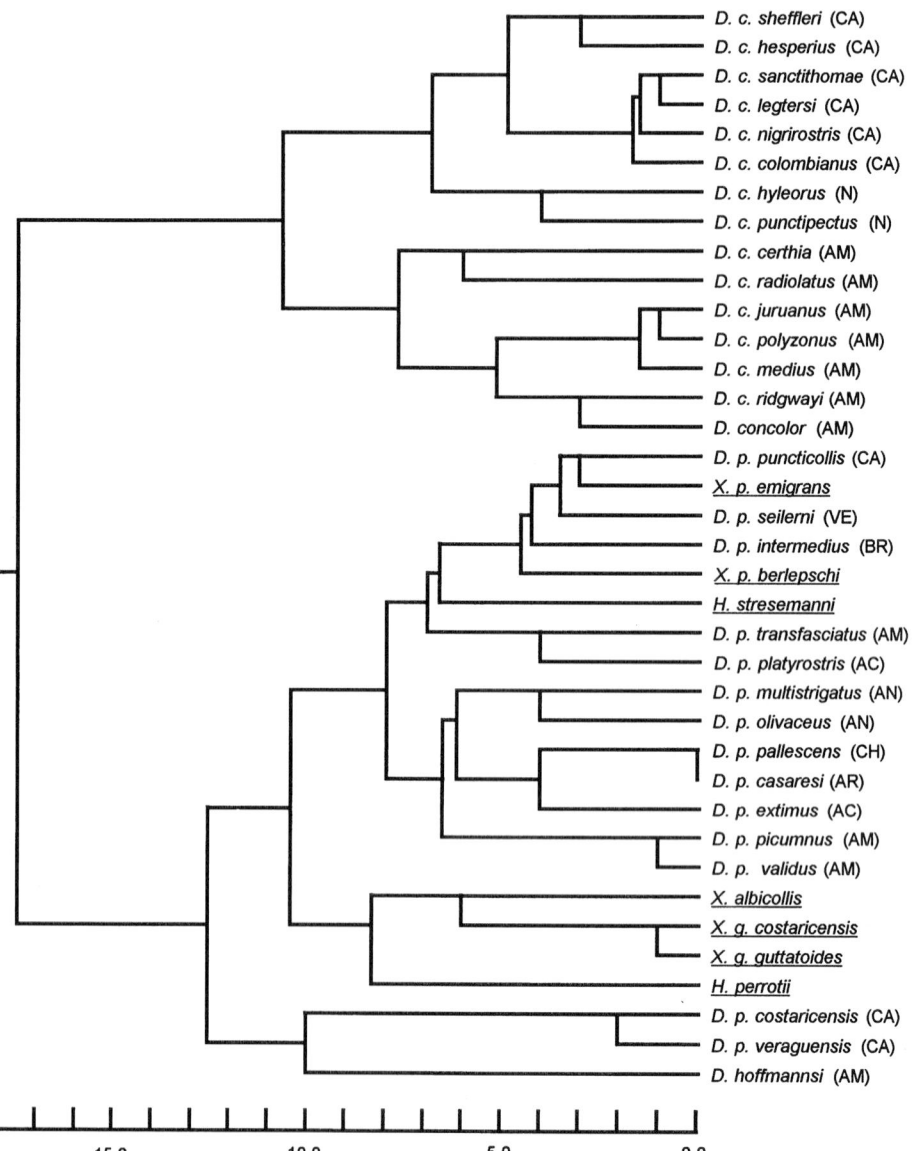

FIG. 11. UPGMA phenogram for *Dendrocolaptes* plumage characters. Outgroup taxa (underlined) and abbreviations are the same as in Figure 10. This phenogram was generated using a matrix computed using Gower's (1971) General Similarity Coefficient. The Cophenetic Correlation Coefficient of 0.796 suggests a relatively good fit to the data.

America, the Caribbean-Slope subspecies traversed the central ranges through low passes in northwestern Costa Rica and now occurs on the Pacific Slope in the northern half of the country. That this subspecies only recently arrived on the Pacific coast is further supported by its apparently limited introgression with a form endemic to the coastal forests of the Chiriquí area of endemism. The phenomenon of Caribbean-Slope taxa colonizing the Pacific Slope though the low passes in northern Costa Rica has been documented in other species as well (Stiles and Skutch 1989: 7).

The Nechí *D. certhia* are closely allied with the Central American taxa, differing in only a single plumage character. Even so, my cladistic and phenetic analyses produced contradictory hypotheses of relationship for these two populations. The cladistic analysis produced an unre-

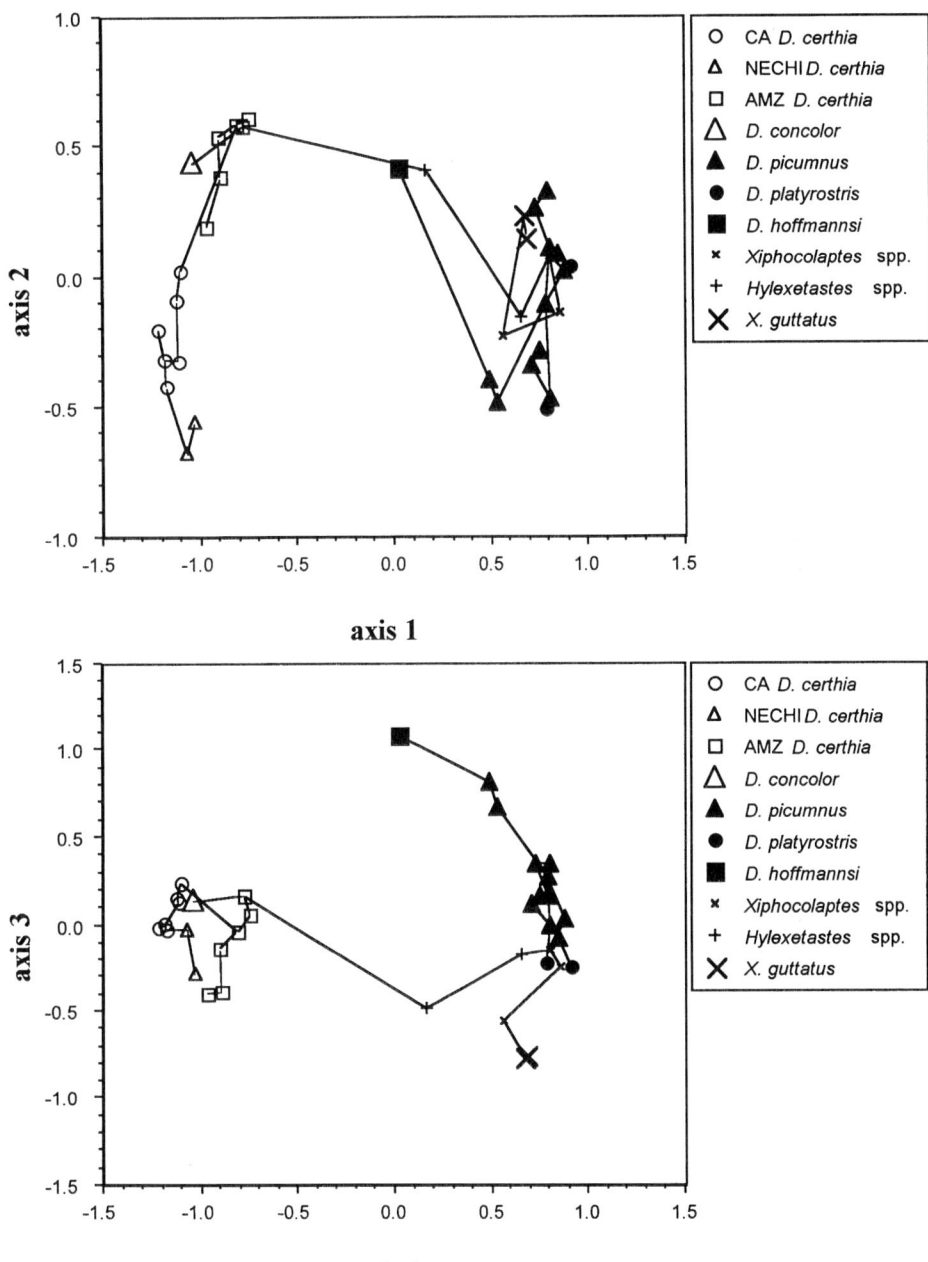

FIG. 12. Graphical representation of the analysis using non-metric, multidimensional scaling (MDS) on 22 plumage and bill characters for *Dendrocolaptes* and selected outgroup taxa. The species studied formed three distinct groups apparently based on degree of barring versus streaking. The extensively barred *D. certhia* and *D. concolor* were segregated to the far left of the graphs, whereas the heavily streaked *D. picumnus*, *D. platyrostris*, *Xiphocolaptes* spp., *Xiphorhynchus guttatus*, and *Hylexetastes stresemanni* were placed at the right. The minimally streaked *D. hoffmannsi* and unmarked *H. perrotii* were found near the center of the graph. The three song-types of *D. certhia* were separated along axis 2 (upper graph), whereas *D. picumnus* and *D. platyrostris* were divided along axis 3 (lower graph). The minimal distortion of the minimum-spanning tree indicates that this analysis accurately portrayed true group similarity.

solved group of trans-Andean taxa that allied the Nechí birds with the geographically adjacent Chocó representative of the Central American song-type, whereas the phenetic analyses separated the Central American and Nechí taxa. The former relationship suggests that the Nechí birds were separated from the Central American taxa after differentiation of most Central American forms, whereas the similarity depicted by the phenogram suggests that the Nechí and Chocó regions were isolated soon after the initial invasion of the trans-Andean region from Amazonia, but prior to the differentiation of Middle American forms. This latter hypothesis is further supported by the striking vocal differences between the Nechí, *D. p. punctipectus*, and the Chocó representative of the Central American song type (Marantz 1992, Willis 1992). Although I could not quantify the degree of reproductive isolation between the Central American and Nechí taxa based on the limited number of available specimens, hybridization between these forms was apparent in the Magdalena Valley of central Colombia. Haffer (1967) presented two additional examples of secondary contact and hybridization by birds in a nearby area just east of the Gulf of Urabá; unfortunately, few specimens of *D. certhia* have been collected in this region.

In contrast to the complex variation shown by trans-Andean *D. certhia*, biogeographic patterns of the Amazonian taxa are relatively simple and corroborate patterns documented for other Amazonian birds by Haffer (1985), Cracraft and Prum (1988), and Hackett (1993). Among Haffer's areas of endemism, five of six regions contain distinct, barred woodcreeper taxa. The correspondence between the distributions of these birds and areas of endemism is illustrated by comparing my Figures 5 and 6 with Haffer's Figure 4. The Napo region forms the western center of the range of *D. c. radiolatus*, the Inambari region corresponds with *D. c. juruanus* and the "Balta" population, *D. concolor* (and *D. certhia "ridgwayi"*) are concentrated in the Rondônian area of endemism, *D. c. certhia* the Guianas area, and *D. c. medius* the Belém region. In sum, Haffer's only area of endemism not corresponding to the range of a barred woodcreeper, the Imerí region, is located approximately in the region where *D. c. radiolatus* and *D. c. certhia* meet along the Rio Negro (which also divides *Gymnopithys rufigula* and *G. leucaspis*; Hackett 1993). Unfortunately, my cladistic analysis poorly resolved relationships among Amazonian barred woodcreepers. Even so, the relationship between *D. concolor* and *D. certhia "ridgwayi"* closely matched that of sympatric *Pteroglossus* aracaris (Cracraft and Prum 1988). Likewise, the clade uniting the Guiana and Napo representatives matched the relationships documented in the genus *Gymnopithys* (Hackett 1993), despite countering the relationships in four clades in which the Guiana taxon was the sister-species to a clade uniting the Inambari and Napo taxa (Cracraft and Prum 1988).

Like those of many Amazonian bird species (Capparella 1988, 1991), the ranges of barred woodcreepers in this region are now delimited primarily by major Amazon tributaries. This "river effect" has long been known in many Amazonian taxa in which plumage variation is limited within a taxon over a large geographic area, but differs sharply among specimens collected in close proximity on opposite sides of major rivers (reviewed by Haffer 1992). Even though rivers do not form boundaries between some avian taxa (Haffer 1985; Haffer and Fitzpatrick 1985; Haffer 1997), they do separate nearly all Amazonian *Dendrocolaptes*. Furthermore, those rivers delimiting *Dendrocolaptes* are the ones suggested by Haffer (1992) to present major barriers to avian dispersal (Rio Amazonas, Rio Madeira, Rio Negro, and Rio Tocantins); by contrast, the Rio Tapajós and the Rio Xingu do little more than modify the extent of the apparent hybrid zone between *D. certhia* and *D. concolor*. Moreover, those taxa that have crossed rivers appear to have done so in the headwaters regions, where river size may present less of an obstacle (Haffer 1985, 1992; Haffer and Fitzpatrick 1985; Haffer 1997). Although the crossing of rivers in their headwaters region has been presented as an argument against the rivers initially delimiting populations (Haffer 1985, 1992), Cracraft and Prum (1988) pointed out that this is not necessarily true. In fact, only a correspondence between the age of the taxa and that of the proposed vicariance event can be used to choose between events responsible for dividing populations (Capparella 1988, 1991; Cracraft and Prum 1988). As such, Capparella's molecular data for manakins (Pipridae), which suggest that at least some Amazonian taxa are substantially older than the Pleistocene forest refugia, argues strongly for the development of major Amazonian tributaries as the principal isolating mechanism separating populations of some taxa.

Unlike the barred *Dendrocolaptes*, few streaked taxa are found in lowland regions; however, those that do conform to the patterns described by Haffer (1985). Of the three Amazonian subspecies of *D. picumnus*, one occurs in the Guianas region, another occupies both the Napo and Inambari areas of endemism, and the last is known from a small portion of the Rondônian region. Although Cracraft and Prum (1988) presented two additional examples of taxa not differentiated across the Napo and Inambari regions, the barred woodcreepers occurring in these areas were distinct. My cladistic analysis further indicated that the southern forms of *D. picumnus* evolved from lowland Amazonian taxa. The former comprise both a subspecies occurring in the foothills adjacent to southern Amazonia (*D. p. olivaceus*) and members of the *D. p. pallescens* species-group, which occupy both the "non-forested" Chaco region and a forested site within the southeastern Brazil region. Finally, the two subspecies of *D. platyrostris* occur in Haffer's (1985) forested, southeastern Brazil and his "nonforested," Caatinga and Campos Cerrado areas of endemism, respectively.

Like the Amazonian *D. picumnus*, which are separated by river systems, the montane subspecies presently occupy disjunct regions of the Neotropics separated by expanses of unsuitable habitat. This insular distribution has facilitated formation of several distinct taxa. The similarity of plumage patterns in forms from the coastal ranges of Venezuela and from northern Central America suggests the possibility of an early montane invasion of Central America (in times preceding the uplift of the northern Andes) by ancestral populations allied with the Venezuelan form. This latter taxon now occurs in disjunct populations in the coastal ranges of northern Venezuela and in the Santa Marta region of Colombia (Peters 1951). Such a distribution suggests that this form can either disperse between isolated islands of suitable habitat (Chapman [1917: 154–157] presented a detailed discussion regarding the unlikelihood of such an event), or more likely, that it is an old taxon that has failed to differentiate since the isolation of these mountain ranges by the uplift of the Mérida Andes (which are now occupied by the northern Andean subspecies, *D. p. multistrigatus*). Striking differences, both zoological and geological, between the Santa Marta Mountains and the Andes suggest that these two mountain ranges, though only 65 km apart, are of independent origin and have never been connected (Chapman 1917). In contrast to the results of my cladistic analysis, the phenetic similarity of the northern Andean subspecies to Amazonian taxa suggests that the former evolved more recently from Amazonian ancestors as a result of the uplift of the northern Andes. This hypothesis is additionally supported both by Chapman's (1917) discussion regarding the close relationship between faunas of the northern Andes and upper Amazonia, and by the fact that the northern Andean subspecies (*D. p. multistrigatus*) occurs in an area that geographically separates disjunct populations of *D. p. seilerni* from the coastal ranges of northern Venezuela and the Santa Marta region of Colombia. That the Middle American *D. picumnus* evolved recently from southern South American forms, as suggested by my cladistic analysis, is probably no more likely than their placement by my phenetic analysis with *D. hoffmannsi* (as the outgroup to a mixed-genus group of streaked woodcreepers). By contrast, the close similarity in general appearance between the northern Andean and Middle American taxa supports Chapman's hypothesis that the northern Andes of Colombia and the mountains of Costa Rica and western Panama were connected in the recent past.

Relative to variation shown by northern populations of *D. picumnus*, my cladistic and phenetic analyses allied the southern Andean *D. p. olivaceus* with either the Chaco and Amazonian subspecies, or the northern Andean taxon, respectively. The first of these hypotheses indicates an independent evolution of two montane populations from a lowland form, whereas the second indicates two relictual, foothill populations presently separated by approximately 2,000 km. In either case, the apparent similarity between *D. p. olivaceus* and *D. p. seilerni* (Zimmer 1934), from the coastal ranges of northern Venezuela, appears to be the result of convergence. In conclusion, my cladistic analysis indicates multiple invasions of Central America by populations of *D. picumnus* derived from northern South America, as well as differentiation of lowland forms into the Andean foothills in both the northern and southern parts of the species' range.

The remaining streaked species are *D. hoffmannsi*, a Rondônian endemic, and *D. platyrostris*, a widespread Brazilian species that occurs in both the southeastern Brazil and the "non-forested" Caatinga and Campos Cerrados areas of endemism. Although this latter taxon occurs in "non-forested" regions throughout much of its range, it generally occupies locally occurring woodland habitats, such as gallery forest, within these areas (e.g., Olmos 1993). The apparent effects of convergence between the sympatric, southern Amazonian *D. hoffmansi* and *D. concolor*, and between the Atlantic coastal *D. p. platyrostris* and the Amazonian *D. picumnus transfasciatus* (see also Willis 1982), complicate the biogeographic interpretations of their ranges.

Ecological correlates of plumage variation.—Much of the difficulty in resolving biogeographic patterns in *Dendrocolaptes* appears to result from subtle examples of convergence. In some instances this variation follows patterns predicted by ecological factors. For example, taxa occupying drier and more open habitats are often paler than birds occurring in humid rainforest (i.e., Gloger's rule). However, because most *Dendrocolaptes* frequent forested regions, evolutionary patterns must be inferred from less precise factors such as rainfall and other climatic variables. It should also be stressed that even though Gloger's rule has been well established empirically within birds (e.g., Zink and Remsen 1986), the adaptive bases of such plumage variation are still poorly understood (Selander 1971).

Plumage variation in *Dendrocolaptes* conforms to Gloger's rule in that taxa occurring in humid areas generally appear darker and more brightly colored than do those in drier regions. In Amazonian *D. certhia*, the darkest and most brightly patterned taxa occur in the humid forests at the base of the Andes from eastern Colombia south through northern Bolivia, whereas birds found farther to the east in the basin have paler plumage and less distinct barring on the underparts. Similarly, in eastern Amazonia, birds from the humid Amazon delta (Ratisbona 1976; Haffer and Fitzpatrick 1985) are more noticeably barred than are those occurring in the drier region to the west of the Rio Tocantins. Amazonian *D. picumnus* share the trends in *D. certhia* plumage patterns, albeit to a lesser degree. The plumage patterns of *Dendrocolaptes* in other regions follow Gloger's rule as well. Mexican *D. certhia*, which occur in an area somewhat drier than that found farther south (Mosiño Alemán and García 1974; Portig 1976), are relatively pale, whereas the subspecies that occurs on the Yucatan Peninsula, which is drier than nearby mainland sites, is even paler (Paynter 1954). In southern South America, the *D. picumnus* subspecies found in the dry Chaco region is pale overall and has indistinct barring on its underparts. Similarly, nominate *D. platyrostris*, which occurs in the humid Atlantic coastal forests, is more rufescent and has a more striking head pattern than does the duller subspecies found in drier and more open, woodland habitats. Although bright plumage and bold barring is correlated with rainfall in woodcreepers, within this framework, plumage patterns are uniform in some taxa over large regions in which climate is variable. Such variation may indicate that the effects of climate have been homogenized to some extent by gene flow. Furthermore, it is a lack of gene flow that most likely accounts for the striking differences among taxa now found in areas with similar rainfall patterns on opposite sides of large river systems.

Plumage patterns in the outgroup taxa provide support for those evident in sympatric *Dendrocolaptes*. For example, *Hylexetastes perrotii* and sympatric barred woodcreepers share similar plumage patterns on both sides of the Amazon (Willis 1992), whereas *H. stresemanni*, which occurs in humid forests nearer the Andes, is heavily barred like sympatric forms of both *D. certhia* and *D. picumnus*. Likewise, *Xiphocolaptes albicollis*, of the Atlantic coastal rainforests, has plumage patterns remarkably similar to the sympatric *D. platyrostris*. Even though selection by similar environmental factors would be expected to produce birds of similar appearance, especially given the close phylogenetic relationship of the taxa studied here, the similarity between sympatric species-pairs like the Amazonian *D. concolor* and *Hylexetastes perrotii uniformis*, and *D. c. certhia* and *H. p. perrotii*, and the Atlantic coastal forms, *D. p. platyrostris* and *Xiphocolaptes albicollis*, is remarkable.

Willis (1992) proposed that competitive mimicry, rather than Gloger's rule, may account for the plumage similarity in sympatric woodcreeper species. By mimicking the larger *H. perrotii*, it was suggested that *D. certhia* may gain a competitive advantage in interactions with *D. picumnus*. Because there is no evidence of interspecific territoriality between these three species (Willis 1992), however, I suspect that the only advantage *D. certhia* may obtain from mimicking the larger *Hylexetastes* may be a reduced incidence of attack by *D. picumnus* over ant swarms. Reduced predation when predators confuse *D. certhia* with the larger *Hylexetastes* is another possible explanation (Willis 1992), but this hypothesis has its flaws. First, it seems unlikely that predators strong enough to take large woodcreepers would distinguish between species of roughly similar size, even though *Hylexetastes* are stronger, and possibly more effective fighters, than are *Dendrocolaptes* (A. P. Capparella *in litt.*). Also, because *Dendrocolaptes* are more abundant and widespread than are sympatric *Hylexetastes*, it would seem difficult for both potential competitors and predators to learn which birds to avoid. Furthermore, even though *D. certhia* is distributed over a larger geographic range than are the two *Hylexetastes*, its plumage patterns vary little over this region. Even in the absence of sympatric, intermediate-sized competitors, convergence is apparent between *D. platyrostris* and *Xiphocolaptes albicollis*. In this case the only possible selection for mimicry would have to be exerted by predation. In conclusion, competitive mimicry seems to provide a poor explanation for convergent plumage patterns in sympatric woodcreeper species.

Preliminary comparison of character sets.—Although only limited vocal and morphometric data are presently available (Willis 1982, 1992; Marantz 1992), preliminary comparisons between plumage characters and other data sets can be made. Vocal characters supported both the division of trans-Andean and cis-Andean barred woodcreepers, and the separation of Central American and Nechí groups within the trans-Andean region. Within the trans-Andean region, the Chocó representative was allied with the Central American *D. certhia* both by song-type (based on a recording from Hardy et al. 1991) and overall plumage similarity. Vocally, *D. c. punctipectus*, a representative of the Nechí group, has a song that is in some ways intermediate between those of the Central American and the Amazonian forms, although it is strikingly different from both (Marantz 1992). Within Amazonia, advertising songs of barred woodcreepers varied little, with the plumage similarity between *D. concolor* and Amazonian *D. certhia* supported by similar songs. Despite the striking plumage differences between *D. hoffmannsi* and *D. picumnus*, the single, brief recording of *D. hoffmannsi* contained songs much like those of *D. picumnus*. In a counter example, the similar plumage patterns of *D. picumnus* and *D. platyrostris* contrasted with recognizable song differences. Within *D. picumnus*, however, songs varied little geographically despite most subspecies being well-supported by plumage patterns.

In contrast to the results of the plumage analyses, generic groups were well-supported morphometrically, even though patterns within *Dendrocolaptes* were not as well defined. In general, there was extensive overlap between the streaked and barred *Dendrocolaptes*; however, within the barred woodcreepers, bill morphology supported the separation of trans-Andean and cis-Andean clades (Marantz 1992). These differences in bill morphology appear to correspond to modes of foraging in that Amazonian birds rely to a greater degree on aerial sallying, whereas Central American birds forage more by gleaning, probing, and following ant swarms (Willis 1982, 1992). Although *Dendrocolaptes hoffmannsi* was as distinct morphometrically as it was using plumage characters, the similar plumage patterns of *D. picumnus* and *D. platyrostris* contrasted with different bill shapes in addition to the above noted song differences.

ACKNOWLEDGMENTS

I thank my advisor, J. V. Remsen, for his help through all phases of the project. Robert M. Zink and Mark S. Hafner also contributed valuable suggestions towards the completion of this phase of the project. The analysis of plumage variation was greatly facilitated by the help of Bret M. Whitney, J. V. Remsen, and David A. Wiedenfeld, each of whom examined numerous skins for me and shared with me their ideas on classification of these birds. I am especially indebted to Cerise L. Cauthron for proof-reading and support. Mario Cohn-Haft kindly translated Willis's (1992) paper for me. Reviewers Angelo P. Capparella and David A. Wiedenfeld provided many valuable suggestions in the course of reviewing the manuscript. John M. Bates, Michael A. Patten, Donald E. Kroodsma, and Peter Houlihan also commented on earlier versions of this manuscript. Steven W. Cardiff helped immeasurably in the processing of numerous loans of specimens to LSU. Shannon J. Hackett assisted me with the cladistic analysis. Members of the sections of Ornithology and Biochemical Systematics at the Museum of Natural Science, Louisiana State University (especially John M. Bates, Andrew W. Kratter, Shannon J. Hackett, and Manuel Marín A.), provided a stimulating atmosphere for the development and exchange of ideas. The hospitality of John M. Bates and Shannon J. Hackett greatly facilitated my work at AMNH. I thank the following for either loaning specimens to me or granting me access to their collections: François Vuilleumier, Allison V. Andors, and Mary LeCroy (American Museum of Natural History; AMNH); Kimball L. Garrett (Los Angeles County Museum of Natural History; LACM); Kenneth C. Parkes and James Loughlin (Carnegie Museum of Natural History; CM); Charles R. Brown (Peabody Museum, Yale University; YPM); Scott M. Lanyon, David E. Willard, and Thomas S. Schulenberg (Field Museum of Natural History; FMNH); Gary R. Graves and Philip Angle (United States National Museum of Natural History; USNM); Robert B. Payne and Janet Hinshaw (Museum of Zoology, University of Michigan; UMMZ); Frank B. Gill and Mark B. Robbins (Academy of Natural Sciences, Philadelphia; ANSP); Lloyd F. Kiff and Jon C. Fisher (Western Foundation of Vertebrate Zoology); Ned K. Johnson (Museum of Vertebrate Zoology, University of California, Berkeley); John C. Hafner (Moore Laboratory of Zoology, Occidental College); James R. Northern (University of California, Los Angeles); Richard O. Prum (Museum of Natural History, University of Kansas); the late David J. Klingener (Department of Biology, University of Massachusetts); and Gene K. Hess (Delaware Museum of Natural History; DMNH). Finally, I wish to acknowledge the late Theodore A. Parker III, for initially suggesting that I work on the members of the genus *Dendrocolaptes* and for providing insight and guidance through the early phases of this project.

Literature Cited

Baker, A. J. 1985. Museum collections and the study of geographic variation. Pp. 55–77 *in* Museum Collections: Their Roles and Future in Biological Research (E. H. Miller, Ed.). Occas. Pap. Brit. Columbia Prov. Mus. 25.

Bates, J. M., M. C. Garvin, D. C. Schmitt, and C. G. Schmitt. 1989. Notes on bird distribution in

northeastern Dpto. Santa Cruz, Bolivia, with 15 species new to Bolivia. Bull. Br. Ornithol. Club 109: 236–244.
BATES, J. M., T. A. PARKER III, A. P. CAPPARELLA, AND T. J. DAVIS. 1992. Observations on the *campo, cerrado* and forest avifaunas of eastern Dpto. Santa Cruz, Bolivia, including 21 species new to the country. Bull. Br. Ornithol. Club 112:86–98.
BARROWCLOUGH, G. F. 1983. Biochemical studies of microevolutionary processes. Pp. 223–261 *in* Perspectives in Ornithology (A. H. Brush and G. A. Clark, Jr., Eds.). Cambridge Univ. Press, Cambridge, U.K.
BINFORD, L. C. 1965. Two new subspecies of birds from Oaxaca, Mexico. Occas. Pap. Mus. Zool. Louis. State Univ. 30:1–6.
BRODKORB, P. 1941. A race of woodhewer from the Alto Parana. Occas. Pap. Mus. Zool. Univ. Mich. 453: 1–3.
CAPPARELLA, A. P. 1988. Genetic variation in Neotropical birds: implications for the speciation process. Proc. Int. Ornithol. Congr. 19:1658–1664.
CAPPARELLA, A. P. 1991. Neotropical avian diversity and riverine barriers. Proc. Int. Ornithol. Congr. 20: 307–316.
CHAPMAN, F. M. 1917. The distribution of bird-life in Colombia; a contribution to a biological survey of South America. Bull. Am. Mus. Nat. Hist. 36:1–729.
CORY, C. B., AND C. E. HELLMAYR. 1925. Catalogue of the Birds of the Americas and the Adjacent Islands, Part 4. Publ. Field Mus. Nat. Hist., zool. ser. 13.
CRACRAFT, J. 1985. Historical biogeography and patterns of differentiation within the South American avifauna: areas of endemism. Pp. 49–84 *in* Neotropical Ornithology (P. A. Buckley, M. S. Foster, E. S. Morton, R. S. Ridgely, and F. G. Buckley, Eds.). Ornithol. Monogr. No. 36.
CRACRAFT, J., AND R. O. PRUM. 1988. Patterns and processes of diversification: speciation and historical congruence in some Neotropical birds. Evolution 42:603–620.
ENDLER, J. A. 1983. Testing causal hypotheses in the study of geographic variation. Pp. 424–443 *in* Numerical Taxonomy (J. Felsenstein, Ed.). Springer-Verlag, Berlin.
FEDUCCIA, A. 1973. Evolutionary Trends in the Neotropical Ovenbirds and Woodhewers. Ornithol. Monogr. No. 13.
FJELDSÅ J. 1985. Subspecies recognition in ornithology: history and the current rationale. Fauna Norv. Ser. C, Cinclus 8:57–63.
GOWER, J. C. 1971. A general coefficient of similarity and some of its properties. Biometrics 27:857–871.
GRISCOM, L. 1927. Undescribed or little-known birds from Panama. Am. Mus. Novit. 280.
GRISCOM, L., AND J. C. GEENWAY, JR. 1941. Birds of lower Amazonia. Bull. Mus. Comp. Zool. 88:85–344.
HACKETT, S. J. 1993. Phylogenetic and biogeographic relationships in the Neotropical genus *Gymnopithys* (Formicariidae). Wilson Bull. 105:301–315.
HACKETT, S. J., AND K. V. ROSENBERG. 1990. Comparison of phenotypic and genetic differentiation in South American antwrens (Formicariidae). Auk 107:473–489.
HAFFER, J. 1967. Speciation in Colombian forest birds west of the Andes. Am. Mus. Novit. 2294.
HAFFER, J. 1978. Distribution of Amazon forest birds. Bonn. zool. Beitr. 29:38–78.
HAFFER, J. 1985. Avian zoogeography of the Neotropical lowlands. Pp. 113–146 *in* Neotropical Ornithology (P. A. Buckley, M. S. Foster, E. S. Morton, R. S. Ridgely, and F. G. Buckley, Eds.). Ornithol. Monogr. No. 36.
HAFFER, J. 1992. On the "river effect" in some forest birds of southern Amazonia. Bol. Mus. Paraense Emílio Goeldi, sér Zool. 8:217–245.
HAFFER, J., AND J. W. FITZPATRICK. 1985. Geographic variation in some Amazonian forest birds. Pp. 147–168 *in* Neotropical Ornithology (P. A. Buckley, M. S. Foster, E. S. Morton, R. S. Ridgely, and F. G. Buckley, Eds.). Ornithol. Monogr. No. 36.
HARDY, J. W., T. A. PARKER III, AND B. B. COFFEY, JR. 1991. Voices of the Woodcreepers: Neotropical Family Dendrocolaptidae. Cassette Sound Recording distributed by ARA Records, Gainesville, Florida.
HILTY, S. L., AND W. L. BROWN. 1986. A Guide to the Birds of Colombia. Princeton Univ. Press, Princeton, New Jersey.
IHERING, H. V. 1915. The classification of the family Dendrocolaptidae. Auk 32:145–153.
MARANTZ, C. A. 1992. Evolutionary implications of vocal and morphological variation in the woodcreeper genus *Dendrocolaptes* (Aves: Dendrocolaptidae). M.S. thesis, Louisiana State University, Baton Rouge, Louisiana.
MOSIÑO ALEMÁN, P. A., AND E. GARCÍA. 1974. The climate of Mexico. Pp. 345–404 *in* Climates of North America, vol. 11. (R. A. Bryson and F. K. Hare, Eds.). Elsevier Scientific Publishing Co., Amsterdam.
MUNSELL COLOR. 1988. Munsell Soil Color Charts. Munsell Color, Baltimore.
OLMOS, F. 1993. Birds of Serra da Capivara National Park, in the "caatinga" of north-eastern Brazil. Bird Conserv. Internat. 3:21–36.
PAYNTER, R. A., JR. 1954. Three new birds from the Yucatan Peninsula. Postilla 18:1–4.
PETERS, J. L. 1951. Check-list of Birds of the World, vol. 7. Museum of Comparative Zoology, Cambridge, Massachusetts.

Phelps, W. H., and E. T. Gilliard. 1940. Six new birds from the Perijá Mountains of Venezuela. Am. Mus. Novit. 1100.
Pimentel, R. A. 1979. Morphometrics: The Multivariate Analysis of Biological Data. Kendall/Hunt Publishing Co., Dubuque, Iowa.
Pimentel, R. A., and J. D. Smith. 1986. BioΣtat II: A Multivariate Statistical Toolbox. Sigma-Soft, Placentia, California.
Portig, W. H. 1976. The climate of Central America. Pp. 405–478 in Climates of Central and South America, vol. 12 (W. Schwerdtfeger, Ed.). Elsevier Scientific Publishing Co., Amsterdam.
Raikow, R. J. 1994. A phylogeny of the woodcreepers (Dendrocolaptinae). Auk 111:104–114.
Ratisbona, L. R. 1976. The climate of Brazil. Pp. 219–293 in Climates of Central and South America, vol. 12 (W. Schwerdtfeger, Ed.). Elsevier Scientific Publishing Co., Amsterdam.
Remsen, J. V., Jr., O. Rocha O., C. G. Schmitt, and D. C. Schmitt. 1991. Zoogeography and geographic variation of *Platyrinchus mystaceus* in Bolivia and Peru, and the circum-Amazonian distribution pattern. Ornithol. Neotropical 2:77–83.
Ridgley, R. S., and G. Tudor. 1994. The Birds of South America, vol 2: The Suboscine Passerines. University of Texas Press, Austin.
Selander, R. K. 1971. Systematics and speciation in birds. Pp. 57–147 in Avian Biology, vol. 1. (D. S. Farner and J. R. King, Eds.). Academic Press, New York.
Sneath, P. H. A., and R. R. Sokal. 1973. Numerical Taxonomy: The Principles and Practice of Numerical Classification. W. H. Freeman and Co., San Francisco.
Steullet, A., and E. Deautier. 1950. Una nueva subespecie de *Dendrocolaptes pallescens* Pelzeln. Hornero 9:175–177.
Stiles, F. G., and A. F. Skutch. 1989. A Guide to the Birds of Costa Rica. Comstock Publishing Associates, Ithaca, New York.
Swofford, D. L. 1991. PAUP: Phylogenetic Analysis Using Parsimony, Version 3.0s. Illinois Natural History Survey, Champaign, Illinois.
Todd, W. E. C. 1925. Descriptions of new Furnariidae and Dendrocolaptidae. Proc. Biol. Soc. Wash. 38: 79–82.
Todd, W. E. C. 1948. Critical remarks on the wood-hewers. Ann. Carnegie Mus. 31:5–18.
Todd, W. E. C. 1950. The northern races of *Dendrocolaptes certhia*. Journ. Washington Acad. Sci. 40:236–238.
Vaurie, C. 1980. Taxonomy and geographical distribution of the Furnariidae (Aves, Passeriformes). Bull. Am. Mus. Nat. Hist. 166:1–357.
Wetmore, A. 1942. New forms of birds from Mexico and Colombia. Auk 59:265–268.
Willis, E. O. 1982. The behavior of Black-banded Woodcreepers (*Dendrocolaptes picumnus*). Condor 84: 272–285.
Willis, E. O. 1992. Comportamento e ecologia do arapaçu-barrado *Dendrocolaptes certhia* (Aves, Dendrocolaptidae). Bol. Mus. Paraense Emílio Goeldi, sér Zool. 8:151–216.
Zimmer, J. T. 1934. Studies on Peruvian birds. No. 14. Notes on the genera *Dendrocolaptes*, *Hylexetastes*, *Xiphocolaptes*, *Dendroplex*, and *Lepidocolaptes*. Am. Mus. Novit. 753.
Zink, R. M., and J. V. Remsen, Jr. 1986. Evolutionary processes and patterns of geographic variation in birds. Curr. Ornithol. 4:1–69.

APPENDIX

Qualitative Review of Plumage Variation

Whereas the barred *Dendrocolaptes* are treated below in three, biogeographically defined groups, the streaked species are discussed together under one heading. *Dendrocolaptes hoffmannsi* is treated separately because of its intermediacy between the barred and streaked birds based on plumage characters. Despite making taxonomic recommendations (see Table 2), I have maintained the taxa used by Peters (1951) throughout this discussion for consistency. For more precise color definitions, many specimens were compared using soil color charts (Munsell Color 1988); codes (e.g., 10 YR 6/6), and capitalized color names refer to colors from these charts.

Trans-Andean *D. certhia*

Central American D. certhia.—Specimens in this group were characterized by the combination of a mostly blackish bill, rufescent crown, dark loral region, well-defined barring on the underparts, and the lack of contrast between the chin, throat, and breast (i.e., these birds lack the distinctive pale throat of the Amazonian birds). These characters, several of which were noted by Cory and Hellmayr (1925), combine to give the Central American birds a distinctive appearance that is quite unlike that of the Amazonian populations of the species. Furthermore, the distinction is a discrete one that corresponds closely with the break between the cis-Andean and trans-Andean regions. No intergradation nor clinal variation is evident in the plumage patterns of these presently allopatric groups.

Within the Central American *D. certhia*, birds were separated into three, distinct clusters based on plumage patterns: (1) *D. c. sheffleri*; (2) a group that combined *D. c. sanctithomae*, *D. c. nigrirostris*, *D. c. colombianus*, and possibly, *D. c. legtersi*; and (3) *D. c. hesperius*. Although only two specimens (including the holotype) of *D. c. sheffleri* were examined, the distinctive bill coloration and plumage patterns of this subspecies easily distinguish it from all other taxa. Because my findings matched those of Binford (1965), additional comment is unnecessary.

Compared to those defining *D. c. sheffleri*, no discrete plumage differences exist between the subspecies *D. c. sanctithomae*, *D. c. nigrirostris*, and *D. c. colombianus*. Although birds from Mexico and Belize have a more extensive pale base to the lower mandible than those collected farther south and east, Binford (1965) noted that topotypical specimens of *D. c. sanctithomae* were more similar to *D. c. nigrirostris* than to the western populations of *D. c. sanctithomae* examined by Todd (1950). Moreover, variation in the extent of the pale base to the bill appears to be clinal, with birds having gradually darker bills from Mexico to Nicaragua. Likewise, plumage variation appears to vary clinally, with northern birds appearing more golden below (Brownish-Yellow, 10 YR 6/8 versus 10 YR 6/6), and a little more rufescent on the crown (Yellowish-Red (5 YR 5/8) versus Light Yellowish-Brown (10 YR 6/4), or Dark Yellowish-Brown (10 YR 4/4 and 10 YR 4/6)), compared with those collected along the Pacific coast of South America. Even so, some southern birds examined were identical in both respects to those taken in Mexico. For example, a brightly colored bird from prov. Esmeraldas, Ecuador (ANSP 182408), was identical to birds taken in Belize with respect to coloration of both the underparts (Brownish-Yellow, 10 YR 6/8) and crown (Yellowish-Red, 5 YR 5/8). I therefore agree with Binford (1965) that neither bill coloration nor plumage variation consistently separate *D. c. sanctithomae* and *D. c. nigrirostris*. Furthermore, *D. c. colombianus* was equally difficult to distinguish from *D. c. nigrirostris* based on my examination.

In addition to his assessment of plumage characters, Binford (1965) pointed out a discrepancy in Todd's mensural data and indicated that he found little difference in bill length between *D. c. sanctithomae* and *D. c. nigrirostris*. By making this same correction in the measurements supplied by Todd (1950), it appears that bill lengths of *D. c. nigrirostris* and *D. c. colombianus* are likewise similar. Todd's corrected measurements, along with those supplied by Willis (1992), further suggest that culmen length varies clinally, with the bill becoming progressively longer from north to south. It therefore appears that neither plumage patterns nor bill characters support the recognition of either *D. c. nigrirostris* or *D. c. colombianus*.

A taxonomic assessment of *D. c. legtersi* is complicated by the limited number of available specimens. All specimens examined were collected at two nearby sites located approximately half-way up the east coast of the Yucatan Peninsula (Carrillo Puerto and Tabi, edo. Quintana Roo, Mexico). Other specimens of *D. certhia* taken to the north, west, and south of this region, however, appear typical of *D. c. sanctithomae*. Of 17 *D. certhia* specimens collected in edos. Campeche and Quintana Roo, Mexico (13 identified as *D. c. sanctithomae* and four labeled *D. c. legtersi*), there was a trend towards paler overall coloration, and less golden underparts, as one moved north onto the peninsula. The pale, nearly gray (Light Yellowish-Brown, 10 YR 6/4, to Brownish-Yellow, 10 YR 6/6) underparts of the four *D. c. legtersi* specimens made them easily distinguishable from all of the *D. c. sanctithomae* with which they were compared. Within *D. c. sanctithomae*, birds collected at the base of the peninsula at La Tuxpena, edo. Campeche, were identical to those taken in edo. Tabasco (underparts, Brownish-Yellow 10 YR 6/8), whereas two birds from Chompoton, edo. Campeche, were slightly less golden below (Brownish-Yellow, 10 YR 6/6) than were the birds noted above. These Chompoton specimens matched closely two birds collected west of Chetumal, edo. Quintana Roo. The differing dates of collection for the Chompoton and Chetumal specimens (1 and 24 May [UMMZ 137995 and UMMZ 137996] versus 18 and 14 February [YPM 8473 and YPM 8474], respectively) suggest that the pale coloration of *D. c. legtersi* is not an artifact of feather wear. Although some fading may have resulted from feather wear (e.g., a bird collected on 21 March 1949 [YPM 8472] was slightly brighter and more golden (Brownish-Yellow, 10 YR 6/6, versus Light Yellowish-Brown, 10 YR 6/4) below than were the three taken 6-20 June 1950 [YPM 13507-13509]), this effect appeared to be minor. Although only a limited number of *D. certhia* specimens have been collected on the Yucatan Peninsula, the small distances separating these collecting sites suggest that the plumage variation leading to the recognition of *D. c. legtersi* is the result of a step-cline rather than a smooth transition. When describing *D. c. legtersi*, Paynter (1954) suggested that it probably occurred on the northern portion of the peninsula to the limit of rain forest; however, a single specimen of *D. certhia* taken at the northeast tip of the peninsula (11 km. S of Puerto Juarez, Quintana Roo; DMNH 34872) matched perfectly *D. c. sanctithomae* collected near Chetumal. This specimen, if labeled correctly, indicates that even if *D. c. legtersi* is considered valid, its range must be restricted to the immediate vicinity of the type locality in the east-central portion of the peninsula.

In contrast to the above amalgamation of taxa, *D. c. hesperius* appears to be a valid taxon. Specimens from throughout the limited range of this subspecies (22 from the Osa Peninsula and the Térraba Valley, prov. Puntarenas, Costa Rica; one from prov. Chiriqui, Panama; and two from Pigres, prov. Puntarenas, Costa Rica; Fig. 4) were consistently separable from both *D. c. nigrirostris* and *D. c. sanctithomae* based on the combination of narrow barring on the underparts of *D. c. hesperius* (dark bars generally ≤1 mm, versus bars >1 mm in *D. c. sanctithomae* and *D. c. nigrirostris*) and the more grayish-brown (Brownish-Yellow, 10 YR 6/6), less golden cast to its underparts (especially the breast). Specimens of this subspecies were further distinguished from *D. c. nigrirostris* collected on the Caribbean slope of Costa Rica, based on a slightly more extensive, pale region at the base of their lower mandibles. In this regard, specimens of *D. c. hesperius* matched those of *D. c. sanctithomae* collected in Honduras or Guatemala. In general, the crown of *D. c. hesperius* is slightly less rufescent (Yellowish-Brown, 10 YR 5/6, versus Strong Brown, 7.5 YR 5/8) than in *D. c. nigrirostris* (appearing similar to *D. c. colombianus* in this regard); however, brighter-than-average *D. c. hesperius* were inseparable from typical, Mexican *D. c. sanctithomae* based on crown color.

The two above noted specimens from the mouth of the Río Tarcoles at Pigres, prov. Puntarenas (09°47′

N, 84°39′ W), which appear to represent *D. c. hesperius*, suggest that the range of this subspecies may extend farther north than previously believed. One of the two birds collected at this locality (USNM 198702) matched specimens of *D. c. hesperius* taken both on the Osa Peninsula and in the Térraba Valley, whereas the second individual from this site (USNM 198703) appeared intermediate between typical representatives of both *D. c. hesperius* and *D. c. nigrirostris*. This latter bird had more golden-yellow underparts than did typical *D. c. hesperius* and the underpart barring was wider relative to that of the other specimen taken at this locality. These two specimens suggest both that the range of *D. c. hesperius* extends up the coast of southwestern Costa Rica to the mouth of the Río Tarcoles, and that it may intergrade with *D. c. nigrirostris* in this region. It should be noted, however, that three specimens collected approximately eight kilometers to the *south* of Pigres at Las Agujas, prov. Puntarenas (09°43′ N, 84°39′ W; LACM 16263-16265), appeared typical of *D. c. nigrirostris* (K. L. Garrett *in litt.*). In general, plumage variation in *D. c. hesperius* did not appear to be clinal in that birds collected throughout its limited range differed little with respect to plumage patterns; only a single specimen, collected outside the presently recognized range of the subspecies, showed intermediate characters.

In sum, the Central American *D. certhia* are distinguishable from the Amazonian taxa based on the combination of a blackish bill, dark loral mark, rufescent crown, and the lack of a pale throat. Within this region, taxa appear to be divided into discrete groups as follows: *D. c. sheffleri*, *D. c. hesperius*, *D. c. sanctithomae* (including both *D. c. nigrirostris* and *D. c. colombianus*), and apparently, *D. c. legtersi*. Although it is possible that mensural characters may support the recognition of *D. c. nigrirostris* and *D. c. colombianus*, preliminary morphometric analyses (Marantz 1992, Willis 1992) argue against this possibility. To better assess the validity of *D. c. legtersi*, and its present distribution, additional material will be necessary.

Nechí D. certhia.—Two subspecies of *D. certhia* are restricted to Haffer's (1967) Nechí region of endemism (Fig. 5). Although the Nechí taxa appear to be closely allied with members of the Central American song-type in that they share the dark lores, rufous crown, and dark throat with the Central American birds, they are readily distinguished by the presence of rounded (or triangular) spots along the shafts of numerous breast feathers (Phelps and Gilliard 1940, Wetmore 1942). This spotting gives the birds a superficial appearance of being intermediate between *D. certhia* and *D. picumnus* (Willis 1992). Although the presence of breast spotting distinguished the Nechí representatives from Central American birds, intergradation between *D. c. hyleorus* (a Nechí form) and *D. c. colombianus* (a Chocó representative of the Central American group) is apparent in five specimens taken in central Colombia. These specimens (collected at Regeneración, depto. Bolívar, and Tarazá and Hacienda Belén, depto. Antioquia [USNM 402001, 401992–401994, and 401996, respectively]), labeled by Wetmore as being intermediate between *D. c. hyleorus* and *D. c. colombianus*, are intermediate between the "pure" populations of these taxa both in the extent and pattern of breast spotting and in the contrast of the dusky breast-band. The pale, central spot characteristic of the breast feathers of *D. c. hyleorus* is bordered distally by a dark subterminal-band and, on each side, by dark, vertical lines that extend from the feather base to this subterminal band. The pale, central spot is therefore completely enclosed by blackish lines. By contrast, breast feathers of *D. c. colombianus* have three concentric, black bands on otherwise golden-brown (Brownish-Yellow, 10 YR 6/6) feathers (i.e., no pale, central marks). Relative to "pure" birds of both taxa, breast feathers of intermediate specimens also show a varying degree of duskiness on either side of the central spot, but generally not a dark border. Even though the spot tends to be present in these birds, often on a reduced number of breast feathers, the pattern created is not as striking as on topotypical *D. c. hyleorus*. Similarly, two of these specimens (USNM 401992 and 401994) had slightly paler breasts, and thus less distinct breast-bands, than did topotypical *D. c. hyleorus* (USNM 369010). In this regard, the band of USNM 402001 was even less distinct, in that it appeared similar to typical *D. c. colombianus* (e.g., USNM 443104).

Although the advertising song of *D. c. hyleorus* is unknown, I have tentatively placed this subspecies with *D. c. punctipectus* based on the similarity of plumage patterns in these taxa (they differ only subtly in coloration). Within the Nechí region, *D. c. punctipectus* is found exclusively in the Lake Maracaibo Basin in edo. Zulia, northwestern Venezuela, whereas *D. c. hyleorus* is restricted to lowlands in northeastern Colombia (Peters 1951). Variation within my sample of Nechí birds was minimal and not great enough to merit the recognition of *D. c. hyleorus*. Patterns in these 15 specimens (which included both topotypical *D. c. hyleorus* and the holotype of *D. c. punctipectus*) matched closely the description presented by Wetmore (1942). In general, *D. c. hyleorus* was paler and grayer overall than was *D. c. punctipectus*; however, a specimen of *D. c. hyleorus* collected at Petrólea, depto. Norte de Santander, Colombia (USNM 373293), a locality geographically intermediate between the type localities of the two taxa, was nearly identical to the holotype of *D. c. punctipectus* with respect to the pattern on the breast, crown, and back. The holotype of *D. c. punctipectus* differed only by its more rufescent (Yellowish-Brown, 10 YR 5/6) crown, slightly brighter throat (closest to Brownish-Yellow, 10 YR 6/8), and darker, more golden (Brownish-Yellow, 10 YR 6/8) belly; however, these differences in coloration were subtle. The subtlety of differences between *D. c. hyleorus* and *D. c. punctipectus*, combined with the small sample size and apparently clinal nature of plumage variation in these taxa, argues against the recognition of *D. c. hyleorus*. A more thorough examination of an extensive series of skins, however, may yield a discrete break between the two Nechí populations. The assessment of *D. c. hyleorus* is additionally complicated by its apparent intergradation with *D. c. colombianus*, and the fact that *D. c. punctipectus* and *D. c. colombianus* are similar to each other with respect to plumage brightness (the same character that separates *D. c. punctipectus* and *D. c. hyleorus*). This suggests that intergrades between *D. c. hyleorus* and *D. c. colombianus* may appear similar to *D. c. punctipectus*.

Not only do the Nechí *D. certhia* appear to be closely allied with the Central American forms based on plumage patterns, but lack of reproductive isolation is suggested by intergrade specimens taken where members of these two groups meet in northern Colombia. Additional work is necessary to evaluate both the degree of intergradation between the Nechí and Central American forms, and patterns of vocal variation in this region. Within the Nechí region, however, only a single taxon appears to be present, *D. c. punctipectus* (including *D. c. hyleorus*). Although specimens of *D. c. hyleorus* averaged somewhat paler and grayer than the type of *D. c. punctipectus*, the variation appeared to be clinal in the small number of specimens examined.

CIS-ANDEAN *D. CERTHIA* AND *D. CONCOLOR*

Relative to the trans-Andean *D. certhia*, the Amazonian subspecies are characterized by a pale, often whitish, throat; the lack of a distinctive, dark loral-mark; and barring on the underparts that is less crisply defined. With few exceptions, taxa in this complex fit into discrete groups treated by Cory and Hellmayr (1925) as *D. certhia concolor*, *D. c. certhia*, *D. c. radiolatus*, *D. c. juruanus*, *D. c. medius*, and *D. c. polyzonus* (Figs. 5 and 6). Cory and Hellmayr recognized *D. concolor* as a subspecies of *D. certhia*, and they subsumed *D. c. ridgwayi* into this subspecies; my examination supported their conclusions. I have therefore combined the Amazonian *D. certhia* with *D. concolor* into an Amazonian "barred" woodcreeper complex.

Northern and western Amazonian populations.—Most Amazonian barred woodcreeper taxa are delimited by major river systems and can be diagnosed by one or more discrete plumage characters. For example, even though *D. c. certhia* and *D. c. juruanus* appear remarkably similar, both with respect to underpart coloration and the character of barring thereon, *D. c. juruanus* was readily distinguished by its duller, more obviously barred crown with feathers that lacked both the Yellowish-Brown (10 YR 5/6 to 10 YR 5/8) shaft streaks and the black tips that Cory and Hellmayr (1925) noted were conspicuous on most *D. c. certhia*. These two taxa are separated geographically by *D. c. radiolatus* of northwestern Amazonia. In contrast to the Light Brownish-Yellow to Brownish-Yellow (10 YR 6/4 and 10 YR 6/6, respectively) *D. c. certhia* and *D. c. juruanus*, which also have finely barred underparts (bars less than 2 mm wide and often *ca* 1 mm in width), *D. c. radiolatus* is generally a richer, more ochraceous, brown (Dark Yellowish-Brown, 10 YR 4/6) in color with broader (2–2½ mm wide) and noticeably blacker barring on the underparts, and has distinct, dark subterminal-bars on the greater secondary-coverts. These characters combined to give *D. c. radiolatus* a distinctive appearance that varied little from central Amazonia towards the base of the Andes.

It is unclear how *D. certhia* subspecies are distributed in the western portion of Amazonia based on published accounts (e.g., Peters 1951). For example, Peters indicated that, in the northern portion of its range, *D. c. radiolatus* occurs east to the right bank of the Rio Negro; however, he made no mention of the southeastern boundary of this subspecies' range relative to that of *D. c. juruanus*. By contrast, my analysis (Fig. 9) indicated that *D. c. radiolatus* occurs south in Peru along both banks of the Río Ucayali to at least Lagarto, depto. Ucayali (AMNH 239377–239379), and Yurinaqui Alto, depto. Junín (FMNH 278325). That *D. c. radiolatus* has been collected on both banks of both the Río Ucayali and the Río Amazonas further indicates that neither of these rivers now delimit the range of this subspecies. This pattern is confused somewhat by two specimens, apparently representing *D. c. juruanus*, from the south bank of the Amazon at Orosa (AMNH 232049, 232050); the remaining specimens from this region (LSUMZ) all appear to represent *D. c. radiolatus*. Unfortunately, no specimens of *D. certhia* that I examined were collected on the upper portion of the Río Javarí, the next major river to the east of the Río Ucayali, and the most likely barrier between *D. c. radiolatus* and *D. c. juruanus*. Furthermore, these southernmost sites for *D. c. radiolatus* are only 850 kilometers to the northwest of known, Bolivian localities for *D. c. polyzonus* (the closest site for the latter being north of Puerto Linares, depto. La Paz, Bolivia; LSUMZ 101895, 101896).

Of the 22 specimens that I examined from the intervening region of southeastern Peru and northwestern Bolivia, 14 showed both the fine, underpart barring and the deep, ochraceous coloration (Dark Yellowish-Brown, 10 YR 4/6, to Strong Brown, 7.5 YR 5/6) of *D. c. polyzonus*. For example, LSUMZ 31346 (collected at Balta, Río Curanja, depto. Ucayali, Peru) appeared identical to the few available specimens of *D. c. polyzonus* collected in northern Bolivia (e.g., ANSP 120654, Santa Ana, depto. La Paz); unfortunately, I was not able to compare topotypical *D. c. polyzonus*, from central Bolivia, directly with birds from southern Peru. By contrast, the remaining eight specimens that I examined, from southeastern Peru and northwestern Bolivia, appeared either typical of *D. c. juruanus*, or with only a slight indication of ochraceous coloration (Dark Yellowish-Brown, 10 YR 4/6). For example, a specimen collected along Quebrada Juliaca on the Río Heath, depto. Madre de Dios, Peru (LSUMZ 84637), was nearly identical to another collected at the opposite end of the range of *D. c. juruanus* at the confluence of the Rios Madeira and Amazonas at Rosarinho, edo. Amazonas, Brazil (AMNH 282305). In general, the series of specimens collected both along Quebrada Juliaca and south of Cobija, depto. Pando, Bolivia (LSUMZ 132437–132441), although including some individuals intermediate between *D. c. juruanus* and *D. c. polyzonus*, appeared more typical of *D. c. juruanus* in that most birds showed less ochraceous underparts than did specimens collected farther to the west. Because *D. c. polyzonus* shares the narrow, underpart barring with *D. c. juruanus*, but the deep ochraceous coloration with *D. c. radiolatus*, it may not be possible to determine if specimens from southern Peru and northwestern Bolivia represent *D. c. polyzonus* or, as was indicated by J. P. O'Neill on the labels of LSUMZ specimens from southeastern Peru, intergrades between *D. c. radiolatus* and *D. c. juruanus*. For convenience, I will refer to birds from this region as the "Balta" population, named after the locality in which most of the specimens have been collected.

To provide support for my interpretation of plumage patterns, three additional ornithologists independently examined a subset of the *D. certhia* specimens from northern and western Amazonia. Overall, the percentage of specimens identified correctly was high within this group of birds. Whitney correctly identified 91% of 122 skins, whereas Remsen correctly classified 82% of 67 skins. Wiedenfeld correctly classified 95% of 68 skins into groups of *D. c. radiolatus* versus the other taxa; however, time became a limiting factor in his classification within this latter group. Most mis-identifications by all three observers were between *D. c. juruanus* (or sometimes *D. c. radiolatus*) and birds from the Balta population; however, several *D. c. certhia* were additionally placed with *D. c. juruanus*. The former classification was because the Balta birds shared the narrowly barred underparts with *D. c. juruanus*, but the bright ochraceous coloration with *D. c. radiolatus*. Because the width of barring was less variable than was overall coloration, confusion between specimens from the Balta population and *D. c. juruanus* was more frequent than with *D. c. radiolatus*. The second group of incorrectly classified birds was also expected because *D. c. certhia* and *D. c. juruanus* are nearly identical when viewed from below. The mis-identifications likely resulted from the fact that a small percentage of the *D. c. certhia* specimens had duller than average crown patterns (e.g., CM 83571 and YPM 39301), possibly as a result of feather wear, and therefore appeared remarkably similar to typical *D. c. juruanus*.

In conclusion, most northern and western Amazonian populations of *D. certhia* (*D. c. certhia*, *D. c. radiolatus*, and *D. c. juruanus*) are separated by major river systems that form discrete barriers to intergradation. Only in southeastern Peru, near the confluence of the headwaters of several river systems that otherwise delimit barred woodcreeper ranges, do these barriers break down, apparently allowing populations of *D. c. radiolatus*, *D. c. juruanus*, and *D. c. polyzonus* to come into contact. In this region, birds showing characters of *D. c. polyzonus* were found to occur well to the north of the Bolivian range of this subspecies as recognized by Cory and Hellmayr (1925) and Peters (1951). Based on a sample of only 28 birds from both Bolivia and southern Peru (only six of which were *D. c. polyzonus*), however, it was not possible for me to assess the degree of apparent intergradation between *D. c. radiolatus*, *D. c. juruanus*, and *D. c. polyzonus*.

Southern Amazonian populations of D. certhia *and* D. concolor.—Plumage variation of southern Amazonian barred woodcreepers presents the most confusing geographic pattern within the genus *Dendrocolaptes*. Previous authors have recognized as many as three taxa (*D. concolor*, *D. certhia ridgwayi*, and *D. certhia medius*) from the region east of the Rio Madeira, and these taxa have been treated as either one or two species (Cory and Hellmayr 1925; Zimmer 1934; Griscom and Greenway 1941; Todd 1948; Willis 1992; Ridgely and Tudor 1994). Although Cory and Hellmayr (1925) merged *D. certhia ridgwayi* with *D. concolor* due to plumage variability in the latter taxon, Zimmer (1934) found that the variation was clinal in nature, even though some apparently pure *D. concolor* specimens were taken as far east as the Rio Tocantins. Combining Zimmer's comments with an independent examination of skins, Griscom and Greenway (1941) suggested that *D. c. ridgwayi* was valid because most birds collected around Santarém were both easily separable from *D. concolor* and strikingly distinct from *D. c. medius* (although three individuals from this location were of "the *concolor* type"). Finally, Todd (1948) split *D. certhia* and *D. concolor* stating that: "The very fact the two forms behave as they do, as described by Zimmer [presumably in reference to the sympatric occurrence of pure *D. c. ridgwayi* and *D. concolor*], indicates that they are really distinct species." Todd was additionally unable to distinguish Griscom and Greenway's three *D. concolor* specimens from the rest of the series, and suggested that all specimens he examined could be referred with certainty to either *D. concolor* or *D. c. ridgwayi*. More recently, Willis (1992) found specimens collected between the Rio Tapajós and the Rio Tocantins to be intermediate between *D. concolor* and *D. c. medius* in the degree of barring. Willis attributed these specimens to a hybrid population under the name *D. c. ridgwayi*, and indicated that no *D. concolor* specimens that he examined were collected east of the Rio Tapajós.

Because my combined sample included most specimens examined by previous workers, it was not surprising that the variation I noted paralleled that already described. The initial impression gained from examining this series matched the comments of Cory and Hellmayr in that these birds showed an extreme degree of intrapopulation variation in overall plumage coloration; however, by plotting the localities on a map, certain patterns emerged. These patterns were concordant with those noted by Zimmer (1934), who indicated that specimens varied from typical *D. concolor* on the right bank of the lower Rio Madeira to *D. certhia medius* around Belém, edo. Pará, Brazil, but that specimens collected in the intervening region included both typical *D. concolor* and individuals that showed a combination of characters. Although plumage variation was clinal to some degree, this was not universally so.

I found two conflicting patterns of plumage variation within the southern Amazonian populations, one corresponding to the degree of barring, and the other the coloration of the underparts. Although *D. c. medius* from Belém were strikingly different from *D. c. ridgwayi*/*D. concolor* collected farther to the west based on the boldness of barring, the degree of barring varied substantially within the latter group. Unfortunately, because most specimens that I examined failed to include age-related information such as skull ossification, and few showed obvious signs of immaturity (e.g., juvenal plumage), I had no way to confirm Willis' (1992) finding that immature birds are more obviously barred than adults from the same population. Whereas barring generally became more prevalent from west to east, underpart coloration darkened from south to north. For example, specimens from the southwestern portion of the region were pale-gray overall (Brown, 10 YR 5/3, to Dark Yellowish-Brown, 10 YR 4/4), whereas those collected nearer the Rio Amazonas were, on average, darker and more ochraceous (a darker Dark Yellowish-Brown, 10 YR 3/4). Within this group, however, there was substantial individual variation with birds almost separable into pale and dark "morphs." This pattern was additionally complicated by a tendency for unbarred birds to occur farther east than expected based on the generally clinal nature of variation. In this regard, a series of skins collected by the Olalla brothers, reportedly along the right bank of the Rio Tocantins, is especially problematical in that it contains both barred and unbarred birds collected at the same localities. Two skins in this series from Baião, edo. Pará, Brazil, appeared identical to *D. concolor* from the right bank of the Rio Xingu at Porto de Moz, edo. Pará, Brazil, or the right bank of the Rio Tapajós at Mirituba, edo. Pará, Brazil. Four additional Olalla specimens collected at Mocajuba, edo. Pará, Brazil, included one typical *D. concolor* (AMNH 431012), another intermediate between *D. concolor* and *D. c. medius* (AMNH 431011), and two (AMNH 431010 and 431014) similar to *D. c. medius* collected around Belém. Although the above data suggests that *D. c. medius* and *D. concolor* occur sympatrically, it is possible that labeling problems may account for this phenomenon, given that *D. c. medius* occurs on the right bank of the Rio Tocantins, whereas *D. concolor* is found on the left bank. Even if the two taxa really were collected at the same sites, the degree of true syntopy is unknown.

The complexity of patterns shown by birds from southern Amazonia was confirmed by an independent examination of the skins by three other ornithologists. Whereas these observers had little difficulty separating the barred taxa from northern and western Amazonia, they were unable to divide birds from southern Amazonia into three, well-defined clusters that corresponded to the presently defined taxonomic groups. Instead, they separated birds into groups that comprised *D. c. medius*, *D. concolor* from northeastern Bolivia (the extreme southwestern limit of its range), and between two and ten intermediate groups that combined *D. concolor* and *D. c. ridgwayi* to varying degrees. Furthermore, Whitney noted a characteristic pattern of scaling in the facial region of *D. c. medius* that, in addition to simplifying the identification of *D. c. medius*, was also useful for identifying apparent hybrids between *D. c. medius* and *D. concolor*. These apparent hybrids were collected east of the Rio Xingu (52 km. SSW Altamira; USNM 572539–572540) and on the east bank of the Rio Tocantins (specimens collected in edo. Pará, Brazil, at Mocajuba [AMNH 431011] and at "Rodovia Belém-Brasilia Km 86", edo. Pará [LSUMZ 67194]). These specimens, taken on both sides of the Rio Tocantins, suggest both that *D. concolor* and *D. certhia medius* have crossed this river, and that hybridization does occur where the two forms meet.

In conclusion, plumage variation in the south-bank barred woodcreepers is difficult to assess. Making this assessment even more complex is the possibility of incorrect specimen localities for key specimens that suggest the sympatric occurrence of "pure" *D. certhia* and *D. concolor*. By contrast, other specimens from the east bank of the Rio Xingu and the east bank of the Rio Tocantins show signs of hybridization between *D. certhia medius* and *D. concolor*. Although plumage patterns support the recognition of both *D. concolor* and *D. certhia medius*, *D. certhia ridgwayi* appears to represent a hybrid population. Such an extensive hybrid zone, in addition to the high degree of phenotypic variance within this zone, argues for subsuming *D. concolor* into *D. certhia*.

Dendrocolaptes picumnus and *D. platyrostris*

Like the barred *Dendrocolaptes*, the streaked species occur throughout much of both Central America and South America. Unlike the *D. certhia* complex, which occurs exclusively in humid lowland forests, the streaked *Dendrocolaptes* are more generalized in their habitat preference. *Dendrocolaptes picumnus* occurs in foothill and montane forests in both Central America and South America, lowland rainforests of the Amazon basin, the Chaco, and the Alto Paraná in Paraguay (Figs. 2 and 3; Peters 1951). Although multiple subspecies of *D. picumnus* have been recognized within some regions, taxa within these groups generally share similar plumage patterns and therefore will be discussed together under biogeographically defined subheadings. Whereas *D. picumnus* occurs in the northern and western portion of South America, *D. platyrostris* is found in central, eastern, and southern Brazil and adjacent portions of Argentina and Paraguay (Fig. 8). Within this range are found two well-marked subspecies, *D. p. intermedius* and *D. p. platyrostris*. These two subspecies occur in Haffer's (1985) Caatinga and Campos Cerrados areas of endemism of east-central Brazil and adjacent Paraguay, and in the Atlantic coastal forests of southeastern Brazil, northeastern Argentina, and eastern Paraguay, respectively (Peters 1951).

Although *D. picumnus* and *D. platyrostris* traditionally have been considered separate species (Cory and Hellmayr 1925, Peters 1951), the plumage differences separating them are no more than those distinguishing Central American and Amazonian populations of *D. certhia*. Both are boldly streaked *Dendrocolaptes* united by a combination of plumage characters. They differ in the degree of streaking and in the shade of the crown. Willis (1982) further suggested that all three streaked *Dendrocolaptes* (including *D. hoffmannsi*) may be conspecific because he believed that *D. platyrostris* from edo. Maranhão, Brazil, were intermediate between *D. picumnus* and nominate *D. platyrostris* with respect to bill coloration and wing and tail measurements. As additional evidence for conspecificity, Willis stated that these three taxa are only "... marginally sympatric, perhaps [occurring] in adjacent territories in differing habitats." In fact, the only evidence for the sympatric occurrence of *D. picumnus* and *D. platyrostris*, was presented by Brodkorb (1941), who noted that the type localities of *D. picumnus extimus* and *D. tarefero* [= *D. platyrostris platyrostris*] refer to the same site. Although the specimens I examined confirmed that these two species occur in close proximity to each other in eastern Paraguay (Figs. 3 and 8), my series included no cases in which both species were collected at the same site. Therefore, even though the true degree of sympatric occurrence in this region has yet to be studied in detail, it appears that Willis was correct in suggesting that the two are at most marginally syntopic. The parapatry of these two apparent sister-species, combined with striking plumage differences between *D.*

picumnus pallescens and *D. platyrostris platyrostris*, and the lack of phenotypically intermediate specimens, suggests that the two are valid, biological species.

Montane D. picumnus.—The montane subspecies of *D. picumnus* were divided into four categories based on a combination of biogeography and plumage patterns: 1) *D. p. puncticollis*; 2) a group containing *D. p. costaricensis*, *D. p. veraguensis*, and *D. p. multistrigatus*; 3) *D. p. seilerni*; and 4) *D. p. olivaceus*. The northernmost subspecies, *D. p. puncticollis*, was characterized by narrow but distinct streaking on an otherwise unmarked, Olive (5 Y 4/4) breast; moreover, each streak was bordered by a Very Dark Brown (10 YR 2/2) margin. This distinct subspecies has a blackish (Dark Yellowish-Brown, 10 YR 3/6) crown with fine, Yellow (10 YR 7/6) streaking, and relatively indistinct belly-barring.

The two Middle American taxa (*D. p. costaricensis* and *D. p. veraguensis*) were similar to *D. p. multistrigatus* of the northern Andes, with some aspects of their plumage variation appearing to form a cline with the Andean form. Relative to the northern *D. p. puncticollis*, all three subspecies had less extensive streaking on what appeared to be an indistinctly barred breast, dark and narrow barring on the belly, and a golden (e.g., Yellowish-Brown, 10 YR 5/8, on an especially bright *D. p. costaricensis*; UMMZ 132556) cast to the underparts. In all, *D. p. costaricensis* appeared more similar to *D. p. multistrigatus* than to the Venezuelan *D. p. seilerni*, contra Cory and Hellmayr (1925). Separating *D. p. costaricensis* from *D. p. multistrigatus* were the reduced extent of breast streaking of *D. p. costaricensis* and the irregular, dark spots along the edges of the middle-breast feathers. These spots gave the breast an indistinctly barred appearance in *D. p. costaricensis*, whereas the few *D. p. multistrigatus* that showed any barring had it restricted to the lower throat.

Despite suggestions to the contrary (Griscom 1927), *D. p. veraguensis* is not a "well-marked" subspecies when compared with *D. p. costaricensis*. Direct comparison of the same four *D. p. veraguensis* examined by Griscom (including the holotype) with six *D. p. costaricensis* (including the two specimens examined by Griscom) yielded no striking differences. My examination suggested that the pattern and coloration of the crown, back, wings, and tail were identical in the two subspecies. In *D. p. veraguensis*, the breast was generally brighter and slightly more reddish-brown (Dark Yellowish-Brown, 10 YR 3/6 to 10 YR 4/6) than in the grayish-olive (Olive, 5 Y 4/4) *D. p. costaricensis*; however, one specimen labeled as *D. p. costaricensis* from geographically intermediate prov. Chiriquí, Panama (AMNH 524601), was identical to the series of *D. p. veraguensis*. Although Griscom noted that the breast streaking of *D. p. veraguensis* was slightly more reduced than in *D. p. costaricensis*, the more limited extent of streaking in the specimens of *D. p. veraguensis* that I examined appeared to be an artifact of specimen preparation. Furthermore, the holotype had the most limited amount of streaking of any in the series I examined, whereas the two *D. p. costaricensis* examined by Griscom (AMNH 390657 and 390658; collected at La Hondura, prov. Cartago, Costa Rica) were among the most extensively streaked of the birds examined. Furthermore, the pattern of the plumage underlying the breast streaking was identical in the two taxa, with a trend toward more rounded spots on the upper throat and more linear streaks on the upper breast. Finally, although one bird (AMNH 77624) was slightly brighter and more rufous (Yellowish-Brown, 10 YR 5/6) on the underparts than others, this variation did not appear to have a geographic basis.

Although not geographically linked with each other (Figs. 2 and 3), the Venezuelan *D. p. seilerni* appeared most similar to *D. p. puncticollis* (from northern Central America) based on its combination of narrow bellybarring and narrow streaking on a nearly unbarred breast. In contrast, *D. p. seilerni* did not appear as similar to the northern Andean *D. p. multistrigatus* as was suggested by Cory and Hellmayr (1925). Relative to *D. p. multistrigatus*, the breast streaking of *D. p. seilerni* was more striking due to its greater contrast with an otherwise unmarked breast. Furthermore, the streaking of *D. p. seilerni* was more extensive than that of *D. p. multistrigatus* (extending to the mid-belly rather than the lower breast), with the streaks themselves appearing noticeably narrower, and the throat appearing more distinctly streaked.

In the type description for the southern Andean *D. p. olivaceus*, Zimmer (1934) commented on its close plumage similarity to *D. p. seilerni*, a highland Venezuelan form occurring well to the north of the range occupied by *D. p. olivaceus*, and to *D. p. pallescens*, a southern subspecies that occurs primarily in the lowlands. Zimmer's diagnosis of *D. p. olivaceus*, and his comparison of it with both *D. p. seilerni* and *D. p. pallescens*, applied equally well to the additional specimens that I examined; however, he presented little information on the separation of *D. p. olivaceus* from *D. p. validus* of the nearby Amazonian lowlands. These two subspecies were readily separable based on a combination of narrower streaking across a plainer, less noticeably barred, breast in *D. p. olivaceus*. In this respect, *D. p. olivaceus* was much like *D. p. pallescens*. In contrast, belly barring on *D. p. olivaceus* was slightly narrower than on *D. p. validus* but, unlike that of *D. p. pallescens*, it was nearly as dark as that of *D. p. validus*. As implied by its name, *D. p. olivaceus* averaged more olivaceous (Brownish-Yellow, 10 YR 6/6) and less rufescent or golden (Yellowish-Brown, 10 YR 5/8) below compared to *D. p. validus*. Although intermediate between *D. p. pallescens* and *D. p. validus* with respect to some aspects of plumage variation, *D. p. olivaceus* was easily separable from each of these taxa.

In conclusion, most montane *D. picumnus* appear to form discrete populations that are geographically isolated. With the exception of *D. p. veraguensis*, which was not consistently separable from *D. p. costaricensis* based on the small sample that I examined, all taxa appear to be valid. Additional specimens from southern Central America will be required to assess better the relationships between *D. p. costaricensis*, *D. p. veraguensis*, and *D. p. multistrigatus*.

Central Amazonian D. picumnus.—Three, well-marked subspecies of *D. picumnus* occur in the lowlands of the central Amazon Basin: *D. p. picumnus*, *D. p. validus*, and *D. [p.] transfasciatus* (Cory and Hellmayr 1925). The two most similar forms, *D. p. picumnus* and *D. p. validus*, are distributed on opposite sides of the Rio Negro in the northern and western portions of the Amazon basin, respectively. Whereas Cory and Hellmayr (1925: 270) noted that the throat feathers of *D. p. validus* were "... laterally edged rather than spotted with dark brown [as in nominate *D. picumnus*]," I found that a few *D. p. validus* had irregular, dark borders to the pale shaft-streaks of the upper breast feathers. These atypical individuals had an overall pattern similar to the dark spotting that borders the pale shaft-streaks in *D. p. picumnus*. Like those for *D. p. picumnus* and *D. p. validus*, the diagnosis of *D. [p.] transfasciatus* (Todd 1925) matched the only specimen of this taxon that I examined (AMNH 286964). This striking individual showed a combination of a Very Dark Brown (10 YR 2/2) crown with golden (Very Pale Brown, 10 YR 8/4) shaft-streaking that extended to the lower back; an underlying pattern of blackish barring on the back; broad, White (5 Y 8/1) streaking on the throat and breast; and coarsely barred underparts with each feather having the blackish (Very Dark Brown, 10 YR 2/2) cross-bars broken by a pale shaft-streak. Although Willis (1982) suggested that *D. p. transfasciatus* may be more closely allied with *D. platyrostris* than with nominate *D. picumnus*, the single specimen of *D. p. transfasciatus* that I examined was similar to other Amazonian *D. p. picumnus*, with the exception of its blackish crown and the unusual pattern of underpart barring.

The "*southern*" *D. picumnus*.—The southern forms of *D. picumnus* comprise *D. p. pallescens*, *D. p. casaresi*, and *D. p. extimus*. These subspecies all had similar plumage patterns, so similar in fact that I found no consistent differences among them. Characters used by Cory and Hellmayr (1925) to diagnose *D. [p.] pallescens* matched all three taxa in this group. Past confusion in the taxonomy of these birds likely resulted from their highly variable underpart color. For example, this coloration varied from an intense rusty (closest to Yellowish-Red, 5 YR 4/6) to plain brown (Brown, 10 YR 5/3) in the series of *D. p. pallescens* examined. Although *D. p. extimus* was diagnosed based on its paler, less tawny coloration relative to *D. p. pallescens* (Brodkorb 1941), topotypical *D. p. extimus* that I examined were identical to *D. p. pallescens* in every respect. Whereas three specimens in this series (UMMZ 109758–109760) were slightly duller and less rufescent (closest to Brownish-Yellow, 10 YR 6/6) than some *D. p. pallescens*, two other birds (UMMZ 109755, 109756) were as bright rufous below (Dark Yellowish-Brown, 10 YR 4/6) as were any *D. p. pallescens*. Similarly, their crowns were the same Dark Yellowish-Brown (between 10 YR 3/6 and 10 YR 4/6) as in some *D. p. pallescens* (e.g., LSUMZ 123868). Like *D. p. extimus*, *D. p. casaresi* was inseparable from *D. p. pallescens* based on plumage patterns. Although the *D. p. casaresi* specimens that I examined were worn (as were the *D. p. pallescens* to which they were compared), both taxa varied in both upperpart and underpart coloration, as well as the pattern of breast streaking. In sum, the variation that I noted in *D. p. casaresi* was within the range exhibited by *D. p. pallescens*. Despite the similarity in plumage patterns between *D. p. casaresi* and *D. p. pallescens*, they appear to be separable on the basis of mensural characters (Steullet and Deautier 1950). Unfortunately, the measurements published in the type description of *D. p. casaresi* were insufficient to rule out the effect of clinal variation in these characters.

In sum, the southern forms of *D. picumnus* comprise a single group based on plumage patterns. Because *D. p. casaresi* was described based on mensural characters, however, additional work will be necessary to determine the influence of clinal variation in size.

Dendrocolaptes platyrostris.—Two well-marked subspecies have been recognized for *D. platyrostris*. These subspecies, *D. p. platyrostris* and *D. p. intermedius*, are easily distinguished based on the characters discussed by Cory and Hellmayr (1925). Nominate *D. platyrostris* has a Black (10 YR 2/1) crown with bold, Yellow (5 Y 7/6) streaking; is slightly darker (Dark Yellowish-Brown, 10 YR 3/6) on the back; and the wings (secondaries Dark Reddish-Brown, 2.5 YR 3/4), tail (central rectrices Very Dusky Red, 2.5 YR 2.5/2), and uppertail coverts (Dark Yellowish-Brown, 10 YR 4/6) are darker and more chestnut than in *D. p. intermedius*. In contrast, *D. p. intermedius* has a much duller, browner (Dark Yellowish-Brown, 10 YR 3/6) crown; a more olive (Dark Yellowish-Brown, between 10 YR 3/4 and 10 YR 3/6) back; and noticeably paler (Red, 2.5 YR 4/6) uppertail coverts. I also confirmed that the wings (secondaries Red, 2.5 YR 4/6) and tail (central rectrices Dark Reddish-Brown, 5 YR 3/4) were slightly paler in *D. p. intermedius* than in *D. p. platyrostris*. Of note, was the fact that specimens collected in central Paraguay (deptos. Concepción and Amambay; UMMZ) showed signs of intergradation between these two forms. In these specimens, the crown was generally darker than in *D. p. intermedius*, but not quite as black as that of *D. p. platyrostris*. Intermediate specimens have also been noted from São Paulo, Brazil (Willis 1992).

Dendrocolaptes hoffmannsi

Previous workers (e.g., Willis 1982) have placed the little-known *D. hoffmannsi* with the streaked *Dendrocolaptes*; however, both morphology (Raikow 1994) and plumage patterns suggest that it is either intermediate between the streaked and barred forms, or more closely related to the barred species. This species, endemic to the Rondônian center of endemism, is known from only a few specimens collected between the Rio Madeira and the Rio Xingu in central Brazil (Fig. 7). Furthermore, the accuracy of locality data for specimens collected by the Olalla brothers on the right bank of the Rio Xingu has been questioned (Haffer 1985). The description by Cory and Hellmayr (1925) was accurate for most specimens that I examined in that they had indistinct, pale shaft-streaks on the feathers of the throat and breast that were comparable to the "inconspicuous buffy hair streaks" described by Cory and Hellmayr (1925: 268). The overall impression given by the fine shaft-streaks of this species was different from the boldly marked underparts of both *D. picumnus* and *D. platyrostris*. In addition to the distinct pattern of breast streaking, the crown pattern of *D.*

hoffmannsi is also unlike that of the other streaked species in that the feathers are quite rusty (Yellowish-Red, 5 YR 5/8) in color with well-defined, black tips. Pale shaft-streaks on these feathers are narrow and indistinct, if present at all. The overall appearance of the crown was therefore similar to the Central American *D. certhia*; however, the crown feathers of those taxa have multiple, black bars extending across each feather. The distinctive plumage of *D. hoffmannsi* actually makes it appear more similar to the sympatric *D. concolor* than to the streaked species (cf. Willis 1992).

Whereas most specimens of *D. hoffmannsi* matched Cory and Hellmayr's (1925) description, the holotype (AMNH 524562) differed in some regards (see also Hellmayr. 1909. *Bull. Br. Ornithol. Club*: 66). Relative to most other specimens I examined, this bird had relatively broad, whitish streaking on the throat and upper breast. Most of these streaks had blackish borders that ended near the tip of each feather. One other specimen from the type locality (Calama, edo. Rondônia; AMNH 524563) had dark spots adjacent to an irregular, pale shaft-stripe on each breast feather. This pattern, possibly a result of immaturity, created an appearance of barring underlying the otherwise streaked breast that was also shown by another Rondônian specimen (FMNH 343854).

SPECIES LIMITS AND DISTRIBUTION OF SOME NEW WORLD SPINE-TAILED SWIFTS (*CHAETURA* SPP.)

Manuel Marín

Museum of Natural Science, Foster Hall 119, and Department of Zoology and Physiology, Louisiana State University, Baton Rouge, Louisiana 70803, USA

ABSTRACT.—The "brown" assemblage of New World spine-tailed swifts consists of three groups distinguishable by size: small *C. vauxi*, medium *C. chapmani*, and large *C. pelagica*. Two of the three species groups are migratory: the *vauxi* group undertakes short movements from northern temperate to tropical latitudes, and some populations in tropical highlands may move to lowlands. The *pelagica* group makes long-distance movements. *Chaetura andrei* is not a valid taxon; the nominate form pertains to the *vauxi* species group. It is proposed that the taxa *C. andrei meridionalis* and *C. chapmani viridipennis* be elevated to species rank. The latter two taxa are probably more closely related to *C. pelagica* than to the species in which they are currently placed. All three groups, small, medium, and large, constitute the *pelagica* superspecies.

RESUMEN.—El grupo "café" de los vencejos de cola de espina del Nuevo Mundo esta constituido por tres subgrupos diferenciados por el tamaño: el pequeño, *C. vauxi*, el mediano, *C. chapmani* y el grande, *C. pelagica*. Dos de los tres subgrupos emigran después del período de nidificación; dentro del subgrupo de *vauxi*, los que viven en la zona templada norte emigran a latitudes mas trópicales y al parecer los que viven en las zonas templadas tropicales emigran desde las partes altas a las bajas. El subgrupo *pelagica* emigra transequatorialmente. Se encontró que *C. andrei* no es un taxón válido, la especie pertenece al subgrupo *C. vauxi*. También se encontró que las subespecies *C. andrei meridionalis* y *C. chapmani viridipennis* están taxonómicamente mas cercanas a *C. pelagica* que sus respectivas especies. Los tres subgrupos, el pequeño, el mediano y el grande pertenecen a la superespecie *pelagica*.

Swifts (Apodidae) as a group provide a special difficulty for the taxonomist. They are remarkably uniform in shape, partly because of their extreme adaptation to an aerial life, leaving few characters as indices to phylogenetic relationships. At the family, subfamily, and genus level, taxonomists have used foot structure (Sclater 1865), tail spines (Hartert 1892; Hellmayr 1908), nest structure and placement (Lack 1956a), transpalatine processes (Orr 1963), echolocation (Medway and Pye 1977), and clutch size, growth rate and ecological similarities (Marín and Stiles 1992) as characters. At the species level, however, such characters are of limited use (e.g., Brooke 1970; Marín and Stiles 1992; Browning 1993).

The New World spine-tailed swifts (*Chaetura* spp.) represent the most widespread group of swifts in the New World. Three factors underlie taxonomic problems in swifts, especially in the genus *Chaetura*. The first is the shortage of museum specimens. Many species and subspecies were described from single specimens, e.g., *C. egregia* (Todd 1916) and *C. chapmani viridipennis* (Cherrie 1916b), or only from a small series of specimens. Although some species contain many subspecies, e.g., six in *C. vauxi* (Wetmore 1957; Bull and Collins 1993) and eight in *C. cinereiventris* (Peters 1940), most subspecies have not been evaluated using large sample sizes.

Second, contrasts in coloration have been used in studies of species limits but have generally not been evaluated with respect to geographic variation. These areas of contrast involve the throat versus the breast, the rump versus the mantle, or both. These are some of the few characters that distinguish some species. However, a high degree of plumage similarity among species is almost a rule in *Chaetura* and has generated much taxonomic confusion. For example, Hellmayr (1907, 1908) first brought attention to the similarity of *Chaetura chapmani*, *C. pelagica*, and *C. v. vauxi*, but noted that the latter two differed from the first by having a distinct whitish throat. He also noted that *C. c. chapmani* differs from the other two by having the glossy color of the head and mantle blue instead of green. Nevertheless, he pointed out that *Chaetura gaumeri* (currently classified as *C. vauxi richmondi*) from Costa Rica was similar in pileum and mantle coloration to *C. chapmani*. Subsequently, Lack (1956b) suggested that *C. pelagica*, *C. chapmani*, and *C. vauxi* should be considered under a single species. Wetmore (1957) refuted Lack's arguments and suggested that all should be different species. Wetmore acknowledged that coloration was similar but that body sizes differed between *C. v. vauxi* and *C. pelagica*, and among *C. v. (ochropygia) richmondi*, *C. v. aphanes*, and *C. chapmani*.

For some swift species, color varies along environmental gradients in accordance with Gloger's Rule (in *Apus* [Lack 1956b] and in *Chaetura* [Meise 1964]). This environmental color correlation is known in several other bird groups (e.g., tyrant flycatchers [Smith and Vuilleumier 1971]) and in many North American birds (Zink and Remsen 1986).

The third problem in swift taxonomy is that swifts are so mobile that the limits of breeding distribution and timing of migrations are uncertain in many species. Recent detailed analyses of migratory bird species in the Neotropics have shown that reference works may misrepresent patterns of seasonal distribution (e.g., Remsen and Parker 1990; Marantz and Remsen 1991; Chesser and Marín 1994).

The aim of this study was to investigate the taxonomic position of the Ashy-tailed Swift (*C. andrei*) and its closest relatives. The present paper gives a new interpretation and adds further information on seasonal movements, morphometrics, and species limits for some species of New World spine-tailed swifts.

METHODS

This analysis is based primarily on personal examination of several hundred museum specimens (see Acknowledgments). When used, sight records or specimens from the literature that I did not examine were regarded with extreme caution, because I have encountered many misidentified specimens (also see below). Detailed localities (including latitude and longitude) are given in some cases to assist in defining seasonal boundaries, areas of possible intergradation, or general distributional ranges. The latitudes and longitudes were taken from the gazetteers of Naumburg (1935), Fairchild and Handley (1966), Paynter (1982, 1985, 1989, 1992), Paynter and Traylor (1981, 1991) and Vanzolini (1992). Colors are described with reference to Smithe (1975, 1981) by giving the name and number (e.g., Brownish Olive 29) of the color or colors found closest to that of the museum specimens. Measurements follow Baldwin et al. (1931) except that wing measurements were taken flattened, because it gives the longest possible measurement of the wing and because it is less variable, from person to person, than measuring the chord (pers. observ.). Wing length and tarsus length were used as body size indices for the morphometric analysis. Tail measurements were not analyzed because in *Chaetura* they may vary by 9 mm or more owing to the degree of tail wear (tail-spines) (see Table 2). Body masses were not analyzed because sample sizes were inadequate for most species.

In most New World swifts the molt of the inner primaries occurs during the breeding season (Snow 1962; Collins 1968a; Marín and Stiles 1992, 1993; Bull and Collins 1993; Marín, unpubl.), although the Black Swift (*Nephoecetes niger*) is an exception (Marín and Stiles 1992). Therefore, the timing of primary molt was used to distinguish breeding versus winter range in most species. Nomenclature, but not sequence of species and subspecies, follows Peters (1940).

RESULTS AND DISCUSSION

Chaetura andrei andrei.—The nominate form of the Ashy-tailed Swift is known from only five specimens from three localities in eastern Venezuela. Two of the localities are in edo. Bolivar, at Caicara (7°37'N, 66°10'W) and Altagracia (7°52'N and 65°33'W), and one is in edo. Sucre, at San Felix (10°15'N and 63°55'W) (Berlepsch and Hartert 1902). The only published accounts on the species are those of Cherrie (1916a) and Berlepsch and Hartert (1902), who reported those specimens collected by Cherrie in Venezuela. Since then, virtually nothing has been written about *C. andrei*, except on the specimens claimed by Phelps and Phelps (1958), but their identification needs to be confirmed (see also Fig. 1.)

Two of the known specimens of *C. a. andrei*, including the holotype (AMNH 477325), show strong contrast between the throat and chest, whereas the other two show less contrast. Phenotypically, the two individuals of *C. a. andrei* that have less contrast are identical to the southernmost form of *C. vauxi* [*richmondi*] *aphanes*. Indeed, there is a great deal of confusion between *C. vauxi* and *C. andrei*. For example, two specimens of *C. vauxi aphanes* collected in Venezuela (edos. de Carabobo and Falcón, in the western and central part of its range) by G. Cherrie (AMNH 150209, 150211) were originally labeled as *C. a. andrei* by the collector (who collected

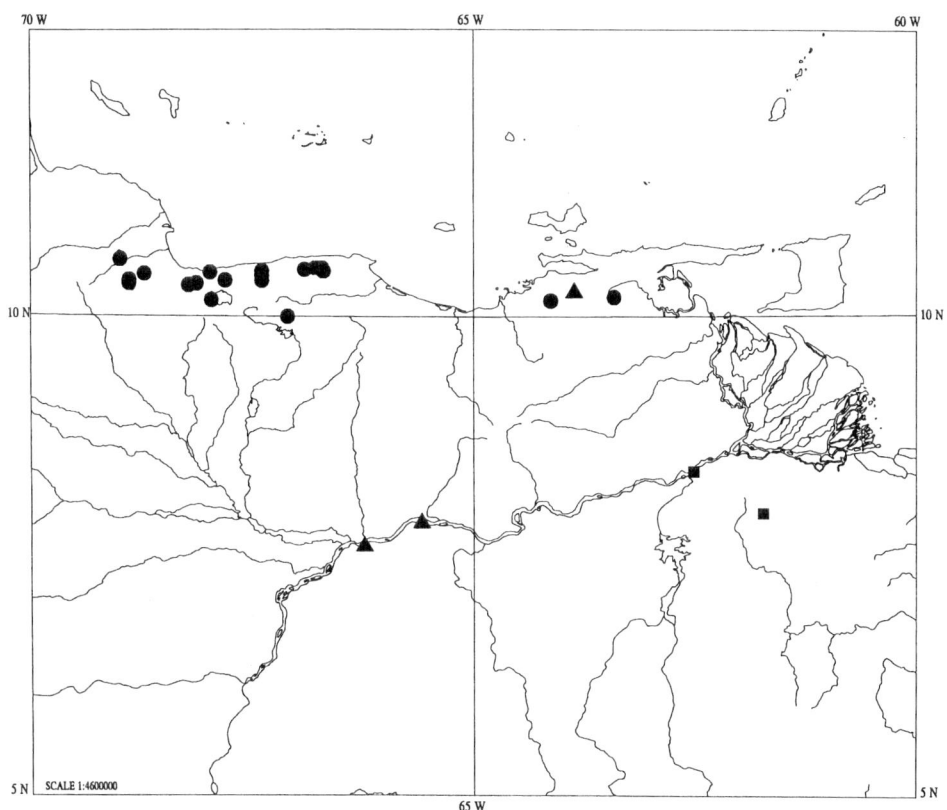

FIG. 1. Map of Venezuela showing the only known localities for *Chaetura a. andrei* (triangles). Squares are the two specimens alleged of this species reported by Phelps and Phelps (1958) (see text); and circles are localities for *Chaetura vauxi aphanes*. Based on museum specimens, Sutton and Phelps (1948), Wetmore and Phelps (1956), and Phelps and Phelps (1958).

the type series of *C. andrei*). George M. Sutton re-identified these and two other specimens labeled as *C. andrei* and collected in the edos. de Lara and Discrito Federal (also both inside the range of *C. v. aphanes*) in the AMNH collection as *C. vauxi richmondi* (Sutton and Phelps 1948). Most specimens from Venezuela examined and labeled as *C. vauxi aphanes* or *C. v. richmondi* are from the western Coastal Range, and some have a paler breast and belly (less contrast with the throat) than those from the eastern Coastal Range than some of those labeled *C. a. andrei* from the south or eastern Coastal Range (Fig. 1). However, there are more differences in coloration between *C. v. vauxi* and *C. v. richmondi* than between *C. a. andrei* and *C. v. aphanes*. Furthermore, Wetmore and Phelps (1956) described *C. vauxi aphanes* within the same geographical range. This only added to the confusion because eight years earlier Sutton and Phelps (1948) reported that the newly discovered *C. vauxi* population in Venezuela did not differ from *C. vauxi richmondi* from Central America. The type locality for *C. v. aphanes* is Caripe (10°12′N, 63°29′W), in the eastern Coastal Range. This locality is only 30 km southeast of a paratype of *C. andrei* (see Fig. 1). These specimens are all very similar if not identical (Table 1), and by size, all *C. a. andrei* fall into the *C. vauxi* group (Fig. 2). I found no major differences in wing length (wl), tarsus length (tl), or culmen length (cl) between *C. a. andrei* and *C. vauxi aphanes*. (Student's t_{wl} = 0.43, P = 0.69, d.f. = 11; t_{tl} = 0.92, P = 0.38, d.f. = 11; t_{cl} = 2.35, P = 0.04, d.f. = 10; see also Fig. 2 and Table 2). Four of the five known specimens of *C. andrei* have on average smaller tails (Table 2). However, I excluded tail length from the analysis because the tails are very worn, and the degree of variation combined with small sample size can give a false impression. Furthermore, all known specimens are from the same dates, and length of tailspines (hence tail length) varies seasonally (Sutton and Phelps 1948; this study).

Cherrie (1916a), the only person to see these birds in the wild, encountered them first on 2 February and collected two on 3 February 1898 in Altagracia and one each on 19 and 21 March

TABLE 1
COLORS OF SOME IMPORTANT BODY REGIONS THAT DISTINGUISH SOME NEW WORLD *Chaetura* SWIFTS. CAPITALIZED AND NUMBERED COLOURS ARE FROM SMITHE (1975, 1981)

Group	Species or subspecies	Throat (vary from)	Chest-belly (intensity and darkness increases toward lower belly)	Crown-mantle	Rump (vary from)
Small	*C. a. andrei*	From whitish to near Drab-Gray (119D), Light Drab (119C).	Near Hair Brown (119A).	Near Brownish Olive (29) (greenish-blue gloss ?).	From Grayish Horn Color (91), Light Drab (119C).
Small	*C. v. vauxi*	From white to near Smoke Gray (45).	Near Light Drab (119C).	Near Brownish Olive (29) (greenish gloss).	From Grayish Horn Color (91), Light Drab (119C), Drab-Gray (119D).
Small	*C. vauxi aphanes*	From whitish to near Light Drab (119C).	Near Drab (27).	Near Brownish Olive (29) (greenish-blue gloss).	From Grayish Horn Color (91), Dark Drab (119B, Drab (27), Light Drab (119C).
Small	*C. v. richmondi*	From whitish to near Smoke Gray (44/45).	From a brownish-gray tone to near Brownish Olive (29).	Near Brownish Olive (29) (greenish-blue gloss).	From Grayish Horn Color (91), Dark Drab (119B, Drab (27).
Intermediate	*C. c. chapmani*	Near Brownish Olive (29), sometimes with an almost unperceivable paler tone than chest.	Near Brownish Olive (29).	A rich near Brownish Olive (29) (greenish-blue gloss).	Between Grayish Horn Color (91), and Brownish Olive (29).
Large	*C. c. viridipennis*	From near Brownish Olive (29), some specimens with feathers tipped with near Light Drab (119C).	Near Brownish Olive (29).	A rich near Brownish Olive (29) (greenish-blue gloss).	Darker than Grayish Horn Color (91), also near Brownish Olive (29).
Large	*C. pelagica*	From white to near Light Drab (119C), some can be near Dark Drab (119B).	From Smoke Gray (45) to Brownish Olive (29).	Near Brownish Olive (29) (greenish gloss).	Closer to Grayish Horn Color (91).
Large	*C. a. meridionalis*	From white to near Light Drab (119C), some are darker near Dark Drab (119B).	From Light Drab (119C) to Brownish Olive (29).	Near Brownish Olive (29) (greenish gloss).	Closer to Grayish Horn Color (91).

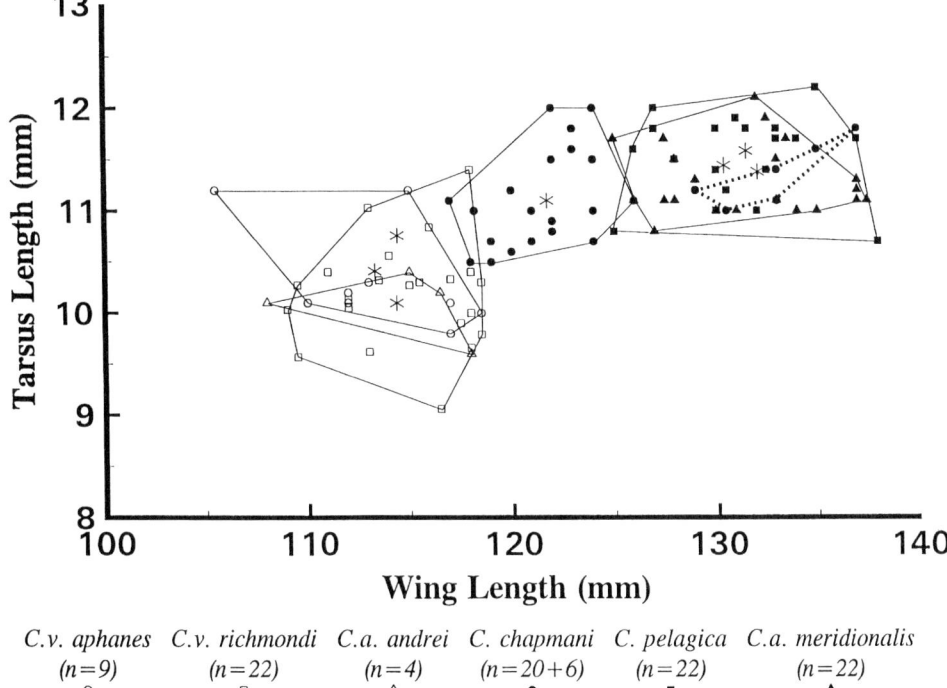

FIG. 2. Scatter plot of tarsus versus wing, showing the separation of three species groups, small, medium, and large (see text). Dashed line represents *Chaetura viridipennis* specimens. The asterisks are the species or subspecies means from Table 2.

at Caicara. Cherrie arrived in the area on 1 November 1897, but no specimens were seen or secured until February; he found that *C. a. andrei* was "not uncommon at Caicara during March and April." Either this would suggest some seasonal movement or confusion with other species. None of the specimens examined showed any primary molt, which would be an indication of breeding in the area. Furthermore, Sutton and Phelps (1948) reported specimens of *C. vauxi aphanes* collected in the Coastal Range from all months except January, February, March, and April. This either indicates seasonal movement or a lack of collecting.

I conclude that *C. a. andrei* is inseparable from the *vauxi* complex. It is the same size as and color as if not identical to, *C. vauxi aphanes*. Therefore, I propose that *C. a. andrei* be treated as a subspecies of *C. vauxi*; however, the name *C. a. andrei* has priority over *C. vauxi aphanes*. Thus, *C. vauxi aphanes* should be regarded as a synonym of *Chaetura vauxi andrei*. The other possibility would be to have two subspecies of the *vauxi* complex in Venezuela, one within the wet western Coastal Range (*C. v. aphanes*) and the other within the deciduous seasonal forest in the southern and eastern lowlands, and eastern Coastal Range (*C. v. andrei*). However, the westernmost population in the western Costal Range is phenotypically and morphologically identical to the Middle American form *C. vauxi richmondi* (Sutton and Phelps 1948; this paper). The small number of specimens from the eastern population makes it difficult to assess adequately the degree of intraspecific variation and seasonal movements. Those specimens labeled as *C. a. andrei* were either post-breeding migrants of a *C. vauxi* population, from the (eastern ?) Coastal Range, or they might represent a resident form in the hilly lowlands of southern Venezuela. This cannot yet be resolved because, for the last 80 years, no specimens have been collected.

Chaetura "andrei" meridionalis.—Before Hellmayr (1907) described the subspecies *C. a. meridionalis*, the Brazilian population of this taxon was treated for over 87 years as a synonym of *Hirundo pelasgia*, *Cypselus pelasgius*, *Chaetura poliura*, and *Chaetura pelasgia* (e.g., Wied, 1820, 1830; Pelzeln, 1871; Ihering, 1899). Thus, through those years, this taxon was classified as the Chimney Swift (*Chaetura pelagica*). However, when Hellmayr described *C. andrei meridionalis*, from northwestern Argentina, he considered it a southern population of *C. andrei*. He noted that both populations had the crown and mantle sooty brown with a light bronzy greenish

TABLE 2
Measurements (Mean ± 1 SD) of Some Species and Subspecies of *Chaetura*. Sample Sizes Are Given in Parentheses

Wing[1] (mm)	Tail[2] (mm)	Tail spines[3] (mm)		Exposed culmen (mm)	Tarsus (m)	Mass (g)
		Min	Max			
Chaetura pelagica						
130.40	39.10	0	7	5.00	11.44	21.33
±2.83	±2.67			±0.22	±0.59	±2.71
(50)	(50)			(50)	(50)	(20)
Chaetura vauxi richmondi						
114.41	36.80	1	8	4.94	10.76	16.25
±3.06	±2.09			±0.32	±0.53	±1.47
(34)	(35)			(35)	(33)	(35)
Chaetura vauxi aphanes						
113.30	36.60	5	8	4.75	10.41	16.00
±4.07	±2.90			±0.25	±0.76	±1.80
(9)	(9)			(9)	(9)	(3)
Chaetura andrei andrei						
114.37	30.25	2	3	4.36	10.10	n/d
±4.42	±1.71			±0.20	±0.34	
(4)	(4)			(3)	(4)	
Chaetura andrei meridionalis						
131.54	37.05	1	5	4.77	11.54	21.56
±3.71	±1.40			±0.27	±0.17	±2.15
(50)	(49)			(45)	(47)	(8)
Chaetura chapmani chapmani						
121.75	40.00	0	9	4.95	11.10	24.69
±2.36	±2.09			±0.16	±0.49	±1.58
(20)	(20)			(16)	(20)	(6)
Chaetura chapmani viridipennis						
132.07	41.50	5	8	5.16	11.38	23.00
±3.79	±0.95			±0.19	±0.29	n/d
(7)	(7)			(6)	(7)	(1)

[1] Wing flat.
[2] Tail spines included.
[3] Sample size, same as for the tail.

sheen, but that the southern differed from the northern population mainly by its larger size and in being "perhaps a shade lighter" (Hellmayr 1907, 1908).

The type specimen (AMNH 477329) was collected on 2 February 1906 in (Santiago?) prov. Santiago del Estero, Argentina. The type, unfortunately an immature, shows a sharp contrast between the throat and chest, but as with *C. pelagica*, individual variation is great. Most specimens show contrast between the throat and chest-belly region, whereas others have only a gradual increase in darkness. In young birds as well as adults with fresh feathers, the greenish gloss in the primaries is intense, contrasting with adult birds with worn primaries that have a steel-blue gloss. Both young and adults have a sooty brown mantle with greenish gloss, but this is more pronounced in young birds. In coloration *C. a. meridionalis* is basically identical to *C. pelagica* (Table 1), and I found no significant difference in wing length (wl)(Student's $t_{wl} = 0.40$, $P = 0.70$, d.f. = 98). However, tarsus length (tl) and culmen length (cl) differed significantly (Student's $t_{tl} = 4.01$, $P < 0.001$ d.f. = 95; $t_{cl} = 4.68$, $P = < 0.001$ d.f. = 93) (see also, Fig. 2 and Table 2). The only way that I found to distinguish these taxa is by the difference between the two outermost primaries; in *C. a. meridionalis* the 9th is 3–8 mm longer than the 10th, whereas in *C. pelagica* the 9th and 10th primary are nearly equal in length. Lack (1957) reported size differences between the two outermost primaries for several species of Apodidae and considered them aerodynamic adaptations rather than taxonomic characters. Both species belong to the large size class (Fig. 2).

Color and size differences between *C. "a." andrei* (= *C. vauxi andrei*) and *C. "a." meri-

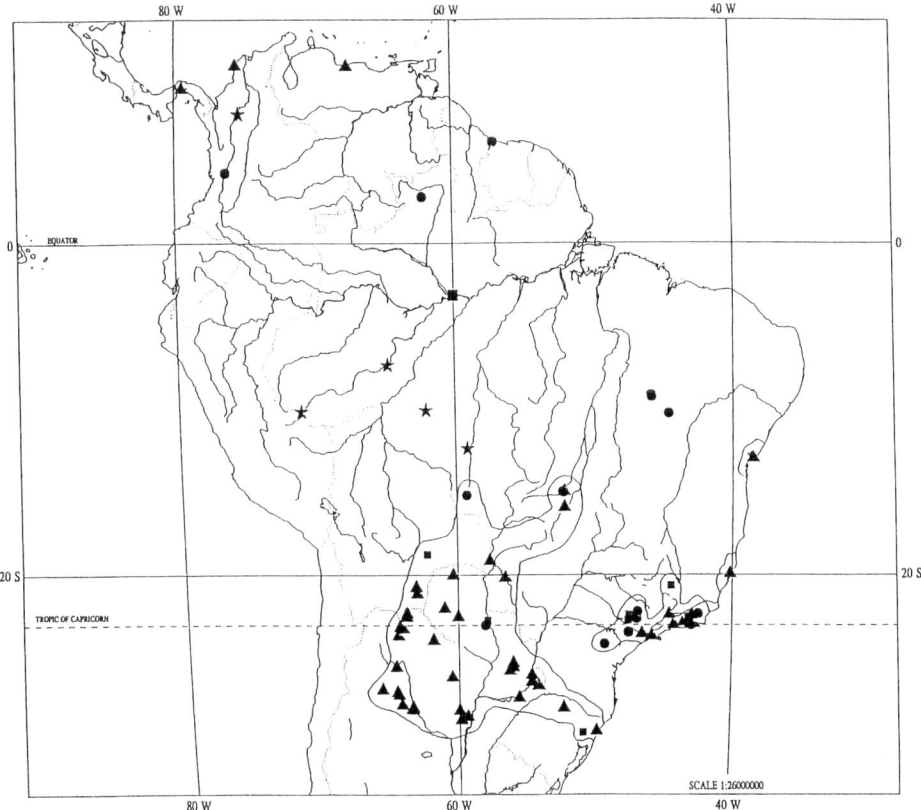

FIG. 3. Distribution of *Chaetura "andrei" meridionalis*. Triangles = examined specimens; circles = reliable specimens reported in literature cited; and squares = reliable sight records. Stars represent specimens of *Chaetura "chapmani" viridipennis*. Additional specimens from Belton (1984) and Schifter (1992); and additional sight records from Parker et al. (1993).

dionalis are nearly the same as the differences found between *C. v. vauxi* and *C. pelagica*. Therefore, for consistency in assigning species rank to allopatric taxa in this genus, I recommend that the taxon *meridionalis* be considered as separate species from *C. vauxi andrei*. I found no phenotypic and mensural differences between *C. meridionalis* and *C. pelagica* (except wing tip, see above). These two taxa are phenotypically so similar that they can be treated as a disjunct populations of the same species. Their relationships and biogeography might be paralleled by that between *Vireo o. olivaceus* and *V. o. chivi* (see Johnson and Zink 1985). It also parallels the situation of many other taxonomic groups that have a resident form in the tropics and migratory forms north and south of the tropics, e.g., Mississippi and Plumbeous kites (*Ictinia mississippiensis* / *I. plumbea*), Tropical Kingbird (*Tyrannus melancholicus*), and martins (*Progne subis* / *P. modesta*), etc. This biogeographical pattern merits further inquiry.

I suggest, however, that *C. meridionalis* should be treated as a different species from *C. pelagica* pending further analyses (e.g., biochemical, vocal, etc.). I suggest that the English name be Sick's Swift in recognition of Helmut Sick's studies of this species and many other South American birds.

Chaetura meridionalis is one of over 200 bird taxa that are austral migrants (Chesser 1994). This species winters as far north as Colombia, Venezuela, and central Panama (Fig. 3). During the austral spring it reaches the southern portion of the tropics and subtropics and is found mainly in the lowlands up to 800—900 m, occasionally over 1,000 m, from about 10° to 31°S and roughly 40° to 65°W. (Fig. 3). It is a single-brooded species with a clutch of 4-5 eggs (Sick 1948, 1959). Molt and breeding commences around October–November and breeding ends by January (Fig. 4). There are no published breeding records from Argentina, Surinam, or Venezuela (contra AOU [1983]; Oniki et al. [1992]). However, based on museum specimen data, gonad

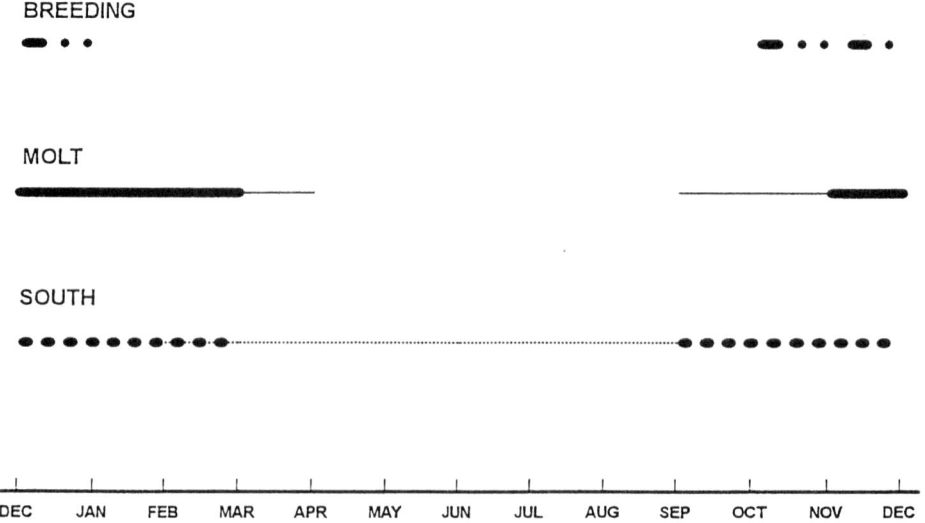

FIG. 4. Annual cycle of *Chaetura meridionalis* in reference to molt, breeding and its presence in the Southern Hemisphere, based on specimen data. Thicker lines represent the southern breeding season and were <95% of recorded individuals from specimens or literature fell, and thinner lines represents the non breeding season and few recorded individuals (see text).

size, and inner primary molt, it evidently does breed in Argentina, as well as in Bolivia, Paraguay, and Brazil.

The earliest arrival dates in the southern breeding range are: 1 September, near Puerto Pinasco (ca. 23°S, 58°W), along the Río Paraguay, Paraguay (Wetmore 1926); 5 September, Niteroi (ca. 23°S, 43°W), near Rio de Janeiro, Brazil (Sick 1959); 10 September, Ocampo (28°S, 59°W), prov. Santa Fe, Argentina (AMNH 477332–33); and 13 September at Corral (28°S, 63°W), prov. Santiago del Estero, Argentina (MACN). Among 28 specimens from southeastern Brazil, the earliest arrival date is 30 August (MZUSP; *fide* D.F. Stotz, *in litt.* and R.T. Chesser, pers. comm.).

Movements away from their breeding grounds are irregular and extend for about three months. This irregularity seems to be common in austral migrants (Chesser 1994). Some birds began to move late in January to early February, shortly after breeding. Movement seems to end by mid-April. Observations from a flock of non-breeders roosting throughout the season at a site in the state of Minas Gerais (20°35'S, 44°30'W), Brazil, indicate that the numbers began to decline by February and that the birds were completely absent by 10 April (Andrade and Freitas 1987). Mitchell (1957) at Lajes (22°40'S, 43°52'W), Rio de Janeiro, Brazil, noted particularly large numbers on 30 March and 5 April 1953. Sick's (1959) latest observations at Niteroi (22°53'S, 43°07'W) were in mid-April. A specimen collected at Yacuiba (22°S), depto. Tarija, Bolivia, on 10 February (Remsen and Traylor 1983) was probably a bird moving north early. Two specimens were taken in April 1958, one (MZUSP 42894) taken at Petrópolis (22°31'S), and the second (MZUSP 42895) at Moricá (22°55'S), both from the vicinity of Rio de Janeiro (D. F. Stotz, *in litt.* and R. T. Chesser, pers. comm.). I examined two late specimens taken on 14 March, both taken by the same collector at Puerto Segundo (ca. 25°59'S), prov. Misiones, Argentina (FMNH 57760; MACN). The latest specimens that I examined from the breeding range are from 26 March, both collected by Kaempfer at Erebango (ca. 28°S), Rio Grande do Sul, Brazil (AMNH 314037-38). The latest specimen in the São Paulo Museum is from 27 March (D.F. Stotz, *in litt.*). Also, there is one extralimital record from 1 March 1959 of a specimen found dead in the Islas Malvinas (Falkland Islands), ca. 51°40'S and 58°W (Woods 1988).

The distribution of *C. meridionalis* during the post-breeding season is difficult to ascertain. The problems are: 1) it seems that at least part of the breeding population in coastal Brazil (at ca. 23°S) occasionally winters there (Sick 1958) or leaves very late and for only a short period of time; 2) few reliable winter records exist from anywhere; and 3) it overlaps geographically with similar species, thus limiting the utility of many sight records. Among the odd records, there are two winter specimens from 22 June from Salvador (Bahia), Brazil, 13° S (AMNH 163159-60). Both specimens were molting their 10th primary; thus it is probable that they were late migrants. Reiser collected three birds in the southern part of the state of Piaui (Piauhy, 9–10°S, 45°W, see also Fig. 4) on 10 May, 10 July, and 14 August 1903 (Hellmayr 1929; NMW 42546, 42548, 42549; M. Cohn-Haft, *in litt.*).

With the present data it is difficult to assess the status of the species in northeastern Brazil. One possibility is that these birds were at the southern terminus of their winter range or northern end of their breeding area. It also might be that those birds were just passing through the area, because Reiser's birds were all collected in the same year and because there are no other records for the area. Alternatively, as reported by Sick (1958), part of the population may remain in the breeding range in climatically favorable years.

Few specimens have been collected during the post-breeding months. The only specimens that I found anywhere are as follows: 13 September, from Venezuela (Aragua, Rancho Grande; 10°22'S 67°41'W; USNM 595566); 27 August, from Colombia (Bolívar, Cartagena; 10°25'N, 75°32'W; USNM 372685); and 4 August, from Panama, (prov. Panamá, Tapia; 9°04'N, 79°25'W; FMNH 302511). Other specimens reported in the literature include: 5 August, from Surinam (Nickerie; ca. 6°N, 57°W; Haverschmidt 1968); 6 May, from Brazil (Roraima, Mucajai; ca. 2°45'N, 62°10'W; São Paulo Museum; D. F. Stotz, in litt. and R. T. Chesser, pers. comm.); and 8 August, from Colombia (Magdalena, Río Frío, 4°09'N, 76°18'W; Darlington 1931). I agree with Sick (1986) and Stotz et al. (1992) that the birds observed in March by Gilliard (1944) in Manaus and identified as possible *C. pelagica* might also have been migrants of *C. meridionalis* from the south.

The scarcity of winter records is surprising, especially considering the thousands of birds that leave the breeding grounds. Why are there so few records and what happens to those birds? The lack of specimens might reflect lack of collecting on those dates or specific areas, or difficulty in collecting swifts in general. For example, few specimens of the abundant *C. pelagica* have been collected in South America, where the entire population must winter. But another possibility is confusion with other species, as was the case with the previous taxon (see above) particularly with *Chaetura pelagica*, *C. chapmani*, and *C. vauxi aphanes*, because all overlap in northern South America. For example, a photograph, p. 8 in Collins (1968b; Fig. 4) does not seem to correspond to *C. chapmani*, as labeled, and may actually be *C. meridionalis* because *C. chapmani* lacks a distinct pale throat (see below). Furthermore, the dates given in the text "August 23 to October 28" are well within possible dates for migrants of *Chaetura meridionalis* from the south or *C. pelagica* from the north, but the bird was probably *C. meridionalis*, because *C. pelagica* seems to migrate through western South America. Additionally, *C. pelagica* and *C. meridionalis* possibly overlap in Bolivia and Argentina, because, *C. pelagica* reaches 23°S (Calama, II Region, Chile) west of the Andes on a regular basis during December-January (Demetrio 1993). Such dates and latitudes are well within the breeding range and dates of *C. meridionalis* east of the Andes. But the extent of overlap cannot be assessed without specimens. The Colombian-Venezuelan llanos and the Orinoco basin should be investigated as potential wintering grounds for *C. meridionalis*.

Chaetura c. chapmani.—Hellmayr (1907) described this subspecies from a series collected by F. M. Chapman in Trinidad. He described it as similar to *C. pelagica*, but differing in having "the pileum and mantle black glossed with steel-blue (instead of sooty brown); the throat smoky brown, like the rest of the underparts (not clear whitish), and the rump and the upper tail-coverts rather paler." Hellmayr (1908) emphasized the similarity of *Chaetura chapmani* to *Chaetura pelagica* and *C. v. vauxi*, but the latter two differ from *C. chapmani* by having a distinct whitish throat. *Chaetura chapmani* also differs from the other two by having a blue glossy color in head and mantle instead of green gloss. Additionally, Hellmayr (1907, 1908) pointed out that *Chaetura gaumeri* (currently known as *C. vauxi richmondi*) from Costa Rica was similar in pileum and mantle coloration to *C. chapmani*. However, *C. chapmani* is larger and lacks a distinct whitish throat.

No seasonal movements are known for *C. c. chapmani*, which seems to be a resident in the lowlands of northern South America, primarily north of the Amazon river. The southernmost specimen examined of this subspecies (M.C-H. # 324) is from Manaus, Brazil, where resident year-round (M. Cohn-Haft, pers. comm.). There are two apparently disjunct populations, one in northeastern South America, eastern Venezuela, Trinidad, Guyana, Surinam, French Guiana, and Brazil, and the other in the lowlands of northern Colombia to central Panama. Several specimens from Panama are molting their mid-primaries, a good indication of breeding in the area. The gap in distribution between the two populations, which is in western and northern Venezuela, is occupied by *C. vauxi*.

Chaetura c. chapmani is intermediate in size between the small *vauxi* group and the large

TABLE 3
CURRENT AND PROPOSED SCIENTIFIC AND ENGLISH NAME FOR SOME SPECIES OF NEW WORLD
Chaetura

Current	Proposed	Old or proposed English name
Chaetura v. vauxi	Same	Vaux's Swift
Chaetura vauxi richmondi	Same	Vaux's Swift
Chaetura vauxi aphanes	*Chaetura vauxi andrei*	Vaux's Swift
Chaetura andrei andrei	*Chaetura vauxi andrei*	Vaux's Swift
Chaetura c. chapmani	Same	Chapman's Swift
Chaetura chapmani viridipennis	*Chaetura viridipennis*	Amazonian Swift
Chaetura andrei meridionalis	*Chaetura meridionalis*	Sick's Swift
Chaetura pelagica	Same	Chimney Swift

pelagica group; the morphological differences between the small and the intermediate forms are as great as the differences between the large and the intermediate (see Table 2 and Fig. 2). Currently, *C. chapmani* contains a second form, *C. chapmani viridipennis*, which differs from the nominate form only in size (also see below). However, it is likely that the disjunct population from the lowlands of central and southern Panama and northwestern Colombia represents a different subspecies. Also in the following section I will show why *viridipennis* should be considered a separate species from *chapmani*.

Chaetura chapmani viridipennis.—Cherrie (1916b) described this taxon from a single specimen (AMNH 127383) collected on 17 February 1914 in Doze Octobre (12°22'S and 59°08'), Matto Grosso, Brazil. He specified that it differs from the nominate species by being larger and "having the pileum, mantle and wings glossed with greenish (instead of steel blue)." As with other *Chaetura* swifts, this color difference in the primaries is actually only a different stage of the molt process (Sutton and Phelps 1948; Collins 1968b, present work). When the primaries are fresh, they have a greenish iridescence; the intermediate stage is a deep, slightly glossy blue, and when worn, the primaries became a dull blackish-brown coloration. By size, locality, and date, the type specimen falls within the range of *C. meridionalis* (see Figs. 3 and 4). Moreover, the bird is an immature, and the locality is not far from Rio das Mortes (15°S, 52°15'W), where nesting of *C. meridionalis* has been reported (Sick 1948); if any specimens accompanied the nesting record, they should be re-examined. Most specimens identified as *C. c. viridipennis* that I have examined fall within the mensural range given for *C. meridionalis*. I found no significant difference between the two taxa in wing and tarsus length (Student's $t_{wl} = 0.40$, $P = 0.70$, d.f. = 55; $t_{tl} = 0.80$, $P = 0.43$, d.f. = 52), but the culmen length differs significantly ($t_{cl} = 3.5$, $P < 0.001$, d.f. = 50; see also, Table 2, and Fig. 2). The two taxa also differ in that *C. c. viridipennis* has a less contrasting and more uniform throat coloration. Also, *C. c. viridipennis* has the two outermost primaries of about equal length, with the 10th primary a maximum of 1–2 mm longer, whereas the 9th primary of *C. meridionalis* is longer than the 10th by about 3–8 mm. In this respect, *C. c. viridipennis* is similar to *C. pelagica*: in both species the 9th and 10th primary are about equal in length. The difference between these two is the throat coloration and color intensity. I also found no significant difference between the two in body mass (Student's $t_{wl} = 1.56$, $P = 0.12$; $t_{tl} = 0.73$, $P = 0.47$; $t_{cl} = 1.50$, $P = 0.14$; all d.f. = 81).

Some museum specimens of *C. c. viridipennis* from northern South America overlap in distribution with *C. meridionalis*; consequently, all specimens previously identified as *Chaetura chapmani viridipennis* should be re-examined using the wing formula, particularly those from northern South America. Furthermore, because of migration, this taxon overlaps the geographic range of a population of the nominate form in northern South America. However, *C. c. viridipennis* differs significantly by wing length from *C. c. chapmani* (Student's $t_{wl} = 8.74$, d.f. = 25, $P < 0.001$; see also Table 2, and Fig. 2). The differences between *C. c. chapmani* and *C. c. viridipennis* are equivalent to those between *C. v. vauxi* and *C. pelagica* in North America; all have the same wing formula, with size being the primary difference (see Table 2). Therefore, for consistency in species limits within *Chaetura*, I recommend treating *C. c. viridipennis* as a full species, *C. viridipennis*. For an English name, I suggest Amazonian Swift (see Table 3).

As suggested by the A.O.U. (1983), the species complex that consists of *C. vauxi*, *C. chapmani*, and *C. pelagica* probably constitute a superspecies. I suggest that five species in the "brown" group of the New World spine-tailed swifts should be maintained, but with some modifications (see Table 3).

ACKNOWLEDGMENTS

I thank the personnel from the following museums for providing specimens or data deposited therein: M. Robbins, Academy of Natural Sciences of Philadelphia; F. Vuilleumier, M. LeCroy, P. Sweet, and J. Bates, American Museum of Natural History, New York (AMNH); R. Prys-Jones, M. Walters, M. Adam, and F. E. Warr, Natural History Museum [ex-British Museum of Natural History], Tring (BMNH); S. Bailey, California Academy of Sciences, San Francisco; E. Alabarce, Fundación Miguel Lillo, Tucumán; S. Lanyon and D. Willard, Field Museum of Natural History, Chicago (FMNH); F. G. Stiles, Instituto de Ciencias Naturales, Universidad Nacional de Colombia, Bogotá (ICN); K. Garrett, Los Angeles County Museum of Natural History (LACM); M. Morton, Moore Laboratory of Zoology, Occidental College, Los Angeles (MLZ); J. Navas, and A. Camperi, Museo Argentino de Ciencias Naturales, Buenos Aires (MACN); J. Sanchez, Museo Nacional de Costa Rica, San José; J. V. Remsen and S. Cardiff, Museum of Natural Science, Louisiana State University, Baton Rouge; N. Johnson and C. Cicero, Museum of Vertebrate Zoology University of California, Berkeley (MVZ); F. G. Stiles, Museo Zoologia Universidad de Costa Rica, San José; Museo Zoologia Universidad de São Paulo, São Paulo (MZUSP); H. Schifter, Naturhistorisches Museum Wien, Wien, Austria (NMW); P. Unitt, San Diego Natural History Museum, San Diego; G. Graves, National Museum of Natural History, Washington D.C. (USNM); J. Hinshaw, University of Michigan Museum of Zoology; and L. Kiff, S. Sumida, R. Corado, and W. Wehtje, Western Foundation of Vertebrate Zoology, Camarillo, California. I am very grateful to G. Barrowclough, M. LeCroy, P. Sweet, J. Bates, and F. Vuilleumier for help at the AMNH, and to R. Prys-Jones and M. Walters for help at the BMNH. I thank J. Bates, R. T. Chesser, K. Garrett, T. Schulenberg, and F. G. Stiles for additional specimen information. Mary Dolack, Geshe Moller, and Heiko Schoenfuss translated papers from German to English. This paper benefitted from comments by M. R. Browing, R. T. Chesser, M. Cohn-Haft, C. T. Collins, J. V. Remsen, and F. G. Stiles. This work benefited greatly by a Collection Study Grants from the American Museum of Natural History, research funds from the Museum of Natural Science, Louisiana State University, and a research grant from the Western Foundation of Vertebrate Zoology. I am particularly indebted to all those collectors whose work seldom receives recognition. This paper is dedicated to the memory of Ted Parker for his great contributions to New World ornithology.

LITERATURE CITED

AMERICAN ORNITHOLOGISTS' UNION. 1983. Check-list of North American Birds, 6th Edition. American Ornithologists' Union, Washington, D.C.

ANDRADE, M. A., AND M. V. FREITAS. 1987. Notas sobre o anilhamento de *Chaetura andrei* (Aves: Apodidae) no estado de Minas Gerais, Brasil. Pages 192–193 in ANAIS II Encontro Nacional de Anilhadores de Aves. Universidade Federal do Rio de Janeiro, Brazil.

BALDWIN, S. P., H. C. OBERHOLSER, AND L. G. WORLEY. 1931. Measurements of birds. Sci. Pub. Cleveland Mus. Nat. Hist. 2:1–165.

BELTON, W. 1984. Birds of Rio Grande do Sul, Brazil. Part 1. Rheidae through Furnariidae. Bull. Amer. Mus. Nat. Hist. 178:369–636.

BERLEPSCH, H. V., AND E. HARTERT. 1902. On the birds of the Orinoco region. Novitates Zoologicae. 9:1–91.

BROOKE, R. K. 1970. Taxonomic and evolutionary notes on the subfamilies, tribes, genera and subgenera of the swifts (Aves: Apodidae). Durban Mus. Novitates 9:1–24.

BROWING, M. R. 1993. Species limits of the cave swiftlets (*Collocallia*) in Micronesia. Pages 101–106 in Recent advances in the study of the Apodidae. (G. Malacarne, C. T. Collins, and M. Cucco, Eds.). Avocetta Vol. 17. (Special issue).

BULL, E. L, AND C. T. COLLINS. 1993. Vaux's Swift *Chaetura vauxi*. in The Birds of North America, No. 77 (A. Poole and F. Gill, Eds.). Academy of Natural Sciences, Philadelphia, and American Ornithologists' Union, Washington, D.C.

CHERRIE, G. K. 1916a. A contribution to the ornithology of the Orinoco region. Sci. Bull. Mus. Brooklyn Inst. Arts Sci. 2:133–374.

CHERRIE, G. K. 1916b. Some apparently undescribed birds from the collection of the Roosevelt South American expedition. Bull. Amer. Mus. Nat. Hist. 35:183–190.

CHESSER, R. T. 1994. Migration in South America: an overview of the austral system. Bird Conserv. Intern. 4:163–179.

CHESSER, R. T, AND M. MARÍN. 1994. Seasonal distribution and natural history of the Patagonian Tyrant (*Colorhamphus parvirostris*). Wilson Bull. 106:649–667.

COLLINS, C. T. 1968a. The comparative biology of two species of swifts in Trinidad, West Indies. Bull. Florida State Mus. 11:257–320.
COLLINS, C. T. 1968b. Notes on the biology of Chapman's Swift *Chaetura chapmani* (Aves, Apodidae). Amer. Mus. Novitates 2320:1–15.
DARLINGTON, P. J. 1931. Notes on the birds of the Río Frío (near Santa Marta), Magdalena, Colombia. Bull. Mus. Comp. Zool. 71:349–421.
DEMETRIO, L. 1993. Aclaraciones sobre la presencia del Vencejo de Chimenea (*Chaetura pelagica*) en el valle de Calama. Bol. Unorch. 15:16–17.
FAIRCHILD, G. B., AND C. O. HANDLEY. 1966. Gazetter of collecting localities in Panama. Pages 9–22 *in* Ectoparasites of Panama (R. L. Wenzel and V. J. Tipton, Eds.) Field Museum of Natural History, Chicago, Illinois.
GILLIARD, E. T. 1944. Chimney Swifts (?) at Manaus, Brazil. Auk 61:143–144.
HARTERT, E. 1892. Catalogue of the Birds in the British Museum. vol. 16:434–518. Taylor and Francis, London, United Kingdom.
HARVERSHMIDT, F. 1968. Birds of Surinam. Oliver and Boyd, Edinburg, U.K.
HELLMAYR, C. E. 1907. [No title]. Bull. Brit. Ornith. Club. 19:62–63.
HELLMAYR, C. E. 1908. Übersicht der südamerikanischen Arten der Gattung *Chaetura*. (s. str.). Verhandlungen Ornithologischer Gesellschaft Bayern 8:144–161.
HELLMAYR, C. E. 1929. A contribution to the ornithology of northeastern Brazil. Field Mus. Nat. Hist., Zool. series, 12:234–501.
IHERING, H. V. 1899. As aves de estado de S. Paulo. Rev. Mus. Paulista 3:113–476.
JOHNSON, N. K., AND R. M. ZINK. 1985. Genetic evidence for relationships among the Red-eyed, Yellow-green, and Chivi vireos. Wilson Bull. 97:421–435.
LACK, D. 1956a. A review of the genera and nesting habits of swifts. Auk 73:1–32.
LACK, D. 1956b. The species of *Apus*. Ibis 98:34–61.
LACK, D. 1957. The first primary in swifts. Auk 74:385–386.
MARANTZ, C. A., AND J. V. REMSEN, JR. 1991. Seasonal distribution of the Slaty Elaenia, a little-known austral migrant of South America. J. Field Ornithol. 62:162–172.
MARÍN, M., AND F. G. STILES. 1992. On the biology of five species of swifts (Apodidae, Cypseloidinae) in Costa Rica. Proc. West. Foundation Vert. Zool. 4:287–351.
MARÍN, M., AND F. G. STILES. 1993. Notes on the biology of the Spot-fronted Swift. Condor 95:479–483.
MEDWAY, L., AND J. D. PYE. 1977. Echolocation and the systematics of swiftlets. Pages 225–238 *in* Evolutionary Ecology (B. Stonehouse and C. Perrins, Eds.). C. M. Macmillan, London.
MEISE, V. W. 1964. Zur systematik und evolution amerikanischer stachelschwanzsegler der gattung *Chaetura*. Mitt. Hamburg. Zool. Mus. Inst. 61, Taf. IV, Suppl., 95–106.
MITCHELL, M. H. 1957. Observations on Birds of Southeastern Brazil. University of Toronto Press, Canada.
NAUMBURG, E. M. B. 1935. Gazetteer and maps showing stations visited by Emil Kaempfer in eastern Brazil and Paraguay. Bull. Amer. Mus. Natur. Hist. 68:449–469.
ONIKI, Y., E. O. WILLIS, AND M. M. WILLIS. 1992. *Chaetura andrei* (Apodiformes, Apodidae): Aspects of nesting. Ornitologia Neotropical 3:65–68.
ORR, R. T. 1963. Comments on the classification of swifts of the subfamily Chaeturinae. Proc. XIIIth International Ornithol. Congress: 126–134.
PARKER, T. A., III, A. H. GENTRY, R. B. FOSTER, L. H. EMMONS, AND J. V. REMSEN, JR. 1993. The Lowland Dry Forests of Santa Cruz, Bolivia: A Global Conservation Priority. RAP Working Papers 4, Conservation International, Washington, D.C.
PAYNTER, R. A., JR. 1982. Ornithological Gazetteer of Venezuela. Museum of Comparative Zoology, Cambridge, Massachusetts.
PAYNTER, R. A., JR. 1985. Ornithological Gazetteer of Argentina. Museum of Comparative Zoology, Cambridge, Massachusetts.
PAYNTER, R. A., JR. 1989. Ornithological Gazetteer of Paraguay. (second edition). Museum of Comparative Zoology, Cambridge, Massachusetts.
PAYNTER, R. A., JR. 1992. Ornithological Gazetteer of Bolivia (second edition). Museum of Comparative Zoology University, Cambridge, Massachusetts.
PAYNTER, R. A., JR., AND M. A. TRAYLOR, JR. 1981. Ornithological Gazetteer of Colombia. Museum of Comparative Zoology, Cambridge, Massachusetts.
PAYNTER, R. A., JR., AND M. A. TRAYLOR, JR. 1991. Ornithological Gazetteer of Brazil. Museum of Comparative Zoology, Cambridge, Massachusetts.
PELZELN, A. V. 1871. Zur ornithologie Brasiliens. (resultate Von Johann Natterers reisen den jahren 1817 bis 1835.) Druck und verlag Von A. Pichler's Witwe and Sohn, Wien, Austria.
PETERS, J. L. 1940. Check-list of Birds of the World, Vol. 4. Museum of Comparative Zoology, Harvard University Press, Cambridge, Massachusetts.
PHELPS, W. H., AND W. H. PHELPS, JR. 1958. Lista de las Aves de Venezuela con su Distribución. Tomo 2, Parte 1, No Passeriformes. Editorial Sucre, Caracas, Venezuela.
REMSEN, J. V., JR., AND T. A. PARKER III. 1990. Seasonal distribution of the Azure Gallinule (*Porphyrula flavirostris*), with comments on vagrancy in rails and gallinules. Wilson Bull. 102:380–399.

REMSEN, J. V., JR., AND M. A. TAYLOR, JR. 1983. Additions to the avifauna of Bolivia, part 2. Condor 85: 95–98.
SCHIFTER, H. 1992. Von Johann Natterer in Brasilien gesammelte Segler (Apodidae) und die darunter befindlichen Typen. Mitt. Zool. Mus. Berl. 68 (1992) Suppl.: Ann. Orn. 16:157–165.
SCLATER, P. L. 1865. Notes on the genera and species of Cypselidae. Proc. Zool. Soc. London 30:593–617.
SICK, H. 1948. The nesting of *Chaetura andrei meridionalis*. Auk 65:515–520.
SICK, H. 1950. Apontamientos sôbre a ecologia de *Chaetura andrei meridionalis* Hellmayr, no estado do Rio de Janeiro (Micropodidae, Aves). Rev. Brasil. Biol. 10:425–436.
SICK, H. 1958. Geselligkeit, schorntein-benutzung und überwinterung beim brasilianishen stalchelschwanzsegler *Chaetura andrei*. Die Vogelwarte 19:248–253.
SICK, H. 1959. Notes on the biology of two Brazilian swifts, *Chaetura andrei* and *Chaetura cinereiventris*. Auk 76:471–477.
SICK, H. 1986. Ornitología Brasileira, Uma Introdução. 2ª Edição Brasília, Ed. Universidade de Brasília, Brasília, Brazil.
SMITH, W. J., AND F. VUILLEUMIER. 1971. Evolutionary relationships of some South American ground tyrants. Bull. Mus. Comp. Zool. 141:181–268.
SMITHE, F. B. 1975. Naturalist's Color Guide. American Museum of Natural History, New York.
SMITHE, F. B. 1981. Naturalist's Color Guide, part III. American Museum of Natural History, New York.
SNOW, D. W. 1962. Notes on the biology of some Trinidad swifts. Zoologica 47:129–139.
STOTZ, D. F., R. O. BIERREGAARD, M. COHN-HAFT, P. PETERMANN, J. SMITH, A. WHITTAKER, AND S. V. WILSON. 1992. The status of North American migrants in central Amazonian Brazil. Condor 94:608–621.
SUTTON, G. M., AND W. H. PHELPS. 1948. Richmond's Swift in Venezuela. Occ. Papers Mus. Zool. Univ. Mich. 505:1–6.
TODD, W. E. C. 1916. Preliminary diagnosis of fifteen apparently new Neotropical birds. Proc. Biol. Soc. Wash. 29:95–98.
VANZOLINI, P. E. 1992. A Supplement to the Ornithological Gazetteer of Brazil. Mus. Zool. Univ. de São Paulo, Brazil.
WETMORE, A. 1926. Observations on the birds of Argentina, Paraguay, Uruguay, and Chile. Bull. U.S. Nat. Mus. 133:1–448.
WETMORE, A. 1957. Species limitations in certain groups of the swifts genus *Chaetura*. Auk 74:383–384.
WETMORE, A., AND W. H. PHELPS. 1956. Further additions to the list of Venezuela. Proc. Biol. Soc. Wash. 69:1–12.
WIED-NEUWIED, M. PRINZ, ZU. 1820. Reise nach Brasilien in den Jahren 1815 bis 1817. Vol. 2. Erster Band, Frankfurt, Germany.
WIED-NEUWIED, M. PRINZ, ZU. 1830. Beitrage zur Naturgeschichte von Brasilien. III band. Erste Abtheilung. In verlage des Gr. H.S. priv. Landes-Industrie-Comptoirs, Weimar, Germany.
WOODS, R. W. 1988. Guide to Birds of the Falkland Islands. Nelson Press, Shropshire, England.
ZINK, R. M., AND J. V. REMSEN, JR. 1986. Evolutionary processes and patterns of geographic variation in birds. Pages 1–69 *in* Current Ornithology, vol. 4 (R. F. Johnston, Ed.). Plenum Press, New York and London.

INSIGHTS INTO THE MAINTENANCE OF HIGH SPECIES DIVERSITY IN THE NEOTROPICS: HABITAT SELECTION AND FORAGING BEHAVIOR IN UNDERSTORY BIRDS OF TROPICAL AND TEMPERATE FORESTS

PETER P. MARRA[1] AND J. V. REMSEN, JR.
Museum of Natural Science, Foster Hall 119, Louisiana State University,
Baton Rouge, Louisiana 70803, USA;
[1]*present address: Department of Biology, Dartmouth College,*
Hanover, New Hampshire 03755, USA

ABSTRACT.—For almost three decades, structural habitat complexity has been regarded as a primary ecological factor responsible for maintaining high species diversity of tropical bird communities. However, differences in habitat complexity between temperate and tropical forests have not been documented sufficiently. Differences between temperate and tropical forests were quantified by measuring 36 variables related to understory habitat structure. Structural components of temperate and tropical habitats differed significantly in several features. However, no differences were found in overall habitat heterogeneity (i.e. complexity) between temperate and tropical forests. To search for "bird-related" factors important in maintaining high tropical species diversity, we compared habitat selection between foliage-gleaning insectivorous birds in temperate and tropical forest understories. The tropical species were more specialized in horizontal and vertical habitat selection, and had lower "niche breadths" in foraging substrate and in foraging height. Tropical species also showed less interspecific overlap in most foraging variables than did temperate species. Therefore, higher species diversity, at least within this guild of birds in tropical and temperate forest understories, can be attributed at the proximate level to greater specialization and "tighter species packing," and may be more independent of greater habitat complexity than previously thought.

RESUMEN.—Por casi tres decadas, la complejidad estructural del habitat se ha considerado como el principal factor ecologico responsable de mantener la alta variedad de especies encontrada en comunidades de pajaros tropicales. Sin embargo, la diferencia en complejidad del habitat entre bosques templados y bosques tropicales no ha sido suficientemente quantificada. Nosotros quantificamos diferencias entre bosques templados y bosques tropicales midiendo 36 variables relacionadas a la estructura del habitat bajo el dosel y encontramos varias diferencias significativas en componentes estructurales entre habitats tropicales y templados. Sin embargo, no encontramos diferencias en la heterogeneidad del habitat (complejidad) entre bosques templados y tropicales. Buscando factores que fueran importantes en mantener la alta diversidad de las especies de pajaros tropicales, comparamos la seleccion del habitat entre pajaros insectivoros que usan el follaje bajo el dosel en bosques templados y tropicales. Las especies tropicales estaban mas especializadas en su seleccion de habitats al nivel horizontal y vertical, teniendo pequenos nichos en el substrato y altitud del area donde forrajeaban. En la mayoria de las variables relacionadas al forraje, las especies tropicales tambien demonstraron menos traslapo entre-especies que las especies templadas. Concluimos pues que la alta variedad de especies, por lo menos dentro de este grupo de pajaros que viven bajo el dosel en bosques templados y tropicales, se puede atribuir inicialmente a mayor especializacion y concentracion de especies y puede que sea mas independiente de la total complejidad del habitat de lo que actualmente se cree.

Why tropical rainforests, particularly those of South America, harbor the greatest number of bird species in the world is a question that continues to intrigue ornithologists. The striking difference in species richness stimulates such intrigue: over 300 species of resident forest birds can be found in an area of about 3 km^2 of Amazonian rainforest (Terborgh et al. 1984), whereas in a comparable area of temperate forest, no more than 50 species typically coexist (and usually no more than 30). As many as 40 species in a single family, the antbirds (Formicariidae), can be found at a single Amazonian forest site (e.g., Terborgh et al. 1984); thus, single-site species richness of one family alone may exceed that of all forest birds combined at many temperate localities. No consensus exists concerning explanations of high species richness of the tropics because the answer involves complex interactions among ecological and historical factors; such factors, at the ultimate level of causation, are intrinsically difficult to investigate.

Testing hypotheses concerning proximate causes of diversity, or factors that maintain diversity, however, is potentially more feasible, as outlined by Remsen (1990). Three hypotheses have been proposed to explain single-point (alpha) diversity at the proximate level in tropical forests: (1) *Increased Resource Diversity*—This hypothesis suggests that more species are able to coexist because of greater resource diversity in the tropics, such as greater structural habitat complexity (MacArthur and MacArthur 1961; MacArthur et al. 1962; MacArthur 1969; Orians 1969; Karr 1971; Willson 1974; Cody 1975; Terborgh 1980a) or unique tropical resources such as bamboo (Parker 1982; Kratter 1995, and in press) or dead leaves suspended in vegetation (Remsen and Parker 1984; K. Rosenberg 1990 1997); (2) *Increased Specialization*—This hypothesis proposes that species are more specialized in the tropics, and therefore more species can "pack" into available habitats (Klopfer and MacArthur 1960); (3) *Increased Ecological Overlap*—This hypothesis proposes that there is greater niche overlap among tropical species, permitting more species to coexist in a given area (Klopfer and MacArthur 1961; May and MacArthur 1972).

These hypotheses are not mutually exclusive, and they all predict a direct causal relationship between diversity and resource use. In this study we tested these three hypotheses by comparing (1) structural habitat complexity and (2) the specificity of macrohabitat, microhabitat, and foraging site selection between a similar guild of tropical and temperate zone birds.

Despite the obvious need for comparative data on habitat selection in tropical and temperate bird species, few data are available. By habitat selection, we mean comparing habitat availability to habitat use by birds. So far, research quantifying habitat preference in tropical birds has consisted only of describing habitat use at the macrohabitat scale by comparing species composition among different habitats or forest types (e.g., Orians 1969; Terborgh and Weske 1969; Karr 1971; Willson et al.. 1973; Karr and Freemark 1983; Lovejoy 1974; Remsen and Parker 1983; Terborgh et al. 1984; Terborgh 1985; De Visscher 1984; Willson and Moriarty 1976; Silva and Constantino 1988; Thiollay 1988a; G. Rosenberg 1990) or at the scale of foraging substrates within a habitat (e.g., Stiles 1978; Askins 1983; Remsen 1985; Thiollay 1988b; K. Rosenberg 1990).

With few exceptions, studies that have quantified microhabitat selection within a forest habitat have been restricted to the temperate zone (e.g., James et al. 1984; Cody 1985; Morse 1985). The exceptions (Kikkawa et al. 1980; Schemske and Brokaw 1981; Karr and Freemark 1983; Wunderle et al. 1987; Levey 1988) relied exclusively on mist-net capture data. We believe that the use of mist nets for determining microhabitat selection of birds is unsound. General problems in interpreting data from mist-net captures have been discussed elsewhere (Karr 1971, 1981; Lovejoy 1974; Remsen and Parker 1983; Terborgh 1985; Remsen and Good 1996). Specifically with respect to microhabitat selection, mist-net data have the following problems: (1) Capture in a mist net means only that the bird was *flying through* that air space, not necessarily that it was using the microhabitat near the mist net (Remsen and Good 1996). In fact, there might be an inverse relationship between use of microhabitat within a few meters of the net and likelihood of capture, because those individuals foraging in the vegetation near the net are probably more likely to detect the net than are those only passing through in flight. This problem is particularly severe for those species with mobile spacing systems, nonterritorial social systems, or relatively large distances between foraging sites, which make them especially prone to mist-net capture (Remsen and Parker 1983; Remsen and Good 1996), e.g., *Pipra* spp., *Manacus* spp., *Mionectes* spp., *Phaethornis* spp., *Glyphorynchus spirurus*, dead-leaf-searching specialists such as certain *Myrmotherula* spp. and *Automolus* spp., and ant-following antbirds such as *Pithys albifrons*, *Hylophylax* spp., *Phlegopsis* spp., and *Rhegmatorhina* spp. (2) Probability of capture in a mist net varies with mist net placement, vegetation density, and incidence of light. For example, a species might actually prefer a certain habitat but would be unlikely to be captured there because the net is more easily detected where vegetation density is low or sunlight frequently exposes the net. (3) Few species of forest undergrowth birds are completely restricted to the first 2–3 m above ground that is the upper limit sampled by mist nets; whenever undergrowth species use microhabitat higher than 2–3 m, this use is "invisible" to mist-net sampling.

In addition to the problems of using mist net data, previous studies have not included detailed habitat measurements to determine habitat availability. Typically, habitat structure has been measured solely by foliage-height-diversity (FHD) profiles (MacArthur and MacArthur 1961), and a species' habitat use has been quantified only with respect to the vertical range within a FHD profile. Although a general relationship between FHD profiles and avian species diversity exists, it is not able to account for differences in species diversity among tropical areas with similar FHD profiles (Orians 1969; Pearson 1975) or similar bird diversities among tropical areas with very different FHD profiles (Terborgh and Weske 1969). Moreover, FHD profiles do not measure specific structural features that might be critical to bird species diversity (Lovejoy 1974). We are not aware of any studies that have compared habitat use to habitat availability at the "horizontal" level. Studies of habitat selection by tropical birds using actual observations rather than mist-netting, combined with detailed information on habitat availability, are nonexistent.

We compared habitat selection between temperate and tropical forest birds using standard methods at both sites and by studying a group of ecologically similar species, namely understory foliage-gleaning insectivores. Because precise counterparts are naturally difficult to identify, we focused on the most common species in this guild at our temperate and tropical sites. By choosing only the most common species, we are unable to determine whether our results are biased by lack of data on the uncommon species. Although between-biome comparisons involving phylogenetically unrelated organisms can be inappropriate because of uncontrollable variables, these problems are inherent and unavoidable. Nevertheless, we regard our approach as a logical first step.

METHODS

Our temperate zone study site was at the Tensas River National Wildlife Refuge (hereafter Tensas), in Madison, Franklin, and Tensas parishes (32°12'N, 91°25'W), northeastern Louisiana. The refuge contains approximately 40,500 ha of bottomland hardwood forest and as such is one of the largest tracts of relatively undisturbed deciduous forests remaining in the southeastern United States. Data were collected from 22 May to 3 June 1988.

Our study area, known locally as Fairfield Woods, is considered "high floodplain" forest. Little, or no standing water was present on the study site while data were collected. The forest is dominated by sweetgum (*Liquidambar styraciflora*) and willow oak (*Quercus phellos*); also present are overcup oak (*Quercus lyrata*), water hickory (*Carya aquatica*), cedar elm (*Ulmus crassifolia*), red maple (*Acer rubrum*), and bald cypress (*Taxodium distichum*). The primary understory vegetation consists of greenbriar (*Smilax*), swamp palmetto (*Sabal minor*), and common buttonbush (*Cephalanthus occidentalis*) (Barrow 1990).

The tropical forest study site was at the Tambopata Reserved Zone (hereafter Tambopata), southeastern Peru, approximately 30 km southwest of Puerto Maldonado on the south bank of the Río Tambopata, depto. Madre de Dios (12°50'S, 69°17'W). Data were collected from 25 June to 7 August 1988.

The biota of this 5,500-ha. reserve is among the richest in the world. So far, 550 species of birds have been recorded since 1977 (Foster et al. 1994), whereas only 244 species have been recorded on the temperate study site (R. B. Hamilton, unpubl. data). At Tambopata on a 1.0-ha plot, 153 species of trees represented by individuals 10 cm or more in diameter at breast height (DBH) were recorded (G. Hartshorn pers comm. in Parker 1982). In contrast, only 26 species of trees have been recorded in all of Tensas N.W.R. (Barrow 1990). For more details on the reserve and its habitats, see Parker (1982) and Erwin (1984).

At Tambopata, data were collected in three forest types (following Remsen and Parker 1983): (1) river-edge forest, (2) transitional forest, and (3) *terra firme* forest. Below we give a brief qualitative description of these forest types; for a more thorough and quantitative analysis see Marra (1989).

Because of annual or historical disturbance by river action, river-edge forest has a more open canopy, allowing for more light penetration and consequently a denser forest understory. Large, almost impenetrable thickets of *Heliconia* spp. and bamboo (*Guadua* sp.) were common. Two km of trail and adjacent forest formed our river-edge forest site. Although the river mostly parallels the trail, our study site did not include the riverbank itself.

The second forest type, "transitional forest," was studied along 3.6 km of trails and adjacent forest. Transitional forest is poorly drained, low-lying forest with occasional knolls or small hills. It is flooded seasonally by rainfall or by rain-swollen streams. Transitional forest differs most importantly from river-edge forest in its more continuous and taller canopy. Bamboo thickets and *Heliconia* spp. were present in this habitat, but to a much lesser degree than in river-edge forest.

TABLE 1
Study Species in Temperate and Tropical Forests

Scientific name (abbreviation)	English name
TEMPERATE SPECIES	
Cardinalis cardinalis (Cc)	Northern Cardinal
Thryothorus ludovicianus (Tl)	Carolina Wren
Vireo griseus (Vg)	White-eyed Vireo
Wilsonia citrina (Wc)	Wilson's Warbler
Oporornis formosus (Of)	Kentucky Warbler
TROPICAL SPECIES	
Myrmoborus myotherinus (Mm)	Black-faced Antbird
Myrmoborus leucophrys (Ml)	White-browed Antbird
Formicarius analis (Fa)	Black-faced Antthrush
Formicarius colma (Fc)	Rufous-capped Antthrush
Myrmeciza hemimelaena (Mh)	Chestnut-tailed Antbird
Hypocnemis cantator (Hc)	Warbling Antbird
Corythopis torquata (Ct)	Ringed Antpipit

The third forest type, *terra firme* forest, is on slightly higher ground and is not flooded. Two km of trail and adjacent forest formed our study area in this habitat type. Only an occasional sprig of bamboo or small patch of *Heliconia* was present.

The target birds at Tambopata (Table 1) were six species of understory antbirds (Formicariidae) and one understory flycatcher (Tyrannidae); all seven species are common, widespread species in Amazonia. They were selected because they were fairly common at the sites and relatively easy to find and follow. To minimize the influence of social factors on habitat selection, none of the seven target species was an army-ant-follower or member of mixed-species flocks; individuals forage predominantly alone, in pairs, or in small family groups. To minimize the direct influence of plant resources on habitat selection (Terborgh 1985), only insectivores were chosen (vs. frugivores or nectarivores). Species chosen for comparison from the temperate zone differ phylogenetically from the tropical species, but are ecologically similar in being common passerine insectivores of forest undergrowth (Table 1).

To quantify habitat availability, and to compare vegetation heterogeneity within tropical and between tropical and temperate forests, 36 variables were measured at the tropical study sites and 32 at the temperate study site (Table 2). A total of 200 random samples was taken at the temperate forest site and 204 at the tropical site (80 in river-edge forest, 144 in transitional forest, and 80 in *terra firme* forest). At each sample point, variables were quantified in two concentric, cylindrical plots, 2 m and 10 m in diameter, extending from the ground to 5 m above the forest floor. This technique was first developed by Moser et al. (1990) and was modified for this study. Only three variables were not shared between temperate and tropical sites (Table 2): bamboo, *Heliconia,* and palms were unique to the tropical sites, and palmetto was unique to the temperate site.

The target bird species were censused at Tambopata to determine habitat preferences at the macrohabitat level (river-edge, transitional, or *terra firme* forest). Censuses began at sunrise and lasted approximately 1 hr. Three censuses were conducted in river-edge, five in transitional, and five in *terra firme* forest. All singing individuals (as well as individuals seen at other times of the day) within 50 m to either side of the trail were spot-mapped and considered to represent one pair. Final relative abundance estimates (pairs per km of trail) were based on the maximum number of pairs seen in each forest type. To assess the relative abundances of the target species at the temperate forest site, the total number of pairs seen or heard was estimated daily. By the end of the study at Tensas, Marra was familiar with the territories of all target species within the study site and was able to estimate the number of individuals of each target species.

For data on habitat use and foraging, birds were located by either sight or sound, more typically by the latter in the tropics. Upon locating a bird, it was watched until a prey attack was seen. The exact point of the attack was used as the center point of the 2-m and 10-m-cylinders (modified after Moser et al.. 1988). A minimum of 60 s elapsed between marking of foraging points. A maximum of five foraging points was flagged for any one individual in a given day. The tropical species were extremely difficult to find and observe, so when an indi-

TABLE 2
Description of Variables and Measurement Techniques

Canopy Height (CH)	Rangefinder; measurement taken directly above foraging point or random sample point.
Percent Canopy Cover (CC)	Spherical densiometer (Lemmon 1956, 1957). Measurement taken directly above foraging point or random sample point.
Gap Association	Frequency of gap occurrence within 10-m cylinder. A gap is considered any open area in understory larger than 3 m in diameter or an area where large amounts of sunlight penetrate below the 2-m ceiling of the 10-m cylinder in the undergrowth (modified after Brokaw 1982). Classified as: (a) Artificial—manmade (i.e. trail) or (b) Natural—treefall; or bamboo.
% Ground Cover (GC)	Estimate (2- and 10-m) of all foliage 0–30 cm above ground.
% Shrub Cover (SC)	Estimate (2- and 10-m) of all shrub foliage 30 cm–3 m
Stems	Counts of stems in 2- and 10-m cylinders.
Fallen Logs (FL)	Counts of fallen logs in 2- and 10-m cylinders.
Mean Leaf-litter (LL)	Mean number of leaves on ground at five point-samples in both 2- and 10-m cylinders.
% Volume Vines (VV)	Estimate of vine volume from total composition in 2- and 10-m cylinders.
% Volume Palms (VP)	Estimate of palm tree volume from total composition in 2- and 10-m cylinders (tropics only).
% Volume Palmetto (VP)	Estimate of palmetto volume from total composition in 2- and 10-m cylinders (temperate only).
% Volume Bamboo	Estimate of bamboo volume from total composition in 2- and 10-m cylinders (tropics only).
% Volume Heliconia	Estimate of *Heliconia* volume from total composition in 2- and 10-m cylinders (tropics only).
% Suspended Dead Leaves	Estimate of suspended-dead-leaf volume from total composition in 2-m cylinder.
% Dead Stems	Estimate of dead-stem volume from total composition in 2-m cylinder.
% Live Leaves	Estimate of suspended live-leaf volume from total composition in 2-m cylinder.
% Live Stems	Estimate of live-stem volume from total composition in 2-m cylinder.
% Air Space	Estimate of open-space availability in 2-m cylinder.

vidual was found, it was necessary to maximize the number observations. If a bird seemed disturbed or altered its behavior, observations were not taken on that individual. To avoid biasing the data in favor of readily observed individuals, no more than 10 observations were taken on an individual bird or pair during the study period, although without color-banded individuals we could not be sure. In both localities, the size of the study sites was large enough to include many pairs.

Three foraging-site variables were recorded at each observation: foraging maneuver, foraging substrate, and foraging height following the terminology of Remsen and Robinson (1990). Each foraging substrate was classified as one of the following: live leaf, dead leaf, live stem, dead stem, vine, palm, palmetto, leaf litter, bare ground, bamboo stem, bamboo leaf, bark, spider web, or air. Heights of foraging observations were measured either with a meter tape or, for foraging points higher than 3 m, a range-finder.

Data analysis.—Rotated orthogonal and nonrotated factor analyses were performed to identify patterns and search for multicolinearity among variables for each habitat type (SAS 1982). To assess overall habitat heterogeneity, general variances were calculated for each forest type. The general variance is the determinant of the variance-covariance matrix computed from the measured variables in each habitat type. It provides a way of consolidating the information on all variances and covariances into a single number (Johnson and Wichern 1982). The general variance is calculated by summing the products of the covariances and subtracting the products of the variances (Johnson and Wichern 1982; SAS Institute 1985). The final value for the general variance is reported as the natural log of the calculated determinant from each matrix. To determine whether a bird species was preferring or avoiding a particular type of microhabitat (as measured by our microhabitat variables; Table 2), we compared the frequency distributions of a bird species' use of a particular variable (as measured by our observational data) with that variable's availability (as measured by our random microhabitat sampling). If these frequency distributions differed significantly (Kolmogorov-Smirnoff tests, $P \leq .05$), we classified species to be preferring (i.e., selecting) or avoiding that microhabitat variable.

To determine the degree of specialization in foraging, niche breadths were calculated for each species in each habitat type using the formula:

$$B = 1/\Sigma\, P_i^2$$

where P_i = proportion in category "i" (Levins 1968). Niche breadths were calculated for foraging height, foraging maneuver, and foraging substrate. Differences in niche breadths by habitat for each foraging variable were determined with a one-way analyses of variance (ANOVA) using JMP (SAS 1995). Niche overlaps were calculated using the formula:

$$O_{jk} = \Sigma\, P_{ia}P_{ja} / \sqrt{(\Sigma\, P_{ia}^2)(\Sigma\, P_{ja}^2)}$$

where P_{ia} & P_{ja} = proportional use of resource state "a" by species "i" and "j" (Pianka 1974). Overlap values were calculated for the same foraging variables analyzed for the niche breadth calculations. A one-way analyses of variance (ANOVA) was then used to test for differences in overlap values between species grouped by forest type for each foraging variable using JMP; a Welch analysis of variance was calculated when variances were unequal (SAS 1995).

To assess the overall specialization of temperate and tropical bird species within each forest type, general variances were also calculated for each bird species in each habitat type. The general variance for each species was divided by the general variance for each respective forest type. This new value estimates the degree to which a species is specialized. It compares the variation available to a species in a forest type (the general variance of the random samples) to a species' use (the general variance for a species) in that same forest type. A species with a general variance equal to its respective forest would be an extreme generalist (Degree of generalization, hereafter DoG = 1.0), whereas a species with a general variance much lower than that of its respective forest would be a specialist (DoG near 0.0). A one-way analyses of variance was then performed between each group of species from each forest type using the derived DoG value using JMP (SAS 1995). When a bird species' mean was zero for a given variable, that variable had to be eliminated in the calculation of the general variance for that species within that habitat because zero values yielded singular matrices, from which it is impossible to calculate a general variance.

RESULTS

Habitat analyses.—For detailed quantification of habitat types by variable, see Marra (1989). Principal component analyses were unable to explain any significant amount of variance in the data. The first five factors explained only 42% of the variance with rotation and 37% without rotation. These results suggest that little multicolinearity exists among variables; therefore, none of the variables were combined or removed for other analyses. Furthermore, no other interpretations from these analyses were made due to dangers in interpreting factor analyses in which so little of the variance can be explained (Gibson et al. 1984).

To evaluate the complexity (= heterogeneity) of the understory within each forest type, the general variance was calculated from the variance-covariance matrix constructed from the random measurements. Among tropical forests, river-edge forest was highest at 79, then transitional forest at 65, and *terra firme* forest at 60 (Table 3). When data from tropical forests were combined, the general variance was 80. The temperate forest general variance fell between transitional and *terra firme* forest types with a value of 63 (Table 4). Therefore, assuming that the general variance is indeed a measure of habitat complexity, the understory of the combined tropical forests appear to be more complex structurally than the temperate forest. When analyzed separately, river-edge forest is the most complex, and the understories of transitional, temperate and *terra firme* forests are similar. Statistical tests between general variances are problematic, so values presented above for each habitat type are meant to be used as a relative index of overall habitat heterogeneity.

Macrohabitat selection.—At the temperate site, *Cardinalis cardinalis* and *Thryothorus ludovicianus* were the most common at 5.0 pairs per km, and *Wilsonia citrina* and *Oporornis formosus* were less common at 3.4 and 3.3 respectively per km in bottomland hardwood forest (Table 5).

In tropical forest, six of the seven target species showed marked differences in abundance among forest types (Table 5). *Myrmoborus leucophrys* was more common in river-edge forest

TABLE 3

GENERAL VARIANCE (DETERMINANT) AND "DEGREE OF GENERALIZATION" (DoG) VALUES FOR TROPICAL SPECIES WITHIN EACH FOREST TYPE AND FORESTS COMBINED. DoG VALUES ARE CALCULATED BY DIVIDING THE DETERMINANT OF A GIVEN SPECIES INTO THAT FOREST TYPE. DoG VALUES NEAR 1.00 WOULD BE A GENERALIST, WHEREAS THOSE NEAR 0.0 A SPECIALIST. RANDOM VALUES ARE THE GENERAL VARIANCES FOR A GIVEN FOREST TYPE. * ONE OR MORE VALUES WERE ELIMINATED IN THE CALCULATION OF THE GENERAL VARIANCE (SEE TEXT)

	River-edge		Transitional		Terra Firme		Combined	
	Determinant	DoG	Determinant	DoG	Determinant	DoG	Determinant	DoG
Mb. myotherinus	—	—	—	—	46	0.77	50	0.63
Mb. leucophrys	56	0.71	63	0.97	—	—	74	0.93
F. analis	41	0.52	52	0.80	—	—	69	0.86
F. colma	—	—	*38	0.58	*48	0.80	57	0.79
Mc. hemimelaena	*49	0.62	62	0.95	28	0.47	73	0.91
H. cantator	49	0.62	45	0.70	—	—	66	0.83
C. torquata	16	0.20	*27	0.42	*31	0.52	44	0.55
MEAN (± SE)		0.53 ± 0.08		0.73 ± 0.07		0.64 ± 0.09		0.78 ± 0.07
RANDOM	79		65		60		80	

than in transitional forest, and absent from *terra firme* forest. The congener *M. myotherinus* showed the opposite habitat preferences of *M. leucophrys*; it was common in *terra firme* forest, rare in transitional, and absent in river-edge forest. *Formicarius analis* was equally abundant in both river-edge and transitional forest, but rare in *terra firme* forest, where *F. colma* was common. The latter was uncommon in transitional forest, and rare in river-edge forest. *Myrmeciza hemimelaena* was common in all three forest types, although slightly more common in river-edge forest. *Hypocnemis cantator* was more common in river-edge forest than in transitional, and absent from *terra firme* forest. *Corythopis torquata* was present in all three forest types but most abundant in transitional forest.

Microhabitat selection.—For those variables in which selectivity was demonstrated, we plotted selectivity (use minus availability) for each category of microhabitat variable. Only six of these graphs will be presented for each species in each forest type; the remainder will be described and summarized in the text. A summary of these comparisons shows that temperate species significantly selected 45% (14/31) of the habitat variables. In contrast, selection was more frequent in the tropical forest birds. River-edge forest species selected 69% (24/35) of the microhabitat variables, transitional forest species 63% (22/35), and *terra firme* forest species 40% (14/35).

TEMPERATE FOREST BIRDS

In general, temperate species selected microhabitat with high densities of live and dead stems, and avoided areas with large amounts of airspace. Each species also exhibited a preference for some unique structural component (i.e., vines, suspended dead leaves, leaf litter) of bottomland

TABLE 4

GENERAL VARIANCE (DETERMINANT) AND "DEGREE OF GENERALIZATION" (DoG) VALUES FOR TEMPERATE FOREST SPECIES. DoG VALUES NEAR 1.00 WOULD BE GENERALIST, WHEREAS THOSE NEAR 0.0, SPECIALIST. RANDOM VALUE IS THE GENERAL VARIANCE FOR TEMPERATE FOREST

	Determinant	DoG
Cardinalis cardinalis	59	0.94
Thryothorus ludovicianus	71	1.13
Vireo griseus	70	1.11
Wilsonia citrina	57	0.90
Oporornis formosus	56	0.89
MEAN (± SE)		0.99 ± 0.08
RANDOM	63	

TABLE 5
Relative Abundance of Tropical and Temperate Target Species within Each Forest Type (Pairs/km)

	Tropical species		
	River-edge	Transitional	Terra firme
Myrmoborus myotherinus	0.0	0.3	4.0
Mrymoborus leucophrys	5.0	2.3	0.0
Formicarius analis	3.0	3.3	1.0
Formicarius colma	0.5	2.3	3.0
Myrmeciza hemimelaena	5.0	4.0	3.5
Hypocnemis cantator	5.0	3.0	0.0
Corythopis torquata	1.0	3.7	1.5
TOTAL	19.5	18.9	13.0
	Temperate species		
Cardinalis cardinalis	5.0		
Thryothorus ludovicianus	5.0		
Vireo griseus	4.3		
Wilsonia citrina	3.4		
Oporornis formosus	3.3		
TOTAL	21.0		

hardwood forest. Therefore, at least when foraging, temperate species preferred areas with dense vegetation and did show signs of some specialization. Presumably, denser vegetation provides a higher density and diversity of prospective foraging substrates and greater concealment from predators.

Cardinalis cardinalis.—This species showed selectivity for 16% (5/31) of the microhabitat variables. It preferred 2-m cylinders with relatively low vine density (Fig. 1, $P \leq 0.025$) and avoided dense vine tangles. It also selected microhabitat containing palmetto in densities of 1—25% in 10-m cylinders, and it avoided areas where the volume of palmetto exceeded 25% in density ($P \leq 0.025$). A significant preference was also shown for 2-m cylinders with 0–40% air volume, and it avoided areas with more air space than this ($P \leq 0.001$).

Thryothorus ludovicianus.—This species selected 26% (8/31) of the microhabitat variables, more than any other target species in temperate forest (Fig. 2). It avoided microhabitat without any small stems (DBH 0–16 cm), preferring areas with at least 3–4 small stems in the 2-m cylinders ($P \leq 0.05$). It also avoided microhabitat that contained more than one stem in the DBH class of 32–50 cm (Fig. 2, $P \leq 0.025$) and strongly avoided those areas with stems >152 cm in the 10-m cylinders ($P \leq 0.01$). This species was the only temperate species to select microhabitat (2-m-cylinders) containing suspended dead leaves ($P \leq 0.001$) and avoided microhabitat without vines, preferring areas where the volume of vines was in the 11–50% range (Fig. 2, $P \leq 0.001$).

Vireo griseus.—This species selected 19% (6/31) of the microhabitat variables (Fig. 3). It avoided microhabitat with stems in the 32–50 cm DBH class ($P \leq 0.001$). It selected 2-($P \leq 0.001$) and 10-m ($P \leq 0.005$) cylinders where vine volume composed 11–75% of the cylinders. It also avoided microhabitat containing greater than 40% air space, preferring areas with denser vegetation ($P \leq 0.001$).

Wilsonia citrina.—This species selected 16% (5/31) of the habitat variables (Fig. 4). It preferred microhabitat with a taller canopy in the 15–22 m range and avoided microhabitat with a canopy lower than 15 m ($P \leq 0.001$). Within 2-m cylinders, areas with more than 40% air space were avoided and those with less air space were selected ($P \leq 0.001$). At the 10-m cylinder level of measurement, it avoided areas with small stems (DBH 0–16 cm) in densities of 30 or more ($P \leq 0.05$).

Oporornis formosus.—This species selected 19% (6/31) of the variables (Fig. 5). It preferred microhabitat with 3–6 leaves per point sample in the leaf litter and avoided areas with more than this both within 2- ($P \leq 0.01$) and 10-m cylinders ($P \leq 0.001$). It preferred 2-m cylinders with a volume of vines of 1–50% and avoided areas with no vines and where vine composition exceeded 50% ($P \leq 0.01$). Microhabitat (2-m cylinders) that exceeded 65% in open air space was avoided and areas less than this selected ($P \leq 0.001$).

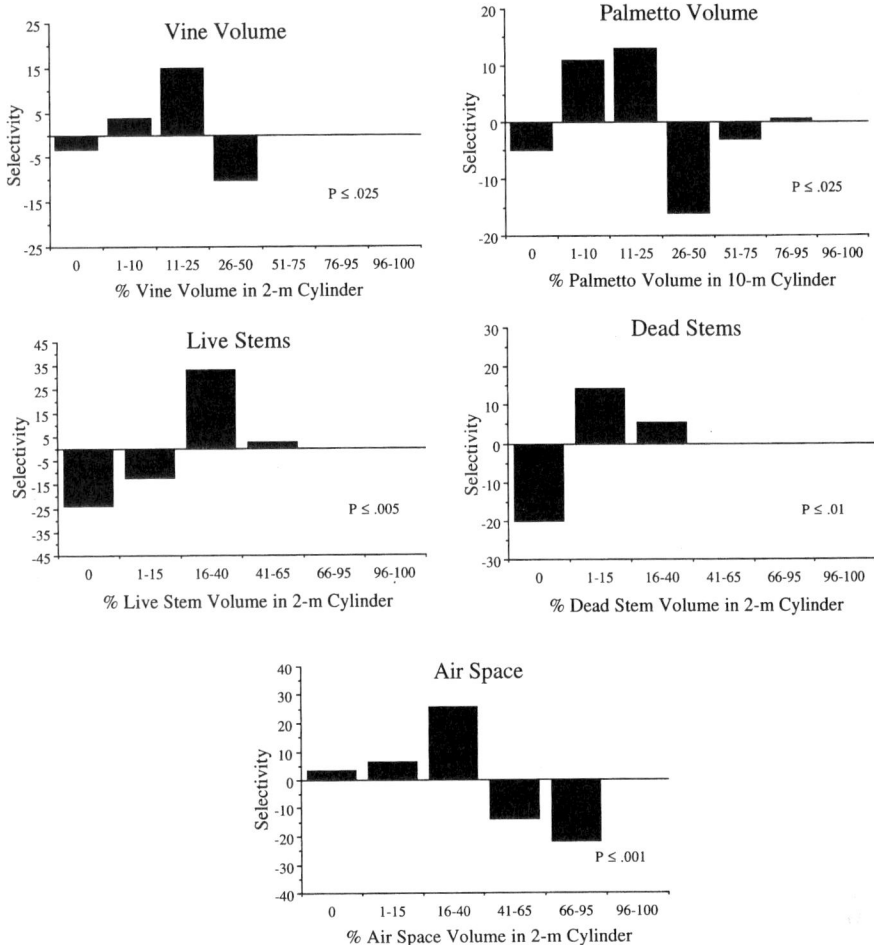

FIG. 1. Microhabitat selectivity in *Cardinalis cardinalis*. Positive values indicate preference (use > availability), whereas negative values indicate avoidance for the habitat variables depicted. Significance according to Kolmogorov-Smirnoff test.

RIVER-EDGE FOREST BIRDS

Species of river-edge forest differed strongly in the microhabitat each selected. *Hypocnemis cantator* and *M. leucophrys* were similar in their selection of bamboo, although *H. cantator* showed a much greater preference. *Myrmeciza hemimelaena* generally preferred areas more typical of transitional forest than river-edge forest, whereas *C. torquata* selected microhabitat within river-edge forest more similar to *terra firme*.

Myrmoborus leucophrys.—This species selected 37% (13/35) of the habitat variables (Fig. 6). It avoided areas with dense (> 26%) ground cover, preferring areas with 11–25% cover ($P \leq 0.025$). Microhabitats with 50–95% shrub cover were preferred and areas with less than 50% avoided in 10-m cylinders ($P \leq 0.001$). Bamboo was selected at the 10-m cylinder level of measurement at densities greater than 15% ($P \leq 0.005$). Accordingly, microhabitat associated with bamboo was shown to be selected; 2-m cylinders were selected with three to four small stems and 10-m cylinders with densities of 61–100 small stems (0–16 cm DBH; $P \leq 0.005$). All larger DBH classes (from 32 cm) in 10-m cylinders, with densities greater than one, were avoided ($P \leq 0.001$). Dense leaf-litter was preferred (2- and 10-m; $P \leq 0.005$), whereas fallen logs (in 10-m cylinder; $P \leq 0.005$) and palms were avoided 10-m; $P \leq 0.001$).

Formicarius analis.—This species selected 11% (4/34) of the habitat variables (Fig. 7). It preferred 10-m cylinders with more than 40 small stems (DBH 0-16 cm; $P \leq 0.001$) and avoided areas with lower densities of stems in this size class (DBH 0-16 cm; $P \leq 0.001$). It avoided

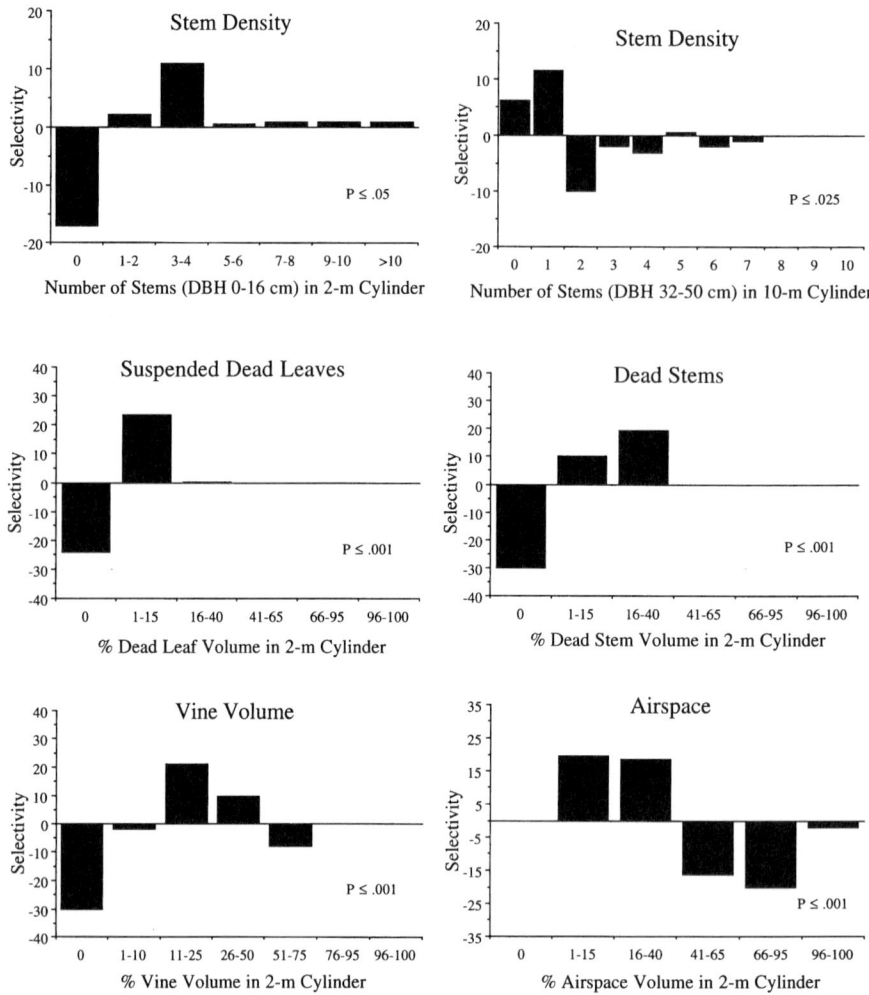

FIG. 2. Microhabitat selectivity in *Thryothorus ludovicianus*. Positive values indicate preference (use > availability), whereas negative values indicate avoidance for the habitat variables depicted. Significance according to Kolmogorov-Smirnoff test.

10-m cylinders with larger stems (DBH 50–76 cm $P \leq 0.025$; DBH 76–152 cm $P \leq 0.005$) and with fallen logs ($P \leq 0.001$).

Myrmeciza hemimelaena.—This species selected 20% (7/34) of the habitat variables (Fig. 8). It avoided areas with canopy cover greater than 80% preferring areas with canopy cover between 30 and 70% ($P \leq 0.005$). Ground cover between 25–95% was preferred in the 2-m cylinders (Fig. 8, $P \leq 0.01$) and 26–75% in the 10-m cylinders ($P \leq 0.001$). This species avoided microhabitat in 2- and 10-m cylinders with bamboo ($P \leq 0.001$) or *Heliconia* ($P \leq 0.001$).

Hypocnemis cantator.—This species selected 37% (13/34) of the habitat variables (Fig. 9). The selection of bamboo was significant in both 2- and 10-m cylinders ($P \leq 0.001$). Similar to the habitat associations of *M. leucophrys,* variables positively correlated with bamboo were preferred (shrub cover, $P \leq 0.001$, and leaf litter [2-m cylinder, $P \leq 0.001$; 10-m cylinder, $P \leq 0.01$]) and variables not associated with bamboo avoided (stems 16–50 cm DBH in 10-m cylinder, $P \leq 0.001$; stems 50–76 cm DBH in 10-m cylinder, $P \leq 0.005$; stems 76–152 cm DBH in 10-m cylinder, $P \leq 0.025$; fallen logs in 10-m cylinder, $P \leq 0.001$, and palms in 2- [$P \leq 0.025$] and 10-m cylinders [$P \leq 0.005$]).

Corythopis torquata.—This species selected 29% (10/34) of the habitat variables (Fig. 10). It preferred areas where the canopy exceeded 22 m in height and avoided areas with a shorter

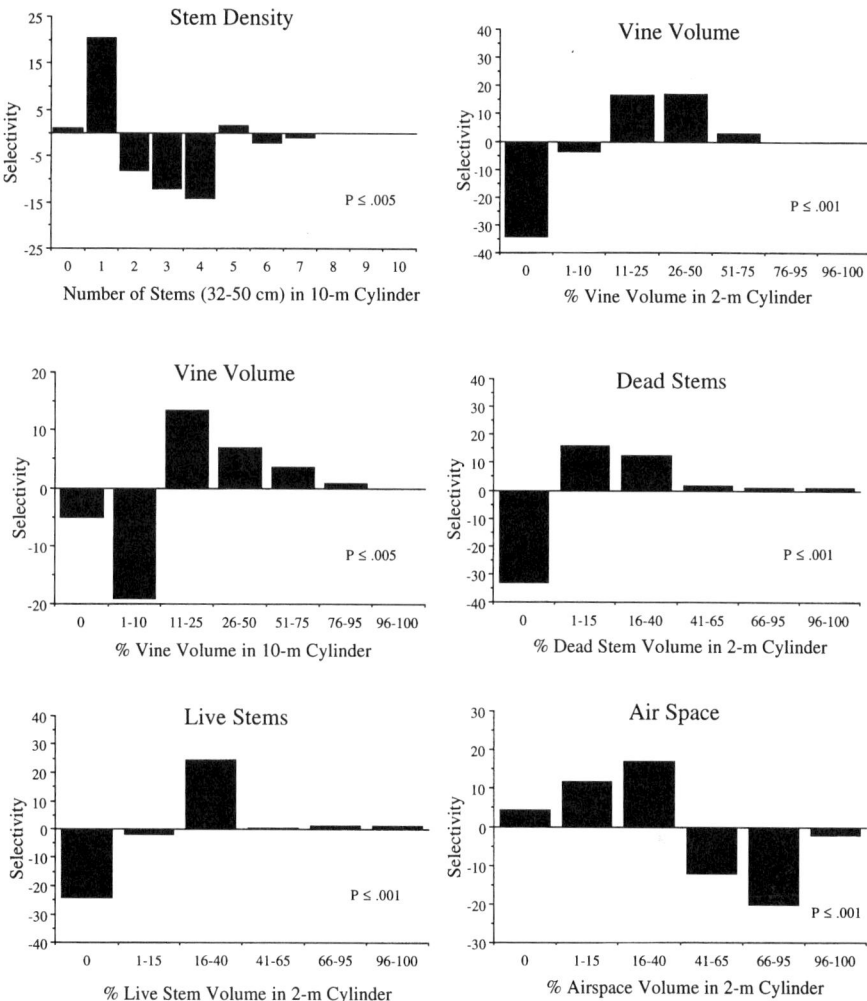

FIG. 3. Microhabitat selectivity in *Vireo griseus*. Positive values indicate preference (use > availability), whereas negative values indicate avoidance for the habitat variables depicted. Significance according to Kolmogorov-Smirnoff test.

canopy ($P \leq 0.001$). Accordingly, areas with large stems (DBH > 152 cm; $P \leq 0.05$) were preferred, and it avoided microhabitat not associated with mature forest, such as small stems (DBH 0–16 cm; $P \leq 0.005$), shrub density greater than 25% (2-m cylinders, $P \leq 0.05$), and vine volume in excess of 10% (10-m cylinders, $P \leq 0.01$). This species preferred areas where the leaf litter was less than two leaves thick ($P \leq 0.001$). Areas containing bamboo were strongly avoided in both the 2- and 10-m cylinders ($P \leq 0.001$).

TRANSITIONAL FOREST BIRDS

In transitional forest, *H. cantator* and *M. leucophrys* showed even stronger selection for bamboo, which is much less extensive there than in river-edge forest. *Myrmeciza hemimelaena* and *C. torquata* were more generalized in microhabitat preference in transitional forest than in river-edge forest. This is probably because transitional forest is much more homogeneous compared to the structurally more heterogeneous river-edge forest (Table 3).

Myrmoborus leucophrys.—This species selected 29% (10/35) of the habitat variables (Fig 11). It showed a stronger selection for bamboo in transitional forest than it did in river-edge forest, both within 2-m and 10-m-cylinders ($P \leq 0.001$). Variables positively associated with bamboo were again also preferred by *M. leucophrys* (stems, DBH 0–16 cm in 2-m [$P \leq 0.01$], and 10-m

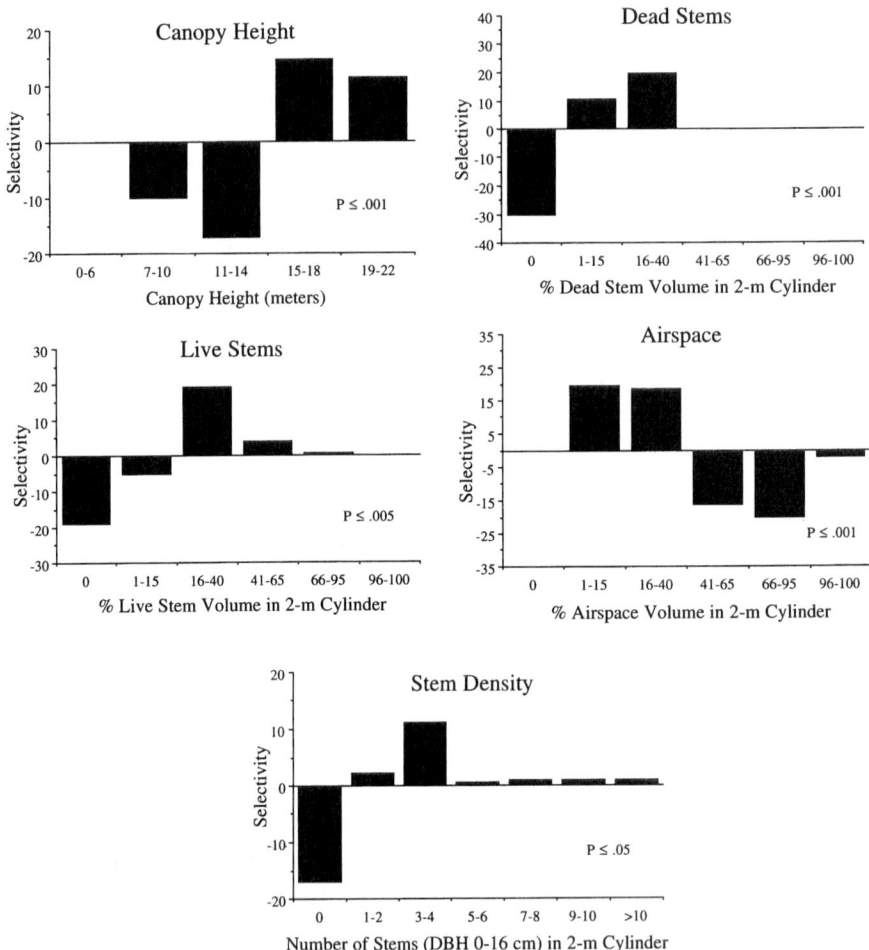

FIG. 4. Microhabitat selectivity in *Wilsonia citrina*. Positive values indicate preference (use > availability), whereas negative values indicate avoidance for the habitat variables depicted. Significance according to Kolmogorov-Smirnoff test.

cylinders [$P \leq 0.001$], dead stems [$P \leq 0.05$], and leaf litter [$P \leq 0.001$]), and it avoided variables negatively associated with bamboo (large stems, DBH > 50 cm, $P \leq 0.005$).

Formicarius analis.—This species selected 14% (5/35) of the habitat variables (Fig. 12). It preferred microhabitat with 70% canopy cover and avoided areas with less cover than this ($P \leq 0.001$). It preferred areas having a canopy of trees from 23 m to over 30 m ($P \leq 0.025$). Stems 32–50 cm in diameter were avoided when their densities were greater than 2 in 10-m cylinders ($P \leq 0.05$). Live stems in densities greater than 15% ($P \leq 0.01$) and shrub cover in densities greater than 25% ($P \leq 0.025$) were also avoided.

Formicarius colma.—This species selected 14 (5/35) of the habitat variables (Fig 13). It avoided areas containing trees with DBH's between 50–152 cm. This species avoided areas with bamboo ($P \leq 0.005$), whereas it preferred areas with *Heliconia* ($P \leq 0.05$). *Formicarius colma*, like *F. analis*, preferred areas with a canopy cover of 70–100% and avoided all areas with less than 70% ($P \leq 0.025$).

Myrmeciza hemimelaena.—This species showed little selectivity in this forest type. Only 9% (3/35) of the variables were selected (Fig 14). It preferred microhabitat with a canopy composed of trees of at least 23 m ($P \leq 0.005$). It also preferred 2-m ($P \leq 0.025$) and 10-m cylinders ($P \leq 0.01$) with high densities of stems in the 0–16 cm DBH class.

Hypocnemis cantator.—In contrast to *M. hemimelaena*, *H. cantator* was highly selective in transitional forest. It selected 43% (15/35) of the habitat variables (Fig. 15). There was again

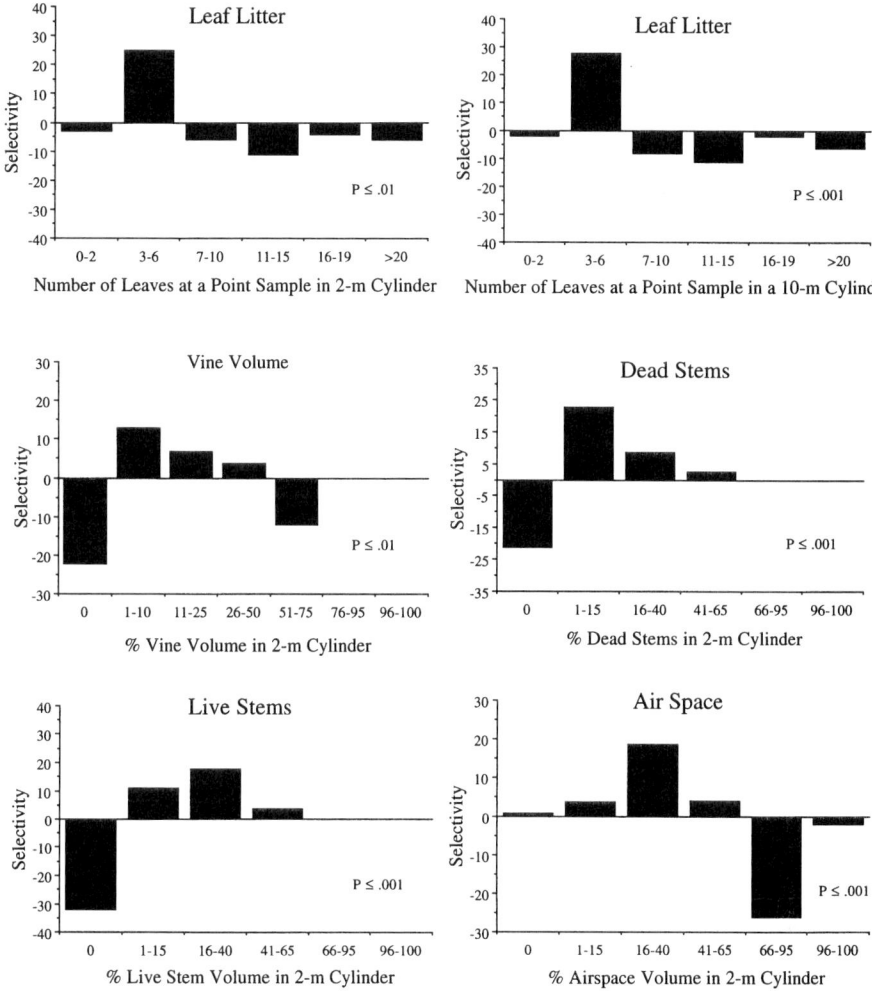

FIG. 5. Microhabitat selectivity in *Oporornis formosus*. Positive values indicate preference (use > availability), whereas negative values indicate avoidance for the habitat variables depicted. Significance according to Kolmogorov-Smirnoff test.

strong statistical significance in its selection of microhabitat dominated by bamboo in 2-m ($P \leq 0.005$) and 10-m cylinders ($P \leq 0.001$). Accordingly, it preferred structural attributes associated with bamboo, such as small stems (DBH 0–16 cm: in 2-m cylinders, $P \leq 0.001$, and 10-m-cylinders, $P \leq 0.01$) and vines in 2-m cylinders ($P \leq 0.001$) and 10-m cylinders ($P \leq 0.01$), and it avoided variables not associated with bamboo, such as stems with DBH's between 16 and 152 cm ($P \leq 0.01$), fallen logs ($P \leq 0.01$), palms ($P \leq 0.01$), and airspace ($P \leq 0.01$). This species demonstrated preference for short trees (<10 m in height), avoided trees between 11 and 22 m, and preference for 23–30 m trees ($P \leq 0.001$). This preference corresponds to this species affinity for bamboo and treefall gaps.

Corythopis torquata.—This species selected 17% (6/35) of the habitat variables (Fig 16). It preferred areas with shrub cover between 1 and 25% and avoided areas with greater than 25% ($P \leq 0.005$). It avoided areas with a leaf litter of more than 2 leaves deep ($P \leq 0.01$), with vine volume more than 10% ($P \leq 0.001$), palms ($P \leq 0.025$), and bamboo ($P \leq 0.001$) in 10-m cylinders.

TERRA FIRME FOREST BIRDS

Species in *terra firme* forest tended to avoid areas with a tall canopy and preferred areas with more canopy cover. Areas with a thick leaf-litter were also avoided by all species.

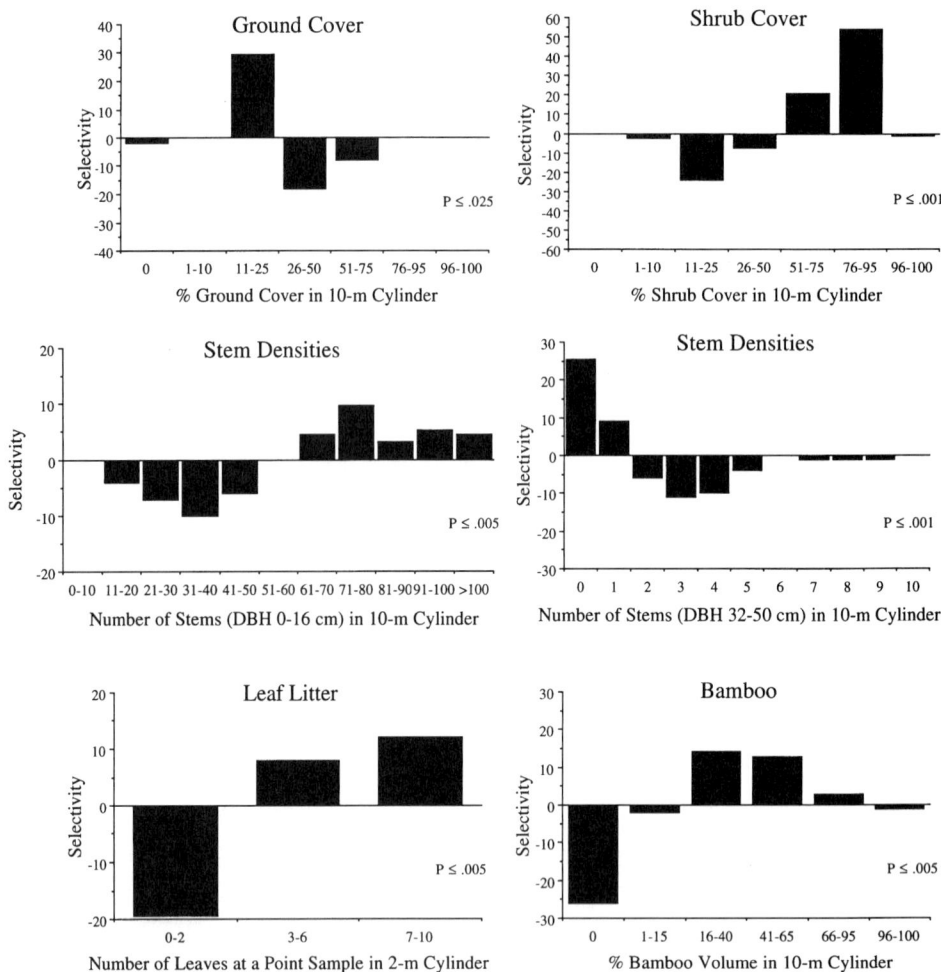

FIG. 6. Microhabitat selectivity in *Myrmoborus leucophrys* in river-edge forest. Positive values indicate preference (use > availability), whereas negative values indicate avoidance for the habitat variables depicted. Significance according to Kolmogorov-Smirnoff test.

Myrmoborus myotherinus.—This species selected 23% (8/35) of the habitat variables (Fig. 17). It avoided areas where the canopy exceeded 27 m in height, preferring areas with a canopy 7–18 m ($P \leq 0.01$). Areas with canopy cover between 70 and 90% were also preferred ($P \leq 0.001$). This species avoided areas of shrub cover that exceeded 50%, preferring shrub cover in the 11–50% range ($P \leq 0.001$), and it preferred areas where the leaf litter was no more than 2 leaves thick ($P \leq 0.005$). This species also demonstrated preference for 2-m cylinders containing no suspended dead leaves ($P \leq 0.05$) and with live leaves in low densities (1–40%, $P \leq 0.001$), but it avoided areas with a densities of live leaves greater than 40%.

Formicarius colma.—This species selected 29% (8/35) of the habitat variables (Fig. 18). It preferred areas where the canopy was 7–22 m in height and avoided areas where the canopy height exceeded 22 m ($P \leq 0.001$). Shrub cover in 2-m cylinders greater than 25% was avoided ($P \leq 0.005$), and shrub cover in 10-m-cylinders greater than 50% was avoided ($P \leq 0.025$). It avoided areas containing stems ranging in DBH from 32 to 76 cm ($P \leq 0.005$). It preferred areas with leaf litter only 0–2 leaves thick in both 2-m and 10-m cylinders ($P \leq 0.001$). It also preferred sites containing less then 40% live leaf volume at the 2-m cylinder scale of measurement ($P \leq 0.001$).

Myrmeciza hemimelaena.—This species selected 20% (7/35) of the habitat variables (Fig. 19). In contrast to its selectivity in river-edge forest, it preferred areas where canopy cover was 80–

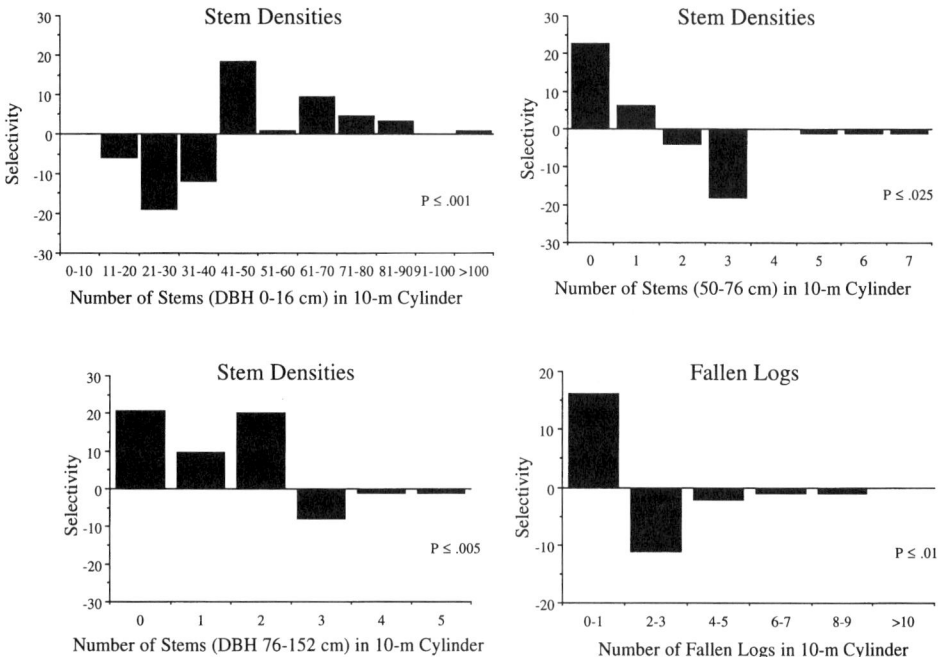

FIG. 7. Microhabitat selectivity in *Formicarius analis* in river-edge forest. Positive values indicate preference (use > availability), whereas negative values indicate avoidance for the habitat variables depicted. Significance according to Kolmogorov-Smirnoff test.

90% ($P \leq 0.001$). This species preferred microhabitat having a density of 0–16 cm stems, ranging from 50 to 100 stems in 10-m cylinders ($P \leq 0.001$). Large stems with DBH's 32–76 cm were avoided in 10-m cylinders ($P \leq 0.001$). Areas with leaf litter 0–2 leaves in depth were preferred, and those areas where leaf litter exceeded this were avoided ($P \leq 0.001$). Palms were also avoided at the 10-m cylinder scale of measurement ($P \leq 0.001$).

Corythopis torquata.—This species selected 29% (10/35) of the habitat variables (Fig 20). In contrast to its preference in river-edge forest, *C. torquata* preferred forest with a shorter canopy, ranging between 15 and 22 m and with a canopy cover between 70 and 90% ($P \leq 0.001$). This species consistently foraged in areas with 11–25% ground cover, and avoided places with denser ground cover ($P \leq 0.005$). It avoided areas with shrubs exceeding 25% in 2-m cylinders and avoided areas exceeding 50% shrub cover in 10-m cylinders ($P \leq 0.001$). Like the other species in this habitat, it preferred thin leaf-litter (0–2 leaves) and avoided areas with a deep leaf litter ($P \leq 0.01$). Other sites avoided were those that included dense vines (>10%, $P \leq 0.025$), suspended dead leaves ($P \leq 0.001$), and dense live leaves (>40%; $P \leq 0.001$).

FORAGING BEHAVIOR

TEMPERATE BIRD SPECIES

Foraging maneuver and substrate.—The primary foraging maneuver used by *C. cardinalis* was gleaning (Fig. 21). Sally-strikes were the next most frequent maneuver, followed by flaking. *Cardinalis cardinalis* searched primarily live leaves and secondarily palmetto (Fig. 22). This corresponds to the results showing its preference for microhabitat containing palmetto (Fig. 1). All ten substrates used as foraging substrates by other temperate species were used by *C. cardinalis*.

Thryothorus ludovicianus gleaned substrates for prey almost as much as it probed them (Fig. 21). It flaked leaves in the leaf litter as well, but at a lower frequency. *Thryothorus ludovicianus* foraged on eight of ten substrates used by all species (Fig. 22) with almost equal frequency in each substrate. Dead leaves were the most frequently searched substrate, but only slightly more frequently than live leaves. Remsen and Parker (1984) and Rosenberg (1990) reported dead leaves as the predominant foraging substrate for tropical members of the genus *Thryothorus*, but

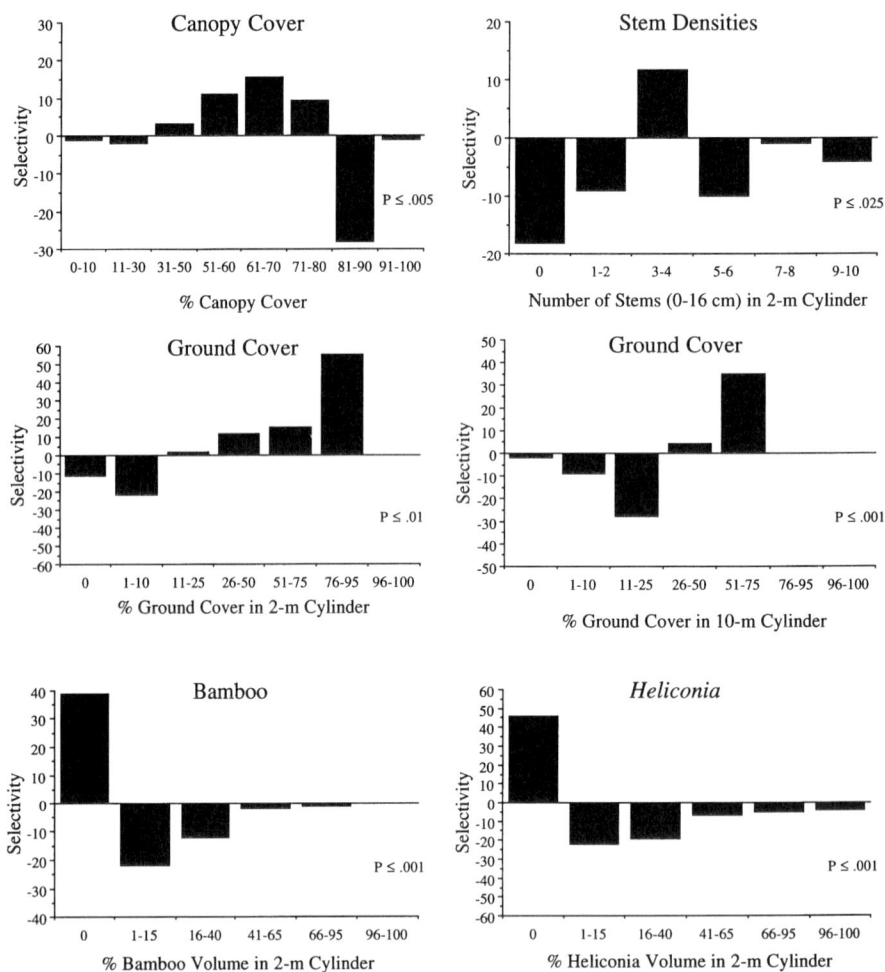

FIG. 8. Microhabitat selectivity in *Myrmeciza hemimelaena* in river-edge forest. Positive values indicate preference (use > availability), whereas negative values indicate avoidance for the habitat variables depicted. Significance according to Kolmogorov-Smirnoff test.

these are the first data to document this for *T. ludovicianus*. This is also in accordance with our results from microhabitat use that found *T. ludovicianus* to prefer 2-m cylinders containing suspended dead leaves (Fig. 2).

Vireo griseus used only a narrow range of foraging maneuvers (Fig. 21). It used predominantly sally-strikes followed by gleans. Thus its foraging maneuvers are similar to those of three other vireos of eastern North America, *V. solitarius*, *V. olivaceus*, and *V. philadelphicus*, studied by Robinson and Holmes (1982). *Vireo griseus* foraged primarily on live leaves and secondarily on vines (Fig. 22). Microhabitat (2-m and 10-m cylinders) with high vine volume was preferred by *V. griseus* (Fig. 3).

Wilsonia citrina used predominantly gleans and sally-strikes (Fig. 21). It occasionally used sally-hovers and was the only species that used flutter-chases. Foraging by *W. citrina* was directed primarily towards live leaves and much less frequently at vines, dead stems, air, live stems, bark, and dead leaves (Fig. 22).

Oporornis formosus, like other temperate species, used predominantly gleans (Fig. 21). Sally-striking, flaking, and leaping were also observed at lower frequencies. *Oporornis formosus* was somewhat generalized in substrate use (Fig. 22). All ten substrates used by the other temperate species were also used by *O. formosus*. Although most foraging maneuvers were directed toward

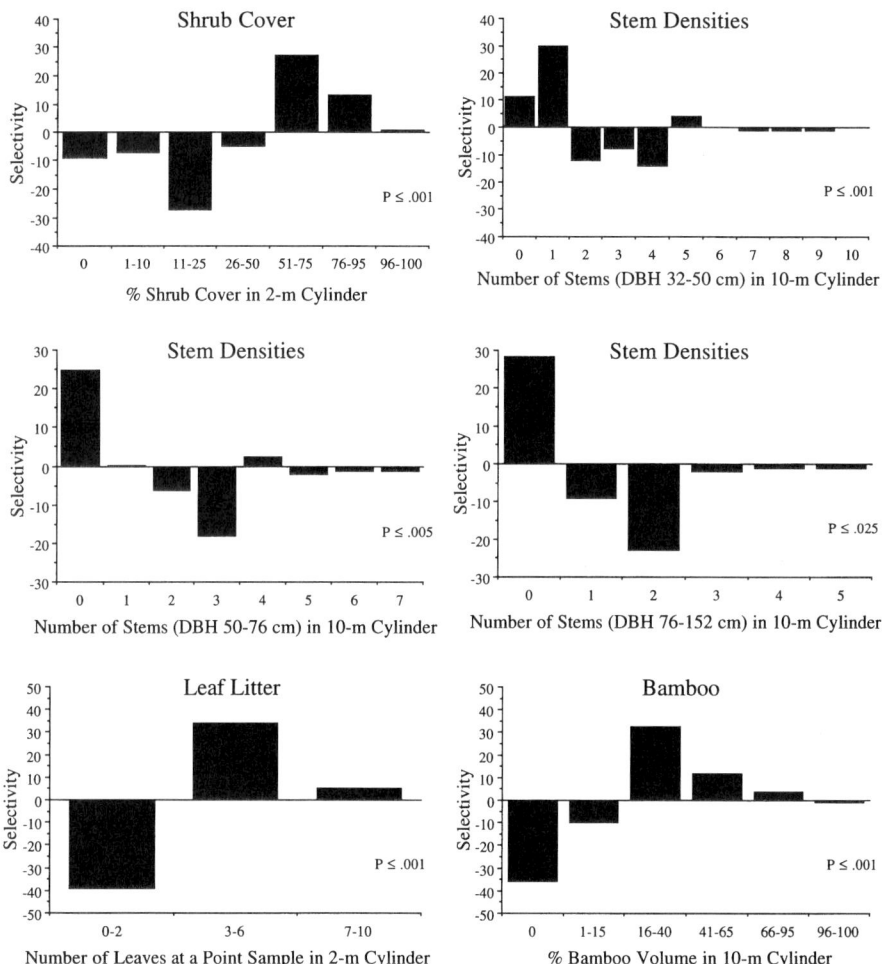

FIG. 9. Microhabitat selectivity in *Hypocnemis cantator* in river-edge forest. Positive values indicate preference (use > availability), whereas negative values indicate avoidance for the habitat variables depicted. Significance according to Kolmogorov-Smirnoff test.

live leaves, it used vines, palmettos, live stems, and dead stems each in more than 10% of the observations.

Foraging height.—*Cardinalis cardinalis* foraged from the ground to the canopy (mean foraging height 3.3 m ± 3.2; Fig. 23). *Thryothorus ludovicianus* foraged from the ground to the midstory (mean foraging height 1.7 m ± 1.8; Fig. 23). *Vireo griseus* was also highly variable in its foraging height (mean foraging height 4.6 m ± 2.9; Fig. 23). It most often foraged by working its way up vine tangles into the canopy. The mean foraging height for *W. citrina* was the same as that for *V. griseus* (4.6 m ± 2.9; Fig. 23); *Wilsonia citrina* foraged from the ground to the subcanopy. *Oporornis formosus* foraged lower and was less variable than other temperate forest species (mean foraging height of 1.5 m ± 1.5; Fig. 23); much of its foraging was on the ground or in understory saplings.

TROPICAL FOREST SPECIES

Foraging maneuver.—Gleaning was the most frequently used maneuver by *H. cantator* (Fig. 24). It used all maneuver types with essentially the same frequency in river-edge and transitional forests except for hanging, which it used only in river-edge forest. Gleaning was the maneuver

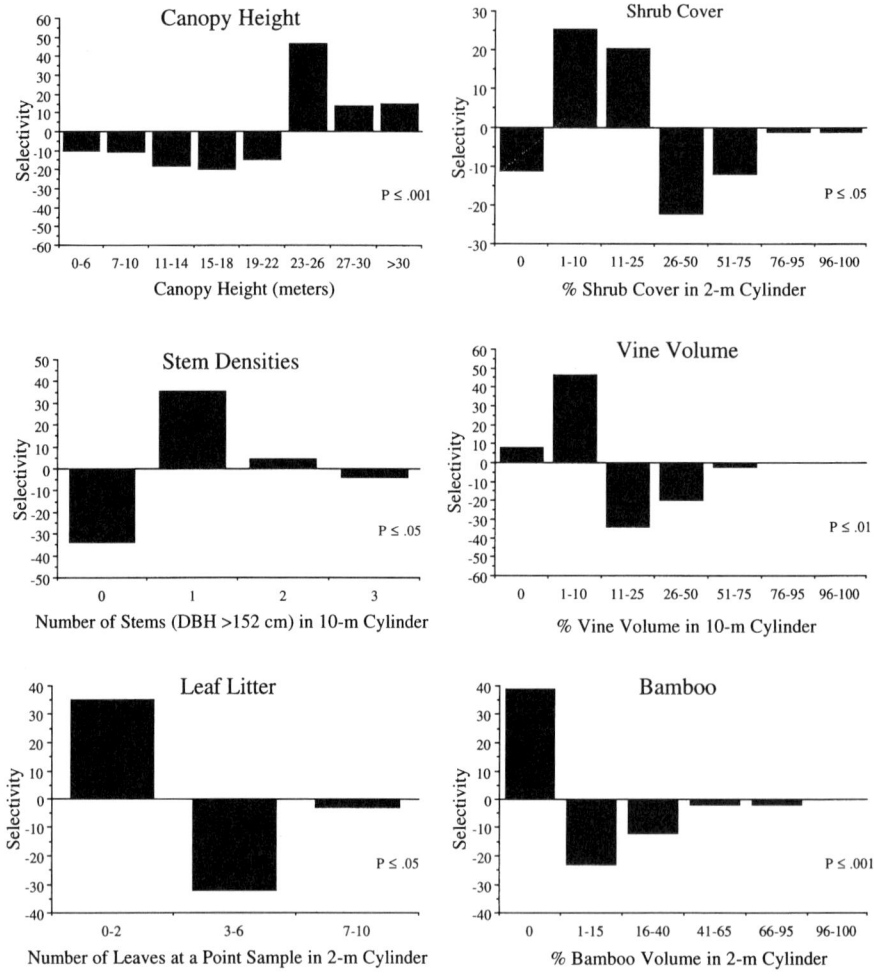

FIG. 10. Microhabitat selectivity in *Corythopis torquata* in river-edge forest. Positive values indicate preference (use > availability), whereas negative values indicate avoidance for the habitat variables depicted. Significance according to Kolmogorov-Smirnoff test.

used most frequently by *M. hemimelaena* (Fig. 24), with leaping (from the forest floor up to foliage) used slightly less frequently. In general, it used all maneuvers with the same frequency in each forest type. The primary foraging maneuver of *M. leucophrys* also was gleaning, with nearly the same frequency in both river-edge and transitional forests (Fig. 24). It used sally-strikes substantially more in river-edge than in transitional forest. The primary foraging maneuvers of *M. myotherinus* were gleaning and flaking (Fig. 24). It also used leaping, sally-striking, and flutter-chasing. Gleaning and flaking were the primary maneuvers used by *F. analis* in river-edge forest (Fig. 24). In transitional forest, however, it used gleaning about twice as frequently as flaking. Gleaning was the primary foraging maneuver used by *F. colma,* and flaking secondarily, in both transitional and *terra firme* forests (Fig. 24). Most foraging maneuvers used by *C. torquata* were either gleans, leaps, or flutter-chases (Fig. 24), as found by Fitzpatrick (1980). Most were used with the same frequency in all three forest types. Leaps were typically launched from the forest floor up to ground cover and even to low shrubs. Flutter-chasing was used less frequently in *terra firme* forest, where gleaning was used more often.

Foraging substrate.—Live leaves and bamboo leaves were the most frequently used substrates by *H. cantator* in both forest types; suspended dead leaves were used much more frequently in river-edge forest (Fig. 25). *Myrmeciza hemimelaena* used seven types of substrates in both river-

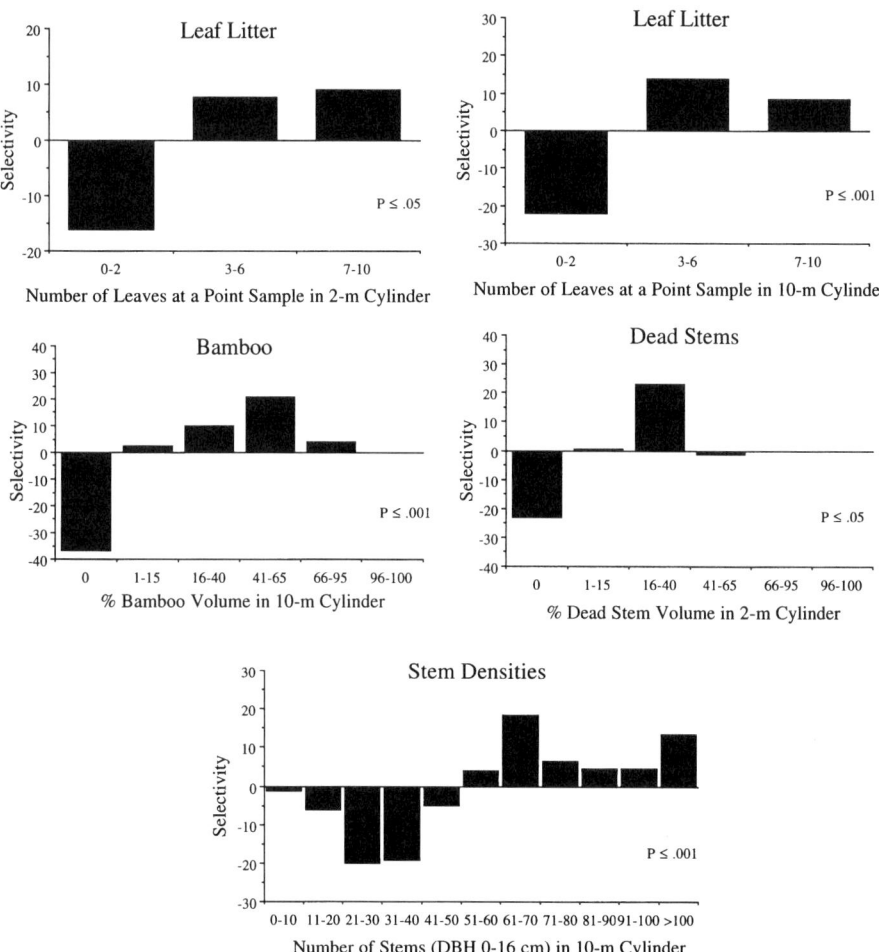

FIG. 11. Microhabitat selectivity in *Myrmoborus leucophrys* in transitional forest. Positive values indicate preference (use > availability), whereas negative values indicate avoidance for the habitat variables depicted. Significance according to Kolmogorov-Smirnoff test.

edge and transitional forest, but only four in *terra firme* forest (Fig. 25). *Myrmeciza hemimelaena* used most of the foraging substrates, each with less than 15% of the observations, and with roughly the same frequency in all forest types. *Myrmoborus leucophrys* used ten substrate types in river-edge forest and seven in transitional forest (Fig. 25). Live leaves and leaf litter were the predominant foraging substrates in river-edge forest, and live leaves, dead stems, leaf litter, and bamboo stems were used most frequently in transitional forest. *Myrmoborus myotherinus* foraged primarily in the leaf litter and on live leaves (Fig. 25). Nearly 100% of the foraging observations of *F. analis* in both forest types were in leaf litter (Fig. 25). *Formicarius colma* used predominantly leaf litter in both forest types (Fig. 25). *Corythopis torquata* most frequently used live leaves, leaf litter, and air in all three forest types (Fig. 25). Air was used substantially less frequently in *terra firme* forest than in any other forest type.

Foraging height.—*Myrmoborus myotherinus* foraged primarily on the ground, with a mean foraging height of 0.3 m (Fig. 23). The mean foraging height of *M. leucophrys* was 0.8 m ± 0.8 in river-edge forest but almost twice as high in transitional forest, with a mean of 1.4 m ± 1.5 (Fig. 23). *Formicarius analis* and *F. colma* were exclusively ground-foragers (mean height 0.0; Fig. 23). *Myrmeciza hemimelaena* had nearly the same mean foraging heights in river-edge (0.6 m) and transitional forests (0.5 m). In *terra firme* forest, however, it foraged much higher

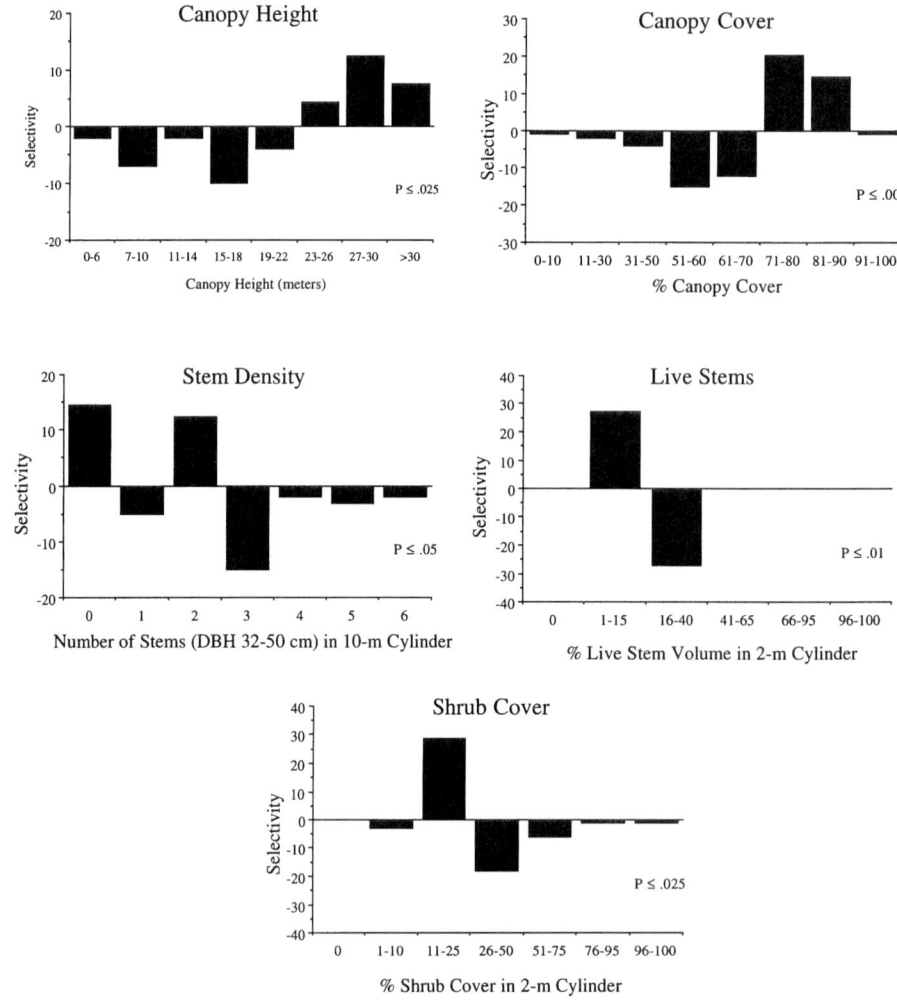

FIG. 12. Microhabitat selectivity in *Formicarius analis* in transitional forest. Positive values indicate preference (use > availability), whereas negative values indicate avoidance for the habitat variables depicted. Significance according to Kolmogorov-Smirnoff test.

(mean 1.2 m ± 1.2; Fig. 23). *Hypocnemis cantator* had mean foraging heights of 3.5 m ± 1.4 in river-edge forest and 4.1 m ± 1.6 in transitional forest (Fig. 23). There was high variation associated with each of these. This species was most commonly found in bamboo patches (Figs. 9, 15), where it had an especially broad vertical foraging range. *Corythopis torquata* had mean foraging heights of 0.02 m ± 0.05 in river-edge and 0.02 m ± 0.04 in transitional forests, and 0.01 m in *terra firme* forest (Fig. 23). Although this species is almost exclusively a ground-dwelling flycatcher, it occasionally forages from low perches to above-ground foliage.

TEMPERATE-TROPICAL COMPARISONS

NICHE BREADTHS

Comparisons of mean niche breadths for species of temperate and tropical forests for each foraging category (Fig. 26) reveal both similarities and differences. For all species-specific niche breadths, see Marra (1989). No differences were found between habitats in foraging maneuver breadth (ANOVA $F_{3,16} = 0.95$, $P = 0.44$). Therefore, these temperate and tropical species use a similar range of maneuvers in each of their respective forest types. No differences were found between habitats in breadth of substrates either (ANOVA $F_{3,16} = 1.3$, $P = 0.31$). However,

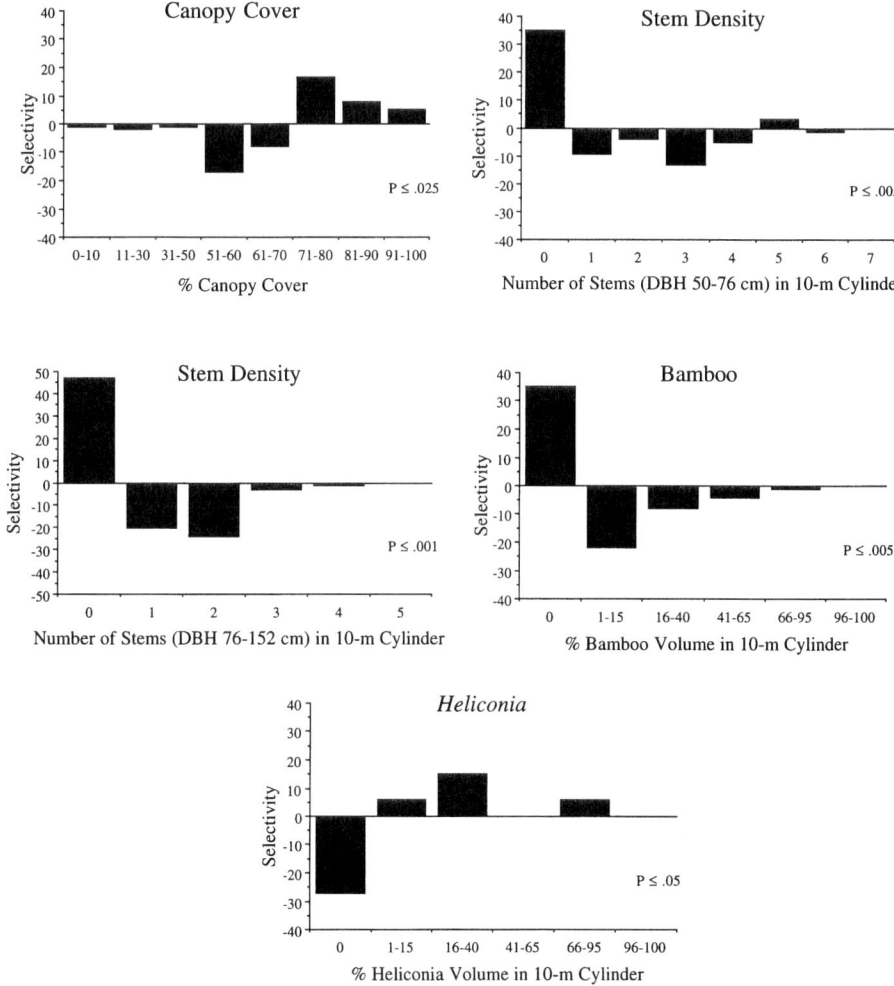

FIG. 13. Microhabitat selectivity in *Formicarius colma* in transitional forest. Positive values indicate preference (use > availability), whereas negative values indicate avoidance for the habitat variables depicted. Significance according to Kolmogorov-Smirnoff test.

temperate forest species did tend to use a broader range of substrates: their mean (3.8 ± 0.8) was higher than those of any of the tropical forest species (river-edge 2.8 ± 0.5, transitional 2.8 ± 0.5, *terra firme* 2.1 ± 0.3), suggesting greater substrate specialization by the tropical species (Fig 26).

With regards to foraging height, temperate forest birds had a significantly broader mean breadth (10.7 ± 1.1; ANOVA $F_{3,16}$ = 9.5, P = 0.0008) than birds of any of the tropical forest types (Fig 24). *Terra firme* forest species had the lowest mean breadth in foraging height (2.7 ± 1.1), followed by river-edge (3.4 ± 1.3), and then transitional forest (3.6 ± 1.2) species. Certain tropical species had comparatively high foraging height breadths. Two bamboo specialists, *Myrmoborus leucophrys* and *Hypocnemis cantator,* species specializing horizontally (in this case on bamboo) are not comparably specialized vertically. This can also be seen with *Myrmeciza hemimelaena*, a species strongly associated with treefall gaps (another form of horizontal specialization; Marra 1989). Its foraging-height breadths were slightly higher than species not specializing on a horizontal component of the forest. This is especially clear in *terra firme* forest, where its foraging height breadth was broader, and its association with treefall gaps stronger, than in any other forest type (Marra 1989).

To summarize results on niche breadths, temperate understory species used a broader range of the vertical dimension of the habitat, relative to tropical understory species. Only tropical

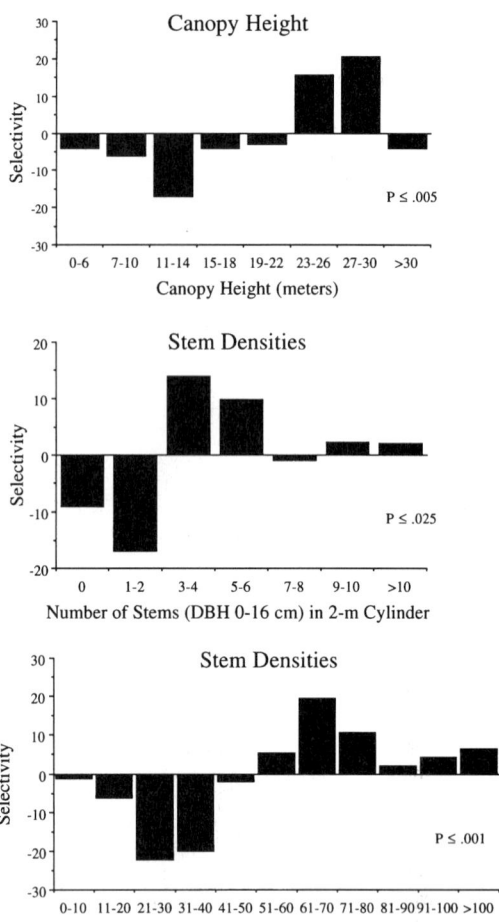

FIG. 14. Microhabitat selectivity in *Myrmeciza hemimelaena* in transitional forest. Positive values indicate preference (use > availability), whereas negative values indicate avoidance for the habitat variables depicted. Significance according to Kolmogorov-Smirnoff test.

species specializing "horizontally" have foraging height breadths as broad as those of temperate species.

NICHE OVERLAPS

Comparing overlap values from different forest types is fraught with interpretational problems. Nevertheless, we provide some cautious interpretations. The degree to which species overlap within each forest type with respect to foraging maneuver was marginally significant (ANOVA $F_{3,37} = 2.6$, $P = 0.06$; Fig 27). River-edge species showed less overlap in all types of foraging maneuvers used (0.65 ± 0.05) compared to species of temperate (0.79 ± 0.05), transitional (0.82 ± 0.03), and *terra firme* (0.85 ± 0.05) forests.

Temperate forest species (0.83 ± 0.05) showed significantly more overlap than tropical species (river-edge 0.61 ± 0.09, transitional 0.55 ± 0.09, *terra firme* 0.71 ± 0.10) in their use of foraging substrates (Welch ANOVA $F_{3,37} = 3.4$, $P = 0.04$; Fig 27). This result combined with those on niche breadth (Fig. 26) suggest greater substrate specialization by tropical forest species.

With respect to overlap in foraging height, species of temperate (0.66 ± 0.05) and *terra firme* (0.75 ± 0.09) forest overlapped slightly more in foraging height than did species of river-edge (0.43 ± 0.12) and transitional (0.47 ± 0.10) forest (Welch ANOVA $F_{3,37} = 2.4$, $P = 0.10$; Fig 27).

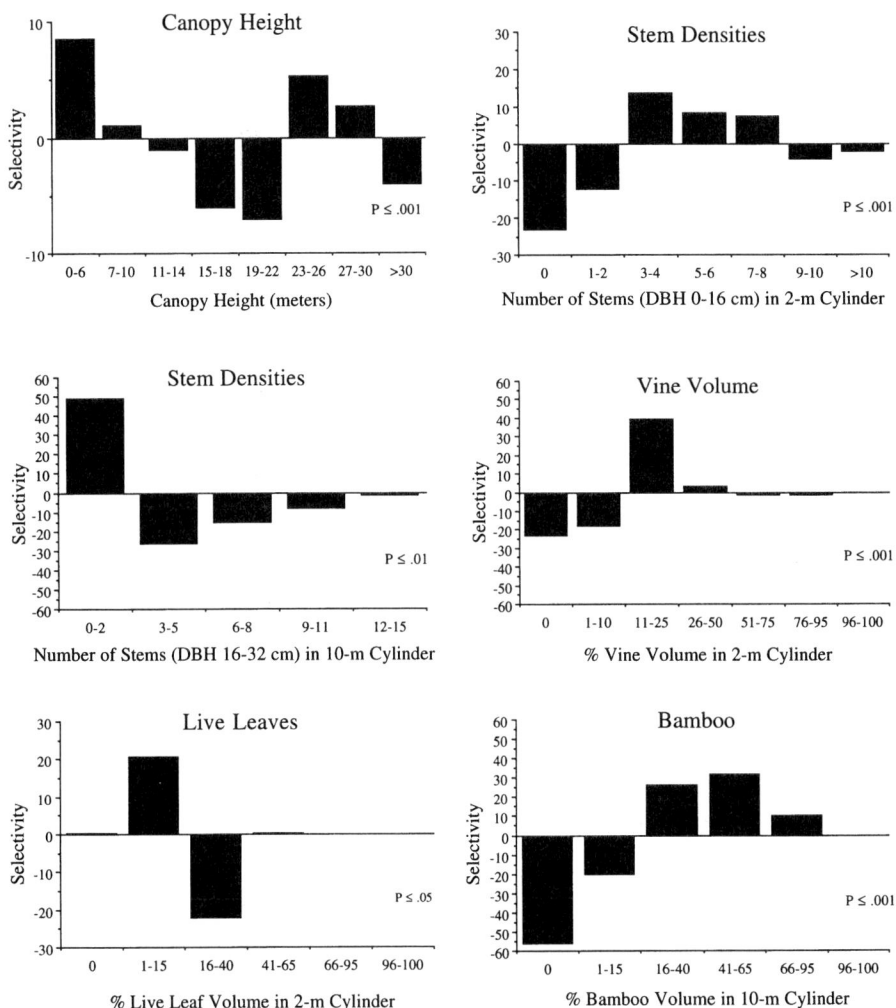

FIG. 15. Microhabitat selectivity in *Hypocnemis cantator* in transitional forest. Positive values indicate preference (use > availability), whereas negative values indicate avoidance for the habitat variables depicted. Significance according to Kolmogorov-Smirnoff test.

MULTIVARIATE ANALYSIS

All temperate species had general variances exceeding 0.89, and two species had values greater than 1 (Table 4). The latter result was due to these species having higher variances for some variables than recorded in the random samples. In river-edge and transitional forests, *C. torquata* and *M. leucophrys* had the lowest and highest DoG values respectively (ranging from 0.20 to 0.97; Table 3). Species in *terra firme* forest had DoG values ranging from 0.47 in *M. hemimelaena* to 0.80 in *F. colma* (Table 3). All DoG values for tropical forest species were significantly lower than those of temperate forest birds (ANOVA $F_{3,16} = 5.8$, $P = 0.007$). Therefore, these understory bird species of tropical forests are more specialized in their habitat selection than the temperate forest species.

A final illustration of the overall degree of specialization by both the temperate and tropical species can be demonstrated by plotting the DoG, an estimate of horizontal use of space, against both (1) the breadth of the foraging site (Σ of foraging maneuver breadth and foraging substrate breadth for each species; Fig. 28) and (2) breadth of foraging height, which is an estimate of vertical use of space (Fig. 28). The first figure illustrates how temperate and tropical species overlap heavily in those niche dimensions related to the breadth of their foraging sites (i.e.,

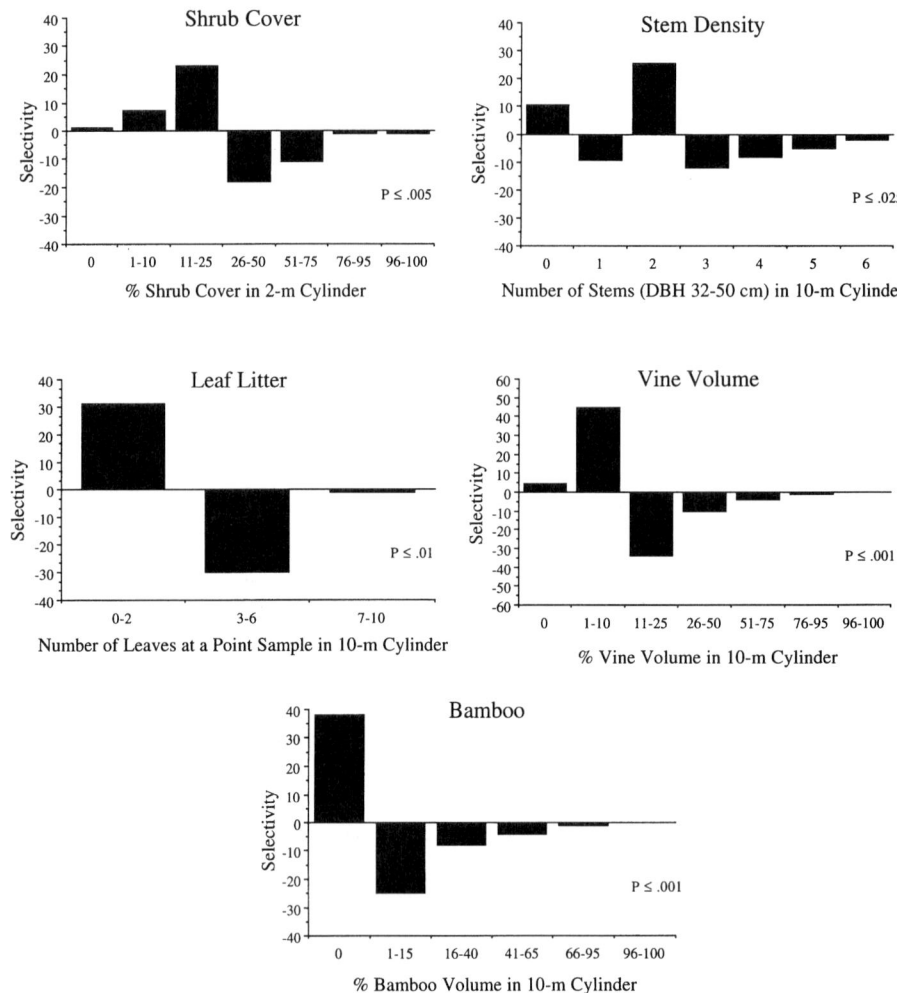

Fig. 16. Microhabitat selectivity in *Corythopis torquata* in transitional forest. Positive values indicate preference (use > availability), whereas negative values indicate avoidance for the habitat variables depicted. Significance according to Kolmogorov-Smirnoff test.

maneuvers and substrates), whereas the two groups still differ strongly in degree of horizontal generalization, with temperate species significantly more generalized than tropical. This second comparison (Fig. 28) demonstrates the high degree of specialization by understory tropical species relative to temperate species in both a vertical (i.e., breadth of foraging height) and horizontal (i.e., degree of generalization) use of space.

DISCUSSION

Our goal was to investigate the proximate, ecological factors rather than the ultimate, evolutionary causes of high tropical species diversity. To do this, comparisons to less diverse areas such as temperate zone forests are necessary. Tropical-temperate comparisons have been conducted between birds of alder forests (Stiles 1978), woodpeckers (Askins 1983), wading birds (Kushlan et al. 1985), and understory birds in general (Karr 1971). That there are a myriad of complications inherent in any intercommunity comparison such as these is well known (Terborgh and Robinson 1986). However, we see no alternative to these comparisons for increasing our understanding of tropical diversity.

Are tropical habitats structurally more complex?—Few quantitative analyses have attempted to measure the complexity of tropical habitats. Until now, either foliage-height-diversity profiles

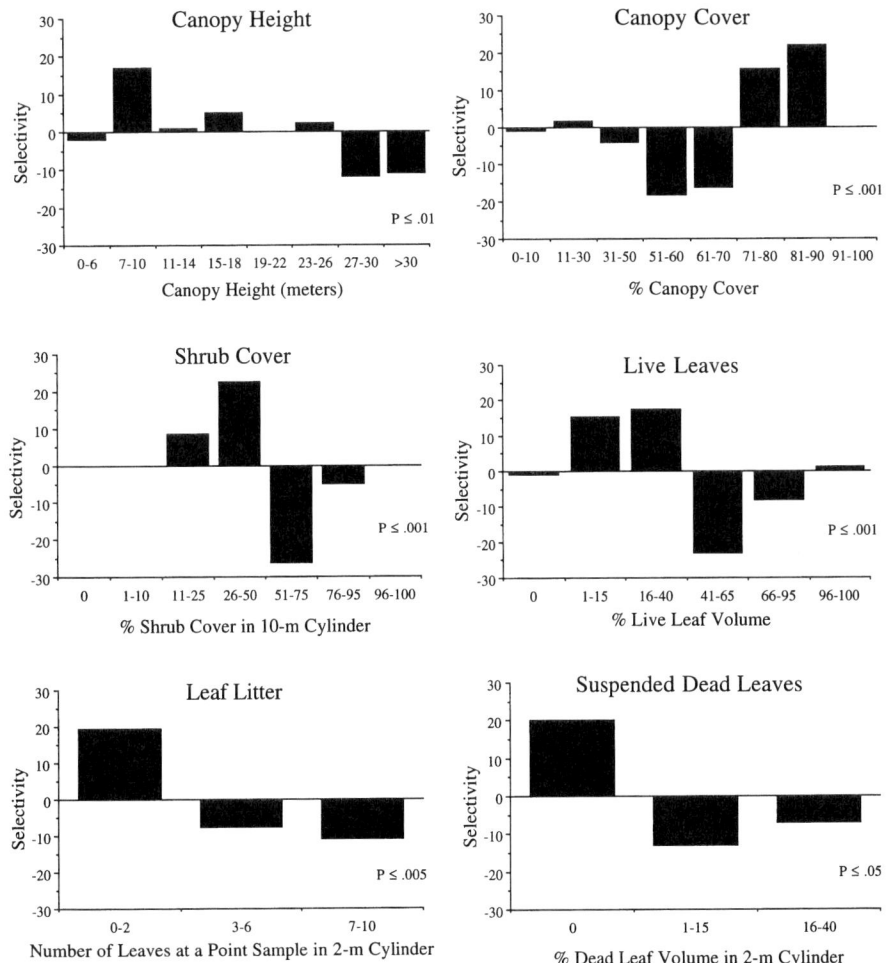

FIG. 17. Microhabitat selectivity in *Myrmoborus myotherinus* in *terra firme* forest. Positive values indicate preference (use > availability), whereas negative values indicate avoidance for the habitat variables depicted. Significance according to Kolmogorov-Smirnoff test.

have been used to gauge the complexity by looking at forest stratification (MacArthur et al. 1966; Smith 1973), or tropical habitats have simply been accepted as more complex based upon qualitative descriptions. A foliage-height-diversity profile is an index of the vertical foliage density from the ground to the canopy. Studies using these indices have shown only a correlation between increasing vertical foliage density and higher species diversity (MacArthur and MacArthur 1961; MacArthur et al. 1962; MacArthur et al. 1966; Karr 1971; Karr and Roth 1971); others have found no such correlation (Orians 1969; Terborgh and Weske 1969; Howell 1971; Pearson 1975). Moreover, the effect that this type of complexity has on bird species restricted to the understory is unknown.

Before understanding the effects that increased habitat complexity might have on bird species diversity, it is necessary to define what is meant by habitat complexity. Habitat complexity can be examined at two different scales: (1) between habitats, i.e., more distinct kinds of habitats available, and (2) within-habitats, i.e., either more different layers of vegetation, or unique structural features, such as bamboo, palms, suspended dead leaves, epiphytes, or lianas. We focused on within-habitat complexity, whether tropical habitats are structurally more complex, and whether bird species diversity is related to this complexity.

We analyzed the structural complexity of just the understory of each forest type separately (temperate and tropical). Many variables differed significantly between temperate and tropical

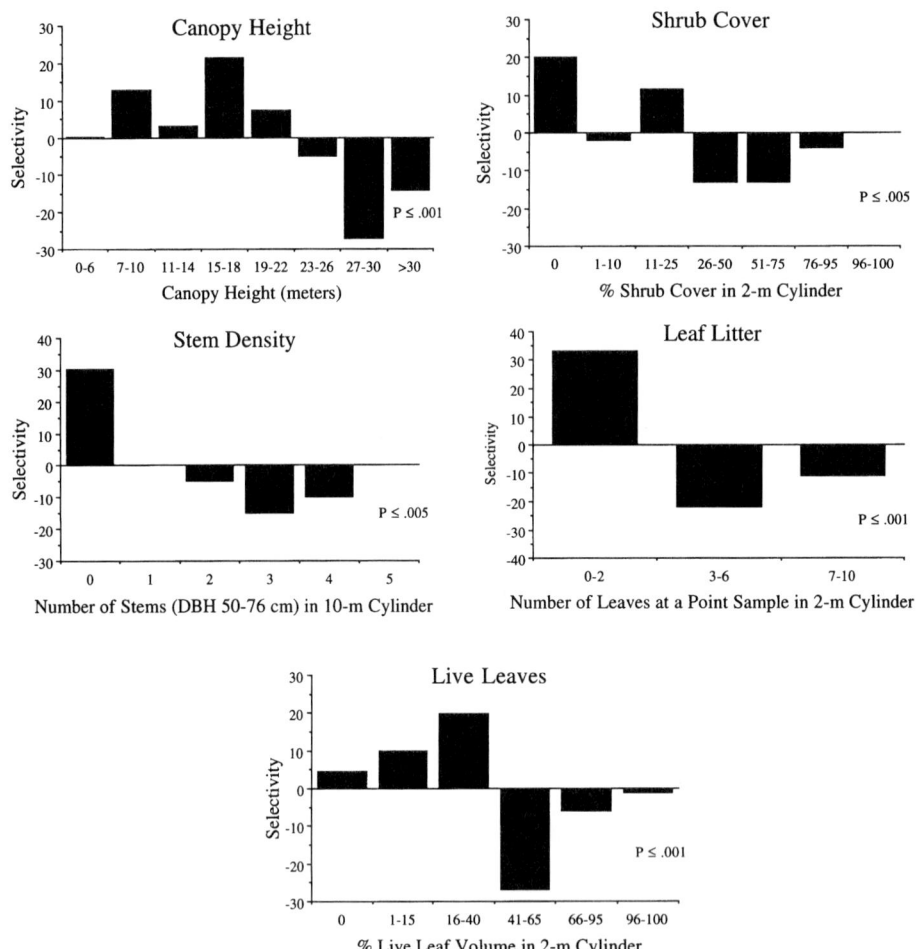

FIG. 18. Microhabitat selectivity in *Formicarius colma* in *terra firme* forest. Positive values indicate preference (use > availability), whereas negative values indicate avoidance for the habitat variables depicted. Significance according to Kolmogorov-Smirnoff test.

forests (see Marra 1989). However, these differences do not really provide insight into complexity *per se*. To compare complexity, it is necessary to analyze the structural heterogeneity of each forest type. Under the assumption that the variability of the vegetation (as measured by the habitat variables) is a measure of complexity (using the general variance), we found no large differences between forest types, except for the river-edge forest, which seemed to have greater structural heterogeneity. Therefore, we agree with Terborgh (1985) that the widely assumed differences in structural habitat heterogeneity, at least as far as the understory is concerned, between temperate and tropical forests need reevaluation, and that we must look elsewhere for the major factors governing species diversity of understory birds. Although structural diversity is similar between temperate and tropical forest types, the diversity of understory birds differs greatly. Terborgh and Weske (1969), working in human-altered habitats, and Pearson (1975), studying forest birds in three Amazon localities, were also unable to explain total species diversity by habitat complexity alone.

Can unique tropical resources explain greater species diversity?—An alternative way for tropical forests to be more complex is through addition of unique resources, such as army ants, bamboo, palms, epiphytes, lianas, or year-round supplies of fruit, nectar, or suspended dead leaves. Although we did measure some of the above, the measurements and analyses conducted in this study were not designed to quantify or integrate *all* of these identifiable unique resources. Unique resources unquestionably add complexity to tropical forests and can explain the presence

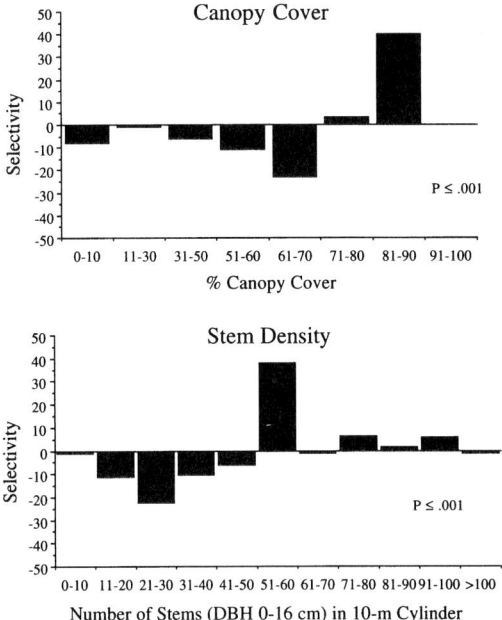

FIG. 19. Microhabitat selectivity in *Myrmeciza hemimelaena* in *terra firme* forest. Positive values indicate preference (use > availability), whereas negative values indicate avoidance for the habitat variables depicted. Significance according to Kolmogorov-Smirnoff test.

of some specialized bird species (e.g., Orians 1969). However, not all the additional bird species found in tropical forests are this easily explained (Terborgh 1980a, 1985).

To compare species numbers between temperate and tropical understory forest types, it is essential to distinguish those species that are members of specialized guilds (frugivores, ant-followers, dead-leaf-searchers, vine specialists, bamboo specialists, nectarivores, and treefall specialists) from the more "generalized" species, i.e., species not obviously associated with one of these guilds added to a tropical forest by special habitats or resources. As a first approach, we examined two tropical forest types for a tropical-temperate comparison: low-lying forest (the combined river-edge and transitional forest; this combination was necessary due to difficulties in distinguishing exact boundaries between the forest types) and *terra firme* forest (Table 6). We used Foster et al. (1994) to classify species according to forest type. Of the 73 species restricted to the understory (0–5 m) of low-lying forest (Table 6), 46 (63%) can be placed into one of the above "unique tropical guilds," whereas 27 (37%) cannot. In *terra firme* forest, 20 (47%) of the 43 species (Table 6) are members of specialized guilds, whereas 23 (53%) species are not. The difference between the two forest types can be attributed to the absence from or decrease in *terra firme* forest of some of these special structural features, e.g., bamboo and vines. Nevertheless, the overall comparison to the temperate forest is dramatic. In the tropical forests, 23–27 species not associated with "tropical" resources coexist in the understory, whereas, at most only five do so at the temperate site. Such strong differences suggest that "species-packing" is much greater in the tropics. Therefore, our data provide strong support for the Klopfer and MacArthur hypothesis (1960).

Few bird species at our temperate study site are as restricted to the understory as those at the tropical site. With the exception of *Oporornis formosus,* all our temperate species spent much of their time in the subcanopy and canopy; therefore, it is even questionable that they can be classified as purely understory species (Fig. 28). Other than *O. formosus,* only *Meleagris gallopavo,* (Wild Turkey), *Pipilo erythrophthalmus* (Eastern Towhee), and *Hylocichla mustelina* (Wood Thrush) could be classified as purely understory species at Tensas, but none of these is widespread in forests there. In fact, temperate forests of North America in general have few species that could be classified as purely understory species (e.g., Ruffed Grouse, *Bonasa umbellus;* thrushes, *Catharus* spp., and Ovenbird, *Seiurus aurocapillus*). Greater vertical stratification of tropical forest birds relative to temperate zone forest birds has been proposed by many

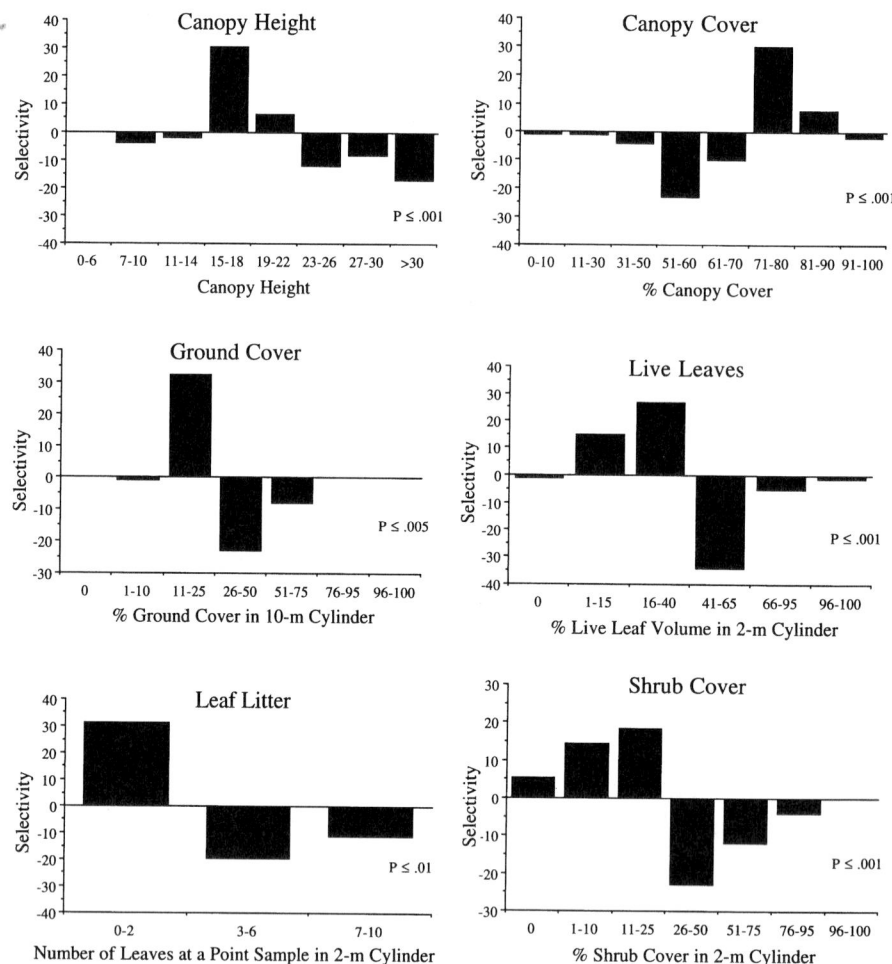

FIG. 20. Microhabitat selectivity in *Corythopis torquata* in *terra firme* forest. Positive values indicate preference (use > availability), whereas negative values indicate avoidance for the habitat variables depicted. Significance according to Kolmogorov-Smirnoff test.

investigators as one of the proximate mechanisms to explain the greater species richness of tropical forests (MacArthur 1966; Orians 1969; Terborgh and Weske 1969; Pearson 1971, 1977; Crome 1978; Terborgh 1980b; Bell 1982).

The presence of specialized guilds of birds on "unique" tropical resources (Orians 1969; Karr 1971, 1975, 1976, 1980; Lovejoy 1974; Pearson 1977; Terborgh 1980a; Remsen 1985) clearly increases species richness substantially, as illustrated above. Remsen (1985) found that most differences in species richness between tropical and temperate montane communities could be attributed to the addition of new "tropical" resources, such as the year-round availability of fruiting trees, nectar, epiphytic vegetation, vines, bamboo thickets, and suspended dead leaves. Terborgh (1980a) found that in lowland forest 34% of the "extra" tropical species were attributable to these additional guilds and attributed the remainder to greater species-packing. Terborgh's figures, however, were based upon species in the entire vertical range of a lowland forest, not just the understory, and so this could explain the differences between our calculations (63%) for low-lying forest.

Many of the special resources thought to be unique to the tropics are actually also found to a lesser degree in many forests of the southeastern United States. For instance, bamboo is present in southeastern forests, and although it has undoubtedly been drastically reduced in historical times, it has no species of birds restricted to it (with the possible exception of the virtually extinct Bachman's Warbler, *Vermivora bachmanii;* Remsen 1986). In contrast, in tropical forests,

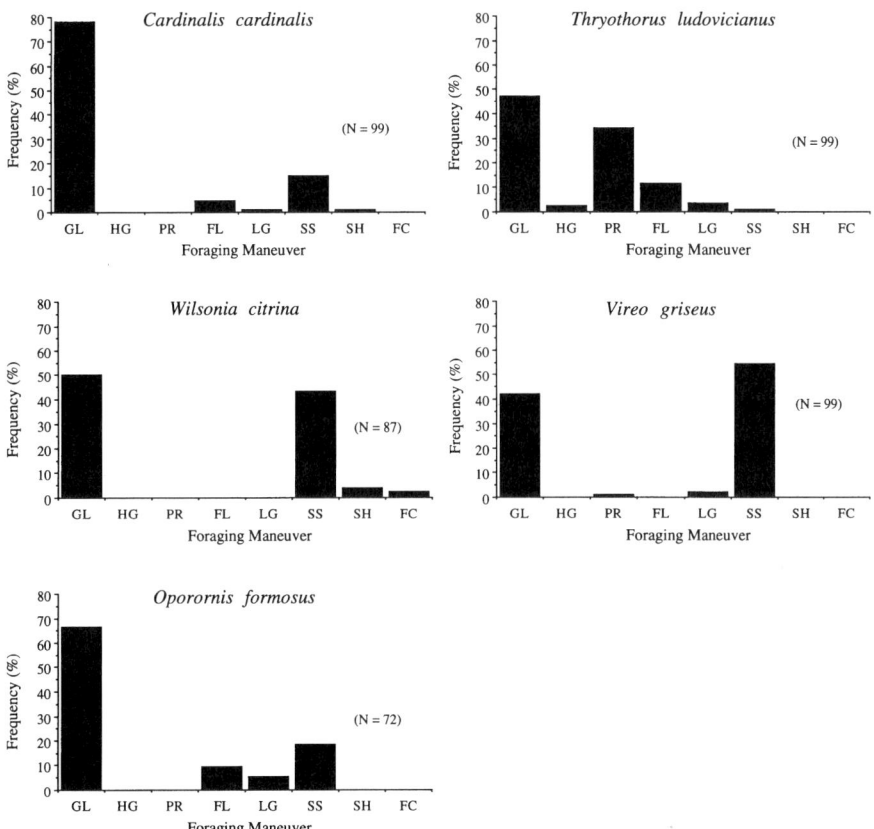

FIG. 21. The frequency (% of samples) of use of different foraging maneuvers by temperate forest species; GL = glean; HG = hang-glean; PR = probe; FL = flake; LG = leap-glean; SS = sally-strike; SH = sally-hover; FC = flutter-chase.

bamboo supports many specialized species (e.g., 16 species in Tambopata; Table 12; Kratter 1995, and in press). Two species used in this study preferred microhabitat containing bamboo but are not considered bamboo specialists. Suspended dead leaves are present in both temperate and tropical forests, although more abundantly in tropical forests, and support a number of specialists in the tropics (Remsen and Parker 1984; ≤6 species in Tambopata, Rosenberg 1990 1997). Dead-leaf foraging has been recorded in some temperate species (Remsen and Parker 1984; Remsen et al. 1989), including *T. ludovicianus* in this study, but these temperate species seem to be facultative users of dead leaves, whereas many tropical species are obligate dead-leaf foragers. Vines tangles, specialized on by some tropical species (Terborgh 1980a), were surprisingly similar in abundance between tropical and temperate forests. Interestingly, four of five temperate study species in our study preferred sites with abundant vines, whereas only one tropical study species did so. This difference is probably because our target tropical species did not include vine specialists (e.g., *Cymbilaimus fasciatus*, *Cercomacra cinerascens*, *Thryothorus* spp., and *Microbates* spp.).

The addition of new resources (structural and food related) can account for some of the increase and maintenance of higher bird species diversity in the tropics, but not for all of it. Below, we try to show that the remainder of this diversity in lowland tropics is primarily maintained by factors only secondarily related to structural habitat complexity.

Are tropical species more specialized?—Our data show that even though habitat complexity was similar in temperate and tropical forests, the target tropical species were indeed more selective and specialized in terms of microhabitat and foraging site than were the temperate species. Therefore, much of the increase in tropical species in lowland forest is due to tighter species-packing.

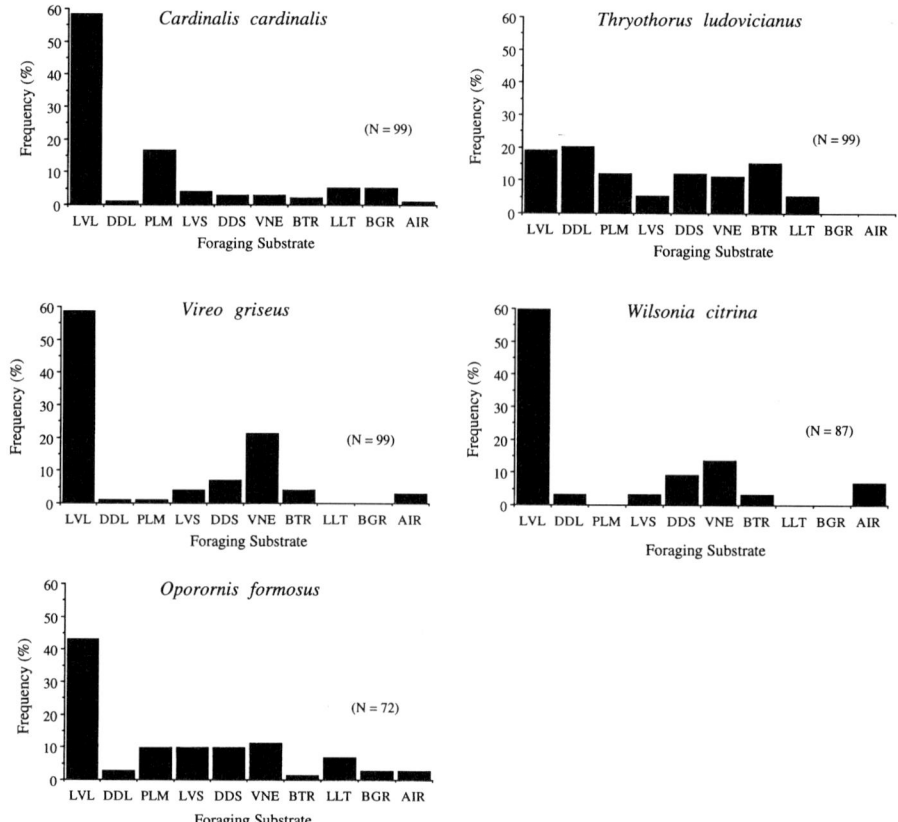

FIG. 22. The frequency (% of samples) of use of foraging substrates by temperate forest species; LVL = live leaves; DDL = dead leaves; PLM = palm; LVS = live stems; DDS = dead stems; VNE = vines; BTR = bark/trunk; LLT = leaf litter; BGR = bare ground; SWB = spider web.

The tropical species in this study specialized in a variety of ways at different levels of scale. *Myrmoborus myotherinus* was specialized at the macrohabitat level, using primarily *terra firme* forest, but was also restricted vertically in foraging height. *Myrmoborus leucophrys* was not as specialized in its selection of macrohabitat, but was more specialized in microhabitat selection, being restricted to some degree to bamboo. Thus, it was a "horizontal" specialist and was not restricted vertically. This was even more true for *H. cantator*, which was more restricted to bamboo than was *M. leucophrys*. *Formicarius analis* and *F. colma* are both "vertical" specialists in that they are entirely terrestrial. They are somewhat selective at the macrohabitat level, although not particularly selective at the microhabitat level. Both are specialized in their foraging maneuvers and substrate use relative to other target species. However, because they both forage only from the ground, the kinds of maneuvers and substrates that they can use is probably limited. The last two species, *M. hemimelaena* and *C. torquata*, neither of which are selective at the macrohabitat level, are the most generalized and most specialized (according to the general variance calculations), respectively, of all target species at the tropical sites. *Myrmeciza hemimelaena* was somewhat selective in that it preferred gaps and was fairly specialized vertically. *Corythopis torquata* can be classified as a "selecting generalist" (after Rosenzweig 1985). It was common in all three forest types; it was not selecting at the macrohabitat level, but was restricted at the microhabitat level, varying little in its selection of microhabitat variables, thus explaining its high "degree of generalization."

Overall, the temperate species were not as horizontally or vertically specialized as the tropical species. Nevertheless, some temperate species were specific in their selection of some of the microhabitat variables, whereas most were not selective of macrohabitat (although we did not use a range of sites comparable to that used in the tropics). *Cardinalis cardinalis*, *T. ludovicianus*, and to some degree *V. griseus* all expand into secondary forest habitats, and the first two species

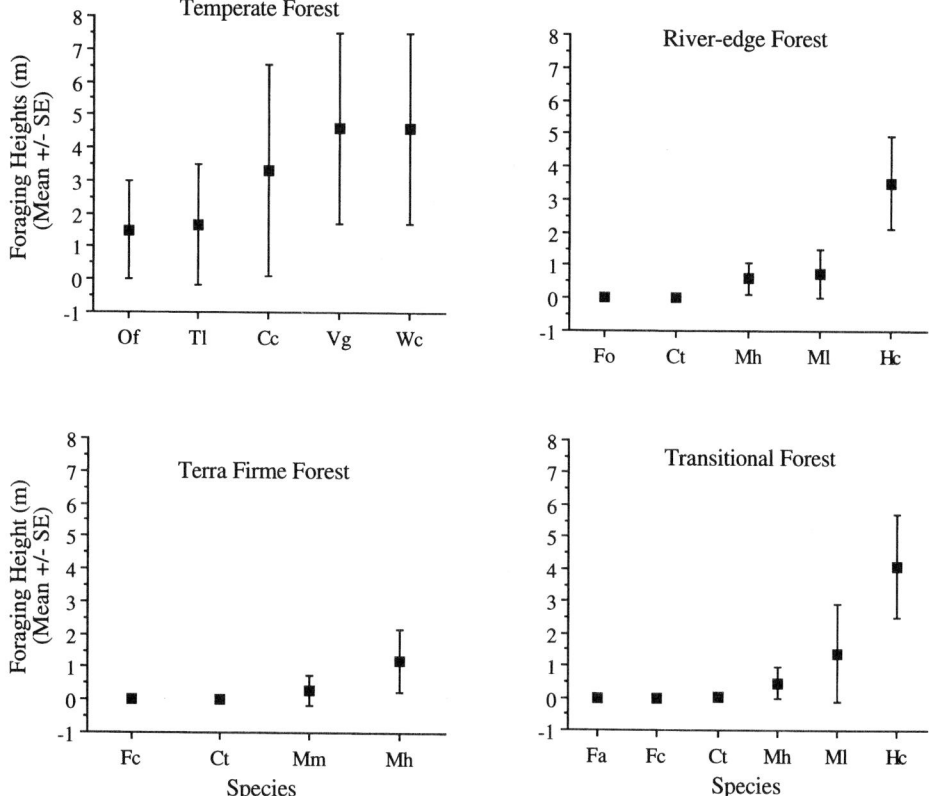

FIG. 23. Mean foraging heights with standard error bars of temperate and tropical forest species. See Table 1 for list of abbreviations.

are common in suburban habitats. However, neither of these habitat types were studied. At the microhabitat level, *V. griseus* and *T. ludovicianus* were somewhat restricted horizontally in their strong preference for microhabitat containing vines. All in all, these temperate understory species were true generalists in almost every sense of the word; they foraged throughout the vertical and to a lesser degree the horizontal range of the forest. These species were using a wide range of the habitat spectrum, unlike the tropical species.

Do tropical species overlap more in niche space?—Our calculations of niche overlap showed that the tropical species generally overlapped less among themselves than did the temperate species, but this varied depending on the foraging variable and forest type (Fig 27). Little overlap was found among tropical species in foraging height (except in *terra firme* forest), which may be a major axis of segregation among tropical species. This was not the case for temperate species. Therefore, our results do not support the hypothesis of Klopfer and MacArthur (1961) concerning the existence of greater overlap among tropical species and its importance as a mechanism contributing to the maintenance of high tropical species diversity.

We are aware of only two other studies that have attempted to quantify differences in niche metrics between guilds or communities of temperate and tropical forest birds. Stiles (1978) found that the differences in species richness between bird communities of alder forests in Washington and Costa Rica could be explained by a combination of greater specialization in foraging behavior and greater diversity of resources in the tropics. Askins (1983) found that the differences in species richness of woodpeckers in Minnesota and Guatemala was best explained by greater diversity of resources in the tropics, with some weak support for greater specialization in foraging behavior. The results of both studies are in general concordance with our results concerning the proximate mechanisms that effect species richness.

What accounts for tighter species-packing in the tropics?—We have shown that niches are more finely divided among certain tropical birds and that neither greater structural habitat com-

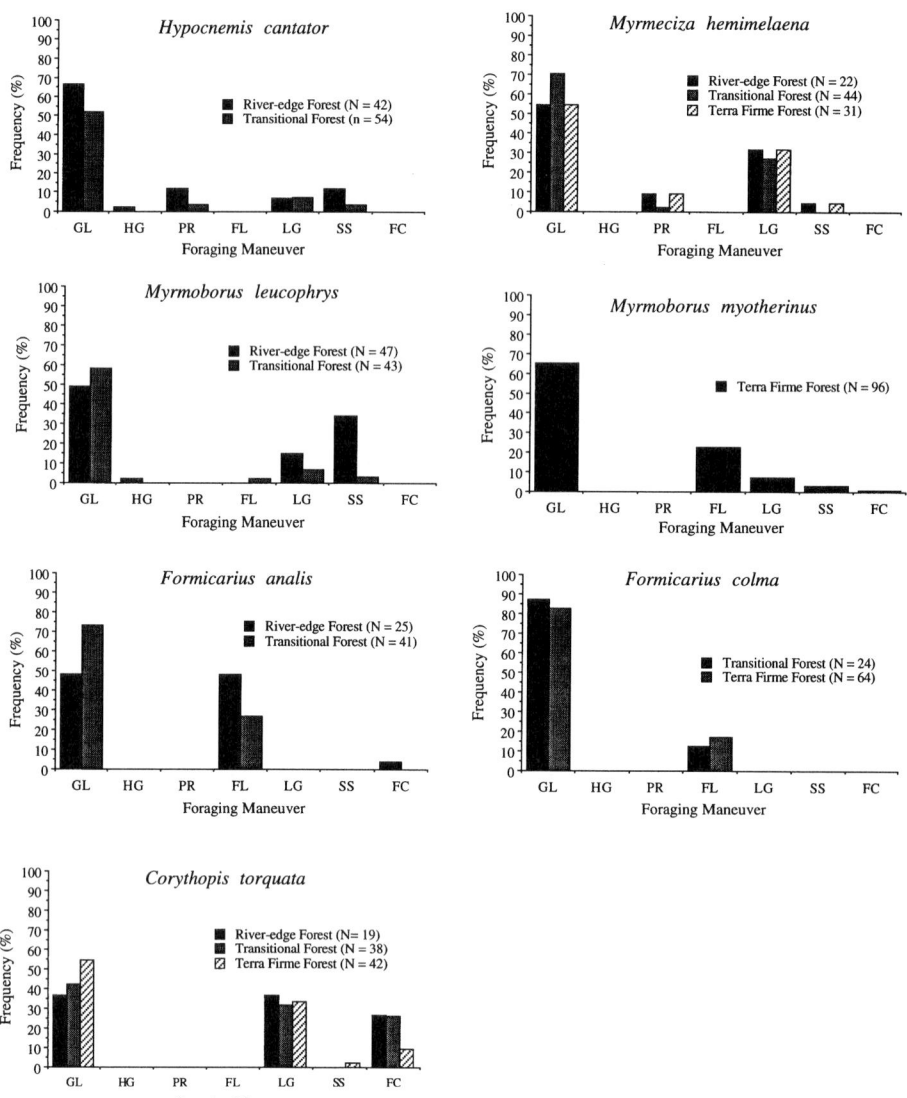

FIG. 24. The frequency (% of samples) of use of different foraging maneuvers by tropical forest species in each forest type; GL = glean; HG = hang-glean; PR = probe; FL = flake; LG = leap-glean; SS = sally-strike; SH = sally-hover; FC = flutter-chase.

plexity or greater niche overlap accounts for the high diversity at the tropical sites. The presence of "unique" tropical resources can account for many of the additional tropical species, but not all. Therefore, the remainder are able to coexist because of greater species-packing. What factors could be responsible for creating or maintaining this tighter species-packing? One hypothesis is that greater habitat and foraging-site specialization, and the higher species diversity in tropical understory birds, is maintained primarily by characteristics of the resource base (arthropods) created by greater climatic stability in the tropics (Stiles 1978). Pianka (1970) first suggested that tropical arthropods were "K-strategists" relative to their temperate counterparts. Below, we will explain how this can potentially influence habitat selection in tropical birds.

First, are tropical arthropods K-strategists, as suggested by Pianka (1970)? One characteristic of K-strategists is that they have stable populations of long-lived individuals. Unfortunately, data documenting year-round insect abundance or on insect longevity in tropical areas are meager and probably insufficient at this time for firm conclusions. Although not yet rigorously tested,

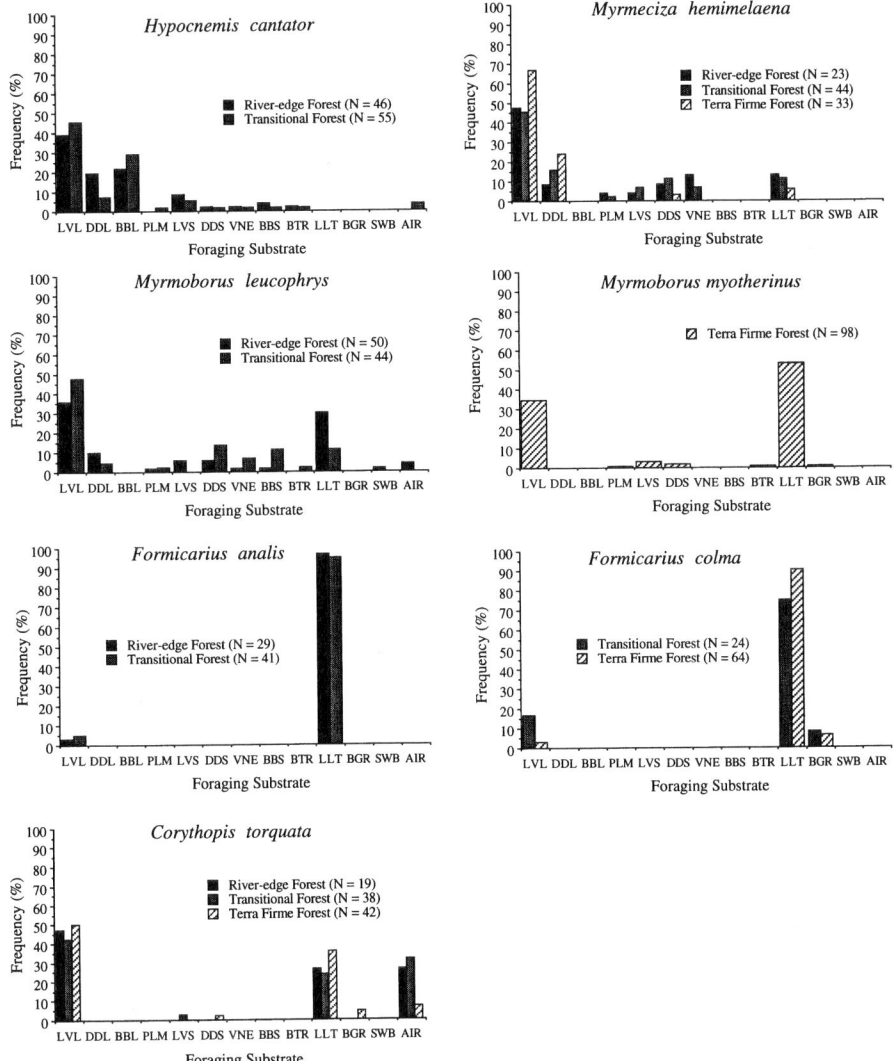

FIG. 25. The frequency (% of samples) of use of foraging substrates by tropical forest species; LVL = live leaves; DDL = dead leaves; BBL = bamboo leaves; PLM = palm; LVS = live stems; DDS = dead stems; VNE = vines; BBS = bamboo stems; BTR = bark/trunk; LLT = leaf litter; BGR = bare ground; SWB = spider web; AIR = air.

the idea maintains that tropical birds have a stable, year-round supply of insects. Recently, however, this hypothesis of a stable, year-round insect supply has been increasingly challenged (Leigh et al. 1982).

Sampling of arthropods is problematic. Many studies that have sampled insects year-round in the tropics have relied primarily on light traps and have shown some seasonality in insect abundance. Unfortunately, light traps have many shortcomings, mainly in that they only attract certain types of insects (Smythe 1982; Wolda 1978, 1982). For instance, light traps do not sample adequately two major groups used by birds, orthopterans and larval insect forms (i.e., caterpillars). These have been shown to be the major food items taken by furnariids, formicariids, and dendrocolaptids, both by field observation (Greenberg 1981; Thiollay 1988b) and by stomach analysis (mainly orthopterans; Chapman and Rosenberg 1991; Rosenberg 1997). In addition, Smythe (1982) found that the lepidopterans (adult forms), which may be effectively sampled with light traps, exhibited less seasonality than the other types of insects that can be sampled

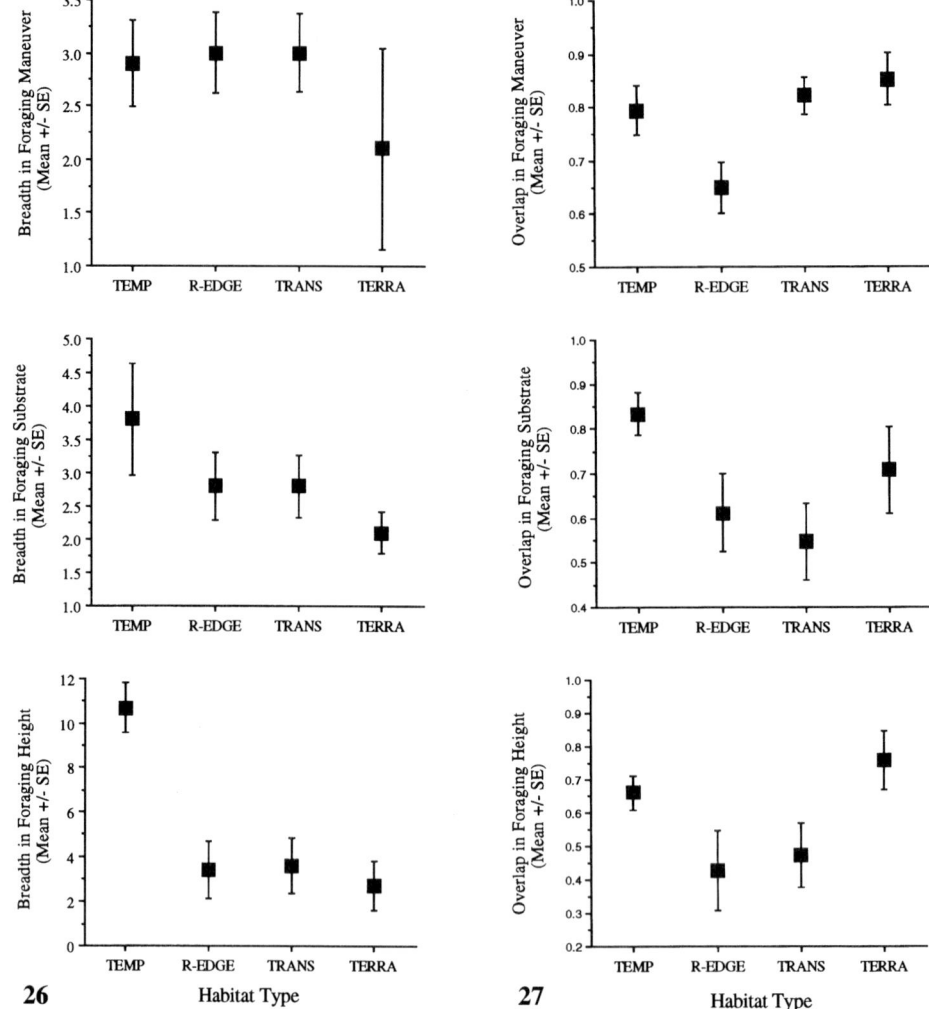

Fig. 26. Mean (±SE) foraging maneuver, substrate and height breadths for species in each habitat type; TEMP = temperate, R-EDGE = river-edge, TRANS = transitional, TERRA = *terra firme*.

Fig. 27. Mean (±SE) foraging maneuver, substrate and height overlaps for species in each habitat type; TEMP = temperate, R-EDGE = river-edge, TRANS = transitional, TERRA = *terra firme*.

efficiently. In addition, Penny and Arias (1982) found that orthopterans make up the majority of the arthropod biomass near ground level, whereas lepidopterans do so near the canopy. Therefore, data on year-round abundance of arthropods important to birds (i.e., arthropods and lepidopterans) are inadequate at this time to test the hypothesis that tropical resource bases in the tropics are more stable. Despite these recent claims that tropical insect populations arc as seasonal as temperate insect populations, extreme differences in the year-round insect abundance between tropical and temperate areas are obvious. Wolda (1988) stated that "activity seasons of tropical insects tend to be large, the percent of species active around the year higher, and the seasonal peaks less well defined, relative to counterparts at higher latitudes."

In addition to having populations lacking seasonal fluctuations, K-strategists are more likely to live longer (Pianka 1970). However, data on longevity for tropical insects is nonexistent, although lack of severe temperature fluctuations makes it seem likely that tropical insects would live longer. If so, one would predict that there would be greater selection for them to develop more anti-predator strategies, such as camouflage, mimicry systems, nocturnal behavior, and chemical defenses, than there is for temperate insects. The life-history strategies of temperate

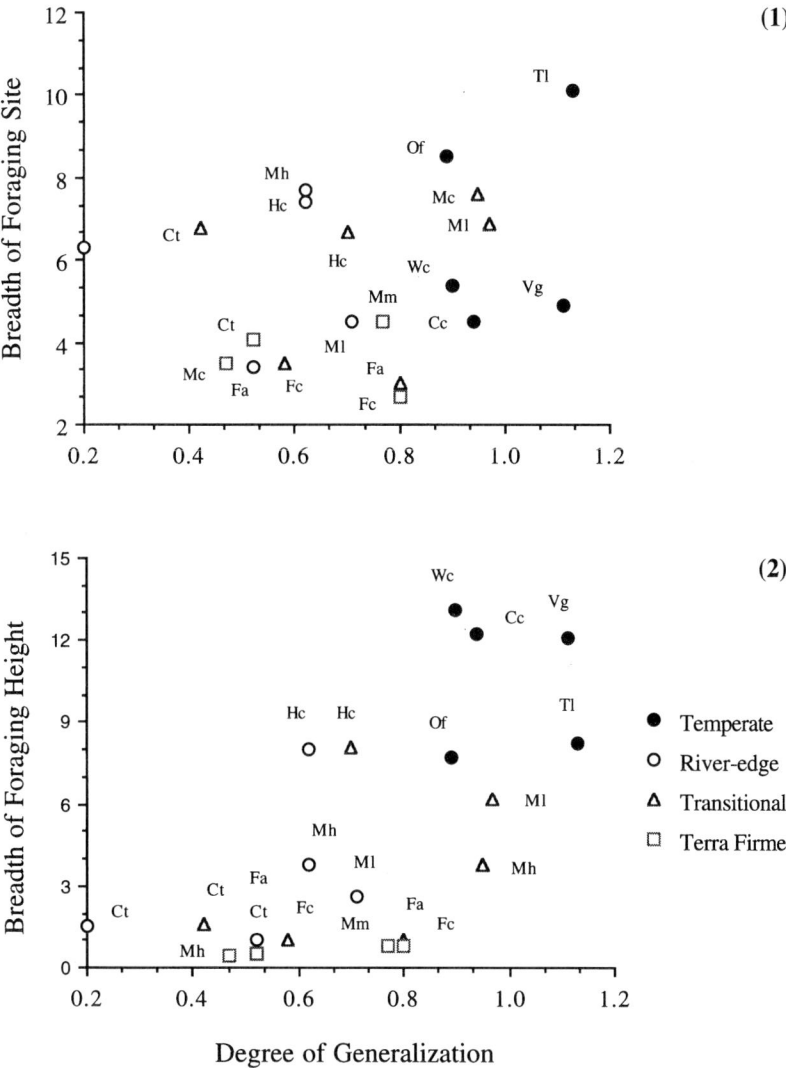

FIG. 28. 1) Breadth of foraging site (maneuver + substrate breadth for each species, see Marra 1989 for specific breadth values) and 2) breadth of foraging height vs. degree of generalization of temperate and tropical bird species. See Table 1 for list of abbreviations.

insects seem to be aimed at predator satiation (Lloyd and Dybas 1966) rather than predator avoidance.

How then would the different life-history strategies of tropical insects cause tropical birds to become more specialized and more tightly packed? It is possible that tropical birds are forced to become specialized because their resource is specialized. The key to this hypothesis would be in the lower average detectability of tropical insects compared to temperate insects (Thiollay 1988b). It may be that tropical birds have to spend more time searching substrates for prey that have developed anti-predator strategies. Support from this comes from Thiollay (1988b), who found that the mean foraging attack rate in tropical insectivorous birds was 4–6 times lower than recorded for temperate birds. Therefore, tropical birds may spend more time searching for well-hidden prey. However, it is also possible that birds are getting a better yield (i.e., larger insects; Schoener 1965; Greenberg 1981; Thiollay 1988b) from each prey item, which would mean that they do not have to feed as often, or alternatively, that metabolic demands differ between temperate and tropical birds. In any case, although there are trends in the predicted direction, our data do not demonstrate strong specialization at the foraging site by tropical birds

TABLE 6

Total Number of Understory Bird Species Listed According to Their Most Conspicuous Resource Specialization (if Any) in Two Forest Types at the Tambopata Reserve, Southeastern Peru. Some of These Categories Are Not Mutually Exclusive, and Species Can Occur in Both Types. River-edge and Transitional Forests Have Been Combined and Are Referred to as Low-lying Forest. Numbers of Species Summarized from Marra (1989)

	Low-lying forest	Terra firme forest
Resource specialization		
Frugivores	12	7
Nectarivores	4	3
Bamboo Thickets	16	0
Vine Tangles	2	0
Treefall Gaps	2	4
Army-ant Followers	4	4
Dead-leaf searchers	6	2
Total Specialists	46 (63%)	20 (47%)
Total species without a resource specialization	27 (37%)	23 (53%)

with respect to maneuver or substrate breadths. Instead, understory tropical birds become specialized vertically (in terms of foraging height) and horizontally (in terms of microhabitat selection) relative to their temperate counterparts, and use whatever means necessary to obtain arthropods within these restricted zones. However, more data on foraging behavior combined with analyses of stomach contents is needed to conclusively determine the degree that tropical birds are specialized on specific types of arthropods.

Another factor that potentially influences the selection of habitat, either primarily, secondarily, or in combination with resources, is interspecific competition. It is possible that greater interspecific competition among the tropical study species causes them to have lower niche breadths, lower general variances, and to be overall more specialized than the temperate birds. With so many species occurring together (27 in the understory of low-lying forest), it is difficult to imagine that interspecific competition would not influence habitat selection to some degree. Recent experimental work by Robinson and Terborgh (1995) has provided further support for the importance of interspecific competition in shaping habitat selection in Amazonian birds. More data such as these are critical to increasing our understanding of what factors allow for the maintenance of high species diversity in tropical regions.

In conclusion, our comparison of the understories of temperate and tropical forests has revealed that structural differences between forest types are slight and almost certainly not sufficient to account for all differences in bird species richness. Our data support the findings of others in that the majority of additional bird species in tropical forests can be explained by two factors: (1) the presence of resources "unique" to the tropics and (2) "tighter species-packing."

ACKNOWLEDGMENTS

This work is dedicated to the memory of our dear friend and colleague Ted Parker. His dedication to and passion for understanding tropical biology inspired our own efforts. This research would not have been possible without financial support provided by the Fugler Fellowship for Tropical Vertebrate Biology, J. S. McIlhenny, and the LSU chapter of Sigma Xi. We thank Max Gunther of Peruvian Safaris S. A., Peru, and all the people at the Tensas National Wildlife Refuge for providing assistance with logistics during the research period.

J. M. Bates, R. B. Hamilton, T. A. Parker III, S. K. Robinson, K. V. Rosenberg, and G. B. Williamson provided valuable assistance with the design of the project. J. M. Bates, R. Bierregaard, J. M. Hagan, and K. V. Rosenberg made important contributions to the development of this paper. We are grateful to F. Rutledge, V. Lancaster, B. Moser, T. S. Sillett and M. P. Ayres for assistance with data analysis. Additional support was provided by the following: W. Barrow, A. P. Capparella, A. Fogg, S. J. Hackett, M. L. and P. R. Isler, G. Kinney, J. P. O'Neill, T. S. Schulenberg, P. E. Scott, and R. M. Zink.

LITERATURE CITED

Askins, R. A. 1983. Foraging ecology of temperate-zone and tropical woodpeckers. Ecology 64:945–956.

Barrow, W. C. 1990. Ecology of small insectivorous birds in a bottomland forest. Ph.D. dissertation, Louisiana State University. Baton Rouge, Louisiana.

Bell, H. L. 1982. A bird community of lowland rainforest in New Guinea; vertical distribution of the avifauna. Emu 82:143–162.

Brokaw, N. V. L. 1982. The definition of treefall gap and its effects on measures of forest dynamics. Biotropica 14:158–160.

Chapman, A., and K. V. Rosenberg. 1991. Dietary relationships of four sympatric Amazonian woodcreepers (Dendrocolaptidae). Condor 93:904–915.

Cody, M. L. 1975. Towards a theory of continental species diversity. Pages 214–257 in Ecology and Evolution of Communities (M. L. Cody and J. M. Diamond, Eds.). Belknap Press, Cambridge, Massachusetts.

Cody, M. L. 1985. Habitat selection in the sylviine warblers of western Europe and North Africa. Pages 85–129 in Habitat Selection in Birds (M. L. Cody, Ed.). Academic Press, Orlando, Florida.

Crome, F. H. J. 1978. Foraging ecology of an assemblage of birds in lowland rainforest in northern Queensland. Austral. J. Ecol. 3:195–212,

De Visscher, M. N. 1984. Une analyse multifactorielle des relations "oiseaux-végétation" pour décrire l'habitat dans une région du piémont des Andes vénézuéliennes. Gerfaut 74:127–137.

Erwin, T. L. 1984. Tambopata Reserved Zone, Madre de Dios, Peru: History and description of the reserve. Revista Peruviana Entomol. 27:1–8.

Fitzpatrick, J. W. 1980. Foraging behavior of Neotropical tyrant flycatchers. Condor 82:43–57.

Foster, R. B., T. A. Parker III, A. H. Gentry, L. H. Emmons, A. Chicchon, T. S. Schulenberg, L. Rodriguez, G. Lamas, H. Ortega, J. Icochea, W. Wust, M. Roma, J. A. Castillo, O. Phillips, C. Reynel, A. Kratter, P. K. Donahue, and L. J. Barkley. 1994. The Tambopata-Candamo Reserved Zone of Southeastern Peru: A Biological Assessment. RAP working papers 6, Conservation International, Washington, D.C.

Gibson, R. A., Baker, A. J., and A. Moeed. 1984. Morphometric variation in introduced populations of the Common Myna (*Acridotheres tristis*): An application of the jackknife to principal component analysis. Syst. Zool. 33:408–421.

Greenberg, R. 1981. Dissimilar bill shapes in tropical versus temperate forest foliage gleaning birds. Oecologia 49:143–147.

Howell, T. R. 1971. An ecological study of the birds of the lowland pine savanna and adjacent rain forest in northeastern Nicaragua. Living Bird 10:185–242.

James, F. C., R. F. Johnston, N. O. Wamer, G. J. Niemi, and W. J. Boecklen. 1984. The Grinnellian niche of the Wood Thrush. Am. Nat. 124:17–47.

Johnson, R. A., and D. W. Wichern. 1982. Applied Multivariate Statistical Analysis. Prentice Hall, Englewood Cliffs, New Jersey.

Karr, J. R. 1971. The structure of avian communities in selected Panama and Illinois habitats. Ecol. Monogr. 41:207–233.

Karr, J. R. 1975. Production, energy pathways, and community diversity in forest birds. Pages 161–176 in Tropical Ecological Systems: Trends in Terrestrial and Aquatic Research. (F. B. Golley and E. Medina, Eds.). Springer-Verlag, New York, New York.

Karr, J. R. 1976. Within- and between-habitat avian diversity in African and Neotropical lowland habitats. Ecol. Monogr. 46:457–481.

Karr, J. R. 1980. Geographical variation in the avifaunas of tropical forest undergrowth. Auk 97:283–298.

Karr, J R. 1981. Surveying birds with mist nets. Stud. Avian Biol. 6:62–67.

Karr, J. R., and R. R. Roth. 1971. Vegetation structure and avian diversity in several New World areas. Am. Nat. 105:423–435.

Karr, J. R., and K. E. Freemark. 1983. Habitat selection and environmental gradients: dynamics in the stable tropics. Ecology 64:1481–1494.

Kikkawa, J., T. E. Lovejoy, and P. S. Humphrey. 1980. Structural Complexity and Species Clustering in Tropical Rainforests. Acta XVII Congressus Intern. Ornith. (1978), Pages 962–967. Berlin, West Germany.

Klopfer, P. H., and R. H. MacArthur. 1960. Niche size and faunal diversity. Am. Nat. 104:293–300.

Klopfer, P. H., and R. H. MacArthur. 1961. On the causes of tropical species diversity: niche overlap. Am. Nat. 95:223–226.

Kratter, A. W. 1995. Bamboo specialization in Amazonian birds. Ph.D. Dissertation. Louisiana State University, Baton Rouge, Louisiana.

Kratter, A. W. 1997. Bamboo specialization in Amazonian birds. Biotropica 29:100–110.

Kushlan, J. A., G. Morales, and P. C. Frohring. 1985. Foraging niche relations of wading birds in tropical wet savannas. Pages 663–682 in Neotropical Ornithology (Buckley, P. A., M. S. Foster, E. S. Morton, R. S. Ridgely, and F. G. Buckley, Eds.). Ornithol. Monogr. No. 36.

Leigh, E. G., A. S. Rand and D. M. Windsor. (Eds.) 1982. The Ecology of a Tropical Forest. Seasonal Rhythms and Long Term Changes. Oxford Univ. Press, Oxford, U.K.

LEMMON, P. E. 1956. A spherical densiometer for estimating forest overstory density. Forest Science 2:314–320.
LEMMON, P. E. 1957. A new instrument for measuring forest overstory density. J. Forestry 55:667–668.
LEVEY, D. J. 1988. Treefall gaps and the distribution of understory birds and shrubs in a tropical wet forest. Ecology 69:1076–1089.
LEVINS, R. 1968. Evolution in Changing Environments. Princeton Univ. Press, Princeton, New Jersey.
LLOYD, M., AND H. S. DYBAS. 1966. The periodical cicada problem. Evolution 20:466–505.
LOVEJOY, T. E. 1974. Bird diversity and abundance in Amazon forest communities. Living Bird 13:127–191.
MACARTHUR, R. H. 1969. Patterns of communities in the tropics. Biol. J. Linn. Soc. 1:19–30.
MACARTHUR, R. H., AND J. W. MACARTHUR. 1961. On bird species diversity. Ecology 42:594–598.
MACARTHUR, R. H., J. W. MACARTHUR, AND J. PREER. 1962. On bird species diversity II. Prediction of bird census from habitat measurements. Am. Nat. 96:167–174.
MACARTHUR, R. H., H. RECHER, AND M. L. CODY. 1966. On the relation between habitat selection and species diversity. Am. Nat. 100:319–332.
MARRA, P. P. 1989. Habitat selection in understory birds of tropical and temperate forests. Master's Thesis, Louisiana State University, Baton Rouge, Louisiana.
MAY, R. M., AND R. H. MACARTHUR. 1972. Niche overlap as a function of environmental variability. Proc. Natl. Acad. Sci. U.S.A. 69:1109–1113.
MORSE, D. H. 1985. Habitat selection in North American parulid warblers. Pages 131–157 in Habitat Selection in Birds (M. L. Cody, Ed.). Academic Press, Orlando, Florida.
MOSER, E. B., W. C. BARROW, AND R. B. HAMILTON. 1990. An exploratory correspondence analysis of avian foraging behavior-habitat relationships. Pages 309–317 in Avian Foraging: Theory, Methodology, and Applications (M. L. Morrison, C. J. Ralph, J. Verner, and J. R. Jehl, Jr., Eds.). Stud. Avian Biol. No. 13.
ORIANS, G. H. 1969. The number of bird species in some tropical forests. Ecology 50:783–801.
PARKER, T. A., III. 1982. Observations of some unusual rainforest and marsh birds in southeastern Peru. Wilson Bull 94:477–493.
PEARSON, D. L. 1971. Vertical stratification of birds in a tropical dry forest. Condor 73:46–53.
PEARSON, D. L. 1975. The relation of foliage complexity to ecological diversity of three Amazonian bird communities. Condor 77:453–466.
PEARSON, D. L. 1977. Ecological relationships of small antbirds in Amazonian bird communities. Auk 94:283–292.
PENNY, N. D., AND J. R. ARIAS. 1982. Insects of an Amazon Forest. Columbia Univ. Press, New York, New York.
PIANKA, E. R. 1970. On r- and k- selection. Am. Nat. 104:592–596.
PIANKA, E. R. 1974. Niche overlap and diffuse competition. Proc. Natl. Acad. Sci. U.S.A. 71:2141–2145.
REMSEN, J. V., JR. 1985. Community organization and ecology of birds of high elevation humid forest of the Bolivian Andes. Pages 733–756 in Neotropical Ornithology (Buckley, P. A., M. S. Foster, E. S. Morton, R. S. Ridgely, and F. G. Buckley, Eds.). Ornithol. Monogr. No. 36.
REMSEN, J. V., JR. 1986. Was Bachman's Warbler a bamboo specialist? Auk 103:216–219.
REMSEN, J. V., JR. 1990. Community ecology of Neotropical kingfishers. Univ. Calif. Publ. Zool. 124:1–116.
REMSEN, J. V., JR., M. ELLERMAN, AND J. COLE. 1989. Dead-leaf searching by the Orange-crowned Warbler in Louisiana in winter. Wilson Bull. 101:645–648.
REMSEN, J. V., JR., AND D. A. GOOD. 1996. Misuse of data from mist-net captures to assess relative abundance in bird populations. Auk 113:381–398.
REMSEN, J. V., JR., AND T. A. PARKER III. 1983. Contribution of river-created habitats to bird species richness in Amazonia. Biotropica 15:223–231.
REMSEN, J. V., JR., AND T. A. PARKER III. 1984. Arboreal dead-leaf-searching birds of the Neotropics. Condor 86:36–41.
REMSEN, J. V., JR., AND S. K. ROBINSON. 1990. A classification scheme of avian foraging behavior in terrestrial habitats. Pages 144–160 in Avian Foraging: Theory, Methodology, and Applications (M. L. Morrison C. John Ralph, J. Verner, and J. R. Jehl, Jr., Eds.). Stud. Avian Biol. No. 13.
ROBINSON, S. K., AND R. T. HOLMES. 1982. Foraging behavior of forest birds: the relationships among search tactics, diet, and habitat structure. Ecology 63:1918–1931.
ROBINSON, S. K., AND J. TERBORGH. 1995. Interspecific aggression and habitat selection by Amazonian birds. J. Animal Ecology 64:1–11.
ROSENBERG, G. H. 1990. Habitat specialization and foraging behavior by birds of Amazonian river islands in northeastern Peru. Condor 92:427–443.
ROSENBERG, K. V. 1990. Dead-leaf foraging specialization in tropical forest birds: Measuring resource availability and use. Pages 360–368 in Avian Foraging: Theory, Methodology and Applications. (Morrison, M. L., C. J. Ralph, J. Verner, and J. R. Jehl, Jr., Eds.). Stud. Avian Biol. No. 13.
ROSENBERG, K. V. 1997. Ecology of dead-leaf-searching specialists and their contribution to Amazonian bird diversity. Pages 673–700 in Studies in Neotropical Ornithology Honoring Ted Parker (J. V. Remsen, Jr., Ed.). Ornithol. Monogr. No. 48.

ROSENZWEIG, M. L. 1985. Some theoretical aspects of habitat selection. Pages 517–540 *in* Habitat Selection in Birds (M. L. Cody, Ed.). Academic Press, Orlando, Florida.
SAS INSTITUTE. 1985. SAS User's Guide: Statistics, version 5 edition. SAS Institute, Inc., Cary, North Carolina.
SAS INSTITUTE. 1995. JMP: Statistics and Graphics Guide, version 3.1. SAS Institute, Inc., Cary, North Carolina.
SCHEMSKE, D. W., AND N. BROKAW. 1981. Treefalls and the distribution of understory birds in a tropical forest. Ecology 62:938–945.
SCHOENER, T. W. 1965. The evolution of bill size differences among sympatric congeneric species of birds. Evolution 19:189–213.
SILVA, J. M. C. DA, AND R. CONSTANTINO. 1988. Aves de um trecho de mata no baixo Rio Guamá —uma reanálise: riqueza, raridade, diversidade, similaridade e preferências ecológicas. Bol. Mus. Paraense Emilio Goeldi, sér. Zool. 4:201–210.
SMITH, A. P. 1973. Stratification of temperate and tropical forests. Am. Nat. 107:671–683.
SMYTHE, N. 1982. The seasonal abundance of night-flying insects in a Neotropical forest. Pages 309–318 *in* The Ecology of a Tropical Forest. Seasonal Rhythms and Long Term Changes (Leigh, E., Jr., A. S. Rand, and D. Windsor, Eds.). Oxford Univ. Press, Oxford, U.K.
STILES, E. W. 1978. Avian communities in temperate and tropical alder forests. Auk 80:276–284.
TERBORGH, J. W. 1980a. Causes of Tropical Species Diversity. Actis Congr. Int. Ornithol., XVII (1978): 955–961. Berlin, West Germany.
TERBORGH, J. 1980b. Vertical Stratification of a Neotropical Forest Bird Community. Actis Congr. Int. Ornithol. XVII (1978):1005–1012. Berlin, West Germany.
TERBORGH, J. 1985. Habitat Selection in Amazonian Birds. Pages 311–338 *in* Habitat Selection in Birds (M. L. Cody, Ed.). Academic Press, Orlando, Florida.
TERBORGH, J. W., AND J. S. WESKE. 1969. Colonization of secondary habitats by Peruvian birds. Ecology 50:765–782.
TERBORGH, J. W., AND S. K. ROBINSON. 1986. Guilds and their utility in ecology. Pages 65–90 *in* Community Ecology: Pattern and Process (J. Kikkawa and D. J. Anderson, Eds.). Blackwell Scientific Publications, Oxford, U.K.
TERBORGH, J. W., J. W. FITZPATRICK, AND L. H. EMMONS. 1984. Annotated checklist of bird and mammal species of Cocha Cashu Biological Station, Manu National Park, Peru. Fieldiana (Zool.) 21:1–29.
THIOLLAY, J. M. 1988a. Diversité spécifique et coexistence en forêt tropicale: L'occupation des mileux par une guilde d'oiseaux insectivores en Guyane. Rev. Ecol. (Terre Vie) 43:59–92.
THIOLLAY, J. M. 1988b. Comparative foraging success of insectivorous birds in tropical and temperate forests: ecological implications. Oikos 53:17–30.
WILLSON, M. F. 1974. Avian community organization and habitat structure. Ecology 55:1017–1029.
WILLSON, M. F., S. H. ANDERSON, AND B. G. MURRAY. 1973. Tropical and temperate bird species diversity: Within-habitat and between-habitat comparisons. Carib. J. Sci. 13:81–90.
WILLSON, M. F., AND D. J. MORIARTY. 1976. Bird species diversity in forest understory: analysis of mist-net samples. Oecologia 25:373–379.
WOLDA, H. 1978. Fluctuations in abundance of tropical insects. Am. Nat. 112:1017–1045.
WOLDA, H. 1982. Seasonality of Homoptera on Barro Colorado Island. Pages 319–330 *in* The Ecology of a Tropical Forest. Seasonal Rhythms and Long Term Changes (Leigh, E., Jr., A. S. Rand, and D. Windsor, Eds.). Oxford Univ. Press, Oxford, U.K.
WOLDA, H. 1988. Insect seasonality: Why?. Ann. Rev. Ecol. Syst. 19:1–18.
WUNDERLE, J. M., JR., A. DIAZ, I. VELAZQUEZ, AND R. SCHARRON. 1987. Forest openings and the distribution of understory birds in a Puerto Rican rainforest. Wilson Bull. 99:22–37.

NEW SUBSPECIES OF *MYRMOBORUS LEUCOPHRYS* (FORMICARIIDAE) AND *PHRYGILUS ALAUDINUS* (EMBERIZIDAE) FROM THE UPPER HUALLAGA VALLEY, PERU

JOHN P. O'NEILL AND THEODORE A. PARKER III

Museum of Natural Science, Foster Hall 119, Louisiana State University, Baton Rouge, Louisiana, 70803-3216, USA
[1]*Deceased*

ABSTRACT.—The ecologically diverse valley of the upper middle Río Huallaga, depto. Huánuco, Perú, is isolated by mountains on the south, east and west and supports several endemic subspecies of birds. We herein describe a new subspecies of *Myrmoborus leucophrys* from the humid forest near Tingo María, and of *Phrygilus alaudinus* from the rain-shadow desert near the town of Huánuco.

RESUMEN.—El valle del Río Huallaga medio en el depto. Huánuco, Perú, es ecologicamente diverso y esta aislado por cordilleras al sur, este, y oeste. En este valle viven varias subespecies de aves endemicas. Aqui publicamos la descripción de una nueva subespecie de *Myrmoborus leucophrys* del bosque húmedo de las cercanías de la ciudad de Tingo María, y de una nueva subespcie de *Phrygilus alaudinus* de la zona arida de las cercanías de la ciudad de Huánuco.

During the past 30 years personnel from the Louisiana State University Museum of Natural Science (LSUMZ) have visited repeatedly the Department of Huánuco in central Peru. This department is geologically and ecologically complex (see O'Neill and Parker 1976), with the humid lowlands along the upper middle Huallaga river isolated by mountains to the east and west and the uppermost part of the valley isolated in a rain-shadow desert surrounded by the high Andes. It is not surprising that the differentiation of populations of certain species of birds has occurred in both the humid and dry areas.

The Cordillera Azul, a major north-south spur of the eastern Andes, originates just southeast of the city of Tingo María as an extension of the mountains on the east bank of the Río Huallaga (See Fig. 1). This range rarely exceeds 1,500 m in elevation, but it continues north-northwestward from the Tingo María region more than 300 km, flanked on the west by the Río Huallaga, with the only major break being the point at which the Huallaga cuts through it on its northeastward flow to join the Río Marañón. South of Tingo María, the Huallaga flows east-northeastward through the easternmost main Andes in a deep, narrow canyon that is the only low-elevation connection between the humid eastern slopes of the mountains and the arid, rain-shadow-affected Huánuco valley. The lowland "Amazonian" forest in the valley between the Cordillera Azul and main Andes is isolated by high elevations on the east and west and by the dry canyon and valley to the southwest. The tropical fauna in this area is in an ecological *cul-de-sac* that connects with the main Amazonian forest only far to the north around the northern extension of the Cordillera Azul, and thus it is not surprising that the area supports endemic, isolated taxa such as *Brotogeris cyanoptera gustavi*, *Ramphocelus melanogaster* and *Pyrrhura (melanura) berlepschi*, and that certain species, such as *Arremon aurantiirostris,* reach their southern limit there (Meyer de Schauensee, 1966). The widespread antbird *Myrmoborus leucophrys* is another species represented in the area by a well-marked previously undescribed form that may be known as:

Myrmoborus leucophrys koenigorum subsp. nov.

Holotype.—LSUMZ 28431, adult male, collected on the Río Azul near junction of Río Azul and Río Tulumayo NE of Tingo María, depto. Huánuco, Peru (elevation not recorded, but approximately 800 m), 22 June 1962; collected by W. T. Van Velzen (original number 414).

Diagnosis.—Differs in male plumage from all known forms of *Myrmoborus leucophrys* by

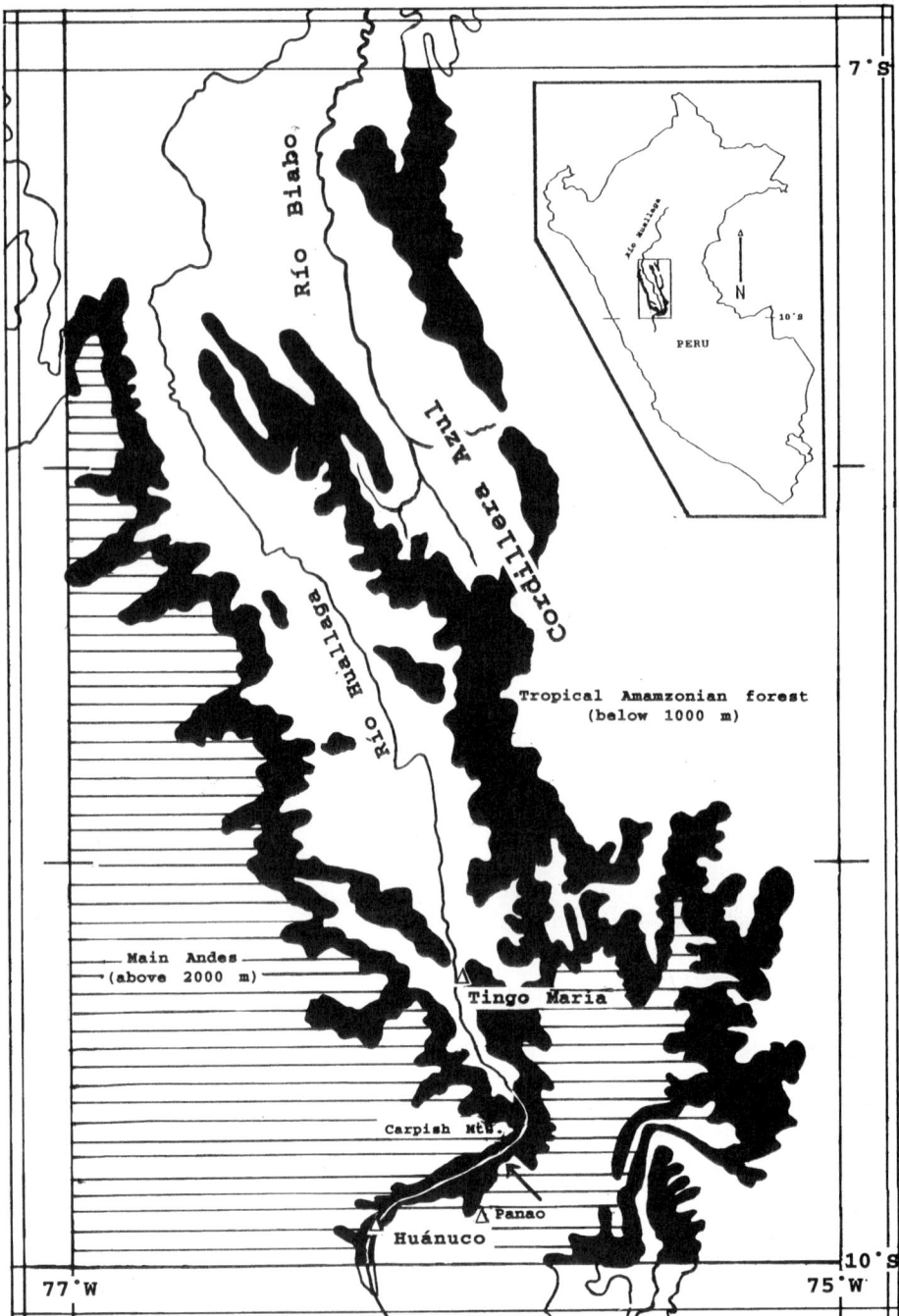

FIG. 1. Map of the upper middle Huallaga river valley. Areas in black represent elevations between 1,000 and 2,000 m; areas covered by horizontal parallel lines reach elevations of more than 2,000 m. The small arrow northeast of Panao in the lower center of the map indicates the approximate point at which humid forest ends and rain shadow desert begins.

FIG. 2. Dorsal (left) and lateral views of the heads of *Myrmoborus leucophrys koenigorum*, ssp. nov. (upper) and *M. l. leucophrys* to illustrate the nearly solid white crown of the new form.

having the crown solid pale grayish white instead of gray with a white forehead and superciliaries (see Figure 2), and from *M. l. leucophrys*, the only other form known in Peru, by also being somewhat larger (see Table 1); although the series is small, males tend to all be darker midventrally and to show less demarcation between the black throat and the charcoal gray belly than do those of the nominate subspecies; the female differs from the females of *M. l. leucophrys* in its somewhat larger size, by having the crown slightly paler cinnamon-rufous with pale cinnamon shaft streaks on the feathers of the mid-crown, and the forehead and superciliaries not strongly demarcated from the crown, thus approaching the condition found in the male.

Description of the holotype.—Forehead and crown pale grayish white, closest to #10 Gray (Pale Gull Gray) (capitalized colors from Ridgway [1912]), this color also extending onto the nape. Feathers in central part of crown show blackish shaft streaks and dusky centers. Lores, ear coverts, chin, and throat black, this color extending onto the upper breast where it blends imperceptibly into the dark bluish gray (between #5 Slate Gray and #4 Slate Color) of the lower breast, belly, sides, dorsum, rump, and rectrices. Lesser, median, and greater secondary coverts warm black (near Fuscous Black), with bluish gray present to varying degrees on the outer webs of the feathers. Primaries, secondaries, and tertials Blackish Brown (3), edged on the outer web with bluish gray. Soft part colors not recorded, but bill, feet and tarsi are black in dried specimens. Testes 2×1 mm.

Measurements of type (mm).—Wing (chord) 68.6, tail 47.5, tarsus 27.0, culmen from base 19.3.

Range.—So far as known, restricted to the tropical forest of the upper Huallaga River valley near the junction of the Cordillera Azul and the Cordillera Carpish (main eastern Andes) in depto. Huánuco, Peru, but likely found throughout most of the upper and middle Huallaga valley.

Specimens examined.—(all LSUMZ except as noted): *Myrmoborus l. leucophrys* (24). PERU: depto. Amazonas; Huampami, 4 ♂♂, 3 ♀♀, Kigkis, 1 ♂. depto. Loreto (now Ucayali); Balta, 4 ♂♂, 2 ♂♂. depto Madre de Dios; km 105 W of Puerto Maldonado, 1 ♂, 1 ♀, km 40 E of Quincemil, 1 ♂, Quebrada Juliaca, 1 ♀. BOLIVIA: depto. Beni; 25 km S Riberalta, 1 ♂, depto. La Paz; Ixiamas, 2 ♂♂, 1 ♀, depto. Cochabamba; Todos Santos, 1 ♂, 1 ♀. *M. l. koenigorum* (10). PERU: all depto. Huánuco; Vista Alegre, 2 ♂♂ (Field Museum of Natural History, hereafter FMNH), Río Azul, 3 ♂♂, 2 ♀♀, 40 km E Tingo María, 1 ♂, Sta. Elena, 1 ♀, Divisoria, 1 ♀.

Remarks.—We take great pleasure in naming this new form for Helen and Arturo Koenig of Lima, Peru, who have been of tremendous help to LSUMZ personnel for more than three de-

TABLE 1
SELECTED MEASUREMENTS IN MM OF TWO SUBSPECIES EACH OF *Myrmoborus leucophrys* AND *Phrygilus alaudinus*

	M. l. leucophrys		M. l. koenigorum		P. a. bipartitus		P. a. bracki	
	Males	Females	Males	Females	Males	Females	Males	Females
Wing	15 (63.0–68.4) 65.4*	9 (61.0–67.4) 64.7	6 (66.9–70.0) 68.8	4 (64.4–66.3) 65.4	5 (71.1–74.2) 73.0	1 (—) 72.0	5 (66.1–71.1) 67.7	1 (—) 64.6
Tail	14 (42.7–50.4) 45.6	8 (43.1–48.3) 44.4	6 (45.5–50.5) 48.0	4 (45.4–49.6) 46.9	5 (54.8–59.0) 56.3	1 (—) 54.0	4 (50.0–53.1) 52.0	1 (—) 49.5
Culmen from base	15 (17.3–19.5) 18.6	9 (17.5–19.6) 18.6	6 (18.7–19.9) 19.4	4 (18.7–20.1) 19.4	5 (13.8–15.2) 14.4	1 (—) 14.3	5 (13.7–15.1) 14.0	1 (—) 13.7
Tarsus	15 (20.1–28.5) 25.4	9 (23.3–27.5) 25.2	6 (25.2–28.5) 26.7	4 (24.8–27.9) 26.4	5 (20.2–22.0) 21.3	1 (—) 21.7	5 (20.3–22.6) 22.5	1 (—) 20.9

* n (range) mean.

cades. The new bird is resident at Sta. Elena in the foothills of the Cordillera Azul, where the Koenigs previously raised coffee, and where LSUMZ personnel collected vertebrates from 1967 to 1972.

Myrmoborus leucophrys is a widespread and common antbird of forest edge, river-edge forest, second-growth forest, and other non-mature forest habitats. Although seemingly suitable areas in Peru from which we have found it to be apparently absent include the middle Río Santiago valley, depto. Amazonas, the region of the junction of the Marañón and Ucayali rivers north and east in depto. Loreto to the Ecuadorean, Colombian and Brazilian borders, and from all of the middle and lower Río Huallaga valley except for the Tingo María region. LSUMZ personnel have made collections along the middle Río Santiago, along the lower Río Napo, and on the south bank of the Amazon without finding the species, but there are so few collections from most of the length of the Huallaga that it could have been missed. Because the lower Huallaga valley, especially at the point where it cuts through the Cordillera Azul, is somewhat drier, it may not provide suitable habitat for the species. For some birds we might argue that the extreme habitat destruction caused by most of the middle Huallaga valley's conversion to coca plantations would make the area uninhabitable, but this activity should create more habitat for *Myrmoborus leucophrys*, not less.

It is somewhat surprising to find an isolated population of *Myrmoborus leucophrys* in the southernmost part of the *cul-de-sac* of the upper Huallaga valley. We might be tempted to give the new form species status, but we would like to have more specimens, including tissue for studies of the genetic relatedness of the Huallaga birds and the typical white-browed birds, as well as recordings of vocalizations and detailed notes on microhabitat use. We also are hesitant to give the new form species status because some of the males in the known series do show a crown somewhat grayer than the forehead and superciliaries, possibly indicating recent or present contact with typical white-browed birds. The closest white-browed population is in the tropical forest east of the Cordillera Azul. Contact could possibly take place in the headwaters of the Río Biabo where the main ridge of the Cordillera Azul seems to be less than 1,000 m. More extensive fieldwork in the middle Huallaga and Cordillera Azul areas may also turn up additional localities for *Myrmoborus leucophrys*, as well as examples of other new subspecies of Huallaga valley birds.

The valley of the upper Huallaga River, in the area of the city of Huánuco, on the southern side of the Carpish Mountains varies in elevation from approximately 1,800–2,000 m and lies in a rain-shadow vegetated with *Acacia, Prosopis,* cactus, and other xerophytic plants. The avifauna reflects this xeric habitat in harboring such species as *Columbina minuta, Tapera naevia, Thraupis bonariensis, Campylorhynchus fasciatus,* and *Phrygilus alaudinus*. The six known specimens of the last-mentioned species from this valley are extremely small and represent a new form that may be known as:

Phrygilus alaudinus bracki, subsp. nov.

Holotype.—LSUMZ 75430, Adult male collected on the west bank of the Río Huallaga below (= south of) Acomayo, depto. Huánuco, Peru (no elevation recorded, but approximately 2,000 m on the bank of the river), 18 November 1973; collected by Dan A. and Erika J. Tallman (Dan A. Tallman original number 2404).

Diagnosis.—Differs in both sexes from the widespread Peruvian form *P. a. bipartitus* and from the subspecies of southern Peru and Bolivia, *P. a. excelsus,* in being distinctly smaller in wing and tail length and in body weight, but not in culmen or tarsus length (Table 1). Although colors in this species are subject to seasonal variation due to wear, specimens of the new form in fresh plumage have the brown edges to the dorsal feathers a richer, rustier tone than do specimens of *bipartitus* of comparable freshness, and the gray of the breast of the males is also somewhat paler, less blue than is the same area in *bipartitus*. *P. a. humboldti* of the isolated hills in the coastal desert in depto. Piura, Peru is also small in size, and similar to *P. a. bracki*. Only one specimen of *humboldti* was examined, but measurements of the holotype are taken from Koepcke (1963). The bird we looked at had a wing chord of 66.7 mm (67 mm for the type) and a tail of 47.0 mm (51 mm for the type), and had the freshest wing coverts edged with a grayish brown similar to that found in *P. a. bipartitus,* unlike the rich rusty color bordering the same feathers in *P. a. bracki*. The nominate subspecies is distinctly larger and darker in all aspects than is either the new form, *humboldti,* or *bipartitus,* and is known only from the temperate zone of Chile (Peters, 1970).

Description of holotype.—Forehead and anterior half of superciliary gray (nearest #6 Gray

[Dark Gull Gray]). Lores blackish grizzled with dirty white. Posterior superciliary, ear-coverts, throat, breast and sides of neck a slightly paler gray than the forehead (nearest #7 Gray [Deep Gull Gray]), this color extending across the upper mantle where it mixes with the dusky-streaked plumage of that area. Posterior crown, mantle, and scapulars near Cinnamon-Drab, with dull blackish centers, giving an overall streaked effect. Rump pale gray, washed with Cinnamon-Drab, but without streaking. Lesser and median wing coverts pale gray; greater secondary coverts, tertials, and secondaries dusky black, edged, especially on outer web, with Verona Brown; primaries dusky black, edged on outer web with pale brownish gray. Central pair of rectrices dusky black, edged narrowly on both webs with pale Cinnamon-Drab; outer four pairs of rectrices blackish with large white tear-drop shaped mid-feather spot occupying center half of inner web, and forming a prominent mid-caudal band on each side of the spread tail; the outer web of these feathers blackish with narrow whitish edging, especially distally. Mid-belly dull white; flanks and under tail coverts buffy (between Sayal Brown and Cinnamon). Bill, feet, and tarsi bright yellow; iris dark chestnut. Wt 17.0 g; left testis only (right missing) 7 × 5 mm; skull ossified; little fat; no molt.

Measurements of type (mm).—Wing (chord) 66.1, tail 51.9, tarsus 20.5, culmen from base 13.7.

Range.—So far as known, the arid inter-Andean valley of the upper Río Huallaga in the vicinity of the city of Huánuco, depto. Huánuco, Peru.

Specimens examined.—(all LSUMZ except as noted): *Phrygilus a. excelsus* (8). PERU: depto. Puno; 2 km N Juliaca, 1 ♂, 5 km from Puno on road to Arequipa, 1 ♂. BOLIVIA: depto. La Paz; Finca Capiri, 12 km W Viacha, 3900 m, 4 ♂♂, 2 ♀♀. *P. a. bracki* (6). PERU: depto. Huánuco; below Acomayo, 5,300 ft., 1 ♂, below Acomayo, Río Huallaga, 1 ♂, Huánuco, 3 ♂♂, 1 ♀ (FMNH). *P. a. bipartitus* (6). depto. Lima; La Molina, 1 ♂, Chosica Mt., 1 ♂, 1 ♀ (FMNH), depto. Ancash; Yánac, 8,500 ft., 1 ♂, *P. a. humboldti* (1). depto. Piura; Quebrada Chorillo, Cerro Illescas, 1 ♂ (FMNH).

Remarks.—We take great pleasure in naming this new form for our friend and colleague Antonio Brack Egg of Lima, Peru. Dr. Brack has been a driving force in the establishment of a conservation ethic in Peru as well as being a scientist who has a deep love for his country and its flora and fauna.

Phrygilus alaudinus is widespread and common in the drier parts of Peru. It is found more often at middle elevations and locally along the coast, but does occasionally occur into the puna zone, mainly in agricultural areas. The new form seems to be isolated in the arid upper Huallaga valley near the city of Huánuco. It is isolated on the east by humid temperate forest and on the west, south, and north by the high elevations of the main Andes, where *P. a. bipartitus* would be expected. The valley of the Río Marañón, the headwaters of which come very close to those of the upper Huallaga, is extremely difficult of access. It is not surprising that there seem to be no specimens of *Phrygilus alaudinus* from the valley, especially from the middle and lower reaches, the regions that are most like the arid Huánuco area in climate and elevation. If the species does occur in the middle and lower Marañón valley, we would predict that it would be represented by a smaller, paler population like *bracki* or *humboldti*. If there are small birds in the Marañón valley, then it would be tempting to lump them with *bracki* and *humboldti* and put the three groups under a single name. At present there is not enough material available to look closely at this possibility, so we prefer to consider *P. a. humboldti* and *P. a. bracki* as two widely allopatric subspecies.

The new form and *humboldti* (for which no weight is available) are both small in body size [average of 19 g ($n = 2$) for *bracki* vs 21.5 g ($n = 2$) for *bipartitus*, and 27.4 ($n = 6$) for *excelsus*] and wing and tail length, but their bill and tarsal lengths are essentially the same as those of *P. a. bipartitus*. As pointed out by Grant (1965), a reduction in body size may be related to greater cooling efficiency by birds in warmer temperatures such as those of the desert-like Huánuco valley, or the coastal desert inhabited by *humboldti*, for which no weight is available. Grant (1965) also pointed out that birds with *relatively* larger bills and tarsi can, in an area of less competition, obtain a wider range of food. Since we know little of the competition for food faced by any population of *Phrygilus alaudinus*, a discussion of such is not possible.

ACKNOWLEDGMENTS

We gratefully acknowledge the financial assistance of the late Babette M. Odom, the late Laura R. Schweppe, H. Irving Schweppe, and John S. McIlhenny, the last two of whom continue to have a strong interest in the LSUMZ and its research program in Latin America. John W.

Fitzpatrick and David E. Willard kindly loaned material from the Field Museum. Willet T. Van Velzen and Dan A. and Erika J. Tallman assisted with the fieldwork that resulted in the forms herein described. We also thank Eric Cardich B., Susana Moller-Hergt, and José Purisaca P. of the Ministerio de Agriculture, Lima, Peru, for their interest in the work of the LSUMZ and for the issuance of necessary permits. Gary R. Graves and Thomas S. Schulenberg reviewed the manuscript and offered many helpful suggestions.

LITERATURE CITED

GRANT, P. R. 1965. The adaptive significance of some size trends in island birds. Evolution 19:355—367.

KOEPCKE, M. 1963. Finken des peruanischen Kustengebietes. Beitrage zur Neotropischen Fauna III Band, Heft 1.

MEYER DE SCHAUENSEE, R. 1966. The Species of Birds of South America with their distribution. Narberth, Pennsylvania, Livingston Publ. Co., 577 pp.

O'NEILL, J. P., AND T. A. PARKER III. 1976. New subspecies of *Schizoeaca fuliginosa* and *Uromyias agraphia* from Peru. Bull. British Ornith. Club 96 (4): 136–141.

PETERS, J. L. 1970. Check-list of the Birds of the World, Vol XIII. *R. A. Paynter, Jr, Ed.,* Cambridge, Massachusetts, Mus. Comparative Zoology.

RIDGWAY, R. 1912. Color Standards and Nomenclature. Washington, D.C., published by the author.

AVIFAUNA OF THE TAPAJÓS NATIONAL PARK AND VICINITY, AMAZONIAN BRAZIL

DAVID C. OREN[1] AND THEODORE A. PARKER III[2,3]

[1]Departamento de Zoologia, Museu Paraense Emílio Goeldi,
Caixa Postal 399, 66.017-970 Belém, Pará, Brazil; and
[2]Foster Hall 119, Museum of Natural Science,
Louisiana State University, Baton Rouge, Louisiana 70803, USA
[3]Deceased

ABSTRACT.—The Tapajós National Park in central Brazilian Amazonia is one of the world's largest tropical rainforest reserves. Studies of the region's avifauna began almost 90 years ago, when Emilie Snethlage made collections at Vila Braga and Ilha de Goyana. More than 2,000 bird specimens have been collected in the region; important observational studies have also been carried out, in particular in the last two decades. This paper presents a summary of the relevant collections at the Museu Paraense Emílio Goeldi (Belém, Brazil), Carnegie Museum (Pittsburgh, USA), and Museu de Zoologia da Universidade de São Paulo (São Paulo, Brazil), and of observations made by T. A. Parker, III, T. S. Schulenberg, H. Sick, and D. C. Oren in visits to the Park. The number of bird species recorded totals 448 for the region and 387 for the park. With 19 species of Dendrocolaptidae, the park has more woodcreeper species than any other Neotropical site.

RESUMEN.—O Parque Nacional do Tapajós, localizado na parte central da Amazônia brasileira, é uma das maiores reservas de floresta tropical do mundo. Estudos da avifauna da região se iniciaram há quase 90 anos, quando Emilie Snethlage coletou espécimes ornitológicos em Vila Braga e na ilha de Goyana. No total, mais de 2000 espécimes de aves foram coletados na região; foram elaborados também estudos baseados em observações, especialmente durante as últimas duas décadas. O presente trabalho apresenta um compêndio dos dados das coleções relevantes do Museu Paraense Emílio Goeldi (Belém, Brasil), Carnegie Museum (Pittsburgh, EUA), e Museu de Zoologia da Universidade de São Paulo (São Paulo, Brasil), bem como observações de T. A. Parker, III, T. S. Schulenberg, H. Sick e D. C. Oren durante visitas ao parque. O número de espécies de aves registrado na região totaliza 448, sendo 387 para o parque. Com 19 espécies de dendrocolaptídeos, o parque sustenta a fauna mais rica de arapaçus conhecida no Neotrópico.

The Tapajós National Park in Amazonian Brazil is one of the largest forest reserves in the world, encompassing 994,000 ha of gently undulating terrain along the west bank of the Rio Tapajós. Most of the park lies within the state of Pará, although the far western section is in Amazonas State (Fig. 1). Created in 1974 during the construction of the Transamazonian Highway, when Brazil was the object of strong international pressure regarding regional environmental issues, it received the official designation of "Parque Nacional da Amazônia." At the time, the name was more logical than it is today, for then it was Brazil's only national park in the vast Amazon region. Although officially still designated as "Amazonia National Park," most researchers, environmental administrators, and visitors prefer to call the protected area the "Tapajós National Park" to distinguish it from Brazil's 32 other conservation units (National Parks, Biological Reserves, and Ecological Stations) in the Amazon region. The national park should not be confused with the Tapajós National Forest, a 631,000 ha federally administered area on the east bank of the river where controlled resource exploitation and forestry are allowed. In this paper we use the expression "vicinity of the park" to mean the region west of the Rio Tapajós from Itaituba, in the north, to km 212 of the Transamazonian Highway, in the south, and include Ilha de Goyana (Fig. 1).

The vegetation of the Park is primarily high *terra firme* forest (approximately 92%, according to RADAM-Brasil (1975)). There are important areas of seasonally flooded habitats along the

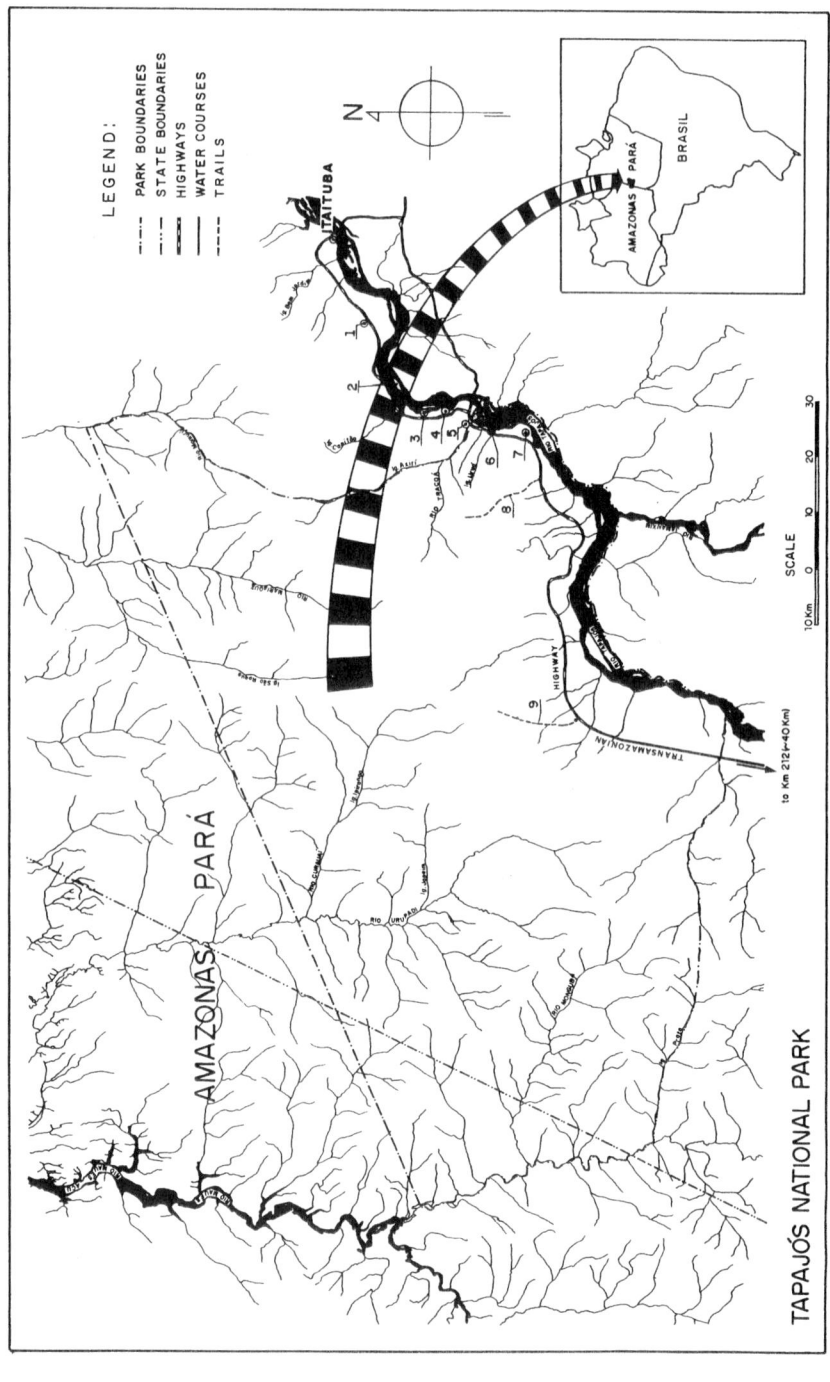

FIG 1: Tapajós National Park. 1 = Km 19 Transamazonian Highway; 2 = Ilha de Goyana; 3 = Vila Braga; 4 = Vila Nova; 5 = Tracoá; 6 = Uruá; 7 = Buburé; 8 = Capelinha Trail; 9 = Km 120 Trail.

Rio Tapajós and other water courses (4%), with the remaining 4% occupied by second growth (primarily along the Transamazonian highway) in various stages of ecological succession.

Brazilian Amazonia in general can be considered very poorly explored ornithologically (Oren and Albuquerque 1991). Within this context, the region of the Tapajós National Park deserves to be considered relatively well-known, having received visits by ornithological collectors and researchers for almost a century. Part of the available information, however, is in the literature in various languages (English, Portuguese and German), or exists as unpublished manuscripts by Parker, Oren and H. Sick, and an even larger part resides almost forgotten in museum collection catalogues and drawers. Even some of the published information has been systematically overlooked; an example is Griscom and Greenway's (1941) record of *Hyloctistes subulatus* from Vila Braga, missed by Peters (1951) and subsequently by every other compilation-style ornithological work on South America, as well as in the specialized literature on furnariids. The diverse available data for the region of the park have never been brought together in a single document, leading to a belief that this area, like so many others in the region, is almost unknown. The truth is quite different, however, as well over 2,000 avian specimens (including the Types of 16 taxa, 15 still recognized) have been collected in the general vicinity of the park, which has been visited by some of the region's foremost ornithologists and bird collectors over the past 90 years.

The first ornithological explorer of the region was Emilie Snethlage, who made three major expeditions there to collect birds (December 1906 to January 1907; October to November 1908, end of December 1908; June to July 1917), primarily at Vila Braga and Ilha de Goyana (Fig. 1). Of her three collections, originally deposited at the Museu Paraense Emílio Goeldi, Belém, Brazil, only the first, consisting of 209 specimens of 119 species, was published (Snethlage 1907, 1908, 1914). The last collection she made in the region of today's park was the largest, consisting of over 700 specimens. Many specimens were later exchanged and donated to other institutions, including the Berlepsch Museum, American Museum of Natural History, Bern Natural History Museum, the Museu Nacional, Rio de Janeiro and Zoological Museum, Berlin). Among the Museu Goeldi specimens with known destination are 1050 that Snethlage sent to the Museu Nacional do Rio de Janeiro in 1921. She herself left the Goeldi and formally transferred to the Museu Nacional on 15 January 1922 to work as itinerate researcher (Cunha 1989). The material sent to Rio de Janeiro included not only a good part of the specimens from the Tapajós region, but almost all the Types of the taxa she had described (see Gonzaga 1989a, 1989b).

Samuel Klages, professional collector for the Carnegie Museum, Pittsburgh, USA, visited Vila Braga from Nov 1919 to January 1920 and spent February to March 1920 in Itaituba; he also made a short visit to Ilha de Goyana in December 1919. The ornithological specimens he sent to Pittsburgh were the subject of several publications (Todd 1925, 1927, 1937, 1943; Griscom and Greenway 1941); Griscom and Greenway's article lists the specimens taken by Klages at the three localities near the Park, but a good part of the paper is difficult to interpret, because it reports such things as "Rio Tapajoz, various localities" or "a long series from the Rio Tapajoz"; also, in the case of multiple localities from the basin, these are lumped, making it impossible to distinguish which specimens, in fact, came from Itaituba, Vila Braga, or Ilha de Goyana, when one or more of these localities and others are cited. Given these problems, a new study of this material would be highly desirable.

In late 1920 and early 1921, Ernst Garbe visited the Tapajós region, spending most of his time near Santarém but also some at Itaituba. The fewer than 10 specimens collected at Itaituba were deposited at the Zoological Museum of the University of São Paulo (MZUSP), São Paulo, Brazil.

In 1955, Brazilian professional collector José Hidasi visited Itaituba, where he collected 49 birds purchased by the Museu Paraense Emílio Goeldi. Other specimens were in all likelihood sold to other institutions, but we were not able to locate their whereabouts in time for this paper.

Edwin O. Willis traveled in the region west of the Tapajós near and in the park twice to study principally ant-following birds, visiting Maloquinha, Pará (4°18'S, 56°04'W) from 20 February to 1 March 1966 and along the Transamazonian Highway in September 1974 (Willis 1969; Oren and Willis 1981).

As part of its field epidemiological work to help prevent diseases of importance to human populations in 1972 and 1973, during the construction of the Transamazonian Highway, the Evandro Chagas Institute, Belém, Brazil, collected hundreds of birds to check for evidence of human arboviruses. Evandro Chagas' technician Geraldo P. Silva was responsible for the field work, carried out at a site east of the Park (km 19 of the Transamazonian Highway, measured as distance from Itaituba) and at another site some 40 km west of the Park (a locality called Flexal, km 212). All the specimens were bled for arbovirus testing and preserved in formalin.

Some of the most important specimens were made into skins after arrival in Belém. The Evandro Chagas Institute and Silva later donated the entire collection of anatomical specimens and skins to the Museu Paraense Emílio Goeldi.

From August to December 1978, D. Oren visited the Park as part of the then Instituto Brasileiro de Desenvolvimento Florestal's efforts to develop a management plan for the conservation unit. He made observations and a small reference collection, primarily of *terra firme* forest birds within the Park along a trail at km 120 of the Transamazonian Highway, but also of seasonally flooded sand vegetation on the east bank of the Rio Tapajós. All specimens were deposited at the Museu Paraense Emílio Goeldi.

In August 1979, Oren returned to the Park with Helmut Sick and Julie Zickefoose. Emphasis during this excursion was on observations, but a small collection, mostly of relatively well-developed second growth (*capoeira alta*) birds, was also made.

T. Parker and T. S. Schulenberg visited the park for seven days in November 1985 (the 8th to 14th). They spent most of their time in primary forest on high ground and river edge forest along the west bank of the Tapajós, concentrating their efforts in four areas: (1) along the Transamazonian Highway between the guard stations at Tracoá and Uruá (mostly regenerating forest and second-growth along the road edge); (2) forest on the slopes above the Tapajós at Uruá and a forest trail that leads from Uruá to the south along and high above the river ("Uruá Trail"); (3) river-edge forest at Buburé; (4) primary forest along the Capelinha ("Little Chapel") Trail southwest of Buburé; this forest path traverses more than 30 km of rainforest before ending at a religious shrine periodically visited by local people. They spent 16 hours along this trail over a period of three days.

IMPORTANCE OF THE PARK FOR AMAZONIAN CONSERVATION

This national park is extremely important for conservation in the Amazon, and one of the only parks in Brazil with the necessary infrastructure to handle visitors. It is highly threatened by gold miners, squatters, and (perhaps most dangerously) by hydroelectric development of the Tapajós basin. Clearly this park is a national treasure that deserves more attention from biologists and tourists alike. It lies within a region of great biogeographic importance: the uplands between the Rio Madeira and the Rio Tapajós. Several bird species endemic to this faunal area occur there: White-crested Guan (*Penelope pileata*), Red-necked Aracari (*Pteroglossus bitorquatus*), Hoffmanns' Woodcreeper (*Dendrocolaptes hoffmannsi*), Harlequin Antbird (*Rhegmatorhina berlepschi*), Pale-faced Bare-eye (*Skutchia borbae*) and Snow-capped Manakin (*Pipra nattererii*). The survival of these and countless other species depends on the continued protection of this national park.

SPECIES LIST

The following list for the park and vicinity cites a total of 448 species. The list of species reported specifically for the Tapajós National Park is 387, even though only a small portion of it has been visited (a few hundred hectares, at most, and along approximately 100 km of the Transamazonian Highway), and for relatively short periods; this makes it one of the richest sites known anywhere. The 19 species of Dendrocolaptidae (all confirmed for the park) represent the richest assemblage of woodcreepers known (cf. Haffer 1990). The 60 tyrannids, 51 antbirds and 18 furnariids are further evidence of the region's exceptional diversity. In this context it is important to note that the list for these last three groups is definitely incomplete.

Some data for the Snethlage collections have already been published (Snethlage 1914), the outstanding exception being the 1917 collection, which was the largest, amounting to over 700 specimens. Since many of these were loaned or were destroyed by inadequate curatorial conditions in the past, the information presented for her material comes largely from the Museu Paraense Emílio Goeldi (MPEG) accession catalogues. The other MPEG collections for the region (Hidasi, Silva and Oren) were individually verified in the serial collections by the first author.

Data from the Carnegie Museum (CM) come from Griscom and Greenway (1941); individual specimens are cited only if their precedence is unambiguous.

The few specimens from the region at the Museu de Zoologia da Universidade de São Paulo (MZUSP) are from Pinto's (1938, 1944) two catalogues of those holdings.

The comments in the species accounts below are principally from Parker's unpublished manuscript, based on his and Schulenberg's observations, and include information on behavior, vocalizations, taxonomy and other aspects of the biology of not only the birds of the Tapajós

National Park, but also of related taxa in the western Amazon. Comments with dates of 1978 and 1979, and those regarding specimens, are of the first author.

Species names preceded with an asterisk indicate those which have been registered within the boundaries of the park. The number following each family name indicates the number of species registered for the park followed by the total known for the park and vicinity. For example, THAMNOPHILIDAE (40/44 species) means that 44 species are known from the park and vicinity, while 40 have been confirmed for the park itself.

Locality designations (principally for Parker's and Schulenberg's visit):

TS = area around Tracoá guard station.
TAH = forest and second-growth bordering the Transamazonian Highway between Tracoá and the entrance to the Capelinha Trail. For specimens, TAH is the abbreviation for the Transamazonian Highway, and these locality designations include the specific km where the trails used for collecting in *terra firme* forest begin.
UT = forest along the "Uruá Trail" behind Uruá.
BA = area around the settlement of Buburé, on the west bank of the Tapajós, primarily a small stretch of river-edge forest upstream from buildings.
CT = the "Capelinha Trail"; a 6-km section of this 30(+)-km forest trail.

Localities for the specimens are standardized here to a single spelling. For example, we have used "Ilha de Goyana" (the spelling used by Snethlage) for the island in the Tapajós, though it is called Goyana Island by Griscom and Greenway (1941), and modern Portuguese orthographic rules require "ilha de Goiana."

TINAMIDAE (8/8 SPECIES)

*Tinamus tao (Gray Tinamou). 1 imm. ♀, MPEG 14845, Itaituba, 8 November 1955, col. J. Hidasi; forest; iris: brown; stomach contents: seeds. Two heard in disturbed forest along the TAH at the beginning of Ramal Terra Preta road. Not recorded in mature forest along the CT.

*Tinamus major (Great Tinamou). 1 ♀, CM, Vila Braga (Griscom and Greenway 1941:98). Several heard just before dusk along the TAH midway between Tracoá and Uruá; in forest bordering a swamp and (probably) the river. Also heard at dawn along the TAH near the beginning of the CT.

*Tinamus guttatus (White-throated Tinamou). 2 ♀♀, MPEG 13764–65, Vila Braga, 4/17 July 1917, col. E. Snethlage and F. Lima. The common forest *Tinamus*; heard in all areas, especially along the CT, where at least three were noted daily. As in other Amazonian forests, this species and *Crypturellus variegatus* characterize mature *terra firme*. Also heard from along the TAH, including the beginning of the Ramal Terra Preta road where *T. major* and *T. tao* were also heard.

*Crypturellus cinereus (Cinereous Tinamou). 1 ♀, MPEG 13767, Vila Braga, 6 July 1917, col. E. Snethlage and F. Lima. Apparently common in second-growth woodland and forest; heard early and late at several points along the TAH, including around TS. Also heard in mature forest along the CT.

*Crypturellus soui (Little Tinamou). 1 ♂, MPEG 13766, Vila Braga, 24 July 1917, col. E. Snethlage and F. Lima. Common in regenerating forest along the TAH (up to 10 heard daily), and also in forest along the CT (probably areas disturbed long ago). Up to four singing at once along the TAH at the beginning of the CT at dawn.

*Crypturellus undulatus (Undulated Tinamou). 1 ♀, MPEG 5226, Ilha de Goyana, 21 December 1906, col. E. Snethlage, 1 ♀, CM, Ilha de Goyana (Griscom and Greenway 1941:100). Several heard late in the afternoon at Uruá in stunted forest on sandy soil. Also a few heard in disturbed forest along the TAH between Tracoá and Uruá (again, probably in scrubby woodland on sandy soil—but heard at a distance).

*Crypturellus variegatus (Variegated Tinamou). 1 ♀, 1 ♂, MPEG 13768–69, Vila Braga, 8 June/20 July 1917, col. E. Snethlage and F. Lima. Heard at least two on both mornings in mature forest along the CT. Also heard once late in the afternoon from the TAH west of Tracoá.

*Crypturellus strigulosus Brazilian Tinamou. Heard in high forest at km 120 of TAH; voice identified by H. Sick in August 1979.

PODICIPEDIDAE (1/1 SPECIES)

*Podiceps dominicus (Least Grebe). Up to 10 individuals were seen on small ponds NE of Tracoá just outside of the park in 1985. Similarly common on the numerous ponds within the park in 1978 and 1979.

PHALACROCORACIDAE (1/1 SPECIES)

Phalacrocorax brasilianus (Neotropic Cormorant). As many as 60 noted at once on rocks in the river at Uruá; also common (up to 30) along the river at Buburé.

ANHINGIDAE (1/1 SPECIES)

Anhinga anhinga (Anhinga). One or two noted at three roadside ponds surrounded by forest along the TAH, and several more seen east of the park.

PELECANIDAE (0/1 SPECIES)

Pelecanus occidentalis (Brown Pelican). Itaituba (Snethlage, 1914). This is one of only two Brazilian records for the species, represented in the country's fauna as a rare vagrant.

ARDEIDAE (8/8 SPECIES)

Ardea cocoi (Cocoi Heron). Seen occasionally on the Tapajós and in ponds along the TAH, always singly in 1978 and 1979.
Ardea alba (Great Egret). Observed in grassy swamps at the edge of the Tapajós, usually singly but on one occasion five sighted together in 1978.
Egretta thula (Snowy Egret). Seen in small numbers along the Tapajós, typically near rapids, and in roadside ponds in 1978 and 1979.
Butorides striatus (Striated Heron). Singles seen twice along the river edge at Buburé.
Agamia agami (Agami Heron). 1 ♂, MPEG 5225, Itaituba, 14 January 1907, col. E. Snethlage. One seen inside high forest along a small stream in 1979.
Pilherodius pileatus (Capped Heron). Apparently common, at ponds and pools along streams along TAH; seen at TS. Up to six daily; usually alone, but once three seen in close proximity.
Tigrisoma lineatum (Rufescent Tiger Heron). One or two seen/heard at least three different marshes along TAH; two heard (protracted mooing calls) in swampy woods bordering stream at the beginning of the CT.
Cochlearius cochlearius (Boat-billed Heron). Seen at ponds along the TAH on several nights in 1978.

CICONIIDAE (1/1 SPECIES)

Jabiru mycteria (Jabiru Stork). A single individual sighted soaring over the TAH in August 1979.

THRESKIORNITHIDAE (1/1 SPECIES)

Mesembrinibis cayennensis (Green Ibis). 1 ♀, MPEG 5224, Ilha de Goyana, 20 December 1906, col. E. Snethlage.

ANHIMIDAE (1/1 species)

Ahima cornuta (Horned Screamer). A pair sighted in a grassy pond at the edge of the Tapajós in September 1978.

ANATIDAE (2/2 species)

Cairina moschata (Muscovy Duck). Two (females?) observed on a pond in the cleared land just NE of the park along the road to Itaituba. Probably common in the park as well.
Dendrocygna autumnalis (Black-bellied Whistling Duck). Five sighted once in a roadside pond along the TAH.

CATHARTIDAE (5/5 SPECIES)

Sarcoramphus papa (King Vulture). One adult seen circling low over forest along the TAH near Buburé.
Coragyps atratus (Black Vulture). Very scarce within the park (to three seen in a day around TS and along the TAH). Quite common (300+) over cleared lands to the NE along the TAH.
Cathartes aura (Turkey Vulture). Frequently sighted east of the park in cleared areas and several times along second-growth areas of the TAH within the park.
Cathartes melambrotos (Greater Yellow-headed Vulture). Common over forest within the park and cleared lands to the NE.

Cathartes burrovianus (Lesser Yellow-headed Vulture). Frequently sighted in small numbers (two to five) perched at the edge of ponds and swamps along the TAH in 1979.

Accipitridae (16/17 Species)

Gampsonyx swainsonii (Pearl Kite). Three individuals seen atop dead trees in widely separated places along the TAH east of the park.

Elanoides forficatus (Swallow-tailed Kite). Large flocks (20+) regularly seen perched in isolated trees in the late afternoon along the TAH and soaring during the day by Oren. The species' absence during Parker's visit suggests seasonal movements.

Leptodon cayanensis (Gray-headed Kite). Two (a pair?) observed performing a flight display over forest at Uruá, and another seen circling over a large cleared area just east of the park.

Chondrohierax uncinatus (Hook-billed Kite). Sighted once in swampy woods near a stream dammed by the TAH in 1979.

Harpagus bidentatus (Double-toothed Kite). 1 imm. ♂, MPEG 5216, Itaituba, 15 January 1907, col. E. Snethlage. One glimpsed in forest middlestory along the UT: the bird was probably following a group of capuchin monkeys.

Ictinea plumbea (Plumbeous Kite). One of the commonest raptors, seen regularly both soaring and perched during Oren's visits, but absent during Parker's, suggesting seasonal movements.

Rostrhamus sociabilis (Snail Kite). An immature seen in a small marsh east of the park.

Buteo magnirostris (Roadside Hawk). As in most other places in the Neotropics, this was the only common and frequently seen bird of prey. Daily counts ranged from 6 to 20, these being single birds perched at the forest edge along the TAH. None heard or seen inside mature forest.

Buteo brachyurus (Short-tailed Hawk). One noted over forest at Tracoá is the only record (Schulenberg).

Asturina nitida (Gray Hawk). A pair of these hawks flew from dead tree to tree in the cleared land east of the park in 1985. Seen regularly in pairs or alone near the Uruá station in 1978 and 1979.

Leucopternis albicollis (White Hawk). 1 ♀, MZUSP, Itaituba (Pinto 1938:75). One seen perched on a snag along the forest edge just west of Buburé, and another heard (as it circled over the forest?) along the CT.

Leucopternis kuhli (White-browed Hawk). 1 ♀, CM, Vila Braga (Griscom and Greenway 1941:113); 1 ♂, MPEG 47658, km 19 TAH, 9 August 1972, col. G. P. Silva.

Busarellus nigricollis (Black-collared Hawk). An adult circling low over a marsh east of the park is the only record from 1985. Much more common along the TAH and in marshes in 1978 and 1979.

Buteogallus urubitinga (Great Black Hawk). One adult seen on a rock in the middle of the Tapajós above Uruá.

Harpia harpyja (Harpy Eagle). Seen on all visits to the park, typically perched high (30 m) at the edge of the TAH in emergent trees; the Harpy Eagles in the park were remarkably fearless of human presence.

Spizaetus tyrannus (Black Hawk-Eagle). 1 ♂, CM, Vila Braga (Griscom and Greenway 1941: 115). One heard calling (performing a flight display) over forest at Uruá.

Pandion haliaetus (Osprey). Single individuals seen on three separate occasions on the Rio Tapajós in November and December 1978.

Falconidae (8/8 Species)

Herpetotheres cachinnans (Laughing Falcon). At least two different adults seen along the TAH, and one heard at dawn on both mornings at the beginning of the CT.

Micrastur semitorquatus (Collared Forest-Falcon). Vocalizations identified by H. Sick in forest next to abandoned banana plantations along the TAH in 1979.

Micrastur mirandollei (Slaty-backed Forest-Falcon). 1 ♀, CM, Vila Braga (Griscom and Greenway 1941:116). Heard on several occasions in high second growth and *terra firme* forest.

Micrastur gilvicollis (Lined Forest-Falcon). 2 ♀♀, CM, Vila Braga (Griscom and Greenway 1941:116); 1 sex?, MPEG-A 508, Flexal = km 212 TAH, 29 October-6 December 1973, col. G. P. Silva; 1 sex?, MPEG 34064, km 120 TAH, 18 November 1978, col. D. C. Oren, *cipoal* (vine forest); 205 g; total length 315 mm.

Daptrius ater (Black Caracara). Surprisingly scarce; noted only once: three together over river-edge forest at Buburé.

Daptrius americanus (Red-throated Caracara). One (alone?) seen in a dead tree west of Buburé, and groups heard in the distance from the CT and TS.

Milvago chimachima (Yellow-headed Caracara). Uncommon along the Tapajós, where seen singly flying along the shore in 1978.

Falco rufigularis (Bat Falcon). 1 ♂, CM, Vila Braga (Griscom and Greenway 1941:116); 1 ♀, MPEG 47663, Flexal = km 212 TAH, November 1973, col. G. P. Silva. Single birds seen twice along the TAH west of TS, and once near the station. All perched high atop dead trees along the road edge.

CRACIDAE (5/5 SPECIES)

Ortalis guttata (Speckled Chachalaca). 1 ♂ CM, Vila Braga (Griscom and Greenway 1941: 121); 1 ♂ CM, Itaituba (Griscom and Greenway 1941:121). Apparently common, although infrequently seen. Several small groups encountered along the TAH, and pairs/groups heard almost daily near Tracoá and Uruá.

Penelope superciliaris (Rusty-margined Guan). 1 ♂, 1 ♀, CM, Vila Braga (Griscom and Greenway 1941:120); 1 ♂, MPEG 6557, Vila Braga, 28 October 1908, col. E. Snethlage. This small guan noted only once; three (including a recently-fledged young bird) seen along the CT. Wing drumming heard at dawn in this and other localities may have been produced by either this or the next species.

Penelope pileata (White-crested Guan). 1 ♀, MPEG 13763, Vila Braga, 10 July 1917, col. E. Snethlage and F. Lima. Seen only twice: two together in forest canopy at Uruá, and two in fruiting trees at the forest edge along the TAH. Several others heard in the latter area.

Pipile nattereri (Red-throated Piping-Guan). According to park guards, this species regularly seen in river-edge forest within the park. We flushed what was probably a piping-guan from a treetop at Buburé. Pairs seen twice in August 1979.

Mitu tuberosa (Razor-billed Curassow). 1 ♂, CM (Griscom and Greenway 1941:116). Two seen and one or two more heard along the CT; another heard at dawn along the TAH east of Uruá. The presence of this large gamebird in such an accessible part of the park indicates that poaching is infrequent, if it occurs at all.

ODONTOPHORIDAE (1/1 SPECIES)

Odontophorus gujanensis (Marbled Wood-Quail). 1 ♀, MPEG 5220, Vila Braga, 10 January 1907, col. E. Snethlage. Two pairs counter-singing (loud duets) one morning well before dawn (03:30) at Tracoá, and two more heard at dawn near the beginning of the CT; others heard at/near dawn from the TAH.

PSOPHIIDAE (1/1 SPECIES)

Psophia viridis (Dark-winged Trumpeter). 2 ♂♂, 1 ♀, MPEG 13749–51, Vila Braga, 16/29 June, 19/14 July 1917, col. E. Snethlage and F. Lima; 1 ♂, 1 ♀, CM, Vila Braga (Griscom and Greenway 1941:123). At least three flushed (one seen briefly as it ran along the trail) near a swarm of army ants along the CT.

RALLIDAE (2/2 SPECIES)

Laterallus melanophaius (Rufous-sided Crake). 1 ♀, MPEG 13752, Vila Braga, 1 July 1917, col. E. Snethlage and F. Lima. Reported as captured in mammal traps in 1978.

Aramides cajanea (Gray-necked Wood-Rail). 1 ♀, CM, Ilha do Goyana (Griscom and Greenway 1941:1). Heard evenings along the Tapajós near Uruá and at the pond near the Chapel of the Capelinha Trail.

HELIORNITHIDAE (1/1 SPECIES)

Heliornis fulica (Sungrebe). One seen on a pond just east of Buburé.

EURYPYGIDAE (1/1 SPECIES)

Eurypyga helias (Sunbittern). Two (a pair?) singing at dawn along the stream at the beginning of the CT.

JACANIDAE (1/1 SPECIES)

Jacana jacana (Wattled Jacana). Several noted in a marsh just east of Uruá, and a few others seen in marshes east of the park.

CHARADRIIDAE (3/3 SPECIES)

Hoploxypterus cayanus (Pied Lapwing). 1 ♀, MPEG 5221, Ilha de Goyana, 23 December 1906, col. E. Snethlage. One observed on a small sandbar in the river at Uruá. Nesting on sandbars in the river in August 1979 (recently hatched chicks).

Pluvialis dominica (American Golden-Plover). A group of 6-7 seen on 12 November 1978 in low grasses where bulldozers had cleared at Uruá.

Charadrius collaris (Collared Plover). 1 ♂, MPEG 5222, Ilha de Goyana, 20 December 1906, col. E. Snethlage.

SCOLOPACIDAE (2/2 SPECIES)

Tringa solitaria (Solitary Sandpiper). 1 ♀, MPEG 5223, Ilha de Goyana, 23 December 1906, col. E. Snethlage.

Actitis macularia (Spotted Sandpiper). Several sighted on rocks, sandbars, and stranded wood in the Tapajós from October to December 1978.

LARIDAE (2/2 SPECIES)

Phaetusa simplex (Large-billed Tern). One or two seen over the river on each visit to Uruá and Buburé.

Sterna superciliaris (Yellow-billed Tern). Rather common (30+ seen) along the Tapajós; breeding on sandbars in the river in August 1979.

RYNCHOPIDAE (1/1 SPECIES)

Rynchops nigra (Black Skimmer). 1 sex?, CM, Vila Braga (Griscom and Greenway, 1941: 134).

COLUMBIDAE (9/10 SPECIES)

Columba speciosa (Scaled Pigeon). 2 ♂♂, MPEG 5217/6552, Ilha de Goyana, 11 December 1906/21 October 1908, col. E. Snethlage; 2 ♂♂, MPEG 13753–54, Vila Braga, 10 July 1917 (both), col. E. Snethlage and F. Lima; 1 ♂, CM, Vila Braga, (Griscom and Greenway 1941:134). Two flying high over TS late one afternoon, the only ones observed.

Columba cayennensis (Pale-vented Pigeon). 1 ♂, MPEG 5218, Ilha de Goyana, 6 January 1907, col. E. Snethlage; 1 ♂, MPEG 13755, Vila Braga, 21 July 1917, col. E. Snethlage and F. Lima. Up to three pairs noted in a day, all perched in trees at the edges of marshes along the TAH, or flying nearby. One to several individuals also seen in river-edge trees and bushes at Buburé and Uruá.

Columba subvinacea (Ruddy Pigeon). 1 ♂, MPEG 13756, Vila Braga, 3 July 1917, col. E. Snethlage and F. Lima; 1 ♂, 1 ♀, CM, Vila Braga (Griscom and Greenway 1941:135). Common and vocal in forest canopy in all areas visited; up to a dozen heard daily, but only a few seen.

Columba plumbea (Plumbeous Pigeon). 1 ♀, MPEG 13757, Vila Braga, 11 June 1917, col. E. Snethlage and F. Lima; 2 ♂♂, 2 ♀♀, CM, Vila Braga (Griscom and Greenway 1941:135).

Columbina passerina (Common Ground-Dove). Common along the road edge and in clearings at Tracoá, Uruá and Buburé; usually in small groups of 4–8, up to 40 seen in a day; mostly flushed from grassy or weedy places not far from taller cover. Also common in the recently cleared land east of the park, but outnumbered by Ruddy Ground Doves (*C. talpacoti*) near Itaituba. The latter seemingly more common around the town and airport.

Columbina talpacoti (Ruddy Ground-Dove). 1 ♀, MPEG 5219, Ilha de Goyana, 9 January 1907, col. E. Snethlage; 1 ♀, MPEG 13758, Vila Braga, 5 June 1917, col. E. Snethlage and F. Lima. Very scarce in the park, where a few seen along the TAH; common in open land to the east.

Claravis pretiosa (Blue Ground-Dove). 1 ♂, CM, Vila Braga (Griscom and Greenway 1941: 137). One male glimpsed as it flew low across the TAH between Tracoá and Uruá.

Leptotila rufaxilla (Gray-fronted Dove). 1 ♂, MPEG 6554, Ilha de Goyana, 20 October 1908, col. E. Snethlage. A few heard in all areas within the park, mostly at the forest edge, but also in disturbed areas (the open palm forest along the CT) within mature forest; up to six heard daily along the TAH, and several flushed at the beginning of trails and just inside the forest along the road.

Leptotila verreauxi (White-tipped Dove). 1 ♂, MPEG 13759, Vila Braga, 15 June 1917, col. E. Snethlage and F. Lima. Several heard and one seen in low woodland at the airport in Itaituba. Seen regularly in open vegetation near Uruá in 1978 and 1979.

Geotrygon montana (Ruddy Quail-Dove). 1 ♀, MPEG 13760, Vila Braga, 18 July 1917, col. E. Snethlage and F. Lima, 1 ♂, MPEG 15297, Itaituba, 17 November 1955, col. J. Hidasi. This species flushed daily from the ground in mature forest along the CT, and another seen briefly in forest at the beginning of the Ramal Terra Preta. Also heard at dawn along the CT. Probably common, but easily overlooked.

PSITTACIDAE (18/19 SPECIES)

Ara ararauna (Blue-and-yellow Macaw). One pair probably heard over forest along the CT. Recorded in small numbers (usually in pairs) in 1978 and 1979, but not observed in 1985.

Ara macao (Scarlet Macaw). A pair seen just west of Buburé.

Ara chloroptera (Red-and-green Macaw). Seen daily but in small numbers (two to four pairs); observed in all areas, mainly early and late in the day over the TAH, but also seen in forest canopy along the CT.

Ara manilata (Red-bellied Macaw). Up to 40 roosted in tall, dead trees in a wooded swamp along the TAH between Tracoá and Uruá; these birds arrived late in the afternoon from the direction of the river. None seen in other areas.

Aratinga guarouba (Golden Parakeet). Flocks of 6–10 (6, 6, 8, 8, 10) seen almost daily at several points along the TAH between Tracoá and Buburé. The park's Golden Parakeet population has already been reported by Oren and Willis (1981) and recently has been found as far west as the Rio Jamari in Rondônia (Yamashita and França 1991).

Aratinga leucophthalmus (White-eyed Parakeet). This the commonest psittacid observed during our stay in the park. Large flocks of up to 80 seen several times along the TAH between Tracoá and Buburé, and up to 200 (a roost?) seen at once in trees along the airstrip at Uruá. More normal flock size 12–20, with nearly all birds within any aggregation being paired. The species not heard in/over mature forest along the CT, thus, as in other areas, it seems to prefer disturbed forest and edges.

Pyrrhura picta (Painted Parakeet). 2 ♂♂, MPEG 6537–38, Vila Braga, 6 November 1908 (both), col. E. Snethlage; 1 ♂, 1 ♀, MPEG 13736–37, Vila Braga, 5 June 1917, col. E. Snethlage and F. Lima. Like other members of its genus, this canopy-dwelling species regularly heard but infrequently seen. Uncommon, being noted primarily in the taller forest at Uruá (a pair, 6+, 6+) and Buburé (6); surprisingly, a flock of five observed in isolated trees of a clearing east of the park. None heard or seen in good forest along the CT.

Forpus sp. (*sclateri ?*) (Dusky-billed Parrotlet ?). Several small parrots heard at a great distance over the TAH probably this species. Snethlage collected seven individuals of *F. sclateri* along the Rio Jamauchim (a right-bank tributary of the Tapajós), some 20 km east of the park.

Brotogeris versicolurus (Canary-winged Parakeet). A few heard in the cleared lands just east of the park, then common around Itaituba, particularly at the airport, where flocks of 10, 10+, 12, 20 and 40 seen flying high overhead. Found inside the park boundaries flying over secondary forest in August 1979.

Brotogeris chrysopterus (Golden-winged Parakeet). 1 ♂, 1 ♀, MPEG 13740–41, Vila Braga and Ilha de Goyana (resp.), 30 June 19/27 July 1917, col. E. Snethlage and F. Lima. Common; recorded in all areas; usually in small, tight flocks of 6–10 (range = 4–20+) flying high over forest and clearings. The flock of 20+ observed in the crowns of melastome trees along the airstrip at Uruá; they appeared to be feeding on flowers and small (mostly green?) fruits. Also heard in the canopy of mature forest along the CT.

Touit purpurata (Sapphire-rumped Parrotlet). 1 ♀, MPEG 13747, Vila Braga, 17 July 1917, col. E. Snethlage and F. Lima. Seen regularly in small flocks (six to 15) in high forest in 1979.

Pionopsitta vulturina (Vulturine Parrot). 1 ♀, MPEG 6551, Vila Braga, 7 November 1908, col. E. Snethlage; 1 ♀, CM, Vila Braga (Griscom and Greenway 1941:148); 1 ♂, MPEG 14809, Itaituba, 3 November 1955, col. J. Hidasi. Uncommon and difficult to observe; seen or heard flying rapidly just above the treetops as follows: 1, 1, 1, 2, 2–3. Remarkably like *P. barrabandi* in color (head orange and black, not yellow as shown in Forshaw) and voice. Half of the above birds heard along the CT in mature forest.

Pionus menstruus (Blue-headed Parrot). 1 ♀, MPEG 13743, Vila Braga, 2 July 1917, col. E. Snethlage and F. Lima. The second-commonest parrot observed during our visit, being outnumbered only by the White-eyed Parakeet. Up to 80 of these parrots seen in a day, with group size ranging from four to 20; several pairs also noted each day. Seen or heard in all areas, though like the *Aratinga*, this species seemed to prefer the forest borders.

Pionus fuscus (Dusky Parrot). 2 ♂♂, 1 ♀, MPEG 13744-46, Vila Braga, 18 June 1917 (all), col. E. Snethlage and F. Lima.

Amazona ochrocephala (Yellow-headed Parrot). Four seen in flight (and tape-recorded) over forest at Tracoá the only ones observed.

Amazona amazonica (Orange-winged Parrot). 1 imm. ♀, CM, Vila Braga (Griscom and Greenway 1941:145). At least four (two pairs) roosted in the wooded swamp between Tracoá and Uruá (see under Red-bellied Macaw). None heard or seen in other areas.

Amazona kawalli (White-cheeked Parrot). 1 ♀, MPEG 14804, Itaituba, 7 November 1955, col. J. Hidasi. This species only recently described (Grantsau and Camargo 1989). It is probably sympatric with *A. farinosa* in most of its range. The above specimen was incorrectly reported as being from "Santarém" in the original description.

Amazona farinosa (Mealy Parrot). This species noted daily in small numbers (up to 6 pairs), usually one or two pairs seen at once, these flying low over mature forest (over the TAH or along the CT).

Deroptyus accipitrinus (Hawk-headed Parrot). 1 ♀, MPEG 14805, Itaituba, 7 November 1955, col. J. Hidasi. Apparently scarce; two (a pair?) seen just west of Tracoá in scrubby forest on sandy soil. These perched atop adjacent trees and uttered their characteristically peculiar calls, which included several notes well-described in Forshaw (1973): "ee-yah" and "chak-chak." One of the birds flew across the road at treetop level, about 200 feet from us. It looked quite *Accipiter*-like, employing series of four to five shallow flaps followed by glides of 7 to 10 m.

Cuculidae (5/7 Species)

Coccyzus melacoryphus (Dark-billed Cuckoo). 1 ♀, MPEG 13713, Ilha de Goyana, 30 June 1917, col. E. Snethlage and F. Lima.

Piaya cayana (Squirrel Cuckoo). Fairly common in forest and forest edge; up to 6 seen daily, and along the CT in mature forest (middlestory and canopy). One or two noted at the periphery of canopy flocks along the UT and the CT. Otherwise seen alone.

Piaya melanogaster (Black-bellied Cuckoo). 3 ♀, CM, Vila Braga (Griscom and Greenway 1941:152). Seen regularly as singles in high forest in 1978 and 1979.

Crotophaga major (Greater Ani). Seen in small flocks at ponds along the TAH and in vegetation at the edge of the Tapajós in 1978 and 1979.

Crotophaga ani (Smooth-billed Ani). Fairly common in second-growth along the TAH, with daily counts of up to 40, but mostly 6–12 (several small groups); more common in cleared land east of park.

Tapera naevia (Striped Cuckoo). 1 ♂, CM, Itaituba (Griscom and Greenway 1941:153). One heard in second-growth at the airport in Itaituba; probably occurs in clearings within the park.

Neomorphus geoffroyi squamiger (Rufous-vented Ground-Cuckoo). 1 ♀, 1 imm. ♀, MPEG 13714-15, Vila Braga, 16 June 19/6 July 1917, col. E. Snethlage and F. Lima.

Opisthocomidae (1/1 Species)

Opisthocomus hoazin (Hoatzin). A few seen in a marsh just east of Buburé, and one seen in a wooded swamp between Tracoá and Uruá.

Strigidae (6/7 Species)

Otus choliba (Tropical Screech-Owl). Voice identified by H. Sick almost nightly at Uruá and in second growth forest along the Tapajós.

Otus watsonii (Tawny-bellied Screech-Owl). Two heard singing just before dawn in two areas, at the beginning of the Ramal Terra Preta and along the TAH near the beginning of the CT; the birds at mid-levels in mature forest. Songs consisted of a series of closely-spaced, soft hoots delivered on one pitch; more rapid series given in response to playback.

Lophostrix cristata (Crested Owl). 1 imm. ♂, MPEG 13748, Vila Braga, 1 July 1917, col. E. Snethlage and F. Lima.

Pulsatrix perspicillata (Spectacled Owl). One heard just before dawn at Tracoá.

Glaucidium hardyi (Amazonian Pygmy-Owl). Two heard just before dawn in forest at the beginning of the CT, and another heard along the CT late in the morning, our only records. These birds gave slightly descending series of short, soft whistles (in pattern and quality somewhat like the songs of Eastern Screech Owls (*Otus asio*)).

Glaucidium brasilianum (Ferruginous Pygmy-Owl). Heard and then sighted twice during the day at km 120 in 1978.

Ciccaba sp. (*virgata* or *huhula*) (Mottled Owl or Black-banded Owl). One heard before dawn near the beginning of the CT. The emphatic hoots of one of these two species.

NYCTIBIIDAE (1/1 SPECIES)

Nyctibius griseus (Common Potoo). One heard along the road-edge at the beginning of the CT.

CAPRIMULGIDAE (8/9 SPECIES)

Lurocalis semitorquatus (Semicollared Nighthawk). Common near ponds along the TAH and along the Tapajós in 1978.
Chordeiles rupestris (Sand-colored Nighthawk). 3 ♂♂, 2 ♀♀, MPEG 6490–94, Rio Tapajós, 15 November 1908 (all), col. E. Snethlage. Nesting on islands in the Tapajós in August 1979 (recently hatched chicks).
Chordeiles acutipennis (Lesser Nighthawk). Identified by H. Sick at Uruá in August 1979.
Nyctiprogne leucopyga (Band-tailed Nighthawk). Heard regularly along the Tapajós.
Podager nacunda (Nacunda Nighthawk). Sighted nightly in small numbers at Uruá in October 1978.
Nyctidromus albicollis (Pauraque). 2 ♂♂, MPEG 13732–33, Vila Braga, 14 June 19/4 July 1917, col. E. Snethlage and F. Lima; 1 ♂, MPEG 47675, km 19 TAH, 27 July 1972, col. G. P. Silva. Regularly seen in open habitats and along the TAH 1978 and 1979.
Caprimulgus rufus (Rufous Nightjar). The commonest caprimulgid in open habitats at Uruá in 1978 and 1979.
Caprimulgus nigrescens (Blackish Nightjar). 1 ♂, MPEG 6498, Vila Braga, 23 October 1908, col. E. Snethlage; 1 ♀, CM, Vila Braga (Griscom and Greenway 1941:164). Seen just before dawn in several places along the TAH between Tracoá and Uruá (mainly forest on sandy soil); and on 9 November one flushed from two eggs on the ground under bushes at the road-edge several kilometers west of Tracoá; the eggs buff-colored with blackish and brownish scrawls over their entire surface.
Hydropsalis brasiliana (Scissor-tailed Nightjar). 1 ♂, CM, Itaituba (Griscom and Greenway 1941:163). 1 ♀, MPEG 14975, Itaituba, 25 October 1955, col. J. Hidasi.

APODIDAE (3/4 SPECIES)

Chaetura chapmani (Chapman's Swift). Several swifts seen by Schulenberg at Tracoá were probably Chapman's Swifts.
Chaetura spinicauda (Band-rumped Swift). This relatively distinctive species common at Tracoá (up to 30 seen daily around the station) and along the TAH.
Chaetura brachyura (Short-tailed Swift). A flock of 10+ observed at Buburé; the distinctive shape of this species (wide-looking wings, very short tails) makes field identification easy.
Reinarda squamata (Fork-tailed Palm-Swift). One seen flying low over a recently-burned clearing east of the park in 1985. Regularly seen in small flocks in 1978 and 1979.

TROCHILIDAE (10/14 SPECIES)

Threnetes leucurus (Pale-tailed Barbthroat). One seen in the overgrown banana plantation along the UT.
Phaethornis superciliosus (Long-tailed Hermit). 5 ♂♂, 1 ♀, CM, Vila Braga and Itaituba (Griscom and Greenway 1941:168) (includes the Type of *P. s. insignis* Todd, 1937, CM 77,635, from Itaituba); 1 sex?, MPEG-A 3292, Flexal = km 212 TAH, 29 October–6 December 1973, col. G. P. Silva. The common forest *Phaethornis*; several seen daily along the trails and road-edges.
Phaethornis philippi (Needle-billed Hermit). 1 ♂, MPEG 34082, km 120 TAH, 1 November 1978, col. D. C. Oren. One seen well and several heard in undergrowth along the CT (Parker).
Phaethornis squalidus (Dusky-throated Hermit). 1 ♂, MPEG 6482, Ilha de Goyana, 30 December 1908, col. E. Snethlage.
Phaethornis ruber (Reddish Hermit). 1 ♂, MPEG 5197, Vila Braga, 8 January 1907, col. E. Snethlage; 1 ♀, MPEG 34083, 2 November 1978, col. D. C. Oren. A few noted daily, low at forest edge along the TAH and in undergrowth along the forest trails (UT, CT).
Campylopterus largipennis (Gray-breasted Sabrewing). 1 sex?, MPEG-A 3391, Flexal = km 212 TAH, 29 October-6 December 1973, col. G. P. Silva; 1 ♂, MPEG 34086, 18 November 1978, col. D. C. Oren. One observed in forest along the CT (Schulenberg).
Thalurania furcata (Fork-tailed Woodnymph). 1 ♂, MPEG 5200, Vila Braga, 7 January 1907, col. E. Snethlage; 1 ♂, 1 ♀, CM, Vila Braga (Griscom and Greenway 1941:176); 1 ♀, MPEG 34088, km 120 TAH, 27 August 1979, col. D. C. Oren. Fairly common in forest undergrowth; up to four seen in a morning along the CT. Also seen along forest edges at Tracoá and Uruá.

Heliomaster longirostris (Long-billed Starthroat). As many as four found within a relatively small area around Tracoá.

Hylocharis cyanus (White-chinned Sapphire). 1 ♂, CM, Itaituba (Griscom and Greenway 1941:173). Several probable females glimpsed in second-growth at Tracoá (Parker).

Hylocharis sapphirina (Rufous-throated Sapphire). 1 ♂, 2 ♀♀, CM, Itaituba (Griscom and Greenway 1941:173).

Chlorestes notatus (Blue-chinned Sapphire). 1 ♂, CM, Itaituba (Griscom and Greenway 1941: 174).

Polytmus theresiae (Green-tailed Goldenthroat). 1 ♀, MPEG 5201, Ilha de Goyana, 6 January 1907, col. E. Snethlage; 1 ♂, CM, Vila Braga (Griscom and Greenway 1941:178).

Amazilia fimbriata (Glittering-throated Emerald). 1 ♂, MPEG 6484, Ilha de Goyana, 20 October 1908, col. E. Snethlage. Fairly common in second-growth along the TAH, especially at Tracoá and Uruá. One fed at banana flowers along the UT.

Heliothryx aurita (Black-eared Fairy). 1 ♀(?), MPEG 13730, Vila Braga, 8 June 1917, col. E. Snethlage and F. Lima; 1 ♂, MPEG 15525, Itaituba, 16 November 1955, col. J. Hidasi. One seen several times at mid-heights in forest along the UT.

TROGONIDAE (4/6 SPECIES)

Trogon melanurus (Black-tailed Trogon). 1 ♂, MPEG 13719, Vila Braga, 26 June 1917, col. E. Snethlage and F. Lima; 5 ♂, CM, Vila Braga (Griscom and Greenway 1941:183). A few heard daily along the forest trails (UT, CT).

Trogon viridis (White-tailed Trogon). 2 ♂♂, MPEG 13716–17, Vila Braga, 6\14 June 1917, col. E. Snethlage and F. Lima; 1 ♂, 1 ♀, CM, Vila Braga (Griscom and Greenway 1941:181). This the most frequently recorded *Trogon* (by voice); heard in forest in all areas, especially along the CT (up to three each day).

Trogon collaris (Collared Trogon). Specimen(s), CM, Vila Braga (Griscom and Greenway 1941:180). Females seen in the upper undergrowth in forest along the UT and CT.

Trogon rufus (Black-throated Trogon). 1 ♂, MPEG 6501, Vila Braga, 3 November 1908, col. E. Snethlage. The Type of *T. r. amazonicus* Todd, 1943 (adult ♂, CM 75,244) is from Vila Braga.

Trogon curucui (Blue-crowned Trogon). 1 ♂, MPEG 5204, Ilha de Goyana, 6 January 1907, col. E. Snethlage; 1 ♀, MPEG 13718, Vila Braga, 18 June 1917, col. E. Snethlage and F. Lima.

Trogon violaceus (Violaceous Trogon). 1 ♀, 1 ♂, MPEG 6502–03, 29 October/7 November 1908, col. E. Snethlage. Two pairs heard high in the canopy of mature forest along the CT.

ALCEDINIDAE (4/4 SPECIES)

Ceryle torquata (Ringed Kingfisher). Scarce; singles observed over the river at Uruá and at a wooded pond east of the park.

Chloroceryle amazona (Amazon Kingfisher). One seen along the river at Buburé.

Chloroceryle americana (Green Kingfisher). Uncommon in ponds along the TAH in 1978 and 1979.

Chloroceryle aenea (American Pygmy Kingfisher). 1 ♂, MPEG 13735, Ilha de Goyana, 15 June 1917, col. E. Snethlage and F. Lima; 1 ♂, MPEG 34103, km 120 TAH, 24 November 1978, col. D. C. Oren.

MOMOTIDAE (2/2 SPECIES)

Electron platyrhynchum (Broad-billed Motmot). 1 ♂, 1 ♀, CM, Vila Braga (Griscom and Greenway 1941:186) (includes the Type of *E. p. orientale* Todd, 1937, CM 75,219). Fairly common in forest along the CT.

Baryphthengus martii (Rufous Motmot). 1 imm. ♀, 1 sex?, MPEG 5202–03, Vila Braga, 3 January 1907 (both), col. E. Snethlage; 1 ♀, MPEG 15136, Itaituba, 5 November 1955, col. J. Hidasi; 1 ♂, MPEG 34104, km 120 TAH, 18 November 1978, col. D. C. Oren. Heard at dawn in several places along the TAH, and also along the CT in forest (to four daily). The songs of Amazonian populations of this species and *Momotus momota* are very similar.

GALBULIDAE (4/5 SPECIES)

Galbula cyanicollis (Blue-cheeked Jacamar). 1 ♀, MPEG 5211, Vila Braga, 10 January 1907, col. E. Snethlage; 1 ♂, MPEG 34105, km 120 TAH, 27 October 1978, col. D. C. Oren; 2 sex?, MPEG-A 353–54, Flexal = km 212 TAH, 29 October-6 December 1973, col. G. P. Silva.

Galbula galbula (Green-tailed Jacamar). A pair seen in August 1979 along the bank of the Rio Tapajós near Uruá. According to Haffer (*in litt.*), Griscom and Greenway's (1941) "Rio Tapajoz" specimens include seven from Vila Braga.

Galbula leucogastra (Bronzy Jacamar). 1 ♂, MPEG 13682, Vila Braga, 3 July 1917, col. E. Snethlage and F. Lima.

Galbula dea (Paradise Jacamar). 1 sex?, MPEG 5210, 15 January 1907, col. E. Snethlage; 1 ♂, MPEG 6519, Vila Braga, 7 November 1908, col. E. Snethlage. Seen in canopy regularly in 1978 and 1979.

Jacamerops aurea (Great Jacamar). 1 ♀, MPEG 5212, Vila Braga, 8 January 1907, col. E. Snethlage. 1 ♀, 1 ♂, MPEG 47684–85, km 19 TAH, 11 July 1972 and 2 August 1972, col. G. P. Silva. One to two heard daily in the canopy of mature forest along the CT.

BUCCONIDAE (9/10)

Notharchus macrorhynchus (White-necked Puffbird). 1 ♂, MPEG 13664, Vila Braga, 19 June 1917, col. E. Snethlage and F. Lima; 1 ♂, 1 ♀, CM, Itaituba (Griscom and Greenway 1941: 189). A pair observed in the open, bare limbs of a forest-edge tree at Tracoá, and one heard in forest canopy along the CT.

Notharchus tectus (Pied Puffbird). 1 ♂, 2 ♀♀, CM, Vila Braga (Griscom and Greenway 1941:190). Apparently fairly common; two pairs and a single bird seen in treetop branches along the TAH west of Tracoá.

Bucco capensis (Collared Puffbird). Specimen(s), CM, Vila Braga (Griscom and Greenway 1941:189). Seen at km 120 TAH in 1978.

Bucco tamatia (Spotted Puffbird). 1 ♂, MPEG 13665, Vila Braga, 21 June 1917, col. E. Snethlage and F. Lima. Seen in mature second growth in August 1979.

Malacoptila rufa (Rufous-necked Puffbird). 1 ♂, 3 ♀♀, MPEG 13666–69, Vila Braga, 4/4/15 June 19/13 July 1917, col. E. Snethlage and F. Lima; 1 ♀, MZUSP, Itaituba (Pinto 1938:314); 1 ♀, MPEG 15017, Itaituba, 11 November 1955, col. J. Hidasi; 3 sex?, MPEG-A 388–90, Flexal = km 212 TAH, 29 October-6 December 1973, col. G. P. Silva; 7 sex?, MPEG-A 392–98, km 19 TAH, 6 Jul/15 Aug. 72, col. G. P. Silva. One captured in mist-nets at km 120 TAH in 1978 and released.

Nonnula rubecula (Rusty-breasted Nunlet). 2 ♂♂, 1 ♀, MPEG 13670–72, Vila Braga, 28 June 19/5/5 July 1917, col. E. Snethlage and F. Lima; 1 ♂, CM, Vila Braga (Griscom and Greenway 1941:193) (this is the Type of *N. r. simplex* Todd, 1937, CM 76, 152); 1 ♀, MPEG 47676, km 19 TAH, 6 July 1972, col. G. P. Silva; 1 imm. ♀, MPEG 34108, km 120 TAH, 25 November 1978, col. D. C. Oren.

Nonnula ruficapilla (Gray-cheeked Nunlet). 1 ♂, MPEG 47677, km 19 TAH, 6 July 1972, col. G. P. Silva.

Monasa nigrifrons (Black-fronted Nunbird). 1 ♂, MPEG 5213, Ilha de Goyana, 7 January 1907, col. E. Snethlage; 1 ♀, MPEG 6534, Vila Braga, 22 October 1908, col. E. Snethlage. Common along forest borders (TAH) and in river-edge forest at Uruá and Buburé; groups of 3–5+ individuals observed several times each day.

Monasa morphoeus (White-fronted Nunbird). 2 ♀♀, MPEG 13673–74, Vila Braga, 6 June 19/12 July 1917, col. E. Snethlage and F. Lima. Common at mid-levels and in the canopy of mature forest, where it replaces the last species, though some overlap occurs, especially on the slopes above the river.

Chelidoptera tenebrosa (Swallow-wing). 2 sex?, MPEG 5214–15, Itaituba, 12 January 1907 (both), col. E. Snethlage. Abundant all along the TAH, where one to four perched conspicuously atop trees every few hundred meters (or so it seemed); 30–50 noted daily. Many flushed from burrows in sandy banks along the road.

CAPITONIDAE (1/1 SPECIES)

Capito brunneipectus (Brown-breasted Barbet). 1 ♂, 3 ♀♀, MPEG 13709–12, Vila Braga, 16/19/19 June 19/3 July 1917, col. E. Snethlage and F. Lima (two of these specimens, including the Type of *brunneipectus* Chapman, 1921, sent by Snethlage to the MNRJ, later taken to Europe on one of her study trips, and apparently never returned to the MNRJ: see Gonzaga 1989a; the other two specimens have disappeared from the Goeldi); 1 ♀, CM, Vila Braga (Griscom and Greenway 1941:194). A male seen on two mornings along the UT our only record; the bird associated with a large mixed-species flock in the canopy, where it hopped along vines and bare limbs and branches.

Ramphastidae (6/6 Species)

Pteroglossus aracari (Black-necked Araçari). 2 ♂♂, 1 ♀, MPEG 13692–94, Vila Braga, 3/27 June 19/11 July 1917, col. E. Snethlage and F. Lima. Seen only three times; always two birds together in the canopy of trees along the TAH. We were surprised to see so few, and no larger groups.
Pteroglossus inscriptus (Lettered Araçari). 2 ♂♂, 2 ♀♀, MPEG 13699–702, Vila Braga, 4/28/29 June 19/9 July 1917, col. E. Snethlage and F. Lima; 1 ♂, CM, Vila Braga (Griscom and Greenway, 1941:198). One seen in the canopy at Uruá our only observation.
Pteroglossus bitorquatus (Red-necked Araçari). 1 ♀. MPED 13698, Vila Braga, 14 June 1917, col. E. Snethlage and F. Lima; 2 ♂, 1 ♀, CM, Vila Braga (Griscom and Greenway 1941:198). Our only records of this aracari as follows: two (a pair?) along the TAH between Tracoá and Uruá, and one just west of Tracoá.
Selenidera maculirostris (Spot-billed Toucanet). 2 ♂♂, 2 ♀♀, MPEG 13705–08, 19 June 19/17/17/2 July 1917, col. E. Snethlage and F. Lima; 1 ♀, MZUSP, Itaituba (Pinto 1938:334). Single birds heard in the canopy along the CT. The hoarse, croaking song (gu-ow gu-ow gu-ow etc.; up to five notes) of this toucanet is very like that of the Golden-collared Toucanet (*Selenidera reinwardtii*) of western Amazonia.
Ramphastos vitellinus (Channel-billed Toucan). 1 ♀, MPEG 6514, Vila Braga, 5 November 1908, col. E. Snethlage; 1 ♂, 4 ♀♀, MPEG 13687–91, Vila Braga, 2/6/9/11/18 June 1917, col. E. Snethlage and F. Lima. Common in forest canopy all along the TAH and forest trails; up to 30 seen/heard in a day along the TAH (once 7 seen in the top of one tree); 8+ heard in a morning along the CT.
Ramphastos (tucanus) cuvieri (White-throated Toucan). 2 ♂♂, MPEG 13685–86, Vila Braga, 18 June 19/3 July 1917, col. E. Snethlage and F. Lima. Nearly as common as the last (and perhaps just less vocal); found in the same places, nearly in the same trees.

Picidae (10/11 Species)

Picumnus (aurifrons) borbae (Bar-breasted Piculet). 3 ♀♀, MPEG 5209/6510–11, Vila Braga, 7 January 1907/October/6 November 1908, col. E. Snethlage; 7 ♂♂, 2 ♀♀, CM, Vila Braga (Griscom and Greenway 1941:207); 3 ♂♂, CM, Itaituba (Griscom and Greenway 1941:207); 1 imm. ♂, MPEG 34111, Uruá, 27 August 1979, col. D. C. Oren. Relative to other forest-dwelling members of its genus, this species common; single birds or pairs seen 5 times in or near mixed-species flocks in the forest middlestory or lower canopy along the UT and CT; a pair also seen (again with other species) in bushes and small trees along the road-edge at Tracoá.
Piculus flavigula (Yellow-throated Woodpecker). 1 ♂, MPEG 6505, Ilha de Goyana, 23 October 1908, col. E. Snethlage; 1 ♀, MPEG 13649, Vila Braga, 6 June 1917, col. E. Snethlage and F. Lima; 2 ♂♂, 1 ♀, CM, Vila Braga (Griscom and Greenway 1941:200). Heard (one to three) daily along both trails.
Piculus chrysochloros (Golden-green Woodpecker). 2 ♂♂, 1 ♀, CM, Vila Braga Griscom and Greenway 1941:199); 1 ♀, MPEG 14920, Itaituba, 29 November 1955, col. J. Hidasi. Regularly seen in high forest in 1978.
Celeus elegans (Chestnut Woodpecker). 1 ♂, MPEG 5208, Bela Vista, 22 December 1906, col. E. Snethlage. Single birds seen in flight at Tracoá and Buburé.
Celeus grammicus (Scale-breasted Woodpecker). 2 ♂♂, 2 ♀♀, CM, Vila Braga (Griscom and Greenway 1941:204). This series includes the Type of *C. g. subcervinus* Todd, 1937, CM 76,440.
Celeus flavus (Cream-colored Woodpecker). 1 ♀, MPEG 6509, Ilha de Goyana, 21 October 1908, col. E. Snethlage; 2 ♂♂, 2 ♀♀, CM, Vila Braga (Griscom and Greenway 1941:204). A male seen in scrubby forest midway between Tracoá and Uruá (rarely have I seen solitary birds of this species).
Dryocopus lineatus (Lineated Woodpecker). Fairly common along the TAH, where several pairs encountered each day.
Melanerpes cruentatus (Yellow-tufted Woodpecker). 2 ♂♂, MPEG 6506–07, Vila Braga, 5 November 1908 (both), col. E. Snethlage. Fairly common along forest borders (TAH, TS), with daily counts of up to 10, but considering the extensive available habitat (secondary forest with scattered, tall trees) not nearly as common as in western Amazonia. Mostly recorded singly and in pairs (no larger family groups).
Veniliornis affinis (Red-stained Woodpecker). 1 imm. ♂, 1 ♀, MPEG 5206/6508, Vila Braga, 19 December 1906/28 October 1908, col. E. Snethlage. Common in forest middlestory and

canopy; pairs found with several mixed-species flocks along both trails, and also in river-edge forest at Buburé. Mostly seen on slender vines and branches in the lower canopy.

Campephilus melanoleucos (Crimson-crested Woodpecker). 1 ♂, CM, Vila Braga (Grimson and Greenway 1941:206). A pair observed in tall trees around TS.

Campephilus rubricollis (Red-necked Woodpecker). One or two pairs (families?) detected daily at mid-heights (on trunks) in mature forest along the CT.

DENDROCOLATIDAE (19/19 SPECIES)

Dendrocincla fuliginosa (Plain-brown Woodcreeper). 3 ♀♀, MPEG 13362-64, Vila Braga, 23/23/26 June 1917, col. E. Snethlage and F. Lima; 1 ♀, MPEG 47718, km 19 TAH, 14 July 1972, col. G. P. Silva; 9 sex?, MPEG-A 900, 907–14, Flexal = km 212 TAH, 29 October–6 December 1973, col. G. P. Silva. Single birds glimpsed over two army ant (*Eciton*) swarms, one along each trail.

Dendrocincla merula (White-chinned Woodcreeper). 16 sex?, MPEG-A 920–25, 929–38, Flexal = km 212 TAH, 29 October–6 December 1973, col. G. P. Silva; 1 ♀, MPEG 34114, km 120 TAH, 27 October 1978, col. D. C. Oren. One at a large swarm of ants along the CT; it clung to trunks from 1–7 m above the ground.

Deconychura longicauda (Long-tailed Woodcreeper). 1 ♀, MPEG 13367, Vila Braga, 12 July 1917, col. E. Snethlage and F. Lima; 2 ♂♂, CM, Vila Braga (Griscom and Greenway 1941: 217); 1 sex?, MPEG-A 899, Flexal = km 212 TAH, 29 October–6 December 1973, col. G. P. Silva; 1 ♂, MPEG 34115, km 120 TAH, 7 November 1978, col. D. C. Oren.

Deconychura stictolaema (Spot-throated Woodcreeper). 1 ♂, CM, Vila Braga (Griscom and Greenway 1941:218); 8 sex?, MPEG-A 871–73, 888–92, Flexal = km 212 TAH, 29 October-6 December 1973, col. G. P. Silva. This inconspicuous species seen twice with undergrowth and middlestory mixed-species flocks of *Thamnomanes* antshrikes and *Myrmotherula* antwrens.

Sittasomus griseicapillus (Olivaceous Woodcreeper). 3 ♂♂, MPEG 13343–45, Vila Braga, 23/26 June 19/3 July 1917, col. E. Snethlage and F. Lima. One or two seen in forest along the UT; another heard along the CT.

Glyphorynchus spirurus (Wedge-billed Woodcreeper). 1 ♀, MPEG 6307, Vila Nova, 22 October 1908, col. E. Snethlage; 2 ♂♂, 1 ♀, CM, Vila Braga (Griscom and Greenway 1941:216); 36 sex?, MPEG-A 646–47, 649–652, 663, 674–75, 689–90, 696, 713–15, 719–30, 751–65, 801–05, Flexal = km 212 TAH, 29 October–6 December 1973, col. G. P. Silva; 1 ♀, MPEG 34116, km 120 TAH, 29 October 1978, col. D. C. Oren; 1 ♀, MPEG 34116, Uruá, 27 August 1979, col. D. C. Oren. Fairly common in forest along both trails, where 1–2 detected with most understory flocks.

Nasica longirostris (Long-billed Woodcreeper). 1 ♂, MPEG 6321, Ilha de Goyana, 19 October 1908, col. E. Snethlage. At least one of these woodcreepers associated with a canopy flock along the UT; another seen with a flock in river-edge forest at Buburé, and a third heard in mature forest at the beginning of the CT. Although usually associated with varzea, the last mentioned individual several miles from the river in hilly *terra firme* forest.

Dendrexetastes rufigula (Cinnamon-throated Woodcreeper). Our only record of one heard at dawn in tall trees along the TAH just west of Tracoá (Parker).

Hylexetastes perrotii uniformis (Red-billed Woodcreeper). 2 ♂♂, 2 ♀♀, CM, Vila Braga (Griscom and Greenway 1941:210); 3 sex?, MPEG-A 807–09, Flexal = km 212 TAH, 29 October–6 December 1973, col. G. P. Silva; 1 ♂, MPEG 34119, km 120 TAH, 25 November 1978, col. D. C. Oren.

Xiphocolaptes promeropirhynchus (Strong-billed Woodcreeper). One seen briefly in forest near the beginning of the CT. At dawn on the following day one singing in the same area (taped).

Dendrocolaptes certhia concolor (Amazonian Barred-Woodcreeper). 1 ♂, MPEG 6330, Vila Braga, 5 November 1908, col. E. Snethlage; 4 sex?, MPEG-A 863–66, Flexal = km 212 TAH, 29 October–6 December 1973, col. G. P. Silva; 1 ♂, 1 ♀, MPEG 34120–21, km 120 TAH, 4 November 1978 (both), col. D. C. Oren.

Dendrocolaptes hoffmannsi (Hoffmann's Woodcreeper). Specimen(s), CM, Vila Braga (Griscom and Greenway 1941:209); 1 ♂, MPEG 14966, Itaituba, 16 October 1955, col. J. Hidasi. Single birds seen over two of three antswarms found along the CT; these stayed fairly high (6–13 m) on trunks, and sallied to adjacent foliage.

Xiphorhynchus picus (Straight-billed Woodcreeper). 1 ♀, MPEG 6319, Ilha de Goyana, 21 October 1908, col. E. Snethlage; 1 ♂, MPEG 13357, Vila Braga, 5 June 1917, col. E. Snethlage and F. Lima. A pair regularly seen in second-growth along the road and stream at Tracoá; also heard in several places to the west along the TAH.

Xiphorhynchus obsoletus (Striped Woodcreeper). 1 ♂, MPEG 6312, Vila Braga, 6 November 1908, col. E. Snethlage. One or two heard at dawn in forest along the road at Tracoá, and another watched for several minutes as it followed a mixed-species flock in river-edge forest at Buburé.

Xiphorhynchus ocellatus (Ocellated Woodcreeper). 2 ♂♂, MPEG 13350-51, Vila Braga, 9/16 June 1917, col. E. Snethlage and F. Lima; specimen(s), CM, Vila Braga (Griscom and Greenway 1941:213); 4 sex?, MPEG-A 850–53, Flexal = km 212 TAH, 29 October–6 December 1973, col. G. P. Silva; 1 ♂, MPEG 34125, km 120 TAH, 27 August 1979, col. D. C. Oren.

Xiphorhynchus elegans (Elegant Woodcreeper). 1 ♀, MPEG 6318, Ilha de Goyana, 21 October 1908, col. E. Snethlage; 1 ♀, MPEG 47703, km 19 TAH, 12 July 1972, col. G. P. Silva; 10 sex?, MPEG-A 811–13, 817–21, 841–42, Flexal = km 212 TAH, 29 October–6 December 1973, col. G. P. Silva; 1 ♀, MPEG 34126, km 120 TAH, 27 August 1979, col. D. C. Oren.

Xiphorhynchus (guttatus) eytoni (Dusky-billed Woodcreeper). 2 ♂♂, 2 ♀♀, MPEG 6308-11, Vila Braga, 23/27/28/31 October 1908, col. E. Snethlage; 3 sex?, MPEG-A 854, 860–61, Flexal = km 212 TAH, 29 October-6 December 1973, col. G. P. Silva. This species fairly common in all forest types, being found in both secondary woodland bordering the TAH and mature forest along both trails. Seen singly and in pairs, usually in association with other birds, especially canopy flocks. Up to 10 recorded in a morning along the CT, where observed high on trunks and limbs (17–23+ m), at mid-heights on trunks (in association with mixed-species flocks led by *Thamnomanes caesius*), and low on trunks (3–10 m) over an antswarm.

Lepidocolaptes albolineatus (Lineated Woodcreeper). 1 ♂, CM, Vila Braga (Griscom and Greenway 1941:214). A pair observed several times with a large canopy flock along the UT; the birds foraged primarily on large, horizontal limbs in the upper parts of tall trees; they searched smooth bark and rougher surfaces, probing in crevices and knot-holes. Another individual seen alone high in the branches of a *Mimosa* near the river-edge at Buburé.

Campylorhamphus procurvoides (Curve-billed Scythebill). 2 ♂♂, MPEG 63–25, Vila Braga, 27/31 October 1908, col. E. Snethlage; specimens, CM, Vila Braga and Itaituba (Griscom and Greenway 1941:215); 3 sex?, MPEG-A 593–95, Flexal = km 212 TAH, 29 October–6 December 1973, col. G. P. Silva; 1 ♂, MPEG 34127, km 120 TAH, 26 November 1978, col. D. C. Oren.

Furnariidare (10/18 Species)

Synallaxis gujanensis (Plain-crowned Spinetail). Apparently common in second-growth along the TAH, at Tracoá, in secondary forest at beginning of CT, and riverside vegetation at Buburé.

Synallaxis rutilans (Ruddy Spinetail). 2 ♂♂, MPEG 6288-89, Vila Braga, 28 October/4 November 1908, col. E. Snethlage; 3 sex?, MPEG-A 499–501, Flexal = km 212 TAH, 29 October-6 December 1973, col. G. P. Silva.

Cranioleuca vulpina (Rusty-backed Spinetail). 1 ♀, MPEG 6287, Ilha de Goyana, 19 October 1908, col. E. Snethlage; 1 ♂, 1 ♀, MPEG 13326–27, Vila Braga, 5/8 June 1917, col. E. Snethlage and F. Lima.

Cranioleuca gutturata (Speckled Spinetail). 1 ♀, MPEG 5113, Vila Braga, 10 January 1907, col. E. Snethlage.

Berlepschia rikeri (Point-tailed Palmcreeper). One or two (the species usually duets) heard from a distance in a wooded marsh along the TAH midway between Tracoá and Uruá, and in a swampy area in cleared land east of the park.

Hyloctistes subulatus (Striped Woodhaunter). 1 ♂, MPEG 13328, Vila Braga, 6 June 1917, col. E. Snethlage and F. Lima; 3 ♂♂, 2 ♀♀, CM, Vila Braga (Griscom and Greenway 1941:2); 3 sex?, MPEG-A 481–83, Flexal = km 212 TAH, 29 October-6 December 1973, col. G. P. Silva; 1 sex?, MPEG 34130, km 120 TAH, 9 November 1978, col. D. C. Oren. One to three of these inconspicuous furnariids found in mature forest along the CT; one seen about 10 m above the ground in palm fronds, amidst a large mixed-species flock that also contained several Rufous-rumped Foliage-gleaners (*Philydor erythrocercus*) and both species of *Thamnomanes* antshrikes (see below); later two observed together, again at mid-heights near a mixed flock. The latter birds tape-recorded.

Ancistrops strigilatus (Chestnut-winged Hookbill). 1 ♀, MPEG 13337, Vila Braga, 3 June 1917, col. E. Snethlage and F. Lima; 2 ♂♂, 1 ♀, CM, Vila Braga (Griscom and Greenway 1941: 2). One or two heard in forest canopy along the CT.

Philydor erythrocercus (Rufous-rumped Foliage-gleaner). 2 ♂♂, 1 ♀, MPEG 6297–99, Vila Braga, 27 October/4/7 November 1908, col. E. Snethlage; 1 ♂, MZUSP, Itaituba (Pinto 1938: 431); 8 sex?, MPEG-A 949–55, 967, Flexal = km 212 TAH, 29 October-6 December 1973, col. G. P. Silva; 1 ♀, MPEG 34131, km 120 TAH, 4 November 1978, col. D. C. Oren. Common in middlestory and lower canopy mixed-species flocks along the CT; noted in three flocks on one morning (three individuals in one and two in another).

Philydor ruficaudatus (Rufous-tailed Foliage-gleaner). 1 ♂, CM, Vila Braga (Griscom and Greenway 1941:225).

Philydor pyrrhodes (Cinnamon-rumped Foliage-gleaner). Seen twice along the CT.

Philydor erythropterus (Chestnut-winged Foliage-gleaner). 3 ♂♂, CM, Vila Braga (Griscom and Greenway 1941:225). Common in forest along both trails, where one to two seen or heard in nearly all mixed-species flocks encountered in the upper middlestory and canopy; up to 8 heard/seen in a morning along the CT. Most of these searching large live leaves at or near the ends of slender branches in the upper and outer portions of trees (more than 16 m above ground).

Automolus ochrolaemus (Buff-throated Foliage-gleaner). 1 ♂, MPEG 6293, Vila Braga, 3 November 1908, col. E. Snethlage; 2 sex?, MPEG-A 580 and 585, Flexal = km 212 TAH, 29 October-6 December 1973, col. G. P. Silva. Fairly common in forest undergrowth along the CT, and also in secondary forest along the TAH. Up to 4 found daily along the CT, including two individuals that followed *Thamnomanes/Myrmotherula* flocks in the undergrowth (up to 5 m).

Automolus infuscatus (Olive-backed Foliage-gleaner). 5 sex?, MPEG-A 460–64, Flexal = km 212 TAH, 29 October-6 December 1973, col. G. P. Silva. Detected regularly at km 120 TAH in 1978.

Xenops tenuirostris (Slender-billed Xenops). 1 ♀, CM, Itaituba (Griscom and Greenway 1941:227). One seen in a mixed-species flock in river-edge (seasonally-flooded?) forest at Buburé.

Xenops minutus (Plain Xenops). 1 ♀, MPEG 6302, Vila Braga, 23 October 1908, col. E. Snethlage; 6 sex?, MPEG-A 539–40, 553–54 and 567–68, Flexal = km 212 TAH, 29 October-6 December 1973, col. G. P. Silva; 1 sex?, MPEG 34133, km 120 TAH, 27 August 1979, col. D. C. Oren. Single birds observed with three undergrowth flocks (containing both *Thamnomanes* antshrikes, several *Myrmotherula* antwrens, and ant-tanagers (*Habia rubica*) found along the CT).

Sclerurus mexicanus (Tawny-throated Leafscraper). 1 sex? MPEG-A 536, Flexal = km 212 TAH, 29 October-6 December 1973, col. G. P. Silva.

Sclerurus rufigularis (Short-billed Leafscraper). 3 sex?, MPEG-A 512–4, Flexal = km 212 TAH, 29 October-6 December 1973, col. G. P. Silva.

Sclerurus caudacutus (Black-tailed Leafscraper). 1 ♂, MPEG 6305, Vila Braga, 29 October 1908, col. E. Snethlage; 5 sex?, MPEG-A 526–30, Flexal = km 212 TAH, 29 October-6 December 1973, col. G. P. Silva; 2 ♂♂, MPEG 34135/36, km 120 TAH, 2 and 9 November 1978, col. D. C. Oren. Two (a pair?) glimpsed and tape-recorded on and near the ground in mature forest along the CT.

THAMNOPHILIDAE (40/44 SPECIES)

Cymbilaimus lineatus (Fasciated Antshrike). 3 ♀♀, MPEG 6332–34, Vila Braga, 29 October/6 November 1908, col. E. Snethlage; 1 sex?, MPEG-A 977, Flexal = km 212 TAH, 29 October-6 December 1973, col. G. P. Silva. Fairly common at mid-heights and in the canopy of secondary and primary forest along the TAH (where heard) and both trails; up to four heard in a morning (CT).

Taraba major (Great Antshrike). Recorded daily in small numbers in secondary woodland at TS and along the TAH.

Sakesphorus luctuosus (Glossy Antshrike). 2 ♂♂, MPEG 6352–53, Ilha de Goyana, 19/21 October 1908, col. E. Snethlage; 1 ♂, MPEG 34138, east bank Tapajós, in front of Uruá, 22 November 1978, col. D. C. Oren. A pair in forest along the river at Buburé foraged in dense foliage and vine tangles from within a few meters of the water up to mid-heights (16–20 m).

Thamnophilus palliatus (Lined Antshrike). 1 imm. ♂, MPEG 5142, Vila Braga, 8 January 1907, col. E. Snethlage. A pair nest-building in shrubbery along the TAH just west of Tracoá (11 November); others heard in second-growth in the cleared land east of the park. Here this species seems to occupy the niche of the widespread Barred Antshrike (*Thamnophilus doliatus*).

Thamnophilus nigrocinereus (Blackish-gray Antshrike). 1 ♀, 1 ♂ (Types of *T. n. huberi* Snethlage, 1907), MPEG 5139–40, Ilha de Goyana, 26/31 December 1907 (both Type specimens sent to the Zoological Museum, Berlin); 4 ♂♂, 3 ♀♀, 1 sex?, MPEG 6337–44, Ilha de Goyana, 19/20/20/20/20/20/21 October 1908, col. E. Snethlage (this series donated to the Zoological Museum, Berlin; Berlepsch Museum and British Museum); 6 ♂♂, 5 ♀♀, MPEG 13379–89, Ilha de Goyana, 5/5/5/8/8/8/15/15/20 June/14/14 July 1917; 1 ♂, CM, Itaituba (Griscom and Greenway, 1941:231).

Thamnophilus aethiops (White-shouldered Antshrike). 2 ♂♂, 2 ♀♀, MPEG 6345–48, Vila Braga, 4 November 19/7/26 October/4 November 1908, col. E. Snethlage; 11 ♂♂, 13 ♀♀, CM, Vila Braga (Griscom and Greenway 1941:231); 2 ♂♂, MPEG 47756–57, km 19 TAH, 3 and 17 July 1972, col. G. P. Silva; 11 sex?, MPEG-A 1011–19, 1031–2, Flexal = km 212 TAH, 29 October-6 December 1973, col. G. P. Silva; 1 ♂, 1 ♀, MPEG 34139/40, km 120 TAH, 18 November 1978 (both), col. D. C. Oren. This secretive antshrike heard in disturbed forest undergrowth along the Ramal Terra Preta.

Thamnophilus schistaceus (Black-capped Antshrike). 1 ♀, MPEG 5152, Vila Braga, 3 January 1907, col. E. Snethlage; 1 ♂, MPEG 34141, km 120 TAH, 2 November 1978, col. D. C. Oren. Fairly common (up to five daily).

Thamnophilus stictocephalus (Natterer's Slaty-Antshrike). 1 ♀, CM 69,260, Vila Braga, 6 September 1918, col. S. M. Klages. This is the type of *Thamnophilus punctatus saturatus* (Todd 1927).

Thamnophilus amazonicus (Amazonian Antshrike). 1 ♀, 1 ♂, MPEG 6349–50, Vila Braga, 27 October/6 November 1908, col. E. Snethlage; 3 ♂♂, 6 ♀♀, Vila Braga (Griscom and Greenway 1941:233); 1 sex?, MPEG-A 1028, Flexal = km 212 TAH, 29 October-6 December 1973, col. G. P. Silva. Common in roadside woods along the TAH (including several pairs around TS); also found in river-edge forest at Buburé, where a pair consistently remained in the middlestory (10–20 m).

Pygiptila stellaris (Spot-winged Antshrike). 2 ♂♂, 1 ♀, MPEG 5143–45, Vila Braga, Ilha de Goyana and Bela Vista (resp.), 8/9 January 1907/22 December 1906, col. E. Snethlage. Probably fairly common, but easily overlooked in the darkened vine tangles and leafy branches of the lower canopy, where they spend most of their time; one to three observed in several mixed-species flocks, one along the UT and at least 3 along the CT.

Thamnomanes saturninus (Saturnine Antshrike). 1 ♀, 1 ♂, MPEG 5153/6359, Vila Braga, 19 December 1906/26 October 1908, col. E. Snethlage; specimen(s), CM, Vila Braga (Griscom and Greenway, 1941:234); 2 ♂♂, MPEG 47764–65, km 19 TAH, 6/8 July 1972, col. G. P. Silva; 4 sex?, MPEG-A 1035, 1038–40, Flexal = km 212 TAH, 29 October–6 December 1973, col. G. P. Silva; 1 ♂, MPEG 34146, km 120 TAH, 27 October 1978, col. D. C. Oren. This species found in almost every undergrowth flock along both trails, and probably serves as a flock leader to some degree (along with closely related *Thamnomanes caesius*); usually three to four noted together. Other flock associates included the Stipple-throated Antwren (*Myrmotherula haematonota*), Tawny-crowned Greenlet (*Hylophilus ochraceiceps*), and Red-crowned Ant Tanager (*Habia rubica*).

Thamnomanes caesius (Cinereous Antshrike). 5 ♀♀, 8 ♂♂, MPEG 6362–74, Vila Braga, 23/26/26/26/27/28/28/29/31 October/3/5/5/7 November 1908, col. E. Snethlage; 1 ♂, 1 ♀, MPEG 47758–59, km 19 TAH, 14 July 1972 and 26 July 1972, col. G. P. Silva; 25 sex?, MPEG-A 1054–61, 1078–95, Flexal = km 212 TAH, 29 October–6 December 1973, col. G. P. Silva; 1 ♂, MPEG 34147, km 120 TAH, 2 November 1978, col. D. C. Oren. A common leader of mixed-species flocks at all levels in mature forest; two to five seen in as many as 5 mixed-species flocks encountered along 6 km of the CT in one morning.

Myrmotherula brachyura (Pygmy Antwren). 1 ♂, 1 ♀, MPEG 13431–32, Vila Braga, 29 June 19/11 July 1917, col. E. Snethlage and F. Lima. Heard along the TAH in disturbed forest, and also (once) in primary forest along the CT. As in western Amazonia, this species overlaps ecologically with the similar-looking Sclater's Antwren (*M. sclateri*).

Myrmotherula sclateri (Sclater's Antwren). 2 ♂♂, MPEG 13435–36, Vila Braga, 3 June 1917 (both), col. E. Snethlage and F. Lima. 1 ♂, CM, Vila Braga (Griscom and Greenway 1941:236). Common in the canopy and upper middlestory of forest along both trails; singles and pairs noted in vine tangles and foliage along slender branches from 13–23+ m above the ground; all (with the possible exception of a few singing individuals) associating with mixed-species flocks.

Myrmotherula surinamensis (Streaked Antwren). 3 ♀♀, MPEG 5157–58/6380, Ilha de Goyana, 1/3 January 1907/21 October 1908, col. E. Snethlage. Pairs recorded in two areas: in shrubby tangles bordering a stream along the TAH and at mid-heights in river-edge forest (BA); the latter birds (a male and two female-plumaged individuals) searched small leaves in tree crowns of the lower canopy at 20 m. Songs and calls very like those of west Amazonian birds.

Myrmotherula hauxwelli (Plain-throated Antwren). 1 ♀, MPEG 6394, Vila Braga, 31 October 1908, col. E. Snethlage; 1 ♀, MZUSP, Itaituba (Pinto, 1938:471); 13 sex?, MPEG-A 1132–37, 1156–62, Flexal = km 212 TAH, 29 October–6 December 1973, col. G. P. Silva; 2 ♀♀, 1 ♂, MPEG 34150–52, km 120 TAH, 27 August 1979 (1st ♀ and ♂) and 2 November 1978 (2nd ♀), col. D. C. Oren. Pairs encountered low in forest undergrowth, twice along the UT and twice along the CT; observed in association with *Thamnomanes saturninus*-led mixed-species flocks and alone.

*Myrmotherula leucophthalma (White-eyed Antwren). 1 ♂, 2 ♀♀, MPEG 6389–91, Vila Braga, 31/31 October/3 November 1908, col. E. Snethlage; specimen(s), CM, Vila Braga (Griscom and Greenway, 1941:237) (includes the Type of *M. l. phaeonota* Todd, 1927, CM 75, 173); 1 ♂, MPEG 47784, Flexal = km 212 TAH, 27 November 1973, col. G. P. Silva; 11 sex?, MPEG-A 1188, 1196–97, 1207–14, Flexal = km 212 TAH, 29 October–6 December 1973, col. G. P. Silva; 2 ♂♂, MPEG 34153–54, km 120 TAH, 29 Oct/4 November 1978, col. D. C. Oren. A red-backed, checker-throated antwren found in most undergrowth mixed-species flocks along both trails; in pairs (probably family groups), with up to 8 individuals seen in a morning along the CT. All within 4 m of the ground (mostly at 1–1.3 m), where they searched curled, dead leaves suspended from branches and trapped in small palms and vine tangles. The taxonomy of the *M. leucophthalma*/*M. haematonota* group is confusing; probable examples of *M. l. phaeonota*, although red-backed *haematonota* are known not far to the west and north. The birds vocally (calls and song) quite like *leucophthalma* of southeastern Peru.

*Myrmotherula ornata (Ornate Antwren). 1 ♂, MPEG 5163, Vila Braga, 8 January 1907, col. E. Snethlage; 1 ♂, MZUSP, Itaituba (Pinto 1938:473); 10 sex?, MPEG-A 1237-46, Flexal = km 212 TAH, 29 October–6 December 1973, col. G. P. Silva. Pairs found with *Thamnomanes caesius*-led mixed-species flocks along both trails; in dense tangles at mid-heights (10–17 m; once at 5 m), where they mostly probed trapped dead leaves. Similar in behavior and appearance to last species, but consistently seen higher.

*Myrmotherula axillaris (White-flanked Antwren). 2 ♂♂, MPEG 6397–98, Ilha de Goyana and Vila Braga (resp.), 19 October 1908, col. E. Snethlage; 7 sex?, MPEG-A 1253–59, Flexal = km 212 TAH, 29 October–6 December 1973, col. G. P. Silva; 1 ♂, MPEG 34155, km 120 TAH, 13 November 1978, col. D. C. Oren. Common; found in secondary forest along the TAH, river-edge forest at Buburé, and primary forest along both trails. Always found with or near other species, particularly the *Thamnomanes*-led flocks of the undergrowth and middlestory; typically foraged in the slender branches of small trees at 5–10 m, but observations ranged from 3–23+ m. Pairs also seen with pairs of Dot-winged Antwrens (*Microrhopias quixensis*).

*Myrmotherula longipennis (Long-winged Antwren). 3 ♂♂, MPEG 6403–04, Vila Braga, 23 October/3 November 1908, col. E. Snethlage; 11 ♂♂, 5 ♀♀, CM, Vila Braga (Griscom and Greenway, 1941:238) (includes the Type of *M. l. ochrogyna* Todd, 1937, CM 69,230); 20 sex?, MPEG-A 1308–10, 1322–35, 1374–76, Flexal = km 212 TAH, 29 October-6 December 1973, col. G. P. Silva; 1 ♂, MPEG 34156, km 120 TAH, 1 November 1978, col. D. C. Oren. Common in undergrowth mixed-species flocks (always near *Thamnomanes saturninus* or *T. caesius*) forest along both trails; as in western Amazonia, this species seems to overlap ecologically with the White-flanked Antwren (*M. axillaris*); both glean green leaves on slender branches.

*Myrmotherula iheringi (Ihering's Antwren). 5 ♂♂, 2 ♀♀, MPEG 13507–13, Vila Braga, 9/13/13/18/21 June–19/7/4/Jul. 17, col. E. Snethlage and F. Lima; 8 ♂♂, 4 ♀♀, CM, Vila Braga (Griscom and Greenway, 1941:239). Fairly common in forest at mid-heights; pairs found with mixed flocks along both trails (1 and 2 pairs, respectively). Seen in close association with Ornate (*M. ornata*) and Dot-winged Antwrens (*Microrhopias quixensis*). Usually noted on slender branches and vines.

*Myrmotherula menetriesii (Gray Antwren). 1 ♂, MPEG 6414, Vila Braga, 23 October 1908, col. E. Snethlage; 1 sex?, MPEG-A 1321, Flexal = km 212 TAH, 29 October–6 December 1973, col. G. P. Silva; 1 ♀, 1 ♂, MPEG 34157–58, km 120 TAH, 27 October 1978 and 27 August 1979, col. D. C. Oren. Probably common in mixed-species flocks; seen with at least two flocks along the CT, and another on the UT.

*Dichrozona cincta (Banded Antbird). 1 ♀, 1 imm. ♀, MPEG 13540–41, 29 June–19/July 1917, col. E. Snethlage and F. Lima; 9 sex?, MPEG-A 1390–96, 1402–03, Flexal = km 212 TAH, 29 October–6 December 1973, col. G. P. Silva; 1 ♂, 1 ♀, MPEG 34159–60, km 120 TAH, 17 November 1978 (both), col. D. C. Oren. Encountered only once: one territorial bird (probably a male) tape-recorded and lured briefly into view as it foraged in leaf litter on the ground under tall forest along the CT. In other areas in Amazonia *D. cincta* seems to be uncommon and unevenly distributed in *terra firme* forest. Silent individuals are almost impossible to detect.

*Microrhopias quixensis (Dot-winged Antwren). 4 ♂♂, MPEG 6420–23, Vila Braga, 22/22/27/28 October 1908, col. E. Snethlage; 1 ♂, MPEG 34161, Island in Rio Tapajós 12 km above Uruá, 24 August 1979, col. H. Sick. Common in secondary and primary forest; in pairs from mid-levels up into the canopy, especially with mixed-species flocks (pairs+ observed in up to

5 flocks during one morning) but also alone (although not far from at least one or two other species, such as White-flanked Antwrens (*M. axillaris*). Nearly all *Microrhopias quixensis* found in vine tangles and dense foliage from 10–17 m above the ground.

Formicivora grisea (White-fringed Antwren). 1 ♂, MPEG 13521, Vila Braga, 22 June 1917, col. E. Snethlage and F. Lima.

**Terenura spodioptila* (Ash-winged Antwren). 2 ♂♂, 2 ♀♀, CM, Vila Braga (Griscom and Greenway 1941:2). Pairs located with canopy flocks along both trails; though mostly seen in foliage near the ends of slender branches (where from below they looked much like *Hylophilus* greenlets), one pair gathered bits of moss from limbs and the trunk of a tree about 12 m above the ground.

**Cercomacra cinerascens* (Gray Antbird). 1 ♂, MPEG 5176, 19 December 1906, col. E. Snethlage; 1 sex?, MPEG-A 1457, Flexal = km 212 TAH, 29 October–6 December 1973, col. G. P. Silva. Common (by voice; up to 12 daily) in mid-level vine tangles in tall forest along the TAH and both trails.

**Cercomacra nigrescens* (Blackish Antbird). 2 ♀♀, MPEG 13598–99, Vila Braga, 4/14 June 1917, col. E. Snethlage and F. Lima. Fairly common (2–3 pairs daily) in forest edge second-growth along the TAH and side trails; found in pairs, usually away from other birds, in dense cover from 0.6–10 m above the ground. Those seen brought into view with the aid of a tape-recorder.

**Myrmoborus myotherinus* (Black-faced Antbird). 1 ♂, 1 ♀, MPEG 5183–84, Vila Braga, 8 January 1907 (both), col. E. Snethlage; 1 ♀, MPEG 6441, Vila Nova, 22 October 1908, col. E. Snethlage; 34 sex?, MPEG-A 1548, 1550–62, 1565–70, 1579–90, 1768, Flexal = km 212 TAH, 29 October–6 December 1973, col. G. P. Silva. Quite common in the undergrowth of both forest types in all areas (three pairs daily (UT), up to 10 pairs daily (CT); perhaps the most numerous antbird in this part of the park (along with the White-flanked Antwren).

**Hypocnemis cantator* (Warbling Antbird). 1 ♀, 1 imm. ♂, MPEG 6428–29, Vila Braga, 23/27 October 1908, col. E. Snethlage; 22 sex?, MPEG-A 1488–89, 1493–1511, 1525, Flexal = km 212 TAH, 29 October–6 December 1973, col. G. P. Silva; 1 ♀, MPEG 34168, km 120 TAH, 17 November 1978, col. D. C. Oren. Common in the dense tangles of disturbed forest, along edges (TAH) and locally within mature forest (regenerating garden plots (?) along both trails).

**Hypocnemoides maculicauda* (Band-tailed Antbird). 4 ♂♂, 3 ♀♀, MPEG 5185–89/6447–48, Ilha de Goyana, 21/23/26 December 1906/1/1 January 1907/20/20 October 1908, col. E. Snethlage. Pairs found in river-edge forest at Uruá and Buburé foraged in branches near or overhanging the water, rarely moving higher than 1 m.

**Schistocichla leucostigma* (Spot-winged Antbird). 4 sex?, MPEG-A 1594–97, Flexal = km 212 TAH, 29 October-6 December 1973, col. G. P. Silva. A pair observed both low in forest undergrowth and on the ground along a small stream (CT).

*Sclateria naevia (Silvered Antbird). 1 ♂, MPEG 34170, km 120 TAH, 25 November 1978, col. D. C. Oren.

**Myrmeciza ferruginea* (Ferruginous-backed Antbird). 1 ♂, 1 ♀, MPEG 13537–38, Vila Braga, 8 June 19/13 July 1917, col. E. Snethlage and F. Lima; 9 ♂♂, 6 ♀♀, CM, Vila Braga (Griscom and Greenway 1941:247) (includes the Type of *M. f. eluta* [Todd, 1937], CM 76,019); 1 ♂, MPEG 47808, km 19 TAH, 12 July 1972, col. G. P. Silva; 2 sex?, MPEG-A 1606–07, Flexal = km 212 TAH, 29 October-6 December 1973, col. G. P. Silva; 1 ♀, MPEG 47809, Flexal = km 212 TAH, 1 December 1973, col. G. P. Silva. This antbird heard once in forest at Uruá, and three seen together walking through leaf litter on the ground in mature forest along the CT.

**Rhegmatorhina berlepschi* (Harlequin Antbird). 1 ♂ (Type), MPEG 5191, Vila Braga, 5 January 1907, col. E. Snethlage; 1 ♀, 1 ♂, MPEG 6464–65, Vila Braga, 3/4 November 1908, col. E. Snethlage; (MPEG 6465 sent by Snethlage to the Berlepsch Museum; the Type and the other 1908 Snethlage specimen of *R. berlepschi* have disappeared from the Museu Goeldi); 4 ♂♂, 8 ♀♀, MPEG 13604–15, Vila Braga, 14/25/25/26/27/27/27 June/1/1/6/11 July 1917, col. E. Snethlage and F. Lima (of these, only 13608 and 13612 are still in the Museu Goeldi); 1 ♂, 1 ♀, CM, Vila Braga (Griscom and Greenway 1941:250); 3 sex?, MPEG 31947–49, km 19 TAH, b/n 6 Jul/15 August 1972, col. G. P. Silva; 1 ♂, 2 ♀♀, MPEG 47820–22, km 19 TAH, 10 July 1972 (first 2) and 20 July 1972, col. G. P. Silva; 1 ♂, MPEG 34171, km 120 TAH, 29 October 1978, col. D. C. Oren. A male seen at one (the largest) of four antswarms (CT); it stayed close to the ground in undergrowth over the columns of ants; nearby other ant-following species, including Black-spotted Bare-eyes (*Phlegopsis nigromaculata*), Pale-faced Bare-eyes (*Skutchia borbae*) and Scale-backed Antbirds (*Hylophylax poecilinota*). The bare orbital skin of the *Rhegmatorhina* bright yellow, in contrast to the dull green portrayed in one of the only illustrations of this rare species (see Willis 1969).

*Hylophylax naevia (Spot-backed Antbird). 10 sex?, MPEG-A 1690–91, 1710–14, 1716–18, Flexal = km 212 TAH, 29 October-6 December 1973, col. G. P. Silva; 1 ♂, MPEG 34172, km 120 TAH, 7 November 1978, col. D. C. Oren. Two pairs found in forest undergrowth (CT).

*Hylophylax punctulata (Dot-backed Antbird). 9 ♂♂, 2 ♀♀, MPEG 13575–86, Vila Braga, 4/14/21/21/25 June 19/4/6/6/6/20/20 July 1917, col. E. Snethlage and F. Lima; 2 sex?, MPEG-A 1692, 1715, Flexal = km 212 TAH, 29 October–6 December 1973, col. G. P. Silva. Pairs found in two places along the CT.

*Hylophylax poecilinota (Scale-backed Antbird). 1 ♀, MPEG 6438, Vila Braga, 31 October 1908, col. E. Snethlage; 18 sex?, MPEG-A 1753–67, 1790–92, Flexal = km 212 TAH, 29 October–6 December 1973, col. G. P. Silva; 1 ♂, 2 juv. ♂♂, MPEG 34173–75, km 120 TAH, 2 November 19/27 October/7 November 1978, col. D. C. Oren. Pairs (+) observed at all four antswarms (UT,CT).

*Skutchia borbae (Pale-faced Antbird). 2 sex?, MPEG 30210–11, km 19 TAH, 8 July 1972 (both), col. G. P. Silva; 1 sex?, MPEG 31945, km 19 TAH, b/n 6/Jul/15 August 1972, col. G. P. Silva; 1 ♀, MPEG 47848, km 19 TAH, 25 July 1972, col. G. P. Silva; 1 sex?, MPEG-A 1793, Flexal = km 212 TAH, 29 October–6 December 1973, col. G. P. Silva. This interesting species recorded at three different antswarms along the CT; at least two seen repeatedly at the two main swarms (described earlier), and the species heard at a third. The birds perched low (30–100 cm) on slender saplings and occasionally hopped on the ground, often within centimeters of the ants.

*Phlegopsis nigromaculata (Black-spotted Bare-eye). 1 ♂, 1 sex?, MPEG 5192–93, Vila Braga, 3/10 January 1907, col. E. Snethlage; 5 sex?, MPEG-A 1794–97, 1804, Flexal = km 212 TAH, 29 October–6 December 1973, col. G. P. Silva; 1 ♂, MPEG 34176, km 120 TAH, 2 November 1978, col. D. C. Oren. Heard and seen at all four antswarms (UT,CT), and also at several points along the TAH (wandering birds?); at least 3 (and perhaps as many as 6) present at the two large swarms along the CT. These active and aggressive, but usually stayed well concealed in undergrowth close to the ground; seen to frequently displace each other as well as Pale-faced Bare-eyes and Scale-backed Antbirds. In western Amazonia, this species occurs mainly in seasonally-flooded forest, so it was surprising to find it so common in *terra firme*.

FORMICARIIDAE (7/7 SPECIES)

*Chamaeza nobilis (Striated Antthrush). Vocalizations identified by H. Sick in high forest in 1979.

*Formicarius colma (Rufous-capped Antthrush). 1 ♀, MPEG 6472, Vila Braga, 7 November 1908, col. E. Snethlage; 5 sex?, MPEG-A 1814–18, Flexal = km 212 TAH, 29 October-6 December 1973, col. G. P. Silva; 1 ♀, MPEG 34178, km 120 TAH, 27 October 1978, col. D. C. Oren. Up to three singing birds noted daily along the CT.

*Formicarius analis (Black-faced Antthrush). 1 ♂, MPEG 5194, Itaituba, 17 January 1907, col. E. Snethlage; 2 sex?, MPEG-A 1805–06, Flexal = km 212 TAH, 29 October-6 December 1973, col. G. P. Silva; 1 ♂, MPEG 34179, km 120 TAH, 27 August 1979, col. D. C. Oren. Common in both secondary and primary forest in all areas, with up to six recorded (mainly heard) in a morning along the CT. In most well-known Amazonian localities this species prefers disturbed/younger forests while *F. colma* prefers undisturbed/older forests. Within this park, however, this pattern not as clearly defined.

*Myrmornis torquata (Wing-banded Antbird). 1 ♂, MPEG 5190, Vila Braga, 2 January 1907, col. E. Snethlage; 4 ♂♂, 1 ♀, MPEG 13616–20, Vila Braga, 2/9/22 June/16/18 July 1917, col. E. Snethlage and F. Lima; 1 ♀, MPEG 34181, km 120 TAH, 29 October 1978, col. D. C. Oren.

*Myrmothera campanisona (Thrush-like Antpitta). Fairly common, on the ground in well-shaded forest of both main types; heard in all areas (up to 4 in a morning along the TAH and CT).

*Grallaria varia (Variegated Antpitta). Specimen(s), CM, Vila Braga (Griscom and Greenway 1941:252) (includes the Type of *G. varia distincta* Todd, 1927, CM 75,444). One of these large antpittas flushed along the CT; none heard vocalizing during our stay in the park.

*Hylopezus berlepschi (Amazonian Antpitta). Specimen(s), CM, Vila Braga (Griscom and Greenway 1941:253). One heard in dense secondary forest bordering a marsh along the TAH.

CONOPOPHAGIDAE (1/2 SPECIES)

Conopophaga melanogaster (Black-bellied Gnateater). 1 ♂, 1 ♀, MPEG 5195–96, Vila Braga, 5 January 1907 (both), col. E. Snethlage; 1 ♀, CM, Vila Braga (Griscom and Greenway 1941: 254); 1 ♂, 1 ♀, MPEG 47853–54, km 19 TAH, 6 July 1972 (both), col. G. P. Silva.

*_Conopophaga aurita_ (Chestnut-bellied Gnateater). 2 ♂♂, 3 ♀♀, MPEG 13637–41, Vila Braga, 4/8/10/12/13 July 1917, col. E. Snethlage and F. Lima; 1 ♂, MPEG 34180, km 120 TAH, 27 October 1978, col. D. C. Oren; 5 sex?, MPEG-A 1840–44, Flexal = km 212 TAH, 29 October–6 December 1973, col. G. P. Silva.

Rhinocryptidae (1/1 Species)

*_Liosceles thoracicus_ (Rusty-belted Tapaculo). 1 ♂, MPEG 13647, Vila Braga, 19 June 1917, col. E. Snethlage and F. Lima; 1 ♂, 1 ♀, CM, Vila Braga (Griscom and Greenway 1941:254). Up to four heard in a morning in primary forest along the CT.

Cotingidae (6/9 Species)

Cotinga cotinga (Purple-breasted Cotinga). 1 ♂, MPEG 13310, Vila Braga, 16 July 1917, col. E. Snethlage and F. Lima; 1 ♂, 1 ♀, MPEG 15135/232, Itaituba, 2 and 9 November 1955, col. J. Hidasi.
*_Cotinga cayana_ (Spangled Cotinga). 4 ♂♂, 2 ♀♀, MPEG 13311–16, Vila Braga, 17 June 19/14/16/16/19/19 July 1917, col. E. Snethlage and F. Lima. Several males and females noted almost daily perched conspicuously atop trees along the road edges (TAH).
*_Xipholena lamellipennis_ (White-tailed Cotinga). 2 ♂♂, 1 imm. ♂, 1 ♀, MPEG 13317–20, Vila Braga, 14/16/16/16 July 1917, col. E. Snethlage and F. Lima; 1 ♂, MZUSP, Itaituba (Pinto 1944:13).
*_Iodopleura isabellae_ (White-browed Purpletuft). Three seen around a large clump of mistletoe on the top of a tall, road-edge tree near Uruá.
*_Lipaugus vociferans_ (Screaming Piha). 1 ♀, MPEG 6282, Vila Braga, 7 November 1908, col. E. Snethlage; 1 ♀, MPEG 34182, km 120 TAH, 25 November 1978, col. D. C. Oren. A few heard (singing) in the distance from several points along the TAH; also heard (calling) along the CT.
*_Querula purpurata_ (Purple-throated Fruit-crow). Probably heard once at a great distance from the TAH. Surprisingly scarce.
*_Gymnoderus foetidus_ (Bare-necked Fruit-crow). Seen singly flying over the TAH regularly in 1978 and 1979.
Phoenicircus carnifex (Guianan Red-Cotinga). 1 ♂, 1 ♀, CM, Vila Braga (Griscom and Greenway 1941:255).
Phoenicircus nigricollis (Black-necked Red-Cotinga). 1 imm. ♂, MPEG 13309, Vila Braga, 8 June 1917, col. E. Snethlage and F. Lima.

Pipridae (10/12 Species)

*_Pipra rubrocapilla_ (Red-headed Manakin). 2 ♀♀, 1 ♂, MPEG 6262–64, Vila Braga, 26/27/27 October 1908, col. E. Snethlage; 10 sex?, MPEG-A 2005, 2012–19, 2028, Flexal = km 212 TAH, 29 October-6 December 1973, col. G. P. Silva; 1 ♀, MPEG 34185, km 120 TAH, 25 November 1978, col. D. C. Oren. A few detected daily; recorded in secondary and primary forest in all areas.
*_Pipra pipra_ (White-crowned Manakin). A female glimpsed in forest undergrowth near the beginning of the UT.
*_Pipra nattererii_ (Snow-capped Manakin). 3 ♂♂, 1 ♀, MPEG 5088–91, Vila Braga, 2/2/3/4 January 1907, col. E. Snethlage; 26 ♂♂, ♀♀, CM, Vila Braga (Griscom and Greenway 1941:265); 1 ♂, 1 ♀, MPEG 30213–14, km 19 TAH, 20/14 July 1972, col. G. P. Silva; 26 sex?, MPEG-A 1946–49, 1966–81, 1984–89, Flexal = km 212 TAH, 29 October–6 December 1973, col. G. P. Silva; 1 ♂, 2 ♀♀, MPEG 34200–02, km 120 TAH, 27/29/29 October 1978, col. D. C. Oren. A female-plumaged individual seen in undergrowth over an antswarm along the CT.
*_Pipra fasciicauda_ (Band-tailed Manakin). 1 sex?, MPEG-A 1876, Flexal = km 212 TAH, 29 October-6 December 1973, col. G. P. Silva; 1 ♀, MPEG 34198, right margin Tapajós, in front of Uruá, 22 November 1978, col. D. C. Oren.
*_Chiroxiphia pareola regina_ (Blue-backed Manakin). 2 ♂♂, MPEG 6270–71, Vila Braga, 31/7 October 1908, col. E. Snethlage; 4 ♂♂, 6 ♀♀, CM, Vila Braga (Griscom and Greenway 1941:267); 1 ♂, MPEG 47885, Flexal = km 212 TAH, November 1973, col. G. P. Silva. At least one male heard (one seen) in three places along the CT; all in the crowns of small trees at 6–13 m. Display calls of this yellow-crowned form are very similar to those of red-crowned birds (_napensis_) found north of the Amazon northeast of Iquitos.

Manacus manacus (White-bearded Manakin). 1 ♂, MPEG 13277, Vila Braga, 7 July 1917, col. E. Snethlage and F. Lima.

Xenopipo atronitens (Black Manakin). 1 ♂, MPEG 34221, east margin Tapajós, in front of Uruá, 22 November 1978, col. D. C. Oren.

*Machaeropterus pyrocephalus (Fiery-capped Manakin). Vocalizations heard in mature second growth forest identified by H. Sick in 1979.

*Heterocercus linteatus (Flame-crowned Manakin). 1 imm. ♂, MPEG 6272, Ilha de Goyana, October 1908, col. E. Snethlage; 2 ♀♀, MPEG 30196–97, Itaituba, 7/8 December 1971, col. G. P. Silva; 1 imm. ♂, MPEG 34226, right margin Tapajós, in front of Uruá, 22 November 1978, col. D. C. Oren. Two adult males and two probable immature males seen in river-edge forest at Buburé perched on open, slender branches in the crowns of trees at mid-heights and in the lower canopy (13–20 m up). Although quiet while perched, these birds uttered sharp, buzzy calls during intraspecific chases. The birds probably in a display area of some kind, but they gave no visual display while perched.

*Tyranneutes stolzmanni (Dwarf Tyrant-Manakin). 1 ♀, MPEG 34230, km 120 TAH, 18 November 1978, col. D. C. Oren. As many as six singing males heard (and several seen) at mid-levels (3–15 m) along the CT, and a few also heard in all other areas (UT, TS, TAH). As in other parts of Amazonia, they perched in the open on slender branches of middlestory trees, but motionless and extremely hard to see.

*Piprites chloris (Wing-barred Manakin). 1 ♂, MPEG 5086, Vila Braga, 3 January 1907, col. E. Snethlage; 1 ♂, MPEG 6261, Ilha de Goyana, 20 October 1908, col. E. Snethlage; 1 ♂, CM, Vila Braga (Griscom and Greenway 1941:263). Heard once along the Ramal Terra Preta and twice along the CT.

*Schiffornis turdinus (Thrush-like Manakin). 1 ♀, MPEG 13278, Vila Braga, 2 July 1917, col. E. Snethlage and F. Lima; 6 ♂♂, 2 ♀♀, CM, Vila Braga (Griscom and Greenway 1941: 269); 1 ♂, MPEG 15154, Itaituba, 17 November 1955, col. J. Hidasi; 1 sex?, MPEG-A 2096, Flexal = km 212 TAH, 29 October-6 December 1973, col. G. P. Silva. One in undergrowth of mature forest along the CT.

TYRANNIDAE (51/60 SPECIES)

Pachyramphus rufus (Cinereous Becard). 1 ♂, 1 ♀, MPEG 13297–98, Vila Braga, 15 June 1917 (both), col. E. Snethlage and F. Lima.

*Pachyramphus castaneus (Chestnut-crowned Becard). Two heard in secondary forest near Tracoá.

*Pachyramphus polychopterus (White-winged Becard). Uncommon; a few seen or heard daily in secondary forest along the TAH.

*Pachyramphus marginatus (Black-capped Becard). 1 ♀, MPEG 6275, Vila Braga, 31 October 1908, col. E. Snethlage. At least one or two (usually a pair) found each day with canopy mixed-species flocks along both trails; also noted in the canopy of disturbed forest along the TAH (where it overlaps with the last species).

*Pachyramphus minor (Pink-throated Becard). 1 ♀, MPEG 6273, Vila Braga, 31 October 1908, col. E. Snethlage; 1 ♀, CM, Vila Braga (Griscom and Greenway 1941:261). A female seen in a tall treetop along the TAH Between Tracoá and Uruá.

*Tityra semifasciata (Masked Tityra). A few single birds and pairs seen in treetops along the TAH. Surprisingly, no other member of the genus encountered.

*Knipolegus poecilocercus (Amazonian Black-Tyrant). 1 ♂, 1 imm. ♂, MPEG 13175–76, Vila Braga, 5 June 1917 (both), col. E. Snethlage and F. Lima; 2 ♂♂, 3 ♀♀, MPEG 13177–81, Ilha de Goyana, 15/20/20/20 June 19/14 July 1917, col. E. Snethlage and F. Lima. A male seen in riverine vegetation along the Tapajós in August 1979.

*Fluvicola albiventer (Black-backed Water-Tyrant). A pair seen in a pond along the TAH in 1979.

*Pyrocephalus rubinus (Vermilion Flycatcher). 1 imm. ♂, MPEG 13182, Ilha de Goyana, 20 June 1917, col. E. Snethlage and F. Lima.

*Ochthornis littoralis (Drab Water-Tyrant). 1 ♂, 1 ♀, MPEG 13183–84, Vila Braga, 24 June 1917, col. E. Snethlage and F. Lima; 2 ♂♂, 1 ♀, CM, Vila Braga (Griscom and Greenway 1941: 272). Seen along the banks of the Tapajós regularly in 1978 and 1979.

*Tyrannus savana (Fork-tailed Flycatcher). A regional migrant, common in open areas in August 1979 only.

*Tyrannus melancholicus (Tropical Kingbird). Common in clearings, around marshes, and along forest borders; 50+ noted daily along the TAH and in adjacent areas.

Myiodynastes maculatus (Streaked Flycatcher). Seen several times in high trees at forest edge near Uruá in August 1979.

Empidonomus varius (Variegated Flycatcher). 2 ♀♀, MPEG 5084–85, Itaituba and Ilha de Goyana, 19 December 1906/13 January 1907, col. E. Snethlage. A pair frequented the trees around the guard station at Uruá; the birds appeared to be territorial and may have been nesting.

Legatus leucophaius (Piratic Flycatcher). 1 ♂, MPEG 5071, Ilha do Papagaio, 28 December 1906, col. E. Snethlage. Fairly common, in canopy of disturbed forest along the TAH and both forest trails. Single singing birds observed around two colonies of Olive Oropendolas (*Gymnostinops yuracares*) and Yellow-rumped Caciques (*Cacicus cela*) at TS and to the west (TAH).

Conopias trivirgata (Three-striped Flycatcher). Small (family?), noisy groups of this flycatcher observed two or three times in road-edge canopy (TAH); these associated with a variety of other species in loosely formed mixed-species flocks.

Megarynchus pitangua (Boat-billed Flycatcher). 1 ♀, MPEG 6227, Ilha de Goyana, 23 October 1908, col. E. Snethlage. Uncommon (2–3 pairs daily), but found in all localities.

Myiozetetes luteiventris (Dusky-chested Flycatcher). Two pairs and two family groups (3–4) observed perched atop low trees and in bushes along the edge of the TAH, and the families in the canopy of forest along the CT. This supposedly local species is easily overlooked, particularly in the forest where often found in the treetops. The nasal and buzzy calls are quite like those of other *Myiozetetes* species, especially the Gray-capped Flycatcher (*M. granadensis*).

Myiozetetes cayanensis (Rusty-margined Flycatcher). 1 ♂, 1 ♀, MPEG 13185–86, Vila Braga, 26 June 1917, col. E. Snethlage and F. Lima. Very common in roadside marshes and adjacent forest-edge, in second-growth along streams, and in clearings (TS, UT); in pairs (+), usually atop bushes and low trees. As in southeastern Venezuela and the Guianas, this species seems to replace the widespread Social Flycatcher (*M. similis*), although the Rusty-margined seems more tied to marshes and streams; the two species do co-occur in several parts of Amazonia.

Pitangus sulphuratus (Great Kiskadee). Fairly common.

Pitangus lictor (Lesser Kiskadee). Less common than the two previous species, but pairs found in several marshes along the TAH and also along the stream at TS.

Attila spadiceus (Bright-rumped Attila). 1 ♀, MPEG 5104, Itaituba, 15 January 1907, col. E. Snethlage; 1 ♀, MPEG 13308, Vila Braga, 4 June 1917, col. E. Snethlage and F. Lima; 1 sex?, MPEG-A 1864, Flexal = km 212 TAH, 29 October-6 December 1973, col. G. P. Silva; 1 ♀, MPEG 34237, km 120 TAH, 29 October 1978, col. D. C. Oren. Fairly common; one or two heard in the canopy in all areas.

Attila cinnamomeus (Cinnamon Attila). 1 ♂, MPEG 5105, Ilha de Goyana, 21 December 1906, col. E. Snethlage; 1 sex?, MPEG-A 1865, Flexal = km 212 TAH, 29 October-6 December 1973, col. G. P. Silva. One seen in river-edge forest at Buburé.

Laniocera hypopyrra (Cinereous Mourner). 1 sex?, MPEG 5108, Itaituba, 12 January 1907, col. E. Snethlage; 1 ♂, 1 ♀, MPEG 13306–07, Vila Braga, 7 June 19/18 July 1917, col. E. Snethlage and F. Lima; 1 ♀, MPEG 47865, km 19 TAH, 14 July 1972, col. G. P. Silva.

Rhytipterna simplex (Grayish Mourner). 1 ♀, 1 sex?, MPEG 6284–85, Vila Nova and Vila Braga, 22 October/6 November 1908, col. E. Snethlage; 1 ♀, MPEG 47866, km 19 TAH, 25 July 1972, col. G. P. Silva; 1 sex?, MPEG 340, km 120 TAH, 27 August 1979, col. D. C. Oren. A few heard (several also seen) in all areas.

Myiarchus ferox (Short-crested Flycatcher). A few recorded daily in second-growth along the TAH and in clearings at TS and UT.

Myiarchus tuberculifer (Dusky-capped Flycatcher). 1 ♀, MPEG 6228, Vila Braga, 23 October 1908, col. E. Snethlage. One heard in forest bordering the TAH.

Lathrotriccus euleri (Euler's Flycatcher). One singing at dawn in the middlestory of forest along the CT (Parker).

Cnemotriccus fuscatus (Fuscous Flycatcher). 1 sex?, 1 ♀, MPEG 6229–30, Ilha de Goyana, 20 October 1908, col. E. Snethlage; 3 ♂♂, MPEG 13188–90, Vila Braga, 5/5/15 June 1917, col. E. Snethlage and F. Lima; 3♂♂, 2♀♀, 1 sex?, CM, Ilha de Goyana (Griscom and Greenway 1941:280); 1 ♂, MPEG 15430, Itaituba, 12 October 1955, col. J. Hidasi.

Terenotriccus erythrurus (Ruddy-tailed Flycatcher). 1 sex?, MPEG 6231, Vila Braga, 31 October 1908, col. E. Snethlage; 1 ♂, MPEG 15431, Itaituba, 3 November 1955, col. J. Hidasi; 4 sex?, MPEG-A 29–32, Flexal = km 212 TAH, 29 October–6 December 1973, col. G. P. Silva. Heard once in mid-level of forest along UT.

Neopipo cinnamomea (Cinnamon Tyrant). 1 ♂, CM, Vila Braga (Griscom and Greenway 1941:268). Mist-netted, banded and released by Oren and Zickefoose in tall second-growth forest at Uruá in August 1979.

Myiobius barbatus (Sulphur-rumped Flycatcher). 3 ♂♂, MPEG 6233–35, Vila Braga, 26/29 October/4 November 1908, col. E. Snethlage; 1 ♀, MPEG 15436, Itaituba, 2 November 1955, col. J. Hidasi; 3 sex?, MPEG-A 2372, 2374–75, Flexal = km 212 TAH, 29 October-6 December 1973, col. G. P. Silva. One seen briefly with a large mixed flock in the lower canopy of mature forest.

Onychorhynchus coronatus (Royal Flycatcher). 1 ♂, MPEG 6575, Vila Braga, 3 November 1908, col. E. Snethlage; 1 sex?, MPEG-A 55, Flexal = km 212 TAH, 29 October-6 December 1973, col. G. P. Silva; 1 ♂, MPEG 34258, km 120 TAH, 26 November 1978, col. D. C. Oren.

Platyrinchus saturatus (Cinnamon-crested Spadebill). 1 ♀, CM, Vila Braga (Griscom and Greenway, 1941:282).

Platyrinchus platyrhynchos (White-crested Spadebill). 2 ♂♂, MPEG 13218–19, Vila Braga, 10/17 June 1917, col. E. Snethlage and F. Lima; 8 sex?, MPEG-A 2272–76, 2284–86, Flexal = km 212 TAH, 29 October–6 December 1973, col. G. P. Silva; 2 ♂♂, MPEG 34259–60, km 120 TAH, 29 October/1 November 1978, col. D. C. Oren.

Platyrinchus coronatus (Golden-crowned Spadebill). 13 sex?, MPEG-A 2251–52, 2261–71, Flexal = km 212 TAH, 29 October-6 December 1973, col. G. P. Silva; 1 ♂, MPEG 34261, km 120 TAH, 7 November 1978, col. D. C. Oren. Fairly common (by voice), but quite inconspicuous, in the upper undergrowth of forest along the CT (two or more noted in three places).

Tolmomyias assimilis (Yellow-margined Flycatcher). 1 ♀, MPEG 13200, Vila Braga, 28 June 1917, col. E. Snethlage and F. Lima.

Tolmomyias poliocephalus (Gray-crowned Flycatcher). 1 ♂, 2 ♀♀, MPEG 13207–09, Vila Braga, 3/6/11 June 1917, col. E. Snethlage and F. Lima. Common in secondary and primary forest canopy in all areas; single birds heard (singing) in numerous places along the TAH and both trails (up to six daily along the TAH); all high (13–23+ m) in taller trees (in contrast to the next species).

Tolmomyias flaviventris (Yellow-breasted Flycatcher). 1 ♂, MPEG 13214, Ilha de Goyana, 15 June 1917, col. E. Snethlage and F. Lima. Very common (20+ recorded in a day) in scrubby forest on sandy soil and along the edge of mature forest in all areas; mostly singly or in pairs apart from other birds, in the foliage of small trees.

Rhynchocyclus olivaceus (Olivaceus Flatbill). 1 ♀, 1 ♂, MPEG 6237–38, Vila Braga, 3 November 1908 (both), col. E. Snethlage; 1 sex?, MPEG-A 2392, Flexal = km 212 TAH, 29 October/6 December 1973, col. G. P. Silva.

Ramphotrigon ruficauda (Rufous-tailed Flatbill). 2 ♂♂, 1 ♀, MPEG 13215–17, 19 June 1917 (all), col. E. Snethlage and F. Lima; 1 ♀, MPEG 15448, Itaituba, 10 October 1955, col. J. Hidasi; 1 ♀, MPEG 30212, km 19 TAH, 26 July 1972, col. G. P. Silva. Fairly common at mid-levels (6–20 m) in mature forest along both trails; 2–4 (2,2+,4) observed apart from other species in at least three places along the CT.

Todirostrum chrysocrotaphum (Painted Tody-Flycatcher). 1 ♂, 1 ♀, CM, Itaituba and Vila Braga (Griscom and Greenway 1941:286). One or two seen in the middle and upper branches of tall trees along the road-edge at Tracoá; one lured down into the canopies of smaller trees through playback of the bird's song: a series of evenly-spaced tic notes much like those of Peruvian *chrysocrotaphum*. This individual, apparently an example of the little-known subspecies *T. c. similis*, had white lores and a white throat, but otherwise looked like the western form *T. c. neglectum*.

Todirostrum maculatum (Spotted Tody-Flycatcher). 1 ♂, MPEG 5054, Itaituba, 16 January 1907, col. E. Snethlage; 1 ♂, 3 ♀♀, MPEG 5051–53/64, Ilha de Goyana, 19 December 1906/1 January 1907/6 November 1908, col. E. Snethlage. Pairs observed in second-growth along the stream and road at Tracoá, and in river-edge trees at Buburé.

Hemitriccus striaticollis (Stripe-necked Tody-Tyrant). 1 ♀, CM, Vila Braga (Griscom and Greenway 1941:289). Several pairs found in shrubbery and low trees along the road-edge and clearing at Tracoá. These birds seen from near the ground up into the upper branches of small trees (up to 7–10 m). They found alone and in the company of other second-growth birds. The species fairly common in scrubby woods opposite the airport in Itaituba.

Hemitriccus minimus (Zimmer's Tody-Tyrant). 1 ♂, 1 ♀, MPEG 5055/68, Vila Braga, 19 December 1906/28 October 1908, col. E. Snethlage; specimens, CM, Vila Braga and Itaituba (Griscom and Greenway 1941:289) (includes the Type, ♂, CM 77,080, Itaituba, 26 February 1926, col. S. M. Klages); 1 sex?, MPEG-A 47, Flexal = km 212 TAH, 29 October-6 December 1973, col. G. P. Silva. A pair found at mid-heights (10–13 m) in dense vines against large trunks and in the crowns of middlestory trees in mature forest along the CT; the birds initially associating with a mixed flock that included woodcreepers, foliage-gleaners and antbirds. Their primary call a protracted, grating churr, somewhat *Oncostoma*-like, but thinner; they also uttered soft tic notes.

Poecilotriccus andrei (Black-chested Tyrant). 1 ♀, CM, Itaituba (Griscom and Greenway 1941:290). This specimen is the Type of *P. klagesi* Todd, 1925, a junior synonym of *andrei*.

**Lophotriccus galeatus* (Helmeted Pygmy-Tyrant). 1 sex?, MPEG-A 00, Flexal = km 212 TAH, 29 October-6 December 1973, col. G. P. Silva. This species heard in disturbed forest in several places along the TAH and UT, and also along the CT; always in the upper understory or at mid-heights in shaded, viny vegetation. Not seen well, but its *Lophotriccus*-like calls are unmistakable (but reminiscent of those of last species).

**Myiornis ecaudatus* (Short-tailed Pygmy-Tyrant). 1 ♀, MPEG 5056, Ilha do Papagaio, 28 December 1906, col. E. Snethlage; 1 ♀, 1 sex?, MPEG 5057/6250, Vila Braga, 4 January 1907/ 28 October 1908, col. E. Snethlage. A common, but very inconspicuous bird of the canopy of all forest types; its cricket-like calls heard from the road-edges and along both forest trails. Usually encountered in pairs apart from other birds; often in the upper, sunlit branches and vines of forest trees, but also lower (down to 7 m) in viny growth along the edges.

**Phylloscartes flaveolus* (Yellow Tyrannulet). 1 ♂, 1 ♀, MPEG 5059–60, Ilha de Goyana, 23 December 1906/3 January 1907, col. E. Snethlage; 1 ♀, MPEG 34268, right bank Tapajós, in front of Uruá, 22 November 1978, col. D. C. Oren. Fairly common in road-edge scrubby forest; encountered in small groups of 3–4 individuals in vines and crowns of small trees (up to 13 m) along TAH and side roads; usually with small mixed-species flocks. None seen or heard inside mature forest.

*Inezia subflava (Pale-tipped Tyrannulet). 1 ♀, MPEG 5058, Vila Braga, 19 December 1906, col. E. Snethlage; 1 ♀, MPEG 6253, Ilha de Goyana, October 1908, col. E. Snethlage.

**Myiopagis gaimardii* (Forest Elaenia). 2 ♂♂, MPEG 5068–69, Itaituba, 12/15 January 1907, col. E. Snethlage; 1 ♂, MPEG 6257, Ilha de Goyana, 22 October 1908, col. E. Snethlage. Very common in the canopy of secondary and primary forest; usually one to two heard with mixed-species flocks; as many as 30 heard in a day along the TAH, and up to six along the CT in forest.

Myiopagis viridicata (Greenish Elaenia). 1 ♂, MPEG 13239, Vila Braga, 23 June 1917, col. E. Snethlage and F. Lima.

**Phaeomyias murina* (Mouse-colored Tyrannulet). 1 ♀, MPEG 5062, Itaituba, 13 January 1907, col. E. Snethlage; 1 ♀, MPEG 130, Ilha de Goyana, 8 June 1917, col. E. Snethlage and F. Lima.ND pairs seen in second-growth (bushes and low trees) at Tracoá and Buburé; the latter quite territorial. Also observed in scrub opposite the airport in Itaituba.

**Camptostoma obsoletum* (Southern Beardless-Tyrannulet). 1 ♂, MPEG 5064, Itaituba, 13 January 1907, col. E. Snethlage. A few noted daily in second-growth at TS, along the TAH, and at Buburé.

**Zimmerius gracilipes* (Slender-footed Tyrannulet). At least 2 pairs observed at TS, in the tops of vine-covered trees along the road and clearing edge. This easily overlooked species probably occurs in the forest canopy as well.

**Tyrannulus elatus* (Yellow-crowned Tyrannulet). 1 ♂, MPEG 5061, Ilha de Goyana, 23 December 1906, col. E. Snethlage; 1 ♂, MPEG 132, Vila Braga, 3 June 1917, col. E. Snethlage and F. Lima. Commonly heard and occasionally seen in the canopy of secondary forest along the TAH; up to 10 recorded daily. A few also heard in forest canopy along the UT.

*Ornithion inerme (White-lored Tyrannulet). A few heard daily in the canopy of both forest types. This species is very inconspicuous; it is usually seen in sunlit, vine-covered tree crowns.

**Mionectes oleagineus* (Ochre-bellied Flycatcher). 2 ♂♂, MPEG 133–44, Vila Braga, 22 June 19/16 July 1917, col. E. Snethlage and F. Lima; 10 sex?, MPEG-A 2336–45, Flexal = km 212 TAH, 29 October-6 December 1973, col. G. P. Silva. Two or three heard (displaying) at mid-height in shaded forest along the CT.

**Mionectes macconnelli* (McConnell's Flycatcher). 2 sex?, MPEG-A 23–25, Flexal = km 212 TAH, 29 October-6 December 1973, col. G. P. Silva. Seen in high forest with H. Sick in August 1979.

**Corythopis torquata* (Ringed Antpipit). 4 sex?, MPEG-A 2310–13, Flexal = km 212 TAH, 29 October-6 December 1973, col. G. P. Silva. One walking on the ground (and singing) in the deep shade of mature forest (CT) our only record. Whether the species is uncommon locally or relatively silent during our stay remains to be determined.

HIRUNIDINIDAE (8/9 SPECIES)

**Tachycineta albiventer* (White-winged Swallow). 1 ♂, 1 ♀, CM, Vila Braga (Griscom and Greenway 1941:299). Common over the river at Uruá and Buburé; a few over marshes along the TAH.

Progne tapera (Brown-chested Martin). 1 ♂, MPEG 5022, Vila Braga, 4 January 1907, col. E. Snethlage. Three seen over the airstrip at Uruá.

Progne subis (Purple Martin). 1 ♂, CM, Ilha de Goyana, 23 December 1919 (Griscom and Greenway 1941:297).

Progne chalybea (Gray-breasted Martin). 1 ♂, CM, Ilha de Goyana (Griscom and Greenway 1941:297). Uncommon; six or more perched atop snags over a marsh along the TAH west of Tracoá, and a few others seen over Tracoá and Buburé. This species common over cleared lands east of the park and very common over residential areas in Itaituba.

Notiochelidon cyanoleuca (Blue-and-white Swallow). Large flocks numbering up to 100 seen daily at Uruá in August and September 1978; presumably southern migrants.

Atticora fasciata (White-banded Swallow). Common along the stream and over the clearing at TS (as many as 20 at once); also low over the river at Uruá and Buburé; many on river rocks.

Atticora melanoleuca (Black-collared Swallow). Seen regularly in rapids in the Rio Tapajós in 1978, sometimes in the company of the previous species.

Stelgidopteryx ruficollis (Southern Rough-winged Swallow). The most common swallow in the park, occurring in all open areas, especially along the TAH; dozens seen every day.

Hirundo rustica (Barn Swallow). 1 ♀, CM, Itaituba, 13 March 1920 (Griscom and Greenway 1941:299). One seen over the river at Buburé on 10 November, and noted over the airport at Itaituba on 14 November.

CORVIDAE (0/1 SPECIES)

Cyanocorax chrysops (Plush-crested Jay). 1 sex?, MPEG 14882, Itaituba, 5 November 1955, col. J. Hidasi.

TROGLODYTIDAE (6/6 SPECIES)

Campylorhynchus turdinus (Thrush-like Wren). Uncommon in forest edge in 1978 and 1979.

Thryothorus genibarbis (Moustached Wren). 1 ♂, MPEG 6151, Ilha de Goyana, 21 October 1908, col. E. Snethlage; 3 ♂♂, MPEG 13049–51, Vila Braga, 5/13 June 19/12 July 1917, col. E. Snethlage and F. Lima. This species common and vocal in low, dense second-growth along the forest edges in all areas; also found in disturbed forest (in tangles as high as 13 m) along the UT and CT.

Thryothorus leucotis (Buff-breasted Wren). 2 ♂♂, MPEG 5017–18, Ilha de Goyana, 23/25 December 1906, col. E. Snethlage; 1 sex?, MPEG 34305, Uruá, 27 August 1979, col. H. Sick; 1 ♂, MPEG 34506, right bank Tapajós, in front of Uruá, 22 November 1978, col. D. C. Oren. A few pairs heard in tangled vegetation along streams crossing the TAH; also in river-edge shrubbery and seasonally flooded forest at Buburé and in scrubby woods opposite the airport in Itaituba.

Troglodytes aedon (House Wren). 1 ♂, MPEG 6158, Vila Braga, 7 November 1908, col. E. Snethlage. Uncommon in the park; heard (1–2) in the clearing edges at TS and US; also several in recently cleared land east of the park and in Itaituba.

Microcerculus marginatus (Nightingale Wren). Specimen(s), CM, Vila Braga (Griscom and Greenway 1941:302). 3 sex?, MPEG-A 2633–35, Flexal = km 212 TAH, 29 October–6 December 1973, col. G. P. Silva; 1 ♂, MPEG 34307, km 120 TAH, 17 November 1978, col. D. C. Oren. Heard once, in mature forest along the CT.

Cyphorhinus aradus (Musician Wren). 6 ♂♂, 4 ♀♀, MPEG 13035–44, 12/16/16/16 June 19/7/13/13/16/22 July 1917, col. E. Snethlage and F. Lima; 5 sex?, MPEG-A 2705–09, Flexal = km 212 TAH, 29 October–6 December 1973, col. G. P. Silva. One to two groups of 2–4 individuals encountered daily low in forest undergrowth and on the ground along the CT.

TURDIDAE (3/4 SPECIES)

Catharus fuscescens (Veery). 1 ♀, MPEG 15461, Itaituba, 2 November 1955, col. J. Hidasi.

Turdus leucomelas (Pale-breasted Thrush). 1 sex?, MPEG 5016, Ilha de Goyana, 3 January 1907, col. E. Snethlage; 1 sex?, MPEG 15469, Itaituba, 15 November 1955, col. J. Hidasi. Regularly detected around the Uruá station in 1978 and 1979.

Turdus fumigatus (Cocoa Thrush). One seen well and tape-recorded as it sang from mid-heights in river-edge trees at Buburé; the songs (two types heard—one in response to playback) very like those of Peruvian *T. hauxwelli*.

Turdus albicollis (White-necked Thrush). 1 ♂, 1 imm. ♀., MPEG 13033–34, Vila Braga, 26

June/19/11 July 1917, col. E. Snethlage and F. Lima; 1 ♂, MPEG 34311, km 120 TAH, 29 October 1978, col. D. C. Oren.

SYLVIIDAE (1/1 SPECIES)

Ramphocaenus melanurus (Long-billed Gnatwren). 4 ♀♀, MPEG 13531–34, Vila Braga, 6/6 June/19/12/21 July 1917, col. E. Snethlage. Pairs observed in vine tangles in the middlestory of disturbed forest at TS, and in forest along the CT.

VIREONIDAE (6/7 SPECIES)

Cyclarhis gujanensis (Rufous-browed Peppershrike). 1 ♀, MPEG 13081, Vila Braga, 4 July 1917, col. E. Snethlage and F. Lima. Our only record of this species of one heard (songs and calls tape-recorded) in the canopy of forest along the CT.
Vireolanius leucotis (Slaty-capped Shrike-Vireo). Single birds heard (one also seen) with lower canopy flocks along the UT and CT.
Vireo olivaceus chivi (Red-eyed Vireo). 1 ♂, MPEG 6156, Ilha de Goyana, 22 October 1908, col. E. Snethlage. This vireo common in scrubby woods at Buburé, and a few noted daily in canopy flocks in mature forest (UT,CT); also common in scrub at Itaituba.
Vireo altiloquus (Black-whiskered Vireo). 1 ♂, CM, Vila Braga, 10 December 1919 (Griscom and Greenway 1941:309).
Hylophilus semicinereus (Gray-naped Greenlet). 1 ♂, MPEG 5020, Itaituba, 12 January 1907, col. E. Snethlage; 1 ♂, 1 ♀, MPEG 13060–61, Ilha de Goyana, 30 June 19/14 July 1917, col. E. Snethlage and F. Lima; specimen(s), CM, Vila Braga (Griscom and Greenway 1941:311). Common and vocal in road-edge woods (all along TAH, and at TS, UT); mainly found in pairs in the leafy crowns and tangled branches of small to medium-sized trees; often with other small birds.
Hylophilus muscicapinus (Buff-cheeked Greenlet). 1 ♀ (Type of *H. m. griseifrons* (Snethlage 1907), MPEG 5021, Vila Braga, 10 January 1907, col. E. Snethlage (the Type specimen sent by Snethlage to the Zoological Museum, Berlin after she left the Museu Goeldi; it is registered as ZMB-30.3173 (Haffer *in litt.*)); 1 ♀, MPEG 6159, Vila Braga, 23 October 1908, col. E. Snethlage. Apparently fairly common in the canopy of mature forest. One or two detected with mixed-species flocks from the upper middlestory up into the lower canopy (UT, one pair; CT, up to three pairs). These seen primarily amidst small leaves near the ends of slender branches, often in association with other small insectivores such as Ash-winged Antwrens.
Hylophilus ochraceiceps (Tawny-crowned Greenlet). 1 ♂, MPEG 6158, Vila Braga, 29 October 1908, col. E. Snethlage; 1 sex?, MPEG-A 2558, Flexal = km 212 TAH, 29 October–6 December 1973, col. G. P. Silva; 1 ♀, MPEG 34314, km 120 TAH, 26 November 1978, col. D. C. Oren. A common but inconspicuous member of mixed-species flocks in mature forest undergrowth; mainly seen from 1 to 4 m, occasionally as high as 8 m. Associated closely with *Thamnomanes* antshrikes (especially *T. saturninus*) and *Myrmotherula* antwrens; pairs seen with one to two mixed-species flocks along the UT, and up to four along the CT. Owing to morphological and vocal differences, this population (*H. o. lutescens*) might best be regarded as a distinct species.

ICTERIDAE (6/8 SPECIES)

Scaphidura oryzivora (Giant Cowbird). This species uncommon; a few solitary birds noted daily, usually flying high over the TAH.
Molothrus bonariensis (Shiny Cowbird). 2 ♂♂, 1 ♀, CM, Ilha de Goyana (Griscom and Greenway 1941:318).
Psarocolius viridis (Green Oropendola). 2 ♂♂, 1 ♀, MPEG 13171–73, Vila Braga, 14/18/19 July 1917, col. E. Snethlage and F. Lima; 1 ♂, 2 ♀♀, CM, Vila Braga (Griscom and Greenway, 1941:317). Scarce; an apparently wandering bird seen west of TS (it sang repeatedly as it moved from one tall tree to another). Others heard at a great distance as we walked the CT.
Psarocolius yuracares (Olive Oropendola). Common, particularly at TS and along the TAH mid-way to Uruá where there are active colonies in tall, isolated trees (30+ nests in both).
Cacicus cela (Yellow-rumped Cacique). Common and widespread; nesting in trees used by last species, but in smaller numbers.
Cacicus haemorrhous (Red-rumped Cacique). 1 ♀, MPEG 5042, Vila Braga, 5 January 1907, col. E. Snethlage; 7 ♂♂, 2 ♀♀, CM, Vila Braga (Griscom and Greenway 1941:317).

Icterus cayanensis (Epaulet Oriole). Several (1–2 pairs?) heard daily in mature forest canopy along the CT associating with large mixed-species flocks.

Leistes militaris (Red-breasted Blackbird). 1 ♂, CM, Itaituba (Griscom and Greenway 1941: 319).

PARULIDAE (2/2 SPECIES)

*Granatellus pelzelni (Rose-breasted Chat). 2 ♂♂, MPEG 5023-24, Vila Braga and Itaituba (resp.), 10/15 January 1907, col. E. Snethlage. Single males observed with small mixed-species flocks in viny tangles along the road-edge opposite TS (from 3 to 8 m above the ground) and in forest along the CT (12+ m).

*Basileuterus rivularis (River Warbler). 1 ♂, CM, Vila Braga (Griscom and Greenway 1941: 315). A pair in vegetation overhanging a small stream on the CT; these birds also seen in nearby undergrowth over army ants.

THRAUPIDAE (25/36 SPECIES)

*Coereba flaveola (Bananaquit). 1 ♂, MPEG 5030, Ilha de Goyana, December 1906, col. E. Snethlage; 1 ♂, MPEG 13105, Vila Braga, 11 July 1917, col. E. Snethlage and F. Lima. Common in second-growth along the TAH. Mostly scattered, solitary individuals.

Conirostrum speciosum (Chestnut-vented Conebill). 1 ♀, MPEG 5028, Ilha de Goyana, 21 December 1906, col. E. Snethlage.

*Cyanerpes caeruleus (Purple Honeycreeper). 1 ♀, MPEG 5029, Vila Braga, 5 January 1907, col. E. Snethlage; 2 ♂♂, 2 ♀♀, CM, Vila Braga (Griscom and Greenway 1941:312). A few observed daily in treetops at the forest edge (TAH).

*Cyanerpes cyaneus (Red-legged Honeycreeper). 1 imm. ♂, MPEG 13101, Ilha de Goyana, 30 June 1917, col. E. Snethlage and F. Lima; 2 ♂♂, MPEG 15499/500, Itaituba, 13/14 November 1955, col. J. Hidasi. Seen regularly in mixed-species flocks at flowering trees in 1978 and 1979.

*Chlorophanes spiza (Green Honeycreeper). 1 imm. ♂, MPEG 13100, Vila Braga, 16 July 1917, col. E. Snethlage and F. Lima; 1 ♂, CM, Vila Braga (Griscom and Greenway 1941:312). A pair seen once at Terra Preta in August 1979.

*Dacnis cayana (Blue Dacnis). 3 ♀♀, MPEG 13083–85, Ilha de Goyana, 30/30 June/19/14 July 1917, col. E. Snethlage and F. Lima; 1 ♂, 1 ♀, MPEG 13086–87, Vila Braga, 17 July 1917, col. E. Snethlage and F. Lima. Fairly common; one to several pairs and a few solitary birds seen daily in second-growth and canopy of road-edge forest along TAH.

*Dacnis flaviventer (Yellow-bellied Dacnis). 2 ♂♂, 1 ♀, MPEG 5025–27, Ilha de Goyana, 20 December 1906 (all), col. E. Snethlage; specimen(s), CM, Ilha de Goyana (Griscom and Greenway, 1941:313). A pair in the tops of river-edge trees at Buburé represent our only record.

*Tersina viridis (Swallow Tanager). At least 20 (an equal number of males and females) counted in one tree along the TAH east of Uruá.

Euphonia minuta (White-vented Euphonia). 1 ♂, 1 ♀, MPEG 13106–07, Ilha de Goyana, 20 June 1917 (both), col. E. Snethlage and F. Lima; 1 ♂, MPEG 15148, Itaituba, 7 November 1955, col. J. Hidasi.

*Euphonia chlorotica (Purple-throated Euphonia). 1 ♂, 1 ♀, MPEG 13108–09, 7 June 1917, col. E. Snethlage and F. Lima. Uncommon; a few heard daily in second-growth along the TAH and in riverside woods at Buburé.

*Euphonia violacea (Violaceous Euphonia). 1 ♂, MPEG 5032, Ilha de Goyana, 31 December 1906, col. E. Snethlage; 2 ♂♂, MPEG 13112, Vila Braga, 18 June 19/22 July 1917, col. E. Snethlage and F. Lima. One or two pairs seen each day along the TAH; also noted (a few) at Buburé. All in treetops, often in bare branches, at the forest edge.

*Euphonia ruficentris (Rufous-bellied Euphonia). 1 ♂, MPEG 13114, Vila Braga, 17 June 1917, col. E. Snethlage and F. Lima; 1 ♂, CM, Vila Braga (Griscom and Greenway 1941:322). One or two of these euphonias heard daily in the canopy of disturbed (TAH) and mature (CT) forest.

*Euphonia chrysopasta (Golden-bellied Euphonia). 1 ♂, CM, Vila Braga (Griscom and Greenway 1941:323). A small flock seen at Terra Preta in 1979.

Tangara velia (Opal-rumped Tanager). 2 ♂♂, MPEG 13115–16, Vila Braga, 16/22 July 1917, col. E. Snethlage and F. Lima; 1 ♂, MPEG 15054, Itaituba, 9 November 1955, col. J. Hidasi.

*Tangara chilensis (Paradise Tanager). 2 ♀♀, MPEG 13117–18, Vila Braga, 16/25 July 1917, col. E. Snethlage and F. Lima; 2 ♂♂, CM, Vila Braga (Griscom and Greenway 1941:323); 1 ♂, MPEG 15116, Itaituba, 4 November 1955, col. J. Hidasi. Five birds seen in a large *Cecropia*

along the TAH just west of TS (Parker). Others may have been heard along the CT. Small tanagers curiously scarce, and few high-canopy flocks encountered.

Tangara punctata (Spotted Tanager). 1 ♂, CM, Vila Braga (Griscom and Greenway, 1941:3).

*Tangara varia (Dotted Tanager). 1 ♀, MPEG 5033, Vila Braga, 5 January 1907, col. E. Snethlage; 1 ♀, CM, Vila Braga (Griscom and Greenway 1941:3); 1 ♂, MPEG 15133, Itaituba, 22 November 1955, col. J. Hidasi.

**Tangara nigrocincta* (Masked Tanager). Seen occasionally in fruiting and flowering trees in 1978.

Tangara mexicana (Turquoise Tanager). 1 ♀, MPEG 13123, Ilha de Goyana, 15 June 1917, col. E. Snethlage and F. Lima; 1 ♂, 2 ♀♀, MPEG 131–26, 25/25/10 July 1917, col. E. Snethlage and F. Lima.

Tangara gyrola (Bay-headed Tanager). 2 ♂♂, 1 ♀, MPEG 13128–30, Vila Braga, 25 July 1917 (all), col. E. Snethlage and F. Lima.

**Thraupis episcopus* (Blue-gray Tanager). 1 ♀, MPEG 5035, Itaituba, 13 January 1907, col. E. Snethlage; 1 ♂, MPEG 13131, Ilha de Goyana, 30 June 1917, col. E. Snethlage and F. Lima. Common in open habitats, including road-edge and clearing-edge second-growth in all localities visited.

**Thraupis palmarum* (Palm Tanager). Almost as common as last, and in similar habitats; seemed to prefer taller (disturbed) forest, and also seen in the canopy of mature forest.

**Ramphocelus carbo* (Silver-beaked Tanager). 1 imm. ♂, MPEG 5036, Vila Braga, December 1906, col. E. Snethlage; 4 sex?, MPEG-A 2937–40, Flexal = km 212 TAH, 29 October–6 December 1973, col. G. P. Silva. Common to abundant in edge habitats, where found from low in shrubby growth to the canopy.

Piranga flava (Hepatic Tanager). 1 ♂, MPEG 15100, Itaituba, 19 October 1955, col. J. Hidasi.

*Habia rubica (Red-crowned Ant-Tanager). 2 ♂♂, 2 ♀♀, MPEG 6192–95, Vila Braga, 26/26 October/3/4 November 1908, col. E. Snethlage; specimen(s), CM, Vila Braga (Griscom and Greenway 1941:327); 1 ♂, 1 ♀, MPEG 34333–34, km 120 TAH, 7 November 19/26 November 1978, col. D. C. Oren. Fairly common in undergrowth and middlestory (1 to 13 m) of mature forest along CT.

*Lanio versicolor (White-winged Shrike-Tanager). 1 ♂, MPEG 15095, Itaituba, 16 November 1955, col. J. Hidasi; 1 sex?, MPEG-A 2728, Flexal = km 212 TAH, 29 October–6 December 1973, col. G. P. Silva. A pair seen at 23 m in the canopy of mature forest along the CT.

**Tachyphonus rufus* (White-lined Tanager). Single males seen in scrub at Uruá and along the TAH east of the park.

**Tachyphonus cristatus* (Flame-crested Tanager). 1 ♀, 1 ♂, MPEG 6205–06, Vila Braga, 5/7 November 1908, col. E. Snethlage; 5 ♂♂, 2 ♀♀, CM, Vila Braga (Griscom and Greenway 1941: 329); 1 ♀, MPEG 15050, Itaituba, 7 November 1955, col. J. Hidasi; 1 sex?, MPEG-A 2943, Flexal = km 212 TAH, 29 October–6 December 1973, col. G. P. Silva; 1 ♀, MPEG 34336, km 120 TAH, 7 November 1978, col. D. C. Oren. Several observed with one lower canopy-flock along the UT (two males and a female), and with two or three flocks along the CT.

**Tachyphonus surinamus* (Fulvous-crested Tanager). 1 ♂, MPEG 13154, Vila Braga, 18 June 1917, col. E. Snethlage and F. Lima; 1 ♀, MPEG 34337, km 120, Trans-Amazonian Highway, 25 November 1978, col. D. C. Oren.

Tachyphonus phoenicius (Red-shouldered Tanager). 1 ♂, MPEG 15048, Itaituba, 8 November 1955, col. J. Hidasi.

**Tachyphonus luctuosus* (White-shouldered Tanager). 1 ♂, MPEG 6203, Vila Braga, October 1908, col. E. Snethlage. Pairs (+) observed in most of the lower-canopy flocks along the UT, but also with middlestory flocks along both trails.

Hemithraupis guira (Guira Tanager). 1 ♂, 1 ♀, MPEG 13155–56, Vila Braga, 23 June 1917 (both), col. E. Snethlage and F. Lima.

Thlypopsis sordida (Orange-headed Tanager). 1 ♂, MPEG 13157, Vila Braga, 20 June 1917, col. E. Snethlage and F. Lima.

**Lamprospiza melanoleuca* (Red-billed Pied Tanager). 1 ♀, MPEG 6215, Vila Braga, 7 November 1908, col. E. Snethlage. Quite common in high forest in August 1979.

**Cissopis leveriana* (Magpie Tanager). 2 ♂♂, CM, Itaituba (Griscom and Greenway 1941: 331). Fairly common (to 3 or 4 pairs+ daily) in forest edge along the TAH, and in all other localities; not seen in groups as in some other places.

Schistochlamys melanopis (Black-faced Tanager). 1 ♀, MPEG 15032, Itaituba, 11 November 1955, col. J. Hidasi.

EMBERIZIDAE (11/11 SPECIES)

Saltator maximus (Buff-throated Saltator). 1 ♂, MPEG 6212, Ilha de Goyana, October 1908, col. E. Snethlage; 1 ♂, MPEG 34358, right bank Tapajós, in front of Uruá, 22 November 1978, col. D. C. Oren. Quite common (to 10+ in a morning) in forest edge along the TAH and side roads.

Saltator coerulescens (Grayish Saltator). 1 ♂, MPEG 13159, Ilha de Goyana, 20 June 1917, col. E. Snethlage and F. Lima; 1 ♂, 1 ♀, CM, Itaituba (Griscom and Greenway 1941:337). A few in riverside shrubbery and woods at Buburé.

Saltator grossus (Slate-colored Grosbeak). 2 ♂♂, MPEG 6216–17, Vila Braga, 31 October/4 November 1908, col. E. Snethlage. Fairly common in upper middlestory and canopy of disturbed forest along TAH and side roads; also several heard daily along UT and CT.

Paroaria gularis (Red-capped Cardinal). 1 ♂, MPEG 5050, Ilha de Goyana, December 1906, col. E. Snethlage. One seen at Buburé in scrub (not near water); another in roadside second-growth at TS.

Cyanocompsa cyanoides (Blue-black Grosbeak). 1 imm. ♂, MPEG 13166, Vila Braga, 12 July 1917, col. E. Snethlage and F. Lima; 1 ♂, MPEG 47978, km 19 TAH, 14 July 1972, col. G. P. Silva; 1 sex?, MPEG-A 2957, Flexal = km 212 TAH, 29 October–6 December 1973, col. G. P. Silva. One or two heard in disturbed forest undergrowth at TS.

Volatinia jacarina (Blue-black Grassquit). 1 ♂, MPEG 6220, Ilha de Goyana, 20 October 1908, col. E. Snethlage. Common in weeds in the clearings at TS and US, also along the TAH in several places.

Sporophila lineola (Lined Seedeater). 1 ♂, 1 imm. ♂, MPEG 5047–48, Ilha de Goyana, 1 January 1907 (both), col. E. Snethlage; 1 sex?, MPEG 13169, Vila Braga, 16 June 1917, col. E. Snethlage and F. Lima; 1 imm. ♂, CM, Itaituba (Griscom and Greenway 1941:335). Several (at least three) males and 12+ (probable) females observed in weeds and grasses at UT. Also recorded at this same locality in August 1979.

Sporophila castaneiventris (Chestnut-bellied Seedeater). 1 ♂, MPEG 5046, Itaituba, 13 January 1907, col. E. Snethlage; 1 ♂, CM, Ilha de Goyana (Griscom and Greenway 1941:333). Single males seen in river-edge shrubbery at Buburé and in a marsh-edge east of Tracoá.

Oryzoborus angolensis (Lesser Seed-Finch). 1 ♂, MPEG 5043, Ilha de Goyana, 21 December 1906, col. E. Snethlage. Common in road-edge second-growth along TAH, including areas of TS and UT; numerous singing males heard (up to 12 in a day).

Arremon taciturnus (Pectoral Sparrow). 1 ♂, MPEG 5040, Ilha Campinho, 28 December 1906, col. E. Snethlage; 2 ♂♂, CM, Vila Braga (Griscom and Greenway 1941:339); 1 sex?, MPEG-A 2956, Flexal = km 212 TAH, 29 October–6 December 1973, col. G. P. Silva; 1 ♂, MPEG 34367, km 120 TAH, 25 November 1978, col. D. C. Oren. Common (by voice) in dense undergrowth along the forest edge and in disturbed forest (TAH and side roads); up to 12 heard in a morning (six or more pairs).

Ammodramus aurifrons (Yellow-browed Sparrow). 1 ♂, 1 ♀, MPEG 5049/6221, Ilha de Goyana, 19 December 1906/23 October 08, col. E. Snethlage; 1 ♂, MPEG 13170, Vila Braga, 5 June 1917, col. E. Snethlage and F. Lima. Several seen in sparse, low vegetation on the riverbank at Buburé.

ACKNOWLEDGMENTS

We thank park director Egydio Castro and park guards Carlos Melo, Joaquim da Silva and José Salles de Souza of IBDF for making our visits possible. Parker extends gratitude to Christoph and Ina Hrdina of André Safaris in Brasília, and Victor Emanuel of Victor Emanuel Nature Tours for providing generous logistic and financial assistance for his travels. Oren thanks Lynn Branch, at the time working under a Peace Corps (USA) agreement with IBDF, for the invitation to work in the Park and for support when there. We are extremely grateful for the many helpful comments made by J. V. Remsen, J. Haffer, R. O. Bierregaard, and T. S. Schulenberg on earlier versions of this paper.

LITERATURE CITED

CHAPMAN, F. M. 1921. Descriptions of some apparently new birds from Bolivia, Brazil, and Venezuela. Amer. Mus. Novit. 2:1–8.

GRISCOM, L., AND J. C. GREENWAY, JR. 1941. Birds of Lower Amazonia. Bull. Mus. Comp. Zool. 88:81–344.

CUNHA, O. R. DA. 1989. Maria Elizabeth Emília Snethlage (1868–1929). Pages 83–102 in Talento e Atitude: Estudos Biográficos do Museu Emílio Goeldi. Museu Paraense Emílio Goeldi, Belém.

GONZAGA, L. P. 1989a. Catálogo dos Tipos na coleção ornitológica do Museu Nacional. I—Não Passeriformes. Bol. Mus. Para. Emílio Goeldi, sér. Zool. 5:9–40.
GONZAGA, L. P. 1989b. Catálogo dos Tipos na coleção ornitológica do Museu Nacional. II -/Passeriformes. Bol. Mus. Para. Emílio Goeldi, sér. Zool. 5:41–69.
GRANTSAU, R., AND H. F. DE ALMEIDA CAMARGO. 1989. Nova espécie brasileira de *Amazona* (Aves, Psittacidae). Rev. Brasil. Biol. 49:1017–1020.
HAFFER, J. 1990. Avian species richness in tropical South America. Studies Neotrop. Fauna Environ., Ecol. Syst. 25:157–183.
OREN, D. C., AND H. G. ALBUQUERQUE. 1991. Priority areas for new avian collections in Brazilian Amazonia. Goeldiana Zool. 6:1–11.
OREN, D. C., AND E. O. WILLIS. 1981. New Brazilian records for the Golden Parakeet *Aratinga guarouba*. Auk 98:394-396.
PETERS, J. L. 1951. Catalogue of Birds of the World, vol. 7. Museum of Comparative Zoology, Cambridge, Mass.
PINTO, O. M. O. 1938. Catálogo das aves do Brasil, primeira parte. Rev. Mus. Paulista 22:1–566.
PINTO, O. M. O. 1944. Catálogo das aves do Brasil, segunda parte. Departamento de Zoologia, Secretaria de Agricultura, São Paulo.
RADAM-BRASIL. 1975. Projeto RADAM-Brasil, vol. 8: Tapajós. DNPM, Rio de Janeiro.
SICK, H. 1985. Ornitologia Brasileira: Uma Introdução. Editora Universidade de Brasília, Brasília.
SNETHLAGE, E. 1907. Neue Vogelarten aus Südamerika. Orn. Monatber. 15:160–164.
SNETHLAGE, E. 1908. Ornithologisches vom Tapajoz und Tocantins. J. f. Orn. 56:493–539.
SNETHLAGE, E. 1909. Novas espécies de aves amazônicas das collecções do Museu Goeldi. Bol. Mus. Goeldi 5:437–448.
SNETHLAGE, E. 1914. Catálogo das aves amazônicas. Bol. Mus. Goeldi 8:1–465.
TODD, W. E. C. 1925. Sixteen new birds from Brazil and Guiana. Proc. Biol. Soc. Washington 38:91–100.
TODD, W. E. C. 1927. New gnateaters and antbirds from tropical America, with a revision of the genus *Myrmeciza* and its allies. Proc. Biol. Soc. Washington 40:149–178.
TODD, W. E. C. 1937. New South American birds. Ann. Carnegie Mus. 25(19):3–255.
TODD, W. E. C. 1943. Critical remarks on the trogons. Proc. Biol. Soc. Wash. 56:3–16.
WILLIS, E. O. 1969. On the behavior of five species of *Rhegmatorhina*, ant-following antbirds of the Amazon basin. Wilson Bull. 81:363–395.
YAMASHITA, C., AND J. T. FRANÇA. 1991. A range extension of the Golden Parakeet *Aratinga guarouba* to Rondonia state, western Amazonia (Psittaciformes: Psittacidae). Ararajuba 2:91–92.

THE IMPORTANCE OF NATIONAL PARKS AND BIOLOGICAL RESERVES TO BIRD CONSERVATION IN THE ATLANTIC FOREST REGION OF BRAZIL

THEODORE A. PARKER III[1,3] AND JAQUELINE M. GOERCK[2]

[1]*Museum of Natural Science, Louisiana State University, Baton Rouge, Louisiana 70803, USA, and Conservation International, Washington, DC 20036, USA*
[2]*Department of Biology, University of Missouri–St. Louis, St. Louis, MO 63121, USA*
[3]*Deceased*

ABSTRACT.—Deforestation of Brazil's Atlantic Forest has threatened many plant and animal species with extinction. Here, the importance of protected areas in the Atlantic Forest Region of Brazil for the conservation of many of the most endangered bird species in the region is emphasized. A list of species is presented for four national parks and two biological reserves within the Atlantic Forest domain.

RESUMEN.—O desmatamento da Mata Atlântica no Brasil tem ameaçado de extinção muitas espécies de plantas e animais. Este trabalho enfatiza a importância das áreas protegidas na região da Mata Atlântica para a conservação de aves consideradas ameaçadas de extinção. Listas de espécies de aves são apresentadas para quatro parques nacionais e duas reservas biológicas pertencentes ao Domínio da Mata Atlântica.

The Atlantic Forest of Brazil consists primarily of humid evergreen forest that is geographically isolated from similar vegetation (i.e., the Amazon) by drier (caatinga, cerrado, and chaco) habitats. This distinct biogeographic area contains a unique combination of ecosystems in which as many as 75% of the plant species (Gentry 1992) and 29% of the bird species (Parker et al. 1996) are endemic. The avifauna contains species with close relatives in both Andean and Amazonian regions (Sick 1985; Willis 1992).

The original forest covered approximately 1 million km^2 (12% of the Brazilian territory) principally along 4,000 km of the Brazilian coast, from Ceará in the north to Rio Grande do Sul in the south. The Atlantic Forest also extended inland within the states of São Paulo, Paraná and Santa Catarina, as well as into Paraguay and Argentina. The entire region has had a long history of destruction (see review in Sick and Teixeira 1979), and only a few small remnants remain relatively intact in Brazil (Mittermeier 1988; Myers 1988). In 1992, the Atlantic Forest Biosphere Reserve was established (Lino 1992), which will likely enhance protection of these habitats by bringing international attention to the region. With increasing human population in Brazil, however, little hope remains for saving areas that do not lie within national parks, biological reserves, or other conservation units. Even such protected areas still suffer exploitation of timber, orchids, bromeliads, and *Euterpe* palms, as well as hunting, bird trapping, and other threats to the flora and fauna of the region (JMG, pers. obs.).

We present here a preliminary list of birds and their relative abundances at two biological reserves (BR)—Sooretama, and Augusto Ruschi (formerly Nova Lombardia)—and four national parks (NP)—Tijuca, Itatiaia, Iguaçu, and Aparados da Serra—within Atlantic Forest ecosystems (Appendix; see IBAMA 1989 for description of sites). Together these areas contain 533 species (78% of the total Atlantic Forest avifauna), with 157 endemic (cf. Parker et al. 1996), 27 threatened and 44 near-threatened species (Table 1). Following Collar et al. (1992), threatened refers to species that are 'at risk' or 'in peril' of extinctions and encompasses endangered, vulnerable, rare, or species for which the status is indeterminate or insufficiently known. Near-threatened species are those that, while apparently not seriously in danger of global extinction, give cause for concern (cf. Collar et al. 1992). Some of these areas have already been thoroughly studied

TABLE 1

Total Numbers of All Species, Endemics, Threatened and Near-Threatened Species of Four National Parks (NP) and Two Biological Reserves (BR) in the Atlantic Forest Region of Brazil: Sooretama SBR, Augusto Ruschi ARBR, Itatiaia ItNP, Tijuca TiNP, Iguaçu IgNP, Aparados da Serra ASNP. Endemism to the Atlantic Forest Region Follows Parker et al. (1996). Threat Status Follows Collar et al. (1992)

Status	SBR	ARBR	ItNP	TiNP	IgNP	ASNP	Total
All species	286	212	251	127	226	129	533
Endemic species	55	71	101	38	61	31	157
Near-threatened species	10	13	21	6	10	10	44
Threatened species	12	8	5	2	5	2	27

and published species accounts exist (e.g., Sooretama, Ruschi 1980; Scott and Brooke 1985), but data on abundance are generally not reported (but see Scott and Brooke 1985 for Sooretama). Consequently we hope that this paper will provide an important source for further studies urgently needed in the entire Atlantic Forest region.

Bird observations were done mainly by one of us (TAP) during a series of brief (2–5 day) visits to each area during the breeding season (August–November) every other year from 1980 to 1989. Intensive surveys lasted from just before dawn until late morning (0430–1200 h). Surveys were undertaken during the peak of vocal activity only; therefore, these lists are not a complete nor exhaustive account for each area.

1. Sooretama Biological Reserve (STBR) (18°53'–19°05'S, 39°55'–40°15'W) is located in the state of Espírito Santo and encompasses 24,000 ha. This state has been almost entirely deforested, and Sooretama, with the adjoining Linhares Reserve of the Companhia Vale do Rio Doce, represents perhaps the most important forest remnant at low elevations (80 m) in the region. The avifauna consists of at least 286 species, and contains approximately one-third of the Atlantic Forest threatened or near-threatened birds. Of special interest is the unusual number of cotingid species, *Carpornis melanocephalus*, *Cotinga maculata*, *Xipholena atropurpurea*, and *Procnias nudicollis*. Tinamous, particularly *Tinamus solitarius* and *Crypturellus noctivagus*, and the heavily poached curassow *Crax blumenbachii*, apparently are well protected in this area (Collar et al. 1992). This group of large frugivores has been shown to be especially vulnerable to local extinction (cf. Willis 1979). Several large raptors have been observed in the reserve as well, for example, *Accipiter poliogaster*, *Spizastur melanoleucus*; notably a single sighting of *Morphnus guianensis* perched along the main road occurred in October 1989 (TAP and JMG, pers. obs.).

2. Augusto Ruschi Biological Reserve (ARBR) (19°45'–20°00'S, 40°27'–40°38'W), formerly known as Nova Lombardia Biological Reserve, also lies in the state of Espírito Santo, and encompasses a small area (4,000 ha) of fairly humid tropical hill-forest within an elevational range of 700 to 900 m. This reserve contains an unusual number of threatened and near threatened southeastern Brazilian endemics (see Table 1). Among the most important of these are highly endangered psittacids, namely *Touit surda*, *Amazona rhodocorytha*, and *Triclaria malachitacea*. Additionally, Parker (1983) reported a new record of a migrant from North America, *Dendroica fusca*, in the reserve. Because forest cover has become so patchy in this region, forest islands like ARBR will undoubtedly increase in importance as refuges for both resident and migratory birds. Moreover, ARBR is an important haven not only for birds, but also for three, perhaps four, species of endangered primates (cf. Fonseca et al. 1994), and for the endangered Maned Three-toed Sloth, *Bradypus torquatus* (JMG, pers. obs., cf. Fonseca et al. 1994).

3. Itatiaia National Park (ItNP) (22°16'–22°28'S, 44°34'–44°42'W), established in 1937 and located in the States of Rio de Janeiro and Minas Gerais, was Brazil's first national park. This 30,000 ha area, which spans a gradient from evergreen forest (starting at 300 m elevation) to high altitude grasslands with sparse shrub cover in the Serra da Mantiqueira (2,800 m), harbors a minimum of 251 bird species. The complex vegetation structure, including several species of bamboo, provides essential habitat for rare bamboo specialists, such as *Claravis godefrida*, *Biatas nigropectus*, *Sporophila frontalis*, *S. falcirostris*, and *Amaurospiza moesta*. Important populations of some of the rarer and more restricted large frugivores (e.g., *Baillonius bailloni*, *Tijuca atra*), also occur in this area. For example, as many as eight individuals of *Phibalura flavirostris*, a near-threatened cotinga, were seen feeding together in a single fruiting tree (JMG, pers. obs., August 1989). Itatiaia National Park is essential to the conservation of many Atlantic Forest

species, because this national park is one of the few localities where certain species (e.g., cotingas) presently survive in fairly large numbers, or where some species (e.g., *Piprites pileatus*) are known to occur at all (Collar et al. 1992).

4. Tijuca National Park (TiNP) (22°55′–23°00′S, 43°11′–43°19′W), in the heart of Rio de Janeiro city (Rio de Janeiro state), encompasses an area of 3,200 ha. Although this area suffered anthropogenic occupation in the 19th Century, and some segments were converted to coffee, tea, and sugar plantations, the forest in Tijuca is presently in an advanced state of regeneration. At lower altitudes (80–1,020 m) than ItNP, Tijuca National Park lacks high-elevation grassland and consists primarily of humid evergreen forest. Although apparently not very rich in species numbers (127), TiNP harbors some rare species (e.g., *Touit melanonota*). Perhaps the major purpose of this national park should be as a source of environmental education to one of the world's most populated cities.

5. Iguaçu National Park (IgNP) (25°04′–25°41′S, 53°38′–54°28′W), in the state of Paraná in Brazil along the border with Argentina, is a fairly large protected area (170,086 ha) at mid elevations (300–600 m). Established in 1939, Iguaçu, together with adjacent remaining forests in Argentina, surrounds the spectacular Iguaçu Falls. Humid forests rich in epiphytes (orchids and bromeliads), lianas, ferns, palms, and bamboo, alternate with areas of evergreen Brazilian Pine (*Araucaria angustifolia*) forests. These forests harbor at least 226 species, including the endangered *Pipile jacutinga*, which has been overhunted throughout most of its range. *Dryocopus galeatus*, perhaps the rarest woodpecker in South America, is also found in this park.

6. Aparados da Serra National Park (ASNP) (29°07′–29°15′S, 50°01′–50°10′W), which comprises an area of 10,250 ha, lies in the more temperate states of Rio Grande do Sul and Santa Catarina. A 5.8 km long canyon that drops down 500 m is surrounded by evergreen Brazilian Pine forests and by natural grassland vegetation. These grasslands provide excellent pasture for cattle, and as late as 1990 the park was still occupied by ranchers (JMG, pers. obs.). This park includes important habitat for *Amazona vinacea*, a formerly abundant and widespread species, which has now declined substantially in numbers throughout its range (Collar et al. 1992). Because of the existence of natural grassland areas and marshes, some rare, threatened, or near-threatened birds associated with this type of vegetation are found in relatively large numbers here, e.g., *Limnornis rectirostris*, *Heteroxolmis dominicana*, *Emberizoides ypiranganus*, and *Xanthopsar flavus*, the latter now experiencing considerable reduction in numbers (Collar et al. 1992; see also Collar et al. 1994). The presence of several species that generally occur in the grasslands of southern South America contributes to the overall number of species present in this park (129).

The total numbers of bird species for these parks and reserves certainly exceed those presented here, and care should be taken when comparing these lists to others (cf. Remsen 1994). Contrary to what one would expect for such threatened ecosystems with easy access, very few long-term surveys have been undertaken in the region, and the distribution and natural history of rare endemic species remain poorly known. We encourage those interested in Atlantic Forest birds to use these preliminary lists to initiate more intensive studies at these and other nearby sites.

ACKNOWLEDGMENTS

We are deeply grateful to B. A. Loiselle and J. V. Remsen for encouragement and comments at many stages of this paper. The paper also benefitted from the comments and suggestions of N. J. Collar and an anonymous reviewer from Brazil. D. F. Stotz kindly provided a list with data on endemism for the birds of the Atlantic Forest Region. We would also like to thank IBAMA for allowing us to enjoy the beauty of the national parks and biological reserves in Brazil, and Christoph and Ina Hrdina for so enthusiastically helping with every phase of this work. Special thanks are due Mr. Alberto Hoffman, an IBAMA park guard, who cheerfully introduced T. A. Parker III to Augusto Ruschi Biological Reserve in October 1980, and subsequently accompanied him on each visit. Finally, T. A. Parker III thanks Victor Emanuel Nature Tours for supporting his field-work in Brazil. The list for ARBR is dedicated to the memory of Augusto Ruschi, whose intense interest in all living things led to the establishment of this and other biological reserves in Brazil, as well as to a growing conservation movement within the country.

LITERATURE CITED

BELTON, W. 1984. Birds of Rio Grande do Sul, Brazil, Part 1: Rheidae through Furnariidae. Bull. Amer. Mus. Nat. Hist. 178:369–631.

COLLAR, N. J., L. P. GONZAGA, N. KRABBE, A. MADROÑO NIETO, L. G. NARANJO, T. A. PARKER III, AND D.

C. WEGE. 1992. Threatened Birds of the Americas. International Council for Bird Preservation, Cambridge, United Kingdom.

COLLAR, N. J., M. J. CROSBY, AND A. J. STATTERSFIELD. 1994. Birds to watch 2—The world list of threatened birds. Birdlife Conservation Series No. 4, Smithsonian Institution Press, Washington, D.C.

FONSECA, G. A. B. DA, A. B. RYLANDS, C. M. R. COSTA, R. B. MACHADO, AND Y. L. R. LEITE. 1994. Livro vermelho dos mamíferos brasileiros ameaçados de extinção. Fundação Biodiversitas, Belo Horizonte, Brazil.

GENTRY, A. 1992. Tropical forest biodiversity: distributional patterns and their conservational significance. Oikos 63:19–28.

IBAMA. 1989. Unidades de Conservação do Brasil. Ministério do Interior, Brasília, Brazil.

ISLER, M. P., P. R. ISLER, AND B. M. WHITNEY. 1997. Biogeography and species limits in the *Thamnophilus punctatus* group. Pp. 355–381 *in* Studies in Neotropical Ornithology Honoring Ted Parker (J. V. Remsen, Ed.). Ornithol. Monogr. No. 48.

LINO, C. F. (ED.). 1992. Reserva da Biosfera da Mata Atlântica. Plano de ação. Volume 1: referências básicas. Consórcio Mata Atlântica, Universidade Estadual de Campinas, São Paulo, Brazil.

MITTERMEIER, R. A. 1988. Primate diversity and the tropical forest: case studies from Brazil and the importance of megadiversity countries. Pages 145–154 *in* Biodiversity (E. O. Wilson, Ed.). National Academy Press, Washington, D.C.

MYERS, N. 1988. Threatened biotas: "Hotspots" in tropical forests. Environmentalist 8:1–20.

PARKER, T. A., III. 1983. A record of Blackburnian Warbler (*Dendroica fusca*) for southeastern Brazil. Amer. Birds 3:274.

PARKER, T. A., III, D. F. STOTZ, AND J. W. FITZPATRICK. 1996. Ecological and distributional databases. *In* Neotropical Bird Ecology and Conservation (D. F. Stotz, J. W. Fitzpatrick, T. A. Parker III, and D. K. Moskovits, Eds.). University of Chicago Press, Chicago, Illinois.

REMSEN, J. V., JR. 1994. Use and misuse of bird lists in community ecology and conservation. Auk 111: 225–227.

RUSCHI, A. 1980. A fauna e a flora da Estacao Biológica de Sooretama. Bol. Museu Biol. Prof. Mello Leitão, Série Zool. 98:1–24.

SCOTT, D. A., AND M. L. BROOKE. 1985. The endangered avifauna of southeastern Brazil: a report on the BOU/WWF expedition of 1980/81 and 1981/82. Pages 115–139 *in* Conservation of Tropical Forest Birds (A. W. Diamond, and T. E. Lovejoy, Eds.). International Council for Bird Preservation (Techn. Publ. 4), Cambridge, United Kingdom.

SICK, H. 1985. Observations on the Andean-Patagonian component of southeastern Brazil's avifauna. Ornithol. Monogr. 36:233–237.

SICK, H. 1993. Birds in Brazil. Princeton University Press, Princeton, New Jersey.

SICK, H., AND D. M. TEIXEIRA. 1979. Notas sobre aves brasileiras raras ou ameaçadas de extinção. Publ. Avulsas Museu Nacional 62.

WILLIS, E. O. 1979. The composition of avian communities in remanescent woodlots in southern Brazil. Pap. Avulsos Zool. 33:1–25.

WILLIS, E. O. 1992. Zoogeographical origins of eastern Brazilian birds. Ornitologia Neotropical 3:1–15.

APPENDIX

Preliminary List of the Birds of Four National Parks and Two Biological Reserves in the Atlantic Forest Region of Brazil. Relative Abundance of Species Is Given for Each Locality as Follows: Sooretama SBR, Augusto Ruschi ARBR, Itatiaia ItNP, Tijuca TiNP, Iguaçu IgNP, Aparados da Serra ASNP. Endemism to the Atlantic Forest Region (According to Parker et al. 1996), and Threat Status (According to Collar et al. 1992) Are Also Presented. Taxonomic Order Mostly Follows Sick (1993), with Some Changes According to Other Authors (e.g., Collar et al. 1992, Parker et al. 1996)

Species	SBR	ARBR	ItNP	TiNP	IgNP	ASNP	Endemism	Threat Status
Tinamidae								
Tinamus solitarius	F	R	R		U		AF	NT
Crypturellus soui	U							
Crypturellus obsoletus		F	C	F	F	F		
Crypturellus variegatus	U							
Crypturellus noctivagus	U						AF	NT
Crypturellus parvirostris	R	R?						
Crypturellus tataupa	?				U			
Rhynchotus rufescens						C		
Nothura maculosa						C		
Phalacrocoracidae								
Phalacrocorax brasilianus	X				C			
Ardeidae								
Ardea alba	U				U			
Egretta thula					U			
Bubulcus ibis					U			
Butorides striatus	U	U						
Syrigma sibilatrix						F		
Pilherodius pileatus	U							
Nycticorax nycticorax	X				U			
Tigrisoma lineatum	U							
Ixobrychus involucris	R							
Threskiornithidae								
Theristicus caudatus						F		
Mesembrinibis cayennensis					U			
Ciconiidae								
Ciconia maguari						R		
Cathartidae								
Sarcoramphus papa	X				R	X		
Coragyps atratus	C	F	F	F	C	C		
Cathartes aura	F	F	F	F	F	F		
Cathartes burrovianus	U							
Anatidae								
Anas flavirostris						F		
Anas georgica						F		
Amazonetta brasiliensis	F					U		
Cairina moschata	F				F			
Accipitridae								
Elanoides forficatus	F	F	U		U			
Leptodon cayanensis	U	R						
Chondrohierax uncinatus	U							
Harpagus diodon	F	U	U					
Harpagus bidentatus	?							
Ictinia plumbea	C	U			C			
Rostrhamus sociabilis	X				U			
Accipiter poliogaster	R							NT
Accipiter striatus			U					
Geranoaetus melanoleucus						R		
Buteo albicaudatus				X		R		

APPENDIX
CONTINUED

Species	SBR	ARBR	ItNP	TiNP	IgNP	ASNP	Endemism	Threat Status
Buteo leucorrhous			U					
Buteo brachyurus		R	U					
Buteo magnirostris	F	F	F	F	F	U		
Leucopternis polionota	U?		R		R		AF	NT
Leucopternis lacernulata		R					AF	T
Buteogallus meridionalis						U		
Buteogallus urubitinga					U	U		
Morphnus guianensis	R							NT
Spizastur melanoleucus	R					X		NT
Spizaetus tyrannus	U	R			R	X		
Falconidae								
Herpetotheres cachinnans	F	R						
Micrastur semitorquatus	X	R			R			
Micrastur ruficollis		U	F		F	F		
Milvago chimachima	F	U	U		U	F		
Milvago chimango						F		
Caracara plancus	U		U			F		
Falco rufigularis	X				R			
Falco femoralis	R					F		
Falco sparverius	U		U		U	U		
Cracidae								
Ortalis (motmot) araucuan	U						AF	
Penelope superciliaris	F			U				
Penelope obscura		R?	F		U	F		
Pipile jacutinga					R		AF	T
Crax blumenbachii	R						AF	T
Odontophoridae								
Odontophorus capueira	U	U	F		U		AF	
Aramidae								
Aramus guarauna	U				U			
Rallidae								
Rallus sanguinolentus						U		
Rallus nigricans	F	U	U		U			
Aramides cajanea	U			X				
Aramides saracura			F		U	F	AF	
Porzana albicollis	F					U		
Laterallus melanophaius	U	R		U				
Laterallus leucopyrrhus						U		
Porphyriops melanops						R		
Gallinula chloropus	X				U			
Porphyrula martinica	F							
Cariamidae								
Cariama cristata			X			F		
Jacanidae								
Jacana jacana	F				U	R		
Charadriidae								
Vanellus chilensis	F				U	C		
Scolopacidae								
Gallinago (gallinago) paraguaiae	U	R				U		
Gallinago undulata						F		
Tringa solitaria	X							
Actitis macularia						U		
Laridae								
Sterna superciliaris						U		

APPENDIX
CONTINUED

Species	SBR	ARBR	ItNP	TiNP	IgNP	ASNP	Endemism	Threat Status
Columbidae								
Columba speciosa	F							
Columba picazuro	R				F			
Columba cayennensis	F				C			
Columba plumbea		F	C			U		
Columbina talpacoti	C	R		F	F			
Claravis pretiosa	F				U			
Claravis godefrida			R				AF	T
Scardafella squammata	F							
Leptotila verreauxi	X	U?			F	F		
Leptotila rufaxilla	F		F		?			
Geotrygon montana	U	U						
Geotrygon violacea					R			
Psittacidae								
Propyrrhura maracana	U		R					NT
Diopsittaca nobilis	U?							
Aratinga leucophthalmus			R		F			
Aratinga aurea	U							
Pyrrhura cruentata	C						AF	T
Pyrrhura frontalis		F	C	F	F	F	AF	
Pyrrhura (leucotis) leucotis	C						AF	
Forpus (xanthopterygius) crassirostris	F	U	F	U	F			
Brotogeris versicolurus	?			U?				
Brotogeris tirica	C		U	F			AF	
Touit melanonota				R			AF	T
Touit surda	R	R					AF	T
Pionopsitta pileata		F	F		F	F	AF	NT
Pionus menstruus	U							
Pionus maximiliani	F	U	C	F	F			
Amazona rhodocorytha	F	R					AF	T
Amazona aestiva	R							
Amazona amazonica	U							
Amazona farinosa	C							
Amazona vinacea						U	AF	T
Triclaria malachitacea		R					AF	T
Cuculidae								
Coccyzus melacoryphus	U				U			
Coccyzus euleri			R	R				
Coccyzus americanus	X							
Piaya cayana	C	F	F	F	F			
Crotophaga ani	C	U	U	U	F			
Crotophaga major	F				F			
Guira guira	F				F	U		
Tapera naevia	F	U			U			
Dromococcyx phasianellus					U			
Neomorphus geoffroyi	R?							
Strigidae								
Otus choliba	F				U			
Otus atricapillus	X?		U		?	X	AF	
Pulsatrix perspicillata	U							
Pulsatrix koeniswaldiana		?	F				AF	
Glaucidium brasilianum	U	U?	U		F			
Athene cunicularia	U					U		
Ciccaba virgata	R		U		U			
Strix hylophila			U		R	U?	AF	
Rhinoptynx clamator	R							
Nyctibiidae								
Nyctibius aethereus			X					
Nyctibius griseus	F	U	F		C			

APPENDIX
CONTINUED

Species	SBR	ARBR	ItNP	TiNP	IgNP	ASNP	Endemism	Threat Status
Caprimulgidae								
Lurocalis semitorquatus	X	F	C		F	U		
Nyctidromus albicollis	C	U	F		U			
Nyctiphrynus ocellatus	?	F						
Caprimulgus longirostris			U	X				
Hydropsalis brasiliana		U						
Macropsalis creagra			R				AF	NT
Apodidae								
Streptoprocne zonaris		U	C	F		U		
Streptoprocne biscutata			R		R	C		
Cypseloides senex					C			
Cypseloides fumigatus			F			C		
Chaetura cinereiventris	F	C	?	X				
Chaetura andrei	F	?	C	F	F			
Panyptila cayennensis	X (R)							
Reinarda squamata	R							
Trochilidae								
Ramphodon naevius		R		U			AF	NT
Glaucis hirsuta	X							
Phaethornis eurynome		F	F	?	U		AF	
Phaethornis squalidus			U					
Phaethornis pretrei		R						
Phaethornis ruber				X				
Phaethornis idaliae	U						AF	
Eupetomena macroura	X		U	X				
Melanotrochilus fuscus	X (R)	F	F				AF	
Colibri serrirostris		R?	C					
Anthracothorax nigricollis	U		U		U			
Chrysolampis mosquitus	R							
Stephanoxis lalandi			F		?	F	AF	
Lophornis magnifica		U	U					
Popelairia langsdorffi		R						
Chlorostilbon aureoventris		U	U		U	C		
Thalurania glaucopis	F	F	C	F	U		AF	
Hylocharis sapphirina	F							
Hylocharis cyanus	C							
Hylocharis chrysura					U			
Leucochloris albicollis		R	C			C	AF	
Polytmus guainumbi	U							
Amazilia versicolor	X	R	U		U			
Aphantochroa cirrochloris	R						AF	
Clytolaema rubricauda		U	C				AF	
Calliphlox amethystina		U	R					
Trogonidae								
Trogon viridis	C							
Trogon rufus	U	F	U		F			
Trogon surrucura		F[1]	F[1]		C[2]		AF	
Alcedinidae								
Ceryle torquata	F				F	U		
Chloroceryle amazona	U							
Chloroceryle americana	F	R						
Momotidae								
Baryphthengus ruficapillus	F		F		F		AF	
Galbulidae								
Galbula ruficauda	C							

APPENDIX
Continued

Species	SBR	ARBR	ItNP	TiNP	IgNP	ASNP	Endemism	Threat Status
Bucconidae								
Notharchus macrorhynchus	U	R			U			
Nystalus chacuru					U			
Malacoptila striata	U	U	U				AF	
Nonnula rubecula					R			
Monasa morphoeus	F							
Chelidoptera tenebrosa	C							
Ramphastidae								
Pteroglossus castanotis					F			
Pteroglossus aracari	F							
Baillonius bailloni			F	R	F		AF	NT
Selenidera maculirostris	X	F	U		F		AF	
Ramphastos vitellinus	F	C		?	F?			
Ramphastos dicolorus			C	?	C	R	AF	
Picidae								
Picumnus (cirratus) cirratus	U	U	F	F				
Picumnus (cirratus) temminckii					F		AF	
Picumnus exilis	R							
Picumnus nebulosus						R		NT
Colaptes campestris	F³		U³			C⁴		
Colaptes melanochloros		F	F		F			
Piculus flavigula	F	F						
Piculus aurulentus			U			U	AF	NT
Piculus chrysochloros	U							
Celeus flavescens	F				F			
Celeus flavus	X							
Dryocopus lineatus					F			
Dryocopus galeatus					R		AF	T
Melanerpes flavifrons	U		R?		F		AF	
Melanerpes candidus	R							
Veniliornis spilogaster			C		F	F	AF	
Veniliornis maculifrons	X?	U		F			AF	
Veniliornis affinis	F							
Campephilus robustus	R	R	U		U		AF	
Rhinocryptidae								
Psilorhamphus guttatus					X		AF	NT
Merulaxis ater			U				AF	NT
Scytalopus speluncae			C		?	U	AF	
Scytalopus indigoticus			R				AF	
Formicariidae								
Hypoedaleus guttatus		F			U		AF	
Batara cinerea		R	U		R	U		
Mackenziaena leachii			F		R	R	AF	
Mackenziaena severa		R	F		R		AF	
Taraba major	F							
Biatas nigropectus			R		R		AF	T
Thamnophilus palliatus	F			X				
*Thamnophilus (punctatus) ambiguus*⁵	C			X				
Thamnophilus caerulescens		U	C		F	F		
Thamnophilus ruficapillus		U	F			F		
Dysithamnus stictothorax		F	F	F			AF	NT
Dysithamnus mentalis		C	C	F	C			
Dysithamnus xanthopterus			U				AF	
Dysithamnus plumbeus	U						AF	T
Thamnomanes caesius	F							
Myrmotherula gularis		R	F				AF	
Myrmotherula axillaris	C							
Myrmotherula minor		R					AF	NT
Myrmotherula unicolor	X?			X			AF	NT

APPENDIX
CONTINUED

Species	SBR	ARBR	ItNP	TiNP	IgNP	ASNP	Endemism	Threat Status
Myrmotherula urosticta	F						AF	NT
Herpsilochmus rufimarginatus	C			C	C			
Drymophila ferruginea		U	C				AF	
Drymophila rubricollis					C	R	AF	
Drymophila genei			C				AF	NT
Drymophila ochropyga			C				AF	NT
Drymophila malura					R	U	AF	
Drymophila squamata	C			C			AF	
Terenura maculata	F	F	F	C	F		AF	
Pyriglena leucoptera	C	U	C	C	U		AF	
Myrmeciza loricata		F	F				AF	
Myrmeciza ruficauda	U						AF	T
Chamaeza campanisona		C	F		F			
Chamaeza ruficauda			C			U	AF	
Formicarius colma	C							
Grallaria varia		F	F		U	U		
Hylopezus nattereri			F				AF	
Conopophaga melanops	F						AF	
Conopophaga (lineata) lineata		U	C		U		AF	

Furnariidae

Species	SBR	ARBR	ItNP	TiNP	IgNP	ASNP	Endemism	Threat Status
Cinclodes pabsti						U	AF	
Furnarius rufus	F	U				U		
Furnarius figulus	U							
Limnornis rectirostris						F		NT
Leptasthenura setaria						C	AF	
Leptasthenura striolata						U	AF	
Schizoeaca moreirae			C				AF	
Synallaxis spixi	F	F	F		U	F		
Synallaxis (ruficapilla) ruficapilla		R	C		U	U	AF	
Synallaxis albescens	F		U					
Synallaxis cinerascens			R		U	U		
Certhiaxis cinnamomea	U				U			
Cranioleuca pallida		U	F				AF	
Cranioleuca obsoleta					U	C	AF	
Thripophaga macroura	F						AF	T
Anumbius annumbi						U		
Anabazenops fuscus		F	F				AF	
Syndactyla rufosuperciliata		U	C		U	F		
Anabacerthia amaurotis			R				AF	
Philydor atricapillus	F	F	F	F?	F		AF	
Philydor lichtensteini		U	R	F	C		AF	
Philydor rufus		C	C	F	U			
Automolus leucophthalmus	F	C	C	U	F		AF	
Cichlocolaptes leucophrus	F	F	U				AF	
Heliobletus contaminatus			U			F	AF	
Xenops minutus	F	R	?	X	R			
Xenops rutilans	U	F	F	F	F			
Sclerurus scansor		R		X	R		AF	
Sclerurus caudacutus	R?							
Lochmias nematura		U	F		U	F		

Dendrocolaptidae

Species	SBR	ARBR	ItNP	TiNP	IgNP	ASNP	Endemism	Threat Status
Dendrocincla turdina	F	F	F	X	F		AF	
Sittasomus griseicapillus		C	C	C	C			
Glyphorynchus spirurus	X							
Xiphocolaptes albicollis		U	F		U	F		
Dendrocolaptes platyrostris	U	U	F	U	F			
Xiphorhynchus guttatus	F							
Lepidocolaptes squamatus	R	F	C	F		F	AF	

APPENDIX
CONTINUED

Species	SBR	ARBR	ItNP	TiNP	IgNP	ASNP	Endemism	Threat Status
Lepidocolaptes fuscus	U	C	F	F	F		AF	
Campylorhamphus falcularius	U	F	F				AF	
Tyrannidae								
Phyllomyias fasciatus			C			F		
Phyllomyias burmeisteri		F	F	U				
Phyllomyias virescens			?				AF	
Phyllomyias griseocapilla		U	U				AF	NT
Ornithion inerme	R							
Camptostoma obsoletum	F	U	F	F	F	F		
Phaeomyias murina					U			
Myiopagis viridicata					C			
Myiopagis gaimardii	X?							
Myiopagis caniceps	F		U		U			
Elaenia flavogaster	C	F		X	U			
Elaenia mesoleuca			C		F	C		
Elaenia obscura			?					
Serpophaga nigricans					U	F		
Serpophaga subcristata		U	U	U		F		
Mionectes oleagineus	X							
Mionectes rufiventris		F	F	F			AF	
Leptopogon amaurocephalus	U	F	F	F	C			
Phylloscartes eximius			?		C		AF	NT
Phylloscartes sylviolus					F		AF	NT
Phylloscartes ventralis			U					
Phylloscartes paulistus					U		AF	T
Phylloscartes oustaleti		F					AF	NT
Phylloscartes difficilis			U			?	AF	NT
Capsiempis flaveola	X		F		U			
Corythopis delalandi					C			
Myiornis auricularis	C	F	F	F	F		AF	
Hemitriccus diops		U	F	U	U		AF	
Hemitriccus obsoletus			C			?	AF	
Hemitriccus orbitatus				X			AF	NT
Hemitriccus nidipendulus		U	R				AF	NT
Todirostrum poliocephalum		U	F	U			AF	
Todirostrum plumbeiceps		U	F		U			
Ramphotrigon megacephala		R?	F					
Rhynchocyclus olivaceus	U							
Tolmomyias sulphurescens		F	C	C	F			
Tolmomyias poliocephalus	F							
Tolmomyias flaviventris	U							
Platyrinchus mystaceus		F	F	F	F			
Platyrinchus leucoryphus		R			R		AF	T
Myiobius barbatus	U	F		X				
Myiobius atricaudus			U					
Myiophobus fasciatus	F	U		F	U			
Contopus cooperi			R					
Contopus cinereus	U	U	F	U	U			
Lathrotriccus euleri		F	C	F	F	U		
Pyrocephalus rubinus					U			
Xolmis cinerea						U		
Heteroxolmis dominicana						F		NT
Knipolegus lophotes			U					
Knipolegus nigerrimus			U	U		F	AF	
Knipolegus cyanirostris			F			F		
Fluvicola nengeta	U	U		X				
Arundinicola leucocephala	F							
Colonia colonus		U	F		U			
Hirundinea ferruginea			F	F		F		
Machetornis rixosus	F							

APPENDIX
CONTINUED

Species	SBR	ARBR	ItNP	TiNP	IgNP	ASNP	Endemism	Threat Status
Muscipipra vetula		R	U				AF	NT
Attila rufus	F	F	C	F			AF	
Attila spadiceus	U							
Attila phoenicurus			U			R		
Rhytipterna simplex	C	F						
Sirystes sibilator	F	F	U	?	F			
Myiarchus ferox	U	U			?			
Myiarchus swainsoni		U	F		F	U		
Myiarchus tuberculifer	U	U	?					
Pitangus sulphuratus	C	F	F	F	C			
Pitangus lictor	U							
Megarynchus pitangua	F	F	F	?	F			
Myiozetetes similis	F	U	F	F				
Conopias trivirgata					F			
Myiodynastes maculatus	C	U	C	F	C			
Legatus leucophaius			F		F			
Empidonomus varius	U	U	U		F			
Tyrannus savana	U		U			U		
Tyrannus melancholicus	C	F	F	C	F	U		
Laniocera hypopyrra	X							
Pachyramphus viridis	F	U	U		U			
Pachyramphus castaneus	F	F	C	F	U			
Pachyramphus polychopterus	F	R	F	F	C			
Pachyramphus marginatus	C	F						
Pachyramphus validus	U	R	U	U				
Tityra cayana	F	F	F		?			
Tityra inquisitor	U		U		F			
Pipridae								
Pipra rubrocapilla	F							
Pipra pipra	R							
Pipra fasciicauda					R			
Chiroxiphia caudata		C	C	C	F		AF	
Ilicura militaris		F	U	F			AF	
Manacus manacus	U	U	U	X	U			
Neopelma aurifrons		U	R				AF	
Schiffornis virescens		F	F		F		AF	
Schiffornis turdinus	C							
Piprites pileatus			R				AF	T
Piprites chloris		U			F			
Cotingidae								
Laniisoma (elegans) elegans		R		R?			AF	T
Phibalura flavirostris			U		R			NT
Tijuca atra			F				AF	NT
Carpornis cucullatus		F					AF	NT
Carpornis melanocephlus	R						AF	T
Cotinga maculata	R						AF	T
Xipholena atropurpurea	R						AF	T
Lipaugus vociferans	U							
Lipaugus lanioides	R?	F					AF	T
Pyroderus (scutatus) scutatus	R	R	R		F		AF	
Procnias nudicollis	U	F		X	R		AF	NT
Oxyruncus cristatus		F	U	U	R			
Hirundinidae								
Tachycineta albiventer					F			
Tachycineta leucorrhoa	R					C		
Progne tapera	U							
Progne subis	R							
Progne chalybea	C	U						
Atticora melanoleuca					U			

APPENDIX
CONTINUED

Species	SBR	ARBR	ItNP	TiNP	IgNP	ASNP	Endemism	Threat Status
Notiochelidon cyanoleuca	X	U	C	C		F		
Neochelidon tibialis			R?					
Stelgidopteryx ruficollis	C	F	U	F	F			
Hirundo rustica	F							
Corvidae								
Cyanocorax caeruleus						U	AF	NT
Cyanocorax chrysops					C			
Troglodytidae								
Campylorhynchus turdinus	C							
Thryothorus genibarbis	F							
Thryothorus longirostris				F				
Troglodytes aedon	F	F	F	C	F	F		
Donacobius atricapillus	F							
Muscicapidae								
Polioptilinae								
Ramphocaenus melanurus	C	?						
Polioptila lactea					F		AF	
Turdinae								
Myadestes leucogenys		U						NT
Platycichla flavipes		U	F	X				
Turdus (nigriceps) subalaris			U		U	F	AF	
Turdus rufiventris		U	C	C	F	F		
Turdus leucomelas	X				C			
Turdus amaurochalinus	U	U	F	X	U	F		
Turdus fumigatus	U							
Turdus albicollis	X	F	C	F?	C			
Mimidae								
Mimus saturninus	F	F?						
Motacillidae								
Anthus hellmayri			X			C		
Anthus lutescens	X							
Vireonidae								
Cyclarhis gujanensis	C	C	C	C	F	C		
Vireo (olivaceus) chivi	C	U	C	C	C	U		
Hylophilus (poicilotis) poicilotis		U	F		F		AF	
Hylophilus thoracicus			R	U				
Emberizidae								
Parulinae								
Parula pitiayumi	F	F	F	F	C	U		
Geothlypis aequinoctialis	U	U			F			
Basileuterus culicivorus		F	C	C	F	U		
Basileuterus leucoblepharus			C		U	C	AF	
Phaeothlypis rivularis					U			
Dendroica fusca		R						
Coerebinae								
Coereba flaveola	U	U	C	C				
Thraupinae								
Orchesticus abeillei			F				AF	NT
Schistochlamys ruficapillus		U	R					
Cissopis leveriana			F		U			
Pyrrhocoma ruficeps			U		U		AF	
Hemithraupis guira					C			
Hemithraupis ruficapilla	X	F	C	F			AF	

APPENDIX
CONTINUED

Species	SBR	ARBR	ItNP	TiNP	IgNP	ASNP	Endemism	Threat Status
Hemithraupis flavicollis	C			X				
Nemosia pileata	U							
Orthogonys chloricterus			U				AF	
Tachyphonus cristatus	C	F		F				
Tachyphonus coronatus		U	C	X	F		AF	
Trichothraupis melanops		F	F	F	F			
Habia rubica	F	C	F	F	C			
Ramphocelus bresilius	U						AF	
Thraupis sayaca	C	F	C	C	F	U		
Thraupis cyanoptera		F					AF	NT
Thraupis ornata	F	F	C	U			AF	
Thraupis palmarum	F	?		F				
Stephanophorus diadematus			F			C		
Pipraeidea melanonota			F					
Euphonia chlorotica	F			F	U			
Euphonia violacea	F	F	F	F	F			
Euphonia chalybea					U			NT
Euphonia musica			U					
Euphonia pectoralis		F	F	X	F		AF	
Chlorophonia cyanea		F	F	F	F			
Tangara (mexicana) brasiliensis	F						AF	
Tangara seledon	F	U	F	C	F		AF	
Tangara cyanocephala		F		C			AF	
Tangara desmaresti			C				AF	
Tangara cyanoventris		C	C				AF	
Tangara cayana			U					
Tangara preciosa						U		
Tangara (velia) cyanomelaena	U						AF	
Dacnis nigripes		R?					AF	T
Dacnis cayana	F	F	C	F	C			
Chlorophanes spiza		F						
Conirostrum speciosum			U	X	C			
Tersina viridis			U		F			
Emberizinae								
Zonotrichia capensis		F	C	C		C		
Ammodramus humeralis	F							
Haplospiza unicolor			C			U	AF	
Donacospiza albifrons			U			F		
Poospiza thoracica			F			R	AF	
Poospiza lateralis			C			C		
Sicalis flaveola					F			
Emberizoides herbicola	U							
Emberizoides ypiranganus						F		NT
Embernagra platensis			U			C		
Volatinia jacarina	F	F	U	U	U			
Sporophila frontalis	R		C				AF	T
Sporophila falcirostris			R				AF	T
Sporophila (nigricollis) nigricollis	U							
Sporophila (nigricollis) ardesiaca	R						AF	
Sporophila caerulescens	F	U	U	U	?			
Sporophila leucoptera	F							
Sporophila bouvreuil	R							
Sporophila melanogaster						U	AF	NT
Oryzoborus angolensis					U			
Amaurospiza moesta			R				AF	NT
Tiaris fuliginosa	R	R						
Arremon (taciturnus) taciturnus	C		R					
Arremon (taciturnus) semitorquatus			F				AF	
Arremon flavirostris					R			
Coryphospingus pileatus	U							
Coryphospingus cucullatus					C			

APPENDIX
CONTINUED

Species	SBR	ARBR	ItNP	TiNP	IgNP	ASNP	Endemism	Threat Status
Cardinalinae								
Caryothraustes canadensis	C	C						
Saltator fuliginosus	U	F	F	U	U		AF	
Saltator maximus	F	U		X				
Saltator similis		C	C	C	F	?		
Saltator maxillosus			F			U	AF	NT
Cyanocompsa brissonii	X				U			
Icterinae								
Psarocolius decumanus	U					U		
Cacicus haemorrhous	F	?	C		C?			
Cacicus chrysopterus			U			U		
Icterus cayanensis					C?			
Xanthopsar flavus						U		T
Agelaius cyanopus	R							
Agelaius ruficapillus	U							
Sturnella superciliaris	F							
Pseudoleistes guirahuro						U		
Gnorimopsar chopi	F					F		
Molothrus bonariensis	F	U	U	U	U	U		
Scaphidura oryzivora	X				U			
Fringillidae								
Carduelis magellanica			U					
Passeridae								
Passer domesticus	F			F	F			

Note: [1] *Trogon (surrucura) aurantius*; [2] *Trogon (surrucura) surrucura*; [3] *Colaptes (campestris) campestris*; [4] *Colaptes (campestris) campestroides*; [5] (cf. Isler et al. 1997).

Relative abundance: C: common; more than 10 individuals recorded daily (by sight *or sound*), in the preferred habitat. F: fairly common; up to 10 individuals recorded daily. U: uncommon; one or several individuals recorded at least every other day. R: rare; recorded fewer than 3 times. X: species recorded in park or reserve by other observers [esp. Belton 1984, Scott and Brooke 1985, David Stemple (Augusto Ruschi BR) and Allen Altman (Tijuca NP), pers. comm.]. ?: indicates uncertainty, further evidence needed (either for record, or abundance status).

Endemism and Threat Status: AF: endemic to the Atlantic Forest region. T: threatened. NT: near-threatened.

NOTES ON AVIAN BAMBOO SPECIALISTS IN SOUTHWESTERN AMAZONIAN BRAZIL

THEODORE A. PARKER III,[1,2] DOUGLAS F. STOTZ,[3,5] AND
JOHN W. FITZPATRICK[4,6]

[1]*Museum of Natural Science, Foster Hall 119, Louisiana State University,*
Baton Rouge, Louisiana 70803, USA;
[2]*Deceased;*
[3]*Museu de Zoologia da Universidade de São Paulo,*
Caixa Postal 7172-01051, São Paulo, Brazil; and
[4]*Division of Birds, Field Museum of Natural History,*
Roosevelt Road, at Lake Shore Drive, Chicago, Illinois 60605, USA
[5]*present address: Environmental and Conservation Programs,*
Field Museum of Natural History,
Roosevelt Rd. at Lake Shore Dr., Chicago, Illinois 60605, USA;
[6]*present address: Laboratory of Ornithology, Cornell University,*
Ithaca, New York 14850, USA

ABSTRACT.—A number of bird species in southwestern Amazonia are associated with stands of bamboo. We studied the birds in bamboo in two regions, eastern Rondônia near the Rio Jiparaná where a single small patch was located, and near Alta Floresta, in northern Mato Grosso, where several extensive bamboo stands were explored. We found a moderately large bamboo avifauna at each site, with eight species in Rondônia and ten at Alta Floresta. In species accounts, we discuss the significance of these records and provide notes on the ecology and behavior of these bamboo species. For most of these species, these records represented significant range extensions. Four species were recorded in Brazil for the first time. We also located a population of *Poecilotriccus tricolor*, previously known from a unique type, in Rondônia. We determined that it should be treated as a weakly differentiated subspecies of *Poecilotriccus capitalis*.

RESUMO.—Várias espécies de aves no sudoeste da Amazônia associam-se com bambuzais. Investigamos as aves de bambú em duas regiões: na região leste de Rondônia, perto ao rio Jiparaná onde encontramos um pequeno bambuzal, e perto de Alta Floresta, no norte de Mato Grosso, onde exploramos vários bambuzias. Encontramos uma avifauna bastante diversa associada com bambús, composta de oito espécies em Rondônia, e de dez em Atla Floresta. No relato de espécies, discutimos a importância destes registros, e fornecemos imformações sobre a ecologia e o comportamento destas espécies. Para a maioria destas espécies, estes registros representam um aumento significante na área de distribuição. Quatro espécies foram registradas no Brasil pela primeira vez. Também, localizamos uma população de *Poecilotriccus tricolor*, antes conhecido sómente do holotipo, em Rondônia. Determinamos que esta forma é uma subespécie pouco diferenciada de *Poecilotriccus capitalis*.

During ornithological surveys in Rondônia and Mato Grosso, Brazil, in 1986 and 1989, we made observations on birds inhabiting tall, dense thickets of spiny bamboo (*Guadua* spp.) within both disturbed and undisturbed rainforests. Large species of bamboo occur patchily in southern Amazonia, primarily in floodplain forests along riverbanks or on natural landslides in hilly terrain. *Guadua* bamboos quickly invade clearings, and some bamboo-dwelling bird species may be spreading because of deforestation in SW Amazonia (Parker and Remsen 1987).

On 6 and 11 November 1986, during an avifaunal survey in Rondônia conducted by the Field Museum of Natural History (FMNH) and the Museu de Zoologia, Universidade de São Paulo (MZUSP), Stotz and Fitzpatrick, together with T. S. Schulenberg, D. E. Willard, S. M. Lanyon and A. T. Peterson surveyed an isolated bamboo area of ca. 4 ha in partly disturbed *terra firme* forest at the edge of a recently planted corn field, along the road from Jaru to Maçadinho, ca. 40 km W of Cachoeira Nazaré (9°44'S, 61°53'W), Rio Jiparaná. This was the only substantial patch of bamboo found along more than 50 km of road through forest in the region. From 26 October to 9 November 1989, Parker, M. Isler, and P. Isler studied birds in bamboo thickets in upland *terra firme* forests 28–30 km north of Alta Floresta, Mato Grosso (9°41'S, 55°54'W), ca. 5 km south of the Rio Teles Pires. *Guadua* bamboo was common and widespread in the latter region, and we located several large (2–5 ha) areas of bamboo, both inside forest and along forest edges. A more detailed description of the area, as well as a discussion of the entire avifauna of the Alta Floresta area can be found in Zimmer et al. (1997).

Specimens from Rondônia mentioned herein were collected under the auspices of the MZUSP, but presently are housed in the Museu Goeldi (Belem) except as noted. Tape-recordings are archived in the Library of Natural Sounds (LNS), Cornell Laboratory of Ornithology, Ithaca, New York.

We here report the first Brazilian records of Cabanis' Spinetail (*Synallaxis cabanisi*), Crested Foliage-gleaner (*Automolus dorsalis*), Manu Antbird (*Cercomacra manu*), and Dusky-tailed Flatbill (*Ramphotrigon fuscicauda*), all birds found primarily or exclusively in bamboo thickets. These birds had previously been known only from southwestern Amazonia near the base of the Andes, no further east than Pando, Bolivia (Parker and Remsen 1987). Since our studies, *Ramphotrigon fuscicauda* has been found in northeastern dpto. Santa Cruz, Bolivia (Kratter et al. 1992).

Amazonian birds associated with bamboo have been discussed by Parker (1982), Pierpont and Fitzpatrick (1983), Terborgh et al. (1984), Parker and Remsen (1987), Fitzpatrick and Willard (1990) and Kratter (in press). The highest diversity of bamboo birds is found in southwestern Amazonian forests. In southeastern Peruvian forests as many as 19 bird species can be restricted to bamboo thickets at a single site (Kratter, in press). Bamboo patches are extensive in southwestern Amazonia, east to the state of Acre, Brazil. East of there in Amazonia, bamboo patches are more scattered and smaller. Along with the extent of bamboo, the diversity of bamboo specialist birds declines eastward in Amazonia, but we document here a diverse community of bamboo birds as far east in Amazonia as north-central Mato Grosso, with eight species in Rondônia and ten at Alta Floresta (two bamboo inhabiting species, *Nonnula ruficapilla* and *Myrmotherula ornata*, that we recorded are not discussed below). Recent work in northeastern Santa Cruz, Bolivia has also demonstrated the presence of some bamboo birds in lowland forest, with up to four species at one site (Kratter et al. 1992; Parker and Bates unpubl. manuscript). However, the species found at all of these sites are a subset of those in southeastern Peru (see Kratter, in press). Of the bamboo species we recorded, only *Poecilotriccus capitalis*, replaced by its congener, *P. albifacies* in southeastern Peru, does not occur in southeastern Peru.

SPECIES ACCOUNTS

Curve-billed Scythebill *Campyloramphus procurvoides*. This species was common in extensive bamboo thickets north of Alta Floresta and also was noted in adjacent vine-rich forests. It was rare in *terra firme* forests at Cachoeira Nazaré, where two specimens were collected. These are the first records of *C. procurvoides* for Mato Grosso and Rondônia, and the southernmost ever for the species. As with southwestern Amazonian populations of the Red-billed Scythebill, *C. trochilirostris*, *C. procurvoides* appears to be strongly associated with bamboo, but is not restricted to it. Both species frequently probe cavities in dead bamboo stalks, including those excavated by woodpeckers (e.g., *Celeus spectabilis* in the case of *C. trochilirostris* in southeastern Peru) and enlarged by Peruvian Recurvebills, *Simoxenops ucayalae* (Parker, pers. obs.).

Cabanis' Spinetail *Synallaxis cabanisi*. A pair observed and tape-recorded in forest-edge bamboo and second-growth ca. 25 km north of Alta Floresta represents the first record for Brazil, and an easterly range extension of 1,200 km. *Synallaxis cabanisi* was considered a distinct species from *S. macconnelli* and *S. moesta* by Vaurie (1980), and later by Ridgely and Tudor (1994). Although these three taxa may be conspecific, we here follow their treatment, considering each as separate species. *Synallaxis cabanisi* was known previously from foothill forest and second-growth along the lower slopes of the Andes from dpto. San Martín, Peru, to dpto. Cochabamba, Bolivia (Vaurie 1980), and in bamboo thickets (Parker et al. 1994a, 1994b) or *Gynerium* cane (Terborgh et al. 1984) along rivers near the base of the Andes. It probably occurs locally in appropriate habitat between the large area between Madre de Dios and north-central Mato Grosso.

Peruvian Recurvebill *Simoxenops ucayalae*. This furnariid was fairly common in the larger patches of bamboo in forest north of Alta Floresta, where up to six individuals were noted each morning. These foraged singly and in pairs, from 1-6 m above ground in dense, dark thickets of bamboo stalks intermixed with vines and tree branches. They enlarged small cavities in bamboo stalks by hammering and prying with their large bills. A variety of arthropods, including spiders, roaches, katydids, and crickets, and small vertebrates such as frogs, use these cavities as refuges. Pairs of *S. ucayalae* sang and called regularly as they moved through their large territories (>1.0 ha).

Although until recently known only from southeastern Peru (Parker 1982) and adjacent northern Bolivia (Parker and Remsen 1987), *S. ucayalae* apparently occurs widely, if locally, in bamboo thickets across southern Amazonia. The first Brazilian record was a specimen from an unknown locality; Novaes (1978) assumed that the specimen was taken in Acre, because, at the time *S. ucayalae* was known only from adjacent Peru. However, the collector (G. P. Silva) believed that it came from eastern Amazonia, near Santarem, Pará. The first Brazilian record from a definite locality was a male collected ca. 52 km south-southwest of Altamira, along the lower Rio Xingu, Pará (Graves and Zusi 1989). The species has also been found recently in the Carajás region of eastern Pará (specimen at Museu Goeldi; D. Oren, pers. comm.).

Crested Foliage-gleaner *Automolus dorsalis*. Two specimens collected in bamboo 40 km west of Cachoeira Nazaré, and many individuals observed and tape-recorded in bamboo understory of forests north of Alta Floresta, represent the first records for Brazil. The species was known previously from a handful of widely scattered localities along the base of the Andes from southeastern Colombia to southeastern Peru and northern Bolivia (Parker and Bailey 1991).

At both sites, Crested Foliage-gleaners were fairly common and occurred primarily in the crowns of tall bamboo thickets, 4–8 m above the ground. They also were seen occasionally at similar heights in middlestory vine tangles in adjacent forest at Alta Floresta. Singly or in pairs, they usually associated with understory mixed-species flocks. They typically searched curled dead leaves and other debris trapped in bamboo branches and foliage. Their varied vocalizations included a staccato series (*krek-krek-krek-krek-krek*) about 5 s long, sometimes delivered antiphonally by both members of a pair, and at least two types of calls, including a loud *jejejejejeje* and a single *klek* or *kep* notes. Only one other foliage-gleaner, *Anabazenops fuscus*, a bamboo specialist of southeastern Brazil (Rodrigues et al. 1994, is known to duet antiphonally. This and other behavioral similarities (Kratter 1994), as well as plumage similarities, suggest a close relationship between these little-known taxa (see Kratter and Parker 1997).

Bamboo Antshrike *Cymbilaimus sanctaemariae*. This southwestern Amazonian endemic was recently recognized as a species distinct from the sympatric Fasciated Antshrike (*Cymbilaimus lineatus*) (Pierpont and Fitzpatrick 1983). The FMNH group collected eight specimens at the bamboo site west of Cachoeira Nazaré. The species was previously known in Brazil from one specimen collected in 1968 by J. Hidasi at Empresa Nova along the Rio Acre (Pierpont and Fitzpatrick 1983, LSUMZ 68109). In addition, MZUSP has two specimens (66060, 66061) taken by Hidasi in 1968 at Rio Branco, Acre.

Cymbilaimus sanctaemariae regularly foraged in the crowns of tall bamboo thickets (especially in hill forest), but it also sometimes inhabits dense vine tangles from mid-heights up into the lower canopy of tall floodplain forests, including some lacking bamboo. This species may occur in suitable habitat east of Rondônia, although it appears to be absent in suitable habitat at Alta Floresta.

Striated Antbird *Drymophila devillei*. Four specimens collected in bamboo west of Cachoeira Nazaré are referable to the nominate race *devillei*, previously unrecorded from Brazil. Novaes (1976) reported specimens of the race *subochracea* from the Rio Aripuanã, and only *subochracea* was found at Alta Floresta by Parker. Where these morphologically distinct forms come into contact is unknown. Surprisingly, the Aripuanã and Jiparaná localities are less than 250 km apart, with no apparent geographical barrier between them. However, several other species (e.g., *Conopophaga aurita* and *Pipra nattereri*) also are represented by different subspecies at these two sites. The songs of *D. d. devillei* and *D. d. subochracea* appear identical to us (pers. obs., LNS), suggesting that the two forms are indeed conspecific, despite distinctive male plumages.

Drymophila devillei appears more restricted to *Guadua* bamboo than any of the other bird species mentioned here, except perhaps *Cercomacra manu*. Pairs defend small territories (of ca. 0.3 ha) within large areas of bamboo. They usually foraged in pairs or families, gleaning bamboo foliage and stems in the upper parts of thickets, regularly associating with mixed-species flocks as they move through their territories. They sometimes follow flocks into middlestory vine tangles in adjacent forest.

Manu Antbird *Cercomacra manu*. This recently described species, until now known from a few localities in southeast Peru and adjacent northern Bolivia (Parker and Remsen 1987; Fitzpatrick and Willard 1990), was surprisingly common in extensive bamboo thickets north and south of Alta Floresta, although absent from Rondônìa. In one area of ca. 4 ha (28 km north) at least four and perhaps five pairs held contiguous, small territories of ca. 0.3 to 0.5 ha, and M. Isler noted five territorial pairs along ca. 100 m of road through disturbed forest with bamboo understory ca. 46 km southeast of the town. Smaller *Guadua* thickets scattered through forest and along forest-edges supported one to three pairs. Pairs foraged mainly in bamboo foliage and branches from 1–9 m above ground (mainly at 4–6 m), where they gleaned arthropods, especially green lepidopteran larva, from leaves and stems.

This species should be looked for in bamboo areas all across the southern periphery of Amazonian Brazil. Like *Simoxenops ucayalae*, it probably occurs far to the east of north-central Mato Grosso. Interestingly, it was absent from the small, isolated bamboo thicket we surveyed in Rondônia, where *Cercomacra nigrescens* occurred instead.

Dusky-tailed Flatbill *Ramphotrigon fuscicauda*. Several seen and tape-recorded (LNS) in tall bamboo inside forest ca. 28 km north of Alta Floresta apparently represent the first records for Brazil. These individuals perched from 3 to 6 m up in the tangled middle to upper portions of thickets. Unlike *Ramphotrigon megacephala*, this species is not confined to bamboo throughout its range, but occurs regularly in some vine-rich floodplain forests without bamboo (Parker 1984; Terborgh et al. 1984).

Large-headed Flatbill *Ramphotrigon megacephala*. This bamboo specialist was common at both Alto Floresta and west of Cachoeira Nazaré, where five specimens were collected. Birds from southern Amazonia are referable to *R. m. boliviana*. This species was previously known in Amazonian Brazil only from the Rio Juruá, Amazonas (Novaes 1960), although the nominate subspecies is widespread in bamboo in southeastern Brazil. For more information on this and the previous species see Parker (1984).

Black-and-white Tody-Flycatcher *Poecilotriccus capitalis*. This species was fairly common in bamboo 40 km west of Cachoeira Nazaré, where four males and two females were collected (housed at FMNH and MZUSP). Several pairs also were observed and tape-recorded outside of bamboo in dense, roadside second-growth vegetation within 2 km of the bamboo site. Stotz observed a male in similar dense second-growth at Cachoeira Nazaré (where there is no bamboo) in March 1988, and the Islers observed a female in bamboo ca. 28 km north of Alta Floresta.

Birds in Rondônia belong to the subspecies *tricolor*, described by Berla (1946) as *Todirostrum tricolor* from a single male taken on the Rio Jamari, Rondônia about 100 km west of Cachoeira Nazaré. It can be distinguished from *capitalis* in both sexes by the pale outer web of the outermost primary. Berla distinguished *tricolor* from *capitalis* by its complete black pectoral band. Fitzpatrick and Stotz examined the holotype at the Museu Nacional (Rio de Janeiro) and found that the complete pectoral band is an illusion resulting from poor skin preparation. Males of *tricolor* may have slightly more black on the sides of the breast and generally have a darker wash across the white of the center of the breast than nominate *capitalis*, but these are not well-defined characters. Females of *tricolor* appear identical to those of *capitalis*, except for the pattern of the outermost primary.

Songs of *tricolor* and *capitalis* are extremely similar, and Rondônia *tricolor* responded to tape-recordings of *capitalis* from northern Peru. In view of the similarity of song and plumage in both sexes, we recommend that *tricolor* be treated as a weakly differentiated subspecies of *P. capitalis*.

Fitzpatrick (1976) and Traylor (1977) merged *P. albifacies* of southeastern Peru into *P. tricolor*, erroneously assuming the unique male holotype of *albifacies* (Blake 1959) to be a missexed example of the female, at that time unknown, of *tricolor*. On the basis of new specimens of *P. albifacies* (both sexes) collected near the type locality, they later (Traylor and Fitzpatrick 1982) treated *albifacies* as a distinct species from *tricolor*, a conclusion overlooked by some recent authors. The songs and calls of *albifacies* differ in pattern and quality from those of *capitalis*, including *tricolor* (pers. obs., LNS). *P. albifacies* is also quite distinct in both male and female plumage. It is clear, therefore, that Blake (1959) and Traylor and Fitzpatrick (1982) were correct in considering *capitalis* and *albifacies* to be distinct species.

In addition to the Brazilian records of *Poecilotriccus capitalis* reported above, this species has also been found recently in the Carajás region of eastern Pará (D. Oren, pers. comm.).

ACKNOWLEDGMENTS

We thank P. Vanzolini (Museu de Zoologia, Universidade de São Paulo, MSUSP) for his help in organizing the FMNH fieldwork in Rondônia. This work was undertaken for Eletronorte, through a contract between the Academia Brasileira de Ciencias and the Consorcio Nacional de Engenheiros Consultores. We are also grateful for logistical help provided by the Museu de Zoologia da Universidade de São Paulo, the Conselho Nacional de Desenvolvimento Cientifico e Tecnologico, and the Consorcio Nacional de Engenheiros Consultores. Victor Emanuel Nature Tours, Inc. funded Parker's work at Alta Floresta. We thank our field companions L. da Silva, C. Hrdina, M.L. and P.R. Isler, S.M. Lanyon, A.T. Peterson, T.S. Schulenberg, and D.E. Willard; the Islers, D. Oren, and T.S. Schulenberg generously provided data on some of the species discussed herein. Finally, we thank A.W. Kratter, J.V. Remsen, Jr., K. Rosenberg, and T.S. Schulenberg for their reviews of the manuscript.

LITERATURE CITED

BERLA, H. F. 1946. Uma nova espécie do gênero *Todirostrum* Lesson, 1831 (Passeriformes, Tyrannidae). Summa Brasilia Biol. 1:125–127.

BLAKE, E. R. 1959. A new species of *Todirostrum* from Peru. Nat. Hist. Misc. 171.

FITZPATRICK, J. W. 1976. Systematics and biogeography of the tyrannid genus *Todirostrum* and related genera (Aves). Bull. Mus. Comp. Zool. 147:435–463.

FITZPATRICK, J. W., AND D. E. WILLARD. 1990. *Cercomacra manu*, a new species of antbird from southwestern Amazonia. Auk 107:239–245.

GRAVES, G. R., AND R. L. ZUSI. 1989. Avian body weights from the lower Rio Xingu, Brazil. Bull. Brit. Ornithol. Club 110:20–25.

KRATTER, A. W. 1994. The nest of the Crested Foliage-gleaner *Automolus dorsalis*. Orn. Neotrop. 5:105–107.

KRATTER, A. W. In press. Bamboo specialization by Amazonian birds. Biotropica.

KRATTER, A. W., M. D. CARREÑO, R. T. CHESSER, J. P. O'NEILL, AND T. S. SILLETT. 1992. Further notes on bird distribution in northeastern Dpto. Santa Cruz, Bolivia, with two species new to Bolivia. Bull. Brit. Ornithol. Club 112:143–150.

KRATTER, A. W., AND T. A. PARKER III. 1997. Relationship of two bamboo-specialized foliage-gleaners: *Automolus dorsalis* and *Anabazenops fuscus* (Furnariidae). Pp. 383–397 *in* Studies in Neotropical Ornithology Honoring Ted Parker (J. V. Remsen, Jr., Ed.). Ornithol. Monogr. No. 48.

NOVAES, F. C. 1960. Sobre *"Ramphotrigon megacephala"* (Swainson)(Tyrannidae, Aves). Rev. Brasil. Biol. 20: 217–221.

NOVAES, F. C. 1976. As aves do rio Aripuanã, Estados do Mato Grosso e Amazonas. Acta Amazonica 6:61–85.

NOVAES, F. C. 1978. Sobre algumas aves pouco conhecidas da Amazonia Brasileira II. Bol. Mus. Para. Emílio Goeldi, Zool., Nova Ser., no. 9.

PARKER, T. A., III. 1982. Observations of some unusual rainforest and marsh birds in southeastern Peru. Wilson Bull. 94:477–493.

PARKER, T. A., III. 1984. Notes on the behavior of *Ramphotrigon* flycatchers. Auk 101:186–189.

PARKER, T. A., III, AND B. BAILEY. (EDS.). 1991. A Biological Assessment of the Alto Madidi Region. RAP Working Papers 1. Conservation International, Washington, D. C.

PARKER, T. A., III, P. K. DONAHUE, AND T. S. SCHULENBERG. 1994a. Birds of the Tambopata Reserve (Explorer's Inn Reserve). Pages 106–124. *in* The Tambopata Reserved Zone of Southeastern Perú: A Biological Assessment (R. B. Foster, T. A. Parker III, A. H. Gentry, L. H. Emmons, A. Chicchón, T. Schulenberg, L. Rodríguez, G. Lamas, H. Ortega, J. Icochea, W. Wust, M. Romo, J. A. Castillo, O. Phillips, C. Reynal, A. Kratter, P. K. Donahue, and L. J. Barkley). RAP Working Papers 6. Conservation International, Washington, D. C.

PARKER, T. A., III, A. W. KRATTER, AND W. WUST. 1994b. Birds of the Ccolpa de Guacamayos, Madre de Dios. Pages 91–105. *in* The Tambopata Reserved Zone of Southeastern Perú: A Biological Assessment (R. B. Foster, T. A. Parker III, A. H. Gentry, L. H. Emmons, A. Chicchón, T. Schulenberg, L. Rodríguez, G. Lamas, H. Ortega, J. Icochea, W. Wust, M. Romo, J. A. Castillo, O. Phillips, C. Reynal, A. Kratter, P. K. Donahue, and L. J. Barkley). RAP Working Papers 6. Conservation International, Washington, D. C.

PARKER, T. A., III, AND J. V. REMSEN, JR. 1987. Fifty-two Amazonian bird species new to Bolivia. Bull. Brit. Ornith. Club 107:94–107.

PIERPONT, N., AND J. W. FITZPATRICK. 1983. Specific status and behavior of *Cymbilaimus sanctaemariae*, the Bamboo Antshrike, from southeastern Peru. Auk 100:645–652.

RIDGELY, R. S., AND G. TUDOR. 1994. The Birds of South America. Vol. 2 The Suboscine Passerines. University of Texas Press, Austin, Texas.

RODRIGUES, M. R., S. M. R. ALVARES, AND C. G. MACHADO. 1994. Foraging behavior of the White-collared Foliage-gleaner: a bamboo specialist. Orn. Neotrop. 5:65–67.

TERBORGH, J. W., J. W. FITZPATRICK, AND L. EMMONS. 1984. Annotated checklist of bird and mammal species of Cocha Cashu Biological Station, Manu National Park, Peru. Fieldiana, Zool., N. S. 21.

TRAYLOR, M. A., JR. 1977. A classification of the tyrant flycatchers (Tyrannidae). Bull. Mus. Comp. Zool. 148: 129–184.

TRAYLOR, M. A., JR., AND J. W. FITZPATRICK. 1982. A survey of the tyrant flycatchers. Living Bird 19:7–50.

VAURIE, C. 1980. Taxonomy and distribution of the Furnariidae (Aves, Passeriformes). Bull. Amer. Mus. Nat. Hist. 166:1–357.

ZIMMER, K. T., T. A. PARKER III, M. L. ISLER, AND P. R. ISLER. 1997. Survey of a southern Amazonia avifauna: the Alta Floresta region, Mato Grosso, Brazil. Pp. 887–918 *in* Studies in Neotropical Ornithology Honoring Ted Parker (J. V. Remsen, Jr., Ed.). Ornithol. Monogr. No. 48.

NOTES ON THREE TINY GRASSLAND FLYCATCHERS, WITH COMMENTS ON THE DISAPPEARANCE OF SOUTH AMERICAN FIRE-DIVERSIFIED SAVANNAS

THEODORE A. PARKER III[1,3] AND EDWIN O. WILLIS[2]

[1]Museum of Natural Science, Louisiana State University,
Baton Rouge, Louisiana 70803, USA; and
[2]Departamento de Zoologia, UNESP, 13506-900 Rio Claro, SP, Brazil
[3]Deceased

ABSTRACT.—The Rufous-sided Pygmy-Tyrant (*Euscarthmus rufomarginatus*, Tyrannidae) was rediscovered in high-grass bushy savannas ("campo-cerrados") in western Mato Grosso and nearby Bolivia. Pairs of this species hop wrenlike low in the vegetation, at times in mixed-species flocks, eating insects and fruits; a twittery song and January fledgling were noted. Two other small grassland flycatchers, the Bearded Tachuri (*Polystictus pectoralis*) and Sharp-tailed Tyrant (*Culicivora caudacuta*), are in more open remnant grasslands ("campos") of interior South America. Single adults care for tachuri fledglings, pairs (with family groups until winter) in Sharp-tails. All three flycatcher species are now rare, owing to destruction of huge areas by agribusiness, and annual burning of many remnant savannas. Other rare savanna species needing research include Blue-eyed Ground-Dove (*Columbina cyanopis*), encountered only once in our studies, and Ochre-breasted Pipit (*Anthus nattereri*), one of several species that prefer lightly grazed or burned savanna. Birds of natural savannas shift every few years with local fires: tall-grass species move to older grassland and fire-followers to the new burns. Large or connected reserves are needed to provide both types of habitat; small reserves protected from fire turn to scrub, while annually burned ranches save few birds.

RESUMO.—Notas sobre três pequenos tiranídeos dos campos nativos, com comentários sobre o desaparecimento das savanas diversificadas pelo fogo na América do Sul. A maria-corruíra (*Euscarthmus rufimarginatus*, Tyrannidae) foi redescoberta nos campos cerrados arbustivos com gramineas altas a oeste de Mato Grosso e cercanias de Bolivia. Pares desta espécie saltitam como as corruíras, baixo na vegetação e por vezes juntam-se aos bandos mistos, alimentando-se de insetos e frutos; foi notado um filhote em janeiro. Dois outros tiranideos pequenos, o tricolino-canela, *Polystictus pectoralis* e a maria-do-campo, *Culicivora caudacuta*, são encontrados mais comumente em remanescentes campestres mais abertos do interior da América do Sul. Apenas um dos adultos cuida dos filhotes em *P. pectoralis*, e o casal (com grupos familiares até o inverno) em *C. caudacuta*. Atualmente, todas as três espécies são raras, devido à destruição de grandes áreas para agricultura e o fogo anual em muitos dos cerrados remanescentes. Outras espécies raras de cerrado que necessitam ser pesquisados incluem a rolinha-brasileira, *Columbina cyanopis*, encontrada somente uma vez durante nossos estudos e, o caminheiro-grande, *Anthus nattereri*, uma das várias espécies mais comumente preferem pastos levemente utilizados ou os cerrados queimados. As aves dos cerrados naturais mudam a cada poucos anos com os fogos locais: espécies de gramas altas mudam-se para os campos mais desenvolvidos e as seguidoras-de-fogo para as áreas recentemente queimadas. Reservas grandes ou interligadas são necessárias para proporcionar ambos os tipos de habitat; pequenas reservas protegidas do fogo transformam-se em arbustos densos, enquanto os pastos queimados anualmente retêm poucas aves.

Recent field work in the campos, cerrados and gallery forests of western Brazil and nearby Bolivia has revealed several bird species once considered endemic to central Brazil (Silva and Oniki 1988; Willis and Oniki 1990; Bates et al. 1989, 1982). These open habitats are rapidly being converted into agricultural and cattle lands (Cavalcanti 1988; Willis and Oniki 1988; Silva 1995), and not many undisturbed examples survive outside of the few national parks and biological reserves. Savanna regions, although large, are eclipsed by the more speciose Amazon and Atlantic rain forests in the public mind with respect to conservation, and are almost ignored internationally.

Here we draw attention to several grassland flycatchers (Tyrannidae) and other birds that may be on the verge of extinction, despite having large historic ranges. Small and cryptically colored organisms, even vertebrates, tend not to stir the emotions of conservationists the way larger, more glamorous ones do, such as jaguar and giant otter. Nonetheless, the decline of several species of small, nondescript grassland birds in central South America reflects an alarming situation that urgently requires attention. In the five years since we first drafted this paper more records of these birds have come to light, and more attention has been directed to grassland ecosystems, but the situation is still serious.

RUFOUS-SIDED PYGMY-TYRANT

The Rufous-sided Pygmy-Tyrant (*Euscarthmus rufomarginatus*) is a small (6 g) brownish flycatcher with a white throat, pale yellowish breast, and extensively rufous sides. It was described by Pelzeln (1868) from specimens J. Natterer collected at Calção de Couro (20°20'S, 47°47'W) and Rio das Pedras (20°15'S, 47°54'W) in the state of São Paulo, Brazil. Calção de Couro Creek is now a concrete storm sewer in modern Ituverava. Rio das Pedras has been plowed except for a small buriti-palm marsh and cerrado at its junction with the Rio do Carmo. Recent field work by Willis and Oniki (1988) failed to reveal any significant unmodified "warm campo-cerrado" vegetation, let alone *E. rufomarginatus*, in the state of São Paulo, where 1484 km^2 were said to be present in 1971–1973 (Serra-Filho 1974). The species is also known from a few old specimens from Maranhão, Piaui, and northern Mato Grosso (Traylor 1979), while Olrog (1979) reports it from Paraguay. A population in Surinam is known only from the Sipaliwini Savanna (Mees 1968), which should be resurveyed. Silva et al. (1997) register the species from Amapa savannas. Repeated visits to seemingly suitable habitats in the three most important campo reserves in central Brazil (Parque Nacional Brasilia, Parque Nacional Serra da Canastra, and Parque Nacional das Emas) and work of many ornithologists (Collar et al. 1992; R. Cavalcanti; A. Negret; H. Sick; R. Ridgely, pers. comms.) have failed to yield new records.

In July 1987 and January 1988 Willis rediscovered *E. rufomarginatus* in high-grass campo-cerrado, or bushy prairie, in the Reserva Ecológica Serra das Araras, Mato Grosso (Willis and Oniki 1990). (We recommend using "campo-cerrado" instead of the "campo sujo" of Eiten 1972, for the latter means "dirty prairie" and perpetuates old prejudices.) In June 1989, John Bates and Curtis Marantz (Bates et al. 1992) discovered a second population in campo-cerrado on the Bolivian side of the Serrania de Huanchaca or (Brazilian name) Serra de Ricardo Franco, a previously unexplored plateau on the Brazilian border just west. Parker, Hermes Justiniano, and Omar Rocha studied *E. rufomarginatus* in the same area on 23–24 August 1990.

The flycatchers in the Serra das Araras were in tall-grass bushy areas both at 700 m atop the serra and at 200 m near a river in the central valley of the serra. The campo-cerrado at "Huanchaca Uno," a remote airstrip on the northern end of the Huanchaca and about 21 km south of Catarata Arco Iris, covered a level area at 600 m of some 2 km^2. The tall grass and bushes with small trees were almost too dense to walk through. The area had not been burned since at least October 1986 (H. Justiniano, pers. comm.) and appeared to be shielded from fire by a large island of forest to the south, west and north, and by rocky, cerrado-covered slopes to the east. The diversity of shrubs seemed great in this unburned zone, including shrubby melastomes, dwarf palms, terrestrial bromeliads, and various small trees, such as *Byrsonima* sp., *Curatella americana*, *Eriotheca* (*gracilipes*?), and *Tabebuia ochracea*.

Within an area of about 300 by 300 m in the Huanchaca, Parker and coworkers found at least four pairs of *E. rufomarginatus*. They were among the most conspicuous birds by voice, especially at dawn and late in the afternoon. Single birds, presumably males, sang persistently from the cover of grasses and bushes, but presumed mates were never far away and responded to playback as quickly and aggressively as their more vocal counterparts, with similar vocalizations. The territorial song was a loud, accelerating series of 3 "tic" notes followed by a "tiker" note and a long, emphatic buzz (e.g., "tic-tic-tic-tiker-trrrik," Fig. 1, A); this was given, at short intervals, for up to 30 seconds at a time. In response to playback, one or both members of a pair flew to within a few meters of the sound source and uttered a series of similar but longer calls ("trrrika-trrrika-trrrika," Fig. 1, B). When both called in unison, the effect was antiphonal. Individuals were heard frequently until about 8:00 (on a clear morning), but then became silent and difficult to locate. One could easily overlook them when not calling.

In Brazil, songs matched the recordings from Bolivia. On 26 January, 7–10 fast buzzy trills

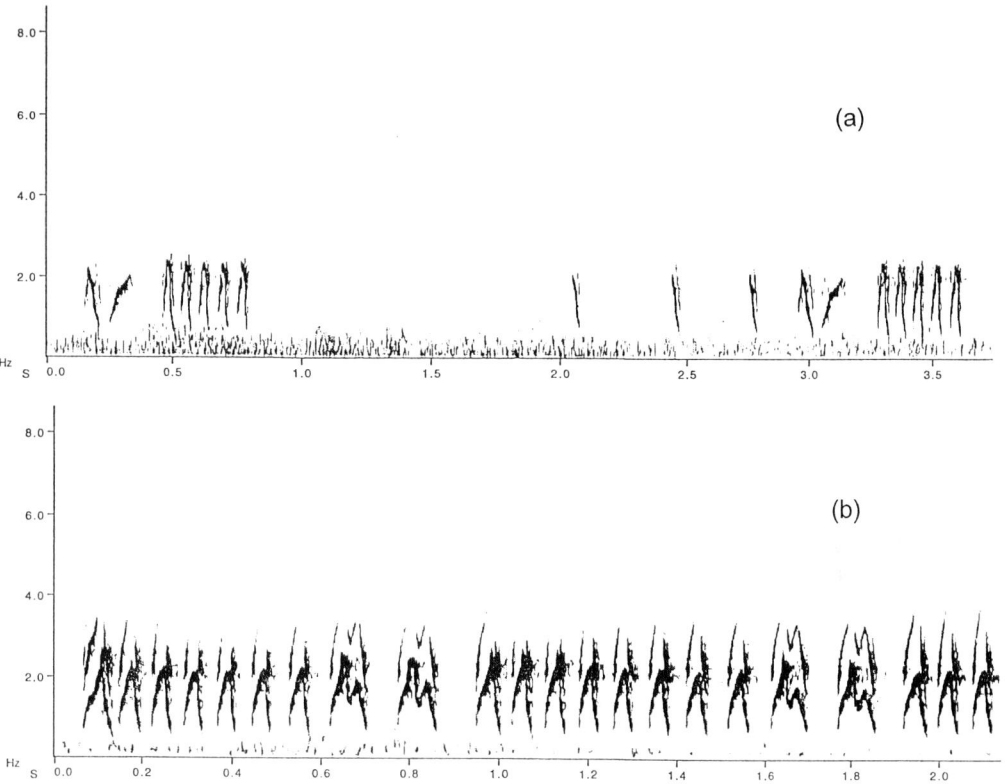

FIG. 1. (a) "tik-tik-tik-tiker-trrrik" songs of *Euscarthmus rufomarginatus,* and (b) "trrrika" long series in response to playback. Recordings by T. A. Parker from Serra de Huanchaca (Library of Natural Sounds, Cornell Lab of Ornithology).

("pe'e'eert, . . .") seemed to be alarm notes as a pair fled in a mixed-species flock of Rusty-backed Antwrens (*Formicivora rufa*) and House Wrens (*Troglodytes aedon*). The pair led a large young with gray-blotched chest and once fed it a rather large green grasshopper; the fledgling gave a "pee'eert" as it looked at Willis before fleeing. When the pair met, they engaged in duets of "trrrika" vocalizations.

In Brazil and Bolivia, foraging birds usually remained within a few centimeters of the ground and were difficult to observe. They perched on grass stalks and gleaned small arthropods from grasses and foliage of small bushes, occasionally making short forward or upward sallies to vegetation. They jerked their tails upward, wrenlike, and fled nervously if approached. Three different birds in the Huanchaca fed on the small, dark fruits of an unidentified shrub, and later regurgitated the seeds.

M. and P. Isler (Library of Natural Sounds, Cornell University) recorded the song from an extensive area of dense bamboo scrub south of Alta Floresta, Mato Grosso. It is not certain whether the unseen bird was along the road or in the bamboo, but they noted no grassland (pers. comm.). It may be that bamboo scrub can be used by some birds, perhaps temporarily after fires in natural habitat.

BEARDED TACHURI

Several other small grassland flycatchers may be nearly as threatened as *E. rufomarginatus.* Several old specimens of the Bearded Tachuri (*Polystictus pectoralis*) exist from Bolivia (Traylor 1979). We know of only a few populations of this species in central Brazil, notably in fairly open shrubby grassland in Emas (Parker and Willis, pers. obs.; R. Ridgely, pers. comm.) and at Itirapina in São Paulo (Willis 1992, 1993). The species is similarly scarce in northern South America (e.g., Gran Sabana of southeastern Venezuela; Parker, pers. obs.), and the Bogotá Savanna population may be endangered (Hilty and Brown 1986). Collar and Wege (1995) summarize records, noting that it is a summer visitor to northern Argentina and that it deserves near-threatened status.

At Itirapina, *P. pectoralis* is commonest in campo-cerrado, some of which is protected in an agricultural experiment station of the state Instituto Florestal and in a reserve of the Universidade de São Paulo. In adjoining private property, Willis had to go to newspapers and television to stop a chicken company from planting eucalyptus and a paper company from planting pines, where the rare Lesser Nothura (*Nothura minor*) and Ochre-breasted Pipit (*Anthus nattereri*) also occurred. Unfortunately, this continuous tract of campo and campo-cerrado, several kilometers across and one of the largest in the state, is not even registered on recent maps of native vegetation of the state because native savannas look like pastures in aerial and satellite photos; we hope techniques of geographic analysis can be improved rapidly.

The male tachuri gives an odd little flight song on warm mornings, especially in spring (September-December), a rising "tee-tee-tee-teet" followed by a short grasshopper-like "wing-buzz" at the apex of a 20–100 m flight to 10 m above the grass and bushes. The song can be heard on the cassette of Straneck and Carrizo (1990). It is also given, without the buzz, inside or atop a low bush before sunrise in springtime. Earlier still, in the near-dark, unseen birds circle slowly in the sky, repeating their songs and buzzes every 3 s. They seem to be advertising territories up to 200 m across (04:32 to 04:45 on 21 November 1993; as early as 04:21 but mostly 04:30–04:50 on 4 and 11 December 1994; later, 05:02–05:15, on 11 October 1994, and 04:55–05:10 on 11 November 1995).

Single birds or pairs wander from bush to bush or through low grass and herbs, fluttering short distances or gleaning insects. With their long legs, they perch at times between two upright grass blades. Single birds also pass through disturbed bushy pastures or abandoned fields nearby, especially in winter. A short "peewee" is given when birds are together, the tail jerking up before the bird flees from the observer. On 9 December 1984, a sharp "pee" of a parent bird caused two fledglings to stop their thin "peeeh" begging whistles, but they resumed "peewee" exchanges with it as they followed low in weeds and sparse grass. Parker (*in litt.* to Collar and Wege 1995) incorrectly reported this as a pair caring for fledglings. We even doubt Holland's (1893) suggestion of pair formation, and think it possible that males display while females care for nests and fledglings. Two similar young, reddish on the primaries and with buffy wing bars and yellowish underparts, followed one adult on 25 December 1994; it gave a short rattle and a faint "chup-chup, seet-seet" as well as "wheebee" or "peewee" calls before it fed one fledgling.

Contrary to Collar and Wege (1995), we have noted more use of moist grasslands for the following species. Wet areas occur in most grasslands, but the tachuris stay away from them unless dry areas have been destroyed (as is often the case).

SHARP-TAILED TYRANT

The Sharp-tailed Tyrant (*Culicivora caudacuta*) also is endangered by conversion of moist campos to agriculture. We have found the species at Brasilia, Emas, and a few localities in São Paulo State. At Itirapina and other sites, these tiny flycatchers live in pairs, members of which fly alternately low over the grass and sally for seeds or insects in foliage or flowers of bushes, tall grass or air (see Fitzpatrick 1980; Traylor and Fitzpatrick 1982; Parker and Rocha 1991). Cup nests and three buffy eggs are reported (Hartert and Venturi 1909). Young are buffy and can be mistaken for Rufous-sided Pygmy-Tyrants. On 17 February 1991 at Itirapina, two fledglings gave "t-t-t-t-t-t-t-twee" stutters or brief "twee" notes. By 7 April and 1 May, these two had partly white superciliaries and, at the long "wheef, wheef, wheef, ..." song series of an adult, gave a single "wheef" then "t-t-t-t-t-t-t-t-t-t," which sounded like a pigeon taking off, before following. Duets of stutters alternating with "wheef" notes are frequent for pairs, also. Family groups seem to last into the winter, perhaps forming the wandering groups of 3-10 birds one sees in that season. They can join mixed-species flocks, often including Chalk-browed Mockingbirds (*Mimus saturninus*), Cock-tailed Tyrants (*Alectrurus tricolor*) and two campo-cerrado tanagers (*Neothraupis fasciata, Cypsnagra hirundinacea*), as can lone individuals of *P. pectoralis*.

OTHER CAMPO-CERRADO BIRDS

The Lesser Nothura, although common at Emas, barely survives at Itirapina and is otherwise nearly gone in much of central Brazil (Collar et al. 1992). *Taoniscus nanus*, the Dwarf Tinamou, is of uncertain status near Brasilia (Teixeira and Negret 1984) and is often confused vocally with Ocellated Crakes (*Micropygia schomburgkii*), so new studies are needed.

Ochre-breasted Pipits have disappeared from several campo-cerrados in São Paulo in recent years, and are rarely recorded in other areas south to Argentina (Collar et al. 1992). Because they prefer burned areas in wet campos (Willis and Oniki 1993), they disappear both in over-protected reserves like Itirapina and in places where agriculture and cattle take over. P. Martuscelli (pers. comm.) found it recently in one large remnant of campo-cerrado on a private ranch near Santa Cruz do Rio Pardo. In September 1995, Willis found a few singing males in a campo pasture just a few kilometers southwest of the Itirapina reserve, where the few cattle feeding on native grasses next to moist campo left the area rather like a recent burn. Coutinho (1982) reported that artificial removal of terminal sprouts (hoeing, light grazing) increased herbaceous productivity and flowering much like fire, perhaps explaining pipit use of either burned or lightly grazed zones. Native grazing mammals do not seem to have been common in southeastern Brazil, suggesting that the pipit was using burned zones rather than lightly grazed ones in the past. Itirapina ranchlands are usually planted to more productive introduced grasses, or converted to valuable orange or eucalyptus groves, so it is not certain this site can be protected.

We hardly know the Blue-eyed Ground Dove (*Columbina cyanopis*), known from eight specimens (Collar et al. 1992) and recently rediscovered in campos at the Serra das Araras (Silva and Oniki 1988). Willis briefly noted one low at the edge of a gallery wood near the hospital of the University of Mato Grosso do Sul, Campo Grande (barely 100 m from the second Congress of the Brazilian Ornithological Society) in October, 1992, among Ruddy Ground-Doves (*C. talpacoti*). He only confirmed the identification after checking specimens at the Museum of Zoology of the University of São Paulo, for it seemed to have a bluish face, forming small pale triangles before and behind the dark eyes; it was rusty on head and back, a pattern never found in Ruddy Ground-Doves of any age. In November 1993, Willis and Oniki could not find the species. We suspect that it is overlooked, but it may be rare.

DISCUSSION

Euscarthmus rufomarginatus probably was once widely distributed in the campos of inland South America. We suspect that it began to decline with the increase of cattle ranching in the last century. The combined effects of overgrazing, introduced grasses and annual burning lead to dramatic changes in the structure and floristic composition of campos. Protected areas of campo-cerrado vegetation, especially areas not burned for at least three years, are rare. In such Brazilian national parks as Emas, extensive fires occur almost annually, despite efforts to control or prevent them. Many fires are set by local hunters to aid in poaching game and by people unhappy with such large areas being "unproductive." The continuing survival of a variety of grassland plants and animals may depend on staging burns at intervals of several years within a large area, preferably connected to other areas by corridor zones, to provide a variety of successional habitats.

The three small flycatchers, and species that need natural occasionally-burned or lightly-grazed bushy prairies, are only some of the birds threatened by further conversion of South American savannas in this century, principally into eucalyptus, soybeans and pastures for exportable crops produced by wealthy ranchers. Poor farmers have rarely used such areas, but recently government-aided land-reform projects in Brazil have fostered group invasion of "unproductive latifundios" or large ranches, incidentally stimulating other ranchers to quickly plow under remaining patches of natural habitats. Some disappearing birds such as Coal-crested Finch (*Charitospiza eucosma*) and Campo Miner (*Geobates poecilopterus*) use burned zones. Other species, such as Cock-tailed Tyrants and Collared Crescentchests (*Melanopareia torquata*), use taller grass. When fire is completely suppressed in open savannas, bushes close in to a "cerradão" or woodland (Coutinho 1982). With annual fires not even pipits or miners survive. Under natural conditions, the differing birds engaged in local migrations or seminomadism at intervals of more than one year, in response to fire succession.

Only birds of northeastern Atlantic forests are as endangered as campo and campo-cerrado species (Willis and Oniki 1992), but Brazilian, Bolivian and international attention has only recently started to consider the conservation problems in these economically valuable habitats. As Willis emphasized in a presentation at the International Ecological Congress in Japan in 1990, the situation of two million square kilometers of former savanna in South America is an ecological disaster, fully comparable to problems in North American grasslands, now that liming of large areas as an antidote to aluminum toxicity and acid soils has allowed agribusinesses to dominate vast areas. Recently, eucalyptus-forest managers have been completely suppressing fire in narrow reserves around and through their plantations, allowing bushes to take over. These habitats will not protect grassland birds, nor will numerous cattle trampling up natural or artificial grasslands on so-called "sustainable use" areas. Mares (1992) emphasized mammalian diversity in these open ecosystems, and Chesser and Hackett (1992) agreed, though they noted that diversity in forested regions is even greater. Botanists, zoologists and others at a congress on cerrados of the state of São Paulo (Pirassununga, October 1995) suggested complete protection for the scattered 1.8% of the original areas of this habitat spectrum still present there. A congress on grassland birds (Association of Field Ornithologists, Tulsa, Oklahoma) later the same month also agreed that research is urgently needed in South American savannas, especially on the ecology of fires at landscape scales.

ACKNOWLEDGMENTS

H. Justiniano and the staff of the Parque Nacional Noel Kempff Mercado made possible Parker's field work in the Huanchaca. The Conselho Nacional de Desenvolvimento Cientifico e Tecnológico (CNPq) of Brazil and the Fundação de Amparo à Pesquisa do Estado de São Paulo (FAPESP) helped Willis' studies, as well as personnel of the Instituto Florestal in Itirapina. Recordings and sonograms were provided by the Library of Natural Sounds, Laboratory of Ornithology, Cornell University, Ithaca, New York. We appreciate suggestions from J. Bates, J. Fitzpatrick, and J. M. C. Silva.

LITERATURE CITED

BATES, J. M., M. C. GARVIN, D. C. SCHMITT, AND C. G. SCHMITT. 1989. Notes on bird distribution in northeastern Depto. Santa Cruz, Bolivia, with 15 species new to Bolivia. Bull. Brit. Orn. Club 109: 236–244.

BATES, J. M., T. A. PARKER III, A. P. CAPPARELLA, AND T. J. DAVIS. 1992. Notes on bird distribution in northeastern Santa Cruz, Bolivia II, with 21 species new to the country. Bull. Brit. Orn. Club 112: 86–98.

CAVALCANTI, R. B. 1988. Conservation of birds in the *cerrado* of central Brazil. Pages 59–66 in Ecology and Conservation of Grassland Birds (P. D. Goriup, Ed.). ICBP Technical Publication 7, Cambridge, UK.

CHESSER, R. T., AND S. J. HACKETT. 1992. Mammalian diversity in South America (Letter). Science 256: 1502–1504.

COLLAR, N. J., L. P. GONZAGA, N. KRABBE, A. MADROÑO-NIETO, L. G. NARANJO, T. A. PARKER III, AND D. C. WEGE. 1992. Threatened Birds of the Americas. Int. Council Bird Preserv., Cambridge, U.K.

COLLAR, N. J., AND D. C. WEGE. 1995. The distribution and conservation status of the Bearded Tachuri *Polystictus pectoralis*. Bird Cons. International 5:367–390.

COUTINHO, L. M. 1982. Ecological effects of fire in Brazilian *cerrado*. Pages 273–291 in Ecology of Tropical Savannas (B. J. Huntley and B. H. Walker, Eds.). Springer-Verlag.

EITEN, G. 1972. The cerrado vegetation of Brazil. Bot. Rev. 38:201–341.

FITZPATRICK, J. W. 1980. Foraging behavior of neotropical tyrant flycatchers. Condor 82:43–57.

HARTERT, E., AND S. VENTURI. 1909. Notes sur les oiseaux de la République Argentine. Novit. Zool. 16:43–57.

HILTY, S. L., AND W. L. BROWN, JR. 1986. A Guide to the Birds of Colombia. Princeton Univ. Press, Princeton, New Jersey.

HOLLAND, A. H. 1893. Field-notes on the birds of Estancia Sta. Elena, Argentine Republic. Ibis (6)5:483–488.

MARES, M. 1992. Neotropical mammals and the myth of Amazonian biodiversity. Science 255:976–979.

MEES, G. F. 1968. Enige voor de avifauna van Suriname nieuwe vogelsoorten. Giervalk 58:101–107.

OLROG, C. C. 1979. Notas ornitológicas, XI: sobre la colección del Instituto Miguel Lillo. Acta Zool. Lilloana 33:5–7.

PARKER, T. A., III, AND O. ROCHA. 1991. La avifauna del Cerro San Simon, una localidad de campo rupestre aislado en el Depto. Beni, noreste boliviano. Ecol. en Bolivia 17:15–29.

PELZELN, A. 1868. Zur Ornithologie Brasiliens. A. Pichler's Witwe & Sohn, Vienna.

SERRA-FILHO, R. (ED.) 1974. Levantamento da cobertura vegetal natural e do reflorestamento do Estado de São Paulo. Bol. Técnico Inst. Florestal, S. Paulo 11:1–55.

SILVA, J. M. C. DA. 1995. Avian inventory of the cerrado region, South America: implications for biological conservation. Bird Cons. International 5:291–304.

SILVA, J. M. C. DA, AND Y. ONIKI. 1988. Lista preliminar da avifauna da Estação Ecológica Serra das Araras, Mato Grosso, Brasil. Bol. Museu Paraense "Emilio Goeldi", sér. Zool. 4:123–143.

SILVA, J. M. C. DA, D. C. OREN, J. C. ROMA, AND L. M. P. HENRIQUES. 1997. Composition and distribution patterns of the avifauna of an Amazonian [upland] savanna, Amapá, Brazil. Pp. 743–762 in Studies in Neotropical Ornithology Honoring Ted Parker (J. V. Remsen, Jr., Ed.), Ornithol. Monogr. No. 48.

STRANECK, R., AND G. CARRIZO. 1990. Canto de Las Aves de los Esteros y Palmares. Library of Latin America (LOLA), Buenos Aires.

TEIXEIRA, D. M., AND A. NEGRET. 1984. The Dwarf Tinamou (*Taoniscus nanus*) of central Brazil. Auk 101: 188–189.

TRAYLOR, M. A., JR. (ED.) 1979. Checklist of Birds of the World. VIII: Tyrannidae through Phytotomidae. Museum of Comparative Zoology, Cambridge, Massachusetts.

TRAYLOR, M. A., JR. AND J. W. FITZPATRICK. 1982. A survey of the tyrant flycatchers. Living Bird 19:7–50.

WILLIS, E. O. 1992. Itirapina, São Paulo, Brazil [Christmas Bird Count]. Am. Birds 46:1021.

WILLIS, E. O. 1993. Itirapina, São Paulo, Brazil [Christmas Bird Count]. Am. Birds 47:995.

WILLIS, E. O., AND Y. ONIKI. 1988. Bird conservation in open vegetation of São Paulo, Brazil. Pages 67–70 in Ecology and conservation of grassland birds (P. D. Goriup, Ed.). ICBP Technical Publication 7, Cambridge, UK.

WILLIS, E. O., AND Y. ONIKI. 1990. Levantamento preliminar das aves de inverno em dez áreas do sudoeste de Mato Grosso, Brasil. Ararajuba 1:19–38.

WILLIS, E. O., AND Y. ONIKI. 1992. Losses of São Paulo birds are worse in the interior than in Atlantic forests. Ciência e Cultura 44:326–328.

WILLIS, E. O., AND Y. ONIKI. 1993. New and reconfirmed birds from the state of São Paulo, Brazil, with notes on disappearing species. Bull. Brit. Orn. Club 113:23–34.

BIRDS OF THE CENTRAL RÍO TUICHI VALLEY, WITH EMPHASIS ON DRY FOREST, *PARQUE NACIONAL MADIDI*, DEPTO. LA PAZ, BOLIVIA

ALAN PERRY[1], MICHAEL KESSLER[2], AND NICHOLAS HELME[3]

[1]*Foundation for Tropical Research and Exploration (TREX),*
Casilla 11215, La Paz, Bolivia;
[2]*Foundation for Tropical Research and Exploration (TREX),*
Systematisch-Geobotanisches Institut,
Untere Karspuele 2, 37073 Goettingen, Germany;
[3]*Foundation for Tropical Research and Exploration (TREX),*
189 Main Road, Kalk Bay, 7975 South Africa

ABSTRACT.—During eight weeks of fieldwork between July and September 1993 we inventoried the avifauna of the central Río Tuichi valley, depto. La Paz, northwestern Bolivia. The valley supports a complex mosaic of habitat types, including dry forest, lower montane rainforest, and floodplain forest. This study emphasized bird surveys of the 1,200 km² tract of inter-Andean dry forest. Most of this dry forest valley falls within the boundaries of the newly created *Parque Nacional Madidi*. A list of the 275 bird species that were recorded is presented, along with notes on preferred habitat, relative abundance, and evidence. Species accounts are provided for noteworthy records, including one species new to Bolivia, (*Basileuterus chrysogaster*), four species new to depto. La Paz, (*Dendrocygna viduata*, *Buteo albonotatus*, *Micrastur semitorquatus*, and *Cyanerpes cyaneus*), and five additional species (*Neomorphus geoffroyi*, *Myrmotherula grisea*, *Herpsilochmus* sp., *Cranioleuca* sp., and *Oxyruncus cristatus*). Biogeographically, the avifauna represents an interesting mixture of species from dry and open habitats, lowland Amazonian rainforest, lower montane rainforest, as well as a number of species with wide elevational ranges. The central Río Tuichi dry forest is a high priority for conservation. Its large pristine area, low human population, close proximity and dependence on surrounding humid ecosystems, and isolated geographical position make it ideally situated for preservation. Additionally, no other Andean or lowland dry forests are represented in Bolivian protected areas, and Neotropical dry forests are threatened throughout their Neotropical range. The conservation importance is further underscored by the high bird diversity for such dry forests, the unique biogeographical bird species mixture, and the presence of rare, poorly known, undescribed, threatened, endemic, and range-restricted bird species. *Parque Nacional Madidi* represents one of the best opportunities in South America for the conservation of inter-Andean dry forests and a number of other globally threatened ecosystems.

RESUMEN.—Durante ocho semanas de trabajo de campo entre Julio y Septiembre de 1993 realizamos un inventario de la avifauna de una región de 1,200 km² de bosque seco interandino del valle central del Río Tuichi, depto. de La Paz, noroeste de Bolivia. Este valle mantiene un mosaico complejo de diferentes tipos de habitats, incluyendo bosque seco, bosque lluvioso de baja montaña, y bosque estacionalmente inundado. La mayor parte del bosque seco de este valle se encuentra dentro de los límites de la recien-declarado *Parque Nacional Madidi*. Una lista de las 275 especies de aves que fueron registradas es presentada, junto con notas sobre las preferencias de habitas, abundancia relativa, y evidencia. Algunos registros son proporcionadas para una especie nueva para Bolivia, (*Basileuterus chrysogaster*), cuatro especies nuevas para el depto. de La Paz, (*Dendrocygna viduata*, *Buteo albonotatus*, *Micrastur semitorquatus*, y *Cyanerpes cyaneus*) y cinco especies adicionales (*Neomorphus geoffroyi*, *Myrmotherula grisea*, *Herpsilochmus sp.*, *Cranioleuca sp.*, y *Oxyruncus cristatus*), para los cuales obtuvimos notables registros. Biogeográficamente, la avifauna representa una interesante mezcla de especies de habitats secos y abiertos, bosque lluviosos de tierras bajas amazonicas, bosques lluviosos de baja montaña, asi también un número de especies con amplio rango altitudinal. El bosque deciduo de este valle representa uno de los mejores ejemplos de su tipo en los Neotropicos, y esta idealmente situado para la preservación debido a su gran dimension y estatus pristino, la baja población humana, y aislamiento del valle. La importancia de su conservación es más valiosa por la alta diversidad de especies de aves para tales bosques secos, la singular mezcla biogeográfica de especies de aves, y la presencia de especies de aves raras, poco conocidas, no descritas, amenazadas, y de rango restringido. *Parque Nacional Madidi* representa una de las mejores oportunidades en Sud America para la conservación del bosques secos interandinos y un sinnúmero de otras ecosistémas globalmente amenazados.

Bolivia has an exceptional diversity of species and habitats, with well preserved tracts of many major Neotropical habitats. Over 1,340 bird species have been recorded from within its boundaries (Remsen and Traylor 1989; Armonia 1995). The relatively small human population and vast uninhabited areas allow many opportunities to integrate conservation and sustainable development. One of the potential biodiversity showcases is *Parque Nacional Madidi* (Fig. 1). This vast area of over two million hectares harbours nearly the full array of Amazonian and Andean ecosystems ranging from 200 m to 5,500 m. It is likely to contain over 1,000 species of birds and has the highest biodiversity of any protected area in the world (Parker and Bailey 1991; Remsen and Parker 1995).

Distributional information for birds in the region is available for lowland rainforest (Foster et al. 1994; Parker unpubl. data; Parker and Bailey 1991), foothill forest (Foster et al. 1994; Friggens et al unpubl. data; Parker unpubl. data; Perry et al. unpubl. data), lower montane rainforest (Cardiff et al. unpubl. data; Parker unpubl. data; Parker and Bailey 1991; Perry and Helme unpubl. data), cloud forest (Perry et al. unpubl. data), savanna (Foster et al. 1994; Parker unpubl. data; Parker and Bailey 1991; Helme and Perry unpubl. data), and dry forest (Pearman 1993; Parker and Bailey 1991). The Bolivian *Centro de Datos para la Conservación* (unpubl. data) and *Veterenarios sin Fronteras* (unpubl. data) have summarized the published flora and fauna information for Madidi and Pilon Lajas respectively.

Dry forest, which occurs in scattered areas throughout South America (Fig. 2), is considered one of the most threatened Neotropical ecosystems (Gentry 1977, 1993; Janzen 1986, 1988; Parker and Bailey 1991; Collar et al. 1997). Lowland dry forest occurs at elevations of up to about 800 m throughout the Neotropics, and is particularly widespread in western Central America, northern South America, and south-central South America. Inter-Andean dry forests occur at elevations of 500 to 2,000 m in isolated "islands," mainly in rain-shadowed valleys of the eastern slope of the Andes.

Lowland dry forest birds have been surveyed in Central America (Stiles 1983), the Tumbes region of northwestern Peru and southwestern Ecuador (Wiedenfeld et al. 1985; Best et al. 1992; Parker and Carr 1992), and southeastern Bolivia (Parker et al. 1993). But for inter-Andean dry forest avifauna, very little information exists (Chapman 1921, and various bird records from depto. Huanuco, Peru, and depto. Guaillabamba, Ecuador). Because of the information from this study and preliminary studies that briefly visited the Río Macharipo valley (Figs. 1, 3, and 4, Parker and Bailey 1991; Pearman 1993), the dry forest of the Río Tuichi and Machariapo valleys is the best-surveyed in Bolivia and Peru. In this paper we present results of avifaunal surveys in the isolated inter-Andean dry forest of the central Río Tuichi valley (Figs. 3 and 4), the first in a series of TREX expeditions to explore and survey remote and biologically important sites within the *Parque Nacional Madidi*.

STUDY SITES

Fieldwork in 1993 was conducted from 9 to 23 July (A. Perry; M. Kessler; N. Helme) and from 7 to 30 September (A. Perry; N. Helme), and included intensive surveys of dry forest, brief surveys of surrounding habitats, and exploration of most of the central Río Tuichi drainage between 500 and 1,500 m (Fig. 3–5). Leaving from Apolo, the nearest town, we hiked with mules loaded with gear and equipment for three days to the study sites. On the July expedition, the team worked out of Hac. Ubito, and explored the dry and floodplain forests of the Río Ubito and Río San Juan valleys, as well as lower montane rainforest near Nogal and Rabiana. During the September expedition, we set up camps in the Río Ubito valley (Campamento Cerro Yanomayo and Campamento Tapir Pool) and Río Recina valley (Campamento Agreyoyo), and explored the Río San Juan, Río Recina, and Río Tuichi valleys. The primary objectives were to, (1) obtain an overview of the avifaunal composition and its biogeographical affinities, (2) to collect data useful to organizations promoting conservation and sustainable development in the region, (3) to assess the conservation status of the dry forest and rare or threatened bird species, and (4) to provide training and fieldwork opportunities for Bolivian student biologists.

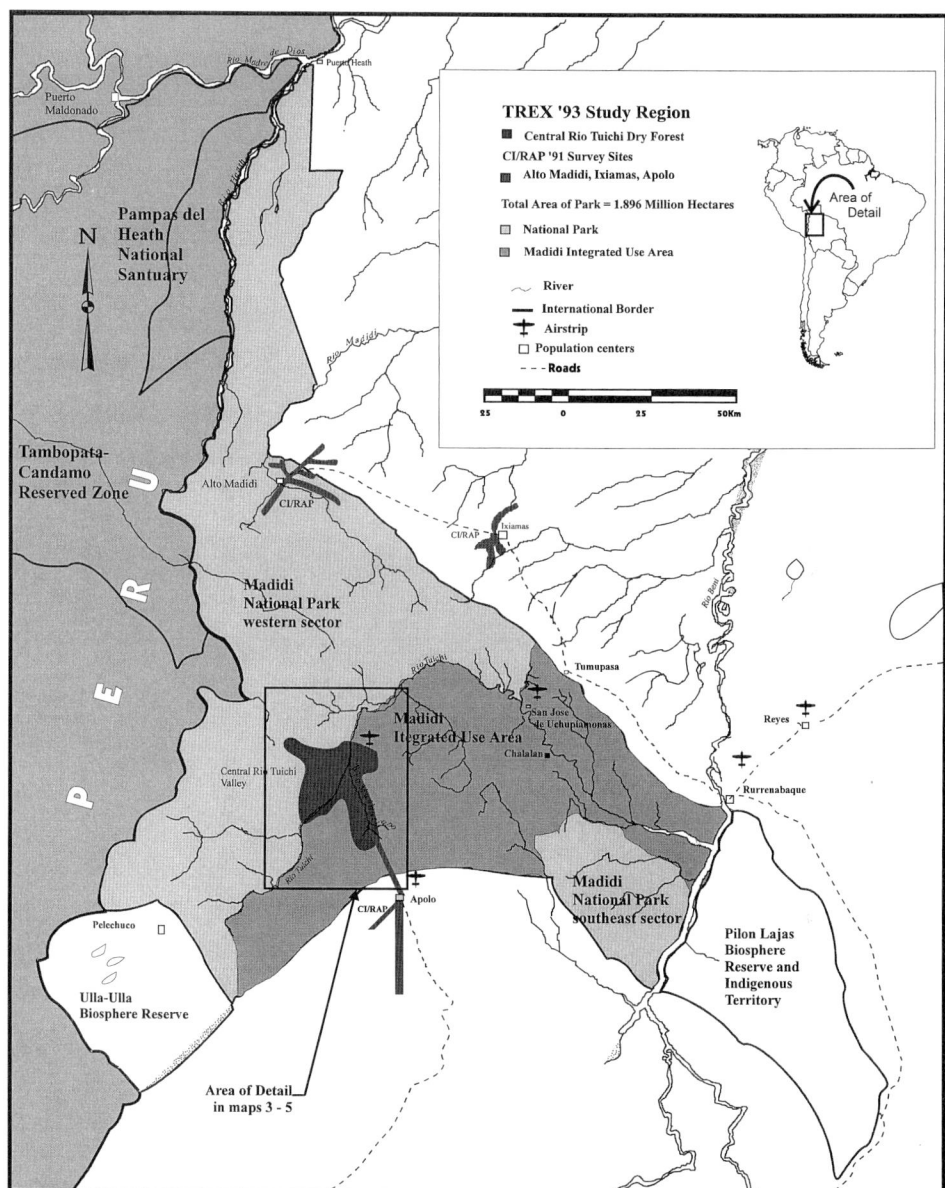

FIG. 1. Map of the Madidi National Park and Integrated Use Area, depto. La Paz, Bolivia, showing adjacent conservation units in Bolivia and Peru, TREX study region (the central Río Tuichi valley) and the survey sites of the Conservation International-Rapid Assessment Program (CI-RAP, Parker and Bailey 1991).

The central Río Tuichi and associated valleys occur between 68°20′W and 68°50′W and 14°10′S and 14°40′S (Figs. 3–5). Average rainfall in Apolo, about 12 km SSE of the Tuichi dry forest, is 1,360 mm, most of which falls during the wet season months from October to March. The dry season, from April to September, has a monthly rainfall average of less than 40 mm. Although there are no rainfall data from the Tuichi dry forest area, it is certainly less than in Apolo (pers. obs., Ramos, pers. comm.). We estimate it to be from 900–1,200 mm per year. The mean annual temperature in Apolo is 20°C, with an extreme range of 5°C to 35°C (Soux et al. 1991).

The Tuichi valley supports a complex mosaic of forest and second-growth habitats. Figure 4 shows the present-day distribution of vegetation types, and Figure 5 shows the hypothetical

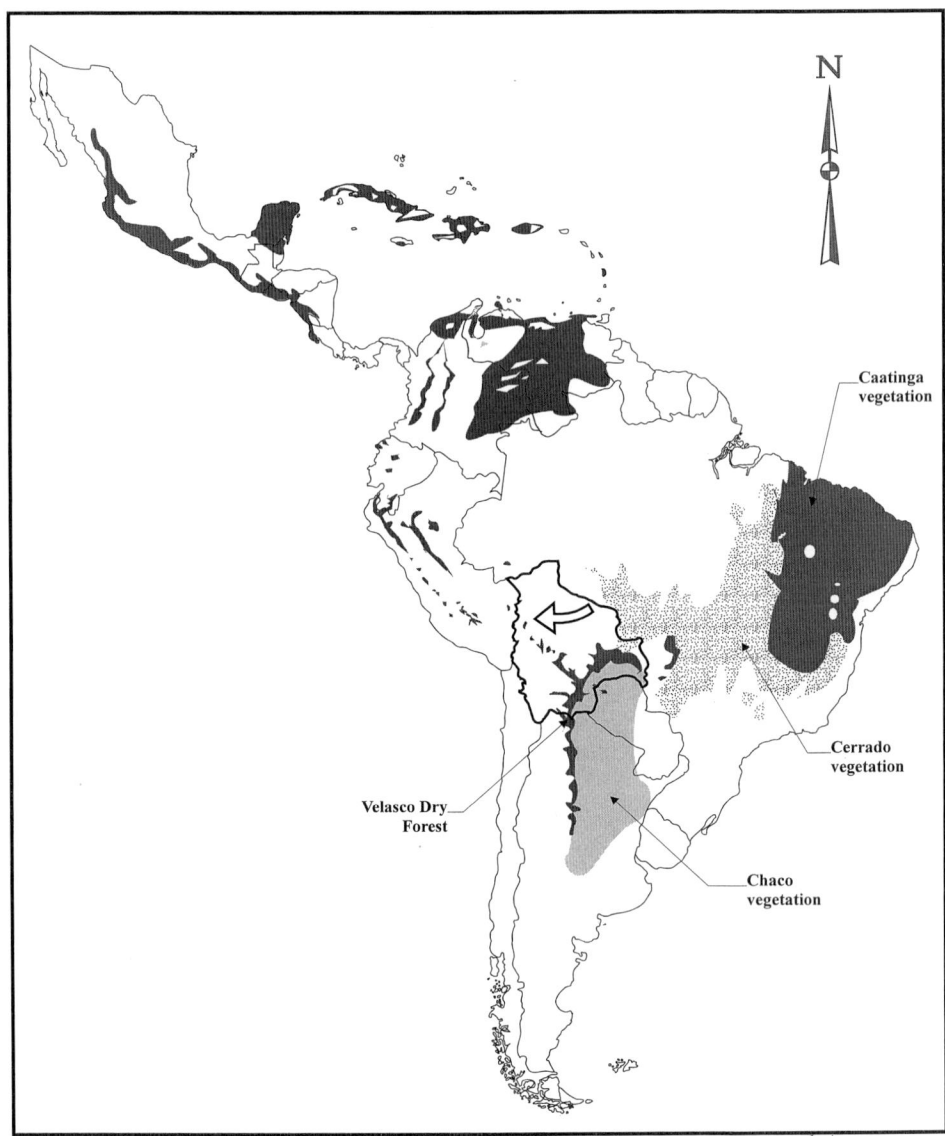

FIG. 2. Map of the dry vegetation of the Neotropical region, highlighting Bolivia (bold line) and the central Río Tuichi valley dry forest (arrow). Shaded areas represent dry vegetation; dark shading in Bolivia and northern Argentina represents the Velasco dry forest; light shading in southern Bolivia, Paraguay, and Argentina represents chaco vegetation; stipling in eastern Bolivia and Brazil represents cerrado vegetation; dark shading in eastern Brazil represents caatinga vegetation. Map based on Hueck and Seibert (1981), Gentry (1995), and Kessler (1995a).

distribution of vegetation types without human impact. Dry forest is restricted to rain-shadow foothill valleys between 500 and 1,250 m (usually between 700 and 1,000 m) and is defined by the presence of columnar cacti (*Cereus* spp., *Echinocereus* spp.). The canopy, which is mainly deciduous, reaches 25 m and is dominated by *Anadenanthera colubrina*, *Cochlospermum orinocense*, *Schinopsis brasiliensis*, and *Ceiba* spp. Typical understory species include *Vriesia amazonica*, *Agonandra* spp., *Trichilia* spp., *Peperomia* spp., and *Clavija tarapotana*. "Semi-deciduous forest," a transition forest between dry forest and lower montane rainforest (usually between 1,000 and 1,300 m), is defined by the absence of terrestrial cacti, other than *Opuntia brasiliensis*, a diverse canopy with both deciduous and evergreen species, and a mostly evergreen understory dominated by *Capparis* spp., *Prunus* aff. *tucumanensis*, and *Eugenia* spp.

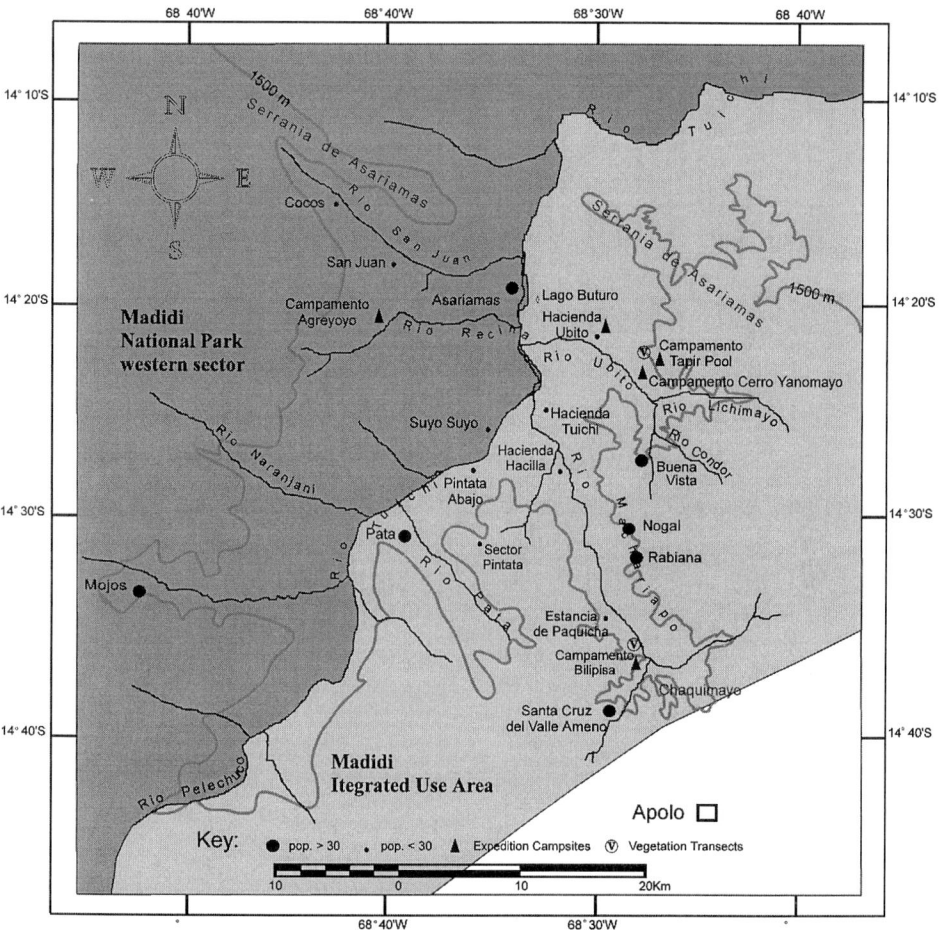

FIG. 3. Map of the central Río Tuichi and associated valleys, showing physical features, communities, and the boundary between the Madidi National Park, western sector, and the Madidi Area of Integrated Use. The topographic line represents 1,500 m. Dark shading represents Madidi National Park, western sector, and light shading represents Madidi Integrated Use Area.

Floodplain forest (Fig. 4 and 5) occurs both on floodplains with year-round access to groundwater, and in areas seasonally inundated by summer rainfall. The canopy, mainly evergreen, reaches 40 m, and is characterized by large numbers of lowland rainforest tree species (e.g., Apocynaceae spp., *Hura crepitans*, *Inga* spp., *Ficus* spp., *Attalea phalerata*, and *Astrocaryum murumuru*). The understory is dominated by *Guarea guidonia*, *Costus* spp., *Heliconia* spp., *Aphelandra* spp., and *Olyra latifolia*. Where disturbed, this forest may include dry forest taxa such as *Opuntia brasiliensis*, *Clavija tarapotana*, and *Anadenanthera colubrina*.

Lower montane rainforest (Fig. 4 and 5) is restricted to humid mountain slopes above the elevational limit of dry forest, i.e., generally higher than 1,250 m, and is defined by the presence of understory Melastomataceae (especially *Miconia* spp.), which generally require constantly humid conditions (Richter and Lauer 1987). The canopy, mainly evergreen, reaches 40 m. Floristic composition and structure are notably different from dry forest, being essentially evergreen with abundant epiphytes. The canopy reaches 30 m, and is characterized by *Myroxylon balsamum*, *Clusia* spp., and *Ficus* spp., and the understory has numerous bamboos.

Riverbank forest and associated second-growth (Fig. 4 and 5; part of floodplain forest) occur along rivers, where landslides and river floodings have created disturbances. The 20-m canopy is characterized by trees such as *Cecropia* spp., *Ochroma pyramidale*, *Inga* spp., and *Erythrina* sp., and the undergrowth is dominated by *Gynerium sagittatum*, *Costus* spp., *Heliconia* spp., and various Asteraceae.

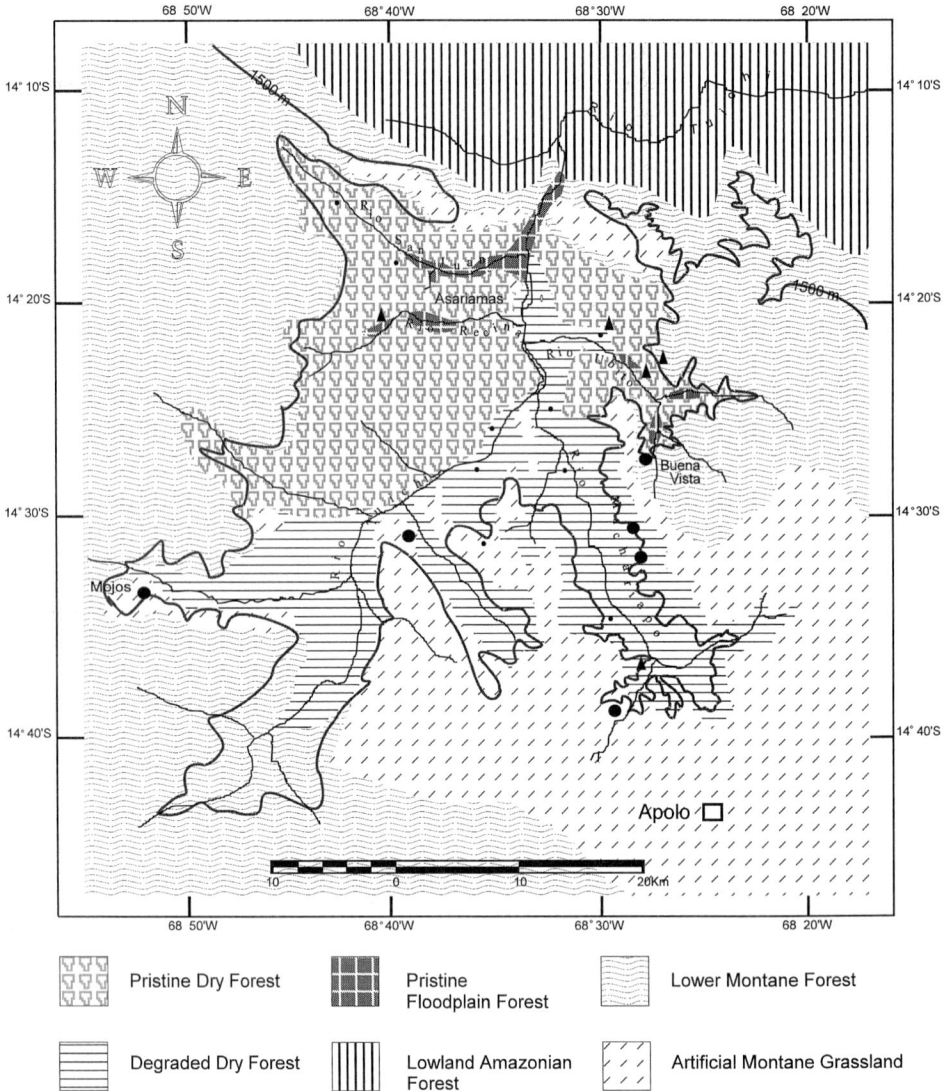

FIG. 4. Map of the central Río Tuichi and associated valleys, showing the present-day distribution of vegetation types. Note that "dry forest" includes "deciduous" and "semi-deciduous" forests (as described in the Flora and Vegetation survey).

A variety of other habitats occur in the study area. Clearings (Fig. 4 and 5) are found wherever forest has been cleared for agriculture, or burned to produce suitable grazing areas. They range in size from small clearings near settlements to relatively large grasslands such as those of the Apolo valley and Serranía de Asariamas, and are usually dominated by grasses and second growth (Asteraceae, *Inga* spp., *Alchornea* sp.) with scattered trees and bushes bordered by a fringe of secondary woodland. Seasonally flooded marshes of the Apolo valley are filled with grasses (Poaceae), sedges (Cyperaceae), and other water-adapted plants. Lago Buturo (Fig. 3), an isolated oxbow of the Río Tuichi, is a shallow, permanent lake dominated by 30-m tall *Hura crepitans*, and supports abundant populations of the floating aquatic plant *Lemna* sp. and small islands of sedges.

METHODS

Birds were surveyed mainly in dry forest, with limited time spent in floodplain forest and lower montane rainforest. Birds were tape-recorded by AHP, using a Marantz PMD 221 portable

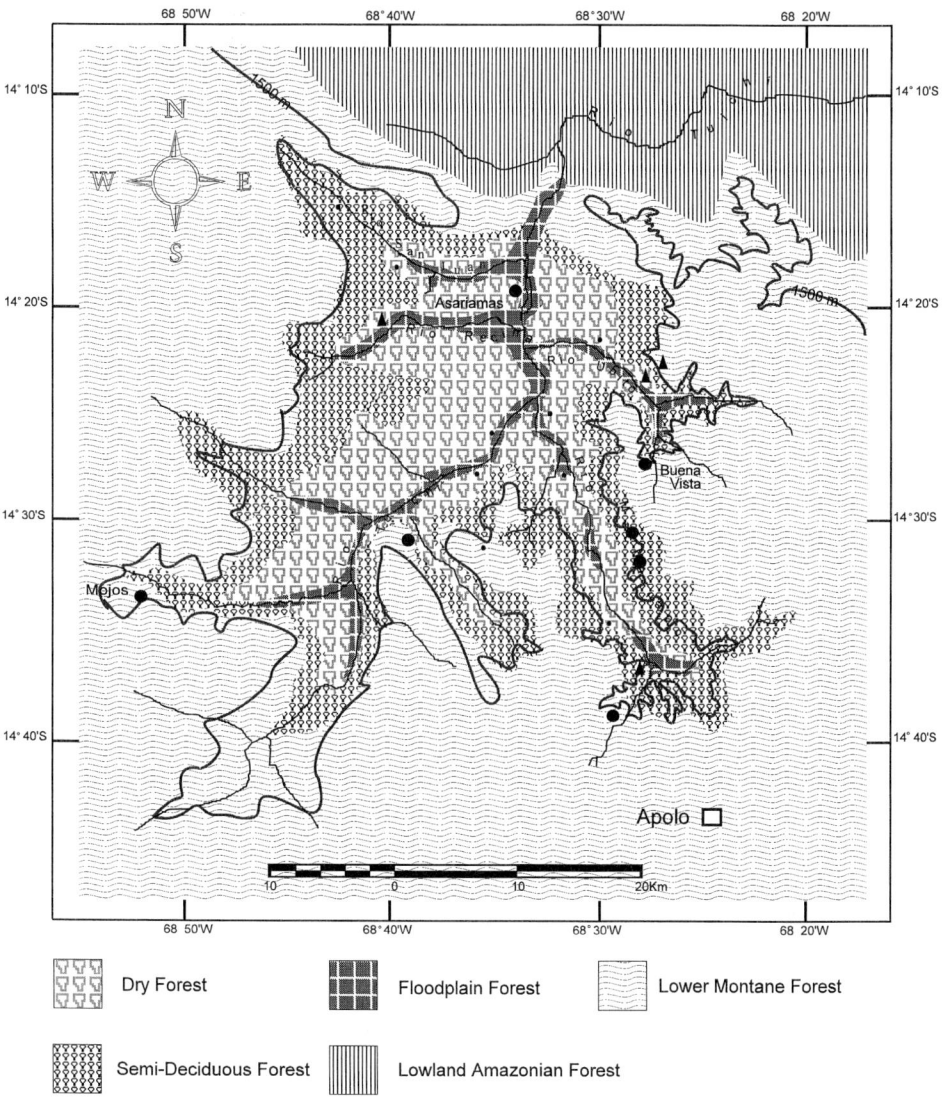

FIG. 5. Map of the central Río Tuichi and associated valleys, showing the hypothetical distribution of vegetation types, as they probably were before human impact.

tape-recorder and a Sennheiser ME-88 directional microphone. Identifications were made in the field through visual and acoustical information. Voice playbacks were used to better visually identify certain species. Song recordings were later verified by comparing them to tape-recordings that were commercially available or made by AHP at other sites. Tapes were also analyzed by Bret Whitney, who verified identifications and added species from the background of recordings. Specimens, representing thirteen species, were collected and prepared as study skins by NAH and AHP. A limited number of mist nets were placed in a variety of habitats. Photographs were taken of all birds captured in mist-nets. Field identifications of study skins and photographs were verified by comparing them to material at Coleccíon Boliviana de Fauna (CBF, La Paz, Bolivia) and Louisiana State University Museum of Natural Science (LSUMNS). Tape-recordings were deposited at the Library of Natural Sounds, Cornell Laboratory of Ornithology. Study skins were deposited at CBF. Observations made by particular individuals are indicated by their initials. Bolivian departments are abbreviated as follows: LP = La Paz, CO = Cochabamba, BE = Beni, SC = Santa Cruz, PA = Pando, TA = Tarija.

The biogeographical analysis was limited to only species recorded in (but not necessarily limited to) dry forest. Species were classified into biogeographical elements according to the life zones in which they occur in Bolivia (Remsen and Traylor 1989): the Amazonian rainforest element is made up of species from life zone "a" (Amazonian lowlands) and "l" (lowlands); the humid montane element is made up of species from any combination of life zones "f" (foothill forests; 500–1,100 m), "u" (upper tropical zone forests; 1,100–1,700 m), "s" (subtropical zone forests; 1,700–2,600 m), or h (humid Andean forests; 500–3,600 m); and the dry and open habitat element is made up of species from one or two of life zones "l" (lowlands), "n" (non-Amazonian lowlands; savannas, grasslands, deciduous and partly deciduous forest, cerrado, and ranches), and "v" (valleys; semi-arid and arid inter-montane valleys from 1,000 to 3,600 m). Note that species known from life zone "l" were included in both the lowland Amazonian and dry and open group. Unidentified species were excluded from these calculations.

RESULTS

A total of 262 species (Appendix) was recorded and documented by sight records (149, 54%), voice recordings (97, 35%), study skins (12, 4%), and photographs (10, 4%), and preferred habitat and relative abundance were noted for each species. The following important records are treated separately:

Dendrocygna viduata.—White-faced Whistling-Duck.—One was seen by MK on 5 July near marshes of the Apolo valley. These marshes are seasonally flooded wetlands surrounded by extensive secondary grasslands that are regularly burned. The marshes are shallow (usually less than 0.5 m deep) and are dominated by grasses (Poaceae) and sedges (Cyperaceae). This is the first record for LP, but was not unexpected, as the species is known from non-Amazonian lowlands of BE and SC (Remsen and Traylor 1989).

Buteo albonotatus.—Zone-tailed Hawk.—One was seen by MK on 17 July south of Hac. Ubito over dry forest circling high with one *Spizaetus tyrannus* and three *Cathartus aura*. [*Buteo albonotatus* was seen on the first hot and sunny day after a week of rain.] The individual was identified by its long wings, tilting flight, two-toned underwings, and at least two white tailbands. This is the first record for LP; the species was previously known only from the Amazonian lowlands of PA and SC (Remsen and Traylor 1989), and a sight record from the subtropical zone of CO (Whitney et al. 1994).

Micrastur semitorquatus.—Collared Forest-Falcon.—An adult bird was seen on 15 July by NAH and AHP perched at the edge of riverbank second-growth along the Río Ubito near Hac. Ubito. During the same month this species was heard calling on two consecutive evenings from the edge of lower montane rainforest near Buena Vista. *Micrastur semitorquatus* was also tape-recorded by AHP in April 1993 and October 1994 near Laguna Chalalan (500 m, foothill rainforest, lower Río Tuichi, depto. La Paz) in the subcanopy of a 40-m-tall lone tree in a clearing. These records constitute the first for LP; it was previously known in Bolivia from the lowlands of BE, CO, SC, and TA (Remsen and Traylor 1989).

Neomorphus geoffroyi.—Rufous-vented Ground-Cuckoo.—This species was seen every other day in relatively undisturbed dry forest throughout the entire study area. Observations were either of solitary individuals or of single birds in the company of *Cyanocorax cyanomelas* and *Psarocolius decumanus* following army ants. Birds were almost entirely terrestrial, but were also seen hopping on to low branches (up to 1 m). The only sound given by the species was an irregular series of bill-claps, which could be heard up to 30 m away. In Bolivia, *N. geoffroyi* is known from humid Amazonian and upper tropical zone forests of PA, LP, SC (Remsen and Traylor 1989). It is noteworthy that we recorded the species almost exclusively in dry forest, because previous records place the species only in humid forest. (Hilty and Brown 1986; Parker et al. 1982).

Cranioleuca sp.—An unidentified *Cranioleuca* spinetail was identified by Bret Whitney in tape-recordings made by AHP. The high, thin, fast descending song was recorded on 9 and 17 September. These vocalizations most likely refer to a *Cranioleuca* being described by J. Fjeldså and S. Mayer (unpubl.), but could also represent *C. pyrrhophia* (Whitney, pers. comm.). Recorded vocalizations were compared to those of *C.* sp. and *C. pyrrhophia*, but the three could not be distinguished (Mayer, pers. comm.). A record of either *C.* sp. or *C. pyrrhophia* would represent a northern range extension, from the type locality near Inquisivi, eastern LP (for *C.* sp.), or from near Coroico, eastern LP (for *C. pyrrhophia*).

Myrmotherula grisea.—Ashy Antwren.—A pair was seen by NAH near Hac. Ubito (elev. 850 m) on 11 July, foraging actively in creeper tangles at the border of dry forest and riverbank forest, at a height of 8–12 m. The pair was part of a mixed-species flock that included *Sittasomus griseicapillus, Herpsilochmus* sp., *Hemithraupis guira,* and *Cissopis leveriana.* During the short period of observation the birds scanned clusters of dry leaves caught in vine tangles, a behavior similar to that reported in Parker et al. (1992). Additional pairs were also seen by NAH on 29 July at 950 m and MK on 23 July at 1,150 m in the Río Machariapo valley where semi-deciduous forest meets a stream edge (dominated by *Anadenanthera colubrina* intermixed with *Cecropia* sp. and *Eugenia* sp., with a dense understory of shrubs and vines). The pair was part of a mixed-species flock, foraging in the understory at 1–5 m, with additional species *Arremon flavirostris, Thamnophilus aroyae, Herpsilochmus* sp., *Dysithamnus mentalis, Basileuterus* sp., and *Poecilurus scutatus.* The pair moved actively from bush to bush, searching both sides of both green and dead leaves, but no prey was taken. Calls were not recorded during either of these observations. This species is known from only a few forest localities from 600 to 1400 m on the lower slopes of the Bolivian Andes in LP, CO, SC (Parker et al. 1992; Remsen and Traylor 1989) and is considered threatened (Collar et al. 1992).

Herpsilochmus sp. nov.—One male of this species was collected by AHP (CBF 02679), and numerous tape-recordings were made of its vocalizations. This species was common (>10 recorded daily) in dry forest of the study area, and was not recorded in either floodplain or lower montane rainforest. At 950 m in the Río Ubito valley, densities of one pair or singing male along every 100 m of trail were documented. Nearly one-half of our observations of this species consisted of a single pair foraging solitarily, whereas the other half were of pairs traveling with mixed-species flocks that included *Myiornis albiventris, Sittasomus griseicapillus,* and *Casiornis rufa.* Pairs and mixed-flocks foraged actively in shrubs and small trees at 3–8 m, most commonly at 5–6 m. Birds were noted hanging upside down, and often hover-gleaning from leaves. Contact calls were given almost continually and were very similar to those of *H. atricapillus.* (pers. obs., Whitney, pers. comm.). This antwren was discovered in 1990 at this site by T. Parker (Parker and Bailey 1991), and subsequently tape-recorded and collected in 1993 by M. Pearman (Pearman 1993), and may prove to be new to science (Pearman unpubl. data).

Cyanerpes cyaneus.—Red-legged Honeycreeper.—One individual was collected by NAH and AHP (CBF 02674) on 15 September in dry forest (with a canopy dominated by *Anadenathera colubrina*) near the Río Recina, and four individuals were observed in tall riverbank forest along the Río Ubito. Birds favored the canopy (15–20 m) and foraged in restless monospecific flocks of 4–5. The birds foraged in the large flowers of *Inga* sp. (Leguminosae) and hover-gleaned for insects on nearby leaves. This apparently represents the first record for LP, although the species was previously known from the Amazonian lowlands of PA, BE, and SC (Remsen and Traylor 1989). Ridgely and Tudor (1994) reported *C. cyaneus* for LP, but no published reference to this record was found.

Basileuterus chrysogaster.—Golden-bellied Warbler.—An individual was collected (CBF #02670) on 15 September from the border of floodplain forest and riverbank second-growth near the Río Recina. An additional pair was observed nearby by NAH the following day, foraging from 1-5 m above ground in riverbank second-growth along the Río Recina. No tape-recording was obtained. This represents the first record of the species for Bolivia, previously known from Peru, Ecuador, and Colombia, and the southern-most record for the species.

Oxyruncus cristatus.—Sharpbill.—A single individual of this species was seen by NAH on 15 September in riverbank second-growth along the Río Ubito near the Yanomayo camp at 950 m. The bird was immediately recognized by its pointed bill, the distinct black scaling on the yellowish belly and chest, and the orange iris. The bird was foraging in a mixed-species flock of *Veniliornis* sp. and other unidentified species. The bird was seen for less than 20 s, and was noted probing, possibly for insects, the flowers of *Inga* sp. and *Ochroma pyramidale* ca. 6 m above the ground. No vocalization was noted. This is the second published report for Bolivia; the species was previously known only from southwest of Apolo, depto. La Paz in stunted montane rainforest on sandy soil at 1,400–1,600 m (Parker and Bailey 1991). The Sharpbill is found in lower and upper montane rainforest along the eastern slopes of the Andes.

DISCUSSION

A breakdown of the 275 species by habitat (Appendix) shows that 102 (38.9%) were found in dry forest (40, or 15.3% of these restricted to it), 51 (19.4%) in floodplain forest (10, or 3.8% of these restricted to it), 50 (19.1%) were found in lower montane rainforest (18, or 6.9% restricted to it), 33 (12.6%) were found in all three forest types (dry, floodplain, and lower montane), 46 (17.6%) were found in riverbank forest and second growth (3, 1.1% of these restricted), 89 (34.0%) in clearings (46, 17.6% restricted to them). The most species-rich families

TABLE 1

THE INFLUENCE OF EACH BIOGEOGRAPHICAL ELEMENT ON THE AVIFAUNA OF EACH HABITAT IN THE STUDY AREA. THIS ANALYSIS WAS LIMITED TO THE 133 SPECIES THAT WERE FOUND IN DRY FOREST

	Biogeographical element		
Habitat	Lowland Amazonian	Humid montane	Dry and open
Dry forest (Fd)	67 (52.8%)	27 (21.3%)	33 (26.0%)
Floodplain (Ft)	20 (64.5%)	3 (9.7%)	8 (25.8%)
Humid lower montane (Fm)	7 (28.0%)	14 (56.0%)	4 (16.0%)
All forest types (Fd, Ft, Fm)	20 (50.0%)	10 (25.0%)	10 (25.0%)
Disturbed (Fr, C)	25 (50.0%)	4 (8.0%)	21 (42.0%)

were the Tyrannidae (56 species), Thraupidae (31 species), Formicariidae (15 species), Accipitridae (11 species), Emberizidae (11 species), Trochilidae (9 species), and Furnariidae (9 species). Pearman (1933) recorded 106 species in the Río Machariapo valley, a tributary of the Río Tuichi. Of these, 75 (71%) occurred in dry forest, and 35 (33%) were restricted to dry forest.

Direct comparison of the species list and relative abundance indices (Appendix) should be treated with caution for reasons outlined by Remsen (1994) and Terborgh et al. (1990). Although we believe that most resident bird species were detected during our survey, some undoubtedly went unrecorded, particularly secretive undergrowth species (limited mist net use) and non-dry forest species (limited survey time spent away from dry forest). Relative abundance is difficult to assess when there are discrepancies between numbers of observed individuals and the frequency of recorded vocalizations of these species (e.g., canopy, mixed-species flock, and secretive species), and when species are identified solely as background vocalizations on tape recordings.

In terms of biogeography, the avifauna of the central Río Tuichi Valley represents an interesting mixture of species from the lowland Amazonian rainforest element, humid montane element, dry and open habitat elements, as well as a number of species with wide altitudinal ranges. This combination of species is found elsewhere only in a few dry valleys of Peru and Bolivia. Table 1 shows the influence of each biogeographical element on the avifauna of each habitat in the study area. It was limited to the 133 species that were found in dry forest. The Amazonian element is the largest and includes 87 species (65.4% of total), of which 67 were found only in dry forest (Fd), with an additional 20 found in floodplain forest (Ft), 7 in lower montane rainforest (Fm), and 20 in all forest types (Fd, Ft, Fm), and 25 in disturbed habitats (Fr, C). This suggests that the floodplain forests of the valley, despite their small total area, play an important role in connecting bird populations of the dry forest with those of the extensive tracts of lowland Amazonian rainforest to the north and east. Changes in abundance (or detectability) in our surveys suggest that some species (e.g. *Aratinga leucopthalmus*, *Amazona farinosa*, *Trogon melanurus*, *Galbula ruficauda*, *Megaryncus pitangua*, and some foliage gleaners and tanagers) undergo seasonal movements from the dry forest in the wet season to the Amazonian lowland rainforest in the dry season, but the extent of these movements remains unknown.

The humid montane element includes 37 species (27.8% of total), of which 27 were found only in dry forest, with an additional three found in floodplain forest, 14 in lower montane rainforest, and 10 in all forest types. Some species of this element (e.g., *Phyllomyias burmeisteri*, *Myioborus miniatus*, *Myiarchus cephalotes*) were found in dry forest at lower elevations than normal (Parker and Bailey 1991).

The dry and open habitat element includes 43 species (32.3% of total), of which 33 were found only in dry forest (Fd), with an additional eight found in floodplain forest (Ft), four in lower montane rainforest (Fm), and 10 in all forest types (Fd, Ft, Fm). The number of typical dry forest species is low compared to other dry forest sites (e.g., the lowlands of depto. Santa Cruz; Parker et al. 1993), possibly because of the comparatively small area of dry habitat in the valley and its isolated geographical position. The low number of dry forest species that have been able to successfully colonize the Río Tuichi valley might open up a number of ecological niches to invading species from surrounding humid habitats. This could also explain the unusually low elevational distribution of some humid montane species (see last paragraph). The remaining species found are less specific in their habitat requirements and are found in a variety of habitats, or are typical of forest edge and disturbed habitats (e.g., *Otus choliba, Leptotila verreauxi, Synallaxis albescens, Pachyramphus polychopterus*).

In contrast to the avifauna of the dry forest, which is made up of species from a variety of faunal elements, the floodplain and humid lower montane bird communities are composed mostly of species belonging to their respective faunal elements. Twenty species found in floodplain forests (64.5%) pertain to the lowland Amazonian faunal element, and 14 species found in lower montane rainforest (56.0%) pertain to the humid montane element. Floodplain and lower montane bird communities include relatively few species that pertain to other faunal elements (11 (35.5%) and 11 (44%) respectively). Amazonian element species form the largest component (67, or 52.8%), while humid montane and dry and open elements represent 27 species (21.3%) and 33 species (26.0%) respectively. It seems remarkable that the dry forest bird community is the only one in which species typical of its own faunal element are not the predominant members.

The Tuichi dry forests represent one link in a chain of isolated inter-Andean dry forest patches. Recent studies suggest that each of these patches has a unique avifaunal composition and endemic taxa. For example, the Río Tuichi valley contains what probably represents a new, endemic species (or subspecies) of *Herpsilochmus* antwren (Parker and Bailey 1991; Pearman 1993) and either a hybrid (as has been collected in depto. Puno (M. Pearman, pers. comm), or undescribed, form of *Pyrrhura molinae* (Pearman 1993). *Tangara meyerdeschauenseei* occurs only in the Sandia Valley (Peru) to the north (Schulenberg and Binford 1985), and possibly near Apolo, depto. La Paz (Bolivia), very near the study site (Parker and Bailey 1991) To the south, *Asthenes berlepschi* is endemic to the Consata valley, and *Cranioleuca* sp. nov. (S. Mayer, pers. comm.) and a possibly undescribed form of *Pyrrhura molinae* (different from that from Río Tuichi valley, S. Mayer and M. Friggens, pers. comm.) are only found in the Río La Paz/Inquisivi valley (Bolivia). All of these taxa are found in dry forests.

The low number of typical dry and open habitat species species indigenous to the area probably results in there being a variety of ecological niches available to species from adjacent habitats. Some lower montane rainforest species (e.g. *Phyllomyias burmeisteri*) occur in the Tuichi dry forests at lower elevations than usual (Parker and Bailey 1991). This suggests that competition determines the elevational limits of certain montane rainforest birds (cf. Terborgh and Weske 1975). Dry and open element species may be limited in the valley perhaps because their "source habitats" are relatively remote (i.e., southern Peru and central and southern Bolivia).

The dry forest of the Río Tuichi valley represents a major distributional boundary for dry forest birds: no fewer than eight species (*Leptotila megalura, Pyrrhura molinae, Colibri serrirostris, Poecilurus scutatus, Phacellodomus ruber, Formicivora melanogaster, Sporophila ruficollis,* and *Arremon flavirostris*) of dry forest or open habitats reach their northwestern limits here. Furthermore, several genera with Amazonian and dry forest representatives (e.g., *Arremon flavirostris/taciturnus, Cyanocompsa cyanoides/brissonii*) meet in this area. Some montane species from surrounding humid habitats reach their northern (e.g., *Myrmotherula grisea, Hemitriccus spodiops*) or southern (e.g., *Basileuterus chrysogaster*) limits here. It becomes clear that the patches of inter-Andean dry forest in northern Bolivia and adjacent Peru represent an interesting, largely under appreciated, example of isolated habitat islands where location and isolation influenced the avian community structure and lead to local speciation and endemism.

The central Río Tuichi dry forest is a high priority for conservation. Its large pristine area, low human population, close proximity and dependence on surrounding humid ecosystems, and isolated geographical position make it ideally situated for preservation. Additionally, no other Andean or lowland dry forests are represented in Bolivian protected areas, and Neotropical dry forest are threatened throughout their Neotropical range. It is the largest continuous tract of well preserved inter-Andean dry forest anywhere in the Andes, and one of the largest in the entire Neotropics (Kessler 1992; Gentry in press). Because of its geographical position, it represents a unique opportunity to study the effect of isolation on the faunal composition and differentiation of taxa.

In terms of avifauna, the conservation importance of the valley is further underscored by the high species diversity for such an area and its unique biogeographical species mixture. As well, it supports numerous threatened (*Mitu tuberosa, Ara militaris*, and *Myrmotherula grisea*) and near-threatened species (*Tigrisoma fasciatum, Harpyhaliaetus solitarius, Spizastur melanoleucus, Myiornis albiventris,* and *Ampeliodes tschudii*) (Appendix 4; Collar et al. 1992; Remsen and Quintela unpubl. data). Recently though, some of these species (*T. fasciatum, H. solitarius, S. melanoleucus,* and *A. tschudii*) have been found to be more abundant than previously thought, and may not represent threatened species (M. Robbins, pers. comm.). Other noteworthy records include new species for Bolivia and depto. La Paz, as well as a possibly undescribed *Herpsilochmus* antwren and *Cranioleuca* spinetail, and the rare species *Neomorphus geoffroyi* and *Ara militaris*. Restricted-range species include *Myrmotherula grisea*, *Thamnophilus aroyae*, and *Chiroxiphia boliviana* (Lower Yungas Endemic Bird Area; Bibby et al. 1992; Stattersfield et al. unpubl. data). These noteworthy records also represent bird species of potential ecotourism interest that can be used to promote the area's ecotourism value, and show that dry forests is relatively intact and ecosystem integrity is still high. It is also likely that rare and threatened species from surrounding humid habitats (e.g., *Hemitriccus spodiops*) probably use, at least seasonally, the dry forest of the Río Tuichi valley. These species are also of potential ecotourism interest and can be used to promote the area's ecotourism value. And finally, the Tuichi dry forests form an integral part of the larger *Parque Nacional Madidi*, an area that has been predicted to contain over 1,000 bird species, and the world's most biodiverse protected area (Remsen and Parker 1995).

The major threat to the dry forest is the planned La Paz-Cobija road, which has been constructed from La Paz to Apolo. The existing road would continue north, bisecting the Río Machariapo and Tuichi valleys. Uncontrolled colonization and destruction of large areas of dry forest would inevitably follow. Hunting would have a severe impact on the populations of birds and mammals, and would be difficult to control. Road construction must be prevented, or rigorous laws enacted to prevent the normal habitat destruction that follows road building.

We further recommend developing an integrated management plan for *Parque Nacional Madidi* (Fig. 1) that is community-based and economically sustainable. It undoubtedly represents the best (and possibly only such) opportunity for the conservation of inter-Andean dry forests in Peru and Bolivia. Over 80% of the least disturbed dry forest in the valley lies west of the Río Tuichi, i.e., within the boundaries of the *Parque Nacional Madidi*. Furthermore, dry forest is but one example of the variety of globally threatened habitats, including Neotropical savannas and humid montane rainforest, that are represented by large, pristine, examples within the boundaries of *Parque Nacional Madidi*. The conservation of the entire valley and its representative habitats is clearly a high conservation priority.

ACKNOWLEDGEMENTS

We thank the staff of the Dirección Nacional de Conservación de la Biodiversidad (DNCB), with special thanks to Alejandra Sanchez de Lozada (National Director), and Marco Octavio Ribera (Coordinator, Unidad de Planificación e Información), and Abel Castillo (Coordinator, Unidad de Manejo de Recursos Naturales) for their enthusiastic support and encouragement, and for their commitment to apply the results of this project to conservation strategies and management plans in Bolivia. Special thanks go to individuals and institutions in La Paz who actively promote conservation and sustainable development in the Madidi region, especially Mario Baudoin Weeks (former National Director, Dirección Nacional de la Conservación de la Biodiversidad), Guillermo Ríoja and Ana Martinet de Mollinedo (Conservation International Bolivia Program), Michael Jackson, Elizabeth Coloma, and David Ridgeway (British Embassy), Gary Hunnisett (World Bank), Rosa Maria Ruiz (EcoBolivia), Eduardo Forno (Centro de Datos para la Conservación), Charlie Munn (Wildlife Conservation International), Leon Merlot (Veterenarios sin Fronteras), Martin Proctor (ProEco), and Oscar Sainz (Colibri Tours).

We are grateful for the collaboration of Conservation International (CI) staff in Washington, especially Enrique Ortiz, Karen Ziffer, and Adrian Forsyth, and Kim Awbrey. Special thanks go to members of the CI–Rapid Assessment Program, the late Ted Parker, the late Alwyn Gentry, Robin Foster, and Louise Emmons, for exploring and documenting the biological richness of the Madidi region, and for guiding TREX activities there.

We acknowledge the collaboration of the people of Apolo, Asariamas, Buena Vista and Nogal, especially Cesár Gariazu Gamez, Anival Delaney, Enrique Koehnke, Augusto Machicau, Darian Delaney, Don Jorge, Saturnino Ramos, Felix Valencia, Mario Albarez, and Angel Ubano (Rabiana). Bolivian biology students Boris Ríos and Enzo Aliaga, and volunteers Ruth Maier, Ulrike Switzer, Geoff Pease, and Catherine Fowler, provided invaluable support in the field. Additional thanks go to the staff and pilots of Transporte Aereo Militar, and to Liam O'Brien and Manuel Herrera of the United States Defense Mapping Agency.

We thank the staff of the Colección Boliviana de Fauna, Museo Nacional de Historia Natural, and Herbarío Nacional for arranging access to biological collections, coordinating student training opportunities, and providing work space and field equipment. Special thanks go to Stephan Beck, Emilia García, Monica Moraes, Jaime Sarmiento, Teresa Tarifa, James Aparicio, Liliana

Villalba, Michelle Blair, and Omar Rocha. We thank those who have given advice and reviewed manuscripts, especially Bret Whitney, J. V. Remsen, Jr., Tom Schulenberg, Robin Foster, Mark Robbins, Phil Ward, Mark Pearman, Susan Davis, and Bennett Hennessey. We thank Lucho Jammes, David Wege, Adrian Long, and Martin Kelsey for valuable discussions on Bolivian bird conservation.

We are extremely grateful to those who provided funding for this project: British Overseas Development Agency; Conservation, Food and Health Foundation; Sophie Dansforth Conservation Biology Fund; and the World Nature Association. We also thank our corporate sponsors: American Airlines, Mountainsmith, Mountain Safety Research, LowePro, TEVA, Crazycreek, Best American Duffel, Flatland Mountain Works, Osprey, Trimble Navigation, Pur, Pelican Products, and Sevylor. Lastly, we thank our family and friends who provided an endless stream of support and inspiration to continue working for the goal of a sustainable *Parque Nacional Madidi*.

LITERATURE CITED

BEST, B., C. CLARKE, M. CHECKER, A. BROOM, R. THEWLIS, W. DUCKWORTH, AND A. MCNAB. 1993. Distributional records, natural history notes, and conservation of some poorly known birds from southwestern Ecuador and northwestern Peru. Bull. Brit. Ornith. Club 113:108–255.

BIBBY, C. J., N. J. COLLAR, M. J. CROSBY, M. F. HEATH, C. H. IMBODEN, T. H. JOHNSON, A. J. LONG, A. J. STATTERSFIELD AND S. J. THIRGOOD. 1992. Putting Biodiversity on the Map: Priority Areas for Global Conservation. International Council for Bird Preservation, Cambridge, England.

CHAPMAN, F. 1921. The distribution of bird life in the Urubamba valley of Peru. US Nat. Mus. Bulletin 117: 27–29.

COLLAR, N. J., L. P. GONZAGA, N. KRABBE, A. MADROÑO NIETO, L. G. NARANJO, T. A. PARKER III, AND D. G. WEGE. 1992. Threatened Birds of the Americas: The ICBP/IUCN Red Data Book. Third edition, part 2. International Council for Bird Preservation, Cambridge, England.

COLLAR, N. J., D. C. WEGE, AND A. J. LONG. 1997. Patterns and causes of endangerment in the New World avifauna. Pp. 237–260 *in* Studies in Neotropical Ornithology Honoring Ted Parker (J. V. Remsen, Jr., Ed.), Ornithol. Monogr. No. 48.

FJELDSÅ, J., AND N. KRABBE. 1990. Birds of the High Andes. Apollo Books, Svendborg, Denmark.

FOSTER, R. B., T. A. PARKER III, A. H. GENTRY, L. H. EMMONS, A. CHICCHÓN, T. SCHULENBERG, L. RODRÍGUEZ, G. LAMAS, H. ORTEGA, J. ICOCHEA, W. WUST, MÓNICA ROMO, J. A. CASTILLO, O. PHILLIPS, C. REYNEL, A. KRATTER, P. K. DONAHUE, AND L. J. BARKLEY. 1994. The Tambopata-Candamo Reserved Zone of southeastern Peru: a biological assessment. RAP Working Papers 6, Conservation International, Washington, D.C.

GENTRY, A. H. 1995. Diversity and floristic composition of Neotropical dry forests. *In* H. Mooney, S. Bullock, and E. Medina (Eds.), Tropical Deciduous Forests. Cambridge Univ. Press, Cambridge, England.

HILTY, S. L., AND W. L. BROWN. 1986. A Guide to the Birds of Colombia. Princeton Univ. Press, Princeton, New Jersey.

HUECK, K. AND P. SEIBERT. 1981. Vegetationskarte von Südamerika, with map 1:8,000,000. Gustav Fischer Verlag, Stuttgart, Germany.

KESSLER, M. 1995a. Vegetation. Pp. 53–118, *In* B. Best and M. Kessler (Eds.), Biodiversity and Conservation in Tumbesian Ecuador and Peru. Birdlife International, Cambridge, England.

PARKER, T. A., III, AND B. BAILEY. (EDS.). 1991. A biological assessment of the Alto Madidi region and adjacent areas of northwest Bolivia, May 18–June 15, 1990. RAP Working Papers 1, Conservation International, Washington, D.C.

PARKER, T. A., III, J. M. BATES, AND G. COX. 1992. Rediscovery of the Bolivian Recurvebill with notes on other little-known species of the Bolivian Andes. Wilson Bull. 104:173–178.

PARKER, T. A., III, AND J. L. CARR. (EDS.). 1992. Status of forest remnants in the Cordillera de la Costa and adjacent areas of southwestern Ecuador. RAP Working Papers 2, Conservation International, Washington, D.C.

PARKER, T. A., III, A. CASTILLO U., M. GELL-MANN, AND O. ROCHA. 1991. Records of new and unusual birds from northern Bolivia. Bull. Brit. Ornith. Club. 111:120–138.

PARKER, T. A., III, A. H. GENTRY, R. B. FOSTER, L. H. EMMONS, AND J. V. REMSEN, JR. (EDS.). 1993. The lowland dry forests of Santa Cruz, Bolivia: A global conservation priority. RAP Working Papers 4, Conservation International, Washington, D.C.

PARKER, T. A., III, S. A. PARKER, M. A. PLENGE. 1982. An Annotated Checklist of Peruvian Birds. Buteo Books, Vermillion, South Dakota.

PARKER, T. A., III, T. S. SCHULENBERG, M. KESSLER, AND W. W. WUST. 1995. Natural history and conservation of the endemic avifauna in north-west Peru. Bird Conserv. Intern. 5:201–213.

PEARMAN, M. 1993. The avifauna of the Río Machariapo dry forest, northern depto. La Paz, Bolivia: a preliminary investigation. Bird Conserv. Intern. 3:105–117.

PERRY, A., M. KESSLER, N. HELME, AND J. MITTON (EDS.). In press. Biological and environmental assessment of the inter-Andean dry tropical forests of the central Río Tuichi valley, *Parque Nacional Madidi*, depto. La Paz, Bolivia. Quipus, La Paz, Bolivia.

REMSEN, J. V., JR. 1994. Use and misuse of bird lists in community ecology and conservation. Auk 11:225–227.

REMSEN, J. V., JR. 1995. The importance of continued collecting of bird specimens to ornithology and bird conservation. Bird Conserv. Intern. 5:145–180.

REMSEN, J. V., AND T. A. PARKER III. 1995. Bolivia has the opportunity to create the planet's richest park for terrestrial biota. Bird Conserv. Intern. 5:181–189.

REMSEN, J. V., JR., AND M. A. TRAYLOR. 1989. An Annotated List of the Birds of Bolivia. Buteo Books, Vermillion, South Dakota.

RICHTER, M., AND W. LAUER. 1987. Pflanzenmorphologische Merkmale der hygrischen Klimavielfalt in der Ostkordillere Boliviens. Aachener Geogr. Arb. 19:71–108.

RIDGELY, R. S., AND G. TUDOR. 1994. The Birds of South America, Volume II. Univ. of Texas Press, Austin, Texas.

SIBLEY, C. G., AND MONROE, B. L. 1991. Distribution and Taxonomy of Birds of the World. Yale Univ. Press, New Haven, Connecticut.

SOUX, M. L., M. G. GISMONDI, R. JIMENEZ, L. DE LOS ANGELES CARDENAS, AND R. HILARI. 1991. Apolobamba, Caupolican, Franz Tamayo-Historia de una region Paceña. Prefectura del Departamento de La Paz, La Paz, Bolivia.

STILES, F. G. 1983. Birds. Pp. 502–544. *In* D. H. Janzen (Ed.). Costa Rican Natural History. Univ. of Chicago Press, Chicago.

TERBORGH, J. W., J. W. FITZPATRICK, AND L. EMMONS. 1984. Annotated checklist of bird and mammal species of Cocha Cashu biological station, Manu National Park, Peru. Fieldiana (Zoology, New Ser.) No. 21.

TERBORGH, J., S. K. ROBINSON, T. A. PARKER III, C. A. MUNN, AND N. PEIRPONT. 1990. Structure and organization of an Amazonian forest bird community. Ecol. Monogr. 60:213–238.

WHITNEY, B. M., J. L. ROWLETT, AND R. A. ROWLETT. 1994. Distributional and other noteworthy records for some Bolivian birds. Bull. Brit. Ornith. Club. 114:149–162.

WIEDENFIELD, D. A., T. S. SCHULENBERG, AND M. B. ROBBINS. 1985. Birds of a tropical deciduous forest in extreme northwestern Peru. Pp. 305–315, *In* P. A. Buckley et al. (Eds). Neotropical Ornithology. Ornithol. Monogr. No. 36. American Ornithologists' Union, Washington, D.C.

APPENDIX

BIRD SPECIES RECORDED IN THE CENTRAL RÍO TUICHI VALLEY DRY FOREST, DEPTO. LA PAZ, BOLIVIA, JULY AND SEPTEMBER 1993. TAXONOMY AND SEQUENCE FOLLOWS REMSEN AND TRAYLOR (1989) AND REMSEN AND PARKER (1995), WITH THE EXCEPTION OF *Streptoprocne rutilus* FOLLOWING MARÍN AND STILES (1992), *Picumnus albosquamatus* FOLLOWING SIBLEY AND MONROE (1991) AND *Lepidocolaptes lacrymiger* FOLLOWING RIDGELY AND TUDOR (1994). SEE TEXT FOR DETAILED HABITAT DESCRIPTIONS AND LIMITATIONS OF RELATIVE ABUNDANCE INDICES. THE NUMBER OF CODES IS LIMITED TO THREE FOR HABITAT, AND ONE (THE MOST RELIABLE) FOR EVIDENCE, AND THE ORDER FOLLOWS THAT OUTLINED BELOW

Species	Habitats	Abundance	Evidence
TINAMIDAE			
Tinamus tao	Fd, Fm	F	t
Tinamus major	Fd, Ft	U	t
Nothocercus nigrocapillus	Fm	?	t
Crypturellus obsoletus	Fm	?	t
Crypturellus atrocapillus	F	C	t
Crypturellus tataupa	Fd, Fr	F	t
PHALACROCORACIDAE			
Phalacrocorax brasilianus	R	F	si
ARDEIDAE			
Tigrisoma fasciatum	Sh	R	si
Tigrisoma lineatum	Sh	U	si
Bubulcus ibis	Sh	R	si
Egretta thula	Sh	R	si
Butorides striatus	Sh	R	si
CATHARTIDAE			
Coragyps atratus	Fd, Fr, C	C	si
Cathartes aura	Fd, Fr	F	si
Sarcoramphus papa	Fd	R	ph
ANATIDAE			
Dendrocygna viduata	M	R	si*
ACCIPITRIDAE			
Chondrohierax uncinatus	Fd	R	si
Elanoides forficatus	Fd, Fm	R	si
Ictinia plumbea	Fd	U	si
Buteogallus urubitinga	Fd, Ft	R	si
Geranoaetus melanoleucos	Fd, Fm	R	si
Harpyhaliaetus solitarius	Fd	R	si
Buteo magnirostris	Fr, C, Rm	F	si
Buteo brachyurus	Fd	R	si
Buteo albonotatus	Fd	R	si*
Spizastur melanoleucus	Fd, Fm	R	si
Spizaetus tyrannus	Fd	R	si
FALCONIDAE			
Micrastur ruficollis	Fd	U	t
Micrastur semitorquatus	Fm, Fr	R	si*
Falco sparverius	C	C	si
Falco rufigularis	C	U	si
CRACIDAE			
Ortalis motmot	Fd, Ft, Fr, C	F	t
Penelope jacquacu	Fm	U	t
Pipile pipile	Ft, Fm	F	si
Mitu tuberosa	Ft	U	si
RALLIDAE			
Aramides cajanea	L	?	si
Laterallus melanophaius	L	?	si
Porphyrula martinica	L	?	si

APPENDIX
CONTINUED

Species	Habitats	Abundance	Evidence
SCOLOPACIDAE			
Bartramia longicauda	C	R	si
Actitis macularia	S	U	si
COLUMBIDAE			
Columba cayennensis	Fr, C	R?	si
Columba plumbea	F, C	C	si
Columba subvinacea	C	?	si
Zenaida auriculata	C	F	si
Columbina talpacoti	Fr, C	F	si
Columbina picui	Fr, C	F	si
Claravis pretiosa	Fd, C	U	t
Leptotila verreauxi	Fd, C	C	t
PSITTACIDAE			
Ara militaris	F	F	t
Aratinga leucophthalmus	Fd, Ft	F	t
Pyrrhura molinae	F	C	t
Pionus menstruus	F	C	t
Amazona farinosa	Fd, Ft	F	t
CUCULIDAE			
Piaya cayana	F, C	F	t
Crotophaga ani	Fr, C	C	si
Dromococcyx phasianellus	Fm C	?	t
Neomorphus geoffroyi	Fd	F	si+
TYTONIDAE			
Tyto alba	T	F	si
STRIGIDAE			
Otus choliba	Fm, C	?	si
Pulsatrix perspicillata	Fd	U	t
NYCTIBIIDAE			
Nyctibius griseus	C	R	t
CAPRIMULGIDAE			
Nyctidromus albicollis	Fr, C	F	t
Nyctiphrynus ocellatus	Fd	U	t
Caprimulgus rufus	Fd	R	t
Hydropsalis climacocerca	C, Sh	R	si
APODIDAE			
Streptoprocne rutila	O	F	si
Streptoprocne zonaris	O	F	si
Chaetura brachyura	O	R	si
TROCHILIDAE			
Phaethornis superciliosus	Fd, Ft	?	sp
Colibri delphinae	Ft	R	si
Colibri serrirostris	C	C	sp
Chlorostilbon mellisugus	Fd, Fr, C	F	ph
Thalurania furcata	Fd, Ft	F	si
Hylocharis cyanus	Ft	U	si
Polytmus guainumbi	C	R	si
Amazilia lactea	Fd, C	R	si
Heliomaster longirostris	Fd	R	si
TROGONIDAE			
Trogon melanurus	Ft	U	si
Trogon curucui	Fd, Ft	C	t
Trogon sp.	Fd	?	t

APPENDIX
Continued

Species	Habitats	Abundance	Evidence
MOMOTIDAE			
Electron platyrhynchum	Ft	F	t
Momotus momota	Fd, Ft	U	t
ALCEDIDAE			
Ceryle torquata	Rm	R	si
Chloroceryle americana	Rm	R	si
BUCCONIDAE			
Nystalus chacuru	C	?	si
Malacoptila fulvogularis	Fd	R	si
Monasa nigrofrons	Fd, Ft, Fr	F	t
GALBULIDAE			
Galbula ruficauda	Fd, Fr	F	t
CAPITONIDAE			
Eubucco richardsoni	Fd	R	si
Eubucco versicolor	Fd, Fm	U	si
RAMPHASTIDAE			
Aulacorhynchus prasinus	Fd, Fm	U	si
Pteroglossus castanotis	Fd, Ft	F	t
Ramphastos vitellinus	F	F	t
PICIDAE			
Picumnus aurifrons	Fd	?	si
Picumnus albosquamatus	Fr, C	F	sp
Melanerpes cruentatus	Fd, Ft, Fr	U	t
Veniliornis affinis	F	F	si
Piculus leucolaemus	Fd	F	si
Dryocopus lineatus	F	F	t
Campephilus melanoleucos	F	U	t
Campephilus rubricollis	Fd	R	si
DENDROCOLAPTIDAE			
Dendrocincla fuliginosa	Ft	?	si
Deconychura longicauda	Fd	?	sp
Sittasomus griseicapillus	F	C	t
Dendrocolaptes picumnus	Fd	R?	si
Xiphorhyncus ocellatus	Fd, Ft	?	sp
Xiphorhyncus guttatus	F	C	t
Lepidocolaptes lacrymiger	Fd, Fm	?	si
Lepidocolaptes albolineatus	Fd	?	sp
FURNARIIDAE			
Synallaxis azarae	Fm, C	F	si
Synallaxis albescens	C	C	n
Poecilurus scutatus	Fd	C	t
Crainoleuca sp.	Fd	U	t+
Phacellodomus ruber	C	F	si
Premnoplex brunnescens	Fm	?	si
Philydor ?erythrocercus	Fm	F	si
Automolus ochrolaemus	F	F	t
Xenops minutus	F	F	si
FORMICARIIDAE			
Taraba major		F	t
Thamnophilus palliatus	Fd, Fr, C	F	t
Thamnophilus aroyae	C	F	t
Thamnophilus caerulescens	Fd, Fm	F	t
Dysithamnus mentalis	Fd, Fm	F	t
Myrmotherula surinamensis	Fd, Fm	U	t
Myrmotherula longicauda	C	F	t

APPENDIX
CONTINUED

Species	Habitats	Abundance	Evidence
Myrmotherula grisea	Fd, Fr	R	si+
Herpsilochmus sp.	Fd	C	sp+
Formicivora melanogaster	Fd, C	F	si
Pyriglena leuconota	Fd, Fm	U	si
Hypocnemis cantator	C	R	t
Myrmeciza hemimelaena	Fd, Ft	F	t
Formicarius analis	Fd, Ft	F	sp
Chamaeza campanisona	Fm	F	t
CONOPOPHAGIDAE			
Conopophaga ardesiaca	Fd	R	si
TYRANNIDAE			
Phyllomyias burmeisteri	Fd, Fm	?	t
Camptostoma obsoletum	Fd, Fr, C	C	t
Phaeomyias murina	Fd	?	t
Myiopagis gaimardii	Ft, Fr	?	t
Myiopagis caniceps	F	F	t
Elaenia flavogaster	C	?	si
Elaenia chiriquensis	C	?	si
Elaenia sp.	C	?	sp
Serpophaga cinerea	Sh	R	si
Inezia inornata	C	F	si
Mionectes macconelli	Fd	R	t
Leptopogon superciliaris	C	F	ph
Phylloscartes ophthalmicus	F	F	si
Corythopis torquata	Fd, Ft	F	t
Myiornis albiventris	F	F	t
Myiornis ecaudatus	Fr	F	si
Hemitriccus margaritaceiventer	C	?	si
Hemitriccus sp.	Fd	?	t
Todirostrum sp.	Ft, Fr	?	si
Tolmomyias flaviventris	Ft, Fr	?	ph
Tolmomyias sp.	Fd	?	t
Onychorhynchus coronatus	C	R	ph
Terenotriccus erythrurus	Ft	U	ph
Myoiphobus fasciatus	C	U	ph
Pyrrhomyias cinnamomea	Fd, Fm	F	si
Mitrephanes olivaceus	Ft, Fr	R	si
Lathrotriccus euleri	Fr, C	F	t
Cnemotriccus sp.	Fd	?	t
Sayornis nigricans	Rm	F	si
Pyrocephalus rubinus	C	C	si
Ochthoeca cinnamomeiventris	Fd	R	si
Ochthornis littoralis	Sh	F	si
Muscisaxicola fluviatilis	C, Rm, Sh	F	si
Knipolegus aterrimus	C	F	si
Colonia colonus	C	U	si
Satrapa icterophrys	C	R	si
Hirundinea ferruginea	C	U	si
Attila spadiceus	Fd	?	t
Casiornis rufa	F	C	t
Rhytipterna simplex	Fd, Ft	F	t
Sirystes sibilator	Fd, Ft	F	t
Myiarchus tuberculifer	F	F	t
Myiarchus swainsoni	Fd	R	si
Myiarchus ferox	Fd, C	F	n
Myiarchus cephalotes	C	U	t
Myiarchus tyrannulus	F	C	t
Megarynchus pitangua	Fd, C	U	t
Myiozetetes similis	C	F	si

APPENDIX
Continued

Species	Habitats	Abundance	Evidence
Myiodynastes maculatus	F, C	F	t
Legatus leucophaius	C	?	t
Tyrannus melancholicus	Fr, Rm	C	si
Pachyramphus castaneus	Fd	?	t
Pachyramphus polychopterus	Fd, C	F	t
Pachyramphus marginatus	Fd	U	si
Tityra semifasciata	Fd, Fm	R	si
Tityra inquisitor	Fd	R	si
COTINGIDAE			
Oryruncus cristatus	Fr	R	si
Pipreola frontalis	Fm	?	si
Ampelioides tschudii	Fm	U	si
Cephalopterus ornatus	Fd, Ft	U	t
Rupicola peruviana	Fm	?	t
PIPRIDAE			
Schiffornis turdinus	F	F	t
Piprites chloris	Fd, Fm	F	t
Tyranneutes stolzmanni	Fm	?	si
Chiroxiphia boliviana	Fm, C	F	si
Pipra chloromeros	Ft	U	t
HIRUNDINIDAE			
Tachycineta albiventer	R	C	si
Notiochelidon cyanoleuca	C	C	n
Atticora fasciata	R	F	si
Stelgidopteryx ruficollis	C	C	n
TROGLODYTIDAE			
Campylorhynchus turdinus	Fr, C	U	si
Odontorchilus branickii	Fd, Fm	F	si
Thryothorus genibarbis	Fr, C	F	t
Troglodytes aedon	C	C	n
Microcerculus marginatus	Fd	F	t
TURDIDAE			
Catharus ustulatus	Fm	R	si
Turdus serranus	Fd, Fm	R	si
Turdus amaurochalinus	Fm, C	F	si
Turdus albicollis	F	F	si
CORVIDAE			
Cyanocorax cyanomelas	Fd, Ft, Fr	C	t
VIREONIDAE			
Cyclarhis gujanensis	F, C	F	t
Vireolanius leucotis	Fd	F	t
Vireo olivaceus	F, Fr	C	n
MOTACILLIDAE			
Anthus sp.	C	?	si
EMBERIZIDAE			
Zonotrichia capensis	C	C	si
Ammodramus aurifrons	Fr, C	F	t
Emberizoides herbicola	C	?	si
Volatinia jacarina	C	F	si
Sporophila schistacea	C	R	sp
Sporophila caerulescens	C	U	si
Sporophila ruficollis	C	C	si
Arremon flavirostris	Fd, C	F	st
Pheucticus aureoventris	Fd, Fm	F	si
Saltator maximus	Fd, Ft, Fr	F	si
Cyanocompsa cyanoides	C	?	t

APPENDIX
CONTINUED

Species	Habitats	Abundance	Evidence
THRAUPIDAE			
Schistochlamys melanopis	C	C	ph
Cissopis leveriana	Fr	F	si
Hemithraupis guira	F, Fr	C	si
Nemosia pileata	Fd, Ft, Fr	F	si
Tachyphonus rufiventer	Fd, Fm, Fr	F	si
Trichothraupis melanops	F	F	ph
Habia rubica	Fd, Ft	U	t
Piranga flava	Fd, Fm	U	t
Piranga leucoptera	Fm	U	t
Ramphocelus carbo	Fr, C	C	t
Thraupis sayaca	C	F	si
Thraupis palmarum	C	F	si
Anisognathus flavinucha	Fm	F	si
Pipraeidea melanonota	F	U	si
Euphonia laniirostris	Fd, Ft, Fr	F	si
Euphonia musica	Fm	U	si
Euphonia xanthogaster	Fm	F	si
Euphonia sp.	Fd	?	t
Chlorophonia cyanea	Ft	U	t
Tangara chilensis	F	F	si
Tangara xanthocephala	Fm	?	si
Tangara punctata	Fm	?	si
Tangara gyrola	Fm, Fd	?	si
Tangara cyanotis	Fm	?	si
Tangara cyanicollis	Fm, C	U	si
Dacnis cayana	Fm, C	F	si
Chlorophanes spiza	Fm, C	F	si
Cyanerpes cyaneus	Fd, C	F	sp*
Diglossa sittoides	C	F	si
Tersina viridis	Ft, Fr	U	si
Coereba flaveola	Fd, C	F	si
PARULIDAE			
Parula pitiayumi	F, Fr	C	t
Geothlypis aequinoctialis	Fr, C	F	t
Myioborus miniatus	F	F	t
Basileuterus bivittata	Fd, Fm	F	ph
Basileuterus chrysogaster	Ft, Fr	U	sp**
Phaeothlypis rivularis	Rm, Sh	?	t
Conirostrum speciosum	F, C	C	si
ICTERIDAE			
Psarocolius decumanus	F, Fr	C	t
Psarocolius atrovirens	Ft	U	si
Psarocolius angustifrons	Ft, Fr, Sh	F	si
Molothrus bonariensis	Rm	R	si
Scaphidura oryzivora	Rm	R	si
FRINGILLIDAE			
Carduelis xanthogastra	C	F	si

Codes for habitat are as follows: "Fd" = Dry and "semi-deciduous" forest; "Ft" = Floodplain forest; "Fm" = Lower montane rainforest; "F" = All forest types, species recorded in dry forest (Fd), floodplain forest (Ft), and lower montane rainforest (Fm); "Fr" = Riverbank forest and associated second-growth; "C" = Clearings; "Ma" = Marshes, limited to Apollo valley; "Sh" = Shores, sandbars, and rock outcrops along the Río Tuichi; "L" = Lake, Lago Buturo; "R" = River, the open water of the Rão Tuichi, Río Ubito, Río San Juan, and Río Recina; "Rm" = River margins; "T" = Town, in gardens or on buildings in Apollo; "O" = Overhead, for aerial foragers.

Codes for abundance are based on species abundance in preferred habitats. As a general rule, only common (C) and fairly common (F) are valid estimations. We do not feel confident of uncommon (U) or rare (R) designations, and they are included only as a rough indication of abundance status. Codes for abundance are as follows: "C" = Common, recorded daily in moderate to large numbers (more than 10 individuals); "F" = Fairly common, recorded almost daily in small numbers (fewer than 10 individuals), or irregularly in moderate to large numbers (more than 10 individuals); "U" = Uncommon, recorded once every third or fourth day in small numbers (fewer than 10 individuals); "R" = Rare; recorded less than once a week or fewer than five times during the study; "?" = Uncertain, see discussion in text.

Codes for evidence of occurrence are as follows: "sp" = specimen obtained; "t" = tape-recording of voice obtained; "ph" = photo obtained; "n" = mist-netted (and no photo taken); "si" = sight or sound; "*" = first record for depto. La Paz; "**" = first record for Bolivia; "+" = noteworthy record (see species accounts in text).

PATTERNS OF SPECIES COMPOSITION AND ENDEMISM IN THE NORTHERN NEOTROPICS: A CASE FOR CONSERVATION OF MONTANE AVIFAUNAS

Luis Miguel Renjifo, Grace P. Servat, Jaqueline M. Goerck, Bette A. Loiselle, and John G. Blake

Department of Biology and International Center for Tropical Ecology, University of Missouri–St. Louis, St. Louis, Missouri 63121-4499, USA

ABSTRACT.—A review of the composition of five montane avifaunas in northwestern South America and southern Central America confirmed the distinctness of these communities from adjacent lowland areas. Excluding species that depend on aquatic resources, 1,800 bird species from 52 families were classified according to principal life zones in this review. There were 1,366 species associated with lowland areas, whereas 877 species occurred in montane areas (i.e., generally above 1,200 or 1,500 m elevation). Nearly one-half of these montane species are restricted to these high elevations, with the greatest diversity found within subtropical zones, followed by temperate and páramo zones, respectively. Comparisons with lowland avifaunas revealed that montane communities differed in trophic structure as well as familial composition. Specifically, montane communities had proportionately more nectarivores than expected by chance. When compared to randomly generated montane communities, nearly 30% of the families differed significantly in number of species from that expected if these communities were randomly assembled; eight families were more species-rich and seven families less species-rich in montane areas than expected. Moreover, montane areas had a greater number and percentage of species with restricted ranges than did lowland areas. Overall, 217 montane species (24.7%) had small geographic ranges; 142 of these were restricted to one of the five montane regions reviewed here. As evidence of the endangerment of these montane communities, nearly 10% (82) of the species are listed as threatened or near-threatened. Montane habitats are under extreme pressure from human activities. Most urban centers are located in or close to the mountains in the regions reviewed here. Given the high diversity and singularity of these avifaunas, together with high levels of habitat alteration, protection of montane ecosystems should become a priority for conservation efforts in the Neotropics.

RESUMEN.—Una revisión de la composición de cinco avifaunas de montaña en el Noroeste de Sur América y el sur de Centro América confirmó la diferencia de estas comunidades con las de las tierras bajas. Sin incluir especies que dependen de recursos acuáticos, 1,800 especies de aves de 52 familias fueron clasificadas de acuerdo a las principales zonas de vida. 1,366 especies fueron asociadas con areas de tierras bajas, de las cuales 877 especies se encontraron en areas montañosas (generalmente encima de 1,200–1,500 m de elevación). Casi la mitad de estas especies de areas montañosas están restringidas a estas elevaciones altas, encontrándose la mayor diversidad en zonas subtropicales, seguidas por zonas templadas y de páramo respectivamente. Comparaciones con avifaunas de tierras bajas revelan que las comunidades de montaña difieren en estructura trófica así como en la composición de las Familias. Específicamente, las comunidades de montaña tuvieron proporcionalmente mayor número de nectarívoros que los esperados al azar. Se compararon avifaunas de montañas con comunidades ensambladas al azar usando simulaciones de Monte Carlo. En estas comparaciones, cerca del 30% de las familias se diferenciaron significativamente en el número de especies; ocho fueron mas ricas y siete fueron menos ricas en areas montañosas de lo esperado. Además, las areas de montaña tuvieron un mayor número y porcentaje de especies con rangos restringidos que las de tierras bajas. Un total de 217 especies de montaña (24.7%) tuvieron rangos geográficos pequeños; 142 de estas fueron restringidas a una de cinco regiones de montaña que fueron revisadas en este trabajo. Cerca del 10% (82) de estas especies se encuentran enlistadas como amenazadas o casi-amenazadas, lo que evidencia el peligro de extinción de estas comunidades de montaña. Los areas de montaña se encuentran bajo extrema presión por actividades humanas. Muchas de los centros urbanos se encuentran localizados en ó cerca de las montañas que fueron revisadas. La protección de los ecosistemas de montaña debe ser una prioridad para los esfuerzos de conservación en los Neotrópicos debido a la alta diversidad y singularidad de estas avifaunas, así como a los altos niveles de alteración en los habitats.

The avifauna of the Neotropics has long been recognized by biologists as being exceedingly rich. Colombia, with 1,758 species (Hernández 1993), alone contains approximately 19% of the world's bird species, and even the relatively small country of Costa Rica (ca. 50,000 km^2) contains about 9% of the world's avifauna (840 species; Stiles and Skutch 1989). Several factors promote high diversity in the Neotropics, including historical, geological, biotic, and environmental factors (e.g., Haffer 1967, 1974; Karr 1976; Pearson 1982; Cracraft 1985; Ricklefs and Schluter 1993). Indeed, the great diversity of habitat types, the complex topographies, and range of elevations found in Neotropical areas, together with historical events, undoubtedly have been important influences on patterns of bird species composition and richness.

Many Neotropical regions have, in recent times, experienced considerable deforestation (e.g., Houghton et al. 1991; Sayer and Whitmore 1991; Reid 1992; Whitmore and Sayer 1992), and much attention by conservation groups, media, and the public has been directed to the deforestation crisis in Neotropical rain forests (especially the Amazon basin) (e.g., Hecht and Cockburn 1990; Miller and Tangley 1991; Myers 1992). Much less attention, however, has been given to montane forests in the Neotropics (T. Parker, pers. comm.), despite the fact that these forests harbor many endemic species and have been subject to habitat alteration for centuries (see Long 1993). In fact, recent evidence reveals that forest loss is greatest in countries with high endemism levels (Balmford and Long 1994). Andean forests, which contain 6.3% of the world's bird species (Fjeldså and Krabbe 1990), have only 10 to 27% of forest remaining in the Colombian Andes (Henderson et al. 1991; Etter 1993). The situation is even more grim in Ecuador, where less than 4% of cloud forests remain intact (Dodson and Gentry 1991). In contrast, only 6 to 10% of Amazonia has been deforested, although certain Amazonian regions experience much greater pressure than others (Skole and Tucker 1993; see also Balmford and Long 1994).

Here we draw attention to Neotropical montane avifaunas, with the intent of documenting some of their differences from lowland counterparts and their contribution to the region's biodiversity. We focus on montane birds of northwestern South America and southern Central America, describing general community patterns and taxonomic affiliations of montane and lowland faunas. Additionally, we summarize patterns of restricted-range species and status of threatened species in this region. Many of the patterns that we report here are well known to ornithologists who conduct research in the Neotropics. By quantifying these patterns, however, we aim to highlight and better document the importance of montane forest avifaunas to biodiversity conservation.

METHODS

We examined the structure and composition of the regional avifauna of northwestern South America and southern Central America by compiling information on the distribution, endemism, body size, and feeding behavior of landbird species using general sources (Mayr and Phelps 1967; Delacour and Amadon 1973; Ridgely 1976; Meyer de Schauensee and Phelps 1978; Stiles 1983; Hilty and Brown 1986; Isler and Isler 1987; Ridgely and Gwynne 1989; Ridgely and Tudor 1989, 1994; Stiles and Skutch 1989; Fjeldså and Krabbe 1990; Karr et al. 1990; Willard et al. 1991; Collar et al. 1992; ICBP 1992; Dunning 1993; Remsen et al. 1993; BirdLife International, unpubl. data), supplemented by specific accounts (Fitzpatrick and O'Neill 1986; Graves 1988; Graves and Uribe 1989; Renjifo 1991, 1994; Stiles 1992; Vuilleumier et al. 1992) and personal observations of the authors. We generally followed the taxonomic classification of Clements (1991); use of a different classification system (e.g., Sibley and Monroe 1990) would have led to slightly different results, but would not be expected to change the major conclusions of this review.

We compiled distributional data for birds over an area of approximately 2,180,000 km^2, which included continental Costa Rica, Panama, Colombia, and Venezuela (Fig. 1). We used geographic rather than political criteria to separate the five montane and two lowland regions used in this review. Broad regional boundaries were defined by elevation or habitat discontinuities between regions. Lowland areas were divided into cis-Andean and trans-Andean regions by the Andes (cf. Haffer 1967). Cis-Andean encompassed lowland areas east and south of the Andes, including the Orinoco Basin and a sector of the Amazon Basin; trans-Andean encompassed lowlands north

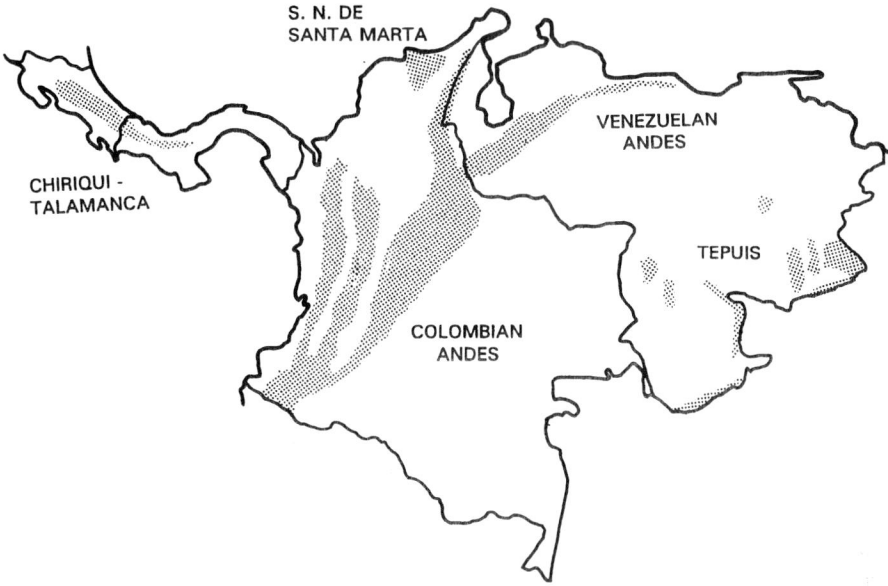

FIG. 1. Schematic map showing location of the five montane systems in southern Central America and northwestern South America used in this review. Stippled areas indicate montane regions. Heavy black lines indicate coastlines and country borders.

and west of the Andes, including the Pacific and Caribbean drainages of these four countries. We did not include avifauna of the dry northwestern Costa Rica in this review because it has strong affinities to northern Central America and differs markedly from avifaunas of southern Central America (Stiles 1983; Stiles and Skutch 1989; Howell and Webb 1995).

We analyzed the five largest mountain systems of southern Central America and northwestern South America (Fig. 1). Three of these montane regions—Chiriquí-Talamanca mountains, Sierra Nevada de Santa Marta, and Tepuis—are totally isolated from other mountains. Indeed, Tepuis are a series of relict table top mountains in the Guiana Shield which are largely isolated from each other (Mayr and Phelps 1967; Meyer de Schauensee and Phelps 1978). Colombian and Venezuelan Andes are relatively more connected, but are divided by the Táchira depression, an important biogeographic barrier forming a gap of 40 km (elevation from 800–900 m) between the two systems (Fjeldså and Krabbe 1990; Hernández 1990). Although the Colombian Andes are divided into three distinct cordilleras separated by two large and deep Inter-Andean valleys of ca. 500 m and ca. 1,000 m elevation, these cordilleras merge in the south (Hilty and Brown 1986). We treat the Colombian Andes as a single unit for this analysis, but it should be recognized that the distributions of a number of bird species are restricted to certain regions (e.g., Central Andes of Colombia and Ecuador, Eastern Andes of Colombia; Hilty and Brown 1986; Long 1993). The eastern slope of northern Perijá, as well as the eastern slope of the Tamá massif, were included in biogeographic Colombian Andes, even though they are politically within Venezuela. Venezuelan Andes, as used here, encompassed high mountains of the west, as well as lower coastal cordilleras. The oldest mountain system in the area is the Tepuis, which has been above sea level since Precambrian and was uplifted in the Tertiary. In contrast, Sierra Nevada de Santa Marta and Chiriquí-Talamanca mountain systems in Colombia and southern Central America, respectively, were uplifted during the Pliocene and Pleistocene (Haffer 1970, 1974). The highest mountains among these five regions are in Sierra Nevada de Santa Marta (hereafter Santa Marta), with peaks at nearly 5,800 m; other regions decrease in elevation in the following order: Colombian Andes, Venezuelan Andes, Chiriquí-Talamanca, Tepuis.

We did not include avifaunas from Cordillera de Caripe and Paria Peninsula in northeastern Venezuela, Serranía de la Macarena in central Colombia, and Serranía del Darién and Cerro Pirré along the border between Colombia and Panama. At least 17 bird species are restricted to these mountain ranges (Haffer 1974; Robbins et al. 1985; Hilty and Brown 1986; Long 1993; BirdLife International, unpubl. data). Inclusion of these areas would be valuable, but lack of specific details regarding elevational distributions would contribute additional sources of error to this study.

When a species was present in at least one region, we added the following information to the database: (1) trophic group (fruit or fruit and seeds [FR]; large invertebrates and small vertebrates [LI]; large invertebrates, small vertebrates, and fruit (i.e., large omnivores) [LO]; small invertebrates [SI]; small invertebrates and fruits (i.e., small omnivores) [SO]; nectar and insects [NI]; nectar, fruit, and insects [NF]; vertebrates [PR]; carrion [CA]; seeds and insects [SE]; herbivore [HE]; (2) body weight; (3) restricted-range species (i.e., species that occupy a geographic range $<50,000$ km^2, BirdLife International, unpubl. data); (4) "endemic" species (i.e., species restricted to a single biogeographic region); (5) minimum and (6) maximum elevations regularly used by species in each montane region. We limited our analyses to resident and migrant landbirds, and excluded birds that take primarily aquatic foods.

Species were assigned to only one trophic group based on their primary food habits. If no data were available on a species' diet and we had no personal knowledge, then we placed the species in the same category as other members of that genus. Placement of species in either SI or LI (or SO, LO) categories often was somewhat arbitrary because no information about prey size was available in any reference that we examined. In those cases, decisions about trophic group were based on bird's overall size and bill size.

We assigned birds to one of seven size classes based on body weights: $\leq$ 10 g, 10–25 g, 25–50 g, 50–100 g, 100–300 g, 300–1,000 g, $>$1,000 g. We used body weight of a similar-sized congener when no data were available for a species and average weight when sexes differed.

We used four elevational zones defined by Chapman (1917, 1926, 1931) and followed in recent treatments of Neotropical avifaunas (Meyer de Schauensee and Phelps 1978; Hilty and Brown 1986; Fjeldså and Krabbe 1990). These zones are tropical (equivalent to lowland), subtropical, temperate, and páramo. We adopt the original terms used by Chapman because of their widespread use by later authors and for ease of presentation; we recognize, however, that use of some terms (e.g., temperate) is misleading. Stiles and Skutch (1989) did not separate the subtropical and temperate avifaunas in the Chiriquí-Talamanca; however, we acknowledge these avifaunas here based on similar vegetation and elevational zones. In the relatively low lying Tepuis, a distinction cannot be drawn between temperate and páramo avifaunas (Chapman 1931; Mayr and Phelps 1967). The actual elevation range of these zones changes among regions (as well as within regions) because of variation in rainfall, height of mountain range, and local topography. We assigned species to zones based on their elevational distribution and the characterization of their habitats. We could not control for fine-scale differences in elevational limits of these zones within regions and, consequently, some errors likely occur in our classification. We feel, however, that these errors are uncommon given the broad elevational categories used. Moreover, such misclassifications would tend to obscure rather than overemphasize patterns described here. It would likely be wrong to assume, however, that these elevational zones control the distributional limits of most bird species (e.g., Terborgh 1971, 1985), but such zones are used here to facilitate broad scale comparisons among regions.

The limits for tropical, subtropical, temperate, and páramo were set to $<$1,500 m, 1,500–2,500 m, $\geq$2,500 m to timberline (ca. 3,500 m), respectively, for the Colombian and Venezuelan Andes, and Santa Marta; $<$1,200 m (Caribbean slope) or 1,500 m (Pacific slope), 1,200 (or 1,500 m)–2,000 m, and $\geq$2,000 m to timberline (ca. 3,000 m), respectively, for Chiriquí-Talamanca. In the Tepuis, we defined only two elevational zones: $<$1,000 m, tropical; $\geq$1,000 m subtropical.

We compared avifaunas among montane regions, elevational zones, and between montane and lowland areas. In addition to a general description of montane and lowland avifaunas, we focused our comparisons on: distribution of species among species-rich families and genera; trophic group composition; number and composition of restricted-range and "endemic" species; and size class distribution. Species-rich families and genera for the entire lowland and montane avifauna (n = 1,800 species) were defined as those families and genera with equal to or more than 45 and 10 species, respectively. We used Sorensen's similarity coefficient (Pielou 1977) to evaluate similarity in species composition among regions.

To determine whether montane avifaunas were simply chance assemblages of the combined lowland and montane avifaunas, we generated random montane communities using Monte Carlo simulations (n = 100 runs); the total source pool consisted of all bird species in the data set. Selection of the source pool for simulations can be problematic (e.g., Connor and Simberloff 1978; Diamond and Gilpin 1982; Graves and Gotelli 1983; Harvey et al. 1983; Gilpin and Diamond 1984; see also review in Wiens 1989) and may greatly affect interpretation of results. For this study, we used the entire data set (i.e., bird species occurring in lowland or montane

TABLE 1

Numbers of Species, Genera, Families, and Restricted-range (i.e., <50,000 km²) Species (RRS) Contained in Different Biogeographic Areas and Elevational Zones. Species Restricted to a Biogeographic Region ("Endemics") Follow RRS in Parentheses. For "Restricted Lowland" and "Restricted Montane" Areas, Genera and Family Numbers Represent Those That Have Species Limited to Lowland or Montane Areas, and Do Not Represent Genera or Families Totally Restricted to Those Areas (See Text for Description of Regions)

	Species	Genera	Families	RRS
Areas				
Total lowland	1,366	502	52	148
Restricted lowland	923	387	49	96
Total montane	877	372	48	217
Restricted montane	434	208	38	165
Shared montane-lowland	443	260	44	52
Regions				
Cis-Andean	975	445	49	32 (16)
Trans-Andean	932	412	46	120 (79)
Colombian Andes	634	324	46	95 (47)
Venezuelan Andes	350	215	39	37 (17)
Chiriquí-Talamanca	253	174	39	51 (47)
Santa Marta	224	165	36	21 (14)
Tepuis	137	106	28	37 (17)
Montane Elevational Zones				
Subtropical	807	346	47	186
Temperate	465	241	43	99
Páramo	117	86	26	32
Totals	1,800	596	52	313

areas) and assumed that if communities were assembled randomly, then each of these species would have an equivalent chance of occurring in montane habitats. This source pool, however, was biased because it only contained birds that occurred in the countries and montane regions selected for this study, and thus did not include all bird species found in montane or cis-Andean and trans-Andean lowlands.

We used chi-square tests for independence to compare distributions of species between areas with respect to trophic groups, families, or size classes. Area comparisons included (1) lowland vs. montane areas, (2) among montane regions, and (3) among elevational zones in montane areas. In all cases, the null hypothesis was that distributions did not differ between areas, among regions, or among zones. When necessary, categories were excluded or merged if expected cell frequencies did not meet assumptions of the chi-square test (see results). Differences in number of species within taxonomic or trophic groups were compared between randomly generated and observed montane communities using t-tests (Zar 1984:100).

Similarly, chi-square tests were used to examine the composition of restricted-range species relative to widespread species (i.e., distribution among elevational zones, lowland vs. montane areas, trophic groups, etc.). To compare number of species within genera between lowland and montane areas, we used Fisher's Exact Test (Sokal and Rohlf 1981); only genera with 10 or more total species were examined.

RESULTS

General description of avifauna.—The avifauna of montane and lowland regions combined was composed of 1,800 species representing 52 families and 596 genera (Table 1); this is almost 20% of the world's bird species. The total species list contained 673 non-Passeriformes, 590 suboscine Passeriformes, and 537 oscine Passeriformes. Six families (Tyrannidae, Thraupidae, Trochilidae, Formicariidae, Emberizidae, and Furnariidae) accounted for 51.8% of the 1,800 species (Table 2).

A total of 877 species representing 48 families was present in all montane regions combined; 1,366 species from 52 families were present in lowland areas (Table 1). Within the lowlands,

TABLE 2

NUMBER OF TOTAL AND RESTRICTED-RANGE (RRS) SPECIES CONTAINED WITHIN FAMILIES IN LOWLAND AND MONTANE AREAS OF NORTHWESTERN SOUTH AMERICA AND SOUTHERN CENTRAL AMERICA. TOTAL NUMBER OF SPECIES FOR THESE FAMILIES (I.E., MONTANE/LOWLAND COMBINED) ARE PROVIDED. IN ALL CASES EXPECTED VALUES FOR CHI-SQUARE TESTS USED IN TEXT FOLLOW OBSERVED VALUES IN PARENTHESES. ONLY FAMILIES WITH MORE THAN 45 SPECIES TOTAL ARE INCLUDED IN THE TABLE

Family	Lowland	Montane	RRS	Total
Accipitridae	44 (40.0)	24 (28.0)	0 (9.8)	52
Psittacidae	53 (48.9)	30 (34.1)	14 (12.9)	69
Trochilidae	114 (133.1)	112 (92.9)	51 (34.1)	182
Picidae	41 (33.0)	15 (23.0)	7 (9.2)	49
Furnariidae	48 (58.3)	51 (40.7)	18 (15.8)	84
Formicariidae	128 (101.3)	44 (70.7)	34 (29.1)	155
Cotingidae	37 (34.7)	22 (24.3)	12 (8.8)	47
Tyrannidae	178 (167.8)	107 (117.2)	19 (42.9)	229
Parulidae	48 (57.1)	49 (39.9)	13 (12.8)	68
Emberizidae	64 (71.2)	57 (49.8)	20 (18.6)	99
Thraupidae	120 (129.5)	100 (90.5)	40 (34.1)	182
Total	875	611	228	1,216

the cis-Andean region contained slightly more species, genera, and families than did the trans-Andean. Among montane regions, greater numbers of species, genera, and families were found in the Colombian Andes, followed in decreasing order by Venezuelan Andes, Chiriquí-Talamanca, Santa Marta, and Tepuis (Table 1). We suggest caution when interpreting and comparing among regions the species richness values reported here as these regions differ in total land area (an important determinant of species richness patterns), as well as in a number of environmental factors (e.g., rainfall, seasonality, etc.). Our purpose here is not to highlight the effects of area on species richness patterns, but rather compositional differences between montane and lowland avifaunas. The latter comparisons are particularly important from a conservation perspective, and that, together with information on number of total and "endemic" species, will facilitate setting of conservation priorities.

Much of the region's avifauna was restricted to either montane (435 species, 24.2%) or lowland (923 species, 51.3%) areas; 442 species (24.6%) were found in both areas (Table 1). Four families, Opisthocomidae, Psophiidae, Burhinidae, and Sylviidae, only occurred in the lowlands. Both Psophiidae and Opisthocomidae had one species in the cis-Andean region. Burhinidae had one species in both cis-Andean and trans-Andean regions, whereas, Sylviidae had five and four species in cis-Andean and trans-Andean regions, respectively.

Three families were restricted to higher elevations: Ptilogonatidae, Alaudidae, and Oxyruncidae. The two species of Ptilogonatidae are only found in Chiriquí-Talamanca highlands. A very localized distribution in the Colombian Andes of one species of Alaudidae is a marginal isolate of a widespread northern hemisphere species. Oxyruncidae (*Oxyruncus cristatus*) was more problematic because it is found in the subtropical zone, as well as the upper tropical zone. We regarded it here solely as a montane taxon because it is always associated with mountains in our study area (note: other classification systems would include this species within the subfamily Cotinginae; e.g., Sibley and Monroe 1990).

One-half of the bird species present in montane regions were strictly montane; species restricted to montane areas were represented by 209 genera of which 93 were confined to the mountains. Some of these genera, however, are found at lower elevations in other regions (Sick 1985; Fjeldså and Krabbe 1990; Willis 1992; Parker and Goerck 1997).

Nearly 20% (170 species) of the bird species found in montane areas were restricted to a single elevational zone; 127 species and 13 genera were restricted to the subtropical zone, 30 species and nine genera to the temperate zone, and 13 species and nine genera to páramo (Appendix 1). Two families, Trochilidae (12 genera) and Furnariidae (seven genera), accounted for most of the genera restricted to a particular elevational zone. The proportion of montane species restricted to specific elevational zones was notably similar among Chiriquí-Talamanca, Colombian Andes, and Venezuelan Andes (i.e., 72% subtropical zone, 19% temperate zone, and 9% páramo). Santa Marta diverged from these regions by having a greater proportion of montane

TABLE 3
Sorensen's Similarity Coefficients Comparing Species Composition Between Regions Using All Species (A) and Species Restricted to Montane Regions (B)

	Cis-Andean	Trans-Andean	Colombian Andes	Venezuelan Andes	Chiriquí-Talamanca	Santa Marta
A. ALL SPECIES						
Trans-Andean	0.56	—	—	—	—	—
Colombian Andes	0.30	0.46	—	—	—	—
Venezuelan Andes	0.24	0.29	0.61	—	—	—
Chiriquí-Talamanca	0.18	0.27	0.29	0.32	—	—
Santa Marta	0.21	0.27	0.48	0.53	0.37	—
Tepuis	0.15	0.11	0.16	0.24	0.16	0.24
B. RESTRICTED MONTANE SPECIES						
Venezuelan Andes	—	—	0.59	—	—	—
Chiriquí-Talamanca	—	—	0.09	0.12	—	—
Santa Marta	—	—	0.29	0.40	0.10	—
Tepuis	—	—	0.07	0.10	0.05	0.11

species restricted to páramo (43%). The relatively small area but high elevation of Santa Marta may account for this departure.

The two lowland regions generally shared more species and had a higher similarity coefficient than did most montane regions (Table 3); Colombian and Venezuelan Andes, however, displayed the greatest similarity of all pair-wise comparisons, which likely corresponds to their geographic proximity. Tepuis had the most distinct avifauna as indicated by consistently low similarity coefficient scores with other regions (Table 3). When similarities in species compositions were compared among montane regions excluding those taxa that also occur in lowland areas, a two-fold or greater reduction in similarity coefficients generally occurred (Table 3). To a large extent, then, species shared among montane regions also are shared with lowland avifaunas.

Composition of lowland and montane avifaunas.—Despite similarities in family composition, the distribution of species within the most species-rich families (i.e., 11 families with ≥45 species) differed between lowland and montane areas ($\chi^2 = 43.7$, $P < 0.0001$; Table 2). In particular, montane areas had more Trochilidae and Parulidae, but fewer Picidae and Formicariidae than did lowlands when based on expected values (Table 2).

When numbers of species contained within species-rich genera (i.e., ≥10 species) were compared between lowland and montane areas, seven genera showed significantly different distributions than expected (Table 4); 1.4 of 29 genera would have been expected to show differences by chance at $P < 0.05$ significance criteria. Four of the seven genera (*Grallaria, Myioborus, Basileuterus, Atlapetes*) were more species-rich in montane areas, whereas three (*Pteroglossus, Myrmotherula,* and *Myrmeciza*) had more species in lowlands (Table 4).

Distribution of species among major trophic categories differed between lowland and montane areas ($\chi^2 = 27.5$, $P < 0.001$; Table 5). Differences in trophic composition were influenced primarily by nectarivores, as expected given the importance of Trochilidae in montane areas. Indeed, when nectarivores were removed from the analysis, differences in distribution of species among trophic groups did not differ significantly between lowland and montane areas ($\chi^2 = 10.5$, $P > 0.15$). In contrast, no significant differences in trophic composition were found either among the five montane regions ($\chi^2 = 28.6$, $P > 0.4$; NF combined with SO due to low numbers; CA and HE excluded from analysis due to low numbers) or among elevational zones ($\chi^2 = 19.5$, $P > 0.15$; Table 5).

In both montane and lowland areas, the majority of the avifauna occurred in the three smaller size classes (68% and 69% of lowland and montane birds below 50 g, respectively). There was a trend for larger size classes to be poorly represented in montane bird communities ($\chi^2 = 11.3$, $P < 0.08$).

Comparison of observed and randomly generated montane avifaunas.—We found that montane communities were not simply chance assemblages; nearly 30% of bird families (15 of 52 families) had a significantly different number of species in montane communities than expected if patterns were random (Table 6) (note: 2.6 of 52 families would be expected to show differences by chance alone using $P < 0.05$ criteria). Moreover, when the number of species occurring in

TABLE 4

Number of Species Contained within Genera in Lowland and Montane Areas for Those Genera with More Than 10 Species. Results of Fisher's Exact Test Are Shown (P Value); Genera Showing Significant Differences in Number of Species between Lowland and Montane Areas Are Bold. Also Shown Are the Number and Proportion of Species with Restricted Ranges (RRS)

Genera	Lowland	Montane	P value	Number of RRS species	Proportion of RRS species
Crypturellus	10	3	0.239	5	0.42
Buteo	8	10	0.138	0	0.00
Caprimulgus	7	6	0.381	2	0.20
Phaethornis	12	6	0.345	0	0.00
Amazilia	18	7	0.178	5	0.28
Trogon	12	4	0.221	3	0.23
Pteroglossus	10	0	0.014	2	0.20
Picumnus	10	3	0.239	3	0.30
Xiphorhynchus	10	3	0.239	0	0.00
Synallaxis	8	8	0.266	4	0.31
Thamnophilus	12	3	0.141	2	0.14
Myrmotherula	19	2	0.005	2	0.10
Myrmeciza	14	1	0.011	4	0.28
Grallaria	3	17	0.001	10	0.53
Elaenia	9	8	0.306	1	0.08
Phylloscartes	6	5	0.431	6	0.60
Pachyramphus	9	4	0.385	0	0.00
Hylophilus	12	3	0.141	1	0.08
Catharus	8	8	0.266	1	0.10
Turdus	10	12	0.116	1	0.07
Thryothorus	15	6	0.228	4	0.25
Dendroica	14	10	0.316	0	0.00
Myioborus	1	10	0.002	6	0.60
Basileuterus	8	13	0.038	5	0.31
Atlapetes	3	14	0.001	6	0.40
Sporophila	13	5	0.248	0	0.00
Icterus	10	2	0.141	0	0.00
Euphonia	20	6	0.081	2	0.09
Tangara	26	22	0.147	3	0.08

TABLE 5

Number of Species Contained within Trophic Groups by Area and Elevational Zones. Only Montane Zones are Shown. See Text for Descriptions of Trophic Groups

	Trophic group								
	SI	SO	LI	LO	FR	SE	PR	NI	NF
Areas									
Lowland	444	242	135	73	214	59	62	114	16
Montane	280	177	60	34	115	37	43	122	6
Montane Regions									
Colombian Andes	199	127	49	24	79	26	38	82	6
Venezuelan Andes	118	64	23	11	47	16	23	42	3
Chiriquí-Talamanca	84	64	13	13	26	18	15	18	0
Santa Marta	80	41	15	6	23	17	18	18	1
Tepuis	47	31	5	4	17	8	7	17	1
Montane Elevational Zones									
Subtropical	255	167	58	31	109	33	41	104	6
Temperate	161	88	28	15	53	21	30	64	2
Páramo	46	18	2	3	9	7	6	24	1

TABLE 6

NUMBER OF SPECIES CONTAINED WITHIN SPECIES RICH MONTANE FAMILIES FOR OBSERVED AND RANDOMLY GENERATED MONTANE COMMUNITIES (SEE TEXT). MEAN NUMBER AND STANDARD DEVIATION OF SPECIES ARE SHOWN FOR RANDOM COMMUNITIES (N = 100 RUNS). RESULTS OF T-TEST COMPARING DIFFERENCES BETWEEN NUMBER OF SPECIES IN OBSERVED AND RANDOM COMMUNITIES ARE SHOWN. ONLY FAMILIES SHOWING SIGNIFICANT PATTERNS ARE INCLUDED

Family	Montane species		t-value	P value
	Observed	Random		
More Species-rich in Montane Areas:				
Odontophoridae	10	5.6 ± 1.8	−2.5	<0.05
Trochilidae	112	88.1 ± 7.3	−3.3	<0.01
Furnariidae	51	41.7 ± 4.4	−2.1	<0.05
Rhinocryptidae	8	4.5 ± 1.4	−2.4	<0.05
Ptiligonatidae	3	1.3 ± 0.8	−2.0	<0.05
Turdidae	26	15.0 ± 2.7	−4.1	<0.001
Hirundinidae	13	8.6 ± 2.0	−2.2	<0.05
Parulidae	49	32.8 ± 3.8	−4.3	<0.001
Less Species-rich in Montane Areas:				
Galbulidae	1	5.9 ± 1.7	2.9	<0.01
Bucconidae	4	12.0 ± 2.5	3.2	<0.01
Ramphastidae	8	14.8 ± 2.7	2.5	<0.05
Picidae	15	24.3 ± 3.7	2.5	<0.05
Formicariidae	44	75.6 ± 6.2	5.1	<0.001
Pipridae	9	15.9 ± 2.5	2.8	<0.01
Sylviidae	0	3.1 ± 1.2	2.5	<0.05

trophic groups was compared between observed and randomly generated montane communities, we found that three of nine trophic groups showed non-random patterns (CA and HE excluded). Specifically, more small omnivores and nectarivores, and fewer large insectivores were recorded in the montane community than expected if those communities were randomly assembled subsets of the region's avifauna ($t \geq 2.0$, $P < 0.05$).

Patterns of restricted-range species.—Based on distributional data made available by BirdLife International (unpubl. data), 313 species (17.4% of the total) were restricted-range species (<50,000 km^2); most occurred in only a single biogeographical region (142 in montane and 95 in lowland) (Table 1). Restricted-range species occurred in a wide range of families (32 of 52) and genera (172 of 596), but they were not distributed equally among species-rich families ($\chi^2 = 35.50$, $P < 0.001$) (Table 2). Tyrannidae, as well as Accipitridae, had a low number of small range species. When these two families were removed from the analysis, the frequency of restricted-range species no longer differed ($\chi^2 = 6.3$, $P > 0.60$). Although Terborgh and Winter (1983) suggested that species with small geographic ranges tend to be of small size, we found no difference in body size between bird species with restricted ranges and other bird species ($\chi^2 = 10.4$, $P > 0.1$).

Montane areas had both a greater number and percentage of species with small geographic ranges than did lowland areas (Table 1). In fact, of those birds only occurring in either montane or lowland areas, more restricted-range species than expected occurred in the montane avifaunas ($\chi^2 = 172.6$, $P < 0.0001$). Moreover, restricted-range species were particularly prevalent among those genera that were significantly associated with montane areas (Table 4). Such differences, however, may be artifacts of current classification systems; further analysis may reveal that many wide-ranging species are actually morphologically similar allospecies (e.g., see Isler et al. 1997).

DISCUSSION

Montane avifaunas of northwestern South America and southern Central America contribute substantially to the region's diverse avifauna. As this review has confirmed, montane avifaunas are not simply a random subset of the rich lowland communities. Within the study region, 93 genera and 435 species occur exclusively in the mountains. Genera restricted to montane areas represent evolutionary lineages that would be lost if montane habitats are not effectively protected (cf. Humphries et al. 1991; Williams et al. 1991). Moreover, the Colombian and Venezuelan Andes and Santa Marta were found to have some of the Neotropics' greatest density of recently evolved species, and thus may represent critical regions for evolutionary change (Fjeldså 1994). Consequently, the evidence strongly supports the fact that montane avifaunas are not only distinct bird communities but also centers of radiation that deserve special attention (see also Erwin 1991).

One of the most important ecological differences between lowland and highland areas is the greater importance (i.e., proportionally more species than expected) of nectarivores in montane avifaunas. In fact, the most species-rich genera restricted to the mountains are nectarivores, namely *Eriocnemis*, *Heliangelus*, and *Diglossa*. This increased importance of nectarivores in montane areas also has been documented elsewhere at different geographic scales of analysis (Stiles 1985; Graham 1990). Indeed, differences between montane and lowland avian communities match major changes in plant communities in terms of plant family composition and physiognomy (Gentry 1988, 1992). For example, the greatest number of bird-pollinated plants occurs in middle elevations, and bird-pollinated families such as Ericaceae and Gesneriaceae become more important with increasing elevation (i.e., abundance and species richness) (Stiles 1981; Gentry 1988, 1992).

Differences in land surface area may largely account for observed differences in diversity of lowland and montane bird communities (Table 1). Yet at the scale of our biogeographic units, montane areas were more distinct from each other than were the two lowland regions; this highlights the importance of regional diversity at high elevations. The low species overlap among montane avifaunas calls for a system of protected areas covering a wide range of geographic locations.

Restricted-range species are of particular concern from a conservation perspective. Given their small geographic ranges, these species are highly vulnerable to habitat destruction and other forms of human pressure. Consequently, degree of overlap in distribution of restricted-range species is considered a useful criterion for assigning conservation priorities. The majority of bird species with restricted-ranges in the regions reviewed here are found in the mountains (Table 1; see also Terborgh and Winter 1983; ICBP 1992; BirdLife International, unpubl. data). Thus, the greater total number and proportion of restricted-range species in montane avifaunas further highlight these regions as areas of utmost conservation concern.

As we have attempted to demonstrate in this review, the distinctiveness and richness of the montane avifauna provide a strong argument for establishment of montane regions as priority areas for conservation. When taken together with human activities in these regions, however, the urgency of conservation action becomes substantially more obvious and pressing. The Andean region of Colombia, with the highest diversity of montane birds and the greatest number of "endemics", is considered one of the most human-modified regions in South America (Hilty 1985; Myers 1988). Similarly, major population centers of Costa Rica and Venezuela are located, to a large extent, in or close to the mountains. Although Panama's population is centered in the lowlands, deforestation in the Chiriquí Mountains is extensive. Worldwide trends reveal an alarming correlation between avian endemism patterns and deforestation, and the countries represented in our study were among those with the highest degrees of endemism and deforestation rates (see Fig. 1b in Balmford and Long 1994). As a result, 82 species of these montane regions are currently listed as threatened or near threatened (Collar et al. 1992), and as many as 50 are restricted to the mountains (Appendix 2).

Despite the richness, distinctiveness and endangerment of the montane avifaunas, conservation of Neotropical mountain ecosystems has been largely neglected. More emphasis towards conservation of montane ecosystems is sorely needed.

ACKNOWLEDGMENTS

We all strongly share Ted Parker's enthusiasm for montane birds and his deep concern for their conservation. We are profoundly indebted to Ted for his willingness to share so freely with us his immense knowledge of birds, conservation, and natural history. We thank David Wege and BirdLife International for providing us with unpublished information on the birds with small geographic ranges for the countries included in this review. This paper would not have been possible without their help. We thank J. V. Remsen, G. Barrantes, M. Kelsey, and an anonymous reviewer for their comments and suggestions on the manuscript. Any errors contained within this paper rest solely with the authors. Due to space limitations, we have not printed a full appendix here, but this is available from the authors upon request by sending a blank disk.

LITERATURE CITED

BALMFORD, A., AND A. LONG. 1994. Avian endemism and forest loss. Nature 372:623–624.

CHAPMAN, F. M. 1917. The distribution of birdlife in Colombia: a contribution to a biological survey of South America. Bull. Amer. Mus. Nat. Hist. 37:1–784.

CHAPMAN, F. M. 1926. The distribution of bird-life in Ecuador: a contribution to the study of the origin of Andean bird-life. Bull. Amer. Mus. Nat. Hist. 55:1–133.

CHAPMAN, F. M. 1931. The upper zonal bird-life of Mts. Roraima and Duida. Bull. Amer. Mus. Nat. Hist. 63:1–135.

CLEMENTS, J. F. 1991. Birds of the World: a Checklist. The Two Continents Publishing Group, Ltd., New York, New York.

COLLAR, N. J., L. P. GONZAGA, N. KRABBE, A. MADRONO NIETO, L. G. NARANJO, T. A. PARKER III, AND D. C. WEGE. 1992. Threatened Birds of the Americas. ICBP/IUCN Red Data Book. International Council for Bird Preservation, Cambridge, United Kingdom.

CONNOR, E. F., AND D. SIMBERLOFF. 1978. Species number and compositional similarity of the Galápagos flora and avifauna. Ecol. Monogr. 48:219–248.

CRACRAFT, J. 1985. Historical biogeography and patterns of differentiation within the South American avifauna: areas of endemism. Pages 49–84 in Neotropical Ornithology (P. A. Buckley, M. S. Foster, E. S. Morton, R. S. Ridgely, and F. G. Buckley, Eds.). Ornithological Monographs No. 36, American Ornithologists' Union, Lawrence, Kansas.

DELACOUR, J., AND D. AMADON. 1973. Curassows and Related Birds. American Museum of Natural History, New York, New York.

DIAMOND, J. R., AND M. E. GILPIN. 1982. Examination of the "null" model of Connor and Simberloff for species co-occurences on islands. Oecologia 52:64–74.

DODSON, C. H., AND A. H. GENTRY. 1991. Biological extinction in western Ecuador. Ann. Missouri Bot. Gard. 78:273–295.

DUNNING, J. B., JR. (Ed.). 1993. CRC Handbook of Avian Body Masses. CRC Press, Boca Raton, Florida.

ERWIN, T. L. 1991. An evolutionary basis for conservation strategies. Science 253:750–752.

ETTER, A. 1993. Diversidad ecosistémica en Colombia hoy. Pages 43–61 in Nuestra Diversidad Biológica. (S. Cárdenas and H. D. Correa, Eds.). CEREC, Bogotá, Colombia.

FITZPATRICK, J. W., AND J. P. O'NEILL. 1986. *Otus petersoni*, a new screech-owl from the eastern Andes, with systematic notes on *O. columbianus* and *O. ingens*. Wilson Bull. 98:1–14.

FJELDSÅ, J. 1994. Geographical patterns for relict and young species of birds in Africa and South America and implications for conservation priorities. Biodiversity and Conservation 3:207–226.

FJELDSÅ, J., AND N. KRABBE. 1990. Birds of the High Andes. Zool. Mus. Univ. Copenhagen and Apollo Books Publ., Svendborg, Denmark.

GENTRY, A. H. 1988. Changes in plant community diversity and floristic composition on environmental and geographical gradients. Ann. Missouri Bot. Gard. 75:1–34.

GENTRY, A. H. 1992. Diversity and floristic composition of Andean forest of Peru and adjacent countries: implications for their conservation. Pages 11–29 in Biogeografía, Ecología y Conservación del Bosque Andino en el Perú (K. R. Young, and N. Valencia, Eds.). Memorias del Museo de Historia Natural No. 21. Universidad Nacional Mayor de San Marcos, Lima, Perú.

GILPIN, M. E., AND J. R. DIAMOND. 1984. Are species co-occurences on islands non-random, and are null hypotheses useful in community ecology? Pages 297–315 in Ecological Communities: Conceptual Issues and the Evidence. (D. R. Strong Jr., D. Simberloff, L. G. Abele, and A. B. Thistle, Eds.). Princeton Univ. Press, Princeton, New Jersey.

GRAHAM, G. L. 1990. Bats versus birds: comparisons among Peruvian volant vertebrate faunas along an elevational gradient. J. Biogeogr. 17:657–668.

GRAVES, G. R. 1988. *Phylloscartes lanyoni*, a new species of bristle-tyrant (Tyrannidae) from the lower Cauca Valley of Colombia. Wilson Bull. 100:529–534.

GRAVES, G. R., AND N. J. GOTELLI. 1983. Neotropical land bridge avifaunas: new approaches to null hypotheses in biogeography. Oikos 41:322–333.

GRAVES, G. R., AND D. URIBE. 1989. A new allopatric taxon in the *Hapalopsittaca amazonina* (Psittacidae) superspecies from Colombia. Wilson Bull. 101:369–376.

HAFFER, J. 1967. Speciation in Colombian forest birds west of the Andes. Amer. Mus. Novitates 2294:1–57.

HAFFER, J. 1970. Geologic-climatic history and zoogeographic significance of the Uraba region in northwestern Colombia. Caldasia 10:603–636.

HAFFER, J. 1974. Avian speciation in tropical South America. Publ. Nuttall Ornithological Club, No. 14. Cambridge, Massachusetts.

HARVEY, P. H., R. K. COLWELL, AND J. W. SILVERTON. 1983. Null models in ecology. Annu. Rev. Ecol. Syst. 14:189–211.

HECHT, S., AND A. COCKBURN. 1990. The Fate of the Forest: Developers, Destroyers and Defenders of the Amazon. Harper Collins Publ., New York, New York.

HENDERSON, A., S. CHURCHILL, AND J. LUTEYN. 1991. Neotropical plant diversity. Nature 351:21–22.

HERNANDEZ, J. 1990. La selva en Colombia. Pages 13–47 in Selva y Futuro, Colombia (J. Carrizosa and J. Hernándes, Eds.). El Sello Editorial, Bogotá, Colombia.

HERNANDEZ, J. 1993. Una síntesis de la historia evolutiva de la biodiversidad en Colombia. Pages 270–287 *in* Nuestra diversidad biológica (S. Cárdenas and H. D. Correa, Eds.). CEREC, Bogotá, Colombia.

HILTY, S. L. 1985. Distributional changes in the Colombian avifauna: a preliminary Blue List. Pages 1000–1012 *in* Neotropical Ornithology (P. A. Buckley, M. S. Foster, E. S. Morton, R. S. Ridgely, and F. G. Buckley, Eds.). Ornithological Monographs No. 36, American Ornithologists' Union, Lawrence, Kansas.

HILTY, S. L., AND W. L. BROWN. 1986. A Guide to the Birds of Colombia. Princeton Univ. Press, Princeton, New Jersey.

HOWELL, S. G., AND S. WEBB. 1995. A Guide to the Birds of Mexico and Northern Central America. Oxford University Press, Oxford, United Kingdom.

HOUGHTON, R. A., D. S. LEFKOWITZ, AND D. L. SKOLE. 1991. Changes in the landscape of Latin America between 1850 and 1985. I. Progressive loss of forests. For. Ecol. Manag. 38:143–172.

HUMPHRIES, C., D. VANE-WRIGHT, AND P. WILLIAMS. 1991. Biodiversity reserves: setting new priorities for the conservation of wildlife. Parks 2:34–38.

ICBP. 1992. Putting Biodiversity on the Map: Priority Areas for Global Conservation. International Council for Bird Preservation, Cambridge, United Kingdom.

ISLER, M. L., AND P. R. ISLER. 1987. The Tanagers: Natural History, Distribution and Identification. Smithsonian Institution Press, Washington, D.C.

ISLER, M. L., P. R. ISLER, AND B. M. WHITNEY. 1997. Biogeography and species limits in the *Thamnophilus punctatus* group. Pages 355–381 *in* Studies in Neotropical Ornithology Honoring Ted Parker (J. V. Remsen, Jr., Ed.), Ornithol. Monogr. No. 48.

KARR, J. R. 1976. Within- and between-habitat diversity in African and Neotropical lowland habitats. Ecol. Monogr. 46:457–481.

KARR, J. R., S. K. ROBINSON, J. G. BLAKE, AND R. O. BIERREGAARD, JR. 1990. Birds of four Neotropical rainforests. Pages 237–269 *in* Four Neotropical Rainforests. (A. H. Gentry, Ed.). Yale Univ. Press, New Haven, Connecticut.

LONG, A. J. 1993. Restricted-range and threatened bird species in tropical montane cloud forests. Pages 47–65 *in* Proc. Intl. Symp. on Tropical Montane Cloud Forests (L. S. Hamilton, J. O. Juvik, and N. F. Scatena, Eds.). East-West Center, Honolulu, Hawaii.

MAYR, E., AND W. H. PHELPS, JR. 1967. The origin of the bird fauna of the south Venezuelan highlands. Bull. Amer. Mus. Nat. Hist. 136:269–328.

MEYER DE SCHAUENSEE, R., AND W. H. PHELPS, JR. 1978. A Guide to the Birds of Venezuela. Princeton Univ. Press, Princeton, New Jersey.

MILLER, K. E., AND L. TANGLEY. 1991. Trees of Life: Saving Tropical Forests and Their Biological Wealth. Beacon Press, Boston, Massachusetts.

MYERS, N. 1988. Threatened biotas: "Hotspots" in tropical forests. Environmentalist 8:1–20.

MYERS, N. 1992. The Primary Source: Tropical Forests and Our Future. W. W. Norton and Co., New York, New York.

PARKER, T. A., III, AND J. M. GOERCK. 1997. The importance of national parks and biological reserves to bird conservation in the Atlantic Forest Region of Brazil. Pages 527–541 *in* Neotropical Ornithology Honoring Ted Parker (J. V. Remsen, Jr., Ed.), Ornithol. Monogr. No. 48.

PEARSON, D. L. 1982. Historical factors and bird species richness. Pages 441–452 *in* Biological Diversification in the Tropics (G. T. Prance, Ed.). Columbia Univ. Press, New York, New York.

PIELOU, E. 1977. Mathematical Ecology. J. S. Wiley Publ., New York, New York.

REID, W. V. 1992. How many species will there be? Pages 55–71 *in* Tropical Deforestation and Species Extinction (T. C. Whitmore and J. A. Sayer, Eds.). Chapman and Hall, London.

REMSEN, J. V., JR., M. A. HYDE, AND A. CHAPMAN. 1993. The diets of Neotropical trogons, motmots, barbets and toucans. Condor 95:178–192.

RENJIFO, L. M. 1991. Discovery of the Masked Saltator in Colombia, with notes on its ecology and behavior. Wilson Bull. 103:685–690.

RENJIFO, L. M. 1994. First records of the Bay-vented Cotinga (*Doliornis sclateri*) in Colombia. Bull. Brit. Ornith. Club 114:101–103.

RICKLEFS, R. E., AND D. SCHLUTER, (EDS.). 1993. Species Diversity in Ecological Communities: Historical and Geographical Perspectives. Univ. Chicago Press, Chicago, Illinois.

RIDGELY, R. S. 1976. A Guide to the Birds of Panama. Princeton Univ. Press., Princeton, New Jersey.

RIDGELY, R. S., AND J. A. GWYNNE. 1989. A Guide to the Birds of Panama with Costa Rica, Nicaragua, and Honduras. Princeton Univ. Press, Princeton, New Jersey.

RIDGELY, R. S., AND G. TUDOR. 1989. The Birds of South America. Vol I: Oscine Passerines. Univ. Texas Press, Austin, Texas.

RIDGELY, R. S., AND G. TUDOR. 1994. The Birds of South America. Vol II: Suboscine Passerines. Univ. Texas Press, Austin, Texas.

ROBBINS, M., T. A. PARKER III, AND S. E. ALLEN. 1985. The avifauna of Cerro Pirré, Darién, Eastern Panama. Pages 198–232 *in* Neotropical Ornithology (P. A. Buckley, M. S. Foster, E. S. Morton, R. S. Ridgely, and F. G. Buckley, Eds.). Ornithological Monographs No. 36, American Ornithologists' Union, Lawrence, Kansas.

SAYER, J. A., AND T. C. WHITMORE. 1991. Tropical moist forests: destruction and species extinction. Biol. Conserv. 55:199–214.

SIBLEY, C. G., AND B. L. MONROE, JR. 1990. Distribution and Taxonomy of Birds of the World. Yale Univ. Press, New Haven, Connecticut.

SICK, H. 1985. Observations on the Andean-Patagonian component of southeastern Brazil's avifauna. Pages 233–237 in Neotropical Ornithology (P. A. Buckley, M. S. Foster, E. S. Morton, R. S. Ridgely, and F. G. Buckley, Eds.). Ornithological Monographs No. 36, American Ornithologists' Union, Lawrence, Kansas.

SKOLE, D., AND C. TUCKER. 1993. Tropical deforestation and habitat fragmentation in the Amazon: satellite data from 1978 to 1988. Science 260:1905–1909.

SOKAL, R. R., AND F. J. ROHLF. 1981. Biometry. W. H. Freeman, San Francisco, California.

STILES, F. G. 1981. Geographical aspects of bird-flower coevolution, with particular reference to Central America. Ann. Missouri Bot. Gard. 68:323–351.

STILES, F. G. 1983. Birds: introduction. Pages 502–530 in Costa Rican Natural History (D. H. Janzen, Ed.). Univ. Chicago Press, Chicago, Illinois.

STILES, F. G. 1985. On the role of birds in the dynamics of Neotropical forests. ICBP Tech. Publ. 4:49–59.

STILES, F. G. 1992. A new species of antpitta (Formicariidae: *Grallaria*) from the eastern Andes of Colombia. Wilson Bull. 104:389–399.

STILES, F. G., AND A. SKUTCH. 1989. A Guide to the Birds of Costa Rica. Cornell Univ. Press, Ithaca, New York.

TERBORGH, J. 1971. Distribution on environmental gradients: theory and a preliminary interpretation of distributional patterns in the avifauna of the Cordillera Vilcabamba, Peru. Ecology 52:23–40.

TERBORGH, J. 1985. The role of ecotones in the distribution of Andean birds. Ecology 66:1237–1246.

TERBORGH, J., AND B. WINTER. 1983. A method for siting parks and reserves with special reference to Colombia and Ecuador. Biol. Conserv. 27:45–58.

VUILLEUMIER, F., M. LECROY, AND E. MAYR. 1992. New species of birds described from 1981 to 1990. Bull. Brit. Onithol. Club Centenary Supp. 112A:267–309.

WHITMORE, T. C., AND J. A. SAYER. 1992. Deforestation and species extinction in tropical moist forests. Pages 1–14 in Tropical Deforestation and Species Extinction (T. C. Whitmore and J. A. Sayer, Eds.). Chapman and Hall, London.

WIENS, J. A. 1989. The Ecology of Bird Communities. Vol I. Cambridge Univ. Press, Cambridge, United Kingdom.

WILLIAMS, P. H., C. J. HUMPHRIES, AND R. I. VANE-WRIGHT. 1991. Measuring biodiversity: taxonomic relatedness for conservation priorities. Aust. Syst. Bot. 4:665–679.

WILLARD, D. E., M. S. FOSTER, G. F. BARROWCLOUGH, R. W. DICKERMAN, P. F. CANNELL, S. L. COATS, J. L. CRACRAFT, AND J. P. O'NEILL. 1991. The birds of Cerro de la Neblina, Territorio Federal Amazonas, Venezuela. Fieldiana (Zoology) 65:1–80.

WILLIS, E. O. 1992. Zoogeographical origins of eastern Brazilian birds. Ornitologia Neotropical 3:1–15.

ZAR, J. H. 1984. Biostatistical Analysis. Second edition. Prentice-Hall, Inc., Englewood Cliffs, New Jersey.

APPENDIX 1
Species Restricted to a Single Elevational Zone within Montane Areas. Species of Genera Restricted to a Particular Elevational Zone Are Indicated by an Asterisk

SUBTROPICAL ZONE

Nothocercus bonapartei; Crypturellus ptaritepui; Crypturellus obsoletus; Penelope perspicax; Odontophorus melanonotus; Odontophorus strophium; Odontophorus columbianus; Odontophorus leucolaemus; Leptotila conoveri; Pyrrhura viridicata; Pyrrhura hoematotis; Touit batavica; Amazona dufresniana; Otus colombianus; Otus guatemalae; Nyctibius leucopterus; Caprimulgus whitelyi; Campylopterus hyperythrus; Campylopterus duidae; Chlorostilbon alice; Polytmus milleri; Eupherusa eximia*; Eupherusa nigriventris*; Elvira chionura*; Elvira cupriceps*; Lampornis hemileucus; Adelomyia melanogenys*; Heliodoxa rubinoides; Sternoclyta cyanopectus*; Urochroa bougueri*; Boissonneaua matthewsii; Eriocnemis godini; Eriocnemis mirabilis; Urosticte ruficrissa; Schistes geoffroyi*; Selasphorus ardens; Pharomachrus fulgidus; Trogon aurantiiventris; Veniliornis kirkii; Cranioleuca demissa; Siptornis striaticollis*; Roraimia adusta*; Premnornis guttuligera*; Premnoplex brunnescens*; Premnoplex tatei*; Margarornis stellatus; Syndactyla guttulata; Anabacerthia striaticollis; Thripadectes virgaticeps; Thripadectes melanorhynchus; Automolus roraimae; Sclerurus albigularis; Thamnophilus unicolor; Thamnophilus insignis; Thamnomanes plumbeus; Myrmotherula behni; Herpsilochmus roraimae; Chamaeza ruficauda; Chamaeza turdina; Grallaria excelsa; Grallaria alleni; Grallaria chthonia; Grallaria bangsi; Grallaria kaestneri; Grallaria flavotincta; Grallaria hypoleuca; Grallaricula loricata; Pipreola lubomirskii; Pipreola whitelyi; Lipaugus streptophorus; Rupicola peruviana; Oxyruncus cristatus*; Chloropipo flavicapilla; Todirostrum russatum; Phyllomyias zeledoni; Phyllomyias plumbeiceps; Zimmerius viridiflavus; Elaenia dayi; Elaenia pallatangae; Mecocerculus minor; Phylloscartes poecilotis; Phylloscartes nigrifrons; Phylloscartes chapmani; Phylloscartes superciliaris; Platyrinchus flavigularis; Myiophobus flavicans; Myiophobus roraimae; Myiophobus pulcher; Contopus lugubris; Myiarchus cephalotes; Cyanolyca cucullata; Troglodytes rufulus; Myioborus castaneocapillus; Myioborus brunniceps; Myioborus albifacies; Myioborus cardonai; Basileuterus bivittatus; Basileuterus griseiceps; Melozone biarcuatum*; Melozone leucotis*; Atlapetes gutturalis; Atlapetes flaviceps; Atlapetes fuscoolivaceus; Atlapetes albofrenatus; Atlapetes personatus; Pselliophorus luteoviridis; Oreothraupis arremonops*; Pheucticus chrysopeplus; Chlorospingus parvirostris; Thlypopsis fulviceps; Mitrospingus oleagineus; Piranga leucoptera; Bangsia melanochlamys; Bangsia aureocincta; Iridosornis porphyrocephala; Euphonia elegantissima; Euphonia musica; Chlorochrysa nitidissima; Tangara xanthocephala; Tangara rufigenis; Tangara ruficervix; Tangara labradorides; Tangara cyanotis; Diglossa duidae; Diglossa major; Diglossopis glauca.

TEMPERATE ZONE

Metriopelia melanoptera*; Hapalopsittaca fuertesi; Aegolius ridgwayi; Caprimulgus saturatus; Panterpe insignis*; Eugenes fulgens*; Aglaeactis cupripennis*; Coeligena orina; Coeligena lutetiae; Ensifera ensifera*; Ramphomicron microrhynchum; Metallura iracunda; Andigena hypoglauca; Veniliornis nigriceps; Grallaria milleri; Grallaricula lineifrons; Acropternis orthonyx*; Contopus ochraceus; Muscisaxicola maculirostris; Cyanolyca turcosa; Eremophila alpestris*; Urothraupis stolzmanni*; Acanthidops bairdii*; Saltator cinctus; Conirostrum rufum; Conirostrum sitticolor; Hemispingus goeringi; Hemispingus verticalis; Buthraupis wetmorei; Anisognathus igniventris.

PÁRAMO

Phalcoboenus carunculatus*; Vanellus resplendens; Chalcostigma heteropogon*; Chalcostigma herrani*; Oxypogon guerinii*; Cinclodes fuscus*; Cinclodes excelsior*; Leptasthenura andicola*; Asthenes wyatti*; Asthenes flammulata*; Cnemarchus erythropygius*; Junco vulcani*; Phrygilus unicolor*

APPENDIX 2

List of Threatened and Near Threatened Bird Species Occurring in Five Montane Regions of Northern South America and Southern Central America. Status Is Based on Recent Information from Collar et al. (1992); t = Threatened, nt = Near Threatened. Approximate Minimum and Maximum Elevations (Meters) for Each Species in Each Region Are Provided and Are Based on Literature Accounts (See Methods for Description of Study Sites and References). Species Restricted to Montane Areas (i.e., Do Not Occur in Tropical Zone, See Text) Are Indicated by an Asterisk

Species	Status	Venezuelan Andes		Tepuis		Colombian Andes		Santa Marta		Chiriquí-Talamanca	
		Min.	Max.	Min.	Max.	Min.	Max.	Min.	Max.	Min.	Max.
ACCIPITRIDAE											
Accipiter collaris	nt	1,300	1,800			600	1,800	600	1,800		
Oroaetus isidori	nt	600	2,500			1,600	3,300	1,600	3,300		
FALCONIDAE											
Falco deiroleucus	nt	0	1,700			100	2,400				
CRACIDAE											
*Penelope perspicax	t					1,300	2,000				
Aburria aburri	nt					600	2,500	600	2,500		
Chamaepetes unicolor	nt									1,000	3,000
Pauxi pauxi	t	0	2,200			900	1,800				
ODONTOPHORIDAE											
Odontophorus atrifrons	nt					1,200	2,800	1,200	2,800		
*Odontophorus hyperythrus	nt					1,600	2,700				
*Odontophorus melanonotus	nt					1,200	1,800				
*Odontophorus strophium	t					1,500	1,800				
*Odontophorus columbianus	nt	1,300	2,400								
*Odontophorus leucolaemus	nt									750	1,850
RALLIDAE											
Neocrex colombianus	nt					0	2,100	1,500	2,100		
COLUMBIDAE											
*Leptotila conoveri	t					1,800	2,500				

APPENDIX 2
Continued

Species	Status	Venezuelan Andes Min.	Venezuelan Andes Max.	Tepuis Min.	Tepuis Max.	Colombian Andes Min.	Colombian Andes Max.	Santa Marta Min.	Santa Marta Max.	Chiriquí-Talamanca Min.	Chiriquí-Talamanca Max.
PSITTACIDAE											
*Leptosittaca branickii	t					1,800	3,500				
*Ognorhynchus icterotis	t					2,000	3,400				
*Pyrrhura viridicata	nt							2,000	2,500		
*Pyrrhura calliptera	t					1,700	3,400				
*Pyrrhura rhodocephala	nt	800	3,050								
*Bolborhynchus ferrugineifrons	t					3,000	3,800				
*Touit costaricensis	nt									500	3,000
Touit stictoptera	t					600	1,700				
*Hapalopsittaca amazonina	t	2,300	3,000			1,900	3,000				
*Hapalopsittaca fuertesi	t					3,100	3,650				
STRIGIDAE											
*Aegolius ridgwayi	nt									2,500	3,000
*Aegolius harrisii	nt	0	3,800								
CAPRIMULGIDAE											
*Caprimulgus whitelyi	nt			1,300	1,800	1,700	2,000				
TROCHILIDAE											
*Campylopterus phainopeplus	nt							1,200	4,800		
Amazilia castaneiventris	t					850	2,045				
Anthocephala floriceps	nt					600	2,300				
*Coeligena prunellei	t					1,400	2,600				
*Eriocnemis mirabilis	t					2,195	2,440				
*Eriocnemis derbyi	nt					2,500	3,600				
Haplophaedia lugens	t					1,100	2,500				
*Metallura iracunda	nt					2,800	3,100				
TROGONIDAE											
*Pharomachrus mocinno	nt									1,500	3,000
GALBULIDAE											
Galbula pastazae	t					1,000	2,100				

APPENDIX 2
CONTINUED

Species	Status	Venezuelan Andes Min.	Venezuelan Andes Max.	Tepuis Min.	Tepuis Max.	Colombian Andes Min.	Colombian Andes Max.	Santa Marta Min.	Santa Marta Max.	Chiriquí-Talamanca Min.	Chiriquí-Talamanca Max.
CAPITONIDAE											
Semnornis ramphastinus	nt					1,000	2,400				
RAMPHASTIDAE											
Andigena laminirostris	nt					300	3,200				
*Andigena hypoglauca	nt					2,700	3,400				
*Andigena nigrirostris	nt	1,800	2,700			1,600	3,200				
DENDROCOLAPTIDAE											
Campylorhamphus pucherani	nt					900	2,500				
FURNARIIDAE											
*Schizoeaca perijana	nt					3,000	3,400				
FORMICARIIDAE											
*Grallaria gigantea	t					2,300	3,000				
*Grallaria excelsa	nt	1,700	2,300			1,700	2,300				
*Grallaria alleni	t					2,000	2,100				
*Grallaria chthonia	t					1,800	2,100				
*Grallaria rufocinerea	t					2,100	3,100				
*Grallaria milleri	t					2,700	3,100				
*Grallaricula loricata	nt	1,440	2,100								
*Grallaricula lineifrons	nt					3,000	3,200				
*Grallaricula cucullata	t					1,500	2,700				
COTINGIDAE											
*Pipreola lubomirskii	nt					1,600	2,300				
Ampelioides tschudii	nt					650	2,700				
Cotinga ridgwayi	nt									0	1,850
Cephalopterus glabricollis	t									0	2,000
Cephalopterus penduliger	nt					700	1,800				
Procnias tricarunculata	nt									0	3,000

APPENDIX 2
Continued

Species	Status	Venezuelan Andes Min.	Venezuelan Andes Max.	Tepuis Min.	Tepuis Max.	Colombian Andes Min.	Colombian Andes Max.	Santa Marta Min.	Santa Marta Max.	Chiriquí-Talamanca Min.	Chiriquí-Talamanca Max.
TYRANNIDAE											
Polystictus pectoralis	nt					150	2,600				
*Contopus ochraceus	nt									2,200	3,000
*Myiotheretes pernix	nt							2,100	2,900		
CORVIDAE											
Cyanolyca pulchra	nt					900	2,300				
TROGLODYTIDAE											
*Cistothorus apolinari	t					2,500	4,000				
PARULIDAE											
*Basileuterus griseiceps	t	1,200	1,600								
Basileuterus cinereicollis	nt	1,100	2,100			800	1,800				
EMBERIZIDAE											
*Atlapetes leucopis	nt					2,300	3,000				
Atlapetes flaviceps	t					1,300	2,100				
*Atlapetes fuscoolivaceus	nt					1,600	2,400				
*Oreothraupis arremonops	t					1,700	2,300				
*Acanthidops bairdii	nt									2,100	3,000
*Saltator cinctus	nt					2,500	3,000				
THRAUPIDAE											
*Oreomanes fraseri	nt					3,000	4,500				
*Hemispingus goeringi	t	2,600	3,200								
*Buthraupis wetmorei	t					3,300	3,450				
Chlorochrysa nitidissima	t					900	2,000				
ICTERIDAE											
Hypopyrrhus pyrohypogaster	t					1,200	2,700				
*Macroagelaius subalaris	nt					950	3,100				

WHAT IS THE CLOSEST LIVING RELATIVE OF *CATHAROPEZA* (PARULINAE)?

MARK B. ROBBINS[1] AND THEODORE A. PARKER III[2,3]

[1]*Division of Ornithology, Museum of Natural History, University of Kansas, Lawrence, Kansas 66045, USA; and*
[2]*Museum of Natural Science, Louisiana State University, Baton Rouge, Louisiana 70803, USA*
[3]*Deceased*

ABSTRACT.—It has been proposed that the closest relatives of the endemic Whistling Warbler (*Catharopeza bishopi*) are the Plumbeous (*Dendroica plumbea*), Arrow-headed (*Dendroica pharetra*), and the Elfin Woods (*Dendroica angelae*) warblers. We present vocal, plumage, structural, and behavioral data that indicate that *Catharopeza* is not closely related to *Dendroica*; these data suggest that the closest living relative of *Catharopeza* is the *Phaeothlypis* group of the basileuterine warblers.

RESUMEN.—Se ha propuesto que los grupos filogenéticamente más cercanos a *Catharopeza bishopi* son *Dendroica plumbea*, *D. pharetra* y *D. angelae*. En este escrito, mostramos caracteres de vocalización, plumaje, estructura y comportamiento sugieren que el grupo más cercano a *Catharopeza* es el grupo *Phaeothlypis*, que se considera parte de *Basileuterus*. Se presenta la sugerencia que *D. plumbea* no es un miembro de la misma superespecie como *D. pharetra* y *D. angelae*.

The Whistling Warbler (*Catharopeza bishopi*), endemic to the island of St. Vincent in the Lesser Antilles, traditionally has been considered to be closely related to the genus *Dendroica*, with several authors suggesting that it be merged into that genus (Ridgway 1902; Bond 1956; Kepler and Parkes 1972). Although historically placed at the end of the *Dendroica* following the Plumbeous (*D. plumbea*) and Arrow-headed (*D. pharetra*) warblers, Kepler and Parkes (1972), in their description of the Elfin Woods Warbler (*D. angelae*), were the first to state specifically that *Catharopeza* is most closely allied with the *Dendroica plumbea* complex (*D. plumbea*, *D. pharetra*, and *D. angelae*).

Kepler and Parkes (1972) based their recommendation primarily on adult plumage patterns in *D. angelae* and *C. bishopi* claiming that the then newly discovered *angelae* bridged the gap between *D. pharetra* and *C. bishopi*. In particular, they underscored what they felt was a similar facial pattern shared by *D. angelae* and *C. bishopi*, as well as similarities in song patterns. Nevertheless, they noted that *C. bishopi* has a unique ventral pattern, and that it is brown in the first basic plumage, rather than green as in *plumbea*, *pharetra*, and *angelae*. Although Bond (1972) endorsed Kepler and Parkes' recommendation, he pointed out that the white eye-ring of *C. bishopi* was the only character shared with *D. angelae*. Earlier, Bond (1928) had noted how different the voice of *Catharopeza* was from other *Dendroica*, and he even remarked that components of the song were similar to the song of the Mourning Warbler (*Oporornis philadelphia*). Andrle and Andrle (1976) disagreed with Kepler and Parkes' assertion that the song of *C. bishopi* was similar to that of *D. angelae*, stating that "no sound basis exists for their assertion that the songs of the two species have a remarkably similar pattern." Because of its unique morphological and vocal attributes, Andrle and Andrle (1976) retained *Catharopeza* as a separate genus, a recommendation followed by the A.O.U. (1983).

We first became familiar with *Catharopeza bishopi* while leading a tour to St. Vincent on 9 May 1988. During the course of listening to an apparent adult male sing for ca. 20 minutes, Parker noted how similar the song was to that of the River Warbler (*Phaeothlypis* [*Basileuterus*] *rivularis*). Although recognition of the genus *Phaeothlypis* is controversial (Ridgely and Tudor 1989), for convenience and to underscore what may be the closest living relatives of *C. bishopi* (i.e., the sibling species pair *P. fulvicauda* and *P. rivularis*), we follow the A.O.U. (1983) in recognizing *Phaeothlypis*. We believe that the following vocal, plumage, structural, behavioral, and distributional characters refute the notion that *C. bishopi* is allied with the *D. plumbea*

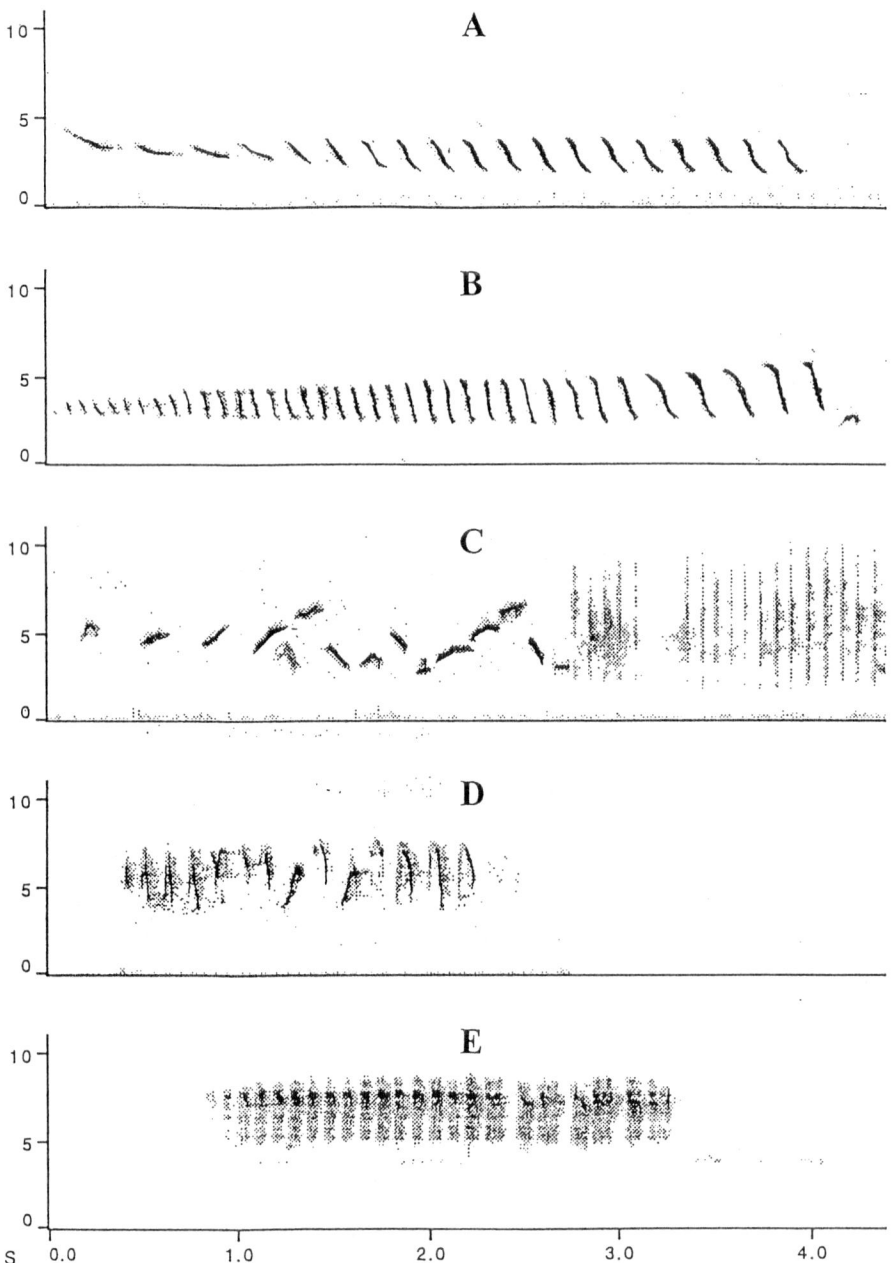

FIG. 1. Sound spectrograms of songs of *Catharopeza bishopi* and possible close relatives. Y-axis represents frequency in kilohertz (kHz). X-axis represents time in seconds (S). (a) *Phaeothlypis rivularis*, Las Claritas, Bolivia; recorded by B. Behrstock; (b) *Catharopeza bishopi*, St. Vincent; recorded by R. Andrle; (c) *Dendroica plumbea*, Dominica; recorded by Robbins; (d) *D. pharetra*, Jamaica; recorded by R. Sutton; (e) *D. angelae*, Puerto Rico; recorded by C. Kepler. All recordings from Hardy et al. (1994).

complex, and better support the hypothesis that the closest living relative of *Catharopeza* is the *Phaeothlypis* assemblage.

Vocalizations.—Parker's assessment was confirmed when the song of *Catharopeza bishopi* was compared to that of its purported closest *Dendroica* relatives (*D. plumbea* complex) and with *Phaeothlypis rivularis* and *P. fulvicauda*. As can be heard on Hardy et al. (1994) and seen in Figure 1, the song of *C. bishopi* is strikingly similar to songs of *Phaeothlypis*, but bears little or no resemblance to those of any member of the *D. plumbea* complex or to any other member of *Dendroica*. Both *C. bishopi* and the *Phaeothlypis* group have loud, ringing songs, with the rapidly delivered whistled notes increasing in volume and frequency (Andrle and Andrle 1976; Ridgely and Tudor 1989; Hardy et al. 1994; Fig. 1). The shape and frequency range of each element is remarkably similar between the songs of *Catharopeza* and both species of *Phaeothlypis*.

We recognize the possibility that the vocal similarities between *Catharopeza* and *Phaeothlypsis* may be convergent in response to features of sound transmission in similar habitats, given that *Catharopeza* is often found on slopes of mountain streams, and the *Phaeothlypis* complex primarily inhabits riparian areas. There are certainly other examples of apparent song convergence in unrelated taxa within the Parulinae. For example, the distantly related Cerulean Warbler (*Dendroica caerulea*) and Northern Parula (*Parula americana*) coexist in the canopy of riparian forest and have confusingly similar songs (compare song of *D. caerulea* and Type 2 song [Borror and Gunn 1985; referred to as Type B by Moldenhauer 1992] of *P. americana*). As demonstrated in Figure 1, we propose that the frequency, tempo, and the structure of the songs of *Catharopeza bishopi* and the *Phaeothlypis* group indicate a common ancestor, and thus are not the result of convergent evolution. The following additional characters support this conclusion.

Morphology and coloration.—The brown first basic plumage of *C. bishopi* is totally unlike the green first basic plumage of the *D. plumbea* complex or any other *Dendroica* (see Curson et al. [1994] for renditions of these plumages). Birds that appear in the brown first basic plumage of *C. bishopi* lack the wing bars and narrow superciliums in first basic plumage of *D. plumbea*, *D. pharetra*, and *D. angelae*. Immature *C. bishopi* appear most similar to adult *Phaeothlypis fulvicauda leucopygia* and juveniles of the entire basileuterine assemblage. Dorsally, these taxa are unmarked, and uniformily dark in color. Immature *C. bishopi* are chocolate brown, whereas adults of *P. f. leucopygia* are brownish-olive. Adult and immature *C. bishopi* lack wing bars as do all age classes of the basileuterine assemblage. With the single exception of the Black-throated Blue Warbler (*Dendroica caerulescens*), all age classes of *Dendroica* species possess wing bars.

Contrary to Kepler and Parkes (1972), we do not believe that even the facial pattern in adult *Catharopeza bishopi* is similar to that of species in the purported *D. plumbea* complex. Like virtually all features of its adult plumage, the eye-ring of *Catharopeza* is unique in the Parulinae in its width and completeness. In comparison to the *D. plumbea* group, the white loral spot of *Catharopeza* is indistinct. Moreover, loral spots are present throughout the majority of paruline genera, so we do not consider it a phylogenetically informative character.

The relatively long, yellow tarsi of *C. bishopi* are notable even under field conditions (pers. obs.; Table 1). Both these features are characteristic of *Phaeothlypis*, but not of a single member of *Dendroica*. Unfortunately, no unique skeletal characters offer obvious clues as to the relationships of *Catharopeza* (J. D. Webster, in litt.;, pers. obs).

Kepler and Parkes (1972) also noted that *C. bishopi* was larger in body mass than all species of *Dendroica*. Dunning's (1993) summarized body mass data underscore the dramatic differences between *Catharopeza* and its purported close relatives: *Catharopeza* ($\bar{x}$ = 17.2 g), *D. plumbea* ($\bar{x}$ = 11 g), *D. pharetra* ($\bar{x}$ = 10.3 g), and *D. angelae* ($\bar{x}$ = 8.4 g). No members of a paruline superspecies complex exhibit a two-fold difference in body mass. Interestingly, body mass of *P. fulvicauda* ($\bar{x}$ = 14.9 g) and *P. rivularis* ($\bar{x}$ = 13.5 g) are closer to *Catharopeza*.

Behavior.—Andrle and Andrle (1976) documented that *Catharopeza bishopi* spends at least 90% of its time within 4 m of the ground. Because of density of the vegetation, they were unable to determine how much time it spends on the ground, but Andrle and Andrle (1976) commented that "it is likely that they are on or close to it [the ground] much of the time," and they observed it foraging "about boulders and large rocks along streams." Lister (1880) commented that it foraged frequently along streams. The propensity of *C. bishopi* for foraging near the ground and at least occasionally along streams certainly is more characteristic of *Phaeothlypis* (and most other members of the basileuterine assemblage) than of any species of *Dendroica* (Ridgely 1976; Hilty and Brown 1986; Curson et al. 1994;, pers. obs.).

Catharopeza bishopi typically sings and forages with its tail cocked above the plane of the body. Andrle and Andrle (1976) described the tail movements of *Catharopeza* as follows: "In the typical 'cocked tail' position, the warbler holds its tail about 75–85° above the horizontal. Frequently, but at varying intervals depending on its position, the bird flicks its tail in a rapid up-down movement thorough an arc of 20–30° so that the tail often reaches the vertical." Although tail flicking and tail wagging are not uncommon in the *Dendroica*, no species, including any in the *D. plumbea* complex (all of which constantly flick their tails), has tail movements approaching the exaggerated motions of *C. bishopi*.

TABLE 1
Means (±SD) of Morphogical Measurements of Adults of Selected Warbler Species

Species	n	Exposed culmen	Bill width[1]	Tarsi	Wing chord	Tail[2]
Phaeothlypis fulvicauda						
cis-Andean[3]						
male	10	10.3 (0.4)	4.9 (2.9)	20.6 (0.6)	63.9 (2.1)	48.8 (2.0)
female	6	10.4 (0.5)	4.9 (2.5)	21.1 (0.6)	62.1 (2.1)	49.2 (2.4)
trans-Andean[4]						
male	6	10.5 (0.4)	4.8 (0.2)	22.2 (1.1)	63.5 (1.9)	49.1 (2.2)
female	6	10.2 (0.2)	4.8 (0.1)	21.7 (0.4)	60.2 (1.5)	47.2 (1.4)
P. rivularis[5]						
male	8	10.9 (0.4)	4.6 (0.2)	23.2 (0.8)	63.8 (1.9)	57.2 (1.4)
female	5	10.5 (0.3)	4.9 (0.2)	22.7 (0.9)	60.1 (0.5)	54.4 (1.5)
Catharopeza bishopi						
male	7	12.0 (0.4)	4.5 (0.1)	22.9 (2.0)	67.9 (1.2)	52.4 (1.6)
female	1	11.6	4.6	23.1	69.5	53.3
Dendroica plumbea[6]	24	10.1 (0.4)	3.8 (0.2)	19.6 (0.7)	62.4 (1.7)	53.3 (1.7)
D. pharetra						
male	2	10.9 (0.6)	3.5 (0.0)	18.4 (0.8)	65.2 (1.4)	52.0 (1.4)
female	1	10.0	3.4	18.3	61.9	50.7
D. angelae						
male	3	9.2 (0.2)	3.4 (0.1)	15.7 (0.2)	54.6 (0.6)	43.1 (1.7)
female	1	10.0	3.2	15.9	50.7	40.4

[1] Midpoint of nares.
[2] For central rectrices.
[3] For birds from western Amazonia.
[4] For birds from Central America and west of the Andes.
[5] For birds from southeastern Brazil.
[6] Sexes pooled; no significant differences ($P < 0.05$).

Biogeography.—A *Catharopeza*/*Phaeothlypis* relationship coincides better with biogeographic patterns seen in other extant West Indian taxa than a *Catharopeza*/*Dendroica* relationship. We are aware of no West Indian bird species or species group sharing the distribution pattern of the latter proposed relationship. However, several taxa that have clearly originated in South America and dispersed into the Lesser Antilles have populations or sister taxa that have their northernmost limits on St. Vincent. This distributional pattern is shared by ecologically distinct taxa representing five avian families: Short-tailed Swift (*Chaetura brachyura*), Yellow-bellied Elaenia (*Elaenia flavogaster*), Bare-eyed Thrush (*Turdus nudigenis*; although this species has recently expanded to islands north of St. Vincent), Cocoa Thrush (*Turdus fumigatus*), and the Lesser Antillean Tanager (*Tangara cucullata*). Moreover, the South American affinities of the southern Lesser Antillean fauna have been documented in several non-avian groups including butterflies (Brown 1978; Miller and Miller 1989), bats (Baker and Genoways 1978; Jones 1989), and reptiles and amphibians (Schwartz 1978). Although this biogeographic pattern is not positive evidence for the relationship between *Catharopeza* and *Phaeothlypis*, it demonstrates that this proposed relationship is certainly biogeographically plausible.

Conclusion.—We believe that there is no evidence that indicates *Catharopeza* is a member of *Dendroica*. We consider that the above similarities of song, plumage, structure, body mass, and behavior indicate a close relationship between *Catharopeza* and *Phaeothlypis*. This hypothesis is also congruent with biogeography. It is hoped that the above evidence will catalyze researchers to test our hypotheses and to determine generic limits within the basileuterine assemblage (that is surely not monophyletic). In particular, future work should focus on whether the expanded *Phaeothlypis*, as outlined by Ridgely and Tudor (1989), is monophyletic. Given the present evidence, we recommend that the genus *Catharopeza* continue to be recognized, and we suggest that it would best be placed *genus incertae sedis* in the taxonomic sequence.

ACKNOWLEDGMENTS

We thank the following people and institutions for the loan of material under their care: Frank Gill and David Agro (Academy of Natural Sciences, Philadelphia), Van Remsen and Steve

Cardiff (Louisiana State Museum of Natural Science), and Gary Graves and James Dean (National Museum of Natural History, Smithsonian). Dan Webster kindly shared his manuscript on paruline osteology. Kimball Garrett, Kenneth Parkes, Town Peterson, Rick Prum, and Van Remsen made helpful comments on the manuscript.

LITERATURE CITED

AMERICAN ORNITHOLOGISTS' UNION. 1983. Check-list of North American Birds. 6th edition. Allen Press, Lawrence, Kansas.
ANDRLE, R. F. AND P. R. ANDRLE. 1976. The Whistling Warbler of St. Vincent, West. Indies. Condor 78: 236–243.
BAKER, R. J. AND H. H. GENOWAYS. 1978. Zoogeography of Antillean bats. Pages. 53–98 in Zoogeography in the Carribean (F. B. Gill, Ed.). Academy of Natural Sciences of Philadelphia.
BOND, J. 1928. On birds of Dominica, St. Lucia, St. Vincent, and Barbados, B.W.I. Proc. Academy of Natural Sciences of Philadelphia. 80:523–545.
BOND, J. 1956. Check-list of Birds of the West Indies, 4th ed. Academy Natural Sciences of Philadelphia.
BOND, J. 1972. Seventeenth Supplement to the Check-list of Birds of the West Indies (1956). Academy of Natural Sciences of Philadelphia.
BORROR, D. J. AND W. W. GUNN. 1985. Songs of the Warblers of North America. Cornell Library of Natural Sounds, Cornell University, Ithaca, New York.
BROWN, F. M. 1978. The origins of the West Indian butterfly fauna. Pages 5–30, in Zoogeography in the Caribbean (F. B. Gill, Ed.). Academy of Natural Sciences of Philadelphia.
CURSON, J., D. QUINN, AND D. BEADLE. 1994. Warblers of the Americas. Houghton Mifflin Co., New York.
DUNNING, J. B., JR. 1993. CRC Handbook of Avian Body Masses. CRC Press, Ann Arbor, Michigan.
HARDY, J. W., B. B. COFFEY, AND G. B. REYNARD. 1994. Voices of the Neotropical Wood Warblers. ARA Records, Gainesville, Florida.
HILTY, S. L. AND W. L. BROWN. 1986. A Guide to the Birds of Colombia. Princeton University Press, Princeton, New Jersey.
JONES, J. K., JR. 1989. Distribution and systematics of bats in the Lesser Antilles. Pages 545–660 in Biogeography of the West Indies, Past, Present, and Future (C. A. Woods, Ed.). Florida Museum of Natural History, Gainesville, Florida.
KEPLER, C. B. AND K. C. PARKES. 1972. A new species of warbler (Parulidae) from Puerto Rico. Auk 89: 1–18.
LISTER, C. E. 1880. Field-notes on the birds of St. Vincent, West Indies. Ibis 4:38–44.
MILLER, L. D. AND J. Y. MILLER. 1989. The biogeography of West Indian butterflies (Lepidoptera: Papilionoidea, Hesperioidea): a vicariance model. Pages 229–262 in Biogeography of the West Indies, Past, Present, and Future (C. A. Woods, Ed.). Florida Museum of Nat.Hist., Gainesville, Florida.
MOLDENHAUER, R. R. 1992. Two song populations of the Northern Parula. Auk 109:215–222.
RIDGELY, R. S. 1976. A Guide to the Birds of Panama. Princeton University Press, Princeton, New Jersey.
RIDGELY, R. S. AND G. TUDOR. 1989. The Birds of South America. Vol. 1. The Oscine Passerines. University of Texas Press, Austin.
RIDGWAY, R. 1902. The birds of North and Middle America, part 2. U.S. Natl. Mus., Bull. 50.
SCHWARTZ, A. 1978. Some aspects of the herpetogeography of the West Indies. Pages 31–51 in Zoogeography in the Caribbean (F. B. Gill, Ed.). Academy of Natural Sciences of Philadelphia.

VOICE AND TAXONOMY OF *CAPRIMULGUS* (*RUFUS*) *OTIOSUS* (CAPRIMULGIDAE), WITH A REEVALUATION OF *CAPRIMULGUS RUFUS* SUBSPECIES

MARK B. ROBBINS[1,2] AND THEODORE A. PARKER III[3,4]

[1]*Department of Ornithology, Academy of Natural Sciences,*
1900 Benjamin Franklin Parkway, Philadelphia, PA 19103, USA
[2]*Present address: Division of Ornithology, Museum of Natural History,*
University of Kansas, Lawrence, Kansas 66045, USA
and [3]*Museum of Natural Science,*
Louisiana State University, Baton Rouge, LA 70803, USA
[4]*Deceased*

ABSTRACT.—The St. Lucia Nightjar (*Caprimulgus otiosus*) exhibits minimal vocal and plumage differentiation from mainland forms of the Rufous Nightjar (*C. rufus*), and, therefore, we consider *otiosus* conspecific with *rufus*. *Caprimulgus rufus otiosus* appears to be resident on the northeastern section of St. Lucia, Lesser Antilles; specimen records from Venezuela are considered erroneous. We recognize the following described subspecies: *otiosus* (St. Lucia), *minimus* (Costa Rica to N. Venezuela), *rufus* (NW Brazil, S. Venezuela, east through the Guianas, and south to south of the Río Amazonas), and *rutilus* (S. Brazil, N. Argentina, E. Bolivia).

RESUMEN.—Las vocalizaciones y el plumaje del tapacaminos de Santa Lucía (*Caprimulgus otiosus*) y de la forma continental *C. rufus*, son muy parecidas, por lo que se les considera como conespecíficos. *Caprimulgus rufus otiosus* aparentemente es residente de la sección noreste de Santa Lucía en las Antillas Menores; consideramos que los registros de Venezuela son erróneos. Nosotros reconocemos las siguientes subespecies descritas: *otiosus* (Santa Lucía), *minimus* (de Costa Rica al N de Venezuela), *rufus* (NW Brazil, S Venezuela, E de las Guyanas hasta el lado sur del Río Amazonas), y *rutilus* (S Brazil, N Argentina y E Bolivia).

In the latest A.O.U. Check-list (1983), the Lesser Antillean form (*otiosus*) of the wide-ranging Rufous Nightjar (*Caprimulgus rufus*) was elevated to species status. Previously, the St. Lucia Nightjar (*C. otiosus*) had been regarded as an isolated subspecies of *C. rufus* by most authors (Bangs 1911; Cory 1918; Griscom and Greenway 1937; Peters 1940; Bond 1947, 1959, 1977), although Wetmore and Phelps (1953) believed that *otiosus* deserved specific recognition. The A.O.U. (1983) gave the following rationale for treatment of *otiosus* as a species, "... it seems best to retain *C. otiosus* as specifically distinct until its status is determined." Herein we demonstrate that *otiosus* shows minimal vocal and plumage differentiation from mainland forms of *C. rufus*, and is best treated as a subspecies of *C. rufus*.

METHODS AND MATERIALS

Study skins were accumulated at the Academy of Natural Sciences, Philadelphia (ANSP) for plumage color and pattern comparisons. Wing chord and length of central rectrices were taken using dial calipers to the nearest 0.1 mm. All tape recordings of *otiosus* were made on the windward side of St. Lucia, West Indies on 7 May 1988 and 29 April 1989. Tape recordings were made using a Sony TCM 5000 cassette recorder, with a Sennheiser ME 80 shotgun microphone. Additional recordings were obtained from Hardy et al. (1988). Sonagrams were produced with "SoundEdit" of Farallon Computing, Inc., Emeryville, California, and "Canary" of the Bioacoustics Research Program at the Cornell Laboratory of Ornithology, Ithaca, New York.

TABLE 1

Mean (±SD) for Wing and Tail Measurements of *Caprimulgus rufus* Subspecies (Excludes Specimens from Central Brazil)

Subspecies	Sex	N	Wing (chord)	Tail (central rectrix)
otiosus				
	male	8	187.9 (±4.0)	122.4 (±3.3)
	female	4	183.9 (±6.0)	119.1 (±2.7)
rufus				
	male	14	170.1 (±4.4)	116.8 (±3.1)
	female	7	167.4 (±4.4)	112.4 (±4.3)
minimus				
	male	6	171.8 (±6.3)	110.6 (±5.4)
	female	13	175.5 (±3.5)	112.9 (±3.9)
rutilus				
	male	7	185.5 (±3.8)	126.8 (±3.3)
	female	9	180.6 (±2.8)	128.1 (±3.8)

RESULTS

Voice.—Bond (1947) was the first to describe the voice of *otiosus*; he stated that its voice was very similar to that of the Chuck-will's-widow (*Caprimulgus carolinensis*). Apparently no sound-recordings were available of *otiosus* until we recorded it on the windward side of St. Lucia in May 1988. Comparison of these recordings to those of all other described forms of the Rufous Nightjar demonstrates the striking similarity in this aspect among all populations; we cannot distinguish the song of *otiosus* from those of other *C. rufus* populations (Fig. 1; listen to voice of *C. rufus* on Hardy et al. 1988).

Morphometrics and plumage coloration.—Wing length and plumage color and pattern have been used to characterize subspecies within this complex. But, as Griscom and Greenway (1937) pointed out, much overlap exists in wing length, and purported plumage characters are not diagnostic among some subspecies. Indeed, our comparison of material from most of the range of *C. rufus* revealed confusion on the characterization, distribution, and nomenclature of the described forms.

Our examination of 71 specimens reveals that St. Lucian *otiosus* and southern *rutilus* (S. Brazil, N. Argentina, E. Bolivia) have significantly longer wings and tails than the nominate (NW Brazil, S. Venezuela east through the Guianas, and south of the Río Amazonas) subspecies and *C. r. minimus* (Costa Rica to N. Venezuela) (Table 1). Two morphometric characters separate *otiosus* from *rutilus*: length of the buffy spot on the outer rectrix in males ($x = 58.7 \pm 4.4$ in *otiosus*, $x = 43.6 \pm 5.1$ in *rutilus*; two-tailed, $P < 0.001$) and tail length (two-tailed, $P < 0.05$ in males; $P < 0.001$ in females; Table 1). Male *rufus* and *minimus* differ in tail length (two-tailed, $P < 0.01$; Table 1). However, females of *minimus* and *rufus* are indistinguishable based on tail length, but their wing lengths differ significantly (two-tailed, $P < 0.01$; Table 1).

Griscom and Greenway (1937) noted that birds from the Guianas and northeastern and central Brazil, considered by them to represent the nominate race, are more rufous than other subspecies. Since Griscom and Greenway's review, birds from the southern states of Amazonas and Bolívar in Venezuela have been described (*C. r. noctivigularis*) as even more rufous than the nominate race (Wetmore and Phelps 1953). However, Wetmore and Phelps (1953) did not compare their southern Venezuelan specimens with the holotype of *rufus*; their comparison was limited to birds presumed to be of the nominate race from south of the Río Amazonas in Brazil (Pará, Goias, and Bahia).

While examining *C. rufus* specimens in the British Museum of Natural History, Tring (BMNH), Nigel Cleere (pers. comm.) found a male (BMNH 95.11.27.134) from the Rupununi River, Guyana, identified as *C. r. noctivigularis*; he noted that it was "much darker and more rufescent" than specimens labeled as the nominate race from south of the Río Amazonas, Brazil (Pará, Bahia, Pernambuco). We compared this Guyana specimen and the only specimen known from Surinam (National Natuurhistorisch Museum, Leiden; NNML 7964; Mees 1968) to *C. rufus* specimens in United States museums. Both the Guyana and Surinam specimens are virtually indistinguishable from those of the upper tributaries of the Río Negro, extreme northwestern

Brazil (American Museum of Natural History, New York; AMNH), and one from Bolívar, Venezuela (United States National Museum of Natural History, USNM 389293; a paratype of *C. r. noctivigularis*). A series of specimens from Santarém, Brazil, at the mouth of the Río Tabajos (Carnegie Museum of Natural History, CMNH), is referable to the nominate race, although these birds average slightly larger in both wing and tail measurements than birds north of the Amazon. All these birds differ from specimens farther south of the Amazon (Bahia, Goias) by being much darker and having a much darker rufous coloration.

Apparently the holotype of *C. r. rufus* is no longer extant. Richard Prum (pers. comm.) could not locate it during a visit to the Paris Museum in February 1994, and Gerlof Mees (*in litt.*, January 1994) concurred stating, "Chances that the type still exists as a specimen are virtually nil . . .". Examination of the crude color plate (copy of Daubenton [1783] at the ANSP) that depicts the holotype of *C. r. rufus* indicates a very rufescent bird that seems to match better the specimens from Surinam, Guyana, southern Venezuela, and northern Brazil than specimens from south of the Amazon. Indeed, the bird depicted is so rufescent that it has incorrectly been identified as the Rufous Potoo (*Nyctibius bracteatus*) (see comments in Sclater 1866). Had Wetmore and Phelps (1953) compared their southern Venezuelan birds to some of the above material they surely would have realized that the Venezuelan Amazonas and Bolívar birds were typical examples of the nominate race. Therefore, we consider *C. r. noctivigularis* a synonym of *C. r. rufus*.

Specimens from central Brazil (Goias, Bahia) are not assignable either to nominate or to southern *rutilus*. For example, we examined five specimens (AMNH 163156, 477303; BMNH 90.2.18.110, 52.3.8.5; ANSP 21903) that have the plumage color and pattern characters of *rutilus*, but measurements that fall within the range of variation of the nominate race. Three of these are labelled as coming from "Bahia", whereas one is from Goias, and the other is without locality information. More confusing is another "Bahia" specimen (BMNH 88.8.1.73) that both in color and in measurements is clearly referable to the nominate race. This latter specimen may be mislabelled as to locality.

Another clarification that needs to be made is the identity of two types that were overlooked by Peters (1940). Pelzeln's (1871) description of *Antrostomus cortapau* was based on at least nine specimens (fide E. Bauernfeind, *in litt.*). The designated types were two males collected by J. Natterer at Engenho do Gama, Mato Grosso (southeast of the town of Mato Grosso; Paynter and Traylor 1991). Bauernfeind's measurements and plumage descriptions of the two cotypes indicate that *cortapau* should be synonymized with Burmeister's (1856) *rutilus*. Griscom and Greenway (1937) considered Sclater's (1866) *Antrostomus ornatus* to be a synonym of *rutilus*. Measurements and comparison of Sclater's type with other BMNH specimens confirm Griscom and Greenway's treatment (fide N. Cleere, *in litt.*).

Finally, the convoluted taxonomic status of *Caprimulgus saltarius* is germane to *C. rufus*. Olrog (1979) originally described this taxon as a race of *C. sericocaudatus*, not *C. rufus* as stated by Hardy et al. (1988). Nores and Yzurieta (1984) were convinced that *saltarius* was more closely allied to *rufus*, initially thinking that it was no more than a color phase. Nonetheless, because they believed *saltarius* was sympatric with *C. rufus* in northwestern Argentina they elevated *saltarius* to species rank. At that time the voice of *saltarius* was still unknown. Later, Robert Straneck determined that the voice and morphometrics of *saltarius* were very similar to *rufus*; he concluded that *saltarius* was no more than a gray color phase of *rufus* (*in litt.* to N. Cleere, pers. comm.). Therefore, we consider *saltarius* to be a synonym of *rutilus*.

Status, distribution, and conservation of otiosus.—All but two records of *C. r. otiosus* are from the dry, scrubby woodland of the northeastern coast of St. Lucia. Danforth (1935) mentioned that Bond heard this species at Anse La Raye in the southwestern part of the island in arid scrub in early June 1929. The only other non-windward side record is of "many" heard in humid forest above the valley between Piton Flore and La Sorciére in the "spring" of 1976 by W. Gladfelter (Bond 1977). The latter record is odd, because *otiosus* otherwise is known only from dry, scrubby forest. There are no other records from this area, despite it having been surveyed by others.

Babbs et al. (1988) searched unsuccessfully for *C. r. otiosus* between mid-June and mid-August 1987, along the leeward side of St. Lucia at Anse Mamin, Savannes Bay, and the Edmund Forest Reserve, and we failed to find it in dry scrubby forest on 5 May 1988, just northwest of Soufriere in the southwestern part of the island. The survey by Babbs et al. (1988) found that *otiosus* was restricted to the lowland, dry scrub woodland on the windward side of the island between Petite Anse and the Louvet River, a distance of only ca. 6 km. At dusk on 7 May 1988, between the Louvet River and Caille Des, we heard and recorded a minimum of 12 birds. Our roadside census undoubtedly underestimated the number of birds in this area. Parker returned to the latter area at dusk on 29 April 1989 and heard at least 8 birds, with an additional 2–3 birds heard below Desbarra. One bird approached the group after a prerecorded tape, ca. 30 s in duration, of mainland *C. rufus* was played. Another individual also responded strongly to tapes of *C. r. otiosus*, but less so to tapes of mainland *C. rufus*. During 1992, just south of Grande Anse, Wayne Burke (pers. comm.) heard three birds on the evening of 11 April, and on 1 May 1992 Armas Hill (*in litt.*) heard a minimum of 8 birds singing there at twilight (ca. 18:00).

The principal vocal period of *C. r. otiosus* appears to be from late February through June (J. Semper in Danforth 1935; W. Burke, pers. comm.;, pers. obs.). It has been heard near Louvet in February (no date specified; Bond 1986). At least five birds were heard on 27 February 1992, near Caille Des by Burke (*in litt.*), and six responded to tape playback under a full moon near Caille Des on 4 February 1993 (F. Vilella, C. Cox, D. Anthony and L. Jones; Burke, pers. comm.). The only known nest, with two eggs, was found on 26 June 1932 near Petite Anse (S. John in Danforth 1935). No information exists on the vocal activity of *otiosus* from July through mid-October (see below).

Much speculation still remains on whether *C. r. otiosus* is a permanent resident. Our queries of the local people in the Louvet area brought conflicting reports on the year-round presence of this population. Intriguingly, of the few specimens collected, a female was taken on 12 December 1934 (ANSP 108389), and a male was collected on 25 February 1901 (AMNH 477311); there are also the above February records of vocalizing birds. If *otiosus* is migratory, it would be extraordinary for it to migrate to St. Lucia from December through February, given that no other West Indian bird is known to migrate during those months. On the other hand, Burke (*in litt.*) made two unsuccessful attempts to find *otiosus* between Desbarra and Louvet in fall 1992. He used prerecorded tapes of *otiosus* under seemingly ideal conditions (light wind, no precipitation) on 16 October and 6 November. Whether *otiosus* would respond to prerecorded tapes at this time of year is unknown. We underscore that care must be exercised in identifying non-vocalizing nightjars on St. Lucia from September through April, as it would be almost impossible to distinguish migrating or wintering Chuck-will's-widows (*C. carolinensis*) from *C. r. otiosus* in the field.

Wetmore and Phelps (1953) stated that three specimens (two were taken in different months at the same locality) collected in northern Venezuela were referable to *C. r. otiosus*. We have not examined these specimens as they are deposited in the Phelps Collection in Caracas; hence, the following interpretation is based solely on what they wrote. Unfortunately, diagnostic plumage measurements, wing and tail length, could not be taken as the two males were in molt (one male's tail was not in molt) and the female lost both remiges and rectrices during preparation. Therefore, Wetmore and Phelps relied on plumage coloration and the tail length of the one male to separate these three specimens from *C. r. minimus*; they believed that these specimens were duller than *minimus*. Yet, they also commented that the three specimens appeared darker than St. Lucian specimens. The female was taken on 22 April with enlarged ova, which lead Wetmore and Phelps to conclude that *otiosus* was a resident in northern Venezuela. It was this latter specimen that convinced them that *otiosus* should be considered a species, based on the assumption that both *otiosus* and *minimus* breed sympatrically in northern Venezuela. However, as stated above, the lack of flight feathers of this specimen precludes subspecific identification.

We believe that a more plausible explanation of the identity of the three purported "*otiosus*" Venezuelan specimens is that they are either *minimus* or less likely an undescribed resident subspecies. It is inconceivable that *otiosus* and *minimus* could breed sympatrically in Venezuela because their voices are indistinguishable (Fig. 1). This lack of a mechanism for assorted mating would presumably result in extensive hybridization between *otiosus* and *minimus* that would swamp out average differences in wing and tail length (Table 1). As mentioned above, Wetmore and Phelps recognized that the three Venezuelan birds did not match the St. Lucian birds. Furthermore, from the limited samples of *minimus* that we have examined, it appears that the Venezuelan population may be longer-winged and longer-tailed than Costa Rica and Panama populations (the Colombia birds seem intermediate in size). Wetmore and Phelps (1953) even acknowledged that the Venezuelan birds may be separable from other *minimus*. Thus, there is no conclusive evidence for *otiosus* occurring in Venezuela.

Finally, although St. Lucia *otiosus* has relatively long wings and tail like southern *rutilus* (Table 1), we are unable to find convincing evidence that *otiosus* is migratory. However, southern Brazilian and Argentinian populations of *rutilus* are apparently migratory (Belton 1984; Parker, pers. obs.), and it is conceivable that some *rutilus* may spend the months of May through August as far north as Venezuela, not unlike other austral migrants, such as the Fork-tailed Flycatcher

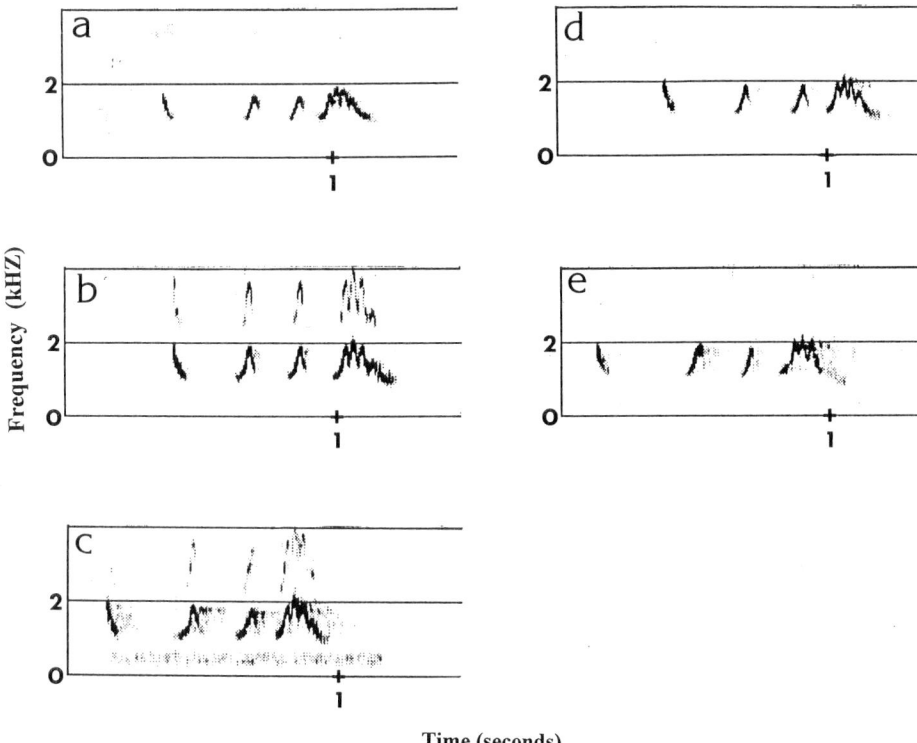

FIG. 1. Spectrograms of voices of subspecies of *Caprimulgus rufus*. a) *otiosus*, 7 May 1988, Louvet, St. Lucia; recorded by Robbins (LNS 43447); b) *minimus*, 30 May 1956, near Caracas, Venezuela; recorded by P. Schwartz (LNS 4923); c) *minimus*, 2 February 1982, near Tocumen, Panama; recorded by Parker (LNS 25600); d) *rufus*, 20 November 1973, Iguarassú, Pernambuco, Brazil; recorded by J. Vielliard (from Hardy et al. 1988); e) *rutilus*, 31 October 1975, Fazenda Barreiro Rico, São Paulo, Brazil; recorded by W. Belton (LNS 19023). Y-axis represents frequency in kilohertz. X-axis represents time in seconds.

(*Tyrannus savana savana*), Streaked Flycatcher (*Myiodynastes maculatus solitarius*) (Zimmer 1938), and the Slaty Elaena (*Elaenia strepera*) (Marantz and Remsen 1991). Should *rutilus* overlap temporally with *minimus* in Venezuela, specimens would be required to distinguish them.

St. Lucian *otiosus* is certainly vulnerable given that its potential habitat is limited to no more than 30 sq. km on the windward, northeastern section of the island. Catastrophic events, such as a hurricane, could easily eliminate or severely reduce this subspecies to the point of extinction; however, this threat is present to virtually all West Indian taxa. Presumably, mortality of incubating adults, eggs, and young is high from the ubiquitous mongoose, but the fact that this nightjar is still common in much of its historical range, even after a hundred years of association with the mongoose, suggests that the mongoose's impact is not threatening. The most immediate, predictable threat to this population is the conversion of the dry, scrubby forest for charcoal and real estate development. To ensure that the nightjar, the highly endangered White-breasted Thrasher (*Ramphocinclus brachyurus*), and the nearly extinct St. Lucian subspecies of the House Wren (*Troglodytes aedon mesoleucus*) will continue to survive we recommend that a large tract of forest be preserved between Petite Anse and the Louvet River area.

DISCUSSION

Species limits.—Voice is the primary taxonomic character for ascertaining species limits within the Caprimulgidae, although song differences are usually associated with rather pronounced plumage pattern differences (Reynard 1962; Robbins and Ridgely 1992, Robbins et al. 1994). Relatively minor differences in plumage coloration and measurements, as exemplified by subspecies of the Lesser (*Chordeiles acutipennis*) and Common (*C. minor*) nighthawks (Ridgway 1914, Dickerman 1981, 1985), are not considered sufficient to warrant species level treatment.

Plumage differences among the *C. rufus* populations are comparable to variation observed in the above nighthawk subspecies. This lack of pronounced plumage differentiation coupled with the lack of audible and spectrographic song differences among the *C. rufus* populations, argues for treating *otiosus* as a subspecies. Although the Chuck-will's-widow (*C. carolinesis*) is undeniably very closely related to the *C. rufus* complex, it does have a distinct voice and morphological characters that merit it being treated as a species. Spectrographically, the first element of *carolinesis*' song appears quite similar to the initial element of *rufus*' song, but *carolinensis* lacks the second and third elements of *rufus*' song (Fig. 1). The "will's" and "widow" elements of *carolinensis*' song appear very similar to the final element of *rufus*' song. Morphologically, *carolinensis* has lateral pinnae on the rictal bristles that are apparently present only in juvenile *C. rufus* (Ridgway 1914, Wetmore and Phelps 1953), and a different wing formula (Ridgway 1914). Not surprisingly, given that it is highly migratory, *C. carolinensis* has longer wings than all *C. rufus* subspecies.

Distributional summary of C. rufus subspecies.—The following summary should be considered preliminary as much more work is needed to clarify subspecific distributions in *C. rufus*. It is likely that there are yet undescribed forms, as this species has recently been found, but not collected, at several Peruvian localities (depto. Junín, Río Ene, J. Weske and J. Terborgh, J.P. O'Neill, pers. comm.; depto. San Martín, W of Rioja, Parker and G. Graves; depto. San Martín, near Jirillo, T. Schulenberg, pers. comm.) and in extreme southern Ecuador (Prov. Zamora-Chinchipe, Zumba, R. Ridgely and Robbins, ANSP unpubl. data).

C. r. minimus: Costa Rica, Panama, northern Colombia, and northern Venezuela (east to Sucre).
C. r. otiosus: St. Lucia, Lesser Antilles.
C. r. rufus (includes synonym *noctivigularis*): northwest Brazil (upper tributaries of the Rio Negro, Amazonas) and southern Venezuela (western Amazonas, western Bolívar) east across Guyana and Surinam to French Guiana, and south to tributaries on the south bank of the Rio Amazonas.
C. r. rutilus (includes synonyms *ornatus*, *cortapau*, and *saltarius*): southern Brazil (Mato Grosso, Rio de Janeiro) and eastern Bolivia south to northern Argentina (Tucuman, Corrientes) and Rio Grande do Sul, Brazil.

ACKNOWLEDGMENTS

This paper is dedicated to Jaqueline Goerck.

We thank Greg Budney and the Library of Natural Sounds, Cornell University, Ithaca, New York, for providing recordings of the Rufous Nightjar. R. Wayne Burke, Armas Hill, Francisco Vilella, Christopher Cox, Donald Anthony, Lindon Jones, and Allan Keith kindly shared unpublished material on St. Lucian birds. Nigel Cleere and Michael Walters provided information on specimens housed in the British Museum of Natural History, Tring; Thomas Schulenberg provided information on specimens in the Field Museum of Natural History. Gerlof Mees graciously shared information on the specimen he collected in Surinam, and we thank Richard Prum for attempting to locate the type of *C. r. rufus* in the Paris Museum. Mario Cohn-Haft, Gary Graves, Bill Hardy, Van Remsen, and Tom Schulenberg provided valuable comments on the manuscript. We are grateful to the following people and institutions (listed alphabetically by institution) for loans and information of specimens under their care: George Barrowclough and Mary LeCroy, American Museum of Natural History (AMNH); Michael Walters, British Museum of Natural History, Tring (BMNH); Kenneth Parkes, Carnegie Museum of Natural History (CMNH); R. Piechocki, Martin-Luther-Universitat Hall Institut fur Zoologie, Germany; Raymond Paynter, Jr., Museum of Comparative Zoology, Cambridge; Gary Graves, National Museum of Natural History, Smithsonian (USNM); Rene Dekker, National Natuurhistorisch Museum, Leiden (NNML); Ernst Bauernfeind, Naturhistorisches Museum Wien, Austria; and Fred Sibley, Peabody Museum of Natural History, Yale University.

LITERATURE CITED

AMERICAN ORNITHOLOGISTS' UNION. 1983. Check-list of North American Birds, 6th ed. Allen Press, Lawrence, Kansas.
BABBS, S., S. BUCKTON, P. ROBERTSON, AND P. WOOD. 1988. Report of the 1987 University of East Anglia-ICBP St. Lucia expedition. International Council for Bird Preservation Study Report 33. Cambridge, United Kingdom.
BANGS, O. 1911. Descriptions of new American birds. Proc. Biol. Soc. Wash. 24:187–190.
BELTON, W. 1984. Birds of Rio Grande do Sul, Brazil. Pt. 1. Rheidae through Furnariidae. Bull. Am. Mus. Nat. Hist. 178:1–631.

BOND, J. 1947. Field Guide to the Birds of the West Indies. MacMillan Co., New York, New York.
BOND, J. 1959. Fourth Supplement to the Check-list of Birds of the West Indies (1956). Acad. Nat. Sci., Philadelphia, Pennsylvania.
BOND, J. 1977. Twenty-first Supplement to the Checklist of Birds of the West Indies (1956). Acad. Nat. Sci., Philadelphia, Pennsylvania.
BOND, J. 1986. Twenty-sixth Supplement to the Check-list of Birds of the West Indies (1956). Acad. Nat. Sci., Philadelphia, Pennsylvania.
BURMEISTER, H. 1856. Systematische Ubersicht der tiere Brasiliens welche wahrend einer Reise durch die Provinzen von Rio de Janeiro und Minas Gerais gesammelt und beobachtet wurden. Vol. 2. G. Reimer, Berlin.
CORY, C. B. 1918. Catalogue of Birds of the Americas and the Adjacent Islands. Pt. 2. Publ. Field Mus. Nat. Hist., Zool. Ser. no. 197. Chicago, Illinois.
DANFORTH, S. T. 1935. The Birds of Saint Lucia. Univ. of Puerto Rico, San Juan.
DAUBENTON, E. 1783. Tables planches enluminees pour servir a l'histoire naturelle de M. Le Comte de Buffon. Paris.
DICKERMAN, R. W. 1981. Geographic variation in the juvenal plumage of the Lesser Nighthawk (*Chordeiles acutipennis*). Auk 98:619–621.
DICKERMAN, R. W. 1985. Taxonomy of the Lesser Nighthawks (*Chordeiles acutipennis*) of North and Central America. Pages 356–359 *in* P. Buckley et al. (eds.), Neotropical Ornithology. Ornithol. Monogr. No. 36.
GRISCOM, L. AND J. C. GREENWAY, JR. 1937. Critical notes on new Neotropical birds. Bull. Mus. Comp. Zool. 81:416–437.
HARDY, J. W., G. B. REYNARD, AND B. B. COFFEY. 1988. Voices of the New World Nightjars and their Allies. ARA Records, Gainesville, Florida.
MARANTZ, C. A., AND J. V. REMSEN, JR. 1991. Seasonal distribution of the Slaty Elaenia, a little-known austral migrant of South America. J. Field Ornithol. 62:162–172.
MEES, G. F. 1968. Enige voor de avifauna van Suriname nieuwe vogelsoorten. Gerfaut 58:101–107.
PAYNTER, R. A., JR., AND M. A. TRAYLOR, JR. 1991. Ornithological Gazetteer of Brazil. Museum of Comparative Zoology, Cambridge, Massachusetts.
PELZELN, A. 1871. Zur ornithologie Brasiliens. Resultate von Johann atterers Reisen 1817 bis 1835. Pichlers, Vienna.
PETERS, J. L. 1940. Check-list of Birds of the World. Vol. 4. Museum of Comparative Zoology, Cambridge, Massachusetts.
REYNARD, G. B. 1962. The rediscovery of the Puerto Rican Whip-poor-will. Living Bird 1:51–60.
RIDGWAY, R. 1914. The Birds of North and Middle America, Pt. 6. Bull. U.S. Natl. Mus. No. 50.
ROBBINS, M. B., AND R. S. RIDGELY. 1992. Taxonomy and natural history of *Nyctiphrynus rosenbergi* (Caprimulgidae). Condor 94:984–987.
ROBBINS, M. B., R. S. RIDGELY, AND S. W. CARDIFF. 1994. Voice, plumage and natural history of Anthony's Nightjar (*Caprimulgus anthonyi*). Condor 96:224–228.
NORES, M., AND D. YZURIETA. 1984. Consideriones acerca de la taxonomia y distribucion de *Caprimulgus sericocaudatus saltarius* Olrog 1979. V Renunion Argentina de ornitologia. Buenos Aires, Argentina.
OLROG, C. C. 1979. Notas ornitologicas. XI. Sobre la coleccion del Instituto Miguel Lillo. Acta Zoologica Lilloana 33:2.
SCLATER, M. A. 1866. Additional notes on the Caprimulgidae. Proc. Zool. Soc., London 38:581–590.
WETMORE, A., AND W. H. PHELPS, JR. 1953. Notes on the Rufous Goatsuckers of Venezuela. Proc. Biol. Soc. Wash. 66:15–19.
ZIMMER, J. T. 1938. Notes on migrations of South American birds. Auk 55:405–410.

TAXONOMY AND NEST DESCRIPTION OF THE TUMBES SWALLOW (*TACHYCINETA [ALBILINEA] STOLZMANNI*)

MARK B. ROBBINS,[1,2] GARY H. ROSENBERG,[3]
FRANCISCO SORNOZA MOLINA,[4] AND MARCO A. JÁCOME[4]

[1]*Department of Ornithology, Academy of Natural Sciences,
1900 Benjamin Franklin Parkway, Philadelphia, Pennsylvania 19103, USA;*
[2]*Present address: Division of Ornithology, Museum of Natural History,
University of Kansas, Lawrence, Kansas 66045, USA*
[3]*8101 N. Wheatfield Drive, Tucson, Arizona 85741, USA;*
[4]*Museo Ecuatoriano de Ciencias Naturales, Casilla 8976-Suc. 7, Quito, Ecuador*

ABSTRACT.—Until recently the endemic Tumbesan swallow, *Tachycineta [albilinea] stolzmanni*, was known from only a handful of specimens, with nothing known about its natural history. Because of this dearth of information, *stolzmanni* has been treated for the past sixty years as a subspecies of the Mangrove Swallow (*Tachycineta albilinea*) of Central America. Newly acquired data on the vocalizations, nest structure, and morphology of *stolzmanni* indicate that it deserves specific status.

RESUMEN.—Hasta recientemente la golondrina de Tumbes, *Tachycineta [albilinea] stolzmanni*, era conocida por pocos especímenes y no se conocía algo de su historia natural. Debido a la escasez de información, *stolzmanni* ha sido tratada en los últimos sesenta años como una subespecie de la golondrina manglera (*Tachycineta albilinea*) de Centroamérica. Información adquirida recientemente sobre las vocalizaciones, estructura del nido y morfología de *stolzmanni* indica que merece reconocimiento como especie.

The taxonomic status of *Tachycineta [albilinea] stolzmanni*, a swallow restricted to the arid coastal region of northwest Peru, has been in question since Hellmayr (1935) examined one of the few extant specimens and concluded that it was a subspecies of the Mangrove Swallow (*Tachycineta albilinea*) of Middle America. Every subsequent author has followed Hellmayr's revision, but a few have doubted that treatment (Zimmer 1955; Peters 1960; Schulenberg and Parker 1981; Parker et al. 1982; Ridgely and Tudor 1989; Turner and Rose 1989; Sibley and Monroe 1990). Herein we present new information on morphology, nest structure, and voice that supports treating *stolzmanni* as a species. We first give an overview of the nomenclatural history of this taxon followed by our data.

In September, 1878, along the Peruvian coast at Chepén, depto. La Libertad, M. Stolzmann collected four specimens of an undescribed swallow. Taczanowski (1880) described the Peruvian birds as a new species, *Hirundo leucopygia*, but he was unaware that this name was preoccupied by an earlier *H. leucopyga* (Meyen 1834); Gould (1841) had changed Meyen's *H. leucopyga* to *leucopygia*, which created the nomenclatural conflict when Taczanowski described the Peruvian bird. Philippi (1902) corrected the problem by renaming the Peruvian bird after the collector. Hellmayr (1935) examined one of the four original specimens (the type in the Warsaw Museum was lost), and although he noted that *stolzmanni* was morphologically distinct from the Mangrove Swallow (*Tachycineta albilinea*), he nonetheless treated it as a race of *albilinea*.

Morphology.—The recent acquisition of eight specimens (ANSP, Louisiana State Museum of Natural Science [LSUMNS], MECN) corroborates Hellmayr's (1935) comparison of a single *stolzmanni* to a series of *albilinea*. As Hellmayr pointed out, *stolzmanni* differs from *albilinea* in a number of aspects of plumage pattern and coloration. Some of the more prominent differences include the lack of white supraloral streak in *stolzmanni*. The underparts of most *albilinea* are nearly immaculate white, whereas those of *stolzmanni* are washed with light gray, especially on the breast. The dusky shaft streaks on the underparts and the upper tail coverts are more pronounced in *stolzmanni* than in *albilinea*. This latter feature is most noticeable on the throat,

TABLE 1
SELECTED MEASUREMENTS ($X \pm SD$) OF *Tachycineta albilinea* AND *T. stolzmanni*

	n	Wing (chord)	Tail (lg. rectrix)	Bill (base)
T. albilinea				
Male	23	97.4 ($\pm$1.8)	40.5 ($\pm$1.4)	11.1 ($\pm$0.5)
Female	22	93.6 ($\pm$1.7)	39.8 ($\pm$1.5)	10.8 ($\pm$0.6)
T. stolzmanni				
Male	5	93.8 ($\pm$1.7)	48.5 ($\pm$2.5)	8.8 ($\pm$0.2)
Female	2	88.5 (88.1–89.0)	44.9 (44.4–45.5)	8.6 (8.6–8.7)

where *stolzmanni* has streaking and *albilinea* has none. Hellmayr (1935) also correctly noted that the under wing coverts and axillaries of *stolzmanni* are smoky gray, but white in *albilinea*, and the duller, less bluish coloration to the dorsum. Two immatures (both males with unossified skulls, bursae, and small gonads; ANSP 186248-9) that were collected with adults on 21 April 1993, southwest of Zapotillo, Ecuador, prov. Loja, by R. S. Ridgely and Sornoza differ from adult *stolzmanni* in the following manner: dorsally, especially on the pileum, they are sooty brown. The throat, abdomen and crissum are white with only the crissum having a hint of streaking. Although the breast streaking is present, it is much more indistinct than that of the adult.

Because Hellmayr could only examine a single specimen of *stolzmanni*, he was unable to fully appreciate the morphometric differences between these two taxa. Male *albilinea* have significantly longer wings than female *albilinea* (Mann Whitney, $U = 753.0$; $P = 0.000$), but not in tail and bill length (Table 1). Sample sizes are not sufficient to determine if significant size differences exist between male and female *stolzmanni*. Because only two adult female *stolzmanni* were available, between taxon comparisons are restricted to males. Male *stolzmanni* have significantly shorter wings (Mann Whitney, $U = 385.0$, $P < 0.002$), and bills (Mann Whitney, $U = 368.0$, $P < 0.001$), but longer tails (Mann Whitney, $U = 276.0$, $P < 0.000$) than male *albilinea*. Although larger body mass sample sizes are needed, it appears that *stolzmanni* may average less than *albilinea*. The mean masses of four adult male and two adult female *stolzmanni* were 11.0 g ($\pm$0.7) and 12.0 g ($\pm$1.4), respectively. Pooling sexes for *albilinea*, Turner and Rose (1989) gave a mean of 13.9 g (range 10–16 g; no n provided), and Dunning (1993) gave the same mean ($n = 9$) but no range.

Nest.—At about 14:00 on 6 April 1992, ca. 3 road km north of Zapotillo, prov. Loja, Ecuador, we discovered a pair of *stolzmanni* sitting on a telephone wire that paralleled the unsealed, primary road. On numerous occasions, between 14:00 and 15:30, both birds visited a somewhat isolated acacia or mesquite that was on the flat part of a small knoll in the desert (VIREO r08/12/001–002; the pair was subsequently collected [ANSP 184817; MECN uncatalogued]). Upon closer inspection of the tree, we discovered a nearly round hole, ca. 2.5 cm in diameter, just under 2 m above the ground on the northwest section of the main trunk. Prior to the swallow occupation, we suspect that this hole had been made, or at least enlarged, by either an Ecuadorean Piculet (*Picumnus sclateri*) or a Scarlet-backed Woodpecker (*Veniliornis callonotus*). The piculet seems most likely given the hole's diameter. Between 15 and 20 cm directly below the entrance hole was a natural, narrow crevice ($<$ 3 cm at its widest point) that was vertically oriented. At the bottom of this crevice were many white, irregularly shaped pebbles (all $<$ 1 cm in diameter). We presume that the adults had gathered these from a nearby stream bed or the road. Although no other nest material was present, it is conceivable that other materials may have been eventually added, given that the female (ANSP 184817) would not lay the first egg for at least two days (her two largest ova were 10×8 mm and 5×5 mm). By contrast, Mangrove Swallow (*T. albilinea*) cavity nests (usually in a tree) in Panama are loosely constructed of grass stems and moss mixed with leaves and small sticks, with an interior lining of feathers (Wetmore et al. 1984; Dyrcz 1984). Sheppard (1977) found that Tree Swallows (*Tachycineta bicolor*) add feathers to their cavity nests after the clutch is complete, whereas Winkler (1993) reported that this species usually added feathers before the eggs were laid, and on occasion before the nest was complete.

The nest locality of *stolzmanni* was at 350 m in elevation in relatively flat desert with low, rolling hills nearby. Typically this area is quite barren, as a result of the general aridity coupled

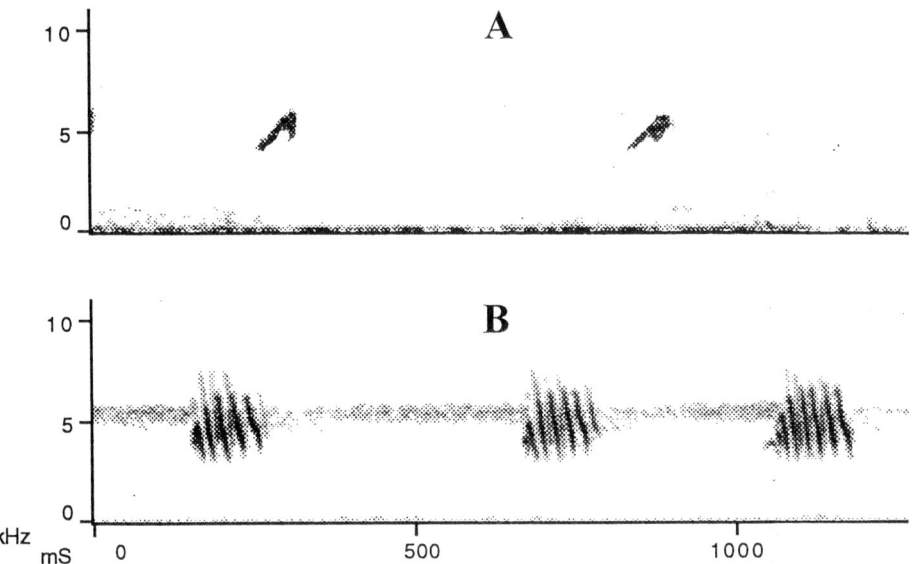

FIG. 1. Spectrograms of flight calls of a) *Tachycineta stolzmanni*, April 1992, near Zapotillo, prov. Loja, Ecuador; recorded by G. H. Rosenberg; and of b) *T. albilinea*, February 1994, Yucatan, Mexico; recorded by S. N. G. Howell. Y-axis represents frequency in kilohertz (kHz). X-axis represents time in milliseconds (mS).

with severe overgrazing by goats and cattle. Even the acacia/mesquite are badly pruned by the goats; at the nest site several goats were seen foraging in trees 2–3 meters above the ground. However, during our visit, the desert had been transformed by heavy rains that began about three weeks before our arrival. A relatively lush carpet of vines and bushes had appeared with the advent of the rains. The rains in 1992 were particularly heavy, as a result of the effects of "El Niño."

Voice.—At the nest we heard both adults giving a single-noted buzzy call primarily in flight (Fig. 1). This call is analogous to the flight calls of *T. albilinea* (pers. obs.), with the latter's flight call described as a "*dzreet, dzreet*" or "*jeet* or *jrrt*" (Ridgely 1976, Stiles and Skutch 1989). Although the flight calls of *stolzmanni* and *albilinea* (S. Howell recording from Yucatan, Mexico) sound superficially similar, spectrographically they are quite different (Fig. 1). The flight call of *stolzmanni* is structurally simple, ranging from ca. 4–6 kHz in frequency, whereas the more complex flight call of *albilinea* ranges from 3.5–6.5 kHZ. We are unaware of any sound recordings of the song of either *stolzmanni* or *albilinea*.

Distribution and status.—With a single exception, no new information has been published on the distribution and status of this swallow since it was discovered. In Zimmer's review (1955) of Peruvian swallows, he stated that *stolzmanni* had not been encountered since the original discovery. It was not until the mid-/to late 1970s that field work by the staff and students of the LSUMNS obtained information on this swallow. Schulenberg and Parker (1981) summarized those observations by indicating that the species was locally common in irrigated land and mesquite groves from the depto. Tumbes south to the depto. La Libertad. Our records from Ecuador barely extend the range of this swallow across the Peruvian border. The closest population of *T. albilinea* is in eastern Panama (Ridgely 1976; Wetmore et al. 1984), ca. 1,300 km to the north of the range of *stolzmanni*, and neither swallow is known to be migratory.

Discussion.—We believe that the differences in plumage color and pattern, morphometrics, nest construction, vocalizations, and distribution between *T. stolzmanni* and *T. albilinea* equal or exceed differences in these same attributes found between the following swallow species pairs: White-rumped (*Tachycineta leucorrhoa*)/Chilean (*T. leucopyga*), Northern Rough-winged (*Stelgidopteryx serripennis*)/Southern Rough-winged (*S. ruficollis*), and the recently split Chestnut-collared (*Hirundo rufocollaris*)/Cave (*Hirundo fulva*) pair (Ridgely and Tudor 1989). We predict that molecular data will demonstrate that *stolzmanni* and *albilinea* differ genetically as much as the other swallow species pairs mentioned above. We suggest Tumbes Swallow as an

English name for *stolzmanni*, because its range is entirely contained within the endemic Tumbesan Faunal Center (see Cracraft 1985).

ACKNOWLEDGMENTS

Our work in Ecuador was supported by a grant from the MacArthur Foundation. We thank our friends at the Museo Ecuatoriano de Ciencias Naturales, Quito, for logistical and field assistance. The Ministerio de Agricultura, Quito, kindly provided authorization for our work in Ecuador. Steve N. G. Howell kindly obtained vocalizations of *Tachycineta albilinea*. We thank the following people and museums (listed alphabetically by institution) for specimen loans or information from specimens under their care: Mary LeCroy (American Museum of Natural History, New York), Michael Walters (British Museum of Natural History, Tring), Van Remsen and Steve Cardiff (Louisiana State Museum of Natural Science, Baton Rouge), and Gary Graves and Jim Dean (National Museum of Natural History, Smithsonian Institution). John O'Neill kindly took measurements of a specimen in the collection at the Museo de la Universidad de San Marcos, Lima, Peru. We thank the following reviewers for improving the manuscript: Steve Cardiff, Van Remsen, Fred Sheldon, and Linda Whittingham.

LITERATURE CITED

CRACRAFT, J. 1985. Historical Biogeography and patterns of differentiation within the South American avifauna: areas of endemism. Pages 49–84 *in* P. Buckley et al. (eds.), Neotropical Ornithology. Ornithol. Monogr. 36.

DUNNING, J. B., JR. 1993. CRC Handbook of Avian Body Masses. CRC Press, Ann Arbor, Michigan.

DYRCZ, A. 1984. Breeding biology of the Mangrove Swallow (*Tachycineta albilinea*) and the Grey-breasted Martin (*Progne chalybea*) at Barro Colorado Island, Panama. Ibis 126:59–66.

GOULD, J. 1841. Part 3, Birds Described by J. Gould, with a Notice on Their Habits and Ranges. Smith and Elder, London.

HELLMAYR, C. E. 1935. Catalogue of the birds of the Americas. Part 8. Field Mus. Nat. Hist., Zool. Ser. 13: 1–541.

MEYEN, F. J. F. 1834. Beitrage zur Zoologie, gesammelt auf einer Reise un die Erde. Ueber Vogel. Nova Acta Acad. Caes. Leop. Carol. 16:1–125.

PARKER, T. A., III, S. A. PARKER, AND M. PLENGE. 1982. An Annotated Checklist of Peruvian birds. Buteo Books, Vermillion, South Dakota.

PETERS, J. L. 1960. Check-list of Birds of the World. Museum of Comparative Zoology, Cambridge, Massachusetts.

PHILIPPI, R. A. 1902. Figuras i descripciones de aves Chilenas. Anal. Mus. Nac. Chile, Zool. 15:1–114.

RIDGELY, R. S. 1976. A Guide to the Birds of Panama. Princeton Univ. Press, New Jersey.

RIDGELY, R. S., AND G. TUDOR. 1989. The Birds of South America. Vol. 1, the Oscine Passerines. Univ. Texas Press, Austin.

SCHULENBERG, T. S., AND T. A. PARKER III. 1981. Status and distribution of some northwest Peruvian birds. Condor 83:209–216.

SHEPPARD, C. D. 1977. Breeding in the Tree Swallow, *Iridoprocne bicolor*, and its implications for the evolution of coloniality. Ph.D. dissertation, Cornell Univ., Ithaca, New York.

SIBLEY, C. G., AND B. L. MONROE, JR. 1990. Distribution and Taxonomy of Birds of the World. Yale Univ. Press, New Haven, Connecticut.

STILES, F. G., AND A. F. SKUTCH. 1989. A Guide to the Birds of Costa Rica. Cornell Univ. Press, Ithaca, New York.

TACZANOWSKI, M. L. 1880. Liste des oiseaux recueillis au nord du Péron par M. Stolzmann pendentles derniers mois de 1878 et dans la premiere moitie de 1879. Proc. Zool. Soc. London 189–215.

TURNER, A., AND C. ROSE. 1989. Swallows and Martins. An Identification Guide and Handbook. Houghton Mifflin Co., Boston, Massachusetts.

WETMORE, A., R. F. PASQUIER, AND S. L. OLSON. 1984. The Birds of the Republic of Panama. Pt. 4. Smithsonian Institution Press, Washington, D.C.

WINKLER, D. W. 1993. Use and importance of feathers as nest lining in Tree Swallows (*Tachycineta bicolor*). Auk 110:29–36.

ZIMMER, J. T. 1955. Studies of Peruvian birds. 66. The swallows (Hirundinidae). Amer. Mus. Novit., no. 1723, 35 pp.

BIRDS OF A PERUVIAN OXBOW LAKE: POPULATIONS, RESOURCES, PREDATION, AND SOCIAL BEHAVIOR

SCOTT K. ROBINSON

Illinois Natural History Survey, 607 East Peabody Drive, Champaign, Illinois 61820, USA

ABSTRACT.—The bird community of a small (22-ha) oxbow lake, Cocha Cashu, of the Manu River in the Amazon basin of southeastern Peru was studied during 11 field seasons, 1979–1989. Here, field observations on the population status, interactions with predators, and social systems are summarized for many of the 186 species that regularly occurred there.

Oxbow lakes such as Cocha Cashu are characterized by narrow but very productive strips of marsh, shrubs, isolated trees, and vines along their borders. These habitats attract high populations of resident and nonterritorial birds, but also attract predators. In addition to species confined to the lake margins, Cocha Cashu attracted many forest birds to abundant flowering and fruiting trees (especially figs and Lauraceae) and isolated nest sites that provided some protection from mammalian predators. Birds respond to the constraints of limited habitat, high population density, and intense predation through various kinds of group living, including coloniality, cooperative breeding, and mono- and multi-species flocking. The various kinds of sociality further influence the kinds of mating systems observed on the lake, with several species showing polygynous mating behavior. Anti-predator adaptations included mobbing of some but not all nest predators, and vigilance coupled with alarm calling against raptors that attack adult birds. *Cacicus cela* used different escape tactics when faced with different kinds of predators. Mobbing and group vigilance were effective at deterring most, but not all, avian predators. Interspecific aggression appeared to be most intense around cavity nests and some fruiting trees. Possible cooperative breeding was documented for two species in which this behavior has not previously been described (*Ramphocelus carbo* and *R. nigrogularis*). Relative to forest habitats, lake-margin birds showed a stronger tendency to form monospecific than multi-species flocks, perhaps because variable resource availability and high population densities of some species precluded the formation of stable, multi-species flocks. Oxbow lakes strongly affect local patterns of species richness and abundance, but appear to have few specialists that do not occur in other aquatic, wetland, or second-growth habitats associated with riverine systems. The combination of rich foraging and nesting resources concentrated in a small area and high predation pressure influence life histories, population dynamics, and community structure of birds of this oxbow lake.

RESUMEN.—Durante 11 estaciones de campo (1979–1989) se estudió a la comunidad de aves de un pequeño recodo de río (22 ha) en Cocha Cashu en el río Manu, Perú. En éste estudio se resumen las observaciones sobre el estado de la población, interacciones con depredadores, y sistemas sociales para las 168 especies que regularmente ocurren en la zona de estudio.

Los recodos tales como Cocha Cashu se caracterizan por contener lineas muy angostas, pero muy productivas de pantanos, arbustos, árboles aislados y vainas a lo largo de sus bordes; éstos tipos de hábitats atraen grandes poblaciones de aves residentes y no territoriales, sin embargo también atraen depredadores. Además de las especies confinadas a los márgenes del lago, Cocha Cashu atrajo muchas aves de bosque hacia los árboles florales y frutales que son muy abundantes (especialmente higos y Lauraceae), y hacia sitios aislados de anidación que proveyeron de protección contra los mamíferos depredadores. La respuesta de las aves a las restricciones de la poca disponibilidad del hábitat, altas densidades de población, y altos niveles de depredación consiste en distintos tipos de vida en grupo los cuales incluyen colonialidad, anidación cooperativa, y la formación de parvadas mono y multiespecíficas. Los distintos tipos de vida social influyen a su vez las distintas variantes en los sistemas reproductivos que se observan en el lego, existiendo asi distintas especies que muestran un comportamiento reproductivo de tipo poligénico. Las adaptaciones contra los depredadores incluyeron el asalto de algunos, pero no todos los depredadores de nidos, y la vigilancia asociada con llamados de alarma contra las rapaces que atacan a las aves adultas; *Cacius cela* utilizó diferentes tácticas contra distintos depredadores. Los asaltos a depredadores, y la vigilancia en grupo fueron efectivos para auyentar a la mayoría, pero no a todas las aves depredadoras. La agresión interespecífica aparentó ser más intensa al rededor de nidos en cavidades y algunos árboles frutales. La posibilidad de anidación cooperativa se documentó para dos especies en las cuales no se había descrito anteriormente (*Ramphocelus carbo* y *R. nigrogularis*). Las aves de los márgenes de los lagos mostraton una mayor tendencia a formar parvadas mono-específicas, en comparación a las parvadas multi-específicas de las aves de bosque; ésto se debe probablemente a que la variable disponibilidad de recursos, y las altas densidades de población de algunas especies imposibilitaron la formación de parvadas multiespecíficas estables. Los recodos de los ríos afectan fuertemente los patrones de riqueza específica y abundancia, éstos recodos sin embargo, parecen contener pocos especialistas que no ocurren en otros hábitats acuáticos y de humedales de crecimiento secundario asociados a sistemas riverinos. La combinación de recursos ricos para el forrajeo y la anidación concentrados en una área pequeña, asi como las altas presiones de depredación influyen las historias de vida, las dinámicas de población, y la estructura de la comunidad de aves en éste recodo del río.

The meandering of Amazonian rivers creates numerous oxbow lakes (Salo et al. 1986) that provide diverse habitats and resources for birds (Parker 1982; Terborgh et al. 1984; Terborgh 1985; Willard 1985) and has contributed to species richness in Amazonia (Remsen and Parker 1983). Once oxbow lakes have been pinched off from rivers, they begin to undergo complex successional patterns as the lakes fill with silt and younger vegetation along the lake margins begins to mature (Terborgh 1985). The viny lake edges, marshes, shrubs, and open water provide habitat for marsh-dwelling species (Parker 1982; Kiltie and Fitzpatrick 1984), piscivores (Willard 1985), migrant shorebirds (Bolster and Robinson 1990), migrant landbirds (Fitzpatrick 1980; Robinson et al. 1988), colonial icterines (Robinson 1985a, 1986a), omnivores (Eason 1992), and raptors (Robinson 1995). Many lake-margin specialists defend essentially linear territories on one or both sides of the lake (Kiltie and Fitzpatrick 1984; Eason 1992). Isolated lake-margin trees and shrubs provide nest sites that are safe from monkeys (Robinson 1985a). Fruiting trees along lake margins also attract large but variable populations of frugivores (Fitzpatrick 1980, 1985; Janson 1987). The open space around oxbow lakes also attracts some species of raptors and makes them relatively easy to observe (Robinson 1994a). Although most lake-margin birds are generally thought to be behaviorally monogamous (Terborgh et al. 1984, 1990), several other breeding systems have been documented, including cooperative and communal breeding (Kiltie and Fitzpatrick 1984) and polygyny (Robinson 1986a,b). Many aquatic species vary in abundance seasonally (Willard 1979, 1985) and with water levels (Bolster and Robinson 1990). Populations of colonial icterines vary tenfold depending upon the availability of safe sites and unpredictable attacks by some predators (Robinson 1985a). Taken together, these studies suggest that oxbow lakes contain both a core of lake-margin specialists, some of which are abundant, and others whose populations fluctuate in response to varying resource availability.

The purpose of this paper is to describe new aspects of the natural history of lake-margin birds in an oxbow lake in Amazonian Peru and to use these data to increase our understanding of selective forces that shape the community ecology, population dynamics, and breeding behavior of lake-margin birds and of the well-studied Manu bird community in general (e.g., Terborgh et al. 1990; Robinson and Terborgh 1997). Specifically, I quantify the abundance of lake-margin birds, sources of population variation, and interactions with predators that attack nests and adults. I also describe several breeding systems and unusual foraging tactics of lake-margin species, several of which never have been described. The paper concludes with a discussion of how resource availability has shaped community organization of this oxbow lake.

STUDY AREA

Cocha Cashu (Fig. 1) is a 2.2-km-long oxbow lake of the Manu River in Manu National Park, southeastern Peru (400 m elevation, 11°55′S, 77°18′W). The lake is surrounded by undisturbed forest in various stages of floodplain succession (Terborgh 1983), including forests in the late transitional stage (sensu Terborgh and Petren 1991) along the western edge (the former inside of the meander loop) and mature floodplain forest along the eastern and northern edge (Terborgh et al. 1990). An intermittent stream flowing into the lake on the east side of the lake only carries water after heavy wet-season rains. Two seasonal swamps to the north and east of Cocha Cashu, which may be old oxbow lakes, fill with water during the wet season. During major floods,

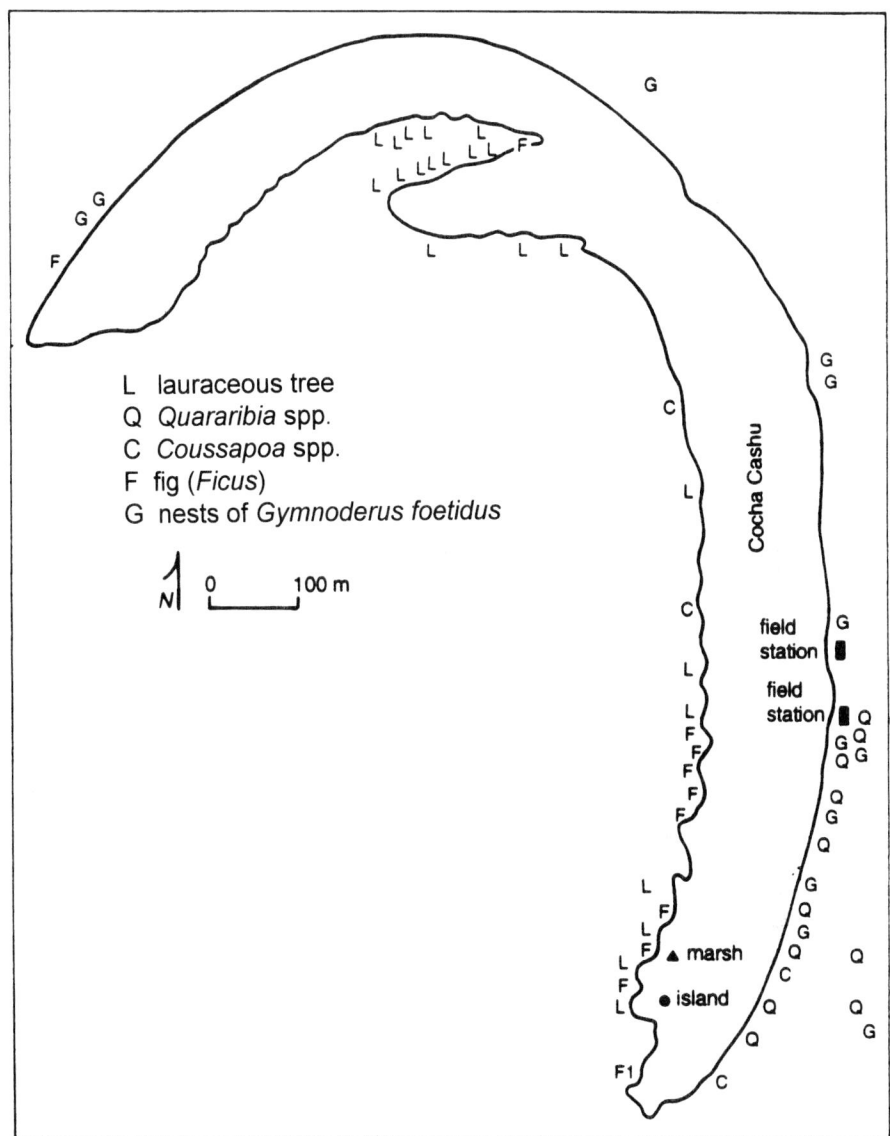

FIG. 1. Map of Cocha Cashu showing locations of key landmarks, food resources, and nest sites.

water flows into Cocha Cashu through these swamps. Direct flow from the river is cut off by natural levees except during the highest floods (see below). Because it is protected from direct river flow and lacks a permanent stream, Cocha Cashu is filling in slowly with silt and may be at least 200 years old (J. Terborgh, pers. comm.).

Cocha Cashu contains the following habitat types. (1) Open water on Cocha Cashu covers only about 20 ha during the dry season. Trees routinely fall into the water from the lake margin and provide perches and nest sites isolated from the forest. Perhaps because of the lack of a permanent stream providing nutrients, Cocha Cashu lacks the dense mats of floating *Pistia stratiotes* (Araceae) that cover much of the surface of some Manu oxbow lakes. (2) Marshes cover the ends of the three arms of the lake. The two largest marshes on the southern end of the lake (1.5 ha) and the "thumb" (1.2 ha) also contain many shrubs, chiefly *Annona tessmanii* (Annonaceae). During a major October flood in 1982, water flowed through the lake and opened up extensive areas of both marshes. The marsh at the far western end of the lake consists of a grassy border no more than 15 m wide and 125 m long. Marshy vegetation is dominated by grasses and contains shrubs and saplings (e.g., *Heura crepitans, Ficus trigona*). In some areas *Heliconia* spp. have invaded the marsh and provide a source of nectar. At low water, large sections of the marsh become dry, and mudflats form along the edges (Bolster and Robinson 1990). (3) Shrub thickets dominated by *Annona tessmanii* form as a natural part of lake-bed succession in the older sections of marshes. These shrubby areas provide low (<10m), dense cover until invaded by trees. The shrub thickets on Cocha Cashu were largely destroyed by the flood of 1982, when the lake level rose 5 m (S. Robinson, unpubl. data). (4) Lake-bed trees include several flood-tolerant species that grow in shrubby areas of the marsh: 11 large *Ficus* trees (at least eight are *F. trigona*, Fig. 1), several *Heura crepitans* (Bombacaceae) and numerous *Cecropia* (Moraceae). In the older sections of the lake bed and on the inside of the former meander loop (the west shore), large *Cedrela odorata* (Meliaceae) and various species of Lauraceae grow. The land between the two western arms of the lake, a former river island, is covered with lauraceous trees. (5) Lake-margin trees and vines cover the eastern and northern edges of the lake, where an abrupt 3–4 m tall bank (levee) separates the lake from mature floodplain forest. This abrupt edge is often covered with vines, including *Combretum* spp. (Combretaceae), which also provide a source of dry-season nectar for lake-margin birds (Terborgh 1983). Two large *Ficus insipida* have grown out over the lake from the lake edge. At least nine *Quararibea* (Bombacaceae) trees grow within 20 m of the lake margin (Fig. 1) and provide a rich source of nectar during the late dry season.

METHODS

Observations of lake-margin birds were made primarily during 11 years of field work (1979–1989) on a population of Yellow-rumped Caciques (*Cacicus cela*) centered on Cocha Cashu (e.g., Robinson 1985a). During this period, colonies of caciques on the lake were visited daily for 2–12 h/day during 33 months in residence (1–7 months per year). Short visits in 1988 (1 month) and 1989 (3 weeks) enabled me to make only brief observations of lake-margin birds. During my study of caciques, over 1,200 yellow-rumped caciques and 1,000 individuals of other lake-margin species were netted and color banded. Nets were located at 10 sites in the southern section of Cocha Cashu (Fig. 1). Additional netting by Willard and Fitzpatrick (e.g., Willard 1985) in 1975–1976 provided a few banded birds still present on Cocha Cashu at the beginning of my study. The 1975–1976 netting included sites scattered throughout the entire lake margin.

Lake-margin censuses were conducted in 1980 and 1985 to quantify population sizes of most species. During these censuses (eight in 1980, 10 in 1985), I paddled a canoe or kayak slowly around the perimeter of the entire lake and mapped the locations of any birds detected. These censuses, in combination with netting and incidental observations, were used to estimate population sizes for all species confined to lake margins. Relative abundances of species that only visited the lake were estimated from visitation rates to fruiting or flowering trees and mist-net captures.

In this paper, I included only bird species that regularly used lake-margin vegetation and open water. Population data were not included for forest species occasionally caught in mist-nets or observed crossing the lake, although some of these species are discussed in the context of interactions with lake-margin birds (e.g., raptors crossing the lake). I also included primarily forest-dwelling species that regularly visited fruiting and flowering trees along the lake (e.g., *Tangara callophrys, Dacnis flaviventer*) or that appeared to show a preference for foraging along the lake margin (e.g., *Nasica longirostris, Dendrexetastes rufigularis*). I only estimated population sizes of species mostly restricted to lake-margin vegetation, which excluded species that foraged regularly along lake margins, but that also had territories extending into the forest.

RESULTS

I observed 186 species of birds that regularly used Cocha Cashu, its border, and the airspace immediately (<100 m) overhead for foraging (Appendix). An additional 117 species have been observed or captured along the lake, but not regularly enough to be considered in this paper.

Population Dynamics

Territorial species.—Some species restricted to lake-margin vegetation and aquatic habitats, which totaled only 22 ha, were abundant in their limited habitat and included birds from many different guilds (Appendix). Typically, however, most territorial species had two to three pairs on the lake (Appendix). Population densities of piscivores and other aquatic species reported in Willard (1985) were also generally high, as were marsh-nesting species (Appendix).

Several lake-margin birds also foraged in mid-successional forests bordering the lake and in seasonal swamps and other aquatic habitats. Marked individuals of *Sclateria naevia* and *Hypocnemoides maculicauda* have been observed on the edge of the lake and in a seasonal fig swamp 400–500 m east of Cocha Cashu (see also Parker 1982). Individual *Agamia agami, Cochlearius cochlearia, Tigrisoma lineatum, Eurypyga helias, Chloroceryle inda,* and *C. aenea* have been observed entering the forest along the intermittent stream (see also Willard 1979, 1985, Remsen 1990). Characteristic birds of lake-margin vegetation that also foraged extensively and defended territories in adjacent mid-successional forests included *Ortalis motmot, Aramides cajanea, Leptotila rufaxilla, Glaucidium brasilianum, Eubucco tucinkae, Pteroglossus castanotis, P. inscriptus, Dryocopos lineatus, Campephilus melanoleucos, Trogon melanurus, T. curucui, Taraba major, Percnostola lophotes, Myrmeciza hyperythra, M. goeldii, Nasica longirostris, Myiozetetes granadensis, Legatus leucophaius,* all three species of *Tityra, Saltator maximus, Thraupis palmarum, Vireo olivaceus,* and *Cacicus solitarius.* Only a few species regularly foraged in both lake-margin and mature floodplain forest (*Melanerpes cruentatus, Galbula cyanescens, Myrmeciza hyperythra, Tityra cayana,* and *T. semifasciata*). A family group of *Daptrius ater* divided its time between Cocha Cashu and the river. A color-banded *Myiozetetes granadensis* from Cocha Cashu was later observed 600 m away in early successional vegetation along the river.

Some of the territorial residents defended linear territories on one side of the lake only (*Myrmotherula surinamensis*), whereas others included both sides (*Donacobius atricapillus*: Kiltie and Fitzpatrick 1984; *Paroaria gularis*: Eason 1992). One pair of *M. surinamensis* had a home range 500–700 m long and less than 10 m wide.

Small backwaters that form behind natural levees in mid-sucessional forests (Robinson and Terborgh 1997) were sometimes visited by lake-margin birds. Individually marked birds of the following species caught on Cocha Cashu have been observed or caught in forest backwaters at least 150 m from the lake: *Jacana jacana, Chloroceryle inda, Galbalcyrhynchus purusianus, Pitangus lictor, P. sulphuratus, Myiozetetes similis, Donacobius atricapillus, Paroaria gularis,* and *Ramphocelus nigrogularis.* Unmarked family groups of *Cissopis leveriana* and *Ramphocelus carbo,* most likely from Cocha Cashu, have also been observed foraging in backwaters in mid-successional habitats.

Loosely organized multi-species flocks were sometimes formed by species that forage in shrubs, vines, and *Heliconia* thickets that overhang the water. Pairs of *Sclateria naevia, Myrmeciza hyperythra, Hypocnemoides maculicauda* and, occasionally, *Myrmotherula surinamensis* foraged close together within 1–2 m of the water (see also Parker 1982). Forest-canopy flocks described by Munn (1985) often foraged high (<30 m) along the lake margin, particularly after heavy rains.

Some non-territorial species that live primarily on Cocha Cashu were also abundant. *Cairina moschata* sometimes occurred in flocks of up to 25 individuals; at times, a flock of up to 200 *Brotogeris sanctithomae* foraged in fruiting figs (especially *Ficus trigona*) along the lake (see also Forshaw 1973). Among colonial icterids, populations of *Cacicus cela* (up to 112 simultaneously breeding females) and *Psarocolius angustifrons* (up to 28 simultaneously breeding females) varied with recent histories of predator attacks (Robinson 1985a, 1988). Flocks of 50–200 *Chordeiles rupestris,* 5–50 *Stelgidopteryx ruficollis,* and 16 *Rynchops nigra* regularly visited Cocha Cashu from their nesting areas on river-edge beaches (Groom 1992).

Seasonality.—Robinson et al. (1988, 1995), Bolster and Robinson (1990) and Fitzpatrick (1980) provided data on use of the lake margin by austral and Nearctic migrants. Among the arrival dates for southward-migrating *Hirundo rustica* were Sept. 14 (1981), 13 (1982), 9 (1983) and 29 (1985). Flocks of *Myiodynastes luteiventris* and *Tyrannus tyrannus* foraged in fruiting trees from September through December (at least). Most Nearctic migrants other than shorebirds (Bolster and Robinson 1990) were only rarely observed (Appendix).

Austral migrants were also present in small numbers only (Appendix). The most abundant species was *Pyrocephalus rubinus,* which appeared to have territories or small home ranges on the lake margin. One banded individual foraged in a canopy opening for most of August and early September before disappearing. *Fluvicola pica* appeared in only one of the years (1987) that I visited Cocha Cashu, and was only present until mid-September. Elsewhere, *F. pica* is a marsh dweller (Wetmore et al. 1984).

Several species were only present on Cocha Cashu during the wet season. *Crotophaga major* began appearing in late August and September (arrival dates: 26 August, 1982; 30 August, 1983; 5 September, 1981; 11 September, 1979; 16 September, 1980; 19 September, 1984; 22 September, 1987), but did not appear to reach full populations (up to 32 individuals) until breeding began in December. Marked birds were site-faithful; one banded individual returned to Cocha Cashu for at least six consecutive seasons. *C. major* is seasonal elsewhere in its range (Wetmore 1968). Willard (1985) described seasonality in aquatic species. *Ardea alba* did not appear in 1979–1983 until August (arrival dates: 4 August, 1979; 15 August, 1981; 10 August, 1982; 2 August, 1983), and remained at least until December. *Bubulcus ibis* primarily occurred as a non-foraging wanderer in October-December (10 of 11 records). *Heliornis fulica* was mostly present during the wet season, but a few individuals remained during the dry season, a pattern also typical of *Phalacrocorax brasilianus* (Willard 1985), in which individuals are known to wander after the breeding season (del Hoyo et al. 1993).

Irregular species.—Observations of some species were strongly associated with dry-season cold fronts. Flocks of up to 150 *Notiochelidon cyanoleuca* visited the lake during cold spells in July-October, possibly coming from higher elevations. Of the 23 records of foraging flocks of *N. cyanoleuca* on Cocha Cashu, all followed cold fronts when the temperature dropped below 18°C. *N cyanoleuca* also shows evidence of seasonal movements in Central America (Skutch 1952). Cold, wet weather also attracted flocks of *Stelgidopteryx ruficollis*, *Progne tapera*, and *Atticora fasciata*, which usually foraged along the river. *Falco rufigularis* regularly attacked these flocks, sometimes successfully (Robinson 1994a).

Food-resource based aggregations.—During the dry season, two sources of nectar were heavily used by lake-margin birds (Appendix). The vine *Combretum alnifolium* flowered profusely in many locations along the lake margin from July through early August, and *Quararibea* sp. trees provided abundant nectar along the outer levee bordering the lake from August through November (Fig. 1, see also Terborgh 1983). Although many species of mammals also visited these flowers, birds were the most frequent visitors (Janson et al. 1980). During a two-hour observation period (10:00–12:00 AM, 5 August 1979) at a *Quararibea*, I recorded 50 visits by *Cacicus cela*, 20 by *Paroaria gularis*, four by *Psarocolius angustifrons*, *Melanerpes cruentatus*, and *Ramphocelus nigrogularis*, three by *Aratinga weddillii* and *Capito niger*, and two by *Celeus elegans* and *Brotogeris sanctithomae*. A flock of 14 *Ara ararauna* ate flowers in the same *Quararibea* the next day.

The epiphytic *Coussapoa* (Moraceae) provided a rich food resource for lake margin birds. At least six fruiting *Coussapoa* grew along the lake and were heavily used by many lake-margin birds (Appendix). All of my observations of *Porphyrolaema porphyrolaema* in the Cocha Cashu area ($n = 7$ total) are from *Coussapoas* along Cocha Cashu ($n = 4$) and on the edge of a seasonal swamp ($n = 3$). *Cacicus cela* and *Psarocolius angustifrons* fed *Coussapoa* to their nestlings.

Especially during the dry season (June–August), lake-margin figs (*Ficus spp.*) provided abundant fruit eaten by many lake-margin and forest birds (Appendix). Flocks of up to 200 *Brotogeris sanctithomae* (possibly with a few *B. cyanoptera*) and 40 *Columba cayennensis* descended on single fruiting trees. Seasonal dependence on figs has been documented in these species in Central America, as well (Skutch 1964, Forshaw 1973). *Porphyrula martinica* were observed eating figs on at least six occasions.

Lauraceous trees that lined the lake on the younger (western) side and on the former river island appeared to be particularly critical to the life histories of many bird species. Annual and seasonal fluctuations in populations of *Gymnoderus foetidus* (see below), *Cotinga maynana*, (see also Snow 1982) and *Tersina viridis* may have been closely tied to the availability of lauraceous fruits. Flocks of migrant *Tyrannus tyrannus* and *Myiodynastes luteiventris* descended on fruiting lauraceous and aggressively competed with resident flycatchers (especially *Myiozetetes similis*, *Tyrannus melancholicus*, and *Myiodynastes maculatus*) (Fitzpatrick 1980). Flocks of *Crotophaga major* also fed in fruiting Lauraceous trees. Large flocks or aggregations of forest species such as *Ramphastos culminatus* (up to 22 individuals at a time), *R. cuvieri* (up to 20 individuals at a time), and *Pipile pipile* (up to 24 individuals at a time) gathered in lake-margin lauraceous trees, as did smaller groups of *Mitu tuberosa* (1–7 individuals) and *Penelope jacquacu* (1–14 individuals). Other forest birds that visited lake-margin lauraceous trees included *Trogon melanurus*, *T. curucui*, and *Turdus hauxwelli*.

During the wet season, major emergences of mayflies (Ephemeroptera) were frequent over the lake, especially at dusk. Large flocks of up to 200 *Chordeiles rupestris* fed in these swarms. After heavy rains early in the wet season, termite swarms provided a short-lived but abundant resource. On 10 October 1985, one swarm attracted an estimated 150 *Tyrannus tyrannus*, 20 *Cacicus cela*, six *Monasa nigrifrons*, five *Gymnoderus foetidus*, three *Scaphidura oryzivora*, three *Crotophaga major*, six *Paroaria gularis*, two *Myiozetetes similis*, two *Pitangus sulphuratus*, two *P. lictor*, one *Falco rufigularis*, and one *Ramphastos cuvieri*. The fruitcrow and the toucan were flycatching 20–50 m above the forest in continuous flight. On 18 October 1985, at one swarm I counted at least eight *Gymnoderus*, two *Crotophaga ani*, >50 *Cacicus cela*, >10 *Psarocolius*

TABLE 1
ATTACKS BY NEST PREDATORS ON COCHA CASHU. DATA FOR *Daptrius ater* AND *Buteogallus urubitinga* FROM ROBINSON (1994)

Predator	Prey	Attacks	% Successful
Pteroglossus castanotis	*Cacicus cela*	33	24.2
($\bar{X}$ group size = 2.3 ± 0.45 N = 12)	*Tachycineta albiventer*	1	100.0
		1	0.0
P. inscriptus	*Cacicus cela*	4	0.0
(group sizes = 4, 5)	*Columba cayennensis*	2	50.0
	Paroaria gularis	1	100.0
Ramphastos cuvieri	*Psarocolius angustifrons*	5	0.0
(group sizes = 1, 2)	*Cacicus cela*	44	36.4
	Megarynchus pitangua	3	33.3
	Pitangus sulphuratus	10	20.0
	Columba cayennensis	1	100.0
	Myiozetetes similis	2	50.0
Daptrius ater	*Psarocolius angustifrons*	12	0.0
($\bar{X}$ group size: 3.0 ± 0.8 SD N = 21)	*Cacicus cela*	32	34.4
	Myiozetetes similis	6	16.7
	Pitangus sulphuratus	6	0.0
	Megarynchus pitangua	3	0.0
	Columba cayennensis	1	100.0
Buteogallus urubitinga	*Psarocolius angustifrons*	9	44.4
(group size = 1)	*Cacicus cela*	53	58.4
	Opisthocomus hoazin	5	100.0
	Myiozetetes similis	3	33.3
	Pitangus sulphuratus	6	0.0
	Megarynchus pitangua	3	0.0
Cebus apella	*Cacicus cela*	11	100.0
(group sizes = 3–12)	*Megarynchus pitangua*	1	100.0
	Pitangus sulphuratus	2	50.0

angustifrons, *Conioptilon mcilhennyi*, >150 *Streptoprocne zonaris*, *Ictinia plumbea*, *Myiozetetes granadensis*, *Megarynchus pitangua*, and a pair of *Phaetusa simplex*. From October-December 1981, I noted termite swarms attended by birds on 12 evenings. Additional species observed at other swarms were *Pteroglossus castanotis*, *Chaetura brachyura*, *Lurocalis semitorquatus*, *Tityra cayana*, *T. semifasciata*, *T. inquisitor*, *Cissopis leveriana*, *Melanerpes cruentatus*, and *Cotinga maynana*, all of which flew back and forth through the swarm flycatching at least 20 m above the top of the canopy (see also Robinson 1994a for an account of use of termites by raptors).

NEST SITES AND NEST PREDATION

In addition to providing abundant food resources, lake margins provided many potential nest sites isolated enough to provide some protection from forest-dwelling nest predators such as mammals. Nevertheless, many predators attacked these colonies (Table 1). Forest-dwelling nest predators such as *Ramphastos cuvieri* and *Cebus apella* and lake-margin species such as *Daptrius ater*, *Pteroglossus castanotis* and *P. inscriptus* attacked concentrations of nests frequently. Because such attacks were easy to observe in open lake-margin habitats, I was able to obtain detailed data on the relative success of attacks by different predators, the advantages of different nest sites, and the behavior of both predators and prey. In this section, I describe the interactions between nest predators and prey, beginning with selection of "safe" nest sites and concluding with active defense against predators.

Isolated trees.—The edges of Cocha Cashu provided birds with many opportunities to nest in isolated or semi-isolated trees that provide protection against the mammalian nest predators that are abundant in the forest (Robinson 1985a, Table 1). In addition to a colony of up to 112 simultaneously active nests of *Cacicus cela* and up to 28 active nests of *Psarocolius angustifrons*, the tiny one-tree island on the southern edge of Cocha Cashu (Fig. 1) has been used for nesting by *Tigrisoma lineatum* (five nests in 11 breeding seasons, 1979–1989), *Opisthocomus hoazin* (seven nests), *Columba cayennensis* (two nests), *Glaucis hirsuta* (one nest), *Furnarius leucopus* (six nests), *Pitangus sulphuratus* (eight nests), *Megarynchus pitangua* (one nest), *Myiozetetes similis* (16 nests, five in one season [1981]), *Legatus leucophaius* (three nests pirated from *Cacicus cela*, Robinson 1985b), *Tyrannus melancholicus* (four nests), *Tityra cayana* (two nests), *Donacobius atricapillus* (three nests), *Turdus ignobilis* (two nests), *Paroaria gularis*, (13 nests), *Thraupis episcopus* (seven nests), *Cacicus solitarius* (one nest), *Icterus icterus* (two nests pirated from *C. cela*: Robinson 1985b), and *Scaphidura oryzivora* (at least three parasitized nests of *Psarocolius angustifrons*: Robinson 1988). None of these nests was depredated by mammals, but many were attacked and depredated by raptors (Robinson 1994) and *Ramphastos cuvieri*, (Robinson 1985a).

The various fig trees growing along the edge of the lake were also heavily used as nest sites by many birds even though they were accessible to mammals. When a troop of *Cebus apella* monkeys crossed the shrubby marsh to attack a huge fig at the southern end of the lake on 22 August 1985, they depredated 12 nests of *Cacicus cela* and one nest each of *Paroaria gularis, Opisthocomus hoazin, Dryocopus lineatus, Columba cayennensis, Myiozetetes similis*, and *Thraupis episcopus*. Two younger *Cebus apella* repeatedly moved towards a nest of *Pitangus sulphuratus*, but the flycatcher was able to drive them away by repeatedly diving at them and even pecking one of them on the face.

From 1979–1989, I also observed *Cebus apella* depredating the nests of the following species in lake-margin figs and trees: *Opisthocomus hoazin* (six nests), *Columba cayennensis* (three nests), *Aratinga weddellii* (one nest in a cavity chewed open by an adult male), *Ara severa* (one nest also in a cavity), *Pitangus sulphuratus* (three nests), *Myiozetetes similis* (two nests), *Megarynchus pitangua* (one nest taken by an adult male), *Legatus leucophaius* (one pirated nest of *Cacicus cela*), *Tyrannus melancholicus* (one nest), *Paroaria gularis* (two nests), *Saltator coerulescens* (one nest), *Thraupis episcopus* (one nest), *Icterus icterus* (one pirated nest of *C. cela*), and *C. cela* (214 nests in 12 colonies). *Cebus apella* may also have been responsible for the four oven-like *Furnarius leucopus* nests depredated by an animal that excavated a hole in the side. *Psarocolius angustifrons* colonies were always placed in trees where monkeys could not reach them. I never witnessed a successful attack by monkeys on any lake-margin oropendola colony.

Cavity nests.—The high incidence of snags (standing dead trees) in marshes of the lake bed created ideal sites for cavity-nesting birds. Many snags could not be reached by monkeys, which otherwise seemed able to depredate cavity nests. One large male *C. apella* attacked a nest of *Dryocopus lineatus* by chewing a hole in the dead wood 5–10 cm below the nest entrance. When the new hole was large enough, the monkey pulled out a relatively large but still mostly featherless nestling. Two large dead *Cedrela odorata* that remained standing in the southern Cashu lake bed at least through 1989 were used by the following cavity nesters: *Ara severa* (two nests), *Aratinga weddellii* (six nests), *Pteroglossus castanotis* (two nests), *P. inscriptus* (one nest), *Dryocopus lineatus* (one nest), *Campephilus melanoleucos* (one nest), *Melanerpes cruentatus* (active nest sites all 11 years), *Tityra inquisitor* (two nests), *T. semifasciata* (nine nests), *T. cayana* (11 nests). As far as I know, none of these nests was depredated. A third *C. odorata* snag west of the field station was used all 11 years by *Melanerpes cruentatus*.

A small, old snag of ca. 35 cm dbh and 6-m-tall protruding from the marsh was the site of intensive interspecific competition for the two active cavities. During 1979–1986 before the top rotted away, the cavities in this stump were visited by *Dryocopus lineatus* (nested once), *Nasica longirostris, Xiphorhynchus picus, Dendrexetastes rufigula, Tityra semifasciata* (nested three times), *T. cayana* (nested twice), *Amazona ochrocephala* (nested once), *Ara severa* (nested once), and *Aratina weddellii* (nested once). Prolonged aggressive interactions were observed between *Ara severa* and *Amazona ochrocephala, Tityra semifasciata* and *T. cayana*, and *T. cayana* and *Melanerpes cruentatus* when both species appeared to be competing for one of the nest sites. *Amazona ochrocephala* appeared to "win" the largest nest site in July of 1981; *Ara severa* moved into the nest when *Amazona* was finished in November 1981. Similarly, after several days of intermittent aggression, *Tityra semifasciata* occupied the cavity, apparently having "won" its interaction with *T. cayana*. *Tityra cayana* occupied a nest cavity in a lake-bed tree after a long aggressive interaction with *T. inquisitor*. *Ara severa* may also have displaced a pair of *Aratinga weddellii* nesting in the snag west of the field station. A group of five *Pteroglossus castanotis* was also observed fighting with a group of four *P. inscriptus* in the lake-bed *Cedrela* snag. *Pteroglossus castanotis* occupied the cavity at the end of the fight.

Tachycineta albiventer nested in shallow cavities in the still-exposed branches of trees fallen into the water. I observed depredation of three of their nests by *Pteroglossus inscriptus*, and one by *P. castanotis*. Of the *Paroaria gularis* that also occasionally nested in these cavities, at least one lost its nest to a group of *P. inscriptus* and another to a pair of *P. castanotis*.

Interactions with avian nest predators.—Monkeys could not reach all nest sites, but once they approached a nest, mobbing rarely deterred them (Table 1). Avian predators, on the other hand, could reach all nests (except perhaps for some cavities), but most could be deterred by mobbing and other active forms of defense. The effectiveness of mobbing by *Cacicus cela* against toucans and caracaras depended upon the number of birds mobbing (Robinson 1985a); the relatively low success rate of attacks by toucans and caracaras (Table 1) reflected the effectiveness of mobbing. *Pteroglossus inscriptus* is apparently too small to attack colonies of *C. cela* successfully (Table 1), and most toucans and caracaras did not attack colonies with more than 50 active nests (Robinson 1985a). Similarly, *Psarocolius angustifrons* are large enough (200–500 g) to deter predator attacks by all toucans and caracaras (Table 1: $n = 14$ attacks). I also saw a group of eight *Gymnoderus foetidus* (including two males) repeatedly dive at and chase away three *Ramphastos cuvieri* that were within 100 m of two known nests near the field station (Fig. 1).

The only defense caciques and oropendolas had against the large *Buteogallus urubitinga* was to hide their active nests amid abandoned nests; in general, neither *Cacicus cela* nor *Psarocolius angustifrons* ever mobbed *B. urubitinga* (Robinson 1985a). Interestingly, however, the much smaller *Megarynchus pitangua*, *Myiozetetes similis*, and *Pitangus sulphuratus* succeeded in chasing *B. urubitinga* away from the vicinity of their nests on 11 of 12 approaches (Table 2). These flycatchers dived on the hawks and pecked them on the back of the head, which appeared to distract the much larger hawk. The greater aerial maneuverability of the smaller flycatchers might enable them to chase away predators that the larger colonial icterids cannot deter. Alternatively, the value of small flycatcher eggs and nestlings to *B. urubitinga* may not have been high enough to make it worth the cost of being mobbed.

Mobbing by the "yellow-bellied" flycatchers appeared to be particularly effective in defense of nest sites (Table 1). In addition to chasing away larger predators such as *B. urubitinga* and *C. apella*, the majority of attacks by caracaras and toucans on their nests were unsuccessful (Table 1). "Yellow-bellies" sometimes mobbed in multi-species groups and even attacked predators of adult birds (*Micrastur semitorquatus* and *Accipiter bicolor*). These flycatchers sometimes attacked *Ramphastos cuvieri* as they flew across the lake, even riding on the backs of their necks for short distances. On two occasions, I saw *Myiozetetes similis* attack a flying *Crotophaga major* and drive it into the water. Other species benefit from this mobbing as well; I have on eight occasions seen yellow-bellied flycatchers chasing toucans and caracaras out of cacique colony trees, even though only cacique nests were under attack. Similarly, I twice saw *Pitangus sulphuratus* and *M. similis* chase parasitic *Scaphidura oryzivora* out of trees where oropendolas were nesting.

One way that some nest predators circumvented prey defenses was by attacking in groups. I twice observed groups of five and six *P. inscriptus* attacking the nests of *Columba cayennensis* nesting in dense shrubs 1–2 m above the water. In both cases, the araçaris surrounded the nests. Each time the pigeon lunged toward an araçari, another araçari would attack the nest from behind. The defense of one nest was successful; after roughly 10 min of attacking, the araçaris left the nest. After about 3 min the two small nestlings in the other nest were carried off by araçaris. These attacks appeared to represent coordinated group attacks on the part of *P. inscriptus*. In both attacks, only a single pigeon was present to defend the nest. *Pteroglossus inscriptus* was also observed carrying an advanced nestling of *Paroaria gularis* on 31 August 1982. In September 1982, I also saw four *P. inscriptus* attacking a nest of *Donacobius atricapillus*. Although I could see few details, one *P. inscriptus* flew away carrying something about the size of a small nestling and the nest was no longer active afterwards.

Groups of three to four *Pteroglossus castanotis* also occasionally attacked colonies of *Cacicus cela* 40 m up in a tall tree bordering Cocha Cashu. Two different groups of three *P. castanotis* attacked this colony at least 18 times in September and October 1986 and 23 times in September 1987. A group of three *P. castanotis* also depredated a nest of *Columba cayennensis* in a dense shrub about a meter above water level. The foliage, however, was too dense to allow detailed observations of the tactics used by *P. castanotis*.

Ramphastos cuvieri, on the other hand, rarely foraged in groups when attacking colonies of *Cacicus cela*, and nests of *Columba cayennensis*, *Pitangus sulphuratus*, and *Myiozetetes similis*. The latter attacks on the flycatchers were carried out in spite of intensive mobbing. In other areas, however, *Ramphastos* toucans have been observed in coordinated attacks on nests (Mindell and Black 1984).

TABLE 2
Attacks by Predators of Adult Birds Witnessed along the Lake Margin, Including a Subset of Data from Robinson (1994a)

Predator	Prey	Attacks	% Successful
Accipiter bicolor	Flocks of *Brotogeris sanctithomae* in figs	6	16.7
	Birds at *Cissus* vine (*Turdus ignobilis*?)	1	100.0
	Birds at flowering trees (*Cacicus cela*)	10	10.0
	Cacicus cela colonies	13	0
	Porphyrula martinica	1	0
	Birds at *Coussapoa* (*Cacicus cela*)	3	33.0
Spizastur melanoleucus	*Psarocolius angustifrons*	1	0
Spizaetus ornatus	*Psarocolius angustifrons*	4	0
	Porphyrula martinica	3	100
	Cacicus cela	1	0
Falco rufigularis	Swallows	27	0
	Cacicus cela	6	0
	Chordeiles rupestris	4	0
Buteo albonotatus	*Cacicus cela* colonies	3	0
	Tyrannus melanocholicus	1	0
Micrastur semitorquatus	*Cacicus cela*	14	0
	Other birds	12	0

Interactions with Predators of Adult Birds

Black Caimans (*Melanosuchus niger*) were abundant on Cocha Cashu and ate birds at least occasionally. One Black Caiman appeared to catch a Sungrebe (*Heliornis fulica*) under overhanging vegetation in November 1986. Although I could not be certain, a sudden lunge by a White Caiman (*Caiman crocodilus*) at a group of four *Jacana jacana* appeared to be successful—I only saw three fly away. Predator attacks may explain why few adult jacanas at Cocha Cashu have all of their toes. An attack by a Black Caiman on a *Laterallus melanophaius* also appeared to be successful, but again, I could not be certain. Predation by caimans on lake-margin birds may explain why most birds of marshes and open water give loud alarm calls every time they hear a loud splash.

Other occasional predators of birds of the lake edge include giant otters (*Pteroneura brasiliensis*) and *Ardea alba*. Giant otters twice lunged at *Laterallus melanophaius* feeding at the edge of a marsh; one attack may have been successful. On three separate occasions, I also saw *A. alba* eating *Laterallus melanophaius*. On one of these occasions, a group of four *Porphyrula martinica*, two *Gallinula chloropus*, a *Jacana jacana*, and an *A. alba* were feeding together on the edge of a marsh. The *L. melanophaius* approached to within about 0.5 m of the egret, which then quickly grabbed it. The heron flew off to a low branch, shook the crake, and ate it after about 10 min. On the other two instances, I saw *A. alba* carrying adult *L. melanophaius*. All three observations of *A. alba* eating *L. melanophaius* were late in the dry season (26 September 1979; 29 September 1981; 13 October 1984) when low water levels may have forced many marsh-nesting species to feed on exposed mudflats. Evens and Page (1986) also observed *Ardea alba* eating *Laterallus jamaicensis* in San Francisco Bay, U.S.A., during high tides when *L. jamaicensis* was forced to feed in the open.

Although colonies of *Cacicus cela* and *Psarocolius angustifrons* were fully exposed to predators, I never saw a successful attack on an adult (Table 2). Their vigilance and frequent alarm calls appeared to provide ample warning and gave them time to escape. Of 68 attacks by six species of raptors (*Accipiter bicolor*, *Micrastur semitorquatus*, *Falco rufigularis*, *Buteo albonotatus*, *Spizastur melanoleucos*, and *Spizaetus ornatus*), none was successful. Only twice did a predator come even as close as 2 m from a fleeing bird (one by *B. albonotatus* and the other by *A. bicolor*).

Cacique colonies were attacked at different rates depending on their position relative to the forest (Table 3). Forest-dwelling raptors such as *A. bicolor* and *M. semitorquatus* primarily attacked colonies within 25 m of the shore, whereas colonies in the middle of the lake were attacked at a higher rate by *B. albonotatus*. Both attacks by *S. ornatus* were also on colonies close to the forest. Attack rates at all colonies, however, were low (less than one attack/40 hours of observation). Caciques on the large island colony spotted *B. albonotatus* on the island long

TABLE 3
ATTACK RATES BY PREDATORS AT DIFFERENT COLONIES OF *Cacicus cela* DURING ALL FIELD
SEASONS, 1979–1987

Predator	No. attacks/100 observation hours[1] (total no. witnessed) Cocha Cashu			
	Island colony	Colonies <25 m from forest	Marsh colonies >25 m from forest	Forest colonies[2]
Accipiter bicolor (adult)	0.13 (3)	1.18 (9)	0.00 (0)	0.71 (1)
Buteo albonotatus	0.00 (0)	0.00 (0)	2.50 (3)	0.00 (0)
Micrastur semitorquatus	0.09 (2)	1.58 (12)	0.00 (0)	0.00 (0)
Falco deiroleucus	0.26 (6)	0.00 (0)	0.00 (0)	0.00 (0)

[1] Observations hours: Island colony, 2,300; Colonies <25 m from forest, 760; Small colonies >25 m from forest, 120; Forest wasp nest colonies, 140.
[2] Forest colonies were those high in trees around wasp nests at least 100 m from Cocha Cashu.

before they got close enough to launch an attack (see below). The attacks on the smaller (<10 nests) colonies in marsh shrubs may have been possible because of reduced vigilance in small colonies.

Caciques showed a high degree of discrimination among potential predators at the island cacique colony (Table 4). Approaches by such potentially dangerous predators as *A. bicolor*, *B. albonotatus*, *B. brachyurus*, and *M. semitorquatus* elicited strong reactions (abandonment of the colony) when predators were still far (>50 m) from the colony. *Buteo albonotatus* elicited reactions at the greatest distances, which indicates that the caciques were not being fooled by its resemblance to *Cathartes aura* or *C. melambrotus* (Willis 1963, 1965). Often, I would hear the distinctive, high-pitched alarm calls of *Monasa nigrifrons* first and then hear the alarm calls

TABLE 4
REACTIONS OF *Cacicus cela* TO APPROACHES OR FLY-BYS BY POTENTIAL PREDATORS AT THE
LARGE (20–100 ACTIVE NEST) ISLAND CACIQUE COLONY, 1980–1985, DURING 1200
OBSERVATIONS HOURS

	$X \pm SD$ distance at first reaction by Caciques (n)[1]	Median % of birds that left[2]	% of reactions in which birds		No. approaches
			Dived for cover[3]	Flew to treetops[3]	
Dangerous predators					
Accipiter bicolor	110 ± 70 (44)	100	2	98	103
Buteo albonotatus	150 ± 60 (20)	100	97	3	31
B. brachyurus	110 ± 50 (7)	100	86	14	7
Spizaetus ornatus	50 ± 30 (5)	100	17	83	12
Spizastur melanoleucos	80 (1)	100	17	83	6
Micrastur semitorquatus	100 ± 60 (6)	100	0	100	17
Other raptors					
Leptodon cayannensis	70 ± 80 (9)	25–50	100	0	5
Harpagus bidentatus	30 ± 20 (7)	25–50	100	0	4
Ictinia plumbea	70 ± 50 (5)	25–50	100	0	6
Buteo magnirostris	30 ± 30 (7)	1–25	100	0	2
Leucopternis schistacea	30 ± 10 (12)	1–25	100	0	8
Pandion haliaetus	30 ± 30 (15)	0	100	0	12
Herpetotheres cachinnans	10 ± 10 (4)	25–50	100	0	5
Falco deiroleucus	40 ± 30 (13)	1–25	100	0	12
Non-Raptors					
Sarcoramphus papa	170 ± 90 (5)	25–50	100	0	5
Coragyps atratus	100 ± 40 (10)	25–50	100	0	10
Cathartes melambrotus	60 ± 34 (66)	50–75	100	0	68
Anhima cornuta	20 ± 20 (6)	25–50	100	0	6

[1] Distance from colony when I first heard an alarm call or saw birds leave the tree.
[2] Estimates in 25% increments of the proportion of birds that left the tree in response to the approach of that potential predator.
[3] Birds tended either to dive for cover or fly up to exposed treetops at the edge of the lake.

of caciques and oropendolas. A loud alarm call by *Psarocolius angustifrons* was almost always ($n = 21$ of 22 records) followed by abandonment of the colony by both caciques and oropendolas. Reactions to less-threatening species (e.g., *Leptodon cayanensis, Ictinia plumbea*) were typically much less extreme (<50% of the birds left the colony) and only occurred when the predators approached the colony closely (Table 4). Nevertheless, caciques did panic occasionally when harmless species (e.g., *Cathartes melambrotus, Sarcorhamphus papa*) such as vultures flew by. Perhaps at a distance, *Cathartes* vultures resemble *Buteo albonotatus* enough to fool a few individuals and provoke alarm calls.

Caciques also reacted differently to raptors that attacked from the forest than to those that attacked from above (Table 4). When approached by soaring buteos and other species flying over the colony, caciques nearly always dived for cover under the colony in low, dense shrubs (Table 4). In contrast, caciques flew to tree tops when attacked by raptors that attacked from the forest interior (Table 4: *A. bicolor. M. semitorquatus*). This latter tactic may provide better visibility against predators that ambush their prey. Diving into dense cover, on the other hand, may be the most effective way to escape predators that hunt in the open. Caciques seemed to use different alarm calls for the two kinds of predators, but I lack the recordings to test this hypothesis.

Breeding Systems

Monogamy.—Species for which at least two marked pairs were observed nesting without assistance from any other birds include: *Columba cayennensis, Eubucco tucinkae, Colaptes punctigula, Furnarius leucopus, Pitangus sulphuratus, Megarynchus pitangua, Myiozetetes similis, Tyrannus melancholicus, Tityra semifasciata, Tachycineta albiventer, Turdus ignobilis, Paroaria gularis, Saltator coerulescens, Thraupis episcopus,* and *Icterus icterus.* Although *P. gularis* was behaviorally monogamous and territorial, flocks of floaters, mostly in juvenal plumage, moved around the lake and fed in the territories of breeding adults (Eason 1992).

Polygyny.—The best-documented cases of polygyny were in the two colonial icterids, *Cacicus cela* and *Psarocolius angustifrons* (Robinson 1986a). Groups of *Scaphidura oryzivora* numerically dominated by females foraged and visited oropendola colonies together (Robinson 1988). In these flocks, males displayed to females foraging on beaches.

Circumstantial evidence suggests that *Cairina moschata* may also have a male dominance-based mating system. On two separate occasions (Sept. 1979 and October 1981) I observed long, violent fights between males. Both fights lasted at least 20 min. and involved face-to-face butting of breasts while they loudly splashed their wings in the water, pecking each other on the head and breast, grabbing the feathers on each other's napes and hitting each other with the crook of their wings (Sick 1993). The fights were audible at least 700 m away. At the end of both fights, the apparent loser was held underwater and escaped only after swimming about 5 m underwater. These fights were considerably more violent than those described in the limited literature on this species (Bolen 1983). Because male *C. moschata* have distinctive patterns of red warts around their bare faces, I was able to confirm that the winners of both fights were later associated with small groups of 2–4 females accompanied by no other males. What appeared to be one of these males (I could not be sure) was observed copulating with a female 30-35 m up on a large horizontal branch of a *Ficus insipida*. I have also observed *C. moschata* copulating in trees (a *Cedrela odorata* and a *Ficus perforata*) on two other occasions (Sept., 1980 and Nov. 1983).

Group-living species.—Many typical lake-margin birds either bred cooperatively or lived in groups (possibly extended families) for much of the year. In addition to cooperative breeding previously documented in *Donacobius atricapillus* (Kiltie and Fitzpatrick 1984), I observed cooperative breeding in two other species in which such behavior had never before been documented (Brown 1987): *Ramphocelus carbo* and *R. nigrogularis. Ramphocelus nigrogularis,* for which I found two nests (the first nests of this species located: Isler and Isler 1987), bred in groups as large as eight individuals. One nest, 2 m up at the base of a broken *Astrocaryum* palm frond in the field-station clearing, was attended by five adults. The two eggs in this nest were depredated before they hatched. The second nest was located 1 m above the water in a tree that had fallen into the lake. The nest was situated in a mixture of living vines and dead leaves and contained two pale blue eggs with brown speckling. Six individual *R. nigrogularis* attended this nest, including two marked adults. Two adults had bright red plumage and two had duller plumage. At least four different adults fed the nestlings soon after they hatched. The nestlings disappeared four to five days after hatching. Cooperative breeding in *R. carbo* was documented at a nest adjacent to the field station in September 1988. This nest was 1 m up in an unidentified sapling and was closely attended by five adults, all of which fed the nestlings repeatedly. Only one adult was banded, but at one point, all five adults visited the nest in quick succession. All three of these species were usually observed in groups; during late December 1981, a group of 12 *R. nigrogularis* was repeatedly observed together. A closely related species in Central America (*R. passerinii*), however, does not breed cooperatively; even though they feed in groups, they nest separately (Skutch 1954).

Several other species may have bred cooperatively. *Pteroglossus inscriptus* nested roughly 20 m up in a cavity of an unidentified tree species. A group of four birds was observed repeatedly in the vicinity of the tree and at one point, three birds brought food to the nest hole at the same time. Each bird delivered its food, and then all three flew off together. Otherwise, I only saw single birds bringing food to the nest. Other *Pteroglossus* spp. are not known to breed cooperatively (Skutch 1958). *Pteroglossus castanotis* often foraged in groups of 3–4, but seemed to form pairs when breeding. *Opisthocomus hoazin* also occurred in groups and defended nests against predators as a group; elsewhere, it is known to nest cooperatively (Strahl 1988).

Crotophaga major lived in communal groups of up to 10 individuals and at times foraged in groups as large as 24 individuals. *Crotophaga major* also displayed in groups. When I approached a group, they often formed a circle with their beaks pointed toward the center, their bodies held horizontally with tails pointing outwards, and made a low vocalization that sounded like an idling motor. When I left in 1981, two nests were being constructed by groups of 8 individuals. I strongly suspect but could not confirm that this species nested communally as do the other two species of *Crotophaga* (Vehrencamp 1978).

MISCELLANEOUS NATURAL HISTORY OBSERVATIONS

Cocha Cashu as a refuge during floods.—Because Cocha Cashu was protected from the river by levees, it provided a refuge for riverine birds during normal wet-season floods. Species that foraged on Cocha Cashu during floods included *Ardea cocoi*, *Egretta thula*, *Pilherodias pileata*, and *Hydropsalis climacocerca*.

Effects of the flood of 1982.—During the "El Niño" year of 1982, a major flood swept through Cocha Cashu in late October. After the floodwaters receded, the two major marshes (Fig. 1) were much smaller and more open. Population declines between 1980 and 1985 of many bird species characteristic of marshes (e.g., *Crotophaga ani*, *Anhima cornuta*, *Gallinula chloropus*, and *Laterallus melanophaius*) (Appendix) were most likely caused by this habitat alteration. This flood also drowned the nests of 18 *Cacicus cela*, 5 *Myiozetetes similis*, 1 *Furnarius leucopus* among others.

Insects flushed by mammals.—When troops of monkeys foraged along the lake margin, they were often accompanied by several lake-margin birds, including *Aramides cajanea*, *Furnarius leucopus*, *Sclateria naevia*, *Crotophaga major*, *C. ani*, *Pitangus sulphuratus*, *P. lictor*, and *Donacobius atricapillus*. One group of 8 *C. major* caught three green katydids (Orthoptera: Tettigoniidae), a mantid, and a green lizard flushed by a mixed troop of *Cebus apella* and *Saimiri sciureus* on 16 October 1981. When Capybaras walked through dense marshy vegetation, they were sometimes followed by *Pitangus lictor*, *P. sulphuratus*, *Myiozetetes similis*, *Crotophaga ani*, *Laterallus melanophaius*, *Donacobius atricapillus*, and *Jacana jacana*. These species caught insects flushed by the Capybaras and on the trail of open water that formed behind the moving Capybaras. *Laterallus melanophaius* was twice observed catching and eating blood-sucking tabanid flies that were feeding on Capybaras. The crakes would dart towards the Capybara's flank, pluck a fly, and then quickly move back into the grass. Robinson (1988) described how *Scaphidura oryzivora* perched on the backs of Capybaras and Tapirs to catch Tabanids.

Nectar robbing of Heliconia flowers.—Two icterids, *Agelaius xanthophthalmus* ($n = 3$ observations) and *Icterus icterus* ($n = 7$ observations) were observed robbing *Heliconia* flowers by pecking holes in the base. Both species have sharp beaks that may be well-adapted for this purpose.

Ecology and behavior of Gymnoderus foetidus.—The poorly known (Snow 1982) Bare-necked Fruitcrow (*Gymnoderus foetidus*) occurred very erratically over southern Cocha Cashu. Counts of *G. foetidus* flying over cacique colonies showed no consistent seasonal pattern (Table 1). In 1980, their year of greatest abundance, *Gymnoderus* did not become abundant until late September and October, when I found six nests (for locations of nests, see Fig. 1). In 1982 and 1983, fruitcrows were present in all months. The other six nests located were in October and November of 1982 (four nests) and 1983 (two nests). The high abundance in November 1988 (Table 1) reflected heavy use of a fruiting lauraceous tree behind the island colony where I conducted my

TABLE 5
SEASONAL AND ANNUAL OCCURRENCE OF *Gymnoderus foetidus* FLYING OVER COCHA CASHU, AN OXBOW LAKE OF THE MANU RIVER

Year	July	August	September	October	November
			No. sightings/No. individuals per 10 census hours (No. census hours)		
1980	—	0 (12)	0.7/2.3 (60)	3.8/11.2 (50)	2.5/55.0 (4)
1981	1.3/1.3 (32)	1.0/2.1 (62)	0 (16)	0 (8)	0 (12)
1982	2.1/2.1 (58)	3.5/5.2 (62)	2.7/4.2 (60)	7.7/16.0 (62)	3.8/3.8 (24)
1983	—	1.1/4.2 (62)	0.3/0.5 (60)	1.1/1.5 (62)	3.6/4.3 (14)
1984	—	0 (10)	0 (44)	0 (4)	—
1985	—	—	1.0/1.0 (10)	0.4/2.9 (28)	0 (14)
1986	—	—	0 (12)	0 (4)	0 (6)
1987	—	—	0.4/0.4 (26)	0 (10)	0 (4)
1988	—	0 (4)	1.5/3.0 (40)	1.4/3.6 (28)	1.7/21.1 (18)
1989	—	0 (24)	0 (12)	—	—

censuses. Eleven of 12 nests were within 25 m of the lake (Fig. 1), eight in emergent *Dipteryx pentandra* (Leguminosae) trees at heights of 40-50 m. Two other nests were located in *Cedrela odorata* (Meliaceae) at heights of 15 m and 17 m. The remaining two nests were 20 m up in a *Ficus insipida* (Moraceae) and 25 m up in an unidentified tree. I only saw females visit these nests.

The life history of this species in the Cocha Cashu area appears be closely tied to its primary food, trees of the family Lauraceae. The high populations of *Gymnoderus* in October and November 1980 (Table 5) were centered around at least six fruiting lauraceous trees on the peninsula between the two western arms of the lake (Fig. 1). Nesting females sometimes joined others in flocks of as many as 25 birds when visiting clusters of fruiting lauraceous trees.

Gymnoderus foetidus displayed at fruiting lauraceous trees. In 1980, at least eight different males with "adult" plumage (extensive folds of blue skin on neck and grayish wings) flew conspicuously back and forth between fruiting trees in an area of intense feeding activity. On at least three occasions one male supplanted another at the end of a flight of approximately 10 m. Males were observed displaying in this way on all four visits in 1980. In 1982, three males with differing amounts of folded bare skin on their necks established a dominance hierarchy in a fruiting lauraceous tree. The male with the most extensive and bluest folds of the neck skin supplanted a male with darker, less folded skin, which in turn supplanted a male with only a small dark patch of bare skin on the neck. The most dominant male supplanted the middle-ranking male four times, and the middle-ranking male supplanted the subordinate eight times in a period of approximately 10 min. All three males left the tree together at the end of the bout of supplanting.

Foraging ecology and behavior of Galbalcyrhynchus purusianus.—*G. purusianus* appeared to be a wasp specialist: all 21 recorded prey items were wasps or bees caught during aerial sallies. Wasps were smashed against branches repeatedly before being swallowed whole. The very long bill of this species may aid in catching the wasps and provide protection against being stung. *Galbalcyrhynchus purusiana* displayed in groups of four to six. I once saw two groups of five individuals interacting aggressively. Each group perched together and sang while two individuals locked their long beaks together lengthwise and pulled back and forth. I never found a nest cavity, however, and was unable to determine if more than two adults fed young.

Mitu tuberosa.—In addition to the fruit regularly consumed by this species, I have observed *M. tuberosa* eating fish scales surrounding the den entrance of giant otters (*Pteroneura brasiliensis*). *M. tuberosa* occasionally was observed feeding on mudflats exposed during the dry season.

Opisthocomus hoazin.—In addition to its normal diet of young leaves, I also once saw *O. hoazin* eating snail (*Pomacea*) eggs off a blade of grass 15–20 cm above the water.

Longevity in Xiphorhynchus picus.—The small population of *X. picus* on Cocha Cashu seemed to contain very long-lived individuals. One bird was caught seven years after its initial capture. The annual return rate of marked adults was 86% ($n = 29$ bird-years of five individuals).

Nest site of Eubucco tucinkae.—A presumed nest hole of this species was located in a horizontal, 8–10-cm diameter dead fig branch 1.5 m above the water in September 1979. A pair was observed excavating the hole and entering it for long (>10-min) periods during the day. It was abandoned or depredated before I looked inside to check its contents.

Nests of Colaptes punctigula.—Of eight nests of this species located on Cocha Cashu, six were in cavities excavated in small (<20 cm) dbh *Cecropia* trees at heights of 4–6 m. The other two were in dead branches of *Ficus trigona* at 5–6 m. Four of the *Cecropia* trees later broke at the site of the nest suggesting that these nests significantly weakened the trees. One long-lived marked individual (at least six years old) nested in four consecutive *Cecropias* during the 1981 breeding season (July–December).

Foraging substrate use in Nasica longirostris.—*N. longirostris* probed in knotholes of trunks and large (>15 cm diameter) branches on three separate foraging bouts. However, 16 of 18 substrates searched during three other foraging bouts were bromeliad epiphytes. The birds used their long bills to probe in the base of the leaves. Hilty and Brown (1986) described similar foraging behavior. One probe resulted in the capture of a 3–4-cm brown katydid (Orthoptera: Tettigoniidae).

Donacobius atricapillus *foraging.*—On windy days, I twice observed this species perched motionless on moving branches roughly 20 cm above the water. On both occasions, the bird was perched with its bill pointed toward the water where several dragonflies were sitting on grass blades. When the branch dipped low enough, the bird would lunge and grab one of the dragonflies. The bird appeared to be using the motion of the branch to approach the prey without moving its body.

Turdus ignobilis.—This species was an abundant breeder in some years, but may be somewhat nomadic. Of 83 individuals captured in mist nets, only five were recaptured in subsequent years and I only resighted one of 46 individuals that were color-banded in subsequent years. Their movements may be tied to the availability of fruit (virtually all fecal material left during the banding process contained purple fruit), but I do not know which species they ate. At least one bird that nested along the lake was observed feeding in an unidentified fruiting tree along the Manu River 700 m away from its nest site in November 1982.

DISCUSSION

Oxbow lakes offer dramatically different conditions for foraging and nesting than the surrounding forests. The marshes, shrubby borders, open water, and isolated trees of oxbow lakes provide habitat for birds of early successional vegetation and opportunities for forest birds to take advantage of food resources and nest sites rarely available in forests. Cocha Cashu was used regularly by at least 186 species of birds and was visited by at least 117 additional species during the 1979-1989 study period. Some of these species were remarkably abundant in the limited amount of lake-margin vegetation, a result consistent with other studies showing high population densities in early successional Amazonian vegetation (Rosenberg 1988, Robinson and Terborgh 1997). These species can be roughly divided into: (1) a core group of territorial species, some of which were far more abundant than typical forest species (Terborgh et al. 1990), mostly restricted to the lake margin; (2) species that bred in lake-margin vegetation but foraged primary in the surrounding forest (mostly in mid-successional vegetation); (3) species from other habitats that aggregated at particular food resources such as fruiting trees (*Gymnoderus foetidus*) and insect emergences (e.g., termite swarms); and (4) purely aquatic and wetland species, many of which moved about seasonally as water levels fluctuated and many of which suffered population declines as a result of the flood of 1982 (Willard 1985). Because most birds in the latter three groups of species had extremely variable populations, the bird communities of oxbow lakes seemed far more loosely organized than those of mature forest (Terborgh et al. 1990). Indeed, the community was more like an aggregation of species than a coevolved community in which heterospecific social groups were a conspicuous feature (e.g., Munn 1985). Bird communities of aquatic habitats of the Venezuelan llanos also seem to be dominated by many seasonal, erratic species (Thomas 1979). In the following section, I first discuss the contribution of oxbow lakes to regional species diversity and then discuss how sociality helps lake-margin birds deal with the resources available and cope with predation. I end with a discussion of the implications of sociality for the evolution of mating systems.

CONTRIBUTION TO SPECIES DIVERSITY IN THE MANU REGION

Oxbow lakes contained few species not found in rivers, river-edge successional vegetation, streams, and other backwater habitats elsewhere in Amazonia (see also Remsen and Parker 1983, Parker and Remsen 1987). Exceptions included *Opisthocomus hoazin*, *Aramus guarauna*, and

Gallinula chloropus, which were only found on oxbow lakes in the Cocha Cashu area. A large (>100 ha) marshy backwater four km from Cocha Cashu was the only other place in the Manu region where *Busarellus nigricollis* (see also Remsen 1990), *Colaptes punctigula*, *Ixobrychus exilis*, *Icterus icterus*, *Crotophaga ani*, *Porphyrula martinica*, and *Rallus nigricans* were found. *Myrmotherula surinamensis* was found along one 10-m wide stream flowing through *terra firme* forest; otherwise, I only found them along oxbows. *Cissopis leveriana* was mostly restricted to oxbow lakes, but also occurred in a flood-damaged section of forest and in a marshy backwater. *Xiphorhynchus picus* and *Crotophaga major* occasionally nested along rivers, but were mostly found along oxbow lakes. Many piscivores also foraged along forest streams and, occasionally, along rivers. Other species rare away from oxbow lakes were *Anhinga anhinga* (rivers also), *Cochlearius cochlearius* (wooded swamps; also see Willard 1979), *Heliornis fulica* (rivers also), *Piaya minuta* (occasionally large areas of river-edge early successional vegetation), *Donacobius atricapillus* (marshy backwaters also), *Ramphocelus nigrogularis* (shrub swamps; see also Parker and Remsen 1987), and *Agelaius xanthophthalmus* (marshy backwater; see also Parker 1982). Many characteristic birds of lake margins also occupied the early to mid-successional forest that typically grows on the inside of oxbow lakes, which were formerly the inside of meander loops (Robinson and Terborgh 1997). Bird communities of oxbow lakes therefore consist of a mixture of species from various successional stages, backwaters, and other aquatic habitats rather than a distinct community with its own specialized species. Small oxbow lakes, such as Cocha Cashu, therefore contribute more to local than to regional species richness. Even lake-margin specialists in the Manu Region, however, occur in other habitats elsewhere in their ranges (e.g., Hilty and Brown 1986).

ADVANTAGES OF GROUP LIVING

Group living offers many advantages to lake-margin birds. Groups of *Opisthocomus hoazin* defended their nests against predators, although often to no avail: I have seen *Buteogallus urubitinga* (Robinson 1994a) and *Cebus apella* eating eggs in nests surrounded by as many as eight calling Hoatzins. The advantages of group living in defense against predators have been demonstrated several times (reviewed by Koenig et al. 1992) and seems particularly plausible in an environment in which the densities of both nests and nest predators are so high.

Another advantage of group living is cooperative hunting. Groups of *Pteroglossus inscriptus*, *P. castanotis*, and *Daptrius ater* were observed attacking nests of birds that defend their nests by mobbing. Larger groups seemed to be better able to depredate nests because some birds can distract the defending parents while others grab the eggs or young. Bednarz (1988) also hypothesized that cooperative hunting was an important ecological factor underlying cooperative breeding in Harris' Hawk (*Parabuteo unicinctus*).

Competition for nest sites and food resources creates another situation in which group living would be advantageous (Emlen 1982, Emlen 1994). Intense inter-and intraspecific competition for limited cavity nests might select for group defense in *Melanerpes cruentatus* and the two species of *Pteroglossus*. Brawn (1990) hypothesized that groups of *Tachycineta thalassina* are more successful in interspecific competition for nest sites, and Koenig and Mumme (1987) found that another cooperative species of *Melanerpes* (*M. formicivora*) benefits from the advantages of defense of trees used for food storage. *Tyrannus tyrannus* gains access to fruiting trees by overwhelming other species through numerical dominance (Fitzpatrick 1980). Leighton (1986) hypothesized that cooperative breeding in some hornbills (Bucerotidae) in Borneo evolved because of selection for group defense of fruiting trees.

Interspecific competition for safe nest sites only seems well developed in cavity nesters, most of which use the lake margin for nest sites rather than foraging (e.g., *Tityra*, parrots). In every case in which two species were observed fighting over nest cavities, the larger species occupied the nest site after the fight ended. Robinson and Terborgh (1995) found that larger congeners also tended to be socially dominant to smaller ones.

Non-cavity nesters can benefit from interspecific nesting associations. The aggressive mobbing of predators by some of the "yellow-bellied" flycatchers (e.g., *Myiozetetes, Pitangus sulphuratus, Megarynchus*) can benefit nearby nests as well. Such advantages of multi-species nesting aggregations may be a secondary benefit derived from nesting in the sites best protected from the mammals.

In the most abundant territorial species, *Paroaria gularis*, nonbreeders lived in groups and territorial birds lived in monogamous pairs (Eason 1992). Floater flocks moved freely around the lake by overwhelming territory holders, which sometimes joined the flock when it passed through their territory after unsuccessfully defending their territory. The young floaters therefore have the advantages of group living and access to all the resources (fruiting and flowering trees) of the territory holders. As floaters mature, they eventually replace territory holders (Eason 1992).

Although lake margins lack multi-species flocks of the complexity shown by forest species (Munn 1985), a few territorial species foraged at least occasionally in mixed-species flocks. By foraging together, flocks of *Sclateria naevia, Hypocnemoides maculicauda, Myrmeciza hyperythra,* and *Myrmotherula surinamensis* might be able to benefit from predator protection without increasing food competition for food significantly, because each of these species forages differently (Parker 1982). Shorebirds on Cocha Cashu often form multi-species flocks consisting of just one individual from as many as seven species (Bolster and Robinson 1990). Many species respond interspecifically to alarm calls (S. Robinson, pers. obs.). Where avian predators are frequently present (Robinson 1994) and potential prey occur at high densities, it should benefit prey individuals to respond interspecifically to alarm calls.

The interactions among resources, nest predators and social organization are particularly evident among colonial Icterids (Robinson 1985a, b, 1986a,b). Isolated trees provide nest sites safe from mammalian predators of caciques and oropendolas. In addition, group defense provides good protection against most avian predators. The effectiveness of the caciques' anti-predator adaptations (e.g., Tables 2, 4) may explain why three other species parasitize or pirate their nesting behavior *(Legatus leucophaius, Icterus icterus,* and *Scaphidura oryzivora*: Robinson 1985a, 1988).

Nonterritorial monogamous birds that nest separately but forage gregariously include most parrots, *Gymnoderus foetidus,* and *Columba cayennensis*. Their monospecific flocks should provide protection against predators at fruiting trees, whereas their nests are either protected in cavities (parrots) or scattered about the lake in dense shrubs *(C. cayennensis)*. *Gymnoderus foetidus* also scattered their nests in emergent canopy trees but gathered in monospecific flocks when visiting fruiting trees. Such monospecific flocks would be particularly advantageous given the high incidence of predator attacks by *Accipiter bicolor* and *Micrastur semitorquatus* on lake-margin fruiting trees (Table 2, see also Robinson 1994a).

SOCIALITY AND BREEDING SYSTEMS

The consequences of high population densities and advantages of group living may be reflected in the variety of social systems of lake-margin birds. The high population density of several species relative to the limited availability of habitat might be a key selective factor favoring cooperative breeding in *Donacobius atricapillus* (Kiltie and Fitzpatrick 1984), *Ramphocelus carbo, R. nigrogularis, Opisthocomus hoazin,* and, possibly, *Pteroglossus inscriptus*. Koenig et al. (1992) proposed that the most important ecological constraint favoring cooperative breeding is not population density *per se*, but rather the population density relative to high-quality habitat. With lake-margin territories saturated, it may pay for young of many species to remain together with their parents and await nearby vacancies in territories or inherit their natal territory. Kiltie and Fitzpatrick (1984) proposed that habitat saturation was a crucial ecological constraint underlying the evolution of cooperative breeding in *D. atricapillus*. The existence of at least three other cooperative species in the same environment strongly supports this hypothesis. More studies of the breeding systems of other group-living species are needed to test this hypothesis.

The high concentrations of females made possible by localized resources such as those of oxbow lakes may have been key factors in the evolution of polygynous mating systems (Emlen and Oring 1977). *Cacicus cela* and *Psarocolius angustifrons,* two of the most abundant lake-margin birds, both have male-dominance-based polygynous mating systems (Robinson 1986a,b; Webster 1994). In these species, the only form of male parental care is mobbing nest predators. *Cairina moschata* appears to have a true harem polygynous mating system, which is exceedingly rare in birds (reviewed in Webster 1994). In the frugivorous *Gymnoderus foetidus,* males appear to form mobile display areas at their lauraceous fruiting trees and engage in interactions suggestive of male dominance. The occasional abundance of fruiting lauraceous trees may free males from parental care. Sociality in *Scaphidura oryzivorus* may also reflect a polygynous mating system (Robinson 1988). The synchrony of nesting in oropendolas, their chief host, would reduce competition among *S. oryzivorus* females, enabling them to form multi-female flocks. In contrast, *Molothrus ater*, which mostly searches for scattered nests in North America, searches alone and is primarily monogamous (reviewed in Robinson et al. 1995). The males that accompany *S. oryzivorus* flocks help distract defending oropendolas at colonies and display to females in the afternoon when they are foraging (Robinson 1988).

TABLE 6
Contrasting Breeding and Social Systems of Species That Nest on the Lake Margin (Appendix A, Not Including Seasonal Visitors or Species Found Only on the Edge ["e" of the Lake]) and Species That Breed in the Mature Floodplain Forest Bordering Cocha Cashu (Terborgh et al. 1990)

Mating system	Sociality	Lake-margin	Mature forest
Monogamous	Territorial	.57 (65)	.44 (105)
	Multi-species flocks	.06 (7)	.30 (71)
	Monospecific groups or flocks	.29 (33)	.13 (30)
Subtotal		.92 (105)	.87 (206)
Polygamous[a]		.07 (8)	.10 (23)
Polyandrous[b]		.01 (1)	.04 (10)
Total		(114)	(239)

[a] Includes lek-breeding and male-dominance based mating systems.
[b] Includes tinamous.

CONCLUSIONS

Cocha Cashu and other oxbow lakes add resource-rich, but variable habitats to the Manu landscape. Oxbow lakes provide foraging habitat for birds of riverine habitats (both aquatic and early successional vegetation on meander loops: Robinson and Terborgh 1997) and foraging and nesting habitat for forest birds (primarily from mid-successional stages). The availability of many of these resources varies daily (e.g., termite swarms), seasonally (e.g., mudflats: Bolster and Robinson 1990), annually (e.g., lauraceous trees: Table 5), over longer cycles (e.g., el niño flooding), or even stochastically (availability of colony sites that are not attacked by *Buteogallus urubitinga*: Robinson 1985a). Perhaps because of this variability and the low diversity of territorial, permanent resident species, complex multi-species flocks such as those found in mature forests (Munn 1985), rarely form along lake margins (Table 6). Instead, social organizations are dominated by monospecific groups or flocks that provide the same anti-predator functions as multi-species flocks (Munn 1985). Group living also benefits lake-margin species in competition for nest sites, feeding young, hunting, and defending food resources. Perhaps because of the rich, but variable resources of oxbow lakes, species with complex, monospecific social organizations can thrive (e.g., cooperatively breeding *Opisthocomus hoazin*, *Donacobius atricapillus* and *Ramphocelus* tanagers, communally breeding *Crotophaga* anis, floater flocks in *Paroaria gularis*, and colonially breeding icterids and their associated brood parasites and nest pirates). In mature forests where resources may be more predictable and evenly spaced, most territorial species occur at very low population densities (Terborgh et al. 1990; Robinson and Terborgh 1997). Such low population densities may have favored the evolution of complex multi-species groupings (Munn 1985) rather than the monospecific social organizations that dominate lake-margin species.

ACKNOWLEDGMENTS

I would like to thank J. Terborgh for the opportunity to work in the Manu National Park. E. J. Heske, J. D. Brawn, D. Willard, and an anonymous reviewer provided many helpful suggestions for improving the flow of this manuscript. The Peruvian Ministerio de Agricultura, Direccion General Forestal y Fauna, graciously allowed me to work in the Park. I am particularly grateful to K. Wehr for logistical help throughout the study. Partial funding for the work came from the National Science Foundation (DEB 8025975), the Frank M. Chapman Memorial Fund, the Society of Sigma Xi and Princeton University. I also thank J. Terborgh for guidance and J. Fitzpatrick and D. Willard for access to their mist-netting data.

LITERATURE CITED

Bednarz, J. 1988. Cooperative hunting in Harris' Hawks (*Parabuteo unicinctus*). Science 239:1525–1527.
Bolen, E. G. 1983. *Cairina moschata* (Pato Real Aliblanco, Pato Real, Muscovy Duck). Pages 554–556 *in* Costa Rican Natural History (D. H. Janzen, ed.). University of Chicago Press, Chicago, Illinois.
Bolster, D., and S. K. Robinson. 1990. Habitat use and relative abundance of migrant shorebirds in Amazonian Peru. Condor 92:239–242.
Brawn, J. D. 1990. Interspecific competition and social behavior in Violet-green Swallows. Auk 107:606–608.

BROWN, J. L. 1987. Helping and Communal Breeding in Birds. Princeton Univ. Press, Princeton, New Jersey.
DEL HOYO, J., A. ELLIOTT, AND J. SARGATAL. 1992. Handbook of the Birds of the World. Volume I: Ostrich to Ducks. Lynx Edicions, Barcelona, Spain.
EASON, P. 1992. Optimization of territory shape in heterogeneous habitats: a field study of the Red-capped Cardinal. J. Anim. Ecol. 61:411–424.
EMLEN, S. T. 1982. The evolution of helping. I. An ecological constraints model. Amer. Natur. 119:29–39.
EMLEN, S. T. 1994. Benefits, constraints, and the evolution of the family. Trends Ecol. Evol. 9:282–285.
EMLEN, S. T., AND L. W. ORING. 1977. Ecology, sexual selection, and the evolution of mating systems. Science 197:215–223.
EVENS, J., AND G. W. PAGE. 1986. Predation on Black Rails during high tides in salt marshes. Condor 88: 107–109.
FITZPATRICK, J. W. 1980. Wintering of North American tyrant flycatchers in the Neotropics. Pages 67–78 in Migrant Birds in the Neotropics (A. Keast and E. S. Morton, Eds.). Smithsonian Institution Press, Washington, D.C.
FITZPATRICK, J. W. 1982. *Conioptilon mcilhennyi*—The Black-faced Cotinga. Pages 125–127 in the Cotingas (D. W. Snow, Ed.) British Museum and Oxford University Press, London.
FITZPATRICK, J. W. 1985. Form, foraging behavior, and adaptive radiation in the Tyrannidae. Pages 447–470 in Neotropical Ornithology (P. A. Buckley et al., eds.). Ornithol. Monogr. No. 36.
FORSHAW, J. M. 1973. Parrots of the World. Doubleday Press, Garden City, New York.
GROOM, M. J. 1992. Sand-colored Nighthawks parasitize the antipredator behavior of three nesting bird species. Ecology 73:785–793.
HAVERSCHMIDT, F. 1968. Birds of Surinam. Oliver and Boyd, Edinburgh, Scotland.
HILTY, S. L., AND W. L. BROWN. 1986. A Guide to the Birds of Columbia. Princeton University Press, Princeton, New Jersey.
ISLER, M. L., AND P. R. ISLER. 1987. The Tanagers: Natural History, Distribution, and Identification. Smithsonian Institution Press, Washington, D.C.
JANSON, C. H., J. TERBORGH, AND L. H. EMMONS. 1980. Non-flying mammals as pollinating agents in the Amazonian forest. Biotropica Suppl. 13–1:1–6.
JANSON, C. H. 1987. Bird consumption of bicolored fruit displays. Amer. Natur. 130:788–792.
KILTIE, R. A., AND J. W. FITZPATRICK. 1984. Reproduction and social organization of the Black-capped Donacobius (*Donacobius atricapillus*). Auk 101:804–811.
KOENIG, W. D., AND R. L. MUMME. 1987. Population Ecology of the Cooperatively Breeding Acorn Woodpecker. Princeton Univ. Press, Princeton, New Jersey.
KOENIG, W. D., F. A. PITELKA, W. J. CARMEN, R. L. MUMME, AND M. T. STANBACK. 1992. The evolution of delayed dispersal in cooperative breeders. Q. Rev. Biol. 67:111–150.
LEIGHTON, M. 1986. Hornbill social dispersion: Variations on a monogamous theme. Pages 108–130 in Ecological Aspects of Social Evolution. (D.I. Rubenstein and R.W. Wrangham, Eds.). Oxford University Press, Oxford, U.K.
LIMA, S. L., AND L. M. DILL. 1990. Behavioral decisions made under the risk of predation: A review and prospectus. Can. J. Zool. 68:619–640.
MINDELL, D. P., AND H. L. BLACK. 1984. Combined-effort hunting by a pair of Chestnut-mandibled Toucans. Wilson Bull. 96:319–321.
MUNN, C. A. 1985. Permanent canopy and understory flocks in Amazonia: Species composition and population density. Pages 683–712 in Neotropical Ornithology (P. A. Buckley, M. S. Foster, E. S. Morton, R. S. Ridgely, and F. G. Buckley, Eds.). Ornithol. Monogr. No. 36. American Ornithologists' Union, Washington, D.C.
PARKER, T. A., III. 1982. Observations of some unusual rainforest and marsh birds in southeastern Peru. Wilson Bull. 94:477–493.
PARKER, T. A., III, AND J. V. REMSEN, JR. 1987. Fifty-two Amazonian bird species new to Bolivia. Bull. Brit. Ornithol. Club 107:94–106.
RABENOLD, K. N. 1990. *Campylorhynchus* wrens: The ecology of delayed dispersal and cooperation in the Venezuelan savanna. Pages 157–196 in Cooperative Breeding in Birds (P. B. Stacey and W. D. Koenig, Eds.). Cambirdge University Press, New York.
REMSEN, J. V., JR. 1990. Community ecology of Neotropical kingfishers. Univ. Calif. Publ. Zool. 124:1–116.
REMSEN, J. V., JR., F. G. STILES, AND P. E. SCOTT. 1986. Frequency of arthropods in stomachs of tropical hummingbirds. Auk 103:436–441.
REMSEN, J. V., JR., M. A. HYDE, AND A. CHAPMAN. 1993. The diets of Neotropical trogons, motmots, barbets, and toucans. Condor 95:178–192.
REMSEN, J. V., JR., AND T. A. PARKER III. 1983. Contribution of river-created habitats to bird species richness in Amazonia. Biotropica 15:223–231.
ROBINSON, S. K. 1985a. Coloniality as a defense against nest predators of the Yellow-rumped Cacique. Auk 102:509–519.
ROBINSON, S. K. 1985b. The Yellow-rumped Cacique and its associated nest pirates. Pages 896–907 in Neotropical Ornithology (P. A. Buckley, M. S. Foster, E. S. Morton, R. S. Ridgely, F. G. Buckley, Eds.). Ornithol. Monogr. 36, American Ornithologist Union, Washington, D.C.
ROBINSON, S. K. 1986a. The evolution of social behavior and mating systems in the blackbirds (Icterinae).

Pages 175–200 *in* Ecological Aspects of Social Evolution (D. I. Rubenstein and R. A. Wrangham, Eds.). Princeton University Press, Princeton, New Jersey.

ROBINSON, S. K. 1986b. Benefits, costs, and determinants of dominance in a polygynous Neotropical oriole. Anim. Behav. 34:241–255.

ROBINSON, S. K. 1986c. Three-speed foraging during the breeding cycle of the Yellow-rumped Cacique (Icterinae: *Cacicus cela*). Ecology 67:394–405.

ROBINSON, S. K. 1988. Ecology and host relationships of Giant Cowbirds in southeastern Peru. Wilson Bull. 100:224–235.

ROBINSON, S. K. 1994a. Habitat selection and foraging ecology of raptors in Amazonian Peru. Biotropica 26:443–458.

ROBINSON, S. K. 1994b. Use of bait and lures by Green-backed Herons in Amazonian Peru. Wilson Bull. 106-3:567–569.

ROBINSON, S. K., J. TERBORGH, AND J. W. FITZPATRICK. 1988. Habitat selection and relative abundance of migrants in southeastern Peru. Proc. XIX Intern. Ornithol. Congr:2298–2307.

ROBINSON, S. K., J. TERBORGH, AND J. W. FITZPATRICK. 1995. Distribution and Abundance of Neotropical Migrant Land Birds in the Amazon Basin and Andes. Bird Conserv. Int.

ROBINSON, S. K., AND J. TERBORGH. 1990. Bird communities of the Cocha Cashu Biological Station in Amazonian Peru. Pages 199–216 *in* Four Neotropical Forests (A. Gentry, Ed.). Yale. University Press, New Haven, Connecticut.

ROBINSON, S. K., AND J. TERBORGH. 1995. Effects of interspecific aggression on habitat selection in Amazonian birds. J. Anim. Ecol. 64:1–11.

ROBINSON, S. K., AND J. TERBORGH. 1997. Bird community dynamics along primary successional gradients of an Amazonian whitewater river. Pages 641–672 *in* Neotropical Ornithology Honoring Ted Parker (J. V. Remsen, Jr., Ed.), Ornithol. Monogr. No. 48.

ROBINSON, S. K., J. TERBORGH, AND C. A. MUNN. 1990. Lowland tropical forest bird communities of a site in western Amazonia. Pages 229–258 *in* Biogeography and Ecology of Forest Bird Communities (A. Keast, Ed.). SPB Acad. Publ., The Hague, The Netherlands.

ROBINSON, S. K., S. I. ROTHSTEIN, M. C. BRITTINGHAM, L. J. PETIT, AND J. A. GRZYBOWSKI. In press. Ecology of cowbirds and their impact on host population dynamics. *In* Ecology and Conservation of Neotropical Migratory Birds (T. E. Martin and D. Finch, Eds.) Oxford University Press, Oxford, U.K.

ROSENBERG, G. H. 1990. Habitat specialization and foraging behavior by birds of Amazonian river islands in northeastern Peru. Condor 92:427–443.

SALO, J., R. KALLIOLA, I. HAKKINEN, Y. MAKINEN, P. NIEMELA, M. PUHAKKA, AND P. D. COLEY. 1986. River dynamics and the diversity of Amazon lowland forest. Nature 322:254–258.

SHORT, L. L. 1970. Notes on the habits of some Argentine and Peruvian woodpeckers. Amer. Mus. Novit. 2413:1–37.

SICK, H. 1993. Birds in Brazil. Princeton University Press, Princeton, New Jersey.

SKUTCH, A. F. 1952. Life history of the Blue-and-white Swallow. Auk 69:392–406.

SKUTCH, A. F. 1954. Life Histories of Central American Birds. Pacific Coast Avifauna, No. 31. Cooper Ornithological Society, Berkeley, California.

SKUTCH, A. F. 1958. Roosting and nesting of Aracari toucans. Condor 60:201–219.

SKUTCH, A. F. 1964. Life histories of Central American pigeons. Wilson Bull. 76:211–247.

SNOW, D. W. 1982. The Cotingas. Cornell University Press, Ithaca, New York.

STEARNS, S. C. 1992. The Evolution of Life Histories. Oxford University Press, Oxford, U.K.

STRAHL, S. D. 1988. The social organization and behavior of the Hoatzin *Opisthocomus hoazin* in central Venezuela. Ibis 130:483–502.

TERBORGH, J. 1983. Five New World Primates. Princeton Univ. Press, Princeton, New Jersey.

TERBORGH, J. 1985. Habitat selection in Amazonian birds. Pages 311–338 *in* Habitat Selection in Birds. (M. L. Cody, Ed.). Academic Press, New York.

TERBORGH, J., J. W. FITZPATRICK, AND L. H. EMMONS. 1984. Annotated checklist of bird and mammal species of Cocha Cashu Biological Station, Manu National Park. Fieldiana: Zool., no. 21:1–29.

TERBORGH, J., S. K. ROBINSON, T. A. PARKER III, C. A. MUNN, AND N. PIERPONT. 1990. Structure and organization of an Amazonian forest bird community. Ecolog. Monogr. 60:213–238.

TERBORGH, J., AND K. PETREN. 1991. Development of habitat structure through succession in an Amazonian floodplain forest. Pages 28–46 *in* Habitat Structure: The Physical Arrangements of Objects in Space (S. S. Bell, E. D. McCoy, and H. R. Mushinsky, Eds.). Chapman and Hall, New York.

THOMAS, B. T. 1979. The birds of a ranch in the Venezuelan llanos. Pages 213–232 *in* Vertebrate Ecology of the Northern Neotropics (J. F. Eisenberg, ed.). Smithsonian Institution Press, Washington, D.C.

VEHRENCAMP, S. L. 1978. The adaptive significance of communal nesting in Groove-billed Anis (*Crotophaga sulcirostris*). Behav. Ecol. Sociobiol. 4:1–33.

WEBSTER, M. S. 1994. The spatial and temporal distribution of breeding female Montezuma Oropendolas: Effects on male mating strategies. Condor 96:722–733.

WETMORE, A. 1968. The Birds of the Republic of Panamá. Part 2.—Columbidae (Pigeons) to Picidae (Woodpeckers). Smithsonian Institution Press, Washington, D.C.

WETMORE, A., R. F., PASQUIER, AND S. L. OLSON. 1984. The Birds of the Republic of Panamá. Part 4.

Passeriformes: Hirundinidae (Swallows) to Fringillidae (Finches). Smithsonian Institution Press, Washington, D.C.

WILLARD, D. E. 1979. Comments on the feeding of the Boat-billed Heron (*Cochlearius cochlearius*). Biotropica 11:158.

WILLARD, D. E. 1985. Comparative feeding ecology of twenty-two tropical piscivores. Pages 788–797 *in* Neotropical Ornithology (P. Buckley, M. S. Foster, E. S. Morton, R. S. Ridgely and F. G. Buckley, Eds.). Ornithol. Monogr. No. 36, American Ornithologist Union, Washington, D.C.

WILLIS, E. O. 1965. A prey capture by the zone-tailed hawk. Condor 68:104–105.

WILLIS, E. O. 1963. Is the Zone-tailed Hawk a mimic of the Turkey Vulture? Condor 65:313–317.

ZACK, S. AND B. J. STUTCHBURY. 1992. Delayed breeding in avian social systems: The role of territory quality and "floater" tactics. Behaviour 123:194–219.

APPENDIX

POPULATION SIZE, RESOURCE USE, MICROHABITAT USE, POSSIBLE MATING SYSTEMS, AND SEASONALITY OF BIRDS OF COCHA CASHU. DATA FROM PERSONAL OBSERVATIONS OR FROM THE GENERAL LITERATURE

Species	Population[1] 1980	Population[1] 1985	Range[2]	Resource use[3]	Microhabitats[4]	Breeding system[5] (nests found)	Seasonality[6]
Phalacrocorax brasilianus	22i	17i	0–29i	F	AQ	—	LD-W(D)
Anhinga anhinga	3i	4i	0–5	F	AQ	—	W(D)
Ardea cocoi	0–1i	0–1i	0–3	F	AQ, MA	—	W(D), F
Ardea alba	3i	4i	0–7	F, B	AQ, MA, MF	—	LD-W
Egretta thula	0	0	0–3	F	AQ, MA, MF	M*	F
Butorides striatus	8–14i	9–13i	8–17	F	AQ, MF, LM	M?	R
Agamia agami	4p	4p	—	F	LM	—	R
Bubulcus ibis	0	0	0–13	T, I	S	—	LD-W(Nov)
Pilherodius pileatus	0	0	0–3	F	MF, AQ	—	F
Nycticorax nycticorax	0	0	0–1	F	LM	—	V
Tigrisoma lineatum	5p	6p	—	F	LM	M*	R
Ixobrychus exilis	0	0	0–1	F, I	MA	—	V
I. involucris	0	0	0–1	F, I	MA	—	V
Cochlearius cochlearius	5	—	—	AI	LM	—	R
Jabiru mycteria	0	0	0–3	F, C	MF, MA	—	V
Mycteria americana	0	0	0–5	F, C	MF, MA	—	V
Mesembrinibis cayennensis	3p	4p	—	AQI	LM	—	R
Anhima cornuta	2p	1p	0–8i	V	MA	M*	R, F
Cairina moschata	8i	16i	4–26	V	MA, LM	PY	R, S
Oxyura dominica	0	0	0–1	V	MA	—	V
Elanoides forficatus	—	—	0–25	AEI	AE	M*	R, S
Ictinia plumbea	—	—	—	AEI, TE	AE	M*	R
Rostrhamus sociabilis	0	0	0–13	C	AQ, MA	—	S
Accipiter bicolor	e	e	—	C(B)	LT, S	—	R
Buteo albonatatus	—	—	0–1	C(B)	AE	—	R(S)
Busarellus nigricollis	1p	1p	0–1p	C, F	MA, AQ, S	M*	R
Buteogallus urubitinga	1p	1p	0–1p	C	MA, S, AQ, LT, MF	—	R
Rostrhamus sociabilis	0	0	0–13	SN	AQ, MA	—	S
Spizaetus ornatus	e	e	—	C	LT	—	R
Pandion haliaetus	li	li	0–2	F	AQ	—	LD-W(D)
Micrastur semitorquatus	e	e	—	C	LT, S	—	R, S
Daptrius ater	1f	1f	—	C, FI	LT, S, MF	F(C?)	R

APPENDIX
CONTINUED

Species	Population[1] 1980	Population[1] 1985	Range[2]	Resource use[3]	Microhabitats[4]	Breeding system[5] (nests found)	Seasonality[6]
Falco rufigularis	1p	1p		C, AEI, TE	AE	M*	R
Ortalis motmot	3.0e	4.0e		FF, CF, LF, S	LT, S	F(C?)	R
Pipile pipile	e	e	0–22i	LF	LT	U	R
Penelope jacquacu	e	e	0–13i	LF	LT	U(M?)	R
Mitu tuberosa	e	e	0–8i	LF	LT, LM	M	R
Opisthocomus hoazin	83i	91i	0–93i	V	LT, V, S	C*	R
Aramus guarauna	11i	6i	2–12i	SN	LM, M	U	R, S?
Rallus nigricans	6m	3m	2–6m	AQI	MA	U	R
R. maculatus	0	0	0–1	AQI	MA	U	V
Aramides cajanea	5p	6p	7–11p	AQI, I	LM, H	M	R
Laterallus melanophaius	30p	19p	17–30p	AQI	MA, H	M	R
Gallinula chloropus	13ad	4a	2–15i	AQI	MA, H, S, AQ	M?	R
Porphyrula martinica	25i	14i	10–31i	AQI	MA, H, S, LM, AQ	M?	R
Heliornis fulica	24i	18i	0–24	AQI	AQ, LM	U	LD-W
Eurypyga helias	3p	4p	3–4	AQI	LM	M	R
Jacana jacana	2li	9i	4–28i	AQI	MA, AQ	PA	R, S
Phalaropus tricolor	+	+	0–6	AQI	MF, AQ	—	LD
Tringa solitaria	+	+	0–11i	AQI	MF	—	LD
T. flavipes	+	+	0–16i	AQI	MF	—	LD
T. melanoleuca	+	+	0–6i	AQI	MF	—	LD
Actitis macularia	+	+	0–8i	AQI	MF, LM	—	LD-W(D)
Calidris himantopus	+	+	0–4i	AQI	MF	—	LD
Phaetusa simplex	+	+	0–8i	F	AQ, TE	—	D-EW(F)
Sterna superciliosus	+	+	0–11i	F, AQI	AQ	—	D-EW(F)
Rynchops nigra	+	+	0–18i	F	AQ	—	D-EW(F)
Columba cayennensis	33i	30i	20–36i	FF	LT, S	M	R
Leptotila rufaxilla	4p(e)	5p(e)		S	LM, S	M*	R
Ara ararauna	+	+	0–28	LF, Q/C	LT	M	R, S
A. macao	+	+	0–32	Q/C, OF	LT	M	R, S
A. severa	+	+	0–116i	Q/C, FF	LT, S	M*	R, S
Aratinga weddellii	4–34i	10–26i	18–34i	FF, Q/C, OF	LT, S	M*	R
Brotogeris cyanoptera	e	e	0–80i	FF, Q/C	LT	M*	R, S
B. sanctithomae	10–200i	30–200i	0–275i	FF, OF, Q/C	LT, S	M*	R, S

APPENDIX
CONTINUED

Species	Population[1] 1980	Population[1] 1985	Range[2]	Resource use[3]	Microhabitats[4]	Breeding system[5] (nests found)	Seasonality[6]
Amazona ochrocephala	+	+	0–18i	OF	LT	M*	R, S
Coccyzus melacoryphus	+	+	0–1i	FI	LT, V, S	U	?
Piaya cayana	e	e		FI	LT, V	M*	R
P. minuta	1p	2p	1–2p	FI	S	M	R
Crotophaga major	16p	14p	12–16p	FI, TE, LF	LT, V, S	CM*	LD-W
C. ani	6p	2p	0–7p	FI, TE, LF	MA, S, V	CM*	R
Pulsatrix perspicillata	e	e		C	LT	—	R
Glaucidium brasilianum	3p	3p	3p	C, FI, TI	S, LT	M*	R
Nyctibius griseus	0–1	0	0–1i	AEI	S, LT	U	S
Lurocalis semitorquatus	e	e	0–2i	AEI, TE	AE	U	R
Chordeiles rupestris	0–120	0–140	0–200i	AEI	AE	M	R
C. minor	0–1	0–1	0–2i	AEI	AE	—	LD-W
Nyctidromus albicollis	7.0m	3.0m	2–7m	AEI	AE, S	M*	R
Hydropsalis climacocerca	0–2	0–3	0–4i	AEI	AE, S	M	R, F
Tachornis squamata	0–8	0–6	0–10i	AEI	AE	U	S, R
Chaetura brachyura	0–10	0–8	0–30i	AEI, TE	AE	U	R
Streptoprocne zonaris	0–150	0–200	0–350i	AEI, TE	AE	U	R, S
Glaucis hirsuta	+	+	0–6i	HN, N, I	H	U	R
Threnetes leucurus	+	+	0–2i	HN, N, I	H	U	R
Phaethornis superciliosus	+	+	0–8i	HN, I	H	U	R
P. hispidus	+	+	0–7i	HN, I	H	U	R
Anthracothorax nigricollis	+	+	0–1i	N, I	LT, S, V	U	LD, W
Heliomaster longirostris	+	+	0–2i	N, I	LT, V, S	U	D
Ceryle torquata	1–2i	1–2i	1–3i	F	AQ, LT, S	M?	R
Chloroceryle americana	1.0p	1.0p	1–5i	F	AQ	M?	R
C. amazonica	1.0p	1.0p	2–5i	F	AQ	M?	R
C. inda	5.0p	4.0p		F	AQ	M?	R
C. aenea	4.0p	4.0p		F	AQ	M?	R
Momotus momota	e	e		FI, OF	LT, V	M*	R
Galbalcyrhynchus purusianus	2.0f	2.0f		AEI	LT, S	F	R
Galbula cyanescens	4.0e	3.0e		AEI	LT, V, S	M	R
Monasa nigrifrons	e	e		AEI, FI, TE	LT	F	R
Chelidoptera tenebrosa	0–1i	0	0–3i	AEI, TE	LT, AE	U	S

APPENDIX
Continued

Species	Population[1] 1980	Population[1] 1985	Range[2]	Resource use[3]	Microhabitats[4]	Breeding system[5] (nests found)	Seasonality[6]
Eubucco tucinkae	3.0e	3.0e		FI, Q/C, CF, FF, LF, TE	LT, V, S	M*	R
Pteroglossus castanotis	2.0f	3.0f		FI, Q/C, CF, FF, OF, C, LF	LT, V, S	F*	R
P. inscriptus	2.0f	1.0f		FI, Q/C, CF, FF, OF, C	LT, V, S	F*	R
Ramphastos cuvieri	e	e	3–22i	OF, FF, CF, LF, FI, Q/C, C, TE	LT, S	F?	R
R. culminatus	e	e	0–19i	FI, Q/C, TE, OF, FF, CF, LF		F?	R
Veniliornis passerinus	2.0	1.0p		BI			
Colaptes punctigula	3.0p	3.0p		BI, Q/C	S, LT	M*	R
Celeus elegans	e	o		T, BI, Q/C	LT	F	R
Dryocopus lineatus	2.0e	2.0e		BI, Q/C	LT, S	M*	R
Melanerpes cruentatus	4.0e	4.0e		OF, Q/C, BI, TE	LT, S	CO*	R
Campephilus melanoleucos	2.0e	1.0e		BI, Q/C	LT	M*	R
Trogon melanurus	e	e		F, I	LT	M	R
T. curucui	e	e		F, I	LT	M	R
Nasica longirostris	e	e		BI	LT	M?	R
Xiphorhynchus picus	3.0p	4.0p		BI	LT, V, S	M*	R
Furnarius leucopus	4.0p	5.0p		TI	LM	M*	R
Taraba major	4.0p	4.0p		FI	V, S	M	R
Myrmotherula surinamensis	3.0p	3.0p		FI	V, S	M*	R
Hypocnemoides maculicauda	3.0p	3.0p		FI, TI, AQI	V, S, LM	M	R
Percnostola lophotes	0	1.0p(e)		FI	V, S	M?	R
Sclateria naevia	3.0p	3.0p		FI, TI, AQI	V, S, LM	M?	R
Myrmeciza hyperythra	5.0p	4.0p		FI, TI	S, H, V	M?	R
M. goeldii	1.0p	1.0p		FI	S	M?	R
Elaenia spectabilis	0	0	0–2	F	S, MA	—	D
Leptopogon amaurocephalus	0	1.0p		FI	LT, V	M*	R
Myiophobus fasciatus	0–1	0–1	0–2	FI	S	—	D
Pyrocephalus rubinus	2i	3i	0–5	AEI	LT, S	—	D
Fluvicola pica	0	0	0–1	AEI, AQI	S, MA	—	D, s
Colonia colonus	0	0	0–1	AEI	S, MA	—	V
Pitangus lictor	4.0p	5.0p	3–5p	LF, OF, AQI, AEI, TE, F	S,MA, V	M*	R
P. sulphuratus	4.0p	5.0p		AQI, AEI, LF, TE	S, MA, V, LT	M*	R
Megarynchus pitangua	1.0	1.0p		AQI, AEI, LF, TE	S, MA, V, LT	M*	R
Myiozetetes similis	12.0p	13.0p	7.0–9.0p	AQI, AEI, LF, TE	S, MA, V, LT	M*	R

APPENDIX
CONTINUED

Species	Population[1] 1980	Population[1] 1985	Range[2]	Resource use[3]	Microhabitats[4]	Breeding system[5] (nests found)	Seasonality[6]
M. granadensis	e	e	0–5i	AQI, AEI, FI, LF, TE	LT, V, S	M*	R
Myiodynastes maculatus	e	e		LF, FI	LT	—	R?
M. luteiventris	0–32	0–16	0–32	LF, FI	LT	—	LD-W, S
Legatus leucophaius	3.0p	2.0p	0–4.0p	LF, AEI	LT, S	M*	R?
Tyrannus melancholicus	3.0p	3.0p	2–4p	AEI, LF, TE	LT, S	M*	R
T. tyrannus	0–3,500i	0–450i	0–5,200i	LF, AIE, TE	LT, S	—	LD-W, S
Schiffornis major	e	e	0–1	FI, F?	V	U	R, S
Tityra cayana	3.0p	3.0p	3–4p	CF, FI, BI, LF, OF, TE	LT, S, V	M*	R
T. semifasciata	2.0p	3.0p	2–3p	CF, FI, BI, LF, OF, TE	LT, S, V	M*	R
T. inquisitor	1.0p	1.0p	0–1.0p	CF, FI, LF, OF, TE	LT	M?	R
Porphyrolaema porphyrolaema	0	0–3	0–3	CF	LT, V	U	R, S
Cotinga maynana	3m	1m	0–5m	LF, TE	LT	PY?	S
C. cayana	0–2	0	0–2	LF	LT	PY?	S
Conioptilon mcilhennyi	e	e	0–3	LF, OF, BI?, TE	LT	F	R
Gymnoderus foetidus	12p	8p	0–28i	LF, BI, TE	LT	PY*	R, S
Tachycineta albiventer	7.0p	10.0p	12–90i	AEI	AE	M*	R
Progne tapera	+	+	0–12i	AEI	AE	M	R, S
Notiochelidon tibialis	0–150i	0–60i	0–200i	AEI	AE	—	D, EW, C
Atticora fasciata	0–22i	0–17i	0–35i	AEI	AE	—	R
Stelgidopteryx ruficollis	4–30i	4–40i	0–40i	AEI	AE	—	R
Riparia riparia	0–8i	0–6i	0–20i	AEI	AE	—	LD-W
Hirundo rustica	0–5i	0–20i	0–20i	AEI	AE	—	LD-W
Donacobius atricapillus	12f	10f	10–14f	FI, AQI, TE	MA, S	CO*	R
Turdus ignobilis	8.0p	16.0p	4–20.0p	LF, OF	S, V	M*	R, S
Sporophila caerulescens	0–12i	0	0–12i	S	S, MA	—	D
Oryzoborus angolensis	1.0p	0	0–1.0p	S	S, MA	—	U, S
Paroaria gularis	26.0p	27.0p	70–103i	FI, CF, FF, Q/C, AQI, TE	S, MA, V, LT	M?	R
Saltator maximus	0p	2p	0–2p	FF, FI, LF, CF	LT, V, S	M*	R
S. coerulescens	2.0p	2.0p		FF, FI, LF, CF	S, V	M	R
Cyanocompsa cyanoides	e	e	0–1p	S, OF	V	M*	R
Cissopis leveriana	2.0f	2.0f		CF, FF, OF, Q/C, TE	S, MA, V	M?	R
Ramphocelus nigrogularis	3.0f	3.0f		FF, CF, OF, Q/C, FI	S, V	CO*	R
R. carbo	4.0f	5.0f		FF, CF, OF, Q/C, FI	S, V, MA	CO*	R

APPENDIX
CONTINUED

Species	Population[1] 1980	Population[1] 1985	Range[2]	Resource use[3]	Microhabitats[4]	Breeding system[5] (nests found)	Seasonality[6]
Thraupis episcopus	3.0p	4.0p	2–4p	FF, CF, OF, Q/C, FI	S, V, LT	M*	R
T. palmarum	e	e		FF, CF, OF, Q/C, FI	LT, S	M*	R
Euphonia laniirostris	0	1p	0–1p	F	LT, S	M	R
E. chrysoposta	e	e		FF	LT	M?	R, S
Tangara mexicana	e	e		FF, CF, OF, FI, Q/C	LT	F	R
T. chilensis	e	e		FF, CF, OF, FI, Q/C	LT	F	R
T. schrankii	e	e		FF, CF, OF, FI, Q/C	LT	F*	R
T. callophrys	e	e		FF, FI, Q/C, CF	LT	M/F	R
Dacnis lineata	e	e		FF, CF, OF, FI, Q/C	LT	M	R
D. flaviventer	e	e		FF, CF, OF, FI, Q/C	LT	M	R
D. cayana	e	e		FF, CF, OF, FI, Q/C	LT	M	R
Tersina viridis	0–3i	0–6i	0–6i	LF, AEI	LT	U	S
Geothlypis aequinoctialis	0	0–1i	0–1i	FI	MA	U	U
Cyclarhis gujanensis	0	0	0–1p	FI	S	M?	S
Vireo olivaceus chivi	4.0p	4.0p		FI, CF	LT, V	M	R
Psarocolius angustifrons	18f	17f	12–24f	TE, FI, CF, FF, OF, Q/C	LT, V, S	PY*	R, S
Cacicus cela	98f	88f	38–124f	FI, CF, FF, OF, Q/C, TE	LT, V, S	PY*	R, S
C. solitarius	5.0m	4.0m	66–750i	FI, CF, FF, OF, Q/C	HNH, S	U	R
Icterus icterus	1.0p	1.0p	0–1.0p	FI, CF, FF, OF, Q/C, HN, H	H, LT, V, S	M*	R, S
Agelaius xanthophthalmus	6.0f	5.0f	4–6f	FF, OF, FI, HN	S, MA	F*	R
Scaphidura oryzivora	0–16i	0–14i	0–18i	TI, TE, MF, CF, FF, Q/C	LM, LT, V	PY*	R, S
Dolichonyx oryzivorus	0	0–1i	0–1i	FI, S	MA	—	LD, W

[1] p = pairs, fa = families, i = individuals, e = edge only.
[2] Maximum and minimum population counts from other years.
[3] Resources (listed in approximate order of use): F = fish, I = insects, AQI = aquatic arthropods, FI = foliage insects, Q/C = *Quararibea/Combretum* nectar or flowers, AI = aerial insects, FF = fig fruits, LF = Lauraceous fruits, CF = Coussapoa fruits, T = termites, TE = termite emergences, OF = other fruits, Hn = *Heliconia* nectar, C = carnivore, B = birds, H = herptiles, V = vegetation, S = seeds, BI = bark insects, TI = terrestrial insects, Sn = snails. Blank = never observed feeding in the Cocha Cashu area. Data from direct observations at Cocha Cashu supplemented by data from other sources (Hilty and Brown 1986, Short 1970, Remsen et al. 1986, 1993, del Hoyo et al. 1993, Haverschmidt 1968, Sick 1993, Wetmore et al. 1984).
[4] Microhabitats: AQ = aquatic, MF = mudflats, MA = marshes, H = *Heliconia* thickets, V = vines along edges, S = shrubs along lake margin, LT = lake-margin trees, A = aerial, LM = lake margin water/soil interface, IT = isolated trees.
[5] Breeding Systems: M = monogamous, M/F = monogamous with floaters, F = extended families (possibly cooperative), CO = cooperative, CM = communal, PY = polygynous, PA = polyandrous, U = unknown. * = nesting documented.
[6] Seasonality: W = wet only, D = dry only, E = early, L = late, parentheses indicates occasional records, C = cold fronts, F = river floods, R = year-round resident, V = vagrant (<10 records total), S = sporadic.

BIRD COMMUNITY DYNAMICS ALONG PRIMARY SUCCESSIONAL GRADIENTS OF AN AMAZONIAN WHITEWATER RIVER

SCOTT K. ROBINSON[1] AND JOHN TERBORGH[2]

[1]*Illinois Natural History Survey, 607 East Peabody Drive, Champaign, Illinois 61820, USA;*
[2]*Center for Tropical Conservation, Duke University, 3705 Erwin Road, Durham, North Carolina 27705, USA*

ABSTRACT.—We used spot-map methods to census birds in two plots (40 ha and 80 ha) representing the early and middle stages of primary succession generated by the meandering of the Manu River in southeastern Peru in 1983 and 1985. We distinguished seven distinct successional stages beginning with the first plant communities growing on open beaches (*Tessaria*: stage 1) and proceeding through mature floodplain forest (stage 7). Each successional stage was more structurally and floristically complex. One plot was dominated by early successional vegetation (stages 1–5) and the other was dominated by middle-to-late stages (5–6). We supplemented spot-map data with mist netting to census nonterritorial species, principally nectarivores and frugivores. Because the earliest successional stage (*Tessaria*) covered only small (<3 ha) areas on any given meander tongue, we censused six additional stands to characterize their bird communities.

Species richness increased along the successional gradient, but not uniformly. Structurally simple *Tessaria* stands (stage 1) contained more diverse communities of breeding birds than the next two successional stages, which appeared to have few resources available to birds. After stage 3, breeding species richness increased by 31–71 species per successional stage as a distinct canopy layer formed (stage 4), fruiting trees became more available (stages 5–6) and the vertical structure of the forest became fully developed (stage 7: 71 more species than stage 6). Early successional bird communities (stages 1–3) were dominated by a few very abundant insectivores and omnivores, but wanderers from the adjacent forest plus Nearctic and austral migrants seasonally outnumbered breeding residents. Many species that breed in *Tessaria* stands were restricted to just one or two of the seven stands censused. The middle stages of succession had the highest estimated richness, abundance, and biomasses of nectarivores and frugivores. Nectarivore abundance reflected the huge stands of *Heliconia* that dominate the understory of mid-successional stages. Bark foragers reached their highest abundance in late successional stages as the dominant canopy tree species of middle successional stages (e.g., *Cedrela odorata*) began to die off, but remained standing.

Middle successional stages lacked many of the insectivores characteristic of more structurally complex mature forest. Among the missing species were understory flockers, ant-followers, ground foragers, and many arboreal insectivores that participate in canopy flocks. Species of early and middle stages tended to be more abundant on average than those of mature forest. Early-/and late-successional bird communities had few species in common, and many congeners segregated along this successional gradient. Most species characteristic of early successional vegetation, however, were also found in many other kinds of natural and anthropogenic disturbances. The *Tessaria* beaches of the Manu lacked most species endemic to the same successional stage on the larger rivers in the Amazon system. Mid-successional stages had relatively fewer species restricted to them, but some of these species are rare and endemic to southwestern Amazonia. Many characteristic species of floodplains reached their peak abundances in stages 4–6. Mid-successional stages, therefore, may play a greater role in contributing to the unusually high species richness of western Amazonia than had previously been recognized.

These results point to the importance of preserving natural, undammed meandering rivers in conserving the avian diversity of Amazonia. Logging and agricultural development of mid-successional floodplain forests probably has a negative impact on regional bird communities. Not all species of the early and middle stages of succession have adapted to human-generated secondary succession. Recovery of the full diversity and richness of tropical bird communities following disturbance may take several hundred years, although care must be taken when extrapolating results from primary to secondary successional gradients.

RESUMEN.—Durante 1983 y 1985, se utilizó el método de mapeo de territorios para censar aves en dos parcelas (40 ha y 80 ha), las cuales representan las etapas temprana y media de sucesión primaria generadas por los meandros del río Manú en el sureste de Perú. Se pudieron distinguir siete etapas de sucesión que van desde las primeras comunidades vegetales que crecen en las playas abiertas (*Tessaria*: etapa 1), hasta el bosque maduro en la llanura aluvial (etapa 7); siendo cada una de éstas etapas más compleja estructural y florísticamente que las anteriores. Una de las parcelas estuvo compuesta por vegetación temprana (etapas 1–5), mientras que la otra fue representada por etapas medias a tardías de sucesión (5–7). Para censar especies no territoriales, principalmente aves nectarívoras y frugívoras, se suplmentó el método de mapeo de territorios con el uso de redes de viento. Debido a que la etapa más temprana de sucesión (*Tessaria*) abarcó solamente áreas muy paqueñas (<3 ha) en cada uno de los meandros del río, se censaron seis zonas adicionales para poder caracterizar a las communidades de aves que éste hábitat sostiene.

La riqueza de especies de aves se incrementó a lo largo del gradiente de sucesión florística, éste incremento sin embargo, no fue uniforme. Las zonas estructuralmente más simples, compuestas por *Tessaria* (etapa 1) soportaron comunidades más diversas de aves anidando que las dos etapas siguientes de sucesión; aparentemente éstas últimas etapas contienen menos recursos disponibles para las aves. A partir de la tercera etapa, la riqueza de especies de aves se incrementó entre 31 y 71 especies por etapa como resultado de la formación de una capa adicional de la vegetación (etapa 4), el aumento en la disponibilidad de árboles frutales (etapas 5 y 6), y el desarrollo completo de la estructura vertical del bosque (la etapa 7 presentó 71 especies más que la etapa 6). Las comunidades de aves en las etapas tempranas de sucesión (etapas 1–3), estuvieron compuestas principalmente por unas cuantas especies muy abundantes de insectívoros y omnívoros; las especies errantes del bosque adyacente y las migratorias neotropicales y australes, sin embargo, excedieron estacionalmente en número a las residentes anidantes. Muchas de las especies que anidan en zonas dominadas por *Tessaria* estuvieron restringidas únicamente a una o dos de las siete etapas censadas. Las etapas medias de sucesión tuvieron los niveles más altos de riqueza, abundancia y biomasa estimadas de frugívoros y nectarívoros; la alta abundancia de éstos últimos refleja la gran cantidad de áreas dominadas por *Helicorina* que ocupan la mayor parte del sotobosque y las etapas medias de sucesión. A medida que las especies de árboles dominantes en las etapas medias de sucesión (e.j. *Cedrela odorata*) comenzaron a morir permaneciendo de pie, especies que forrajean en la corteza de los árboles alcanzaron su mayor abundancia en las etapas de sucesión tardía.

Las etapas medias de sucesión se caracterizaron por la ausencia de muchos de los insectívoros característicos de bosques maduros estructuralmente más complejos; entre las especies ausentes destacan especies de parvadas del sotobosque, especies que se alimentan de hormigas, especies que forrajean en el suelo, y muchas especies de insectívoros arboreos que forman parte de parvadas del dosel. En promedio las especies de las etapas mediana y temprana tendieron a ser más abundantes que las del bosque maduro. Las comunidades mediana y tardía de aves tuvieron pocas especies en común, y muchos congéneres fueron segregados a lo largo de éste gradiente de sucesión; la mayoría de las especies características de vegetación temprana fueron sin embargo registradas en muchos otros tipos de perturbaciones naturales y antropogénicas. En las playas de *Tessaria* de Manu no se registraron la mayoría de las especies endémicas de la misma etapa de sucesión en rios mayores en el sistema Amazónico. Las etapas medias de sucesión presentaron relativamente menos especies restringidas a ellas, algunas de éstas sin embargo, son raras y endémicas de la Amazonia del suroeste. Muchas especies características de llanuras aluviales alcanzaron sus mayores abundancias en las etapas 4-6; las etapas de sucesión media por consiguiente, probablemente cumplen un papel más importante al que se había reconocido previamente en su contribución a la inusual riqueza de especies de la Amazonia del oeste.

Estos resultados sugieren la importancia para conservar la diversidad de aves de la Amazonia, de preservar rios naturales que tengan meandros, y que no formen parte de presas; la tala y el desarrollo de agricultura afectan probablemente a las comunidades regionales de aves de una manera negativa. No todas las especies de aves de etapas tempranas y medias de sucesión se han adaptado a la sucesión secundaria generada por el hombre, la recuperación de la total diversidad y riqueza de especies de comunidades de aves tropicales después de perturbaciones puede tomar varios cientos de años, se debe tener cuidado sin embargo al extrapolar resultados de gradientes de sucesión primaria a secundaria.

Understanding the successional dynamics of bird communities is central to understanding their ability to recover from human-caused disturbances, as well as evaluating the habitat requirements of vulnerable species within communities (Andrade and Rubio-Torgler 1994). An extensive literature exists on the dynamics of bird communities along secondary successional gradients following logging and abandonment of agricultural land in the north temperate zone (e.g., Adams 1908; Odum 1950; Johnston and Odum 1956; Martin 1960; Shugart and James 1973; May 1984; Helle and Mönkkönen 1990; Glowacinski and Järvinen 1975). In general, bird community richness, diversity, density and biomass increase along secondary successional gradients in the temperate zone, although in some situations, avian diversity and biomass pass through a mid-successional trough (e.g., Johnston and Odum 1956; Robinson 1988). Few species are restricted to mid-successional stages in the temperate zone (Shugart and James 1973). Bird communities at the early and late ends of secondary successional gradients have few species in common, whereas mid-successional bird communities tend to consist of a mixture of early and late-successional species. Bird community composition continues to change even in relatively old (>70 year) temperate forests (e.g., Holmes et al. 1986; Litwin and Smith 1992). Primary successional gradients, however, have rarely been studied in the temperate zone because of a paucity of currently unconstrained rivers that meander actively and create new lands where succession can occur (but see Zimmerman and Tatschl 1975).

The extensive river systems of Amazonia offer opportunities to study primary successional gradients in a region where rivers remain unconstrained by levees and dams. Consequently, the literature on the plant communities of Amazonian primary successional gradients is growing (Foster et al. 1986; Salo et al. 1986; Rasanen et al. 1987; Gentry 1988; Foster 1990a, b; Kalliola et al. 1991; Terborgh and Petren 1991). Tree species diversity and complexity of vertical structure increase progressively through time for at least 300 years after the onset of succession (Terborgh and Petren 1991).

Bird community composition also changes dramatically along Amazonian successional gradients, with many species confined to the early and later ends (Terborgh and Weske 1969; Remsen and Parker 1983; Terborgh et al. 1984; Terborgh 1985; Robinson et al. 1990; Robinson and Terborgh 1990; Dyrcz 1990; Terborgh et al. 1990; Rosenberg 1990). The existence of successional gradients in close proximity to mature forest may explain much of the increased species richness of Amazonian bird communities relative to those elsewhere in the Neotropics (Remsen and Parker 1983; Salo et al. 1986; Thiollay 1986; Robinson et al. 1990; Karr et al. 1990) and in other parts of the world (e.g., Bell 1982; Brosset 1990). Many congeners, for example, segregate along riparian successional gradients and some defend interspecific territories where they come in contact (Terborgh 1985; Robinson and Terborgh 1995).

A number of species are apparently restricted to early successional vegetation on mid-channel islands of major Amazonian tributaries (Remsen and Parker 1983; Rosenberg 1990). Many species that colonize secondary successional habitat created by humans derive from the earliest stages of riverine successional gradients (Terborgh and Weske 1969; Remsen and Parker 1983). Austral and neotropical migrants also occur most abundantly in early successional habitats in Amazonia (Fitzpatrick 1980; Robinson et al. 1988, 1995).

Despite the acknowledged importance of successional gradients in contributing to Amazonian bird species richness, few quantitative studies document bird communities of early and mid-successional habitats. Terborgh et al. (1990) censused birds on a continuous successional gradient that extended from open beach to mature floodplain forest in southeastern Peru. Species richness increased dramatically with age of the vegetation. Few species were restricted to mid-successional forests. The successional gradient included within the census plot, however, was a relatively narrow strip which might have caused an underestimation of the number of mid-successional species.

Mist-net samples of early successional communities show higher species richness and abundance than those in certain later successional stages (Dyrcz 1990, Robinson and Terborgh 1990). This finding, however, reflects several factors unrelated to the diversity of breeding species. First, mist nets sample most of the community under the low canopy of early successional habitats, rather than just understory birds. Second, juveniles and transients of many forest birds commonly wander through early successional habitats. Third, austral and Nearctic migrants show strong preferences for early successional habitats (Robinson et al. 1988, 1995).

The purpose of this paper was to present the results of censuses of early and mid-successional bird communities along the Manu River, a whitewater tributary of the Madre de Dios River. Our goal was to expand upon the preliminary analysis of successional bird communities presented in Terborgh et al. (1990), Robinson et al. (1990) and Robinson and Terborgh (1990) by presenting new results from censuses of two relatively large (40–80 ha) plots consisting entirely of early and mid-successional vegetation. We compared these results with censuses of adjacent mature floodplain forest presented in Terborgh et al. (1990) and related changes in bird communities to those in plant communities. Specifically, we examined patterns of avian species richness, relative abundance, and guild structure on plots containing a mixture of different successional stages. We also compared the successional bird communities of the Manu River with those of larger tracts of early successional vegetation occurring along the Amazon and its major tributaries (Remsen and Parker 1983; Rosenberg 1990).

STUDY AREA AND METHODS

Our study was conducted within 25 km of the Cocha Cashu Biological Station in the Manu National Park, depto. Madre de Dios, southeastern Peru (11°55' south and 77°18' west). The park contains the entire watershed of the Manu River, which is thereby protected from any human perturbation. During the November-May rainy season, the river regularly overflows its banks, but floods last only a few days. At the level of Cocha Cashu, floods crest 24–30 hours after rains stop. Because rains are intermittent, the river rises and falls continuously during the wet season (Terborgh and Petren 1991). Downstream, owing to the confluence of independent watersheds, the Amazon and its larger tributaries experience much more gradual, longer-lasting, and predictable flooding. A distinct forest community (várzea) specifically adapted to season-long flooding is characteristic of the central Amazonian floodplain (Remsen and Parker 1983), but not extreme western Amazonia.

The Manu is a typical whitewater river that meanders within a broad floodplain, creating a complex mosaic of successional and mature stands, backwaters, swamps, marshes, and oxbow lakes (Salo et al. 1986, 1991). As the river erodes the outside of a meander bend (Fig. 1), it deposits sediment on the inside (point bar) where primary succession begins. The spatial distribution of these habitats in the Cocha Cashu area is shown in Terborgh (1983, 1985), and in the photographs in Robinson et al. (1990). Salo et al. (1986) estimated that 26% of the Amazonian lowlands of Peru was in active floodplains, and that 12% consisted of early and middle successional stages. Salo et al. (1986) hypothesized that the variety and extent of these early successional plant communities contribute to the high species richness of Amazonian plant and animal communities.

Terborgh and Petren (1991) measured the annual erosional loss caused by meandering along one bend of the river near the field station (Fig. 1). At the point of maximum curvature, the loss averaged 25 m/yr. At this rate, the Manu River can meander nearly halfway across its 6-km wide floodplain in a century. Other bends in the vicinity appear to be extending at twice this rate (Kalliola et al. 1991). In principle, the river could sweep out the floodplain every several hundred years. As a result of local vagaries of meandering, however, many areas are recycled within a few decades, whereas others may escape disturbance for many centuries.

Successional stages.—Here, we provide a synopsis of the detailed descriptions of the Manu primary succession provided by Terborgh and Petren (1991) and Kalliola et al. (1991). For purposes of this paper, we recognize seven successional stages, although distinctions between them may not always be easily discernible to an inexperienced observer.

(1) The first successional stage forms on the highest sections of point bars, and consists of nearly pure monocultures of *Tessaria integrefolia* (hereafter *Tessaria*), a woody composite that grows rapidly on exposed beaches during the dry season (Kalliola et al. 1991). *Tessaria* grows to heights of 8–10 m and reaches reproductive maturity in 3–4 years. Most incipient *Tessaria* stands are washed away or buried in sediment during the wet season. Stands that resist flooding are on the highest sections of beaches. Sediment deposition around surviving *Tessaria* stems results in the formation of abrupt levees, behind which backwater depressions form in areas of reduced sedimentation. Coarse, knee-high stands of grass (*Paspalum* sp.) form in the backwaters and along the upper margins of beaches. The characteristic topography of levee/backwater units remains obvious for at least a century, until siltation gradually levels the landscape.

(2) The second stage begins when a stout cane (*Gynerium sagittatum*, hereafter referred to as cane) invades 3–5 year-old *Tessaria* stands via underground rhizomes (Kalliola et al. 1991). Cane grows to 8 m or higher, and eventually replaces the short-lived *Tessaria* stands. The cane stage persists for 15–20 years. Many tree seedlings become established in open canebreaks in the early part of the cane stage. Later, the understory is invaded by broadleafed monocots (Musaceae, Zingaberaceae, Marantaceae), which collectively cast a deep shade and suppress further establishment of tree seedlings.

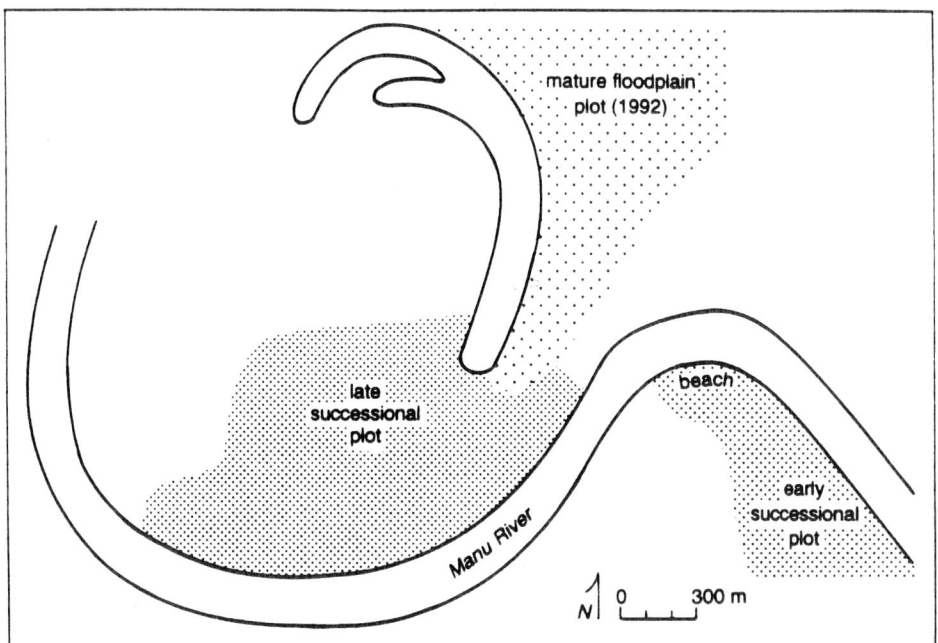

FIG. 1. Location of study plots relative to the Manu River and Cocha Cashu (the oxbow lake bordering mature floodplain forest).

(3) Succession enters the third stage when the canebreaks are overtopped by fast-growing trees (*Cecropia, Erythrina, Sapium, Inga*, and others). Because *Cecropia* grows faster than most other early successional trees, Terborgh (1983, 1985) recognized the *Cecropia/Gynerium* association as a separate stage.

(4) *Cecropia* trees are generally short-lived, and soon are overtopped by other species, which, in turn, shade out the cane and lead to the next successional stage. Broad-leafed monocots invade to monopolize the understory, and a mixed canopy, composed of *Guarea* (Meliaceae), *Guatteria* (Anonaceae), *Ficus* (Moraceae), *Sapium* (Euphorbiaceae) and others, replaces the *Cecropia* canopy. These trees may persist for two or more decades. Between the broad-leafed monocots and the canopy is a sparse and irregular mid-story composed of additional tree genera (e.g., *Rheedia* (Guttiferae), *Trichilia* (Meliaceae), *Sorocea* (Moraceae)).

(5) Eventually, two species, *Ficus insipida* (Moraceae) and *Cedrela odorata* (Meliaceae), begin to overtop the mixed canopy of the previous stage and establish a uniform and long-lasting upper layer at 35–40 m. Some trees from the previous stage remain in the subcanopy (e.g., *Guarea, Inga, Sorocea, Trichilia*). The *Ficus-Cedrela* stage likely persists for a century or more and some individual trees attain large girths (>1.5 m dbh). Because it is so much longer-lasting, the area occupied by this stage exceeds those of all previous stages combined.

(6) The *Ficus-Cedrela* stage gradually blends into the next stage as the trees composing the canopy die and are replaced by other species. During this period, a more coherent mid-story begins to fill in the space beneath the canopy. This "transitional" forest (Terborgh 1985) contains many characteristic mid-successional species, and lacks many others that occur in the mature phase, implying that further compositional change takes place before the forest attains mature status. During the transitional phase, the levee/backwater topography gradually vanishes, as repeated floods and rains deposit silt disproportionately in the backwaters.

(7) The final, "mature" stage, possibly attained after 300 or more years, is characterized by high plant species richness and a complex vertical structure consisting of five vertically superimposed layers (Gentry 1990, Terborgh and Petren 1991).

Backwaters.—Sections of both census plots (see below) contain backwaters and abandoned river channels, where water collects during the wet season, forming swamps, canebreaks, and small open marshes (<0.5 ha). Because several bird species are restricted to these habitats, we include a separate "backwater" category in our analysis of successional bird communities.

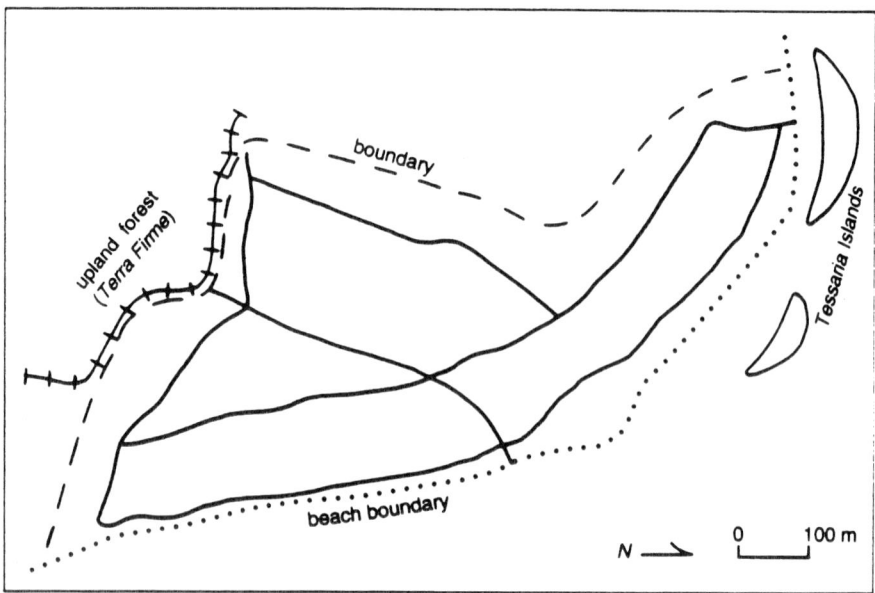

FIG. 2. The trail system (solid lines) used for censusing in the early-successional plot, 1985. Censuses were also conducted along the beach boundary and around the edges of the *Tessaria* islands.

Study plots.—We censused two large tracts containing successional gradients (Figs. 2 and 3). The early successional plot (Figs. 2 and 3) contained the first five stages and a backwater comprised of an old river channel. This plot abutted a steep, 30-m bank atop which occurred upland, or *terra firme* forest outside the floodplain. This study plot was chosen because it contained extensive stands of the first four stages. Only a limited area of stage 5 was included, and later stages are not represented in the plot or on adjacent areas. All six successional stages were

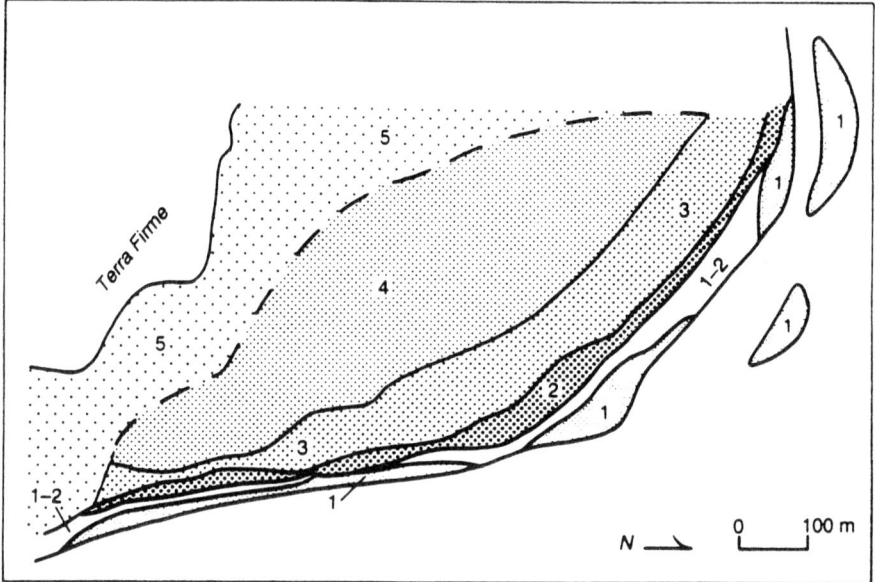

FIG. 3. Successional stages represented in the early-successional plot. Stage 1–2 consists of a mixture of old *Tessaria* and young cane. Stage 1 = *Tessaria*; Stage 2 = cane; Stage 3 = *Cecropia*; Stage 4 = *Guarea*; Stage 5 = *Ficus-Cedrela*.

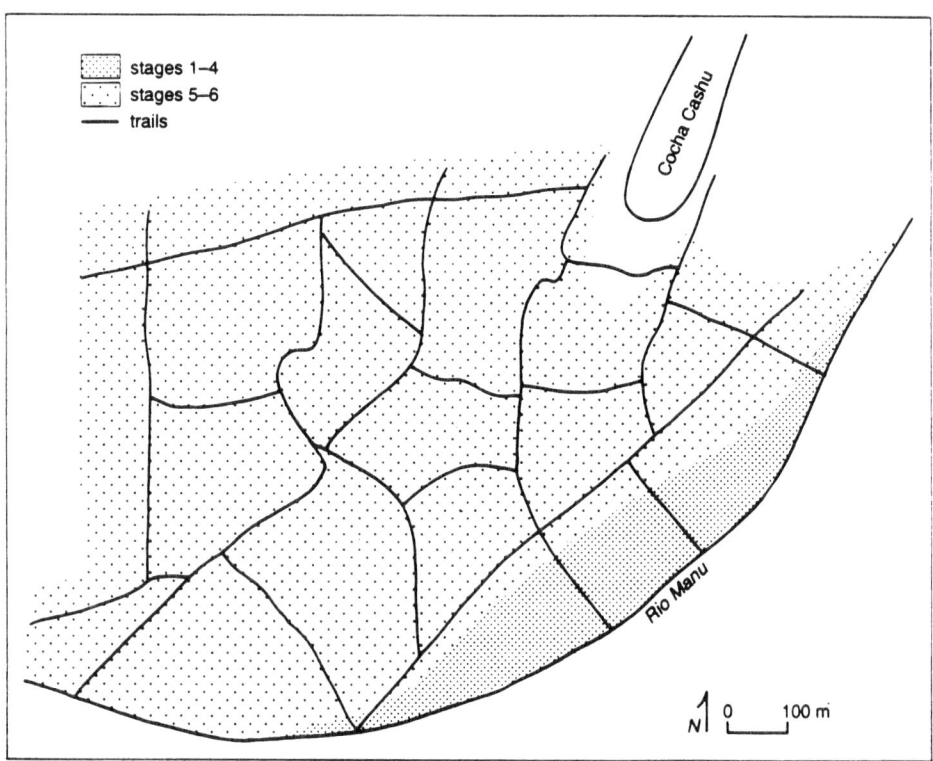

FIG. 4. Successional stages of the late-successional plot and locations of census trails. Censuses were also conducted along the edge of the river.

included within the late successional plot (Fig. 4), but most of the area (86 hectares) was occupied by stages 5 and 6. In addition to censusing these two plots, we conducted intensive censuses of seven *Tessaria* stands along the Manu River.

Song censuses.—We employed a combination of mist-netting and spot-mapping to census birds in the study plots shown in Figures 2–4. The methods are described in detail in Terborgh et al. (1990); here we provide only essential detail. We constructed trail systems, such that no point in any plot was more than 100 m from a trail. In the early successional plot, we cut parallel trails along the interfaces between successional stages to improve the resolution of bird locations within the narrow vegetation zones (Fig. 2).

All trails were censused an equal number of times by each of us, and the locations of every bird heard or seen were recorded with reference to surveyed locations on a trail. When two or more conspecifics were heard countersinging, we noted this and the locations of the singing birds. After every field season, we mapped registrations by species, and estimated the number of territories or home ranges. For nonterritorial species, we tallied the registrations by successional stage. Population densities of territorial species were estimated by dividing the number of territories by the area occupied on each plot. For stages 1–2, we estimated population density for the early successional plot only, because the area occupied by these stages in the late successional plot was too small to give reliable figures. In all, we tallied over 4,000 registrations on the early successional plot and 8,000 registrations on the late successional plot.

Population density estimates for the two successional plots are considerably less precise than those reported for mature floodplain forest by Terborgh et al. (1990). Many species were represented in the relatively small early successional plot by only a single pair that probably occupied areas outside the plot as well (e.g., *Bucco macrodactylus*, *Buteo magnirostris*). Moreover, we did not color-band the populations of either plot, so we had no back-up method to confirm the estimates derived from song censuses. Furthermore, we did not employ transect counts or estimates of group sizes of canopy frugivores (pigeons, parrots, toucans), so we made no attempt to estimate population densities of these or other nonterritorial species such as hummingbirds and manakins.

All censuses were conducted during early morning sessions between 15 August and 15 November 1983 (late-successional plot) and 1985 (early-successional plot). On any given day, each of us covered one of the routes (four in the early-successional plot and seven in the late-successional plot). Coverage of the routes and the direction walked were rotated systematically. We began censuses 15–30 minutes before first light to census nocturnal birds and those with a predawn singing period. Because censuses were conducted during different years, we cannot be certain that differences in population density were not caused by year effects. Logistical constraints, however, precluded censusing more than one plot per year.

Mist-netting.—We supplemented our song censuses with understory mist-netting conducted in a variety of successional stages. Variously, from eight to 40 mist nets were opened for four (usually consecutive) days from dawn to dusk. Netting was curtailed during rains and severe mid-day heat in the weakly shaded *Tessaria* habitat. Further details of our netting procedures are available in Robinson and Terborgh (1990). We used mist-netting results solely to provide an index of relative abundance of nonterritorial understory species in different successional stages, thereby avoiding some, but not all of the pitfalls of extracting population estimates from mist-net captures (e.g., Remsen and Parker 1983; Graves et al. 1983; Terborgh 1985; Remsen and Good 1996). Mist-netting also helped in detecting non-vocal species that were under-represented in song censuses (Terborgh et al. 1990).

Additional Tessaria *censuses.*—In addition to the *Tessaria* stands censused in the two main study plots (Figs. 2 and 3), we censused birds on six more beaches as part of a different project (Robinson and Terborgh 1995). Because these beaches were visited only 3–10 times each, we only include estimates of presence/absence of the most vocal territorial species. The purpose of these censuses was to increase our sample of this narrow, but bird-rich habitat (see below), and to provide replicates that would enable us to detect some of the rarer and more patchily distributed species.

Guild characterizations.—Here, we use a simple stratified system to characterize guilds based on foraging stratum (terrestrial, understory, arboreal) and diet (insectivore, frugivore, omnivore, granivore, and nectarivore) (Terborgh and Robinson 1986). We defined birds that forage predominantly below 5 m as understory guild members. Birds of taller saplings and the subcanopy were included in the arboreal guilds. The distinction between understory and arboreal guilds in early successional habitats was occasionally somewhat arbitrary, but most species either foraged high in the *Tessaria* crowns, or beneath the crowns in vines and other invading plants. We recognize, however, that understory birds in stages 1–2 may be more ecologically similar to canopy species in stages 6–7 than to understory species in these well-shaded, mature forests.

Based on our observations of birds in the Manu region, we classified species as omnivores only if they regularly feed at two trophic levels (e.g., fruit and insects). Genera, such as *Celeus* and *Campephilus*, that feed primarily on insects and only occasionally consume fruit and nectar were classified as insectivores. *Melanerpes cruentatus* was classified as an omnivore because it seemed to consume as much fruit and nectar as it does insects. Conversely, primarily frugivorous species, such as *Gymnoderus foetidus* and various toucans, which occasionally consumed insects and small vertebrates, were classified as frugivores (Remsen et al. 1993). Parrots, most of which consume seeds while discarding fruit pulp, were classified as granivores.

Plot classification.—We calculated population densities for most species by successional stage. This required us to classify the plots by area. The early-successional plot was divided into two sections. The first comprised a band of nearly pure *Tessaria* and an adjacent band of older *Tessaria* that was being invaded by cane. This habitat harbors a distinctive set of species, most of which are restricted to the earliest successional stages, a convenience that makes it possible to calculate population densities with reasonable accuracy. Species that extended into later-successional stages (e.g., *Leptotila rufaxilla* and *Crypturellus undulatus*) were assigned an average population density for all zones occupied. We made no attempt to estimate population densities of the many nonterritorial species that occurred in the earliest-successional stages.

We used the older portions of the early-successional plot for estimating population densities in Stages 4 and 5 combined. Included within the area was an abandoned river channel with its associated backwaters. We used the combined area of all habitats occupied to estimate the densities of species whose territories extended into earlier stages.

We similarly subdivided the late-successional plot into two sections, an early-successional section (stages 1–4), and a late-successional section (stages 5–6). Population estimates were derived only for the late-successional section, because bird locations could not be confidently

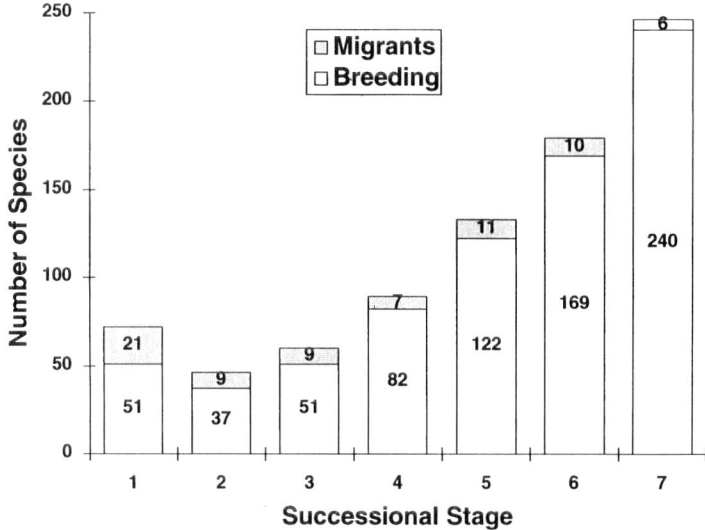

FIG. 5. Species richness of breeding species (including species that regularly visited stage 1) and migrants (Nearctic and austral) in each successional stage.

assigned to the narrow bands of vegetation that comprised the early stages. Within the area occupied by Stages 5 and 6 were several backwaters and zones of jumbled vegetation created by numerous treefalls where the *Ficus/Cedrela* stage was giving way to transitional forest. As in the early-successional plot, all areas occupied by species that extended into earlier-successional stages were included in the estimates of population densities.

Occupancy of successional stages.—The territory maps for each species were inspected to determine the stages in which they occurred. We noted the distribution of sightings and mist-net captures of non-territorial species in different successional stages.

Biomasses.—To estimate biomasses of breeding, territorial species, we multiplied the estimated number of adults by their body masses. Most weights were obtained from birds captured near Cocha Cashu (many are included in Dunning 1992). To fill the gaps in our data, we obtained weights from specimens from Peru in the Louisiana State University Museum of Natural Science and the Field Museum of Natural History. We used average group size (three for toucans, four for all others [S. Robinson, pers. obs.]) to estimate the biomasses of species that live in extended families or communal groups. Because biomass estimates depend so critically on the values for a few large species within each guild (Terborgh et al. 1990), they should only be regarded as approximations. Moreover, because we did not estimate the populations of nonterritorial species and nonbreeding floaters, our estimates presumably represent minimum levels. This applies especially to parrots and colonial icterines, which were large and common on our plots.

RESULTS

Species richness.—Species richness of both territorial and nonterritorial birds (Appendices 1, 2, and 3) increased along the successional gradient, but not uniformly (Fig. 5). The number of species that used the first stage was remarkably high, given the nearly monoculture structure of the *Tessaria* zone. If one includes regular visitors and migrants, 90 species occurred regularly in this stage, and many others occurred as vagrants (see below). Species richness declined in the cane stage (stage 2). Few species were added in stage 2, but at least 30 new resident species appeared in each successive stage, with the largest jump occurring between stages 6 and 7 (see Appendices 1, 2, and 3). As *Cecropia* began overtopping the cane (stage 3), avian species richness began a steady increase that continued until the mature phase was reached (stage 7).

Species turnover.—Although the overwhelming trend was toward additions of resident species along the successional gradient (Fig. 5), considerable species turnover occurred both early and late in the successional gradient (Table 1). Early-successional species dropped out when the cane was overtopped by *Cecropia* and mid-successional species dropped out between stages 6 and 7 (Table 1). Overall, 66 species (22% of the 306 total year-round residents recorded in the area) dropped out of the community between stages 1 and 7.

TABLE 1
BREEDING SPECIES TURNOVER BETWEEN DIFFERENT ADJACENT SUCCESSIONAL STAGES

Stages	Number of species	
	Lost (%)[a]	Gained (%)[b]
1–2	12 (23.5)	3 (8.1)
2–3	16 (43.2)	30 (58.8)
3–4	4 (7.8)	35 (42.6)
4–5	4 (4.9)	40 (32.8)
5–6	3 (2.5)	44 (26.0)
6–7	27 (16.0)	89[c] (37.1)
Total	66	241

[a] Percentage of species from the earlier successional stage that were absent from the later stage.
[b] Percentage of species from the later stage that first appeared in that stage.
[c] Includes species listed in Appendix B and all species listed in Appendix A with only a "+" in stage 6 (i.e., they were not regular enough in this stage to estimate population densities).

Abundances of territorial species.—Rank-abundance curves for the four subplots differed conspicuously (Fig. 6). The *Tessaria* community contained several abundant species, some of which had territories of less than one hectare in contrast to the 4–6 ha territories of the most abundant mature-forest species (Terborgh et al. 1990). Stage 1 had significantly more abundant (>20 pairs/20 ha) and significantly fewer rare (<5 pairs/100 ha) species than stages 4–5 ($\chi^2 = 29.9$, $df = 3$, $P < 0.001$), stages 5–6 ($\chi^2 = 42.0$, $df = 3$, $P < 0.001$) and stage 7 ($\chi^2 = 87.4$, $df = 3$, $P < 0.001$). Subsequent successional stages contained progressively fewer common species (> 20 pairs or breeding units/100 ha), and more rare species (< 5 pair/100 ha). The abundances of species in stages 4–5 and 5–6 (Fig. 6) did not differ significantly. Stage 7 differed in the proportion of species in four categories (>20 pairs/100 ha, 10-20 pairs/100 ha, 5–10 pairs/100 ha, or <5 pairs/100 ha) from stage 4–5 ($\chi^2 = 13.5$, $df = 3$, $P < 0.001$) and stage 5–6 ($\chi^2 = 10.83$, $df = 3$, $P < 0.02$); most of these differences resulted from a much higher proportion of rare

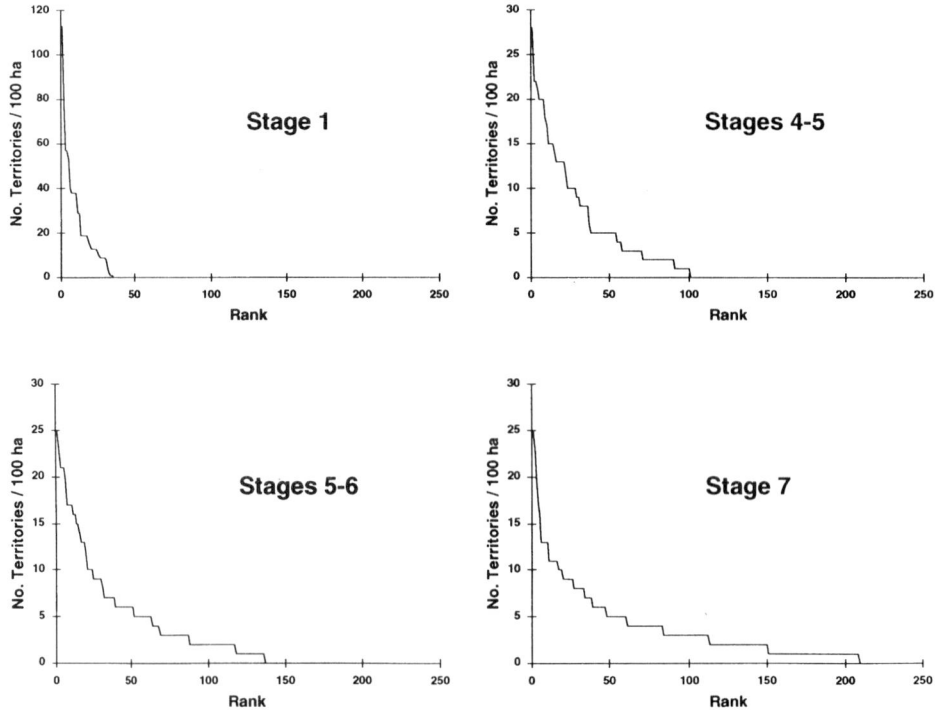

FIG. 6. Rank order of abundance/100 ha of territorial species at four different successional stages. Data for stage 7 from Terborgh et al. (1990).

TABLE 2
Species Richness, Population Densities/100 ha, and Biomass/100 ha of Guilds along the Successional Gradient. Stage 3 is Not Included Because We Do Not Have Reliable Abundance Estimates (See Methods)

		Successional stage			
		1–2	4–5	5–6	7
Terrestrial Granivore	No. Species	3	5	7	10
	No. Pairs/100 ha	34	34.5	56	42
	Biomass (kg)/100 ha	18.0	29.0	76.7	61.4
Arboreal Frugivore	No. Species	0	14	18	18
	No. Pairs/100 ha	0	45	59.5	60.5
	Biomass (kg)/100 ha	0	18.3	42.5	20.2
Arboreal Insectivore	No. Species	7	23	33	47
	No. Pairs/100 ha	176	193	217	193.5
	Biomass (kg)/100 ha	7.2	16.0	18.3	13.2
Arboreal Omnivore	No. Species	7	22	28	29
	No. Pairs/100 ha	228	177.5	200	109
	Biomass (kg)/100 ha	18.7	23.0	21.7	10.8
Understory Insectivore	No. Species	11	11	18	39
	No. Pairs/100 ha	330	70.5	95	178
	Biomass (kg)/100 ha	14.6	3.6	5.7	7.8
Bark Insectivore	No. Species	3	7	19	20
	No. Pairs/100 ha	37	50.5	79	57
	Biomass (kg)/100 ha	1.1	5.2	8.9	5.3
Terrestrial Insectivore	No. Species	3	6	6	14
	No. pairs/100 ha	57	56.5	30.5	71.3
	Biomass (kg)/100 ha	6.1	4.2	2.5	5.5
Terrestrial Frugivore	No. Species	0	1	2	2
	No. Pairs/100 ha	0	5.5	6	6.5
	Biomass (kg)/100 ha	0	1.3	3.2	9.8
Understory Omnivore	No. Species	0	2	2	2
	No. Pairs/100 ha	0	15	11	2
	Biomass (kg)/100 ha	0	2.2	1.7	2

species (<5 pairs/100 ha). Overall abundance of territorial species, however, was similar among the plots (stages 1–2: 862 pairs or breeding units/100 ha; stages 4–5: 648; stages 5–6: 745; stage 7: 720).

Biomass of territorial species.—Estimated biomass was higher in the late-successional plot (stages 5–6: 181.2 kg/100 ha) than in the plots representing stages 1 and 2 (65.7 kg/100 ha), stages 4 and 5 (102.8 kg/100 ha), and mature forest (stage 7: 134.2 kg/100 ha). The relatively high biomass estimates of mid-successional stages resulted from higher abundances of a few large-bodied frugivores and granivores.

Guilds of territorial species.—Successional patterns of richness, abundance, and biomass varied among guilds (Table 2). Species richness of some guilds increased with each successional stage (e.g., terrestrial granivores; arboreal, terrestrial and understory insectivores), whereas others increased little after mid-to-late successional stages (5–6) (e.g., arboreal frugivores, arboreal omnivores, bark insectivores). Among territorial species, most of the increase in species richness between stages 6 and 7 was caused by a 42% increase in richness of arboreal insectivores, a 116% increase in understory insectivores, and a 133% increase in terrestrial insectivores. In contrast, the number of frugivores and omnivores (all strata combined) increased only 2% between the late-successional and mature phases.

The increased number of insectivores in mature forest (stage 7) is associated with the presence of more diverse mixed-species flocks (both canopy and understory), ant-followers, and ground-foraging insectivores (Table 1). Army ant-followers were essentially absent in early successional stages, and most species were found only in the mature phase (Appendix 2). Understory flocks were notably scarce and species poor in early successional habitats, including just 6–8 species

TABLE 3
SPECIES RICHNESS OF GUILDS OF NONTERRITORIAL SPECIES ALONG THE SUCCESSIONAL GRADIENT

	Number of species/stage Stages			
	1–2	4–5	5–6	7
Canopy Nectarivore	3	2	2	4
Understory Nectarivore	6	7	8	8
Understory Frugivore	4	3	5	5
Canopy Frugivore	1	3	4	2
Canopy Granivore	0	9	9	12
Understory Granivore	3	0	0	0
Canopy Omnivore	0	2	3	4
Aquatic	0	2	6	8

in stages 4–6. Understory flocks were absent altogether from stages 1–3. Several of the most abundant understory flock species of the mature phase were absent even from the late successional plot (stages 5–6: e.g., *Thamnomanes ardesiacus*, *Myrmotherula longipennis*, *M. leucophthalma*). Similarly, canopy flocks of mid-successional habitats, although relatively diverse (21 species), lacked the two core "sentinel" species of the mature phase (*Lanio versicolor* and *Tachyphonus rufiventer*, Munn 1985). Most terrestrial insectivores restricted to the mature phase were rare (<5 pairs/100 ha) and had large (>20 ha) territories (e.g., *Chamaeza nobilis*, *Myrmothera campanisona*, *Dichrozona cincta*, *Sclerurus caudacutus*, *Conopophaga peruviana*, *Liosceles thoracicus*). The terrestrial insectivores of stages 1–2 (*Hydropsalis climacocerca*, *Nyctidromus albicollis*, and *Furnarius leucopus*) all foraged extensively on beaches as well as along the edges of *Tessaria* stands, so our estimates of their population densities are probably inflated, as were estimates of *Buteo magnirostris*.

Early-successional breeding communities were dominated by a few abundant species of arboreal and understory insectivores and arboreal omnivores (Table 2). *Turdus ignobilis*, the most abundant species, also foraged across the river in flood-damaged forest; our estimates therefore exaggerate its true density. The biomass of understory insectivores in stages 1–2 was the highest of any successional stage (Table 2). In contrast, frugivore and granivore abundance and biomass reached peak levels in the late-successional plot. Similarly, the abundance and biomass of omnivores was higher in mid-successional than in mature forest.

Bark-foragers represent the only group of insectivores that did not increase in richness, abundance, or biomass between late-successional and mature forest.

Guilds of non-territorial species.—We restrict our analysis of non-territorial species to patterns of species richness and relative abundance (Tables 3, 4). Species richness of nectarivores varied little along the successional gradient, because both canopy and understory species regularly visited the earliest successional stages. Abundance of understory hummingbirds was greatest in stages 4–6 (Table 4), undoubtedly in response to the abundance of *Heliconia* and other nectar-producing understory plants.

Understory frugivores, which include lek-breeding manakins and *Mionectes* flycatchers, were abundant in mist-net samples (Table 4), even though no known leks occurred in either study area. Among arboreal frugivores that lack defined territories, several appeared to be confined largely to mid-to-late successional stages, including *Gymnoderus foetidus*, *Cotinga maynana*, *Iodopleura isabellae*, and *Tersina viridis*. These species were observed only rarely during the study.

Aquatic species routinely entered the mature phase plot along a small, intermittent stream that flowed into an oxbow lake (Terborgh et al. 1990). Backwaters included within the late-successional plot attracted aquatic species such as *Anhima cornuta* and *Cairina moschata*. The lack of permanent water in the early-successional plot likely accounted for the scarcity of aquatic species on that study area (Table 3).

Parrots, the only arboreal granivores, were commonly observed in stages 3–7 (Appendix 3), but reached their greatest richness in the mature phase (Table 2). Flocks of *Ara severa* and *Brotogeris cyanoptera* fed in flowering *Erythrina* trees in stage 3, along with occasional *Ara couloni*, *A. manilata*, and *Aratinga weddellii*. *Pionus menstruus* and *Forpus sclateri* appeared to reach their peak abundance in mid-successional vegetation, whereas *Ara chloroptera*, *Aratinga*

TABLE 4
MIST-NET CAPTURE RATES OF NON-TERRITORIAL BIRDS ALONG THE MANU SUCCESSIONAL GRADIENT

Species	1[a]	2[b]	1–3[c]	1–4[d]	5[e]	6[f]	7[g]	7[h]	7[i]
Geotrygon montana	0	0	0.0 (1)	0.0 (2)	0.5 (9)	1.1 (7)	0.1 (12)	0	0.3 (28)
Glaucis hirsuta	0	0.2 (2)	0.4 (22)	0.6 (26)	2.8 (55)	3.3 (1)	0.1 (8)	0.1 (2)	0.2 (19)
Threnetes leucurus	0.2 (1)	0	0.1 (4)	0.11 (5)	0.2 (4)	1.4 (9)	0.1 (14)	0.6 (13)	0.3 (26)
Phaethornis superciliosus	3.7 (17)	0.2 (1)	0.8 (39)	0.4 (18)	1.9 (38)	3.3 (21)	0.2 (32)	0.1 (3)	0.1 (11)
P. hispidus	5.9 (27)	0.8 (4)	0.4 (22)	0.4 (20)	0.7 (13)	5.4 (34)	0.1 (15)	0.1 (3)	0.4 (39)
Campylopterus largipennis	2.6 (12)	0	0.1 (4)	0.5 (22)	0.7 (14)	1.4 (9)	0.0 (3)	0	0
Polyplancta aurescens	0	0	0	0.0 (1)	0	0.5 (3)	0.0 (1)	0	0
Thalurania furcata	0.4 (2)	0.2 (1)	0.3 (15)	0.5 (22)	0.5 (10)	3.9 (25)	0.0 (12)	0.1 (2)	0.2 (13)
Mionectes olivaceus	0.2 (1)	0	0.0 (1)	0.0 (2)	0.1 (1)	0	0.0 (7)	0	0.0 (1)
M. oleagineus	0.7 (3)	0	0.1 (7)	0.1 (6)	0.2 (3)	0	0.0 (6)	0	0.1 (5)
M. macconnelli	0	0	0.0 (1)	0.0 (2)	0.1 (1)	0	0	0	0
Cnemotriccus fuscatus	1.1 (5)	0.6 (3)	0	0.0 (2)	0	0	0	0	0
Myiophobus fasciatus	1.1 (5)	0	0	0	0	0	0	0	0
Neopelma sulphureiventris	0	0	0	0.2 (7)	0.0 (2)	0	0	0	0
Machaeropterus pyrocephalus	0	0	0	0	0	0	0	0	0
Pipra coronata	0	0	0	0	0	0.2 (1)	0.1 (19)	0.2 (5)	0.1 (11)
P. fasciicauda	1.7 (8)	2.5 (13)	0.5 (25)	1.0 (48)	2.3 (55)	3.3 (21)	0.9 (134)	1.6 (33)	1.6 (139)
P. chloromeros	0.2 (1)	0.4 (2)	0.0 (1)	0.3 (15)	0.2 (3)	0.3 (2)	0.1 (14)	0.1 (1)	0.0 (1)
Sporophila caerulescens	8.9 (41)	0	0.1 (3)	0.0 (2)	0	0	0	0	0
S. castaneiventris	1.7 (8)	0	0	0	0	0	0	0	0

[a] 12 nets opened in August 1983 and November 1985 (460 total net-hrs) on a rapidly growing beach.
[b] 16 nets opened in August 1986 (512 net-hrs) in the early-successional plot.
[c] 39 nets opened in August 1974, September 1976, and November 1980 (5,150 net-hrs) in a mixture of the first three successional stages on a slowly growing beach (4,500 net-hrs).
[d] 36 nets opened in August 1974, September 1975, and November 1976 in a cross section from the river edge through late stage 4 vegetation in the early-successional plot.
[e] 20 nets opened in August 1976 and November 1980 (2,000 net-hrs).
[f] 29 nets opened in November 1986 (638 net-hrs).
[g] 36 nets opened 1973–1982 (15,840 net-hrs).
[h] 24 nets opened in 1981–1982 (2,112 net-hrs).
[i] 40 nets opened in 1974, 1975, 1976, 1979, 1981 (8,800 net-hrs).

leucophthalma, Pyrrhura rupicola, Pionopsitta barrabandi, and *Amazona farinosa* were only observed in the mature phase (Appendix 2).

Understory granivores, including *Sporophila* spp. and a territorial ground-running sparrow (*Ammodramus aurifrons*), were restricted to the grasses and sedges that grow along the beach/ *Tessaria* interface.

Omnivorous, arboreal caciques and oropendolas were abundant in virtually all habitats. The oropendola *Psarocolius angustifrons* fed and roosted in cane stands (stage 2), whereas *P. oseryi* and *Cacicus cela* fed primarily in later stages (3–7). The local abundance of these species reflected the availability of safe colony sites; they can decline sharply during years when their colonies have been destroyed by predators (Robinson 1985). The consistent presence of 1–3 forest colonies of *C. cela* in the mature-phase plot (Robinson 1985, unpubl. data), however, suggests a possible preference for mature forest. In contrast, the early-successional plot had just one small (8-nest) colony during the study, and the late-successional plot had a colony of 12–15 nests during the study. *Psarocolius oseryi*, on the other hand, seemed to prefer late-successional forest and swamps (Leak and Robinson 1989); a colony of 20 nests was built on the plot at the end of the census period.

Migrants.—The abundance of migrants in early-successional vegetation at Cocha Cashu has already been documented (Fitzpatrick 1980, Robinson et al. 1988, 1995.). It is worth noting, however, that the population densities of some migrants were sometimes very high. Flocks of *Tyrannus tyrannus*, numbering in the hundreds, foraged in successional vegetation during the late September–early December southward migration, and a few flocks remained through December. During this period, *T. tyrannus* may be the most abundant species in all successional stages. Flocks of *Myiodynastes luteiventris* also foraged in mid-to-late successional vegetation. Censuses of two *Tessaria* stands in early November revealed 10–17 *Empidonax alnorum* singing and, presumably, defending territories, which would indicate population densities in excess of 200 birds/100 ha, although this figure may have been inflated by migrants. November censuses suggested that 8–15 *Contopus virens* wintered in the early successional plot, and 10–20 in the late successional plot. *Myiarchus swainsoni*, an austral migrant (Chesser 1997), was comparably abundant in the same habitat during the June–September dry season. Other migrants that occupy *Tessaria* also maintained high population densities (10–100 individuals/100 ha): *Cnemotriccus fuscatus, Pyrocephalus rubinus, Myiophobus fasciatus, Catharus ustulatus, Turdus amaurochalinus,* and *Vireo flavoviridis*. Although *Catharus ustulatus* occurred regularly in all successional stages, it was not abundant in any. Other migrants recorded only in stages 1–3 included *Coccyzus erythropthalmus, C. americanus, Sublegatus* sp., *Inezia inornata*, and *Dolichonyx oryzivorus*. Two additional migrants, *Casiornis rufa* and *Empidonomus aurantioatrocristatus*, were only recorded in stages 5–6.

Patchiness of Tessaria *bird communities.*—Because the areas of *Tessaria* (stage 1) tend to be small on any given meander tongue, we censused six additional beaches to document the occurrence of patchily distributed species (Appendix 4). Of the species found only in *Tessaria* in the Manu area, several were found on only one or two of the beaches, including *Claravis pretiosa, Columbina talpacoti, Cercomacra nigrescens, Todirostrum latirostre, Thlypopsis sordida,* and *Conirostrum speciosum*. The largest of the surveyed *Tessaria* stands, which included areas of cane intermingled with *Tessaria*, supported breeding populations of species more characteristic of later-successional stages, including *Anabazenops* (formerly *Automolus*; see Kratter 1997) *dorsalis* (a bamboo specialist that occasionally occurs in cane), *Camptostoma obsoletum, Thryothorus genibarbis, Ramphocaenus melanurus, Saltator coerulescens,* and *Cyanocompsa cyanoides*. The presence of *Synallaxis albigularis* appeared to depend upon the presence of grassy backwaters.

The earliest-successional stages attracted the greatest numbers of wanderers from other habitats and regions outside of the Manu. The following species have only been recorded in stages 1–2 (fewer than 3 records each): *Columbina picui, Dromococcyx phasianellus, Otus choliba, Hydropsalis brasilianum, Caprimulgus sericocaudatus, Amazilia viridicauda, Celeus spectabilis, Synallaxis cabanisi, Thamnophilus amazonicus, Neochelidon tibialis, Sporophila nigricollis, Schistochlamys melanopis, Euphonia chlorotica,* and *Molothrus bonariensis* (on the beach). Less than 10 other extralimital wanderers have been recorded from later-successional stages (Terborgh et al. 1984; S. Robinson, unpubl. data).

During censuses and mist-netting of the *Tessaria* and cane stages, we recorded many species that were much more abundant in later-successional and even *terra firme* forest. Among these forest species detected more than once in stages 1–2 are the following: *Florisuga mellivora, Glyphorynchus spirurus, Simoxenops ucayalae, Philydor pyrrhodes, Thamnomanes ardesiacus,*

Myrmotherula hauxwelli, M. axillaris, M. menetriesii, M. longipennis, Cercomacra serva, Hypocnemis cantator, Ramphotrigon ruficauda, Attila bolivianus, and *Tachyphonus rufiventer.*

Other successional species.—Only six species known to occupy successional vegetation in the vicinity of Cashu Cocha were not detected on any of our study areas. One species with a vocalization unknown to us, *Hemitriccus iohannis,* has been recorded from other early successional areas in the Manu area and elsewhere (A. W. Kratter, pers. comm.). *Piaya minuta,* which is restricted to lake margins in the Cocha Cashu area, has been recorded in stages 1–2 areas closer to the mouth of the Manu River than Cocha Cashu (Robinson, unpubl. data). Similarly, *Xiphorhynchus picus,* primarily a bird of lake and river edges, occasionally foraged in these *Tessaria* in other years (J. W. Fitzpatrick, pers. comm.). *Cercomacra manu* occupied a patch of bamboo in disturbed river-edge vegetation 300 m north of the early successional plot. We did not record *Pachyramphus castaneus* in mid-successional forest, even though both previous records were from this successional stage (Terborgh et al. 1984).

DISCUSSION

Bird communities change considerably along the successional gradients of the Manu River. The surprisingly rich, insectivore-dominated community of structurally simple *Tessaria* stands (stage 1) gives way to less species-rich assemblages in the cane and *Cecropia* stages (2–3) after which species richness increases steadily as the forest becomes more structurally complex and new kinds of resources become available. In the vertically complex mature phase (stage 7), more than 160 breeding species can coexist over a single point (Terborgh et al. 1990). Here, we first review the contribution of primary successional gradients to regional species richness, then discuss the community dynamics of each successional stage separately. We conclude with discussion of the conservation implications of these results.

Contribution to regional species richness in the Cocha Cashu area.—Our results support the hypothesis that the disturbances created by meandering rivers contribute to the high species richness of Amazonian bird communities, which are rich even by Neotropical standards (Karr et al. 1990). Many species were restricted to the earliest and latest stages of floodplain succession, especially among congeners (see Table 7; Robinson and Terborgh 1995). Many species found in the earliest successional stages are also found in other low, dense habitats such as the margins of oxbow lakes, swamps and marshes (e.g., *Daptrius ater* [Robinson 1994], *Synallaxis gujanensis, Turdus ignobilis*) and bamboo thickets (e.g., *Percnostola lophotes, Myrmeciza goeldii* [Terborgh et al. 1984]). Similarly, many successional species readily colonize secondary-successional areas created by human activities (e.g., *Buteo magnirostris, Synallaxis albigularis, Thamnophilus doliatus, Cercomacra nigrescens*) (Terborgh and Weske 1969, Remsen and Parker 1983). Perhaps because the Manu River is a relatively small upper tributary of the Amazon, it lacks most of the species restricted to the huge *Tessaria*-dominated islands and seasonally inundated forests (varzea) of the Amazon and its major tributaries (Remsen and Parker 1983; Rosenberg 1990). In the Manu region (including human settlements), only a few breeding species are restricted to this successional stage (e.g., *Automolus rufipileatus* [also in bamboo elsewhere in Peru: A. W. Kratter, pers. comm.], *Todirostrum maculatum*). Early successional stages contribute many migrants and wanderers to the Manu list (Robinson et al. 1988, 1995, see below).

The mature phase of floodplain succession adds many species, primarily insectivores, as a result of the increased vertical complexity of the vegetation (Pearson 1971, 1975, 1977; Terborgh 1980). Nevertheless, most species of the mature phase of floodplain succession also occur in upland, or *terra firme* forest (Terborgh et al. 1984, Terborgh and Robinson, unpubl. data).

The contribution of the middle states of succession to regional species richness in southwestern Amazonia is difficult to evaluate from just a single site. Many species that reach their peak abundance in mid-successional stages were also found at low population densities in mature or early successional stages (e.g., *Columba subvinacea*). Many other species were so rare that we could not characterize habitats reliably (e.g., *Coccyzus melacoryphus, Polyplancta aurescens, Formicarius rufifrons* [see Kratter 1995], *Iodopleura isabellae, Ara couloni, Casiornis rufa, Myiarchus tyrannulus, Empidonomus aurantioatrocristatus, Neopelma sulphureiventer*). Two of these species, *Ara couloni* and *Polyplancta aurescens,* in fact, may be wanderers from higher elevations where they occupy *terra firme* forest (J. W. Fitzpatrick, pers. comm.). Other species we recorded only in mid-successional stages are found in mature forest at higher elevations (*Eubucco tucinkae, Xiphorhynchus ocellatus*) (J. W. Fitzpatrick, pers. comm.). Two other species that occur in montane forests (*Aulacorhynchus prasinus* and *Philydor rufus*) only occur in mid-successional vegetation in the Cocha Cashu area. Perhaps the relatively rich, but variable resources, and low diversity of forest species in mid-successional habitats enable many species of higher elevations to invade these habitats.

Many other species we recorded only in mid-successional stages on our plots were also found in overgrown lake beds (successional swamp forest: Terborgh et al. 1984), successional vegetation along the margins of lakes (Robinson 1997), flood disturbances, and marshes (e.g., *Chondrohierax uncinatus, Ortalis motmot, Forpus sclateri, Psarocolius oseryi, Cyanocorax violaceus, Aratinga weddellii, Galbalcyrhynchus purusianus, Pteroglossus castanotis, P. inscriptus, Celeus elegans, Dryocopus lineatus, Nasica longirostris, Formicarius rufifrons, Myiopagis viridicata, Myiodynastes luteiventris, Tityra inquisitor, Gymnoderus foetidus, Conioptilon mcilhennyi, Cissopis leveriana, Ramphocelus carbo, Nemosia pileata, Euphonia laniirostris*, and *Dacnis flaviventer*). This group of generalized mid-successional species is equivalent to the flood-dependent species described in Remsen and Parker (1983). Given the complex mosaic of floodplain habitats available in Amazonia (e.g., Salo et al. 1986), it is perhaps not surprising that few, if any species were entirely confined to the vegetation that characterizes mid-successional vegetation in the Cocha Cashu area.

BIRD COMMUNITY DYNAMICS IN DIFFERENT SUCCESSIONAL STAGES

In this section, we discuss some characteristic features of bird communities of each successional stage. Species richness increased along this gradient (Fig. 5), but the entire organization of the community changed as well.

Tessaria.—Considering their structural simplicity and general lack of abundant fruit and nectar, *Tessaria* stands supported remarkably rich bird communities. The foliage height diversity of a *Tessaria* stand would be expected to support fewer than the 34 breeding species recorded in the censuses based on regressions in MacArthur and MacArthur (1960). The breeding bird community is dominated by insectivores and a few omnivores. Flycatchers were particularly abundant and sometimes formed loosely organized mixed-species flocks (Robinson and Terborgh, pers. obs.). The population densities of some *Tessaria* specialists were extraordinarily high with some reaching 50–125 pairs/100 ha. Rosenberg (1990) documented even higher population densities of river-island *Tessaria* specialists along the Amazon. The biomass and abundance of breeding insectivores in *Tessaria* were comparable to or higher than those of later successional stages, despite the greater foliage volume of the latter (Terborgh and Petren 1991).

Tessaria is remarkably rich in arthropod resources, as confirmed in both sweep-net and Malaise-trap samples (J. Terborgh, unpubl. data). *Tessaria* foliage is often riddled with holes left by folivorous insects and first-year saplings are commonly defoliated by outbreaks of caterpillars and orthopterans. *Tessaria* thus conforms to the Resource Limitation Hypothesis that rapidly growing plants should invest little in anti-folivore defenses (Coley et al. 1985). It is likely that many of the frugivores (e.g., *Pipra, Mionectes*), and nectarivores (e.g., *Glaucis, Phaethornis*) that visit this habitat in large numbers are supplementing their diets with insects. Similarly, the migrants that abound in these stages (Robinson et al. 1988) may be attracted by high arthropod densities. Additionally, some vines that invade *Tessaria* may provide limited amounts of fruit (e.g., *Cissus* spp. vitaceae) or nectar on a seasonal basis.

The high diversity of birds in mist-net samples in *Tessaria* (Robinson and Terborgh 1990) reflects a combination of sampling artifacts and ecological factors. First, mist nets sample a higher proportion of canopy species in habitats that lack the tall canopy of forests (Karr 1981). Second, mist nets also disproportionately sample the non-territorial frugivores, nectarivores, and migrants that visit these habitats (Remsen and Parker 1983). Third, *Tessaria* may also act as a dispersal corridor and holding area for non-territorial floaters. The many vagrant species that we only recorded in *Tessaria* suggest that dispersing birds may move along rivers and take advantage of abundant resources where encountered. Similarly, "floaters" of forest species may benefit from high resource availability without the threat of conspecific interference (see also Dyrcz 1989). We have no data on where birds of this community go when wet-season floods inundate this habitat.

Compared with the large *Tessaria* islands of the Amazon, however, the bird communities of the small (<3 ha) stands of *Tessaria* along the Manu lack specialists (Remsen and Parker 1983; Rosenberg 1990). None of the Amazonian *Tessaria*-island specialists were found in the small (<3 ha) *Tessaria* stands found along the Manu. Even within the Manu area, the distribution of early-successional species was extremely patchy (Appendix 4) and may have depended at least partially on the size of the stand, which in turn reflected the speed with which each meander loop was eroding away the forest on the outer loop.

Cane (stage 2).—Bird communities of cane stands were depauperate and contained few, if any, specialists. Several species characteristic of bamboo stands occasionally have territories in cane stands (e.g., *Campylorhamphus trochilirostris, Anabazenops dorsalis, Thryothorus genibarbis, Simoxenops ucayale* [Appendix 4]), perhaps because of some similarities in vegetation structure (cane and bamboo are both tall, hollow-stemmed woody monocots). *Picumnus rufiventris*, another species that is common in bamboo, also feeds extensively in cane breaks in addition to forest vine tangles. Otherwise, cane stand bird communities consisted of species that were much more abundant in *Tessaria* or in later stages. Canebreaks offer unusual foliage, but no nectar. Cane leaves emerge in a tuft from a short length of stem, are notably tough, and show little evidence of herbivory. Canebreaks may therefore offer relatively few food resources to birds. Dense canebreaks, however, were used as roosting sites for icterids (mostly *Psarocolius angustifrons* and *Scaphidura oryzivora*) and *Columba cayennensis*. A large cane stand on the Madre de Dios River (ca. 250 km downriver from Cocha Cashu) contained an estimated 20,000–40,000 *Psarocolius angustifrons*, 1,000–2,000 *Scaphidura oryzivora,* and 200–500 *Psarocolius oseryi* as well as a few other species of icterids (S. Robinson, pers. obs.).

Cecropia *(stage 3).*—The *Cecropia* stage also has a poorly defined bird community in our study area. When *Cecropia* fruits (March–April) and *Erythrina* flowers (July–August), these common trees attract many frugivores, nectarivores, and omnivores, but few birds breed in this stage. Only *Nemosia pileata* appears to be restricted to this zone in the Manu, although this conclusion is based on observations of just three pairs (see also Isler and Isler 1987). The large *Cecropia*-dominated islands of faster-flowing rivers such as the nearby Madre de Dios would offer better opportunities to study birds of this stage.

Mixed forest and Ficus/Cedrela *(stages 4–5).*—The *Cecropia* stage is superseded by a taller forest containing more than two dozen tree species (Terborgh and Petren 1991). In association with the increasing plant diversity, the bird community rapidly increases in species richness with the addition of many arboreal frugivores, omnivores, insectivores, and granivores (parrots). The *Heliconia*-dominated understory is rich in nectarivores and omnivores (e.g., *Cacicus solitarius*), but poor in understory insectivores. *Heliconia* and other broadleafed monocots produce abundant nectar, but do not appear to hold high arthropod densities. Among understory insectivores, only *Myrmeciza hyperythra* appeared to specialize on *Heliconia* habitats. Understory flocks of insectivores were mostly lacking in these stages, and canopy flocks inhabiting these stages tended to have few species and to lack the complex organization of mature-forest flocks (Munn 1985). Perhaps a threshold of flock size must be reached before they can support "sentinel" species such as *Thamnomananes schistogynus* and *Lanio versicolor*. Pearson (1977) also found that structurally simple forests lacked many species of antwrens. In contrast, many common floodplain species reached their peak abundance in these stages (e.g., *Crypturellus undulatus*). Perhaps because flood duration is short in the Manu relative to the Amazon, the mid-successional bird communities we studied had quite different species composition from those of varzea described by Remsen and Parker (1983).

Ficus/Cedrela *through transitional forest (stages 5–6).*—As the large *Ficus* and *Cedrela* trees die out and are replaced by a more diverse late-successional forest, species richness continued to increase, reaching levels comparable to mature forest for some guilds (e.g., bark insectivores, arboreal frugivores and omnivores). The estimated total biomasses of territorial species attained maximal levels in late-successional forest. The abundance of bark insectivores may reflect the many standing dead and dying *Cedrela* and *Ficus* trees. Mid-successional forests along secondary-successional gradients are also rich in frugivores (Martin and Karr 1986, Loiselle and Blake 1994).

Late-successional forest was rich in frugivorous, granivorous, and omnivorous birds, but was nevertheless deficient in most insectivorous guilds (except bark foragers). Arboreal insectivores and omnivores formed large multi-species flocks, but generally lacked sentinel species. Understory flocks were also scarce and species-poor relative to mature forest (see also Pearson 1977). Ant-followers were mostly absent, in keeping with our impression that ant swarms rarely forage in late-successional forests, at least in our study area. Many insectivores characteristic of the middle vertical strata were also rare or absent in the late successional forests (e.g., *Jacamerops aurea, Microrhopias quixensis, Myiornis ecaudatus, Hemitriccus zosterops, Ramphotrigon ruficauda*). The low diversity of understory insectivores in our late-successional plot may reflect the continued dominance of broadleafed monocots in contrast to the mature forest understory, which is composed of saplings, palms and a diverse assemblage of low treelets.

One caveat about late-successional forests is that their structure may be strongly affected by local variations in drainage and topographic levels. Late-successional forest on the higher terraces north of the late-successional plot had a more uniform canopy and diverse understory. Several mature-forest species characteristic of mixed-species flocks occurred in this area just outside of our plot, including *Lario versicolor, Glyphorynchus spirurus, Hyloctistes subulatus, Myrmotherula longipennis, Terenura humeralis,* and *Hylophilus ochraceiceps*. Better-drained areas may also regain faster the full diversity of terrestrial insectivores that were conspicuously lacking in the flood-prone successional forests. Our study plot may therefore under-represent the variability of late-successional bird communities.

Mature floodplain forest (stage 7).—The bird community of this successional stage has already been described in detail (Terborgh et al. 1990). In this well-drained, vertically complex forest, many insectivores specialize on particular vertical strata (Pearson 1971, 1975, 1977; Terborgh 1980), mixed-species flocks reach their greatest diversity (Munn 1985), ant swarms are typically attended by 5–10 species, and many terrestrial insectivores coexist in a forest that is only rarely inundated.

CONGENERIC REPLACEMENTS

Another way in which successional gradients contribute to regional diversity is through congeneric habitat segregation. Many closely related species occupy different ends of successional gradients (Table 5). In many cases, habitat segregation is mediated through interspecific aggression, which sometimes takes the form of interspecific territoriality (Robinson and Terborgh 1995). Interestingly, understory insectivores of early-successional habitats tend to be larger and more aggressive than congeners occupying later-successional habitats (Robinson and Terborgh 1995). The high insect abundance and dense understories of early-successional stages may make them richer foraging habitats for understory insectivores. The high population densities and small territories of insectivores in early-successional stages supports this argument. Dominant congeners may be able to occupy these rich habitats by excluding subordinate congeners (Robinson and Terborgh 1995). Alternatively, population densities may be higher because there is a higher density of foraging substrate for the species that specialize on the foraging substrates provided by the limited number of plant species in early successional habitats.

IMPLICATIONS FOR CONSERVATION

The meandering of Amazonian whitewater rivers generates a great diversity of habitats in their floodplains (Kalliola et al. 1991). Because these habitats are successional and dynamic, they are severely altered when rivers are dammed, levees are constructed, or if deforestation alters the hydrology (see also Remsen and Parker 1983). Fortunately, birds of early-successional habitats, which include many migrants, often appear to be preadapted to secondary-successional gradients created by human activities (Terborgh and Weske 1969, Remsen and Parker 1983). Even many southwestern Amazonian endemics appear to persist in human-altered landscapes adjacent to the Manu park (e.g., *Crypturellus atricapillus, Ara couloni, Formicarius rufifrons* [see also Kratter 1995]). Among these endemics, only *Conioptilon mcilhennyi* appears to be confined entirely to successional vegetation along floodplains. We know little, however, about the habitat requirements of some species of early successional habitats such as the northern migrant *Oporornis agilis*, which might be more specialized than we realize. Nor do we know anything about the spatial requirements or nesting success of most Amazonian species. Undoubtedly, the community of predators differs between primary and secondary successional habitats.

The apparent importance of mid-successional habitats for maintenance of high species richness in the Manu region poses potential problems for conservation of regional biodiversity. Mid-successional forests contain the valuable *Cedrela odorata*, a mahogany relative, and are ideal sites for some kinds of agriculture. The logging and conversion to agriculture of mid-successional forests presumably would have a major impact on mid-successional specialists (e.g., *Conioptilon mcilhennyi*) and would reduce the regional abundance of many species that reach their peak abundance in this habitat. Because many mid-successional species are frugivorous, a strategy of leaving behind stands of fruiting trees (e.g., Lauraceae) might reduce the impact of human settlement.

Perhaps the most sobering of all is the finding that the recovery of diversity after disturbance may be prolonged (Terborgh and Petren 1991). Even though mid-successional bird communities are reasonably rich, they lack a great many species of the mature phase. Terborgh and Petren (1991) estimated that it takes at least 300 years to reach the mature phase characterized by a multi-layered vertical structure consisting of approximately five layers. Recovery of avian diversity in secondary temperate forests appears to occur more rapidly (Johnston and Odum 1956;

TABLE 5
HABITAT SELECTION BY CLOSELY RELATED SPECIES PAIRS AND TRIOS ALONG A SUCCESSIONAL GRADIENT IN THE MANU NATIONAL PARK, SOUTHEASTERN PERU

Genus	Plant successional stage					
	1	2–3	4–5	5–6	7	Terra firme
Crypturellus (<300 g)					soui or bartletti	soui
Crypturellus (>300 g)	undulatus	undulatus	undulatus	undulatus	undulatus or variegatus	variegatus
Leucopternis						
Odontophorus						kuhli
Columba	gujanensis	gujanensis	schistacea	schistacea	schistacea	stellatus
Otus	cayennensis	cayennensis	overlap	plumbea	stellatus	plumbea
Glaucidium	choliba	choliba	watsonii	watsonii	plumbea	watsonii
Momotus/Baryphthengus	brasilianum	brasilianum	overlap	hardyi or overlap	watsonii	hardyi
Monasa			M. momota	M. momota	hardyi	B. martii
			nigrifrons	nigrifrons	M. momota	morphoeus
					nigrifrons or morphoeus	
Eubucco		tucinkae	overlap	richardsoni	richardsoni	richardsoni
Celeus		elegans	elegans	elegans	torquatus	torquatus
Veniliornis	passerinus	passerinus	passerinus	overlap	affinis	affinis
Campephilus	melanoleucos	melanoleucos	melanoleucos	overlap	overlap	overlap
Xiphorhynchus (<40 g)	picus	picus	ocellatus	ocellatus	spixii	spixii
Synallaxis	albigularis or gujanensis	gujanensis				
Philydor	rufus	rufus	overlap	overlap	erythropterus	erythropterus
Automolus	rufipileatus	rufipileatus			infuscatus	infuscatus
Taraba/Thamnophilus	Th. doliatus or Ta. major	Ta. major or Th. doliatus	Ta. major	Ta. major	Th. aethiops	Th. aethiops
Cercomacra	nigrescens	nigrescens				
Myrmoborus	leucophrys	leucophrys			serva	serva
Myrmeciza (<30 g)	atrothorax	atrothorax or hemimelaena	overlap	overlap	myotherinus	myotherinus
			hemimelaena	hemimelaena	hemimelaena	hemimelaena
Formicarius		rufifrons	rufifrons	overlap	colma	colma
Hemitriccus			iohannis	iohannis	zosterops	zosterops
Todirostrum	maculatum	latirostre	chrysocrotaphum	chrysocrotaphum	chrysocrotaphum	chrysocrotaphum
Tolmomyias	flaviventris	flaviventris	flaviventris or assimilis	overlap	assimilis	assimilis
Pachyramphus	polychopterus	polychopterus	polychopterus	polychopterus	marginatus	marginatus
Coniopitlon/Querula		C. mcilhennyi	C. mcilhennyi	overlap	Q. purpurata	Q. purpurata
Turdus	ignobilis	ignobilis	ignobilis	albicollis or hauxwelli	albicollis or hauxwelli	albicollis
				maximus	maximus	
Saltator	coerulescens	coerulescens	overlap			[*Pitylus*] grossus

Shugart and James 1973). In temperate forests, species richness increases little after the 20–60 years required for a distinct canopy and shrub layer to form. As we found here (Fig. 5), species richness actually may decline in the early-to-middle phases of succession (Johnston and Odum 1956). In contrast with successional gradients in tropical forests, however, "mature" temperate forests contain few species not also present in mid-successional forests (Shugart and James 1973). Tropical forests probably are far less resilient than temperate forests to large-scale disturbances such as agriculture and clearcutting (e.g., Johns 1991).

Secondary succession following logging (Johns 1991) or abandonment of agricultural fields may follow different patterns from the primary successional gradients that we describe here. Recovery of soil fertility accompanies secondary succession, whereas sediment deposition and gradual building of the topographic level accompany primary succession. Secondary succession might proceed more rapidly in small openings surrounded by mature forests that could act as a seed source (Andrade and Rubio-Torgler 1994). Alternatively, the soils of secondary successional gradients might be much less nutrient-rich as a result of the infrequent siltation during floods.

Recovery of forests and their bird communities from anthropogenic disturbance depends upon the size of the disturbance, its location with respect to sources of seeds and vertebrate colonists, and past history of land use. Relatively light use of floodplains, such as selective logging and low density slash-and-burn agriculture, might have little long-term consequence for biodiversity, because such small-scale disturbances may mimic the damage caused by the natural flooding and meandering of rivers (Andrade and Rubio-Torgler 1994). Extensive clearing of land, however, whether for timber harvest or creation of pasture or cropland, creates disturbance on a scale that precludes the rapid recovery of biodiversity. Deforestation of whole landscapes, as occurred during the development of the eastern United States, would therefore have catastrophic consequences for biodiversity in Amazonia.

ACKNOWLEDGMENTS

We thank Ken Petren, Ernesto Raez, and Walter Wust for assistance with mist netting. Melisse Reichman and Mike Riley mapped most spot-map records. John Fitzpatrick, Andy Kratter, Ed Heske, and Jeff Brown all provided many useful comments on the first draft. We are indebted to the Direccion General Forestal y Fauna of the Peruvian Ministerio de Agricultura for permission to work in the Manu National Park. Financial support was provided by NSF grants DEB-8207002 and DEB-025975. Last, and not least, we honor the memory of Ted Parker, who taught us several dozen new songs and made this research possible. In just six weeks, Ted added 19 species to the Manu list.

LITERATURE CITED

ADAMS, C. C. 1908. The ecological succession of birds. Auk 25:109–153.
ANDRADE, G. I., AND H. RUBIO-TORGLER. 1994. Sustainable use of the tropical rain forest: Evidence from the avifauna in a shifting cultivation habitat mosaic in the Colombian Amazon. Conserv. Biol. 8:545–554.
BELL, H. L. 1982. A bird community of lowland rainforest in New Guinea. 4. Birds of secondary vegetation. Emu 82:217–224.
BROSSET, A. 1990. A long-term study of the rain forest birds in M'Passa (Gabon). Pages 299–318 in Biogeography and Ecology of Forest Bird Communities (A. Keast, Ed.). SPB Acad. Publ., The Hague, The Netherlands.
CHESSER, R. T. 1997. Patterns of seasonal and geographical distribution of austral migrant flycatchers (Tyrannidae) in Bolivia. Pp. 171–204 in Studies in Neotropical Ornithology Honoring Ted Parker (J. V. Remsen, Jr., Ed.), Ornithol. Monogr. No. 48.
COLEY, P. D., J. P. BRYANT, AND F. S. CHAPIN III. 1985. Resource availability and plant antiherbivore defense. Science 230:895–898.
DUNNING, J. B., JR. 1992. CRC Handbook of Avian Body Masses. CRC Press, Boca Raton, Florida.
DYRCZ, A. 1990. Understory bird assemblages in various types of lowland tropical forest in Tambopata Reserve, SE Peru (with faunistic notes). Acta Zool. Cracov. 33:215–233.
FITZPATRICK, J. W. 1980. Wintering or North American tyrant flycatchers in the Neotropics. Pages 67–78 in Migrant Birds in the Neotropics: Ecology, Behavior, Distribution and Conservation (A. Keast and E. S. Morton, Eds.). Smithsonian Inst. Press, Washington, D.C.
FOSTER, R. B., J. B. ARCE, AND T. S. WACHTER. 1986. Dispersal and the sequential plant communities in an Amazonian Peru floodplain. Pages 357–370 in Frugivores and Seed Dispersal (A. Estrada and T. H. Fleming, Eds.). Dr. W. Junk, Dordrecht, The Netherlands.
FOSTER, R. B., 1990a. The floristic composition of the Rio Manu floodplain forest. Pages 99–111 in Four Neotropical Rainforests (A. H. Gentry, Ed.). Yale Univ. Press, New Haven, Connecticut.
FOSTER, R. B. 1990b. Long-term changes in the successional forest community of the Rio Manu floodplain.

Pages 565–572 *in* Four Neotropical Rainforests (A. H. Gentry, Ed.). Yale Univ. Press, New Haven, Connecticut.

GENTRY, A. 1988. Changes in plant community diversity and floristic composition on environmental and geographical gradients. Ann. Missouri Bot. Garden 75:1–34.

GENTRY A. H., AND J. TERBORGH. 1990. Composition and dynamics of the Cocha Cashu "mature" floodplain forest. Pages 542–564 *in* Four Neotropical Rainforests (A. H. Gentry, Ed.). Yale Univ. Press, New Haven, Connecticut.

GLOWACINSKI, Z., AND O. JÄRVINEN. 1975. Rate of secondary succession in forest bird communities. Ornis Scand. 6:33–40.

GRAVES, G. R., M. B. ROBBINS, AND J. V. REMSEN, JR. 1983. Age and sexual differences in spatial distribution and mobility in manakins (Pipridae): Inferences from mist-netting. J. Field Ornithol. 54:407–412.

HELLE, P., AND M. MÖNKKÖNEN. 1990. Forest succession and bird communities: Theoretical aspects and practical implications. Pages 299–318 *in* Biogeography and Ecology of Forest Bird Communities (A. Keast, Ed.). SPB Acad. Publ., The Hague, The Netherlands.

HOLMES, R. T., T. W. SHERRY, AND F. W. STURGES. 1986. Bird community dynamics in a temperate deciduous forest: Long-term trends at Hubbard Brook. Ecol. Monogr. 56:201–220.

ISLER, M. L., AND P. R. ISLER. 1987. The Tanagers: Natural History, Distribution, and Identification. Smithsonian Institution Press, Washington, D.C.

JOHNS, A. D. 1991. Responses of Amazonian rain forest birds to habitat modification. J. Trop. Ecol. 7:417–437.

JOHNSTON, D. W., AND E. P. ODUM. 1956. Breeding bird populations in relation to plant succession on the Piedmont of Georgia. Ecology 37:50–61.

KALLIOLA, R., J. SALO, M. PUHAKKA, AND M. RAJASITTA. 1991. New site formation and colonizing vegetation in primary succession on the Western Amazon floodplains. J. Ecol. 79:877–901.

KARR, J. R. 1981. Surveying birds with mist nets. Stud. Avian Biol. 6:62–67.

KARR, J. R., S. K. ROBINSON, J. G. BLAKE, AND R. O. BIERREGAARD, JR. 1990. Birds of four Neotropical forests. Pages 237–269 *in* Four Neotropical Rainforests (A. H. Gentry, Ed.). Yale Univ. Press, New Haven, Connecticut.

KRATTER, A. W. 1995. Status, habitat, and conservation of the Rufous-fronted Antthrush *Formicarius rufifrons*. Bird Conserv. Inst. 5:391–404.

KRATTER, A. W. AND T. A. PARKER III. 1997. Relationship of two bamboo-specialized foliage-gleaners: *Automolus dorsalis* and *Anabazenops fuscus* (Furnariidae). Pp. 383–397 *in* Studies in Neotropical Ornithology Honoring Ted Parker (J. V. Remsen, Jr., Ed.), Ornithol. Monogr. No. 48.

LEAK, J. A., AND S. K. ROBINSON. 1989. Notes on the social behavior and mating systems of the Casqued Oropendola. Wilson Bull. 101:134–137.

LITWIN, J. S, AND C. R. SMITH. 1992. Factors influencing the decline of Neotropical migrants in a northeastern forest fragment: Isolation, fragmentation, or mosaic effects? Pages 483–496 *in* Ecology and Conservation of Neotropical Migrant Landbirds. (J. M. Hagan III and D. W. Johnston, Eds.). Smithsonian Inst. Press, Washington, D.C.

LOISELLE, B. A, AND J. G. BLAKE. 1994. Annual variation in birds and plants of a tropical second-growth woodland. Condor 96:368–380.

MACARTHUR, R. H., AND J. W. MACARTHUR. 1961. On bird species diversity. Ecology 42:594–598.

MARTIN, N. D. 1960. An analysis of bird populations in relation to forest succession in Algonquin Provincial Park, Ontario. Ecol. Monogr. 41:126–140.

MARTIN, T. E., AND J. R. KARR. 1986. Temporal dynamics of Neotropical birds with special reference to frugivores in second-growth woods. Wilson Bull. 98:38–60.

MAY, P. G. 1984. Avian reproductive output in early and late successional habitats. Oikos 43:277–281.

MUNN, C. A. 1985. Permanent canopy and understory flocks in Amazonia: Species composition and population density. Pages 683–712 *in* Neotropical Ornithology (P. A. Buckley, M. S. Foster, E. S. Morton, R. S. Ridgely, and F. G. Buckley, Eds.). Ornithol. Monogr. 36.

ODUM, E. P. 1950. Bird populations of the highlands (North Carolina) plateau in relation to plant succession and avian invasion. Ecology 31:587–605.

PEARSON, D. L. 1971. Vertical stratification of birds in a tropical forest. Condor 77:453–466.

PEARSON, D. L. 1975. The relation of foliage complexity to ecological diversity of three Amazonian bird communities. Condor 77:453–466.

PEARSON, D. L. 1977. Ecological relationships of small antbirds in Amazonian bird communities. Auk 94:283–292.

RASANEN, M. E., J. S. SALO, AND R. J. KALLIOLA. 1987. Fluvial perturbance in the western Amazon basin: Regulation by long-term sub-Andean tectonics. Science 238:1398–1401.

REMSEN, J. V., JR, AND D. A. GOOD. 1996. Misuse of data from mist-net captures to assess relative abundance in bird populations. Auk 113:381–398.

REMSEN, J. V., JR, AND T. A. PARKER III. 1983. Contribution of river-created habitats to bird species richness in Amazonia. Biotropica 15:223–231.

REMSEN, J. V., JR., M. A. HYDE, AND A. CHAPMAN. 1993. The diets of Neotropical trogons, motmots, barbets and toucans. Condor 95:178–192.

ROBINSON, S. K. 1985. Coloniality as a defense against nest predators of the Yellow-rumped Carique. Auk 102:509–519.
ROBINSON, S. K. 1988. Reappraisal of the costs and benefits of habitat heterogeneity for nongame wildlife. Trans. N. Amer. Wildl. Nat. Res. Conf. 53:145–155.
ROBINSON, S. K. 1995. Habitat selection and foraging ecology of raptors in Amazonian Peru. Biotropica 26: 443–458.
ROBINSON, S. K., J. W. FITZPATRICK, AND J. TERBORGH. 1995. Distribution and abundance of Neotropical migrant land birds in the Amazon basin and Andes. Bird Conserv. Intern 5:305–323.
ROBINSON, S. K., J. TERBORGH, AND J. W. FITZPATRICK. 1988. Habitat selection and relative abundance of migrants in southeastern Peru. Acta XIX Int. Congr. Orn. 2298–2307.
ROBINSON, S. K., AND J. TERBORGH. 1990. Bird communities of the Cocha Cashu Biological Station in Amazonian Peru. Pages 199–216 in Four Neotropical Rainforests (A. H. Gentry, Ed.). Yale Univ. Press, New Haven, Connecticut.
ROBINSON, S. K., AND J. TERBORGH. 1995. Interspecific aggression and habitat selection in Amazonian birds. J. Anim. Ecol. 64:1–11.
ROBINSON, S. K., J. TERBORGH, AND C. A. MUNN. 1990. Lowland tropical forest bird communities in a site in western Amazonia. Pages 229–258 in Biogeography and Ecology of Forest Bird Communities (A. Keast, Ed.). SPB Acad. Publ., The Hague, The Netherlands.
ROSENBERG, G. H. 1990. Habitat selection and foraging behavior by birds of Amazonian river islands in northeastern Peru. Condor 92:427–443.
SALO, J., R. KALLIOLA, I. HAKKINEN, Y. MAKINEN, P. NIEMELA, M. PUHAKKA, AND P. D. COLEY. 1986. River dynamics and the diversity of Amazonian lowland forest. Nature 322:254–258.
SHUGART, H., JR, AND D. JAMES. 1973. Ecological succession of breeding bird populations in northwestern Arkansas. Auk 90:62–77.
TERBORGH, J. 1980. Vertical stratification of a Neotropical forest bird community. Acta XVII Congr. Int. Orn. (Berlin):1005–1012.
TERBORGH, J. 1983. Five New World Primates. Princeton Univ. Press, Princeton, New Jersey.
TERBORGH, J. 1985. Habitat selection in Amazonian birds. Pages 311–318 in Habitat Selection in Birds (M. L. Cody, Ed.). Academic Press, New York, New York.
TERBORGH, J., AND S. K. ROBINSON. 1986. Guilds and their utility in ecology. Pages 65–90 in Community Ecology: Pattern and Process (J. Kikkawa and B. Anderson, Eds.). Blackwell, Oxford.
TERBORGH, J., J. W. FITZPATRICK, AND L. E. EMMONS. 1984. Annotated checklist of bird and mammal species of the Cocha Cashu Biological Station, Manu National Park, Peru. Fieldiana, New Series 21:1–29.
TERBORGH, J., AND K. PETREN. 1991. Development of habitat structure through succession in an Amazonian floodplain forest. Pages 28–46 in Habitat Structure: The Physical Arrangement of Objects in Space. (S. S. Bell, E. D. McCoy, and H. R. Mushinsky, Eds.). Chapman and Hall, New York.
TERBORGH, J., S. K. ROBINSON, T. A. PARKER III, C. A. MUNN, AND N. PIERPONT. 1990. Organization of an Amazonian bird community. Ecol. Monogr. 60:213–238.
TERBORGH, J., AND J. S. WESKE. 1969. Colonization of secondary habitats by Peruvian birds. Ecology 50: 765–782.
THIOLLAY, J. M. 1986. Structure comparée du peuplement avien don trois sites de foret primaire en Guyane. Rev. Ecol. 41:59–105.
UHL, C., K. CLARK, H. CLARK, AND P. MURPHY. 1981. Early plant succession after cutting and burning in the upper Rio Negro region of the Amazon Basis. J. Ecol. 69:631–649.
ZIMMERMAN, J. L., AND J. L. TATSCHL. 1975. Floodplain birds of Weston Bend, Missouri River. Wilson Bull. 87:196–206.

APPENDIX 1
ESTIMATES OF POPULATION DENSITIES OF TERRITORIAL SPECIES ALONG THE SUCCESSIONAL GRADIENT OF THE MANU RIVER

Species (mass[g])	No. territories/100 ha Successional stages[a]						Guild[f]	Other habitats[g]
	1[b]	2[b]	3[b]	4–5[c]	5–6[d]	7[e]		
Tinamus major (1,170)				2.5	8.5	8.5	G,T	TF
Crypturellus cinereus (450)				5.0(B)[h]	8.5	1.5	G,T	TF
C. soui (250)				2.0	3.0	1.0	G,T	TF, BA
C. bartletti (241)					4.0	13.0	G,T	
C. atrocapillus (453)				+[i]			G,T	BA
C. undulatus (540)	12.5			12.5	20.0	5.0	G,T	
Agamia agami (609)					+(B)	+	AQ	AQ
Anhima cornuta					+(B)		AQ	AQ
Cairina moschata					+(B)		AQ	AQ
Leptodon cayanensis (550)	+				+	+	R,D	TF
Chondrohierax uncinatus		+			+		R,D	
Harpagus bidentatus (260)					+	1.0	R,D	TF
Accipiter bicolor (210)					+	0.25	R,D	TF
Leucopternis schistaceus (355)					+(B)	0.25	R,D	
Buteo magnirostris (265)	0.3	0.1						
Spizaetus ornatus (925)					+	0.25	R,D	TF
Herpetotheres cachinnans (650)					+	1.0	R,D	
Micrastur semitorquatus (550)					0.1	0.25	R,D	TF
M. ruficollis (230)			+		2.0	1.5	R,D	
Daptrius ater (370)	+	+		+	+	+	R,D	
D. americanus (583)					+	+	R,D	TF
Ortalis motmot (415)				1.0	2.0	1.0	F,A	
Penelope jacquacu (1,280)				+	3.0	2.5	F,A	TF
Pipile pipile (1,200)				1.5	5.5	2.5	F,A	TF
Mitu tuberosa (3,060)	+		+	+	3.5	2.5	G,T	TF
Odontophorus gujanensis (315)	+					G,T		
Psophia leucoptera (990)					0.5	2.0	F,T	TF
Aramides cajanea (515)	+	+	+	+(B)	3.5(B)	1.5	AQ	AQ
Columba cayennensis (260)			+				F,A	LM
C. subvinacea (125)				1.5	6.5	+	F,A	TF
C. plumbea (210)				10.0	4.5	8.5	F,A	TF
Leptotila rufaxilla (175)	12.5	12.5	12.5	12.5	8.5	1.5	G,T	

APPENDIX 1
CONTINUED

Species (mass[g])	No. territories/100 ha Successional stages[a]						Guild[f]	Other habitats[g]
	1[b]	2[b]	3[b]	4–5[c]	5–6[d]	7[e]		
Geotrygon montana (115)	+			5.5	5.5	4.0	F,T	TF
Coccyzus melacoryphus (52)		+	+				I,A,G	
Piaya cayana (105)				2.5	7.0	4.0	I,A,G	TF
Neomorphus geoffroyi (340)					+	0.25	I,AF	TF
Otus watsonii (145)					3.0	5.5	R,N	TF
Glaucidium hardyi (60)				2.5	4.5	5.0	R,N	TF
G. brasilianum (67)				5.0	6.0	1.0	R,N	LM
Ciccaba virgata (320)					0.25	1.5	R,N	TF
Lurocalis semitorquatus (87)					+	1.5	I,Aer.	TF
Nyctiphrynus ocellatus (43)	19				+	2.5	I,T,S	TF
Nyctidromus albicollis (65)	38						I,T,S	LM, BE
Hydropsalis climacocera (48)							I,T,S	LM, BE
Trogon melanurus (122)			+	20.0	17.0	13.0	O,A,S	TF
T. collaris (59)			+	10.0	14.5	8.0	IF,A,S	TF(?)
T. curucui (61)			+	10.0	8.5	8.5	IF,A,S	
T. violaceus (44)					0.25	3.5	IF,A,S	TF
Chloroceryle aenea (14)				+(B)	3.0(B)	0.5	AQ	AQ
C. inda (54)				+(B)	4(B)	1.5	AQ	AQ
Electron platyrhynchum (65)				2.5	1.5	2.0	I,A,S	TF
Momotus momota (111)				22.0	21.0	6.0	O,A	
Galbalcyrhynchus purusianus (500)				+(B)	+(B)		I,A,S	AQ
Galbula cyanescens (24)				13.0	6.5	3.0	I,U,S	TF, BA
Notharcus macrorhynchus (120)					0.5	2.0	I,A,S	TF
Bucco macrodactylus (25)	19.0				+(B)	1.0	I,A,S	
Nystalus striolatus (47)				5.0	13.0	2.5	I,A,S	
Nonnula ruficapilla (22)	19.0		1.0		3.0(B)	+	I,U,S	
Monasa nigrifrons (85)			+	22.0	21.0	7.0	I,A,S	
Chelidoptera tenebrosa (47)	+	+		+(B)			I,Aer	
Capito niger (64)				20.0	17.0	10.5	O,A	TF
Eubucco richardsoni (35)				5.0	4.5	4.0	O,A	TF
E. tucinkae (41)				1.0	3.0	0.5	O,A	LM
Aulacorhynchus prasinus (132)					2.0	1.5		F,A
Pteroglossus castanotis (310)				0.5	+	F,A	LM	

APPENDIX 1
Continued

Species (mass[g])	No. territories/100 ha Successional stages[a]						Guild[f]	Other habitats[g]
	1[b]	2[b]	3[b]	4–5[c]	5–6[d]	7[e]		
P. inscriptus (126)			+	1.0	0.5	0.25	F,A	LM
P. flavirostris (135)			2.0	2.0	1.5	0.5	F,A	TF
P. beauharnaesii (203)				1.5	2.0	2.0	F,A	
Selenidera reinwardtii (138)					+	1.5	F,A	TF
Ramphastos culminatus (369)					2.0	1.5	F,A	TF
R. cuvieri (734)			2.0	2.0	3.5	2.0	F,A	TF
Picumnus rufiventris (21)	19.0	19.0	19.0	++	1.5(6)	4.0	I,B,I	BA, TF
Colaptes punctigula (78)					0.5(B)		I,B,I	LM
Piculus chrysochloros (88)				+	+	1.0	I,B,I	TF
P. leucolaemus (69)					+	+	I,B,I	TF
Celeus elegans (136)			1.0	1.0	0.5	2.0	I,B,I	
C. grammicus (79)					0.5	1.0	I,B,I	TF
C. flavus (101)				1.5	1.5	+	I,B,I	TF
Dryocopus lineatus (209)				+	2.0		I,B,I	
Melanerpes cruentatus (59)			9.0	9.0	4.5	3.0	O,A	TF
Veniliornis passerinus (36)	9.0	9.0	9.0	9.0	5.5		I,B,I	
V. affinis (36)					5.5	4.0	I,B,I	TF
Campephilus melanoleucos (231)			+	+	3.0	0.25	I,B,I	SW, TF, LM
C. rubricollis (220)					0.25	1.00	I,B,I	TF
Dendrocincla fuliginosa (31)				5.0	6.0	4.0	I,A,S	TF
Deconychura longicauda (23)					+	2.0	I,A,S	TF
Sittasomus griseicapillus (16)			7.5	7.5	8.5	6.5	I,B,S	TF
Glyphorynchus spirurus (14)					+	3.0	I,B,S	TF
Nasica longirostris (92)					3.0	0.25	I,B,I	TF
Dendrexetastes rufigula (70)				1.5	6.5	3.0	I,A,G	TF
Xiphocolaptes promeropirhynchus (36)				5.0	1.5	1.0	I,B,I	TF
Dendrocolaptes certhia (73)					3.0	2.5	I,A,S	TF
D. picumnus (80)				5.0	0.25	2.0	I,A,F	TF
Xiphorhynchus obsoletus (39)				1.5(B)	+	0.25	I,B,S	SW
X. ocellatus (32)					0.5	1.5	I,B,S	
X. spixii (40)					+	3.5	I,B,S	TF
X. guttatus (65)			27.5	27.5	22.5	11.0	I,B,S	TF

APPENDIX 1
Continued

Species (mass[g])	No. territories/100 ha Successional stages[a]						Guild[f]	Other habitats[g]
	1[b]	2[b]	3[b]	4–5[c]	5–6[d]	7[e]		
Lepidocolaptes albolineatus (33)	9.0	9.0		+	5.5	5.0	I,B,S	TF
Campylorhamphus trochilirostris (38)	+	+	+	+(B)	2.0(B)	0.5(B)	I,B,I	BA
Furnarius leucopus (44)	+				+(B)	+	I,T,G	LM
Synallaxis albigularis (17)	53.0						I,U,G	M
S. gujanensis (19)							I,U,G	M
Hyloctistes subulatus (29)				+		2.5	I,U,G	TF
Philydor erythrocercus (25)				2.5	2.0	4.5	I,U,DL	TF
P. rufus (34)	12.0	12.0	12.0	12.0	8.0	+	I,A,G	
P. erythropterus (30)					2.0	2.5	I,A,G	TF
P. ruficaudatus (29)				5.0	5.5	3.0	I,A,DL	
Automolus ochrolaemus (34)	38.0	3.5	3.5	3.5	2.0	2.5	I,U,DL	TF
A. rufipileatus (37)		38.0					I,U,DL	
Xenops rutilans (13)					0.5	4.0	I,B,S	TF
X. minutus (12)					1.5	7.5	I,B,S	TF
Cymbilaimus lineatus (40)	8.5	8.5	8.5	2.5	1.5	7.5	I,U,G	TF
Taraba major (60)	38.0	38.0		+	6.5(B)	+	I,U,G	BA
Thamnophilus doliatus (29)							I,U,G	
T. schistaceus (21)				0.5	21.0	10.5	I,A,G	TF
Pygiptila stellaris (25)				2.5	13.0	7.5	I,A	TF
Thamnomanes schistogynus (17)		+		5.0	5.0	11.0	I,U,S	TF; BA
Myrmotherula brachyura (8)				15.0	13.0	13.0	I,A,G	TF
M. sclateri (8)					0.25	4.0	I,A,G	TF
M. axillaris (8)				7.5	3.0	16.0	I,U,G	TF
M. longipennis (9)					+	11.0	I,U,G	TF
M. menetriesii (9)				5.0	10.0	15.0	I,A,G	TF
Terenura humeralis (13)					+	6.0	I,A,G	TF
Cercomacra cinerascens (20)				5.0	5.5	17.0	I,A,G	TF
C. nigrescens (18)	19.0	19.0	3.0	3.0	1.5	+	I,U,G	
Myrmoborus leucophrys (19)				17.5	3.0	20.0	I,T,G	BA
M. myotherinus (20)				2.5(B)	2.0(B)	0.5	I,T,G	TF
Hypocnemoides maculicauda (13)				+(B)	10.0	0.5	I,U,G	SW
Percnostola lophotes (28)	8.0	8.0				0.5	I,U,G	BA
Sclateria naevia (22)					3.5(B)		I,U,G	SW

APPENDIX 1
CONTINUED

Species (mass[g])	1[b]	2[b]	3[b]	4–5[c]	5–6[d]	7[e]	Guild[f]	Other habitats[g]
Myrmeciza hemimelaena (15)			15.0	15.0	10.0	9.0	I,U,G	TF
M. hyperyrha (41)				7.5	4.5	6.5	I,U,G	TF, SW
M. goeldii (42)	9.5	9.5	9.5	9.5	16.5	1.5	I,U,G	BA
M. atrothorax (18)	56.5	56.5					I,U,G	
Hylophylax naevia (13)			+		7.0	1.5	I,T,G	TF
Phlegopsis nigromaculata (45)	+	+	+	2.0	7.0	4.5	I,AF	TF
Formicarius colma (49)			+		1.5	5.0	I,T,G	TF
F. analis (58)		+	20.0	20.0	16.5	13.0	I,T,G	TF
F. rufifrons (57)					+		I,T,G	
Hylopezus berlepschi (48)		+	7.5	7.5(B)	1.0	+	I,T,G	BA, TF
Camptostoma obsoletum (9)		+			4.5		I,A,G	
Phaeomyias murinus (13)	38.0	38.0					O,A	
Todirostrum latirostre (8)	+	+					I,U,S	
Tyrannulus elatus (8)			15.0	15.0	5.5	2.5	O,A	TF
Myiopagis gaimardii (12)			20.5	20.5	16.0	4.0	I,A,S	
M. viridicata (11)			5.0	5.0		+	I,A,S	
Terenotriccus erythrurus (7)					+	2.0	I,U,S	TF
Leptopogon amaurocephalus (11)				3.0	2.5		I,U,S	BA
T. maculatum (7)	40.0	40.0					I,A,S	
T. chrysocrotaphum (7)				7.5	5.5	3.5	I,A,S	TF
Tolmomyias assimilis (17)				15.0	5.0	4.0	I,A,S	TF
T. poliocephalus (11)				10.0	16.0	3.0	I,A,S	TF
T. flaviventris (14)	16.5	16.5	16.5		4.5	+	I,A,S	
Ochthornis littoralis (13)	+						I,U,S	BE
Muscisaxicola fluviatilis (14)	+						I,T,S	BE
Attila bolivianus (45)				12.5	14.0	4.0	I,A,S	
A. spadiceus (35)					2.0	4.0	I,A,S	TF
Sirystes sibilator (38)				1.0	0.5	2.0	I,A,S	TF
Myiarchus tuberculifer (20)			+	+	+	+	I,A,S	TF
M. ferox (28)	56.5	56.5					I,A,S	
Pitangus sulphuratus (57)	+	+	+	+(B)	0.5(B)		O,A	LM
Megarynchus pitangua (66)	1.5	1.5	1.5		+(B)		O,A	LM
Myiozetetes similis (33)	28.5	28.5		1.5(B)	+(B)		O,A	LM

APPENDIX 1
CONTINUED

Species (mass[g])	No. territories/100 ha Successional stages[a]						Guild[f]	Other habitats[g]
	1[b]	2[b]	3[b]	4–5[c]	5–6[d]	7[e]		
M. granadensis (31)	14.0	14.0	14.0	4.0(B)	4.0		O,A	LM
Myiodynastes maculatus (51)				+(B)	3.0		O,A	LM
Legatus leucophaius (28)				7.5(B)	5.5(B)	1.5	O,A	TF, LM
Rhytipterna simplex (36)					1.5	4.5	I,A	TF
Tyrannus melanocholicus (47)	+	+					I,A,S	BE
Pachyramphus polychopterus (22)	+	+	+	13.0		+	I,A,S	
P. marginatus (18)					1.5	4.0	I,A,S	TF
P. minor (37)					4.0	5.5	I,A,S	
Tityra cayana (66)				2.5(B)	3.0	1.5	O,A	TF
T. semifasciata (88)				2.5(B)	+	4.0	O,A	TF
T. inquisitor (45)					1.5(B)	+	O,A	
Piprites chloris (20)					5.5	5.5	I,U,S	TF
Schiffornis major (31)					1.5(B)	0.5	F,U	SW
Lipaugus vociferans (81)			+		3.0(B)	10.0	F,A	SW, TF
Conioptilon mcilhennyi (40)				2.5	2.0	0.25	F,A	SW
Querula purpurata (120)				1.5	+	2.0	F,A	TF
Campylorhynchus turdinus (25)					2.0	1.5	I,A,G	TF
Thryothorus genibarbis (19)					+	0.5	I,U,DL	TF, BA
Troglodytes aedon (10)	+						I,T,G	BE, RI
Microcerculus marginatus (18)				5.0	+	2.0	I,T,G	TF
Cyphorhinus aradus (30)				3.5	+	2.5	I,T,G	TF
Turdus ignobilis (58)	113.0	+	+	+			O,A	LM
T. hauxwelli (72)				20.0	24.5	8.5	O,A	
T. albicollis (52)					.0	3.0	O,A	TF
Ramphocaenus melanurus (12)	+	+	+		4.5(B)		I,U,G	TF
Cyanocorax violaceus (262)	+			1.5	+	+	O,A	
Ammodramus aurifrons (18)	+						O,A	BE
Sporophila castaneiventris (8)	9.0						T,G	BE
Arremon taciturnus (28)					1.0	0.5	T,G	SW, TF
Paroaria gularis (31)	+				+(B)	+(B)	O,T	BE, LM
Saltator maximus (46)					1.5(B)	+	O,U	TF, SW
Cyanocompsa cyanoides (27)	+	+	+	2.5	3.0	1.5	O,A	TF, SW, BA
Lanio versicolor (19)					+	4.5	I,A,S	TF
Cissopis leveriana (66)			2.0(B)	+(B)			O,A	LM

APPENDIX 1
CONTINUED

No. territories/100 ha Successional stages[a]

Species (mass[g])	1[b]	2[b]	3[b]	4–5[c]	5–6[d]	7[e]	Guild[f]	Other habitats[g]
Nemosia pileata (14)		+	1.5				O,A	TF
Tachyphonus luctuosus (13)	28.5			1.5	5.5	6.0	O,A	TF
Ramphocelus carbo (27)		28.5					O,U	LM
Thraupis episcopus (42)	+						O,A	RI, LM
Euphonia laniirostris (16)			3.5	3.5			F,A	
E. chrysopasta (15)				10.0	5.5	4.0	F,A	TF
E. minuta (10)					1.5	2.0	F,A	TF
E. xanthogaster (14)					3.0	3.0	F,A	TF
E. rufiventris (15)				5.0	11.5	9.0	F,A	TF
Tangara mexicana (19)				1.5	3.0	2.5	O,A	TF
T. chilensis (24)			+		1.5	4.5	O,A	TF
T. schrankii (20)			+		1.5	4.0	O,A	TF
Dacnis lineata (13)			+		2.0	4.0	O,A	TF
D. cayana (14)	+	+	+	4.5	3.0	4.0	O,A	TF
D. flaviventer W					+(B)		O,A	TF
Chlorophanes spiza (18)				+	+	3.0	O,A	TF
Tersina viridis (32)				+	+		F,A	TF, LM
Geothlypis aequinoctialis (10)	77.0						I,U,G	M
Vireolanius leucotis (26)					1.0	5.5	I,A,G	TF?
Vireo olivaceus chivi (15)				5.0	8.5	3.0	O,A	LM
Hylophilus hypoxanthus (13)					2.0	4.5	I,A,G	TF
H. ochraceiceps (11)					+	.5	I,U,G	TF
Psarocolius oseryi (132)					15.0	6.0	O,A	
Cacicus solitarius (84)				12.5(B)	10.0(B)	0.25	O,U	LM
Icterus cayanensis (39)					0.5	1.0	O,A	TF

[a] Successional stages: 1 = *Tessaria*; 2 = *Gynerium* (cane); 3 = *Cecropia*; 4 = Mixed Canopy; 5 = *Ficus/Cedrela*; 6 = Transitional Forest; 7 = Mature Phase.
[b] Data from stages 1–3 from the early successional plots only.
[c] Data include the older end of the early successional plot (stage 4 through early stage 5).
[d] Data include the mature end of the late successional plot (stages 5 and 6).
[e] From Terborgh et al. (1990).
[f] Guilds defined as follows: G,T = Terrestrial Granivore; AQ = Aquatic; R,n = Nocturnal Raptor; R,D = Diurnal Raptor; F,A = Arboreal Frugivore; FT = Terrestrial Frugivore; I,A,G = Arboreal Insect Gleaner; I,AF = Insectivorous Ant Follower; I,Aer = Aerial Insectivore; I,A,S = Arboreal Sallying insectivore; I,U,S = Understory Sallying Insectivore; I,T,S = Terrestrial Sallying Insectivore; O,A = Arboreal Omnivore; I,B,I = Bark-dwelling Insectivore feeding in Trunk Interiors; I,B,S = Bark-dwelling Insectivores feeding on surfaces; I,T,G = Terrestrial Gleaning Insectivore; I,U,G = Understory Gleaning Insectivore; I,U,DL = Dead-leaf searching Understory Insectivore; I,A,DL = Arboreal dead-leaf searching Insectivore; I,U,S = Sallying Understory Insectivore; IF,A,S = Arboreal Sallying Insectivore and Frugivore.
[g] Other habitats occupied by this species: TF = *Terra firme*; BA = bamboo; BE ⟩ beach; SW = swamp; M = marsh; LM = Lake Morgin; AQ = aquatic.
[h] (B) = found in backwaters.
[i] + = Present, but too infrequently to estimate abundance.
[j] All population densities for birds in stages 2 and 3 are cumulative population densities for the entire early successional plot for species that occupied both early (1–3) and later (4–5) successional stages.

APPENDIX 2

SPECIES RESTRICTED TO THE MATURE PHASE OF THE SUCCESSIONAL GRADIENT. SEE APPENDIX 1 FOR CODES FOR GUILD ASSIGNMENTS

Species (mass [g])	Population density pairs or groups/100 ha	Guild
Tinamus tao (2,000)	0.5	G,T
Crypturellus variegatus (350)[a]	0.5	G,T
Tigrisoma lineatum (840)[a]	0.25	AQ
Cochlearius cochlearius (550)[a]	0.25	AQ
Mesembrinibis cayennensis (670)[a]	0.25	AQ
Accipiter superciliosus (120)	0.25	R,D
Morphnus guianensis (1,750)	+	R,D
Harpia harpyja (4,500)	+	R,D
Spizaetus tyrannus (1,025)	+	R,D
Micrastur gilvicollis (220)	1.5	R,D
Odontophorus stellatus (310)	8.0	T,G
Eurypyga helias (190)[a]	0.25	AQ
Ara chloroptera (1,250)	1.0	A,G
Aratinga leucophthalma (190)	3.0	A,G
Pyrrhura rupicola (75)	3.5	A,G
Pionopsitta barrabandi (140)	0.5	A,G
Amazona farinosa (800)	2.0	A,G
Pulsatrix perspicillata (795)	0.75	R,N
Lophostrix cristata (510)	2.0	R,N
Ciccaba huhula (370)	1.5	R,N
Nyctibius grandis (575)	1.0	I,A,N
N. bracteatus (125)	1.0	I,A,N
Popelairia popelairii (2.5)	0.5	N,A
Chrysuronia oenone (4.0)	2.5	N,A
Trogon viridis (91)	+	O,A,S
Jacamerops aurea (79)	1.0	I,A,S
Malacoptila semicincta (44)	2.0	I,U,S
Monasa morphoeus (74)	1.0	I,A,S
Celeus torquatus (134)	1.0	I,B,I
Dendrocincla merula (46)	8.0	I,A,F
Automolus infuscatus (39)	1.5	I,U,DL
Sclerurus caudacutus (36)	3.0	I,T,G
Thamnophilus aethiops (27)	1.0	I,U,G
Thamnomanes ardesiacus (18)	13.0	I,U,S
Myrmotherula hauxwelli (11)	11.0	I,U,G
M. leucophthalma (10)	7.0	I,U,DL
M. iheringi (8)	3.0	I,U,G
Dichrozona cincta (16)	4.5	I,T,G
Microrhopias quixensis (11)[b]	2.0	I,U,G
Hypocnemis cantator (13)[b]	1.0	I,U,G
Myrmeciza fortis (46)	1.0	I,AF
Gymnopithys salvini (25)	0.5	I,AF
Rhegmatorhina melanosticta (32)	2.0	I,AF
Hylophylax poecilinota (18)	0.25	I,AF
Chamaeza nobilis (123)	2.0	I,T,G
Myrmothera campanisona (47)	4.5	I,T,G
Conopophaga peruviana (23)	3.0	I,T,G
Liosceles thoracicus (81)	2.0	I,T,G
Zimmerius gracilipes (9)[b]	2.0	O,A,S
Ornithion inerme (7)	2.0	I,A,S
Corythopis torquata (17)	7.0	I,T,G
Myiornis ecaudatus (6)	10.0	I,A,S
Hemitriccus zosterops (9)	5.0	I,A,S
Ramphotrigon fuscicauda (19)	1.0	I,A,S
R. ruficauda (19)	2.5	I,A,S
Platyrinchus coronatus (10)	7.5	I,U,S

APPENDIX 2
CONTINUED

Species (mass [g])	Population density pairs or groups/100 ha	Guild
P. platyrhynchos (12)	6.0	I,U,S
Onychorhynchus coronatus (14)	0.5	I,U,S
Laniocera hypopyrra W	1.0	O,A,S
Tyranneutes stolzmanni (9)	9.0	F,A,S
Pipra coronata (9)	3.5	F,U,S
Porphyrolaema porphyrolaema (60)	1.0	F,A,S
Lamprospiza melanoleuca (39)	1.5	O,A,G
Hemithraupis guira (13)	0.5	I,A,G
H. flavicollis (17)	0.5	I,A,G
Tachyphonus rufiventer (17)	4.5	I,A,G
Habia rubica (33)	8.5	I,A,S
Tangara nigrocincta (17)	+	O,A,G
T. velia (21)	0.5	O,A,G
T. callophrys (23)	1.0	O,A,G
Cyanerpes caeruleus (16)[b]	+	O,A,G
Psarocolius yuracares (360)	1.0	O,A,G

[a] Aquatic species recorded along the margin of Cocha Cashu that occasionally entered the forest along an intermittent stream; not necessarily associated with mature floodplain forest.
[b] Species recorded from mid-successional vegetation elsewhere in the Manu region (J. W. Fitzpatrick, pers. comm.).

APPENDIX 3
REGISTRATIONS OF NON-TERRITORIAL OR MIGRANT BIRDS DURING SONG CENSUSES (ONLY INCLUDES SPECIES WITH ≥5 REGISTRATIONS)

Species	Early successional plot				Late successional plot		
	1	2	3	4–5	1–3	4–5	6
Columba cayennensis	2	3	2				
Ara macao					1	1	4
A. severa		1	12		4	2	0
Aratinga weddellii				3	1	4	3
Pyrrhura picta					0	8	22
Forpus sclateri					3	8	7
Brotogeris sp.		1	16	30	2	5	11
Pionites leucogaster				4	1	0	13
Pionus menstruus			1	2	0	3	6
Amazona ochrocephala				1	3	3	14
Glaucis hirsuta	1	1	6	16	1	3	1
Phaethornis superciliosus	1			8	0	0	3
P. hispidus	1	2	6	12	0	16	9
P. stuarti	1	7	9	19	0	4	1
Campylopterus largipennis	1	3	6		0	1	0
Florisuga mellivora	3	1			0	0	3
Thalurania furcata		2	9	8	1	2	10
Elaenia spp.	6			1			
Contopus virens				7	0	8	14
Pyrocephalus rubinus	8			1			
Myiarchus swainsoni			2	2	2	9	9
Myiodynastes luteiventris			1	7	0	2	0
Lipaugus vociferans				1	0	1	12
Cotinga maynana				2	0	4	2
Gymnoderus foetidus				1	0	1	4
Catharus ustulatus	2		1		0	0	2
Turdus amaurochalinus	4				2		
Sporophila caerulescens	6						
Psarocolius angustifrons	2	1	6	1	2	4	14
Cacicus cela			2	20	3	15	26

APPENDIX 4
Distribution of Birds in Seven *Tessaria* Stands on Beaches Upriver and Downriver of the Manu

	No. beaches occupied		No. beaches occupied
Crypturellus undulatus	7	*Ochthornis littoralis*	7
Buteo magnirostris	7	*Muscisaxicola fluviatilis*	7
Odontophorus gujanensis	2	*Myiarchus ferox*	7
Claravis pretiosa	1	*Myiozetetes granadensis*	7
Columbina talpacoti	1	*Myiophobus fasciatus*	4
Coccyzus melacoryphus	1	*Tyrannus melancholicus*	7
Nyctidromus albicollis	6	*Satrapa icterophrys*	1
Hydropsalis climacocerca	7	*Phaeomyias murina*	3
Bucco macrodactylus	4+	*Thryothorus genibarbis*	1
Picumnus rufiventris	5+	*Turdus amaurochalinus*	7
Veniliornis passerinus	6+	*T. ignobilis*	7
V. affinis	5	*Ramphocaenus melanurus*	1
Furnarius leucopus	7	*Ammodramus aurifrons*	7
Synallaxis albigularis	3	*Sporophila caerulescens*	5
S. gujanensis	7	*S. castaneiventris*	3
Philydor rufus	7	*Poroaria gularis*	4
Anabazenops dorsalis	1	*Saltator coerulescens*	1
Automolus rufipileatus	7	*S. maximus*	1
Taraba major	5+	*Cyanocompsa cyanoides*	1
Cercomacra nigrescens	2	*Cissopis leveriana*	1
Percnostola lophotes	6	*Thlypopsis sordida*	2
Myrmeciza atrothorax	7	*Nemosia pileata*	1
Camptostoma obsoletum	1	*Ramphocelus carbo*	7
Todirostrum latirostre	2	*Geothlypis aequinoctialis*	2
Todirostrum maculatum	7	*Conirostrum speciosum*	1

ECOLOGY OF DEAD-LEAF FORAGING SPECIALISTS AND THEIR CONTRIBUTION TO AMAZONIAN BIRD DIVERSITY

KENNETH V. ROSENBERG

Museum of Natural Science, Foster Hall 119, Louisiana State University, Baton Rouge, Louisiana 70803, USA
Present address: Cornell Lab of Ornithology 159 Sapsucker Woods Rd. Ithaca, New York 14850, USA

ABSTRACT.—One reason suggested for the high avian species diversity in tropical forests is increased specialization on resources that are absent in temperate habitats. This study investigates in detail one such specialization, namely foraging for arthropods in suspended aerial leaf-litter in lowland tropical rainforest. Up to 16 species at two southwestern Amazonian sites constitute a guild of specialized dead-leaf foragers that make up roughly 11% of the region's insectivorous bird species. Most dead-leaf specialists are ovenbirds (Furnariidae) or antbirds (Formicariidae) that are characteristic members of mixed-species foraging flocks in the understory or canopy. These specialists, compared with other insectivores, tended to use more acrobatic postures and manipulated foraging substrates with the bill or feet. These species segregated to some extent by habitat, including several congeneric replacements. The guild reaches its highest diversity in a belt across southwestern Amazonia and along the base of the Andes, where bamboo and other disturbance-related microhabitats add to forest heterogeneity. Individual dead leaves, as resources for birds, were abundant in all forest types and supported higher prey densities (number per leaf) than adjacent live foliage. Prey density was highest in larger leaves, especially in large, crumpled *Cecropia* leaves. The arthropod fauna of aerial leaf-litter was similar among seasons, habitats, and sites, being dominated by spiders, roaches, other orthopterans, and small beetles. This contrasts greatly with arthropods available on live foliage. Guild members differed significantly from each other in either foraging height, size or type of leaves searched, diet composition, or prey size, although overlaps between species pairs were usually high (≥ 0.900). Although twice as many species were supported in low-lying forest than in upland forest, ecological overlaps among species in each habitat were usually similar. Behavioral similarity among species was not related to dietary overlap. Size of prey taken, however, was correlated with bill size, except that the largest species, *Xiphorhynchus guttatus*, ate surprisingly small prey. Diet composition of all species differed significantly from prey availability in dead leaves, with orthopterans selected by all species and small roaches and spiders often avoided. Censuses of 92 mixed-species flocks revealed no negative and only two positive associations between species, suggesting that birds join flocks independent of other species present. Co-occurring *Myrmotherula leucophthalma* and *M. ornata* in the same flocks were not aggressive and converged in foraging height and substrate use. In contrast, co-occurring *Automolus* foliage-gleaners tended to diverge in foraging height and exhibited overt aggression. Niche segregation among dead-leaf foragers therefore represents a balance between benefits and constraints imposed by feeding in a mixed-species flock; that is, increased vigilence and group defense of territories versus feeding close to potential competitors. Dead-leaf specialization evolved independently in several bird families but shows strong phylogenetic constraints among genera. Phylogenetic study of *Myrmotherula* antwrens revealed that all specialist species were related and that they have been evolving separately from other antwrens perhaps for as long as nine million years. Foraging specialization is therefore a primitive trait within this group (and probably others), appearing before the radiation of modern species. Study of present-day ecology may not elucidate factors leading to the evolution of such specialization, especially without concurrent phylogenetic analyses.

RESUMEN.—Una de las razones sugeridas para la gran diversidad de aves en los bosques tropicales es el aumento en la especialización sobre recursos que están ausentes en hábitats templados. Este estudio investiga en detalle una de tales especializaciones, esto es, el forrajeo para artrópodos en hojarasca suspendida en bosques tropicales de tierras bajas. Hasta 16 especies de aves en dos sitios en el suroeste de Amazonia constituyen un tipo de forrajeadores especializados en la hojarasca, los cuales forman aproximadamente el 11% de las especies de pájaros insectívoros de la región. La mayoría de los especialistas en hojarasca son pizpitas (Furnariidae) o pájaros comehormigas (Formicariidae), que son miembros característicos de bandadas mixtas de forrajeo en el sotobosque o el dosel. Estos especialistas, comparados con otros insectívoros, tendieron a colocarse en posturas más acrobáticas y a manipular el sustrato de forrajeo con las patas o el pico. Estas especies se segregaban por hábitat, hasta cierto punto, incluyendo varios congéneres que les reemplazaban. Esta asociación alimentaria alcanzó su mayor diversidad en la faja a través del suroeste de Amazonia y a lo largo de la base de los Andes, donde las bambúas y otros microhábitats de áreas perturbadas le añaden heterogeneidad al bosque. Hojas muertas individuales, como recurso para los pájaros, eran abundantes en todos los tipos de bosque y sostuvieron una mayor densidad de presas (número por hoja) que el follaje vivo adyacente. La densidad de presas fue mayor en hojas más grandes, especialmente en las hojas arrugadas grandes de *Cecropia*. La fauna artrópoda de la hojarasca suspendida fue similar a través de las estaciones del año, hábitats, y sitios, siendo dominada por arañas, cucarachas, otros ortópteros y escarabajos pequeños. Esto contrasta grandemente con los artrópodos disponibles en el follaje vivo. Miembros de las distintas asociaciones alimentarias difirieron significativamente uno del otro en altura de forrajeo, tamaño o tipo de hojas forrajeadas, composición de la dieta, o el tamaño de la presa, aunque los solapamientos ecológicos entre parejas de especies fueron usualmente altos (≥ 0.900). Aunque se encontró el doble de especies en terrenos bajos comparados con bosques en terrenos altos, el solapamiento ecológico entre las especies en cada hábitat fue usualmente similar. La similaridad en el comportamiento entre especies no estuvo relacionada con el solapamiento dietético. El tamaño de las presas cazadas, sin embargo, estuvo correlacionado con el tamaño del pico, excepto que la especie más grande *Xiphorhynchus guttatus*, consumió presas sorpresivamente pequeñas. La composición dietética de todas las especies difirió significativamente de la disponibilidad de presas en hojas muertas, siendo los ortópteros seleccionados por todas las especies, mientras que cucarachas y arañas pequeñas fueron evitadas. Censos de 92 bandadas de especies mixtas no revelaron asociaciones negativas y solamente dos asociaciones positivas entre especies, lo que sugiere que las aves se asocian con bandadas independientemente de qué otras especies estén presentes. Individuos de *Myrmotherula leucophthalma* y *M. ornata* que se encontraban en las mismas bandadas no eran agresivos y convergían en la altura y substrato utilizado. En contraste, *Automolus* tendió a divergir en la altura de forrajeo y exhibió agresión abiertamente. La segregación de nichos entre forrajeros de hojas muertas, por lo tanto, representa un balance entre beneficios y costos impuestos por alimentarse en bandadas de especies mixtas; esto es, aumento en la vigilancia y la defensa grupal de territorios versus la alimentación cerca de competidores potenciales. La especialización en hojas muertas evolucionó independientemente en varias Familias de aves, pero muestra fuertes restricciones filogenéticas entre Géneros. El estudio filogenético de *Myrmotherula* reveló que todas las especies especialistas estaban relacionadas y han estado evolucionando separadamente de otros del mismo Género, quizás por tanto como nueve millones de años. La especialización en el forrajeo, por lo tanto, es un rasgo primitivo dentro de este grupo (y probablemente otros), que aparece antes de la radiación evolutiva de especies modernas. El estudio de la ecología actual puede no elucidar factores que conlleven a la evolución de tal especialización, especialmente sin los análisis filogenéticos correspondientes.

Specialization on food resources that are unique to tropical habitats has been suggested as a major mechanism promoting high avian diversity in tropical versus temperate regions (Orians 1969, Karr 1971, 1976; Terborgh 1980, Remsen 1985). Examples of specialized tropical birds include those restricted to localized habitats such as bamboo and river-edge forests, those foraging exclusively on novel substrates such as epiphytic plants, vine tangles, and suspended dead foliage, those relying year-round on nectar or fruit, and those species that rely on other organisms such as army ants or monkey troops to flush their prey. This study investigates one of these novel specializations, namely the extraction of arthropods from curled dead leaves suspended above the ground in tropical forests.

In many tropical forests, leaves falling from the canopy are trapped before reaching the ground by vines and other understory vegetation, forming an aerial leaf-litter. These suspended dead leaves are used as diurnal hiding places for nocturnal arthropods, such as roaches, katydids, beetles, and spiders. A number of bird species have been shown to forage exclusively by searching for arthropods in these dead leaves (Gradwohl and Greenberg 1984, Remsen and Parker 1984, Rosenberg 1990a, 1993).

This specialized foraging system is of interest for two reasons. First, it is virtually absent outside of Neotropical forest communities and therefore may contribute significantly to increased avian species diversity in these communities. Second, because the dead leaves represent discrete resource patches that are easily quantified and sampled for arthropod prey, resource availability and use can be directly measured and compared (Gradwohl and Greenberg 1982a,b; Rosenberg 1990a). This contrasts with many other situations in which arthropods may move among a variety of microhabitats. Studies of insectivorous bird communities often have been hampered by the difficulties in measuring such mobile prey resources.

The 11 species of dead-leaf specialists listed by Remsen and Parker (1984) are members of two exclusively Neotropical families, the ovenbirds (Furnariidae) and antbirds (Formicariidae). Members of these families have been observed foraging in dead leaves as far north as southern Mexico (Slud 1964; Skutch 1972, 1982; Alvarez del Toro 1980), but their degree of substrate specialization has not been studied. In addition, a variety of other species may regularly use dead-leaf substrates, including barbets (Capitonidae), woodcreepers (Dendrocolaptidae), wrens (Troglodytidae), tanagers (Thraupinae), and blackbirds (Icterinae) (Remsen and Parker 1984). Some North American wood-warblers (Parulinae) use dead leaves to some extent on their wintering grounds (Morton 1980, Remsen and Parker 1984, Remsen et al. 1989), and one species, *Helmitheros vermivorus*, is a specialist on dead leaves in winter (Greenberg 1987a,b).

Detailed studies of *Myrmotherula fulviventris* in Panama, where it is possibly the only member of this guild (Gradwohl and Greenberg 1980, 1982a, 1982b, 1984), concluded that this species: (1) spent 98% of its foraging time searching curled dead leaves; (2) was able to reduce populations of its preferred prey (orthopterans and spiders) by 50% over a 6-wk period; and (3) was most successful at longer, highly curled leaves, which contained significantly more arthropods. Closely related *Myrmotherula* species in South America show similar degrees of specialization (98–99% of observations; Rosenberg 1993).

In Amazonia, where up to 10–15 dead-leaf foraging species may co-occur locally, the potential for interactions among guild members is enhanced because these birds are characteristic members of mixed-species foraging flocks in the forest understory or sub-canopy (Munn and Terborgh 1979, Munn 1985). Some species are known to defend year-round territories against conspecifics in neighboring flocks, and they frequently travel and forage together with congeners or other flock members that search live foliage or other substrates. Comparisons among dead- and live-leaf foraging antwrens (Rosenberg 1993) revealed that substrate specialization resulted from fundamental differences in search behavior. Remsen and Parker (1984) suggested that dead-leaf specialists may further subdivide the aerial leaf-litter resource by segregating with respect to habitat, foraging height, leaf size, or prey type.

In this paper I describe in detail the ecology of dead-leaf foraging birds at two southwestern Amazonian sites. This study addresses four questions: (1) what contribution does this specialized guild make to overall insectivorous bird diversity in Amazonia? (2) are species-specific behaviors associated with dead-leaf specialization? (3) what aspects of resource availability (including arthropod prey) serve to promote specialization? and (4) do members of this specialized foraging guild further partition the aerial leaf-litter resource?

STUDY AREA AND METHODS

Study area.—I worked at two lowland sites in southwestern Amazonia. The first was the 5,500-ha Tambopata Reserve in depto. Madre de Dios, southeastern Peru (12°50'S, 69°17'W), at 290 m. General aspects of the reserve are described by Erwin (1985). This region is characterized by a distinct dry season, corresponding to the austral winter, usually from June to October. Rainfall during this period frequently accompanies southern cold fronts (*friajes*), which also usually bring high winds and temperatures as low as 10°C.

At Tambopata, I worked in three habitat types, all in primary rainforest. Upland forest (Upland type II of Erwin 1985, *terra firme* of Marra 1989) occurred on high, ancient alluvial terraces on relatively well-drained, sandy soils. Low-lying forest (Upland type I of Erwin 1985, transitional forest of Parker 1982, Marra 1989) occurred throughout the reserve on poorly drained soils; these flooded locally from high rainfall but were above the influence of fluctuating river levels. Vegetation in these forests is described further in Rosenberg (1990a), Erwin (1985), and Marra (1989). Locally within the low-lying forest, and along rivers, the understory is dominated by nearly pure stands of bamboo (*Guadua* spp.), which I consider a third habitat type. Over 20 km of trails traverse the reserve, allowing easy access to each forest type. The avifauna of Tambopata is relatively well known (Parker 1982, unpubl. data). I worked at Tambopata for 231 field-days

(5 May–20 July 1987, 28 June–15 October 1988, and 5 September–23 October 1989), covering the period from late rainy season to late dry season.

The second study site was in depto. Pando, northwestern Bolivia, 12 km SW Cobija (11°9'S, 68°51'W), at 325 m. This site was in hilly forest in the Acre-Purus drainage, about 200 km NNE Tambopata. At Pando, I worked in two habitats, upland forest and bamboo. The upland forest was similar to that at Tambopata, with a relatively open understory consisting mostly of shrub-like palms (e.g., *Geonoma* spp.) and a canopy of 30–40 m. This forest was dissected by streams, along which grew dense thickets of bamboo. The bamboo here was spineless and structurally different from that at Tambopata. I worked at the Pando site from 9 June to 8 August 1986 (mid-dry season) as part of a general avifaunal survey conducted by Louisiana State University Museum of Natural Science (LSUMNS; Parker and Remsen 1987).

Resource use and degree of specialization.—To ascertain degree of specialization and associated behaviors, I attempted to quantify the foraging behavior of all arboreal insectivorous species at each study site. I observed foraging birds primarily by first locating mixed-species foraging flocks in each habitat and then following these for as long as possible. The vast majority of observations were made before 1200 h. I recorded data on all species and noted flock compositions, but I concentrated my observations on dead-leaf foraging species. To minimize consecutive observations, I rotated my attention among flock members.

For each foraging individual, I recorded onto microcassette: height above ground, canopy height, foraging site (e.g., vine tangle, live branch), relative foliage density (scale, 0–5) in a 1-m-radius sphere around the bird, mode of searching or prey attack (including associated postures, such as hanging), substrate (including specific characteristics, such as leaf size and type), and perch type. Because dead-leaf searching species sought prey primarily hidden inside substrates, it was often impossible to distinguish between searching maneuvers and prey captures. I therefore recorded all unambiguous visual searches and included these in analyses of substrate use. Otherwise, my categorization of behaviors closely followed that of Remsen and Robinson (1990). I also noted associated bird species and any interactions among flock members.

I assessed the arthropod diet of each species directly to determine degree of prey selectivity and resource partitioning among dead-leaf specialists. Birds were collected for stomach analysis using mist-nets and shotguns, primarily at the Pando study site and on the Río Shesha, depto. Ucayali, Peru. The Río Shesha site was in hilly lowland rainforest with an avifauna typical of western Amazonia and similar to that of both Pando and Tambopata (LSUMNS unpubl. data). These samples were supplemented with a few birds taken elsewhere in eastern Peru and northern Bolivia (LSUMNS Stomach Contents Collection). All stomach samples were preserved directly in 70% ethanol as soon as possible after collection.

Stomach contents were sorted and identified to lowest taxonomic category possible under a 6X–25X dissecting microscope. Minimum number of prey items in each category was determined from diagnostic fragments, such as mouthparts, heads and wings. Identification of arthropod fragments was facilitated by dissecting voucher specimens collected at the study sites and by illustrations in Ralph et al. (1985), Moreby (1987), and Chapman and Rosenberg (1991). Prey sizes were estimated by measuring characteristic parts with an optical micrometer. Fragment size was then converted to prey length using regression equations in Calver and Wooller (1982), Diaz and Diaz (1990), or those determined in the present study. Each individual stomach was considered as a sample, and the diet of each species was determined by averaging the proportions of each prey category across individuals (i.e., samples were not pooled).

Resource subdivision and partitioning.—To assess relative specialization along finer axes of foraging height, substrate type, dead-leaf size, diet composition, and prey size, I calculated niche breadth as $B = 1/\Sigma p_i^2$, where p_i is the proportion of category i in the sample (Levins 1968). Each niche measure was divided into 10 categories to allow for comparisons of breadth across variables. Overlaps between each species pair were then calculated for each variable as $O_a = \Sigma P_{ia}P_{ja}\sqrt{(\Sigma P_{ia}^2)(\Sigma P_{ja}^2)}$, where P_{ia} and P_{ja} are the proportional uses of resource state "a" by species i and j respectively (Pianka 1974, May 1975). In addition, I calculated two combined measures of overlap: foraging space, equal to the product of O_{height} and $O_{substrate}$; and diet, equal to the product of $O_{prey-type}$ and $O_{prey-size}$. I used "product-alpha" (Cody 1974) in these cases because the measures being combined represented relatively independent niche parameters.

Differences between species were tested for continuous measures (foraging height, leaf size, prey size) using the Kolmogorov-Smirnoff test and for categorical measures (substrate, prey type) using the G-test. Details of pairwise comparisons may be found in Rosenberg (1990b).

Resource availability.—Numbers of suspended dead leaves within 10 m of the ground were assessed on randomly placed 10-m line transects perpendicular to existing trails, as described in

TABLE 1
CHARACTERISTICS OF 16 DEAD-LEAF FORAGING BIRD SPECIES IN SOUTHWESTERN AMAZONIA. HABITAT AND FLOCK TYPES ARE LISTED IN ORDER OF IMPORTANCE

Species	Code	Body weight[1]	Habitat[2]	Flock type[3]	% Dead leaves[4]	Number of obs.
CAPITONIDAE						
Capito niger	CN	62.5	U, L	C, SC	73	121[5]
Eubucco richardsoni	ER	31.8	L	C, SC	97	136[5]
DENDROCOLAPTIDAE						
Xiphorhynchus guttatus	XG	57.8	L, U	U, SC, C	63	331
FURNARIIDAE						
Cranioleuca gutturata	CG	14.9	L	SC, U	70	96
Philydor erythrocercus	PE	26.3	U	SC, U	80	122
Philydor ruficaudatus	PR	30.1	L	SC, C	92	36
Automolus rufipileatus	AR	36.5	L	U, SC	100	107
Automolus ochrolaemus	AO	33.8	L	U	94	236
Automolus melanopezus	AM	30.7	L	U	97	283
Automolus infuscatus	AI	38.8	U, L	U	88	201
Hyloctistes subulatus	HS	27.1	U, L	SC, U	85	20
FORMICARIIDAE						
Pygiptila stellaris	PS	24.1	L, U	C, SC, U	58	338
Myrmotherula leucophthalma	ML	9.3	L, U	U	99	1,137
Myrmotherula haematonota	MH	8.7	U	U	94	81
Myrmotherula ornata	MO	9.5	L	U	98	538
TROGLODYTIDAE						
Thryothorus genibarbis	TG	18.5	L	U	95	116

[1] Mean of 5 male and 5 female specimens (grams).
[2] U = Upland forest; L = Low-lying forest.
[3] Type of mixed-species foraging flock: C = Canopy; SC = Subcanopy; U = Understory.
[4] Percent of foraging observations at dead-leaf substrates.
[5] Includes only insectivorous foraging.

Rosenberg (1990a). I established 10 transects in each habitat at each site; leaves were censused at Pando in July 1986 and at Tambopata in May and July 1987 and in July and October 1988. During each census, I counted all dead leaves and clusters in a 1-m wide strip and recorded the distribution of leaf sizes (to the nearest 1 cm) and types, especially palms, bamboo, and *Cecropia* leaves. I also measured the accumulation, persistence, and turnover of individual leaves in each habitat at Tambopata, as described by Rosenberg (1990a).

The arthropod fauna of aerial litter was sampled by placing individual dead leaves in zip-lock plastic bags and spraying them with insecticide; arthropods exited the leaves and were easily separable. All arthropods were identified to the lowest taxonomic level possible, measured to the nearest 1 mm, and preserved in 70% ethanol. Characteristics of each leaf (e.g., size, type) were also recorded at the time of collection. I selected leaves in two ways. At Pando and at Tambopata in 1987, samples consisted of the first 30–50 dead leaves encountered 1–2 m above ground, along transects from randomly determined starting points along a trail. Some leaves proved impossible to collect without disturbing their arthropod inhabitants; therefore, these samples may be somewhat biased towards more exposed leaves. At Tambopata in 1988, I established 30 1-m^3 plots, 1–2 m above ground in low-lying forest. Within each plot I searched for arthropods on every substrate surface, including all live and dead leaves. In this way, I determined arthropod density on live vs. dead leaves, in addition to number per leaf. Arthropods on live foliage were also assessed by visually searching leaf surfaces in areas adjacent to the dead-leaf samples described above.

RESULTS

Dead-leaf foraging guild.—Sixteen bird species were found to feed most frequently at suspended dead leaves at either the Pando or Tambopata sites (Table 1). Ten species occurred at both sites and showed little or no variation in degree of specialization between areas. Fifteen of these 16 species were regular members of mixed-species feeding flocks, based on censuses of

92 flocks at Tambopata (82 in low-lying forest, 10 in upland; Rosenberg unpubl. data). Two barbets (Capitonidae) were regular members of canopy feeding flocks, searching for both insects and fruit. *Eubucco richardsoni* was common only in low-lying forest at Tambopata (33% of canopy flocks) and was very rare at Pando; virtually all of its insectivorous foraging was at dead leaves. *Capito niger* occurred in most forest types at both sites. It was a more generalized forager, searching branch and trunk surfaces in addition to dead leaves. One woodcreeper, *Xiphorhynchus guttatus*, foraged at dead leaves more than on any other substrate and was often the most conspicuous dead-leaf forager in any particular mixed-species flock. It occurred widely in most forest types, joining flocks in the canopy or understory (39% of all flocks).

As noted by Remsen and Parker (1984), most dead-leaf specialists belong to the families Furnariidae or Formicariidae (Table 1). Seventy-six percent of understory flocks in both forest types contained at least one species of *Automolus* foliage-gleaner. *Automolus rufipileatus* was restricted to river-edge forest at Tambopata, usually with extensive thickets of bamboo in the understory. *Automolus ochrolaemus* and *A. melanopezus* occurred in low-lying forest with bamboo at both sites; *A. ochrolaemus* was more widespread at Tambopata in dense, low-lying forest away from bamboo (40% of flocks). *Automolus infuscatus* was the common species in upland forest at both sites (55% of flocks) and also in more open areas of low-lying forest at Tambopata far from bamboo (22% of flocks). Of the five species of *Philydor* foliage-gleaners in this region, only *P. erythrocercus* and *P. ruficaudatus* are apparently dead-leaf specialists. *Philydor erythrocercus* was fairly common in upland forest at both sites (30% of flocks); it was often the only specialist in canopy flocks. *Philydor ruficaudatus* was rare, occurring only in a few canopy flocks in low-lying forest at Tambopata. The other *Philydor* species (*erythropterus*, *pyrrhodes*, and *rufus*) may use dead leaves regularly, but concentrate their foraging more on live foliage (especially palms). *Hyloctistes subulatus* is only tentatively listed as a dead-leaf specialist because of my small sample of observations. It was uncommon in upland forest at Pando and was inexplicably rare at Tambopata during the study period. Its inclusion is supported, however, by observations of this species in lowland forest in Costa Rica, where 70% of 37 foraging attempts were at dead leaves (pers. obs.). *Cranioleuca gutturata* foraged in dense parts of the subcanopy in low-lying forest at Tambopata, travelling with either understory or low canopy flocks (16% of flocks). It foraged along branches and vines in addition to searching trapped dead leaves. An additional furnariid, *Thripophaga fusciceps*, was listed as a specialist by Remsen and Parker (1984); although it is recorded from Tambopata, I did not observe it.

Of the antbirds, *Pygiptila stellaris* was the only canopy-flocking species that used dead leaves to a large extent. Although only 58% of its foraging was at dead leaves, it is included here because when feeding at these leaves this species employed many of the same behaviors (see below) exhibited by other specialists. Individual *P. stellaris* were observed to switch between bouts of dead-leaf foraging and searching live foliage, and this was the only species for which dead-leaf foraging appeared to be height-dependent; they searched dead leaves significantly more when in understory or sub-canopy flocks (i.e., ≤ 10 m) than in the upper canopy ($X^2 = 20.4$; $P < 0.001$).

Three small antwrens are extreme specialists in this region (94%–99% of foraging; Table 1). *Myrmotherula leucophthalma* was the most widespread, occurring in most (73%) forest understory flocks at Tambopata, but restricted to streamside bamboo and disturbed forest at Pando. In upland forest at Pando, this species was replaced by *M. haematonota* (Parker and Remsen 1987). *Myrmotherula ornata* was common in the vicinity of bamboo thickets at Tambopata and often occurred in the same mixed flocks as *M. leucophthalma* in this habitat (41% of flocks). The remaining specialist is a wren, *Thryothorus genibarbis*, which lived primarily in bamboo thickets at both sites, as well as in other disturbed and river-edge forest. This species foraged in solitary pairs or family groups and only occasionally joined understory flocks that passed through their territories.

Thus, each forest type supported a distinct assemblage of dead-leafing birds. In upland forest, understory flocks contained two species, and one to three species occurred in the canopy. In low-lying forest, especially with bamboo, many more species were present, with up to five species in understory flocks and five or six in some canopy flocks. When, on occasion, an understory flock temporarily joins with a sub-canopy flock, as many as nine dead-leafing species may forage in close proximity. The importance of bamboo to certain guild members will be discussed further below. Note also that the most specialized members of the guild (i.e., >75%) were nearly all understory species; canopy foragers (except *Philydor* spp.) were either partly frugivorous or regularly searched live foliage, branches or vines.

The 16 species included in this study represent roughly 20% of the 84 insectivorous species

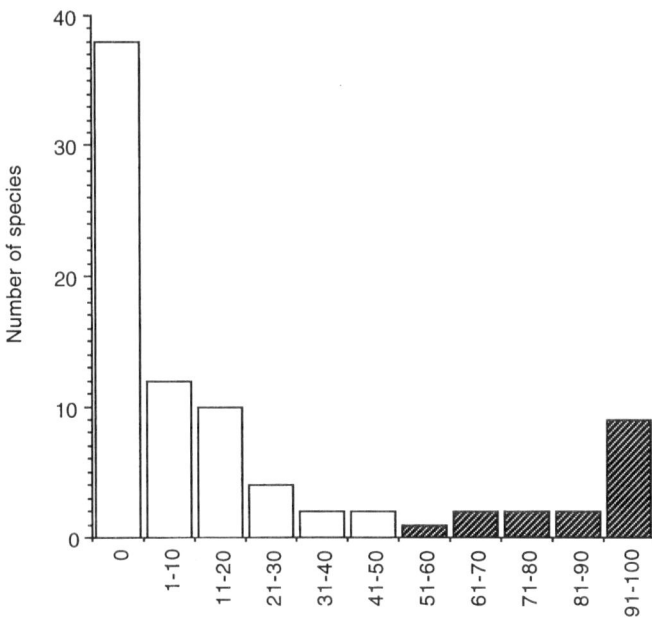

FIG. 1. Distribution of dead-leaf foraging among 84 insectivorous bird species in southwestern Amazonia. Based on 6962 observations at Pando, Bolivia and Tambopata, Peru. Shaded species are specialists considered in this study.

for which I collected foraging data (Fig. 1). An additional 38 species never were observed to forage on dead-leaf substrates. The remaining insectivorous birds at these forest sites inspected dead leaves opportunistically, but this usually accounted for <20% of their foraging (Fig. 1). A few species, however, regularly included dead leaves in their repertoires, although these substrates were not specifically sought out and the birds rarely if ever used specialized behaviors to inspect leaves or extract prey (see below). Remsen and Parker (1984) listed several species as "Regular Users" of dead leaves (25%–75% of observations). A few of these I have placed in the specialist guild above, and my observations suggest several others also are regular users; *Myrmotherula hauxwelli* (50% of 70 observations), *M. iheringi* (47% of 219 obs.), *Ancistrops strigilatus* (32% of 134 obs.), *Hypocnemis cantator* (31% of 54 obs.), *Philydor pyrrhodes* (29% of 41 obs.), *Microrhopias quixensis* (26% of 34 obs.), and *Dendrexetastes rufigula* (23% of 52 obs.). One additional regular user listed by Remsen and Parker (1984) was the ant-tanager *Habia rubica*. This species proved to be an extreme generalist, using dead-leaf substrates in 29% of 78 observations. Finally, my limited observations of a few other species included bouts of systematic dead-leaf searching; these were *Philydor rufus, Myrmeciza hyperythra, Thryothorus leucotis, Campylorhynchus turdinus, Paroaria gularis, Icterus icterus, Cacicus cela,* and *Psarocolius decumanus*. The extent to which these species may be specialized remains to be quantified.

Species-specific behaviors.—Dead-leaf specialists typically moved directly from leaf to leaf, inspecting them for hidden arthropod prey and ignoring intervening areas of live foliage or other substrates. Because dead leaves were often suspended in difficult to reach places or on flimsy substrates, the birds often employed acrobatic postures or behaviors to inspect them. Extending the body or neck (reaching) or hanging with legs extended was observed frequently in all species (Table 2). The "hanging" category includes clinging directly to the dead leaf and (especially in *E. richardsoni* and *Philydor* spp.) the completely vertical suspension of the body to reach leaves directly below a perch. These behaviors were not unique to dead-leaf foraging species, however; reaching was commonly observed in other arboreal foliage-gleaning species, and hanging is a characteristic behavior in certain genera (e.g., *Xenops, Terenura, Hylophilus*; pers. observ.).

In general, what separated guild members from other birds that occasionally inspected dead leaves was their tendency to manipulate these substrates physically with their bills or feet. All species studied picked at dead leaves with their bills on at least 50% of their foraging attempts

TABLE 2
Percent Use of Unusual Postures and Behaviors Associated with Dead-Leaf Foraging in 16 Amazonian Birds

Species (N)	Reach/ lean	Hang	Use bill (pick)	Pull	Hold	Tear	Thrash
Capito niger (88)	44	17	81	17	11	9	0
Eubucco richardsoni (124)	37	52	81	11	7	13	0
Xiphorhynchus guttatus (200)	8	21	83	0	0	2	20
Cranioleuca gutturata (62)	13	42	57	0	0	0	0
Philydor erythrocercus (104)	18	48	68	2	3	0	1
P. ruficaudatus (28)	25	50	89	0	0	0	0
Automolus rufipileatus (94)	16	22	72	3	3	4	6
A. ochrolaemus (230)	16	30	56	1	1	1	4
A. melanopezus (271)	22	27	61	2	2	2	6
A. infuscatus (208)	24	15	56	1	2	1	4
Hyloctistes subulatus (19)	0	68	74	0	0	0	5
Pygiptila stellaris (191)	27	10	59	2	0	1	5
Myrmotherula leucophthalma (741)	34	31	58	0	0	0	1
M. haematonota (78)	23	44	68	0	0	0	0
M. ornata (512)	26	21	52	0	0	0	1
Thryothorus genibarbis (94)	34	15	65	0	0	0	6

(Table 2). This behavior was often associated with cocking the head to listen, or peering inside the leaf, and served to jostle or flush otherwise immobile and hidden prey. Non-specialists visually inspected dead leaves but rarely disturbed the leaves to facilitate prey detection. This fundamental difference in behavior was confirmed with close observations of captive birds (Rosenberg 1993).

In addition to simply picking at a leaf, some species exhibited more complex behaviors to aid in prey capture. One such tactic was to pull a suspended leaf closer to the bird with the bill and (usually) grab or hold the leaf with the foot. This technique was used most frequently by the barbets and was observed consistently in nearly all the larger furnariid species (Table 2). Typically, a leaf was held next to a branch with the foot, and the prey was extracted with the bill from beneath the feet. A variation was seen in *P. erythrocercus*, in which the birds hung acrobatically with one foot, reached out and grabbed a leaf (usually undehisced) with the other foot and pulled it to the face to peer inside or extract prey, much in the manner of a parrot. The antbirds, as well as *Thryothorus genibarbis* and *Cranioleuca gutturata*, did not exhibit these additional behaviors. In particular, no antbird was seen to use its feet to manipulate substrates or prey, in the wild or during feeding experiments (Rosenberg 1993). *Pygiptila stellaris* occasionally tugged on a leaf with its bill, and the *Myrmotherula* antwrens rarely used the bill to bite down on curled leaves to test for hidden prey.

Another behavior distinguishing the barbets and large furnariids (*Automolus*) from the other species was their tendency to tear apart large leaves in search of prey. As pointed out by Remsen and Parker (1984), this behavior often destroys the leaf as a future hiding place for arthropods. Another "destructive" searching technique, used most frequently by *Xiphorhynchus guttatus*, was to thrash and toss leaves from clusters, often knocking them to the ground. Overall, the behaviors of *Myrmotherula* spp., *Cranioleuca gutturata*, and *Philydor* spp. were least destructive to the leaves and allowed them to serve as potentially renewable resource patches (see below).

Habitat associations.—Birds already specialized on dead-leaf substrates apparently subdivide this resource in a number of ways. As noted above (Table 1), many species were restricted to only one major habitat (forest type), and different combinations of species coexisted in each forest. Segregation by habitat was most evident among congeners. For example, *Philydor*, *Automolus*, and *Myrmotherula* all showed species replacements between upland and lowland forests at one or both sites.

The presence of bamboo has been recognized as an important habitat component for birds in southwestern Amazonia (Parker 1982, Pierpont and Fitzpatrick 1983, Parker and Remsen 1987, Kratter 1993). On my study sites, *A. melanopezus*, *A. rufipileatus*, *M. ornata*, and *T. genibarbis* were absent in forests without at least some bamboo, although segregation among congeners was far from absolute. For example, *M. ornata* joined some flocks containing *M. leucophthalma* at Tambopata, *A. melanopezus* and *A. ochrolaemus* occasionally occurred in the same flocks at both sites, and *A. rufipileatus* occurred with either of these species at Tambopata. However, even in low forest, *A. infuscatus* avoided areas with any bamboo and therefore rarely overlapped with other *Automolus* species.

Except for a few well-defined dense thickets, bamboo was distributed patchily throughout the low-lying forest at Tambopata. Consequently, flocks containing dead-leafing species encountered a gradient of bamboo densities, making it difficult to assign specific observations as either "bamboo" or "nonbamboo." Therefore, I consider only two habitat types, upland and low-lying forest, in the following comparisons.

I tested for spatial associations among dead-leafing species in 92 mixed-species flocks at Tambopata using Cramer's V (Pielou 1977:201); this is essentially a correlation coefficient between two species, based on their pattern of co-occurence in individual flocks. No significant negative associations existed between any species pair in either forest type (Appendix). The strongest positive associations (only two significant) were among species sharing an affinity for bamboo (listed above) and those foraging at similar heights in the canopy (e.g., *Eubucco richardsoni* with *Capito niger* and *Philyder* spp.; *Xiphorhynchus guttatus* with *Pygiptila stellaris*). In fact, overlap in foraging height was the only variable significantly (although weakly) correlated with this measure of association ($r = 0.284$; $P < 0.01$). The strongest negative associations (none significant) were between canopy and low understory species, which tended to travel in separate flocks, and between *A. infuscatus* and the above-listed bamboo species.

Foraging height and leaf size.—Within each habitat, most species differed significantly from each other in either their foraging height distributions or the average sizes of leaves searched (Appendix; for details of pairwise comparisons, see Rosenberg 1990b). Species overlapped more (i.e., were more densely packed) in low-lying forest than in upland forest (Fig. 2). The additional species in low-lying forest were primarily understory foragers, and more guild members searched larger leaves in low-lying forest than in upland. Among the understory species in both habitats, size of leaves searched was highly correlated with body size (weight) ($r = 0.897$; $P < 0.001$), but this relationship disappeared if canopy birds were included.

The breadths of foraging heights and leaf sizes used were similar among most species (Table 3). *Myrmotherula haematonota* exhibited the most restricted height range, and *X. guttatus* the broadest. On average, species in upland forest used a narrower range of heights than species in low forest. *Pygiptila stellaris* and *C. gutturata* showed the greatest diversity of leaf-sizes used, and *T. genibarbis* and *X. guttatus* showed the lowest.

Foraging site.—Because dead leaves could become trapped above ground in a variety of situations in the forest, guild members had the opportunity to concentrate their foraging efforts in particular microhabitats (Table 4). Barbets, for example, along with *Philydor erythrocercus*, foraged more than other species on bare twigs and branches; these species also consequently were seen in more open, exposed areas, as reflected by their lower average foliage density measures than all other species. *Xiphorhynchus guttatus* searched for dead leaves relatively frequently along trunks and on large canopy palm fronds. Many species foraged in dense vine-tangles, where leaves often gathered in large clusters. These areas were particularly important to *C. gutturata* and *T. genibarbis*, and probably *P. ruficaudatus* and *H. subulatus*. *Myrmotherula haematonota* at Pando, and *M. leucophthalma* in upland forest at Tambopata, fed frequently in understory palm vegetation (especially *Geonoma* spp.). Although six species commonly occurred in bamboo habitats, only *A. rufipileatus* and *M. ornata* foraged often within bamboo foliage.

Most dead-leafing species showed a tendency to perch directly on the leaves being searched (Table 4), another behavior rarely seen in other species that only occasionally inspected dead leaves. This was most evident in *X. guttatus*, which hung on dead palm fronds and clung to large *Cecropia* leaves, and in *M. haematonota*, which routinely clung to the tips of understory palm leaflets. Other species, such as *C. gutturata*, *C. niger*, and *P. erythrocercus*, most often inspected leaves from adjacent perches.

Substrate types.—Perhaps the most important way in which guild members differed was in their use of various types of dead leaves and other substrates (Fig. 3). Based on 10 substrate categories (including live foliage and branches), nearly every species pair in both habitats differed significantly in substrate use (Appendix). Among the canopy species, *C. niger* and *X. guttatus* often inspected large, suspended *Cecropia* leaves. *Xiphorhynchus guttatus* also inspected large dead palm fronds and clusters of leaves, which were not exploited by *C. niger*. *Eubucco richardsoni* and *P. erythrocercus* rarely used these distinctive leaf-types or clusters but instead concentrated their foraging at relatively small (10–12 cm) leaves that were often undehisced at the tips of dead branches (48% and 30% respectively). In my small sample of observations for *P. ruficaudatus*, 39% of the leaves searched also were undehisced on branch-tips. This contrasts

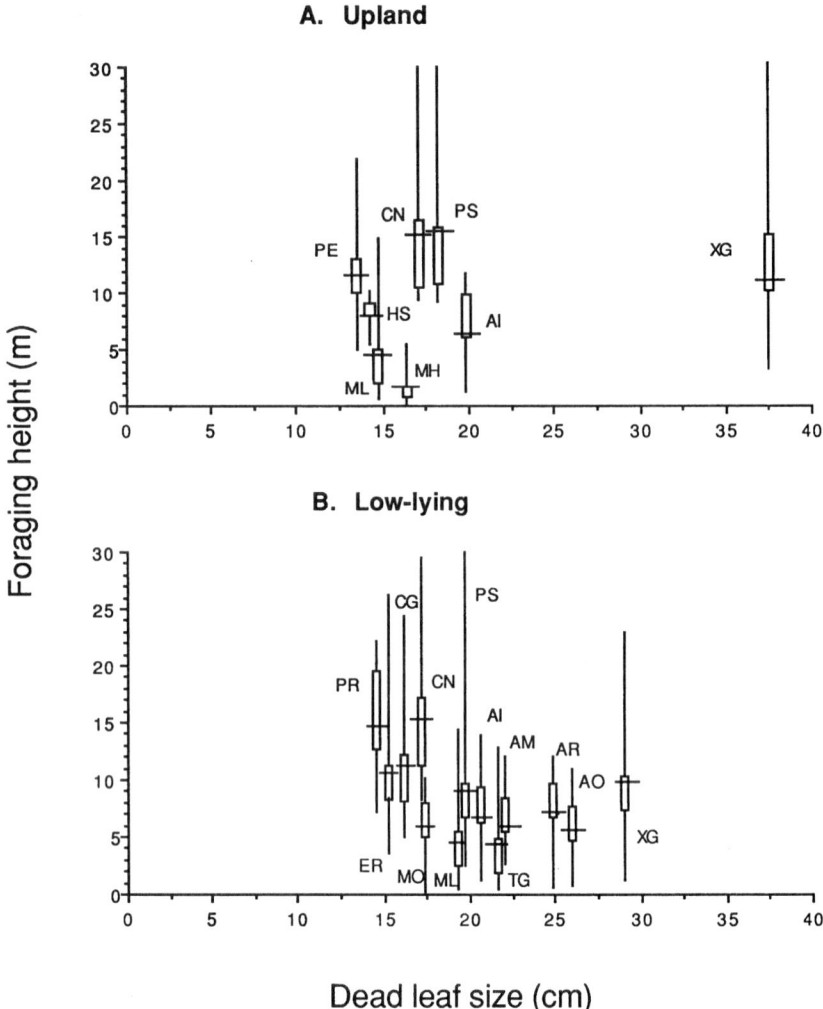

FIG. 2. Foraging heights and average leaf sizes used by dead-leaf foraging birds in Upland (A) and Low-lying (B) Amazonian forest. Horizontal line = mean height; vertical bar = modal 50% of observations; vertical line = height range. Bird species codes from Table 1.

with nearly every other species, which searched mostly leaves that had fallen and lodged on branches or vines.

In the understory, all *Automolus* spp. exploited large and distinctive leaf types, such as *Cecropia* and palms, as well as dead-leaf clusters (Fig. 3) Among the smaller antwrens, *M. leucophthalma* in upland forest at Tambopata, and *M. haematonota* in upland forest at Pando, both fed frequently at understory palm leaflets (e.g., *Geonoma* spp.). In low-lying forest at Tambopata, *M. ornata* differed greatly from *M. leucophthalma* in its heavy use of dead bamboo leaves. At Pando, however, where *leucophthalma* was the only *Myrmotherula* in bamboo, it often fed in dead bamboo foliage. *Thryothorus genibarbis* at both sites fed most often at dead bamboo and *Cecropia* leaves and in large clusters.

Diversity of substrates used ranged from 2.21 (*E. richardsoni*) to 6.44 (*T. genibarbis*), out of a possible 10.00 (Table 3). Both *A. infuscatus* and *M. leucophthalma* used a narrower range of foraging substrates in upland forest than in low-lying forest, but the average breadth for all upland species combined was only slightly less than for low-lying forest species. Substrate diversity was not related to foraging height, leaf size, or body size in these birds.

Substrate availability.—Suspended dead leaves were abundant in all forest habitats at each

TABLE 3

FORAGING AND DIETARY DIVERSITY (NICHE BREADTH) IN 16 DEAD-LEAF SEARCHING BIRDS. NICHE BREADTH = $1/\Sigma P_{iA}^2$ (SEE TEXT); EACH MEASURE BASED ON 10 CATEGORIES

Species	Height	Leaf size	Substrate	Prey type	Prey size
Capito niger	4.92	5.52	6.28	4.64	3.57
Eubucco richardsoni	4.09	4.93	2.21	4.20	3.76
Xiphorhynchus guttatus	5.11	4.48	5.52	3.80	4.83
Cranioleuca gutturata	4.85	6.74	5.74	3.18	1.97
Philydor erythrocercus	3.46	5.05	3.43	4.35	5.08
P. ruficaudatus	2.99	6.22	4.00	4.45	4.31
Automolus rufipileatus	4.78	5.39	5.44	5.98	5.02
A. ochrolaemus	4.80	5.66	6.09	3.65	6.17
A. melanopezus	4.16	5.29	4.64	3.32	5.00
A. infuscatus (upland)	4.62	5.39	3.74	4.70	2.96
A. infuscatus (low-lying)	4.84	5.53	5.16	—	—
Hyloctistes subulatus	2.37	—	4.76	3.37	5.77
Pygiptila stellaris	4.04	7.45	4.08	4.39	5.97
Myrmotherula leucophthalma (upland)	4.89	6.19	3.72	—	—
M. leucophthalma (low-lying)	4.16	5.53	4.62	—	—
M. leucophthalma (Pando)	3.12	—	4.47	4.08	4.39
M. haematonota	2.22	4.58	3.85	2.60	3.59
M. ornata	4.80	4.91	4.98	4.64	2.57
Thryothorus genibarbis	4.36	4.30	6.44	5.26	3.52
Upland (ave.)	3.85	5.52	4.42	3.98	4.49
Low-lying (ave.)	4.32	5.53	4.98	4.30	4.66

site. Rosenberg (1990a) reported a seasonal change in leaf abundance in two of three habitats at Tambopata in 1987. Repeated sampling in 1988 revealed that local variation among transects was greater than seasonal changes within a habitat (Rosenberg 1990b). Greatest concentrations of dead leaves were in the vicinity of tree-fall gaps or dense vine-tangles; also, local variation in leaf drop from particular tree species (e.g., *Cecropia*) contributed greatly to changes in leaf abundance.

At Tambopata, the density of dead leaves was consistently lower in upland forest (x = 3.9/m^3) than in low-lying forest with (5.1/m^3) or without (4.7/m^3) bamboo. At Pando, leaf density was higher, averaging 6.2/m^3 in upland forest and 6.7/m^3 in bamboo. In all areas, dead leaves

TABLE 4

PERCENT USE OF PERCH AND FORAGING SITES IN 16 DEAD-LEAF SEARCHING BIRDS. FD = FOLIAGE DENSITY IN A 1-M RADIUS SPHERE AROUND BIRD (SCALE, 1–5)

Species (N)	On leaf	Vine tangle	Live branch	Dead branch	Trunk	Palm	Bamboo	FD (ave.)
Capito niger (47)	9	10	47	32	2	0	0	2.2
Eubucco richardsoni (97)	23	13	20	34	0	0	0	2.2
Xiphorhynchus guttatus (175)	42	13	10	0	21	14	0	3.0
Cranioleuca gutturata (115)	3	52	23	6	2	4	0	3.4
Philydor erythrocercus (85)	11	20	26	32	5	7	0	2.1
P. ruficaudatus (33)	21	55	6	18	0	0	0	3.3
Automolus rufipileatus (99)	23	28	22	1	0	2	36	3.6
A. ochrolaemus (266)	13	36	28	6	1	5	9	3.3
A. melanopezus (261)	20	38	26	3	0	3	10	3.4
A. infuscatus (210)	14	34	36	4	0	6	0	3.1
Hyloctistes subulatus (20)	30	55	0	15	0	0	0	3.3
Pygiptila stellaris (155)	14	29	43	6	0	5	4	3.3
Myrmotherula leucophthalma[upland] (154)	22	17	21	16	0	30	0	2.8
M. leucophthalma[low] (572)	20	31	30	7	0	8	4	3.0
M. haematonota (71)	39	1	18	8	0	38	0	2.7
M. ornata (494)	10	32	15	6	0	4	38	3.0
Thryothorus genibarbis (77)	25	53	0	0	0	0	0	3.8

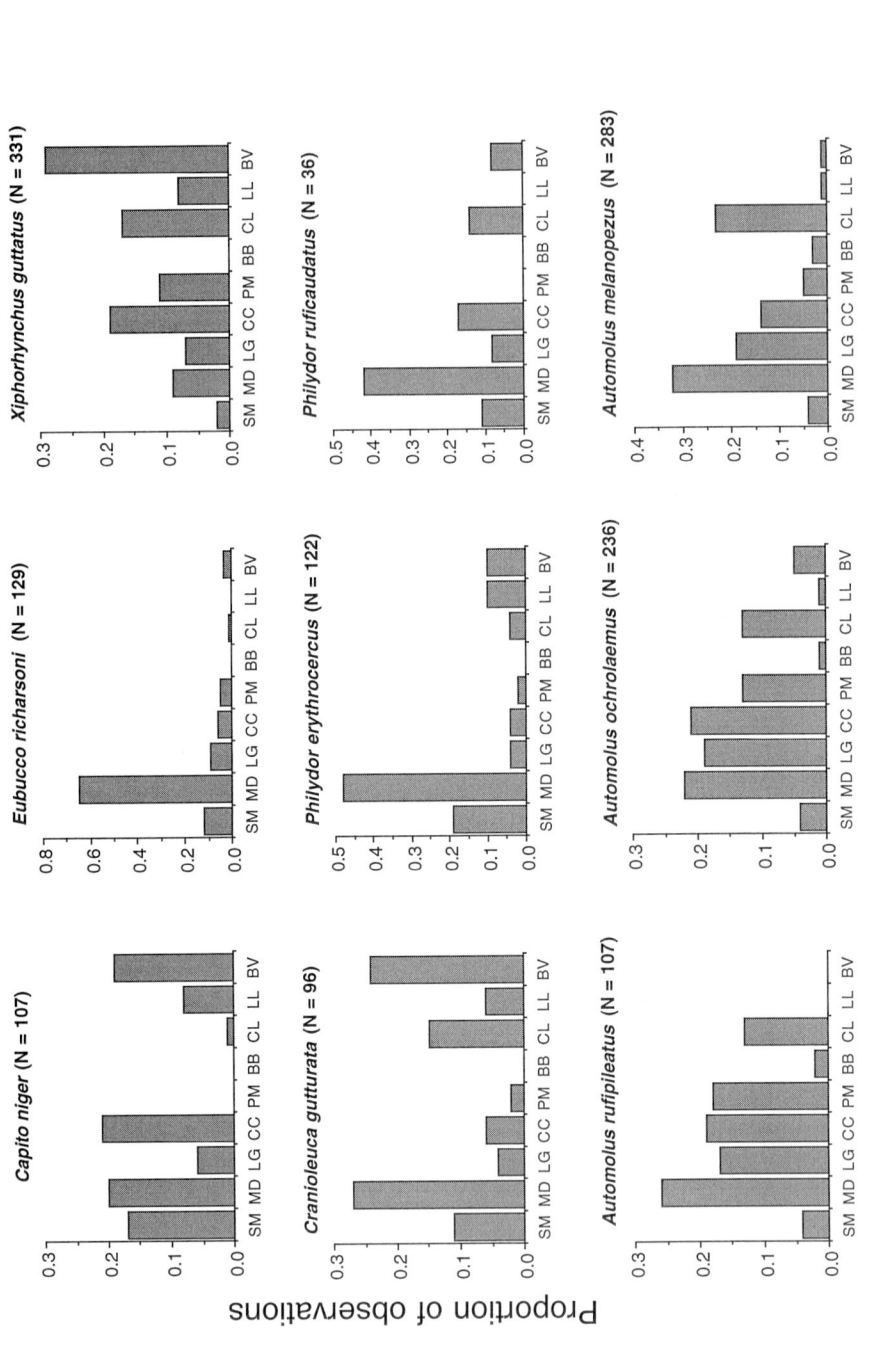

FIG. 3. Use of substrate types by dead-leaf foraging birds in sw Amazonia. SM, MD, LG = small (3–8 cm), medium (10–15 cm), and large (≥20 cm) curled dead leaves; CC = dead *Cecropia* leaves; PM = dead palm fronds; BB = dead bamboo leaves; CL = dead leaf clusters; LL = live leaves; BV = branches and vinestems.

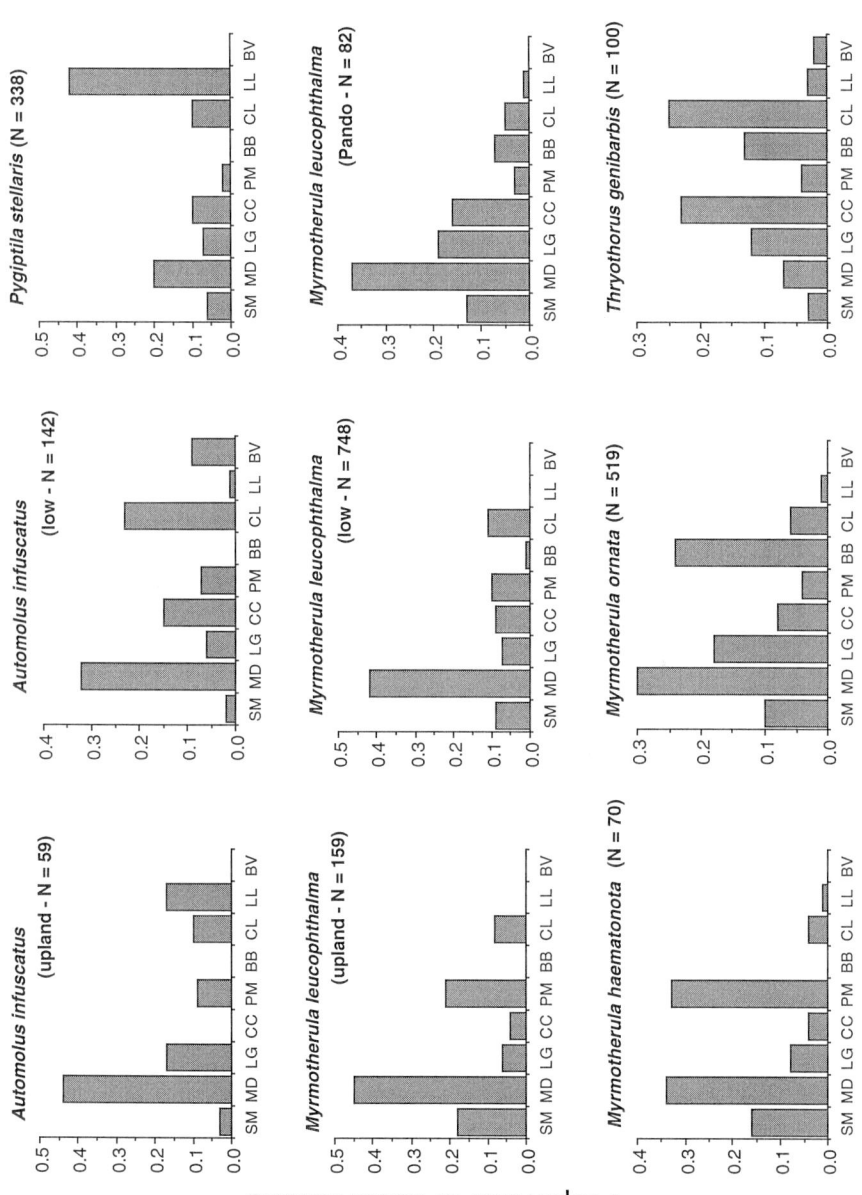

Fig. 3. Continued.

were concentrated in the first 3 m above the ground (Rosenberg 1990a). At Tambopata, overall height distribution of dead leaves was similar in all habitats, but with a tendency for bamboo sites to have more dead leaves at mid-levels (3–5 m) than did upland forest. At Pando, this was more pronounced, with a much greater proportion of leaves above 3 m in bamboo (51%) than in upland forest (25%).

Distribution of dead-leaf sizes and types differed among habitats, with each habitat offering particular distinctive leaf-types, such as palms, bamboo, or *Cecropia* leaves (Fig. 4). In upland forest at Tambopata, understory palm leaflets (mostly *Geonoma* spp.) made up 39% of the available dead leaves below 2 m. Above that height, small or medium-sized curled or entire leaves were most abundant. In low-lying forest, 44% of the leaves above 5 m were large palm fronds (*Iriartea* spp., *Socratea* spp.), whereas at lower levels, curled leaves ≤ 15 cm predominated. In bamboo thickets, dead bamboo leaves were most abundant at all levels, and *Cecropia* leaves made up 13% of the available dead leaves above 5 m. *Cecropia* leaves were patchily distributed, mostly in the vicinity of light-gaps and close to rivers (up to 8 leaves/ m^3).

At Tambopata, overall average dead-leaf size ranged from 14.5 cm in upland forest to 17.3 cm in low forest without bamboo. In each habitat, average leaf size was highest above 5 m (Fig. 4). At Pando, dead leaves averaged smaller in both habitats (11.7 cm in upland; 11.3 cm in bamboo), primarily because of the scarcity of large palm or *Cecropia* leaves. In upland forest, 16% of the leaves below 2 m were palms (mostly *Geonoma* spp.). Bamboo at Pando was structurally quite different from that at Tambopata; leaves were shorter (15 cm vs. 18–20 cm) and thinner (< 1 cm), and formed dense mats after dying, rather than lodging and curling individually. Because of difficulty enumerating dead bamboo leaves at Pando, leaf densities underestimated the total number of dead-leaf substrates available in this habitat.

Considering only nonentire leaves ≥8 cm as a closer measure of leaf availability to birds, differences between habitats were more marked. Upland forest supported nearly 50% fewer suitable leaves in 1987 than did bamboo thickets (2.6/m^3 versus 3.8/m^3), with low forest being intermediate (3.4/m^3). In addition, the proportion of total dead leaves considered suitable was greater in bamboo (84%) than in upland (69%) or low forest (72%).

Diet.—Diets of nearly all species were qualitatively similar, differing only in the proportions of major prey taxa (Fig. 5). Because of low variability among individuals, samples of five to six stomachs were adequate to represent the diets of most species (see Rosenberg and Cooper 1990). In six species, samples of three or more stomachs from each collecting locality allowed a geographic comparison in diets. In no case were there significant differences in prey types eaten between sites (G-tests; P's > 0.27). Therefore, I believe that pooling samples from several Amazonian localities is justified.

The barbets were largely frugivorous; on average, 37% of the food items in stomachs of *Eubucco richardsoni* and 18% of those in *Capito niger* were arthropods. These percentages support the estimates of frugivory found for these genera in a broader diet survey of barbets and other tropical bird families (Remsen et al. 1993). Those arthropods I identified in barbet stomachs indicate that the animal portion of these species' diets is similar to those of other dead-leaf foraging species (Fig. 5). In all remaining species except *Cranioleuca gutturata*, most of the diet (64%–94%) consisted of Orthoptera (including roaches), beetles, and spiders. The large woodcreeper, *X. guttatus*, ate more beetles and fewer orthopterans than most other species, whereas in *A. melanopezus*, *H. subulatus*, and *M. haematonota*, 50% or more of the prey consumed were katydids or crickets. In general, smaller species (e.g., *Myrmotherula* spp.) ate more roaches and spiders than did larger species (e.g., *Automolus* spp.). Species feeding in bamboo tended to eat more Heteroptera (mostly Pentatomidae) than species restricted to upland forest. Remains of vertebrate prey were found in eight species, including the relatively small-billed *T. genibarbis*. All identifiable bones were of iguanid lizards (probably *Anolis* spp.), except for two tree-frogs eaten by *A. melanopezus* and *A. rufipileatus*. Finally, the diet of *Cranioleuca gutturata* was notably different from all other species studied, consisting mostly of small Homopterans (leafhoppers) and ants. Unfortunately, my small sample of stomachs was from northeastern Peru; therefore, the uniqueness of this species' diet must be considered tentative.

Dietary diversity was lowest in *M. haematonota* (2.92), which ate the highest proportion of orthoptera, and *C. gutturata* (3.18) and was highest in *A. rufipileatus* (5.98) and *X. guttatus* (5.72) (Table 3). All other species ranged from 3.32 to 4.70 (out of a possible 10.00). On average, species in upland forest exhibited a narrower dietary breadth than those in low forest. This measure was otherwise not related to a species' body or bill size, taxonomic affinity, foraging height, substrate diversity, or number of stomachs examined.

Average size of prey eaten was positively correlated with bill size in these bird species (Fig.

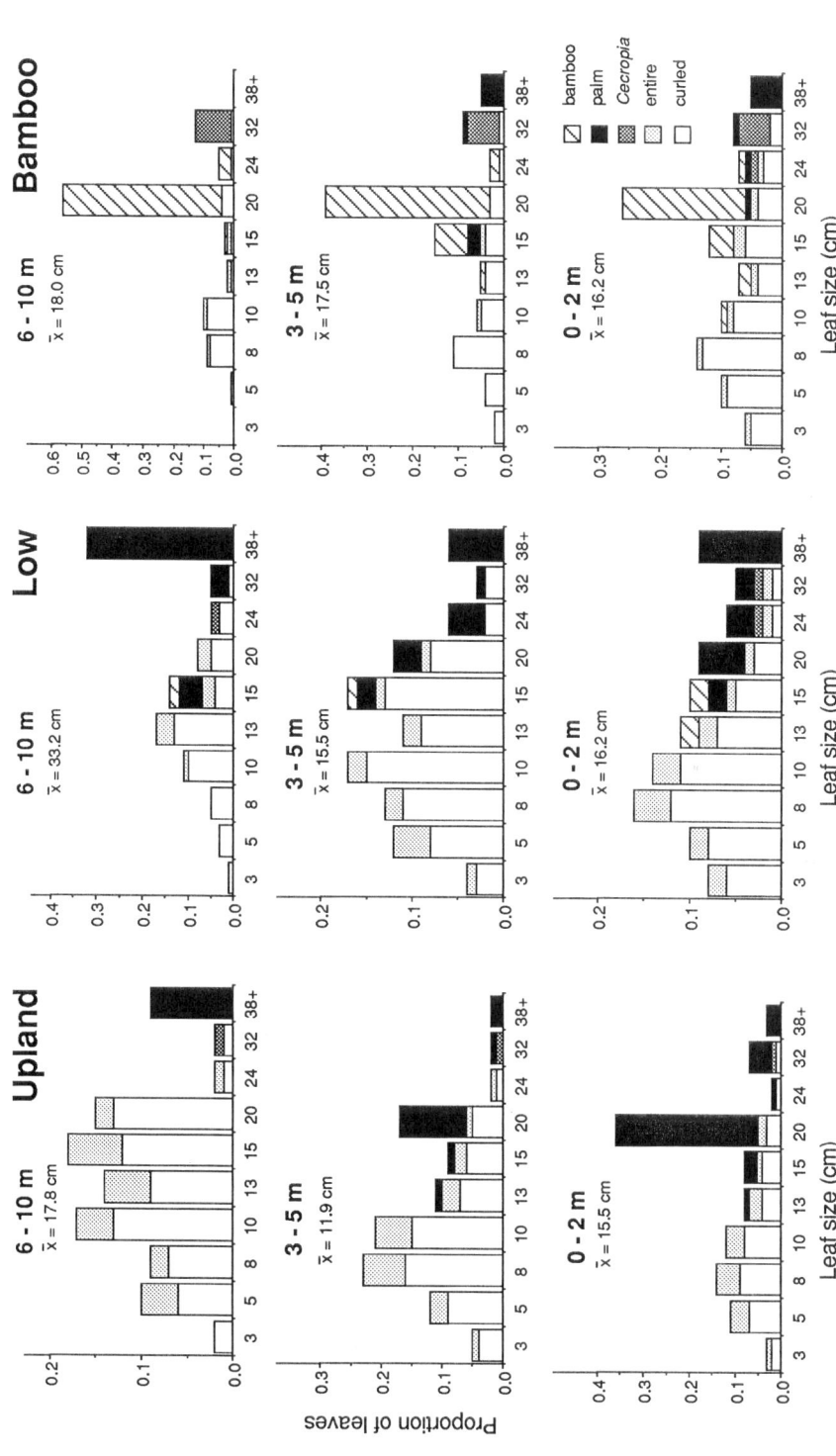

FIG. 4. Distribution of dead-leaf sizes and types at three height levels (0–2 m, 3–5 m, 6–10 m) in three Amazonian forest habitats. Based on 7417 leaves in Upland forest, 7794 leaves in Low-lying forest, and 9025 leaves in Bamboo. Average leaf size at each level is shown.

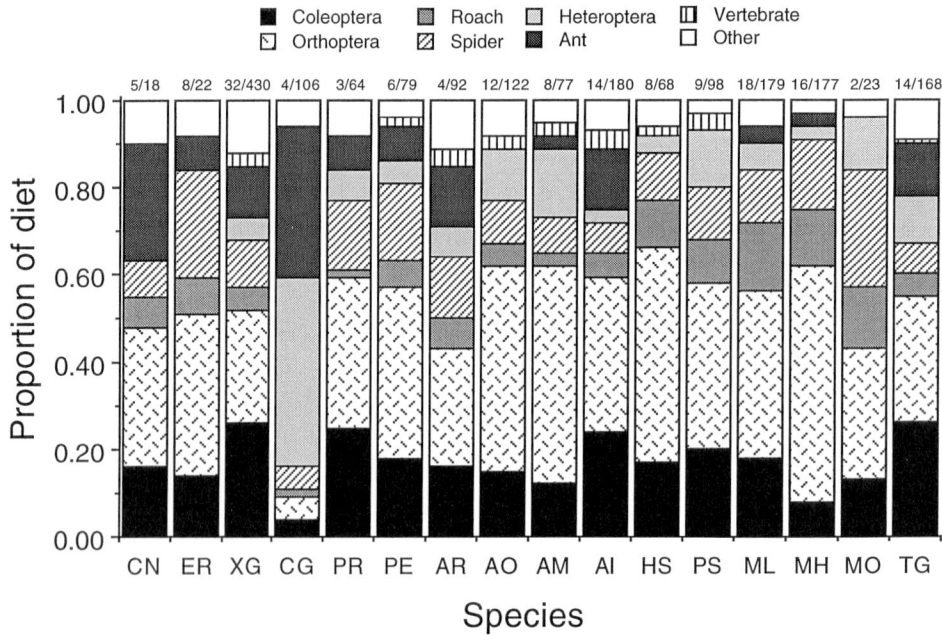

FIG. 5. Diets of 16 dead-leaf foraging birds in sw Amazonia. Bird species codes from Table 1. Sample sizes (above bars) are number of stomachs/number of prey items examined.

6), although much overlap existed for each prey category. Small antwrens generally did not take orthopterans larger than 20 mm, whereas the larger species ate many large as well as small orthopterans. *Xiphorhynchus guttatus* preyed on relatively small orthopterans for its size, however, overlapping greatly with the much smaller *Myrmotherula* spp.. All species preyed on relatively small spiders and beetles (Fig. 5b, c); in *Myrmotherula* spp. and *T. genibarbis*, virtually all beetles eaten were < 10 mm. The larger species ate beetles up to 18 mm long; in *X. guttatus*, which ate the highest proportion of beetles, nearly half were > 10 mm. *Cranioleuca gutturata*, which ate mostly homopterans and ants, had the smallest overall prey size.

Prey-size diversity, based on the frequency distribution among 5-mm size classes, was lowest in *C. gutturata* (1.97) and highest in *A. ochrolaemus* and *H. subularis* (5.60-5.63); all other species ranged from 3.28 to 4.98 (Table 3). Average prey-size diversity was similar in the two forest types (4.14 vs. 4.36).

Prey availability.—The arthropod fauna found in suspended dead leaves consisted mostly (>70%) of roaches, other orthopterans, spiders, and small beetles (Fig. 7). Also consistently present were ants (mostly colonial nesters), heteropterans, parasitic wasps, tiny flies, and a few moths and larvae. Four tree-frogs were also found inside curled dead leaves. Samples were similar among habitats, study sites, seasons, and years (Rosenberg 1990b).

Arthropods found in live foliage differed considerably from those in dead leaves (see Rosenberg 1990a, 1993). No roaches were found on live leaves, and fewer orthopterans and beetles. Spiders were about equally abundant on dead and live leaves, but ants, bugs, flies, and wasps were more numerous on live vegetation.

About one-half of the dead-leaf arthropods were ≤ 5 mm in length, including most beetles and nearly all flies, wasps, and ants (Fig. 7). Medium-sized (6–10 mm) prey consisted of orthopterans, roaches, spiders, and some beetles, whereas only orthopterans and roaches were among the larger available prey. Virtually all arthropods > 20 mm were katydids (Tettigoniidae).

The number of arthropods per dead leaf varied according to leaf type and size (Fig. 8). Abundance was highly correlated with leaf size in every habitat and seasonal sample. *Cecropia* leaves nearly always contained at least some arthropods (x = 1.34/leaf), whereas entire leaves supported virtually none (0.04/leaf). Number of prey in palm leaflets was slightly above the overall average, whereas number in bamboo leaves was slightly below average.

I estimated overall arthropod density for each habitat at Tambopata by multiplying the density of prey in each leaf type by the abundance of that leaf type in each habitat, excluding entire

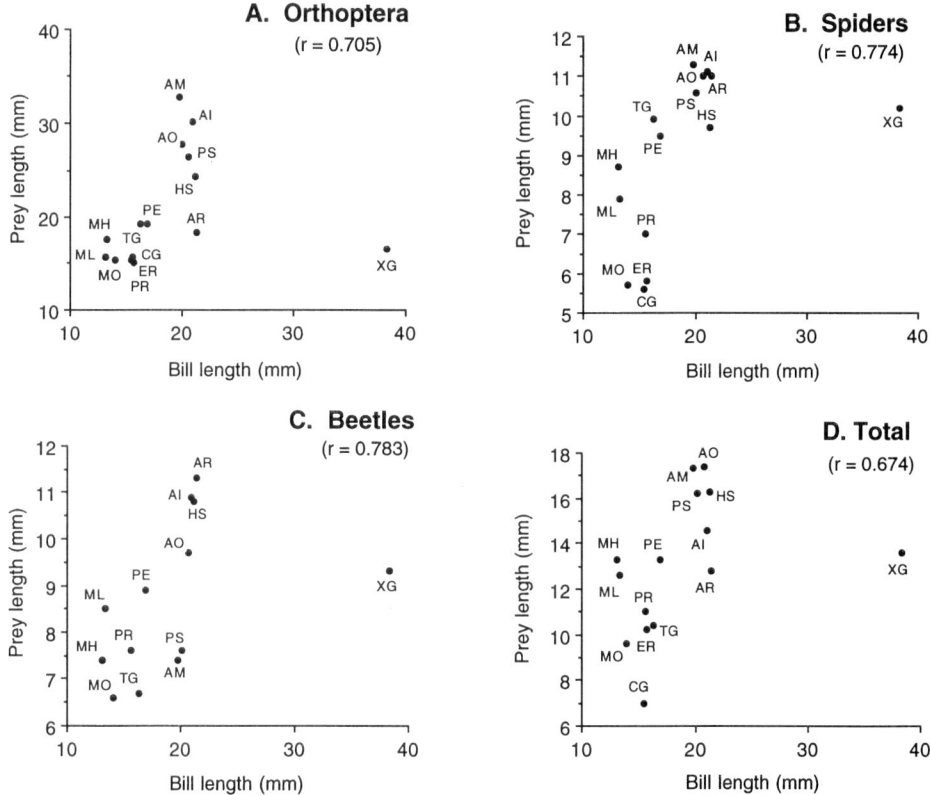

FIG. 6. Relationship between prey size and bill size among dead-leaf foraging birds. Data are the average lengths of prey eaten by each species. Bill length is the exposed culmen, averaged for five male and five female specimens. Correlations based on all species except *Xiphorhynchus guttatus*. Bird species codes from Table 1.

leaves. In May 1987, prey density was identical in upland and low-lying forest (0.29/leaf), but higher in bamboo (0.40/leaf). In July 1987, arthropod density in low forest had not changed (0.27/leaf), even though leaf abundance increased by 50%. In July 1988, however, prey density in this habitat increased to 0.37/leaf. Density of prey on live foliage at Tambopata (based on 3,155 leaves) averaged 0.10/leaf. Despite this 3–4 fold increase in number of arthropods from live to dead leaves, density per cubic meter of space was nearly identical for prey in live (6.3/m^3) and dead (6.1/m^3) foliage.

Diets versus prey availability.—The dietary composition of all bird species sampled differed significantly from prey availability in the dead leaves (Fig. 9). In nearly all species, orthopterans were highly selected, and except in the small antwrens, roaches were seemingly avoided. Other prey categories were usually eaten roughly in proportion (+ or − 10%) to their availability. The diets of *A. rufipileatus* and *X. guttatus* most closely matched the arthropods present in dead leaves, whereas that of *C. gutturata* was highly divergent (not shown).

Overall niche relationships.—A summary of the ecological similarities within the dead-leaf foraging guild is illustrated by combined measures of overlap in resource space and diet (Fig. 10). Although roughly twice as many species were supported in low-lying forest, average overlaps in the two forest types were nearly identical. Considering overlap in resource space, the primary division in each habitat was between understory and canopy species, with greater overall separation among the latter species due to foraging on substrates other than dead leaves. A close ecological similarity among congeneric *Automolus* and *Myrmotherula* species was apparent in low-lying forest.

Combined overlaps in diet were generally much greater than overlaps in resource space. In each forest type, the most similar species consisted of most or all of the highly specialized antbirds and ovenbirds, as well as the more generalized *P. stellaris* and *X. guttatus*. Also, no

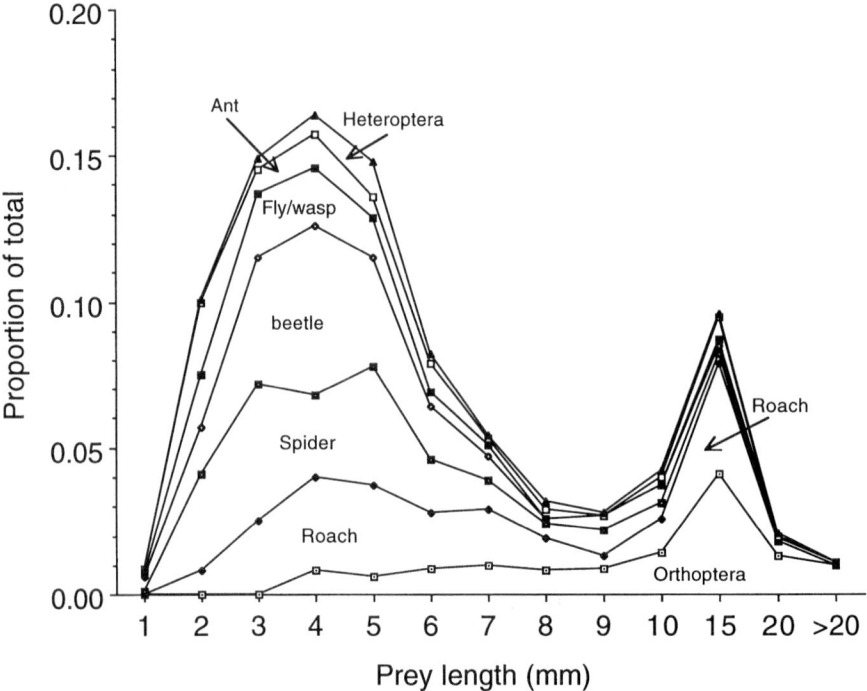

FIG. 7. Composition and size distribution of arthropod prey in suspended dead leaves in Amazonian forest. Based on 1025 arthropods collected from dead leaves at Pando, Bolivia and Tambopata, Peru.

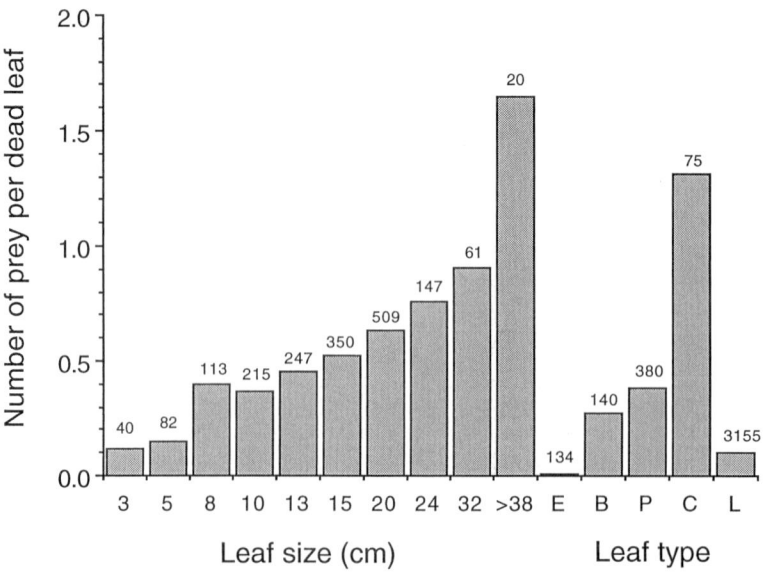

FIG. 8. Abundance of arthropod prey in dead leaves of different sizes and types. E = entire (uncurled); B = bamboo; P = palm; C = *Cecropia*; L = live leaves. Numbers above each bar indicates number of leaves sampled.

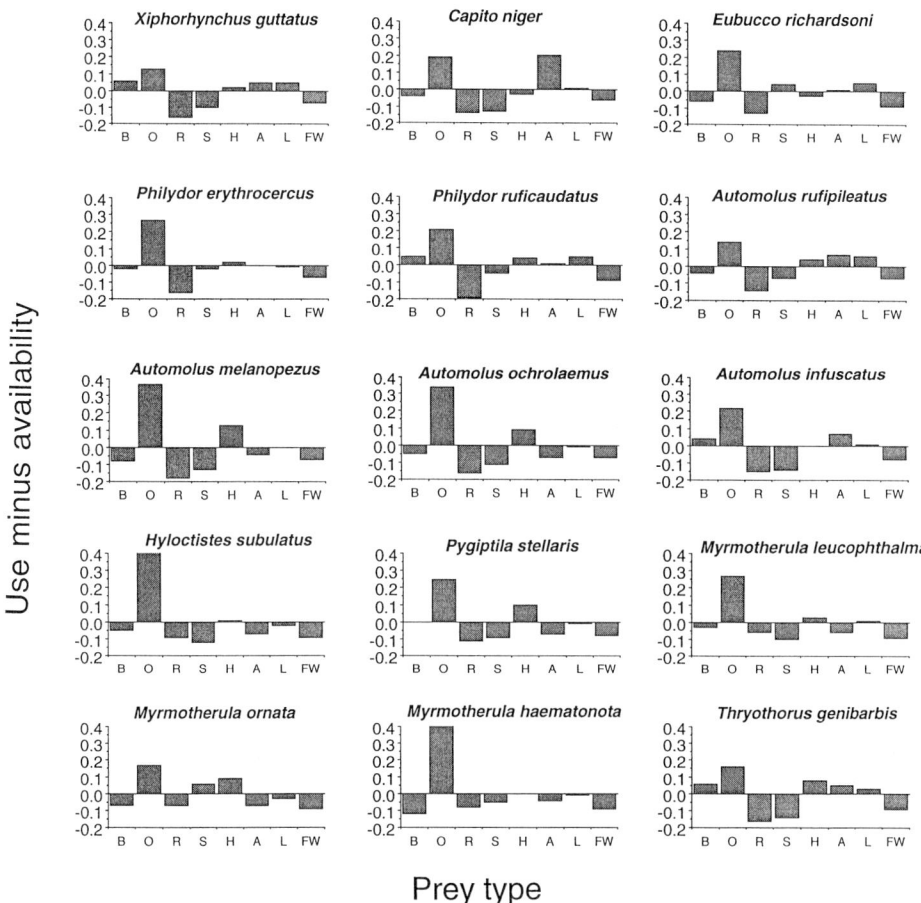

FIG. 9. Diets of 15 dead-leaf foraging birds compared with prey availability in suspended dead leaves. Horizontal line (0.0) indicates use = availability; bars above the horizontal indicate selection of prey, bars below indicate avoidance.

consistent segregation in diet was apparent between understory and canopy species. The most clearly divergent member of the guild was *Cranioleuca gutturata*.

Similar conclusions can be reached by summarizing significant differences in foraging and diet measures (Appendix). The fewest differences overall were among *Automolus*, *Myrmotherula*, and *Philydor* species, whereas the most divergent species were *X. guttatus*, *T. genibarbis*, and especially *C. gutturata*.

DISCUSSION

Contribution to avian species richness.—The species that I studied, although extremely specialized in their foraging, are by no means rare or restricted geographically. In fact, dead-leaf searching species are a common and conspicuous component of Amazonian forest avifauna. At Tambopata, of roughly 150 insectivorous bird species, 17 species (11%) are dead-leaf foragers (11 are specialists by Remsen and Parker's criterion). These represent 18% of the 93 species that regularly join mixed-species foraging flocks. In 77 understory flocks that I censused at Tambopata (4–20 species per flock), an average of 36% of the species in each flock were dead-leaf foragers, with up to seven species in a single large flock. Thus, not only do these species contribute to the overall regional species diversity, but they also comprise a substantial proportion of the flocking insectivores at any point and time.

These results probably apply equally to the avifauna at Pando and at other Bolivian sites sampled by Remsen and Parker (1984). In 25 understory flocks censused by Munn (1985) at Manu National Park in southern Peru, an average of 23% of the species in each flock were dead-

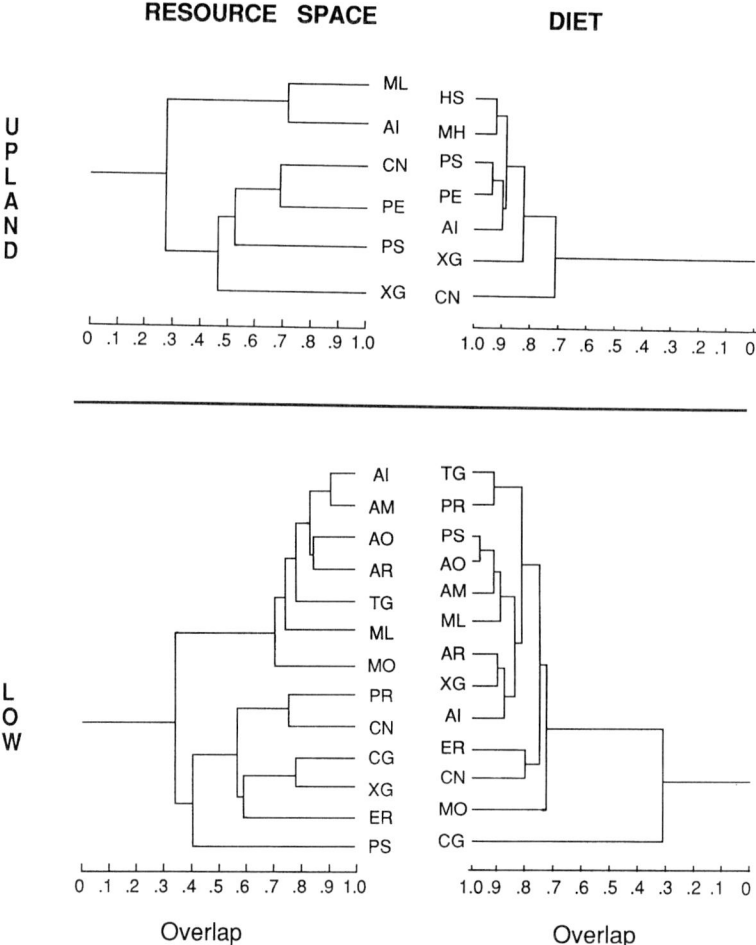

FIG. 10. Ecological overlap among dead-leaf foraging birds in two forest habitats in sw Amazonia. Overlap in resource space = O (foraging height) multiplied by O (substrate); Overlap in diet = O (diet composition) multiplied by O (prey size). All original overlaps based on proportional representation of 10 categories for each measure (see text). Phenograms constructed by computer program "PHYLIP" (Felsenstein 1987). Species codes from Table 1.

leaf foragers; this value is lower perhaps because the censuses were over longer periods and included many species that only occasionally joined a particular flock. Munn's data (1985) also indicated that up to seven dead-leafing species occurred in a single flock. Thiollay (1988), working in the opposite corner of the Amazon Basin in French Guiana, calculated that 23% of the prey attacks by 13 common, understory, foliage-gleaning species were at dead leaves. Four of the 13 species in that study were probably dead-leaf specialists.

In lowland forests of Central America, dead-leafing *M. fulviventris* helps to form the nucleus of typical mixed-species foraging flocks (Gradwohl and Greenberg 1980). Other dead-leaf specialist genera (e.g. *Eubucco*, *Automolus*) are represented by at least one species in Panama and Costa Rica, in some cases by the same species that is widespread throughout Amazonia. In southwestern Costa Rica, as many as five dead-leafing species may join a single understory flock (Rosenberg, pers. obs.). Therefore, in spite of geographic differences in total species number, the presence of a dead-leaf foraging guild that is a core component of mixed-species flocks appears to be constant throughout most Neotropical lowland forests. A decline in the importance of dead-leaf foraging is seen only with increasing latitude outside the tropics and with increasing elevation in the South American Andes (Remsen and Parker 1984). The extent to which this phenomenon is restricted to Neotropical as opposed to Asian or African forest communities is unknown.

Niche segregation among dead-leaf specialists.—Two factors may influence the subdivision of this already specialized resource: the distribution and productivity of specific resource types and interactions with potentially competing guild members. As shown by Rosenberg (1990a), most understory birds selected dead-leaf types nonrandomly, avoiding very small leaves and selecting large and distinctive leaf types in each habitat. Avoidance of small leaves was explained by their low productivity in terms of prey availability. To some extent, most species showed evidence of exploiting the most abundant substrate types available. In spite of this general tendency for the guild to track resources in each habitat, virtually every species pair differed significantly with respect to substrate type, leaf size, or foraging height (Appendix). This pattern initially suggests that, in addition to responses to resource availability, species-specific foraging niches in this guild are influenced by co-occurring species.

The only consistent pattern of dietary differences among these birds was in prey size. Each major prey group was partitioned, at least by small versus large bird species. Large orthopterans were the most important prey for all species, and size of orthopterans eaten was highly correlated with bill size (except in *X. guttatus*). Thus, of all the niche parameters measured, only the difference in bill size was a useful predictor of dietary differences within this guild. This contrasts with a study of four woodcreepers from the same study sites (including *X. guttatus*), in which bill size was unrelated to prey size (Chapman and Rosenberg 1991). It is possible, however, that many small prey represent nymphal stages of the same katydid species eaten as adults by large birds, such that predation by the smaller species may potentially reduce prey availability for larger species at a later time.

Because the arthropod fauna of dead leaves was relatively uniform, even across habitats and sites, it is perhaps no surprise that all species searching dead leaves would have similar diets. Subtle differences in foraging height or behavior may be important, however, in reducing spatial overlap among birds in the same flock, thus affecting interference rather than exploitative competition. If so, then species with the most similar foraging niches might be expected to avoid feeding in the same flocks, thus contributing to a "checkerboard" distribution of species among flocks (Graves and Gotelli 1993). This was partially true, to the extent that several pairs of congeners segregated completely by habitat and did not occur syntopically. However, within each habitat, I found no negative spatial associations between any species pairs, suggesting that these birds joined flocks independent of the other species present and independent of their ecological similarity. This result is partly in conflict with Graves and Gotelli's (1993) analysis of Munn's (1985) flock data, which found a significant checkerboard distribution pattern in several ecological groups of congeners, including dead-leaf foragers.

Of particular relevance, then, is the degree of interaction among ecologically similar congeners. In low forest at Tambopata, *M. leucophthalma* shared the understory with *M. ornata*, and a potential further segregation in habitat did not take place. The two antwrens occurred together in 23 of 45 understory flocks in areas with at least some bamboo. In this habitat, *leucophthalma* and *ornata* differed significantly in average foraging height, leaf size, and substrate use (Appendix). The most striking difference was the avoidance of bamboo leaves by *leucophthalma*, even though this species frequently used bamboo leaves where it occurred alone at Pando. Further south in Bolivia, in the absence of *leucophthalma*, *ornata* becomes a habitat generalist in forest with little bamboo (Remsen and Parker 1984; J. V. Remsen and S. M. Lanyon, unpubl. data).

It is tempting to cite this as evidence for ecological release and competitive interactions among these antwren species. However, when foraging together, the two species, on average, converged in foraging height and overlapped more in substrate use than when apart (Rosenberg 1990b). This convergence may indicate a shared response to local resource conditions and is consistent with my observation that most members of any mixed-species flock forage at approximately the same heights at any particular time. Differences between the two species when together were still significant, however, suggesting that subtle, relative height differences are maintained. It is noteworthy that in 598 observations of these two antwrens together at Tambopata, I noted only a single mildly aggressive interspecific encounter, a female *ornata* briefly chasing a *leucophthalma*. On many occasions, however, the two species fed in close proximity (sometimes on the same branch) without interacting in any way.

A different situation existed among the *Automolus* foliage-gleaners. As noted above, because of habitat segregation these species rarely occurred together in the same flock. When two species did forage together, however, divergence in foraging height was striking (Rosenberg 1990b); in four flocks their foraging heights were completely nonoverlapping. Furthermore, in one flock, I observed prolonged physical aggression between an individual *A. ochrolaemus* and *rufipileatus*, and in a second flock both birds vocalized frequently, giving calls normally used in intraspecific encounters. The more aggressive nature of *Automolus* species, compared with *Myrmotherula* species, is supported by additional observations of individuals fighting with *Pygiptila stellaris*, and displacing smaller antwrens from specific foraging sites. At nearby Manu National Park, Robinson and Terborgh (1995) found *Automolus* species pairs to exhibit a range of interspecific territorial behavior, from nonoverlapping to complete overlap. Among overlapping species pairs, assymetrical aggression was noted, with the species occupying the later-successional stage of habitat (i.e., more mature forest) usually dominating.

Niche segregation within the dead-leaf foraging guild appears to represent a dynamic balance between the constraints imposed by feeding in a mixed-species flock and those imposed by feeding close to potential competitors. The former may lead to convergence in foraging height and substrate use, both as a shared response to resource conditions and as a means of deriving the maximum benefit from group vigilance. At the same time, divergence should be expected if competition for shared resources is important.

Each species (or set of species) may solve this apparent dilemma in different ways. In *Myrmotherula* antwrens, local syntopy at Tambopata is tolerated without overt aggression or divergence of niches within individual flocks. In contrast, the same level of niche segregation in *Automolus* foliage-gleaners is apparently rarely sufficient to allow syntopy. Near total segregation is maintained through habitat differences, niche divergence within flocks, or aggression. Possibly, the relative rarity of large leaves and large orthopterans within those leaves increases the potential for competition among the larger species. This also supports Powell's (1989) suggestion that smaller species may underutilize food resources and coexist with greater niche overlap in flocks because the flock's territory size is determined by the needs of the larger species.

Maintenance of dead-leaf specialization.—The ubiquity of dead-leaf foraging in lowland tropical forests is certainly related to the abundance and high productivity of aerial leaf litter. Although leaf distribution was extremely patchy, it is likely that the scale of that patchiness affected the movement patterns of individual birds within their flocks' territories more than the distribution of flocks. Overall availability of dead leaves did differ among forest types, however.

In upland forest at Tambopata, the lower average dead-leaf density and relative scarcity of high density sites, along with the smaller average leaf size and relative scarcity of novel leaves such as *Cecropia*, probably resulted in fewer dead-leafing species using this habitat and a lower density of understory flocks (based on encounter rate along trails). Whether individual flocks had larger home ranges in upland forest, or whether portions of the forest were not occupied by flocking birds, was not determined. Where understory flocks occurred, however, the proportion of dead-leaf specialist species was similar to that in the other habitats.

In contrast, bamboo thickets offered the highest density of leaf-types preferred by birds (nearly 50% more than in upland), the greatest number of *Cecropia* leaves, and the highest average density of prey per leaf. The addition of 3 or 4 species to the dead-leafing guild at Tambopata is probably a result of this added productivity of bamboo. The largest understory flocks at Tambopata, including all flocks containing pairs of congeners (see below), were in the vicinity of bamboo thickets. It is possible that resource availability for dead-leafing species that form the nucleus for many flocks determines, in part, the formation and distribution of these flocks in lowland forests.

Taxonomic composition of prey available in dead leaves, at least at the ordinal level, appears to vary little geographically, perhaps contributing to the uniformity of dead-leafing behavior from site to site. Gradwohl and Greenberg (1982a) found that 68% of the arthropods in dead leaves in Panama were roaches, other orthopterans, and spiders, and suggested that the species involved were unique to aerial leaf-litter. Similarly, samples from Belize contained 62% orthopterans (including roaches), 17% spiders, and 14% beetles (Greenberg 1987b). That prey abundance is far greater and average prey size higher in dead leaves than on live foliage is also supported by several studies (Gradwohl and Greenberg 1982b, Greenberg 1987b, Rosenberg 1990a). Greenberg (1987b) calculated a 153:1 difference in arthropod biomass in dead versus live leaves at several sites in the West Indes and Belize. In particular, the consistent availability of large orthopterans, especially katydids, seems important in supporting the dead-leaf foraging birds.

For dead-leaf specialists to remain resident in tropical forests, resource availability must remain relatively stable year-round. Boinski and Fowler (1989) found that the accumulation of aerial leaf-litter was the least seasonal aspect of forest phenology measured in a Costa Rican rainforest. Furthermore, arthropods in dead leaves were the only subset of the arthropod fauna not to decrease during the mid-wet season, when other arthropods may be limiting to their predators (Boinski and Fowler 1989). Although my seasonal sampling was limited, arthropod availability in dead leaves at Tambopata seemed similar between late rainy season and mid-dry season, as well as between years and sites. Exclosure experiments in Panama indicated that the dead-leaf foraging birds themselves may deplete the arthropod resource by as much as 50% over a 6-wk period (Gradwohl and Greenberg 1982a). I found a high degree of turnover of arthropods at individual leaves, however, with a leaf recolonized, on average, 3–4 days after prey removal (Rosenberg 1990a). It is likely that birds may re-visit individual leaves every few days with reasonable probability of success.

Seasonal variation in dead-leaf abundance existed at many sites at Tambopata, but this variation was generally less than that among sites within a season. New leaves accumulated locally at the start of the dry season as deciduous canopy trees became bare and as high winds associated with austral winter storms redistributed leaf clusters and opened up areas with new treefalls. The short-term effect of this seasonal change was to superimpose a temporal aspect on the already highly patchy spatial distribution of dead leaves. The long-term effect, particularly of high winds, is to maintain a broken canopy and promote the growth of dense vine tangles and bamboo thickets (Erwin 1985), which in turn enhances the accumulation of aerial leaf litter. On a regional basis, it is perhaps no surprise that the highest diversity within several dead-leaf foraging genera (e.g., *Automolus, Myrmotherula*) lies in a belt across southern Amazonia and along the base of the Andes, where exposure to windstorms and presence of bamboo is most pronounced.

Costs of dead-leaf specialization.—Although the benefits of specializing on an abundant, predictable resource are relatively easy to quantify, the possible costs that may constrain such behavior are more difficult to evaluate. One likely constraint is the apparent dependency of most specialists on mixed-species foraging flocks. Several qualitative lines of evidence suggest that flocking by dead-leaf foragers is related to their reduced opportunities for vigilance while feeding. Searching dry, dead leaves frequently involves noisy rummaging that is audible to an observer (and presumably a predator) beyond the range of visual contact. In addition, the birds frequently insert their bills and heads inside leaves or dark clusters, sometimes for relatively long time periods, and scan more distant areas only when travelling between leaves. This contrasts with most other species, which constantly search more distant live-leaf surfaces and adjacent airspace (Rosenberg, pers. obs.). These latter species, because of their tendency to spot predators, sometimes serve as sentinels and are usually the first to give alarm calls that potentially warn other flock members (Munn 1986). This tendency is especially well-developed in certain species (e.g., *Thamnomanes* antshrikes; Schulenberg 1983) whose vocal repertoire includes specific alarm calls that elicit immediate and often dramatic anti-predator responses in other flock members, including dead-leaf foraging birds. The dead-leaf foragers are often vocally silent while foraging and do not frequently give alarm calls.

These observations, although largely anecdotal, suggest that dead-leaf specialists may benefit directly from increased vigilance provided by mixed-species foraging flocks. Foraging in mixed-species flocks, however, may impose costs on dead-leaf foragers in several ways. First, the rate at which flocks normally travel may be greater than that most efficient for exploiting dead leaves. Thiollay (1988) recorded lower foraging rates (moves/minute) in dead-leafing *M. gutturalis* (12.0) than in 3 live-leafing *Myrmotherula* species (18.3–18.9). Birds foraging rapidly must flit from leaf to leaf and only cursorily inspect each one. In 23 leaves previously inspected by 9 different *M. leucophthalma* at Tambopata, I found 4 potential prey items remaining, suggesting that the hidden prey sought by these birds may be difficult to detect. On several occasions I observed dead-leaf foragers that lagged behind a flock to extract large prey from tightly curled leaves or to manipulate and eat prey after capture, efforts that sometimes required several minutes. Typically in these cases, the birds would then fly directly to join the distant flock, which was often still audible. Pairs with dependent young may find it particularly difficult to forage efficiently with a flock. Most of my observations of antwrens feeding away from flocks were of family groups, and in several instances families moved temporarily with a flock but then lagged behind. Whether these constraints affect dead-leaf specialists more than other species is unknown.

Another potential cost of joining a mixed-species flock is the close proximity of possibly competing species, especially other dead-leaf foragers. The presence of more than one specialist in most flocks may necessitate the subdivision of an already restricted source (as documented in this study), and may require the retention of at least some plasticity in resource selection. Intraspecific competition may be reduced in flocks, however, because for most species membership is restricted to one pair or family group per flock. The flock thus serves as a basis for spacing and territorial establishment (Powell 1989).

An obvious consequence of extreme resource specialization is the potential vulnerability to a decrease in resource abundance or availability. Although dead leaves are a seemingly ubiquitous resource, their accumulation as aerial litter is in part dependent on the structure of the forest. It is noteworthy that nearly all dead-leaf specialist birds are restricted to primary lowland rainforests, a trait shared with most other insectivorous, mixed-flocking species. Some species inhabit naturally disturbed sites within forests, such as bamboo thickets, but only nonflocking *T. genibarbis* (and other *Thryothorus* spp.) occur in severely disturbed, nonforest habitats. Thus, the mutual dependency by dead-leaf specialists and other flocking species on a variety of resource types in intact forests may signal a shared vulnerability to human-induced disturbances.

Finally, a potential cost of extracting large prey hidden in curled leaves is danger from the prey itself. Large katydids and spiders in particular often have sharp spines on their legs or body and can give a nasty bite (pers. obs.). In cage experiments with *Myrmotherula* antwrens (Rosenberg 1993), dead-leaf specialist species were more willing and able to attack large orthopterans (>30mm) than were live-foliage-gleaning species. The largest prey (up to two-thirds the length of the bird's body) were often killed and eaten with great difficulty over long time periods (up to 45 min). Ability to handle these large prey may be an additional specialization of small dead-leaf foragers.

Evolution of dead-leaf specialization.—Most of this paper has been devoted to the ecological factors affecting species-specific behaviors and niche segregation among species. Those factors important in maintaining specialization, however, may not be the same as the evolutionary forces shaping long-term, genetically based foraging niches (Futuyma and Moreno 1988, Sherry 1990). By comparing species under variable ecological conditions, such as in different habitats or geographic locations, I hoped to identify the most stereotyped and flexible aspects of their behavior. Highly stereotyped behaviors are probably evolutionarily fixed, whereas behavioral flexibility may still be available for modification by natural selection or genetic drift.

Most dead-leaf foragers appear to be highly stereotyped in a qualitative sense; that is, their overall degree of specialization (dead vs. live leaves), modes of searching dead leaves (e.g., hanging postures, manipulative use of the bill), general foraging strata (understory vs. canopy), and diet composition do not vary substantially among individuals or populations. However, quantitative differences may exist in exact foraging heights or types of leaves searched. The ability (or need) to fine-tune these behaviors in response to subtly variable resource conditions or competitive regimes may have prevented further specialization over evolutionary time. In addition, more or less continuous gene flow among populations may prevent local specialization from occurring (Fox and Morrow 1981).

Specialized and stereotyped behaviors are thought to evolve most easily when resources are highly predictable. Aerial leaf litter and its component arthropod fauna appear to represent an extreme case of resource predictability, and dead-leaf foraging birds may be extreme in their stereotypy, even among tropical organisms. This system offers a stark contrast to many temperate-zone studies in which unpredictability and opportunism may be common (e.g., Wiens and Rotenberry 1979, Rosenberg et al. 1982). The extent to which this contrast reflects a true latitudinal gradient in resource stability remains to be shown with further comparative studies.

Important to the evolution of foraging specialization in these birds are psychological adaptations involving search images, learning, or memory. Greenberg (1987a, 1990) showed that the tendency to investigate novel substrates (including dead leaves) was inversely related to degree of neophobia, which was innate and varied among species. Greenberg (1987a) suggested that dead-leaf searching may represent a neotenic retention of curiosity towards novel substrates, which is usually more prevalent in young rather than in adult birds. Learning and the development of search images may enable individual organisms to become resource specialists and may be an important step in the fixation of these behaviors in populations. For example, Werner and Sherry (1987) documented that a "generalist" species, the Cocos Finch (*Pinaroloxias inornata*), was actually composed of specialized individuals (including dead-leaf specialists).

Ultimately, degrees of ecological specialization must be traced through phylogenetic lineages of species. Dead-leaf foraging obviously evolved independently in several families of birds; however, within each family only one or a few genera exhibit this behavior, suggesting strong phylogenetic constraints. Hackett and Rosenberg (1990) studied the genetic relationships among *Myrmotherula* and other small antwrens, including the dead-leaf specialists considered in this study. Genetic data clearly indicate that the "checker-throated" group of *Myrmotherula* (all dead-leaf specialists) represent a distinct clade (six species distributed throughout the Amazon basin,

Andean foothills, and southern Central America) not closely related to other *Myrmotherula*. In fact, the only other antbirds (of 12 genera tested) in the same clade as the dead-leafing *Myrmotherula* were *Pygiptila stellaris* and *Microrhopias quixensis*. The former species is a habitual dead-leaf forager, as documented in this study, and the latter is one of few other antbirds thought to be a regular user of dead leaves. Thus, we must conclude that this particular behavioral specialization represents a primitive trait that arose early in the history of this lineage, possibly more than nine million years before present (Hackett and Rosenberg 1990).

That dead-leaf specialization evolved before the radiation of the current species in independent lineages, and has remained qualitatively unchanged through these radiations, implies that the present-day ecology of these species may be irrelevant to the question of what originally led to the evolution of this specialization. In groups in which not all species are specialized (e.g., *Philydor*), the evolution of dead-leaf foraging may be more recent, or even ongoing. In such cases, current ecologies may represent the range of conditions that led to specialization in other lineages, and phylogenetic analyses may enable us to track these avenues of change. Also, comparative studies of geographic variation (if any) in less specialized species, such as *Pygiptila stellaris* and especially *Xiphorhynchus guttatus*, may illuminate conditions under which specialization was most likely to have evolved.

ACKNOWLEDGMENTS

I thank J. V. Remsen, Jr. for suggesting this area of study, and for advice and support through all phases of my research. T. A. Parker III and J. P. O'Neill also shared their extensive knowledge of tropical birds and along with the other staff and students at Louisiana State University Museum of Natural Science (LSUMNS) provided an unparalleled setting in which to conduct this research. I also benefitted from discussions about tropical biology or birds with J. Belwood, T. L. Erwin, R. Greenberg, C. Munn, T. W. Sherry, and D. F. Stotz, among others.

Fieldwork in Pando, Bolivia was made possible through the expedition funds of the LSUMNS and a grant by the National Geographic Society to J. V. Remsen, Jr. Members of this and other expeditions helped collect specimens for dietary analysis. For the opportunity to work at the Tambopata Reserve in Peru, I thank Dr. Max Gunther D. for a reduced rate at the Explorer's Inn. Chris Canaday and Robb Brumfield were invaluable field assistants at Tambopata and helped with stomach analysis and insect identifications. The LSU Entomology Museum loaned collecting equipment and also helped with identifications. Grace Servat kindly provided additional stomach samples, and A. Chapman identified the stomach contents of *X. guttatus*.

Various drafts of this manuscript benefited from comments by J. M. Bates, K. M. Brown, A. Chapman, S. J. Hackett, R. Hamilton, P. P. Marra, C. A. Munn, T. A. Parker III, D. P. Pashley, J. V. Remsen, Jr., T. S. Sillett, D. F. Stotz, G. B. Williamson, R. M. Zink, and an anonymous reviewer. Jorge Saliva provided the Spanish (translation of the) abstract. My research was supported by a National Science Foundation dissertation improvement grant (BSR-8800905), an LSU Alumni Federation Fellowship, a Charles M. Fugler fellowship in tropical biology, and research funds from LSUMNS and LSU Department of Zoology and Physiology.

LITERATURE CITED

ALVAREZ DEL TORO, M. 1980. Las Aves de Chiapas. Universidad Autonoma de Chiapas, Tuxtla Guttierrez, Mexico.

BOINSKI, S., AND N. L. FOWLER. 1989. Seasonal patterns in a tropical lowland forest. Biotropica 21:223–233.

CALVER, M. C., AND R. D. WOOLLER. 1982. A technique for assessing the taxa, length, dry weight, and energy content of the arthropod prey of birds. Aust. Wildl. Res. 9:293–301.

CHAPMAN, A., AND K. V. ROSENBERG. 1991. Diets of four sympatric Amazonian woodcreepers (Dendrocolaptidae). Condor 93:904–915.

CODY, M. L. 1974. Competition and the Structure of Bird Communities. Princeton Univ. Press, Princeton, New Jersey.

DIAZ, J. A., AND M. DIAZ. 1990. Estimas de tamaños y biomasas de arthropodos aplicables al estudio de la alimentacion de vertebrados insectivoros. Doñana, Acta Vertebr. 17:67–71.

ERWIN, T. L. 1985. Tambopata Reserve Zone, Madre de Dios, Peru: history and description of the reserve. Rev. Peru. Entomol. 27:1–8.

FELSENSTEIN, J. 1987. Documentation of Computer Program "Phylip." Univ. of Washington.

FOX, L. R., AND P. A. MORROW. 1981. Specialization: species property or local phenomenon? Science 211: 887–893.

FUTUYMA, D. J., AND G. MORENO. 1988. The evolution of ecological specialization. Ann. Rev. Ecol. Syst. 19:207–233.

GRADWOHL, J., AND R. GREENBERG. 1980. The formation of antwren flocks on Barro Colorado Island, Panama. Auk 97:385–395.
GRADWOHL, J., AND R. GREENBERG. 1982a. The effects of a single species of avian predator on the arthropods of aerial leaf-litter. Ecology 63:581–583.
GRADWOHL, J., AND R. GREENBERG. 1982b. The breeding seasons of antwrens on Barro Colorado Island. Pages 345–351 in The Ecology of a Tropical Forest: Seasonal Rhythms and Long-term Changes, (E. G. Leigh, A. S. Rand, and D. Windsor, Eds.). Smithsonian Institution. Press, Washington, D. C.
GRADWOHL, J., AND R. GREENBERG. 1984. Search behavior of the Checker-throated Antwren foraging in aerial leaf-litter. Beh. Ecol. Sociobiol. 15:281–285.
GRAVES, G. R. AND N. J. GOTELLI. 1993. Assembly of avian mixed-species flocks in Amazonia. Proc. Natl. Acad. Sci. 90:1388–1391.
GREENBERG, R. 1987a. Development of dead-leaf foraging in a tropical migrant warbler. Ecology 68:130–141.
GREENBERG, R. 1987b. Seasonal foraging specialization in the Worm-eating Warbler. Condor 89:158–168.
GREENBERG, R. 1990. Ecological plasticity, neophobia, and resource use in birds. Pages 431–437 in Avian foraging: theory, methodology, and applications, (M. L. Morrison, C. J. Ralph, J. Verner, and J. R. Jehl, Jr., Eds.). Stud. Avian Biol. 13.
HACKETT, S. J., AND K. V. ROSENBERG. 1990. Comparison of phenotypic and genetic differentiation in South American antwrens (Formicariidae). Auk 107:473–489.
KARR, J. R. 1971. Structure of avian communities in selected Panama and Illinois habitats. Ecol. Monogr. 41:207–233.
KARR, J. R. 1976. Seasonality, resource availability, and community diversity in tropical bird communities. Am. Nat. 111:817–832.
LEVINS, R. 1968. Evolution in Changing Environments. Monogr. Pop. Biol. No. 2. Princeton University Press, Princeton, New Jersey.
MARRA, P. P. 1989. Habitat selection in understory birds of tropical and temperate forests. Thesis. Louisiana State University, Baton Rouge, Louisiana.
MAY, R. M. 1975. Some notes on estimating the competition matrix, alpha. Ecology 56:737–741.
MOREBY, S. J. 1987. An aid to the identification of arthropod fragments in the faeces of gamebird chicks (Galliformes). Ibis 130:519–526.
MORTON, E. S. 1980. Adaptations to seasonal changes by migrant land birds in the Panama Canal Zone. Pages 437–453 in Migrant Birds in the Neotropics: Ecology, Behavior, Distribution, and Conservation, (A. Keast, and E. S. Morton, Eds.). Smithsonian Institution Press, Washington, D. C.
MUNN, C. A. 1985. Permanent canopy and understory flocks in Amazonia: species composition and population density. Pages 683–712 in Neotropical Ornithology, (P. A. Buckley, M. S. Foster, E. S. Morton, R. S. Ridgely, and F. G. Buckley, Eds.). Ornithol. Monogr. 36.
MUNN, C. A. 1986. Birds that cry wolf. Nature 319:143–145.
MUNN, C. A., AND J. W. TERBORGH. 1979. Multispecies territoriality in Neotropical foraging flocks. Condor 81:338–347.
ORIANS, G. H. 1969. The number of bird species in some tropical forests. Ecology 50:783–801.
PARKER, T. A., III. 1982. Observations of some unusual rainforest and marsh birds in southeastern Peru. Wilson Bull. 94:477–493.
PARKER, T. A., III, AND J. V. REMSEN, JR. 1987. Fifty-two Amazonian bird species new to Bolivia. Bull. Br. Ornithol. Club 107:94–107.
PIANKA, E. R. 1974. Niche overlap and diffuse competition. Proc. Nat. Acad. Sci. USA 71:2141–2145.
PIELOU, E. C. 1977. Mathematical Ecology. New York, Wiley-Interscience, Inc.
PIERPONT, N., AND J. W. FITZPATRICK. 1983. Specific status and behavior of *Cymbilaimus sanctaemariae*, the Bamboo Antshrike, from southwestern Amazonia. Auk 100:645–652.
POWELL, G. V. N. 1989. On the possible contribution of mixed species flocks to species richness in Neotropical avifaunas. Behav. Ecol. Sociobiol. 24:387–393.
RALPH, C. P., S. E. NAGATA, AND C. J. RALPH. 1985. Analysis of droppings to describe diets of small birds. J. Field Ornithol. 56:165–174.
REMSEN, J. V., JR. 1985. Community organization and ecology of birds of high elevation humid forest of the Bolivian Andes. Pages 733–756 in Neotropical Ornithology, (P. A Buckley, M. S. Foster, E. S. Morton, R. S. Ridgely, and F. G. Buckley, Eds.). Ornithol. Monogr. 36.
REMSEN, J. V., JR., M. ELLERMAN, AND J. COLE. 1989. Dead-leaf searching by the Orange-crowned Warbler in Louisiana in winter. Wilson Bull. 101:645–648.
REMSEN, J. V., JR., M. A. HYDE, AND A. CHAPMAN. 1993. The diets of Neotropical trogons, motmots, barbets, and toucans. Condor 95:178–192.
REMSEN, J. V., JR., AND T. A. PARKER III. 1984. Arboreal dead-leaf searching birds of the Neotropics. Condor 86:36–41.
REMSEN, J. V., JR., AND S. K. ROBINSON. 1990. A classification scheme for foraging behavior of birds in terrestrial habitats. Pages 144–160 in Avian foraging: theory, methodology, and applications, (M. L. Morrison, C. J. Ralph, J. Verner, and J. R. Jehl, Jr., Eds.). Stud. Avian Biol. 13.
ROBINSON, S. K., AND J. TERBORGH. 1995. Interspecific aggression and habitat selection by Amazonian birds. J. Anim. Ecol. 64:1–11.

ROSENBERG, K. V. 1990a. Dead-leaf foraging specialization in tropical forest birds: measuring resource availability and use. Pages 360–368 in Avian foraging: theory, methodology, and applications, (M. L. Morrison, C. J. Ralph, J. Verner, and J. R. Jehl, Jr., Eds.). Stud. Avian Biol. 13.

ROSENBERG, K. V. 1990b. Dead-leaf foraging specialization in tropical forest birds. Dissertation. Louisiana State University, Baton Rouge, Louisiana.

ROSENBERG, K. V. 1993. Diet selection in Amazonian antwrens: Consequences of substrate specialization. Auk 110:361–375.

ROSENBERG, K. V., AND R. J. COOPER. 1990. Approaches to avian diet analyses. Pages 80–90 in Avian foraging: theory, methodology, and applications, (M. L. Morrison, C. J. Ralph, J. Verner, and J. R. Jehl, Jr., Eds.). Stud. in Avian Biol. 13.

ROSENBERG, K. V, R. D. OHMART, AND B. W. ANDERSON. 1982. Community organization of riparian breeding birds: response to an annual resource peak. Auk 99:260–274.

SCHULENBERG, T. S. 1983. Foraging behavior, eco-morphology, and systematics of some antshrikes (Formicariidae: *Thamnomanes*). Wilson Bull. 95:505–521.

SHERRY, T. W. 1990. When are birds dietarily specialized? Distinguishing ecological from evolutionary approaches. Pages 337–352 in Avian foraging: theory, methodology, and applications, (M. L. Morrison, C. J. Ralph, J. Verner, and J. R. Jehl, Jr., Eds.). Stud. Avian Biol. 13.

SKUTCH, A. F. 1972. Studies of tropical American birds. Publ. Nuttall Ornithol. Club. 10.

SKUTCH, A. F. 1982. New studies of tropical American birds. Publ. Nuttall Ornithol. Club. 19.

SLUD, P. 1964. The birds of Costa Rica. Distribution and ecology. Bull. Am. Mus. Nat. Hist. 128:1–430.

TERBORGH, J. W. 1980. Causes of tropical species diversity. Proc. XVII Internat. Ornithol. Congr. (1978): 955–961.

THIOLLAY, J. M. 1988. Comparative foraging success of insectivorous birds in tropical and temperate forests: ecological implications. Oikos 53:17–30.

WERNER, T. K., AND T. W. SHERRY. 1987. Behavioral feeding specializations by *Pinaroloxias inornata*, the "Darwin's Finch" of Cocos Island, Costa Rica. Proc. Nat. Acad. Sci. 84:5506–5510.

WIENS, J. A., AND J. T. ROTENBERRY. 1979. Diet niche relationships among North American grassland and shrubsteppe birds. Oecologia 42:253–292.

APPENDIX

Summary of Niche Differences and Spatial Overlaps among Dead-leafing Species in Two Amazonian Forest Habitats. Above Diagonal Are Measures That Differed Significantly between Species Pairs (K-S or G-tests; $P < 0.01$; See Rosenberg 1990b): H = Foraging Height; S = Substrate; L = Leaf Size; D = Diet Composition; P = Prey Size. Below Diagonal Are Measures of Spatial Association (Cramer's V); * = $P < 0.05$. Species Codes from Table 1. Note that Dietary Comparisons Are Not Possible for CG, PR, or AR; Partly Frugivorous Diet of CN and ER Are Assumed to Differ from Other Species

A. Upland forest

	CN	XG	PE	AI	PS	ML	MH
CN		HSLD	HSLD	HSD	SD	HSLD	HSLD
XG	.000		HSLP	HSLP	HSLDP	HSLD	HSLD
PE	.000	.218		HSP	HSL	HSL	HSLP
AI	.000	.000	.218		HSD	HSDP	HSDP
PS	.000	.667	−.005	−.004		HSP	HSLDP
ML	.000	.000	.000	.000	.000		—

B. Low-lying forest

	ER	XG	CG	PR	AR	AM	AO	AI	PS	ML	MO	TG
CN	HS	HSL	HSD	S	HSL	HSL	HSL	HS	HSD	HSL	HSL	HSL
ER		SL	SD	HS	HSL	HSL	HSLD	HSL	SL	HSL	HSL	HSL
XG	−.005		SLDP	HSLP	HSL	HSLD	HSLD	HSLDP	SLD	HSLDP	HSL	HSLP
CG	.054	−.055		HSDP	HSLDP	HSLDP	HSLDP	HSLDP	HSLDP	HSLDP	HSD	HSLDP
PR	.532*	−.179	.163		HL	H	HLD	H	HS	HSD	HS	HSL
AR	−.046	−.121	.036	−.115		HSD	HD	SL	HSLD	HSLD	HSL	HSLP
AM	.025	.052	.091	−.009	−.138		HS	SD	HS	HSLD	SL	HSD
AO	−.009	.105	.017	−.224	.264	−.038		HSD	HSL	HSL	HSL	HSLDP
AI	−.119	−.185	−.061	.054	−.166	−.186	−.187		HSD	HSL	S	HSP
PS	.100	.250	−.032	.000	.039	.107	.000	−.158		HS	HSL	HSLD
ML	−.300	.033	−.102	−.149	−.091	.195	.164	.153	−.123		HSL	SLDP
MO	−.115	.199	−.291	−.125	.240	.157	.348*	−.164	.000	.118		HSL
TG	−.122	−.129	−.133	−.093	.367	.027	.110	−.133	.000	−.136	.318	

CN	.308
ER	−.099
XG	.011
CG	.088
PR	−.089
AR	−.100
AM	.045
AO	−.121
AI	−.005
PS	−.089
ML	−.225
MO	−.072

BIRDS OF THE TAMBO AREA, AN ARID VALLEY IN THE BOLIVIAN ANDES

C. GREGORY SCHMITT[1], DONNA C. SCHMITT[1], AND
J. V. REMSEN, JR[2]

[1]*Box 15818, Santa Fe, New Mexico 87506, USA, and*
[2]*Museum of Natural Science, 119 Foster Hall, Louisiana State University, Baton Rouge, Louisiana 70803, USA*

ABSTRACT.—We surveyed birds in an arid valley at 1,500 m elevation in the Andes of western depto. Santa Cruz, Bolivia, in January and February, and in June and July. This is one of the only localities with xeric vegetation to be surveyed in an arid intermontane valley of the Eastern Andes. The 50 species breeding or presumed to be breeding in the arid scrub provides a striking example of high species richness at tropical latitudes. Differences between the surveys in January–February ("summer" wet season) and the one in June ("winter" dry season) show strong seasonality in the presence/absence or abundance in 32 of the 92 regularly occurring species. Whereas at least 51 species were breeding or probably breeding during the January–February surveys, only five were breeding or probably breeding during the June–July survey. Subcutaneous fat levels were scored as "no fat" or "low" in 95% of all individuals collected. Most species occurring at this Andean locality have primarily lowland, rather than Andean, geographic distributions.

RESUMEN.—Investigamos las aves en un valle árido a 1,500 m de altura en los Andes al oeste del Departamento de Santa Cruz, Bolivia, en Enero y Febrero, y en Junio y Julio. Esta es una de las únicas localidades con vegetación xerofítica a ser investigadas en un valle árido intermontano en los Andes Orientales. Las 50 especies que se reproducen o que se presume que se reproducen en arbustos áridos nos dá un notable ejemplo de una gran riqueza de especies en latitudes tropicales. Las differencias entre las investigaciones en Enero–Febrero ("verano" estación húmeda) y la de Junio ("invierno" estación seca) muestran una fuerte estacionalidad en la presencia/ausencia o abundancia en 32 de las 92 especies que regularmente están presentes. Considerando que por lo menos 51 especies se estaban reproduciendo o probablemente reproduciendo durante la investigación de Enero-Febrero, solamente seis se estaban reproduciendo o probablemente reproduciendo durante la investigación de Junio–Julio. Los niveles de grasa subcutánea fueron calificados como "sin grasa" o "bajo" en 95% de todos los individuos colectados. La mayoria de las especies que ocurren en esta localidad andina tienen primordialmente distribución geográfica de tierras bajas, en vez de andina.

The eastern slope of the Eastern Andes of South America is best known for its humid cloud-forest. Rain-shadows in some intermontane valleys, however, produce arid conditions that create isolated areas of xerophytic vegetation. Perhaps the best-studied of these areas are the Marañón Valley of northern Peru (Dorst 1957) and the Urubamba Valley of central Peru (Chapman 1921), both of which possess many endemic bird taxa and form important biogeographic barriers to humid forest biota (e.g., Vuilleumier 1984; O'Neill 1992). Other intermontane Andean valleys that do not transect the Eastern Andes have been studied by Miller (1947, 1952).

Although also rich in endemic bird taxa and important as a biogeographic barrier, the Río Grande Valley of central Bolivia has received less attention. This dry, intermontane valley system forms a major barrier to dispersal for humid forest birds of the Andes, few of which occur south of the Río Grande Valley (Remsen et al. 1986, 1987). Within the valley, a mosaic of vegetation types ranges from semi-humid forest to barren scrub. The Río Grande Valley has been disturbed heavily by humans, and so the character of the original vegetation is unknown. Two bird species are endemic to the valley system: *Ara rubrogenys* (Red-fronted Macaw) and *Poospiza garleppi* (Cochabamba Mountain-Finch); five other species whose main distribution is the Río Grande Valley system but which also occur in the nearby Río La Paz or Río Pilcomayo valleys are: *Oreotrochilus adela* (Wedge-tailed Hillstar), *Upucerthia harterti* (Bolivian Earthcreeper), *Sicalis luteocephala* (Citron-headed Yellowfinch), *Poospiza boliviana* (Bolivian Warbling-Finch), and *Oreopsar bolivianus* (Bolivian Blackbird) (Remsen and Traylor 1989, Fjeldså and Krabbe 1990).

METHODS

We conducted intensive avifaunal surveys at site near the Tambo school and the village of San Isidro, a small community of about 30 dwellings at 1,500 m on the Río San Isidro (known locally as the Río Pulquina), roughly 150 km west of the city of Santa Cruz de la Sierra, depto. Santa Cruz. Our study site (hereafter "Tambo"; ca. 18°02'S, 64°25'W; Paynter 1992) was 2.5 km north of the Tambo school, roughly 0.5 km northeast of the El Rancho church.

The area surveyed covered roughly 2 km². Approximately 75% of this area consisted of xeromorphic scrub vegetation (Fig. 1), 25% of cultivated fields bordered by hedgerows, and less than 1% of riparian woodland. The river, estimated to be 15–30 m wide, had created a flat floodplain valley in an otherwise hilly area. The floodplain has been converted completely to cultivated fields (mainly tomatoes, corn, and beans) except for (a) a very narrow (usually just a few meters wide) riparian strip of trees 10–20 m tall and (b) bushes and small trees along hedgerows and irrigation ditches, often with bordering patches of grasses and weeds. The hillsides were covered with a variety of xeromorphic shrubs, small trees, and many columnar cacti; vegetation height was roughly 2–5 m, slightly higher in the arroyos. The vegetation only locally formed dense thickets. Virtually every tree and shrub was thorny. Epiphytes and hemiparasites were present but not conspicuous. Terrestrial bromeliads were abundant on the steepest slopes. Numerous cattle and goats made trails everywhere through the hillsides. Although no rainfall data are available from the region, the similarity of the vegetation physiognomy (Fig. 1) to that found in the Sonoran Desert of North America suggests that annual rainfall is roughly equivalent.

We did not collect plant specimens for identification. However, Michael Nee, who is familiar with the vicinity of the site and with the flora of depto. Santa Cruz, provided us (pers. comm.) with the following predictions on the identifications of the most common arborescent species: (1) riparian trees and shrubs: *Prosopis alba, P. kuntzei,* and *Acacia aroma* (Mimosaceae); *Tipuana tipu* (Fabaceae); *Schinopsis haenkeana* and *Schinus fasciculatus* (Anacardiaceae); and *Vassobia breviflora* (Solanaceae); (2) xeromorphic hillside and dry-wash scrub: the columnar cacti *Harrisia tephracantha* and *Neoraimondia herzogiana*; *Jodina rhombifolia* (Santalaceae); *Condalia* sp. (Rhamnaceae); *Celtis chichape* and *C. spinosa* (Ulmaceae); *Aloysia gratissima* (Verbenaceae); *Capparis retusa, C. speciosa, C. atamisquea,* and *Koeberlinia spinosa* (Capparidaceae); *Gymnosporia spinosa* (Celastraceae); *Achatocarpus praecox* (Achatocarpaceae); and *Porlieria microphylla* (Zygophyllaceae). The terrestrial bromeliad is almost certainly *Deuterocohnia longipetala*.

We surveyed birds during the wet season from 8 to 20 February 1979 and from 15 January to 3 February 1984, and during the dry season from 23 June to 2 July 1984. The general aspect of the vegetation differed strongly between the two study periods, with many trees and shrubs losing their leaves in the dry season. We observed and collected birds daily from dawn to noon, and often again late in the afternoon until dusk. In 1979, we operated 3–6 mist nets in xeromorphic habitats and 1–2 nets in riparian habitats for a total of ca. 530 daytime net-h. In January and February 1984, we operated 3–5 nets in xeromorphic habitats and 1–2 in riparian habitats for a total of ca. 850 daytime net-h. In June–July 1984, we operated 3–5 mist nets in xeromorphic habitats and one in riparian habitats for a total of 441 daytime net-h. All nets were 12-m long.

We placed each species into one of six categories of relative abundance based on the average number of individuals detected (sight, voice, and mist-net capture) per day: "Abundant" (> 50/day), "Common (15–49/day), "Fairly Common" (3–14/day), "Uncommon" (1–2/day), "Rare" (not seen daily, but regularly enough to be assumed to be part of the "core" avifauna, as defined by Remsen [1994]), and "Visitor" (1–2 records, probably not part of core avifauna). No substantial differences in abundance were noted between the January and February (wet season) samples, and so these were combined. Within our limited temporal framework, we doubt that our data would allow us to detect differences in relative abundance of only one of the above categories between samples (January-February vs. June-July); therefore, we only consider between-season differences in abundance of at least two categories to represent real differences. Because our visits were relatively brief and sampled only portions of four of the twelve months of the year, and because we were not intimately familiar with the region's avifauna, we predict that some of our assignments to categories are incorrect.

Specimens were inspected during preparation for evidence of breeding. We considered as "strong" evidence for breeding the presence of an egg with shell in the oviduct, a brood patch,

FIG. 1. Arid scrub on hillsides at study site near Tambo, 1,500 m, depto. Santa Cruz, Bolivia.

a cloacal protuberance, or enlarged ova. We considered enlarged testes as "weak" evidence for breeding. See Appendix 1 for details. Body mass (Appendix 2) was measured using Pesola balances. Specimens are deposited at the Delaware Museum of Natural History, the Museum of Natural Science, Louisiana State University, and the Museo Nacional de Historia Natural, La Paz. Miscellaneous natural history observations of little-known species are summarized in Appendix 3.

RESULTS AND DISCUSSION

We recorded 103 bird species during our three surveys (Table 1), 11 of which, were detected only 1–2 times. One of these 11 species, *Elaenia strepera*, is a long-distance migrant that had presumably just begun migrating from its breeding towards its winter range (Marantz and Remsen 1991). Seven other species may have been wanderers from adjacent regions. The status of three nocturnal species was uncertain. The remaining 92 species were considered the core avifauna for analysis (Table 2).

Of the 92 core species, 50 (54%) were breeding-season or permanent residents in xeromorphic vegetation (scrub and washes; Table 1). Therefore, this structurally simple habitat contains more breeding species than the richest forests at temperate latitudes in North America or elsewhere. For example, the number of species recorded on 43 Breeding Bird Census plots in "broadleaf forests" in 1994 in North America ranged from 11 to 47 with a mean of 27 species (*J. Field Ornith.* 66:S5–S6, 1995); although the plot size of these censuses is smaller than the area that we sampled, we found 47 of the 50 species on our most intensively sampled area, roughly 20 ha. Thus, the species diversity of the Tambo avifauna provides another example of the extraordinary richness of bird communities at tropical latitudes.

Of the core avifauna, 69 species (75%) were detected during both the wet and dry season surveys (Table 1) and were presumably permanent residents. However, 14 (20%) of these showed substantial change in relative abundance between seasons: 13 were classified only as Visitors during the dry season, and one species (*Thraupis bonariensis*) decreased in relative abundance from wet to dry season by two ranks on our scale. Of these 14 species, four changed in status from Rare to Visitor, and so whether these truly changed in abundance is open to question. Four of the 14 species are known to be migratory elsewhere in South America (*Crotophaga ani, Myiophobus fasciatus, Vireo olivaceus, Thraupis bonariensis*). That leaves four species for which we can find no previous documentation of seasonal movements: *Coragyps atratus, Turdus rufiventris, Poospiza nigrorufa,* and *Molothrus badius*.

Of the 23 core species detected during only one season, 15 (16%) were wet-season residents that were breeding or presumed breeding. Of these 15, 13 are known to be migratory elsewhere in South America (*Phalacrocorax brasilianus, Tapera naevia, Hydropsalis brasiliana, Elaenia parvirostris, Myiodynastes maculatus, Tyrannus melancholicus, Empidonomus aurantioatrocristatus, Pachyramphus polychopterus, P. validus, Phytotoma rutila, Catamenia analis, Tiaris obscura, Sporophila caerulescens*; Remsen and Hunn 1979; Fjeldså and Krabbe 1990; Chesser 1994, 1995, 1997; Bates 1997). The two for which we can find no previous report of seasonal movements are *Phaethornis pretrei* and *Euphonia musica*. Concerning the latter, T. A. Parker (in Isler and Isler 1987) had suspected that it moved seasonally in the Andes. Two species, *Empidonax alnorum* and *Catharus ustulatus*, are Nearctic breeding species present at our Bolivian site only during the boreal winter. Six species (*Accipiter striatus, Pyrrhura molinae, Sublegatus modestus, Knipolegus aterrimus, Pygochelidon cyanoleuca,* and *Myioborus brunniceps*) were present only during the dry season; all are known to be make seasonal movements elsewhere in South America (Fjeldså and Krabbe 1990; Chesser 1994, 1995, 1997).

From specimen data, we found strong evidence (see Methods) for wet-season breeding for 37 species and weak evidence for another 14 species (Appendix 1). Because we did not search actively for nests, and because relatively small numbers of specimens (none for many species) were obtained, we regard this as a conservative estimate of the proportion (51 of 83 core species, 62%) of the wet-season avifauna that was breeding. Other than the two Nearctic wintering species, we suspect that virtually all species were breeding during the wet season. In the dry season, the proportion of species that was breeding was significantly lower (Chi-square = 348, $P = 0.0001$): we found strong evidence for breeding for only two (3%) of the 62 core species (*Picoides lignarius, Columbina picui*) and weak evidence for three other species (*Nothoprocta pentlandii, Chlorostilbon aureoventris, Amazilia chionogaster*). Breeding during the dry season is well-known for pigeons and hummingbirds (Skutch 1950). In contrast to the wet season, only a few species (*Amazilia chionogaster, Cranioleuca pyrrhophia, Stigmatura budytoides, Polioptila*

TABLE 1

Birds of the Tambo Area, 1,500 m, Depto. Santa Cruz, Bolivia. Status Refers to the Typical Number Found in Appropriate Habitat within the Study Area in a Morning of Fieldwork, as Follows: "A" = abundant (> 50), "C" = Common (15–49), "FC" = Fairly Common (3–14), "U" = Uncommon (1–2), and "R" = Rare (Not Seen Daily, but Regularly Enough to be Assumed to be Part of the Core Avifauna), and "V" = Visitor (1–2 Records, Presumably Not Part of Core Avifauna). Habitat Codes Are: "Scrub" = Arid Hillside Scrub; "Washes" = Mainly Washes in Desert Scrub (Used for Species Restricted to, or in Much Higher Densities in, Washes Than on Surrounding Hillsides); "Riparian" = Riparian Woods, Thickets; "Fields" = Agricultural Fields or Their Weedy Edges; "Aerial" = Foraging Overhead; "River" = Aquatic Habitat; "Overhead" = Seen Flying over Only, Not Foraging in the Study Area; and "?" = Habitat Not Recorded. Habitat with Highest Density for That Bird Species Is Listed First. Distribution Patterns Are as Follows: "Widespread" = Includes Lowlands and Highlands; "Lowlands" = Widespread in Neotropical Lowlands; "Dry Andes" = Primarily in Dry Montane Regions; "S. Dry Andes" = Dry Andes, Mainly Southern Peru to Northern Argentina; "S. Lowlands" = Lowlands of South America Generally South of Amazonia; "Humid Andes" = Primarily Humid or Semi-Humid Andes; "HA-SL" = Humid Andes and Southern Lowlands; "DA-SL" = Dry Andes and Southern Lowlands; "SDA-SL" = Southern Dry Andes and Southern Lowlands; "N.A. Migrant" = Migrant from North America; "Endemic" = Endemic to Río Grande Valley System

Family/Species	Status Jan.–Feb.	Status June–July	Habitat	Distribution
TINAMIDAE				
Nothoprocta pentlandii	R	U	Washes	Dry Andes
PHALACROCORACIDAE				
Phalacrocorax brasilianus	R	—	River	Lowlands
CATHARTIDAE				
Cathartes aura	FC	U	Aerial	Widespread
Coragyps atratus	U	V	Aerial	Lowlands
Vultur gryphus	R	R	Aerial	Dry Andes
ACCIPITRIDAE				
Accipiter striatus	—	R	Washes	HA-SL
Buteo magnirostris	R	U	Fields, scrub	Lowlands
COLUMBIDAE				
Zenaida auriculata	FC	U	Fields	Lowlands
Columbina picui	C	A	Fields	S. lowlands
Leptotila verreauxi	FC	FC	Washes, riparian	Lowlands
PSITTACIDAE				
Ara rubrogenys	R	V	Overhead	Endemic
Aratinga acuticauda	C	A	Riparian, washes	Lowlands
Aratinga mitrata	U	C	Overhead	Dry Andes
Bolborhynchus aymara	R	R	Scrub	S. dry Andes
Brotogeris chiriri	FC	FC	Overhead	Lowlands
Pyrrhura molinae	—	R	Scrub	S. lowlands
Amazona aestiva	FC	U	Scrub	S. lowlands
CUCULIDAE				
Crotophaga ani	U	V	Fields	Lowlands
Guira guira	U	FC	Fields	S. lowlands
Tapera naevia	R	—	Riparian	Lowlands
STRIGIDAE				
Glaucidium brasilianum	—	V?	?	Lowlands
CAPRIMULGIDAE				
Caprimulgus parvulus	—	R?	Scrub	Lowlands
Hydropsalis brasiliana	U	—	Riparian, scrub	S. lowlands

TABLE 1
CONTINUED

Family/Species	Status Jan.–Feb.	Status June–July	Habitat	Distribution
APODIDAE				
Aeronautes montivagus	—	V?	Aerial	Humid Andes
Streptoprocne zonaris	—	V?	Aerial	Widespread
TROCHILIDAE				
Phaethornis pretrei	U	—	Riparian	S. lowlands
Chlorostilbon aureoventris	R	U	Scrub	S. lowlands
Amazilia chionogaster	FC	C	Scrub	SDA-SL
Lesbia nuna	—	V	Washes	Dry Andes
BUCCONIDAE				
Nystalus chacuru	U	U	Riparian, washes	S. lowlands
PICIDAE				
Picumnus dorbignianus	U	U	Riparian, washes	Dry Andes
Melanerpes cactorum	C	C	Scrub	S. lowlands
Picoides lignarius	U	R	Washes	S. dry Andes
Piculus rubiginosus	V	—	Riparian, washes	Humid Andes
Colaptes melanochloros	R	U	Scrub	S. lowlands
Campephilus leucopogon	R	U	Washes	S. lowlands
DENDROCOLAPTIDAE				
Lepidocolaptes angustirostris	C	FC	Washes	S. lowlands
FURNARIIDAE				
Upucerthia harterti	FC?	FC	Scrub	Endemic
Furnarius rufus	C	FC	Fields, washes	S. lowlands
Synallaxis frontalis	U	U	Riparian	S. lowlands
Cranioleuca pyrrhophia	FC	C	Scrub	S. lowlands
Phacellodomus striaticeps	U?	U	Washes	S. dry Andes
FORMICARIIDAE				
Thamnophilus caerulescens	U	FC	Riparian, washes	Widespread
Thamnophilus ruficapillus	U	R	Riparian, washes	HA-SL
TYRANNIDAE				
Phaeomyias murina	U	V?	Washes	Lowlands
Camptostoma obsoletum	—	V?	Washes	Lowlands
Sublegatus modestus	—	R	Washes	Lowlands
Suiriri suiriri	FC	C	Washes	S. lowlands
Elaenia flavogaster	V	—	Washes	Lowlands
Elaenia parvirostris	C	—	Riparian, scrub	S. lowlands
Elaenia strepera	V	—	Riparian	Humid Andes
Serpophaga munda	U	FC	Riparian, washes	S. lowlands
Stigmatura budytoides	C	C	Scrub	S. lowlands
Hemitriccus margaritaceiventer	U	FC	Scrub, riparian	Lowlands
Tolmomyias sulphurescens	R	—	Washes	Lowlands
Myiophobus fasciatus	FC	V	Fields	Lowlands
Empidonax alnorum	R	—	Riparian	N.A. migrant
Knipolegus aterrimus	—	U	Washes	SDA-SL
Sayornis nigricans	—	V?	Riparian	Humid Andes
Satrapa icterophrys	U	V	Riparian	S. lowlands
Hirundinea ferruginea	U	U	Washes (steep banks)	DA-SL
Pitangus sulphuratus	U	U	Fields	Lowlands
Myiodynastes maculatus	U	—	Washes	Lowlands
Tyrannus melancholicus	U	—	Riparian	Lowlands
Empidonomus aurantioatrocristatus	U	—	Scrub	S. lowlands
Pachyramphus polychopterus	U	—	Riparian	Lowlands
Pachyramphus validus	U	—	Riparian	HA-SL
COTINGIDAE				
Phytotoma rutila	U	—	Washes	SDA-SL

TABLE 1
Continued

Family/Species	Status Jan.–Feb.	Status June–July	Habitat	Distribution
HIRUNDINIDAE				
Pygochelidon cyanoleuca	—	FC	Aerial	Widespread
TROGLODYTIDAE				
Troglodytes aedon	FC	FC	Fields, scrub	Widespread
POLIOPTILIDAE				
Polioptila dumicola	C	FC	Scrub	S. lowlands
TURDIDAE				
Catharus ustulatus	R	—	Riparian	N.A. migrant
Turdus chiguanco	R	R	Riparian, scrub	S. dry Andes
Turdus rufiventris	U	V	?	S. lowlands
Turdus amaurochalinus	C	U	Washes	S. lowlands
VIREONIDAE				
Cyclarhis gujanensis	U	U	Riparian, washes	Lowlands
Vireo olivaceus	FC	V	Riparian	Lowlands
EMBERIZIDAE				
Zonotrichia capensis	A	A	Fields, scrub	Widespread
Lophospingus griseocristatus	A	A	Scrub, fields	S. dry Andes
Poospiza nigrorufa	FC	V	Riparian	S. lowlands
Poospiza torquata	U	FC	Fields, scrub	SDA-SL
Poospiza melanoleuca	FC	FC	Riparian, scrub	S. lowlands
Sicalis flaveola	C	A	Fields, scrub	Lowlands
Sporophila caerulescens	FC	—	Riparian	S. lowlands
Tiaris obscura	C	—	Riparian	Dry Andes
Catamenia analis	U	—	Riparian	Dry Andes
Arremon flavirostris	U	U	Riparian, washes	S. lowlands
Coryphospingus cucullatus	FC	U	Fields, riparian, scrub	S. lowlands
Pheucticus aureoventris	C	FC	Riparian	HA-SL
Saltator aurantiirostris	C	C	Riparian, scrub	SDA-SL
Cyanocompsa brissonii	U	U	Riparian, washes	Lowlands
THRAUPIDAE				
Piranga flava	U	U	Riparian, washes	Widespread
Thraupis sayaca	FC	FC	Riparian	Lowlands
Thraupis bonariensis	A	FC	Washes	DA-SL
Euphonia chlorotica	FC	FC	Riparian	Lowlands
Euphonia musica	R	—	Riparian	Lowlands
PARULIDAE				
Parula pitiayumi	U	FC	Riparian, washes	Widespread
Geothlypis aequinoctialis	R	V	Fields	Lowlands
Myioborus brunniceps	—	U	Riparian	Humid Andes
ICTERIDAE				
Icterus cayanensis	R	V	Washes	Lowlands
Molothrus badius	C	V	Fields	SDA-SL
Molothrus bonariensis	R	V	Fields	Lowlands
FRINGILLIDAE				
Carduelis magellanica	U	U	Riparian	Widespread

dumicola, Saltator aurantiirostris, Parula pitiayumi) sang regularly during the dry season, when dawn was strikingly devoid of bird song.

Of the 88 species for which body mass data were obtained, individuals with subcutaneous fat levels scored as "moderate" (following McCabe 1943) were found in 25 (28 %) species, and as "heavy" in eight (9 %) species (Appendix 2). Of the 708 adult individuals collected, sub-

TABLE 2
SUMMARY OF SPECIES RICHNESS, AND SEASONAL CHANGES OF THE AVIFAUNA AT TAMBO, DEPTO. SANTA CRUZ, BOLIVIA

	Seasons combined	Wet season	Dry season
Total species	103	90	82
Core avifauna	92	83	62

cutaneous fat levels were generally low: only 34 (5%) had moderate fat and only eight (1%) had heavy fat (Appendix 2). No substantial seasonal differences were evident in the proportion of individuals with moderate or heavy fat. Unfortunately, we are not aware of any community-wide samples of tropical birds for comparison. The eight individuals having heavy subcutaneous fat deposits were: *Hirundinea ferruginea pallidior* (1 July), *Catharus ustulatus* (17 February), *Turdus amaurochalinus* (27 June), *Vireo olivaceus chivi* (19 January), *Sporophila caerulescens* (26 January), *Catamenia analis* (26 January), *Euphonia chlorotica* (1 July), and *Carduelis magellanica* (27 January). Five of these species were strongly seasonal at our site, and so the elevated fat levels were presumably associated with migration, but three species (*H. ferruginea, E. chlorotica, C. magellanica*) did not differ in status between our samples.

Of the 88 species with body mass data (Appendix 2), no previous data were listed for 13 species by Dunning (1993), although body mass data had already been published for one (*Upucerthia harterti*) by Remsen et al. (1988). For the vast majority of the other 75 species, however, our data are the first that include differences due to sex and to subcutaneous fat levels. The wide ranges in masses reported by many species by Dunning (1993) are almost certainly due to differences in fat levels, not true individual variation.

Of the 83 species in the breeding avifauna, the majority of species are primarily lowland forms; of these, 27 (33%) are widespread in South American lowlands, and 26 (31%) are found primarily south of Amazonia (Table 1). Although our study site, at 1,500 m, is definitely in the Andes, only 11 core species (13%) are found largely in the Andes, all in dry regions. (The remaining species cannot be assigned easily to exclusively montane or lowland distribution patterns.) Thus, the Tambo avifauna of central Bolivia is primarily derived from the lowlands rather than the adjacent montane regions at higher elevations.

Only two species (*Ara rubrogenys* and *Upucerthia harterti*; see Remsen et al. [1988] for summary of our natural history observations of the latter at Tambo) of the seven endemic to the Río Grande Valley or adjacent dry valleys are found at Tambo; the other five are restricted to higher elevations (see Remsen and Traylor 1989; Fjeldså and Krabbe 1990), as are many other species of birds of the dry Andes whose elevational ranges do not extend as low as 1,500 m. At the subspecies level, however, 11 (14%) of the 81 nonendemic, core breeding species are represented at Tambo by subspecies evidently endemic to the Río Grande Valley system or adjacent dry valleys: *Lepidocolaptes angustirostris hellmayri, Cranioleuca pyrrhophia striaticeps, Thamnophilus caerulescens connectens, Stigmatura b. budytoides, Pitangus sulphuratus bolivianus, Phytotoma rutila angustirostris, Polioptila dumicola saturata, Cyclarhis gujanensis dorsalis, Poospiza h. hypochondriaca, Thraupis bonariensis composita,* and *Molothrus badius bolivianus*.

The results from the few other locality surveys from comparable elevations in Andean dry valleys suggest that derivation of these intermontane avifaunas from lowland avifaunas is a general pattern. Chapman (1917), whose pioneering work on Andean biogeography included the first analyses of avifaunas of arid montane valleys, and Miller (1947, 1952) showed that the avifauna of the upper Magdalena Valley was derived from lowland avifaunas, but the localities that they studied were much lower than Tambo. Chapman's (1921) analysis of the Urubamba Valley avifauna of Peru, which included several localities, and only one as high as 1,500 m, showed that the avifauna of dry habitats there consisted of 58% widespread lowland forms and 29% of species found in lowlands primarily south of Amazonia, i.e. roughly comparable to the affinities of the Tambo avifauna.

Similarly, absence of locality surveys from higher elevations within the Río Grande Valley prevents us from determining whether the avifauna in this valley system becomes more montane in nature gradually with increasing elevation, or whether some threshold elevation is crossed from primarily lowland to primarily montane avifaunas.

ACKNOWLEDGMENTS

We thank the Academia Nacional de Ciencias, La Paz, Gastón Bejarano, and James Solomon for their aid and cooperation in Bolivia. The Delaware Museum of Natural History sponsored the Schmitts' field-work in 1979. Babette C. Odom, John S. McIlhenny, and H. Irving and Laura Schweppe sponsored our field-work in Bolivia in 1984. We are grateful to Bruce D. Glick and Helen Glick for assistance with field-work at Tambo and logistic support. We are grateful to Michael Nee for the plant names and to Manuel Plenge for the Spanish abstract. John Hagan, Steve Hilty, and Francois Vuilleumier provided many helpful comments on the manuscript.

LITERATURE CITED

BATES, J. M. 1997. Distribution and geographic variation in three South American grassquits (Emberizinae, *Tiaris*). Pages 91–110 *in* Studies in Neotropical Ornithology Honoring Ted Parker (J. V. Remsen, Jr., Ed.), Ornithol. Monogr. No. 48.

CHAPMAN, F. M. 1917. The distribution of bird life in Colombia. Bull. Amer. Mus. Nat. Hist. 36:1–169.

CHAPMAN, F. M. 1921. The distribution of bird life in the Urubamba Valley of Peru. Bull. U. S. Nat. Mus. 117:1–134.

CHESSER, R. T. 1994. Migration in South America: an overview of the austral system. Bird Conservation International 4:91–107.

CHESSER, R. T. 1995. Biogeographic patterns of seasonal distribution of austral migrant tyrannid flycatchers. Unpublished Ph.D. thesis, Louisiana State University, Baton Rouge, Louisiana.

CHESSER, R. T. 1997. Patterns of seasonal distribution of austral migrant flycatchers (Tyrannidae) in Bolivia. Pages 171–204 *in* Studies in Neotropical Ornithology Honoring Ted Parker (J. V. Remsen, Jr., Ed.), Ornithol. Monogr. No. 48.

CURSON, J., D. QUINN, AND D. BEADLE. 1994. Warblers of the Americas. Houghton Mifflin, Boston, Massachusetts.

DORST, J. 1957. Contribution a l'étude écologique des oiseaux du haut Marañon (Pérou septentrional). L'Oiseau Revue Fran. Orn. 27:235–269.

DUNNING, J. B. (ED.). 1993. CRC Handbook of Avian Body Masses. CRC Press, Boca Raton, Florida.

FJELDSÅ, J., AND N. KRABBE. 1990. Birds of the High Andes. Zoological Museum, Univ. of Copenhagen, Denmark.

ISLER, M., AND P. ISLER. 1987. The Tanagers. Smithsonian Institution Press, Washington, D.C.

MARANTZ, C. A., AND J. V. REMSEN, JR. 1991. Seasonal distribution of the Slaty Elaenia (*Elaenia strepera*), a little-known austral migrant of South America. J. Field Ornithol. 62:162–172.

MCCABE, T. E. 1943. An aspect of collectors' technique. Auk 60:550–557.

MILLER, A. H. 1947. The tropical avifauna of the upper Magdalena Valley, Colombia. Auk 64:351–381.

MILLER, A. H. 1952. Supplementary data on the tropical avifauna of the arid upper Magdalena Valley of Colombia. Auk 69:450–457.

O'NEILL. J. P. 1992. A general overview of the montane avifauna of Peru. Mem. Mus. Hist. Natural, U.N.M.S.M. (Lima) 21:47–55.

PAYNTER, R. A., JR. 1992. Ornithological Gazetteer of Bolivia. Second edition. Museum of Comparative Zoology, Harvard University, Cambridge, Massachusetts.

REMSEN, J. V., JR. 1994. Use and misuse of bird lists in community ecology and conservation. Auk 111: 225–227.

REMSEN, J. V., JR., AND E. S. HUNN. 1979. First records of *Sporophila caerulescens* from Colombia; a probable long distance migrant from southern South America. Bull. Brit. Orn. Club 99:24–26.

REMSEN, J. V., JR., C. G. SCHMITT, AND D. C. SCHMITT. 1988. Natural history notes on some poorly known Bolivian birds, Part 3. Gerfaut 78:363–381.

REMSEN, J. V., JR., AND M. A. TRAYLOR, JR. 1989. An Annotated List of the Birds of Bolivia. Buteo Books, Vermillion, South Dakota.

REMSEN, J. V., JR., M. A. TRAYLOR, JR., AND K. C. PARKES. 1986. Range extensions for some Bolivian birds, 2 (Columbidae to Rhinocryptidae) Bull. Brit. Orn. Club 106:22–32.

REMSEN, J. V., JR., M. A. TRAYLOR, JR., AND K. C. PARKES. 1987. Range extensions for some Bolivian birds, 3 (Tyrannidae to Passeridae). Bull. Brit. Orn. Club 107:6–16.

RIDGELY, R. S., AND G. TUDOR. 1989. The Birds of South America. Vol. I. Univ. Texas Press, Austin, Texas.

RIDGELY, R. S., AND G. TUDOR. 1994. The Birds of South America. Vol. II. Univ. Texas Press, Austin, Texas.

SHORT, L. L. 1982. Woodpeckers of the World. Delaware Museum Natural History, Greenville, Delaware.

SKUTCH, A. F. 1950. The nesting seasons of Central American birds in relation to climate and food supply. Ibis 92:185–222.

VUILLEUMIER, F. 1984. Zoogeography of Andean birds: two major barriers; and speciation and taxonomy of the *Diglossa carbonaria* superspecies. Nat. Geogr. Soc. Res. Rep. 16:713–731.

WINKLER, H., D. A. CHRISTIE, AND D. NURNEY. 1995. Woodpeckers. An identification guide to the woodpeckers of the world. Houghton Mifflin, Boston, Massachusetts.

APPENDIX 1

Breeding Condition of Birds at Tambo, Depto. Santa Cruz, Bolivia. Breeding Condition: "+" = Enlarged Gonads (Testes ≥2 × 2 mm for Birds <8 g; >7 × 4 mm for Birds <50 g, or >10 × 5 mm for Birds >50 g; Largest Ova >1 mm for Birds <50 g or >2 mm for Birds >50 g; "−" = Nonreproductive Gonads (Testes <2 × 2 mm for Birds <8 g, <4 × 3 mm for Birds <50 g, or <7 × 4 mm for Birds >50 g; Largest Ova <1 mm; "±" = Intermediate Testes Size or for Largest Ovum = 1 mm for Females <50 g, or = 2 mm for Females >50 g. Numbers in Parentheses Refer to Number of Specimens From Each Sex in the Given Condition; ★ = Presence of a Cloacal Protuberance; ● = Presence of a Brood Patch; E = Egg with Shell in Oviduct

Species	Breeding condition	
	January–February	June–July
Nothoprocta pentlandii	−(1♂, 1♀)	+(1♂), ±(1♀), −(1♂)
Phalacrocorax brasilianus	+(5♂♂)	
Columbina picui	+(1♂), −(1♂)	+(2♂♂, 1♀), ±(1♂, 1♀), −(2♂♂, 1♀)
Aratinga acuticauda	±(1♂)	−(1♂)
Aratinga mitrata		
Bolborhynchus aymara		−(2♂♂, 1♀)
Brotogeris versicolurus	−(1♀)	
Amazona aestiva	±(1♀), −(1♀)	
Crotophaga ani	+(1♂), ±1(♀●)	
Guira guira	+(2♀♀), −(1♀)	
Tapera naevia	−(2♂♂)	
Caprimulgus parvulus	−(1♀)	±(1♀)
Hydropsalis brasiliana	+(2♂♂), −(1♀)	
Phaethornis pretrei	±(1♀), −(1♂, 1♀)	+(1♂), −(2♂♂)
Chlorostilbon aureoventris	+(2♂♂), ±(1♂), −(2♀♀)	+(8♂♂), ±(2♀♀), −(2♂♂, 4♀♀)
Amazilia chionogaster	+(2♀♀; 1●), ±(1♂, 1♀)	−(3♂♂, 1♀)
Nystalus maculatus	−(5♂♂, 2♀♀)	−(4♂♂, 1♀)
Picumnus cirratus	+(1♀), ±(6♀♀), −(3♀♀)	±(2♀♀), −(9♂♂)
Melanerpes cactorum	±(2♀♀), −(2♂♂, 1♀)	+(1♀), −(1♂)
Picoides lignarius		−(1♀)
Piculus rubiginosus	−(1♂)	
Colaptes melanochloros		±(1♀), −(1♀)
Campephilus leucopogon		±(2♀♀), −(4♂♂)
Lepidocolaptes angustirostris	+(1♂), ±(2♀♀), −(4♂♂, 2♀♀)	±1(♀), −(2♂♂, 5♀♀)
Upucerthia harterti		

APPENDIX 1
Continued

Species	Breeding condition	
	January–February	June–July
Furnarius rufus	+(1♂), ±(1♀), −(1♂, 2♀♀)	
Synallaxis frontalis	+(3♂♂, 1♀), ±(2♀♀), −(2♀♀)	−(3♂♂)
Cranioleuca pyrrhophia	+(1♂, 2♀♀; 1●), −(2♂♂, 2♀♀)	−(1♂, 1♀)
Phacellodomus striaticeps		−(6♂♂, 3♀♀)
Thamnophilus caerulescens	+(1♂, 2♀♀●), −(1♀)	−(1♀)
Thamnophilus ruficapillus	+(2♀♀; 1●), ±(2♂♂), −(4♂♂)	−(1♂, 1♀)
Camptostoma obsoletum		±(1♀), −(1♂)
Phaeomyias murina	+(1♂), ±(4♂♂), −(1♂, 3♀♀)	−(1♂)
Sublegatus modestus		
Suiriri suiriri	+(1♂), −(1♀)	−(1♂)
Elaenia flavogaster	−(1♀)	−(3♂♂, 2♀♀)
Elaenia parvirostris	+(19♂♂; 2★, 12♀♀, 6●), ±(6♀♀; 1●), −(2♀♀)	
Elaenia strepera	+(1♂)	
Serpophaga munda	−(1♀)	
Stigmatura budytoides	+(1♂, 2♀♀; 1●), ±(2♂♂, 2♀♀; 1●), −(1♂, 5♀♀)	−(4♂♂, 8♀♀)
Hemitriccus margaritaceiventer	−(1♂)	−(1♂, 1♀)
Tolmomyias sulphurescens	−(1♀)	
Myiophobus fasciatus	+(3♂♂, 2♀♀●), ±(5♂♂, 1♀), −(4♂♂, 6♀♀)	
Empidonax alnorum	−(2♀♀)	
Knipolegus aterrimus		−(1♀)
Satrapa icterophrys	+(2♂♂), ±(1♀), −(2♂♂, 1♀)	
Hirundinea ferruginea		−(3♂♂, 1♀)
Pitangus sulphuratus		−(1♂)
Myiodynastes maculatus	+(1♂)	
Empidonomus aurantioatrocristatus	+(1♂)	
Tyrannus melancholicus	+(1♂, 2♀♀●), ±(1♀●), −(1♂)	
Pachyramphus polychopterus	+(1♂★)	
Pachyramphus validus	+(1♂★)	
Phytotoma rutila	+(2♂♂)	
Troglodytes aedon	+(1♂)	−(1♂, 1♀)
Polioptila dumicola	+(1♂), −(1♀)	−(3♂♂, 1♀)
Catharus ustulatus	−(3♂♂, 2♀♀)	
Turdus chiguanco		−(1♀)

APPENDIX 1
CONTINUED

Species	Breeding condition	
	January–February	June–July
Turdus rufiventris	+(4♂♂; 3★, 1♀●)	−(1♀)
Turdus amaurochalinus	+(13♂♂; 4★, 4♀♀; 3●), ±(1♂, 4♀♀), −(4♀♀)	
Cyclarhis gujanensis	±(1♂★)	±(1♀)
Vireo olivaceus		−(1♀)
Zonotrichia capensis	+(17♂♂; 6★, 2♀♀●), ±(1♂, 2♀♀), −(1♂, 2♀♀)	−(9♂♂, 7♀♀)
Lophospingus griseocristatus	+(13♂♂; 7★, 6♀♀●; 1E), ±(2♀♀●)	−(3♂♂, 6♀♀)
Poospiza nigrorufa	+(5♂♂, 5♀♀; 3●), ±(2♂♂, 1♀), −(2♂♂, 2♀♀)	
Poospiza torquata	+(6♂♂; 2★, 1♀●)	
Poospiza melanoleuca	+(4♂♂; 1★), ±(1♂, 1♀), −(1♂)	−(4♂♂)
Sicalis flaveola	+(10♂♂; 5★, 4♀♀; 1●), ±(5♀♀; 3●), −(5♀♀)	
Sporophila caerulescens	+(11♂♂; 4★, 5♀♀; 2●), ±(1♂, 7♀♀), −(4♀♀)	−(4♂♂, 3♀♀)
Catamenia analis	−(3♂♂, 1♀)	
Tiaris obscura	+(5♂♂, 6♀♀), ±(5♂♂), −(1♂)	
Arremon flavirostris	+(2♂♂; 1★, 4♀♀; 2●)	−(1♂, 1♀)
Coryphospingus cucullatus	+(3♂♂, 2♀♀; 1●)	−(1♂, 1♀)
Pheucticus aureoventris	+(11♂♂; 3★, 5♀♀●), ±(2♀♀)	−(1♂, 1♀)
Saltator aurantiirostris	+(7♂♂, 6♀♀; 5●), ±(2♀♀)	−(3♀♀)
Cyanocompsa brissonii	+(4♂♂; 2★, 2♀♀; 1●)	−(2♂♂)
Piranga flava	−(1♀)	
Thraupis sayaca	+(3♂♂; 1★)	−(1♂)
Thraupis bonariensis	+(17♂♂; 8★, 3♀♀; 3●; 1E), −(1♂, 1♀)	−(1♂, 2♀♀)
Euphonia chlorotica	±(2♀; 1●), −(1♂, 2♀♀)	−(3♂♂, 2♀♀)
Euphonia musica	−(1♂)	
Parula pitiayumi		−(1♂)
Geothlypis aequinoctialis	+(2♂♂; 1★, 1♀●), −(2♂♂)	−(1♂)
Myioborus brunniceps		−(2♂♂, 2♀♀)
Icterus cayanensis	±(1♂)	−(1♀)
Molothrus badius	+(11♂♂; 1★, 1♀●), ±(7♀♀)	−(2♀♀)
Molothrus bonariensis		−(1♂)
Carduelis magellanica	+(1♂), ±(1♀), −(2♀♀)	−(1♂)

APPENDIX 2

Body Mass (g) of Birds at Tambo, Depto. Santa Cruz, Bolivia. Body Fat Classified as: "+" = Heavy or Very Heavy Fat (Deep Fat in Feather Tracts, Furcula Area, and Throughout Intestinal Tract); "±" = Moderate Fat (Furcula Area Almost Filled with Fat; Fat Present in Feather Tracts); "—" = No Fat to Only Trace or Light Fat. Individuals in Juvenal Plumage Not Included

Species	♂♂ (N; range) —	±	+	Mean mass (g) —	±	+	♀♀ (N; range) —	±	+
Nothoprocta pentlandii	186 (2; 164–208)	—	—	192 (1)	—	—	—	—	—
Phalacrocorax brasilianus	1400 (1)	—	—	1520 (1)	—	—	—	—	—
Zenaida auriculata	—	—	—	78.2 (1)	—	—	—	—	—
Columbina picui	44 (9; 40.3–51)	—	—	42 (1)	—	—	45 (1)	—	—
Aratinga acuticauda	168 (3; 167.7–172)	—	—	—	—	—	—	—	—
Aratinga mitrata	273.5 (1)	—	—	—	—	—	—	—	—
Bolborhynchus aymara	30.0 (2; 29–31)	—	—	30.5 (1)	—	—	—	—	—
Brotogeris versicolorus	—	—	—	60.0 (1)	—	—	—	—	—
Amazona aestiva	—	—	—	405.0 (2; 402–408)	—	—	—	—	—
Crotophaga ani	108.9 (1)	—	—	84.0 (1)	—	—	—	—	—
Guira guira	—	—	—	122.6 (3; 108.7–139.5)	—	—	—	—	—
Tapera naevia	43.8 (2; 40.0–47.7)	—	—	—	—	—	—	—	—
Caprimulgus parvulus	—	—	—	45.5 (1)	—	—	31.5 (1)	—	—
Hydropsalis brasiliana	—	—	—	4.3 (1)	—	—	—	—	—
Phaethornis pretrei	4.9 (2; 4.8–5.1)	—	—	3.2 (2; 2.8–3.6)	—	—	—	—	—
Chlorostilbon aureoventris	3.4 (3; 3.4–3.5)	3.4 (1)	—	—	—	—	—	—	—
Amazilia chionogaster	5.3 (9; 4.9–6.7)	5.8 (3; 5.6–6.0)	—	4.8 (7; 4.6–5.2)	—	—	4.9 (1)	—	—
Nystalus maculatus	35.3 (4; 32.6–38.3)	—	—	38.7 (4; 38.7–41.7)	—	—	—	—	—
Picumnus cirratus	10.1 (9; 8.5–11.5)	—	—	10.3 (3; 10.1–10.5)	—	—	—	—	—
Melanerpes cactorum	37.8 (10; 31.0–42.0)	—	—	32.3 (14; 26.4–35.2)	—	—	—	—	—
Picoides lignarius	28.2 (2; 28.1–28.4)	—	—	31.2 (4; 28.6–33.2)	—	—	—	—	—
Piculus rubiginosus	—	—	—	66.5 (1)	—	—	—	—	—
Colaptes melanochloros	106.4 (1)	—	—	117.3 (2; 116.8–117.9)	—	—	—	—	—
Campephilus leucopogon	—	—	—	191 (2; 185–198)	—	—	—	—	—
Lepidocolaptes angustirostris	26.7 (9; 23.6–30.0)	—	—	30.3 (4; 24.7–45.1)	—	—	—	—	—
Upucerthia harterti	24.9 (3; 23.6–26.2)	—	—	23 (6; 21.6–23.8)	—	—	—	—	—
Furnarius rufus	39 (5; 31.4–44)	—	—	41.3 (3; 38.0–43.2)	—	—	—	—	—
Synallaxis frontalis	15.1 (4; 13.1–16.8)	—	—	13.3 (6; 12.1–15.3)	—	—	—	—	—

APPENDIX 2
CONTINUED

Species	♂♂ (N; range)		±		Mean mass (g)	♀♀ (N; range)		±		+
Carnioleuca pyrrhophia	13 (9; 12.6–14.5)	—	—	—	13.4 (8; 12.2–14.5)	—	—	—	—
Phacellodomus striaticeps	23.3 (1)	—	—	—	24 (1)	—	—	—	—
Thamnophilus caerulescens	19 (2; 17.2–20)	—	—	—	19 (4; 15.2–20.9)	—	—	—	—
Thamnophilus ruficapillus	20.8 (7; 19.9–21.5)	—	—	—	19.5 (3; 16.3–21.3)	—	—	—	—
Phaeomyias murina	10.4 (6; 7.8–11.7)	—	—	—	8.3 (2; 7.5–9.2)	8.0 (1)	—	—	—
Sublegatus modestus	—	14.4 (1)	—	—	—	—	—	—	—
Suiriri suiriri	15.1 (3; 15.0–15.3)	—	—	—	13.5 (2; 13.0–14.0)	—	—	—	—
Elaenia flavogaster	23.4 (1)	—	—	—	—	—	—	—	—
Elaenia parvirostris	14.0 (19; 11.4–16.8)	—	—	—	13.5 (18; 10.9–16.6)	15.3 (2; 14.3–16.3)	—	±	—
Elaenia strepera	20.5 (1)	—	—	—	—	—	—	—	—
Serpophaga munda	6.8 (1)	—	—	—	5.5 (1)	—	—	—	—
Stigmatura budytoides	11.7 (9; 10.9–12.3)	—	—	—	10.5 (14; 9.4–12.6)	11.7 (1)	—	—	—
Hemitriccus margaritaceiventer	8.1 (2; 7.1–9.2)	—	—	—	—	—	—	—	—
Tolmomyias sulphurescens	—	—	—	—	13.1 (1)	—	—	—	—
Myiophobus fasciatus	10.5 (12; 9.0–17.0)	—	—	—	9.1 (8; 7.1–10.6)	—	—	—	—
Empidonax alnorum	11.0 (2; 10.9–11.2)	—	—	—	—	—	—	—	—
Knipolegus aterrimus	—	—	—	—	21 (1)	—	—	—	—
Satrapa icterophrys	16.8 (4; 15.5–17.8)	—	—	—	17.0 (1)	21.0 (1)	—	—	—
Hirundinea ferruginea	21.2 (1)	24.6 (1)	—	—	—	23.1 (1)	—	—	—
Pitangus sulphuratus	64.5 (1)	—	—	21.5 (1)	—	—	—	—	—
Myiodynastes maculatus	—	55.4 (1)	—	—	—	—	—	—	—
Empidonomus aurantioatrocristatus	22.9 (1)	—	—	—	—	—	—	—	—
Tyrannus melancholicus	38.8 (2; 38.1–39.5)	—	—	—	42.2 (3; 39.0–44.6)	—	—	—	—
Pachyramphus polychopterus	20.9 (1)	—	—	—	—	—	—	—	—
Pachyramphus validus	39.1 (1)	—	—	—	—	—	—	—	—
Phytotoma rutila	40.5 (2; 36.0–45.1)	—	—	—	—	—	—	—	—
Troglodytes aedon	10.0 (2; 9.0–11)	—	—	—	10.3 (1)	—	—	—	—
Polioptila dumicola	6.6 (4; 5.8–7.7)	—	—	—	6.6 (2; 6.1–7.1)	—	—	—	—
Catharus ustulatus	27.5 (2; 26.8–28.3)	28.6 (1)	—	—	24.1 (1)	—	—	—	—
Turdus chiguanco	—	—	—	—	80 (1)	—	—	—	32.4 (1)
Turdus rufiventris	63.9 (3; 63.1–64.6)	—	—	—	77 (2; 68.2–86)	—	—	—	—

APPENDIX 2
Continued

Species	♂♂ (N; range)			Mean mass (g)			♀♀ (N; range)		
		±				+		±	+
Turdus amaurochalinus	57.5 (11; 46.5–64.6)	61.2 (1)		60.8 (9; 54.3–70)			64.2 (1)		58 (1)
Cyclarhis gujanensis	36.1 (1)			32 (1)					
Vireo olivaceus	13.1 (19; 10.1–15.0)	14.6 (1)		12.8 (6; 11.4–14.7)					15.5 (1)
Zonotrichia capensis	20.8 (22; 16.8–24.0)			20 (14; 17.1–31)			24.1 (1)		
Lophospingus griseocristatus	17.8 (11; 15.6–20.0)			18 (14; 13.1–20.2)					
Poospiza nigrorufa	16.0 (6; 13.0–17.4)			18.3 (1)					
Poospiza torquata	10.6 (6; 10.0–11.0)								
Poospiza melanoleuca	12.9 (6; 11.7–13.7)			13.3 (1)					
Sicalis flaveola	15.7 (14; 13.1–17.8)			15 (14; 13.8–17.2)			18.6 (1)		
Sporophila caerulescens	10.2 (11; 9.1–11.6)			9.3 (16; 7.4–11.5)	11.1 (1)				
Catamenia analis	11.6 (2; 11.3–11.9)				13.1 (1)		12.4 (1)		
Tiaris obscura	10.4 (5; 9.5–11.9)			10.9 (10; 8.5–18.3)			11.2 (2; 11.0–11.3)		
Arremon flavirostris	23.1 (3; 20.3–26.0)			25 (4; 23–28.0)					
Coryphospingus cucullatus	12 (3; 11.7–13)			14.2 (3; 12.9–14.9)					
Pheucticus aureoventris	49 (11; 44.8–53)			50 (8; 43.8–57.2)			61.4 (1)		
Saltator aurantiirostris	39 (9; 33.1–44.8)			45 (11; 38–54.3)					
Cyanocompsa brissonii	23.8 (6; 20.0–26.1)			24.6 (1)			29.8 (1)		
Piranga flava	34.5 (1)			27.7 (1)					
Thraupis sayaca	32.2 (4; 29.7–36.0)								
Thraupis bonariensis	30.8 (18; 24.9–36.3)	12.1 (2; 12.0–12.3)		38 (5; 33–45.1)			30.0 (1)		
Euphonia chlorotica	12.5 (1)			11.4 (4; 8.4–13.6)			12.9 (1)		14.3 (1)
Euphonia musica	17.1 (1)								
Parula pitiayumi	7.9 (1)								
Geothlypis aequinoctialis	10.2 (3; 9.6–11.0)	10.9 (1)		13.5 (1)					
Myioborus brunniceps	8.8 (2; 8.1–9.6)			8.5 (2; 8.2–8.9)					
Icterus cayanensis	31.5 (1)			29.5 (1)					
Molothrus badius	47.4 (11; 39.3–61.5)			43.2 (10; 39.9–45.6)					
Molothrus bonariensis	59 (1)								
Carduelis magellanica	10.2 (1)	13.3 (1)					12.2 (1)		12.6 (1)

APPENDIX 3

Natural History Observations on Some Little-known Species of the Tambo Area, Depto. Santa Cruz, Bolivia. Although We Think That Sonagrams Are the Only Valid Means for Comparing Vocalizations, We Did Not Obtain Recordings of Birds from the Study Site Because of Mechanical Problems with Our Tape-recorder. In the Absence of Sonagrams, We Offer Some Transliterations of Voices and Comparisons to Other Descriptions, But We Caution Against Over-Interpretation of Perceived Differences

Amazilia chionogaster. This hummingbird was one of the most common bird species in xeromorphic scrub. Chases and fights were noted frequently among individuals competing for their primary nectar source, the red flowers of an epiphytic or hemiparasitic plant that grew in dense clusters in the scrub. (The plant was most likely a mistletoe, *Ligaria cuneifolia,* or possibly *Tristerix penduliflorus* [M. Nee, pers. comm.].) The vocalizations of this hummingbird were variable and complex. We rarely heard what we though was the "song": a series of about five syllables of hoarse, grating, complex buzzy, "churring" notes sometimes introduced by a few high-pitched, clear, finchlike notes "tseep, tseep, tseep, tseep." This species sometimes also gave low, harsh, buzzy notes at regular intervals, ca. 1/sec, that also might have been a "song." Similar notes were also given more irregularly, as well as a thin "tsip." The differences between our descriptions of the voice and those given by Fjeldså and Krabbe (1990) seem to be more than merely differences in interpretation and transliteration.

Picumnus dorbygnianus. Winkler et al. (1995) found no published descriptions of the voice of this taxon. The "song" is a series of sharp notes, with bouncing-ball rhythm, that is remarkably similar to a faint version of the "song" of *Picoides pubescens* (Downy Woodpecker).

Melanerpes cactorum. The population density of this species seemed to be extraordinarily high for a woodpecker, with 20 to 35 individuals regularly tallied in a morning on the hillsides with columnar cacti near camp. The density was the highest for any woodpecker species in our collective experience. The tops of columnar cacti were favored perches. The typical call was a "throaty," hoarse *whu-hu, whi-hu, whi-huh, whi,* although often only one syllable was given. Also frequently heard was a hoarse *ji-ji* that was similar in quality to calls of many melanerpine woodpeckers. These descriptions are similar to those given by Short (1982).

Picoides lignarius. The descending rattled "song" is remarkably similar to that of North American *P. scalaris.*

Lepidocolaptes angustirostris. The most frequently heard vocalization was a descending, bouncing-ball-rhythm trill: *zee, zew-zew-zew-zew-zew-zew.* It also gave a "liquid" call reminiscent of that of *Myiozetetes similis,* but lower-pitched, and a mournful, tyrannidlike, whistled *wheeeeaa.*

Stigmatura budytoides. Ridgely and Tudor (1994) described the vocalizations and displays performed by duetting pairs. We add that up to 20 phrases are delivered, in total lasting roughly 5 s. The phrases in the songs of birds at Tambo consisted of only four notes, with the second one accented: *ja-JE-je-je.* Duetting birds pointed their tails downwards in contrast to the characteristic horizontal posture during foraging. Individuals occasionally gave a loud *wheep-wheep-wheep-WHEEP?*

Hemitriccus margaritaceiventer: Birds at Tambo gave vocalizations that evidently differ somewhat from those described by Ridgely and Tudor (1994). The vocalization that we heard most frequently was a hoarse, three-noted *chew-chew-chew?* (the last note inflected upwards). We also heard it give a soft, descending, low-pitched, trilled *tsu, TSEE-tse-tse-tsu-tsu.*

Lophospingus griseocristatus. Little has been published on the natural history or voice of this species (Ridgely and Tudor 1989). At Tambo, this was the most abundant bird species, both in arid scrub, where they frequently perched on the tops of columnar cacti, and in weedy fields; more than 100 were tallied most mornings. Flocks typically numbered 10–20 individuals; these were often joined by *Sicalis flaveola, Zonotrichia capensis,* and *Poospiza torquata.* The most frequent call note, difficult to describe, was *tsew,* reminiscent of a *Thraupis* tanager; less frequently given was a more sparrowlike *pick?*

Myioborus brunniceps. The call note of this species is evidently undescribed (Ridgely and Tudor 1989, Curson et al. 1994). Birds at Tambo gave a high-pitched *tsew.*

NOTES ON THE YELLOW-BROWED TOUCANET *AULACORHYNCHUS HUALLAGAE*

THOMAS S. SCHULENBERG[1,2] AND THEODORE A. PARKER III[1,3,4]

[1]*Museum of Natural Science, Foster Hall 119, Louisiana State University, Baton Rouge, Louisiana 70803, USA*
[2]*Present address: Environmental and Conservation Programs, Field Museum, Roosevelt Road at Lake Shore Drive, Chicago, Illinois 60605, USA and Conservation International, 2501 M Street NW, Suite 200, Washington D.C. 20037, USA;*
[3]*Conservation International, 2501 M Street NW, Suite 200, Washington D.C. 20037, USA*
[4]*Deceased*

ABSTRACT.—The Yellow-browed Toucanet *Aulacorhynchus huallagae*, previously known only from one specimen, was rediscovered at the type locality in 1979. Here the species occupies a narrow elevational band (2,100–2,350 m), and there is little elevational overlap with sympatric species of toucan. The species probably occurs both north and south of the type locality, but nonetheless has a very restricted distribution; we are not aware of any immediate threat, however, to the forests in which *A. huallagae* is found. *Aulacorhynchus huallagae* forms a superspecies with two taxa, *A. haematopygus* and *A. coeruleicinctus*, found to the north and south, respectively.

RESUMEN.—El Toucancito de Ceja Amarilla (*Aulacorhynchus huallagae*), que anteriormente solo se conocía de un espécimen, fue redescubierto en la localidad del tipo en 1979. Aquí la especie habita una zona altitudinal muy estrecha (2,100–2,300 m), y hay muy poco traslapo altitudinal con especies simpátricas de toucanes. La especie probablemente se encuentra al norte y al sur de la localidad tipo, sin embargo tiene una distribución muy restringida. No obstante, no estamos enterados si hay alguna amenaza inmediata a los bosques donde *A. huallagae* se encuentra. *Aulacorhynchus huallagae* forma una superespecie con dos especies, *A. haematopygus* y *A. coeruleicinctus*, que se encuentran al norte y al sur, respectivamente.

The Yellow-browed Toucanet *Aulacorhynchus huallagae* was described from a single male specimen collected "on the trail to Utcubamba, in the Huallaga Valley, east of Tayabamba," Peru, on 3 May 1932 (Carriker 1933). The specimen was collected by M. A. Carriker, Jr. from "a small band," the only individuals he saw in the area. The species was not seen again for 47 years, until October–November 1979 when, as members of a Louisiana State University Museum of Natural Science (LSUMZ) expedition, we retraced Carriker's route along the mule trail from Tayabamba to Utcubamba, Depto. La Libertad. Camps were established in relatively undisturbed cloud forest along the trail at "Cumpang" (08°12'S, 77°10'W; 2,625 m) and on a ridge just above the village of Utcubamba (08°13'S, 77°08'W; 2,100 m; the local spelling of the village is now "Uctubamba").

We found *A. huallagae* in the canopy of lush, epiphyte-laden cloud forest dominated by 12–15 m tall trees of the genus *Clusia* (Guttiferae). This species was recorded almost daily by some member of our group from 5 October–9 November. Although usually noted in pairs, the species was also observed on several occasions in small groups of 3–4 individuals. Rugged terrain and prolonged periods of inclement weather prevented us from gathering detailed behavioral data. We twice saw pairs feeding on medium-sized, purple melastome fruits (a *Miconia*?) and once observed an individual probing a cluster of *Clusia* flowers, possibly in search of fruits.

The call of *A. huallagae* typically is a series of 20 to 30 frog-like "*krik*" notes delivered at a rate of slightly more than one note per second (recordings by Parker housed in Library of Natural Sounds, Cornell Laboratory of Ornithology). In response to playback of this type of vocalization, a Yellow-browed Toucanet approached closely and began rhythmically and emphatically moving its head and tail from side to side, with the head and tail moving in opposite directions. This may be a territorial display. The call of *A. huallagae* is very similar to that of *A. coeruleicinctus* (Blue-banded Toucanet), which occupies a similar elevational range in cloud forest from Depto. Pasco, Peru, south to Depto. Cochabamba, Bolivia.

Along the Tayabamba-Utcubamba trail, *A. huallagae* occupied a rather narrow elevational range from 2,100 to 2,350 m, with little overlap in elevational distribution with other sympatric toucans. The larger *Andigena hypoglauca* (Gray-breasted Mountain-Toucan) was fairly common above 2,300 m, and the smaller *Aulacorhynchus prasinus* (Emerald Toucanet) was found only below 2,120 m, where it was uncommon. M. Robbins observed three *A. huallagae* feeding on the purplish fruits of a large fruiting tree in which *Andigena* was seen feeding on another occasion.

We have no explanation for the restricted geographic range of *A. huallagae*. Seemingly suitable habitat within its narrow elevational range occurs along a largely unexplored 550 km-long section of the eastern Andes, interrupted only by the arid inter-montane valleys of the Marañón and Huallaga rivers. The species has not been found in the Cordillera Colán, in depto. Amazonas, just to the southeast of the Río Marañón (pers. obs. T.S.S.). Subsequent to our field work in La Libertad, it has been found in the Río Abiseo National Park, Depto. San Martín (E. Ortíz in litt., Collar et al. 1992). To the south, *A. huallagae* has not been found in the relatively well-surveyed Carpish Mountains in depto. Huánuco just to the north of the Río Huallaga (Tallman 1975;, pers. obs. T.A.P.). Although the species thus does not appear to occupy all of the potentially available forest, little of the area between Colán and the Carpish region is accessible; the geographic limits of its distribution are unknown, and we predict that the species will eventually be found both north and south of the two known localities. The geographic distribution of this bird may be similar to that of other north Peruvian endemic birds, such as *Xenoglaux loweryi* (Long-whiskered Owlet; O'Neill and Graves 1977), *Thripophaga berlepschi* (Russet-mantled Softtail), *Grallaria carrikeri* (Pale-billed Antpitta; Schulenberg and Williams 1982), and *Grallaricula ochraceifrons* (Ochre-fronted Antpitta; Graves et al. 1983), as well as to that of the threatened Yellow-tailed Woolly Monkey *Lagothrix flavicauda* (Parker and Barkley 1980). Most of the area within the range of *Aulacorhynchus huallagae* is relatively lightly settled by humans, and much forest remains; most of the rapidly ongoing deforestation of the Huallaga Valley is occurring below 2,000 m, the lower limit of this toucanet's range. Consequently, unless there is a significant change in human settlement patterns, we see no immediate conservation threat to this species.

We have not compared our four specimens (three males, one female) of *A. huallagae* directly with the type; they agree closely, however, with descriptions of that specimen (Carriker 1933, Haffer 1974). As noted by Haffer (1974), *A. huallagae* combines features of its two presumed closest relatives, *A. haematopygus* (Crimson-rumped Toucanet) and *A. coeruleicinctus* (see Table 1). Even the two "unique" features of *A. huallagae*, the yellow superciliary and bright yellow undertail coverts, represent only modifications of the character states of these features in *A. coeruleicinctus*.

The salient feature of our series of *A. huallagae* is its uniformity. The only character that shows any individual variation is the shape of the maxilla in cross-section. In most specimens, there is a slight "buckling" of the side at the base of the maxilla. One male specimen (LSUMZ 92032) also has a short, shallow groove along the side of the maxilla anterior to the nares. This groove, and to a lesser extent the "buckling," are reminiscent of the maxillary groove of *A. coeruleicinctus*.

The irides of the four recent specimens of *A. huallagae* were described on specimen labels as some shade of "red" (three) or "reddish-brown" (one) (see also the color photograph reproduced in Parker [1990]). The irides of LSUMZ specimens (n = nine) of *A. coeruleicinctus* from Deptos. Pasco, Junín, and Puno, Peru, and Depto. La Paz, Bolivia, were described as pale or creamy yellow. There appears, however, to be a population of *A. coeruleicinctus* in southern Peru with dark irides. The only specimens that we have seen with soft-part color information from the region between Junín and Puno (one from Huanhuachayo, Depto. Ayacucho [LSUMZ 69411] and another from Bosque Aputinye, Depto. Cuzco [LSUMZ 78173]) were described as having "dark brown" and "reddish-brown" irides, respectively. Iris coloration thus appears to show geographic variation independent of present species limits in these toucanets. Additional soft-part color data from *A. huallagae* specimen labels describe the facial skin of two males and one female as "black", and mention a pale spot or area below the eye as being "yellow-white", "blue", or (in the female), "white."

The *A. huallagae* males (LSUMZ) weighed 250 g, 253 g, and 278 g, and had testes measuring 9×4 mm, 6×4 mm, and 8×3 mm, respectively; one male had a brood patch. The female

TABLE 1
COMPARISON OF MORPHOLOGICAL FEATURES OF THREE SPECIES OF *Aulacorhynchus* TOUCANS

Character	A. haematopygus	A. huallagae	A. coeruleicinctus
Bill color	Dark red, mixed with black; basal band white	Basally black or dark gray, shading to whitish-gray tip; basal band white	Basally black or dark gray; shading to whitish-gray tip; no contrasting basal band
Bill shape	Nares on top of maxilla; maxilla slightly concave in cross-section	Nares on side of maxilla; maxilla generally smooth in cross-section	Nares on side of maxilla; maxilla "stepped" in cross-section, caused by longitudinal groove at height of nares
Iris color	Reddish-brown	Red or brown	Pale yellow or reddish-brown (see text)
Superciliary color	Short blue streak; almost lacking	Pale or bright yellow	Bluish-white to blue
Sides of face	Green with blue area at base of mandible	Green	Light blue behind eye, white at base of mandible
Throat color	Green	Dull white washed with green	White
Breast color	Green with faint blue	Green with blue chest band	Green with blue chest band
Color of undertail coverts	Green	Bright yellow	Pale green or yellowish-green
Number of rectrices tipped chestnut	4	4	2–4

weighed 278 g, and its ovary measured 14 × 8 mm; several ova were slightly enlarged. Stomach contents included "green fruit pulp and white seeds 4 mm long" and "small greenish fruits."

Although *Aulacorhynchus huallagae*, *A. coeruleicinctus*, and *A. haematopygus* show relatively slight morphological differences (see Table 1), we recommend that these allopatric taxa be retained as full species within a superspecies until biochemical or behavioral studies clarify their relationships. Future fieldwork may greatly reduce the present distributional gaps between populations of these species. We are unable to explain, for example, the absence of *A. haematopygus* on the eastern slope of the Andes from Colombia to northern Peru, nor of *A. huallagae* from deptos. La Libertad to Pasco, Peru.

ACKNOWLEDGMENTS

We are indebted to John S. McIlhenny, the late Babette M. Odom, H. Irving Schweppe, and the late Laura R. Schweppe for their generous support of the LSUMZ expedition to La Libertad. We thank the Dirección General Forestal y de Fauna of the Ministerio de Agricultura, Lima, for its continuing interest in and support of LSUMZ field studies. Our expedition would not have been possible without the aid of our field assistants, Manuel Sánchez S. and Reyes Rivera A., and the support and companionship of fellow expedition members Linda J. Barkley, Paul K. Donahue, J. William Eley, Mark B. Robbins, and David A. Wiedenfeld. Arturo and Helen Koenig, Manuel and Isabel Plenge, and Gustavo del Solar R. provided invaluable logistic support in Peru. M. Robbins and B. Eley shared their information on *A. huallagae* and other species of *Aulacorhynchus*. The Spanish abstract was prepared by Tyana Wachter and Manuel Plenge.

LITERATURE CITED

CARRIKER, M. A., JR. 1933. Descriptions of new birds from Peru, with notes on other little-known species. Proc. Acad. Nat. Sci. Philadelphia 85:1–38.

COLLAR, N. J., L. P. GONZAGA, N. KRABBE, A. MADROÑO NIETO, L. G. NARANJO, T. A. PARKER III, AND D. C. WEGE. 1992. Threatened Birds of the Americas. International Council for Bird Preservation, Cambridge, United Kingdom.

GRAVES, G. R., J. P. O'NEILL, AND T. A. PARKER III. 1983. *Grallaricula ochraceifrons*, a new species of antpitta from northern Peru. Wilson Bull. 95:1–6.

HAFFER, J. 1974. Avian speciation in tropical South America. Publ. Nuttall Ornithol. Club No. 14.

O'NEILL, J. P., AND G. R. GRAVES. 1977. A new genus and species of owl (Aves: Strigidae) from Peru. Auk 94:409–416.

PARKER, T. A., III. 1990. La Libertad revisited. Birding 22:16–22.

PARKER, T. A., III AND L. J. BARKLEY. 1981. New locality for the yellow-tailed woolly monkey. Oryx 16:71–72.

SCHULENBERG, T. S., AND M. D. WILLIAMS. 1982. A new species of antpitta (*Grallaria*) from northern Peru. Wilson Bull. 94:105–113.

TALLMAN, D. A. 1974. Colonization of a semi-isolated temperate cloud forest: preliminary interpretation of distributional patterns of birds in the Carpish region of the department of Huánuco, Peru. M.S. thesis, Louisiana State University, Baton Rouge, Louisiana.

Orange-eyed Flycatcher, a new species of tyrant-flycatcher (Tyrannidae: *Tolmomyias*) from the western Amazon Basin. From an acrylic painting by John P. O'Neill.

A NEW SPECIES OF TYRANT-FLYCATCHER (TYRANNIDAE: *TOLMOMYIAS*) FROM THE WESTERN AMAZON BASIN

THOMAS S. SCHULENBERG[1] AND THEODORE A. PARKER III[2,3]

[1]*Conservation International, 2501 M Street NW, Suite 200, Washington D.C. 20037, USA and Environmental and Conservation Programs, Field Museum, Roosevelt Road at Lake Shore Drive, Chicago, Illinois 60605, USA;*
[2]*Museum of Natural Science, 119 Foster Hall, Louisiana State University, Baton Rouge, Louisiana 70803, USA and Conservation International, 2501 M Street NW, Suite 200, Washington D.C. 20037, USA*
[3]*Deceased*

ABSTRACT.—A new species of tyrant-flycatcher (*Tolmomyias*) is described from the western Amazon basin. The species occurs in the sub-canopy of river-edge and *várzea* forest, and is syntopic with three other members of the genus (*T. assimilis*, *T. poliocephalus*, and *T. flaviventris*). The plumage of the new species is quite distinctive, but the wing-formula resembles that of *T. sulphurescens*, and the vocalizations of the new species are similar to those of some populations of *T. sulphurescens* as well. Furthermore, the new species appears to be parapatric with *T. sulphurescens insignis*, with which it may show an unusual pattern of "opposite-bank" replacement in *várzea* and river-edge habitats.

Resumen.—Se describe una nueva especie de atrapamoscas (Tyrannidae: *Tolmomyias*) de la parte occidental de la cuenca amazónica. La nueva especie se encuentra en el sub-dosel del bosque del margen de río y *várzea*, y es simpátrica con tres otros miembros del género (*T. assimilis*, *T. poliocephalus*, y *T. flaviventris*). El plumaje de la nueva especie es muy distinctivo, pero la formula de la ala se parece a *T. sulphurescens*, y las vocalizaciones de la nueva especie son parecidas a algunas de las poblaciones de *T. sulphurescens* también. Además, la nueva especie parece ser parapátrica con *T. sulphurescens insignis*, con la cual podría demostrar un patrón poco común de reemplazo de "rivera opuesta" en *várzea* y hábitats de la margen de río.

The remarkable story of the late Ted Parker's role in the discovery of a previously unrecognized species of tyrant flycatcher is by now a well-known part of field ornithologists' folklore. A detailed account has been published elsewhere (Stap 1990), so only a synopsis of the story will be repeated here. On 14 June 1983, along the Río Napo in eastern Peru, Parker noticed two adult *Tolmomyias* feeding young out of the nest. According to his field notes, these birds had "... [a] pale eye, diff[erent] calls (taped); a bird I've never seen"; at the time he was unable to identify them beyond genus. Eight months later, in February 1984, the late John S. Dunning captured and photographed an unusual tyrannid at a site along the Amazon River, downstream from the bustling port of Iquitos. Parker again was not able to name the bird photographed by Dunning, but recognized it as matching the unidentified *Tolmomyias* that he had seen and taped the year before. Subsequently, specimens of this tyrannid were located, and it was determined that these represented a previously unknown species.

The press of other engagements always kept Parker from completing the formal description of this species before his untimely death. We can be sure that what is published here is not quite the paper that Ted would have written, because Parker knew the new species in life as did no other naturalist. Although this description draws on his field notes and tapes, no doubt Ted would have been able to amplify many aspects of the following account.

In accordance with Ted's oft-expressed wishes, we propose to name this new species

Tolmomyias traylori sp. nov.
Orange-eyed Flycatcher

Holotype.—Louisiana State University Museum of Natural Science [LZUMZ] no. 120175; male, taken on the north bank of the Amazon River, 5 km ESE Orán (approximately 03°25′S,

TABLE 1
Measurements (mm) of Five Species of *Tolmomyias* Flycatchers[a]

Species	Wing (chord)	Tail	Culmen (from base)	Tarsus
MALES				
T. traylori (4)	60.3 ± 1.8	49.3 ± 0.8	13.2 ± 0.5	15.8 ± 0.4
	(58.5–62.6)	(48.6–50.2)	(12.6–13.9)	(15.6–16.4)
T. sulphurescens insignis (8)	60.9 ± 2.1	48.0 ± 1.6	14.6 ± 0.6	16.2 ± 0.4
	(59.1–63.4)	(45.3–48.9)	(13.9–15.5)	(15.5–16.6)
T. assimilis clarus, obscuriceps (16)	64.9 ± 1.3	50.5 ± 1.8	14.4 ± 0.5	16.8 ± 0.5
	(61.3–66.9)	(47.5–53.8)	(13.5–15.3)	(16.2–17.7)
T. poliocephalus poliocephalus (11)	56.5 ± 1.7	44.8 ± 2.8	12.6 ± 0.4	15.1 ± 0.5
	(54.3–59.0)	(41.2–49.4)	(12.0–13.5)	(14.2–15.9)
T. flaviventris viridiceps (14)	57.4 ± 1.8	45.3 ± 1.3	12.8 ± 0.2	15.7 ± 0.5
	(54.3–60.5)	(43.1–48.1)	(12.4–13.2)	(15.1–16.9)
FEMALES				
T. traylori (2)	56.3	46.3	13.3	15.4
	(54.9–57.7)	(45.4–47.2)	(13.0–13.6)	(15.1–15.7)
T. sulphurescens insignis (14)	59.7 ± 1.8	48.0 ± 1.8	14.9 ± 0.5	16.8 ± 0.5
	(56.9–63.4)	(45.6–50.8)	(14.3–15.8)	(15.7–17.6)
T. assimilis clarus, obscuriceps (14)	63.2 ± 2.1	48.9 ± 1.9	15.1 ± 0.6	16.6 ± 0.7
	(58.9–66.6)	(45.3–51.3)	(14.0–15.8)	(15.5–17.7)
T. poliocephalus poliocephalus (12)	54.7 ± 2.1	43.3 ± 1.4	12.9 ± 0.6	15.0 ± 0.4
	(52.3–58.1)	(41.1–45.9)	(11.1–13.4)	(14.4–16.0)
T. flaviventris viridiceps (13)	56.2 ± 1.6	44.6 ± 1.7	13.4 ± 0.5	15.5 ± 0.4
	(53.0–58.5)	(42.1–47.5)	(12.7–14.6)	(14.6–15.8)

[a] Mean ± standard deviation, range in parentheses; sample sizes for each sex follow species name.

72°30′W), approximately 85 km NE Iquitos, depto. Loreto, Peru, elevation 80 m; collected 30 July 1984 by A. P. Capparella, field number 2618.

Diagnosis.—A tyrannid, assignable to the genus *Tolmomyias* by the combination of relatively small size; broad, flat bill; lack of strong wing-bars; and lack of "roughening" of the outer web of the outermost primary (Hellmayr 1927). Separable from all known species of *Tolmomyias* by the prominent buff wash across the chest, auriculars, and forecrown. Further separable from *T. flaviventris* by the dusky or grayish crown that contrasts with the color of the back. Wing formula as in *T. sulphurescens* (primary 4 > 10); wing speculum reduced or lacking, as in *T. sulphurescens* and *T. poliocephalus*. Auriculars plain, lacking the dusky tips of *T. sulphurescens*.

Because of the small number of specimens of *T. traylori*, only limited morphometric comparisons can be made (Table 1). Males of *T. traylori* are similar to the two largest species of the genus, *T. sulphurescens* and *T. assimilis*, in body size, as indexed by wing and tail length, with little or no overlap in these characters between *T. traylori* and the two smaller species, *T. poliocephalus* and *T. flaviventris*. *Tolmomyias traylori* is relatively small-billed, however, showing more overlap in this character with *T. poliocephalus* and *T. flaviventris* than with the two larger species.

Description of holotype.—Feathers of forehead cinnamon, tipped olive-gray. Lores buff, near Pinkish-Cinnamon (capitalized color names from Ridgway [1912]), feathers narrowly tipped dusky. Crown Olive-Green. Auriculars and sides of nape yellowish buff, near Empire Yellow or Cinnamon Buff. Center of nape, scapulars, lesser wing coverts, back, rump, and uppertail coverts pale olive-green, near Yellowish Oil Green. Greater wing coverts and remiges Fuscous; outer web of greater wing coverts edged greenish-yellow, near Citron Yellow; outer web of remiges narrowly edged olive-green, near Citron Green; on the innermost secondary, the greenish-yellow outer margin is bordered medially by a narrow stripe of yellowish-white, near Pale Chalcedony Yellow. Inner web of inner primaries and all but innermost two secondaries edged cream. Rectrices fuscous, outer web of all but outermost rectrix edged olive-green, near Olive-Yellow. Chin and throat whitish-buff. Center of breast Ochraceous-Buff, sides of breast slightly duller. Center of belly bright yellow, near Picric Yellow. Lower breast and flanks pale yellow, near Light Green-Yellow or Primrose Yellow. Undertail coverts cream.

Soft part colors in life: irides yellowish-flesh; maxilla black; mandible pale purple; tarsi, toes dull purple.

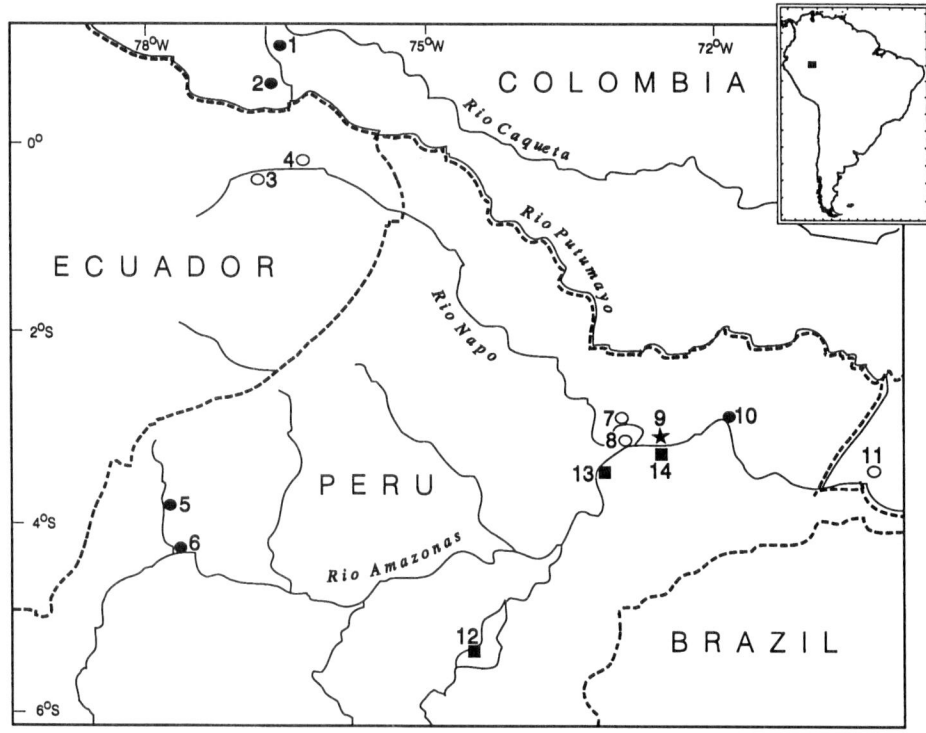

FIG. 1. Northwestern Amazon Basin, showing localities at which *Tolmomyias traylori* has been recorded. Black circles indicate specimen records of *T. traylori*; black star, type locality; open circles, sites documented by tape-recordings or photographs. Localities are: (1) Umbria, (2) San Antonio, (3) Taracoa, (4) Río Mandiyacu, (5) Caterpiza, (6) *boca* Santiago, (7) Sucusari, (8) Explorama, (9) Quebrada Orán, (10) Pebas, and (11) Amacayacu. Localities at which *T. sulphurescens insignis* has been collected are shown by black squares: (12) Victoria, (13) Padre Isla, and (14) Mayococha and Isla Aysana.

Measurements of holotype (mm).—Wing chord 58.5, tail 48.6, culmen from base 13.1, tarsus 15.6, testes 7 × 3 ½; mass 11 ½ g.

Distribution.—*Tolmomyias traylori* occurs (Fig. 1) from the base of the Andes Mountains in southern Colombia and northern Peru (north of the Amazon River) out into the western Amazon basin, east across northernmost depto. Loreto, Peru. In addition to the six localities at which the new species has been collected (see below), the presence of *T. traylori* has been documented at the following sites: ECUADOR—prov. Napo: Taracoa, south bank of the Río Napo, ca. 00°31′S, 76°48′W (tape-recording by Paul Greenfield); prov. Napo: Río Mandiyacu (at La Selva Lodge), north bank of the Río Napo, ca. 00°28′S, 76°21′W (tape recording by David Michael). PERU—depto. Loreto: Explorama Lodge, north bank of the Amazon River, ca. 03°26′S, 72°48′W (photographs of captive bird by John Dunning, archived at Visual Resources for Ornithology [VIREO, Academy of Natural Sciences of Philadelphia] as d01/25/056 and d01/25/057); depto. Loreto: along the lower Río Sucusari, on the north bank of the Río Napo, ca. 03°16′S, 72°54′W (tape recordings by Parker and others, archived at the Library of Natural Sounds [LNS], Cornell Laboratory of Ornithology). COLOMBIA—depto. Amazonas: Parque Nacional Amacayacu, ca. 03°49′S, 70°15′W (tape recording by Bret M. Whitney, pers. comm. to T.S.S.). The limits of the distribution of the new species are not known, and we strongly suspect that the new species will be found at additional localities beyond the currently known range, especially to the north and to the east (see below).

Etymology.—We take great pleasure in naming this species after Melvin A. Traylor, Jr., in recognition of his outstanding contributions to ornithology. Although Mel's interests range widely, our own work in Neotropical ornithology benefitted in particular from his interest in, and revisions of, the tyrannid flycatchers, and from his collaboration with Raymond J. Paynter, Jr. in the production of a series of Neotropical gazetteers. As a sign of Mel's attention to detail and thoroughness, we also note that long ago he singled out as unusual the Field Museum's two specimens of the species that now bears his name.

The English name calls attention to the pale irides of this species, which often have a distinctive orangish tinge in life.

Specimens examined.—T. traylori. We know of a total of seven specimens: COLOMBIA—depto. Putumayo: Umbria, 1000 ft. (= 300 m), 1 ♂ (Academy of Natural Sciences of Philadelphia [ANSP] 160166), collected 7 December 1947 by Kjell von Sneidern; depto. Putumayo: San Antonio, Guamuez, 400 m, two ♂♂ (Field Museum of Natural History [FMNH] 287310, 287311), collected 1 November and 30 October 1969 by Kjell von Sneidern. PERU—depto. Amazonas: Caterpiza, 200 m, 1 ♀ (LSUMZ 99192), collected 27 December 1979 by Nicolás Chigkun N.; depto. Amazonas: *boca* [Río] Santiago, 1 ♀ (American Museum of Natural History [AMNH] 407180), collected 14 January 1929 by J. Schunke; depto. Loreto: 5 km ESE Orán, 1 ♂ (holotype); depto. Loreto: Pebas, 1, sex undetermined (British Museum [Natural History] 1888.1.13.720), collected in 1867 by John Hauxwell.

We have examined hundreds of specimens of *Tolmomyias* (representing the combined series of AMNH, FMNH, and LSUMZ), including representatives of all named taxa in the genus. Listed here are only those specimens from taxa that are sympatric or parapatric with *T. traylori*, and which formed the series on which Table 1 and Fig. 1 are based.

T. sulphurescens insignis (8 ♂♂, 14 ♀♀). PERU—depto. Loreto: Victoria (Canal de Puinahua) (2, FMNH); depto. Loreto: Sarayacu (1, AMNH); depto. Loreto: Padre Isla, 8 km E Iquitos (1, ANSP); depto. Loreto: Mayococha (1, LSUMZ); depto. Loreto: Isla Aysana (1, LSUMZ). BRAZIL—Amazonas: Santo Antonio, Rio Eiru (2, FMNH); Amazonas: Tefé (1, AMNH); Amazonas: Rosarinho, Rio Madeira (6, including holotype, AMNH); Amazonas: Borba (2 AMNH); Amazonas: Rio Madeira (2, AMNH); Amazonas: Rio Negro (1, AMNH); Amazonas: Mirapinima (1, AMNH); Amazonas: Faro (1, AMNH).

T. assimilis (*obscuriceps* and *clarus*) (16 ♂♂, 14 ♀♀). COLOMBIA—depto. Nariño: La Guayacana (1, FMNH). ECUADOR—prov. Napo: Limoncocha (2, LSUMZ); prov. Napo: lower Río Suno (2, AMNH, including holotype of *obscuriceps*); below San José (1, AMNH). PERU—depto. Amazonas: Pomará (3, AMNH); depto. Amazonas: Caterpiza (1, LSUMZ); depto. San Martín: Río Seco (1, AMNH, holotype of *clarus*); depto. Loreto: Libertad (3, LSUMZ); depto. Loreto: Sucusari (1, LSUMZ); depto. Loreto: Río Yanayacu (1, LSUMZ); depto. Loreto: Quebrada Orán (1, LSUMZ); depto. Loreto: Apayacu (2, AMNH); depto. Loreto: Quebrada Vainilla (1, LSUMZ); depto. Loreto: Jeberos (1, AMNH); depto. Loreto: Rio Saimiria (2, ANSP); depto. Loreto: Chamicuros (1, AMNH); depto. Huánuco: Tingo Maria (1, LSUMZ); depto. Ucayali: Yarinacocha (1, LSUMZ); depto. Ucayali: Lagarto (2, AMNH); depto. Ucayali: Santa Rosa (2, AMNH).

T. poliocephalus poliocephalus (11 ♂♂, 12 ♀♀). ECUADOR—prov. Napo: Limoncocha (1, LSUMZ); prov. Napo: Río Suno (2, AMNH). PERU—depto. Amazonas: Pomará (1, FMNH); depto. Loreto: Oceania, Canal de Puinahua (1, FMNH); depto. Loreto: Puerto Indiana (1, AMNH); depto. Loreto: Quebrada Vainilla (3, LSUMZ); depto. Huánuco: Tingo María (1, LSUMZ); depto. Pasco: Cacazú (1, FMNH); depto. Pasco: Puellas (1, FMNH); depto. Ucayali: Balta (4, LSUMZ); depto. Madre de Dios: 5 km NNE Shintuya (1, FMNH); depto. Madre de Dios: Hda. Amazonia (2, FMNH). BOLIVIA—depto. Pando: Camino Mucden (1, LSUMZ). BRAZIL—Amazonas: Iauretê (1, AMNH); Amazonas: Tahuapunto (1, AMNH); Amazonas: Rio Negro (1, AMNH).

T. flaviventris viridiceps (14 ♂♂, 13 ♀♀). COLOMBIA—depto. Putumayo: San Antonio, Guamez (5, FMNH). ECUADOR—prov. Napo: Limoncocha (2, LSUMZ); prov. Napo: below San José (2, AMNH); prov. Napo: upper Río Suno (1, AMNH); prov. Zamora-Chinchipe: Zamora (2, AMNH). PERU—depto. Amazonas: Río Cenepa (1, LSUMZ); depto. Amazonas: La Poza (1, LSUMZ); depto. Amazonas: Caterpiza (2, LSUMZ); depto. Amazonas: *boca* Santiago (1, AMNH); depto. Amazonas: Nazareth (1, LSUMZ); depto. Amazonas: Chiriaco (1, LSUMZ); depto. Loreto: *boca* Río Curaray (1, AMNH); depto. Loreto: Puerto Indiana (2, AMNH); depto. Loreto: Libertad (1, LSUMZ); depto. Loreto: Quebrada Vainilla (1, LSUMZ); depto. Loreto: Sucusari (1, LSUMZ); depto. Loreto: Isla Nesaria (1, LSUMZ); depto. Loreto: 40 km E Iquitos (1, FMNH).

Variation in the paratypes.—The known specimens differ little in most features. In two specimens (AMNH 407180, LSUMZ 99192) the edgings of the greater wing coverts are slightly more ochraceous than in the other specimens. In other members of the genus, immature birds typically are characterized by such wing markings, and it is possible that these two specimens

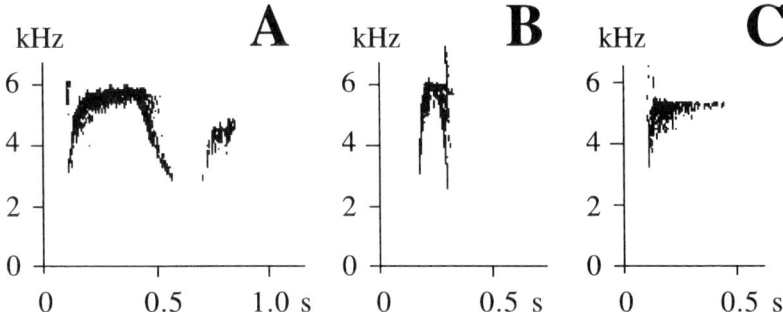

FIG. 2. Vocalizations of *Tolmomyias traylori* and *T. sulphurescens*. (A) Typical call of *T. traylori*, recorded at Sucusari, depto. Loreto, Peru by T. A. Parker (LNS 34313). (B) Call of *T. traylori*, recorded at Sucusari, depto. Loreto, Peru, by T. A. Parker (LNS 30903). (C) Call of *T. sulphurescens* (*insignis*?), recorded at Anavilhanas, Amazonas, Brazil by K. J. Zimmer.

possess juvenal wing coverts. The crown of the AMNH specimen is browner and less gray than the other specimens, which also may be an age-related distinction.

Soft-part colors.—The two color transparencies of a captive individual by John Dunning (VIREO) clearly show the distinctive color of the irides. Although field observers often have referred to this color as "orange," it is closer in color to a pale orangish-brown, possibly paler and more yellowish around the pupil. This is consistent with the description of iris color recorded for the holotype. The iris of the specimen from Umbria apparently also was pale, although it was described as "whitish-gray." In contrast, the iris of the specimen from the mouth of the Río Santiago was recorded as "black." The dark iris on the latter specimen may be another indication that this bird is a juvenile; juvenile *T. poliocephalus* also have dark irides (specimens, FMNH), although the irides of the adults of that species often are pale.

In the transparencies, the tarsi and toes are dull bluish-gray, as in other species in the genus, and the mandible is pale pinkish-brown, possibly more dusky distally.

REMARKS

Ecology and behavior.—Four other species of *Tolmomyias* currently are recognized. Geographic patterns of variation in voice and plumage suggest that at least two of these, *T. flaviventris* (Bates et al. 1992; Ridgely and Tudor 1994) and *T. sulphurescens* (Ridgely and Tudor 1994, pers. obs. T.A.P. and T.S.S.), each contain more than one biological species, but reviews of these taxa are beyond the scope of the present paper. Within much of Amazonia, three or four species of *Tolmomyias* occur syntopically, with a certain amount of segregation by habitat (e.g., Terborgh et al. 1990; Foster et al. 1994; Stotz et al. 1997; Zimmer et al. 1997; pers. obs. T.A.P. and T.S.S.).

Three other species of *Tolmomyias* are found within the known range of *T. traylori*: *T. assimilis*, *T. poliocephalus*, and *T. flaviventris*. At each of the sites from which there are specimens of *traylori*, one or two of the other species also have been collected (AMNH, ANSP, BMNH, FMNH, LSUMZ). None of these sites has been well-surveyed, however, and the sets of species collected syntopically with *traylori* (*assimilis*; *assimilis* and *poliocephalus*; *poliocephalus*; *flaviventris*; *poliocephalus* and *flaviventris*) may not represent the full range of *Tolmomyias* species present at any of these sites. In the vicinity of Sucusari, along the lower Río Napo, eastern Peru, which is the site at which *T. traylori* is best known in life, it is found syntopically with all three other species. Here, *T. assimilis* is found in upland (*terra firme*) forest; *T. poliocephalus* in the canopy of seasonally or permanently flooded forest (*várzea*) bordering rivers and streams, and in advanced second-growth; *T. flaviventris* in *Cecropia*-dominated forest on river islands and in advanced second-growth; and *T. traylori* is largely restricted to the middle-story (subcanopy) of *várzea*, where found in relatively low densities.

Tolmomyias traylori forages in a manner similar to other members of the genus, primarily with sallies (*sensu* Remsen and Robinson 1990) directed at leaves. It sometimes joins mixed-species flocks, although it regularly is found away from flocks as well.

Vocalizations.—Perhaps the most frequently-heard vocalization of *T. traylori* is a distinctive, two-parted call (Fig. 2A). The first note has the shape of an inverted "U," with a very rapid rise and fall, and is followed immediately by a shorter, lower note, giving an emphatic end to

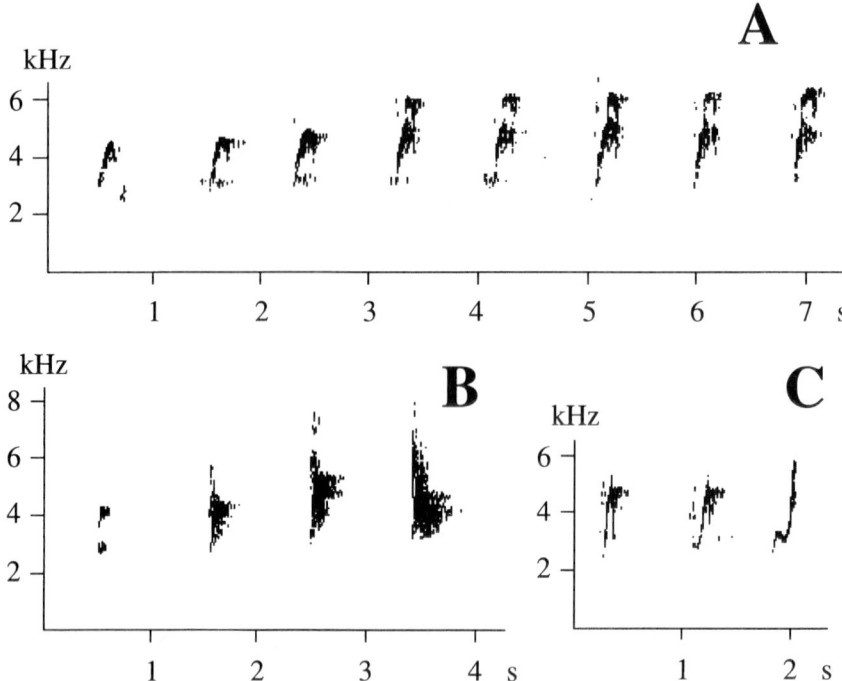

FIG. 3. Vocalizations of *Tolmomyias traylori* and *T. sulphurescens*. (A) Song of *T. traylori*, recorded at Sucusari, depto. Loreto, Peru by T. A. Parker (LNS 35366). (B) Song of *T. sulphurescens* (*insignis*?), recorded at Anavilhanas, Amazonas, Brazil by K. J. Zimmer. (C) Song of *T. sulphurescens pallescens*, recorded at Río Tucavaca, depto. Santa Cruz, Bolivia, by T. S. Schulenberg.

the vocalization. A less frequently heard call is a sharp inverted "V" (Fig. 2B), like a more rapid version of the first part of the preceding vocalization.

What we presume is a song of *T. traylori* is a series of several (three to eight) notes (Fig. 3A). Each note is short and rises sharply, and the entire series increases in intensity and, slightly, in pitch. Both *T. assimilis* and *T. poliocephalus* have comparable vocalizations, but in both species this song differs from that of *T. traylori* in a number of features. The song of *T. assimilis* in much of Peru and southwestern Amazonia (Fig. 4A) consists of a fast series of very buzzy notes. The song of nominate *T. poliocephalus* in western Amazonia (Fig. 4B) contains a series of notes, each of which is relatively long, and later notes in the series typically vary in pitch, 'dipping' slightly following a short initial rise, and often terminating with yet another increase in frequency. The sounds and shape of the notes in the songs of both species are distinctly

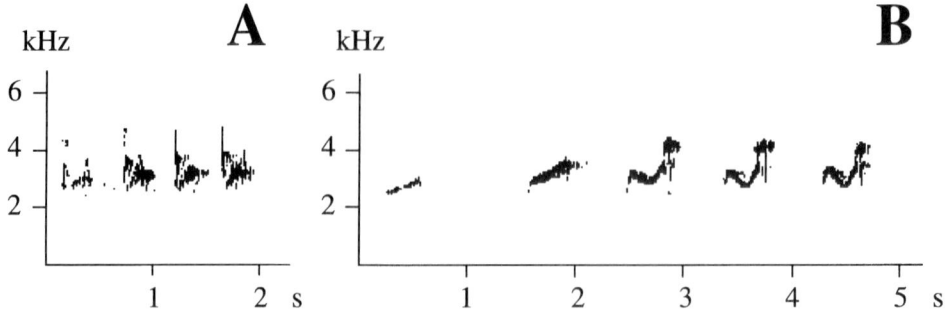

FIG. 4. Vocalizations of *Tolmomyias assimilis* and *T. poliocephalus*. (A) Song of *T. assimilis calamae* recorded at Cachoeira Nazaré, Rôndonia, Brazil by T. S. Schulenberg (LNS 43303). (B) Song of *T. p. poliocephalus*, recorded at Sucusari, depto. Loreto, Peru, by T. A. Parker (LNS 34323).

different from the swift upward sweep of the notes of *T. traylori*, and the songs of both species also are generally lower in pitch than is the song of *T. traylori*.

The songs of *Tolmomyias sulphurescens* show tremendous geographic variation. In some populations, the songs contain series of notes that share some similarities to those in the song of *T. traylori*, in that the notes rise sharply in pitch and reach relatively high frequencies. One such population is *T. sulphurescens pallescens* (Fig. 3C), which enters southern Amazonia in northern Bolivia (Gyldenstolpe 1945) and in Rôndonia, Brazil (specimens, FMNH); birds with vocalizations similar to those of *T. sulphurescens pallescens* also have been recorded at Alto Floresta, Brazil (Zimmer et al. 1997). In contrast, apparent songs of *T. sulphurescens* from the Anavilhanas archipelago on the lower Rio Negro in Brazil (presumably *T. sulphurescens insignis*) are a short series of relatively high, broad amplitude notes (Fig. 3B), which do not "sweep" up from start to finish. Some calls of these birds somewhat resemble a truncated version of the first note of the call of *T. traylori* (Fig. 2C).

Biogeography and systematics.—It may seem remarkable for a new species of bird to be described at such a late date from forests along the Amazon River and its tributaries; these rivers were explored by early collectors as long ago as the mid-nineteenth century, and investigations in the region have continued up to the present day. In fact, only one of the seven known specimens of *T. traylori* was collected after Parker and Dunning had discovered the new species, and the earliest known specimen was taken along the Amazon River 130 years ago (Sclater and Salvin 1867).

We discovered that several institutions with large collections from the western Amazon basin have specimens of the new species. In many cases, the identification of these specimens clearly presented "problems". In his revision of *Tolmomyias*, for example, Zimmer (1939:15) puzzled over the specimen of *T. traylori* from the mouth of the Río Santiago. Zimmer provided an excellent description of the characters of this specimen, but concluded that "For the present I am unable to give it a name"; elsewhere (Zimmer 1939:21) he suggested that this bird might represent a hybrid *T. assimilis* × *T. flaviventris*. Evidently Zimmer's description of this specimen, which serves as a good diagnosis of *T. traylori*, was overlooked, and the common features of these specimens, scattered among different institutions, were not appreciated until after the bird was discovered in life. We expect that more specimens of *T. traylori* may exist in other museums.

All currently known localities are from the north bank of the Amazon River. As presently known, the distribution of the species closely conforms to the outlines of the proposed Napo Refuge (Haffer 1974) or North Amazon (Napo) center of endemism (Cracraft 1985). *Tolmomyias traylori* is unusual, however, in that it is one of the few Napo endemics that inhabits, or is largely restricted to, *várzea* and other riverine habitats. One would expect that *T. traylori* would be found in comparable habitats along the south bank of the Amazon River, but even recent collectors have not found it there. Instead, *T. sulphurescens insignis* has been collected in *várzea* forest and on old river islands in northern Peru along the south bank of the Amazon River; some of these sites are very near to, if not directly across the Amazon from, localities at which *T. traylori* has been found in comparable habitats (Fig. 1).

Interestingly, *Tolmomyias sulphurescens* is absent from the regions on the north bank of the Amazon River from which *T. traylori* is known. Populations of *T. sulphurescens* (subspecies *peruvianus*) are found on the eastern slopes of the Andes, at higher elevations than those from which *T. traylori* is known, in Ecuador and northern Peru. Unlike other members of the genus, however, *T. sulphurescens* is not found in the Amazonian lowlands of eastern Ecuador and northern Peru. There also are very few records of *T. sulphurescens* from southeastern Colombia. One specimen, identified as *T. sulphurescens confusus*, was collected at Aserrío, depto. Caquetá (Nicéforo and Olivares 1975); this locality is about 110 km NE of Umbria, the northernmost Colombian collecting site for *T. traylori*. This specimen, and other possible specimens from eastern Colombia (e.g., Olivares 1962:335), should be re-examined.

This pattern of distribution raises the intriguing possibility that *T. sulphurescens insignis* and *T. traylori* replace each other in river-edge habitats. Although "opposite-bank" species replacements across large rivers are a well-known phenomena in Amazonia (Snethlage 1913, Capparella 1987, Haffer 1992), the species involved usually are understory species of upland forest. Examples of pairs of Amazonian river-edge and *várzea* species that replace each other across a river are quite rare. In northern Peru, for example, there are indications (specimens, LSUMZ, FMNH) that the two species of *Hypocnemoides* (*H. melanopogon* and *H. maculicauda*), both of which are closely tied to riverine habitats, replace each other across the Amazon River, in the same region in which *T. traylori* and *T. sulphurescens insignis* do so. Elsewhere, however, the distributional patterns of replacement between the two *Hypocnemoides* species may be more complicated, or are at least less clear (Gyldenstolpe 1951).

In any event, in the case of the *Tolmomyias*, the pattern of distribution at both the small scale, with each taxon found at sites only a few kilometers apart, but on opposite banks of the Amazon, and at the large scale, with *T. traylori* occupying a geographic region in Amazonia in which *T. sulphurescens* is absent, is consistent with an interpretation of "opposite-bank", river-edge replacement. If this pattern of replacement between these taxa across the Amazon River in northern Peru holds true elsewhere in Amazonia, then it might be possible to make some inferences about the extent of the area in which *T. traylori* is found. Unfortunately, the north bank of the western Amazon River is a very poorly-collected region (e.g., Haffer 1974:fig. 7.1; Oren and Albuquerque 1991). In addition to the Peruvian localities mentioned above, *T. sulphurescens insignis* has been collected at several sites along the south bank of the Amazon in western Brazil (São Paulo de Olivença; opposite Tonantins; Tefé; Zimmer 1939, specimens, Carnegie Museum of Natural History [CM]). Although *T. traylori* has been recorded as far east as Parque Nacional Amacayacu, in extreme southeastern Colombia, neither species has been collected along the north bank of the Amazon River between Pebas, Peru (*T. traylori*) and Manacapurú, Brazil (*T. sulphurescens insignis*; specimens, CM). The two species may replace each other somewhere between these two sites, possibly across the lower Río Caquetá (Rio Japurá); this river appears to be the site of "opposite-bank" replacements between some primate taxa (Hershkovitz 1977, 1988).

The distinctiveness of the plumage features and vocalizations of *Tolmomyias traylori* leave us in no doubt that it should be recognized as a species. The similarities of *T. traylori* to *T. sulphurescens* in wing formula, the pattern of distribution of *T. traylori* (i.e., at least in part replacing *T. sulphurescens*), and the similarities in voice between *T. traylori* and some populations of *T. sulphurescens*, all lead us to conclude that *T. traylori* is closely related to what may be called the "*T. sulphurescens* complex". Further resolution of the affinities of *T. traylori* must await a full revision of *T. sulphurescens*, and of other members of the genus.

ACKNOWLEDGMENTS

We are grateful to the curators and staff at the following museums, all of whom allowed their specimens to be under our care for an extended time during the preparation of the manuscript: Academy of Natural Sciences, Philadelphia; American Museum of Natural History, New York; British Museum (Natural History), Tring, United Kingdom; Field Museum of Natural History, Chicago; and Louisiana State University Museum of Natural Science, Baton Rouge. Doug Wechsler (VIREO) and Greg Budney (Library of Natural Sounds, Cornell Laboratory of Ornithology) kindly provided additional material and information from their collections, as did Mario Cohn-Haft, David Michael, Paul Greenfield, Bob Ridgely, Bret M. Whitney, and Kevin Zimmer. Angelo Capparella, Kenneth C. Parkes (Carnegie Museum of Natural History), Mark Robbins, and Michael Walters (British Museum [Natural History]) all responded to requests for information. The map was prepared by Jodi Sedlock and Clara Simpson, Shannon Hackett prepared the final versions of the remaining figures, and Tyana Wachter provided the Spanish abstract. Jaqueline Goerck made Parker's original field notes available to the senior author, which greatly facilitated the production of the manuscript. The manuscript benefitted from reviews by John Bates, Laurie Binford, A. Capparella, M. Cohn-Haft, S. Hackett, Van Remsen, Doug Stotz, and K. Zimmer. Parker's field work at Sucusari, where most observations of the new species were made, was supported in part by Peter Jenson, Explorama Tours, and Victor Emanuel Nature Tours.

LITERATURE CITED

BATES, J. M., T. A. PARKER III, A. P. CAPPARELLA, AND T. J. DAVIS. 1992. Observations on the *campo, cerrado* and forest avifaunas of eastern Dpto. Santo Cruz, Bolivia, including 21 species new to the country. Bull. Brit. Ornithol. Club 112:86–98.

CAPPARELLA, A. P. 1987. Effects of riverine barriers on genetic differentiation of Amazonian forest undergrowth birds. Ph.D. dissertation, Louisiana St. Univ., Baton Rouge.

CRACRAFT, J. 1985. Historical biogeography and patterns of differentiation within the South American avifauna: areas of endemism. Pages 49–84 in Neotropical Ornithology (P. A. Buckley, M. S. Foster, E. S. Morton, R. S. Ridgely, and F. G. Buckley, Eds.). Ornithol. Monogr. No. 36.

FOSTER, R. B., T. A. PARKER III, A. H. GENTRY, L. H. EMMONS, A. CHICCHÓN, T. SCHULENBERG, L. RODGRÍGUEZ, G. LAMAS, H. ORTEGA, J. ICOCHEA, W. WUST, M. ROMO, J. A. CASTILLO, O. PHILLIPS, C. REYNAL, A. KRATTER, P. K. DONAHUE, AND L. J. BARKLEY. 1994. The Tambopata-Candamo Reserved Zone of southeastern Perú: a biological assessment. Rapid Assessment Program Working Papers No. 6. Conservation International, Washington D.C.

GYLDENSTOLPE, N. 1945. A contribution to the ornithology of northern Bolivia. Kungl. Svenska Vet.-Akad. Handl.(3) 23:1–300.
GYLDENSTOLPE, N. 1951. The ornithology of the Rio Purús region in western Brazil. Arkiv Zool. 2:1–320.
HAFFER, J. 1974. Avian speciation in tropical South America. Publ. Nuttall Ornithol. Club 14.
HAFFER, J. 1992. On the "river effect" in some forest birds of southern Amazonia. Bol. Mus. Paraense Emílio Goeldi 8:217–245.
HELLMAYR, C. E. 1927. Catalogue of birds of the Americas. Part V. Tyrannidae. Field Mus. Nat. Hist., Zool. Ser., 13.
HERSHKOVITZ, P. 1977. Living New World Monkeys (Playrrhini). University of Chicago Press, Chicago.
HERSHKOVITZ, P. 1988. Origin, speciation, and distribution of South American titi monkeys, genus *Callicebus* (Family Cebidae, Platyrrhini). Proc. Acad. Nat. Sci. Philadelphia 140:240–272.
NICÉFORO M., H., AND A. OLIVARES. 1975. Adiciones a la avifauna Colombiana, VI (Tyrannidae-Bombicillidae). Entrega A. Lozania (Acta Zool. Colombiana) No. 19.
OLIVARES, A. 1962. Aves de la region sur de la Sierra de la Macarena, Meta, Colombia. Rev. Acad. Colombiana Cienc. Exactas, Fisicas Nat. 11:305–346.
OREN, D. C., AND H. G. DE ALBUQUERQUE. 1991. Priority areas for new avian collections in Brazilian Amazonia. Goeldiana (Zoologia) No. 6.
REMSEN, J. V., JR., AND S. K. ROBINSON. 1990. A classification scheme for foraging behavior of birds in terrestrial habitats. Pages 144–160 *in* Avian foraging: theory, methodology and applications (M. L. Morrison, C. J. Ralph, J. Verner, and J. R. Jehl, Jr., Eds.). Stud. Avian Biol. 13.
RIDGELY, R. S., AND G. TUDOR. 1994. The Birds of South America, vol 2. University of Texas Press, Austin.
RIDGWAY, R. 1912. Color Standards and Color Nomenclature. Washington, D. C.
SCLATER, P. L., AND O. SALVIN. 1867. List of birds collected at Pebas, upper Amazons, by Mr. John Hauxwell, with notes and descriptions of new species. Proc. Zool. Soc. London 1867:977–981.
SNETHLAGE, E. 1913. Über die Verbreitung der Vogelarten in Unteramazonien. J. Ornithol. 61:469–539.
STAP, D. 1990. A Parrot Without a Name: The Search for the Last Unknown Birds on Earth. Alfred A. Knopf, New York.
STOTZ, D. F., S. M. LANYON, T. S. SCHULENBERG, D. E. WILLARD, A. T. PETERSON, AND J. W. FITZPATRICK. 1997. Results of an avifaunal survey of the middle Rio Jiparaná. Rondônia, Brazil. Pages 763–781 *in* Studies in Neotropical Ornithology Honoring Ted Parker (J. V. Remsen, Jr., Ed.), Ornithol. Monogr. No. 48.
TERBORGH, J., S. K. ROBINSON, T. A. PARKER III, C. A. MUNN, AND N. PIERPONT. 1990. Structure and organization of an Amazonian forest bird community. Ecol. Monogr. 60:213–238.
ZIMMER, J. T. 1939. Studies of Peruvian birds. No. XXXIII. The genera *Tolmomyias* and *Rhynchocyclus* with further notes on *Ramphotrigon*. Amer. Mus. Novitates 1045.
ZIMMER, K. J., T. A. PARKER III, M. L. ISLER, AND P. R. ISLER. 1997. The avifauna of the Alta Floresta region, Mato Grosso, Brazil. Pages 887–918 *in* Studies in Neotropical Ornithology Honoring Ted Parker (J. V. Remsen, Jr., Ed.), Ornithol. Monogr. No. 48.

NOTE ADDED TO PROOF:

Tolmomyias traylori recently has been reported at a third site in Ecuador, in the Kapawi Ecological Reserve (ca. 02°33′S, 76°51′W) on the upper Río Pastaza (David Michael, pers. comm.).

BROMELIAD FORAGING SPECIALIZATION AND DIET SELECTION OF *PSEUDOCOLAPTES LAWRENCII* (FURNARIIDAE)

T. SCOTT SILLETT[1,3], ANNE JAMES[2,4], AND KRISTINE B. SILLETT[1,3]

[1]*Museum of Natural Science, Foster Hall 119, and Department of Zoology and Physiology, Louisiana State University, Baton Rouge, Louisiana 70803, USA*
[2]*Department of Ecology, Evolution, and Organismal Biology, Tulane University, New Orleans, Louisiana 70118, USA*
[3]*Present address: Department of Biological Sciences, Dartmouth College, Hanover, New Hampshire 03755, USA*
[4]*Present address: Cornell Lab of Ornithology, 159 Sapsucker Woods Road, Ithaca, New York 14850, USA*

ABSTRACT.—Over 50 species of Neotropical birds have been recorded foraging for animal prey in bromeliads. Of these bird species, *Pseudocolaptes lawrencii* is one of the most specialized. At a montane rainforest site in Costa Rica, 74% of its documented foraging efforts were in epiphytic bromeliads. *P. lawrencii* selected large bromeliads and foraged for arthropods within leaf litter and organic debris trapped in the plants. Based on our analyses of the bromeliad prey base and bird stomach contents, *P. lawrencii* was an opportunistic predator of the litter-inhabiting arthropods. Birds consumed dermapterans, orthopterans, arachnids, and coleopterans in proportions equal to the prey's availability and did not select for prey size. However, *P. lawrencii* avoided isopods. *P. lawrencii* did not consume aquatic insect larvae, which were the largest component of the bromeliad prey base and occurred in 80% of bromeliads sampled.

RESUMEN.—Mas de 50 especies de aves Neotropicales han sido estudiadas mientras forrajean por animales en bromelias. De estas especies de aves, *Pseudocolaptes lawrencii* es una de las mas especializadas. En un bosque tropical montañoso en Costa Rica, se encontró que el 74% de los atentos de forrajeo fueron en bromelias epifiticas. *Pseudocolaptes lawrencii* selecciono bromelias grandes y forrajeo en las hojas y los escombros orgánicos atrapados en las bromelias. Basado en nuestro análisis de los animales encontrados en las bromelias y el contenido estomacal de las aves, concluimos que *P. lawrencii* es un predator oportunistico de los artrópodos que viven en los escombros orgánicos. Las aves consumieron dermapteros, orthopteros, aracnidos y coleopteros en proporciones idénticas a la disponibilidad de estas presas. No se encontró selección de presas, sin embargo, *P. lawrencii* evito isopodos. No consumió larvas de insectos acuáticos, la presa mas abundante y se encontró en un 80% de las bromelias estudiadas.

Among the masses of epiphytes that give Neotropical montane forests their "fantastic appearance" (Slud 1964:205), bromeliads are often the most conspicuous plants. Bromeliads increase the structural complexity of forests and create additional microhabitats for birds and their animal prey. Indeed, a diverse fauna exists within the impounded water and detritus of tank bromeliads (e.g., Picado 1911, Pittendrigh 1948, Laessle 1961, Diesel 1989, Paoletti et al. 1991), consisting of two primary components: animals living within the aquatic medium (e.g. dipteran larvae, frogs) and animals typically associated with soil and organic debris (e.g., earwigs [Dermaptera], roaches [Orthoptera], isopods). Thus, bromeliads can enhance opportunities for resource subdivision and specialization by birds in Neotropical forests. Foraging specialization on unique tropical resources, such as bromeliads, is thought to be one mechanism responsible for the high bird species diversity of the Neotropics relative to the Temperate Zone (Schoener 1968; Orians 1969; Karr 1971; Terborgh 1980; Remsen 1985). At least 51 Neotropical bird species have been recorded foraging for animal prey in bromeliads (Appendix). Nine species appear to be specialized on bromeliad foraging; most of these belong to the Dendrocolaptidae and Furnariidae.

Tuftedcheeks (Furnariidae: *Pseudocolaptes lawrencii*, *P. boissonneautii*, and *P. johsoni*) are among the most specialized of bromeliad-foraging birds. They occur in wet montane forests of southern Central America and the Andes (Slud 1964; Hilty and Brown 1986; Fjeldså and Krabbe 1990; Ridgely and Tudor 1994; Sillett 1994). At a montane rainforest site in Costa Rica, 74% of foraging observations of *P. lawrencii* were in arboreal bromeliads, and nearly 99% of its foraging efforts were in epiphytes of one type or another (Sillett 1994).

In this paper, we present further data on the natural history and foraging ecology of *P. lawrencii*. We focus on the bird's use of and selectivity for the bromeliad resource base to determine if the bird specializes on particular prey types, prey sizes, and bromeliad sizes. The null hypothesis we test is that *P. lawrencii* uses bromeliad resources in proportion to their availability.

STUDY AREA AND METHODS

Our research was conducted in the Cordillera de Talamanca, Costa Rica, near Villa Mills and the Pensión La Georgina (83°40'W longitude, 9°30'N latitude; hereafter "La Georgina"), approximately 95 km south of San José along the Pan American Highway. All data were collected from 3 July to 11 August 1991. We worked in a 4-km^2 area of montane rainforest (Holdridge 1967) between 2,800 and 3,100 m elevation, near the transition zone from oak forest to páramo vegetation. Trees are covered with diverse epiphytic vegetation, including bryophytes, lichens, and tank bromeliads (species of *Guzmania*, *Vriesea*, and *Tillandsia* [Burt-Utley and Utley 1977]). *Quercus costaricensis* is the dominant canopy tree. A more complete description of the study site is given by Sillett (1994).

To compare *P. lawrencii*'s selection of bromeliad sizes to the bromeliad size distribution available at La Georgina, we collected foraging observations and conducted vegetation surveys. Foraging data were gathered on opportunistically encountered birds. We took only one observation per individual bird per day to minimize sequential observations and to avoid serial correlation problems (Martin and Bateson 1986; Hejl et al. 1990). Bromeliad size (diameter across the top of each plant's rosette of leaves) was estimated for every bromeliad in which *P. lawrencii* was observed foraging. To quantify the available bromeliad size distribution, we randomly selected 120 points in the oak forest at La Georgina. At each point, imaginary 1-m diameter cylinders were delineated, extending from ground to forest canopy. We estimated sizes of all bromeliads encompassed by the cylinders. Bromeliads were classified into three size categories before data analyses: small (1–30 cm), medium (31–60 cm), and large (>60 cm). More detailed descriptions of methods used to gather foraging and vegetation data are given in Sillett (1994).

Ten foraging *P. lawrencii* were collected with shotguns in the vicinity of La Georgina for analysis of stomach contents. Birds were prepared as either study skins or skeletons; tissue samples from each bird were preserved in liquid nitrogen. Stomach samples were preserved in 70% ethanol as soon as possible after collection. All specimens, as well as tissue and stomach samples, were deposited in the Louisiana State University Museum of Natural Science. Stomach contents were sorted and identified to Class or Order under a dissecting microscope. Minimum numbers of prey items in each category were determined from diagnostic fragments (e.g. mouthparts, heads, and wings). Arthropod fragments were identified using illustrations in Ralph et al. (1985), Moreby (1987), Borer et al. (1989), and Chapman and Rosenberg (1991), and then measured with the microscope's optical micrometer. Fragment size was converted to prey size using regression equations in Calver and Wooller (1982), K. V. Rosenberg (unpublished data), and an equation determined for Dermaptera in the present study (body length = 0 + 3.02 × [cercus length]; $R^2 = 0.93$; 20 animals measured). We believe that with knowledge of the particular fragments representing different types of arthropods, we were able to detect hard-bodied and soft-bodied prey equally well. However, the potential biases associated with differential digestion of hard-versus soft-bodied prey are poorly understood (Rosenberg and Cooper 1990 and references therein).

We collected 45 tank bromeliads from randomly selected locations in the study site to quantify the bromeliad prey base. All bromeliads collected were attached to trees and within 2.5 m of the ground. To quantify the bromeliad prey base encountered by foraging *P. lawrencii*, we sampled a size distribution of bromeliads comparable to the size distribution selected by the bird. Bromeliads were placed in plastic bags immediately upon collection to minimize escape of arthropods. Before sealing the bags, a small amount of insecticide was sprayed inside to kill any flying insects. We opened bags in a large wash tub within 24 hr of collection and measured each bromeliad across the top of the rosette of leaves. We then carefully dissected the bromeliads, collected all animals encountered and preserved them in 70% ethanol. Arthropods were identified to Class or Order and measured under a dissecting microscope. Insect larvae were classified as either terrestrial (larvae found in impounded dry leaf litter and detritus, most of which were Coleoptera and Lepidoptera) or aquatic (larvae found in impounded water and wet detritus, primarily Diptera, e.g., Syrphidae, Ceratopogonidae).

We used the Brillouin diversity index, H (Hurtubia 1973; Pielou 1975; Sherry 1984), to assess if our samples of *P. lawrencii* stomachs and bromeliad contents adequately represented the diversity of prey types consumed by the bird and available at La Georgina.

$$H = \left(\frac{1}{P}\right) \times \ln\left(\frac{P!}{(p_1! \times p_2! \times \ldots \times p_n!)}\right)$$

where there are p prey items in each of n different prey categories, with P total prey items per sample (Pielou 1975). To calculate H, samples were taken in random order, and the diversity of prey items was computed for sample 1, then for samples 1 + 2 (contents pooled), and so on through the total number of stomach or bromeliad samples. The saturation curves generated by these calculations become asymptotic if enough samples exist to characterize prey composition (Sherry 1984).

We conducted a series of statistical tests to measure specialization by *P. lawrencii* on bromeliad resources. Statistics were calculated using JMP (SAS Institute 1994). The null hypothesis for all tests was that use of bromeliad resources by *P. lawrencii* equaled resource availability. We considered the bird to specialize on, or be selective of, a resource when use was significantly greater than availability by 10%. A resource was classified as avoided by *P. lawrencii* when use was significantly less than availability by 10%. We tested for a difference between selection of bromeliad size classes by *P. lawrencii* and available size classes at La Georgina with a Pearson χ^2 test. Multiple analysis of variance (MANOVA) was used to test first for an overall difference between bird diet and available bromeliad-inhabiting prey, comparing prey composition and prey size. Four prey types (i.e. Dermaptera, Orthoptera, Arachnida, and Coleoptera) were sufficiently common in both stomachs and bromeliad samples to use in assessing prey size-selectivity by *P. lawrencii*. Bird use of individual prey types was compared to those available in La Georgina bromeliads with one-way analyses of variance (ANOVA). Individual bird stomachs and bromeliad samples were treated as replicates, and proportion data were arcsine-transformed before statistical analysis. Homogeneity of treatment variances (i.e. "used" by *P. lawrencii* and "available" in bromeliads) was assessed with Levene's test (Milliken and Johnson 1984). We used Welch's ANOVA (Welch 1951; Milliken and Johnson 1984) when treatment variances were heterogeneous.

RESULTS

Diversity of available prey generally increased with increasing bromeliad size (Fig. 1), as did mean number of prey per bromeliad (Welch ANOVA, $F_{2, 11.49} = 15.25$, $P = 0.0006$). Mean prey size, however, did not change with bromeliad size (ANOVA, $F_{2, 67} = 0.0046$, $P = 0.99$). Prey-type diversity saturation-curves became asymptotic for stomachs of *P. lawrencii* and medium and large bromeliads (Fig. 1). Therefore, our samples were adequate to characterize the range of prey items consumed by *P. lawrencii* and available at La Georgina, given the level of taxonomic resolution used in this study.

Use of bromeliad size classes by *P. lawrencii* differed from the available size distribution at La Georgina ($\chi^2_2 = 19.51$, $P = 0.0001$). The birds avoided small bromeliads (≤ 30 cm diameter) and specialized on the largest size class (<60 cm diameter, Fig. 2). Although we did not quantify sequential foraging behavior and substrate selection of individual *P. lawrencii*, we typically observed birds moving deliberately among large bromeliads and ignoring most small plants as they foraged. The size distribution of the 45 bromeliads collected for prey base analysis did not differ from use of bromeliad sizes by *P. lawrencii* ($\chi^2_2 = 2.84$, $P = 0.24$).

Proportional use of all prey types by *P. lawrencii* differed from prey availability (MANOVA, Wilks' $\lambda = 0.43$, $F_{9, 45} = 6.51$, $P < 0.0001$). In contrast, mean sizes of prey types consumed by *P. lawrencii* did not differ from available prey sizes (MANOVA, Wilks' $\lambda = 0.80$, $F_{3, 14} = 1.15$, $P = 0.36$). *Pseudocolaptes lawrencii* primarily fed on dermapterans, orthopterans (mainly roaches), coleopterans, and insect egg cases (Fig. 3). Nearly all egg cases in stomachs of *P. lawrencii* were from roaches. We considered insect egg cases to be a separate prey type because stomachs of several other species of epiphyte-searching insectivorous birds at La Georgina contained roach egg cases without any evidence that the birds consumed roaches (Sillett 1994). However, less than five percent of documented foraging observations of these species were in bromeliads (Sillett 1994). In addition, only a small fraction of roaches collected from bromeliads were

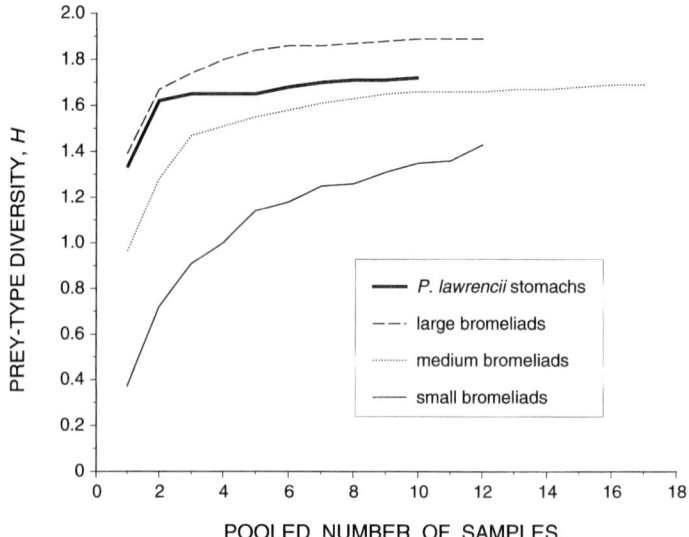

FIG. 1. Brillouin diversity saturation curves as a function of number of samples examined. Curves were produced by randomly sampling bromeliad and stomach data, with replacement.

carrying egg cases (personal observations). We concluded that *P. lawrencii* probably obtains most egg cases from substrates other than bromeliads, such as mats of epiphytic bryophytes (see Sillett 1994), and therefore did not include insect egg cases in further analyses.

Aquatic insect larvae were the largest component of the bromeliad prey base, and occurred in 80% of bromeliads sampled; yet, *P. lawrencii* did not use this resource, based on stomach contents (Fig. 3). Aquatic insect larvae, especially dipterans, have few sclerotized body parts and thus might be underrepresented in stomach samples. However, we have additional evidence suggesting that *P. lawrencii* did not feed on aquatic larvae. First, while in bromeliads, *P. lawrencii* primarily forages in leaf-litter trapped among the plants' outer leaves. One can usually find the birds by listening for their noisy rummaging in bromeliads and then by looking for the falling leaves and detritus tossed out as they forage. We never saw *P. lawrencii* visibly foraging in impounded water. Second, none of the 10 specimens we collected had wet or soiled feathers around the face, throat, or breast that would have been expected if the birds were foraging in water and wet debris.

We concluded that only terrestrial bromeliad-inhabiting prey were available to *P. lawrencii* at La Georgina, and removed all aquatic animals from further analyses. Considering only terrestrial

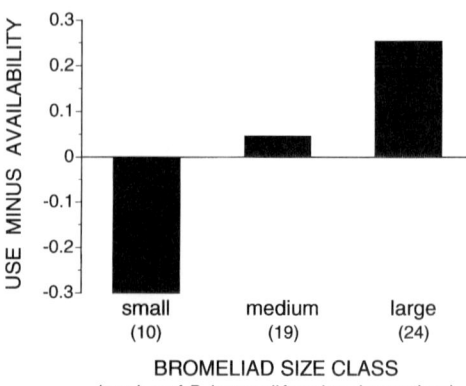

FIG. 2. Use of bromeliad size classes by *P. lawrencii* compared to available bromeliad sizes. Bars above 0.0 horizontal axis indicate selection; bars below indicate avoidance.

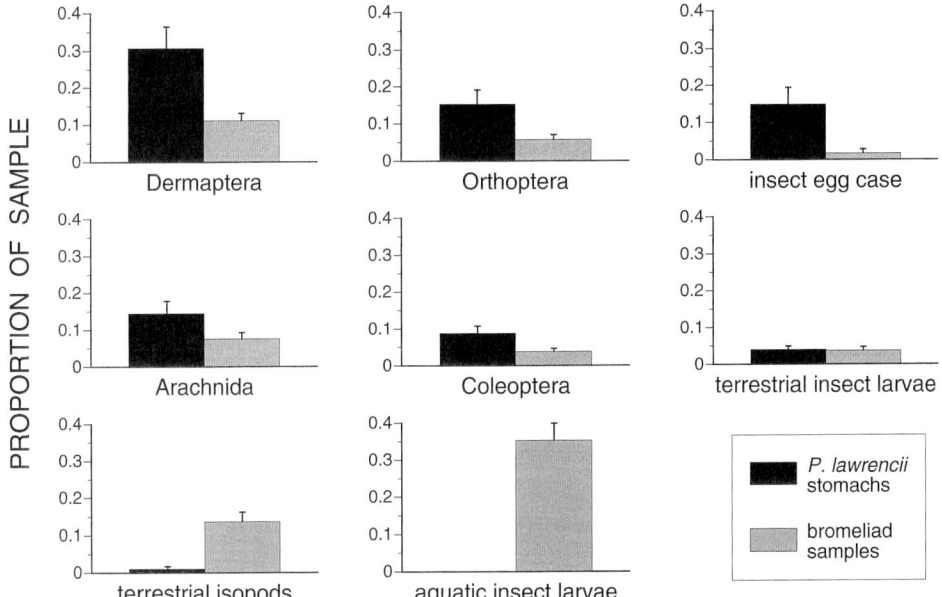

FIG. 3. Comparison of mean proportions of nine prey types in the diet of *P. lawrencii* with the bromeliad prey base. Error bars represent one standard error of mean.

prey, the bird's proportional use of prey types still differed from prey availability (MANOVA, Wilks' $\lambda = 0.69$, $F_{7, 40} = 2.52$, $P = 0.03$). With the exception of isopods, however, which were avoided by *P. lawrencii*, use did not differ from availability for all other prey types (Table 1).

DISCUSSION

The avoidance of isopods by *P. lawrencii* suggests that these crustaceans may be difficult to catch or unpalatable to the bird. The former explanation is unlikely because bromeliad-inhabiting isopods at La Georgina are not fast-moving (personal observations). Isopods are frequently consumed by some land and aquatic bird species (e.g., Weller 1975; Reinecke 1979; Sakai et al. 1986). However, isopods are dorsoventrally flattened and covered by a heavy, calcified exoskeleton (Siefert 1961 as cited in Graveland and Van Gijzen 1994). They may thus present less of an energy reward to *P. lawrencii*, relative to bromeliad-inhabiting insects, causing the birds to spend the majority of their foraging efforts on more profitable prey. A third explanation for the absence of isopods in the diet of *P. lawrencii* is that isopods are not prevalent in canopy bromeliads at La Georgina. Nadkarni and Longino (1990) documented significantly fewer crus-

TABLE 1

RESULTS OF ANOVAS COMPARING PROPORTIONAL USE OF EIGHT PREY TYPES BY *P. lawrencii* TO PREY AVAILABILITY IN LA GEORGINA BROMELIADS

Prey type	DF[a]	F	P-value	Power[b]
Dermaptera	1, 46	3.00	0.09	0.92
Orthoptera	1, 46	2.01	0.16	0.99
Arachnida	1, 46	0.02	0.90	0.97
Coleoptera	1, 46	0.02	0.89	0.84
Terrestrial insect larvae	1, 45.99	2.10[c]	0.15	1.00
Terrestrial isopods	1, 38.14	21.29[c]	<0.0001	—

[a] DF = degrees of freedom.
[b] Statistical power $(1 - \beta)$ is given for all tests that failed to reject the null hypothesis that *P. lawrencii* use of prey did not differ from prey availability. Power was computed as the probability of an ANOVA to detect an actual difference (δ) of 10% between use and availability (i.e. $\delta = 0.1$) at the $\alpha = 0.05$ level.
[c] Welch ANOVA (see Methods).

taceans (isopods and amphipods) in canopy organic matter relative to the forest floor in a Costa Rican cloud forest. All of our bromeliad samples were collected within 2.5 m of the ground.

Pseudocolaptes lawrencii is highly stereotyped in its foraging behavior and selection of foraging substrates (Sillett 1994). This stereotypy may explain why *P. lawrencii* rarely consumes aquatic prey. Searching for prey in detritus-filled water probably requires different behaviors than foraging in drier, impounded leaf litter and organic matter. Leaf litter-inhabiting insects and spiders with an active predator-avoidance response would quickly move to seek cover if suddenly exposed by a rummaging bird. Such mobile prey would be rapidly detected by an actively foraging bird. In contrast, *P. lawrencii* could not easily remove impounded water to expose aquatic prey, given the bird's pointed and relatively stout bill. Water collects in the bases of bromeliad leaf axils and occurs at a greater depth in the center of plants, where leaves are younger, denser, and more erect (Laessle 1962). The majority of aquatic animals we sampled occurred toward bromeliad centers. It may be more difficult for *P. lawrencii* to probe and rummage among dense, young leaves than in more widely spread, older leaves.

Little is known about what components of the bromeliad prey base are exploited by other specialist bird species (see Appendix). Some dendrocolaptids, especially *Nasica longirostris* and scythebills (*Camplylorhamphus* spp.), have long bills, and may be better able to exploit aquatic prey. There are anecdotal accounts of some species, including *P. lawrencii*, taking aquatic vertebrates, such as salamanders and frogs, from bromeliads (e.g., Todd and Carriker 1992; Stiles and Skutch 1989). Only one of 45 bromeliads sampled for this study contained a vertebrate (a small frog), suggesting that vertebrate prey are rare in bromeliads at La Georgina.

Pseudocolaptes lawrencii is a substrate-restricted forager (*sensu* Robinson and Holmes 1982) whose foraging behavior and prey choice are mediated by the nature of its foraging substrate. This species selectively forages in leaf-litter and organic debris trapped in large arboreal bromeliads, which have the greatest diversity and quantity of prey items. When *P. lawrencii* find suitable substrates, they opportunistically consume prey, in terms of both prey size and prey composition, as it is encountered. Rosenberg (1993) documented a similar phenomenon among *Myrmotherula* antwrens specialized on foraging in suspended aerial leaf-litter. Specialist antwrens foraged in curled dead leaves over 90% of the time but took prey roughly in proportion to availability. The existence of highly specialized and stereotyped behaviors that limit foraging to a narrow range of substrates implies that these substrates have been predictable and productive sources of food over evolutionary time (Rosenberg 1993). Arthropods associated with leaf-litter should therefore be predictable and abundant year-round in arboreal bromeliads at La Georgina. In contrast, bromeliad-inhabiting aquatic invertebrates may be highly ephemeral and thus unpredictable from the bird's perspective. However, the seasonality of the bromeliad prey base remains to be quantified.

ACKNOWLEDGMENTS

This paper benefitted from reviews and comments by M. P. Ayres, R. T. Holmes, A. W. Kratter, P. P. Marra, J. V. Remsen, Jr., G. H. Rosenberg, K. V. Rosenberg, S. C. Sillett, and one anonymous reviewer. We are also grateful to J. V. Remsen, Jr. for help in sifting through the ornithological literature, and to M. Marin for translating the abstract into Spanish. D. A. Good wrote the computer program we used to calculate Brillouin diversity indices. The field work for this research was supported by the generous donations of John S. McIlhenny and the patrons of the LSU Museum of Natural Science, and by a grant from the Frank M. Chapman Memorial Fund of the American Museum of Natural History. A. W. Kratter and K. V. Rosenberg helped collect and prepare bird specimens. J. Sanchez and D. Hernandez of the Instituto Nacional de Biodiversidad in Costa Rica allowed us to accompany them during their field work in the La Georgina area and helped us find study sites. We thank the Herrera family for their hospitality and use of their cabin at La Georgina. We also thank Lic. M. Rodriguez R. and J. Hernández B. of the Ministerio de Recursos Naturales, Energía y Minas, Dirección General de Vida Silvestre, and Dr. J. Jimenez A. of the Ministerio de Agricultura y Ganaderia, Dirección de Salud Animal, for permitting us to collect specimens in Costa Rica.

LITERATURE CITED

Askins, R. A. 1983. Foraging ecology of temperate-zone and tropical woodpeckers. Ecology 64:945–956.

Borror, D. J., C. A. Triplehorn, and N. F. Johnson. 1989. An Introduction to the Study of Insects. 6th edition. Saunders College Publishing, Philadelphia, Pennsylvania.

Burt-Utley, K., and J. F. Utley. 1977. Phytogeography, physiological ecology, and the Costa Rican genera of Bromeliaceae. Hist. Nat. de Costa Rica 1:9–29.

CALVER, M. C., AND R. D. WOOLLER. 1982. A technique for assessing the taxa, length, dry weight, and energy content of the arthropod prey of birds. Aust. Wildl. Res. 9:293–301.
CHAPMAN, A., AND K. V. ROSENBERG. 1991. Diets of four sympatric Amazonian woodcreepers (Dendrocolaptidae). Condor 93:904–915.
CRUZ, A. 1978. Adaptive evolution in the Jamaican Blackbird *Nesopsar nigerrimus*. Ornis Scand. 9:130–137.
DIESEL, R. 1989. Parental care in an unusual environment: *Metopaulias depressus* (Decapoda: Grapsidae), a crab that lives in epiphytic bromeliads. Anim. Behav. 38:561–575.
FJELDSÅ, J., AND N. KRABBE. 1990. Birds of the High Andes. Zoological Museum, University of Copenhagen. Copenhagen, Denmark.
FRIEDMANN, H., AND F. D. SMITH, JR. 1950. A contribution to the ornithology of northeastern Venezuela. Proc. U.S. Natl. Mus. 100:411–538.
GRAVELAND, J., AND T. VAN GIJZEN. 1994. Arthropods and seeds are not sufficient as calcium sources for shell formation and skeletal growth in passerines. Ardea 82:299–314.
HEJL, S. J., J. VERNER, AND G. W. BELL. 1990. Sequential versus initial observations in studies of avian foraging. Stud. Avian Biol. 13:166–173.
HOLDRIDGE, L. R. 1967. Life Zone Ecology. Revised edition. Tropical Science Center, San José, Costa Rica.
HURTURBIA, J. 1973. Trophic diversity measurement in sympatric predatory species. Ecology 54:885–890.
KARR, J. 1971. Structure of avian communities in selected Panama and Illinois habitats. Ecol. Monogr. 41:207–233.
LACK, D. 1976. Island Biology Illustrated by the Land Birds of Jamaica. University of California Press. Berkeley, California
LACK, D., AND P. LACK. 1972. Wintering warblers in Jamaica. Living Bird 11:129–153.
LAESSLE, A. 1961. A microlimnological study of Jamaican bromeliads. Ecology 42:499–517.
MARTIN, P., AND P. BATESON. 1986. Measuring Behavior: an Introductory Guide. Cambridge University Press, Cambridge, U.K.
MILLIKEN, G. A., AND D. E. JOHNSON. 1984. Analysis of Messy Data. Vol. 1. Designed Experiments. Van Nostrand Reinhold, New York.
MOREBY, S. J. 1987. An aid to the identification of arthropod fragments in the faeces of gamebird chicks (Galliformes). Ibis 130:519–526.
NADKARNI, N. M., AND J. T. LONGINO. 1990. Invertebrates in canopy and ground organic matter in a Neotropical montane forest, Costa Rica. Biotropica 22:286–289.
NADKARNI, N. M., AND T. J. MATELSON. 1989. Bird use of epiphyte resources in Neotropical trees. Condor 91:891–907.
ORIANS, G. H. 1969. The number of bird species in some tropical forests. Ecology 50:783–801.
PAOLETTI, M. G., R. A. J. TAYLOR, B. R. STINNER, D. H. STINNER, AND D. H. BENZING. 1991. Diversity of soil fauna in the canopy and forest floor of a Venezuelan cloud forest. J. Trop. Ecol. 7:373–383.
PARKER, T. A., III, T. S. SCHULENBERG, G. R. GRAVES, AND M. J. BRAUN. 1985. The avifauna of the Huancabamba region, northern Peru. Pp. 169–197 *in* Neotropical Ornithology (P. A. Buckley et al., Eds.). Ornithol. Monogr. No. 36.
PARKER, T. A., III, AND J. P. O'NEILL. 1980. Notes on little known birds of the Upper Urubamba Valley, southern Peru. Auk 97:167–176.
PICADO, C. 1911. Les Bromeliacees epiphytes comme milieu biologique. Bull. Sci. France Belg. 45:215–360. Reprinted in Hist. Nat. de Costa Rica 1:9–29 (1977).
PITTENDRIGH, C. S. 1946. Bromeliad malaria in Trinidad, B. W. I. Am. Jour. Trop. Med. 26:47–66.
RALPH, C. P., S. E. NAGATA, AND C. J. RALPH. 1985. Analysis of droppings to describe diets of small birds. J. Field Ornith. 56:165–174.
REINECKE, K. J. 1979. Feeding ecology and development of juvenile Black Ducks in Maine. Auk 96:737–745.
REMSEN, J. V., JR. 1985. Community organization and ecology of birds of high elevation humid forests of the Bolivian Andes. Pp. 733–756 *in* Neotropical Ornithology (P. A. Buckley et al., Eds.). Ornithol. Monogr. No. 36.
RIDGELY, R. S., AND G. TUDOR. 1989. The Birds of South America. Vol. 1. The Oscine Passerines. University of Texas Press, Austin, Texas.
RIDGELY, R. S., AND G. TUDOR. 1994. The Birds of South America. Vol. 2. The Suboscine Passerines. University of Texas Press, Austin, Texas.
RIDGELY, R. S., AND J. A. GWYNNE, JR. 1989. A Guide to the Birds of Panama. 2nd ed. Princeton University Press, Princeton, New Jersey.
ROBBINS, M. B., T. A. PARKER III, AND S. E. ALLEN. 1985. The avifauna of Cerro Pirre, Darien, eastern Panama. Pp. 198–232 *in* Neotropical Ornithology (P. A. Buckley et al., Eds.). Ornithol. Monogr. No. 36.
ROSENBERG, K. V., AND R. J. COOPER. 1990. Approaches to avian diet analysis. Stud. Avian Biol. 13:80–90.
SAKAI, H. F., C. J. RALPH, AND C. D. JENKINS. 1986. Foraging ecology of the Hawaiian Crow, an endangered generalist. Condor 88:211–219.
SAS INSTITUTE. 1994. JMP. Version 3.1 [Macintosh OS]. SAS Institute, Inc., Cary, SC.

SCHOENER, T. W. 1968. The *Anolis* lizards of Bimini: resource partitioning in a complex fauna. Ecology 49: 704–726.
SEIFERT, G. 1961. Die Tausendfüssler. Die Neue Brehm-bücherei. A. Ziemsen Verlag, Wittenberg Lutherstad, Germany.
SHERRY, T. W. 1984. Comparative dietary ecology of sympatric insectivorous Neotropical flycatchers. Ecol. Monogr. 54:313–338.
SILLETT, T. S. 1994. Foraging ecology of epiphyte-searching insectivorous birds in Costa Rica. Condor 96: 863–877.
SKUTCH, A. F. 1967. Life histories of Central American highland birds. Publ. Nuttall Ornithol. Club No. 7.
SKUTCH, A. F. 1972. Studies of tropical American birds. Publ. Nuttall Ornithol. Club No. 10.
SLUD, P. 1964. The birds of Costa Rica: distribution and ecology. Bull. Amer. Mus. Nat. Hist. 128:1–403.
STILES, F. G. AND A. F. SKUTCH. 1989. A Guide to the Birds of Costa Rica. Cornell University Press, Ithaca, New York.
TERBORGH, J. W. 1980. Causes of tropical species diversity. Proc. XVII Inter. Ornithol. Congr. 955–961.
TODD, W. E. C., AND M. A. CARRIKER, JR. 1922. The birds of the Santa Marta region of Colombia. Ann. Carnegie Mus. 14:3–582.
WELCH, B. L. 1951. On the comparison of several means: an alternative approach. Biometrika 38:330–336.
WELLER, M. W. 1975. Ecological studies of the Auckland Islands Flightless Teal. Auk 92:280–297.
WETMORE, A. 1926. Observations on the birds of Argentina, Paraguay, Uruguay, and Chile. U.S. Natl. Mus. Bull. 133.
WILLIS, E. 1960. A study of the foraging behavior of two species of ant-tanagers. Auk 77:150–170.
WOLF, L. L. 1976. Avifauna of the Cerro de la Muerte region, Costa Rica. Am. Mus. Novitates 2606:1–37.
ZUSI, R. L. 1969. Ecology and adaptations of the Trembler on the island of Dominica. Living Bird 8:137–164.

APPENDIX

Fifty-one Neotropical Bird Species Recorded Foraging for Animal Prey in Bromeliads. SP = Probable Foraging Specialist; DL = Recorded Foraging in Dead, Curled Bromeliad Leaves or Inflorescences, Not Plant Interior

Family	Scientific name	Common name	Notes	Reference
Accipitridae	*Geranospiza caerulescens*	Crane Hawk		21, 22
Picidae	*Picoides villosus*	Hairy Woodpecker		17
	Piculus rubiginosus	Golden-olive Woodpecker		1
Furnariidae	*Cranioleuca marcapatae*	Marcapata Spinetail		11
	Margarornis stellatus	Star-chested Treerunner		15
	M. bellulus	Beautiful Treerunner		16
	Anabacerthia striaticollis	Montane Foliage-gleaner		23, 5
	A. variegaticeps	Spectacled Foliage-gleaner		13
	Pseudocolaptes lawrencii	Buffy Tuftedcheek	SP	21, 19, 28, 5, 17
	P. boissonneautii	Streaked Tuftedcheek	SP	5, 3, 15
	P. johnsoni	Pacific Tuftedcheek	SP	15
	Hyloctistes subulatus	Striped Foliage-gleaner		5, 22
	Syndactyla (Automolus) ruficollis	Rufous-necked Foliage-gleaner		12, 15
	Automolus ochrolaemus	Buff-throated Foliage-gleaner		21
	Cichlocolaptes leucophrus	Pale-browed Treehunter	SP	15
	Thripadectes rufobrunneus	Streak-breasted Treehunter		21, 19, 22
Dendrocolaptidae	*Nasica longirostris*	Long-billed Woodcreeper	SP	5, 15
	Xiphocolaptes albicollis	White-throated Woodcreeper	SP	15
	X. falcirostris	Moustached Woodcreeper	SP	15
	X. promeropirhynchus	Strong-billed Woodcreeper	SP	23, 5
	Xiphorhynchus flavigaster	Ivory-billed Woodcreeper	DL	27
	Campylorhamphus trochilirostris	Red-billed Scythebill		26, 4, 15
	C. procurvoides	Curve-billed Scythebill		5, 15
	C. falcularius	Black-billed Scythebill		15
	C. pusillus	Brown-billed Scythebill		5, 22
Corvidae	*Cyanolyca argentigula*	Silvery-throated Jay		22, 18
Troglodytidae	*Campylorhynchus zonatus*	Banded-backed Wren		19, 22
	Troglodytes aedon	House Wren		9
	T. ochraceus	Ochraceous Wren		9
Mimidae	*Cinclocerthia ruficauda*	Trembler	SP	29
Parulidae	*Vermivora gutturalis*	Flame-throated Warbler	DL	17
	Dendroica dominica	Yellow-throated Warbler		8

APPENDIX
CONTINUED

Family	Scientific name	Common name	Notes	Reference
Thraupidae	*Orchesticus abeillei*	Brown Tanager		6
	Orthogonys chloricterus	Olive-green Tanager		6
	Buthraupis arcaei	Blue-and-gold Tanager		6
	Wetmorethraupis sterrhopteron	Orange-throated Tanager		6
	Delothraupis castaneoventris	Chestnut-bellied Mountain-Tanager		6
	Chlorochrysa phoenicotis	Glistening-green Tanager		6
	Tangara chilensis	Paradise Tanager		6
	T. icterocephala	Silver-throated Tanager		6
	T. vassorii	Blue-and-black Tanager		6
	Dacnis hartlaubi	Turquoise Dacnis		6
	D. venusta	Scarlet-thighed Dacnis		9
	D. cayana	Blue Dacnis		6
Cardinalidae	*Caryothraustes poliogaster*	Black-faced Grosbeak		20
Icteridae	*Nesopsar nigerrimus*	Jamaican Blackbird	SP	7, 2, 10
	Oreopsar bolivianus	Bolivian Blackbird		14
	Icterus chrysater	Yellow-backed Oriole		10
	Cacicus uropygialis	Scarlet-rumped Cacique		20, 14
	Psarocolius wagleri	Chestnut-headed Oropendola		24
	P. montezuma	Montezuma Oropendola		25

(1) Askins 1983; (2) Cruz 1978; (3) Fjeldså and Krabbe 1990; (4) Friedmann and Smith 1950; (5) Hilty and Brown 1986; (6) Isler and Isler 1987; (7) Lack 1976; (8) Lack and Lack 1972; (9) Nadkarni and Matelson 1989; (10) Orians 1985; (11) Parker and O'Neill 1980; (12) Parker et al. 1985; (13) Ridgely and Gwynne 1989; (14) Ridgely and Tudor 1994; (15) Ridgely and Tudor 1994; (16) Robbins et al. 1985; (17) Sillett 1994; (18) Sillett, pers. observation; (19) Skutch 1969; (20) Skutch 1972; (21) Slud 1964; (22) Stiles and Skutch 1989; (23) Todd and Carriker 1922; (24) M. S. Webster *in litt.*; (25) M. S. Webster unpubl. ms; (26) Wetmore 1927; (27) Willis 1960; (28) Wolf 1976; (29) Zusi 1969.

COMPOSITION AND DISTRIBUTION PATTERNS OF THE AVIFAUNA OF AN AMAZONIAN UPLAND SAVANNA, AMAPÁ, BRAZIL

José Maria Cardoso Da Silva[1], David Conway Oren[2], Júlio César Roma,[2] and Luiza Magalli Pinto Henriques[2]

[1]*Zoological Museum, University of Copenhagen, Universitetsparken 15, DK-2100, Copenhagen, Denmark*; and
[2]*Museu Paraense Emílio Goeldi, Departamento de Zoologia, C. P. 399, 66017-940, Belém, Pará, Brazil*

ABSTRACT.—The avifauna of an isolated patch of Amazonian upland savanna in the Brazilian state of Amapá was studied during October–November 1990. A total of 179 species was recorded, some of which were found for the first time in Amapá or were known from this state only by old sight records. Forty-six species are known to inhabit savanna habitats in Amapá. They may be grouped into four major distribution categories: (a) fourteen species with widespread distribution; (b) three species with distribution restricted to the savannas of northern South America; (c) two species distributed in a patchy fashion almost all the way around Amazonia; and (d) twenty-seven species with ranges centered in the open vegetation in Central Brazil. North of the Amazon, the high density of Central Brazilian species is found in Amapá and Marajó savannas. This may indicate that possibly the most recent connections between savannas of the Cerrado Region in central Brazil and the islands of Amazonian savannas occurred along the Atlantic coast instead of across central Amazonia. This contrasts with what has been generally suggested by some proponents of the refuge theory. An urgent plan of conservation for Amazonian savannas is necessary, as large areas of this unique environment have been drastically modified in the last few decades.

RESUMO.—A avifauna de uma localidade de uma ilha de campo amazônico de *terra firme* no Estado do Amapá, Brasil, foi estudada entre Outubro e Novembro de 1990. 179 espécies foram registradas para esta localidade, algumas das quais foram encontradas pela primeira primeira vez para o Amapá ou foram conhecidas deste Estado somente com base em registros visuais antigos. Um total de 46 espécies registradas para o Amapá são associadas primariamente com campos de *terra firme*. A distribuição destas espécies podem ser agrupadas em quatro grandes categorias: (a) quatorze espécies largamente distribuidas na América do Sul; (b) três espécies com distribuição centrada nas savanas do norte da América do Sul; (c) duas espécies com distribuição em ilhas de vegetações abertas em torno da Amazônia; e (d) vinte e sete espécies com o centro da distribuição no Brasil central. Nos enclaves de campos de *terra firme* localizados ao norte do Rio Amazonas, a mais alta densidade de espécies com distribuições centradas no Brasil Central está nas savanas do Amapá e Marajó. Isto pode indicar que talvez as conexões mais recentes entre estes enclaves e a região do cerrado no Brasil Central ocorreram ao longo da costa Atlântica e não pelo centro da Amazônia, como tem sido geralmente proposto por alguns proponentes da teoria dos refúgios. Um plano urgente de conservação para as ilhas de campos de *terra firme* dentro da Amazônia é necessário, por causa que grandes áreas deste tipo de ambiente singular tem sido drasticamente modificadas durante as últimas décadas.

The dominant paradigm for explaining avian diversification in Amazonia is the refuge theory (Haffer 1969; but see Haffer 1985 for alternatives). It suggests that most of the speciation in Amazonian forest birds was caused by major climatic fluctuations and associated vegetational changes during the Quaternary (Haffer 1969, 1974, 1985).

One of the most important pieces of biogeographic evidence used to support the refuge theory (Haffer 1969, 1974) is the presence of several islands of upland savanna vegetation within Amazonia (Pires 1973; Sarmiento 1983). These savanna patches have been interpreted as relicts of a vegetation that was widespread during the Quaternary glacial periods (Haffer 1967, 1974,

FIG. 1. Distribution of the major Amazonian upland savanna islands. Area of the savannas of the Paru-Trombetas region (including the Sipaliwini savannas in southern Surinam) has been overestimated, because it actually consists of several small patches with little or no connection. Division of South America in morphoclimatic domains follows Ab'Saber (1977). Several small domains have been grouped into a large Andean Region.

1985). Currently, Amazonian savannas are regarded as "present-day refuges" (Prance 1987), where populations of savanna-restricted species may be passing through a process of isolation and differentiation.

Despite their importance of Amazonian avian biogeography, little is known about the flora and fauna of the major patches of Amazonian savannas. The little information available regarding the ranges of savanna birds within Amazonia is scattered in general catalogues and books (e.g., Snethlage 1914; Griscom and Greenway 1941; Pinto 1944, 1978). So far, no synthetic treatment of the patterns of distribution of savanna-restricted species of any of the largest patches of Amazonian savannas has been attempted.

In this paper we first present a list of species recorded at a savanna site in Amapá, one of the largest savanna islands in Brazilian Amazonia; we also include information about the taxonomy, ecology and distribution of selected species. We then delineate the major patterns of distribution of savanna-restricted species recorded in Amapá and discuss how these data can be used to evaluate one of the main assumptions of the refuge theory. Finally, we present recommendations to ensure the conservation of *terra firme* savanna islands in Amazonia.

STUDY AREA

The Amapá savannas occupy 17,000 km^2 (Pires 1973) and are on a narrow and mainly flat, north-south oriented belt composed of Tertiary sediments ("Barreiras Formation") parallel to the Atlantic coast (Fig. 1). They are bordered on the west by tropical forests on Pre-Cambrian crystalline terrains, and on the east by a combination of seasonally flooded grasslands and mangroves on Holocene sediments. The climate is hot (average temperature 26–27°C) and humid (relative humidity 75–90%). The average annual precipitation in Porto Planton, a savanna site, is 1,900 mm, compared to 2,400 mm in neighboring sites covered by forests (IBGE 1966). In

Porto Planton, there is a well-marked dry season (total rainfall <100 mm) between August and December (Azevedo 1967).

Typical Amapá savanna has an herbaceous stratum dominated by grass species (e.g., *Trachypogon plumosus*, *Bulbostylis conifera*, *B. spadicens*, *Scleria* spp. and *Paspalum* spp.) and a 3–10 m tall arboreal stratum composed mainly of the following trees: *Curatella americana*, *Byrsonina* sp., *Himanthoantus obovata*, *Pallicourea rigida* and *Hancornia speciosa*. In the bottom of some narrow valleys, where soils are shallow and permanently inundated, there is a humid grassland habitat with narrow belts of *Mauritia martiniana* palms; this special type of vegetation is locally known as "buritizal". In the wide valleys, there are tall (15–25 m) gallery forest patches of variable size, dominated by tree species such as *Jacaranda copaia* and *Symphonia globulifera*, and palm species such as *Euterpe oleracea*. Magnanini (1953), Azevedo (1967), and Berg and Rabelo (1982) described in detail the Amapá savanna vegetation.

METHODS

We intensively sampled the avifauna of one of the best conserved tracts of savanna in Amapá, on the Campus Experimental da EMBRAPA, 48 km north of Macapá (00°02'N, 51°03'W) between 10 October and 2 November 1990. In addition, one of us (Silva) made observations during three days on the avifauna of other savanna sites located between Macapá and Calçoene. We collected specimens by mist-netting (4,000 net-hours) and with shotguns. All specimens collected are housed at the Museu Paraense Emílio Goeldi (MPEG), Belém, Pará, Brazil. Because we studied the avifauna only during a season, our data on species abundance and composition are not representative of other parts of the year, as savanna avifaunas generally exhibit important seasonal changes caused by migratory movements of some species (Silva 1995b). While sampling the avifauna, we also collected general life-history data. We estimated abundance of species using the following categories: (C) common, (FC) fairly common, (U) uncommon, and (R) rare (see Appendix). Body mass was obtained from collected specimens using Pesola balances 2–3 hours after the collecting time; we did not collect any bird with a high quantity of fat in the body. We also present data on the gonadal development using the terminology proposed by Willard et al. (1991). Taxonomic sequence generally follows Howard and Moore (1980) whereas English names (in the Appendix) are those from Sibley and Monroe (1990). In addition to specimens at MPEG, we mention in the species account some important specimens housed at the American Museum of Natural History (AMNH), New York, Field Museum of Natural History (FMNH), Chicago, and Rijksmuseum van Natuurlijke Historie (RMNH), Leiden.

Patterns of distribution and differentiation of savanna-restricted birds recorded in Amapá were evaluated based on museum specimens, recent fieldwork in some of the other large savannas islands (e.g., Roraima and Marajó), and a literature survey. General patterns of distribution were determined by evaluating the spatial congruence of the species' ranges. We used as reference for describing the distribution patterns a modified version (Fig. 1) of the map of the South American morphoclimatic domains (Ab'Saber 1977).

RESULTS

Species accounts.—Amapá state is one of two administrative regions (the other is Roraima) in Brazilian Amazonia for which a complete and detailed list of bird species is available (Novaes 1974, 1978). Therefore, the following species accounts include only species recorded for Amapá for the first time, species known from Amapá only from old sight records, and species whose taxonomy or population status deserve comments. A full list of species recorded during our expedition is given in the Appendix.

Elanoides forficatus. For five consecutive days (14–19 October), we observed two individuals of this species flying over the savanna at our base camp. Because we did not see this species during the rest of our study, we suspect that both individuals were long-distance migrants from other regions. This is the first record for Amapá.

Buteo albicaudatus. This species had been recorded on Amapá based only on an old sight record by E. Goeldi at Lago Grande do Amapá, 2°03'N, 50°42'W (Novaes 1974). We observed this species twice (14 and 24 October 1990). On both occasions, one single individual was seen flying over savanna.

Zenaida auriculata. Novaes (1974) recorded this species for Amapá based on a single sight record by E. Goeldi at Lago Grande do Amapá. We collected 13 individuals, including three fledglings.

Uropelia campestris. This species is known from Amapá from only a single sight record by

E. Goeldi at Lago Grande do Amapá (Novaes 1974). We observed this species once (two individuals) in an open tract of savanna.

Amazona ochrocephala. Three specimens that we collected are the first records for Amapá.

Glaucidium brasilianum. A specimen that we collected is the first record for Amapá.

Tachornis squamata. As Novaes (1974) pointed out, this species' occurence in Amapá is based only on Meyer de Schauensee's (1966) compilation. We observed this species fairly commonly along the "buritizais" in our study area as well as other savannas close to Porto Platon (00°42'N, 51°27'W).

Heliactin bilopha. Two specimens that we collected are the first records for Amapá. This species was known in northern South America only from the Sipaliwini savannas, Surinam (02°06'N, 56°02'W) (Mees 1968). Specimens were collected in a savanna tract close to a gallery forest. Mees (1985) presented the reasons for using *H. bilopha* (Temminck 1820) instead of the commonly used *H. cornuta* (Wied 1821).

Ramphastos toco. A specimen that we collected is the first for the Amapá savannas; the previous records listed by Novaes (1974) are based only on sight records made in the last century.

Picumnus cirratus. Novaes (1974) recorded this species from Amapá (02°03'N, 50°48'W) and Macapá (00°02'N, 51°03'W). We collected one specimen. Two distinct populations of this species occur in eastern Amazonian Brazil. The MPEG contains specimens referable to both *P. c. macconnelli* and another subspecies, perhaps *P. c. confusus*. Specimens from Amapá at Macapá, Rio Gurijuba (Fazenda Itaporan, 0°52'N, 50°19'W), and our specimen; and from Pará at Monte Alegre (2°01'S, 54°04'W), Caviana Island (0°10'S, 49°00'W), and several localities on Marajó Island (Chaves, 0°10'S, 49°55'W; Fazenda Tapereba, Rio Guará, Municipality of Cachoeira do Arari, approx. 1°10'S, 48°45'W; Ilha da Roça, 1°11'S, 48°53'W; and Fazenda Cedro, Municipality of Ponta de Pedras, approx. 1°23'S, 48°52'W) are assignable to *macconnelli*. All are darkly marked below and are rich brown with no or obsolete markings above. Three specimens collected in coastal Pará at Vista Alegre, Marapanim, 00°36'S, 47°42'W (MPEG 33, 386–88) in mangroves and previously reported without detailed comments by Novaes (1981) are very different, much lighter both below and above, showing characters to those given for *confusus* by Todd (1946). A female and one male have no markings on the back, whereas the other male is strongly barred above. This subspecies is a challenging one to work with, because Kinnear's (1927) original description of *confusus* is rather ambiguous, and its geographic distribution is not clear. If the Vista Alegre birds are not *confusus*, then they represent an undescribed form.

Colaptes campestris. Three specimens that we collected are the first records for the Amapá savannas. Our specimens have the following wing length (chord): males 149.0 and 147.5 mm, and female 144.0 mm. Specimens from the Sipaliwini savannas (at RMNH) have the following measurements: male 156.0 mm and females 145.0, 151.0, 151.0 and 152.0 mm. On average, these small series from Amazonian savanna islands have wing lengths shorter than those recorded for populations of *C. c. campestris* from regions south and east of Amazonia by Short (1972: 69, Tables 17 and 18): means 154.00–163.89 mm for males; 156.91–162.95 mm for females.

Lepidocolaptes angustirostris. We collected eight specimens of this species, which is common in the Amapá savannas. *Lepidocolaptes angustirostris* was found frequently in mixed-species flocks with *Synallaxis albescens, Suiriri suiriri, Elaenia cristata, E. chiriquensis, Myiarchus swainsoni, M. tyrannulus, Myospiza humeralis, Emberizoides herbicola, Neothraupis fasciata,* and *Cypsnagra hirundinacea*. The species composition of these flocks in Amapá is similar to those recorded in some sites of cerrado vegetation in Central Brazil (Silva, unpubl. data).

There is great individual plumage variation in some populations of *L. angustirostris*, depending on age and plumage wear (Silva and Oren, unpubl.). At the MPEG, the birds in the Amapá series are much lighter on average than the majority of individuals assigned to *L. a. bivittatus*, the subspecies from Central Brazil, and to *L. a. coronatus*, the subspecies from much of northeastern Brazil. The Amapá birds have undertail coverts concolor with the rest of the underparts, and in this character are more similar to *bivittatus* than to *coronatus*.

Mees (1974) described *L. a. griseiceps* based on a single specimen collected in the Sipaliwini savannas, Surinam. According to Mees (1974) *L. a. griseiceps* differs from all other subspecies of *L. angustirostris* by the following characters: (a) crown brownish gray, with broad (around 2 mm) and not well-defined white streaks; (b) throat white; and (c) under-tail coverts cream, similar to the remainder of the underparts. The characters pointed out by Mees as diagnostic of *L. a. griseiceps* are also found in two specimens collected in Amapá (MPEG 28,649, male, Amapá, Igarapé Ariramba; MPEG 46.452, male, Campus Experimental da EMBRAPA), which are much lighter above and with grayer heads with broader streaks when compared to other specimens in the same series. With only one specimen from Surinam and the high level of individual variation observed within the populations of *L. angustirostris*, we are reluctant to make a decision on the proper assignment of the Amapá birds, and on the validity of *L. a. griseiceps*. The geographic variation of *L. angustirostris*, one of the few woodcreeper species that inhabits open-vegetation, is badly in need of revision.

Formicivora rufa. Novaes (1974) recorded this species from Porto Platon and Rio Tartarugal (01°32'N, 50°54'W). We collected two specimens. This species was found more commonly in dense thickets on the border between savanna and "buritizais". In the more dense riverine forest, it was replaced by *Formicivora grisea*.

In Amapá *F. rufa* is represented by the subspecies *chapmani*, characterized by females with much darker streaking on the underparts and a black (rather than brown in *rufa*) tail, tipped white. Males have darker tails and more (although not always) heavy shaft streaks on the head than in *rufa*. As far as is known, the distribution of *chapmani* includes, in addition to Amapá: Alter do Chão (02°31'S, 54°57'W), the type locality (Cherrie 1916); Santarém (02°24'S, 54°42'W; 7 males, 7 females in AMNH); the Sipaliwini savannas, Surinam (RMNH, 4 males and 4 females); and possibly the following localities listed by Snethlage (1914): Rio Acará (01°40'S, 48°25'W), Monte Alegre (02°01'S, 54°04'W), Serra de Ereré (01°53'S, 54°11'W) and Rio Maicurú (02°14'S, 54°17'W). The subspecies found in the savannas of Humaitá (07°31'S, 63°02'W), Amazonas, Brazil (6 males and 2 females in MPEG; 4 males, 2 females in AMNH), close to the border between Amazonia and the Cerrado Region (Fig. 1), is *F. r. rufa*. Birds from Gilbués (ca. 09°50'S, 45°21'W), Piauí (2 males in AMNH), and Estiva (9°16'S, 46°35'W), Municipality of Alto Parnaíba, Maranhão (4 males, 2 females in MPEG) are intermediate between *chapmani* and *rufa*, the females showing heavy streaking below and males with prominent scaling on the head, but with the tails mostly brown in both sexes.

Sublegatus modestus. Traylor (1982) suggested that *S. obscurior*, with distribution centered on Amazonia, and *S. modestus*, found mostly south of the Amazon from Central Brazil and Bolivia to central Argentina (Buenos Aires, La Pampa, and Mendoza), could be regarded as subspecies of a single biological species. The main reason for that was the presence of putative hybrids between these two taxa in central Amazonia. Nevertheless, Traylor (1982) also indicated that these two taxa overlap their ranges without any evidence of hybridization in southeastern Peru and eastern Amazonia (Mexiana Island). Ridgely and Tudor (1994) described the voices of *modestus* and *obscurior*, and indicated that they are different. Then, they suggested that these two taxa could be regarded as distinct biological species. Regardless of the discussion on the taxonomic status of *modestus* and *obscurior*, we believe that the ranges of these two taxa in eastern Amazonia require clarification.

Novaes (1978:32) recorded *S. obscurior* for Mazagão, Igarapé Novo (00°07'S, 51°17'W), the only previous record of this species group for Amapá. The MPEG also has specimens of *obscurior* from: Jauareté, Rio Uaupés, Amazonas (00°36'N, 69°12'W); Belém, Pará; Vista Alegre, Mun. Marapanim, Pará; and Rodovia Belém-Brasília, km 36, Pará (approx. 01°47'S, 47°29'W) (wing length of 4 males = 72.0 (2), 73.0, 76.0 mm; 1 female = 69.5 mm). At FMNH, there is a male specimen (FMNH 344091) collected in Roraima, BU-8, 6 km east on Venezuelan border by L. Silva and D. F. Stotz (wing length = 76.0 mm). Information from the labels of these specimens indicate that they were collected in forest edges, forest clearings, forest-savanna edge and mangroves.

We collected two specimens of *S. modestus* in the Amapá savannas (MPEG 46.621–22; wing length of one male = 66.5 mm, one female = 65.0 mm), which gave every indication of approaching the breeding season (Appendix). Traylor (1982) also mentioned specimens of *modestus* from Roraima (Serra da Lua, 02°15'N, 60°45'W, one female taken 12 March; and Boa Vista, 02°49'N, 60°40'W, one male taken 11 December) and Mexiana Island (00°02'S, 49°35'W; specimen taken 13 November), whose dates are too late so that these individuals may be regarded as austral migrants. Interestingly, all these three localities listed by Traylor (1982) are dominated by savannas (Silva and Oren, pers. obs.).

Evidence assembled so far indicates that *modestus* and *obscurior* are separated by habitat in at least some parts of Amazonia: *obscurior* being found in forest clearings and edges, while Amazonian populations of *modestus* are found in savanna patches. In addition, records of *modestus* for the savanna patches north of the Amazon (Mexiana, Amapá, and Roraima) indicate that the breeding distribution of this taxon is larger than that indicated by Traylor (1982) and Ridgely and Tudor (1994). The taxonomic status of the isolated Amazonian populations of *modestus* requires further study.

Elaenia chiriquensis. According to Novaes (1978), this species is recorded for Amapá based exclusively on one old record for Oyapoc (= Oiapoque). Haffer (1978) pointed out that many records from this locality are not specified with certainty as being inside the borders of Amapá. Considering this, our 18 specimens are the first confirmed records for Amapá. Marini and Cavalcanti (1990) studied the pattern of seasonal distribution of this species in South America. They found that in Amazonia two populations occur that cannot be distinguished by plumage: one that breeds in the open vegetation of this region and another that migrates into Amazonia during the austral autumn and winter.

Euscarthmus rufomarginatus. This species is known in northern South America only from the Sipiliwini savannas, Surinam (Mees 1968). We recorded two individuals of this rare species in an open tract of savanna close to a "buritizal" on 11 October 1990. This is the first record for Amapá. Perhaps this species has been overlooked by ornithologists working in savanna regions because of both its low populations and generally inconspicuous habits outside the breeding season.

Todirostrum fumifrons. Two specimens that we collected are the first records for Amapá. We observed this species generally in pairs in the forest edge or in dense growth in the understory of gallery forests.

Xolmis velata. The northernmost known record of this species is Marajó Island (Traylor 1979). Two individuals seen in an open tract of savanna close to a "buritizal" on 18 October 1988 are the first reports for Amapá. We failed in relocating these birds despite repeated attempts; possibly both were migrants or vagrants from another region.

Myiarchus swainsoni. Novaes (1978) recorded this species from Rio Tartarugal and Amapá. Lanyon (1978) suggested that populations of this species in Amapá and Guianas are intermediate between *M. s. pelzelni* and *M. s. phaeonotus*. However, Mees (1985) challenged this conclusion based on specimens collected in Surinam. He pointed out that at least three different taxa included in *M. swainsoni* occur in Surinam: (1) *pelzelni*, found only in the Sipiliwini savannas, southern Surinam; (b) migrant nominate *swainsoni* from southern South America; and (c) *albomarginatus*, described by Mees (1985) as a distinct subspecies known only from the northern sand savanna belt along the Surinam coast. One of us (Silva) examined all specimens (housed at RMNH) used by Mees in his study and all of them support the pattern described by Mees (1985). The 14 specimens that we collected in Amapá are all referred to *pelzelni*.

Mees (1985) also suggested that *pelzelni* could be regarded as a species. We do not agree that *pelzelni* can be ranked as a species under the biological species concept, because Lanyon (1978) indicated that *pelzelni* and *swainsoni* intergrade in the southern part of their ranges.

We observed *M. swainsoni* singing frequently and defending territories, suggesting that the population in Amapá was starting the breeding season.

Myiarchus tyrannulus. Novaes (1978) recorded this species for Amapá and Rio Tartarugal. We collected five specimens which we identified as *M. t. bahiae*. We observed displays and territorial defenses. These observations, combined with the enlarged gonads of most specimens (Appendix), indicate that the population was starting the breeding season. Lanyon (1978) did not mention this species for Amapá.

Tyrannus albogularis. Novaes (1978) recorded this species from Porto Platon, Amapá, Rio Maruanum, and Macapá. We collected seven specimens. Observation of territorial defense and the gonad size of these specimens suggest that at least part of the Amapá population had started the breeding season in October-November. Based on MPEG specimens, *T. albogularis* has been recorded for the Amapá savannas in the following months: February, August, October, November, and December. Some of these months are too late for the population of *T. albogularis* in Amapá to be regarded as composed only of austral migrants.

Davis (1993) showed that *T. albogularis* occurs in central Bolivia only from September to April. Thus, records for this species from Colombia in May-late Aug may be of austral migrants, as suggested by Hilty and Brown (1986). However, the records of this species for other places in northern South America also indicate the presence of resident populations there.

Tostain et al. (1992) recorded two nests under construction in French Guiana in February, as well as sight records for the species in November. Haverschmidt & Mees (1994) recorded specimens of *T. albogularis* for Surinam in February, June, and November. Hilty (1997) listed records from Amazonas and Bolivar, Venezuela, from January to February.

In summary, the evidence available so far supports the hypothesis of the existence of a breeding population of *T. albogularis* in the savannas of Amapá, Surinam, and French Guiana. The status of the populations found in the Venezuelan savannas remains to be defined.

Tyrannus savana. Two subspecies have been recorded in Amapá. Novaes (1978) recorded *T. s. circumdatus* at Amapá, Rio Tartarugal, Rio Maruanum, and Macapá, and only a single specimen (6 December 1951) of *T. s. monachus* for Macapá. We collected nine specimens in our study area, all identifiable as *circumdatus*. We observed males displaying and defending territories, suggesting that the breeding season was beginning. If so, this is the first evidence for breeding of *T. savana* in Amapá. The presence of *circumdatus* in the Amapá savannas is documented by specimens and our observations for the following months: February, March, July, August, September, October and November. This suggests the possibility that this subspecies is a year-round resident in this area. Besides the Amapá savannas, *circumdatus* has been recorded in the following localities along the Tapajós and Amazon Rivers: Santarém (March, AMNH), Urucurituba (February, AMNH), Parintins (August, AMNH), Tauari (April, AMNH), Igarapé Brabo (May and June, AMNH), Monte Alegre (February, September, December; MPEG), Faro (January, MPEG), and Belém (June, MPEG).

Anthus lutescens. This species had not been recorded previously from Amapá (Novaes 1978). We observed solitary individuals and pairs of *A. lutescens* several times in open savannas during our study period.

Sporophila leucoptera. This species is known in northern South America only from Mexiana Island (Paynter 1970) and the Sipaliwini savannas, Surinam (Renseen 1974). We observed a small group (one male and three individuals with female-like plumage) foraging in an open savanna near a "buritizal" on 23 October 1990. This is the first report of this species for Amapá.

Sporophila bouvreuil. We observed a large flock (three males and 10 individuals with female-like plumage) foraging in a dense tract of savanna on 14 October 1990. This is the first report of this species for Amapá. *Sporophila bouvreuil* was previously known in northern South America only from Marajó and Mexiana Islands and the Sipaliwini savannas, Surinam (Paynter 1970).

Neothraupis fasciata. Cavalcanti (1989) was the first to report this species for Amapá, based on a sight report between kms 50 and 70 of the BR-156 highway, north of Macapá. The 10 specimens that we collected are the first for Amapá. The nearest record is Barra do Corda, Maranhão (Pinto 1944), around 900 km southeastern from Amapá. *Neothraupis fasciata* had been thought to be endemic to the Cerrado Region (Cracraft 1985, Haffer 1985).

Hylophilus pectoralis. The specimen that we collected in the canopy of gallery forest is the first record for Amapá.

Patterns of distribution of savanna species.—We recorded 179 species of birds for our study area (Appendix). Of these, 45 (30.2%) species were recorded only in savanna vegetation. Fourteen (*Elanoides forficatus, Gampsonyx swainsonii, Zenaida auriculata, Columbina passerina, Tapera naevia, Bubo virginianus, Picumnus cirratus, Synallaxis albescens, Elaenia chiriquensis, Myiarchus swainsoni, Tyrannus melancholicus, Progne chalybea, Troglodytes aedon,* and *Volatinia jacarina*) are known also to inhabit other open-vegetation habitats within Amazonia (e.g., várzea grasslands, open second-growth forest, open riverine forests, mangroves).

To the remaining 31 savanna species, we add *Polystictus pectoralis* (Bearded Tachuri), the only savanna species listed by Novaes (1974, 1978) not recorded in our study area. The ranges of these 32 savanna species may be grouped into three major categories: (a) three species whose distribution in South America is mostly restricted to the savannas north of the Amazon (*Colinus cristatus, Burhinus bistriatus,* and *Sturnella magna*; only the latter has been recorded south of the Amazon, in a small savanna patch near the mouth of the Tocantins river [Snethlage 1914]); (b) Twenty-seven species that have ranges centered in the open vegetations south and east of Amazonia (e.g., Caatinga and Cerrado) and which have isolated populations in patches of Amazonian savannas (Table 1); (c) two species distributed in a patchy fashion almost all the way around Amazonia, with isolated populations in both Amazonian savannas and open vegetation in the Andean Region (*Piranga flava* and *Tangara cayana*).

DISCUSSION

Patterns of distribution of savanna birds and their implications for the refuge theory.—The patterns of distribution of the savanna-restricted avifauna of Amapá show biogeographic influences from the two largest South American savanna regions: the Llanos and the Cerrado. However, because the number of elements that have ranges centered in Central Brazil surpasses that of elements recorded mainly for savannas in northern South America by nine times, it is obvious that Amapá savannas received much more influence from the Cerrado avifauna than from the Llanos avifauna.

Silva (1995a) showed that the number of Central Brazilian species (his Peri-Atlantic elements) present in savannas of northern South America decreases from Amapá both to the southeast (Marajó Island and other smaller islands in the Amazon estuary) and west (Parú-Trombetas and Santarém towards Roraima and the Llanos). This indicates that Amapá savannas have been much

TABLE 1

Central Brazilian Species Recorded for the *terra firme* Savannas of Amapá. Species Marked with an Asterisk Occur in Central or North America as Well. However, Their Ranges in South America Are Mainly Centered on Regions Located South of Amazonia (*e.g.*, Caatinga, Cerrado, and Chaco)

Elanus leucurus*	Anthus lutescens
Circus buffoni	Cypsnagra hirundinacea
Buteo albicaudatus*	Neothraupis fasciata
Uropelia campestris	Sporophila leucoptera
Aratinga aurea	Sporophila bouvreuil
Chordeiles pusillus	Myiospiza humeralis
Chrysolampis mosquitus	Emberizoides herbicola*
Heliactin bilopha	
Colaptes campestris	
Lepidocolaptes angustirostris	
Formicivora rufa	
Suiriri suiriri	
Elaenia cristata	
Polystictus pectoralis	
Euscarthmus rufomarginatus	
Xolmis cinerea	
Xolmis velata	
Tyrannus savanna*	
Tyrannus albogularis	
Mimus saturninus	

more important than any other patch of Amazonian savanna in maintaining viable populations of savanna-restricted species that have their centers of distribution in Central Brazil.

Silva (1995a) also suggested that the high density of species with ranges centered in Central Brazil in the savannas near the Atlantic coast (Amapá and Marajó) may indicate that the most recent biotic connections between savanna patches north of Amazon and the Cerrado Region occurred along this path (Haffer 1967, Veloso et al. 1975) rather than directly across central Amazonia, as has been generally suggested under the framework of the refuge theory (Haffer 1967, 1974).

The coastal savanna connection hypothesis is also supported by the fact that all Central Brazilian species recorded in the savanna islands in central Amazonia also occur in the coastal savannas (Silva 1995a). Based on this evidence, we suggest that Quaternary biotic connections between central Amazonian savannas and the Cerrado Region occured only through the coastal savannas. Small patches of upland savannas located along the Amazon (FIBGE 1988) can represent relicts of a narrow, ancient savanna belt that linked the coastal (Amapá and Marajó) with the central Amazonian savannas (Santarém and Parú-Trombetas).

One could suggest that the high number of Central Brazilian species found in the coastal savannas is a consequence of the strong ecological similarities between these savannas and those of Central Brazil rather than an indication of recent biotic connections. However, this seems not to be the case, as coastal savannas differ from those of the Cerrado Region by a number of ecological features, such as soils, climate, altitude, and floristic composition (Veloso et al. 1975, Eiten 1972).

Ávila-Pires (1995) mapped the ranges of all savanna lizards recorded in Amazonia and concluded that there is no evidence for the existence of a broad savanna corridor across central Amazonia. She also suggested that a savanna connection along the Atlantic coast better explains the patterns of distribution observed.

The presence of a savanna corridor along the Atlantic coast also has support from the geosciences. Paleopalinological studies in Surinam and Guyana indicated that coastal savannas expanded their distribution, occupying the exposed terrains when sea level was around 70–100 m lower than at present during the Late Pleistocene-Holocene (Van der Hammen 1974). In contrast, there is not any unequivocal paleoecological evidence supporting the hypothesis of a broad savanna corridor right across central Amazonia during the Quaternary (Salo 1987, Colinveaux 1987). The paleopalinological studies in Rondônia (Van der Hammen and Absy 1994) and Carajás (Absy et al., 1991) that are generally used as support for postulating a central Amazonian savanna corridor are in the transition zone between Amazonia and the Cerrado Region. Therefore, the documented pollen history of these two sites may be explained by ecotonal changes during the Quaternary climatic cycles.

Based on phytogeographical data, Prance (1987) also questioned the existence of a broad savanna corridor in central Amazonia during the Quaternary. Alternatively, he suggested that most probably dry forests rather than savanna-like vegetation replaced humid forests in central Amazonia during the Quaternary glacial periods. If Prance's suggestion is correct, then one can deduce that perhaps Quaternary climatic-vegetational fluctuations had less influence on the speciation of most Amazonian forest birds than it has been usually suggested (Haffer 1969, 1974, 1985). The main reason for this is that the barrier effect of dry forests would be limited, because several Amazonian forest bird species are able to survive in dry forests that are found along the borders of Amazonia (Fry 1970; Willis 1976; Willis and Oniki 1990; Silva and Oren, unpubl. data).

Conservation.—Amazonian savannas are probably one of the most threatened ecosystems in Brazilian Amazonia. Two main factors contribute to this: (1) Amazonian savannas are distributed as islands in a "sea" of forest, and their total area is only 150,000 km^2, or 3.0% of the region (Pires 1973); (2) their occupation and use by humans is somewhat easier and less expensive than areas covered by dense forest, so some of the principal regional population centers with rapid growth (e.g., Boa Vista, Macapá, Humaitá, Santarém) are in or near the main patches of the Amazonian savannas.

During the last decades, Amazonian conservation planning and policies have focused almost exclusively on forested ecosystems. As a result, the Amazonian savanna islands and other non-forest ecosystems have been entirely neglected, and none has any representative portion protected in a conservation unit. All major savanna islands in Brazilian Amazonia are threatened, in particular those in Amapá and Roraima and near the cities of Santarém and Monte Alegre, Pará, and Humaitá, Amazonas. Ironically, although much of Amazonia is threatened by deforestation, one of the major threats to the Amapá savannas is a forestation program utilizing Caribbean Pine (*Pinus caribea*) for production of paper pulp.

Both as natural laboratories for the study of isolation processes and as reservoirs of important components of regional biological diversity, Amazonian upland savannas merit specific conservation programs. We recommend that significant sections (on the order of at least several thousands of hectares) of each of the major islands be demarcated for conservation. An immediate moratorium on new development projects in the islands, an assessment of the alterations they have already suffered, detailed biological surveys, and designation of representative portions of the best-preserved sections as reserves are called for to guarantee the conservation of this important part of Amazonian biodiversity.

ACKNOWLEDGMENTS

Our expedition to Amapá was funded by a grant of the J. D. and C. T. MacArthur Foundation to the Museu Paraense Emílio Goeldi, administered through the World Wildlife Fund-US. We are very grateful to B. Rabelo, former director of the Museu de História Natural A. Costa Lima (MHNACL), for his support. We thank M. Santa Brígida, N. Santa Brígida and technicians of the MHNACL for field assistance. We thank J. V. Remsen, Jr., M. Robbins, K. J. Zimmer, J. Fjeldså, and W. Belton for reviewing the manuscript. We thank M. LeCroy and G. Barrowclough (AMNH), J. Bates (FMNH), and R. Dekker (RMNH) for allowing one of us (Silva) to use the collections under their care. We thank G. F. Mees for his information on the taxonomy and distribution of savanna birds in Surinam. Collection studies were supported by the Frank M. Chapman Memorial Fund and the Danish Natural Science Research Council (Grant J.no. 11-0390). The authors are supported by fellowships (Oren) or scholarships (Silva, Roma and Henriques) from the Conselho Nacional de Desenvolvimento Científico e Tecnológico (CNPq), Brasília, Brazil. We dedicate this paper to the memory of Ted Parker, for his great contribution to South American ornithology.

LITERATURE CITED

AB' SABER, A. N. 1977. Os Domínios morfoclimáticos da América do Sul. Primeira Aproximação. Geomorfologia 52:1–21.

ABSY, M. L., A. CLEFF, M. FOURNIER, L. MARTIN, M. SERVANT, M. SIFFEDINE, M. FERREIRA DA SILVA, F. SOUBIES, K. SEGUIO, B. TURCQ, AND T. VAN DER HAMMEN. 1991. Mise en évidence de quatre phases d'ouverture de la forest dense dans le sud-est de l'Amazonie au cours des 60 000 dernières annés. Première comparaison avec d'austres régions tropicales. C. R. Acad. Sci. Paris 312, sér. II:673–678.

ÁVILA-PIRES, T. C. S. 1995. Lizards of Brazilian Amazonia (Reptilia: Squamata). Zool. Verh. Leiden 299:1–706.
AZEVEDO, L. G. 1967. Tipos eco-fisionômicos de vegetação do Território Federal do Amapá. Rev. Brasil. Geografia 29:25–51.
BERG, M. E. VAN DEN, AND B. RABELO. 1982. Nota prévia sobre o estudo dos cerrados do Amapá. Anais Congr. Brasil. Bot. 32:134–140.
CAVALCANTI, R. B. 1989. Ocorrência de *Neothraupis fasciata* (Passeriformes: Thraupidae) no Estado do Amapá. Anais Encontro Nac. de Anilhadores Aves 5:8.
CHERRIE, G. K. 1916. New birds from the collection of the Collins-Day expedition to South America. Bull. Amer. Mus. Nat. Hist. 35:391–397.
COLINVEAUX, P. 1987. Amazonian diversity in light of the paleoecological record. Quat. Sci. Rev. 6:93–114.
CRACRAFT, J. 1985. Historical biogeography and patterns of differentiation within the South American avifauna. Areas of endemism. Pages 49–84 *in* Neotropical Ornithology (P. A. Buckley, M. S. Foster, E. S. Morton, R. S. Ridgely, and F. G. Buckley, Eds.). Ornithol. Monogr. 36.
DAVIS, S. E. 1993. Seasonal status, relative abundance, and behavior of the birds of Concepción, Departamento Santa Cruz, Bolivia. Fieldiana Zoologia 71:1–32.
EITEN, G. 1972. The cerrado vegetation of Brazil. Bot. Rev. 38:201–341.
FIBGE. 1988. Mapa da vegetação do Brasil. Fundação Instituto Brasileiro de Geografia e Estatística, Rio de Janeiro.
FRY, C. H. 1970. Ecological distribution of birds in the northeastern Mato Grosso state, Brazil. Anais Acad. Bras. Ciências 42:275–318.
GRISCOM, L., AND J. C. GREENWAY, JR. 1941. Birds of lower Amazonia. Bull. Mus. Comp. Zool. 88:83–344.
HAFFER, J. 1967. Zoogeographical notes on the 'nonforest' lowland bird faunas of northwestern South America. Hornero 10:315–333.
HAFFER, J. 1969. Speciation in Amazonian forest birds. Science 165:131–137.
HAFFER, J. 1974. Avian speciation in South America. Publ. Nuttall Ornith. Club 14:1–390.
HAFFER, J. 1978. Distribution of Amazonian birds. Bonn. zool. Beitr. 29:38–78.
HAFFER, J. 1985. Avian zoogeography of the Neotropical lowlands. Pages 113–146 *in* Neotropical Ornithology (P. A. Buckley, M. S. Foster, E. S. Morton, R. S. Ridgely, and F. G. Buckley, Eds.). Ornithol. Monogr. 36.
HAVERSCHMIDT, F., AND G. F. MEES. 1994. Birds of Suriname. Vaco, Paramaribo.
HILTY, S. L., AND W. L. BROWN. 1986. A Guide to the Birds of Colombia. Princeton Univ. Press, Princeton, New Jersey.
HILTY, S. L. 1997. Seasonal distribution of birds of a cloud-forest locality, the Anchicayá Valley, in western Colombia. Pp. 321–343 *in* Studies in Neotropical Ornithology Honoring Ted Parker (J. V. Remsen, Jr., Ed.), Ornithol. Monogr. No. 48.
HOWARD, R., AND A. MOORE. 1980. A Complete Checklist of Birds of the World. Oxford University Press, Oxford.
IBGE. 1966. Atlas do Amapá. Instituto Brasileiro de Geografia e Estatística, Rio de Janeiro.
KINNEAR, N. B. 1927. [*Picumnus cirrhatus* (sic) *confusus*, subsp. nov]. Bull. Brit. Orn. Club 47:112–113.
LANYON, W. E. 1978. Revision of the *Myiarchus* flycatchers of South America. Bull. Amer. Mus. Nat. Hist. 161:427–628.
MAGNANINI, A. 1953. As regiões naturais do Amapá. Rev. Brasil. Geografia 14:243–304.
MARINI, M. A., AND R. B. CAVALCANTI. 1990. Migrações de *Elaenia albiceps chilensis* e *Elaenia chiriquensis albivertex* (Aves: Tyrannidae). Bol. Mus. Para. Emilio Goeldi, ser. zool. 6:59–67.
MEES, G. F. 1968. Enige voor de avifauna van Suriname nieuwe vogelsoorten. Gerfaut 58:101–107.
MEES, G. F. 1974. Additions to the avifauna of Suriname. Zool. Mededelingen 48:55–67.
MEES, G. F. 1985. Nomenclature and systematics of birds from Suriname. Proc. Neth. Acad. Sci., C 88:75–91.
MEYER DE SCHAUENSEE, R. 1966. The Species of Birds of South America and Their Distribution. Livingston Publ. Co., Narberth, Pennsylvania.
NOVAES, F. C. 1974. Ornitologia do Território do Amapá. I. Publ. Avulsas Mus. Para. Emílio Goeldi 25:1–121.
NOVAES, F. C. 1978. Ornitologia do Território do Amapá. II. Publ. Avulsas Mus. Para. Emílio Goeldi 29:1–75.
NOVAES, F. C. 1981. Sobre algumas aves do litoral do Estado do Pará. Anais Soc. Sul-Riograndense Ornitol. 2:5–8.
PAYNTER, R. A., JR. 1970. Emberizinae. Pages 1–214 *in* Check-list of Birds of the World, vol. 12 (R. A. Paynter, Jr. and R. W. Storer, Eds.). Museum of Comparative Zoology, Cambridge, Massachusetts.
PINTO, O. M. O. 1944. Catálogo das aves do Brasil e lista dos exemplares existentes na coleção do departamento de zoologia, 2ª parte. Secretaria de Agricultura, Indústria e Comércio, Depto. Zoologia, São Paulo.
PINTO, O. M. O. 1978. Novo catálogo das aves do Brasil. Editora Gráfica dos Tribunais, São Paulo.
PIRES, J. M. 1973. Tipos de vegetação da Amazônia. Publ. Avulsas Mus. Para. Emílio Goeldi 20:179–202.
PRANCE, G. T. 1987. Vegetation. Pages 28–45 *in* Biogeography and Quaternary History in Tropical America (T. C. Whitmore and G. T. Prance, Eds.). Clarendon Press, Oxford, U.K.
REENSEN, T. A. 1974. Twelve bird species new for Surinam. Ardea 62:118–122.

RIDGELY, R. S., AND G. TUDOR. 1994. The Birds of South America, Vol. 2. Oxford University Press, Oxford, U.K.
SALO, J. 1987. Pleistocene forest refuges in the Amazon: evaluation of the biostratigraphical, lithostratigraphical and geomorphological data. Ann. Zool. Fennici 24:203–211.
SARMIENTO, G. 1983. The savannas of tropical America. Pages 245–288 in Ecosystems of the World 13. Tropical Savannas (F. Bourliére, Ed.). Elsevier, Amsterdan.
SHORT, L. L. 1972. Systematics and behavior of South American Flickers (Aves, *Colaptes*). Bull. Amer. Mus. Nat. Hist. 149:1–109.
SIBLEY, C. G., AND B. L. MONROE. 1990. Distribution and Taxonomy of Birds of the World. Yale University Press, New Haven, Connecticut.
SILVA, J. M. C. 1995a. Biogeographic analysis of the South American cerrado avifauna. Steenstrupia 21: 49–67.
SILVA, J. M. C. 1995b. Birds of the Cerrado Region, South America. Steenstrupia 21:69–92.
SNETHLAGE, E. 1914. Catálogo das aves amazônicas. Bol. Mus. Goeldi 8:1–530.
TODD, W. E. C. 1946. Critical notes on the woodpeckers. Ann. Carnegie Mus. 30:297–317.
TRAYLOR, M. A., JR. 1979. Tyrannidae. Pages 1–245 in Check-list of Birds of the World, Vol. 8 (M. A. Traylor, Ed.). Museum of Comparative Zoology, Cambridge, Massachusetts.
TRAYLOR, M. A., JR. 1982. Notes on tyrant flycatchers (Aves, Tyrannidae). Fieldiana Zoology 13:1–22.
TOSTAIN, O., J. L. DUJARDIN, C. H. ERARD, AND J. M. THIOLLAY. 1992. Oiseaux de Guyane. Societe d'Etudes Ornithologiques. Maxéville, France.
VAN DER HAMMEN, T. 1974. The Pleistocene changes of vegetation and climate in tropical South America. J. Biogeogr. 1:3–26.
VAN DER HAMMEN, T., AND M. L. ABSY. 1994. Amazonia during the last glacial. Palaeogeo. Palaeoclimat. Palaeoeco. 109:247–261.
VELOSO, H. P., L. GOÉS FILHO, P. F. LEITE, S. B. SILVA, H. DE C. FERREIRA, R. L. LOUREIRO, AND E. F. DE M. TEREZO. 1975. Vegetação: as regiões fitoecológicas, sua natureza e seus recursos econômicos-estudo fitogeográfico. Pages 305–404 in Ministério das Minas e Energia. Departamento Nacional de Produção Mineral. Projeto RADAMBRASIL, Folha NA.20 Boa Vista e partes das folhas NA. 21 Tumucumaque, NB. 20 Roraima e NB. 21. Rio de Janeiro, Brazil.
WILLARD, D. E., M. S. FOSTER, G. L. BARROWCLOUGH, P. F. CANNELL, S. L. COATS, J. L. CRACRAFT, AND J. P. O'NEILL. 1991. The birds of Cerro de la Neblina, Territorio Federal Amazonas, Venezuela. Fieldiana Zoology 65:1–80.
WILLIS, E. O. 1976. Effects of a cold wave on an Amazonian avifauna in the upper Paraguay drainage, western Mato Grosso, and suggestions on oscine-suboscine relationships. Acta Amazonica 6:379–394.
WILLIS, E. O., AND Y. ONIKI, Y. 1990. Levantamento preliminar das aves de inverno em dez áreas do sudeste do Mato Grosso, Brasil. Ararajuba 1:19–38.

APPENDIX

LIST OF SPECIES RECORDED IN CAMPUS EXPERIMENTAL OF EMBRAPA, MACAPÁ, AMAPÁ, BRAZIL. HABITAT: (F) GALLERY FOREST, (S) TERRA FIRME SAVANNAS, (B) BURITI PALMS AND WET GRASSLANDS ALONG STREAM VALLEYS. ABUNDANCE: (R) RARE, (U) UNCOMMON, (FC) FAIRLY COMMON, AND (C) COMMON. GONAD SIZE[1]: (S) SMALL, (M) MODERATE, (L) LARGE

	Habitat	Abundance	Sex	Body mass (g)	Gonads
TINAMIDAE					
Little Tinamou (*Crypturellus soui*)	F	FC	—	—	—
Black-capped Tinamou (*Crypturellus atrocapillus*)	F	R	—	—	—
ARDEIDAE					
Great Egret (*Ardea alba*)	S, B	U	—	—	—
Snowy Egret (*Egretta thula*)	S, B	U	—	—	—
Striated Heron (*Butorides striatus*)	B	U	—	—	—
Rufescent Tiger-Heron (*Tigrisoma lineatum*)	B, F	R	—	—	—
Zigzag Heron (*Zebrilus undulatus*)	F	R	—	—	—
CATHARTIDAE					
Black Vulture (*Coragyps atratus*)	S, B	C	—	—	—
King Vulture (*Sarcoramphus papa*)	F	R	—	—	—
ACCIPITRIDAE					
Hook-billed Kite (*Chondrohierax uncinatus*)	B	R	—	—	—
Swallow-tailed Kite (*Elanoides forficatus*)	S	U	—	—	—
Pearl Kite (*Gampsonyx swainsonii*)	S	R	—	—	—
White-tailed Kite (*Elanus leucurus*)	S	U	—	—	—
Plumbeous Kite (*Ictinia plumbea*)	F	U	—	—	—
Long-winged Harrier (*Circus buffoni*)	S	R	—	—	—
Savanna Hawk (*Heterospizias meridionalis*)	S, B	U	—	—	—
Black-collared Hawk (*Busarellus nigricollis*)	S, B	R	—	—	—
Gray Hawk (*Asturina nitida*)	F	U	—	—	—
Roadside Hawk (*Buteo magnirostris*)	S, F, B	C	—	—	—
White-tailed Hawk (*Buteo albicaudatus*)	S	R	—	—	—
FALCONIDAE					
Red-throated Caracara (*Daptrius americanus*)	F	R	—	—	—
Crested Caracara (*Polyborus plancus*)	S, B	F	—	—	—
Yellow-headed Caracara (*Milvago chimachima*)	S, B	C	—	—	—
Laughing Falcon (*Herpetotheres cachinnans*)	F	U	—	—	—

APPENDIX
Continued

	Habitat	Abundance	Sex	Body mass (g)	Gonads
CRACIDAE					
Little Chachalaca (*Ortalis motmot*)	F	U	—	—	—
ODONTOPHORIDAE					
Crested Bobwhite (*Colinus cristatus*)	S	F	4 ♀♀ 4 j.	142.5 (20.6) 5.8 (0.8)	L S
RALLIDAE					
Gray-necked Wood-Rail (*Aramides cajanea*)	F	F	—	—	—
Russet-crowned Crake (*Laterallus viridis*)	B	U	—	—	—
JACANIDAE					
Wattled Jacana (*Jacana jacana*)	B	U	—	—	—
BURHINIDAE					
Double-striped Thick-knee (*Burhinus bistriatus*)	S	R	—	—	—
CHARADRIIDAE					
Southern Lapwing (*Vanellus chilensis*)	S, B	U	—	—	—
COLUMBIDAE					
Scaled Pigeon (*Columba speciosa*)	F	R	—	—	—
Pale-vented Pigeon (*Columba cayennensis*)	S, F	F	3 ♀♀	213.3 (15.2)	L
Plumbeous Pigeon (*Columba plumbea*)	F	R	—	—	—
Eared Dove (*Zenaida auriculata*)	S	C	5 ♂♂ 5 ♀♀ 4 ♂j. 4 ♀♀	117.0 (6.7) 109.6 (10.5) 78.0 (15.4) 30.5 (3.1)	L L(3), S(2) M L(3), S(1)
Common Ground-Dove (*Columbina passerina*)	S	C	—	—	—
Ruddy Ground-Dove (*Columbina talpacoti*)	S, B	U	—	—	—
Blue Ground-Dove (*Claravis pretiosa*)	F	R	—	—	—
Long-tailed Ground-Dove (*Uropelia campestris*)	S	R	—	—	—
White-tipped Dove (*Leptotila verreauxi*)	F	F	1 ♂	130.0	L
PSITTACIDAE					
Blue-and-yellow Macaw (*Ara ararauna*)	F, B	R	—	—	—
Red-bellied Macaw (*Ara manilata*)	F, B	U	—	—	—
Red-shouldered Macaw (*Ara nobilis*)	F, B	F	1 ♂	160.0	S

APPENDIX
CONTINUED

	Habitat	Abundance	Sex	Body mass (g)	Gonads
Peach-fronted Parakeet (*Aratinga aurea*)	S	C	8 ♂♂ 5 ♀♀	86.5 (6.7) 82.8 (5.6)	M(7), L(1) L(4), S(1)
Green-rumped Parrotlet (*Forpus passerinus*)	S, B	U	—	—	—
Canary-winged Parakeet (*Brotogeris versicolurus*)	F	F	—	—	—
Tui Parakeet (*Brotogeris sanctithomae*)	B	R	—	—	—
Yellow-crowned Parrot (*Amazona ochrocephala*)	F	F	1 ♂ 2 ♀♀	530.0 365.0, 370.0	L L
Orange-winged Parrot (*Amazona amazonica*)	F	U	1 ♀	310.0	L
CUCULIDAE					
Squirrel Cuckoo (*Piaya cayana*)	F, S	U	1 ♂ 1 ♀	84.0 94.0	M L
Little Cuckoo (*Piaya minuta*)	F	R	—	—	—
Smooth-billed Ani (*Crotophaga ani*)	S, B, F	C	—	—	—
Guira Cuckoo (*Guira guira*)	S, B	F	—	—	—
Striped Cuckoo (*Tapera naevia*)	S	U	—	—	—
STRIGIDAE					
Tawny-bellied Screech-Owl (*Otus watsonii*)	F	U	—	—	—
Great Horned Owl (*Bubo virginianus*)	S	R	—	—	—
Ferruginous Pygmy-Owl (*Glaucidium brasilianum*)	S, F	U	1 ♂	54.0	S
NYCTIBIIDAE					
Common Potoo (*Nyctibius griseus*)	F	R	—	—	—
CAPRIMULGIDAE					
Least Nighthawk (*Chordeiles pusillus*)	S	R	—	—	—
Pauraque (*Nyctidromus albicollis*)	F	R	—	—	—
Ladder-tailed Nightjar (*Hydropsalis climacocerca*)	B	U	—	—	—
APODIDAE					
Band-rumped Swift (*Chaetura spinicauda*)	F	R	—	—	—
Short-tailed Swift (*Chaetura brachyura*)	F	U	1 ♀	—	—
Fork-tailed Palm-Swift (*Tachornis squamata*)	S, B	F		9.5	S

APPENDIX
CONTINUED

	Habitat	Abundance	Sex	Body mass (g)	Gonads
TROCHILIDAE					
Rufous-breasted Hermit (*Glaucis hirsuta*)	F	R	—	—	—
Long-tailed Hermit (*Phaethornis superciliosus*)	F	U	—	—	—
Reddish Hermit (*Phaethornis ruber*)	F	U	1 ?	2.2	?
Swallow-tailed Hummingbird (*Eupetomena macroura*)	S, F	F	2 ♂♂	7.2, 8.0	L, S
			1 ♂	7.8	L
Black-throated Mango (*Anthracothorax nigricollis*)	F	R	—	—	—
Ruby-topaz Hummingbird (*Chrysolampis mosquitus*)	S	U	2 ♂♂	3.6, 3.8	S
Fork-tailed Woodnymph (*Thalurania furcata*)	F	U	—	—	—
White-tailed Goldenthroat (*Polytmus guainumbi*)	S, B	U	4 ♂♂	4.2 (0.6)	L
Glittering-throated Emerald (*Amazilia fimbriata*)	B, S	F	2 ♀♀	4.6, 5.0	L, S
Horned Sungem (*Heliactin bilopha*)	S	U	1 ♂	2.8	L
			1 ♀	2.8	L
TROGONIDAE					
White-tailed Trogon (*Trogon viridis*)	F	R	—	—	—
ALCEDINIDAE					
Ringed Kingfisher (*Ceryle torquata*)	S, B	F	—	—	—
Green Kingfisher (*Chloroceryle americana*)	B	U	—	—	—
Green-and-rufous Kingfisher (*Chloroceryle inda*)	F	R	—	—	—
American Pygmy Kingfisher (*Chloroceryle aenea*)	F	U	—	—	—
MOMOTIDAE					
Blue-crowned Motmot (*Momotus momota*)	F	U	—	—	—
GALBULIDAE					
Yellow-billed Jacamar (*Galbula albirostris*)	F	R	—	—	—
BUCCONIDAE					
Swallow-wing (*Chelidoptera tenebrosa*)	F	U	—	—	—
RAMPHASTIDAE					
Toco Toucan (*Ramphastos toco*)	S, F	U	1 ♀	640.0	S

APPENDIX
CONTINUED

	Habitat	Abundance	Sex	Body mass (g)	Gonads
PICIDAE					
White-barred Piculet (*Picumnus cirratus*)	S	U	1 ♀	14.0	S
White Woodpecker (*Melanerpes candidus*)	S, F	R	—	—	—
Little Woodpecker (*Veniliornis passerinus*)	F	F	1 ♂	27.0	S
			1 ♀	30.5	M
Spot-breasted Woodpecker (*Colaptes punctigula*)	F	U	—	—	—
Campo Flicker (*Colaptes campestris*)	S	F	2 ♂♂	150.0	S, L
			1 ♀	150.0	L
Chestnut Woodpecker (*Celeus elegans*)	F	R	—	—	—
Crimson-crested Woodpecker (*Campephilus melanoleucus*)	F	U	—	—	—
DENDROCOLAPTIDAE					
Straight-billed Woodcreeper (*Xiphorhynchus picus*)	F	F	2 ♂♂	39.5, 41.0	L
			1 ♀	38.0	L
Buff-throated Woodcreeper (*Xiphorhynchus guttatus*)	F	U	1 ♀	54.0	S
Narrow-billed Woodcreeper (*Lepidocolaptes angustirostris*)	S	C	5 ♂♂	31.5 (1.1)	L(3), M(2)
			3 ♀♀	31.6 (1.0)	S, M, L
FURNARIIDAE					
Pale-breasted Spinetail (*Synallaxis albescens*)	S	U	—	—	—
Plain-crowned Spinetail (*Synallaxis gujanensis*)	F	R	—	—	—
Plain Xenops (*Xenops minutus*)	F	U	—	—	—
FORMICARIIDAE					
Great Antshrike (*Taraba major*)	F	F	1 ♂	56.0	M
Barred Antshrike (*Thamnophilus doliatus*)	F	C	4 ♂♂	27.1 (1.6)	M(3), S(1)
			7 ♀♀	28.1 (1.5)	M(4), S(3)
Guianan Slaty-Antshrike (*Thamnophilus punctatus*)	F	F	2 ♂♂	19.5, 20.0	M, L
			1 ♀	20.0	M
White-flanked Antwren (*Myrmotherula axillaris*)	F	U	1 ♂	7.0	L
			1 ♀	8.5	M
White-fringed Antwren (*Formicivora grisea*)	F	F	1 ♂	10.5	L
			2 ♀	10.0, 11.0	S
Rusty-backed Antwren (*Formicivora rufa*)	S	U	1 ♂	11.0	M
			1 ♀	12.5	S

APPENDIX
CONTINUED

	Habitat	Abundance	Sex	Body mass (g)	Gonads
Dusky Antbird (*Cercomacra tyrannina*)	F	U	2 ♂♂	15.0, 17.0	M, S
Warbling Antbird (*Hypocnemis cantator*)	F	R	—	—	—
Black-chinned Antbird (*Hypocnemoides melanopogon*)	F	R	—	—	—
TYRANNIDAE					
Southern Beardless-Tyrannulet (*Camptostoma obsoletum*)	F, S	F	—	—	—
Mouse-colored Tyrannulet (*Phaeomyias murina*)	S, F	F	1 ♂	10.5	L
Southern Scrub-Flycatcher (*Sublegatus modestus*)	S, F	U	1 ♀	9.6	L
Campo Suiriri (*Suiriri suiriri*)	S	C	8 ♂♂	23.3 (2.2)	L(5), S(3)
			6 ♀♀	19.2 (2.1)	S(3), M(3)
Forest Elaenia (*Myiopagis gaimardii*)	F	F	2 ♂♂	10.5, 11.5	M, L
			2 ♀♀	9.4, 10.5	S, M
Yellow-bellied Elaenia (*Elaenia flavogaster*)	S, F	F	5 ♂♂	21.3 (2.6)	L(4), M(1)
			2 ♀♀	21.5, 24.0	M, L
Plain-crested Elaenia (*Elaenia cristata*)	S	F	4 ♂♂	18.3 (0.2)	L
			2 ♀♀	14.0, 18.0	M, L
Lesser Elaenia (*Elaenia chiriquensis*)	S	C	14 ♂♂	16.3 (1.3)	L(11), M(3)
			4 ♀♀	14.1 (0.5)	S(2), M(2)
			1 ?	11.5	?
Rufous-sided Pygmy-Tyrant (*Euscarthmus rufomarginatus*)	S	R	—	—	—
Helmeted Pygmy-Tyrant (*Lophotriccus galeatus*)	F	F	4 ♂♂	6.5 (0.4)	L(4), S(1)
			1 ♀	5.8	M
Smoky-fronted Tody-Flycatcher (*Todirostrum fumifrons*)	F	U	2 ♂♂	6.0, 6.0	S, M
Spotted Tody-Flycatcher (*Todirostrum maculatum*)	F	U	1 ♀	6.6	S
Common Tody-Flycatcher (*Todirostrum cinereum*)	S, B	C	6 ♂♂	6.5 (0.5)	L
			3 ♀♀	6.4 (0.5)	S(2), M(1)
Yellow-breasted Flycatcher (*Tolmomyias flaviventris*)	F	F	6 ♂♂	11.5 (0.3)	L
			5 ♀♀	11.9 (0.5)	S(2), M(2)
			1 ?	11.0	?
Sulphur-rumped Flycatcher (*Myiobius barbatus*)	F	R	—	—	—
Gray Monjita (*Xolmis cinerea*)	S	F	1 ♂	54.0	L
			2 ♀♀	56.0, 58.0	S, M
White-rumped Monjita (*Xolmis velata*)	S	R	—	—	—

APPENDIX
Continued

	Habitat	Abundance	Sex	Body mass (g)	Gonads
Swainson's Flycatcher (*Myiarchus swainsoni*)	S	C	11 ♂♂	23.5 (1.7)	L(2), S(1)
			3 ♀♀	21.2 (0.7)	L(2), S(1)
Short-crested Flycatcher (*Myiarchus ferox*)	F	U	—	—	L
Brown-crested Flycatcher (*Myiarchus tyrannulus*)	F	F	—	—	S, L
Great Kiskadee (*Pitangus sulphuratus*)	S, F, B	C	3 ♂♂	29.8 (2.2)	—
Boat-billed Flycatcher (*Megarynchus pitangua*)	F	U	2 ♀♀	28.0, 28.5	—
Rusty-margined Flycatcher (*Myiozetetes cayanensis*)	F	U	—	—	—
Sulphury Flycatcher (*Tyrannopsis sulphurea*)	F	U	2 ♂♂	52.0, 54.0	L
			2 ♀♀	56.0, 62.0	M, L
White-throated Kingbird (*Tyrannus albogularis*)	S	F	3 ♂♂	36.1 (0.7)	L
Tropical Kingbird (*Tyrannus melancholicus*)	S	F	4 ♀♀	36.5 (2.0)	S(1), L(3)
			3 ♂♂	40.3 (2.3)	L
			1 ♂	39.5	L
Fork-tailed Flycatcher (*Tyrannus savana*)	S	C	6 ♂♂	29.5 (0.6)	L(5), M(1)
			2 ♀♀	29.5, 28.0	S
			1 ?	28.0	?
White-winged Becard (*Pachyramphus polychopterus*)	F	U	1 ♂	19.0	L
Masked Tityra (*Tityra semifasciata*)	F	U	1 ♀	74.0	M
PIPRIDAE					
Pale-bellied Tyrant-Manakin (*Neopelma pallescens*)	F	R	—	—	—
White-bearded Manakin (*Manacus manacus*)	F	R	—	—	—
Crimson-hooded Manakin (*Pipra aureola*)	F	F	4 ♂♂	15.8 (0.7)	L(3), S(3)
			6 ♀♀	15.9 (0.8)	
HIRUNDINIDAE					
Gray-breasted Martin (*Progne chalybea*)	S	U	1 ♀	42.0	L
MOTACILLIDAE					
Yellowish Pipit (*Anthus lutescens*)	S	U	—	—	—
TROGLODYTIDAE					
Black-capped Donacobius (*Donacobius atricapillus*)	B	U	—	—	—
Buff-breasted Wren (*Thryothorus leucotis*)	F	F	2 ♂	18.0, 20.0	L, M
			2 ♀	14.5, 17.5	M
			1 ?	18.5	?
House Wren (*Troglodytes aedon*)	S	C	1 ♂	13.0	L

APPENDIX
CONTINUED

	Habitat	Abundance	Sex	Body mass (g)	Gonads
MIMIDAE					
Chalk-browed Mockingbird (*Mimus saturninus*)	S	F	3 ♂♂	64.6 (6.4)	M(2), L(1)
MUSCICAPIDAE					
Pale-breasted Thrush (*Turdus leucomelas*)	F	C	5 ♂♂	64.4 (4.5)	L(4), M(1)
			4 ♀♀	65.5 (3.4)	M(3), S(1)
White-necked Thrush (*Turdus albicollis*)	F	R	—	—	—
Tropical Gnatcatcher (*Polioptila plumbea*)	F, S	F	3 ♂♂	6.4 (0.4)	L
			2 ♀♀	7.2, 8.0	S, M
EMBERIZIDAE					
Rufous-collared Sparrow (*Zonotrichia capensis*)	S, F	F	4 ♂♂	20.4 (0.7)	L(3), S(1)
			3 ♀♀	18.0 (1.0)	M(2), S(1)
Grassland Sparrow (*Myospiza humeralis*)	S	C	7 ♂♂	15.1 (0.3)	L(4), M(3)
			8 ♀♀	13.6 (1.5)	M(5), S(3)
Wedge-tailed Grass-Finch (*Emberizoides herbicola*)	S	C	6 ♂♂	24.2 (1.6)	M(4), L(1)
			4 ♀♀	24.8 (1.1)	S(3), M(1)
Blue-black Grassquit (*Volatinia jacarina*)	S	U	—	—	L
Plumbeous Seedeater (*Sporophila plumbea*)	S, B	F	5 ♂♂	10.3 (0.3)	S
			1 ♀	10.5	
Variable Seedeater (*Sporophila americana*)	F, S	U	—	—	—
White-bellied Seedeater (*Sporophila leucoptera*)	S	R	—	—	—
Capped Seedeater (*Sporophila bouvreuil*)	S	R	—	—	—
Lesser Seed-Finch (*Oryzoborus angolensis*)	F	U	2 ♂♂	11.5, 13.5	S, M
			1 ♀	11.5	S
Pectoral Sparrow (*Arremon taciturnus*)	F	R	—	—	—
Buff-throated Saltator (*Saltator maximus*)	F	U	—	—	—
THRAUPIDAE					
Black-faced Tanager (*Schistochlamys melanopis*)	F, S	U	—	—	—
White-banded Tanager (*Neothraupis fasciata*)	S	F	8 ♂♂	27.1 (1.1)	L(4), M(4)
			3 ♀♀	26.1 (1.5)	M(2), L(1)
White-rumped Tanager (*Cypsnagra hirundinacea*)	S	C	6 ♂♂	31.6 (2.6)	M(3), L(3)
			8 ♀♀	31.3 (1.5)	M(5), S(3)
White-lined Tanager (*Tachyphonus rufus*)	F	U	—	—	—

APPENDIX
CONTINUED

	Habitat	Abundance	Sex	Body mass (g)	Gonads
Hepatic Tanager (*Piranga flava*)	S	F	5 ♂♂	41.1 (2.0)	L
Silver-beaked Tanager (*Ramphocelus carbo*)	F	C	6 ♂♂	23.7 (1.3)	L(4), M(2)
			6 ♀♀	22.6 (1.8)	M(3), S(3)
Blue-gray Tanager (*Thraupis episcopus*)	F, S	U	1 ♀	35.0	M
Palm Tanager (*Thraupis palmarum*)	B, F	U	1 ♂	36.0	L
			2 ♀♀	33.5, 38.5	M
Purple-throated Euphonia (*Euphonia chlorotica*)	S, F	U	1 ♂	10.5	M
Turquoise Tanager (*Tangara mexicana*)	F	R	—	—	—
Burnished-buff Tanager (*Tangara cayana*)	S	F	2 ♂♂	18.5, 19.5	M, L
			3 ♀♀	19.1 (0.6)	M(2), S(1)
Blue Dacnis (*Dacnis cayana*)	F	R	1 ♂	13.5	L
Red-legged Honeycreeper (*Cyanerpes cyaneus*)	F	R	—	—	—
PARULIDAE					
Bananaquit (*Coereba flaveola*)	F, S	F	5 ♂♂	9.8 (0.4)	L(4), S(1)
			1 ♀	8.0	L
			1 ?	10.0	?
Masked Yellowthroat (*Geothlypis aequinoctialis*)	B, S	R	—	—	—
VIREONIDAE					
Rufous-browed Peppershrike (*Cyclarhis gujanensis*)	F, S	F	4 ♂♂	27.7 (1.8)	L(3), M(1)
			1 ♀	24.0	L
			1 ?	23.0	?
Ashy-headed Greenlet (*Hylophilus pectoralis*)	F	U	1 ♀	11.5	M
ICTERIDAE					
Epaulet Oriole (*Icterus cayanensis*)	B, S	F	2 ♂	42.5, 44.5	L
			1 ♀	40.0	M
Unicolored Blackbird (*Agelaius cyanopus*)	B	R	—	—	—
Red-breasted Blackbird (*Leistes militaris*)	B	U	—	—	—
Eastern Meadowlark (*Sturnella magna*)	S	F	6 ♂♂	97.0 (6.5)	L
			2 ♀♀	72.0, 72.0	M, L
Shiny Cowbird (*Molothrus bonariensis*)	S, B	F	—	—	—

[1] Conventions for gonad developments (from Willard et al. 1991): Female, largest ovum (<1 × 1 mm = small; 1 × 1 mm = moderate; >1 × 1 mm = large). Males, under 10 g, testes (small, 1 × 1 mm; moderate, 2 × 1 mm; large, >2 × 2 mm); 10–50 g (small, 2 × 2 mm; moderate, <2 × 2 to 4 × 4 mm; large, >4 × 4 mm); 50–100 g (small, 3 × 3 mm; moderate, 4 × 3 to 7 × 4 mm; large, >7 × 4 mm); >100 g (small, 6 × 4 mm; moderate, >6 × 4 to >8 × 5 mm; large, >8 × 5 mm).

AN AVIFAUNAL SURVEY OF TWO TROPICAL FOREST LOCALITIES ON THE MIDDLE RIO JIPARANÁ, RONDÔNIA, BRAZIL

DOUGLAS F. STOTZ[1], SCOTT M. LANYON[2], THOMAS S. SCHULENBERG,
DAVID E. WILLARD, A. TOWNSEND PETERSON[3], AND
JOHN W. FITZPATRICK[4]
Division of Birds, Field Museum of Natural History,
Roosevelt Road at Lake Shore Drive,
Chicago, Illinois 60605, USA
[1]*Present address: Environmental and Conservation Programs,*
The Field Museum, Chicago, Illinois 60605, USA
[2]*Present address: Bell Museum of Natural History, 100 Ecology Building,*
University of Minnesota, St. Paul, Minnesota 55108, USA
[3]*Present address: Division of Ornithology, Museum of Natural History,*
University of Kansas, Lawrence, Kansas 66045, USA
[4]*Present address: Laboratory of Ornithology, Cornell University, Ithaca,*
New York 14850, USA

ABSTRACT.—We surveyed the avifauna of two humid lowland forest sites in eastern Rondônia, Brazil. Although Rondônia has been suggested as a site of a Pleistocene refugium for birds and contains a number of species and subspecies endemic to it, the state has attracted little field work by modern ornithologists. We provide locality lists with information on relative abundance, habitat use and migratory status for Cachoeira Nazaré, on the Rio Jiparaná, and Pedra Branca, on the Rio Anari. The species list of 459 at Cachoeira Nazaré is the largest for any single locality in Brazil. We also report in more detail on 30 species that are poorly known or for which eastern Rondônia represented a significant range extension. One species of bird (Rondonia Bushbird, *Clytoctantes atrogularis*) was recently described from the site (Lanyon et al. 1990), and an additional species (Gray-throated Leaftosser, *Sclerurus albigularis*) was recorded from Brazil for the first time.

RESUMO.—Apesar de seu grande número de espécies e subespécies de aves endêmicas, o estado de Rondônia é mal explorado por ornitólogos. Fizemos um levantamento de aves em duas localidades no leste do estado de Rondônia, Brasil, e fornecemos aqui listas com informações sobre a abundância relativa, o uso de habitat, e a categoria migratória de cada ave encontrada em Cachoeira Nazaré, no Rio Jiparaná, e em Pedra Branca, no Rio Anari. Com 459 espécies, a lista de Cachoeira Nazaré é a maior existente para uma única localidade no Brasil. Fornecemos mais detalhes sobre trinta espécies que são pouco conhecidas, ou que mostram uma grande extensão em área de distribuição. Uma espécie de ave (choca-de-garganta-preta, *Clytoctantes atrogularis*) foi recentemente decrita nesta localidade (Lanyon et al. 1990). Uma outra espécie (vira-folha-de-garganta-cinza, *Sclerurus albigularis*) foi registrada pela primeira vez no Brasil.

Since 1970, the state of Rondônia has experienced extensive deforestation. The rate of habitat loss accelerated in 1984 with the paving of the highway to Porto Velho in western Rondônia from the southeastern part of Brazil (Fearnside 1987, 1989). Because the Brazilian government also actively encouraged settlement in the state, the forests of Rondônia, along with their dependent fauna, are among the most threatened in Amazonia. The region between the Rio Madeira and the Rio Tapajós, including virtually the entire state of Rondônia, is recognized as "one of the main areas of endemism [among birds] within the southern Amazonian basin" (Cracraft 1985:71). Haffer (1974) postulated a Pleistocene refugium for birds located within this region, and similar refugia in this region have been proposed for other animal (Müller 1973, Brown 1977) and plant (Prance 1973) groups. However, the birds of Rondônia remain poorly known,

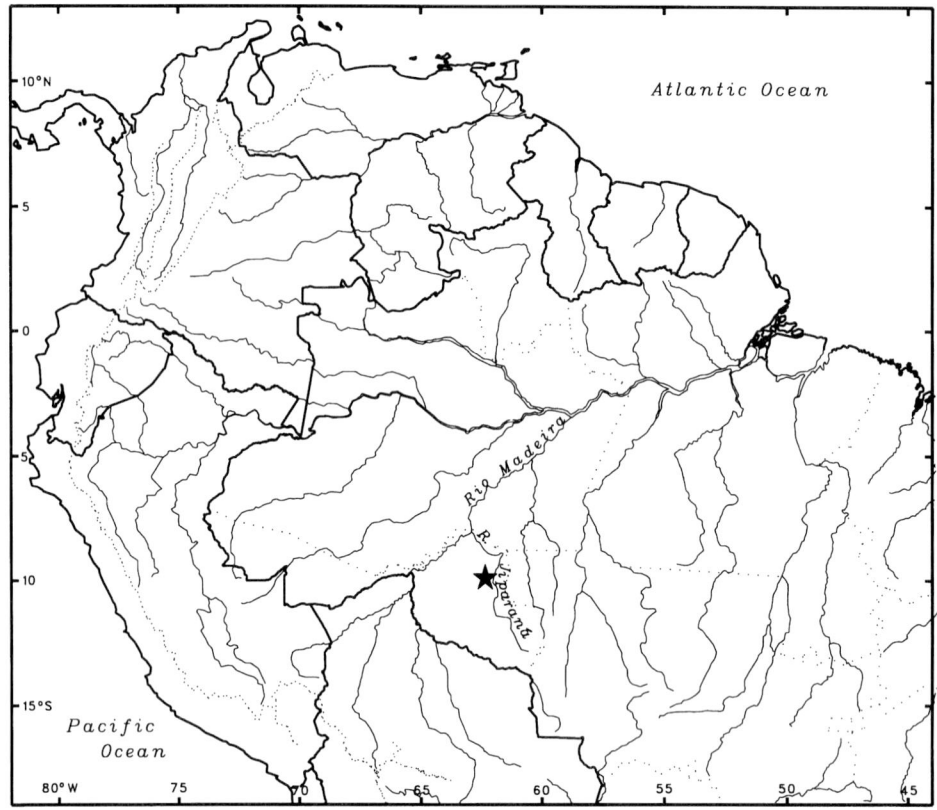

FIG. 1. Black star marks the location of Cachoeira Nazaré, Rondônia, the main study site. Our other study site, Pedra Branca, too close to Cachoeira Nazaré to distinguish at this scale, is 40 km. southwest of Cachoeira Nazaré.

more so than in most states in Brazil. Prior to our work, the only major collection from the state was that of Hoffmanns, who worked primarily along the lower Rio Jiparaná and the Rio Madeira (Hellmayr 1910). Natterer also collected within the region along the Rio Madeira and the Rio Guaporé (Pelzeln 1868–1870). The combination of rapid deforestation and incomplete knowledge of the avifauna of Rondônia heightens the need for basic survey work in this region.

In 1986 and 1988, the Museu de Zoologia of the Universidade de São Paulo (MZUSP) and the Field Museum of Natural History (FMNH) surveyed the avifauna of the middle Rio Jiparaná in eastern Rondônia. The sites surveyed are to be flooded by a dam to provide hydroelectric power for Rondônia's burgeoning human population. We surveyed two main sites (Figure 1):

Cachoeira Nazaré (9°44′S, 61°53′W) on the Rio Jiparaná is the proposed location for the dam. Using the camp created for workers surveying the dam site as our base, we covered an area of about 20 km² on an extensive trail system constructed for topographic survey work by the Consórcio Nacional de Engenheiros Consultores (CNEC). See Lanyon et al. (1990) for further details on Cachoeira Nazaré. Stotz was alone at this site from 20 May to 14 July 1986 and from 1 to 13 March 1988. All six authors, along with Leopoldo da Silva from MZUSP, worked at this site between 10 October and 21 November 1986. We accumulated about 1,400 field hours observing and collecting birds. Schulenberg made extensive tape-recordings, which are deposited at the Library of Natural Sounds, Cornell Laboratory of Ornithology (LNS). We ran seven mist net lines consisting of 10 to 40 twelve meter nets for a total of 1,450 net days.

Pedra Branca (10°02′S, 62°06′W) is along the road from Jaru to Machadinho where it crosses the Rio Anari, a tributary that enters the Rio Jiparaná about 8 km upstream from Cachoeira Nazaré. Stotz and da Silva worked here from 4 to 28 February 1988. Coverage totaled 250 observer hours and 375 net days in three net lines.

Both sites were primarily *terra firme* forest at about 100 m elevation. The forest canopy reached about 35 m with emergents up to 50 m. Both rivers were bounded by bluffs that rose about eight meters from the river's edge at low water in late October. At Cachoeira Nazaré, these bluffs were covered by open forest dominated by *Cecropia*, which graded into more typical *terra firme* forest about 500 m away from the river. At Pedra Branca, this *Cecropia*-dominated vegetation was only a few meters wide. Both sites had limited areas of floodplain forest along the river edge in places where the bluffs disappeared. These areas were largely inaccessible on the 1988 trip because of seasonal flooding. At Cachoeira Nazaré, an extensive area of sandy soil about a kilometer east of the river supported a different forest referred to herein as "low forest." The canopy of this low forest reached only about 25 m and emergents were almost nonexistent. Stotz ran a net line here and made extensive observations in July 1986, but coverage on the other trips was limited. A 5-ha marsh at Pedra Branca lay along the river course; no marsh vegetation existed within the study site at Cachoeira Nazaré. Second growth at Cachoeira Nazaré was limited to a 10-ha clearing for the camp about 1.5 km west of the river, an overgrown clearing of several ha on the east bank of the river, and narrow bands of regrowth bordering two dirt roads through the site. At Pedra Branca, a highway running through the study site was bordered by a band roughly one km wide cleared for agriculture. Most of the clearing had been abandoned several years earlier, but a field of about 50 ha planted in rice was ready for harvest in February 1988. Neither of the study sites held stands of bamboo.

In the appendix, we list all species of birds recorded at each site, with habitat use, migratory status, and subjective estimates of their abundance. The 459 species recorded at Cachoeira Nazaré make it the most diverse single site known in Brazil. However, it remains about 100 species below the most diverse sites in Amazonian Peru (Terborgh et al. 1984; Haffer 1990; Parker et al. 1994). Diversity at this site is high despite rather poor riverine habitats and the lack of oxbow lakes, which can add substantially to the species richness of a site (Remsen and Parker 1983). It seems likely that other sites in southwestern Amazonian Brazil with good *terra firme* forest and riverine habitats together will surpass 500 species.

The much lower total of 306 species recorded at Pedra Branca in part reflects the more limited coverage of this site. In addition, however, this site also had less habitat diversity than Cachoeira Nazaré, and it had lower diversity even inside *terra firme* forest. Several species relatively common in *terra firme* forest at Cachoeira Nazaré, such as Gray Tinamou (*Tinamus tao*), Paradise Jacamar (*Galbula dea*), Yellow-crowned Tyrannulet (*Tyrannulus elatus*), and White-thighed Swallow (*Neochelidon tibialis*), were not recorded at Pedra Branca, and probably were absent.

Specimens from our work in Rondônia are currently deposited at FMNH, MZUSP, and the Museu Paraense Emílio Goeldi. One new species has been described from our collections at Cachoeira Nazaré (Lanyon et al. 1990). The status of North American migrants at Cachoeira Nazaré is discussed elsewhere (Stotz et al. 1992). Here, we provide information on significant range extensions and some poorly known species. Since we began work in Rondônia, ornithologists have increased their attention to this state. In particular, a number of observers have worked at Fazenda Rancho Grande, roughly 150 km southwest of Cachoeira Nazaré. The following species for which we provide accounts below also have been recorded at Fazenda Rancho Grande (R. S. Ridgely, pers. comm.): *Brachygalba lugubris, Deconychura stictolaema, Sclerurus mexicanus, Leptopogon amaurocephalus, Platyrinchus saturatus, Odontorchilus cinereus*, and *Parkerthraustes humeralis*. Most range extensions discussed below involve species not previously known as far south or east as Rondônia. A series of range extensions of species known primarily from northeastern Amazonia, but also south of the lower Amazon, was perhaps the most striking new distributional pattern encountered.

The results of this survey are important not only for the added knowledge they provide concerning the avifauna of the region, but also for what they tell us about the general state of knowledge of bird distributions in Amazonia. As a result of our 1,650 observer hours in one relatively small geographic area, one species new to science (*Clytoctantes atrogularis*) was discovered (Lanyon et al 1990), and significant range extensions for many additional taxa were documented. Seven taxa found in our survey were identified by Cracraft (1985) as endemics of areas other than the Rondonian center in which our study was conducted (Napo/Inambari Center—*Nothocrax urumutum, Nonnula rubecula cineracea, Sclerurus mexicanus peruvianus, Terenura humeralis, Parkerthraustes humeralis*; Belem Center—*Topaza pella microrhyncha*; Guyanan Center—*Haematoderus militaris*).

Therefore, knowledge of faunal distributions in Amazonia clearly remains in a primitive state. Unfortunately, conservation organizations are currently in need of accurate assessments of single site faunal diversity and of broad-scale geographic patterns. Additional basic survey work is hence urgently needed. Until this region of the world is more adequately surveyed, caution must be exercised when interpreting distribution patterns based on currently available information and when formulating policy founded on these patterns.

SPECIES ACCOUNTS

Nocturnal Currasow (*Nothocrax urumutum*). Several were heard and tape-recorded calling at night in *terra firme* forest at Cachoeira Nazaré. This is the southeasternmost record of the species, which previously was known from the base of the Andes east to the Rio Purus, and along the southern margin of the Amazon east to the east bank of the Rio Madeira (Blake 1977). Paynter and Traylor (1991) were unable to determine whether Lago Andirá referred to the lake of this name along the Rio Juruá or to Lago Açu Andirá east of the Rio Madeira. Vanzolini (1992) associated the locality with the lake along the Rio Juruá. However, notes and maps of A. M. Olalla, archived at the American Museum of Natural History, make it clear that specimens he collected at Lago Andirá mentioned in the literature were all taken at Lago Açu Andirá.

Blackish Rail (*Rallus nigricans*). One was seen and heard on 11 February in a small marsh at Pedra Branca. This is the only record from Amazonian Brazil, although it occurs locally in eastern Peru (Parker et al. 1982; Terbough et al. 1984; Parker et al. 1994) and in Pando (Parker and Remsen 1987) and La Paz (Parker et al. 1991) in Amazonian Bolivia. This record extends the range of Blackish Rail about 700 km eastward in Amazonia, although the species is also found in eastern Brazil.

Ash-colored Cuckoo (*Coccyzus cinereus*). Two specimens from Cachoeira Nazaré, one collected 30 June from along the river and another collected 6 July in low forest, are our only records. Elsewhere in Brazil, this species is a summer resident, apparently rare, in Rio Grande do Sul (Belton 1984) and São Paulo (Willis and Oniki 1985). There are also records from Mato Grosso, Goiás, and Bahia (Sick 1985). In Amazonia, it is a rare austral migrant. There are sight records from Peru (Parker 1982) and Colombia (Hilty and Brown 1986), and a specimen from the depto. Beni, Bolivia (Gyldenstolpe 1945b), besides our records.

Fiery-tailed Awlbill (*Avocettula recurvirostris*). We netted one specimen in a tree-fall gap in *terra firme* forest at Cachoeira Nazaré. This species was primarily known from the Guianas, Amazonia north of the Amazon west to Manaus (Cohn-Haft et al. 1997) and south of the Amazon to the west bank of the Rio Tapajós (Pinto 1938; and a sight record SW of Itaituba, Schulenberg and Ridgely, pers. obs.). In addition, the species is known from eastern Ecuador, where its status is uncertain, but there are at least three specimen records (Berlioz 1938; Ridgely, pers. comm.). Our specimen extends the known range about 850 km to the southwest.

Crimson Topaz (*Topaza pella*). We observed *Topaza* infrequently at flowering vines in riverine forest and along forest streams at Cachoeira Nazaré, and we collected several specimens. South of the Amazon, this species previously was known only from the east bank of the lower Rio Tapajós east to the Belém region. The specimens from Rondônia are referable to the weakly characterized race *microrhyncha* described from near Belém. A recent sight record of a *Topaza* from Acre (B. Forrester, in litt.) cannot be assigned unequivocally to species, because *T. pyra* has been collected south of the Amazon at the Rio Urucu (Peres and Whittaker 1991).

Brown Jacamar (*Brachygalba lugubris*). This species was fairly common in river-edge forest at Cachoeira Nazaré and uncommon in the same habitat at Pedra Branca. The nearest and previously westernmost localities are in northern Mato Grosso (Haffer 1974). Specimens are referable to *melanosternum*, the subspecies of southcentral Brazil.

Collared Puffbird (*Bucco capensis*). This species was uncommon at Cachoeira Nazaré and rare at Pedra Branca in the understory of *terra firme* forest. The nearest previous records come from ca. 600 km to the north at Borba on the Rio Madeira (Hellmayr 1910) and ca. 550 km northwest at Jaburu on the Rio Purus (Gyldenstolpe 1951).

Rusty-breasted Nunlet (*Nonnula rubecula*). One collected at Cachoeira Nazaré on 10 July 1986 in low-lying river-edge forest, and preserved in alcohol, is our only record. The specimen represents the Amazonian form, *cineracea*, previously known south only to Borba, Rio Madeira (Hellmayr 1910) and Huitanaã, Rio Purus, 450 km west-northwest of Cachoeira Nazaré (Gyldenstolpe 1951).

Spot-throated Woodcreeper (*Deconychura stictolaema*). Common in understory at both sites, this species joined mixed-species flocks, but was nearly as often encountered alone. It was previously known south only as far as Borba, and remains unrecorded from Bolivia (Remsen and Traylor 1989).

Brown-rumped Foliage-gleaner (*Automolus melanopezus*). One individual was seen and subsequently collected in an area of vine tangles in *terra firme* forest at Cachoeira Nazaré. This species, usually associated with bamboo (Parker 1982), had previously been recorded from southeastern Colombia south to Huitanaã on the Rio Purus (Glydenstolpe 1951), and Pando, Bolivia (Parker and Remsen 1987). This record extends the range of *A. melanopezus* ca. 400 km to the east.

Rufous-tailed Xenops (*Xenops milleri*). This species was common in mixed-species flocks in the canopy of *terra firme* forest at Cachoeira Nazaré. In riverine forest there, *X. tenuirostris* instead of *milleri* was a member of the canopy flocks, whereas at Pedra Branca these flocks in *terra firme* forest contained *X. rutilans*, but not *milleri*. The nearest localities for *milleri* are along the Rio Purus (Huitanaã, Nova Olinda, and Arimã; Gyldenstolpe 1951), all about 450 km to the northwest. The only previous report of this species east of the Rio Madeira, south of the Amazon, is that of Sick (1960) from along the Rio Cururú, Pará.

Gray-throated Leaftosser (*Sclerurus albigularis*). One specimen was collected at Cachoeira Nazaré on 7 November 1986. We netted it in riverine forest along the bluffs on the east side of the river. This is the first record of this species from Brazil. It is known from northwestern Bolivia, but is generally found in Andean foothills, rather than in lowland forest. However, this species recently has been encountered in extreme eastern depto. Santa Cruz, Bolivia, in lowland forest as well (Kratter et al. 1992). These Bolivian specimens apparently represent an undescribed subspecies (Kratter, pers. comm.). It seems likely that our Rondônian specimen would pertain to this subspecies.

Tawny-throated Leaftosser (*Sclerurus mexicanus*). We netted three individuals of this species at Cachoeira Nazaré. The closest records come from near San Borja, dpto. Beni, Bolivia, ca. 700 km to the southwest (Schmitt and Schmitt 1987), and Alta Floresta, 600 km east (Zimmer et al. 1997). Our specimens are referable to the subspecies *peruvianus*, not previously known from Brazil (Pinto 1978), although records of *mexicanus* from the Rio Urucu (Peres and Whittaker 1991) should belong to *peruvianus*. Four species of *Sclerurus* (see Appendix) occurred together at this site, although only *rufigularis* and *caudacutus* were common. Generally, only two *Sclerurus* occur together, although three species coexist near Manaus (Stotz and Bierregaard 1989).

Rondonia Bushbird (*Clytoctantes atrogularis*). A single female of this species, designated as the type, was netted, and a male was observed twice at Cachoeira Nazaré (see Lanyon et al. 1990).

Black-capped Antwren (*Herpsilochmus atricapillus*). In the low forest at Cachoeira Nazaré, this species was common in pairs in the canopy. We obtained specimens on both 1986 trips and sound recordings in October 1986. *Herpsilochmus atricapillus* has long been considered a subspecies of *H. pileatus* (e.g., Hellmayr 1908, 1924; Peters 1951; Meyer de Schauensee 1970; Sick 1985), but Davis and O'Neill (1986) showed that it is better regarded as a separate species. This is the first record of *H. atricapillus* from Amazonia. The closest known localities for *H. atricapillus* are in western Mato Grosso, ca. 550 km southeast of our site. We have examined specimens at MZUSP from Serra do Cachimbo in southern Pará reported by Pinto and Camargo (1957) as *atricapillus* and found that these specimens represent *H. pileatus*. This represents a major range extension for that species, previously known only from northeastern Brazil (Davis and O'Neill 1986), with records from only as far west as Maranhão (Ridgely, pers. comm.).

Chestnut-shouldered Antwren (*Terenura humeralis*). This species was a common member of mixed-species canopy flocks in *terra firme* forest at both sites, with specimens taken at Cachoeira Nazaré. In addition, MZUSP has an unpublished specimen taken at Pimenta Bueno, farther up the Rio Jiparaná, ca. 250 km south of Cachoeira Nazaré. This species was previously known in Brazil east only to the Rio Purus (Pinto 1978). Recently, it has been recorded in Bolivia for the first time in Pando (Parker and Remsen 1987). Another member of the same superspecies, *T. spodioptila*, occurs on the west bank of the Rio Tapajós (Zimmer 1932). Where these species come into contact remains to be determined.

Plain Tyrannulet (*Inezia inornata*). Stotz found this species commonly in the low forest during May through July, when three specimens were collected. One observed on 21 October in low forest was our only record in the second trip. This is an austral migrant into southwestern Amazonia, from which there are records in southern Peru (Parker 1982) and northern Bolivia (Gyldenstolpe 1945b). There are three previous records from Amazonian Brazil: an October specimen from the Rio Juruá (Gyldenstolpe 1945a), a November specimen from the Rio Iténey [=Rio Guaporé], Rondônia (Ridgely and Tudor 1994), and a previously unpublished May specimen from Rio Branco, Acre (MZUSP 65900). In Peru, this species ordinarily occurs in early successional vegetation along rivers (Parker 1982), but we did not find it in this habitat at Cachoeira Nazaré.

Sepia-capped Flycatcher (*Leptopogon amaurocephalus*). This species was common in river-edge forest at Cachoeira Nazaré. Specimens are referable to the Amazonian race *peruvianus*, previously recorded in Brazil only from along the Rio Juruá (Pinto 1944), 950 km northwest of Cachoeira Nazaré. However, specimens reported by Novaes (1976) from Vista Alegre, Amazonas, ca. 250 km northeast of Cachoeira Nazaré, as nominate *amaurocephalus* may belong to *peruvianus*. The nominate race of southeastern Brazil occurs west to Mato Grosso, south of the Amazonian forest.

Cinnamon-crested Spadebill (*Platyrinchus saturatus*). This species occurred at both sites, with specimens collected at each. It was difficult to observe, but occasionally we captured individuals in mist nets in *terra firme* forest. The species is widespread north of the Amazon, but south of the Amazon, previously was known only east of the Rio Tapajós, including Alta Floresta (Zimmer et al. 1997).

Rufous Casiornis (*Casiornis rufa*). This is apparently a migrant from the south. It was fairly common in low forest at Cachoeira Nazaré in May and June, when specimens were obtained, but unrecorded in October and November. The nearest breeding sites are in southern Mato Grosso. Its status in Amazonia is unclear. It is recorded from Monte Alegre, Pará, north of the Amazon (Snethlage 1914), Acre (unpublished specimen in MZUSP), southern Peru (Parker and O'Neill 1980) north to Junin (Weske 1972) and northern Bolivia (Gyldenstolpe 1945b). Except in northern Bolivia, Amazonian records fall between May and September, consistent with a migrant origin for these birds.

Brown-crested Flycatcher (*Myiarchus tyrannulus*). One was collected and another observed in June 1986 in the low forest at Cachoeira Nazaré. We failed to find it in October and November suggesting that the earlier birds may have been migrants. There are only scattered records of this species in western Amazonia (Lanyon 1978), where its status remains unclear. Records from western Amazonia in Acre (recorded in August and September, Pinto and Camargo 1954), in eastern Peru at Río Tambopata (where considered an austral migrant; Parker et al. 1994) and Manu (Terborgh et al. 1984), and in northern Bolivia (recorded in May and October, Gyldenstolpe 1945b) are consistent with occurrence as an austral migrant in this region. However, Lanyon (1978) considered the species not to be migratory in South America and observed birds in southeastern Peru in late November. Similarly, Traylor (1979) considered the species resident in eastern Peru and Amazonian Brazil.

Dusky-chested Flycatcher (*Myiozetetes luteiventris*). We found this species common in small groups around edges and treefalls in *terra firme* forest at both sites, and collected specimens at Cachoeira Nazaré. It has long been considered rare (Remsen 1977), but recently has been found at numerous new localities and has a much broader range than generally appreciated, including Pando in northern Bolivia (Parker and Remsen 1987). Prior to these records, the species was known in southern Amazonian Brazil only from along the lower Rio Tapajós (Traylor 1979) and along the Rio Iquiri in Acre (Pinto and Camargo 1954). A previously unreported specimen (MZUSP 68904) from Açailandia, Maranhão, is far east of the previously known range. It has also recently been recorded at Alta Floresta, Mato Grosso (Zimmer et al. 1997).

Crested Becard (*Pachyramphus validus*). Stotz obtained two specimens and observed this species several times during June and July 1985. All observations and both specimens were of female-plumaged individuals. The distribution of records, along with the status of nominate *validus* as only a summer resident in the southern part of its breeding range (from depto. Santa Cruz, Bolivia and São Paulo, Brazil [Stotz, pers. obs.], southward) suggests that these birds represented wintering birds. Although this species is known to be migratory, it had not previously been recorded in western Amazonia. However, beyond this specimen, the Field Museum has a previously unpublished specimen of a male *P. v. validus* from the Rio Tambopata, Madre de Dios, Peru, collected on 5 September 1958, and the species has been reported based on sight records at Explorer's Inn in Madre de Dios, Peru (Parker et al. 1994).

Crimson Fruitcrow (*Haematoderus militaris*). A female collected from the canopy of *terra firme* forest at Cachoeira Nazaré on 5 March 1988 was our only record for this species. *Haematoderus militaris* was previously known only from the Guianas, eastern Amazonia east of the Rio Negro north of the Amazon and the east of the Rio Tocantins south of the Amazon, 1400 km northeast of Cachoeira Nazaré (Snow 1982). This species is easily overlooked owing to its quiet behavior in the canopy, so recently there have been substantial extensions to its known range (Bierregaard et al. 1987; Willard et al. 1991). Despite being a member of the predominantly frugivorous Cotingidae, *H. militaris* may be largely insectivorous. The stomach of our specimen contained remains of scarab beetles and large Orthoptera. Contents of two stomachs reported previously (Snow 1982), foraging observations from near Manaus (Bierregaard et al. 1987;

Whittaker 1993), and the well-developed bristles in the loral region are all consistent with this view. Other large cotingas, such as *Cephalopterus* and *Pyroderus*, which are known to take some large insects (Snow 1982), also show bristles, but these are much less extensive than in *H. militaris*.

Snow-capped Manakin (*Pipra nattereri*). This species was common in the understory of *terra firme* forest at both sites. This species is currently treated as monotypic (Snow 1979), although Hellmayr (1903) named *gracilis* based on a green-crowned female from the Rio Guaporé. He later concluded (Hellmayr 1910) that green and blue-crowned females represented only individual variants. In fact, two subspecies can be recognized based on female plumage: *P. n. nattereri* with blue-crowned females, and *P. n. gracilis* with green-crowned females.

All females of nominate *nattereri*, which ranges south at least to Calama, on the north bank of the Rio Jiparaná at its mouth, and Praínha, on the Rio Aripuanã, have a blue tinge to the crown, although this varies both in intensity and extent. In *gracilis*, which ranges from Aliança, on the east bank of the Rio Madeira 95 km south of Calama, south to include our study sites and the recently discovered Bolivian population (Bates et al. 1989), females have pure green crowns, concolor with the back. Males of the two subspecies appear to be indistinguishable.

Tooth-billed Wren (*Odontorchilus cinereus*). Previously known from only four specimens collected between the Rio Madeira and the Rio Tapajós, this species was common in canopy flocks at Cachoeira Nazaré and uncommon at Pedra Branca. We obtained specimens at both sites. Since our work began at Cachoeira Nazaré, *O. cinereus* has been found in northeastern depto. Santa Cruz, Bolivia (Bates et al. 1989, 1992) and Alta Floresta (Zimmer et al. 1997). This species behaves much as does *O. branickii* of the lower slopes of the Andes (Fitzpatrick, Stotz and Willard, pers. obs.). It forages by slowly moving high in the canopy along bare limbs from which it gleans insects, while a member of mixed-species flocks. The song of *O. cinereus* is a long dry trill (see Bates et al. 1992), very similar to that of *O. branickii* (Hardy and Coffey 1988). In June 1986, Stotz observed a pair investigating a hole in a dead limb, 25 m above the ground in *terra firme* forest. They may have been considering the site for nesting, but were not seen subsequently. In addition, on 25 April 1990, R. Ridgely (pers. comm.) observed a pair of this species bringing twigs to a dead limb 20 m up in *terra firme* forest edge at Fazenda Rancho Grande, 150 km southwest of Cachoeira Nazaré. There appear to be no previous observations of nesting behavior in either species in this genus.

Guianan Gnatcatcher (*Polioptila guianensis*). Stotz observed a pair in a canopy flock at Cachoeira Nazaré on 5 and 12 March, and collected the female on 12 March. This specimen is referable to the subspecies *paraensis*, otherwise known only from east of lower Rio Tapajós; recent sight records from the Rio Urucu (Peres and Whittaker 1991) and Alta Floresta (Zimmer et al. 1997) probably represent this subspecies as well. The irides on the Cachoeira Nazaré specimen were pale gray; on a male of the race *facilis* collected by Stotz at Colonia Apiaú, Roraima, the iris was bright orange-yellow, and birds belonging to the nominate race at Manaus have dark irides (M. Cohn-Haft, pers. comm.). Recently collected specimens of *P. schistaceigula*, sometimes treated as conspecific with *guianensis*, had dark red irides (Ridgely, pers. comm.). In addition, there is substantial plumage diversity within the three disjunctly distributed subspecies of *guianensis*. The populations vary most prominently in: the tone of the gray on the chest, darkest in nominate *guianensis*, palest in *paraensis*; the extent of white in the tail, greatest in *guianensis*, least in *facilis*; and the presence of white lores and eyering, occurring in *guianensis* and *paraensis*, absent in *facilis*. All populations share the habit of foraging almost exclusively high in the canopy of *terra firme* forest, where they accompany mixed-species flocks.

Fulvous-crested Tanager (*Tachyphonus surinamus*). Previously known south only to the Rio Juruá and Borba, this species was uncommon in small monospecific flocks, mainly along small streams through *terra firme* forest. Specimens are referable to the eastern race *insignis*, rather than *napensis*, which is known from the Rio Juruá.

Yellow-shouldered Grosbeak (*Parkerthraustes humeralis*). We found this species to be uncommon in mixed-species canopy flocks at Cachoeira Nazaré at edges in *terra firme* forest. The only record from Pedra Branca was a female collected on 24 February. Although occurring from Colombia south to Bolivia, this species is usually rare at any locality; its relative abundance at Cachoeira Nazaré is notable. Until recently, *P. humeralis* was known in Brazil only from the Rio Purus (Pinto 1944). Recently it has been found to be reasonably widespread in southern Amazonian Brazil (Ridgely and Tudor 1989), east to Serra de Carajás, Pará, and in northern Bolivia with records southeast to depto. Cochabamba (Remsen and Traylor 1989).

Slate-colored Seedeater (*Sporophila schistacea*). This species was fairly common in February at Pedra Branca where males sang from perches 3 to 8 m up at the forest edge. Females were occasionally encountered with mixed-species flocks in the forest interior. Most stomachs examined from the nine specimens collected contained rice, which was being harvested in the region at that time. The only two records at Cachoeira Nazaré were obtained in March despite extensive coverage at other times of year. This suggests that the species is nomadic in the region, as found by Parker (1982) in southeastern Peru. Unlike in Peru, however, this species was not restricted to bamboo, which is absent at both sites. South of the Amazon in Brazil, *S. schistacea* was previously known only from near the lower Amazon east of the lower Rio Xingu (Ridgely and Tudor 1989, Vielliard, pers. comm.).

ACKNOWLEDGMENTS

We thank Paulo Vanzolini, Director of the Museu de Zoologia, Universidade de São Paulo (MZUSP), for his help both in São Paulo and in Rondônia. This survey was undertaken for Eletronorte, through a contract between the Academia Brasileira de Ciências and the Consórcio Nacional de Engenheiros Consultores (CNEC), as part of a biological survey of the site of a planned hydroelectric dam. We received logistical support from MZUSP, the Conselho Nacional de Desenvolvimento Científico e Tecnológico, and CNEC, especially from the people at their camp JP-14. J. V. Remsen and S. Cardiff (Louisiana State University Museum of Natural Science) provided access to specimens under their care. Mary LeCroy (AMNH) made available copies of A. M. Ollala's notes with respect to Lago Andirá. This paper benefitted from a careful review of an earlier version by Robert Ridgely. Finally, we thank L. da Silva, A. L. Gardner, B. D. Patterson, and M. A. Rogers for assistance in the field. The second field trip was supported financially by the Eppley Foundation for Research.

LITERATURE CITED

ARNDT, T. 1983. Neue Erkenntnisse über den Artstatus des Blausteiβsittich *Pyrrhura perlata perlata* Spix 1824. Spixiana Suppl. 9:425–428.

BATES, J. M., M. C. GARVIN, D. C. SCHMITT, AND C. G. SCHMITT. 1989. Notes on bird distribution in northeastern Dpto. Santa Cruz, Bolivia, with 15 species new to Bolivia. Bull. Brit. Ornithol. Club 109:236–244.

BATES, J. M., T. A. PARKER III, A. P. CAPPARELLA, AND T. J. DAVIS. 1992. Observations on the *campo*, *cerrado* and forest avifaunas of eastern dpto. Santa Cruz, Bolivia, including 21 species new to the country. Bull. Brit. Ornithol. Club 112:86–98.

BELTON, W. 1984. Birds of Rio Grande do Sul, Brazil. Part 1. Rheidae through Furnariidae. Bull. Amer. Mus. Nat. Hist. 178:369–636.

BERLIOZ, J. 1938. Notes critiques sur des Trochilidés. Oiseau (n.ser.) 8:3–19.

BIERREGAARD, R. O., JR., D. F. STOTZ, L. H. HARPER, AND G. V. N. POWELL. 1987. Observations on the occurrence and behavior of the Crimson Fruitcrow *Haematoderus militaris* in central Amazonia. Bull. Brit. Ornithol. Club 107:134–137.

BLAKE, E. R. 1977. Manual of Neotropical Birds. Vol. 1. University of Chicago Press, Chicago.

BROWN, K. S., JR. 1977. Centros de evolução, refugios quaternários e conservação de patrimínios genéticos na região neotropical: padrões de diferenciação em Ithomiinae (Lepidoptera: Nymphalidae). Acta Amazonica 7:75–137.

COHN-HAFT, M., A. WHITTAKER, AND P. C. STOUFFER. 1997. A new look at the "species-poor" central Amazon: the avifauna north of Manaus, Brazil. Pp. 205–235 *in* Studies in Neotropical Ornithology Honoring Ted Parker (J. V. Remsen, Jr., Ed.), Ornithol. Monogr. No. 48.

CRACRAFT, J. 1985. Historical biogeography and patterns of differentiation within the South American avifauna: areas of endemism. Pages 49–84 *in* Neotropical Ornithology (P. A. Buckley, M. S. Foster, E. S. Morton, R. S. Ridgely, and F. G. Buckley, Eds.). Ornithol. Monogr. 36.

DAVIS, T. J., AND J. P. O'NEILL. 1986. A new species of antwren (Formicariidae: *Herpsilochmus*) from Peru, with comments on the systematics of other members of the genus. Wilson Bull. 98:337–352.

FEARNSIDE, P. M. 1987. Deforestation and international economic development projects in Brazilian Amazonia. Conserv. Biol. 1:214–221.

FEARNSIDE, P. M. 1989. A ocupação humana de Rondônia: impactos, limites e planejamento. Assessoria Editorial e Divulgação Científica, Brasília, Brazil.

GYLDENSTOLPE, N. 1945a. The bird fauna of Rio Juruá in western Brazil. Kungl. Sven. Vet. Apsak. Handl. 22(3):1–338.

GYLDENSTOLPE, N. 1945b. A contribution to the ornithology of northern Bolivia. Kungl. Sven. Vet. Apsak. Handl. 23(1):1–300.

GYLDENSTOLPE, N. 1951. The ornithology of the Rio Purús region in western Brazil. Arkiv. Zoolog. ser. 2, 2:1–320.

HAFFER, J. 1974. Avian speciation in tropical South America. Publ. Nuttall Ornithol. Club 14:1–390.

HAFFER, J. 1990. Avian species richness in tropical South America. Stud. Neotropical Fauna Environment 25:157–183.

HARDY, J. W., AND B. B. COFFEY, JR. 1988. Voices of the Wrens (cassette tape). Ara Records, Gainesville, Florida.
HELLMAYR, C. E. 1903. Über neue und wenig bekannte südamerikanische Vögel. Verh. Ges. Wien 53:199–223.
HELLMAYR, C. E. 1908. An account of the birds collected by Mons. G. A. Baer in the state of Goyaz, Brazil. Nov. Zool. 15:13–102.
HELLMAYR, C. E. 1910. The birds of the Rio Madeira. Nov. Zool. 17:257–428.
HELLMAYR, C. E. 1924. Catalogue of birds of the Americas. Part III. Field Mus. Nat. Hist., Zool. Ser. 13.
HILTY, S. L., AND W. L. BROWN. 1986. A Guide to the Birds of Colombia. Princeton Univ. Press, Princeton, New Jersey.
KRATTER, A. W., M. D. CARREÑO, R. T. CHESSER, J. P. O'NEILL, AND T. S. SILLETT. 1992. Further notes on bird distribution in northeastern Dpto. Santa Cruz, Bolivia, with two species new to Bolivia. Bull. Brit. Ornithol. Club 112:143–150.
LANYON, S. M., D. F. STOTZ, AND D. E. WILLARD. 1990. *Clytoctanctes atrogularis*, a new species of antbird from western Brazil. Wilson Bull. 102:571–580.
LANYON, W. E. 1978. Revision of the *Myiarchus* flycatchers of South America. Bull. Amer. Mus. Nat. Hist. 161:427–628.
MEYER DE SCHAUENSEE, R. 1970. A Guide to the Birds of South America. Livingston, Wynnewood, Pennsylvania.
MÜLLER, P. 1973. The dispersal centers of terrestrial vertebrates in the Neotropical Realm. W. Junk, The Hague, Netherlands.
NOVAES, F. C. 1976. As aves do rio Aripuanã, Estados de Mato Grosso e Amazonas. Acta Amazonica 6 (suppl.):61–85.
PARKER, T. A., III. 1982. Observations of some unusual rainforest and marsh birds in southeastern Peru. Wilson Bull. 94:477–493.
PARKER, T. A., III, A. CASTILLO U., M. GELL-MANN, AND O. ROCHA O. 1991. Records of new and unusual birds from northern Bolivia. Bull. Brit. Ornithol. Club 111:120–138.
PARKER, T. A., III, P. K. DONAHUE, AND T. S. SCHULENBERG. 1994. Birds of the Tambopata Reserve (Explorer's Inn Reserve). Pages 106–124. *in* The Tambopata Reserved Zone of Southeastern Perú: A Biological Assessment (R. B. Foster, T. A. Parker III, A. H. Gentry, L. H. Emmons, A. Chicchón, T. Schulenberg, L. Rodríguez, G. Lamas, H. Ortega, J. Icochea, W. Wust, M. Romo, J. A Castillo, O. Phillips, C. Reynal, A. Kratter, P. K. Donahue, and L. J. Barkley). RAP Working Papers 6. Conservation International, Washington, D. C.
PARKER, T. A., III, AND J. P. O'NEILL. 1980. Notes on little known birds of the upper Urubamba Valley, southern Peru. Auk 97:167–176.
PARKER, T. A., III, S. A. PARKER, AND M. A. PLENGE. 1982. An annotated checklist of Peruvian birds. Buteo Books, Vermillion, South Dakota.
PARKER, T. A., III, AND J. V. REMSEN, JR. 1987. Fifty-two Amazonian bird species new to Bolivia. Bull. Brit. Ornithol. Club 107:94–107.
PAYNTER, R. A., JR., AND M. A. TRAYLOR, JR. 1991. Ornithological Gazetteer of Brazil. Museum of Comparative Zoology, Cambridge, Massachusetts.
PELZELN, A. 1868–1870. Zur Ornithologie Brasiliens. Resultate von Johann Natterers Reisen in den Jahren 1817 bis 1835. Druck und Verlag von A. Pichler's Witwe & Sohn, Vienna, Austria.
PERES, C. A., AND A. WHITTAKER. 1991. Annotated checklist of the bird species of the upper Rio Urucu, Amazonas, Brazil. Bull. Brit. Ornithol. Club 111:156–171.
PETERS, J. L. 1951. Check-list of Birds of the World, Vol VII. Museum of Comparative Zoology, Cambridge, Massachusetts.
PINTO, O. M. O. 1938. Catálogo das aves do Brasil (1ª parte). Rev. Mus. Paulista 23:1–566.
PINTO, O. M. O. 1944. Catálogo das aves do Brasil (2ª parte). Secretário Agri., Ind., e Com., São Paulo, Brazil.
PINTO, O. M. O. 1978. Novo catálogo das aves do Brasil. 1ª Parte. São Paulo, Brazil.
PINTO, O. M. O., AND E. A. CAMARGO 1954. Resultados ornithológicos de uma expedição ao território de Acre pelo Departamento de Zoologia. Papéis Avulsos de Zoologia (São Paulo) 11:371–418.
PINTO, O. M. O., AND E. A. CAMARGO. 1957. Sobre uma coleção de aves da região de Cachimbo (sul do estado do Pará). Papéis Avulsos de Zoologia (São Paulo) 13:51–69.
PRANCE, G. T. 1973. Phytogeographic support for the theory of Pleistocene forest refuges in the Amazon basin, based on evidence from distribution patterns in Caryocaraceae, Chrysobalanaceae, Dichapetalaceae and Lecythidaceae. Acta Amazonica 3:5–28.
REMSEN, J. V., JR. 1977. A third locality in Colombia for the Dusky-chested Flycatcher *Tyrannopsis luteiventris*. Bull. Brit. Ornithol. Club 97:93–94.
REMSEN, J. V., JR., AND T. A. PARKER III. 1983. Contribution of river-created habitats to bird species richness in Amazonia. Biotropica 15:223–231.
REMSEN, J. V., JR., AND M. A. TRAYLOR, JR. 1989. An Annotated List of the Birds of Bolivia. Buteo Books, Vermillion, South Dakota.
RIDGELY, R. S., AND G. TUDOR. 1989. The Birds of South America. Vol. I. The Oscine Passerines. University of Texas Press, Austin, Texas.

RIDGELY, R. S., AND G. TUDOR. 1994. The Birds of South America. Vol. II. The Suboscine Passerines. University of Texas Press, Austin, Texas.
SCHMITT, C. G., AND D. C. SCHMITT. 1987. Extensions of range of some Bolivian birds. Bull. Brit. Ornithol. Club 107:129–134.
SICK, H. 1960. The honeycreeper *Dacnis albiventris* in Brazil. Condor 62:66–67.
SICK, H. 1985. Ornitologia brasileira, uma introdução. Vols. 1 and 2. Editora Universidade de Brasília, Brasília, Brazil.
SNETHLAGE, E. 1914. Catálogo das aves amazonicas. Bol. Mus. Goeldi. 8:1–530.
SNOW, D. 1979. Family Pipridae. Pages 245–280 *in* Check-list of Birds of the World, Vol VIII (M. A. Traylor, Jr., Ed.). Museum of Comparative Zoology, Cambridge, Massachusetts.
SNOW, D. 1982. The Cotingas. Cornell University Press, Ithaca, New York.
STOTZ, D. F., AND R. O. BIERREGAARD, JR. 1989. The birds of the Fazendas Porto Alegre, Esteio and Dimona north of Manaus, Amazonas, Brazil. Rev. Bras. Biol. 49:861–872.
STOTZ, D. F., R. O. BIERREGAARD, JR., M. COHN-HAFT, J. SMITH, P. PETERMANN, A. WHITTAKER, AND S. WILSON. 1992. The status of North American migrants in central Amazonian Brazil. Condor 94:608–621.
TERBORGH, J. W., J. W. FITZPATRICK, AND L. H. EMMONS. 1984. Annotated checklist of bird and mammal species of Cocha Cashu Biological Station, Manu National Park, Peru. Fieldiana, Zool., N.S. 21.
TRAYLOR, M. A., JR. 1979. Family Tyrannidae. Pages 1–245 *in* Check-list of Birds of the World, Vol VIII (M. A. Traylor, Jr., Ed.). Museum of Comparative Zoology, Cambridge, Massachusetts.
VANZOLINI, P. E. 1992. A Supplement to the Ornithological Gazetteer of Brazil. Museu de Zoologia, Universidade de São Paulo, São Paulo, Brazil.
WESKE, J. S. 1972. The distribution of the avifauna in the Apurimac Valley of Peru with respect to environmental gradients, habitat, and related species. Unpubl. Ph.D. dissertation, University of Oklahoma.
WHITTAKER, A. 1993. Notes on the behaviour of the Crimson Fruitcrow *Haematoderus militaris* near Manaus, Brazil, with the first nesting record for this species. Bull. Brit. Ornithol. Club 113:93–96.
WILLARD, D. E., M. S. FOSTER, G. F. BARROWCLOUGH, R. W. DICKERMAN, P. F. CANNELL, S. L. COATS, J. L. CRACRAFT, AND J. P. O'NEILL. 1991. The birds of Cerro de la Neblina, Territorio Federal Amazonas, Venezuela. Fieldiana, Zool., N.S. 65.
WILLIS, E. O., AND Y. ONIKI. 1985. Bird specimens new for the state of São Paulo, Brazil. Rev. Bras. Biol. 45:105–108.
ZIMMER, J. T. 1932. Studies of Peruvian birds. VIII. The Formicarian genera *Cymbilaimus, Thamnistes, Terenura, Percnostola, Formicarius, Chamaeza,* and *Rhegmatorhina.* Am. Mus. Novit. 584.
ZIMMER, K. T., T. A. PARKER, III, M. L. ISLER, AND P. R. ISLER. 1997. Survey of a southern Amazonia avifauna: the Alta Floresta region, Mato Grosso, Brazil. Pp. 887–918 *in* Studies in Neotropical Ornithology Honoring Ted Parker (J. V. Remsen, Jr., Ed.), Ornithol. Monogr. No. 48.

APPENDIX
Birds Recorded at Cachoeira Nazaré and Pedra Branca.
Abundance Status: C = Common—seen, heard or netted more than once every other day; U = Uncommon; R = Rare—seen, heard or netted less than once a week. Abundances based on the habitat in which the species is most abundant as follows: a bird that is common in river-edge forest, uncommon in second growth and not present in terra firme forest would be listed as common, as would a bird common in all those habitats.
Habitats: B = River-bluff forest; F = Seasonally-flooded forest; I = Forest streams (Igarapés); L = Low forest; M = Marsh; R = River, including its banks and edges; S = Second-growth; T = *Terra firme* Forest. Habitats are listed in order of importance for species.
Migratory Status: AP = Austral Migrant, present only during passage migration; AS = Austral Migrant, present during summer and breeding; AW = Austral Migrant, wintering; NP = Nearctic Migrant, present only during passage migration; NW = Nearctic Migrant, wintering; R = Resident *—Species not represented in our collection by specimens.
+—Species not collected by Hoffmans or Natterer (see Hellmayr 1910) in Rondônia.
Note: Following Arndt (1983) we treat the name *Pyrrhura rhodogaster* as a synonym of *P. perlata*

Species	Cachoeira Nazaré	Pedra Branca	Habitats	Migratory Status
Tinamidae				
Tinamus tao	C	—	T, B, L	R
Tinamus major	C	U	T, B, F	R
Crypturellus cinereus	C	C	F, B, T	R
Crypturellus soui	U	C	B, T, S	R
Crypturellus obsoletus+	C	U	T	R
Crypturellus variegatus*	C	C	T, B, L	R
Crypturellus strigulosus+	U	C	T	R
Crypturellus parvirostris*+	—	C	S	R
Phalacrocoracidae				
Phalacrocorax brasilianus*+	R	—	R	R
Anhingidae				
Anhinga anhinga+	R	—	R	R
Ardeidae				
Tigrisoma lineatum	R	—	F	R
Zebrilus undulatus+	R	—	F	R
Pilherodius pileatus*	R	—	R	R
Bubulcus ibis+	R	—	R	R
Butorides striatus+	R	R	R, M	R
Egretta thula*+	R	—	R	R
Ardea alba*+	R	—	R	R
Ardea cocoi*	R	—	R	R
Agamia agami	R	—	F	R
Ciconiidae				
Mycteria americana*+	R	—	R	R
Threskiornithidae				
Mesembrinibis cayennensis*	R	R	R, M	R
Cathartidae				
Cathartes melambrotus+	C	C	T, B	R
Coragyps atratus*+	C	U	S	R
Sarcoramphus papa*+	U	R	T	R
Anhimidae				
Anhima cornuta*	R	—	F	R
Anatidae				
Neochen jubatus*	R	—	R	R
Accipitridae				
Pandion haliaetus*+	R	—	R	NP
Leptodon cayanensis*+	R	—	R	R
Chondrohierax uncinatus*	R	—	S	R
Elanoides forficatus*	R	U	T, S	R?
Gampsonyx swainsonii*	—	R	S	R
Harpagus bidentatus	R	R	T	R

APPENDIX
CONTINUED

Species	Cachoeira Nazaré	Pedra Branca	Habitats	Migratory Status
Ictinia plumbea*	C	U	B, S, T, F	AS
Accipiter superciliosus	R	—	T	R
Accipiter bicolor*+	R	—	T	R
Leucopternis schistacea*+	R	—	F, T	R
Leucopternis kuhli*	R	R	T	R
Leucopternis albicollis*	R	—	S	R
Buteogallus urubitinga*+	U	R	F, R	R
Asturina nitida*+	—	R	S	R
Buteo magnirostris*	R	—	S, R	R
Buteo platypterus+	U	R	S, T, B, F	NW
Buteo brachyurus*+	—	R	S	R
Morphnus guianensis*+	R	—	B	R
Spizaetus ornatus*+	R	R	T	R
Spizaetus tyrannus*+	R	—	R, T	R
Falconidae				
Daptrius ater	C	U	B, F, R	R
Daptrius americanus	U	C	T	R
Herpetotheres cachinnans*+	R	U	S	R
Micrastur ruficollis+	R	—	T, B	R
Micrastur gilvicollis	R	R	T	R
Falco rufigularis+	U	C	S	R
Cracidae				
Penelope jacquacu	U	C	T, B, F	R
Pipile cujubi*+	R	—	F	R
Nothocrax urumutum*+	R	—	T	R
Crax mitu	R	R	T	R
Odontophoridae				
Odontophorus gujanensis*	R	—	S	R
Odontophorus stellatus	C	U	T, B	R
Psophiidae				
Psophia viridis	U	R	T	R
Rallidae				
Rallus nigricans*+	—	R	M	R
Aramides cajanea*	—	R	F	R
Laterallus melanophaius*+	R	C	M, S	R
Porphyrula martinica+	—	C	M	R
Heliornithidae				
Heliornis fulica*	R	R	R	R
Eurypygidae				
Eurypyga helias*	R	—	R	R
Charadriidae				
Vanellus cayanus	C	—	R	R
Pluvialis dominica	R	—	S	NP
Scolopacidae				
Bartramia longicauda*	R	—	S	NW?
Tringa melanoleuca*+	R	—	R	NP
Tringa flavipes*+	R	—	R	NP
Tringa solitaria	U	—	R	NW
Actitis macularia	U	—	R	NP
Calidris fuscicollis+	R	—	R, S	NP
Calidris melanotos+	C	—	R, S	NP
Laridae				
Phaetusa simplex*+	R	—	R	R
Sterna superciliaris*+	R	—	R	R
Columbidae				
Columba plumbea	U	U	T, B, L	R
Columba subvinacea+	C	C	B, S, F, T, L	R
Columbina talpacoti	C	C	S	R
Claravis pretiosa*+	R	C	S, B	R
Leptotila verreauxi*+	R	R	S	R
Leptotila rufaxilla	C	U	S, B, F	R

APPENDIX
Continued

Species	Cachoeira Nazaré	Pedra Branca	Habitats	Migratory Status
Geotrygon montana	U	C	T, B, F	R
Psittacidae				
Ara ararauna*	R	—	T, B, F, R	R
Ara macao*	C	C	T, B, F	R
Ara chloroptera*+	U	R	T	R
Ara severa*	C	U	T, B, F, R, S	R
Ara manilata*+	R	U	F, R	R
Aratinga weddellii	—	C	S	R
Pyrrhura perlata	C	U	S, T, B, F, L	R
Pyrrhura picta	C	C	T, S, B, F, L	R
Brotogeris chrysopterus	C	C	T, S, B, F, L	R
Pionites leucogaster	C	C	T	R
Pionopsitta barrabandi	C	C	T, S	R
Graydidascalus brachyurus*	R	—	F, R	R
Pionus menstruus	C	C	T, S, B, F, L	R
Amazona ochrocephala*	U	U	T	R
Amazona farinosa*	U	R	T, B, F, S	R
Deroptyus accipitrinus+	C	R	B, F, S, T, L	R
Opisthocomidae				
Opisthocomus hoazin*+	—	U	M	R
Cuculidae				
Coccyzus cinereus+	R	—	L, R	AW
Coccyzus americanus+	R	—	L	NP
Coccyzus euleri*+	R	—	B	AW?
Coccyzus melacoryphus	U	—	L, R	R
Piaya cayana	C	C	T, B, F, S, L	R
Piaya melanogaster+	C	U	T, L	R
Piaya minuta*+	R	—	S	R
Crotophaga major+	U	U	F, M	R?
Crotophaga ani*+	R	C	S	R
Tapera naevia*+	R	—	S, L	R
Dromococcyx phasianellus*+	R	—	F	R
Neomorphus geoffroyi*	R	—	B	R
Tytonidae				
Tyto alba*+	—	R	S	R
Strigidae				
Otus watsonii+	C	R	T	R
Lophostrix cristata*+	C	R	T	R
Pulsatrix perspicillata*	C	—	T	R
Glaucidium hardyi+	C	U	T	R
Glaucidium brasilianum*+	R	—	L	R
Ciccaba virgata	R	R	T	R
Nyctibiidae				
Nyctibius grandis+	R	—	T	R
Nyctibius aethereus*+	R	—	B	R
Nyctibius griseus*+	R	—	S	R
Caprimulgidae				
Lurocalis semitorquatus*	R	—	S	R
Chordeiles minor*+	U	—	S, R	NP
Podager nacunda*	R	—	R	R
Nyctidromus albicollis	U	—	S, R	R
Nyctiphrynus ocellatus	R	—	B, F, T	R
Caprimulgus nigrescens	C	—	B	R
Hydropsalis climacocerca+	C	R	R	R
Apodidae				
Chaetura cinereiventris*+	C	R	R, B, S, T, F	R
Chaetura egregia*+	R	R	R, B, S, T, F	R
Chaetura chapmani+	R	C	T, S, B, F	R
Chaetura brachyura*+	—	R	S	R
Tachornis squamata*+	R	U	R, S	R
Panyptila cayennensis*+	R	R	R	R

APPENDIX
CONTINUED

Species	Cachoeira Nazaré	Pedra Branca	Habitats	Migratory Status
Trochilidae				
Glaucis hirsuta	—	U	F, S	R
Threnetes leucurus	U	R	B, F, T	R
Phaethornis superciliosus	C	U	T, B, F, L	R
Phaethornis philippii	C	C	T, B, F, L	R
Phaethornis ruber	C	C	L, T, B	R
Campylopterus largipennis	C	U	S, T, B	R
Florisuga mellivora+	C	R	T, S, B	R
Avocettula recurvirostris+	R	—	T	R
Lophornis chalybea*+	R	—	S	R
Popelairia langsdorffi*	R	—	B	R
Chlorestes notatus*	—	R	S	R
Thalurania furcata	C	C	T, B, F, L, S	R
Hylocharis sapphirina*+	R	R	B, F	R
Hylocharis cyanus*+	R	—	S	R
Amazilia fimbriata*+	R	—	S	R
Polyplancta aurescens+	R	—	B	R
Topaza pella+	U	—	I, F, S	R
Heliothryx aurita	U	R	T, B	R
Heliomaster longirostris+	U	R	B, S	R
Trogonidae				
Pharomachrus pavoninus*+	U	R	T	R
Trogon melanurus+	C	C	T, B	R
Trogon viridis+	C	C	T, B, L	R
Trogon collaris+	U	R	B, F, T	R
Trogon rufus+	C	U	T, L	R
Trogon curucui+	C	R	B, F	R
Trogon violaceus	U	—	T, B, L, F	R
Alcedinidae				
Ceryle torquata*	C	C	R	R
Chloroceryle amazona*+	U	C	R	R
Chloroceryle americana	U	R	R, I	R
Chloroceryle inda+	R	R	R, I, M	R
Chloroceryle aenea	U	—	F, I	R
Momotidae				
Electron platyrhynchum	C	U	T, B, F	R
Baryphthengus martii	C	C	T	R
Momotus momota	C	U	B, F, T	R
Galbulidae				
Brachygalba lugubris+	C	U	B, F, S	R
Galbula cyanicollis	C	C	T	R
Galbula ruficauda	C	U	B, S	R
Galbula leucogastra+	U	—	L	R
Galbula dea	C	—	B, T, S, L, F	R
Jacamerops aurea	C	U	T, B, F, L	R
Bucconidae				
Notharchus macrorhynchos	C	R	T	R
Notharchus tectus+	R	R	T	R
Bucco macrodactylus	U	—	S	R
Bucco capensis+	U	R	T, L	R
Nystalus striolatus*+	R	U	T, B	R
Malacoptila rufa	C	C	T, B, F, L	R
Nonnula rubecula+	R	—	L	R
Monasa nigrifrons	C	C	B, F	R
Monasa morphoeus+	C	C	T	R
Chelidoptera tenebrosa	C	U	S, R	R
Capitonidae				
Capito dayi+	C	C	T, B, F	R
Ramphastidae				
Pteroglossus inscriptus	C	R	B, F, S, T	R
Pteroglossus bitorquatus	C	U	T, B, L	R

APPENDIX
Continued

Species	Cachoeira Nazaré	Pedra Branca	Habitats	Migratory Status
Pteroglossus castanotis	C	—	B, F, S	R
Selenidera gouldii[+]	C	U	T, B, F	R
Ramphastos vitellinus	C	U	T, B, F, L	R
Ramphastos tucanus	C	C	B, T, F, L	R
Picidae				
Picumnus aurifrons	U	U	T, B, S, F	R
Melanerpes cruentatus	C	C	B, S, F, T	R
Veniliornis passerinus[*+]	R	—	B	R
Veniliornis affinis	C	C	T, B, L, F, S	R
Piculus flavigula	C	U	T, B, F, L	R
Piculus chrysochloros	C	U	B, F, T	R
Celeus grammicus	C	U	T, L	R
Celeus elegans	U	—	T, B, F, L	R
Celeus flavus	R	—	B, L, T	R
Celeus torquatus[*]	R	—	B	R
Dryocopus lineatus[+]	R	C	S	R
Campephilus melanoleucos[+]	U	R	B, F	R
Campephilus rubricollis	C	C	T, B, F	R
Dendrocolaptidae				
Dendrocincla fuliginosa[+]	R	U	T	R
Dendrocincla merula	C	C	T, L, B	R
Deconychura longicauda	R	—	T, L	R
Deconychura stictolaema[+]	C	U	T, L	R
Sittasomus griseicapillus	C	U	T, B, F	R
Glyphorynchus spirurus	C	C	T, B, F, L	R
Nasica longirostris	U	R	F, B	R
Dendrexetastes rufigula	C	U	S, B, T	R
Hylexetastes perrotii	U	R	T, B, F, L	R
Xiphocolaptes promeropirhynchus[+]	C	—	T, L	R
Dendrocolaptes concolor	U	R	T, L	R
Dendrocolaptes hoffmannsi	R	R	T	R
Xiphorhynchus picus	R	R	S, B	R
Xiphorhynchus obsoletus	U	—	F	R
Xiphorhynchus elegans	C	C	T, L, B	R
Xiphorhynchus guttatus	C	C	T, B, F, L	R
Lepidocolaptes albolineatus[+]	C	U	T, B, F	R
Campylorhamphus procurvoides[+]	R	R	T	R
Furnariidae				
Furnarius sp.[*]	R	—	B	R
Synallaxis gujanensis[*]	R	—	S	R
Synallaxis rutilans	C	R	L, T	R
Cranioleuca gutturata	U	R	T, F	R
Hyloctistes subulatus	C	R	T, B, F	R
Ancistrops strigilatus[+]	C	U	T, B, F, L	R
Philydor erythrocercus	C	C	T, L, B, F	R
Philydor pyrrhodes	R	—	T	R
Philydor erythropterus[+]	C	U	T	R
Philydor ruficaudatus[+]	U	R	T	R
Automolus infuscatus	C	U	T	R
Automolus ochrolaemus	C	R	T, F, B	R
Automolus melanopezus[+]	R	—	T	R
Xenops milleri[+]	C	—	T, B, L	R
Xenops tenuirostris	R	—	B, F	R
Xenops rutilans[+]	—	R	T	R
Xenops minutus	C	C	T, F, B, L	R
Sclerurus albigularis[+]	R	—	B	R
Sclerurus mexicanus[+]	R	—	B, F, T	R
Sclerurus rufigularis	U	U	T, L	R
Sclerurus caudacutus	U	R	T, L	R
Formicariidae				
Cymbilaimus lineatus	C	U	T, B, F, L, S	R

APPENDIX
CONTINUED

Species	Cachoeira Nazaré	Pedra Branca	Habitats	Migratory Status
Sakesphorus luctuosus	C	—	R, F, S	R
Thamnophilus doliatus*	—	R	S	R
Thamnophilus palliatus+	C	C	S	R
Thamnophilus aethiops	C	C	T	R
Thamnophilus schistaceus	C	C	T, B, F, L	R
Thamnophilus murinus+	R	—	T	R
Thamnophilus amazonicus	C	—	B, F	R
Pygiptila stellaris	C	U	T	R
Megastictus margaritatus	U	—	L	R
Clytoctantes atrogularis+	R	—	T	R
Thamnomanes saturninus	C	C	T, B, L	R
Thamnomanes caesius	C	C	T, L, B, F	R
Myrmotherula brachyura+	C	C	T, B, F	R
Myrmotherula sclateri	C	C	T, L, B	R
Myrmotherula surinamensis	C	R	F, B, R	R
Myrmotherula hauxwelli	C	U	T, L	R
Myrmotherula leucophthalma	C	C	T, F	R
Myrmotherula haematonota	C	—	T, L	R
Myrmotherula ornata	R	R	F, B, S	R
Myrmotherula axillaris	C	—	B, F, S, L	R
Myrmotherula longipennis	C	C	T, L, B, F	R
Myrmotherula iheringi+	R	U	T	R
Myrmotherula menetriesii	C	C	T, B, L, F	R
Dichrozona cincta+	C	U	T, L	R
Herpsilochmus atricapillus+	C	—	L	R
Herpsilochmus rufimarginatus+	R	—	F	R
Microrhopias quixensis	C	C	B, T, L, F, S	R
Formicivora grisea	R	R	S	R
Terenura humeralis+	C	R	T	R
Cercomacra cinerascens	C	C	T, B, F, L	R
Cercomacra nigrescens	U	C	S, I	R
Myrmoborus leucophrys	C	U	B, S, F	R
Myrmoborus myotherinus	C	C	T, L	R
Hypocnemis cantator	C	C	S, T, B, F	R
Hypocnemoides maculicauda	R	R	F	R
Sclateria naevia+	R	R	I	R
Percnostola leucostigma	U	R	I	R
Myrmeciza hemimelaena	C	C	T, B, S, F, L	R
Myrmeciza atrothorax*+	R	—	S	R
Rhegmatorhina hoffmannsi	C	U	T, L, B	R
Hylophylax naevia	C	C	T	R
Hylophylax punctulata	U	R	F, I	R
Hylophylax poecilinota	C	C	T, L	R
Phlegopsis nigromaculatus	C	C	T	R
Formicarius colma	C	U	T, L	R
Formicarius analis	C	C	F, B, T	R
Chamaeza nobilis+	U	C	T	R
Myrmornis torquata	U	R	T	R
Myrmothera campanisoma	C	C	T, S	R
Grallaria varia	R	R	T	R
Hylopezus macularius	C	R	T	R
Conopophaga melanogaster	U	R	T	R
Conopophaga aurita	U	R	T	R
Rhinocryptidae				
Liosceles thoracicus	C	C	T	R
Tyrannidae				
Zimmerius gracilipes	U	—	T, B	R
Camptostoma obsoletum+	R	—	S	AW?
Phaeomyias murina+	R	—	L	AW?
Sublegatus modestus*+	R	—	S	AW
Tyrannulus elatus*+	C	—	B, T, F, L	R

APPENDIX
Continued

Species	Cachoeira Nazaré	Pedra Branca	Habitats	Migratory Status
Ornithion inerme+	U	R	S, T, B	R
Myiopagis gaimardii	C	C	T, B, F, L, S	R
*Myiopagis caniceps**+	R	—	T	AW?
Myiopagis viridicata+	U	—	L, S	AW
Elaenia parvirostris	R	—	S, B, T	AW
*Elaenia spectabilis**	R	—	S	AP
Inezia inornata+	C	—	L	AW
Leptopogon amaurocephalus+	U	—	B, F, T	R
Mionectes oleagineus	C	R	L, S, B	R
Corythopis torquata	U	R	L, T, F, B	R
Myiornis ecaudatus	C	U	B, S, T, F	R
Poecilotriccus capitalis+	R	—	S	R
Hemitriccus minor	C	C	T, B, F, L	R
*Hemitriccus striaticollis**	R	—	S	R
*Todirostrum latirostre**+	R	—	S	R
Todirostrum maculatum	C	—	R, B, S	R
*Todirostrum chrysocrotaphum**+	R	—	T	R
Ramphotrigon ruficauda	C	R	T, B, L	R
Rhynchocyclus olivaceus	R	—	T	R
Tolmomyias sulphurescens+	C	U	B, F	R
Tolmomyias assimilis	C	C	T, L	R
Tolmomyias poliocephalus	U	R	T, B, F, L	R
*Tolmomyias flaviventris**+	R	—	S	R
Platyrinchus saturatus+	R	R	T	R
Platyrinchus coronatus	C	R	F	R
Platyrinchus platyrhynchos	U	U	T	R
Onychorhynchus coronatus+	R	U	T	R
Terenotriccus erythrurus	C	C	T, B, F, L	R
*Myiophobus fasciatus**+	R	—	S	AP
Myiobius barbatus	C	U	T, F, B, L	R
Contopus virens+	U	R	S	NW
Cnemotriccus fuscatus	R	—	L	R
*Pyrocephalus rubinus**	R	—	S	AW
Ochthornis littoralis	C	—	R	R
Muscisaxicola fluviatilis	R	—	R	R
Colonia colonus+	C	—	S, T, B	R
*Attila bolivianus**+	R	—	F	R
Attila spadiceus	C	R	T, B, F, L, S	R
Casiornis rufa+	C	—	L	AW
Rhytipterna simplex	C	R	T, B, F, L	R
Laniocera hypopyrra	U	R	T	R
*Sirystes sibilator**+	R	—	T	AW?
Myiarchus tuberculifer+	U	—	T	R
Myiarchus swainsoni+	U	—	T, B	AW
Myiarchus ferox	C	C	S	R
Myiarchus tyrannulus+	R	—	L	AW?
Pitangus sulphuratus+	U	—	S	R
*Megarynchus pitangua**+	U	—	S, B, F	R
Myiozetetes cayanensis+	R	C	S	R
*Myiozetetes similis**+	R	R	S, B	R
*Myiozetetes granadensis**+	R	—	S, B	R
Myiozetetes luteiventris+	C	C	T	R
Myiodynastes maculatus+	R	—	S, T	AW
Legatus leucophaius	R	R	S, B, T	R
Tyrannopsis sulphurea+	R	R	S	R
*Empidonomus varius**+	R	—	T	AP
Empidonomus aurantioatrocristatus+	C	R	T, S	AW
Tyrannus melancholicus	C	C	S	R
Tyrannus savana	C	U	S	AP
Pachyramphus castaneus+	R	R	S	R
Pachyramphus polychopterus	R	—	B	R?

APPENDIX
Continued

Species	Cachoeira Nazaré	Pedra Branca	Habitats	Migratory Status
Pachyramphus marginatus	C	R	T	R
Pachyramphus minor	U	R	T	R
Pachyramphus validus[+]	U	—	B, T	AW
Tityra cayana[*+]	R	—	B	R
Tityra semifasciata	C	C	T, B, S, F, L	R
Tityra inquisitor[*]	R	—	B, F, T	R
Pipridae				
Schiffornis major	—	R	F	R
Schiffornis turdinus	C	C	T, B, F, L	R
Piprites chloris	C	C	T, B, L, F	R
Tyranneutes stolzmanni	C	C	T, F, B, L	R
Heterocercus linteatus	R	—	F	R
Manacus manacus	C	R	S, B	R
Chiroxiphia pareola	C	C	F, L	R
Pipra nattereri	C	U	T	R
Pipra fasciicauda	C	—	F	R
Pipra rubrocapilla	C	C	L, T, B, F, S	R
Cotingidae				
Phoenicircus nigricollis	C	R	T, L	R
Iodopleura isabellae[+]	U	—	T, B, F	R
Lipaugus vociferans	C	C	T, F, B	R
Cotinga maynana[*+]	R	R	B, S	R
Cotinga cayana	C	U	T, F, B	R
Xipholena punicea[+]	R	R	T, F, B	R
Gymnoderus foetidus	U	R	F, B, T	R
Haematoderus militaris[+]	R	—	T	R
Querula purpurata[+]	R	U	T	R
Hirundinidae				
Tachycineta albiventer[+]	C	—	R	R
Progne tapera[*]	R	—	R, S	AP
Progne chalybea[*]	U	C	S	R
Notiochelidon cyanoleuca[*+]	R	–	R	AP
Atticora fasciata	C	C	R	R
Atticora melanoleuca	C	—	R	R
Neochelidon tibialis[+]	C	—	B, T	R
Stelgidopteryx ruficollis[+]	C	R	S, R	R
Riparia riparia[*+]	R	—	R	NP
Hirundo rustica[*+]	R	—	R	NP
Troglodytidae				
Campylorhynchus turdinus	U	U	S	R
Odontorchilus cinereus	C	U	T, B, F	R
Thryothorus genibarbis	C	C	B, F, S	R
Thryothorus leucotis	C	U	S, R	R
Troglodytes aedon	—	C	S	R
Microcerculus marginatus[+]	C	C	T	R
Cyphorhinus arada	R	U	T	R
Muscicapidae, Turdinae				
Catharus fuscescens[+]	U	R	T	NP
Turdus amaurochalinus	U	—	S, B	AW
Turdus lawrencii[+]	R	R	T	R
Turdus fumigatus	U	—	F, B	R
Turdus albicollis	U	U	T, B, F, L	R
Muscicapidae, Polioptilinae				
Ramphocaenus melanurus	R	—	T, B	R
Polioptila guianensis[+]	R	—	T	R
Emberizidae, Emberizinae				
Ammodramus aurifrons[+]	—	U	S	R
Volatinia jacarina[+]	U	C	S	R
Sporophila schistacea[+]	R	C	T, S	R?
Sporophila lineola	U	—	S	AP
Sporophila caerulescens	R	—	S	AW

APPENDIX
Continued

Species	Cachoeira Nazaré	Pedra Branca	Habitats	Migratory Status
Sporophila castaneiventris[+]	R	C	S	R
Oryzoborus angolensis	—	C	S	R
Arremon taciturnus	R	—	T, S	R
Paroaria gularis[+]	C	C	S, R	R
Emberizidae, Cardinalinae				
Parkerthraustes humeralis[+]	U	R	T	R
Saltator grossus	C	C	T, B, L	R
Saltator maximus	U	U	B, S, T	R
Saltator coerulescens[*]	R	C	S, B	R
Cyanocompsa cyanoides	C	C	T, F, B	R
Emberizidae, Thraupinae				
Lamprospiza melanoleuca[+]	C	C	T	R
Cissopis leveriana[+]	R	C	S	R
Hemithraupis guira[+]	R	—	B	R
Hemithraupis flavicollis	C	R	T, B, F, L	R
Tachyphonus cristatus	C	U	T, B, F, L	R
Tachyphonus surinamus[+]	U	—	T	R
Tachyphonus luctuosus	C	—	B, F, T	R
Habia rubica	C	U	T	R
Ramphocelus carbo	C	C	S, B, R	R
Thraupis episcopus	C	C	S	R
Thraupis palmarum	C	C	S, B, T	R
Euphonia laniirostris	R	—	T	R
Euphonia chrysopasta	C	U	S, B, T, F	R
Euphonia minuta[+]	U	R	B, T, F	R
Euphonia xanthogaster[+]	R	—	T	R
Euphonia rufiventris	C	U	T, B, F, L	R
Tangara mexicana	C	R	B, T, S, F	R
Tangara chilensis	C	C	T, B, S, F, L	R
Tangara gyrola	C	R	T, B, F, L	R
Tangara cayana[*+]	R	—	L, S	R
Tangara nigrocincta[+]	U	—	T, B, F	R
Tangara velia[+]	C	U	T, B	R
Dacnis lineata	C	—	T, B, F, L	R
Dacnis flaviventer	R	—	B	R
Dacnis cayana	C	R	T, B, F, L, S	R
Chlorophanes spiza	U	—	T, B, F	R
Cyanerpes nitidus[+]	U	—	B, F	R
Cyanerpes caeruleus	U	R	T, B, F	R
Emberizidae, Tersininae				
Tersina viridis	U	R	B, T, S	AW?
Emberizidae, Parulinae				
Phaeothlypis fulvicauda	C	—	I	R
Granatellus pelzelni	U	R	F, B, I	R
Coereba flaveola[*+]	R	R	F	R
Vireonidae				
Cyclarhis gujanensis[+]	C	U	L, T, B	R
Vireolanius leucotis[+]	C	U	T, B, L	R
Vireo olivaceus	C	R	S, B, T	R
Vireo altiloquus[*+]	R	—	T	NW?
Hylophilus semicinereus	C	C	L, S, B	R
Hylophilus muscicapinus	C	C	T, L, B	R
Hylophilus ochraceiceps	U	—	T	R
Icteridae				
Psarocolius decumanus[*+]	U	U	B, F, S	R
Psarocolius viridis	U	R	T	R
Psarocolius yuracares	C	C	T, B	R
Cacicus cela	C	C	S, L, B, T	R
Cacicus haemorrhous[*+]	U	R	T, F, B	R
Icterus cayanensis[+]	C	U	T, B, S	R
Lampropsar tanagrinus[*+]	R	—	S	R
Scaphidura oryzivora[*+]	U	U	R	R
Dolichonyx oryzivorus[*+]	R	—	S	NP

TIMING OF BREEDING BY ANTBIRDS (FORMICARIIDAE) IN AN ASEASONAL ENVIRONMENT IN AMAZONIAN ECUADOR

DAN A. TALLMAN AND ERIKA J. TALLMAN
Department of Mathematics and Natural Sciences, Northern State University, Aberdeen, South Dakota 57401, USA

ABSTRACT.—From September 1975 through November 1976, we studied breeding of 26 species of antbirds (Formicariidae) at Limoncocha, in equatorial Amazonian Ecuador. At Limoncocha, where daylength, rainfall, temperature, and insect population sizes showed almost no seasonality, the antbirds appeared to breed all year. Evidence for breeding included the finding of nests and the use of dissections on both male and female birds. A high overlap of molting and breeding also was noted. Aseasonal breeding may be a result of an unfluctuating environment, a situation infrequently encountered, even in the tropics. Therefore, aseasonal breeding in birds is rare.

RESUMEN.—Desde de Septiembre de 1.975 hasta Noviembre de 1.976, nosotros estudimos la crianza de 26 especies de hormigueros (Formicariidae) en Limococha, en el Ecuador Amazónico. Donde la luz del día, la lluvia, temperatura y una gran cantidad de insectos, no se encontró en una estación especial, los hormigueros aparecieron para criar durante todo el año. Evidencias de las crianzas incluyeron el hallazgo de nidos y el uso de la disecación en el caso de los pajaros macho y hembra. Tambien se noto un cambio de plumaje en el mismo tiempo que se criaron. La crianza durante todo el año puede ser un resultado de un unfluctuante medío ambiente, una situación no frecuentemente encontrada en los trópicos. Por esta razon, este tipo de crianza de pajaros es muy dificil de encontrar.

For many birds of temperate and Arctic latitudes, not only is the seasonal pattern of breeding well-documented, but the proximate and ultimate causes underlying the timing are reasonably well understood (Gill 1994). Although the majority of bird species breed at tropical latitudes, little or nothing has been published for most on their annual pattern of breeding. Even for those tropical species that have been studied, most come from areas where rainfall showed strong seasonality (e.g., Skutch 1950; Snow and Snow 1964; Fogden 1972).

To investigate whether absence of seasonality in temperature and rainfall might also be associated with an absence of seasonality in breeding of birds, we studied annual breeding and molting cycles of antbirds (Formicariidae) at a site virtually on the equator in Amazonian Ecuador, where temperature and rainfall showed little or no seasonal fluctuation.

METHODS

Study area.—The study area was near Limoncocha (0°24'S, 76°37'W; 300 m elev.), a small village bordered by the Río Jivino on the west and an oxbow lake formed by the Río Napo on the east. The area was relatively undisturbed, moist tropical forest (sensu Holdridge 1967) at the time of the study.

To avoid over-collecting, which could possibly affect breeding cycles, we divided the study area into ten 1 km^2 regions. Ten to 20 mist nets were set on a monthly basis in each region and were open for roughly 8 h per day, 6 days per week. Each month a few antbirds also were shot in forest areas outside the regions within the study area.

Definition of breeding activity.—Breeding activity by antbirds was defined by the presence of active nests, eggs in the ovary or oviduct, ruptured follicles, oviducts more than 3-mm-wide, testis size over 30 mm^2, and by the presence of juvenals with completely unossified skulls and with the Bursa of Fabricius (following Nero 1951; Naik and Andrews 1966; and Benson 1962; for testis size as an indication of breeding time see Moreau 1966 and Foster 1975). In the case of testes size, we arbitrarily chose 30 mm^2 because this seemed to be especially large; similarly, the 3 mm wide oviducts also were arbitrarily chosen as being enlarged. Data on testis size from males were not included in our assessment of the timing of breeding because tropical species may retain large testis size well past their breeding season (Foster 1975). No data were available for the rate of skull ossification or atrophy of Bursa of Fabricius in antbirds. Therefore, we assumed these are valid indicators of age in Formicariidae, as has been demonstrated in other passerines.

Flowering and leaf cycles.—To find indicators of seasonality in the forest, we studied various aspects of the forest. During monthly aerial surveys of the forest, notes and photographs were taken to ascertain the relative flowering condition of the forest canopy. Erika Tallman recorded the flowering time for 46 species of herbs and low trees in the primary forest from October 1975 through September 1976.

Insect cycles.—From November 1975 through October 1976, insect samples were taken on sunny afternoons in one of the study area sectors. The insects were captured with 300 sweeps of a 48-cm-diameter net mounted on a 122-cm pole.

Weather.—We reviewed a compilation of 14 years of data (1961 to 1975) collected by the Jungle Aviation and Radio Service.

Antbirds.—Twenty-six species of antbirds have been recorded at Limoncocha (Tallman and Tallman 1977).

RESULTS

Weather.—Except that June and July tended to be more overcast than the other months of the year, our data support Pearson's (1977) assertion that Limoncocha is climatically aseasonal. Indeed, Limoncocha has one of the most unfluctuating climates in the world. Average monthly temperatures ranged from only 24.1 to 25.6°C (Pearson 1977). Rainfall reports showed no pronounced dry or wet seasons. Average total rainfall was 3,100 mm with an average monthly total of 258.5 mm. An average minimum rain (191.6 mm) fell in February and an average maximum (313.1) occurred in May. However, any month is potentially the wettest of the year.

Flowering and leaf seasonality.—Over one-half of the 46 species of herbs and low trees that we monitored flowered all year. Because of the patchy distribution of tropical plants, other species that flowered all year may have been overlooked. Because they are usually shaded by upper levels, lower strata plants are less affected by cloudiness than are canopy species (Tallman 1979). Pearson (1977) described the Limoncocha canopy as resembling that of many tropical forests, with a mixture of trees that fruited either year-round or synchronously but acyclically throughout the year (Richards 1952; Miller 1963; Moreau 1966). However, we recorded a large number of trees in flower from October through December and a peak of leafless trees from June through September. At least during our study, leaflessness and flowering corresponded with periods of overcast skies (for leaflessness) and a return to clear skies (for flowering). These associations agree with Janzen's (1967) observations in Central America.

Insect seasonality.—Except for dipterans, insects remained at fairly constant numbers at Limoncocha throughout the year, in contrast to other tropical areas that have greatly reduced insect numbers during dry seasons (Davis 1971; Janzen 1973; Wolda 1977, 1978). Pearson (1977) made similar observations regarding damselflies, dragonflies and tiger beetles at Limoncocha. Dipterans, however, tend to have patchy distributions near carrion and dung. Tabanids may be truly cyclic at Limoncocha: we noticed more of these biting flies in both Octobers of the study.

Timing of breeding and molting in antbirds.—The forest-inhabiting antbirds at Limoncocha appeared to breed and/or molt all year without showing peaks of reproductive activity (Fig. 1; Fig. 2; Appendix). Among the female antbirds collected, the monthly average of those in breeding condition (as described above) was 55% with a range from 42% in July to 76% in April.

DISCUSSION AND CONCLUSIONS

Coupled with the breeding data, the aseasonality of molt and the high incidence (33%) of individual birds molting and breeding simultaneously indicate noncyclic breeding by Limoncocha antbirds. In most other birds, the onset of prebasic molt is the most dependable evidence of the cessation of breeding cycles, even in species showing a low year-round incidence of breeding (Tordoff and Dawson 1965). The physiological demands of molt usually preclude simultaneous molting and breeding. Low molt frequency has been used to determine peaks of breeding activity in other bird populations (Miller 1963; Snow and Snow 1964; Ward 1969; Davis 1971; Snow 1976; Dowsett and Dowsett-Lemaire 1984; Dittami and Knauer 1986). These later authors mention that they found little evidence for individuals molting and breeding simultaneously. Stiles (1980) found most hummingbirds in Costa Rica showed discrete molting and breeding seasons,

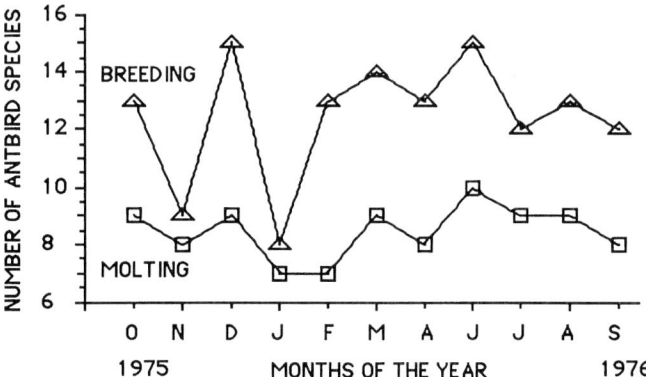

FIG. 1. Composite molt and breeding activity of Limoncocha antbirds.

although some individuals bred and molted at the same time. Chapman (1995) found Little Greenbuls (*Andropadus virens*), a bird of both forest and cultivated areas, breeding all year but did not find individuals breeding and molting simultaneously. Foster (1975) suggested that breeding and molting may coincide in some tropical birds because of cyclic shifts in food abundance; as noted in this paper, we have no evidence for such shifts at Limoncocha.

Aseasonal breeding by antbirds at Limoncocha in the primary forest was an unexpected discovery. Almost all forest bird species, including antbirds in other geographic locations, are seasonal breeders. For example, the antbird, *Gymnopithys leucaspis*, is a seasonal breeder in Panama, where dry seasons are well defined (Willis 1967), but it breeds throughout the year at Limoncocha. Hackett (1993), suggested substantial genetic differences exist between these populations. Around the world, species known to breed all year are primarily second-growth inhabitants or waterbirds (Burger 1949; Thompson 1950; Skutch 1950; Betts 1952; Davis 1953; Miller 1955, 1958; Marchant 1960; Benson 1962; Snow and Snow 1964; Ricklefs 1966; Willis 1967, 1968, 1969; Haverschmidt 1968; Harris 1974; Snow 1966, 1976; Diamond 1974). Even in Sarawak, another aseasonal equatorial location, birds showed marked breeding seasons (Fogden

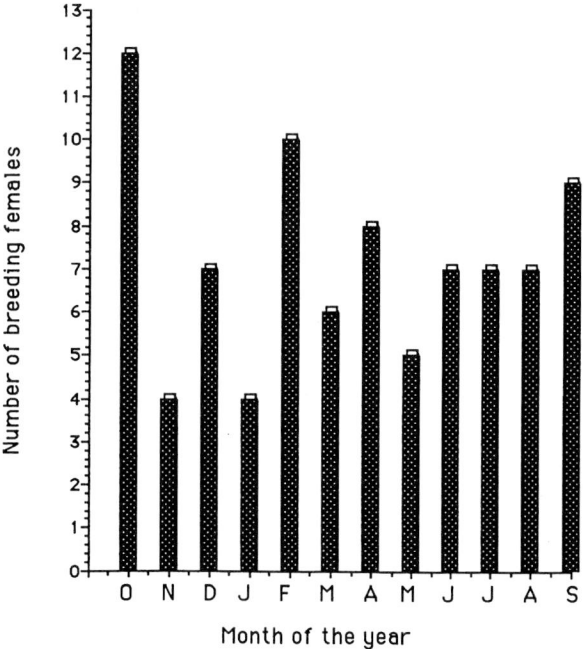

FIG. 2. Summation of evidence by month for breeding by female antbirds at Limoncocha.

1972). An exception may have been found in West Cameroon, were Serle (1981) found evidence for breeding aseasonality in some individuals, although groups of genera at the family level usually showed preferred breeding seasons. Thus there appears to be a tendency in aseasonal environments for birds to show a relaxation of breeding seasonality. However, because truly aseasonal localities are comparatively rare, acyclical breeding seasons appear to be equally uncommon.

ACKNOWLEDGMENTS

We are grateful to the many people who assisted us in Ecuador and the United States. The Ecuadorian Ministry of Agriculture and Dr. Fernando Ortiz-Crespo of the Catholic University, Quito, assisted us in obtaining permits. For logistical support and hospitality in Ecuador, we thank the personnel of the Summer Institute of Linguistics. Gary Lester and Carol Foil assisted at Limoncocha. Nat and Genie Wheelwright, Lindsey Hammond and Doug Wysham, all from the Peabody Museum of Natural History, Yale University, collected birds at Limoncocha and shared their data. Insect samples were counted and identified by Joseph Egers, Louisiana State University Department of Entomology. Financial support was provided by the Organization of American States (Fellowship 48830 PRA) and the Louisiana State University Museum of Natural Science. John Hagan, Mort and Phyllis Isler, Van Remsen, and Lloyd Kiff provided helpful suggestions during the preparation of this manuscript.

LITERATURE CITED

BENSON, C. W. 1962. The breeding seasons of birds in the Rhodesias and Nayasaland. Proc. XIII Inter. Ornith. Congr. 623–639.
BETTS, F. N. 1952. The breeding seasons of birds in the hills of south India. Ibis 94:621–628.
BURGER, J. W. 1949. A review of experimental investigations of seasonal reproduction in birds. Wilson Bull. 61:211–230.
CHAPMAN, A. 1995. Breeding and moult of four bird species in tropical West Africa. Tropical Zoology 8: 1–12.
DAVIS, J. 1971. Breeding and molt schedules of the Rufous-collared Sparrow in coastal Peru. Condor 73: 127–146.
DAVIS, T. A. W. 1953. An outline of the ecology and breeding seasons of birds of the lowland forest region of British Guinea. Ibis 95:450–467.
DIAMOND, A. W. 1974. Annual cycles in Jamaican forest birds. J. Zool., Lond. 173:277–301.
DITTAMI, J. P., AND B. KNAUER. 1986. Seasonal organization of breeding and molting in the Fiscal Shrike (*Lanius collaris*). J. Ornithol. 127:79–84.
DOWSETT, R. A., AND F. DOWSETT-LEMAIRE. 1984. Breeding and moult cycles of some montane forest birds in south-central Africa. Rev. Biol. (Terre Vie) 39:89–111.
FOGDEN, M. P. L. 1972. The seasonality and population dynamics of equatorial forest birds in Sarawak. Ibis 114:113–343.
FOSTER, M. S. 1975. The overlap of molting and breeding in some tropical birds. Condor 77:304–314.
GILL, F. B. 1994. Ornithology. Second Edition. W. H. Freeman, New York.
HACKETT, S. J. 1993. Pylogenetic and biogeographic relationships in the Neotropical genus *Gymnopithys* (Formicariidae). Wilson Bull. 105:301–315.
HARRIS, M. 1974. A field Guide to the Birds of Galapagos. Collins, London.
HAVERSCHMIDT, F. 1968. Birds of Surinam. Oliver and Boyd, London.
HOLDRIDGE, L. R. 1967. Life Zone Ecology. Trop. Science Center, San José, Costa Rica.
JANZEN, D. H. 1967. Synchronization of sexual reproduction of trees within the dry season in Central America. Ecology 21:620–637.
JANZEN, D. H. 1973. Sweep samples of tropical foliage insects: effects of seasons, vegetation types, elevation, time of day and insularity. Ecology 54:687–708.
MARCHANT, S. 1960. The breeding of some SW Ecuadorian birds. Ibis 102:349–382, 584–599.
MILLER, A. H. 1955. Breeding cycles in a constant equatorial environment in Colombia, South America. Acta XI Inter. Ornith. Congr. 495–503.
MILLER, A. H. 1958. Adaptation of breeding schedule to latitude. Proc. XII Inter. Ornith. Congr. 513–522.
MILLER, A. H. 1963. Seasonal activity and ecology of the avifauna of an American equatorial cloud forest. Univ. Calif. Publ. Zool. 66:1–78.
MOREAU, R. E. 1966. The Bird Faunas of Africa and Its Islands. Academic Press, New York, New York.
NAIK, R. M., AND M. I. ANDREWS. 1966. Pterylosis, age determination and moult in the Jungle Babbler. Pavo 4:21–47.
NERO, R. W. 1951. Pattern and rate of cranial ossification in the House Sparrow. Wilson Bull. 63:84–88.
PEARSON, D. L. 1977. A pantropical comparison of bird community structure on six lowland forest sites. Condor 79:232–244.
RICHARDS, P. W. 1952. The Tropical Rainforest. Cambridge Univ. Press, Cambridge, United Kingdom.
RICKLEFS, R. E. 1966. The temporal component of diversity among species of birds. Evolution 20:235–242.

SERLE, W. 1981. The breeding season of birds in the lowland rainforest and in the Montane forest of West Cameroon. Ibis 123:6274.

SKUTCH, A. F. 1950. The nesting seasons of Central American birds in relation to climate and food supply. Ibis 92:185–222.

SNOW, D. W. 1966. Moult and the breeding cycle in Darwin's Finches. J. Orn. 107:283–291.

SNOW, D. W. 1976. The relationship between climate and annual cycles in the Cotingidae. Ibis 118:366–401.

SNOW, D. W., AND B. K. SNOW. 1964. Breeding seasons and annual cycles of Trinidad landbirds. Zoologica 49:1–39.

STILES, F. G. 1980. The annual cycle in a tropical wet forest hummingbird community. Ibis 122:322–343.

TALLMAN, D. A. 1979. Ecological Partitioning by Antbirds of a Moist Tropical Forest in Amazonian Ecuador. Unpubl. Diss. Lousiana State University (Baton Rouge).

TALLMAN, D. A., AND E. J. DE TALLMAN. 1977. Adiciones y revisiones a la lista de la avifauna de Limoncocha, Provincia de Napo, Ecuador. Rev. Univ. Catolica (Quito) 16:217–224.

THOMPSON, A. L. 1950. Factors determining the breeding seasons of birds: an introductory review. Ibis 92:173–184.

TORDOFF, H. B., AND W. R. DAWSON. 1965. The influence of day length on the reproductive timing in the Red Crossbill. Condor 67:416–422.

WARD, P. 1969. The annual cycle of the Yellow-vented bulbul *Pycnonotus goiavier* in a humid equatorial environment. J. Zool., Lond. 157-25-45.

WILLIS, E. O. 1967. The behavior of Bicolored Antbirds. Univ. Calif. Publ. Zool. 79:1–132.

WILLIS, E. O. 1968. Studies of the behavior of Lunulated and Salvin's antbirds. Condor 70:128–148.

WILLIS, E. O. 1969. On the behavior of five species of *Rhegmatorhina*, ant-following antbirds of the Amazon basin. Wilson Bull. 81:363–395.

WOLDA, H. 1977. Fluctuations in abundance of some Homoptera in a Neotropical forest. Geo-Eco-Trop. 3:229–257.

WOLDA, H. 1978. Seasonal fluctuations in rainfall, food and abundance of tropical insects. J. Anim. Ecol. 47:369–381.

APPENDIX

BREEDING DATA FOR ANTBIRDS SUSPECTED TO BREED YEAR-ROUND. THE FOLLOWING SUBSCRIPTS ARE USED: ♀ = BREEDING FEMALE OR ACTIVE NEST; ♂ = TESTIS[1] GREATER THAN 30 MM[2]; ? = BIRD POSSIBLY BREEDING; i = FLEDGLING OR IMMATURE; * = MOLT PRESENT; † = MOLTING AND BREEDING SIMULTANEOUSLY

Species	Type of evidence	Oct	Nov	Dec	Jan	Feb	Mar	Apr	May	Jun	Jul	Aug	Sep
Cymbilaimus lineatus	Breeding	?♀				X♀				X♀			
	Molting									†			
Thamnophilus schistaceus	Breeding	X♀	X♀	X♀		X♀				X♀	X♀		
	Molting			*						*			
Thamnomanes ardesiacus	Breeding	X♀		?♀	X♀	X♀		X♀	X♀	X♀	X♀	X♀	X♀
	Molting	*	*			*		†	†	†		†	
Thamnomanes caesius	Breeding	X♀ᵢ		Xᵢ			Xᵢ	Xᵢ	X♀		X♀		
	Molting			*		*	*				*		
Myrmotherula hauxwelli	Breeding		Xᵢ		X♀	Xᵢ							
	Molting		*		*	*							
Myrmotherula ornata	Breeding			X♀	Xᵢ	X♀	Xᵢ			?♀	X♀	X♀	?♀
	Molting	*		†	*	†	*			*	*	†	*
Myrmotherula axillaris	Breeding	X♀			Xᵢ	X♀	X♀	Xᵢ		X♀		X♀	X♀
	Molting	*	*	*	*			*		*		*	
Myrmotherula menetriesii	Breeding						Xᵢ			Xᵢ	Xᵢ		Xᵢ
	Molting	*		*		*	*	*				*	*
Myrmoborus myotherinus	Breeding		?♀	Xᵢ		X♀		*	X♀	Xᵢ	Xᵢ	X♀	X♀
	Molting			*		*			†			†	†
Hypocnemis cantator	Breeding	Xᵢ		X♀				X♀				X♀	X♀
	Molting				*		*	*					†
Schistocichla leucostigma	Breeding	Xᵢ					Xᵢ	Xᵢ	Xᵢ	X♀	X♀	X♀	
	Molting	*	*				*			†	†		
Myrmeciza hyperythra	Breeding	X♀		X♀				X♀			Xᵢ	*	
	Molting			*				†			*		
Myrmeciza melanoceps	Breeding	X♀	Xᵢ			X♀					Xᵢ	Xᵢ	X♀
	Molting										*		*
Myrmeciza fortis	Breeding	X♀	Xᵢ	Xᵢ	Xᵢ		Xᵢ	X♀	Xᵢ	X♂			
	Molting		*	*						*			
Gymnopithys leucaspis	Breeding	X♀		Xᵢ	Xᵢ		X♀	X♀	Xᵢ	Xᵢ		Xᵢ	
	Molting			*		*			*		*		*

APPENDIX
CONTINUED

Species	Type of evidence	Oct	Nov	Dec	Jan	Feb	Mar	Apr	May	Jun	Jul	Aug	Sep
Hylophylax naevia	Breeding	$X_♀$		X_i	X_i		X_i	$X_♀$	X_i	X_i	$X_♀$	$X_♀$	$X_♀$
	Molting	*				*			*				
Hylophylax poecilinota	Breeding			$X_♀$		$X_♀$ †				X_i	$X_♀$		
	Molting		*				*		*				
Phlegopsis nigromaculata	Breeding	$X_♀$ †	$X_{♀♂}$ †		$X_♂$ †	$X_♂$	$X_{♀♂i}$			X_i	?	$X_♂$	$X_♂$
	Molting	$X_♀$ †	$X_♀$	*				*		*	*	*	*
Phlegopsis erythroptera	Breeding	X_i		$X_♀$			$X_♀$ †		$X_♀$	$X_♀$ †			
	Molting						X_i						
Chamaeza nobilis	Breeding		$X_{♂i}$	X_i	$X_♀$			$X_{♂i}$					$X_♀$
	Molting							*					
Formicarius colma	Breeding	$X_{♂i}$	$X_♂$	$X_{♂i}$	$X_{♀♂}$ †	$X_{♀♂}$	$X_♀$	$X_♂$ †			$X_{♂i}$ †	X_i	
	Molting	*											
Formicarius analis	Breeding	$X_♂$	X_i	$X_{♂i}$ †		$X_{♀♂}$	$X_♀$	$X_♀$		X_i	X_i	X_i	$X_{♀♂}$
	Molting	*							*			*	

[1] See text for discussion of testes size and breeding.

STATUS AND DISTRIBUTION OF *ASTHENES ANTHOIDES* (FURNARIIDAE), A SPECIES ENDEMIC TO FUEGO-PATAGONIA, WITH NOTES ON ITS SYSTEMATIC RELATIONSHIPS AND CONSERVATION

FRANÇOIS VUILLEUMIER

Department of Ornithology,
American Museum of Natural History, Central Park West at 79th Street,
New York, New York 10024-5192, USA

ABSTRACT.—The Austral Canastero *Asthenes anthoides* (King, 1831; Furnariidae), endemic to Patagonia and Tierra del Fuego (southern South America south of about 37°S), was studied in 1985–1988. Fieldwork showed the Austral Canastero to be (a) relatively widespread and locally common in mesic shrubsteppes, (b) absent from grasslands, and (c) more common in Tierra del Fuego than it was 90–100 years ago. These observations contrast with reports stating (a) that *A. anthoides* is vanishing and threatened because its formerly extensive grassland habitat has been virtually destroyed by overgrazing, and (b) that its occurrence in shrublands today reflects a relictual distribution in a marginal habitat. To resolve these divergent points of view, the habitat preferences, status, and distribution of *A. anthoides* are reviewed. Recent field surveys revealed that *A. anthoides* is not rare and occurs both on the Patagonian mainland and in Tierra del Fuego. *A. anthoides* is most frequently found in mesic shrubsteppe receiving 350-450 mm of rain/year, with an upper stratum of shrubs (*Chiliotrichum* and *Berberis*), and a ground stratum of tussock-grasses (*Festuca*). *A. anthoides* is also present in three other habitats: (1) more humid, tall and dense woodland with small trees (*Nothofagus antarctica*), (2) drier, low and open shrubsteppe with shrubs (*Mulinum spinosum*) and some grass tussocks, and (3) drier, low shrubsteppe with uniform stands of *Lepidophyllum cupressiforme* without grass cover. In its mesic shrubsteppe habitat *A. anthoides* reaches densities of up to one singer per ha. *A. anthoides* was looked for, but not found in spite of repeated visits, in the extensive grasslands of Magallanes receiving 200–300 mm of rain/year. The geographical range of *A. anthoides* seems to be disjunct. A population in northwestern Patagonia (western Neuquén, western Río Negro, and western Chubut, Argentina; and adjacent Chile in Ñuble, Malleco, and Aysén) may be geographically isolated from a southern population inhabiting the Patagonian mainland (southern Santa Cruz, Argentina, and Magallanes, Chile) and northern Tierra del Fuego (in both Argentina and Chile). Nineteenth and early twentieth century reports mentioned that *A. anthoides* was rare in Tierra del Fuego, whereas the species is common there today. *A. anthoides* has thus increased numerically in the last 90–100 years and is not vanishing on Tierra del Fuego. On the Patagonian mainland this species is relatively localized and patchily distributed but is not rare where found. There is no evidence suggesting that *A. anthoides* occupied the formerly more widespread grasslands of Patagonia, that its alleged rarity is due to grassland destruction from overgrazing by sheep, or that its modern shrubland habitat is marginal. All steppe habitats, including grasslands where *A. anthoides* does not occur, and shrubsteppes, including the *Chiliotrichum-Festuca* shrubsteppe where it is found most commonly, are overgrazed. In Chilean Fuego-Patagonia, grasslands support on average one sheep per 2 ha, and shrubsteppes twice as many sheep, on average one per hectare. Overgrazing has degraded both grasslands and shrubsteppes, and the shrubsteppe habitat where *A. anthoides* lives is no less impacted by grazing than the grasslands where it does not occur. The apparent distributional gap between northern and southern populations of *A. anthoides* is located in a low rainfall area where the climate is too arid for the mesic shrubsteppe (degraded or not) favored by this species. Careful exploration of mesic shrubsteppe pockets along the eastern Andean foothills of western Santa Cruz may therefore reveal the presence of *A. anthoides* there. Even though the southern population of *A. anthoides* does not seem threatened, its shrubsteppe habitat is not only subject to intensive sheep grazing pressure (too many sheep/ha), but also to overgrazing on a large geographical scale. In Chilean Fuego-Patagonia over 80% of grazing land is now eroded because of overgrazing (55% of it moderately to severely). Increased deterioration of shrubsteppe could therefore alter the status of *A. anthoides* for the worse. Consequently the long-term status of this species remains uncertain. *A. anthoides* is the only member of a group of closely related allopatric and parapatric taxa that consistently occupies a shrubby habitat, the others being reported to live in grasslands. Considered by some authors to be a subspecies of a group of canasteros including *A. wyatti, A. punensis,* and *A. sclateri,* allopatric *A. anthoides* is sufficiently distinct morphologically to be treated as a species. This Patagonian endemic could serve as an indicator species, whose density in shrubsteppes of Fuego-Patagonia might give an easily obtainable measure of the health of its habitat.

RESUMEN.—*Asthenes anthoides* (King, 1831; Furnariidae), llamado Espartillero Austral o Canastero manchado chico en Argentina y Canastero del sur en Chile, especie endémica de la Patagonia y Tierra del Fuego (América austral al sur del 37°S), fue estudiada en 1985–1988. El trabajo de campo mostró que (a) *A. anthoides* tiene una distribución relativamente amplia, siendo común, localmente, en la estepa arbustiva mésica, (b) se encuentra ausente en pajonales o estepas con dominancia de gramíneas, y (c) ha aumentado numéricamente en Tierra del Fuego en los últimos 90–100 años. Dichas observaciones no concuerdan con informes previos según los cuales (a) *A. anthoides* estaría desapareciendo y se encontraría amenazada, debido a que el sobrepastoreo por ovinos practicamente destruyó los pajonales, originariamente su ambiente preferencial, y (b) que su presencia actual en estepa arbustiva correspondería a una distribución relictual en un ambiente marginal. Para tratar de resolver estos puntos de vista divergentes, se han analizado las preferencias ecológicas, status, y distribución geográfica de *A. anthoides*. El trabajo de campo del autor reveló que *A. anthoides* no es una especie rara. La especie fue encontrada en la Patagonia continental y en Tierra del Fuego. *A. anthoides* habita más frecuentemente en estepa arbustiva mésica alta con 350–450 mm de lluvia/año, con un estrato superior de arbustos (*Chiliotrichum* y *Berberis*) y un estrato inferior de gramíneas (*Festuca*). *A. anthoides* se encuentra presente en otros tres ambientes: (1) matorral alto y denso con pequeños árboles (*Nothofagus antarctica*) y mayor pluviosidad anual, (2) estepa arbustiva más baja y abierta, con arbustos (*Mulinum spinosum*) y algunas gramíneas, y bajo regimen pluviométrico, y (3) estepa arbustiva baja de *Lepidophyllum cupressiforme,* uniformemente distribuido, sín gramíneas, y con baja pluviosidad. En su ambiente preferencial de estepa arbustiva mésica la densidad de *A. anthoides* puede alcanzar hasta un indivíduo por ha. *A. anthoides* no fue encontrada en los pajonales extensos de Magallanes, con precipitación anual de 200–300 mm, más allá de que este ambiente fue visitado durante varias ocasiones. El rango de distribución de *A. anthoides* parece ser disyuncto, con una población en el noroeste de la Patagonia (oeste de Neuquén, oeste de Río Negro, y oeste de Chubut, Argentina; y zonas adyacentes de Chile en Ñuble, Malleco y Aysén) que probablemente se encuentra geográficamente aislada de una población austral que ocupa la parte continental de la Patagonia austral (sur de Santa Cruz, Argentina, y Magallanes, Chile), y la zona norte de Tierra del Fuego (Argentina y Chile). Informes del siglo pasado y del comienzo del siglo veinte señalan la rareza de *A. anthoides* en Tierra del Fuego, sín embargo la especie es actualmente común en esta isla. Estos datos sugieren que *A. anthoides* ha aumentado numéricamente en los últimos 90–100 años, y no indican que la especie esta desapareciendo de Tierra del Fuego. En la Patagonia continental esta especie esta más localizada y su distribución es disyunta, pero donde se encuentra no es rara. No hay información que indique que (a) *A. anthoides* ocupaba los extensos pajonales de la Patagonia de antaño, (b) su supuesta rareza actual resultaría de la destrucción de pajonales por sobrepastoreo ovino, o (c) su habitat arbustivo actual es marginal. Todos los ambientes de la estepa, incluyendo los pajonales donde *A. anthoides* no habita actualmente, y las estepas arbustivas, incluyendo la estepa arbustiva con *Chiliotrichum-Festuca* donde la especie se encuentra con más frecuencia, estan sobrepastoreados. En Fuego-Patagonia Chilena, los pajonales tienen una carga media de 1 oveja/2 ha y las estepas arbustivas dos veces más, con un promedio de 1 oveja/ha. Por lo tanto, el sobrepastoreo ha degradado ambos pajonales y estepas arbustivas y la estepa arbustiva donde habita *A. anthoides* ha sido tan deteriorada por sobrecarga ovina como lo han sido los pajonales donde esta especie no se encuentra. La distribución aparente disyunta entre las poblaciones septentrionales y meridionales de *A. anthoides* se encuentra localizada en una zona con poca precipitación anual, que corresponde a una zona donde el clima es demasiado árido para el desarrollo de las estepas arbustivas mésicas ocupadas por *A. anthoides,* sea tanto en áreas con o sin degradación por sobrepastoreo. Exploraciones de campo en zonas de estepa arbustiva mésica, a lo largo de la franja oriental del piedemonte de los Andes en el oeste de Santa Cruz, deberían revelar la presencia de *A. anthoides*. Aunque la población austral de *A. anthoides* no parece amenazada, su ambiente de estepa arbustiva esta sometido no solo a pastoreo ovino intensivo (demasiado ovinos/ha) sino también a sobrepastoreo a gran escala geográfica. En Fuego-Patagonia Chilena, más del 80% del área de uso agropecuario esta erosionado por sobrepastoreo (55% de esto de manera moderada a severa). Un aumento de la deterioración de las estepas arbustivas empeoraría el status de *A. anthoides*. Esto significa que el status futuro de *A. anthoides* no esta asegurado. *Asthenes anthoides* es el único miembro de un grupo de taxones alopátridos o parapátridos estrechamente emparentados, que ocupa un ambiente arbustivo de manera exclusiva. Las otras formas se consideran como ecológicamente ligadas a pajonales. Considerado por algunos autores como una subespecie de un grupo de canasteros o espartilleros que incluyen *A. wyatti*, *A. punensis*, y *A. sclateri*, la forma alopátrida *A. anthoides* es suficientemente distinta morfológicamente, como para ser considerada una especie. Esta especie endémica podría servir como especie indicadora cuya densidad en la estepa arbustiva de Tierra del Fuego y Patagonia podría proporcionar una medida de la salud relativa de dichas estepas de facil obtención.

Patagonia and the archipelago of Tierra del Fuego across the Strait of Magellan in southern Argentina and southern Chile, a region called "Fuego-Patagonia" by Auer (1956, 1960) and Pisano (1977, 1981), has been subjected to environmental perturbations including Pleistocene glacial-interglacial cycles that have led to much faunal extinction (e.g., Humphrey and Péfaur 1979; Markgraf 1985; Vuilleumier 1991). In the last hundred years or so human activities, especially sheep ranching, resulting in wide scale overgrazing and subsequent soil erosion, have greatly altered the natural steppe vegetation throughout Fuego-Patagonia, and have locally scarred it dramatically, and perhaps irrevocably. It is becoming increasingly clear that these landscape degradations have also modified the composition of the avifauna studied by early authors such as Crawshay (1907).

A major question is whether these human influences have changed the status of some endemic species to the point where they are at risk of significant population decline and eventual extinction. In another paper I examined the status of the Ruddy-headed Goose *Chloephaga rubidiceps* (Anatidae), which I considered to be seriously endangered (Vuilleumier 1994b). In the present report I analyze the status of the Austral Canastero *Asthenes anthoides* (King 1831) (Furnariidae), a species endemic to Fuego-Patagonia.

In their book on Andean and Patagonian birds Fjeldså and Krabbe (1990:377–378) wrote: (1) that *Asthenes anthoides* was "Perhaps characteristic of the Patagonian long-grass prairie, until this habitat was destroyed by sheep-grazing," (2) that *A. anthoides* was "Now apparently vanishing, but still locally common near Cabo Virgenes, s Sta Cruz," and (3) that this species was "Uncommon or rare on Isla Grande [of Tierra del Fuego]." Earlier, Fjeldså (1988:93) had stated that "Today, most observers look in vain for [*Asthenes anthoides*], and the few recent observations have been in brush vegetation near the base of the southern Andes." He added: "My own interpretation is that [*Asthenes anthoides*] disappeared together with the tall-grass vegetation, but survives locally in a marginal habitat." Collar et al. (1992:609–612) listed *A. anthoides* as "threatened" and stated: "The species appears to be very local and decreasing in numbers." Ridgely and Tudor (1994:4) wrote: "Found only on the steppes of southern Argentina and Chile, the Austral Canastero, though evidently formerly numerous, is now a very scarce and local resident of less disturbed areas, its numbers having declined because of the effects of severe overgrazing across almost its entire range. No substantial population receives formal protection." Ridgely and Tudor (1994:112) also stated that the Austral Canastero "Certainly deserves formal threatened status." In other words, according to the view expressed by these authors, *A. anthoides* used to be widespread and common in tall grassland, but now that this habitat has been largely destroyed by human interference, this species occupies shrubby habitats, which are marginal, and is diminishing in numbers.

During field work in Fuego-Patagonia I found that *A. anthoides* does not occur in grasslands but in shrubsteppes, and that it is relatively widespread and locally common in southern Fuego-Patagonia. The species appears to be absent from central Patagonia, but reappears in shrubsteppes of northwestern Patagonia.

These observations, which seem to contradict some of the statements quoted above, raise several related questions. Was *A. anthoides* living in grasslands in the past? Is it now reduced to a marginal existence in suboptimal shrublands, since the original grasslands have been largely destroyed by overgrazing? Is *A. anthoides* declining and is it endangered? In order to give preliminary answers to these questions, I first present observations resulting from the field work I carried out in Chilean Fuego-Patagonia, and next summarize the literature on the status and distribution of *A. anthoides*. Given the importance of the systematic status of taxa in conservation biology, I finally discuss the taxonomic affinities of this species. In this report I use the name *A. anthoides* rather than *Thripophaga anthoides*, as did also Fjeldså and Krabbe (1990) and Ridgely and Tudor (1994), *contra* Vaurie (1971:25-28, 1980:181-182). The similarities in color and pattern shared by *Asthenes* and *Thripophaga* that led Vaurie (1971, 1980) to merge them are probably superficial and in my opinion do not reflect a common phylogenetic origin.

METHODS

My field observations are based on a total of 103 field-days in Chilean Fuego-Patagonia at 52°–54°S (Magallanes and Tierra del Fuego Archipelago) in November–December 1985 (35 days), February–March 1987 (15 days), October 1987 (20 days), January 1988 (8 days), November 1988 (22 days), and November 1993 (3 days). Field work covered the austral spring and summer period, from October to March, when most species of Fuego-Patagonian birds breed, but not the autumn, winter, or early spring (April to September). Earlier field observations are from February–March 1965 near Bariloche, about 41°10′S, Río Negro, Argentina, and the Lonquimay Valley, about 38°30′S, Malleco (IX Región), Chile.

In Chilean Fuego-Patagonia I traveled on the mainland from San Juán, Brunswick Peninsula, northward, northeastward, and eastward along the Strait of Magellan all the way to Punta Dungeness at the eastern entrance of the Strait; from Punta Arenas northwestward to Seno Otway, Seno Skyring, Puerto Natales, and the Torres del Paine National Park; and inland along the Chile-Argentina border eastward from Morro Chico to Gallegos Chico, Pali-Aike, and Monte Aymond. On Isla Grande of Tierra del Fuego I traveled on all the main roads between Punta Espora in the north and Estancia Vicuña in the south; from Porvenir in the west to San Sebastián in the east; and inland northward from Onaisín to China Creek, Cullen, Sombrero, and Bahía Azul-Punta Espora.

During these travels I looked for *A. anthoides* in all steppe habitats encountered during the transects, ranging from pure grassland steppes to dense shrubsteppes, and including forest-steppe ecotones and man-modified shrublands and open woodlands. Contacts with *A. anthoides* were visual, vocal, or both. Because the *kik*, *tik*, *pik*, or *tick* calls and the songs (rapid trills) of *A. anthoides* are relatively loud, emitted frequently, and very similar to those of its high Andean relative *A. wyatti* (with which I am familiar; e.g., Vuilleumier and Ewert 1978:71–72), it was easy to detect *A. anthoides* by its voice. Furthermore, *A. anthoides* is relatively conspicuous because it often sings while perched on or near the top of shrubs. Fjeldså and Krabbe (1990: 375–378) described the vocalizations of Patagonian *A. anthoides* and of its allopatric Andean relatives *A. wyatti* and *A. punensis*, but stated (p. 377) that there were "No data" about the voice of its Córdoba Mountains relative *A. sclateri*, which I will therefore describe later.

Besides *A. anthoides*, I have had extensive field experience with seven of the other eleven taxa (species or subspecies) of this group of canasteros (see discussion of systematic position): *A. wyatti* (subspecies *mucuchiesi* in Venezuela, *aequatorialis* in Ecuador, and *graminicola* in Peru); *A. punensis* (subspecies *punensis* in Peru and Bolivia, *cuchacanchae* in Bolivia, and *lilloi* in Argentina); and *A. sclateri* (subspecies *sclateri* in Argentina).

RESULTS

I observed or heard *A. anthoides* on a total of 26 out of 103 field-days (25%): three days (of 35, 9%) in November-December 1985; six days (of 15, 40%) in February–March 1987; zero day (of 8, 0%) in January 1988; seven days (of 20, 35%) in October 1987; nine days (of 22, 41%) in November 1988; and one day (of three, 33%) in November 1993. I saw this species at 22 sites: nine on the Patagonian mainland north of the Strait of Magellan, and 13 on the island of Tierra del Fuego.

The habitat where I found *A. anthoides* most often was a relatively dense mesic shrubsteppe containing a mixture of shrubs, mostly *Chiliotrichum diffusum* (Compositae), forming an upper stratum 1 to 1.5 m tall, and of low, 50 cm tall tussock grasses (especially *Festuca gracillima*, Gramineae), making up an uneven ground cover (Fig. 1, top). This shrubby habitat was studied most intensively on Isla Grande of Tierra del Fuego, and is part of what Pisano (1977:162–167) named the Patagonian Mesic Shrubsteppe or the *Festuca gracillima–Chiliotrichum diffusum* Association (see also description in Lara and Cruz 1987a:17, and their photo no. 5, p. 18).

I observed *A. anthoides* in three other habitats: (1) tall, dense, and uniform woodland composed mostly of 2–3 m high trees (especially *Nothofagus antarctica*, Fagaceae) on the mainland north of Puerto Natales (this habitat forms part of the Deciduous Magellanic Woodland of Pisano 1977:181–197, including his *Nothofagetum antarcticae* Association; see also Lara and Cruz

FIG. 1. Top. Dense shrubsteppe habitat of *Asthenes anthoides* in Fuego-Patagonia, with a mixture of shrubs (*Chiliotrichum diffusum*, Compositae; *Berberis* sp., Berberidaceae), and tussock-grasses (*Festuca* sp., Gramineae) on morainic soil; northwestern Isla Grande de Tierra del Fuego near Bahía San Felipe, Chile, November 1985; photograph F. Vuilleumier. Bottom. Open shrubsteppe habitat of *Asthenes anthoides* in Fuego-Patagonia, with nearly uniform, low cover of *Lepidophyllum cupressiforme* (Compositae) on sandy soil; mainland near Punta Dungeness, Magallanes, Chile, February 1987; photograph F. Vuilleumier.

1987a:15); (2) low, open shrubsteppe with sparse vegetation of 50–100 cm high shrubs (especially *Mulinum spinosum*, Umbelliferae), and some grass tussocks (especially *Festuca gracillima*) on the mainland near the Torres del Paine National Park (this habitat belongs to the *Mulinetum spinosum* or Pre-Andean Xerophytic Matorral Association of Pisano 1974:80–81; see also Lara and Cruz 1987a:18–19, and their photo no. 6, p. 19); and (3) low, open, and relatively uniform, 50–120 cm high scrub of *Lepidophyllum cupressiforme* (Compositae) with little or no ground stratum of tussock grasses on the mainland near Punta Dungeness (Fig. 1, bottom), a habitat that forms part of the Psammophytic Matorral or *Lepidophylletum cupressiforme* Association of Pisano (1977:168–170; see also Lara and Cruz 1987a:18).

At no locality did I see or hear *A. anthoides* in pure grassland, grassy steppe, or grassy meadows, although I spent considerable time in 1985–1988 sampling the avifauna of grasslands in Chilean Patagonia, where they cover large areas, especially on the mainland. Thus I did not encounter *A. anthoides* in the two most extensive grassland types, respectively called the Mesic Patagonian Steppe or *Festucetum gracillimae* Association (Pisano 1977:154–159) and the Xeric Patagonian Steppe or *Festuca gracillima-Stipetum* Association (Pisano 1977:159–162). The first of these two associations is illustrated by Pisano (1977:160, his fig. 8), by Moore (1983:his plate 2a between pages 30 and 31), and by Lara and Cruz (1987a: 21, their photo no. 8), and the second by Pisano (1977:160, his fig. 10) and Vuilleumier (1991:10, his fig. 6, top).

All sites where I observed *A. anthoides* had dry soil, and the shrubby vegetation in which this species occurred was not located near water, either pools, ponds, or streams. Soils varied from hard and pebbly or even rocky on well-drained, gently sloping morainic hillsides or hilltops (*Chiliotrichum-Festuca* shrubsteppe; Fig. 1, top), to soft and sandy on poorly drained flat ground or dunes (*Lepidophyllum* scrub; Fig. 1, bottom).

At one site, Estancia Los Tehuelches, about 4 km inland from Puerto Nuevo along the coast of Bahía Inútil, Tierra del Fuego, Chile, *A. anthoides* was observed daily during three sojourns, 8–13 October 1987, 20–23 October 1987, and 16–20 November 1988. At Estancia Los Tehuelches, *A. anthoides* did not occur in the moist grassy meadows covering the valley floor near the small seasonal stream in the center of my study area (*vega*, see Pisano 1977:174–175, his fig. 20; Lara and Cruz 1987a:20, their photo no. 7; Vuilleumier 1991:22, his fig. 17), but was found along the slopes and on the hilltops in relatively dense *Chiliotrichum* shrubsteppe grazed by sheep and horses and growing on gravelly or pebbly soils of fluvio-glacial origin (till). Territorial singers were regularly seen while displaying vocally from the top or near the top of 50–120 cm tall *Chiliotrichum* shrubs. From three to five different singing individuals were encountered in this habitat on any given day in October 1987 and November 1988. Because this habitat in my study area covered about 5 ha, the maximum density was therefore of one singing individual (and presumably pair) per hectare.

Densities of *A. anthoides* similar to those recorded in *Chiliotrichum-Festuca* shrubsteppe at Estancia Los Tehuelches were also noted elsewhere. Thus on 11 February 1987, along a transect from Punta Espora to Porvenir, Chilean Tierra del Fuego, I recorded *A. anthoides* at each of six study sites (km 100, km 69, km 55, km 49, km 43, and km 5 northeast of Porvenir). At each of these six sites two to three singing individuals could be heard simultaneously, emitting their characteristic trills within an area of about 3 ha. On 28 February 1987, at three sites along the road from Posesión to Punta Dungeness on the mainland of Magallanes north of the Strait of Magellan, *A. anthoides* was found in similar densities in *Berberis* or *Lepidophyllum* scrub on sandy soil.

In spite of my repeated encounters with *A. anthoides* in Fuego-Patagonia I did not succeed in discovering any nest. On 16 November 1988 at Estancia Los Tehuelches a bird was seen carrying food (insects?) in its bill, perhaps to nestlings, but I did not find a nest although I searched diligently for it. To verify whether the Los Tehuelches population was breeding, one specimen was collected on 16 November 1988 (American Museum of Natural History [AMNH] 826141). It was a male with enlarged testes (left testis 11 × 8 mm, right testis 10.5 × 7.5 mm), but with skull only about 25% ossified. It weighed 24.5 g, and had no subcutaneous fat. Some tail molt was noted, but no body or wing molt. Its stomach contained insects. The iris was dark brown, the upper mandible black, the lower mandible gray, tipped black, the tarsi dusky-gray, and the toes grayish black. The large size of the gonads, together with the fact that this bird had a well developed brood patch, strongly suggest that it was (or had been) incubating.

The two most striking facts emerging from my field work in Fuego-Patagonia in 1985, 1987, 1988, and 1993 were (1) that I did not find *Asthenes anthoides* in pure grassland habitats, but only in habitats with much, and often dense, shrubbery, and (2) that whereas the species was widespread and common in 1987 and 1988, I observed it only three times in 1985 (at only two sites, 7 and 9 km east of Porvenir, Isla Grande of Tierra del Fuego). Thus along the transect from Punta Espora to Porvenir *A. anthoides* was noted at the six sites censused on 11 February 1987, as mentioned above, but was not detected on 11 November 1985. This observation suggests either that the species shows year-to-year fluctuations in relative abundance, or that it has some seasonal movements, or else that it increased in the period 1985–1988. Clearly, my observations call for an quantitative study of the habitat preferences and relative abundance of the species, but since this does not appear to be forthcoming, and given the view of several authors that *A. anthoides* may be vanishing, I compare below my observations with published data, in order to provide a basis for future work.

DISCUSSION

Status and distribution of Asthenes anthoides.—In this section I address the two related problems, whether *A. anthoides* is rare and vanishing, and whether its geographical range is more restricted today than it was in the past. Different authors have held varying opinions concerning the status of this species. Some writers found the species to be common. Thus, Venegas and Jory (1979:153) wrote that it was "común localmente" in Chilean Fuego-Patagonia, and Olrog (1948:510) that *A. anthoides* was "a common bird in the *dense thickets* [italics mine] of the rolling plains near Porvenir [Tierra del Fuego] and Cabeza del Mar [Chilean mainland north of the Strait of Magellan]." By contrast, Clark (1986:222) called it "poco común" in Tierra del Fuego, and Johansen (1966:249) did not even record it on that island. Other authors have written what seem to be contradictory statements. For instance Humphrey et al. (1970:261–262) wrote that in Tierra del Fuego, *A. anthoides* had been "recorded rarely in the northeastern part of the Island," but elsewhere called it "a common species . . . in the vicinity of Porvenir . . . " Much earlier, Crawshay (1907:79–80) had written: "Few expeditions record this Spine-Tail, and none from the island [of Tierra del Fuego]." Crawshay (1907) only saw it once during several months of field work in Tierra del Fuego (one bird, which he collected in *Chiliotrichum* shrubbery on 29 November 1907 at Rio McClelland settlement south of Bahía Inútil). Crawshay (1907:80) returned to his collecting site repeatedly over a period of two months but never saw another bird. The status of *A. anthoides* on Staten Island (Isla de los Estados, Argentina), southeast of Tierra del Fuego, is uncertain. Chebez and Bertonatti (1994:55) did not observe it and cited Salvadori's (1900) record as the only one for that island. Salvadori (1900) reported the results of a collecting trip to southern South America from December 1881 to October 1882, during which the expedition visited Staten Island. *A. anthoides* was collected at a locality called "Penguin Rookery," Puerto Roca (see Paynter 1985:334), which was visited on 10, 16, 21, and 28 February 1882 (Salvadori 1900:9). The specimen, presumably still in the Museo Civico di Genova, might be the only record of *A. anthoides* for Staten Island, and its identification should be verified. If confirmed, the occurrence of this species on Staten Island would be very unusual, in view of the fact that shrubsteppe with *Chiliotrichum* or other similar shrubs is not at all widespread there, because of the high rainfall prevailing in the area. Instead, Staten Island is covered mostly by dense forests dominated by evergreen beeches (*Nothofagus betuloides*). Non-forest vegetation consists of isolated patches of wet grasslands, especially along the coasts, of montane grasslands with cushion plants at higher elevations, and of bogs with mosses (*Sphagnum*) or *Marsippospermum grandiflorum* (Juncaceae) (Chebez and Bertonatti 1994). None of these vegetation types appear suitable for *A. anthoides*.

Interesting in the context of possible fluctuations in numbers mentioned earlier are the remarks published by Philippi et al. (1954:52) about the relative abundance of *A. anthoides* in Tierra del Fuego. They saw the species commonly in 1946 at Estancia Ekewern and in 1952 along the road from China Creek to Cerro MacPherson but did not find it in similar habitats near Porvenir in 1952. Information summarized by Collar et al. (1992:610) gave the status of *A. anthoides* in Tierra del Fuego variably as "uncommon breeder," or "occurring in fair numbers," or else as "truly common." Howell and Webb (1995:63–64) found *A. anthoides* "fairly common" in northern Tierra del Fuego and on the adjacent mainland in Magallanes in 1992 and 1994. Their remarks about the relative abundance of this species in Chilean Fuego-Patagonia agree more closely with my own observations than those of any other authors. Philip S. Humphrey collected a series of 21 specimens in Santa Cruz province (housed at the Yale University Peabody Museum of Natural History [YPM], Nos. 82920–82940; Fred Sibley, pers. comm.) and Rollo H. Beck three (at AMNH, Nos. 165169, 165170, 165173) in "Santa Cruz, Argentina," and one (AMNH 165872) along the "Straits of Magellan," Chile.

Besides my observations from Chilean Fuego-Patagonia reported above, I have seen *A. an-*

thoides elsewhere in Patagonia only in the northern part of its range, near Bariloche, Río Negro, Argentina, on 6, 11, and 19 February 1965, and in the upper Lonquimay Valley, Malleco (IX Región IX), Chile, on 31 March 1965. *A. anthoides* was stated to be "not rare near Bariloche" by Collar et al. (1992:610). I did not see or hear *A. anthoides*, even though I was on the lookout for it, during extensive fieldwork in Argentine Patagonia in 1991, 1992, and 1993, when I traveled from the Río Colorado and Río Negro in the north southward to the Beagle Channel in southern Tierra del Fuego, and from the Atlantic coast westward to the Andean foothills.

Whether *A. anthoides* migrates out of its breeding range in the southern hemisphere winter, and if it does, how far north it goes, remains entirely unclear. Philippi (1964:132) stated that in Chile, records from Concepción northward to Aconcagua referred to winter visitors. Fjeldså and Krabbe (1990:378) wrote that *A. anthoides* was "partly migratory, in winter n to Aconcagua." I suspect that these statements had their origin with comments such as Hellmayr's (1932:211) remark that "In Santiago Province [*A. anthoides*] appears to be only a winter visitor." Venegas and Jory (1979:153) stated that there was "probably local migration" in Chilean Fuego-Patagonia. Rollo H. Beck collected 3 specimens in Santa Cruz on 17 May 1915 (AMNH 165170 and 165173) and 20 May 1915 (AMNH 165169) and one along the Strait of Magellan on 23 July 1914 (AMNH 165872). Bernath (1965:100) recorded one *A. anthoides* on 24 April 1959 (autumn) at Gente Grande, Chilean Tierra del Fuego. Humphrey et al. (1970:261) cited specimens from Tierra del Fuego collected by Percival W. Reynolds in April (autumn) and July (winter), and classified the species as a "breeding resident." Humphrey et al. (1970:261) further wrote that *A. anthoides* "almost certainly [is] a non-migratory, breeding resident of the northern, non-forested part of the Island [of Tierra del Fuego] in appropriate habitat." These various records show the presence of *A. anthoides* in southern Fuego-Patagonia there in the southern hemisphere autumn (April–May) and winter (July), thus suggesting year-round residency of at least part of the population. The most detailed statements on the migration of *A. anthoides* remain the remarks written by Hellmayr (1932:211), based on a few specimens and field observations from the 1860s–1890s, some specimens from the 1920s, and some field observations from the 1870s and 1920s–1930s. Altogether the evidence is weak. Even though Chesser (1994:104) lists *A. anthoides* as an austral migrant, I believe that the migratory status of this species needs further critical scrutiny.

Various reports in the literature have mentioned the presence of *A. anthoides* in Buenos Aires province, including, most recently, Collar et al. (1992:609). Narosky and Di Giacomo (1993:73) cited these records, qualified them all as being erroneous, and attributed them to misidentification with the superficially similar *Asthenes hudsoni*. Ridgely and Tudor (1994:113) wrote that they regarded "the two published specimens from Buenos Aires as unverified." Once again, it would be good if all these alleged records from Buenos Aires could be verified by direct examination of the specimens and critical reading of the published sources.

Many authors have indicated or implied that the distribution of *A. anthoides* is continuous from northern to southern Patagonia, from south-central Chile and adjacent Argentina at about 37°S southward to the Strait of Magellan (e.g., Hellmayr, 1925:148–149, 1932:210–211; Peters 1951:108; Johnson 1967:195–196; Olrog 1979:181; Vaurie 1980:160; de la Peña 1988: map p. 32; Narosky and Yzurieta 1989: 188; Collar et al. 1992:609–610; Ridgely and Tudor 1994:map p. 112), but a few other workers have suggested a range disjunction. Thus Fjeldså and Krabbe's (1990:378) detailed map shows what I would interpret as two breeding distribution centers, one in northwestern and west-central Patagonia and a second much farther south in southern Santa Cruz-Magallanes and Tierra del Fuego. Other species of Patagonian birds exhibit similar, although not identical, apparent disjunctions in their breeding ranges, for instance the tyrannids *Neoxolmis rufiventris* and *Muscisaxicola capistrata* (Vuilleumier 1994a:maps on pp. 29 and 36). Neither Pemberton (*in* Wetmore 1926:443), Peters (1923:317), Bettinelli and Chébez (1986), nor I (Vuilleumier 1993:41, and unpublished field work in 1992) have found *A. anthoides* on plateaus of the central part of northern Patagonia (central Neuquén, central Río Negro, and central Chubut provinces), even though shrubsteppes are found there, at least locally. In these three provinces, the species seems confined to the base of the Andean foothills in the west. For example, *A. anthoides* occurs at such locations in Neuquén, near Bariloche (Río Negro), where several authors have reported it, and where I observed it myself, at Valle del Lago Blanco (Chubut, four specimens at AMNH, Nos. 523768–523771; note that one of these birds was collected in March, one in August, and two in September: are they migrants?), at El Hoyo (Chubut, specimens at LSUMZ, Nos. 53318 collected in May, and 69987 collected in June: are they migrants?; Remsen, pers. comm.), and at Lago General Vintter (Chubut, Rasmussen et al. 1992:8). Farther south, as mentioned earlier, it has been collected in Santa Cruz Province (specimens at the AMNH collected by Rollo H. Beck, and at the YPM collected by Philip S. Humphrey; Fred Sibley, pers. comm.). There is no clear evidence that *A. anthoides* did, or does, occur in the vast area between the western part of north-central Patagonia (south of Chubut province) southward through western Santa Cruz to southern Santa Cruz and adjacent Magallanes.

The status of *A. anthoides* in the Falkland (Malvinas) Islands was dealt with briefly by Hellmayr (1925:149, footnote a): "No reliable record exists for its occurrence on the Falkland Islands." Much later, this status was summarized by Woods (1988:212) at greater length: "Uncertain: possibly breeding species or vagrant apparently recorded once. Darwin collected an adult specimen in 1833 or 1834 (Sclater and Sharpe 1885–98). It was reported as originating in the Falklands. Darwin does not comment on this species (Gould and Darwin 1841) and Barlow (1963) records the series of skins, including this one, as originating in Chile." More recently, Ridgely and Tudor (1994:113), basing themselves, I suspect, on Woods' (1988) remarks, wrote that "One very early specimen from Falkland Is., taken by Darwin in the 1830s, is now usually considered as having been mislabeled." There is thus no substantiated record of *A. anthoides* for the Falklands. As *A. anthoides* lives in shrubsteppes, its absence from the Falklands is not surprising, because the shrubsteppe habitat it favors in Fuego-Patagonia is itself largely lacking on these islands. However, *Chiliotrichum diffusum*, one of the important structural components of the shrubsteppe habitat of *A. anthoides* in southern Fuego-Patagonia, used to be widespread in the Falklands until the introduction of stock, which contributed to its demise (Strange 1992: 26–27). During a visit to the Falklands in February 1996, I saw a small relictual patch of shrubsteppe with *Chiliotrichum* and grasses near Port Stanley, East Falkland, that looked much like southern Fuego-Patagonian shrubsteppes where I had observed *A. anthoides* in 1985–1988. It is therefore not impossible that Austral Canasteros occurred in the Falklands in earlier times, when "many ... valleys supported Fachine Bush [*Chiliotrichum diffusum*]" (Strange 1992:26). Some hillsides in the Falklands, including some "stone runs" (Strange 1992:26–27, fig. I, p. 27), are covered today with a shrublike vegetation ("dwarf shrub heath community," Davies and McAdam 1989: 5, Plate 3) dominated by Diddle-dee (*Empetrum rubrum*, Empetraceae) and Tall Ferns (*Blechnum magellanicum*, Blechnaceae). In some areas where the *Empetrum-Blechnum* shrub is intermixed with a grassland ("grass heath community," Davies and MacAdam 1989:5, plate 2) dominated by White Grass (*Cortaderia pilosa*, Gramineae), the vegetation looks somewhat like a shrubsteppe and this is where I would look for the Austral Canastero in the Falklands. For an illustration of this shrub-grassland mosaic see middle-ground of photograph in Strange (1987:122). For a more detailed review of Falkland Islands biogeography see Vuilleumier (1966).

Sixty-five years ago, Hellmayr (1932:211) had written that "The breeding range of [*A. anthoides*] cannot be properly defined at present." Sadly, we do not have a much clearer idea of this range today. A thorough and critical review of the literature, combined with a thorough and critical examination of available specimens in many museums will be necessary before the uncertainties concerning the geographical distribution of *A. anthoides*, past and present, breeding and non-breeding, are fully clarified. This research was beyond the scope of this study. Hence I have refrained in this paper from drawing a distributional map of *A. anthoides* that would show both where the species has been seen or collected, and localities where it has been searched for by competent observers, but not found. At present such a map would be based on many unverified records, and would merely repeat what is found in already published maps, without a critical assessment.

Pending such a review, my current understanding of the status of *A. anthoides* in Fuego-Patagonia, based on my own field work and on the literature (as summarized for instance by Fjeldså and Krabbe 1990; Collar et al. 1992; Ridgely and Tudor 1994) is that the species is present regularly in the southern part of this region, where it occurs at a number of both mainland (Santa Cruz, Magallanes) and island (Tierra del Fuego) sites. Ridgely and Tudor (1994:112) independently reached a similar conclusion: "Small numbers persist in a few areas northeast of Punta Arenas in Chile and at Punta Dungeness in Santa Cruz, Argentina [sic]." [Note that Punta Dungeness is actually in Magallanes, Chile, and that it is the adjacent Cabo Vírgenes which is in Santa Cruz, Argentina. Note also that it is not clear to me why Ridgely and Tudor (1994:112) wrote that the species "persists" in these areas.] Farther north in Patagonia, after an apparent distributional hiatus, *A. anthoides* occurs along the Andean foothills and neighboring pre-Andean shrubsteppes of western Chubut (Argentina) and adjacent Aysén (Chile; e.g. Hellmayr 1932:210 [but the locality was erroneously said to be in Llanquihue, see Paynter 1988:34, 240]; Olrog 1948:510; Philippi 1938:11, however, did not find this species in Aysén), and in similar habitats of western Río Negro (pers. obs.) and western Neuquén (Argentina) and neighboring Chilean provinces (e.g., Ñuble, Estades et al. 1994; Malleco, pers. obs.). In the Chilean province of Ñuble at least (Estades et al. 1994), *A. anthoides* does not appear to be rare in its shrubsteppe habitat.

Anecdotal reports in the literature (including the present paper) suggest that in southern Fuego-Patagonia, especially perhaps in Tierra del Fuego, *A. anthoides* may be more common in some years than in others. That similar population fluctuations may also occur elsewhere in the species' range is worth noting. Thus Pässler (1922:468) found the species nesting at Coronel, near Concepción, Chile, whereas Johnson (1967:195–196) wrote that "Many years later we searched this area . . . but found no trace of either bird or nests."

Given Crawshay's (1907) observations about the rarity of *A. anthoides* in Tierra del Fuego in the early 1900s or earlier and mine (this report) or Howell and Webb's (1995) that the species is not rare in the 1980s–1990s on that island, as well as the reports that in northern Patagonia *A. anthoides* is not rare near Bariloche (Collar et al. 1992) and reaches a density of about 2 individuals per hectare in Ñuble (Estades et al. 1994), I conclude that *A. anthoides* is not vanishing as suggested by Fjeldså and Krabbe (1990:378), but instead has increased numerically in the last 90–100 years. Nevertheless, the status of *A. anthoides* needs to be monitored closely in the future.

The shrubsteppes that *A. anthoides* favor are not uniformly distributed in Patagonia, nor, for that matter, are the grasslands that they do not occupy. Overgrazing by sheep and horses has locally resulted in severe land erosion and desertification of both shrubsteppe and grassy steppe habitats. An idea of the vast geographic scale of this degradation is given by the figures published by Cruz and Lara (1987a) for Chilean Fuego-Patagonia. Out of a total of 3,525,525 ha surveyed by Cruz and Lara (1987a), 2,903,990 ha (about 82%) were affected by erosion, and of the latter surface, no fewer than 2,836,145 ha (about 80% of total land surveyed, and about 98% of eroded area) were found to be eroded because of human activities (in other words overgrazing by sheep). Furthermore, of the area eroded through human activities, 693,366 ha (about 24%) were considered to be "severely" or "very severely" eroded, another 1,240,395 ha (about 44%) were considered to be "moderately" eroded, and 887,284 ha (about 31%) "lightly" eroded. Only 501,330 ha, or 14% of the total surface, were considered by Cruz and Lara (1987a) to have "no apparent erosion." An example of "severe" erosion was illustrated for a *Festuca gracillima* grassland (Cruz and Lara 1987a: 13, photo No. 2). Unfortunately, these authors did not publish a photograph of a degraded *Chiliotrichum* shrubsteppe. Cruz and Lara (1987a:15) concluded that "In general one may say that erosion in the XIIth Region [Chilean Fuego-Patagonia] is serious because of the generalized state of this phenomenon." Indeed!

Erosion due to overgrazing by sheep has most likely been as severe for grasslands as for shrubsteppes, given the geographical scale of overgrazing. Thus a crucial question in regard to the status of *A. anthoides* is: did populations of *A. anthoides* decrease as a result of overgrazing by sheep and subsequent degradation of the original vegetation in the last 100 years or so? In spite of statements in recent reviews of this species' status (Fjeldså and Krabbe 1990, Collar et al. 1992, Ridgely and Tudor 1994), to the effect that populations of this species have declined in the last several decades, I have found no evidence to suggest that a decline has occurred. On the contrary, as I pointed out earlier, the reverse seems to be true in Tierra del Fuego, where the species has apparently increased. I now turn to the question of whether today's shrubsteppe habitat favored by *A. anthoides* represents, as Fjeldså (1988) concluded, a marginal habitat, to which the species retreated after the disappearance of its former grassland habitat.

Habitat preferences.—Fjeldså and Krabbe (1990:377) stated that *A. anthoides* lives in "Open meadows, damp grassland, well-watered hillsides, dense thorny thickets in open terrain." This habitat description is very similar to or even paraphrases those published earlier by Hellmayr (1932:211) and Johnson (1967:195–196). These two latter authors, in turn, had cited Landbeck (1877:238). It is therefore worthwhile quoting Landbeck's statement in full, so that readers can appreciate the extent to which various authors quote the same basic remarks, which over the years acquire the value of "facts." Landbeck (1877:238) had written the following about *Synallaxis rufogularis* Gould, 1839 (= *Asthenes anthoides*, see Hellmayr 1925:149): "Er lebt stets auf Pampa's, auf Wiesen, auch auf Bergen, aber nicht hoch hinauf und wo möglich an feuchten Stellen." ("It lives mostly in the pampas, in meadows, and in mountains, although not high up and wherever possible in damp [or wet] places.") The term "pampas" is ambiguous. There is no way of telling what kind of grassland Landbeck implied by "pampas." Landbeck did mention "wet places," however. There seems to be a discrepancy between Landbeck's (1877:238) or Fjeldså and Krabbe's (1990:332) descriptions of wet, grassy habitats for *A. anthoides* and those of other authors. Thus, Johnson (1967:196) found this species "among the *bush-covered* [italics mine] slopes of a small hill" at Estancia Ekewern on Tierra del Fuego, and I observed it (this report) in a variety of shrubsteppes, but not in "open meadows," "open pastures," "wet meadows," or "damp grassland."

As I have argued earlier in this report, my extensive experience with *A. anthoides* in Fuego-Patagonia strongly suggests that this canastero *lives in shrubsteppes but does not occur in grasslands*. This preference for shrubby habitats is also described by Humphrey et al. (1970:261), Ralph (1985), de la Peña (1988:33), Collar et al (1992:610), Rasmussen et al. (1992), Ridgely and Tudor (1994:112), Estades et al. (1994), and Howell and Webb (1995:63). It is thus not at all clear to me why Fjeldså (1988:93) suggested that this species lived in tall-grass vegetation in the past and why he considered its present brush habitat to be "a marginal habitat" where it survives. Unfortunately, this idea was adopted uncritically by Collar et al. (1992:611), even though, paradoxically, they described the species' actual habitat correctly. I know of no evidence to suggest that *A. anthoides* now lives in, or formerly inhabited, tall-grass vegetation or any kind of pure grassland, grassy steppe, or grassy meadow. Let me give an example based on recent evidence. The large areas of dry grassland of Chilean Patagonia (over 570,000 ha, Cruz and Lara 1987b) do not have Austral Canasteros in them, at least according to my extensive surveys of the avifauna in this vegetation. In eastern Magallanes, only a few kilometers from extensive grasslands covering several thousand ha, the Austral Canastero occurs in small patches (covering perhaps a total of 200–300 ha) of slightly more mesic shrubsteppes dominated by *Lepidophyllum cupressiforme*. It seems odd that a species alleged to inhabit, or to have inhabited, grasslands should be absent from extensive stands of this vegetation, but present nearby in small areas of uniform shrubsteppe.

It is worth making a comparison here between the case of *A. anthoides* and that of another species of Patagonian bird, the Black-throated Finch *Melanodera melanodera* (Emberizinae), that does occur in grassland and that also has been considered threatened because of grassland degradation due to overgrazing by sheep. Thus Ridgely and Tudor (1989:37) wrote: "Until recently [*Melanodera melanodera*] appears to have been quite numerous on the Fuegian grasslands to which it is restricted, as indeed it still is on the Falkland Islands; however, on the mainland [of southern Fuego-Patagonia] Black-throated Finches have declined precipitously of late. The reasons for this decline remain uncertain, but we suspect that it may be correlated with severe overgrazing by sheep." Interestingly, as in the case of *A. anthoides*, Fjeldså (1988:93) had written that *Melanodera melanodera* "also seems to have become very local." And Fjeldså and Krabbe (1990:668) gave the Black-throated Finch's status as follows: "N Isla Grande [of Tierra del Fuego] (where now extinct?) and Sta Cruz, Arg., vanishing due to overgrazing by sheep, although recently rec. as fairly common at Cabo Vírgenes . . . ." I have had personal experience with *Melanodera melanodera* in Fuego-Patagonia in 1985–1993, and in the Falklands in 1996. On this basis I conclude that (a) *Melanodera melanodera* is not extinct on Tierra del Fuego (it occurs, for example, in the grassy moorlands of the Sierra Boquerón; see Vuilleumier 1991:9, his fig. 5, bottom), (b) it is common in dry grasslands of the *Festuca gracillima-Stipa* spp. type on the Patagonian mainland in Magallanes, where I have found it at every visit (Fig. 2; see also fig. 6, top in Vuilleumier 1991:10, which illustrates the most open of this kind of grassland, and fig. 10 in Pisano 1977:160, which shows a denser aspect of the same grassland type), and (c) there is no evidence for or against a decline as described by Fjeldså and Krabbe (1990) or Ridgely and Tudor (1989). Thus in Fuego-Patagonia *Melanodera melanodera* is found precisely in the grassland habitats from which *A. anthoides* is missing (Fig. 2). In the Falkland Islands, again, the situation of *M. melanodera* is unlike that of *A. anthoides*. Whereas there is no substantiated record of *A. anthoides* for these islands (as I argued earlier), the Black-throated Finch is common there in grassland ("grass heath community" dominated by *Cortaderia pilosa*; Vuilleumier 1996) that looks remarkably similar to the *Festuca-Stipa* grassland of Magallanes (compare plate 2 in Davies and McAdam 1989:5 for the Falklands grassland, with figs. 8–9 in Pisano 1977:160 for the Chilean one). In the case of *M. melanodera* I agree with the view that it is a grassland-inhabiting species, but I disagree with the idea that this species is vanishing or declining on the mainland. I am not saying that *M. melanodera* is *not* declining or vanishing there, only that I find no evidence about its status one way of the other. However, if indeed further research shows that *M. melanodera* has a relictual range and that its status is threatened, then I would argue that this situation is due to the severe degradation and local destruction of its grassland habitat by sheep overgrazing. The main point here is that I see no such parallel in the case of the distribution pattern of *A. anthoides*.

I conclude from all the foregoing arguments that *A. anthoides* is not now, and probably never has been, a grassland-inhabiting species, but instead that its habitat preference is, and always

FIG. 2. Dry grassland of the *Festuca gracillima-Stipa* spp. type (Pisano 1977: 159–162) on clayey and organic-rich soil, habitat of *Melanodera melanodera*, on the Patagonian mainland in Magallanes, Chile about 3 km north of O'Higgins (Punta Delgada on maps), 23 November 1993; photograph Allison V. Andors. Note that *Asthenes anthoides*, a putative prairie inhabitant (see text for details), was never recorded in this habitat during repeated visits in 1985–1993.

has been, for shrubsteppes, especially mesic ones with an annual rainfall of about 350–500 mm (Pisano 1977; Thomasson 1979). Such shrubsteppes are found on either side of the Strait of Magellan under a relatively moist, maritime climate, and northward on the Patagonian mainland along the eastern Andean foothills, immediately to the east of the forest-steppe ecotone, in the lee of the mountains where rainfall is low yet not desert-like. Further east still along the decreasing west-east precipitation gradient, the climate becomes too dry and the vegetation too low and too open to be suitable for the Austral Canastero. Hence, I believe that thorough exploration of a narrow zone of mesic shrubsteppe areas along the Andean foothills from Chubut southward to Santa Cruz, where *A. anthoides* appears to be absent at present, may yet reveal its presence in the future. The apparent distributional gap between the northern and the southern populations thus corresponds largely to an area that, even though severely degraded by sheep grazing and subsequent erosion, was unlikely for climatic reasons to have been covered with suitable shrubsteppe habitat for *A. anthoides* in the pre-sheep grazing past.

If *A. anthoides* is primarily a shrubsteppe-inhabiting species, it may be the only taxon to do so among its several close relatives. Thus Ridgely and Tudor (1994:112) stated that "All forms [of the *punensis-sclateri* group] are found in puna grasslands." I well remember my surprise, when I first encountered *A. anthoides* near Bariloche in 1965, to observe these birds in rather dense shrubbery and not in grassland, where I had expected to find them on the basis of previous experience with related taxa like *A. wyatti* in the Andes of Ecuador. This apparent ecological difference makes it worthwhile to analyze the systematic position of *A. anthoides*, in order to place this species in its taxonomic context, and so to provide systematic background information that would be useful to define more precisely this species' conservation status.

Systematic position.—*Asthenes anthoides* belongs to a group of six streaked species-level taxa in the genus *Asthenes* Reichenbach, 1853. The term "species-group" is used here loosely (in the sense of Mayr 1942:290–291, 1963:672, and of Cain 1954:270), since no phylogenetic study has been carried out on these birds yet. In these six forms, the center of feathers, for example on the dorsum, is dark, almost black, on either side of the rachis, whereas the sides (the external margins of the webs) are pale yellowish, brownish, or reddish. Even though these six taxa may not all belong to a monophyletic assemblage, they are uniform morphologically within the streaked species of *Asthenes* in that their pattern of streaking is quite different from that in the taxa that I earlier had called the *A. flammulata* superspecies (Vuilleumier 1968; superspecies used *sensu* Amadon 1966). The most markedly streaked species in this group of six taxa is *A. hudsoni* and the least-streaked one is *A. humilis*. Whether these two species are closely related to the remaining four remains to be established, and this point, being outside the scope of this paper, will not be discussed further here. The four remaining taxa are similarly patterned and constitute a complex of forms that replace each other from north to south along the Andes, in the Sierras Pampeanas, and in Patagonia. Although they have not been studied from a phylogenetic point of view, it is likely that such a study will reveal these four taxa to belong to a monophyletic assemblage. Hopefully the notes below will spur someone to undertake this work.

Fuego-Patagonian *A. anthoides* (King, 1831) is clearly closely related to two high Andean taxa, *A. wyatti* (Sclater and Salvin, 1871) and *A. punensis* (Berlepsch and Stolzmann, 1901), and to *A. sclateri* (Cabanis, 1878) from the Sierras Pampeanas of Argentina (Nores 1995:65). The references to the original descriptions of these taxa can be found in Hellmayr (1925) and Peters (1951). *A. anthoides* and *A. sclateri* are entirely allopatric within this complex. *A. sclateri* is entirely allopatric from Andean *punensis*. *A. wyatti* has several disjunct, morphologically differentiated, and subspecifically named populations in the Andes of Venezuela, Colombia, Ecuador, and central Peru, and meets *A. punensis* in the Andes of southern Peru and western Bolivia (see below). Whether these four taxa should be treated as one, two, three, or four species would seem to depend largely on one's species concept in these parapatric and allopatric forms.

A. sclateri, originally described in the genus *Synallaxis*, was synonymized in *A. hudsoni* by Hellmayr (1925:150), but later was recognized as a distinct species (e.g., by Peters 1951:108). Later, *A. sclateri* was discussed by Hoy [1965, 1973; see also appendum to Hoy's (1965) paper by Stresemann (1965), and Stresemann (1948)]. Hoy (1973) and Stresemann (1965) recognized that *A. sclateri* was not related to the superficially similar *A. hudsoni*, but Stresemann (1965) stated that because he lacked comparative material he could not determine the nearest relatives of *sclateri* within what he called the heterogeneous genus *Asthenes*. Although Olrog (1979: 181–182) considered *hudsoni* and *sclateri* to be closely related to one another, color and pattern ally *sclateri* unequivocally with the *wyatti-punensis-anthoides* complex, a treatment adopted by Meyer de Schauensee (1966), Vaurie (1980), and Fjeldså and Krabbe (1990), and with which I concur, on the basis of my museum and field studies of these taxa, including *A. hudsoni*. Navas and Bó (1982) even merged Andean *punensis* and Córdoba Mountains *sclateri* (the species name would become *sclateri*), a treatment recently adopted by Ridgely and Tudor (1994:111–112). Independently of Navas and Bó (1982), Nores and Yzurieta (1983), in their description of a new subspecies, *A. punensis brunnescens*, from the isolated Sierras de San Luis, part of the Sierras Pampeanas, stated that *punensis* and *sclateri* were probably conspecific, because morphological differences between these two allopatric forms are small (Nores, pers. comm.). *Asthenes sclateri* is found on each of the several isolated ranges called Sierras Pampeanas by Nores (1995), a much broader geographical distribution than that usually given in the literature (e.g., "mts of Córdoba," Fjeldså and Krabbe 1990:377). The taxonomic question that remains is whether *sclateri* should be merged with Andean *punensis* (as in Ridgely and Tudor 1994:111–112) or be left as a separate species but included in a superspecies with *wyatti*, *anthoides*, and *punensis* (as in Fjeldså and Krabbe 1990:377).

Bond (1945:32) thought that there was "no reason why *punensis* should be retained as a distinct species [from *wyatti*], since characters by which it differs from *wyatti* are merely those of degree." Olrog (1962:117–118) similarly concluded that geographical variability in the *wyatti-punensis-anthoides* complex was such that all three forms should be considered conspecific under the oldest available name *A. anthoides*. Other authors, however, for example Vaurie (1980: 181–182), treated the three taxa as three separate species. Still other authors have held yet a different opinion. Thus, Meyer de Schauensee (1966:253), 1982:215–216) combined *punensis* and *anthoides* in one species (*anthoides*), but kept *wyatti* separate. The discussion above shows that almost any combination of species-level taxa has been suggested in the literature for four taxa (*wyatti*, *punensis*, *anthoides*, and *sclateri*), which are to a large extent similar in color, pattern, and vocalizations.

The vocalizations of *A. sclateri* were not described by Fjeldså and Krabbe (1990:377: "No data"). *Asthenes sclateri* studied in the Pampa de Achala, Sierra de Córdoba, on 28–29 October 1996, had two kinds of vocalizations: soft *tzup*, *tsup*, or *chup* calls, and simple trilled songs. Single trills, slightly accelerating and increasing in pitch toward the end, each lasting between 1 and 2 seconds, were emitted from a perch on a low shrub, and repeated at a rate of one every 8–9 seconds. Trills of *A. sclateri* sounded much like those of *A. wyatti mucuchiesi* (Vuilleumier and Ewert 1978), *A. wyatti aequatorialis, A. punensis punensis,* and *A. anthoides* (pers. obs.; see also Fjeldså and Krabbe 1990; Ridgely and Tudor 1994).

The montane forms (*wyatti-punensis-sclateri*) are often stated to live in grassland (e.g., Ridgely and Tudor 1994:112, writing about enlarged *sclateri*: "all forms are found in puna grasslands"). The situation is not that simple. In the high Andes, *Asthenes wyatti* and *A. punensis* are indeed found in high Andean tussock-grassland growing in plains, on rolling hills, or along mountain slopes (páramo or puna grasslands; Dorst 1963, pers. obs.), but the grass tussocks are not always dominant and are often mixed with other plants, including especially shrubs and terrestrial bromeliads. Furthermore, rocks are often an important component of this "grassland" habitat. In the Sierra de Córdoba, *Asthenes sclateri* occurs on slopes with large rock outcroppings where the vegetation consists of scattered grass tussocks, low shrubs, and occasional *Polylepis australis* trees near small cliffs, the soil being covered with dense matted turf (pers. obs.; see also Hoy 1965:206). *Asthenes wyatti* occupies shrubby habitats rather than grasslands at the northern edge of its range in the Santa Marta Mountains, northern Colombia (Todd and Carriker 1922:290) and in the Mérida Andes, Venezuela (Vuilleumier and Ewert 1978:71, 83). Indeed in Venezuela *A. wyatti* lives in a shrubsteppe-like vegetation consisting of rosettes of *Espeletia* spp. (Compositae), shrubs of *Hypericum* sp. (Compositae), and grass tussocks, that resembles the *Chiliotrichum-Festuca* shrubsteppe occupied by *A. anthoides* at the southern extremity of the distribution of the complex. In parts of its range, *A. punensis* occurs in pure or nearly pure tussock-grassland (puna-grassland; Dorst 1963, pers. obs.), but elsewhere (e.g., Bolivia) I have found it in rocky areas with more varied and denser vegetation including tussock-grass, shrubs, and the gigantic, tree-like Bromeliaceae *Puya Raimondii*. Thus, even though *A. anthoides* appears to live in a different habitat from that of its relatives in this complex, geographical variation in ecological preferences clearly shows that in the group of taxa including *wyatti, punensis, sclateri,* and *anthoides* a grassland habitat cannot be considered "typical" of the montane taxa. I find it intriguing that the northernmost populations of the complex, *wyatti* in the Venezuelan Andes, should live in a shrubsteppe-like habitat that does not look that different, although botanically distinct, from that occupied by the southernmost populations of *anthoides* in Fuego-Patagonia.

What about the species versus subspecies status of these forms, including especially *anthoides*? Several years ago I challenged the concept of geographical overlap proposed by Vaurie (1980: 173; and later accepted by Ridgely and Tudor 1994:111) between *wyatti* and *punensis* (Vuilleumier *in* Vaurie 1980:340, footnote no. 103), on the basis of an examination of series of skins at the Museum of Comparative Zoology and at the AMNH, and reached a different interpretation based on geographical variation in tail pattern.

Birds from the disjunct populations found in Venezuela, Colombia, and northern and southern Ecuador generally have the three outermost rectrices largely or entirely rufous, the tip of the next (fourth) rectrix rufous, and the innermost two rectrices largely or entirely dark brown. These birds are traditionally assigned to *A. wyatti*. Birds from central Bolivia and northwestern Argentina do not have any rectrix wholly or almost wholly rufous. Instead, the tips of the four outermost rectrices are rufous and the innermost two rectrices dark brown. These birds are usually placed in *A. punensis*. Birds from intermediate areas in central and southern Peru and in northern and western Bolivia have intermediate tail patterns. The outermost rectrix is entirely or almost entirely rufous, the second rectrix has much rufous along the outer vane, the third and fourth rectrices have a rufous tip, and the fifth and sixth are dark brown. In these populations with intermediate tail patterns, in addition, individual variation is extensive. My conclusion from this study (in Vaurie 1980:340, footnote no. 103) was that there is a fairly broad zone of secondary intergradation or hybridization (Mayr 1963), at least with respect to tail pattern, in southern Peru and western Bolivia. If there is hybridization between *wyatti* and *punensis*, then a case could be made for these two taxa to be considered a single biological species, *A. wyatti* (Sclater and Salvin, 1871). Only a detailed character analysis based on series of well-labelled specimens and further field study in the overlap or hybridization zone will clarify this situation.

A. sclateri and *A. anthoides* are entirely allopatric from such an enlarged *A. wyatti* and are morphologically distinct from each other. *A. sclateri* and *A. anthoides* can thus be considered separate species. *A. anthoides* differs from the other taxa in this complex by having, among other differences, less rufous in the carpal area, a narrower and less richly rufescent wing-bar, and yellowish rather than rufous or cinnamon tips to the outer rectrices. According to Mayr's (1963) biological species concept and Amadon's (1966) superspecies concept, the evidence presented above would suggest that these four taxa should be included in a single superspecies and grouped into three species or allospecies: *A. wyatti* (including *punensis*) from the high Andes, *A. sclateri* from the Sierras Pampeanas, and *A. anthoides* from Patagonia. Although Fjeldså and

Krabbe (1990) kept *wyatti* and *punensis* separate and included all four species in a superspecies, Fjeldså identified two LSUMZ specimens from Departamento Puno, southern Peru, as intergrades between *wyatti graminicola* and *punensis*, and hence would seem to concur with my treatment (Remsen, in litt.). According to Cracraft's (1983), McKitrick and Zink's (1988), and Zink and McKitrick's (1995) phylogenetic species concept, no fewer than twelve species could potentially be recognized in this complex, because they are all more or less diagnosable as morphological units, namely: *wyatti* (Sclater and Salvin 1870); *sanctaemartae* (Todd 1950); *mucuchiesi* (Phelps and Gilliard 1941); *aequatorialis* (Chapman 1921); *azuay* (Chapman 1923); *graminicola* (Sclater 1874) [these six taxa are usually treated as subspecies of *A. wyatti*]; *punensis* (Berlepsch and Stolzmann 1901); *cuchacanchae* (Chapman 1921); *lilloi* (Oustalet 1904) [these three taxa are usually considered as subspecies of *A. punensis*]; *sclateri* (Cabanis 1878); *brunnescens* (Nores and Yzurieta 1983 [described as a subspecies of *A. punensis*]); and *anthoides* (King 1831). However, a thorough study of geographical variation based on a detailed character analysis in the three taxa *punensis*, *cuchacanchae*, and *lilloi* will probably reveal intermediacy, and will establish whether that variation is smoothly clinal, or stepped. It seems to me that the use of a phylogenetic species concept in the case of these canasteros and the recognition of about twelve taxa would obscure, rather than illuminate, the complex evolutionary relationships among all these populations, especially those that are geographically isolated, and those that are not equally differentiated from each other. For the time being, I recommend that *A. wyatti* (including *A. punensis*), *A. sclateri*, and *A. anthoides* be recognized as three separate species or allospecies in a superspecies. Considering *A. anthoides* as a separate species is thus justifiable on morphological grounds.

Conservation.—From the point of view of conservation biology, treating *A. anthoides* as a separate species will simplify the task of conservationists in Chile and Argentina, who must now undertake further studies of this still poorly known Patagonian endemic species of canastero or espartillero. Indeed, *A. anthoides* could well serve as an indicator species for Fuego-Patagonian shrubsteppes. Even though the status of *A. anthoides* appears to be relatively stable at present, at least in southern Fuego-Patagonia where shrubsteppes are extensive (about 570,000 ha according to Cruz and Lara 1987b), there is no room for complacency. Sheep grazing pressure in these shrubsteppes is intense. For example, in the shrubsteppes east of the Useless Bay area of Tierra del Fuego, where I found a density of about one territorial *A. anthoides* per ha in 1987-1988, the average number of sheep according to Lara and Cruz (1987b) is of one per hectare or one per two hectares. My own visual estimates of sheep density in these shrubsteppes, however, are that there are probably two sheep per hectare, and locally perhaps even more. If this is the case, then some ranchers do not follow the generally recommended average density of one sheep per hectare in this type of landscape (the "1:1 rule"). In my opinion, one sheep per hectare is already too many sheep in Chilean Fuego-Patagonian shrubsteppes. These shrubsteppes are now markedly and visibly degraded and cannot sustain for long the further erosion that this overgrazing will inevitably cause. I recommend that future monitoring of the density of *A. anthoides* in shrubsteppes take into account the sheep density there as well. Sheep ranching will only survive if shrubsteppes are healthy and are allowed to regenerate. Estimating the density of an indicator species like the endemic *A. anthoides* might give an easily measurable evaluation of the relative health of shrubsteppes, providing a quicker measure than the rather complicated techniques explained in Lara and Cruz (1987b).

ACKNOWLEDGMENTS

The financial help of the National Geographic Society and the continuing support of the Sanford family and of the Leonard C. Sanford Fund is most gratefully acknowledged. I thank Allison Andors, Angelo Capparella, Gladys Garay, Linda Gregory, Ivan Lazo, Luis Palma, Tobias Salathé, Claudio Venegas, and Jeanie Vesall for help in the field, and Azize Atalah, Leonardo Guzmán, Jorge Jordan, Herman Núñez, Edmundo Pisano, Eduardo Scott, Claudio Venegas, Carlos Weber, and José Yáñez for logistical and other assistance in Chile. I thank the authorities of the Servicio Agrícola y Ganadero (S.A.G.), División de Protección de los Recursos Naturales Renovables, Ministerio de Agricultura, in Santiago and Punta Arenas for having granted me the necessary permits in their respective administrative regions. I am grateful to Fred Sibley for information about specimens in the collection of the Peabody Museum of Natural History at Yale University and to J. Van Remsen who allowed me to cite information from specimens at LSUMZ. I wish to thank Jorge Jordan and Claudio Venegas for the many courtesies they extended to me during my expeditions to southern Chile. Jorge Jordan in particular helped make the sojourns at Estancia Los Tehuelches very productive. I am very grateful to Maria and Ian Strange for their help before and during my visit to the Falkland Islands. Luis M. Chiappe kindly corrected the Spanish abstract. I thank J. Van Remsen, Manuel Nores, and an anonymous reviewer for their many helpful comments on earlier versions of the manuscript, and Allison V. Andors for invaluable help in its preparation and for allowing me to use Fig. 2.

LITERATURE CITED

AMADON, D. 1966. The superspecies concept. Syst. Zool. 15:245–249.

AUER, V. 1956. The Pleistocene of Fuego-Patagonia. Part I. The ice and interglacial ages. Ann. Acad. Sci. Fennicae Ser. A III 45:1–226.

AUER, V. 1960. The Quaternary history of Fuego-Patagonia. Proc. Roy. Soc. London Ser. B 152:507–516.

BARLOW, N. 1963. Darwin's ornithological notes. Bull. Brit. Mus. (Nat. Hist.) Histor. Ser. 2:203–278.

BETTINELLI, M. D., AND J. C. CHEBEZ. 1986. Notas sobre aves de la Meseta de Somuncurá, Río Negro, Argentina. Hornero 12:230–234.

BERNATH, E. L. 1965. Observations in southern Chile in the southern hemisphere autumn. Auk 82:95–101.

BOND, J. 1945. Notes on Peruvian Furnariidae. Proc. Acad. Nat. Sci. Philadelphia 97:17–39.

CAIN, A. J. 1954. Subdivisions of the genus *Ptilinopus* (Aves, Columbae). Bull. Brit. Mus. (Nat. Hist.) 2: 265–284.

CHEBEZ, J. C., AND C. C. BERTONATTI. 1994. La avifauna de la Isla de los Estados, Islas de Año Nuevo y mar circundante. L.O.L.A. Buenos Aires, Monogr. Especial No. 1:1–64.

CHESSER, R. T. 1994. Migration in South America: an overview of the austral system. Bird Conserv. Intern. 4:91–107.

CLARK, R. 1986. Aves de Tierra del Fuego y Cabo de Hornos. Guía de campo. L.O.L.A., Buenos Aires.

COLLAR, N. J., L. P. GONZAGA, N. KRABBE, A. MADROÑO NIETO, L. G. NARANJO, T. A. PARKER III, AND D. C. WEGE. 1992. Threatened Birds of the Americas. The ICBP/IUCN Red Data Book. Third edition, part 2. Smithsonian Inst. Press, Washington, D. C.

CRACRAFT, J. 1983. Species concepts and speciation analysis. Pages 159–187 in Current Ornithology vol. 1 (R.F. Johnston, Ed.). Plenum Press, New York.

CRAWSHAY, R. 1907. The birds of Tierra del Fuego. Bernard Quaritch, London.

CRUZ M., G., AND A. LARA A. 1987a. Evaluación de la erosión del área de uso agropecuario de la XII Región, Magallanes y de la Antártica Chilena. Inst. Invest. Agropec. (INIA), Est. Exper. Kampenaike, Santiago.

CRUZ M., G., AND A. LARA A. 1987b. Regiones naturales del área de uso agropecuario de la XII Región, Magallanes y de la Antártica Chilena. Inst. Invest. Agropec. (INIA), Est. Exper. Kampenaike, Santiago.

DAVIES, T. H., AND J. H. MCADAM. 1989. Wild Flowers of the Falkland Islands. Bluntisham Books, Bluntisham, Cambridgeshire, U. K.

DE LA PEÑA, M. R. 1988. Guía de las Aves Argentinas. Tomo V. Passeriformes: Dendrocolaptidae-Furnariidae-Formicariidae-Tyrannidae. L.O.L.A., Buenos Aires.

DORST, J. 1963. Note sur la nidification et le comportement acoustique du jeune *Asthenes wyatti punensis* (Furnariidae) au Pérou. L'Oiseau et R. f. O. 33:1–6.

ESTADES, C., J. P. GABELLA, AND J. ROTTMANN. 1994. Nota sobre el Canastero del Sur (*Asthenes anthoides*) en la Reserva Nacional Ñuble. Bol. Chileno Ornitol. 1:31–32.

FJELDSÅ, J. 1988. Status of birds of steppe habitats of the Andean zone and Patagonia. ICBP Techn. Publ. No. 7:81–95.

FJELDSÅ, J., AND N. KRABBE. 1990. Birds of the High Andes. A Manual to the Birds of the Temperate Zone of the Andes and Patagonia, South America. Zool. Mus. Univ. Copenhagen and Apollo Books, Svendborg, Denmark.

GOULD, J., AND C. R. DARWIN. 1841. The zoology of the voyage of HMS 'Beagle.' Part 3 Birds:8–145.

HELLMAYR, C. E. 1925. Catalogue of birds of the Americas. Part IV. Furnariidae-Dendrocolaptidae. Field Mus. Nat. Hist. Publ. 234. Zool. Ser. XIII:i–iv, 1–390.

HELLMAYR, C. E. 1932. The birds of Chile. Field Mus. Nat. Hist. Publ. 308. Zool. Ser. XIX:1–472.

HOWELL, S. N. G., AND S. WEBB. 1995. Noteworthy bird observations from Chile. Bull. Brit. Ornithol. Club. 115:57–66.

HOY, G. 1965. Wiederentdeckung von *Asthenes sclateri* (Cabanis) in der Sierra de Cordoba. J. Ornithol. 106:204–206.

HOY, G. 1973. Sobre la taxonomía de *Asthenes sclateri* Cabanis (Aves, Furnariidae). Physis Sec. C 32, 84: 219–221.

HUMPHREY, P. S., AND J. E. PEFAUR. 1979. Glaciation and species richness of birds on austral South American islands. Occ. Papers Mus. Nat. Hist. Univ. Kansas No. 80:1–9.

HUMPHREY, P. S., D. BRIDGE, P. W. REYNOLDS, AND R. T. PETERSON. 1970. Birds of Isla Grande (Tierra del Fuego). Published and distributed for Smithsonian Institution by Univ. Kansas Mus. Nat. Hist.

JOHANSEN, H. 1966. Die Vögel Feuerlands (Tierra del Fuego). Vidensk. Medd. fra Dansk naturh. Foren. 129: 215–260.

JOHNSON, A. W. 1967. The Birds of Chile and Adjacent Regions of Argentina, Bolivia and Peru. Vol. 2. Platt Establ. Gráficos S.A., Buenos Aires.

LARA A., A., AND G. CRUZ M. 1987a. Vegetación del área de uso agropecuario de la XII Región, Magallanes y de la Antártica Chilena. Inst. Invest. Agropec. (INIA), Est. Exper. Kampenaike, Santiago.

LARA A., A., AND G. CRUZ M. 1987b. Evaluación del potencial de pastoreo del área de uso agropecuario de la XII Región, Magallanes y de la Antártica Chilena. Inst. Invest. Agropec. (INIA), Est. Exper. Kampenaike, Santiago.

LANDBECK, C. L. 1877. Bemerkungen über die Singvögel Chile's. Der zoologische Garten 18:233–261.

MARKGRAF, V. 1985. Late Pleistocene faunal extinctions in southern Patagonia. Science 228:1110–1112.

MAYR, E. 1942. Systematics and the Origin of Species. Columbia Univ. Press, New York.

MAYR, E. 1963. Animal Species and Evolution. Belknap Press of Harvard Univ. Press, Cambridge, Massachusetts.

MCKITRICK, M. C., AND R. M. ZINK. 1988. Species concepts in ornithology. Condor 90:1–14.

MEYER DE SCHAUENSEE, R. 1966. The Species of Birds of South America and Their Distribution. Livingston Publ. Co., Narberth, Pennsylvania.

MEYER DE SCHAUENSEE, R. 1982. A Guide to the Birds of South America. 2nd revised edition. Intercollegiate Press (no city given).

MOORE, D. M. 1983. Flora of Tierra del Fuego. Anthony Nelson, Oswestry, United Kingdom, and Missouri Botanical Garden, Saint Louis, USA.

NAROSKY, T., AND A. G. DI GIACOMO. 1993. Las Aves de la Provincia de Buenos Aires: Distribución y Estatus. Asoc. Ornit. del Plata, Vázquez Mazzini Editores, and L. O. L. A., Buenos Aires.

NAROSKY, T., AND D. YZURIETA. 1989. Birds of Argentina and Uruguay, a Field Guide. Vázquez Mazzini, Buenos Aires.

NAVAS, J. R., AND N. A. BO. 1982. La posición taxonómica de "Thripophaga sclateri" y "T. punensis" (Aves, Furnariidae). Com. Museo Argentino Cienc. Nat. "Bernardino Rivadavia" 4:85–93.

NORES, M. 1995. Insular biogeography of birds on mountain-tops in north western Argentina. J. Biogeogr. 22:61–70.

NORES, M., AND D. YZURIETA. 1983. Especiación en las Sierras Pampeanas de Córdoba y San Luis (Argentina), con descripción de siete nuevas subespecies de aves. Hornero No. Extraordinario: 88–102.

OLROG, C. C. 1948. Observaciones sobre la avifauna de Tierra del Fuego y Chile. Acta Zool. Lilloana 5: 437–531.

OLROG, C. C. 1962. Notas ornitológicas sobre la colección del Instituto Miguel Lillo (Tucumán). VI. Acta Zool. Lilloana 18:111–120.

OLROG, C. C. 1979. Nueva lista de la avifauna argentina. Opera Lilloana 27:1–324.

PÄSSLER, R. 1922. In der Umgebung Coronel's (Chile) beobachtete Vögel. Beschreibung der Nester und Eier der Brutvögel. J. Ornithol.:430–482.

PAYNTER, R. A., JR. 1985. Ornithological Gazetteer of Argentina. Museum of Comparative Zoology, Cambridge, Massachusetts.

PAYNTER, R. A., JR. 1988. Ornithological Gazetteer of Chile. Museum of Comparative Zoology, Cambridge, Massachusetts.

PETERS, J. L. 1923. Notes on some summer birds of northern Patagonia. Bull. Mus. Comp. Zool. 65:277–337.

PETERS, J. L. 1951. Check-list of Birds of the World. Volume VII. Mus. Comp. Zoöl., Cambridge, Massachusetts.

PHILIPPI B., R. A. 1938. Contribución al conocimiento de la ornitología de la provincia de Aysén (Chile). Rev. Chilena Hist. Nat. 42:4–20.

PHILIPPI B., R. A. 1964. Catálogo de las aves chilenas con su distribución geográfica. Invest. Zool. Chilenas 11:1–179.

PHILIPPI B., R. A., A. W. JOHNSON, J. D. GOODALL, AND F. BEHN. 1954. Notas sobre aves de Magallanes y Tierra del Fuego. Bol. Mus. Nac. Hist. Nat. Santiago 26:1–63.

PISANO V., E. 1974. Estudio ecológico de la región continental sur del área andino-patagónica. II. Contribución a la fitogeografía de la zona del Parque Nacional "Torres del Paine." Anal. Inst. Patagonia Punta Arenas 5:59–104.

PISANO V., E. 1977. Fitogeografía de Fuego-Patagonia chilena. I.-Comunidades vegetales entre las latitudes 52° y 56°S. Anal. Inst. Patagonia Punta Arenas 8:121–250.

PISANO V., E. 1981. Bosquejo fitogeográfico de Fuego-Patagonia. Anal. Inst. Patagonia Punta Arenas 12: 159–171.

RALPH, C. J. 1985. Habitat association patterns of forest and steppe birds of northern Patagonia, Argentina. Condor 87:471–483.

RASMUSSEN, P. C., P. S. HUMPHREY, AND J. MUNIZ-SAAVEDRA. 1992. Imperial Shags and other birds of the Lago General Vintter area, Chubut Province, Argentina. Occas. Papers Mus. Nat. Hist. Univ. Kansas No. 146:1–16.

RIDGELY, R. S., AND G. TUDOR. 1989. The Birds of South America. Vol. I. The Oscine Passerines. Univ. Texas Press, Austin.

RIDGELY, R. S., AND G. TUDOR. 1994. The Birds of South America. Vol. II. The Suboscine Passerines. Univ. Texas Press, Austin.

SALVADORI, T. 1900. Contribuzione all'avifauna dell'America australe (Patagonia, Terra del Fuoco, Isola degli Stati, Isole Falkland). Ann. Mus. Civico Storia Nat. Genova Ser. 2, 20:610–634.

SCLATER, P. L., AND R. B. SHARPE. 1885–1898. Catalogue of Birds in the British Museum. Vols. X–XV and XXIII–XXVII. London.
STRESEMANN, E. 1948. The status of *Synallaxis sclateri* Cabanis. Auk 65:445–446.
STRESEMANN, E. 1965. Zusatz [addendum to Hoy 1965]. J. Ornithol. 106:206–207.
STRANGE, I. J. 1987. The Falkland Islands and Their Natural History. Hippocrene Books Inc., New York.
STRANGE, I. J. 1992. A Field Guide to the Wildlife of the Falkland Islands and South Georgia. HarperCollins, London.
THOMASSON, K. 1959. Nahuel Huapi: plankton of some lakes in an Argentine National Park, with notes on terrestrial vegetation. Acta Phytogeogr. Suecica 42:83.
TODD, W. E. C., AND M. A. CARRIKER, JR. 1922. The birds of the Santa Marta region of Colombia: a study in altitudinal distribution. Ann. Carnegie Mus. 14:i–vii + 1–611.
VAURIE, C. 1971. Classification of the Ovenbirds (Furnariidae). H. F. and G. Witherby Ltd., London.
VAURIE, C. 1980. Taxonomy and geographical distribution of the Furnariidae (Aves, Passeriformes). Bull. Amer. Mus. Nat. Hist. 166:1–357.
VENEGAS C., C. 1986. Aves de Patagonia y Tierra del Fuego Chileno-Argentina. Univ. Magallanes, Punta Arenas.
VENEGAS C., C., AND J. JORY H. 1979. Guía de Campo Para las Aves de Magallanes. Instituto de la Patagonia, Punta Arenas.
VUILLEUMIER, F. 1968. Population structure of the *Asthenes flammulata* superspecies (Aves: Furnariidae). Breviora, Mus. Comp. Zool. No. 297:1–21.
VUILLEUMIER, F. 1980. Notes. Pages 333-342 *in* Vaurie, C. 1980. Taxonomy and geographical distribution of the Furnariidae (Aves, Passeriformes). Bull. Amer. Mus. Nat. Hist. 166:1–357.
VUILLEUMIER, F. 1991. A quantitative survey of speciation phenomena in Patagonian birds. Orn. Neotrop. 2:5–28.
VUILLEUMIER, F. 1993. Field study of allopatry, sympatry, parapatry, and reproductive isolation in steppe birds of Patagonia. Orn. Neotrop. 4:1–41.
VUILLEUMIER, F. 1994a. Nesting, behavior, distribution, and speciation of Patagonian and Andean ground tyrants (*Myiotheretes*, *Xolmis*, *Neoxolmis*, *Agriornis*, and *Muscisaxicola*). Orn. Neotrop. 5:1–55.
VUILLEUMIER, F. 1994b. Status of the Ruddy-headed Goose *Chloephaga rubidiceps* (Aves, Anatidae): a species in serious danger of extinction in Fuego-Patagonia. Rev. Chilena Hist. Nat. 67:341–349.
VUILLEUMIER, F. 1996. Is the avifauna of the Falkland (Malvinas) Islands impoverished or at equilibrium? Southern Connection Newsletter 10:22–33.
VUILLEUMIER, F., AND D. EWERT. 1978. The distribution of birds in Venezuelan páramos. Bull. Amer. Mus. Nat. Hist. 162:47–90.
WETMORE, A. 1926. Report on a collection of birds made by J. R. Pemberton in Patagonia. Univ. Calif. Publ. Zool. 24:395–474.
WOODS, R. W. 1988. Guide to Birds of the Falkland Islands. Anthony Nelson, Oswestry, U. K.
ZINK, R. M., AND M. C. MCKITRICK. 1995. The debate over species concepts and its implications for ornithology. Auk 112:701–719.

BEHAVIOR, VOCALIZATIONS, AND RELATIONSHIPS OF SOME *MYRMOTHERULA* ANTWRENS (THAMNOPHILIDAE) IN EASTERN BRAZIL, WITH COMMENTS ON THE "PLAIN-WINGED" GROUP

BRET M. WHITNEY[1,2] AND JOSÉ FERNANDO PACHECO[1]

[1]*Laboratorio de Ornitologia, Departamento de Zoologia, Instituto de Biologia-CCS, Cidade Universitária, 21944-970 - Rio de Janeiro - RJ Brasil*

[2]*Museum of Natural Science, Foster Hall 119, Louisiana State University, Baton Rouge, Louisiana 70803, USA*

ABSTRACT.—We describe the foraging behavior and vocalizations of four *Myrmotherula* antwrens endemic to the Atlantic Forest biome of Brazil: *minor* (Salvadori's Antwren), *unicolor* (Unicolored Antwren), *snowi* (Alagoas Antwren), and *urosticta* (Band-tailed Antwren). Application of these data in intrageneric comparisons suggests that *minor* is most closely related to *M. schisticolor* (Slaty Antwren) and *M. sunensis* (Rio Suno Antwren) and that *unicolor* is a member of the "plain-winged" group, which we suggest is monophyletic, including *M. behni* (Plain-winged Antwren), *M. grisea* (Ashy Antwren), and *M. snowi*. Based on significant heterogynism and external morphological distinctions and, to a lesser extent (owing to inadequate and ambiguous samples of songs), vocal differences, we recommend that *snowi*, originally described as a subspecies of *unicolor*, be elevated to species rank. *M. urosticta* is the Atlantic Forest representative of the widespread *urosticta/longipennis* (Long-winged Antwren) complex. We discuss the concept that foraging data sets must be quantified to be comparable or applicable in systematic revision.

RESUMO.—Descrevemos os hábitos de forrageamento e vocalizações de quatro espécies do gênero *Myrmotherula* endêmicas ao bioma brasileiro da Mata Atlântica: *minor*, *unicolor*, *snowi* e *urosticta*. A aplicação desses dados em comparações intragenéricas sugere que *M. minor* é mais proximamente relacionada a *M. schisticolor* e *M. sunensis* e que *M. unicolor* é membro do grupo de espécies de "asas-lisas" - o qual sugerimos considerar monofilético - incluindo *M. behni*, *M. grisea* e *M. snowi*. Baseado na significante heteroginia e diferenças morfológicas externas e, em menor extensão (considerando que a amostra existente é inadequada e ambígua), diferenças no repertório vocal, recomendamos que *M. snowi*, originalmente descrita como subespécie de *M. unicolor*, seja elevada ao nível de espécie. *M. urosticta* é o representante na Mata Atlântica do - bem espalhado - complexo *urosticta/longipennis*. Discutimos o conceito que dados de forrageamento devem ser quantificados para ser comparáveis ou aplicáveis em revisões sistemáticas.

As currently treated (e.g., Sibley and Monroe 1990), the thamnophilid genus *Myrmotherula* is represented in the Atlantic Forest of Brazil by six species: *gularis* (Star-throated Antwren), *axillaris* (White-flanked Antwren), *fluminensis* (Rio de Janeiro Antwren, known only from the unique type), *minor* (Salvadori's Antwren), *unicolor* (Unicolored Antwren), and *urosticta* (Band-tailed Antwren). All but *axillaris* and *minor* are considered to be endemic to the region. However, clear distinctions in the morphology and vocalizations of the disjunct, endemic subspecies *luctuosa* of *M. axillaris* indicate that it is best regarded as a separate species (Isler et al. 1997), and Whitney and Pacheco (1995) restricted *M. minor* to the Atlantic Forest as well. Additionally, the isolated subspecies *snowi* of *M. unicolor* appears to merit recognition as a distinct species (see below). In this paper, we present the first substantive descriptions of the foraging behavior and vocalizations of four of these endemic *Myrmotherula* (*minor*, *unicolor*, *snowi*, and *urosticta*), and apply these data to suggest species relationships. We provide evidence to suggest that the "plain-winged" antwrens (as defined below) form a monophyletic group.

The distribution, habitat, and conservation status of these species were reported by Whitney and Pacheco (1995). *Myrmotherula minor*, *M. unicolor*, and *M. urosticta* are lowland species, with distributions centered in undisturbed forest between sea level and about 500 m elevation. *Myrmotherula snowi* is known only from the type locality, currently incorporated in the Murici

Biological Reserve above Murici, Alagoas, at about 550 m elevation. As species restricted to lowland forest, virtually the entire populations of all of them are highly threatened by habitat loss despite the fact that one or more species is known from some officially designated (but inadequately protected) forest reserves (Whitney and Pacheco 1995).

Methods and terminology: Most of our behavioral observations were conducted in coastal habitats of southern (around Parati) Rio de Janeiro and northern (around Ubatuba) São Paulo states (*minor* and *unicolor*); at the Sooretama Biological Reserve in Espírito Santo, and the forest reserve of the Companhia Vale do Rio Doce (CVRD) at Porto Seguro in southern Bahia (*urosticta*); and at the Murici Biological Reserve (*snowi*). We also made observations on all of these species except the last in other localities on a less regular basis. Antwrens were almost always first located by their distinctive vocalizations. We observed foraging behavior of individual birds continuously for a minimum of 5 min and for up to approximately 2 h irrespective of time of year or time of day. We tended to follow those species less easily or regularly observed (*M. minor*, *M. snowi*) for longer than we did others for which data were more readily gathered. Numbers of individuals and of foraging maneuvers observed are provided at the beginning of each species account. Foraging and other behaviors observed in the field were described verbally on cassette tape. Terminology for foraging behavior follows Remsen and Robinson (1990). All measurements given below (heights, distances, times, etc.) are estimates. Tape recordings were made using Nagra 4.2, and Sony TCM-5000 recorders, and Sennheiser ME-80 shotgun microphones. All recordings have been or will be archived at the Library of Natural Sounds (LNS), Cornell Laboratory of Ornithology, Ithaca, New York, and the Arquivo Sonoro Prof. Elias P. Coelho (ASEC) at the University Federal do Rio de Janeiro, Rio de Janeiro. Sound spectrograms were produced with "SoundEdit" of Farallon Computing, Inc., Emeryville, California, "Canary" 1.0 of the Bioacoustics Research Program at the Cornell Laboratory of Ornithology, Ithaca, New York, and "Canvas" of Deneba Software, Miami, Florida.

MYRMOTHERULA MINOR (SALVADORI'S ANTWREN)

Myrmotherula minor occurs from at least as far north as north-central Espírito Santo south to central São Paulo, with single old records in extreme northern Santa Catarina and the Rio Doce valley in eastern Minas Gerais (Whitney and Pacheco 1995). It inhabits the interior of tall, humid forest (moss coating trunks and limbs and numerous bromeliads) and old second-growth, and is almost always near running water.

Behavior.—We observed the foraging behavior of at least 18 individual *M. minor* over a cumulative time of approximately 4 h. A conservative estimate of 150 foraging maneuvers was observed. *M. minor* almost always traveled with mixed-species flocks. Within flocks, pairs or family groups (up to four individuals) foraged from practically on the ground to as high as about 12 m, generally in the 2–8 m range, most often at approximately 4 to 5 m above ground. The only instance in which we have observed *M. minor* foraging apart from a mixed-species flock (except early in the morning, before flocks had formed) involved a single immature male (probable) that foraged between 1.5 and 4 m above ground around the edges of a largely regrown light-gap inside forest. Members of a pair foraged close to each other, usually not more than about 3 m apart.

M. minor moved rapidly through the peripheral portions of trees and shrubs, less often through herbaceous vegetation, with short hops, hitches (as defined by Whitney 1994a), flutters, and flights, using both horizontally and vertically oriented perches. Foraging maneuvers were often accompanied by a high, rapid wing-flick (above the back, not laterally) at short intervals; the tail was also flicked, but we were not able to determine whether the emphasis of this movement was upward or downward. Movements were generally quite acrobatic, and the birds frequently hung sideways or upside-down for up to about 3 s. on terminal twigs or leaf margins or strands of *Usnea* sp. lichens, sometimes fluttering the wings to maintain balance before dropping or fluttering to a stable perch nearby.

The birds searched in live foliage by leaning over and craning the neck and often stretching the legs to their full extent to reach and glean, sometimes with a short lunging motion, from the upper-/and undersides of petioles and leaves. The solitary individual mentioned above spent several minutes exploring a 2 m-tall shrub with 0.5 cm-diameter red flowers, rapidly hitching and hopping along each limb to the end, then methodically gleaning at each flower for tiny arthropod prey. Vines were inspected at close range as the birds hitched up them, and they occasionally probed moss-coated trunks and limbs and *Usnea* hanging from limbs. Prey items sometimes seemed to be flushed or dislodged from limbs or leaves by the sheer energetic movements of the birds, and these were actively pursued with downward flutter-chases. However, one prey item that flushed in front of a male *M. minor* and flew directly away was not pursued. Sally-hovers of less than 1 s duration also were performed, although not often. A pair of *M. minor* associated with a slow-moving, understory mixed-species flock spent about 15 min foraging in low shrubs and herbaceous growth within 1 m of the ground.

Dead foliage, including leaf clusters and small, isolated, curled dead leaves, was approached specifically and inspected carefully in the manner described above for live foliage. Dead foliage usually was not manipulated (although birds sometimes hung on dead leaf margins), but we noted that an individual once poked an isolated dead leaf hard with the bill, and another time pulled on a dead leaf, both movements producing audible crunching sounds that are probably designed to startle arthropods hiding within. We observed *M. minor* using dead-leaf foraging substrates about ⅓ of the time.

We noted prey captures only about 20 times. All items were small (apparently adult) arthropods and caterpillars less than about 1.5 cm long. We suspect that many more small prey items were consumed than we could see.

In the humid coastal forests of southern Rio de Janeiro and northern São Paulo states, we have not observed *M. minor* traveling with any mixed-species flock that also included another *Myrmotherula* species. We have observed it within about 200 m of a flock containing *M. unicolor*, however, and D. Stotz (pers. comm.) has seen the two species in the same mixed-species flock once near Ubatuba, São Paulo. In the above-mentioned region, we have detected no difference in flock composition for the two species, both of which regularly associate with such species as *Lepidocolaptes fuscus* (Lesser Woodcreeper), *Philydor atricapillus* (Black-capped Foliage-gleaner), *Hypoedaleus guttatus* (Spot-backed Antshrike), *Drymophila squamata* (Scaled Antbird; when it forages with flocks), *Dysithamnus stictothorax* (Spot-breasted Antvireo), *D. mentalis* (Plain Antvireo), *Terenura maculata* (Streak-capped Antwren), and a variety of other small insectivores. The extent to which *M. minor* may forage in the same mixed-species flocks with *M. axillaris* or *M. urosticta* within the relatively narrow zone of known overlap with these species needs further investigation.

Vocalizations.—We have examined tape-recordings of songs of 3 individual *M. minor*, and calls of 7 individuals (7 recordings by Whitney, 4 by J. L. Rowlett). The song is complex and highly distinctive. It consists of one or two, occasionally even three multi-syllabic parts. The first part is a series of four to six sharply whistled syllables that descends evenly from about 5.5 kHz to about 3.0 kHz, ending with a single, even sharper syllable at about 4.5 kHz; the series is sometimes introduced with one or more quiet chipping notes. After a pause of about 0.25 s, the second part of the song is delivered much like the first, but consists of only three or four syllables, descends more rapidly to end at about 2.5 kHz, and lacks the sharp terminal syllable of the first part. An entire two-parted song lasts about 2 s (Fig. 1A). In some songs, the third part repeats the second part after another pause of about 0.25 s. A single individual is capable of delivering one-, two-, and three-parted songs.

Members of a pair frequently gave several types of quiet calls as they foraged, which were usually the first sign of the birds' presence. These notes seemed to function as pair contact calls (especially the rapid series of three syllables on the right side Fig. 2A) and possibly as alarm or heightened awareness notes (two single-syllable calls on the left side of Fig. 2A, which were delivered in succession by one male in response to tape playback of the song, but which also are given in a natural context). It was not possible to determine with confidence the function of these single-syllable calls.

MYRMOTHERULA UNICOLOR (UNICOLORED ANTWREN)

Myrmotherula unicolor occurs from northern Rio de Janeiro south to extreme northern Rio Grande do Sul, inhabiting the interior of tall, undisturbed forest and well developed second growth along the base of the Serra do Mar. In coastal São Paulo, it also inhabits restinga, which may be defined broadly as dense coastal woodland with a canopy height of less than about 15 m growing on white-sand soil (Whitney and Pacheco 1995).

Behavior.—A brief description of the behavior of *M. unicolor* was provided by Belton (1985), who reported that it was "continually searching leaves in the midstory," and "when standing still it peers around inspecting leaves and branches." We observed the foraging behavior of approximately 35 individual *M. unicolor* over a cumulative time of about 5 h. A conservative estimate of 200 foraging maneuvers was observed. *M. unicolor* usually traveled in pairs or family groups of up to four individuals with or apart from mixed-species flocks. We noted pairs foraging

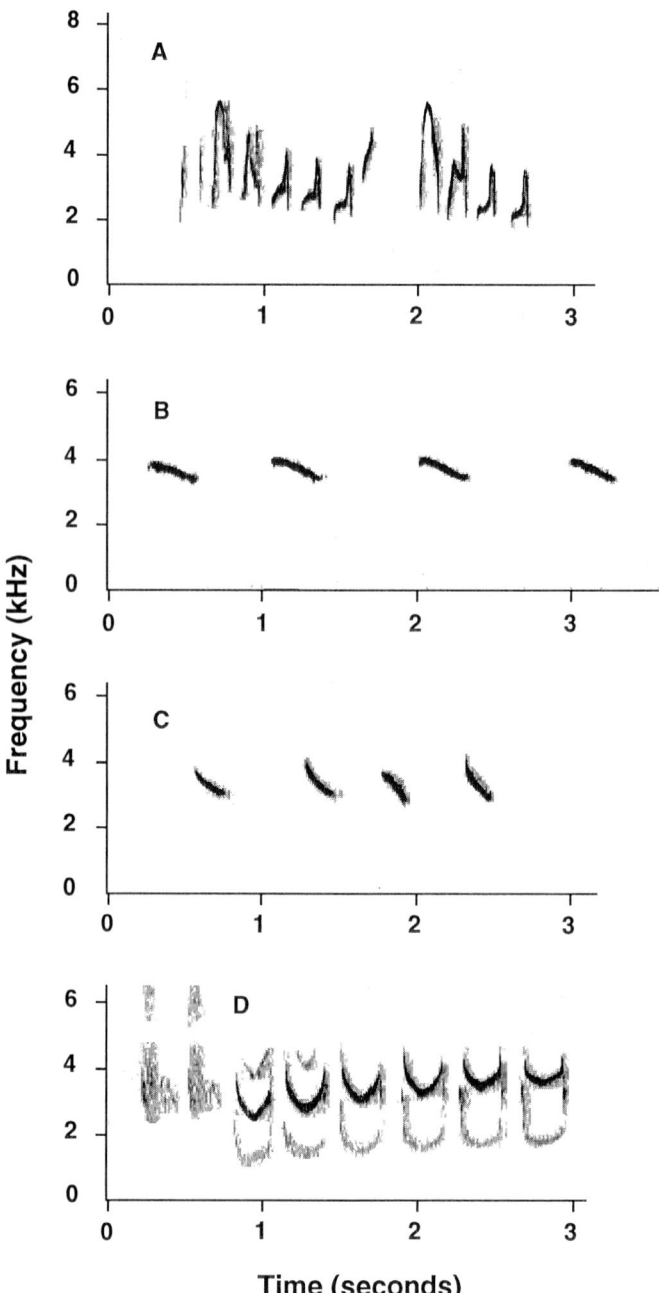

FIG. 1. Loudsongs of four species of *Myrmotherula* antwrens endemic to the Atlantic Forest of Brazil. A. *minor*, two-parted song, October 1992 near Paratí, Rio de Janeiro, recorded by J. L. Rowlett; B. *unicolor*, October 1991 near Paratí, Rio de Janeiro, recorded by BMW (number of syllables variable from two to about ten); C. *snowi*, February 1979 above Muricí ("Pedra Branca"), Alagoas, recorded by L. P. Gonzaga (range of variation unknown); D. *urosticta*, six-syllable song introduced by two "djéer" syllables, February 1988 at the CVRD forest reserve at Porto Seguro, Bahia, recorded by BMW (number of syllables variable from three to six).

apart from mixed-species flocks early in the morning, then observed them joining flocks as these formed within about two hours of dawn. We also noted pairs foraging solitarily late in the morning on at least three occasions. Members of a pair usually foraged within about 4 m of each other, but were occasionally as far apart as 10 m. Foraging heights ranged from 0.5 m (in restinga woodland) to 11 m in tall forest, with all observations in restinga between 0.5 and 4 m, and most observations in forest between about 4 and 7 m. We did not observe *M. unicolor* below about 2.5 m inside forest or humid second-growth. We noted *M. unicolor* in the upper end of the foraging range (above about 8 m) most often when with large mixed-species flocks.

While foraging, the birds regularly flicked the tail up rather sharply and shallowly, and lowered it more slowly (sometimes the emphasis of the movement seems to be reversed); the wings were not moved often, but were occasionally flicked shallowly. These tail and wing movements were sometimes exaggerated greatly after playback of songs and calls.

Whether with a mixed-species flock or not, *M. unicolor* moved through trees with short hops and flights, and often hitched upward on vertical vines and other vertical substrates such as *Philodendron* plants to inspect vegetation and bark surfaces at close range. The birds moved outward along more horizontal limbs from near the interior of trees to the tips with short advances in which they frequently reversed the head-tail orientation about 60°, almost constantly leaning over, extending the legs, and craning the neck up and sideways to scan the undersurfaces of leaves. After checking a limb for prey, individuals usually flew a short distance to another limb to land somewhere along its length (usually not the base), and then proceeded outward toward the tip. *M. unicolor* attacked prey primarily with gleans and reaches, and occasionally performed brief sally-hovers at the periphery of trees and leaf undersides. "Hangs" on live vegetation were infrequent and brief, although an immature male once hung on a leaf margin for about 4 s as it pecked intently at something on the underside of the leaf tip.

Myrmotherula unicolor foraged in dead leaves, even small, isolated ones, by peering intently and from various angles into the folds and crevices of the leaf, sometimes poking or grasping and tugging the leaf with the bill, and occasionally rummaging audibly in clusters of dead leaves. They sometimes briefly hung on dead leaf margins as they peered into otherwise inaccessible crevices. We noted much variation in the extent to which individuals or pairs of *M. unicolor* searched dead leaves and leaf clusters versus live leafy vegetation: some pairs searched little (about 25% of the time) in dead leaves, but others foraged principally (about 50–75% of the time) in dead leaves. We suspect that this variation has more to do with the density of dead leaves in the specific place in the forest in which we encountered the birds than preference of the birds. Thus, we observed *M. unicolor* foraging more in dead leaves in places where there seemed to be more suspended dead leaves available. We estimated, however, that in no place in the forest in which we found *M. unicolor* were dead leaves more abundant than live ones (either in terms of absolute numbers or surface areas). This suggests that *M. unicolor* does select dead leaf substrates more often than live ones. High-resolution definition of substrate-specific foraging behavior requires that data be extensively quantified (see comment at the end of this paper).

Prey items included small moths and caterpillars less than about 2 cm in length obtained in live vegetation, and a variety of arthropods up to about 4 cm in length found in dead leaves (including on three separate occasions a pale-colored, short-winged type of katydid with long legs and greatly elongated antennae) that were thrashed into submission on limbs before being swallowed whole.

In northern Rio de Janeiro state, where the ranges of *M. unicolor* and *M. axillaris* overlap, the two regularly forage in the same mixed-species flocks. Some common flock associates of *M. unicolor* in tall forest are listed in the *M. minor* account above, and in Teixeira and Gonzaga (1985). In restinga on Ilha do Cardoso, São Paulo, D. Stotz (pers. comm.) listed the most common flock associates of *M. unicolor* as *Sittasomus griseicapillus* (Olivaceous Woodcreeper), *Herpsilochmus rufimarginatus* (Rufous-winged Antwren), *Leptopogon amaurocephalus* (Sepia-capped Flycatcher), *Phylloscartes kronei* (Restinga Tyrannulet), and *Parula pitiayumi* (Tropical Parula Warbler).

Vocalizations.—We have examined tape-recordings of songs of 18 individual *M. unicolor*, and calls of 18 individuals, not all the same as those for which songs were recorded (20 recordings by Whitney, 1 recording by P. S. Fonseca, 1 recording by W. Belton). Belton (1985) described the voice of *M. unicolor* as "High, plaintive, short: 'eeeeeeeee' trending slightly downscale. Also very short, weak: 'whee'." Similarly, Sick (1993) described the voice as a "relatively long, descending whistle." The song of *M. unicolor* is actually a more or less evenly paced series of two to about 10 of these lengthened, descending syllables at slightly below 4 kHz (Fig. 1B). The syllables are usually delivered at intervals of about 0.75 s, but intervals vary somewhat,

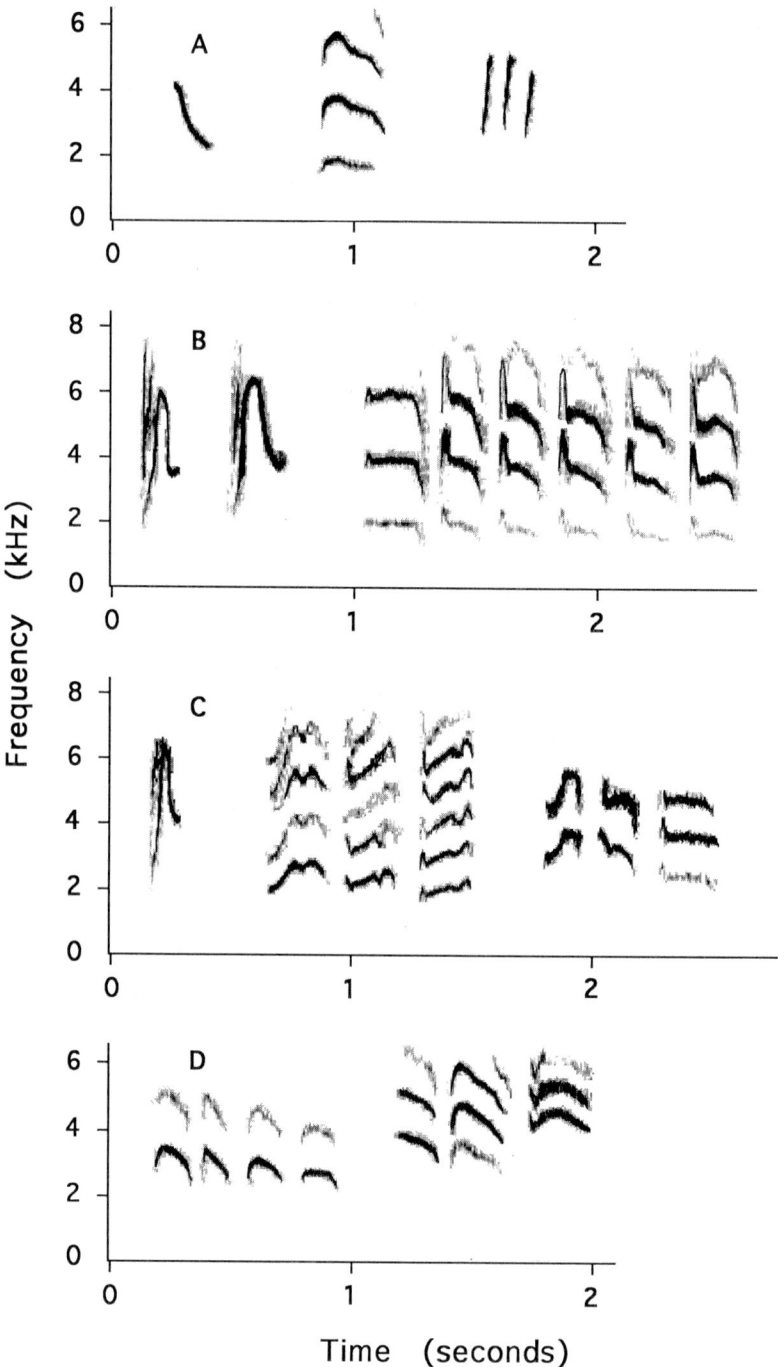

FIG. 2. Representative calls of the same four *Myrmotherula* species designated in Fig. 1 (all recorded by BMW). A. *minor*, possible alarm or heightened awareness calls (left and center) delivered by a single male in response to tape playback but also delivered in an unsolicited context, and apparent contact calls typically delivered in couplets or triplets very quietly (right), October 1993 near Ubatuba, São Paulo; B. *unicolor*, two examples (left) of the "kleek" or "wheet" pair-contact call (given in succession by a single male), which is shared by the four presently described "plain-winged" *Myrmotherula* species (compare with 2C and Fig. 2A of Whitney 1994a), and apparent scold or mobbing vocalization (multi-syllabic, right side) that usually comprises four to six syllables, October 1992 near Ubatuba, São Paulo; C. *snowi*, the "kleek" pair-contact call (left), and two examples (center and right) of the three-syllable vocalization that apparently functions as

and irregular pauses are sometimes interspersed in the sequence, especially after tape playback. Additionally, the song is sometimes introduced with two or three distinctive syllables sounding like "djéer, djéer. . . ." A series of raspy syllables lasting from 1 to 2 s at about 4 kHz, the first of which is structurally and qualitatively different from the rest (Fig. 2B, right side), seems to be a scold or a mobbing vocalization. It was given infrequently (by few individuals), sometimes in an agitated manner with attention directed at the observer (both in a natural context and in response to tape playback) and once in response to the presence of a *Glaucidium brasilianum* (Ferruginous Pygmy-Owl), on which occasion numerous other birds were giving homologous mobbing vocalizations. The most common contact vocalization between members of a pair is a single-syllable "kleek" or "wheet" (Fig 2B, two examples on left side, delivered in succession by one male), which was described as "plee-e" by Willis and Oniki (1992).

MYRMOTHERULA SNOWI (ALAGOAS ANTWREN)

We consider *snowi*, originally described as a subspecies of *M. unicolor* (Teixeira and Gonzaga 1985), a distinct species (evidence provided below). It remains known only from the type locality above Murici in Alagoas (Collar et al. 1992), which is a semi-humid forest remnant (probably historically maintained by moisture-laden winds off the Atlantic Ocean) on the easternmost extension of the Borborema Plateau (Whitney and Pacheco 1995). *Myrmotherula snowi* inhabits the interior of tall forest, and is a rare bird.

Behavior.—To augment the ecological information provided by Collar et al. (1992), it is noteworthy that Teixeira and Gonzaga (1985) reported that *M. snowi* searched for prey with great liveliness in both live and dead, trapped leaves, although foraging behavior was not described. A good list of flock associates was provided by Teixeira and Gonzaga (1985).

We observed the foraging and other behaviors of four individuals for a period of about 2 h in October 1990 and about 2 h in November 1993. At least 40 foraging maneuvers were observed. *Myrmotherula snowi* foraged in pairs with small mixed-species flocks in the forest understory, ranging from about 1.5 m to as high as about 9 m, keeping mostly 5-8 m above ground. This may vary somewhat seasonally, as others have observed pairs and family groups mostly nearer the ground, with all observations in the period February (when breeding probably takes place) to May (Collar et al. 1992). The birds moved actively through the vegetation with short hops and flights, regularly pumping the tail slightly downward but seldom moving the wings. We observed *M. snowi* searching both live and dead foliage, with more than half of our observations involving short reaches and gleans in dead foliage. Prey items were located primarily with close-range visual inspections in which the neck was craned and the legs were often extended. The birds sometimes hung briefly on dead leaf margins (more rarely on live leaves) to probe into curls, and sometimes poked dead leaves with the bill. Prey items included small orthopterans and greenish and whitish caterpillars less than 2 cm long. On one occasion, a male obtained from a large, curled dead leaf a 4-cm-long, pale-colored katydid with long legs and very long antennae (appearing to us the same as those taken by *unicolor* in southern Rio de Janeiro), grasping it and shaking it for several seconds as it kicked and struggled, then flew off with it.

Vocalizations.—Including vocalizations of both sexes, we have examined two recordings of songs and eight recordings of calls of *M. snowi* (eight recordings by Whitney, two recordings by L. P. Gonzaga). In the type description of *M. unicolor snowi*, Teixeira and Gonzaga (1985) described its vocalizations as seeming to be identical to those of the nominate form, the song being a low, mournful whistle: "üüüuuu . . ." extremely like the voice of *Myiarchus tuberculifer* (Dusky-capped Flycatcher) but repeated less consecutively than the flycatcher gives its call (translation from Portuguese). We have only two recordings of the song of *snowi*, which may be described as a series of 3–6 downslurred, clear-whistled syllables at about 3.5 kHz delivered in about 2–4 s (Fig. 1C; thus, whistled syllables delivered much more frequently than in *Myiarchus tuberculifer*). The single-syllable "kleek" pair-contact call is shown in Fig. 2C (far left side). Both sexes of *M. snowi* frequently gave a three-syllable (sometimes two-syllable), slightly descending vocalization, "nyiih-nyeeh-nyaah," that apparently functions as an alarm or height-

---

←

an alarm or heightened awareness call, and possibly as a scold, November 1993 above Muricí ("Pedra Branca"), Alagoas; *D. urosticta*, three-syllable vocalization (female left, male right) that apparently functions as an alarm or heightened awareness call, and possibly as a scold, October 1992 at Sooretama Biological Reserve near Linhares, Espírito Santo.

ened awareness call, and as a scold (Fig. 2C, two examples, center and right). It was given when the presence of an observer was detected, in the same context in which *Thamnomanes caesius* (Cinereous Antshrike), which is usually in the same flocks as *M. snowi*, begins giving its alarm "rattle," and it was sometimes repeated several times in response to tape playback.

MYRMOTHERULA UROSTICTA (BAND-TAILED ANTWREN)

Myrmotherula urosticta occurs from just south of Salvador, Bahia south to northern Rio de Janeiro in tall, undisturbed forest (Whitney and Pacheco 1995).

Behavior.—We observed a minimum of 50 foraging maneuvers of about 15 individuals of *M. urosticta* over a cumulative period of about 3 h. *Myrmotherula urosticta* traveled almost exclusively with mixed-species flocks. Pairs or small family groups foraged from 3–9 m above ground, mostly 5–7 m. The birds foraged primarily in live vegetation and vines, hopping and hitching along both in the interior and at the periphery of trees, frequently flicking the wings rapidly and simultaneously out from the sides in the manner of *M. axillaris*. Most foraging attacks were gleans and reaches, with occasional flutters and flutter-chases and brief upside-down hangs. Vines and the undersides of leaves were carefully searched as the birds craned the neck in various positions while leaning over or hanging from leaf petioles or margins. We have not noted dead-leaf searching by *M. urosticta*.

All flocks with *M. urosticta* also contained *M. axillaris*, and *Thamnomanes caesius* was present in most. In the few places where overlap is known (see Whitney and Pacheco 1995), we have not found *M. urosticta* in the same flock as either *M. minor* or *M. unicolor*. Our opportunities to observe this have been brief and seasonally patchy, however, and the extent to which competetive exclusion or other factors might be involved warrants further study.

Vocalizations: We have examined tape-recordings of songs of eight individual *M. urosticta*, and calls of nine individuals (16 recordings by Whitney, one recording by P. S. Fonseca). The song of *M. urosticta* is a steady series of 3–6 "U"-shaped syllables delivered at a rate of approximately 3/sec. that rises from about 2.5 to 3.5 kHz (Fig. 1D). Each syllable of the song has a distinctly burry quality (i.e., not a pure, whistled sound). The song is sometimes introduced with 2–3 harsh "djéer" syllables, much like that of *M. unicolor*. The most conspicuous call of both sexes (female is lower frequency) of *M. urosticta* is like the three-syllable, descending "nyiih-nyeeh-nyaah" of *M. snowi* described above, and is given in similar context, with variation ranging from 1–4 syllables (Fig. 2D).

TAXONOMIC IMPLICATIONS

In the structure of the syllables of its individual parts and in overall quality, the song of *M. minor* suggests a closer relationship to *M. schisticolor* and *M. sunensis* (Slaty and Rio Suno antwrens, respectively; see Whitney [1994a] for voice comparisons) than to any other members of the genus. From his morphological studies, Zimmer (1932) suspected a close relationship between *minor* and *sunensis*. Limited observations of the foraging behavior of both species suggests that *minor* forages more acrobatically, and changes foraging stations more often, than does the relatively lethargic (Whitney 1994a) *sunensis*. Foraging behavior of *M. schisticolor interior* in eastern Ecuador (described by Whitney 1994a) is quite similar to that of *M. minor* (i.e., both seem to be active, "generalist foragers" and "regular users" of dead-leaf foraging substrates).

Collar et al. (1992) treated *snowi* as a species based on vocal differences reported in part by us, which treatment was followed by Ridgely and Tudor (1994). Whitney and Pacheco (1995, citing the present paper) also treated *snowi* as a species. However, there has been no corroborated justification for the elevation of *snowi* to species status. Although Teixeira and Gonzaga (1985) included discussion of the morphology and some aspects of the voice and natural history of *M. unicolor snowi* in the type description, the level of its relationship to nominate *unicolor* merits further investigation. We have not compared skins of these two, but the description of a single male *snowi* (the holotype) by Teixeira and Gonzaga (1985), and our field observations, indicate that male *snowi* are practically indistinguishable from male *unicolor*. Female plumages were diagnosed by Teixeira and Gonzaga (1985) as notably different, with *snowi* ($n = 2$ adults) being "much more rufous" in the upperparts and the entirety of the underparts, and "considerably brighter" in color than nominate *unicolor* ($n = 22$), which we have also noted in the field (notwithstanding the impossibility of direct comparison). Hellmayr (1929) coined the term "heterogynism" to emphasize the taxonomic importance of geographic differences in the female plumage of formicariids.

Comparison of external morphology of *snowi* ($n =$ one male and two females) and *unicolor*

($n = 30$ males and 22 females) presented by Teixeira and Gonzaga (1985, their Table 1) showed that, for both sexes, there is no overlap in wing-length, tail-length, or length of culmen. More importantly, perhaps, the two taxa appear to be proportioned differently, with *snowi* having a significantly longer wing and longer bill, but a shorter tail.

The songs of *unicolor* and *snowi* are structurally similar, and individual syllables are of similar shape, although inter-syllable intervals appear to be significantly longer in *unicolor*. Songs of the two taxa also sound quite similar. We hesitate to offer any analysis based on songs, however, because the two *snowi* we have (both from the only locality known for the taxon) are themselves somewhat different, and a larger sample is particularly desirable in this case. The alarm or scolding type of vocalization recorded for both taxa, although similar in appearance (frequency, spacing of syllables; compare right sides of fig's. 2B and 2C), does sound diagnostically distinct. This is reflected in the level frequency of the series of *unicolor*, and the consistency of shape of the individual syllables, which open sharply then quickly rise and fall. The series of *snowi* tends to fall in frequency overall, and the individual syllables are more variable in shape and frequency range. These differences become particularly apparent when several successive vocalizations of a single individual are considered.

Our limited observations of foraging behavior indicate that *unicolor* and *snowi* search for prey in closely similar fashion, and they show roughly equal preference for dead- and live-leaf substrates.

Given the appreciably different female plumages and external morphologies of *unicolor* and *snowi*, and the reasonably well-documented distinctions in their alarm or scold type vocalizations, at least, we conservatively recommend that *snowi* be treated as specifically distinct from *unicolor*. *Myrmotherula snowi* appears to be about as clearly differentiated from *M. unicolor* as are any of the other "plain-winged" (see below) *Myrmotherula* from each other. The degree of differentiation of another thamnophilid species pair sharing a basically similar but even less disjunct distribution than *snowi* and *unicolor*, *Terenura sicki* (Alagoas Antwren) and *T. maculata* (Streak-capped Antwren), is comparable, as they show significant heterogynism and slight but consistent vocal differences, albeit no distinctions in external morphology (Teixera and Gonzaga 1983, pers. observ.).

The "nyiih-nyeeh-nyaah" vocalization of *M. urosticta* appears to be the "beer bin" of Willis (1984). He reported (and we can confirm) that this call is shared by *M. longipennis* (Long-winged Antwren) and suggested that *urosticta* "is perhaps a well-marked subspecies of *longipennis*." Despite the fact that the two species never have been placed together in taxonomic lists (and that *urosticta* has priority over *longipennis*), we agree with Willis (1984) that *longipennis* is close to *urosticta*, just not that close. Species limits within the *urosticta/longipennis* group are complex, and will be treated in detail in a future study (in prep.).

Relationships within the "Plain-Winged" Myrmotherula Group

The "plain-winged" group of *Myrmotherula* may be distinguished from all congeners by close similarity in structure and pattern of the known songs and the dominant "kleek" calls of the four presently described species: *unicolor*, *snowi*, *behni* (Plain-winged Antwren), and *grisea* (Ashy Antwren). Additionally, their basically all-gray male plumage lacks any marking on the wing-coverts. Some species have restricted and intraspecifically variable distribution of black feathering on the throat. The apparently relictual distribution of the plain-winged group lies in elevated ranges primarily east of the main cordillera of the Andes (with *unicolor* being restricted to the coastal foot of the Serra do Mar), forming a broken ring around Amazonia and the dry habitats of eastern Brazil.

The songs of *M. unicolor*, *M. snowi*, and *M. behni* (of the Sumaco region of eastern Ecuador, for which Whitney [1994a] provided sound spectrograms of "kleek" call and song), are similar except that the interval between syllables in the song of *unicolor* may prove to be longer than those of *snowi* and *behni* (very few songs of *snowi* and *behni* have been recorded). Whitney also has tape-recorded the "kleek" call for *M. grisea*. The full song of *grisea* remains unknown, but it is likely to be much like those of the others. Whitney once heard it give the introductory "djéer, djéer," as given by *unicolor*, in response to his whistled approximation of a basic "plain-winged" antwren song.

The "nyiih-nyeeh-nyaah" alarm or scold vocalization shared by *unicolor* and *snowi* does not appear to be given by *grisea* (but opportunity to establish this has been limited), or by *behni* in eastern Ecuador. A functionally and structurally similar vocalization is in the repertoire of some populations of *M. axillaris* and all populations of the *M. urosticta/longipennis* complex, as mentioned above (pers. observ.). A possibly homologous vocalization is given in single-syllable (usually) form by *M. schisticolor* (Whitney 1994a.). In fact, this stereotyped vocalization appears to be given by at least one (and usually only one where multiple taxa are syntopic) representative of the "gray" group of *Myrmotherula* (as defined by Hackett and Rosenberg 1990, which included the "plain-winged" group) at every locality over its vast distribution, and is not shared by any member of the genus outside the "gray" group (pers. observ.). It is apparently of ancestral origin, as opposed to having evolved independently several times, and only, within the "gray" group. There may have been selective pressure to maintain this warning or alert message across two or more subgroups of the "gray" group if such lack of variation promotes interspecific identification of a potential threat among members of the avian understory community, as suggested generally by Thorpe (1956), and for scolds of *Scytalopus* tapaculos by Whitney (1994b).

The "djéer, djéer" syllables often introducing the songs of *M. unicolor*, and once heard by Whitney from *M. grisea*, are probably homologous to the practically identical "djéer" syllables often introducing the songs of members of the *urosticta/longipennis* complex, and do not appear to be given by any other *Myrmotherula* (pers. observ.). Thus, with regard to vocalizations, the "plain-winged" and "long-winged" groups, the distributions of which are essentially complementary and non-overlapping, appear to be related in a number of respects. If wing-spots are removed from consideration, plumage of the two groups is also quite similar. Members of the "plain-winged" group appear to forage in dead leaves much more than do "long-winged" antwrens, which rarely do (pers. observ.).

It has always been assumed that *snowi*, isolated in semi-humid, "hilltop" forest in northeastern Brazil, is most closely related to *unicolor*, which occurs much further south along the seaward base of the Serra do Mar. This is certainly a reasonable assumption, especially when one considers the obvious links of other isolates roughly sharing the distribution of *snowi* with southern forms (such as *Philydor novaesi/atricapillus* and *Terenura sicki/maculata*). However, recognizing that there is in northeastern Brazil at least one upland isolate that almost certainly colonized from the tepuis region (*Procnias averano*), it also seems plausible to us that *snowi* could be more closely related to some form of *M. behni*.

In their comparison of phenotypic and genetic differentiation of *Myrmotherula* species, Hackett and Rosenberg (1990, and K. Rosenberg, pers. comm.) reported that *M. grisea*, and particularly *M. behni* (sample from Cerro de la Neblina, Venezuela) and *M. schisticolor* (Slaty Antwren, sample from Panama) are closely related. *Myrmotherula longipennis* (sample from Pando, Bolivia) did not cluster with this group, but much closer to *M. menetriesii* (Gray Antwren), with which it is syntopic over its wide distribution. Hackett and Rosenberg (1990) lacked samples of *unicolor*, *snowi*, *urosticta*, and several other of their "gray" *Myrmotherula*. A complete geographic sampling (including all named "subspecies," at least) would probably reveal several subgroups of the broadly defined "gray" group, one of which would be our "plain-winged." Remsen et al. (1991) included *M. unicolor* as a "sister taxon" to *behni* and *grisea* without explanation. A phylogenetic analysis incorporating all available data, including more rigorous samples of voices and a much more complete geographical representation of biochemical samples than have been available to date, is needed to corroborate relationships and shed light on the evolutionary history of these antwrens.

Unquantified Foraging Behavior: How Bad Can It Be?

Our foraging data are not quantified. This greatly diminishes their analytical application, we have been told, because only quantitative data sets may be objectively and statistically compared to each other. It is important to note at the same time, however, that we did not make any assumptions concerning such variables as, to mention a few examples: seasonal, temporal, or weather-related fluctuations in composition, abundance, or behavior of potential prey species; variation in numbers, density, and species-composition of live and dead leaves at different strata in the forest, across structurally different habitats, or at different localities at different seasons; differential composition of mixed-species flocks; effects of observer-related distraction; or possible dietary shifts (which might involve shifts in foraging height and/or search method) correlated with reproductive condition of the antwrens involved. None of these factors is easily measured. The fact remains, however, that without concurrent quantification of such variables, and without stringent uniformity of data-collection methodology (this has been recommended, but in reality, published papers on the subject vary widely in methodology) and adherence to a standard terminology for foraging behaviors (a reasonable one has been published for birds in terrestrial habitats, but to date it has been applied by very few), it is unlikely that any quantified foraging data sets from any independent sources are justifiably comparable—except perhaps for data collected in exactly the same manner by observers of equal skill level at a single locality in a single habitat (and definition of this is complicated) during a single season, or for documentation of gross substrate use (i.e., bamboo, bark, dead leaves, live foliage). This is not to suggest that "selectively quantified" foraging data are not useful. Our point is that *de facto* acceptance of the applicability and comparability of incompletely or differently quantified data sets on the foraging behavior of birds could actually be misleading in systematic revision or comparative studies of behavioral ecology. To argue that it could not is to accept that it would suffice, perhaps even prove to be most conservative, to define differences or similarities in foraging behaviors of birds on only a carefully descriptive level after all.

ACKNOWLEDGMENTS

This paper is written for and in the spirit of Ted Parker, who truly knew the importance of tape-recording and observing the behaviors of birds and other organisms, from the perspectives both of a scientist investigating the relationships of species and a conservationist working to preserve the diversity that he had the capacity to appreciate in its multi-layered depths as have few others. We are grateful to L. P. Gonzaga and J. L. Rowlett for allowing us to make sound spectrograms of their rare recordings of *M. snowi* and *M. minor*, respectively, and to C. Bauer for assistance with tape recordings housed at ASEC. We benefited from discussion with M. Isler and P. Isler concerning distribution and vocalizations of *Myrmotherula* species, and we thank them also for their helpful comments on an early draft of the manuscript. We are particularly grateful to D. Stotz, K. Rosenberg, and J. V. Remsen, Jr. for their well-informed and constructive criticism.

LITERATURE CITED

BELTON, W. 1985. Birds of Rio Grande do Sul, Brazil. part 2. Formicariidae through Corvidae. Bull. Amer. Mus. Nat. Hist. 180:1–242.

COLLAR, N. J., L. P. GONZAGA, N. KRABBB, A. MADROÑO NIETO, L. NARANJO, T. A. PARKER III, AND D. C. WEGE. 1992. Threatened Birds of the Americas. Cambridge, U.K. International Council for Bird Preservation, Cambridge, United Kingdom.

HACKETT, S. J., AND K. V. ROSENBERG. 1990. Comparison of phenotypic and genetic differentiation in South American antwrens (Formicariidae). Auk 107:473–489.

HELLMAYR, C. E. 1929. On heterogynism in formicarian birds. J. Ornithol., Festschr. Hartert, pages 41–70.

ISLER, M. L., P. R. ISLER, AND B. M. WHITNEY. 1997. Biogeography and systematics of the *Thamnophilus punctatus* complex. pp. 355–381 *in* Studies in Neotropical Ornithology Honoring Ted Parker (J. V. Remsen, Jr., Ed.), Ornithol. Monogr. No. 48.

REMSEN, J. V., JR., AND S. K. ROBINSON. 1990. A classification scheme for foraging behavior of birds in terrestrial habitats. pages 144–160 *in* Avian foraging: theory, methodology, and applications (M. L. Morrison, C. J. Ralph, J. Verner, and J. R. Jehl, Jr., Eds.). Stud. Avian Biol. 13.

REMSEN, J. V., JR., O. ROCHA O., C. G. SCHMITT, AND D. C. SCHMITT. 1991. Zoogeography and geographic variation of *Platyrinchus mystaceus* in Bolivia and Peru, and the circum-Amazonian distribution pattern. Ornit. Neotrop. 2:77–83.

RIDGELY, R. S., AND G. TUDOR. 1994. The Birds of South America, Vol II, the Suboscine Passerines. Univ. Texas Press, Austin, Texas.

SIBLEY, C. G., AND B. L. MONROE, JR. 1990. Distribution and Taxonomy of Birds of the World. Yale Univ. Press, New Haven, Connecticut.

SICK, H. 1993. Birds in Brazil. A Natural History. Princeton Univ. Press, Princeton, New Jersey.

TEIXEIRA, D. M., AND L. P. GONZAGA. 1983. A new antwren from northeastern Brazil. Bull. Brit. Orn. Club 103:133–135.

TEIXEIRA, D. M., AND L. P. GONZAGA. 1985. Uma nova subespécie de *Myrmotherula unicolor* (Ménétriès, 1835) (Passeriformes, Formicariidae) do nordeste do Brasil. Bol. Mus. Nac. Rio de Janeiro. n.s. 310:1–16.

THORPE, W. H. 1956. The language of birds. Scientific Amer. 195:128–138.

WHITNEY, B. M. 1994a. Behavior, vocalizations, and possible relationships of four *Myrmotherula* antwrens (Formicariidae) from eastern Ecuador. Auk 111:469–475.

WHITNEY, B. M. 1994b. A new *Scytalopus* tapaculo (Rhinocryptidae) from Bolivia, with notes on other Bolivian members of the genus and the *magellanicus* complex. Wilson Bull. 106:585–614.

WHITNEY, B. M., AND J. F. PACHECO. 1995. Distribution and conservation status of four *Myrmotherula* antwrens (Formicariidae) in the Atlantic Forest of Brazil. Bird Conservation International. 5:295–313.

WILLIS, E. O. 1984. *Myrmotherula* antwrens (Aves, Formicariidae) as army ant followers. Rev. Bras. Zool. 2:153–158.

WILLIS, E. O., AND Y. ONIKI. 1992. A new *Phylloscartes* (Tyrannidae) from southeastern Brazil. Bull. Brit. Orn. Club. 112:158–165.

ZIMMER, J. T. 1932. Studies of Peruvian birds. IV. The genus *Myrmotherula* in Peru, with notes on extralimital forms. part 2. Amer. Mus. Novit., 524:1–16.

LAND OF MAGNIFICENT ISOLATION: M. A. CARRIKER'S EXPLORATIONS IN BOLIVIA

David A. Wiedenfeld
Museum of Natural Science, 119 Foster Hall, Louisiana State University, Baton Rouge, Louisiana 70803, USA

ABSTRACT.—Melbourne Armstrong Carriker, Jr. (1879–1965) traveled and collected birds and mallophaga in Central America and South America from Colombia to Bolivia. He collected more than 53,000 bird specimens and described many new species and subspecies from his collections. From 1934–1938 he worked in Bolivia, and afterwards wrote a manuscript of his adventures and travails that was never published. I have excerpted sections of that manuscript describing travel on the rivers, and work in the Andes and Chaco.

RESUMEN.—Melbourne Armstrong Carriker, Jr. (1879–1965) viajó y colectó ejemplares de aves y de mallaphagos en América Central y tambien en América Sur, de Colombia a Bolivia. Obtuvó mas de 53,000 muestras de aves, y de sus colecciones describió muchas especies y subespecies nuevas. Desde 1934–1938, Carriker trabajó en Bolivia y despues escribió sus aventuras, pero el manuscrito nunca fue publicado. Yo he extraido partes del manuscrito describiendo viajes por los rios bolivianos, y trabajos en los Andes y chaco boliviano.

M. A. Carriker, Jr. (1879–1965) was one of the outstanding ornithological collectors and explorers in the Neotropics in the early part of this century. He traveled and collected bird specimens throughout much of Spanish Latin America, spending many years in Colombia, Peru, and Bolivia. By 1944, his efforts had produced about 53,000 bird specimens (Phelps 1944). Because of this accomplishment, he is often remembered as an ornithological collector, but his study and collecting of mallophaga (feather mites) were at least as great as his work on birds, and he produced many more articles on mallophaga systematics than on birds. Excerpts from Carriker's own writing of his adventures in Bolivia follow this biography.

Born in Sullivan, Illinois, in 1879, Melbourne Armstrong Carriker, Jr. (Fig. 1) was the son of a Nebraska doctor. As a young man, he attended college for two years, but soon decided his career lay elsewhere. He had always liked hunting and the out-of-doors, so he gradually became involved in bird collecting. At the age of 23 he made his first of several collecting trips to Costa Rica. He then collected for two years in Venezuela, and in 1911 began collecting in Colombia (Table 1).

On a visit to Santa Marta, Colombia, Carriker met Myrtle Carmelita Flye, the daughter of an American engineer who had come to Santa Marta to help with electrification of the city and had stayed on to become a coffee planter. Carriker and Miss Flye were married in Wayland, Michigan, on 22 June 1912, and moved back to Colombia to start a coffee plantation adjacent to Cincinnati Plantation, which belonged to Mrs. Carriker's father. Their own plantation they named Vista Nieve ("View of the Snow"). The finca had a pleasant, cool climate and a spectacular view; down the valley sparkled the Caribbean Sea, and behind could be seen the eponymous snow-covered peaks of the Santa Marta Mountains.

The Carrikers lived at Vista Nieve until 1927 and had five children there, the eldest, Melbourne Romaine, born in 1915. This son, referred to as Mel, accompanied his father on the first of the expeditions to Bolivia. The Carriker's younger children were Myrtle Florence, Howard Holland, Frederick Ruthven, and Alva Marie. Throughout this period Carriker continued to collect, primarily in the area around Vista Nieve, but also elsewhere in Colombia and nearby Venezuela. The specimens that he collected he sold to American museums, primarily the Academy of Natural Sciences of Philadelphia, for $5 apiece, a good price for the time. He continued to write during this period, too, and in 1922 coauthored a landmark book with W. E. Clyde Todd of the Carnegie Museum on the birds of the Santa Marta Mountains.

In 1927, when Mel was 12, the Carrikers decided to return to the United States to further

Fig. 1. Melbourne Armstrong Carriker. Photo from Phelps (1944).

TABLE 1
Carriker's Major Expeditions in the Neotropics, Compiled from Phelps (1944), Stephens and Traylor (1983), Paynter (1992), and from Carriker (ms.), and from Melbourne R. Carriker (pers. comm.). Most Expeditions Seem to Have Been Six to Eight Months Long

Years	Expedition
1902–1907	Costa Rica, for Carnegie Museum
1909–1911	Venezuela
1910?	Panama
1911–1917	Colombia
1922	Venezuela
1929–1933	Four times to Peru, for Academy of Natural Sciences of Philadelphia
1934–1938	Three times to Bolivia, for Academy of Natural Sciences of Philadelphia
1940	Veracruz, Mexico, for U.S. National Museum
1942–1952	Colombia, for U.S. National Museum
1952–1963	Colombia, for Yale Peabody Museum and other museums

their children's education and to alleviate concern that they would "grow up Colombian." Carriker received a job offer from the Academy in Philadelphia, but when it came time to leave for the States, the Academy discovered that it had no money for his position. Nevertheless, the Academy told him to come anyway, as the funding would come through in a year or two. So the Carriker family moved to Beachwood, New Jersey, in 1927, after selling Vista Nieve to the in-laws next door. Once in the U.S., times were hard for the family for the next year and a half. Carriker was a skilled carpenter and worked in that trade as a contractor, and in local politics, and Mrs. Carriker took in sewing to cover expenses.

Finally, in fall 1929, the Academy did come up with a paid curator position for Carriker. Then began a busy collecting period in his life; between 1929 and 1938 he made four expeditions to Peru, then three to Bolivia (Table 1), each longer than six months. Immediately after returning from Bolivia in May 1938, he left the Academy. It is not clear why. Mel Carriker thinks it may have had to do with his concerns about soon reaching retirement age with little in the way of savings. He again began working as a contractor and carpenter, and worked on the manuscript of his Bolivian adventures.

The long years of separation and constant expeditioning had taken a toll on Carriker's marriage. In 1941 the Carrikers were divorced, and Carriker moved back to Colombia permanently, but this time to Popayán, in the south of the country. It would have been cheaper to retire in Colombia, and Carriker had always held a deep love for Latin America. He continued to collect birds and mallophaga, mainly in Colombia. He was made a Research Associate of the Natural History Museum of the Universidad del Cauca by its director, the Colombian naturalist Carlos Lehmann V. He later was remarried, to Felisa Quintano Ropero. He died on 27 July 1965, at age 86, and was buried in Bucaramanga, Colombia. During his later years, he seems to have published almost strictly on mallophaga. His last ornithological article was published in 1959, but he continued to publish on mallophaga, and at the time of his death had several publications on mallophaga systematics in preparation. Most of his papers and his collection of mallophaga were given to the U. S. National Museum Division of Entomology.

Ornithological expeditions and collecting during the period of Carriker's Bolivian trips (1934–1938), from which the excerpts following the biography were taken, were very different from what they would later become. It was only at this time that ornithologists were discovering the usefulness of mist nets, which were then still exotic, expensive silk meshes imported from Japan. During these three trips, Carriker did not use the "newfangled" nets. In Bolivia during the 1934–1935 expedition, Mel Carriker says he and his father would shoot 15 to 18 birds apiece per day. They prepared the specimens on an assembly line: Carriker would collect the mallophaga feather mites from the bird and hand it to Mel, who would skin it and hand it back to his father to be stuffed.

Carriker's expeditions, usually consisting of himself and one or two assistants, did not carry sufficient supply of most consumables to last the entire expedition. The expeditioners reloaded their shotgun shells on days when they could not go into the field. They carried only a small amount of food, sufficient for only a couple of days. Once they had reached a working area, they would hire a local woman to prepare midday and evening meals for them; for their breakfast, Carriker and his assistants would usually fry some batter to make "doughboys."

Even though the expeditions carried little in the way of consumable supplies, they did carry a prodigious amount of other equipment. Travel, at least in Bolivia, was mostly by truck, train, and boat, with a smaller fraction of travel by mule or horseback. This may have been required by the fact that the gear added up to about 500 kg, including collecting equipment, tents, cots, bedding and camping supplies, but also items such as Carriker's own saddle, which he took everywhere even when not traveling by horseback.

Rarely did they remain for long periods in a field camp on the Bolivian trips. Usually the group stayed in spare rooms or hotels in small towns or villages, or occasionally at a ranch or mine headquarters. They would arise very early and walk far to their collecting sites, returning to their lodgings around midday.

The Bolivian expeditions were difficult and hazardous. They took place in an era before effective insect repellents or penicillin. Quinine was then known as an effective treatment for malaria, and it proved a necessary item of the expeditions' pharmacopeia. On two of the trips, his assistants, one being his son Mel, contracted malaria and were treated with the drug. Even with all of these hazards and hardships, during the three Bolivian expeditions Carriker discovered some 50 species and subspecies of birds new to science, from about 8,700 specimens that he and his assistants collected and prepared.

M. A. Carriker was a hero of Ted Parker's, as the man who had gone many places Ted later went in Peru, and collected many birds Ted later sought with difficulty. But Carriker had done it 50 years earlier, without using mist nets and without the advantages of lightweight modern fabrics and camping equipment. Because he did not use mist nets, Carriker did miss a few species, including one named for him much later, *Grallaria carrikeri*, the Pale-billed Antpitta (Schulenberg and Williams 1982). Although first discovered elsewhere, the antpitta was found in 1979 to occur along the trail between Tayabamba and Ongón, a trail on which Carriker must have passed twice in the 1930s.

Nevertheless, Carriker's efforts and successes were extraordinary. I remember Ted relating his delight in meeting in Yanac, Peru, the son of a man who had guided Carriker in 1932 to the site of his discovery of Tawny Tit-Spinetail (*Leptasthenura yanacensis*) and Ash-breasted Tit-Tyrant (*Anairetes alpinus*). Ted followed the same route above Yanac across a 5000 m pass, to the forest where these two rare and poorly known species were discovered by Carriker. Ted later told me how difficult the climb was, crossing a pass at such high elevation that his nose bled. His admiration for Carriker only grew.

Soon after his last Bolivian expedition Carriker wrote the manuscript "Experiences of an Ornithologist along the Highways and Byways of Bolivia," from which the excerpts following this biography were taken. He sent the manuscript to several publishers, but all turned it down. He became discouraged and put the manuscript aside. After his death his sons Mel and Frederick privately circulated a few photocopies of the manuscript, but none of it has appeared in print for general distribution.

Although Carriker may have been a consummate bird collector, he was less of an adventure writer. The writing of his Bolivia story is flat, reading as a list of daily activities and travel, with little to say of his assistants or description of the wonderful countryside through which he must have traveled. For example, on the day he must have collected the first specimen of a spectacular new species of curassow (Horned Curassow, *Pauxi unicornis*), he wrote only that he had obtained a species of "turkey" that he thought might be new to science—but said nothing about his experience of finding and stalking the bird, nothing of his feelings when he had realized it must be something spectacularly new.

His field assistant on the 1936–1937 and 1937–1938 expeditions is never provided with more than a first name, Berto, and all we know of his background is that he was Peruvian and had worked with Carriker in Peru. In the manuscript, his American field assistant on the 1937–1938 trip is also present in first name only. ("Mike" in the manuscript was Mike Howes, the son of friends in Toms River, New Jersey.) In fact, the assistants are rarely even mentioned, although Carriker does occasionally praise Berto's hunting skills and disparage Mike. Mike must have turned out to be a less than satisfactory assistant, for Carriker does mention that Mike scarcely learned the rudiments of Spanish, that he was often ill (sometimes with malaria), and that he lacked Carriker's and Berto's stamina to spend the day out hunting and hiking in the mountains. We will likely never know who Berto was, for Carriker did not put his assistants' names on his collecting tags.

The excerpts from the Bolivia manuscript are a small part of its 340 pages. They represent portions of all three Bolivian expeditions and some of his best descriptions of expedition life and collecting. Within the excerpts of Carriker's writing, my comments are closed in brackets. The excerpts have been lightly edited, mainly for grammar and typographical errors. I have changed the order of some of the excerpted sections, to make them more coherent without the intervening sections in the original manuscript. I have also added subheadings where Carriker had chapters. Carriker prepared a map of his travels in Bolivia as the frontispiece for his manuscript (Fig. 2).

EXCERPTS FROM "EXPERIENCES OF AN ORNITHOLOGIST ALONG THE HIGHWAYS AND BYWAYS OF BOLIVIA" BY M. A. CARRIKER, JR.

Along the Coroico and Beni Rivers.—[In 1934–1935, Carriker and his 19-year-old son Mel collected mostly in Depto. La Paz, on the eastern slope of the Andes. They arrived on the upper Río Coroico at Santa Ana, in the humid foothills at 794 m elevation, about 10 July 1934.]

There were many species of exquisitely colored birds at Santa Ana, the most beautiful being the small tanagers. One of these bore the highly descriptive name of "*Siete Colores*" [possibly *Tangara chilensis*] and actually did possess seven distinct colors. But with all their vivid hues, very few were gifted with the powers of song, their repertoires being limited to faint chirps, a few whistling call notes, harsh cries or guttural croaks. The gift of song seems to have been developed inversely to high specialization in color as though Mother Nature had endowed her

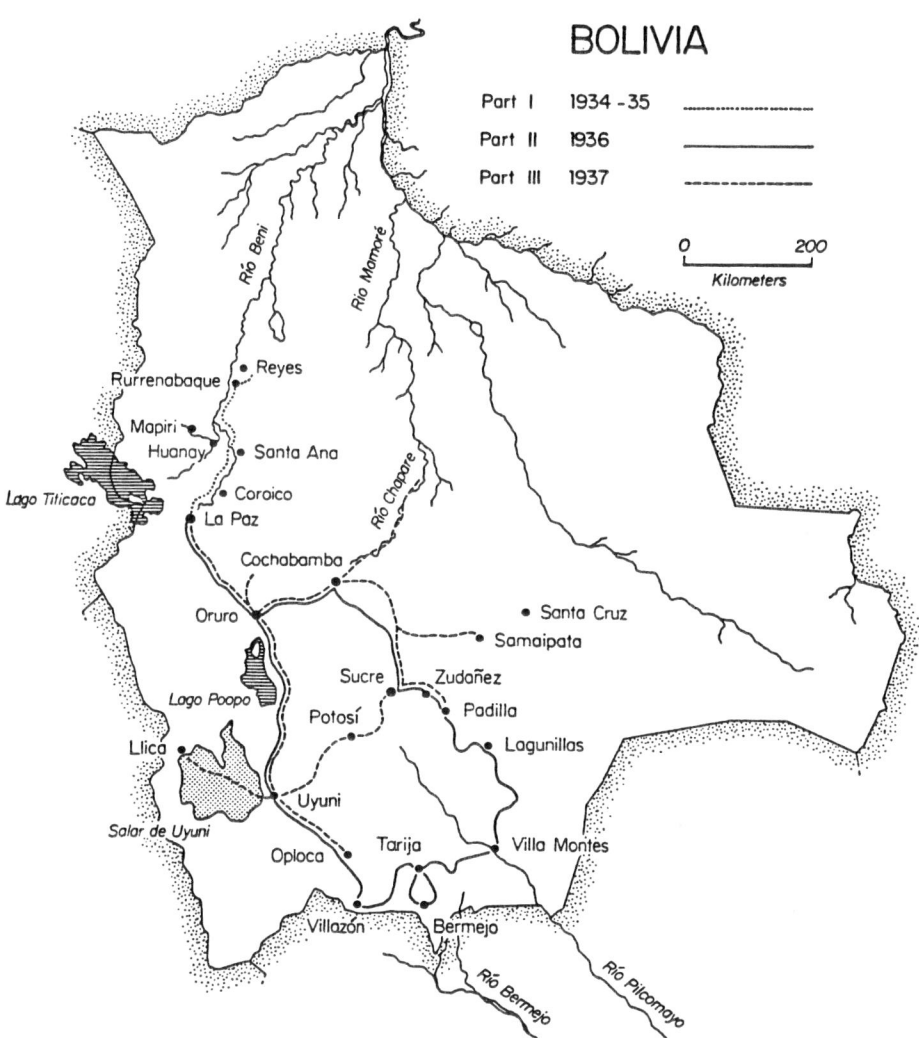

Fig. 2. Map tracing the three expeditions made by the author in Bolivia in the period 1934–1937 [actually ending in 1938].

homeliest children with that enchanting gift as a recompense for their ugliness. The gorgeously attired little tanagers could do no more than utter faint chirps, while the call-note of the Cock-of-the-Rock [*Rupicola peruviana*] was extremely harsh and unmusical. On the other hand some of the wrens and thrushes, inconspicuous in their somber dress of browns and grays, poured forth a torrent of clear, liquid notes or little haunting melodies that held the listener spellbound.

Bird life in the tropical jungles presents many curious aspects, one of which is the gregarious nature of most of the smaller species. A few of them are true hermits shunning their own kind and living apart in gloomy solitude, but by far the greater number are just the opposite. They gather together in wandering bands of varying size, and the heterogeneous lot of species composing the bands seem to have tacitly divided the forest between them in vertical planes as they move through it in search of food. The terrestrial antbirds, forest finches, and some of the ovenbirds [Furnariidae] search the ground and shrubbery for insect life and small berries; other ovenbirds, arboreal antbirds, tree creepers [Dendrocolaptidae], woodpeckers, flycatchers, and many tanagers work through the foliage of the lower limbs of the larger trees and the tops of the smaller ones, up and down their trunks and branches, while other species of tanagers, warblers [probably honeycreepers], cotingas and flycatchers pursue their quest through the tops of the forest.

One may wander through the jungle for an hour or more and scarcely see or hear a single bird when suddenly, out of thin air, will appear such a bewildering galaxy of feathered creatures that one hardly knows which to shoot at first. When one of these wandering bands is encountered, it is necessary to work fast, picking out the most desirable species first before they can escape. Even during the nesting season, when the females are incubating, the unattached males will flock together in large "stag" parties in their search for food.

Apparently it is a matter of mere luck to encounter these flocks, because they will be met with in the most unexpected places. A certain procedure is generally followed by them. In the early morning they usually hunt on the sunny side of the ridges, even in the low foothills, but at midday they may be anywhere. I have often had the experience of walking from daylight until near noon without finding a half dozen desirable birds and then, discouraged and almost on the point of returning to camp, have suddenly found myself in the midst of a huge band. By following them for a half hour, I secured more specimens than could be cared for during the afternoon.

There were very few mosquitoes at Santa Ana, but sometimes the tiny blackflies were very troublesome especially as we sat skinning the birds in the afternoon. Fortunately, these little imps of Satan were never found in the jungle. They occurred only in the open, along the margins of streams, and around houses and in the fields. In some places they made life insupportable. Sometimes while working, we rubbed our hands and faces with "fly dope," usually quite effective although it had to be renewed from time to time. Just before sunset the blackflies always became increasingly vicious, as though frantic over the idea of going to bed supperless; then, with the last rays of the setting sun, they likewise disappeared.

It required more than a month to secure a good representation of the amazing number of beautiful and rare birds present at Santa Ana. A total of 141 species was taken, quite a number of which later proved to be new to science [including two new species, the Bolivian Recurvebill *Simoxenops striatus*, and Ashy Antwren *Myrmotherula grisea*].

When starting out on one of these long collecting trips I rarely, if ever, carried much in the way of food supplies, preferring to "live off the country" whenever possible, going on the theory that wherever any sort of human inhabitants lived, I, too, could always subsist for a short time at least on whatever was used by them for food. This custom enabled me to travel with less baggage, but sometimes resulted in rather severe hardship and a monotonous diet. On the whole the food provided in those primitive regions is wholesome and nourishing, although not of a great variety, and frequently suffers from the crude culinary efforts of the native women.

I always tried to keep on hand a supply of flour, baking powder, lard, coffee, tea and sugar, and also bacon when possible, from which we prepared our morning meal and enjoyed a cup of tea in the late afternoon.

In all of these out-of-the-way places the greatest problem was to secure meat. Although fresh beef was an extremely rare luxury, pork was often obtainable. However, in our work we shot many birds and often some mammals whose flesh could be utilized. The meat thus secured was all we had for long intervals, and the bodies of such birds and beasts were always carefully saved after the skin had been removed and preserved. Nevertheless there were many meatless days.

[In early August 1934, Carriker and Mel moved camp towards the lowlands. From Santa Ana they traveled by balsa rafts down the Coroico and Beni rivers towards Rurrenabaque.]

These *balsas* are curious crafts and deserve a description. They are constructed of the peeled dried trunks of the *balsa* tree, which is so light that two men may easily carry on their shoulders a dry log 30 ft. long and a foot in diameter at the butt. For the construction of the raft, or "*balsa*" as they are familiarly termed, seven logs are used, the size of the timbers depending on the capacity desired for the raft or the nature of the stream in which it is to be used. The usual size used on the Río Beni are logs having a diameter at the butt of ten inches to a foot, whereas those used on the Coroico and Mapiri [rivers] are more slender. The middle log is selected with a sharply curved front end, and the ones on each side of it are less upturned at the tip. The two outer ones are straight, and the lower sides are beveled off at the end. The logs are fastened together by at least four pieces of black palm wood driven through them from one side to the other through holes previously cut with a chisel. A light platform, covered with wild cane stalks, is constructed amidships. It is about 10 ft. long, nearly the width of the craft, and about 15 in. in height. On this platform all of the cargo is carefully loaded, covered with a tarpaulin and lashed down tightly with ropes. When loaded to capacity, the logs of a *balsa* will be just about two inches above the surface of the water, and the capacity of a Beni *balsa* is around 1,000 lbs. [454 kg] including the crew of three men.

Fig. 3. The start of the journey down the Río Beni from Santa Ana on *balsas*. The *balseros* are at their stations. Mel is sitting on the luggage at the right. The little girl is saying goodbye.

It was Mel's first experience a-voyaging on a raft (Fig. 3), while I had never before navigated one on such turbulent water. The whole trip was packed full of thrills, especially the first day. It was a fearsome sight when the craft entered one of the long rapids, dancing up and down like an animated cork, great waves constantly rushing over the raft from bow to stern, boiling under the platform on which the baggage was stowed and on which we sat. The men were tense and excited, their long paddles poised, leaping from side to side to ward off some half-submerged boulder, yelling, cursing joyously at some hair-raising escape from destruction.

Once, on the [Río] Coroico, we narrowly escaped disaster. The river had widened, and then ran steeply over a shallow riffle at right angles to its course and struck against a rocky wall where a huge whirlpool formed. By a miracle the *balsa* escaped crashing against the rocks but was caught in the whirlpool, which sucked down the stern until the after half of our cargo was completely submerged. In that one frenzied moment I envisioned the loss of all of our precious cargo and the end of the year's work, because most of our materials could not have been replaced in Bolivia. Then with the men paddling furiously, the one at the stern in water to his waist, we gradually pulled out of the deadly whirlpool and rose to the surface. It was a narrow squeak all right, and even the boatmen admitted it.

[About the 22nd of August, 1934, Carriker and Mel arrived at their collecting site downriver, Chiñiri, Depto. La Paz, elevation *ca* 1,000 m.]

Although birds were fairly abundant around Chiñiri, there were fewer species and individuals than at Santa Ana, so that we were forced to work harder and cover more ground in our search for them. The most interesting birds taken there were the tinamous. There were four species present, and two of these we had not taken before. We have no birds in North America (north of Mexico) which even slightly approach the tinamou in appearance. They resemble a miniature ostrich except that the neck and legs are proportionately shorter. The bill is rather long, curves slightly downward, is slender toward the tip and is comparatively soft. The head is small and the neck slender, and both are rather sparsely covered with minute rounded feathers, except on the top of the head where they are long and thick. The breast is massive, the wings are short but very powerful, and the tail is diminutive, the upper and lower coverts being as long or longer than the somewhat rudimentary rectrices. The body plumage is long and dense but rather soft, and the feathers have thickened ribs and a soft texture somewhat like that of pigeons. Tinamous range in size from smaller than a quail to that of a large domestic fowl.

They are solitary in habit and never more than a pair or family group are ever seen together. They feed exclusively on the ground, but apparently the forest species roost in low trees. Their

Fig. 4. The author and Mel in camp at Chiñiri.

food consists largely, if not entirely, of vegetable matter, small nuts, seeds and tuberous roots, wild figs, and other fruit. In all our museums no large family of birds is so poorly represented by adequate series of specimens as the tinamous. The range of many species is purely conjectural.

They all utter a rather loud, characteristic call note, differing decidedly in the different genera. The call is heard frequently during the breeding season but rarely at other times of the year. When the call note is skillfully imitated, the birds will often reply and will sometimes slowly and carefully approach the person who is calling provided he is well concealed and motionless. Many times I have attempted this feat but rarely with success. I have also spent many hours stalking these birds, but the thick jungle that they invariably select for their haunts makes this an impossible task.

The best time to hunt tinamous is early in the morning, especially after a rain, when they will often be encountered in some footpath or trail through the jungle. When approached under such conditions, the bird will do one of two things, either slip like a shadow into the adjacent undergrowth or squat behind some slight bit of cover. Rarely will they run ahead along the trail. When this happens, it is just one of those heaven-sent gifts from the Red Gods to the hunter who has not failed to offer libations at their shrine (Fig. 4).

If they decide to hide, they will remain motionless until the hunter has passed, and then go quietly about their business. If approached too closely, a feathered bomb will suddenly explode at your feet and rocket upward through the trees until an opening is reached, then glide down again and disappear. The initial speed developed as the bird leaves the ground is incredible, while the line of flight is almost vertical. When the bird flushes in this manner, there is just one possible shot: when the bird reaches the zenith of its upward rush and is almost stationary for the fraction of a second before the downward plunge. However, when that instant arrives, the bird is usually safely sheltered behind some friendly tree or limb. The sudden explosive flight is so disconcerting and usually happens at such unexpected times and places, that a successful wing shot becomes one of those soul-satisfying events that makes life endurable for the hard-working and constantly harassed collector.

The jungle-inhabiting tinamous are not only masters of the art of concealment, noiseless and invisible locomotion, and quick escape, but are greatly aided by Nature in their grim battle for existence by the character of their habitat and their own highly protective coloration. The fiery rays of the tropical sun rarely penetrate the dense leafy canopy overhead, and even at midday dusk lurks in the cloistered aisles below so that the tinamou, as it glides with noiseless tread through the tangled undergrowth, becomes only another flickering shadow, elusive and ghostly. Any attempt to stalk the bird through such a maze is fore-doomed to failure.

Slender lianas, tough as wire, entangle the surest foot, wickedly hooked briers clutch at clothing and flesh alike while the unwary hand, thrust suddenly against a tree for support, will more likely meet with a mat of slender thorns, sharp as needles and brittle as glass. Hidden crevices between jagged, moss-covered rocks clutch the groping feet and harmless appearing leaves pushed gently aside may prove to be a poisonous nettle that will leave the skin swollen and racked with burning, itching pain.

Huge black ants, little red ants, stinging and biting ants, ants of every conceivable size and color are legion. Mosquitoes, little biting flies, hard-hitting and vicious horse-flies, deer flies, wood ticks, scorpions and tarantulas all add their quota to the troubles of the stalker.

But underneath all these petty annoyances there lurks the one and only real danger for the hunter in the South American jungles, the silent deadly strike of the fer-de-lance [a snake in the genus *Bothrops*, probably *B. atrox*] whose bite, next only to that of the Hooded Cobra of India, is the deadliest of all serpents. He gives no warning, has no pity, and horrible beyond words is the death that he carries in his bloated poison sacs. Six to seven ft. [about 2 m] in length, the girth of a man's forearm, and fangs three-fourths of an inch [1.9 cm] long are not uncommon for this snake. Fortunately for us, these reptiles are not abundant in Bolivia, their range of greatest abundance being in northern South America and around the Caribbean Sea.

Although there were four species of tinamous at Chiñiri, we secured only five specimens in two weeks of hunting. Of the four forms taken there, two proved to be undescribed [sub]species [*Crypturellus soui inconspicuus* and *Crypturellus cinereus cinerascens*. The latter proved to be synonymous with *Crypturellus c. cinereus*.], and another was known from a single specimen taken in Peru. [The other two species were *Crypturellus noctivagus garleppi* and *Tinamus major peruvianus*.] Capturing these birds seemed to be entirely a matter of luck. After hunting for days without success, using every conceivable precaution, we would encounter and secure a bird when least expected. One of our rarest of the Chiñiri tinamous was shot by Mel while following a narrow trail we had cut through the jungle. He saw the leaves of the low undergrowth moving slightly as though some animal were moving rapidly through it. Without waiting to see what it might be, he fired at the moving leaves and secured one of the rarest birds of the year!

Another species we nicknamed the "ghost." It proved to be entirely sooty gray in color and about the size of a very large pigeon. A pair was heard calling almost every day in different parts of the surrounding jungle, and we had both spent literally hours in stalking them without even a glimpse of the birds. One afternoon a "ghost" was heard quite close to the clearing, and with high hopes I once again took up the chase. It always remained just ahead of me but always out of sight, circling around, calling at intervals and seemingly enjoying our little game of tag. Finally I returned to camp, called Mel, and we planned a campaign. I would stalk the bird, while he would circle ahead and try for a shot when it approached him. It all sounded fine and we were already gloating over our "ghostie." For a full half hour I kept up the chase, moving about as silently as I knew how, straining every nerve, and streaming with perspiration. The bird continued calling and circling and I after it with Mel somewhere ahead. I glimpsed it just once, but before I could put gun to shoulder it had disappeared. In desperation I resolved to close in rapidly and flush the bird and risk a wing shot. When it next called I rushed up to the spot with gun ready. Nothing happened. Carefully I quartered over the surrounding jungle but without result. It had vanished into thin air, and more than ever I began to think that our little "ghost bird," applied in jest, might have some foundation in fact. As I stood ruminating over this idea, the bird suddenly burst out of a scant bit of cover not three ft. [1 m] away. I threw up the gun, but just as I was going to pull the trigger, it swerved behind a small tree and disappeared.

The last day I spent in the field at Chiñiri, I was returning to camp along a narrow trail that we often used, when a sudden rustling in the undergrowth at one side froze me in my tracks. I glimpsed a slender neck and head not 15 ft. [4.6 m] away and shot with the auxiliary barrel. Walking over, I picked up—the "ghost." [This may have been the specimen which he used to describe *Crypturellus soui inconspicuus*.]

We took 110 species of birds at Chiñiri and out of that number were 59 that we had not previously taken. For only 14 days spent in the field, I considered this a very satisfactory performance (Fig. 5).

In the Andes east of La Paz.—[On 19 December 1934, Carriker and Mel moved their operations to Hichuloma, Depto. La Paz, elevation 3,150 m, a locality in the Andes just across the first pass east of La Paz. They remained there until mid-January 1935.]

Our room at Hichuloma resembled a tomb more than a human habitation. The walls and roof

FIG. 5. The author and Mel skinning and mounting birds after a good morning's hunt at Chiñiri.

were of corrugated iron and the floor was of concrete. When the sun shone warmly upon the roof, the heat was rapidly absorbed into the room through the thin metal shell, but unfortunately the cold entered equally fast. Even with our little "Primus" [brand kerosene] stove going full blast, the cold air entered faster than we could heat the room. In any event we slept warmly and in comfort, and the meals in the little restaurant were much superior to those we had been getting in Sandillani. However, we were still forced to fall back on our own resources for breakfast, because we were invariably off shooting before the hotel cook even thought of getting up. But now breakfast was better than formerly because we had bacon and a bottle of milk, in addition to our old standby, with oranges and tangerines besides.

From the view off the edge of the great stone retaining wall in front of the little hotel, it seemed as though half of Bolivia was spread out before us. The steep-walled Unduavi Valley dropped rapidly down at the right and swung in a sharp curve to the left more than a 1,000 ft. [305 m] below, where it was joined by another gash in the huge range which descended from everlasting snow. To the left towered a series of stupendous sheer cliffs supporting a flat-topped plateau. From this poured several foaming cascades, all dissolving in mist long before they reached the valley below. To the southeast stretched the Unduavi Valley, still imprisoned between grim-walled mountains that steadily lessened in height as the river bored deeper and deeper into the earth and finally became lost in a shadowy purple haze.

There were many routes over which we could hunt: the railway and the auto highway up the valley, the road toward Coroico and the one down the valley, and the mule trail up over the divide to Sandillani. We were just below timber-line so that after a climb upward of 500 ft. [about 150 m] one could leave behind the dense thickets and low trees and emerge on the grass covered *puna*. [Puna is the name given in Peru and Bolivia to fairly arid grassland above tree-line.]

After four days of work with such success that it only whetted our appetites for more, we left for La Paz on the afternoon of December 24th. Two short days there were spent in a continual round of gayeties, and then we returned to the same old grind [at Hichuloma].

Weather permitting, life once more settled down to the routine of hunting in the morning and skinning in the afternoon. Long hunting trips were made in every direction. We often returned cold, wet, hungry and exhausted from these, but almost always with something worthwhile. I made several trips far up the valley toward La Cumbre, following the railway for several miles, then climbing the long slope above it to the foot of the great naked crags which formed the backbone of the range. The slope at the foot of the cliffs was littered with masses of broken rock that had been gradually breaking away from the crags above for past ages. Here grew clumps of the curious *quinúa* trees [*Polylepis* spp., Rosaceae], which are found only above the limit of ordinary arboreal vegetation. The pale chestnut-colored bark covers the trunk in paper-thin layers an inch [about 2.5 cm] in thickness or more, like shaggy tattered garments. The trunk and limbs are gnarled and twisted, and the tiny gray-green leaves are thick and harsh.

It was in this type of woodland that I had discovered two most interesting new species of birds in northern Peru [Carriker is probably referring to the Ash-breasted Tit-Tyrant, *Anairetes alpinus* and Tawny Tit-Spinetail, *Leptasthenura yanacensis*, which he discovered at Yanac, Peru, in March and April 1932.], and I now searched carefully amongst them for something of unusual interest. The large rocks were often so nicely balanced that the weight of a person springing lightly from one to another would send them crashing down the slope, and their lichen-covered surfaces rendered footing most insecure.

Vizcachas [*Lagidium* sp., Chinchillidae] scurried about leaping nimbly from rock to rock, or loped with astonishing ease up almost perpendicular walls of the cliffs and disappeared in the crevices. The few birds seen were very shy, flitting from clump to clump of the scattered *quinúa* trees, making their rapid pursuit over the loose rock impossible. Finally I saw a pair of birds which seemed vaguely familiar, and after a long chase shot them. One was lost, having fallen into a deep crevice between the great boulders, but the other proved to be one of the rare new species [either the *Anairetes* or *Leptasthenura*] of northern Peru. I subsequently made other trips to that place, but never saw another either there or anywhere else in Bolivia.

Export permits.—[Carriker and Mel returned to La Paz in mid January 1935 to prepare their return to the United States.]

At the time of my last visit to La Paz, I had learned that we would have to return home on the *Santa Barbara*, sailing from Mollendo [Peru] on February 3rd [1935]. This would give me a week to arrange for our departure and a week to do a bit of collecting down the valley below La Paz. I decided to obtain at once the necessary government permits for the shipment of the collection and then spend the last week collecting. There seemed to be a million things to do, and it was difficult to know which should be done first.

At all events the first link in the seemingly endless chain of red-tape was the presentation of a petition to the Minister of Public Instruction, asking for permission to export the collection. The petition must clearly state the number of specimens in the consignment, exactly when and where they were taken and for what institution, and for what specific purpose. This petition, if favorably received, was turned over to the *Directora del Museo Nacional de Tiahuanáco* [Director of the Tiahuanáco National Museum] in La Paz, who was *supposed* to inspect the collection and give her written approval for its removal from the country.

El Ministro de Instrucción y Asuntos Indígenes [Minister of Instruction and Indian Affairs], Col. Peñaranda, received me very courteously. After listening attentively to my lengthy discourse, he advised me that as far as he was concerned, there would be no obstacle placed in the way of the exportation of my collection, but that it remained for the *Señora* Urióste, *Directora* of the museum, to give it her approval.

The task of convincing that estimable lady that it was wholly unnecessary to make a minute inspection of the collection was a delicate one and required not a little tact and diplomacy before it was finally accomplished. Armed with my original petition, its written approval by Col. Peñaranda and the additional approbation by the *Señora* Urióste, I called on *El Ministro de Hacienda* (Finance and Industries). After two days delay he added another signed and stamped sheet to my growing sheaf of official-looking documents, which waived all export duties and granted permission for passing the collection *gratis* through the customs. Then followed a visit to the *Ministro de Defensa Nacional*, who likewise attached an imposing and sweeping ratification of all and sundry resolutions by the other ministers.

With this check and double check on the activities of the various Cabinet members, I loaded the numerous boxes and chests on a truck and went up the long hill to the custom house.

Armed with this imposing array of official signatures, seals and stamps, I entered the office of the *Administrador de la Aduanas*, the gentleman who has complete control over all of the customs of the country. That gentleman was absent but his secretary, after a careful perusal of my papers and a couple of telephone calls, decided that he could find no flaw in the mystic maze and graciously affixed signature and official stamp at the foot of the last page.

I then wended my weary way to the sanctum of the *Administrador de la Aduana de La Paz*. Now would come the acid test. By this time my nerves had reached the jittery stage. As I sat waiting the pleasure of this demigod, my thoughts turned longingly to the club bar and a tall cold glass, while I feverishly planned how I was ever to persuade this grim-looking gorgon to allow my mountain of baggage to pass the customs without the inspection they were supposed

FIG. 6. The author's camp at La Merced [en route to Monte Bello].

to make. The furtive glances which he bestowed upon me from time to time as he sat writing at his desk were not reassuring, and I only hoped that I was wearing my best poker face.

Suddenly I was startled from my pessimistic dreaming by hearing, "*Bueno, Señor, en que puedo servirlo?*" ["Well, sir, how can I help you?], and I snapped into action. I presented my sheaf of documents, and after he had run rapidly through them, I put on the high pressure. I described some of the many wonderful places we had visited, the gorgeous birds, the majesty of the vast mountain ranges, the kindness of the people, and the great courtesy of the government officials and their efforts to assist us in spite of the fact that they were so intensely occupied with the terrible war against the *bárbaros* [barbarians] *de Paraguay*. I ended by telling him that I should be most delighted to show him some of the choicest of my specimens, so that he might see for himself what beautiful birds were to be found in his great and glorious land.

By that time I noticed that the surrounding atmosphere seemed to have changed from frigid hostility to balmy interest. I wound up with a final burst of eloquence as this now friendly and courteous gentleman rose to accompany me downstairs, where I opened a chest at random and showed him some of the birds. As each was removed from its wrapping of tissue paper, I explained why they had been packed so carefully and how long it would take to open them all for an inspection and that such a task would greatly interfere with the routine of the *aduana* [Customs]. To all of this he readily agreed, and to my infinite relief he called a subordinate and instructed him to place stickers on all of the packages at once without further inspection, and that I had full authority of the President and all of his Ministers to pass them through the *aduana, libre de impuestos* [tax-free].

I thanked him profusely, from the bottom of my heart, as he warmly shook my hand in parting and wished me a pleasant voyage and a safe return for the continuance of my explorations.

After the nerve-wracking experience of the past three hours, I was a physical and mental wreck. Hailing a passing taxi, I hurried to the Strangers Club, where Mel anxiously awaited my arrival. As I entered the lounge I suppose that my face told the good news, for Mel jumped up and yelled, "Hot-diggity-dog, let's celebrate," and celebrate we did.

Tough trail from Monte Bello to Bermejo.—[In May 1936, Carriker returned to Bolivia. In place of his son Mel as assistant, this time Berto came with him, a Peruvian with whom Carriker had worked during his expeditions to that country a few years before. Around the beginning of September 1936, they were at Monte Bello, in Depto. Tarija (Fig. 6) in the arid southern part of the country, at 600 m elevation. Monte Bello was the sugar cane plantation of expatriate American Collie Manatt, who had loaned Carriker a horse named Alec, and with whom he was seeking to rent more pack animals for a trip to Bermejo.]

On Sunday evening Collie and I rode up the Lipeito River and brought down two horses, and I paid in advance for another that was to be caught and delivered to us the next day. Carlos [guide for the trip] arrived that evening riding the horse I had rented for him. We were still short the one that I had paid for the previous day and another belonging to a chap named Pedro living across the river form Monte Bello.

Monday I began packing early while waiting for the two horses to be brought, but noon arrived and no sign of them. While we were eating, a boy arrived with the sad news that the horse from up the river could not be found and returned to me the money I had paid for it. After lunch Collie and I went to *Don* Pedro's house to get his horse and at the same time try for another nearby. We found that it was a *día de fiesta* [holiday] and Pedro had gone down the river spearing fish, while his horse still roamed in the woods. Collie went to look for Pedro and found him dead drunk. There was another small still nearby whose vile concoction was sold largely to the local inhabitants, and this being a day of fiesta, there was certain to be a crowd there. We rode over to see if by any chance we might pick up a horse.

There was a crowd of some twenty persons, men and women, most of them considerably worse for the wear. I saw a good-looking mule tied to a tree and after inquiring to whom it belonged, I asked the owner to rent it to me. He was half "overseas," and much to my surprise he readily assented. Before he could change his mind, I paid over the money and led off the mule.

Tuesday morning I rode over to see Pedro and found him sober and very apologetic, and he returned with me to hunt for his horse, which was loose in the forest on our side of the river. About two o'clock he again appeared, drunk as a tick, but leading a little buckskin horse. I suppose he secured the liquor up at the Monte Bello still, since he came from that direction. We finished shoeing all the animals that afternoon and were ready to begin another uncertain adventure the next morning.

We decided to load the three packs on two large strong-looking horses and the mule, and we picked out a big black for the two duffel bags, which were the largest of the loads. Collie had loaned me three pack-saddles of a somewhat revised sawbuck style, and we had an abundance of rawhide ropes. No sooner had the two bags been hoisted to the back of the big black horse than he began to buck furiously, and men and bags flew in all directions. After another hard struggle, we replaced and partly lashed the bags, and the horse went into another spasm. The landscape was again strewn with wreckage, this time including the pack-saddle and pads.

I always suspected that Berto inherited his great beak of a nose from some swashbuckling conquistador, and if he did, the old leaven was working strong that day. He picked up a club, seized the halter rope of the horse and proceeded to give him an exceedingly thorough beating. I would have gladly assisted in the operation, but no help was needed.

After the battle with the black, everything went smoothly with the loading, and we were soon ready to start with Carlos, the guide and general factotum, taking the lead mule in tow. I followed leading the roan, and Berto brought up the rear mounted on Pedro's little buckskin and leading the big black. I think that he was still hoping that the horse would start something to give him an excuse for beating him up again.

We crossed the [Río] Lipeo and followed it to its junction with the Candado, the two forming the Río Bermejo, then crossed the latter and headed down the left bank. We were to pick up another man, named Juan Tapia, a few miles below at Campo de Vaca near the home of Carlos. Arriving at Tapia's place, we found that he was not ready, having expected us on Tuesday. He tried to persuade us to spend the night there, which I refused, deciding to continue the journey and let Juan overtake us by evening, which he could easily do. They were making *chancáca* [crude brown sugar], very clean-looking and clear golden yellow in color, so I bought an *arroba* (25 lbs.) [11.4 kg] to take with us to sweeten our coffee and munch on the road.

We planned to arrive at the first bad spot on the trail that afternoon, where the hills closed in on both sides of the river making a narrow gorge and a tremendously deep pool. We hoped to pass that obstacle the same evening and make camp a short distance below. Juan sent one of his boys with us to assist in passing the *pozo* [deep pool], and we set out leaving him to follow.

Juan overtook us about three o'clock before we reached the big pool, and upon arriving there we rapidly began our task of passing it. The animals were all completely stripped, excepting their halters, driven into the river and forced to swim down through the deep pool between the cliffs to a small beach on the opposite shore. Next came the real work. Everything had to be packed up a steep path to the top of a rocky ridge and across a narrow ledge, which ran diagonally down the face of the cliff to a boulder-strewn playa below. At this point Juan suddenly, and most opportunely, developed "*un ataque de corazón*" [a "heart attack"], and began to wheeze and heave his chest up and down like a bellows. I think he did carry over a few pack pads and saddle blankets, but it fell to Berto, Carlos and me to do the lion's share. I have seldom seen so much strength and dynamic energy as had been packed in the small body of Berto. Carlos was a hulk of a man, well over six ft. [1.8 m], but slow and clumsy. He claimed to be an Australian, but spoke with a strong teutonic accent, and I afterwards learned that he had been a tropical tramp and beachcomber previous to appearing on the Bermejo. There he lived alone, cultivated a little patch of sugarcane, cotton, corn, and *yuca* [*Manihot esculenta*, a starchy root also called cassava], and eked out a miserable existence.

Juan's boy carried over some of the lighter gear, while I took the saddles, saddle-bags, etc., leaving the six heavy packages for Berto and Carlos. The first big duffel bag that Carlos attempted to bring across the cliff bumped into the rock, threw him off balance, and to save himself, he dropped the bag, which rolled down the almost perpendicular slope and was caught on a point of rock a foot [30 cm] under water. It was impossible to descend the slope, and equally impossible to ascend it, so I sent Juan's boy to swim over to the bag and slip a rope around it while we held the other end above. In this manner we snaked the bag up the cliff and hauled it to safety. After that episode Berto carried over the remainder of the cargo alone, while Carlos and Juan moved the stuff further downstream past the big boulders, where it could be loaded, while the boy swam across the river and brought back the animals.

We had scarcely gotten well started the next morning when the trail entered a maze of rocky ledges along the edge of the river. It was not broken rock or loose boulders, but *solid* rock that had been cut and gouged and scoured by countless floods, leaving sharp ridges, points and gullies. We picked our way carefully for several hundred yards without accident until a deep gully cut across the path at right angles, easy enough to enter but almost impossible to escape from on the other side.

From the bottom of the gully there was a sheer rise of perhaps four ft. [1.2 m] into another V-shaped cleft in the rock with smooth slippery sides that left no footing at all along the bottom. Carlos was ahead, and after dismounting, managed to get his horse through. Then he and Berto started the big black, one ahead with the lead rope, the other behind. He attempted to make the leap from the gully to the cleft, reached it, fell, rose and fell again, rolling completely over. Wedged in the narrow cleft the unfortunate brute could not be hoisted to his feet without removing the pack, and then it required the combined efforts of three men to accomplish it. The pack was damaged, the saddle smashed, and the rocks were spattered with blood. The other two pack animals were more fortunate, getting through it without a fall, one man ahead and two pushing from the rear. I came last with Alec, and as the men were all ahead repairing the damages, I foolishly attempted to drive him through alone.

As he landed in the cleft his feet flew from under him, he lost his balance completely and crashed back into the gully, landing on his rump and knocking me head over heels, but I managed to roll clear of his flailing hooves as he turned completely over. Poor Alec! He was badly shaken by the fall and deathly afraid to make another attempt, so that it required the combined efforts of the four of us to finally get him through.

We had barely reorganized after this catastrophe when we reached a very steep rocky ascent of perhaps 60 ft. [18 m]. Half way up this slope the black again went to pieces, reared, plunged, fell over backwards and rolled to the bottom of the slope. He was again unloaded and taken up the hill empty, while Carlos and Berto carried up his load.

Eventually, after a three-hour battle, we again reached the river bank. We were bathed in perspiration from the heat, which rose in shimmering waves from the boulder-strewn playas. The air, too, had become filled with a blue haze so that objects but a short distance away were indistinct. I had been too much preoccupied with other matters to pay much attention to this phenomenon, but now noticed that the blue haze was smoke. Undoubtedly the forest fires which had been raging in the south for the past two weeks, and of which we had received rumors, were now working their way north, up the Bermejo valley. The prospect looked pretty grim.

The next morning dawned murky and oppressive, the sun rising malignantly through the ever thickening pall of smoke. The day was one long nightmare of heat, smoke, vicious flies and fording rivers. The floor of the stream was covered everywhere with a jumble of boulders of all sizes, worn smooth by their ceaseless battle with rushing floods and grinding sands. The sediment and slime had left the boulders so slippery that it was a miracle the animals kept their footing. Alec was particularly adept at fording these rocky streams, never putting his weight on a foot until it was firmly set, often feeling about for several seconds before a firm footing could be found. The other horses slipped, stumbled and plunged through, often falling to their knees and wetting the packs, but luckily none ever fell down completely. We crossed the river 22 times on the third day and 41 times on the whole trip.

At about 2 pm we left the river and followed up a small creek for several miles, wading in the water most of the time, then struck off through the forest and up over a ridge perhaps 500 ft. in height. This detour was made to avoid a series of pools and cliffs, and after crossing the ridge we again descended to the shores of the river.

The faint trail through the forest was badly choked with fallen trees, limbs and masses of vines, so that frequent detours and much cutting were necessary to get through. There was no serious trouble until we had nearly reached the top of the range. A large fallen tree blocked the path at a point where it was impossible to pass except by making a nasty detour. We were forced to ascend a very steep slope for about 50 ft. [15 m], then make an abrupt turn around a large tree to a more level stretch. With the aid of two men all the animals had been safely passed except the black horse, which we always left for the last hoping that he would gain some meager speck of courage by watching the others pass first. Carlos went ahead with the lead rope while Berto was pushing behind. Half way to the top the horse, from sheer cowardice, refused to make further effort, reared and went over backwards, knocking Berto down and narrowly missing him when he [the horse] twice rolled completely over and came up in a heap at the bottom of the hill. Berto was not badly hurt but very severely shaken up, and the palm of his right hand was filled with thorns. Again we were forced to unload the pack and carry it over the hill. Once over the top, there followed a sharp descent into a deep ravine, after which the trail followed down a wide ridge. The path dipped into the ravine, crossed it, and then followed along its brink for a short distance before turning away from it. After the pack of the black had been carried over the ridge and since it was now getting late, I sent Carlos and Juan on ahead while Berto and I loaded the pack.

The ravine dropped abruptly below the path and was choked with high brush. When the black horse reached that point he stumbled, seemed to let go all holds and crashed down into the bottom of the ravine, the pack under him and his four legs in the air. It was impossible to extricate the horse until pack and saddle had been removed, and to accomplish that we needed more assistance. I left Berto to loosen the ropes as best he could while I ran down the hill and called Carlos and Juan. I think that we could have cheerfully shot that horse then and there had there been any other possible manner of transporting his pack.

It was dark night by the time we reached a camp site, and so dark from the pall of smoke and clouds that had rolled up that we were forced to use a flashlight for unloading the horses and making camp. Supper that night was a half-raw, miserable mess that we had little desire to eat after it had been dished out. Berto and I finished a half bottle of Monte Bello rum, the last we had, called it a day and turned in. That night there arrived in camp literally millions of tiny gnats, which were attracted by the light of our fire and which added further to our complete misery. But the end was not yet. About midnight hordes of *manta blanca* arrived on the scene, probably advised by the gnats that there were good pickings. Our only escape from their vicious attacks was to cover our heads completely with our blankets. I know of nothing worse than these tiny almost invisible insects whose bite is like the prick of a red-hot needle. They are the "no-see-ums" of the Canadian Ojibwas, the "*ala blanca*" of Central America and the "*jejeng*" of Peru. The only thing that will keep them out is a net made of fine muslin or cheese-cloth. Fortunately they usually make their appearance sporadically and do not as a rule last long, sometimes an hour or two just at dusk or before dawn.

Once I was travelling alone down the beach from the Río Sicsola [Sixaola] to Bocas del Toro on the Caribbean shores of northern Panama. Night overtook me on the beach, far from any house, so I lay down on the soft sand and prepared to spend the night there. I was quite comfortable until millions of these little demons appeared. I was without blanket or covering of any sort, and they soon crawled inside my clothing at every possible opening and bit me until I was nearly frantic. I finally buried myself completely in the sand, except for my face which I covered with a kerchief, and thus managed to pass the remainder of the night.

The sand flies drove us out at the first break of dawn, but by keeping in the smoke of the campfire we managed to eat breakfast in comparative comfort. The shores of the river were now almost free of boulders, so that we traveled much more rapidly and easily, our only annoyance being the thick smoke and black-flies.

At noon I suddenly became aware of a curious noise, a muffled "thump, thump, thump," and finally identified it as the exhaust of an internal combustion motor. Ten minutes later we rounded a sharp bend and I gazed at two tall oil derricks, from one of which issued the noise we had been hearing.

FIG. 7. The author's rented truck stopping temporarily at a Bolivian outpost near the Paraguayan boundary.

It was a most curious sensation to come suddenly out of that God-forsaken wilderness into this bit of modern civilization. For three days we had battled with nature in her most perverse moods, and had not seen a single human being or anything to indicate that one had recently passed that way, and then, as if dropped from the heavens, there were huge petroleum derricks and gas motors.

Travails of collecting in Tarija.—[From 1932–1935, Bolivia had been at war with Paraguay over the arid Chaco region, with Bolivia losing a large portion of its Chaco lands to the Paraguayans. Carriker and Berto spent several months collecting in the former war zone in 1936 (Fig. 7). In mid-October they were at Entre Rios, at 1,400 m elevation about 58 km east of the departmental capital, Tarija.]

Although very little rain had fallen there, the trees were putting out leaves, especially in the valleys, where many were in bloom. It was a delightful change to work again in woodland, odorous of flowers and new leaves. The birds, too, were singing, chirping or screaming harshly, according to their kind, all excited by the approach of the nesting season.

Only common birds were seen that day, and few specimens reposed in the old hunting bag. As a result I had wandered far up the valley searching through the fields and prowling in the scattered groves of *algarrobas* [*Prosopis* sp.] along the sides of the valley. One such grove was quite extensive, including many unusually large trees behind which a man could easily be concealed. Several birds had been shot in the grove, and I was kneeling on the ground inserting cotton into their throats, when suddenly a twig snapped behind me.

From force of habit my head turned slowly in the direction of the sound. The half-wrapped bird dropped to the ground as with a single rapid motion I seized the gun at my side, leaped to my feet, and threw up the weapon. In the same second the safety catch had been released. Immediately before pulling the trigger, I had yelled, "*Alto,*" putting every atom of force and authority that I could muster in the command. As though galvanized by an electric shock, the half-naked creature that leaped toward me stopped in mid-stride. Tangled strings of dirty hair framed a dark face, distorted with rage, from which gleamed the eyes of an untamed beast. A huge, sharp-pointed knife gleamed in his hand, held high in the air, ready to strike.

We stood thus for an appreciable moment, unquestionably a striking tableau. Then the fire seemed to dim in the glaring eyes, and his hand dropped to his side. "Why would you stab me?" I quietly asked, and in reply he mumbled something about killing all of the "*canalla de Paraguayos*" [Paraguayan dogs]. Then I realized he was not in reality a blood-thirsty assassin but only the pitiful wreck of what had once been a man, driven to insanity by some unknown shock or torture suffered in the recent war. I told him sternly to drop the knife and walk ahead of me, which he did quietly enough.

I escorted him to a house that I had seen not far away, and the people there confirmed my theory, and told me they would take care of him. I reported the matter to the Sub-prefect, who apologized profusely for the incident. He said that he had heard about the man, who was a Guaraní Indian laborer in the army during the war and had been sent back to the Chaco in his present condition. Just what had caused his insane hatred for Paraguayans I never learned, but he doubtless mistook me for one of his mortal enemies. It was a most unpleasant experience, one which might easily have ended in tragedy, either for me or for the unfortunate creature who had attempted to take my life.

[On 28 October 1936, Carriker and Berto moved their collecting camp to the bank of the Río Pilcomayo at Villa Montes, depto. Tarija, at 500 m elevation. They were still in the dry Chaco, and summer was coming on. At Villa Montes they lodged on a military base.]

It was impossible to go out that afternoon. The mercury stood at 110° Fahrenheit [43.3°C] in the shade under the tile roof of the veranda, and what vagrant breeze found its way into our room seemed to have issued from a blast furnace. There was no ice, and the lukewarm water available for drinking purposes was brought by ourselves from an iron pipe back of the hospital in our aluminum kettles. We stripped to our underwear and spent the afternoon sweltering on our cots. At 5 pm the wind swept up from the south in hot waves, filling the air with clouds of dust and sand. By seven o'clock it was blowing a gale, and sheets of corrugated roofing rattled and banged and were hurled through the air until they crashed against another building or fell to the earth. Just when it seemed that the whole place was going to be blown to pieces, the gale died away as suddenly as it began.

The following morning it was much cooler. We arose before dawn, prepared our own breakfast, and at 6 am were in the field. The vegetation had been entirely removed for a considerable distance on all sides of the town, there being a large airport on the west side. The nearest woodland was back toward the foothills beyond the airport. We made our way toward this. By nine o'clock I felt like a sponge from which the last drop of moisture had been sucked by the blazing sun, and was forced to make two trips to the river for water. Although the water was muddy and warm, it tasted like nectar. At 10 am I found it impossible to continue and returned to the house, meeting Berto as I entered. We consumed impossible quantities of tepid water, which oozed from our pores as fast as it entered our parched stomachs.

We had both found birds abundant and between the two of us shot 30, many of which were new to the collection. In such a high temperature birds decompose very rapidly, and no time was lost getting to work. The meager lunch was hastily disposed of and consisted more of water than of food. By 2 pm the thermometer hanging on the veranda on the east side of the house registered 115° Fahrenheit [46.1°C], but in spite of the heat we finished our specimens at 6:30 pm.

The second day was much cooler and slightly overcast, so that we did not suffer so much from the heat. We had started out with the first streaks of dawn and reached the heavy woodland of the foothills by sunrise. Again we were very successful and were able to return at 9:30 am with 32 specimens. The day continued cool and we again finished before dark.

On the upper reaches of the Río Mamoré.—[Carriker left Bolivia for the U.S. in early January of 1937. While he was away, he left Berto to collect on his own around Incachaca, depto. Cochabamba, 2,350 m elevation. In May 1937 Carriker again returned to Bolivia. This time he brought an assistant from the United States, Mike Howes. They rejoined Berto, and in mid-June the group of three was at Palmar, on the Río San Matéo, whose waters eventually flow into the Mamoré. The following section begins the day after their arrival at a road-builders' camp just below the town of Palmar.]

The next morning water oozed dismally from the ragged cloud bank that filled the valley. Below Palmar the hills again closed in, rising abruptly from the margins of the small river that roared and foamed over a jumbled mass of boulders, gouging out infrequent pools in the sharp bends. A small flat had been left at one side of the valley on which stood the buildings of the camp, and the highway, swinging sharply across in front of it, descended to the bridge that spanned the river.

It was impossible to think of the three of us remaining in the small, crowded camp, so after breakfast I cast about for a spot where we might pitch our tent. I found an old thatch-roofed camp down close to the river that had been built for the use of the men who erected the bridge. The sides and both gables were entirely open, and outside of some leaks along the ridge, the roof seemed to be in good condition. *Don* Orlando [Chief Engineer of the Road-building Commission] thought that the timbers that supported it were so badly decayed that it was dangerous to use it, but after thorough examination I decided that it was safe enough, at least for a few weeks. The building was about 22 by 45 ft. [6.7 by 13.7 m], so we pitched our tent under one side, about 12 ft. [3.7 m] back from the front. The tent was barely large enough to take our three cots and a couple of chests, so we arranged the remainder of the baggage in front on the other side of the building, and stretched the big tent-fly over it with the lower edge to the front. The fly reached entirely across the house, so that by tying up the lower corner on the side in front of our tent, we had fine shelter under which to erect our tables and prepare our specimens.

It required the greater part of the day to make all of these arrangements and make ready for the work to come. I arranged with *Don* Orlando to furnish us with lunch and dinner at the camp, while, as usual, we prepared our own breakfast at an early hour. The members of the Commission who lived at the camp were delighted at the idea of having company for several weeks, and everyone turned out to assist me.

The weather during our stay at Palmar was simply abominable, and we lost so much time from rain that it required four weeks to finish the work there. Under less adverse conditions, it would easily have been completed in three. Not only were we handicapped by the weather but by the nature of the jungle and almost complete absence of any sort of trails or paths penetrating it. We spent much time in cutting narrow hunting trails, selecting the less precipitous and broken section. Had it not been for these trails, shooting would have been almost impossible. The highway between the camp and Palmar ran along the base of almost perpendicular slopes and close alongside the river, so that very few birds were ever taken there. The large flat at Palmar had been almost entirely denuded of primeval forest, the rocky land being either planted with coca and a few patches of *plátanos* [bananas] and *yuca* or else allowed to revert to almost impenetrable second-growth scrub of varying heights. Then, too, it was a three mile [about 5 km] walk to reach those flats, and a hunting trip up there meant covering never less than 10 or 12 miles [16 km or 19 km].

Beyond the bridge the highway skirted the slopes of the high hills that diverged from the river and cut far across to another larger stream, the Ibirízu. It was fairly good hunting along this road, and one of us usually took that route. The new road which was being built continued on down the left bank of the San Mateo, there having been an old mule trail there that eventually reached San Antonio at the junction of the Río San Mateo and Espíritu Santo. This old trail had been abandoned for some years, and except for a narrow passageway of eight to 10 ft. [2.4 to 3 m], had grown up in a dense tangle of brush, and the road itself, having lately been reopened to pack trains, had been churned into a ribbon of deep mud.

The almost daily or nightly rains kept the jungle dripping with water, so that one always emerged from it with saturated clothing. Heavy downpours frequently caught us far from camp, and there was nothing to do but turn back and endure the discomfort of the deluge as best we might until camp might be reached. On the second morning *Don* Orlando returned to Cochabamba, leaving his brother, Oscar, the Doctor, Rios, and García, the truck driver. There was little or no excitement for some days, merely the monotonous grind of each succeeding day, roaming through the tangled, moisture-laden undergrowth of the jungle, pricked by thorns, stung by nettles, bitten by vicious ants and venomous mosquitoes, and often returning with little to show for the labor and discomfort endured. Then, after we changed to dry clothing, perhaps after a dip in the icy pool below the bridge, came a short respite when we all gathered around the table for the midday meal. After that until 7 pm we toiled at skinning and stuffing the birds and animals secured.

It was tough going for Mike, but Berto and I were too much accustomed to that sort of thing to mind and took it all in stride, although we did curse the weather often and fluently. But with Mike it was different. It was his first experience in the tropics, or anywhere else outside of southern New Jersey, and while willing and fairly enthusiastic, he seemed to make no headway at learning to collect birds but was getting along fairly well at skinning them. He was always falling over banks or into holes and scratching and bruising himself a great deal more than was necessary.

I had located a good hunting ground about two miles down the river, back in a valley that debouched on the old mule road, and had cut a little trail back to the low ridge that formed the watershed between it and some other creek. On the morning of July 4th, I decided to return to this valley and perhaps continue my exploration over into the next one.

After hunting with just slight success up to the low ridge where my trail ended, I crossed over and dropped down into another rivulet, which seemed to be running more or less parallel to the Río San Mateo, onto the old trail that followed its left bank. Thinking that it must, of course, eventually swing around and flow into the San Mateo, I continued following it. The valley opened out, and except for the wicked, curving thorns of the great clumps of bamboo scattered about, the going was fairly good. After an hour or so I began to feel uneasy about the direction that the little stream, now rapidly increasing in size, was following because it showed no signs of turning across toward the San Mateo.

I should have turned back then and retraced my steps over the same route by which I had entered, but I am a stubborn cuss sometimes, and also was becoming curious to see where the creek would lead me. After another hour of fairly fast walking (I was getting uneasy by then), slashing the undergrowth and vines that obstructed my way with my machete, I suddenly came out on the banks of another larger stream. This stream wound back and forth across a fairly level valley a mile [1.6 km] or more in width, flowing over a sandy or pebbly bottom and was nowhere more than knee deep. The general direction of its course seemed to be toward the San Mateo, but at an angle that would reach it many miles below our camp. There was now no question of turning back. The day was too far spent and I had traveled much too far away to do anything but continue on with the hope of eventually reaching the San Mateo, where I knew that a right-of-way had been cleared for the new road. Once on that I would have little difficulty in following it up to the other construction camp about four miles [6.4 km] below our own.

My clothing was saturated and I was tired and very hungry, but I dared not stop to rest for I had no idea when I should get out to the new road, or when there, how far it might be back to camp. I had matches, but all dead wood that could be cut with my light machete was so saturated with water that it would have been almost impossible to kindle a fire, and to spend a night in the jungle, without a fire, in wet clothing and a temperature of about 52° Fahrenheit [11°C] was madness and not to be considered, except as a last resort. So I doggedly continued, walking as rapidly as possible, tearing my way ruthlessly through brush and vines, only cutting when it was impossible to proceed without it. My flight through the jungle was eased at times by walking some distances in the shallow stream.

Ultimately, about 2 pm, I crossed a wide flat and suddenly came to the edge of a plateau, heard the roar of a river, and soon found a spot where I could look out over a great valley, and I knew that it *should be* the San Mateo. I tumbled headlong down the long, steep slope, slipping, sliding, leaping from tree to tree, in a mad frenzy to reach the bottom and learn the truth. If the road was there, everything was all right, but if not—but I dared not let my thoughts dwell on that idea. At last I reached the bottom, bruised and battered, clothing in tatters and my hands and face scratched and gouged by the unbreakable cat-claw thorns of the long, slender and often invisible branches of the bamboo.

With a great surge of relief rising in my breast, I burst through a thicket of low palms and fell headlong into the newly cleared right-of-way. Then, and only then, did I realize my condition and sat down on a nearby log to rest and gather together my scattered wits. The mad rush of that last hour had been my undoing, and while greatly relieved that I was not irretrievably lost, it was not pleasant to think of the long miles that still separated me from food and shelter.

To add to my troubles, the high hills rose steeply from the very water's edge, the clearing for the new highway had been cut along the face of those steep slopes, and all the logs and debris had been thrown down along the lower edge. There was no sign of a path, only the bare slope thickly strewn with stumps of trees and the sharp-cut stubs of the underbrush. At frequent intervals, deep ravines descended from the hills, cutting across the way. Long detours had been made into and out of these deep, narrow valleys, following the grade of the road. The great masses of fallen trees and debris that choked these intersecting ravines below the roadway made it extremely difficult to descend into them and out onto the road opposite, so that after a couple of futile attempts I gave up the idea and doggedly followed the cleared route. The sun had broken through the pall of clouds shortly after noon, and its broiling rays again drenched my clothing with perspiration. The strap of my shooting bag seemed to be slowly eating its way through my left collar bone, and my hands and arms were stiff and sore from carrying my gun and the long hours of swinging a machete. Every 100 yards or so [100 m] I had to stop and rest for a few minutes, then force myself to continue.

The sun was settling down to the summits of the distant mountains when I finally reached the roadway, merely a mule trail 2 m in width, but following the grade of the highway. It being Sunday, there were no men working, so I still had no idea where I was. In any event, there was no more clambering up and down slippery slopes or crawling painfully around the faces of muddy, crumbling bluffs. From some far hidden recess of my being there now surged new strength and hope, and I trudged rapidly on over the smooth pathway. It was still a long three miles [4.8 km] (I learned afterwards) before the new construction camp was reached.

Fortunately, I found *Don* Julio, the foreman, in the camp and rapidly related my tale of woe.

I have no idea how I looked, but I guess it must have been pretty bad for the first thing he did was to pour me a stiff drink of cane liquor and then shout to the cook to bring some hot food as rapidly as possible. He soon came with a large bowl filled with a stew of rice, *chárqui* [dried beef], *chuña* [potato flour] and a few carrots, but it seemed that I had never tasted anything so good. A large cup of coffee and a piece of dry bread completed the repast. It was nearly 5 pm when I arrived at the camp, and there still remained three miles [4.8 km] of very muddy trail ahead of me. My clothing was still wet, and I now began to feel cold, so thanking *Don* Julio, I forced my stiffened limbs to continue the march. A mile [1.6 km] up the road I met Berto, the Major [in charge of conscript labor], and a *peón* [unskilled laborer].

When I failed to return at 1:30 pm, Berto had become alarmed, and knowing the direction I had taken, followed, locating my tracks in the wet earth where I had left the trail and entered the forest. He made a long circuit up the valley, shouting and firing his gun but getting no response, started back to camp where he met the Major, who had also started out to look for me. The three were then on their way to *Don* Julio's camp, where they meant to take all of his available men and scour the jungle in search of me.

It was a joyful meeting, and I gladly turned over my bag and gun to the *peón* and returned with them to the camp where *Don* Oscar [brother of Orlando] and Rios [the storekeeper] anxiously awaited news of the lost one. I do not know how many miles I traveled on that ill-fated day, but since I had been on the move almost continuously for nearly 12 hours, it could not have been much less than 30 [48 km]. A bath, clean clothes, a couple of drinks and dinner did wonders for me, and beyond a deep lassitude and a bruised hip, I felt pretty good.

[On 31 July 1937 Carriker, Berto, and Mike moved their operations further downriver to Todos Santos, depto. Cochabamba, elevation 300 m, lodging in a house in the town. At Todos Santos there was weekly air service, arriving on Monday. The first day Carriker arranged camp, while Berto and Mike collected, bringing in 27 specimens of 19 species, 11 of which were new for the year and five new for Carriker's Bolivian collections.]

The second morning also I left the two boys to do the shooting while I built a platform on which to dry our specimens in the sun, set up the artificial drier, and wrote some letters for the plane that was due to arrive the following day. The boys were again successful and brought in 16 additional species, of which four were new to my list.

After supper that evening, I went shooting with the electric headlight down the road, across the flying field to the river, and then up the shore to the end of the cleared area. There were numerous nightjars of which I secured eight specimens including two species, one of them having the outer tail feathers more than a foot [30 cm] in length [possibly Scissor-tailed Nightjar, *Hydropsalis brasiliana*?]. It was a fascinating sport to hunt them. I walked along slowly, swinging the slender beam of light from side to side, up and down, searching over every yard of surface for the tell-tale spot of glowing, fiery red that would reveal the presence of one of the elusive birds. It was very difficult to judge accurately the distance of the creature solely by the glowing eyes, so that sometimes I shot them too close, and then again a wounded bird would escape. Once the open eye was closed or turned from the ray of light, the bird instantly disappeared from view, and there was no way of telling which way it flew or how far. A wounded bird might flit only a few yards and then drop, but without knowing just where to search for it in the thick grass, it would be impossible to locate.

For several days thereafter our affairs moved along in their usual, monotonous routine. We all went shooting very early in the morning, returning about 11:30 as a rule, had lunch at twelve, and then spent the afternoon until more or less 7 pm preparing the specimens which had been secured. Each day many new species were added to the collection; our food was plentiful and fairly good; and there were no blackflies at all to ruffle our tempers as we sat working in the somewhat dim light within the house. However, after the cool temperatures at Palmar, we did suffer rather severely from the unaccustomed heat and high humidity. There was no drinking water to be had in the jungle unless we happened to be near the river, which was not often, whereas at Palmar there was always an abundance of clear, cold water wherever we went. The mosquitoes, too, were very bad in the jungle, a swarm of them being always at hand to settle on hands and face whenever we stopped for a moment or when we slowly and carefully stalked some particularly elusive bird.

A goodly number of the species of birds taken at Palmar were also present around Todos Santos, but perhaps two-thirds of all those taken at Todos Santos were different. Some of them had been taken on the [Río] Beni by Mel and me in 1934, but others had not, so that on the whole we added many new forms to the collection, some of which had never been recorded from Bolivia by any previous collector. Several species of monkeys were common, and tracks of wild hogs, deer and tapir were common sights within a mile [1.6 km] or two of the town.

There had not been many species of large birds at Palmar, but here they abounded, the most conspicuous being the two huge, gaudily colored macaws [possibly Red-and-green (*Ara chloroptera*) and Blue-and-yellow (*A. ararauna*) macaws]. The little green toucan [Emerald Toucanet, *Aulacorhynchus prasinus*] of Palmar was replaced by two larger species, one black with white throat and crimson upper and under tail coverts [*Ramphastos* sp.], the other with alternate bands of scarlet and gold across the under parts and a black throat [*Pteroglossus* sp.]. There were two forms of large oropendolas [*Psarocolius* spp.], those huge orioles which nest in colonies and swing their long, purse-shaped nests from the tips of the branches of some jungle monarch rising high above the average trees or standing solitary in some dense cane-brake along the river.

There were tinamous and wild turkeys [probably guans or chachalacas], great hawks and silent herons, the latter wading in the shallows of the river or in the big lagoon in the jungle back of the town. Tiny woodpeckers [*Picumnus* spp.], no larger than a chickadee [*Parus* spp.], prowled among the branches of the tall underbrush, while crimson-headed giants [*Campephilus* spp.] filled the jungle with the resounding roll of a snare drum. However, the families most largely represented were the flycatchers, the antbirds, and the tree creepers [woodcreepers]. There were scarcely any sparrows at all.

[In mid August, 1937, the expedition moved further downriver from Todos Santos to the mouth of the Río Chaparé, depto. Cochabamba, elevation 250 m, where they lodged in a decrepit room in the only building at the settlement. They hired the other inhabitants of the house to prepare their meals.]

It was surprising what a decided change took place between Todos Santos and the mouth of the Río Chaparé, not only in the bird fauna but also in the flora. Many species of birds and trees, abundant at the former place, were not seen at the latter and vice versa.

As we retreated further and further from the base of the foothills, the dry season became more pronounced, and there were fewer cloudy days and increasing heat. During the 15 days spent at the *boca* [mouth of the river], we were visited by one terrific storm accompanied by the usual heavy electric discharges. This was later followed by a *surrazo* [cold front coming north from the Antarctic] of 24 hours duration. The wind that accompanied it was not of great violence, but the cold drizzle and "scotch mist" made life very uncomfortable and forced us to work in the little room at the end of the house. The only blessing brought by it was the temporary cessation of hostilities by the countless hordes of mosquitoes that otherwise made life an unadulterated misery at the approach of dusk.

We always continued skinning birds until the absence of light forced us to suspend the work, after which the lamp was lighted and the evening meal was served on a stationary table under the large tree. It was at this time that the shock troops of our winged enemies emerged in overwhelming numbers from the gloom of the encircling thickets. With appetites sharpened from the long day in hiding, they set about their task of merciless slaughter with fanatic zeal. Not satisfied with attacking every vulnerable portion of our long-suffering anatomy, they fell in the soup, explored the rice, and gnawed at the *plátanos*. Supper was not what could be termed a pleasant interlude and was terminated with utmost dispatch. Clouds of tobacco smoke only seemed to whet their appetites, and our only respite was to seek the shelter of the nets that covered our cots.

On moonlit nights the mosquitoes were less abundant, seeming to shun the brilliantly lighted patio as though fearful of being caught at their nefarious occupation. After the first rush from cover in the early evening, their numbers usually diminished so that we were often able to come out from hiding after an hour in bed and stroll for a time back and forth along the edge of the high bank, swishing kerchiefs about our heads to discourage the near approach of the persistent pests.

One of the few enjoyable events that broke the round of monotony was the daily plunge into the cool waters of the swiftly flowing river. Sticky, clay mud margined the river, but there was always a canoe moored at the bank in which we could dry ourselves and dress. It was not possible to stand quietly in the water without being surrounded by a swarm of small fish, whose sharp teeth invariably searched out the tenderest portions of our bodies. They were not capable of doing serious damage, but were very annoying, always returning quickly to the attack as soon as our splashing ceased.

Our host [Pedro] was an ardent disciple of Izaak Walton, spending long hours in his canoe in pursuit of the finny denizens of the river. His delight in fishing was not untinged with a touch of commercialism. *Chárqui* cost him real money, whereas fresh fish were free for the taking and decidedly increased the profits derived from catering to the healthy appetites of three guests. It was also a most satisfactory arrangement for the guests, to whom fresh fish was a welcome change from the tough and not always savory beef.

Berto and I made several long trips up the [Río] Chimoré in Pedro's light dugout canoe, and because paddling was not one of his expert accomplishments, I always managed the stern. Paddling up the swift current was a man-sized job, so that it was with much satisfaction that I found myself still able to stand in the stern of the canoe and pole it upstream through the shallows that bordered the long stretches of sandy beach. I had learned the trick many years before from the Indians of Central America, acquiring sufficient skill to make a small canoe, single hand, up through a hundred yard [100 m] riffle. It is a task requiring considerable skill, much practice, and a nice sense of balance. It is not easy to stand erect in the stern of a little, cranky, round-bottomed dugout, and force it up through a 10-mile [16 km/h] current with no one in the bow to prevent the swift water from swinging it completely around and out of control.

A week after our arrival we had an unexpected visitor. A stern-wheeled steamer, much larger than the one that meets the launch from Todos Santos, made its appearance shortly after noon. The scantily clad roustabouts swarmed ashore and disappeared into the grove of *plátanos*, each carrying a long machete. Soon we could hear the crashes of the stalks as the big crown of leaves toppled over, leaving the bunches of fruit hanging within reach of the cutters. Several hundred bunches were conveyed to the steamer. Then they cast off, dropped down to the Chimoré, and ascended that stream to the logging camp, where they loaded a month's accumulation of lumber. The following day they returned to Trinidad.

A few days before the bimonthly steamer was due from Trinidad, huge cargo canoes began to arrive from Todos Santos loaded with freight that would be transferred to the steamer. Some of these canoes, patiently hewn by hand from a single tree-trunk, were capable of carrying two and a half tons of cargo, in addition to the crew of seven men with their food and meager possessions. One such craft arrived with a load of potatoes in bags. They had made the long journey from Cochabamba on mule-back to Todos Santos, where all of the decayed tubers were carefully sorted out and the remainder repacked. Four or five days on the river, exposed to the broiling sun with perhaps a rain for good measure, did not improve their condition. It was usually necessary to again pick out the damaged ones when they reached the *boca*. The combined loss by decay and petty pilfering all the way from Cochabamba to Trinidad must have reduced the original shipment by half and doubled the cost. As a result the people of Trinidad looked upon the tubers as one of their greatest luxuries.

Beer and salt were the principal articles of merchandise that made this long and expensive trip down from the mountains, and there were always many tons of both carried by the Trinidad steamer on every trip, not only to supply the inhabitants of that city, but all of the straggling population along the banks of the endless miles of river between the Boca Chaparé and the far distant frontier of Brazil.

The cost entailed and time lost on the trip to the mouth of the Chaparé were amply repaid by the number of species of birds added to the collection there. Forty-three forms were secured which had not been taken that year, nor were they found at any other point visited later. Some few, but not many, had been taken in 1934 on the upper Río Beni.

On the Salar de Uyuni.—[By January 1938 the expedition had made its return to the high Andes (Fig. 8) and was quartered at Uyuni, depto. Potosí, on the shores of the vast, high Salar de Uyuni, elevation 3,669 m. There were few birds in the arid, high lands around the salt lake, and Carriker wished to visit other areas around in it in his search for more birds.]

I also wanted to get over to the western cordillera of the Andes, along the Chilean frontier. I had thought of going on the Antofagasta Railway as far as the border and do some work there, but learned that the only town near enough to the mountains was across the frontier in Chilean territory, which made the matter too difficult. Also they told me that the mountains were very dry, barren, and contained much sulphur, and that animal life was almost non-existent. Then I got an unexpected break.

Mr. Carr of Oruro had given me a card of introduction to a merchant of Uyuni by the name of Ñieto. He was a Peruvian but had lived there for many years. He was greatly interested in the Indians of the region and had spent much time traveling in the remoter districts in order to study their customs and mode of life. There was a village of these Aymara *indígenes* on the northwest side of the salt basin not far from the Chilean border. It was an isolated spot and difficult of access. The people of the village did most of their trading with a town just inside Chile and seldom came to Uyuni. Señor Ñieto had also made careful studies of the famous Salar de Uyuni, that vast lake of pure salt that lies north of the town, and he had also been one of

Fig. 8. The author and an Indian girl above Zudáñez [Depto. Chuquisaca, ca. 2,500 m elevation].

the pioneers of automotive traffic over its glistening surface. He had explored and mapped a route across the Salar to this village of Llica [depto. Potosí] and had been instrumental in getting a causeway of earth and stone built out into the Salar a sufficient distance to allow a motor vehicle to reach a point where the salt would bear its weight. A good road continued on from the causeway to the village, a distance of several miles.

He now offered to take me over in his automobile because there was no regular motor service to Llica. I could return in the truck that belonged to the government school for *indígenes* located at Llica. The truck was expected to come to Uyuni sometime within a week or 10 days, but traffic over the Salar at that time of year was uncertain because a heavy rain might fall at any day and cover the salt with water, rendering its passage impossible or very dangerous. There was not room in the car of *Señor* Ñieto for all of us, so I left Mike in Uyuni to amuse himself hunting rheas [Puna Rhea, *Pterocnemia pennata*], while Berto and I took our beds and a small outfit and started out on a Sunday after lunch.

The salt [in the lake bed] is so pure that it is universally used without any refinement, and is undoubtedly rich in iodine, because the inhabitants of the whole region where that salt is used are unusually free from goiter, a disease prevalent in some of the mountainous region of South America where ordinary rock salt is used.

Heavy banks of clouds were hovering at the base of the Chilean mountains, and the outlook was anything but cheerful. Once the car had crossed the thin edge of salt near shore and reached a firm base, we traveled over the grandest speed-way it would be possible to imagine. Can you visualize such a vast plain, level as a floor and as hard as concrete with its surface just sufficiently roughened to prevent skidding and give a splendid grip to the tires? After traveling about 30 miles [48 km] we met the onrushing storm, and 45 miles [72 km] of salt still remained to be covered. Soon we were running through a half-inch [1.2 cm] of water, then an inch [2.5 cm], then two inches [5.1 cm], which the rushing car threw up in a shower that drenched the whole car. We were compelled to stop and wrap some rags around the distributor to prevent the entrance of the salt water, and attempted to shield the battery in the same manner. However, in spite of our precautions, the strong brine seeped into the battery cells, and coming in contact with the acid, generated chlorine gas that soon filled the interior of the car and almost drove us frantic. The rain continued to fall, driven before a lateral gale of wind. We were forced to reduce our speed to 25 miles per hour [40 km/h], but could not open the windows except for a mere crack on the lee side. Our eyes were smarting and our lungs were half asphyxiated, but we dared not stop. Nothing could be done except to continue as rapidly as possible with the hope of reaching land before the early dusk would shut out all landmarks and we should be irretrievably lost in that wilderness of salt. If we missed the entrance to the causeway, it would be impossible to regain the shore, and we would have no means of knowing which way to turn to find it.

Fortunately, after an hour or more we ran out of the rain, and soon reached a stretch of almost dry surface because the gale of the wind had blown the water away from the shore for which we were headed. We were then able to open the windows and breathe again the pure, cold, air, but it was several days before we fully recovered from the effects of the gas. The whole car was covered with a thin film of salt, and the brine even penetrated around the badly fitting door, flooding much of the interior, including myself, so that everything it touched was soon encrusted with salt. The clouds lifted sufficiently for Señor Ñieto to recognize certain landmarks by which he steered our course safely to the causeway, and we reached the village just at dusk.

Except for the few springs of brackish water at the village, and at other places along the shore line of the ancient sea, there was not a drop of moisture for many miles. As a result bird life was extremely scarce. We wandered long distances back among the hills, but found almost nothing. Most of the birds taken were found along the shores where water was present. Although badly disappointed with the few feathered creatures which we found, nevertheless we did take several worthwhile species, especially a nice series of the beautiful Andean Curlew [Andean Avocet, *Recurvirostra andina*], which had always eluded me.

But of all the weird things we saw around Llica, I think that the most impossible of them all were the little fish that I found in the tiny brooks that issued from the earth and were quickly swallowed up again. The largest were about five inches [12.7 cm] in length, dark in color, and thick of body with somewhat the shape of a sucker, and belonged, I think, to two species. Quite a number were preserved, but they have not as yet been reported by the museum.

There is no permanent stream of fresh water within many, many miles of that spot. How did the fish ever reach their present habitat? Are they relics of species that inhabited the ancient sea and ascended the little streams that flowed into it, remaining there when the sea turned to salt? Their careful study will doubtless reveal who are their nearest relatives, and thus throw some light on their origin.

We were allowed only a week at this fascinating spot, being forced to take advantage of the early departure of the truck. Because there was little prospect of it making another trip in the near future, we could not take a chance on being marooned in that isolated spot.

All trace of water had disappeared from the Salar when we returned to Uyuni. The sun threw dazzling reflections from the shimmering surface of the salt that soon became painful to unprotected eyes. There was just one sign of life on that vast expanse—flamingos [*Phoenicopterus* sp. or *Phoenicoparrus* spp.]. We had seen them on the first trip, but dared not stop on account of the storm and the lateness of the hour. Now, I tried hard to approach within shooting distance of them, but it was quite impossible. Neither by truck nor on foot would they allow us to approach nearer than three or four hundred yards [300 or 400 m] before they took flight. They formed blotches and long lines of crimson on the snowy surface and were the only break in the monotony of the gleaming surface between us and the distant mountains.

[Carriker ended the manuscript at the point when he and Mike returned to the United States in April 1938. Carriker never returned to Bolivia.]

ACKNOWLEDGMENTS

Much of the information in the biography is taken from interviews and correspondence with Melbourne R. Carriker. Without his assistance, there would yet be many gaps in our knowledge about the man who was his father. J. Van Remsen and François Vuilleumier both reviewed and greatly improved the manuscript. Melissa Wiedenfeld helped with the historical research for the biography.

LITERATURE CITED

EMERSON, K. C., ED. 1967. Carriker on mallophaga. Posthumous papers, catalog of forms described as new, and bibliography. Melbourne R. Carriker, Jr., 1879–1965. U. S. National Museum Bull. 248.

PAYNTER, R. A., JR. 1992. Ornithological Gazetteer of Bolivia, second edition. Museum of Comparative Zoology, Harvard University, Cambridge, Massachusetts.

PHELPS, W. H. 1944. Resumen de las colecciones ornitologicas hechas en Venezuela. Boletín de la Sociedad Venezolana Ciencias Naturales 61:325–444.

SCHULENBERG, T. S., AND M. D. WILLIAMS. 1982. A new species of antpitta (*Grallaria*) from northeastern Peru. Wilson Bull. 94:105–113.

STEPHENS, L., AND M. A. TRAYLOR, JR. 1983. Ornithological Gazetteer of Peru. Museum of Comparative Zoology, Cambridge, Massachusetts.

APPENDIX 1

Carriker's ornithological publications. Long gaps, especially 1910–1930, appear to correspond to a time when Carriker was living in Colombia and doing little ornithologically but collecting. A listing of his entomological publications is given by Emerson (1967).

Books:

CARRIKER, M. A., JR. 1910. An annotated list of the birds of Costa Rica including Cocos Island. Annals Carnegie Mus. 6:314–915.

TODD, W. E. C., AND M. A. CARRIKER, JR. 1922. The birds of the Santa Marta region of Colombia, a study in altitudinal distribution. Annals Carnegie Mus. 14:1–611.

Articles:

CARRIKER, M. A., JR. 1908. Notes on Costa Rican Formicariidae. Annals Carnegie Mus. 5:8–10.

CARRIKER, M. A., JR. 1930. Descriptions of new birds from Peru and Ecuador. Proc. Acad. Nat. Sci. Philadelphia 82:367–376.

CARRIKER, M. A., JR. 1931. Descriptions of new birds from Peru and Bolivia. Proc. Acad. Nat. Sciences of Philadelphia 83:455–467.

CARRIKER, M. A., JR. 1932. Additional new birds from Peru with a synopsis of the races of *Hylophylax naevia*. Proc. Acad. Nat. Sciences of Philadelphia 84:1–7.

CARRIKER, M. A., JR. 1933. Descriptions of new birds from Peru, with notes on other little-known species. Proc. Acad. Nat. Sciences of Philadelphia 85:1–38.

CARRIKER, M. A., JR. 1934. Descriptions of new birds from Peru, with notes on the nomenclature and status of other little-known species. Proc. Acad. Nat. Sciences of Philadelphia 86:317–334.

CARRIKER, M. A., JR. 1934. Rediscovery of *Conothraupis speculigera* (Gould). Auk 51:497–499.

CARRIKER, M. A., JR. 1935. Descriptions of new birds from Bolivia, with notes on other little-known species. Proc. Acad. Nat. Sciences of Philadelphia 87:313–341.

CARRIKER, M. A., JR. 1935. Description of new birds from Peru and Ecuador, with critical notes on other little-known species. Proc. Acad. Nat. Sciences of Philadelphia 87:343–359.

CARRIKER, M. A., JR., AND R. MEYER DE SCHAUENSEE. 1935. An annotated list of two collections of Guatemalan birds in the Academy of Natural Science of Philadelphia. Proc. Acad. Nat. Sciences of Philadelphia 87:411–455.

CARRIKER, M. A., JR. 1936. Glimpses of Bolivia by a naturalist. Bolivia vol. 5, no. 8 (January–February), pp. 9–14, 22–23.

CARRIKER, M. A., JR. 1936. A new antbird from Venezuela. Auk 53:316–317.

CARRIKER, M. A., JR. 1954. Additions to the avifauna of Colombia. Novedades Colombianas 1:14–19.

CARRIKER, M. A., JR. 1955. Notes on the occurrence and distribution of certain species of Colombian birds. Novedades Colombianas 2:48–64.

CARRIKER, M. A., JR. 1959. New records of rare birds from Nariño and Cauca and notes on others. Novedades Colombianas 1(4):196–199.

CARRIKER, M. A., JR. 1959. Itinerario del autor durante sus recolecciones en la región de Santa Marta, Colombia de junio de 1911 a octubre de 1918. Novedades Colombianas 1(4):214–222.

CARRIKER, M. A., JR. 1960. Itinerario del autor durante sus recolecciones en la región de Santa Marta, Colombia de junio de 1911 a octubre de 1918 (continuacíon). Novedades Colombianas 1(5):330–335.

APPENDIX 2

CARRIKER'S ITINERARY IN BOLIVIA. THE FOLLOWING BOLIVIAN ITINERARY IS MAINLY TAKEN FROM CARRIKER'S MANUSCRIPT OF "EXPERIENCES OF AN ORNITHOLOGIST ALONG THE HIGHWAYS AND BYWAYS OF BOLIVIA," BUT AUGMENTED WITH INFORMATION FROM PAYNTER (1992). DATES, WHEN GIVEN, ARE DATES ON WHICH CARRIKER IS KNOWN TO HAVE BEEN AT THE LOCALITY. ABBREVIATIONS ARE: ARR. = ARRIVE; LV. = LEAVE

Locality	Day	Month Year
Lv. New York	early	June 1934
Arr. La Paz		June 1934
La Paz to Sandillani	5	July 1934
Passes Coroico	6	July 1934
Calabatea	9	July 1934
Santa Ana (Depto. La Paz)	11, 15, 19, 26, 31	July 1934
Santa Ana	1	Aug. 1934
Lv. Santa Ana	2	Aug. 1934
Guanay	6–18, 10	Aug. 1934
Teoponte	15–20, 17	Aug. 1934
Chiñiri	23	Aug. 1934
Lv. Chiñiri	6	Sept. 1934
Down Río Beni to Soza farm (above Rurrenabaque)		Sept. 1934
Soza farm	1st week	Sept. 1934
Rurrenabaque	10	Sept. 1934
Susi	11, 14, 15	Sept. 1934
Chatarona	16, 18, 19, 20, 24, 30	Sept. 1934
Chatarona	1	Oct. 1934
To Soza farm again	2	Oct. 1934
Lv. Soza farm for upriver	15	Oct. 1934
Arr. Sipiápo	19?	Oct. 1934
Guanay	25	Oct. 1934
Caranavi	30?	Oct. 1934
Playa Ancha (between Caranavi & Santa Ana)	1	Nov. 1934
Calabatea	7, 9, 18	Nov. 1934
Calabatea to Sandillani	21	Nov. 1934
Sandillani	22, 26	Nov. 1934
Sandillani to Hichuloma to La Paz	27	Nov. 1934
La Paz back to Hichuloma	1?	Dec.? 1934
Hichuloma to Sandillani	2?	Dec.? 1934
Sandillani	10, 11	Dec. 1934
Sandillani to Hichuloma again	19	Dec. 1934
Hichuloma	20	Dec. 1934
To La Paz	24	Dec. 1934
Back to Hichuloma again	26?	Dec. 1934
Hichuloma	29	Dec. 1934
Hichuloma	1, 2	Jan. 1935
Overnight to La Cumbre	4	Jan. 1935
Back to Hichuloma	5	Jan. 1935
Hichuloma	8	Jan. 1935
Km. 50 Yungas Railway	9	Jan. 1935
Hichuloma to La Cumbre again	10, 11	Jan. 1935
Km. 34 Yungas Railway	13	Jan. 1935
La Cumbre	14	Jan. 1935
Hichuloma to La Paz	16?	Jan. 1935
La Paz to Irpaví	23?	Jan. 1935
Irpaví to La Paz	30	Jan. 1935
La Paz to Mollendo, Peru	1?	Feb.? 1935
Lv. Mollendo for New York	3	Feb. 1935

APPENDIX 2
Continued

Locality	Day	Month Year
La Paz to Oruro	1	May 1936
Oruro	2–3	May 1936
Oruro to Llallagua/Cataví	4	May 1936
Llallagua/Cataví	5–25	May 1936
Llallagua/Cataví to Pazña	29	May 1936
Pazña to Callipampa	30	May 1936
Callipampa-Oruro-Callipampa	7–9	June 1936
Callipampa	10	June 1936
Callipampa to Chocaya	12	June 1936
Chocaya	14–18	June 1936
Chocaya to Oploca	20	June 1936
Oploca to El Salo	30	June 1936
Back to Oploca	3	July 1936
Oploca to Villazón	4	July 1936
Villazón to Tarija	5	July 1936
Tarija to San Lorenzo (Depto. Tarija)	7?	July 1936
San Lorenzo	8–19	July 1936
San Lorenzo back to Tarija	21?	July 1936
Tarija to La Merced	26	July 1936
La Merced	1	Aug. 1936
La Merced to La Capilla	3	Aug. 1936
La Capilla to Coyombulla	5	Aug. 1936
Coyombulla to Río Lipeo/Monte Bello	6	Aug. 1936
Río Lipeo/Monte Bello	9, 10, 11–28	Aug. 1936
To Bermejo	1?	Sept.? 1936
Bermejo	7–15	Sept. 1936
Bermejo to Fortín Campero	16	Sept. 1936
Fortín Campero	23	Sept. 1936
Fortín Campero to Agua Blanca via Bermejo	24	Sept. 1936
Arr. back at Río Lipeo/Monte Bello	1	Oct. 1936
Río Lipeo/Monte Bello back to La Capilla	3	Oct. 1936
La Capilla to La Merced	8?	Oct. 1936
La Merced to Tarija	10?	Oct. 1936
Tarija to Entre Rios	17	Oct. 1936
Entre Rios	23	Oct. 1936
Entre Rios to village on road	27	Oct. 1936
To Villa Montes	28	Oct. 1936
Villa Montes	29	Oct. 1936
Villa Montes	8, 9	Nov. 1936
Villa Montes to Camiri	10?	Nov. 1936
Camiri to Laguníllas	11	Nov. 1936
Laguníllas	16	Nov. 1936
Lv. Laguníllas, on road all night	19	Nov. 1936
Arr. Río Azuero	20	Nov. 1936
Río Azuero	21	Nov. 1936
Río Azuero to Padilla	28	Nov. 1936
Padilla to Zudáñez	?	Dec.? 1936
Zudáñez to Sucre	?	Dec. 1936
Arr. la Paz	24	Dec. 1936
Lv. La Paz for Mollendo, Peru	29?	Dec. 1936
Lv. Mollendo for New York	4?	Jan. 1937

APPENDIX 2
Continued

Locality	Day	Month Year
[Note: Dates for "Carriker" specimens from 8–23 April and 7–29 May, 1937, mostly from Incachaca, must have been collected by his field assistant known only as "Berto." At that time, Carriker himself was in the United States, and he did not return to Bolivia until late May, 1937. Berto had been left to collect in the Incachaca area back in December 1936.]		
Arr. La Paz	?	May? 1937
La Paz to Oruro	?	May 1937
Oruro	23	May 1937
Oruro to Cochabamba	?	May 1937
Cochabamba to Incachaca	29	May 1937
Incachaca	10	June 1937
Incachaca to Cochabamba	13	June 1937
Cochabamba to Palmar	1	July 1937
Palmar	4	July 1937
Day trip to Río Ibirízu	11	July 1937
Palmar	28	July 1937
Palmar to Todos Santos	31	July 1937
Todos Santos	6, 8	Aug. 1937
Lv. Todos Santos on launch	12	Aug. 1937
Arr. mouth Río Chaparé	16	Aug. 1937
Lv. mouth Río Chaparé	1	Sept. 1937
Arr. Todos Santos	6	Sept. 1937
Todos Santos to Palmar	11	Sept. 1937
Palmar to Cochabamba	18?	Sept. 1937
Cochabamba to Tiraque/El Juno	26	Sept. 1937
Lv. Tiraque	4	Oct. 1937
Ele Ele	8	Oct. 1937
Ele Ele to Mataral	17	Oct. 1937
Mataral to Samaipata	26	Oct. 1937
Samaipata	3, 10	Nov. 1937
Samaipata to Cochabamba	27	Nov. 1937
Cochabamba to Oruro	12	Dec. 1937
Oruro to Sucre	14–15	Dec. 1937
Sucre to Tomina	20	Dec. 1937
Tomina to Padilla	30	Dec. 1937
Padilla	31	Dec. 1937
Padilla	1–8, 12	Jan. 1938
Potosí	19–27	Jan. 1938
Uyuni	29	Jan. 1938
Uyuni	5	Feb. 1938
Llica	6–10	Feb. 1938
To Chocaya	17, 18	Feb. 1938
Oploca	24 Feb.–2	Mar. 1938
Callipampa	6–13	Mar. 1938
Viloco	28–30	Mar. 1938

SPECIES LIMITS IN *CRANIOLEUCA VULPINA*

KEVIN J. ZIMMER
1665 Garcia Rd., Atascadero, California 93422, USA

ABSTRACT.—New information on the vocalizations, behavior, and habitat of populations of the Rusty-backed Spinetail (*Cranioleuca vulpina*) reveal that it actually consists of two species-level taxa: a wide-ranging, polytypic species (*C. vulpina*) that occupies *várzea* and riverine forests and flooded savanna woodlands north and south of the Amazon, and a more specialized form (*C. vulpecula*) restricted to successional habitats on islands in the Amazon and its major white-water tributaries. Long considered a subspecies of *C. vulpina*, *C. vulpecula* is shown to be morphologically, vocally, and ecologically distinct from sympatrically distributed populations, with no evidence of intergradation. I found *C. vulpecula* to be common on Ilha Marchantaria in the Rio Solimões near Manaus, Brazil, a major eastward extension of its known range. One form or another of *C. vulpina* occupies the adjacent "mainland" banks of the river north and south of Marchantaria, as well as black-water islands in the nearby Rio Negro.

RESUMO.—Novas informações sobre vocalizações, comportamento, e habitat de populações do joão-do-rio (*Cranioleuca vulpina*) revelam que este taxon contém duas espécies: uma (*C. vulpina*) politípica, de ampla distribuição, habitando várzea, mata de galeria, e savanas inundáveis ao norte e sul do Rio Solimões-Amazonas, e a outra (*C. vulpecula*) especializada em ambientes sucessionais de ilhas no Rio Solimões e seus principais afluentes de água branca. *Cranioleuca vulpecula*, por muito tempo tratada como subespécie de *C. vulpina*, é morfologicamente, vocalmente, e ecologicamente distinta de populações simpátricas de *C. vulpina*, sem evidência de intergradação. *Cranioleuca vulpecula* é comúm na Ilha Marchantaria no Rio Solimões em frente a Manaus, Amazonas, Brasil, ou outra de *C. vulpina* ocorre em áreas adjacentes, nas margens do rio e nas ilhas de água preta do Rio Negro.

In the past decade careful attention to vocalizations, habitats, and other field data has revealed the presence of many previously overlooked sibling species (Pierpont and Fitzpatrick 1983; Stiles 1983; Graves 1987; Groth 1988; Johnson and Marten 1988; Willis 1988, 1991; Fitzpatrick and Willard 1990). During the same period, more intensive study of a previously largely overlooked habitat, islands in the larger rivers of Amazonia, has shown that the avifaunas of these islands may differ almost completely from those of nearby *terra firme* forest (Remsen and Parker 1983; Rosenberg 1990). Indeed, it has been shown that many species inhabiting these river islands are obligate river-island specialists that are seldom, if ever, found on adjacent "mainland." These specialists occupy successional plant communities created by periodic river flooding and by the constant erosion of islands at their upstream ends with the concomitant deposition of silt and sand at their downstreams ends (Remsen and Parker 1983; Rosenberg 1990). With such highly specialized avifaunas the Amazonian river islands would seem a likely habitat in which to find cryptic sibling species that are ecologically differentiated from their "mainland" counterparts (e.g., *Furnarius torridus*, *Knipolegus orenocensis*, and *Conirostrum margaritae*).

While conducting recent field work on Ilha Marchantaria in the Rio Solimões (= Amazon) near Manaus, Brazil, I found that resident island populations of *Cranioleuca vulpina*, the Rusty-backed Spinetail, differed strikingly in voice, plumage, and habitat from other populations with which I was familiar from both north and south of the Amazon. They also differed just as noticeably in these respects from river-island populations in a nearby black-water river, the Rio Negro. Subsequent analysis of tape-recorded vocalizations and museum specimens has revealed that *Cranioleuca vulpina* comprises two species: a widespread form that occurs north and south of the Amazon in various *várzea* and riverine habitats (including river-islands of black-water rivers), and a second species that is geographically widespread through Amazonia but restricted to islands in the Amazon and its larger white-water tributaries. This second species was originally described as a separate species, *Synallaxis vulpecula* (Sclater and Salvin 1866), but shortly thereafter (Sclater 1874) was treated as a subspecies of *vulpina*, where it has remained ever since.

NOMENCLATURAL HISTORY

In 1866 Sclater and Salvin described *Synallaxis vulpecula* from three specimens taken from the upper and lower Río Ucayali, Peru, collected by Bartlett. They noted that the specimens were most similar to *S. vulpina* (Von Pelzeln), but differed in being longer-billed, with more uniformly rusty upperparts, and with more spotting on the underparts. Upon further examination of specimens, Sclater (1874) synonymized *S. vulpecula* and *S. alopecias* (Von Pelzeln) of northern Brazil (mostly north of the Amazon) and Venezuela, with *S. vulpina* (Von Pelzeln) of interior Brazil south of the Amazon. Taczanowski (1884) also treated the three forms as synonomous.

Hellmayr (1925) treated the three forms as separate subspecies and placed them in the genus *Cranioleuca*. He also recognized a fourth subspecies, *C. vulpina reiseri* Reichenberger, from eastern Brazil. Hellmayr gave the range of *vulpecula* as "Eastern Peru (Río Ucayali; Pebas, Iquitos, and Nauta, Río Marañon) and western Brazil (Rio Purús)." He noted that *vulpecula* was "insufficiently known" and appeared to be intermediate between *vulpina* and *alopecias*. He also pointed out that one specimen of *vulpecula* was very long-billed.

Peters (1951) followed Hellmayr's treatment of *C. vulpina*, but recognized two additional subspecies, *C. v. apurensis* Zimmer and Phelps from the state of Apure, Venezuela, and *C. v. foxi* Bond and de Schauensee from the junction of the Río Chaparé and the Río Mamoré in Bolivia. Peters listed the range of *C. v. vulpecula* as "Northeastern Perú from the Napo, the Marañón and the Ucayali, east to western Brazil on the upper Rio Madeira (São Antonio de Guajará) and northeastern Bolivia on the lower Río Beni (Victoria)."

Wetmore (1957) described a new subspecies, *C. v. dissita*, from Coiba Island, Panama, a location separated from the nearest known populations of *C. vulpina* by the width of Colombia and most of Panama, as well as by the Andean cordillera. This unexpected find led Wetmore (1957) to examine more closely the entire *Cranioleuca vulpina* complex. He particularly noted that *vulpecula* differed greatly from other subspecies in having a "decidedly heavier bill, and in the much more distinct pattern of spotting and streaking on the under surface." He believed that "the sum of these differences warrants recognition of *vulpecula* as a separate species, distinct from any others of the genus" (Wetmore 1957).

Vaurie (1980) continued to treat *vulpecula* as conspecific with *vulpina*. Vaurie noted Wetmore's (1957) assertion that *vulpecula* should be considered specifically distinct, but dismissed this notion without further discussion. He did observe, however, that *vulpecula* was "better differentiated than the other populations," pointing out the heavier and larger bill, paler throat, and more distinctly mottled underparts.

Current workers have continued to treat *vulpecula* as a subspecies of *Cranioleuca vulpina*, under the English name of "Rusty-backed Spinetail" (Meyer de Schauensee 1970; Meyer de Schauensee and Phelps 1978; Hilty and Brown 1986; Sibley and Monroe 1990; Monroe and Sibley 1993; Ridgely and Tudor 1994). The highly disjunct form *dissita* is now considered by some workers a distinct species, *C. dissita*, the Coiba Spinetail, based on morphological, vocal, and habitat differences (Ridgely and Gwynne 1989; Sibley and Monroe 1990). Although this may be the correct course, published documentation of vocal differences between *dissita* and other taxa in the complex is limited to a brief qualitative description (Ridgely and Gwynne 1989), and no analysis has been published. Concerning reported habitat differences (Ridgely and Gwynne 1989), habitat expansion in insular forms is a regular phenomenon (Lack 1971), and the described "forest habitat" of *dissita* is not clearly different from the *várzea* woodlands inhabited by many other "Rusty-backed" Spinetails.

METHODS

I made field observations of "Rusty-backed" Spinetails in edo. Cojedes, Venezuela (February 1993, 1994 and 1995); terr. Amazonas, Venezuela (March 1993 and 1994); in the Anavilhanas Archipelago of the Rio Negro, Amazonas, Brazil (October 1992, January 1993, November 1994, and January 1995); along the Rio Branco near Boa Vista, Roraima, Brazil (November 1994); at various sites in Mato Grosso, Brazil (October 1992, September 1993, 1994 and 1995); along the Rio Juma, Amazonas, Brazil (January 1995); and on Ilha Marchantaria, Rio Solimões, Amazonas, Brazil (October 1992, January 1993, November 1994, and January 1995). Observations were made using Zeiss 10 × 40 binoculars. All behavioral data were recorded on cassette or micro-cassette in the field, and later transcribed.

Mapped distributions (as they appear in this paper) are based largely on ranges reported in Hellmayr (1925), Peters (1951), Meyer de Schauensee and Phelps (1978), Hilty and Brown

(1986), and Remsen and Traylor (1989), supplemented by label data from specimens which I examined, and by more recent records documented by tape recordings.

I assembled 77 recordings of *Cranioleuca vulpina* from various localities for comparison to my 11 recordings of Marchantaria birds. Included among these recordings are vocalizations of *C. v. vulpina* (Mato Grosso, Brazil; depto. Santa Cruz, Bolivia), *C. v. alopecias* (edo. Cojedes, edo. Bolivar, terr. Amazonas, Venezuela; Rio Negro, Amazonas, Brazil; Rio Branco, Roraima, Brazil), *C. v. reiseri* (Minas Gerais, Brazil), and *C. v. vulpecula* (Río Napo and Río Amazonas, Peru; and Río Napo, Ecuador). Locations and recordists for all recordings examined are listed in Appendix 1. I could not locate recordings of *C. v. apurensis* or *C. v. foxi*. For comparison I categorized vocalizations as songs, calls, or duets. "Songs" were unsolicited longer vocalizations given by an individual bird, seemingly in the context of territorial advertisement. All vocalizations categorized as "calls" involved foraging birds and seemingly were given in the context of contact vocalizations between members of a pair or family group. "Duets" involved simultaneous singing by a pair of birds, often (but not always) in response to playback of another vocalization. My recordings were made with a Sony TCM-5000 recorder and Sennheiser ME-80 and MKH-70 shotgun microphones. Sonagrams used in illustrations were made on a Macintosh Centris 650 computer using Canary version 1.1 (Bioacoustics Research Program, Cornell Laboratory of Ornithology, Ithaca, New York), Canvas version 3.0.6 graphics software (Deneba Software, Miami, Florida) and a LaserWriter Pro 630 printer.

To confirm my field impressions of morphological differences, I examined representative specimens of *vulpina* (17), *alopecias* (24), and *vulpecula* (27), and *reiseri* (2). These specimens are housed at the Carnegie Museum, Pittsburgh (CM), Field Museum of Natural History, Chicago (FMNH), Los Angeles County Museum of Natural History, Los Angeles (LACM), and the Louisiana State University Museum of Natural Science, Baton Rouge (LSUMZ). A list of specimens examined is provided in Appendix 2. I used calipers to measure (1) height of bill at the base, (2) length of exposed culmen, and (3) length of closed wing (= chord of folded wing). Tail measurements were not taken because this character is strongly influenced by molt and wear in spinetails. Measurement terminology conforms with that used by Baldwin et al., (1931). Analysis of variance (ANOVA) and Duncan's Multiple Range Tests were used for the statistical comparison of these measurements (Steel and Torrie 1980).

RESULTS

Distribution.—The distributions of the various forms of "Rusty-backed" Spinetails (Fig. 1) shows that *vulpecula* is largely allopatric to the other taxa. Gaps in distribution may be real or artifacts of under-sampling, as relatively little work has been done in western Amazonian Brazil (Oren and Albuquerque 1991). Sympatry of *vulpecula* and *alopecias* (and possibly also *vulpina*) is now confirmed from at least three areas in Amazonas, Brazil. Near Manaus, only the water barrier of the Rio Solimões separates *vulpecula* on Ilha Marchantaria from mainland sites occupied by *alopecias* to the north, and either *alopecias* or *vulpina* (see below) to the south. On 28 August 1995, A. Whittaker (pers. comm.) observed a duetting pair of *alopecias/vulpina* in mature *várzea* woodland on the upstream end of Ilha Marchantaria, less than 400 m from localities where we have recorded *vulpecula* in younger successional habitats. In September 1995, Mario Cohn-Haft visited this same *várzea* woodland on the upstream end of Ilha Marchantaria, and collected an apparently mated pair of *alopecias/vulpina* (collector's catalog numbers: female = MCH 385, male = MCH 386, specimens to be deposited with the Museu Paraense Emílio Goeldi in Belém [MPEG]). Cohn-Haft also collected a male *vulpecula* (MCH 384, to be deposited at MPEG) from a patch of cane in the center of the island. The latter bird provides the first specimen corroboration of *vulpecula* from Ilha Marchantaria, and the three birds together provide the first specimen corroboration of sympatry. On 22 September 1993, J. F. Pacheco tape-recorded *vulpecula* (Fig. 2d) on Ilha do Mateiro (ca. 10 km northwest of 30°07'S, 64°47'W), Mamirauá Ecological Station (north bank of the Rio Solimões). On 1 September 1994, less than 1 km away from Ilha do Mateiro, he tape-recorded *alopecias* (Fig. 2e) in *várzea* on the bank of the mainland. On 1 April 1995, 5–6 km west of Benjamin Constant, Bret Whitney and Mort and Phyllis Isler observed and tape-recorded a pair of either *vulpina* or *alopecias* in *várzea* along the Rio Javarí, just a few kms distant from river islands occupied by *vulpecula*. At each of these localities *vulpecula* and either *alopecias* or *vulpina* occur within a few kms to 400 m of one another, but are separated by a habitat barrier, and thus are sympatric but not syntopic. Whitney (*in litt.* 1994) has reported *vulpecula* from the same mature *várzea* woodland at the upstream end of Ilha Marchantaria where Whittaker and Cohn-Haft recently found *alopecias/vulpina*, suggesting that

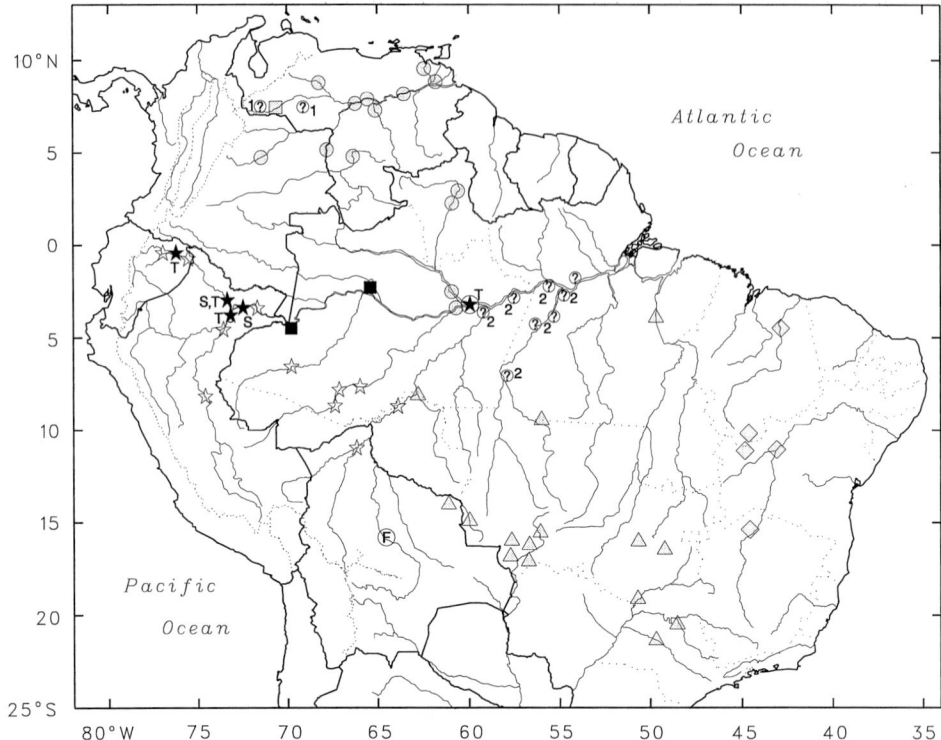

FIG. 1. Distribution of "Rusty-backed" Spinetails (*Cranioleuca vulpina*): black stars = confirmed sites for *C. v. vulpecula* (S indicates specimens examined by the author, T indicates confirmation by tape recordings); open stars are unconfirmed *vulpecula* from the literature or sight records; the black star within the circle locates Ilha Marchantaria on the Rio Solimões where *vulpecula* has recently been discovered; black squares locate (from left to right) Benjamin Constant and the Mamirauá Ecological Station (respectively), two sites in Amazonas, Brazil where sympatry between *C. v. vulpecula* and either *C. v. alopecias* or *C. v. vulpina* has recently been confirmed. gray circles = *C. v. alopecias*; gray triangles = *C. v. vulpina*; gray squares = *C. v. apurensis*; gray diamonds = *C. v. reiseri*; and "F" in a circle indicates the lone site for *C. v. foxi*. Question marks enclosed by circles = taxon uncertain. A "1" next to the question mark indicates that the taxon may be either *apurensis* or *alopecias*. A "2" next to the question mark indicates that the taxon is either *alopecias* or *vulpina*.

two forms may even exist syntopically at that spot. Specimen evidence suggests that *vulpecula* and *vulpina* may also come into contact along the upper Rio Madeira (Fig. 1).

Pacheco's observations and recordings from Mamirauá and the Whitney/Isler record from Benjamin Constant are important not only because they confirm additional sites of sympatry between *vulpecula* and *alopecias/vulpina*, but also because they document a significant westward range extension for *alopecias/vulpina*, which were not previously known from west of the lower Rio Madeira or lower Rio Negro drainages (Fig. 1).

There are some areas where there is uncertainty over the form of "Rusty-backed" Spinetail present (Fig. 1). I have tape-recorded spinetails from a locality in edo. Apure, Venezuela, that is equidistant from the known ranges of both *apurensis* and *alopecias*. Without corroborative specimens, it is impossible to know which taxon these birds represent. Additionally, it is not clear which form inhabits the immediate south bank of the Amazon in Brazil. Hellmayr (1925), listed *alopecias* as occurring only north of the Amazon in Brazil, with *vulpina* occurring south of the Amazon from the Madeira to the Tocantins. According to Peters (1951), *alopecias* occurs south of the Amazon from the mouth of the Madeira east to the Xingú, and *vulpina* occurs farther south, from the upper Río Madeira eastward to the Araguaya, and south into Mato Grosso, Goias, and western São Paulo. Specimens from the east bank of the Tapajós near Santarém, Pará, have been variously assigned to both *alopecias* (LACM) and *vulpina* (CM). Furthermore, I have tape-recorded birds from the south bank of the Amazon and just west of the Madeira, a locality that again falls between the known ranges of the two taxa. In the absence of solid criteria

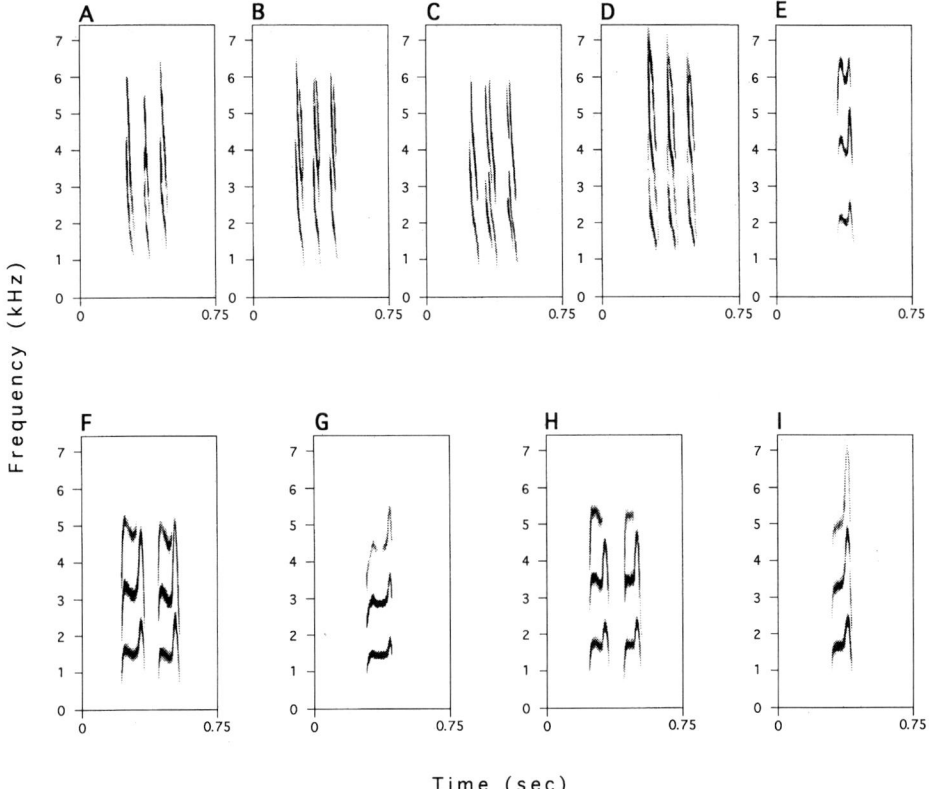

FIG. 2. Sonagrams of calls of "Rusty-backed" Spinetails, *Cranioleuca vulpina* (a. *C. v. vulpecula*, Isla Timicurillo, Río Amazonas, depto. Loreto, Perú [T. A. Parker, recordist (c) 1994 All rights reserved. Library of Natural Sounds, Cornell Laboratory of Ornithology, Ithaca, New York]; b. *C. v. vulpecula*, Ilha Marchantaria, Rio Solimões, Amazonas, Brazil; c. *C. v. vulpecula*, Río Napo, Ecuador [G. H. Rosenberg, recordist]; d. *C. v. vulpecula*, Mamirauá Ecol. Station, Amazonas, Brazil [J. F. Pacheco, recordist]; e. *C. v. alopecias*, Mamirauá Ecol. Station, Amazonas, Brazil [J. F. Pacheco, recordist]; f. *C. v. alopecias*, edo. Cojedes, Venezuela; g. *C. v. alopecias*, Anavilhanas Archipelago, Rio Negro, Amazonas, Brazil; h. *C. v. vulpina*, Mato Grosso, Brazil; i. *C. v. reiseri*, Minas Gerais, Brazil [B. Whitney, recordist]). Recordists other than the author are cited in brackets.

(either morphological or vocal) for separating these forms in the field, and given the difficulty of visually identifying even specimens, I believe that it is best to make no assumptions regarding the subspecific identity of birds from near the south bank of the Amazon.

Voice.—On 8 October 1992, while observing and tape recording birds on Ilha Marchantaria (Rio Solimões, Brazil) with Andrew Whittaker, I heard an unfamiliar vocalization that Whittaker identified as *Cranioleuca vulpina*. This vocalization was unlike any that I knew from *vulpina*, a species with which I had extensive field experience from both Venezuela and from Brazil south of the Amazon. I tape-recorded the vocalization, and in response to playback, a pair of spinetails flew in and responded with a noisy duet. The duet and subsequent calls were equally unfamiliar to me, and the birds exhibited several morphological differences from Rusty-backed Spinetails with which I was familiar. On 11 October 1992, Whittaker and I were conducting field work on some of the many islands in the Anavilhanas Archipelago of the Rio Negro, northwest of Manaus. Rusty-backed Spinetails were common on these islands and were vocally and visually similar to those that I had encountered in Venezuela and Mato Grosso, Brazil.

"Calls" of Marchantaria birds (Fig. 2b) match those of *C. v. vulpecula* from Peru (Fig. 2a) and Ecuador (Fig. 2c). Calls of all *vulpecula* differ markedly from those of *alopecias* (north of the Amazon), and from those of nominate *vulpina* and *reiseri* (south of the Amazon). However, calls of *alopecias* (including Anavilhanas birds from the Rio Negro near Manaus), *reiseri*, and *vulpina* are very similar to one another. Calls of the latter three forms may be one-noted (Fig.

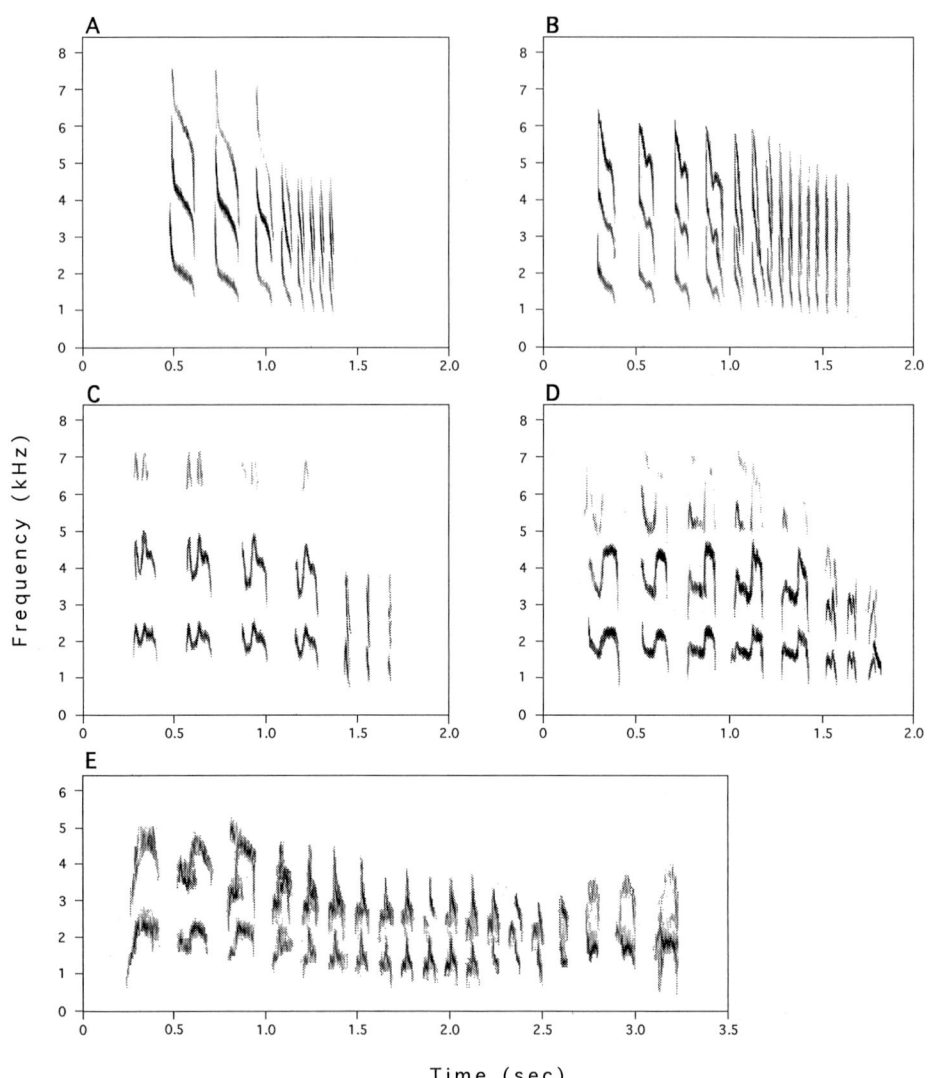

FIG. 3. Sonagrams of songs of "Rusty-backed" Spinetails, *Cranioleuca vulpina* (a. *C. v. vulpecula*, Isla Timicurillo, Río Amazonas, depto. Loreto, Perú [T. A. Parker, recordist (c) 1994 All rights reserved. Library of Natural Sounds, Cornell Laboratory of Ornithology, Ithaca, New York]; b. *C. v. vulpecula*, Ilha Marchantaria, Rio Solimões, Amazonas, Brazil; c. *C. v. alopecias*, Anavilhanas Archipelago, Rio Negro, Amazonas, Brazil; d. *C. v. vulpina*, Mato Grosso, Brazil; and e. *C. v. vulpina*, Mato Grosso, Brazil). Recordists other than the author are cited in brackets.

2e, 2g and 2i) or two-noted (Fig. 2f and 2h). The differences between the higher, thinner calls of *vulpecula* and the richer, more nasal calls of *alopecias*, *vulpina*, and *reiseri* are strikingly evident in the field. These calls are commonly given by foraging spinetails, and are the most frequently heard vocalizations.

The songs in all members of this complex (Fig. 3) seem more subject to individual variation than do the calls, but this variation primarily involves song length (Fig. 3d and 3e). The pattern of the song remains largely the same but the number of notes at the end is variable. Such differences can be noted in successive songs of a single bird. Spinetails subjected to tape playback frequently respond with longer songs (pers. obs.). Differences among taxa involve rates of delivery, different patterns of frequency shifts, and individual note shape within the song. Once again, the similarities between *vulpecula* from Peru (Fig. 3a) and birds from Marchantaria (Fig. 3b) are obvious, as are the differences between the songs of these birds and those of *alopecias*

(Fig. 3c) and *vulpina* (Fig. 3d and 3e). The songs of *alopecias* and *vulpina* differ somewhat from one another, but are more similar in pattern, delivery rate, frequency band, and the shape of individual notes to one another than either is to *vulpecula*. There are few recordings of individual songs of *alopecias* and *vulpina*, largely because the beginnings of a song from one member of a pair typically elicits an immediate duet response from its mate. Therefore, recordings of duets and calls greatly outnumber those of individual songs.

Differences in duets among these three taxa parallel those in the calls and songs, with Marchantaria birds sounding like *vulpecula* from Peru and Ecuador, and unlike *alopecias* and *vulpina*, which resemble one another closely. Unfortunately, without recording in stereo with two microphones, it is nearly impossible to get a clean sonagram of the duets of these birds because the individual notes are too difficult to pick out. These difficulties are compounded by the problem of the male and female routinely being different distances from the microphone. Therefore, no sonagrams of duets are presented here. Typical duets for all Rusty-backed Spinetails seem to involve both members of the pair singing similar phrases, with one bird initiating the song and the other immediately joining in. As with individual songs, duets of *vulpecula*, *alopecias*, and *vulpina* vary substantially in length (2 to 5 seconds), and tend to become longer when the birds are clearly agitated (as when responding to tape playback).

Playback trials.—Playback experiments with *Cranioleuca* recordings offer further evidence of the significance of vocal differences among the taxa, and of their role as potential isolating mechanisms.

On 22 January 1993, I returned to Marchantaria with Whittaker. After locating a pair of spinetails, I played recordings of songs, calls, and duets of nominate *vulpina* from Mato Grosso, Brazil. Several playbacks over 1–2 minutes elicited no response. When I switched to recordings of spinetails made at Marchantaria the previous October, the birds moved aggressively toward us and responded with repeated, loud duets. Bret Whitney and Mario Cohn-Haft visited Marchantaria on 7 October 1993. They located a pair of spinetails that had never been exposed to tape playback and presented them with a recording of a duet from Anavilhanas (Rio Negro) birds (*alopecias*). The birds showed no response over a period of 2–3 minutes, during which the tape was played 4–5 times. When Whitney presented the birds with a tape from their own (Marchantaria) population, they responded within 30 seconds by flying in and immediately vocalizing. Whitney returned to Marchantaria on 22 October 1993 and performed a similar experiment. Once again the birds did not respond to tape of *alopecias*, but did fly in immediately to tape of Marchantaria birds (Whitney *in litt.*).

On 6 November 1994, I returned to Ilha Marchantaria and performed several playback experiments with three widely separated pairs of spinetails. Each pair of birds was initially presented with recordings of *vulpecula* from Peru. The response of all three pairs was identical and dramatic. Immediately after the first playback, both members of the pair responded with a loud duet. A second playback caused both the male and female to approach me rapidly and in a highly agitated state, with wings drooped and quivering, and tail spread. This was followed by a sustained period of vocalizing (mostly duets, but with some individual songs and calls), which continued for several minutes without further playback from me. Pairs were allowed to calm down and resume foraging for a period of 2–3 minutes after their last vocalizations. I then presented the birds with tapes of nominate *vulpina* from Mato Grosso, Brazil, and *alopecias* from Venezuela and from the Anavilhanas Archipelago, Brazil. Repeated playback of songs, calls, and duets from these populations elicited no response from the spinetails. I then reverted to playback of *vulpecula* from Ecuador, which drew an immediate loud vocal response and aggressive approach by both members of the pair.

On 3 March 1993, along a small tributary of the Río Ventuari in Amazonas, Venezuela, I played tape of songs, calls, and duets of Marchantaria birds (*vulpecula*) to a pair of already vocalizing spinetails (*alopecias*). These birds showed no interest in the recordings, which I presented several times over the course of 5+ minutes, with 30–60 second intervals between playbacks. I then switched to a recording of calls and duets from Rio Negro birds (also *alopecias*). The spinetails responded to the first playback of these calls by immediately approaching my canoe and delivering repeated duets while hopping about in the shrubbery in a clearly agitated fashion. In that same month, Whitney (*in litt.* 1994) performed a similar experiment at Hato Piñero (Cojedes, Venezuela). He located an undisturbed spinetail (*alopecias*) and presented it with a recording of the call and duet from birds (*vulpecula*) recorded near Iquitos, Peru. Whitney played this tape several times over the course of 1–2 minutes, with the bird showing no interest in the tape.

On 23 February 1994 I presented tape of calls, songs, and duets of Marchantaria birds and of *vulpecula* recorded in Peru and Ecuador to spinetails (*alopecias*) along a small caño at Hato Piñero (Cojedes, Venezuela). Repeated playback attempts elicited no response from either of two pairs of spinetails. On 17 January 1995, in the Anavilhanas Archipelago of the Rio Negro, I presented recordings of nominate *vulpina* (recorded in the Pantanal of Mato Grosso) to two pairs of spinetails (*alopecias*), both of which responded immediately and aggressively to recordings of calls, songs, and duets.

On 9 September 1994 I presented tape of spinetails from Marchantaria to two pairs of foraging *vulpina* near Pixaim, Mato Grosso, Brazil. Neither pair of spinetails exhibited any interest in the tape. When I switched to playback of vocalizations of *alopecias* (from edo. Cojedes, Venezuela, and from the Anavilhanas Archipelago, Rio Negro, Brazil), the spinetails immediately responded with rapid and nearly continuous calls, punctuated by frequent duets, and they approached me in a clearly agitated manner. A pair of *vulpina* along the Rio Cristalino (Mato Grosso, Brazil) on 20 September 1994 displayed a similar lack of response to playback of vocalizations of Marchantaria birds, but responded vigorously to tape of *alopecias* from Venezuela and the Rio Negro. On 8 and 9 September 1995, near Pixaim (Mato Grosso, Brazil), I presented recordings of individual songs of *alopecias* (from the Anavilhanas Archipelago, Rio Negro) to each of three different pairs of nominate *vulpina*. Each pair responded immediately with rapidly repeated loud calls and duets followed by close approach.

On 21 January 1995, along the Rio Juma (a west-bank tributary of the Rio Madeira, south of the Amazon), I presented recordings of calls, songs, and duets of both Anavilhanas birds (*alopecias*) and Pantanal birds (*vulpina*) to what appeared to be a family group of Rusty-backed Spinetails (either *alopecias* or *vulpina*). Both sets of recordings elicited vocal and aggressive responses from the group of spinetails. On 1 April 1995, along the Rio Javarí near Benjamin Constant (Amazonas, Brazil), Whitney (pers. comm.) presented tape of *vulpecula* recorded from Sucusari, Peru (Río Napo) to a foraging pair of spinetails (either *vulpina* or *alopecias*). They responded by singing two duets (typical *vulpina/alopecias* type). When Whitney presented the pair with recordings of their own voices the birds approached closely and sang several times. A second playback of *vulpecula* song from Peru elicited no further response.

On 9 November 1994, in Minas Gerais, Brazil, Whitney (pers. comm.) presented recordings of nominate *vulpina* from Alta Floresta (Mato Grosso) to a pair of *reiseri*. The pair responded vigorously by immediately approaching Whitney closely and vocalizing.

The results of these playback trials are summarized in Table 1.

Morphology.—Although vocalizations are perhaps the most conspicuous difference between *vulpecula* and other taxa in the *Cranioleuca vulpina* complex, morphological differences are also pronounced.

Specimens of *vulpecula* have significantly longer and heavier bills than either *alopecias* or *vulpina*, with only minimal overlap in bill height or length of exposed culmen (Table 2). Based on this sample, *alopecias* and *vulpina* do not appear to differ in bill size. The wing chord of *vulpecula* is also significantly longer than that of *alopecias*, which, in turn, is longer than that of *vulpina*. The sample size for *reiseri* is too small for statistical comparison, but bill and wing measurements for the two specimens examined fall within the middle of the range for both *alopecias* and *vulpina*, and at the extreme low end for *vulpecula*.

Plumage differences are less easily quantified, but are more striking (Fig. 4). Compared to all other forms, *vulpecula* is extensively clean white on the chin, throat, and upper breast (giving an overall much paler look to the underparts), contrasting with gray-buff on the lower half of the abdomen. The dividing region on the lower breast of *vulpecula* is also conspicuously flammulated or mottled (this sometimes less evident on older specimens). Both *alopecias* and *vulpina* are very similar to one another below, with nearly uniform dingy gray-buff underparts (darker than the lower abdomen of *vulpecula*), with only vague flammulation on the breast, and a marginally whiter throat (less extensive and duller white than in *vulpecula*). The two specimens of *reiseri* examined are warmer ochraceous-buff below, with little evidence of breast flammulation, and only a slightly whiter throat.

The dorsal plumage of *vulpecula* is brighter and more uniformly rust-colored (a paler, "livelier" rust), whereas *alopecias* and *vulpina* are a darker rust color, with the crown, wings, and tail brighter than the back (which is suffused with brown), and the rump usually duller than the back. In *vulpecula* the forehead tends toward a mildly contrasting buffy color which is overlain with visibly dark streaking. No such effect is noticeable in most *alopecias* and *vulpina*, both of which are typically uniformly dark rust from the forehead to the rear crown.

Hellmayr (1925) commented that *alopecias* was nearest to *vulpina* but with the "upper back more or less suffused with brownish (male) or entirely brown (female), contrasting with the

TABLE 1

Summary of Playback Trials (as Detailed in the Text) Involving Various Taxa within the *Cranioleuca vulpina* Complex. A "Strong" Response Involved Immediate and Repeated Vocalizations (from Previously Non-vocalizing Birds) as Well as Approach toward the Sound Source. Only One Response Was Considered Anything Other Than "Strong", That Involving Only Two Songs (without Approach or Subsequent Vocalizations). "Subject taxon/Locality" Identifies the Taxon of the Bird on Which the Trial Was Being Conducted, and the Site. Locations Where These Trials Were Conducted Were: Ilha Marchantaria, Amazonas, Brazil; Edo. Amazonas, Venezuela; Edo. Cojedes, Venezuela; Anavilhanas Archipelago, Rio Negro, Brazil; Mato Grosso, Brazil; Benjamin Constant, Amazonas, Brazil; Amazonas, Brazil; and Minas Gerais, Brazil. Codes for the Various Playback Variables Are as Follows: Pec = *vulpecula*, Alo = *alopecias*, Vul = *vulpina*, (M) = Marchantaria, (P) = Peru, (E) = Ecuador, (V) = Venezuela, (AA) = Anavilhanas Archipelago, and (MG) = Mato Grosso. Blank Lines Indicate Tests That Were Not Performed

Subject Taxon/Locality	Response to playback of:					
	Pec (M)	Pec (P)	Pec (E)	Alo (V)	Alo (AA)	Vul (MG)
vulpecula (Marchantaria)	strong	strong	strong	none	none	none
alopecias (Amaz, VEN)	none	—	—	—	strong	—
alopecias (Cojedes, VEN)	none	none	none	—	—	—
alopecias (Anavil., BRA)	—	—	—	—	—	strong
vulpina (Mato Grosso, BRA)	none	—	—	strong	strong	—
vulpina/alopecias (Benjamin Con., BRA)	—	weak	—	—	—	—
vulpina/alopecias (Amaz., BRA)	—	—	—	—	strong	strong
reiseri (Minas Gerais, BRA)	—	—	—	—	strong	—

TABLE 2

Mean Measurements (± Std. Error, with Ranges beneath the Means) in mm of "Rusty-backed" Spinetails (*Cranioleuca vulpina*), Followed by Statistical Comparisons of the Means. Values of P (= Probability of a Greater Value of F) Derived from Analysis of Variance (ANOVA). Duncan's Multiple Range Tests Were Used to Group Taxa by Character, Based on Significant Differences in the Means. There Were Significant Between-taxa Differences for All Three Characters Measured. *C. v. vulpecula* Differed Significantly from *C. v. vulpina* and *C. v. alopecias* over All Three Measurements. *C. v. vulpina* and *C. v. alopecias* Differed from One Another Only in Wing Chord Length. Sample Size for *C. v. reiseri* Was Too Small to Permit Statistical Comparisons

	N	Bill height at base	Culmen	Wing chord
C. v. vulpecula	27	4.5 (.05) [3.9–5.1]	13.6 (.14) [11.7–14.7]	68.3 (.40) [64.3–72.8]
C. v. alopecias	23	3.8 (.04) [3.4–4.1]	11.2 (.11) [10.0–11.9]	66.3 (.44) [62.1–70.5]
C. v. vulpina	17	3.9 (.05) [3.5–4.2]	11.3 (.11) [10.4–12.0]	65.0 (.54) [60.9–68.8]
*C. v. reiseri**	2	3.8 [3.8–3.9]	11.4 [11.2–11.6]	65.0 [64.6–65.5]

Statistical Comparisons of Means

Measurement	P	Groupings
1) Bill height	<0.0001	Group A = *C. v. vulpecula*
		Group B = *C. v. vulpina* and *C. v. alopecias*
2) Culmen	<0.0001	Group A = *C. v. vulpecula*
		Group B = *C. v. vulpina* and *C. v. alopecias*
3) Wing Chord	<0.0001	Group A = *C. v. vulpecula*
		Group B = *C. v. alopecias*
		Group C = *C. v. vulpina*

FIG. 4. Ventral views of "Rusty-backed" Spinetail (*Cranioleuca vulpina*) specimens. Left to right: *C. v. alopecias, C. v. alopecias, C. v. alopecias, C. v. vulpecula, C. v. vulpecula, C. v. vulpecula, C. v. vulpina, C. v. vulpina, C. v. vulpina, C. v. reiseri, C. v. reiseri*. (photo by K. J. Zimmer).

hazel crown; rump and lower parts darker, more tinged with ochraceous brown." I could find no such consistent differences between *alopecias* and *vulpina* in my examination of specimens, although subtle distinctions could have been obscured by the uncertainty of the racial allocation of specimens from the lower Rio Tapajós. Hellmayr (1925) did note that although specimens of *alopecias* from Venezuela closely matched the type (from Forte do Rio Branco, Brazil), three birds from the north bank of the lower Amazon closely approached *vulpina*. This suggests that any variation in the two forms may be clinal, which would explain the confusion surrounding birds from the immediate south bank of the Amazon.

In the field, Marchantaria birds (as compared to both *alopecias* and *vulpina*) appear to be a brighter, paler rust color above, with larger bills, and a conspicuously white throat and upper breast that contrast with a mottled, streaky, gray-buff chest. Overall they appear brighter and show more contrast than do other taxa in the complex. These observations match those made by Sclater and Salvin (1866), Wetmore (1957), and Vaurie (1980) regarding specimens of *vulpecula* from Peru and western Brazil, as well as my own observations of *vulpecula* specimens. The specimen of *vulpecula* (#MCH 384) from Marchantaria also matches others in the LSUMZ taken from Peru (M. Cohn-Haft, pers. comm.).

Habitat and behavior.—The literature, specimen label data, and my own field experience (combined with that of many other workers) suggest marked differences in habitat occupied by *vulpecula* versus both *alopecias* and *vulpina*. The following accounts of habitat and behavior are based on my own field observations, except where otherwise noted.

In the llanos of Venezuela, *alopecias* occurs in gallery forest and riverine woodlands bordering small streams or larger rivers that retain water during the severe dry season. Farther south, in terr. Amazonas, Venezuela, *alopecias* is found in a narrow band of seasonally flooded *várzea* forest adjacent to rivers, streams, and oxbows. In each of these regions the habitats occupied are well-wooded, with largely closed canopies (in the llanos, degree of canopy closure is subject to the drought-deciduous nature of many of the trees) that may exceed 20–30 m in height. The understory is shaded but fairly open, and is characterized by abundant woody vines, thickets, and aerial roots. Typically, many trees and shrubs bordering the watercourses form dense tangles that overhang the banks. Spinetails in each of these regions are generalized gleaners of bark and foliage, spending much time creeping along upper and under sides of branches and along woody vines (typically with the body held parallel to the foraging substrate). Crevices in bark, bases of stems, and clusters of dead leaves are all important foraging substrates. Prey are less frequently taken from live foliage. These spinetails are encountered as individuals, pairs, or in small, noisy (presumably family) groups. In the llanos, individuals and pairs often accompany mixed-species flocks of woodcreepers, tyrant-flycatchers, and greenlets (*Hylophilus*).

In the Anavilhanas Archipelago on the Rio Negro (Brazil), *alopecias* occupies habitats similar to those used in Amazonas, Venezuela. The larger islands support mature woodland that is seasonally flooded and similar in profile to *várzea* forest along the rivers and streams of Amazonas, Venezuela. Spinetails observed there remained in woody tangles near the river edge, and in shrubbier thickets that border lagoons in island centers. *Várzea* woodlands bordering the "mainland" banks of the Rio Negro are also occupied by spinetails. Foraging behavior and degree of sociality is as described above.

My observations of nominate *vulpina* come mostly from Mato Grosso, Brazil. In the Amazonian lowlands of northern Mato Grosso, it occupies forest thickets and vine tangles along streams and rivers (similar to habitats used by *alopecias* in southern Venezuela). In the south it is also found in the seasonally flooded savanna woodlands of the Pantanal. In the Pantanal these spinetails are most common in riverine woodlands along the larger (or at least more permanent) watercourses, where a mostly closed canopy (10–20 m) shades a fairly open understory that has an abundance of woody vines and thickets. Also occupied are shrubbier edges of savanna woodlands where these border ditches and depressions that act as catch basins for water during the dry season. Again, foraging behavior and degree of sociality are as described for *alopecias* in Venezuela.

In contrast, *C. v. vulpecula* appears to be limited to islands in the larger "white-water" rivers of Amazonia. Rosenberg (1990) listed *Cranioleuca vulpina* (= *vulpecula*) as one of 18 obligate river-island species found in northeastern Peru along the Río Napo and the Río Amazon. In eight months of field work and extensive collecting in all habitats in the region, neither Rosenberg nor other LSUMNS personnel recorded *vulpecula* or any other "obligate river-island species" away from the islands (Rosenberg 1990). Rosenberg found *vulpecula* to be something of a habitat generalist on young-to-medium-aged islands, where it used a variety of successional habitats, from scrub dominated by *Tessaria integrifolia* to various-aged stands of *Cecropia*. Within these habitats *vulpecula* used a variety of foraging substrates, including *Tessaria*, *Cecropia*, *Mimosa*, and vines, mostly at lower and middle heights in the canopy (Rosenberg 1990).

Precise localities where spinetails were collected were frequently difficult to determine from specimen labels, particularly for older specimens. However, of the specimens examined, all *vulpecula* that could be pinpointed were collected on islands in white-water rivers (among these the Ríos Amazon, Napo, and Madeira). Those with more ambiguous label data were at least referable to sites located along white-water rivers. Similarly, I could find no specimens of *alopecias*, and only three of *vulpina* (all from one island location, Obidos Island in the Amazon—these are called *alopecias* by Peters 1951) that were referable to islands of white-water rivers. Where locality data were specific, all other specimens of *alopecias* and *vulpina* appeared to have been taken from "mainland" localities on one bank or another of a given river, or on islands of "black-water" rivers.

"Rusty-backed" Spinetails on Ilha Marchantaria occupy habitats similar to those inhabited by *vulpecula* on islands in the Napo and Amazon in Peru and Ecuador. As with other islands in large white-water rivers of Amazonia, Marchantaria is constantly eroded on the older upstream end, which is forested with broad-leaved woodland similar to "mainland" varzea. Continual new silt and sand deposition at the younger downstream end creates favorable conditions for a dynamic succession of plant communities, from canegrass and sandbar scrub to stands of *Salix*, *Mimosa*, and *Cecropia*. "Rusty-backed" Spinetails occupy virtually all undisturbed habitats on Marchantaria. Whitney (*in litt.*) has found them in the interior of tall woodland at the upstream end, where the dominant trees are *Ficus* and *Cecropia*, and where the mostly open canopy ranges from 6–15 m in height, although spinetails appear to be much more common in younger successional habitats (canegrass, *Salix*, and younger stands of *Cecropia*) nearer the downstream end.

My observations of these spinetails on Marchantaria have all involved pairs of birds foraging between 1–4 m above the ground in shrubby, successional habitats, particularly in tall (3–5 m) canegrass. Here, they gleaned a variety of small arthropod prey from stems, branches, masses of organic debris suspended in the vegetation by changing water levels, and especially from the bases of leaves, usually moving outward from the center of the shrub or tree with a series of short hitching motions (Zimmer and Whitney, pers. obs.) typical of the genus. Whitney (*in litt.*)

has also observed them foraging to heights of 12–14 m in *Ficus* at the upstream end of the island. One such bird foraged on thin limbs and twigs in the subcanopy of the tree, mostly away from the trunk (closer to the periphery), crawling and hitching along while gleaning from bark and twigs only (not from foliage). It paid particular attention to fissures and cavities in the limbs, moving along limbs and twigs with its body parallel to the substrate, and often progressing outward from the trunk along the undersides of limbs (Whitney *in litt.*).

DISCUSSION

"Rusty-backed" Spinetails inhabiting successional habitats on Ilha Marchantaria are clearly referable to *C. v. vulpecula*. Songs and calls of Marchantaria birds closely match those of *vulpecula* recorded in Peru and Ecuador (Figures 2 and 3), as do duet vocalizations (pers. obs.). Plumage characteristics of Marchantaria birds (as observed by Zimmer, Whittaker, Whitney, and Cohn-Haft, and as corroborated by specimen #MCH 384) also match those of specimens of *vulpecula* collected in Peru, Ecuador, and western Brazil. Habitat characteristics (white-water river islands with a variety of successional habitats) also conform to those of places occupied by *vulpecula* in Peru and Ecuador. Perhaps most significantly, Marchantaria birds respond vigorously to playback of recordings of calls, songs, and duets of *vulpecula* from both Peru and Ecuador, and do not respond to vocalizations of geographically proximate *alopecias* and *vulpina*. This represents a significant eastward range extension for *vulpecula*, which was previously known in Brazil only from the Rio Purús (Hellmayr 1925), the Rio Juruá (Gyldenstolpe 1945), and the upper Rio Madeira at São Antonio de Guajará (Peters 1951). That *vulpecula* occurs on Marchantaria should not be surprising. Of the 18 obligate river-island species that Rosenberg (1990) found on his study sites on the Napo, at least 14 (*Leucippus chlorocercus, Furnarius minor, Synallaxis propinqua, Cranioleuca* [*vulpina*] *vulpecula, Certhiaxis mustelina, Thamnophilus cryptoleucus, Myrmochanes hemileucus, Elaenia pelzelni, Serpophaga hypoleuca, Stigmatura napensis, Cnemotriccus fuscatus, Knipolegus orenocensis, Conirostrum margaritae* and *Conirostrum bicolor*) also occur on Ilha Marchantaria (Zimmer, unpublished ms.). It would appear that *vulpecula*, like many other obligate river-island species, has a widespread but essentially linear distribution along the Amazon and its major tributaries.

It also seems clear that *vulpecula* should be treated as a species, not as a subspecies of *vulpina*. Wetmore previously (1957) recommended it be accorded full specific rank based on morphological differences alone. Plumage differences (particularly in ventral coloration) are striking even in the field, and *vulpecula* is probably diagnosable solely by mensural characters. All known vocalizations of *vulpecula* are diagnostic, and tape-playback experiments indicate that *vulpecula* does not respond to the voice of either nominate *vulpina* or *alopecias*, nor do the latter forms respond to the voice of *vulpecula*. "Rusty-backed" Spinetails north of the Amazon (*alopecias* and *apurensis*) and south of the Amazon (*vulpina*) are nearly identical to one another morphologically, vocally, and ecologically, and respond to tape playback of one anothers' voices, but the range of the morphologically, vocally, and ecologically distinct *vulpecula* lies between them. Additionally, *vulpecula* is sympatric (although not syntopic) with both *alopecias* and *vulpina* in parts of Amazonian Brazil.

The distinctions between *alopecias* and *vulpina* appear tenuous at best. The two forms are continuously distributed across the Amazon, with specimens from near the river on the north bank showing signs of intermediacy (Hellmayr 1925). Populations from the lower Rio Tapajós have been variously assigned to either *alopecias* or *vulpina*. Although there may be slight morphological differences between Venezuelan populations of *alopecias* and populations of *vulpina* from well south of the Amazon, these differences appear to me to be clinal, with no obvious discontinuities. Similarly, analysis of vocalizations of north-bank *alopecias* and south-bank *vulpina* reveals that the two forms have nearly identical calls and similar songs. Current sample sizes of tape-recorded songs are insufficient to determine whether noted differences in songs of *alopecias* versus *vulpina* are merely attributable to individual variation, if they are diagnosably different, or, if they may vary clinally. Playback trials have shown that *alopecias* responds vigorously to calls and songs of *vulpina*, as does the latter to vocalizations of *alopecias*. On current evidence, I recommend that *alopecias* be treated as a synonym of *vulpina*.

It is not known how far east along the Amazon *vulpecula* ranges, or whether the many islands near the mouth provide suitable habitat. Contact in northeast Brazil between *vulpecula* and *reiseri* seems unlikely, but the latter form is morphologically and vocally similar to *vulpina* and *alopecias*, and there is no reason at present to suggest that its taxonomic affinities lie outside of the Rusty-backed Spinetail complex.

With respect to English names, the name "Rusty-backed Spinetail" is well established and should be retained for *vulpina, alopecias, apurensis, foxi,* and *reiseri* (even though these birds are no rustier dorsally than *vulpecula*). The English name of "Peruvian Rusty-backed Spinetail" used for *vulpecula* by Hellmayr (1925) is no longer appropriate now that the bird is known to occur widely in Ecuador and Brazil. It is tempting to suggest a name for *vulpecula* that will reflect the unique nature of its habitat preferences, but "River-Island Spinetail" is not exclusive enough (other species of spinetails also inhabit river islands), and anything more specific is unwieldy. In light of these difficulties, I suggest the English name of "Parker's Spinetail", in honor of the late Theodore A. Parker III. The elucidation of the status of *vulpecula* as a distinct species was initially based on field recognition of vocal differences, an area in which Parker excelled above all others. Furthermore, he was among the first to recognize the unique nature of Amazonian river-island avifaunas and bring them to the attention of the scientific community.

The ecological gap between *vulpecula* and Rusty-backed Spinetails to the north and south is of particular interest. Some closely related species-pairs replace one another at the transition from *terra firme* to *várzea* in Amazonia, among them *Monasa morphoeus* and *M. nigrifrons*, *Myrmoborus myotherinus* and *M. leucophrys*, and *Schiffornis turdinus* and *S. major*. Remsen and Parker (1983) and Rosenberg (1990) have also documented the almost complete difference in species composition between "mainland" *terra firme* forest and young and medium-aged Amazonian river islands. Therefore, that sibling species of spinetails could replace one another from "mainland" *várzea* to Amazonian river-islands dominated by successional habitats is not a novel idea. It is interesting however, that nearby (<70 kms distant) river islands in a major black-water tributary (the Rio Negro) would be populated by *alopecias* rather than by *vulpecula*. This suggests that the distribution of *vulpecula* is not determined by the presence of river islands per se, but by specific habitats not available on all islands. Rosenberg (1990) reached similar conclusions regarding the distributions of some *Tessaria*-scrub specialists. Certainly the islands that I have visited in the Anavilhanas Archipelago of the Rio Negro bear little or no similarity to Marchantaria or other islands in the Solimões (= Amazon). The islands in the Rio Negro have more mature, stable forests structurally similar to the *várzea* on either bank of the adjacent "mainland." Among the suboscines that breed in close proximity to *alopecias* in this habitat are *Nasica longirostris, Thamnophilus nigrocinereus, Myrmotherula assimillis, M. klagesi, Hypocnemoides melanopogon, Myrmoborus lugubris, Lathrotriccus euleri,* and *Hemitriccus minor*, none of which occur with *vulpecula* in the successional habitats on Marchantaria dominated by canegrass, scrub, *Cecropia*, and *Salix*. *M. assimilis* does occur with *vulpecula* in more wooded habitats at the upstream end of Marchantaria. Remsen and Parker (1983) and Rosenberg (1990) also found that older islands with more mature woodlands did share many species with nearby "mainland" *várzea*.

Current evidence suggests that some factor or combination of factors specific to islands in white-water rivers determines the presence of *vulpecula*. Nearby islands on black-water (e.g., the Rio Negro) or clear-water (e.g., the Rio Tapajós) rivers are instead populated by either *alopecias* or *vulpina*. A parallel white-water island to black-water island replacement within a closely related species pair can be found among antshrikes of the genus *Thamnophilus*. Near Manaus, Brazil, *Thamnophilus nigrocinereus*, the Blackish-gray Antshrike, is common on river islands in the Río Negro, but is replaced by the closely related *T. cryptoleucus* (Castlenau's Antshrike) on the white-water islands of the Solimões (including Ilha Marchantaria). Remsen and Parker (1983) noted the primitive state of our knowledge of habitat preferences of Amazonian birds and the likelihood that differences existed between white-water and black-water river-created habitats. Many plant ecologists distinguish between *várzea* (seasonally inundated forests bordering sediment-filled white-water rivers) and *igapó* (inundated forests flooded by sediment-free, clear-water or black-water rivers), and have noted edaphic, physiognomic, and floristic peculiarities of each (Janzen 1974; Prance 1979; Anderson 1981). Species replacements from white-water river-islands to black-water river-islands or *igapó* within sibling-species pairs of spinetails and antshrikes provide further evidence of the complexities of river-created habitats in Amazonia, and of their effect on species distributions.

With its linear and somewhat fragmented distribution, and its restriction to successional habitats on river islands, *vulpecula* seems particularly vulnerable to habitat perturbation. As noted by Rosenberg (1990), the overall population size of most obligate river-island birds is probably relatively small, and because of the dynamic nature of their habitats, the continued existence of these birds is dependent on the perpetual formation of new islands. Although the significant contribution of river-created habitats to overall avian diversity in Amazonia has been well documented (Remsen and Parker 1983), few existing reserves or parks include extensive river-island habitats within their boundaries. Any changes in water flow in the Amazon and its major tributaries, as could result from damming or from increased flooding and erosion resulting from deforestation (Gentry and Lopez-Parodi 1980), could place *Cranioleuca vulpecula* and many other river-island specialists at extreme risk.

ACKNOWLEDGMENTS

This paper is dedicated to the memory of Theodore A. Parker III, my friend, colleague, and mentor, who shared so freely of his knowledge of neotropical birds. Some of my most cherished memories are of time spent in the field with Ted in Brazil, and in particular in sharing his excitement over the many discoveries from Alta Floresta (this volume).

Andrew Whittaker has been my enthusiastic field partner in much of Brazil, and first introduced me to the avian richness of the Manaus region. Tom Schulenberg, Mario Cohn-Haft, Mort and Phyllis Isler, and Bret Whitney have provided much advice and stimulating conversation regarding this and other taxonomic problems. Cohn-Haft and Rita Mesquita were also most gracious hosts during my visit to Baton Rouge. Whitney provided tape recordings of spinetails, as well as valuable field observations of Marchantaria birds. Mort and Phyllis generously donated their time and talents to providing the map and sonograms (respectively) used in this paper. J. V. Remsen and his students and staff at the Louisiana State University Museum of Natural Science were most helpful in coordinating specimen loans from the various institutions, as well as in providing access to resources at LSU. I must also thank Kenneth Parkes (Carnegie Museum), Kimball Garrett (Los Angeles County Museum), and Tom Schulenberg (Field Museum of Natural History) for the loan of specimens from their respective institutions. Dave Lightfoot was a valuable consultant on the statistical aspects of this work. Greg Budney and the staff at the Library of Natural Sounds, Cornell Laboratory of Ornithology, were of inestimable help in allowing access to the LNS collection of recordings, including 25 cuts of *Cranioleuca vulpina*. Additional recordings were provided by Jose Fernando Pacheco, Gary Rosenberg and Andrew Whittaker. J. V. Remsen, Douglas Stotz, and Bret Whitney reviewed earlier drafts of this manuscript and made numerous helpful suggestions for improvement. I also thank Victor Emanuel, for providing me with the travel opportunities that made this research possible.

LITERATURE CITED

ANDERSON, A. B. 1981. White-sand vegetation of Brazilian Amazonia. Biotropica 13:199–210.
BALDWIN, S. P., H. C. OBERHOLSER, AND L. G. WORLEY. 1931. Measurements of Birds. Scientific Publ. Cleveland Mus. Nat. Hist. Vol. II, 1–165. Cleveland, Ohio.
FITZPATRICK, J. W., AND D. E. WILLARD. 1990. *Cercomacra manu*, a new species of antbird from southwestern Amazonia. Auk 107:239–245.
GENTRY, A., AND J. LOPEZ-PARODI. 1980. Deforestation and increased flooding of the upper Amazon. Science 210:1354–1356.
GRAVES, G. R. 1987. A cryptic new species of antpitta (Formicariidae: *Grallaria*) from the Peruvian Andes. Wilson Bull. 99:313–321.
GROTH, J. G. 1988. Resolution of cryptic species in Appalachian red crossbills. Condor 90:745–760.
GYLDENSTOLPE, N. 1945. The bird fauna of Rio Juruá in western Brazil. K. Svenska Vetensk. Akad. Handl., ser. 3, 22:1–338.
HELLMAYR, C. E. 1925. Catalogue of birds of the Americas. Field Mus. Nat. Hist. Publ. Zool. Ser. Vol. 13, part IV.
HILTY, S. L., AND W. L. BROWN. 1986. A Guide to the Birds of Colombia. Princeton University Press, Princeton, New Jersey.
JANZEN, D. H. 1974. Tropical blackwater rivers, animals, and mast fruiting by the Dipterocarpaceae. Biotropica 6:69–103.
JOHNSON, N. K., AND J. A. MARTEN. 1988. Evolutionary genetics of flycatchers. II. Differentiation in the *Empidonax difficilis* complex. Auk 105:177–191.
LACK, D. 1971. Ecological Isolation in Birds. Harvard University Press, Cambridge, Massachusetts.
MEYER DE SCHAUENSEE, R. 1970. A Guide to the Birds of South America. Livingston Press, Narbeth, Pennsylvania. (Reprinted by International Council for Bird Preservation with new addenda, 1982.)
MEYER DE SCHAUENSEE, R., AND W. H. PHELPS, JR. 1978. A Guide to the Birds of Venezuela. Princeton University Press, Princeton, New Jersey.
MONROE, B. L., JR., AND C. G. SIBLEY. 1993. A World Checklist of Birds. Yale University Press, New Haven and London.
OREN, D. C., AND H. G. ALBUQUERQUE. 1991. Priority areas for new avian collections in Brazilian Amazonia. Goeldiana (Zool.) 6:1–11.
PETERS, J. L. 1951. Checklist of Birds of the World. Vol. VII. Harvard University Press, Cambridge, Massachusetts.

PIERPONT, N., AND J. W. FITZPATRICK. 1983. Specific status and behavior of *Cymbilaimus sanctaemariae*, the Bamboo Antshrike, from southwestern Amazonia. Auk 100:645–652.
PRANCE, G. T. 1979. Notes of the vegetation of Amazonia III. The terminology of Amazonian forest types subject to inundation. Brittonia 31:26–38.
REMSEN, J. V., JR., AND T. A. PARKER III. 1983. Contribution of river-created habitats to bird species richness in Amazonia. Biotropica 15:223–231.
REMSEN, J. V., JR., AND M. A. TRAYLOR, JR. 1989. An Annotated List of the Birds of Bolivia. Buteo Books, Vermillion, South Dakota.
RIDGELY, R. S., AND J. A. GWYNNE. 1989. A Guide to the Birds of Panama (with Costa Rica, Nicaragua and Honduras). Princeton University Press, Princeton, New Jersey.
RIDGELY, R. S., AND G. TUDOR. 1994. The Birds of South America, Vol. 2. Univ. of Texas Press, Austin, Texas.
ROSENBERG, G. H. 1990. Habitat specialization and foraging behavior by birds of Amazonian river islands in northeastern Peru. Condor 92:427–443.
SCLATER, P. L. 1874. On the species of the genus *Synallaxis* of the family Dendrocolaptidae. Proc. Zool. Soc. London 1874:2–28.
SCLATER, P. L., AND O. SALVIN. 1866. Catalogue of birds collected by Mr. E. Bartlett on the River Ucayali, with notes and descriptions of new species. Proc. Zool. Soc. London 1866:175–201.
SIBLEY, C. G., AND B. L. MONROE, JR. 1990. Distribution and Taxonomy of Birds of the World. Yale University Press, New Haven and London.
STEEL, R. G. D., AND J. H. TORRIE. 1980. Principles and Procedures of Statistics. 2nd ed. McGraw-Hill, New York.
STILES, F. G. 1983. The taxonomy of *Microcerculus* wrens (Troglodytidae) in Central America. Wilson Bull. 95:169–183.
TACZANOWSKI, L. 1884. Ornithologie du Pérou. Vol. I. Friedlander and Sohn, Berlin.
VAURIE, C. 1980. Taxonomy and geographical distribution of the Furnariidae (Aves, Passeriformes). Bull. Amer. Mus. Nat. Hist. 166:1–357.
WETMORE, A. 1957. The birds of Isla Coiba, Panama. Smithsonian Miscellaneous Collections 134:1–105.
WILLIS, E. O. 1988. *Drymophila rubricollis* (Bertoni, 1901) is a valid species (Aves, Formicariidae). Rev. Brasil. Biol. 48:431–438.
WILLIS, E. O. 1991. Sibling species of greenlets (Vireonidae) in southern Brazil. Wilson Bull. 103:559–567.

APPENDIX 1

Recording locations and recordists. Numbers following each name represent the number of recordings from the recordist at each site.

alopecias—VENEZUELA: Hato Piñero, edo. Cojedes (T. A. Parker 3, K. J. Zimmer 12); Junglaven Camp, Río Ventuari, terr. Amazonas (K. J. Zimmer 1); lower Río Caura near Maripa, edo. Bolivar (B. Whitney 1). BRAZIL: Anavilhanas Archipelago, Rio Negro, Amazonas (M. Cohn-Haft 1, K. J. Zimmer 6); Baixo Japurá, Boca do Japurá, Estação Ecológica Mamirauá, Amazonas (J. F. Pacheco 1); Rio Branco, Boa Vista, Roraima (K. J. Zimmer 3).

vulpina—BRAZIL: Pixaim to Porto Jofre, Mato Grosso (K. J. Zimmer 18); Rio Cristalino, Mato Grosso (K. J. Zimmer 3). BOLIVIA: Flor de Oro, Parque Nacional Noel Kempff Mercado, depto. Santa Cruz (B. Whitney 1).

vulpina/alopecias—BRAZIL: Amazon Lodge, Rio Juma, Amazonas (A. Whittaker 1, K. J. Zimmer 1); Rio Javarí, 5–6 km west of Benjamin Constant, Amazonas (B. Whitney 1).

vulpecula—BRAZIL: Ilha Marchantaria, Rio Solimões, Amazonas (K. J. Zimmer 11); Ilha do Mateiro, Estação Ecológica Mamirauá, Amazonas (J. F. Pacheco 1). ECUADOR: La Selva Lodge, Río Napo (G. H. Rosenberg 2). PERU: Isla Ronsoco, Río Napo, depto. Loreto (T. A. Parker 5); Isla Llachapa, Río Napo, depto. Loreto (T. A. Parker 5, G. Budney and T. A. Parker 1); Isla Timicurillo, Río Amazonas, depto. Loreto (T. A. Parker 7); Isla de Iquitos, depto. Loreto (M. B. Robbins 1).

reiseri—BRAZIL: Rio São Francisco S. Januária, Minas Gerais (B. Whitney 2).

APPENDIX 2

List of localities and lending instituitions for specimens examined. All specimens were from one of the following institutions: Carnegie Museum, Pittsburgh (CM), Field Museum of Natural History, Chicago (FMNH), Los Angeles County Museum of Natural History, Los Angeles (LACM), and the Louisiana State University Museum of Natural Science, Baton Rouge (LSUMZ).

vulpecula (15 males, 12 females)—BRAZIL: Amazonas, Rio Juruá (FMNH, 1 male, 1 female). PERU: Dpto. Loreto, Isla Tuhayo and Isla Ronsoco, Río Napo, ca. 80 km N of Iquitos and Isla Pasto, Río Amazonas, ca. 80 km NE of Iquitos (LSUMZ, 14 males, 10 females); Ucayali, Yarina-Cocha, Río Ucayali (FMNH, 1 female).

vulpina (7 males, 10 females)—BRAZIL: São Paulo, Barra do Rio Dourado (FMNH, 1 female); Mato Grosso, Descalvados, Rio Paraguay (FMNH, 1 male); Pará, Santarem, Amazon River (CM, 3 males, 3 females); Pará, Obidos Island, Amazon River (CM, 1 male, 2 females); Amazonas, Rio Purus (CM, 2 males, 2 females); Amazon River (CM, 1 female); Tapajós River (CM, 1 female).

alopecias (13 males, 10 females)—BRAZIL: Amazonas, Caviana, Rio Solimões (CM, 1 male, 1 female); Amazonas, Manacaparu, Rio Solimões (2 males, 2 females); Pará, Monte Alegre area, N of Amazon River (LACM, 3 males, 1 female); Pará, Ilha de Urucurituba, Rio Amazonas (FMNH, 1 male, 1 female); Pará, Rio Cururu, Tapajós River (LACM, 2 males, 1 female); Pará, Rio Tapajós (LACM, 1 female); Pará, Santarém (LACM, 1 male); Roraima, Rio Mucajai S of Boa Vista (LACM, 1 male, 1 female). COLOMBIA: Boyaca, Trinidad (LACM, 1 male); Meta, Carimagua (FMNH, 1 male, 1 female). VENEZUELA: Delta Amacuro, Caño Mariusa, Barrancos, 150 km NE of Orinoco River Delta (LACM, 1 female).

reiseri (1 male, 1 female)—BRAZIL: Bahia, Santa Rita de Cassia, Rio Preto (LACM).

AVIFAUNA OF A LOCALITY IN THE UPPER ORINOCO DRAINAGE OF AMAZONAS, VENEZUELA

KEVIN J. ZIMMER[1] AND STEVEN L. HILTY[2]

[1]1665 Garcia Rd., Atascadero, California 93422, USA; and
[2]6316 West 102nd St., Shawnee Mission, Kansas 66212, USA

ABSTRACT.—This paper reports the findings of several surveys of the avifauna of the region surrounding two remote fishing camps in the upper Río Orinoco region of Amazonas, Venezuela. This area is of particular interest because it has been little-explored by ornithologists, and because of the predominance in the region of habitats derived from white-sand soils. We provide information on a number of range extensions, including one species (Pale-bellied Mourner, *Rhytipterna immunda*) for which there were no previous records for Venezuela. We also provide information on the vocalizations, foraging ecology, and habitat preferences for many rare or poorly known species, including *Mitu tomentosa, Neomorphus rufipennis, Notharchus ordii, Myrmotherula cherriei*, and *Hylophilus brunneiceps*. Our observations of *Hylophilus brunneiceps* also have taxonomic implications. The soil-based insularity of the local *terra firme* habitat as a factor limiting local bird distributions is also discussed.

RESUMÉN.—Este papel reporta los encuentros de varios reconocimientos de la avifauna de la region rodeando dos campamentos de pezca remotos en la parte superior del Río Orinoco en la region de la Amazonas, Venezuela. Esta área es de interes particular porque se ha explorado poco departe de ornitólogos y porque la predominancia de habitacíones en la region se diriven de terrenos de arena-blanca. Proveemos informacíon sobre varias extensíones, incluyendo un especie (*Rhytipterna immunda*) por el cual no se encuentran archives anteriores en Venezuela. Tambien proveemos informacíon sobre las vocalizacíones, forraje ecologico, y habitacíones preferidas departe de muchos especies raros ó poco conocidos incluyendo *Mitu tomentosa, Neomorphus rufipennis, Notharchus ordii, Myrmotherula cherriei*, y *Hylophilus brunneiceps*. Nuestras observacíones de *Hylophilus brunneiceps* tambien tiene implicacíones taxonomicas. La insulacíon de la base terrena cercano, habitacíones *terra firme*, como punto que limita la distribucíon de pajaros tipicos tambien se discute.

In February 1990 we began surveying the birds in the area surrounding a small fishing camp, "Campamento Junglaven," in the interior of the Territorio Federal de Amazonas, Venezuela, at approximately 05°06′N, 66°44′W (ca. 156 km E-SE from Pto. Ayacucho), south and east of the Río Orinoco and northwest of the Río Ventuari (Fig. 1). The camp is located on the left (east) bank of Camani Creek, a small tributary flowing south into the Río Ventuari. A second nearby fishing camp "Camani Camp" is on the north bank of the middle Río Ventuari, between the Indian village of Camani and the mouth of the Río Manapiare. The area is completely undeveloped except for the two camps and ca. 15 km of sandy roads connecting them to the shared airstrip. Access is limited to boat or small aircraft. Annual precipitation is ca. 2,550–3,000 mm, much of it concentrated in June through October (Schwerdtfeger 1976). There is a pronounced November–May dry season.

The surrounding area contains a mosaic of different soil types that support a natural patchwork of distinct vegetation types. Much of the area is dominated by white-sand soils covered by grassy savanna with scattered shrubs (1–3 m in height) and small stands of *Mauritia* palms. Bordering these savannas are scrubby, low-canopy (3–10 m) woodlands (= savanna woodlands) that are also found on white-sand soils. These woodlands are dense, with an understory that is often nearly impenetrable. They are somewhat deciduous, with a partially open canopy, and are seasonally flooded (June–October). Scattered lagoons and oxbows within these woodlands retain standing water throughout the dry season (November–May). The Ventuari and its many small tributaries are flanked by wide bands of taller (15–25 m) forest that grows on yellow-clay soils and is seasonally flooded. This *várzea* forest is characterized by a more closed canopy with a fairly open understory and an abundance of vines and lianas. Farther from the river, large isolated

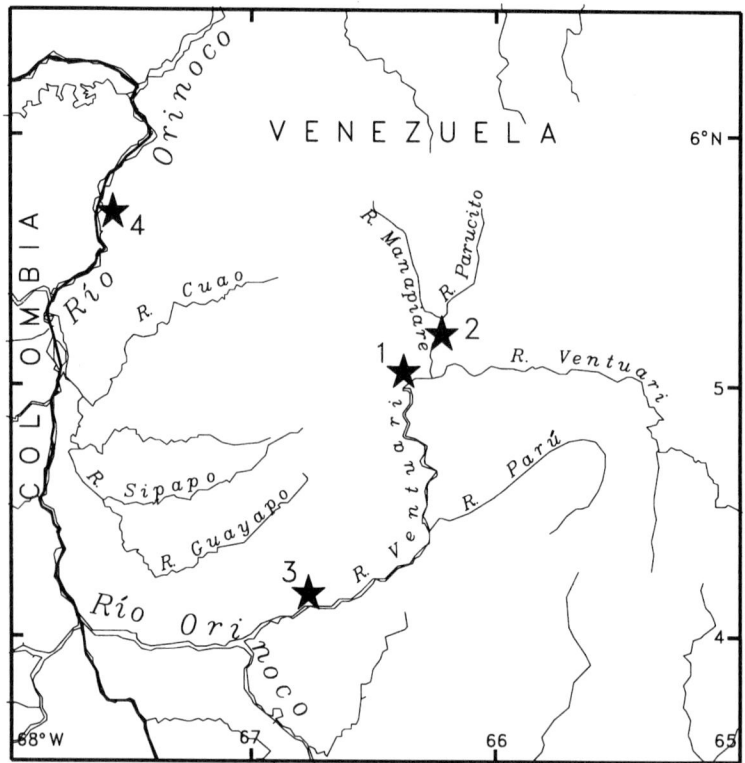

FIG. 1. Location of Junglaven Camp in relation to Pto. Ayacucho, major rivers, and to the two nearest Phelps expedition collecting sites. Black stars and their numbers locate: 1 = the area of this survey surrounding Junglaven and Camani Camps, 2 = San Juan de Manapiare (Phelps and Phelps 1952), 3 = Las Carmelitas (Phelps, Jr. 1947), and 4 = Pto. Ayacucho, Amazonas, Venezuela.

patches of red-clay soils support tall (>30 m), lush "islands" of humid tropical forest that are not seasonally flooded (= *terra firme* forest). The boundaries of these habitat types are clearly delineated by the underlying soil mosaic. Thus, local bird distributions are often sharply defined. Sandstone tabletop mountains (tepuis) and quartzite dome-like sandstone hills (cerros) are scattered throughout the region as a whole, but none are readily accessible from the Junglaven and Camani camps, and thus were not surveyed.

Access to these habitats is limited. The road system (ca. 15 km) traverses mostly savanna habitats. About 7 km of road winds through the large patch of *terra firme* forest adjacent to Junglaven Camp. This narrow track is shaded by the forest canopy, and has served as our only real trail into the *terra firme* forest. Two additional trails figured prominently in our surveys. These were the "Laguna Galapago Trail" (ca. 2 km), which traverses seasonally flooded savanna woodland and ends at a shallow lagoon, and the "Manaca Trail" (ca. 1 km), which dissects high-canopy *várzea* forest along the Río Ventuari. Additional coverage of *várzea* and riverine habitats has been by boat along Camani Creek (ca. 2–3 km north and south from Junglaven Camp), the Río Ventuari (ca. 5 km east and west from Camani Camp), and Caño Guayaje (ca. 3–5 km upstream from its confluence with the Ventuari).

Dates of our visits are as follows: 10–14 February 1990 (Hilty), 29 December 1990–4 January 1991 (Hilty and Zimmer), 29 December 1991–6 January 1992 (Hilty), 28 February–5 March 1992 (Zimmer and R. S. Ridgely), 29 December 1992–5 January 1993 (Hilty), and 26 February–5 March 1993 (Zimmer). Additionally, the area has been surveyed by R. O. Prum and J. D. Kaplan, who spent 10–11 March and 22 March–28 April, 1990 at Camani Camp. They summarized their sight observations in a field checklist titled *Preliminary Checklist to the Birds of Campamento Camani, Territorio Amazonas, Venezuela* (*in litt.*), which was later revised by R. A. Rowlett and J. Coons (*in litt.*) following their visit to Junglaven Camp from 23 February–2 March 1991. Further visits have been limited to occasional small parties of birders. By their own estimation, Prum and Kaplan spent the "vast majority" of their time in the *várzea* forest, and this bias is reflected in their checklist. Our time has been more evenly distributed (10% savanna, 20% *várzea*, 30% savanna woodland, 40% *terra firme*), but still has been biased toward the more diverse *terra firme* forest, with relatively little time spent in the savanna. All of our observations from *terra firme* forest were made along the ca. 7-km entrance road into Junglaven Camp. Savanna woodland was surveyed primarily along the Laguna Galapago trail and from the edges of the Junglaven Camp clearing. The Phelps Foundation from Caracas has sent numerous collecting expeditions into Amazonas, both upriver and downriver of Junglaven. The closest collecting sites (Fig. 1) are San Juan de Manapiare on the Río Manapiare (Phelps and Phelps 1952), an affluent of the upper Río Ventuari (Paynter 1982) <50 km upstream from Junglaven, and Las Carmelitas (ca. 04°10'N, 66°45'W), on the left bank of the Río Ventuari, about 100 km south-southwest of Junglaven. A collection at Las Carmelitas was made in February 1947 (Phelps, Jr. 1947; Paynter 1982).

The avifauna of this region contains an interesting blend of species derived from several biogeographic regions. Many species are widespread in western Amazonia. Others are characteristic of the Guianan Shield or Pantepui region (Mayr and Phelps 1967; Meyer de Schauensee and Phelps 1978; Cracraft 1985) that encompasses the tablelands and surrounding lowlands of Amazonas territory and Bolívar state in Venezuela, as well as adjacent areas in Guyana and northern Brazil. A few species are endemic to the upper Río Orinoco and Río Negro regions. Cracraft (1985) called this the "Imeri Center" of endemism and defined it as including the lowlands of southern Amazonas (Venezuela), Brazil north of the upper Río Negro and Rio Vaupés, and the eastern portions of Vaupés, Guainia, and southeastern Vichada, Colombia. Many species endemic to this region seem to be specialists of white-sandy-soil forests (Hilty and Brown 1986; Hilty and Zimmer, pers. obs.) and are very locally distributed.

Many widespread *terra firme* species seem to be curiously absent from this region. Several of these are common in lowland humid forest to the east in the Venezuelan state of Bolívar, and to the south in southern Amazonas and northern Brazil. We would note particularly the lack of records of Guianan Slaty-Antshrike (*Thamnophilus punctatus*, Isler et al. 1997), Mouse-colored Antshrike (*T. murinus*), Dusky-throated Antshrike (*Thamnomanes ardesiacus*), Rufous-winged Antwren (*Herpsilochmus rufimarginatus*), any species of antpitta, and Guianan Cock-of-the-Rock (*Rupicola rupicola*), as well as a general lack of *Automolus* foliage-gleaners, *Myrmotherula* antwrens, and wrens (Troglodytidae).

The Guianan Cock-of-the-Rock is unrecorded at Junglaven, probably due to lack of suitable nesting sites. The surrounding region is spotted with cerros and tepuis whose steep sides and forested lower slopes appear to be ideal nesting sites (Snow 1982), but these are not readily accessible from Junglaven. The species is fairly common near Pto. Ayacucho, where nesting habitat is accessible (pers. obs.).

More difficult to explain are the apparent distributional gaps in the ranges of the previously mentioned antbirds. *Thamnophilus murinus* is common in *terra firme* forest farther east in Bolívar, and occurs in smaller numbers in the same habitat south of Pto. Ayacucho (pers. obs.), but there are no Junglaven records. Willard et al. (1991) found it fairly common in the lowlands around Cerro Neblina. *Thamnomanes ardesiacus* is common in *terra firme* forests to the east in Bolívar and to the south in Amazonas (Willard et al. 1991) but has not been recorded at Junglaven. The absence of *Thamnophilus punctatus* is particularly interesting, given that the species is fairly common in second-growth edges of *terra firme* forest to the north and west in Amazonas near Pto. Ayacucho (pers. obs.). To the south, in northern Brazil, *T. punctatus* seems to be a specialist in sandy-soil campina woodlands (Zimmer, pers. obs.) that are structurally similar to the savanna woodlands at Junglaven (where this habitat is occupied by both *Thamnophilus nigrocinereus* and *Thamnophilus amazonicus*). *Herpsilochmus rufimarginatus* has not been found at Junglaven, although it is common in *terra firme* forest at Pto. Ayacucho (pers. obs.) and is known generally from northern and central Amazonas (Meyer de Schauensee and Phelps 1978). No species of antpittas have been found at Junglaven, although *Grallaria varia* and *Myrmothera campanisona* are known from *terra firme* forest in Amazonas (Meyer de Schauensee and Phelps 1978; Willard et al. 1991).

Both *Automolus infuscatus* (Olive-backed Foliage-gleaner) and *A. ochrolaemus* (Buff-throated Foliage-gleaner) were unusually scarce in the *terra firme* forests near Junglaven, but are more common in the more humid forests of eastern Bolívar (pers. obs.) and to the south in Amazonas (Willard et al. 1991). Similarly, the *Myrmotherula* assemblage in the *terra firme* forests at Junglaven seemed notably depauperate. Only *M. axillaris* and *M. menetriesii* were common (*M.*

haematonota was uncommon and there was only one record of *M. guttata*), and there were no records of *M. longipennis* (which is known from central Amazonas).

The near complete absence of wrens (Troglodytidae), particularly from *terra firme* forest, was notable. *Thryothorus leucotis* was common in *várzea*, and *Troglodytes aedon* occurred in clearings around the camps, but these were the only wrens recorded. The absence of *Thryothorus coraya*, *Henicorhina leucosticta*, and *Microcerculus bambla*, all known from central and southern Amazonas and Bolívar, is puzzling.

In many cases the seeming absence of these birds may result from the insular nature of *terra firme* forest in the Junglaven area. The only accessible tall, humid forest at Junglaven, although many square kilometers in extent, is isolated from other contiguous tall forest by several kilometers of low, dense *várzea* scrub, or by some type of savanna or savanna woodland. These drier habitats may serve as effective barriers to many forest species, preventing them from colonizing such forest "islands." Alternatively, the current insular nature of these forest patches could be the result of climatic changes that isolated forest patches that began with a full complement of *terra firme* species from the surrounding region. Over time, species diversity in the newly isolated patches would then have dropped (through differential extirpation) until some "island" equilibrium was reached. Another possible explanation is that the entire Junglaven area is somewhat drier than surrounding regions to the east in Bolívar and to the south in Amazonas. Perhaps more importantly, the Junglaven region has a more seasonal climate (with a more pronounced dry season) than the aforementioned areas (Mayr and Phelps 1967), and this seasonality may produce physiological or resource bottlenecks that exclude some humid-forest birds. If the forest "islands" at Junglaven were indeed formerly connected to larger expanses of forest, then climate-induced physiological or resource bottlenecks could have been the agents of differential extirpation. In the absence of regular immigration across unsuitable habitat, populations not well-adapted to the drier conditions would have eventually died out.

In the accounts that follow we provide notes on some little-known species (including observations on behavior) from the vicinity of Junglaven and Camani camps. We supplement these notes (where relevant) by additional observations from a structurally similar site 20–60 km south of Puerto Ayacucho (= Campamento Camturama), Amazonas, Venezuela. Some observations detailed here represent range extensions within the country or in the territory of Amazonas; for one species, the Pale-bellied Mourner *Rhytipterna immunda*, we report the first documented records for Venezuela. Because several of the species recorded from this region are locally distributed and poorly known, we have included notes on vocalizations, behaviors, and habitat. For one of these species, the Brown-headed Greenlet *Hylophilus brunneiceps*, the information presented in this paper has taxonomic significance.

Our observations should be viewed as a preliminary survey of the Junglaven avifauna. Visits in other seasons may yield additional information on the presence or absence of many species. Our understanding of the regional avifauna in general, and, in particular, of certain taxa, could be greatly enhanced by additional surveys that employed systematic mist-netting and collecting procedures. The 394 species of birds known from the vicinity of Junglaven and Camani camps (Appendix) were found primarily on our surveys, but we have included supplemental records reported in Prum and Kaplan (*in litt.*) and Rowlett and Coons (*in litt.*), as well as a few records submitted by other observers (most of these from P. Coopmans). Birds not recorded by us are attributed to the observers. Many records reported in this paper are sight observations; others are substantiated by tape-recordings or photographs. All sound recordings referred to in this paper by the authors, and by R. O. Prum and P. Coopmans, have been (or will be) deposited at the Library of Natural Sounds (LNS), Cornell University, Ithaca, New York. Photographs referred to in this paper are housed at VIREO (Visual Resources for Ornithology), Academy of Natural Sciences of Philadelphia. Terminology used in describing foraging behaviors follows Remsen and Robinson (1990).

SPECIES ACCOUNTS

Agami Heron (*Agamia agami*).—We have consistently found small numbers of these rarely seen forest herons along small tributary streams of the Ventuari. Virtually all records have been of lone individuals foraging quietly beneath tangles of dead or live branches drooping down over the water from adjacent banks. On 31 December 1990, we observed two adults and an immature bird fishing together (within 10 m) in a shaded, backwater stretch of the Caño Guayaje. These birds foraged in shallow water by walking among fallen dead branches and large exposed root buttresses that concentrated small fish. The birds were relatively undisturbed by our presence. During a 45-minute observation period, all three birds periodically uttered a low, gutteral "uur'r'r'r'r'" (Zimmer tape-recording). R. O. Prum (pers. comm.) saw and tape-recorded two adults and a juvenal several times at the same site in March and April 1990.

Zigzag Heron (*Zebrilus undulatus*).—We have found this poorly known, small heron on several occasions, in *várzea* forest along small tributaries of the Ventuari or in dense savanna woodland near swampy depressions with much decaying leaf litter. On 3 March 1992, Zimmer photographed an adult that was feeding uncharacteristically in the open (under overcast skies) along the Caño Guayaje. This bird flew at the initial approach of the boat, but then came bounding back down the bank in a series of antpitta-like hops. It hopped up in some dead branches hanging out low over the water and began fishing like a Striated Heron (*Butorides striatus*), clinging to the branch by its feet and extending its neck slowly toward the water before making a sudden jab. It also hopped quickly from a perch into the shallow water to spear prey and then hopped back out. Prey items were believed to be small fishes or tadpoles. The heron made several abrupt perch changes, many of which involved a jumping "about-face." For the 15 minutes that we observed it, the bird flicked its tail constantly, jerking it from a resting position downward at an angle to one side, then back to normal, and then jerking it downward at an angle to the other side.

In our experience this species is largely crepuscular, becoming vocal shortly before dusk and again in the pre-dawn. At these times individuals are often responsive to tape playback, sometimes flying in from more than 100 m distant. The advertising call is a deep, resonant "whoooo" or "whooah" repeated approximately once every 5 seconds. This call is similar to calls of this species south of the Amazon in Brazil (Zimmer, pers. obs.) and similar to that described from Ecuador by English and Bodenhorst (1991). In response to tape playback one individual repeatedly made a barely audible call reminiscent of a grumbling stomach.

These records extend the known range of this species in Amazonas (Meyer de Schauensee and Phelps 1978), where it was known previously only from El Carmen on the left bank of the Río Negro near the Brazilian border (Paynter 1982). However, this secretive bird is easily overlooked in many areas where it occurs. The only published records of *Zebrilus* from north of the Orinoco in Venezuela are from the east in Delta Amacuro and Sucre (Meyer de Schauensee and Phelps 1978), yet Zimmer (unpubl. data) and D. Wolf (pers. comm.) have recently found Zigzag Herons along small streams bordered by deciduous gallery forest near El Baul (Cojedes, Venezuela) in the llanos, a major extension of the known range.

Lesser Razor-billed Curassow (*Mitu tomentosa*).—This species was present in *várzea* forest and in savanna woodlands that bordered lagoons or oxbows. It was usually encountered singly or in groups of up to five, rarely to 10, birds. In Colombia this species is reported to occupy gallery forest in the llanos, and *terra firme* forest elsewhere (Hilty and Brown 1986). At Junglaven we did not find it more than 200 m inside *terra firme* forest. *Mitu* may be replaced in that habitat by another large curassow, *Crax alector*, which was fairly common in *terra firme* forest, but which was not found in other habitats.

Upland Sandpiper (*Bartramia longicauda*).—This Nearctic migrant was seen twice in savanna in April 1990 by Prum and Kaplan (*in litt.*). Zimmer and R. S. Ridgely observed a single bird in savanna near the airstrip on 2 March 1992 and again on 5 March 1992. Records in Amazonas are few, being reported only in March at San Fernando de Atabapo (Phelps and Phelps 1958).

White-rumped Sandpiper (*Calidris fuscicollis*).—A single individual of this Nearctic migrant was reported by P. Coopmans from 10 to 14 October 1993. There are no published records from Amazonas.

Scarlet-shouldered Parrotlet (*Touit huettii*).—Our only record was of a pair perched high in *várzea* canopy on 3 January 1991. Prum and Kaplan (*in litt.*) recorded a single pair in flight over the Río Ventuari. The nearest known localities in Venezuela are northwestern Bolívar on the upper Río Parguaza, and in Amazonas along the Río Orinoco at San Fernando de Atabapo (Meyer de Schauensee and Phelps 1978). Parker and Remsen (1987) suggested that this species may be somewhat nomadic, which may account for the paucity of records from this site.

Orange-cheeked Parrot (*Pionopsitta barrabandi*).—This widespread parrot is probably more numerous in the savanna forest around Junglaven than in any place where we have encountered it in western Amazonia. We regularly found this species venturing onto the savanna to feed, particularly in the early mornings. In March 1993, Zimmer found 12 birds feeding in close proximity on small fruits of an unidentified savanna shrub 1.5–3 m above the ground.

Yellow-billed Cuckoo (*Coccyzus americanus*).—Sightings of this Nearctic migrant were reported by Prum and Kaplan (*in litt.*), and also by Paul Coopmans, 10–14 October 1993. Previously published records in Amazonas are from Nericagua and Puerto Yapacana, both on the Río Orinoco (Phelps and Phelps 1958).

Rufous-winged Ground-Cuckoo (*Neomorphus rufipennis*).—Zimmer recorded this poorly known Pantepui endemic at Junglaven on four occasions during the February–March 1993 trip, and Hilty recorded it on one occasion during the 1991–1992 trip. All of Zimmer's records were from *terra firme* forest. Hilty's record was from *várzea*. On 2 March 1993 we encountered an army ant swarm (*Eciton* sp.) attended by *Dendrocolaptes certhia*, *Dendrocincla fuliginosa*, and numbers of *Gymnopithys rufigula*. Repeated broadcasting of tape of the voice of Rufous-winged Ground-Cuckoo (obtained in Bolívar, Venezuela) induced a vocal response from a bird that called several times from more than 100 m away and then approached silently. It ran rapidly toward us on a zigzagging course, pausing occasionally to hop onto logs or run up low, leaning branches to peer about. During this time the bird partially erected and then lowered its crest, and dipped its tail up and down. On 3 March (at 1130 hrs), about 2 km west of the previous day's antswarm, we saw a ground-cuckoo dust-bathing in the road. The bird ran into the forest upon our approach, but tape playback induced several loud mandible clackings similar to those described for *N. geoffroyi* (Slud 1964; Hilty and Brown 1986; Ridgely and Gwynne 1989).

In all instances calling birds gave a resonating "whooop" repeated at 3–6 second intervals (Fig. 2). This call is soft and somewhat dove-like at close range, but is far-carrying. After playback, the call occasionally sounds slightly tremulous in quality, more like "hrroop." These vocalizations are very similar to those recorded from *N. rufipennis* by both Hilty and Zimmer in the state of Bolívar, Venezuela. *Neomorphus rufipennis* seems unusually vocal compared to *N. geoffroyi*. Observers familiar with that species rarely report any sound other than mandible clacking, although Slud (1964) on one occasion found a bird in Costa Rica giving a "low muffled 'woof'."

Long-tailed Potoo (*Nyctibius aethereus*).—This species was recorded by R. Rowlett and J. Coons (*in litt.*) on 2 March 1991, when they taped two birds, and saw one in the pre-dawn in *terra firme*. This species was also recorded by R. Behrstock, B. Finch, and J. Kingery on 12–13 May 1992, when they heard three birds and saw one (R. Berhstock, pers. comm.). Observers in both parties had previous field experience with this species and the other species of potoos found in Venezuela. The Long-tailed Potoo is known from scattered records from Colombia and the Guianas south through eastern Ecuador and Peru to Amazonian Brazil, southeastern Brazil, and Paraguay (Hilty and Brown 1986). It was previously known in Venezuela only from El Dorado in northeastern Bolívar (Meyer de Schauensee and Phelps 1978). The Long-tailed Potoo has also been recorded from Vaupés in eastern Colombia (Hilty and Brown 1986).

Rufous Nightjar (*Caprimulgus rufus*).—This species was heard and tape-recorded by Hilty on 30 December 1992 and 3 January 1993 in *várzea* scrub, and by P. Coopmans 10–14 October 1993. The only published records of this species for Venezuela south of the Orinoco are at Caño Cataniapo, Río Orinoco, Amazonas, and at El Carmen, Alto Río Paraguaza, Bolívar (Phelps and Phelps 1958).

Brown-banded Puffbird (*Notharchus ordii*).—On 31 December 1990, at 15:30, B. Masters and R. Komuniecki (members of our party) found a Brown-banded Puffbird perched quietly in scrubby vegetation at the edge of *terra firme* forest. When we arrived at the site, we discovered a pair of birds excavating a nest cavity in a termitarium 4–5 m up in a mostly bare tree. The pair was observed at this same spot on two subsequent days, both times in the early-mid afternoon. During each observation one member of the pair typically perched silently outside the nest (within 5 m), while its mate worked inside the termitarium for up to 15 minutes. The only vocalization heard was a soft, nasal "yank" (similar to a subdued White-breasted Nuthatch [*Sitta carolinensis*]), given several times by one of the birds when inside the termitarium. Whenever the working member of the pair emerged, the other bird would either take its place inside the nest, or the two birds would sit side by side for a few minutes before both flying off. This is

→

FIG. 2. Sonograms of voices of some poorly known bird species recorded near Junglaven Camp, Amazonas, Venezuela [K. J. Zimmer, recordist (c) 1994 All rights reserved. Library of Natural Sounds, Cornell Laboratory of Ornithology, Ithaca, New York]. 1 = "whooop" call of *Neomorphus rufipennis* (LNS #64590). 2 = song of male *Myrmotherula cherriei* followed by contact call of a female *M. cherriei* (LNS #64585). 3 = display call of male *Heterocercus flavivertex* (LNS #64573). 4 = song of *Rhytipterna immunda*. 5a = typical song ("peern peern peern") of *Hylophilus brunneiceps* (LNS #64591). 5b = complex song of *H. brunneiceps*. This vocalization may be given separately, or as a prelude to the typical song.

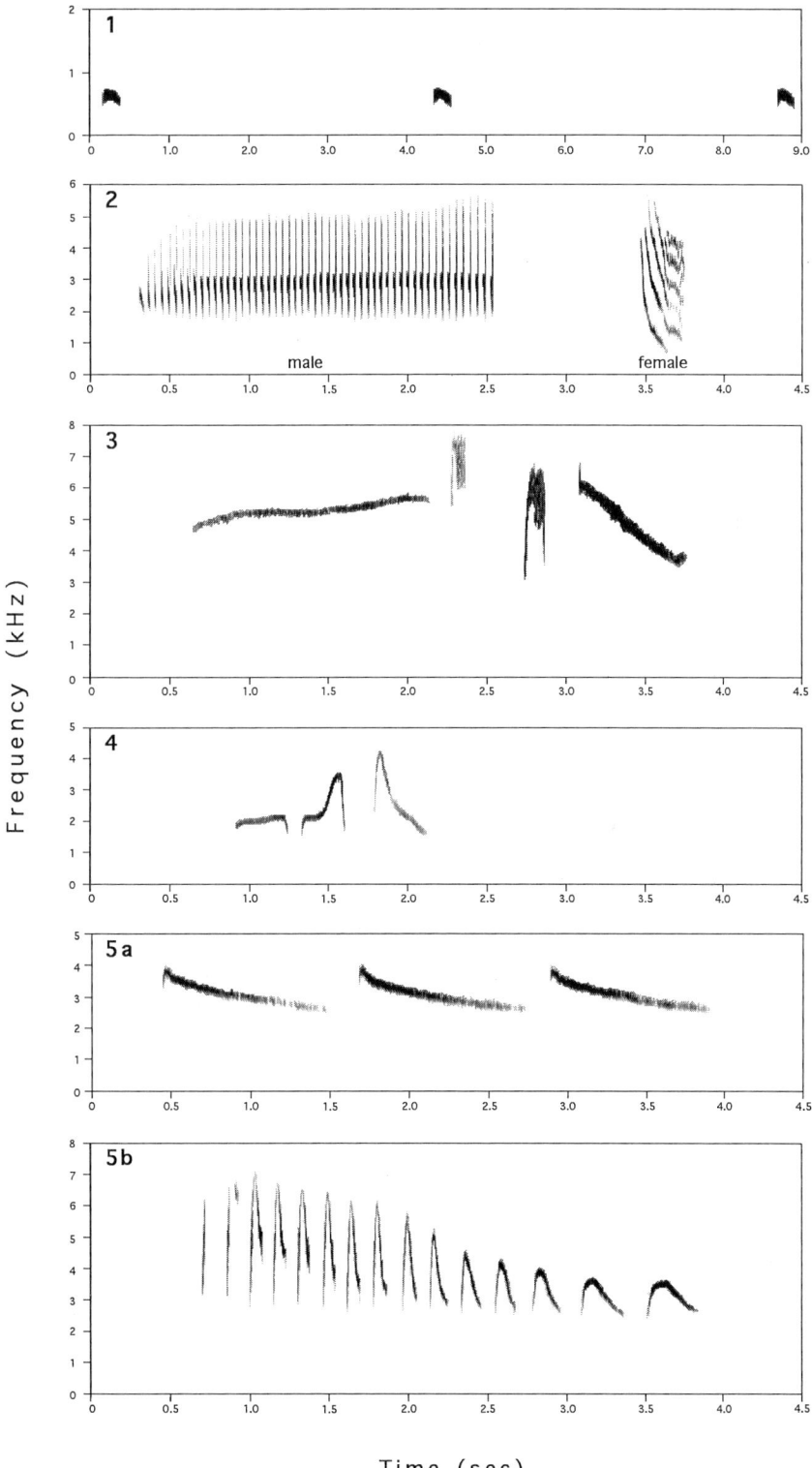

the first described nest of *Notharchus ordii*, although other species of *Notharchus* are known to nest in arboreal termitaria (Hilty and Brown 1986; Sick 1993). Zimmer photographs are on file at VIREO.

Prior to the last 10 years, this species was known only from study skins collected in Venezuela and Brazil prior to the middle part of this century. Meyer de Schauensee (1982) listed the known range as "Venezuela in Amazonas, adjacent Brazil and south of the Amazon between ríos Xingú and Tapajós." In Venezuela *N. ordii* is known only from Amazonas south of the Río Ventuari (Meyer de Schauensee and Phelps 1978). In the past several years, single records from Peru (Foster et al. 1994) and Bolivia (Parker and Remsen 1987), and several from Amazonian Brazil (Péres and Whittaker 1990; Zimmer et al. 1997, M. Cohn-Haft, pers. comm.) suggest that this species may be widely distributed in low densities throughout much of western and central Amazonia and the upper Orinoco river drainage. The single records from Peru and Bolivia and the recent record from near the upper Urucú river, Tefé, Amazonas, Brazil (Péres and Whittaker 1990) were of birds in the canopy of *terra firme* forest, as were many of the recent records from the Rio Cristalino, Mato Grosso, Brazil (Zimmer et al. 1997). Willard et al. (1991) collected an individual at forest edge near Cerro Neblina, Amazonas. Zimmer and A. Whittaker have found *N. ordii* in stunted, deciduous forest growing on rocky outcroppings near the Rio Teles Pires, northern Mato Grosso, Brazil (Zimmer et al. 1997). Zimmer (unpubl. data) has also recently found this species in somewhat stunted forest growing on white-sand soils in the upper Rio Negro region of northwestern Brazil. Clearly, this species should be watched for in the canopy of both tall humid forest and scrubby forest habitats.

Rusty-breasted Nunlet (*Nonnula rubecula*).—We have found this infrequently observed nunlet on several occasions in both *terra firme* and *várzea* forest. On 31 December 1990, Zimmer found a bird perched 0.5 m above the ground carrying a 3–4 cm unidentified arthropod in its bill. It sat quietly, nervously swinging its tail in an arc for several seconds before flying off. We returned to the site the next day, and Hilty played a tape of *N. frontalis* from Panama. This elicited an instant response from a *N. rubecula*, which flew in from at least 20 m away and gave several excited squealing calls before flying off again. On 2 January 1991, we encountered another individual perched 5 m above the ground on an open limb. This bird sat quietly in front of us for several minutes and repeatedly flicked one or both wings high above the back in the manner of *Leptopogon*, *Mionectes*, and *Pogonotriccus* flycatchers.

Tawny-tufted Toucanet (*Selenidera nattereri*).—Our records of this poorly known Imeri endemic (Cracraft 1985) were all from *terra firme* forest, but R. A. Rowlett and J. Coons (*in litt.*) reported seeing a female in dense savanna woodland on 25 February 1991, as did R. S. Ridgely in March 1992. This species is locally distributed from the Guianas through southern Venezuela to eastern Colombia and northwestern Amazonian Brazil (Meyer de Schauensee and Phelps 1978; Hilty and Brown 1986).

Blackish-gray Antshrike (*Thamnophilus nigrocinereus*).—This species was common in dense scrub and seasonally flooded savanna woodland. The areas of highest abundance have low canopies (<12 m) with thick understories that contain many small, spiny-trunked palms (*Bactris* sp.?) that grow in standing water for much of the May-November rainy season. This species also occurs, but in lower densities, in higher-canopy *várzea* along the Ventuari and its tributaries. Behavior and vocalizations noted were similar to those described by Ridgely and Tudor (1994).

Amazonian Antshrike (*Thamnophilus amazonicus*).—This species was common in a variety of habitats, occupying *terra firme* forest, *várzea* forest, and seasonally flooded savanna woodlands. It was encountered most frequently in pairs (occasionally in family groups), either alone or associated with mixed-species flocks. In the high-canopy *terra firme*, *T. amazonicus* foraged mostly 6–15 m above ground. In savanna woodlands, where the canopy was much lower, Amazonian Antshrikes foraged from the canopy (13 m) to about 1.5 m above ground. The foraging pace was similar in style but faster than in most congeners (perhaps most similar to *T. caerulescens*; Zimmer, pers. obs.). *T. amazonicus* punctuated short hops with pauses to scan foliage. They perched on substrates ranging from woody vine tangles to fairly open branches. Prey items were gleaned from both upper and under surfaces of vegetation, usually by reaching. Leaps were often used to obtain prey from overhanging foliage. Prey items identified at Junglaven include lepidopteran and coleopteran larvae, katydids, spiders, mantids, and walking-sticks.

White-shouldered Antshrike (*Thamnophilus aethiops*).—This species was regularly encountered in *terra firme* forest. It occurred in dense patches of second-growth vegetation within light gaps in the primary forest. They were encountered frequently near *Heliconia* thickets. This furtive species foraged alone or in pairs, and independent of mixed-species flocks (as also noted by Ridgely and Tudor 1994), about 0.5 to 5 m above ground. Foraging movements were deliberate, with short hops separated by long pauses to peer about; the birds frequently kept to dense cover.

Singing males sat with an upright posture and frequently erected the crown feathers into a bushy frontal crest. A singing bird's chest expanded in and out to the rhythm of the song, but the birds did not seem to pump or vibrate their tails in the manner of many congeners (Zimmer, pers. obs.). The typical song was 8–10 trogon-like "cah" notes strung together (3 s), the female sounding more nasal and plaintive.

Cherrie's Antwren (*Myrmotherula cherriei*).—This little-known Imeri endemic (Cracraft 1985) was fairly common in three different forest types at Junglaven: 1) *várzea* scrub forest—a dense, low-canopy (<10 m) scrub belt on white-sand soil adjacent to rivers and oxbows; 2) poorly drained savanna woodland; and 3) high-canopy (20 m) river-edge *várzea*. In this region Streaked Antwren (*M. surinamensis*) seemed to be local and restricted to the immediate vicinity of rivers (within 20 m).

Cherrie's Antwrens were typically encountered in pairs, either alone or less often in association with mixed-species flocks that included other antwrens, greenlets, gnatcatchers, honeycreepers, and tanagers. Members of a pair generally foraged within 3–7 m of one another. The contact note is a frequently uttered, somewhat nasal "choop" (Fig. 2). The song is a harsh rattle that starts softly and builds gradually in volume, lasting 2–3 s (Fig. 2). The tail is vibrated as the bird sings.

Foraging was conducted mostly 1–7 m above ground, although in higher-canopy *várzea*, birds often ascended to 10 m. They gleaned foliage in a generalized manner, working rapidly (but not hyperactively) through thin leafy branches as well as hanging dead vines. They hopped from side-to-side as they foraged, taking prey either by gleaning from leaf or vine surfaces, or, more frequently by leaping (to 15 cm) to the underside of leaves. Small green or brown lepidopteran larvae were frequent prey.

Spot-backed Antwren (*Herpsilochmus dorsimaculatus*).—This poorly known species of southern Venezuela, southeastern Colombia, and northern Brazil was fairly common in the vine-rich, leafy canopy of *várzea* forest at Junglaven. Its soft, descending, woodcreeper-like trill (2 s) was often the only indication that it was present. Individuals and pairs foraged alone or associated with canopy mixed-species flocks.

White-plumed Antbird (*Pithys albifrons*).—Although known from the Pto. Ayacucho area (sightings by Hilty, D. Delaney, and J. Langham) and from elsewhere in Amazonas and Bolívar (Meyer de Schauensee and Phelps 1978; Hilty and Zimmer, pers. obs.), this obligate ant-follower has not been recorded at any of the numerous ant swarms that we've encountered at Junglaven. The only Junglaven record is a sighting by P. Coopmans 10–14 October 1993. Willard et al. (1991) found it "common in lowland forest" near Cerro Neblina, Amazonas.

Amazonian Umbrellabird (*Cephalopterus ornatus*).—This species was uncommon but regularly recorded in *várzea* along the Río Ventuari and its tributaries, as well as along scrubbier forest bordering the creek adjacent to Junglaven camp. Most of our records involve lone birds perched high above the river or in undulating flight over the river. Twice we have seen groups of at least five individuals in *várzea* scrub bordering the river near camp. At Pto. Ayacucho, this species also occurs in savanna woodlands along small streams. Zimmer observed a pair carrying nesting material in such habitat 30 km south of Pto. Ayacucho on 28 February 1994.

Bare-necked Fruitcrow (*Gymnoderus foetidus*).—We have recorded this species on several occasions along the Río Ventuari near Camani Camp. All our records have been of lone birds perched high in trees overlooking the river, or in flight over the river. We have also recorded it from savanna woodland (white sand soil) 30 km south of Pto. Ayacucho, a minor northward range extension in Amazonas (Meyer de Schauensee and Phelps 1978).

Yellow-crowned Manakin (*Heterocercus flavivertex*).—This Imeri endemic (Cracraft 1985) was fairly common in *várzea* scrub and savanna woodland at Junglaven. We found males occupying the same display perches (usually ends of broken bare branches or looping woody vines 2–4 m above the ground and beneath the shaded canopy) in consecutive years. The arresting song, repeated at irregular intervals, is a long (1.5–2 s) sliding whistle, followed by an explosive 2-note call, "pseeeeeeee CHIK-KYEE!" (Fig. 2). Males frequently engaged in extended noisy chases, dashing through the dense scrub and uttering a number of loud squeals and sharp, excited notes. R. O. Prum and J. D. Kaplan collected more than 100 hours of behavioral data on male Yellow-crowned Manakins from the Junglaven-Camani area, the results of which are in preparation (R. Prum, pers. comm.).

Rufous-crowned Elaenia (*Elaenia ruficeps*).—This species occurred in bushy savanna habitats near Junglaven. We have also found it in similar brush-dotted savanna 60 km south of Pto.

Ayacucho. These records represent a minor northward extension of the known range in the territory of Amazonas (Meyer de Schauensee and Phelps 1978).

Cinnamon-crested Spadebill (*Platyrinchus saturatus*).—Our only record of this inconspicuous understory flycatcher from Junglaven is an individual seen and taped (distinctive two-note "chip-it") by Zimmer in *terra firme* forest on 28 February 1992. This record represents a minor northward extension of the known range in the territory of Amazonas (Meyer de Schauensee and Phelps 1978).

Pale-bellied Mourner (*Rhytipterna immunda*).—This little-known bird is a fairly common inhabitant of savanna woodland and *várzea* scrub in the Junglaven area. It was first found by us on 31 December 1990, when Zimmer and Hilty taped a singing bird at the edge of Junglaven camp. On 1 March 1992 Zimmer and R. S. Ridgely found and tape-recorded several birds foraging along the borders of savanna woodland. These individuals were sally-hovering for small fruits and calling sporadically. Ridgely compared our recordings to a tape of *R. immunda* made by W. E. Lanyon for confirmation of identity. In February-March 1993 Zimmer recorded this species on 6 of 8 days in the area, made additional tape recordings, and photographed (VIREO) a bird near camp. These constitute the first records for Venezuela. The closest previously known localities were from the mouth of the Río Guainía and the lower Río Inírida, northeastern Guainía, Colombia (Hilty and Brown 1986). Sick (1993) found it in "savanna bordering Amazonian forest, including campinas of white sand" in northern Brazil. It occurs locally from the Guianas to northern Mato Grosso in Brazil.

Rhytipterna immunda may be overlooked because of its striking resemblance to *Myiarchus* flycatchers. In comparison to *Myiarchus* species, the posture of *R. immunda* is more horizontal, its head grayer and more rounded (no apparent crest), and its eyes proportionately larger. It also tends to behave more sluggishly, sitting still in heavy vegetation for long periods. Calling birds can be difficult to locate, because they remain in dense scrub, usually within 5 m of the ground. The usual response to tape playback is to simply maintain position and continue calling, with no change in vocalization. The only vocalization that we have recorded with certainty is an emphatic, whistled "duuuuu-WEET'you" (Fig. 2), although Zimmer has tape-recorded a dawn song (singer unseen) thought to be of this species. The latter vocalization contained elements similar to the known songs, but was much longer and more complex.

Foraging birds moved sluggishly, peering about somewhat like other *Rhytipterna* (pers. obs.). They obtained insect prey by gleans, short leaps, or sally-hovers. Small fruits were regularly obtained by sally-hovering. They foraged alone or in pairs and sometimes joined mixed-species flocks that included *Galbula galbula*, *Thamnophilus amazonicus*, *Myrmotherula cherriei*, *Myiopagis gaimardii*, *Tolmomyias flaviventris*, *Tyrannulus elatus*, *Xenopipo atronitens*, *Polioptila plumbea*, and *Hylophilus brunneiceps*.

(White-throated Kingbird (*Tyrannus albogularis*)).—Not recorded from Junglaven, but on 18 February 1994 Zimmer found a pair ca. 60 km south of Pto. Ayacucho in savanna with scattered shrubs and moriche palms. This represents the first known record from Amazonas (Meyer de Schauensee and Phelps 1978). White-throated Kingbirds are thought to be austral migrants (Chesser 1994); however, the February date of the Pto. Ayacucho record does not fall within a pattern expected of austral migrants. We have found *T. albogularis* in February in palm-savannas near Santa Elena, Bolívar. This species may be overlooked due to its resemblance to Tropical Kingbird (*T. melancholicus*).

Great Kiskadee (*Pitangus sulphuratus*).—Interestingly, there are only two Junglaven records of this species (February 1990 by Hilty, and March 1992 by R. S. Ridgely, both times along Camani Creek near Junglaven camp) although it is common at Pto. Ayacucho (pers. obs.).

Gray-cheeked Thrush (*Catharus minimus*).—There are four records of this Nearctic migrant from Junglaven: single birds seen twice (foraging alone on the forest road and on the ground inside the tall humid forest) in December 1991 by SLH; another bird on 3 January 1993 by Hilty; and one on 1 February 1991 by J. Coons. Although there are scattered records through most of the country (Meyer de Schauensee and Phelps 1978), this species is infrequently encountered on its tropical non-breeding grounds (pers. obs.), and little is known about its areas of concentration during the boreal winter (Ridgely and Tudor 1989). The forests of Amazonas may be an important wintering area for this species.

Black-whiskered Vireo (*Vireo altiloquus*).—One individual, seen by Zimmer and R. S. Ridgely as it foraged with a mid-level/canopy mixed-species flock in *terra firme* forest on 5 March 1992 is the only record. There are few records of Black-whiskered Vireo from Venezuela. All but one of the specimens taken from the Venezuelan mainland has been of *V. a. altiloquus*, a migrant from the Greater Antilles (Meyer de Schauensee and Phelps 1978). The only published Amazonas records, from two locations near Cerro Duida (Boca de Sina, and Cunucunuma), also pertained to the nominate subspecies, and were also from March (Phelps and Phelps 1963).

Brown-headed Greenlet (*Hylophilus brunneiceps*).—This Imeri endemic (Cracraft 1985) was uncommon in savanna woodlands and low-canopy *várzea* scrub. It is known only from blackwater areas and associated sandy-belt forests in southwestern Venezuela and adjacent Colombia and Brazil (Ridgely and Tudor 1989). Confusion has long surrounded this species, largely because *H. hypoxanthus inornatus* (found in Amazonian Brazil south of the Amazon between the Rio Tapajós and the Rio Tocantins) is in many respects phenotypically intermediate between *H. brunneiceps* and nominate *H. h. hypoxanthus*. As noted in Ridgely and Tudor (1989), *inornatus* was originally considered (incorrectly) by Hellmayr (1918) to be a disjunct race of *H. brunneiceps*. Vocal and morphological differences between *H. brunneiceps* and *H. hypoxanthus* confirm that *inornatus* belongs in at least the same superspecies with the latter. Songs of both *H. h. hypoxanthus* (LNS recordings) and *H. h. inornatus* (Zimmer recordings from Mato Grosso, Brazil) are similar: a complex phrase that Ridgely and Tudor (1989) interpreted as "itsochuweet" or "purcheechoweer." This phrase is, in turn, similar to the songs of *H. muscicapinus* (Zimmer recordings). In contrast, Brown-headed Greenlets tape-recorded at Junglaven sing a simple, loud, descending "PEERN PEERN PEERN" repeated up to 1/s (Fig. 2). They also give a longer song that begins with 5-8 twittering notes that lead into 4–5 soft "peer" notes on a level pitch, which in turn leads into the loud, descending notes described above ("swe-swe-swe-swe-swe-swe-peer-peer-peer-peer PEERN PEERN PEERN") [Fig. 2].

Iris color in *H. brunneiceps* has long been another point of confusion. Meyer de Schauensee and Phelps (1978) illustrate the iris of Brown-headed Greenlet as dark, but indicate uncertainty on this point in the caption on the page facing the illustration. Hilty and Brown (1986) describe and illustrate the iris color as dark. Ridgely and Tudor (1989) do not illustrate *H. brunneiceps*, nor do they describe its iris color, but they do include the species in their dark-eyed species grouping. Field observations from Junglaven (R. Prum and J. Kaplan, pers. comm.) have shown that the iris color of *H. brunneiceps* is actually pale, and that the legs are fleshy-pink (not brown as described in Hilty and Brown 1978). Specimens with soft part color descriptions in the Phelps Collection confirm these observations (R. Prum, pers. comm.). Iris color in *H. h. inornatus* is dark (Zimmer, pers. obs.) as it is in nominate *hypoxanthus* (Ridgely and Tudor 1989). Therefore, in voice, morphology and habitat preference, *H. brunneiceps* is more closely allied with the pale-eyed, scrub or forest-border inhabiting *H. flavipes*, *H. thoracicus*, *H. pectoralis*, and *H. semicinereus* group than with the dark-eyed, forest-interior-inhabiting *H. hypoxanthus/H. muscicapinus* group. Ridgely and Tudor (1989) placed *H. brunneiceps* in the latter group, noting that the voice of *brunneiceps* was unrecorded but "probably similar" to that of other members of the *hypoxanthus/muscicapinus* group. This assumption was probably based on the lack of accurate information regarding iris color of *brunneiceps*.

At Junglaven this species foraged in the canopy of savanna woodlands (5–12 m above the ground) and *várzea* forest (usually in the scrubbier, lower-canopied forest). It was usually encountered in pairs, alone or foraging in association with mixed-species flocks. It took small insects by gleaning from leafy outer branches, hanging from the edges of leaves, or occasionally sally-hovered or sally-stalled to take prey from the undersides of overhanging leaves.

Gray-chested Greenlet (*Hylophilus semicinereus*).—Singing individuals were heard from *várzea* scrub along the river near Junglaven camp on 2 March 1992 (Zimmer) and 4 March 1992 (R. S. Ridgely). Paul Coopmans saw and tape-recorded a bird 10–14 October 1993. This represents a small northern extension of the range in the state of Amazonas for this poorly known bird. It was previously known from San Fernando de Atabapo (Río Orinoco) and Las Carmelitas (located on a tributary of the lower Río Ventuari) southward to the Río Casiquiare (Phelps and Phelps 1963).

Short-billed Honeycreeper (*Cyanerpes nitidus*).—This species was fairly common in *terra firme* and *várzea* forests and savanna woodlands, usually in mixed-species flocks with other *Cyanerpes* spp., *Dacnis* spp., and small tanagers. As noted by Ridgely and Tudor (1989), this is usually the least numerous species of *Cyanerpes*. At Junglaven and Pto. Ayacucho it is the most common, outnumbering both *C. caeruleus* and *C. cyaneus*.

White-bellied Dacnis (*Dacnis albiventris*).—There is only one Junglaven record for this poorly known Amazonian species. On 13 May 1992, R. Behrstock and B. Finch observed an adult male in *terra firme* forest near Junglaven camp (Behrstock, pers. comm.). The bird was feeding with a mixed-species flock (that included *Tachyphonus cristatus*, *Dacnis cayana*, and *Hemithraupis flavicollis*) among "lavender-colored flowers of a vine 6–10 m up on a tree at a treefall clearing adjacent to the forest road."

Plumbeous Euphonia (*Euphonia plumbea*).—This species was uncommon around the camp clearing in *várzea* scrub and in savanna woodland. Our records constitute a minor range extension, as it was previously known in Amazonas only from San Fernando de Atabapo (04°03′N, 67°28′W), farther south in Amazonas (Phelps and Phelps 1963).

The call is a whistled "PEE-DEE," usually repeated many times. In pattern it is similar to the calls of *E. chlorotica/E. trinitatus*. The song is jumbled and complex, and may or may not incorporate the "pee-dee" notes. Similar complex songs are given by several other species in the genus, including *E. chrysopasta* and *E. fulviventris* (pers. obs.).

Bobolink (*Dolichonyx oryzivorus*).—A single bird seen by P. Coopmans 10–14 October 1993 in savanna is the only record. Large numbers migrate through Venezuela and are regularly recorded in the llanos and northern areas of the country during each northward and southward migration (Meyer de Schauensee and Phelps 1978; Hilty, pers. obs). This is the first published record from Amazonas.

ACKNOWLEDGMENTS

We would like to thank R. S. Ridgely, who accompanied Zimmer on one of his visits to Junglaven, as well as Robert Berhstock, Paul Coopmans, Dale Delaney, and Jeri Langham for sharing their sightings from Junglaven and Pto. Ayacucho with us. Captain Lorenzo Rodriguez and the Junglaven Camp staff were most pleasant and accomodating hosts during our many visits. We would also like to thank Mort Isler for contributing the map (Fig. 1) and Phyllis Isler for preparing the sonograms (Fig. 2). Andrew Kratter, Richard Prum and J. V. Remsen, Jr. reviewed earlier drafts of this paper and made numerous helpful suggestions for its improvement. We are particularly grateful to Victor Emanuel Nature Tours, Inc., for supporting our work at Junglaven.

LITERATURE CITED

CHESSER, R. T. 1994. Migration in South America: an overview of the austral system. Bird Cons. Intl. 4: 91–107.

CRACRAFT, J. 1985. Historical biogeography and patterns of differentiation within the South American avifauna: areas of endemism. Pp. 49–84 *in* Neotropical Ornithology (P. A. Buckley, M. S. Foster, E. S. Morton, R. S. Ridgely, and F. G. Buckley, Eds.). Ornithological Monograph No. 36.

ENGLISH, P., AND C. BODENHORST. 1991. The voice and first nesting records of the Zigzag Heron in Ecuador. Wilson Bull. 103:661–664.

FOSTER, R. B., T. A. PARKER III, A. H. GENTRY, L. H. EMMONS, A. CHICCHÓN, T. SCHULENBERG, L. RODRÍGUEZ, G. LAMAS, H. ORTEGA, J. ICOCHEA, W. WUST, M. ROMO, J. A. CASTILLO, O. PHILLIPS, C. REYNEL, A. KRATTER, P. K. DONAHUE, AND L. J. BARKLEY. 1994. The Tambopata-Candamo Reserved Zone of Southeastern Peru: a biological assessment. RAP Working Pap. No. 6, Conservation International, Washington, D.C.

HELLMAYR, C. E. 1918. Catalogue of birds of the Americas. Field Mus. Nat. Hist. Publ. Zool. Ser., vol. 13.

HILTY, S. L., AND W. L. BROWN. 1986. A Guide to the Birds of Colombia. Princeton University Press, Princeton, New Jersey.

ISLER, M. L., P. R. ISLER, AND B. M. WHITNEY. 1997. Biogeography and systematics of the *Thamnophilus punctatus* complex. Ornithol. Monogr. pages 355–381 *in* Studies in Neotropical Ornithology Honoring Ted Parker (J. V. Remsen, Jr., Ed.), Ornithol. Monogr. No. 48.

MAYR, E., AND W. H. PHELPS, JR. 1967. The origin of the bird fauna of the south Venezuelan highlands. Bull. Amer. Mus. Nat. Hist. 136:269–328.

MEYER DE SCHAUENSEE, R. 1982. A Guide to the Birds of South America. Livingston Publishing Co., Wynnewood, Pennsylvania.

MEYER DE SCHAUENSEE, R., AND W. H. PHELPS, JR. 1978. A Guide to the Birds of Venezuela. Princeton University Press, Princeton, New Jersey.

PARKER, T. A., III, AND J. V. REMSEN, JR. 1987. Fifty-two Amazonian bird species new to Bolivia. Bull. Brit. Ornithol. Club 107:94–107.

PAYNTER, R. A. 1982. Ornithological Gazetteer of Venezuela. Museum of Comparative Zoology, Cambridge, Massachussetts.

PERES, C. A., AND A. WHITTAKER. 1991. Annotated checklist of the bird species of the upper Rio Urucu, Amazonas, Brazil. Bull. Brit. Ornithol. Club 111:156–171.

PHELPS, W. H., AND W. H. PHELPS, JR. 1952. Nine new subspecies of birds from Venezuela. Proc. Biol. Soc. Washington 65:39–54.

PHELPS, W. H., AND W. H. PHELPS, JR. 1958. Lista de las aves de Venezuela con su distribución. Bol. Soc. Venezolana Cienc. Natur. 19:1–317.

PHELPS, W. H., AND W. H. PHELPS, JR. 1963. Lista de las aves de Venezuela con su distribución. Bol. Soc. Venezolana Cienc. Natur. 24:1–479.

PHELPS, W. H., JR. 1947. The ornithological collections. Pp. 559–560 *in* The Orinoco-Ventuari Region, Venezuela (C. B. Hitchcock, Ed.). Geogr. Rev. 37:525–566.
REMSEN, J. V., JR., AND S. K. ROBINSON. 1990. A classification scheme for foraging behavior of birds in terrestrial habitats. Pp. 144–160 *in* Avian Foraging: Theory, Methodology, and Applications (M. L. Morrison, C. J. Ralph, J. Verner, and J. R. Jehl, Jr., Eds.). Stud. Avian Biol. 13.
RIDGLEY, R. S., AND J. A. GWYNNE. 1989. A Guide to the Birds of Panama, with Costa Rica, Nicaragua, and Honduras. 2nd ed. Princeton University Press, Princeton, New Jersey.
RIDGELY, R. S., AND G. TUDOR. 1989. The Birds of South America, Vol. 1. University of Texas Press, Austin, Texas.
RIDGELY, R. S., AND G. TUDOR. 1994. The Birds of South America, Vol. 2. University of Texas Press, Austin, Texas.
SCHWERDTFEGER, W. (ED.). 1976. Climates of Central and South America. World Survey of Climatology. Vol. 12. Elsevier Scientific Publishing Company, Amsterdam, Netherlands.
SICK, H. 1993. Birds in Brazil, A Natural History. Princeton University Press, Princeton, New Jersey.
SLUD, P. 1964. The birds of Costa Rica: distribution and ecology. Bull. Amer. Mus. Nat. Hist. 121:1–430.
SNOW, D. 1982. The Cotingas: Bellbirds, Umbrellabirds, and Other Species. Cornell University Press, Ithaca, New York.
WILLARD, D. E., M. S. FOSTER, G. F. BARROWCLOUGH, R. W. DICKERMAN, P. F. CANNELL, S. L. COATS, J. L. CRACRAFT, AND J. P. O'NEILL. 1991. The birds of Cerro de la Neblina, Territorio Federal Amazonas, Venezuela. Fieldiana Zool. 65:1–90.
ZIMMER, K. J., T. A. PARKER III., M. L. ISLER, AND P. R. ISLER. 1997. Survey of a southern Amazonian avifauna: the Alta Floresta region, Mato Grosso, Brazil. pages 887–918 *in* Studies in Neotropical Ornithology Honoring Ted Parker (J. V. Remsen, Jr., Ed.), Ornithol. Monogr. No. 48.

APPENDIX

Following is a complete list of the 398 species of birds recorded to date from the vicinity of Junglaven Camp, Amazonas, Venezuela. Observers and their initials (as used here and in the text) are: Robert Behrstock (RB), John Coons (JC), Paul Coopmans (PC), Brian Finch (BF), Steven L. Hilty (SLH), Joseph D. Kaplan (JDK), Jeff Kingery (JK), Richard O. Prum (ROP), Robert S. Ridgely (RSR), Rose Ann Rowlett (RAR), and Kevin J. Zimmer (KJZ). In instances in which a species has been recorded by only one or two parties, the initials of the observers are included following the species name. Species whose occurrence in the area is known to be documented by tape recordings are indicated with a "(+)". The primary recordists are K. J. Zimmer, S. L. Hilty, and P. Coopmans. Species whose names are enclosed in "[ ]" are judged "hypothetical". These species have been recorded at least once, but are judged to be sufficiently rare or unlikely in this region (and are often more easily confused with a regularly occurring species) as to require more documentation. Migrant species are indicated by an "M." Dates of occurrence are included for some migrants. Habitat is recorded for each species: "a" is aerial (often observed flying high over any habitat); "s" is savanna (on white-sand soil); "sw" is savanna woodland (stunted, scrubby forest growing on sandy soil); "v" is *várzea* (seasonally flooded forest along streams and rivers); "t" is *terra firme* forest (high-canopy, moist forest growing on lateritic soils, and not seasonally flooded); "r" is river edge (observed on rivers, sandbars, marshy edges, ponds, or on perches or in vegetation immediately above or adjacent to the water); "fe" is forest edge (refers to species found at the margins of any type of forest, including in and around clearings within the forest).

TINAMIDAE

Tinamus major (+) t, v
Crypturellus cinereus (+) v, sw
C. soui (+) v, fe
C. undulatus (+) v, sw
C. variegatus (+) t
Crypturellus sp. ? fe, v heard (RSR,KJZ), heard (SLH) (+)

PHALACROCORACIDAE

Phalacrocorax brasilianus r

ANHINGIDAE

Anhinga anhinga r

ARDEIDAE

Ardea cocoi r
A. alba r
Egretta thula r
E. caerulea r
Butorides striatus r
B. virescens M, r (KJZ) (PC)
Agamia agami (+, photos, video) r

Bubulcus ibis M, s
Pilherodius pileatus r
Nycticorax nycticorax r
Tigrisoma lineatus r
Zebrilus undulatus (+, photos) r, v, sw
Cochlearius cochlearius r

CICONIIDAE

Mycteria americana M, r, v
Jabiru mycteria M, r

THRESKIORNITHIDAE

Mesembrinibis cayennensis r

CATHARTIDAE

Sarcoramphus papa a
Coragyps atratus a
Cathartes aura a
C. melambrotus a,t

ANATIDAE

Dendrocygna autumnalis M, r
Cairina moschata M, r

ACCIPITRIDAE

Elanoides forficatus a,s (ROP, JDK) (RAR, JC)
Leptodon cayanensis a, t, sw, v
Harpagus bidentatus a, t, v
Ictinia plumbea a, fe, v, sw
Accipiter superciliosus t (KJZ, RSR) (PC)
Buteo albicaudatus s (PC)
B. magnirostris (+) fe
B. brachyurus a (SLH, KJZ)
Leucopternis albicollis a, sw (ROP, JDK)
Busarellus nigricollis r
Heterospizias meridionalis s
Buteogallus urubitinga a, r, t, v, sw
Spizastur melanoleucus a, v (RAR, JC-1 ad. Feb. 91)
Spizaetus ornatus a, t, v (SLH, KJZ) (SLH) (KJZ) (+)
S. tyrannus a, t, v (KJZ, RSR) (SLH) (+)
Geranospiza caerulescens r, sw, v
Pandion haliaetus M, a, r

FALCONIDAE

Herpetotheres cachinnans s, fe, r (SLH) (KJZ)
Micrastur semitorquatus (+) t
M. gilvicollis t, v (RAR, JC-heard) (KJZ, RSR-heard) (+)
Daptrius ater (+) r, v, sw
D. americanus (+) t, v
Milvago chimachima s, sw, v, r
Polyborus plancus s
Falco rufigularis r, fe

CRACIDAE

Ortalis motmot sw (ROP, JDK) (RAR, JC)
Penelope jacquacu (+) t, v
Pipile pipile r, v, t
Mitu tomentosa (+) sw, v
Crax alector t

ODONTOPHORIDAE

Odontophorus gujanensis t, v

PSOPHIIDAE

Psophia crepitans (+) t, v

RALLIDAE

Aramides cajanea (photos) r, v
Laterallus sp. (ROP, JDK-heard) (+) (SLH-heard) (+) s

EURYPYGIDAE

Eurypyga helias (+) r, v

HELIORNITHIDAE

Heliornis fulica r (SLH, KJZ) (RAR, JC)

JACANIDAE

Jacana jacana r

CHARADRIIDAE

Vanellus chilensis r
Hoploxypterus cayanus r
Pluvialis dominica M, r, s (ROP, JDK) (RAR, JC)
Charadrius collaris r (SLH, Jan. 93)

SCOLOPACIDAE

Tringa solitaria M, r
Actitis macularia M, r
Calidris fuscicollis M, r (PC)
Bartramia longicauda M, s (ROP, JDK) (KJZ, RSR) (RAR, JC)
Gallinago paraguaiae M, s
G. undulata M, s (RAR, JC)

LARIDAE

Phaetusa simplex r
Sterna superciliaris r
Rynchops niger M, r

COLUMBIDAE

Columba speciosa (+) v, t
C. cayennensis (+) r, fe, v, sw
C. subvinacea (+) v
Columbina passerina s (KJZ, RSR)
C. minuta s, sw (ROP, JDK) (RAR, JC)
C. talpacoti s,
Claravis pretiosa v, r
Leptotila verreauxi fe (KJZ)
L. rufaxilla v, r, sw
Geotrygon montana t

PSITTACIDAE

Ara macao (+) a, v, t, r
A. chloroptera (+) a, v, t, r
A. severa (+) a, s, sw, r
Aratinga leucophthalmus t, fe (SLH-Dec/Jan 92–93, daily)
A. pertinax (+) s, sw
Pyrrhura melanura (+) t, v, r
Forpus sp. a (heard SLH-Dec./Jan 91–92)
Brotogeris cyanoptera (+) a, v, t, sw
Touit purpurata a (SLH-Dec./Jan. 92–93, daily flocks 45+)
T. huetii v (ROP, JDK) (SLH, KJZ)
Pionites melanocephala (+) t, sw, v
Pionopsitta barrabandi (+) s, sw, v, t
Pionus menstruus (+) a, t, v, r
Amazona ochrocephala sw, v, s
A. amazonica s, sw, v,
A. farinosa (+) t

CUCULIDAE

Coccyzus americanus-M, v (ROP, JDK) (PC)
Coccyzus sp. fe (SLH-*euleri* or *americanus*)

Piaya cayana (+) t, v, sw
P. melanogaster t, v, sw
P. minuta sw, fe (SLH, KJZ) (KJZ, RSR)
Crotophaga major r
C. ani r
Neomorphus rufipennis t, v (KJZ) (SLH) (+, photo)

STRIGIDAE

[*Otus guatemalae*] v, t (reported heard only by ROP, JDK, RAR, JC, RB)
O. choliba v, t, sw
O. watsonii v, t
Lophostrix cristata v, t
Pulsatrix perspicillata t, v
Glaucidium brasilianum v, sw

NYCTIBIIDAE

Nyctibius grandis (+) t, v
N. aethereus (+) t, v (RAR, JC) (RB, BF, JK)
N. griseus (+) v, t, sw

CAPRIMULGIDAE

Lurocalis semitorquatus fe (KJZ, RSR) (SLH)
Chordeiles pusillus s
C. rupestris r (ROP, JDK) (SLH)
C. acutipennis s
Nyctiprogne leucopyga r
Nyctidromus albicollis (+) fe, sw, v
Caprimulgus rufus (+) sw (SLH) (PC)
C. cayennensis s
C. nigrescens t

APDODIDAE

Streptoprocne zonaris a
Chaetura cinereiventris a
C. spinicauda a, r
C. brachyura a, s
Panyptila cayennensis a
Reinarda squamata a, s

TROCHILIDAE

Phaethornis superciliosus t, v
P. hispidus sw, v
P. squalidus sw, v
P. ruber t, v
Campylopterus largipennis t (SLH, Feb. 90)
Florisuga mellivora-t (KJZ, RSR) (PC)
Anthracothorax nigricollis fe
Chrysolampis mosquitus fe (PC)
Lophornis chalybea v (SLH-3 Jan. 93, a male)
Popelairia langsdorfii fe (PC)
[*Discosura longicauda*] fe, sw (SLH, KJZ) (KJZ, RSR) (SLH)
Chlorestes notatus v, sw
Chlorostilbon mellisugus fe, v, sw
Thalurania furcata t (SLH)
Hylocharis sapphirina sw (SLH, 29 Dec. 92)
H. cyanus t, sw, v
Polytmus guainumbi s, sw (ROP, JDK) (RAR, JC)
P. theresiae s (SLH)
Amazilia chionopectus fe
A. versicolor fe
Heliothryx aurita v, t, r
Heliomaster longirostris sw, fe, r

TROGONIDAE

Trogon melanurus (+) t, v
T. viridis (+) t, v, sw
T. violaceus t, v

ALCEDINIDAE

Ceryle torquata r
Chloroceryle amazona r
C. americana r
C. inda r
C. aenea r

MOMOTIDAE

Momotus momota (+) t, v

GALBULIDAE

Brachygalba lugubris (+) r, v
Galbula albirostris t
G. galbula (+) v, r, sw, fe, t
G. leucogastra r, sw
G. dea (+) t, v
Jacamerops aurea (+) t

BUCCONIDAE

Notharchus macrorhynchus (+) t, v
N. ordii fe (SLH, KJZ, photos)
N. tectus t (SLH-4 Jan. 93)
Bucco tamatia (+) sw, v
Nonnula rubecula t, v, sw
Monasa atra (+) t, v, sw
Chelidoptera tenebrosa (+) fe, r, s, v

CAPITONIDAE

Capito niger (+) t, v

RAMPHASTIDAE

Pteroglossus pluricinctus v, t, fe
P. viridis v, t, fe
P. flavirostris v, t, fe
Selenidera nattereri (+) t, sw
Ramphastos culminatus x *vitellinus* (mostly intergrades) (+) t, v, r, sw
R. cuvieri (+) v, t, r

PICIDAE

Picumnus exilis (+) t, v, sw
Piculus rubiginosus t (SLH -/30 Dec. 92)
P. flavigula t, v
P. chrysochloros t, v
Celeus elegans t, v, sw
C. grammicus (+) v, t, sw
C. flavus (+) v, r, t
C. torquatus (+) t, v, sw
Dryocopus lineatus (+) fe, sw
Melanerpes cruentatus (+) fe, r, sw
Veniliornis sp. (*cassini* or *affinis*) (+) t, v, sw
Campephilus melanoleucos (+) t, v, sw
C. rubricollis t, sw (KJZ, RSR)

DENDROCOLAPTIDAE

Dendrocincla fuliginosa t, v, sw
D. merula (+) t (PC)
Sittasomus griseicapillus t, v
Glyphorynchus spirurus t, v
Nasica longirostris (+) v, r, t
Dendrocolaptes picumnus (+) t, v

D. certhia (+) t
Xiphorhynchus picus (+) v, r
X. obsoletus (+) v, sw, r
X. pardalotus t? (PC)
X. guttatus (+) t
Lepidocolaptes albolineatus (plain-crowned *duidae* race) (+) t, v

FURNARIIDAE

Synallaxis albescens s (heard, PC)
S. rutilans (+) t, v
Cranioleuca vulpina (+) v, r
Philydor pyrrhodes (+) t
Automolus infuscatus t (RAR, JC)
A. ochrolaemus (+) t
Xenops tenuirostris t (KJZ, RSR) (SLH)
X. minutus (+) t, v

FORMICARIIDAE

Cymbilaimus lineatus (+) t
Frederickena viridis t (RAR, JC)
Sakesphorus canadensis (+) v
Thamnophilus nigrocinereus (+) sw, v
T. aethiops (+) t
T. amazonicus (+) t, sw, v
Pygiptila stellaris (+) t, v
Thamnomanes caesius (+) t
Myrmotherula brachyura (+) t, v
M. surinamensis (+) r
M. cherriei (+) sw, v, fe
M. guttata t (SLH)
M. haematonota (+) t
M. axillaris (+) t, v, sw
M. menetriesii (+) t
Herpsilochmus dorsimaculatus (+) v
Cercomacra cinerascens (+) t
C. tyrannina (+) fe
Myrmoborus leucophrys (+) v, t, sw
Hypocnemis cantator (+) t, v
Hypocnemoides melanopogon (+) sw, v
Schistocichla leucostigma v, sw (ROP, JDK) (SLH, KJZ)
Sclateria naevia (+) r
Myrmeciza longipes (+) fe, sw, v
M. atrothorax (+) fe, sw, v
Pithys albifrons t (PC)
Gymnopithys rufigula (+) t, sw
Hylophylax punctulata (+) v (KJZ, RSR)
H. poecilinota (+) t
Formicarius colma (+) t

COTINGIDAE

Cotinga cayana v, fe, sw, r
Xipholena punicea fe, sw
Iodopleura isabellae (+) sw, v
Lipaugus vociferans (+) v
Cephalopterus ornatus r, v, sw
Gymnoderus foetidus r, v

PIPRIDAE

Pipra erythrocephala (+) t, v
P. pipra sw, v, t
Manacus manacus (+) sw
Xenopipo atronitens (+) sw, v
Heterocercus flavivertex (+, photos, video) sw, v
Tyranneutes stolzmanni (+) t, v
Piprites chloris t (heard SLH) (heard PC)
Schiffornis turdinus (+) sw, t, v

TYRANNIDAE

Laniocera hypopyrra t, sw, v
Pachyramphus polychopterus v, sw, t
P. marginatus (+) t, v, sw
P. minor t
Tityra cayana t, v, sw, fe
[*T. semifasciata*] (ROP, JDK)
Zimmerius gracilipes t, v
Ornithion inerme (+) t
Camptostoma obsoletum s, sw
Tyrannulus elatus (+) t, sw, fe
Myiopagis gaimardii (+) t, v, sw
M. caniceps t
M. flavivertex v (RSR-heard)
[*Elaenia parvirostris*] M, s (ROP, JDK-March–April, 1990)
E. chiriquensis s (KJZ, RSR)
E. cristata M (?) s, sw
E. ruficeps s
Inezia subflava v, r
Mionectes oleagineus (+) t, v, sw
Capsiempis flaveola (+) sw (KJZ, RSR) (KJZ)
Corythopis torquata t (RAR, JC)
Colopteryx galeatus (+) t, v
Todirostrum pictum (+) t, v
T. sylvia (+) sw
Ramphotrigon ruficauda (+) t, v
Tolmomyias assimilis (+) t
T. poliocephalus (+) v
T. flaviventris (+) sw, v
Platyrinchus platyrhynchos (+) t
P. coronatus t (heard, SLH)
P. saturatus (+) t (KJZ, RSR)
Terenotriccus erythurus (+) t, v
Myiophobus fasciatus fe (SLH)
Contopus virens M, fe (KJZ, RSR) (SLH)
Lathrotriccus euleri (+) v (ROP, JDK)
Ochthornis littoralis r
Knipolegus poecilocercus v, sw
Fluvicola pica r
Attila spadiceus (+) t, v
A. cinnamomeus (+) v, sw
Rhytipterna simplex (+) t, v, sw
R. immunda (+) sw
Myiarchus ferox v, sw, fe
M. tyrannulus fe (KJZ, RSR)
M. tuberculifer t, v
Pitangus sulphuratus r (SLH)
P. lictor (+) r
Megarynchus pitangua-r (SLH, KJZ)
Myiozetetes luteiventris (+) t (SLH)
M. cayanensis (+) fe, s, sw, v
M. similis s, v, fe
Conopias parva v (PC)
Myiodynastes maculatus M?, t, v
Legatus leucophaius M (?), t, fe, v, sw
Empidonomus varius M (?), fe (KJZ)
Tyrannopsis sulphurea (+) s
Tyrannus savana M, s, v, r
T. melancholicus s, fe, r, v

HIRUNDINIDAE

Tachycineta albiventer r, s
Progne tapera r
P. chalybea a, s, r
P. subis M, r (KJZ-5 March 93, 2 males)
Atticora fasciata r

A. melanoleuca r
Stelgidopteryx ruficollis s, r, fe
Hirundo rustica M, a, s, r

CORVIDAE

Cyanocorax violaceus (+) sw, v, fe

TROGLODYTIDAE

Thryothorus leucotis (+) r, v, sw
Troglodytes aedon fe (KJZ) (SLH)

MIMIDAE

Mimus gilvus s (KJZ, RSR)

TURDIDAE

Catharus fuscescens M, v, sw (ROP, JDK: 22 Mar–28 Apr)
C. minimus M, t (SLH: 28 Dec–5 Jan, 3 Jan) (JC: 1 Feb)
C. ustulatus M, v, sw (ROP, JDK: 22 Mar–28 Apr) (RAR, JC:23 Feb–02 Mar)
Turdus ignobilis s (KJZ, RSR)
T. fumigatus v, t
T. albicollis t (SLH, KJZ)

POLIOPTILINAE

Ramphocaenus melanurus (+) t, v
Polioptila plumbea (+) sw, v, fe

VIREONIDAE

Cyclarhis gujanensis v, sw (ROP, JDK)
Vireolanius leucotis (+) t (SLH)
Vireo olivaceus M, t, v, sw
V. altiloquus M, t (KJZ, RSR)
Hylophilus semicinereus (+) r (KJZ, RSR) (KJZ) (PC)
H. brunneiceps (+) sw
H. ochraceiceps (+) t

PARULIDAE

Dendroica petechia M, v, fe, sw
D. striata M, t, v (dates span the period of 28 Dec–27 Feb)
Seiurus noveboracensis M, r, v
Oporornis agilis M, sw (ROP, JDK: 22 Mar–28 Apr, 1990) (photos)
Setophaga ruticilla M, fe, v

THRAUPIDAE

Coereba flaveola (+) t, fe, v, sw
Dacnis cayana v, t, fe
D. lineata fe (SLH, Feb. 90)
D. albiventris t (RB, BF, JK)
D. flaviventer v, sw, t
Cyanerpes caeruleus t
C. nitidus t, fe, v, sw
C. cyaneus t, fe, sw
Chlorophanes spiza t
Hemithraupis flavicollis t, v
Tangara cayana s
T. mexicana v, t
T. velia t, v, sw
T. chilensis t, v
Euphonia cyanocephala sw (ROP, JDK) (RAR, JC)
E. violacea v, sw
E. minuta t, fe, v, sw
E. chlorotica/trinitatis (+) s, sw
E. rufiventris t
E. plumbea (+) sw
E. chrysopasta t, v, sw
Tersina viridis v, sw, r, t
Thraupis palmarum sw, s

T. episcopus (+) fe, sw, v
Ramphocelus carbo sw, v
Tachyphonus luctuosus t (SLH)
T. cristatus t, v, sw
T. surinamus t, v (ROP, JDK)
T. phoenicius s, sw
Schistochlamys melanopis s, sw

ICTERIDAE

Scaphidura oryzivora r (KJZ, RSR)
Psarocolius yuracares (+) t, v
Cacicus cela (+) v, r, fe, sw
Agelaius icterocephalus s (SLH)
Icterus chrysocephalus s (SLH)
Icterus icterus r (SLH, KJZ)
Dolichonyx oryzivorus M, s

EMBERIZIDAE

Saltator maximus sw
Cyanocompsa cyanoides (+) t, v
Oryzoborus angolensis r, v (ROP, JDK)
Sporophila lineola s (PC)
S. nigricollis s, r
Zonotrichia capensis s
Ammodramus humeralis s
Emberizoides herbicola s
Sicalis columbiana sw, r, s (ROP, JDK) (SLH, KJZ)

SURVEY OF A SOUTHERN AMAZONIAN AVIFAUNA: THE ALTA FLORESTA REGION, MATO GROSSO, BRAZIL

KEVIN J. ZIMMER[1], THEODORE A. PARKER III,[2,4] MORTON L. ISLER[3], AND
PHYLLIS R. ISLER[3]

[1]1665 Garcia Rd., Atascadero, California 93422, USA; [2]Museum of Natural Science,
Louisiana State University, Baton Rouge, Louisiana 70803, USA; and
[3]Division of Birds, National Museum of Natural History, Smithsonian Instituition,
Washington, D.C. 20560, USA
[4]Deceased

ABSTRACT.—This paper reports the findings of several surveys of the avifauna of the region surrounding the town of Alta Floresta, Mato Grosso, Brazil. We recorded a total of 474 species, many of which are regional endemics, which makes Alta Floresta one of the richest known sites in Brazil and all of eastern Amazonia. We provide information for a number of species range extensions and for species for which there were few previous records for Brazil. We also provide information on the vocalizations, foraging ecology, and habitat preferences for many rare or localized species that are poorly known. Of particular interest to us were a number of birds that, at least locally, are restricted in distribution to stands of spiny bamboo (*Guadua* sp.). The relevance of our findings to previously conducted surveys, current biogeographic hypotheses, and the many urgent regional conservation problems (including conservation priorities), are discussed.

RESUMÉN.—Apresentam-se os resultados de vários levantamentos ornitológicos da região próxima a Alta Floresta, Mato Grosso, Brasil. Tendo um total de 474 espécies, inclusive muitas endêmicas da região, Alta Floresta é um dos sítios mais ricos do Brasil e de toda a Amazônica oriental. Descrevemos extensões de distribuição de várias espécies, muitas dessas tendo puocos registros no país. Também descrevemos vocalizações, ecologia de forrageamento, e preferência de habitat em algumas espécies raras ou mal conhecidas. Particularlmente interessantes são espécies que, pelo menos localmente, são restritas a tabocais (bambuzais de *Guadua* sp.). Discutimos os resultados no contexto de levantamentos anteriores, hipóteses biogeográficas, e problemas urgentes regionais de conservação.

The Alta Floresta region (centered at 9°41'S, 55°54'W) sits astride the Rio Teles Pires, a principal tributary of the Rio Tapajós, in extreme north-central Mato Grosso near the border with Pará (Fig. 1). It is located in the southern fringe of forested Amazonia, an area suffering alarmingly from rapid deforestation and environmental degradation (Oren and Albuquerque 1991). The accumulating impact of this devastation readily could be seen in return visits to Alta Floresta during the few years of this avifaunal survey. The threat to the region creates an urgent need to inventory its biological resources and to identify representative habitats for conservation.

Knowledge of the avifauna of the Alta Floresta region also affords an opportunity to explore important biogeographical issues. Alta Floresta lies in the contact zone between two main areas of avian endemism south of the Amazon (Fig. 1)—the Belém and Rondônia areas as defined by Haffer (1985), and the Pará and Rondônia centers of Cracraft (1985). To the north, these regions of endemism are clearly delimited by the Amazon, and in the region immediately south of the Amazon, by its major tributaries, the Tocantins, the Tapajós, and the Madeira. A key problem for biogeographers is to determine the extent to which avian distributions are limited by the physical barriers of the rivers or by the presence of an ecologically competitive taxon on the other side (Haffer 1992). Avifaunal inventories of regions such as Alta Floresta that are nearer the headwaters of major rivers (where physical barriers become narrow) should help answer this question.

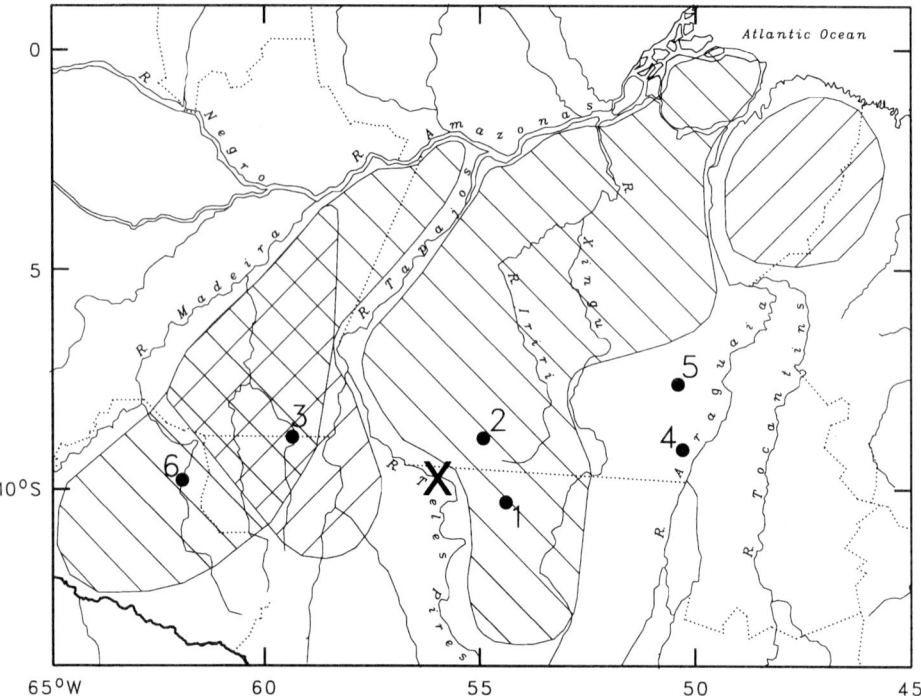

FIG. 1. Location of Alta Floresta in relation to modern collecting stations and to defined areas of endemism of forest birds. The large "X" locates the Alta Floresta region. The filled circles locate collecting stations. 1 = Rio Peixoto de Azevedo (Novaes and Lima 1991). 2 = Cachimbo (Pinto and Camargo 1957). 3 = Rio Aripuanã (Novaes 1976). 4 = Santana do Araguaia (Oren, unpubl. data). 5 = SE Pará. dot is approx. center of 3 sites (Novaes 1960). 6 = Cachoeira Nazaré (Stotz et al. 1997). Areas of endemism shaded downward to the left are those of Haffer (1985); the Belém area of Haffer lies to the east and the Rondônia area to the west. Areas of endemism shaded downward to the right are those of Cracraft (1985); the Pará center of Cracraft lies to the east and the Rondônia center to the west. Dotted lines are state boundaries.

In spite of the immediate threat to its avifauna and its importance to the study of avian evolution in the Amazon basin, the southern tier is one of the least-known sectors of Amazonia (Oren and Albuquerque 1991). Figure 1 locates sites of recent collections of birds along the southeastern edge of the forested region of Amazonia, the results of which are published or in preparation for publication. Some of these sites encompass non-forest as well as forest habitat, because the broad ecotone between forested and nonforested regions is not a straight line, but rather an intricate array of "tongues" of vegetation, as shown in simplified form by Haffer (1985; Fig. 1). Two collecting localities are especially pertinent: the Rio Peixoto de Azevedo (Novaes and Lima 1991), on the east side of the Teles Pires ca. 100–150 km from Alta Floresta, and Cachimbo (Pinto and Camargo 1957), even closer to Alta Floresta, but on the other side of a possible barrier to avian distributions, the Serra do Cachimbo. Not shown are collecting stations at substantial distances from this broad ecotone between forest and nonforest, whether within the forested regions or in the Campos Cerrados region that begins close to Alta Floresta to the south. Although farther removed from Alta Floresta, the Cachoeira Nazaré site (Fig. 1) on the west bank of the Rio Jiparaná, Rondônia, is also significant because it harbors many of the same bamboo-inhabiting birds found in our survey. This site was surveyed in 1986 through a joint venture of the Field Museum of Natural History and the Museu de Zoologia da Universidade de São Paulo (Lanyon et al. 1990; D.F. Stotz et al. 1997).

In October–November 1989, Parker and the Islers made the first ornithological investigation of the region surrounding the town of Alta Floresta (elevation 296 m) in northern Mato Grosso. From 26 October to 9 November they worked tracts of mostly upland (*terra firme*) forest and bamboo thickets along the road leading north from Alta Floresta to the Rio Teles Pires (hereafter the "Teles Pires Road"), concentrating their efforts in the area 25–30 km north of town (ca. 5 km south of the Rio Teles Pires). The Islers also visited another site 45–50 km southeast of Alta

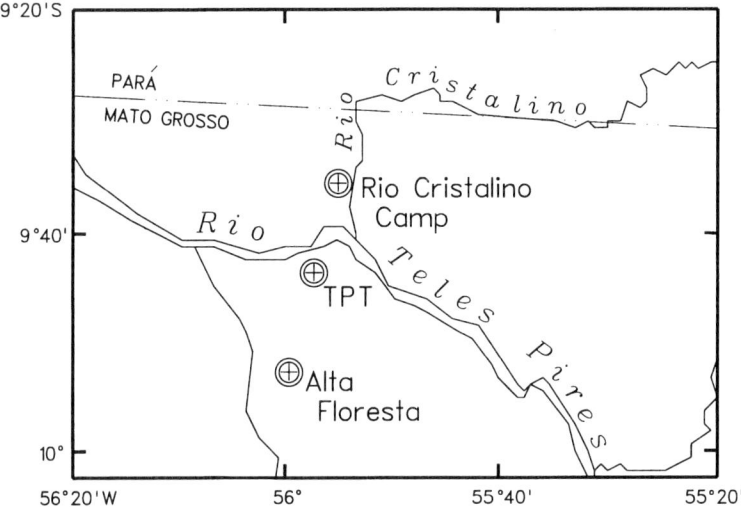

FIG. 2. Map of the Alta Floresta region, Mato Grosso, Brazil. Circles locate the city of Alta Floresta and the two primary sites at which this survey was conducted: the Teles Pires Trail (TPT) on the west (south) bank of the Rio Teles Pires, and the Rio Cristalino Camp (the center of the Reserva Florestal Cristalino) on the east (north) bank of the Rio Teles Pires. The dashed line near the top of the map locates the boundary between the states of Mato Grosso and Pará.

Floresta. Parker and Zimmer returned to Alta Floresta in August 1991, dividing their time between the forests along the Teles Pires Road and a privately owned forest reserve along the Rio Cristalino (= Reserva Florestal Cristalino). Subsequent visits by Zimmer in October 1992, September 1993, September 1994, and September 1995 have followed a similar strategy, but with more time being spent in the Reserva Florestal Cristalino. Bret M. Whitney (*in litt.*) also worked the area in September 1993 and made a number of important observations. A complete list of observers and dates of visits is found in the Appendix.

Additionally, in late November and early December 1993, an expedition of the Museo Paraense Emílio Goeldi led by Jürgen Haffer collected specimens (skins and tissue) on both sides of the Rio Teles Pires in the Alta Floresta region. Materials from this expedition are now being analyzed, and the results are not included in this paper. Materials collected by the Goeldi Museum expedition will provide a basis for studying differences in bird populations between the west (south) and east (north) banks of the Rio Teles Pires, a subject that this paper does not address.

METHODS

The Alta Floresta "region" (as defined simply by the total area encompassed by the perimeter of all sites visited) is ca. 1,800 km². The actual area surveyed (viewed as multiple belt transects encompassing ca. 40 km of roads, 10 km of river, and 10 km of forest trails, with an effective width of 500 m [based on the average maximum distances from either side of the transect at which birds could be detected by the observers]) is ca. 30 km². An estimated 90% of our work has been concentrated in an area slightly less than half of that size.

Most of our fieldwork was conducted at two sites (Fig. 2). The first of these, which we refer to as the "Teles Pires Trail" (hereafter the "TPT"), is a gold-miners' trail that traverses ca. 2 km of upland forest located ca. 28 km north of Alta Floresta off the Teles Pires Road. This forest has a canopy height of 25–30 m, a relatively open understory, and several large (1–5 ha) patches of spiny bamboo (*Guadua* spp.). The largest of these patches are nearly pure stands of bamboo (average stem dbh. ca. 8–10 cm), with relatively few (scattered) trees. The stands are dense, but the canopy is open (few overtopping trees), and during mid-day much sunlight penetrates to the floor. In spite of this, the understory is relatively open except for the bamboo and bamboo litter. In August 1991, this site comprised more than 1,000 ha of upland forest, bounded on the south and east by pastures, and transected by a small east-west-oriented logging road. By October 1992, all forest north of the logging road had been clear-cut for a distance of more than 1 km, and the western boundaries were being similarly eroded. As of September 1994, the TPT and its ambient forest and bamboo stands remained intact, although the site as a whole was surrounded by pasture on three sides and had been reduced by at least 30% of its 1991 area.

Our other primary site is the Reserva Florestal Cristalino (hereafter "RFC"), a privately owned forest reserve just upstream along the Rio Cristalino from its confluence with the Rio Teles Pires. The Cristalino is a small, black-water river, as yet uncontaminated by either mercury from mining operations, or from sediments resulting from logging and subsequent erosion. The RFC is situated along the right bank of the river, and encompasses 1,500 ha of primary forest. It is contiguous with vast areas of virgin forest along the opposite bank and farther up the Cristalino watershed that remain unprotected. The RFC is administered by a non-profit ecological institute, the Cristalino Ecological Institute (IEC). The only development in the reserve consists of a small research/tourism lodge (The Rio Cristalino Camp, hereafter the "RCC") and several 1–1.5 km trails. Access is solely by boat. Hunting and trapping pressure is minimal to non-existent, as indicated by large numbers of cracids, macaws, and monkeys (seven species recorded within a few kms of the RCC).

The forest in the RFC (at least that accessible by existing trails) differs in structure from that along the TPT. Canopy height is 20–30 m with occasional emergents reaching 40 m. The forest here is extremely rich in vines from the canopy to the floor, with an abundance of lianas and creepers throughout all strata. The understory vegetation is dense. Two of the trails, the "Rio Cristalino Trail" (RCT) and the "Taboca Trail" (TT), transect multiple small (0.25–0.50 ha) stands of bamboo. With few exceptions these stands are less homogeneous than those along the TPT, having an abundance of other trees and vines intermixed. The canopy height is ca. 15 m and largely closed by mats of overlapping bamboo and tree crowns. Woody vine tangles extend to the floor, where little light penetrates. The bamboo in these patches is smaller-stemmed (6–8 cm dbh), with slender (0.6 cm diameter) leaves 16–21 cm long. Similar stands of bamboo are frequent at the edge of primary and secondary forest along the Teles Pires Road and the Fazenda Cristalino Road north of Alta Floresta. Whether the bamboo in these patches is a different species of *Guadua* from that along the TPT, or simply a younger stage of the same species, is not clear. The Taboca Trail also transects two mostly homogeneous stands of thicker-stemmed bamboo similar in structure (although smaller in extent) to those along the TPT.

Some of the small hilltops and bluffs scattered along the Rio Cristalino are topped by rocky outcroppings that support a drier, stunted forest quite different from surrounding low areas. These forest patches have an open canopy (15–20 m above ground) and an abundance of large terrestrial bromeliads. Most of the trees within these patches are drought-deciduous, and have been mostly or entirely leafless during recent September visits. One such patch is accessible at the top of the "Serra Trail" (ca. 360 m elevation) and is the only place in the Alta Floresta region where Streaked Xenops (*Xenops rutilans*), Natterer's Slaty-Antshrike (*Thamnophilus stictocephalus*; see Isler et al. 1997), White-fringed Antwren (*Formicivora grisea*), Rufous Casiornis (*Casiornis rufa*), Xenopsaris (*Xenopsaris albinucha*), Chestnut-vented Conebill (*Conirostrum speciosum*), and Hooded Tanager (*Nemosia pileata*) have been found.

We spent less time surveying roadside forests along the Teles Pires Road and the Fazenda Cristalino Road (which leads from the Teles Pires Road to the IEC boat-launch on the Rio Teles Pires). As recently as 1992 there were still good stands of high-canopied primary forest along the Teles Pires Road 15–25 km north of Alta Floresta. These are now largely gone. Similarly, much of the primary forest between the TPT and the Rio Teles Pires has been cut, although some areas of swampy secondary forest remain. The Fazenda Cristalino Road is now largely protected, and transects good secondary forest with an abundance of *Cecropia*, as well as some smaller stands of primary forest. Other "satellite" sites that have received minor attention from us include a small stand of protected forest adjacent to the Floresta Amazonica Hotel in Alta Floresta, and the aforementioned extensive stands of bamboo ca. 45–50 km southeast of Alta Floresta, which were surveyed briefly by the Islers in 1989. This latter site has not been visited since the initial survey, but given the almost total deforestation south of Alta Floresta (as judged from airplane flights between Alta Floresta and Cuiabá), it is unlikely that the site is still forested.

Voucher tape-recordings were made whenever possible to document species occurrence. These recordings were made using Nagra 4.2 and Sony TCM-5000 and D5M recorders, and Sennheiser ME-80, ME-88 and MKH-70 shotgun microphones. All recordings have been or will be archived at the Library of Natural Sounds, Cornell University, Ithaca, New York. Behavioral data were summarized verbally on micro-cassette or cassette at the time of observation. All measurements used in such data (distances, heights, etc.) are estimates. Foraging terminology follows Remsen and Robinson (1990). Two additional foraging terms used here are "hitching" and "side-to-side hitching." As used in this paper, "hitching" is a slight modification of the term as used by

Whitney (1994) and is defined as movement in which both feet are moved sideways either simultaneously or in rapid succession, such that the axis of the body is held at an obvious angle (if not perpendicular) to the substrate (stem, vine, branch, etc.) and progression is made by a series of short advances. "Side-to-side" indicates hitching progression in which the orientation of the body relative to the substrate changes from one motion to the next, such that the bird leads first with one side, then with the other.

RESULTS

The Appendix lists all species of birds recorded by us and colleagues from the Alta Floresta region, along with a subjective estimate of relative abundance. Species thought to be migrants (either austral or boreal) are labelled as such. Because much of our attention was focused on bamboo-inhabiting birds, we have indicated in our list those species that could be considered bamboo specialists in the Alta Floresta region. Many of these species appear to be obligate bamboo specialists throughout their range, whereas others are typically associated with other habitats.

Because the micro-distribution of birds in the region may have broader implications concerning biogeographic theories, we have also noted in the Appendix those species recorded from only one side of the Rio Teles Pires. A total of 63 taxa has been found only on the west (south) side of the river. At least 28 of these are open-country species (e.g., *Amazonetta brasiliensis*, *Caprimulgus parvulus*, *Columbina talpacoti*) for which no suitable habitat exists in the RFC east of the Teles Pires. Similarly, of the 47 taxa currently recorded only from the east (north) side of the Teles Pires, 12 are species closely tied to the rivers or immediately adjacent forest (e.g., *Heliornis fulica*, *Chloroceryle aenea*, *Cranioleuca vulpina*), habitats that have received scant coverage on the west bank of the Teles Pires. Many other species recorded from only one side of the river are known locally from only one or two records (e.g., several raptors), and thus their seeming distributional limits may be a sampling artifact. Excluding raptors and species for which suitable habitat is lacking on one side or the other of the Teles Pires, several forest species remain that are known from only one side of the river. Many of these are also known (in this region) from only a few records (these are marked by an "*" in the lists that follow), and are secretive birds that could easily escape detection. Many will undoubtedly be found on both sides of the river as the area is investigated more thoroughly. West (south) bank taxa as yet unrecorded from the RFC are *Pyrrhura rhodogaster*, *Synallaxis cabanisi**, *Cranioleuca gutturata*, *Sclerurus rufigularis*, *Sclerurus albigularis**, *Microrhopias quixensis bicolor*, *Myrmornis torquata**, *Conopophaga melanogaster**, *Schiffornis major**, *Colonia colonus*, *Onychorhynchus coronatus**, *Myiobius barbatus*, *Rhynchocyclus olivaceus*, *Poecilotriccus capitalis**, *Polioptila guianensis**, *Vireo olivaceus*, *Vireo altiloquus**, *Dacnis albiventris**, *Tangara cyanicollis*, *Piranga flava*, *Cacicus haemorrhous*, and *Cyanocompsa cyanoides*. East (north) bank taxa as yet unrecorded from the west bank include *Mitu tuberosa*, *Touit purpurata*, *Dromococcyx pavoninus**, *Neomorphus* sp.*, *Popelairia langsdorffi**, *Hylocharis sapphirina**, *Brachygalba lugubris*, *Malacoptila rufa*, *Notharchus ordii*, *Pteroglossus beauharnaesii*, *Deconychura stictolaema**, *Xenops milleri**, *Microrhopias quixensis emiliae*, *Pyriglena leuconota*, *Chamaeza nobilis*, *Conopophaga aurita**, *Xipholena* sp., *Cephalopterus ornatus*, *Ramphotrigon ruficauda*, *Platyrinchus coronatus*, *Conirostrum speciosum**, *Nemosia pileata**, and *Tangara punctata**.

The total of 474 species makes the Alta Floresta region one of the richest in Brazil. Of these, only 21 species (4%) are thought to be wholly or partially migratory (noted in the Appendix). Of particular note is the area's richness of antbirds (families Thamnophilidae and Formicariidae), with 50 species present. At least 309 species (65%) are substantiated by voucher tape-recordings. Voucher-tape documentation is particularly strong for forest species and passerines, with over 75% of the latter group (including 83% of all suboscines and 98% of all antbirds) so documented.

The richest sites in Amazonian Peru have locality lists of well over 500 species of birds (Terborgh et al. 1984; Haffer 1990; Foster et al. 1994), but the difference may be partly attributable to differences in sampling intensity and procedures, rather than entirely to intrinsic regional differences in species richness (Remsen 1994). Up to the end of the period covered by this report, no mist-netting had been done at Alta Floresta (except for that done by J. Haffer in November–December 1993, the results of which are not included in this paper) in contrast to other localities in Amazonia (Remsen 1994), and the total number of observer-hours and especially seasonal spread of coverage for the area is unquestionably low compared to such areas as Cocha Cashu Biological Station (Manu National Park, Peru), Tambopata (Peru), Sucusari (Peru), and Limoncocha (Ecuador), all of which have higher species totals (Haffer 1990). Site differences in the number of habitats surveyed can also distort comparisons of species richness (Remsen and Parker 1983; Remsen 1994). The Alta Floresta region encompasses both sides of a major river, and includes black-water, bamboo, and deciduous forest habitats lacking from some of the upper Amazonian sites. Conversely, Alta Floresta lacks the extra dimensions provided by an abundance of oxbow lakes, river islands, and river-edge second-growth (e.g., *Tessaria*, *Cecropia*, and *Gynerium*) found at many of the other sites. Increased coverage, particularly of secondary and marshy habitats peripheral to the TPT and the RFC (which have received scant coverage to date), will undoubtedly add many species to the Alta Floresta list, as will coverage during seasons outside of August–November. We echo the concerns of Remsen (1994) regarding comparisons of relative species diversity between sites. When differences in (1) the number of habitats surveyed; (2) sampling procedure (with or without mist-netting or tape-recording), seasonality, and intensity; and (3) the relative contribution of migrants or wanderers (as opposed to 'core' or resident species) to the total species list are taken into account, Alta Floresta seems closer in species richness to the aforementioned sites than a straight comparison of species totals would suggest. Comparisons that fail to take these factors into account could result in important reservoirs of biodiversity being overlooked when critical conservation decisions are made.

Deforestation in this part of Amazonian Brazil has progressed at a rapid pace, spurred on by huge international economic development projects and concomitant human immigration (Fearnside 1987). The rapid conversion of primary forest to pastures for cattle grazing, combined with the destruction of forest from hydroelectric projects and mining operations, has placed some of the richest-known bird communities on earth at extreme risk (Collar et al. 1992: 649). Construction on a large hydroelectric project at Cachoeira Nazaré began almost as soon as the bird survey was completed (Lanyon et al. 1990), and the rapid rate of deforestation around Alta Floresta has been described above. A further environmental threat to the Alta Floresta region comes from the pollution of the Rio Teles Pires by upstream gold-mining operations along its tributaries, particularly the Rio Peixoto de Azevedo (E. Carvalho, pers. comm.). In light of this, surveys such as this one are critical to assess properly regional biodiversity and to prioritize sites for the creation of forest reserves. Collar et al. (1992: 650) have noted the "urgent need for the establishment of various kinds of forest reserves in the biologically rich areas along the northern edge of the Brazilian Shield, . . . for example along the rios Aripuanã, Juruena, and Teles Pires." The most pressing need is for the establishment of reserves on the west (south) side of the Teles Pires. Expansion of the existing 1,500-ha Reserva Florestal Cristalino, or the creation of a second contiguous reserve, to protect as much of the still-pristine Cristalino watershed as possible would seem to be an equally important step.

In the species accounts that follow we provide information on a number of range extensions, including many species for which there were few previous records for Brazil. We also provide information on the vocalizations, foraging ecology, and habitat preferences for many rare or localized species which are poorly known. When specific records cited are our own, we have used only the last names of the author(s) involved. We have also incorporated significant observations made by a few colleagues who have either accompanied us during our field work or who have visited the region on their own. Citations involving these individuals include first initials. All observers and the dates of their visits to Alta Floresta are listed in the appendix.

SPECIES ACCOUNTS

Snail Kite (*Rostrhamus sociabilis*).—The first Alta Floresta record of this species was of 16 birds kettling over the Río Teles Pires on the afternoon of 22 September 1993 (Zimmer, A. Whittaker). Five days later, B. Whitney (pers. comm.) observed 175 individuals (mostly in subadult plumage) at dawn, flying steadily down the Río Cristalino about 20 m above the forest canopy. They passed in a period of 3–4 minutes. This event was interpreted as a large social roost (Sick 1993) lifting off to continue a migratory movement. Zimmer has witnessed similar spectacular dawn lift-offs of this species in the Pantanal of southern Mato Grosso during September–October. Snail Kites undergo dramatic population movements, seemingly in response to drought and changing food availability (del Hoyo et al. 1994), although the pattern and seasonality of these movements are not completely understood. For example, in some years the species is abundant in September–October at Iguacu Falls (Brazil), whereas in other years it is completely absent during the same time period (Zimmer, pers. obs.). Two specimens were collected in the Rio Peixoto de Azevedo in March (Novaes and Lima 1991).

Gray-bellied Hawk (*Accipiter poliogaster*).—Almost nothing is known of the food habits or movements of this rarely recorded raptor, although at least in some areas it is thought to be migratory (del Hoyo et al. 1994). There are two records from Alta Floresta. On 17 September 1993 Zimmer observed an adult at close range along the TPT. The bird seemed to be tracking a mixed-species understory flock (consisting primarily of *Thamnomanes* spp., and *Myrmotherula* spp.) in an extensive stand of *Guadua* bamboo. It landed on a bent-over bamboo stalk 2–3 m above the ground less than 25 m away. It sat in that spot, peering about for at least two minutes, and then flew off in the direction of the still moving flock. In the same week B. Whitney (pers. comm.) observed a distinctively plumaged immature bird (plumage like a miniature adult Ornate Hawk-Eagle *Spizaetus ornatus*) perched at the forest edge on the TPT near the top of a mostly bare tree. It perched in a vertical posture (tail straight down, back erect) for 3–4 minutes. A Cuvier's Toucan *Ramphastos cuvieri* perched about 2 m away the entire time, and it eventually lunged at the hawk, which rapidly flew outward and down out of view.

Roadside Hawk (*Buteo magnirostris*).—There are only two records of this widespread and normally common species, a single individual seen by A. Whittaker at the RCC clearing on 21 September 1993, and two birds seen in second-growth pastures <50 km south of Alta Floresta in September 1994. Four specimens were collected at Cachimbo (Pinto and Camargo 1957) and one at Rio Peixoto de Azevedo (Novaes and Lima 1991). The apparent rarity of this species in the Alta Floresta area suggests that it did not occur in the nearly continuous forests that until recently covered the region. Its absence from river-edges (a habitat commonly occupied in other forested parts of Amazonia) of the Teles Pires and Cristalino is puzzling. With suitable pasture and second-growth habitats now in abundance around Alta Floresta, this species should be expected to colonize the area in the coming years.

White-browed Hawk (*Leucopternis kuhli*).—This rarely seen forest hawk has been recorded on three occasions. On 20 August 1991, Parker and Zimmer observed two birds along the logging road bounding the north edge of the TPT forest. One of these was a typical adult, and the other was similar in appearance to *L. melanops* (whitish crown, dark mask, pale-spotted mantle) and was believed to have been an immature bird. The two birds called back and forth to one another, both uttering similar whistled "pseeoo" calls resembling those described by Parker and Remsen (1987) from Bolivia. On 21 September 1993 Zimmer observed an adult along the RCT after inadvertently playing a single pre-recorded taped call (recorded in Rondônia). The hawk, whose presence prior to this moment was unknown, responded immediately by descending from the canopy to 20 m above ground (perching close to the trunk on a large tree) and giving several calls of its own (tape-recorded). This bird was observed off and on over the next 30 minutes, while it seemed to be following a mixed-species canopy flock of woodcreepers, furnariids, tanagers, etc. On 19 September 1995 Zimmer recorded a single individual vocalizing repeatedly from the canopy along the Taboca Trail. All vocalizations of this species recorded at Alta Floresta are identical to those tape-recorded by Zimmer in Rondônia. This species has also been collected along the nearby Rio Peixoto de Azevedo (Novaes and Lima 1991).

Harpy Eagle (*Harpia harpyja*).—A single tertial and whitewash on the ground (at the TPT stream crossing) found near the decomposing skin of a large mammal (possibly an anteater) on 20 August 1991 are evidence that this species still occurs in the region. However, in September 1993, local settlers told Zimmer and A. Whittaker of having recently shot an enormous bird of prey (described as being 1.5 m in length) in the immediate vicinity of the TPT. Habitat destruction and fragmentation between the town of Alta Floresta and the Rio Teles Pires probably precludes the continued existence of this species within that area, but the RFC would seem to have both sufficient habitat and healthy prey populations to support this species. It was also collected in the Rio Peixoto de Avezedo region (Novaes and Lima 1991).

Crimson-bellied Parakeet (*Pyrrhura rhodogaster*).—This species, endemic to the Rondônia Center (Haffer 1978; Cracraft 1985), is uncommon here near the eastern edge of its range. In October 1989, Parker and the Islers found it common at the Floresta Amazonica Hotel and in the TPT area and to the north and south along the main road in patches of tall forest. In August 1991, Parker and Zimmer found it common in the same areas, but the species was not found along the Rio Cristalino. Groups of six to eight (maximum of 12) individuals were commonly heard or seen flying through or low over the TPT canopy (up to 30 birds/day). One small group fed on *Cecropia* catkins at the TPT forest edge; others were observed feeding on *Erythrina* flowers in the same area. Birds in the interior forest canopy at TPT were feeding on a small *Ficus*, and a group seen in swamp forest canopy farther north along the Teles Pires Road were feeding on seeds taken from 20 cm-long pods of a leguminous tree. On subsequent visits *P. rhodogaster* has been decidedly less common. Zimmer and B. Schram missed it entirely in October 1992; Zimmer and A. Whittaker saw only two groups in September 1993, one at the Floresta Amazonica Hotel forest, and the other a group of eight in disturbed tall forest north of the TPT on the Teles Pires Road; and the only ones recorded in September 1994 and September 1995 were three pairs (1994) and a group of five birds (1995) at the Floresta Amazonica Hotel. Whether the decreased numbers recorded during recent visits indicate declines due to habitat loss, or merely reflect seasonal movements between August and September/October, is not known. The apparent absence of *P. rhodogaster* from the pristine RFC is puzzling. *Pyrrhura picta* is common along the Cristalino, as it is throughout the area.

The Crimson-bellied Parakeet was previously considered a Brazilian endemic (Meyer de Schauensee 1982), but has recently been found at several sites in Bolivia (Bates et al. 1989; Parker et al. 1991). It is replaced to the east by its sister taxon, the Pearly Parakeet (*P. perlata*). Haffer (1992) hypothesized a contact zone between the two forms somewhere between the Tapajós and the Xingú. As neither *P. rhodogaster* nor *P. perlata* have yet been found along the Rio Cristalino, along the Rio Peixoto de Azevedo to the east (Novaes and Lima 1991) or at Cachimbo to the northeast (Pinto and Camargo 1957), it is possible that in this region the two species may be separated by an area in which neither occurs.

Sapphire-rumped Parrotlet (*Touit purpurata*).—This small parrot has been recorded on six occasions. Parker heard and tape-recorded the distinctive nasal flight calls of two or more birds as they flew over the forest canopy above the RCT one morning in August 1991. Zimmer and B. Schram had excellent views of a single calling bird as it flew low over the RCC boat launch area on the morning of 20 October 1992. Bret Whitney (pers. comm.) heard flight calls of one or more birds over the forest canopy along the RCT on 20 September 1993. Zimmer heard flight calls of 1–2 birds passing over the RCC clearing on two different days in September 1994. On 21 September 1995 Zimmer and Andrew Whittaker had good views of three birds flying past the top of the Serra Trail (RFC). These records represent a significant southerly range extension. The other Brazilian localities south of the Amazon from which this species was previously known are from along the Río Capim in northeastern Pará (Sibley and Monroe 1990); the Javarí and Kayapó Indigenous Reserves, Pará (B. Whitney, *in litt.*); the Belém area, Pará (R. Ridgely, pers. comm.); the Rio Urucu, Amazonas (Peres and Whittaker 1991); and from Quebrada Socó, lower Rio Javarí, Amazonas (J. V. Remsen, Jr., pers. comm.).

Neomorphus sp.—The only ground-cuckoo record is of an individual glimpsed by Parker shortly after dawn (August 1991) along the RCT; it was not seen well enough to discern whether it was *N. squamiger* or *N. geoffroyi*. The bird was seen in an area where a large antswarm (*Eciton* sp.) attended by *Phlegopsis nigromaculata* and *Rhegmatorhina gymnops* was located late the previous afternoon, suggesting that army ants were still somewhere nearby.

Brown-banded Puffbird (*Notharchus ordii*).—This little-known puffbird is uncommon in the Alta Floresta region, although it remained undetected until 1993. The first record was on 19 September 1993, when W. Maynard, Zimmer, and A. Whittaker found a pair of birds in open stunted forest on a rocky dome at the top of the Serra Trail (RFC). As noted in the site descriptions, the forest atop this dome has an open canopy and an abundance of large, terrestrial bromeliads. Several prominent arboreal termitaria were also nearby. Zimmer and Hilty (1997) have found *N. ordii* nesting in arboreal termitaria in Venezuela, as has been recorded for other species of *Notharchus* (Hilty and Brown 1986; Sick 1993). When first encountered, only a single puffbird was in view. It was perched ca. 9 m above the ground in a bare tree, where it remained quietly for several minutes. As we were watching it, a second individual flew in and landed in the same tree. The two birds sat quietly for another few minutes and then flew suddenly into the forest and out of sight. On 27 September 1993 Bret Whitney (pers. comm.) heard and tape-recorded one bird along a different trail in the RFC, and later used that tape to lure in a pair of puffbirds at the site where Maynard, Zimmer, and Whittaker had seen them a week earlier. The only foraging move observed was a spectacular sally-strike performed by one of the birds at Serra. It launched from a branch on which it had been perched for more than five minutes, flying downward at about 20° below horizontal for a distance of about 8 m across an opening to another tree, swooped sharply laterally as it hit the outer leaves on the tree and landed in another tree about 5 m away, without having visibly captured anything (B. Whitney, pers. comm.). In September 1994 Zimmer and A. Whittaker recorded individuals or pairs of these puffbirds at the same two localities where birds were found in 1993, and heard either a single song of *N. ordii*, or an imitation by *Turdus lawrencii*, at a third location. In September 1995 Zimmer and A. Whittaker saw three pairs of Brown-banded Puffbirds and heard at least three or four others along the Rio Cristalino. One pair perched 5 m apart in the bare upper branches of an emergent tree, ca. 25 m above the left bank of the Cristalino at about 16:30.

The songs recorded by Whitney are 11–12 s in length, and consist of several clear, loud introductory whistles followed by cadenced couplets and triplets to finish. Zimmer and A. Whittaker have subsequently recorded songs of similar pattern and quality, but lasting 15–16 s. These fit well the songs recorded by K. Rosenberg on 11 September 1989 from Madre de Dios, Peru (LNS #46284). Rosenberg's recording is, to our knowledge, the first recording made of *N. ordii*, and also documents the only Peruvian record of the species (Foster et al. 1994). This song could be confused with a seldom-heard variant song of *Notharchus macrorhynchos* that also begins with loud whistles and ends with couplets (K. Rosenberg, B. Whitney, A. Whittaker, Zimmer, pers. obs.). In response to playback one pair of Brown-banded Puffbirds responded with a duet in which one bird sang a typical song and the second bird began with a subdued version of the typical song but ended with a series of 8 low whistles (1/s) inserted within and between the ending triplets of its mate (Zimmer and A. Whittaker recordings). Whitney (pers. comm.) suggested that *N. ordii* may be closely related to Black-breasted Puffbird (*N. pectoralis*) of Panama and west of the Andes, pointing out that it shares both a morphological (size, proportions, shape of the breastband) and vocal resemblance.

There are few records of this species, which is known largely from specimens collected in northern Amazonian Brazil and southwestern Venezuela. The first record for southeastern Amazonia was of a male collected 17 August 1978 from nearby Rio Peixoto de Azevedo (Novaes and Lima 1991). The species has recently been recorded for the first time in Peru (Foster et al. 1994) and depto. Pando, Bolivia (Parker and Remsen 1987).

Collared Puffbird (*Bucco capenis*).—Three records. Parker heard one singing at dawn on 25 August 1991 at the beginning of the TPT. On 22 September 1993 Zimmer saw one perched ca. 10 m above the ground for 2 minutes along the TT. In the same week B. Whitney (pers. comm.) heard a pair (pre-dawn) from the forest canopy along the RCT. Aside from a single bird collected in the Rio Peixoto de Azevedo area (Novaes and Lima 1991), these may constitute the only records for Mato Grosso.

Rufous-necked Puffbird (*Malacoptila rufa*).—Our only records of this poorly known, Amazonian puffbird are from the vine-rich forest along the Rio Cristalino. In August 1991 Zimmer saw one perched in a vine tangle 2 m above ground. In October 1992, Zimmer and B. Schram found a pair of these puffbirds following a mixed-species understory flock composed of *Thamnomanes caesius*, *T. saturninus*, *Thamnophilus schistaceus*, *Myrmotherula* spp., *Hylophilus ochraceiceps*, and *Habia rubica*. The puffbirds changed perches frequently (perching 2.5–5.0 m above ground for ca. 10–180 s at a time), remaining close to the flock as it moved through two small stands of bamboo and beyond (linear distance of 70+ m) before being lost from sight. In September 1994 Zimmer found lone birds (apart from flocks) along the Taboca Trail (perched quietly 2–3 m above ground) and the Serra Trail (perched on an open branch beneath a large vine tangle 10+ m above ground). In September 1995 Zimmer found a lone bird perched 10 m above the ground along the Taboca Trail. It sally-pounced from its perch to the top of an understory palm (ca. 6 m in height), where it remained for a moment before darting quickly away with an unidentified object in its bill. The following day Zimmer found a lone bird in a stand of bamboo farther along the same trail. It perched at heights of 1.5–3 m for 15–150 s. The general behavior of all of these individuals has been similar to that of other species of *Malacoptila* (Hilty and Brown 1986; Ridgely and Gwynne 1989; R. Ridgely, pers. comm.; Zimmer, pers. obs.) No vocalizations have been noted from Alta Floresta, but B. Whitney (pers. comm.) describes the song from the lower Rio Tapajós, extreme eastern Pará, and Serra dos Carajás as "a high, thin descending, slightly modulated whistle lasting about 1.5–3.0 s." He describes the call as "a similarly high, thin, slightly descending 'seeee' lasting about 0.5 s." Similar calls have been described for this species from western Amazonia by Hilty and Brown (1986). Three specimens were collected from the Rio Peixoto de Azevedo area (Novaes and Lima 1991). This species is seemingly more common in Rondônia (D. F. Stotz et al. 1997), where Zimmer (unpubl. data) has also found it following *Thamnomanes*-led mixed-species flocks, as well as attending a swarm of small (not *Eciton burchelli*) army ants.

Gray-cheeked Nunlet (*Nonnula ruficapilla*).—Parker and Zimmer observed one in the upper branches (10 m above ground) of a bamboo stand along the TPT (August 1991) and heard another in dense, low forest along the RCT stream, for the only area records. Emilio Dente collected a single female in the Rio Peixoto de Azevedo area (Novaes and Lima 1991). This species is known in Brazil from the Rio Juruá east to the upper Rio Xingú (Sick 1993) but within that range it may be only locally distributed. At Alta Floresta, near the eastern edge of its known range, it appears to be rare. In most other parts of its range *N. ruficapilla* is associated with bamboo (Parker and Zimmer, pers. obs., B. Whitney, pers. comm.) However, at Fazenda Rancho Grande near Ariquemes (Rondônia), where *N. ruficapilla* is uncommon, Zimmer (unpubl.

data) has found several territories linearly distributed along roadside forest edge with an abundance of viny second-growth, but without bamboo.

Black-girdled Barbet (*Capito dayi*).—This little-known barbet is found in Brazil south of the Amazon from the Rio Javari to the Tocantins and south to the Rio Teles Pires (Sick 1993) and Rio Peixoto de Azevedo (Novaes and Lima 1991). It has also recently been found in northeastern Bolivia (Bates et al. 1989; Kratter et al. 1992). This species is uncommon to fairly common at Alta Floresta, most often seen in the company of large mixed-species canopy flocks of foliage-gleaners, tanagers, and dacnises. Pairs and small groups (family?) of 4–6 barbets are also encountered independent of other species. We have recorded them at heights of 4–30 m (most often 15–25 m) above the ground, searching dead leaves (particularly *Cecropia*) and feeding on fruits of *Ficus*, *Cecropia*, and a small tree with small, reddish (rubiac-like) fruits. Bret Whitney (pers. comm.) observed one female along the Serra Trail foraging just 2 m above the ground, where it was searching both live vegetation and trapped dead leaves in an *Acacia*. They were most common in the canopy and at borders of tall primary forest (although they seemed strangely absent from the canopy along the TPT), but were also seen feeding in heavily disturbed secondary forest (usually in *Cecropia*). Zimmer (unpubl. data) has found them regularly venturing from primary forest into adjacent abandoned cacao plantations in Rondônia. Although these barbets have not been particularly vocal during the time of our visits (August–October), we have noted three types of vocalizations. The song is a steadily accelerating series of hollow "boo" notes (ca. 7-8 s) that is similar to that of Striated Antthrush (*Chamaeza nobilis*) but lacking the distinctly slowed "woop woop woop woop . . ." terminal notes (Zimmer recording; B. Whitney, pers. obs.). Another vocalization is a staccato series of rough notes that could be translated as "rerk erk erk erk erk," or "rok rok ock ock" and which is frequently given by foraging birds. Of the vocalizations that we have recorded, this one is most similar to the "song" described by Bates et al. (1989) from Bolivia. The third vocalization is a peculiar grating rattle (with the quality of tearing thick cardboard) often given by foraging birds when they make longer flights (as in crossing the road or in leaving a given area).

Red-billed Woodcreeper (*Hylexetastes perrotii uniformis*).—Rare. Our only records are of a pair (presumably the same ones) recorded at the edge of tall forest along the Teles Pires Road (south of the TPT) in 1989 and again in 1991, and of a single bird heard by Parker from the canopy along the RCT in 1991. In each instance these birds were vocal for only 15+ minutes at first light (during which time the Teles Pires Road pair responded strongly to tape playback) and then not seen or heard subsequently. The dawn song was a penetrating series of 2–4 drawn-out (almost two-syllabled) "ERWEEÉK" notes (slightly faster than 1 note/s), each series being given several seconds apart (Zimmer recording). This is similar to songs described for this form from Rondônia (Ridgely and Tudor 1994). Each member of the pair gave similar calls on slightly different pitches. In response to playback both birds gave faster versions of the song, as well as several types of nasal, whining snarls, among them a "nnnyeah" and a "nyip, nyeek, nyeek, weeweweweip" (Zimmer recording). The "nnnyeah" call is similar to the distress calls given in response to playback by some *Xiphocolaptes* woodcreepers, notably *X. major* and *X. promeropirhynchus* (Zimmer, pers. obs.). The resident form of Red-billed Woodcreeper is the unbarred, brownish-red-billed *H. p. uniformis*. This form is replaced somewhere to the east by a conspicuously barred, undescribed subspecies (Haffer 1992), but the exact area of this replacement must be considerably farther east, because specimens collected at Cachimbo (Pinto and Camargo 1957) and Rio Peixoto de Azevedo (Novaes and Lima 1991) were identified as *uniformis*.

Curve-billed Scythebill (*Campylorhamphus procurvoides*).—Uncommon to fairly common in extensive stands of bamboo and in adjacent vine-rich forests. Often seen with mixed-species understory and mid-level flocks in upland forest along the TPT and RCT, where in 1991–1994 up to 3–4/day were seen. In 1989 and 1991 this species was also conspicuous in the larger stands of bamboo along the TPT, where it foraged on small trunks, large vines, and bamboo stalks at 5–12 m. No scythebills were found in these stands of bamboo in 1992–1993, but there was a resident pair there in 1994, and other pairs were found inhabiting stands of bamboo along the TT and RCT in 1993 and 1994. These birds were also seen in the company of *Thamnomanes*-led mixed-species flocks that passed through the territories. Alta Floresta is close to the southernmost locality at which *C. procurvoides* has been recorded, along the Rio Peixoto de Azevedo (Novaes and Lima 1991).

Cabanis's Spinetail (*Synallaxis cabanisi*).—A pair observed and tape-recorded (October 1989) in low, spreading thickets of bamboo in cut-over forest ca. 25 km north of Alta Floresta by Parker and the Islers represented the first record for Brazil and an easterly range extension of

1,200 km (Parker et al. 1997). A single tape-responsive bird (presumably of this same pair) was seen by Parker and Zimmer in the same location in August 1991. This species has only recently been recognized as distinct from *S. moesta* (lower slopes of the eastern Andes and adjacent lowlands in Colombia, Ecuador, and Peru) and *S. macconnelli* (slopes of the tepuis and adjacent lowlands in southern Venezuela and the Guianas) (Vaurie 1980: 99–100; Sibley and Monroe 1990). *Synallaxis cabanisi* was known previously only from the foothills and adjacent lowlands bordering the base of the Andes from northern depto. San Martin, Peru, south to depto. Cochabamba, Bolivia (Vaurie 1980), but more recently has been found in river-edge bamboo thickets near Puerto Maldonado (Parker et al. 1997), and along the upper Río Tambopata, Madre de Dios, Peru (Foster et al. 1994).

Chestnut-throated Spinetail (*Synallaxis cherriei*).—At Alta Floresta this poorly known spinetail inhabits the larger, more mature and homogeneous stands of bamboo within the forest interior. It does not appear to be present in younger second-growth thickets of bamboo prevalent at the forest edge along the Teles Pires and Fazenda Cristalino roads. In 1989 Parker and the Islers found four pairs along ca. 400 m of the TPT in the largest stand of bamboo. Three pairs were found in the same area in 1991, and their territory size was estimated to be less than 0.5 ha. At least 1–2 pairs were present in this same bamboo patch in 1992–1995, and an additional two pairs were in smaller stands of bamboo at the RCC (along the RCT and TT) in 1993 and 1995. Pairs of these spinetails foraged in close proximity to one another (within 10 m), usually in dense, dark tangles of fallen bamboo stalks and suspended debris. They worked from the ground up to 3 m (but usually much lower), occasionally ascending along hanging vine tangles close to the trunks of trees, and picking at vine surfaces and dead leaves. They spent most of their time probing in dead leaves and other suspended debris trapped in the tangles of dead bamboo stalks near the ground. On occasion we have seen them hopping on the ground beneath these tangles, picking at the leaf litter and flipping dead leaves to uncover prey. In response to tape playback one individual climbed to 12 m above the ground, where it sat and called continuously for several minutes (Zimmer and A. Whittaker, pers. obs.).

The Chestnut-throated Spinetail disjunctly occupies a large range, from the foothills of the Andes in Colombia, Ecuador, and Peru east to the drainage of the lower Rio Xingú (Collar et al. 1992). It is currently known in Brazil from only six scattered sites in Rondônia, Mato Grosso, and Pará (Collar et al. 1992). Within its known range *S. cherriei* occupies a variety of habitats; only at Alta Floresta does it seem specialized on bamboo (Collar et al. 1992). Competitive relationships with sympatric congeners have been implicated as determining factors in local habitat selection of *S. cherriei* (Collar et al. 1992). At Alta Floresta this species is sympatric with *S. rutilans*, which occupies dense tangles in interior forest light-gaps, and *S. gujanensis*, a rare inhabitant of thickets at the forest edge. In 1991 a pair of *S. rutilans* occupied a streamside territory adjacent to the largest stand of bamboo on the TPT, and within earshot of a pair of *S. cherriei*.

Peruvian Recurvebill (*Simoxenops ucayalae*).—This poorly known bamboo specialist (Parker 1982; Parker et al. 1997; Kratter, in press) occurs in the larger stands of mature bamboo (generally homogeneous stands with relatively open canopy) along the TPT and the TT. At least two pairs have maintained territories in the largest bamboo stand (ca. 4–5 ha) along the TPT from 1989–1993. Two more territories were located in separate patches of large bamboo along the newly opened Taboca Trail in 1993, and an additional territory was found along this trail in 1995. This species seems to maintain large territories (responding to tape playback along a linear trail segment of more than 100 m) and is absent from the many smaller thickets of bamboo along the RCT and TT, as well as from more extensive second-growth bamboo thickets along the forest edge.

Recurvebills foraged from 1–10 m above ground, most commonly below 6 m, working on small cavities in bamboo stalks by hammering and prying with their heavy bills (Parker et al. 1997). They frequently probed and further fractured the ends of broken and dead stalks. They also ascended higher to forage in woody vine tangles beneath the crowns of trees, hitching vertically through the vines like a woodcreeper, and hammering at vine surfaces. They were fairly vocal, particularly at first light, giving sporadic nasal "chak" notes (presumably contact notes between foraging members of a pair) as well as the song, which is a loud, *Syndactyla*-like rattle that accelerates as it rises in pitch and then slows at the end (Parker 1982). When agitated by playback, the typical response was to deliver multiple "chak" notes in a staccato rhythm, accelerating into the song.

Until recently, this species was known only from disjunct localities in eastern Peru (Parker 1982), but it is now known to occur widely (but patchily) over much of southern Amazonia

(Parker et al. 1997). The initial Alta Floresta records (1989) by Parker and the Islers represented only the second known locality for Brazil (Parker et al. 1997); the first was along the lower Rio Xingú in 1986 (Graves and Zusi 1989). The species has also recently been found in depto. Pando, Bolivia (Parker and Remsen 1987).

Olive-backed Foliage-gleaner (*Automolus infuscatus*).—This species was frequently found in association with *Thamnomanes*-led understory flocks, foraging from 1–8 m (typically from 1–5 m) above ground. One individual moving with a mixed-species flock along the TPT was foraging unusually high at 15 m above the ground. This species often foraged in the tops of small understory palms, in lower vine tangles, and in the forks and crotches of lower trees, where it probed in trapped dead leaves (especially of large leaves such as *Cecropia*) and other suspended debris (as found elsewhere by Remsen and Parker [1984]; Rosenberg 1997). Loud single "QUIT" notes and occasional "jureet-reetreetreetreet" calls (songs ?) differ from corresponding vocalizations of birds from north of the Amazon (Parker and Zimmer, pers. obs.; T. Schulenberg, pers. comm.).

Crested Foliage-gleaner (*Anabazenops dorsalis*).—This foliage-gleaner was a fairly common inhabitant of bamboo along the TPT, RCT, and TT. As of 1993 at least three pairs occupied the largest stand of big bamboo on the TPT, and virtually every patch of bamboo along the RCT (3+ pairs) and the TT (8 pairs) had its own pair of these foliage-gleaners. They foraged from 5–10 m above ground in the crowns of bamboo, but also to 15+ m in vine tangles of adjacent trees. When foraging in bamboo, they probed curled dead leaves and other debris trapped in the mats of bamboo foliage. They also sat for periods of 30 seconds or more picking at joints in the bamboo where there are clusters of spines. When foraging in trees, they concentrated on woody vine tangles, moving up through the tangles vertically while hitching from side-to-side, but also hitching horizontally along branches, using the tail as a brace by wrapping or tucking it around or under branches. Here they concentrated on dead leaves trapped in the vines. Birds were encountered singly or in pairs, either alone or in the company of mixed-species, mid-level flocks moving through their territories.

The most common vocalization (given frequently) is a loud "clock-clock-clock-clock," the individual notes repeated 5–10 times in a somewhat staccato delivery. Members of a pair give this call antiphonally, perching near one another with throat feathers flared prominently and pumping their tails up and down (each downstroke bringing the tail underneath the branch or vine on which the bird is perched). They also utter a long (up to 15 s) *Simoxenops*-like rattle, which may lead into the stacatto "clock" notes. More subdued "kek" notes are given as apparent contact notes while foraging.

This species was previously considered to be a bird of foothill forest and adjacent lowlands (particularly along rivers) from southeastern Colombia south through eastern Peru (Hilty and Brown 1986). Throughout its range it is restricted to *Guadua* bamboo thickets inside floodplain or disturbed upland forest, to 800 m (Parker et al. 1991). The first Brazilian record was of two specimens collected west of Cachoeira Nazaré, Rondônia, in 1986 (D. F. Stotz et al. 1997; Parker et al. 1997). The Alta Floresta records represent only the second documented occurrence for the country (Parker et al. 1997). The species has also recently been found at Alto Madidi, depto. La Paz, Bolivia (Parker et al. 1991). Recent research (Kratter 1997) indicates that this species is allied to the disjunct *Anabazenops fuscus* of southeast Brazil, and not to *Automolus*, where it was formerly placed (e.g., Sibley and Monroe 1990; Ridgely and Tudor 1994).

Chestnut-crowned Foliage-gleaner (*Automolus rufipileatus*).—Fairly common, particularly in stands of bamboo. It is strangely rare along the Rio Cristalino in river-edge forest, the habitat in which it is usually most common at other localities (Remsen and Parker 1983; Hilty and Brown 1986; Ridgely and Tudor 1994; Parker and Zimmer, pers. obs.). Although not confined to bamboo (it is also in overgrown treefall tangles in the forest interior, and in dense low forest above *Heliconia* thickets along small forest streams), most stands of bamboo along the TPT, RCT, and TT have a territorial pair of *A. rufipileatus*. These forage from 2–20 m above ground in dense tangles of vegetation, particularly where the crowns of bamboo form interlacing mats with vine tangles and branches of overstory trees. Along the aforementioned trails, territories of *A. rufipileatus* frequently overlapped with those of *Anabazenops dorsalis*.

Gray-throated Leaftosser (*Sclerurus albigularis*).—One bird seen and tape-recorded by Parker and Zimmer in shaded forest beyond the TPT bamboo in August 1991, and heard in the same area by Zimmer and B. Schram in October 1992, represent the only records. The bird was lured in with playbacks of its song (a long series of 8–12 emphatic "week" notes, similar to songs of Central American populations of *S. albigularis* but longer and faster) and perched a few inches above the ground on a slanted sapling. In response to playback, it increased the pace of its song and added a terminal rattle (again, similar to Central American birds; Zimmer, pers. obs.) before moving off and becoming unresponsive to playbacks (although continuing to sing incessantly from far back in the undergrowth). This bird had a rich rufous wash on the breast, contrasting strongly with a pale whitish-gray throat, brownish upperparts, and a rufous rump, appearing similar to other populations of *S. albigularis* (and more distinctly pale-throated than the otherwise similar *S. scansor*). We have recorded three other species of *Sclerurus* (*rufigularis*, *mexicanus*, and *caudacutus*) within a few hundred meters of this site.

The only previous Brazilian record of Gray-throated Leaftosser was an individual netted at Cachoeira Nazaré in 1986 (D. F. Stotz et al. 1997). The Amazonian distribution of this bird was thought to be restricted to the Andean foothills (Sibley and Monroe 1990), but it has recently been found to be fairly common in lowlands at the base of the Serranía Huanchaca, depto. Santa Cruz, Bolivia, some 350 km east of known Andean sites (Kratter et al. 1992). Specimens taken by Kratter et al. (1992) were thought possibly to represent a previously undescribed subspecies of *S. albigularis*. An individual *S. albigularis* singing a song similar to that of the Alta Floresta bird was tape-recorded by Zimmer (unpubl. data) in January 1993 at Fazenda Rancho Grande, Rondônia, another site with four sympatric species of leaftossers.

Glossy Antshrike (*Sakesphorus luctuosus*).—This antshrike was uncommon in riparian thickets along the Rio Cristalino and the Rio Teles Pires, in spite of what seems to be an abundance of suitable habitat. It foraged from 1–8 m above ground in viny or leafy thickets, progressing by heavy hops of ca. 5–30 cm and pausing for 2–20 s at each station to peer about. Prey were usually perch-gleaned from leaf or vine surfaces. Prey items that we were able to identify included a 4-cm-long brownish caterpillar, and a 5–6-cm-long brown katydid beaten several times against a branch and then fed by an adult male to a fledgling (September). The crest is usually folded back when foraging, but erected when singing. In seeming response to tape playback of songs of *Thamnophilus palliatus*, a male Glossy Antshrike displayed a partially erect to fully erect crest and a fanned tail held cocked at 45° above horizontal. The fanned tail was dramatic, revealing the bold white tips on the rectrices. The wings were held slightly drooped and were frequently twitched. No vocalizations or other signs of agitation were obvious.

((bushbird sp. ?)) ((*Clytoctantes/Neoctantes*)).—Judged hypothetical on the basis of a brief sighting by Parker in November 1989. While working alone along a small stream inside tall *terra firme* forest ca. 1 km south of the Rio Teles Pires, Parker saw a large, all-black antbird that he felt was a male bushbird, 0.5–1.0 m above the ground in dense tangles of a tree-fall. The bird flew before it could be studied (Collar et al. 1992; Parker, pers. comm.). Parker felt that the bird was probably the male of the recently described Rondônia Bushbird *Clytoctantes atrogularis*, (Lanyon et al. 1990), but Black Bushbird *Neoctantes niger*, whose patchy distribution in Amazonia and Brazil is poorly understood (Lanyon et al. 1990), could not be ruled out. *Contra* Lanyon et al. (1990), the Islers did not observe a bushbird at Alta Floresta, and the "different song" alluded to was actually a variant song of *Myrmornis torquata*, recorded at a completely different site along the TPT.

Ornate Antwren (*Myrmotherula ornata*).—Fairly common in upland forest, usually in stands of bamboo, and almost invariably in the company of mixed-species understory/mid-level flocks. Although clearly not restricted to bamboo, Ornate Antwrens in this region seem strongly tied to it. Most mixed-species flocks in stands of bamboo included a pair of *M. ornata*, as did some flocks in adjacent *terra firme* forest that lacked bamboo. This species foraged from 5–20 m above ground (most often 6–10 m), typically concentrating on vine tangles, particularly in the crowns of bamboo, where it searched single dead leaves (especially large curled leaves) and clusters of dead leaves that were suspended in the vegetation (as found elsewhere by Remsen and Parker [1984], Rosenberg 1997), seldom spending more than a few seconds at a given station. It flicked its wings frequently while foraging, and members of a pair typically remained within 5–10 m of one another. The closely related (Hackett and Rosenberg 1990) *M. leucophthalma*, another dead-leaf specialist (Remsen and Parker 1984; Rosenberg 1993; Zimmer, pers. obs.), is also fairly common at Alta Floresta, and occurred in many of the same flocks with *M. ornata*, but only outside of bamboo. Ornate Antwren is the only "checker-throated" *Myrmotherula* associated with bamboo at Alta Floresta. Where the two species occurred in the same flocks (*terra firme* forest), *M. leucophthalma* was invariably lower (1–6 m above ground) and usually foraged on slender branches of small understory trees (especially palms), where it systematically inspected suspended dead leaves, often spending 15–30 s at a single leaf. Common flock-mates of *M. ornata* (in bamboo) were *Anabazenops dorsalis*, *Automolus infuscatus*, *A. rufipileatus*, *Thamnophilus amazonicus*, *Thamnomanes caesius*, *Drymophila devillei*, *Microrho-*

pias quixensis, Ramphotrigon megacephala, Ramphocaenus melanurus, Hylophilus semicinereus, and *Granatellus pelzelni.*

White-flanked Antwren (*Myrmotherula axillaris*).—This species is noteworthy at Alta Floresta mainly for its scarcity there (in much of its extensive range *M. axillaris* is the most common *Myrmotherula* [Zimmer, pers. obs.]) and also for its apparent restriction at this site to bamboo. Only a single territorial male or pair was recorded in each visit (1989 and 1991-95) to the TPT and always from the largest stand of bamboo. Similarly, single pairs found along the RCT and TT in 1992–1995 were also in patches of bamboo.

Dot-winged Antwren (*Microrhopias quixensis*).—This species is of local interest primarily because it is represented by two distinct forms that apparently replace one another across both the Rio Tapajós to the north (Peters 1951), and the Rio Teles Pires in the south (Zimmer and A. Whittaker, pers. obs.). West (south) of the Teles Pires we have found *M. q. bicolor* (females slaty above and entirely rufous below, both sexes with extensive white tail tipping) to be fairly common but patchily distributed in denser forests with numerous large, vine-covered trees. Along the TPT these birds favor dense tangles 5–10 m above ground, particularly at the edges of treefall gaps. East (north) of the Teles Pires, in the RFC, we have found only *M. q. emiliae* (females blacker above with throat and chest deep chestnut-maroon and belly black, both sexes with reduced white tail-tipping). These birds also frequented light gaps, particularly favoring tangles and mats of vegetation in the crowns of bamboo (places also used by *Drymophila devillei* and *Granatellus pelzelni*), and foraged at heights of 5–10 m. Both *M. q. bicolor* and *M. q. emiliae* were encountered in pairs or family groups, frequently alone, but also joining mixed-species flocks for short distances.

Vocalizations differ between the two forms. The song of *emiliae* begins with 2–4 high, clipped notes, accelerating into a short terminal rattle ("EE EE EE CHECHCHCHCH"), the entire song lasting 1.5–2.0 s (Zimmer tape recordings, A. Whittaker tape recordings). Songs of *bicolor* are longer, with a more musical quality, and lack the terminal rattle although they also accelerate toward the end ("SIP SIP SIP SIP SEE-SEE-SEE-SESESE") (Zimmer tape recordings). Both forms have a similar loud, sharp "CHEEP!" call (given frequently when foraging), and *emiliae* also utters a differently pitched "CHOOP" repeated 4–8 times in rapid succession when excited (Zimmer tape recordings). In September 1995 Zimmer and A. Whittaker conducted numerous tape playback experiments with the two forms of Dot-winged Antwrens. In more than 15 trials (conducted over several days and involving 5–6 pairs of birds along various trails in the RFC) no individual of *emiliae* was found to respond in any way to playback of songs of *bicolor*. These same individuals and pairs responded strongly to playback of songs of other *emiliae*. Unfortunately, reciprocal experiments could be performed with only one pair of *bicolor* along the TPT, but they similarly failed to respond to playback of songs of *emiliae*. Interestingly, playback of songs of either form of *Microrhopias* in patches of bamboo almost always drew an immediate vocal response and aggressive approach from *Drymophila devillei*, which continued to sing until several minutes after playback of *Microrhopias* vocalizations ceased.

A detailed analysis of vocal and ecological differences, as well as examination of the exact nature of this zone of parapatry/contact between the morphologically distinct *emiliae* and *bicolor*, is needed.

Striated Antbird (*Drymophila devillei subochracea*).—This species was found in shaded stands of slender bamboo intermixed with trees. In all years but 1995 it was absent from the larger, more homogeneous stands (which have a relatively open, sunlit canopy) found along the TPT (and from two smaller but similar stands along the TT), all of which were occupied by *Cercomacra manu*. We have not found the two species occupying the same patches of bamboo, except in September 1995, when Zimmer found one territorial male *Drymophila* in one of the large stands of bamboo along the TPT. In 1991–1995 a pair of *D. devillei* occupied each of three small (ca. 0.5 ha) stands of bamboo along the 1.0–1.5 km RCT. In 1993 Zimmer found eight pairs scattered over several patches of bamboo along the 1.5 km Taboca Trail. Each of these patches was occupied by at least one pair of *Drymophila* in September 1995. All occupied areas were dense, shaded stands of slender bamboo intermixed with numerous taller trees, and the overlapping canopies contained many vine tangles. Similar stands of bamboo just inside the forest edge along the Fazenda Cristalino Road were also occupied by *D. devillei* in 1993. In 1989 Parker and the Islers located three territories along 500 m of roadside north of the TPT, each 15–50 m or more inside the forest from the road edge. That entire area had been clear-cut by October of 1992. All *D. devillei* found at Alta Floresta have been of the morphologically distinctive subspecies *subochracea*, previously known only from the Rio Curvá (a tributary of the Iriri), Pará (Pinto 1978). *Drymophila devillei* found in bamboo west of Cachoeira Nazaré, Rondônia, were of the more widespread nominate subspecies (Parker et al. 1997). The contact zone, if any, between the two forms is unknown.

Pairs of these antbirds foraged in close contact to one another, remaining mostly in the shaded crowns of the bamboo and adjacent vine tangles 5–13 m above ground. They hopped rapidly along vines and horizontal limbs, and hitched from side-to-side as they made their way vertically along bamboo stalks, picking at the bases of clusters of bamboo spines and leaves. Prey were gleaned directly from stem, vine, and leaf surfaces. These birds often flicked their wings and wagged their tail up and down as they moved along limbs. Both members of a pair sang periodically (often in a responsive manner), becoming more vocal when a mixed-species flock passed through the territory. Singing birds erected their crown feathers into a bushy crest and vibrated the tail to the rhythm of the song. They occasionally joined mixed-species flocks passing through. On one occasion Zimmer saw a male *D. devillei* aggressively displace an Ornate Antwren (that was moving with a flock) from two consecutive perches.

Manu Antbird (*Cercomacra manu*).—This recently described species (Fitzpatrick and Willard 1990) was locally common at Alta Floresta, where it occupied essentially pure stands of *Guadua* bamboo (canopy ca. 10–15 m), with fairly open canopy with a sunlit interior. It was absent from shaded thickets of bamboo inside tall forest, in which the bamboo was thoroughly intermixed with other vegetation (these being occupied by *Drymophila devillei*). Territories seemed to be very small, at least in the optimal areas, and probably average 0.2 ha or less. At least 10 pairs have consistently occupied the largest stand (4–5 ha) of bamboo on the TPT, and on occasion we have noted four pairs countersinging within 50 m of us. The Islers found 5 pairs along ca. 100 m of a new road cut through dense bamboo thickets roughly 45+ km southeast of Alta Floresta. Scattered individuals were also heard during random stops in disturbed roadside forest with extensive bamboo along the Teles Pires Road (1989 and 1991), but most of these areas have since been clear-cut. Four pairs found in 1993 along the TT were the first for the RFC. The sum of these records represents the first for Brazil (Parker et al. 1997). Since its initial discovery in Manu National Park, Peru (Terborgh et al. 1984; Fitzpatrick and Willard 1990), *C. manu* has also been found in northwestern Bolivia (Parker and Remsen 1987) and near Porangaba, Acre, Brazil (A. Whittaker, pers. comm.).

Pairs foraged in close proximity, frequently engaging in syncopated antiphonal duets, with one such duet frequently setting off countersinging bouts by two or more neighboring pairs. Vocalizations were as described by Fitzpatrick and Willard (1990). This species foraged mostly in the crowns of the bamboo, as well as in vine tangles and shaded canopies of adjacent overstory trees, from heights of 4–15+ m. In the morning hours individuals seemed to concentrate their searches in the sunlit crowns, descending to lower heights in mid-day. They moved along branches in a rapid series of short hops, swinging their tails (often kept partially fanned) from side-to-side, and stopping frequently to peer about. Arthropods were gleaned directly (sometimes hover-gleaned) from stem and leaf surfaces of bamboo and interlaced foliage of vine tangles. Small, green lepidopteran larvae seemed to be an important prey item.

Bare-eyed Antbird (*Rhegmatorhina gymnops*).—This localized antbird is known only from Brazil south of the Amazon between the Rios Tapajós and Xingú (Meyer de Schauensee 1982). It seems to be uncommon in the RFC (rarer in upland forest along the TPT), but none were recorded during the 1989, 1992, or 1994 surveys. Pairs or groups of 3–5 individuals have been found at army-ant (*Eciton burchelli*) swarms along most of the trails in the RFC, and occasional "cruising" individuals and possible family groups have been lured in from some distance by tape playback when no ants were obviously present (K. Zimmer 1991, 1993; B. Whitney 1993, pers. comm.). In September 1995, Zimmer and A. Whittaker found three groups of Bare-eyed Antbirds attending *Eciton* swarms along various trails at the RCC. Two of these swarms were static, remaining in the same spot for at least three days, although the ants appeared to raid only in the afternoons for a few hours. At least 4–6 Bare-eyed Antbirds attended each of these static swarms for three consecutive days, and remained nearby (vocalizing in alarm whenever we approached and responding to tape playback) even when the ants were not actively raiding. Both swarms were sporadically attended by a pair of White-backed Fire-eyes (*Pyriglena leuconota*) and 2–6 Black-spotted Bare-eyes (*Phlegopsis nigromaculata*), but unlike *Rhegmatorhina*, these species were not obviously present at times when the ants were inactive. One of these two ant swarms was in sunlit second-growth (mostly *Heliconia*) at the edge of the RCC clearing. The other was in shaded stream-bottom forest along the Taboca Trail. At this latter spot the antbirds could be easily observed. Foraging behaviors and perch heights (0.2 m to 1 m) were similar to those described by Willis (1969). A spider and a brown orthopteran (both taken from the leaf litter) were the only prey items identified. Zimmer observed three brief agonistic encounters in which one *Rhegmatorhina* supplanted another from its perch, but noticed no such conflicts between *Rhegmatorhina* and nearby *Phlegopsis*. On 20 September 1995, Zimmer and A. Whittaker also heard a Bare-eyed Antbird singing from cut-over *várzea* forest on the south bank of the Rio Teles Pires. Vocalizations include harsh, churring "shooo" or "shrew" calls given frequently at antswarms, a loud staccato chatter and hard chips given in alarm, and the song, a descending series of 7–10 notes, the first protracted (almost two-syllabled) and the last couple buzzier, "EEYOU YOU YOU YOU YOU ZHEW ZHEW" (Zimmer tape recording, LNS). *Rhegmatorhina gymnops* has been collected along the Rio Peixoto de Azevedo (Novaes and Lima 1991).

Striated Antthrush (*Chamaeza nobilis*).—There are only a few records of this species from Alta Floresta. One was seen and its alarm calls tape-recorded by Parker and Zimmer along the RCT in August 1991, and another was heard once along the TPT that same week. None were recorded in 1989 or 1992, but in September 1993, 1–3 singing birds were noted (and taperecorded) daily along the RCT and TT (K. Zimmer; B. Whitney, pers. comm.) indicating that the species may be more common than previously thought. In September 1994 Zimmer and A. Whittaker found a pair of birds singing repeatedly along the lower portion of the Serra Trail (RFC) in shaded forest with a relatively open understory. A lone bird was heard by Zimmer and A. Whittaker in similar habitat along the Taboca Trail (RFC) in September 1995. Parker has noticed the tendency for this species to remain silent for weeks at a time (Parker et al. 1991), a habit that could explain the sporadic pattern of Alta Floresta records. Zimmer found three other terrestrial formicariids (*Myrmothera campanisona*, *Hylopezus macularius*, and *Grallaria varia*) to be much more vocal, and thus seemingly more common, in September 1993, than in previous visits in mid-August and late October. Songs of Striated Antthrush are similar to those of Blackgirdled Barbet (see discussion of that species), but end with several distinctive "woop" or "wah" notes (Zimmer, pers. obs.; B. Whitney, pers comm.). Observers too far away to hear these terminal notes may have difficulty distinguishing the voices of the two species. These records and an individual tape-recorded by the Islers in early October 1989 near Rurópolis, Pará, east of the Tapajós, constitute a minor eastward extension of the known range of *C. nobilis*, which was previously known only from west of the lower Rio Tapajós (Sibley and Monroe 1990; Sick 1993).

Wing-banded Antbird (*Myrmornis torquata*).—Known at Alta Floresta only from one territorial pair found in upland forest along the TPT, first tape-recorded by the Islers in 1989, identified in 1990 by R. Ridgely after playback of the Islers' tape, and relocated by Parker and Zimmer in 1991. The song of Alta Floresta birds was noticeably different from that of birds recorded in southern Venezuela (which sing a series of 12–15+ loud "whew" notes strung together and building in intensity to the end), being a longer series of 10–20 notes with the first several notes sounding 2-syllabled "er-whee er-whee er-whee . . . whee whee whee whee." This song was given by the male immediately following tape playback as well as more than 20 min after any playback. A straight series of 12+ "whee" notes was also recorded. In response to tape playback both the male and female approached the recorder and clung to vertical saplings while wing-flicking in exaggerated fashion. The tail was flicked upwards simultaneously with the flick of the wings. The male perched 1.0–1.5 m above the ground in a hunched posture with his back feathers fluffed up, revealing a white interscapular patch as a thin vertical streak down the center of the back. Several vocalizations were given in response to playback. Both birds uttered a low chirring sound. The male gave a short whiny call "werup wereup wheee wheee," a high and rising "wheee wheee churr" (the latter note low and harsh), a harsh "shrooo," several isolated angry "churrs," short whisper versions of the song, 3–4 "whee" notes (but softer) ending with a "churr," a loud "squee'ah" (similar to calls of Royal Flycatcher, *Onychorhynchus coronatus*), and isolated "er-whee" notes, whereas the female gave one very long song with 45+ notes in which the intensity faded and the pace slowed beyond 20 notes (Zimmer taperecordings). The apparent scarcity and localized distribution of this species in Amazonia is puzzling. Wing-banded Antbirds are extremely difficult to detect when they are not vocalizing, and thus, some of their apparent rarity may be an artifact of under-sampling.

Amazonian Antpitta (*Hylopezus berlepschi*).—Apparently uncommon and locally mostly restricted to dense stands of essentially pure bamboo (although Zimmer and B. Schram recorded one in dense *Heliconia* thickets in swamp forest along the Teles Pires Road). In 1989, in the area 45+ km southeast of Alta Floresta (which was more extensively covered with bamboo), the Islers found this species to be considerably more common. One or two pairs inhabited the largest bamboo thicket along the TPT from 1989–1993, and two more were heard in extensive stands of bamboo along the TT in 1993. One territorial bird was seen in a small, dense stand of bamboo and vine tangles in a forest light-gap along the RCT in September 1995. On four separate occasions we have observed singing birds (perched to heights of 2 m) dramatically inflate large, dull-pink air sacs on either side of the neck (documented by Zimmer photographs), the effect being similar to that seen in displays of some North American grouse and prairie-chickens. We have not seen this phenomenon in any other antpitta, nor have we seen it described in print. Other species of *Hylopezus* commonly inflate air sacs in the throat and neck region when calling (Zimmer, pers. obs.), but do not reveal prominent patches of bare skin when doing so.

Black-bellied Gnateater (*Conopophaga melanogaster*).—This species is known in this area from a single sighting of a male by Parker from along the TPT in October 1989. This individual was seen clearly, but was not tape-recorded. Next-day attempts by the Islers to find this species at this location were unsuccessful, as were attempts by other observers in later years. The nearest localities from which this species has been recorded are along the Rio Jiparaná, Rondônia (Pinto 1978; Stotz et al. 1997) and middle Rio Xingú, Pará (Graves and Zusi 1990).

Purple-throated Cotinga (*Porphyrolaema porphyrolaema*).—There are five records of this cotinga, which has been called "one of the least known members of its family" (Snow 1982). It was first reported by R. Ridgely and V. Emanuel (pers. comm.), who saw a pair (male and female) on the grounds of the Floresta Amazonica Hotel in Alta Floresta (5–9 October 1990). On 18 September 1993 Zimmer and A. Whittaker observed a male and female at length as they perched (but were not seen to feed) 20+ m above the ground in a fruiting tree that was also attracting various ramphastids. Zimmer tape-recorded several calls, most of which were similar to the plaintive whistles of Dusky-capped Flycatcher *Myiarchus tuberculifer*, and could be transcribed as "wheeeuw" (similar to those described in Hilty and Brown, 1986). Bret Whitney (pers. comm.) has noted similar vocalizations from this species in Peru, as has R. Ridgely (pers. comm.) from Ecuador. A second type of call (interspersed with the plaintive whistle) was similar in quality but with a tremulous element, so as to sound three-syllabled, "werleeyooo" (Zimmer, tape recordings). Zimmer and Whittaker observed a pair in this same area in September 1994. Whitney (pers. comm.) observed two males perched quietly (no vocalizations or interactions) in the canopy above the Serra Trail (RFC) on 27 September 1993. In September 1994 1–2 pairs were seen repeatedly over a three-day period as they visited a fruiting tree along the Rio Cristalino (Zimmer and A. Whittaker). These records constitute a significant easterly range extension for *Porphyrolaema*, which was previously known from eastern Colombia, Ecuador, and Peru east to the Rios Negro and Purús in western Amazonian Brazil (Snow 1982; Sibley and Monroe 1990; Sick 1993). R. Ridgely (*in litt.*) has also recorded this species at Fazenda Rancho Grande near Ariquemes, Rondônia, Brazil.

Snow-capped Manakin (*Pipra nattereri*).—This Rondônia Center endemic (Cracraft 1985) is uncommon in moderately dense understory of the forest interior. It is seemingly much more common in the RFC than along the TPT and in other forest between the Rio Teles Pires and Alta Floresta. Parker and Zimmer found 2–4 males displaying in an area of ca. 0.5 ha along the RCT and heard another bird along the TPT in August 1991. Zimmer and B. Schram recorded 1–5 birds daily along the RCT in October 1992, including an adult male and an immature male on separate song perches less than 10 m apart with a female sitting nearby. They also heard single birds along the TPT and the Castanheira Trail (RFC). Zimmer and A. Whittaker recorded 1–2 daily from along the RCT and saw or heard one male on multiple mornings along the TPT in September 1993. These birds were surprisingly difficult to observe compared to other members of the *P. coronata* complex (Zimmer, pers. obs.). The males called from thin, open branches 1–6 m (most often 3–4 m) above the ground, giving a thin upward "seeeep" and a frog-like "chur-reep," both of which can be ventriloquial. Males seem to call continuously from a given song perch for several minutes and then to move directly to another one 5–25 m away and resume calling. Females and immature birds often sit on a perch for extended periods, giving the "seeeep" call incessantly.

Sick (1993) considered Snow-capped Manakin to be part of a superspecies, along with Opal-crowned Manakin (*P. iris*) and Golden-crowned Manakin (*P. vilasboasi*). The latter species is known only from the upper Rio Cururu, southwestern Pará (Sick 1993), whereas the former is known from both the right (east) bank of the lower Rio Tapajós and from the upper Rio Xingu (Sick 1993), but apparently not from intervening areas (Ridgely and Tudor 1994). The three members of this complex are thought to replace one another geographically (Sick 1993), but two or more of the species could yet prove sympatric in this under-sampled region (Oren and Albuquerque 1991). Observers should be alert to the possibility that Opal-crowned Manakin (in particular) could occur in the Alta Floresta region (particularly east of the Teles Pires). It is essentially vocally identical to Snow-capped Manakin (Ridgely and Tudor 1994; B. Whitney, pers. comm.), and the two forms could prove difficult to distinguish visually in the dimly lit forest interior.

Cinnamon-crested Spadebill (*Platyrinchus saturatus*).—This inconspicuous understory flycatcher has been recorded at Alta Floresta on only a few occasions. Zimmer and B. Schram watched one foraging from 0.3–1.5 m above the ground in moderately dense understory along the RCT on 21 October 1992, and Zimmer taped the distinctive two-note "chip-it" calls of another bird along the TPT on 24 October 1992. Bret Whitney (pers. comm.) recorded one pair (and on one occasion, probably a second pair) from along the TPT on each of three days in September 1993. In September 1994, A. Whittaker tape-recorded a singing bird in upland forest along the TPT. In response to playback of the song, the bird exposed and raised its crest in dramatic fashion. Whittaker found a territorial bird in this same location in September 1995. These are the first records for Mato Grosso and quite probably represent the most southerly locality for the species (Traylor 1979; Meyer de Schauensee 1982; Sibley and Monroe 1990).

Dusky-tailed Flatbill (*Ramphotrigon fuscicauda*).—This tyrannid is known from scattered lowland and foothill localities in southeastern Colombia, eastern Ecuador and Peru, and northern Bolivia (Sibley and Monroe 1990). Few specimens exist in museums (Parker 1984, Terborgh et al. 1984). It was first recorded from Brazil in October 1989, when Parker and the Islers found two pairs inhabiting the largest bamboo stand on the TPT. One or two pairs were found in the same area on subsequent visits in 1991–1994. Parker heard another individual calling from extensive bamboo in cut-over forest 5 km south of the TPT in August 1991. This species has not been found in the small, shaded stands of bamboo along the RCT, but Zimmer found three to four pairs in partially sunlit stands of pure bamboo along the newly opened TT in September 1993, two pairs there in 1994, and at least four pairs there in September 1995. These were the first records for the RFC. This species is also found in vine tangles and other second-growth habitats (with and without bamboo) in Peru (Parker 1984) and Ecuador (R. Ridgely, pers. comm.), but appears to be a bamboo specialist at Alta Floresta. Vocalizations recorded are similar to those described by Parker (1984) from Peru. Both males and females have responded vigorously to tape playback, darting back and forth overhead at heights of 6–10 m, but rarely perching in view. When not disturbed by playback, this species tends to sit quietly 3–8 m above the ground, making occasional sallies to leaves and stems to glean prey. Unlike *R. megacephala*, Dusky-tailed Flatbills do not seem to associate with mixed-species flocks that pass through the bamboo. They call regularly at first light, but only sporadically thereafter.

Large-headed Flatbill (*Ramphotrigon megacephala*).—This bamboo specialist (Parker 1984) is present in virtually every stand of bamboo along the TPT, RCT, and TT, the larger stands (such as along the TPT) having multiple territorial pairs. Roadside surveys along the Teles Pires Road and the Fazenda Cristalino Road have also revealed these flatbills to be present in extensive bamboo thickets at the edge of disturbed forest. This species was much more vocal than *R. fuscicauda*, and its "whu-hoo" calls were repeated incessantly throughout the day. Individuals foraged from 3–12 m up in the bamboo, making short (usually <1.5 m), typically upward-directed sally-gleans to bamboo foliage and stems. They often associated with mixed-species flocks that passed through their territories, such flocks commonly including *Thamnophilus amazonicus, Myrmotherula ornata, Microrhopias quixensis, Drymophila devillei, Ramphocaenus melanurus, Hylophilus semicinereus*, and *Granatellus pelzelni*. This species is disjunctly distributed through Amazonia, southeastern Brazil, eastern Paraguay, and northeastern Argentina (Parker 1984; Sibley and Monroe 1990; Ridgely and Tudor 1994). It was previously known in Amazonian Brazil only from the Rio Juruá (Amazonas), and from west of Cachoeira Nazaré, Rondônia (Parker et al. 1997; D. F. Stotz et al. 1997).

Black-and-white Tody-Tyrant (*Poecilotriccus capitalis*).—This species is known at Alta Floresta from a single record, a female observed by the Islers moving through the large stand of bamboo along the TPT at ca. 3–5 m above the ground on 7 November 1989. Its rufous crown, olive-green back, and contrasting yellow tertial edgings were seen clearly. Its distinctive calls were recorded for ca. 1 minute (verified by Parker, LNS #48203). This species was previously known in Brazil from a specimen (the type for *P. c. tricolor*) collected on the Rio Jamari in western Rondônia (Sibley and Monroe 1990, Sick 1993), and from several individuals (also referable to *P. c. tricolor*) found near Cachoeira Nazaré, Rondônia (Parker et al. 1997; D. F. Stotz et al. 1997).

Zimmer's Tody-Tyrant (*Hemitriccus minimus = aenigma*).—Parker tape-recorded three different individuals high in the canopy of upland forest along the TPT in August 1991: Two were 15–20+ m above ground in or near vine tangles near the centers of large tree crowns in almost continuous canopy with only small light gaps, and the third was in lower canopy 12–14 m up in vine-filled treetops. All fell silent in response to playback of their calls. In 1992 Zimmer heard this same voice (a short, somewhat musical trill) from near the end of the RCT. In September 1993 Zimmer and A. Whittaker heard and tape-recorded 1–2 individuals on each of three days from the same part of the TPT where Parker and Zimmer had found birds in 1991. Exhaustive effort was expended to see these calling birds, but the ventriloquial nature of the vocalization, combined with the density of foliage and vine tangles at that level (15–20+ m above ground) of the canopy, resulted in nothing more than a naked-eye glimpse of a small bird flitting from one vine tangle to another. The tendency for these birds to fall silent for extended periods following tape playback also hampered efforts to see them. In September 1995 Zimmer and Whittaker tape-recorded 1–2 birds in evergreen forest bordering the stunted, deciduous forest on the rocky dome at the top of the Serra Trail (RFC).

Until recently, this species was known only from the type series collected at 4 localities along the east bank of the Rio Tapajós, Pará, Brazil (D. F. Stotz, pers. comm.). Parker and colleagues have recently found this species in sandy-soil, stunted forest in the Serranía de Huanchaca, depto. Santa Cruz, and in the canopy of taller, vine-rich forest on lateritic soils near Versalles, depto. Beni, in Bolivia (Parker et al. 1991; Bates et al. 1992). Bret Whitney (pers. comm.) has recently found *minimus* to be common along both banks of the lower Rio Tapajós.

Helmeted Pygmy-Tyrant (*Lophotriccus galeatus*).—Bret Whitney (pers. comm.) reported multiple birds along the various trails of the RFC and along the Rio Cristalino road in September 1993. The only previous record for Mato Grosso was a male collected near the Rio Peixoto de Azevedo (Novaes and Lima 1991). This species was previously undetected in the Alta Floresta region because of vocal similarities with *Hemitriccus minor*. Populations of *Lophotriccus galeatus* from southern Venezuela typically utter a series (sometimes short and intermittent, sometimes repeated almost interminably) of dry "tic" notes, only occasionally punctuated by a short trill (Zimmer tape recordings, Ridgely and Tudor 1994). These trills take on a harsher, more gravelly quality, and may be lengthened when birds are agitated by tape playback (Zimmer, pers. obs.). At such times the calls of *L. galeatus* sound similar to those of *H. minor*, which, near Manaus (Amazonas, Brazil), sings a variable number (most often three) of well-spaced, long, harsh trills, which may be preceded by several staccato "tic" notes (Zimmer tape recordings; A. Whittaker tape recordings; Ridgely and Tudor 1994).

In September 1994 Zimmer and A. Whittaker tape-recorded and closely observed two Helmeted Pygmy-Tyrants (each lacking wingbars and having obvious crests) along the TT, and tape-recorded another in upland forest along the TPT. The primary vocalization of each of these birds was a series of long, harsh trills (punctuated by "tic" notes), similar to those described earlier for *Hemitriccus minor*. In September 1995 Zimmer and Whittaker attempted to visually confirm the identity of all birds in the RFC giving vocalizations similar to those of *Hemitriccus minor*. All four birds that could be visually confirmed were *Lophotriccus galeatus*. It is apparent that local populations of *Lophotriccus* differ vocally from those with which we are familiar from north of the Amazon. Because of this, and because directed searches in 1994 and 1995 have yielded only *Lophotriccus*, we are now uncertain as to the true status of *Hemitriccus minor* (previously recorded [largely on the basis of heard vocalizations] by every party to visit Alta Floresta, and considered to be "fairly common") in the Alta Floresta region. Closer examination of the local status of *H. minor* and further investigation into possible geographic variation in *Lophotriccus galeatus* are needed.

Xenopsaris (*Xenopsaris albinucha*).—There are two local records of this poorly known species. A. Whittaker (pers. comm) observed one in September 1994 from the rocky dome atop the Serra Trail (RFC). On 21 and 22 September 1995 Whittaker and Zimmer observed an individual in the same location. This species is locally distributed in interior northeastern and central Brazil (Ridgely and Tudor 1994), and is more typical of savanna, marsh, and caatinga habitats (Hilty and Brown 1986; Sick 1993; Ridgely and Tudor 1994). As noted in the site description, the forest at the top of the Serra Trail is stunted and more xeric than surrounding areas, and is highly deciduous. It is the only spot in the Alta Floresta region where several characteristic species of savanna woodlands and gallery forests have been recorded. This species usually forages near the ground (Ridgely and Tudor 1994; Zimmer, pers. obs.). The 1995 bird was observed at length, foraging at heights of 8–20 m above the ground, in the crowns and outer branches of leafy (mostly a single large flowering legume) and leafless trees. It occasionally perch-gleaned from the tops of green leaves, but most foraging maneuvers were sally-stalls or sally-pounces to live foliage and bare branch and stem surfaces. These sallies were generally of 1–5 m, with a few longer sallies reaching 8 m. This bird was briefly associated with a mixed-species flock which included *Ornithion inerme*, *Todirostrum chrysocrotaphum*, *Tolmomyias poliocephalus*, *Odontorchilus cinereus*, *Coereba flaveola*, *Cyanerpes nitidus*, and

Cyanerpes caeruleus, but was later observed foraging apart from other birds. The only vocalizations noted were sporadic, thin, wispy notes.

Tooth-billed Wren (*Odontorchilus cinereus*).—This poorly known wren seems to be uncommon at Alta Floresta; however, experience with it in Rondônia suggests a marked degree of seasonality in the period of vocal activity (Zimmer, unpubl. data), and hence, it may be overlooked. During each of our visits from 1989–1992, we recorded one to three pairs, but in September 1993 Zimmer and A. Whittaker recorded 1–2 pairs daily from each of three different trails in the RFC, as well as a pair along the TPT. In all instances these pairs moved with mixed-species canopy flocks at heights of 15–30+ m, where they actively gleaned prey from foliage, vines, and limbs. They seemed to frequent vine tangles close to trunks, as well as thicker, more open, horizontal branches of the inner canopy. In behavior, they are almost gnatcatcher-like, holding their long tail partially cocked upward and flipping it about. A frequently observed foraging maneuver is a "reach-down" or "duck-under" (Remsen and Robinson 1990), in which the bird leans far forward to peer under the branch it is sitting on, a maneuver commonly used by many *Tangara* tanagers. So frequent are these maneuvers, that it is one of the best ways to spot *O. cinereus* in the flock (Zimmer, pers. obs.). This behavior has also been reported for the sister taxon *O. branickii* of the Andes (Parker et al. 1980). Typical flock associates at Alta Floresta are *Capito dayi, Lepidocolaptes albolineatus, Ancistrops strigilatus, Philydor* spp., *Xenops* spp., *Cymbilaimus lineatus, Pygiptila stellaris, Myrmotherula sclateri, Vireolanius leucotis, Hylophilus hypoxanthus, Hylophilus semicinereus, Euphonia rufiventris, Euphonia chrysopasta, Lanio versicolor, Tachyphonus cristatus* and *Lamprospiza melanoleuca*.

In September 1994 Zimmer and A. Whittaker found two different family groups of four birds each in the RFC. These groups were associated with mixed-species flocks and foraged at heights of 10–30 m above ground. The juvenal birds periodically gave harsh, thin begging calls accompanied by posturing and wing flutter, which continued until they were fed by one of the adults. In September 1995 Zimmer and A. Whittaker observed a Tooth-billed Wren pass an arthropod prey item to a second bird (presumed to be its mate) which in turn carried the prey item into what was assumed to be a nest cavity (a knot-hole ca. 17 m above the ground in the main trunk of a large, deciduous [and nearly leafless] tree). The second bird emerged from the hole a few seconds later and wiped its bill on a branch before flying off.

Virtually all pairs observed have been vocal, with both males and females singing persistently. Vocalizations are as described by Bates et al. (1992), most commonly, a reverberating trill similar to the song of a Pine Warbler (*Dendroica pinus*) or a Chipping Sparrow (*Spizella passerina*), and less frequently (but a common response to tape playback) a ringing series of 8–15+ "swee" notes (ca. 3–6 s). Zimmer (unpubl. data) has found birds at Fazenda Rancho Grande, Rondônia, to be much more vocal in mid-October than in November or January.

Until recently, this species was known only from Brazil south of the Amazon between the Rios Madeira and Xingú (Meyer de Schauensee 1982; Ridgely and Tudor 1989), but it has now been found to be fairly common in some parts of depto. Santa Cruz, Bolivia (Bates et al. 1992; Kratter et al. 1992).

Guianan Gnatcatcher (*Polioptila guianensis*).—Parker observed an individual of this species moving with a mixed-species canopy flock (October 1989) for the only area record. This enigmatic species, known primarily from the Guianas, southern Venezuela, and Brazil north of the Amazon (Meyer de Schauensee 1982), was until recently thought to occur south of the Amazon only from the lower Rio Tapajós east to Belém (Ridgely and Tudor 1989). In 1986 a pair was observed and the female collected at Cachoeira Nazaré, Rondônia (Stotz et al. 1997), and a survey of a site near the upper Rio Urucu, Tefé, Amazonas, Brazil found *P. guianensis* to be a common member of canopy flocks (Peres and Whittaker 1990). It is locally fairly common in tall *terra firme* forest (sometimes lightly disturbed) in the Serra dos Carajas, central Pará, where up to six in one day have been recorded, always with canopy mixed-species flocks (B. Whitney and J. F. Pacheco, pers. comms.). The Alta Floresta sight record, along with that from Cachoeira Nazaré, extends the known range south.

White-bellied Dacnis (*Dacnis albiventris*).—Known from a single record, a pair well-seen by Parker as they perched in the bare upper branches of a tall tree in swamp forest along the Teles Pires Road (just south of the Rio Teles Pires) on 24 August 1991. These birds were moving with a mixed-species flock that included three other species of *Dacnis* (*lineata, cayana, flaviventer*), *Tachyphonus cristatus, Tangara chilensis, Tangara schrankii,* and *Tangara mexicana*. This species is little-known and considered rare throughout its rather wide range in Amazonia (Isler and Isler 1987). The nearest known previous record is from the upper Rio Cururú in southeastern Pará (Ridgely and Tudor 1989).

Yellow-shouldered Grosbeak (*Parkerthraustes humeralis*).—There are only a few Alta Floresta records of this poorly known species. Robert Ridgely and V. Emanuel (pers. comm.) observed two birds eating flowers in the forest canopy along the Teles Pires Road (October 1990). Zimmer and B. Schram saw two birds moving with a large canopy mixed-species flock along the Teles Pires Road on 25 October 1992. Zimmer and A. Whittaker found 2–4 birds moving with a canopy mixed-species flock (composed mostly of *Tangara velia*, *T. mexicana*, *T. chilensis*, *Thraupis palmarum*, *Dacnis cayana*, and *Chlorophanes spiza*) in cut-over forest along the Teles Pires Road on 16 September 1993. Bret Whitney (pers. comm.) also tape-recorded a pair of birds on the grounds of the Floresta Amazonica Hotel in Alta Floresta in late September 1993, and heard or saw birds along the TPT, the Fazenda Cristalino road, and along the Serra and Taboca trails (RFC). Andrew Whittaker observed 1–2 birds moving with a mixed-species canopy flock along the TPT in September 1994. This wide-ranging species is thought to be rare throughout much of its range, but has recently proven to be widespread and fairly common in eastern Ecuador (R. Ridgely, pers. comm.), and fairly common at elevations of 500–600 m in foothills in depto. La Paz, Bolivia (B. Whitney, pers. comm.). It was also found to be relatively common at Cachoeira Nazaré in Rondônia (D. F. Stotz et al. 1997), and Zimmer (unpubl. data) has likewise found it to be more common at Fazenda Rancho Grande, near Ariquemes, Rondônia.

ACKNOWLEDGMENTS

We would like to thank J. M. Goerck, B. Schram, and A. Whittaker, all of whom accompanied Parker or Zimmer during portions of the field work at Alta Floresta and who shared in many of the finds. Special thanks in this regard to Whittaker, who has accompanied Zimmer on three of his visits to the region. Bret M. Whitney was most helpful in providing details of many important records from his work at Alta Floresta, as well as in giving much useful advice regarding this paper. We appreciate access to the unpublished trip reports of R. S. Ridgely and V. Emanuel, and comments from J. Haffer on an earlier draft of this manuscript. J. V. Remsen, R. S. Ridgely, and B. M. Whitney reviewed earlier drafts of this manuscript and made numerous helpful suggestions for improvement. Special thanks are due V. Emanuel and Victor Emanuel Nature Tours, Inc., for supporting Zimmer's field work at Alta Floresta, as well as one of Parker's trips. We are also extremely grateful to Vitoria and Edson de Carvalho of the Cristalino Ecological Institute, both for their continued support, and for their dedication to conserving the biological diversity of the Rio Cristalino watershed.

LITERATURE CITED

BATES, J. M., M. C. GARVIN, D. C. SCHMITT, AND C. G. SCHMITT. 1989. Notes on bird distribution in northeastern Dpto. Santa Cruz, Bolivia, with 15 species new to Bolivia. Bull. Brit. Ornithol. Club 109:236–244.

BATES, J. M., T. A. PARKER III, A. P. CAPPARELLA, AND T. J. DAVIS. 1992. Observations on the *campo, cerrado*, and forest avifaunas of eastern Dpto. Santa Cruz, Bolivia, including 21 species new to the country. Bull. Brit. Ornithol. Club 112:86–98.

COLLAR, N. J., L. P. GONZAGA, N. KRABBE, A. MADROÑO NEITO, L. G. NARANJO, T. A. PARKER III, AND D. C. WEGE. 1992. Threatened Birds of the Americas: The ICBP/IUCN Red Data Book. International Council for Bird Preservation, Cambridge, United Kingdom.

CRACRAFT, J. 1985. Historical biogeography and patterns of differentiation within the South American avifauna: areas of endemism. Pp. 49–84 *in* Neotropical Ornithology (P. A. Buckley, M. S. Foster, E. S. Morton, R. S. Ridgely, and F. G. Buckley, eds.). Ornithol. Monogr. No. 36.

DEL HOYO, J., A. ELLIOTT, AND J. SARGATAL (EDS.). 1994. Handbook of Birds of the World. Vol. 2. New world vultures to guineafowl. Lynx Editions, Barcelona.

FEARNSIDE, P. M. 1987. Deforestation and international economic development projects in Brazilian Amazonia. Conserv. Biol. 1:214–221.

FITZPATRICK, J. W., AND D. E. WILLARD. 1990. *Cercomacra manu*, a new species of antbird from southwestern Amazonia. Auk 107:239–245.

FOSTER, R. B., T. A. PARKER III, A. H. GENTRY, L. H. EMMONS, A. CHICCHÓN, T. SCHULENBERG, L. RODRÍGUEZ, G. LAMAS, H. ORTEGA, J. ICOCHEA, W. WUST, M. ROMO. J. A. CASTILLO, O. PHILLIPS, C. REYNEL, A. KRATTER, P. K. DONAHUE, AND L. J. BARKLEY. 1994. The Tambopata-Candamo Reserved Zone of Southeastern Peru: a biological assessment. RAP Working Pap. No. 6, Conservation International, Washington, D.C.

GRAVES, G. R., AND R. L. ZUSI. 1990. Avian body weights from the lower Rio Xingu, Brazil. Bull. Brit. Ornithol. Club. 110:20–25.

HACKETT, S. J., AND K. V. ROSENBERG. 1990. Comparison of phenotypic and genetic differentiation in South American antwrens (Formicariidae). Auk 107:473–489.

HAFFER, J. 1978. Distribution of Amazonian forest birds. Bonn. zool. Beitr. 29:38–78.

HAFFER, J. 1990. Avian species richness in tropical South America. Stud. Neotrop. Fauna Environment 25: 157–183.
HAFFER, J. 1992. On the "river effect" in some forest birds of southern Amazonia. Bol. Mus. Para. Emílio Goeldi, sér. Zool. 8:207–245.
HILTY, S. L., AND W. L. BROWN. 1986. A Guide to the Birds of Colombia. Princeton Univ. Press, Princeton, New Jersey.
ISLER, M. L., AND P. R. ISLER. 1987. The Tanagers. Smithsonian Institution Press. Washington, D.C.
KRATTER, A. W. Bamboo specialization by Amazonian birds. Biotropica (in press).
KRATTER, A. W., M. D. CARREÑO, R. T. CHESSER, J. P. O'NEILL, AND T. S. SILLETT. 1992. Further notes on bird distribution in northeastern Dpto. Santa Cruz, Bolivia, with two species new to Bolivia. Bull. Brit. Ornithol. Club. 112:143–150.
KRATTER, A. W., AND T. A. PARKER III. 1997. Relationship of two bamboo-specialized foliage gleaners: *Automolus dorsalis* and *Anabazenops fuscus* (Furnariidae). Pp. 383–397 *in* Studies in Neotropical Ornithology Honoring Ted Parker (J. V. Remsen, Jr., Ed.), Ornithol. Monogr. No. 48.
LANYON, S. M., D. F. STOTZ, AND D. E. WILLARD. 1990. *Clytoctantes atrogularis*, a new species of antbird from western Brazil. Wilson Bull. 102:571–580.
MEYER DE SCHAUENSEE, R. 1982. A Guide to the Birds of South America. Livingston Publishing Company, Wynnewood, Pennsylvania.
NOVAES, F. C. 1976. As aves de rio Aripuaña, Estados de Mato Grosso e Amazonas. Acta Amazonica 6:61–85.
NOVAES, F. C., AND M. F. C. LIMA. 1991. As aves do rio Peixoto de Azevedo, Mato Grosso, Brasil. Rev. Bras. Zool. 7:351–381.
OREN, D. C., AND H. G. ALBUQUERQUE. 1991. Priority areas for new avian collections in Brazilian Amazonia. Goeldiana (Zool.) 6:1–11.
PARKER, T. A., III. 1982. Observations of some unusual rainforest and marsh birds in southeastern Peru. Wilson Bull. 94:477–493.
PARKER, T. A., III. 1984. Notes on the behavior of *Ramphotrigon* flycatchers. Auk. 101:186–188.
PARKER, T. A., III, J. V. REMSEN, JR., AND J. A. HEINDEL. 1980. Seven bird species new to Bolivia. Bull. Brit. Ornithol. Club 100:160–162.
PARKER, T. A., III, A. CASTILLO U., M. GELL-MANN, AND O. ROCHA O. 1991. Records of new and unusual birds from northern Bolivia. Bull. Brit. Ornithol. Club 111:120–138.
PARKER, T. A., III, J. W. FITZPATRICK, AND D. F. STOTZ. 1997. Notes on avian bamboo specialists in southwest Amazonian Brazil. Pages 543–547 *in* Studies in Neotropical Ornithology Honoring Ted Parker, (J. V. Remsen, Jr., Ed.), Ornithol. Monogr. No. 48.
PARKER, T. A., III, AND J. V. REMSEN, JR. 1987. Fifty-two Amazonian bird species new to Bolivia. Bull. Brit. Ornithol. Club 107:94–107.
PARKER, T. A., III, J. V. REMSEN, JR., AND J. A. HEINDEL. 1980. Seven bird species new to Bolivia. Bull. Brit. Ornithol. Club 100:160–162.
PERES, C. A., AND A. WHITTAKER. 1991. Annotated checklist of the bird species of the upper Rio Urucu, Amazonas, Brazil. Bull. Brit. Ornithol. Club 111:156–171.
PINTO, O. M. O. 1978. Novo catálogo das aves do Brasil; premeira parte. Emprea Gáfica do Revista dos Tribunais, São Paulo, Brazil.
PINTO, O. M. O., AND E. A. CAMARGO. 1957. Sôbre uma coleção de aves da região de Cachimbo (sul do Estado do Pará). Pap. Avul. Dept. Zool. São Paulo 13:51–69.
REMSEN, J. V., JR. 1994. Use and misuse of bird lists in community ecology and conservation. Auk 111:225–227.
REMSEN, J. V., JR., AND T. A. PARKER III. 1983. Contribution of river-created habitats to bird species richness in Amazonia. Biotropica 15:223–231.
REMSEN, J. V., JR., AND T. A. PARKER III. 1984. Arboreal dead-leaf searching birds of the Neotropics. Condor 86:36–41.
REMSEN, J. V., JR., AND S. K. ROBINSON. 1990. A classification scheme for foraging behavior of birds in terrestrial habitats. Pp. 144–160 *in* Avian Foraging: Theory, Methodology, and Applications (M. L. Morrison, C. J. Ralph, J. Verner, and J. R. Jehl, Jr., Eds.). Stud. Avian Biol. 13.
RIDGELY, R. S., AND J. A. GWYNNE. 1989. A Guide to the Birds of Panama (with Costa Rica, Nicaragua and Honduras). Princeton University Press, Princeton, New Jersey.
RIDGELY, R. S., AND G. TUDOR. 1989. The Birds of South America, Vol. 1. Univ. of Texas Press, Austin, Texas.
RIDGELY, R. S., AND G. TUDOR. 1994. The Birds of South America, Vol. 2. Univ. of Texas Press, Austin, Texas.
ROSENBERG, K. V. 1997. Ecology of dead-leaf foraging specialists and their contribution to Amazonian bird diversity Pp. 673–700 *in* Studies in Neotropical Ornithology Honoring Ted Parker (J. V. Remsen, Jr., Ed.), Ornithol. Monogr. No. 48.
SIBLEY, C. G., AND B. L. MONROE, JR. 1990. Distribution and Taxonomy of Birds of the World. Yale University Press, New Haven, Connecticut.
SICK, H. 1993. Birds in Brazil, A Natural History. Princeton University Press, Princeton, New Jersey.
SNOW, D. 1982. The Cotingas: Bellbirds, Umbrellabirds, and Other Species. Cornell University Press, Ithaca, New York.
STOTZ, D. F., S. M. LANYON, T. S. SCHULENBERG, D. E. WILLARD, A. T. PETERSON, AND J. W. FITZPATRICK. 1997. An avifaunal survey of two tropical forest localities on the middle Rio Jiparaná, Rondônia,

Brazil. Pages 763–781 *in* Studies in Neotropical Ornithology Honoring Ted Parker (J. V. Remsen, Jr., Ed.), Ornithol. Monogr. No. 48.

TERBORGH, J. W., J. W. FITZPATRICK, AND L. EMMONS. 1984. Annotated checklist of bird and mammal species of Cocha Cashu Biol. Station, Manu National Park, Peru. Fieldiana Zool. 21:1–29.

TRAYLOR, M. A., JR. 1979. Subfamily Elaeniinae. Pp. 3–112 *in* Check-list of Birds of the World. Vol. 7 (M. A. Traylor, Jr., Ed.). Museum of Comparative Zoology, Cambridge, Massachusetts.

VAURIE, C. 1980. Taxonomy and geographical distribution of the Furnariidae (Aves, Passeriformes). Bull. Amer. Mus. Nat. Hist. 166:1–357.

WHITNEY, B. M. 1994. Behavior, vocalizations, and possible relationships of four *Myrmotherula* antwrens (Formicariidae) from eastern Ecuador. Auk 111:469–475.

WILLIS, E. O. 1969. On the behavior of five species of *Rhegmatorhina*, ant-following antbirds of the Amazon basin. Wilson Bull. 81:363–395.

ZIMMER, K. J., AND S. L. HILTY. 1997. Avifauna of a locality in the upper Orinoco drainage of Amazonas, Venezuela. Pages 865–885 *in* Studies in Neotropical Ornithology Honoring Ted Parker (J. V. Remsen, Jr., Ed.), Ornithol. Monogr. No. 48.

APPENDIX

Bird species recorded from the Alta Floresta region (Mato Grosso, Brazil). Relative abundance of each species in suitable habitat is estimated. As with any such list, our abundance designations are subjective. Observers with differing levels of experience with Amazonian bird vocalizations, or who visit the region in other seasons, will no doubt draw different conclusions regarding the abundance of many species. Also, because the region has been visited so few times, and because most workers have concentrated their efforts in forest habitats, some species restricted to second-growth and marshy habitats may have been dramatically under-sampled. We have included the initials of primary observers after many species to provide more complete documentation (an absence of observer initials indicates that the species has been recorded by at least four parties). In a few cases we have indicated uncertainty of a specific identification within a sibling-species pair by separating the specific names with a diagonal bar (e.g., *Xiphorhynchus spixii/elegans*). Many species have been documented by tape recordings of their calls, songs, displays, or drums (woodpeckers); these are noted in the list below. The principal recordists are the Islers (MPI), Parker (TAP), Bret Whitney (BW), Andrew Whittaker (AW), and Zimmer (KJZ). All recordings of note are (or will be) deposited in the Library of Natural Sounds (LNS) at The Laboratory of Ornithology, Cornell University, Ithaca, New York. Bret Whitney's recordings will also be deposited at the Arquivo Sonoro Elias P. Coelho (ASEC), Universidade Federal do Rio de Janeiro, Rio de Janeiro, Brazil.

Key:

Status/Distribution:

    C = common, several-to-many individuals encountered per day in appropriate habitat.
    FC = fairly common, encountered irregularly in numbers, or at least a few almost every day.
    U = uncommon, not encountered every day
    R = rare, only a few records
    (( )) = hypothetical, reported at least once, but judged to be sufficiently rare or unlikely in the region as to require more documentation
    b = bamboo, denotes those species which at least regionally appear to be largely confined to stands of bamboo within the forest
    m = migrant or probable migrant, either austral or boreal (these species may be present only during the austral or boreal winters, or during species-specific migratory periods)
    e = indicates that the species has been found only on the east (north) side of the Rio Teles Pires, in the Reserva Florestal Cristalino (RFC)
    w = indicates that the species has been found only on the west (south) side of the Rio Teles Pires (ie. the Alta Floresta side of the river)
    (+) = indicates that the species has been tape-recorded in the region

Observers/Dates:

    TAP = Parker (26–31 Oct. 1989)
    MPI = Islers (26 Oct.–9 Nov. 1989)
    RET = Robert S. Ridgely, Victor Emanuel, and Guy Tudor (5–9 Oct. 1990); these initials are given only for species not recorded by TAP and MPI
    P/Z = Parker and Zimmer (18–26 Aug. 1991)
    Z/S = Zimmer and Brad Schram (20–26 Oct. 1992)
    Z/W = Zimmer and Andrew Whittaker (16–24 Sept. 1993)
    BW = Bret Whitney (17–22 and 24–30 Sept. 1993)
    ZW2 = Zimmer and Andrew Whittaker (15–24 Sept. 1994)
    AW = indicates species observed only by Andrew Whittaker during the Sept. 1993 Z/W trip.
    AW2 = indicates species observed only by Andrew Whittaker during the Sept. 1994 ZW2 trip.
    ZW3 = Zimmer and Andrew Whittaker (15–23 Sept. 1995)

TINAMIDAE

Tinamus tao (+) U
Tinamus major (+) FC
Tinamus guttatus Rw (TAP)
Crypturellus cinereus (+) C
Crypturellus soui (+) FCw
Crypturellus obsoletus (+) Ub
Crypturellus variegatus (+) FC
Crypturellus strigulosus (+) FC-C
Crypturellus tataupa (+) U
Crypturellus parvirostris Rw (BW)

PHALACROCORACIDAE

Phalacrocorax brasilianus C

ANHINGIDAE

Anhinga anhinga FC

ARDEIDAE

Ardea cocoi U
Ardea alba U (ZS, ZW, ZW2)
Egretta thula R (ZW, ZW2)
Butorides striatus FC
Bubulcus ibis C
Pilherodius pileatus U
Tigrisoma lineatum (+) FC

CICONIIDAE

Mycteria americana U

THRESKIORNITHIDAE

Mesembrinibis cayennensis (+) FC

CATHARTIDAE

Sarcoramphus papa U
Coragyps atratus C
Cathartes burrovianus Rw (PZ)
Cathartes melambrotus FC

ANATIDAE

Amazonetta brasiliensis Rw (BW, ZW2, ZW3)
Cairina moschata U

ACCIPITRIDAE

Gampsonyx swainsonii Rm (PZ)
Elanoides forficatus Um (how many resident versus migrants??)
Elanus leucurus Rw (ZW2)
Leptodon cayanensis U
Chondrohierax uncinatus U
Harpagus bidentatus U
Ictinia plumbea (+) FC
Rostrhamus sociabilis U (ZW, BW)
Accipiter bicolor Re (AW2)
Accipiter superciliosus Rw (ZW)
Accipiter poliogaster Rw (ZW, BW)
Buteo albicaudatus Rw (ZW)
Buteo magnirostris R (ZW, ZW2)
Buteo brachyurus Rw (ZW, ZW2, ZW3)
Leucopternis kuhli (+) R (PZ, ZW, ZW3)
Buteogallus urubitinga Ue
Morphnus guianensis Re (ZW3)
Harpia harpyja Rw (PZ)
Spizastur melanoleucus Rw (RET)
Spizaetus ornatus (+) R (ZS, ZW, BW)
Spizaetus tyrannus (+) Re (ZW2, ZW3)

Geranospiza caerulescens Re (ZW)
Pandion haliaetus Rm (RET, BW)

FALCONIDAE

Herpetotheres cachinnans Uw
Micrastur semitorquatus Re (ZW3)
Micrastur mirandollei (+) U-R (TAP, ZW, BW)
Micrastur ruficollis (+) U
Micrastur gilvicollis (+) Rw (PZ, BW, ZW2)
Daptrius ater (+) U
Daptrius americanus (+) U
Milvago chimachima Rw (ZW)
Polyborus plancus Uw
Falco rufigularis FC-U
Falco sparverius Rw (ZW2)

CRACIDAE

Penelope jacquacu (+) FC
Pipile cujubi (+) FC
Mitu tuberosa (+) Ue

ODONTOPHORIDAE

Odontophorus gujanensis (+) FC-U

ARAMIDAE

Aramus guarauna U

PSOPHIIDAE

Psophia viridis (+) U

RALLIDAE

Aramides cajanea Uw (RET, BW)
Porzana albicollis Rw (AW)
Laterallus exilis (+) Uw
Laterallus melanophaius Uw (TAP, MPI, BW)
Porphyrula martinica Uw

HELIORNITHIDAE

Heliornis fulica Re (PZ, ZW, ZW3)

EURYPIGIDAE

Eurypyga helias U

JACANIDAE

Jacana jacana FCw

CHARADRIIDAE

Vanellus chilensis Uw
Hoploxypterus cayanus R

SCOLOPACIDAE

Tringa solitaria Um
Actitis macularia Rme (BW, ZW2, ZW3)

LARIDAE

Phaetusa simplex R (ZW, BW)
Sterna superciliaris R (ZW2)
Rynchops niger R (ZW, BW, ZW2)

COLUMBIDAE

Columba cayennensis (+) R (BW, ZW2, ZW3)
Columba subvinacea (+) FC-C
Columba plumbea (+) FC
Columbina talpacoti Cw
Claravis pretiosa (+) U

Leptotila rufaxilla (+) FC
Geotrygon montana (+) FC

PSITTACIDAE

Ara ararauna (+) FC
Ara macao (+) FC
Ara chloroptera (+) U
Ara severa (+) FC
Ara manilata (+) Uw
Aratinga leucophthalmus U
Pyrrhura rhodogaster (+) Uw
Pyrrhura picta (+) C
Forpus sclateri (+) U
Brotogeris chrysopterus (+) C
Touit purpurata (+) U-Re
Pionites leucogaster (+) FC
Pionopsitta barrabandi (+) U
Pionus menstruus (+) C
Amazona ochrocephala (+) FC
Amazona amazonica (+) U
Amazona farinosa (+) U
Deroptyus accipitrinus (+) U-FC

CUCULIDAE

Coccyzus melacoryphus Um
Piaya cayana (+) FC
Piaya melanogaster U
Piaya minuta (+) U
Crotophaga major U
Crotophaga ani (+) C
Dromococcyx phasianellus (+) U
Dromococcyx pavoninus Re (PZ)
Neomorphus sp. Re (PZ)

TYTONIDAE

Tyto alba Uw (TAP, BW)

STRIGIDAE

Otus watsonii (+) FC-C
Lophostrix cristata (+) U
Pulsatrix perspicillata U (TAP, MPI, BW)
Glaucidium hardyi (+) U
Glaucidium brasilianum Rw (ZW3, and possibly PZ)
Athene cunicularia Rw (PZ, ZW2)
Ciccaba huhula Re (ZS)
Ciccaba virgata (+) Uw (MPI, BW)

NYCTIBIIDAE

Nyctibius grandis U (TAP, MPI, BW)

CAPRIMULGIDAE

Lurocalis semitorquatus (+) FC
Chordeiles minor Rmw (TAP, MPI, ZW3)
Chordeiles sp. Rw (BW)
Nyctidromus albicollis U
Nyctiphrynus ocellatus (+) U-FC
Caprimulgus maculicaudus U-Rw (PZ, BW)
Caprimulgus parvulus (+) Uw (PZ, BW)
Caprimulgus nigrescens (+) C
Hydropsalis climacocerca Re (ZW, BW)

APODIDAE

Streptoprocne zonaris U-R
Cypseloides senex Rw (ZW2)
Chaetura cinereiventris (+) C
Chaetura egregia U-FC

Chaetura brachyura U-FC
Tachornis (=Reinarda) squamata U-FC

TROCHILIDAE

Glaucis hirsuta (+) U
Threnetes leucurus U
Phaethornis superciliosus FC
((*Phaethornis bourcieri*)) (TAP, PZ, BW)
Phaethornis ruber (+) FC
Campylopterus largipennis FC
Florisuga mellivora U
Anthracothorax nigricollis U
Popelaria langsdorffi Re (ZW3)
Popelaria langsdorffi/Discosura longicauda Re (BW) (1 record of a female)
Thalurania furcata (+) C
Hylocharis sapphirina Re (BW, AW2)
Hylocharis cyanus (+) U-R (PZ, ZW2, ZW3)
Amazilia versicolor FC-U
Amazilia fimbriata R (AW, AW2, AW3)
Heliothryx aurita U (PZ, ZS, ZW, BW, ZW2)
Heliomaster longirostris U
Calliphlox amethystina U

TROGONIDAE

Pharomachrus pavoninus (+) U-R
Trogon melanurus (+) FC-C
Trogon viridis (+) FC-C
Trogon collaris (+) FC
Trogon rufus (+) U
Trogon curucui (+) U-FC
Trogon violaceus (+) U-FC

ALCEDINIDAE

Ceryle torquata (+) FC-U
Chloroceryle amazona FC-C
Chloroceryle americana U
Chloroceryle inda Ue
Chloroceryle aenea Re (PZ, ZW3)

MOMOTIDAE

Electron platyrhynchum (+) FC
Momotus momota (+) FC

GALBULIDAE

Brachygalbus lugubris Ue
Galbula cyanicollis (+) U
Galbula dea (+) FC-C
Galbula ruficauda (+) FC
Galbula leucogastra Re (AW2)
Jacamerops aurea (+) U-FC

BUCCONIDAE

Notharchus macrorhynchos (+) FC
Notharchus ordii (+) Ue
Notharchus tectus (+) U
Bucco capensis (+) R (PZ, ZW, BW)
Nystalus striolatus (+) FC
Malacoptila rufa Re
Nonnula ruficapilla Rb (PZ)
Monasa nigrifrons (+) C
Monasa morphoeus (+) C
Chelidoptera tenebrosa (+) C

CAPITONIDAE

Capito dayi (+) FC-U

RAMPHASTIDAE

Pteroglossus castanotis (+) FC
Pteroglossus inscriptus (+) FC
Pteroglossus bitorquatus (+) U
Pteroglossus beauharnaesii (+) Ue
Selenidera gouldii (+) FC
Ramphastos vitellinus/culminatus (+) FC (many phenotypic intermediates)
Ramphastos cuvieri (+) C

PICIDAE

Picumnus aurifrons (+) FC
Piculus flavigula (+) FC
Piculus chrysochloros U
Celeus elegans U
Celeus grammicus (+) FC
Celeus flavus (+) FC
Celeus torquatus (+) U
Dryocopus lineatus (+) FC
Melanerpes cruentatus (+) C
Veniliornis affinis (+) FC
Campephilus melanoleucos (+) FC
Campephilus rubricollis (+) U

DENDROCOLAPTIDAE

Dendrocincla fuliginosa (+) U
Deconychura longicauda (+) U (PZ, BW)
Deconychura stictolaema (+) Re (ZS, ZW)
Sittasomus griseicapillus U
Glyphorynchus spirurus (+) FC
Nasica longirostris (+) U-FC
Dendrexetastes rufigula (+) U-FC
Hylexetastes perrotii (+) R (TAP, MPI, PZ)
Xiphocolaptes promeropirhynchus (+) U (PZ, BW, ZW2)
Dendrocolaptes certhia (+) U
Dendrocolaptes picumnus (+) Re (BW, ZW3)
Xiphorhynchus picus (+) FC-Ce
Xiphorhynchus obsoletus (+) U-FC
Xiphorhynchus spixii/elegans (+) C
Xiphorhynchus guttatus/eytoni (+) C
Lepidocolaptes albolineatus (+) FC
Campylorhamphus procurvoides (+) U-FC

FURNARIIDAE

Synallaxis cabanisi (+) Rbw (TAP, MPI, PZ)
Synallaxis gujanensis (+) R (ZW, ZW3)
Synallaxis rutilans (+) U
Synallaxis cherriei (+) Ub
Cranioleuca gutturata Uw
Cranioleuca vulpina (+) Ue
Berlepschia rikeri (+) Rw (ZW3)
Simoxenops ucayalae (+) Ub
Hyloctistes subulatus (+) U
Ancistrops strigilatus (+) FC
Philydor erythrocercus (+) FC
Philydor erythropterus (+) FC
Philydor ruficaudatus (+) U
Automolus infuscatus (+) U-FC
Automolus ochrolaemus (+) U-FC
Automolus rufipileatus (+) FCb
Anabazenops dorsalis (+) FCb
Xenops rutilans (+) R (ZW3)
Xenops rutilans/tenuirostris (+) Rw (PZ, BW)
Xenops milleri Re (BW, AW2, AW3)
Xenops minutus (+) FC
Sclerurus albigularis (+) Rw (PZ, ZS)
Sclerurus mexicanus (+) U (MPI, PZ, ZS)

Sclerurus rufigularis (+) Uw
Sclerurus caudacutus (+) U

FORMICARIIDAE

Cymbilaimus lineatus (+) FC-C
Taraba major (+) FC
Sakesphorus luctousus (+) U
Thamnophilus palliatus (+) FC
Thamnophilus aethiops (+) Ub
Thamnophilus schistaceus (+) FC-C
Thamnophilus stictocephalus (+) FCe (only in deciduous forest on rocky outcroppings) (BW, ZW2, ZW3)
Thamnophilus amazonicus (+) U-FC
Pygiptila stellaris (+) FC
((*Megastictus margaritatus*)) w (RET)
((*Clytoctantes/Neoctantes* sp.)) w (TAP)
Thamnomanes saturninus (+) U
Thamnomanes caesius (+) C
Myrmotherula brachyura (+) C
Myrmotherula sclateri (+) FC
Myrmotherula surinamensis (+) FC
Myrmotherula hauxwelli (+) C
Myrmotherula leucophthalma (+) FC
Myrmotherula ornata (+) FCb
Myrmotherula axillaris (+) U-Rb
Myrmotherula longipennis (+) FC
Myrmotherula menetriesii (+) FC-C
Herpsilochmus rufimarginatus (+) U-FC
Microrhopias quixensis (+) FC
Formicivora grisea (+) Re (only in deciduous forest on rocky outcroppings) (ZW2, ZW3)
Drymophila devillei (+) FCb
Cercomacra cinerascens (+) C
Cercomacra manu (+) FCb
Cercomacra nigrescens (+) FC-C
Pyriglena leuconota (+) Ue
Myrmoborus leucophrys (+) FC
Myrmoborus myotherinus (+) U
Hypocnemis cantator (+) FC-C
Hypocnemoides maculicauda (+) U-FC
Sclateria naevia (+) U
Myrmeciza hemimelaena (+) Ub
Mrymeciza atrothorax (+) Cw, Re
Myrmornis torquata (+) Rw (MPI, PZ)
Rhegmatorhina gymnops (+) U
Hylophylax naevia (+) FC
Hylophylax punctulata (+) U-R
Hylophylax poecilinota (+) U-FC
Phlegopsis nigromaculata (+) FC
Chamaeza nobilis (+) R-Ue
Formicarius colma (+) FC
Formicarius analis (+) FC
Grallaria varia (+) U-FC
Hylopezus macularius (+) U-FC
Hylopezus berlepschi (+) Ub
Myrmothera campanisona (+) U

CONOPOPHAGIDAE

Conopophaga melanogaster Rw (TAP)
Conopophaga aurita (+) Re (BW, ZW)

COTINGIDAE

Porphyrolaema porphyrolaema (+) R
Cotinga cayana U
Xipholena sp. Ue
Iodopleura isabellae (+) R-U
Lipaugus vociferans (+) C
((*Querula purpurata*)) (RET)

Cephalopterus ornatus Ue
Gymnoderus foetidus U-FC
Phoenicircus sp. R (PZ, ZS)

PIPRIDAE

Pipra rubrocapilla (+) FC
Pipra nattereri (+) U
Pipra fasciicauda (+) FC
Chiroxiphia pareola (+) FC
Manacus manacus R-Uw (TAP, MPI)
Machaeropterus pyrocephalus (+) R-U
Heterocercus linteatus R (PZ, ZW2)
Tyranneutes stolzmanni (+) U
Piprites chloris (+) FC
Schiffornis major (+) Rw (ZW, ZW2, ZW3)
Schiffornis turdinus (+) FC

TYRANNIDAE

Zimmerius gracilipes (+) U-FC
Ornithion inerme (+) FC
Camptostoma obsoletum (+) U
Phaeomyias murina (+) R (BW)
Sublegatus sp. (probably *brevirostris*) Re m (?) (PZ, ZW3)
Tyrannulus elatus (+) C
Myiopagis gaimardii (+) C
Myiopagis caniceps (+) U
Myiopagis viridicata U-R m (?) (PZ)
Elaenia spectabilis Rmw (PZ)
Elaenia parvirostris (+) Rm (BW)
Euscarthmus meloryphus U-Rw (TAP, MPI, PZ)
Mionectes oleagineus (+) U
Leptopogon amaurocephalus (+) U (TAP, ZS, BW)
Corythopis torquata (+) U
Capsiempis flaveola (+) Ub
Myiornis ecaudatus (+) FC
Lophotriccus (= *Colopteryx*) *galeatus* (+) U (BW, ZW2, ZW3)
Poecilotriccus capitalis (+) Rw (MPI)
Hemitriccus minimus (= *aenigma*) (+) R-U
Hemitriccus zosterops (+) U
Hemitriccus minor (+?) FC
Todirostrum chrysocrotaphum (+) U
Todirostrum maculatum R
Todirostrum latirostre (+) R-U
Ramphotrigon fuscicauda (+) Ub
Ramphotrigon megacephala (+) FCb
Ramphotrigon ruficauda (+) Re (AW2, ZW3)
Rhynchocyclus olivaceus (+) Rw (ZS, ZW)
Tolmomyias sulphurescens (+) FC
Tolmomyias poliocephalus (+) U
Tolmomyias flaviventris (+) U-R (TAP, BW, ZW2)
Platyrinchus platyrhynchos (+) U
Platyrinchus coronatus (+) Re (ZW2, ZW3)
Platyrinchus saturatus (+) R
Onychorhynchus coronatus (+) Rw (ZW, BW)
Terenotriccus erythrurus (+) U
Myiobius barbatus Uw (RET, ZW, BW)
Myiophobus fasciatus Rw (PZ)
Lathrotriccus euleri (+) U
Pyrocephalus rubinus R-U (PZ, BW, ZW2)
Ochthornis littoralis (+) U
Fluvicola albiventris Re (ZW2, ZW3)
Colonia colonus (+) Uw
Attila spadiceus (+) U
Attila cinnamomeus (+) U-FC
Casiornis rufa Re m (?) (only in deciduous forest on rocky outcroppings) (BW, ZW3)
Rhytipterna simplex (+) FC

Laniocera hypopyrra (+) U
Sirystes sibilator (+) U (PZ, BW, ZW2)
Myiarchus ferox (+) U
Myiarchus tyrannulus (+) R m (?) (PZ, ZW, ZW3)
Myiarchus tuberculifer (+) U
Pitangus sulphuratus U
Pitangus lictor (+) U
Myiozetetes luteiventris (+) U
Myiozetetes cayanensis (+) FC
Myiozetetes granadensis Rw (TAP, MPI)
Myiodynastes maculatus (+) R-Um (ZW, BW, ZW2)
Legatus leucophaius (+) FCm
Empidonomus varius (+) U m (in part?)
Empidonomus aurantioatrocristatus Rm (PZ, ZW2, ZW3)
Tyrannopsis sulphurea (+) Rw (BW, ZW3)
Tyrannus savana Um
Tyrannus melancholicus FC
Tyrannus albogularis R-Um
Xenopsaris albinucha Re (only in deciduous forest on rocky outcroppings) (ZW2, ZW3)
Pachyramphus castaneus (+) U
Pachyramphus polychopterus (+) U
Pachyramphus marginatus (+) FC
Pachyramphus minor (+) U
Tityra cayana (+) U
Tityra semifasciata FC
Tityra inquisitor R-U (ZW, BW)

HIRUNDINIDAE

Tachycineta albiventer FC
Progne tapera U-FC
Progne chalybea U-FC
Atticora fasciata FCe
Neochelidon tibialis (+) R
Stelgidopteryx ruficollis C

TROGLODYTIDAE

Campylorhynchus turdinus (+) FC
Odontorchilus cinereus (+) U
Thryothorus genibarbis (+) FCb
Thryothorus leucotis (+) FC-C
Troglodytes aedon C
Microcerculus marginatus (+) FC
Cyphorhinus aradus (+) R
Donacobius atricapillus Re (ZW2)

TURDIDAE

Turdus amaurochalinus U-Rw m (?) (PZ, BW)
Turdus lawrencii (+) U
Turdus fumigatus/hauxwelli (+) U
Turdus albicollis U (PZ, ZW, ZW2)

POLIOPTILINAE

Ramphocaenus melanurus (+) FC
Polioptila guianensis Rw (TAP)

VIREONIDAE

Cyclarhis gujanensis (+) U
Vireolanius leucotis (+) FC-C
Vireo olivaceus Uw
Vireo altiloquus Rw m (RET)
Hylophilus semicinereus (+) FC
Hylophilus hypoxanthus (+) C
Hylophilus ochraceiceps (+) FC

ICTERIDAE

Scaphidura oryzivora R (ZW2, ZW3)

Psarocolius decumanus (+) FC
Psarocolius bifasciatus (+) FC-C
Cacicus cela (+) U-FC
Cacicus haemorrhous Rw (MPI)
Cacicus solitarius Rw (TAP)
Icterus cayanensis (+) U

PARULIDAE

Granatellus pelzelni (+) FC
Basileuterus culicivorus (+) U
Phaeothlypis rivularis (+) U

THRAUPIDAE

Coereba flaveola (+) U
Conirostrum speciosum Re (ZW2, ZW3)
Cyanerpes caeruleus (+) U
Cyanerpes nitidus U-R
Chlorophanes spiza U
Dacnis cayana C
Dacnis lineata (+) FC
Dacnis flaviventer (+) FC
Dacnis albiventris (+) Rw (PZ)
Euphonia minuta U-R (MPI, ZW)
Euphonia violacea/laniirostris (+) R (PZ, BW, ZW3)
Euphonia rufiventris (+) FC
Euphonia chrysopasta (+) FC
Euphonia xanthogaster (+) U
Tangara velia FC
Tangara chilensis (+) C
Tangara schrankii (+) U
Tangara punctata Re (BW, ZW3)
Tangara cyanicollis (+) Uw
Tangara nigrocincta U-R (TAP, BW, ZW2)
Tangara mexicana (+) FC
Tangara gyrola U
Thraupis episcopus/sayaca Uw
Thraupis palmarum (+) FC-C
Ramphocelus carbo (+) FC
Piranga flava Rw (ZS)
Habia rubica (+) U-FC
Lanio versicolor (+) FC
Tachyphonus cristatus (+) C
Tachyphonus luctuosus (+) FC
Hemithraupis flavicollis U-FC
Nemosia pileata Re (ZW3)
Lamprospiza melanoleuca (+) FC
Cissopis leveriana (+) U
Tersina viridis (+) U-FC

EMBERIZIDAE

Saltator maximus (+) FC-C
Saltator grossus (+) FC-C
Parkerthraustes humeralis (+) U
Paroaria gularis U-FCe
Cyanocompsa cyanoides (+) FCw
Volatinia jacarina C
Oryzoborus angolensis (+) Rw (ZS)
Sporophila caerulescens Rw m (?) (TAP, PZ, ZW3)
Arremon taciturnus (+) FC

PASSERIDAE

Passer domesticus Rw (ZW)